APCOM Conference 2025

Application of Computers and Operations Research in the Minerals Industry

10–13 August 2025
Perth, Australia

The Australasian Institute of Mining and Metallurgy
Publication Series No 4/2025

AusIMM

Published by:
The Australasian Institute of Mining and Metallurgy
Ground Floor, 204 Lygon Street, Carlton Victoria 3053, Australia

ISBN 978-1-922395-51-1

Advisory Committee

Tony Tang
FAusIMM(CP)
Conference Advisory Committee Chair
Curtin University

Chaoshui Xu
MAusIMM

Ernest Baafi
MAusIMM

Mehmet Kizil
MAusIMM

Farjad Ather
MAusIMM

Gamini Senanayake
MAusIMM

Oliver Mowbray
AAusIMM

Winfred Assibey-Bonsu

Robert Solomon

AusIMM

Julie Allen
Head of Events

Amelia Lundstrom
Event Manager

Raha Karimi
Program Manager

Reviewers

We would like to thank the following people for their contribution towards enhancing the quality of the papers included in this volume:

Siddhartha Agarwal

Angelina Anani

Heath Arvidson

Winfred Assibey-Bonsu

Baris Ates

Farjad Ather

Ernest Baafi

Marcel Antonio Arcari Bassani

Filipe Beretta

Pedro Campos

Liam Findlay

Willem Fourie

Matin Ghasempour

Ryan Goodfellow

Nasib Al Habib

Alireza Kamrani

Mehmet Kizil

Fabian Manriquez

Ahlam Maremi

Jorge Mariz

Douglas Mazzinghy

Rudrajit Mitra

Ehsan Moosavi

Oliver Mowbray

Cuthbert Musingwini

Michael Owusu-Tweneboah

Kate Willa Brown Requist

Roberto Rolo

Oscar Rondon

Christopher Roos

Scott Rosenthal

Gamini Senanayake

Tony Tang

Augusto Toledo

Chaoshui Xu

Guang Xu

Foreword

Welcome, on behalf of the Conference Advisory Committee, we are delighted to welcome you to the AusIMM's Application of Computers and Operations Research in the Minerals Industry – APCOM 2025, hosted in Perth, Western Australia.

The APCOM 2025 technical conference is bringing together researchers, industry professionals, innovators, technology practitioners, and government regulators to share innovations, best practices, and challenges within mining operations. The conference is unique in its ability to create a global discussion platform, attracting multidisciplinary groups.

The program will explore current and emerging innovations and technologies to forward thinking for our modern mining practices, strategies, risk management and critical controls, as well as highlight effective leadership strategies and management that promote leading practice for the mining industry.

Delegates will have the opportunity to hear from an innovative range of papers, sharing insights on thought-provoking themes such as Artificial Intelligence (AI), Machine Learning (ML), Robotics, and Remote Operations. AusIMM, along with the Conference Advisory Committee and the APCOM International Council, have secured a strong list of industry experts to present keynotes on global rapid advancement with innovations and technologies and approaches, including Michelle Keegan, Dr Christoph Mueller, Gustavo Pilger, Eduardo Coloma, Dr Penny Stewart, Professor Rousos Dimitrakopoulos, Professor Peter Dowd, Professor Shirong Ge, Ben Cabanas and Cam Stevens.

We look forward to welcoming you to the AusIMM's APCOM 2025 in Perth.

Yours faithfully,

Prof Tony Tang FAusIMM(CP)
APCOM 2025 Conference Advisory Committee Chair

About APCOM

In 1961 The University of Arizona, Tucson, USA hosted a modest conference and workshop designed to demonstrate the advantage of digital computers in solving problems in the minerals industry. The meeting was a huge success and served to launch a continuing symposium series on this topic that soon became international in scope.

At the 11th Symposium, again at the University of Arizona, in 1973, a formal steering committee for the symposium was established which is now named the Council for the Application of Computers and Operations Research in the Mineral Industry. The meeting, however, has become widely known as the APCOM Symposium, an acronym coined by the South Africans at the 10th Symposium in Johannesburg in 1972.

The APCOM symposia also has an educational component – the first symposium was called a 'short course' and four of the five organisers were professors at the University of Arizona. The educational component was reinforced by the fact that the next seven symposia were organised by universities in the USA.

The APCOM series of events has truly become a major driver of innovation in the minerals industry. The APCOM organisation continues to operate on an informal basis, and its continuing success is largely partly due to the personal commitment of dedicated individuals, many of whom are present at the 42nd APCOM Symposium in Perth, Australia. This is the fourth time in its 64-year history that APCOM has been held in Australia. The 42nd APCOM is aimed at maintaining the APCOM goal of playing a significant role in contributing to effective decision-making processes in the broad minerals industries.

The APCOM Council is currently comprised of the following members:

- The University of Arizona, USA
- Colorado School of Mines, USA
- The Pennsylvania State University, USA
- The Society for Mining, Metallurgy and Exploration, USA
- The Canadian Institute of Mining and Metallurgy, Canada
- The Institution of Mining and Metallurgy, UK
- The South African Institute of Mining and Metallurgy, Republic of South Africa
- The Australasian Institute of Mining and Metallurgy, Australia
- China University of Mining and Technology, China
- Federal University of Rio Grande do Sul, Brazil
- University of Utah, USA
- McGill University, Canada
- The Wroclaw University of Technology, Poland

The current Chairperson of the Council is Andrea Brickey, Professor, Mining and Management Department, South Dakota Mines, Rapid City, USA. She is also the representative of the Society for Mining, Metallurgy and Exploration, USA.

To ensure greater continuity and control, the APCOM Council has entered into a Letter of Agreement with the Society for Mining, Metallurgy and Exploration (SME), USA to provide custodial management of the APCOM Council funds. As part of the APCOM agreement, SME receives AUD$40 per paid registration towards the APCOM scholarship fund named after the late Professor Daniel G Krige for the best higher degree research student paper presented at APCOM 2025.

APCOM International Council

CURRENT COUNCIL MEMBERS

Professor Angelina Anani, The University of Arizona, USA

Professor Ernest Baafi, The University of Wollongong, Australia

Professor Andrea Brickey, South Dakota Mines, USA

Dr Winfred Assibey-Bonsu, Gold Fields Ltd, Australia

Professor Kadri Dagdelen, Colorado School of Mines, USA

Professor Roussos Dimitrakopoulos, McGill University, Canada

Professor Shirong Ge, China University of Mining and Technology, China

Professor Jose Filipe, Federal University of Rio Grande do Sul, Brazil

Dr Christoph Mueller, Mobile Tronics, Germany

Professor Nelson Morales, Polytechnique Montréal, Canada

Professor Cuthbert Musingwini, University of Witwatersrand, South Africa

Professor Antonio Nieto, University of Utah, USA

Professor Julian Ortiz, Camborne School of Mines, University of Exeter, UK

EMERITUS COUNCIL MEMBERS

Professor Don McKee, formerly with University of Queensland, Australia

Professor Thys Johnson, formerly with Colorado School of Mines, USA

Professor Tim O'Neil, formerly with University of Arizona, USA

Emeritus Professor Raja Ramani, Pennsylvania State University, USA

Professor Sukumar Bandopadhyay, formerly with University of Alaska, USA

Sponsors

Major Conference Sponsor

Platinum Sponsor

Contents

Advanced technologies

Exploration and geology

Investment and risk management

Mine planning and operations

Sustainable practices

Advanced technologies

Advances in vibration monitoring of mining equipment with deep machine learning methods

C Aldrich[1] and X Liu[2]

1. Professor, Western Australian School of Mines, Curtin University, Perth WA 6845.
   Email: chris.aldrich@curtin.edu.au
2. Research Fellow, Western Australian School of Mines, Curtin University, Perth WA 6845.
   Email: xiu.liu@curtin.edu.au

ABSTRACT

In this study an emerging novel approach for real-time visual monitoring and analysis of vibrational signals from process equipment is explored. To this end, features were extracted from image-encoded simulated vibrational signals from an industrial screen. More specifically, local binary patterns (LBP) and a deep convolutional neural network, GoogleNet were used to extract features from wavelet spectrograms of signal segments related to changes in the operation of the vibrational excitor of the screen. These features were projected onto a process monitoring map generated with a t-distributed stochastic neighbour embedding (t-SNE) algorithm. The results indicate that this method provides a reliable basis for visual real-time monitoring, that could enhance the operational efficiency and longevity of industrial screens and other mineral processing equipment.

INTRODUCTION

Vibrational monitoring of mineral processing equipment, such as grinding mills (Zeng, Zheng and Forssberg, 1993; Tang *et al*, 2010; Mohanty, Gupta and Raju, 2015), fluidised beds (Ikonen *et al*, 2023; Xiao, Ma and Liu, 2024), and hydrocyclones (Tyeb *et al*, 2024), is essential for maintaining optimal performance and preventing costly breakdowns. By continuously measuring the vibrations of these machines, operators can detect early signs of wear and tear, misalignment, or imbalance. This proactive approach allows for timely maintenance and repairs, minimising downtime and ensuring that the equipment operates at peak efficiency.

Essentially, all these methods depend on feature extraction from the signals in one form or another. These methods can be categorised as belonging to one of four different domains, namely time domain methods, such as autoregressive modelling and statistical analysis, frequency domain methods, such as Fourier and harmonic analysis, time-frequency domain methods, such as short-time Fourier transforms and wavelet transforms and complex domain methods based on machine learning models, recurrence quantification analysis and complex wavelet transforms, among other. The analytical approach used in this investigation can be considered to belong to the latter domain.

ANALYTICAL METHODOLOGY

The overall analytical methodology (Aldrich, 2019) consists of the acquisition of vibrational signals, segmentation of the signals, image encoding of the signal segments, feature extraction from the images, dimensionality reduction of the features and mapping of the dimensionally reduced features to a process monitoring chart or map, as outlined in Figure 1.

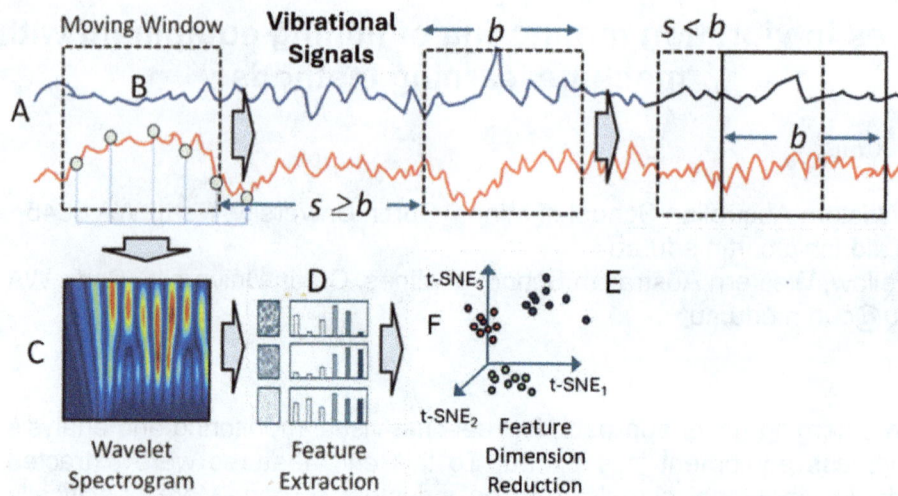

FIG 1 – Analytical methodology for vibrational monitoring of equipment based on signal acquisition (A), signal segmentation (B), imaging of signal segments (C), extracting features from the images (D), reducing the image features with the t-distributed stochastic neighbour (t-SNE) algorithm to generate a visual process control chart.

The reason for employing image-encoding techniques for time series data is that it enables the application of well-established image processing and computer vision methods for analysis. This is an approach that was proposed by, among others, Wang and Oates (2015) as a novel framework that enables the use of convolutional neural networks and other state-of-the-art deep learning models. Since then, these methods have attracted considerable and growing interest in time series analysis across multiple technical domains and applications. The analytical steps are described in more detail below.

Signal acquisition and segmentation (A, B)

In practice, vibrational signals would be acquired over a period when the equipment is considered to operate normally. If the period of observation is relative long, eg in the order of hours, short signals could be collected intermittently over time. Such a signal is segmented by use of a moving window. Two parameters define segmentation, ie the window size, b and the step size or stride of the moving window, s. When $b = s$, the windows are contiguous, when $b > s$, the windows overlap and when $b < s$, the segments are taken from the raw signal are non-contiguous and non-overlapping. The size of the window, b, has to be determined empirically, but as a general principle, it should be sufficiently large to capture repeated periodic behaviour of the signal, if this is present, but not so large its ability to detect changes in the signal is compromised. The selection of the step size would have to determined similarly as well, generally in tandem with the selection of the window size.

Encoding signal segments into images (C)

Encoding signals as images is a powerful technique that leverages visual analysis to interpret complex time series data. These methods include global recurrence plots (Hou *et al*, 2017), wavelet spectrograms representing the signal in both time and frequency domains, as well as Gramian angular fields that transform time series into polar coordinate systems with the angular information encoded in a matrix that can be visualised as an image (Memarian *et al*, 2024).

Image-encoded signals enable the application of advanced image processing and computer vision to time series analysis, which enhances the ability of the algorithms to detect and interpret complex patterns in the data. By converting signals into visual representations, these techniques also make it easier for humans to understand and analyse the underlying dynamics of the system.

In this investigation, wavelet spectrograms were used to encode the signals. A wavelet spectrogram is a time-frequency representation of a signal, created by applying wavelet transforms to the signal. Specifically, when using Morlet wavelets, the process involves convolving the signal with a series of Morlet wavelets at different scales. Morlet wavelets are sinusoidal functions modulated by a

Gaussian envelope, which allows for the analysis of both the frequency and temporal characteristics of the signal.

The Morlet wavelet is defined by a central frequency and a Gaussian window, which provides a balance between time and frequency resolution. This balance is controlled by the width of the Gaussian, often referred to as the number of cycles. The wavelet transform decomposes the signal into components at various scales, capturing both high-frequency details and low-frequency trends.

More formally, the Morlet wavelet is defined by Equation 1:

$$\psi(t) = \pi^{-\frac{1}{4}} e^{-i\omega_0 t} e^{-\frac{t^2}{2}} \tag{1}$$

where ω_0 is the central frequency of the wavelet. The continuous wavelet transform of a signal $x(t)$ using the Morlet wavelet, is given by Equation 2:

$$W_x(a, b) = \int_{-\infty}^{\infty} x(t)\, \psi^*\left(\frac{t-b}{a}\right) dt \tag{2}$$

where a and b are scale and translation parameters respectively and ψ^* denotes the complex conjugate of the Morlet wavelet.

The wavelet spectrogram is obtained by plotting the magnitude of the wavelet transform ($\lceil W_x(a, b)\rceil$) as a function of time (b) and scale (a). This representation or wavelet scalogram provides a detailed view of the signal's frequency content over time, allowing for the identification of transient events and localised features.

Feature extraction from images (D)

Local binary patterns (LBP)

Features can subsequently be extracted from wavelet scalograms by use of any of several methods. In this investigation, local binary patterns were used. Local binary patterns are highly suitable for extracting features from textural images, owing to their ability to effectively capture local texture information and their robustness to changes in illumination and rotation.

The method involves comparing each pixel in an image to its neighbouring pixels. For each pixel, a binary code is generated based on whether the neighbouring pixels have higher or lower intensity values than the central pixel. This binary code is then converted into a decimal value, which represents the local texture pattern.

By applying this process across the entire image, LBP creates a histogram of these decimal values, capturing the distribution of local texture patterns. This histogram serves as a feature vector that can be used for various image analysis tasks, such as classification, segmentation, and recognition. LBP is highly effective due to its simplicity, computational efficiency, and robustness to changes in illumination and rotation.

More formally, given a pixel at location (x_c, y_c) in an image, consider a neighbourhood of P pixels around it, which are position on a circle of radius R, the set-up of which is denoted as (P, R). The LBP value of the pixel is calculated as follows:

- *Thresholding*: For each neighbour p (where $p = 0, 1, \ldots P - 1$), compare the intensity of the neighbouring pixel g_p with the intensity of the centre pixel, g_c. Create a binary pattern by thresholding:

$$S(g_p - g_c) = \begin{cases} 1, if\ g_p \geq g_c \\ 0, otherwise \end{cases} \tag{3}$$

- *Binary pattern formation*: Form a binary number by concatenating the binary results of the thresholding step for the neighbouring pixels.

- *Calculation of LBP code*: Convert the binary number to a decimal one to get the LBP code for the centre pixel:

$$LBP_{P,R}(x_c, y_c) = \sum_{p=0}^{P-1} S(g_p - g_c) 2^p \tag{4}$$

This yields an LBP code for each pixel in the image, which captures local texture around that pixel. The histogram of these LBP code can then be used as a feature vector for further analysis of the image.

GoogleNet

GoogleNet, also known as Inception v1, is a deep convolutional neural network (CNN) that was designed by Google researchers and made publicly available. It was introduced in 2014 and won the ImageNet Large Scale Visual Recognition Challenge (ILSVRC) that year. The core innovation of GoogleNet is the Inception module, which allows the network to use multiple convolutional filter sizes (1×1, 3×3, 5×5) and max-pooling operations simultaneously. This helps the network capture different levels of detail and spatial information. GoogleNet is significantly deeper than its predecessors, with 22 layers in total. This depth allows the network to learn more complex features. Despite its depth, GoogleNet is computationally efficient due to the use of 1×1 convolutions that reduce the number of parameters and computational cost. Moreover, to combat the so-called vanishing gradient problem, GoogleNet includes auxiliary classifiers at intermediate layers. These classifiers provide additional gradient signals during training, improving convergence.

Process monitoring (E)

Dimensional reduction of the imaged signal features

The features are reduced to two dimensions by use of the t-distributed stochastic neighbour embedding (t-SNE) algorithm (van der Maaten and Hinton, 2008). The t-SNE algorithm reduces the dimensionality of data, while preserving the relationships between data points, making it easier to visualise complex data sets. It does so by calculating the pairwise similarities between data points in the high-dimensional space and converting them into probability distributions. t-SNE then maps the data to a lower-dimensional space as specified by the user by minimising the divergence between the probability distributions of the high-dimensional and low-dimensional spaces. The result is a scatter plot where similar data points are clustered together, revealing patterns and structures in the data.

More formally, t-SNE maps are computed as follows. Given a set of N samples over M variables, $X \in \mathbb{R}^{N \times M}$, the t-SNE algorithm computes the values of the conditional probabilities $p_{j|i}$ based on the similarities of the samples $\mathbf{x}_i$ and $\mathbf{x}_j$ by defining Equation 5.

$$p_{j|i} = \frac{\exp\left(-\|\mathbf{x}_i - \mathbf{x}_j\|^2 / 2\sigma_i^2\right)}{\sum_{k \neq i} \exp\left(-\|\mathbf{x}_i - \mathbf{x}_k\|^2 / 2\sigma_i^2\right)}, \text{ for all } i \neq j \qquad (5)$$

and setting $p_{j|i} = 0$. $p_{j|i}$ is the conditional probability that the j'th sample ($\mathbf{x}_j$) is a neighbour of the i'th sample ($\mathbf{x}_i$), if neighbours were selected based on their probability density under a Gaussian distribution centred on $\mathbf{x}_i$. Note that $\sum p_{j|i} = 1$, for all i.

Given that he probabilities of p_i and p_j from N samples can be approximated by $1/N$, the conditional probabilities can be expressed as $p_{j|i} = Np_{ji}$ and given that $p_{ji} = p_{ij}$, p_{ji} can be expressed in terms of conditional probabilities as in Equation 6, noting that that $p_{ii} = 0$ and $\sum p_{ij} = 1$.

$$p_{ji} = \frac{p_{j|i} + p_{i|j}}{2N}, \text{ for all } i \neq j \qquad (6)$$

The kernel bases are adjusted to match the data density by equating a predefined entropy to the entropy of the conditional distribution. To create a lower-dimensional map (P <M), similarities between points in the P-dimensional space are measured similarly to those in the M-dimensional space. The probabilities (q_{ij}) are defined accordingly by setting $q_{ii} = 0$.

$$q_{ij} = \frac{(1 + \|\mathbf{y}_i - \mathbf{y}_j\|^2)^{-1}}{\sum_k \sum_{l \neq k} (1 + \|\mathbf{y}_i - \mathbf{y}_j\|^2)^{-1}}, \text{ for all } i \neq j \qquad (7)$$

A student t-distribution with one degree of freedom is employed to estimate the similarity between points in the low-dimensional space. The map is then generated by minimising the Kullback-Leibler

(KL) divergence using an appropriate gradient method with respect to the mapped points y_i in Equation 7.

Mapping of features for monitoring

Ab initio optimisation was used to find the t-SNE features every time a new signal segment became available. In practice, this would require the collection of s signal samples to populate part of a new map, ie equal to the step size of the moving window. The main impact of this is the computational cost, which could be a problem when large feature sets need to be processed rapidly. However, this can be readily circumvented by training a machine learning model on the data associated with normal operations to map the features of new signals onto the map.

CASE STUDY 1 – MONITORING OF INSTRUMENT FAILURE IN A SCREEN

In mineral processing plants, screens play a crucial role in separating and classifying materials based on size. The operation of these screens involves the use of vibrational energy to move and sort the material. The vibration of these screens is often generated by two eccentric shafts, which are positioned either on top of or in the middle of the screen, as shown diagrammatically in Figure 2.

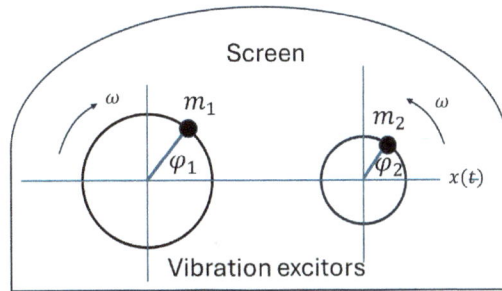

FIG 2 – Simulation of vibration exciter signals in a screen (after Chen, Tong and Li, 2020).

These eccentrically weighted shafts are designed to create an elliptical motion, which is highly effective for screening purposes. When the shafts rotate, they produce an unbalanced force that causes the screen to vibrate. The elliptical motion generated by the vibrational exciter ensures that the material moves in a controlled manner across the screen surface. This motion helps in stratifying the material, allowing finer particles to pass through the screen openings while larger particles are retained on the surface.

The elliptical trace created by the vibrational exciter is particularly advantageous, as it combines both horizontal and vertical motion. This dual motion enhances the screening efficiency by ensuring that the material is evenly distributed across the screen and has multiple opportunities to come into contact with the screen openings. As a result, the screening process becomes more effective, leading to better separation and classification of materials.

Modelling of the vibration of screen boxes can be complicated, but can be approximated by a simplified elliptical trace of the screen deck, as described by Equations 8 and 9. This model is similar to the one studied by Chen, Tong and Li (2020), where $x(t)$ and $y(t)$ are the horizontal and vertical displacements of the screen deck at time t, A_x and A_y are the amplitudes of the harmonic motion, ω is the angular velocity and φ_x and φ_y are the initial phase angles of the motion. When the phase angle difference, $\varphi_y - \varphi_x = \frac{\pi}{2}$, it is a right elliptical trace with clockwise motion, and conversely, it is a right elliptical trace with anticlockwise motion, if the phase difference $\varphi_y - \varphi_x = -\frac{\pi}{2}$. e_x and e_y represent additive Gaussian noise components:

$$x(t) = A_x \cos(\omega t + \varphi_x) + e_x \tag{8}$$

$$y(t) = A_y \cos(\omega t + \varphi_y) + e_y \tag{9}$$

Equations 8 and 9 were used to simulated the signal with $A_x = A_y = 1$, $\omega = 10\pi$ and the phase difference $\varphi_y - \varphi_x = \frac{\pi}{4}$. Zero mean Gaussians with a variance of 0.001, ie $e_x = e_y \sim N(0, 0.001)$ were added to these signals; 5000 such samples were generated. Instrument failure was subsequently generated by Gaussian noise with a zero mean and a variance of 0.05, as shown by samples 5001 to 6000 in Figure 3.

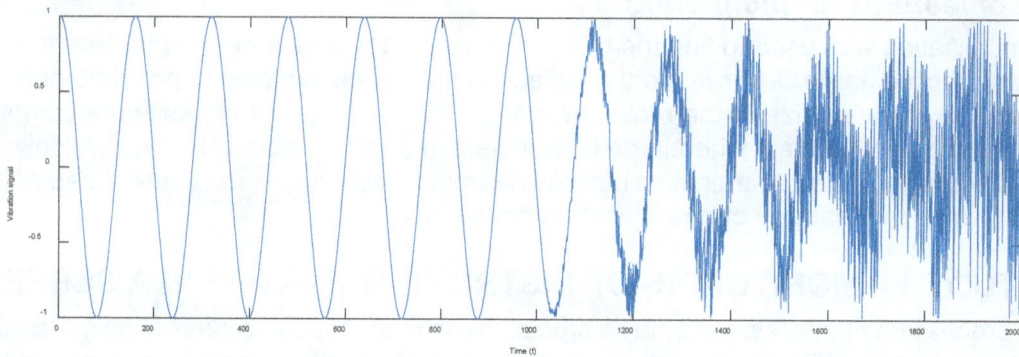

FIG 3 – Simulated signal followed by incremental instrument failure starting at index 1001.

The entire signal was segmented with parameters b = 100 and step size s = 10, which resulted in overlapping signal segments and images. Each image was in turn converted to a Morlet wavelet scalogram, examples of which are shown in Figure 4. The top row in Figure 4 shows images of the noiseless harmonic vibration, which are very similar in appearance. The bottom row shows segments 97, 107 and 127, bearing in mind that signal corruption becomes progressively worse from segment 97 onwards.

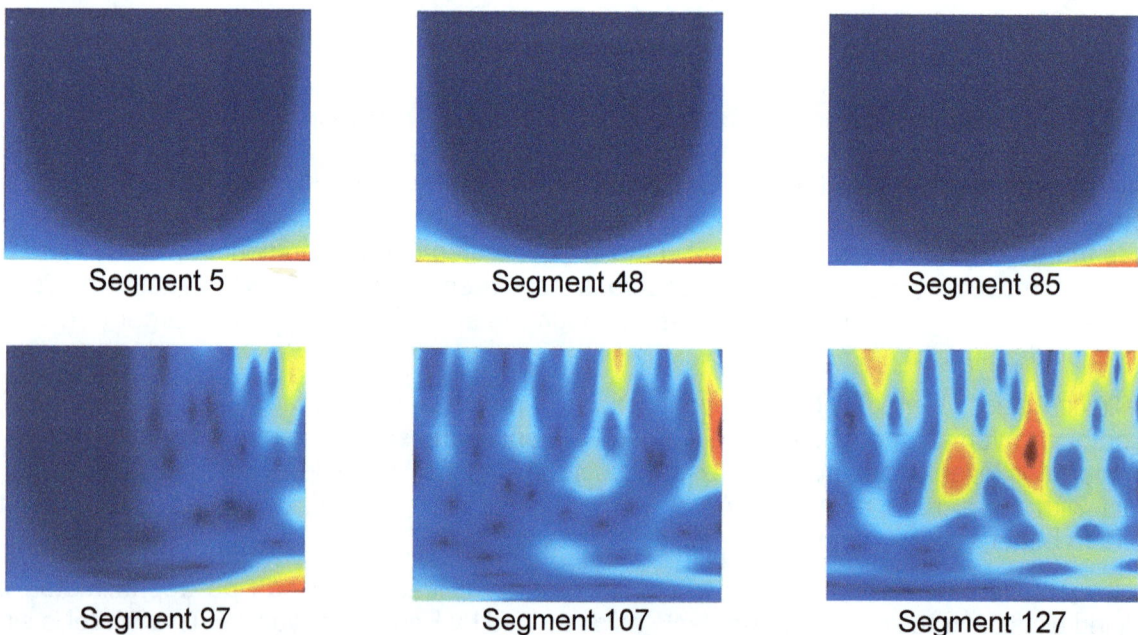

Segment 5   Segment 48   Segment 85

Segment 97   Segment 107   Segment 127

FIG 4 – Examples of wavelet scalograms of vibration image segments. The top row shows random samples from the simulated noiseless elliptical motion signal (time indices 1 to 1000) and the bottom row shows random samples from simulated progressive instrumentation failure (time indices 1001 to 2000).

LBP and GoogleNet features were extracted from all 191 images and the 96 image segments corresponding with the signal at time indices 1 to 1000 were projected to a t-SNE score plot (black circles). 95 per cent confidence ellipses (blue dashed lines) were fitted to the features representing normal process conditions, as shown in Figure 5. The t-SNE features associated with the changing conditions (time indices 1001 to 2000) are shown on the charts as red stars. As can be seen from Figure 5, the process control charts based on both the LBP and GoogleNet features were able to detect the change in the signal very reliably within ten samples (the step s).

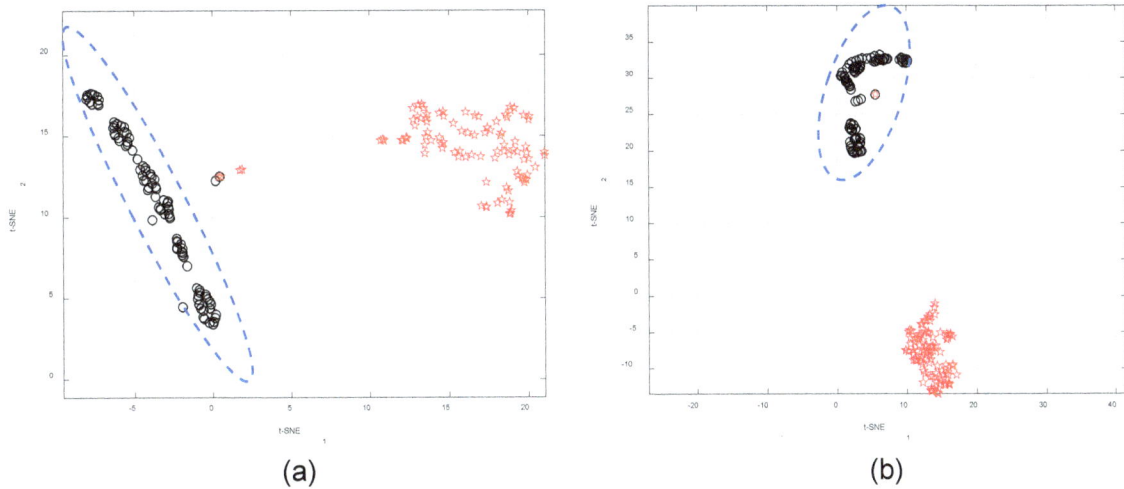

(a)   (b)

FIG 5 – Process monitoring charts in Case Study 1 with: (a) Vibration signal monitoring with LBP features; (b) Vibration signal monitoring with GoogleNet features.

CASE STUDY 2 – MONITORING OF CHANGES IN VIBRATION FREQUENCY

In the second case study, the ability of the systems to detect an increase in the vibration frequency of the excitor on the screen is investigated. Simulated data are again used for this purpose, similar to those in Case Study 1. 10 000 samples were generated, the first 5000 of which had a stable frequency generated by the synchronised angular shaft velocity of 10π rad/s. This velocity was subsequently uniformly increased to reach 20π rad/s at the 10 000th sample. The signal is shown in Figure 6.

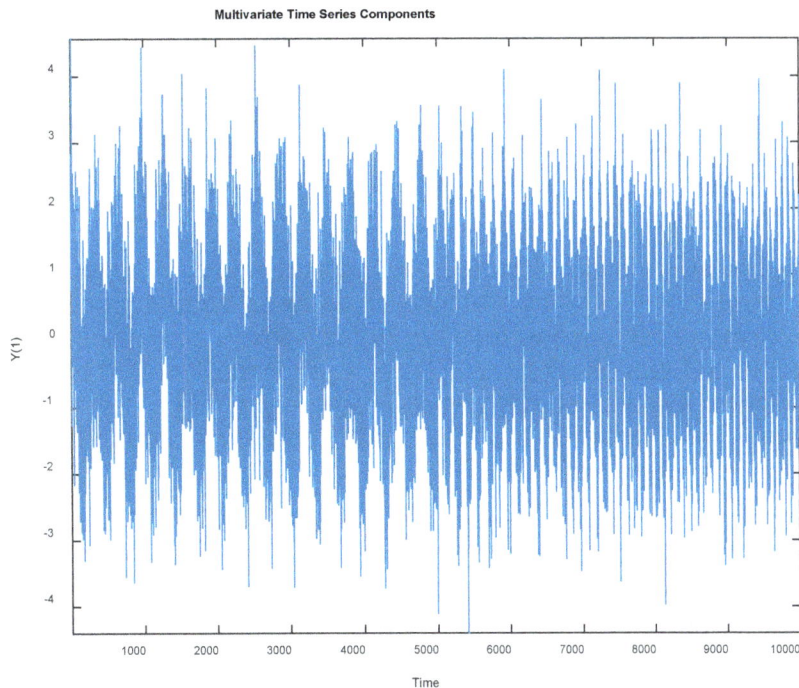

FIG 6 – Vibration signal investigated in Case Study 2. The first 5000 samples represent a signal generated with an angular shaft velocity of 10π rad/s, which increases uniformly from time index 5001 to reach 20π rad/s at time index 10 000.

The parameters of the moving window were set at b = 1000 and s = 10, which generated 901 segments. These were again imaged with the Morlet wavelet scalogram, as was used in Case Study 1. LBP and GoogleNet features were extracted from all the images and the 451 image segments corresponding with the signal at time indices 1 to 5000 were projected to a t-SNE score plot (black circles), to which a 95 per cent confidence ellipse was fitted, as shown in Figure 7. The t-

SNE features associated with the changing conditions (time indices 5001 to 10 000) are shown on the charts as red stars.

With the GoogleNet process monitoring chart, approximately 77 per cent of the signal segments could be identified as abnormal. With the LBP process monitoring chart, only approximately 13 per cent of the segments could be detected as such.

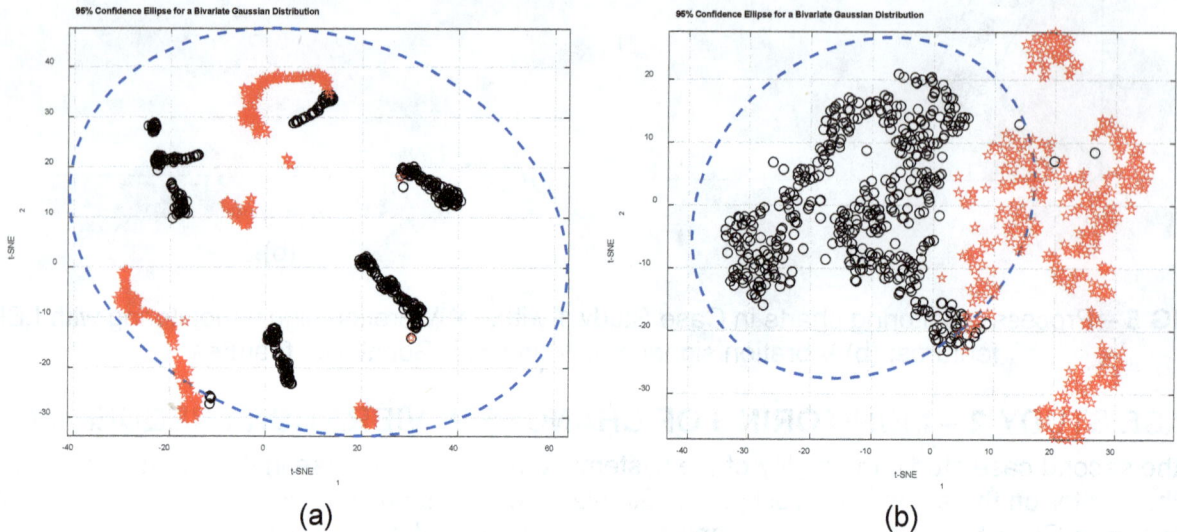

(a)   (b)

FIG 7 – Process monitoring charts in Case Study 2: (a) Vibration signal monitoring with LBP features; (b) Vibration signal monitoring with GoogleNet features.

DISCUSSION AND CONCLUSIONS

Vibrational signal processing based on signal segmentation and imaging of the segments can serve as a flexible framework to capture the signal dynamics. When used for process monitoring, features extracted from the imaged signals can be projected onto process charts with suitable machine learning models. In this study, 2D t-SNE maps were used to construct process monitoring charts. Such charts have previously been proposed for visual process monitoring, but based on supervised process monitoring, not unsupervised, as with this study.

Unsupervised monitoring is more complicated than supervised monitoring, since it also requires identification of a representation of what are considered to be normal process features. Machine learning models that can map the image features to the t-SNE coordinates would be an obvious approach to accomplish this (Lu and Yan, 2022), except for the fact that most machine learning models are not reliable when presented with data that are very different from the data on which they had been trained. In general, such models would tend to generate features that are poorly related to the imaged signals they are presented with.

The approach followed here was to add new data to the existing training data and to recompute the t-SNE features ab initio. Automated detection of changes in the vibration signals was accomplished by fitting 95 per cent confidence ellipses to the t-SNE features associated with normal process conditions. However, in general, a nonlinear approach would be required, as the distributions of the t-SNE features are generally not normal and can deviate considerably from Gaussian distributions. This can be accomplished by a number of different approaches, such as support vector data description methods (Tax and Duin, 2004), one-class support vector machines (Schölkopf et al, 1999; Bounsiar and Madden, 2014), kernel-based methods or multimodal Gaussian models.

Feature extraction with local binary patterns was compared with that of a convolutional neural network, GoogleNet. It is important to note that GoogleNet had been pretrained on ImageNet features only and not trained on any of the imaged vibration signal features. It was therefore used purely as a feature extractor, similar to algorithms such as local binary patterns.

In principle, it would be possible to enhance these features by any of a number of approaches that could be referred to as self-supervised learning or contrastive learning. This would require the set-up of artificial classes that would enable supervised learning of the network and will be the focus of future work.

In summary, the following can be concluded from this study:

- The study highlights the efficacy of image-encoded signal processing as a framework for the development of process monitoring systems based on vibration signals, as well as the direct use of pretrained convolutional neural networks without any training on the signal data themselves.

- Visual process monitoring based on the use of t-distributed stochastic neighbour embeddings provides a useful and information-rich monitoring framework when high-dimensional image and signal features are used.

- Used directly as image feature extractors, pretrained deep learning models, such as GoogleNet, can provide more sensitive monitoring that what can be attained with traditional methods, such as local binary patterns.

ACKNOWLEDGEMENTS

The authors acknowledge the funding support from the Australian Research Council for the ARC Centre of Excellence for Enabling Eco-Efficient Beneficiation of Minerals, Grant Number CE200100009.

REFERENCES

Aldrich, C, 2019. Process fault diagnosis for continuous dynamic systems over multivariate time series, *Time Series Analysis – Data, Methods and Applications*. https://doi.org/10.5772/intechopen.78491

Bounsiar, A and Madden, M G, 2014. One-class support vector machines revisited, in *Proceedings of the 2014 International Conference on Information Science and Applications (ICISA)*, pp 1–4. https://doi/org/10.1109/ICISA.2014.6847442

Chen, Z, Tong, X and Li, Z, 2020. Numerical investigation on the sieving performance of elliptical vibrating screen, *Processes*, 8(9):1151. https://doi.org/10.3390/pr8091151

Hou, Y, Aldrich, C, Lepkova, K, Machuca, L L and Kinsella, B, 2017. Effect of electrode size on the electrochemical noise measured in different corrosion systems, *Electrochimica Acta*, 256:337–347. https://doi.org/10.1016/j.electacta.2017.09.169

Ikonen, E, Liukkonen, M, Hansen, A H, Edelborg, M, Kjos, O, Selek, I and Kettunen, A, 2023. Fouling monitoring in a circulating fluidized bed boiler using direct and indirect model-based analytics, *Fuel*, 346:128341. https://doi.org/10.1016/j.fuel.2023.128341

Lu, W and Yan, X, 2022. Variable-weighted FDA combined with t-SNE and multiple extreme learning machines for visual industrial process monitoring, *ISA Transactions*, 122:163–171. https://doi.org/10.1016/j.isatra.2021.04.030

Memarian, A, Damarla, S K, Memarian, A and Huang, B, 2024. Detection of poor controller tuning with Gramian Angular Field (GAF) and StackAutoencoder (SAE), *Computers and Chemical Engineering*, 185:108652. https://doi.org/10.1016/j.compchemeng.2024.108652

Mohanty, S, Gupta, K K and Raju, K S, 2015. Vibration feature extraction and analysis of industrial ball mill using mems accelerometer sensor and synchronized data analysis technique, *Procedia Computer Science*, 58:217–224. https://doi.org/10.1016/j.procs.2015.08.058

Schölkopf, B, Williamson, R, Smola, A and Shawe-Taylor, J, 1999. Single-class support vector machines, in *Unsupervised Learning* (eds: J Buhmann, W Maass, H Ritter and N Tishby), Dagstuhl-Seminar-Report, 235:9–20.

Tang, J, Zhao, L-J, Zhou, J-W, Yue, H and Chai, T-Y, 2010. Experimental analysis of wet mill load based on vibration signals of laboratory-scale ball mill shell, *Minerals Engineering*, 23(9):720–730. https://doi.org/10.1016/j.mineng.2010.05.001

Tax, D M and Duin, R P, 2004. Support Vector Data Description, *Machine Learning*, 54:45–66. https://doi.org/10.1023/B:MACH.0000008084.60811.49

Tyeb, M H, Mishra, S, Singh, A and Majumder, A K, 2024. Prediction of operating state of hydrocyclones using vibrometry and 1D convolutional neural networks, *Advanced Powder Technology*, 35(2):104337. https://doi.org/10.1016/j.apt.2024.104337

van der Maaten, L J P and Hinton, G E, 2008. Visualizing high-dimensional data using t-SNE, *Journal of Machine Learning Research*, 9:2579–2605.

Wang, Z and Oates, T, 2015. Imaging time-series to improve classification and imputation, arXiv.1506.00327 [cs.LG], https://doi.org/10.48550/arXiv.1506.00327

Xiao, H, Ma, Y and Liu, M, 2024. Diagnoses of flow behaviors in gas-liquid-solid circulating fluidized beds using vibration acceleration signals, *Powder Technology*, 445:120070. https://doi.org/10.1016/j.powtec.2024.120070

Zeng, Y, Zheng, M and Forssberg, E, 1993. Monitoring jaw crushing parameters via vibration signal measurement, *International Journal of Mineral Processing*, 39(3–4):199–208. https://doi.org/10.1016/0301-7516(93)90015-3

Comparative analysis of underground geological mapping – traditional methods versus LiDAR scanning with Apple devices and survey-grade laser scanners

C Birch[1], H Grobler[2] and A Olivier[3]

1. Snr Lecturer, University of the Witwatersrand, Johannesburg Gauteng, South Africa.
   Email: clinton.birch@wits.ac.za
2. Head of Department, University of Johannesburg Gauteng, South Africa.
   Email: hgrobler@uj.ac.za
3. MSc student (deceased), University of the Witwatersrand, Johannesburg Gauteng, South Africa

ABSTRACT

Geological mapping in underground mining has traditionally relied on manual tools such as tapes, clinorules, compasses, and notebooks. However, advancements in digital technology present alternatives that improve efficiency, accuracy, and integration with mine planning systems. This study evaluates three geological mapping methods, traditional manual techniques, Apple iPad Pro and iPhone Pro LiDAR scanning, and survey-grade laser scanners, in underground mining environments. Testing was undertaken at Mponeng Gold Mine, Maseve Platinum Mine, and simulated mining tunnels at the University of the Witwatersrand and the University of Johannesburg. Findings reveal that while survey-grade laser scanners provide the highest accuracy, their cost and operational complexity limit their widespread use for geological mapping especially in narrow, confined stopes. Apple's LiDAR-enabled devices, though constrained by a 5 m scanning range, offer a cost-effective and practical alternative for geological mapping, enabling geologists to rapidly capture 3D face maps with minimal equipment. The study also presents workflows for integrating LiDAR-derived geological and production data into mine planning software. Overall, the research highlights the potential for consumer-grade LiDAR to modernise underground geological mapping, offering a balance between accuracy, cost, and ease of use. There are also safety benefits gained by the geologist being able to remain back from the immediate face vicinity. The findings have led to the adoption of iPhone Pro Maxes for geological mapping at Mponeng Mine.

INTRODUCTION

Background information

Geologists have traditionally mapped underground workings to identify geological features such as lithology and structural elements like dykes and faults. The primary purpose of this mapping is orebody delineation, which is crucial for mine planning and resource estimation. However, lithology mapping extends beyond just the orebody – mine planners also need to understand the lithology of access tunnels to inform excavation designs.

In most cases a geologist or geotechnical engineer will visits working 'faces' and map structures of interest that correlates the mining excavation with known stratigraphy and where a displacement of reef is observed and structures associated with such displacement (faults, dykes, joints) and structures with potential safety implications are mapped. To meet these requirements, geologists regularly map both tunnels and stopes. The thickness of the reef, faults and dykes are recorded and plotted on mine plans. The Mine Health and Safety Act Chapter 17 (22) prescribes a Geological Plan *'… drawn to a legible scale, depicting geological features that could affect mining, or these features may be shown on the plan(s) referred to in regulation 17(23)'*, plans of the workings showing *'…outlines and dips of the workings; …faults; dykes and water plugs.'* (Department of Mineral Resources of South Africa, 2011; Grobler and Pienaar, 2021).

In the past, this mapping process involved manual tools like tapes, clinorules, and compasses, with geologists recording their findings in wet-strength notebooks using pencils (Figure 1). However, as the industry transitioned to a digital 3D environment, traditional paper maps were replaced with 3D models. This shift necessitated that the underground geological maps be georeferenced and digitised to integrate seamlessly into the 3D digital environment.

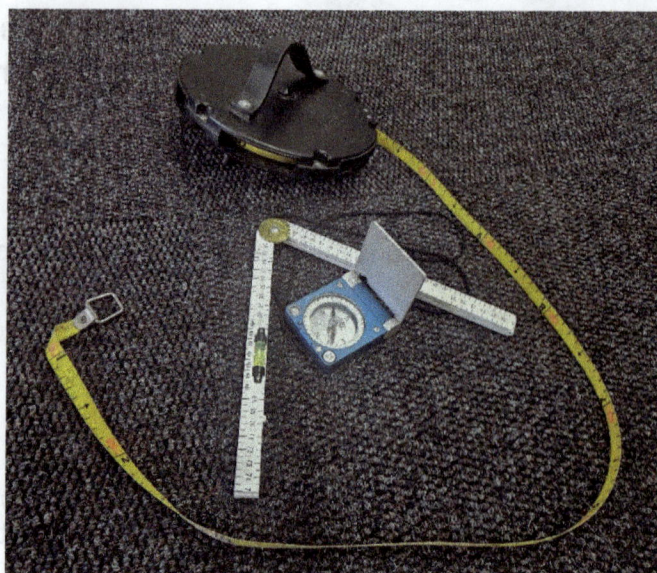

FIG 1 – Tape, clinorule and compass for underground geological mapping.

Mine surveyors have led the way in using 3D laser scanners in underground environments to measure excavation volumes. These advanced scanners provide surveyors with the ability to capture highly accurate scans, which can then be georeferenced using survey pegs.

There are two primary methods for achieving this georeferencing:

1. Pre-scan georeferencing: The 3D laser scanner is mounted on a survey tripod in the place of the total station directly under the survey peg, allowing the scanner's position to be known before the scan begins.

2. Post-scan georeferencing: The scanner captures the area first, and the survey pegs visible in the scan are then used to georeference the data afterward.

These scanners collect millions of data points per second, generating detailed point clouds with colour information that allows for the visualisation of full-colour images. Additionally, built-in cameras capture digital photos during the scanning process, further enhancing the data collected. Despite their many advantages, 3D laser scanners are bulky and come with a high price tag, making them a significant investment for survey teams.

Problem statement

Traditional geological mapping methods are time-consuming, and their 2D outputs, typically paper maps, are not easily integrated into modern digital 3D environments. While 3D laser scanners offer a way to capture geological information digitally, they are often bulky, expensive, and primarily used by mine surveyors to generate detailed scans of excavations. This limited access to digital geological data can create challenges. For instance, at Mponeng Mine, geologists' mapping were not transferred onto the plans used during mine planning sessions where teams collaborated using digitally generated maps (Olivier, 2023).

Research aim and objectives

In 2022, a research project was identified at Mponeng Mine to explore effective methods for digitally capturing geological mapping and integrating it into survey plans for mine planning purposes. The high cost and bulkiness of 3D laser scanners limited their use by the geology department, especially during mapping in the tunnels and narrow, tabular stopes. Since geologists typically work without assistants, carrying a survey tripod along with a bulky 3D laser scanner was impractical (Birch and Olivier, 2022). This initial study developed a technique to use Apple iPad Pro and iPhone Pro devices to capture 3D geological information in narrow, confined ultra-deep stopes. Furthermore, the project proposed workflows to incorporate both geological and production 3D scans information into a mine's block model as well as routine reporting (Olivier, 2023).

To expand on this initial research project, additional testing has been conducted at the Maseve Platinum Mine and in the University of Johannesburg (UJ) mine tunnel. This testing is aimed to qualitatively and quantitatively compare the accuracy and effectiveness of iPad Pro and iPhone Pro devices with dedicated 3D laser scanners for capturing geological information.

LITERATURE REVIEW

Traditional geological mapping methods

Traditionally, geologists have recorded their field observations—including sketches, measurements (such as the angle of tilted strata), and narratives—in field notebooks. While the validity of these observations remains unchanged, digital tools have enhanced the process. Digital photographs now often supplement sketches, and advanced instrumentation improves measurement accuracy. For instance, global positioning system (GPS) instruments provide more precise location data compared to relying solely on topographic and cultural features (US Geological Survey, 2025).

Traditional underground geological mapping involves creating detailed representations of subsurface geological features, often using manual observations, sketches, and measurements within tunnels or mines. This process is crucial for understanding geological structures, identifying potential hazards, and managing underground spaces in mines. However, traditional mapping methods have several shortcomings, including:

- Safety risks: Operating close to the working face exposes geologists to significant safety hazards.

- Environmental challenges: darkness, dust, and confined spaces make accurate observations difficult, especially when production crews are present.

- Time and labour intensive: The process is both time-consuming and requires substantial manual effort.

- High costs: Accurate observations necessitate costly expertise.

- Limited accuracy and detail: Observations are often captured using pencil sketches and field notes, reducing precision.

- Integration difficulties: Manual observations are challenging to incorporate into the digital 3D models commonly used in modern mines.

Overall, traditional underground geological mapping is constrained by its reliance on manual techniques, which are often risky, costly, and less accurate than modern technologies. Contemporary tools such as geographic information systems (GIS), computer-aided design (CAD) software, LiDAR, and virtual reality (VR) visualisation offer safer and more efficient alternatives (Berg *et al,* 2011; Minrom, 2023).

Limitations of traditional geological maps and 2D models

Traditional geological mapping and 2D geological maps have several limitations, with the most critical being their inability to effectively represent the third dimension. These maps typically provide plan views or single-plane cross-sections of geological features. While cross-sections offer some insights into geological structures, they only present information along a specific vertical slice, leaving significant gaps in understanding the complete 3D nature of the geology. This limitation can lead to inaccuracies in geological interpretation and hinder effective decision-making. One attempt to address this issue is through stack-unit maps. However, these maps can quickly become overly complex when deeper or more detailed geological features are included, reducing their practicality and usability (Berg *et al,* 2011).

Printed maps also face limitations due to their static nature, lacking the flexibility to adapt to different scales or provide dynamic interaction with the data. This restricts the ability to zoom in or out and explore geological features at varying levels of detail. The issue is exacerbated by the irregular distribution of geological data, which is particularly challenging in subsurface mapping where data density typically decreases with depth. Traditional 2D maps often struggle to convey geological complexity, especially in structurally intricate regions. They may not effectively visualise critical geological structures such as faulting and folding – features that are essential for effective mine planning and resource estimation. Moreover, modern geological exploration techniques generate significantly more data than in the past, including borehole logs, geophysical surveys, and digital terrain models. Unfortunately, traditional 2D mapping methods are not well-suited to integrate and visualise this multidimensional data, limiting their effectiveness in supporting robust planning and decision-making processes (Berg *et al,* 2011).

Benefits of 3D geological models

3D geological models offer significant advantages, particularly in visualising complex geological structures. Unlike traditional 2D maps, they provide clear and detailed representations of subsurface geology through multiple map views in various formats, enhancing the communication and understanding of intricate geological processes (Berg *et al,* 2011). One of the key benefits of 3D models is their flexibility. They can be easily updated with new data and customised to meet the specific needs of different users. These models integrate diverse geological and geophysical data, enabling more accurate analysis and interpretation of subsurface conditions. They also support advanced modelling techniques, such as geostatistical methods, which are vital for resource assessments. By offering a comprehensive geological perspective, 3D models improve our understanding of geological deposits and serve as effective teaching tools across all educational levels (Birch, 2018). Additionally, they facilitate the prediction of geological features and material distributions in data-sparse regions, benefiting scientific research and practical applications in industries like construction, mining, and environmental management (Berg *et al,* 2011). However, challenges remain, particularly when integrating traditional non-digital geological mapping into digital 3D models, as highlighted by the example of Mponeng Mine (Olivier, 2023).

Laser scanning methods in underground mining

3D laser scanning is transforming underground mining by delivering highly accurate and detailed maps of mine environments, leading to safer and more efficient operations (Davey, 2023). Dusza-Pilarz, Kirej and Jasiołek (2024) examined the integration of laser scanning with 3D software as a powerful tool for improving mining design. Their study focused on the use of laser scanning measurements in both horizontal and vertical headings of copper ore mines in Poland. Laser scanners play a crucial role in inventorying large and hard-to-access chambers or mine workings. They generate detailed point clouds, which are then processed using CAD programs to create precise digital models of mining environments. These digital models support modernisation projects and the planning of new mine workings. The benefits of this technology extend to various aspects of mining operations, including enhanced geological modelling, improved slope stability analysis, optimised ventilation planning, and better equipment management (Dusza-Pilarz, Kirej and Jasiołek, 2024).

The primary laser scanning methods used in underground mining are terrestrial laser scanning (TLS) and mobile laser scanning (MLS). TLS offers exceptional accuracy, achieving precision down to the millimetre level, while MLS provides greater efficiency and mobility for surveying larger areas. The use of simultaneous localisation and mapping (SLAM) algorithms further enhances mobility and mapping accuracy. However, these algorithms can encounter challenges in environments with limited features or highly symmetrical structures (Singh, Banerjee and Simit, 2023).

From a geological and geotechnical perspective, laser scanning technology holds significant potential for tasks such as estimating the geological strength index (GSI), characterising rock masses, mapping discontinuities, and analysing deformations. While current technology effectively supports applications like change detection and structural mapping, further advancements are needed in areas such as lithology identification and autonomous navigation. Safety remains a critical consideration, particularly in coalmines and other safety-sensitive environments. Developing intrinsically safe laser scanning systems is essential to mitigate risks. Additionally, improving georeferencing and data registration techniques, integrating deep learning methods, utilising multi-imaging sensors, and implementing robust spatial referencing approaches – such as ground control tags (GCTs) – could significantly enhance the performance of laser scanning technology in underground mining (Singh, Banerjee and Simit, 2023).

Manual underground geological mapping requires geologists to measure dips and strikes of geological contacts using a compass and clinorule, which is only effective when the geologist is positioned directly at the working face. This traditional method involves significant risks, as it necessitates physical presence in potentially hazardous environments. Laser scanning technology offers a safer and more efficient alternative. It allows geologists and geotechnicians to assess underground areas remotely, minimising the need to be near dangerous working faces. This technology enhances safety by enabling operators to maintain a safe distance while still obtaining precise geological data (Birch and Olivier, 2022).

While laser scanning holds great promise for improving safety, efficiency, and automation in underground mining, addressing technological and infrastructural challenges is crucial for broader adoption and achieving real-time operational capabilities. Integrating georeferencing techniques ensures that laser scans are accurately aligned with the mine's coordinate system, improving the precision of mining operations (Dusza-Pilarz, Kirej and Jasiołek, 2024). However, several challenges remain. Collecting coherent and consistent data can be difficult due to feature-deficient environments, dynamic conditions, and external factors such as dust and water. Additionally, processing large volumes of spatial data requires advanced graphical computing power. With continued advancements, laser scanning technology could become a vital tool in modernising underground mining practices, driving progress toward safer and more efficient operations (Singh, Banerjee and Simit, 2023).

Using iPad Pro and iPhone Pro LiDAR scanning for mining and geoscience surveys

Traditional topographic surveying for geosciences often involves high costs, complex logistics, and extensive training requirements. While drones equipped with optical sensors have helped reduce expenses, surveying remains a challenging and costly endeavour. The integration of LiDAR sensors into consumer devices like the iPad Pro 2020 and iPhone 12 Pro offers a promising, low-cost alternative for mining and geoscientific applications (Luetzenburg, Kroon and Bjørk, 2021).

Apple's LiDAR sensor uses direct time-of-flight (dToF) technology with single photon avalanche diode (SPAD) sensors and vertical-cavity surface-emitting laser (VCSEL) lasers to create detailed depth maps. This method calculates depth by measuring the time it takes for light pulses to bounce back from a target. In normal mode, 576 points are used, while power-saving mode reduces it to 144 points. The iPad's dToF sensor has a 60° × 48° field of view (shown in Figure 2). Apple's LiDAR fuses data from other sensors (camera, accelerometer, gyroscope and magnetometer) to improve depth mapping (Marc, 2021).

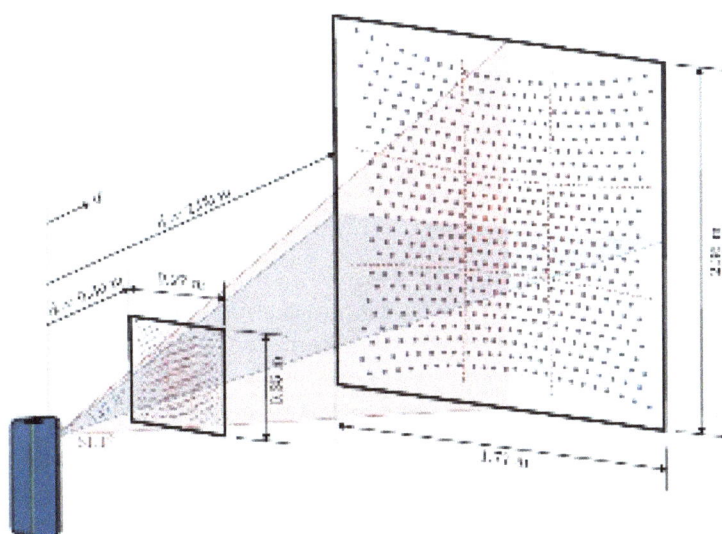

FIG 2 – Apple iPhone Pro LiDAR sensor field of view (Tondo, Riley and Morgenthal, 2023).

Treccani, Adami, and Fregonese (2024) evaluated the effectiveness of five iOS LiDAR-based apps for surveying medium-sized indoor and outdoor environments using a second-generation iPad Pro.

The study focused on comparing the accuracy, precision, usability, and practicality of these apps against traditional TLS methods. The testing was conducted in two distinct environments: Indoor: A university corridor measuring 40 m × 2 m; Outdoor: A narrow street in the historical centre of Mantua, Italy, stretching 72 m.

The five iOS apps assessed included: 3D Scanner App; Dot3D; Polycam; RTAB-Map; and Scaniverse.

Two survey path strategies were tested: a closed loop and a zigzag path. The accuracy of the apps was measured against a reference data set created using a TLS Leica RTC360.

The study found that apps utilising loop-closure algorithms, such as Dot3D and RTAB-Map, with marker assistance, delivered the best performance – especially when following the closed-loop path. Dot3D (designed for commercial use) and RTAB-Map (research-oriented) exhibited the smallest deviations from the TLS data, achieving mean error values of 4 to 7 cm with standard deviations of 4 to 8 cm. In contrast, 3D Scanner App, Polycam, and Scaniverse performed better with zigzag paths, but displayed higher error margins overall (Treccani, Adami and Fregonese, 2024).

Recent studies have explored the efficiency and accuracy of Apple LiDAR-equipped devices in challenging underground conditions for both production and geoscientific purposes (Świerczyńska, Kurdek and Jankowska, 2024). One such study conducted at the Kłodawa Salt Mine in Poland compared the performance of the Leica Disto laser rangefinder, the Leica RTC360 3D laser scanner, and the iPad Pro using the 3D Scanner app. The study concluded that while the iPad Pro was not suitable for high-precision tasks in salt mine conditions, it could still be valuable for preliminary or non-critical measurements (Świerczyńska, Kurdek and Jankowska, 2024). This finding suggests that consumer-grade LiDAR technology, despite its limitations, may offer a cost-effective solution geoscientific application, particularly in situations where high precision is not required.

Rutkowski (2023) evaluated the performance of the iPhone 13 Pro's LiDAR sensor in a surface model of a mine shaft. The study found that the sensor has a maximum effective range of 5 m and is well suited for scanning large objects that do not require high precision, offering an accuracy of approximately ±10 cm.

Building on this research, Ordóñez et al (2024) examined the potential of the iPad Pro 11 inch with a LiDAR sensor for geospatial surveying in Ecuador's artisanal and small-scale mining (ASM) sector. The study was conducted at 'La Zamorana', an underground gold mine, comparing the LiDAR sensor's performance against traditional surveying methods. The primary objective was to assess both accuracy and practicality in real-world mining conditions.

The results indicated that the iPad Pro's LiDAR sensor outperformed traditional surveying methods, particularly in delivering detailed and accurate elevation measurements. Most discrepancies were within 5 cm, highlighting the sensor's precision. Additionally, the LiDAR-based approach demonstrated superior alignment with control points compared to traditional techniques, emphasizing its potential to enhance geospatial data collection in the ASM sector (Ordóñez et al, 2024). The study followed a structured approach across five phases: Establishing control points; Conducting traditional surveys; Conducting LiDAR surveys; Post-processing data; and Evaluating ASM's readiness for technological adoption.

The results showed that 87.5 per cent of the iPad Pro's measurements were within 0.7 m of the control points, significantly outperforming the traditional compass and tape methods, which achieved only 37.5 per cent accuracy. The LiDAR sensor also demonstrated superior elevation accuracy, with 75 per cent of its measurements averaging just 3 cm higher than the control points. Moreover, the iPad Pro produced detailed 3D models in only 50 mins, compared to the four days required by traditional total station methods, underscoring its remarkable efficiency (Ordóñez et al, 2024).

The study concluded that the iPad Pro's LiDAR sensor provides a cost-effective, efficient, and accurate alternative to traditional surveying methods in artisanal and small-scale mining (ASM) environments. The strong positive feedback received highlighted the ASM sector's readiness to adopt advanced technologies. However, the authors recommended further research to assess the technology's effectiveness across diverse mining environments to validate its broader applicability. Overall, the research underscores the transformative potential of consumer-grade LiDAR technology in modernising geospatial-surveying practices within the ASM sector. It suggests that adopting this technology could enhance safety, operational efficiency, and data accuracy, contributing significantly to the industry's modernisation efforts (Ordóñez *et al,* 2024).

Rutkowski and Lipecki (2023) investigated the potential of the low-cost iPhone 13 Pro LiDAR scanner for inspections during the mineshaft sinking process. The study aimed to evaluate the accuracy and practicality of 3D models generated by the smartphone scanner compared to TLS methods. The research took place at the GG-1 ventilation shaft in Kwielice, Poland, at depths ranging from 1320–1350 m. Measurements were recorded at seven concrete lining intervals before the final lining was installed. Mineshaft sinking involves complex tasks such as drilling, blasting, mucking, concrete placement, and infrastructure installation, all of which require regular inspection and measurement. Traditionally, these measurements are performed manually using mechanical verticals and tape measures – methods that, while reliable, offer room for improvement (Rutkowski and Lipecki, 2023).

The iPhone 13 Pro LiDAR scanner, paired with the PolyCam app, provided an intuitive and efficient scanning method. Initial surface tests showed a geometric error of no more than 2 cm when scanning a simple steel cube. In underground conditions, the device maintained performance, capturing over 2.7 million points per scan and generating detailed 3D models with reasonable accuracy. The optimal scanning distance was between 0.5 and 2 m from the rock face (Rutkowski and Lipecki, 2023).

The study highlighted several practical applications of the LiDAR scanner in mining operations, including: daily inspections and shift handovers; bolting pattern pre-inspections and safety assessments; and quick geometry control and concrete volume calculations.

For instance, volume calculations using the LiDAR scanner showed discrepancies of less than 2 m^3 compared to actual concrete volumes, demonstrating its potential for resource management and cost savings. When compared to a professional FARO FOCUS 130 scanner, the iPhone 13 Pro exhibited a lower accuracy range (5–15 cm versus 3 mm). However, it offered substantial advantages in cost (approximately $1000 compared to $34 000) and durability in harsh underground environments. The iPhone's IP68 rating for water and dust resistance enhanced its robustness compared to more delicate professional equipment (Rutkowski and Lipecki, 2023).

While the iPhone 13 Pro LiDAR scanner cannot fully replace high-precision TLS devices, the authors concluded that it provides a viable, cost-effective solution for tasks where operational efficiency is more important than extreme accuracy. The scanner could be particularly useful for quick, agile measurements throughout the mineshaft sinking process. Additionally, the technology could enhance safety protocols by creating 3D models of accident sites and visualising hazardous areas. Overall, the study underscores the growing potential of consumer-grade LiDAR technology in industrial applications. It suggests that devices like the iPhone 13 Pro could play a pivotal role in modernising mining operations, enhancing safety, and reducing costs (Rutkowski and Lipecki, 2023).

Luetzenburg, Kroon, and Bjørk (2021) assessed the technical capabilities of Apple's LiDAR sensor on the iPhone 12 Pro by comparing its performance against traditional structure-from-motion multi-view stereo (SfM MVS) point clouds. The study aimed to evaluate the sensor's accuracy and precision under both controlled and field conditions. Small objects were scanned to test precision, and a large-scale model of a coastal cliff at Roneklint, Denmark, was generated using the '3D Scanner App' on the iPhone. The resulting LiDAR models were compared to SfM MVS reference models using the multi-scale model-to-model cloud comparison (M3C2) method in CloudCompare software (Luetzenburg, Kroon and Bjørk, 2021). CloudCompare is 3D point cloud and mesh processing software Open Source Project.

The LiDAR sensor demonstrated high accuracy for small objects with dimensions greater than 10 cm, achieving an absolute accuracy of ±1 cm. For the large-scale coastal cliff model (130 × 15 × 10 m), the sensor maintained an accuracy of ±10 cm. The device efficiently produced realistic 3D models within 15 mins, displaying strong performance on flat surfaces and un-vegetated cliff areas. The average M3C2 distance between the LiDAR and reference clouds was -0.11 m, with most discrepancies within 15 cm. The sensor exhibited high precision, with 92 per cent of points showing a mean distance of less than 5 cm between repeated scans (Luetzenburg, Kroon and Bjørk, 2021).

The authors concluded that the iPhone 12 Pro's LiDAR sensor offers promising potential for geoscientific applications, providing a practical tool for high-resolution topographic modelling. While it does not yet match the accuracy of state-of-the-art SfM MVS standards, it offers notable advantages in accessibility, ease of use, and integrated data processing. With continued advancements in software and hardware, consumer-grade LiDAR devices could play a more prominent role in scientific and educational settings (Luetzenburg, Kroon and Bjørk, 2021).

These findings of these various studies focused on mining and the geosciences indicate that consumer-grade LiDAR technology, such as the sensors in iPhone and iPad devices, could significantly contribute to modernising surveying practices in the mining industry. This technology is particularly valuable in environments where traditional methods are impractical or too costly.

METHODOLOGY

Study design

The study incorporates both qualitative and quantitative components. Qualitative approaches are used to compare the outputs of various mapping and scanning techniques, assessing whether the iPad Pro and iPhone Pro scans are suitable for their intended purposes. A quantitative approach is employed to evaluate the accuracy of the iPad Pro and iPhone Pro scans in comparison to those obtained from a TLS, as well as comparisons with surveyed ground control points.

Site selection

Several sites were used in this study. Initial testing of the iPad Pro was conducted at the Sibanye Stillwater Digital Mining Laboratory (DigiMine) tunnel and stope to develop a technique suitable for underground scanning. To assess volume and distance accuracy in a controlled environment, closed offices were scanned with the iPad Pro and compared to those obtained using a laser distometer. Underground scanning was then performed in the ultra-deep stopes of Mponeng Gold Mine, followed by additional underground scanning at Maseve Mine.

Finally, scanning was conducted using an MLS and two TLSs, as well as the iPad Pro and iPhone Pro, in the simulated mine tunnel at UJ. These tests were carried out to measure the time required for different scanning approaches and to compare their cost, efficiency and accuracy with traditional geological mapping methods.

Data collection

The data collected at the DigiMine and Mponeng Mine was primarily focused on developing a method for scanning using an iPad Pro and, later, an iPhone Pro. Various scanning apps were tested to determine which is most suitable for capturing scans and integrating them into software packages used for geological analysis, such as Leapfrog Geo or Deswick.CAD.

Scanning conducted at Maseve Mine compared and contrasted the quality of scans obtained using the iPhone Pro and iPad Pro LiDAR scanners. This process was repeated in the UJ mine tunnel, where different scanning techniques and settings in the 3D Scanning App was also be evaluated.

The scanning exercise in the UJ mine tunnel assessed the time required to map the geology using traditional methods versus the various scanning technologies. The simulated geology represents 18 m of Ventersdorp Contact Reef (VCR) as well as Merensky Reef lithology. The scanners tested include the iPad Pro, iPhone Pro, two TLSs, and an MLS.

Seven ground control points have been installed in the tunnel, with their coordinates determined to millimetre accuracy using a Leica MS50 MultiStation. These points are marked with a black-and-white pattern, making them easily identifiable in the scans. Three points are located on the western side of the tunnel, three on the eastern side, and one on the tunnel roof.

Data analysis

The initial study conducted at the DigiMine and Mponeng Mine was a qualitative investigation aimed at establishing a method for scanning mine stopes and tunnels using an iPad Pro. Various iOS apps were tested, and the most suitable one for scanning in the ultra-deep-level VCR mine stopes at

Mponeng Mine was identified. Additionally, an optimal technique for scanning in highly constricted stope environments was developed. The benefits and limitations of the scanning technique and equipment were evaluated, and a workflow was designed to integrate iPad Pro and iPhone Pro scans into Mponeng Mine's digital databases for use in the mine design software, Deswick.CAD.

The study at Maseve Platinum Mine focused on assessing the effectiveness of 3D laser scanning for geological mapping. A Zoller and Fröhlich (Z&F) 5010X TLS was tested on a Merensky Reef pillar. The iPad Pro and iPhone Pro were also tested in the same pillar to evaluate the quality of the geological information they provided in comparison to the Z&F 5010X scanner. These tests did not assess the time efficiency of 3D scanning versus traditional geological mapping, and no comparative time studies were conducted. The findings demonstrated that LiDAR technology offers significant advantages for geological mapping, including the ability to capture detailed 3D point clouds, high-resolution images, and thermal data. These capabilities enhance the understanding of geological structures and potential hazards. Additionally, the study confirmed that the strikes and dips of geological features could be derived directly from the LiDAR point cloud and incorporated into a 3D geological model (Grobler and Pienaar, 2021).

The third part of the study took place in the UJ mine tunnel and compared traditional geological mapping techniques (tape, compass, and clinorule) with iPad Pro and iPhone Pro scanning using the 3D Scanner App, along with an MLS and two TLS dedicated 3D scanners. These scanners included:

- FjD Trion P1 – a handheld (mobile) LiDAR scanner equipped with a scanner and a colour image-capturing device.

- Riegl VZ600 – a terrestrial (tripod-mounted) LiDAR scanner.

- Z&F 5010X – a terrestrial (tripod-mounted) LiDAR scanner.

This study evaluated both the time required to capture geological information and the quality of the resulting output. Additionally, equipment costs were compared. The scanning accuracy of the iPad Pro and iPhone Pro was quantitatively assessed using total station survey points, as well as through visual comparisons with the Z&F 5010 scanner. The scanning outputs were analysed using CloudCompare and Hexigon Cyclone 3DR software.

RESULTS

Developing and testing underground scanning using Apple LiDAR

An iPad Pro 11 inch (2nd generation) was tested at the DigiMine in 2021. This test provided valuable insights into its potential as a 3D LiDAR scanning device for mining applications. The primary goal was to evaluate the device's accuracy, usability, and practicality for capturing geological data and supporting production activities in conditions that closely resemble real underground mining environments. The DigiMine offers a controlled yet realistic setting, featuring narrow tabular stopes and simulated Bushveld Complex geological formations, making it an ideal testing ground. The tests involved scanning both the tunnel cross-cut and the stope panel, allowing for a comprehensive assessment of the device's performance across various underground scenarios. Four 3D scanning applications were tested: Scaniverse (Niantic, Inc.); 3D Scanner App (Laan Labs); Polycam; and Sitescape (FARO Solution).

These applications were selected to compare their effectiveness in generating accurate and practical 3D scans of geological features.

The findings indicated that the iPad Pro is highly user-friendly, requiring minimal training for operators to initiate and manage scans efficiently. The touch interface and intuitive design of the applications contributed to a smooth user experience. In terms of scanning performance, the device produced accurate 3D scans within an acceptable error margin, particularly when using the Sitescape application. This app excelled at capturing detailed geological features of the stope panel. Scaniverse and 3D Scanner App also performed well, providing quick photometric scans. It was found that the 3D Scanner App was the most flexible for the number of file formats it was able to output, and there are no restrictions to the size of the scans. Considering the typical lengths of the narrow, tabular stopes in South Africa, this aspect was considered critical and thus the 3D Scanner App was selected as being the most suitable for the later tests conducted at the Mponeng Mine. The scans captured in the DigiMine stope and tunnel are shown in Figures 3 and 4.

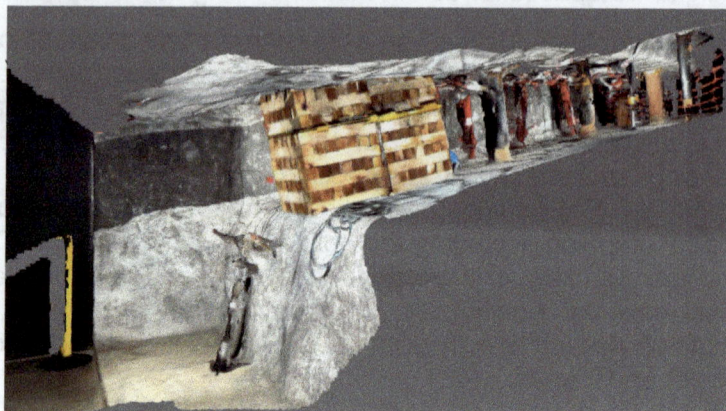

FIG 3 – DigiMine stope (scanned with iPad Pro using 3D Scanner App) (Birch and Olivier, 2022).

FIG 4 – DigiMine tunnel (scanned with iPad Pro using 3D Scanner App) (Birch and Olivier, 2022).

The geo-referencing process using CloudCompare was successful, with scans then seamlessly imported into Deswick.CAD software. Georeferencing directly in the Deswick.CAD program is also possible. This integration enabled precise tracing of geological features and enhanced the development of a 3D geological model. Such capability is crucial for translating raw scan data into actionable insights for mine planning and production control. Additionally, the testing demonstrated significant time efficiency. The iPad Pro allowed for rapid capture of geological features within minutes, greatly improving productivity over traditional mapping methods and facilitating more frequent updates to geological data.

However, some challenges were noted during testing. While the iPad Pro performed well in the controlled DigiMine environment, its size and weight could pose issues in tighter underground spaces. More critically, the device's lack of inherent dust and water resistance raises concerns about its durability in harsh mining conditions. Without adequate protective measures, the iPad Pro may not withstand prolonged exposure to dust, moisture, and potential physical impacts underground. The low-light conditions in the tunnel and stope when the lights were turned off (to simulate the expected underground lighting conditions) also meant the scanning photogrammetry was ineffective. To remedy this and ensure that the scanning surfaces were properly illuminated, a suitable LED light was. A 3D printed holder for the iPad Pro was designed using Shapr3D to hold the light and provide some protection to the device (Figure 5).

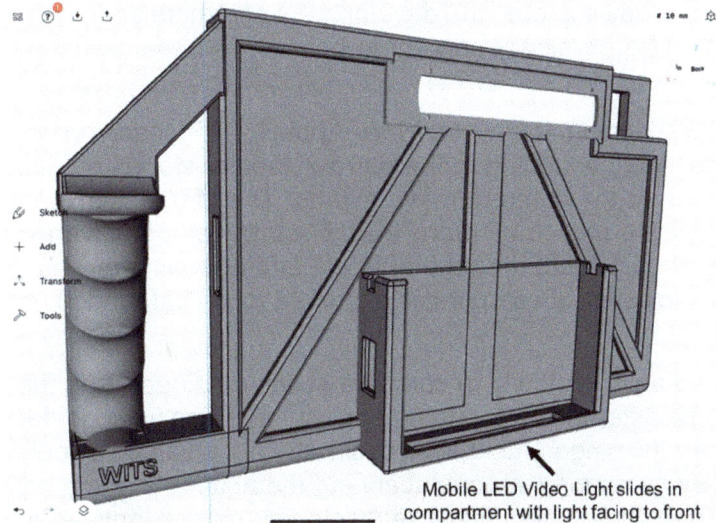

FIG 5 – Apple iPad Pro 11 inch holder designed using Shapr3D software (Birch and Olivier, 2022).

Following the initial testing conducted in the DigiMine, additional testing took place in an active underground production stope at Mponeng Mine, renowned for its ultra-deep mining operations and complex geological formations. The mine's narrow tabular stopes and harsh conditions provided a challenging yet realistic environment for evaluating the iPad Pro. The primary objective was to capture 3D scans of the stope panels using the 3D Scanner App. This app allows the operator to pause scanning, enabling them to move and resume the scan from a different position. In the narrow confines of the Mponeng VCR stopes, this feature was crucial, as movement requires crawling on hands and knees – making continuous scanning impossible.

Two tests were conducted. In the first test, scans were performed in a full 360° manner, capturing both the stope faces and the backfill support. In the second test, the scanning focused solely on the mining face. Both test scans started in the upper gully and ended in the lower gully, with survey pegs clearly marked to ensure visibility in the scans. This marking was essential for geo-referencing the scans when importing them into subsequent image processing software.

The iPad Pro demonstrated strong performance in scanning accuracy, capturing geological features with a point cloud resolution of approximately 50 mm. This represents the lowest scanning resolution, which results in faster data capture and processing times. The resolution is considered suitable for geological mapping, as the app also collects photographs that are draped over the surface during mesh processing. The captured scans were seamlessly imported into Deswick.CAD software, allowing for precise geo-referencing of geological features and enhancing the 3D geological block model used in mine planning. A significant advantage of the iPad Pro was its time efficiency. The scanning process was considerably faster than traditional geological mapping methods, reducing the time required for data capture. This speed is particularly valuable in underground environments, where minimising time at the stope face is crucial for both safety and productivity.

However, the study identified limitations, particularly in terms of durability. The iPad Pro's size and fragility posed challenges in the confined spaces of the mine. Its lack of ruggedisation raised concerns about its resilience to dust, moisture, and physical impacts. Even with a robust 3D-printed protective case, the device did not appear to be adequately shielded for the harsh conditions of the Mponeng stopes.

Testing TLS, iPad Pro and iPhone Pro at Maseve Mine

The TSL scanning study conducted by Grobler and Pienaar (2021) at Maseve Mine primarily focused on identifying the benefits of TLS scanners and integrating this data with other sources, such as thermal imaging measurements. Although thermal imaging was found to be ineffective in detecting geological features due to the stabilisation of rock temperatures in underground environments, the study opened the door to exploring other applications of LiDAR point cloud data. This included the potential use of artificial intelligence for the automated identification of reef contacts and structural features. Additionally, the study explored the development of remote analysis techniques to minimise the need for geologists to be physically present in hazardous underground environments, thereby improving safety (Grobler and Pienaar, 2021).

During the testing of TLS scanning at Maseve Mine, Apple LiDAR scanning was also evaluated. This complemented previous tests conducted in the narrow, tabular stopes of Mponeng Mine, providing valuable comparisons due to the different mine layouts. The Merensky Reef pillar where the tests were conducted resembled a reef drive more than a narrow stope. Given the higher, more open spaces left after mining, the limited 5 m range of Apple LiDAR scanners was initially expected to be a significant drawback. However, these concerns proved to be unfounded, as the areas could still be effectively scanned.

This testing also provided an opportunity to compare scans obtained using the iPad Pro and iPhone Pro, as well as to evaluate different 3D scanning applications. The findings demonstrated that Apple LiDAR scanning, despite its range limitations, could still be a valuable tool for certain geological mapping applications. By comparing various scanning methods, the study contributed to a broader understanding of how different Apple LiDAR technologies can be applied in underground mining environments.

Comparison between traditional geological mapping, MLS, TLS and Apple LiDAR Scanning

Traditional geological mapping

The final phase of testing took place in the UJ mine tunnel, which features 18 m of simulated Witwatersrand Ventersdorp Group conglomerates and lava (the VCR), along with Bushveld Complex Merensky Reef geology, including norite, pegmatoidal pyroxenite, and chromite, represented on the walls. The tunnel also contains a fault and a dyke. Although the Witwatersrand Supergroup and Bushveld Complex are not naturally found in direct contact as depicted in the tunnel, they represent the typical geology that students will encounter when entering the South African mining industry.

Traditional geological mapping in a mine cross-cut or reef drive tunnel involves the use of tapes, clinorules, and a compass (Figure 1). Additional equipment, such as a line level and string, is required to determine the true strike of structures (Figure 6).

FIG 6 – Line level (Chamberlain, 2025).

The geologist lays out the tape along the side of the tunnel. At regular intervals – typically every 2 m in a reef drive tunnel or incline raise – profiles are measured using a clinorule and recorded in a notebook. Generally, only one tunnel wall is mapped (the east sidewall in the case of UJ). Any additional structures, such as a fault or a dyke (as seen in the UJ mine tunnel), are noted, and their dip and strike are determined using a clinorule or compass inclinometer. The true strike is established by using a line level with a string stretched either across the tunnel or by observing the intersection of the structure with the grade line. A second person is needed to hold the string, but only if no grade line is present, as is the case in the UJ mine tunnel.

This process is slow and further complicated by the normal flow of people and equipment through mine tunnels. Mapping the 18 m of geology represented in the UJ mine tunnel using this method took 35 mins, even under ideal, controlled conditions. The collected data then had to be manually plotted onto graph paper at the surface for record-keeping, with the various structures transferred onto the mine survey plans. Plotting the geology for the 18 m tunnel took approximately an hour (see Figure 7).

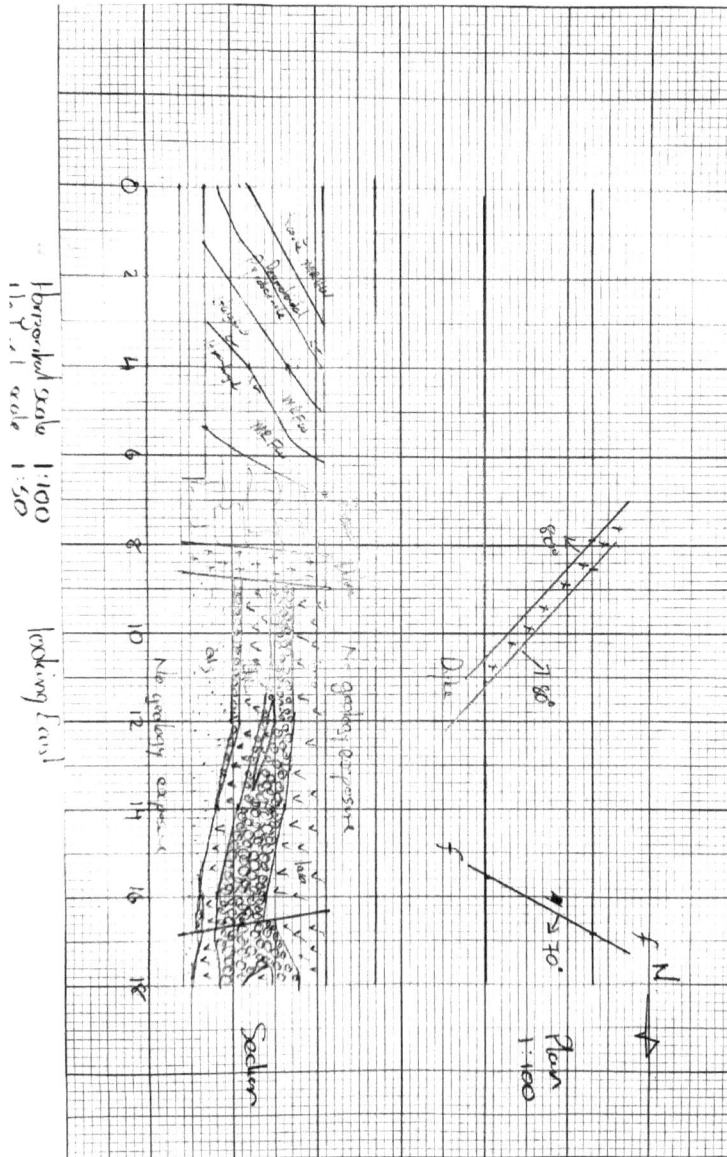

FIG 7 – Manual plotting of the tunnel geological mapping exercise.

The geological compass used for the mapping exercise, a basic Breithaupt Kassel Gekom Pro Compass, currently costs around US$600, while certain specialised Breithaupt Kassel models can reach up to US$1500. The classic Brunton Pocket Transit Geo Compass is priced at approximately US$780. Although inexpensive geological mapping compasses, often Brunton clones, are available for as little as US$20, they are unlikely to withstand the harsh underground conditions. In modern practice, geologists frequently take digital cameras underground to document specific observations. However, these cameras must first be approved for use in the mine.

MLS and TLS scanning

Three different laser scanners were used to scan the tunnel.

The MLS FjD Trion P1 is a handheld scanner that can also be mounted on a backpack or vehicle. It performs continuous scanning using SLAM algorithms. On the surface, automatic georeferencing is achieved through the Global Navigation Satellite System (GNSS). However, since GNSS is unavailable underground, georeferencing must be done using ground control points. Scanning the 18 m tunnel took approximately 4 mins, and the estimated cost of the FjD Trion P1 is around US$15 000.

The TLS Riegl VZ600 is mounted on a lightweight camera tripod. It scans very quickly while simultaneously capturing images using a Sony α7 full-frame camera. The set-up between scans is efficient, as the unit only needs to be roughly levelled. For the scanning exercise in the UJ mine tunnel, six set-ups were used. While automatic GNSS georeferencing is possible on the surface, underground scanning requires ground control points. The files generated are very large, necessitating a computer with a powerful graphics card to open and view the scans. Scanning the tunnel took approximately 7 mins, and the estimated cost of the Riegl VZ600 is US$70 000.

The TLS Z&F 5010X must be mounted on a survey tripod and precisely levelled before scanning. In this exercise, four scans were conducted, and the scanning process was noticeably slower compared to the newer Riegl VZ600. However, the Z&F 5010X produced more precise results, with less scatter in the resulting point clouds. As with the other scanners, GNSS georeferencing is possible on the surface but not underground. Scanning the 18 m tunnel took 17 mins, and the estimated cost of the scanner is US$45 000.

Apple LiDAR scanning

An iPad Pro and iPhone Pro were used to test scanning efficiency and accuracy in the UJ mine tunnel. Due to concerns raised during underground testing at Mponeng Mine regarding the iPad Pro's vulnerability to damage, the iPhone Pro is considered a more suitable option. The iPhone Pro offers similar LiDAR scanning capabilities but is a smaller, more rugged device, making it more portable and durable for underground 3D scanning. These attributes address many of the challenges encountered with the iPad Pro. However, it is important to note that cell phones are not permitted in South African underground fiery mines. The iPad Pro tested at Mponeng Mine was a non-cellular model, which was allowed underground without restrictions.

Both devices were tested using the 3D Scanner App, but the scanning technique and settings differed significantly.

iPad Pro scanning method

For testing the iPad Pro, a video light was attached to illuminate the scanned area. The 3D Scanner App settings were adjusted to the lowest available scanning resolution (using the app's advanced settings) to speed up data acquisition and processing.

The scanning process was highly systematic:

- The operator stood still while scanning, rotating 360° before pausing the scan.
- They then moved forward a few metres, ensuring sufficient overlap with the previous scan.
- The scanning motion resembled spray painting, with slow, consistent vertical movements.
- This process was repeated throughout the entire 18 m tunnel.

Using this method, the 18 m tunnel was scanned in 4 mins. Since non-cellular iPad Pro models lack GPS, the scans required georeferencing using ground control points. While this is not an issue underground, it is a consideration for surface LiDAR scanning.

iPhone Pro scanning method

For testing the iPhone Pro, scanning was conducted using only the available tunnel lighting and the 3D Scanner App was left in its default settings. Unlike the segmented approach used with the iPad Pro, scanning was performed as a single, continuous process.

Key considerations included:

- Ensuring full surface coverage with sufficient resolution to capture details.
- Monitoring the triangulated mesh surface size to confirm scanning accuracy.
- Avoiding duplicate scans, which can cause ghost images in the processed point cloud.

Using this method, scanning the 18 m tunnel took 7 mins. Since iPhones are equipped with GPS, surface scans would be automatically georeferenced. However, the GPS data is recorded in longitude, latitude, and altitude (EPSG:4326 coordinate system), which may not be compatible with the mine's survey system, requiring coordinate conversions.

3D Scanner App output

The 3D Scanner App can process meshes directly on the device and export scans in various 3D shape formats, including: OBJ; GLTF; GLB; and STL.

The OBJ file format is commonly used to transfer meshed shapes into 3D software like Leapfrog Geo. It was observed that colour information may be lost during this process, making meshed shapes unsuitable for integrating geology into 3D geological modelling software. The various options should be tested to see which gives the best results for the modelling software and how the files should be imported to retail the necessary information. The colour is retained when viewing meshed shapes in software like Microsoft 3D Viewer.

The app can also export scans as point clouds in various formats, including: XYZ colour; PLY; PTS; LAS; LAS Geo-referenced; E57; PCD; XYZ colour, space-delimited; and XYZ (no colour).

E57 and LAS point clouds are commonly used for importing laser scans into software such as CloudCompare or Hexigon Cyclone 3DR, both of which were used in this research.

The quality of the Apple LiDAR scans can be viewed in Figure 8.

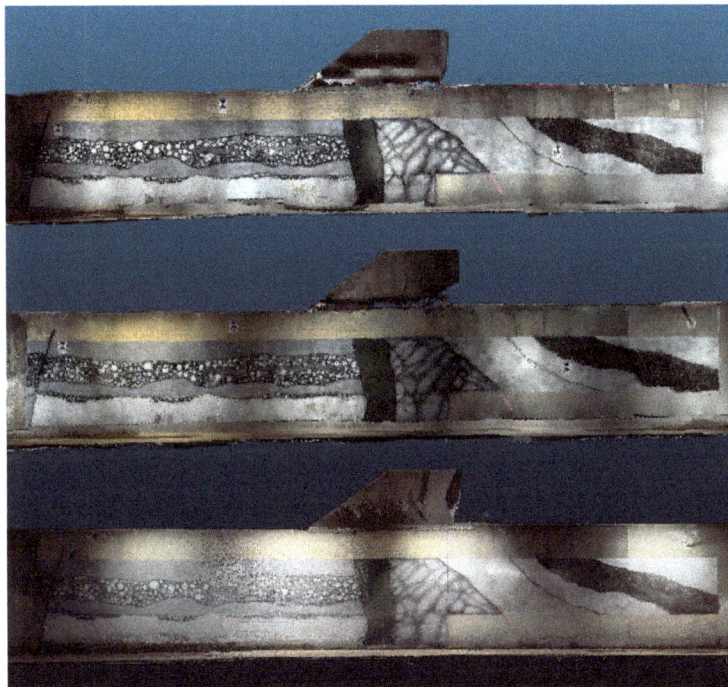

FIG 8 – Apple LiDAR scan quality compared to MLS FjD Trion P1.

The top figure is the E57 point cloud from the iPad Pro (with a video light illuminating the scanned surfaces), the middle figure is the E57 point cloud from the iPhone Pro (using available light). The bottom E57 point cloud is obtained using the MLS FjD Trion P1. It is clear that the bottom image has a higher resolution resulting in a sharper image. However, for the purposes of geological modelling, the two Apple LiDAR E57 point clouds can be clearly digitised in the geological modelling software. If a clearer image is required, the GLTF mesh from 3D Scanner App can be used. However, but it is not able to be incorporated in the geological modelling software used by the university (Leapfrog Geo) with the colour visible.

Accuracy of Apple LiDAR scanning

The accuracy of iPad Pro and iPhone Pro scans was tested in two ways:

1. Distance measurement comparison. The distances between seven surveyed ground control points installed in the tunnel were measured in the LAS or E57 point clouds. These

measurements were then compared to the actual distances between the control point coordinates, which were obtained using a Leica MS50 MultiStation with sub-millimetre accuracy.

2. Point cloud georeferencing and comparison. The LAS point clouds were georeferenced in Hexigon Cyclone 3DR. The cloud comparison tools were then used to compare the Apple LiDAR scans with those obtained from the Z&F 5010X TLS.

Distance measurement comparison

The distances between two ground control survey points (BW02 and BW25) are measured in the GLTF mesh (in the case of the iPad Pro) or E57 point cloud (in the case of the iPhone Pro) to the other ground control survey points (BW11, BW12, BW14, BW15, BW19, and BW25). BW14 was placed on the roof of the tunnel and was not visible in the mesh or point cloud (Figure 9).

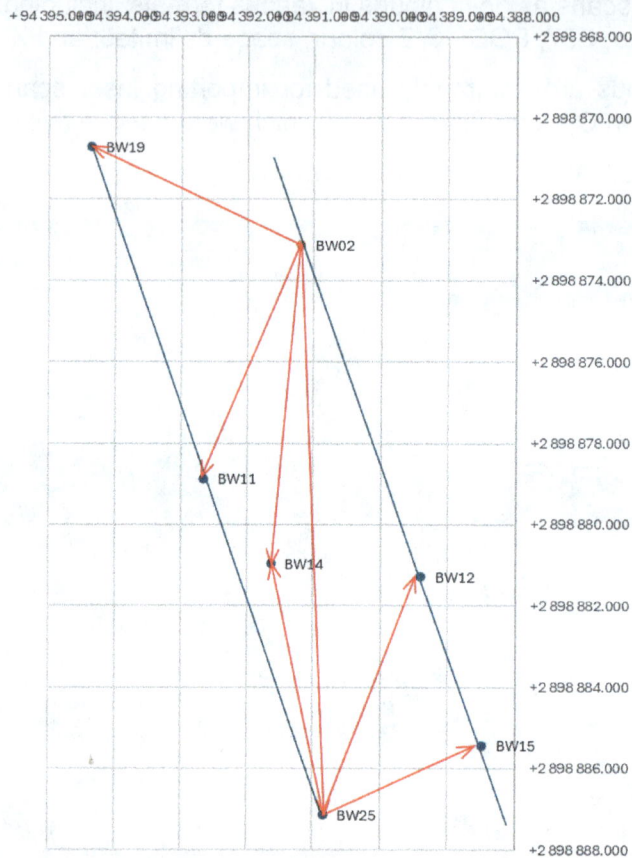

FIG 9 – Position of the ground control survey points in the UJ mine tunnel.

When measuring using the E57 point cloud, the nearest point is not always in the middle of the survey marker. This is not a factor when conducting the measurements using the GLTF mesh. However, the GLTF mesh cannot be opened in geological modelling software like Leapfrog Geo showing the colours. Table 1 shows the distances measured using the Leica MS50 MultiStation with those obtained using the iPad Pro and iPhone Pro.

TABLE 1

Distance differences between ground control points.

From	BW02					
	Leica MS50	iPad GLTF	Difference		iPhone E57	Difference
BW25	14.021	13.889	0.132		14.076	-0.055
BW11	5.938	5.996	-0.058		5.949	-0.011
BW12	8.349	8.318	0.031		8.346	0.003
BW15	12.607	12.607	0.000		12.479	0.128
BW19	3.983	4.034	-0.051		3.996	-0.013
Average difference			**0.011**			**0.010**

From	BW25					
	Leica MS50	iPad GLTF	Difference		iPhone E57	Difference
BW25	14.021	14.048	-0.027		13.935	0.086
BW11	8.459	8.509	-0.050		8.363	0.096
BW12	6.039	6.098	-0.059		5.949	0.090
BW15	2.906	2.895	0.011		2.865	0.041
BW19	16.798	16.939	-0.141		16.782	0.016
Average difference			**-0.053**			**0.066**

It can be observed that the average differences are all below 7 cm. For a surveys for engineering purposes, these errors would be unacceptable. However, for the purposes of geological mapping and modelling, this margin of error would be more than acceptable especially when compared to traditional tape and clinorule mapping.

Point cloud georeferencing and comparison

Hexigon Cyclone 3DR has the function to compare two point clouds or a point cloud and mesh. The two files must be georeferenced (using the ground control survey points) or registered to each other. The TLS Z&F 5010X E57 point cloud has been used as the base reference and it is compared to the iPad Pro GLTF mesh (Figure 10), as well as the iPhone Pro E57 point cloud (Figure 11).

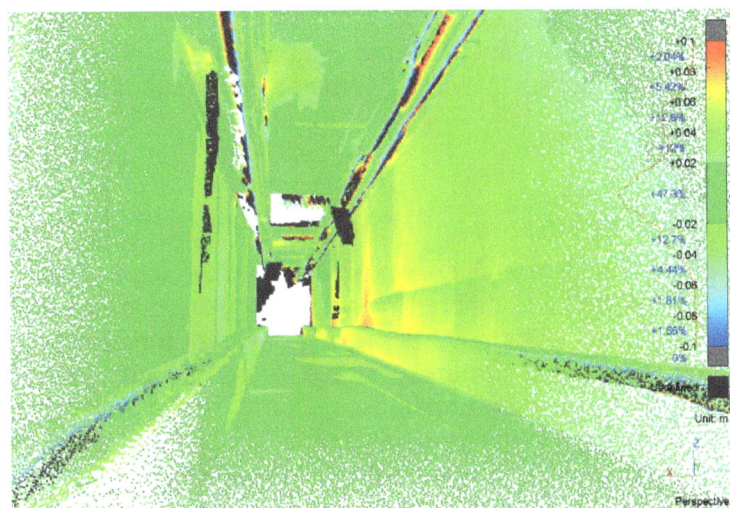

FIG 10 – iPad Pro comparison to reference scan.

FIG 11 – iPhone Pro comparison to reference scan.

In Figure 10, it can be observed that the differences between the TLS Z&F 5010X E57 point cloud and the iPad Pro GLTF mesh ranges between -10 cm and +10 cm, with 47 per cent of the of the points falling within ±2 cm.

The E57 point cloud from the iPhone Pro shows similar differences compared to the TLS Z&F 5010X E57 point cloud. 28 per cent of the points are within ± 2 cm of the reference scan with the same overall spread of ±10 cm.

These results fit in with the accuracies noted by other authors for scanning outcrops with Apple LiDAR devices for geological information (Luetzenburg, Kroon and Bjørk, 2021). They are considered far superior to the traditional methods of geological modelling, especially in underground stopes where little consideration is given to face shapes that are often assumed to be straight.

Apple LiDAR equipped devices are available from about US$1000. A rechargeable video light is about US$100. For the cost of a high-end geological compass, the geologist can be equipped with a LiDAR scanning device that would enable faster and more accurate underground mapping. The mine's protocol regarding cell phones underground may require the adaption of iPad Pros rather than iPhone Pros. If this can be overcome then the iPhone Pro would be the better device. Options could be to physically disable the cellular function completely (for example sealing the SIM card slot).

DISCUSSION

Implications for mining practices

Following the testing of an iPad Pro at Mponeng Mine, the Geology Department has decided to switch to using iPhone Pro Maxes for traditional mapping in the deep-level Ventersdorp Contact Reef (VCR) stopes. The iPhone Pro presents several advantages over the iPad Pro, notably its compact, pocket-sized design and waterproof features. These enhancements effectively resolve issues encountered during the iPad Pro-testing phase, particularly its susceptibility to damage due to its large screen and lack of waterproofing. The short 5 m range of the Apple LiDAR scanners are not considered a restriction in the confined mining conditions found in the narrow-tabular stopes of Mponeng Mine.

The adoption of the iPhone Pro Maxes for geological mapping required the development of a workflow to integrate the 3D scanning technology into the routine data acquisition and dissemination processes of the Geology Department. A second workflow to acquire mining production information using 3D scanning technology and make this available to the production teams has also been developed.

Geological scanning workflow

The geological scanning workflow incorporates 3D scanning into geological mapping and modelling processes through a series of well-defined steps. The process begins with data capture, where 3D LiDAR scans of stope panels are obtained using the iPhone Pro Max paired with the 3D Scanner application. Once the scans are captured, they undergo geo-referencing using Deswick.CAD software to align the spatial data with existing geological models accurately. The next phase involves data processing in CloudCompare software, which is used to clean and enhance the quality of the point cloud data. Geologists then perform a detailed geological interpretation, analysing the geo-referenced scans to accurately and dynamically map geological features. The interpreted geological data is integrated into the 3D geological block model, providing essential insights for short- and medium-term mine planning. Finally, the updated geological models are communicated to the Production team to support decisions related to stoping width control and grade dilution management. The geological scanning workflow is shown in Figure 12.

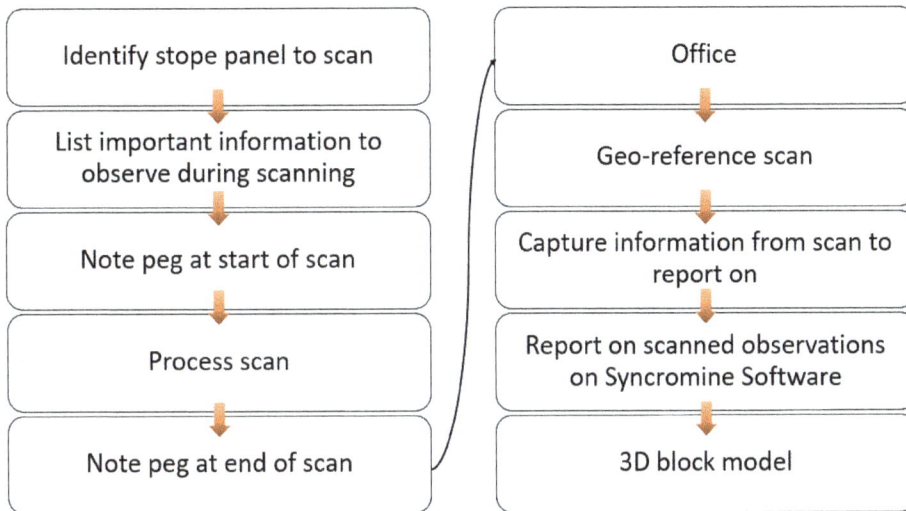

FIG 12 – Geological scanning workflow.

Production scanning workflow

The production scanning workflow is tailored for the operational needs of the Production team, aiming to improve efficiency and accuracy in daily mining activities. The workflow initiates with daily scanning, where production team members use 3D scanners to regularly capture the conditions of underground stopes. Immediate feedback is a critical component of this workflow, as scans are analysed in real-time to assess stope width, equipment positioning, and compliance with the established mining plans. Upon returning to the surface, the team promptly uploads the scanned data to a central database, ensuring all departments have immediate access to the most recent information. This data is instrumental in operational planning, allowing the Production team to make informed decisions regarding equipment movement, shift management, and verification of stope cleaning quality. Additionally, the scans play a vital role in maintaining legal and safety compliance by facilitating accurate documentation and enabling the identification of potential hazards through precise mapping of excavation progress. The production scanning workflow is shown in Figure 13.

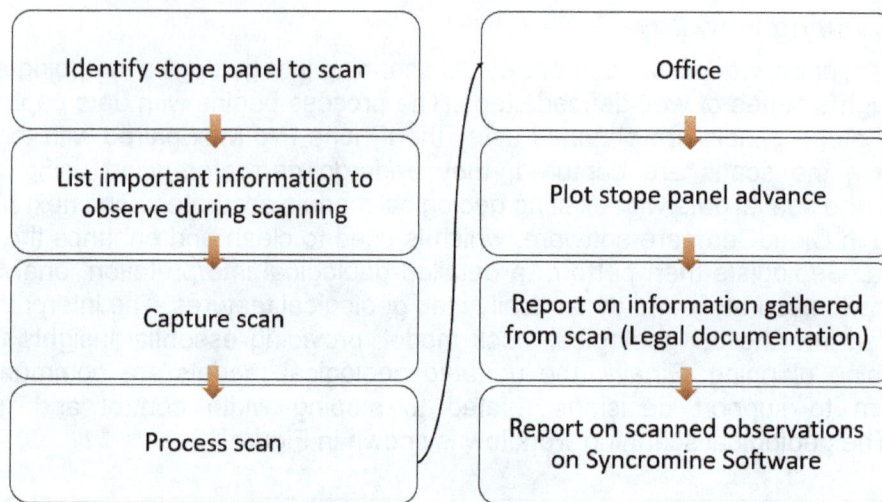

FIG 13 – Production scanning workflow.

Key benefits of including 3D scanning in mine planning and management

The adoption of 3D scanning workflows at Mponeng Mine delivers several key benefits. Efficiency is significantly enhanced as both workflows reduce the manual workload and time required for traditional mapping and surveying methods. The high precision of geo-referenced 3D models boosts accuracy, leading to improved mining outcomes by minimising grade dilution and enhancing resource management. The dynamic control enabled by rapid data capture and processing allows for quick adjustments to mining strategies, particularly in controlling stoping width and minimising the processing of waste materials. The mine is looking to expand the use of Apple LiDAR devices into other disciplines, for example, Rock Mechanics.

CONCLUSIONS AND RECOMMENDATIONS

This study has explored using Apple LiDAR devices to replace traditional geological mapping in underground stopes and mine tunnels for mine planning purposes. The initial study was conducted at the DigiMine (stope and tunnel) to establish a methodology for the scanning. This was followed up with testing in the ultra-deep stopes at Mponeng Mine, which led to the geology department procuring iPhone Pro Max devises. A workflow for geological, as well as production 3D scanning has been established for bringing the digital information into the mine's databases. The scans are digitised and incorporated into the 3D geological models.

The key findings of the comparison scanning exercise in the UJ mine tunnel can be summarised in Table 2 in the Appendix.

In conclusion, Apple LiDAR equipped devices offer a cost-effective and efficient tool for 3D LiDAR scanning in mining, delivering accurate scans quickly to support geological mapping and production planning. The durability of iPad Pro models in the harsh underground environment is a concern with the iPhone Pro might providing a more robust and practical solution if allowed by possible mine restrictions regarding cellular phones.

ACKNOWLEDGEMENTS

The authors would like to acknowledge the following organisations: The University of the Witwatersrand School of Mining Engineering; The University of Johannesburg Department of Mining Engineering and Mine Survey; The Wits Mining Institute Sibanye Stillwater Digital Mining Laboratory (DigiMine); Harmony Mponeng Gold Mine; Mandela Mining Precinct Maseve Mine; Horts Solutions; and Aero Geomatics.

REFERENCES

Berg, R, Mathers, S, Kessler, H and Keefer, D, 2011. Synopsis of current three-dimensional geological mapping and modeling in geological survey organizations, Illinois State Geological Survey.

Birch, C and Olivier, A, 2022. Narrow, tabular stope 3D scanning in deep-level gold mines using an iPad Pro LiDAR, in *Proceedings of the 27th International Mining Congress and Exhibition of Turkey*, TMMOB Maden Mühendisleri Odası, pp 658–669.

Birch, C, 2018. Geological mapping and modelling training in the University of the Witwatersrand Mine Tunnel, South Africa, *Journal of the Southern African Institute of Mining and Metallurgy*, 118(8). https://doi.org/10.17159/2411-9717/2018/v118n8a3

Chamberlain, 2025. Ross Line Level 75mm F7728 LS2. Available from: <https://www.chamberlains.co.za/ross-line-level-75mm-f7728-ls2-1059509>

Davey, R, 2023. 3D Laser scanning in underground mining: Development and prospects. Available from: <https://www.azomining.com/Article.aspx?ArticleID=1765>

Department of Mineral Resources of South Africa, 2011. Mine Health and Safety Act No 29 of 1996 Government Gazette 27 May 2011, Pretoria.

Dusza-Pilarz, K, Kirej, M and Jasiołek, J, 2024. Use of laser scanning and 3D software in mining design, in Mineral Resources and Energy Congress, 526:01012. https://doi.org/10.1051/e3sconf/202452601012

Grobler, H and Pienaar, M, 2021. Enhanced orebody knowledge through scanning technologies and workflows, in *Proceedings of the 5th International Future Mining Conference*, pp 46–68 (The Australasian Institute of Mining and Metallurgy: Melbourne).

Luetzenburg, G, Kroon, A and Bjørk, A A, 2021. Evaluation of the Apple iPhone 12 Pro LiDAR for an application in geosciences, *Scientific Reports*, 11(1):1–9. https://doi.org/10.1038/s41598-021-01763-9

Marc, Y, 2021. Apple LIDAR demystified: SPAD, VCSEL and Fusion. Available from: <https://4sense.medium.com/apple-lidar-demystified-spad-vcsel-and-fusion-aa9c3519d4cb>

Minrom, 2023. Unravel the hidden wonders of geological mapping!. Available from: <https://minrom.com/unravel-the-hidden-wonders-of-geological-mapping/>

Olivier, A, 2023. Optimising the use of three-dimensional data to lower gold grade dilution by controlling stope width in the mining of ultra-deep complex ore bodies, The University of the Witwatersrand.

Ordóñez, C, Calvopiña, J, Toapanta, S, Carranco, A and González, J, 2024. Integrating lidar technology in artisanal and small-scale mining: A comparative study of iPad Pro LiDAR sensor and traditional surveying methods in Ecuador's artisanal gold mine, *Journal of Geodetic Science*, 14(1). https://doi.org/10.1515/jogs-2022-0181/machinereadablecitation/ris

Rutkowski, W and Lipecki, T, 2023. Use of the iPhone 13 Pro LiDAR scanner for inspection and measurement in the mineshaft sinking process, *Remote Sensing*, 15(21):5089. https://doi.org/10.3390/rs15215089

Rutkowski, W, 2023. Possible application of low-cost laser scanners and photogrammetric measurements for inspections in shaft sinking process, *Journal of Konbin*, 53(1):147–156. https://doi.org/10.5604/01.3001.0016.3245

Singh, S, Banerjee, B and Simit, R, 2023. A review of laser scanning for geological and geotechnical applications in underground mining, *International Journal of Mining Science and Technology*, 33(2):133–154. https://doi.org/10.1016/j.ijmst.2022.09.022

Świerczyńska, E J, Kurdek, D and Jankowska, I, 2024. Accuracy of the application of mobile technologies for measurements made in headings of the Kłodawa Salt Mine, *Reports on Geodesy and Geoinformatics*, 117(1):55–68. https://doi.org/10.2478/rgg-2024-0007

Tondo, G R, Riley, C and Morgenthal, G, 2023. Characterization of the iPhone LiDAR-Based Sensing System for Vibration Measurement and Modal Analysis, *Sensors*, 23(18):7832. https://doi.org/10.3390/S23187832

Treccani, D, Adami, A and Fregonese, L, 2024. Assessing the effectiveness of LiDAR-based apps on Apple devices to survey indoor and outdoor medium sized areas, *The International Archives of the Photogrammetry, Remote Sensing and Spatial Information Sciences*, XLVIII-2/W8-2024:431–438. https://doi.org/10.5194/isprs-archives-xlviii-2-w8-2024-431-2024

US Geological Survey, 2025. Introduction to Geologic Mapping. Available: <https://www.usgs.gov/programs/national-cooperative-geologic-mapping-program/introduction-geologic-mapping>

APPENDIX

TABLE 2
Summary of geological mapping and LiDAR scanning approaches.

Method	Cost	Speed and efficiency	Quality	Accuracy	Comments
Traditional Geological Mapping	Equipment costs are low but the labour cost for skilled geologists to do the mapping is very high.	Very slow to physically map the faces, with additional time required afterwards to plot the results onto paper, and then digitise this into the mine's 3D geological model. 35 mins to map the 18 m and an additional hour to plot the mapping onto paper.	While detailed observations of the lithology can be done, the manner in which the UJ mine tunnel was mapped results in a low quality map. Sections every 2 m with the sections linked together results in a blocky sketch.	Low accuracy due to using a compass and clinorule. Dip measurements are often taken on apparent dip. True strike of the structures can be difficult to determine, requiring string and a line level.	Geologists are required to spend time at the mine face exposing them to hazards.
Apple LiDAR	LiDAR equipped iPad Pro and iPad Pro devices start from about US$1000. Labour costs are reduced compared to traditional geological mapping.	Very effective because far less time is spent gathering the information and it is in a digital form allowing rapid inclusion in the mine's 3D geological model. Time to scan the tunnel depends on the application resolution settings and technique used. Four mins for low-resolution scan-pause-scan and 17 mins for high-resolution continuous scanning.	Fair. The images in the point clouds depend on the scanning resolution. The OBJ or GLTF mesh images can be very clear (depending on the lighting available) even for low resolution scans. Both techniques (scan-pause-scan or continuous scanning) gave similar quality scans.	Fair. Accuracy of between 2 and 10 cm was achieved in the UJ mine tunnel for both approaches used (scan-pause-scan or continuous scanning)	A low-cost practical solution to improve the quality of geological mapping underground. Already implemented at Mponeng Mine.
MLS Scanning (FjD Trion P1)	Around US$15 000	Very efficient for gathering LiDAR information (four mins for the tunnel including the areas before and after the 18 m with simulated geology).	High. The image quality is clearly higher resolution than those obtained from the Apple LiDAR scans.	This was not tested due to the scanning taking place before the survey ground control marks being installed in the tunnel.	The equipment is bulky to be used in a narrow, confined stope. However, scanning tunnels can be done very efficiently.
TLS Scanning (Z&F 5010X and Riegl VZ600)	Very high (US$45 000 to US$70 000)	The Z&F 5010X was relatively slow. Each scan required precise levelling and the scan acquisition was slow (17 mins). The newer Riegl VZ600 was very fast (4 mins).	Both scanners take high quality visible light images while obtaining the LiDAR information.	Exceptional. Accurate to 1 mm.	The equipment is bulky and must be placed on a tripod before scanning takes place. Scanning in narrow, confined stopes is not practical. The expense of the equipment would mean most mines would only be able to purchase limited quantities.

Discrete events simulation approach to investigating the impact of electrifying haul trucks – a case study

T Chimbwanda[1], A Anani[2] and N Risso[3]

1. Student, University of Arizona, Tucson Arizona 85721, USA. Email: tchimbwanda@arizona.edu
2. Associate Professor, University of Arizona, Tucson Arizona 85721, USA. Email: angelinaanani@arizona.edu
3. Assistant Professor University of Arizona, Tucson Arizona 85721, USA. Email: nrisso@arizona.edu

ABSTRACT

As near-surface deposits deplete and global demand for mineral resources soars, the mining industry faces the dual challenge of maximising resource extraction while optimising operational efficiency. The adoption of electrically powered haulage solutions presents a potential alternative to conventional diesel-powered trucks, but this transition is highly capital-intensive, requiring reliable data insights to assess feasibility. This study employs stochastic Discrete Event Simulations (DES) to evaluate the operational, energy efficiency, and economic impacts of replacing diesel-powered haul trucks with battery-electric haul trucks in an open pit mining operation in the USA. The simulation model, validated to within 5 per cent of operational data, reveals that transitioning to battery-electric haul trucks reduces productivity by up to 23 per cent. Despite this decline, the shift significantly improves energy efficiency, with potential reductions in fuel costs and overall energy expenditures by approximately 34 per cent. These findings provide critical insights into the trade-offs between productivity and cost efficiency, supporting data-driven decision-making for mining operations.

INTRODUCTION

Energy transition in mining

With depletion of near surface deposits, mines continue to deepen in search of higher grades. This trend has a direct implication on haul distances which continue to rise. Longer hauls, coupled with escalating fuel costs present some of the biggest challenges the mining industry faces today. Haulage costs account for up to 60 per cent of operating costs of a typical large-scale surface mine (Navarro *et al*, 2020). The mining industry stands at an inflection point, marked by a transition from diesel powered equipment to electric alternatives to meet decarbonisation goals through a global commitment to achieve net-zero (Bao *et al*, 2023a). The most common haulage option for open pit mines is the truck-shovel set-up. The biggest advantages of the truck-shovel configuration with diesel powered trucks are flexibility and scalability which makes mine planning easier. However, a critical challenge with such a configuration is the high fuel consumption associated with it. Haul trucks utilise 70–80 per cent of their fuel on ramp inclines, significantly impacting operational costs (Bao *et al*, 2023b).

The work done by Lindgren *et al* (2022) reveals that transitioning to battery-electric alternatives for open pit haulage can substantially reduce not only the carbon footprint, but also operating costs. In this study, it is demonstrated that battery-electric haul trucks can achieve higher production and operating and maintenance costs relative to the diesel powered alternative. Lindgren *et al* (2022) further notes that this transition could be economically viable with the capability of frequents charging opportunities, more efficient and durable batteries. The outcome of this study highlights the energy efficiency and operational gains brought by adoption of fully electric hauling solutions.

Dating back from the 1950s, In-pit Crushing and Conveying (IPCC) systems are an alternative to the truck-shovel configuration see (Figure 1). Perceived benefits of IPCC systems include lower operating costs and reduced energy consumption (Bao *et al*, 2023b). This translates to substantially lower carbon footprint. However, despite the clear benefits of IPCC systems, they suffer from significant limitations. Their lack of flexibility and scalability which presents challenges in mine planning.

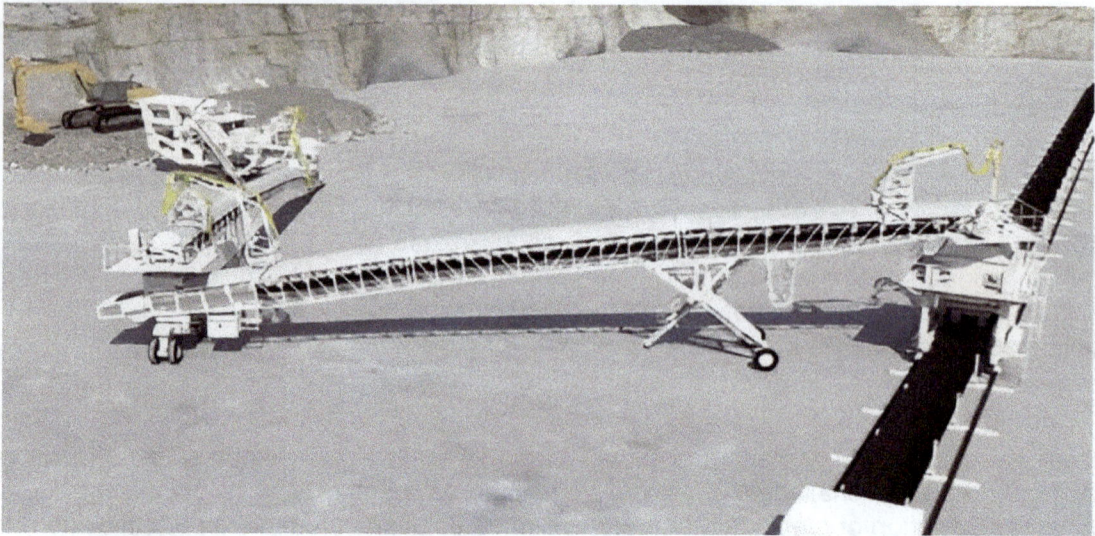

FIG 1 – Mobile in-pit crushing and conveying system (IPPC) (Bao *et al*, 2023b).

Another alternative to the conventional truck-shovel configuration is trolley assist a technology that is backed by historical success (Nuric, Nuric, and Brčaninović, 2008). Trolley assist systems provide external power to diesel-electric equipment. They are most effective on inclines. In most trolley assist configurations; the diesel engine runs idle. This has a direct impact on substantial savings in energy (Cruzat and Valenzuela, 2017). However, trolley assist systems only supplement diesel engine energy requirements for limited haul distances. In other sections of the haulage system, the trucks would still run entirely on diesel. Studies reveal that a decrease in consumption by 19 L/km is achievable with trolley assist implementations (Bao *et al*, 2023b). Despite the gain they bring, trolley assist system still face significant limitations. They are complex and highly capital intensive. Furthermore, they are less flexible relative to conventional truck-shovel configuration with the need to relocate trolley infrastructure as the mine evolves.

Battery Trolley (BT) Systems are another promising technology which continues to advance with breakthroughs in battery technology (Bao *et al*, 2023a). Figure 2 shows a typical configuration of a BT system. BT systems aim to power haulage entirely with electric energy. Electro-mobility is a crucial aspect in the successful adoption of BT systems. Some of the advantages include fewer mechanical components which translates to relatively simple maintenance and less breakdowns. The study notes some of the limitations of BT that needs to be considered. Battery size is one of the biggest challenges currently. Trolley assist has potential to reduce battery size requirement for haul trucks. Another downside of BT systems is the time lost in charging and battery swapping. This can significantly increase cycle times and in turn reduce overall productivity.

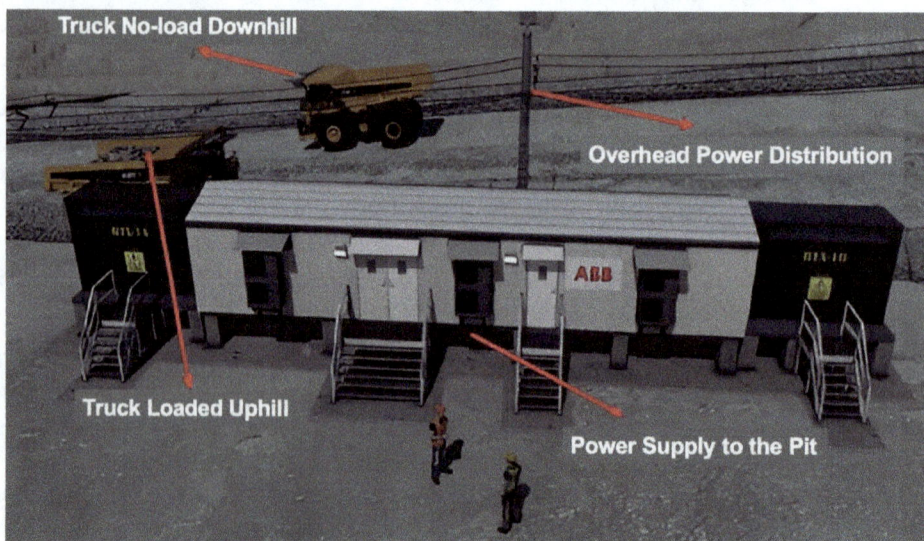

FIG 2 – Battery trolley configuration (Bao *et al*, 2023a).

DES in mining

Discrete-event simulations (DES) relies on modelling systems as a sequence of events occurring at distinct time points, often incorporating stochastic elements. They are used to statistically analyse system performance, optimise processes, and generate actionable insights derived from simulated outcomes (Saïd, Eveno and Villaneau, 2022). In mining, DES play pivotal role in enhancing operational efficiency and decision-making. They find wide application in mining, spanning from underground mining fleet optimisation, analysing Mine to Mill optimisation systems to determining optimum fleet sizes. DES enable modelling of complex mining operations and scenarios like assessment of various alternatives in equipment selection which aids to better mine planning outcomes and improved productivity (Salama and Skawina, 2023).

Fahl (2017) conducts a comparative analysis of DES and other techniques like spreadsheet analysis. A case study example of a stochastic equipment selection framework is used to demonstrate the value and applicability of DES within the mining industry. Emphasis is made on the enabling role played by DES in enhancing effective decision-making.

Failure to account for duty cycles can lead to mismatch, low productivity and increased operating costs (Anani and Awuah-Offei, 2017). The study highlights the difficulty in determining the optimal shuttle car fleet size when varying duty cycles are not accounted for. This has implication in potential overestimation of fleet requirements. To address this, they investigate the impact of varying duty cycles on equipment matching and optimal fleet sizing. In the study, DES simulation methodology is used to model a 17-entry with a shuttle capacity of four cars. Even though, four shuttle cars is optimal for the entire panel, investigations revealed that such a fleet size is only optimal for 80 per cent of the segments when duty cycles are ignored. Data from time and motion studies and equipment monitoring systems is used and fitted to statistical distributions. The Chi-squared tests check the goodness of fit of such distributions like normal, erlang among others. The results of the study were compared and validated with mine data. The study provides insights on how DES simulation is a powerful tool for mining applications.

In another study (Anani, Nyaaba and Córdova, 2022) demonstrates the applicability of DES in tackling mine planning problems. Cognisant of the limitations of prevailing mine planning tools, the study takes an integrated approach to optimisation of fleet size and change-out time. To that end, the researchers proposed a DES simulation approach to maximising production efficiency of a coalmine room and pillar panel. They developed a stochastic simulation model that took into account dynamics like continuous miner continuous miner cutting sequences, loading processes and shuttle car routing. Panel widths, fleet sizes and change-out times were also considered. The shuttle car cycle time data utilised in this study was collected from time and-motion experiments done over two days at an underground room-and-pillar coalmine in Illinois, USA. Spearman correlation test was used to ascertain relationships between variables, which is vital to evaluate relationships that cause complex modelling scenarios. To that end, correlation testing minimises modelling risks. As a result of their study, they found out that by optimising panel width and fleet size, there is potential to increase production by 5 per cent. Moreover, utilising 15-entry panel and 19-entry panel with respectively three shuttle cars and four to five shuttle cars would enhance operational efficiency at the underground coal mining operation. The outcome of this research massively contributes to making underground mining more production efficient.

Huayanca, Bujaico and Delgado (2023) proposed a stochastic approach to estimate truck fleet size considering anticipated production capacity increase from 100 000 t to 140 000 t at a copper mine in Peru. The study recognises the challenge of selecting approaches methodology, pointing out that some methodologies are too complex or require advanced programming expertise. DES as an approach to stochastic methodology was chosen over other methods which are complex. In the study, data was collected using inputs from CAT MineStar System. The data was used to generate probability distribution functions and distribution fits in Arena, a simulation software developed by Rockwell Automation (Patole, 2024). Subsequently, a calibration model was developed. For validation, results from the calibration model were compared with real data from operations. The difference between the half-width indicator and parameter mean values was used as a performance metric. For all parameters, this difference was less than 1 per cent. Following validation of the calibration model, a yearly model was built. Except for planned production capacities, they used the same probability functions as used in the calibration model. Their results projected the need for increasing the number of trucks by 13 units to match the anticipated increase in production. The value in the outcome gives insights to anticipate for future challenges making it a vital ingredient for proactive mine planning.

Foundational work is done by Bao *et al* (2023a), to estimate productivity of battery trolley mining truck fleets. Noting the urgent call for the industry to reduce emissions and improve energy efficiency, three battery trolley configurations are presented. The configurations are namely, dynamic charging; stationary charging; dual trolley systems. In this study, all configurations are evaluated in their strengths, weaknesses and limitations. Mining system theory is used to identify deployment requirements, operational processes and power sources for the three battery trolley configurations. Subsequently, a hypothetical copper mining scenario is created using data from open access databases and Aitik Copper mine project parameters. An equation for estimating trolley system productivity is modelled. Dependent variables used in this study include shovel and truck productivity, maximum number of trolleys, battery trolley productivity. Trolley power and ore type are some of the independent variables used in the equations modelled. Results showed that the stationary dual trolley option offers higher capacity relative to dynamic charging. Moreover, they found out that trolley power determines the capacity of the battery trolley system. Comparatively, stationary trolleys allow for deployment of 68 truck units whereas the dynamic charging option is limited to a fleet size of 13 trucks. This reveals a significant disparity in productivity, a significant factor to consider in mine planning. A configuration of a typical Trolley Battery system is shown in Figure 3.

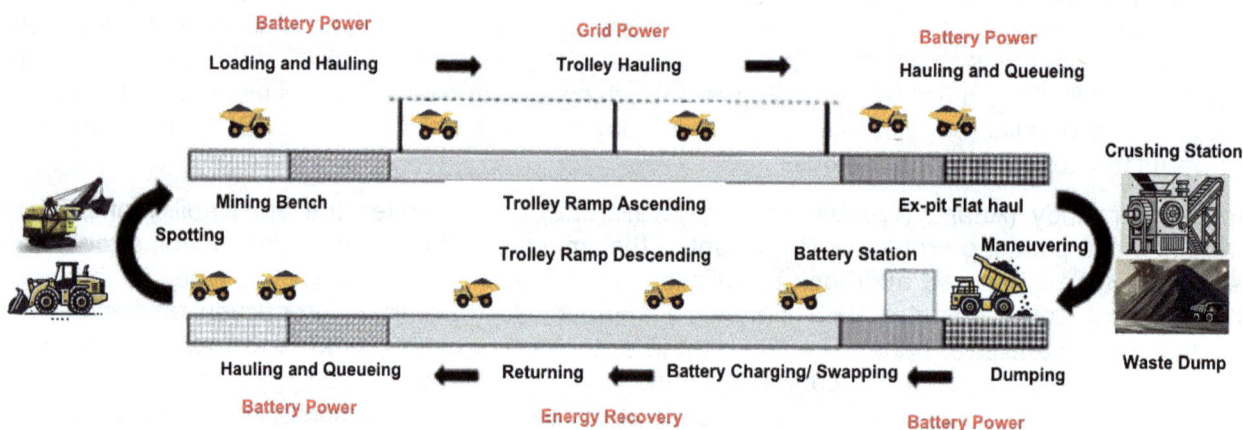

FIG 3 – Production cycle of trolley-battery configuration (Bao *et al*, 2023a).

Gap analysis

The work done by Bao *et al* (2023a) provides foundational base for in-depth research as the battery trolley systems evolve. Although relevant, this work highlights some of the research gaps. Currently battery trolley haulage solutions are still at primitive and conceptual and lacks widespread implementation. Moreover, the study used assumptions to create specific mining scenarios of each configuration. If the assumptions don't hold, simulated results may not reflect the actual reality. Another limitation to the progression of battery trolley haulage systems is that it is bottle-necked by the pace of advancement of battery technology. In light of current limitations, there is need for further research to validate the conceptual designs proposed. Although Bao *et al* (2023a) laid a strong foundation to estimating productivity of battery trolley systems through theoretical modelling and parameter analysis, there is need for advanced simulation and analysis that account for the inherent dynamic and stochastic nature of mining. This study aims to build on the work by Bao *et al* (2023a), and adopt the stochastic approach used by Huayanca, Bujaico and Delgado (2023) for a case study of a copper mine in Peru. Insights provided by Salama and Skawina (2023), Fahl (2017) and Anani and Awuah-Offei (2017) on the applicability of DES for equipment selection and performance analysis reinforce the use of stochastic a methodology for more detailed evaluation of the implications of transitioning to a battery powered haulage solution. Taking the approach of stochastic modelling through DES, this study will provide a more comprehensive and detailed evaluation of the trade-offs between energy efficiency improvements and productivity changes.

METHODOLOGY

This study uses a stochastic Discrete Event Simulations methodology to investigate the impact of transitioning from the conventional diesel-powered truck-shovel configuration to battery-electric haul trucks. In this study, operational data from an open pit copper mine in the USA is used to develop and validate the simulation model. The methodology is divided into four main stages, namely i) data collection, ii) simulation design, iii) scenario analysis, and iv) validation.

Data description

The data set used in this study was collected from an open pit mine in the USA. The operation consists of five shovels and 24 haul trucks, with each truck having a maximum payload capacity of 267 tons. Two shovels were designated for waste stripping while the remainder worked on ore. Fourteen truck units are assigned to hauling ore while nine are dedicated to hauling waste (see Table 1). The data set includes detailed operational data for a single day, capturing key aspects of the haulage cycle:

- Spotting Time: The time taken by a truck to position itself at the shovel.

- Loading Time: The duration required to load a truck at the shovel.

- Full Load Haul Time: The time taken by a fully loaded truck to travel from the shovel to the designated dumping or crusher location.

- Empty Load Haul Time: The duration for an empty truck to return from the dumping or crusher location to the shovel.

- Dumping Time: The time spent unloading material at the crusher.

TABLE 1
Assignment of trucks between ore and waste.

Truck assignment	
Ore	Waste
14	10

While this data set provides valuable insights into fleet performance, and cycle efficiency, it is important to acknowledge that a single day may not fully capture variations in operational conditions, such as fluctuations in ore grade, weather impacts, or changes in haulage patterns.

In addition to these metrics, the data set contains the payload for each individual truck, providing granular details about the haulage process. The primary destinations for the ore are classified as the A-Side and B-Side of the operation, collectively represented as the crusher in the simulation model. This comprehensive data enables accurate modelling and validation of the haulage system, forming the basis for credible scenario analysis and evaluation of the transition to battery-electric haul trucks.

The data used in this study was provided in the form of a spreadsheet containing a comprehensive collection of operational metrics. To ensure the data set's relevance and accuracy, it underwent detailed analysis and preprocessing. From the extensive data set, a subset of relevant information was selected for the simulation, including, but not limited to, spotting time, loading time, full load haul time, empty load haul time, dumping time, and truck payload. This selection process focused on identifying the data most critical for accurately modelling the haulage cycle. Outliers were identified and removed to improve data reliability and ensure precise simulation results. This cleaning process was performed using the Data Analysis tool in Microsoft Excel's, specifically leveraging the Descriptive Statistics feature to systematically review the data set. Table 2 shows an extract of descriptive statistics of full haul time from one of the waste stripping shovels to a designated dumping location.

TABLE 2

Example of descriptive statistics of full haul time from a shovel to a dumping location.

Full haul time S44 shovel	
Mean	864.183
Standard error	4.708
Median	859.5
Mode	841
standard deviation	79.349
sample variance	6296.185
Kurtosis	20.306
Skewness	-0.557
Range	1075
Minimum	361
Maximum	1436
Sum	245428
Count	284

Once refined, the data was further analysed using Arena's Input Analyser to generate appropriate statistical distributions for each variable, such as hauling times, spotting times, and dumping times. These distributions captured the inherent variability within the data and formed the basis for creating a stochastic model that accurately represents the mining operations in the Arena simulation software (see Table 3). Figure 4 shows the distribution representing haul times from one of the waste stripping shovels to a designated dump location.

TABLE 3

Summary of the distribution analysis results produced by Arena's Input Analyser for the selected waste shovel route to the dump location.

Distribution summary	
Distribution	Weibull
Expression	754 + WEIB (0, 0)
Square error	0.01103

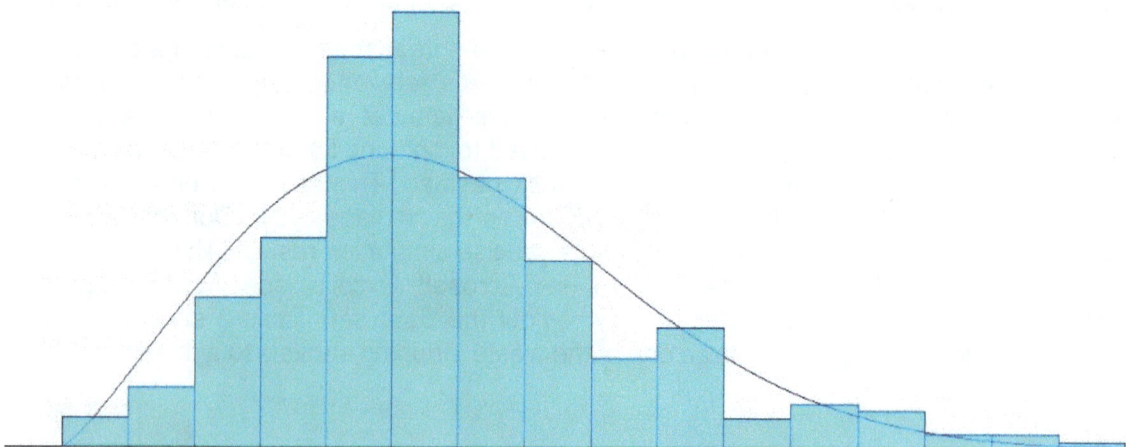

FIG 4 – Full load haul times distribution generated by Arena's Input Analyser tool.

For modelling the battery-trolley alternative, we assume that each haul truck requires at least 20 min of charging to reach a sufficiently high state of charge before resuming operations. This figure is based on the battery capacity and charge rates detailed by Bao *et al* (2023a) as well as typical operational constraints observed in pilot battery-electric mining systems. Consequently, this 20 min charging downtime directly reduces the available operating window for each truck, thereby lowering overall productivity compared to scenarios where no charging is required.

Simulation design

This study draws inspiration from several studies in approaching the simulation design problem. Park, Jung and Choi (2023) develops simulation design a mine haulage system based on the truck cycle times theory. The design included input data such as daily operating conditions, truck IDs, truck capacities, and loading points. Parameters such as time for the simulation were derived from the predicted truck travel times, loading times, and dumping times. The simulation model was verified and validated by comparing the results with actual vehicle operation logs to ensure accurate representation of the haulage system. The simulation design was structured to accommodate changes in haulage routes and operational conditions, reflecting real-world scenarios encountered in the mining process. The value in the simulation design proposed by Park, Jung and Choi (2023) is that it enables analysis of ore production by truck and loading point, facilitating the determination of optimal truck dispatch combinations. Navarra (2023) also shows that the simulation design for discrete event simulation of haulage systems includes using predicted truck cycle times as input data, incorporating operating conditions and time factors, to accurately model ore production and optimise equipment utilisation and dispatch strategies.

Baek and Choi (2019) develops a discrete event simulation for a mine haulage system based on truck cycle time theory to model truck operation times. Input data, including truck travel times and other factors such as loading and dumping times, were obtained through big data analysis from the mine safety management system. The simulation includes a graphical user interface (GUI) for entering simulation factors and time parameters. The simulation engine performs three main tasks: generating input data, executing the simulation, and outputting results, such as total simulation time, ore quantity moved, crusher utilisation, and truck wait times. The design was validated by comparing simulation results with actual vehicle operation records, ensuring accuracy. This approach provides reliable estimates of haulage time, operational cycles, production quantities, and truck wait times, enhancing production efficiency and equipment utilisation.

Our simulation model is developed using Arena software, employing a stochastic Discrete Event Simulation (DES) framework to accurately replicate the operational dynamics of the mine. This approach accounts for the inherent variability and complexities of mining operations. Key aspects of the simulation design include:

- **Random variations:** Variability in operational parameters, such as haul times and loading durations, is modelled to reflect the stochastic nature of the mining environment. This ensures a realistic representation of daily operations.

- **Duty cycles:** Detailed duty cycles are incorporated for both diesel-powered and trolley-assisted battery-electric haul trucks. These include the full sequence of activities—loading, hauling, dumping, and for electric trucks, the additional consideration of charging times.

In the model, trucks are represented as entities, while shovels are modelled as resources, allowing for a realistic depiction of the interaction between equipment. Two scenarios are simulated to evaluate the system: the base case, representing the current configuration with diesel-powered trucks, and the alternative case, which models the transition to battery-electric haul trucks, including the additional requirements of charging infrastructure and its impact on operational cycles.

The simulation incorporates random variations from the generated distributions to reflect the dynamic nature of mining, such as variability in haul times. It also includes detailed duty cycles for each configuration, covering activities like loading, hauling, dumping, and, for the alternative case, charging times. In the base case scenario, fuel refuelling has been assumed to be implicitly captured within the probabilistic distributions generated from the operational data. These distributions reflect real-world variability in haul cycle times, idle periods, and truck availability, which inherently incorporate the impact of refuelling events. Since the data-driven approach accounts for operational delays and cycle disruptions, explicit modelling of refuelling was not separately implemented. The base case is initially developed and validated against ground truths derived from real operational data to ensure accuracy (see Figure 5). Following successful validation, the alternative case introduces a new system component—incorporating charging infrastructure to represent the transition to battery-electric haul trucks. The two scenarios are evaluated on key performance metrics, including truck-shovel system productivity measured in total ore production and energy efficiency. Figure 6 shows a schematic representation of the alternative case.

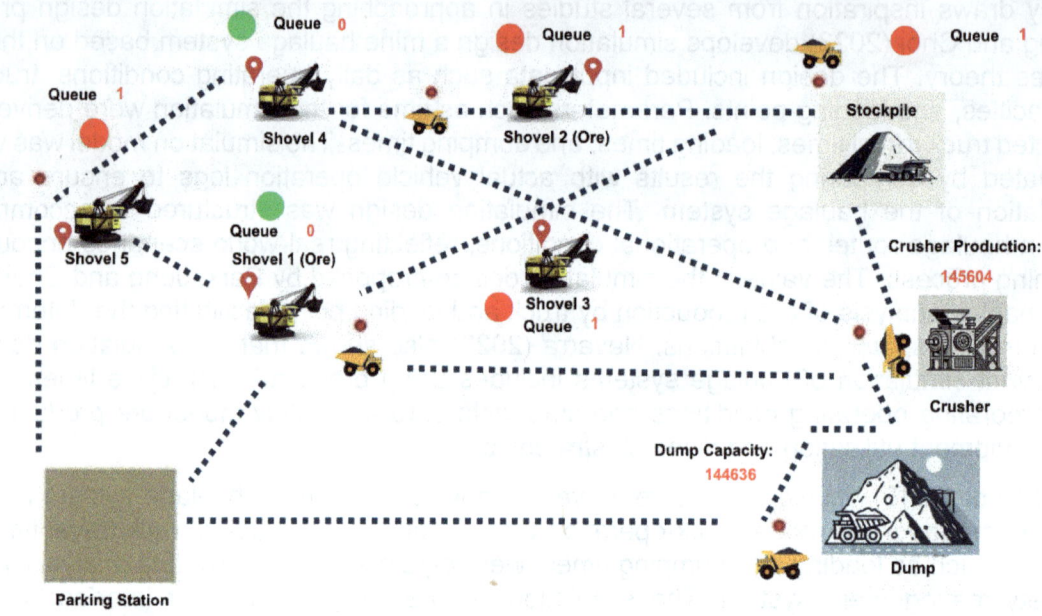

FIG 5 – Schematic representation of the diesel-powered trucks base case scenario.

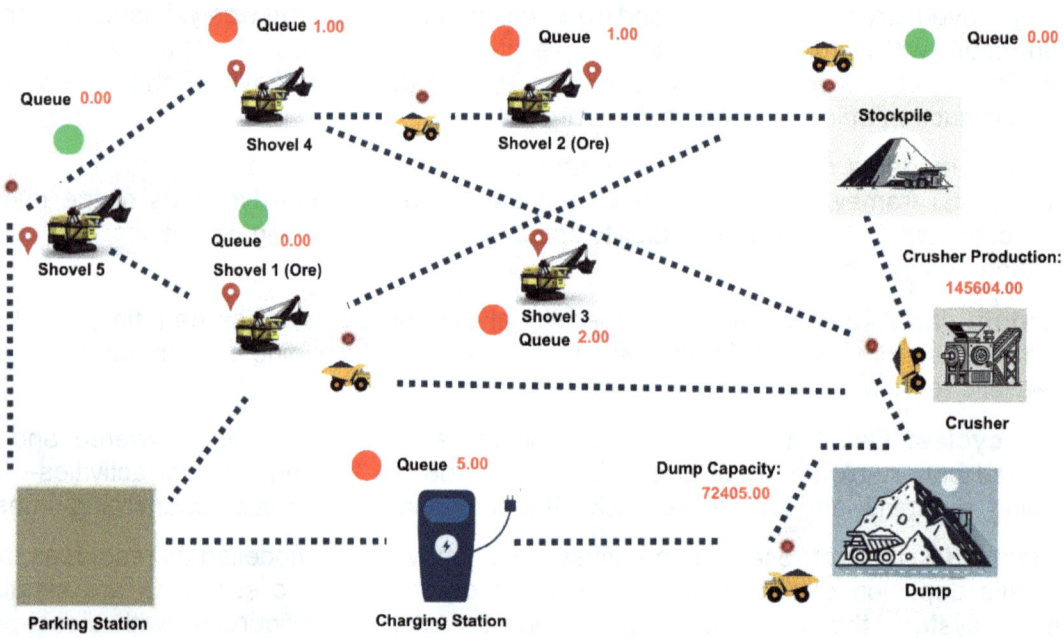

FIG 6 – Schematic representation of the battery-powered haul trucks alternative case.

Base case model validation

To validate the DES model, base case results for the truck-shovel configuration are compared with actual data from the mine's operations to ensure consistency and credibility of the simulation methodology.

Alternative case

Following validation, the alternative case is implemented. Charging is an essential component of battery electric haul truck configuration. State-of-art technology require at least 20 mins for a full charge (Bao *et al*, 2023a). Moreover, for an open pit mine with stationary charging stations, a fully charged electric haul truck cannot complete more than three cycles without requiring recharging (Bao *et al*, 2023b). Guided accordingly, our simulation considers the current realities of the technology. As such, the charging times conservatively are represented as a normal distribution with a mean 25 mins and standard deviation of 2 mins. After completing three cycles, each haul truck is routed to the charging station. Apart from the additional charging logic, the alternative case is identical to the base case scenario.

RESULTS

Validation

Operational data reveal that a total ore production of 97 000 t. The simulation results yield an average total ore production of 91 766 t, which is 5234 t (approximately 5 per cent) lower than the actual production data. This level of deviation is within an acceptable range, suggesting that the simulation model effectively captures the operational dynamics of the truck-shovel haulage system. The slight difference may be attributed to inherent variations in daily operations, random events, or minor simplifications in the simulation framework.

Alternative case

The alternative case achieves an average ore production of 70 301 t. This is approximately 23 per cent less than the base case which achieves 91 766 t. This is approximately 23 per cent lower than the base case production of 91 766 t, reflecting the impact of charging times and reduced operational efficiency associated with the battery-electric configuration (see Table 4).

TABLE 4

Results.

	Base case	Alternative case	% change
Crusher production (tons)	917 767	70 301	-23%
Dump capacity (tons)	85 279	51 450	-39%

Despite the perceived reduced productivity, there are energy cost reduction benefits of the battery electric configuration. Based on findings from (Wirantaya, Biyanto and Satwika, 2019), the average fuel consumption of a 267 t mining haul truck is approximately 147 L/kg. Assuming a diesel price of $1.20/L, the total fuel cost for the base case can be calculated as follows:

$$\text{Fuel Costs} = \text{Trucks} \times \text{Fuel Consumption} \times \text{Fuel Price} \times \text{Daily Hours} \times \text{days}$$

$$\text{Fuel Costs} = 24 \times 147 \times 1.20 \times 24 \times 360 = \$36\,578\,304$$

Energy cost comparison with battery-electric haul trucks

According to Bao *et al* (2024), for a 220 t haul truck operating solely on battery power, the energy consumption is approximately 645 kWh per cycle, with the truck completing roughly 1.44 cycles per hr (around 35 cycles per day). Extrapolating these figures to a 267 t truck suggests that the required energy per cycle would be higher—on the order of an additional 20–25 per cent—due to the truck's increased weight and associated rolling resistance (ie 774–806 kWh per cycle). This assumption can be reasonable, particularly in accounting for not only the direct payload increase (which is roughly 21–22 per cent when going from 220 to 267 t) but also additional inefficiencies like increased rolling resistance and non-linear effects in energy consumption. While this range provides a ballpark estimate, the actual energy demand for a 267 t truck will also depend on specifics such as the haul route, drive system efficiency, and battery chemistry. In the USA, the average electricity cost for industrial consumers is approximately \$0.08 per kWh. To be conservative, however, one might assume a rate of \$0.10 per kWh. Therefore, assuming an electricity price of \$0.10 per kWh and an estimated demand of 806 kWh per cycle, the total electricity cost would be calculated as follows:

$$\text{Electricity Cost} = \text{Trucks} \times \text{Cycles} \times \text{Energy Consumption} \times \text{Electricity Price} \times \text{Daily Hours} \times \text{days}$$

$$\text{Daily Electricity Cost} = 24 \times 1.44 \times 806 \times 0.10 \times 24 \times 360 = \$\,24\,067\,031$$

Contextualising the benefits

While the alternative case experiences a 23 per cent reduction in production compared to the base case, it also significantly improves energy efficiency. As demonstrated from analysis, switching from diesel to electricity for a fleet of 24 trucks results in roughly a 34 per cent reduction in annual energy costs. Moreover, battery-electric haul trucks convert a higher proportion of their input energy into actual hauling work compared to their diesel counterparts, thereby enhancing overall operational efficiency. Although the reduced productivity may require some operational adjustments, the improved energy efficiency and reduced diesel dependency can contribute to better long-term cost stability, particularly if fuel prices rise.

CONCLUSION

Discussion

The results of this study highlight a trade-off between operational productivity and energy efficiency when transitioning from diesel-powered to battery-electric haul trucks. The base case scenario, which represents current operations with diesel haul trucks, achieves an average daily ore production of 91 766 t, closely aligning with the actual figure of 97 000 t and thus validating the reliability of the simulation model. In contrast, the alternative case incorporating battery-electric haul trucks yields an average of 70 301 t daily—approximately 23 per cent less than the base case.

This reduction in production can be attributed primarily to the operational constraints introduced by the battery-electric haul trucks. Current state-of-the-art battery technologies require at least 20 mins to fully charge a truck, with a conservative estimate of 25 mins used in the simulation. Additionally, battery-electric trucks are limited to completing only three cycles before needing to recharge. These charging requirements create significant delays in the haulage process, lowering overall system throughput.

Despite this production loss, the energy efficiency gains are considerable. Battery-electric trucks convert a higher proportion of their input energy into actual hauling work compared to diesel trucks, thereby reducing overall energy costs. As demonstrated in the analysis, switching from diesel to electricity can cut annual haulage energy expenditures by approximately 34 per cent, a figure that can substantially improve operational cost stability—particularly in the face of volatile diesel prices. Over time, these energy savings can offset part of the productivity trade-off, especially as battery technologies advance and charging infrastructure becomes more efficient.

From an operational perspective, the loss in production presents a challenge, especially for mines with high production targets or tight profit margins. However, this loss may be mitigated through faster charging rates, higher energy-density batteries, and extended cycle capabilities. Additionally, optimising operational strategies, such as staggered charging schedules or hybrid fleet configurations, could help narrow the gap between productivity and energy savings.

In summary, the findings demonstrate that while the transition to battery-electric haul trucks introduces certain operational constraints, it also offers notable improvements in energy efficiency and potential cost savings. These results emphasise the importance of a balanced approach, where technological advancements and operational innovations work in tandem to maintain production efficiency while capturing the benefits of lower, more predictable energy costs.

Future work

Future work will explore strategies to enhance the operational efficiency of battery-electric haul truck–shovel configurations. One promising approach is the implementation of battery-swapping systems, which could significantly reduce downtime by eliminating the need for trucks to remain idle during charging. This, in turn, may boost truck utilisation and increase production rates. Additionally, alternative electrified configurations such as Trolley Assist will be investigated for their potential to further improve haulage efficiency and reduce on-board battery requirements. Furthermore, since the current study relied on data collected over a 24-hour period, efforts will be made to collect and analyse data over a longer operational time frame, allowing for a more comprehensive understanding of haulage system performance under varying conditions and improving the robustness of future simulation models. Additionally, future work will explore a more detailed breakdown of refuelling impacts to assess their specific influence on operational efficiency and cycle times. This could involve explicit modelling of refuelling delays to further refine the simulation analysis.

REFERENCES

Anani, A and Awuah-Offei, K, 2017. Incorporating changing duty cycles in CM-shuttle car matching using discrete event simulation: a case study, *Int J Min Miner Eng*, 8:96. https://doi.org/10.1504/IJMME.2017.084202

Anani, A, Nyaaba, W and Córdova, E A, 2022. An integrated approach to panel width, fleet size and change-out time optimization in room-and-pillar mines, *J South Afr Inst Min Metall*, 122. https://doi.org/10.17159/2411-9717/1509/2022

Baek, J and Choi, Y, 2019. Simulation of Truck Haulage Operations in an Underground Mine Using Big Data from an ICT-Based Mine Safety Management System, *Appl Sci*, 9:2639. https://doi.org/10.3390/app9132639

Bao, H, Knights, P F, Kizil, M and Nehring, M, 2024. Energy Consumption and Battery Size of Battery Trolley Electric Trucks in Surface Mines, *Energies*, https://doi.org/10.3390/en17061494

Bao, H, Knights, P, Kizil, M and Nehring, M, 2023a. Productivity estimation of battery trolley mining truck fleets, *Int J Min Reclam Environ*, https://doi.org/10.1080/17480930.2023.2278013

Bao, H, Knights, P, Kizil, M S, Nehring, M, 2023b. Electrification Alternatives for Open Pit Mine Haulage, *Mining*, 3:1–25. https://doi.org/10.3390/mining3010001

Cruzat, J V and Valenzuela, M A, 2017. Modeling and evaluation of benefits of trolley assist system for mining trucks, in *2017 IEEE Industry Applications Society Annual Meeting*, pp 1–10. https://doi.org/10.1109/IAS.2017.8101840

Fahl, S K and Askari-Nasab, H, 2017. Benefits of Discrete Event Simulation in Modeling Mining Processes, Mining Optimization Laboratory (MOL) – Report Eleven, Paper 204, pp 190–195, University of Alberta, Edmonton, Canada. https://doi.org/10.13140/RG.2.2.23709.69604

Huayanca, D, Bujaico, G and Delgado, A, 2023. Application of Discrete-Event Simulation for Truck Fleet Estimation at an Open pit Copper Mine in Peru, *Appl Sci*, 13:4093–4093. https://doi.org/10.3390/app13074093

Lindgren, L, Grauers, A, Ranggård, J and Mäki, R, 2022. Drive-Cycle Simulations of Battery-Electric Large Haul Trucks for Open pit Mining with Electric Roads, *Energies*, 15:4871–4871. https://doi.org/10.3390/en15134871

Navarra, A, 2023. Discrete Event Simulation for the Integrated Management of Mining and Metallurgical Systems, in *Proceedings of the 62nd Conference of Metallurgists, COM 2023*, pp 957–963 (Springer Nature Switzerland, Cham).

Navarro, V F, Mateus, G R, Martins, A G, Carneiro, W and Chaves, L S, 2020. Integrated optimization and simulation models for short-term open-pit mine planning, *J South Afr Inst Min Metall*, 120:311–320.

Nurić, S, Nurić, A and Brčaninović, M, 2009. Haulage solutions with trolley assist diesel-electric AC trucks on the pit mine RMU Banovici, *Journal of Mining and Metallurgy*, 45A(1):78–87.

Park, S, Jung, D and Choi, Y, 2023. Prediction of Ore Production in a Limestone Underground Mine by Combining Machine Learning and Discrete Event Simulation Techniques, *Minerals*, 13:830. https://doi.org/10.3390/min13060830

Patole, S, 2024. Using Simulation software Rockwell Arena for effective teaching of Value Stream Mapping in Undergraduate Lean Six Sigma Class, in *The Future of Engineering Education – 2024 Annual Conference & Exposition*, 14 p (American Society for Engineering Education).

Saïd, F, Eveno, I and Villaneau, J, 2022. Discrete-Events Simulation for Teaching Statistics in Industrial Engineering, *Athens J Technol Eng*, 9:61–76. https://doi.org/10.30958/ajte.9-1-4

Salama, A and Skawina, B, 2023. Selection of Discrete Event Simulation Software for Simulating Mining Operations, Tanzan, *J Engineering Technol*, https://doi.org/10.52339/tjet.v42i2.832

Wirantaya, D, Biyanto, T R and Satwika, N A, 2019. A study of operational deviation control to reduce diesel fuel consumption of a hundred tons mine haultruck, *AIP Conf Proc*, 2088:020009. https://doi.org/10.1063/1.5095261

Enhancing pit optimisation with direct block scheduling (DBS) and the Bienenstock Zuckerberg (BZ) algorithm – maximising NPV and efficiency

J Chung[1] and D Rahal[2]

1. Global Technical Advisor-Optimisation, Deswik Consulting, Perth WA 6000.
   Email: joyce.chung@deswik.com
2. Product Manager, Deswik Consulting, Brisbane Qld 4001. Email: david.rahal@deswik.com

ABSTRACT

Direct Block Scheduling (DBS) has become an essential tool in modern mine planning. It is particularly important in open pit optimisation because it can ensure an optimal net present value (NPV) by modelling intricate mining systems. DBS maximises NPV by leveraging dynamic cut-off grades, blending constraints, and capital expenditure options. This paper presents the application of DBS to integrate the Bienenstock Zuckerberg (BZ) algorithm with mixed-integer linear programming (MILP) and clustering algorithms for generating mining phases. This integration offers a highly efficient solution to complex pit optimisation challenges.

DBS excels in modelling multi-mine and multi-block model systems. The method offers a significant advantage in strategic planning, where ore blending strategies can incorporate multiple pits or regions. The early-stage integration of blending constraints in DBS ensures that ore quality is maintained, enhancing downstream processing efficiency. In addition, the ability to strategically manage capital expenditure phases over time makes DBS a powerful tool for long-term financial planning in mining projects.

The BZ algorithm enhances DBS by allowing it to solve complex scheduling problems within a reasonable time frame. The reduced time required to solve complex linear relaxations is a substantial improvement over the traditional MILP method. The computational efficiency of BZ algorithm allows for the generation of high-quality, optimised strategic mining strategy that aligns with both operational and financial objectives, even in large-scale mining environments.

The use of a clustering algorithm to augment DBS offers significant strategic advantages for schedule optimisation. It does this by grouping similar blocks based on DBS period, ore quality, proximity, or operational characteristics. This clustering of mining tasks into phases/pushbacks transitions the optimised mining sequence into an efficient mining production schedule. This reduces the size of the model to be solved.

This paper discusses the combined use of DBS (with the BZ algorithm) and clustering techniques. It demonstrates the collective impact on optimising pit designs, maximising NPV, managing capital expenditures, and enhancing operational profitability. The current approach offers superior results for mining operations that need a comprehensive solution to the challenge of producing an optimised schedule for complex mining operations.

INTRODUCTION AND PROBLEM STATEMENTS

Optimising the net present value (NPV) of mining projects remains a fundamental goal in open pit mine planning. It involves complex decision-making processes due to the multifaceted nature of mining operations, which require balancing economic efficiency with operational feasibility. In recent decades, the complexity of mining projects has increased significantly due to the growing scale of operations, the need to manage multiple orebodies simultaneously, stringent quality requirements, and heightened environmental considerations. This increased complexity necessitates sophisticated optimisation methodologies that can robustly handle multiple operational, financial, and logistical constraints simultaneously.

Direct Block Scheduling (DBS) has emerged as a crucial strategic tool, providing mine planners with advanced methodologies to address modern challenges. DBS maximises NPV by dynamically managing cut-off grades, blending constraints, and phased capital expenditures. It directly integrates economic and operational parameters, enabling detailed and realistic strategic planning. However, as mining operations scale up and become increasingly complex, traditional optimisation methods, such as Mixed-Integer Linear Programming (MILP), often fall short in computational efficiency, rendering them inadequate for practical, large-scale applications.

The core problem addressed in this research is the inherent limitations of existing DBS methodologies in handling large-scale, multi-pit, multi-block optimisation challenges effectively. Traditional DBS approaches face significant computational constraints, often resulting in impractical processing times in optimised solutions. To address this critical gap, this study proposes the integration of advanced computational techniques, specifically the Bienenstock Zuckerberg (BZ) algorithm and clustering methods, into DBS frameworks. The objective is to efficiently generate optimal schedules that simultaneously satisfy complex operational constraints, strategically manage capital expenditure, and maximise financial returns.

The Bienenstock Zuckerberg (BZ) algorithm, originally developed as a robust optimisation technique, significantly enhances the computational efficiency of DBS by efficiently solving linear relaxation problems common in large-scale optimisation scenarios. Its primary advantage lies in its iterative process, where solutions from previous iterations inform and accelerate subsequent computations. Unlike traditional MILP methods that can struggle with scalability and require excessive computational resources, the BZ algorithm employs a dual decomposition approach, decomposing the large-scale scheduling problem into smaller, more manageable sub-problems. These sub-problems are solved iteratively, converging rapidly to high-quality, feasible solutions. The computational efficiency of the BZ algorithm ensures practical applicability even for extensive and complex mining operations.

Clustering algorithms further augment DBS by providing practical solutions for grouping sequenced mining blocks based on critical operational parameters, such as proximity, ore quality, extraction timelines, and operational compatibility. These algorithms efficiently transform complex and detailed block schedules into simplified mining phases or pushbacks. This approach significantly improves the practicality and operational feasibility of schedules by ensuring smoother transitions from optimised theoretical sequences to executable mining plans.

By integrating the BZ algorithm with MILP and clustering algorithms, this research aims to significantly enhance the computational efficiency, accuracy, and practicality of DBS, providing robust solutions to complex pit optimisation challenges in multi-mine and multi-block scenarios.

LITERATURE REVIEW

Open pit optimisation has long focused on maximising NPV while addressing operational constraints inherent in mining projects. Early optimisation techniques, notably Lerchs-Grossmann and the floating cone methods, primarily emphasised economic viability but did not sufficiently account for practical constraints such as ore blending and phased capital expenditures (Whittle, 1999; Lerchs and Grossmann, 1965; Hustrulid and Kuchta, 2006). These shortcomings necessitated the development of more sophisticated optimisation models capable of addressing multiple operational and economic dimensions simultaneously (Kuchta, Newman and Topal, 2004). Among such developments, the Pseudoflow algorithm (Picard, 1976; Hochbaum and Chen, 2000) has gained attention for its computational efficiency in generating nested pit shells by solving maximum flow problems. While Pseudoflow improves performance over traditional methods, it still assumes immediate mining of pit shells and does not integrate scheduling logic or operational constraints over time.

DBS evolved as an advanced scheduling approach, providing significant advantages over traditional optimisation methods. DBS allows for dynamic determination of cut-off grades, direct integration of blending constraints, and strategic planning of capital expenditures (Morales and Rubio, 2010; Ramazan, 2007; Newman, Rubio and Martinez, 2010). The explicit inclusion of blending constraints into DBS at an early stage has proven essential in maintaining ore quality, thereby improving downstream processing efficiency and overall profitability (Askari-Nasab, Frimpong and Szymanski, 2011; Osanloo and Ataei, 2003).

Despite its benefits, conventional DBS methods employing MILP face computational limitations, particularly when applied to large-scale multi-pit or multi-block model scenarios. Computational complexity often leads to impractical solution times, undermining the applicability of DBS in expansive operational contexts (Lamghari and Dimitrakopoulos, 2012).

Recent advancements have highlighted the effectiveness of heuristic and metaheuristic approaches in overcoming these computational barriers (Muir and Dimitrakopoulos, 2019). Among these, the Bienenstock Zuckerberg (BZ) algorithm has gained prominence due to its ability to rapidly solve linear relaxation problems. Originally developed for solving large-scale precedence constrained linear programming problems, the BZ algorithm leverages an iterative and decomposition-based method to enhance the speed and scalability of solving complex optimisation problems (Bienstock and Zuckerberg, 2010; Chicoisne et al, 2012; Lambert, Brickey and Newman, 2014). Extensive studies have demonstrated its applicability and efficiency in addressing open pit mine scheduling problems, highlighting significant reductions in computational effort compared to traditional MILP solutions (Espinoza, Goycoolea and Moreno, 2013; Lambert, BrickeyA and Newman, 2014).

Further, clustering algorithms have gained attention in translating DBS results into practical mining phases. Clustering methods, including K-means, hierarchical clustering, and fuzzy clustering, significantly reduce the complexity of mine scheduling by aggregating blocks based on parameters such as proximity, ore quality, and operational characteristics (Dagdelen and Kawahata, 2008; Tabesh, Askari-Nasab and Frimpong, 2015; Samavati, Essam and Sarker, 2018). These methods facilitate the transition from theoretical optimisation to practical production schedules, improving operational efficiency and reducing uncertainties in the execution phase.

Overall, integrating these advanced computational techniques, notably the BZ algorithm and clustering methods, represents a significant advancement in DBS capabilities, providing solutions to the complex scheduling challenges faced by modern mining operations.

APPLICATIONS AND DISCUSSIONS

This study employs a block model to evaluate the effectiveness of DBS, enhanced by integrating the BZ algorithm and a fuzzy clustering algorithm. The block models, illustrated in Figure 1, are used as case studies to demonstrate the proposed optimisation methodology. Each block model comprises approximately 1.5 million blocks, with key characteristics summarised in Table 1. Figure 1 presents multiple section views of the deposit, highlighting the distribution of nickel (Ni) grades within the model.

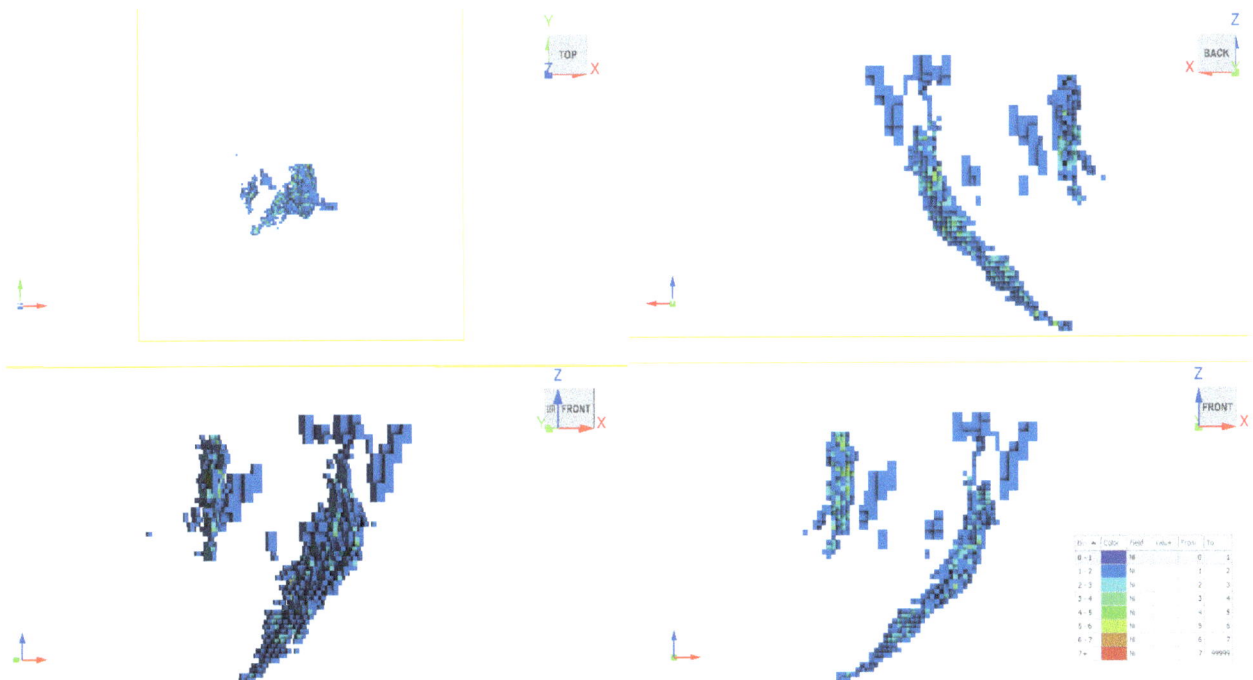

FIG 1 – Deposit section views.

TABLE 1

Block model summary.

Legend (Ni range)	Mass	Ni%	Cu	Co
0 to 1	7 585 599 642	0.01	0.03	0.00
1 to 2	10 862 837	1.36	0.99	0.03
2 to 3	3 460 971	2.46	0.98	0.08
3 to 4	1 734 405	3.44	1.29	0.11
4 to 5	1 053 408	4.47	1.76	0.14
5 to 6	674 832	5.44	2.13	0.18
6 to 7	266 334	6.37	2.50	0.19
7+	120 280	7.27	2.88	0.22

To implement this optimisation, a commercial strategic mine planning tool was used. The software integrates DBS functionalities and leverages the computational efficiency of the BZ algorithm to solve large-scale linear relaxations within a reasonable time frame. Additionally, it incorporates a fuzzy clustering algorithm that transforms block-level DBS outputs into practical, executable mining phases, ensuring alignment with operational constraints.

Table 2 summarises the operational parameters applied in this study. These include mining and milling capacities, grade constraints, and vertical advance limits—factors critical for realistic, implementable mine schedules.

TABLE 2

Operational parameters used in the study.

Mining capacity	50 million tpa
Milling capacity	10 million tpa
Grade constraint	0.55% Ni (Max)
	0.45% Ni (Min)
	0.7% Cu (Max)
Vertical advance rate	Six benches per annum

Conventional method – pit optimisation and phase bench schedule

In conventional open pit optimisation workflows, techniques such as the Pseudoflow algorithm or the Lerchs-Grossmann (LG) method are commonly employed to generate a series of nested revenue pit shells. These revenue shells represent potential pit limits based on varying economic factors, serving as a foundational step in strategic planning. Figure 2 illustrates an example of the revenue shells generated by the Pseudoflow method.

Following the optimisation, a manual phase selection process is typically carried out. This involves selecting specific revenue shells to define practical mining phases that can be implemented in scheduling. In the case presented here, the optimisation results, summarised in Figure 2, guided the selection of revenue factors 0.42, 0.45, and 0.99 to represent the intermediate phases and the ultimate pit limit. The figure highlights the selections as Phases 1, 2, and 3.

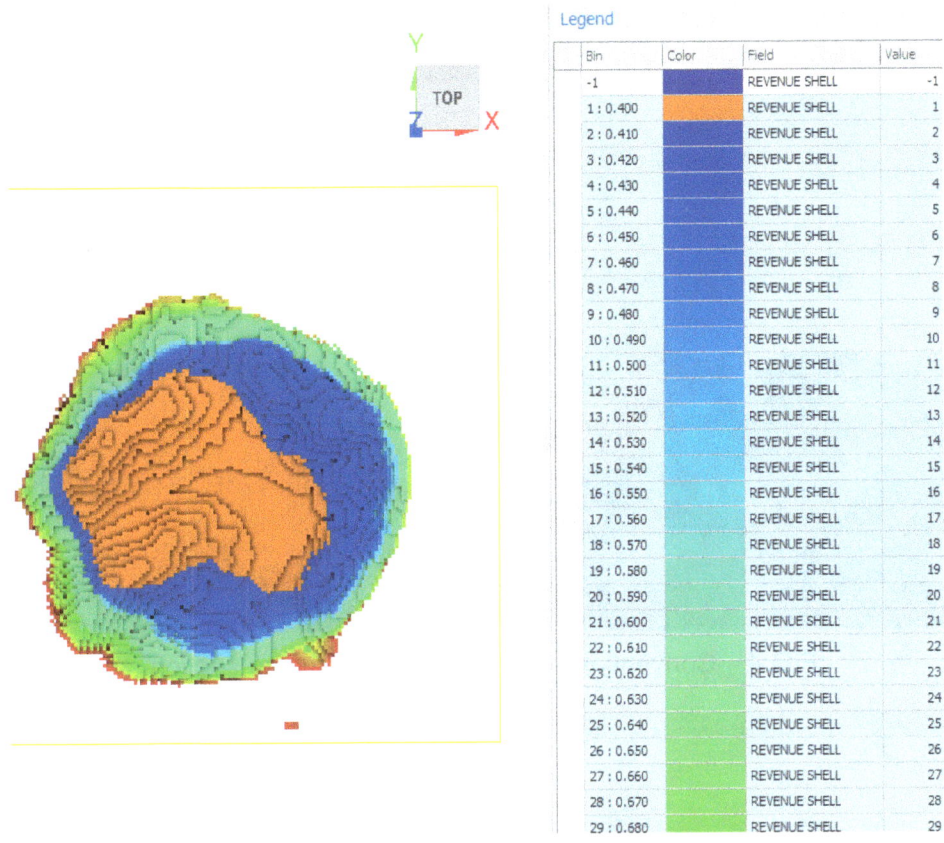

FIG 2 – Example revenue shells generated by Pseudoflow algorithm.

Using these selected phases, a Phase Bench Scheduling (PBS) process was conducted to develop a production schedule over time. The PBS has considered all the operational parameters listed in Table 2. The outcome of this schedule is illustrated in Figure 3, which displays the NPV and individual material NPV contributions. The resulting NPV from this phase-based schedule is $2.70 billion, representing a 38 per cent reduction compared to the best-case NPV of $4.36 billion. The best-case shells achieved in the pit optimisation, as shown in Figure 3, did not include the operational capacity, grade and vertical advance constraints.

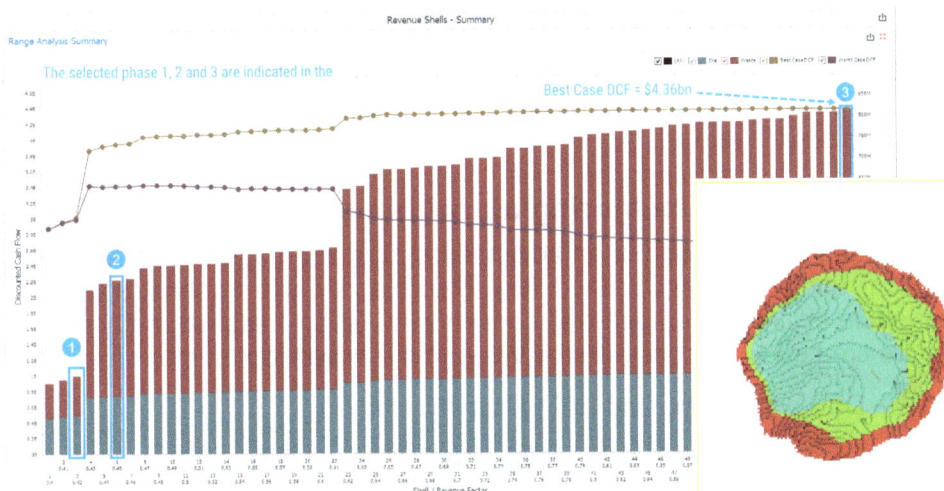

FIG 3 – Example of revenue shells ore and waste curve.

This reduction demonstrates a key limitation of the conventional approach. While pit optimisation identifies economic pit limits, the manual selection of intermediate phases (see Phases in Figure 4) often introduces inefficiencies and rigidities that prevent the full realisation of the optimal NPV. The loss of value stems from a lack of flexibility in integrating downstream constraints, such as blending, cut-off grade dynamics, and capital expenditure phasing, during the initial-phase definition process.

Figure 5 shows that the cumulative DCF decreased to $2.70 billion once time-based constraints were added to the best-case schedule from the revenue shells (Figure 3).

	PHASE	0
	PHASE	1
	PHASE	2
	PHASE	3

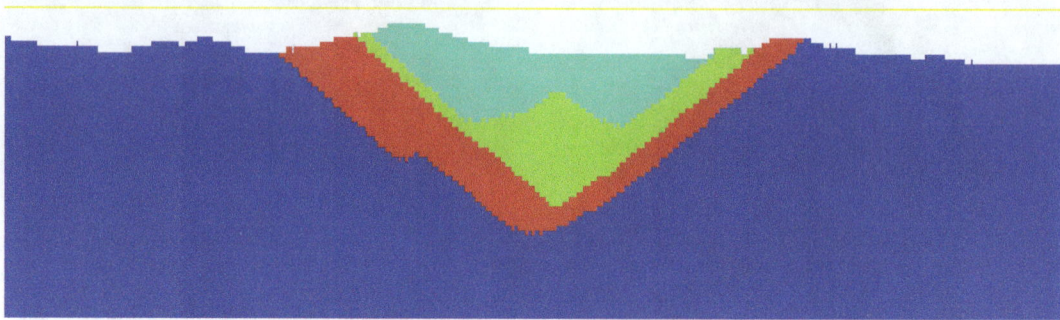

FIG 4 – Conventional method phases – cross-section.

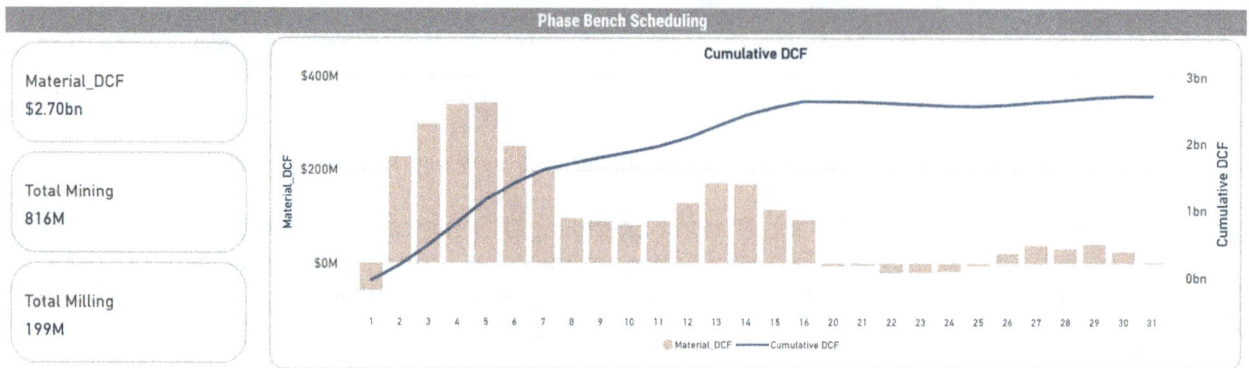

FIG 5 – PBS result for conventional approach.

Additionally, the Conventional Method cannot correctly account for the time value of money, as it doesn't schedule the block model to generate the shells; instead, it solves for a maximum value at a point in time, as if the deposit were mined instantly. The NPVs generated are a post-process calculation, assuming a very basic style of scheduling, ie top-down inner shell to outer shell, which is unlikely to be replicated in reality.

Advanced method – DBS, clustering algorithm and phase bench schedule

This section presents the application of an advanced optimisation approach that integrates DBS, a clustering algorithm, and phase bench scheduling. The objective is to demonstrate the improvements in NPV and schedule efficiency compared to conventional pit optimisation methods. In this section, four scenarios are created to demonstrate the value of the advanced method:

- Scenario 1 – DBS with mining and milling constraints
- Scenario 2 – DBS with mining, milling and blending constraints
- Scenario 3 – DBS with mining and milling constraints with vertical advancement rate limit
- Scenario 4 – DBS with mining, milling and blending constraints with vertical advancement rate limit.

Each of the above scenarios provides a DBS result that will be an input to the clustering algorithm. Once the clustering algorithm generates mining phases, these phases will be used as input to perform the PBS. In the PBS step, all the operational constraints are considered.

Scenario 1 – DBS with mining and milling constraints

In this scenario, DBS is applied with mining and milling capacity constraints without predefining any phases. The system dynamically optimises the block extraction sequence while respecting some of the operational parameters in Table 2. Figure 6 (left) shows the optimised extraction sequence prior to generating mining phases. The optimised block schedule yields a material NPV of $4.07 billion, with 816 Mt mined and 145 Mt milled (Figure 7 top).

A fuzzy clustering algorithm is then used to group the optimised blocks into operational mining phases as shown in Figure 6 (right) and Figure 8. These phases are refined into an executable plan through phase bench scheduling, resulting in an NPV of $2.88 billion (Figure 7 bottom). The phase bench schedule ensures that the optimised plan is translated into a structured, implementable production schedule.

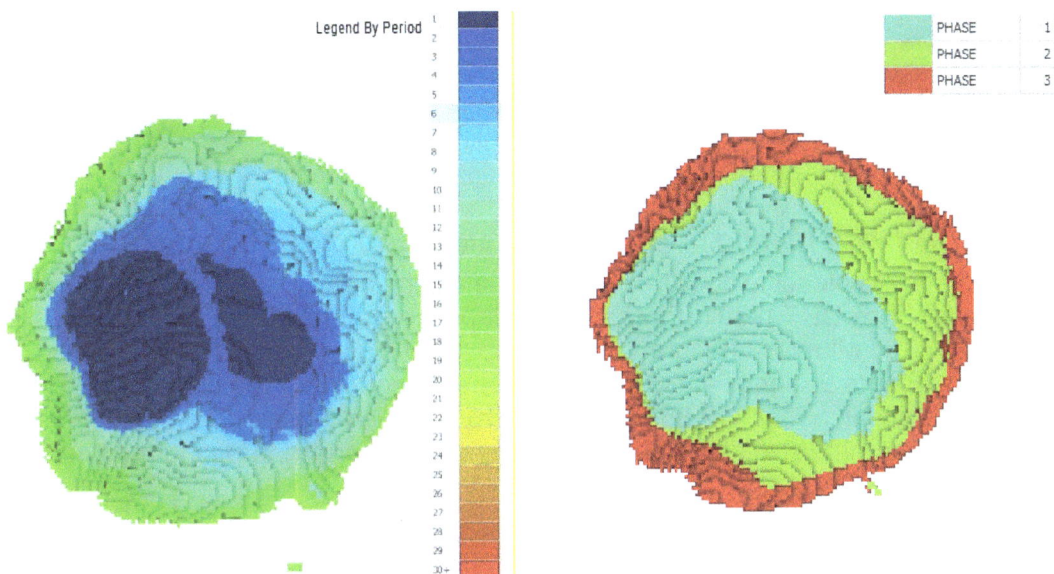

FIG 6 – left: DBS schedule by period; right: clustering algorithm output – defined mining phases.

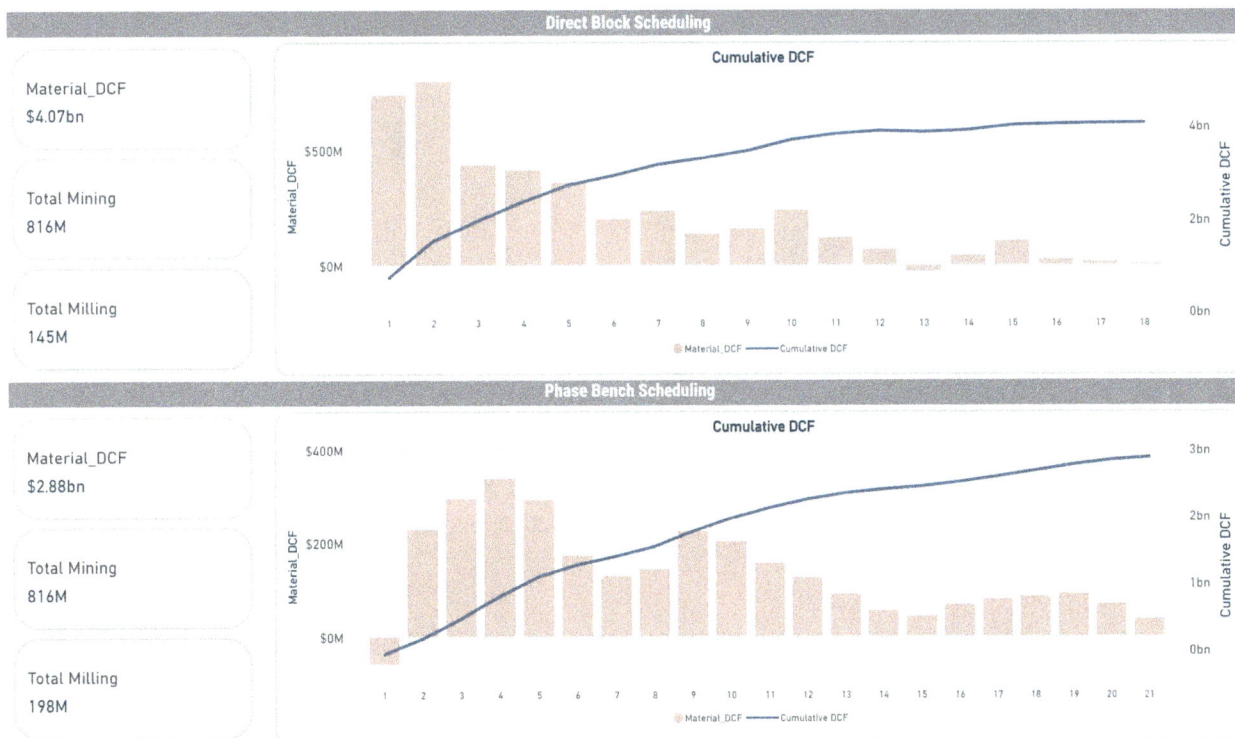

FIG 7 – top: DBS result – material NPV and cumulative NPV; bottom: phase bench scheduling result.

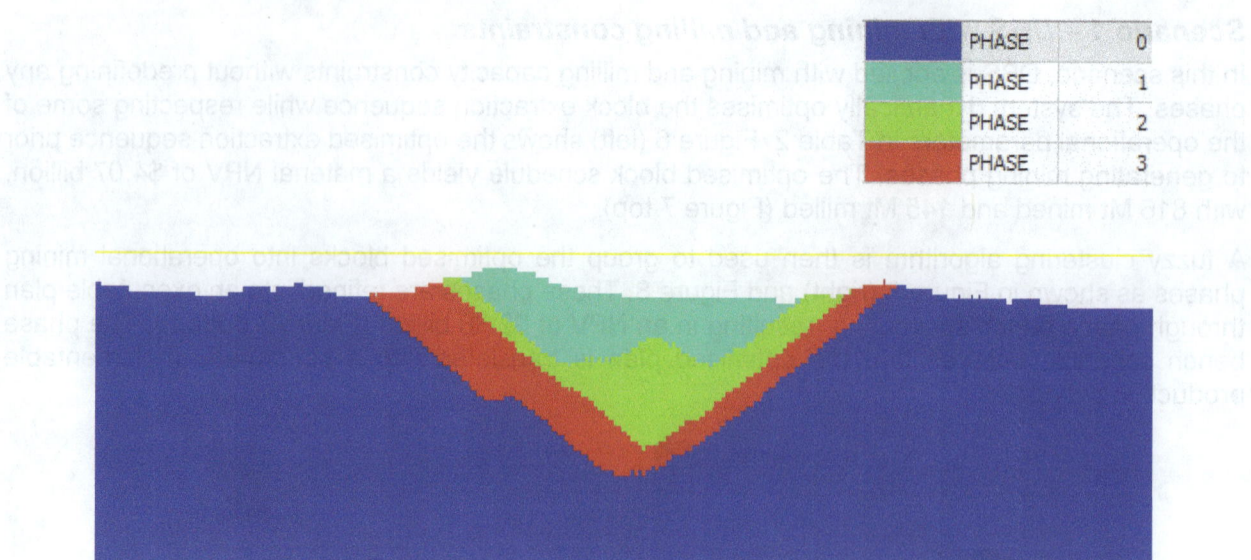

	PHASE	0
	PHASE	1
	PHASE	2
	PHASE	3

FIG 8 – Scenario 1 phases – cross-section.

Scenario 2 – DBS with mining, milling constraints and blending constraints

Scenario 2 incorporates blending constraints alongside mining and milling capacities to manage feed grade consistency. The DBS result (Figure 9, left) produces a material NPV of $3.45 billion, with 816 Mt mined and 136 Mt milled (Figure 10, top). The fuzzy clustering algorithm groups blocks into practical phases (Figure 9, right), and Figure 11. Subsequent phase bench scheduling delivers a feasible schedule with an NPV of $2.88 billion with 816 Mt milled (Figure 10 bottom). This approach balances optimisation objectives with blending requirements, ensuring economic and operational feasibility.

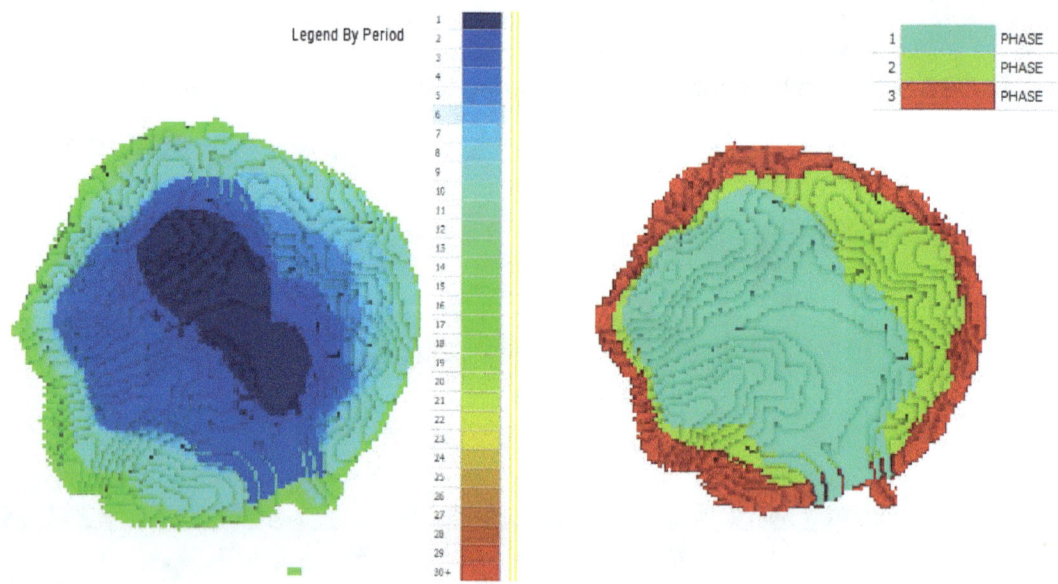

FIG 9 – left: Scenario 2 – DBS output by period (mining, milling and blending constraints); right: Scenario 2 – phase definition via fuzzy clustering.

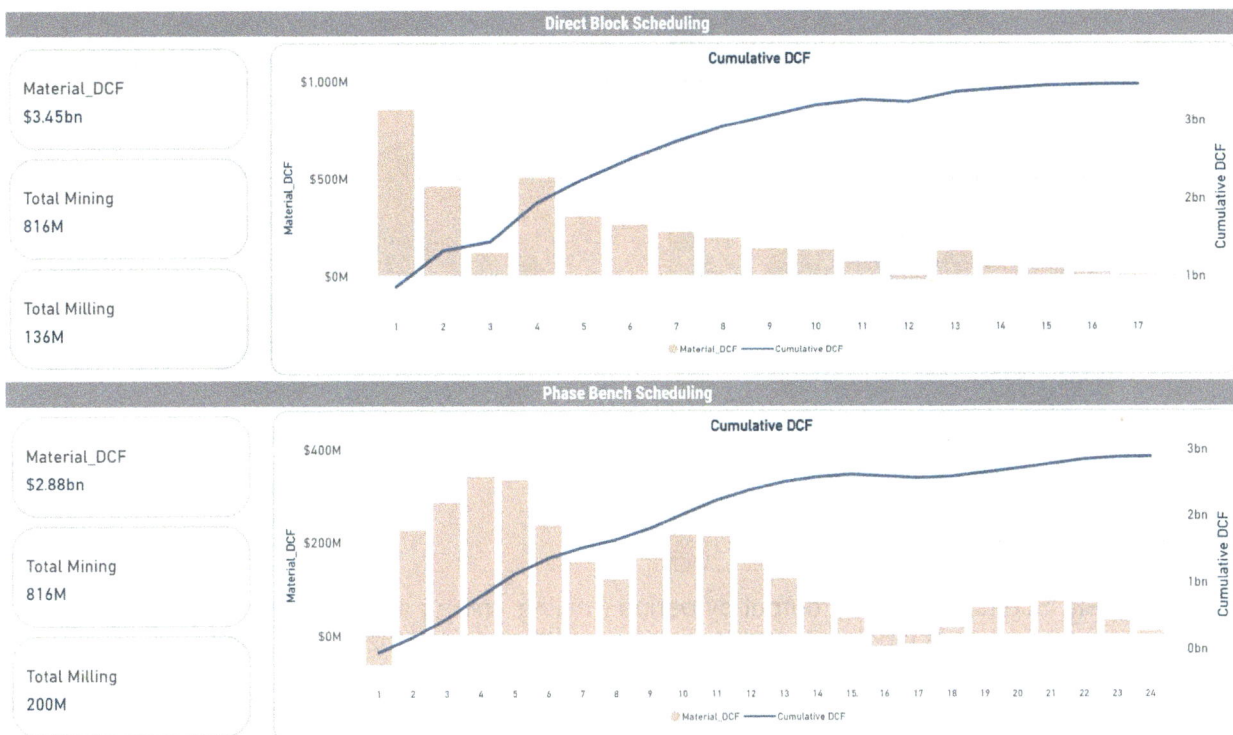

FIG 10 – top: Scenario 2 – DBS result (material NPV and cumulative NPV); bottom: Scenario 2 – phase bench scheduling result.

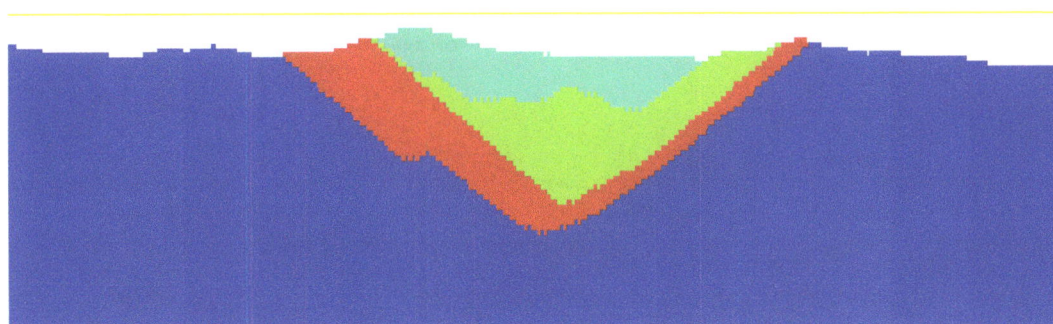

FIG 11 – Scenario 2 phases – cross-section.

Scenario 3 – DBS with mining and milling constraints with vertical advancement rate limit

In Scenario 3, DBS is applied with mining and milling capacity constraints along with a vertical advancement rate limit to control vertical advancement. The DBS result (Figure 12 left) achieves a material NPV of $3.63 billion, with 725 Mt mined and 133 Mt milled (Figure 13 top). Using fuzzy clustering, mining phases are generated (Figure 12 right and Figure 14), and the phase bench schedule finalises an actionable schedule with an NPV of $2.72 billion (Figure 13 bottom). This ensures a realistic schedule while complying with vertical advancement rate limits and capacity constraints.

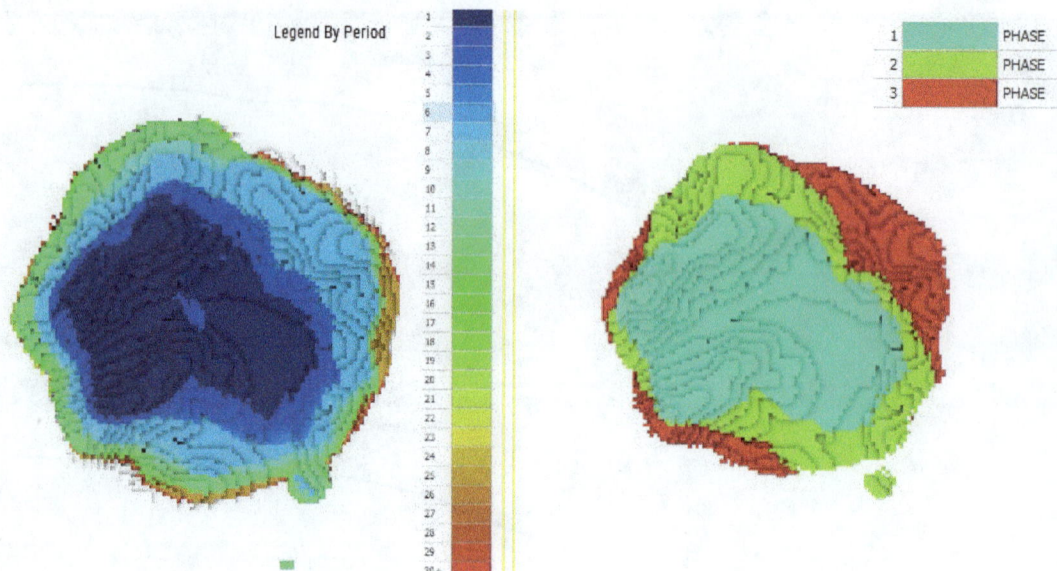

FIG 12 – left: Scenario 3 – DBS output by period (mining, milling and vertical advancement rate constraints); right: Scenario 3 – phase definition via fuzzy clustering.

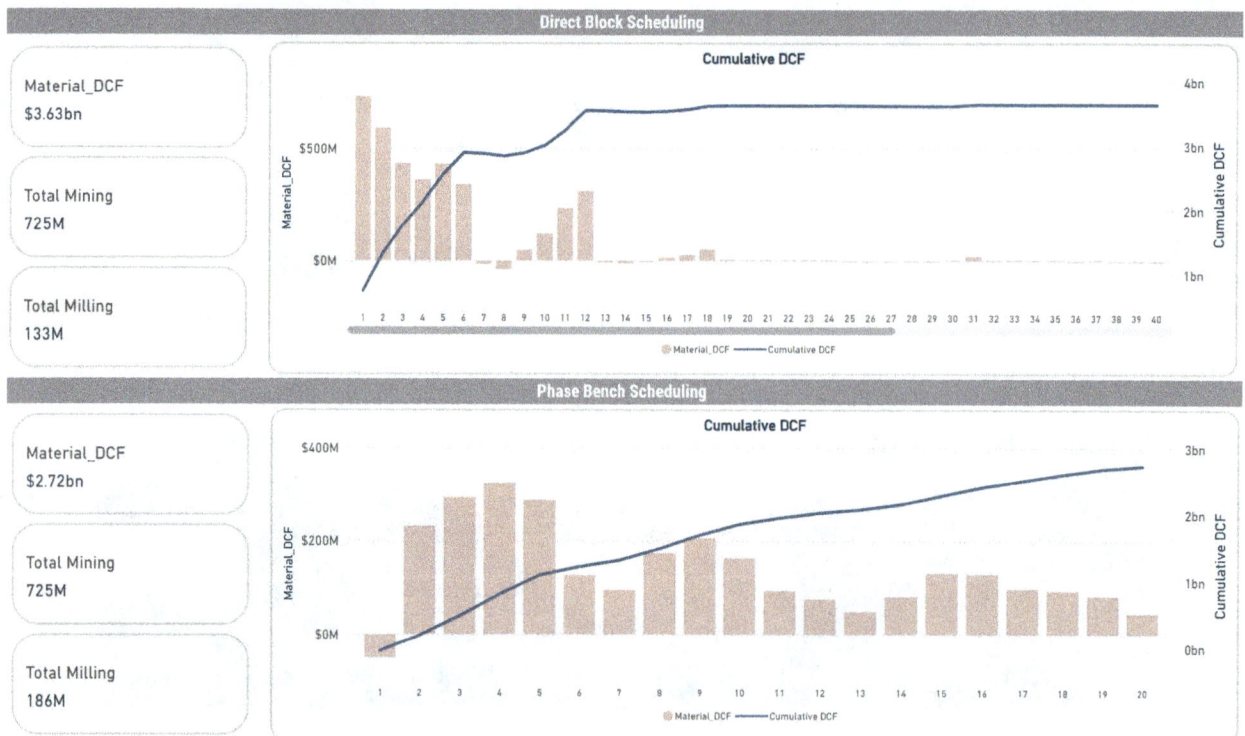

FIG 13 – top: Scenario 3 – DBS result (material NPV and cumulative NPV); bottom: Scenario 3 – phase bench scheduling result.

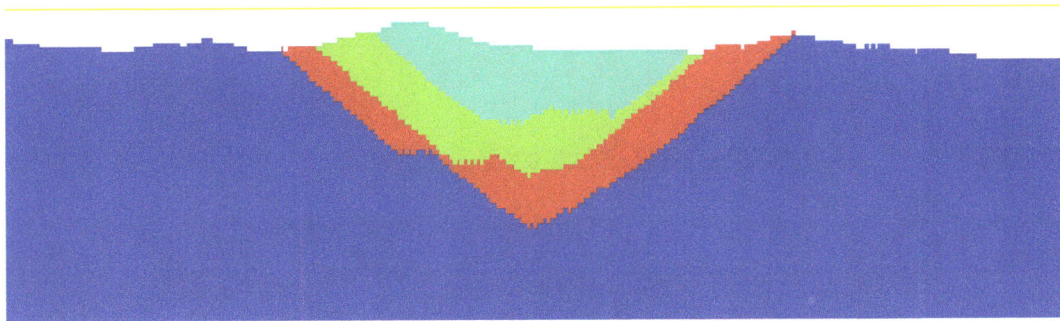

FIG 14 – Scenario 3 phases – cross-section.

Scenario 4 – DBS with mining, milling and blending constraints with vertical advancement rate limit

Scenario 4 applies DBS with the most comprehensive constraint set, incorporating mining, milling, blending constraints, and vertical advancement rate limitations to fully reflect operational complexities.

The DBS optimisation result (Figure 15 left) produces a material NPV of $3.36 billion (Figure 16 top). Blocks are grouped into operational phases using fuzzy clustering (Figure 15 right and Figure 17), followed by phase bench scheduling. The final schedule achieves an NPV of $2.90 billion (Figure 16 bottom). This demonstrates the advanced method's ability to handle complex, multi-dimensional constraints while ensuring the production schedule remains practical, executable, and aligned with operational realities.

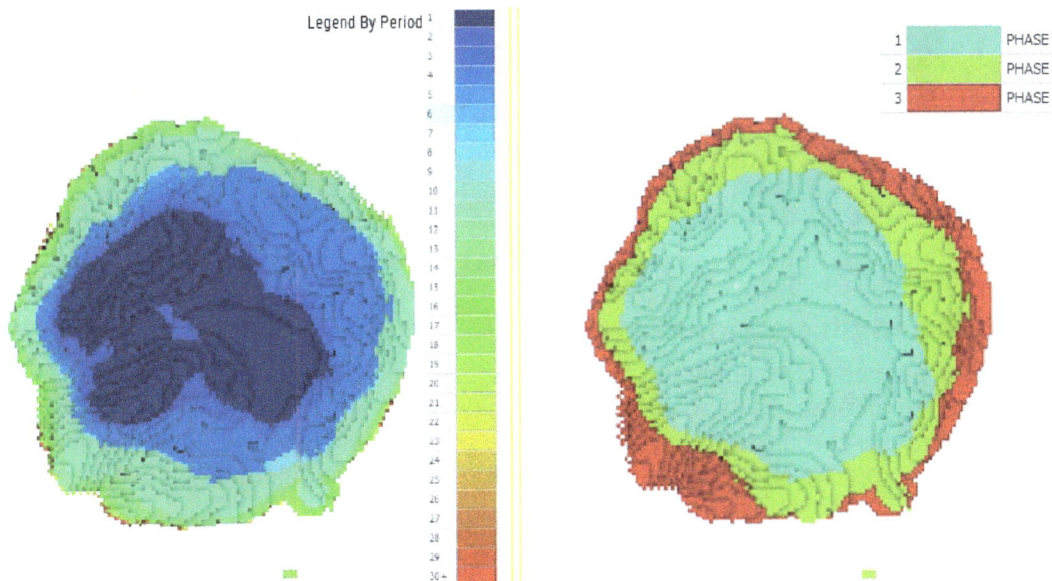

FIG 15 – left: Scenario 4 – DBS output by period (mining, milling, blending and vertical advancement constraints); right: Scenario 4 – phase definition via fuzzy clustering.

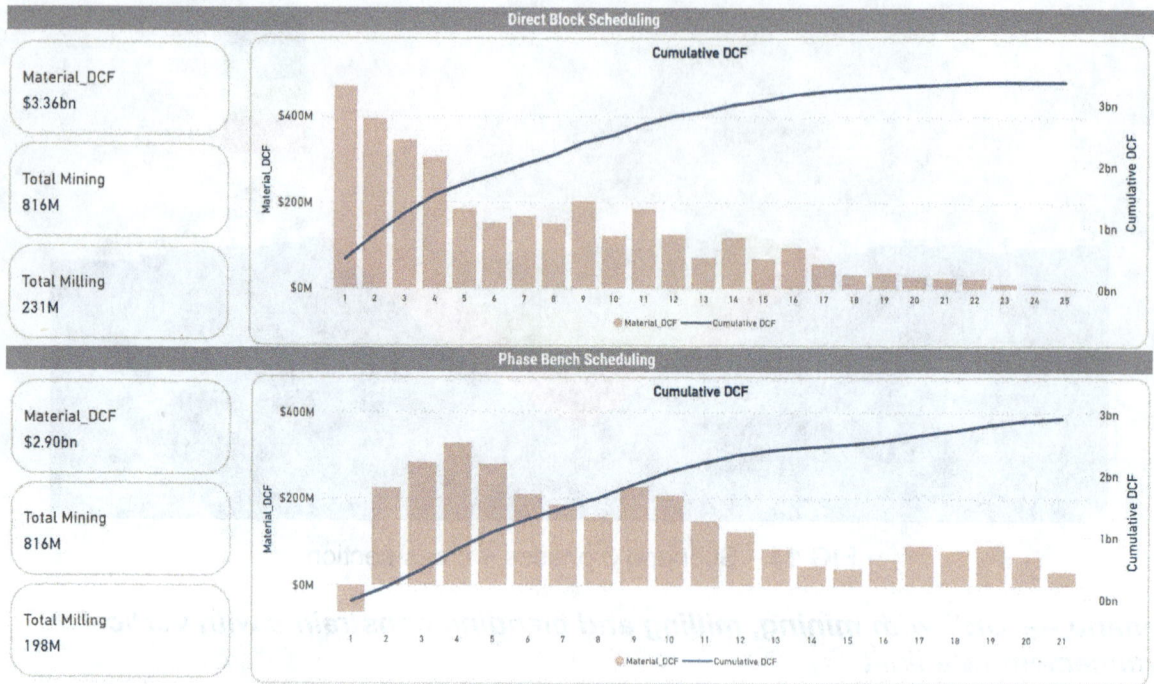

FIG 16 – top: Scenario 4 – DBS result (material NPV and cumulative NPV); bottom: Scenario 4 – phase bench scheduling result.

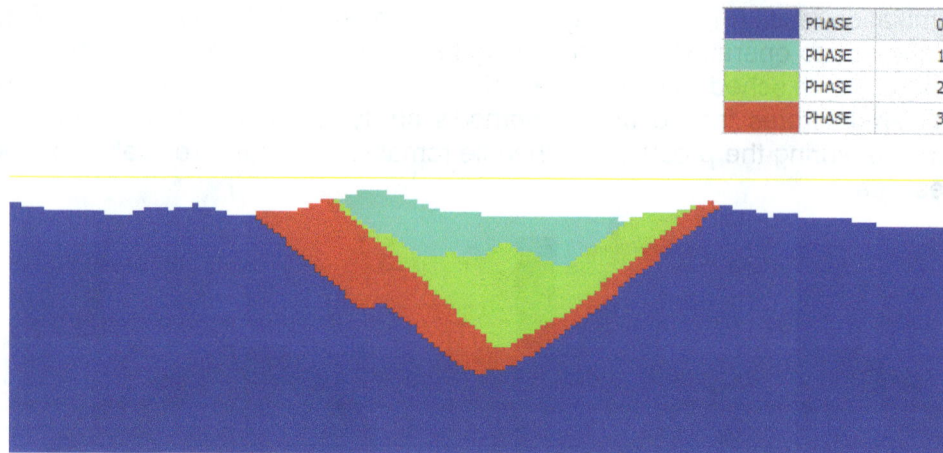

FIG 17 – Scenario 4 phases – cross-section.

Comparisons and discussions

The comparison between conventional pit optimisation methods and advanced DBS approach reveals fundamental differences in how theoretical value translates into executable mine plans. The most striking contrast is the discrepancy between the theoretical NPV produced by pit optimisation techniques and the practical NPV realised through production scheduling.

Conventional workflows rely heavily on pit shell generation algorithms, such as Lerchs-Grossmann (LG) or Pseudoflow—to identify a series of nested pits based on revenue factors. These shells are often interpreted as proxies for mining phases and are subsequently used to construct phase-based production schedules. However, this methodology introduces several structural inefficiencies:

- Overestimation of NPV in Pit Shell Generation

  Pit shell generation assumes that the entire volume within a shell can be mined instantly, ignoring temporal dependencies and constraints such as vertical advance rates, equipment availability, and scheduling logistics. This simplification leads to an inflated NPV that does not reflect the time value of money in a practical sequence. The theoretical NPV of $4.36 billion, derived from pit optimisation in this study, assumes perfect operational execution with no lag, dilution, or phase transition costs—conditions that are unrealistic in actual mining environments.

- Inflexibility in Phase Selection

  The manual selection of shells to define mining phases is inherently rigid and lacks dynamic adaptability to operational constraints. Decisions made at this stage—such as which shells to select and how to split them—are typically driven by visual inspection or heuristics, which may not optimally balance ore quality, blending requirements, or infrastructure timing. This leads to the observed 38 per cent drop in NPV (from $4.36 billion to $2.70 billion) after phase bench scheduling is applied in the conventional method.

- Delayed Incorporation of Constraints

  In conventional methods, the impact of key operational constraints such as blending, mill feed quality, and capital deployment schedules are captured after the pit limits and phases are established. This decoupling of economic optimisation from practical mining constraints introduces inefficiencies, as these constraints may conflict with the geometry or economics of selected phases.

The conventional pit optimisation method, employing revenue shells followed by manual selection of intermediate phases and PBS, resulted in a final NPV of approximately $2.70 billion. This represents a substantial 38 per cent decrease from the theoretically achievable best-case NPV of $4.36 billion identified in the pit optimisation process. This significant discrepancy highlights the inherent inefficiencies associated with traditional optimisation approaches, particularly regarding the manual selection of intermediate mining phases. Such manual selections often lack the flexibility needed to dynamically address key operational factors, such as varying cut-off grades, blending requirements, CAPEX timing, and other operational constraints.

In contrast, the advanced DBS approach integrates economic and operational considerations from the outset. It determines optimal block sequences based on actual scheduling logic, accounting for capacities, cut-off grades, blending windows, and other constraints. Importantly, DBS generates a time-dependent sequence, inherently preserving the time value of money during NPV calculation.

The integration of the BZ algorithm allows for solving the optimisation model efficiently, even with very large data sets (block models). In addition, fuzzy clustering bridges the gap between theoretical block sequences and practical mining phases by intelligently grouping blocks based on multiple criteria. This process reduces model size while maintaining adherence to scheduling logic, ensuring the resulting production schedule remains close to optimal.

The results of this integrated method are summarised in Table 3. Although transitioning from DBS to PBS results in a value loss of 14–29 per cent, this is a significant improvement over the 38 per cent loss observed in conventional methods. Moreover, the absolute NPV values achieved by PBS under DBS scenarios consistently outperform the conventional approach by 10–25 per cent, demonstrating the superiority of the integrated approach.

TABLE 3

Result comparison.

Scenario	DBS NPV (billion USD)	PBS NPV (billion USD)	Δ (DBS – PBS) %
Conventional method (considering all operational parameters)	$4.36	$2.70	38%
Scenario 1	$4.07	$2.88	29%
Scenario 2	$3.45	$2.88	17%
Scenario 3	$3.63	$2.72	25%
Scenario 4	$3.36	$2.90	14%

In summary, these results underscore the limitations of traditional pit shell approaches, where overestimation of value and under-integration of constraints impair decision-making. By contrast, the DBS methodology, with BZ optimisation and clustering algorithm, produces schedules that more accurately reflect the real-world potential of mining projects. This leads to better-informed strategic decisions and a tighter alignment between theoretical models and operational outcomes.

CONCLUSIONS

Integrating DBS with the BZ algorithm and clustering methods significantly enhances open pit mine planning, addressing limitations inherent in conventional optimisation approaches. Traditional methods, which rely on rigid, manually selected mining phases, often result in substantial inefficiencies and a notable reduction in NPV (after PBS), as evidenced by a 38 per cent decrease from the theoretical maximum.

The DBS approach consistently demonstrates superior economic performance by dynamically addressing operational constraints such as blending, dynamic cut-off grades, and capital expenditure phasing early in the scheduling process. Although transitioning from theoretical DBS outcomes to practical PBS introduces a 14–29 per cent reduction in NPV, the resulting plans still outperform conventional methods by 10–25 per cent. Scenario 4, which incorporates comprehensive constraints (mining, milling, blending, and vertical advancement), clearly illustrates this advantage, achieving a practical PBS NPV of $2.90 billion, significantly exceeding the results of conventional methods.

Moreover, the integration of the BZ algorithm dramatically improves computational efficiency, enabling practical optimisation of large-scale, multi-constraint mine planning scenarios. The incorporation of clustering algorithms further enhances operational feasibility by effectively transitioning theoretically optimised schedules into actionable mining phases.

In conclusion, the advanced DBS methodology presented in this paper offers a robust, computationally efficient, and economically advantageous approach to modern open pit mine planning. It enables mining operations to maximise profitability while simultaneously maintaining operational practicality, ensuring competitiveness in today's increasingly complex mining landscape.

ACKNOWLEDGEMENTS

We gratefully acknowledge our colleagues at Deswik for their thorough peer reviews and constructive feedback, which significantly contributed to the quality and clarity of this paper. Additionally, we extend our sincere appreciation to our partner, Alicanto Labs, for their continued commitment and hard work in enhancing the efficiency and effectiveness of the BZ algorithm.

REFERENCES

Askari-Nasab, H, Frimpong, S and Szymanski, J, 2011. Open-pit optimization using mixed integer linear programming, *Mining Technology*, 120(1):2–10.

Bienstock, D and Zuckerberg, M, 2010. Solving LP relaxations of large-scale precedence constrained problems, in *International Conference on Integer Programming and Combinatorial Optimization*, pp 1–14 (Springer: Berlin).

Chicoisne, R, Espinoza, D, Goycoolea, M, Moreno, E and Rubio, E, 2012. A New Algorithm for the Open-Pit Mine Production Scheduling Problem, *Operations Research*, 60(3):517–528.

Dagdelen, K and Kawahata, K, 2008. Strategic mine planning with clustering algorithms, *Mining Technology*, 117(4):134–141.

Espinoza, D, Goycoolea, M and Moreno, E, 2013. Advances in mine optimization algorithms, *Computers and Operations Research*, 40(1):107–118.

Hochbaum, D S and Chen, A, 2000. Performance analysis and best implementations of old and new algorithms for the open-pit mining problem, *Operations Research*, 48(6):894–914. https://doi.org/10.1287/opre.48.6.894.12095

Hustrulid, W and Kuchta, M, 2006. *Open Pit Mine Planning and Design*, 2nd edn (CRC Press).

Kuchta, M, Newman, A M and Topal, E, 2004. Implementing mixed integer linear programming models for production scheduling, *Journal of Mining Science*, 40(6):554–565.

Lambert, W, Brickey, A and Newman, A, 2014. A study of the Bienstock-Zuckerberg algorithm: applications in mining and resource constrained project scheduling, *Computers and Industrial Engineering*, 72:131–140.

Lamghari, A and Dimitrakopoulos, R, 2012. Large-scale open pit mine scheduling with uncertainty, *International Journal of Mining, Reclamation and Environment*, 26(5):345–362.

Lerchs, H and Grossmann, I F, 1965. Optimum design of open-pit mines, *CIM Bulletin*, 58:17–24.

Morales, N and Rubio, E, 2010. Block scheduling with mixed integer programming in mining operations, *European Journal of Operational Research*, 207(3):1443–1451.

Muir, W and Dimitrakopoulos, R, 2019. Metaheuristics for mining optimization, *Computers and Operations Research*, 115:193–205.

Newman, A M, Rubio, E and Martinez, M, 2010. Production Scheduling for Strategic Open Pit Mine Planning: A Mixed-Integer Programming Approach, *Interfaces*, 40(3):222–236.

Osanloo, M and Ataei, M, 2003. Open pit optimization with blending constraints, *Mining Technology*, 112(1):53–57.

Picard, J C, 1976. Maximal closure of a graph and applications to combinatorial problems, *Management Science*, 22(11):1268–1272. https://doi.org/10.1287/mnsc.22.11.1268

Ramazan, S, 2007. The new optimization approach in mine planning, *Mining Technology*, 116(1):34–57.

Samavati, M, Essam, D and Sarker, R, 2018. Clustering methods in mine planning, *Mining Engineering*, 70(8):45–52.

Tabesh, M, Askari-Nasab, H and Frimpong, S, 2015. Application of clustering algorithms in mine planning, *Minerals Engineering*, 81:84–95.

Whittle, J, 1999. The Whittle method, *SME Mining Handbook*, Society for Mining, Metallurgy and Exploration.

From static models to dynamic solutions – the future of resource modelling

H Dillon[1] and J Mackenzie[2]

1. Global Customer Success Manager – Geoscience, Maptek, Christchurch 8025 New Zealand.
   Email: henry.dillon@maptek.com.au
2. Global Strategy Manager – Core Technologies, Maptek, Brisbane Qld 4000.
   Email: james.mackenzie@maptek.com.au

ABSTRACT

Mining businesses must adapt to stay competitive in the face of increasing deposit complexity. Technological advances in automation and machine learning (ML) promise a way forward for smarter resource modelling. These advances deliver more than just efficiency; they represent a paradigm shift in resource estimation (Vespignani and Smyth, 2024). This paper explores how leveraging advanced technologies, cloud-based simulation, and integration with desktop solutions can overcome the inherent limitations of traditional practices, providing unprecedented insights and precision.

Historically, the software tools for resource estimation have often functioned in isolation, creating data silos and inefficiencies. Introducing cloud-integrated simulations and fostering seamless platform interoperability and visualisation at end-user workstations can unlock dynamic solutions that better serve geologists, engineers, and decision-makers. In this narrative, we demonstrate how breaking down these silos can enable real-time collaboration, more accurate data reconciliation, and an overall increase in confidence for block model outputs.

This paper focuses on the transformative potential of artificial intelligence (AI) and machine learning (ML). By feeding real-time data into dynamic models, these technologies can refine estimates, predict, and adjust for uncertainties in ways that traditional methods, such as conditional simulation, struggle to achieve. The ability to run cloud-based simulations through dynamic cloud-predicted domain models offers a future where resource models become truly adaptive, responding dynamically to changes in input conditions.

The authors discuss an automation and software interoperability framework that supports integration across geological, geostatistical, and engineering tools. This approach preserves data integrity while optimising system usability, paving the way for a new era in resource estimation. The paper concludes with strategic recommendations for fostering a culture of innovation and ensuring that industry adoption of these emerging technologies keeps pace with evolving challenges.

INTRODUCTION

The increasing geological and operational complexity of mineral deposits presents significant challenges for resource estimation and mine planning. Traditional approaches to resource modelling rely on static models, which, although effective historically, often struggle to adapt to the dynamic nature of mining data. As deposits become smaller, more geologically complex, and more challenging to mine, precise delineation of ore-waste boundaries is vital to optimising mill recovery and maintaining economic viability.

In response, technological advancements, particularly in artificial intelligence (AI), machine learning (ML), and cloud-based simulations, offer a new paradigm in which resource models can refine and adapt based on real-time data inputs. Static models in resource estimation typically depend on historical data and predefined assumptions to predict orebody geometry, but the process of updating models whenever new geological or operational data emerges is often time-consuming and prone to human error. Furthermore, traditional geostatistical methods (eg kriging, conditional simulation) offer limited ability to incorporate real-time data streams effectively.

This paper illustrates how AI, ML, and cloud-based simulations can move resource modelling from manual processes to dynamic, continuously evolving systems. By leveraging these technologies, resource models become self-updating and predictive, empowering mining professionals to make more informed, data-driven decisions in real time. Although cloud computing, data integration, and workflow automation are essential for modern mining operations, we focus on showing how real-time data can enhance geological and geostatistical modelling, ultimately improving decision-making and resource estimation accuracy.

We also present a high-level framework for integrating AI-driven modelling solutions into existing mining workflows, demonstrating the potential to revolutionise resource estimation, operational efficiency, and long-term planning.

ADVANCEMENTS IN ARTIFICIAL INTELLIGENCE AND MACHINE LEARNING FOR RESOURCE MODELLING

The application of AI and ML in mineral resource estimation has expanded considerably in recent years (Azhari et al, 2023). These technologies excel at identifying complex, non-linear patterns within geological data, often surpassing traditional geostatistical methods in both adaptability and predictive power (Dumakor-Dupey and Arya, 2021). Dynamic modelling approaches (where real-time data integration refines estimates continuously) can reduce reconciliation errors and improve forecasting accuracy.

A systematic review highlights the efficiency gains from ML-driven techniques in mineral exploration, orebody delineation, and resource estimation. Approaches such as neural networks, random forests, and support vector machines have shown greater accuracy than conventional kriging methods (Galetakis et al, 2022). Deep learning algorithms have also played an increasing role in automated geological modelling, expediting the interpretation of complex subsurface structures.

Cloud computing and big data integration

Cloud computing has emerged as a critical enabler of scalable and high-performance simulations for resource estimation. Cloud-native computational frameworks can efficiently process large geological data sets, allowing parallel processing and iterative model updates (Liang, 2024). By enhancing data accessibility across teams, these systems promote real-time collaboration and adaptive modelling techniques.

Big data analytics further amplifies the potential of AI in resource estimation (Mattera, 2023). Cloud-enabled ML workflows can ingest large quantities of exploration and production data, refining models dynamically as new drilling or geophysical information emerges. Integrating cloud-based ML with conventional geostatistical software can thus bridge the gap between static resource modelling and adaptive, real-time decision-making.

Challenges and future directions

Despite its promise, implementing AI and ML in resource modelling presents challenges, including data quality concerns, computational costs, and industry scepticism toward automation. Recent research has underscored the significance of data governance and training data quality in ensuring the reliability of AI-driven predictions (Hoseinzade et al, 2025).

ML algorithms are only as good as the data they ingest, and domain-specific training data must be carefully curated or coupled with validated data to maximise predictive accuracy. Success stories of cloud adoption in the geosciences are being introduced, but a clear business case and strong leadership support are crucial for widespread acceptance. Return on investment (ROI) studies on mining AI deployments suggest that long-term precision gains can offset upfront costs, reduce operational downtime, and improve resource recovery.

METHODOLOGY

Mining resource estimation has traditionally relied on established geostatistical methods such as kriging, conditional simulation, and block modelling (Mahboob, Celik and Genc, 2022). These techniques have effectively quantified geological uncertainty and predicted ore distributions. However, each method has inherent limitations:

- **Block models** represent orebodies as discretised 3D grids, but updates are periodic, manual, and reliant on fixed parameters.

- **Kriging** provides reliable estimates based on spatial correlation but assumes stationary statistical conditions, which limits its effectiveness for highly variable or complex deposits.

- **Conditional simulations** capture variability by generating multiple realisations yet still depend on predefined parameters and lack real-time adaptability.

The static nature of these methods creates inefficiencies and reconciliation errors, particularly as geological and operational conditions evolve. Increasingly, the mining industry acknowledges these limitations (Yoon *et al*, 2024), highlighting issues such as:

- **Reconciliation errors** due to outdated assumptions.

- **Inefficient long-term planning** arising from slow incorporation of new geological insights.

In contrast, dynamic resource modelling leverages machine learning (ML) and artificial intelligence (AI) to facilitate real-time model updates and improved predictive capabilities.

For clarity, we define our methodology at two levels:

1. **Technical framework:** Outlining the end-to-end workflow for data ingestion, automated triggers, cloud-based processing, and real-time updates to domain models.

2. **Practical application (case studies):** Illustrating how this framework applies in real-world mining scenarios.

TECHNICAL FRAMEWORK

The fundamental requirements for our framework are that the user can view and manipulate inputs to the modelling process locally in desktop applications, but the data is stored anywhere in the cloud, processing is completed in the cloud, and the orchestration of moving data to processes follows process requirements or prespecified specifications.

Data ingestion and preparation

Figure 1 illustrates that the workflow begins by gathering all necessary inputs, such as drill holes, topographic surveys, and geophysical logs, into a central repository. Automated connectors or APIs continuously ingest and validate these data sets, ensuring consistent formats and checking for anomalies (eg missing domain codes or invalid coordinates). This cleansing stage standardises the data and establishes a single source of truth, setting the foundation for all subsequent modelling tasks.

Domain definition, exploration data analysis, and domain modelling

Next, the system applies pre-existing domain codes or user-defined rules to delineate geological domains. Before finalising these rules, exploration data analysis (EDA) may be performed to examine distributions, identify outliers, and shape geostatistical parameters. In Figure 1, parallel processes, such as implicit modelling or indicator simulation, translate these inputs into 3D domain surfaces, refining boundaries and quantifying uncertainty. The result is an up-to-date representation of the deposit's geology, ready for visualisation and iterative refinement.

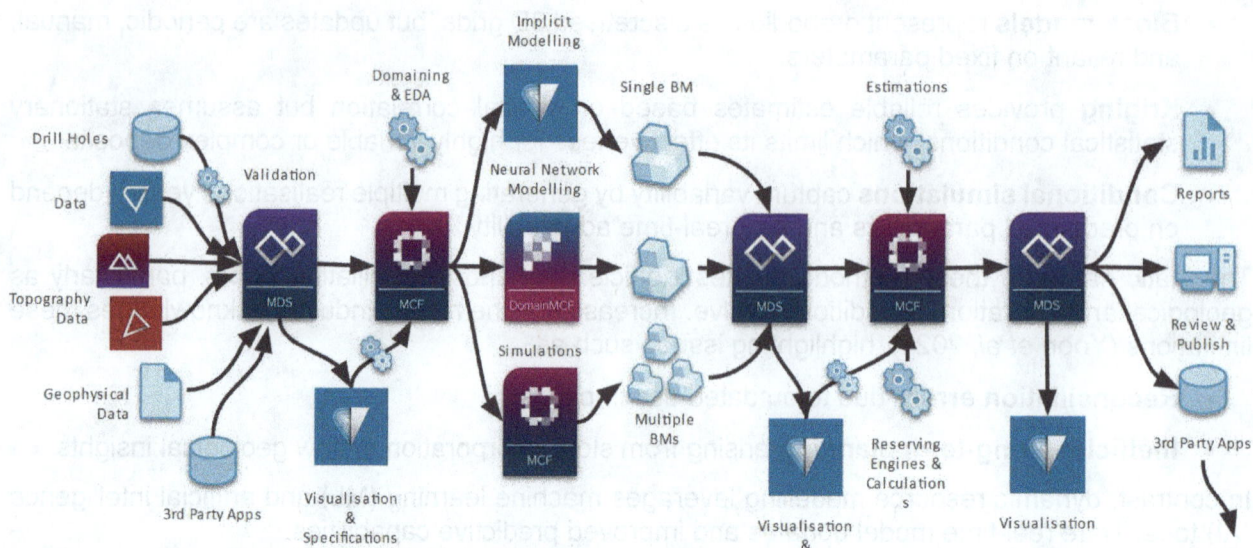

FIG 1 – Conceptual framework illustrating dynamic interactions between desktop applications and cloud-based data management, computation, and orchestration components, enabling automated, real-time updates in resource modelling.

Block model creation and rule-based calculations

Once the domain solids are generated, they feed into block model creation if not automatically done during domain modelling. The system applies rule-based calculations to classify or annotate each block (eg ore–waste classification, rock type, density), ensuring changes to threshold values trigger recalculations. This process keeps the block model in sync with the latest geological information. Users can quickly adjust parameters, validate results, and finalise the block model with additional rule-based steps by maintaining a continuous connection between the desktop environment and the cloud.

Grade simulation, estimation, and reporting

In the final modelling stage, grade simulation and estimation are performed using deterministic methods (eg kriging) and sequential Gaussian simulation to capture uncertainty. Multiple realisations and post-processing statistics are stored for deeper analysis or classification. These outputs flow into a resource reporting phase, including applying economic cut-off criteria (eg net smelter return) and classifying resources as Measured, Indicated, or Inferred. Finally, resource numbers can be compared to previous models to gauge tonnage or grade prediction improvements. The workflow supports generating comparative statistics and visuals (eg swath plots, Q–Q charts) to assess local and global consistency. By keeping all data and results centrally managed, each update or recalculation remains auditable, ensuring real-time adaptability and higher-confidence decisions for mining professionals throughout the operation.

This framework underscores how dynamic connections between a desktop environment and integrated data management, computational, and process orchestration components streamline the entire resource modelling life cycle. By automating routine steps (eg data ingestion, domain modelling, simulation) and maintaining a single source of truth, mining professionals can reduce manual errors, accelerate updates, and derive higher confidence estimates of the subsurface.

In practice, these processes create a transparent and auditable pipeline: each change (such as new drill hole data or updated domain rules) triggers automatic recalculations, ensuring real-time adaptability. The resultant workflow fosters collaboration across geology, engineering, and management teams, empowering them to make informed decisions with the most current data available.

PRACTICAL APPLICATION

Machine learning is increasingly being applied to grade control and resource estimation in mining, enabling faster and more accurate identification of high-risk zones (Mining Events, 2024). We present two case studies of processes completed using Maptek's Vestrex ecosystem. For each of these processes, many of the parameters were predefined. The clients predefined SMU for block size selections and Grade estimation parameters.

Voids project case study

The Voids Project addresses the critical challenge of identifying hidden underground voids that pose significant safety and operational risks in mining. Due to the concealed nature of voids, traditional static modelling methods were insufficient for rapid detection and effective risk management. The following outlines the structured methodology used, highlighting the integration of cloud computing, artificial intelligence (AI), and workflow automation. A graphical representation of this is provided in Figure 2.

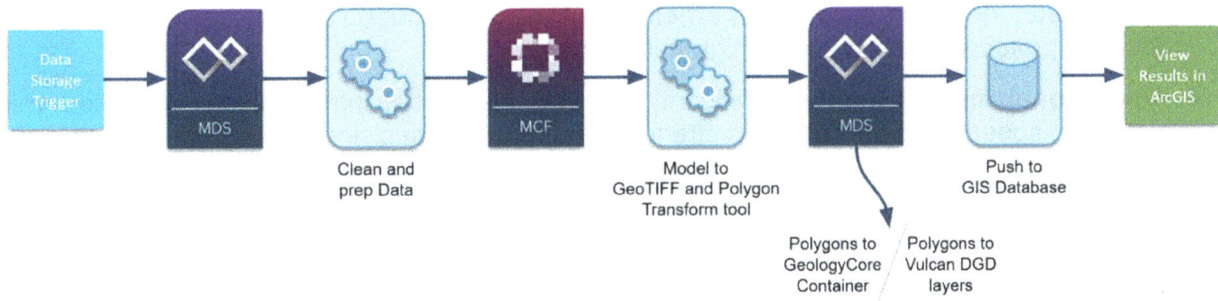

FIG 2 – Illustration of the Voids Project workflow.

Process summary

1. Data ingestion and integration

   o Disparate data sets, including Measurement While Drilling (MWD) and historical drilling data, were securely ingested using open APIs from the client cloud-hosted data environment.

   o Automated transformation pipelines cleansed and standardised these data sets, ensuring consistent and efficient downstream processing.

2. Cloud-based computation using machine learning

   o The integrated data was processed using the Maptek Compute Framework (MCF), which uses cloud-native computational power to generate dynamic domain models.

   o Specifically, Maptek DomainMCF, a proprietary neural network algorithm to provide an ML-driven solution, quickly classified drilling data into void risk zones, significantly reducing computation times from more than a day using the manual method to less than 15 mins.

   o This rapid turnaround allowed production teams to identify and visualise high-risk zones promptly, enhancing real-time risk mitigation strategies.

3. Workflow and data integration

   o Real-time workflows were orchestrated through the Maptek ecosystem, comprising:

   - Maptek Data System (MDS) for centralised data storage.

   - Maptek Compute Framework (MCF) to execute computational algorithms.

   - Maptek Orchestration Environment (MOE) for seamless data transfer.

   o Automated pipelines continuously imported, transformed, and processed data, ensuring each modification (such as adjustments to domain rules or new drill hole data) triggered immediate model updates.

4. Collaborative editing and auditability

   o Geologists and engineers interacted with a single cloud-hosted project, maintaining a single source of truth.

o Edits to domain models or specifications were auditable, with automated triggers ensuring model integrity and transparency throughout the workflow.

Operational benefits and outcomes

This project employed DomainMCF (a Neural Network based deep learning model) to model a problem which had previously not been possible. By automating complex domain detection, the process of domain modelling, polygonisation and delivery to the drills was completed in less than 20 mins, highlighting the transformative potential of ML in unlocking previously inaccessible insights such as:

- **Real-time risk identification:** Rapid visualisation of void risks allowed immediate deployment of targeted risk management protocols.

- **Enhanced safety and production efficiency:** Mining continued safely in low-risk zones, with focused protocols applied only in high-risk areas.

- **Operational improvement:** Significant reduction in manual interventions, increased void detection speed and accuracy, and strengthened real-time decision-making capabilities.

This methodology demonstrates how transitioning from traditional static modelling to dynamic, cloud-powered workflows and AI-enhanced simulations can deliver precise, real-time adaptive resource modelling. The Voids Project highlights how this approach enhances operational safety, resource estimation accuracy, and decision-making speed, laying a foundation for broader industry adoption.

Block modelling and simulation case study

The Block Modelling and Simulation Project demonstrates the transformative potential of applying cloud-based dynamic modelling workflows in mineral resource estimation. This project employed the Maptek Vestrex ecosystem, including Maptek Data Store (MDS), Maptek Compute Framework (MCF) and Maptek Orchestration Environment (MOE) to automate and dynamically refine resource models. The goal was to create adaptive resource models that were continuously updated as new geological and drilling data became available, reducing delays typically associated with manual model recalibration.

Process summary

1. Data ingestion and integration

   o Data from an acQuire database was integrated into the system from three distinct drilling campaigns (Exploration, Resource Definition, and Production).

   o Drill hole data sets containing domain classifications and assay (grade) information were automatically ingested into the MDS, creating a centralised data repository.

2. Dynamic model construction and processing

   o Geological domain shapes, generated through a multi-domain implicit modelling process, were the foundation for creating the initial block model.

   o Using Maptek Vulcan software, preliminary block calculations were performed to prepare the model structure for subsequent estimation.

   o MCF executed grade estimations and conditional simulations across multiple grade domains.

3. Automated workflow and real-time updates

   o The model utilised specification files created in Vulcan, which were automated within the MOE.

   o New validated drill hole data entering the MDS automatically triggered cloud-based processes in the MCF, leading to immediate recalculation of block estimations and simulations.

- o Once completed, Vulcan was used for final calculations and classification, integrating seamlessly with cloud-processed outputs for rapid validation and updating.

4. Validation and reporting

- o A final validation step automatically compared the newly generated model against previous iterations and the latest drill hole data sets, ensuring continuous quality assurance.
- o Results were returned to user desktops in real time, enabling rapid validation and interpretation.

Figure 3 presents a graphical representation of the Block modelling and simulation workflow.

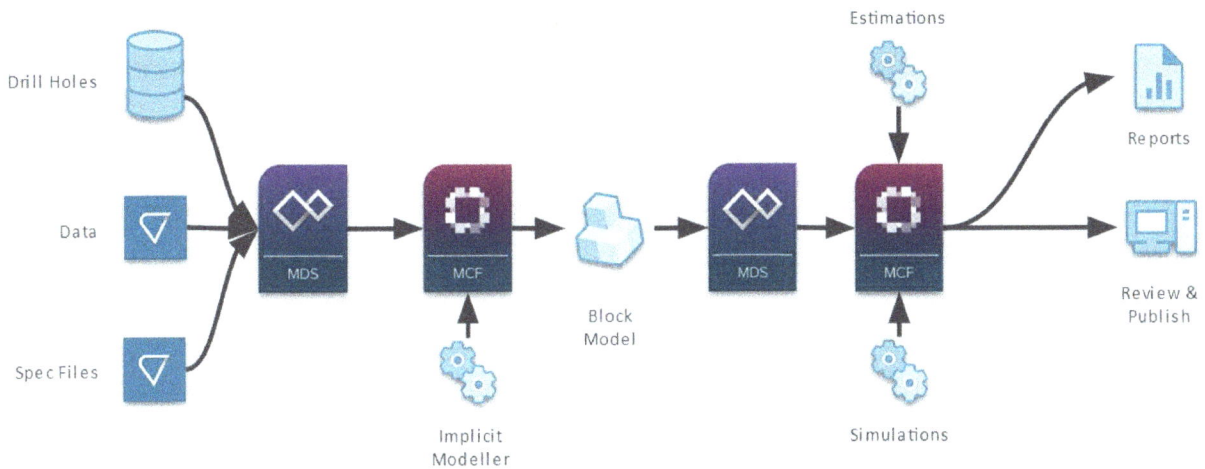

FIG 3 – Illustration of the Block Modelling and Simulation project workflow.

Operational benefits and outcomes

In this case study, cloud based execution of sequential Gaussian simulation resulted in a 60 per cent time reduction, from 1 hr 52 mins to 42.7 mins, demonstrating the computational efficiencies gained through cloud orchestration. Additionally, the following improvements and outcomes were identified:

- **Reduced modelling cycle time:** The automation and real-time triggering of updates significantly reduced the time required for resource model recalibration, allowing rapid responses to newly acquired geological data.

- **Enhanced model accuracy and reconciliation:** Continuous refinement of block models using dynamic data reduced reconciliation errors between predicted and actual production data, increasing confidence in resource estimates.

- **Improved collaboration and auditability:** Centralised cloud hosting ensured geological teams accessed a unified, auditable model version. Changes in the data set and model parameters were transparently tracked, ensuring data integrity and reducing inconsistencies.

- **Efficient use of resources:** Leveraging cloud computing for computationally intensive processes freed local resources for more complex analysis and interpretation tasks. The scalable nature of cloud processing meant computational resources could dynamically adapt to workload requirements, providing cost efficiency and operational agility.

- **Increased confidence in resource estimates:** Continuous model updates and automated recalculations provided teams with greater confidence in the accuracy and precision of resource estimations, improving the reliability of strategic decision-making and long-term planning.

By shifting computationally intensive tasks to the cloud, local resources remained available for advanced analytics and interpretive work. The rapid reinterpretation of resource models reduced reconciliation errors between predicted and actual mining outcomes, increasing overall confidence in resource estimates. Furthermore, collaboration was enhanced as teams shared a unified project with auditable version control, improving both data consistency and transparency.

DISCUSSION

Dynamic cloud-predicted domain models

Unlike traditional static models, AI-driven cloud simulations continuously update domain and grade models based on the latest drilling and geological data. The process works as follows:

1. **Single source of truth** – A cloud-hosted project either hosts, or is connected to an authoritative model, ensuring all team members access and modify the same data set from their geology core desktop solution.

2. **Data updates in real time** – When new drilling data is added (eg a new drill hole), the model dynamically updates to reflect the new geological inputs.

3. **Editable specification files** – Key parameters, such as domain rules and classification criteria, are stored in specification files that team members can modify.

4. **Trigger mechanisms for automatic processing** – When changes are made (eg adjusting domain rules, adding or removing data points), cloud-based triggers initiate a model rebuild, ensuring that the latest information is always reflected.

5. **Result delivery back to desktop** – Once processing is completed, updated resource estimates and simulations are streamed back to the geology desktop solution, allowing for real-time interpretation and decision-making.

This approach ensures that each resource estimate is genuinely adaptive, dynamically responding to changes in input conditions rather than relying on static, manually adjusted models.

Comparison between static and AI-driven models

Table 1 summarises the key differences between traditional static models and AI-enhanced dynamic models, highlighting how advanced technology significantly improves resource modelling processes.

TABLE 1

Comparison of static and AI-driven resource models.

Feature	Traditional static models	AI-enhanced dynamic models
Update frequency	Periodic with manual interpretation	Continuous, real-time updates
Collaboration	Localised, often siloed	Centralised, accessible from multiple locations
Computational power	Limited to desktop/laptop capabilities	Scalable cloud computing
Data integration	Requires manual import/export	Automatic, real-time data ingestion
Model adaptability	Predefined assumptions, static structure	Adjusts dynamically based on new data
Auditability	Manual version tracking	Automated version control and traceability

By leveraging cloud computing, the mining industry can transition from batch-processed, static models to dynamic, continuously evolving resource estimations. This allows teams to make faster, more informed decisions, improving confidence in resource estimates and overall operational efficiency.

Automation for adaptive modelling with interoperable workflows

Transitioning resource modelling to cloud-based solutions involves integrating cloud computing with existing desktop-based workflows rather than replacing established geological and geostatistical tools. Interoperability plays a critical role in enabling adaptive modelling. Building workflows locally on familiar desktop solutions and linking these seamlessly to cloud computing infrastructure allows visualisation and specification tasks to remain accessible at the desktop, while computationally demanding simulations and processing tasks leverage cloud capabilities.

By breaking down traditional operational silos, this approach ensures consistent data management through a centralised 'single source of truth'. Teams maintain full auditability and version control; every change is automatically recorded and traceable. Automation in this context focuses on ensuring model integrity by triggering immediate updates when users modify domain rules, add new drill hole data, or adjust specifications. This ensures that geological models adapt dynamically in real time, enhancing resource estimation accuracy without sacrificing data integrity or process transparency.

Although the case studies presented illustrate specific use cases requested by Maptek clients, the modular structure of the Maptek Compute Framework (MCF) enables broader integration. Users can incorporate Maptek tools, third-party applications, or custom Python scripts to suit their specific requirements. While workflows in this paper do not include processes such as SMU optimisation or QA/QC validation, these can be integrated into the modelling pipeline as required, supporting flexible, organisation-specific standards for quality, compliance, and decision-making.

CONCLUSION

The transition from static to dynamic resource modelling marks a fundamental transformation in resource estimation, representing a shift from periodic, manual updates toward continuous, adaptive, and real-time data integration. The key takeaway is that traditional methods, though effective historically, are inherently limited in addressing the complexities of contemporary geological challenges. Conversely, dynamic models leveraging AI, ML and cloud computing offer the agility, scalability and predictive accuracy essential for modern mineral resource management.

Artificial Intelligence and Machine Learning are not replacements for geologists; they are powerful tools that augment their decision-making capabilities. These technologies enable professionals to rapidly identify patterns, predict uncertainties, and dynamically adapt models based on real-time geological inputs. As such, AI and ML empower geologists, engineers, and decision-makers by enhancing precision, improving resource reconciliation, and facilitating more confident, data-driven decisions (Vespignani and Smyth, 2024).

Key takeaways

- **AI/ML as enablers:** Far from replacing geologists, AI and ML empower them to detect patterns, reduce reconciliation errors, and adapt models in near real time.

- **Cloud infrastructure:** Scalable processing, unified data repositories, and automated workflows facilitate immediate model updates, promoting collaborative decision-making.

- **Return on investment (ROI):** While implementing cloud and AI solutions involves upfront costs, the long-term gains (eg improved safety, reduced downtime, higher accuracy, and better resource recovery) can justify the investment.

- **Data governance:** Upholding data integrity and security is critical, as AI-driven insights hinge on high-quality, well-managed data sets. Establishing rigorous standards and protocols fosters trustworthy predictions.

Strategic recommendations

- **Invest in robust cloud infrastructure** to accommodate dynamic, large-scale modelling.

- **Promote interoperability** between desktop and cloud platforms, ensuring seamless workflow automation.

- **Cultivate a culture of innovation** by providing training and support for AI/ML adoption, encouraging ongoing professional development.

- **Maintain rigorous data quality standards** to guarantee the reliability of AI-driven model updates.

- **Conduct ROI evaluations** to ensure cost-effectiveness and stakeholder buy-in.

By embracing these strategic recommendations, the mineral resource industry can effectively harness the full potential of dynamic resource modelling, positioning itself for sustained operational excellence and long-term competitiveness.

REFERENCES

Azhari, F, Sennersten, C C, Lindley, C A, Genc, B, Kubler, S, Robert, J and Moser, B, 2023. Deep learning implementations in mining applications: A compact critical review, Artificial Intelligence Review, 56:14367–14402. https://doi.org/10.1007/s10462-023-10500-9

Dumakor-Dupey, N K and Arya, S, 2021. Machine learning—A review of applications in mineral resource estimation, Energies, 14(14):4079. https://doi.org/10.3390/en14144079

Galetakis, M, Vasileiou, A, Rogdaki, A, Deligiorgis, V and Raka, S, 2022. Estimation of mineral resources with machine learning techniques, Materials Proceedings, 5(122):122. https://doi.org/10.3390/materproc2021005122

Hoseinzade, Z, Shojaei, M, Khademi, F, Mokhtari, A R and Saremi, M, 2025. Integration of deep learning models for mineral prospectivity mapping: A novel Bayesian index approach to reducing uncertainty in exploration, Modeling Earth Systems and Environment, 11:161. https://doi.org/10.1007/s40808-025-02342-x

Liang, M, 2024. How AI can enhance your resource modeling, AusIMM Bulletin, Bulletin articles. Available from: <https://www.ausimm.com/bulletin/bulletin-articles/how-ai-can-enhance-your-resource-modeling/> [Accessed: 7 March 2025].

Mahboob, M A, Celik, T and Genc, B, 2022. Review of machine learning-based mineral resource estimation, Journal of the Southern African Institute of Mining and Metallurgy, 122(11):655-664. https://doi.org/10.17159/2411-9717/1250/2022

Mattera, M, 2023. Benefits and challenges of using big data in resource estimation, AusIMM Bulletin, Bulletin articles. Available from: <https://www.ausimm.com/bulletin/bulletin-articles/benefits-and-challenges-of-using-big-data-in-resource-estimation/> [Accessed: 7 March 2025]

Mining Events, 2024. The role of machine learning in grade control resource estimation, Mining Events, News and Blog. Available from: <https://mining-events.com/the-role-of-machine-learning-in-grade-control-resource-estimation/> [Accessed: 7 March 2025].

Vespignani, J and Smyth, R, 2024. Artificial intelligence investments reduce risks to critical mineral supply, Nature Communications, 15:7304. https://doi.org/10.1038/s41467-024-51661-7

Yoon, S, Jun, H, Chung, M, and Lee, W J, 2024. Preliminary study for AI application of mineral exploration using machine learning, International Journal of High School Research, 6(12):15–22. https://doi.org/10.36838/v6i12.3

Improving underground mobile mechanised mining productivity with safety in mind

C Domoney[1]

1. MMP Technical Services Manager, Maptek Pty Ltd, Warabrook NSW 2304.
   Email: craig.domoney@maptek.com.au

ABSTRACT

Historically, adverse vehicle interactions have disproportionately contributed to fatalities in mining, with underground operations being particularly affected. Underground mining presents unique challenges for collision avoidance due to global navigation satellite system (GNSS) denial, limited infrastructure, and confined tunnels and junctions. These constraints reduce the effectiveness of traditional surface solutions, which often rely on line-of-sight (LOS) technologies. In response, underground operations have implemented Collision Warning Systems (CWS) and collision avoidance systems (CAS) for low-speed, short-range interactions. However, their performance in high-speed, long-range, non-line-of-sight (NLOS) scenarios is limited and poses significant safety concerns.

This paper evaluates the potential of vehicle-to-everything (V2X) and dedicated short-range communication (DSRC) systems (adapted from the automotive industry) for improving underground safety and productivity. A scenario-based model aligned with ISO 21815-3 (2023) and ISO 19296 (2018) standards is used to calculate minimum safe stopping distances. At 20 km/h, a detection range of at least 80 m is required to allow for hazard recognition, operator reaction, braking, and safety margin. This far exceeds the 50 m range provided by many legacy systems currently in use but is comfortably covered by V2X technology.

The model was validated using data from a mid-sized underground mine. Over 40 per cent of empty haul trips exceeded the safe speed for 50 m NLOS detection, highlighting the operational risk. Additionally, modern underground haul trucks such as the Caterpillar AD30 and Sandvik TH320 are capable of reaching or exceeding 25 km/h, reinforcing the need for extended detection capabilities that reflect real-world vehicle performance.

An illustrative productivity calculation shows that increasing empty travel speed by 2 km/h across 80 per cent of the travel segment can enable one additional trip per vehicle per shift (equating to a ~10 per cent productivity gain). In addition to improved safety outcomes, benefits may include enhanced travel consistency, reduced congestion, and greater operator confidence.

This study demonstrates that standards-aligned, real-time NLOS detection technologies have the potential to significantly enhance both the safety and operational efficiency of underground mobile mechanised mining operations.

INTRODUCTION

Mining is foundational to modern civilisation, supplying the raw materials necessary for infrastructure, technology, and industry. Yet, it remains one of the most hazardous occupations globally. Underground mining, in particular, poses elevated risks due to confined spaces, limited visibility, and reliance on mobile equipment. Trackless mobile machinery (TMM) is a key contributor to productivity in such settings, but also represents a significant hazard. According to the Mine Health and Safety Inspectorate (DMRE, 2023), TMM-related incidents remain among the leading causes of fatalities in South African mines; this trend is consistent across multiple jurisdictions.

As surface resources become depleted, the global shift toward underground mining introduces heightened safety concerns. The operational environment underground is notably complex, characterised by GNSS denial, narrow and irregular tunnels, and minimal infrastructure for real-time communication. Surface mining operations are comparatively less constrained, allowing for the effective application of GNSS and LOS technologies. As a result, the uptake and success rate of CWS/CAS has been significantly higher in surface mining.

In response, many underground operations have adopted CWS and CAS better suited to low-speed, short-range interactions. Generally, these systems have demonstrated limited performance in higher-speed or long-range scenarios, particularly under NLOS conditions (Qian *et al*, 2022; NIOSH, 2024). Current industry initiatives, such as those led by the Earth Moving Equipment Safety Round Table (EMESRT, 2023), have largely concentrated on improving low-speed, close-proximity collision scenarios. While critically important, this focus has resulted in comparatively less attention being paid to high-speed, long-range interaction hazards, particularly in underground mining environments. Furthermore, the lack of unified standards and limited interoperability between vendor solutions inhibits broader adoption.

Scientific and technological context

Existing literature underscores the challenges of collision avoidance in underground mining, particularly under NLOS conditions (Qian *et al*, 2022; NIOSH, 2024). In contrast, wireless technologies such as DSRC and V2X (originally developed for automotive applications) have demonstrated reliable high performance in challenging urban environments. Recent studies and guidelines have explored the application of DSRC for underground vehicle communication, highlighting its robustness in confined underground tunnels and complex operational settings (Chehri *et al*, 2020; Gaber *et al*, 2021; Wikipedia, 2025). These technologies operate using decentralised, real-time communication protocols and may be well-suited for underground adaptation.

This paper builds on these developments, combining industry standards, operational data, and performance modelling to assess the viability of long-range NLOS detection systems in underground hard rock mining. The aim is to demonstrate how such systems may enhance both safety and operational throughput through increased situational awareness and coordinated vehicle interaction.

METHODOLOGY – SCENARIO MODELLING AND STOPPING DISTANCE CALCULATIONS

To evaluate safe interaction distances for underground mobile equipment, this study models a scenario in which two haul trucks approach an intersection (each travelling at 20 km/h) under NLOS conditions (see Figure 1). The analysis is based on ISO 21815-3:2023(E) (ISO, 2023), which outlines discrete intervals for collision warning and avoidance, and ISO 19296:2018 (ISO, 2018), which provides braking performance standards for underground machinery.

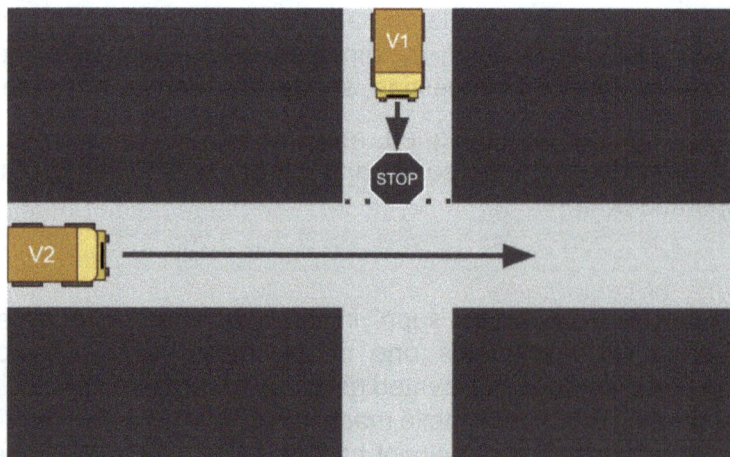

FIG 1 – Two vehicles approaching an intersection underground.

Scenario definition and assumptions

- Both vehicles are travelling underground at 20 km/h.

- The intersection is completely NLOS; drivers do not have visual or infrastructure assisted awareness of each other.

- There is no centralised communication infrastructure; only direct vehicle-to-vehicle communication is considered.

- In this scenario Vehicle 1 gives way to Vehicle 2, that is it will stop before the intersection, while Vehicle 2 continues through uninterrupted.

The goal is to determine the minimum required detection distance to ensure a safe stop for Vehicle 1 before reaching the intersection.

Stopping distance framework (based on ISO 21815-3)

Figure 2 provides a framework for measurement and understanding of a collision avoidance interaction. It illustrates the standard naming conventions for critical points (A–F) and intervals (AB, BC, CD, DE, EF) according to ISO 21815-3 (ISO, 2023).

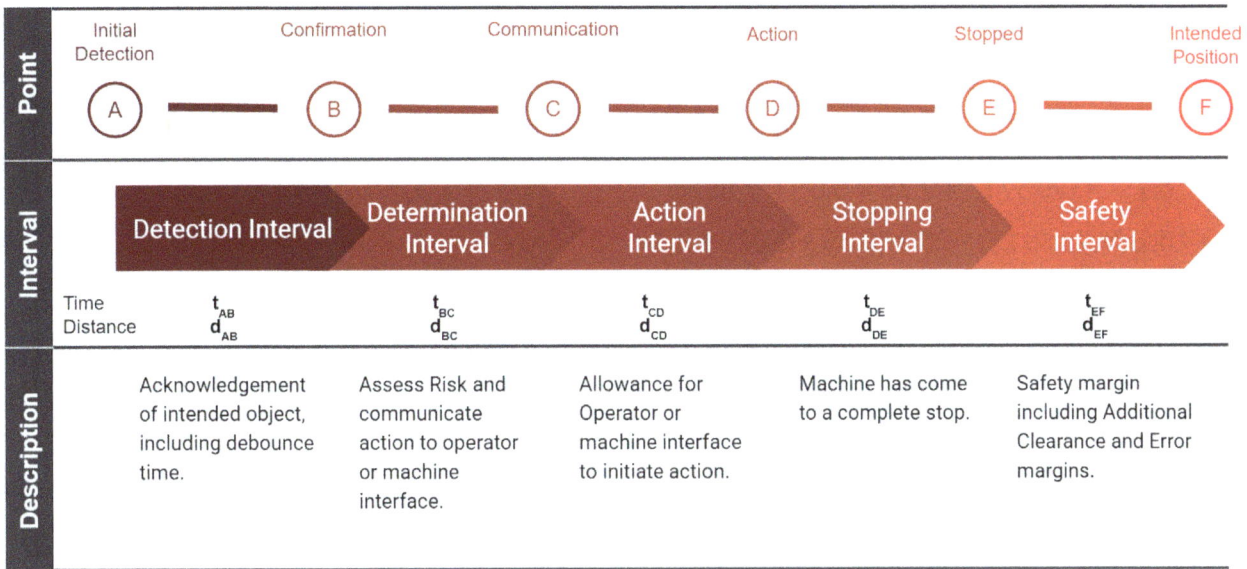

	Point					
	Initial Detection	Confirmation	Communication	Action	Stopped	Intended Position
	A	B	C	D	E	F

Interval					
	Detection Interval	Determination Interval	Action Interval	Stopping Interval	Safety Interval
Time Distance	t_{AB} d_{AB}	t_{BC} d_{BC}	t_{CD} d_{CD}	t_{DE} d_{DE}	t_{EF} d_{EF}

Description					
	Acknowledgement of intended object, including debounce time.	Assess Risk and communicate action to operator or machine interface.	Allowance for Operator or machine interface to initiate action.	Machine has come to a complete stop.	Safety margin including Additional Clearance and Error margins.

FIG 2 – ISO 21815 (2023) points and intervals of interest.

Variable definitions

- t_{AC} time to detect and alert (combined A to C)
- d_{AC} distance during detection and alert
- t_{CD} operator reaction time (C to D)
- d_{CD} distance during operator reaction
- t_{MBS} master brake system response time
- d_{DE} braking distance including master brake system delay
- d_{EF} safety margin
- v vehicle speed (20 km/h)
- a minimum deceleration (2.75 m/s^2)

Distance calculations

Formulas used

- d_{AC} $= (v \times t_{AC}) / 3.6$
- d_{CD} $= (v \times t_{CD}) / 3.6$
- d_{DE} $= (v^2) / (2 \times a \times 3.6^2)$
- d_{EF} $=$ fixed safety margin (4 m)

For Vehicle 2 (constant speed travel during braking):

- d_{DE} (Vehicle 2) = (v × (t_{MBS} + t_{brake})) / 3.6, where t_{brake} = (v / 3.6) / a

Substituted values

- t_{AC} = 1.5 s
- t_{CD} = 2.74 s
- t_{MBS} = 0.35 s
- d_EF = 4 m

Derived values

- t_{brake} = (20 / 3.6) / 2.75 ≈ 2.02 s
- d_{AC} = (20 × 1.5) / 3.6 = 8.33 m
- d_{CD} = (20 × 2.74) / 3.6 = 15.22 m
- d_{DE} = (20^2) / (2 × 2.75 × 3.6^2) = 7.53 m
- d_{DE} (Vehicle 2) = (20 × (0.35 + 2.02)) / 3.6 ≈ 13.17 m

Combined distance summary

Table 1 summarises the calculated distances for Vehicle 1 (stopping) and Vehicle 2 (continuing at constant speed), including detection, operator reaction, braking, and safety margin components.

TABLE 1

Summary of calculated vehicle interaction intervals.

Component	Vehicle 1 – stopping (m)	Vehicle 2 – continuing (m)
Detection and alert distance (d_{AC})	8.33	8.33
Operator reaction distance (d_{CD})	15.22	15.22
Braking distance (d_{DE})	7.53	13.17*
Safety margin (d_EF)	4.00	4.00
Total distance (d_AF)	35.08	40.72

* Distance covered by Vehicle 2 while Vehicle 1 is braking.

Total stopping distance for Vehicle 1

$$d_AF_1 = d_{AC} + d_{CD} + d_{DE} + d_EF$$

$$d_AF_1 = 8.33 + 15.22 + 7.53 + 4 = 35.08 \text{ m}$$

Total distance for Vehicle 2

$$d_AF_2 = d_{AC} + d_{CD} + d_{DE} \text{ (Vehicle 2)} + d_EF$$

$$d_AF_2 = 8.33 + 15.22 + 13.17 + 4 = 40.72 \text{ m}$$

Combined minimum required detection distance

$$d_{tot} = d_AF_1 + d_AF_2$$

$$d_{tot} = 35.08 + 40.72 = 75.80 \text{ m}$$

Rounded up, a minimum detection distance of approximately 80 m is required under NLOS conditions at 20 km/h.

This analysis demonstrates that long-range detection systems capable of reliably identifying other vehicles at distances of at least 80 m are necessary to enable effective and consistent collision avoidance at moderate speeds underground. Systems falling short of this range may not provide sufficient time for operator recognition, reaction, brake system response, and safe stopping, particularly where intersections or blind corners are encountered. These findings highlight the critical need for underground communication and sensing technologies to prioritise not only short-range proximity alerts but also extended non-line-of-sight hazard detection capabilities.

DATA ANALYSIS – SPEED THRESHOLDS, RISK EVALUATION AND PRODUCTIVITY

Safe operational speeds versus detection distance

Using the methodology above, and consistent with thresholds outlined in ISO 21815-3:2023 (ISO, 2023), safe operational speeds can be related to required NLOS detection distances.

Table 2 shows that at 20 km/h a minimum of 76 m NLOS detection is required; a 50 m range is only sufficient for speeds up to 13.5 km/h.

TABLE 2
Safe operational speed versus detection distance.

Safe operational speed (km/h)	Total distance (m)	Vehicle 1 distance (m)	Vehicle 2 distance (m)
30	122	55	67
25	98	45	53
20	76	35	41
13.5*	50	24	26
10	38	18	20
5	22	11	11

* 50 m NLOS is often raised in operational conversations as a safety requirement.

Typical maximum speeds of underground haul trucks

Modern underground haul trucks are designed for both reliability and high productivity in confined mining environments. While operational speeds are often limited by tunnel geometry, safety protocols, and traffic management, many vehicles are technically capable of exceeding 20 km/h.

For example:

- The Caterpillar AD30, a 30 tonne underground truck, has a top speed of 36.7 km/h in its highest gear (Caterpillar, 2022).

- The Sandvik TH320, designed for small to medium hard rock mines, is optimised for long ramp haulage and can exceed 25 km/h under appropriate conditions (Sandvik, 2022).

- The UK XTUT20, a 20 tonne haul truck, is geared to reach up to 25 km/h (Siton, 2023).

These specifications show that 20 km/h is a realistic and conservative benchmark for safety evaluation. For example, at 25 km/h, (well within the capability of the Sandvik TH320 and UK XTUT20) a minimum of 98 m of detection is required to safely manage a NLOS interaction. This far exceeds the capabilities of many currently deployed short-range systems.

Moreover, Table 2 demonstrates that a 50 m detection range, which is common in legacy proximity systems, is only sufficient at speeds below approximately 13.5 km/h.

Evaluating the role of speed

Truck speed plays a critical role in balancing both safety and productivity in underground hard rock mines. As highlighted by Przhedetsky (2010), productivity is influenced by a broad set of variables, including vehicle payload, road surface condition, tyre type and wear, and operator technique. While manufacturers provide Rimpull (speed-on-grade tables or performance curves), actual performance often varies based on site-specific operational and environmental conditions. Przhedetsky also notes that simply increasing truck size is no longer a straightforward strategy to improving productivity due to the need for corresponding changes in mine design and infrastructure, as well as the associated cost implications. Additionally, factors such as equipment availability, utilisation, workforce wages, and other overheads further contribute to operational efficiency, highlighting the multi-dimensional nature of productivity challenges.

Simulation based studies of truckloader haulage systems reinforce this complexity. For instance, Salama and Greberg (2012) demonstrate how the number of trucks, haul distances, and loading rates affect not only overall productivity, but also equipment utilisation and traffic congestion. Although this study does not directly examine truck speed, it illustrates how increasing the number of trucks may reduce truck utilisation and lead to congestion, ultimately causing productivity to plateau as travel distances increase. These findings emphasise the interdependent factors that must be considered when evaluating haulage system performance.

Speed related risk – Maptek data study

To better understand potential operational risk posed by speed within a typical underground hard rock mine, an analysis was conducted on haul truck data spanning two years of production from a mid-sized Australian hard rock mine (Maptek, 2025). The data set includes timestamped travel speeds across full and empty legs over multiple haulage levels.

Figure 3 outlines the typical underground load and haul environment, in which trucks travel from production stopes to the surface via ramps with intersecting levels. This environment is GNSS denied, constrained by narrow tunnels, and often shared with other mobile equipment, complicating real-time tracking and traffic coordination.

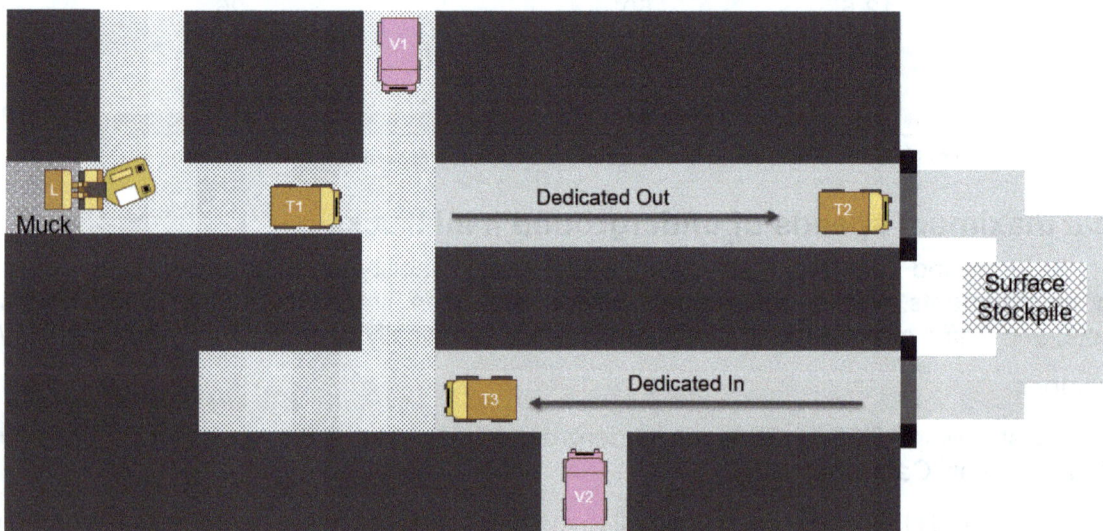

FIG 3 – Load and haul to surface.

Data cleaning was conducted to remove outliers by determining the median and excluding data points beyond two standard deviations. The primary objective of this process was to create a safe, representative data set for analysis.

The data provides insights into the average speed during full travel and the average speed during empty travel for each haul cycle. Although no additional data on instantaneous speed is available, the existing information is sufficient to perform an initial risk evaluation when combined with the safe operational speeds outlined in Table 2.

Key statistics

- Median full-travel speed: 8.6 km/h
- Median empty-travel speed: 13.0 km/h

- 40.6 per cent of empty-travel cycles exceed 13.5 km/h.

The individual haul cycles have been plotted in Figure 4. The x-axis is segmented into intervals of 0.5 km/h, rounded down. Note the graph is shaded with a red background at the 50 m NLOS cut-off of 13.5 km/h as indicated in Table 2.

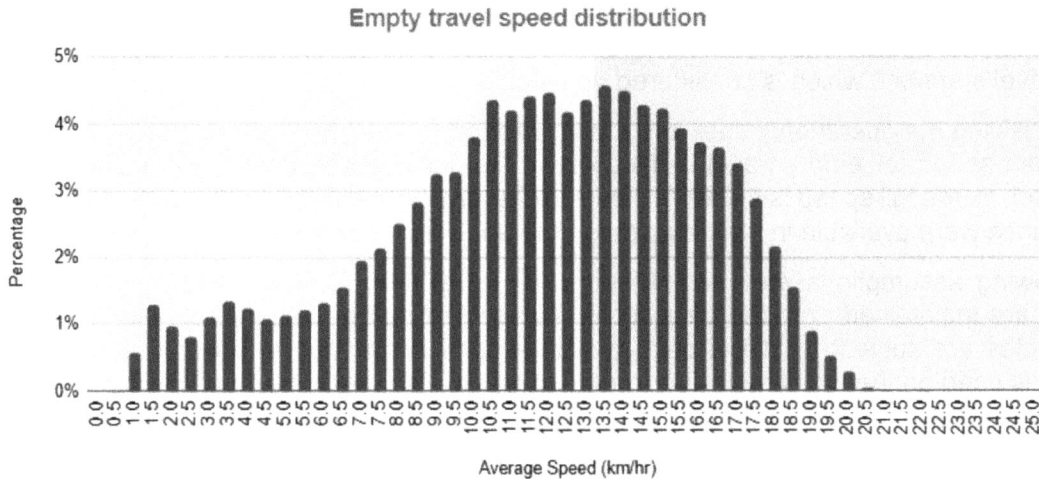

FIG 4 – Empty travel speed distribution.

This analysis highlights a material collision risk under existing short-range NLOS capabilities. Given that many vehicles are capable of operating at or above 20 km/h, there is clear risk if detection range is not matched to vehicle performance.

Potential productivity gains

Field trials, such as those conducted with Sandvik's prototype underground truck, demonstrate that vehicles can achieve maximum haul speeds exceeding 40 km/h under optimal conditions (Sandvik, 2013). However, average operational speeds tend to be lower due to gradient, turning radius, and safety restrictions.

DSRC systems have been shown to achieve consistent detection beyond 100 m even under challenging NLOS conditions in underground tunnels, thereby validating their suitability for enabling earlier hazard recognition and improved traffic flow (Maptek, 2024; Chehri *et al*, 2020; Qureshi *et al*, 2016).

Improved long-range NLOS detection can yield productivity benefits by:

1. Reduced congestion: There is significant research linking traffic congestion in underground mining to productivity losses, particularly around ramp and haulage scheduling. For example, recent studies have examined how improved scheduling algorithms and dispatch systems can alleviate congestion and increase efficiency (Zhang, Xu, and Yu, 2024; Fourie and Minnitt, 2014). Early awareness allows better traffic coordination, minimising idle time at intersections. As highlighted in the previous section, where vehicles are restricted by narrow tunnels or one-way traffic loops, disruption to production vehicles can result in considerable losses. One of the most effective ways to mitigate this is through enhanced real-time coordination, allowing operators to proactively manage priority routes and avoid unnecessary stops. This can reduce the total nonproductive time, allowing vehicles to spend more time in motion and reduce delays.

2. Increased speed confidence: When operators trust that the environment is well managed and risks are mitigated (such as through consistent NLOS hazard detection) they are more likely to maintain efficient, higher speeds within safety limits. While specific peer-reviewed research on this topic in mining is limited, broader studies in occupational safety suggest that improved environmental safety enhances operator confidence and performance. Further research is needed to quantify this relationship in underground mining settings.

Illustrative calculation – impact of increased speed on productivity

Building on the previous section, which identified improved traffic coordination and reduced congestion as key contributors to productivity, this section isolates the potential gains from increased average haulage speed alone. Due to the absence of controlled trials with measured speed changes, the historic data set is revisited to model possible outcomes. Specifically, this analysis calculates the required increase in empty travel speed to enable one additional trip per truck per shift. The scenario assumes that all operational parameters remain constant except for speed over 80 per cent of the empty travel segment, which is considered adjustable.

Upon reviewing the operational data, the median speed for full travel is found to be 8.6 km/h, while the median speed for empty travel is 13.0 km/h. This indicates that vehicles are operating under heavy load, as the full speed is 4.4 km/h slower, reflecting typical uphill full haulage. Only time, state, and distance were available in the data set, so the analysis focuses on median speed.

The following assumptions are used. Median full travel time (0.40 hr) and median loading time (0.09 hr) are maintained from the data set. It is assumed that operations run with 12 hr shifts and that vehicles are subject to 80 per cent availability and 80 per cent utilisation. Empty travel time (0.31 hr) is used as the baseline for performance gains. The median empty and full distances from the data set are 4.05 km and 3.44 km, respectively.

The effective hours per shift are calculated as 12 hrs × 80 per cent availability × 80 per cent utilisation, resulting in 7.68 productive hours per shift.

Using the median cycle time of 0.82 hrs per trip, the number of completed trips per shift can be calculated as the floor of the effective hours divided by the median cycle time: $\lfloor 7.68 / 0.82 \rfloor = 9$ trips.

Let the variables be defined as follows:

- t_N is the new cycle time
- t_F is the median full travel time
- t_L is the median loading time
- t_E is the median empty travel time
- d_E is the median empty travel distance
- p_S is the fixed (unchanged) portion of the empty travel segment (assumed to be 0.2)
- p_C is the adjustable portion of the empty travel segment (assumed to be 0.8)
- a is the required proportional reduction in travel time for the adjustable segment

To increase to ten trips per truck per shift:

$$t_N = 7.68 / 10 = 0.768 \text{ h}$$

Using these definitions:

$$t_N = t_F + t_L + t_E(p_S + p_C \times a)$$

Therefore:

$$a = (t_N - t_F - t_L - t_E p_S) / (t_E \times p_C) \quad a = (0.768 - 0.40 - 0.09 - 0.31 \times 0.2) / (0.31 \times 0.8) \approx 87 \text{ per cent}$$

The new average speed required:

$$v = (d_E \times p_C) / (a \times t_E) = (4.05 \times 0.8) / (0.87 \times 0.31) \approx 15.01 \text{ km/h}$$

The median empty travel speed was 13.04 km/h. This analysis shows that increasing the speed by as little as 2 km/h across 80 per cent of the empty travel segment could enable one additional trip per vehicle per shift. This modest increase appears achievable with improved non-line-of-sight (NLOS) detection, which would enhance operator awareness, reduce traffic congestion, and increase confidence in both personal safety and the safety of others.

This example is illustrative, demonstrating potential performance gains based on the available data set and defined operational assumptions.

CONCLUSION

This paper investigated the relationship between NLOS hazard detection, operating speed, and productivity in underground hard rock mining. Using a standards aligned model based on ISO 21815-3 and ISO 19296, a scenario was developed to calculate the minimum detection distances required to enable safe vehicle interaction at moderate speeds. The analysis showed that at 20 km/h, a total NLOS detection distance of at least 80 m is necessary to accommodate detection, reaction, braking, and safety margins. Systems with shorter detection ranges (such as legacy 50 m solutions) are only suitable to ensure safety for speeds below 13.5 km/h.

These calculations were tested against operational data from a mid-sized underground mine. The data set showed that over 40 per cent of empty travel cycles exceeded the 13.5 km/h threshold, placing them at risk of insufficient stopping distance using short-range CWS and CAS technologies. Additionally, modern underground haul trucks such as the Caterpillar AD30 and Sandvik TH320 are technically capable of reaching or exceeding 25–35 km/h, meaning the potential risk at higher speeds is significant. This reinforces the need for longer-range NLOS detection systems to match the real-world performance capabilities of underground equipment.

Technologies such as DSRC and V2X offer promising solutions. Originally developed for the automotive industry, DSRC has demonstrated the ability to achieve reliable NLOS detection beyond 100 m even in challenging underground conditions. Field trials, supported by published studies, highlight the potential for early hazard recognition and proactive traffic coordination.

This study also examined the potential for improved NLOS awareness to yield operational gains. An illustrative calculation showed that increasing empty haul speeds by just 2 km/h over 80 per cent of the travel segment could enable one additional trip per vehicle per shift representing a ~10 per cent productivity gain. This modest speed increase appears achievable when operators are equipped with better situational awareness and confidence in the surrounding environment.

Beyond immediate safety gains, enhanced consistency in vehicle speeds may also lead to reduced mechanical stress, lower fuel consumption, and more predictable maintenance intervals. These downstream benefits contribute to both safety and productivity.

In conclusion, this study demonstrates that real-time, long-range NLOS detection systems aligned with ISO standards have the potential to significantly improve the safety and efficiency of underground mobile mechanised mining. Continued research and implementation efforts should prioritise interoperability, broader deployment, and operational validation across diverse underground environments.

ACKNOWLEDGEMENTS

The author would like to acknowledge the assistance provided by ChatGPT (OpenAI, 2025) for grammar and text enhancement. The suggestions and edits made by the AI tool has assisted in refining and polishing the document. The author has reviewed and made necessary adjustments to ensure that the final content reflects their own voice and ideas.

The author would like to express gratitude to the individuals whose contributions have been invaluable to the success of this project:

- Gideon Slabbert, Transformation Lead, for insightful discussions regarding underground mining productivity. These insights helped structure the data analysis.

- James Matheson, Development Manager and Gideon Slabbert, Transformation Lead, for proof reading and technical feedback. Their support and guidance enhanced the quality of this work.

- Jane Ball, Manager, Global Marketing Communications, for proof reading and feedback to enhance readability.

Additionally, the author would like to acknowledge Maptek for providing the opportunity, which made this research possible.

Finally, the author extends thanks to all the team members who have collaborated on this project. Their commitment and teamwork have been essential to our success.

REFERENCES

Caterpillar, 2022. AD30 underground mining truck – specifications, Caterpillar Inc. Available from: <https://s7d2.scene7.com/is/content/Caterpillar/CM20220523-b2ea9-e564b> [Accessed: 29 Apr 2025].

Chehri, A, Chehri, H, Hakem, N and Rachid, S, 2020. Realistic 5.9 GHz DSRC Vehicle-to-Vehicle Wireless Communication Protocols for Cooperative Collision Warning in Underground Mining, in Proceedings of the International Conference on Wireless and Mobile Computing, Networking and Communications. Available from: <https://www.researchgate.net/publication/341760133> [Accessed: 6 March 2025].

Department of Mineral Resources and Energy (DMRE), 2023. MHSI Annual Report 2022–2023, DMRE, South Africa. Available from: <https://static.pmg.org.za/MHSI_Annual_Report_2022-2023_pdf.pdf> [Accessed: 6 March 2025].

Earth Moving Equipment Safety Round Table (EMESRT), 2023. Interaction Control: Operational and Design Philosophies for Earthmoving Equipment. Available from: <https://emesrt.org/> [Accessed: 28 April 2025].

Fourie, C and Minnitt, R C A, 2014. Behaviour of production traffic in underground mine ramps, *Journal of the Southern African Institute of Mining and Metallurgy*, 114(12):1015–1023.

Gaber, T, El Jazouli, Y, Eldesouky, E and Ali, A, 2021. Autonomous Haulage Systems in the Mining Industry: Cybersecurity, Communication and Safety Issues and Challenges, *Electronics*, 10(11):1357. https://doi.org/10.3390/electronics10111357

International Organization for Standardization, 2018. ISO 19296:2018 Mining — Mobile machines working underground — Machine safety, ISO: Geneva.

International Organization for Standardization, 2023. ISO 21815-3:2023 Earth-moving machinery — Collision warning and avoidance — Part 3: Risk area and risk level for forward/reverse motion, ISO: Geneva.

Maptek, 2024. Understanding Non-Line-of-Sight Vehicle Detection Underground [online]. Available from: <https://go.maptek.com/l/19542/2024-03-18/44j1zkc/19542/1714546991v9FGcWrN/Maptek_Understanding_Non_Line_of_Sight_Vehicle_Detection_Underground.pdf> [Accessed: 28 April 2025].

Maptek, 2025. Underground Load and Haul Operational Review, confidential report, internal technical dataset, Maptek Pty Ltd.

National Institute for Occupational Safety and Health (NIOSH), 2024. Mining and Machinery Struck-by Injuries [online]. Available from: <https://www.cdc.gov/niosh/mining/topics/machinery-struck-by-injuries.html> [Accessed: 28 April 2025].

OpenAI, 2025. ChatGPT (April 2025 version), OpenAI, Assistance with grammar and text enhancement provided during manuscript development, <https://chat.openai.com/>

Przhedetsky, D, 2010. Productivity of the underground truck – how can it be measured?, Rock Cognition Pty Ltd, Available from: <https://rockcognition.com.au/articles-upload/100513-rc-article-productivity-of-the-underground-truck-by-dmitry-przhedetsky.pdf> [Accessed 29 Apr 2025].

Qian, M, Zhao, K, Li, B, Gong, H and Seneviratne, A, 2022. Survey of Collision Avoidance Systems for Underground Mines: Sensing Protocols, *Sensors*, 22(19):7400.

Qureshi, M A, Noor, R M, Shamim, A, Shamshirband, S and Raymond Choo, K-K, 2016. A Lightweight Radio Propagation Model for Vehicular Communication in Road Tunnels, *PLoS ONE*, 11(3):e0152727. https://doi.org/10.1371/journal.pone.0152727

Salama, A J and Greberg, J, 2012. Optimization of truck-loader haulage system in an underground mine: a simulation approach [online], Luleå University of Technology. Available from: <https://www.diva-portal.org/smash/get/diva2:1011835/FULLTEXT01.pdf> [Accessed 29 Apr 2025].

Sandvik, 2013. Sandvik prototype underground truck sets new standards in Australian operational trials, Sandvik Mining and Rock Technology. Available from: <https://www.rocktechnology.sandvik/en/news-and-media/news-archive/2013/09/sandvik-prototype-underground-truck-sets-new-standards-in-australian-operational-trials-/> [Accessed 29 Apr 2025].

Sandvik, 2022. TH320 underground truck, Sandvik Mining and Rock Technology. Available from: <https://www.geotechpedia.com/Equipment/Show/663/TH320-Underground-truck> [Accessed 29 Apr 2025].

Siton, 2023. XTUT20 underground mining trucks, Siton Machinery. Available from: <https://www.globalsiton.com/Underground-truck/uk-xtut-20-tons-underground-mining-trucks-for-non-coal-mines> [Accessed 29 Apr 2025].

Wikipedia, 2025. IEEE 802.11p, Wikipedia. Available from: <https://en.wikipedia.org/wiki/IEEE_802.11p> [Accessed: 6 March 2025].

Zhang, H, Xu, J and Yu, W, 2024. Traffic Congestion Scheduling for Underground Mine Ramps Based on an Improved Genetic Algorithm, *Applied Sciences*, 13(21):9862. Available from: <https://www.mdpi.com/2076-3417/14/21/9862> [Accessed 29 Apr 2025].

Breaking down silos – enhancing mining operations through Agile collaboration and data-driven decision-making

Elisa[1] and D Farascarina[2]

1. Scientific Analytic Expert, PT. Pamapersada Nusantara, Jakarta 13930.
   Email: elisa@pamapersada.com
2. Operation Research Expert, PT. Pamapersada Nusantara, Jakarta 13930.
   Email: dania.farascarina@pamapersada.com

ABSTRACT

PT. Pamapersada Nusantara (PAMA) embarked on a digital transformation journey to optimise productivity and decision-making at its Berau Coal Binungan (BRCB) mining district. By adopting Scrum, PAMA successfully broke down traditional organisational silos and fostered a data-driven culture.

The implementation of Scrum resulted in significant improvements in the productivity of key machinery. The Excavator PC20008 achieved a 17 per cent increase in output, from 597 Bcm/hrs to 697 Bcm/hrs. Similarly, the Excavator EX2600 experienced an 8 per cent increase, and the Heavy Dumptruck saw a 9.4 per cent improvement in efficiency. These gains can be attributed to the iterative nature of Scrum, which enabled rapid experimentation and continuous improvement.

A real-time performance dashboard, developed as part of the Scrum initiative, provides valuable insights into operational performance. By visualising data from multiple sources, the dashboard empowered decision-makers to make data-driven decisions and identify areas for optimisation.

The success of the Scrum implementation at the BRCB prompted a broader rollout across other PAMA districts, including KIDECO, BRCG, and KPCB. The expanded adoption of Scrum fostered a company-wide culture of innovation and agility.

This case study demonstrates the transformative power of Scrum in the mining industry. By aligning data analytics with operational goals and empowering teams to collaborate, PAMA achieved significant productivity gains and enhanced decision-making capabilities. Scrum has proven to be a valuable framework for organisations seeking to improve operational efficiency and drive sustainable growth.

INTRODUCTION

Operational efficiency is a critical factor in large-scale mining operations, where productivity, resource allocation, and cost control determine overall performance. Many mining companies face challenges in optimising their production systems due to fragmented decision-making, lack of integration between engineering and operations teams, and inconsistent execution strategies. Studies have shown that inefficiencies in equipment utilisation, cycle times, and operational planning contribute significantly to production losses (Beer and Eisenstat, 2000).

PT Pamapersada Nusantara (PAMA) has encountered similar challenges in its Berau Coal Binungan (BRCB) district, where operational inefficiencies have hindered production performance. In Q1 2022, the BRCB achieved only 79 per cent of its planned overburden production. Data from March 2022 indicate that productivity gaps were the largest contributing factor to production loss, accounting for 35 per cent. Key indicators further highlighted inefficiencies in the heavy equipment fleet, with HD 785 trucks operating at 87.4 per cent of the target productivity, EX2600 excavators at 87.41 per cent efficiency, and PC2000 units at 74.6 per cent efficiency. These inefficiencies were further intensified by abnormal cycle times and road conditions, which affected 67 per cent of dump truck rotations.

Despite the BRCB's struggles, it remains a strategic asset for PAMA, containing 66 per cent of Berau Coal's total reserves. To address these challenges, PAMA introduced the MA OPRENG (Operation-Engineering) Strategy, which aims to bridge the gap between operational execution and engineering planning. However, internal assessments revealed that adherence rates to planned activities were only 74.5 per cent in the Engineering (ENG) team and 81.1 per cent in the operations (OPR) team, indicating that traditional management approaches were insufficient.

To enhance coordination and improve operational efficiency, this study explores the application of Scrum methodologies in a mining environment. Scrum, which is widely used in IT and software development, has shown promise in industrial and non-IT sectors (Serrador and Pinto, 2015). By implementing structured, iterative workflows and integrating data-driven decision-making, PAMA aims to improve cross-departmental collaboration to drive operational improvement, specifically fostering a culture of continuous improvement.

Additionally, this research incorporates Evidence-Based Management (EBM) principles to strengthen operational control through leading and lagging indicators (Parmenter, 2015). Leading indicators, such as equipment utilisation rates and operator adherence, help predict potential inefficiencies, allowing for proactive interventions. Moreover, lagging indicators, such as actual production output and fuel efficiency, validate performance trends over time.

This study is based on an improvement project conducted at BRCB from March 2022 to December 2023, focusing on optimising the EX2600 and PC2000 fleets, which contribute 53.64 per cent of total planned production in the CD area. The research aims to answer:

- How can the Scrum framework be applied in mining operations to improve the coordination between operations and engineering teams?

- What measurable impacts can Scrum implementation have on mining production efficiency?

By integrating Scrum-based workflows and EBM-driven performance controls, this research aims to provide a structured approach for improving adaptability, decision-making, and execution in large-scale mining operations.

LITERATURE REVIEW

Scrum, an Agile framework emphasising iterative development, continuous feedback, and adaptive planning, has evolved beyond software development into various industrial applications (Schwaber and Sutherland, 2020). Originally designed to enhance software engineering efficiency, Scrum has been adopted in industries such as manufacturing, oil and gas, and mining, where dynamic decision-making and real-time problem-solving are crucial (Rigby, Sutherland and Takeuchi, 2016). The Stacey Matrix (Stacey, 1996) provides a structured approach to determining when Agile methodologies, including Scrum, are most applicable. In complex environments with high uncertainty, such as mining operations, traditional linear management approaches struggle to accommodate such uncertainty. Agile frameworks offer a more suitable alternative by enabling iterative adaptation and real-time responsiveness (Conforto et al, 2014). As illustrated in Figure 1, the Stacey matrix shows how Scrum is best suited for complex decision-making, where both requirements and technical solutions are highly uncertain.

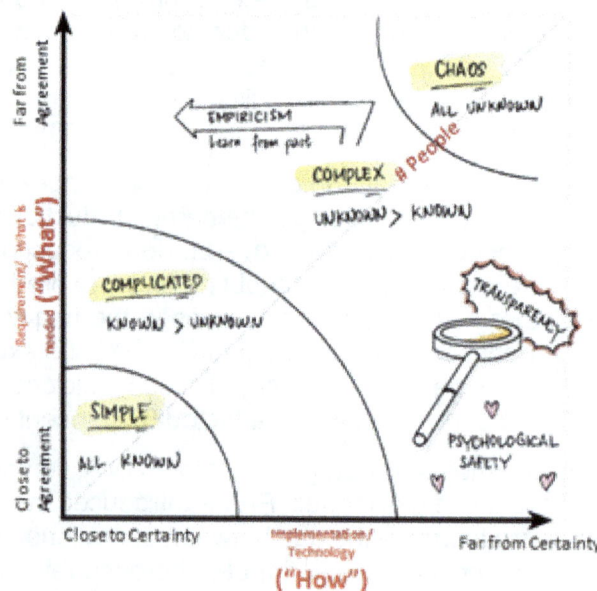

FIG 1 – Stacey complexity matrix (source: Obrutsky, 2016).

Scrum has been successfully implemented across various non-software industries. In the oil and gas sector, BP (British Petroleum) has leveraged Agile methodologies to optimise offshore drilling asset management, improving equipment reliability and reducing downtime through iterative improvements (McKinsey and Company, 2021). Similarly, Rio Tinto has integrated Scrum with predictive maintenance systems to enhance fleet availability and reduce operational delays in mining operations (Deloitte, 2020). Airbus, a key player in the aerospace industry, has adopted Agile principles to streamline aircraft production, reducing project lead times and increasing adaptability to market fluctuations (Lehmann and Reineke, 2019). These case studies demonstrate the viability of Scrum in complex, high-risk industrial environments where rapid adaptation is required to optimise operations (Denning, 2018).

In product development, approaches can vary significantly. **Type A (Sequential)** represents a traditional, linear approach in which each phase of the project is completed before moving on to the next, similar to the approach often used in NASA's systems. Although this approach ensures control and order, it can be slow and rigid. **Type B (Overlapping at Phase Boundaries)** allows certain phases to overlap slightly, meaning that the next phase starts as the previous one is nearing completion. This allows for more efficiency and faster progression while maintaining some level of order. **Type C (Extensive Overlapping)** takes this a step further by overlapping multiple phases simultaneously, enabling rapid iteration and flexibility, which aligns more closely with Agile principles. This approach accelerates the pace of development, adapting quickly to changing needs, as seen in industries like aerospace and mining. Figure 2 further illustrates how overlapping phases in product development, as described by Takeuchi and Nonaka (1986), align with Agile principles to accelerate innovation and efficiency.

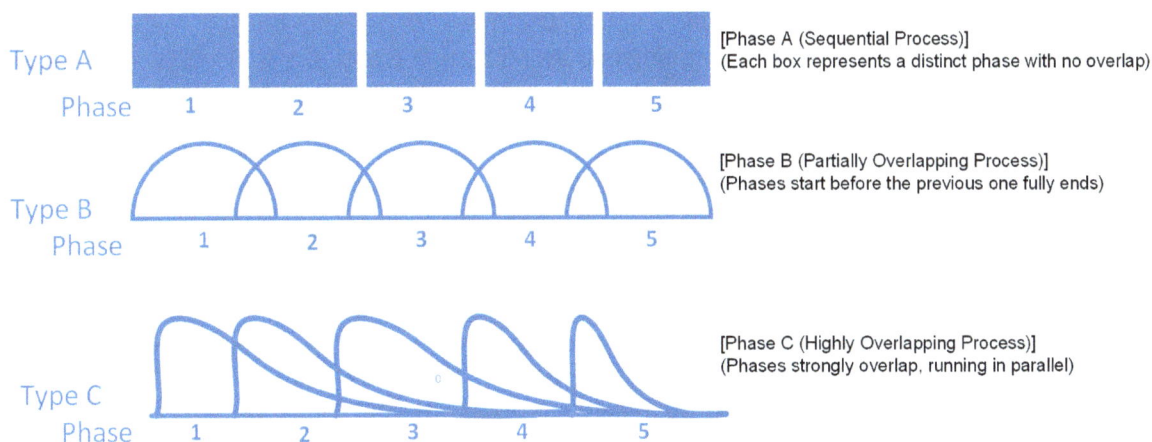

FIG 2 – Comparison of product development processes (source: Takeuchi and Nonaka, 1986).

A critical component of Scrum's effectiveness in industrial operations is the implementation of Evidence-Based Management (EBM), which facilitates performance optimisation through Leading and Lagging Indicators (Scrum.org, 2021). Leading indicators, such as cycle time and backlog refinement frequency, provide insights into potential future performance trends, enabling proactive decision-making. In contrast, lagging indicators, including production output and the mean time between failures, reflect historical performance and inform long-term strategic planning (Nehring et al, 2018). By utilising real-time performance metrics, industrial organisations, including those in the mining sector, can refine their Scrum implementation to enhance operational efficiency and resilience (Topal, Knights and Kizil, 2019).

In Agile-driven industrial environments, statistical disparity, measured using the Coefficient of Variation (CV), serves as a key metric for evaluating variability and process control. CV is widely applied in Six Sigma and Lean methodologies to measure operational performance fluctuations and identify process instability (Antony, 2018). A higher CV value indicates increased variability, necessitating iterative adjustments within Scrum Sprints, whereas a lower CV suggests a stable process, allowing teams to focus on incremental optimisations (Knights, Topal and Kizil, 2017). In mining operations, CV-based statistical models have been employed to monitor excavation rates, equipment performance, and material haulage efficiency, ensuring that Scrum-driven improvements align with measurable operational outcomes (Nehring *et al*, 2018).

The literature confirms that Scrum is highly applicable beyond software development, particularly in industrial sectors characterised by complexity and uncertainty, such as mining, oil and gas, and aerospace. By integrating Agile methodologies, companies can enhance operational resilience, optimise performance through EBM, and effectively manage variability using statistical process control methods such as CV. Future research should focus on refining domain-specific Scrum adaptations to further improve productivity and efficiency in mining operations while addressing industry-specific challenges and operational constraints.

METHODS

In the methodology section, the author will explain the approach starting from the methodology to the implementation of improvement solutions.

Introduction to the methodology

The methodology used in this study involves a mixed-methods approach, combining both quantitative and qualitative techniques. For quantitative data collection, observational methods were employed along with a questionnaire using a Likert scale (1–4) distributed through Google Forms. Secondary data from the year 2022 were also utilised. The target respondents for this study are section heads or individuals in similar roles that can make decisions related to problem-solving and improvement initiatives. In terms of data analysis, the study applied descriptive statistical analysis with Pearson Correlation and Cronbach's Alpha to evaluate the reliability and relationships between variables. In addition, diagnostic analytics were employed to provide further insights. The data analysis was conducted using software tools such as Microsoft Excel, Smart PLS, and MySQL.

Scrum implementation process

Fundamentals of Scrum – awareness and training

The Scrum process operates as a continuous, closed-loop cycle that constantly evolves to address recurring challenges from different perspectives, as shown in Figure 3. Each sprint builds upon the previous one by maintaining focus on the same core problem but exploring it from new angles, ensuring that solutions remain relevant and progressively more effective. This approach allows for constant reflection, feedback, and refinement, creating a dynamic environment in which each iteration strengthens the team's ability to tackle complex issues with increasing precision. The closed-loop system ensures that learning and improvement are embedded in every phase, driving sustained progress while adapting to changing needs and priorities, as illustrated in the figure.

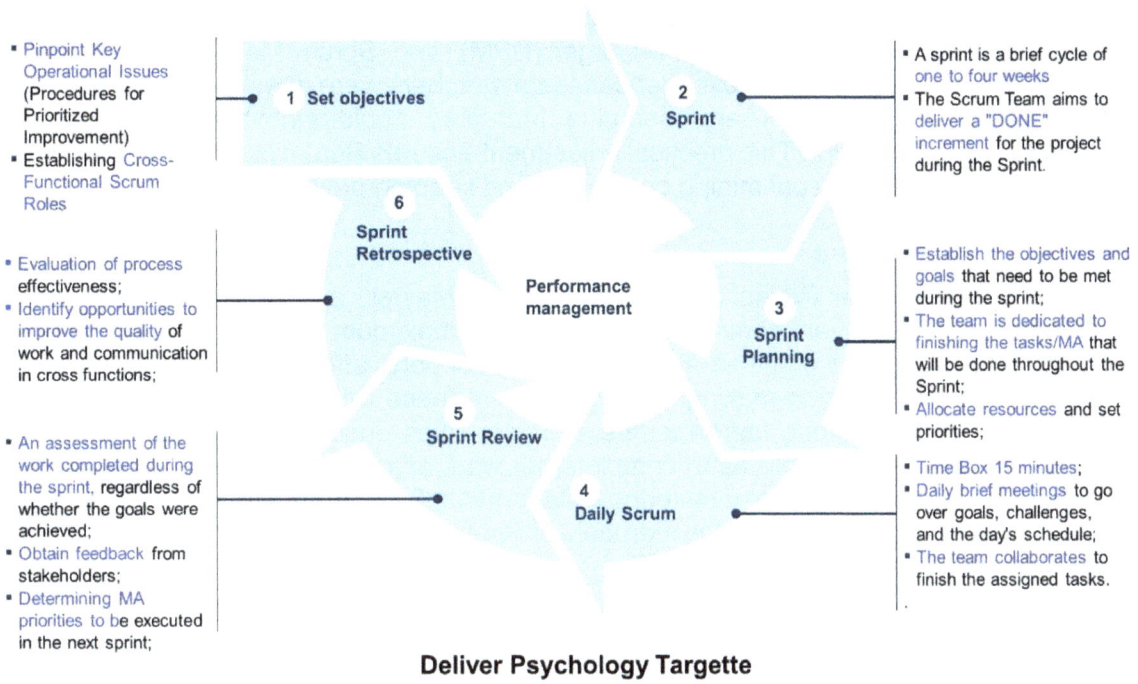

FIG 3 – Scrum in operation workflow (adapted from Schwaber and Sutherland, 2020).

Delivering the psychological target here should be seen not as a 'nice-to-have' goal, but as a target rooted in actual capacity. This approach focuses on setting goals based on the unit's capacity, ensuring that improvements are not just aspirational but achievable within the team's operational capabilities. By selecting specific areas for enhancement, we can focus efforts on areas that are most likely to yield the greatest impact, thus aligning the targets with the unit's true capacity. This method ensures that the psychological targets set are realistic, measurable, and driven by the team's ability to perform, creating a foundation for sustainable progress and continual achievement.

In the application of Scrum within the coal mining project at PAMA BRCB, the standard Scrum Framework, as outlined in the 2020 Scrum Guide, was adapted to fit the unique needs of the industry (Figure 4). Schwaber and Sutherland's (2020) paper provides the foundation for Scrum, but for clarity and alignment with PAMA's operational processes, key changes were made. The evolving nature of coal mining projects, with annual changes in volume and strategy, necessitates a flexible Scrum approach. Unlike software development, where the focus is on Product Backlog Items (PBI) and tasks to achieve deliverables, in mining, the emphasis is on defining tactical activities that can be executed quickly to maintain leading and lagging indicators on target.

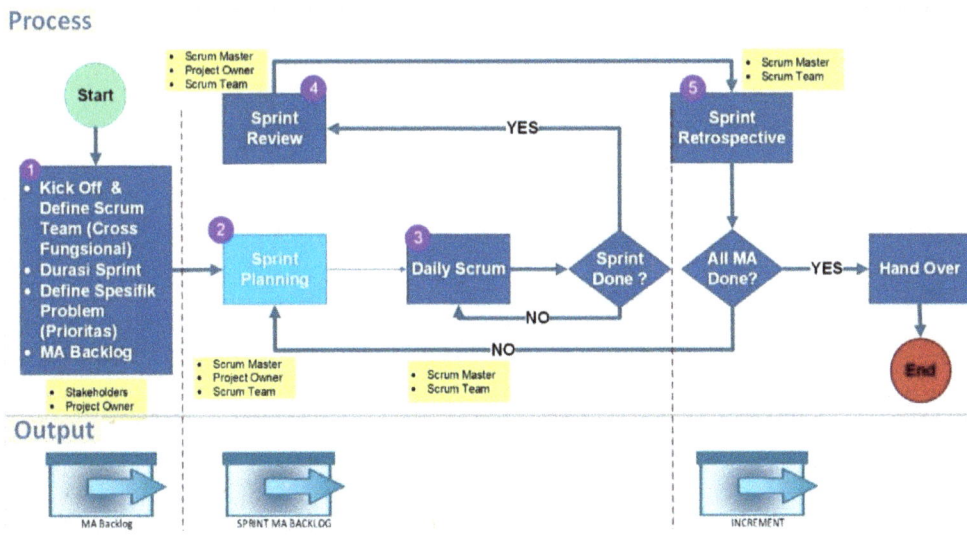

FIG 4 – Scrum Agile project management (source: Streule *et al*, 2016).

As part of this adaptation, 'Product Owner' was renamed to 'Project Owner,' represented by the Project Manager (PM) or Deputy Project Manager (DPM), and 'Scrum Master' and 'Scrum Team' roles remain. To ease the transition, experienced team members were directly appointed to the roles of Scrum Master and Scrum Team, ensuring that they could innovate and drive process improvements within the sprint. This practical adjustment ensures Scrum is not just a framework but a strategic tool for optimising coal mining operations and keeping projects on track.

Formation of Scrum roles

Scrum roles consist of one Project Owner, one Scrum Master, and at least two Scrum Team members, with a minimum team composition of four and a maximum of ten individuals. The primary reason for this structure is to maintain a team size that supports effective communication, efficient coordination, and swift decision-making. The purpose of these roles in Scrum is to ensure clear responsibility distribution among team members. Scrum roles ensure that the team possesses a balanced set of skills and capabilities to complete the work efficiently. Each role is responsible for ensuring smooth information flow and fostering collaboration. Furthermore, these roles provide clear focus and accountability for each team member, promoting project overall success. For a visual representation of this structure, see Figure 5.

Scrum Team size ≤ 10 (this is just recommendation not a rule!)

FIG 5 – Scrum size (source: Schwaber and Sutherland, 2020).

Diagnostic analysis for improvement priorities

The analysis process is conducted across all critical upstream to downstream processes in the Overburden supply chain. Subsequently, responsibilities are assigned to each Scrum team to control these processes, ensuring they remain within normal limits. One example of this analysis, shown in Figure 6 includes the formation of seven streams based on the urgency of processes that need to be controlled (Figure 7).

Penentuan Prioritas Leading Indicator / Sprint Goals

Regression Equation

$$VOL = 1349.6 - 6.17 \text{ avg speed} - 62.5 \text{ achload} - 24.4 \text{ Avg LST} + 0.0 \text{ befrest\_0} - 32.8 \text{ befrest\_1}$$
$$+ 0.0 \text{ afrest\_0} - 129.8 \text{ afrest\_1} + 0.0 \text{ befCS\_0} - 430.8 \text{ befCS\_1} + 0.0 \text{ aftCS\_0}$$
$$- 39.4 \text{ aftCS\_1} + 0.0 \text{ ashrpray\_0} - 108.2 \text{ ashrpray\_1} + 0.0 \text{ ishapray\_0} - 153.7 \text{ ishapray\_1}$$
$$+ 0.0 \text{ subuhpray\_0} - 152.4 \text{ subuhpray\_1} + 0.0 \text{ EGI\_EX2600} - 793.7 \text{ EGI\_PC1250SP8}$$
$$- 475.6 \text{ EGI\_PC20008}$$

✓ **Variable Respon:** Volume Produksi per loader per hour

✓ **Variable Predictor :**

1 Plan Working Geometri (HH, LH, LL, HL)	10 Jam After Change Shift
2 Average Loaded Stop Time	11 Jam Sholat Ashar
3 Average Empty Stop Time	12 Jam Sholat Maghrib
4 Ach. Loading time (Act/Plan)	13 Jam Sholat Isha
5 Average Travel Speed	14 Jam Sholat Subuh
6 Std Dev. Speed	15 EGI Loader
7 Jam Before Rest	16 Categorical Pit
8 Jam After Rest	17 N HD
9 Jam Before Change Shift	18 Distance

Model Summary

S	R-sq	R-sq(adj)	R-sq(pred)
147.388	72.49%	72.31%	72.08%

N Unit	17		Unit
PA	92%		
Avg. Prodty	622,34		Bcm/Hrs

Coefficients

Term	Coef	SE Coef	T-Value	P-Value	VIF
Constant	1349.6	43.7	30.89	0.000	
avg speed	-6.17	1.34	-4.59	0.000	1.21
achload	-62.5	20.6	-3.03	0.002	1.93
Avg LST	-24.4	16.5	-1.48	0.139	1.04
befrest 1	-32.8	12.7	-2.58	0.010	1.08
afrest 1	-129.8	13.4	-9.72	0.000	1.09
befCS 1	-430.8	11.7	-36.81	0.000	1.10
aftCS 1	-39.4	12.3	-3.20	0.001	1.08
ashrpray 1	-108.2	18.0	-6.03	0.000	1.04
ishapray 1	-153.7	17.3	-8.88	0.000	1.04
subuhpray 1	-152.4	17.5	-8.69	0.000	1.04
EGI PC1250SP8	-793.7	19.4	-40.83	0.000	3.52
PC20008	-475.6	11.7	-40.72	0.000	2.25

Factor	Coef	WH/Freq	Opploss/Day
Speed	6,17	16,5	1.592
Loading Time	62,50	16,5	16.129
LST	24,40	16,5	6.297
Before Rest	32,80	2	1.026
After Rest	129,80	2	4.060
Before CS	430,80	2	13.475
After CS	39,40	2	1.232
Ashar Pray	108,20	1	1.692
Isha Pray	153,70	1	2.404
Subuh Pray	152,40	1	2.384

Priority to Improve:

1. Improve loading time;
2. Kontrol jam sebelum Change shift;
3. Improve loaded stop time (LST), cek traffic area disposal / jalan yang menyebabkan HD muatan berhenti;
4. Kontrol produksi pasca rest (start stop loader dan hauler);

FIG 6 – An example of correlation analysis of variables that significantly influence production March 2022.

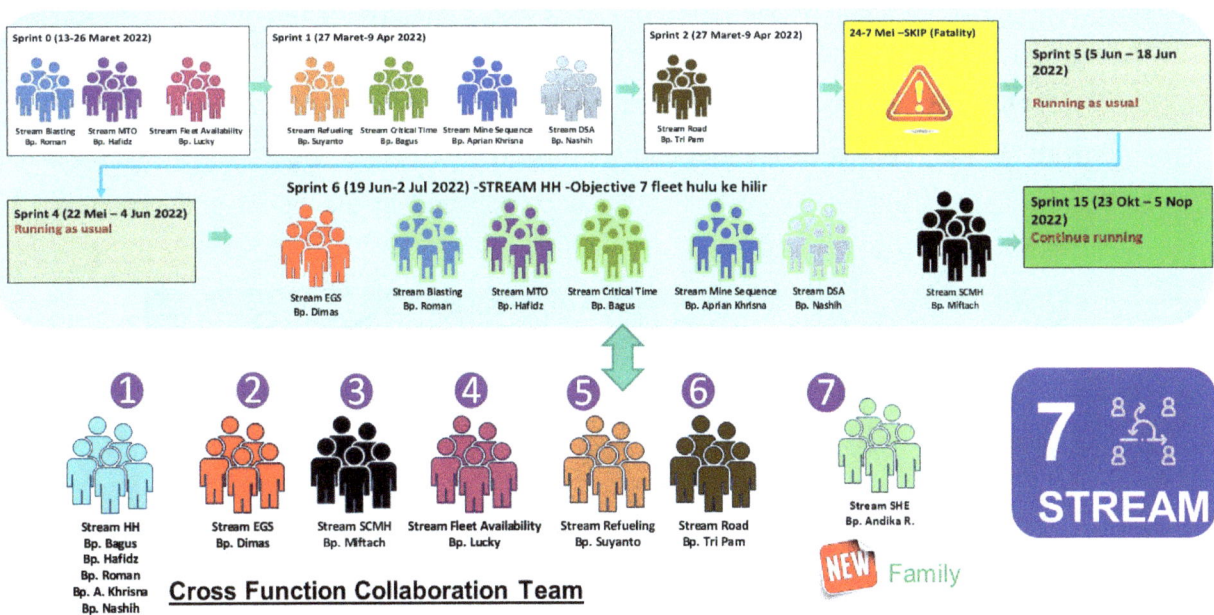

FIG 7 – Scrum team structure at PAMA BRCB (source: author's own work).

Implementation of Scrum events

Scrum Events are key meetings within the Scrum process that form an iterative cycle, ensuring continuous improvement and alignment with project goals. These events are time-boxed to promote efficiency and focus. These include Sprint, Sprint Planning, Daily Scrum, Sprint Review, and Sprint Retrospective. Each event facilitates collaboration, provides regular feedback, and allows for necessary adaptations, ensuring the team remains transparent, accountable, and on track towards achieving project objectives. By consistently reviewing progress and addressing challenges, Scrum events play a crucial role in maintaining momentum and driving project success.

- **Sprint:** A fixed time ranging from 1 to 4 weeks, during which the team focuses on completing the planned work/activities. The mapping can be seen in Figure 8.

- 1-Week sprint:
  - A short duration allows the team to move quickly and receive more frequent feedback.
  - Ideal for small projects with a limited scope.
  - Provides the team to adapt quickly to changing requirements
  - However, a sprint that is too short may overwhelm the team with more frequent meeting overheads.

- 2-Week sprint:
  - The most used duration in agile development.
  - Provides a good balance between iteration speed and the team's ability to complete significant work.
  - Offers enough time for the team to plan, develop, and test features.
  - Allows the team to receive regular feedback.

- 3-Week and 4-Week sprint:
  - Longer durations give the team more time to complete more complex tasks.
  - Suitable for larger projects with a broader scope.
  - This allows the team to focus on tasks over an extended period.
  - However, longer sprints may cause the team to lose momentum and delay quick feedback.

- **Sprint planning**: A meeting held at the beginning of each sprint to plan the work to be done. During Sprint Planning, the following roles must be present:
  - **Project Owner**: Explains and prioritises the issues that need to be addressed by the team.
  - **Scrum Master**: Facilitates the meeting and ensures everyone understands the objectives.
  - **Scrum Team**: Determines what can be accomplished in the sprint and how it will be achieved.

Activity	Sprint	Planning	Review	Retrospective	Daily Scrum
30 days	8 hours	4 hours	3 hours	3 hours	15 minutes
3 weeks	~ 6 hours	~ 3 hours	~ 3 hours	~ 2 hours 15 minutes	15 minutes
2 weeks	~ 4 hours	~ 2 hours	~ 2 hours	~ 1.5 hours	15 minutes
1 week	~ 2 hours	~ 1 hour	~ 45 minutes	~ 45 minutes	15 minutes

FIG 8 – Scrum duration (source: Partogi, 2015).

An example of formulating sprint goals and Milestone Activities (MA) within a sprint can be seen in Figure 9.

- **Daily Scrum**: A brief daily meeting (time-boxed to 15 mins or less) to discuss progress, challenges, and the daily plan. The Daily Scrum helps the team stay focused and coordinated. Each team member answers three questions:

  - What have I done since the last meeting?

  - What will I do today?

  - Are there any obstacles preventing me from moving forward?

FIG 9 – New example of Sprint backlog used in PAMA (Source: Author's own adaptation based on internal PAMA documentation, 2024).

An illustration of this can be seen in Figure 10.

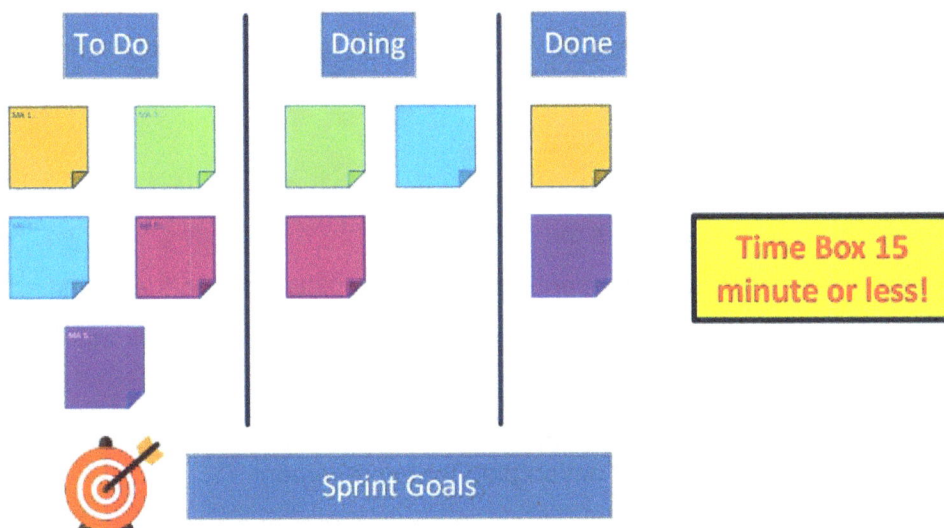

FIG 10 – Daily Scrum illustration (source: created by the author in collaboration with Scrum Master, PAMA, 2024).

During the Daily Scrum, the following roles must be present:

- **Scrum Team:** All team members must be present.
- **Scrum Master:** May attend to assist, but it is not mandatory.
- **Project Owner:** This person can attend but is not required.

Common Agile practices during the Daily Scrum include the Daily Standup and Daily Plank Meeting, as shown in the field practice in Figure 11. The goal is to ensure effective communication, review progress, and focus on the tactical Milestone Activities (MA) to be executed for the day.

Stream HH

Stream SCMH

Daily Standup

✔ 3 peraturan Daily Plank Meeting:
1. Setiap orang bergiliran menjelaskan progres pekerjaannya
2. Kamu harus tetap melakukan plank saat berbicara
3. Kamu bisa beristirahat (tidak melakukan posisi plank) saat tidak berbicara

Daily Plank Meeting

FIG 11 – Daily Scrum practices observed at PAMA BRCB (source: author's own work, based on internal observations at PAMA BRCB, 2022).

- **Sprint Review**: A meeting at the end of the sprint to showcase the work completed to stakeholders. During this meeting, the team receives feedback and discusses what has been achieved. The goals of the Sprint Review are:
  - o Review the work completed during the sprint (MA versus Sprint Goals).
  - o Receive feedback from stakeholders on the results achieved.
  - o Discuss what went well and what needs improvement.
  - o Adjust the direction of the program/project if necessary for corrective action in the next sprint.

Roles that must be present during the Sprint Review:

- **Project Owner:** Provides feedback.
- **Scrum Master:** This role facilitates the meeting.
- **Scrum Team:** The scrum team presents the work completed during the sprint.
- **Stakeholders:** stakeholders may be invited to provide feedback and engage in discussion.

For implementation details, refer to Figure 12.

- **Sprint Retrospective**: A meeting at the end of the sprint to reflect on what went well, what could be improved, and create an action plan for improvement.

SPRINT 15 SPRINT REVIEW (05/11/2022)

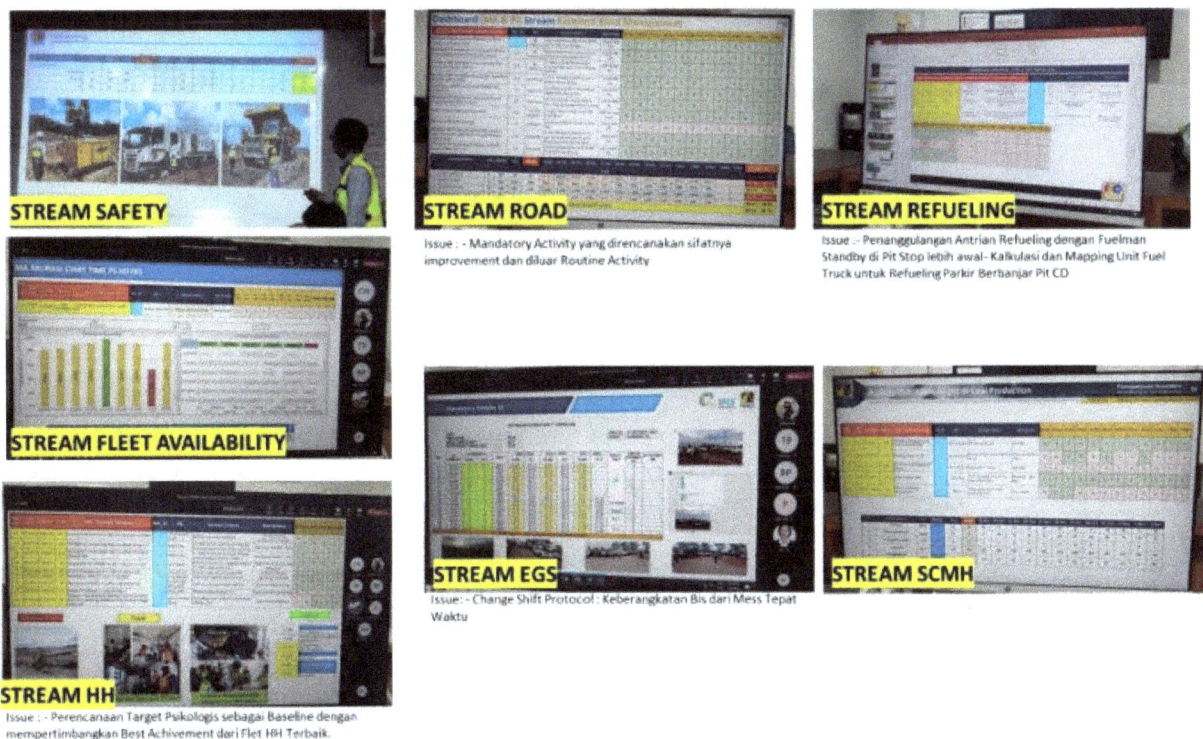

FIG 12 – Sprint review (source: PAMA BRCB, 2022).

Roles required during the Sprint Retrospective:

- **Scrum Master:** This role facilitates the meeting and ensures all voices are heard.
- **Scrum Team:** Discusses what went well, what did not, and how to improve processes.
- **Project Owner:** This person is not required to attend but may join if desired.

The retrospective cycle helps the team analyse and improve performance. It can be visualised as follows Figure 13:

- **Set the Stage**: Create a safe and open environment for discussion, ensuring that all team members feel comfortable sharing their views and experiences.
- **Look Back**: Reflect on the past sprint by considering what happened and how team members felt about it.
- **Gather Data**: Collect concrete information on the work results, challenges, and successes during the sprint.
- **Generate Insights**: Analyse the gathered data to identify patterns or recurring issues. These insights help us understand what went well and what needs improvement.
- **Decide What to Do**: Based on the identified insights, the team discusses and decides on concrete actions to improve future performance.
- **Action plan:** A clear, specific action plan, including responsibilities and deadlines for implementation.
- **Close the Retrospective**: Conclude the session by summarising discussions, reaffirming commitment to the action plan, and allowing team members to provide feedback on the retrospective process itself.

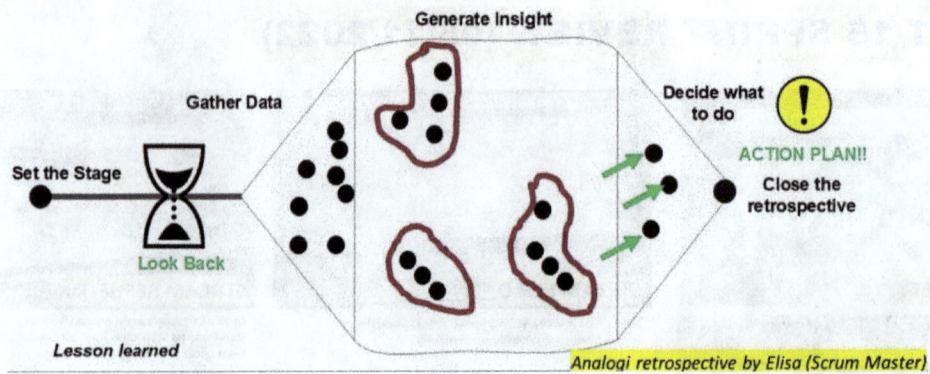

FIG 13 – Sprint retrospective analogy by Scrum Master.

SCRUM artefacts

Scrum Artefacts are objects used to enhance transparency and provide insights into progress within the Scrum framework. These artefacts include the MA Backlog, which represents the Project Goals; the Sprint Backlog, which represents the Sprint Goals; and the Increment, which is defined by the Definition of Done (DoD).

The Project Goals refer to medium-term objectives, typically spanning 3 months, 6 months, or 1–2 years. Sprint Goals, on the other hand, represent short-term objectives to be achieved within a single Sprint, which usually lasts between 1 and 4 weeks. These goals can be aligned with performance indicators such as PI (Performance Indicator), OPRENG (Operational Performance), OKRs, and KPIs. The Increment represents the tangible result of activities completed during the sprint, or a version of the product that is usable and deliverable at the end of the sprint.

The definition of 'done' (DoD) provides a clear and consistent understanding of what is considered 'done' within the context of a project during a sprint. Establishing the DoD criteria is crucial for preventing differing perceptions among team members about what constitutes completion. For example, clear DoD criteria can help ensure team alignment regarding what needs to be achieved before work is deemed finished. For a detailed view of the Scrum artefacts within a single sprint, refer to Figure 14.

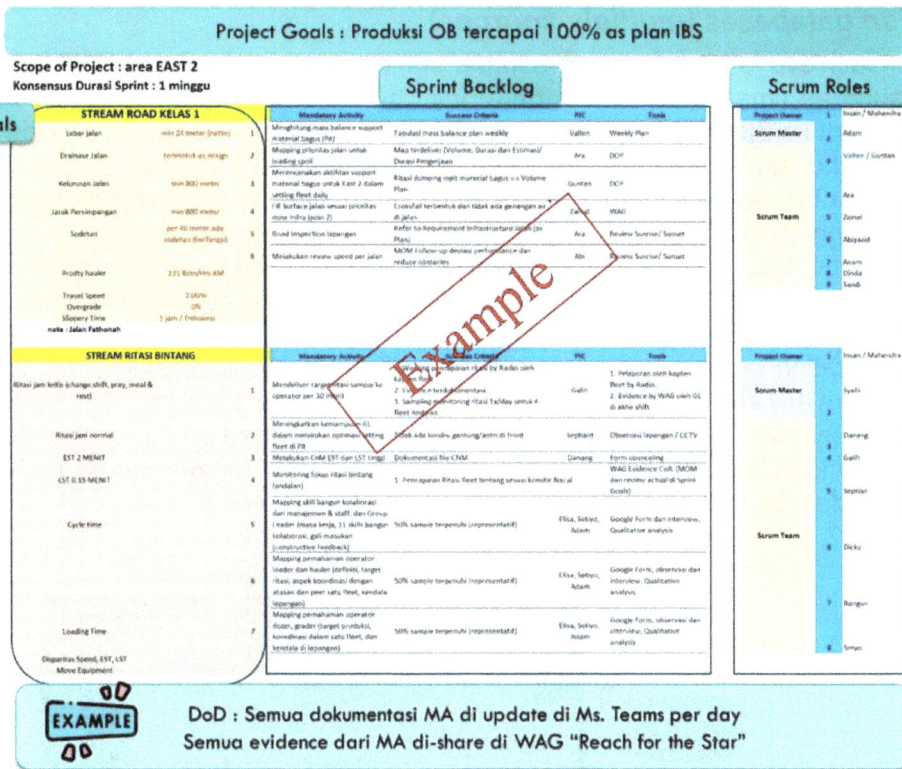

FIG 14 – Scrum artefacts: MA backlog, sprint goals, project goals and team roles at PAMA BRCB (source: author's own work, based on internal practices at PAMA BRCB, 2024).

Scrum measurement (lagging and leading indicator)

Scrum measurement utilises basic knowledge and the collection of measurement processes to assess process changes. Both Lagging and Leading Indicators are defined collaboratively with the Scrum team, covering the entire process from upstream to downstream. These indicators will serve as the foundation for evaluating the success of each sprint. The measurement mapping for the Overburden Process, starting from the critical processes that need to be controlled to ensure the achievement of lagging indicators, is illustrated in Figure 15.

FIG 15 – Lagging and leading indicator's used in PAMA BRCB operations (source: author's own work, based on internal performance metrics at PAMA BRCB, 2024).

Integration databases (multiple sources)

PAMA can achieve a comprehensive understanding of production by integrating various data sources, such as timesheets, hourly data, joint surveys, VHMS, SmartD, Ewacs Pro, and other technologies, into a Big Data Tree Diagram. This integration creates a unified view for both the Head Office (HO) and site operations, enabling the identification of root causes behind production shortfalls. The objective is to empower site decision-makers by providing an integrated view of both leading and lagging indicators, eliminating the need for manual analysis and multiple dashboards. The process involves drafting a tree diagram for production non-attainment, defining key data attributes like metrics, goals, and constraints, and designing a data architecture that supports data normalisation through ETL (Extract, Transform, Load). Continuous monitoring and maintenance of the data integration process ensure accuracy and efficiency. The final result, as visualised in the mapping (Figure 16) and Tableau dashboard (Figure 17), offers a clear, real-time view of production performance, enabling better decision-making. This streamlined process allows decision-makers to quickly identify leading indicators causing issues in a district, all on one page, without the need to open multiple dashboards—saving time and enhancing efficiency.

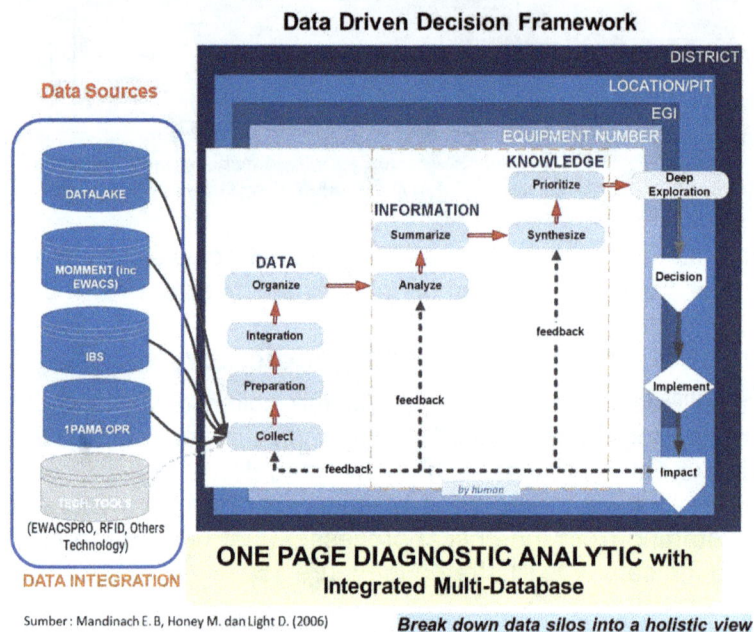

FIG 16 – Data integration.

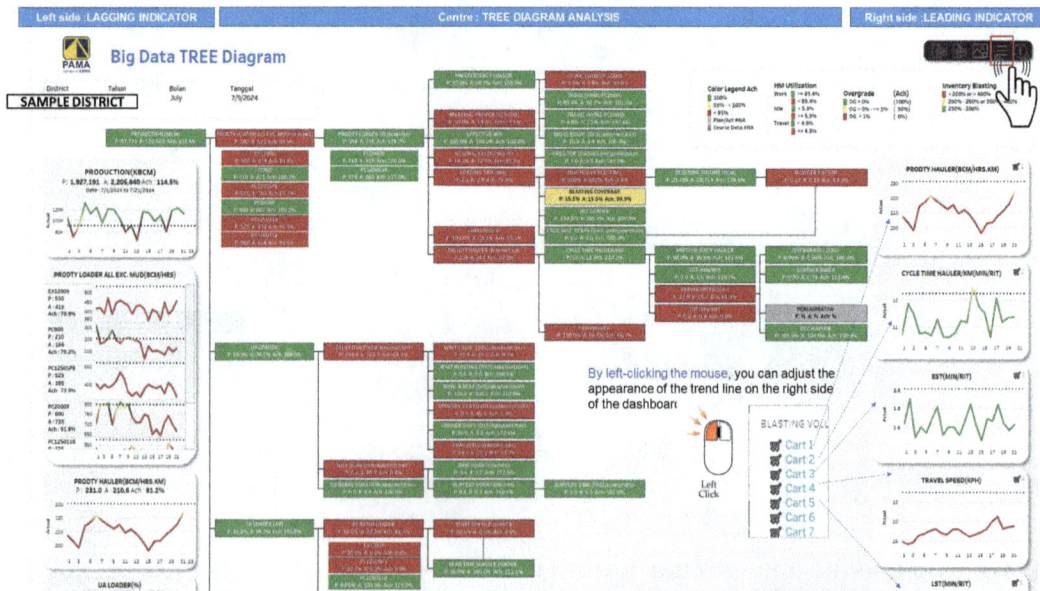

FIG 17 – Dashboard big data tree diagram (lagging and leading indicator) one page dashboard.

Gamification and coaching clinic

After a year of implementing the Coaching Clinic program in 2022, conducted over 20 sprints, the gamification applied at the site in Figure 18 and the Coaching Clinic have shown significant results in improving team performance and the quality of Scrum implementation. A Coaching Clinic is a mentoring session designed to deepen the team's understanding of Scrum principles, enhance skills in executing Scrum roles, and address challenges faced during the Scrum implementation process. In this program, coaches provide hands-on, interactive training to overcome practical difficulties that arise on the ground.

❑ Gamification Scrum in Operation (SiO) ❑ Pemberian Apresiasi dari Manajemen Site

GAMIFIKASI STREAM SCRUM IN OPERATION (SIO BRCB)

MA	H1	H2	H3	H4	H5	H6	H7	H8	H9	H10	H11	H12	H13	H14	ALL
1 HH	84%	83%	58%	80%	82%	86%	82%	76%	84%	84%	79%	83%	91%	89%	84%
2 FA	77%	68%	67%	55%	55%	90%	46%	70%	67%	61%	45%	64%	48%	44%	58%
3 ROAD	80%	79%	81%	83%	79%	82%	79%	79%	80%	81%	84%	82%	80%	82%	81%
4 SCMH	60%	61%	67%	68%	66%	67%	82%	75%	74%	74%	77%	75%	68%	70%	70%
5 REFUELING	71%	73%	71%	68%	70%	70%	70%	71%	73%	71%	73%	73%	73%	93%	71%
6 EGS	95%	100%	93%	100%	94%	95%	93%	91%	96%	95%	91%	96%	93%	92%	95%
7 SAFETY	88%	74%	73%	73%	81%	81%	84%	79%	70%	77%	76%	82%	83%	87%	78%

PI	Sprint 6	Sprint 7	Sprint 8	Sprint 9	Sprint 10	Sprint 11	Sprint 12	Sprint 13	Sprint 14	Sprint 15	Sprint 16	Sprint 17	Sprint 18	Sprint 19	ALL
1 HH	84%	76%	86%	82%	86%	86%	85%	82%	85%	82%	87%	88%	85%	85%	84%
2 FA	67%	58%	57%	60%	59%	58%	63%	61%	83%	86%	93%	84%	87%	88%	73%
3 ROAD	N/A	N/A	N/A	N/A	N/A	N/A	81,9%	84,4%	97,60%	95,10%	72,60%	95%	101,20%	89,60%	95%
4 SCMH	78%	65%	96%	96%	98%	98%	85%	96%	83%	75%	75%	97%	85%	88%	86%
5 REFUELING	86%	77%	87%	83%	93%	91%	98%	87%	94%	94%	94%	N/A	92%	92%	88%
6 EGS	99%	100%	100%	102%	102%	100%	100%	96%	102%	107%	99%	97%	97%	99%	100%
7 SAFETY	N/A	N/A	N/A	N/A	N/A	N/A	100%	75%	N/A	90%	87%	94%	90%	79%	88%

	MA	PI	80%	20%		
HH	84%	84%	67%	17%	83,8%	2
FA	58%	73%	47%	15%	61,3%	
ROAD	81%	95%	65%	19%	83,6%	
SCMH	70%	86%	56%	17%	73,3%	
REFUELING	73%	88%	58%	18%	75,9%	
EGS	95%	100%	76%	20%	95,6%	1
SAFETY	78%	88%	62%	18%	79,9%	

FIG 18 – Gamification SiO implementation (hybrid after completing 20 sprints).

The need for a Coaching Clinic arises for several key reasons: first, to ensure a deeper understanding of Scrum values and principles; second, to support the development of team skills in applying Scrum effectively, particularly in the roles of Scrum Master and Product Owner; third, to improve collaboration and communication among team members; and fourth, to drive continuous improvement in each sprint through reflection and adjustments made during retrospectives. All these points can be observed in the evidence shown in Figure 19, which illustrates improved team performance, better collaboration, and progress in completing sprint tasks more efficiently after the implementation of the Coaching Clinic and gamification.

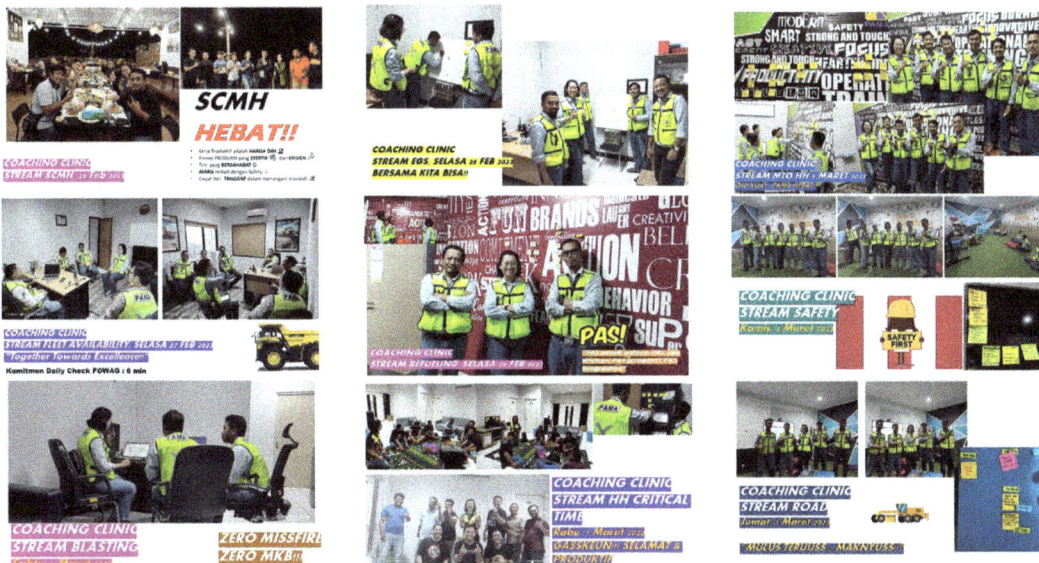

FIG 19 – Coaching clinic about Scrum in operation implementation.

RESULTS

Evaluation of collaboration and performance post-Scrum implementation

After the Scrum teams completed the Scrum in Operation Implementation process from March 13, 2022 to December 31, 2022, spanning 20 sprints, a survey was conducted to measure the effectiveness of the collaboration process through Scrum. A total of 26 respondents from all streams within PAMA BRCB participated in the survey. Most respondents (30.8 per cent) had less than three years of experience at PAMA, with 69.2 per cent having been at PAMA BRCB for under three years. The majority (34.6 per cent) of those involved in Scrum collaboration had participated for 5–8 months. The survey results showed a balanced representation of Scrum teams across all streams, with management also contributing assessments. **The collaboration-building capability improved significantly, with the index rising from 28 (indicating developing collaboration) to 39.3, reflecting effective collaboration after the implemented improvements.** This improvement directly answers the business question by demonstrating how the Scrum framework has enhanced coordination between operations and engineering teams at PAMA BRCB. The comparison between pre-Scrum and post-Scrum collaboration can be seen in Figure 20.

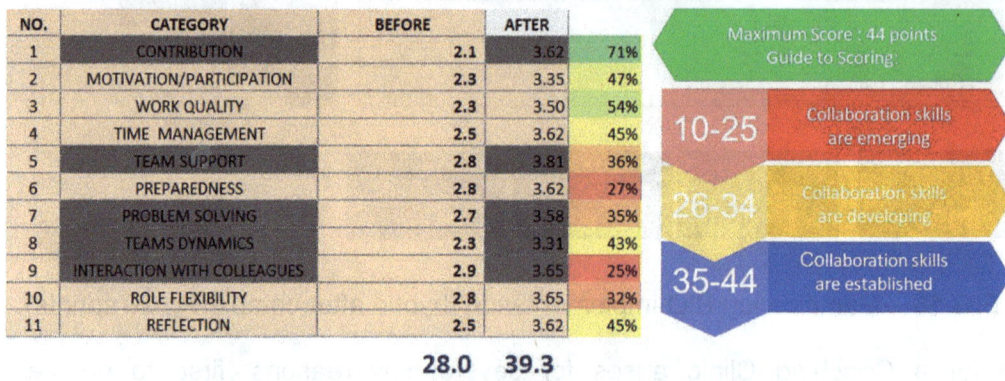

NO.	CATEGORY	BEFORE	AFTER	
1	CONTRIBUTION	2.1	3.62	71%
2	MOTIVATION/PARTICIPATION	2.3	3.35	47%
3	WORK QUALITY	2.3	3.50	54%
4	TIME MANAGEMENT	2.5	3.62	45%
5	TEAM SUPPORT	2.8	3.81	36%
6	PREPAREDNESS	2.8	3.62	27%
7	PROBLEM SOLVING	2.7	3.58	35%
8	TEAMS DYNAMICS	2.3	3.31	43%
9	INTERACTION WITH COLLEAGUES	2.9	3.65	25%
10	ROLE FLEXIBILITY	2.8	3.65	32%
11	REFLECTION	2.5	3.62	45%
		28.0	**39.3**	

Maximum Score : 44 points
Guide to Scoring:

10-25 — Collaboration skills are emerging
26-34 — Collaboration skills are developing
35-44 — Collaboration skills are established

FIG 20 – Survey results based on the collaboration self-assessment tool reference (source: Ofstedal and Dahlberg, 2009).

Evaluation of metric dashboard effectiveness in monitoring Scrum performance

The significant productivity gap observed in the PC20008 model, despite having the largest fleet size of 12 units and an average productivity of only 16 per cent, indicates a potential issue that requires further investigation. While the EX26007 and PC200011R models perform satisfactorily, the low and inconsistent productivity of the PC20008 suggests that factors such as operator skills, equipment condition, and maintenance practices could be influencing its performance. To improve equipment utilisation and overall project efficiency, a thorough evaluation of these factors is needed, including operator training, regular equipment assessments, and optimised maintenance schedules. Addressing these issues will help identify and resolve the root causes of the low productivity, leading to enhanced overall fleet performance. The detailed analysis and productivity trends are presented in Figure 21.

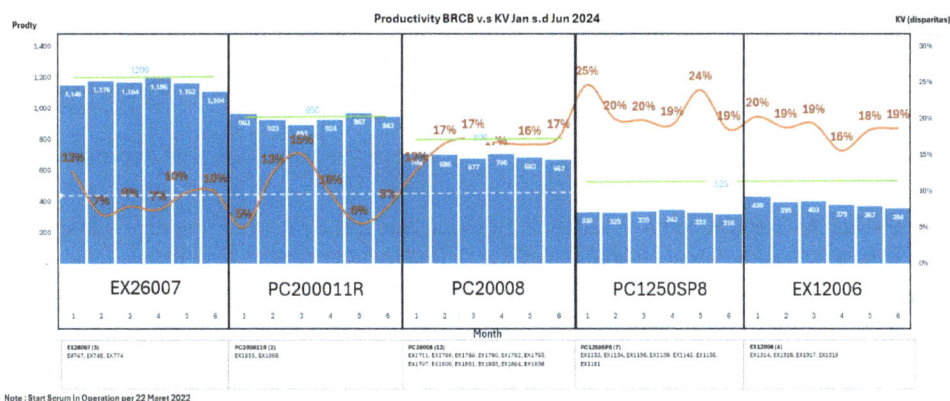

FIG 21 – Productivity performance (source: PAMA BRCB, 2022).

CONCLUSIONS

Scrum can improve coordination between operations and engineering teams in mining by fostering collaboration, continuous improvement, and adaptability. It helps breakdown silos and align teams, ensuring better control over MA OPRENG (Operation-Engineering) strategy. However, it's not a 'one-size-fits-all' solution and must be tailored to the unique needs of each mining operation for maximum effectiveness.

Measurable impacts of Scrum include increased production efficiency through iterative improvements, faster identification of inefficiencies, and enhanced decision-making using data-driven tools like the Big Data Tree Diagram and Key Variance (KV) analysis. These tools help assess process efficiency and predictability, leading to more reliable decisions.

In conclusion, integrating Scrum with analytics improves collaboration, optimises processes, and boosts production performance. It also strengthens control over MA OPRENG strategy. Its success, however, depends on adapting it to the specific needs of each operation. To maximise its impact, it is recommended to leverage Scrum in Operations across other PAMA districts as a mass production approach to control Operational Performance. Future studies could explore scalability and adaptability in varying operational contexts by applying it to other PAMA districts, ensuring broader applicability across the mining industry.

ACKNOWLEDGEMENTS

The authors sincerely appreciate the Project Manager of PT. Pamapersada Nusantara, BRCB District, for granting the necessary approval and support that enabled this research. Special thanks to the Operation Manager of BRCB (Operations Division) for providing critical resources and facilitating improved process control and operational efficiency.

The authors also acknowledge the Technical Service Team (Engineering Division) for their expertise in technical implementation and the Corporate Management Development Division for their strategic facilitation. Their collective contributions were essential in advancing the Scrum framework for mining operations.

REFERENCES

Antony, J, 2018. Lean Six Sigma for the Process Industry: A Comprehensive Review, *Journal of Manufacturing Science and Technology,* 45(3):221–234.

Beer, M and Eisenstat, R A, 2000. The silent killers of strategy implementation and learning, *MIT Sloan Management Review*, 41(4):29–40.

Conforto, E C, Salum, F, Amaral, D C, Silva, S L and Almeida, L F, 2014. Can Agile Project Management Be Adopted by Industries Other than Software Development?, *Project Management Journal*, 45(3):21–34.

Deloitte, 2020. The Future of Mining: Agile Operations and Predictive Maintenance, Deloitte Insights Report.

Denning, S, 2018. The Age of Agile: How Smart Companies Are Transforming the Way Work Gets Done, Harvard Business Review Press.

Knights, P, Topal, E and Kizil, M S, 2017. Data-Driven Optimization in Mining Operations, *Mining Technology Journal*, 128(1):78–95.

Lehmann, S and Reineke, A, 2019. Agile Manufacturing in Aerospace: A Case Study on Airbus, *International Journal of Production Economics*, 210:105–118.

McKinsey and Company, 2021. Agile in Heavy Industries: BP's Offshore Asset Management Approach, McKinsey Industry Insights.

Nehring, M, Topal, E, Kizil, M S and Knights, P, 2018. A Decision Support System for Improving the Efficiency of Mining Operations, *International Journal of Mining Science and Technology*, 28(4):537–548.

Obrutsky, S L, 2016. Comparison and contrast of project management methodologies PMBOK and Comparison and contrast of project management methodologies PMBOK and SCRUM, ResearchGate.

Ofstedal, K and Dahlberg, K, 2009. Collaboration in student teaching: Introducing the collaboration self-assessment tool, *Journal of Early Childhood Teacher Education*, 30(1):37–48. https://doi.org/10.1080/10901020802668043

Parmenter, D, 2015. *Key Performance Indicators: Developing, Implementing and Using Winning KPIs* (John Wiley and Sons).

Partogi, J, 2015. *Manajemen Modern dengan Scrum,* Andi Offset.

Rigby, D K, Sutherland, J and Takeuchi, H, 2016. Embracing Agile, *Harvard Business Review*, 94(5):40–50.

Schwaber, K and Sutherland, J, 2020. The Scrum Guide: The Definitive Guide to Scrum, Scrum.org.

Scrum.org, 2021. Evidence-Based Management: Measuring Agile Performance, Scrum Inc, Whitepaper.

Serrador, P and Pinto, J K, 2015. Does Agile work? A quantitative analysis of Agile project success, *International Journal of Project Management*, 33(5):1040–1051.

Stacey, R D, 1996. *Strategic Management and Organizational Dynamics* (Pitman Publishing).

Streule, T, Miserini, N, Bartlome, O, Klippel, M and Garcia, B, 2016. Implementation of Scrum in the Construction Industry, Procedia Engineering.

Takeuchi, H and Nonaka, I, 1986. The New New Product Development Game, *Harvard Business Review*, 64(1):137–146.

Topal, E, Knights, P and Kizil, M S, 2019. Digital Transformation and Operational Optimization in Mining: The Role of Real-Time Data Analytics, *Mining Technology Journal*, 128(1):78–95.

Fuel efficiency optimisation of HD785–7 trucks at open pit coalmine – an adaptive, clustering-based approach

A H M Fadhil¹, S Andika² and M R Pratama³

1. Operation Research Expert, PT. Pamapersada Nusantara, Jakarta 13930, Indonesia. Email: muhammad.hafiz40917@pamapersada.com
2. Operation Research Expert, PT. Pamapersada Nusantara, Jakarta 13930, Indonesia. Email: satya.andika@pamapersada.com
3. Scientific Analytic Expert, PT. Pamapersada Nusantara, Jakarta 13930, Indonesia. Email: mochamadrp@pamapersada.com

ABSTRACT

Enhancing fuel consumption efficiency is an essential topic for operational sustainability and operational cost in the mining industry. In open pit mining, one of the biggest contributors to fuel consumption is dump trucks (DT). The varying operator driving behaviour and frequently changing mine road environmental conditions, such as road surface and road grade, lead to often unmanageable DT operations and result in sub-optimal fuel consumption. Therefore, the author aims to develop a guidance system that can manage the driving behaviour of DT operators. This study proposes a guidance system that uses a dynamic analytical model with customised algorithms and a near real-time telemetry system to adapt to the changing mine road environment. Using the DBSCAN (Density-Based Spatial Clustering of Applications with Noise) algorithm combined with other data processing techniques, the model can determine fuel-efficient driving behaviour represented by speed, engine speed, and accelerator position and can adapt to changing mine road environmental conditions without compromising DT productivity. In addition, the model runs every three hrs to increase the model's flexibility to changes in the mine road environment. The authors evaluate the model using two parameters, the cycle fuel (L) which is a parameter that calculates fuel consumption in a mining road segment and the normalised productivity metric (m^3/h.km). In other words, the cycle fuel indicates the actual fuel consumption, and the normalised productivity metric indicates the efficiency of the fuel consumption. From the evaluation results, this approach was able to produce an average fuel saving of 2.42 L/h per unit or 75 015.1 L in total. In addition, the higher adherence group has a higher productivity by 11.6 per cent compared to lower adherence groups. This result explains that the model developed by the authors, not only produces fuel saving, but also enhances the fuel consumption efficiency of DT at open pit coalmine sites. This paper details the algorithm specification and model development, deployment of the model, and the closed loop workflow of the model.

INTRODUCTION

In 2016, Indonesia was among the countries that signed the Paris Agreement. The main objective is to keep global temperature rise 'below' 2°C above pre-industrial levels, with a maximum effort to limit temperature rise to 1.5°C (Mutu International, 2022). This aims to reduce the risk of more severe climate change impacts. Indonesia's Nationally Determined Contribution (NDC) outlines targets of 31.89 per cent unconditional and 43.2 per cent conditional greenhouse gas (GHG) emissions reduction by 2030, with a long-term goal of achieving net-zero emissions by 2060 (United Nations Development Programme, 2023).

Based on data from EDGAR (Emissions Database for Global Atmospheric Research) in 2022 Indonesia produces greenhouse gas emissions of 1.15 Gt CO_2e and 44.2 per cent is dominated by CO_2 emissions (EDGAR, 2023). One sector that has a large contribution to these carbon emissions is industrial combustion, which amounted to 27.9 per cent. 'Industrial combustion' refers to the burning of fuels for industrial processes, such as manufacturing, and excludes power generation and buildings. Thus, industrial processes have a large contribution to CO_2 and greenhouse gas emissions in Indonesia.

One of the large and growing industrial sectors in Indonesia is the coal mining sector. This can be seen in 2022 in Indonesia coal still dominates the total energy supply by 36.4 per cent beating petroleum (International Energy Agency, 2023). This value increased from the previous year by 5.1 per cent. This increase is also in line with the increase in coal production in Indonesia in 2022 from 687.43 Mt to 775.18 Mt, an increase of 12.8 per cent (Ministry of Energy and Mineral Resources of Indonesia, 2023). With the increasing need for coal, of course, coal mining activities as a process of extracting coal from the earth will increase.

Coal mining activities have been carried out for centuries. This includes overburden removal, coal getting and coal processing. However, coal mining is as energy intensive as any other mine. 40 per cent of total energy used in surface mines relates to fuel consumption (US Department of Energy (DOE), 2012) and Haul trucks are responsible for most of this fuel consumption (Energy Efficiency Opportunities (EEO), 2010). Therefore, efforts related to improving the fuel performance of hauling activities can significantly reduce the impact of surface mines on the environment.

In this study, the authors propose a new approach in Indonesia to improve the fuel consumption performance of hauling dumptrucks. The authors propose a data-driven approach that uses customised machine learning algorithms to recommend fuel-efficient driving behaviours to dumptruck drivers. In addition, the authors will conduct an analysis of the impact of these recommended driving behaviours on the fuel consumption performance of the dumptruck.

LITERATURE REVIEW

Mine haulage operations are the most fuel-intensive activity in open pit mining, accounting for about 40 per cent of total energy consumption (US DOE, 2012; Ercelebi and Bascetin, 2009). This operation is used for overburden removal and coal transportation in open pit coal mining. Several factors affect fuel consumption in mine haulage, including haul road conditions, truckload, speed, acceleration pattern and terrain slope (Kecojevic and Komljenovic, 2010; Soofastaei et al, 2015, 2018). Fuel consumption in haulage is also affected by operator behaviour, with aggressive acceleration and inconsistent speeds contributing to inefficiencies (Alamdari et al, 2022; Awuah-Offei, 2016). These findings highlight the need for data-driven adaptive approaches to optimise speed and fuel efficiency in mine haulage operations.

Many efforts have been made to improve fuel consumption performance in mine haulage activities in Indonesia. PT Bukit Asam (PTBA) implemented industry 4.0-based technology to monitor engine status on hauling dumptrucks, which was previously done manually using mine cars (Environment Indonesia, nd). PT Borneo Alam Semesta optimised the fuel ratio between the digging and hauling equipment (Ramadhani, Mustofa and Melati, 2022). The implementation of IoT-based fleet management systems has helped mining companies monitor and optimise fuel consumption (Cartrack Indonesia, nd). However, these methods often fail to adapt to dynamic mining conditions, making it necessary to apply machine learning techniques to improve fuel efficiency.

Machine learning (ML) has been widely used in transportation and industrial applications to analyse fuel consumption patterns and develop predictive models for optimisation (Xie et al, 2023). Density-based clustering methods, such as DBSCAN (Density-Based Spatial Clustering of Applications with Noise), have been applied to classify driving behaviours and estimate fuel consumption more accurately. Kwon, Park and Park (2018) introduced a DBSCAN-based approach for estimating fuel consumption in intracity buses, highlighting the effectiveness of clustering methods in identifying efficient driving patterns. How the DBSCAN algorithm works can be seen in Figure 1.

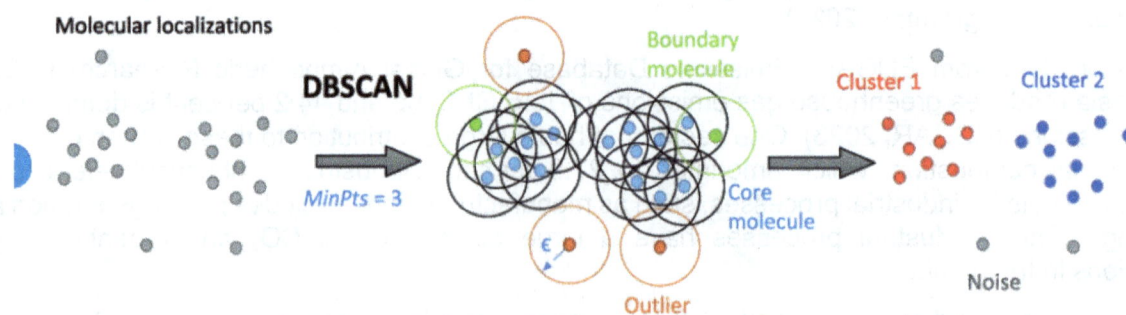

FIG 1 – DBSCAN algorithm.

Incorporating DBSCAN clustering with post-processing techniques, such as filtering and scoring, allows for the identification of optimal speed, engine speed, and accelerator position, which can then be recommended to operators for improved fuel efficiency (Kwon, Park and Park, 2018). This adaptive approach ensures that recommendations are based on actual operational conditions, rather than predefined theoretical models, making them more practical for implementation in active mining sites. Application of DBSCAN for Fuel-Efficient Driving Recommendations.

Building on these methodologies, the proposed approach leverages DBSCAN clustering to analyse Vehicle Health Monitoring System (VHMS) data of haul trucks operating in an open pit mining environment. The model processes real-time telemetry data every three hrs, clustering driving behaviours based on key operational parameters such as speed, engine speed, and accelerator position. Post-processing steps, including filtering and scoring, are applied to identify the most fuel-efficient clusters, which are then used to generate driving recommendations for the subsequent three-hour period.

This adaptive data-driven approach ensures that fuel optimisation strategies evolve dynamically with changes in mining conditions. Unlike static rule-based methods, DBSCAN allows continuous learning from operational data, making it highly suitable for large-scale mining applications where variability in road conditions, payloads, and operator behaviour significantly impacts fuel consumption.

METHODOLOGY

This study proposes a data-driven approach to recommend driving behaviour to Komatsu HD785–7 dumptruck operators. The general workflow can be seen in Figure 2.

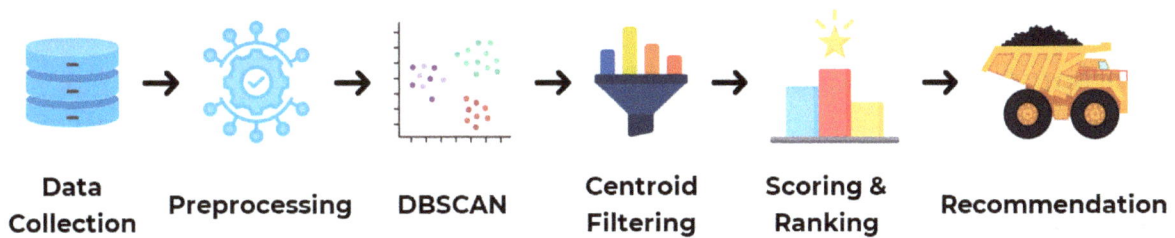

| Data Collection | Preprocessing | DBSCAN | Centroid Filtering | Scoring & Ranking | Recommendation |

FIG 2 – Proposed model workflow.

Data collection

This study uses near-real-time data from the VHMS of a Komatsu HD785–7 dumptruck as the data set of the model. The VHMS data is the telemetry data of various sensors of the dumptruck transmitted every three seconds and stored in the database at the local server of the mine site. The author combined the factors affecting dumptruck fuel consumption obtained from the literature review with the parameters available in the VHMS data. Some of the parameters used in this study are as follows in Table 1.

TABLE 1
VHMS data parameters used as data sets set.

Column name	Details
pos_name	Road segment names in the Auto Dispatch System application
mobileid	Unit number of the dumptruck
mobileactivityid	ID of the activity performed by the dumptruck (hauling, traveling, loading etc)
fuel_rate_01l	Fuel consumption rate of dumptruck (L/h)
plm_speed	Speed of dumptruck unit (km/h) measured from payload metre (PLM) sensor
eng_speed	Rotating speed of dumptruck engine (rev/min)
accel_pos	Accelerator position in percentage
plm_inc	Road inclination of PLM sensor (°)
plm_payload	Weight of the load transported by the dumptruck (tons)

The data set is structured based on road segments (pos_name) and truck activities (mobileactivityid), distinguishing between travelling (unloaded) and hauling (loaded) conditions. The segmentation allows for detailed analysis of fuel efficiency across different haul road conditions and operational scenarios.

Data preprocessing and feature engineering

A rigorous data preprocessing pipeline is implemented to enhance data quality and extract relevant features. The process includes the following steps:

Data cleaning

First, the author conducted data cleaning to improve the quality of the data. The first step is to take VHMS data for hauling (loaded) and traveling (unloaded) activities, and take VHMS data whose pos_name column is on the mine road, not front loading, disposal, or pitstop. Then, the author handled the missing data by using several approaches such as moving average and interpolation. After the missing data was resolved, the author performed data cleaning in the form of removing outliers and data that did not make sense. This was done with two methods, namely by filtering data based on mine site observations and filtering data using the IQR (interquartile range) approach for each road segment and dumptruck activity.

Data aggregation

After the data is cleaned, this VHMS data will be aggregated based on its pos_name and mobileactivityid for each mobileid. This combination of *pos_name* and *mobileactivityid* at a given dumptruck and time is called a traversal. Traversal represents an instance of a dumptruck passing a *pos_name* in a particular activity (loaded or unloaded). The purpose of this aggregation is not only to reduce noise from the sensor data, but also because the recommendations provided will be specific to each combination of *pos_name* and *mobileactivityid*.

Feature engineering

This study currently uses one feature generated from feature engineering, *diff_accel*. This *diff_accel* column is a column that represents the differential of the *accel_pos* column.

Model development

The core component of the proposed methodology is an unsupervised learning approach using Density-Based Spatial Clustering of Applications with Noise (DBSCAN). DBSCAN (Density-Based

Spatial Clustering of Applications with Noise) is a clustering algorithm that groups data based on the density of points in the data space. It identifies clusters as areas with high density of points, while points in areas with low density are considered as noise or outliers (Wang, Lu and Rinaldo, 2019). As can be seen from Figure 1, the preprocessed data set is clustered using DBSCAN. The result of this clustering is a cluster in each pos_name and mobileactivityid. Then, each cluster will have its centroid calculated as the representative value of each cluster. Then, the centroids will be filtered to match the safety rules in the Company, such as the maximum speed limit. In addition, to prevent speed recommendations from driving behaviour recommendations from being low, the author also filters clusters whose speed is below Q2 in each pos_name and mobileactivityid. Then, the centroids will be calculated the fuel score using this formula:

$$score = \left\lceil \frac{fuel\ rate}{speed \times payload} \right\rceil$$

After calculating the score of each centroid, the centroids will be ranked based on the score. It can be seen from the formula above that the lower the score, the better the centroid. The centroid chosen is the centroid with the smallest score in each *pos_name* and *mobileactivity*. The values of the driving behaviour parameters)—speed, engine speed, and accelerator position)—from the centroid will be used as recommendations in the corresponding *pos_name* and *mobileactivityid*.

Deployment

This data-driven model is programmed using the Python language and there are several SQL queries to perform data collection and recommendation updates. There are two Python scripts that will be deployed, namely recommendation script and evaluation script. Recommendation script as the name suggests aims to generate recommendations. This script is deployed on the mine site server as can be seen in Figure 3. The recommendations generated will be displayed on the UMPC of the site's Auto Dispatch System that has been installed on each dumptruck. On the other hand, the Evaluation script aims to evaluate the dumptruck using the evaluation metrics mentioned above. This script is deployed on the head office server. The output of this script will be evaluation data that will be connected to the reporting dashboard (Tableau). Figure 3 is an overview of the data pipeline.

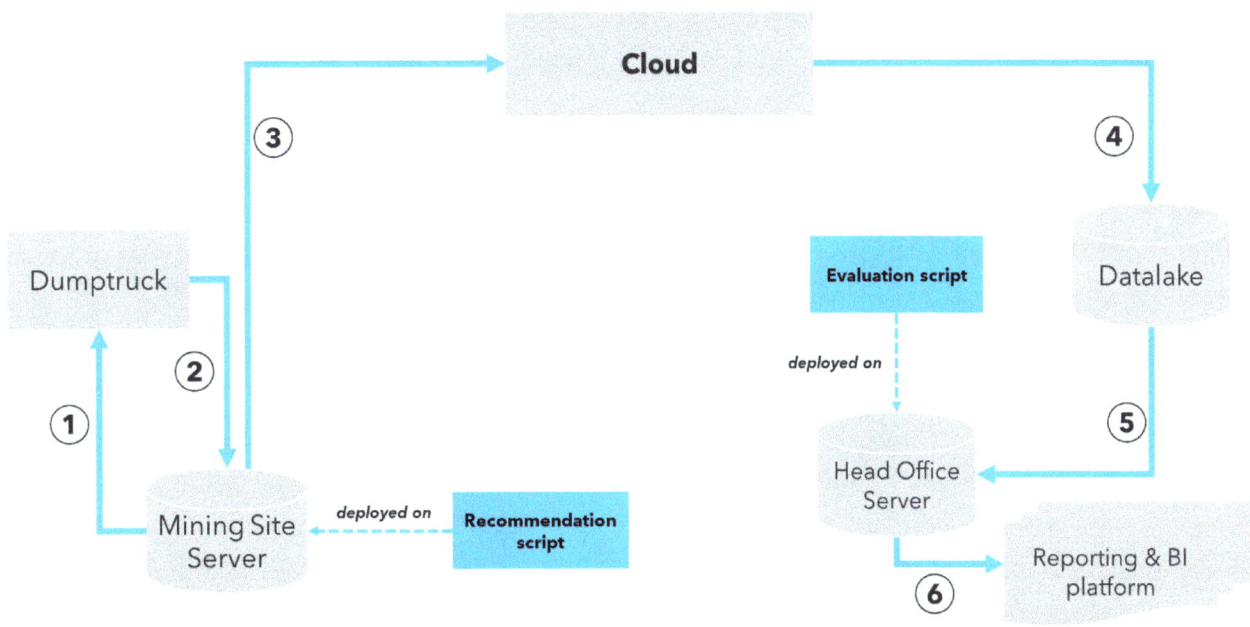

FIG 3 – Data pipeline.

EVALUATION

To evaluate the performance of this approach, the author uses several parameters, including:

- Adherence (%): is calculated by summarising the operator's adherence to recommendations based on his aggregated VHMS data for each traversal. For now, adherence only refers to the speed parameter.

- Fuel Saving (L): is calculated by comparing the average cycle fuel (L)—a parameter that calculates fuel consumption in a mining road segment—of compliant operators and the average cycle fuel of non-compliant operators in each pos_name-mobileactivityid combination. The difference is then multiplied by the number of compliant operators in that pos_name-mobileactivityid combination.

- Productivity (m³/h.km): is a dumptruck productivity parameter that has been normalised to distance (m³/h.km). This value represents the fuel consumption efficiency of the dumptruck.

Based on the data from 16 December 2024 – 12 January 2025, it can be seen that with 124 total dumptruck units and an average adherence rate of 37.04 per cent, this approach can result in a total fuel saving of 75 015.1 L or an average of 2.42 L/h per unit. In addition, in terms of average fuel burn, it can be seen that the top 25 dumptrucks with the highest adherence (average adherence of 41.79 per cent) have higher productivity than the bottom 25 dumptrucks with the lowest adherence (average adherence of 24.51 per cent). More details can be seen in Figures 4 and 5.

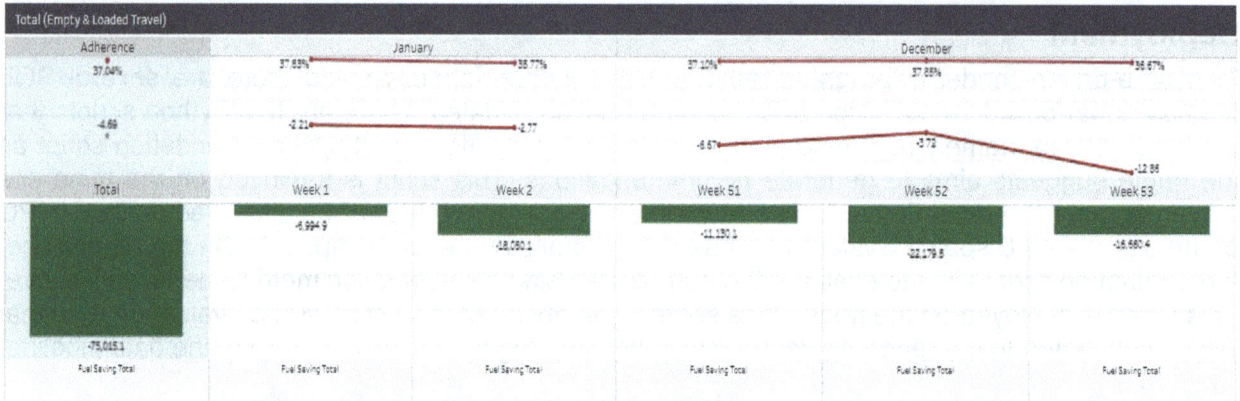

FIG 4 – Fuel saving and adherence trend.

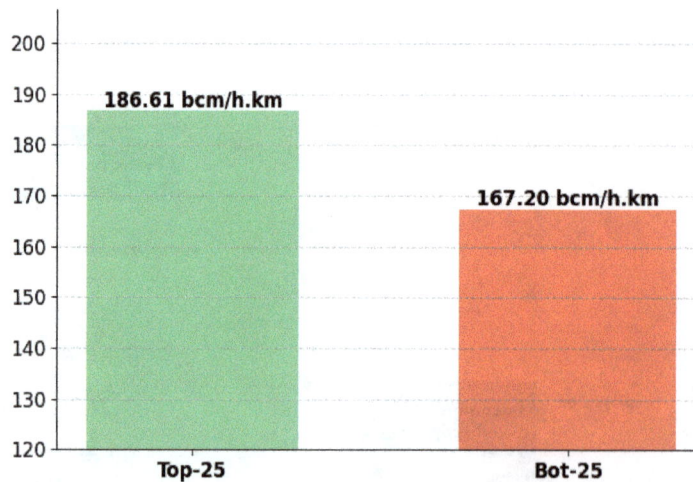

FIG 5 – Top-25 versus Bottom-25 dumptrucks productivity.

CONCLUSIONS

This study presents a data-driven approach to optimising fuel efficiency in open pit mining haul trucks using DBSCAN clustering. By analysing VHMS data in three-hour intervals, the model identifies optimal driving behaviours—speed, engine speed, and accelerator position—that contribute to fuel-efficient operations. The proposed methodology integrates data preprocessing, clustering, and post-processing techniques to generate actionable recommendations for truck operators, ensuring continuous improvement in fuel efficiency.

The findings demonstrate that the clustering approach effectively distinguishes driving patterns associated with low fuel consumption while maintaining operational performance. Implementing this model in real-world mining operations can lead to substantial fuel savings and reduced greenhouse gas emissions, aligning with Indonesia's commitment to the Paris Agreement and its Nationally Determined Contribution (NDC) goals.

Future work should explore refining feature selection, incorporating additional operational variables, and validating the model across different mining sites to enhance generalisability. Additionally, integrating real-time feedback mechanisms for operator adaptation could further improve fuel-saving outcomes.

ACKNOWLEDGEMENTS

The authors express their gratitude to the company for providing the opportunity and resources to conduct this study. In addition, the authors would like to thank all the functional people who contributed to this study. Special thanks to the research team who contributed to the data processing and model development. We also acknowledge the support of industry experts and academic advisors for their valuable discussions and feedback throughout this research.

REFERENCES

Alamdari, S, Basiri, M, Mousavi, A and Soofastaei, A, 2022. Application of Machine Learning Techniques to Predict Haul Truck Fuel Consumption in Open-Pit Mines, *Journal of Mining and Environment*, 13(1):69–85.

Awuah-Offei, K, 2016. Energy Efficiency in Mining: A Review with Emphasis on the Role of Operators in Loading and Hauling Operations, *Journal of Cleaner Production*, 117:89–97.

Cartrack Indonesia, nd. Fuel Management System to Maximize Mining Industry [in Indonesian: Sistem manajemen bahan bakar untuk maksimalkan industri pertambangan] [online]. Available from: <https://cartrack.id/id/sistem-manajemen-bahan-bakar-untuk-maksimalkan-industri-pertambangan> [Accessed: 28 February 2025].

Emissions Database for Global Atmospheric Research (EDGAR), 2023. Indonesia – Country Fact Sheet [online]. Available from: <https://edgar.jrc.ec.europa.eu/country_profile/IDN> [Accessed: 28 February 2025].

Energy Efficiency Opportunities (EEO), 2010. Energy-Mass Balance: Mining, Canberra: Australian Government, Department of Resources Energy and Tourism.

Environment Indonesia, nd. PTBA's Innovation in Reducing Fuel Consumption as an Energy Transformation Effort [in Indonesian: Inovasi PTBA dalam Pengurangan Konsumsi BBM sebagai Upaya Transformasi Energi] [online]. Available from: <https://environment-indonesia.com/inovasi-ptba-dalam-pengurangan-konsumsi-bbm-sebagai-upaya-transformasi-energi/> [Accessed: 28 February 2025].

Ercelebi, S G and Bascetin, A, 2009. Optimization of Shovel-Truck System for Surface Mining, *Journal of the Southern African Institute of Mining and Metallurgy*, 109(7):433–439.

International Energy Agency, 2023. Indonesia – Energy Statics Data Browser [online]. Available from: <https://www.iea.org/data-and-statistics/data-tools/energy-statistics-data-browser?country=INDONESIA> [Accessed: 28 February 2025].

Kecojevic, V and Komljenovic, D, 2010. Haul Truck Fuel Consumption and CO_2 Emission Under Various Operating Conditions, *Mining Engineering*, 62(1):42–48.

Kwon, O H, Park, Y and Park, S H, 2018. Density-based Clustering Methodology for Estimating Fuel Consumption of Intracity Bus by Using DTG Data, in *Advanced Multimedia and Ubiquitous Engineering*, pp 879–885 (Singapore: Springer).

Ministry of Energy and Mineral Resources of Indonesia, 2023. Handbook of Energy and Economic Statistics of Indonesia 2023. Jakarta: Indonesia Government.

Mutu International, 2022. Understanding What is the Paris Agreement and Its Implementation in Indonesia [in Indonesian: Mengenal Apa itu Paris Agreement dan Implementasinya di Indonesia] [online]. Available from: <https://mutucertification.com/mengenal-paris-agreement-indonesia/> [Accessed: 28 February 2025].

Ramadhani, A, Mustofa, A and Melati, S, 2022. Optimization of Fuel Ratio for Excavators and Transport Vehicles at PT Borneo Alam Semesta [in Indonesian: Optimalisasi Fuel Ratio Alat Gali-Muat dan Alat Angkut PT Borneo Alam Semesta], *Jurnal Himasapta*, 7:157.

Soofastaei, A, Aminossadati, S, Kizil, M and Knights, P, 2015. The Effect of Rolling Resistance on Fuel Consumption and Greenhouse Gas Emissions by Haul Trucks in Surface Mines, *Tribology International*.

Soofastaei, A, Karimpour, E, Knights, P and Kizil, M, 2018. Energy-efficient loading and hauling operations, in *Green Energy and Technology* (Cham: Springer).

United Nations Development Programme, 2023. Indonesia – Climate Promise [online]. Available from: <https://climatepromise.undp.org/what-we-do/where-we-work/indonesia> [Accessed: 28 February 2025].

US Department of Energy (DOE), 2012. Mining Industry Energy Bandwidth Study, Washington, DC: US Government.

Wang, D, Lu, X and Rinaldo, A, 2019. DBSCAN: Optimal Rates For Density-Based Cluster Estimation, *Journal of Machine Learning Research*, 20(170):1–50.

Xie, X, Sun, B, Li, X, Olsson, T, Maleki, N and Ahlgren, F, 2023. Fuel Consumption Prediction Models Based on Machine Learning and Mathematical Methods, *Journal of Marine Science and Engineering*, 11(4).

Deep learning for predicting hauling fleet production capacity under uncertainties in open pit mines using real and simulated data

N Guerin[1], M Nakhla[2], A Dehoux[3] and J L Loyer[4]

1. PhD student, Mines Paris – PSL, Paris 75006, France. Email: nicolas.guerin@minesparis.psl.eu
2. Professor, Mines Paris – PSL, Paris 75006, France. Email: michel.nakhla@minesparis.psl.eu
3. Lead Data Scientist, Eramet, Paris 75015, France. Email: anita.dehoux@eramet.com
4. Head of Data and AI, Eramet, Paris 75015, France. Email: jl.loyer@eramet.com

ABSTRACT

Accurate short-term forecasting of hauling-fleet capacity is crucial in open-pit mining, where weather fluctuations, mechanical breakdowns, and variable crew availability introduce significant operational uncertainties. We propose a deep-learning framework that blends real-world operational records (high-resolution rainfall measurements, fleet performance telemetry) with synthetically generated mechanical-breakdown scenarios to enable the model to capture fluctuating high-impact failure events. We evaluate two architectures: an XGBoost regressor achieving a median absolute error (MedAE) of 14.3 per cent and a Long Short-Term Memory network with a MedAE of 15.1 per cent. Shapley Additive exPlanations (SHAP) value analyses identify cumulative rainfall, historical payload trends, and simulated breakdown frequencies as dominant predictors. Integration of simulated breakdown data and shift-planning features notably reduces prediction volatility. Future work will further integrate maintenance-scheduling indicators (Mean Time Between Failures, Mean Time to Repair), detailed human resource data (operator absenteeism, crew efficiency metrics), blast event scheduling, and other operational constraints to enhance forecast robustness and adaptability. This hybrid modelling approach offers a comprehensive decision-support tool for proactive, data-driven fleet management under dynamically uncertain conditions.

INTRODUCTION

The increasing competitiveness and constant economic pressure are compelling the mining industry to rigorously optimise operational efficiency and the accuracy of production planning. Traditionally, value engineering and approaches based on Value Driver Trees (VDT) have enabled operators to identify key levers for improvement and prioritise performance indicators (Miles, 1961; Cambitsis, 2012). However, these methods struggle to accurately quantify the impacts of numerous uncertainties, such as unexpected operational disruptions or sudden weather changes, particularly on short-term production plans (Asif, Bessant and Francis, 2010; Carvalho *et al*, 2017; Sánchez *et al*, 2020).

Effectively accounting for these uncertainties is critical, as they directly influence the performance of haulage fleets in open pit mines, where even minor fluctuations can cause significant delays and additional operational costs. Studies have shown that extreme weather conditions, such as heavy rainfall, can lead to substantial annual production losses, reaching up to 10 per cent in some open pit mines. Furthermore, unforeseen equipment failures represent a major source of downtime and productivity loss.

In light of these limitations, advanced machine learning approaches, particularly Deep Learning, have emerged as promising alternatives to improve forecasting accuracy. Recently, several studies have begun exploring these possibilities. For instance, Baek and Choi (2020) employed deep neural networks to forecast ore production based on truck haulage operational data. Fan *et al* (2022), on the other hand, combined tree-based ensemble models with Gaussian mixture models to predict truck productivity, focusing primarily on internal mining operation variables.

However, these studies generally focus on limited sets of internal operational factors and fail to sufficiently capture external dynamics such as detailed weather conditions, which are nonetheless critical for operational reliability. Our study distinguishes itself by explicitly incorporating these external meteorological variables using historical rainfall data, and by leveraging both real and simulated data to enrich the predictive models. The inclusion of simulated data, generated using methods such as Monte Carlo Tree Search (MCTS), specifically enables the analysis of rare but high-impact scenarios such as sudden mechanical failures or significant fluctuations in resource usage (Browne *et al*, 2012).

To achieve this, we propose the use of two complementary models: the XGBoost model to clearly identify key and interpretable features with strong influence on productivity, and a Long Short-Term Memory (LSTM) network capable of capturing the complex temporal dependencies inherent in sequential operational data. The main advantage of our approach lies in this combination of interpretability and predictive performance, aiming to provide a practical and directly usable tool for operational managers to anticipate and adjust their decisions in real time in the face of uncertainties.

This research directly addresses a need in the mining industry: enhancing operational resilience and improving proactive fleet management. By integrating an advanced analytical perspective and a better understanding of uncertainty dynamics, we lay the foundation for more robust operational management, with the potential to have a direct impact on the economic performance of modern mining operations.

MATERIALS AND METHODS

The following section outlines the foundational components of our study: the data sourced from a mining fleet management system (FMS) and framework designed to predict hauling fleet productivity. In production systems, the performance depends on both operator efficiency and equipment availability. However, due to the lack of human resources data in our study, we have chosen to focus solely on the trucks as our primary resource. This approach enables us to concentrate our analysis on the fleet's impact on operational capacity, which is the critical determining factor. By integrating domain-specific data with advanced modelling techniques, this work addresses the unique challenges of mining operations, balancing interpretability and temporal forecasting accuracy.

Fleet management system

Mining fleet operations generate vast amounts of heterogeneous data, driven by the integration of IoT sensors, GPS modules, and onboard telemetry systems. The data set underpinning this study originates from a fleet of heavy-duty mining equipment operating in a large-scale open pit mine. Several large-scale mining management software packages exist, including HxGN MineOperate, which integrates data communication functions, GPS positioning for shovels, drills, and haul trucks, and automated fleet assignments for open pit mining operations. The system consists of the following key components:

- HxGN MineOperate OP Pro: A computerised field system equipped with an advanced interface, installed on trucks, auxiliary equipment, shovels, and crushers to optimise fleet management and operational workflows.

- Global Positioning System (GPS): Ensuring precise tracking and navigation of mining equipment.

- Radio Data Link: Connected to a central computing hub, facilitating real-time data transmission and enabling automated haul truck assignments through peer-to-peer communication, reducing inefficiencies due to network delays.

The hauling process initiates at the beginning of each shift, where operator data is entered into the system to ensure proper registration of personnel and equipment. This information is transmitted to the central control system, which then enables the dispatcher to assign work to operators. Operators receive real-time notifications regarding equipment utilisation, operational requirements, and mandatory inspections, ensuring continuous workflow efficiency.

The temporal nature of the data provides granular insights into equipment behaviour and operational trends. With over 50 features per equipment unit, the data set combines numerical, categorical, and time-series formats, reflecting the complexity of industrial-scale mining operations.

However, raw industrial data often suffers from inconsistencies inherent to harsh mining environments. To address these issues, preprocessing steps were rigorously applied. Short data gaps were resolved using linear interpolation or forward-fill techniques, while longer gaps were flagged for manual review. García, Luengo and Herrera (2015) provided a comprehensive framework for data preprocessing in complex industrial data sets, advocating for context-aware interpolation strategies to maintain data fidelity.

Models' selection

XGBoost for interpretable feature analysis

XGBoost, a gradient-boosted tree algorithm (Figure 1), was selected for its ability to handle mixed data types, including categorical maintenance codes and numerical sensor readings. Its interpretability is particularly valuable in mining, where stakeholders prioritise understanding the drivers of productivity. Furthermore, its inherent regularisation techniques help mitigate overfitting, which is crucial when dealing with noisy industrial data. In addition, its built-in capabilities to perform embedded feature selection complement the rigorous upstream preprocessing, ensuring robust performance on heterogeneous data sets. For instance, XGBoost's feature importance scores can highlight critical predictors that are subsequently used to inform further feature engineering and operational decision-making. Moreover, the algorithm's scalability and computational efficiency enable rapid model iteration and facilitate real-time deployment in environments with large, continuously streaming data sets.

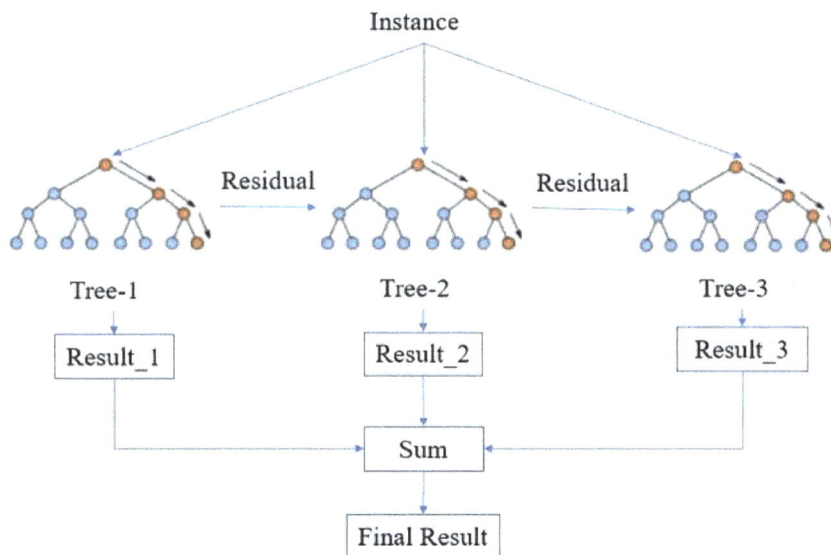

FIG 1 – XGBoost model scheme from Wang, Chakraborty and Chakraborty (2020).

The efficacy of XGBoost in structured data analysis has been well-documented. Chen and Guestrin (2016) demonstrated its scalability and performance in capturing non-linear relationships, while Wang *et al* (2021) proved its capability to model complex interactions in industrial data. Beyond these applications, our choice is primarily driven by XGBoost's strong theoretical foundations and practical advantages, its ability to naturally capture non-linear feature interactions and its built-in mechanisms to control model complexity make it exceptionally well-suited for our predictive tasks.

LSTM networks for temporal dynamics

LSTM networks were implemented to capture the intricate temporal dependencies in equipment behaviour. Their gated architecture (Figure 2) enables effective retention and selective forgetting of information over long sequences. Designed to remember long-term patterns, LSTMs are ideal for mining fleet data, where cyclic production schedules are common. By processing multivariate time-series sequences, the LSTM architecture learns to forecast key metrics, effectively modelling the dynamic evolution of mining operations. Moreover, LSTMs inherently handle variable sequence lengths and are robust to noise and irregular time intervals, common in industrial data sets. This ability to integrate contextual information from previous time steps results in more accurate predictions when past operational conditions heavily influence future outcomes.

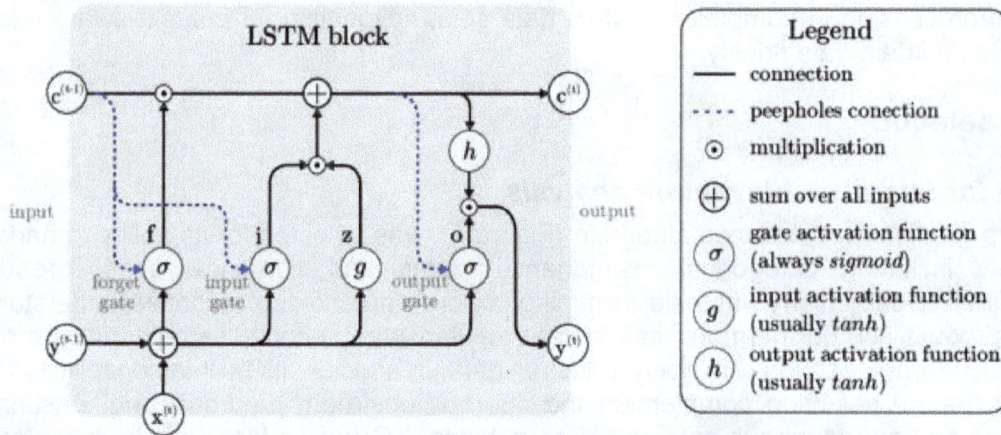

FIG 2 – LSTM model scheme from Van Houdt, Mosquera and Napoles (2020).

The foundational work of Hochreiter and Schmidhuber (1997) established LSTMs as a solution to the vanishing gradient problem, enabling effective learning of long-term dependencies. More recently, Ao, Li and Yang (2023) demonstrated the efficacy of LSTM networks for predicting truck travel time in open pit mines by capturing complex temporal dependencies in historical data, thereby enhancing scheduling and operational efficiency. Our choice is further justified by LSTM's proven performance in handling non-stationarity and noise in time series data, making it an excellent candidate for forecasting in the unpredictable operational environment of open pit mining.

We adopt two distinct modelling approaches to comprehensively analyse mining fleet productivity. The first approach leverages XGBoost, a powerful gradient-boosted tree algorithm, to extract interpretable insights from static, time-windowed data snapshots. The second approach employs LSTM networks, which are adept at capturing the temporal dynamics inherent in sequential equipment behaviour. Together, these methods enable us to dissect both the static and dynamic drivers of operational efficiency in mining fleets.

ANALYSIS AND PROPOSAL

Currently, large-scale mining operations deploy fleets consisting of hundreds of trucks, dozens of shovels, several loaders, and other equipment distributed across multiple pits. As depicted in Figure 3, these operations are integrated across various mining areas, with certain equipment such as shovels and loaders being flexibly repositioned according to daily operational requirements and mining plans, while other assets remain fixed.

FIG 3 – Mining operation of two open pits (image from Google Earth).

A central dispatch system manages the allocation of trucks to ensure that their distribution aligns with the overall operational strategy. However, the actual production capacity of the hauling fleet is subject to significant uncertainties. For instance, truck availability can be disrupted by unexpected breakdowns, and adverse weather conditions—particularly rain—can further impact operational efficiency. These factors introduce variability in the effective production capacity, making accurate forecasting a complex challenge.

In our study, we leverage a multi-faceted and temporally rich data set to address these challenges. The data set captures the dynamic nature of mining operations through continuous, time-stamped records shift-by-shift of 10.5 hrs that reflect real-time fluctuations in performance between shifts. Simultaneously, weather data recorded at matching intervals provides insights into environmental conditions that directly influence operations. Additionally, simulated data augments our observations by modelling scenarios critical for understanding changing conditions and their potential impact on production capacity.

By combining these diverse data sources, our approach aims to develop a robust forecasting model that accounts for both the inherent temporal fluctuations in mining operations and the external factors that impact efficiency. This holistic view is essential for improving decision-making in large-scale, multi-pit mining environments.

Operational data

The operational data utilised in this study are directly sourced from the FMS described in the section earlier. Specifically, the FMS continuously collects detailed, cycle-by-cycle data from individual haul trucks and shovels which are aggregated at the shift level. These data encompass several key performance indicators that reflect the dynamics of our mining operations. Meanwhile, the working trucks metric, which denotes the number of trucks actively engaged in material transport during the shift, averaged 13.8 trucks with a variability of 3.1, while the working shovels feature, indicating the number of shovels in operation, maintained an average of 4.7 with a standard deviation of 1.4, thus ensuring a balanced deployment of equipment. Additionally, the cycle count, signifying the number of complete loading cycles per shift, was observed to average 158 cycles with a standard deviation of 71, directly reflecting the throughput of our operations. The payload, which measures the average weight of material loaded per shift in tons, averaged 13 795 tons with a variability of 6393 tons, underscoring both the efficiency and the inherent variability of material handling.

Furthermore, the average cycle time of a shift, the duration required to complete a full cycle from truck arrival to departure recorded an average of 64 mins with a standard deviation of 17 mins, capturing the relation between operational speed and occasional delays during peak activity. A schematic representation of a hauling truck cycle is provided in Figure 4.

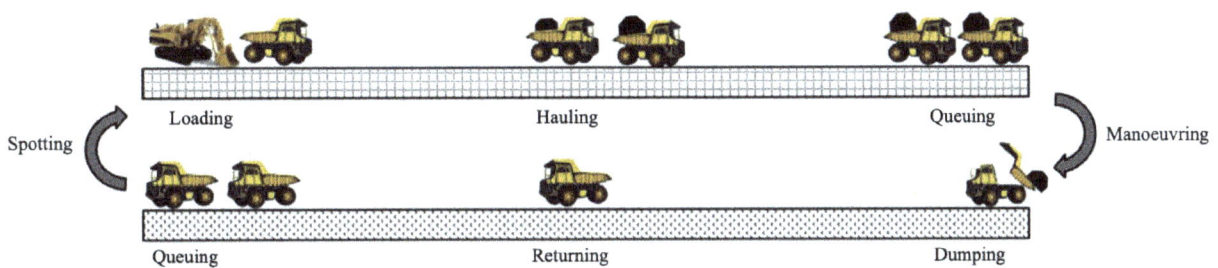

FIG 4 – Schematic of hauling operation in surface mines from Soofastaei *et al* (2015).

Finally, the shift and crew variables categorise the operational periods, revealing that certain shifts experienced slightly faster cycle times, likely a result of differences in crew performance and working conditions. Collectively, these metrics, quantified through their respective means and standard deviations, provide a comprehensive picture of our operational efficiency and variability, and they form the backbone of our model's input data for predictive analysis and simulation.

Weather data

Specifically, we leverage the ERA5-Land hourly data set (spanning from 1950 to the present), which is accessible via the Copernicus platform either through direct download or API queries. The data set provides a high spatial resolution of 0.1° × 0.1° and an hourly temporal resolution, ensuring that even short-term variations in weather are captured.

In order to effectively represent the variability in precipitation across the mining area, we extract data from four key cardinal points that delineate the boundaries of the mine. These points, corresponding to the northern, southern, eastern, and western limits of the site, are selected to encompass the full range of local weather influences. Additionally, we selected the grid point closest to the centre of the mine. After collecting the hourly precipitation data from this location, we aggregate the values at the operational post level (Figure 5). This is achieved by summing the hourly measurements over defined time intervals, thereby yielding cumulative precipitation metrics that not only indicate the weather's impact during each operational shift but also capture the intense rainfall peaks that significantly affect operations and reflect seasonal variations.

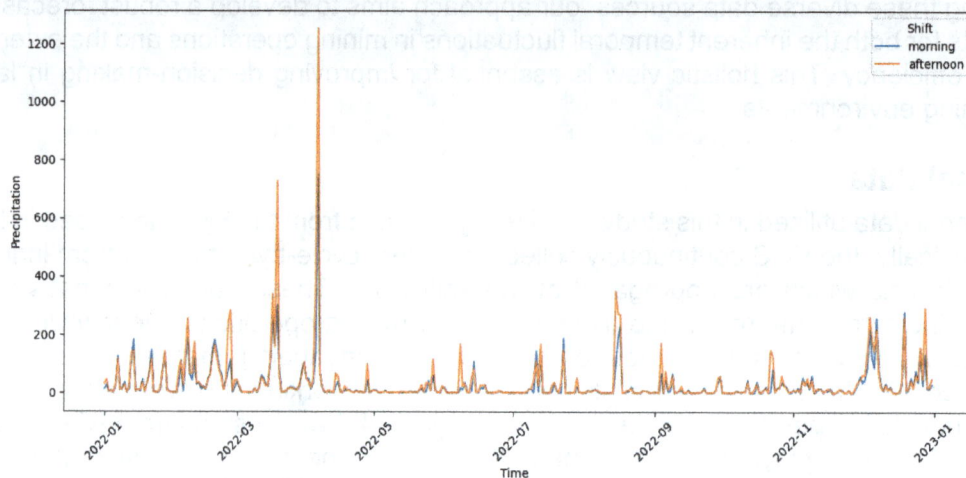

FIG 5 – Rain precipitations variation on mine over one year (2022).

To illustrate the extent of rainfall variability in our tropical context, we analysed precipitation data region from 2021 to 2024. Annual totals ranged from over 2500 mm in 2021—an exceptionally wet year marked by cyclonic activity—to just 1485 mm in 2024, reflecting a 6 per cent deficit relative to the 1991–2020 baseline. The region displays a clear seasonality, with a wet season spanning January to Mars, during which extreme events such as tropical depressions contribute to intense rainfall peaks. For instance, March 2021 recorded up to 320 mm of rain. In contrast, the dry season from August to November is significantly less active, with monthly rainfall sometimes dropping below 30 mm, as observed in September 2022.

Incorporating cumulative precipitation measures into our deep learning model is crucial, as heavy rainfall episodes directly reduce operational efficiency by deteriorating haul road conditions, increasing cycle times, and occasionally forcing temporary equipment downtime due to unsafe conditions. Gonzalez *et al* (2019) evaluated the effects of extreme rainfall events on open-pit mines in Peru, demonstrating marked operational delays during heavy rains. Similarly, Tlhatlhetji and Kolapo (2021) studied the rainy season at the Wescoal Khanyisa Colliery, documenting a significant decrease in equipment availability and cycle counts during wet periods. Note that we exclusively incorporate rainfall metrics, as temperature in our tropical region lacks significant seasonal variability and thus has been excluded from our analysis.

Simulated data

Mechanical breakdowns, significantly impact fleet productivity through sudden equipment downtime. Our study incorporates simulated data designed to estimate the potential utilisation of trucks and shovels for upcoming shifts. The objective of this simulation framework is to realistically model the short-term operational capacity of mining fleets using historical availability and utilisation data collected through the FMS.

A detailed statistical analysis of past shifts revealed that truck availability varies significantly from one shift to the next, with an average fluctuation of approximately 22 per cent, corresponding to about 3.1 trucks around a mean of 13.8 per shift. Similarly, shovel utilisation shows an average variation of around 21 per cent, translating to roughly 1.4 shovels around a mean of 4.7 per shift. These variations, which reflect operational uncertainty, are explicitly embedded into our simulation framework to produce robust and realistic predictions (see Figure 6).

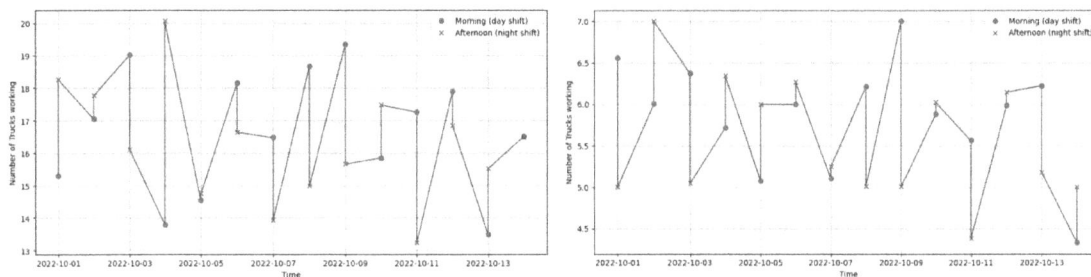

FIG 6 – Trucks (left) and shovels (right) utilisation variation per shift through time.

To generate these simulations, we adopt a hybrid approach that combines regression modelling with Monte Carlo simulation techniques inspired by MCTS. Instead of relying on fixed historical averages, we first train regression models to predict the fleet composition for the next shift based on empirical data. We then inject stochastic noise sampled from the distribution of model residuals to simulate variability. This process yields a range of plausible future scenarios, providing a nuanced representation of operational uncertainty in mining environments (Browne et al, 2012).

Proposed methodology and models

This research focused on applying time series forecasting learning models, specifically XGBoost and LSTM, within the context of trucks and shovels in mining operations. The research utilised the open-source programming language Python, along with relevant libraries that facilitated the simulation of the mining process and training of the learning models. The methodology employed in this research encompassed a comprehensive analysis of the hauling cycle over a three-year period. Through the utilisation of a Transact-SQL script and the Dispatch system, essential data on trucks and shovels' operational performance, as well as weather, were extracted. A wide range of daily operating scenarios was accounted for. This meticulous methodology ensured a comprehensive and reliable foundation for subsequent learning modelling and simulations.

In addition, we integrated temporal lag features into our modelling framework to capture dynamic dependencies and cumulative effects in mining operational data. Each feature was designed to provide specific insights into the system's behaviour. For example, we created two payload-related temporal lags: one that represents the payload recorded during the previous time period—capturing short-term persistence in payload delivery—and another that computes the rolling sum of payloads over the last four shifts, which smooths out short-term fluctuations and highlights sustained trends. Similarly, two lag features related to fleet operations were implemented: one indicates the number of working trucks in the immediate past period, reflecting the current operational state, while the other aggregates the average number of working trucks over four shifts, offering a broader view of fleet performance over time. Finally, we computed the cumulative sum of rainfall over the last six shifts to quantify the overall impact of weather events on operations and road conditions. By incorporating these features, our model leverages both immediate past values and aggregated historical trends, thereby enhancing its ability to predict future operational performance accurately.

Incorporating these temporal lag features is essential in time series analysis, as highlighted in the literature (Hyndman and Athanasopoulos, 2018). They not only enrich the feature set by embedding historical context but also enhance the predictive capabilities of the models by capturing both immediate and aggregated temporal dynamics.

To ensure transparency and reproducibility, Table 1 summarises the input features used in both models, clearly indicating their data sources, nature (empirical, simulated, predictive, temporal lag), and precise definitions. These features were selected based on their operational relevance and predictive value.

TABLE 1
Model input and metadata.

Feature name	Data source	Feature type	Description
Crew_next	Operational	Predictive	Crew scheduled for the next shift
Working_trucks	Operational	Empirical	Number of trucks operating during the current shift
Predicted_working_trucks_next	Simulated	Predictive	Predicted number of trucks for the next shift
Predicted_working_shovels_next	Simulated	Predictive	Predicted number of shovels for the next shift
Working_shovels	Operational	Empirical	Number of shovels operating during the current shift
Cycle_count	Operational	Empirical	Number of loading cycles completed per shift
Payload	Operational	Empirical	Total payload transported during the shift (in tons)
Cycle_time	Operational	Empirical	Average duration of loading cycles per shift
Payload_lag1	Operational	Temporal Lag	Payload from the previous shift
Payload_rolling_sum_4	Operational	Temporal Lag	Sum of payloads over the last four shifts
Shift_next	Operational	Predictive	Next shift scheduled (day or night)
Working_trucks_lag1	Operational	Temporal Lag	Number of trucks operating during the previous shift
Working_trucks_mean4	Operational	Temporal Lag	Mean number of working trucks over the last four shifts
Precipitation	Meteorological	Empirical	Total rainfall during the previous shift
Precipitation_next	Meteorological	Predictive	Historical rainfall for the next shift
Precipitation_sum6	Meteorological	Temporal Lag	Cumulative rainfall over the last six shifts

In this study, two distinct forecasting models were developed to predict the next payload in mining operations: an XGBoost regression model and a Long Short-Term Memory (LSTM) network.

We began by extracting the relevant features from our data set and normalising them using a MinMaxScaler. To preserve the time-dependent structure, the data was split chronologically (80 per cent for training and 20 per cent for testing). The XGBoost regressor was then configured with hyperparameters selected to balance model complexity and generalisation. In particular, we set:

- **n_estimators**: 1000
- **Learning Rate:** 0.01
- **Max Depth:** 3

After training on the training subset and evaluating on the test subset, we conducted a feature importance analysis. To further unpack variable interactions, Shapley Additive Explanations (SHAP) values were computed, offering granular insights into how specific features influence predictions.

In contrast, the LSTM model was tailored to harness the sequential characteristics of our operational data. After removing missing values and scaling both the features and target variable, a sliding window approach with a look-back period of ten-time steps was used to create temporal sequences. The LSTM architecture featured an LSTM layer with 64 units to process these sequences, followed by a dropout layer (with a dropout rate of 0.2) to mitigate overfitting, and a dense layer to output the final prediction. Moreover, to enhance robustness, an attention mechanism was integrated to highlight critical temporal intervals—such as periods of peak engine stress—that could significantly impact performance. The model was compiled using the Adam optimiser (with a learning rate of 0.001 and a clip value of 0.5) and trained using mean squared error as the loss function, with early stopping employed to capture the best model state.

After training, predictions were made on the test set, and the outputs were inverse transformed to their original scale. Performance metrics (MedAE and R^2) were computed for both models, allowing us to draw a direct comparison.

RESULTS

In our study, two forecasting models were developed to predict payload in mining operations. To assess their performance, we selected the median absolute error (MedAE) as our primary evaluation metric due to its robustness against outliers and its interpretability in the context of operational variability.

One of the key innovations in our approach was the integration of rainfall data as a predictor. Rainfall is known to affect operational efficiency, and its inclusion allowed the models to capture weather-induced variations in payload. In the XGBoost model, we computed SHAP values to dissect the influence of each feature. The SHAP analysis revealed that rainfall was among the most significant predictors, with increased rainfall correlating with decreased payload performance as shown in Figure 7. This granular insight not only validates our feature selection but also underscores the practical importance of including weather data in operational forecasts.

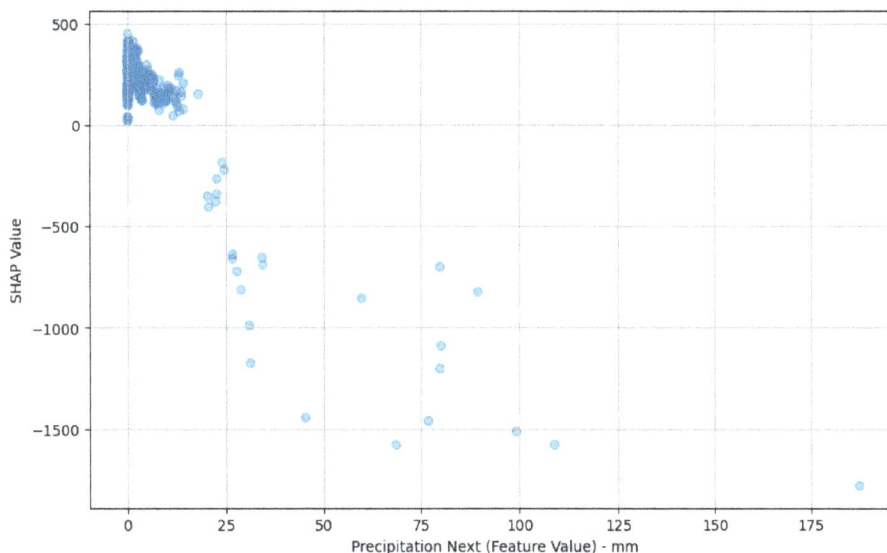

FIG 7 – SHAP analysis: 'upcoming rainfall' feature impact on model predictions.

To better understand the impact of each input feature on the model's predictions, a sensitivity analysis was conducted using SHAP. Figure 8 presents the SHAP summary plot, where each point represents a prediction, and each colour encodes the feature value. The horizontal spread indicates the magnitude and direction of the feature's impact on the output variable.

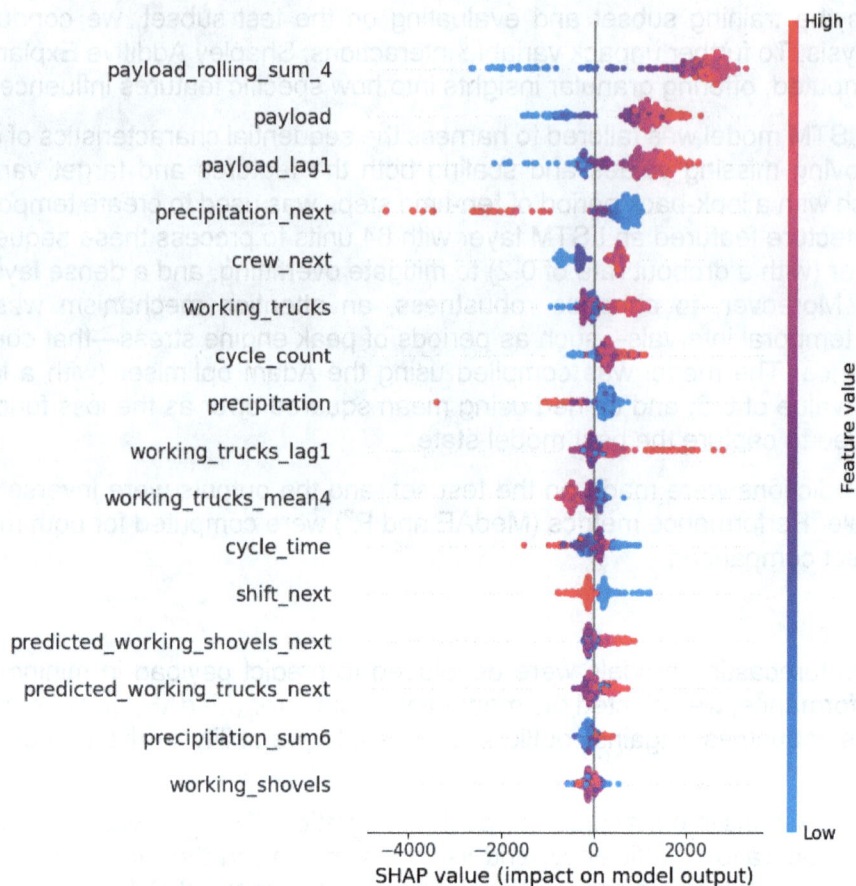

FIG 8 – SHAP summary plot showing the impact of each feature on model output.

The three variables related to historical payload clearly dominate in terms of impact on the model. The observed trend is intuitive: a low payload in previous periods strongly contributes to predicting a low upcoming payload, while a high previous payload tends to indicate a high subsequent level, illustrating strong operational inertia across successive shifts.

Regarding precipitation, as previously discussed, heavy forecasted rainfall has a significant negative impact on productivity, more so than accumulated past rainfall.

We also observe that the variable representing the next shift has a moderate but clear impact. This directly reflects field experience, where the median payload observed during the night is often higher than during the day, mainly due to fewer ancillary activities at night (eg fewer interruptions or preventive maintenance).

Simulated variables show a relatively weak impact, suggesting that integrating maintenance data and HR information (eg actual availability of operators and equipment) would be necessary to significantly improve the predictive relevance of these indicators. Currently, the impact of actual observed availability during the previous shift outweighs that of simulated forecasts, highlighting the room for improvement in integrating real-time information into these predictive variables.

Finally, the next crew to operate has a moderate but structured impact. This regularity highlights its strategic role in understanding overall operational dynamics.

To visualise model performance, we generated several time-series plots comparing the actual payload with the predicted values from each model over a representative period.

For example, the time series analysis presented in Figure 9 compares the actual and predicted payload values over the last 100 shifts of our data set. The blue line represents the observed payload, while the orange line corresponds to the predictions generated by the model, XGBoost and LSTM respectively.

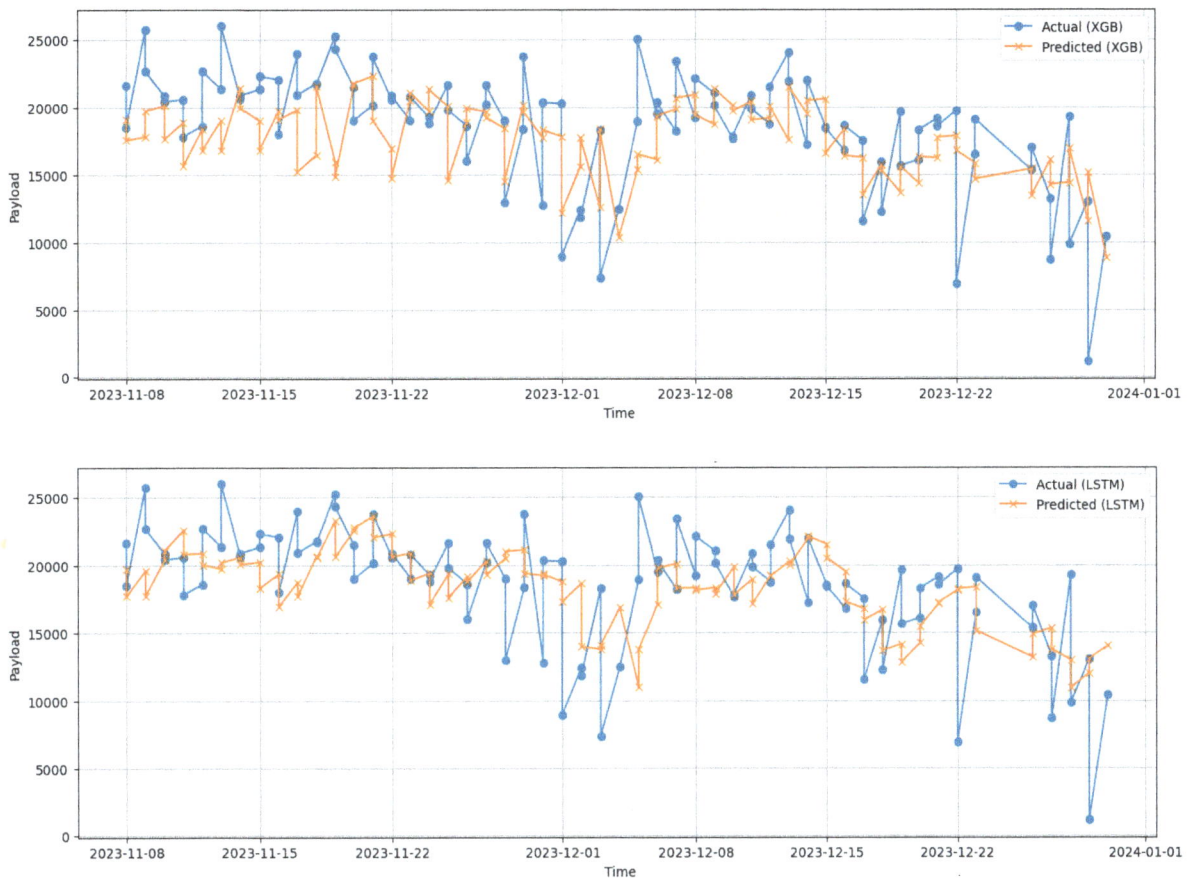

FIG 9 – XGBoost and LSTM next payload forecast respectively versus actual observations over time.

Both models successfully capture the general trend in the data, but they exhibit different behaviours regarding fluctuations and short-term variations. In the current stage of the work, particular attention is being given to the test data to evaluate the robustness and generalisation capabilities of the models. The XGBoost model, being tree-based, shows a greater ability to react to sharp increases and decreases in payload size. This is particularly useful in scenarios where sudden changes occur frequently, as the model can adapt quickly to new patterns. However, this reactivity can sometimes lead to overfitting, where the model becomes too sensitive to noise and short-term variations, potentially reducing its generalisation ability.

In contrast, the LSTM model, due to its sequential learning nature, produces smoother predictions. This stability makes it effective for long-term forecasting and general trend recognition, but it struggles with rapid variations in the payload. The LSTM model tends to exhibit a lagging effect, where it adjusts to changes more slowly than XGBoost. As a result, it is less prone to overfitting but may fail to capture important local fluctuations that influence the overall prediction quality.

Throughout the observed period, both models follow the overall downward trend in payload values towards the end of December 2023. However, the XGBoost model better anticipates sudden peaks and drops, whereas the LSTM model exhibits delay in adapting to these changes. This suggests that while LSTM is better at capturing the broader structure of the time series, it lacks the flexibility needed to track short-term anomalies effectively.

To further evaluate model performance, the percentage error of predictions was analysed for both models over 379 test observations. The evolution of these errors is presented in Figure 10.

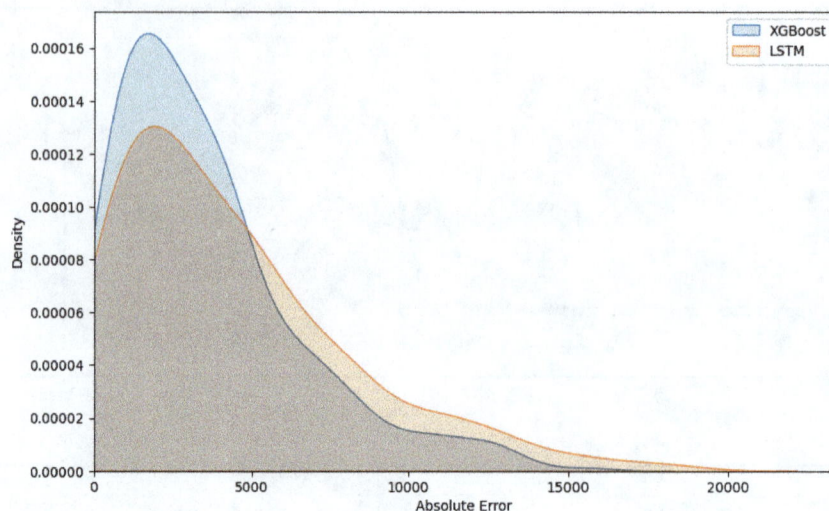

FIG 10 – Kernel Density Estimate (KDE) of Absolute Errors for XGBoost versus LSTM.

XGBoost: Achieved a **MedAE of 14.3 per cent**. This model benefits from its capacity to model non-linear relationships and its enhanced interpretability via SHAP values, which clearly indicate the strong influence of rainfall on predictions.

LSTM: Recorded a **MedAE of 15.1 per cent**. While the LSTM model is particularly adept at capturing sequential dependencies and overall temporal patterns, it occasionally produces larger errors during abrupt operational shifts—often coinciding with heavy impacts events.

The analysis reveals the following key insights: XGBoost produced 33 instances where the error exceeded 50 per cent between the predicted and actual values. LSTM showed a slightly higher count, with 31 instances surpassing the 50 per cent error threshold.

As show in Figure 11 These errors occurred at similar points in time, indicating that both models struggled under the same conditions. A deeper investigation of these high-error occurrences reveals that they do not always coincide with identifiable external rainy events, suggesting that environmental factors were not the primary cause. Instead, these errors are likely due to missing or unrepresented information in the data set that the models were not able to account for.

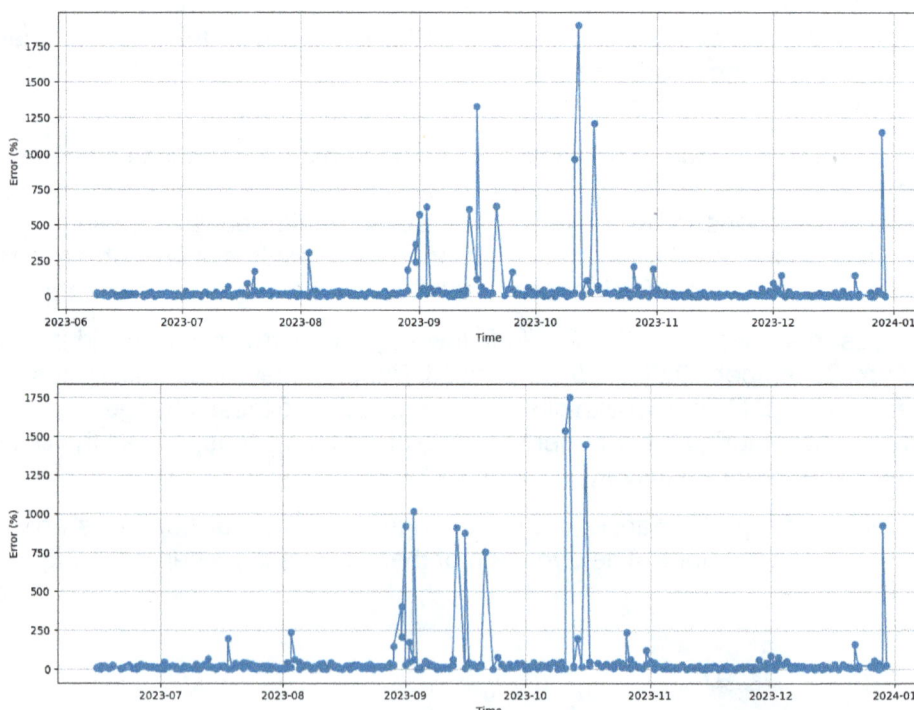

FIG 11 – Temporal evolution of XGBoost (first) and LSTM (second) forecast error percentage.

A further analysis was conducted to assess the operational impact of these high-error occurrences. We observed that when the number of trucks scheduled for the next shift was incorporated as an input feature, the frequency of large deviations (errors exceeding 50 per cent) dropped significantly—down to just eight instances for XGBoost and 27 for LSTM. This demonstrates that integrating key operational planning metrics has a stabilising effect on XGBoost predictions, significantly reducing high-error occurrences. Furthermore, model performance metrics reinforce these findings:

- XGBoost achieved a **MedAE of 8.4 per cent**, highlighting its robustness in most scenarios.

- LSTM showed a **MedAE of 13.5 per cent**, particularly struggling with lower payload values.

- XGBoost attained an **R² score of 0.78**, confirming its strong predictive capabilities, as illustrated in Figure 12.

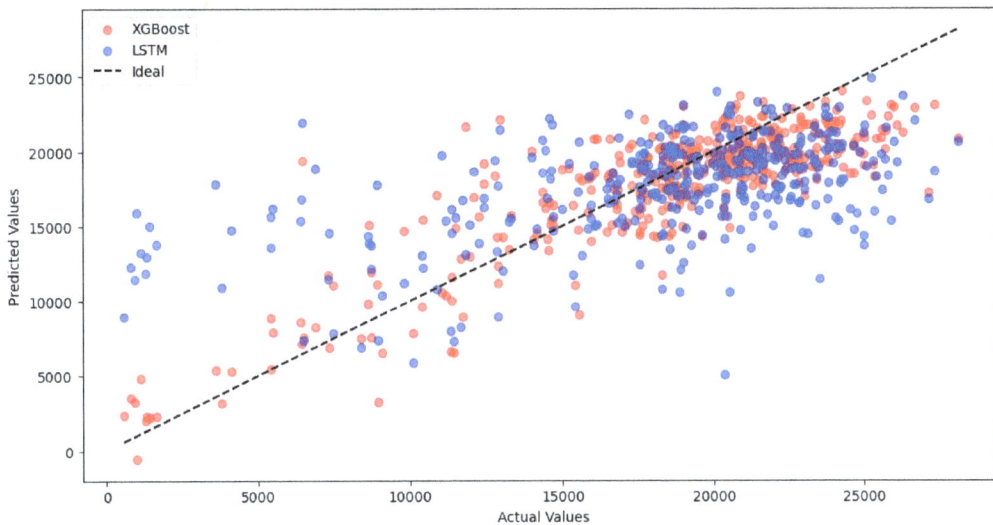

FIG 12 – Model calibration plot (XGBoost and LSTM).

This analysis underscores the importance of integrating operational factors into predictive modelling. It also opens the door for future enhancements, such as incorporating maintenance scheduling and reliability indicators like Mean Time Between Failures (MTBF) and Mean Time to Repair (MTTR), which could further refine forecasting accuracy and improve overall model stability.

Our results yield several important insights, the clear relationship between rainfall and payload performance, as evidenced by both SHAP values and forecast graphs, underscores the necessity of including weather data in operational models. Rainfall not only directly affects payload but also interacts with other operational factors, amplifying its impact.

In summary, both the XGBoost and LSTM models have demonstrated interesting capabilities in forecasting mining payload, with their performance being significantly enhanced by the integration of rainfall data. The median absolute error provides a clear metric for comparison, revealing that while XGBoost offers a slightly better interpretability and handles non-linear relationships effectively. Most importantly, our findings show that incorporating operational planning data—such as the number of trucks scheduled for the next shift—can dramatically reduce major prediction errors. These results lay a strong foundation for future work, where further integration of maintenance planning metrics is expected to drive even greater improvements in forecasting accuracy. Future work should focus on integrating more robust anomaly detection mechanisms and exploring alternative modelling techniques to better address the observed discrepancies.

CONCLUSION

This study has provided compelling evidence that artificial intelligence models offer a promising pathway to accurately forecast hauling fleet production capacity in open pit mining environments fraught with uncertainty. By fusing operational data, meteorological records, and simulated scenarios, our approach captures the complex, dynamic behaviour of mining operations—enabling more nuanced and proactive short-term planning.

The strategic implications of this work are profound. In an industry where operational disruptions ranging from adverse weather conditions to unforeseen equipment downtimes can lead to significant production setbacks, our predictive framework can become an essential decision-support tool. By quantifying the impact of rainfall, on shift operation performance, the model provides actionable insights that allow planners to adjust fleet sequences in short time. This capability not only improves short-term scheduling but also enhances overall operational resilience by aligning production volumes with budget constraints, stock levels, and project timelines.

Looking ahead, several opportunities exist to further refine and extend the predictive accuracy of our models. One promising avenue is the integration of additional operational constraints that are currently underrepresented. For instance, incorporating the effects of blast events could capture the periodic disruptions that frequently slow down operations. Similarly, embedding absenteeism data, especially regarding operator availability, could provide a deeper understanding of the human factors influencing fleet performance. Moreover, including maintenance scheduling indicators such as MTBF and MTTR can address uncertainties arising from equipment reliability, thereby reducing prediction errors even more.

Beyond these enhancements, the integration of our predictive models with real-time decision support systems offers a strategic advantage. By merging forecast outputs with automated dispatch and scheduling tools, mining companies can transition from reactive management to proactive, data-driven planning. Such integration would allow rapid adjustments in operational sequences, mitigating the impact of unpredictable events and ensuring that resources are optimally allocated across various production scenarios. This shift not only has the potential to boost operational efficiency but also to secure a competitive edge in an increasingly digitised industry.

Furthermore, our work underscores a broader vision for the future of mining operations a vision in which advanced analytics and real-time data converge to form a holistic, adaptive management system. As mining companies continue to embrace digital transformation (Shimaponda-Nawa and Nwaila, 2024), the predictive modelling framework presented here can serve as the cornerstone for a comprehensive strategy that encompasses real-time monitoring, anomaly detection, and automated decision-making. This comprehensive approach promises to enhance both economic performance and safety outcomes by minimising disruptions and optimising resource allocation under uncertain conditions.

In summary, this research lays a foundation for the new era of data-driven, short-term planning in open pit mining. By leveraging the strengths of deep learning models and systematically integrating key environmental and operational variables, our approach transforms forecasting from a reactive exercise into a strategic asset. Future work should explore the synergistic integration of diverse data streams and advanced analytics to further refine these models ensuring that mining operations remain agile, resilient, and strategically positioned to navigate the complexities of modern production environments.

REFERENCE

Ao, M, Li, C and Yang, S, 2023. Prediction method of truck travel time in open pit mines based on LSTM model, in *Proceedings of the 42nd Chinese Control Conference (CCC)*, pp 8651–8656. https://doi.org/10.23919/CCC58697.2023.10240705

Asif, M, Bessant, J and Francis, D, 2010. Meta-management of integration of management systems, *The TQM Journal*, 22(6):599–613. https://doi.org/10.1108/17542731011085325

Baek, J and Choi, Y, 2020. Deep neural network for predicting ore production by truck-haulage systems in open-pit mines, *Applied Sciences*, 10(5):1657.

Browne, C B, Powley, E, Whitehouse, D, Lucas, S, Cowling, P I, Rohlfshagen, P, Tavener, S, Perez, D, Samothrakis, S and Colton, S, 2012. A survey of Monte Carlo tree search methods, *IEEE Transactions on Computational Intelligence and AI in Games*, 4(1):1–43. https://doi.org/10.1109/TCIAIG.2012.2186810

Cambitsis, A, 2012. A framework to simplify the management of throughput and constraints, Southern African Institute of Mining and Metallurgy, Johannesburg.

Carvalho, M, Sampaio, P, Rebentisch, E, Carvalho, J Á and Saraiva, P, 2017. Operational excellence, organisational culture and agility: the missing link?, *Total Quality Management and Business Excellence*, 30(13–14):1495–1514. https://doi.org/10.1080/14783363.2017.1374833

Chen, T and Guestrin, C, 2016. XGBoost: A scalable tree boosting system, in *Proceedings of the 22nd ACM SIGKDD International Conference on Knowledge Discovery and Data Mining, KDD '16*, pp 785–794. https://doi.org/10.1145/2939672.2939785

Fan, C, Zhang, N, Jiang, B and Liu, W V, 2022. Prediction of truck productivity at mine sites using tree-based ensemble models combined with Gaussian mixture modelling, *International Journal of Mining, Reclamation and Environment*, 37(1):66–86. https://doi.org/10.1080/17480930.2022.2142425

García, S, Luengo, J and Herrera, F, 2015. *Data preprocessing in data mining* (Springer: Cham).

Gonzalez, F R, Raval, S, Taplin, R and Parsons, M B, 2019. Evaluation of impact of potential extreme rainfall events on mining in Peru, *Natural Resources Research*, 28:393–408. https://doi.org/10.1007/s11053-018-9396-1

Hochreiter, S and Schmidhuber, J, 1997. Long short-term memory, *Neural Computation*, 9(8):1735–1780.

Hyndman, R J and Athanasopoulos, G, 2018. *Forecasting: Principles and Practice*, 2nd edn (OTexts). Available from: <https://otexts.com/fpp2/>

Miles, L D, 1961. *Techniques of value analysis and engineering* (McGraw-Hill, New York).

Sánchez, F, Vargas, J, Pereira, J and Baena, L, 2020. Innovation in the mining industry: Technological trends and a case study of the challenges of disruptive innovation, *Resources Policy*, 65:101569. https://doi.org/10.1016/j.resourpol.2020.101569

Shimaponda-Nawa, M and Nwaila, G T, 2024. Integrated and intelligent remote operation centres (I2ROCs): Assessing the human–machine requirements for 21st century mining operations, *Minerals Engineering*, 207:108565. https://doi.org/10.1016/j.mineng.2023.108565

Soofastaei, A, Aminossadati, S, Kizil, M S and Knights, P, 2015. Simulation of payload variance effects on truck bunching to minimise energy consumption and greenhouse gas emissions, in *Proceedings of the Coal Operators' Conference 2015* (eds: N Aziz and B Kininmonth), pp 337–346 (University of Wollongong – Mining Engineering, the Australasian Institute of Mining and Metallurgy – Illawarra, and Mine Managers Association of Australia).

Tlhatlhetji, M and Kolapo, P, 2021. Investigating the effects of rainy season on open cast mining operation: The case of Wescoal Khanyisa Colliery, *Research Square Preprint*, https://doi.org/10.21203/rs.3.rs-870740/v1

Van Houdt, G, Mosquera, C and Napoles, G, 2020. A review on the long short-term memory model, *Artificial Intelligence Review*, 53:5929–5955. https://doi.org/10.1007/s10462-020-09838-1

Wang, Q, Zhang, R, Lv, S and Wang, Y, 2021. Open pit mine truck fuel consumption pattern and application based on multi-dimensional features and XGBoost, *Sustainable Energy Technologies and Assessments*, 43:100977. https://doi.org/10.1016/j.seta.2020.100977

Wang, W, Chakraborty, G and Chakraborty, B, 2021. Predicting the Risk of Chronic Kidney Disease (CKD) Using Machine Learning Algorithm, *Applied Science*, 11:202. https://doi.org/10.3390/app11010202

Advanced PGNAA Analysers – leveraging machine learning for improved precision in iron making

H Gu[1] and Y Strutz[2]

1. Data Scientist, Scantech, Adelaide SA 5038. Email: h.gu@scantech.com.au
2. Technical Manager, Scantech, Adelaide SA 5038. Email: y.strutz@scantech.com.au

ABSTRACT

This study explores the application of machine learning to enhance the precision of Scantech GEOSCAN Prompt Gamma Neutron Activation Analysis (PGNAA) for on-belt analysis of bulk materials. Focusing on the indirect measurement of Basicity in iron making, we aim to mitigate the uncertainty inherent in traditional analytical methods, which often rely on error-prone, multi-step calculations. Our approach leverages supervised learning, training a model on spectral analysis data, belt loads, statistical measures, and detector temperatures. Batch correlation serves as the loss function, and a defined acceptable response range acts as a filter, ensuring robust network performance.

Traditional PGNAA techniques, including single-peak analysis and Monte Carlo simulations, can be challenging when dealing with elements exhibiting weak or unstable signals. Indirectly calculating these elements using intermediate ratios often leads to cumulative uncertainties. Our machine learning model circumvents this issue by establishing a direct relationship between the spectral analysis data and the Basicity parameter, eliminating the need for intermediate variables.

The network architecture comprises multiple layers with tan-sigmoid and log-sigmoid activation functions, optimised for mapping input features to the target Basicity output. Rigorous filtering criteria were implemented during training to reject networks susceptible to overfitting or convergence to local extrema, ensuring reliable online performance.

Experimental results demonstrate a significant improvement in accuracy, validated by strong correlation with laboratory data and reduced uncertainty. Further validation on independent data sets confirms the model's robustness across varying operational conditions.

This method offers a promising avenue for enhancing the precision and operational efficiency of PGNAA systems, extending their applicability to values previously considered difficult to measure directly. Future work will focus on further network optimisation and validation across a wider range of material compositions.

INTRODUCTION

The mining industry faces escalating economic and environmental challenges, including declining ore grades, stricter sustainability requirements, and the need to reduce processing costs. Pre-concentration through sensor-based bulk ore sorting has the potential to benefit all these areas by enabling mining operations to divert material of low economic value prior to processing, thereby achieving higher feed grades and recovery, while reducing the processing plant and tailings dam footprint and capital required throughout the life of the mine. Of sensing techniques for bulk sorting Prompt Gamma Neutron Activation Analysis (PGNAA) has proven to be a successful measurement technique within many industries, including mining (Ferguson, 2020). PGNAA measures the elemental composition of bulk materials on a conveyor belt by analysing the gamma ray spectrum generated from neutron absorption and has been used in the mining industry since the 1980s (Noble, 2020). By enabling pre-concentration of ores, PGNAA systems help operators divert low-value material early in the process, thereby improving feed grades, reducing waste, and lowering operational costs.

A key challenge lies in deriving parameters that are not directly measurable by PGNAA but are essential for process optimisation. One such parameter is Basicity (B4), a critical metric in iron making, which is defined as:

$$B4 = \frac{[Ca]+[Mg]}{[Si]+[Al]} \tag{1}$$

Where [Ca], [Mg], [Si] and [Al] are the concentrations of calcium, magnesium, silicon, and aluminium respectively.

Traditional PGNAA analysis techniques, such as single peak analysis and Monte Carlo Library Least-Square (MCLLS) fitting (Han, 2005), are widely utilised. Each measured elemental ratio from these models has an associated level of uncertainty resulting from the probabilistic nature of neutron activation, counting statistics and detector measurement uncertainty. For values calculated from the elemental ratios, the uncertainties are cumulative as they propagate through the calculation chain. For the calculation of B4, weak or unstable gamma-ray signals for elements such as Al and Mg amplify uncertainties in their ratios, degrading the accuracy of the resulting B4 value.

To address this limitation, a machine learning (ML) framework to establish a direct mapping between raw full spectrum regression coefficients (full spectrum analysis) and the B4 value was developed. By bypassing error-prone intermediate calculations, this method reduces uncertainty propagation and improves measurement precision. This work focuses on Scantech GEOSCAN PGNAA systems, leveraging our full-spectrum analysis (FSA) capabilities—a technique that aggregates gamma-ray signals across the entire energy spectrum—to train supervised learning models.

The remainder of this paper is structured as follows: Section 2 details the uncertainty propagation in traditional method and the machine learning architecture; Section 3 explains how the neural network was set-up; Section 4 details the selection, evaluation, and optimisation of the neural network model, drawing on multivariance function derivative theory; and Section 5 evaluates the model's performance during training and testing phases, as well as its generalisability to future operational data; Section 6 concludes the study and proposes future research directions to refine and expand the method.

METHODOLOGY

Uncertainty propagation in traditional PGNAA methods

For directly tested element ratios, the traditional method is to do a full spectrum analysis, using regression to attain the contribution coefficient of each element. This is followed by the calibration process where the ratio of each element is calculated via a transformation matrix. Basicity (B4) is then calculated using Equation 1 and depends on the concentrations of Ca, Mg, Si and Al.

This method inherently accumulates errors due to the propagation of uncertainties in intermediate elemental ratios. The total uncertainty in B4 can be quantified using the Law of Propagation of Uncertainty (LPU) (Couto, 2018), these uncertainties can be quantified by expanding the measurement model into a Taylor series and approximating the result by retaining only the first-order terms, thus simplifying the uncertainty propagation. This is shown by:

$$y = f(x_1, \dots, x_N) \tag{2}$$

$$u_y^2 = \sum_{i=1}^{N} \left(\frac{\partial y}{\partial x_i}\right)^2 u_{x_i}^2 + 2\sum_{i=1}^{N}\sum_{j=i+1}^{N} \left(\frac{\partial y}{\partial x_i}\right)\left(\frac{\partial y}{\partial y_i}\right) COV(x_i, x_j) \tag{3}$$

where u_y is the combined standard uncertainty for y and u_{x_i} is the uncertainty for the i input quantity. The second term of equation is due to the correlation between the input quantities. If it is assumed there is no correlation between them, it can be simplified as:

$$u_y^2 = \sum_{i=1}^{N} \left(\frac{\partial y}{\partial x_i}\right)^2 u_{x_i}^2 \tag{4}$$

By substituting Equation 1 into Equation 4, we determine the uncertainty in basicity is given by:

$$u_{B4}^2 = \frac{1}{(u_{Si}+u_{Al})^2} u_{Ca}^2 + \frac{1}{(u_{Si}+u_{Al})^2} u_{Mg}^2 + \frac{(u_{Ca}+u_{Mg})^2}{(u_{Si}+u_{Al})^4} u_{Si}^2 + \frac{(u_{Ca}+u_{Mg})^2}{(u_{Si}+u_{Al})^4} u_{Al}^2 \tag{5}$$

This equation demonstrates how errors in intermediate elemental ratios ($u_{Ca}, u_{Mg}, u_{Si}, u_{Al}$) amplify when combined to compute B4. Weak or unstable signals for elements like Si or Al exacerbate this issue, making traditional methods prone to cumulative inaccuracies.

Considering this, machine learning was used to establish a direct relationship between the FSA and B4, bypassing the need for intermediate element ratios as shown in Figure 1.

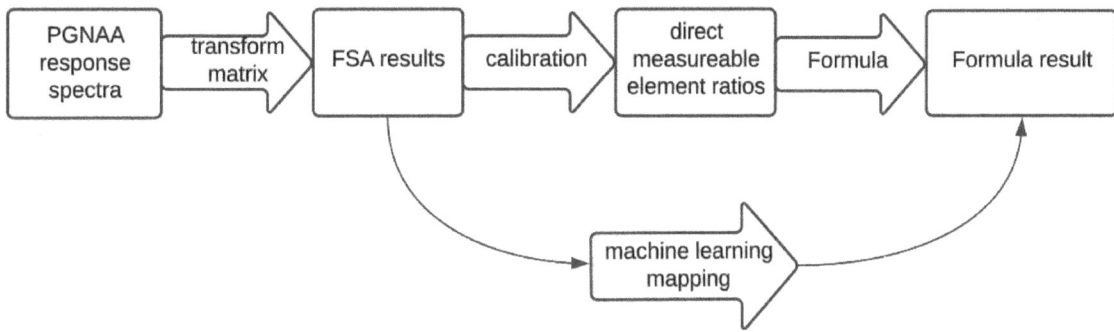

FIG 1 – Flow chart showing how neural network applied in this study.

Backpropagation neural networks, which are widely used for establishing mappings between different groups of data through a guided training process, were employed in this endeavour.

Machine learning structure

The structure of the neural network used is shown in Figure 2. For the network FSA response results are used as the input vectors and Basicity is used as the target.

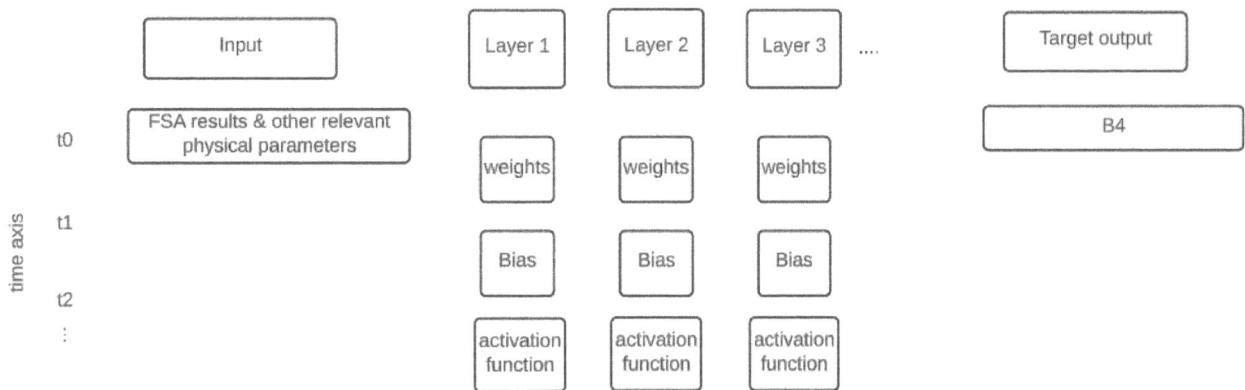

FIG 2 – Network structure.

For each layer, the output can be expressed using Equation 6:

$$y = g(W \cdot x + b) \tag{6}$$

Where:

x represents the input of the layer

W represents the weight matrix of the layer

b represents the bias vector of the layer

g represents the activation function of the layer

Linear, Tan-Sigmoid and Log-Sigmoid functions are commonly used as activation functions (Demuth, 2000). These activation functions can change the scale of the output in a linear or non-linear way. In this study, a sigmoid function was used as the activation function:

$$sigmoid(x) = \frac{1}{1+e^{-x}}.$$

Pearson R and mean squared error were used to provide feedback to the network and adjust its weight, to find the best fit between the network's predicted B4 and lab measurements.

NETWORK SET-UP

The target of the network training is to build a map between the full spectra analysis results of the PGNAA applicable elements and the B4 ratio. The first step of network training is to identify the input vectors of the mapping. This involved identifying parameters related to B4. The selection criteria were as follows:

- Manageable parameter array size: a compact set of parameters to facilitate efficient training and fast computation.

- Representativeness: the chosen parameters should fully represent the factors influencing the target. Any parameter changes outside this scope should not significantly affect the results.

Based on these criteria, the following parameters were selected as inputs to the neural network model:

- FSA Numbers: representing the PGNAA response to the product.

- Chi-square: indicating the discrepancy in FSA simulation.

- Belt load: reflecting neutron and gamma transmission.

After combining these input vectors, a network with three layers was built and trained on the data. The input vectors were preprocessed with normalisation and standardisation to improve the stability and performance of the neural network (Brownlee, 2020). Bayesian Optimisation was applied to tune the hyperparameters for the network. The optimised hyperparameters that were used in the final training and testing are shown in Table 1.

TABLE 1

Network hyperparameters set.

Layers	Number of inputs	First hidden layer nodes	Second hidden layer nodes	Activation of hidden layers	Batch size
3	15 (FSAs and belt load)	14	14	Sigmoid	99

Since the initial weights and biases in the network are random, the networks can be adjusted to different styles through training and finally stopped at a local extremum. Therefore, networks that successfully captured the characteristics of the relationship between input data and target data must be identified. This brings us to the next section – network filter.

NETWORK FILTERING PRINCIPLES IN MODEL TRAINING

The iterative backpropagation process, where network parameters are continuously adjusted to refine the mapping between inputs and outputs, was used to train the neural network. PGNAA analysers sample at a much higher frequency (eg every two mins for SCANTECH) compared to lab-based measurements (eg order of once per hour). This provides a major advantage over traditional lab tests, as continuous online monitoring enables real-time process control without disrupting production. To align online analyser data with lab test results, we take the product throughput weighted mean of online measurements over the lab sampling period.

In the network training process:

- The target vector corresponds to the lab test ratio.

- The input vectors represent the weighted mean of online analyser data over the lab sampling period.

By structuring the training data this way, we ensure that the network learns from representative, temporally aggregated samples rather than being misled by transient fluctuations.

There are two assumptions for a reliable mapping between FSA response values and lab test results when training our model:

- Assumption 1: The short-term relationship between FSA response values and elemental ratios remains unaffected by the product accumulation process. This implies that the mapping function can be approximated as linear or semi-linear.

- Assumption 2: The lab sampling frequency is sufficient to cover the full range of product composition variations.

Due to the random initialisation of network weights, models trained on the same data set may converge to slightly different parameter sets. Some of these models become trapped in local extrema, making them unsuitable for broader applications. To ensure the robustness and generalisability of the trained networks, suboptimal models were excluded using the following two criteria.

Threshold-based evaluation

The first filtering criterion sets a quantitative threshold for network performance. At the completion of training—whether reaching the maximum epoch or an early stopping condition—the predicted outputs are compared against the target values. The network is considered valid only if the difference between the predicted and target values falls within a predefined threshold.

This evaluation is based on two key metrics:

1. Correlation coefficient – measuring the linear association between predicted and target values.
2. Mean Squared Error (MSE) – quantifying the average squared difference between predictions and targets.

These dual objectives ensure that the model not only maintains a high correlation with the target data but also minimises prediction errors.

Online model simulated ratio distribution assessment

This rule aims to exclude networks that are excessively sensitive to minor input fluctuations, leading to disproportionately large variations in output. Networks with this instability often produce results that deviate significantly from the expected range when exposed to inputs slightly outside the training distribution. Introducing Gaussian noise into the input data is a common approach to assessing model stability. However, in this project, we adopt a different approach that leverages the online sampling characteristics of PGNAA analysers.

We treat the original online analyser data as a high-frequency representation of the lab data (assumption 1), but which contains an effective noise relative to the lab data due to the higher online analyser sampling frequency and lab sampling variability. If the trained network accurately captures the fundamental characteristics of the real mapping, it can be analogous a low-resolution sketch of a high-resolution image—the true relationship between FSA response and elemental ratios. The output distribution from the model should therefore closely match the distribution observed from the lab data (assumption 2). An example of the distribution of B4 lab results is shown as the blue distribution in Figure 3.

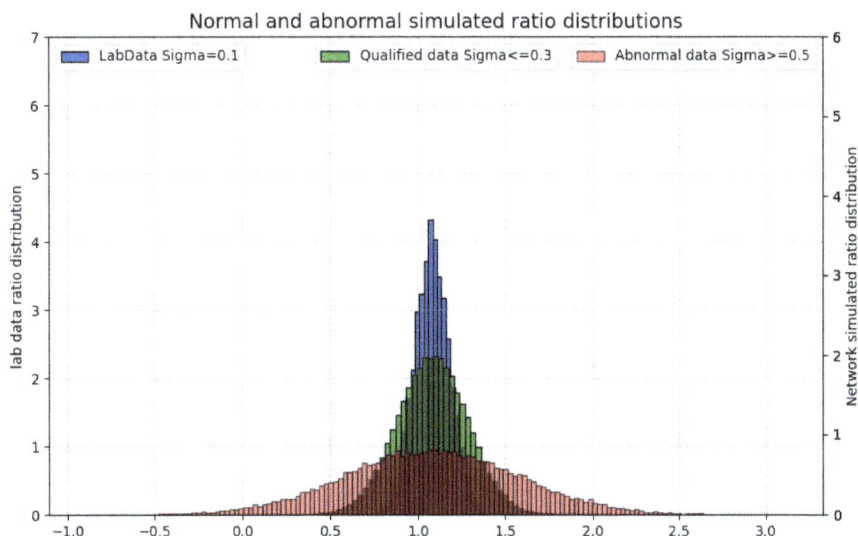

FIG 3 – Lab data distribution of B4 values (Equation 1).

A well-trained network, fed with all PGNAA analyser data (prior to applying the weighted mean), should produce predictions with a distribution comparable to the green distribution shown in Figure 3. Conversely, networks generating irrational predictions—such as distributions that are too narrow, too wide, or with extreme deviations—are flagged for exclusion. An example of an undesirable output distribution is shown by the pink distribution in Figure 3. By systematically comparing network-predicted distributions against expected lab test ratios, unstable and unreliable models are filtered out.

To establish a robust filtering criterion, the original distribution of the lab data was derived and its mean $\bar{x}$ and standard deviation σ were computed. An expanded acceptance range is then defined as:

$$\bar{x} \pm k\sigma$$

where k is an adjustable parameter that depends on the characteristics of the data set and the strictness of the application. In practice, k is typically set below 5. If the predicted output from the network—when given raw analyser data (sampled every 2 mins)—falls outside this range, the network is deemed unreasonable and is filtered out.

NETWORK MESH ANALYSIS AND OPTIMISATION

Once a network is deemed rational (ie it meets the filtering criteria outlined in Section 4), its performance is optimised through mesh analysis. This process involves evaluating how changes in input vectors affect the network's output, enabling identification and refinement of key parameters that drive model behaviour.

Mesh construction and input interdependencies

The neural network can be conceptualised as a multivariate function: $y = f(x_1, x_2, \ldots x_N)$, where x_i represents the ith input vector, and y is the output (eg the predicted B4 value). To analyse this function, a mesh was constructed by discretising each input vector into evenly spaced values across its range. However, since input vectors are often correlated, the effective mesh space is smaller than the Cartesian product of individual ranges.

For example, in Figure 4, the pink bars represent input vectors that are actively varied within predefined ranges (anchors), while the blue bars represent vectors whose ranges are passively constrained due to their interdependencies with the active vectors. This interdependence reduces the dimensionality of the mesh, simplifying the analysis.

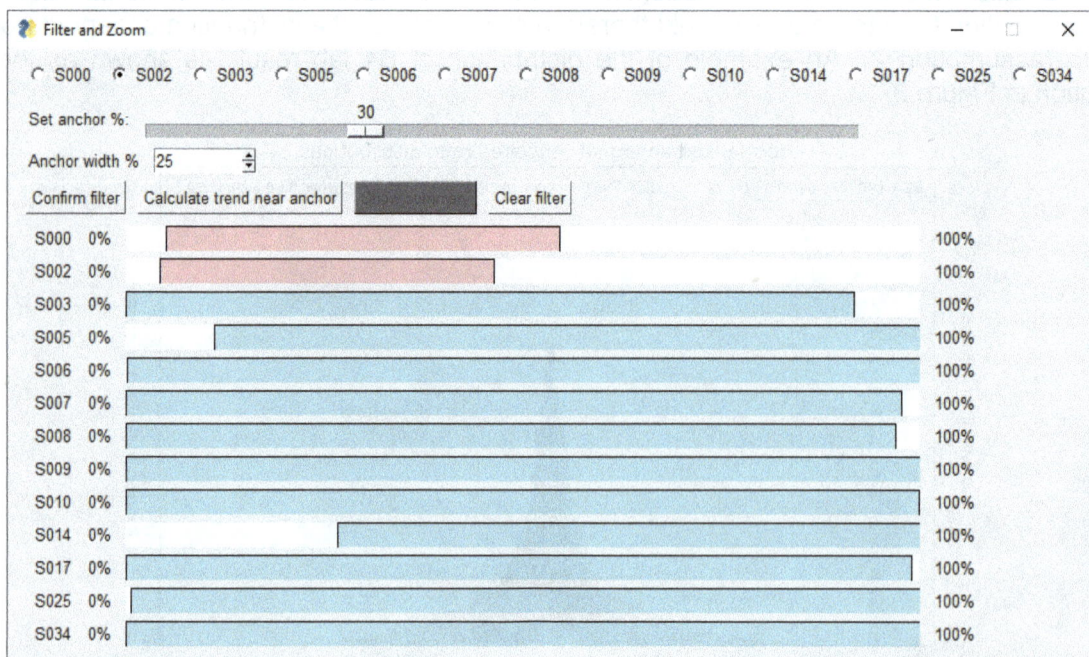

FIG 4 – Mesh construction showing active (pink) and passive (blue) input vectors.

Sensitivity analysis and parameter optimisation

To understand the influence of each input vector on the output, sensitivity analysis was performed as follows:

- Anchor Selection: For each input vector, an anchor point (eg the mean or median value) was selected and a range defined around it.

- Input Filtering: The data set was filtered to include only samples where the remaining input vectors fall within their respective mid-ranges. Refer to Figure 4.

- Output Evaluation: The active input vector was varied across its range while holding other inputs constant at their mid-values. The resulting output changes are plotted to visualise the input-output relationship. Refer to Figures 5 and 6.

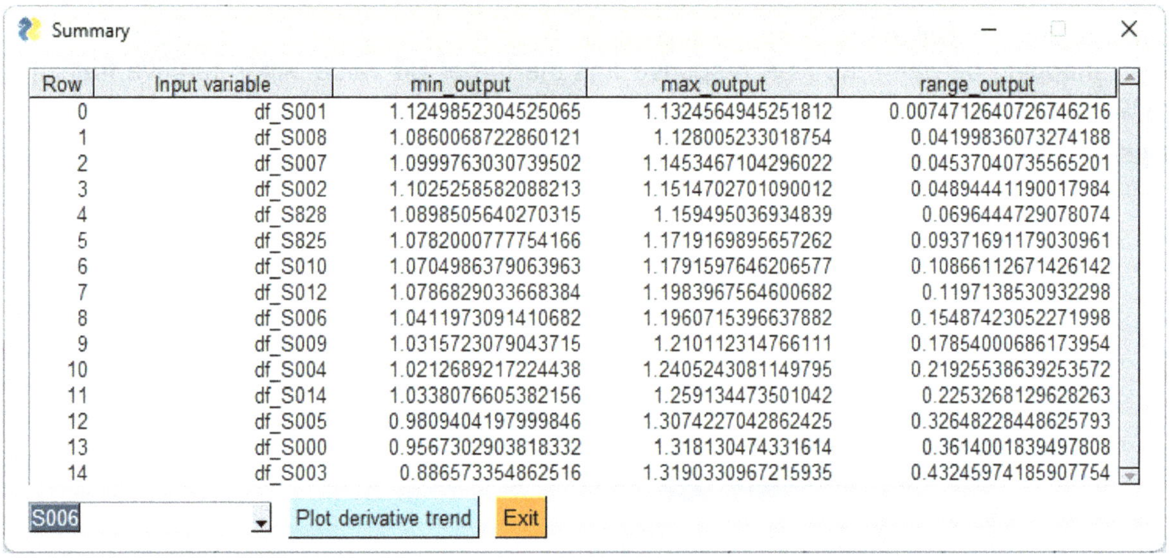

Row	Input variable	min_output	max_output	range_output
0	df_S001	1.1249852304525065	1.1324564945251812	0.0074712640726746216
1	df_S008	1.0860068722860121	1.128005233018754	0.04199836073274188
2	df_S007	1.0999763030739502	1.1453467104296022	0.04537040735565201
3	df_S002	1.1025258582088213	1.1514702701090012	0.04894441190017984
4	df_S828	1.0898505640270315	1.159495036934839	0.0696444729078074
5	df_S825	1.0782000777754166	1.1719169895657262	0.09371691179030961
6	df_S010	1.0704986379063963	1.1791597646206577	0.10866112671426142
7	df_S012	1.0786829033668384	1.1983967564600682	0.1197138530932298
8	df_S006	1.0411973091410682	1.1960715396637882	0.154874423052271998
9	df_S009	1.0315723079043715	1.210112314766111	0.17854000686173954
10	df_S004	1.0212689217224438	1.2405243081149795	0.21925538639253572
11	df_S014	1.0338076605382156	1.259134473501042	0.2253268129628263
12	df_S005	0.9809404197999846	1.3074227042862425	0.32648228448625793
13	df_S000	0.9567302903818332	1.318130474331614	0.3614001839497808
14	df_S003	0.886573354862516	1.3190330967215935	0.43245974185907754

S006 | Plot derivative trend | Exit

FIG 5 – Summary of the impact of each input vector.

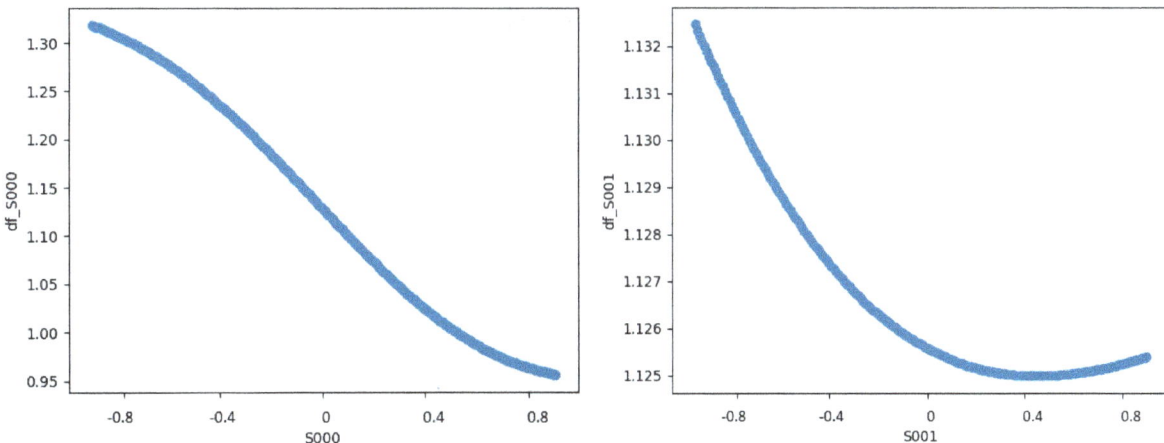

FIG 6 – Examples of how input change affects the output.

This process generates a sensitivity profile for each input vector, revealing its relative contribution to the output. Inputs with minimal influence on the output (eg those producing flat or near-flat sensitivity curves) are flagged for potential removal in subsequent training iterations.

Iterative refinement

Using the insights from sensitivity analysis, the network was iteratively refined by:

- Removing low-impact input vectors to reduce model complexity.
- Retraining the network with the optimised input set.

- Re-evaluating performance metrics (eg Pearson R, mean squared error) to ensure improvements.

This iterative process continues until the network achieves optimal performance, balancing accuracy and computational efficiency.

RESULTS AND DISCUSSION

In this study, the training data covered six months of time, and the whole data set comprised 1078 samples, with each sample consisting of:

- Input features: full-spectrum analysis (FSA) response results, including gamma-ray counts across energy bins, belt load measurements.

- Target value: laboratory-measured Basicity (B4) ratios provided by the mine site.

The neural network was trained using a supervised learning approach, with the goal of establishing a direct mapping between the FSA response and the target B4 value. After iterative training and filtering, the model achieved a strong alignment with the lab results.

The performance of the trained network is illustrated in Figures 7 and 8.

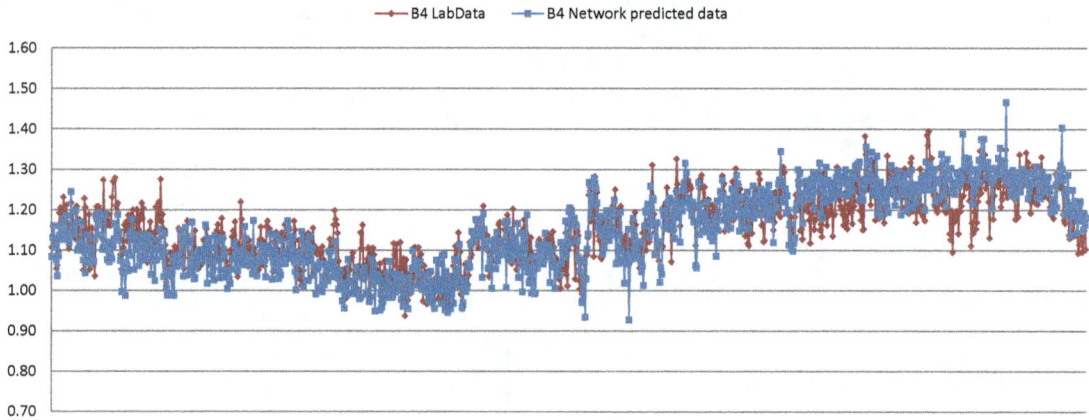

FIG 7 – A time-series plot showing the lab-measured and network-predicted B4 ratios over a representative period. The close overlap between the two curves demonstrates the model's ability to capture temporal variations in B4.

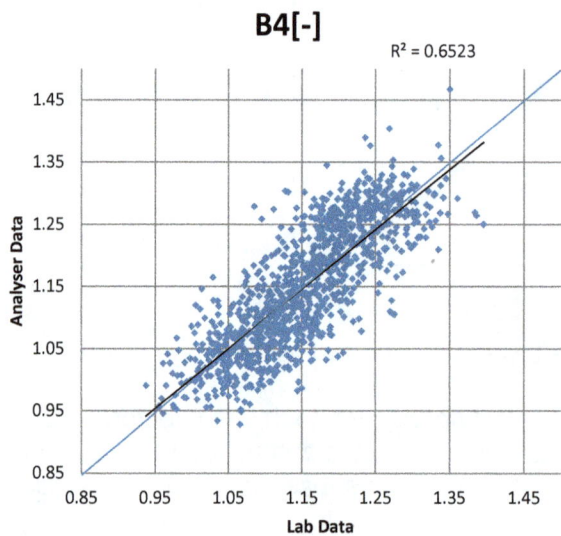

FIG 8 – A scatter plot comparing the lab-measured B4 ratios (x-axis) against the network-predicted B4 ratios (y-axis). The high density of points along the diagonal indicates a strong correlation between predicted and actual values.

To assess the generalisability and robustness of the trained network, its performance was evaluated an independent data set collected one month after the training period. This validation data set consisted of 162 samples, with the same input features (FSA response results and belt load measurements) and target values (lab-measured B4 ratios) as the training set. The network's predictions for the validation data set were compared to the lab-measured B4 ratios in Figures 9 and 10.

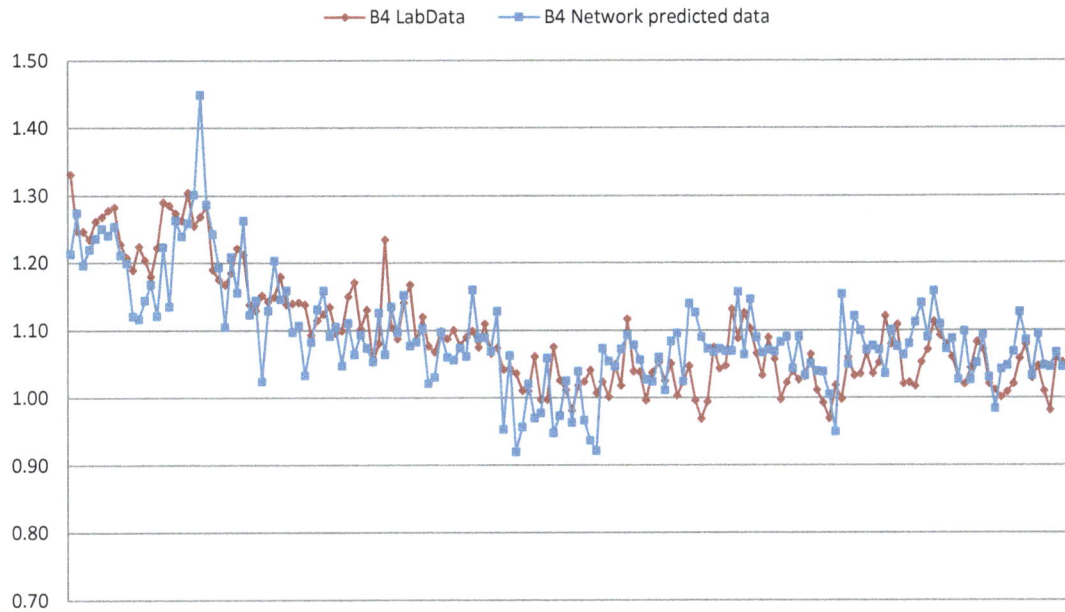

FIG 9 – A time-series plot showing the lab-measured and network-predicted B4 ratios over the validation period. The close alignment between the two curves demonstrates the model's stability and temporal consistency.

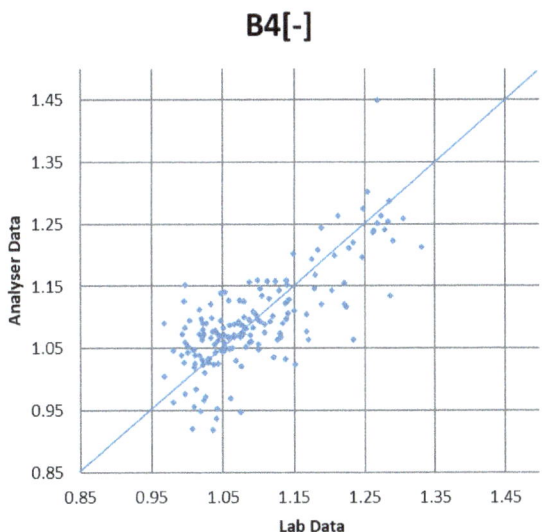

FIG 10 – A scatter plot comparing the lab-measured B4 ratios (x-axis) against the network-predicted B4 ratios (y-axis) for the validation period. The distribution of points along the diagonal indicates the model's ability to maintain accuracy over time.

The performance of the trained network in was evaluated using three categories, the mean absolute error, correlation coefficient and root mean squared error as shown in Table 2. Because the validation data samples are taken after the end of the train data period, so it is actually an extrapolating application in the time series of the attained network. But their correlation are similar, and the RMSE of validation part is a bit smaller than the train data set. So it can validate the attained model from the training.

TABLE 2

Network performance.

	Sample size	Mean Absolute Error (MAE)	Correlation coefficient (R)	Root Mean Squared Error (RMSE)
Training data	1078	0.047	0.81	0.058
Validation data	162	0.044	0.76	0.057

CONCLUSION

In this study, the feasibility of applying machine learning to enhance the indirect measurement of Basicity (B4) in iron making using Scantech GEOSCAN PGNAA analysers was demonstrated. The developed approach establishes a direct mapping between raw full-spectrum analysis (FSA) data and the target B4 value, bypassing error-prone intermediate calculations. This methodology offers two key advantages:

1. Reduction of uncertainty propagation

   Traditional PGNAA methods compute B4 indirectly using elemental ratios (eg Ca, Mg, Si, Al), where uncertainties in individual measurements propagate through the formula Equation 1, amplifying the final error. By training a neural network to directly predict B4 from spectral analysis data, we circumvent this multi-step uncertainty chain. Experimental results showed a correlation coefficient (R) of 0.81 and a root mean squared error (RMSE) of 0.06 on training data, with comparable performance (R = 0.76, RMSE = 0.06) on validation data sets. These metrics indicate a significant improvement over traditional methods, where propagated uncertainties often degrade confidence in B4 estimates.

2. Expansion of PGNAA's applicability

   Our framework extends PGNAA's utility beyond directly measurable elements. For instance, while elements like Lithium or Tin cannot be detected directly via PGNAA, their ratios to measurable elements (eg Ca or Si) can be inferred through machine learning if a stable correlation exists. This approach enables estimation of parameters that would otherwise require costly lab tests. However, such mappings are dependent on ore composition consistency—validation across diverse geological sources remains critical for broader adoption.

Future Directions:

- Model generalisation: validate the framework on ores with varying mineralogical compositions to ensure robustness.

- Real-time optimisation: integrate adaptive learning techniques to handle temporal drift in operational conditions.

- Uncertainty quantification: develop hybrid models that explicitly account for measurement uncertainties in spectral data.

In summary, our work highlights machine learning's potential to improve the precision and scope of PGNAA-based analysis in iron making. By reducing reliance on intermediate calculations and expanding the range of inferable parameters, this approach aligns with industry demands for faster, more reliable process control.

ACKNOWLEDGEMENTS

The author would like to express sincere gratitude to their manager, Yaron Strutz, for his invaluable support in project selection, management, coordination of resources, and reviewing and revising this paper. The author also appreciates their coworker, Dylan Peukert, for his insightful suggestions and for validating my approach of using ML to reduce uncertainty. Additionally, thanks go to Oliver Chew for his early contributions to testing the code and providing valuable feedback for improvement.

Finally, the author acknowledges their supervisor, Lucas Biggins, along with my coworkers Ken Smith, Stiaan Jordaan, and Craig Lockyear, for their assistance in reviewing the paper to ensure compliance with the company's confidentiality regulations.

REFERENCES

Brownlee, J, 2020, August 25. How to use Data Scaling Improve Deep Learning Model Stability and Performance Machine learning mastery. Available from: <https://machinelearningmastery.com/how-to-improve-neural-network-stability-and-modeling-performance-with-data-scaling/>

Damasceno, J C, Couto Paulo, R G, 2018. Methods for evaluation of measurement uncertainty, *Intechopen*, 10.

Demuth, H, 2000. Neural Network Toolbox User's Guide.

Han, X, 2005. Development of Monte Carlo Code for Coincidence Prompt Gamma-ray Neutron Activation analysis, North Carolina State University. Available from <https://www.proquest.com/openview/993bddbe4877acd171f481e033636bf9/1?pq-origsite=gscholar&cbl=18750&diss=y>

Noble, F, 2020. Understanding ore heterogeneity and effective bulk ore sorting using PGNAA/PFTNA, Preconcentration Digital Conference.

Noble, G and Ferguson, S, 2020. Understanding ore heterogeneity and effective bulk ore sorting using PGNAA/PFTNA, Preconcentration Digital Conference, November 2020.

Transforming mining optimisation – from hours to minutes with massively parallel serverless processing

A J Heggart[1]

1. Senior Solutions Architect, Amazon Web Services, Perth WA 6000.
   Email: alexhegg@amazon.com

ABSTRACT

Mining optimisation is an area that can benefit immensely from High Performance Compute (HPC) to provide data driven decisions through the application of novel mathematical, statistical and machine learning approaches. Where these algorithms or approaches scale well with parallel processing we have the ability to apply the massive scale of cloud computing to how quickly we can produce these results. Traditional approaches require large upfront investment in infrastructure and do not scale well with demand. Strategic mine planning optimisation has been evolving since the 1960s, with recent advances in computational power and algorithms enabling more detailed and faster planning (Elkington, 2009).

In this paper the author will discuss an approach that utilises serverless technologies within Amazon Web Services to deliver optimisation of thousands of orebody plans to users within 5 mins. Serverless has emerged as a novel approach to HPC, offering dynamic resource allocation and quick response times (Crouch et al, 2024). This fundamentally changes how this information can be used as the time horizon has shifted from what could be delivered with traditional approaches which may take in the order of tens of hours. It is also cost-effective with no existing infrastructure in place, scaling to meet the demand automatically and requiring the customer to only pay for what they consume.

This approach is in use within a global tier-1 mining organisation. Mechanisms from Amazon's culture of innovation were used to rapidly prototype, develop and implement this solution within the miner's environment. A relentless focus on the customer and working backwards from their needs allowed the team to prioritise experiments and testing to quickly eliminate sub-optimal decisions and approaches and learn quickly from failures. This paper will also discuss the evolution of this architecture and the lessons learned and data points that were taken to deliver the final outcome for the customer.

INTRODUCTION

Optimisation of blasting and dig patterns can yield significant increase in mine productivity. Simulation can be used to iterate though a large number of potential options to find the optimal plan to extract ore. With advances in computational power it is possible to execute these simulations in a reasonable time to optimise mining operations for optimal ore recovery and throughput. Complexity in the variables and uncertainty in the underlying data however means a large number of simulations are required.

A tier-1 global mining company was applying these simulation techniques but was unable to drive effective change within their business. With a traditional computing approach, the required number of simulations would take around 12 hrs to complete. Mine engineers would initiate simulation jobs out of curiosity to compare to their finalised plans as the time taken to get a result would run into the next shift. The simulation software development team at the miner approached Amazon Web Services with a challenge to change the time horizon for simulation results so that mine engineers could incorporate the outputs into their plans.

Amazon has a culture of innovation that encourages teams to experiment at pace, be prepared to fail but to also learn from those mistakes to quickly narrow in on an optimal solution. Adopting these practices, the teams worked closely together to iterate over a number of possible architectures to decrease the total run time, with an ultimate goal of completing a simulation run within 5 mins. Utilising massively parallel serverless processing the team managed to achieve this objective while also delivering it at the same cost as existing solutions.

This paper is structured as follows: We first introduce the simulation process and challenges in reducing the total job time with traditional approaches. We will talk about a number of architectural patterns that were quickly validated, tested and eliminated before selecting the optimal method. We will also talk about how work that is highly parallel can be done for a similar cost in a much shorter time by leveraging the scale of the cloud.

SIMULATION OVERVIEW

A simulation job run consists of four discrete stages with varying levels of parallelism. Stage 3 reaches peak concurrency of up to 10 000 instances of a simulation running at the same time. With a traditional, single server model then the number of concurrent simulations is limited to around ten. This requires simulations to be queued then executed when a slot becomes available. For the existing solutions, this would mean a typical run time of 10–12 hrs.

The simulation job starts with a single instance in Stage 1 which runs for around 15 seconds and does preprocessing of data and allocation of those data sets and hyperparameters to Stage 2 jobs. Stage 2 has up to 1000 concurrent simulations which run for around 15 seconds to further preprocess and allocate simulation jobs for grade blocking. Stage 3 is the longest running stage, taking approximately 2 mins to complete, as well as having the highest level of concurrency with up to 10 000 instances of a simulation.

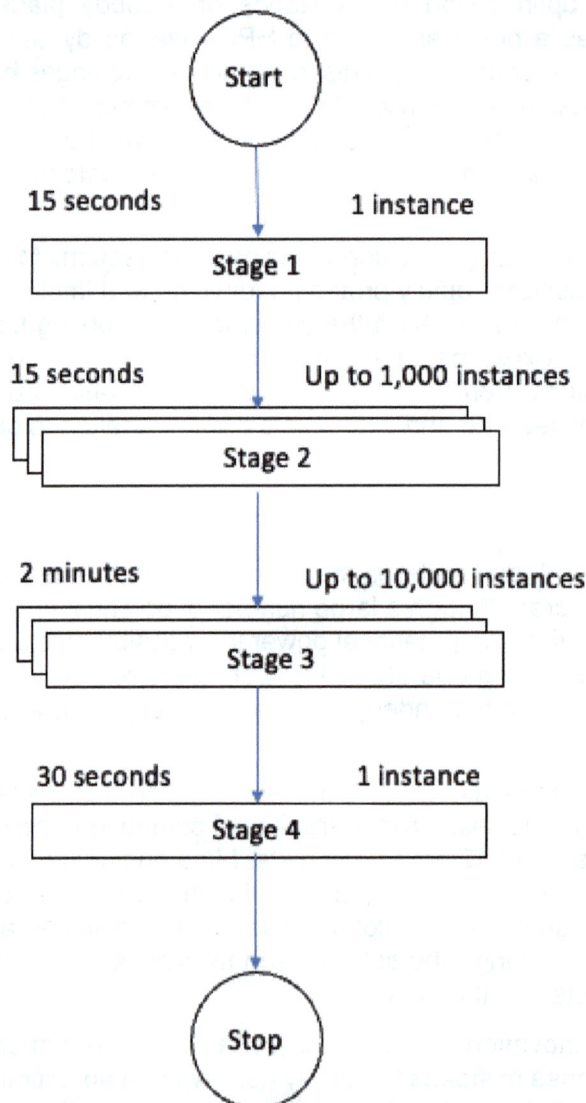

FIG 1 – Overview of the simulation process.

To concurrently execute all of these simulations using traditional servers would require at least 10 000 virtual CPU cores necessitating a large capital investment in computing infrastructure. These servers would also be heavily underutilised when the simulation jobs are not running. Running a single or several serves would bring the cost down significantly but fail to achieve run times close to what is needed to enable near real time decision support.

ARCHITECTURAL OPTIONS

Adopting Amazon's culture of innovation, the team worked to quickly evaluate multiple deployment architectures utilising Amazon Web Services cloud. The jobs fundamentally rely on two components, a scheduler to coordinate and assign work, and compute to execute that work. The team proposed evaluating six architectures across two work scheduling technologies and three compute options. One of the work scheduling technologies does not support one of the compute technologies, so this combination was immediately discounted.

The work scheduling technologies selected were: 1) AWS Batch, a fully managed batch computing service; and 2) AWS Step Functions, a visual workflow service that can be used to orchestrate distributed applications and workloads.

The compute technologies selected were: 1) EC2 instances, a virtual server in the AWS cloud environment; 2) AWS Fargate, a serverless, pay as your go container compute engine; and 3) AWS Lambda, a serverless compute service which allows you to run code with provisioning or managing infrastructure.

The five architectural options were then instantiated with dummy code that were proxies for the real workload, allowing the evaluation to proceed without real data and concurrent to existing development. Each workload was then also timed and averaged over a number of runs give an indicative timing of each architecture.

TABLE 1

Architecture options, including reasons for disqualification for the unselected architectures.

Work scheduler	Compute	Disqualification reason
Batch	EC2	Batch was too slow to initiate and start a job
Batch	Fargate	Batch was too slow to initiate and start a job
Step Functions	EC2	EC2 was too slow to start new instances
Step Functions	Fargate	Fargate was too slow to initiate new containers
Step Functions	Lambda	Selected architecture

The testing identified issues with using Batch and EC2 or Fargate to execute new jobs. Batch is a traditional batch computing service, that is designed to manage pools of compute resources and schedule and queue jobs for execution against scoped compute resources. Traditional high-performance computing (HPC) workloads may execute for hours, days or weeks so the overhead of several minutes to schedule, queue and initiate a new job is minimal, but significantly ate into the targeted 5 min execution time. Fargate also had delays in scheduling new tasks to run, which introduced an overhead that pushed the end-to-end run time outside the targeted window.

Step Functions is a more generalised workflow service, so required a small development overhead to build out the required work scheduling capability. This trade-off was desirable as it provided near instantaneous workload execution to assist with reducing the total simulation job duration. Lambda also typically initiates and start either sub-second or within seconds for larger container workloads, which again provided optimal use of compute time to bring the total simulation job run time down to the 5 min target.

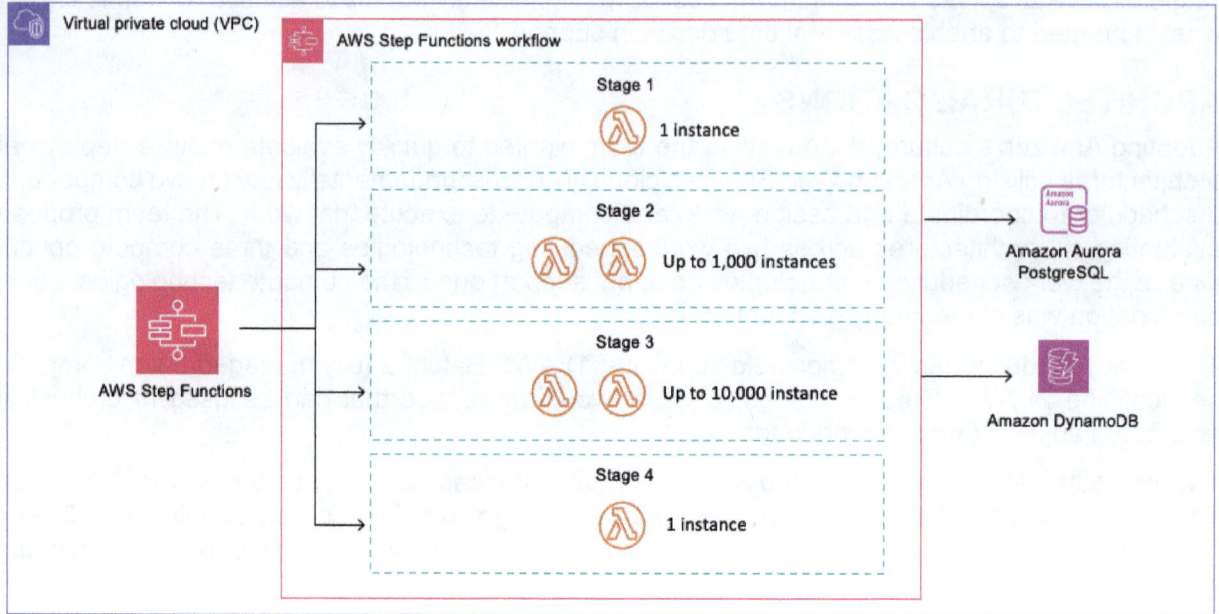

FIG 2 – Selected architecture.

PRICING

Cloud computing enables on-demand access to resources with a pay-as-you-go model. For activities that massively scale in parallel this provides an opportunity to accelerate run time without potentially increasing cost. Running a single Lambda with one virtual server core would take nearly 14 days and cost approximately 10 USD to execute 10 000 simulations that take 2 mins each to run. If these simulations can all be executed concurrently, then for the same cost of 10 USD you can get the result in 2 mins. We observed that there are small overheads in parallelising the simulations and coordinating the workload, though that is a small trade-off to make to change the end-to-end run time.

TABLE 2

Indicative pricing for simulation options. These options are using Amazon Web Services hpc6a.48xl EC2 instances or AWS Lambda serverless compute functions in the ap-south-east-2 region. This pricing is based on an estimate of 100 simulation jobs per month.

Simulation approach	Cost per job	Cost per month	Job time
Traditional server	18 USD	1800 USD	10 hrs
Fixed HPC cluster	1850 USD	185 000 USD	5 mins
Cloud HPC cluster (on-demand)	10 USD	1000 USD	15 mins
Scaled Lambda compute	10 USD	1000 USD	5 mins

CONCLUSIONS

We present here a novel architecture for delivering simulation jobs at scale for the mining industry. For short-lived, highly parallel jobs, this approach can fundamentally change the uptake of these techniques by changing the time horizon from hours to minutes to deliver actionable insights to mining engineers at the same or lower cost than traditional approaches. Like all software architectures, this is not a one size fits all approach, but should have wide applicability to a number of use cases in the optimisation and simulation space.

Further work will look to extend this capability across other areas of the mining and wider industrial domains.

ACKNOWLEDGEMENTS

The author would like to thank Amazon Web Services for providing the opportunity to deliver innovative solutions to customers. Any views or opinions expressed in this paper are solely those of the author and do not necessarily represent those of Amazon Web Services.

REFERENCES

Crouch, M, Chien, A, Dorier, M, Gandhi, R, Jain, R, Kim, J, Kettimuthu, R, Leyffer, S, Munson, T, Papka, M, Vishwanath, V and Foster, I, 2024. Serverless supercomputing: High-performance function as a service for science, in *Proceedings of the SC '24 Workshops of The International Conference on High Performance Computing, Network, Storage, and Analysis*, pp 110–122 (IEEE Computer Society).

Elkington, T, 2009. Optimising mining project value for a given configuration, PhD thesis, University of Western Australia.

Optimising truck fleet selection and sizing in open pit mining using simulation – analysing sustainability impacts of fleet configuration and truck failures

M Kazemi[1], A Moradi[2], J Doucette[3] and H Askari-Nasab[4]

1. MSc Graduate, University of Alberta, Edmonton Alberta T6G 1H9, Canada. Email: mohamad92reza@gmail.com
2. Assistant Professor, Sustainable Intelligent Mining Laboratory, University of Kentucky, Lexington, KY 40506, USA. Email: ali.moradi@uky.edu
3. Professor, University of Alberta, Edmonton Alberta T6G 1H9, Canada. Email: jed3@ualberta.ca
4. Professor, University of Alberta, Edmonton Alberta T6G 1H9, Canada. Email: hooman@ualberta.ca

ABSTRACT

This study introduces a simulation-based model designed to optimise truck fleet type and size configurations in open pit mining operations, focusing on sustainability and emissions reduction. The model evaluates both homogeneous and heterogeneous fleet set-ups to identify optimal strategies that enhance operational efficiency, productivity, and environmental performance. Through comparative analysis, we find that heterogeneous fleets outperform homogeneous configurations by achieving production targets with reduced fuel consumption, supporting an effective balance between productivity and emissions reduction. The simulation results also indicate that, when factoring in potential truck failures, smaller trucks within a homogeneous fleet offer higher flexibility in ore transfer and experience lower average downtimes, thus maintaining operational flow more effectively than heterogeneous fleets under failure scenarios. This insight highlights the advantages of homogeneous fleets with smaller trucks in conditions where truck reliability is variable. Conversely, heterogeneous fleets show higher fuel efficiency in non-failure scenarios, achieving lower fuel consumption per ton moved. The study underscores the critical role of fleet selection, sizing, truck reliability, and emissions in optimising fuel consumption, production rates, and material flow in mining operations. These findings offer practical insights into configuring fleet operations that meet production demands while minimising costs and environmental impacts, contributing to more sustainable mining practices.

INTRODUCTION

Open pit mining presents a variety of substantial challenges requiring solutions. These encompass mine design, road network analysis, infrastructure optimisation, fleet management, truck quantity and type determination, and truck allocation. Achieving an efficient truck fleet size is crucial for a cost-effective hauling system in mining, ensuring production needs are met while minimising expenses. Choosing the quantity and types of trucks for the fleet is a substantial financial commitment, given its non-reversible nature (Salhi and Rand, 1993). Within the mining system, an excess of trucks can lead to over-trucking, with trucks waiting for shovels, while too few trucks result in under-trucking, causing shovels to wait (Ataeepour and Baafi, 1999). Having the right fleet size is crucial for efficient transportation. Having too few or too many trucks can cause delays and underutilisation. A shortage of trucks reduces production, while an excess raises GHG emissions. Achieving the optimum number of trucks maintains a balance between meeting production needs and minimising GHG emissions.

Simulation modelling proves potent for testing alternative actions, offering insights into optimal outcomes. In mining, these models predict the impact of new ideas and policies. Monte Carlo Simulation and specialised languages have simplified discrete event model creation, aiding analysis of production capacities, bottleneck identification, and resource utilisation (Knights and Bonates, 1999). Manríquez et al (2019) highlighted discrete event simulation's role in designing mining systems, including transportation routes and equipment types. For truck allocation, simulation has long been valued. Maran and Topuz (1988) stressed its importance, especially when traditional methods fall short. Discrete event simulation is widely used in optimising truck and shovel systems due to its capacity to model randomness and complexity (Que, Anani and Awuah-Offei, 2016).

This study employs Arena simulation software to develop a discrete event simulation model, evaluating fleet truck sizing and selection's influence on production rates and GHG emissions in open pit mining. Truck allocation in the simulation model relies on a multi-objective optimisation model aiming to minimise deviations from target production, shovel idle time, truck wait time, and fuel consumption. The study also examines the impact of truck failures in production and truck fleet selection.

In the upcoming sections, following a literature review, this research will introduce the simulation model and the integrated optimisation model. Subsequently, a detailed explanation of the case study, including key performance indicators (KPIs), will be provided. Shifting to the results, the paper will analyse and compare the performance of different scenarios related to truck selection, sizing, and the impact of failures within a case study. Lastly, the paper will engage in a comprehensive discussion of the results, drawing conclusions, and outlining potential directions for future research.

LITERATURE REVIEW

The use of simulation techniques is crucial for effectively addressing fleet management and haulage systems within open pit mining. Through the creation of a dynamic virtual environment, simulation empowers researchers and engineers to comprehensively analyse the impact of truck fleet selection and sizing in mining operations, production rate, and GHG emissions. As a result, decision-makers gain the insights needed to make informed choices. Simulation serves as a reliable tool for evaluating trade-offs and alternative scenarios, ultimately providing decision-makers with a clearer perspective, and enhancing efficiency, sustainability, and resource utilisation. In what follows, a series of studies are presented that address the fleet selection and sizing challenge in open pit mining, followed by application of simulation in this context.

Bozorgebrahimi, Hall and Blackwell (2003) reviewed critical parameters of fleet sizing in open pit mining. On the other hand, Burt and Caccetta (2014, 2018) explored fleet selection problem in mining, supported by case studies. They reviewed fleet selection problem challenges, applications, and solution approaches within the context of open pit mining.

Over the years, various techniques have been used in fleet selecting and sizing in open pit mining. Markeset and Kumar (2000) introduced the Life cycle costing technique, followed by Samanta, Sarkar and Mukherjee (2002) who combined the Analytical Hierarchy Process and Life cycle costing. Different approaches like the match factor concept (Douglas, 1964; Burt and Caccetta, 2007), queuing theory (Ercelebi and Bascetin, 2009), linear programming (Edwards, Malekzadeh and Yisa, 2001; Ta, Ingolfsson and Doucette, 2013), and machine repair modelling (Krause and Musingwini, 2007) have been used conventionally. In addition, innovative computer-based algorithms, including expert systems, fuzzy set theory, genetic algorithms, multiple criteria decision-making, and machine learning algorithms have also emerged (Bandopadhyay and Venkatasubramanian, 1987; Marzouk and Moselhi, 2004; Li and Song, 2009; Bazzazi, Osanloo and Karimi, 2011; Nobahar, Pourrahimian and Mollaei Koshki, 2022). To address uncertainties in the selection of surface mining fleet, it is necessary to create a stochastic model. Discrete event simulation, pioneered by Rist (1961) for mine haulage, offers a solution by considering stochastic parameters. Noteworthy applications of this simulation method can be found in various mining studies (Kolonja and Mutmansky, 1994; Baafi and Ataeepour, 1998; Ataeepour and Baafi, 1999; Yuriy and Vayenas, 2008; Que, Anani and Awuah-Offei, 2016; Chaowasakoo *et al,* 2017; Zeng, Baafi and Walker, 2019; Zhang *et al,* 2022). However, these models often lack accuracy with respect to real-world fleet management systems and underestimate production capacity effects. To address these limitations, Moradi-Afrapoli, Tabesh and Askari-Nasab (2019) presented an integrated simulation model encompassing mining, processing, and dispatching systems. Subsequently, several studies utilised the integration of simulation and dispatch optimisation modelling to predict the optimal solution for the fleet selection and sizing problem in the presence of uncertainty (Moradi-Afrapoli and Askari-Nasab, 2020; Mohtasham *et al,* 2021; Moradi-Afrapoli, Upadhyay and Askari-Nasab, 2021; Upadhyay *et al,* 2021; Moradi-Afrapoli, Upadhyay and Askari-Nasab, 2022; Yeganejou *et al,* 2022; Mirzaei-Nasirabad *et al,* 2023). Nevertheless, their model fails to account for energy efficiency, greenhouse gas (GHG) reduction, and truck failure. In this study, an integrated framework is established that integrates simulation and optimisation, considering production capacity, energy efficiency, and GHG mitigation as well as truck failure.

METHODOLOGY

The integrated simulation and optimisation model in this research relies on various input parameters and data, which encompass the short-term production schedule, the mine's road network, specifications for shovels and trucks detailing capacities and performance, information about dumping locations and their capacities, as well as the count of dumping points per dump location. Moreover, fitted probability distributions are necessary for numerous input variables, including loading and dumping times, hauling durations, empty travel times, backing times, spot times, shovel bucket capacities, and truck loading capacities. The majority of these inputs are stochastic, which makes them particularly challenging. Consequently, historical data for such random variables were employed to fit diverse probability density functions.

Multi-objective optimisation model

The truck dispatching optimisation model utilised in this research is centred around four primary objectives: reducing deviations from target path flow rates, minimising fuel consumption (and GHG emissions), minimising shovel idle time, and minimising truck wait time. Since these four objectives exist in varying dimensions, it is necessary to transform them into dimensionless forms. An efficient operation requires the satisfaction of several constraints. Furthermore, certain estimated parameters are utilised within the constraints, and their estimation methods (formulas) are presented in the equalisation constraints of the mathematical model. Several indices, parameters, and decision variables are available within the optimisation model. The indices are as follows:

t Index for set of trucks: $t = \{1, \dots, T\}$

s Index for set of shovels: $s = \{1, \dots, S\}$

d Index for set of dumping points: $d = \{1, \dots, D\}$

d' Index for set of locations where trucks are required to dump their load before traveling to the new shovel: $d' = \{1, \dots, D\}$

w Index for set of weights assigned to individual goals: $w = \{1, 2, 3, 4\}$

g index for the group of trucks that are currently waiting in a queue of the shovel: $g = \{1, \dots, NTWS\}$

The parameters are introduced below:

IT_{tsd} Idle time for shovel s if truck t is assigned to transport material from shovel s to the dumping point d

WT_{tsd} Wait time for truck t if it is assigned to transport material from shovel s to the dumping point d

N_w Normalised weights of individual goals based on priority

AF A factor balancing available trucks with the required capacity of plants

PC_d Capacity of the plant d: $d = \{1, \dots, P\} \subset \{1, \dots, D\}$

SC_s Production capacity of shovel s

MP_{sd} Path flow rate for the path from shovel s to the dumping point d that the production operation has met so far

TC_t Actual capacity of truck t (tonne)

NTC_t Nominal capacity of truck t (tonne)

P_{sd} Path flow rate for the path from shovel s to the dumping point d

TR_{tsd}	Next time truck t reaches shovel s, if truck t is assigned to transport material from shovel s to the dumping point d
SA_{tsd}	Next time shovel s is available to serve truck t, if truck t is assigned to transport material from shovel s to the dumping point d
$TNOW$	Current time of the operation/simulation
$LD_{td'}$	The distance truck t must travel to reach the dumping point d' to dump its load
$ED_{td's}$	The distance truck t must travel from the dumping point d' to the next expected shovel s
ALT_t	Average loading time of truck t
APL_t	Average payload of truck t
$LV_{td's}$	Average loaded velocity of truck t traveling to dumping point d' and will travel to shovel s after dumping its load
$EV_{td's}$	Average empty velocity of truck t traveling from dumping point d' to the next expected shovel s
$DQ_{td'}$	Queue time for truck t in the queue of the dumping point d'
$DT_{td'}$	Dump time for truck t to dump its material in dumping point d'
$NTWS_s$	Number of trucks waiting in queue at shovel s
ST_g	Spotting time for the truck g in the queue
LT_g	Loading time for the truck g in the queue
α_t	Intercept of truck t for the fuel consumption
β_t	Payload coefficient of truck t for the fuel consumption
γ_t	Loading time coefficient of truck t for the fuel consumption
τ_t	Idle time coefficient of truck t for the fuel consumption
ω_t	Empty traveling time coefficient of truck t for the fuel consumption
φ_t	Loaded traveling time coefficient of truck t for the fuel consumption
SIT_{tsd}	Shovel idle time coefficient, by assigning truck t to the path of shovel s to dumping point d
TWT_{tsd}	Truck wait time coefficient, by assigning truck t to the path of shovel s to dumping point d
F_{tsd}	Truck fuel consumption coefficient, by assigning truck t to the path of shovel s to dumping point d

Below are the decision variables:

x_{tsd}	Binary variable equals to 1 if truck t assigns to the path of shovel s to dumping point d, and 0 otherwise
y_{sd}^-	Negative deviation of the met path flow rate and the desired path flow rate for the path between shovel s and dumping point d
y_{sd}^+	Positive deviation of the met path flow rate and the desired path flow rate for the path between shovel s and dumping point d

The model has the following objective functions:

$$f_1 = \sum_{s=1}^{S} \sum_{d=1}^{D} (y_{sd}^- + y_{sd}^+) \tag{1}$$

$$f_2 = \sum_{t=1}^{T}\sum_{s=1}^{S}\sum_{d=1}^{D} F_{tsd}x_{tsd} \tag{2}$$

$$f_3 = \sum_{t=1}^{T}\sum_{s=1}^{S}\sum_{d=1}^{D} SIT_{tsd}x_{tsd} \tag{3}$$

$$f_4 = \sum_{t=1}^{T}\sum_{s=1}^{S}\sum_{d=1}^{D} TWT_{tsd}x_{tsd} \tag{4}$$

The following two formulas are used to normalise the objective functions and to present the normalised weighted sum objective function, respectively.

$$\bar{f}_i = \frac{f_i - U_i}{N_i - U_i} \; \forall i \in \{1,2,3,4\} \tag{5}$$

$$f = N_1\bar{f}_1 + N_2\bar{f}_2 + N_3\bar{f}_3 + N_4\bar{f}_4 \tag{6}$$

The constraints of the model are expressed below:

$$\sum_{s=1}^{S}\sum_{d=1}^{D} TC_t x_{tsd} \leq NTC_t \; \forall t \in \{1, \dots, T\} \tag{7}$$

$$\sum_{t=1}^{T}\sum_{s=1}^{S} TC_t x_{tsd} \geq AF \times PC_d \; \forall d \in \{1, \dots, P\} \tag{8}$$

$$\sum_{t=1}^{T}\sum_{d=1}^{D} TC_t x_{tsd} \leq SC_s \; \forall s \in \{1, \dots, S\} \tag{9}$$

$$\sum_{t=1}^{T} TC_t x_{tsd} + MP_{sd} + y_{sd}^- - y_{sd}^+ = P_{sd} \; \forall s \in \{1, \dots, S\} \text{ and } \forall d \in \{1, \dots, D\} \tag{10}$$

$$AF = \frac{\sum capacity\ of\ available\ trucks}{\sum required\ flow\ rate\ at\ paths} \tag{11}$$

$$TR_{tsd} = TNOW + \frac{LD_{td'}}{LV_{td's}} + DQ_{td'} + DT_{td'} + \frac{ED_{td's}}{EV_{td's}} \tag{12}$$

$$\forall t \in \{1, \dots, T\} \text{ and } \forall s \in \{1, \dots, S\} \text{ and } \forall d \in \{1, \dots, D\} \text{ and } \forall d' \in \{1, \dots, D\}$$

$$SA_{tsd} = TNOW + \sum_{g=1}^{NTWS_s} \left(ST_g + LT_g \right) \tag{13}$$

$$\forall t \in \{1, \dots, T\} \text{ and } \forall s \in \{1, \dots, S\} \text{ and } \forall d \in \{1, \dots, D\}$$

$$SIT_{tsd} = \max\left(0, TR_{tsd} - SA_{tsd}\right) \tag{14}$$

$$\forall t \in \{1, \dots, T\} \text{ and } \forall s \in \{1, \dots, S\} \text{ and } \forall d \in \{1, \dots, D\}$$

$$TWT_{tsd} = \max\left(0, SA_{tsd} - TR_{tsd}\right) \tag{15}$$

$$\forall t \in \{1, \dots, T\} \text{ and } \forall s \in \{1, \dots, S\} \text{ and } \forall d \in \{1, \dots, D\}$$

$$F_{tsd} = \alpha_t + \beta_t \times APL_t + \gamma_t \times ALT_t + \tau_t \times TWT_{tsd} + \omega_t \frac{ED_{td's}}{EV_{td's}} + \varphi_t \frac{LD_{td'}}{LV_{td's}} \tag{16}$$

$$\forall t \in \{1, \dots, T\} \text{ and } \forall s \in \{1, \dots, S\} \text{ and } \forall d \in \{1, \dots, D\} \text{ and } \forall d' \in \{1, \dots, D\}$$

$$x_{tsd} \in \{0,1\} \; \forall t \in \{1, \dots, T\} \text{ and } \forall s \in \{1, \dots, S\} \text{ and } \forall d \in \{1, \dots, D\} \tag{17}$$

$$y_{sd}^- \geq 0 \ \forall s \in \{1, \dots, S\} \text{ and } \forall d \in \{1, \dots, D\} \tag{18}$$

$$y_{sd}^+ \geq 0 \ \forall s \in \{1, \dots, S\} \text{ and } \forall d \in \{1, \dots, D\} \tag{19}$$

The first objective employs a goal programming approach to minimise deviations from path flow rates, which is computed using Equation 1. The second objective function seeks to minimise the total fuel consumption of active trucks using Equation 2. The third objective centres on minimising the idle time of active shovels using Equation 3. The fourth objective is to decrease truck wait time during operations, calculated using Equation 4. To achieve the model's solution, the four objectives are made dimensionless using Nadir and Utopia points (Grodzevich and Romanko, 2006), defining lower and upper limits. This process scales objectives from 0 to 1 using Equation 5. Priority weights for the weighted sum method are based on normalised versions of objectives from Equations 1 to 4 in Equation 6. There are several constraints in the model. Constraint 7 restricts a truck's payload to its nominal capacity for tonnage transport in one cycle. In Constraint 8, the material transported to processing plants via all trucks must fulfill processing targets set by each plant, adjusted by the AF factor (calculated in Equation 11). Only the AF portion of plant requirements can be fulfilled. Constraint 9 limits haulage capacity to a shovel's nominal digging rate, while Constraint 10 calculates path flow rate deviations for paths linking a shovel as a source and a dumping location as a destination. Equation 12 is employed to ascertain the arrival time of each truck for loading by a shovel. Shovel availability is determined using Equation 13, predicting the next time the shovel will be available to load the truck. The coefficients for the three optimisation objectives are computed through Equations 14 15, and 16, corresponding to shovel idle time, truck wait time, and fuel consumption objective functions, respectively. Lastly, Constraint 17 guarantees the binary nature of the first set of decision variables, while Constraints 18 and 19 ensure non-negativity for the goal programming variables.

Integrated simulation and optimisation framework

The simulation section of the framework employs a step-by-step approach, as depicted in Figure 1. Initially, the model identifies trucks awaiting assignment to operational shovels and destinations. Subsequently, the multi-objective optimisation model comes into action, efficiently allocating unassigned trucks. This ensures that all available trucks are effectively assigned their respective tasks. Throughout the simulation, the optimisation model is recalibrated in response to specific events, such as truck initiation, dumping completion, or truck reactivation following a failure. These occurrences trigger a reassessment of the optimal assignment for each truck. The optimisation process for assigning available trucks persists throughout the simulation runtime until the predefined time period for the simulation is reached. The input data for the framework encompasses the quantity and types of trucks present within the system. As a result of this input, the framework generates Key Performance Indicators (KPIs) statistical report, which will be elaborated upon in the subsequent section. Figure 2 depicts the hauling procedure executed by trucks in open pit mining. The sequence commences at the terminal, where trucks are designated to ore or waste shovels, determined by considerations like production goals, travel durations, queue statuses, and processing periods. Following this, the trucks journey to either the waste dump or one of the crushers/plants, contingent on their cargo and the hopper capacities at each plant. Ultimately, the trucks are reassigned to a different shovel based on the timetable and objective functions, and this cycle persists. Finally, Figure 3 presents a flow chart of a truck status during the operation. Once a failed truck has been repaired, it becomes imperative to reassign it to a new loading or unloading point.

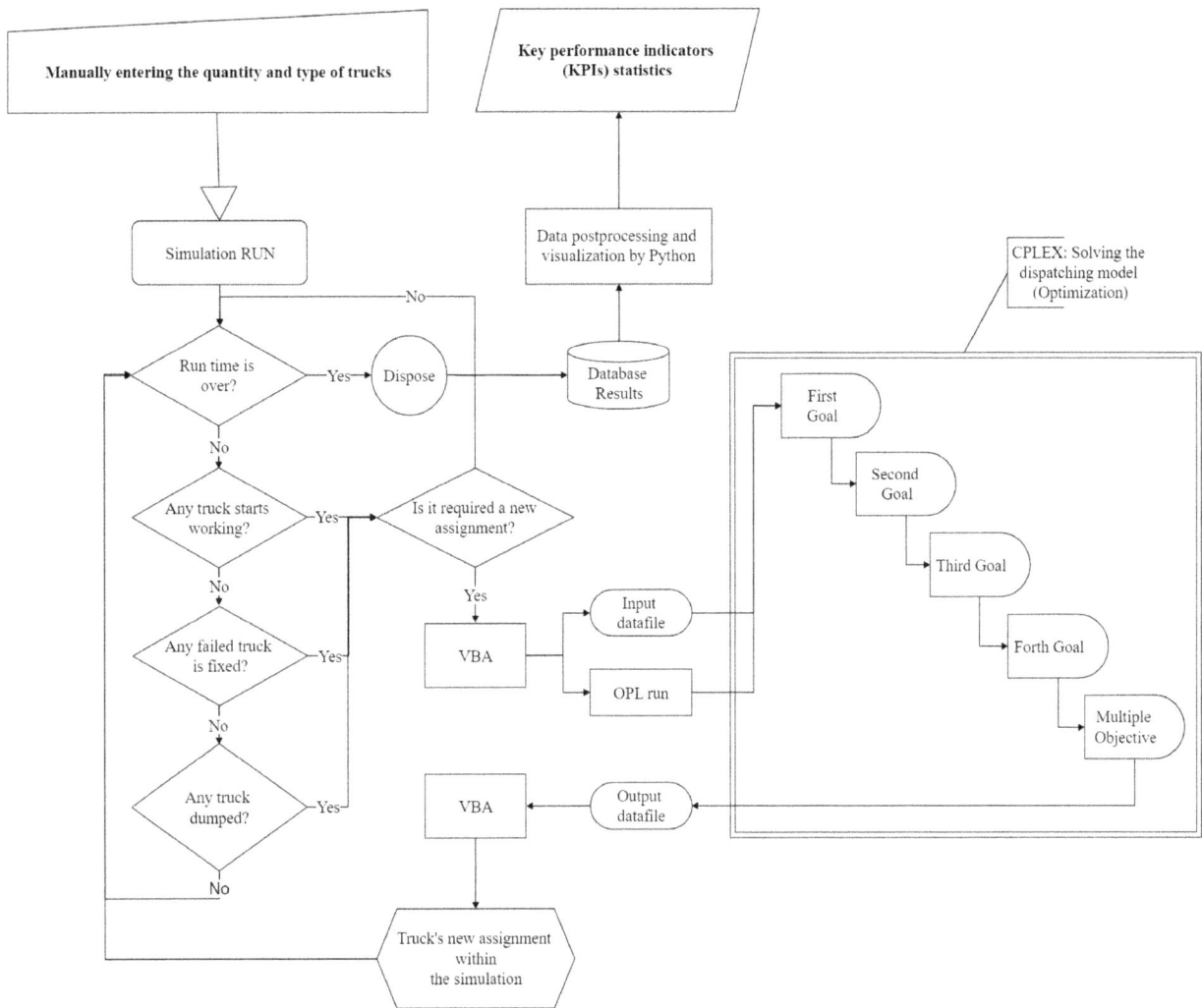

FIG 1 – An overview of the simulation and optimisation integration process.

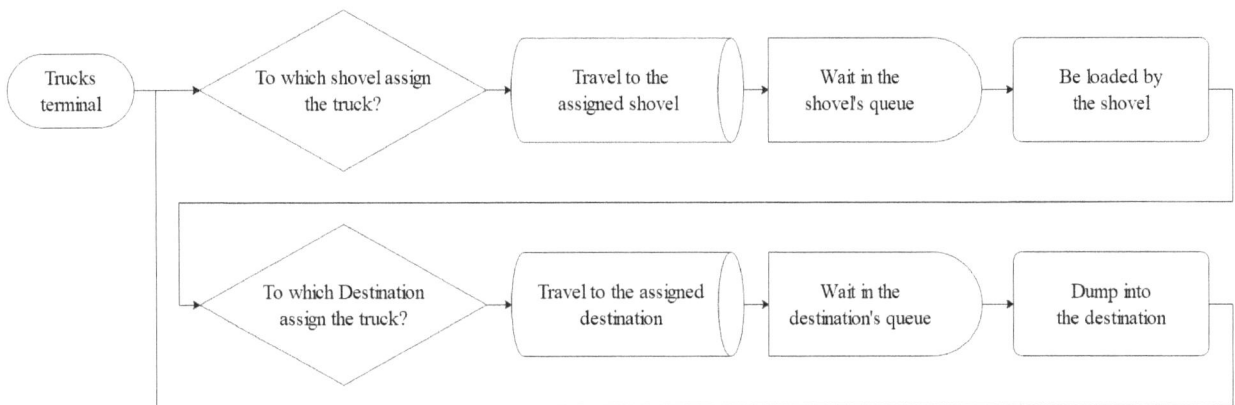

FIG 2 – An active truck's operations.

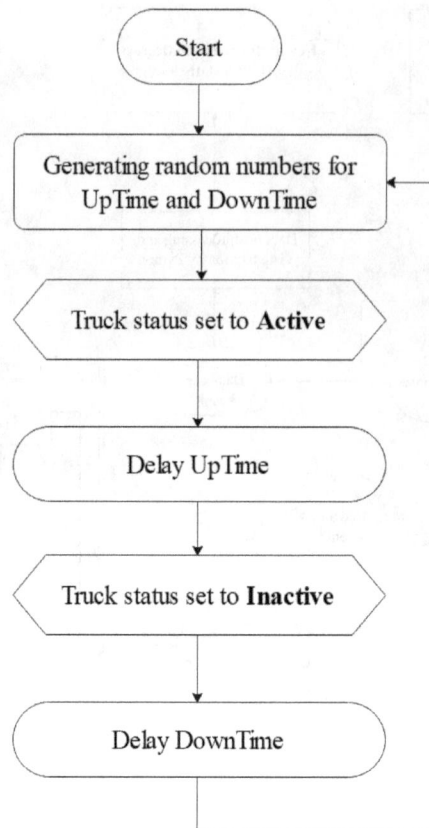

FIG 3 – Truck status flow chart.

Key performance indicators (KPIs)

Variables below are introduced as KPIs in this study. Collectively, these variables wield a substantial influence in evaluating and optimising truck dispatching within mining operations, facilitating improved decision-making, heightened operational efficiency, and enhanced profitability.

- Total ore tonnage production: This metric reflects the overall volume of ore transported to processing plants, directly impacting the mining operation's profitability and productivity.

- Total ore and waste tonnages mined and delivered: Monitoring the total quantities of both ore and waste materials offers insights into mining process efficiency and facilitates resource utilisation optimisation.

- Utilisation of ore and waste shovels: Evaluating the usage of shovels dedicated to ore and waste handling ensures optimal deployment and helps identify potential operational bottlenecks or underutilised equipment.

- Total and average queue times for trucks: Tracking queue times for trucks awaiting loading or unloading provides operational efficiency information and reveals areas where delays might occur.

- Trucks' fuel consumption: Effective fuel management is vital for cost control and environmental sustainability. Monitoring and optimising fuel usage aids in minimising operational expenses and carbon emissions.

- Fuel consumption of a truck per tonne of production: This measure offers insights into trucks' fuel efficiency relative to the amount of material transported. It identifies opportunities for enhancing fuel efficiency and reducing operational costs.

- Ore TPGOH (tonne per gross operating hour): This gauge quantifies mining productivity by calculating extracted ore per equipment operation hour. Higher TPGOH values signify better efficiency and productivity.

- Stripping ratio: This ratio compares waste material removal volume to ore extraction volume, shedding light on the balance between ore production and waste elimination.

- Trucks' availabilities and downtimes: Monitoring truck availability and tracking downtimes identifies potential equipment failures, planned maintenance activities, and minimises mining operation disruptions. Additionally, it can significantly impact TPGOH.

DESIGN OF EXPERIMENTS AND RESULTS

This study includes a case study utilising historical data from the Gol-E-Gohar iron ore open pit mine in Iran to assess the developed framework. The evaluation aims to analyse the performance of different truck fleets in terms of their truck's types and quantities, and also to investigate the impact of truck failures on each fleet scenario. Figure 4 depicts the arrangement of loading and dumping points, as well as the operational road network. At the loading points, there are five active shovels, with two designated for ore extraction and three for waste. At the dumping points, there are three destinations including two processing plants and a waste dump.

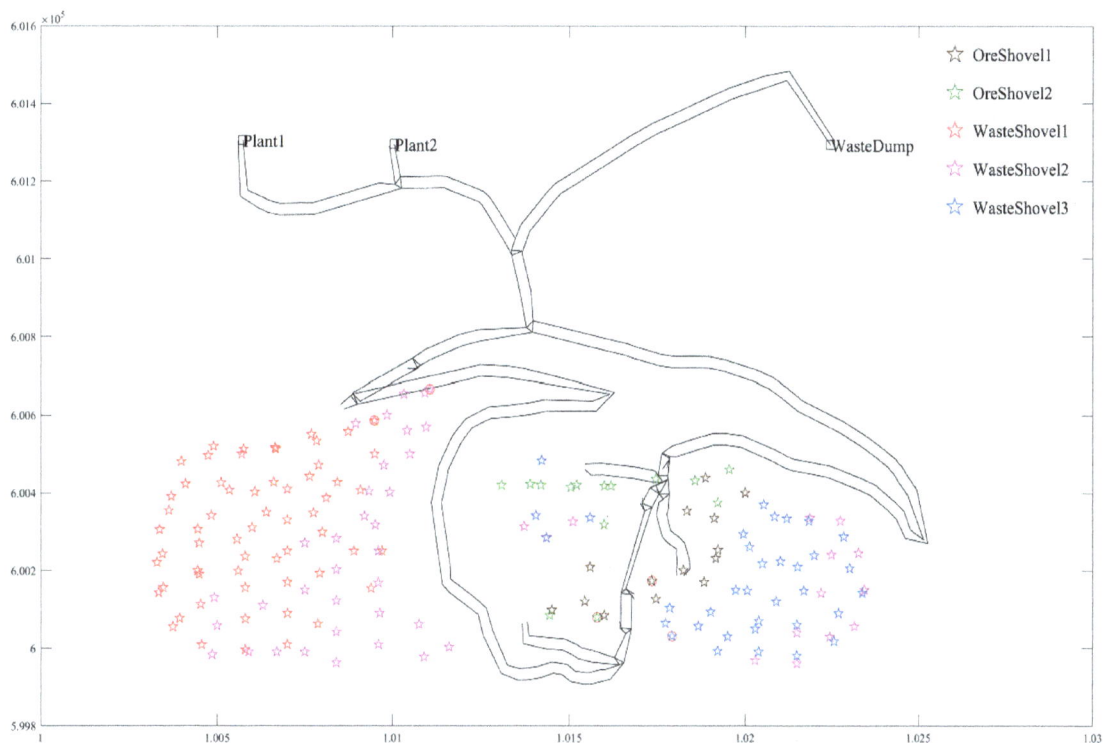

FIG 4 – Gol-E-Gohar iron ore mine network.

The equipment present in the case study comprises Hitachi (HIT) EX2500 and HIT EX5500 shovels, as well as Caterpillar (CAT) 785C and CAT 793C trucks for the transportation operations. The mining activities involve three distinct destinations: two processing plants equipped with two hoppers each, and a waste dump featuring multiple dumping points. The distribution of shovels and trucks to the excavation and dumping locations is outlined in Table 1.

TABLE 1

Equipment distributions.

Origin	Destination	Shovel type	Truck type
Shovel 1	Plant 1 Plant 2	HIT EX2500	CAT 785C CAT 793C
Shovel 2	Plant 1 Plant 2	HIT EX2500	CAT 785C CAT 793C
Shovel 3	Waste dump	HIT EX5500	CAT 785C CAT 793C
Shovel 4	Waste dump	HIT EX5500	CAT 785C CAT 793C
Shovel 5	Waste dump	HIT EX2500	CAT 785C CAT 793C

There is deterministic and stochastic information included in the case study's input data. The Arena Input Analyser tool from Rockwell Automation has been utilised in (Afrapoli, 2018) for the establishment of stochastic input distributions based on historical data. Each processing plant has a feeding rate target (capacity limit) of 2300 t per hr.

The calculation of fuel consumption for individual CAT 785C trucks is performed using Equation 20 obtained from (Dindarloo and Siami-Irdemoosa, 2016):

$$F(\frac{l}{cycle}) = 1.37071 + 0.00483 \times PL + 0.00398 \times LT + 0.00499 \times ES + 0.01471 \times ETR \\ + 0.00278 \times LS + 0.0519 \times LTR \tag{20}$$

F	fuel consumption per cycle (litres)
PL	payload (tonnes)
LT	loading time (seconds)
ES	empty idle time (seconds)
ETR	empty travel time (seconds)
LS	loaded idle time (seconds)
LTR	loaded travel time (seconds)

The fuel consumption for the CAT 793C truck type is determined through Equation 21, where a specific coefficient from the Caterpillar handbook (Caterpillar Inc, 1999) is multiplied with it. This coefficient accounts for factors such as load and haul conditions, road conditions, grades, and rolling resistance. As a result, the CAT 793C's fuel consumption is approximately 1.59 times that of CAT 785C. Thus, Equation 21 outlines the formula utilised to compute the fuel consumption for CAT 793C trucks in each operational cycle.

$$F(\frac{l}{cycle}) = 2.17943 + 0.00768 \times PL + 0.00633 \times LT + 0.00793 \times ES + 0.02339 \times ETR \\ + 0.00442 \times LS + 0.0825 \times LTR \tag{21}$$

The simulation encompassed a duration of ten days, involving 12 hrs of operation per day, with the goal of achieving a satisfactory ore production of 550 000 t for the planned mining operations over this period.

The operational efficiency, productivity, cost-effectiveness, and sustainability of a mining fleet are significantly impacted by the number and varieties of trucks within it. It is crucial to carefully consider the right quantity of trucks, select appropriate truck types, and efficiently manage their dispatch. This plays a vital role in developing a productive and financially viable fleet system. Through analysis of these factors and implementing fleet management approaches, mining companies can streamline operations, boost productivity, and reduce expenses and environmental impacts. This research presents 40 scenarios based on different truck types and quantities. The first nine scenarios focus on a homogenous fleet of CAT 785C trucks. Subsequently, the following eight scenarios involve a homogenous fleet of CAT 793C trucks, each with varying quantities. The remaining scenarios encompass a diverse fleet arrangement, incorporating both CAT 785C and CAT 793C trucks (heterogenous fleet) within the system. Within the Appendix, Table A.1 comprehensively presents the key performance indicators (KPIs) for each distinctive scenario involving diverse combinations of truck types and quantities. Among the scenarios, scenario 6 with a homogenous fleet of 30 CAT 785C trucks, scenario 13 featuring 18 CAT 793C trucks in a homogenous fleet, and scenario 24 that combines 20 CAT 785C trucks with five CAT 793C trucks in a heterogenous fleet, demonstrate the best performance in terms of achieving production goals and reducing fuel usage as shown in Figures 5, 6 and 7.

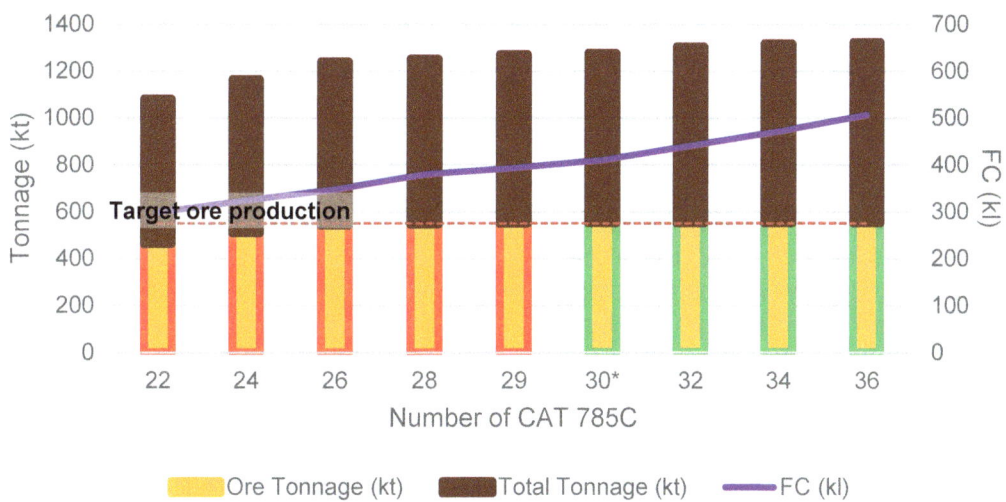

FIG 5 – Production and fuel consumption in homogenous fleet of CAT 785C (FC: fuel consumption).

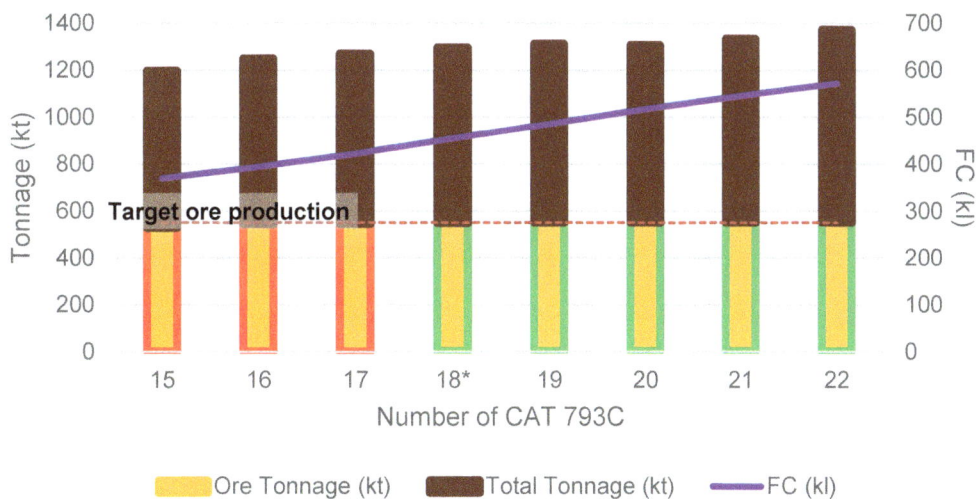

FIG 6 – Production and fuel consumption in homogenous fleet of CAT 793C (FC: fuel consumption).

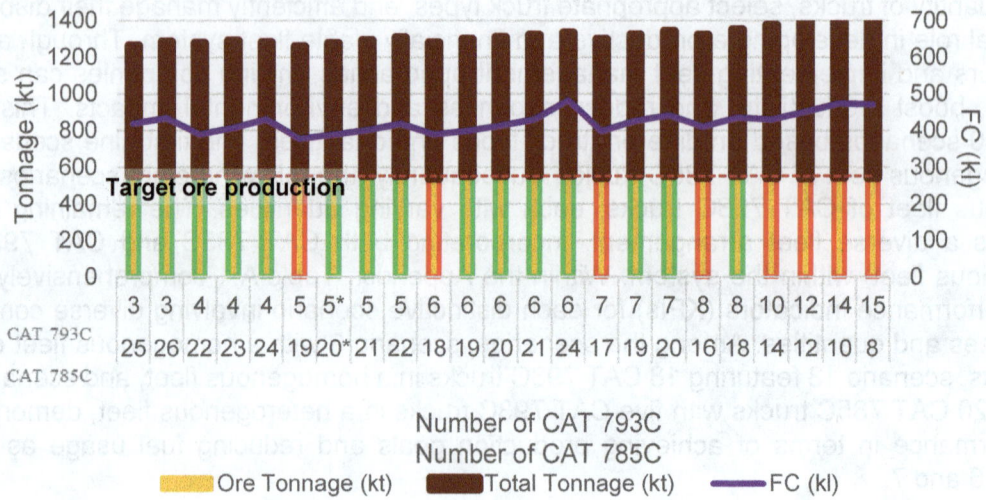

FIG 7 – Production and fuel consumption in heterogenous fleet of CAT 785C and CAT 793C (FC: fuel consumption).

With the central aim being the maximisation of production, scenario 24 emerges as the optimal selection by simultaneously achieving production targets, minimising fuel usage, and reducing carbon emissions. This configuration involves a fleet composition comprising 20 smaller trucks (CAT 785C) and five larger trucks (CAT 793C). This choice effectively strikes a balance between the demand for high productivity and the imperative to curtail fuel consumption and environmental impact, aligning seamlessly with the sustainability objectives outlined in this study. Scenario 6 boasts the highest utilisation of ore and waste shovels, closely followed by scenario 24. Among the three scenarios, scenario 13 demonstrates the lowest utilisation of shovels. When considering average truck queue times, scenario 6 holds the record for the longest, followed by scenario 24. Scenario 13 displays the shortest average truck queue time among the three scenarios. All three scenarios achieve acceptable ore tonnage, with scenario 6 slightly surpassing in ore tonnage, and scenario 13 slightly lagging. Surprisingly, scenario 13 showcases the highest total tonnage, standing notably higher than the other scenarios, followed by scenario 6. In contrast, scenario 24 presents a slightly lower total tonnage when compared to scenario 6. In terms of fuel consumption, a comparison between scenario 6, scenario 13, and the lowest fuel consumption recorded in scenario 24 reveals a clear distinction. Scenario 6 shows a 5.83 per cent higher fuel consumption ratio, while scenario 13 exhibits a significantly higher ratio difference of 17.35 per cent.

Table 2 details the average cycle numbers for each shovel and destination for every truck in scenario 24. Moreover, it provides the OreCycles%, which is the percentage of times a truck transports ore material of the total cycles. Similarly, it presents the WasteCycles%, representing the percentage for waste material transport.

TABLE 2

TABLE 2
Heterogenous fleet cycles of scenario 24.

Truck type	Truck#	SH1	SH2	SH3	SH4	SH5	P1	P2	WD	Ore cycles (%)	Waste cycles (%)
785C	1	84	82.2	50.4	56.6	51.6	82.8	83.4	158.6	51	49
	2	82.8	77.8	54.4	58.8	55.4	79.4	81.2	168.6	49	51
	3	75.4	78.4	60.4	59.8	58	74.8	79	178.2	46	54
	4	79.8	76.4	60.8	63.4	53	81.2	75	177.2	47	53
	5	80.6	74.6	61.6	67	52.6	75.6	79.6	181.2	46	54
	6	81.6	65.4	60.2	76	57.2	68.8	78.2	193.4	43	57
	7	84.2	67.6	65.8	71.8	51.2	69.6	82.2	188.8	45	55
	8	72.6	71.4	71.4	69.2	57.4	66.8	77.2	198	42	58
	9	81.6	65	66.6	83	51	69.4	77.2	200.6	42	58
	10	71.8	68.4	67.2	85.6	56.4	65.2	75	209.2	40	60
	11	72.2	70.8	77.4	82.4	47.6	69.8	73.2	207.4	41	59
	12	73.6	63.4	72	85.4	54	66.4	70.6	211.4	39	61
	13	73.4	65.8	76.6	88	48.2	67	72.2	212.8	40	60
	14	68.4	68	78.2	88.6	53	67.2	69.2	219.8	38	62
	15	71.4	64.6	79.4	82.8	55.2	63.8	72.2	217.4	38	62
	16	77	64.8	84.2	83.2	45.8	65.2	76.6	213.2	40	60
	17	78.4	64	73.2	80.2	50.2	67.6	74.8	203.6	41	59
	18	76.2	76	78.6	75	43.4	74	78.2	197	44	56
	19	70.6	77.6	78	84.4	41.4	67.6	80.6	203.8	42	58
	20	76	78.2	79	75.6	41	74	80.2	195.6	44	56
793C	21	47.8	67.8	48.4	44.8	95.8	67.4	48.2	189	38	62
	22	45.4	60.4	58.4	39	104.2	61.8	44	201.6	34	66
	23	48.2	61	53.8	42.6	98	61.4	47.8	194.4	36	64
	24	52.2	57	54.4	44.4	101.4	58.2	51	200.2	35	65
	25	51.4	61.6	46.8	39.4	103.2	61.8	51.2	189.4	37	63

The findings highlight that CAT 793C trucks transport larger quantities of waste in comparison to CAT 785C trucks. While waste dumping isn't constrained by hourly capacity, plants have specific hourly hopper limits. Trucks with lower capacities offer greater flexibility for transferring ore materials within the system, making them a more suitable choice for assignment to ore shovels. Furthermore, a significant distinction is observed in the assignment of large trucks to waste shovel 5. This difference primarily arises from the fact that shovel 5 boasts a higher digging rate and capacity than the other waste shovels.

In Table 3, the KPIs for the most promising scenarios, accounting for truck failures, are shown. These scenarios, previously discussed without factoring in truck failures. Considering truck failures, scenario 6, featuring a homogeneous CAT 785C truck fleet with 30 trucks, stands out for its remarkable tonnage transportation, production rate, and shovels' utilisation. Although its fuel consumption isn't the lowest, its rate per tonne of production is acceptable. Examining the impact of truck failures, it becomes evident that a fleet with a higher number of smaller trucks holds advantages in achieving the hourly ore production rate. Despite their smaller capacities, the flexibility of smaller trucks enhances their effectiveness Moreover, these smaller trucks (CAT 785C) experience reduced downtime when compared to the larger trucks (CAT 793C), further enhancing their performance in the context of truck failures.

TABLE 3

KPIs of the best scenarios with the trucks failure, and differences' percentages in KPIs.

Scenario	Util ore (%)	Util waste (%)	Average queue time (mins)	Total queue time (hrs)	Fuel consumption (kL)	Ore tonnage (kT)	Total tonnage (kT)	Ore TPGOH (t)	Stripping ratio
6(F)	78.1	53.5	3.4	527	361	531	1222	4421	1.30
13(F)	66.6	43.3	2.3	190	389	512	1180	4264	1.31
24(F)	71.8	50.1	2.9	378	343	502	1171	4181	1.33
(%) Diff 6(F) and 6	-3.8	-4.8	-11.4	-15.3	-12.0	-4.0	-4.4	-4.0	-0.8
(%) Diff 13(F) and 13	-7.0	-10.2	-11.0	-17.4	-14.4	-7.0	-8.9	-7.0	-3.0
(%) Diff 24(F) and 24	-8.8	-7.1	-15.7	-21.8	-11.4	-9.0	-8.3	-9.0	1.5

Scenario 6 stands out with the least variation in KPIs compared to other scenarios. This suggests that incorporating a larger number of smaller trucks in the fleet can minimise production losses in the event of unplanned failures. However, when considering the presence of a stockpile or several stockpiles in the system and a slightly higher number of trucks in both types, a heterogenous fleet still outperforms homogenous fleets.

Figures 8 and 9 illustrate that the daily average TPGOH for scenario 6 and scenario 24, respectively, is significantly impacted by the daily average number of active trucks available in the system. This underscores that a decrease in the active truck count can result in a corresponding reduction in the TPGOH and consequently, total ore production.

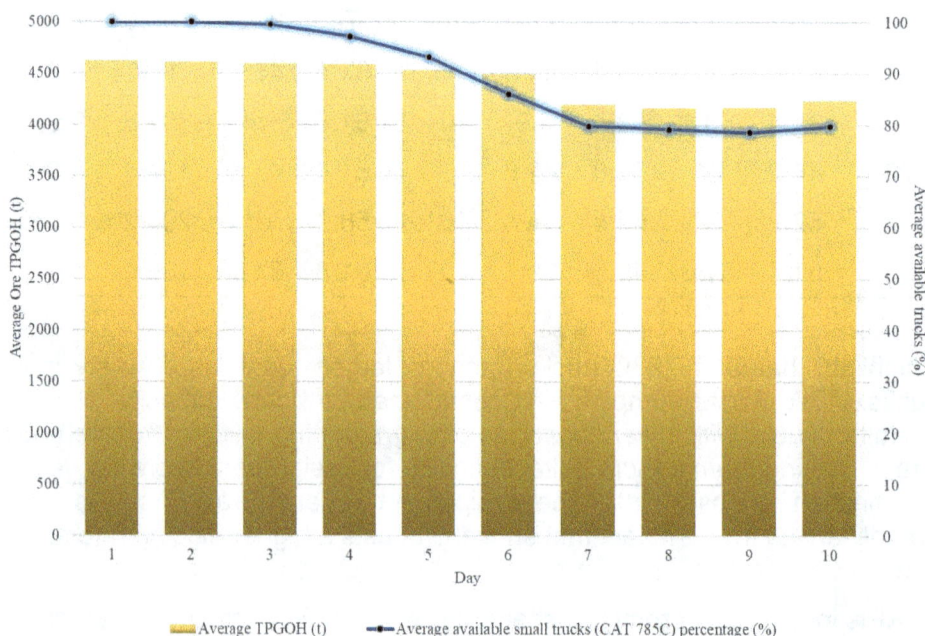

FIG 8 – Scenario 6 (Homogeneous fleet – 30 small trucks) – impact of truck failures on average TPGOH.

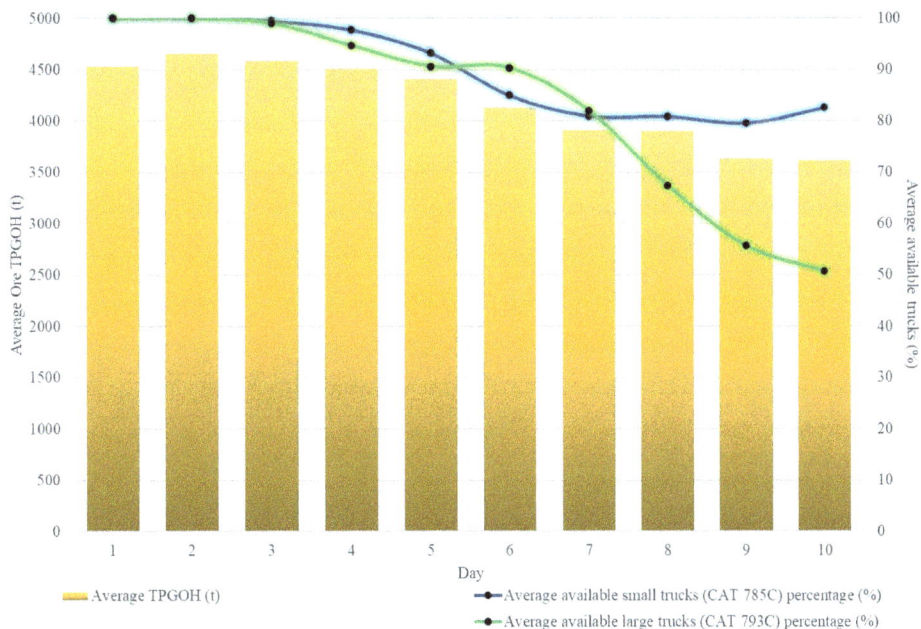

FIG 9 – Scenario 24 (Heterogenous fleet – 25 small and five large trucks) – impact of truck failures on average TPGOH.

CONCLUSIONS

The study primarily focused on determining the optimal quantity and types of trucks required within the system, utilising a developed truck dispatching optimisation model. model's primary objective was to minimise deviations in path flow rates, shovel idle time, truck wait time, and truck fuel consumption. An important contribution of this research was the incorporation of fuel consumption and GHG emissions as criteria for truck fleet selection and sizing. Furthermore, the study enhanced the model's practicality and reliability by considering truck uptime and downtime.

The number and types of trucks within a mining fleet exerted a substantial impact on operational efficiency, productivity, and cost-effectiveness. Apart from an efficient dispatching system, optimising the selection and quantity of available trucks played a pivotal role in establishing a sustainable and productive haulage system for open pit mines. Among the various scenarios explored, scenario 24, a mix of 20 CAT 785C and five CAT 793C trucks in a heterogeneous fleet, displayed optimal performance in meeting production targets and minimising fuel consumption. This configuration effectively balanced high productivity with the imperative to reduce fuel consumption and environmental impact, aligning well with the study's sustainability goals. Scenario 6, consisting of 30 CAT 785C trucks, exhibited a 5.83 per cent higher fuel consumption ratio and 5.45 per cent higher fuel consumption per tonne of production. The CAT 793C trucks transported more waste material per truck due to their higher capacity. Notably, while dumping in the waste disposal area had no hourly capacity restriction, the processing plants had specific hourly hopper capacities. Trucks with lower capacities offered enhanced flexibility in transferring ore materials and were better suited for assignment to ore shovels.

In scenarios accounting for truck failures, a fleet with a higher number of smaller trucks proved advantageous in maintaining the hourly ore production rate due to increased flexibility, despite the smaller average capacity per truck. Additionally, smaller trucks had lower average downtimes compared to larger counterparts, contributing to their superior performance in the context of truck failures. When prioritising fuel consumption per tonne of production, scenario 6 with a homogeneous fleet of 30 small trucks, along with scenarios 20 (comprising 22 small and four large trucks) and 24 (comprising 20 small and five large trucks) with heterogeneous fleets, emerged as the most reliable and efficient choices. However, scenario 6 stood out due to its higher production rate and dispatching flexibility, making it the optimal selection when accounting for unforeseen failures in the model. In conclusion, considering truck failures, fuel consumption, and production rates, scenario 6 with a homogeneous fleet of 30 CAT 785C trucks demonstrated favourable performance. Nonetheless, introducing one or more stockpiles into the system and having a slightly higher number of trucks of both types could potentially lead to a heterogeneous fleet outperforming homogeneous ones. It is important to note that stockpiling wasn't a part of this study's framework. These considerations ensured steady material flow, mitigated truck failure effects, and optimised overall production efficiency in mining operations.

Future research should consider aspects such as truck age, shovel failure impact, and stockpile integration to improve the modelling approach's reliability, realism, and comprehensiveness. Incorporating these factors could lead to more accurate predictions, refined optimisation strategies, and ultimately, greater efficiency and sustainability in mining operations.

REFERENCES

Afrapoli, A M, 2018. A Hybrid Simulation and Optimization Approach towards Truck Dispatching Problem in Surface Mines.

Ataeepour, N and Baafi, E Y, 1999. ARENA simulation model for truck-shovel operation in despatching and non-despatching modes, *International Journal of Surface Mining, Reclamation and Environment*, 13(3):125–129. https://doi.org/10.1080/09208119908944228

Baafi, E Y and Ataeepour, M, 1998. Using ARENA® to simulate truck-shovel operation, *Mineral Resources Engineering*, 7(03):253–266.

Bandopadhyay, S and Venkatasubramanian, P, 1987. Expert systems as decision aid in surface mine equipment selection, *International Journal of Surface Mining, Reclamation and Environment*, 1(2):159–165.

Bazzazi, A A, Osanloo, M and Karimi, B, 2011. Deriving preference order of open pit mines equipment through MADM methods: Application of modified VIKOR method, *Expert Systems with Applications*, 38(3):2550–2556.

Bozorgebrahimi, E, Hall, R A and Blackwell, G H, 2003. Sizing equipment for open pit mining–a review of critical parameters, *Mining Technology*, 112(3):171–179.

Burt, C N and Caccetta, L, 2007. Match factor for heterogeneous truck and loader fleets, *International Journal of Mining, Reclamation and Environment*, 21(4):262–270. https://doi.org/10.1080/17480930701388606

Burt, C N and Caccetta, L, 2014. Equipment selection for surface mining: a review, *Interfaces*, 44(2):143–162.

Burt, C N and Caccetta, L, 2018. *Equipment selection for mining: with case studies* (Springer).

Caterpillar Inc, 1999. *Caterpillar Performance Handbook*, edition 29, Caterpillar Inc.

Chaowasakoo, P, Seppälä, H, Koivo, H and Zhou, Q, 2017. Improving fleet management in mines: The benefit of heterogeneous match factor, *European Journal of Operational Research*, 261(3):1052–1065.

Dindarloo, S R and Siami-Irdemoosa, E, 2016. Determinants of fuel consumption in mining trucks, *Energy*, 112:232–240. https://doi.org/10.1016/j.energy.2016.06.085

Douglas, J and Gunn, J E, 1964. A general formulation of alternating direction methods: Part I, Parabolic and hyperbolic problems, *Numèrische mathèmatik*, 6:428–453.

Edwards, D J, Malekzadeh, H and Yisa, S B, 2001. A linear programming decision tool for selecting the optimum excavator, *Structural Survey*, 19(2):113–120.

Ercelebi, S G and Bascetin, A, 2009. Optimization of shovel-truck system for surface mining, *Journal of the Southern African Institute of Mining and Metallurgy*, 109(7):433–439.

Grodzevich, O and Romanko, O, 2006. *Normalization and other topics in multi-objective optimization Normalization and Other Topics in Multi-Objective Optimization,* Fabien. Available from: <https://www.researchgate.net/publication/233827319>

Knights, P F and Bonates, E J L, 1999. Applications of discrete mine simulation modeling in South America, *International Journal of Surface Mining, Reclamation and Environment*, 13(2):69–72. https://doi.org/10.1080/09208119908944211

Kolonja, B and Mutmansky, J M, 1994. Analysis of truck dispatching strategies for surface mining operations using SIMAN, *Transactions-Society for Mining Metallurgy And Exploration Incorporated*, 296:1845–1851.

Krause, A and Musingwini, C, 2007. Modelling open pit shovel-truck systems using the Machine Repair Model, *Journal of the Southern African Institute of Mining and Metallurgy*, 107(8):469–476.

Li, X and Song, X, 2009. Application of genetic algorithm to optimize the equipment of coal mine, in *2009 International Conference on Computational Intelligence and Software Engineering,* 3 p (IEEE).

Manríquez, F, *et al*, 2019. Discrete event simulation to design open pit mine production policy in the event of snowfall, *International Journal of Mining, Reclamation and Environment*, 33(8):572–588. https://doi.org/10.1080/17480930.2018.1514963

Maran, J and Topuz, E, 1988. Simulation of truck haulage systems in surface mines, *International Journal of Surface Mining, Reclamation and Environment*, 2(1):43–49. https://doi.org/10.1080/09208118808944136

Markeset, T and Kumar, U, 2000. Application of LCC techniques in selection of mining equipment and technology, in *Mine Planning and Equipment Selection 2000*, pp 635–640 (Routledge).

Marzouk, M and Moselhi, O, 2004. Multiobjective optimization of earthmoving operations, *Journal of construction Engineering and Management*, 130(1):105–113.

Mirzaei-Nasirabad, H, *et al*, 2023. An optimization model for the real-time truck dispatching problem in open pit mining operations, *Optimization and Engineering*, pp 1–25.

Mohtasham, M, *et al*, 2021. Truck fleet size selection in open pit mines based on the match factor using a MINLP model, *Mining Technology*, 130(3):159–175.

Moradi-Afrapoli, A and Askari-Nasab, H, 2020. A stochastic integrated simulation and mixed integer linear programming optimisation framework for truck dispatching problem in surface mines, *International Journal of Mining and Mineral Engineering*, 11(4):257–284.

Moradi-Afrapoli, A, Tabesh, M and Askari-Nasab, H, 2019. A stochastic hybrid simulation-optimization approach towards haul fleet sizing in surface mines, *Mining Technology*, 128(1):9–20.

Moradi-Afrapoli, A, Upadhyay, S and Askari-Nasab, H, 2021. Truck dispatching in surface mines-Application of fuzzy linear programming, *Journal of the Southern African Institute of Mining and Metallurgy*, 121(9):505–512.

Moradi-Afrapoli, A, Upadhyay, S P and Askari-Nasab, H, 2022. A nested multiple-objective optimization algorithm for managing production fleets in surface mines, *Engineering Optimization*, pp 1–14.

Nobahar, P, Pourrahimian, Y and Mollaei Koshki, F, 2022. Optimum fleet selection using machine learning algorithms— Case study: Zenouz Kaolin mine, *Mining*, 2(3):528–541.

Que, S, Anani, A and Awuah-Offei, K, 2016. Effect of ignoring input correlation on truck–shovel simulation, *International Journal of Mining, Reclamation and Environment*, 30(5–6):405–421. https://doi.org/10.1080/17480930.2015.1099188

Rist, K, 1961. The solution of a transportation problem by use of a Monte Carlo technique, *Applications for computers and operations research in the minerals industries (APCOM)*, Tucson, US [Preprint].

Salhi, S and Rand, G K, 1993. Incorporating vehicle routing into the vehicle fleet composition problem, *European Journal of Operational Research*, 66(3):313–330. https://doi.org/10.1016/0377-2217(93)90220-H

Samanta, B, Sarkar, B and Mukherjee, S K, 2002. Selection of opencast mining equipment by a multi-criteria decision-making process, *Mining Technology*, 111(2):136–142.

Ta, C H, Ingolfsson, A and Doucette, J, 2013. A linear model for surface mining haul truck allocation incorporating shovel idle probabilities, *European Journal of Operational Research*, 231(3):770–778. https://doi.org/10.1016/j.ejor.2013.06.016

Upadhyay, S P, *et al*, 2021. A simulation-based algorithm for solving surface mines' equipment selection and sizing problem under uncertainty, *CIM Journal*, 12(1):36–46.

Yeganejou, M, *et al*, 2022. Integration of simulation and dispatch modelling to predict fleet productivity: an open pit mining case, *Mining Technology*, 131(2):67–79.

Yuriy, G and Vayenas, N, 2008. Discrete-event simulation of mine equipment systems combined with a reliability assessment model based on genetic algorithms, *International Journal of Mining, Reclamation and Environment*, 22(1):70–83.

Zeng, W, Baafi, E and Walker, D, 2019. A simulation model to study bunching effect of a truck-shovel system, *International Journal of Mining, Reclamation and Environment*, 33(2):102–117.

Zhang, Y, *et al*, 2022. Determination of truck–shovel configuration of open pit mine: a simulation method based on mathematical model, *Sustainability*, 14(19):12338.

APPENDIX

TABLE A.1

KPIs for various types of trucks and number of trucks.

Scenario	Number of trucks 785C	Number of trucks 793C	Utilisation ore (%)	Utilisation waste (%)	Average queue time (mins)	Total queue time (hrs)	Fuel consumption (kl)	Ore tonnage (kt)	Total tonnage (kt)	Ore TPGOH (t)	Stripping ratio
1	22	0	67.84	47.55	2.72	373.29	298.31	461.50	1080.23	3845.87	1.34
2	24	0	74.92	50.84	2.82	416.18	324.12	508.44	1166.06	4237.02	1.29
3	26	0	79.40	54.31	3.12	488.36	346.97	541.50	1241.02	4512.54	1.29
4	28	0	80.20	54.98	3.51	555.85	378.91	545.41	1253.44	4545.11	1.30
5	29	0	80.60	56.15	3.67	591.81	392.78	548.64	1272.59	4572.02	1.32
6*	30	0	81.16	56.22	3.85	622.29	410.04	552.39	1277.51	4603.23	1.31
7	32	0	81.29	58.40	4.16	686.96	440.67	552.36	1304.78	4603.02	1.36
8	34	0	81.18	59.44	4.61	770.65	470.49	552.26	1317.77	4602.15	1.39
9	36	0	81.19	60.14	4.92	825.31	506.12	552.32	1324.50	4602.65	1.40
10	0	15	68.85	43.14	2.13	175.38	369.04	529.10	1196.37	4409.20	1.26
11	0	16	70.85	45.66	2.28	195.68	393.42	544.77	1248.65	4539.73	1.29
12	0	17	70.96	46.96	2.40	213.87	421.65	545.57	1270.10	4546.44	1.33
13*	0	18	71.60	48.20	2.63	229.66	454.54	550.06	1295.28	4583.85	1.35
14	0	19	71.73	49.06	2.66	239.92	485.36	551.08	1312.29	4592.29	1.38
15	0	20	71.95	48.64	2.84	254.45	517.51	552.31	1305.70	4602.59	1.36
16	0	21	71.94	50.61	3.00	274.56	545.11	552.14	1333.44	4601.17	1.42
17	0	22	72.07	52.89	3.18	299.42	572.12	552.27	1372.42	4602.28	1.49
18	25	3	80.05	53.94	3.75	552.06	418.24	552.21	1263.46	4601.73	1.29
19	26	3	80.15	55.59	3.83	577.76	433.62	552.23	1285.20	4601.89	1.33
20	22	4	79.39	54.70	3.59	521.00	388.72	551.17	1277.53	4593.06	1.32

Scenario	Number of trucks 785C	Number of trucks 793C	Utilisation ore (%)	Utilisation waste (%)	Average queue time (mins)	Total queue time (hrs)	Fuel consumption (kl)	Ore tonnage (kt)	Total tonnage (kt)	Ore TPGOH (t)	Stripping ratio
21	23	4	79.70	55.45	3.70	543.67	405.79	551.96	1289.60	4599.71	1.34
22	24	4	79.55	54.89	3.77	549.45	429.54	551.29	1281.59	4594.05	1.32
23	19	5	76.77	53.48	3.14	432.65	376.35	538.51	1258.84	4487.59	1.34
24*	20	5	78.71	53.89	3.45	483.93	387.44	551.52	1276.65	4596.02	1.31
25	21	5	79.12	55.32	3.67	522.96	400.03	551.37	1292.61	4594.76	1.34
26	22	5	79.12	56.31	3.75	543.72	417.23	552.27	1306.81	4602.26	1.37
27	18	6	76.15	54.49	3.17	431.62	388.18	536.69	1279.36	4472.42	1.38
28	19	6	78.55	54.92	3.46	481.03	398.40	551.41	1298.85	4595.12	1.36
29	20	6	78.62	55.95	3.68	518.31	412.42	552.12	1309.17	4601.03	1.37
30	21	6	78.71	56.96	3.82	545.51	428.61	552.37	1322.11	4603.12	1.39
31	24	6	79.14	57.73	4.29	623.72	481.38	552.35	1334.03	4602.92	1.42
32	17	7	76.28	54.88	3.32	445.52	397.33	540.61	1298.82	4505.10	1.40
33	19	7	78.32	56.42	3.70	512.38	424.99	552.04	1322.56	4600.33	1.40
34	20	7	78.52	57.61	3.85	543.51	440.50	552.04	1341.86	4600.34	1.43
35	16	8	76.08	55.82	3.35	442.32	409.77	540.87	1321.28	4507.23	1.44
36	18	8	77.87	57.77	3.80	522.46	433.74	551.64	1351.33	4597.04	1.45
37	14	10	75.36	58.24	3.51	455.19	428.78	541.39	1375.32	4511.56	1.54
38	12	12	74.47	59.72	3.67	462.28	449.87	540.02	1413.49	4500.14	1.62
39	10	14	74.22	60.61	3.78	461.28	472.90	542.37	1449.22	4519.73	1.67
40	8	15	73.56	58.48	3.80	437.64	470.42	541.60	1423.36	4513.37	1.63

Monitoring-based optimisation of horizontal transport in underground mines using machine learning and data integration from NGIMU sensors

W Koperska[1], P Stefaniak[2], A Skoczylas[3] and M Stachowiak[4]

1. Research Specialist, KGHM Cuprum Research and Development Centre Ltd., Wroclaw 53-659, Poland. Email: wioletta.koperska@kghmcuprum.com
2. Head of Analytics Department, KGHM Cuprum Research and Development Centre Ltd., Wroclaw 53-659, Poland. Email: pawel.stefaniak@kghmcuprum.com
3. Research Specialist, KGHM Cuprum Research and Development Centre Ltd., Wroclaw 53-659, Poland. Email: artur.skoczylas@kghmcuprum.com
4. Research Specialist, KGHM Cuprum Research and Development Centre Ltd., Wroclaw 53-659, Poland. Email: maria.stachowiak@kghmcuprum.com

ABSTRACT

Horizontal transport is a critical component of the production process in underground mining operations. With the ever-expanding reach of underground mines, continuous adaptation of equipment and work organisation to dynamically changing operational conditions is essential for optimal equipment usage. This includes selecting appropriate configurations for wheeled loading and haulage machinery responsible for transporting material from production faces to transfer points. Effective process optimisation requires tools that not only facilitate an understanding of current operations but also enable the adjustment of workflows to ensure the targeted tonnage is transported within a specified time frame, while minimising energy expenditure. Key challenges to efficiency, aside from random operational incidents, include equipment downtime caused by traffic bottlenecks on haul roads and machines idling while waiting for loading. Data from existing SCADA systems and IoT sensors such as NGIMU proposed in the article provide essential information, from which equipment utilisation metrics and numerical tracking of material flow within the transport network can be extracted. This article presents an example of the use of data recorded from NGIMU sensors located on machines and applying various techniques, including machine learning to address these challenges. Additionally, the use of reporting tools in a GIS environment is demonstrated to further enhance process visualisation and decision-making.

INTRODUCTION

Horizontal transport in mines plays a key role in ensuring the efficiency of the mining process and provides great opportunities for optimisation. Proper organisation of the transport of ore affects efficiency, operating costs, and work safety. The basis of this system is loading and haulage machines, in particular haulage trucks, which are responsible for the transport of ore underground. However, their suboptimal use or poor organisation can lead to production losses. Therefore, the key issue of optimising horizontal transport is tracking the course of work of these machines. Continuous monitoring of their movement, time and number of work cycles, driving speed, as well as detection of downtimes allows for the identification of problem areas and, consequently, the implementation of improvements. The quality of the road on which the machines move may also be a key issue. Poorer road quality may lead to the need to reduce speed and also affects the comfort of drivers, and sometimes even their safety (Stefaniak *et al*, 2022).

Horizontal transport optimisation is a very broad topic that can be considered at many different levels. At an early stage, optimisation can concern design. For example, in the work of Roumpos *et al* (2014), the important role of optimising the location of the distribution point of the belt conveyor system in surface mines was presented. In order to solve this problem, the authors proposed a computer model based on minimising transportation costs. Optimisation is also usually closely related to the energy issue, which can be considered in the context of individual machines. The article by Feng and Dong (2020) focuses on the development of an energy management strategy for hybrid electric mining trucks. The idea is to balance fuel savings with battery life, which aims to minimise the total operating costs of the vehicle. The authors introduce a method based on the combination of artificial neural networks and fuzzy logic. A very important issue related to transportation optimisation is the appropriate management of the machine fleet. Many studies on this topic can be found in the literature. For example, Mena *et al* (2013), proposed a simulator to optimally allocate trucks in the truck-shovel system by route according to their operational efficiency in an open pit mine. The goal of the optimisation is to maximise the fleet productivity. A similar topic was taken up by Ta *et al* (2005), where the optimisation of truck allocation based on the stochastic optimisation approach was proposed. The possibility of uncertain parameters, such as the truck cycle time and its loading was assumed. In turn, Park, Choi and Park (2016) focussed on the optimisation of the number of trucks used for transport. For this purpose, a program was developed to simulate the transport of trucks and loaders in an underground mine.

However, the key to any optimisation is the proper understanding of the system, which can be influenced by various factors. Only the identification of possible sources of problems and areas for optimisation will allow their implementation. That is why monitoring systems play a crucial role in this issue. In Aguirre-Jofré *et al* (2021) a system based on inexpensive IoT was proposed for automatic collection of information about trucks and surface mining shovels, such as: speed, number of work cycles for trucks, loading time, and machine positioning. The authors suggest that monitoring these tasks allows for optimising the performance of these machines because it is possible, for example, to maximise productivity while reducing the number of required devices. GPS technology is usually used for vehicle monitoring, which allows for obtaining accurate results. Many studies have been conducted on this application in open pit mines as well (Gu *et al*, 2008; Chaowasakoo, Leelasukseree and Wongsurawat, 2014; Peck and Hendricks, 1997). The underground mine is a specific case, where this technology cannot be used, which requires the development of a completely different method, which will be reliable in underground conditions.

In this work, a system is proposed to monitor work progress based on a single NGIMU (Next Generation Inertial Measurement Unit) sensor placed on a vehicle that is performing the haulage of ore. However, inertial data require appropriate processing and development of new methods to extract valuable information about the transport performed. For this reason, a number of methods have been proposed that enable: detection of machine work cycles, estimation of driving speed, assessment of road quality, route segmentation, and determination of the 3D trajectory along which the machine moved. The results section presents sample analyses based on the determined parameters. The methods were developed based on data measured for three haul trucks during their standard operation at the Hellas Gold mine in Greece. The experiments were performed as part of the NetHelix project (2022). The project aims to introduce improvements in the mining sector, including new technologies, automation, and optimisation while minimising the negative impact on the environment.

MATERIAL AND METHODS

One of the basic elements of horizontal transport in the mine is the loading and haulage transport performed by loaders and trucks. These machines perform cyclical work, transporting mined material from point A to B. The course of this process has a significant impact on the course of the entire production. In order to optimise this part of horizontal transport, it is necessary to precisely track the course of work of loaders and trucks. In this part, using the example of trucks, various methods are presented, which are elements aimed at replicating the work of machines as accurately as possible. All methods were developed using only one sensor placed on the machine, namely the NGIMU sensor (Figure 1). The sensor, placed in a protective housing, was mounted on the front of the machine. Three vehicles were used for the experiment, for which data was collected during their standard work. The sensor enables data recording from a three-axis gyroscope and accelerometer at a sampling rate of 400 Hz.

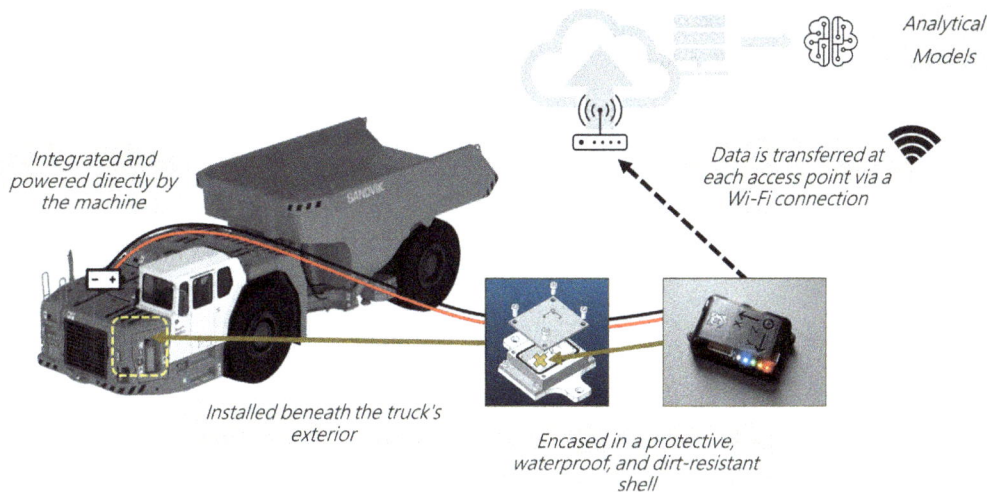

FIG 1 – Schematic diagram of the NGIMU sensor application for monitoring the operation of a haul truck.

Cycle detection

The haul truck performs cyclical work of transporting ore, which can be divided into four operations: loading the box (performed by a loader), driving with a full box to the unloading point, unloading, and returning with an empty box. Detecting the moments when the machine performs these operations is the basis for monitoring the course of the machine's work. In that it is possible to track the basic tasks performed by the machine and their duration. It also provides information on the number of cycles performed during the work shift, which can also be converted into the average value of the transported ore. It was noticed that the raw accelerometer and gyroscope signals have different characteristics for individual operations. However, manual development of rules classifying operations would be difficult and laborious, and other variables such as the route or machine properties could cause the need for continuous updates. Therefore, it was decided to train a neural network model that would learn to recognise machine operations (Skoczylas *et al*, 2025). For this purpose, a neural network architecture based on autoencoders and a modified VGG16 structure was used, and the input parameters were accelerometer and gyroscope data for all three axes. After developing the structure that gives the most accurate result, a validation method was also developed that corrects obvious errors resulting from, for example, incorrect order of operations.

Speed estimation

One of the important parameters regarding the operation of loading and hauling machines, such as trucks, is the driving speed. Theoretically, speed can be determined by integrating the linear acceleration measured by the accelerometer. However, in practice, the result obtained in this way is inaccurate due to noise and drift. In order to obtain a better result, it was decided to use the control points where the speed estimation can be performed in a different way than integrating the accelerometer value (Ustun and Cetin, 2019; Yu *et al*, 2015). At the control points, the speed determined by the additional method is compared with that obtained as a result of integrating the linear acceleration. The estimation error is removed, and the values of the estimated speed between the control points are rescaled in relation to the detected trend. The following were used as control points:

- machine stop moments – speed equal to 0
- machine turn moments – speed determined using values from the gyroscope and accelerometer as centripetal acceleration divided by angular velocity.

The detection of the above moments is also performed based on NGIMU data.

Stops are noticeable in the accelerometer signals as fragments with low value variability. Based on this relationship, it was decided to divide the resultant acceleration signal into fragments at the change locations using the PELT (Pruned Exact Linear Time) method (Killick, Fearnhead and Eckley, 2012). This method aims to optimally divide the time series into fragments with different characteristics, in this case different variance. Then, the IQR (interquartile range) statistics are calculated for each fragment. Using a set threshold, the fragments for which the IQR value is the lowest are separated. These fragments are the moments when the machine stops (Cetin, Ustun and Sahin, 2016).

A very similar method was used to detect the moments of machine turning. The turn can be detected from the gyroscope signal for the vertical axis of the vehicle, usually marked as Z. The integral of this signal is the machine turning angle (yaw). A sudden change in yaw is the moment when the machine turns. Detection of these moments is again possible by dividing the signal into fragments using the PELT method and selecting those with the highest IQR values using the set threshold.

The above two techniques in themselves already give an interesting result, as they provide information on how often the machine stopped (which translates into an increase in the driving time) and how many turning manoeuvres it had to perform.

The last step is to use the above control points to estimate the target machine speed. At the control points, the original velocity determined by integrating the accelerometer values is converted to the velocity indicated by the control point. The original velocity values are also rescaled relative to the error trend determined for each of the two consecutive control points.

Scaling the vibration data

Vibrations recorded by an inertial sensor can be influenced by many different factors. For example, the vibrations measured in the Z-axis are influenced not only by the road quality but also by the travel speed and the current operational status (ie whether the machine is loaded or unloaded). In this case, it is not possible to evaluate road quality (described in the next section) directly from vibrations without taking into account these dependencies. Typically, vibrations and dynamic overloads are significantly higher when the machine is traveling unloaded and at lower speeds.

In order to ensure the analyses are accurate and repeatable, particularly for road quality detection, a method for scaling the vibrations was developed. After several attempts, a Z-Score-based approach was selected as the most appropriate. The following scaling method was applied:

- Vibrations were categorised based on the traveling speed, rounded to the nearest integer. For example, the '15 km/h' category was assigned to all vibration measurements recorded when the machine's speed was in the range of 14.51–15.50 km/h.

- For each speed category, the unloaded vibration readings were normalised to match the levels of the loaded readings within the same speed category.

$$u_i = \left(\frac{u_i - \bar{u}}{\sigma_u}\right) \cdot \sigma_l + \bar{l}$$

- where u_i is the accelerometer reading for the unloaded machine at the current speed category, $\bar{u}$ and σ_u are the mean and standard deviation of unloaded readings, and σ_l and $\bar{l}$ are the mean and standard deviation of loaded readings for the same speed category.

- For each speed group, the accelerometer readings (both for loaded and previously normalised unloaded conditions) were normalised to match the vibration levels of a target. The target was defined as the accelerometer readings recorded when the machine was traveling at a lower speed in a loaded state.

$$x_i = \left(\frac{x_i - \bar{x}}{\sigma_x}\right) \cdot \sigma_t + \bar{t}$$

- where x_i is the accelerometer reading for the machine at the current speed category, $\bar{x}$ and x_u are the mean and standard deviation of loaded and normalised unloaded readings at the current speed category, and σ_t and $\bar{t}$ are the mean and standard deviation of loaded readings for the target speed category.

The optimal target speed category was around 3 km/h. Among all the tested approaches, the Z-score method proved most effective, as it allowed data normalisation without the large modifications observed with methods like quantile mapping.

The normalisation of accelerometer readings enhanced subsequent algorithms by improving the overall comparison between measurements taken during different operational stages. This approach was tested by analysing the cross-correlation of signals recorded on the same roads at different times, under varying operational conditions and speeds. The normalisation increased the mean similarity between signals by approximately 10 per cent and, more importantly, reduced the spread of values, making subsequent analyses more applicable.

Road quality assessment

Another important factor in optimising the transport of mining machines is the quality of the road on which they move. Poorer road quality may require moving at a lower speed and affect the increased frequency of machine failures. In addition, it also affects the comfort of employees and sometimes their safety. For this reason, monitoring the condition of the road surface can significantly contribute to the optimisation of production. Road quality is related to the vibration of the machine, which is recorded by inertial sensors, in particular the Z-axis accelerometer (axis directed vertically to the vehicle). However, as mentioned earlier, the quality of the road is not the only factor influencing the vibration of the machine, these are also the driving speed, machine vibrations, and loading (full and empty box). All these factors mean that despite moving along the same route, machines can record a different level of vibration. The proposed method aims to indicate a three-level of road quality regardless of the above factors. For this reason, as the first step in removing the dependence on some factors, the Z-axis accelerometer data were normalised using the method described in the previous section. The remaining influence of the factors is eliminated by the developed road quality assessment algorithm. This is an extension of the method described in Skoczylas *et al* (2021). The method is based on the analysis of the signal in the frequency domain. First, the spectrum (fast Fourier transform – FFT) is determined for linear acceleration in the Z-axis measured by the accelerometer for moments when the machine is idling. These moments are determined by the same method as during the speed estimation. This will be used to eliminate the influence of machine vibrations. Successively, all signal fragments when the machine was moving are divided into groups according to two criteria:

- load of the box – driving with an empty or full box
- driving speed according to fixed ranges.

An independent road quality classification model is created for each of the groups prepared in this way. The model consists in dividing the signal into several-second windows, for which the spectrum of the Z-axis accelerometer signal is determined, taking into account the subtraction of the spectrum obtained for idle running. For this, a weighted average is determined (higher weight for higher frequencies). The parameter prepared in this way is divided into three groups using the K-means method: good, medium, and bad road quality. The K-means algorithm, as an unsupervised machine learning algorithm, is based on the distance between centroids and allows dividing data into clusters without previously defined rules and labels.

Route segmentation

In order to further improve the monitoring of the machine's route, it was decided to develop a method for segmenting the route into sections. This will allow, for example, to recognise problematic sections, including those with poorer road quality, and will also enable comparison of segments with respect to subsequent cyclic driving. In the case of a route containing a large number of turns, the best solution is to perform the division when they occur. The method of detecting turns was discussed in the speed estimation. However, if the segments are always to correspond to the same road sections, the division at the moment of the turn is not sufficient. This results from the fact that each driving, even on the same route, may be slightly different, additional manoeuvres may occur or some of the turns may not be detected. For this reason, after performing the division, the segments must still be matched to each other. The dynamic time warping method – DTW (Berndt and Clifford, 1994) was used for this purpose. The DTW method, by appropriately shortening or extending two signals, is able to indicate their best match to each other, and thus the similarity between the signals. The matching was based on two variables: the direction of the turn at the end of the segment and the length of the route in each segment.

3D trajectory

Gyroscope data combined with the determined driving speed also make it possible to determine the exact 3D trajectory of the machine's route. This will also allow for visualisation of the results obtained from the above methods directly on the route. For this purpose, two angles are required: yaw and pitch, which are determined by integrating the gyroscope values. However, this can only be done directly from the data when the axes of the mounted sensor coincide with the axes of the vehicle. Otherwise, it is necessary to correct the orientation in order to obtain the proper result. During the analysed experiment, the Z-axis of the sensor was in line with the vertical axis of the vehicle, but the X and Y-axes did not coincide with the axes along and across the machine. Correction of the angles is possible even without knowing the exact orientation of the sensor using the Madgwick orientation filter (Madgwick, 2010). The method is based on the simultaneous use of gyroscope and accelerometer data to determine the correct angles. The Madgwick filter also largely removes the gyroscope drift. Data prepared in this way allows for localisation in 3D space:

$$X = \int v \cdot \cos(\alpha_{yaw}) \cdot \cos(\alpha_{pitch})$$

$$Y = \int v \cdot \sin(\alpha_{yaw}) \cdot \cos(\alpha_{pitch})$$

$$Z = \int v \cdot \sin(\alpha_{pitch})$$

Where:

v	is speed
α_{yaw}	is yaw signal
α_{pitch}	is pitch signal

RESULTS

The described methods contain a number of procedures, by means of which specific results were obtained from the input data from the accelerometer and gyroscope, indicating the course of the machine's operation. An example of a set of input and output data is shown in Figure 2, where (a) is a fragment of the raw signal of the Z-axis accelerometer with the road quality classification marked with appropriate colours, and (b) is the same fragment for the X and Y-axis accelerometers with a colour corresponding to the estimated travel speed. As can be seen, not all dependencies are directly visible in the raw signals. There are moments when the machine stops, where the acceleration in the Z-axis is much lower. Better road quality usually correlates with lower values of acceleration in the Z-axis, but not always, because the dependence is also on the speed. The travel speed is not directly visible in the raw signals.

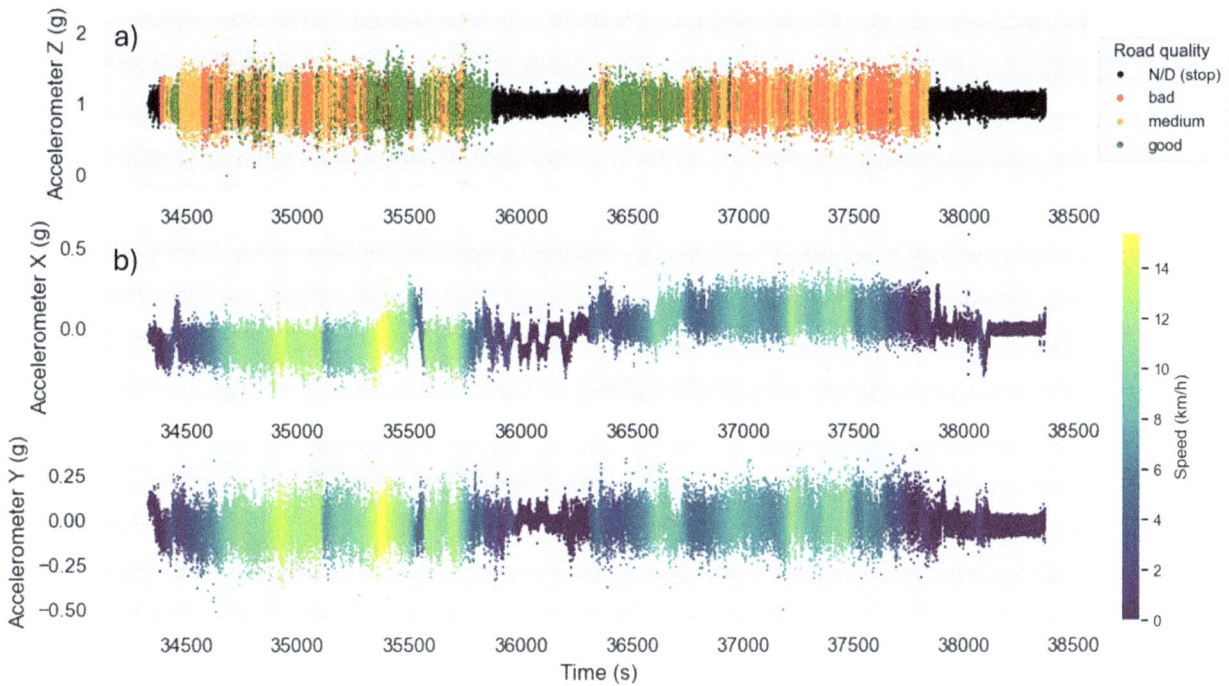

FIG 2 – Fragment of raw accelerometer signals covering one working cycle of the haul truck with: (a) road quality classification for the Z-axis; and (b) travel speed estimation for the X and Y-axes.

The methods described in the previous section provide a lot of new information on the operation of the haul truck, the analysis of which can significantly contribute to the optimisation of this part of horizontal transport. First of all, it is possible to precisely monitor the course of transport performed by the truck and statistically compare individual road segments. Selected analyses are presented below, based exclusively on the results obtained with the described methods using NGIMU data.

Parameters such as road quality can be presented on the three-dimensional trajectory of the machine's route using appropriate colours. Such visualisation is presented in Figure 3, taking into account two different perspectives of the route, which in practice can be interactive. In the case of road quality analysis (Figure 3a), it is easy to recognise where there are better and worse sections. In the case of speed analysis (Figure 3b), it is possible to recognise where the machine can drive at a higher speed and where it must slow down (for example on some sharper bends).

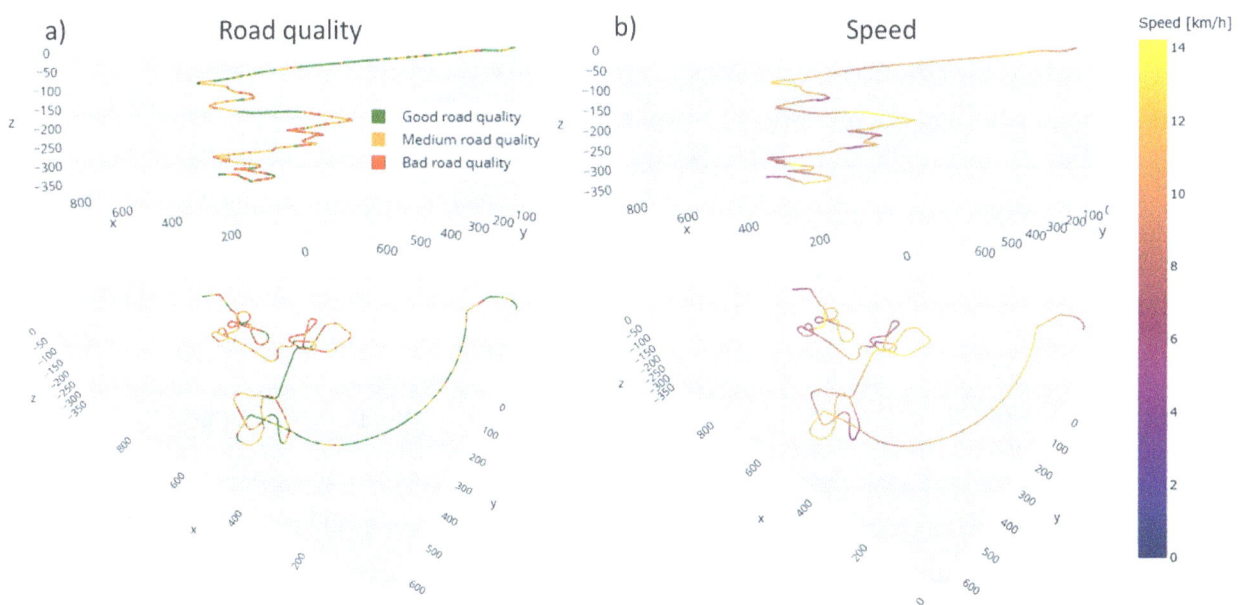

FIG 3 – Road quality (a); and driving speed (b) presented on a 3D trajectory.

It is also possible to perform an interesting comparative analysis of individual road segments, taking into account not only one driving but, for example, the entire work shift or a longer period. Such a comparison for the average speed is presented in Figure 4. It can be seen that some segments (especially those at the beginning and end of the route) are characterised by a lower average speed. There is also a visible difference between driving with an empty and full box. Driving with a full box is usually characterised by a lower speed. On the other hand, for driving with an empty box, there are sometimes values close to zero, which results from the necessity of stopping the machine. Therefore, it is additionally possible to recognise in which segments the machine has to make stops.

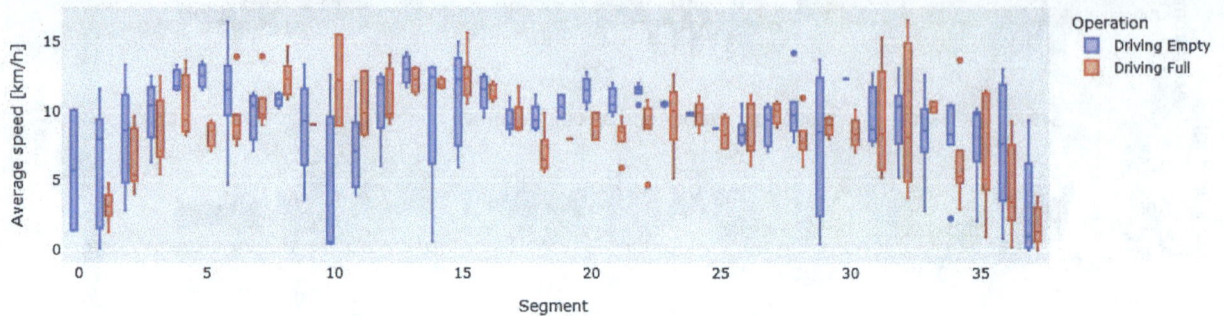

FIG 4 – Comparison of route segments by average machine speed.

Each of the route segments can also be analysed separately. For this purpose, it is worth using an interactive dashboard, in which the road segment is selected directly from the trajectory. The visualisation of this idea is presented in Figure 5. A road segment can be described with statistics such as driving time, average speed, road length, average slope angle, percentage value of good, medium and bad road quality, and downtime.

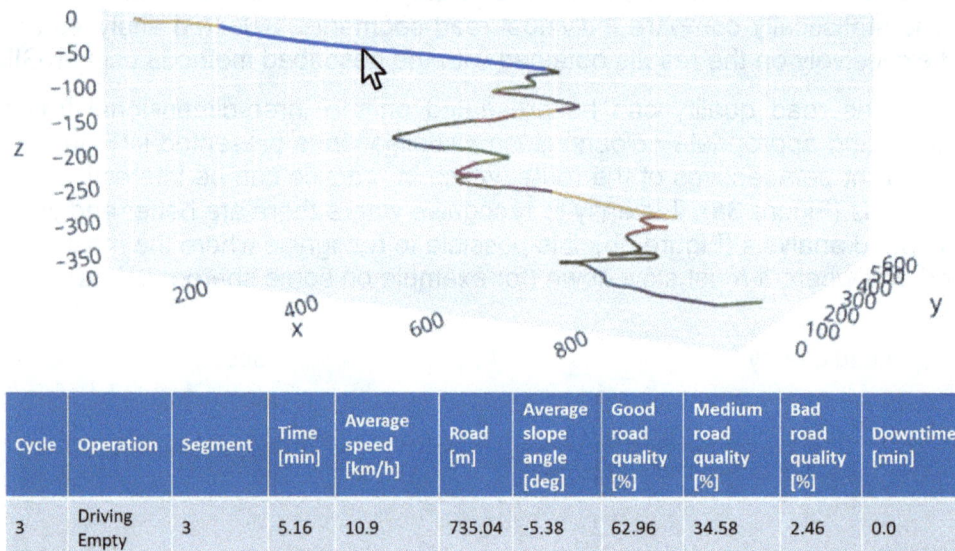

Cycle	Operation	Segment	Time [min]	Average speed [km/h]	Road [m]	Average slope angle [deg]	Good road quality [%]	Medium road quality [%]	Bad road quality [%]	Downtime [min]
3	Driving Empty	3	5.16	10.9	735.04	-5.38	62.96	34.58	2.46	0.0

FIG 5 – Example of an interactive module for analysing the parameters of road segments.

CONCLUSION

The article presents the possibility of using only one NGIMU sensor to monitor the course of the truck work and the quality of the road on which it moves. Methods have been developed to determine the machine's work cycles, travel speed, road quality assessment, route segmentation and 3D trajectory determination. These methods provide very important information about the course of the machine's work and can be used for various types of statistics, analyses and comparisons. Examples of analytical applications of the obtained results have been presented. Such monitoring is the basis for the optimisation of horizontal transport. It allows for a precise understanding of the course of transport and the factors that affect it, which would not be visible without monitoring. Appropriate analyses of historical data and comparisons can contribute to better management and improve decision-making. The basic parameter for monitoring the work of a haul truck is the number of haulage cycles performed during a work shift, which is the basis for assessing the efficiency of the machine, and thus for indicating cases requiring optimisation. Tracking the trajectory of the road and the speed of the vehicle can indicate critical points where travel is difficult and affects the duration of the operation. Here, the quality of the road also plays a key role, the recognition of which can indicate locations that need to be improved. The study of the duration of the machine's unloading and loading operations can provide information on whether these processes are fully optimal or whether a way to shorten their duration should be found. This proves that analyses based on the monitoring of loading and hauling machines are an indirect but key element of the optimisation of horizontal transport in an underground mine.

ACKNOWLEDGEMENTS

This work is a part of the project which has received funding from the European Union's Horizon research and innovation programme under grant agreement No 101092365.

REFERENCES

Aguirre-Jofré, H, Eyre, M, Valerio, S and Vogt, D, 2021. Low-cost internet of things (IoT) for monitoring and optimising mining small-scale trucks and surface mining shovels, *Automation in Construction*, 131:103918.

Berndt, D J and Clifford, J, 1994, July. Using dynamic time warping to find patterns in time series, in *Proceedings of the 3rd International Conference on Knowledge Discovery and Data Mining*, pp 359–370.

Cetin, M, Ustun, I and Sahin, O, 2016. Classification Algorithms for Detecting Vehicle Stops from Smartphone Accelerometer Data, *Transportation Research Board Washington DC,* 2:3.

Chaowasakoo, P, Leelasukseree, C and Wongsurawat, W, 2014. Introducing GPS in fleet management of a mine: Impact on hauling cycle time and hauling capacity, *International Journal of Technology Intelligence and Planning*, 10(1):49–66.

Feng, Y and Dong, Z, 2020. Optimal energy management with balanced fuel economy and battery life for large hybrid electric mining truck, *Journal of Power Sources*, 454:227948.

Gu, Q H, Lu, C W, Li, F B and Wan, C Y, 2008. Monitoring dispatch information system of trucks and shovels in an open pit based on GIS/GPS/GPRS, *Journal of China University of Mining and Technology*, 18(2):288–292.

Killick, R, Fearnhead, P and Eckley, I A, 2012. Optimal detection of changepoints with a linear computational cost, *Journal of the American Statistical Association*, 107(500):1590–1598.

Madgwick, S, 2010. An efficient orientation filter for inertial and inertial/magnetic sensor arrays, *Report x-io and University of Bristol (UK)*, 25:113–118.

Mena, R, Zio, E, Kristjanpoller, F and Arata, A, 2013. Availability-based simulation and optimisation modeling framework for open-pit mine truck allocation under dynamic constraints, *International Journal of Mining Science and Technology*, 23(1):113–119.

NetHelix, 2022. Intelligent Digital Toolbox Towards More Sustainable and Safer Extraction of Mineral Resources (Grant Agreement No. 101092365), European Commission Horizon, Attica, Greece.

Park, S, Choi, Y and Park, H S, 2016. Optimisation of truck-loader haulage systems in an underground mine using simulation methods, *Geosystem Engineering*, 19(5):222–231.

Peck, J and Hendricks, C, 1997. Applications of GPS-based navigation systems on mobile mining equipment in open-pit mines, *CIM Bulletin*, 90.

Roumpos, C, Partsinevelos, P, Agioutantis, Z, Makantasis, K and Vlachou, A, 2014. The optimal location of the distribution point of the belt conveyor system in continuous surface mining operations, *Simulation Modelling Practice and Theory*, 47:19–27.

Skoczylas, A, Stefaniak, P, Anufriiev, S and Jachnik, B, 2021. Road Quality Classification Adaptive to Vehicle Speed Based on Driving Data from Heavy Duty Mining Vehicles, in *International Conference on Intelligent Computing and Optimisation*, pp 777–787 (Cham: Springer International Publishing).

Skoczylas, A, Stefaniak, P, Anufriiev, S and Koperska, W, 2025. Deep Learning for Ore Haulage Monitoring: Vibrational Analysis Using a VGG16 Network, *IEEE Access*.

Stefaniak, P, Stachowiak, M, Koperska, W, Skoczylas, A and Śliwiński, P, 2022. Application of wearable computer and ASR technology in an underground mine to support mine supervision of the heavy machinery chamber, *Sensors*, 22(19):7628.

Ta, C H, Kresta, J V, Forbes, J F and Marquez, H J, 2005. A stochastic optimisation approach to mine truck allocation, *International Journal of Surface Mining, Reclamation and Environment*, 19(3):162–175.

Ustun, I and Cetin, M, 2019. Speed estimation using smartphone accelerometer data, *Transportation Research Record*, 2673(3):65–73.

Yu, J, Zhu, H, Han, H, Chen, Y J, Yang, J, Zhu, Y, Chen, Z, Xue, G and Li, M, 2015. Senspeed: Sensing driving conditions to estimate vehicle speed in urban environments, *IEEE Transactions on Mobile Computing*, 15(1):202–216.

A review of artificial intelligence applications in monitoring slope stability for open pit mines

M C I Madahana[1] and J E D Ekoru[2,3]

1. Lecturer, School of Mining Engineering, University of the Witwatersrand, Braamfontein 2000 Johannesburg, South Africa. Email: milka.madahana@wits.ac.za
2. Lecturer, Academic Development Unit (ADU), University of the Witwatersrand, Braamfontein 2000 Johannesburg, South Africa.
3. School of Electrical and Information Engineering, University of the Witwatersrand, Braamfontein 2000 Johannesburg, South Africa. Email: john.ekoru@wits.ac.za

ABSTRACT

The fundamental objective of any mining operation is to extract the maximum amount of ore in a highly economical way while adhering to safety standards. From an economic and safety perspective, the stability of rock slopes in open pit mine and quarry operations is crucial because unstable slopes have the potential to cause property damage and fatalities. Slope stability in open pit mining operations is crucial for ensuring safety and optimising resource extraction. Traditional monitoring techniques are often labour-intensive and limited in their ability to predict slope failure accurately. Recent advances in Artificial Intelligence (AI) have shown significant potential to enhance slope stability monitoring by providing real-time analysis, predictive capabilities, and optimised decision-making. To ascertain the probability of slope breakdown and how to prevent it, slope stability analysis is a crucial component of mining engineering. There is an immediate need for a method for assessing slope stability that is dependable, affordable, and widely applicable. By examining research work conducted in slope monitoring and testing, there is a need for an alternative approach that makes use of artificial intelligence and machine learning (ML) techniques. The main goal of this scoping review is to determine what has currently been researched and documented about the use of artificial intelligence integrated with machine learning in monitoring slope stability in open pit mines. A scoping review is conducted using electronic bibliographic databases and resources for instance: OneMine, Research Channel Africa, Scopus, ScienceDirect, Web of Science, Taylor and Francis journals, GeoRef, EbscoHost, Proquest, Springer collection, Access World News, World Bank Group and SABINET African Journals, IEEE, were searched to identify peer-reviewed publications, published in English, between January 2011 and February 2025, and related to application of Artificial Intelligence in monitoring of slope stability for open pit mines. The results obtained from the search were treated as follows: The total number of articles obtained and a clear inclusion criterion for the scoping review has been provided. The results show that three main themes have emerged from the previously conducted studies namely; Advanced Machine Learning for Slope Stability Assessment, Optimisation and Hybrid Models for Enhanced Prediction and Real-World Applications and Operational Integration. Future direction of research involves the integration of Artificial Intelligence with real-time monitoring for dynamic slope assessments, to achieve the development of integrated systems, advanced algorithms like ensemble learning and neural networks will be instrumental. There is also a need to develop numerous automated, adaptive machine learning models capable of handling complex data sets from diverse mine environments.

INTRODUCTION

Extracting as much ore as possible in a safe and cost-effective manner is the main goal of any mining operation. Open pit mining operations entails the extraction of valuable minerals from the earth's surface, often requiring the excavation of steep slopes. The stability of these slopes is a critical factor in the success of mining operations (Carlà *et al*, 2018; Qin *et al*, 2022), as slope failures can lead to catastrophic consequences, including the loss of life, significant economic losses, and long-term environmental destruction (Kolapo *et al*, 2022). The dilemma of slope stability for open pit operations does not only affect the safety point of view, it also affects the operations economically (Naghadehi *et al*, 2011). The instability of rock slopes and absence or failure of early warning mechanisms can lead to serious injuries or fatalities and damage to mine's property. Thus, researchers are committed to developing models and systems that can be used for early monitoring and detection of slope failures (Dick *et al*, 2014). Slope failure can lead to loss of profit in mining operations and it can seriously ruin the mine's reputation leading to investors pulling out (Kolapo *et al*, 2022). Consequently, monitoring slope stability is a central aspect of mining safety (Casagli *et al*, 2010), and traditional methods such as visual inspections, geological surveys, and ground-based instrumentation have been the standard approaches for decades (Mohammed, 2021). Slope stability monitoring typically involves a combination of visual inspections, manual data collection, and geological surveys. These methods, while valuable, are often time-consuming and subject to human error. Instruments such as inclinometers, strain gauges, and piezometers are frequently used to monitor ground deformation, pore pressure, and slope movement. However, these methods are typically limited to specific locations within the mine, and real-time data processing is challenging. As a result, traditional methods may miss early signs of instability or provide inadequate warnings, particularly in large mining operations. Traditional monitoring methods have a limitation of capacity to process huge volumes of data, provision of real-time predictions, or ability to comprehensively conduct risk assessments under dynamic environmental conditions (Song *et al*, 2024).

In recent years, Artificial Intelligence (AI) has emerged as a powerful tool for enhancing slope stability monitoring by improving predictive accuracy, automating data analysis, and integrating diverse data sources (Gupta, Sharma and Singh, 2022; Meng, Mattsson and Laue, 2021; Wang, Zhang and Wang, 2023). AI techniques, including machine learning (ML) (Kang *et al*, 2017; Kothari and Momayez, 2018), deep learning (DL), and expert systems, offer significant advantages over conventional methods by enabling more accurate and timely predictions of slope instability (Song *et al*, 2024). The application of AI in this domain involves the integration of various data sources, such as remote sensing data, geotechnical sensors, and environmental monitoring systems, to create comprehensive models for predicting slope failure (Zhou *et al*, 2019; Sahoo *et al*, 2022). This review aims to provide an overview of AI applications in monitoring slope stability in open pit mines, highlighting the methodologies, tools, and challenges involved. The complexity of managing open pit mines, combined with the dynamic nature of slope stability, necessitates more advanced monitoring techniques. AI provides a solution by enabling real-time, data-driven analysis of geological and environmental factors affecting slope stability. Through AI applications, mining companies can integrate data from various sources, predict potential failures, and make informed decisions regarding slope design and operation management.

MATERIALS AND METHODS

The use of a rigorous review process that guaranteed thorough coverage and critical examination of the available information was necessary for a consolidative study on 'Artificial Intelligence Applications in Monitoring Slope Stability' in the context of open pit mines. An integrative assessment of the literature was carried out in order to pinpoint and characterise the existing gaps in artificial intelligence applications in monitoring slope stability. Peer-reviewed papers in identified relevant databases were among the sources.

Search strategy

Systematic review guidelines for database and search engine searches were used to preserve study rigour (Moher *et al*, 2009). OneMine, Research Channel Africa, Scopus, ScienceDirect, Web of Science, Taylor and Francis journals, GeoRef, EbscoHost, Proquest, Springer collection, Access World News, World Bank Group, and SABINET African Journals, IEEE were among the online databases that were searched. The following were among the particular keywords that were used: Using machine learning and artificial intelligence to monitor slope stability in open pit mines. When necessary, asterisks, parentheses, and boolean operators like 'AND', 'OR' and others were employed to further filter the source articles found. According to database requirements, a span between January 2011 and February 2025 was selected in order to guarantee current data within a larger range. The search was then broadened to include 'Application of artificial intelligence and machine learning in Monitoring Slope Stability' when it was unable to find enough papers in the desired field of study. Papers that were closely related to the suggested topic were taken into consideration. To find additional sources, a thorough review of the reference lists of pertinent papers was also carried out. All publications were sourced by the principal investigator, and the final selection was refined through an iterative process. To make sure the final article selection was proper, one secondary researcher examined every article that was initially chosen and those that were included in the final data set.

Inclusion and exclusion strategy

Review papers were not included in order to keep the emphasis on original research. To guarantee quality and accessibility, publications were limited to peer-reviewed, English-language journals. In order to prevent a variety of potentially inconsistent forms and to improve reliability and validity through standardised, high-quality research, grey literature was excluded. The study focused on the use of machine learning and artificial intelligence in open pit mining slope stability monitoring. Figure 1 provides a visual representation of the search technique and its results.

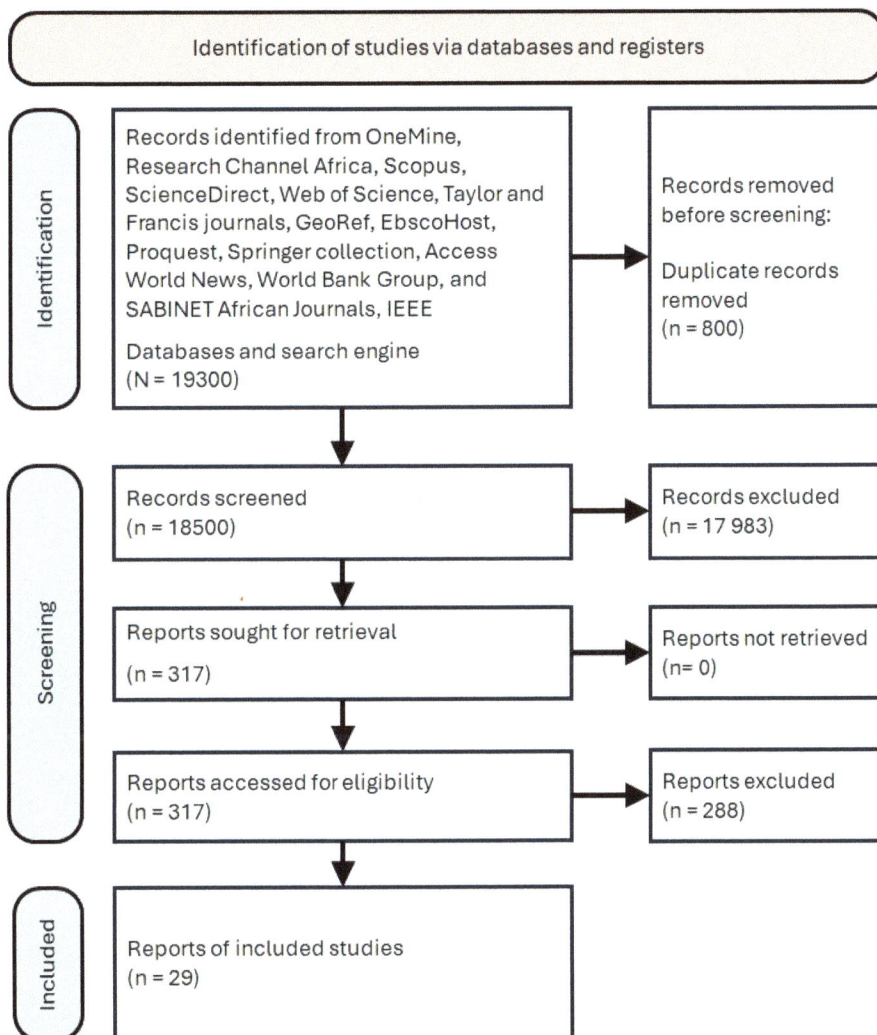

FIG 1 – Prisma diagram showing the inclusion and exclusion criteria.

Data extraction and synthesis

From January 2011 to the beginning of February 2025 was chosen as the review period. Using Boolean operators, 19 300 articles in total were found in relation to the key terms across databases. Of these, 29 journal articles were determined to be suitable for final inclusion using a linear, progressive procedure, as shown in Figure 1. The use of machine learning and artificial intelligence for slope monitoring was the main focus. The six processes of thematic analysis proposed by Braun and Clarke (2006)—data familiarisation, code generation, theme derivation, theme review, theme defining and naming, and report creation—were used to apply deductive thematic analysis. Reviewing topics, defining them precisely, refining them, doing ongoing analysis, and exercising researcher reflexivity were all ways to validate the themes (Braun and Clark, 2006). In particular, the original author highlighted pertinent text passages in actual copies of the article to create the initial codes. After that, codes were examined, trends were found, and related codes were merged into pertinent themes. The authors discussed any disagreements and changed the concepts as needed.

RESULTS AND DISCUSSION

Results and discussion are grouped into a single section for improved coherence, integrated expression, and concept clarity. The reviewed publications are shown in Table 1.

Overview of included studies

The 29 articles are organised into three major themes, each with its respective sub-themes. These are presented and discussed in this section, highlighting their interconnections through specific details, recommendations from the authors, and relevant implications (see Table 1).

TABLE 1

Artificial intelligence applications in monitoring slope stability.

Authors	Field of application	Approach	Models	Input parameters	Findings
Yang et al, 2023	Landslide	Previous studied data	SVM, RF, KNN, GBRT	γ, C, α, H, β, δ	With an accuracy of 85%, the results indicate that RF with K5 cut-off was the best model.
Paliwal et al, 2022	Landslide	Previous studied data	ANN	β, H, W, J1, J2	An android app was developed using ANN with reasonable results for prediction of the factor of safety in rock slopes.
Mahmoodzadeh et al, 2022	Landslide	Previous studied data	(GFB), SVR, DT, LSTM, DNN, KNN	γ, C, α, H, β, δ	XGBoost, predicted slope safety factors, with a good performance with R^2 of 0.8139, RMSE of 0.160893, and MAPE of 7.27%.
Bai et al, 2022	Landslide	Previous studied data	SVM, DT, KNN, ANN, RF, ANN, GCAB, GBDT	H, β, and BD, C, α, δ	FOS, the ANN and RF models performed better.
Ahangari Nanehkaran et al, 2022	Landslide	Field study data	DL, SVM, KNN, DT, RF	H, β, DL, C, α	The MLP, obtained a precision rating of 0.938 and an accuracy rating of 0.90.
Azmoon, Biniyaz and Liu, 2021	Landslide	Field study data	DL	γ, C, α	The model that was trained on 2000 samples outperformed JCM by a factor of 18.
Lin et al, 2021	Landslide	Previous studied data	SVM, ANN, DT, KNN, BRS, GBR, Bagging, RF, DT, RF	γ, C, α, H, β, δ	Best performing algorithms were GBR, SVM, and Bagging.
Kainthura and Sharma, 2021	Landslide	Previous studied data	SVM, Backpropagation, ML	UD, RI, ES	RF performed well Accuracy = 0.84, AUC = 0.93.
Singh and Chakravarty, 2023a	Pit slope	Field study data	RF, LR, SVM, KNN GNB, DT	Cohesion, Phi, Slope angle wind, rain, blast, C_b	A unique way to identify important features that affect dump slope stability was introduced.
Singh and Chakravarty, 2023b	Pit slope	Field study data	RF, LR, SVM, KNN Regression models	C, G, β, t, α, SC, WR, BC, IP	It was found that random forest outperformed other models. The decision tree regressor was the best performer.

Authors	Field of application	Approach	Models	Input parameters	Findings
Jiang et al, 2022	Dump slope	Field study data	SVM, RF, KNN, GBRT	H, β, RI	GBRT was the best model with an accuracy of Factor of Safety (FOS) -1.283, whereas FOS (1.290) was determined via numerical simulation analysis.
Bharati et al, 2022	Dump slope	Field study data	ANN and MBA	Cfn, Sh, Sa	With a coefficient of determination value of 0.996, it was discovered that the ANN model predicts more accurately than the MBA model.
Karir et al, 2022	Dump slope	Previous studied data	SVR, ANN, RF, GBM, XGB	Hs, Hcd, BW, A, B, Hcr, α, C	The effectiveness of various ML models in predicting the safety factor was found to be higher for the artificial dump slope compared to the residual soil slope that occurs naturally.
Nguyen et al, 2019	Open pit	Hybrid model based on clustering and ANN	HKM-ANN, FCM-SVR, ANN, SVR, FCM-ANN, HKM-SVR	Blasting events, Clustered data.	The model improved prediction accuracy by clustering data with the HKM algorithm and using ANN for estimation. It outperformed other models, including FCM–SVR, ANN, and SVR, in predicting PPV from blasting operations.
Nguyen et al, 2021	Open pit	Hybrid model based on nature-inspired optimisation algorithms and DNN	HHOA-DNN, WOA-DNN, PSOA-DNN, DNN	Explosive charge per blast, Monitoring distance, Time delay per blasting group.	Conventional DNN model was outperformed, with the HHOA–DNN model having the best accuracy.
Lyu et al, 2024	Open pit	Multi-parameter sample data set, Bayesian optimisation	BP neural networks, Genetic algorithm-optimised CNN, 1D-CNN, B-1D MCNN	Factor indicators, Training set length.	The B-1D MCNN. outperformed other models, showing improvements of 10.96% to 27.85% in accuracy, 10.26% to 28.55% in F1-score, and 8.98% to 25.05% in precision.
Sun et al, 2024	Open pit	Runge-Kutta optimisation, Machine learning	RUN-XGBoost, PSO-XGBoost, XGBoost, Ridge, LASSO, SVM, SVR	Maximum explosive, Total explosive, Blast centre distance, Blasthole depth, Height difference.	The RUN-XGBoost, achieved high accuracy (R^2 = 0.963, VAF = 96.486). The model's error metrics are also minimal (RMSE = 0.0774, MAE = 0.0552).

Authors	Field of application	Approach	Models	Input parameters	Findings
Fakir, 2017	Open pit	Artificial Neural Networks, Rock Engineering Systems approach	Back Propagation, Self-Organising Maps, Artificial Neuro Fuzzy Inference System	Rock mass structure, *in situ* stress, water flow, construction effect, intact rock quality, rock mass properties, hydraulic conditions, discontinuities properties, geometry, history of instabilities.	the Open Pit Mine Slope Stability Index (OPMSSI) effectively predicts slope stability using a weighted sum of ratings within the Rock Engineering Systems (RES).
Du et al, 2019	Open pit	Machine Learning, Ground-Based Interferometric Radar (GB-SAR)	Ensemble Learning (Super Learner, Weaker Learners)	GB-SAR field data.	The ensemble learning model accurately predicted slope deformation, outperforming individual models. Field data confirmed its reliability, and future work will refine algorithm selection and expand applications.
Alshibani et al, 2024	Open pit	Decision support, sustainability optimisation	Analytical Hierarchy Process (AHP), Multi-Attribute Utility Theory (MAUT), Optimisation Module	Operating cost, Productivity, Repair and maintenance costs, Soil condition, Sustainability variables (social, economic, environmental).	The proposed framework improves decision-making using AHP, MAUT, and SMART, with potential for AI, GIS, GPS, and sensor integration to enhance efficiency, sustainability, and safety.
Wang et al, 2024	Open pit	Field tests and laboratory experiments	Deformation monitoring	Rock mass properties, rainfall, geological conditions (clay mineral content, micro-cracks, water absorption properties).	The SSA-BP model achieves high accuracy.
Filho et al, 2025	Open pit	Failure susceptibility analysis	Global Slope Performance Index with slope classification system (GAMAH-R)	Acceptance criteria for 25 slopes from three open pit ore mines.	The correlation between GAMAH-R and GSPI systems effectively assessed slope risks and guided mine planning. Further validation is needed, especially for hard rock formations.
Song et al, 2024	Open pit	Digital early warning platform	AI, Multi-source information fusion	Data from various monitoring sources, geological disaster data.	The current open pit mine early warning platforms face challenges such as inconsistent data formats, incomplete data types, and fragmented industry practices, which hinder their construction and efficiency.

Authors	Field of application	Approach	Models	Input parameters	Findings
Wang, Dong and Ji, 2025	Rock slope	Sen's slope trend analysis, microseismic monitoring, visual cloud chart	Multi-index fusion early warning model	Cumulative Benioff strain, b-value, S-value, Energy index, Cumulative apparent volume, Hurst exponent, Q-value.	The findings indicate that in high and steep annular slope mining sites, the early warning model based on Sen's slope trend analysis.
Barkhordari et al, 2024	Slope stability	Machine learning	Logistic Regression (LR), Quadratic Discriminant Analysis (QDA), Light Gradient Boosting Machine (LGBM), Linear Discriminant Analysis (LDA)	Peak ground acceleration, friction angle, angle of inclination.	The Light Gradient Boosting Machine (LGBM), are effective for predicting slope stability under seismic conditions.
Yadav et al, 2025	Slope stability	Ensemble machine learning	Decision Tree (DT), Random Forest (RF), Bagging, Boosting, Ensemble Bagging Regression	Seven quantitative parameters (not specifically named in the abstract).	RF and DT, achieved up to 96% accuracy in slope stability prediction. ensemble techniques remained robust, demonstrating a 6–8% improvement over individual models.
Lin et al, 2024	Slope stability	Digital twin database, Bayesian optimisation	CNN, SVM, GBM, ML models	Geometries, Weak layers.	The model, trained on a large DT-generated database, outperforms other machine learning methods, achieving 95.4% accuracy and 0.99 ROC in testing.
Zheng et al, 2024	Slope stability	Machine learning, metaheuristic optimisation	Sparrow Search Algorithm (SSA) + Back Propagation (BP) Neural Network	Unit weight (γ), Cohesion (c), Friction angle (φ), Slope angle (α), Slope height (H).	The SSA-BP neural network model for predicting slope safety coefficients demonstrates high accuracy (regression coefficient of 0.9405 and MAE of 0.1684).
Zhang et al, 2022	Slope stability	Deep Learning, Harris Hawks Optimisation	DMLP, HHO-DMLP, SVM, RF	Shear strength, friction angle, geological and geomorphological data, location-specific observations.	The results showed that RF and XGBoost outperformed SVM and LR in both training and testing accuracy.

Themes

Three clear themes were identified in the articles that were extracted. The themes obtained are presented below.

Theme 1 – advanced machine learning for slope stability assessment

These themes cover the detailed and broad use of numerous Machine learning algorithms and Deep Learning algorithms for slope stability. The subthemes obtained were:

- Application traditional ML methods for instance SVM, RF, ANN.

- Use of ensemble learning techniques for instance Gradient Boosting, XGBoost.

- Application of Deep learning models for instance CNNs.

- Use of Automated Machine Learning (AutoML).

The main objective and focus of these publications were to improve the accuracy and efficiency of slope stability analysis via a data-driven methodologies. The papers that followed these themes have been arranged in Table 2.

TABLE 2

Advanced machine learning for slope stability assessment.

Author and year	Machine learning algorithm(s) used	Purpose
Ma *et al*, 2022	AutoML (various ML models)	Slope stability classification of circular mode failure.
Zhang *et al*, 2022	Ensemble Learning (Random Forest, Gradient Boosting)	Slope stability prediction.
Mahmoodzadeh *et al*, 2022	SVM, ANN, RF, XGBoost	Prediction of safety factors for slope stability.
Lin *et al*, 2021	SVM, RF, ANN	Evaluation and prediction of slope stability.
Yang *et al*, 2023	Optimised ML algorithms	Slope stability prediction.
Ahangari Nanehkaran *et al*, 2022	Various ML algorithms	Estimation of safety factor in slope stability analysis.
Karir *et al*, 2022	Various ML algorithms	Stability prediction of natural and man-made slopes.
Kainthura and Sharma, 2021	Various ML techniques	Prediction of slope failures.
Bai *et al*, 2022	Various ML models	Performance evaluation and engineering verification of slope stability models.
Azmoon, Biniyaz and Liu, 2021	Deep Learning (ANN)	Evaluation of deep learning against limit equilibrium methods.
Paliwal *et al*, 2022	ANN	Stability prediction of residual soil and rock slopes.
Singh and Chakravarty, 2023b	Classification and Regression algorithms	Assessment of slope stability.
Jiang *et al*, 2022	GBRT algorithm	Landslide risk prediction.
Bharati *et al*, 2022	ANN, Multiple Regression	Stability evaluation of dump slope.
Fattahi, 2017	Adaptive Neuro-Fuzzy Inference System	Prediction of slope stability.

Author and year	Machine learning algorithm(s) used	Purpose
Zhang et al, 2022	Deep Neural Network, Harris Hawks Optimisation	Estimating the friction angle of clays in evaluating slope stability.
Fakir, 2017	ANN	Open pit slope stability.
Zheng et al, 2024	Back Propagation, Sparrow Search Algorithm	Prediction of soil slope stability.
Yadav et al, 2025	Ensemble Machine Learning	Enhanced slope stability prediction.
Onyelowe et al, 2025	Advanced Machine Learning Combinations	Geophysical flow prediction of slope behaviour.

Theme 2 – optimisation and hybrid models for enhanced prediction

This theme that emerged demonstrates that some authors over the past ten years have focused on the development and application of optimisation algorithms and hybrid models to improve the performance of ML and DL in slope stability prediction. The use of optimisation algorithms (genetic algorithms, sparrow search algorithm, Harris Hawks optimisation, Bayesian optimisation) to fine-tune model parameters. The researchers have extensively looked at the development of hybrid models that amalgamate various ML/DL methodologies or integrate them with other methods (for instance clustering). The main objective of this type of research work was to improve prediction accuracy, robustness, and efficiency by leveraging optimisation and hybrid approaches (see Table 3).

TABLE 3

Optimisation and hybrid models for enhanced prediction.

Author and year	Machine learning algorithm(s)	Purpose
Yang et al, 2023	Intelligent optimisation algorithms, Machine Learning algorithms	Slope stability prediction method
Zhang et al, 2022	Deep Neural Network, Harris Hawks Optimisation	Estimating the friction angle of clays in evaluating slope stability
Zheng et al, 2024	Sparrow Search Algorithm, Back Propagation	Fast and accurate prediction of soil slope stability
Nguyen et al, 2021	Nature-inspired optimisation algorithms, Deep Neural Network	Predicting blast-induced ground vibration in open pit mines
Nguyen et al, 2019	Clustering, Artificial Neural Network	Hybrid model for predicting blast-induced ground vibration in open pit mines
Lyu et al, 2024	Bayesian Optimisation, 1D-CNN	Stability evaluation of open pit mine slopes

Theme 3 – real-world applications and operational integration

The third theme provides the practical application AI in real-world open pit mining scenarios, focusing on operational integration and decision support. Some of the case studies that were considered were:

- Case studies of slope stability analysis in specific mining regions.

- The development of intelligent early-warning platforms and decision support systems.

- The application of AI to predict blast-induced ground vibrations and optimise blasting parameters.

- The use of AI to create GUI based platforms for practical use.

- The monitoring of large landslides.

Table 4 summarises theme 3 on application of AI in real world open pit mining scenario:

TABLE 4

Real-world applications and operational integration.

Author and year	Machine learning algorithm(s)	Purpose
Alshibani *et al*, 2024	Decision Support Framework (AI-based)	Sustainability and fleet selection in open pit mining construction
Wang *et al*, 2024	Machine Learning Algorithms, Failure Process Monitoring	Failure process and monitoring data analysis of a large landslide
Filho *et al*, 2025	Machine Learning Algorithms	Failure susceptibility analysis of open pit slopes
Barkhordari *et al*, 2024	Machine Learning Algorithms, GUI-based platform	Slope stability prediction under seismic conditions
Song *et al*, 2024	Intelligent Early-warning Platform	Early-warning system for open pit mining stability
Monjezi, Khoshalan and Varjani, 2011	Genetic Algorithm	Optimisation of open pit blast parameters
Nguyen *et al*, 2021	Nature-inspired Optimisation Algorithms, Deep Neural Network	Predicting blast-induced ground vibration in open pit mines
Sun *et al*, 2024	RUN-XGBoost Model	Prediction of Peak Particle Velocity (PPV) in open pit mine
Wang, Dong and Ji, 2025	Sen's Slope Trend Analysis	Early warning model for rock mass instability in mining areas
Du *et al*, 2019	Machine Learning Algorithms	Prediction of slope deformation in an open pit mine

CHALLENGES AND FUTURE DIRECTIONS

The future of Artificial Intelligence (AI) applications in monitoring slope stability for open pit mines holds significant promise, driven by advances in machine learning (ML) and deep learning (DL) algorithms. AI's ability to enhance the accuracy and efficiency of slope stability predictions is becoming increasingly important. Future directions should focus on further integration of AI with real-time monitoring systems to offer dynamic slope stability assessments, leveraging advanced algorithms like ensemble learning, neural networks, and optimisation techniques. Additionally, there is a growing need for the development of automated and adaptive machine learning models that can handle large, complex data sets from diverse mine sites. The use of hybrid models, combining AI with traditional geotechnical methods, can improve predictive capabilities, especially under uncertain conditions. Future research could also explore AI's potential in automating risk assessment, developing early warning systems, and incorporating advanced simulation tools for continuous monitoring, thus increasing safety and reducing environmental impacts in open pit mining operations.

CONCLUSION

Artificial Intelligence offers significant potential for enhancing slope stability monitoring in open pit mines. Through the use of machine learning, deep learning, remote sensing integration, and expert systems, AI can improve prediction accuracy, provide real-time monitoring, and support decision-making. While challenges remain, the continued development of AI technologies holds great promise for improving the safety, efficiency, and sustainability of mining operations.

REFERENCES

Ahangari Nanehkaran, Y, Pusatli, T, Chengyong, J, Chen, J, Cemiloglu, A, Azarafza, M and Derakhshani, R, 2022. Application of machine learning techniques for the estimation of the safety factor in slope stability analysis, *Water*, 14(22).

Alshibani, A, Elmaghraby, B, Bubshait, A, Ghaithan, A M, Mohammed, A and Hassanain, M A, 2024. Advancing sustainability: An integrated decision support framework for fleet selection in open pit mining construction, *Results in Engineering*, 23:102501. https://doi.org/10.1016/j.rineng.2024.102501

Azmoon, B, Biniyaz, A and Liu, Z, 2021. Evaluation of deep learning against conventional limit equilibrium methods for slope stability analysis, *Applied Sciences*, 11(13).

Bai, G, Hou, Y, Wan, B, An, N, Yan, Y, Tang, Z, Yan, M, Zhang, Y and Sun, D, 2022. Performance evaluation and engineering verification of machine learning based prediction models for slope stability, *Applied Sciences*, 12(15).

Barkhordari, M S, Barkhordari, M M, Armaghani, D J, Mohamad, E T and Gordan, B, 2024. GUI-based platform for slope stability prediction under seismic conditions using machine learning algorithms, *Architecture, Structures and Construction*, 4:145–156. https://doi.org/10.1007/s44150-024-00112-4

Bharati, A K, Ray, A, Khandelwal, M, Rai, R and Jaiswal, A, 2022. Stability evaluation of dump slope using artificial neural network and multiple regression, *Engineering with Computers*, 38:1835–1843.

Braun, V and Clarke, V, 2006. Using thematic analysis in psychology, *Qual Res Psychol*, 3:77–101.

Carlà, T, Farina, P, Intrieri, E, Ketizmen, H and Casagli, N, 2018. Integration of ground-based radar and satellite InSAR data for the analysis of an unexpected slope failure in an open pit mine, *Eng Geol,* 235:39–52.

Casagli, N, Catani, F, Del Ventisette, C and Luzi, G, 2010. Monitoring, prediction and early warning using ground-based radar interferometry, *Landslides*, 7:291–301.

Dick, G, Erik, E, Albert, G, Doug, S and Nick, D, 2014. Development of an early-warning time-of-failure analysis methodology for open pit mine slopes utilizing ground-based slope stability radar monitoring data, *Can Geotech J*, 52:515–529.

Du, S, Feng, G, Wang, J, Feng, S, Malekian, R and Li, Z, 2019. A new machine-learning prediction model for slope deformation of an open pit mine: An evaluation of field data, *Energies,* 12(7):1288. https://doi.org/10.3390/en12071288

Fakir, M, and Ferentinou, M, 2017. A Holistic Open Pit Mine Slope Stability Index Using Artificial Neural Networks, paper presented at the ISRM AfriRock - Rock Mechanics for Africa, Cape Town, South Africa.

Fattahi, H, 2017. Prediction of slope stability using adaptive neuro-fuzzy inference system based on clustering methods, *Journal of Mining and Environment*, 8(2):163–177.

Filho, F A F, Bacellar, L d A P, Marques, E A G, de Assis, A P, Gomes, R C and da Costa, T A V, 2025. Failure susceptibility analysis of open pit slopes: A case study from the Quadrilátero Ferrífero Mine, Brazil, *Geotechnical and Geological Engineering*, 43. https://doi.org/10.1007/s10706-024-02992-1

Gupta, G, Sharma, S K and Singh, G, 2022. Dump slope stability analysis using artificial intelligence, *Journal of Mines, Metals and Fuels*, 70(3):129–135. https://doi.org/10.18311/jmmf/2022/30445

Jiang, S, Li, J Y, Zhang, S, Gu, Q H, Lu, C W and Liu, H S, 2022. Landslide risk prediction by using GBRT algorithm: Application of artificial intelligence in disaster prevention of energy mining, *Process Safety and Environmental Protection*, 166:384–392.

Kainthura, P and Sharma, N, 2021. Machine learning techniques to predict slope failures in Uttarkashi, Uttarakhand (India), *Journal of Scientific and Industrial Research*, 80.

Kang, F, Xu, B, Li, J and Zhao, S, 2017. Slope stability evaluation using Gaussian processes with various covariance functions, *Applied Soft Computing*, 60:387–396.

Karir, D, Ray, A, Kumar Bharati, A, Chaturvedi, U, Rai, R and Khandelwal, M, 2022. Stability prediction of a natural and man-made slope using various machine learning algorithms, *Transportation Geotechnics*, 34.

Kolapo, P, Oniyide, G O, Said, K O, Lawal, A I, Onifade, M and Munemo, P, 2022. An Overview of Slope Failure in Mining Operations, *Mining*, 2(2):350–384. https://doi.org/10.3390/mining2020019

Kothari, U C and Momayez, M, 2018. Machine Learning: A Novel Approach to Predicting Slope Instabilities, *International Journal of Geophysics*, 4861254. https://doi.org/10.1155/2018/4861254

Lin, M, Zeng, L, Teng, S, Chen, G and Hu, B, 2024. Prediction of stability of a slope with weak layers using convolutional neural networks, *Natural Hazards*, 120:12081–12105. https://doi.org/10.1007/s11069-024-06674-2

Lin, S, Zheng, H, Han, C, Han, B and Li, W, 2021. Evaluation and prediction of slope stability using machine learning approaches, *Frontiers of Structural and Civil Engineering*, 15(4):821–833.

Lyu, J, Hu, T, Liu, G, Cao, B, Wang, W and Li, S, 2024. Stability evaluation of open pit mine slope based on Bayesian optimisation 1D-CNN, *Scientific Reports*, 14(1):13995.

Ma, J, Jiang, S, Liu, Z, Ren, Z, Lei, D, Tan, C and Guo, H, 2022. Machine learning models for slope stability classification of circular mode failure: An updated database and automated machine learning (AutoML) approach, *Sensors*, 22(23).

Mahmoodzadeh, A, Mohammadi, M, Farid Hama Ali, H, Hashim Ibrahim, H, Nariman Abdulhamid, S and Nejati, H R, 2022. Prediction of safety factors for slope stability: Comparison of machine learning techniques, *Natural Hazards*, 111(2):1771–1799.

Meng, J, Mattsson, H and Laue, J, 2021. Three-dimensional slope stability predictions using artificial neural networks, *International Journal for Numerical and Analytical Methods in Geomechanics,* 45:1988–2000.

Moher, D, Liberati, A, Tetzlaff, J and Altman, D G, 2009. The PRISMA Group, Preferred reporting items for systematic reviews and meta-analyses: The PRISMA statement, *Ann Intern Med,* 151:264–269.

Monjezi, M, Amini Khoshalan, H and Yazdian Varjani, A, 2011. Optimisation of open pit blast parameters using genetic algorithm, *International Journal of Rock Mechanics and Mining Sciences*, 48:864–869.

Naghadehi, M, Jimenez, R, Khalokakaie, R and Jalali, S-M, 2011. A probabilistic systems methodology to analyze the importance of factors affecting the stability of rock slopes, *Eng Geol,* 118:82–92.

Nguyen, H, Bui, X-N, Tran, Q-H, Nguyen, D-A, Hoa, L T T, Le, Q-T and Giang, L T H, 2021. Predicting blast-induced ground vibration in open pit mines using different nature-inspired optimisation algorithms and deep neural network, *Natural Resources Research*, 30(6):4695–4715. https://doi.org/10.1007/s11053-021-09896-4

Nguyen, H, Drebenstedt, C, Bui, X-N and Bui, D T, 2019. Prediction of Blast-Induced Ground Vibration in an Open pit Mine by a Novel Hybrid Model Based on Clustering and Artificial Neural Network, *Sensors*, 20(1):132. https://doi.org/10.3390/s20010132

Onyelowe, K C, Ebid, A M, Hanandeh, S and Kamchoom, V, 2025. Evaluating the slope behavior for geophysical flow prediction with advanced machine learning combinations, *Scientific Reports*, 156531. https://doi.org/10.1038/s41598-025-90882-8

Paliwal, M, Goswami, H, Ray, A, Bharati, A K, Rai, R and Khandelwal, M, 2022. Stability prediction of residual soil and rock slope using artificial neural network, *Advances in Civil Engineering*, 2022:1–14. https://doi.org/10.1155/2022/4121193

Qin, J, Du, S, Ye, J and Yong, R, 2022. SVNN-ANFIS approach for stability evaluation of open pit mine slopes, *Expert Syst Appl*, 198:116816.

Sahoo, A K, Pramanik, J, Jayanthu, S and Samal, A K, 2022. Slope stability predictions using machine learning techniques, in *Proceedings of the 4th International Conference on Advances in Computing, Communication Control and Networking (ICAC3N)*, pp 133–137. https://doi.org/10.1109/ICAC3N56670.2022.10074079

Singh, S and Chakravarty, D, 2023a. Efficient and Reliable Prediction of Dump Slope Stability in Mines using Machine Learning: An in-depth Feature Importance Analysis, *Archives of Mining Sciences*, 68:685–706.

Singh, S K and Chakravarty, D, 2023b. Assessment of slope stability using classification and regression algorithms subjected to internal and external factors, *Archives of Mining Sciences*, pp 87–102.

Song, Z, Li, X, Huo, R and Liu, L, 2024. Intelligent early-warning platform for open pit mining: Current status and prospects, *Rock Mechanics Bulletin*, 3:100098. https://doi.org/10.1016/j.rockmb.2023.100098

Sun, M, Yang, J, Yang, C, Wang, W, Wang, X and Li, H, 2024. Research on prediction of PPV in open pit mine used RUN-XGBoost model, *Heliyon*, 10:e28246. https://doi.org/10.1016/j.heliyon.2024.e28246

Wang, J, Dong, L and Ji, S, 2025. Rock mass instability early warning model: A case study of a high and steep annular slope mining area using Sen's slope trend analysis, *Tunnelling and Underground Space Technology*, 159:106514. https://doi.org/10.1016/j.tust.2025.106514

Wang, J, Yang, X, Tao, Z, He, M and Shen, F, 2024. Failure process and monitoring data of an extra-large landslide at the Nanfen Open pit Iron Mine, *Journal of Mountain Science*, 21(9):2918–2938. https://doi.org/10.1007/s11629-023-8540-5

Wang, S, Zhang, Z and Wang, C, 2023. Prediction of stability coefficient of open pit mine slope based on artificial intelligence deep learning algorithm, *Sci Rep,* 13:12017. https://doi.org/10.1038/s41598-023-38896-y

Yadav, D K, Chattopadhyay, S, Tripathy, D P, Mishra, P and Singh, P, 2025. Enhanced slope stability prediction using ensemble machine learning techniques, *Scientific Reports*, 15:7302. https://doi.org/10.1038/s41598-025-90539-6

Yang, Y, Zhou, W, Jiskani, I M, Lu, X, Wang, Z and Luan, B, 2023. Slope stability prediction method based on intelligent optimisation and machine learning algorithms, *Sustainability*, 15(2):1169.

Zhang, H, Nguyen, H, Bui, X-N, Pradhan, B, Asteris, P G, Costache, R and Aryal, J, 2022. A generalized artificial intelligence model for estimating the friction angle of clays in evaluating slope stability using a deep neural network and Harris Hawks optimisation algorithm, *Engineering with Computers*, 38(5):S3901–S3914. https://doi.org/10.1007/s00366-020-01272-9

Zheng, B, Wang, J, Feng, S, Yang, H, Wang, W, Feng, T and Hu, T, 2024. A new, fast and accurate algorithm for predicting soil slope stability based on sparrow search algorithm-back propagation, *Natural Hazards*, 120:297–319. https://doi.org/10.1007/s11069-023-06210-8

Zhou, J, Li, E, Yang, S, Wang, M, Shi, X, Yao, S and Mitri, H S, 2019. Slope stability prediction for circular mode failure using gradient boosting machine approach based on an updated database of case histories, *Safety Science*, 118:505–518. https://doi.org/10.1016/j.ssci.2019.05.046

Ground vibration prediction using a machine learning approach

M C I Madahana[1] and J E D Ekoru[2]

1. Lecturer, University of the Witwatersrand, Johannesburg, Gauteng, South Africa, 2000. milka.madahana@wits.ac.za
2. Lecturer, University of the Witwatersrand, Johannesburg, Gauteng, South Africa, 2000. john.ekoru@wits.ac.za

ABSTRACT

Mining products play a critical role in our current lives for instance titanium is used in the design of surgical pins and bone plates, renewable energy technologies, use copper for wiring of solar panels. Mobile phones are powered by precious metals for example lithium. To obtain this minerals, explosives are usually applied in rock fragmentation. Ineffective use of explosive energy in an operation may result in excessive ground vibration. Extreme ground vibrations because of blasting activities can result in various problems, for instance damage to property of the nearby residents and ecological damage. The measurement of ground vibration because of blasting is essential to mitigate risks associated with adverse impacts of blasting. The peak particle velocity (PPV) is the most important parameter generally used to evaluate ground vibrations in blasting sites. There are various methods that are currently used in prediction or estimation of PPV. The Empirical approach is limited by few input variables used in the prediction of ground vibration. The statistical and mathematical modelling techniques usually require explicit knowledge and understanding of the progression of the intricate blasting dynamics. With the current paradigm shift towards automated systems in the mining industry and introduction of the fourth industrial revolution concepts, it is imperative that other techniques for the prediction of PPV be applied. The main objectives of this research work is to conduct a study on the application various machine learning algorithms including deep Neural Networks in the prediction of PPV, 799 observations of data sets are used to develop machine learning algorithms. The following input features are used: the powder factor (kg/m^3), spacing (m), stemming length (m), burden (m), maximum charge per delay (kg), blast-face distance to the monitoring point (m) and PPV is considered as the target variable. Random Forest, Artificial Neural Networks and deep Neural networks are some of the machine learning algorithms that are used in this research work. Various criteria, including mean absolute error (MAE), and correlation coefficient (R), are used to evaluate the developed models' accuracy and applicability. The results show the deep Neural Networks outperformed the traditionally known machine learning algorithm. To improve the performance further, the size of the data sets will be increased and various algorithms will be amalgamated.

INTRODUCTION

A wide and economical method to break and displace rocks in the mines is by blasting. Blastholes are charged using explosive material after the rock mass has been drilled. Blasting activities inevitably result in a number of adverse environmental effects, including fly rock, back-break, dust pollution, air overpressure, and ground vibration (Jiang *et al*, 2019; Bakhtavar *et al*, 2021a). The most detrimental of these features are fly rock, air pressure and ground vibration. It is therefore imperative to monitor the impacts of blasting in the mines with the main objective of minimising or eradicating its impact to the mining site and the environment. Figure 1 shows the propagation of vibration waves during rock blasting in open pit mining. Prior to blasting, there should be accurate determination of the impacts of blasting (Bakhtavar *et al*, 2021a, 2021b; Hosseini *et al*, 2021; Hosseini, Mousavi and Monjezi, 2022; Armaghani *et al*, 2018; Nguyen *et al*, 2020; Nguyen and Bui, 2020). Out of the aforementioned effects of blasting, research has shown that ground vibration is the most detrimental. In order to conduct ground vibration prediction, the effective parameters on ground vibration should be determined. Ground vibrations can be measured via Peak Particle Velocity (PPV) or using frequency. PPV is the most popularly used representation for the estimation and evaluation of blast induced ground vibration in surface mines. The number of blastholes, hole depth, burden, spacing, powder factor, charge per delay, and the distance between the blasting bench and the mounted seismograph are the most important factors when determining PPV. Various empirical models have been previously developed to predict PPV in the mines and open pits. For instance,

Davies, Farmer and Attewell (1964), Ambraseys and Hendron (1968), Dowding (1985), Roy (1993) and Rai and Singh (2004). The empirical models have been reported to not be effective. In addition, the accurate estimation of PPV is unachievable using the empirical equations, hence its challenging to minimise or mitigate adverse impacts of blasting. Artificial Intelligence and the current computational techniques have shown capabilities in resolving engineering and scientific problems accurately. The Artificial Intelligence approach and soft computing can be explored in the prediction of ground vibration.

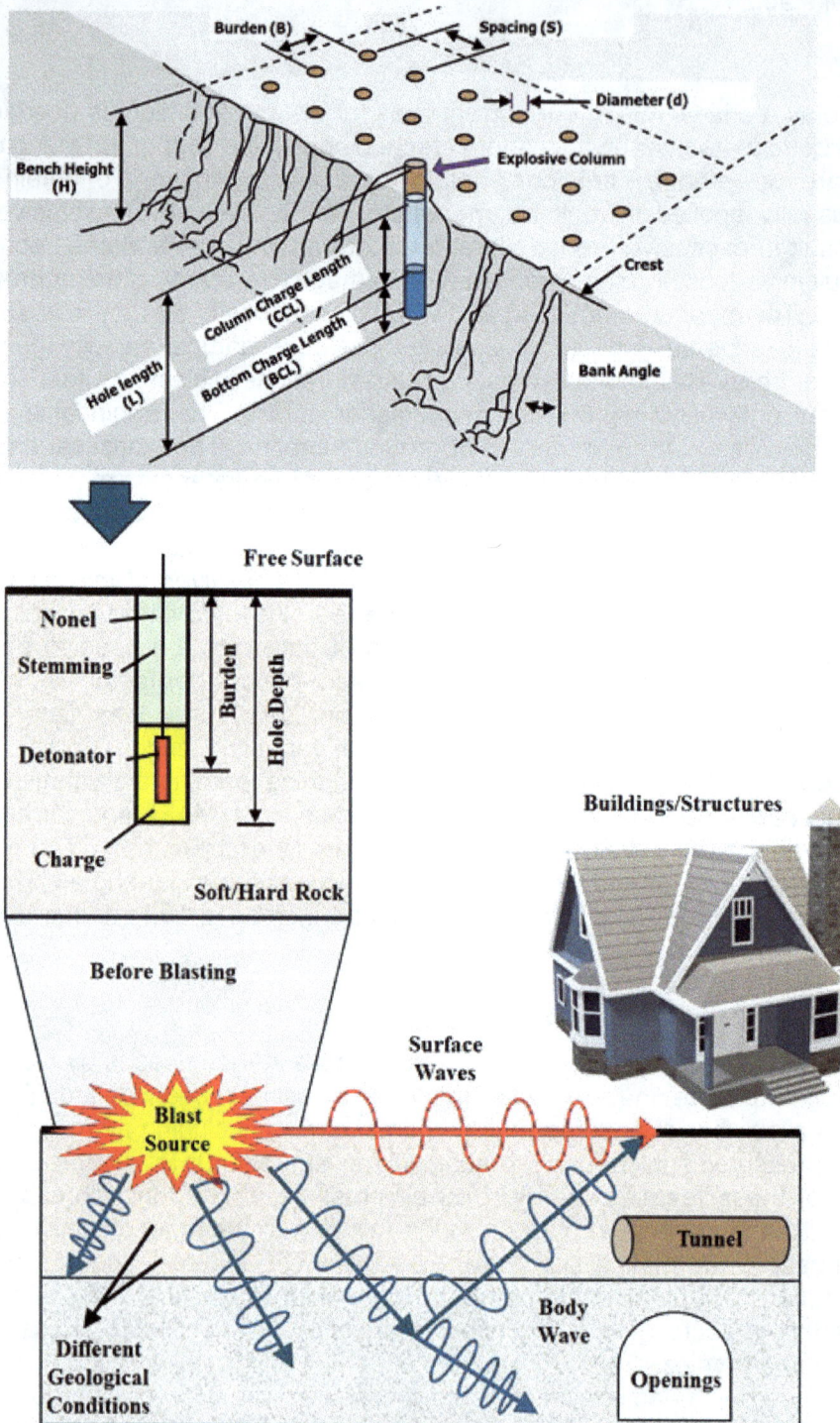

FIG 1 – Illustrates the propagation of vibration waves during rock blasting in open pit mining (Fissha *et al*, 2025; Zhang *et al*, 2023).

LITERATURE REVIEW

Several authors (Hasanipanah *et al*, 2015) have previously conducted research and proposed various techniques for the prediction of PPV using both soft computing and Artificial Intelligence. (Armaghani *et al*, 2014) used Imperialist competitive algorithm (ICA) to estimate PPV. The results showed good performance for the ICA algorithm. Using applying a genetic algorithm, (Hasanipanah *et al*, 2018), predicted PPV, producing results that indicated that the optimisation algorithm can predict with high levels of accuracy. Dehghani and Ataee-Pour (2011), used the ANN to calculate blast induced ground vibration. Taheri *et al* (2017), amalgamated the Artificial Neural Network (ANN) with Artificial Bee Colony (ABC) and compared the results obtained to empirical equations. The results showed that the ANN-ABCA model outperformed the empirical model. Fouladgar, Hasanipanah and Bakhshandeh Amnieh (2017) applied the Cuckoo Search(CS) as a unique swarm intelligence technique for PPV predictions of PPV. Other authors who employed ANN algorithms are Das, Sinha and Ganguly (2019). Hasanipanah *et al* (2017) presented research work that was a hybrid of Fuzzy system (FS) and ICA to predict PPV. (Jiang *et al*, 2019) also researched on a particle Swarm Optimisation (PSO) for prediction of PPV values. Iphar, Yavuz and Ak (2008) using Adaptive Neuro-Fuzzy Inference system (ANFIS) estimated PPV, the results showed a reasonable performance in terms of prediction. A summary of some of the research works that have been conducted since 2005 is provided in Table 1.

TABLE 1

Previously conducted research work (Hosseini *et al*, 2023; Yan *et al*, 2024; Fissha *et al*, 2025).

Author(s)	Year	ML technique	No of parameters	Performance metrics	Model performance
Singh and Singh	2005	ANN	9	R^2	0.82
Iphar, Yavuz and Ak	2008	ANFIS	2	R^2	0.99
Khandelwal, Kankar and Harsha	2010	SVM	2	R^2	0.96
Fişne, Kuzu and Hüdaverdi	2011	FIS	2	R^2	0.92
Khandelwal, Kumar, and Yellishetty	2011	ANN	2	R^2	0.92
Mohamed	2011	ANN, FIS	2	R^2	ANN = 0.94, FIS = 0.90
Monjezi, Ghafurikalajahi and Bahrami	2011	ANN	4	R^2	0.95
Fişne, Kuzu and Hüdaverdi	2011	FIS	2	R^2	0.92
Mohamed	2011	ANN, FIS	2	R^2	0.94
Mohamadnejad, Gholami and Ataei	2012	SVM, ANN	2	R^2	SVM = 0.89, ANN = 0.85
Ghasemi, Ataei and Hashemolhosseini	2013	FIS	6	R^2	0.95
Monjezi, Ghafurikalajahi, and Bahrami	2013	ANN	4	R^2	0.95
Armaghani *et al*	2014	PSO-ANN	9	R^2	0.94
Armaghani *et al*	2014	PSO-ANN	6		0.94
Armaghani *et al*	2015	ANN	2	R^2	0.987
Armaghani *et al*	2015	ANFIS	2	R^2	0.97

Author(s)	Year	ML technique	No of parameters	Performance metrics	Model performance
Dindarloo	2015	SVM	12	R^2	0.99
Hajihassani et al	2015a	ICA-ANN	7	R^2	0.98
Hajihassani et al	2015b	PSO-ANN	8	R^2	0.89
Hasanipanah et al	2015	SVM	6	R^2	0.96
Jahed et al	2015	ANFIS	8	R^2	0.97
Hajihassani et al	2015b	PSO-ANN	7	R^2	0.89
Hajihassani et al	2015a	ICA-ANN	3	R^2	0.98
Amiri	2016	KNN:ANN	2	R^2	0.88
Faradonbeh et al	2016	GEP	8	R^2	0.88
Ghoraba et al	2016	ANN, ANFIS	2	R^2	ANFIS = 0.95, ANN = 0.89
Faradonbeh et al	2016	GEP	2	R^2	0.88
Monjezi, Hasanipanah and Khandelwal	2013	ANN, ANFIS	2	R^2	0.95
Hasanipanah et al	2017	CART	2	R^2	0.95
Koçaslan	2017	ANFIS	5	R^2	1
Hasanipanah et al	2017	CART	2	R^2	0.95
Hasanipanah et al	2017	CART	2	R^2	0.95
Khandelwal et al	2017	CART	2	R^2	0.92
Shahnazar et al	2017	PSO-ANFIS	2	R^2	0.98
Armaghani et al	2018	ICA	2	R^2	0.95
Armaghani et al	2018	ICA	8	R^2	0.95
Nguyen et al	2019	XGBoost	9	R^2	0.952
Nguyen et al	2019	HKM-CA	6	R^2	0.99
Nguyen et al	2019	SVR-GA	2	R^2	0.99
Huang, Koopialipoor, and Armaghani	2020	FA-ANN	9	R^2	0.91
Nguyen et al	2020	SVR-GA	5	R^2	0.99
Zhang et al	2020	PSO-XGBoost	8	R^2	0.968
Zhang et al	2020	RF, CART, CHAID	6	R^2	RF = 0.94, CART = 0.97, CHAID = 0.91
Zhou et al	2020a	FS-RF	5	R^2	0.903
Zhou et al	2020b	RF	5	R^2	0.93
Nguyen et al	2020	HKM-CA	2	R^2	0.99
Huang, Koopialipoor, and Armaghani	2020	FA-ANN	6	R^2	0.91
Zhang et al	2020	RF, CART, CHAID	2	R^2	0.94
Lawal, Kwon and Kim	2021	ANN-MFO	5	R^2	0.97
Qiu et al	2022	(WOA-, GWO-, BO-) XGBoost	13	R^2	0.9757

Author(s)	Year	ML technique	No of parameters	Performance metrics	Model performance
Ragam, Komalla and Kanne	2022	XGBoost-RF	9	R^2	0.95
Hosseini *et al*	2023	BH-LSTM	6	R^2	0.9956
Nguyen, Bui and Topal	2023	(SpaSO-, SaISO-, MFO-)ELM	5	R^2	0.99
Fissha *et al*	2025	PSO-DRVM	2	R^2	0.917

Gaps in the literature

From literature various gaps can be observed to exist. One of the gaps that still exists is the development of rigorous comparative studies to establish models that show good performance with regards to the prediction of PPV. In addition, the current literature shows that Deep Neural Networks (DNNs) have not been extensively applied in the estimation of PPV, however, DNNs, been successfully previously used in other fields of Science, finance and engineering and proven to be effective in prediction.

Research objectives

The objectives of this research are:

- To apply Deep Neural Networks in the prediction of PPV.

- To compare the results to machine learning algorithms that have previously been used.

- To provide recommendations on some of the efficient methods that can be applied in the estimation of PPV.

Research contribution

The main contribution in this research work is the presentation of application of deep neural networks in estimation of PPV. The results obtained are compared to existing results conducted using supervised and un supervised methods.

Research significance

Determination of how the ground vibrates during blasting is time consuming. Therefore, to reduce the impact during blasting, there is a need to come up with more efficient ways to predict the impacts of vibration during blasting. The current study presents application of Deep Neural Network in estimating PPV, the algorithm used to leverage on their ability to handle complex data.

METHODOLOGY AND MATERIALS

Preliminaries to the methods used

Deep Neural Networks are used in this research and their performance is compared to commonly used machine learning methods for instance; Decision tree, Random forest, XGboost. The preliminaries ML algorithms have already been provided in literature in detail by various authors for instance (Yan *et al*, 2024). The sections to come will provide the preliminaries to the Artificial Neural Network (ANN), Long Short-Term Memory Networks (LSTMs), Recurrent Neural Networks(RNN).

Artificial Neural Networks (ANN)

AI techniques were first introduced in the 1970s. Some of the branches of AI are Ann, Case Based Reasoning (CBR), Genetic Algorithm (GA), Expert systems and fuzzy logic. ANN have currently gained traction in various fields of science. The ANN algorithm simulates the structure of the human brain by imitating the way the human brain functions. ANN are made up of various interconnected structures with neurons that are capable of performing parallel computation. A neural network can be defined with three fundamental components: Network architecture, transfer function and learning law. The input layer, output layer and hidden layer form the architecture of a neural network as shown in Figure 2. Each layer has numerous nodes which are connected to each other via weight.

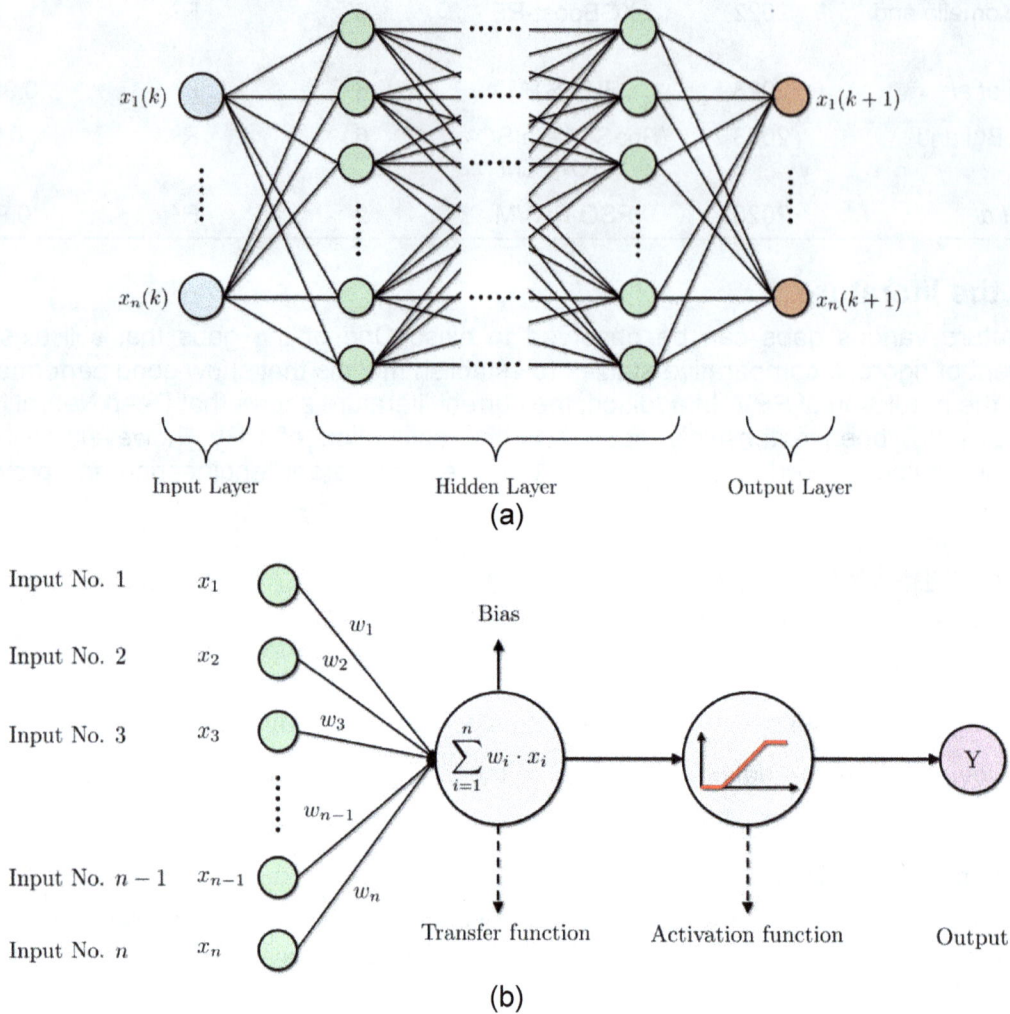

(a)

(b)

FIG 2 – (a) Basic architecture of an ANN; (b) Basic architecture of an individual neuron in an ANN.

Recurrent Neural Networks (RNN)

Recurrent Neural Networks (RNNs) are a class of artificial neural networks that are designed to process sequential data by maintaining an internal memory, enabling them to capture temporal dependencies. Unlike feedforward neural networks, RNNs incorporate feedback loops, allowing information to persist across time steps. This characteristic is achieved through recurrent connections that pass information from one-time step to the next, effectively creating a 'memory' of past inputs. RNNs are particularly well-suited for tasks where the order of information is crucial, such as natural language processing, time series analysis, and speech recognition (Chung *et al*, 2014). However, traditional RNNs can struggle with long-range dependencies due to the vanishing or exploding gradient challenge (Heaton, 2017), which has led to the development of more sophisticated architectures like Long Short-Term Memory (LSTM) networks and Gated Recurrent Units (GRUs) (Hochreiter and Schmidhuber, 1997). These advanced variants introduce gating mechanisms that regulate the flow of information, allowing the network to selectively remember or forget past inputs, thereby mitigating the challenges associated with learning long-term dependencies.

Gated Recurrent Units (GRUs)

Gated Recurrent Units (GRUs) are a form of recurrent neural network (RNN) architecture which were developed specifically to address the vanishing gradient challenge, which hinders the training of traditional RNNs, especially for long sequences. They were introduced by Cho *et al* (2014) as a simplification of the Long Short-Term Memory (LSTM) network. GRUs employ gating mechanisms to control the flow of information within the network. These gates regulate how much of the past information is retained and how much of the new information is incorporated. This allows the network to selectively remember or forget information, which is crucial for capturing long-range dependencies. GRUs have only two gates: the update gate and the reset gate. This simplification reduces the number of parameters and makes GRUs computationally more efficient. The update gate determines how much of the past hidden state is retained. It essentially controls how much of the previous information is carried forward to the next time step. The reset gate determines how much of the past hidden state is ignored. It controls how much of the previous information is used to compute the current candidate hidden state. The hidden state of a GRU represents the memory of the network. It is updated at each time step based on the input and the gating mechanisms. GRUs are computationally less expensive than LSTMs due to their simpler architecture. GRUs are effective for various sequence modelling tasks, including natural language processing, time series analysis, and speech recognition. The gating mechanisms in GRUs help mitigate the vanishing gradient problem, enabling the network to learn long-range dependencies.

Long Short-Term Memory Networks (LSTMs)

LSTM have capabilities of learning long-term dependencies for prediction challenges with sequence. Performance historical data of an LSTM is stored in a memory cell. The memory cell has the following regulating gates; the input, forget and output gates which permit the LSTM network to delete or add information from the cell state. The equations (Hosseini *et al*, 2023) that are used to build the LSTM are provided in Equations 1 to 5:

$$(i_t = \sigma(W_i h_{t-1} + W_i h_t + b_i)) \tag{1}$$

$$f_t = \sigma(W_f h_{t-1} + W_f h_t + b_f) \tag{2}$$

$$\tilde{C}_t = \tanh(W_c h_{t-1} * W_c x_t * b_c) \tag{3}$$

$$o_t = \sigma(W_o h_{t-1} + W_o h_t + b_o) \tag{4}$$

$$h_t = o_t * \tanh(c_t) \tag{5}$$

Performance metrics

This section considers various performance metrics that are generally employed in evaluation of the performance of ML algorithm are provided. For reference purposes the equations (Hosseini *et al*, 2023) that represent this metrics are provided in Equations 6 to 17.

Coefficient of determination (R²)

$$R^2 = \frac{[\sum_{m=1}^{n}(x_m - x_{mm})^2] - [\sum_{m=1}^{n}(x_m - x_p)^2]}{\sum_{m=1}^{n}(x_m - x_{mm})^2} \tag{6}$$

Mean Square Error (MSE)

$$MSE = \frac{1}{n}\sum_{i=1}^{n}(x_m - x_{mm})^2 \tag{7}$$

Root Means Square Error (RSME)

$$RMSE = \sqrt{\frac{1}{n}\sum_{i=1}^{n}(x_m - x_{mm})^2} \tag{8}$$

Variance Account For (VAF)

$$VAF = \left(1 - \frac{var(x_m - x_p)}{var(x_m)}\right) \times 100 \tag{9}$$

Mean Absolute Error (MAE)

$$MAE = \frac{1}{n}\sum_{i=1}^{n}(|x_p - x_m|) \tag{10}$$

Mean Bias Error (NMBE)

$$NMBE = \frac{\frac{1}{N}\sum_{i=-1}^{n}(x_m - x_m)^2}{\frac{1}{N}\sum_{m=1}^{n} x_m} \tag{11}$$

Mean Absolute Percentage Error (MAPE)

$$MAPE = \frac{1}{n}\sum_{i=1}^{n}\left|\frac{x_m - x_p}{x_m}\right| \times 100 \tag{12}$$

Nash–Sutcliffe Efficiency (NS)

$$NS = 1 - \frac{\sum_{i=1}^{n}(x_m - x_p)^2}{\sum_{m=1}^{n}(x_m - x_{mm})^2} \tag{13}$$

Index of Agreement (IOA)

$$IOA = 1 - \frac{\sum_{i=1}^{n}(x_p - x_m)}{2\sum_{i=1}^{n}(x_p - x_{mm})} \tag{14}$$

Index of scatter (IOS)

$$IOS = \frac{RMSE}{\text{Average of actual values}} \tag{15}$$

a20-index (a20)

$$a20index = \frac{m20}{N_D} \tag{16}$$

Performance index (PI)

$$PI = R^2 + \left(\frac{VAF}{100}\right) - RMSE \tag{17}$$

Data preprocessing and feature selection

The general methodology followed in this research work is shown in Figure 3. Open source data set that were obtained from a database undergo the cleaning process which includes checking for missing values. Checking for duplicates and removing them from the data set.

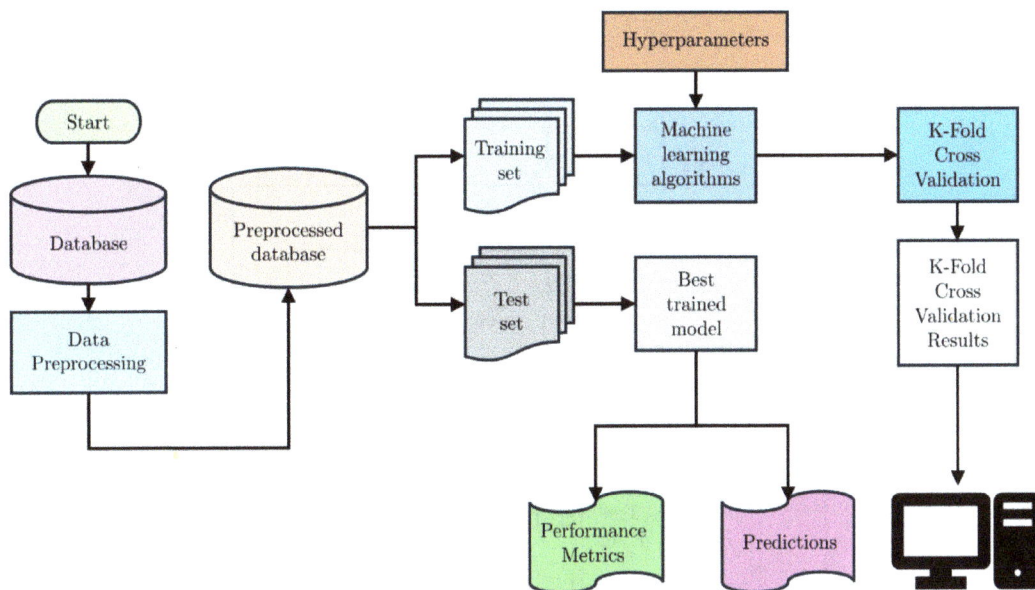

FIG 3 – Overall methodology followed.

Table 2 shows some of the parameters used in the development of the algorithms (Dehghani and Ataee-Pour, 2011).

TABLE 2

Input and output parameters used in the development of the algorithms.

	Parameter	Symbol	Unit
	Burden	B	m
	Spacing	S	m
	Delay between rows	D_e	ms
	Powder factor	q	kg/m³
Input	Number of rows in each blast	n	-
	Distance of monitoring point from blasting face	μ	m
	Maximum hole per delay	θ	-
	Charge per delay	ch	kg
	Point load index	σ	MPa
Output	Peak particle velocity	PPV	mm/s

Data cleaning and preprocessing

A generalised methodology was followed to develop the algorithms, as shown in Figure 4.

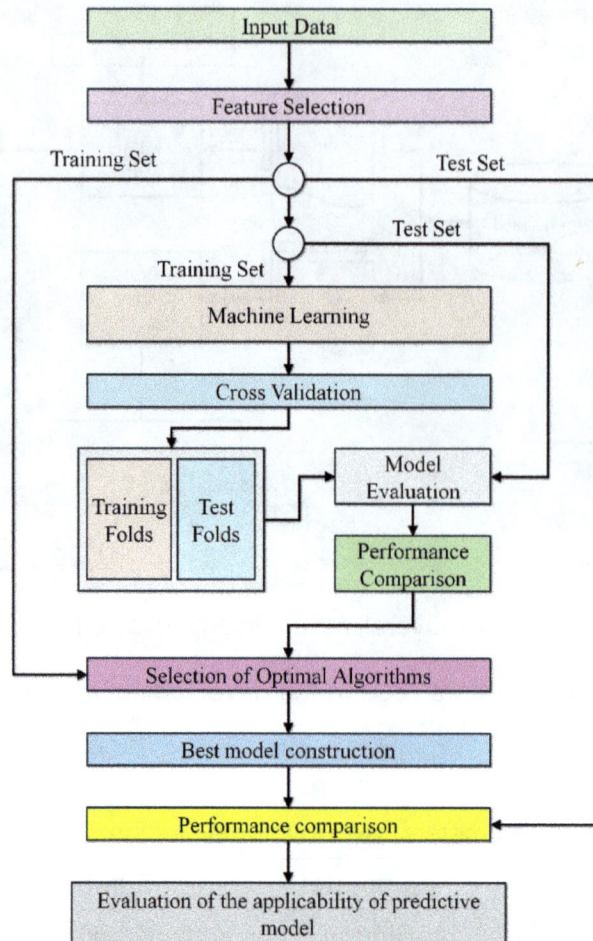

FIG 4 – Model development (Fissha *et al*, 2025).

An open source data set (Morena, 2019) with 17 features and 799 observations was used in the development of the machine learning algorithm used. The data set consists of drilling and blasting parameters for instance: Overburden (m), Hole Size (mm), Charge per hole (kg), Distance (m), and PPV Peak (mm/sec). Spaces and special characters in the data set were removed. Log transformation was applied to reduce the effects of outliers and normalise skewed distributions while maintaining the integrity of the structure of the data. It is important to note that the site number and test are not input variables. The columns with significant outliers were transformed. Figure 5 shows the feature correlation heat map.

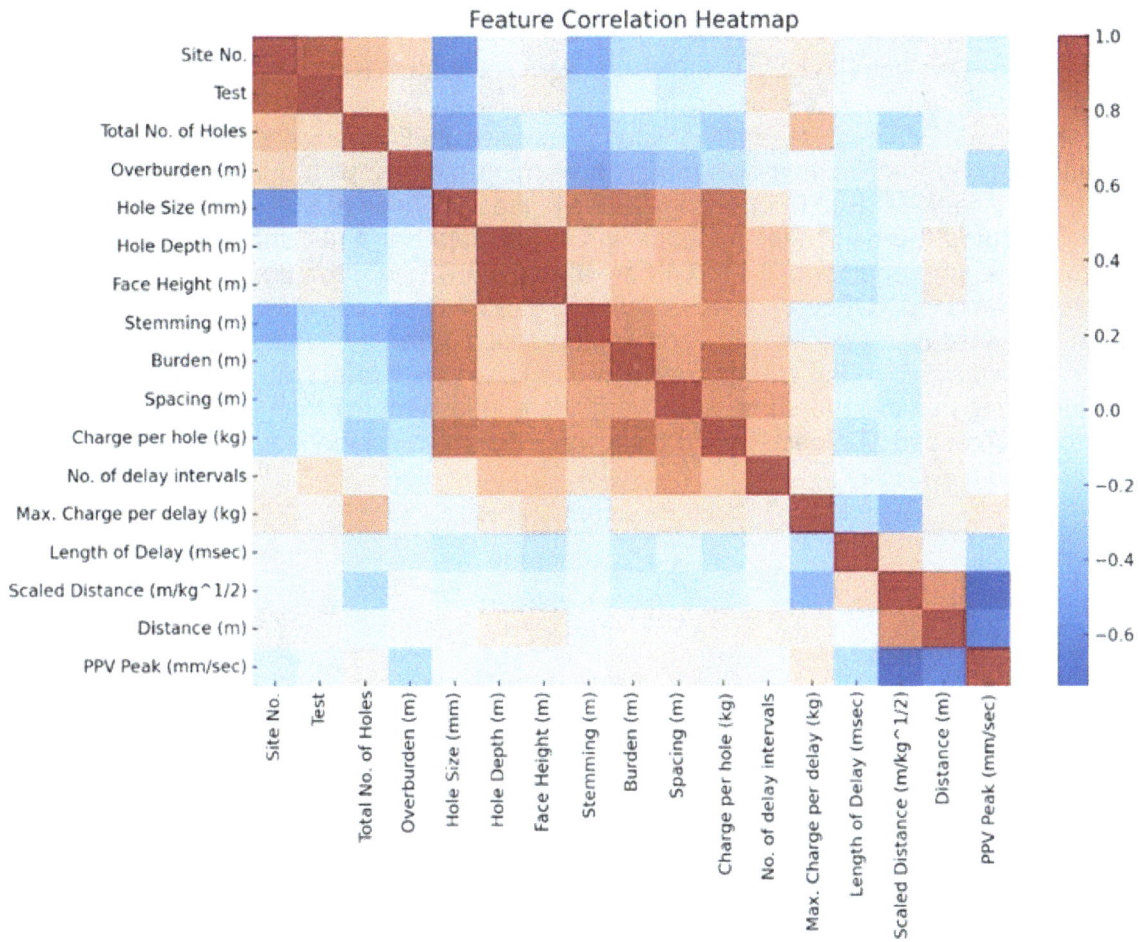

FIG 5 – Shows the feature correlation heat map.

RESULTS AND DISCUSSION

Table 3 shows the performance of ML algorithms in of prediction of PPV.

TABLE 3

Performance of ML algorithms in of prediction of PPV.

Algorithm	MSE	RMSE	NRMSE	MAD	MAPE	R^2
Random Forest	0.0000250	0.015	0.033	0.007	53.731	0.94
Decision Tree	0.000220	0.034	0.012	0.007	30.363	0.95
XGboost	0.00194	0.013	0.030	0.003	18.471	0.96

Table 4 shows the performance of deep learning neural networks algorithms in of prediction of PPV.

TABLE 4

Deep learning neural networks performance in prediction of PPV.

Algorithm	R^2
ANNs	0.96
LSTM	0.99
GRN	0.98
RNN	0.98

Performance of machine learning models

Random Forest, Decision Tree, and XGBoost were evaluated based on the main performance metrics that were previously discussed for instance: Mean Squared Error MSE, RMSE, RMSE, NRMSE, MAD, MAPE, and R^2, which pictorially shown in Figure 5. XGBoost attained a good performance among traditional known machine learning models, with the lowest MSE (0.00194), RMSE (0.013), and NRMSE (0.030), as well as the highest R^2 value (0.96). This shows that XGBoost effectively captured the non-linear relationships in the data set, thus attaining a better accuracy compared to the other methods. Random Forest also showed a good performance, with an R^2 of 0.94, slightly lower than XGBoost. However, its MAPE (53.731 per cent) is significantly higher, which suggests that it had challenges with percentage-based error evaluation. Decision Tree, while achieving the highest R^2 (0.95) among the ML models, had the highest MSE (0.000220) and RMSE (0.034), indicating that it is more prone to overfitting and less robust than ensemble-based models like Random Forest and XGBoost. Figure 6 shows R^2 versus machine learning models.

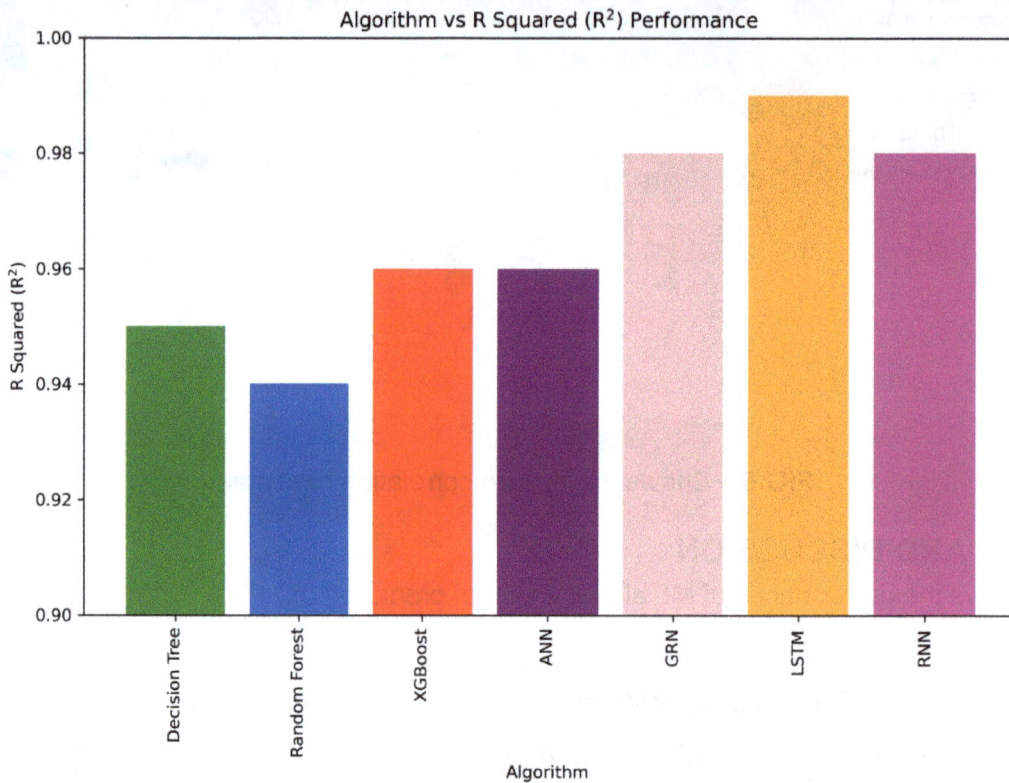

FIG 6 – Shows R^2 verses machine learning models.

Performance of deep learning models

The deep learning models—Artificial Neural Networks (ANNs), Long Short-Term Memory (LSTM), Gated Recurrent Networks (GRN), and Recurrent Neural Networks (RNN)—demonstrated superior predictive performance, as reflected by their higher R^2 values. LSTM ($R^2 = 0.99$) outperformed all other models, indicating that it effectively captures long-term dependencies and sequential patterns in the data set. This suggests that PPV prediction benefits from time-series modelling, where LSTMs excel. GRN and RNN (both $R^2 = 0.98$) also showed strong predictive ability, though slightly below LSTM. These models effectively capture temporal dependencies but may not be as optimised as LSTM for handling complex sequential data. ANNs ($R^2 = 0.96$) performed similarly to XGBoost, highlighting that while ANNs are powerful, they may not fully capture temporal dependencies as well as recurrent models like LSTM and GRN.

Comparison between machine learning and deep learning models

Deep learning models outperformed traditional machine learning models, with LSTM achieving the highest accuracy ($R^2 = 0.99$). While XGBoost was the best machine learning model, it was still outperformed compared to the deep learning techniques. The superiority of LSTM, GRN, and RNN suggests that PPV prediction benefits from sequential modelling, which deep learning models are better suited for.

These results indicate that LSTM is the most effective model for predicting PPV, followed by GRN and RNN, while XGBoost remains the best traditional ML model. The findings highlight the importance of deep learning in improving accuracy when dealing with complex, time-dependent data like blast vibrations.

Implications for real world mining

The results of this study have significant implications for blast vibration monitoring and control in quarries and mining operations. Accurate prediction of Peak Particle Velocity (PPV) is crucial for minimising environmental impact, ensuring regulatory compliance, and optimising blasting efficiency. The superior performance of LSTM, GRN, and RNN models compared to traditional machine learning models suggests that deep learning-based approaches should be prioritised for real-world applications.

CONCLUSIONS AND FUTURE RECOMMENDATIONS

Accurate prediction of PPV is critical in rock blasting. The focus of this research work was on the application of deep learning Neural Networks in the prediction of PPV and thereafter comparing the results to the commonly used ML algorithms. The results show that LSTM, GRN and RNN outperformed the traditional algorithms even though the performance of the traditional machine learning algorithm is still good. Future work will include amalgamation of various machine learning algorithms to improve the accuracy further. The practical integration of the developed algorithms into actual systems will also be explored.

REFERENCES

Ambraseys, N R and Hendron, A J, 1968. Dynamic Behavior of Rock Masses, in *Rock Mechanics in Engineering Practice* (eds: K G Stagg and O C Zienkiewicz).

Amiri, M, Bakhshandeh Amnieh, H, Hasanipanah, M and Khandelwal, M, 2016. A new combination of artificial neural network and K-nearest neighbors models to predict blast-induced ground vibration and air-overpressure, *Engineering with Computers*, 32:631–644. https://doi.org/10.1007/s00366-016-0442-5

Armaghani, D J, Hajihassani, M, Mohamad, E T, Marto, A and Noorani, S A, 2014. Blasting-induced flyrock and ground vibration prediction through an expert artificial neural network based on particle swarm optimization, *Arab J Geosci*, 7:5383–5396.

Armaghani, D J, Hasanipanah, M, Amnieh, H B and Mohamad, E T, 2018. Feasibility of ICA in approximating ground vibration resulting from mine blasting, *Neural Comput Appl*, 29:457–465.

Armaghani, D J, Momeni, E, Abad, S V A N K and Khandelwal, M, 2015. Feasibility of ANFIS model for prediction of ground vibrations resulting from quarry blasting, *Environ Earth Sci*. https://doi.org/10.1007/s12665-015-4305-y

Bakhtavar, E, Hosseini, S, Hewage, K and Sadiq, R, 2021a. Green blasting policy: Simultaneous forecast of vertical and horizontal distribution of dust emissions using artificial causality-weighted neural network, *J Clean Prod*, 283:124562.

Bakhtavar, E, Hosseini, S, Hewage, K and Sadiq, R, 2021b. Air pollution risk assessment using a hybrid fuzzy intelligent probability-based approach: Mine blasting dust impacts, *Nat Resour Res*. https://doi.org/10.1007/s11053-020-09810-4

Cho, K, Van Merriënboer, B, Gulcehre, C, Bahdanau, D, Bougares, F, Schwenk, H and Bengio, Y, 2014. Learning phrase representations using RNN encoder-decoder for statistical machine translation, in *Proceedings of the 2014 Conference on Empirical Methods in Natural Language Processing (EMNLP)*, pp 1724–1734.

Chung, J, Gulcehre, C, Cho, K and Bengio, Y, 2014. Empirical evaluation of gated recurrent neural networks on sequence modeling, in NIPS 2014 Deep Learning Workshop.

Das, A, Sinha, S and Ganguly, S, 2019. Development of a blast-induced vibration prediction model using an artificial neural network, *Journal of the Southern African Institute of Mining and Metallurgy*, 119(2):187–200. https://doi.org/10.17159/2411-9717/2019/v119n2a11

Davies, B, Farmer, I W and Attewell, P B, 1964. Ground vibration from shallow sub-surface blasts, *Engineer*, 217.

Dehghani, H and Ataee-Pour, M, 2011. Development of a model to predict peak particle velocity in a blasting operation, *International Journal of Rock Mechanics and Mining Sciences*, 48(1):51–58. https://doi.org/10.1016/j.ijrmms.2010.08.005

Dindarloo, S R, 2015. Peak particle velocity prediction using support vector machines: A surface blasting case study, *J S Afr Inst Min Metall*, 115(7):637–643. https://doi.org/10.17159/2411-9717/2015/V115N7A10

Dowding, C H, 1985. *Blast Vibration Monitoring and Control* (Prentice-Hall Inc).

Fişne, A, Kuzu, C and Hüdaverdi, T, 2011. Prediction of environmental impacts of quarry blasting operation using fuzzy logic, *Environmental monitoring and assessment*, 174(1):461–470. https://doi.org/10.1007/s10661-010-1470-z

Fissha, Y, Ragam, P, Ikeda, H, Kumar, N K, Adachi, T, Paul, P S and Kawamura, Y, 2025. Data-driven machine learning approaches for simultaneous prediction of peak particle velocity and frequency induced by rock blasting in mining, *Rock Mechanics Bulletin*, 4(1):100166. https://doi.org/10.1016/j.rockmb.2024.100166

Fouladgar, N, Hasanipanah, M and Bakhshandeh Amnieh, H, 2017. Application of cuckoo search algorithm to estimate peak particle velocity in mine blasting, *Engineering with Computers*, 33:181–189. https://doi.org/10.1007/s00366-016-0463-0

Ghasemi, E, Ataei, M and Hashemolhosseini, H, 2013. Development of a fuzzy model for predicting ground vibration caused by rock blasting in surface mining, *JVC/J Vib Control*. https://doi.org/10.1177/1077546312437002

Ghoraba, S, Monjezi, M, Talebi, N, Armaghani, D J and Moghaddam, M R, 2016. Estimation of ground vibration produced by blasting operations through intelligent and empirical models, *Environ Earth Sci*. https://doi.org/10.1007/s12665-016-5961-2

Hajihassani, M, Armaghani, D J, Marto, A and Mohamad, E T, 2015a. Ground vibration prediction in quarry blasting through an artificial neural network optimized by imperialist competitive algorithm, *Bull Eng Geol Environ*, 74:873–886.

Hajihassani, M, Armaghani, D J, Monjezi, M, Mohamad, E T and Marto, A, 2015b. Blast-induced air and ground vibration prediction: A particle swarm optimization-based artificial neural network approach, *Environ Earth Sci*, 74:2799–2817.

Hasanipanah, M, Amnieh, H B, Arab, H and Zamzam, M S, 2018. Feasibility of PSO–ANFIS model to estimate rock fragmentation produced by mine blasting, *Neural Comput and Applic*, 30:1015–1024. https://doi.org/10.1007/s00521-016-2746-1

Hasanipanah, M, Faradonbeh, R S, Amnieh, H B, Armaghani, D J and Monjezi, M, 2017. Forecasting blast-induced ground vibration developing a CART model, *Eng Comput*. https://doi.org/10.1007/s00366-016-0475-9

Hasanipanah, M, Monjezi, M, Shahnazar, A, Jahed Armaghani, D and Farazmand, A, 2015. Feasibility of indirect determination of blast induced ground vibration based on support vector machine, *Meas J Int Meas Confed*. https://doi.org/10.1016/j.measurement.2015.07.019

Heaton, J, 2017. Ian Goodfellow, Yoshua Bengio and Aaron Courville: Deep learning, *Genetic Programming and Evolvable Machines*, 19:305–307.

Hochreiter, S and Schmidhuber, J, 1997. Long short-term memory, *Neural Computation*, 9(8):1735–1780. https://doi.org/10.1162/neco.1997.9.8.1735

Hosseini, S, Monjezi, M, Bakhtavar, E and Mousavi, A, 2021. Prediction of dust emission due to open pit mine blasting using a hybrid artificial neural network, *Nat Resour Res*. https://doi.org/10.1007/s11053-021-09930-5

Hosseini, S, Mousavi, A and Monjezi, M, 2022. Prediction of blast-induced dust emissions in surface mines using integration of dimensional analysis and multivariate regression analysis, *Arab J Geosci*, 15:163.

Hosseini, S, Pourmirzaee, R, Armaghani, D J and Sabri Sabri, M M, , 2023. Prediction of ground vibration due to mine blasting in a surface lead–zinc mine using machine learning ensemble techniques, *Sci Rep*, 13:6591. https://doi.org/10.1038/s41598-023-33796-7

Huang, J, Koopialipoor, M and Armaghani, D J, 2020. A combination of fuzzy Delphi method and hybrid ANN-based systems to forecast ground vibration resulting from blasting, *Sci Rep*, 10:1–21.

Iphar, M, Yavuz, M and Ak, H, 2008. Prediction of ground vibrations resulting from the blasting operations in an open pit mine by adaptive neuro-fuzzy inference system, *Environ Geol*. https://doi.org/10.1007/s00254-007-1143-6

Jahed, D, Ehsan, A, Seyed, M, Alavi, V and Khalil, N, 2015. Feasibility of ANFIS model for prediction of ground vibrations resulting from quarry blasting, *Environ Earth Sci*, pp 2845–2860. https://doi.org/10.1007/s12665-015-4305-y

Jiang, W, Arslan, C A, Tehrani, M S, Khorami, M and Hasanipanah, M, 2019. Simulating the peak particle velocity in rock blasting projects using a neuro-fuzzy inference system, *Engineering with Computers*, 35:1203–1211. https://doi.org/10.1007/s00366-018-0659-6

Khandelwal, M, Armaghani, D J, Faradonbeh, R S, Yellishetty, M, Majid, M Z A and Monjezi, M, 2017. Classification and regression tree technique in estimating peak particle velocity caused by blasting, *Engineering with Computers*, 33(1):45–53. https://doi.org/10.1007/s00366-016-0455-0

Khandelwal, M, Kankar, P K and Harsha, S P, 2010. Evaluation and prediction of blast induced ground vibration using support vector machine, *Mining Science and Technology (China)*, 20(1):64–70. https://doi.org/10.1016/S1674-5264(09)60162-9

Khandelwal, M, Kumar, D L and Yellishetty, M, 2011. Application of soft computing to predict blast-induced ground vibration, *Eng Comput*. https://doi.org/10.1007/s00366-009-0157-y

Koçaslan, A, Yüksek, A G, Görgülü, K and Arpaz, E, 2017. Evaluation of blast-induced ground vibrations in open-pit mines by using adaptive neuro-fuzzy inference systems, *Environ Earth Sci*, 76:57. https://doi.org/10.1007/s12665-016-6306-x

Lawal, A I, Kwon, S and Kim, G Y, 2021. Prediction of the blast-induced ground vibration in tunnel blasting using ANN, moth-fame optimized ANN and gene expression programming, *Acta Geophys*. https://doi.org/10.1007/s11600-020-00532-y

Mohamadnejad, M, Gholami, R and Ataei, M, 2012. Comparison of intelligence science techniques and empirical methods for prediction of blasting vibrations, *Tunn Undergr Sp Technol*, 28:238–244.

Mohamed, M T, 2011. Performance of fuzzy logic and artificial neural network in prediction of ground and air vibrations, *JES J Eng Sci*, 39:425–440.

Monjezi, M, Ghafurikalajahi, M and Bahrami, A, 2011. Prediction of blast-induced ground vibration using artificial neural networks, *Tunn Undergr Sp Technol*, 26:46–50.

Monjezi, M, Hasanipanah, M and Khandelwal, M, 2013. Evaluation and prediction of blast-induced ground vibration at Shur River Dam, Iran, by artificial neural network, *Neur Comput Appl*, 22(7–8):1637–1643. https://doi.org/10.1007/s00521-0120856-y

Morena, B I, 2019. Prediction of blast vibrations from quarries using machine learning algorithms and empirical formulae, Master's dissertation, University of the Witwatersrand.

Nguyen, H and Bui, X-N, 2020. Soft computing models for predicting blast-induced air over-pressure: A novel artificial intelligence approach, *Appl Sof Comput*, 92:106292.

Nguyen, H, Bui, X-N and Topal, E, 2023. Reliability and availability artificial intelligence models for predicting blast-induced ground vibration intensity in open pit mines to ensure the safety of the surroundings, *Reliab Eng Syst Saf*, 231:109032.

Nguyen, H, Bui, X-N, Tran, Q-H and Mai, N-L, 2019. A new soft computing model for estimating and controlling blast-produced ground vibration based on hierarchical K-means clustering and cubist algorithms, *Appl Sof Comput*, 77:376–386.

Nguyen, H, Choi, Y, Bui, X N and Nguyen-Toi, T, 2020. Predicting blast-induced ground vibration in open pit mines using vibration sensors and support vector regression-based optimization algorithms, *Sensors* (Switzerland). https://doi.org/10.3390/s20010132

Qiu, Y, Zhou, J, Khandelwal, M, Yang, H, Yang, P and Li, C, 2022. Performance evaluation of hybrid WOA-XGBoost, GWO-XGBoost and BO-XGBoost models to predict blast-induced ground vibration, *Engineering with Computers*, 38(Suppl 5):4145–4162. https://doi.org/10.1007/s00366-021-01393-9

Ragam, P, Komalla, A R and Kanne, N, 2022. Estimation of blast-induced peak particle velocity using ensemble machine learning algorithms: A case study, *Noise Vib Worldw*, 53:404–413.

Rai, R and Singh, T N, 2004. A new predictor for ground vibration prediction and its comparison with other predictors, *Indian Journal of Engineering and Materials Sciences*, 11:178–184.

Roy, P P, 1993. *Putting ground vibration predictions into practice*, 241 p (Colliery Guard: Kingdom).

Shahnazar, A, Nikafshan Rad, H, Hasanipanah, M, Tahir, M M, Armaghani, D J and Ghoroqi, M, 2017. A new developed approach for the prediction of ground vibration using a hybrid PSO-optimized ANFIS-based model, *Environ Earth Sci*, 76:527. https://doi.org/10.1007/s12665-017-6864-6

Shirani Faradonbeh, R, Jahed Armaghani, D, Abd Majid, M Z, Md Tahir, M, Ramesh Murlidhar, B, Monjezi, M and Wong, H M, 2016. Prediction of ground vibration due to quarry blasting based on gene expression programming: A new model for peak particle velocity prediction, *Int J Environ Sci Technol*, 13(6):1453–1464. https://doi.org/10.1007/s13762-016-0979-2

Singh, T N and Singh, V, 2005. An intelligent approach to prediction and control ground vibration in mines, *Geotech Geol Eng*. https://doi.org/10.1007/s10706-004-7068-x

Taheri, K, Hasanipanah, M, Golzar, S B and Majid, M Z A, 2017. A hybrid artificial bee colony algorithm-artificial neural network for forecasting the blast-produced ground vibration, *Engineering with Computers*, 33:689–700. https://doi.org/10.1007/s00366-016-0497-3

Yan, Y, Guo, J, Bao, S, and Fei, H, 2024. Prediction of peak particle velocity using hybrid random forest approach, *Sci Rep*, 14(1):30793. https://doi.org/10.1038/s41598-024-81218-z

Zhang, H, Zhou, J, Jahed Armaghani, D, Tahir, M M, Pham, B T and Huynh, V V, 2020. A Combination of Feature Selection and Random Forest Techniques to Solve a Problem Related to Blast-Induced Ground Vibration, *Applied Sciences*, 10(3):869. https://doi.org/10.3390/app10030869

Zhang, Y, He, H, Khandelwal, M, Du, K and Zhou, J, 2023. Knowledge mapping of research progress in blast-induced ground vibration from 1990 to 2022 using CiteSpace-based scientometric analysis, Environ Sci Pollut Control Ser, 30(47):103534–103555. https://doi.org/10.1007/s11356-023-29712-1

Zhou, J, Asteris, P G, Armaghani, D J and Pham, B T, 2020a. Prediction of ground vibration induced by blasting operations through the use of the Bayesian Network and random forest models, *Soil Dyn Earthq Eng.* https://doi.org/10.1016/j.soildyn.2020.106390

Zhou, J, Li, C, Koopialipoor, M, Armaghani, D J and Pham, B T, 2020b. Development of a new methodology for estimating the amount of PPV in surface mines based on prediction and probabilistic models (GEP-MC), *Int J Mining Reclam Environ*, 35:48–68.

Using digital twin for production scheduling problems – functions used in mining operations

E Moosavi[1] and K Tolouei[2]

1. Department of Petroleum and Mining Engineering; and Research Center for Modeling and Optimization in Science and Engineering, South Tehran Branch, Islamic Azad University, Tehran, Iran. Email: se_moosavi@azad.ac.ir; se.moosavi@yahoo.com
2. Department of Petroleum and Mining Engineering; and Research Center for Modeling and Optimization in Science and Engineering, South Tehran Branch, Islamic Azad University, Tehran, Iran. Email: kamyar.toloie@gmail.com

ABSTRACT

Production scheduling is concerned about when and where production begins in mining, which is mainly determined by the bench base. According to the processing feed order requirements and the actual situation of the pit, the best scheduling scheme for completing the order can be determined, so that the final product can be accomplished with standard quality and quantity in the specified time that customer stipulated. Digital twin technology allows mining companies to test different initiatives and innovations without risk to operations. Some of the advantages of this technology in the mineral processing industry are: (i) It is used as a decision-analytical tool to avoid unnecessary costs and reducing running, maintenance and environmental inefficiencies; (ii) It tests different scenarios to determine the amount of power; and (iii) Avoid poor process changes. The growth of digital technologies opens the possibility to collect and analyse great amount of field data in real-time, representing a precious opportunity for an improved scheduling activity. Thus, scheduling under uncertain scenarios may benefit from the possibility to grasp the current operating conditions of the mining equipment in real-time and take them into account when elaborating the best production schedules. By digital twin, the work environment miners will be able to create long-term and short-term schedules and create accurate estimates for drilling, crushing and extraction work will be over and what the final product results will be. Moreover, by simulating the equipment, machinery, and the entire work process, on-site workers will be able to test new methodologies on their most crucial work processes in a very cost-effective manner because no capital will be required to accurately find out what works – every test will be executed in a digital simulation, using the same exact machinery and equipment.

INTRODUCTION

Production planning mainly solves the problem of what and how much to produce in a mining operation, which is mainly decided by upstream operations. Production scheduling is concerned about when and where production begins in mining, which is mainly determined by the bench base. According to the processing feed order requirements and the actual situation of the pit, the best scheduling scheme for completing the order can be determined, so that the final product can be accomplished with standard quality and quantity in the specified time that customer stipulated. When planning and scheduling is used to guide the actual production, there will be external changes in customer orders and internal changes in personnel, equipment, materials etc, such as cut-off grade, average grade, equipment fault, production capacity, operational constraints etc.

These changes are collectively referred to as uncertainty factors. Due to the low transparency of the existing mining information, when these uncertain factors appear in the production, it will seriously hinder the formulation and implementation of the scheduling plan in the time dimension. In this aspect, it is the breakthrough point that we can predict the uncertain factors that may appear in the production process through the transparent data of the mining, and take relevant measures in advance, which can form a plan and scheduling scheme to guide the actual production accurately. Digital twin provides technical support for this breakthrough point.

The digital twin, firstly proposed by Dr Michael Grieves (2003), is defined as making full use of the data such as physical model, sensor update, operation history etc and integrating the simulation process with multi-disciplinary, multi-physical quantities, multi-scale, and multi-probability to realise the mapping that can be completed in virtual space, so as to use it to reflect the corresponding entity equipment life cycle process (Glaessgen and Stargel, 2012). Digital twins can be defined as a software system that represents and is interconnected to a specific physical reality (Van Der Horn and Mahadevan, 2021). Beyond this broad definition, digital twins have a plethora of definitions that are usually specific to the characteristic being investigated or described (eg industry and technology) and often have a specific focus on internet of things (IoT) technology. The promise of digital twins is that they will enable new ways of working and solving specific problems. The corollary to this is that specific configuration, development or connections must be performed to make the digital twin specific and valuable to a given use case. Industry use cases must be researched and understood in order for digital twin applications to be valuable. As Van Der Horn and Mahadevan (2021) states, there is a need to move digital twins from idea to practicality in order to understand the actual risks and benefits.

This is an important time in the digital transformation of mining with major miners publicly investing in digital twins and major consultants creating new practices to support adoption. The mining industry needs to better understand the potential of this new technology and direct appropriate research into how best to implement digital transformation initiatives with digital twins.

This study explores the available literature related to digital twins and the mining industry. The mining industry is investing in digital twins, and a review of relevant work is timely. The study reviews the available literature, identifies gaps in the literature and provides recommendations for future research.

DEFINITIONS

Based on the given definitions of a digital twin in any context, one might identify a common understanding of digital twins, as digital counterparts of physical objects. Within these definitions, the terms digital model, digital shadow and digital twin are often used synonymously. However, the given definitions differ in the level of data integration between the physical and digital counterpart. Some digital representations are modelled manually and are not connected with any physical object in existence, while others are fully integrated with real-time data exchange. Therefore, the authors would like to propose a classification of digital twins into three subcategories, according to their level of data integration.

A digital model is a digital representation of an existing or planned physical object that does not use any form of automated data exchange between the physical object and the digital object. The digital representation might include a more or less comprehensive description of the physical object. These models might include, but are not limited to simulation models of planned factories, mathematical models of new products, or any other models of a physical object, which do not use any form of automatic data integration. Digital data of existing physical systems might still be in use for the development of such models, but all data exchange is done in a manual way. A change in state of the physical object has no direct effect on the digital object and *vice versa*.

Digital model

A digital model is a digital representation of an existing or planned physical object that does not use any form of automated data exchange between the physical object and the digital object (see Figure 1). The digital representation might include a more or less comprehensive description of the physical object. These models might include, but are not limited to simulation models of planned factories, mathematical models of new products, or any other models of a physical object, which do not use any form of automatic data integration. Digital data of existing physical systems might still be in use for the development of such models, but all data exchange is done in a manual way. A change in state of the physical object has no direct effect on the digital object and *vice versa*.

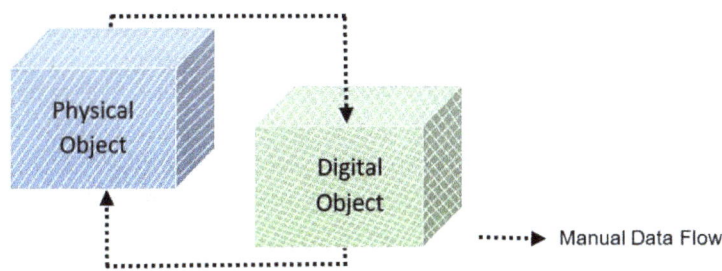

FIG 1 – Data flow in a digital model.

Digital shadow

Based on the definition of a digital model according to the Figure 2, if there further exists an automated one-way data flow between the state of an existing physical object and a digital object, one might refer to such a combination as digital shadow. A change in state of the physical object leads to a change of state in the digital object, but not *vice versa*.

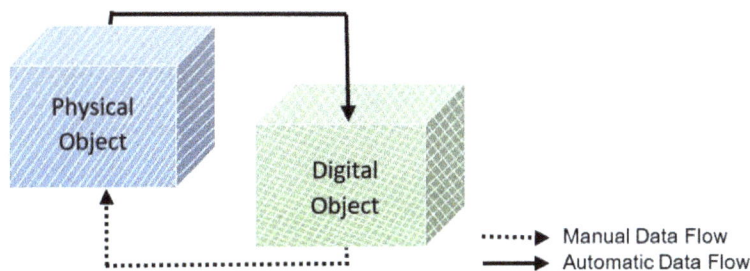

FIG 2 – Data flow in a digital shadow.

Digital twin

If further, the data flows between an existing physical object and a digital object are fully integrated in both directions, one might refer to it as digital twin (Figure 3). In such a combination, the digital object might also act as controlling instance of the physical object. There might also be other objects, physical or digital, which induce changes of state in the digital object. A change in state of the physical object directly leads to a change in state of the digital object and *vice versa*.

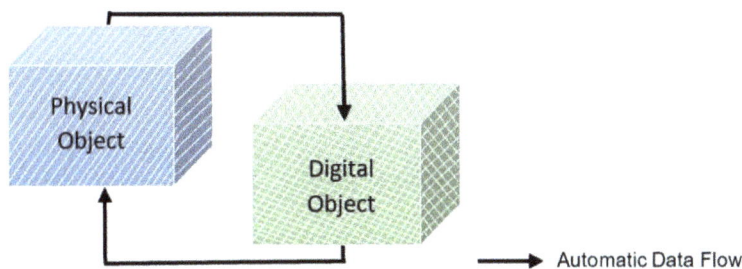

FIG 3 – Data flow in a digital twin.

INTEGRATION OF DIGITAL TWINS AND SCHEDULING

Based on the given definitions of a digital twin (DT) in any context, one might identify a common understanding of digital twins, as digital counterparts of physical objects. Within these definitions, the terms digital model, digital shadow and DT are often used synonymously. However, the given definitions differ in the level of data integration between the physical and digital counterpart. Some digital representations are modelled manually and are not connected with any physical object in existence, while others are fully integrated with real-time data exchange. Therefore, the authors would like to propose a classification of DT into three subcategories, according to their level of data integration.

With the application of digital twins, systematic scheduling can be carried out for the minerals that need to be extracted, as well as the manufacturing methods, resources, and locations etc, and all aspects can be linked together to realise the collaboration between designers and planners (Wang *et al*, 2020). Once a planning design change occurs, the manufacturing process can be easily updated in the digital twin model. In addition to process planning, production layout is also an important problem for intelligent manufacturing systems. With the help of the digital twin model, it is possible to design a production layout that contains all the detailed information, including machinery, automation, equipment, tools, resources, and even operators, and seamlessly relate it to the product design.

Sanchez and Hartlieb (2020) provide an overview of the main functions of the mining industry and the applicability of various digital transformation initiatives, as summarised in Figure 4. Based on the key functions in mining set out by Sanchez and Hartlieb (2020), the relevant literature was critically reviewed in terms of four mining functional areas as follows:

1. Exploration.
2. Mining.
3. Processing.
4. Waste management/tailing.

Further to this, other industries with extensive digital twin applications have investigated functions that are applicable to mining, as shown in Figure 5.

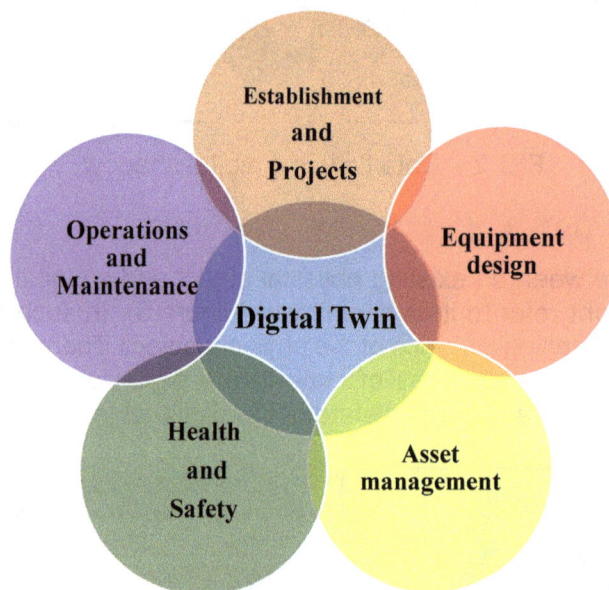

FIG 4 – Functions used in mining.

This is recalled by Figure 5, showing examples of DT-based functionalities for different application domains (production, energy, quality, maintenance etc) in the industrial practice. Moreover, DT represents the virtual counterpart of the physical asset along all its life cycle, then it aids in the management of physical assets by supporting decision-making throughout the life cycle (Macchi *et al*, 2018), also using the integration with an intelligence layer (Negri *et al*, 2020).

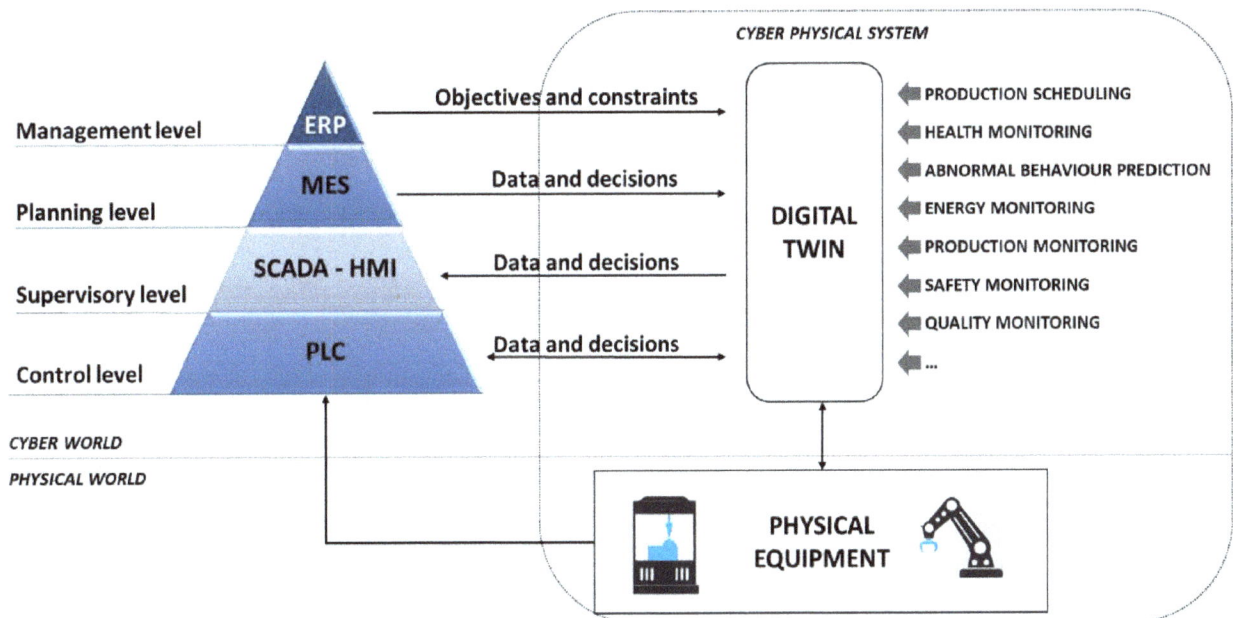

FIG 5 – Role of digital twin to support the control of manufacturing operations in cyber-physical production systems (Negri *et al*, 2021).

CONCLUSIONS

Research on scheduling problems is an evergreen challenge for mining engineers. The growth of digital technologies opens the possibility to collect and analyse great amount of field data in real-time, representing a precious opportunity for an improved scheduling activity. Thus, scheduling under uncertain scenarios may benefit from the possibility to grasp the current operating conditions of the mining equipment in real-time and take them into account when elaborating the best production schedules. Digital twin simulation enables engineers to virtually adjust 'levers' showing future scenarios around mine planning, blasting, metallurgy and process control to guide the best performance going forward. Engineers can conduct risk analysis, cost improvement studies, 'what if' simulations and scenario analysis. They can also show justification for resource allocation and develop strategies for the amelioration of negative events such as breakdowns.

REFERENCES

Glaessgen, E and Stargel, D, 2012. The digital twin paradigm for future NASA and US Air Force vehicles, in *Proceedings of the 53rd Structures Dynamics and Materials Conference*, pp 1–14 (American Institute Of Aeronautics And Astronautics).

Grieves, M, 2006. *Product lifecycle management: Driving the next generation of lean thinking* (New York: McGraw-Hill).

Macchi, M, Roda, I, Negri, E and Fumagalli, L, 2018. Exploring the role of Digital Twin for asset life cycle management, *IFACPapersOnLine*, 51(11):790–795. https://doi.org/10.1016/j.ifacol.2018.08.415

Negri, E, Berardi, S, Fumagalli, L and Macchi, M, 2020. MESintegrated Digital Twin frameworks, *Journal of Manufacturing Systems,* 56:58–71. https://doi.org/10.1016/j.jmsy.2020.05.007

Negri, E, Pandhare, V, Cattaneo, L, Singh, J, Macchi, M and Lee, J, 2021. Field-synchronized Digital Twin framework for production scheduling with uncertainty, *Journal of Intelligent Manufacturing*, 32:1207–1228. https://doi.org/10.1007/s10845-020-01685-9

Sanchez, F and Hartlieb, P, 2020. Innovation in the mining industry: technological trends and a case study of the challenges of disruptive innovation, *Mining, Metallurgy and Exploration*, 37:1385–1399.

Van Der Horn, E and Mahadevan, S, 2021. Digital Twin: Generalization, characterization and implementation, *Decision Support Systems*, 145. https://doi.org/10.1016/j.dss.2021.113524

Wang, H, Li, H, Luo, G and Sun, C, 2020. Digital twin based production design process and work load prediction method, *Computer Integrated Manufacturing Systems*. https://doi.org/10.13196/j.cims.2022.01.002

A novel way to view mining operations

P Muston[1], L Fisher[2], O Siraj[3], N Kashyap[4] and P Mitchell[5]

1. EY Digital Mine Manager, EY, Melbourne Vic 3000. Email: paddy.muston@au.ey.com
2. EY Digital Mine GIS Analyst, EY, Perth WA 6000. Email: lauren.fisher@au.ey.com
3. EY Digital Mine Product Analyst, EY, Perth WA 6000. Email: oatima.siraj@au.ey.com
4. EY Digital Mine Lead, EY, Perth WA 6000. Email: nischith.kashyap@au.ey.com
5. EY Digital Mine Sponsor, EY, Sydney NSW 2000. Email: paul.mitchell@au.ey.com

ABSTRACT

The Spatial Digital Twin (SDT) signifies a major leap forward in the visualisation and management of mining operations. This cutting-edge technology integrates the Internet of Things (IoT), Artificial Intelligence (AI), and data analytics with spatial data and 3D visualisations to create a virtual model that replicates real-world mining assets and systems. The SDT utilises predictive models to forecast potential issues and optimise maintenance schedules, thereby enhancing operational reliability and minimising downtime.

The SDT encompasses a variety of features designed to improve mining operations. The 3D models of each component provide a detailed and accurate representation of the mining site, facilitating improved planning and visualisation. For example, SDT can simulate the entire mine site layout, including haul roads, stockpiles and infrastructure, allowing mine planners to anticipate equipment movements and logistics challenges. Measurement tools within the SDT enable precise calculations of elevation, vegetation, and drainage, which are essential for volumetric calculations and other critical assessments. 3D visualisation tools allow engineers to detect potential geotechnical hazards in open pit mines by analysing slope angles and stress factors in rock formations. These tools assist in monitoring and managing the environmental impact of mining activities, ensuring compliance with regulations and sustainability objectives.

The integration of geospatial and video analytics further enhances site security and operational compliance, mitigating the risk of accidents and ensuring prompt responses to incidents. For instance, AI-powered video analytics can detect unauthorised personnel or unexpected vehicle movements in restricted areas, triggering immediate alerts. This virtual environment allows users to explore a 3D landscape, providing a comprehensive representation of both the static and dynamic aspects of mining operations. Such immersive insights facilitate informed decision-making, operational optimisation, and heightened efficiency.

In conclusion, the Spatial Digital Twin is a transformative tool that enhances the decision-making process for mine planners and engineers by providing a unified, real-time view of mining operations. Its advanced analytics capabilities support enhanced decision-making and operational efficiency, rendering it a valuable asset in the digital transformation of the mining industry.

INTRODUCTION

The mining industry is undergoing a significant transformation driven by advancements in digital technologies that are redefining the operational landscape.

As global demand for minerals and metals surges, the mining industry is increasingly confronted by challenges such as resource depletion and the need for improved operational efficiency, enhanced safety, and reduced environmental impact. Traditional methods of managing mining operations often fall short in addressing these challenges due to their reliance on outdated technologies, siloed data systems and slow adoption of new technology which hinder the effective management of mining operations.

To keep pace with market demands and comply with new regulations, Mipac (2023) estimates the mining industry will invest $1.4 billion by 2030 in data analytics to improve productivity, efficiency, and sustainability. Within data and analytics, Discovery Alert (2024) cites Digital Twins as one of the key areas of growth, with approximately 70 per cent of technology leaders in major corporations are actively pursuing Digital Twin initiatives, signalling a significant shift in operational strategies across the mining sector. Key capabilities include predictive maintenance using real-time data and analytics to forecast equipment failures and optimise maintenance schedules, reducing downtime and costs. Advanced data analytics improve data-driven decisions, while the integration of IoT and AI with Digital Twins allows for real-time monitoring and optimisation of mining operations, enhancing productivity and safety.

Additionally, OT (Operational Technology) systems play a crucial role in deriving value from Digital Twins. OT systems are responsible for controlling and monitoring industrial operations, and they store valuable data that can be integrated with Digital Twins to create a comprehensive and real-time view of mining operations, as well take direction/feedback and control decisions made from the analysis performed in the digital twin. This integration enables predictive maintenance, operational optimisation, and enhanced decision-making, making OT systems an essential component of the Digital Twin ecosystem.

CHALLENGES

However, through our experience in the mining sector, EY has uncovered recurring pitfalls during digital transformations including the following.

Remoteness

As a significant amount of mining activity takes place in remote areas, there are frequently bandwidth and connectivity challenges when streaming data between metropolitan engineering offices and remote mining sites. These remote locations often lack the necessary infrastructure to support high-bandwidth, high-connectivity, high-speed data transmission, leading to delays and potential data loss or service outages. This challenge complicates the real-time monitoring and management of mining operations, making it difficult to maintain seamless communication and data flow between different systems.

Compute power and analytics

OT systems typically lack the compute power and advanced analytics capabilities that modern IT systems, particularly cloud vendor systems, offer. This limitation means that the data collected by OT systems cannot be effectively analysed to generate actionable insights. For example, predictive maintenance, which relies on advanced analytics to forecast equipment failures and optimise maintenance schedules, is challenging to implement with the limited compute power and lack of access to novel modelling approaches on OT systems. As a result, organisations miss out on opportunities to improve operational efficiency and reduce downtime.

OT cybersecurity controls

OT systems, responsible for controlling and monitoring industrial operations, often store valuable data in isolated systems, making it difficult to access for analysis and decision-making. The lack of integration between OT and IT systems exacerbates this issue, preventing data from OT systems from being easily combined with other sources for a comprehensive view of operations. Due to their mission-critical nature, OT systems are designed to operate independently to prevent cybersecurity breaches, limiting connectivity to other systems. This isolation hinders the integration of OT data into higher-level planning functions and collaborations, such as end-to-end value chain optimisation. Without integration, organisations cannot fully leverage their data to drive operational improvements and achieve strategic objectives.

Visualisation limitations

High-fidelity data visualisation is crucial for understanding complex data and making informed decisions. However, deploying visualisations on OT systems to circumvent the challenges above limits the value that can be obtained, as OT systems often lack the advanced visualisation capabilities that modern IT systems provide. This limitation restricts how engineers and planners can interact with the data, even if the analytics were available.

Cost and time of digital transformation

Traditional digital transformation projects in the mining industry are often time-consuming and expensive. These projects can take many years and many millions of dollars to execute, posing the risk of low value ROI and misalignment with business strategy. Custom-built digital products are typically more expensive and require more time from client SMEs, distracting them from their core functions. On the other hand, off-the-shelf solutions lack the flexibility to fully integrate and reflect the operational profile of the organisation.

THE ROLE OF DIGITAL TWINS IN MODERN MINING

The mining industry is at a pivotal moment where the integration of advanced digital technologies can significantly enhance operational efficiency, safety, and sustainability, increasing the viability of increasingly challenged ore reserves. One of the most transformative technologies in this regard is the Digital Twin. By leveraging OT data, Digital Twins provide a comprehensive and real-time view of mining operations, enabling predictive maintenance, operational optimisation, and enhanced decision-making.

Digital Twins are virtual replicas of physical assets, systems, or processes that use real-time data from OT systems to create a dynamic and interactive model. These models are not static; they are continuously updated with data from sensors, equipment, and other sources, providing a real-time representation of the physical world.

Successful deployment of a Digital Twin starts with a robust, integrated data platform to build and implement the structure for data management, leveraging the existing investment in information already captured. Strong foundations allow for uninterrupted, real-time data flow with reduced latency, while future proofing solutions by integrating scalability and security that grow and change with business needs.

EY has extensive experience developing Digital Twins and associated Data Platforms leading to the development of EY's Digital Twin Foundation, an accelerator and a guiding structure for future delivery, and EY's flagship Digital Twin assets.

The Foundation consists of a six-step journey to build a Digital Twin (Figure 1) including:

1. Capture data – Using existing IoT sensors installed within the equipment to collect and monitor metrics such as operating parameters, capacity vibration, temperature etc.

2. Curate data – Process and clean data to ensure data sets are fit-for-purpose and analysis-ready.

3. Publish data products – Provide a unified Data Product consumption layer where data curated products are accessible, governed, adaptive, secure and acts as the single source of truth.

4. Derive analytics – Advanced analytics create high-level data products combining multiple signals to generate actionable insights and indicators, enabling informed decision-making for executive functions in holistic asset management and data prediction.

5. Visualise 3D digital model – An interactive, virtual representation of the asset, overlaid with real-time data and derived analytics data products.

6. Prescribe action – Automated or supported decision-making and prescribed interventions or parameter tuning.

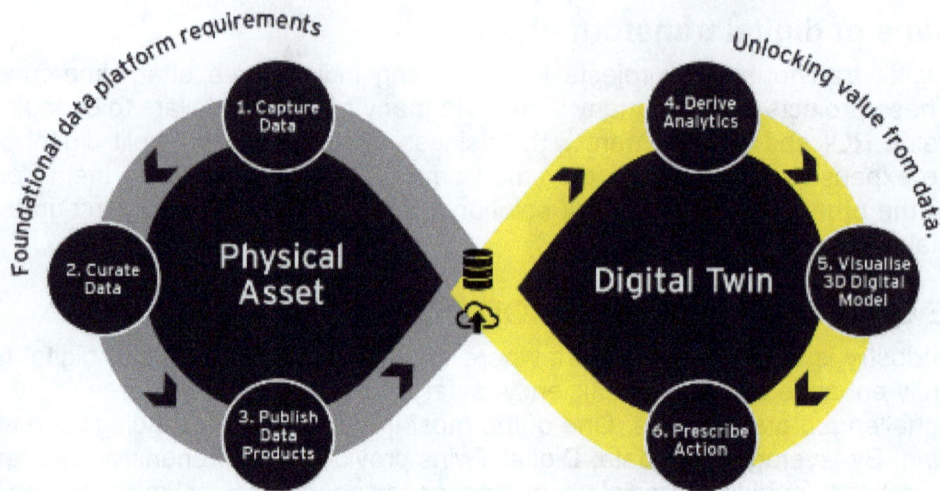

FIG 1 – EY digital twin foundation.

USE CASE ASSESSMENT FRAMEWORK

Use case assessment framework

To effectively identify and address the various requirements associated with different use cases, EY has developed a Use Case Assessment Framework. This Framework helps in determining the specific requirements and constraints for latency, data integration, and other operational parameters, for a given use case. This Framework provides a structured approach to evaluate and prioritise use cases, ensuring that the implementation of Digital Twins is both practical and impactful.

Figure 2 illustrates this Framework, showing how different requirements influence the selection and implementation of use cases, thereby putting the example use cases in the context of the identified requirements and possibilities. Understanding these requirements, provides better assessment for what is feasible and identifies the most suitable solutions for each scenario.

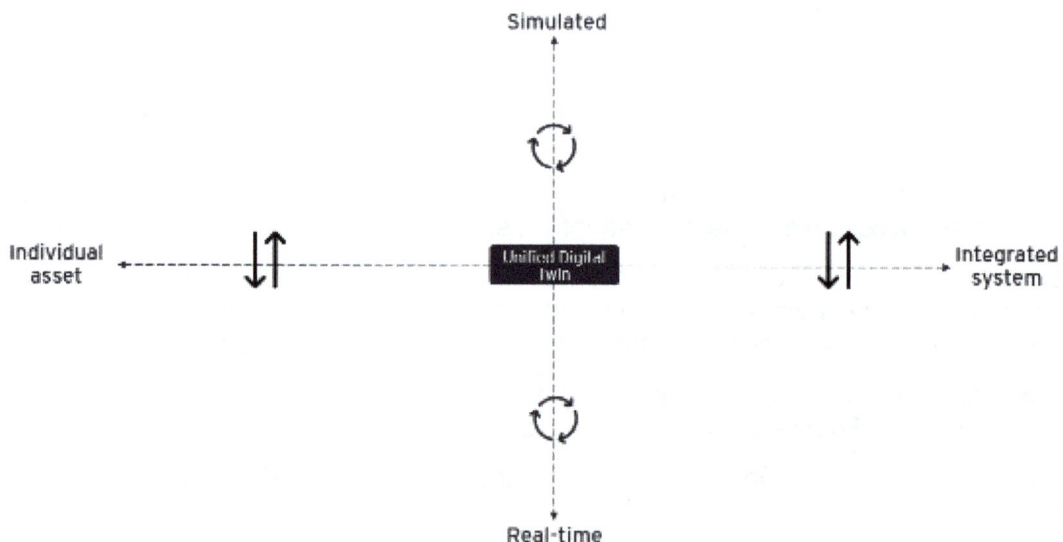

FIG 2 – Use case assessment framework.

Dimension – simulated versus real-time data

Utilising simulated data allows for extensive experimentation and scenario analysis to examine impacts of potential decisions. It is particularly valuable in forecasting and strategic planning, where organisations can model various 'what-if' scenarios to optimise performance.

The complex modelling and scenario analysis often requires significant computational resources, that may not be feasible using constrained on-site systems. This data is typically processed in centralised data centres or cloud environments, where powerful computing capabilities can handle extensive simulations and analytics. By moving the computational load away from on-site Operational Technology (OT) systems, organisations can leverage advanced algorithms and machine learning techniques to generate insights and predictions without the constraints of local hardware limitations. Typically, the latency introduced by data transfer to more compute-abundant environments is not prohibitive for simulated use cases, as these planning and forecasting applications are applicable over longer term timescales and do not require real-time data.

In contrast, real-time data provides an accurate and dynamic reflection of physical assets, enabling immediate insights and responsive actions. The data is typically collected from sensors and devices directly integrated into the on-site OT systems, allowing for continuous monitoring of equipment performance, environmental factors, and operational metrics. The reduced latency is crucial for applications like predictive maintenance, where timely detection of anomalies can prevent costly failures, and control optimisation, maximising efficiency of production assets.

By processing real-time data locally or at the edge, organisations can achieve faster decision-making and enhance operational agility. However, real-time data presents challenges related to data quality, integration, and infrastructure costs and may lack the comprehensive analysis offered by broader simulation problem. Use cases that leverage real-time data should focus on robust data acquisition systems and governance processes to ensure accuracy and reliability.

Dimension – individual asset versus integrated system

Focusing on a single asset allows for detailed monitoring and optimisation of that specific entity, making it ideal for applications such as predictive maintenance or performance tracking. This approach enables organisations to gain deep insights into asset behaviour, leading to improved reliability and reduced downtime. However, the limitation lies in the lack of context provided by surrounding systems; insights derived from an individual asset may not fully capture the broader operational dynamics. Also, standardisation across sites and equipment is a frequent challenge, as legacy assets obtained through acquisition often use different interfaces and systems and have variable performance expectations.

An integrated system approach encompasses multiple assets and their interactions within a broader operational context. This holistic view enables organisations to optimise processes, enhance collaboration across the value chain, and improve overall system performance. Use cases in this category are particularly valuable in mining operations, where the interplay between equipment, personnel, and geological factors significantly impacts productivity. For example, integrating data from drilling rigs, haul trucks, and processing plants can optimise the entire supply chain, from extraction to processing, blending and shipping operations. However, integrating data from various sources and defining realistic system orchestration objectives and policies can be challenging, requiring robust data management and well-defined business rules. Use cases focused on integrated systems should emphasise data integration, real-time analytics, and cross-functional collaboration.

Value lies in dynamic exchange between quadrants

Having both a holistic value-chain level view of mining operations with simulated data for planning, and a real-time view of individual assets simultaneously enables a number of high-value use cases. The holistic perspective enables mining organisations to align their strategic objectives with operational activities, allowing informed decision-making that drives overall efficiency and profitability across the entire operation. While strategic planning is essential, the dynamic nature of mining requires real-time insights into individual assets to ensure operational agility; this allows for immediate responses to changing conditions, such as equipment failures or unexpected environmental factors. By continuously monitoring asset performance, organisations can quickly adapt their operations to maintain productivity and minimise downtime.

Further to the above, integrating both perspectives ensures effective resource management throughout the value chain, as simulated data can identify potential bottlenecks while real-time data provides the necessary feedback to address these issues at the asset level. This synergy enhances decision-making at all levels, empowering executives and managers to understand and adapt to long-term trends while operational teams execute these strategies effectively.

Additionally, the combination of simulated and real-time data helps identify and mitigate risks more effectively, fostering a proactive approach that enhances safety and reduces the likelihood of costly disruptions. Ultimately, this dual perspective enables mining organisations to navigate complexities effectively, driving improved performance and long-term success.

Example use cases

An example set of use cases has been positioned using the Use Case Assessment Framework in Figure 3, to further demonstrate the use of the Framework and the interplay between them.

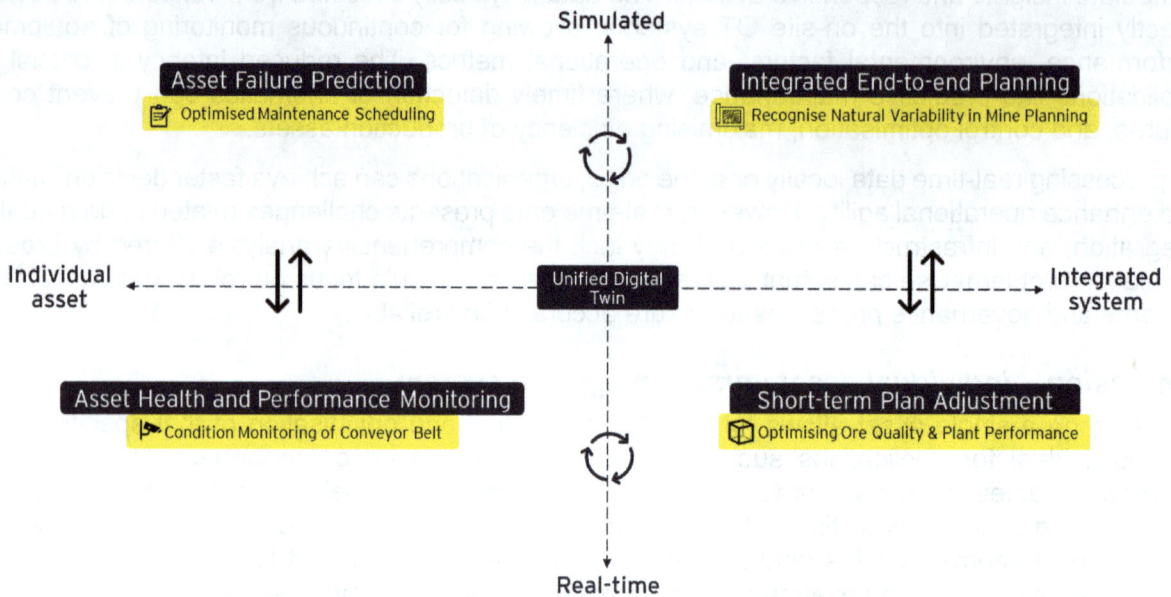

FIG 3 – Use case assessment framework with example use cases.

Example use case – asset failure prediction

Optimised Maintenance Scheduling – Utilise simulations to predict and prioritise maintenance activities, improving asset longevity and reducing downtime. The solution relies on simulation to forecast maintenance needs for individual assets.

By analysing real-time data from OT systems, Digital Twins can predict equipment failures before they occur. This predictive maintenance capability helps reduce downtime, optimise maintenance schedules, and extend the lifespan of assets. For example, sensors on mining equipment can monitor parameters such as vibration, temperature, and pressure, allowing the Digital Twin to identify potential issues and recommend maintenance actions.

Example use case – asset health and performance monitoring

Condition Monitoring of Conveyor Belt – Use video analytics and LiDAR to monitor conveyor belt health in real-time, enhancing safety and operational excellence by detecting carry back and abnormal belt thickness, providing alerts at certain thresholds.

Computer vision and machine learning detect key features on assets or operations such as belt defects, thickness and alignment as well as identifying and quantifying carry-back which can trigger automated maintenance execution, such as online belt cleaning equipment, or act as an early warning system to enact adaptive operations and operational changes.

Example use case – integrated end-to-end planning

Recognise Natural Variability in Mine Planning – Use stochastic scheduling and an orebody productivity model to incorporate uncertainty into mine planning, enhancing accuracy and reliability of production estimates. Update downstream forecasts to inform blending and optimum product grade strategy based on quality variation on the ROM pad. Employ simulated probability or statistical scenarios to account for variability in individual mining blocks.

Modelling material movements and corresponding asset activity across a whole system, combined with the dynamics of how the system responds to product variability, unexpected outages, and changes in asset availability enables powerful scenario planning. For example, new asset management policies or maintenance practices can be tested, or multiple versions of a long-term plan can be run to find the optimal balance between risk and productivity.

Example use case – short-term plan adjustment

Optimising Ore Quality and Plant Performance – Adapt to changing ROM demands and recommend plant settings and stockpile use to maximise output and improve ore recovery. Involves real-time adjustments to integrated production schedules.

Digital Twins can simulate different scenarios to determine the optimal settings for equipment, or material use strategy to target certain objectives, leading to increased productivity and reduced operational costs.

IMPLEMENTATION

The Spatial Digital Twin has been successfully implemented with several clients in the mining industry, which the case study below describes.

Case study

EY has developed a Spatial Digital Twin (SDT) as a core module for clients to customise and implement across numerous detailed use cases (Figure 4).

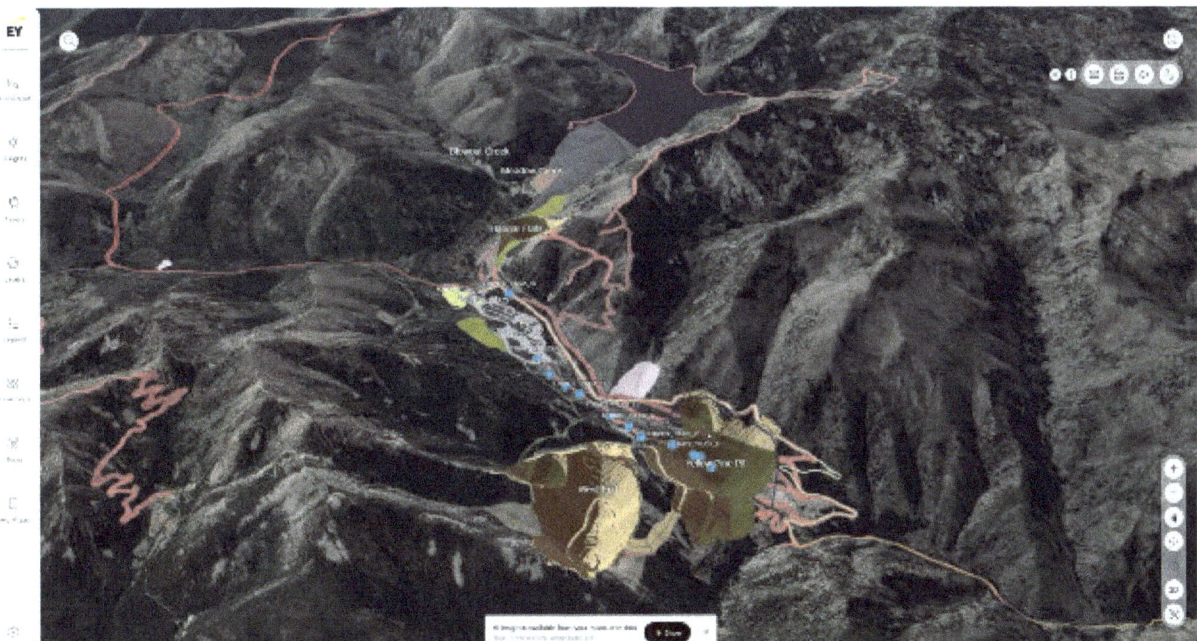

FIG 4 – Overview of EY spatial digital twin.

The SDT serves as a virtual model that mirrors real-world mining assets and systems, offering a cohesive, real-time perspective of mining operations. To host this application, EY first built the Industrial Intelligence Platform (IIP) which supports cutting-edge digital, data and AI capabilities such as geospatial analytics, video and image analytics and machine learning. With these core capabilities the IIP can be implemented for a wide range of applications and solutions, including the Spatial Digital Twin.

Particularly beneficial for the mining industry, the SDT utilises spatial data to facilitate planning, monitoring, and operational optimisation. By combining IoT data, AI, and data analytics with spatial data, the SDT creates a comprehensive representation of mining environments, which aids in informed decision-making and boosts overall operational effectiveness.

One of the standout features of the SDT is the capacity to visualise detailed and accurate 3D models (Figure 5) of mining sites alongside high-resolution imagery and location data providing rich context to assets and operational data, aiding data-driven decisions. The models provide a virtual depiction of the physical landscape, enabling operators to visualise and engage with data meaningfully. For instance, engineers can utilise 3D models to identify potential geotechnical hazards in open pit mines by assessing slope angles and stress factors in rock formations, thereby enhancing planning and risk management for safer, more efficient operations.

FIG 5 – SDT 3D model of a crusher in a mine site.

The SDT also leverages spatial data to deliver a holistic view of mining activities, incorporating information on topography, geology, infrastructure, and environmental conditions. By merging this spatial data with real-time information from OT systems, the SDT facilitates a comprehensive understanding of the mining landscape. This integration also allows for monitoring the environmental impact of mining activities, ensuring compliance with regulations and sustainability goals.

The incorporation of advanced analytics and AI further amplifies the capabilities of the SDT, enhancing the predictive and prescriptive functions. By examining historical and real-time data, SDTs can uncover patterns, forecast potential issues, and suggest actions to streamline operations. For example, AI algorithms can analyse sensor data to predict equipment failures and recommend proactive maintenance, thereby minimising downtime and boosting operational efficiency.

Moreover, the SDT enables real-time monitoring and control of mining operations. By continuously updating with data from OT systems, the SDT provides a dynamic and interactive view of the mining environment (Figure 6). This real-time functionality empowers operators to swiftly respond to changing conditions, optimise processes, and enhance overall efficiency. For instance, monitoring conveyor belt performance in real-time, identifying issues such as carry-back or abnormal belt thickness and triggering maintenance actions to prevent operational disruptions.

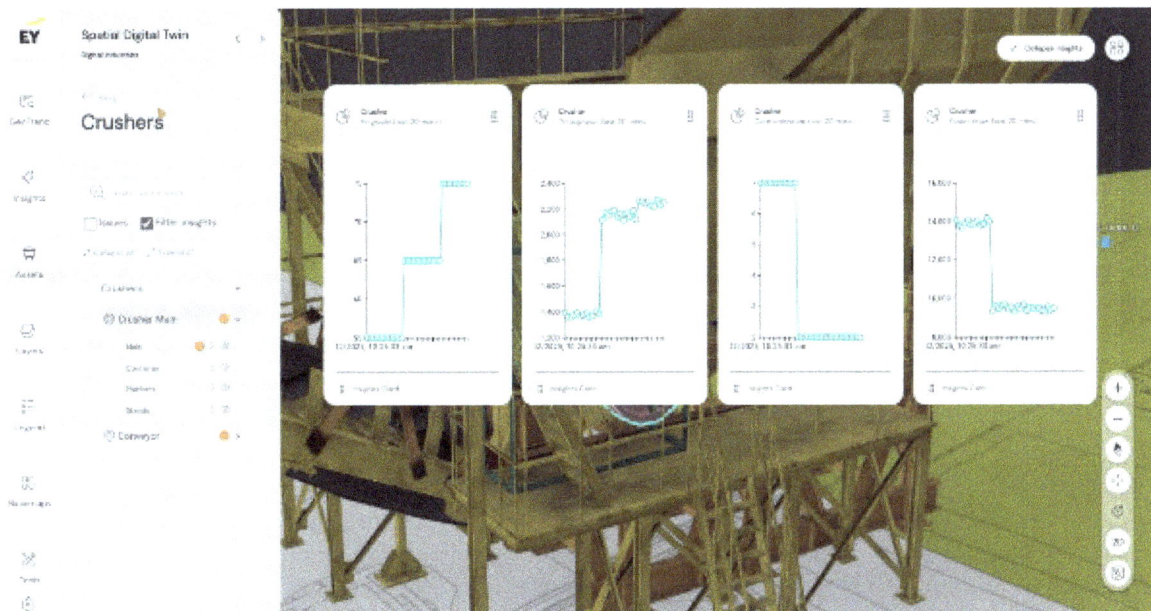

FIG 6 – Real-time analytics of an asset in a mine site.

Advanced analysis of spatial data in EY's GeoTrend function elicits patterns in spatial data which help to focus maintenance or remediate action as well as highlighting repeat issues that need root cause analysis. For example, a maintenance planner can add a query in GeoTrend and look for the areas of a rail network which had the most speed restrictions in the previous year. Areas matching the defined criteria are highlighted on the map and the planner overlays other data layers, including standing water data, which appears to match the areas where the speed restrictions were in place. The planner understands through this visualisation that to stop the re-occurring speed restrictions, drainage in these areas should be improved long-term as this will prevent damage to the rail, thus finding a root-cause solution.

Combining the advanced capabilities with built-in tools such as mark-ups for collaboration, accurate measurements, slope grading and map bookmarking the Spatial Digital Twin allows comprehensive desktop studies to be carried out, reducing the time spent on-site in often remote and harsh conditions (Figure 7). Desktop studies improve safety by providing reliable scoping pre-visit and save cost by reducing the number of site visits necessary and effort wasted travelling into the field and being faced with unforeseen circumstances.

FIG 7 – Detailed view of SDT with measurement and drawing capabilities.

FUTURE WORK ON DIGITAL TWINS

To achieve further innovations and adoption, ongoing future work will be needed to continue to increase the value that Digital Twins can bring to the mining sector.

A number of key challenges to address are below.

Integration challenges between OT and IT systems

Solutions for how to overcome the challenges outlined in section 2.3 will be critical to increasing the breadth of use cases that can be delivered by Digital Twins, and therefore the value that these investments can return. Overcoming the isolation of OT systems will enhance the overall accuracy and reliability of data, and ensuring seamless, continuous connectivity and data flow between IT and OT systems is crucial for leveraging the full potential of digital technologies in mining.

Embedding AI

Implementing advanced analytics and machine learning algorithms to perform more foundational tasks and generate operational intelligence, operators and managers can focus on the most critical functions of their role where they can add the most value. In addition, the advent of qualitative techniques such as Large Language Models (LLMs) supports translation of these quantitative insights into qualitative strategy and decisions. For example, implementing an LLM over the data and insights managed in a Digital Twin can improve system operations by enabling more intuitive and efficient interactions with the operational intelligence delivered by quantitative models.

Broader integration

Broader integration with other service providers will also increase the range of data sources and intelligence services that can be deployed into a Digital Twin. This integration will allow the Digital Twins to cover a wider set of use cases and provide more comprehensive and actionable insights. Ultimately, Digital Twins can act as a one-stop-shop for or single front end for a range of data sources and capabilities managed across the organisation, combining inputs from multiple teams to facilitating better collaboration, decision-making and operational efficiency.

CONCLUSION

The integration of Spatial Digital Twins in the mining industry represents a transformative leap towards enhanced operational efficiency and sustainability. Continual innovation in this technology will be pivotal in supporting growth and meeting the evolving demands of the sector. The adoption of Digital Twins provides tangible benefits, showcasing their ability to optimise monitoring processes, improve decision-making, and ultimately drive productivity while navigating the complexities and challenges of the sector, ensuring a more sustainable and responsible approach to mining.

REFERENCES

Discovery Alert, 2024. Boosting Mining Efficiency with Digital Twin Technology [online]. Available from: <https://discoveryalert.com.au/digital-twins-transform-mining-boosting-efficiency-and-sustainability> [Accessed 6/3/2025].

Mipac, 2023. How data analytics in mining will boost yields and operations [online]. Available from: <https://www.mipac.com.au/insights/harnessing-data-analytics-in-mining> [Accessed 6/3/2025].

Accelerating fleet electrification in mining to meet ESG targets – a simulation and operational analysis approach

A Naik[1], D Kumar[2], K Chhabra[3] and A Kale[4]

1. Industry Consultant, Dassault Systèmes, Perth WA 6000. Email: ajinkya.naik@3ds.com
2. Project Director, TEXMiN and Professor, Mining Engineering, Indian Institute of Technology (Indian School of Mines), Dhanbad Jharkhand 826004, India. Email: dheeraj@iitism.ac.in
3. Industry Consultant Manager, Dassault Systèmes, Perth WA 6000. Email: kriti.chhabra@3ds.com
4. CATIA Systems Modelling and Simulation – Roles Portfolio Manager, Dassault Systèmes, Pune Maharashtra 411057, India. Email: ameya.kale@3ds.com

ABSTRACT

Mining companies want to achieve their Environmental, Social, and Governance (ESG) goals. Adopting electric trucks for haulage operations can help them to significantly reduce carbon emissions. But there are considerable challenges in the adoption of this technology, like constraints with respect to vehicle acquisition and inadequate trial data. Traditional asset acquisition methods are proving inadequate for the unique demands of fleet electrification, leading to delays that jeopardise the timely achievement of ESG goals. This paper presents a dual-layer approach that combines advanced simulation using CATIA DYMOLA with operational analysis to accelerate fleet electrification in the mining sector. Initially, CATIA DYMOLA is employed to develop and contrast various electrification options, including Hybrid Electric Vehicles (HEVs) and Battery Electric Vehicles (BEVs). The simulation outputs include critical data such as range estimations, battery state-of-charge (SOC) across different operational states, optimal charging point locations, and studying the total cost of ownership (TCO) and emissions savings. Following the simulation phase, we transition to the operational layer, where the integration of electric, diesel, and hybrid fleets is analysed in real-world scenarios. How introduction of electric trucks affects operational metrics like total tonnage transported or congestion at charging points is analysed. Several scenarios are explored, such as increasing the number of electric trucks or adjusting the charging infrastructure, to assess their effects on operational efficiency and ESG compliance. Also, the paper explores various operational scenarios as modifying shift schedules, charging schedules, and an increasing number of charging stations, to assess their effects on operational efficiency. This dual-layer approach helps to perform virtual trials rapidly so that mining companies can prioritise investments in electric fleet.

INTRODUCTION

Mining companies are facing regulatory pressure to reach net zero. Replacing the diesel fleet with an all-electric fleet can help them to cut down emissions significantly. Simulation tools that help mining companies to check how electric trucks would operate on mine site in terms of range of the truck on a charge and the requirement of overhead trolley can significantly boost their adoption. Manual trial methods are often insufficient to address the complexities of integrating new technologies.

This paper discusses a Systems Simulation use case, how mining companies can achieve their targets by virtually validating different KPIs such as optimising operational efficiency to meet Environmental, Social, and Governance (ESG) targets, and facilitate informed decision-making. The tool used to perform detailed simulation and analysis is CATIA DYMOLA with its commercial libraries. This paper demonstrates how the battery, electrified powertrain, cooling system, behaviour of the chassis, the terrain etc, can be modelled to analyse the range of the truck, battery state of charge across a typical mining drive cycle, and battery degradation. This leads to the study of the total cost of ownership, planning for infrastructure requirements, and operational efficiency of the Battery Electric Truck, and we can then compare it with diesel-based counterparts.

Several mining companies have defined targets to reduce their emissions significantly by 2035. By using such simulations, mining companies can collate data to maintain reporting standards. CATIA

DYMOLA rapidly enables solving complex multi-disciplinary systems modelling and analysis problems using the MODELICA language. It is a complete environment for model creation, testing, simulation, and post-processing. The libraries and templates inside the tool assist in the rapid modelling of the trucks and the routes; we are then able to simulate the different trucks driving the same/different route. For example, Battery Electric Vehicle with or without Electric Road System (ERS) or Diesel Truck with or without ERS. Further, there is even a possibility to model and simulate hydrogen-based trucks. The simulation is efficient and computes a cycle of ~1600 seconds in less than 1 seconds. To describe a system, most people refer to a diagram that shows process flow with function blocks executing tasks in a certain order. CATIA DYMOLA uses a paradigm that reflects this design process so you can model the same way you think—graphically. The extensible template could be further used to simulate energy supply for electric mine trucks with a buffer battery and grid connection. This simulation considers vehicle dynamics, changing load, power limitation, and recharging on electric load. A full day of operating four trucks is simulated in ~90 seconds. Throughout the modelling process, CATIA DYMOLA allows a block to be used in different circumstances. The next study is the operational schedule assessment, where we assess the charging and refuelling times for electric trucks, hybrid trucks, and diesel trucks. We then analyse the impact on schedules, highlight potential delays, and need for strategic scheduling. This approach also helps us to analyse the critical congestion at charging points or bottlenecks at charging stations, which helps to plan more charging stations or alternative strategies like implementing variable shift patterns to distribute charging demand and optimise fleet utilisation.

METHODOLOGY

The methodology has a two-layer approach where we first start with dynamic modelling of the electric truck, which reduces the need for physical trials, and the next is the operational layer, where we study trade-off analysis for optimal fleet configurations. This dual-layer methodology starts with extensive simulations, assists with implementation plans with data, and helps to evaluate and act on emergent strategies.

Simulation using CATIA DYMOLA

The simulation models created in CATIA DYMOLA are inspired by Lindgren *et al* (2022) with four different truck configurations. The duration for which to run the simulation has to be timed with the overall strategic timeline of the organisation. The focus could be shorter time cycles like few days or weeks. Or it could be at tactical or strategic level. For such a study, the mining organisation should have data related to the terrain, the battery specifications from the battery manufacturer, all the key specifications of the truck like motor torque, gear ratio, weight etc from the truck supplier. More insights are provided in a table in the upcoming pages. Any historical data related to operations like the number of hours of operation of trucks, failure rates, maintenance schedules, fleet usage, production schedules etc is highly relevant. A CATIA DYMOLA specialist and a solution architect can create the simulation study at tactical level or strategic level within a span of 20–30 days, provided all the key input parameters are available:

- Dynamic Modelling:

  o This simulation focuses on the operational cycle of Battery Electric Truck, Battery Electric Truck with/without electric road system (ERS), and Diesel Electric Truck with/without ERS. The energy consumption is analysed. The modelling of components such as electric motors, battery management systems, diesel engines, ERS, battery cooling systems etc enables users to examine their reactions to various operational pressures.

- Drive Cycle Simulation:

  o The truck's operational cycle is modelled to determine how far it can go on a single charge and the dependence on the ERS. The simulation considers energy use during uphill travel, energy regeneration capability during downhill returns, idle energy consumption, etc. The templates enable seamless definition of drive cycles, including the time spent in spot, load, haul, dump, back etc states. In addition, the terrain is modelled by considering the elevation, slope, ERS availability, etc. The amount of time spent by different truck configurations in the different cycle states is analysed and the corresponding fuel consumption, battery usage, and electricity used from the ERS is tracked. The model can even study the impact of environmental factors like temperature or any other similar factors. The model can also compute cost per kilometre for each truck variant, which could assist in choosing the most cost-effective fleet composition.

- State of Charge (SOC) computation and its impact on operational efficiency:
  - The templates and models enable mining companies to incorporate their real operational insights. This helps to analyse the increased/decreased productivity of introducing electric trucks, the cost of ownership, and the right battery configurations, as the detailed state of charge studies enable the analysis of battery degradation and if battery electric trucks can assist with operational schedule assessment. For battery degradation analysis, aging models are used that are based on calendar aging, considering state of charge and temperature. The second approach is cycle aging based on temperature and current, and the last approach is using cycle aging with cycle detection considering charge throughput, mean values of state of charge, and depth of discharge.

Key input parameters are shown in Table 1.

TABLE 1

Input parameters list.

Battery electric truck with / without ERS	Diesel electric truck with / without ERS	Renewable energy sources
Battery capacity	Number of engines	Number of solar panels
Pack voltage	Engine rated power	Peak power of PV system
Motor power	Engine efficiency	Efficiency of solar panels
Motor torque	Fuel tank volume	Price of solar energy
Gear ratio	Mechanical power of generator	Minimum wind speed for production of power
Gradient of terrain	Electrical power of generator	Wind speed above which nominal power is produced
Tyre radius	Efficiency of the generator	Price of wind energy
Mass of tyre	Maximum power of machine	Available energy in the storage system
Grid frequency	Maximum torque of machine	Available power in the storage system
Grid nominal power	Gear ratio	Efficiency for charging and discharging
Efficiency of converters for ERS operations	Gradient of terrain	Nominal frequency of the grid
Price of energy	Tyre radius	Nominal power of the grid
	Mass of tyre	Price of supplied energy from the grid
	Grid frequency	Converter efficiency for ERS operation
	Grid nominal power	
	Efficiency of converters for ERS operations	
	Price of energy	

Simulation scenarios

Haul Truck following a speed/height profile with variable load:

- Simulation model in Figure 1 analyses Battery Electric Trucks with/without ERS and Diesel Electric Trucks with/without ERS. The trucks are subjected to the same route. Their energy consumption, speed, and cycle time are analysed. The energy usage of different trucks is calculated and compared. The single cycle of ~1600 seconds, *viz*, about 26 mins, is simulated in less than 1 second.

- At a certain time in the Figure 1 simulation model, BEV with ERS consumes a total of 472 kWh; 358 kWh is supplied by ERS, reducing battery dependency. The BEV without ERS consumes 490 kWh of energy. The Diesel Truck with ERS consumes 101.7 L of diesel and draws 209 kWh from ERS. The Diesel Truck without the ERS consumes 153.3 L of diesel. Such studies provide an immediate comparison of different truck types. Further, we have the possibility of introducing a hydrogen-powered truck or a hybrid.

- Figure 2 plot depicts that due to the availability of different route templates, the trucks could be subjected to different routes to validate their performance for a different terrain.

- Some key insights can be drawn regarding the usage of ERS and how it reduces fuel and battery consumption. Battery Electric Trucks depend less on onboard energy storage, and diesel trucks rely less on fuel when supplemented with ERS. We can thus test different configurations to determine the optimal configuration, which could lead to an efficient total cost of ownership.

- From Figure 3 plot it can be inferred that the Battery Electric Truck could potentially reach a higher velocity and therefore reduce cycle times.

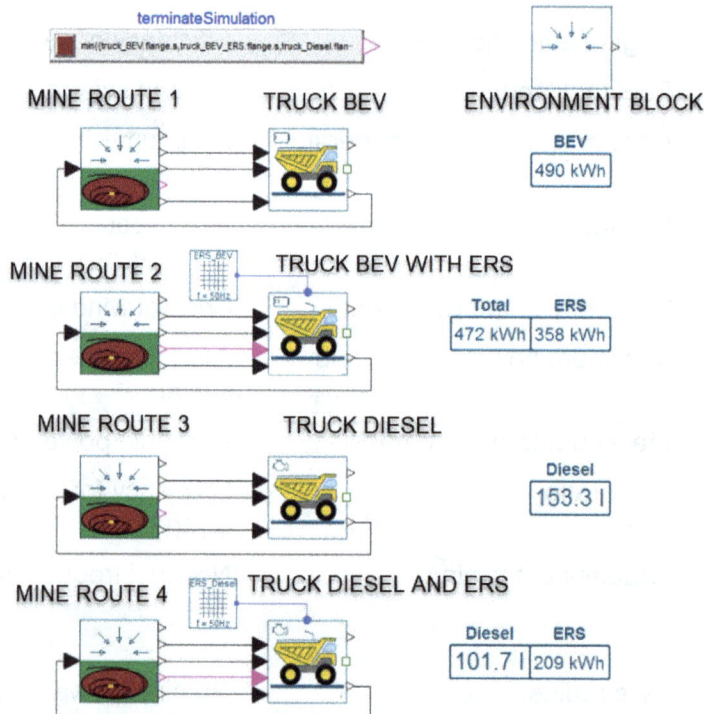

FIG 1 – Comparing different truck configurations.

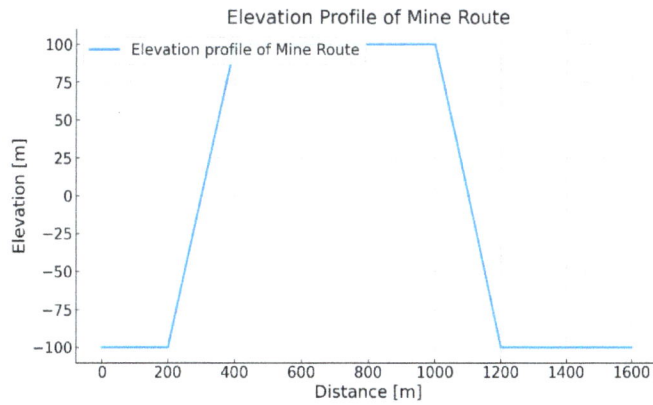

FIG 2 – Elevation profile of the mine route.

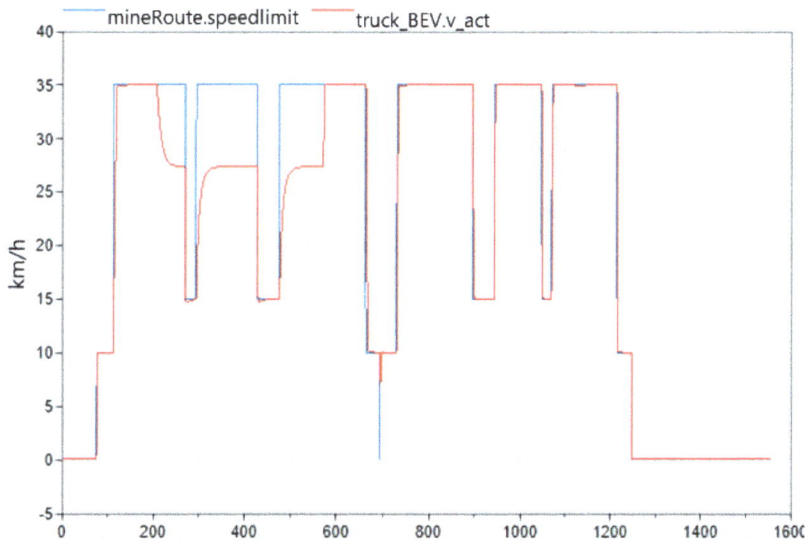

FIG 3 – Maximum allowable speed on mine route (blue) and the actual speed of Battery Electric Truck.

Haul Truck with variable speed on electric road system:

- To ensure continuous operation without requiring charging stops, this approach enables a parameter sweep to find the reduction factor 'k_v_ERS' multiplied with the desired speed for which the battery recharged to the initial state of charge.

- Figure 4 gives the information that a factor of roughly 0.53 will result in a state of charge similar to the initial one, which is set to be 80 per cent.

- In Figure 4, the flat part above roughly 0.78 results from the truck not reaching higher speeds than 27.4 km/h (= 35 km/h × 0.78) due to the limited power of the traction system.

- The simulation helps to determine the optimal reduction factor to balance battery recharge and truck performance. The factor controls truck's speed and regenerative braking efficiency. It ensures energy recuperation and consumption work in tandem to ensure efficient operation. We also take the vehicle load into account, which means we can predict the max. velocity for the truck depending on the vehicle weight.

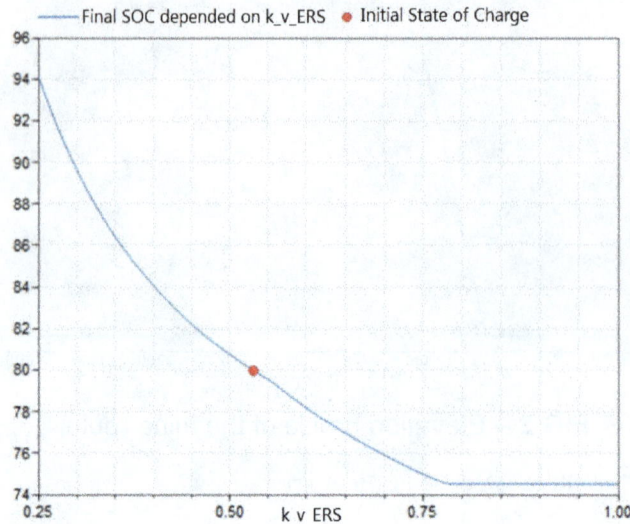

FIG 4 – Final state of charge as function of k_v_ERS reduction factor.

A configurable number of haul trucks operating in a mine operated by renewable energy sources:

- In the Simulation model demonstrated in Figure 5, we simulate energy supply for electric mine trucks using renewables with a buffer battery and grid connection. A full day of operating four trucks is simulated in ~90 seconds. The model considers vehicle dynamics, power limitation, recharging on electric loads etc.

- In this use case, we simulate energy supply for electric mine trucks using renewables with a buffer battery and grid connection. A full day of operating four trucks is simulated in ~90 seconds. The model considers vehicle dynamics, power limitation, recharging on electric loads etc.

- The Energy Mix graph named Figure 6 shows how power from solar (blue), wind (red), and the grid (green) changes over time.

- The Battery State of Charge (SOC) helps to analyse how the battery's charge level changes.

- The Electricity Price Graph represents how the electricity price changes throughout the day. The insights could be used to optimise the charging schedule and to devise a better strategy.

- This holistic approach of merging fleet operations, energy costs, and use of energy mix helps to plan for a sustainable mine.

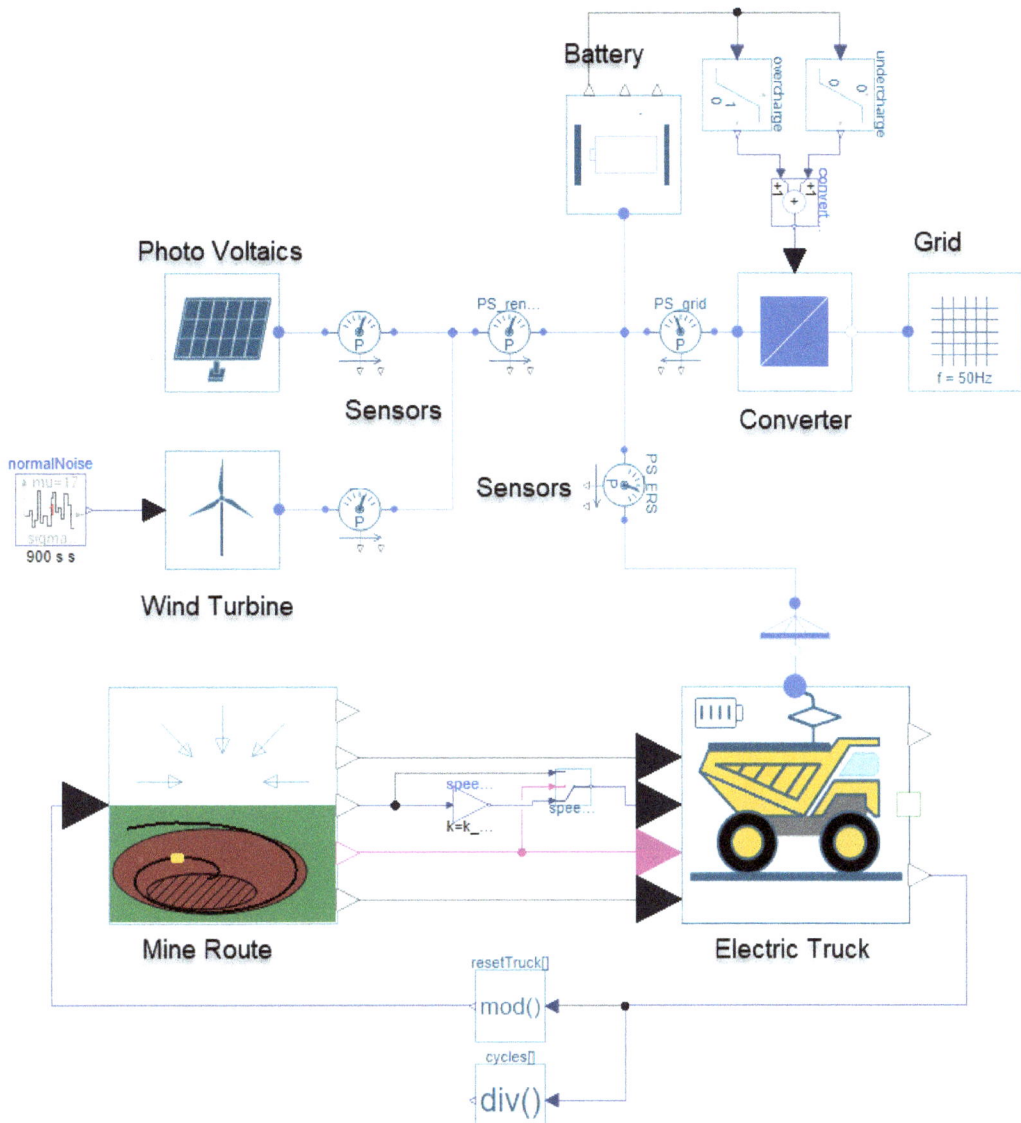

FIG 5 – Electric mine trucks using renewables with a buffer battery and grid connection.

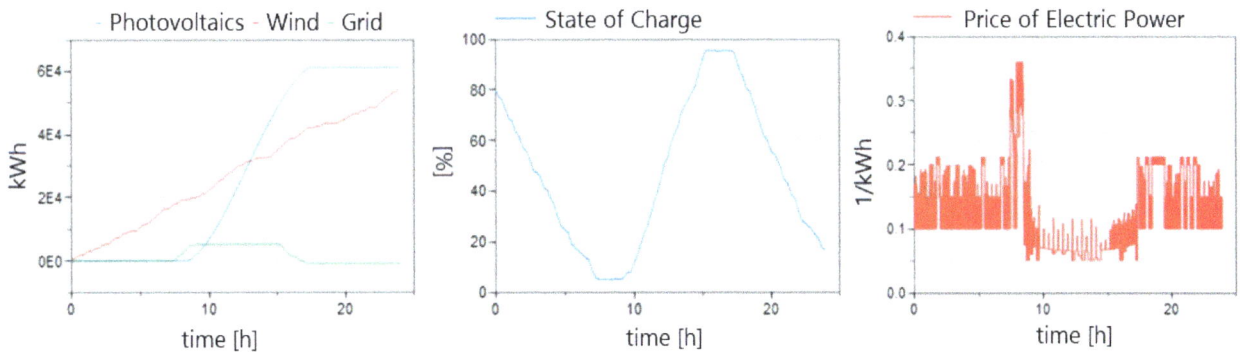

FIG 6 – Energy mix, battery state of charge, and electricity price variations, demonstrating the interaction between renewable generation, storage, and price of electric power.

Operational analysis

The data coming from CATIA DYMOLA related to battery state of charge for electric trucks and the overall mileage of the diesel trucks can now be used to analyse the state times for operational phases like Spot, Load, Haul, Dump, etc. This data can be used to choose the right fleet formations. The operational simulation as seen in Figure 7, helps to perfect factors like recharge time, refuelling time, total tonnage transported, battery size, motor power, ERS length, and operational schedules. This information is key for monitoring operational KPIs, bottleneck identification, and studying the impact of shift patterns. These studies further enable mining companies to analyse the productivity changes due to the introduction of electric trucks. Here we also insert the mine schedule. The results from CATIA DYMOLA and the operational schedule together enable scenario analysis. Some possible scenarios that we can evaluate are how electric charging decreases utilisation of electric trucks. How increasing or decreasing charging points could assist in reducing congestion at charging points. The insights would assist in modifying the shift patterns to distribute charging demand.

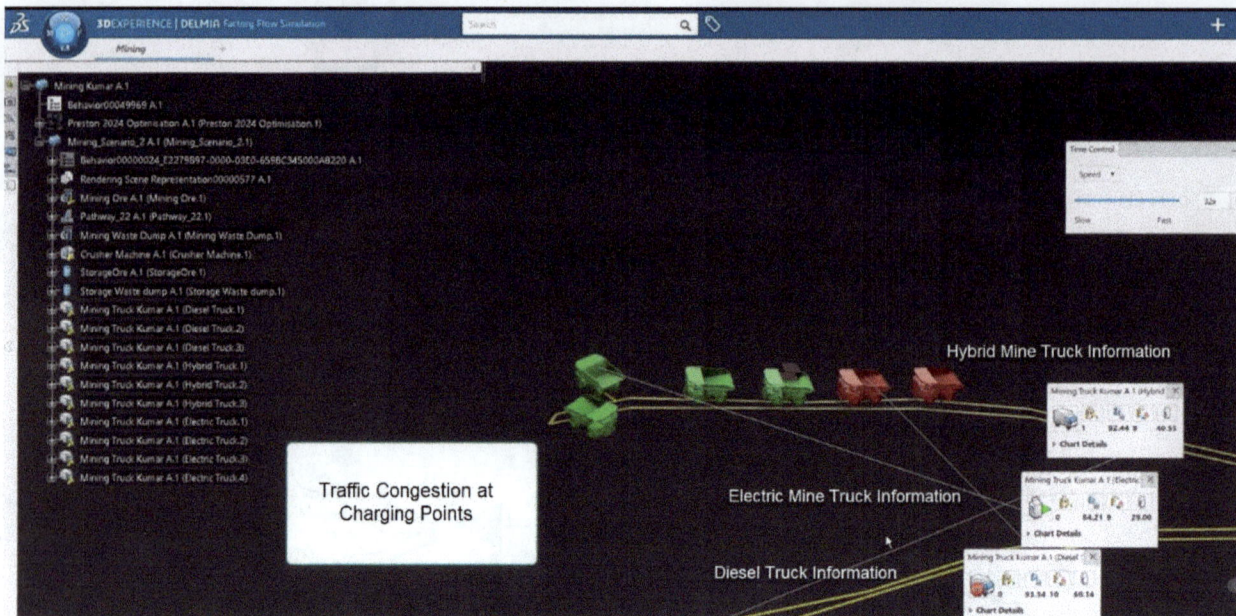

FIG 7 – Analysing congestion at charging points.

As seen in Figure 8, we can generate a report to track the operational efficiency of our entire fleet. The report provides state times for all the trucks in the fleet as total loaded travel, total time spent charging, total idle time, or total unloaded time. The report can help to plan for changes in the operational schedule to achieve the operational KPIs like total ore transported and amount of loaded travel time.

Resource	Products Transferred	State Times (Second)									Total Distance Travelled (Millimeter)	Uti
		Idle	Empty Travel	Loaded Travel	Consumable Replenish	Busy Loading	Busy Unloading	Idle Parked	Travel Blocked	Parking Blocked		
Truck Kumar A.1 (Diesel 1)	10	305	20406.939	15971.588	0	50	50	316.508	2621.089	0	4.002e+007	
Truck Kumar A.1 (Diesel 2)	10	699.137	20469.941	15971.588	0	50	50	324.455	1839.496	316.508	4.009e+007	
Truck Kumar A.1 (Diesel 3)	10	1179.224	20469.149	15971.588	0	50	50	324.455	1035.746	640.963	4.008e+007	
Truck Kumar A.1 (Hybrid 1)	9	1681.466	20488.698	15956.122	0	50	45	324.455	228.872	946.513	4.009e+007	
Truck Kumar A.1 (Hybrid 2)	9	2210.738	20531.852	15837.592	0	50	45	324.455	5.454e-006	721.488	4.001e+007	
Truck Kumar A.1 (Hybrid 3)	9	2783.165	20554.34	15490.19	0	50	45	324.455	[illegible]	[illegible]	e+007	
Truck Kumar A.1 (Electric 1)	9	3378.08	18448.457	14374.43	2902.187	45	45	324.455	[illegible]	[illegible]	e+007	
Truck Kumar A.1 (Electric 2)	9	3995.94	17613.137	15261.448	2526.23	45	45	234.371	[illegible]	[illegible]	e+007	
Truck Kumar A.1 (Electric 3)	9	4628.997	17578.002	15261.448	0	45	45	0	[illegible]	[illegible]	e+007	
Truck Kumar A.1 (Electric 4)	8	5278.421	17591.799	15247.161	0	45	40	0	[illegible]	[illegible]	e+007	

Analyze Loaded Travel Time, Empty Travel Time and other Operational KPIs

ts

Product	Residence Time (Second)				Initial	Input	Consumed	Assembled / Packed	Scrapped	Produced
	Min	Max	Avg	StdDev						
Ore A.1 (Mining Ore.1) [Mining Ore.1]	300	38205.365	11110.243	12152.562	0	3972	70	0	0	70
Waste Dump A.1 (Mining Waste Dump.1) [Mining Waste Dump.1]	1745.172	35874.07	11195.801	11558.641	0	3972	27	0	0	27

FIG 8 – Report for operational KPIs.

CONCLUSIONS

With dual-layer approach, mining companies can perform virtual trials before investing in electric trucks. Based on the energy usage patterns, mining companies can optimise resource allocation and plan for infrastructure upgrades, plan for total cost of ownership, and thereby maximise uptime and decimate costs.

ACKNOWLEDGEMENTS

The authors would like to thank Stephan Diehl (Multi-Physics Modelling Application Director at Dassault Systèmes) and Dag Brück (Modelica Multi-Physics/Control Techno Architecture Director at Dassault Systèmes) for their valuable review of this paper.

REFERENCE

Lindgren, L, Grauers, A, Ranggård, J and Mäki, R, 2022. Drive-Cycle Simulations of Battery-Electric Large Haul Trucks for Open-Pit Mining with Electric Roads, *Energies*, 15(13):4871. https://doi.org/10.3390/en15134871

Triggering a PLC change using a cloud-based machine learning model

C Nethercott[1]

1. Manager Systems and Innovation, SEDGMAN, Brisbane Qld 4104.
   Email: chris.nethercott@sedgman.com

ABSTRACT

Dense medium cyclones (DMCs) are a class of process separation equipment used in coal preparation, iron ore, the pre-concentration of diamonds and in metalliferous and industrial minerals. Under unstable process conditions a symptom called 'surging' occurs disrupting the efficient separation in the DMC causing production losses. This can account for millions of dollars of lost revenue per annum per site. There are different reasons that cause this 'surging' condition, and it is difficult to detect with limited direct sensory feedback from the DMC. Commonly, if surging is suspected it needs to be visually confirmed in the field and a decision needs to be made on what process control set points need to be changed to rectify this behaviour. These changes typically need to be made manually by an operator in the control room.

Although in most cases there is no direct indication from the DMC, symptoms of the surging are evident in downstream equipment trends with distinct patterns across multiple tags. By applying a cloud-based multivariate Machine Learning (ML) model on these trends, it is possible to automatically detect this condition. Furthermore, the output from this model can be used to trigger a response at the control system layer, without the model residing there, like with Advanced Process Control (APC) or Model Predictive Control (MPC).

This method provides a means to validate ML model performance and quantify the actual production value before deploying the model at the site PLC level. The successful hybrid integration of cloud-based machine learning with edge-based control systems offers new possibilities for older equipment, avoiding costly hardware upgrades while still achieving a similar outcome. The knowledge gained from undertaking hybrid PLC testing could prove pivotal in the decision to upgrade a plant control system, reducing risk for the plant owner and operator.

INTRODUCTION

Machine Learning is a class of Artificial Intelligence where algorithms are trained using specific data sets to develop models that can then identify similar patterns in new data. It is a method of data analysis that automates analytical model building. Machine learning provides systems the ability to automatically learn and improve from experience without being explicitly programmed. There are mature products provided by control system manufacturers that perform these functions on edge computers and PLC hardware, but they are typically expensive to implement and maintain. By using Platform as a Service (PaaS) and other Internet of Things (IoT) software services it is possible to train and test machine learning models in the cloud, using data streamed from operating facilities and remote mine sites. Training and testing of models in the cloud can be done without making changes to the control system logic, which reduces the risk of process interruption and impact to productivity. If a high level of certainty can be achieved with a particular model, it can be decided to deploy the model on the edge as part of MPC, or as a trigger to enact a modified control philosophy. The later was done in this circumstance, to test the ability to automatically correct DMC density stability issues on a control system without any advanced control capabilities.

METHODOLOGY

Communications

The challenge was to design a communication loop that would enable the control system to funnel data to the cloud, but also receive feedback on what actions to take. In the interest of cyber security, the decision was made for any action to be preconfigured on the PLC level and just a binary trigger value from the cloud used to activate the change. This was done via an industrial gateway device (Red Lion) that could receive internet messages and communicate to the PLC over industrial protocols, meaning even if the connection was hacked the trigger variable could only be changed from 0 to 1, switching the logic on or off. Additional rules would also be put in place to control the frequency of logic activation. Furthermore, the Red Lion was deployed in an Industrial Demilitarised Zone (IDMZ), preventing other means of a jump to the PLC or other devices via the Red Lion connection.

The continuous communication cycle and flow of data between the PLC and Cloud services is shown in Figure 1:

1. Control Room can select activation status of secondary density control loop (active/not active) in Citect.

2. Data from PLC is written into tags collected by the Red Lion.

3. Red Lion sends the tags to Azure IoT Hub cloud services via the site internet connection.

4. Stream Analytics selects relevant tags for model and groups data into processable chunks.

5. Processed data sent to ML Model which calculates certainty of the data matching surging behaviour.

6. Model output is sent back to a Function that averages to 0 or 1 over a time window.

7. Binary Trigger value is packaged and formatted sent to IoT Hub.

8. IoT Hub Cloud to Device message is sent from the cloud back to Red Lion.

9. PLC watches for trigger value to activate surging density control logic.

10. 'Surging controller Active' shown on Citect.

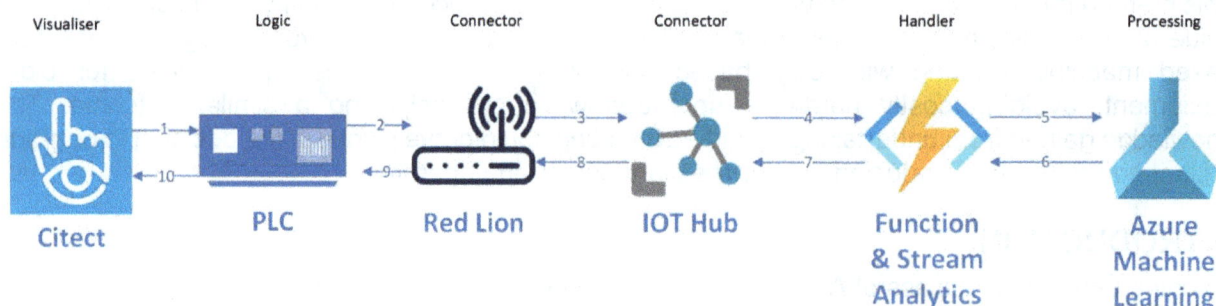

FIG 1 – PLC to cloud data communication cycle.

Controls

The existing control philosophy was a simple closed-loop controller for a slurry density, whereby a control output is used to regulate a measured density process variable at the designated set point density. The plan was to implement a secondary 'surging' control loop in parallel that would take over from the primary density controller should the machine learning model reach the trigger value for density instability. The logic for the two independent controllers are described below.

Primary density controller functional description (existing)

Operator sets density set point (PP424_D_SPY) and water addition (PP424_D_OPY) is used to keep actual density (PP424_D_PVI) close to set point relative density (RD).

Surging density controller functional description (new)

This alternative control loop changes the set point density over a set period of time and then reverts back to the original set point.

Set point change: Increase PP424_D_SPY density set point 0.01 RD every 2 mins for 10 mins until a maximum total offset of 0.05 RD reached (offset to be configurable on Citect details page), then return density set point to original value.

The following activation criteria must be satisfied for the controller to enable:

- Floating trigger value is 1.
- Watchdog tag is increasing meaning communication is flowing successfully.
- Feed has been on for greater than 15 mins.
- Citect page indicator next to density control input showing 'Surging controller Active' when active.

If any of the following criteria occur during activation, the controller will disable and revert back to the Primary controller:

- The five set point changes have been completed.
- The trigger value changes to 0 for 1 min during the step changes.
- Override button selected or density SP moved during Surging controller activation.

The additional conditions were set to prevent excessive activation of the surging controller:

- Maximum of two activations in any given hour.
- 15-minute cool down on surging controller to prevent immediate re-trigger.

The surging controller in action is demonstrated in the results section.

Modelling

A supervised machine learning technique was chosen to classify time series blocks of data into either a surging or non-surging group. The model was trained on labelled surging data to correctly identify the adverse condition. To successfully classify live data, firstly, it is passed through a Principal Component Analysis (PCA) function to reduce the dimensionality of the data into two dimensions which explains most of the data variance. Then, the Mahalanobis distance (Johnson and Wichern, 2007) is calculated, which is a measure of similarity between the simplified points and a distribution of confirmed surging events, providing a measure which indicates a level of confidence between the normal or adverse operating state.

Figure 2 shows a DMC cyclone surging event with a distinct cyclic pattern across multiple weigher, centrifuge and screen tags. Such events in historical data were used to train a multivariate model to detect future events. In this plant, there are centrifuges and desliming screens that are situated at both the underflow and overflow ends of the cyclone. By contrasting the power being drawn by the machines at either side and the weighers, a distinct mirroring of the cyclic pattern is evident during surging.

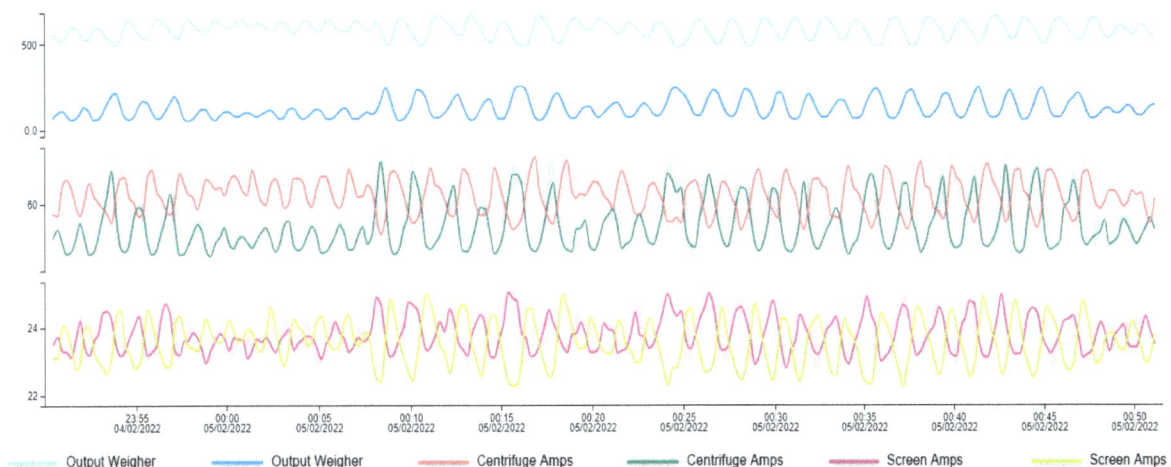

FIG 2 – Plant weigher, centrifuge amp and screen amp trends showing DMC surging event.

In combination, the live tag data is analysed by the ML model and a binary classification variable is created to indicate if a data point is an anomaly (1) or not (0), as shown by the red line in Figure 3. This is used to flag if the DMC is surging or not.

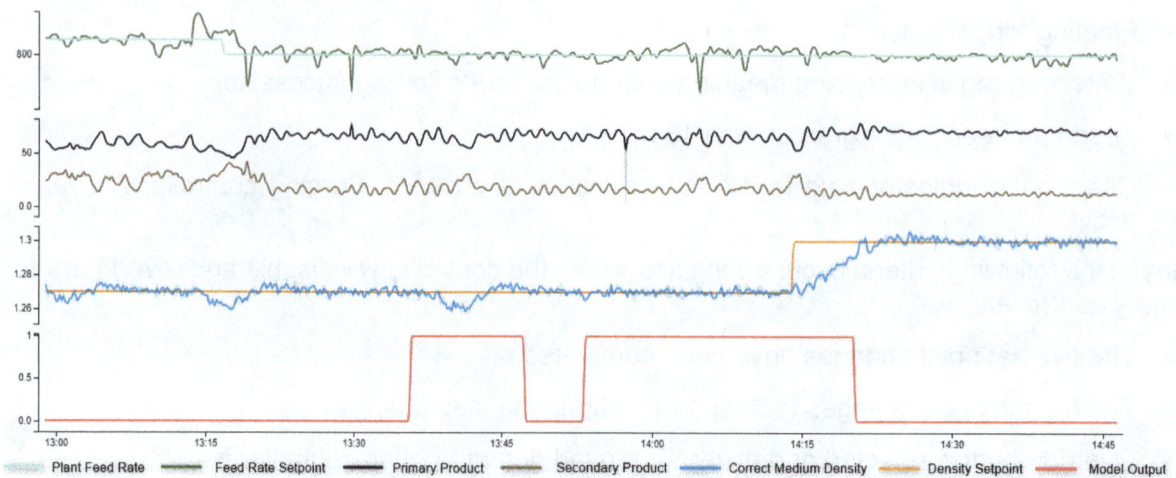

FIG 3 – DMC surging detected by ML Model (red line) corrected with a manual density increase.

RESULTS

The machine learning model was first deployed virtually, to see how the model would perform before using it to trigger the surging controller. It was applied to a two-product coal plant (Coke and Thermal), whereby the primary DMC reject feeds the secondary DMC. If not monitored closely, the primary coking product can be displaced to secondary thermal product. The secondary product price is much lower that the primary product price, which can result in a net financial loss.

Figure 3 demonstrates a DMC surging event that triggered following a decrease in plant feed rate, or solids loading, to the DMC.

Around 13:20, the cyclic behaviour began following a drop in plant feed rate. Following the rolling 10 min average buffer period, the machine learning model output reached a value rounding to '1' (red line) indicating surging of the primary DMC was occurring. The density was then manually raised from 1.27 RD to 1.30 RD around 14:15 alleviating the surging conditions with the model output returning to '0'. A yield uplift in the Primary product was observed by correcting this instability.

After two weeks' validation of the Machine Learning surging model in the cloud, the model output was used to trigger the edge-based PLC surging density control loop automatically. This meant when surging was detected, the surging controller would be triggered to step up the density set point by 0.01 RD every 2 mins for a maximum of five step changes until the surging model deactivated or the sequence had finished. Figure 4 shows a surging event that was rectified automatically, and without human intervention, using this logic. Surging was detected from 12:43 with the model activating at 12:49 sending a '1' to the control system triggering the logic. It can be seen that after the third step change at 12:55 the amps in the Secondary and Primary Product centrifuges inverted indicating the shift of load back into the Primary product. After the fifth step change the surging model deactivated '0' and the density set point returned from 1.29 RD to 1.24 RD.

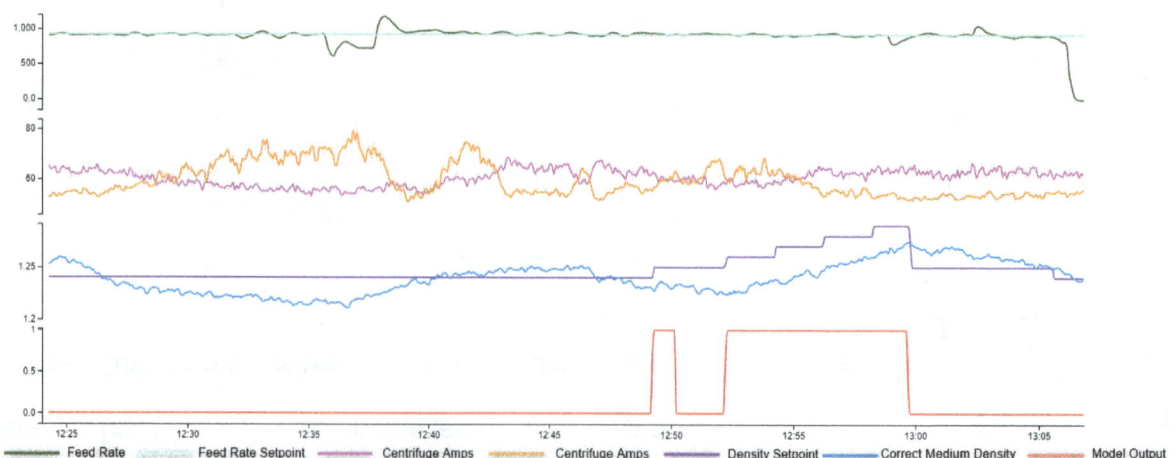

FIG 4 – DMC surging detected by ML Model (red line) corrected with automatic surging controller.

The stabilisation of the DMC trends indicated the surging control logic was successful in correcting the abnormal behaviour.

CONCLUSION

The results above demonstrate how a machine learning model in the cloud can be used to trigger a change in the edge-based control system and validate its effectiveness. Longer term this model would be deployed on the control system, with retraining required should the model accuracy decline due to process changes or equipment maintenance. This method also provides a mechanism for older or less advanced control systems to achieve machine learning control functionality. Most importantly, it shows that complex issues such as DMC surging can be detected and rectified without human intervention, which could be expanded to all manual set points in the plant.

REFERENCES

Johnson, R A and Wichern, D W, 2007. *Applied multivariate statistical analysis*, 6th edn, ch 2 (Pearson: London).

Real-time detection of perimeter breaches using high-precision GPS for enhanced safety in high-risk dumping zones in open pit mining

M R Pratama[1], S Widodo[2] and K P Adiprima[3]

1. Operation Research Engineer, PT Pamapersada Nusantara, Jakarta 13930, Indonesia. Email: mochamadrp@pamapersada.com
2. Software Engineering Manager, PT LAPI ITB, Bandung 40132, Indonesia. Email: sugengwidodo@lapi-itb.com
3. HSE System and Compliance, PT Pamapersada Nusantara, Jakarta 13930, Indonesia. Email: khrisna.protecta@pamapersada.com

ABSTRACT

Mining operations involve high risks, particularly material dumping in designated high-risk disposal areas. A high-risk disposal area is characterised by a five-meter disposal bench height and a base filled with water or mud. The primary concern in these areas is slope stability, which is affected by material properties, bench heights, and water levels. Organisations must establish a safety distance limit for trucks transferring materials to manage safety risks in high-risk dumping areas. This limit is determined through geotechnical assessments. If trucks exceed the designated safe dumping limits, they risk sinking or sliding into water-filled zones, risking safety.

The authors have innovatively developed a Real-Time Detection System with a High-Precision GPS (Global Positioning System) installed in Komatsu HD785-7 trucks. Unlike traditional methods, this system uses two pairs of rovers positioned by a Komatsu DZ375 Bulldozer to define the perimeter of the dumping safety boundary. It also provides real-time audio guidance to the truck operator, ensuring constant awareness of the safe dumping distance. Any violation of the safe dumping perimeter is promptly recorded by CCTV (Closed-Circuit Television) cameras, providing visual evidence of the dumping activity.

The truck operator receives real-time audio guidance in the cab to help maintain a safe dumping distance based on the truck's location relative to the designated safe dump perimeter. The dump truck operator must always be aware of this reference.

INTRODUCTION

In coal mining operations, the disposal area is designated for overburden disposal. The standard specifications for a single-slope disposal area typically have a vertical-to-horizontal ratio of 1:2, corresponding to a slope of 26.57° (Haque and Reza, 2020). The distance from the foot of the in-pit disposal area to the active work zone should be at least three times the total height of the stockpile, or it can be determined based on technical studies. For stockpiling activities involving heights of five metres or less, these can be conducted directly.

Based on regulations issued by the government, it is prohibited to stockpile overburden in areas of former ponds, former river channels, and swamps unless it is based on the results of a technical study (Decree of the Minister of Energy and Mineral Resources, 2018). And if it is forced to do so, it is categorised as a high-risk disposal. High-risk disposal areas have one or more of the following conditions: The height of the dumping bench is more than five metres, and the bottom is filled with water or mud, as shown in Figure 1. An important concern in high-risk dumping areas is slope stability, which is affected by the material's nature, the bench's height, and the water level.

FIG 1 – High-risk disposal area.

To manage the safety risks in a high-risk dumping area, organisations have to set a safety distance limit for trucks dumping the materials in a safe dumping area based on geotechnical assessment (PT Pamapersada Nusantara, 2025) as shown in Figure 2. If trucks exceed the designated safe limit in dumping areas, they could sink or slide into water-filled zones, posing a severe safety risk.

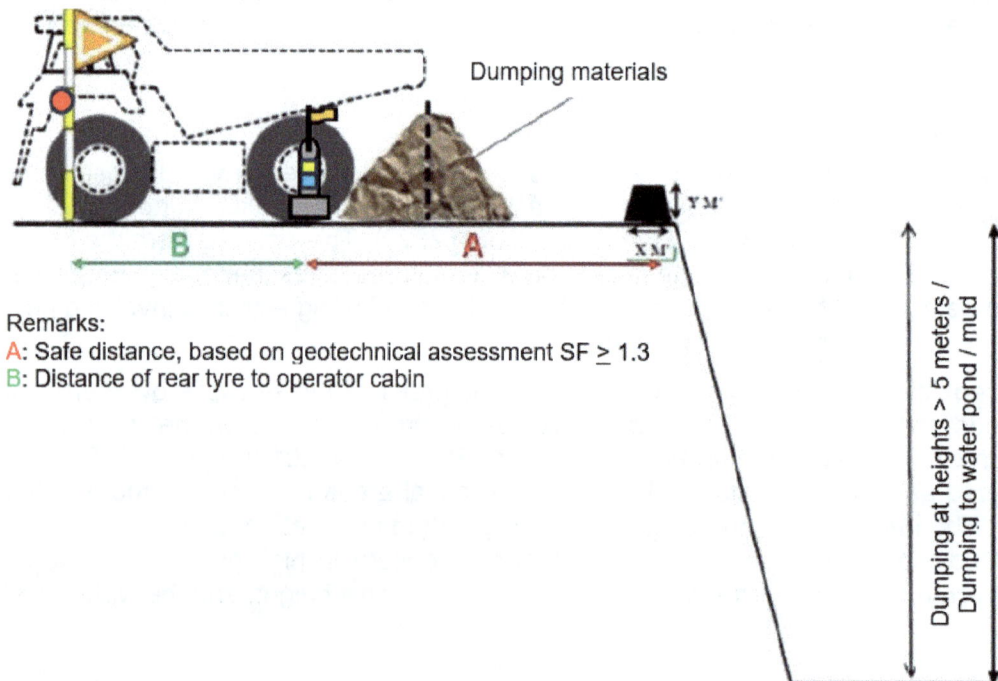

Remarks:
A: Safe distance, based on geotechnical assessment SF $\geq$ 1.3
B: Distance of rear tyre to operator cabin

FIG 2 – Side view illustration of material dumping activities in high-risk disposal.

To anticipate cracks in the dumping point area, at least two pieces pair or more of dumping limiters are installed, with a maximum distance between them of 50 m (PT Pamapersada Nusantara, 2025) as shown in Figure 3.

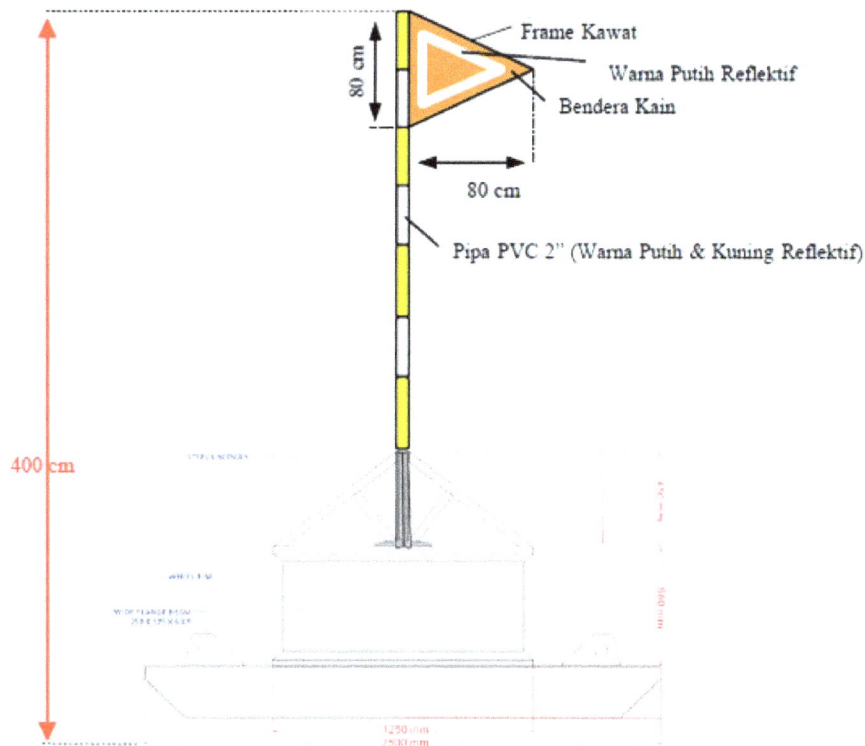

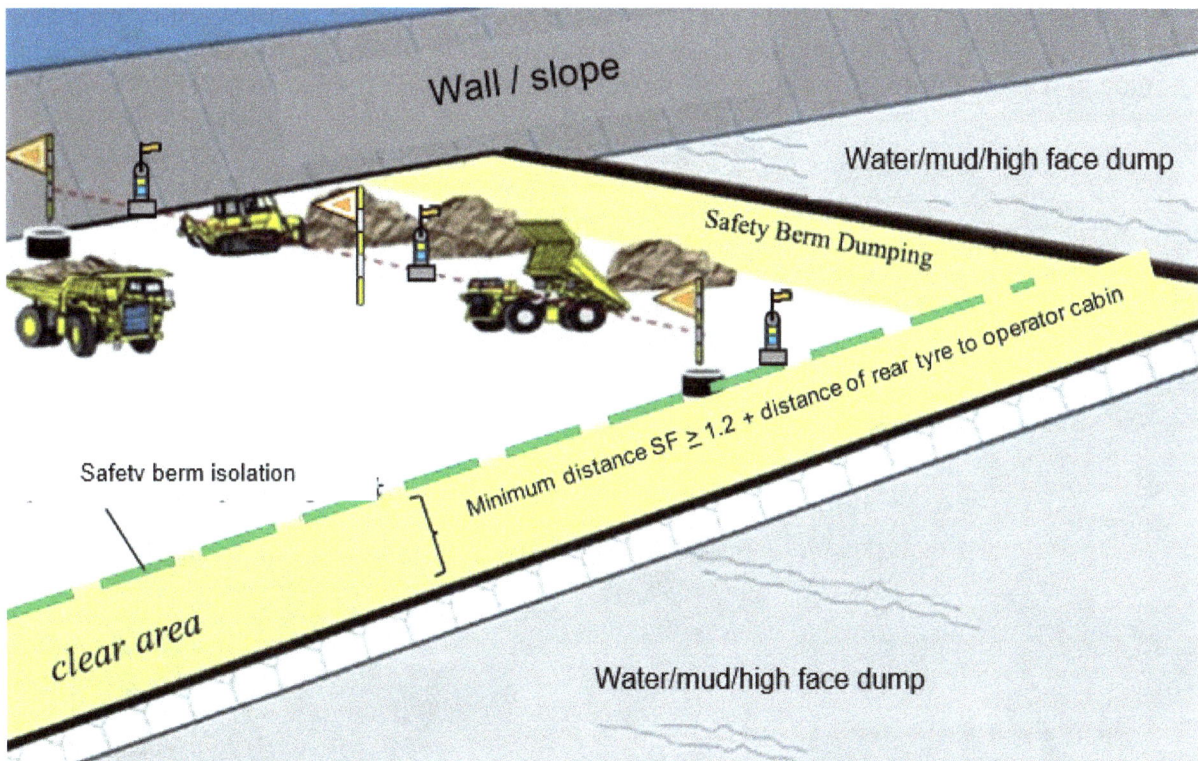

FIG 3 – Standard of dumping limiter placed in dumping activities high-risk disposal.

SYSTEM ARCHITECTURE

The experiments reported in this paper employ advanced safety and monitoring technology utilising RTK-GPS. The system uses RTK-GPS-based devices, including Rovers, Base Stations, and Cabin GPS mounted on Komatsu HD785-7 dump trucks, to monitor dump-truck movements in dangerous dump zones. RTK GPS is a dynamic GPS positioning technique using a short observation time, providing precise results in real-time, thus ensuring the system's accuracy (Lee and Ge, 2006).

The system has several components that work together to ensure the dump truck dumps within the perimeter. Figure 4 represents how a real-time detection component functions by integrating data from multiple GPS-based devices to monitor and control dump-truck operations within designated safe zones.

- Rovers Units

  Rover, shown in Figure 4c, replaces the manual dumping limiter. Two rovers form a straight line as a reference limit for safe dumping, which is referred to as the perimeter. Alerts on the RTK cabin will be generated if the distance to the perimeter starts from 15 m.

- Cabin RTK GPS devices.

  The cabin RTK GPS Device shown in Figure 4d is a high-accuracy GPS device mounted on the dump truck. It communicates with the perimeter established by the rover to determine a safe distance. The base station ensures high-precision RTK GPS signal correction to maintain accuracy.

- Gateway and Mobile Tower CCTV

  Inside the mobile tower, there is a gateway and a CCTV camera, as shown in Figure 4b. The CCTV camera records video and images during crossings. The gateway transmits data transactions to the on-premises server. Additionally, the mobile tower can transmit data to central monitoring systems and provide real-time alerts to supervisors.

- Base station

  The base station were shown in Figure 4a the function is for correcting the position of the Rover and RTK Cabin, ensuring their accuracy. To maintain this accuracy, the base station must be placed high and unobstructed to communicate effectively with the Rover and RTK Cabin.

FIG 4 – Real-time detection component.

The safety distance is established based on the position of the two Rovers, which will align to form a straight line marking the safe dumping limit. The cabin's RTK GPS will perform position corrections with the Rover to continuously verify whether the truck remains within the designated safe distance. If the cabin GPS becomes misaligned with the Rover, it will trigger an alarm and send a notification—containing the unit number and event time—to the gateway. The gateway will then instruct the CCTV in the mobile tower to capture the current conditions and activate a siren from the mobile tower. This will alert the supervisor that a truck has violated the safe distance shown in Figure 5.

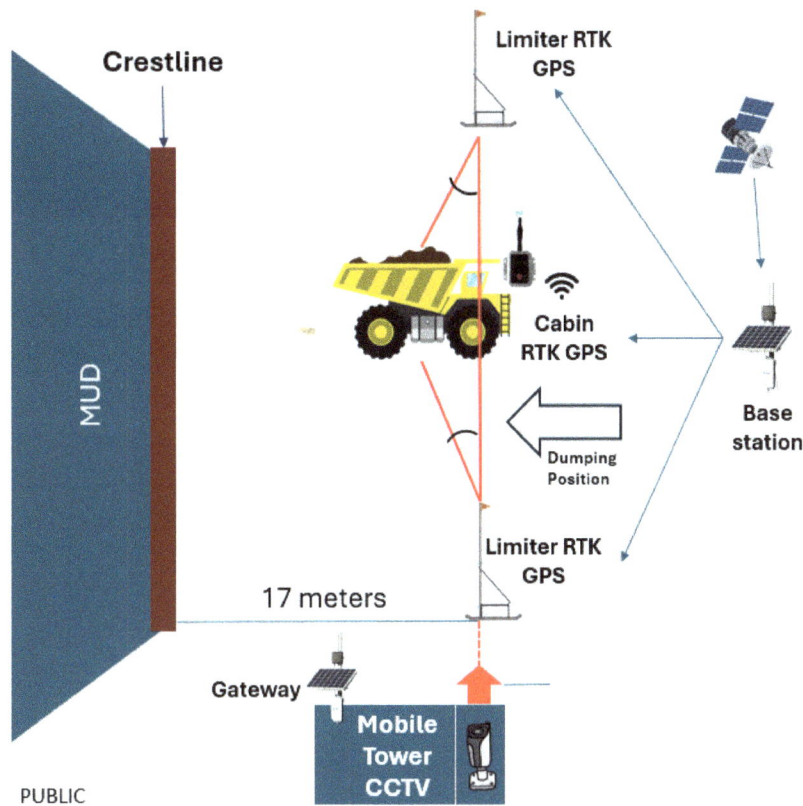

FIG 5 – Real-time detection alert.

DUMPING LIMIT EXPERIMENT

Communication between gateway, rover, and RTK GPS Cabin devices uses specialised wi-fi. This specialised wi-fi is called Long-Range wi-fi, and it is just like other wi-fi, except it uses slower data to achieve long distances. The LR wi-fi could reach a distance of 2 km (Espressif Systems Co, Ltd, 2025), compared to standard wi-fi, which is only 100 m long. Above LR wi-fi is standard TCP/IP protocol, and the authors use the MQTT protocol to communicate between devices. In this case, the gateway works as an Access Point, while the others are wi-fi stations. The authors use RabbitMQ as an MQTT broker at the gateway, which receives and distributes any published message to the subscriber (Spring Source, 2025).

Perimeter determination

Once the rover gets its own coordinates using RTK GPS, it will publish its own coordinates to the broker using the MQTT topic 'dla/pos.' Figure 6 shows when a rover sends its coordinates to the gateway.

```
Position received  b'NODE 007,-1.86056589,115.85724952'
Process coord :   {'name': 'NODE 007', 'lat': -1.86056589, 'lon': 115.85724952, 'distance': 115.8721880857118}
T :  [{'name': 'NODE 007', 'lat': -1.86056589, 'lon': 115.85724952, 'distance': 115.8721880857118, 'to': 24.0}]  R :
[{'name': 'NODE 007', 'lat': -1.86056589, 'lon': 115.85724952, 'distance': 115.8721880857118, 'to': 24.0}]
Coordku :  [{'name': 'NODE 007', 'lat': -1.86056589, 'lon': 115.85724952, 'distance': 115.8721880857118, 'to': 24.0}]
```

FIG 6 – The rover publishes coordinates to the gateway and calculates the distance between rovers.

The gateway serves a critical function in the processing and distributing perimeter data within the system. It initiates its operation by subscribing to the MQTT topic '*dla/pos*' to receive real-time precise coordinate updates from the rovers. The gateway organises and analyses the data to create a clear perimeter when it gets these updates. It arranges the rovers' positions and uses algorithms to understand their spatial relationships. After determining the perimeter, the gateway periodically publishes the updated perimeter coordinates to the MQTT topic '*dla/perimeter*' so that other system components, such as the cabin's RTK GPS device, can be kept informed. In addition, these coordinates are stored in a database for historical tracking analysis and recalibration. Figure 7 describes when the gateway publishes the '*dla/perimeter*'s that is to be received by the RTK cabin.

```
rovercoords : [{'name': 'NODE 007', 'lat': -1.86056589, 'lon': 115.85724952, 'distance': 115.8721880857118, 'to': 24.0},
{'name': 'NODE 008', 'lat': -1.8614091, 'lon': 115.85775376, 'distance': 115.87270580318022, 'to': 24.0}]
Publish peri : 2
Coords : [{'name': 'NODE 007', 'lat': -1.86056589, 'lon': 115.85724952, 'distance': 115.8721880857118, 'to': 23.0},
 {'name': 'NODE 008', 'lat': -1.8614091, 'lon': 115.85775376, 'distance': 115.87270580318022, 'to': 23.0}]
```

FIG 7 – The gateway publishes the dla/perimeter.

When the RTK GPS cabin subscribes to '*dla/perimeter*,' it gets the perimeter coordinates of the safe dumping boundary to inform the truck operator in real-time. Figure 8 divides the dumping safety distance to the perimeter into four segments.

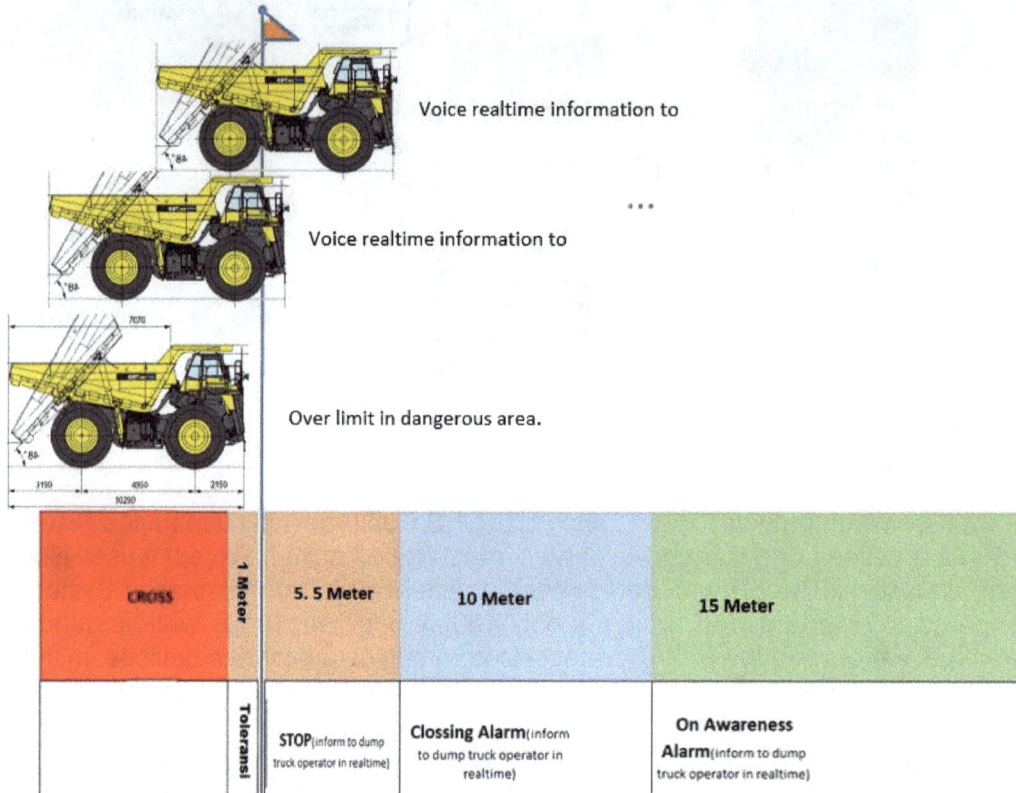

FIG 8 – Real-time alert to the cabin operators.

Unit position segmentation

Find the line equation of the segment

Comparing the RTK coordinates of the GPS Cabin with the rover perimeter, The equation of a line passing through two points (Math Centre, 2009) (x_1, y_1) and (x_2, y_2) is:

$$y = mx + c \tag{1}$$

Where:

$$m = \frac{y_2 - y_1}{x_2 - x_1} \tag{2}$$

(Slope of the line)

$$c = y_1 - mx_1 \tag{3}$$

Find the perpendicular line equation passing through the cabin coordinate

A line perpendicular to another line has a slope that is the negative reciprocal of the original slope. So, if the slope of the segment is m, the perpendicular slope is:

$$m_\perp = -\frac{1}{m} \tag{4}$$

The cabin is at a coordinate. (x_c, y_c) Then, the equation of the perpendicular line passing through this point is:

$$y_c = -\frac{1}{m}x + c_c \tag{5}$$

Find the intersection point of the two lines

The intersection occurs where both equations are equal, meaning:

$$mx + c = -\frac{1}{m}x + c_c \tag{6}$$

This gives the coordinate. (x_{int}, y_{int}) which is the closest point on the segment to the cabin.

The distance between (x_c, y_c) and (x_{int}, y_{int}) calculated using Haversine:

$$d = 2R.\,asin\left(\sqrt{sin^2\left(\frac{\Delta\emptyset}{2}\right) + \cos(\emptyset_1)\cos(\emptyset_2)sin^2\left(\frac{\Delta\lambda}{2}\right)}\right) \tag{7}$$

Where:

R	= Earth's radius (**6371 km**)
$\emptyset_1, \lambda_1$	= Latitude and longitude of the **cabin** (x_c, y_c)
$\emptyset_2, \lambda_2$	= Latitude and longitude of the **closest point on the perimeter** (x_{int}, y_{int})
$\Delta\emptyset$	= $\emptyset_2 - \emptyset_1$ (difference in latitudes in radians)
$\Delta\lambda$	= $\lambda_2 - \lambda_1$ (difference in longitudes in radians)
d	= determines the classification of distance perimeter and RTK GPS Cabin installed in Dump truck

Determine whether the cabin is in a safe area or danger area

If the cabin is close to the perimeter, The authors need to check whether it is in a safe area or a dangerous area:

- Define a parallel line for the safe area:
  - The safe area boundary is a line parallel to the segment line.
  - A parallel line has the same slope m but passes through a known safe coordinate (x_s, y_s).
  - The equation is:

$$y_s = mx + c_s \tag{8}$$

Where:

$$c_s = y_s - mx_s \tag{9}$$

- Compare c_s with c (segment intercept):
  - If $Cs > C$, the save area is above the segment.
  - If $Cs < C$, the save area is below the segment.
- Determine a parallel line passing through the cabin:
  - The equation of a line parallel to the segment and passing through the cabin (x_c, y_c):

$$y_c = mx + c_c \tag{10}$$

Where:

$$c_c = y_c - mx_c \tag{11}$$

- Compare c_c with c:

- If c_c has the same comparison result as c_s. The dump truck cabin is still in a safe area.
- If c_c has the opposite comparison result. The dump truck cabin has entered the danger area.

Data monitoring

If the dump truck enters a hazardous area, it will alert the operator in real-time by sending a message with the topic *'dla/crossing'* via the MQTT message protocol. An alarm will sound in the dump truck operator's cabin, and the supervisor will also be able to hear the alarm from the mobile tower. The mobile tower will also capture real-time images for evidence current conditions using the CCTV camera. Figure 9 illustrates a truck crossing the safe perimeter.

FIG 9 – Crossing the perimeter.

VALIDATION

The validation process involves retrieving dump truck movement log data from the gateway and carefully monitoring the vehicle's approach to the perimeter. This thorough procedure ensures that the alarm in the operator's cabin is activated and that the dumping safety limit violation detection system functions correctly. This is supported by images and video captured from the CCTV camera. The comprehensive dump truck movement log data is presented in Table 1, detailing critical information such as longitude and latitude coordinates, perimeter segments, distance, and speed when the dump truck neared the dumping safety perimeter. To guarantee the accuracy of this vital coordinate data, it was expertly overlaid onto an orthophoto, providing a clear and compelling visualisation, as illustrated in Figure 10. Each perimeter crossing event must be validated to ensure that all systems work correctly. All validation processes are represented in Figure 11.

TABLE 1

Data log RTK GPS cabin.

Latitude	Longitude	GPS	UTC time	Distance (m)	Status	Sound	Speed (kph)
115.855865	-1.864558	RTK FIX	120924 123813.500	15.68	Faraway	0	4.44
115.855857	-1.864546	RTK FIX	120924 123814.500	8.59	On awareness	1	4.63
115.855851	-1.864536	RTK FIX	120924 123815.500	7.33	On awareness	1	4.44
115.855847	-1.864529	RTK FIX	120924 123816.500	6.42	On awareness	1	3.33
115.855844	-1.864524	RTK FIX	120924 123817.500	5.73	On awareness	1	2.41
115.855839	-1.864517	RTK FIX	120924 123818.500	4.76	Closing	1	3.70
115.855834	-1.864509	RTK FIX	120924 123819.500	3.72	Closing	1	3.70
115.855828	-1.864501	RTK FIX	120924 123820.500	2.66	Closing	1	3.89
115.855824	-1.864494	RTK FIX	120924 123821.500	1.78	On perimeter	1	2.96
115.855821	-1.864489	RTK FIX	120924 123822.500	1.1	On perimeter	1	2.41
115.855817	-1.864483	RTK FIX	120924 123823.500	0.26	On perimeter	1	2.96
115.855813	-1.864476	RTK FIX	120924 123824.500	0.54	On perimeter	1	2.78
115.855811	-1.864472	RTK FIX	120924 123825.500	-1.09	Crossing	1	2.04
11.585581	-1.864469	RTK FIX	120924 123826.500	-1.42	Crossing	1	1.11
11.585581	-1864477	RTK FIX	120924 123827.500	-1.39	Crossing	1	0.37
11.585581	-1.864469	RTK FIX	120924 123828.500	-1.4	Crossing	1	0.19
11.585581	-1.864469	RTK FIX	120924 123829.750	-1.41	Crossing	1	0.19
11.585581	-1.864469	RTK FIX	120924 123830.750	-1.42	Crossing	1	0.19
11.585581	-1.864469	RTK FIX	120924 123831.750	-1.41	Crossing	1	0.19
11.585581	-1.864469	RTK FIX	120924 123832.750	-1.43	Crossing	1	0.00
11.585581	-1.864469	RTK FIX	120924 123833.750	-1.43	Crossing	1	0.00
11.585581	-1.864469	RTK FIX	120924 123834.750	-1.44	Crossing	1	0.00
11.585581	-1.864469	RTK FIX	120924 123835.750	-1.44	Crossing	1	0.00
115.855809	-1.864469	RTK FIX	120924 123836.750	-1.45	Crossing	1	0,00
115.855809	-1.864469	RTK FIX	120924 123837.750	-1.49	Crossing	1	0.00

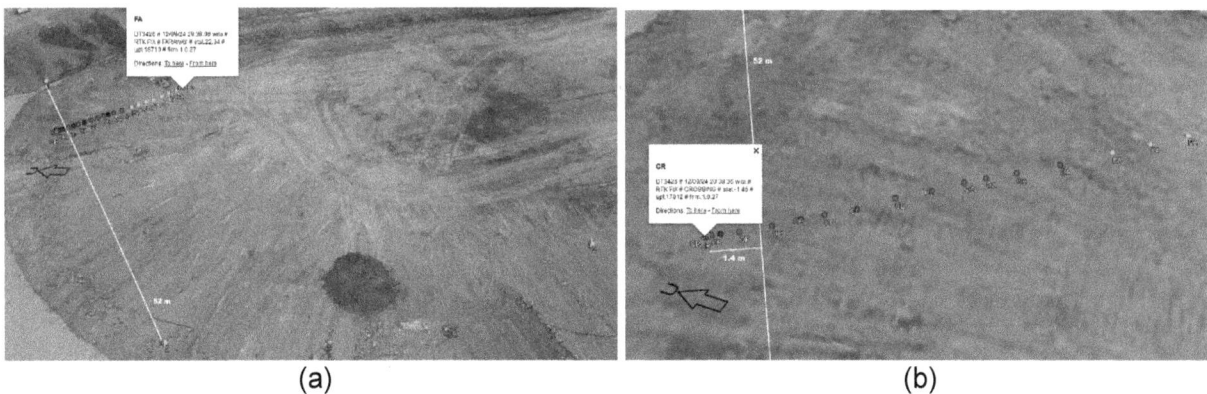

FIG 10 – Coordinate validation by orthophoto: (a) zoomed in; (b) zoomed out.

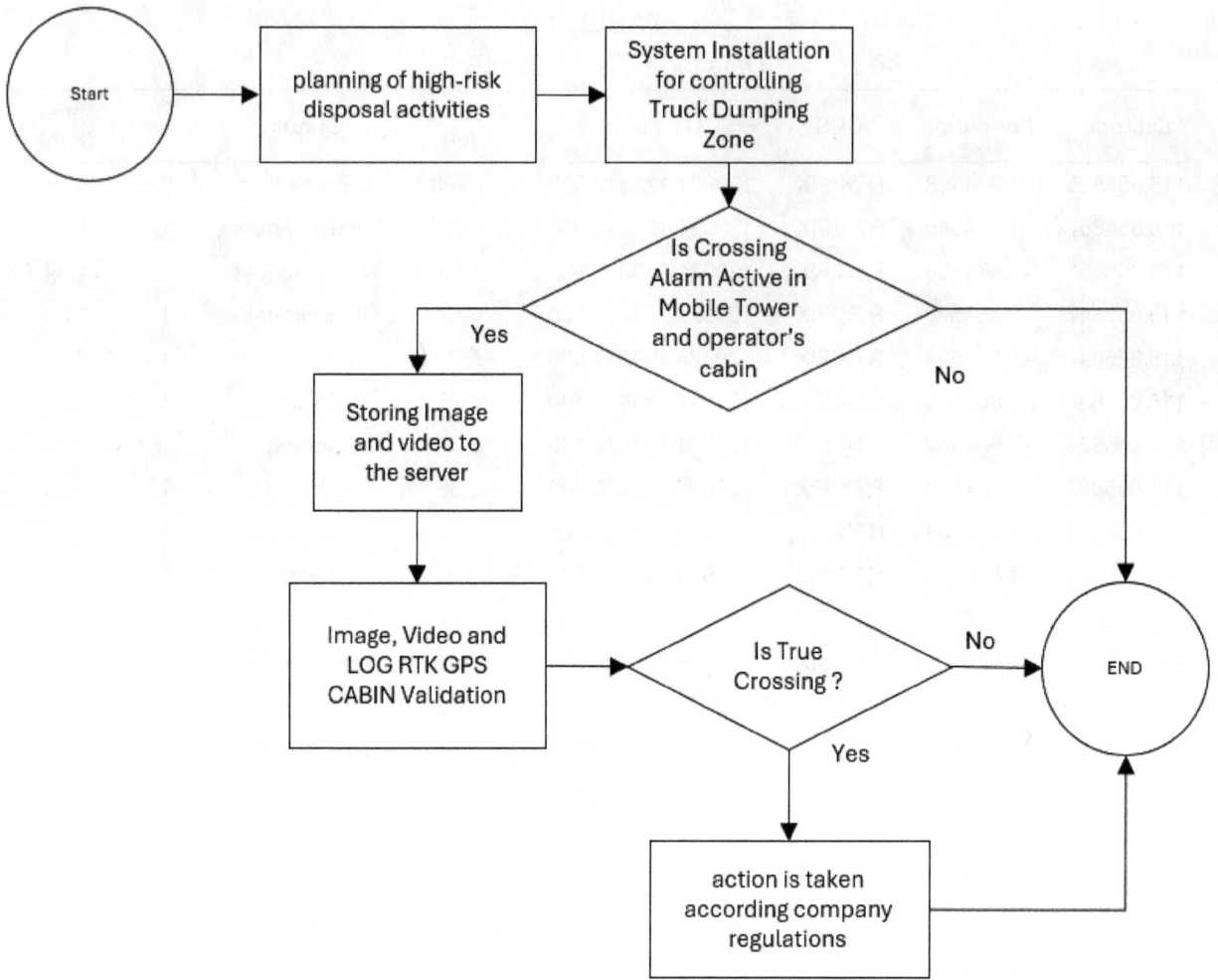

FIG 11 – Data validation of crossing safety perimeter.

CONCLUSIONS

Mining operations carry significant risks, especially in areas where slope stability is critical. This study presents a Dumping Perimeter Warning Procedure that uses a Real-Time Detection System with High Precision RTK GPS technology to monitor safe dumping perimeters, ensuring that dump trucks operate within assessed boundaries to prevent slope failures and accidents. The safe perimeter coming from the two rovers defines the dumping boundary as the RTK GPS installed on the dump truck knows the safe dumping boundary. A mathematical equation then determines whether a truck is safe or at risk. If it enters the danger zone, an automated alert will notify the operator and supervisor while activating CCTV footage for incident analysis. This approach improves safety in mining operations by reinforcing the commitment to protect personnel and equipment, demonstrating that high-precision GPS systems improve safety and risk management in high-risk dump zones and the data can be monitored in real-time by tableau dashboard shown in Figure 12.

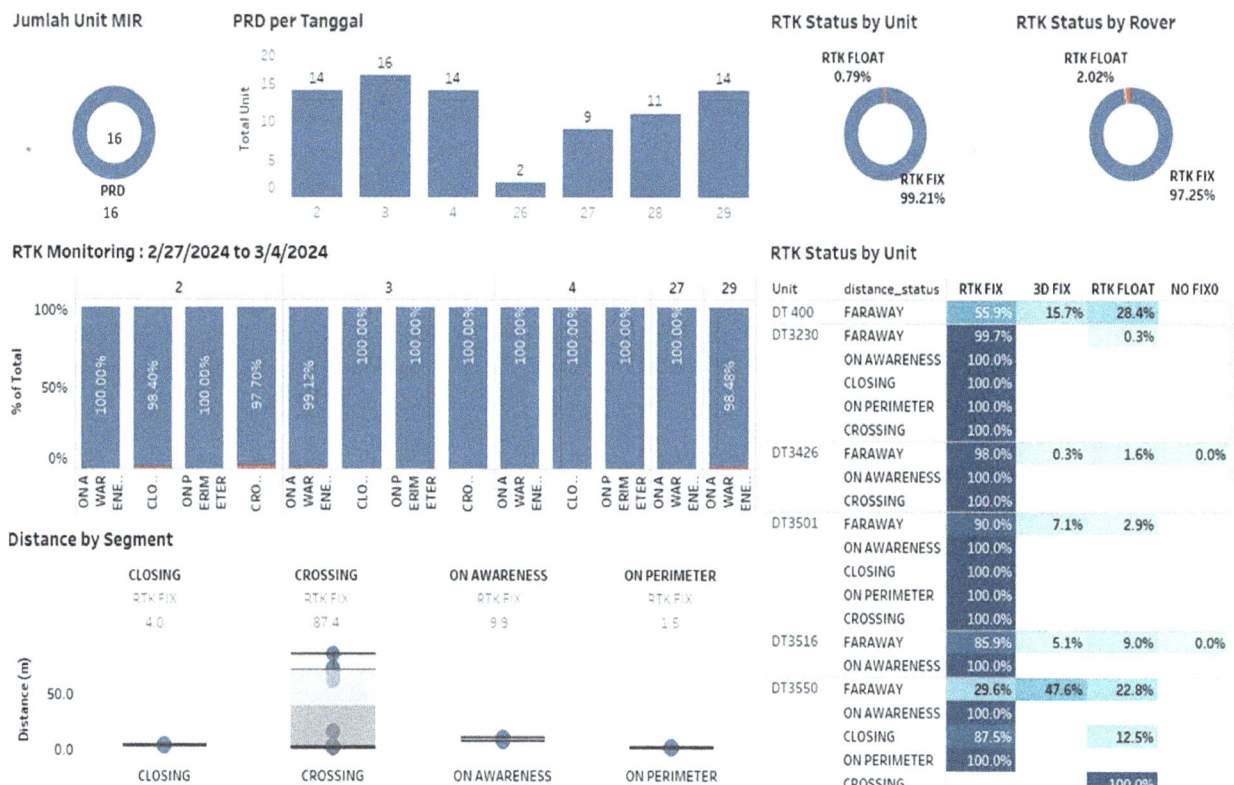

FIG 12 – RTK status monitoring rovers and RTK GPS cabin.

ACKNOWLEDGEMENTS

The authors would like to thank the management of PT Pamapersada Nusantara and PT LAPI ITB for supporting this research. Collaboration in sharing technical expertise and resources ensured the success of this research.

The authors also recognise the contributions of all team members involved in the research, testing, and validation of the Real-Time Detection System for high-risk disposal areas. Special thanks go to the PT Pamapersada Nusantara KIDE District management for their support, input, and critical insights during the research process so that this system can be appropriately implemented at the KIDE District.

REFERENCES

Decree of the Minister of Energy and Mineral Resources, 2018. Pedoman Pelaksanaan Kaidah Teknik Pertambangan yang Baik (Guidelines for the Implementation of Good Mining Engineering Principles), No. 1827 K/30/MEM/2018, 370 p (Jakarta: LL KESDM).

Espressif Systems Co, Ltd, 2025. ESP32 Series Datasheet, version 4.8 (Shanghai).

Haque, D and Reza, M I, 2020. Parametric Analysis of Slope Stability for River Embankment, *Journal of Advanced Engineering and Computation*, 4(3):196. https://doi.org/10.25073/jaec.202043.291

Lee, I-S and Ge, L, 2006. The performance of RTK-GPS for surveying under challenging environmental conditions, *Earth, Planets and Space*, 58:515–522.

Math Centre, 2009. Equation of Straight Lines, Math Centre.

PT Pamapersada Nusantara, 2025. PAMA Production Management System, PAMA/OPRT/21/041/STD STD Disposal. Jakarta.

Spring Source, 2025. Rabbitmq documentation [online]. Available from: <https://www.rabbitmq.com/docs/mqtt>

Enhancing mining operations with digital twin technology through an integrated multi-brand software approach

V Rais[1]

1. Software Expert, Dassault Systèmes, Perth WA 6850. Email: viktor.rais@3ds.com

ABSTRACT

The mining industry faces significant challenges, including economic volatility, stringent environmental regulations, and increasing demands for operational efficiency and sustainability. As the sector embraces digital transformation, Digital Twin technology emerges as a pivotal advancement. Leveraging Dassault Systèmes® advanced software suite—including GEOVIA™, DELMIA™, CATIA™ and SIMULIA™—this technology provides a real-time virtual representation of mining operations, facilitating scenario simulation, challenge prediction, and process optimisation.

This paper explores the role of interoperability and integration in deploying Digital Twin technology across mining operations. By establishing seamless data exchange between various software tools, mining companies can achieve higher levels of integration, ensuring that all components work harmoniously. The Mining Maturity Level framework, inspired by Building Information Modelling (BIM), is proposed to assess and enhance the implementation of digital technologies, automation, and optimisation processes, driving the industry towards greater efficiency and sustainability.

Mining Information Modelling (MIM), inspired by BIM principles, is introduced to advance digital asset representation and improve mining operations. By progressing through defined maturity levels, companies can systematically enhance operational efficiency, safety, and environmental sustainability.

Furthermore, the incorporation of AI and machine learning enhances data security, transparency, and predictive insights, enabling real-time data analytics for predictive maintenance and operational adjustments. Collaborative ecosystems also integrate stakeholders into the decision-making process, leading to improved project outcomes.

This research demonstrates how integrated software solutions and Digital Twin technology can revolutionise mining operations, aligning with the ongoing evolution of digital technologies in the industry. By leveraging these tools and frameworks, mining companies can better adapt to challenges, optimise resource utilisation, and achieve a more sustainable and efficient future.

INTRODUCTION

Dassault Systèmes® Digital Twin approach offers a transformative solution by uniting data acquisition, geological modelling, engineering simulation, operations management, and continuous optimisation.

Dassault Systèmes® supplies a suite of mining-oriented software brands that connect to its cloud-based 3DEXPERIENCE® platform to build a unified Mining Digital Twin. Native 3DEXPERIENCE GEOVIA roles (Earth Engineering Coordinator, Pit Optimizer, Strategic Mine Planner, Surface Mine Designer and Underground Mine Designer) run directly on the platform, whereas established desktop applications—GEOVIA Surpac™, MineSched™ and Whittle™; DELMIA™ solutions (Apriso™, Quintiq™, Ortems™); CATIA™ tools (including CATIA Magic™); and SIMULIA™ applications (Abaqus™ and Isight™)—remain standalone but offer "POWER'BY" connectors or similar interfaces that transfer or synchronise their data with the platform for downstream collaboration, analytics and simulation (Dassault Systèmes, 2025a, 2025b, 2025c, 2025d).

Recent studies converge on the need for an integrated Digital Twin ecosystem in mining: Farrelly and Davies (2021) argue that an interoperable Digital Twin platform is foundational to a network-centric mine; Rathore et al (2021) demonstrate that embedding AI and machine-learning algorithms in Digital Twins yields more adaptive and efficient decision-making; and El Bazi et al (2022, 2023) propose a multi-layered Digital Twin framework covering the full asset life-cycle—from design to decommissioning—to enhance sustainability, reliability, and overall system integration.

LEVELS OF MATURITY AND MINING INFORMATION MODELLING (MIM)

To fully capitalise on these synergies, companies must adopt a structured methodology that not only integrates advanced digital technologies but also systematically measures and guides their implementation. This leads to the concept of Mining Maturity Levels and Mining Information Modelling (MIM)—frameworks designed to provide a clear roadmap for achieving high-value, collaborative, and sustainable digital mining operations.

Parallel to Building Information Modelling (BIM), the Mining Maturity Level framework evaluates the adoption of digital tools, collaboration, and automation in mining operations. Table 1 outlines the incremental journey toward integrated and intelligent ecosystems, while MIM enriches digital asset representation and collaboration.

TABLE 1
Comparison of mining maturity levels.

Level	Description	Key features
0	Traditional methods	Manual or minimal technology, no digital integration
1	Partial automation	Some automation, limited data integration
2	Advanced automation and data integration	Automated processes, partial data exchange
3	Digital integration	Full IT integration, use of Digital Twins and real-time data management
4	Autonomous operation	AI optimisation and risk management, minimal human intervention
5	Fully integrated intelligent ecosystem	Supply chain integration, sustainability emphasis

Progression through these levels enables companies to improve efficiency, safety, and environmental sustainability.

MIM borrows from BIM principles, advancing digital asset representation to enhance mining operations (Table 2).

TABLE 2
Comparison of mining information modelling levels.

Level	Description	Key features
0	No integration	Reliance on traditional methods with minimal digital modelling.
1	Partial integration	Introduction of 2D models and limited 3D modelling for specific operations.
2	Collaborative modelling	Data-sharing among stakeholders with increased model integration.
3	Full modelling integration	Real-time data integration across all operations, promoting seamless interaction among stakeholders.

By advancing through these maturity levels, mining organisations can systematically elevate collaboration, efficiency, and sustainability in alignment with MIM's principles.

METHODOLOGY – WORKFLOW FOR A SINGLE MINING DIGITAL TWIN

Within the Dassault Systèmes® ecosystem, a fully functional Mining Digital Twin integrates all brand solutions and software applications, driving real-time analysis, predictive maintenance, and adaptive operational controls. GEOVIA™, DELMIA™, and SIMULIA™—orchestrated on the 3DEXPERIENCE® platform—form the foundation of seamless digital transformation.

Below is a detailed algorithmic workflow that integrates the main Dassault Systèmes® mining and simulation products into one continuous 'loop,' ensuring a holistic approach to mine design, planning, scheduling, simulation, and operations management (Figure 1).

1. Data acquisition and geoscience management:

   o CATIA™ Magic™ delivers the SysML MBSE layer for 3DEXPERIENCE mining virtual twins; through Teamwork Cloud's version-controlled repository it traces models—via platform APIs—to GEOVIA, DELMIA and SIMULIA data, maintaining a digital thread that links strategic plans to live operations.

   o GEOVIA™ Surpac™ for 3D geology and mine design.

   o GEOVIA™ Earth Modelling for building and managing geological environments.

   o GEOVIA™ Surveyor for updating mine geometry with survey data.

   o Geoscience Referential Manager on 3DEXPERIENCE® for centralising geological data.

2. Strategic mine planning and economic and stope optimisation:

   o GEOVIA™ Whittle™ for pit shells, pushbacks, and strategic plans.

   o Pit Optimizer and Strategic Mine Planner (on 3DEXPERIENCE®) for advanced multi-scenario analysis and optimisation.

   o Stope Optimiser for optimal stope shapes in underground operations.

   o Earth Design and Engineering: Evaluates design viability through integrated pit optimisation and risk analysis for large capital projects.

3. High-fidelity simulation and process optimisation:

   o SIMULIA™ Abaqus™: Executes realistic finite element simulations for rock-mechanics, slope stability, and interactions between mining equipment and geological structures. This analysis ensures that designs uphold safety and performance standards.

   o SIMULIA™ Isight™: Links multiple software components in automated workflows for process integration and design optimisation. By running numerous parameter variations, Isight™ pinpoints optimal solutions, feeding results back into the planning/design pipeline.

4. Tactical scheduling and short-term planning:

   o GEOVIA™ MineSched™ for detailed surface and underground scheduling.

5. Mine design and drill and blast:

   o GEOVIA™ Surface and Underground Mine Designer: Parametrically creates declines (ramps), tunnels, benches, haul roads, stockpiles, dumps etc, for surface and underground projects. It incorporates geotechnical factors such as slope stability and stope geometry.

   o GEOVIA™ Drill and Blast Designer: Plans surface/underground drilling and blast patterns, optimising explosive usage, pattern layout, and detonation sequences.

6. DELMIA™/GEOVIA™ mine operations management (MOM) and supply chain:

   o DELMIA™ Quintiq™, DELMIA™ Apriso™, DELMIA™ Ortems™, DELMIA™/GEOVIA™ MOM™: Provide operational control, scheduling, and supply chain synchronisation across mining sites. They handle real-time production tracking, material flows, and labour management, ensuring that tactical schedules integrate with actual execution.

   o Mine Contributor: Enables multi-user collaboration on models, designs, and data, ensuring continuous alignment between planning and operational teams.

7. Platform collaboration and real-time feedback:

- 3DEXPERIENCE® Platform Roles (eg Geology Modeller, Earth Engineering Coordinator, Pit Optimizer): Maintain a single source of truth, enabling parametric updates, version control, and real-time collaboration.

- DELMIA™ Collaborative Operations on 3DEXPERIENCE®: Aggregates multi-source production data (eg from Apriso™, Ortems™) into a live operational dashboard.

- This continuous feedback ensures iterative updates to the geological model, schedules, designs, and simulations as new data emerges—closing the loop in a truly dynamic Digital Twin.

8. Earth Resources Management and sustainability:

- GEOVIA™ Earth Resources Management: Bridges planning, scheduling, and operations, leveraging the Plan-Do-Check-Act cycle to improve resource management. Allows cross-functional collaboration to ensure alignment with sustainability goals, resource usage tracking, and compliance with environmental standards.

9. Looped feedback and continuous improvement:

- Operational data, simulation insights, and economic evaluations flow back into geological models, strategic plans, and tactical schedules. Machine learning–driven adjustments and predictive maintenance sustain Digital Twin accuracy throughout the mine life cycle.

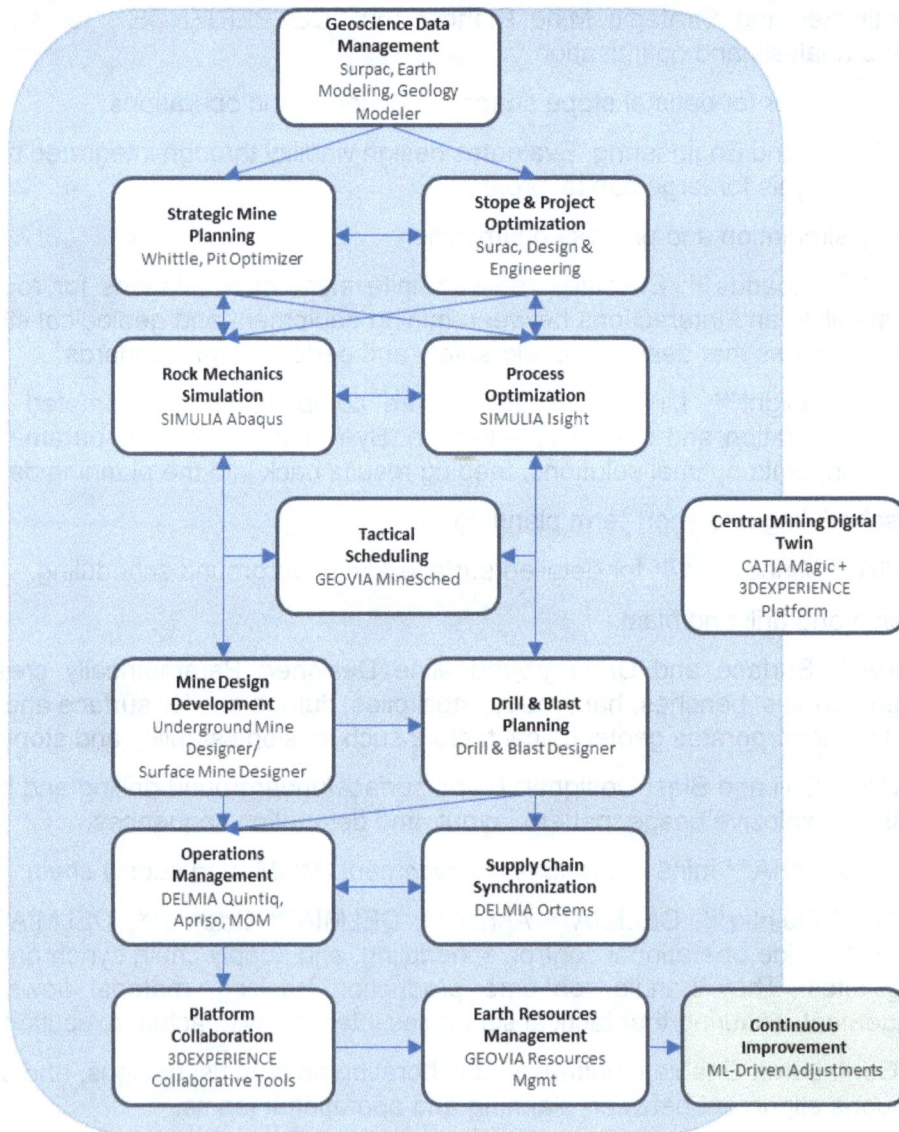

FIG 1 – Mining digital twin — principal workflow.

CONCLUSIONS

By unifying Dassault Systèmes® products—CATIA™, GEOVIA™, DELMIA™, and SIMULIA™—across a single Mining Digital Twin, the industry achieves end-to-end integration from exploration through decommissioning. High-fidelity simulations with SIMULIA™ Abaqus™, advanced planning with GEOVIA™ Whittle™ and MineSched™, collaborative workflows enabled by the 3DEXPERIENCE® platform, and responsive supply chain management via DELMIA™ collectively advance safety, efficiency, and environmental responsibility.

As AI, machine learning, IoT, and other technologies mature, this agile Digital Twin framework helps mining companies adapt swiftly to evolving conditions, extract more value from limited resources, and minimise their ecological impact. The result is a more robust, data-driven, and sustainable mining sector.

REFERENCES

Dassault Systèmes, 2025a. CATIA All Products [online], Dassault Systèmes. Available from: <https://www.3ds.com/products/catia/all-products> [Accessed: March 2025].

Dassault Systèmes, 2025b. DELMIA Portfolio [online], Dassault Systèmes. Available from: <https://www.3ds.com/products/delmia/portfolio/> [Accessed: March 2025].

Dassault Systèmes, 2025c. GEOVIA All Products [online], Dassault Systèmes. Available from: <https://www.3ds.com/products/geovia/all-products> [Accessed: March 2025].

Dassault Systèmes, 2025d. SIMULIA All Products [online], Dassault Systèmes. Available from: <https://www.3ds.com/products/simulia/all-products> [Accessed: March 2025].

El Bazi, N, Mabrouki, M, Chebak, A and Hammouch, F, 2022. Digital twin architecture for mining industry: case study of a stacker machine in an experimental open-pit mine, in *Proceedings of the Fourth Global Power, Energy and Communication Conference (GPECOM) 2022*, pp 232–237. https://doi.org/10.1109/gpecom55404.2022.9815618

El Bazi, N, Mabrouki, M, Laayati, O, Ouhabi, N, El Hadraoui, H, Hammouch, F-E and Chebak, A, 2023. Generic multi-layered digital-twin-framework-enabled asset lifecycle management for the sustainable mining industry, *Sustainability*, 15(4):3470. https://doi.org/10.3390/su15043470

Farrelly, C T and Davies, J, 2021. Interoperability, integration, and digital twins for mining—Part 2: pathways to the network-centric mine, *IEEE Industrial Electronics Magazine*, 15(3):22–31. https://doi.org/10.1109/MIE.2020.3029388

Rathore, M M, Shah, S A, Shukla, D, Bentafat, E and Bakiras, S, 2021. The role of AI, machine learning, and big data in digital twinning: a systematic literature review, challenges, and opportunities, *IEEE Access*, 9:32030–32052. https://doi.org/10.1109/ACCESS.2021.3060863

Automated classification of blast-induced rock fragmentation in underground sublevel caving mine

S Rajpurohit[1], A Gustafson[2], M Tariq[3], I Marin Rodriguez[4], C Quinteiro[5] and H Schunnesson[6]

1. Postdoctoral Researcher, Luleå University of Technology, Luleå Sweden 977 54.
   Email: sohan.singh.rajpurohit@ltu.se
2. Professor, Luleå University of Technology, Luleå Sweden 977 54.
   Email: anna.gustafson@ltu.se
3. Doctoral Student, Luleå University of Technology, Luleå Sweden 977 54.
   Email: muhammad.tariq@ltu.se
4. Doctoral Student, Luleå University of Technology, Luleå Sweden 977 54.
   Email: ivan.ricardo.marin.rodriguez@ltu.se
5. Senior Mining Engineer, LKAB, Kiruna Sweden 981 31. Email: carlos.quinteiro@lkab.com
6. Professor, Luleå University of Technology, Luleå Sweden 977 54.
   Email: hakan.schunnesson@ltu.se

ABSTRACT

In underground mining operations, blast-induced rock fragmentation provides information that can be useful for downstream ore handling operations. Continuous information on rock fragmentation is essential to ensure efficient loading operations using semi-automated and automated LHDs. The manual work of classifying fragmentation is a laborious and time-consuming task. This paper presents an early stage in the development of an AI based image classification model to automate the fragmentation classification process. A camera was installed on the roof of a ramp in an underground sublevel caving mine. The raw images of the truck buckets transporting the ore from underground to surface were captured using a wide angle camera. The raw images were filtered and sorted manually. A subset comprising approximately 1100 sorted images was manually labelled to compile a training data set. The blast-induced fragmented material was classified into five classes: very fine, fine, medium, coarse, and very coarse. The labelled data were used as input to train a deep neural network for image classification. The output of the classification model is the identified material class along with its corresponding confidence score. The developed model was tested on new images from the same mine. The paper also discusses model training on low resolution images, model development and testing, and challenges in adoption and scalability.

INTRODUCTION

In underground mining operations, ore is separated from the *in situ* rock by drilling and blasting. Intact solid ore is blasted and broken into smaller fragments of differing sizes. There is no standardised system to decide the classes of the fragments, but in general, the fragmented material is classified as fine, medium, or coarse. For a higher level of detail, the material can be classified into five classes: very fine, fine, medium, coarse, very coarse.

The fragmented ore is transported from the underground to the surface for further processing. The downstream loading and hauling operations are designed and planned based on the size distribution of the fragmented material, because the size of the fragmented ore influences the efficiency of loading and hauling operations. According to Ur Rehman and Awuah-Offei (2022) the optimum size is medium or medium to coarse for optimal loading and hauling. However, the optimum size is different from mine to mine depending on available machinery and downstream processes. Today, many underground mines are pushing for the use of autonomous vehicles in ore handling. Those automated systems require prior near real time information on the fragmentation to optimise their operating settings, and the blasted ore may contain different fragment size classes other than the desired size.

This paper presents the early stages of the development of an AI based image classification model for five classes of fragmentation. It explains how the model was tested on a new set of images using a graphical user interface (GUI). It also discusses the challenges faced and recommended approach when developing an AI based model for fragmentation classification.

CAMERA SET-UP AND MANUAL DATA COLLECTION

The study was carried out at Loussavaara Kiirunavaaras Konsuln mine, an underground sublevel caving iron ore mine. For the purposes of data collection, a wide-angle camera was installed on the roof of the main ramp (Figure 1) to capture and record short duration videos of each mine truck passing by. The main ramp is used as the primary route for trucks to transport fragmented ore from underground to surface. Flicker free LED lights were installed along with the camera to ensure proper lighting.

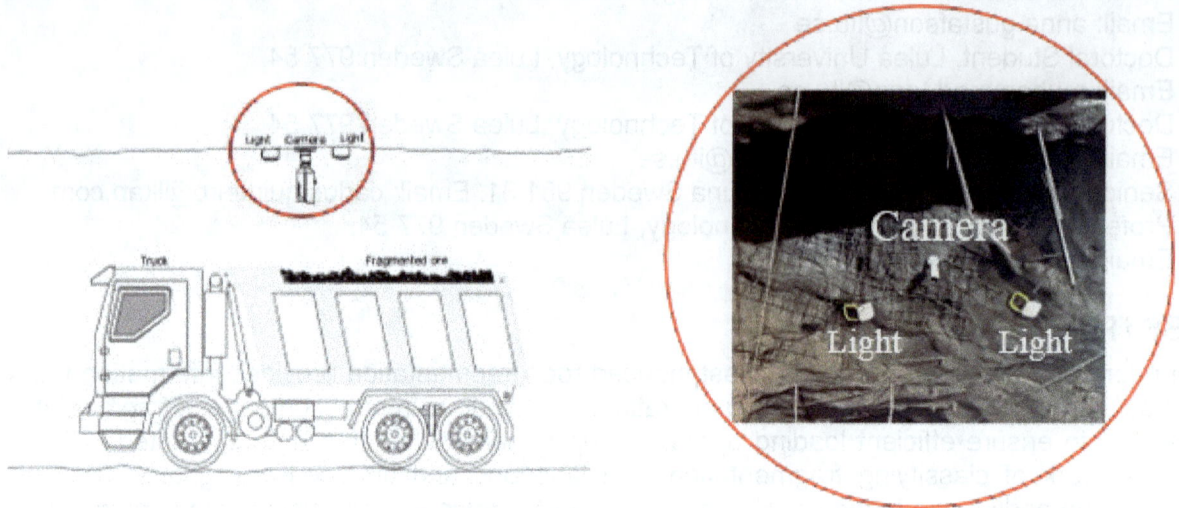

FIG 1 – Installation of camera and LED lights on the roof of main ramp.

For camera control purposes, a radio frequency identification (RFID) tag such as the one seen in Figure 2 was installed on the windshield of each truck. In that way, every time a truck drove through the area where the camera was installed, it triggered a sensor. The sensor collected and logged the identification (ID) number for every mine truck, as well as the date and time it passed through (Marin Rodriguez *et al*, 2024). The digital log containing the raw data was directly uploaded to the network, where data were available for download through remote access.

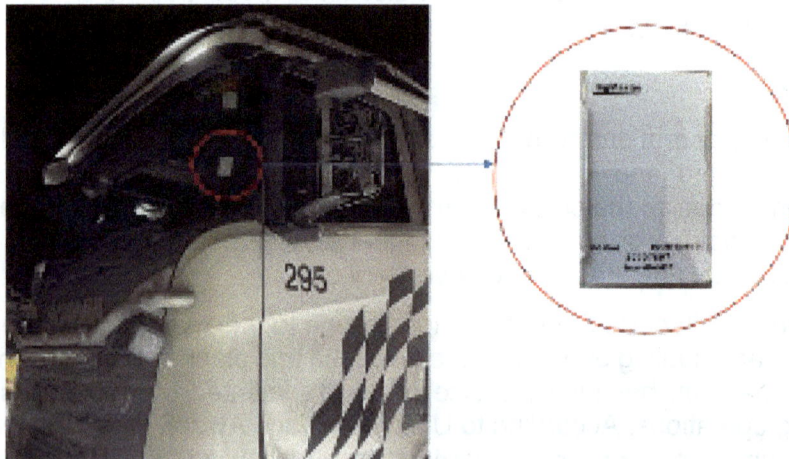

FIG 2 – RDFI tag mounted on mine truck.

To facilitate the correlation of the data generated from the RDFI tags with the video footage, the ID number for each truck was mounted on its roof (Marin Rodriguez *et al*, 2024). Having the number visible meant it was possible to track each truck even when only looking at the video footage. Figure 3 shows an example of an ID number.

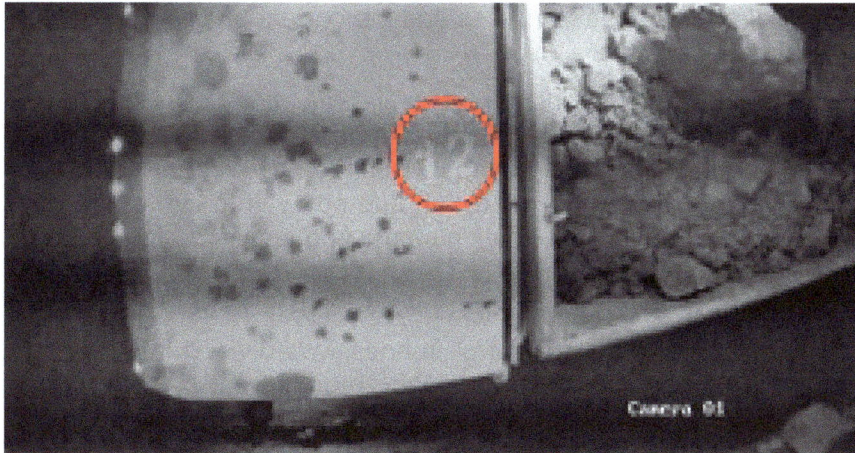

FIG 3 – Identification number on truck.

MANUAL PROCESSING OF IMAGES FOR THE TRAINING DATA SET

To develop a training data set for the automated classification model, the collected data were sorted, filtered, manually classified, and labelled. Raw data containing short videos of the truck box, timestamp, and truck number were downloaded from the remote network drive to the local storage. The videos were sorted and filtered. All videos containing corrupted data, or having no image of fragmented ore, or showing trucks carrying unloaded empty buckets were discarded. The videos of loaded trucks were split into separate image frames. Each video had one folder consisting of all its frames. Out of hundreds of frames in each folder, a suitable image was selected which clearly represented the fragmented ore loaded in the truck bucket. Images with maximum visibility of fragmented ore were preferred even if the quality was lower than in the adjacent frames. All the sorted and filtered images were saved in separate folders for labelling.

The fragmented material was manually subdivided into five classes. The class of *very fine* fragmented ore was assigned as Class 1, *fine* was assigned as Class 2, *medium* was assigned as Class 3, *coarse* was assigned as Class 4, and *very coarse* was assigned as Class 5. The five best images corresponding to each fragment class were selected as visual references (Figure 4). By comparing the reference image with the target image, a fragmentation class was manually assigned to each image using a 'quick rating system' (Manzoor *et al*, 2022).

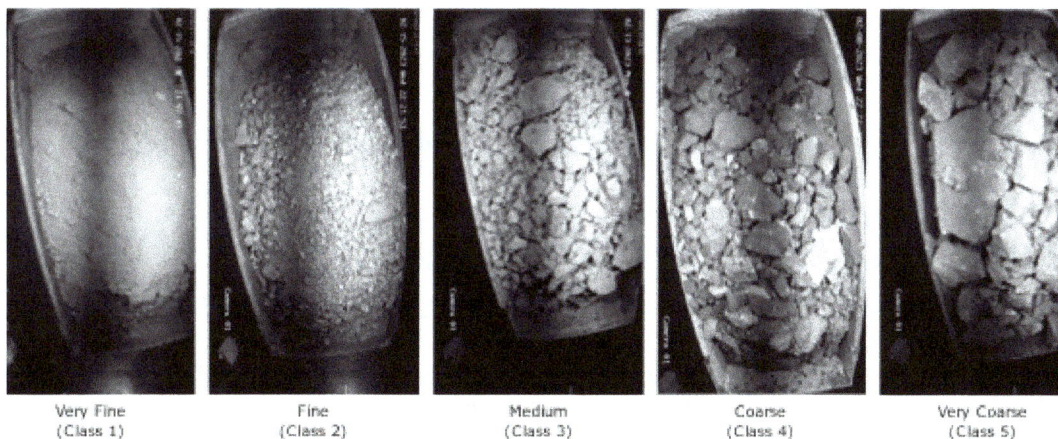

| Very Fine (Class 1) | Fine (Class 2) | Medium (Class 3) | Coarse (Class 4) | Very Coarse (Class 5) |

FIG 4 – Reference particle size of fragmented ore (Marin Rodriguez *et al*, 2024).

To label the image data set, a MATLAB programming script was run to generate the filenames of the images in a comma separated values (csv) file format. The csv file contained two columns called Filename and Class. The second column, Class, contained the class assigned to each individual image. Using the quick rating system, all the images were sorted into their respective classes, and the record was updated in the csv file. All the images were manually labelled in this way. The images and the csv files containing the image filenames and the corresponding fragmentation classes were used as the input for the model.

The image collection process and the fragmentation classification based on the quick rating system were used in previous research by Marin Rodriguez *et al* (2024). The lengthy process of capturing, sorting, filtering, and manually classifying images inspired the idea of automating the process to reduce a significant amount of time. Therefore, an attempt was made to use the same images to train an automated AI based image classification model and test its performance.

TRAINING OF AI BASED FRAGMENTATION CLASSIFICATION MODEL

For model training and testing, the data set of images was divided into train and test data sets (Figure 5). To see precisely how the model responded to different classes of images, fairly equal numbers of images were selected for model training. The typical dimension of the images was 2560 pixels × 1440 pixels. The average size of each image was around 220 kilobytes.

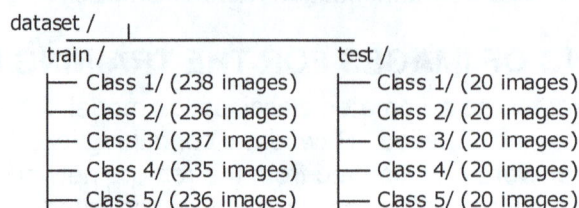

```
dataset /
    train /                              test /
        ├── Class 1/ (238 images)            ├── Class 1/ (20 images)
        ├── Class 2/ (236 images)            ├── Class 2/ (20 images)
        ├── Class 3/ (237 images)            ├── Class 3/ (20 images)
        ├── Class 4/ (235 images)            ├── Class 4/ (20 images)
        └── Class 5/ (236 images)            └── Class 5/ (20 images)
```

FIG 5 – Data set schema of images used in multiclass classification model.

An AI based automated fragmentation classification model was trained using TensorFlow's Deep Neural Network (DNN) architecture (Abadi *et al*, 2015). A pretrained multiclass image classification model ResNet50 (He *et al*, 2016) was used to train the customised model on the image data set of ore fragmentation classes. The model architecture of ResNet50 is shown in Figure 6. The ResNet50 model is efficient in feature extraction and was suitable for fragmentation classification using images (Bamford, Esmaeili and Schoellig, 2021). To optimise the data set, the Adam optimiser (Kingma and Lei Ba, 2014) was used to minimise the cross-entropy loss function. The convergence of the model was assured based on the learning rate scheduling.

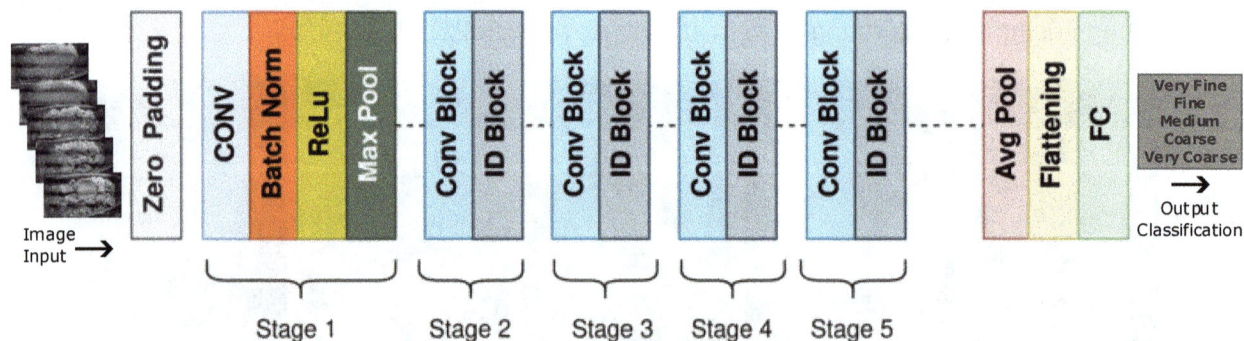

FIG 6 – Resnet50 model architecture (modified after He *et al*, 2016).

The trained model (Figure 7) was exported as an Open Neural Network Exchange (ONNX v1.18.0) file for multiplatform support. An ONNX file can be imported by most software frameworks compatible with TensorFlow's DNN models.

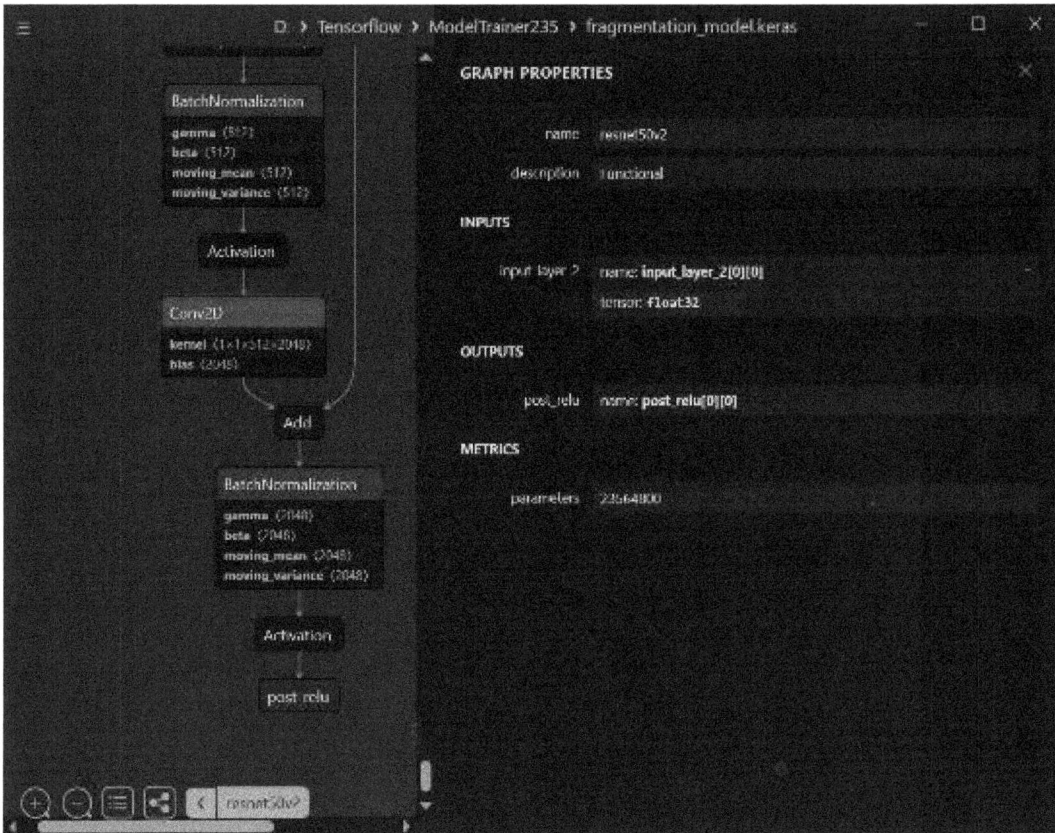

FIG 7 – A screenshot from DNN model visualiser package Netron showing the last part of the model (Roeder, 2025).

The model was trained on 1182 images from the train data set. The TensorFlow's model training algorithm automatically split the train data set into 80 per cent for training and 20 per cent for validation. The model was trained over a number of epochs (iterations) for convergence. During each epoch, the model was trained on 80 per cent of the images and validated on 20 per cent of the images for fine-tuning. The iterative process facilitated the fine-tuning of the model and the extraction of useful features from the images and increased the accuracy of prediction. Model regularisation and early stoppage were used to prevent the overfitting of the model on the training images and ensure the model would have the same level of accuracy on new sets of images. The final accuracy achieved for model training was 81.65 per cent.

Under the iterative process, the model accuracy was calculated based on the accuracy of classification prediction and the associated confidence score. For each input image, the model provided output comprising a predicted class label and a list of five confidence score values. Each value of the confidence score corresponded to a fragment class from 1 to 5. The maximum confidence score belonged to the predicted class. Higher values of the confidence score in image prediction indicated model reliability. The main aim of this trial and iteration was to determine the minimum number of input images required to train the model and ensure the test images were sufficiently accurate. The ultimate goal was to expand the process to other underground material transportation machinery with minimum additional model training.

GRAPHICAL USER INTERFACE

A graphical user interface (GUI) (Figure 8) was developed to test the model. Microsoft C# programming language and Visual Studio integrated development environment (IDE) with.NET framework was used to develop the GUI. Currently, the GUI software package has installation compatibility with Microsoft Windows based operating systems. The C# code and models packaged in ONNX format are compatible with different kinds of software frameworks and operating systems that have compatibility with.NET framework 6.0 or a higher version of it. An ONNX model supports multithreading, parallel computing, and GPU acceleration for processing multiple images simultaneously using ONNX Runtime (version 1.18.0).

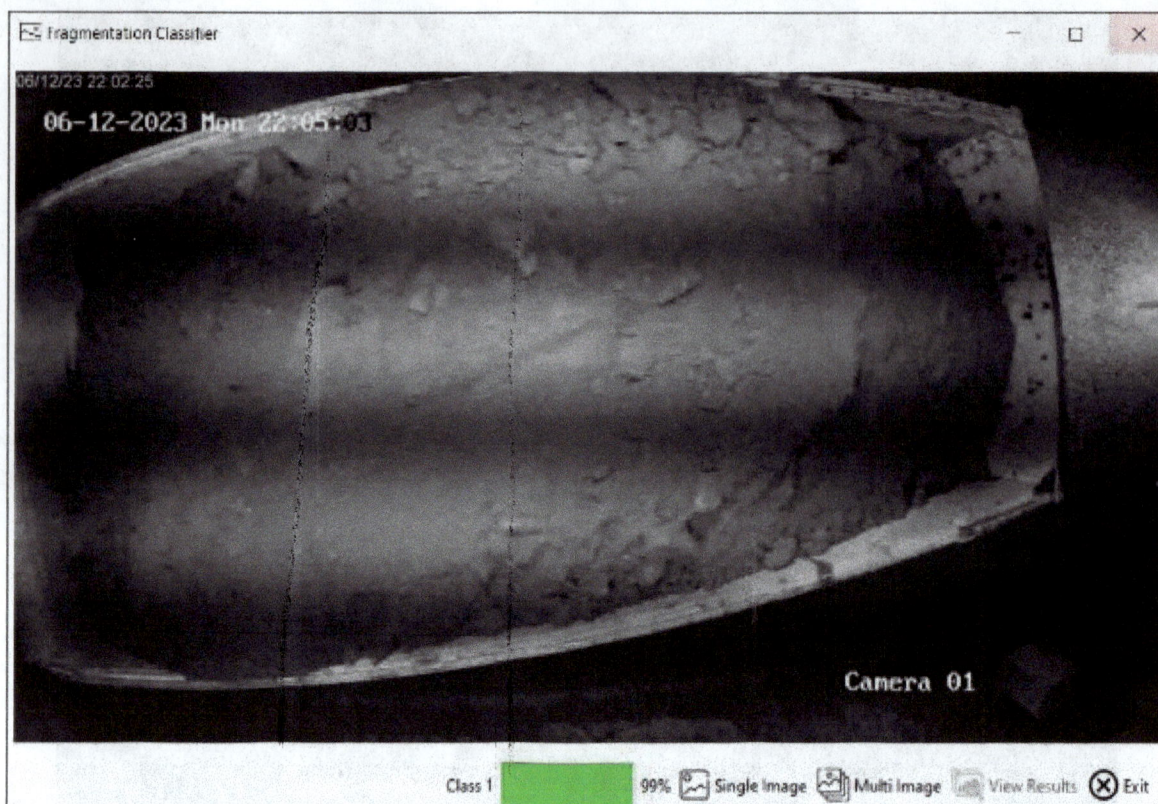

FIG 8 – Developed graphical user interface (GUI) for automated image classification.

The developed GUI is able to check the classification accuracy of the test images of the fragmented ore in two ways: single image classification and batch mode. For classification of a single image, the user can load the image using File Browser. After the image processing is completed by the model, the user will be able to see the image, its classification, and its confidence score. For multi-image classification, a folder containing all the new image files can be selected using the Folder Browser. The GUI will process all the images and generate a csv file containing the file names of the images, their respective fragmentation classes, and the related confidence scores. Additionally, the results of the muti-image classification can be viewed by clicking on the View Results button in the main window. This opens a separate window containing a DataGridView where a list of the images' file names, predicted classes, and confidence scores is displayed. If the user clicks on any row in the DataGridView, the image from that row will be displayed on the GUI. The GUI works in near real time to classify individual images. In the batch mode, the average time for processing 100 images is around 3–4 seconds. The speed may vary depending on the configuration of the computer on which the GUI is used.

MODEL PERFORMANCE

Using the GUI, the trained model was tested on 100 test images of the fragmented ore. Twenty images from each class (*very fine* to *very coarse*) were randomly picked from the manually classified data not used for model training. The test results for the manually classified data and the model-based automated fragmentation classification were compared. The number of manually classified images for each class versus the predicted number of images from each class is presented in a confusion matrix in Figure 9. The green cells along the diagonal of the confusion matrix denote the number of correctly predicted images for each fragmentation class. For example, for *very fine* material, 18/20 images were classified correctly by the model. The total number of correct classified images was 84 out of 100 giving an overall model accuracy of 84 per cent. The upper diagonal section of the matrix denotes the number of images misclassified by the AI model into relatively larger fragment classes than they were assigned in manual classification. A total of ten images were misclassified as belonging to classes with larger fragment sizes. Similarly, the lower diagonal section of the matrix denotes the number of images misclassified by the AI model into relatively smaller fragment sizes. A total of six images were misclassified as belonging to classes with smaller fragment sizes.

	Fragmentation prediction by model					
Class (1–5)	Very Fine (1)	Fine (2)	Medium (3)	Coarse (4)	Very Coarse (5)	Total
Very Fine	18	2	0	0	0	20
Fine	1	17	2	0	0	20
Medium	0	0	15	4ᵃ	1	20
Coarse	0	0	2	17	1	20
Very Coarse	0	0	0	3ᵇ	17	20
Total	19	19	19	24	19	84/100

(Manually classified fragmentation)

FIG 9 – Confusion matrix of the image classification of fragmented ore.

It is evident from the confusion matrix that the model has significant accuracy in predicting fragmentation class. In cases of misprediction, most of the images were misclassified into the next class. No image of a *very fine* class was misclassified as *coarse* or *very coarse*. Similarly, no image of a *coarse* or *very coarse* class was misclassified as *very fine* or *fine*. However, the model slightly misclassified some images of the *medium* ('a' in Figure 9) and *very coarse* classes as *coarse* ('b' in Figure 9).

When the model classifies the images correctly, the confidence score will range between 80 per cent and 99 per cent. When the images are misclassified, the confidence score will range between 40 per cent and 80 per cent. The second-best confidence score given by the model usually corresponds to the correct class of the image. For example, when the fragmentation class is *very coarse*, and misclassified as *coarse*, the confidence score from the model prediction will be 70 per cent for Class 4 and 25 per cent for Class 5.

One *medium* class image in the test set was misclassified as *very coarse* with a confidence score of 73 per cent (Figure 10). When data are manually labelled, the median particle size of the material is used for fragmentation classification. An oversized boulder present in the image is not representative of the size distribution of the material, and the image will be classified as the class of the majority of the material. However, the model was not trained with specific instructions to follow this procedure. Therefore, in this case, the model detected a boulder and misclassified the image as *very coarse*. Work is in progress to allow boulder detection during image processing.

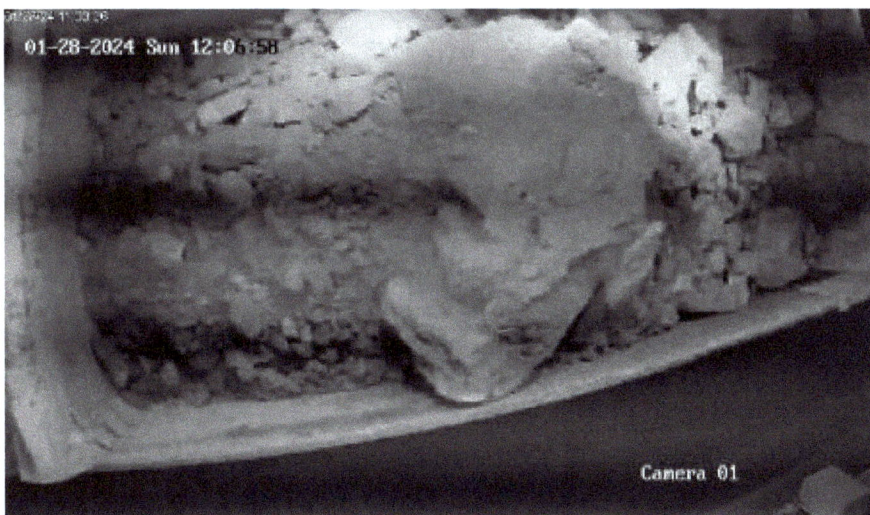

FIG 10 – Fragmented rock of medium class misclassified as very coarse.

For a comprehensive analysis using statistical machine learning matrices, the model performance was determined using the following multiclass classification matrices (Equations 1–4) from Sokolova and Lapalme (2009):

Average accuracy:
$$\frac{\sum_{i=1}^{5} TP_i}{\sum_{i=1}^{5}(TP_i + FP_i + FN_i)} \quad (1)$$

Average precision:
$$\frac{\sum_{i=1}^{5} TP_i}{\sum_{i=1}^{5}(TP_i + FP_i)} \quad (2)$$

Average recall:
$$\frac{\sum_{i=1}^{5} TP_i}{\sum_{i=1}^{5}(TP_i + FN_i)} \quad (3)$$

Average F1-score:
$$\frac{\sum_{i=1}^{5} 2 \times Precision_i \times Recall_i}{\sum_{i=1}^{5} Precision_i + Recall_i} \quad (4)$$

Where:

TP_i = true positive: actual fragmentation class is i, and predicted class is also i

FP_i = false positive: actual class is not i, but predicted as class i

FN_i = false negative: actual class is i, but predicted as class other than i

The matrices given in Equations 1 to 4 were calculated for each class and are presented in Table 1. Additionally, the arithmetic mean values of the matrices are given in the end row of the same table. Precision takes into account the number of true positives and false positives. A higher precision value suggests a smaller number of images from other classes are misclassified as the targeted class. Higher values of recall indicate a smaller number of images are misclassified from the targeted class to other classes. The F1-score is a harmonic mean of both precision and recall values and reflects a balance of precision and recall. The multiplicative factor was assigned based on numeric weighting given to either precision or recall. For this study, equal weight was assigned to precision and recall to calculate the F1-score.

TABLE 1

Performance matrices of multiclass image classification model.

Class (1–5)	TP	FP	FN	Accuracy	Precision	Recall	F1-Score
Very Fine (1)	18	1	2	85.71%	94.74%	90.00%	0.92
Fine (2)	17	2	3	77.27%	89.47%	85.00%	0.87
Medium (3)	15	4	5	62.50%	78.95%	75.00%	0.77
Coarse (4)	17	7	3	62.96%	70.83%	85.00%	0.77
Very Coarse (5)	17	2	3	77.27%	89.47%	85.00%	0.87
			Average	**73.14%**	**84.69%**	**84.00%**	**0.84**

Table 1 shows the model accuracy was relatively balanced in terms of precision, recall, and F1-score, but there was an imbalance in model performance in class prediction. The model performed relatively well for images of *very fine*, *fine*, and *very coarse* classes. Model performance was a bit lower when predicting images of the *medium* class. The highest number of false negatives was observed in the *medium* class, and the highest number of false positives was observed in the *coarse* class.

DISCUSSION

General requirements for a completely automated image classification system include a proper online data collection method with several different components, including instruments (camera, network connectivity), an object detection component to identify the mine truck bucket, an image classification model, and a GUI to visualise the results. APIs are also important to integrate the results with other systems. This paper has assumed a proper data collection system to describe the development of an automated classification model.

IMAGE QUALITY

Image quality is the most important parameter for model accuracy, as images are the main input for model training. The developed model prefers high quality images, as it is more efficient to extract features from images with clear particle boundaries. If the quality of the images is reduced, this adversely affects the efficiency of the feature extraction from the image and ultimately affects the prediction accuracy. In the videos captured during data collection, the image resolution started reducing due to motion blur when vehicles came into the frame. To explain this, all frames from a sample video were split into a number of single frames. A selection of six frames from that video, with frame numbers, is shown in Figure 11.

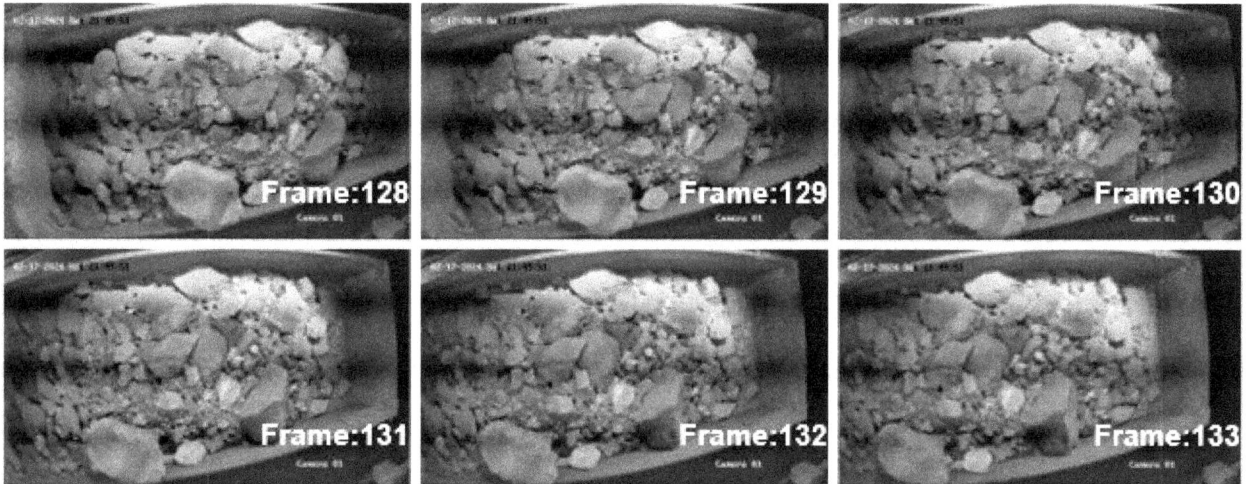

FIG 11 – Frame sequence in a video of a minetruck carrying fragmented rock.

Image brightness, blurriness, contrast, and ortho sharpness were determined for each of the six images as an example of the image quality (Figure 12). In the example shown in Figure 12, the line charts are divided into three zones by vertical lines; the middle zone shows the zone where the vehicle was visible in the camera frames. The graphs show the rapid changes in image quality over each frame. In 4–5 frames, the fragmented ore is clearly visible within the frame. The best strategy is to select the frame with a higher value of sharpness and clear material boundaries. For this sequence of frames from the example video, frame number 131 was selected for image classification because of its relatively higher sharpness compared to the adjacent frames.

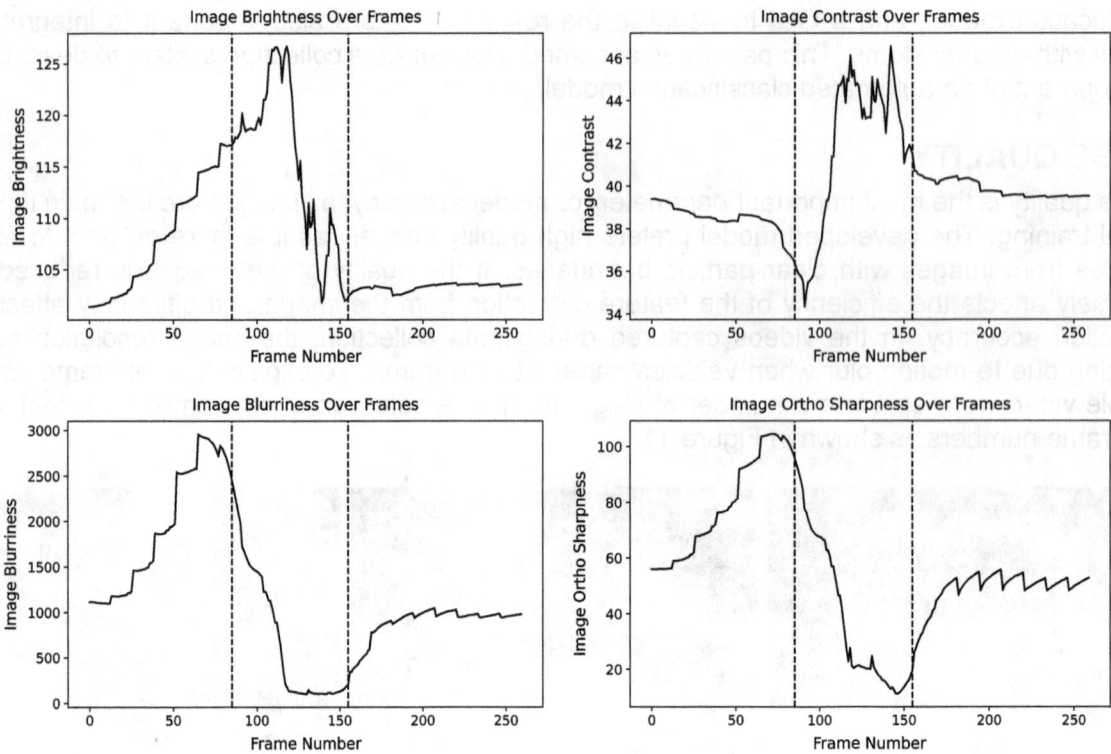

FIG 12 – Quality of images significantly varying frame to frame in a single video.

The most challenging task during the development of an AI-based automated fragmentation classification model was training the model on images with low pixel resolution. Due to noise and blurriness, the pixels at the boundary of finer sized particles tended to merge. This could result in the appearance of a single large size rock fragment in the image instead of the actual fine material. To handle the issue of low-resolution images, the deep residual network based pretrained model ResNet50 was utilised. The ResNet50 model architecture is robust and efficient in extracting features when images have relatively lower resolution or when the boundary surrounding the particles is not clear distinguishable (Arafah, Achmad and Areni, 2020; Bamford, Esmaeili and Schoellig, 2021; He et al, 2016).

The developed model presented in this paper was tested on another set of images (Figure 11) with diverse quality to see how the model handled images with a low resolution. The quality of the images differed in terms of their brightness, blurriness, contrast, and sharpness. The model correctly classified all six images as *coarse*.

MATERIAL VARIATIONS AND SUBJECTIVE BIASES

During model training, the deep convolutional neural network extracted distinct features from the images of each class. When the material present in the image had a uniform size distribution, the model performed with a high accuracy and confidence score. The model clearly differentiated between distinct features of *very fine* and *very coarse* classes of fragmentation. Hence, there was no misclassification of *very fine* as *very coarse* or *vice versa*. Classification was more complex when mixed material was present in the image (Figure 13) and the fragmentation size overlapped between two or more adjacent classes, such as *fine-medium*, *medium-coarse*, or *coarse-very coarse*.

FIG 13 – Examples of mix fragmentation of the fragmented rock loaded into truck bucket.

The decisions made in manual classification could be affected by subjective biases, especially for mixed material. Images having mixed material could potentially be manually classified as *medium* by one person and as *coarse* by another. This could occur when a majority of the material present in the image belongs to the overlapping intersection region between the two classes.

In order to train the model to not be affected by subjective biases, the input images used for model training were manually classified by a single person. The trained model followed the same pattern as it 'learned' from those input images. The model exhibited significant accuracy in classification of the overall test images. However, it misclassified some images with mixed material in the test data set. There is scope for improvement in the model accuracy and consistency in the classification in these particular cases. The issue of subjective bias and its influence on the trained model can be minimised by increasing the number of parameters in training and diversifying the input images corresponding to mixed material. This will result in more consistent prediction by the automated classification model. The advantage of the AI based model over manual classification is that once the model is trained on certain input, it provides consistent results. In addition, the model is very time-efficient in classifying the fragmentation and can be used for a large number of images.

Model generalisation

The model performed well on the unseen test data set showing it was generalised for diversity in the input quality of images. The developed model is machine specific and works for fragmented ore loaded in underground mine trucks. However, its applicability can be extended by training additional images of fragmented material carried by different machines such as LHDs and other kinds of loaders (Mirabedi *et al*, 2018). In such cases, the model does not need to be retrained from scratch; rather, it can be upgraded for the extended applications.

CONCLUSIONS

This paper focuses on the development of a DNN model that requires a relatively small input data set for training. It performs well on unseen data of the same domain and has the potential for generalisation and scalability. The model was trained on a training data set of 1182 images and showed an overall training accuracy of 81.65 per cent which is similar to the model accuracy of 84 per cent achieved during the model testing. The model has a high image processing speed, robustness in correctly classifying low resolution images, and compatibility with most of the ML based software frameworks.

Using the automated classification model is much faster than performing manual classification. The results from the model can provide a larger picture of the fragmentation in the targeted zone within seconds. Other than speed, the main feature of the trained model is consistency in classifying mixed material, thus minimising subjective biases in decision-making. The current version of the model has applicability for underground mine trucks and has the scope to be expanded to underground autonomous LHDs.

The developed model has the potential to be applied to an online system of computer vision to provide near real time information of the size distribution of the blasted material. That near real time information can, for example, be digitally communicated by autonomous LHDs in underground mines.

ACKNOWLEDGEMENTS

The authors acknowledge the staff and management at LKAB's Konsuln mine for their input and support, as well as the company LTH Traktor AB and their operators for facilitating their trucks to be used during data collection.

REFERENCES

Abadi, M, Agarwal, A, Barham, P, Brevdo, E, Chen, Z, Citro, C, Corrado, G S, Davis, A, Dean, J, Devin, M, Ghemawat, S, Goodfellow, I, Harp, A, Irving, G, Isard, M, Jia, Y, Jozefowicz, R, Kaiser, L, Kudlur, M, Levenberg, J, Mane, D, Monga, R, Moore, S, Murray, D, Olah, C, Schuster, M, Shlens, J, Steiner, B, Sutskever, I, Talwar, K, Tucker, P, Vanhoucke, V, Vasudevan, V, Viegas, F, Vinyals, O, Warden, P, Wattenberg, M, Wicke, M, Yu, Y and Zheng, X, 2015. TensorFlow: Large-scale machine learning on heterogeneous systems. Available from: <https://www.tensorflow.org/about/bib>

Bamford, T, Esmaeili, K and Schoellig, A P, 2021. A deep learning approach for rock fragmentation analysis, *International Journal of Rock Mechanics and Mining Sciences*, 145:104839. https://doi.org/10.1016/j.ijrmms.2021.104839

He, K, Zhang, X, Ren, S and Sun, J, 2016. Deep residual learning for image recognition, in *Proceedings of the IEEE conference on computer vision and pattern recognition*, pp 770–778.

Kingma, D P and Lei Ba, J, 2014. Adam: A method for stochastic optimization, arXiv preprint. arXiv:1412.6980

Manzoor, S, Danielsson, M, Söderström, E, Schunnesson, H, Gustafson, A, Fredriksson, H and Johansson, D, 2022. Predicting rock fragmentation based on drill monitoring: A case study from Malmberget mine, Sweden, *Journal of the Southern African Institute of Mining and Metallurgy*, 122(3):155–165. https://doi.org/10.17159/2411-9717/1587/2022

Marin Rodriguez, I R, Gustafson, A, Tariq, M, Rajpurohit, S S, Quinteiro, C and Shunnesson, H, 2024. Comparative analysis of fragmentation sizes in LKAB's Kiirunavara (Konsuln) mine, in *Proceedings of the 9th International Conference and Exhibition on Mass Mining*, pp 703–714.

Mirabedi, S M, Khodaiari, A, Jafari, A and Yavari, M, 2018. The effect of important fragmented rock properties on the penetration rate of loader bucket, *Geotechnical and Geological Engineering*, 36:1295–1307. https://doi.org/10.1007/s10706-017-0393-7

Roeder, L, 2025. Netron: Visualizer for neural network, deep learning and machine learning models (version 8.1.8). https://www.lutzroeder.com/ai

Sokolova, M and Lapalme, G, 2009. A systematic analysis of performance measures for classification tasks, *Information processing and management*, 45(4):427–437.

Ur Rehman, A and Awuah-Offei, K, 2022. Effect of bucket geometry, machine variables and fragmentation size on performance of rubber-tired loaders, *Mining, Metallurgy and Exploration*, 39(1):111–127.

Arafah, M, Achmad, A and Areni, I S, 2020, December. Face Identification System Using Convolutional Neural Network for Low Resolution Image, in *2020 IEEE International Conference on Communication, Networks and Satellite (Comnetsat)*, pp 55–60 (IEEE).

Applying artificial intelligence and the cloud to optimise blast designs

M J Roberts[1]

1. Global Product Strategy Manager, Maptek Pty Ltd, Perth WA 6003.
   Email: mark.roberts@maptek.com.au

ABSTRACT

Key trends in the mining industry, such as automation, data analytics and optimisation, focus on achieving more efficient, productive operations with fewer resources. With the rapid growth of data, the next step is to establish effective processes and systems that leverage this data to support value-driven decision-making.

Significant resources are often allocated to developing accurate simulation models. However, these models are frequently not integrated with existing systems, limiting their utility. In the specialised area of blasting, integration with upstream and downstream processes is essential for driving sustained improvements and value.

This research presents an approach based on artificial intelligence and cloud computing to address complex mining challenges, enabling competing objective optimisation—like good fragmentation and low vibration—through population-based methods that reveal trade-offs across multiple solutions. Secure cloud-based processing accelerates computation, delivering results in timelines not feasible with desktop processing.

A new capability is now available that automates the generation of the three primary components of blast design—drill pattern, charge plan, and timing—in minutes. Integrating seamlessly with leading drill and blast design and reconciliation software, this advanced technology supports fine adjustment for the engineer, and accurate on-bench implementation to achieve reliable downstream performance.

This innovation that incorporates fundamental blasting design principles and empirical models allows mining engineers to focus on higher level thinking and consider how all the factors that influence blast performance can be engineered to better control outcomes. Importantly, this approach will help maximise operational efficiency and contribute to reducing environmental impact and energy costs, promoting a more sustainable drill and blast function.

INTRODUCTION

The digital transformation of the mining industry emphasises efficiency, sustainability and data-driven decision-making. As mining operations generate increasing amounts of data, it becomes essential to employ advanced technology to extract actionable insights.

However, despite significant investments in simulation models, many mining operations struggle to fully utilise these tools due to poor integration with existing processes. Addressing this gap requires comprehensive solutions that align upstream orebody knowledge and mine planning with downstream operational and environmental considerations to achieve improved performance. This is especially true in the context of blast design.

Since 2015, Maptek has pioneered the use of artificial intelligence for mine scheduling using genetic algorithms that have demonstrated their efficacy in solving complex, multi-objective optimisation problems. Genetic algorithms, through their population-based evolutionary approach, can explore vast solution spaces and identify trade-offs between multiple solutions. When combined with cloud computing, this approach provides rapid and scalable scenario analysis needed to address mine planning challenges.

Traditional blast design represents a highly engineering-intensive task, with considerable effort required to generate even a single design. In recent years, the mining industry has faced a steady decline in university enrolments for mining engineering programs, leading to a shrinking talent pool of both graduates and experienced professionals. This scarcity has intensified competition for qualified talent, prompting companies to pursue international recruitment. Additionally, technological advancements now enable certain routine planning tasks, such as blast design, to be performed remotely—an emerging shift that is transforming operational roles (Knights, 2020).

In 2018, Maptek initiated an ambitious project to automate the blast design process using artificial intelligence, with simple design editing tools for fine adjustment and integration with on-bench systems for accurate execution, and blast performance monitoring.

A comprehensive blast design consists of three essential components, generated by engineers with reference to the geological model, topographical terrain and mine design (Wilkinson and Kecojevic, 2005):

1. Drill pattern.

2. Charge plan.

3. Timing design.

Effective blast designs must align these three components with actual localised conditions (Tordoir and Roberts, 2019), and evolve through an optimisation process to better achieve safety, cost and production value drivers (Scott, 1996). Increasingly, mine engineers accomplish this using simulation models for fragmentation, vibration and fly rock as part of their blast design software, while adhering to established design parameters developed specifically for their operation. However, using the current blast design tools to optimise competing blast objectives requires brute force by engineers to generate various scenarios.

Additionally, the ability to review and refine these design parameters is often constrained by production pressures and limited availability of experienced engineering specialists. Consequently, reviews are not routinely conducted and typically occur only as part of larger studies as mining operations advance into new areas.

THE AUTOMATED CREATION OF OPTIMISED BLAST DESIGNS

Comparatively, the genetic algorithm approach can produce detailed results in minutes for competing blast objectives, even for complex requirements such as cost efficiency, optimal fragmentation and vibration control.

The automated creation of optimised and detailed blast designs comprising Drill pattern, Per hole charging and Timing is now available through a new capability called BlastMCF.

This will allow optimal blast design outcomes to be more accessible when compared to manual design tools by providing a cross-section of possible solutions, from which the engineer can better understand the trade-offs between various design parameters.

With the overarching design in place, more time can be spent focusing on fine adjustment which is especially important given the extensive nuance and complexity inherent in the proficiency of drill and blast for localised conditions.

Take for example the interrelated factors that influence blast performance and our greater understanding of them, which is demanding an ever-increasing focus on engineering for better control. There are the fixed details to consider like geology, bench height and drill set-up, and the variable inputs like geometry, explosives and timing that the mine engineer needs to apply to any blast design to achieve the targeted outcome (Salmi and Sellers, 2020).

Therefore, the ability to perform comprehensive scenario analyses to objectively validate design concepts quickly would be highly valued by mine planning teams across the industry.

APPLYING ARTIFICIAL INTELLIGENCE FOR BLAST DESIGN

A Non-dominated Sorting Genetic Algorithm II (GA), commonly known as NSGA-II (Deb et al, 2002), serves as the artificial intelligence foundation for the blast design optimisation. The GA generates a diverse population of blast designs (P) and progresses through successive iterations, with each iteration designated as a generation (Acampora et al, 2013). The algorithm dynamically generates each population of blast designs by strategically manipulating six critical blast design variables within user-defined constraints:

1. **Hole spacing** – The lateral distance between holes in the same row.

2. **Burden distance** – The distance between adjacent rows of holes.

3. **Hole diameter** – The diameter of the drill hole.

4. **Explosive type** – Bulk explosive with fixed density characteristics.

5. **Metres/second timing delay** – The time initiation delay between detonation of blastholes.

6. **Orientation of timing angle** – Where an alteration range is set from the specified orientation.

For configurations involving one or two blast objectives, the GA generates a population of 50 blast designs, with each design consisting of a vector for each of the six blast design variables. To ensure the generation of feasible blast designs, these variables can be constrained by user-specified values, ranges with defined lower and upper bounds, or specific types.

Each individual blast design is encoded to enable modelling and evaluation by specialised blast design objective calculators that quantify their performance relative to the established blast objectives. Five blast design objective calculators have been integrated into the system:

1. **Cost** – The comprehensive estimated cost of all blast components, including drilling, explosive products, primers and associated labour costs.

2. **Fly rock** – The predicted maximum horizontal distance that fly rock may travel as a result of blasting, based on either the McKenzie (2009) or Richards and Moore (2006) models.

3. **Fragmentation** – An estimation of the distribution of rock sizes produced by the blast, utilising the KuzRam model (Cunningham, 2005).

4. **Powder factor** – The ratio of explosive mass per unit mass of rock to be blasted.

5. **Vibration** – An estimation of blast-induced vibration at specified locations of interest, measured as Peak Particle Velocity (PPV).

The performance metrics for each blast design are returned to the GA and sorted according to fitness and non-dominated spread relative to the blast design objectives, enabling the algorithm to determine how the blast design variables influence the blast objectives.

The GA then begins optimising blast designs by randomly selecting two solutions from the population and comparing them through two binary tournaments. Each tournament evaluates two randomly selected blast designs from the population of 50 and determines a winning design. The two tournament winners are paired to generate two offspring (Q) designs.

These offspring are created through a combination of their parent designs' values (crossover). Some values may undergo slight modifications (mutation) to explore a broader search space within each evolutionary cycle. Each offspring represents an alternative blast design to its parents, using probabilistic distribution of the blast design variables while respecting any user-specified constraints. The new offspring designs are evaluated by the blast design objective calculators, and this tournament process continues until an offspring population equal in size to the parent population is generated.

The parent and offspring populations are then combined (R), evaluated, and sorted again according to fitness and non-dominated spread to form a new population of 50 designs. This is accomplished by creating different fronts (F) of non-dominated solutions to complete the first generation. The process repeats for up to 50 generations, unless terminated earlier due to convergence, which is defined as minimal improvement in solutions over five successive generations.

Figure 1 shows the Schematic of the NSGA-II procedure (Acampora et al, 2013).

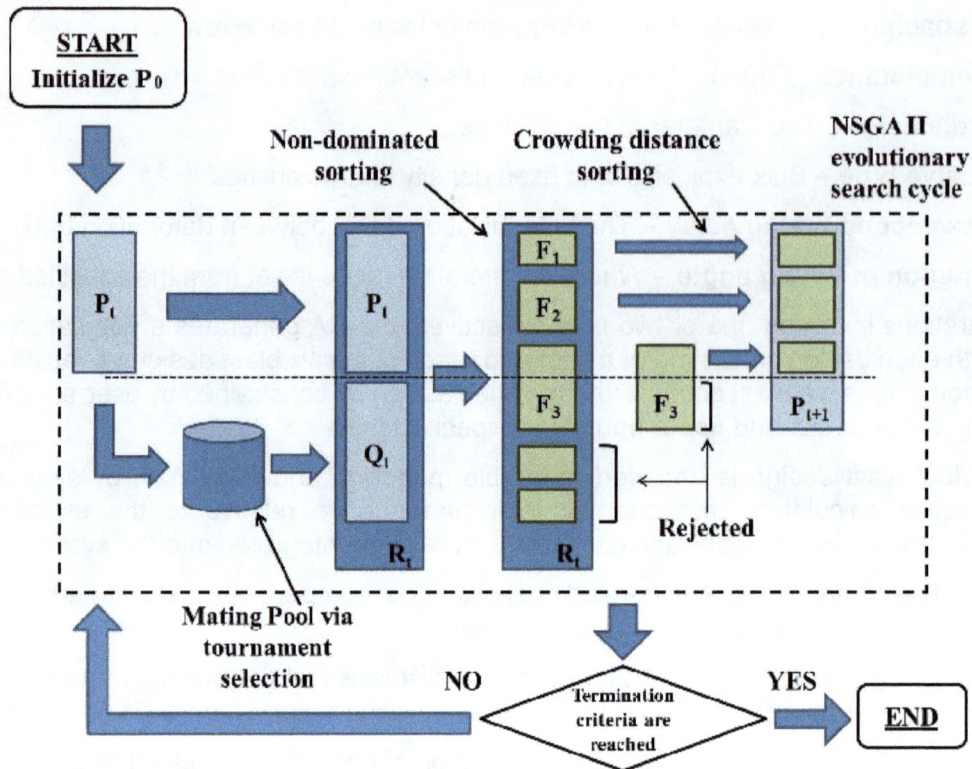

FIG 1 – Schematic of the NSGA-II procedure (Acampora *et al*, 2013).

A crossover operator functions as an exploration mechanism to traverse the search space of possible designs relative to the blast objective and design constraints. With each generation, the results converge toward the optimum for a given blast design objective. The final blast design is decoded into pattern geometry, charge plan and timing design for the user to review, edit and implement.

In the context of evolutionary multi-objective optimisation, there is no single optimal design but rather multiple trade-offs between different objectives. A two-objective scenario is most common and easiest for users to interpret, such as fragmentation versus vibration.

In these cases, the GA generates a successive series of Pareto fronts where non-dominated sorting of blast designs is performed. Similar results form clusters with designs evenly distributed along the Pareto front, with each design considered equivalently optimal. The system returns 15 results for the user to select from, as displayed in Figure 2.

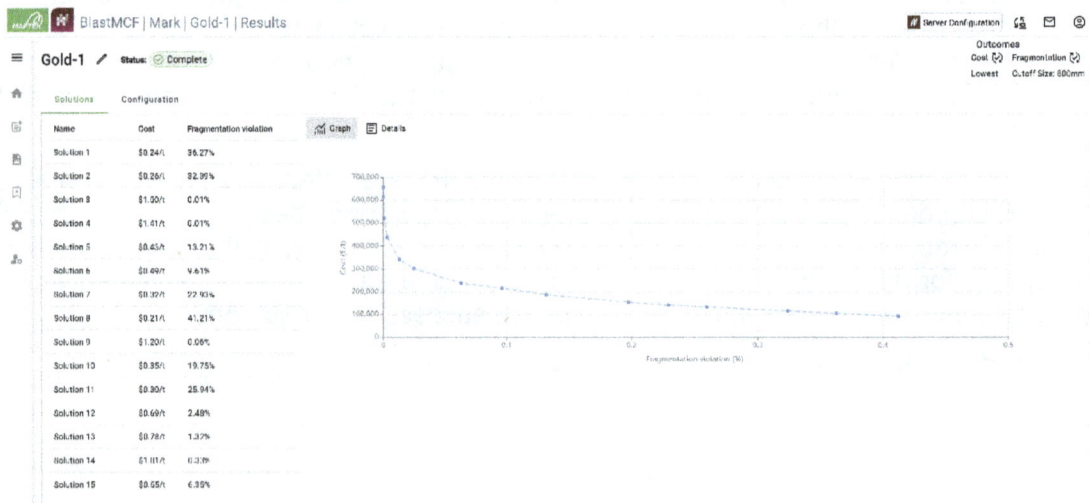

FIG 2 – Two objective blast optimisation result.

APCOM 2025 | Perth, Australia | 10–13 August 2025

The GA can accommodate up to three competing objectives, in which case the Pareto front becomes a three-dimensional surface rather than a two-dimensional curve. When optimising three objectives, a larger population size of 100 designs is employed to ensure greater diversity in ranking the resulting designs, which drives the exploration toward the Pareto surface. An example of results for three competing objectives is provided in Figure 3.

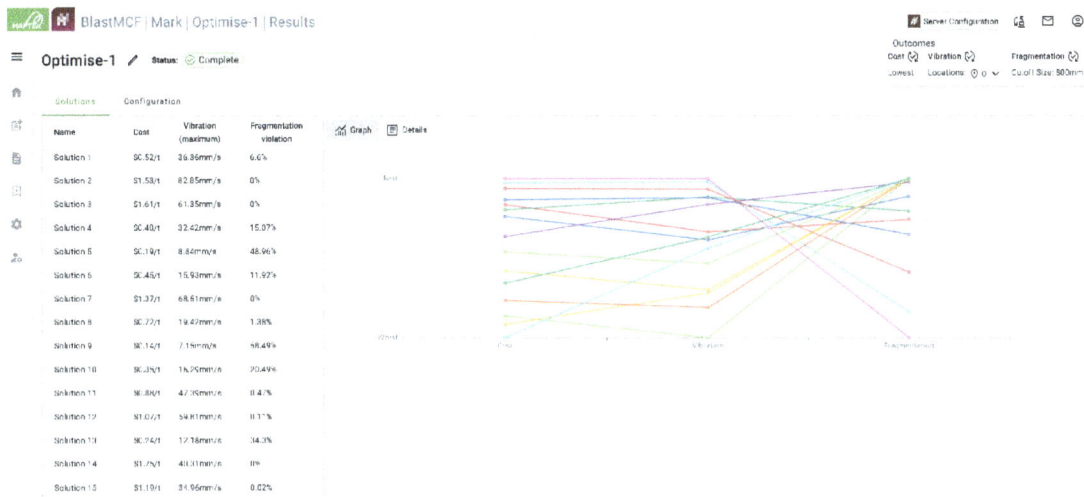

FIG 3 – Three objective blast optimisation result.

A key feature of the GA is its focus on objectives only when all constraints have been satisfied. In scenarios where constraint limits are particularly restrictive, only the design with the smallest constraint violation is returned for user review. This approach is valuable as it allows users to identify which variables violated constraints and adjust input design parameters accordingly, rather than receiving no results at all.

FUNCTIONAL OVERVIEW

The primary input for the software is a blast volume defining the area to be blasted. Engineers can specify this volume by entering a boundary polygon with upper and lower surfaces, specifying a lower RL or depth, or using a solid triangulation to generate the blast volume.

The system employs empirical models for fragmentation, vibration and fly rock assessment, along with inputs for cost and powder factor (energy) calculations. These factors can be configured as either objectives or constraints to optimise the resulting design. Users can further refine the process by specifying values, ranges or types to constrain variables and results based on various parameters including diameter, subdrill, pattern type (square, staggered, or equilateral), and setting minimum/ maximum values and ratios for burden and spacing.

For charge plan creation, the software offers engineers three optimisation approaches:

1. **Optimise charge rule** by adjusting user-selected parameters of an existing charge rule.

2. **Apply charge rule** by allowing the software to select the optimal charge rule from provided options.

3. **Optimise a simple charge plan** with a single explosive deck, designed dynamically based on different bulk explosives and user-defined minimum/maximum stemming heights.

Electronic timing design is configured by selecting and positioning a burden relief tool directly on or near the defined blast volume. Available burden relief options include a four-point timing rose, line or chevron. For chevron configurations, the chevron angle can be specified. The initiation angle can also be optimised within user-defined limits, a particularly valuable feature for vibration-sensitive designs when used as an objective or constraint.

HOLE PATTERN DEFINITION

The software provides sophisticated control over hole pattern definition for common design scenarios. Users can define row direction, set hole perimeter at a specific distance from the blast boundary, and specify additional distances for particular boundary sections, such as representing a free face.

Boundary hole adjustment can be optionally applied, whereby holes within half a burden and/or spacing of the boundary are repositioned onto the boundary in the respective burden or spacing direction. When this adjustment occurs, the next interior hole is also shifted to the midpoint between adjacent holes (including the repositioned boundary hole) as shown in Figure 4.

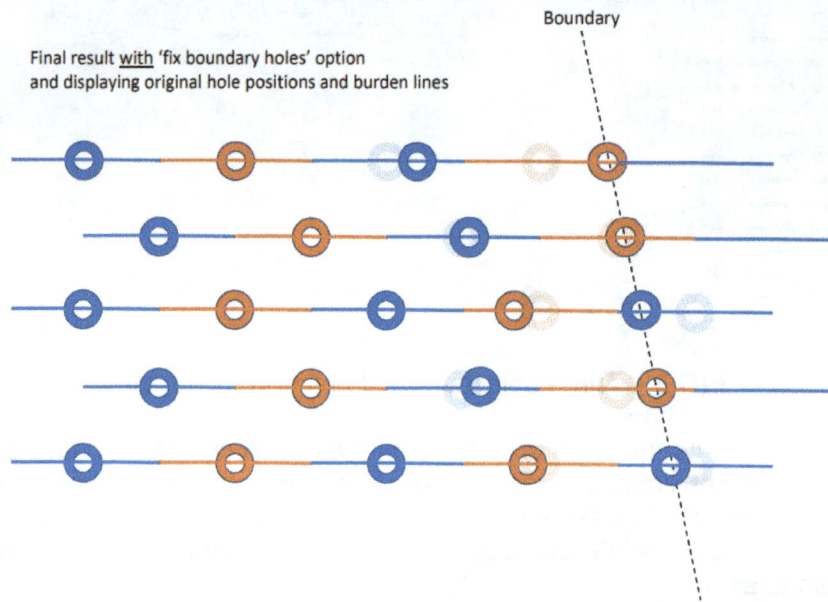

FIG 4 – Boundary hole adjustment.

The initial release of the software creates production-type blast designs where holes share common diameter, spacing and burden, while allowing for variable depth, charge plan and timing.

APPENDING DRILL DESIGNS

The system accommodates pre-existing drill hole designs as input, with the option to generate and apply optimised charge designs to these existing holes. For example, pre-splitting or trim holes with their specific charge plans can be incorporated into the set-up before generating production holes. This ensures that the complete blast design—including specialised and production holes—is considered when evaluating objectives or constraints related to cost, energy, fly rock, fragmentation and vibration.

To develop production blast designs across different lithological domains, the 'Add holes to an existing blast design' feature enables an iterative approach. Separate processing jobs are executed for each domain, with the design output from the first domain serving as input for subsequent domains as illustrated in Figure 5. This workflow can be applied across multiple lithologies, with each processing job contributing to an appended blast design.

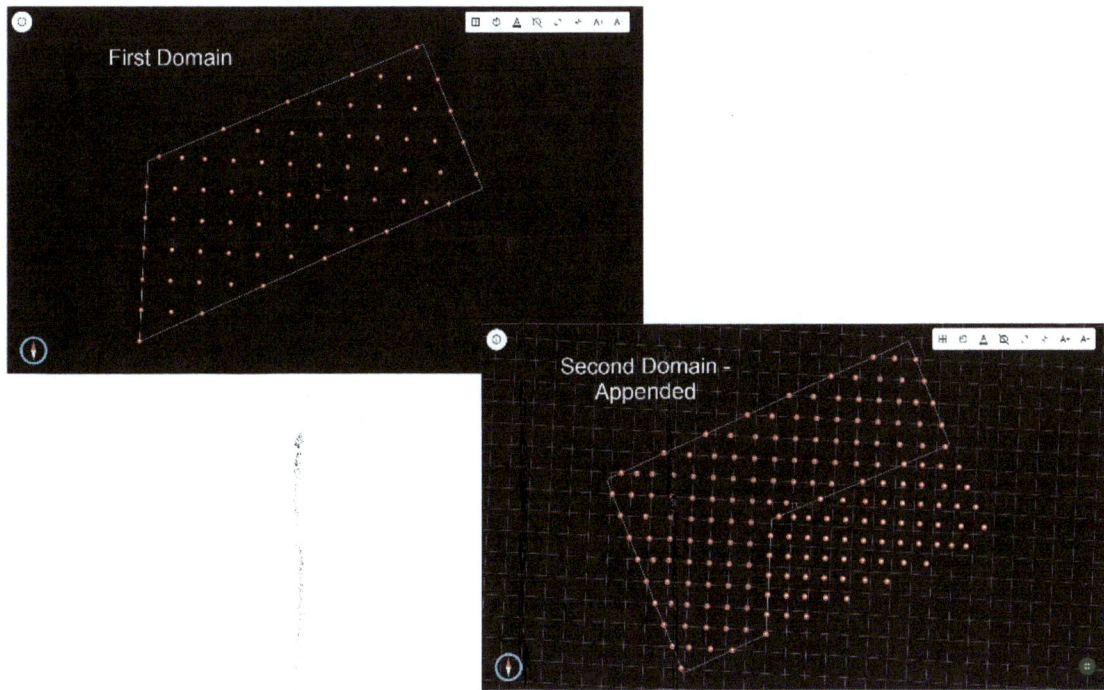

FIG 5 – Appended drill design.

This methodology for addressing blast design across different lithological domains aligns with current engineering practices, where different design parameters (burden, spacing, depth, diameter and explosive mass) are applied based on the specific rock characteristics being blasted.

SET-UP PARAMETERS AND CLOUD PROCESSING

Design parameter set-ups are saved and published to a centrally hosted application server, enabling their use as templates for future designs, facilitating sharing between users, and providing contextual reference alongside resulting blast designs.

Once configured by the user, set-ups undergo validation before their metadata is securely encrypted and transmitted to the Maptek Compute Framework (MCF) for processing. The MCF is a sophisticated technical framework that directs data processing tasks to Amazon Web Services cloud infrastructure, enabling computationally intensive operations to be performed on powerful external systems. All data residing on MCF is encrypted using Amazon Web Services security protocols, while sensitive data remains within the user's network environment.

RESULTS TO DATE

Maptek engaged with a major iron ore mining operation to evaluate the software application, gather user experience feedback, and determine whether the system could deliver on key performance indicators: ease of use, satisfactory results, rapid scenario analysis capabilities, and applicability for both daily design tasks and medium-term planning horizons.

Usability assessments revealed that this new approach is significantly more intuitive and accessible compared to traditional CAD-based drill pattern generation tools, particularly when operating within the mine's established design parameters for various lithologies.

The mine reported that the resulting pattern geometry and charge plans required only minimal post-processing adjustments, primarily consisting of removing, relocating or adding holes.

The MCF's performance enabled rapid scenario testing, with job upload and initialisation typically taking 30–45 secs. The genetic algorithm processing and results calculation generally took around one min for straightforward configurations, though this varies considerably depending on the selected objectives, blast volume size, charge rules complexity, and the number and proximity of vibration and fly rock monitoring locations.

Regarding applicability for both operational and planning purposes, the software supports comprehensive post-processing of results, including editing of drill patterns, charge plans and timing designs as displayed in Figure 6 screenshot. Integration with downstream systems is facilitated through native connectivity with Maptek BlastLogic blast design and reconciliation system, as well as file export capabilities and interfaces with supported third-party systems.

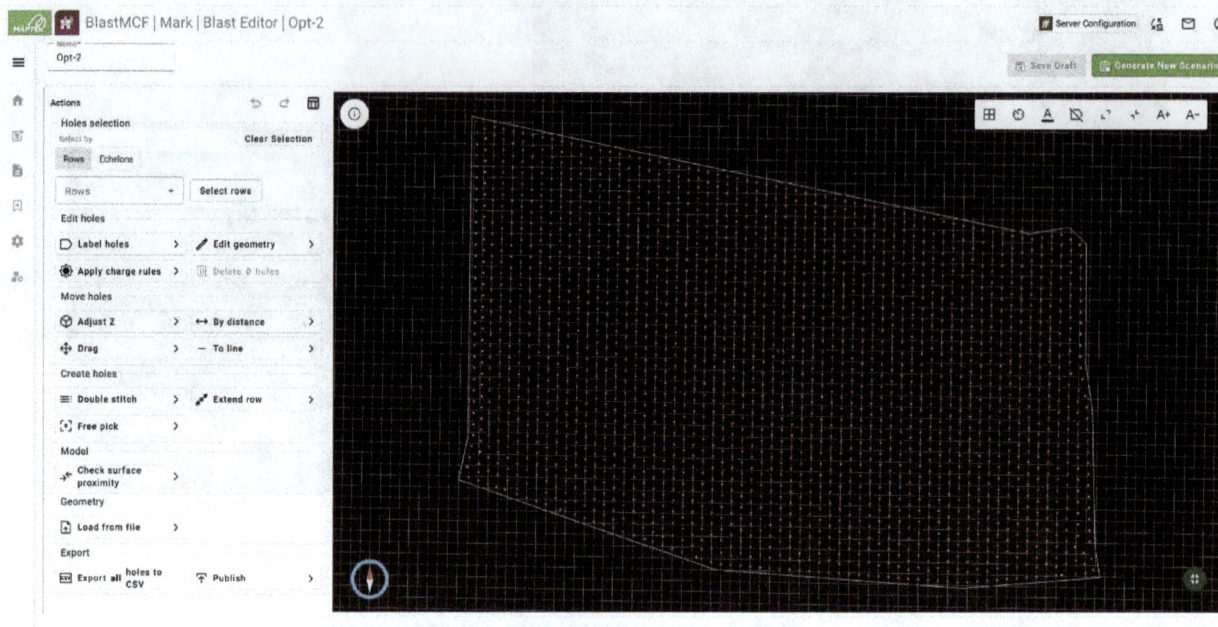

FIG 6 – Draft blast design editing.

ENHANCEMENTS

Ongoing development initiatives include enhanced capabilities for varying timing across a blast design. This involves associating multiple timing tools with different sections of a blast, configuring distinct burden relief parameters at each point of a particular timing tool, and applying different settings across multiple timing tools.

Additional development focuses on optimising hole angles and bearings relative to burden distance to the face and/or wall to improve wall control engineering, as well as expanding charge plan support to include intervals for deck loading required in through-seam and hard-band blasting scenarios.

Another key enhancement area is optimising blast design for different lithological domains within a single processing job. The goal is to optimise blast design geometry as it transitions between domains.

These enhancements will impact processing time, but for specialised and complex large-scale surface mining operations prioritising advanced blast control engineering, longer computation times may be acceptable given the significant benefits.

The ultimate vision for this artificial intelligence based blast design capability is to enable automated creation of multiple blast designs across entire benches or strips, adjusting energy design for each lithological domain within individual blast blocks. This will provide a scalable design solution that brings blast design activities earlier into the mine planning process.

In practical terms, this means blast designs could be generated for blast masters based on lithological domains defined per blast block during medium-term planning. This would enhance scheduling accuracy and resource provisioning for drilling equipment. As blasting approaches, designs can be efficiently updated by re-running the software with current surveyed free-face and terrain data before release to operations.

Implementing variable energy designs for different lithological domains at the blast block level may also make the software suitable for differential blasting design, which targets finer fragmentation for specific ore grades that are subsequently recovered through screening processes (Walters, 2016).

CONCLUSION AND FUTURE WORK

Further investigation is needed to evaluate the software's potential for value-added design applications, particularly in scenarios involving two or three competing objectives related to cost, vibration, fragmentation or powder factor. Initial scenario analyses suggest significant potential for cost savings on a per-blast basis without compromising fragmentation and vibration control.

Additionally, exploring applications where the software optimises only the charge plan, only the timing, or both components after drilling, based on as-drilled geometry—and again after charging based on as-loaded conditions—may yield valuable insights. This research would determine how the tool could rapidly update active blast design components as new information about hole conditions becomes available during implementation. It could also address scenarios requiring blast patterns with variable geometry (through-seam, trim shots, near dewatering wells) that are better suited to traditional CAD design approaches.

Another promising research direction involves validating this new blast design paradigm for remote design work and use by blast designers without formal mine engineering qualifications. This would address the industry's ongoing engineering staffing challenges and the trend toward remote mine planning operations.

ACKNOWLEDGEMENTS

The author extends gratitude to the innovative team of Maptek software engineers for their expertise in applying artificial intelligence to blast design optimisation. Special thanks to the mine engineers and management who participated in the early access program, providing valuable feedback and improvement suggestions. Their support for advancing mining practices and commitment to innovation has been invaluable. Particular appreciation is extended to those who assisted with proofreading the draft manuscript. This paper would not have been possible without these contributions.

REFERENCES

Acampora, G, Kaymak, U, Loia, V and Vitiello, A, 2013. Applying NSGA-II for solving the Ontology Alignment Problem, *IEEE International Conference on Systems, Man and Cybernetics*, pp 1098–1103.

Cunningham, C V B, 2005. The Kuz-Ram fragmentation model – 20 years on, in *Proceedings of the Third EFEE World Conference on Explosives and Blasting,* pp 201–210 (European Federation of Explosives Engineers).

Deb, K, Pratap, A, Agarwal, S and Meyarivan, T A M T, 2002. A fast and elitist multiobjective genetic algorithm: NSGA-II, *IEEE Transactions on Evolutionary Computation,* 6(2):182–197.

Knights, P, 2020. Short-term supply and demand of graduate mining engineers in Australia, *Mineral Economics*, 33:245–251.

McKenzie, C K, 2009. Fly rock range and fragment size prediction, International Society of Explosive Engineers, Denver: ISEE.

Richards, A B and Moore, A J, 2006. Kalgoorlie Consolidated Gold Mines Flyrock Model Calibration Update, report, Terrock Consulting Engineers.

Salmi, E F and Sellers, E J, 2020. A review of the methods to incorporate the geological and geotechnical characteristics of rock masses in blastability assessments for selective blast design, *Engineering Geology*.

Scott, A, (ed), 1996. *Open Pit Blast Design – Analysis and Optimisation*, 338 p (Julius Kruttschnitt Mineral Research Centre, University of Queensland).

Tordoir, A and Roberts, M, 2019. Benchmarking drill and blast compliance to design, a case study, International Society of Explosive Engineers, Nashville: ISEE.

Walters, S, 2016. Driving Productivity by increasing feed quality through application of innovative grade engineering® technologies, CRC ORE.

Wilkinson, W and Kecojevic, V, 2005. Elements of Drill and Blast Design and 3D Visualization in Surface Coal Mines, *Mining Engineering*, 57(9):77–82.

Beyond grade – evolution of automation to provide timely information from blastholes in open cut operations

E Schnetzler[1], J Jackson[2], N Battalgazy[3] and F Blaine[4]

1. Geostatistical Lead, IMDEX Limited, San Francisco CA, USA.
   Email: manu.schnetzler@imdexlimited.com
2. Global Technology Manager – Mining Technologies, Imdex Limited, Brisbane Qld 4000.
   Email: john.jackson@imdexlimited.com
3. Geoscience Data Analyst, Imdex Limited, Brisbane Qld 4000.
   Email: nurassyl.battalgazy@imdexlimited.com
4. Product Manager, Imdex Limited, Perth WA 6000. Email: fred.blaine@imdexlimited.com

ABSTRACT

The development and deployment of BLASTDOG, a semi-autonomous system for acquiring high-resolution, real-time subsurface data from blastholes in open cut mining operations, enables a significant shift in how geological information is used to support mining decisions. The system enhances decision-making across key operational domains—geology, blasting, geotechnical, and processing—by delivering timely, consistent, and high-quality subsurface data.

Integrated with a fully automated surface generation workflow and interactive 3D visualisation via MINEPORTAL, BLASTDOG enables the rapid, repeatable production of actionable insights on a daily basis. Surfaces are generated and version-controlled automatically, with immediate validation tools available to assess data quality and model reliability, ensuring continuous improvement and confidence in operational outcomes.

INTRODUCTION

Modern open cut mining demands timely and precise geological information to inform crucial decisions about blasting, rock mass characterisation, and ore/waste delineation. Historically reliant on assumptions based on one or more of: 1) wide spaced resource drilling, 2) limited samples from geotechnical, metallurgical programs, or 3) assay-based sampling and manual geological logging on grade control or blastholes.

These traditional methods often suffer from substantial delays, poor representativeness of the spatial and parameter variability and significant inconsistencies in the face of increasing orebody complexity plus constraints within the industry associated with the availability of skilled miners and professional and their capacity on sites. These factors can result in poor decisions in geology, blasting, mining, processing types and hence blending/scheduling of material to the plant resulting in non-optimal safety and business outcomes (Jackson, Gaunt and Astorga, 2014).

Although there isn't a single approach or method that addresses the entirety of these factors, providing clarity as to the actual 'in ground' conditions that operations face is a key aspect. A system-based approach to address a number of these factors, BLASTDOG, has and continues to be developed for open pit operations by providing consistent, timely, high resolution 'in ground' orebody and hole physical knowledge that is more relevant process orientated and systemic than traditional assumptions or personal based knowledge. The provision of such knowledge enables operational functional areas (such as geology, drill and blast, geotechnical, mining and processing) to work together to provide a better assessment of the risks of unwanted outcomes, identify and assess options to mitigate such risks, manage the variability and optimise outcomes for the overall operation (Jackson, 2024).

This system consists of acquiring high volume/resolution downhole data, processing and applying advanced analytics, integrating with other site-based data and transforming the data into quality, timely knowledge to improve decision enabling. A key component in providing timely information is the automation of multiple components of the system.

This paper focuses on the evolution and challenges with automation of selected components of the workflows for the generation of material types in both coal and metalliferous operations and coal surfaces in coal operations.

BLASTDOG – SYSTEM OVERVIEW

The BLASTDOG system consists of two major subsystems:

1. Data Acquisition and Ingestion Subsystem which includes a number of components:

   o BLASTDOG's multi-sensor downhole geophysical and hole property data-acquisition including data QA/QC and analysis on a hole basis (BD).

   o Site's Measurement While Drilling Data (MWD).

   o Site's Exploration Drill hole Data.

   o Site's Operational Production data including grade control and mine survey data.

2. Workflows Subsystem to transform the input data and information to knowledge, for which there are three main workflows:

   o Coal surface workflow.

   o 3D material type workflow.

   o 2D process performance and material type workflow.

All the workflows involve visualisation and validation which is via a web-based 3D environment, MINEPORTAL. A schematic overview of the BLASTDOG system is shown in Figure 1.

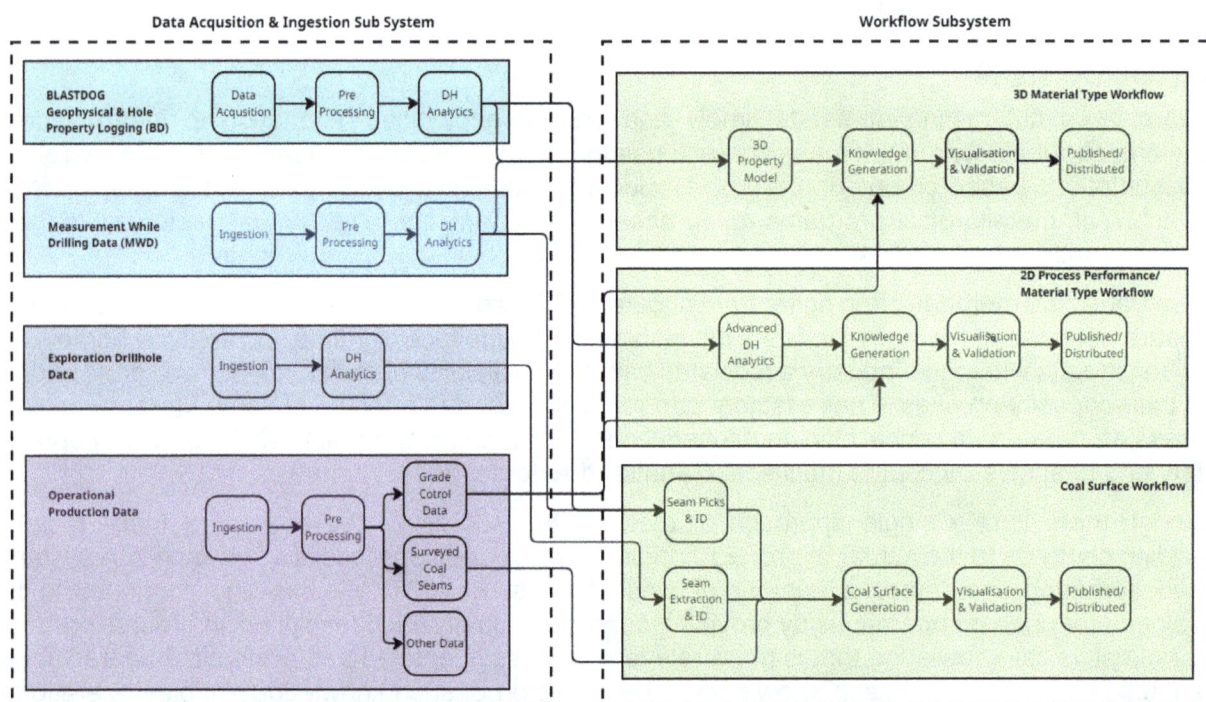

FIG 1 – Schematic overview of the BLASTDOG system.

The BLASTDOG geophysical and hole property logging (BD) and the MWD component of the data acquisition and ingestion subsystem have relatively high data volumes and velocity with data being received in near real time at a high resolution. This is in contrast to the exploration drill hole data which is only updated on a 6 or 12 monthly basis. The operational data falls between these extremes depending on the type and use of the data being ingested.

The system can be considered as semi-autonomous as the end-to-end system has not been fully automated with certain components requiring human in the loop predominately for QA/QC purposes prior to continuation. This was a deliberate choice to ensure that for an immature and evolving system, any issues could be caught and then rectified or mitigated in a time frame that operations could accept.

In this paper, more description will be provided for the following components of the system:

- BLASTDOG geophysical and hole property logging.
- Measurement while drilling.
- Coal surface workflow.

BLASTDOG GEOPHYSICAL AND HOLE PROPERTY – DATA ACQUISITION

The BLASTDOG data acquisition subsystem is a semi-autonomous system to acquire geophysical and hole property data of a blast or grade control hole using a multi-sensor probe. At the normal operating logging speed of 10 m/min, the downhole data is acquired at a sample spacing of 1 cm providing high definition of boundaries, other data is a single set per hole ie collar location. The data acquired is summarised in Table 1.

TABLE 1

Data acquired by BLASTDOG data acquisition subsystem.

	Data/sensors
Rock knowledge	Natural gamma
	Magnetic susceptibility
	Conductivity
	Fractures/fracture frequency
Hole property	Hole depth
	Downhole diameter
	Downhole dip and azimuth
	Water level
	Collar location (RTK GPS)

An example of downhole magnetic susceptibility data from blastholes within a metalliferous open pit showing considerable variability is shown in Figure 2.

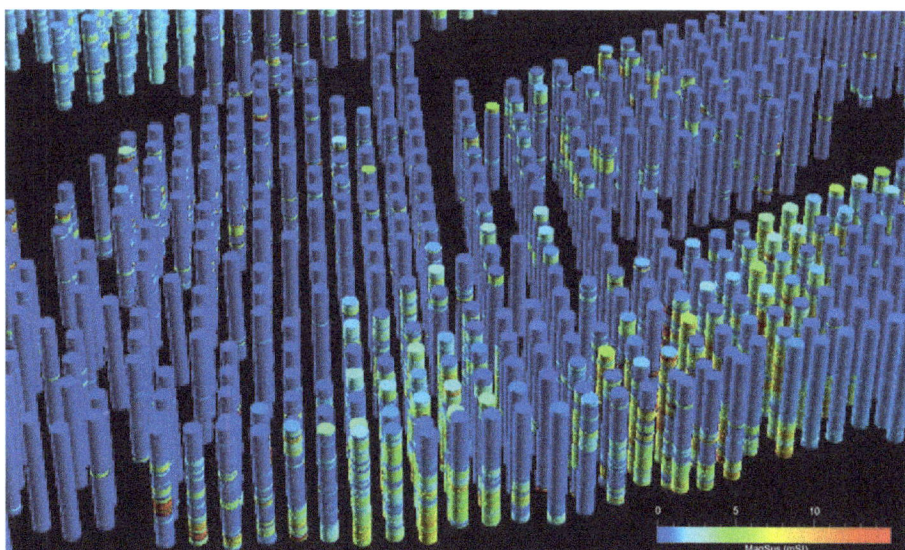

FIG 2 – An example of magnetic susceptibility data from blastholes within a metalliferous open pit deposit.

A block schematic of the data acquisition component is shown in Figure 3 and consists of three major elements:

1. Logging management.
2. Deployment system and multi sensor probe (1AS).
3. Operator control and operator.

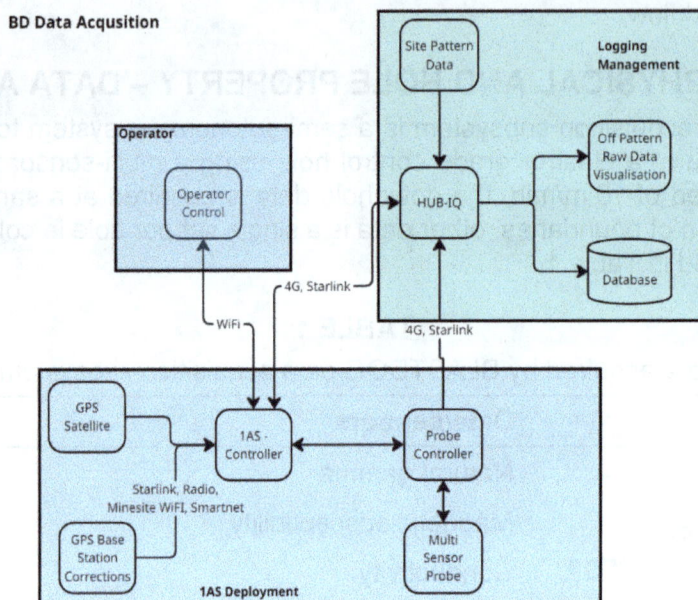

FIG 3 – Block schematic of the BD data acquisition subsystem.

The logging management element itself consists of upload of the site pattern data (name, hole location, EOH depth, azimuth and dip) to the cloud-based HUB-IQ. Once a pattern is made active, HUB-IQ downloads the information to the deployment element.

The deployment element, known as the 1AS, is a robotised CAT 2 tonne excavator currently operating in semi-autonomous mode developed in collaboration with Universal Field Robots in Brisbane. The 1AS is controlled/supervised by a single operator who is located up to 100 m from the machine and can be operated on the same pattern as the drills, subject to exclusion limits (Figure 4). The system can log between 1000–1700 m per shift depending on the depth and spacing of the holes.

FIG 4 – BLASTDOG acquisition subsystem – 1AS Deployment Element on a bench.

The BLASTDOG acquisition system and use of an excavator was developed to address limitations with traditional approaches in order to produce high volume, spatially dense data in a safe manner in open pit operations and particularly coal operations. These included:

- Removing the need for clearing of drill cuttings prior to logging. In the case of deep holes, the cuttings piles can be up to 1–1.5 m high and 3 m in diameter.

- Removing people from having to be near open holes.

- Increasing logging speed by 2–3 times and thus enabling higher volumes of data within the same period of bench access.

- Ability to operate in challenging bench conditions and thus increasing access window and reducing need for further bench preparation.

- Consistent collection of data through automation.

Originally the subsystem was designed to have a high level of autonomy, however as the system was being developed and tested at various operations, a number of limitations were encountered, including:

- Poor communications: poor communications between HUB-IQ and the 1AS were due to IT and logistical issues connecting to mines internal in-pit wi-fi and/or poor 4G coverage. This resulted in: a) having a daily download/upload mechanism via the operator; and b) relying on very local base stations for RTKGPS corrections with the mine grid positions established by mine survey which brought its own problems.

- Safety: although autonomous safety systems within the industry have been established for haul trucks and drill rigs, not every mine has these. Additionally, the bench operations can be very busy with a mix of equipment such as drills, blast crew personnel on foot, and explosive loading trucks. Site preference was for an operator to always maintain situational awareness.

- Productivity: although fully autonomous had been tested in ideal situations, the number of manual interventions due to terrain, difficult access to holes etc, was expected to be initially high and the learning curve too long impacting on the value proposition.

Thus, for these reasons, the system has remained in a semi-autonomous mode with deployment, tramming and movement around the bench being remotely controlled by the operator with alignment to hole and downhole logging being undertaken autonomously.

The advent of Starlink, greatly enhanced the ability for the downhole data, following basic processing on the edge, to be transferred autonomously in near real time data transfer from the 1AS to HUB-IQ enabling fast and efficient QA/QC with any major data issues being flagged to the operator. Starlink has also enabled the ability to single GPS base station broadcasting corrections to the 1AS dramatically reducing the need for roving base stations and mine surveyors.

MEASUREMENT WHILE DRILLING DATA

Measurement While Drilling (MWD) data are considered here as those parameters measured at the drill rig related to the drilling process itself (Paredes Bujes, 2020). The data set includes collar coordinates, downhole survey data, and strata-related metrics such as Specific Energy (SE), Rate of Penetration (ROP), rotary reference, torque, feed pressure, and force on bit. Downhole sample spacing is 10–15 cm, 1/10th of the resolution of that for the BLASTDOG. The data is ideally acquired from the site's MWD database twice daily, coinciding with shift changes and automatically ingested into a MWD database with primary keys to enable data fusion with the BLASTDOG data. Each record is associated with metadata such as hole ID, drill time, and pattern location within the pit, enabling contextual interpretation. In an open pit coalmine each rig can drill in the order of 1000 m per shift with drilling occurring on day and night shift, with larger sites concurrently drilling multiple pits on a 24 hr basis. At one site 4000–6000 m of drilling data was received each shift. Open pit metalliferous operations tend to have significantly lower drilling rates than coal.

MWD has been previously sporadically used for identification of coal, mainly focused on ROP (Partridge, 2019). Here we explored the viability of a number of the parameters with a focus on ROP and SE. Due to inconsistencies and high variability in ROP, SE was selected as the primary indicator owing to its relative consistency and better response to lithological changes. However, raw SE data remained inherently noisy due to multiple factors, including equipment-specific characteristics, functioning sensors (or lack thereof), sensor calibration, part replacements, and operator-dependent drilling behaviour. These variabilities together with the data volumes necessitated a robust automated data cleaning and levelling strategy.

The adopted levelling approach was pit-based, recognising geological variability between pits and ensuring SE profiles were standardised within localised geological contexts. Figure 5 illustrates this transformation, where levelled SE data (RHS) displays improved signal stability compared to its unlevelled counterpart (LHS), enabling more accurate and consistent coal seam delineation.

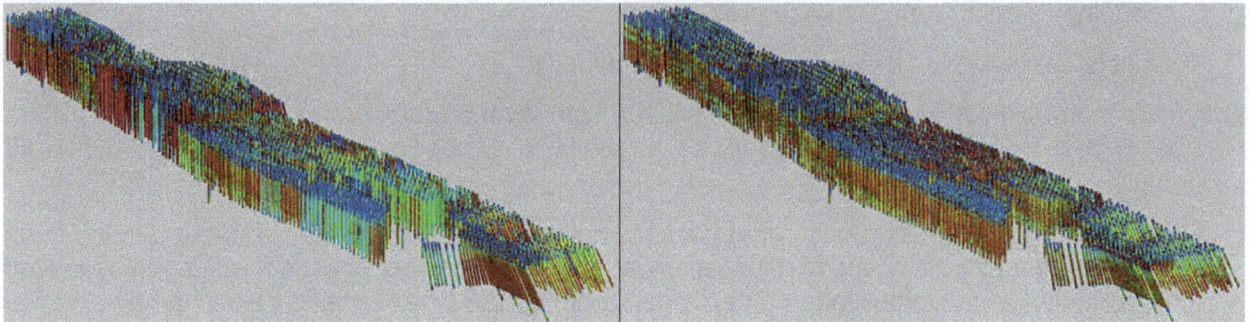

FIG 5 – Comparison of raw SE data from the site's MWD database (LHS) and levelled SE data.

A significant challenge was faced in automating the process in obtaining sufficient expected data to develop data cleaning strategies in parallel with changes within the MWD data due to the rectification of identified issues and/or rigs being added or swapped in/out. All the while still providing quality data that could be used for downstream workflows on a daily basis. This necessitated some manual intervention and the efficient implementation and adaption of the cleansing pipelines through several significant modifications.

COAL SURFACE WORKFLOW

The workflow for the automated generation of coal surfaces consists of five main steps:

1. Seam identification from BLASTDOG geophysical logging ± seam identification from MWD data (if data is available).

2. Conversion into top of coal/bottom of coal de-surveyed XYZ coordinates.

3. Integration of site-based top of coal/bottom of coal XYZ locations from exploration data and/or mine surveyed data.

4. Generation of the coal surface.

5. Validation of the automatically generated surfaces.

The focus here is on the identification of coal seams form BLASTDOG and MWD data, the generation of coal surface and its validation.

BD coal seam identification

The key physical properties of coal seams are low natural gamma, low density, low velocity and low conductivity/high resistivity (Cogswell, 2015). The traditional use of geophysical logging in coal is using natural gamma ± density with coal picking generally undertaken manually by site geologists (Hatherly, 2013). For BLASTDOG geophysical logging, coal seam identification is undertaken using a combination of natural gamma, calliper and conductivity. The difference in tool configuration and sensor selection, means that pre-existing site data and associated manual coal picks cannot be used directly as a training data set for machine learning based coal seam identification. Additionally, initially site geologists generally wish to gain confidence in the new system by comparing seam identification using the same process ie manually.

In parallel, the automation of BLASTDOG coal picks is developed using a two-step process: 1) using a multivariate, multi-scale boundary picking methodology to define units; and 2) application of supervised classification using natural gamma and conductivity validated by manual coal seam picking.

At one site, comparison of traditional logging verses BLASTDOG automated coal seam identification for 71 holes where the logs were through seam. An example of one hole is shown in Figure 6. The automated coal picking algorithm showed a mean thickness difference of 1 cm (StdDev = 13 cm) to that of the traditional manual approach. This was within the expected variation within manual seam identification.

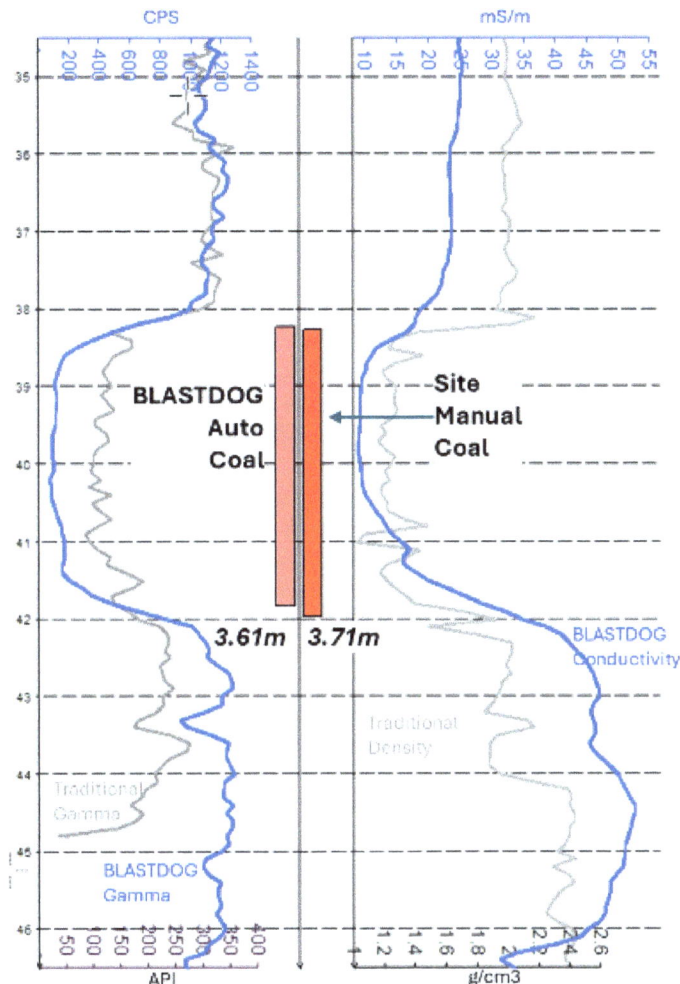

FIG 6 – Comparison of automated coal pick from BLASTDOG with traditional logging and manual coal pick by site.

MWD coal seam picks

Once a reliable preprocessing pipeline has been established, coal picking was initially conducted manually on the cleaned SE profiles as there generally is no training data set available to implement a machine learning approach from the start. Three styles of drilling in relation to the position of the coal seams are need to be accounted for: 1) through seam – where the drill penetrates both the top and bottom of the coal seam; 2) touch coal – where the driller stops drilling when they believe they have just encountered top of coal; and 3) stand off – where the planned hole is not expected to penetrate the coal seam. In all cases, depending on the complexity of the geology the alternatives can occur.

In the example discussed here, most drilling was either stand off or touch coal drilling where the top of coal is identified from as few of three sample points (30–45 cm) or up to ten sample points (100 cm) depending on the driller and geological complexity. As operational consistency improved and more stable MWD data became available, the manual effort enabled the generation of a robust training data set for machine learning. The training data set needed to be sufficiently large to cover the variations in the MWD response due to seam and pit location. Although it was found that the variation in SE response across site was significantly lower than that observed for the geophysical data from BLASTDOG as the variation in lithology types resulted in differing geophysical responses. As the SE is essentially a measure of relative rock hardness there were a number of false positives due to soft rock that was not coal. These had not been identified in the initial training data set which required a detailed review of the drilling data for the patterns selected for the training data set.

During the investigations and development of a machine learning process to automatically identify coal, further preprocessing of the MWD data was found to be required including low pass filtering and stripping of the upper portion of the hole due to soft fill material placed during drill pattern bench preparation.

Once the preprocessing step has been completed, a 1-Dimensional Convolutional Neural Network (1D CNN) was trained on manually labelled data set. Two classes of training data were used for touch coals and no touch coals, respectively. The trained model then is applied to holes processing the last 3 m of each hole to determine if a touch coal is present. For coal through seam cases, similar method is utilised, except longer training intervals are used and rules adjusted to find both top and base of the coal seam.

An example of results of the machine learning in the identification of touch coal holes from patterns other than those used for the training data are shown in Table 2 where the touch coal holes were correctly labelled, 90 per cent were within 30 cm of the manually labelled coal picks which is acceptable. Considering the missed touch coals, these most often occurred where the MWD did not continue far enough. Another cause of errors was artefacts in the MWD. Additional rules were then applied to eliminate some of the false picks.

TABLE 2

Confusion matrix for ML results.

True label	ML result: touch coal	ML result: no touch coal
Touch coal	466	72
No touch coal	5	24

Coal surface generation

The surface generation from labelled picks is a labour-intensive process done manually. It requires importing the points in software and creating the surfaces before exporting them. This is time-consuming and error-prone due to its manual nature.

The automated coal surface generation workflow aims to rapidly deliver a set of accurate surfaces ready for immediate visual inspection in 3D software. The automation integrates coal picks (from BLASTDOG, MWD, exploration and operational production data), blasthole data (from BLASTDOG and MWD), and dynamically generates surfaces. Due to the daily availability of new data and picks, a rigorous version control system ensures consistency and traceability between input data and the corresponding generated surfaces.

The automated workflow includes the following detailed steps:

- **Data Integration:** Read and process coal picks and blasthole data.

- **Surface Generation for Each Coal Seam:**

  o Calculate the general trend for the top of the seam.

  o Estimate residuals (depth minus trend) using kriging.

  o Combine the trend and estimated residuals for accurate top-seam surfaces.

  o Calculate the general trend for the bottom of the seam.

- o Estimate residuals similarly using kriging.
- o Combine trend and estimated residuals to generate bottom-seam surfaces.
- o Alternatively, derive the bottom seam surface by generating a thickness map from the top seam surface.
- o Correct potential crossings.

- **Smoothing:** The surfaces can be optionally smoothed in order to clean local variability in the picks.

- **Priority Rules Application**: Define and apply priority rules between crossing seams.

- **Export and Version Control**: Automatically export generated surfaces to files.

- **QA/QC Integration**: Upload picks, blasthole data, and surfaces into 3D visualisation software for immediate quality assurance and control.

Executed daily with new data inputs, this automated workflow significantly accelerates the surface generation process. Quick and convenient access to version-controlled surfaces and data in 3D visualisation software allows for rapid review and identification of potential discrepancies or anomalies, enhancing data integrity and operational decision-making.

Visualisation and validation

All data and surfaces are automatically imported into MINEPORTAL with version control, enabling immediate visualisation and validation following surface generation. This integrated workflow allows users to quickly assess the quality of both input data and resulting models. Figure 7 illustrates an example of generated top and bottom coal seam surfaces for a drill pattern incorporating picks derived from both BLASTDOG and MWD data.

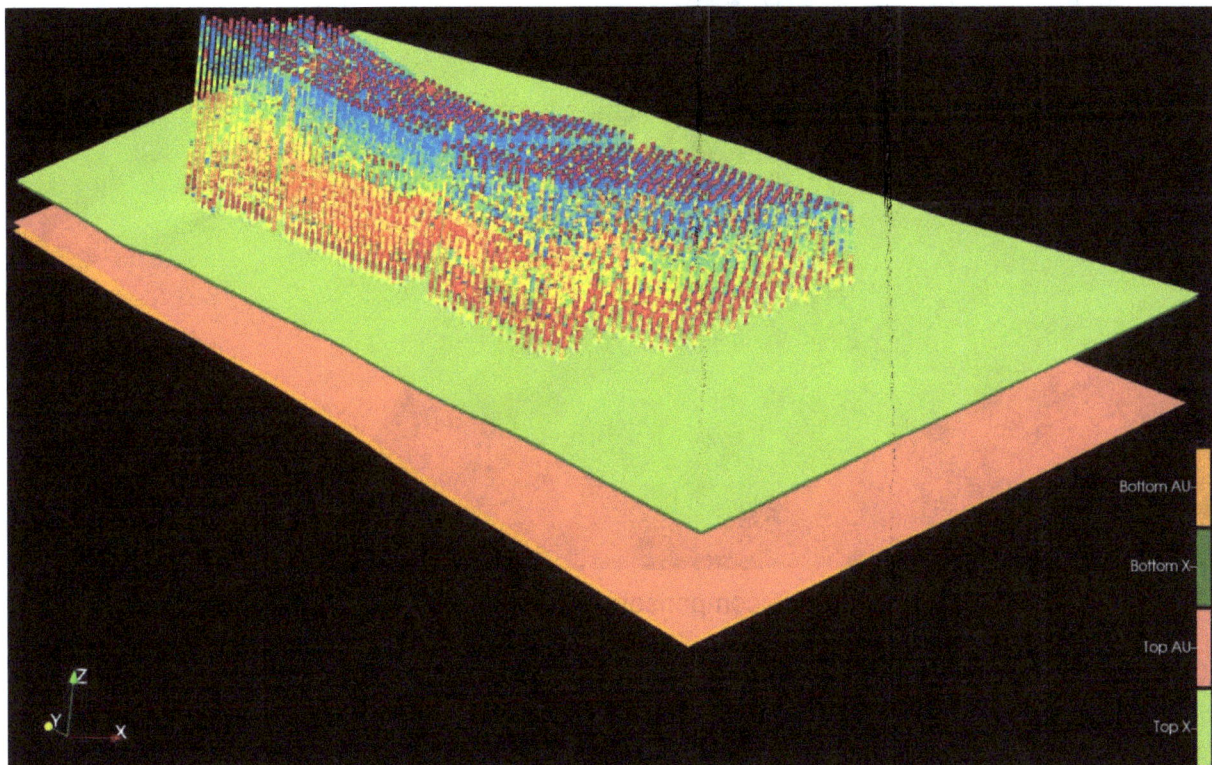

FIG 7 – Example of generated surfaces for two seams.

Validation is supported by tools such as interactive slicing, which allows users to navigate through cross-sections and directly compare the generated surfaces with the underlying picks. This view (Figure 8) effectively highlights discrepancies, enabling the identification of local anomalies or data quality issues. By colour-coding picks based on their distance from the corresponding surfaces, users can visually assess the impact of smoothing and more easily detect problematic picks or geological variability—particularly structural displacements such as faults.

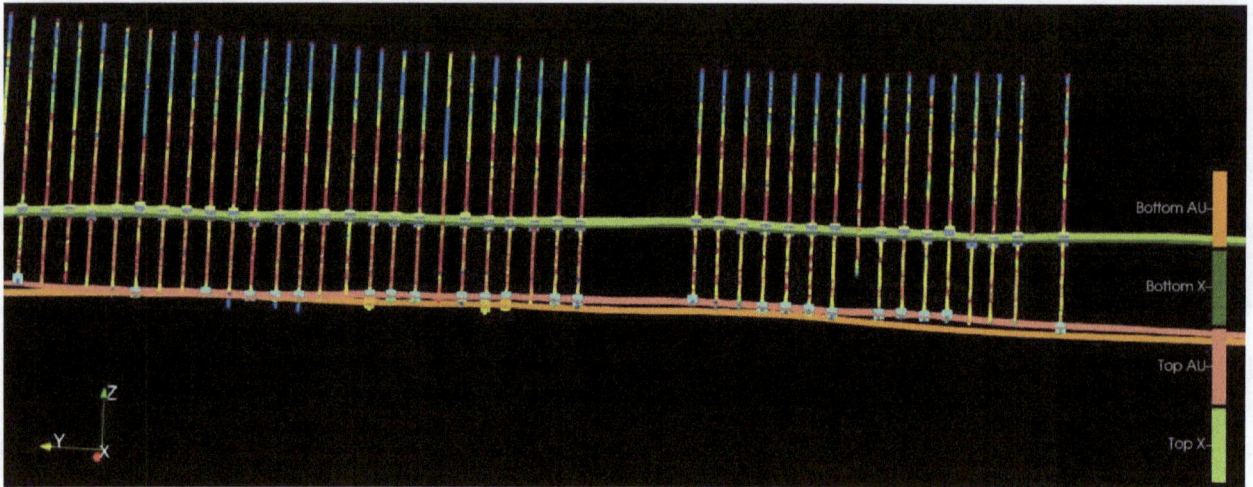

FIG 8 – Slicing through the data sets.

The ability to visualise successive versions of the surfaces as new data becomes available provides valuable insight into the impact of additional information on model quality and continuity. This iterative comparison helps quantify the benefits of increased data density and supports informed decisions about drilling and data acquisition strategies. Figure 9 illustrates this progression: the left panel shows an early surface generated from a limited set of drilled and logged holes, while the right panel displays a more comprehensive surface derived from a denser data set.

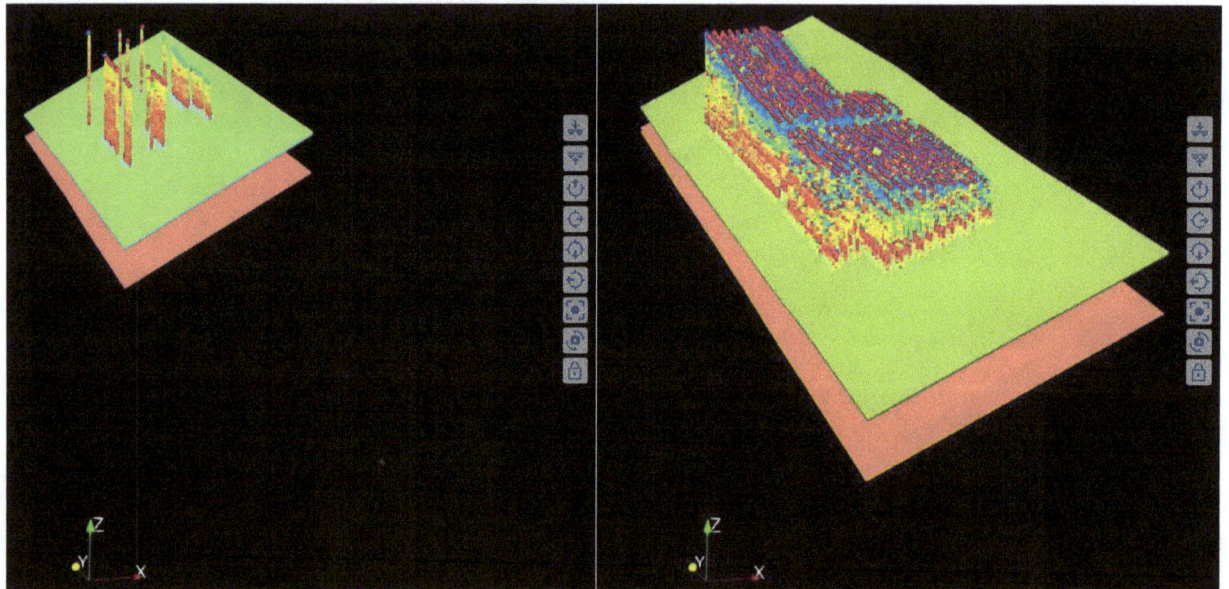

FIG 9 – Side-by-side comparison of different versions of surfaces.

CONCLUSION

The implementation of the BLASTDOG system represents a transformative step in the acquisition, processing, and application of high-resolution geological data in open cut mining operations. By integrating semi-autonomous data acquisition, robust preprocessing workflows, and machine learning for seam detection, the system significantly reduces the time between data collection and decision-making. The combination of geophysical logging, Measurement While Drilling (MWD) data, and real-time processing delivers a more detailed and timely understanding of subsurface conditions than traditional methods can offer.

The automation of the elements of the system uses different approaches, different levels of maturity and evolved at different speeds being highly dependent on: 1) the availability of quality training data; 2) time required to develop training data sets; and 3) operational requirements. In the case of coal seam picks from BLASTDOG geophysical data, a combination of data preprocessing and supervised machine learning based on manually learning the rules has proven to be sufficient whereas a more complex approach was required for coal seam picks from MWD data.

The automation of surface generation and seamless integration with 3D visualisation platforms such as MINEPORTAL further streamline the workflow, enabling daily updates to geological models and immediate validation. This continuous feedback loop supports improved blast design, ore/waste classification, and grade control, ultimately enhancing operational safety, efficiency, and value. While the system remains semi-autonomous by design to accommodate operational realities and QA/QC requirements, its modular architecture and successful field deployments demonstrate the potential for broader adoption and future scaling. The BLASTDOG approach offers a practical and impactful pathway toward more data-informed and agile mining operations.

ACKNOWLEDGEMENTS

A project of this size involves the efforts of many people. Thanks to the teams within IMDEX including engineering, data science, software development, product management, the BLASTDOG operators and support, and management for their support with a special call out to Chris Koplan, Erik Gutterud, Majiga Enkhsaikhan, Brendon Lilly, Arthur Sedek, Glen Casey, Hayden Gray and Terry Mahoney. Also, acknowledgement to Jeff Sterling and Tim Cassell from Universal Field Robots and Katie Silversides from Datarock Pty Ltd.

The authors also thank the support operational teams and technical experts from the various mine sites, particularly those at AngloAmerican and Teck Resources.

REFERENCES

Cogswell, D, 2015. Geophysical logging for coal - current techniques, ASEG Workshop (Perth: Borehole Wireline).

Hatherly, P, 2013. Overview on the application of geophysics in coal mining, *International Journal of Coal Geology*, 74–84.

Jackson, J, 2024. Enabling improved risk management and blasting outcomes through high-fidelity rock knowledge, in 8th Drill and Blast Down Under Conference (Perth: ISEE).

Jackson, J, Gaunt, J and Astorga, M, 2014. Predicting mill ore feed variability using integrated geotechnical/geometallurgical models, in Orebody Modelling and Strategic Mine Planning Conference (The Australasian Institute of Mining and Metallurgy: Melbourne).

Paredes Bujes, C I, 2020. Measure-While-Drilling for Ore Characterisation: Links between drilling and comminution properties of rocks, PhD Thesis, Julius Kruttschnitt Mineral Research Centre, University of Queensland.

Partridge, J, 2019. Measure while drilling: a case study at Daunia Coal Mine, in Mining Geology Conference 2019 (The Australasian Institute of Mining and Metallurgy: Melbourne).

A robust approach for digital mapping of rock masses with a stereo camera

M Shabanimashcool[1]

1. Senior Adviser, Norwegian Geotechnical Institute, Oslo, Østlandet 0484, Norway.
   Email: mahdi.shabanimashcool@ngi.no

ABSTRACT

Digital rock mass mapping has advanced significantly over the last decade, primarily involving the generation of point clouds from exposed rock surfaces. Subsequently, various algorithms have been employed to identify and map the planar segments of these point clouds, representing the surfaces of rock discontinuities. However, this method often overlooks data from the traces of discontinuities that may be present on the exposed rock surfaces. Therefore, utilising stereo cameras could be more effective compared to LiDAR scanning methods for mapping rock masses, as they can capture the rock surface texture (traces of discontinuities on the rock mass exposures) along with the visible planar elements. In this manuscript, we demonstrate a robust mapping algorithm developed for rock mass mapping with a stereo camera, which can be used not only for mapping but also for georeferencing the mapped discontinuities. Furthermore, we provide a demonstrative example where the method is used for rock mass mapping and identifying the rock block size distribution in intact rock.

INTRODUCTION

One of the biggest challenges in mining and rock engineering is the reliably characterisation of rock masses. Traditionally, rock masses are described according to the geometry of their discontinuities which may be measured by scan-lines, scan-windows or oriented boreholes (Priest, 1993). These techniques are time consuming and influenced by the experience and judgment of the geologist conducting it. The mapping outcomes can be used either in rock mass classifications methods to assess the rock mass stability or employed to generate discontinues numerical or analytical calculations. The traditional rock mass mapping becomes more challenging as the size of the projects and geotechnical complexity rises. Furthermore, sometimes safe and accessible zone for a geologist to access and map rock mass is not available. Consequently, the demand for reliable digital tools for rock mass mapping continues to grow.

Digital rock mass mapping techniques can be divided into two categories: traditional point cloud methods that relay on statistical method and geometric constrains; and the powered methods by Artificial Intelligence (AI). AI based approaches are currently less favoured as they typically require large volume of a reliably labelled rock mass training data. However, with utilising the methods which are described for example in this manuscript, it is become possible in the near future to generate standardised labelled databases for training and quantifying the reliability of the AI assisted rock mass mapping methods.

Battulwar et al (2021) critically reviewed all available digital rock mass mapping methods that utilises tradition statistical and geometric methods on point clouds describing their advantages and disadvantages. They concluded that among the different methods of digital rock mass mapping the approaches based on voxels growing are superior compared to the others. In the voxel growing method the user or automatically a small area were selected which is planar, and then the size of area grows as the neighbouring area also satisfies the planarity criteria. Similarly, and more recently Yi et al (2023) applied advanced data analytics methods to implement voxel growing algorithm over point cloud for mapping planar patches, which shows the method of voxel growing has the potential to be the most reliable and used method in the close future. Hence in this manuscript we also implemented a semi-automated similar approach utilising a stereo camera.

Among the literature reviewed by Battulwar et al (2021), they identified that calculating the discontinuity trace length (a measure of discontinuity persistence) utilising Normal tensor voting is a robust method. This approach works only on the meshed surface which is generated from point cloud of the scene. However, such method does not fully consider the three dimensional geometry of the discontinuities.

In conclusion, a robust algorithm for rock mass mapping – considering what an expert geologist might do in the field- should not only map visible planar patches on rock surfaces but also consider discontinuities that lack having visible planar sections in the recorded scene and may have only visible traces. Hence, a robust mapping method should consider both the point cloud and process the images from the rock mass surface to find three-dimensional correspondences on visible discontinuity surfaces. Therefore, this paper demonstrates a method that considers both point cloud data and images from scenes recorded by a stereo camera for comprehensive rock mass mapping. In addition, utilising stereo camera helps us to fuse it with sensor data to automatically georeferenced the mapped scene.

ROCK MASS MAPPING UTILISING STEREO CAMERA

Point cloud is a collection of point coordinates representing surface of a physical object in a specific coordinate system. There are two different techniques for estimating point clouds from images: triangulation (stereo matching) and reconstruction (structure from motion, SFM). In the SFM method both scene geometry (which is called in machine vision terminology as structure) and camera geometry (camera motion) are estimated with a nonlinear optimisation technique (Szeliski, 2011). In SFM method the generated point cloud is not scaled, and we need to know distance between several points within the point cloud to correct the scale.

For the triangulation method, two calibrated cameras are required. Calibrated camera means that we know extrinsic and intrinsic parameters of the camera. Extrinsic properties of the camera define the location and orientation of the camera reference system with respect to a word reference frame (pose of camera in the world coordinate system). Intrinsic properties of the camera link the pixel coordinate of an image with the camera reference frame. Intrinsic properties include focal length, principal point, skew coefficient, and distortion (it shows the radial and tangential distortions in the image). These properties are obtained by mapping several points where their distance from each other is known from forehand. The camera calibration to obtain the intrinsic properties can be done just one time and is not required for the rest of the operation. However, the extrinsic parameters needed to be updated by every frame recorded by the camera. In the stereo matching method two images are used to estimate 3D model (coordinate of points). In this technique the matching pixels of images are find and then the 2D position of them in the images is converted to 3D depth map (like topography map). To make the calculations simpler, two cameras can be parallel with each other while they have only translation type of movement between them. Nowadays there are off-the-shelf stereo cameras with very reasonable price which can be to utilise in digital rock mass mapping. In addition, they might be equipped with Global Navigation Satellite System (GNSS) sensor or Inertia Navigation Systems (INS) which makes estimating of the camera poses per frame quite accurate.

Stereo cameras also provide an ability of storing images from the scene. The images can be used manually by the geologist to quality control the automated mapping results and to perform manual mapping. In this study we are utilising Zed 2i camera from Stereolabs (2023). This stereo camera equipped with its own Software Development Kit (SDK) and an Inertia Measurement Unit (IMU).

Figure 1 presents the algorithm utilised in this study to map rock discontinuities. The movie recorded by the camera is first georeferenced either by using GNSS data from the camera sensor or by using landmark points with known coordinate in the GNSS-denied area (see next section). In addition to mapping the planar patches in the point cloud, discontinuity traces are mapped in the images to identify discontinuities whose surfaces are either not visible for mapping or not captured due to the camera position. After that coplanarity of all the mapped planar elements in the scene are controlled and coplanar elements are merged. Finally, the mapped discontinuities are clustered to obtain their median dip and dip-direction while calculating their normal spacing and rock block size distribution, as denoted as the characterising rock discontinuity network.

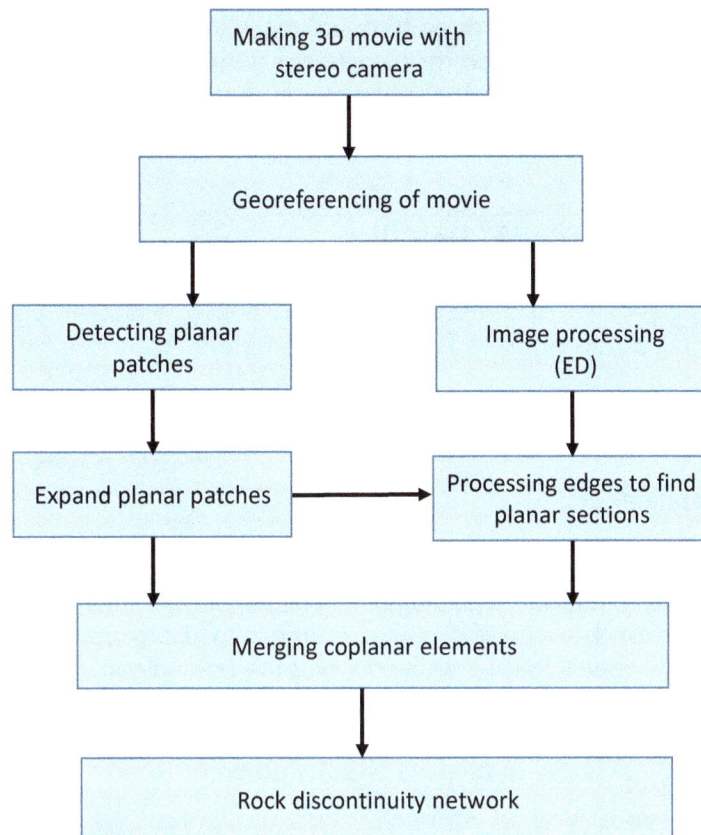

FIG 1 – The algorithm utilised in this manuscript to map rock mass discontinuities with a stereo camera.

Georeferencing

The movies recorded by a stereo camera will be georeferenced to represent the mapped geometry of the discontinuities in a global coordinate. Some of the cameras might be equipped with GNSS sensor to estimate their world coordinate. However, in most of the cases the GNSS pose estimation might not be accurate enough. Therefore, utilising Visual Simultaneous Localisation and Mapping (VSLAM) fused by GNSS sensors data can be used to accurately estimate the pose of the camera (for example see Kudriashov *et al*, 2020).

The VSLAM technique is mostly utilised in robotics for constructing the map of the area where the robot is mobilised and for determining its own location. In VSLAM, initially distinctive visual features from consecutive images are extract and later by matching them a relative mathematical transformation between them is calculated (denoted as homography). The features are a region on an image with unique shape described by specific vectors that are scale invariant (Szeliski, 2011). VSLAM algorithm can also track those features across several consecutive of images. By estimating the 3D local coordinates of those feature points (using their depth estimated by the stere camera) the relative transformation between two consecutive image frames can be obtained. Subsequently, with utilising these consecutive transformations (which are represented in rotation angles and translation vector), it is possible to obtain the absolute pose of the camera compared to the first frame (the start point of the camera, or camera coordinate system). Fusing the global coordinates of the camera from GNSS sensor and the camera pose in the camera coordinate system will allow us to decrease the uncertainties of the estimated poses of the camera and represent the camera pose in the global coordinate. After that, it will be possible to obtain world coordinate of any pixel point with valid depth using the depth estimated per frame of the 3D movie recorded by the stereo camera.

The challenges in VSLAM arise when such technology is intended to implement in GNSS denied area (like underground mine) where light intensity can also vary. To overcome these challenges, we can fuse the relative poses of the camera from VSLAM with measurements by Inertia Navigation System (INS) which might include an accelerometer, inertial measurement unit, magnetic north sensor, gravity acceleration sensors and barometer (for example see Gao and Zhang, 2021;

Kudriashov *et al*, 2020). In addition, we need to have access to several landmark points (minimum three) around the scene that can be used for transforming the local camera coordinates to a world coordinate. As uncertainties from different sensor measurements should be propagate throughout the entire calculation, factor graph based optimisation of the poses can be a helpful tool (Dellaert and Kaess, 2017).

Detecting and expanding planar patches

The planar patches of a point cloud are detected by Random Sample Consensus (RANSAC) algorithm with some adjustment to better suite our objective (Fischler and Bolles, 1981). The implemented RANSAC algorithm requires following parameters:

- approximate diameter of the smallest visible planar patch at the mapping surface (d_{min})
- planarity ratio (λ)
- agglomeration distance (d)
- angular tolerance (θ).

To implement the RANSAC algorithm, a geologist at the site should make observation and estimate the approximate diameter of the smallest planar element in the mapping surface. The selected diameter allows for finding a small planner sections over the point cloud. Moreover, in the developed software tool, it is possible for the user to select a small area on a frame from the 3D movie captured from the scene, if the automated analysis has not managed to capture it. Moreover, in the developed software tool the user has possibility to select a small area on an image from 3D movie.

Planarity of those patches is controlled via principal component analysis in which the eigenvalues and eigenvectors of the covariance matrix of the 3D points located inside the patches are calculated. For a planar patch, the minimum eigenvalue should be much smaller than both the largest and intermediate eigenvalues. The eigenvector corresponding to the minimum eigenvalue is the normal vector of the planar patch. The ratio of the eigenvalues for planarity check can be expressed as a planarity ratio (λ):

$$\lambda = \max \left(\frac{e_3}{e_1} \ and \ \frac{e_3}{e_2} \right)$$

Where e_1, e_2 and e_3 are the maximum, intermediate and minimum eigenvalues of the covariance matrix of the points within the boundaries of a patch contained in a sphere with diameter of d_{min}, respectively. The differences of the coplanarity ration between maximum and intermediate eigenvalues should be less than 10 per cent in a planar patch. This ratio is similar to the roughness or waviness of the joint surface, see Figure 2. Our tests on the different rock surfaces show that the best results is achieved when $\lambda \leq 0.1$. After this step, the algorithm attempts to expand the radius of the planar patch by a ratio of 1–5 per cent of the current radius. The radius of a planar patch is calculated by finding a circumscribed circle that encompasses all points of the patch located on the plane fitted to the patch. If the normal vector of the expanded patch compared to the original patch has angle smaller than the angular tolerance of θ, then the extension of the planar patch is accepted. The process continues until that the expansion of the planar patch radius no longer satisfies the angular tolerance.

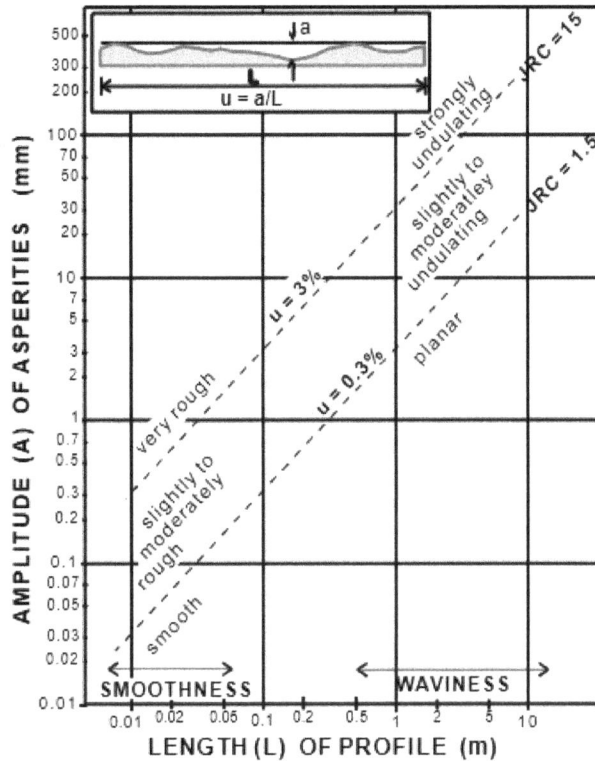

FIG 2 – Rock joint waviness (Palmstrom, 2001).

Mapping discontinuity traces using image processing

To map discontinuities which do not have a visible planar patch in the images, we utilise image processing methods such as edge detection. In the first step, the image transferred to grey scale image, which makes it easy to carry out edge detecting. Then, the image is filtered using Histogram Equaliser Technique (Szeliski, 2011), which is used to remove nonimportant parts of the image and make the intensity of pixels to distribute more uniformly.

Consecutively, after filtering the image, edge detecting algorithm was utilised to find edges in the image. The edges are representing pixels on the images where the intensity of the image pixels changes abruptly. In a simple terminology the edge detector algorithms try to find extrema of the gradient of the image intensity.

Finally, edge pixels which are across from each other, and satisfying the planarity constrains as mentioned before for a planar patch, will be processed. The detected planar patches can be expanded (similar to the planar patches) to include more pixels which are detected as edges by image processing.

Merging coplanar elements

At this stage, all the mapped planar surfaces, whether obtained from the point cloud or through image processing, are presented as planes with a normal vector, centre point and radii. A coplanarity check can be performed across all plane sets with considering by calculating coplanarity angle among them. In this analysis the coplanarity angle threshold is assumed to be slightly larger previous steps. For each set of the combined planes, updated normal vector, centre point and radii will be calculated.

Rock discontinuity network

Eventually, after mapping all visible discontinuity in the scene and storing them as a list of planar circles, it becomes possible to divide them into clusters according to their normal vector orientations. The clustering allows for calculating the best-fitting statistical distributions for each including their orientation (dip and dip-direction), normal spacing and persistence. By utilising Monte Carlo simulation on those previously fitted statistical distribution on the mapping data, user can estimate a statistical distribution of the rock block size in the mapped scene. The detailed mathematical procedure for calculating rock block size distribution is beyond the scope of this manuscript; however, Lu and Latham (1999) provide a comprehensive explanation of those methods.

Example of Implementation

This chapter presents an illustrative example demonstrating the implementation of our developed code for mapping rock mass discontinuities using Zed 2i stereo camera. The study site is a road cut, the fused point cloud from all of the recorded frames from the scene is presented in Figure 3.

FIG 3 – Fused point cloud from all the frames captured by Zed 2i camera from the study site.

The identifying the planar patches revealed numerous planar surfaces across the exposed rock surfaces Figure 4). However, there are several false positives among detected planes (such as planar surfaces which are covered by snow). Therefore, the code provides the user possibility to reject a certain planar patch, or merge several of them manually. Figure 5 shows the traces of the discontinuities identified in an image from the scene. As it is visible, this process might lead to false positives which again user can approve, reject, or merge them.

The rock block size distribution obtained from rock mass mapping is shown in Figure 6. The analysis reveals that rock blocks at the study site have volumes ranging from 2 to 4.9 m³. As a rule of thumb when assessing the rock bolt length sufficient to stabilise the slope, the minimum length can be estimated as:

$$l_{bolt} \cong 0.5 + \sqrt[3]{maximum\ estimated\ block\ volume}$$

This calculation can indicate that rock bolt with minimum length of 2.3 m would be appropriate for the study site.

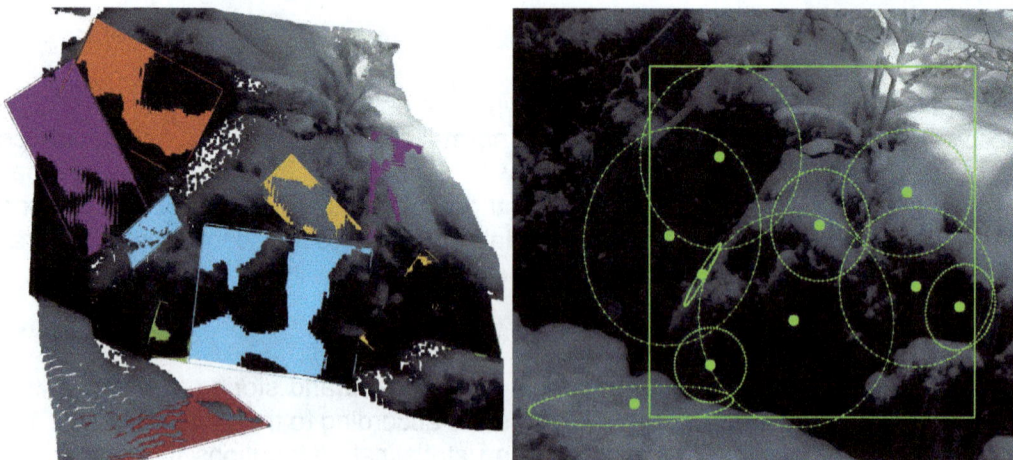

FIG 4 – Mappd planar patches from the point cloud.

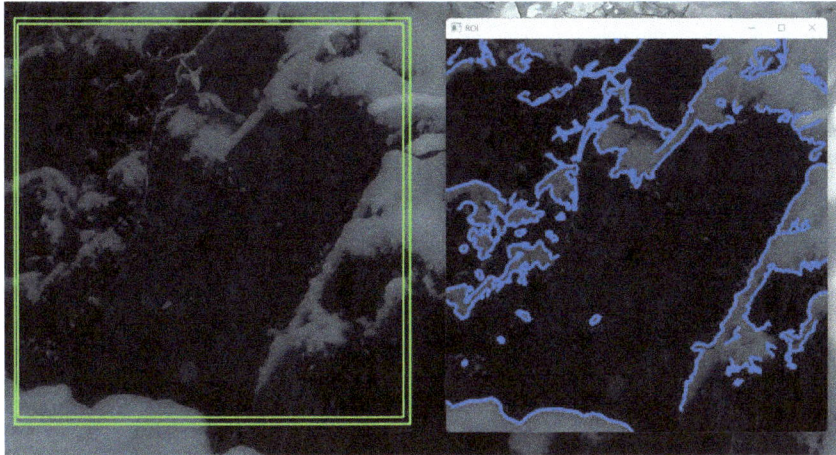

FIG 5 – Mapped traces of the discontinuities.

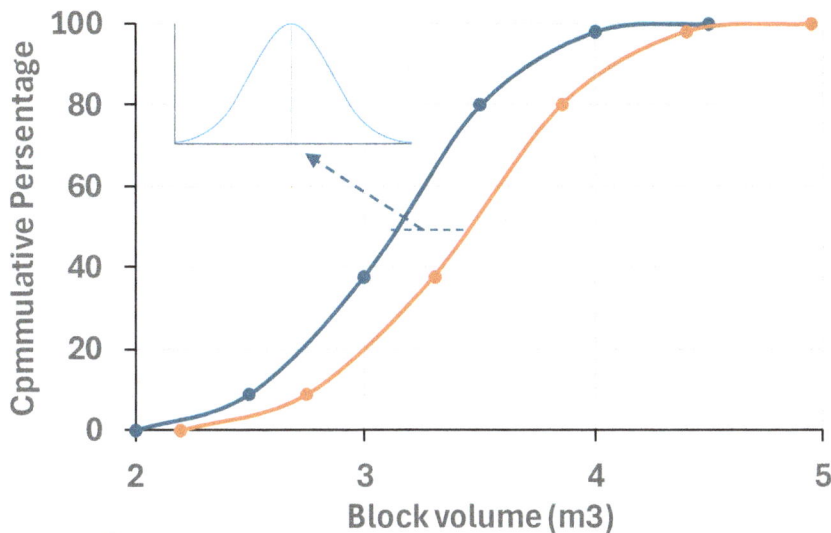

FIG 6 – Estimated distribution of the block sizes at the study site.

CONCLUSIONS

This paper presents a simple and straightforward method for rock mass mapping utilising stereo camera which utilises both the point cloud of the scene and the stored images. The method enables mapping of discontinuities with visible planar patches as well as those without visible planar surfaces in the scene by image processing techniques. In addition, the use of stereo cameras in combination with GNSS sensors or VSLAM allows for simultaneous georeferencing of the mapping results both in the over and underground environments. In addition, the user will have access for manually manipulate the results helping to avoid both false positives and false negatives in the mapping process.

ACKNOWLEDGEMENTS

This work was funded by DINAMINE project. The DINAMINE project is supported by the Horizon Europe research and innovation programme (Grant Agreement No. 101091541). Funded by the European Union. Views and opinions expressed are however those of the author(s) only and do not necessarily reflect those of the European Union or the European Health and Digital Executive Agency (HaDEA). Neither the European Union nor the granting authority can be held responsible for them.

REFERENCES

Battulwar, R, Zare-Naghadehi, M, Emami, E and Sattarvand, J, 2021. A state-of-the-art review of automated extraction of rock mass discontinuity characteristics using three-dimensional surface models, *Journal of Rock Mechanics and Geotechnical Engineering*, 13:920–936.

Dellaert, F and Kaess, M, 2017. Factor graphs for robot perception, *Foundations and Trends in Robotics*, 6(1–2):1–139.

Fischler, M A and Bolles, R C, 1981. A paradigm for model fitting with applications to image analysis and automated cartography, *Communications of the ACM*, 24(6):381–395.

Gao, X and Zhang, T, 2021. *Introduction to Visual SLAM: From Theory to Practice*, 337 p (Springer: Singapore).

Kudriashov, A, Buratoski, T, Gieriegl, M and Malka, P, 2020. *SLAM Techniques Application for Mobile Robot in Rough Terrain*, 131 p (Springer: Cham).

Lu, P and Latham, J P, 1999. Development in the assessment of in-situ block size distribution of rock masses, *Rock Mechanics and Rock Engineering*, 32(1):29–49.

Palmstrøm, A, 2001. Measurement and characterization of rock mass jointing, in *In-situ Characterization of Rocks* (eds: V M Sharma and K R Saxena), pp 49–97 (Balkema: Rotterdam).

Priest, S D, 1993. *Discontinuity Analysis for Rock Engineering*, 473 p (Springer: Dordrecht). https://doi.org/10.1007/978-94-011-1498-1

Stereolabs, 2023. Zed, 2i Stereo Camera [online]. Available from: <https://www.stereolabs.com/en-no> [Accessed: 30 March 2025].

Szeliski, R, 2011. *Computer Vision: Algorithms and Applications* (Springer-Verlag: London).

Yi, X, Feng, W, Wang, D, Yang, R, Hu, Y and Zhou, Y, 2023. An efficient method for extracting and clustering rock mass discontinuities from 3D point clouds, *Acta Geotechnica*, 18:3485–3503.

Real-time monitoring and machine tracking in large-scale underground mines using safety systems and machine learning techniques

A Skoczylas[1], W Koperska[2], P Stefaniak[3] and P Śliwiński[4]

1. Research Specialist, KGHM Cuprum Research and Development Centre Ltd., Wroclaw 53-659, Poland. Email: artur.skoczylas@kghmcuprum.com
2. Research Specialist, KGHM Cuprum Research and Development Centre Ltd., Wroclaw 53-659, Poland. Email: wioletta.koperska@kghmcuprum.com
3. Head of Analytics Department, KGHM Cuprum Research and Development Centre Ltd., Wroclaw 53-659, Poland. Email: pawel.stefaniak@kghmcuprum.com
4. Chief Engineer for Production Analysis and Optimization, KGHM Polska Miedź SA, Lubin 59-301, Poland. Email: pawel.sliwinski@kghm.com

ABSTRACT

Nowadays, one of the main challenges faced by underground mines is achieving real-time monitoring of production processes and machinery movements. Effective monitoring systems are crucial for maintaining situational awareness and ensuring the continuity of operations, particularly in dispersed organisations. This knowledge is essential for safe and efficient extraction, up-to-date production reconciliation, and various operational and planning routines, including the use of specialised simulation environments for optimising production. So far, such monitoring solutions have mostly been implemented in open pit mining or smaller underground operations. This paper introduces a system for tracking and monitoring machinery, utilising data from a safety system specifically designed for underground mining enterprises. Originally, the system was developed to detect potential collisions between machines or between machines and workers. However, the need arose to develop validation algorithms, which included error correction and adaptive filtering. This also required integration with Enterprise Resource Planning (ERP) systems. Additionally, the system's infrastructure was upgraded with extra sensors to allow for the registration of machine locations in specific mining zones, such as the heavy machinery chamber, mining areas, and loading/unloading points. In this study, several analytical models, enhanced with machine learning techniques, were developed to track the movement and interaction of wheeled transport machinery, as well as the overall ore logistics within the mining operation. The paper also covers the system's implementation process in the target environment and provides a description of the user interface, which includes managerial dashboards for production visualisation.

INTRODUCTION

In modern underground mining, achieving real-time awareness of production processes and machinery movements is a key challenge for maintaining operational continuity and optimising productivity. Unlike open pit operations, where visibility and monitoring solutions are more accessible, underground mining requires specialised approaches to track dynamic processes occurring across vast and often inaccessible areas that are difficult to monitor due to structural and technological limitations. Traditional monitoring methods often rely on periodic manual inspections, catalogue-based data, or generalised statistical models, which lack the granularity needed to reflect the real-time state of operations. As a result, decision-making processes remain highly dependent on past performance trends rather than current conditions.

The integration of real-time monitoring solutions plays a crucial role in improving situational awareness, enabling rapid response to emerging issues, and supporting data-driven management strategies. Advanced tracking systems not only provide continuous data acquisition but also serve as the foundation for Decision Support Systems (DSS) and digital twins, which facilitate process optimisation through predictive modelling and simulation. Without such advancements, mining operations are left vulnerable to inefficiencies arising from inaccurate production estimates, unexpected machinery downtime, and suboptimal haulage logistics. In particular, tracking the flow of ore within the transport network over time and space allows for better resource allocation, optimised production cycles, and improved cost control, all of which contribute to the long-term sustainability of underground mining enterprises (Dudycz, Stefaniak and Pyda, 2022; Ranjan, Sahu and Sahu, 2014).

A significant milestone in underground mining was not merely the introduction of control and measurement systems but rather the deployment of broadband telecommunication networks based on fibre-optic technology. This breakthrough enabled a seamless, high-speed exchange of data between underground operations and surface control centres, laying the foundation for modern automation, real-time monitoring, and predictive maintenance. By ensuring continuous data flow, fibre-optic infrastructure has become an essential component of digital transformation in mining, supporting advanced decision-making and enhancing operational control. Before the implementation of real-time data networks, acquiring operational information from haulage machines required extensive manual measurements, often conducted by personnel under challenging conditions. These methods were labour-intensive, prone to human error, and incapable of providing immediate insights necessary for optimising production. The shift toward real-time data acquisition has eliminated these inefficiencies, enabling continuous tracking of key parameters such as machine availability, fuel consumption, transport cycles, and mechanical wear. Additionally, automated data collection has enhanced maintenance scheduling, reducing unplanned machine breakdowns and increasing equipment longevity. Beyond maintenance improvements, real-time monitoring has also revolutionised mining logistics by optimising haulage routes, reducing idle time, and ensuring efficient material flow between excavation points and processing facilities. The ability to dynamically adjust transport strategies based on live data has led to significant reductions in operating costs while improving production output. Furthermore, remote-controlled and semi-autonomous machinery now rely on high-speed data connectivity to ensure safe and efficient operation in hazardous zones where human presence must be minimised. These advancements mark a transformative shift in underground mining, moving towards fully integrated, data-driven management systems that enhance both safety and productivity (Singh, Kumar and Hötzel, 2018; Theissen *et al*, 2023).

Underground mining presents some of the most complex operational challenges due to its confined working conditions, vast underground networks, and the need for precise coordination between multiple processes. Mines often extend for hundreds of kilometres, with excavation zones spread across multiple levels, making effective monitoring and communication essential for safety and efficiency (Clausen *et al*, 2020). Unlike surface operations, where GPS and conventional wireless technologies provide reliable tracking solutions, underground environments introduce significant obstacles to real-time data transmission. Uneven geological structures, poor visibility caused by dust and gas, high humidity, and mechanical vibrations from drilling and blasting all contribute to signal interference and equipment wear. These factors make it crucial to develop monitoring solutions that can withstand harsh mining conditions while maintaining a high level of accuracy and reliability (Skoczylas *et al*, 2023).

Advanced monitoring and communication systems are now essential for ensuring safety, productivity, and cost optimisation in underground mines. Real-time data access has become a fundamental requirement, not only for monitoring personnel and vehicle locations but also for optimising fleet management, scheduling predictive maintenance, and improving decision-making processes. The automation of haulage and excavation processes increasingly depends on uninterrupted data exchange between underground machinery and surface control stations. This trend aligns with the broader digital transformation of the mining industry, which is now centred on Big Data analytics, Artificial Intelligence (AI), automation, robotics, Internet of Things (IoT) applications, and simulation-based modelling. These technologies are redefining how underground mines operate, enabling predictive decision-making, reducing operational risks, and improving overall efficiency (Markham *et al*, 2022). However, despite the progress in digital mining technologies, large-scale integration of real-time monitoring systems remains a challenge. Many solutions focus on small-scale deployments or case studies that do not fully address the complexities of managing large underground operations. There is still a lack of standardised methodologies for obtaining reliable and cost-effective operational data, as well as a need for solutions that automate data fusion from multiple sources. This paper addresses these challenges by introducing an analytical system based on data collected from an anti-collision monitoring framework integrated with self-propelled mining equipment. The system has been tested within KGHM's multi-mine operations, providing an unprecedented scale of research data and operational diversity. The following sections of this article describe the collision-avoidance system, data preprocessing techniques, analytical models developed mainly for assessing operator working time, machine utilisation, and ore haulage cycles. The study also includes a discussion of system outputs and reports developed in collaboration with mining engineers, concluding with recommendations for future advancements in underground mining monitoring technologies.

MATERIAL AND METHODS

The primary objective of this research was to expand the analytical potential of the collision-avoidance (CA) system used in underground mining operations by integrating it with existing ERP systems. This chapter introduces the system, followed by a discussion of data integration methods and the proposed algorithms.

Collision-avoidance system

The research was conducted in the KGHM Polish Copper mines, which are located in the Lower Silesia region of Poland. KGHM operates several mines in this region, which, although treated as separate entities, are interconnected underground and partially overlapping. Collectively, these mines operate hundreds of machines and employ thousands of workers daily. In this environment, a custom-designed collision-avoidance system was implemented to enhance the safety of both machinery and underground personnel.

The collision-avoidance system, launched in those mines, comprised three main modules: passive units, which are integrated into the mandatory headlamps worn by each worker; active units, which are installed on machinery; and a data gateway. In its simplest form, each active unit establishes communication with all passive and active units within its range. The range is determined by the radio signal, which is actively utilised during communication. Through this interaction, the system identifies whether the detected unit is active (mounted on a machine) or passive (worn by a worker). In the case of workers, the system also determines whether they are on foot within the excavation or assigned to another machine, either as an operator or a passenger. The machine operator receives a continuous stream of information regarding all nearby machines and workers who are not assigned to a machine. Additionally, the system offers an option to assign mine personnel as passengers, ensuring they are recognised by other machines as passengers rather than lone workers.

The passive units do not collect any data and serve solely to be detected by active units. In contrast, all active units collect comprehensive data in the form of contact records. Each contact record contains information about the other entity – either the machine's name (for active units) or the worker's ID (for passive units) – along with the contact start and end times. This data is stored within the active unit and transmitted to the database whenever the unit is within range of a data gateway, provided the signal strength allows for successful data transmission.

Data gateways are typically installed in locations that machines visit regularly to ensure a consistent data flow. Two primary types of locations were selected for the placement of data gateways: fuelling stations, where machines refuel and replenish necessary supplies, and heavy machinery chambers (HMC), where machines return after completing servicing or repair operations. The latter location primarily serves as a backup, as not all machines return to the chamber after work, leading to an imbalance between the number of fuelling stations and HMCs. Since there are fewer fuelling stations compared to machine chambers, data collection is more efficient with a smaller number of gateways.

As demonstrated further in this paper, the data collected from this system already provides valuable insights for calculating numerous key performance indicators (KPIs). However, the analytical capabilities can be further enhanced by introducing a third type of unit. In our research, we introduced an additional unit type, referred to as a 'location,' which was implemented by placing passive units in strategically important areas of mining operations. The information generated by the system, combined with the use of this additional type of unit, is presented in the 'Results' section.

Data integration

The primary information collected by the collision-avoidance system is the proximity of contacts occurring within the mine. While this data is highly useful for safety purposes, it is insufficient for comprehensive KPI calculations. Therefore, data integration with the mine's ERP systems was performed. This integration incorporated information regarding the location and type of work being conducted. With these insights, it became possible to establish connections between each machine and its operational environment, such as distinguishing contacts with its operator from those with designated support personnel. Given that each machine logged several hundred to a few thousand events per shift, this integration proved invaluable for creating an efficient management system.

Additional data integration was carried out with systems that collected operational information on feeder activity (crate), CCTV systems installed on some crates, and a system used for haulage cycle detection based on an IMU sensor mounted on the machines. These three sources enhanced the availability and accuracy of the solution used for the automatic detection of haulage cycles over time performed by underground haul trucks and loaders.

Algorithms

The simplified schematic of the system structure, described in the previous subsections, is shown in Figure 1. As mentioned, the machine collects information on contacts with other machines and mining personnel, which is then transferred to the main server via access point connections. On this server, the data is processed daily and integrated with other mine ERP systems. The processed results are made available as reports to the mine management personnel through the company's BI system.

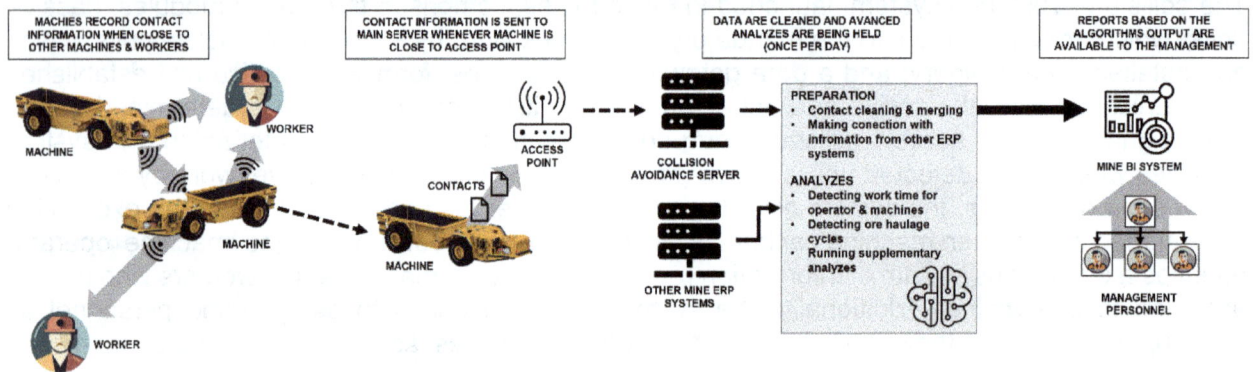

FIG 1 – Simplified schematic of the system's structure, architecture, and operation.

The algorithms implemented in the current stage of the system can be divided into two main modules. The first is responsible for data cleaning and integration with other ERP systems, while the second runs the analyses that generate the desired reports. Since contact data is recorded in a row-based format, the raw readings require processing. During this stage, all events recorded by the machines are sorted, cleaned, and interpolated when necessary. The output of this stage is a column-based data format, where each row represents a contact event between the machine and another object.

The second module runs three main types of analyses. First, the contacts are analysed to establish the relationship between machines and their operators, based on a set of queries and additional implemented logic. The second analysis, regarding the detection of machine work time, is performed similarly, but instead of analysing contacts, the algorithm examines the overall event schema to determine whether the machine was working. The third and most complex analysis involves the detection of ore haulage cycles. During this process, contacts between haul trucks, loaders, and crate sensors are analysed using machine learning algorithms to identify the operations performed during the shift. Whenever possible, data from the CA system is merged with other sources to further enhance detection accuracy.

Additionally, some supplementary analyses are run on user requests or are in the development stage, such as monitoring the fuelling process.

RESULTS

In this section, the currently available reports that demonstrate the potential of the system are presented. Additionally, features that are still in development or planned for future implementation are also described, further highlighting the capabilities and information gain of the approach.

Machine operator work time detection

The most fundamental KPI that can be derived is the total working time of the machine operator. By identifying which operator was assigned to a specific machine, the system can filter relevant contacts and provide precise information on the exact time periods during which the operator was seated in or in close proximity to the machine. Since most operators are required to remain in their machines throughout the shift, their work period typically consists of a single uninterrupted contact, starting near the beginning of the shift and ending close to shift completion. The start and end times of this contact are used to determine the operator's total recorded work time. However, for certain machine types, such as jumbo drills, operators are frequently required to leave the machine to perform external tasks. In such cases, the system detects multiple long-duration contacts with short breaks in between. For these scenarios, the total work time is calculated from the beginning of the first contact to the end of the last.

Instances where operators switched machines or changed assignments were also observed. The system can detect such occurrences by comparing the designated operator's recorded working time with the cumulative contact time of other personnel interacting with the machine. If another worker has a significantly longer contact time with the machine than the assigned operator, it is assumed that an operator switch has likely occurred. This information is highly valuable, as such switches are rarely reported in large mining operations and, therefore, are not reflected in the existing ERP system.

Additionally, if a machine experiences a breakdown or other operational issues occur in the mine, the system may fail to detect any contact with the operator, since the machine is turned off. In such scenarios, the operator's entire working time is typically recorded in the ERP system, while the collision-avoidance system detects significantly shorter working hours. This discrepancy enables the calculation of a wasted time KPI, providing valuable insights into operational inefficiencies.

An example of such a report is shown in Figure 2. All machine-operator interactions are represented as boxes, with width indicating the duration of contact. Additional information is provided for each shift to support managerial tasks. To enhance visibility, alternating shifts are colour-coded in green and blue, while red is used to indicate shifts where a different operator was detected compared to the one assigned in the ERP system.

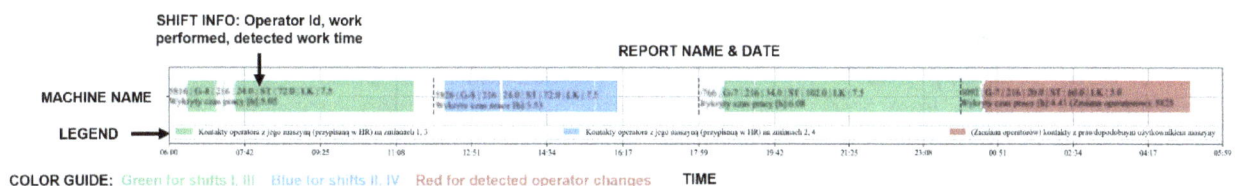

FIG 2 – The structure of the report showing operator work time detection results (for privacy concerns, all information has been blurred out).

Machine work time detection

Machine work-time refers to the period during which the machine is absent from its designated HMC. This can be detected using an additional passive location unit, with work time calculated from the moment the machine loses contact with the sensor at the beginning of the shift until contact is reestablished at the end of the shift.

Furthermore, for certain machines, additional location tags or associations with other machines and personnel can be utilised to refine this KPI. For example, in the case of a haul truck performing ore haulage, the basic work time is calculated from the moment the truck loses contact with the HMC unit until contact is reestablished. However, if a loading machine, a crate operator, or a location unit mounted on the crate is available, the work time can be calculated more precisely – from the beginning of the first ore haulage cycle to the end of the last cycle. By comparing these two work-time measurements, an additional KPI can be established to quantify the travel time from the designated departure point (HMC) to the actual workplace.

Additionally, by tracking contacts with the HMC location unit, it is possible to determine which machines were left at the workplace, as operators sometimes fail to return them to the designated chamber. This tracking capability also enables monitoring of machine routes when their designated department changes, which typically involves a single long-distance transfer – twice if the machine later returns to its original location. Analysing and validating this travel time can provide valuable insights into whether it is more efficient to deploy a machine from another department when no units are immediately available or to accept a temporary loss in productivity.

An example of such a report is shown in Figure 3. Its structure is similar to the operator work time detection report (Figure 2), with the only difference being that the boxes are unicolored. The report does not display contacts directly but instead shows the periods when the machine was working, determined through contact analysis. Since the machine can either be outside in HMC (working) or inside HMC (not working), there is no need for additional colours.

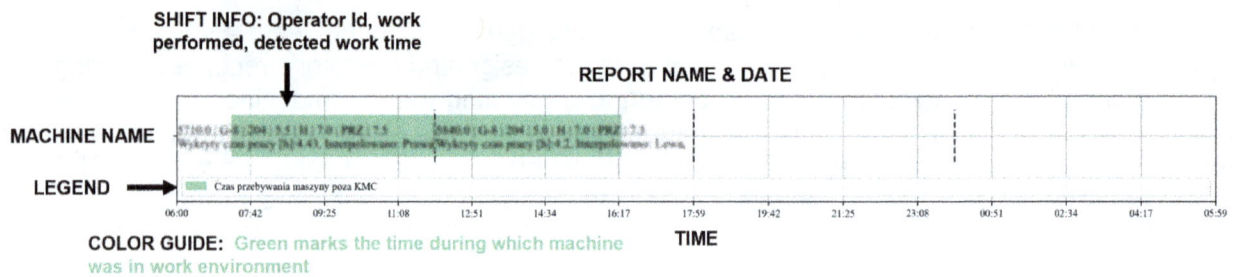

FIG 3 – The structure of the report showing machine work time detection results (for privacy concerns, all information has been blurred out).

Ore haulage cycles detection

The number of cycles performed during ore haulage is a critical KPI for both loader and haul truck operators, as well as their respective management units. Several existing methods are available for cycle detection, primarily utilising data from onboard machine monitoring systems that track parameters such as vehicle speed, engine rev/min, and other operational features – these methods have been implemented using traditional techniques (Gawelski *et al*, 2020) as well as deep learning models (Skoczylas *et al*, 2023).

In addition to onboard monitoring, other sensor-based approaches, such as those utilising IMU sensors, have been explored for cycle detection (Skoczylas *et al*, 2025). Furthermore, computer vision-based methods leveraging camera footage have also been proposed. However, these approaches face limitations, particularly in recognising machinery under challenging underground conditions, where dirt and debris often obscure machine identification numbers, reducing the effectiveness of visual recognition techniques.

Ore haulage cycles can be effectively detected by analysing contacts between specific machines and utilising additional passive localisation tags. For haul trucks, a typical cycle consists of traveling with an empty cargo box to the mining face, being loaded by a loader, transporting the full load to the ore crate, and finally unloading. In contrast, the operation of loaders is generally simpler, as they primarily load haul trucks in a repetitive manner. However, in some regions and mines, loaders are used independently to gather and transport ore without involving haul trucks, a scenario that typically occurs only when the hauling distance is shorter.

In both cases, data from the collision-avoidance system can be leveraged to efficiently measure the ore haulage cycles. The loading process can be detected by analysing contact events between haul trucks and loaders, supplemented by information from the ERP system to establish associations based on the designated work region. However, detecting the loading process in scenarios where loaders operate independently is more challenging and can only be achieved by installing passive location units near the mining face where ore is gathered.

The unloading operation can be identified through two potential methods. If personnel are present at the crate – typically operating the crushing device on the crate – their contact with the haul truck can indicate unloading. Alternatively, in the absence of personnel, passive location units installed at each crate can signal the unloading event. Experimental results indicate that both methods provide comparable accuracy and reliability in detecting unloading operations.

While detecting ore haulage cycles using the CA system is generally efficient and accurate, occasional errors can occur. It is common for haulage machines to encounter multiple loaders or pass near different crates, which can make it challenging to distinguish between actual unloading events and drive-through occurrences. Since the unloading process is typically very brief, misclassification of contacts may arise.

To address these challenges, integrating data from other sources can significantly enhance the accuracy and reliability of cycle detection. For instance, crate feeder operation data can be analysed to identify active unloading periods, helping to eliminate drive-through events. Furthermore, ore haulage contact data can be cross-referenced with additional sources such as CCTV recordings, IMU-based cycle detection, and machine signal-based cycle detection. This multi-source data fusion approach ensures a more robust and precise measurement of haulage cycles, minimising errors and improving operational insights.

An example of such a report is shown in Figure 4. The report is primarily designed from the haul truck's perspective, visualising contacts with loaders (loading operations) in contrast to contacts with the crate (unloading operations). Since a truck may interact with multiple loaders during loading, all contacts are displayed to enhance managerial insight. The plot is colour-coded based on cycle detection results, with green backgrounds indicating detected cycles.

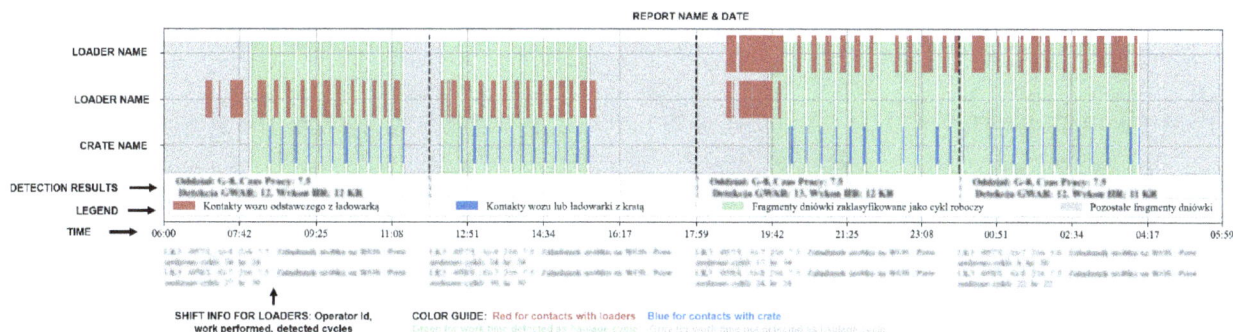

FIG 4 – The structure of the report showing ore haulage cycles from the haul truck perspective (for privacy concerns, all information has been blurred out).

Refuelling time detection

All machines require fuel and other essential resources, such as water and hydraulic fluids, to function properly. These necessities are typically replenished at designated refuelling stations. As previously mentioned, refuelling stations are scarce; therefore, to ensure efficient operations, the time each machine spends at the station should be minimised. By utilising passive location units installed within refuelling stations, it is possible to accurately measure the time each machine spends refuelling. This data can be used to verify refuelling efficiency or serve as a metric to identify inefficiencies within the process.

It is not uncommon to observe queues of machines waiting near refuelling stations for their turn. These queues can be identified by first analysing the contact records of machines within the station using passive tags and then tracing the contact sequences between machines and the station's passive unit. This process can be further extended to detect the entire queue at any given moment.

Such analysis enables the calculation of queue time periods during which machines remain idle while waiting to refuel. This information serves as a valuable tool for optimisation, potentially leading to the development of structured refuelling schedules to reduce downtime and improve operational efficiency.

A distinct case of fuelling operations can be identified in mines equipped with specialised fuelling vehicles that travel throughout the mine to refuel other machines. In this scenario, fuelling times can be determined by analysing contact events between these fuelling vehicles and other machinery. If a contact occurs as a single instance and lasts for several minutes, there is a high probability that it corresponds to a refuelling operation. This approach enables efficient tracking of refuelling activities without requiring additional sensors, providing facilitation to current reporting system and fleet operational efficiency monitoring.

An example of such a report is shown in Figure 5. The plot primarily visualises the contact duration between machines and the sensor/gateway inside the refuelling chamber, indicating close proximity and potential fuelling operations. Additionally, queue detection results are highlighted in orange, representing moments when the machine was waiting outside the refuelling chamber.

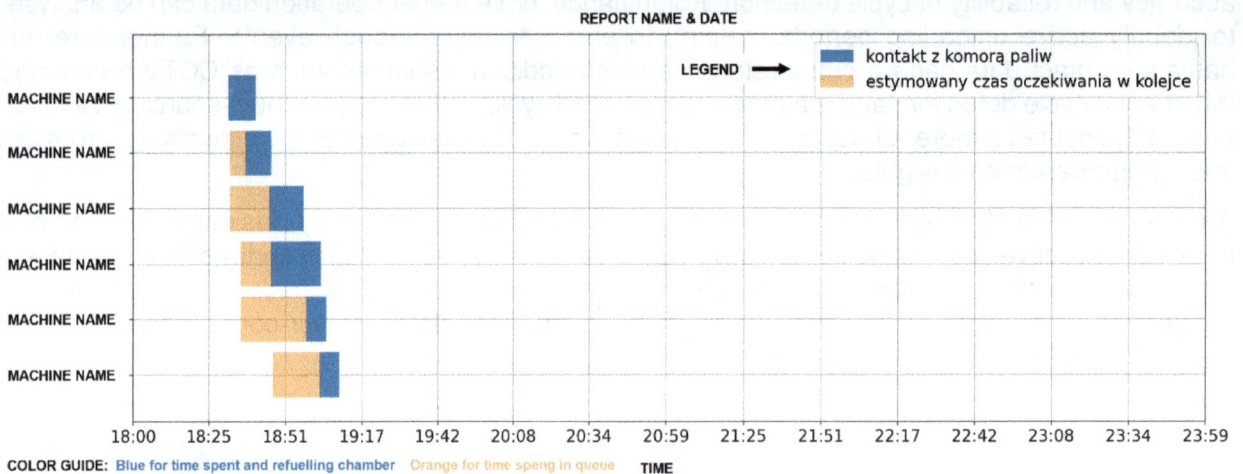

FIG 5 – The structure of the report showing the fuelling process and queues (for privacy concerns, all information has been blurred out).

Personal work time segmentation

Based on the previously described analysis, a comprehensive timesheet of a worker's shift can be generated, which is a common practice in underground mining operations. In an underground mine, the shift begins with the workers descending via the shaft and arriving at the bus stop, where they board personnel transport vehicles. These vehicles are equipped with active units, allowing for the calculation of the time taken to travel from the shaft to the bus stop by measuring the interval between the shaft descent time and the worker-bus connection time.

The subsequent travel to the designated HMC is recorded based on the worker's contact with the bus. Upon arrival at the HMC, personnel typically participate in a start-up meeting where they are briefed on their tasks before preparing their assigned machines for operation. The duration of this contact can be calculated as the time interval between leaving the bus and the first machine start-up event.

The preparation time is then determined from the moment of the machine's first start-up to the point at which it loses contact with the HMC location unit, indicating departure for work. Once the machine leaves for its designated task, as previously described in the work-time detection process, passive location units or links with other machines and personnel can be utilised to differentiate between effective travel and actual work time. Additionally, for certain machines, such as those involved in ore haulage, work can be further divided into operational cycles, as outlined in the ore haulage cycle detection methodology.

Upon task completion, the operator may return to the HMC or request transportation directly to the shaft. In the first scenario, the entire start-up process is reversed, while in the second, long contact durations with other machines can be analysed to identify transportation arrangements. In some cases, personnel may return to the shaft on foot, with the walking time calculated from the point of leaving the machine to the ascent via the shaft.

Personnel work time segmentation is currently in development. The system is already capable of supporting such analyses, as most required modules are operational in other areas. However, since implementation is still in its early stages, no dedicated report is available yet.

Regulations violation detection

Underground mines are considered hazardous environments and are therefore subject to strict regulations imposed by both mining enterprises and regulatory authorities. These regulations clearly define permitted and prohibited activities for personnel. By analysing machine and personnel contact data, it is possible to identify instances of non-compliance with these regulations.

Firstly, most machine operators are required to adhere strict restrictions regarding leaving their vehicles while in the working environment. Contact data between operators and machines can sometimes reveal instances where an operator has exited the vehicle, which appear as interruptions within what should be a continuous contact period, as previously discussed in the machine operator time detection process. Other prohibited activities for operators include, for example, unjustified passing through a crate or unloading waste rock, which can also be easily identified in the data.

Secondly, access to certain areas within the mine is restricted for safety or operational reasons. To ensure compliance with these restrictions, passive location units can be installed in designated areas. If an operator drives a machine into a restricted zone, the system will automatically log the event, allowing for the identification of such incidents through automated queries. This approach provides an effective means of monitoring and enforcing compliance with mine safety protocols.

Currently, no dedicated report exists, but some violations can be identified in other reports. An example is shown in Figure 6, where breaks between machine-operator contacts were detected in two different shifts. These breaks suggest that the machine was either turned off or left unattended by the operator, indicating a potential violation.

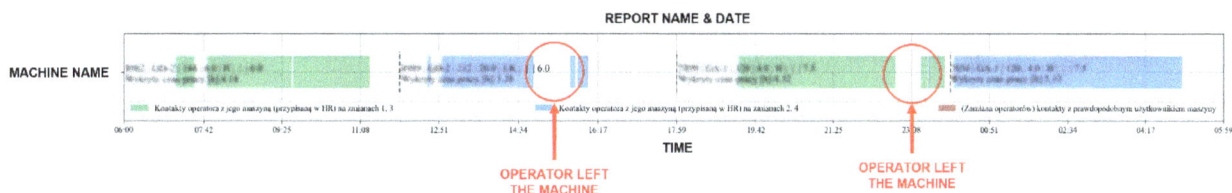

FIG 6 – The structure of the report showing operator work time detection results (for privacy concerns, all information has been blurred out).

Advanced machine localisation

Machine localisation in underground mines is crucial for safety purposes and can also provide valuable insights into operational efficiency. However, accurately determining machine positions in such environments is challenging, as many commonly used localisation technologies are impractical underground. A reliable positioning estimation can be achieved by utilising strategically placed passive location units that interact with machines as they travel. In this approach, localisation accuracy is directly dependent on the number and placement of these sensors. Larger mines with multiple possible routes may require a higher sensor density, making the solution highly site-specific.

By analysing the recorded routes of machines, precise travel time estimations within the mine can be obtained. Additionally, alternative technologies can be integrated to enhance localisation accuracy and provide deeper operational insights. For instance, IMU sensors mounted on machines can be used to assess road conditions by identifying degraded sections that require maintenance (Skoczylas *et al*, 2020). Furthermore, with known travel routes, it becomes possible to evaluate operator driving styles (Stefaniak *et al*, 2023), enabling the identification of individuals who may benefit from additional training to improve driving efficiency and safety.

Advanced machine localisation is considered a potential future improvement for the system, but a dedicated report has not yet been developed. Implementing such analyses is a long-term effort, requiring significant resources for development. However, we believe it is achievable and would provide valuable insights for mine management.

CONCLUSIONS

This study introduced an advanced monitoring system that enhances operational efficiency, safety, and decision-making in underground mining. By integrating data from an anti-collision system with tracking of self-propelled equipment, the research provided insights into machinery movements, ore haulage cycles, and operator efficiency. Tested in KGHM's multi-mine operations, the system demonstrated its ability to improve transport logistics and optimise resource utilisation. A key contribution of this work is the integration of data acquisition with ERP systems, moving beyond traditional catalogue-based estimates. The system enables continuous performance assessment, offering precise analytical models for machine work time, and transport cycle efficiency, supporting predictive maintenance and workflow optimisation. However, large-scale deployment poses challenges, including infrastructure investments in broadband communication networks and the need for robust data validation. The proposed system and algorithms can be deployed in real-time, further enhancing decision support capabilities. In our experimental set-up, data was collected via mine-installed gateways, resulting in a delay of approximately one day. Current efforts aim to reduce this delay to a single shift. However, in newer mines equipped with LTE networks, gateways are unnecessary, enabling real-time collection and processing of machine contact data. Future research should refine data fusion methods, improve localisation accuracy, and explore IoT-based sensors to further enhance system capabilities. The study underscores the importance of digitisation in mining, highlighting the need for real-time, scalable monitoring solutions. As the industry advances towards automation, AI-driven analytics, and IoT connectivity, intelligent monitoring frameworks will be essential for improving safety, efficiency, and sustainability.

REFERENCES

Clausen, E, Sörensen, A M A, Uth, F D, Mitra, R, Schwarze, B and Lehnen, F, 2020. Assessment of the Effects of Global Digitalization Trends on Sustainability in Mining: Part I, Digitalization Processes in the Mining Industry in the Context of Sustainability. Available from: <https://publications.rwth-aachen.de/record/809271> [Accessed: 19 March 2023].

Dudycz, H, Stefaniak, P and Pyda, P, 2022. Problems and challenges related to advanced data analysis in multi-site enterprises, *Vietnam Journal of Computer Science*, 9(01):1–17.

Gawelski, D, Jachnik, B, Stefaniak, P and Skoczylas, A, 2020. Haul truck cycle identification using support vector machine and DBSCAN models, in *Proceedings of the Advances in Computational Collective Intelligence, 12th International Conference (ICCCI 2020)*, pp 338–350 (Springer International Publishing).

Markham, G, Seiler, K M, Balamurali, M and Hill, A J, 2022. Load-Haul Cycle Segmentation with Hidden Semi-Markov Models, in *Proceedings of the 2022 IEEE 18th International Conference on Automation Science and Engineering (CASE)*, pp 447–454.

Ranjan, A, Sahu, H and Sahu, H B, 2014. Communications challenges in underground mines, *Communications*, 5(2):23–29.

Singh, A, Kumar, D and Hötzel, J, 2018. IoT Based information and communication system for enhancing underground mines safety and productivity: Genesis, taxonomy and open issues, *Ad Hoc Networks*, 78:115–129.

Skoczylas, A, Rot, A, Stefaniak, P and Śliwiński, P, 2023. Haulage Cycles Identification for Wheeled Transport in Underground Mine Using Neural Networks, *Sensors*, 23(3):1331.

Skoczylas, A, Stefaniak, P, Anufriiev, S and Jachnik, B, 2020. Road Quality Classification Adaptive to Vehicle Speed Based on Driving Data from Heavy Duty Mining Vehicles, in *International Conference on Intelligent Computing and Optimization*, pp 777–787 (Cham: Springer International Publishing).

Skoczylas, A, Stefaniak, P, Anufriiev, S and Koperska, W, 2025. Deep Learning for Ore Haulage Monitoring: Vibrational Analysis Using a VGG16 Network, *IEEE Access*.

Stefaniak, P, Koperska, W, Skoczylas, A, Witulska, J and Śliwiński, P, 2023. Methods of optimization of mining operations in a deep mine–tracking the dynamic overloads using IoT sensor, *IEEE Access*.

Theissen, M, Kern, L, Hartmann, T and Clausen, E, 2023. Use-Case-Oriented Evaluation of Wireless Communication Technologies for Advanced Underground Mining Operations, *Sensors*, 23(7):3537.

Optimisation of mining machinery maintenance in modern mining enterprises through text mining and machine learning techniques

M Stachowiak[1], W Koperska[2], P Stefaniak[3] and P Śliwiński[4]

1. Research Specialist, KGHM Cuprum Research and Development Centre Ltd., Wroclaw 53-659, Poland. Email: maria.stachowiak@kghmcuprum.com
2. Research Specialist, KGHM Cuprum Research and Development Centre Ltd., Wroclaw 53-659, Poland. Email: wioletta.koperska@kghmcuprum.com
3. Head of Analytics Department, KGHM Cuprum Research and Development Centre Ltd., Wroclaw 53-659, Poland. Email: pawel.stefaniak@kghmcuprum.com
4. Chief Engineer for Production Analysis and Optimization, KGHM Polska Miedz SA, Lubin 59-301, Poland. Email:pawel.sliwinski@kghm.com

ABSTRACT

The efficiency of planning processes is fundamental to the success of modern mining enterprises, particularly in maintaining competitiveness in the raw materials market. This paper addresses the critical challenge of optimising production lines, encompassing strategic arrangement of mining fronts and the management of drilling, transport, and auxiliary machinery. Given the dynamic and often challenging conditions of underground mining, mitigating random disruptions that can lead to machine downtime is crucial for sustaining productivity. This study highlights the importance of leveraging large volumes of diverse data to achieve situational awareness and facilitate informed decision-making under uncertainty. We explore how text mining and machine learning techniques can extract valuable insights from unstructured data, such as equipment failure reports, which often contain complex and ambiguous information. By developing a classification system that categorises failure descriptions into actionable insights, we aim to improve data interpretation and support predictive maintenance strategies. Furthermore, we propose an integrated approach that consolidates data from various sources, including downtime logs and repair records, to establish a comprehensive database for analysis. The proposed methods not only streamline data management but also enhance the accuracy of predictive models, ultimately enabling mining companies to optimise their operations and increase overall productivity.

INTRODUCTION

Given the ever-changing nature of the extraction process and the often challenging operating conditions in most underground mines, one of the key aspects of mine management is addressing the random factors that can slow down or disrupt production. A major challenge in this area is reducing machine downtime caused by equipment failures or incorrect set-up of transport machinery, which can create production bottlenecks (Stefaniak et al, 2023; Skoczylas et al, 2025). For smaller mining operations, achieving situational awareness is not particularly difficult, as data validation, standardisation, and analysis can be handled manually by a small team. However, the situation becomes more complex for large, multi-site operations with extensive fleets of machines.

An intriguing solution was presented in Balaraju, Govinda Raj and Murthy (2020), where the authors aimed to uncover the underlying relationships leading to critical potential failures. They applied statistical reliability analysis to assess the performance of mining trucks, using the Kolmogorov-Smirnov test to identify the best-fitting data distribution. Subsystem reliability metrics for each subsystem of LHD machine were estimated using a Reliability Block Diagram (RBD) approach, structured in a series configuration, to determine the overall reliability of the LHD machine. Another study by Barabady and Kumar (2008) used reliability analysis to identify production bottlenecks and components or subsystems with low reliability compared to their designed expectations. This case study focused on a crusher reliability assessment in a bauxite mine in Iran, revealing that the conveyor subsystem and secondary screen subsystem were critical for the crusher's reliability, while the conveyor subsystem and secondary crusher subsystem were essential for the machine's availability. In Ghodrati, Hoseinie and Kumar (2018), the authors developed a method to estimate the Mean Residual Life (MRL) for mining equipment, aiming to optimise maintenance planning. They employed a statistical modelling approach to estimate the MRL, incorporating a Weibull proportional hazard model (PHM) with time-independent covariates to model the hazard function. The method was verified using historical failure data from the hydraulic system of an LHD machine in a Swedish mine. A similar study in Rahimdel et al (2013) analysed the reliability of drilling machines in a copper mine in Iran, where failures led to delays in blasting operations. The authors used the Markov method for their reliability analysis. In contrast, Paithankar and Chatterjee (2018) highlighted the limitations of Markov models in estimating Remaining Useful Life (RUL), particularly the assumption of a specific statistical distribution for failure time data. The authors proposed a hybrid approach for RUL estimation based on a neural network and genetic algorithm. Their case study focused on an LHD machine, showing the developed method's superiority over traditional techniques. Martinsen et al (2023) explored the reliability assessment of mining process automation using Bayesian networks (BN). The experimental research demonstrated that autonomous haulage route planning can outperformed expert-designed routes. However, it also revealed that autonomous loading and haulage machines had shorter lifespans for tires, brakes, and bearings. A similar study on a fleet of haulage trucks operating in mining environments was presented in Rahimdel (2024). In Jakkula and Ch (2020), reliability analysis results for LHD machines were presented, using a renewal approach to evaluate fleet performance. The Kolmogorov-Smirnov test was applied to match the distribution to the data, and reliability for each subsystem was calculated according to the best-fitting distribution. Preventive maintenance schedules were developed based on these results, targeting a reliability level of 90 per cent. This study focused on evaluating the performance of four highly mechanised LHD systems using reliability, availability, and maintainability (RAM) modelling, identifying the causes of performance decline for each machine and offering recommendations to improve the efficiency of this capital-intensive equipment. In Özfırat, Yetkin and Özfırat (2019), the authors proposed a method for identifying bottlenecks in the circular haulage process due to LHD machine failures. A risk index was used to develop a repair scheduling method, employing Failure Modes and Effects Analysis (FMEA) to assess the associated risks.

Data can be collected in various formats, with different standards applied during the collection process. In the case of data related to failures and repairs, even when recorded using forms in ERP systems, the data often consists of complex, unstructured textual descriptions (Stachowiak et al, 2021). Analysing this type of data is challenging and requires additional processing. In Blanco et al, (2019) the authors propose a method for analysing failure downtimes using wind turbines as an example. In practice, assessing the condition of turbines begins with an expert reviewing the service history, which is a time-consuming task because the expert must examine each entry individually. To automate this process, the authors used text-mining techniques and developed classifiers to assist the expert, allowing them to focus on analysing turbine systems and subsystems to optimise their performance. Meanwhile, the study in Brodny et al (2017) applied the Overall Equipment Effectiveness (OEE) model to assess the availability of selected mining machines. The analysis relied on SCADA logs as the primary data source, from which the authors evaluated the availability of machines in the mechanised longwall system of an underground coalmine. Dindarloo (2016) proposed a novel approach to estimating the Time Between Failures (TBF) for Load-Haul-Dump (LHD) machines using support vector machine regression based on a genetic algorithm. A different approach was presented in Sellami et al (2020), where the authors developed a model to predict operational events and safe operating times based on chronicles identified in the data set using unique sequential patterns over time. A key part of the analysis was identifying frequent event sequences that contributed to the most significant downtimes and then determining the time between these events to predict the one that triggers a failure. The authors suggested using the LSTM deep learning algorithm to predict failures and the remaining time to failure.

Text-mining techniques can extract valuable insights from data sources that were previously considered useless or irrelevant because they were not connected to other data (Stefaniak et al, 2022). The paper highlights an important application of these techniques for analysing text data from failure and repair monitoring of mining machines. By combining information about faults and repairs, a complete history of the machine can be created, enabling the calculation of reliability indicators for the machines and their components based on the classification developed by the authors. These indicators are derived from reliability formulas based on the register of shift states of mining machines.

DATA PREPARATION AND MERGING

The data in this research was collected as part of the NetHelix project (2022). The primary data source consists of textual notes detailing condition irregularities, failures, and service work performed. This data is unstructured and requires thorough analysis, including proper validation and classification based on the specific structure of systems and components. The work involves applying advanced text mining techniques and developing procedures to decompose complex entries into distinct activities, standardise them into a consistent and accurate format for seamless integration with other data sources, and enable further classification. The classification process must encompass critical dimensions such as the system, component, location of the element, symptoms or causes of irregularities, and the type of service work performed.

The data originates from a communication platform that facilitates information exchange within the mine, specifically focusing on operational oversight and production reporting. The system was implemented with the primary objective of reducing workload and enhancing communication across the mine. It serves critical functions across key areas of the mine's operations, including personnel and machinery management, while ensuring seamless electronic information flow-throughout all mining sites. This data contains text notes about the work performed on machines during their operation and technical breaks. In addition, there is a database that contains detailed information about the machine's activities during each work shift, categorised into seven distinct statuses:

1. BS – breakdown stop.
2. ERP – emergency repair.
3. PRP – planned repair.
4. PRD – in production.
5. PR – planned readiness.
6. UPR – unplanned readiness.
7. PS – periodic service.

This can be used to comprehensively track the entire machine history. The next step was to figure out how to connect all these information. Because broken parts can be repaired at different times and for varying durations, it was impossible to connect the databases directly by dates and machine codes. The solution here was to use text mining tools with machine learning. Machine learning was applied to divide the information into categories while text mining tools allowed for proper text preparation. Since the Greek language is characterised by extensive inflection, usually each individual word undergoes case inflection. Additionally, the presence of mining jargon, spelling and stylistic errors, synonyms, and proper names further complicates text processing. For this reason, the results of the qualitative and quantitative analysis were used to develop a specialised dictionary of concepts and to define the target structure of the systems and classification rules. As a result, the following categories were proposed:

- System
- Detailed component/element
- Type of fault
- Work performed
- Cause/symptom
- Location
- Description of fault.

The adopted classification required the creation of five machine learning models for first five categories and two procedures based on logical rules defined for the various combinations of expressions. The seven-stage classification process is preceded by validation algorithms including, among others, lemmatisation, removal of unnecessary words, punctuation correction, capitalisation standardisation,, autocorrection, and text-to-numerical conversion. All corrections and changes of words written in jargon help to bring the words to their correct version. Then, lemmatisation allows to reduce words to one simple form. Words prepared in this way will be properly understood and connected by the model. As a result, a short text note is transformed into categorised information that enables the integration of different databases.

Connecting data sources about failures repair work and shift statuses allows to obtain complete information about history of the machine. The previously proposed categories, assigned to each row of information, allow the combination of information from different databases into a single data set. The linking method involves the following steps:

- Check whether failure and repair entries from the same shift for a machine have been classified in the same way. If so, the cases are merged into single entry.

- Join shift statuses occurring on a given day for a given machine to the corresponding entry.

- Verify whether failure and repair cases are divided into multiple entries or if both occur within the same shift but to different elements. If this happens, the working day is divided into as many segments as there are failures and repairs, distributing time evenly.

Data prepared and combined in this way provides a comprehensive overview of a machine's operational history and enables tracking of all machine-related events, determination of reliability and operational indicators, calculation of the repair duration and assessment of downtime following failure.

RELIABILITY ANALYSIS

The integration of data enabled the calculation of operational and reliability indicators for a selected component, machine system, or group of elements. Using recorded statuses, time spent by the selected group on repairs or stops can be calculated. The formulas are shown below:

$$\text{REPAIRS} = \sum(\text{PRD, ERP}) / \sum(\text{BS, ERP, PRP, PRD, PR, UPR, PS}) \cdot 100\% \tag{1}$$

$$\text{STOP IN BREAKDOWN} = \sum(\text{BS}) / \sum(\text{BS, ERP, PRP, PRD, PR, UPR, PS}) \cdot 100\% \tag{2}$$

$$\text{UNPLANNED DOWNTIME} = \sum(\text{BS, ERP}) / \sum(\text{BS, ERP, PRP, PRD, PR, UPR, PS}) \cdot 100\% \tag{3}$$

By integrating status information with the identification of failure start and end times, derived from the fusion of failure and repair databases, it is possible to determined reliability indicators. The calculated reliability indicators include the Mean Time to Failure (*MTTF*), Mean Time to Repair (*MTTR*), and Mean Time Between Failures (*MTBF*):

$$\text{MTTF} = \sum(\text{PRD}) / \#\text{BLOCK E} \tag{4}$$

$$\text{MTTR} = \sum(\text{ERP,BS}) / \#\text{BLOCK M} \tag{5}$$

$$\text{MTBF} = \text{MTTF} + \text{MTTR} \tag{6}$$

where *BLOCK E* represents the number of periods of continuous operation for a given component or machine system (*PRD* status) and *BLOCK M* represents the number of periods of continuous malfunction for a given component or machine system (*BS* or *ERP* status). The logic behind of determining the elements $BLOCK\ E_i$ and $BLOCK\ M_i$ of the sets is presented schematically in Figure 1.

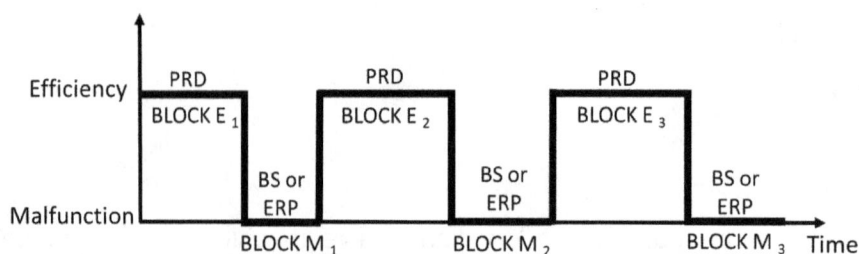

FIG 1 – Machine operation diagram.

Determining the start and end points of blocks was only possible after integrating the data. With the help of the MTBF indicator, the probability of survival and the probability of failure can be determined. They are described by the following formulas:

$$R(t) = e^{-\lambda t} \tag{7}$$

$$F(t) = 1 - e^{-\lambda t} \tag{8}$$

where t is time, λ is failure rate (Meeker, Escobar and Pascual, 2022). The formulas above are based on the assumption of a constant failure rate, which holds true under normal machine operation, excluding the initial break-in and wear-out phases. Therefore, for mining loading and haulage machines, it is reasonable to assume a constant failure rate, which can be estimated as follows:

$$\lambda = 1/MTBF \tag{9}$$

$$P = 1 - e^{-T/MTBF} \tag{10}$$

where T is the failure-free operation time from the last component failure t_0 to the last work shift t_n: $T = \sum_{t_0}^{t_n} PRD$. The *MTBF* indicator value is determined by selecting an appropriate group of machines with a similar expected failure rate for a given component. This group can be defined by machine type and the mining division to which it belongs. To determine the coefficient correctly, it is important to use data from a sufficiently long period to account for the recurrence of failures that occur infrequently.

RESULTS

The following analysis covers data collected over an 11-month period, detailing the failures and repairs of two types of machines across multiple mining divisions. The data has been processed and integrated as described in the previous section. The results presented here offer an evaluation of the reliability indicators for both machine types, highlighting differences across various systems. Additionally, the analysis includes a detailed examination of the Mean Time Between Failures (MTBF) coefficient, observing its fluctuations over time. The analysis also explores the MTBF values for identical machine components, but differentiated by their positions on the machines, offering a deeper understanding of performance variations based on component location. First, Table 1 presents the reliability indicator values for each designated system for both types of machines.

TABLE 1
Reliability indicators for each system.

Machine	System	Repairs	Stop in breakdowns	Unplanned downtime	MTTF	MTTR	MTBF
Type 1	drive system	4.24%	0.31%	0.64%	59.74	0.78	60.52
Type 2		3.28%	0.18%	0.39%	108.92	0.87	109.80
Type 1	construction elements	2.79%	0.18%	0.31%	91.38	0.62	92.00
Type 2		2.85%	0.13%	0.20%	143.12	0.66	143.78
Type 1	fire extinguishing installation	0.18%	0.01%	0.02%	316.60	0.35	316.95
Type 2		0.22%	0.00%	0.02%	378.72	0.64	379.36
Type 1	electrical installation	1.77%	0.07%	0.18%	111.07	0.47	111.55
Type 2		0.84%	0.09%	0.14%	206.00	0.84	206.84
Type 1	cooling system	0.77%	0.05%	0.11%	202.05	0.69	202.74
Type 2		1.06%	0.09%	0.15%	231.15	1.11	232.26
Type 1	brake system	0.59%	0.04%	0.10%	197.10	0.66	197.76
Type 2		0.46%	0.02%	0.07%	270.52	0.69	271.21
Type 1	air conditioning system	0.95%	0.08%	0.17%	131.40	0.56	131.96
Type 2		0.68%	0.05%	0.10%	236.70	0.75	237.45
Type 1	hydraulic system	0.74%	0.05%	0.13%	176.35	0.69	177.04
Type 2		1.19%	0.05%	0.11%	198.52	0.64	199.16
Type 1	actuator system	1.52%	0.13%	0.26%	102.31	0.60	102.91
Type 2		2.00%	0.17%	0.27%	135.26	0.84	136.10
Type 1	wheel system	1.95%	0.11%	0.24%	106.65	0.59	107.24
Type 2		1.28%	0.15%	0.21%	165.77	0.88	166.65

As can be seen in Table 1, the Type 1 machine experiences failures more frequently across each system. In some cases, differences are minor, as in the case of the hydraulic system, while in others, they are significant, such as the electrical system, which fails twice as often. The most critical system from repair perspective appears to be the drive system, which has highest total unplanned downtime. It is also the system with the most frequent failures. Conversely, the fire extinguishing system is the least prone to failures.

Similarly, these statistics can be calculated for more precise elements on machines. Due to the large number of components,, this article does not present a complete table. However, the results indicate that the most common failures in both machines occur in tires and cylinders. Among Type 2 machine-specific components, the bucket is the most failure-prone, while for the Type 1 machine, the most frequently failing components are various covers and joints. The most time-consuming repairs involve tires and the gearbox. Unique components requiring extensive repairs include the cargo box in Type 1 machines and the bucket in Type 2 machines. The analysis can be further refined by adding additional information. In some cases, certain components exist in multiple instances on the machine. Sticking to the case that occurs quite often, ie tires from wheels, they can be further separated by adding information about their position on the machine. Figure 2 illustrates the failure probability function for each wheel as a function of shifts worked.

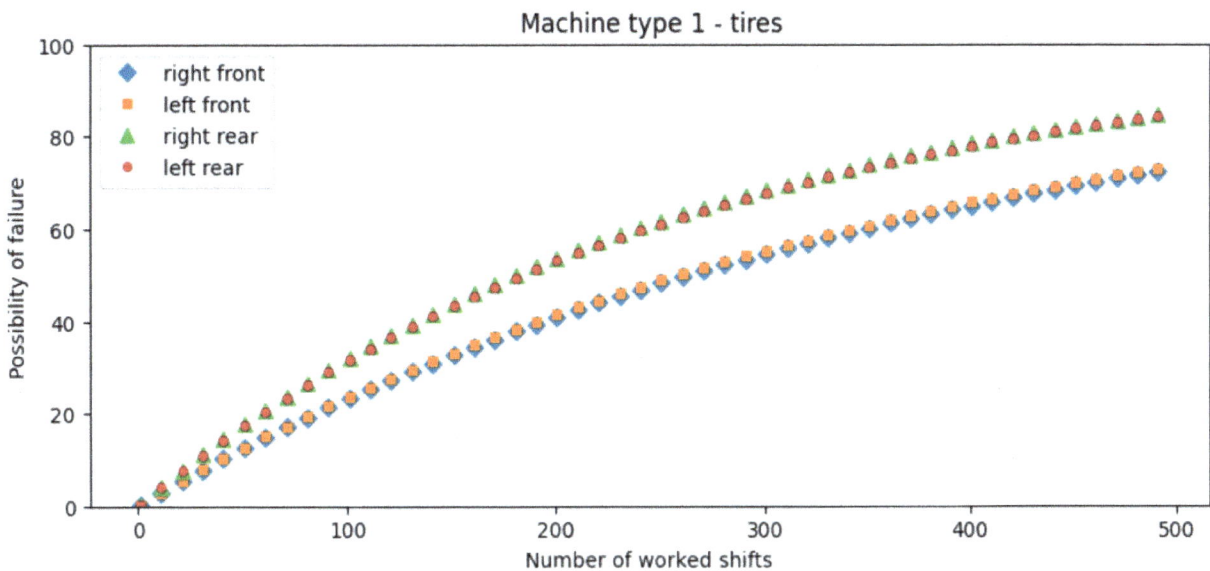

FIG 2 – Likelihood of failure over time for different tires on Type 1 machine.

Figure 2 shows a predictable increase in the likelihood of failure over time for different tires on the machine. Left and right rear tires fails the fastest, which may indicate higher stress on these components. Front tires show a slower increase in failure probability, suggesting better durability or less strain under normal operating conditions. This may be caused by uneven load distribution could place greater strain on the rear tires. The result obtained can be used to increase inspection and maintenance frequency after, for example 200 shifts, with greater focus on rear tires or using by leveraging an existing system to highlight critical areas. The mine conducts periodic tire inspection, and the results can help identify which machines and tires require more thorough examination. This is possible thanks to information about the last recorded failure of a specific component on the machine. Expanding on this idea, a list of potential risk areas can be prepared for mechanics to check during their work shift. This will help distribute the workload more effectively, focusing efforts on high-risk areas.

Table 2 shows three components with the highest failure probability for three selected machines from Machine Group 1. As can be seen, all failures occurred most recently in the first half of last year. A list prepared in this way for all machines within the given departments helps identify components that may require more detailed inspection or additional maintenance. It can be used in implementation of preventive and predictive maintenance, which leads to:

- Significant savings in maintenance costs.

- Reduction in the number of failures and downtimes, which translates into greater equipment availability.

- Extending the service life of components and entire machines.

- Improving work safety through early detection of potential faults.

TABLE 2

The elements with the highest probability of failure for three selected machines.

Name	System	Element	Data of last failure	Likelihood of failure
Machine 1	electrical installation	electrical boxes	2024/04/08	63.17%
	hydraulic system	cap	2024/03/21	69.84%
	drive system	driving gear	2024/03/19	72.75%
Machine 2	cooling system	combustion engine cooling system	2024/04/04	75.67%
	electrical installation	e-gas installation	2024/04/25	76.07%
	drive system	alternator	2024/05/21	76.22%
Machine 3	drive system	multi-ribbed belt	2024/04/25	50.66%
	air conditioning system	compressor-evaporator hose	2024/04/10	52.36%
	actuator system	steering cylinder hose	2024/04/30	57.39%

These benefits confirm that investments in modern technical condition monitoring and predictive analysis systems are profitable and contribute to increased operational efficiency in the mining industry.

It is important to acknowledge that the proposed solution represents an ongoing development effort rather than a finalised system. The dynamic nature of human language, as well as the continuous evolution of mining technologies and equipment, necessitates the regular revision and adaptation of classification models. Over time, new terminology and component types are expected to show, which will need to be reanalysed and classified. There is also the problem of applying the methods to data from other sources. Not all mining departments maintain equally detailed logs. In some cases, equipment shift status or failure cause may be missing, limiting the model's input completeness. These constraints suggest that, while the article demonstrates good results for this kind of database in the case of other type of data, in-depth analysis and appropriate preparation will be required to obtain the desired results.

SUMMARY

The article emphasises the importance of optimising planning processes in modern mining operations to maintain competitiveness. It focuses on enhancing production line efficiency through effective maintenance strategies. Given the unpredictable nature of underground mining, minimising machine downtime is critical. To improve situational awareness, the research employs text mining and machine learning methods to extract insights from unstructured failure reports. A classification system was developed to categorise failure descriptions, facilitating predictive maintenance. Data integration from downtime logs, repair records, and shift statuses creates a comprehensive database, enabling a complete analysis of machine performance. The study analysed 11 months of data on two types of machines, assessing reliability metrics like Mean Time Between Failures (MTBF). The results indicate notable differences between machine types, with more frequent breakdowns observed, particularly in electrical and drive systems – the latter being responsible for the longest downtime. Further analysis revealed that even identical components within the same machine type can exhibit different failure intervals. Tire failure analysis showed that rear tires on Type 1 machines deteriorate faster, suggesting the need for targeted maintenance and inspection strategies. These insights support proactive maintenance scheduling and workload distribution, ultimately enhancing operational efficiency.

ACKNOWLEDGEMENTS

This work is a part of the project which has received funding from the European Union's Horizon research and innovation programme under grant agreement No 101092365.

REFERENCES

Balaraju, J, Govinda Raj, M and Murthy, C S, 2020. Performance evaluation of underground mining machinery: A case study, *Journal of Failure Analysis and Prevention*, 20(5):1726–1737.

Barabady, J and Kumar, U, 2008. Reliability analysis of mining equipment: A case study of a crushing plant at Jajarm Bauxite Mine in Iran, *Reliability Engineering and System Safety*, 93(4):647–653.

Blanco, M A, Marti-Puig, P, Gibert, K, Cusidó, J and Solé-Casals, J, 2019. A text-mining approach to assess the failure condition of wind turbines using maintenance service history, *Energies*, 12(10):1982.

Brodny, J, Alszer, S, Krystek, J and Tutak, M, 2017. Availability analysis of selected mining machinery, *Archives of Control Sciences*, 27(2):197–209.

Dindarloo, S R, 2016. Support vector machine regression analysis of LHD failures, *International Journal of Mining, Reclamation and Environment*, 30(1):64–69.

Ghodrati, B, Hoseinie, S H and Kumar, U, 2018. Context-driven mean residual life estimation of mining machinery, *International Journal of Mining, Reclamation and Environment*, 32(7):486–494.

Jakkula, B and Ch, S N M, 2020. Maintenance management of load haul dumper using reliability analysis, *Journal of Quality in Maintenance Engineering*, 26(2):290–310.

Martinsen, M, Fentaye, A D, Dahlquist, E and Zhou, Y, 2023. Holistic approach promotes failure prevention of smart mining machines based on Bayesian networks, *Machines*, 11(10):940.

Meeker, W Q, Escobar, L A and Pascual, F G, 2022. *Statistical Methods for Reliability Data*, 2nd ed (John Wiley and Sons).

NetHelix, 2022. Intelligent Digital Toolbox Towards More Sustainable and Safer Extraction of Mineral Resources (Grant Agreement No. 101092365. European Commission Horizon, Attica, Greece.

Özfırat, M K, Yetkin, M E and Özfırat, P M, 2019. Risk management for truck-LHD machine operations in underground mines using failure modes and effects analysis, *International Journal of Industrial Operations and Research*, 2(3):1–8.

Paithankar, A and Chatterjee, S, 2018. Forecasting time-to-failure of machine using hybrid Neuro-genetic algorithm–a case study in mining machinery, *International Journal of Mining, Reclamation and Environment*, 32(3):182–195.

Rahimdel, M J, 2024. Bayesian network approach for reliability analysis of mining trucks, *Scientific Reports*, 14(1):3415.

Rahimdel, M J, Ataei, M, Kakaei, R and Hoseinie, S H, 2013. Reliability analysis of drilling operation in open pit mines, *Archives of Mining Sciences*, 58(2):569–578.

Sellami, C, Miranda, C, Samet, A, Bach Tobji, M A and de Beuvron, F, 2020. On mining frequent chronicles for machine failure prediction, *Journal of Intelligent Manufacturing*, 31:1019–1035.

Skoczylas, A, Stefaniak, P, Anufriiev, S and Koperska, W, 2025. Deep Learning for Ore Haulage Monitoring: Vibrational Analysis Using a VGG16 Network, *IEEE Access*.

Stachowiak, M, Skoczylas, A, Stefaniak, P and Śliwiński, P, 2021. Multidimensional Failure Analysis Based on Data Fusion from Various Sources Using TextMining Techniques, in *Intelligent Computing and Optimization: Proceedings of the 3rd International Conference on Intelligent Computing and Optimization 2020* (ICO 2020), pp 766–776 (Springer International Publishing).

Stefaniak, P, Koperska, W, Skoczylas, A, Witulska, J and Śliwiński, P, 2023. Methods of optimization of mining operations in a deep mine–tracking the dynamic overloads using IoT sensor, *IEEE Access*.

Stefaniak, P, Stachowiak, M, Koperska, W, Skoczylas, A and Śliwiński, P, 2022. Application of wearable computer and ASR technology in an underground mine to support mine supervision of the heavy machinery chamber, *Sensors*, 22(19):7628.

Fuel consumption reduction in HD785-7 dump trucks at an open cut coal mining site – a data-driven approach

M Sukma[1], I Maulana[2] and T Yanto[3]

1. Scientific Analytic Expert, PT Pamapersada Nusantara, Jakarta 13930, Indonesia.
   Email: muthia.sukma@pamapersada.com
2. Scientific Analytic Expert, PT Pamapersada Nusantara, Jakarta 13930, Indonesia.
   Email: ikhsan.maulana@pamapersada.com
3. Scientific Analytic Expert, PT Pamapersada Nusantara, Jakarta 13930, Indonesia.
   Email: tri.yanto@pamapersada.com

ABSTRACT

In the mining industry, operational efficiency is crucial for reducing both costs and environmental impact. Fuel consumption, particularly in hauling activities involving dump trucks, is one of the largest contributors to these costs and emissions. This research aims to analyse the factors influencing fuel consumption (L/h) in HD785-7 dump trucks using a data-driven approach. The analysis applies machine learning techniques to predict fuel consumption patterns based on variables such as road conditions, vehicle speed, and operator driving behaviour related to fuel efficiency. The results show that optimising engine speed, acceleration position, and eco mode usage in accordance with actual road conditions significantly improves fuel efficiency. This data-driven approach supports the development of a recommendation system that provides optimal engine speed and acceleration values for each road segment. The system utilises historical fuel-efficient data to generate these recommendations, which are updated hourly using the last five hrs of operational data. Additionally, an alert mechanism is incorporated to optimise eco mode usage under varying travel conditions. All optimisation features are integrated into the Ewacspro Excellent system installed in each HD785-7 unit. The findings indicate that implementing appropriate operational strategies—such as optimising driving techniques for each road segment and improving road conditions—can reduce fuel consumption by up to 6.31 per cent. Furthermore, regular monitoring and maintenance of HD785-7 units contribute significantly to overall fuel efficiency. In conclusion, this data-driven approach not only enhances the fuel efficiency of HD785-7 dump trucks but also provides actionable insights for mine operations management. This research is expected to guide companies in improving both the efficiency and sustainability of equipment operations in the mining industry.

INTRODUCTION

The coal mining industry has a very significant role in the global economy, especially to meet the growing demand for energy. One of the most vital activities in the mining process is the transportation of materials, which generally uses dump trucks to transport materials from the mining site to the processing area or disposal site. The use of dump trucks such as the HD785-7, which is known for its large carrying capacity, is highly dependent on operational efficiency, especially in relation to fuel consumption.

However, Wang *et al* (2021) found the use of heavy vehicles in mining activities often results in high fuel consumption, which not only increases operational costs but also negatively impacts the environment. This environmental impact is increasingly important to consider, given the increasingly stringent regulations related to greenhouse gas emissions and air pollution in the mining industry. Therefore, reducing fuel consumption is one of the main focuses in efforts to improve operational efficiency at mining sites.

This data-driven approach utilises various data sources accessible through sensor technology and vehicle tracking systems. Information related to vehicle speed, operator driving style, terrain travelled, as well as the technical condition of the vehicle, can be collected and analysed to get a clearer picture of the factors affecting fuel consumption. For example, a driver applying an aggressive driving style can lead to significant fuel wastage, while steep or muddy terrain conditions can increase fuel consumption due to the additional load carried by the vehicle. Therefore, with careful data analysis, inefficient operational patterns can be discovered and corrective measures can be implemented to improve them.

With the advancement of technology, data-driven approaches have now become one of the effective methods in monitoring and analysing vehicle performance, including in reducing fuel consumption. This data-driven method enables real-time information collection and analysis of factors that affect fuel efficiency, such as road conditions, vehicle speed, and the operator's driving techniques related to fuel efficiency (Alamsyah, Franto and Andini, 2024). By using appropriate data analysis methods, more measurable and specific solutions can be identified for optimising fuel consumption in dump trucks, including the HD785-7 model.

In addition, Wang *et al* (2021) showed that the current research methods to determine the fuel consumption of dump trucks in the mining process include simulation analysis and regression analysis using a real time databased approach. In this research, utilising real time data from the Ewacspro Excellent features installed on the dump trucks unit as a sensor of all activities carried out, so that it can be very precise and accurate. Where real time data can be analysed for more effective fuel usage by considering environmental aspects, operator's driving techniques and machine. From the results of the analysis using databased approaches such as regression analysis can produce an effective fuel consumption pattern by providing guidance on the use of engine speed, acceleration position, and eco mode according to actual road conditions on the Ewacspro Excellent features.

This research aims to identify and analyse the factors that affect fuel consumption on HD785-7 dump trucks at open pit coal mining site. With a data-driven approach, it is expected that more efficient strategies can be found to reduce fuel consumption and minimise environmental impacts resulting from vehicle operations at mining site.

METHODS

Ewacspro Excellent is a real-time integrated system designed to capture and monitor operational activities occurring in HD785-7 dump trucks. Installed on each unit, the system uses a set of dedicated sensors to record critical parameters such as engine speed, acceleration position, Eco Mode usage, and fuel consumption (fuel burn). The system also features an onboard display, allowing operators to view key performance indicators during operation. In this research, Ewacspro Excellent serves as the main platform for collecting operational data, which is then analysed to generate accurate recommendations tailored to each unit's specific conditions and performance trends.

This research focuses on reducing fuel consumption on HD785-7 dump trucks using a data-driven approach. The data analysis stage is key to gaining insights that can improve operational efficiency. The following is an explanation of the stages of data analysis that are commonly performed in data-driven research (Alhady, 2022).

Data collection

The first stage in data analysis is data collection. At this stage, relevant and quality data needs to be collected to ensure that the analysis results can provide an accurate picture of the problem under study. For the case of reducing fuel consumption on HD785-7 dump trucks, some of the types of data that need to be collected include data related to dump truck performance, such as speed, load, road grade, acceleration position, eco mode, engine speed, road segment, dump trucks activity and fuel burn. Data collection was carried out using sensor devices installed in vehicles, specifically GPS (Global Positioning System), which can collect data in real-time (per 3 seconds).

Data cleaning

This process aims to ensure that the data used in the analysis is free from errors or inconsistencies that could affect the analysis results. Several steps in data cleaning include:

- managing missing data
- removing duplicates
- correcting outliers.

Data exploration

At this stage, an initial exploratory analysis is conducted on the cleaned data to obtain an overview and detect existing patterns or relationships. The techniques used include: data visualisation, descriptive statistics and pattern identification.

Data analysis

Where data processing is carried out to gain deeper insights. Various methods are used such as regression analysis, multivariate analysis, clustering, and descriptive analysis.

Interpretation of result

After analysing the data, the next step is the interpretation of the results. At this stage, the researcher assesses the results of the analysis to draw relevant conclusions. What factors affect the fuel consumption of HD785-7 dump trucks.

Recommendations

After identifying the factors that influence fuel consumption in HD785-7 dump trucks, a recommendation will be given. The recommendations will be presented in the form of visualisations using the Ewacspro Excellent features installed on each dump trucks.

RESULT AND DISCUSSION

This data collection is done through real time data (per 3 seconds) from the performance of each dump truck that has a GPS sensor installed. This test was conducted with one month of data on HD785-7 dump truck activities during hauling and travelling. It is during hauling and travelling activities that dump trucks use high fuel consumption compared to other activities.

Multivariate analysis

After processing the data starting from collecting data to data exploration, then data analysis is carried out with multivariate analysis. Multivariate analysis is conducted to identify the factors that influence fuel consumption in dump trucks, resulting in a specific condition.

As shown in Figure 1, it can be observed that there are several factors that have a moderate relationship (0.4–0.59) and a strong relationship (0.6–0.79), namely eng_speed (engine speed), plm_incl (road gradient), and accel_pos (acceleration position).

Of the three factors identified as having an influence on fuel consumption, two of these are related to the operator's driving techniques (engine speed and acceleration position). Where the driving techniques of each operator is different, so it is necessary to provide guidance in driving to get more efficient fuel consumption. For road gradient, improvements can be made directly on each road segment that has an over-grade.

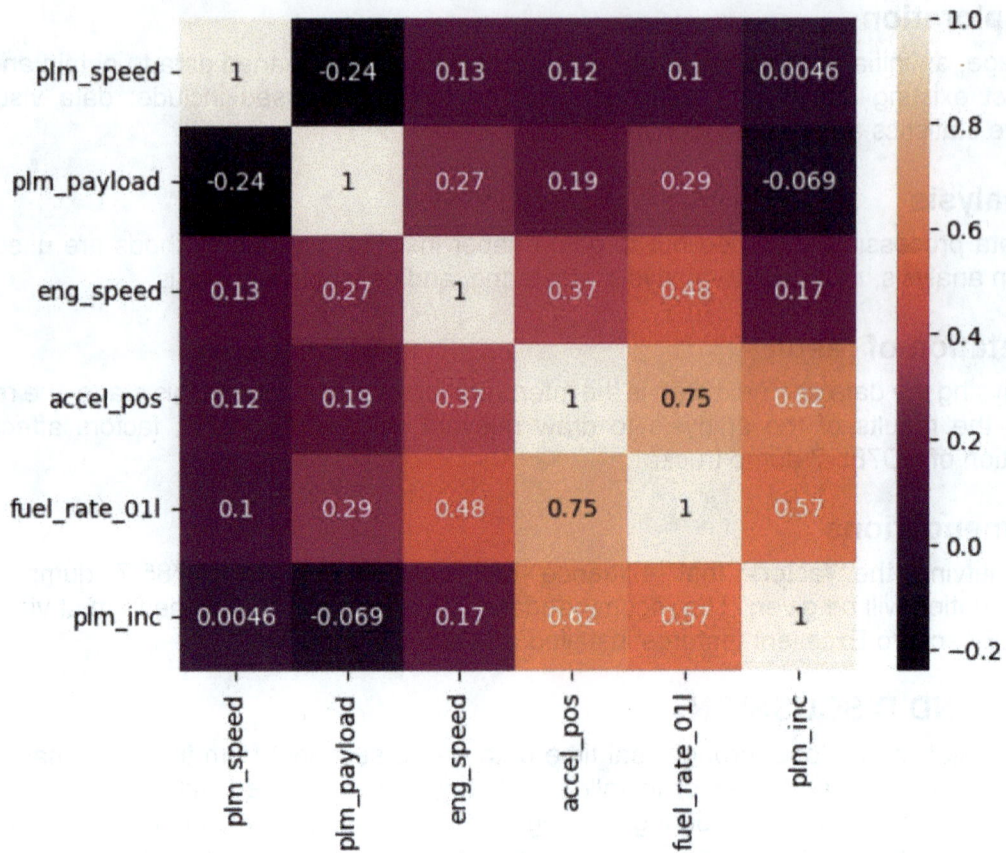

FIG 1 – Correlation matrix of factors affecting fuel consumption.

Guidance for recommendations

Descriptive analysis

A detailed analysis of fuel consumption data was conducted by examining the factors of engine speed and acceleration position. This analysis was carried out for each activity of the HD785-7 dump trucks, specifically during hauling and travelling conditions. Because the operator's driving techniques during hauling and travelling are different. This descriptive analysis is also conducted per road segment as the characteristics of each road segment vary as well.

To provide guidance on driving techniques to each operator, quartile calculations will be used on each data grouped by activity and road segment. This calculation is conducted for data on fuel consumption, engine speed, and acceleration position.

$$Q_1 = X_{[(n+1)/4]} \tag{1}$$

$$Q_2 = X_{[(n+1)/2]} \tag{2}$$

$$Q_3 = X_{[3(n+1)/4]} \tag{3}$$

$$IQR = Q_3 - Q_1 \tag{4}$$

Recommendation

The algorithm used for recommending driving technique guides is based on the quartile calculations of each factor, which will be provided for each activity and road segment. The guidance was built using the most economical of fuel consumption based on fuel rate history. Detection of the most economical database on range value of minimum and quartile 1 of fuel rate.

For the engine speed and acceleration position, based on quartile 2 after data being filtered of the fuel rate that has been generate with less than quartile 1. System will utilise the historical data to generate the recommendation. System will update the recommendation every few hours, with last

5 hrs of historical data. This system will generate the recommendation value for engine speed and acceleration position for each segment road name.

Specifically for the use of Eco Mode, guidance is provided using the algorithm shown in Figure 2. During hauling activities the operator does not have to activate Eco Mode, while during travelling activities the operator must activate Eco Mode.

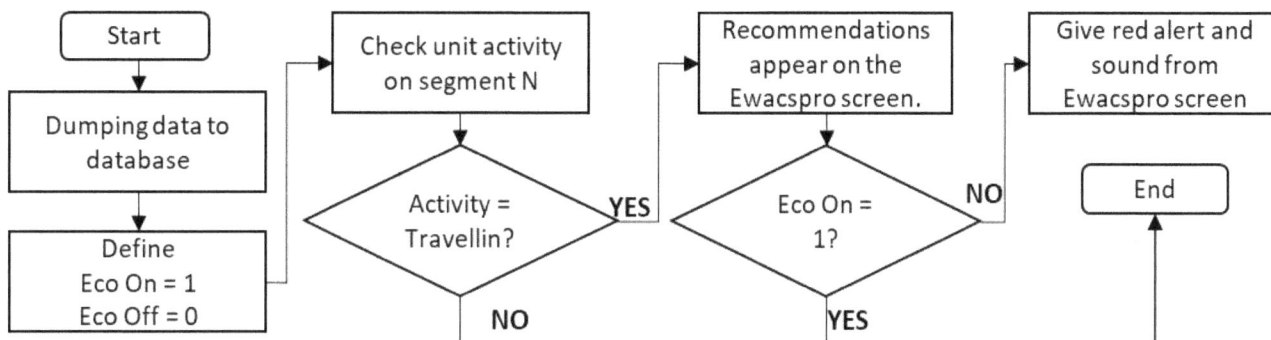

FIG 2 – Algorithm for eco mode guide.

CONCLUSIONS

Fuel consumption in HD785-7 dump trucks is influenced by multiple factors, among which the operator's driving technique plays a significant role. This research highlights the importance of providing data-driven guidance to operators by analysing historical operational data to identify the most fuel-efficient driving patterns. By recommending optimal engine speed and acceleration behaviour for each road segment, and combining this with continuous improvements in road conditions and routine unit maintenance, the research achieved a measurable reduction in fuel consumption. Specifically, the implementation of these integrated management strategies resulted in a decrease in average fuel consumption from 78.32 L/h to 73.37 L/h, representing a 6.31 per cent improvement in fuel efficiency.

ACKNOWLEDGEMENTS

The authors would like to express their heartfelt gratitude to all parties who have provided support and contributions towards the completion of this research. The authors also extends their thanks to PT Pamapersada Nusantara for the assistance provided in terms of facilities, data, and information, which was instrumental in the collection and analysis of the data required for this research.

REFERENCES

Alamsyah, M I, Franto, F and Andini, D E, 2024. Analysis of Fuel Consumption of Loading and Transport Equipment in Overburden Stripping at PT Putra Maga Nanditama North Bengku, *Mining Journal Exploration, Exploitation Georesource Processing and Mine Environmental*, 9(2):75–81.

Alhady, M I, 2022. Analysis of The Implementation of Data Management Processing in PT. Dayamitra Telecommunication Tbk, *Bachelor Internship Report (Unpublished)*, University of Politeknik Negeri Jakarta, Jakarta.

Wang, Q, Zhang, R, Lv, S and Wang, Y, 2021. Open-pit Mine Truck Fuel Consumption Pattern and Application Based on Multi-dimensional Features and XGBoost, *Sustainable Energy Technologies and Assessments*, pp 1–10 (China University of Mining and Technology: Beijing).

Safety berm design – moving beyond rules of thumb

K Thoeni[1]

1. Associate Professor, School of Engineering, The University of Newcastle, Callaghan NSW 2308. Email: klaus.thoeni@newcastle.edu.au

ABSTRACT

Safety berms, also known as safety bunds or windrows, play a critical role in preventing errant vehicles from falling over edges in surface mines and quarries. Despite their importance, the design of these structures often relies on rules of thumb that primarily focus on the berm's height relative to the vehicle's tyre diameter. This oversimplified approach overlooks the complex dynamics of vehicle-berm collisions, potentially undermining safety in high-risk environments. To address this issue, a more rigorous and scientific approach to safety berm design is essential. Effective designs must consider not only the berm's height but also its overall geometry, including the top and base width and batter angle. Incorporating these geometric parameters into a comprehensive analysis allows for a better understanding of the berm's capacity to safely redirect or arrest vehicles during impact. In addition, the vehicle type and the potential approach conditions should be considered. Advanced numerical modelling, combined with full-scale experimental testing, provides a robust framework for improving safety berm designs. This integrated approach enables the simulation of diverse collision scenarios, capturing critical variables such as vehicle type, load, approach angle, velocity, berm geometry, and material properties. By analysing the interplay of these factors, it is possible to identify and optimise berm configurations that offer superior performance under real-world conditions. This paper marks a significant step forward in transitioning from rudimentary design guidelines to simulation-based solutions. By leveraging state-of-the-art numerical modelling and experimental techniques, safety berm designs can be tailored to site-specific conditions. This approach enhances the safety and effectiveness of berm designs, ensuring they meet the demands of modern mining and quarrying operations. Ultimately, this work contributes to safer working environments by advancing the understanding of vehicle-berm interactions and providing a robust foundation for future design standards.

INTRODUCTION

Safety berms, also known as safety bunds or windrows, are critical safety features in surface mining and quarry operations. These structures serve as the primary edge protection along haul roads and dump points, helping prevent potentially fatal accidents involving vehicles running over the edge. Despite their vital role, the design of these structures generally relies on rules of thumb that were established half a century ago. Kaufman and Ault (1977) were the first to provide formal design guidelines for haul roads, including safety berms. They suggested a minimum berm height of 50 per cent of the tyre diameter (mid-axle height) of the largest vehicle using the road. Despite being established so long ago, this rule of thumb is still widely used. Some more recent guidelines suggest that the berm height should be at least 66 per cent of the tyre diameter of the largest truck using the road (Darling, 2011; Thompson, Peroni and Visser, 2019). While these rules of thumb have served the industry for decades, they are not informed by rigorous geotechnical analysis and fail to address modern operational complexities. This oversimplified approach overlooks the complex dynamics of vehicle-berm collisions, potentially undermining safety in high-risk environments. Many guidelines acknowledge this by stating that safety berms are not designed to stop a runaway truck but rather to give drivers a visual and tactile indication of the edge of the road or dump (eg Department of Natural Resources, Mines and Energy (DNRME), 2019; Mine Safety and Health Administration (MSHA), 1999).

Accidents involving ineffective safety berms are very common, with vehicle rollovers, skidding, and inadequate stopping power being primary concerns. Reynoldson (2015) underscores the need for larger berm dimensions in high-risk areas, while Brady (2019) discusses the frequency of vehicle accidents linked to inadequate berm designs. The comprehensive report by Brady (2019) identified several critical factors contributing to incidents, particularly concerning haul trucks and the implementation of safety berms. One of the primary concerns highlighted was the inadequate maintenance of haul roads, which directly impacts the safe operation of haul trucks. Poorly maintained roads can lead to loss of vehicle control, increasing the risk of accidents. Additionally, the report emphasised the importance of properly designed and maintained safety berms. The absence or inadequacy of these safety features was identified as a contributing factor in several fatal incidents. The findings underscore the necessity for stringent design standards and regular maintenance protocols for both haul roads and safety berms. Implementing comprehensive training programs for operators is also crucial to ensure adherence to safety practices. These findings align with data from Dindarloo, Pollard and Siami-Irdemoosa (2016) and Duarte, Marques and Santos Baptista (2021), which indicate a substantial portion of haul truck-related accidents are due to insufficient edge protection which in many cases results in fatalities. Figure 1 shows examples of typical incidents.

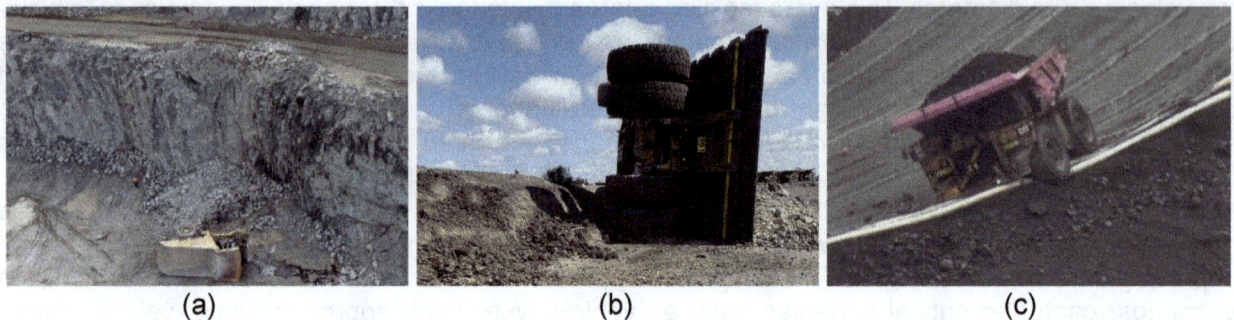

FIG 1 – Example of recent incidents: (a) haul truck run over the edge (Department of Mines, Industry Regulation and Safety (DMIRS), 2019); (b) haul truck rollover due to climbing of safety berm (Resources Safety & Health Queensland (RSHQ), 2023); (c) run-away autonomous mining truck (AMT) stopped by safety berm (Mining Mayhem, 2024).

Autonomous mining trucks (AMTs) have the potential to greatly reduce human error but also introduce new risks related to technological reliability and cybersecurity (Ninnes, 2018). A key concern is software bugs, which may cause unexpected behaviour or system failures in critical situations. Additionally, sensor calibration, crucial for accurate environmental perception and navigation, can degrade over time or be impacted by harsh mining conditions, leading to errors or misjudgements. System security poses another major risk, as autonomous systems are susceptible to cyberattacks that could disrupt operations or even result in accidents. The reliance on Global Navigation Satellite Systems (GNSS) for positioning further increases vulnerability, as signal disruptions can lead to navigation errors. Similarly, wireless mesh networks used to control AMTs are prone to signal loss, which can affect vehicle operation. Another risk factor is the potential loss of traction on slippery roads, which can lead to AMT incidents. Safety berms serve as a critical safety measure by providing a physical barrier that helps prevent vehicles from veering off course or entering hazardous areas in the event of network failures or control system malfunctions. By incorporating safety berms into the site design, the risks associated with technological disruptions can be mitigated, offering an additional layer of protection to ensure the safety of both autonomous and manually operated vehicles. Figure 1c shows an example where a run-away AMT was successfully restrained by the safety berm.

DMIRS (2020) has recently questioned the adequacy of safety berm designed according to a rule of thumb. It is well-known that the proper design of safety berms is very complex as there are several aspects involved including material uncertainty and degradation, environmental effects such as rain and visibility, the behaviour and alertness of the truck driver and finally the type and state of the machine and its breaks. All these aspects influence either the energy absorption capacity of a safety berm or the actual scenario (eg impact speed and approach angle). It is also known that the conventional rule of thumb might not be adequate, and some guidelines suggest increasing the safety berm (Tannant and Regensburg, 2001; Darling, 2011; Thompson, Peroni and Visser, 2019). Other guidelines acknowledge the complexity in such a way that they state that the installation and construction of safety berms must be supported by robust design calculations determined by a competent person (WorkSafe NZ, 2015; NSW Department of Planning and Environment (NSW DPE), 2018).

To address the shortcomings of existing guidelines and industry practice, Thoeni *et al* (2019, 2025) combined full-scale experimental testing with advanced numerical modelling. This integrated approach allows for the simulation of various collision scenarios, capturing critical variables such as vehicle type, approach angle, impact velocity, berm geometry, and material properties. Rigorous full-scale testing provides experimental data on the dynamic impact of haul trucks on safety berms (Giacomini and Thoeni, 2015), allowing for the calibration of the numerical models. Thoeni *et al* (2025) recently validated the calibrated model by back analysing a real accident. This back analysis suggested that a truck travelling at the speed limit in place at the time of the accident, could not have been stopped by the safety berm that was in place constructed based on current guidelines. The numerical model was able to accurately reflect what happened in real life, showcasing its reliability. The current paper discusses how this integrated approach can be used to improve safety across the resource sector. The ultimate aim of this new approach is to move beyond simplistic rules of thumb for safety berm design and develop evidence-based design guidelines and charts that account for the complex dynamics of vehicle-berm collisions.

METHODOLOGY

Full-scale experimental testing

Full-scale experimental testing provides real-world data that is essential for the calibration of the numerical model. The experiments take place in controlled operational mining and quarry environments, where trucks are reversing into purposely built safety berms at low speeds (around 10 km/h). The movement of the truck is captured by video cameras positioned on each side of the truck. Targets on the truck are used to accurately track the motion of the vehicle in the videos. In particular, the motion of the back axle is tracked to provide a clear understanding of the vehicle dynamics upon impact. Fully loaded trucks are generally used, and tests are repeated at least two to three times to account for factors such as variable approach velocity (the truck operator controls the approach velocity) and variation in the material the safety berm is made of (waste rock is generally not very uniform). As shown in Figure 2, tests are conducted at two different approach angles: 90 and 75°. It is important to cover both angles in the testing as both angles will give different responses.

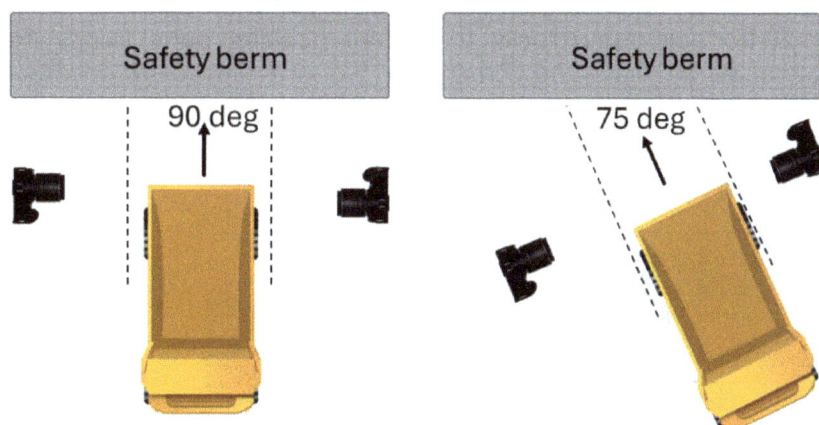

FIG 2 – Full-scale experimental set-up for the reverse impact tests with cameras positioned on each side.

The safety berms are prepared according to current site standards. The particle size distribution of the material is determined by combining data from 3D point clouds for block sizes above 200 mm with sieving of representative samples below 200 mm. Particle shape is characterised by digital photography of representative samples. Additional material characterisation tests are carried out in the form of full-scale tipping tests (Thoeni *et al*, 2025), where a load of material is tipped by the truck and the angle of repose and material footprint is measured, or laboratory-scale tilt box tests (Thoeni *et al*, 2019), where a down-scaled material is used. The former is preferred as it allows using the same material as used for the impact tests.

The primary focus of the experimental testing involves observing the impact dynamics of a truck reversing into a safety berm and the reaction of the safety berm. Detailed data collection from the video footage taken during the tests includes the truck's approach velocity, the dynamic reaction of the truck, the maximum wheel climb and horizontal wheel displacement upon impact. The deformation of the safety berm is estimated by a change detection analysis. This consists of comparing 3D point clouds of the geometry of the safety berm that have been collected before and after the test. The 3D point clouds are obtained by drone-based photogrammetry or laser scanning. When collecting the data it is important to capture the full geometry of the safety berm and avoid any occlusions. Figure 3 shows an example of how a drone survey should be planned.

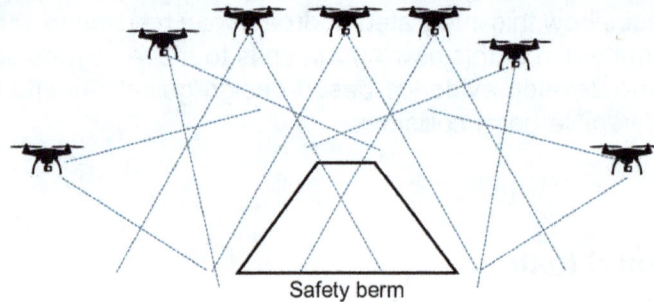

FIG 3 – Example of drone survey for determining the geometry of the safety berm.

Numerical modelling

An advanced numerical model is developed. Within the numerical framework, the waste rock material of the safety berm is represented using the Discrete Element Method (DEM). The DEM is widely used for studying the dynamic behaviour of granular material. The safety berm is simulated as an assembly of spherical particles that interact based on defined contact laws, incorporating parameters like friction, elasticity, and rolling resistance. The latter is used to mimic the behaviour of real, non-spherical rock particles as the modelling of realistic particle shapes would be computationally too expensive. The haul trucks are modelled as a rigid multibody system derived from a simplified 3D CAD representation and parameterised with physical properties estimated from manufacturer specifications. Depending on the complexity of the truck, this multibody system can have anything from 20 (for rigid dump trucks) to 40 (for articulated dump trucks) rigid bodies. These rigid bodies are connected by joints and constraints. Figure 4 shows the concept of the numerical model.

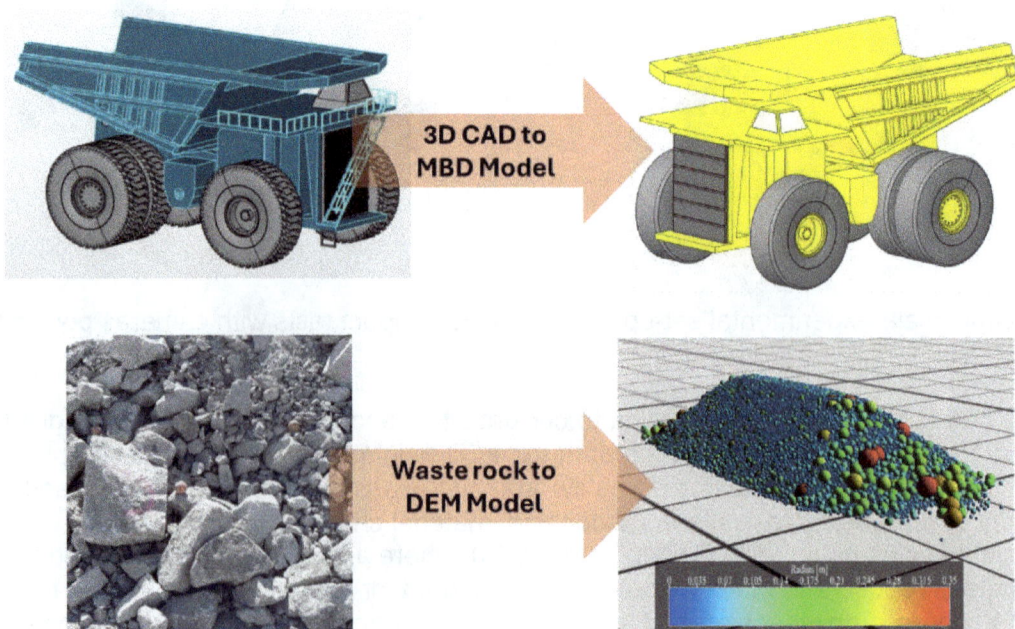

FIG 4 – From a 3D CAD model to a multi-body dynamics model for the truck and from realistic waste rock to spherical discrete elements.

Calibration constitutes the most critical step in the numerical modelling process. The calibration process involves the adjustment of model parameters governing waste rock behaviour and the dynamic response of the truck during impact. This process ensures that simulations accurately reflect real-world behaviour by comparing numerical predictions with full-scale experimental data. The data gathered during the full-scale reverse impact tests is used to calibrate the model parameters. This process is tedious and time-consuming, however, necessary. The calibration is based on reverse impact tests and slow approach velocity. To validate the model's predictive capabilities under real conditions it is recommended to back analyse a specific incident (Thoeni *et al*, 2025).

Once validated, the numerical model can be employed to conduct extensive parametric studies, exploring a wide array of collision scenarios by varying factors such as truck type and load, approach speed and angle, and safety berm geometry, including its height, width, and batter angle. The analysis of simulation results focuses on determining the effectiveness of different berm designs in arresting or redirecting errant vehicles and identifying critical velocities beyond which a safety berm would likely fail.

RESULTS

Figure 5 shows a comparison of a typical snapshot from the full-scale experimental test and the corresponding numerical model after calibration. The snapshots provide a qualitative visual comparison, in this case of a 90° reverse impact test of a CAT 797F on a standard trapezoidal safety berm.

(a)   (b)

FIG 5 – (a) Snapshot from the full-scale experimental test with a CAT 797F; (b) corresponding numerical model after calibration (Thoeni *et al*, 2019).

Figure 6 shows typical quantitative results after calibration for a 75-degree reverse impact test of a CAT 773B. The coloured lines (red, blue, green) show the measured data from the full-scale experimental testing. It can be seen that although the approach velocity *v* is very similar, the measured wheel climb and horizontal wheel displacement data varies. This is mostly due to the variability in the material. Hence, it is important to repeat each test at least two to three times. Overall, it can be seen that the trend of the predicted data from the simulations (black lines) follows the data from the experiments very well. This can be attributed to the careful calibration of the model parameters which is essential for realistic predictions.

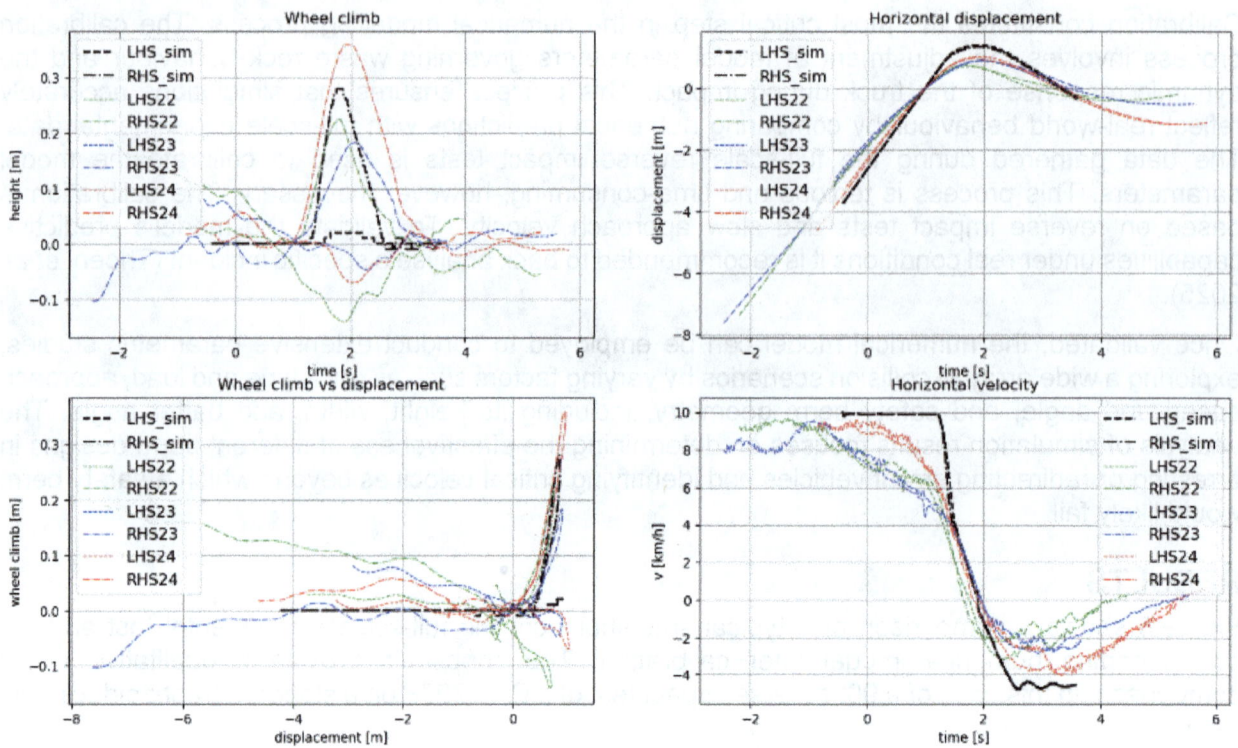

FIG 6 – Example of quantitative comparison of the numerical predictions (black lines) with experimental results (coloured lines) (Thoeni *et al*, 2025).

Figure 7 summarises typical results that can be obtained from a comprehensive parametric study after the numerical model is calibrated and validated. In this particular case, the results represent the critical velocities for a loaded articulated dump truck (ADT) and a rigid dump truck (RDT) of similar size, three different materials and three different safety berm geometries. The height of the trapezoidal safety berm H is increased from 1 to 1.3 and 1.5 m. The top width B of the safety berm is kept the same, ie 1 m. The critical velocity indicates the maximum velocity of the truck at which it can be safely stopped by the safety berm under the given conditions. The results show that the critical velocity decreases with increasing approach angle. A head-on collision at 90° is the worst-case scenario. The results also indicate an increase in critical velocity with an increase in safety berm height. This is expected but the methodology actually allows quantifying this increase. As an example, an increase of the height by 30 per cent from 1 to 1.3 m generally increases the critical velocity by 5–10 km/h. These typical charts allow to either determine a critical velocity for a given berm geometry and, hence, an appropriate speed limit, or to determine the berm geometry for a given speed limit. The data in the charts is scenario-based meaning that it considers berm material, berm geometry, truck type and approach conditions.

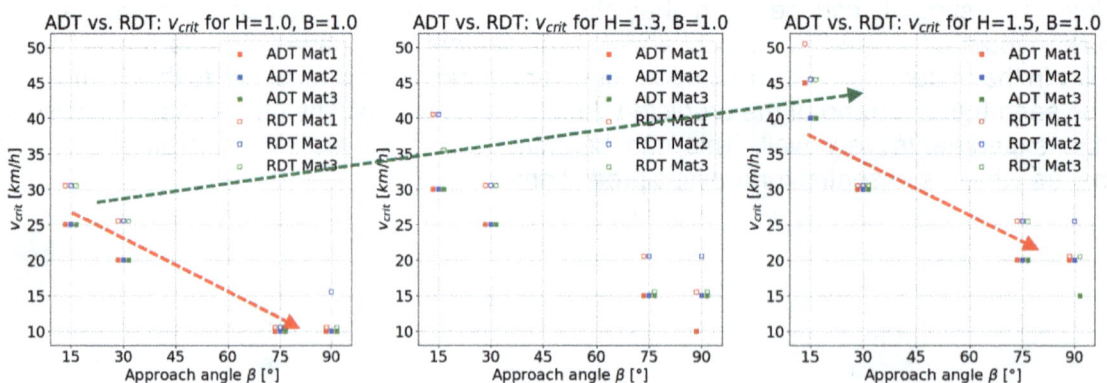

FIG 7 – Typical results from a parametric study (modified after Thoeni *et al*, 2025). The red arrow indicates the performance drop based on the approach angle whereas the green arrow indicates the performance increase due to an increase in berm height.

Figure 8 shows examples of screenshots from simulations where a rigid dump truck (RDT) runs into a safety berm at a shallow approach angle of 30°. The approach velocity considered in the configuration shown in Figure 8a is 30 km/h whereas it is 35 km/h in Figure 8b. The RDT is a CAT 773B with a gross vehicle weight of 85 t. The truck is fully loaded. The considered safety berm has a trapezoidal shape with a top width of 1 m and a base width of 3.4 m. The height is set according to the current rule of thumb, ie 50 per cent of the tyre. For the considered RDT this corresponds to 1 m. Figure 8a shows that the safety berm can safely stop the RDT travelling at 30 km/h. However, Figure 8b shows that if the approach velocity is increased to 35 km/h the safety berm is not effective. The RDT is breaching the barrier and falling off the edge. The consequences in such a case would be catastrophic. Hence, the critical velocity is 30 km/h for this particular case.

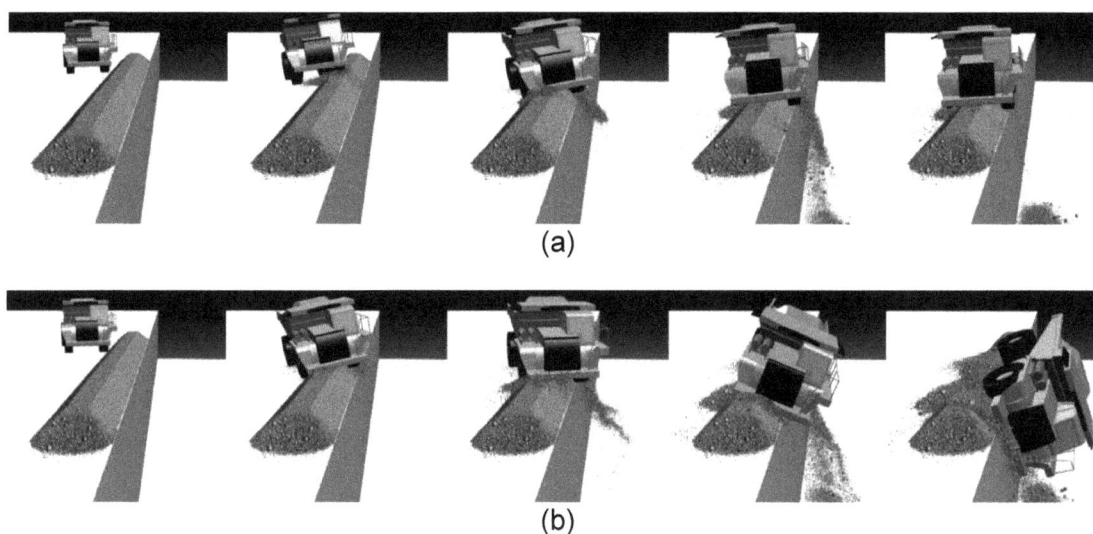

(a)

(b)

FIG 8 – Typical screenshots of a simulation for a CAT 773B approaching a standard safety berm at 15° with an approach velocity of: (a) 30 km/h; (b) 35 km/h.

CONCLUSIONS

Effective safety berm design is crucial for preventing accidents and ensuring safe operations in mines and quarries. The findings from numerical modelling and full-scale testing highlight the complexity of safety berm design, challenging traditional rules of thumb. The results emphasise that berm width is just as critical as height, as a wider berm enhances resistance due to additional mass and provides additional braking distance. The effectiveness of a safety berm also depends on approach conditions, particularly the vehicle's approach conditions upon impact. Head-on collisions present a greater risk compared to shallow-angle impacts. Well-graded materials are essential for optimal performance. Additionally, berm geometry plays a crucial role in redirecting vehicles, with a height of approximately three-quarters of the wheel diameter and a batter angle of 40° identified as particularly effective. These insights demonstrate that a more rigorous approach, integrating advanced numerical modelling and full-scale testing, is necessary to improve safety berm performance and ensure safer operations in mines and quarries. This work provides a robust foundation for future design guidelines.

Given the variability in collision scenarios, berm designs must be tailored to specific site conditions. Ramps, where the risk of accidents is higher, require larger berms, and articulated dump trucks, which are more prone to climbing berms than rigid dump trucks, necessitate additional design considerations. High-risk areas, such as sharp bends and steep inclines, may benefit from increasing both berm height and width, along with additional safety measures like runaway provisions or double berms. Moreover, speed limits should be set based on berm geometry, or alternatively, berms should be designed to accommodate speed constraints for specific machinery. The application of advanced numerical modelling techniques is strongly recommended to evaluate berm performance under different conditions, enabling more precise and reliable design strategies.

To eliminate ambiguity and enhance safety across the resource sector, it is strongly recommended to integrate modern engineering principles based on numerical simulations and full-scale testing in any safety berm design. Such a design approach would provide clear, consistent guidelines tailored to current mining operations, ensuring that safety berms are designed and maintained effectively. However, implementing such a framework requires investment in incident analysis, numerical modelling and large-scale testing. By securing the necessary resources, the industry can ensure that these critical safety enhancements are effectively integrated, leading to long-term operational benefits, reduced risks, and a future-proof haul road infrastructure.

ACKNOWLEDGEMENTS

The author would like to acknowledge the funding support from the Australian Coal Association Research Program (ACARP C21032), Stevenson Aggregates and Fulton Hogan.

REFERENCES

Brady, S, 2019. Review of All Fatal Accidents in Queensland Mines and Quarries from 2000 to 2019, Brady Heywood, Queensland, Australia.

Darling, P, 2011. *SME Mining Engineering Handbook* (Society for Mining, Metallurgy and Exploration: USA).

Department of Mines, Industry Regulation and Safety (DMIRS), 2019. Significant Incident Report No. 277, 22 August 2019, Department of Mines, Industry Regulation and Safety, Western Australia.

Department of Mines, Industry Regulation and Safety (DMIRS), 2020. Mines Safety Bulletin No. 179: Adequacy of windrows (bunds) for vehicle impact, 20 August 2020, Department of Mines, Industry Regulation and Safety, Western Australia.

Department of Natural Resources, Mines and Energy (DNRME), 2019. Recognised Standard 19: Design and Construction of Mine Roads, Department of Natural Resources, Mines and Energy, Queensland Government, Australia.

Dindarloo, S R, Pollard, J P and Siami-Irdemoosa, E, 2016. Off-road truck-related accidents in US mines, *Journal of Safety Research,* 58:79–87.

Duarte, J, Marques, A T and Santos Baptista, J, 2021. Occupational Accidents Related to Heavy Machinery: A Systematic Review, *Safety,* 7(1):21.

Giacomini, A and Thoeni, K, 2015. Full-scale experimental testing of dump-point safety berms in surface mining, *Canadian Geotechnical Journal,* 52(11):1791–1810.

Kaufman, W W and Ault, J C, 1977. Design of Surface Mine Haulage Roads – A Manual, US Department of the Interior, Bureau of Mines, USA.

Mine Safety and Health Administration (MSHA), 1999. Haul Road Inspection Handbook (PH99-I-4), Mine Safety and Health Administration, US Department of Labor.

Mining Mayhem, 2024. AHT lost brakes and travelled 108m before hitting windrow, 26 July 2024. Available from: <https://www.facebook.com/MiningMishaps/posts/aht-lost-brakes-and-travelled-108m-before-hitting-windrow/914914667337256/> [Accessed: 21 July 2025].

Ninnes, J, 2018. Autonomous mining trucks: Are there any limitations? [online], *Australasian Mine Safety Journal*, Available from: <https://www.amsj.com.au/autonomous-mining-truck-safety/> [Accessed: 21 July 2025].

NSW Department of Planning and Environment (NSW DPE), 2018. Health and safety at quarries, State of New South Wales through the NSW Department of Planning and Environment. Available from: <https://www.resources.nsw.gov.au/sites/default/files/2022-07/nsw-resources-regulator-mines-and-quarries-book-complete-v6.pdf> [Accessed: 21 July 2025].

Resources Safety & Health Queensland (RSHQ), 2023. Rear dump truck rollover, Coal Inspectorate, Alert No.434V1, 14 September 2023. Available from: <https://www.rshq.qld.gov.au/safety-notices/mines/rear-dump-truck-rollover> [Accessed: 21 July 2025].

Reynoldson, N, 2015. Specification for design and construction of mine roads, OCE seminar presentation, Department of Natural Resources and Mines, Queensland Australia.

Tannant, D D and Regensburg, B, 2001. *Guidelines for Mine Haul Road Design*, University of British Columbia, Canada.

Thoeni, K, Hartmann, P, Berglund, T and Servin, M, 2025. Edge protection along haul roads in mines and quarries: A rigorous study based on full-scale testing and numerical modelling, *Journal of Rock Mechanics and Geotechnical Engineering,* 17(7):4020–4035.

Thoeni, K, Servin, M, Sloan, S W and Giacomini, A, 2019. Designing waste rock barriers by advanced numerical modelling. *Journal of Rock Mechanics and Geotechnical Engineering,* 11(3):659–675.

Thompson, R J, Peroni, R and Visser, A T, 2019. *Mining Haul Roads* (CRC Press).

WorkSafe NZ, 2015. *Health and Safety at Opencast Mines Alluvial Mines and Quarries: Good Practice Guideline,* WorkSafe New Zealand.

LLM-Nav agent – adaptive ground robot navigation using large language model-based AI agents

T E Vhurumuku[1], C Kuchwa-Dube[2] and P Leeuw[3]

1. MSc Student, University of the Witwatersrand, Johannesburg 2017, South Africa.
   Email: tinashe.erwin@gmail.com
2. Senior Lecturer, School of Mechanical, Industrial Aeronautical Engineering, University of the Witwatersrand, Johannesburg 2017, South Africa. Email: chioniso.kuchwa-dube@wits.ac.za
3. Head of School, School of Mining Engineering, University of the Witwatersrand, Johannesburg 2017, South Africa. Email: paseka.leeuw@wits.ac.za

ABSTRACT

With advancements in robotics and automation, technology is moving toward transforming mine fleets into autonomous robots. While mine fleets engaged in load and haul operations typically follow defined routes, other equipment, such as utility vehicles and robots used for inspection, search, and rescue, will need to operate in undefined environments. This paper explores the application of Large Language Models (LLMs) for autonomous operations in both defined and undefined underground mining environments. LLMs are finding new frontiers in promoting the autonomy of robots, understanding high-level commands, and navigating robots in a way that adheres to the many variations in the environment. This study presents LLM-NavAgent, an AI agent that harnesses the power of LLMs for ground robot navigation with a focus on mining. The mining environment is characterised by moving objects, changing surface levels, and little or no light, making it extremely difficult, thus demanding a high level of adaptability and quick decision-making. With the advancements in LLM-based navigation such as LM-Nav's use of language, vision, and action for autonomous instruction following, also integrated with dynamic navigation in cluttered environments with SayNav, LLM-NavAgent adapts these approaches to address the challenges posed by mining environments. The agent merges LLM-derived natural language processing and ground robot control systems to facilitate the successful execution of high-level commands such as navigation in congested mining environments, obstacle avoidance, and mission completion in a short period of time. This research applies LLM-NavAgent to the minimum viable product in mining operations and addresses scenarios that include performing inspections, face mapping and autonomous exploration. Accordingly, this work contributes to the comprehension of LLM usage in mining robotics, and it conceptually establishes that the LLM-NavAgent represents a step toward adaptable and intuitive robotic navigation in the underground mining environment.

INTRODUCTION

The underground mining environment is typically harsh for both people and equipment, posing constant risks to life, health, and safety, as well as potential damage and incapacity. Equipment designed for underground mining must be robust yet agile enough to operate and manoeuvre in the confined spaces commonly found in underground stopes.

Narrow reef mining, which is typical of gold and Platinum Group Metals (PGM) mining in South Africa, is characterised by long, narrow haulages with tracks on the floor and water accumulation (especially in backfilling mines). The stopes in this type of mining are generally 1 m to 2 m high and face challenges such as steep grades, high humidity, poor lighting, and receding faces after each blast. The mechanised mines with massive stopes also experience the challenges of uneven running surfaces, dust in the haulages and the requirements to comply with collision avoidance systems, which any autonomous robot will have to comply with.

These conditions present harsh conditions for robotic autonomy. According to Wong *et al* (2018), an environment that is challenging for the operation of robots, particularly where robotics and autonomous systems are deployed, can be defined as a harsh environment. This could be exacerbated by the need for multi-agent coordination of mixed agents (Dinelli *et al*, 2023).

Robotic autonomy has advanced significantly in the area of Large Language Model (LLM)-based navigation, for example, the LM-Nav's use of language, vision, and action for autonomous instruction following (Shah *et al,* 2022), also integrated with dynamic navigation in cluttered environments with SayNav (Rajvanshi *et al,* 2022). Despite the integration of artificial intelligence and robotic autonomy, navigating complex and dynamic environments remains a core challenge.

LLM-NavAgent is introduced as a novel approach that leverages large language models (LLMs) to drive ground robot navigation, particularly targeting the demanding conditions found in mining sites. The harsh mining environments present extreme conditions, including moving people and machinery in confined spaces and unpredictable obstacles. These factors make autonomous navigation difficult, requiring a robot to be highly adaptable and capable of quick, informed decision-making to ensure both efficiency and safety.

LLMs offer a promising solution to these challenges by enabling semantic understanding and high-level reasoning based on vast prior knowledge. Unlike traditional navigation algorithms, LLMs provide a more intuitive interface for operators, allowing commands such as *'Inspect the first aid station near the north shaft and proceed to the control room while avoiding heavy machinery'* to be executed seamlessly. By harnessing an LLM, the LLM-NavAgent can translate such natural language directives into a sequence of actionable navigation goals, bridging the gap between human operators and robotic control systems, improving accessibility and ease of use.

RELATED WORK

Several advancements in natural language processing (NLP) based robot navigation serve as the foundation for LLM-NavAgent. Early work, such as LM-Nav (Shah *et al,* 2022), demonstrated how pre-trained models of language, vision, and action can enable robotic navigation by breaking down human instructions into executable waypoints. Similarly, SayCan used an LLM to propose high-level actions constrained by feasibility checks, improving long-horizon planning (Ahn *et al,* 2022).

Recent approaches integrate LLMs with multimodal inputs to enhance spatial reasoning. LangNav treats language as a perceptual representation, converting visual inputs into text descriptions, which are processed by an LLM for navigation planning (Pan *et al,* 2023). Similarly, visual language navigation (VLN) models such as VLMaps and IVLMaps (Huang *et al,* 2024) build spatial maps with open-vocabulary labels, allowing natural language queries to guide robots, with the latter being an improvement of the former. The VLN models are largely lab-based and have not proven themselves in a complex and dynamic real world. For example, Huang *et al* (2023) describe the limitation of VLMaps, among others, as being susceptible to object ambiguity in cluttered scenes during navigation, a common scenario in an underground mining environment. Huang *et al* (2024) acknowledged that IVLMaps could still be enhanced by 3D semantics for object height perception. An avenue of improvement in cluttered environments is also through SayNav by leveraging LLMs to generate real-time adaptive navigation commands.

LLM-driven navigation has been applied in constrained and structured settings, but its potential for underground mining remains largely unexplored. Mining environments present extreme conditions such as GPS-denied spaces, poor lighting, and dynamic obstacles. A conceptual study in Mining LLM envisions LLM-based AI assistants for Mining 5.0, integrating sensor data and mission planning through natural language interfaces (Li *et al,* 2024). Similarly, multi-agent coordination in underground operations, as discussed by Gamache *et al* (2023), could be enhanced by LLMs acting as high-level planners. However, challenges remain, including limited domain-specific data, real-time processing constraints, and safety-critical decision-making.

Compared to traditional navigation methods that rely on SLAM and geometric planners, LLM-driven approaches offer enhanced semantic reasoning (Shah *et al,* 2022), allowing robots to infer contextual relationships and generalise beyond pre-programmed routes. However, they face challenges in precision, real-time reliability, and integration with low-level motion control (Ahn *et al,* 2022). Hybrid architectures combining LLMs for strategic planning with classical control methods for execution have shown promise in mitigating these limitations, enabling more adaptable robotic systems for complex environments (Zhang *et al,* 2024).

These advancements provide a strong basis for our adaptation of LLM-based navigation to mining applications, where dynamic hazards demand a high level of autonomy and reasoning. LLM-

NavAgent builds upon this research by integrating LLM-based semantic understanding with real-time control in challenging subterranean conditions.

METHODOLOGY

The methodology followed is a structured pipeline for processing natural language commands and translating them into executable robotic navigation instructions. Figure 1 illustrates the architecture of the LLM-NavAgent system.

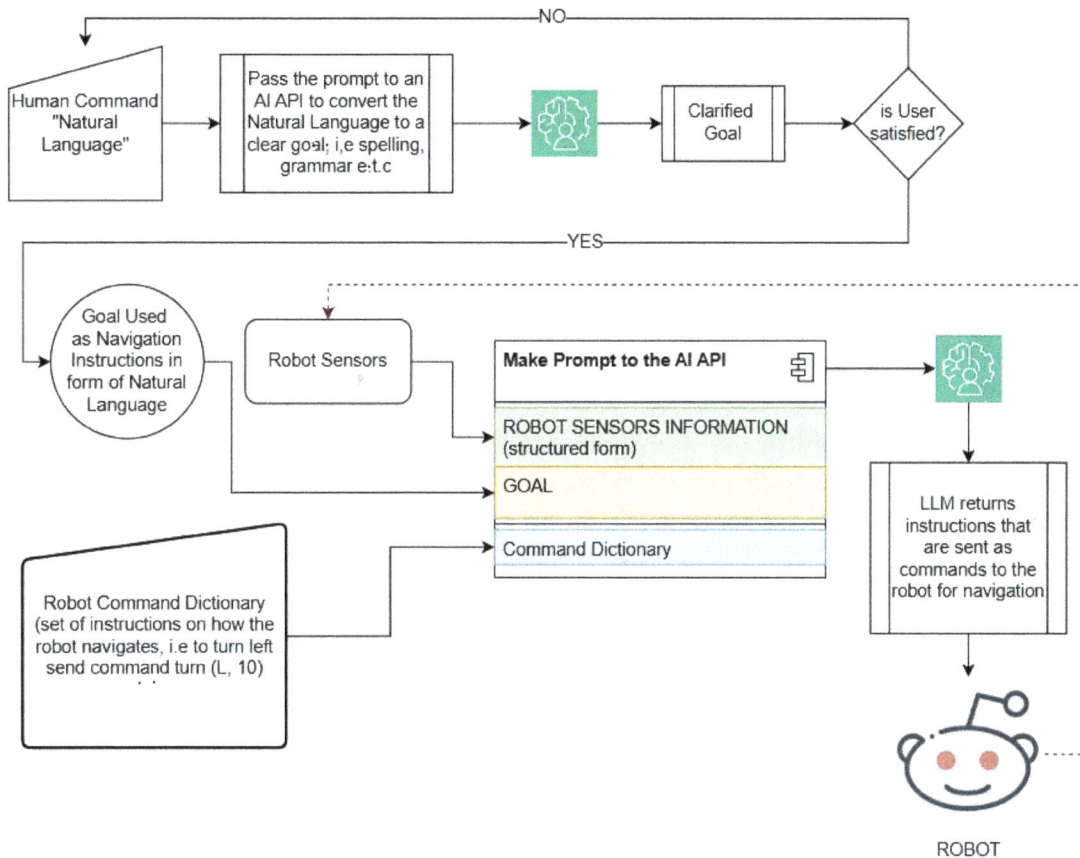

FIG 1 – Proposed LLM-NavAgent framework for translating natural language commands into robotic navigation tasks.

The core process involves:

- Processing natural language commands using an LLM to generate structured navigation tasks.

- Converting high-level commands into robot-compatible navigation instructions.

- Using ROS to manage the robot's motion planning and Gazebo to simulate the mining environment.

- Implementing real-time feedback where the LLM updates its navigation plan based on sensor data.

The robot platform used in this study is equipped with:

- LiDAR: Provides real-time obstacle detection.

- IMU: Assists with motion tracking.

- RGB Camera: Enables object recognition.

- SLAM: Supports localisation in unmapped environments.

- Onboard Control System: Executes low-level motion planning.

The AI agent does not replace the onboard control system but acts as a high-level decision-maker, interpreting human instructions and dynamically adjusting navigation plans. The AI decision-making and navigation involved:

- User Instruction Processing: The AI receives a natural language command.
- Perception and Situation Analysis: Real-time sensor data is used to assess the environment.
- Command Execution: The AI selects relevant robotic commands from a predefined dictionary.
- Real-time Adaptation: The AI continuously refines navigation based on sensor feedback.

ROS AND GAZEBO SIMULATION SET-UP

To validate the feasibility of LLM-NavAgent, we implemented a simulation environment using the Robot Operating System (ROS) and Gazebo. The set-up included:

- A simple ROS environment with basic blocks representing dynamic environments in an abstract way.
- A wheeled robotic platform equipped with LiDAR, cameras, and IMU sensors.
- A navigation stack using the ROS Move Base package for path planning and obstacle avoidance.
- Integration with an LLM API for processing high-level commands and generating movement plans.
- A feedback loop where sensor data is used to update navigation commands dynamically.

This simulation provides a realistic testbed for evaluating LLM-driven navigation and its adaptability to mining environments. As part of the simulation pipeline, LLM-NavAgent utilises a Python code to transform user commands and sensor observations into structured prompts that an LLM can understand and react to. The AI then returns a JSON data packet containing navigation instructions. It thus integrates high level natural language input with physical robotic actions. One example of the prompt template and AI response is found in Appendix A.

RESULTS

LLM-NavAgent was evaluated in a ROS and Gazebo-based environment (Figure 2), testing its ability to process commands while dynamically adapting to obstacles. The evaluation was designed to assess the system's ability to interpret natural language instructions, execute navigation commands, and adapt to real-time environmental changes.

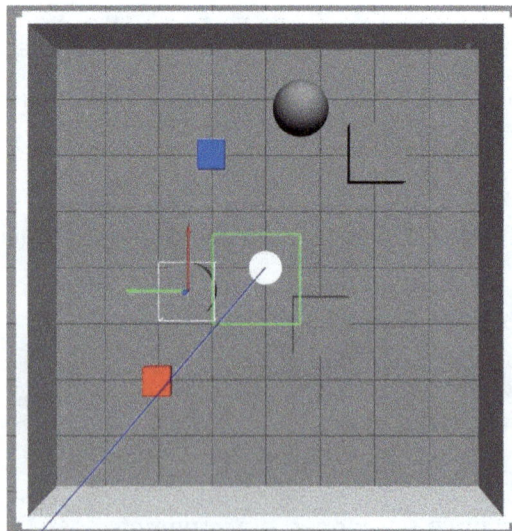

FIG 2 – Minimalistic gazebo set-up

The system was evaluated using:

- Success Rate (%): Percentage of successful task completions, indicating how well the agent executed the given commands.

- Completion Time (s): Time required to execute the command from start to finish, representing the efficiency of the system.

- Path Efficiency (% Deviation from Optimal Path): Measures deviation from the shortest possible route, indicating how well the system maintained an optimal path.

The 'Navigate to Red Block' scenario was the simplest task, requiring the robot to move directly to a fixed target. The system achieved the highest success rate of 96.2 per cent with minimal deviation from the optimal path (4.3 per cent). The completion time of 15.8 seconds suggests that the agent effectively interpreted the instructions and navigated efficiently without encountering major obstacles.

In the 'Avoid Obstacles' scenario, the system's success rate dropped slightly to 92.5 per cent, indicating that while the agent was able to detect and avoid obstacles in most cases, occasional failures occurred due to unexpected environmental variations. The completion time increased to 18.4 seconds, reflecting the additional processing required for dynamic path adjustments. The path deviation of 6.1 per cent suggests that while the agent successfully manoeuvred around obstacles, some detours were longer than necessary.

The 'Explore the Environment' scenario was the most complex, involving real-time mapping and adaptive navigation. The success rate dropped further to 88.3 per cent, with a significantly longer completion time of 32.7 seconds. This was expected, as exploration requires continuous decision-making and recalibration based on sensor inputs. The path deviation of 5.7 per cent indicates that while the system maintained a reasonable level of efficiency, factors such as SLAM-based localisation errors and terrain complexity contributed to deviations from an ideal trajectory.

Table 1 shows the results of the Navigate to Red Block, Avoid Obstacles and Explore the Environment simulations.

TABLE 1

Performance metrics of LLM-NavAgent.

Test scenario	Success rate (%)	Completion time (s)	Path deviation (%)
Navigate to target	96.2	15.8	4.3
Avoid obstacles	92.5	18.4	6.1
Exploration	88.3	32.7	5.7

DISCUSSION

The results demonstrate that LLM-NavAgent effectively integrates natural language-based navigation with real-time sensor feedback, allowing the system to interpret commands and execute tasks with high accuracy. The high success rates across all test scenarios confirm that the agent can perform structured navigation tasks with minimal pre-programming.

One key observation is that direct navigation tasks, such as the 'Navigate to Red Block' test, benefit significantly from the structured approach of LLM-based decision-making. The system follows clear and deterministic paths when there are no significant obstacles, resulting in low completion times and minimal deviation from the optimal route. This highlights the strength of LLM-driven navigation in environments where predefined waypoints or known locations are available.

Obstacle avoidance introduces additional computational challenges, as the agent must continuously update its trajectory in response to changing environmental conditions. The increased completion time and path deviation in the 'Avoid Obstacles' test suggest that real-time adjustments require further optimisation to reduce unnecessary detours. However, the high success rate demonstrates that the system effectively prioritises collision-free navigation, even if it means taking longer routes.

Exploration tasks present the greatest challenge, as they require the agent to navigate unknown terrain without predefined waypoints. The lower success rate in the 'Explore the Environment' scenario suggests that while the system is capable of autonomous mapping, the decision-making process is more complex, leading to longer execution times. The reliance on SLAM for localisation can also introduce errors, particularly in areas with limited visual or geometric features. This highlights the need for improved sensor fusion techniques that combine multiple sources of information, such as LiDAR, inertial measurement units, and depth cameras, to enhance environmental awareness.

A notable factor affecting performance is execution latency. The current implementation relies on sequential processing of sensor data and command execution, which can introduce delays in decision-making, particularly in dynamic environments. Future improvements should focus on reducing processing overhead and optimising real-time control mechanisms to enhance response speed.

Despite these challenges, LLM-NavAgent demonstrates strong potential for real-world mining applications. The ability to interpret high-level instructions and adapt to environmental changes makes it a promising solution for autonomous navigation in underground mining operations. Further testing in real-world settings will provide insights into additional factors, such as terrain stability, lighting conditions, and interaction with human operators, which could influence overall system performance.

CONCLUSIONS

The results of this study confirm that LLM-NavAgent integrates language-based decision-making with real-time navigation, allowing autonomous robots to operate effectively in both structured and unstructured settings. High success rates across all test scenarios demonstrate its ability to interpret and execute commands with minimal manual intervention.

While direct navigation tasks are handled efficiently, obstacle avoidance and exploration still pose computational challenges, evidenced by increased completion times and path deviations in more complex scenarios. Future work will focus on speeding up execution, enhancing sensor fusion, and validating performance under real-world mining conditions.

LLM-NavAgent applies LLM-driven motion planning to the nuanced constraints of underground mining, moving beyond conventional geometric or rule-based approaches. Its language-mediated task decomposition also supports coordinated multi-agent operations, enabling simultaneous inspection, mapping, and material handling under a unified semantic directive.

ACKNOWLEDGEMENTS

The work presented in this paper is based on the MSc research conducted by the first author under the supervision of the second and third authors. The authors would also like to acknowledge the contributions of the Wits Mining Institute's Sibanye-Stillwater Digital Mining Laboratory in the conceptualisation of the MSc research.

REFERENCES

Ahn, M, Brohan, A, Brown, N, Chebotar, Y, Cortes, O, David, B, Finn, C, Fu, C, Gopalakrishnan, K, Hausman, K, Herzog, A, Ho, D, Hsu, J, Ibarz, J, Ichter, B, Irpan, A, Jang, E, Jauregui Ruano, R, Jeffrey, K, Jesmonth, S, Joshi, N J, Julian, R, Kalashnikov, D, Kuang, Y, Lee, K-H, Levine, S, Lu, Y, Luu, L, Parada, C, Pastor, P, Quiambao, J, Rao, K, Rettinghouse, J, Reyes, D, Sermanet, P, Sievers, N, Tan, C, Toshev, A, Vanhoucke, V, Xia, F, Xiao, T, Xu, P, Xu, S, Yan, M, Zeng, A, 2022. Do as I can, not as I say: Grounding language in robotic affordances, *Robotics*. https://doi.org/10.48550/arXiv.2204.01691

Dinelli, C, Racette, J, Escarcega, M, Lotero, S, Gordon, J, Montoya, J, Dunaway, C, Androulakis, V, Khaniani, H and Shao, S, 2023. Configurations and Applications of Multi-Agent Hybrid Drone/Unmanned Ground Vehicle for Underground Environments: A Review, *Drones*, 7:136. https://doi.org/10.3390/drones7020136

Gamache, M, Basilico, G, Frayret, J and Riopel, D, 2023. Real-time multi-agent fleet management strategy for autonomous underground mines vehicles, *International Journal of Mining, Reclamation and Environment*, 37(9):649–666. https://doi.org/10.1080/17480930.2023.2236880

Huang, C, Mees, O, Zeng, A and Burgard, W, 2023. Visual language maps for robot navigation, in *Proceedings of the 2023 IEEE International Conference on Robotics and Automation (ICRA)*, pp 10608–10615 (IEEE).

Huang, J, Zhang, H, Zhao, M and Wu, Z, 2024. IVLMap: Instance-aware visual language grounding for consumer robot navigation. https://doi.org/10.48550/arXiv.2403.19336

Li, Y, Li, L, Wu, Z, Bing, Z, Ai, Y and Tian, B, 2024. MiningLLM: Towards Mining 5.0 via Large Language Models in Autonomous Driving and Smart Mining, *IEEE Transactions on Intelligent Vehicles*. https://doi.org/10.1109/TIV.2024.3382048

Pan, B, Panda, R, Jin, S, Feris, R, Oliva, A, Isola, P and Kim, Y, 2023. Langnav: Language as a perceptual representation for navigation, *Computer Vision and Pattern Recognition*. https://doi.org/10.48550/arXiv.2310.07889

Rajvanshi, A, Sikka, K, Lin, X, Lee, B, Chiu, H and Velasquez, A, 2023. SayNav: Grounding large language models for dynamic planning to navigation in new environments, *International Conference on Automated Planning and Scheduling*. https://doi.org/10.48550/arXiv.2309.04077

Shah, D, Osinski, B, Ichter, B and Levine, S, 2022. Lm-nav: Robotic navigation with large pre-trained models of language, vision and action, *Robotics*. https://doi.org/10.48550/arXiv.2207.04429

Wong, C, Yang, E, Yan, X and Gu, D, 2018. Autonomous robots for harsh environments: A holistic overview of current solutions and ongoing challenges, *Systems Science and Control Engineering*, 6(1):213–219. https://doi.org/10.1080/21642583.2018.1477634

Zhang, Z, Lin, A, Wong, C, Chu, X, Dou, Q and Samuel Au, K, 2024. Interactive navigation in environments with traversable obstacles using large language and vision-language models, IEEE International Conference on Robotics and Automation (ICRA).

APPENDIX A – AI NAVIGATION PROMPT AND EXECUTION CODE

This appendix provides the Python code used to translate natural language navigation commands into step-by-step robotic control actions. The AI processes real-time sensor data and predefined movement commands to generate a structured execution plan.

Python code for AI navigation prompt

The Python script constructs a structured AI prompt, queries an LLM for a navigation plan, and returns a JSON-formatted response. The three data blocks presented in Figure 3 are utilised as raw inputs to the LLM-NavAgent shown in Figure 4. Firstly, the ***user-command*** in natural language gets the intention of the operator ('Navigate to the Red Block avoiding Obstacles'). Secondly, the ***robot-commands*** dictionary specifies all action tokens the robot can perform, enabling the language model to be restricted to known actionable commands. Thirdly, ***sensor-data*** is a snapshot of the environment in real time, which consists of the obstacle flags, the current pose, the porosity, the heading, and the LiDAR ranges.

Figure 4 shows how those inputs are assembled into the final prompt transmitted to GPT-4. The prompt opens with a short role description (that sets up the model as an autonomous ground robot) and then injects a command dictionary and sensor snapshot to ground its reasoning in the capabilities of the robot and the environment. It concludes with an explicit task, a spec for the output, demanding a recursive action plan in JSON. In this manner, the agent provides the LLM with all its context regarding mission intent, authorised actions and awareness of the situation and the LLM is then able to produce a navigation plan.

```python
# Example user command
user_command = "Navigate to the Red Block while avoiding obstacles."

# Predefined robot movement dictionary
robot_commands = {
    "MOVE_FORWARD": "Move forward at 0.5 m/s.",
    "MOVE_BACKWARD": "Move backward at 0.5 m/s.",
    "TURN_LEFT": "Rotate 30 degrees counterclockwise.",
    "TURN_RIGHT": "Rotate 30 degrees clockwise.",
    "STOP": "Halt all movement immediately.",
    "AVOID_OBSTACLE": "Recalculate path and move around detected obstacle.",
    "MAP_ENVIRONMENT": "Use SLAM to explore and generate an occupancy grid.",
    "LOCATE_TARGET": "Use camera object detection to identify target object.",
    "ALIGN_TO_GOAL": "Adjust heading to face target position."
}

# Example real-time sensor data
sensor_data = {
    "front_obstacle": False,
    "left_obstacle": True,
    "right_obstacle": False,
    "position": {"x": 2.1, "y": 3.4},
    "goal_position": {"x": 5.0, "y": 7.2},
    "detected_objects": ["Red Block"],
    "current_heading": 90,
    "lidar_readings": {
        "front_distance": 1.5,
        "left_distance": 0.3,
        "right_distance": 1.7
    }
}
```

FIG 3 – Example of inputs to LLM-NavAgent (command, actions, sensors).

```python
# Construct AI prompt
prompt = f"""
You are an autonomous ground robot equipped with LiDAR, cameras, and motion planning capabilities.
Your goal is to translate user instructions into precise navigation actions based on real-time sensor data.

### Available Robot Commands:
{json.dumps(robot_commands, indent=2)}

### Current Sensor Data:
{json.dumps(sensor_data, indent=2)}

### Task:
- The user has given the following command: "{user_command}"
- Your task is to break down this command into **step-by-step executable robot actions**.
- Ensure safe movement by avoiding obstacles and dynamically adjusting based on real-time feedback.
- Use the predefined robot commands for movement and path planning.

### Output Format:
Return a structured **step-by-step action plan** in JSON format.
"""

# Query the AI model
response = openai.ChatCompletion.create(
    model="gpt-4",
    messages=[{"role": "user", "content": prompt}]
)

# Print AI-generated navigation plan
print(response["choices"][0]["message"]["content"])
```

FIG 4 – Python fragment that builds the LLM prompt and requests the navigation plan from GPT-4.

AI output example

The AI response contains a structured JSON-formatted navigation plan (see Figure 5). An example output for the command 'Navigate to the Red Block while avoiding obstacles' is shown. The AI successfully interprets natural language commands, processes real-time sensor data, and generates a structured step-by-step navigation plan. This approach ensures safe and efficient movement in dynamic environments.

```json
{
    "plan": [
        { "n": 1, "cmd": "LOCATE_TARGET" },
        { "n": 2, "cmd": "ALIGN_TO_GOAL", "p": { "hdg": 42 } },
        { "n": 3, "cmd": "MOVE_FWD",      "p": { "spd": 0.5, "d": 0.8 } },
        { "n": 4, "cmd": "AVOID_OBS",     "p": { "dir": "R",  "rad": 0.6 } },
        { "n": 5, "cmd": "MOVE_FWD",      "p": { "spd": 0.5, "d": 1.5 } },
        { "n": 6, "cmd": "STOP" }
    ],
    "status": "OK",
    "ts": "2025-06-22T06:15:00Z"
}
```

FIG 5 – LLM-generated navigation plan returned as JSON.

Operator behaviour learning to optimise HD785 productivity by adopting modified game theory algorithm

P Vorenza[1] and M Hajjrol Dava[2]

1. Scientific Analytics Engineer, PT. Pamapersada Nusantara, Jakarta 13930, Indonesia.
   Email: pebiria.vorenza@pamapersada.com
2. Mine Geotechnic Engineer, PT. Pamapersada Nusantara, Jakarta 13930, Indonesia.
   Email: m.dava@pamapersada.com

ABSTRACT

The human brain is the most complex object in the known universe. The visible outcomes of brain processes manifest as behaviours that vary from person to person. In this context, human behaviour is significantly influenced by individual capabilities, which develop through interactions with others and the environment. This research conducts an in-depth study of the personal abilities of each employee, specifically focusing on HD785 operators during the production process in open pit coal mining.

To maintain effectiveness, it is crucial to maximise material transport, minimise obstructions, and, most importantly, keep the HD785 operators motivated. The goal is to optimise productivity by studying operator behaviour and focusing on activity parameters within a cycle to form one iteration. This iteration includes the following activity parameters: speed in loaded conditions, load stop time (LST), speed in empty conditions, and empty stop time (EST). These four parameters have been selected to describe historical operator abilities at specific times and locations.

Modified Game Theory offers a robust theoretical framework for analysis, enabling participants (referred to as players) to engage in collective actions that yield mutual benefits without altering the specifications of their abilities. It is used to identify the optimal selection in the optimisation process of HD785 operator assignments, depending on the unique abilities of each operator as determined by the four parameters, allowing them to navigate various road mining conditions. This approach addresses the subsequent assignment problem and reveals the players' bargaining power while assessing the stability of cooperation within a project.

Grounded in the principles of maximising profit (production) and minimising costs (lost opportunities), the results of this implementation aim for an increase in group productivity of approximately 0.5 per cent in bcm per hr continuously. Using this method, the authors would be able to assist all open pit companies facing similar challenges.

INTRODUCTION

Productivity in the mining industry, particularly in open pit mining operations, is significantly influenced by various interrelated factors. The three main factors affecting productivity are equipment condition, environmental conditions, and operator behaviour. Of these three factors, operator behaviour is the most challenging to control. Everyone has different habits, working styles, and driving techniques, ultimately affecting the efficiency and effectiveness of heavy equipment operations (Ramos, Bergstad and Nässén, 2020).

Equipment condition is one factor that is relatively easier to control. Companies can ensure that the equipment remains in optimal condition for peak performance through regular maintenance and proper care. This maintenance includes routine mechanical, hydraulic, and electrical system checks and replacing worn or damaged components. By keeping the equipment in excellent condition, companies can reduce the risk of breakdowns that could lower productivity and extend the equipment's lifespan.

Moreover, good environmental conditions will reduce cycle times and increase productivity (Torres, 2022). Environmental factors include road conditions, front-loading areas, and material disposal sites. Managing and maintaining the roads and infrastructure at the mining site ensures that the equipment can move smoothly without obstacles, which enhances speed and material transportation efficiency. By ensuring that the loading and disposal areas are also in optimal condition, the process of moving material from one point to another can proceed without significant disruptions, thereby supporting the overall mining activities.

However, operator behaviour is a more complex factor and often difficult to predict or control (Abou Elassad *et al,* 2020). This factor remains a challenging topic today (Sagberg *et al,* 2015). Operator behaviour includes driving styles, decisions made during operations, and how they interact with the equipment and the surrounding environment. Each operator may exhibit variations in their driving style, such as the speed they choose when transporting material, the duration of stops during loading, and the speed in empty conditions. All these factors can impact efficiency and productivity, both directly and indirectly.

Given the importance of operator behaviour in supporting productivity, this study aims to identify and categorise the behaviour of HD785 operators under various road conditions at the mining site. By gaining a deeper understanding of operator behaviour patterns in specific contexts, this research will provide recommendations on optimising operator behaviour in certain road conditions to maximise equipment productivity and minimise potential productivity loss.

As part of the research objectives, the analysis will focus on measuring and monitoring four key parameters related to operator activity: speed in full load conditions, load stop time, speed in empty conditions, and empty stop time. By studying the relationship between these parameters and their impact on productivity, ways to manage and adjust operator behaviour to improve operational productivity are expected to be identified.

METHODOLOGY

To achieve the aim of this research, namely, to obtain the best selection combination based on the driving skills of the HD785 operator, which is adapted to the characteristics of road conditions at the mining location. The steps in this study are as follows:

- Conducting a data collection process both qualitatively and quantitatively.

- Conducting experiments on HD785 operator driving patterns on various road characteristics using the Design of Experiments concept.

- Finding the best combination of HD785 operators and road routes to optimise production (overall productivity).

The results at this stage will become recommendations and goals for this research.

Data collection

There are two methods employed in the data collection process: qualitative and quantitative. It is anticipated that these two methods will yield data that is coherent and measurable.

Qualitative approach

The qualitative approach in this study focuses on a comprehensive literature review of the HD785 mining activity cycle, from Front Loading to Disposal, as shown in Figure 1. Generally, this mining activity is divided into three areas: the front-loading area, the road area, and the disposal area, with each area potentially involving multiple activities. After analysing all these activities, the most dominant activity that constitutes one cycle time was identified: transportation by HD785 in the road area. Transportation activities in this area account for up to 60 per cent of the total time in a single cycle. Therefore, the activities in the road area will be evaluated in-depth to uncover the main activities that exert the most influence and serve as parameters in identifying the driving behaviours of HD785 operators. Four key parameters, validated through field studies, were identified, including Load Speed (LS), Load Stop time (LST), Empty Stop time (EST), and Empty Speed (ES).

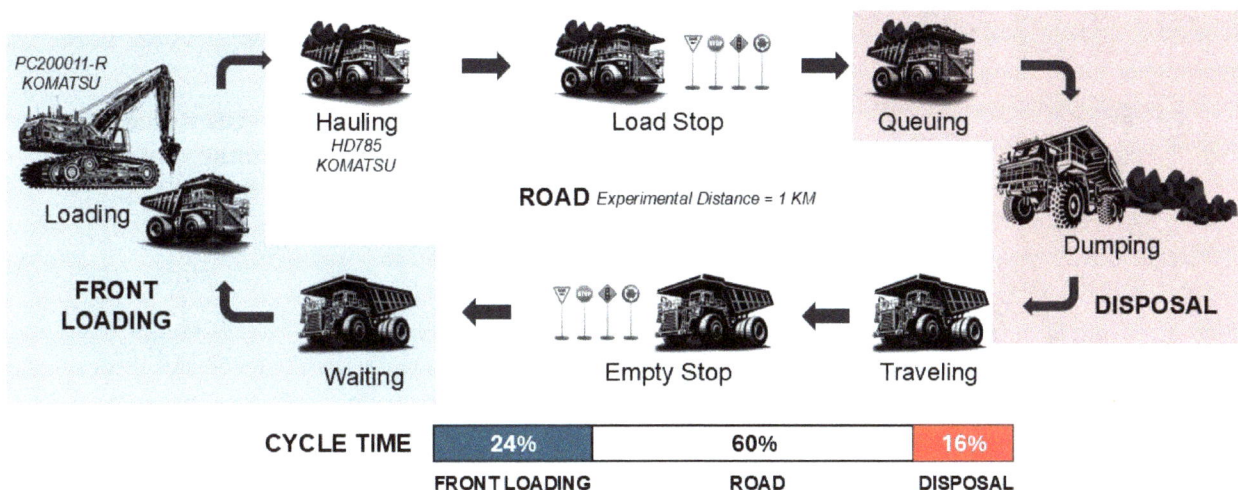

FIG 1 – HD785 mining activity cycle.

Quantitative approach

The quantitative approach in this research aims to measure and analyse objective data related to Road condition and HD785 Productivity. This study uses a field study to collect the necessary data. The following are the steps that will be taken in the quantitative approach:

- Road condition data collection:

  The field study begins with direct observation of the hauling road conditions at the mining site, which includes:

  - Road grade: the maximum road grade is 8 per cent.

  - Road surface: the road surface index must be at least 70 per cent.

  - Road width: The road width must be at least 25 m.

  Based on the road conditions, three types of hauling road categories were created based on the three criteria above, as in Table 1, and documentation of hauling road samples in open pit coal mining based on their categories can be seen in Figure 2.

- Operator data collection:

  At this stage, data collection was carried out on HD785 operators, which were then grouped based on work period, which was broadly divided into three parts, namely:

  - Group 1: operators with entry level experience or probation with 0–5 years of work duration criteria.

  - Group 2: operator with mid-level (intermediate) experience with 5–10 years of work duration criteria.

  - Group 3: operators with senior level (advance) experience with >10 years of work duration criteria.

TABLE 1
Road criterion.

Criterion road condition	Road 1 (excellent)	Road 2 (moderate)	Road 3 (poor)	Standard	Source of assessment information
Road grade	The three criteria must be met for the road conditions	At least two criteria must meet the standard	0–1 criteria meet the standard	<8%	HD785 tilt sensor data
Road surface				>70%	HD785 suspension pressure data
Road width				25 m	Direct measurement in the field

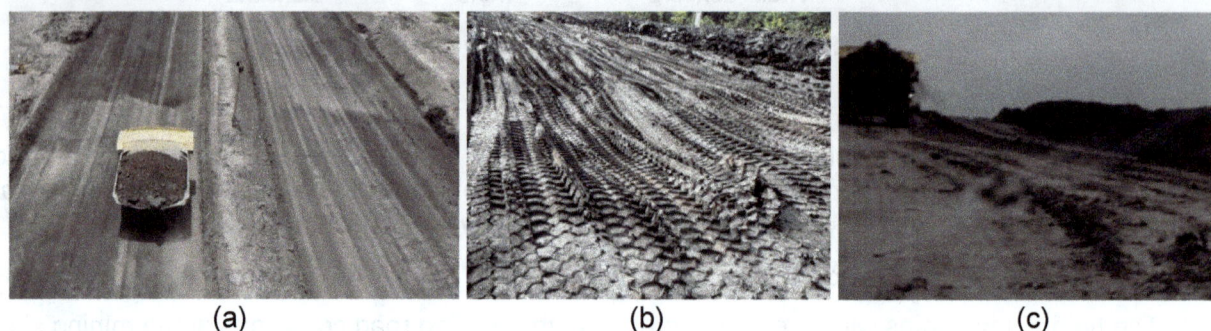

(a)   (b)   (c)

FIG 2 – (a) Good road condition, (b) poor road condition, (c) overgrade and road narrowing.

Design of experiments (DOE)

The design of experiments (DOE) in this study begins with validating key parameters that affect operator behaviours. Furthermore, these parameters are used as benchmarks for each selection of experiments between operator groups on various hauling road categories.

Parameter finalisation

Once the data collection is complete, statistical analysis will be conducted to understand the relationship between haul road conditions and key parameters that produce HD785 productivity values. Linear Regression and Correlation methods demonstrate that the four key parameters selected due to their significant contribution to the value HD785 cycle time are the most dominant influences in one cycle. The results regarding these four key parameters are deemed highly influential, which are:

- **Load Speed (LS)**: This parameter determines how quickly the heavy equipment can move material from the loading point to the disposal point. In other words, it reflects the average speed of the HD785 unit when it is loaded. The standard Load Speed is 23 km/h.

- **Load Stop Time (LST):** The time it takes for the operator to stop on the road while under load can be influenced by several factors, such as the presence of signs, intersections, or other units. The standard for LST is 0.15 mins.

- **Empty Stop Time (EST)**: The time the HD785 unit stops on the road in an empty state. Stopping on the road can be caused by several things, such as signs, intersections, or other units. The standard for LST is 2 mins.

- **Empty Speed (ES)**: This parameter determines how quickly the heavy equipment can transfer material from the disposal point to the loading point. In other words, it represents the average speed of the HD785 unit when it is empty (without a load). The standard set for empty speed is 23 km/h.

Data will be collected in detail for each parameter using the software installed on the HD785 unit. This software operates automatically by reading the sensors in the HD785 and generating databased on the logic of predetermined work activities.

Experiment combination

At this stage, an experimental framework was developed by combining three operator groups with three haul road categories. Measurements were taken from each combination regarding the values of four key parameters, as shown in Table 2.

TABLE 2
Parameter combination.

Parameter	Group name of operator	Criterion (work period)	Road category		
Load Speed (LS)	Group 1 (G₁)	0–5 years	Excellent (E)	Moderate (M)	Poor (P)
	Group 2 (G₂)	5–10 years	Excellent (E)	Moderate (M)	Poor (P)
	Group 3 (G₃)	> 10 years	Excellent (E)	Moderate (M)	Poor (P)
Load Stop time (LST)	Group 1 (G₁)	0–5 years	Excellent (E)	Moderate (M)	Poor (P)
	Group 2 (G₂)	5–10 years	Excellent (E)	Moderate (M)	Poor (P)
	Group 3 (G₃)	> 10 years	Excellent (E)	Moderate (M)	Poor (P)
Empty Stop time (EST)	Group 1 (G₁)	0–5 years	Excellent (E)	Moderate (M)	Poor (P)
	Group 2 (G₂)	5–10 years	Excellent (E)	Moderate (M)	Poor (P)
	Group 3 (G₃)	> 10 years	Excellent (E)	Moderate (M)	Poor (P)
Empty Speed (ES)	Group 1 (G₁)	0–5 years	Excellent (E)	Moderate (M)	Poor (P)
	Group 2 (G₂)	5–10 years	Excellent (E)	Moderate (M)	Poor (P)
	Group 3 (G₃)	> 10 years	Excellent (E)	Moderate (M)	Poor (P)

A total of 125 operators were analysed in each operator group based on years of experience, resulting in a total of 375 data points, each with values for four key parameters. The values of these four parameters were then assessed for achievement against the specified standard. The overall percentage achievement of these parameters was used as the operator's behavioural response to a road category, calculated as follows:

$$\text{Total \% Achievement Experiment-n } (G_n, \ldots) = \text{\% Achievement LS } (G_n, \ldots)$$
$$+ \text{\% Achievement LST } (G_n, \ldots)$$
$$+ \text{\% Achievement EST } (G_n, \ldots)$$
$$+ \text{\% Achievement ES } (G_n, \ldots)$$

Where $(G_n, \ldots)$ is the symbol of Group-n choosing a specific type of road, either Excellent (E), Moderate (M), or Poor (P).

Altruistic game theory

Each player has two traits in the game: partly selfish and partly altruistic (Hoefer and Skopalik, 2013). In this study, the author feels that the Altruistic method is more suitable for applying to HD785 Operators. In altruistic games, players can work together by forming groups called coalitions and can take joint action to achieve their goals better than if they were alone (Rothe, 2021).

This research will utilise an 'Altruistic Game' model to describe the interactions among two or more operators competing under various road conditions. The altruistic concept is employed to select the best aggregate strategy for assigning HD785 operators to road categories based on their driving characteristics, considering not only their benefits but also those of other operators. Generally, each operator will face decisions related to driving behaviour, such as choosing the optimal speed when carrying a full or empty load and determining when to stop or rest under full and empty load conditions. In game theory, each of these decisions will be regarded as a strategy chosen by the operator to maximise productivity. The goal of this game theory approach using the altruistic concept is to optimise total production or joint productivity.

Combination selection strategy

The total value of the percent achievement from the experiment results using the previous Design of Experiments (DOE) method is then combined based on the concept of game theory, where the group of operators is considered a player, and the selection of the haul road category serves as the strategy of choice. Thus, the number of players can be identified as three, with each player having up to three strategies. Table 3 displays the Game Theory combination matrix consisting of 3^3 combination options.

TABLE 3
Game theory matrix.

G_1			G_2		
			E	M	P
	E	E	(G1, E); (G2, E); (G3, E)	(G1, E); (G2, M); (G3, E)	(G1, E); (G2, P); (G3, E)
		M	(G1, M); (G2, E); (G3, E)	(G1, M); (G2, M); (G3, E)	(G1, M); (G2, P); (G3, E)
		P	(G1, P); (G2, E); (G3, E)	(G1, P); (G2, M); (G3, E)	(G1, P); (G2, P); (G3, E)
G_3	M	E	(G1, E); (G2, E); (G3, M)	(G1, E); (G2, M); (G3, M)	(G1, E); (G2, P); (G3, M)
		M	(G1, M); (G2, E); (G3, M)	(G1, M); (G2, M); (G3, M)	(G1, M); (G2, P); (G3, M)
		P	(G1, P); (G2, E); (G3, M)	(G1, P); (G2, M); (G3, M)	(G1, P); (G2, P); (G3, M)
	P	E	(G1, E); (G2, E); (G3, P)	(G1, E); (G2, M); (G3, P)	(G1, E); (G2, P); (G3, P)
		M	(G1, M); (G2, E); (G3, P)	(G1, M); (G2, M); (G3, P)	(G1, M); (G2, P); (G3, P)
		P	(G1, P); (G2, E); (G3, P)	(G1, P); (G2, M); (G3, P)	(G1, P); (G2, P); (G3, P)

Example of a reading in the section with shading, the combination obtained is:

- Group 1 (G_1): Selecting an Excellent Road (E)
- Group 2 (G_2): Selecting an Excellent Road (E)
- Group 3 (G_3): Selecting an Excellent Road (E).

Altruistic game theory calculation

In the dump truck task assignment selection system on the road, achieving cooperation among individuals is a challenge (Riar *et al*, 2024), as this concept encourages HD785 operators to put aside their personal interests and collaborate for reasons that extend beyond their own, specifically driven by altruistic motives. In this study, researcher developed an engineering approach to Altruistic Game Theory tailored for each operator by creating groups that unknowingly focus on their driving habits to optimise overall production or joint productivity. However, this Altruistic Game Theory selection algorithm is confined to group selection. The HD785 operator gamification concept will continue to evolve within each group to enhance its performance.

The applied game theory model will consider:

- **Operator strategy**: The operator can determine their ideal speed or stopping time to maximise the results.

- **Operator interaction**: Implicitly, the interaction between operators serves a common purpose. In other words, the objective of this interaction is to maximise total joint production. This concept does not negate the possibility of one operator forfeiting the chance to achieve greater profits to mitigate the risk of substantial losses for other operators.

- **Environmental factors**: Road conditions can affect the operator's strategies.

- **Combination strategy**: In the mining system, the existing road environment must be utilised. This impacts the combination selection strategy, which mandates that all roads are included in each combination issued.

To recreate the final matrix of game theory based on the existing constraints, see Table 4.

TABLE 4
Final game theory matrix.

G_1			G_2		
			E	M	P
E	E		(G1, E); (G2, E); (G3, E)	(G1, E); (G2, M); (G3, E)	(G1, E); (G2, P); (G3, E)
		M	(G1, M); (G2, E); (G3, E)	(G1, M); (G2, M); (G3, E)	(G1, M); (G2, P); (G3, E)
		P	(G1, P); (G2, E); (G3, E)	(G1, P); (G2, M); (G3, E)	(G1, P); (G2, P); (G3, E)
G_3	M	E	(G1, E); (G2, E); (G3, M)	(G1, E); (G2, M); (G3, M)	(G1, E); (G2, P); (G3, M)
		M	(G1, M); (G2, E); (G3, M)	(G1, M); (G2, M); (G3, M)	(G1, M); (G2, P); (G3, M)
		P	(G1, P); (G2, E); (G3, M)	(G1, P); (G2, M); (G3, M)	(G1, P); (G2, P); (G3, M)
	P	E	(G1, E); (G2, E); (G3, P)	(G1, E); (G2, M); (G3, P)	(G1, E); (G2, P); (G3, P)
		M	(G1, M); (G2, E); (G3, P)	(G1, M); (G2, M); (G3, P)	(G1, M); (G2, P); (G3, P)
		P	(G1, P); (G2, E); (G3, P)	(G1, P); (G2, M); (G3, P)	(G1, P); (G2, P); (G3, P)

This results in six combination options. The concept of calculating Altruistic Game Theory involves summing each selection result. The option with the highest total is the best choice. The total result calculation for the Altruistic Game Theory Combination to N is formulated as follows:

$$Total\ Combination\_n = \sum \% Achievement\ (G_1, ...) $$
$$+ \sum \% Achievement\ (G_2, ...)$$
$$+ \sum \% Achievement\ (G_3, ...)$$

The methodology employed in this research is summarised in the flow diagram in Figure 3.

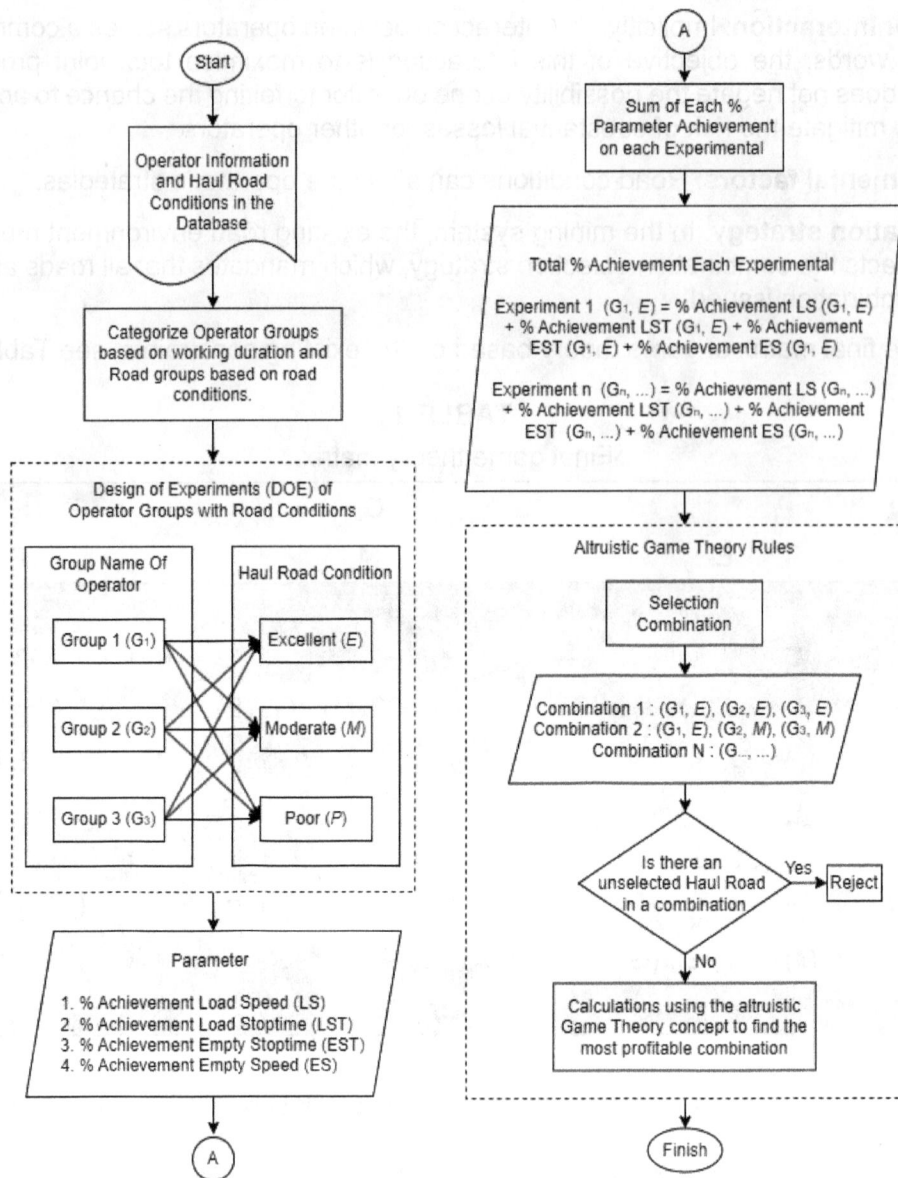

FIG 3 – Methodology flow chart.

RESULT AND DISCUSSION

After identifying and categorising operator behaviours based on previous experiments, the researcher developed a behaviour selection algorithm grounded in modified game theory with altruistic properties. This section will present the results of this study on the impact of implementing the method to increase production or joint productivity on the HD785 unit.

Calculation results for each parameter

In this study, four crucial parameters have been selected to analyse the effect of operator behaviours on the productivity of the HD785 unit in open pit mining operations. The chosen parameters are Load Speed (LS), Load Stop Time (LST), Empty Stop Time (EST), and Empty Speed (ES). Each parameter has predefined standards that serve as benchmarks for measuring operational efficiency. The following are the calculation results for each parameter.

Load Speed (LS)

To calculate the percentage contribution of time savings, the researcher can compare the operator's actual load speed with the established standard. If the operator manages to exceed the standard of 23 km/h, the time required for each loading cycle will be shorter, resulting in time savings. If it is faster than the standard, the percentage achievement obtained is positive, but if it is lower, the percentage achievement is negative. Figure 4 shows the percentage achievement of the Load Speed plot using the Kernel Density Estimates plot.

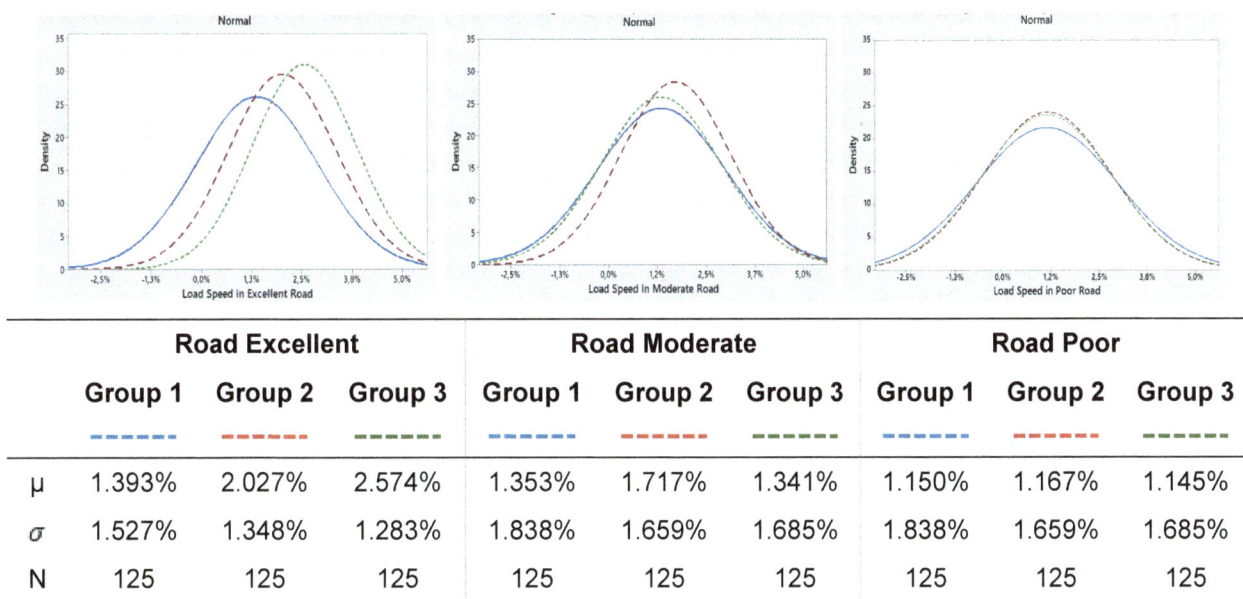

| | Road Excellent | | | Road Moderate | | | Road Poor | | |
	Group 1	Group 2	Group 3	Group 1	Group 2	Group 3	Group 1	Group 2	Group 3
	- - - - -	- - - - -	- - - - -	- - - - -	- - - - -	- - - - -	- - - - -	- - - - -	- - - - -
μ	1.393%	2.027%	2.574%	1.353%	1.717%	1.341%	1.150%	1.167%	1.145%
σ	1.527%	1.348%	1.283%	1.838%	1.659%	1.685%	1.838%	1.659%	1.685%
N	125	125	125	125	125	125	125	125	125

FIG 4 – Kernel density estimates plot of % load speed achievement.

For example, on an excellent road, Operator Group 1 achieved an average time saving of 1.393 per cent faster in minutes for the Load Speed parameter compared to the total travel time from front to disposal at 1 km, with a speed standard of 23 km/h.

Load Stop Time (LST)

The authors will compare the actual LST value with the standard 0.15 mins to calculate the contribution percentage of time savings. Figure 5 shows the percentage achievement of the Load Stop Time plot results using the Kernel Density Estimates plot.

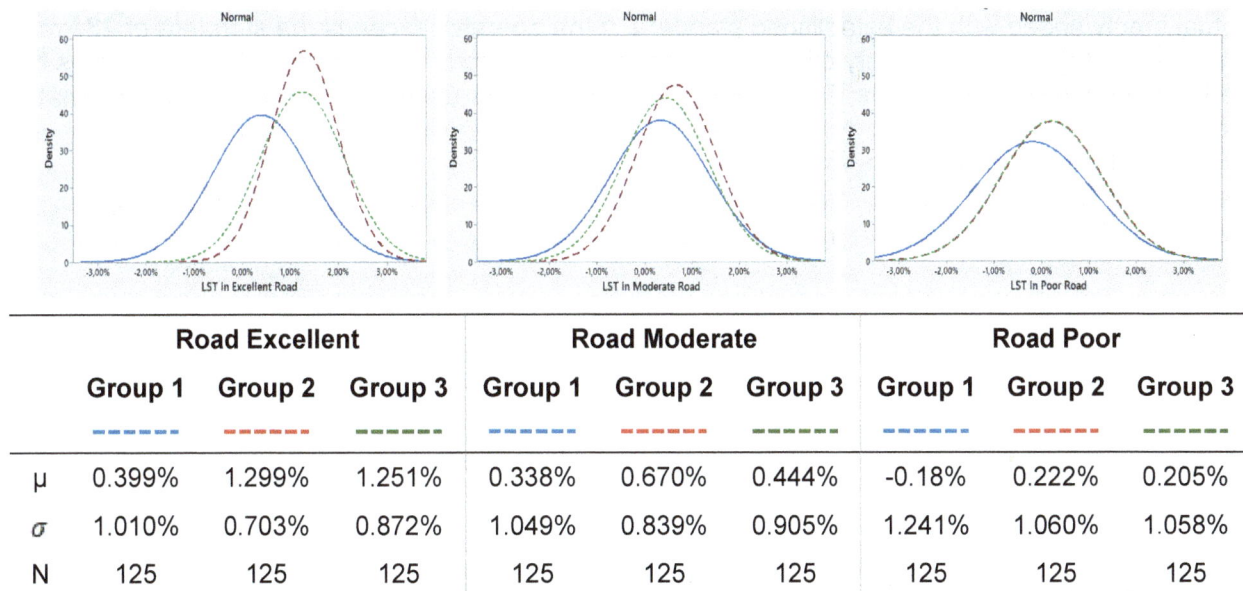

| | Road Excellent | | | Road Moderate | | | Road Poor | | |
	Group 1	Group 2	Group 3	Group 1	Group 2	Group 3	Group 1	Group 2	Group 3
	- - - - -	- - - - -	- - - - -	- - - - -	- - - - -	- - - - -	- - - - -	- - - - -	- - - - -
μ	0.399%	1.299%	1.251%	0.338%	0.670%	0.444%	-0.18%	0.222%	0.205%
σ	1.010%	0.703%	0.872%	1.049%	0.839%	0.905%	1.241%	1.060%	1.058%
N	125	125	125	125	125	125	125	125	125

FIG 5 – Kernel density estimates plot of % LST achievement.

For example, on an excellent road, Operator Group 1 achieved an average time saving of 0.399 per cent faster in minutes compared to the standard allowed stop time of 0.15 mins during a single iteration trip from front loading to disposal.

Empty Stop Time (EST)

To calculate the percentage of time savings, the researcher will compare the actual EST value with the standard 2 mins. Figure 6 shows the percentage achievement of the Empty Stop Time plot using the Kernel Density Estimates plot.

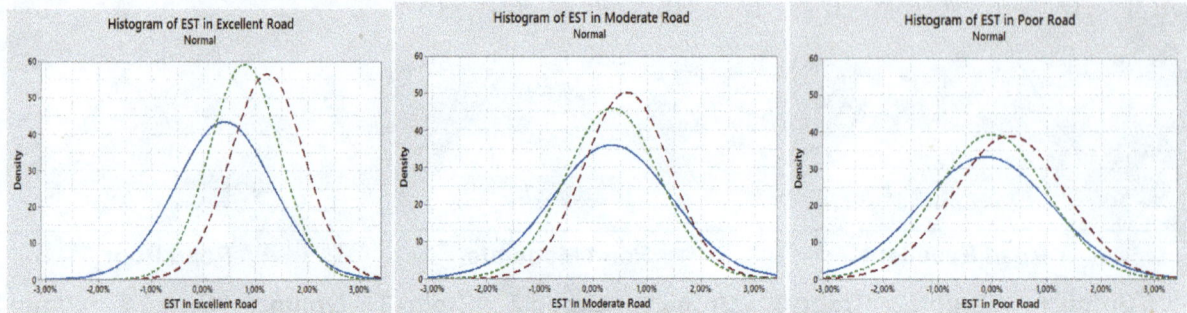

	Road Excellent			Road Moderate			Road Poor		
	Group 1	Group 2	Group 3	Group 1	Group 2	Group 3	Group 1	Group 2	Group 3
	------	------	------	------	------	------	------	------	------
μ	0.407%	1.230%	0.791%	0.344%	0.622%	0.353%	-0.12%	0.314%	-0.01%
σ	0.919%	0.705%	0.673%	1.107%	0.794%	0.869%	1.200%	1.025%	1.013%
N	125	125	125	125	125	125	125	125	125

FIG 6 – Kernel density estimates plot of % EST achievement.

For example, on an excellent road, Operator Group 1 achieved an average time saving of 0.407 per cent faster in minutes compared to the standard allowed stop time of 2 mins during a single iteration trip from disposal to front loading.

Empty Speed (ES)

To calculate the percentage contribution of time savings, the researcher can compare the operator's actual empty speed with the established standard. If the operator exceeds the standard of 23 km/h, the time required for each empty cycle will be shorter, resulting in time savings. Figure 7 shows the percentage achievement of the Empty Stop Time plot results using the Kernel Density Estimates plot.

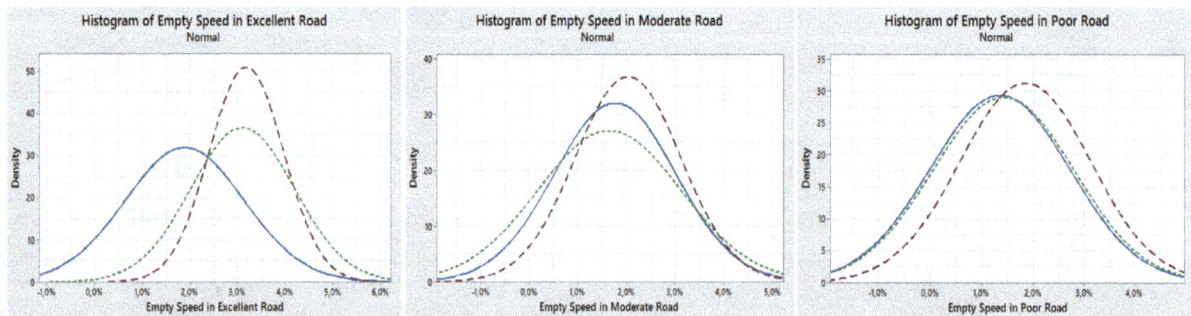

	Road Excellent			Road Moderate			Road Poor		
	Group 1	Group 2	Group 3	Group 1	Group 2	Group 3	Group 1	Group 2	Group 3
	------	------	------	------	------	------	------	------	------
μ	1.892%	3.186%	3.108%	1.771%	2.036%	1.649%	1.360%	1.857%	1.431%
σ	1.249%	0.780%	1.089%	1.243%	1.082%	1.471%	1.358%	1.275%	1.373%
N	125	125	125	125	125	125	125	125	125

FIG 7 – Kernel density estimates plot of % empty speed achievement.

For example, on an excellent road, Operator Group 1 achieved an average time saving of 1.892 per cent faster in minutes for the Empty Speed parameter compared to the total travel time from disposal to front at 1 km, with a speed standard of 23 km/h.

Altruistic game theory selection results

The calculated value used in the selection process is the average value in each experimental group. The average value is obtained from the sum of the total percent achievement of four parameters, namely Load Speed (LS), Load Stop time (LST), Empty Stop time (EST), and Empty Speed (ES). Figure 8 shows the results of the total percent achievement plot using the Kernel Density Estimates plot.

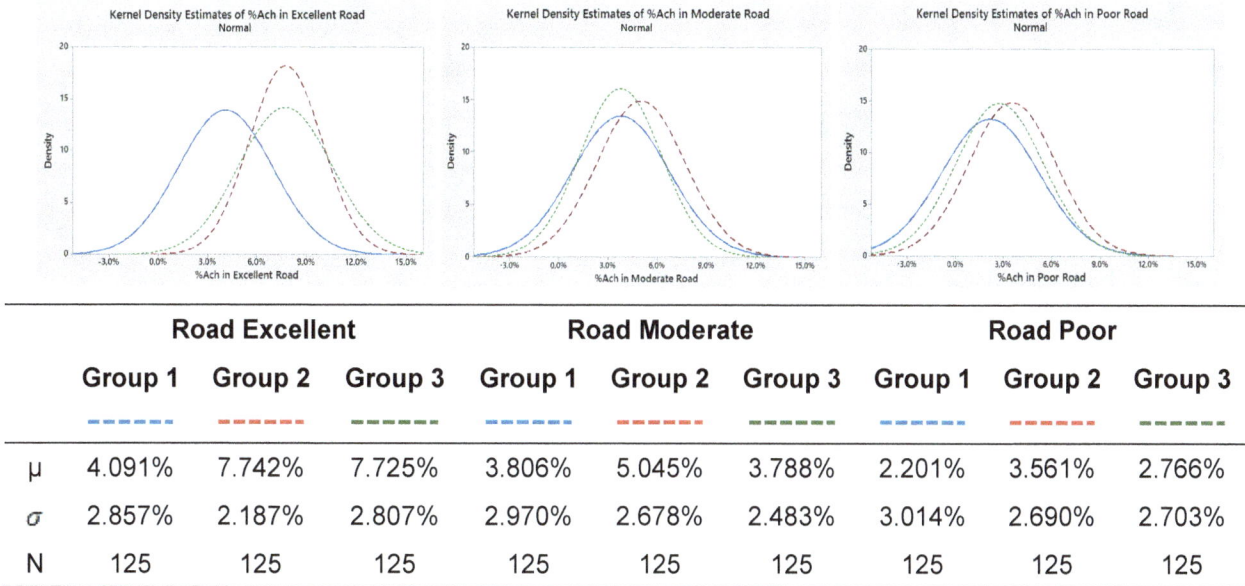

	Road Excellent			Road Moderate			Road Poor		
	Group 1	Group 2	Group 3	Group 1	Group 2	Group 3	Group 1	Group 2	Group 3
	------	------	------	------	------	------	------	------	------
μ	4.091%	7.742%	7.725%	3.806%	5.045%	3.788%	2.201%	3.561%	2.766%
σ	2.857%	2.187%	2.807%	2.970%	2.678%	2.483%	3.014%	2.690%	2.703%
N	125	125	125	125	125	125	125	125	125

FIG 8 – Kernel density estimates plot of % total achievement.

The average value of Percent Total Achievement in Figure 8 is then entered into the Altruistic Game Theory selection matrix. Table 5 shows the matrix of Altruistic Game Theory.

TABLE 5

TABLE 5
Game theory selection matrix.

Iteration 1

		G₁ / G₂ E	M	P
G₃	E	E			
		M			(3.81%); (3.56%); (7.72%)
		P		(2.20%); (5.05%); (7.72%)	
	M	E			(4.09%); (3.56%); (3.79%)
		M			
		P	(2.20%); (7.74%); (3.79%)		
	P	E		(4.09%); (5.04%); (2.77%)	
		M	(3.81%); (7.74%); (2.77%)		
		P			

Iteration 2

		G₁ / G₂ E	M	P
G₃	E	E			
		M			15.09%
		P		14.97%	
	M	E			11.44%
		M			
		P	13.73%		
	P	E		11.90%	
		M	14.32%		
		P			

Based on the calculation results with the matrix in the second iteration, the best selection falls on the following options:

- Group 1 (G₁): Operators with Entry Level (0–5 years working) → Paired with Moderate Road type.

- Group 2 (G₂): Operators with Mid-Level (5–10 years working) → Paired with Poor Road type.

- Group 3 (G₃): Operators with Senior Level (>10 years working) → Paired with Excellent Road type.

With this selection combination, a gain of 15.09 per cent will be obtained in terms of total acceleration time (LS, LST, EST, ES) against the specified standard. Where if described for each Group is the following achievement value is described:

- G₁: +3.81 per cent → Moderate Road.

- G₂: +3.56 per cent → Poor Road.

- G₃: +7.72 per cent → Excellent Road.

In this case, the combination of the above selections is considered the best selection, even though G1 could get a percent achievement of +4.09 per cent if assigned to the Excellent Road, and G2 could get a percent achievement of +7.74 per cent if assigned to the Excellent Road. However, if this selection is made, it will cause tremendous lost opportunities in G3 due to the existing constraints (no roads should not be unselected), It should be paired with a Moderate or Poor type of road, reflecting a percentage loss of opportunities of -4.95 per cent (a drop from +7.72 per cent with an Excellent Road to +2.77 per cent with a Poor Road). Because in the Altruistic Game Theory theorem it adheres to the principle of mutual interest, G1 and G2 are assigned to choose the second best haul road with a total lost opportunity in G1 of -0.28 per cent (from +4.09 per cent with the Excellent Road to +3.81 per cent with the Moderate Road) and in G2 of -4.18 per cent (from +7.74 per cent with the Excellent Road to +3.56 per cent with the Poor Road) so that the total lost opportunity of G1 and G2 is -4.46 per cent smaller than the loss in G3 of -4.95 per cent.

Essentially, all groups will achieve the highest percentage of total accomplishment if they are placed on a good and standardised haul road, which is categorised as an Excellent type of haul road in this study. However, in Altruistic Game Theory, the best solution is pursued by attempting to choose the second option, even though it is not optimal, as it does not lead to significant losses in aggregate or among work groups.

Expert judgement

Based on the selection results obtained from the calculations of Altruistic Game Theory, a study and discussion were held with experts from the Operations, Engineering, and SHE (Safety, Health, and Environment) departments to explore the impact of assignment selection on operator group performance and behaviour. The results obtained were:

- Group 1 (G_1) (Entry Level): This group tends to hesitate to exceed the specified speed limit. Additionally, when traveling on a low-quality road, the operator becomes increasingly cautious due to the road's inconsistency with standards, which leads to reduced productivity. This group proves to be far more profitable on a moderate road compared to others, as the operator is much more stable and efficient on a moderately classified road.

- Group 2 (G_2) (Mid-Level): Compared to other groups, this group tends to seek new challenges in work and dislikes monotonous tasks. Consequently, placing Group 2 on the Poor Road can avoid significant losses, as it allocates other groups to this Poor Road.

- Group 3 (G_3) (Senior Level): They tend to feel demotivated when faced with challenging tasks, as their performance may decline. Consequently, while working, they prefer to remain in a comfort zone. Therefore, with a wealth of work experience and a desire for comfort, Group 3 operators are far more valuable when placed in the excellent category.

CONCLUSIONS

Based on the results of this study, it can be concluded that assigning specific operator behaviours according to road conditions plays a crucial role in determining the productivity of the HD785 unit in open pit mining activities. By grouping operator behaviours and analysing existing road conditions, this research successfully identified behaviour patterns that can be optimised to improve operational efficiency.

As a key outcome, this study developed an algorithm for selecting operator behaviour based on road conditions using the Modified Game Theory approach by adopting the concept of altruistic thinking. This algorithm can provide more accurate recommendations regarding the selection of the best behaviour suitable for specific road conditions, maximising productivity for one operator group without disadvantaging other operator groups. The modified game theory approach with the concept of altruistic thinking provides a strong analytical framework for assessing the bargaining power of individual operators and optimising cooperation among them to achieve more efficient performance.

Additionally, to ensure the accuracy and validity of the results, an in-depth analysis using linear regression methods was conducted on the critical aspects that constitute HD785 unit productivity. This analysis identified significant relationships between the operator activity parameters Load Speed (LS), Load Stop Time (LST), Empty Stop Time (EST), and Empty Speed (ES) with the resulting productivity. To further strengthen these results, data distribution analysis was also carried out, helping to understand the variation and patterns in the collected data and providing a clearer picture of the factors affecting operator performance across various road conditions.

The implementation of this study resulted in a productivity increase of approximately 0.5 per cent bcm per hr continuously. The developed method can significantly contribute to open pit mining companies facing similar challenges, enabling them to optimise operator performance and reduce waste in their operations.

The findings of this research are expected to serve as a reference for the mining industry to enhance human resource management and optimise operator performance, ultimately positively impacting the overall productivity of the company.

ACKNOWLEDGEMENTS

The authors would like to express their sincere gratitude to PT. Pamapersada Nusantara for their invaluable support and resources provided during this research. Their commitment to advancing operational efficiency and safety in the mining industry has significantly contributed to this study's success.

Special thanks to our colleagues and team members at the Mine Optimisation Department, Engineering Department, Operation Department, and SHE (Safety, Health, and Environment) Department for their continuous support, insightful discussions, and collaboration throughout this project. Their expertise and dedication to excellence were instrumental in refining the methodology and ensuring the quality of the research.

REFERENCES

Abou Elassad, Z E, Mousannif, H, Al Moatassime, H and Karkouch, A, 2020. The application of machine learning techniques for driving behavior analysis: A conceptual framework and a systematic literature review, *Engineering Applications of Artificial Intelligence*, 87:103312.

Hoefer, M and Skopalik, A, 2013. Altruism in Atomic Congestion Games, *ACM Transactions on Economics and Computation*, 1(4):21;1–21.

Ramos, É M S, Bergstad, C J and Nässén, J, 2020. Understanding daily car use: Driving habits, motives, attitudes and norms across trip purposes, *Transportation research part F: traffic psychology and behaviour*, 68:306–315.

Riar, M, Morschheuser, B, Zarnekow, R and Hamari, J, 2024. Altruism or egoism – how do game features motivate cooperation? An investigation into user we-intention and I-intention, *Behaviour and Information Technology*, 43(6):1017–1041.

Rothe, J, 2021. Thou shalt love thy neighbor as thyself when thou playest: Altruism in game theory, in *Proceedings of the AAAI conference on artificial intelligence,* 35(17):15070–15077.

Sagberg, F, Selpi, B, Piccinini, G F and Engström, J, 2015. A review of research on driving styles and road safety, *Human factors*, 57(7):1248–1275.

Torres, L C C, 2022. Case study: Simulation and artificial intelligence application for the optimization of the hauling and loading process in an open pit mining system, *IFAC-PapersOnLine*, 55(39):265–269.

Multiphysics modelling of in-place recovery of copper at stope-scale

Z Wang[1], C Xu[2] and P Dowd[3]

1. Grant Funded Researcher, The University of Adelaide, Adelaide SA 5005.
   Email: zhihe.wang@adelaide.edu.au
2. Associate Professor, The University of Adelaide, Adelaide SA 5005.
   Email: chaoshui.xu@adelaide.edu.au
3. Professor, The University of Adelaide, Adelaide SA 5005. Email: peter.dowd@adelaide.edu.au

ABSTRACT

In-Place Recovery (IPR) through underground stope leaching is potentially an effective mining technology to extract copper from low-grade hard deposits that are otherwise considered uneconomic. This study provides a comprehensive modelling of the main physical processes relevant to IPR of copper minerals. This includes adopting Richard's equation with Van Genuchten retention model for unsaturated irrigation, convection-diffusion equation with the Millington and Quirk (1961) effective diffusivity model for transport of species, and reaction zone-based shrinking core model for mineral dissolution. Multiple cases with different material properties and operational parameters were examined to assess how these factors impact the stope leaching performance. It was shown that a good overall performance requires both fast and uniform reaction within the stope. For the tested cases, the results suggest that >50 per cent of the copper in the stope can be dissolved through leaching within one year of operation, given that a sufficiently high irrigation rate and/or lixiviant concentration were applied.

INTRODUCTION

As shallow mineral resources are depleting with years of mining, mining is moving toward deeper and more complex deposits, leading to increased cost and greater technological challenges (Sinclair and Thompson, 2015; Bahamóndez et al, 2016; Sapsford, Cleall and Harbottle, 2017; Dare-Bryan and Hassanvand, 2023). On the other hand, the demand for critical minerals such as copper continues to grow, with a projection of tripling the current market size by 2030 (International Energy Agency (IEA), 2024). Along with an increasing ESG (Environmental, Social, and Governance) requirement worldwide, these have generated huge interest in seeking alternative mining methods that can achieve low-cost and low-environmental-impact mineral extraction. Different from conventional mining methods, In-Place Recovery (IPR) (Figure 1) uses high-intensity blasting to generate connected void space for the intact host rock, creating sufficient contact between lixiviant and ore fragment to dissolve the target minerals (Bahamóndez et al, 2016; Dare-Bryan and Hassanvand, 2023; Estay et al, 2023). With minimal ore transportation and processing, IPR can potentially be an effective mining method for low-grade copper deposits with less cost and lower environmental impacts.

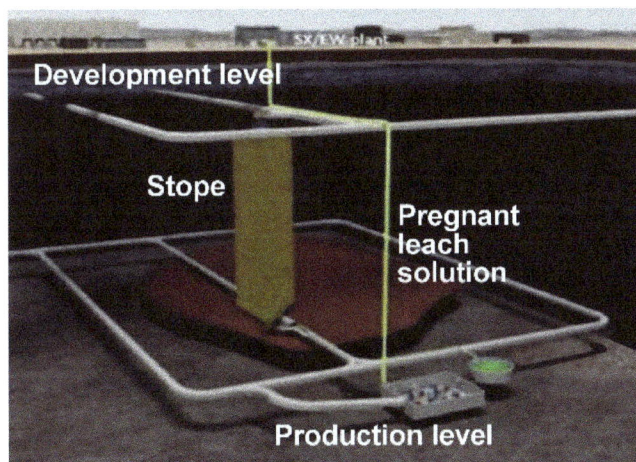

FIG 1 – Illustration of in-place recovery through stope leaching (Dare-Bryan and Hassanvand, 2023).

Many modelling works have been conducted to assess the performance of heap leaching at both lab-scale and field-scale (Leahy, Davidson and Schwarz, 2005; Bennett *et al*, 2012; Ilankoon *et al*, 2017). However, different from heap leaching, in which a uniform material property can be achieved through crushing, grinding, and agglomeration, the stope of IPR contains ore fragments with a wider range of physical size, leading to stronger heterogeneity at various scales. In this work, we considered the ore fragment to have the same size as typical run-of-mine blasted ore fragments, ranging from sub-millimetre to sub-metre. Multiphysics modelling was conducted to simulate IPR operation in a stope, by considering the main physical processes including unsaturated fluid flow, transport of dissolved species, and dissolution of multiple minerals. Various cases with different material properties and leaching parameters were examined to assess their impact on the leaching performance.

METHODOLOGY

Governing equations

Unsaturated flow

Similar to conventional heap leaching, fluid flow in IPR involves applying lixiviant (liquid phase) to the stope through sprinklers/drippers and injecting air (gas phase) to enhance aeration in the stope. Therefore, IPR is expected to also operate under unsaturated conditions. In this work, the unsaturated fluid flow in porous media is considered to be governed by Richard's equation (Bennett *et al*, 2012; Wang *et al*, 2022), as given by:

$$\frac{C_m}{g}\frac{\partial p}{\partial t} + \nabla \cdot \rho[-\frac{K_s K_r}{\mu}(\nabla p + \rho g \nabla D)] = Q_m \tag{1}$$

Where:

C_m	is the specific capacity
g	is the gravitational acceleration
ρ	is the fluid density
t	is the time
K_s	is the saturated permeability
K_r	is the relative permeability
μ	is the dynamic viscosity
D	is the elevation
Q_m	is the flow rate

The van Genuchten retention model (1980) is adopted to link saturation with capillary pressure and relative permeability, as given by:

$$Se = \begin{cases} \frac{1}{[1+|\alpha H_p|^n]^m}, & H_p < 0 \\ 1, & H_p \geq 0 \end{cases} \tag{2}$$

$$K_r = \begin{cases} Se^l\left[1-\left(1-Se^{\frac{1}{m}}\right)^m\right]^2, & H_p < 0 \\ 1, & H_p \geq 0 \end{cases} \tag{3}$$

Where:

Se	is the effective saturation
H_p	is the capillary pressure
K_r	is the relative permeability

α, n, m and l are constants related to material type

Mass transport

Mass transport depicts the movement of dissolved species in solution and is governed by the convection-diffusion equation in porous media (Wang *et al*, 2022), as given by:

$$\frac{\partial(\theta C)}{\partial t} + \boldsymbol{u} \cdot \nabla C = \nabla \cdot (D_e \nabla C) + R + S \tag{4}$$

Where:

θ is the liquid volume fraction

C is the concentration of a certain species

$\boldsymbol{u}$ is the flow velocity

D_e is the effective diffusion coefficient

R is a reaction rate expression

S is a source term

To account for the effect of tortuosity, the Millington and Quirk model (1961) was adopted to derive the effective diffusion coefficient of the stope according to porosity (ie $D_e = \varepsilon^{4/3}D$, where D is the diffusion coefficient of a certain species in free solution and ε is the porosity). Note here we use Equation 4 to solve for mass transport in inter-fragment voids. Mass transport through intra-fragment voids is accounted for by the reaction rate expression (which is a function of ore fragment effective diffusion coefficient) for mineral dissolution in the following section.

Reaction

We consider low-grade copper ores containing chalcocite (Cu_2S), covellite ($Cu_{1.2}S$), and pyrite (FeS_2), which are reacted with lixiviant with ferric sulfate (Fe_2SO_4). The corresponding reactions include:

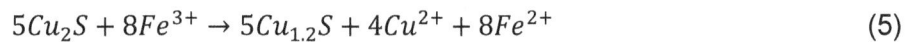

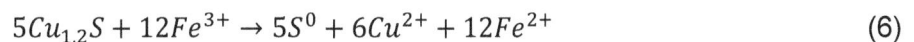

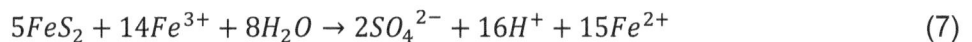

$$5Cu_2S + 8Fe^{3+} \rightarrow 5Cu_{1.2}S + 4Cu^{2+} + 8Fe^{2+} \tag{5}$$

$$5Cu_{1.2}S + 12Fe^{3+} \rightarrow 5S^0 + 6Cu^{2+} + 12Fe^{2+} \tag{6}$$

$$5FeS_2 + 14Fe^{3+} + 8H_2O \rightarrow 2SO_4^{2-} + 16H^+ + 15Fe^{2+} \tag{7}$$

Considering low-grade ores, reaction progresses from the outer surface of the ore toward the centre, which can be described by a reaction zone-based shrinking core model (Braun, Lewis and Wadsworth, 1974; Leahy, Davidson and Schwarz, 2005). The reacted fraction of any mineral in the ore fragment can be expressed as:

$$\frac{da}{dt} = \frac{3(1-a)^{2/3}[Fe^{3+}]}{\tau_c + 6\tau_d(1-a)^{1/3}[1-(1-a)^{1/3}]} \tag{8}$$

Where:

a is the reacted fraction of a certain mineral

τ_c and τ_d are related to ore and mineral properties including reaction rate constant and effective diffusion coefficient, given by Leahy, Davidson and Schwarz (2005).

Oxidation of ferrous ions (Fe^{2+}) back to ferric ions (Fe^{3+}) by dissolved oxygen (DO) in the lixiviant is also considered, and the reaction follows:

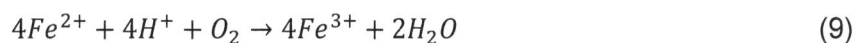

$$4Fe^{2+} + 4H^+ + O_2 \rightarrow 4Fe^{3+} + 2H_2O \tag{9}$$

The corresponding reaction rate (Wermink and Versteeg, 2017) is given by:

$$\frac{d[Fe^{2+}]}{dt} = -k[Fe^{2+}]^2 P_{O_2} \tag{10}$$

Where:

k is the reaction rate constant

P_{O_2} is the oxygen partial pressure

The temperature effect on all reaction rate constants is determined according to the Arrhenius equation given by:

$$k = k_0 exp\left(-\frac{E_A}{RT}\right)$$ (11)

Where:

k_0 is the reaction rate constant at the reference temperature

E_A is the activation energy

R is the gas constant

T is the temperature

The reaction rate constants (at the reference temperature 20°C) and activation energy for the above reactions can be found in previous studies (Leahy, Davidson and Schwarz, 2005; Wermink and Versteeg, 2017).

Numerical settings

All governing equations were implemented and solved with the finite element code COMSOL. For computational efficiency, fluid flow was first solved to reach a steady-state flow field for irrigation and aeration. Interaction between irrigation and aeration is one-way, that is, irrigation settings control the saturation distribution, which determines the air relative permeability (Bennett et al, 2012). Then, the obtained steady-state flow field was used to solve for mass transport and reaction. The lixiviant is considered to possess the same physical properties as water, with a constant concentration of Fe^{3+} and DO when applied from the top of the stope. All dissolved species are set with a constant diffusion coefficient of 2×10^{-9} m^2/s in free solution. Where applicable, aeration is applied at the bottom of the stope. Mass transfer from gas phase oxygen to dissolved oxygen in the liquid phase follows Henry's Law, with a mass transfer coefficient of 1×10^{-5} m/s through the gas-liquid interface. The interface area is assumed to be the same as the surface area of the ore fragments. The 2D stope considered in this work has a height of 30 m and a width of 20 m (as shown in Figure 2), filled with blasted ores of a discrete size distribution. All leaching simulations were run for one year of operation to assess the leaching performance.

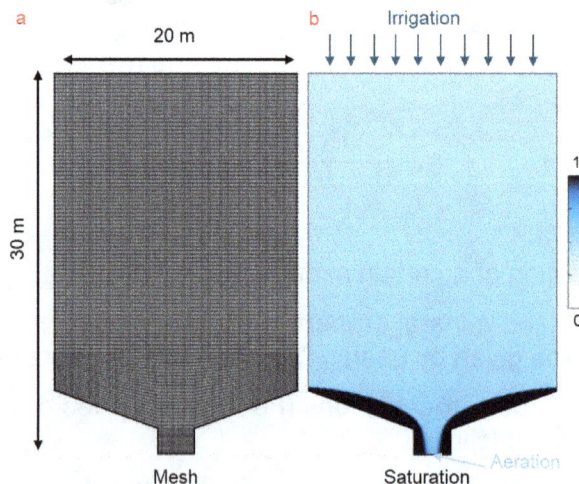

FIG 2 – Dimension of the stope model and associated mesh (a); and simulated steady-state saturation distribution of Base Case (b).

RESULTS

Base case

For the base case, the ore fragment size ranges from 0.5 to 300 mm, with equal mass fraction for each of ten discrete size classes within the size range (ie mass fraction = 0.1 for each of ten size classes). Previous studies showed that the porosity of packed ore structures is generally height-dependent due to potential compaction and particle settlement during operation. To account for this, we adopted the height-dependent hydraulic properties from previous results (McBride *et al,* 2016), including porosity, saturated permeability, and Van Guenchten parameters. The grades (G) of chalcocite and pyrite are set to be 0.9 per cent and 1.0 per cent, respectively. Chalcocite first reacts with Fe^{3+} to generate covellite following Equation 5, and then the generated covellite continues reacting with Fe^{3+} following Equation 6. A uniform irrigation rate of 20 $L/m^2/h$ is applied at the top of the stope. The applied lixiviant has a constant Fe^{3+} concentration of 0.1 mol/L at pH=1 and a maximum amount of DO under atmospheric pressure at an underground temperature of 30°C. The effective diffusion coefficient of ore fragments is set at 1×10^{-12} m^2/s with a porosity of 0.05.

The steady-state saturation distribution is in Figure 2b. It is shown that the majority of the stope has a relatively low saturation of ~0.3, with lower saturation closer to the top of the stope. Evidently, high saturation (>0.9) is achieved near the bottom of the stope, where all pregnant leach solution joins and exits the stope. Figure 3a shows the distribution of Fe^{3+} concentration normalised by its feed concentration of 0.1 mol/L (referred to as Fe^{3+} concentration hereafter), after one year of leaching operation. It can be observed that the Fe^{3+} concentration has barely penetrated halfway into the stope before being fully consumed. This leads to similar distributions of dissolved minerals in Figure 3d–f, where the top regions of the stope show higher percentages of dissolved minerals (Dm). From the reaction rate constants of the three minerals, chalcocite has the fastest reaction, followed by covellite, with pyrite having the slowest reaction. Overall, the base case demonstrates an average Dm of 19.5 per cent (from 0 to 63.6 per cent), 16.5 per cent (from 0 to 58.1 per cent), and 4.9 per cent (from 0 to 21.2 per cent) for chalcocite, and covellite, and pyrite, respectively, and the overall dissolved copper was 18.0 per cent. From the simulation results, the main reason for the low percentage of dissolved copper for the Base Case appears to be insufficient lixiviant supply to the lower half of the stope, as all Fe^{3+} has been consumed in the upper half of the stope. For the upper half with more abundant Fe^{3+} supply, a maximum percentage of dissolved copper of 60.8 per cent can be achieved.

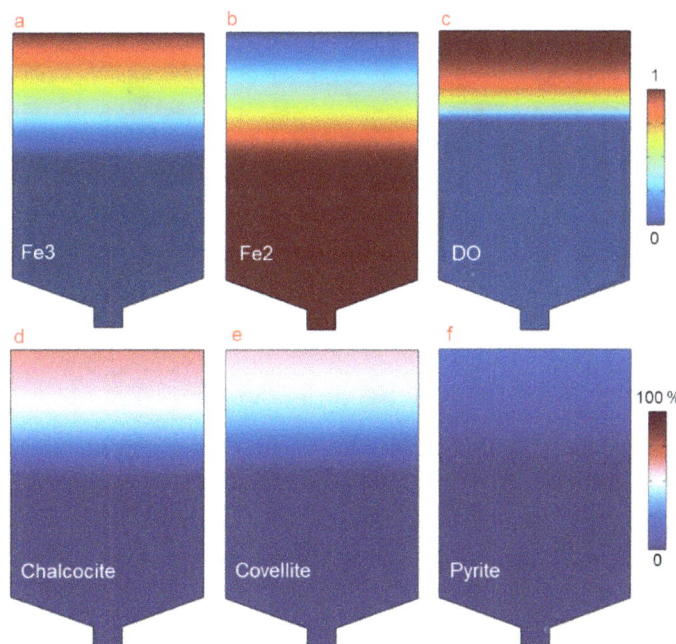

FIG 3 – Base case IPR simulation for one year with distributions of: (a) normalised ferric ion concentration; (b) normalised ferrous ion concentration; (c) normalised dissolved oxygen concentration; (d) percentage of dissolved chalcocite; (e) percentage of dissolved covellite; and (f) percentage of pyrite.

Comparison between cases

Based on the Base Case, the leaching performance of multiple cases was assessed by varying certain material properties or leaching conditions, as listed in Table 1.

TABLE 1

IPR performance of different cases.

Cases	Change compared to base case	Percentage of dissolved copper (relative to base case)
Base case	N/A	18.0% (0%)
Case 1	Higher irrigation at 60 L/m²/h	48.1% (↑30.1%)
Case 2	Higher Fe^{3+} concertation at 0.5 mol/L	63.5% (↑45.5%)
Case 3	Lower diffusion coefficient at 1e-13 m²/s	22.5% (↑4.5%)
Case 4	50% more finer fragments	17.2% (↓0.8%)
Case 5	With aeration of 5 L/m²/h	18.5% (↑0.5%)
Case 6	Optimal aeration	23.0% (↑5.0%)

Case 1: Heap leaching typically has an irrigation rate of 5–20 L/m²/h for a typical lift height of 8–10 m. For a 30 m high stope, a higher irrigation rate appears to be justified to ensure sufficient lixiviant supply. Compared to the Base Case, a higher irrigation rate of 60 L/m²/h was used for Case 1, and the distribution of the percentage of dissolved chalcocite in one year is shown in Figure 4a. A much more uniform distribution can be observed, with >50 per cent of the stope having chalcocite being dissolved by >50 per cent. The overall percentage of dissolved copper in one year is 48.1 per cent, almost triple that of the Base Case.

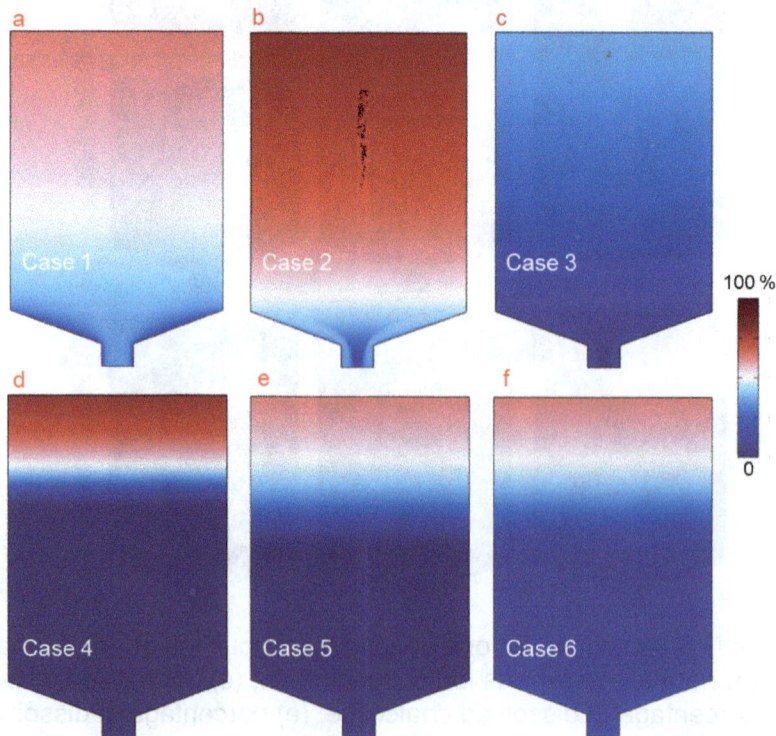

FIG 4 – Distributions of percentage of dissolved chalcocite for Cases 1–6.

Case 2: Apart from increasing the irrigation rate, applying higher concentrations of Fe^{3+} can be another way to enhance lixiviant supply. Here, a higher Fe^{3+} concentration of 0.5 mol/L was applied for the feed lixiviant in Case 2, and the distribution of the percentage of dissolved chalcocite in one year is shown in Figure 4b. It is shown that a majority of the stope demonstrated high percentages of dissolved chalcocite (>70 per cent). The overall percentage of dissolved copper in one year is 63.5 per cent, more than triple that of the Base Case.

Case 3: The diffusion coefficient of ore fragments can be a controlling factor for the transport of lixiviant from inter-fragment voids to the minerals within ore fragments. Diffusion coefficient is mostly determined by the internal pore structure of ore fragments. Although it is an inherent rock property, blasting-induced microcracks can lead to a higher diffusion coefficient. For comparison, a lower effective diffusion coefficient of 1e-13 m^2/s was used for Case 3, and the distribution of the percentage of dissolved chalcocite in one year is shown in Figure 4c. Although the reaction is generally slowed down with a prolonged diffusion process within the ore fragments due to a lower diffusion coefficient, this gives the lixiviant more time to migrate through the inter-fragment voids, leading to a more uniform reaction. The overall percentage of dissolved copper in one year is 22.5 per cent, which is 4.5 per cent higher than that of the Base Case.

Case 4: In addition to the diffusion coefficient, the size of ore fragments is also important for intra-fragment lixiviant diffusion. The fragment size distribution is directly associated with blasting and rock mechanical properties. For Case 4, it is set to contain 50 per cent more finer fragments than Base Case, that is, the lower five size classes of Case 4 with smaller fragments would have a mass fraction of 0.15, whereas the upper 5 size classes with larger fragments would all have a mass fraction of 0.05. The distribution of the percentage of dissolved chalcocite in one year is shown in Figure 4d. Despite having more finer fragments, which resulted in a higher percentage of dissolved chalcocite when there is sufficient lixiviant supply, the overall reaction took place in a narrower region of the stope. This leads to an overall percentage of dissolved copper of 17.2 per cent in one year, which is 0.8 per cent lower than that of the Base Case.

Case 5: Forced aeration is commonly adopted in heap leaching to enhance leaching performance, especially when oxidative bacteria exist. However, here we considered abiotic conditions in this work due to an insufficient understanding of the bacteria in underground stopes. For Case 5, an aeration rate of 5 L/m^2/h was added to the bottom of the stope. The gas phase oxygen will dissolve into the solution when the oxygen partial pressure is higher than its threshold according to Henry's law. The distribution of the percentage of dissolved chalcocite in one year is shown in Figure 4e. In general, due to a relatively slow oxygen dissolution and reaction in Equation 9, Figure 4e is very similar to Figure 3d, showing a limited enhancement. The overall percentage of dissolved copper in one year is 18.5 per cent, which is 0.5 per cent higher than that of the Base Case.

Case 6: As a further examination of the effect of aeration, Case 6 is considered to have an ideal aeration scenario so that the maximum dissolved oxygen is always maintained in the lixiviant regardless of oxygen consumption. The distribution of the percentage of dissolved chalcocite in one year is shown in Figure 4f. It is noticed that compared to Figure 3d and Figure 4e, there is a higher percentage of dissolved chalcocite resulting from minerals reacting with the regenerated Fe^{3+}. The overall percentage of dissolved copper in one year is 23.0 per cent, which is 5.0 per cent higher than that of the Base Case.

From the above case analysis, increasing the irrigation rate or Fe^{3+} concentration was found to have a more evident enhancement to copper dissolution. Therefore, additional cases were examined by increasing the irrigation rate and/or Fe^{3+} concentration as shown in Table 2. It is shown that the cases in the lower-right part of the table, including the three cases along the diagonal, can all achieve an overall percentage of dissolved copper higher than 48 per cent. Considering an estimated copper recovery of 50 per cent in one year for IPR to be potentially economic (Dare-Bryan and Hassanvand, 2023), these cases can all generally achieve this target, without additional engineering effort such as more intensive blasting or adding forced aeration.

TABLE 2

Comparison of IPR performance with different irrigation rates and Fe^{3+} concentrations.

	Fe^{3+} concentration 0.1 mol/L	Fe^{3+} concentration 0.25 mol/L	Fe^{3+} concentration 0.5 mol/L
Irrigation rate 20 L/m²/h	Dissolved Cu: 18.0% (Range: 0–60.8%)	Dissolved Cu: 37.5% (range: 0–70.8%)	Dissolved Cu: 63.5% (Range: 4.5–77.8%)
Irrigation rate 40 L/m²/h	Dissolved Cu: 36.0% (range: 0–60.8%)	Dissolved Cu: 61.5% (range: 38.6–70.8%)	Dissolved Cu: 73.3% (range: 65.3–77.8%)
Irrigation rate 60 L/m²/h	Dissolved Cu: 48.1% (range: 0–60.8%)	Dissolved Cu: 65.6% (range: 51.7–70.8%)	Dissolved Cu: 75.0% (range: 69.2–77.8%)

CONCLUSIONS AND FUTURE WORK

This work has provided stope-scale multiphysics simulations to assess the leaching performance of In-Place Recovery (IPR). Typical material properties and leaching parameters were examined, including ore fragment size, ore fragment effective diffusion coefficient, irrigation rate, lixiviant concentration, and aeration. The results from this work demonstrated the importance of achieving uniform reaction through sufficient lixiviant supply. For constant material properties, increasing irrigation rate and lixiviant concentration are found to evidently enhance the overall leaching performance. By optimising both irrigation rate and lixiviant concentration, a target of dissolving 50 per cent of the copper minerals can be achieved within one year. Considering a minimal or no cost associated with ore transportation and further treatment, the current simulation results demonstrate the potential for IPR to be an effective solution to extract low-grade copper from hard rock.

The cases studied in the current work are all based on available data from previous lab or site tests conducted for heap leaching conditions. Future work will incorporate material properties measured from lab tests and IPR trial tests through a Cooperative Research Centres Projects (CRC-P) grant by the Australian Government to further validate the established multiphysics model. Additional features, where applicable, will be incorporated to consider, eg leaching of other copper minerals (such as chalcopyrite) and gangues.

ACKNOWLEDGEMENTS

The authors would like to acknowledge the support from the Australian Government through a Cooperative Research Centres Projects (CRC-P) grant. We also thank Armineh Hassanvand and Peter Dare-Bryan at Orica Mining Services, David Hunter and Rob Coleman at Core Resources, and BHP Think and Act Differently (TAD) Team for their ongoing support.

REFERENCES

Bahamóndez, C, Castro, R, Vargas, T and Arancibia, E, 2016. In situ mining through leaching: Experimental methodology for evaluating its implementation and economic considerations, *Journal of the Southern African Institute of Mining and Metallurgy*, 116(7):689–698.

Bennett, C R, McBride, D, Cross, M and Gebhardt, J E, 2012. A comprehensive model for copper sulphide heap leaching: Part 1 Basic formulation and validation through column test simulation, *Hydrometallurgy*, 127–128:150–161.

Braun, R L, Lewis, A E and Wadsworth, M E, 1974. in-Place Leaching of Primary Sulfide Ores: Laboratory Leaching Data and Kinetics Model, *Metallurgical Transactions*, 5(8):1717–1726.

Dare-Bryan, P and Hassanvand, A, 2023. Economic and environmental assessment of underground in-situ leaching processes utilising drill and blast to achieve high permeability, in *ALTA2023*, pp 1–23.

Estay, H, Quezada, S D, Arancibia, E and Vargas, T, 2023. Economic Assessment of an In Situ Leaching Operation with Ore Preconditioning Using Sublevel Stoping Techniques, *Mining, Metallurgy and Exploration*, 40(2):493–504.

Ilankoon, I M S K, Neethling, S J, Huang, Z and Cheng, Z, 2017. Improved inter-particle flow models for predicting heap leaching hydrodynamics, *Minerals Engineering*, 111(March):108–115.

International Energy Agency (IEA), 2024. Global Critical Minerals Outlook 2024, Report: 282.

Leahy, M J, Davidson, M R and Schwarz, M P, 2005. A model for heap bioleaching of chalcocite with heat balance: Bacterial temperature dependence, *Minerals Engineering*, 18(13–14):1239–1252.

McBride, D, Gebhardt, J E, Croft, T N and Cross, M, 2016. Modeling the hydrodynamics of heap leaching in sub-zero temperatures, *Minerals Engineering*, 90:77–88.

Millington, R J and Quirk, J P, 1961. Permeability of porous solids, *Transactions of the Faraday Society*, 57:1200–1207.

Sapsford, D, Cleall, P and Harbottle, M, 2017. In Situ Resource Recovery from Waste Repositories: Exploring the Potential for Mobilization and Capture of Metals from Anthropogenic Ores, *Journal of Sustainable Metallurgy*, 3(2):375–392.

Sinclair, L and Thompson, J, 2015. In situ leaching of copper: Challenges and future prospects, *Hydrometallurgy*, 157:306–324.

van Genuchten, M T, 1980. A Closed-form Equation for Predicting the Hydraulic Conductivity of Unsaturated Soils, *Soil Science Society of America Journal*, 44(5):892–898.

Wang, H, Xu, C, Dowd, P A, Wang, Z and Faulkner, L, 2022. Modelling in-situ recovery (ISR) of copper at the Kapunda mine, Australia, *Minerals Engineering*, 186(January):107752.

Wermink, W N and Versteeg, G F, 2017. The Oxidation of Fe(II) in Acidic Sulfate Solutions with Air at Elevated Pressures, Part 1, Kinetics above 1 M H_2SO_4, *Industrial and Engineering Chemistry Research*, 56(14):3775–3788.

Structurally controlled slope stability study using genetic algorithm

G You[1]

1. Senior Lecturer, Federation University Australia, Ballarat Vic 3353.
   Email: g.you@federation.edu.au

ABSTRACT

A genetic algorithm (GA) program has been developed in MATLAB to simultaneously analyse five random variables relevant to slope stability: three structural parameters (joint location, fault location and dip angle) and two strength parameters (cohesion, c and friction angle, φ). The algorithm is implemented within a limit equilibrium framework and applied to a case study of bench-scale slope failure. The results indicate that the failure is structurally controlled, with the GA effectively identifying a near-maximum unstable block that corresponds closely with field observations of the failed mass. The critical block geometry derived by the GA is characterised by Joint 1 located at the toe with a lower-end dip angle and a fault positioned furthest from the bench crest, coupled with minimal cohesion and friction angle values. The analysis concludes that the failure initiated via sliding along Joint 1, where it is intersected at the rear by the fault and progressed through toppling against Joint 2, resulting in a combined sliding-toppling failure mechanism.

INTRODUCTION

Rock slope instability can arise from structurally controlled failure mechanisms, stress-induced failure, or a combination of both. Structurally controlled failures typically manifest as discrete blocks or wedges bounded by discontinuities, whereas stress-induced failures may result in localised spalling, squeezing, or tensile splitting of the rock mass (Barrett, McQueen and Bendtsen, 2008). Common failure mechanisms observed in rock slopes include circular, planar, wedge and toppling failures, as well as combinations of these modes. Circular failures, which have been extensively studied, generally occur in cohesive soils or homogeneous rock masses and are characterised by a rotational failure surface (Gao and Ge, 2024; Manouchehrian, Gholamnejad and Sharifzadeh, 2014; Zolfaghari, Heath and McCombie, 2005). In contrast, open pit slopes are more susceptible to structurally controlled failures—namely planar, wedge and toppling modes—due to the presence of geological discontinuities such as joints, faults and bedding planes. Planar and wedge failures involve sliding along well-defined discontinuity planes, while toppling failure occurs when rock columns rotate forward around a pivot point. Although most rock slope failures tend to exhibit a dominant failure mechanism, combined failure modes are occasionally observed (Al Mandalawi *et al*, 2018; Alejano, Gómez-Márquez and Martínez-Alegría, 2010; Mohtarami, Jafari and Amini, 2014).

Recent years have seen the growing application of artificial intelligence (AI) techniques in slope stability analysis. Gao and Ge (2024) provided a comprehensive review, classifying AI approaches into four categories: simulated evolutionary algorithms, quasi-physical intelligence methods, swarm intelligence methods and hybrid approaches. Among these, genetic algorithms (GAs)—a class of simulated evolutionary algorithms—have proven particularly effective for solving complex optimisation problems in geotechnical engineering. GAs mimic the process of natural selection, employing a population-based approach to iteratively evolve candidate solutions. Each candidate, or chromosome, encodes a set of problem-specific parameters. The fitness of each chromosome is evaluated using a predefined objective function, guiding the selection process for generating new populations through operations such as crossover and mutation. The evolution continues until convergence criteria are met, ideally yielding an optimal or near-optimal solution(Bhandary, Krishnamoorthy and Rao, 2019; Goldberg, 1989; Wu and Utili, 1995). The optimisation process of genetic algorithms can be illustrated in a flow chart (Wijesinghe *et al*, 2022; Zolfaghari, Heath and McCombie, 2005), as shown in Figure 1.

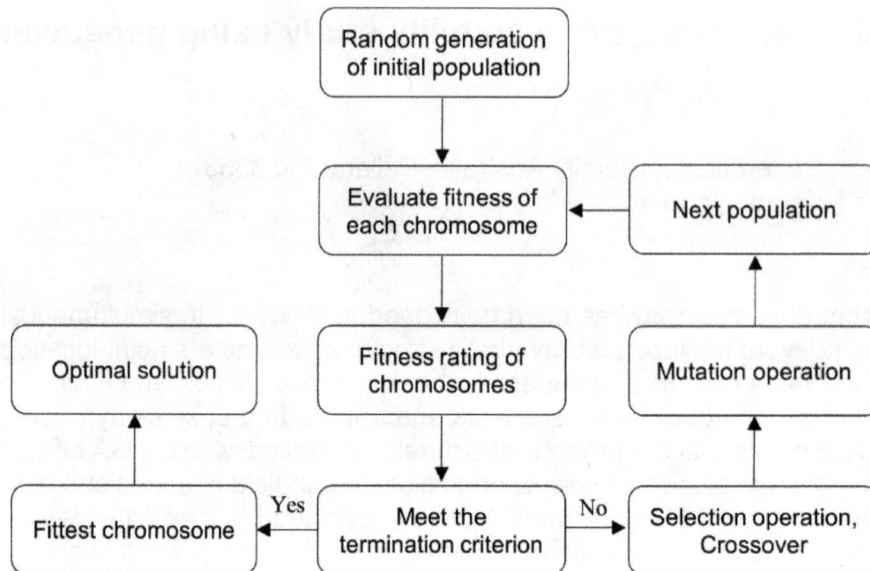

FIG 1 – Flow chart of genetic algorithm optimisation (Wijesinghe *et al*, 2022).

To apply evolutionary computation principles to slope stability analysis, it is essential to define and mathematically formulate the objective function and constraints. Common objective functions include the minimisation of the Factor of Safety (*FS*) (Goshtasbi, Ataei and Kalatehjary, 2008), probability of failure (Zeng, Jimenez and Jurado-Piña, 2015), or geometrical criteria such as bench width (Wijesinghe *et al*, 2022; Wijesinghe and You, 2016). The associated constraints may encompass slope geometry, material properties, geological structures, adherence to design codes, safety and environmental requirements and algorithmic parameters governing selection, crossover, mutation and termination criteria.

In the literature, genetic algorithms have predominantly been employed to identify critical slip surfaces—circular or non-circular—by exploring a wide range of potential failure mechanisms to locate the most critical slip configuration (Cen *et al*, 2019; Gao and Ge, 2024; Manouchehrian, Gholamnejad and Sharifzadeh, 2014; Wang, Moayedi and Kok Foong, 2020; Zolfaghari, Heath and McCombie, 2005). These applications include both the direct implementation of GAs (Bhandary, Krishnamoorthy and Rao, 2019; Cen *et al*, 2019; Goshtasbi, Ataei and Kalatehjary, 2008; McCombie and Wilkinson, 2002; Wu and Utili, 1995; Zeng, Jimenez and Jurado-Piña, 2015; Zolfaghari, Heath and McCombie, 2005) and hybrid approaches that integrate GAs with other artificial intelligence methods, such as artificial neural networks and swarm intelligence techniques (Koopialipoor *et al*, 2018; Wang, Moayedi and Kok Foong, 2020). These stability assessments are often conducted using the Limit Equilibrium Method (LEM) (Goshtasbi, Ataei and Kalatehjary, 2008; Manouchehrian, Gholamnejad and Sharifzadeh, 2014; Zolfaghari, Heath and McCombie, 2005), Finite Element Method (FEM) (Bhandary, Krishnamoorthy and Rao, 2019; Cen *et al*, 2019) or scaled boundary FEM (Wijesinghe *et al*, 2022).

While these studies have primarily focused on identifying slip surfaces, they often overlook the complex structural controls governing rock slope failures. In reality, slope stability in rock masses is influenced by multiple geotechnical factors, including the presence of discontinuities and the uncertainty associated with strength parameters (Fan *et al*, 2020; Zhou *et al*, 2017). his study seeks to advance the application of evolutionary algorithms by systematically identifying the most unfavourable combinations of structural features—such as joints and faults—and strength parameters of the rock mass. The method is demonstrated through a case study of a bench-scale slope failure at the Handlebar Hill Open Pit mine (Al Mandalawi *et al*, 2018), where both sliding and toppling mechanisms were observed.

CASE STUDY OF A BENCH-SCALE SLOPE FAILURE

Site conditions and the bench-scale slope failure

The Handlebar Hill Open Pit Mine operated between 2008 and 2014, ultimately reaching approximately 500 m in length, 300 m in width and 180 m in-depth. Situated 20 km north of Mount Isa in Queensland, the site lies within the up-dip extension of the George Fisher South orebody, part of the Mount Isa Inlier. The geology comprises mid-Proterozoic sedimentary sequences belonging to the Mount Isa Group, characterised by extensive oxidation and sulfide leaching. The site stratigraphy includes, in ascending order, the Magazine Shale, Kennedy Siltstone, Urquhart Shale, Native Bee Siltstone, Judenan Beds and Eastern Creek Volcanics. Mineralisation is predominantly hosted within the Urquhart Shale. The stratigraphic sequence and major fault structures dip steeply (60°–70°) to the west. Regionally significant structural features include the Mount Isa Fault, Paroo Fault and Barkly Shear Zone, in addition to local fault structures within the mine such as the South Barkly Splay (SBS) Fault, 4–6 Fault and the Offset Fault (MIM, 2007).

In July 2008, several bench-scale failures occurred on the fifth bench (3400 level) of the west pit slope (see Figures 2–3). The rock units exposed in this section, from west to east, comprise the Eastern Creek Volcanics, Magazine Shale, Spears Siltstone and Urquhart Shale. The failure zone lies in proximity to two significant fault zones: the Paroo Fault and the Offset Fault. The Paroo Fault, a soft zone measuring 2.0–3.0 m in width and filled with clay gouge, intersects the west slope and marks the contact between the Magazine Shale and the overlying volcanics.

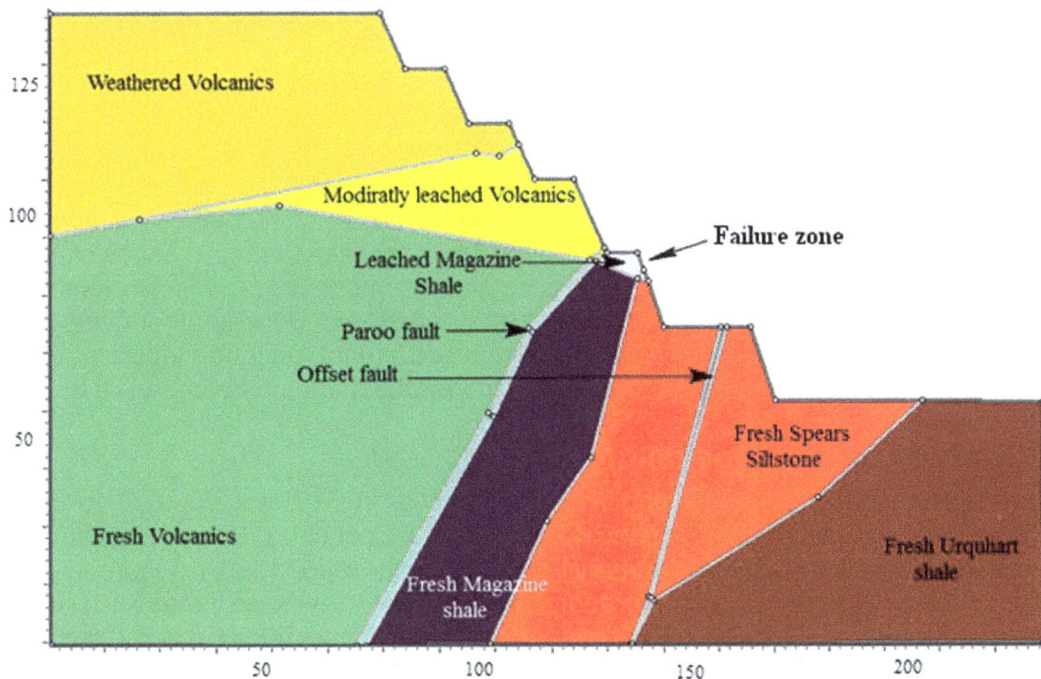

FIG 2 – The west slope profile at HHOC mine prior to the slope failure (Al Mandalawi *et al*, 2018).

FIG 3 – General view of west slope failure – looking south (Rosengren, 2008).

The observed instability was further exacerbated by the interface between leached and unaltered (fresh) rock at the 3400 level, as illustrated in Figure 2. This geological boundary, in combination with nearby fault structures, is considered a key contributing factor to the bench-scale slope failures. Following these failures, the stability of the west slope was assessed through a series of geotechnical investigations, including kinematic analysis, limit equilibrium modelling and finite element analysis. This study revisits the failure to demonstrate the effectiveness of genetic algorithms in analysing structurally controlled slope stability using the limit equilibrium method, incorporating multiple influencing factors as random variables to enable a more robust and comprehensive assessment.

Kinematic study of the slope failure zone

Kinematic analysis was conducted using *RocScience Dips* (RocScience, 2024) to evaluate the structural controls in the failure zone. Among 40 mapped discontinuities, 33 were classified as joints, predominantly grouped into two major sets: J1 (55°/75°) and J2 (67°/289°) (Figure 4). These joint sets produced 767 intersections, with stereographic projection revealing potential for both wedge sliding and flexural toppling. Notably, 32.46 per cent of these intersections fall within the critical zone for wedge sliding (Figure 5), while J2 exhibits a strong influence on flexural toppling failure mechanisms (Figure 6). This study focuses on flexural toppling, in contrast to earlier investigations that emphasised direct toppling (Al Mandalawi *et al*, 2018). Figure 6 shows that 75 per cent of J2 joints lie within the critical flexural toppling zone, highlighting their vulnerability.

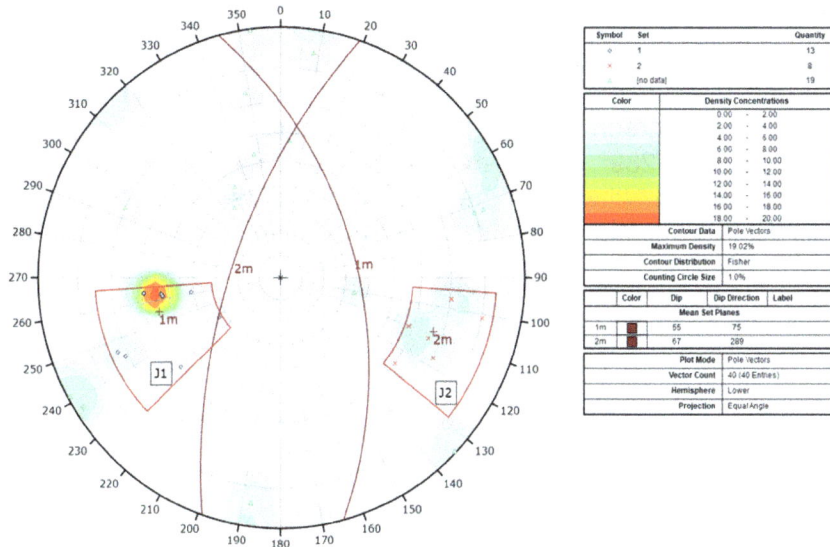

FIG 4 – Pole frequency diagram and orientations of the discontinuities in the zone.

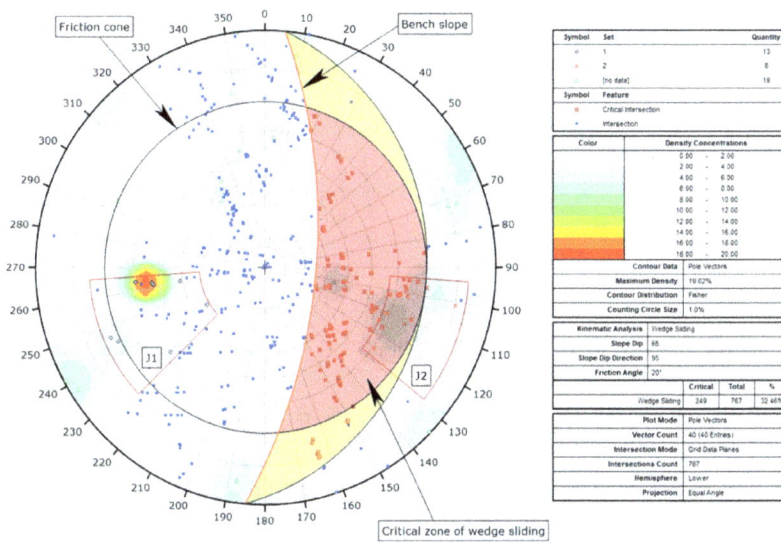

FIG 5 – Kinematic stability analyses of wedge sliding in the zone.

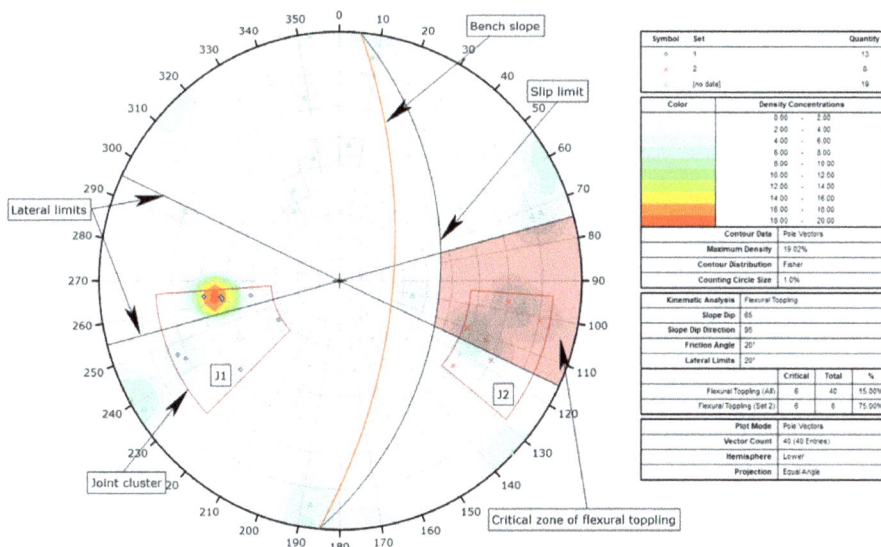

FIG 6 – Kinematic stability analyses of toppling in the zone.

Slope modelling using genetic algorithms

Figure 7 presents the geological structures at the 3400 level (top surface of the fifth bench), focusing on a representative bench segment, *ABCD*, which is 9 m wide and 200 m long. *BC* is the crest of the bench dipping at 65°/95°. The green line (*EJ1*) represents joint set J1 dipping at 55°/75° and the red line (*EJ2*) represents J2 at 67°/289°. The dash line represents the Paroo Fault at 70°/275° (Point E), marking the lithological boundary between the Volcanics and Magazine Shale formations. Both J2 and the Paroo Fault dip in a direction roughly opposite to J1 and the slope surface at this location.

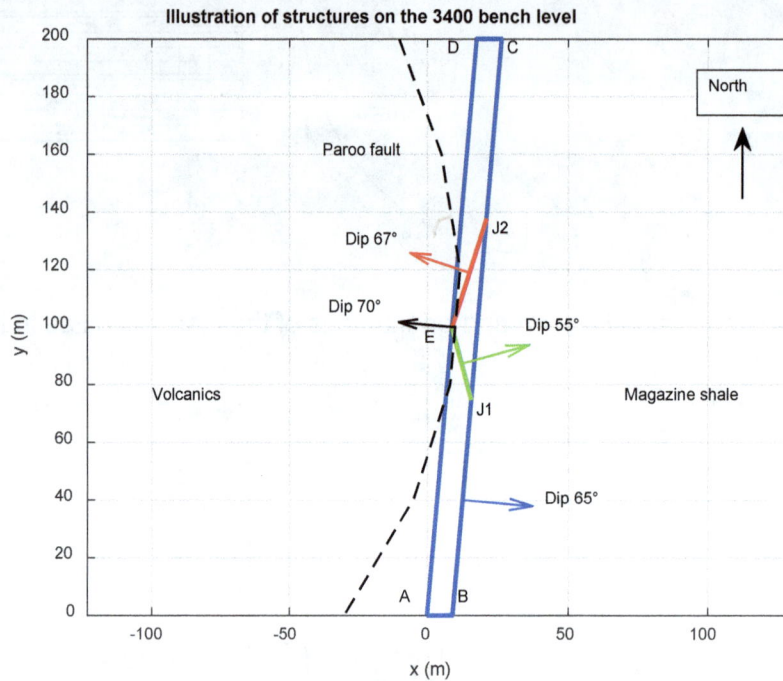

FIG 7 – Illustration of structures on the 3400 bench level.

Based on the kinematic analysis, the slope exhibits a high proportion of critical intersections conducive to wedge failure. Therefore, J1 is selected for a 2D structurally controlled slope stability analysis using LEM in this study. For a detailed analysis of toppling failure associated with J2, refer to Al Mandalawi *et al* (2018).

Figure 8 displays a GA-generated schematic of the fourth and fifth benches. In this cross-section, *EF* represents joint J1. Due to the 2D simplification, J1 appears to have the same dip direction as the slope, although their actual dip directions are 75° and 95°, respectively (as shown by the green and blue arrows in Figure 7). *GI* represents the Paroo Fault, which dips 70° to the west (the black arrow in Figure 7), opposing the slope dip direction. The intersection of *EF* and *GI* at point *H* forms a block, *ECGH*, which is structurally predisposed to instability.

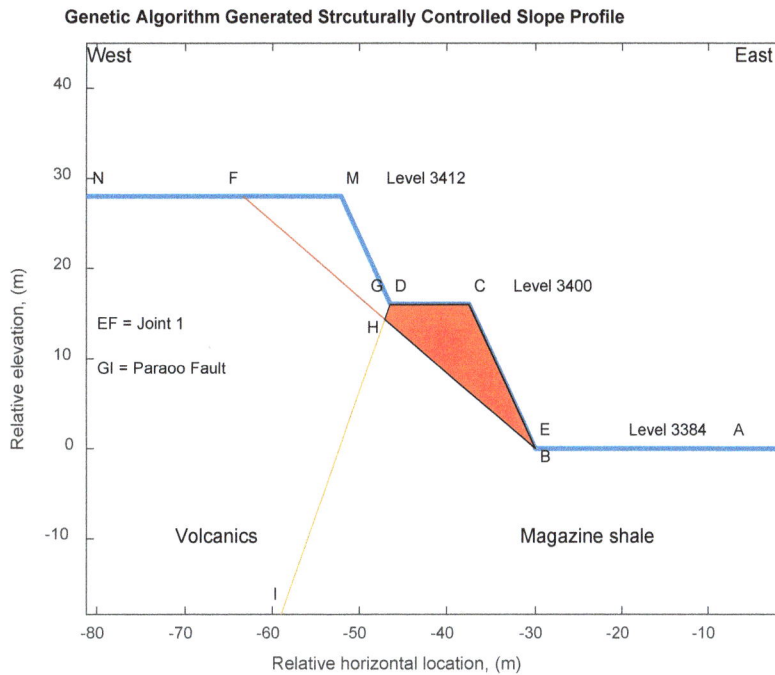

Genetic Algorithm Generated Strcuturally Controlled Slope Profile

FIG 8 – GA-generated schematic configuration of the structurally controlled slope.

The slope failure occurred within the Magazine Shale unit on the fifth bench, located on the right side of the Paroo Fault. This rock mass has a unit weight of 26.5 kN/m³. The bench has a slope angle of 65°, a height of 16 m and a width of 9 m (note that the height of benches 1–4 is 12 m; see Figure 3). The Paroo Fault is exposed near the toe of the fourth bench, with a width of approximately 2–3 m at this location.

For intact rock, cohesion ranges from 0–20 MPa and the friction angle from 6–65° (Seville, 1981). For the rock mass, cohesion ranges from 50–410 kPa and the friction angle from 13–25° (MIM, 2011). In the GA modelling, five rock mass parameters are simultaneously treated as random variables. Their respective value ranges, as summarised in Table 1, are used to identify the most vulnerable configuration of the structurally bounded block *ECGH* in Figure 8.

TABLE 1

Random variables used in the genetic algorithm.

Variable	Unit	Value range	Note
J1 slip plane location (*spl*)	m	0–16	From the toe of the fifth bench (level 3384), represented by Point E in Figure 8
J1 slip plane dip angle (*spa*)	°	40–74	
Paroo Fault location (*pfl*)	m	5–9	From the crest, Point C in Figure 8 of the fifth bench (level 3400), represented by Point G in Figure 8
Cohesion (*c*)	kPa	50–410	Magazine shale rock mass
Frictional angle (*φ*)	°	13–25	Magazine shale rock mass

This study aims to assess the potential instability of the structurally controlled block *ECGH* on the slope. The objective function is defined to identify the minimum *FS* resulting from random combinations of the five variables listed in Table 1. The GA optimisation problem is expressed mathematically as:

$$\text{Min.}\, FS = f(spl, spa, pfl, c, \varphi) \tag{1}$$

Subject to the following constraints:

- Slope geometries: bench slope angle, height and width.

- Rock mass properties: unit weight, cohesion and frictional angle.

- Structural features on the slope, J1 (location and dip angle) and Paroo Fault location and dip angle on the bench.

The corresponding fitness function used in the GA is:

$$Fit = \frac{1}{(FS+Penalty)} \qquad (2)$$

Where, *Fit* is the fitness function to assess the fitness of each chromosome, *FS* is the factor of safety of the chromosome, *Penalty* is a penalty term applied to chromosomes with abnormal genes. The default penalty is zero for valid genes and 1000 for invalid (abnormal) ones.

Five random variables are explored in this study: the J1 slip plane location (*spl*), J1 dip angle (*spa*), Paroo Fault location (*pfl*), cohesion (*c*) and friction angle (*φ*). Each variable is represented as a gene in the chromosome. The GA program is designed to accommodate additional genes if future studies incorporate more variables. The GA settings are summarised in Table 2.

TABLE 2

Parameter settings in the evolutionary genetic algorithm.

Parameter	Value	Note
Type of genes	5	Namely the variables in Table 1
Generation of GA evolution	300	Versatile for the need of convergence
Population size of chromosomes	200	Versatile for the need of convergence
Parent chromosomes selection rate	70%	
Crossover chromosomes	30%	Crossover in pairs
Mutation rate	5%	
Penalty	1000	Applied for an abnormal gene
Termination criteria	300	At the maximum generation

RESULTS AND DISCUSSION

The developed program can perform both deterministic stability analysis when all parameters are assigned fixed values and GA modelling, where one or more parameters are treated as random variables. GA modelling effectively overcomes the limitations of conventional sensitivity analysis, which typically varies one parameter at a time while keeping others constant.

The input conditions and output results of a representative GA simulation are summarised in Table 3, with corresponding graphical outputs shown in Figures 8 to 10. Key results, such as the daylighting of J1 at the slope toe (Figure 8), low cohesion value and a near-minimum friction angle, are consistent with geotechnical expectations. However, some findings diverge from intuitive interpretations—for example, the near-minimum dip angle of J1 combined with the furthest possible location of the Paroo Fault results in a nearly maximum-size structurally controlled block (eg block *ECGH* in Figure 8). This outcome underscores the strength of the GA approach: by randomly sampling within the full range of input variables, it can explore less intuitive yet critical scenarios. Each GA run may produce a different combination of input values, offering a diverse set of possible failure configurations.

Overall, the minimum factor of safety (FS) observed ranges from 0.96 to 1.02, suggesting that the GA-generated block *ECGH* is either unstable or in a state of limiting equilibrium under the input settings listed in Table 3. These findings align well with prior research findings that a cohesion of

64 kPa and a friction angle of 15° would bring the Magazine Shale slope to equilibrium (Al Mandalawi *et al*, 2018).

TABLE 3

MatLab GA modelling conditions and results (0 = fixed parameter, 1 = random variable).

Variable	Slope angle (°)	J1 dip angle (°)	J1 daylight location (m)	Paroo Fault location (m)∗	Cohesion (kPa)	Frictional angle (°)	Factor of safety
Type	0	1	1	1	1	1	LEM
Value	65	40.133	0.062	9	50	13.047	0.98

∗ The Paroo Fault location is the relative horizontal distance from the crest, Point C in Figure 8.

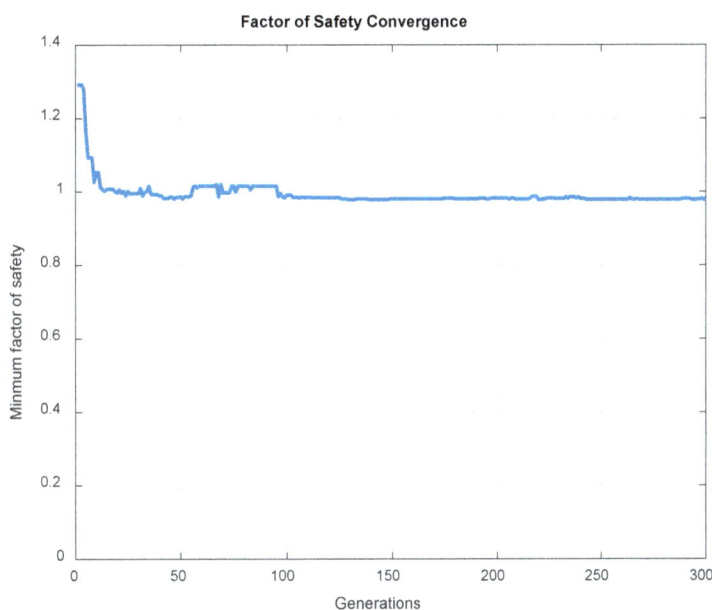

FIG 9 – Convergence of factor of safety during the generations of evolution.

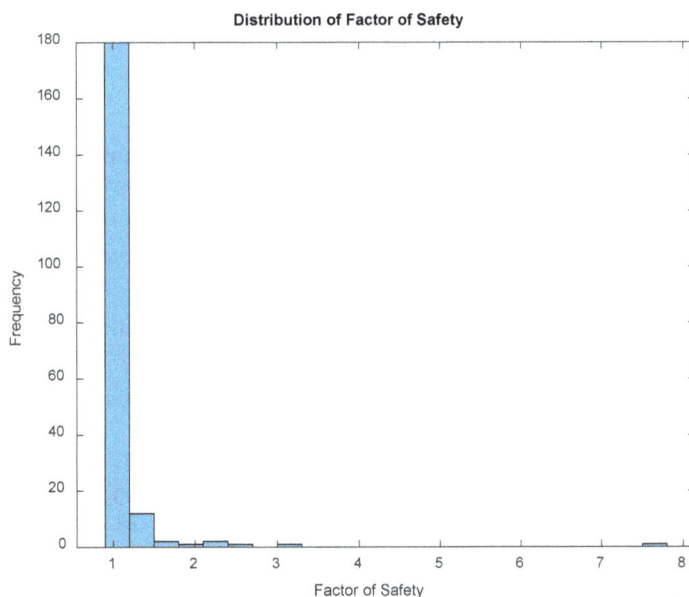

FIG 10 – Histogram of the factor of safety of the population at the 300ᵗʰ generation.

The evolution of the minimum factor of safety across 300 generations is presented in Figure 9. Convergence appears to be achieved after approximately 100 generations, indicating the algorithm's efficiency in identifying critical stability conditions. Figure 10 displays the histogram of *FS* values at the 300th generation, where approximately 90 per cent of the population exhibits *FS* values near 1.0, suggesting a high likelihood of limiting equilibrium under the simulated conditions.

It should be noted that this study does not account for the effects of rainfall. Previous studies on sandstone (Huang *et al,* 2021; Liu *et al,* 2020) have shown that rock strength tends to decrease with increasing degrees of saturation or water content. Therefore, it can be reasonably inferred that under wet conditions, the *FS* would likely decrease due to increases in unit weight and reductions in cohesion and friction angle—potentially leading to slope failure during rainy seasons (Zhao and You, 2020).

CONCLUSIONS

This study re-evaluates the bench-scale slope failure in Magazine Shale at the 3400 level on the west slope at the Handlebar Hill Open Pit Mine, using a genetic algorithm implemented in MATLAB. The analysis incorporates five random variables—three structural parameters (joint and fault geometries) and two strength parameters (cohesion c and friction angle φ)—within a limit equilibrium framework. The results confirm that the failure is structurally controlled and the GA effectively identifies a nearly maximal unstable block configuration, aligning well with field observations of the actual failed mass.

The Paroo Fault is critically positioned near the toe of the fourth bench at the failure zone, where it is exposed and susceptible to infiltration by surface run-off. This exposure can weaken both the fault zone and adjacent discontinuities, further compromising slope stability.

The critical block configuration arises from a combination of adverse conditions: Joint 1 daylights near the toe, cohesion and friction angle approach their minimum values, J1 exhibits its shallowest dip angle and the Paroo Fault extends farthest from the bench crest. Together with field evidence and kinematic analysis, it is concluded that wedge failure occurred along the J1 surface, coalescing with the Paroo Fault at the rear of the block. As sliding progressed, rock bridges within the joints failed, allowing space for blocks to topple along the J2 surfaces at the base of the wedge.

ACKNOWLEDGEMENTS

The author gratefully acknowledges Glencore Zinc for granting permission to conduct the research on slope stability at the Handlebar Hill Open Pit Mine, where Dr Maged Al Mandalawi completed his PhD study on the west slope stability analysis at Federation University Australia. Sincere thanks are also extended to Dr Ahmed Soliman, Principal Geotechnical Advisor, for his valuable support throughout the study.

REFERENCES

Al Mandalawi, M, You, G, Dahlhaus, P, Dowling, K and Sabry, M, 2018. Analysis of a combined circular-toppling slope failure in an open-pit, The 2nd GeoMEast International Congress and Exhibition on Sustainable Civil Infrastructures, Egypt.

Alejano, L R, Gómez-Márquez, I and Martínez-Alegría, R, 2010. Analysis of a complex toppling-circular slope failure, *Engineering Geology*, 114(1):93–104.

Barrett, S, McQueen, L and Bendtsen, B, 2008. Shotcrete Lining Design for Underground Excavations in Rock – The Current State of Practice, in 13th Australian Tunnelling Conference.

Bhandary, R P, Krishnamoorthy, A and Rao, A U, 2019. Stability Analysis of Slopes Using Finite Element Method and Genetic Algorithm, *Geotechnical and Geological Engineering*, 37(3):1877–1889. https://doi.org/10.1007/s10706-018-0730-5

Cen, W, Luo, J, Yu, J and Shamin Rahman, M, 2019. Slope Stability Analysis Using Genetic Simulated Annealing Algorithm in Conjunction with Finite Element Method, *KSCE Journal of Civil Engineering*, 24(1):30–37. https://doi.org/10.1007/s12205-020-2051-5

Fan, B, Wang, L, Gong, W, Wang, C, Jiang, Y and Sun, Z, 2020. Improved robust design of rock wedge slopes with a new robustness measure, *Computers and Geotechnics*, 123. https://doi.org/10.1016/j.compgeo.2020.103548

Gao, W and Ge, S, 2024. A comprehensive review of slope stability analysis based on artificial intelligence methods, *Expert Systems with Applications*, 239. https://doi.org/10.1016/j.eswa.2023.122400

Goldberg, D E, 1989. *Genetic Algorithm in Search, Optimization and Machine Learning* (Addison-Wesley, Massachusetts).

Goshtasbi, K, Ataei, M and Kalatehjary, R, 2008. Slope modification of open pit wall using a genetic algorithm—case study: southern wall of the 6th Golbini Jajarm bauxite mine, *The Journal of The South African Institute of Mining and Metallurgy*, 108:651–656.

Huang, S, He, Y, Liu, G, Lu, Z and Xin, Z, 2021. Effect of water content on the mechanical properties and deformation characteristics of the clay-bearing red sandstone, *Eng Geol Environ*, 80:1767–1790. https://doi.org/10.1007/s10064-020-01994-6Bull

Koopialipoor, M, Jahed Armaghani, D, Hedayat, A, Marto, A and Gordan, B, 2018. Applying various hybrid intelligent systems to evaluate and predict slope stability under static and dynamic conditions, *Soft Computing*, 23(14):5913–5929. https://doi.org/10.1007/s00500-018-3253-3

Liu, H, Zhu, W, Yu, Y, Xu, T, Li, R and Liu, X, 2020. Effect of water imbibition on uniaxial compression strength of sandstone, *Int J Rock Mech Min Sci*, 127. https://doi.org/10.1016/j.ijrmms.2019.104200

Manouchehrian, A, Gholamnejad, J and Sharifzadeh, M, 2014. Development of a model for analysis of slope stability for circular mode failure using genetic algorithm, *Environmental Earth Sciences*, 71(3):1267–1277. https://doi.org/10.1007/s12665-013-2531-8

McCombie, P and Wilkinson, P, 2002. The use of the simple genetic algorithm in finding the critical factor of safety in slope stability analysis, *Computers and Geotechnics*, 29(699–714).

Mohtarami, E, Jafari, A and Amini, M, 2014. Stability analysis of slopes against combined circular–toppling failure, *International Journal of Rock Mechanics and Mining* Sciences, 67:43–56.

Mount Isa Mines Limited (MIM), 2007. Handlebar Hill feasibility study (Geology) (internal technical report).

Mount Isa Mines Limited (MIM), 2011. HHOC Rock Mass Mohr-Coulomb Parameters (internal technical report).

RocScience. 2024. Dips 6.0, Graphical and Statistical Analysis of Orientation Data, RocScience software, Toronto, Canada.

Rosengren, K, 2008. Handlebar Hill Open Cut Pit West Wall, (internal technical report) Mount Isa Mines Limited (MIM).

Seville, R, 1981. Review of Hilton mine rock property data, (internal technical report) Mount Isa Mines Limited (MIM), reference no. RPS/7.5/RES MIN15.1.3.

Wang, H, Moayedi, H and Kok Foong, L, 2020. Genetic algorithm hybridized with multilayer perceptron to have an economical slope stability design, *Engineering with Computers*, 37(4):3067–3078. https://doi.org/10.1007/s00366-020-00957-5

Wijesinghe, D and You, G, 2016. Optimization of the catch bench design using a genetic algorithm, *International Journal of Mining Science and Technology*, 26(6):1011–1016. https://doi.org/10.1016/j.ijmst.2016.09.008

Wijesinghe, D, Dyson, A, You, G, Khandelwal, M, Song, C and Ooi, E, 2022. Simultaneous slope design optimisation and stability assessment using a genetic algorithm and a fully automatic image-based analysis, *International Journal for Numerical and Analytical Methods in Geomechanics*, 46(15).

Wu, W and Utili, S, 1995. On The Optimal Profile Of A Rock Slope, ISRM Congress 2015 Proceedings – Int'l Symposium on Rock Mechanics.

Zeng, P, Jimenez, R and Jurado-Piña, R, 2015. System reliability analysis of layered soil slopes using fully specified slip surfaces and genetic algorithms, *Engineering Geology*, 193:106–117. https://doi.org/10.1016/j.enggeo.2015.04.026

Zhao, L and You, G, 2020. Rainfall Affected Stability Analysis of Maddingley Brown Coal Eastern Batter Using Three-Dimensional FEM, *Arabian Journal of Geosciences*, 13(20). https://doi.org/10.1007/s12517-020-06038-7

Zhou, J-W, Jiao, M-Y, Xing, H-G, Yang, X-G and Yang, Y-C, 2017. A reliability analysis method for rock slope controlled by weak structural surface, *Geosciences Journal*, 21(3):453–467. https://doi.org/10.1007/s12303-016-0058-1

Zolfaghari, A R, Heath, A C and McCombie, P F, 2005. Simple genetic algorithm search for critical non-circular failure surface in slope stability analysis, *Computers and Geotechnics*, 32(3):139–152. https://doi.org/10.1016/j.compgeo.2005.02.001

Study on the influence of surfactants on the wettability of coal dust based on molecular dynamics simulation

Y Zhang[1,2], L Yuan[3], T Ren[4] and J Roberts[5]

1. PhD Student, Anhui University of Science and Technology, Huainan 232001, China.
   Email: ahzhangyi11@163.com
2. Joint PhD Student, University of Wollongong, Wollongong NSW 2500.
   Email: yz062@uowmail.edu.au
3. President, Anhui University of Science and Technology, Huainan 232001, China.
   Email: yuanl_1960@sina.com
4. Professor, University of Wollongong, Wollongong NSW 2500. Email: tren@uow.edu.au
5. Lecturer, University of Wollongong, Wollongong NSW 2500. Email: jon_roberts@uow.edu.au

ABSTRACT

The detailed mechanism by which surfactant molecules interact with water molecules to wet coal powder in solution has been less frequently explored. Advances in computational technologies have enabled the application of mature molecular dynamics (MD) simulation techniques, which have proven to be powerful tools for elucidating intermolecular interactions at the microscopic level. In the field of coalmine dust prevention and control, MD simulations are widely utilised to investigate the wetting effects of surfactants on coal surfaces. These studies often focus on enhancing wettability, and the changes in contact angles between coal particles and liquid droplets. Such research provides deeper insights into the relationship between coal structure and wettability modifications. Overall, molecular dynamics simulations offer a precise microscopic perspective for studying coal wettability. When combined with experimental validation, these simulations help optimise dust removal technologies and surface modification processes. Additionally, recent studies have explored the influence of surfactant-modified nanofluids with drag-reducing properties on coal wettability, revealing potential molecular-level mechanisms through an integrated approach of experimental and computational analysis. This study further explored the effect of surfactant modified nanofluids with drag reducing properties on coal wettability and the potential mechanisms at the molecular level through a combination of experiments and simulations. Molecular simulations were conducted with molecular simulation software to construct H_2O/Coal, H_2O/AOS/Coal, and H_2O/AOS-modified SiO_2/Coal systems. These simulations were intended to explore the synergistic effect between SiO_2 nanoparticles and surfactants, as well as changes in the adsorption configuration, relative concentration distribution, and MSD of water molecules within their composite systems. The research results of this study will provide valuable information for the application of nanofluids in coalmine dust prevention and control.

INTRODUCTION

China's abundant coal resources and limited oil reserves have led to a coal-dominated energy structure that is challenging to alter within a short time frame (Wang *et al*, 2023c; Shi *et al*, 2023; Zhao *et al*, 2022). A high concentration of dust can trigger coal dust explosions and pneumoconiosis (Mo *et al*, 2014; Nie *et al*, 2022; Wang *et al*, 2023b). Additionally, coal dust is the primary culprit that endangers the health of employees and reduces their sense of well-being. Therefore, it is necessary to take effective dust control measures.

Changing the wetting characteristics of the coal body before mining is one such measure. Adding wetting agents is an effective means to improve the wettability of coal seams during coal seam water injection processes (Wang *et al*, 2023a; Zhou *et al*, 2022a; Ma *et al*, 2021). Previous research has shown that a composite system of SiO_2 and surfactants had a better wetting effect on sandstone and materials other than monomers in EOR, indicating that nanofluids have research potential for increasing the wettability of coal (Zou *et al*, 2022; Zhang *et al*, 2022). Studies have consistently indicated that SiO_2 nanofluids have good wetting properties. However, current research on the influence of nanofluids on the wettability of coal is limited mainly to the macroscopic experimental stage. Most researchers have only investigated the effects of water-based nanofluids on coal and have not explored the impacts of surfactant-modified nanofluids (including AOS-modified SiO_2

nanofluids) on the wettability of coal or potential mechanisms at the molecular level. Therefore, further investigations in this field are highly warranted.

EXPERIMENTAL DESIGN AND SIMULATION METHODS

Material

The selected experimental coal sample is bituminous coal. Table 1 presents the coal properties.

TABLE 1

Industrial analysis of coal.

Mad/%	Aad/%	Vdaf/%	FCad/%
1.58	24.13	10.83	63.46

Note: Mad: moisture; Ad: ash; Vdaf: volatile; Fcd: fixed carbon.

The process of preparing nanofluids with different concentrations and compositions using AOS and spherical silica nanoparticles with a diameter of 20 nm as experimental materials includes dispersing them by ultrasound for 1 hr, followed by stirring for 30 mins using a magnetic stirrer to obtain well dispersed nanofluids (Zhang *et al*, 2022; Zhao *et al*, 2023). The chemical structures and molecular models of the experimental materials selected in this research are displayed in Figure 1.

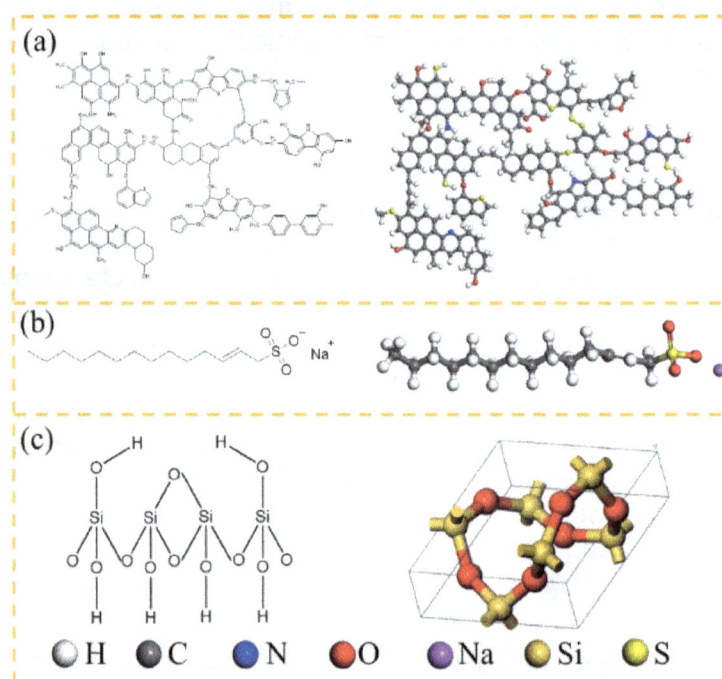

FIG 1 – Chemical structure and molecular model: (a) Bituminous coal, (b) AOS, (c) SiO$_2$.

Wettability experiment

The raw coal was labelled S1, while the coal treated with SiO$_2$ nanofluid, AOS, and modified nanofluids were labelled S2, S3, and S4, respectively. Characterisation experiments were conducted on the wetting characteristics of the coal, including the surface tension, zeta potential, and CA.

Molecular dynamic simulation

Molecular Dynamics (MD) is the process of using computers to analyse molecules at the molecular scale. Perform simulation calculations to simulate molecular motion. The basic principle of molecular dynamics simulation applies Newton's laws of motion to each molecule, and the motion of each molecule follows this law. Therefore, the forces and motion of each molecule in the system can be calculated and solved through Newton's laws of motion, thus obtaining any time. Calculate the relevant parameters such as force, position, and velocity of each atom in the system, and based on these parameters, explore the macroscopic properties of the system through certain methods.

Molecular dynamics simulation should follow periodic boundary conditions and select appropriate force fields. The specific molecular dynamics simulation process is as follows: construct an initial model based on the research object; Given initial conditions and calculation parameters, including system temperature, initial velocity, simulation duration, force field etc; Calculate the forces exerted on each atom or molecule; Substitute the forces acting on each atom or molecule into Newton's equations of motion to calculate the position and velocity of particles; Analyse the macroscopic properties of the system. Molecular dynamics methods have the advantages of low cost, accurate reaction of microscopic processes, and prediction of physical and chemical properties of systems. They can provide reasonable explanations for microscopic phenomena and corresponding mechanisms that cannot be observed in macroscopic experiments. Therefore, exploring the mechanism of the influence of surfactants on the wettability of solid surfaces using molecular dynamics methods has broad application prospects.

An interface model of surfactant solution was constructed using molecular simulation software. The subsequent stage of this process is to optimise the geometric shape of the constructed amorphous cell box; The specific setting parameters are shown in Table 2. Subsequently, a model of the surfactant water interface was constructed using the Build Layers tool and subjected to geometric optimisation. Finally, molecular dynamics simulations were conducted on the surfactant containing solution using the Forcite module, with the parameters set as shown in Table 3.

TABLE 2

Geometry optimisation setting parameters.

Project	Modules	Task	Algorithm	Cut-off distance	Forcefield
Parameter	Forcite	Geometry optimisation	Smart	12.5 Å	Compass II

TABLE 3

Molecular dynamics simulation set-up parameters.

Project	Task	Ensemble	Time step	Total simulation time	Forcefield
Parameter	Geometry optimisation	NVT	1.0 fs	300 ps	Compass II

Construction of bituminous coal model

The Wiser coal chemical structure model was selected for this experiment. It accurately reflects the characteristics of bituminous coal and is a comprehensive and reasonable structure, which is shown in Figure 2a (Zhang, Wang and Yan, 2015; Zhou et al, 2022b; Zhang et al, 2022). Initially, the model had in a planar structure, but after geometric optimisation, some chemical bonds underwent stretching. The total energy decreased from 4644.916 kcal/mol to 1357.403 kcal/mol. The macromolecular model became compact and three-dimensional, as shown in Figure 2b. Finally, a 3D optimal supercell structure containing five Wiser briquette molecules was constructed, as displayed in Figure 2c.

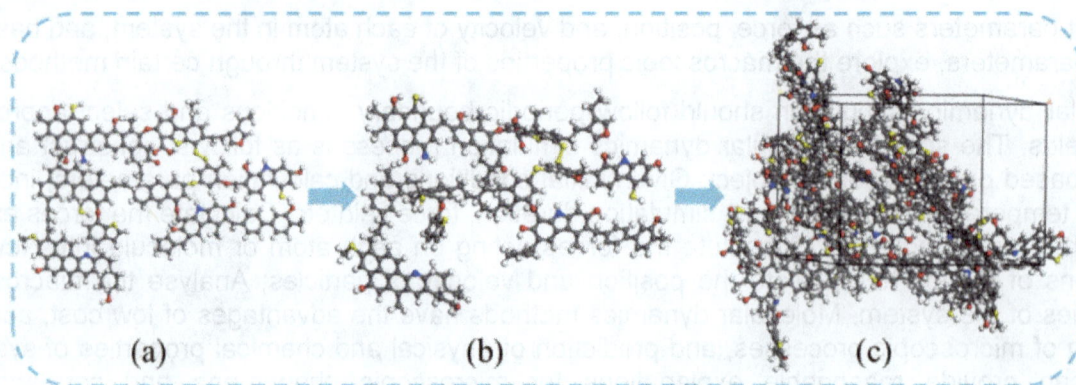

FIG 2 – Molecular model of coal.

Construction of surfactant and nanoparticle models

A nanoparticle model with a radius of 10 Å was built using the Build Nanocluster command, as displayed in Figure 3a. Subsequently, the model was subjected to bond saturation and hydroxylation, as depicted in Figure 3b. Surfactant chains were then modified on both sides of the SiO_2 nanoparticles to construct a composite system of nanoparticles modified by the surfactant agent AOS, namely the AOS-modified SiO_2 model, as shown in Figure 3c.

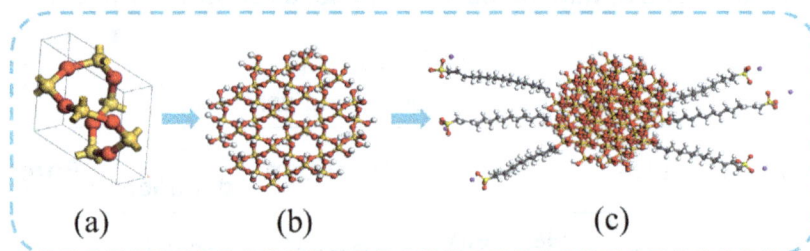

FIG 3 – Surfactant-nanoparticle model.

Construction of simulation system

Two types of simulation systems were built using the Build Layers function. The optimised water molecule model, AOS model, AOS-modified SiO_2 model, and coal molecule model were combined to construct H_2O/Coal, H_2O/AOS/Coal, and H_2O/AOS-modified SiO_2/Coal simulation systems, as shown in Figure 4. Figure 5 depicts the overall experiment and simulation program.

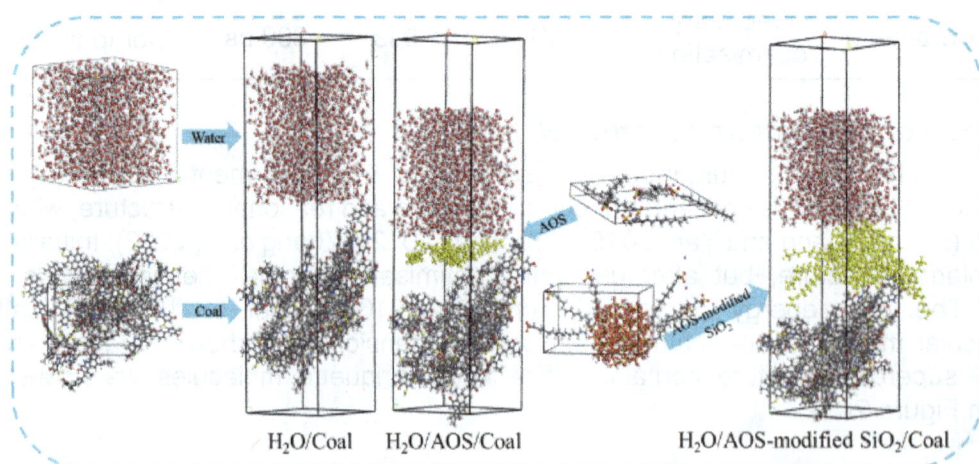

FIG 4 – Construction of different simulation systems.

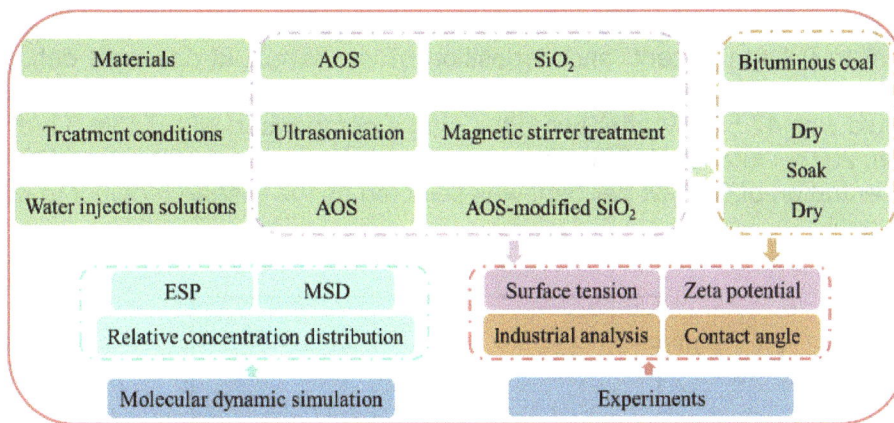

FIG 5 – Overall experimental and simulation program.

RESULTS AND DISCUSSION

Concentration selection of modified SiO₂ nanofluids

The surface tension of water is 72.02 mN/m, while the surface tension of 0.2 wt per cent AOS was 36.10 mN/m under the same conditions. Interestingly, the surface tensions of the NFs modified with AOS were lower than those of AOS and SiO_2 monomers, indicating a synergistic effect between the two components. Figure 6 suggests that the surface tension of the AOS-modified SiO_2 fluid changed very little, but the surface tension first decreased and then increased with increasing SiO_2 concentration. A turning point was observed at 0.01 wt per cent with a surface tension of 27.71 mN/m, indicating that the optimum reduction in intermolecular attraction was achieved at low concentrations. However, in terms of application, the stability of the fluid must be considered, and the zeta potential is a key indicator for characterising the stabilities of colloidal dispersion systems (Mondragon *et al*, 2012).The higher the absolute value of the zeta potential is, the more stable the system. The judgment criteria are shown in Table 4.

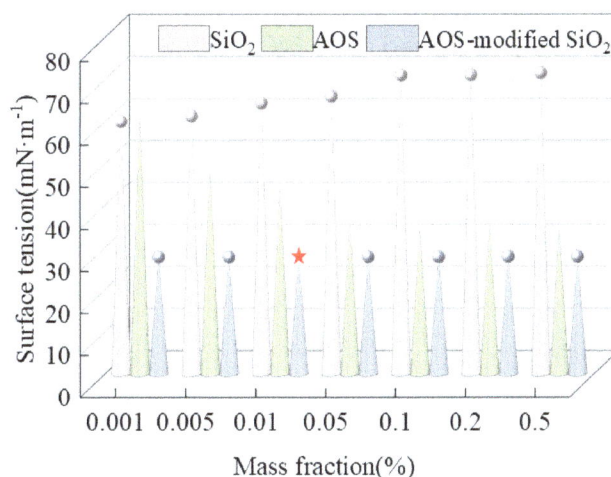

FIG 6 – Surface tension of solutions with different concentrations.

TABLE 4

Criteria for determining solution stability.

Zeta potential (mV)	0 ~ ±5	±10 ~ ±30	±30 ~ ±40	±40 ~ ±60	≥61
Stability	Rapid coagulation	Unstable	Moderate stability	Good stability	Excellent stability

Figure 7 shows that the SiO_2 nanofluids were stable at concentrations, ranging from 0.001 wt per cent to 0.1 wt per cent, and deposition of the nanofluid occurred during the first two days, but it remained stable afterward. The absolute value of zeta potential of 0.01 wt per cent AOS-modified SiO_2 fluid was 47.59 mV after two days, and a comparison with Table 3 indicates its good stability (Xu *et al*, 2023; Liu *et al*, 2023). Therefore, in subsequent experiments, AOS-modified SiO_2 fluid with a concentration of 0.01 wt per cent was selected as the research object to conduct wetting experiments on coal.

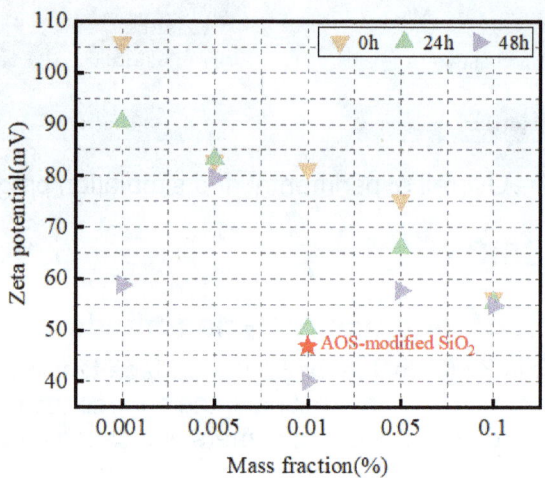

FIG 7 – Zeta potential changes of specific concentration nanofluids.

Analysis of the wettability results

The sizes of the bubble points are used to characterise the CA. Figure 8 describes the initial CA of the four coals did not significantly differ and gradually decreased over time. Notably, the CA formed by deionised water on the S2 surface showed the smallest change, only 8.64 per cent, within 3 sec, as shown in Figure 9. Deionised water was adsorbed completely by the coal treated with the AOS-modified SiO_2 fluid within only 5 sec. The reason may be that AOS infiltrated the interior of the coal body, opened transport channels and carried more SiO_2 nanofluids to immerse the coal body, which increased the hydrophilic area of the coal body from inside to outside.

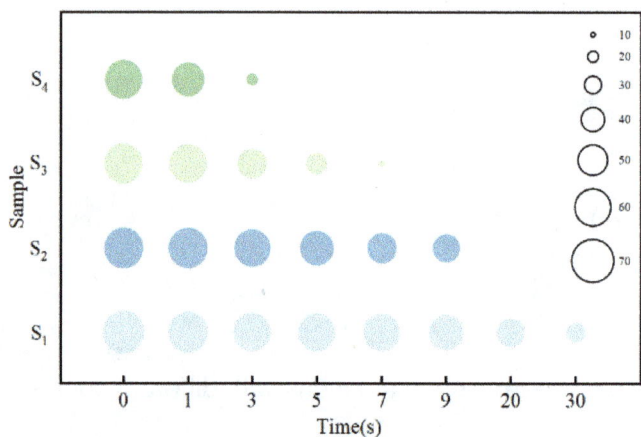

FIG 8 – CA test results and change process.

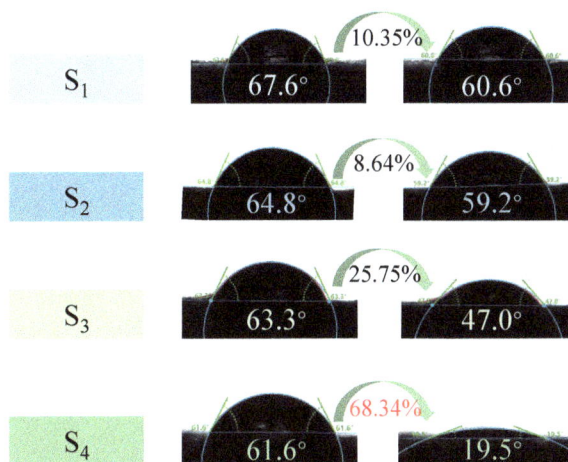

FIG 9 – Change in CA within 3 seconds.

Analysis of MD simulation results

Analysis of MD adsorption state

The equilibrium states of the different adsorption configurations reached after 500 ps of the molecular dynamic simulation are exposed in Figure 10. In the three systems, the substances in the initial state were not in contact. As the simulation reaches equilibrium, water, AOS, and AOS-modified SiO_2 all moved and came into contact with the coal molecular layer. The final equilibrium state was manifested as water molecules penetrating the surfactant from both sides and below to penetrate the coal molecular layer. The three closely adhered to each other. However, the diffusion ranges of water molecules in different systems vary, which may be related to the magnitude of intermolecular forces. It is worth noting that in the H_2O/AOS-modified SiO_2/Coal system, the nanostructure positioned horizontally on the coal-water interface enhances the contact area. This arrangement reduced the surface tension of water and is facilitated migration of the water molecules from all directions toward the coal molecules.

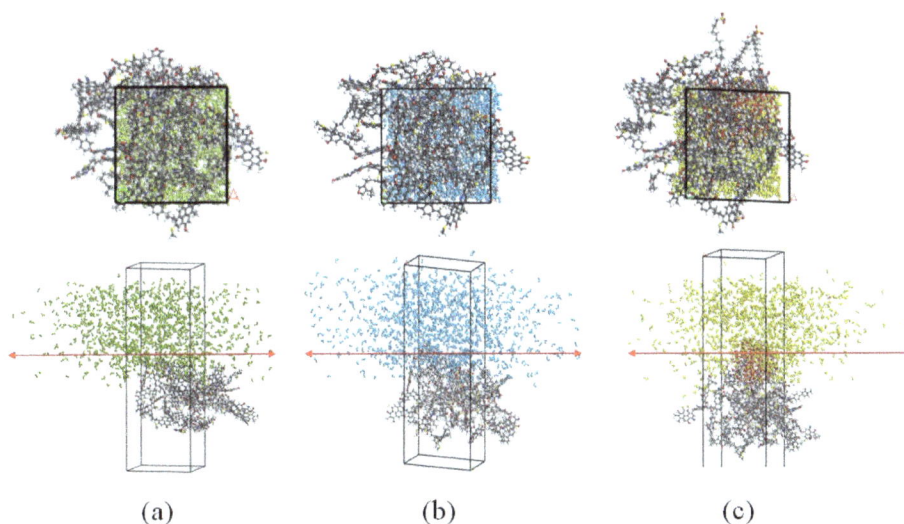

FIG 10 – Adsorption states of water and coal in different simulation systems: (a) H2O/Coal; (b) H2O/AOS/Coal; (c) H2O/AOS-modified SiO_2/Coal.

Analysis of electrostatic potential (ESP)

Molecular ESP can be used to predict the interaction sites of different molecular pairs, which explains the wetting mechanism for the researched substances. In order to investigate the adsorption process of surfactants and nanofluids by coal, the Dmol3 module in MS was accepted to calculate the ESP distributions on bituminous coal, AOS molecules, and SiO_2, as displayed in Figure 11. The ESP distribution on the AOS-modified SiO_2 surface was also predicted and analysed.

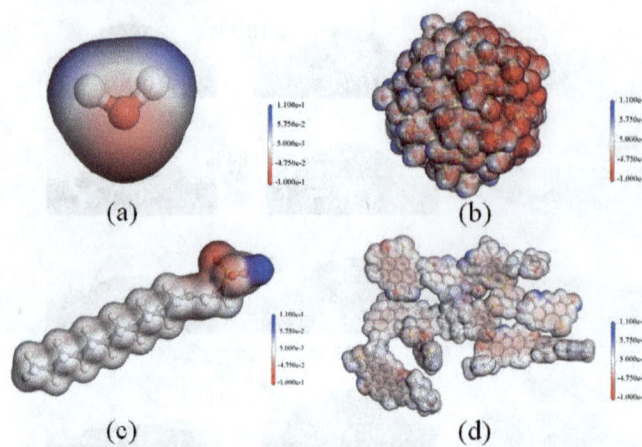

FIG 11 – ESP: (a) H2O; (b) SiO_2; (c) AOS; (d) Bituminous coal.

FIG 12 demonstrates that the potential distribution was uneven at different positions within different molecules. The surface ESP of bituminous coal primarily ranged from -0.475 au to 0.575 au, with positive potential extremes near the oxygen-containing functional groups, which was related to the molecular structure of coal. Observing Figure 12a and 12c, it can be seen that the colour distribution of the ESP on the surface of surfactant molecules was darker than that on the surfaces of water molecules, indicating that hydrogen bonds were more easily formed between AOS and water molecules at extreme positions than between water molecules. Figure 12b, clearly shows that the ESP distribution on the surface of SiO_2 was uniform, with a minimum ESP of -0.1 au and a maximum ESP of 0.11 au, so it was prone to electrostatic attraction with other molecules.

Analysis of relative concentration distribution

The relative concentration distribution is a crucial metric for examining the impacts of different solutions on coal wettability at the microscopic level. By exploring the spatial distributions of water molecules in different systems and analysing the effects of surfactants and modified nanofluids on the diffusion of water molecules, we can gain insight into the factors influencing coal wettability. Figure 12 demonstrates that the concentration distribution of water molecules and coal molecules along the Z-axis varied in different simulation systems. Especially in the system containing AOS-modified SiO_2, the initial distribution point of coal changed from 39.25 Å to 49.45 Å, but the distribution distance of coal was roughly the same. This evidence shows that the presence of AOS-modified SiO_2 affected the distribution of coal and water in the system. In the composite nanoparticles, the AOS tail was in contact with the SiO_2 matrix, and the head was in contact with the surrounding water. Water molecules and coal molecules are strongly attracted to the upper and lower surfaces structures of AOS-modified SiO_2, respectively.

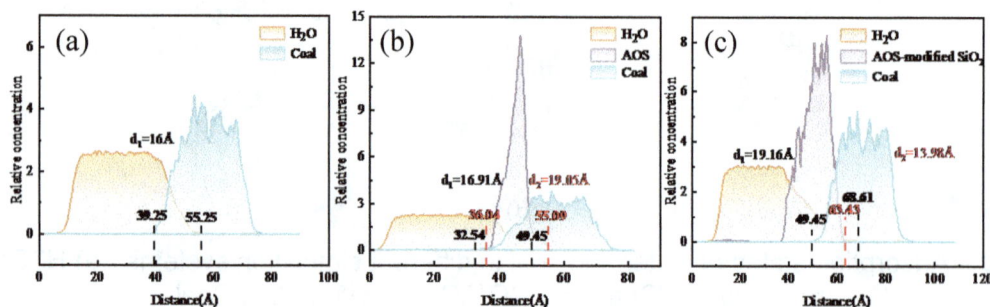

FIG 12 – Relative concentration distribution of three systems.

Mean square displacement of water molecules

The MSD can be used to determine the impact of a surfactant on the migration and aggregation of water molecules. The diffusion coefficient of the system cannot be directly calculated from the trajectory file in MS software. Instead, it was indirectly calculated by analysing the MSD. There is a relationship between the two parameters: one-sixth of the slope of the MSD curve represents the diffusion coefficient of the system.

The MD simulation data for the three systems were all balanced and equilibrated at 300–400 ps, which allowed analyses of the MSD (Zhang *et al*, 2020; Zhou *et al*, 2023). As displayed in Figure 13, the slope of the MSD for water molecules varied in different solutions. The D for the water molecules in the three systems were 0.5022 $Å^2$/ps, 0.5191 $Å^2$/ps, and 0.6684 $Å^2$/ps, respectively. These results indicated that AOS-modified SiO_2 has a superior impact, and the free diffusion rate was 133 per cent for that water molecules. This was attributed to movement of the nano-groups and sodium ions around the water molecules, which resulted in stronger interactions between the coal and water. The movement of water molecules was facilitated, and the probability of collision with the coal molecules was elevated, which was manifested as the higher diffusion coefficient.

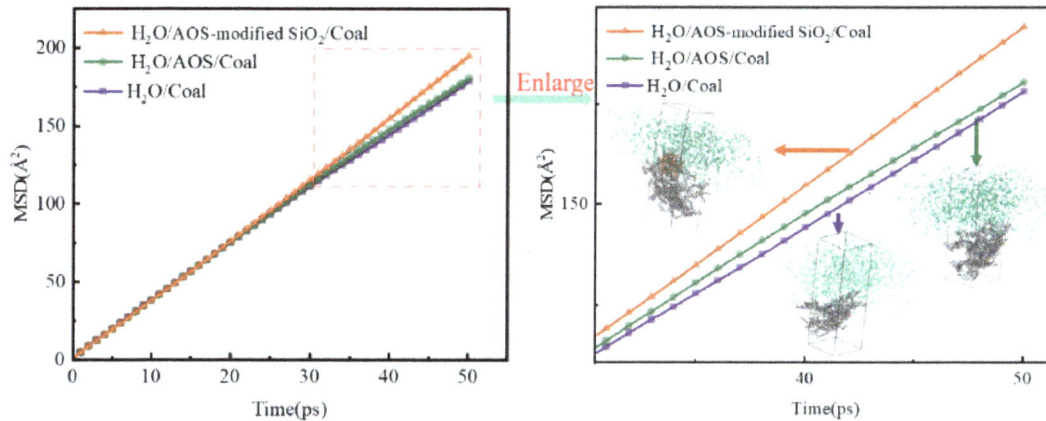

FIG 13 – MSD of water molecules in three systems.

Mechanism for wetting coal with modified SiO_2 nanofluid

Both AOS and SiO_2 have excellent wetting properties. The wetting performance will be more significant by optimising the advantages of the monomer through AOS-modified SiO_2 nanoparticles. Figure 14 displays the corresponding wetting mechanism. Figure 14a indicates that AOS containing sodium ions are adsorbed onto the SiO_2 surface to form amphiphilic hydrophilic head particles containing multiple sodium ions. The accumulation of sodium ions enhances hydration, and the ability to attract water molecules is increased.

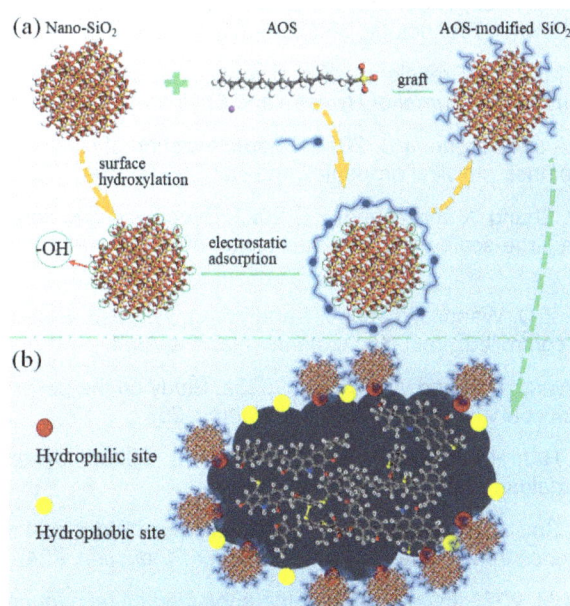

FIG 14 – The wetting mechanism of AOS-modified SiO_2 on coal.

Figure 14b highlights the mechanism for interaction between coal and the AOS-modified SiO_2 composite system. There were many hydrophobic groups, such as aromatic rings on the coal surface. Nevertheless, the SiO_2 nanofluid was highly ductile and adhered and spread on the coal surface to form a strong hydrophilic layer when injected into the coal seam. As a result, much of the dust generated during mining and cutting of the coal bodies was quickly captured by water droplets, wrapped to form large particles and settled. The coal dust concentration rapidly decreased in a short period of time, resulting in a more significant dust reduction effect, which was attributed to the increased wettability of the coal.

CONCLUSIONS

This manuscript explored the mechanism for wetting modified nanoparticles, namely AOS-modified SiO_2 on coal, through a combination of experiments and simulations. The effect of the AOS-modified SiO_2 on the wettability of bituminous coal was explored with surface tension and contact angle experiments, and the adsorption characteristics and wetting mechanism were investigated via simulation methods. By analysing the experimental and simulation results, the following conclusions were drawn.

The wetting experiments showed that there was a synergistic wetting effect between AOS and SiO_2, and the AOS-modified SiO_2 nanocomposite fluid exhibited good wettability and stability.

Molecular dynamics simulation was used to explain the experimental phenomena from a microscopic perspective. Compared with H_2O/Coal, the thickness of adsorbed water molecules increased by 0.91 Å and 3.16 Å, and the diffusion coefficients increased by 3.37 per cent and 33.09 per cent in the H_2O/AOS/Coal and H_2O/AOS-modified SiO_2/Coal simulation systems, respectively.

Simulation results have shown that the AOS-modified SiO_2 composite fluid facilitates the migration of a greater number of water towards coal, thereby enhancing the likelihood of collision. This observation further confirms the superior wetting properties of this fluid on coal.

ACKNOWLEDGEMENTS

This project was financially supported by the Graduate Innovation Fund (Anhui University of Science and Technology) (DPDCM2407).

REFERENCES

Liu, Y-L, Li, Y, Si, Y-F, Fu, J, Dong, H, Sun, S-S, Zhang, F, She, Y-H and Zhang Z-Q, 2023. Synthesis of nanosilver particles mediated by microbial surfactants and its enhancement of crude oil recovery, *Energy*, 272.

Ma, Y L, Sun, J, Ding, J F and Liu, Z Y, 2021. Synthesis and characterization of a penetrating and pre-wetting agent for coal seam water injection, *Powder Technology*, 380:368–376.

Mo, J F, Wang, L, Au, W and Su, M, 2014. Prevalence of coal workers' pneumoconiosis in China: A systematic analysis of 2001–2011 studies, *International Journal of Hygiene and Environmental Health*, 217(1):46–51.

Mondragon, R, Julia, J E, Barba, A and Jarque, J C, 2012. Characterization of silica–water nanofluids dispersed with an ultrasound probe: A study of their physical properties and stability, *Powder Technology*, 224:138–146.

Nie, W, Yang, B, Du, T, Peng, H P, Zhang, X and Zhang, Y L, 2022. Dynamic dispersion and high-rise release of coal dust in the working surface of a large-scale mine and application of a new wet dust reduction technology, *Journal of Cleaner Production*, 351.

Shi, S A, He, J X, Zhang, X L, Yu, Z Q, Wang, J, Yang, T T and Wang, W, 2023. Pore structure evolution of tar-rich coal with temperature-pressure controlled simulation experiments, *Fuel*, 354.

Wang, G, Xie, S L, Huang, Q M, Wang, E M and Wang, S X, 2023a. Study on the performances of fluorescent tracers for the wetting area detection of coal seam water injection, *Energy*, 263.

Wang, H T, Cheng, S S, Wang, H J, He, J, Fan, L and Danilov, A S, 2023b. Synthesis and properties of coal dust suppressant based on microalgae oil extraction, *Fuel*, 338.

Wang, L, Sun, Y W, Zheng, S W, Shu, L Y and Zhang, X L, 2023c. How efficient coal mine methane control can benefit carbon-neutral target: Evidence from China, *Journal of Cleaner Production*, 424.

Xu, L, Li, Q, Myers, M and Cao, X M, 2023. Investigation of the enhanced oil recovery mechanism of CO_2 synergistically with nanofluid in tight glutenite, *Energy*, 273.

Zhang, R, Xing, Y W, Xia, Y C, Luo, J Q, Tan, J L, Rong, G L, Tan, J, Rong, G Q and Gui, X H, 2020. New insight into surface wetting of coal with varying coalification degree: An experimental and molecular dynamics simulation study, *Applied Surface Science*, 511.

Zhang, T C, Zou, Q L, Jia, X Q, Liu, T, Jiang, Z B, Tian, S X, Jiang, C Z and Cheng, Y Y, 2022. Effect of cyclic water injection on the wettability of coal with different SiO_2 nanofluid treatment time, *Fuel*, 312.

Zhang, Z Q, Wang, C L and Yan, K F, 2015. Adsorption of collectors on model surface of Wiser bituminous coal: A molecular dynamics simulation study, *Minerals Engineering*, 79:31–39.

Zhao, B, Li, S G, Lin, H F, Cheng, Y Y, Kong, X G and Ding, Y, 2022. Experimental study on the influence of surfactants in compound solution on the wetting-agglomeration properties of bituminous coal dust, *Powder Technology*, 395:766–775.

Zhao, J J, Tian, S X, Li, P, Xie, H G and Cai, J J, 2023. Molecular dynamics simulation and experimental research on the influence of SiO2-H2O nanofluids on wettability of low-rank coal, *Colloids and Surfaces A: Physicochemical and Engineering Aspects*, 679.

Zhou, G, Wang, C M, Wang, Q, Xu, Y X, Xing, Z Y, Zhang, B Y and Xu, C C, 2022a. Experimental study and analysis on physicochemical properties of coal treated with clean fracturing fluid for coal seam water injection, *Journal of Industrial and Engineering Chemistry*, 108:356–365.

Zhou, G, Xing, M Y, Wang, K L, Wang, Q, Xu, Z, Li, L and Cheng, W M, 2022b. Study on wetting behavior between CTAC and BS-12 with gas coal based on molecular dynamics simulation, *Journal of Molecular Liquids*, 357.

Zhou, G, Yao, J J, Wang, Q W, Tian, Y C and Sun, J, 2023. Synthesis and properties of wettability-increasing agent with multi-layer composite network structure for coal seam water injection, *Process Safety and Environmental Protection*, 172:341–352.

Zou, Q L, Zhang, T C, Ma, T F, Tian, S X, Jia, X Q and Jiang, Z B, 2022. Effect of water-based SiO_2 nanofluid on surface wettability of raw coal, *Energy*, 254.

Zheng S, Xiao P W, Xia Y Z, Liao X C, Yan E, Schulz H, Tian J, Ren F, Cai Y, Ge C, Han H. 2020 New insight into the slow wetting of coal with varying coalification degree: An experimental and molecular dynamics simulation study. Journal Surface Science 511.

Zhang Z B, Zou D L, Xu X, Li S L, Jiang Y P, Shen C Z and Chen B M. 2022 Effect of cyclic wetting on the wettability of coal at each filtration flow nanofluid treatment time. Fuel 352.

Zhang Z, Wang C B and Yan K F. 2015 Adsorption of collectors on model surface of Wiser bituminous coal: A molecular dynamics simulation study. Mineral Engineering 79.

Zhu D, Su S Y, He K F, Cheng J Y, Kong X G and Dong W V. 2022 Experimental study on the influence of surfactant solution on the wetting-agglomeration properties of bituminous coal dust. Power Technology 395.

Zou Q, Liu H, Xu P J, Xia Z X, Chang Gu, Li J J. 2022 Molecular dynamics simulation and experimental potential of the influence of SiO2-H2O nanofluids on wettability of low-rank coal. Colloids and Surfaces A: Physicochemical and Engineering Aspects 579.

Zou Q, Cheng Q M, Wang G M, Kim Y X, Xing Z Y, Sheng B Y and Xu X C. 2024 Experimental study and analysis on the surface tension of coal treated with flash filtration water for coal seam water injection. Advanced Powder Technology 108 300-308.

Optimising drill hole spacing using conditional simulation at the Invincible Gold Mine, Australia

S Zutah[1] and W Assibey-Bonsu[2]

1. Resource Geology Superintendent, Gold Fields Ltd, Perth WA 6000.
   Email: stanley.zutah@goldfields.com
2. Principal Specialist, Geostatistics and Assurance, Gold Fields Ltd, Perth WA 6000.
   Email: winfred.assibeybonsu@goldfields.com

ABSTRACT

The paper presents a probabilistic model for informed decision on mineral resource classification, using conditional simulation.

It includes a case study which investigates the optimisation of drill hole spacing for mineral resource to mineral reserve conversion for an extensional vein package within the Invincible South deposit at Gold Fields' St Ives Gold Mine. The paper explores enhancement of resource classification aimed at reducing exploration costs. The study also investigates the optimal drill hole spacing for grade control.

Results from the study supports a 40 m × 40 m drill spacing to inform an Indicated classification under the assumption that drill angles are optimal. For the grade control, the study showed that a 20 m × 10 m drill spacing is optimal to estimate tonnes, grades and ounces accurately on a quarterly basis. The findings of the study provide valuable insights for exploration strategy, resource definition and classification at Invincible and similar archaean gold mineralisation styles. The study further demonstrates the potential for significant exploration and resource drilling cost savings, and improved resource definition through targeted drilling programs.

INTRODUCTION

Mineral resources and mineral reserves are fundamental assets of mining companies and capital-intensive investments are made with respect to these. One critical risk exists in the uncertainty of the estimation and classification of resources and reserves. If after intensive capital investments, it is subsequently found that the expected mineral resources and mineral reserves were inefficiently estimated or classified, billions of dollars may be lost.

There is also an ever-increasing emphasis on the consideration of mineral resource classification categories by all CRIRSCO-based schemes including, the JORC Code (2012), the CIM Guidelines (2019), the SAMREC Code (2016), and SEC S-K 1300 Rules (SEC, 2021). While all the codes emphasise the prerogative of the individual Competent or Qualified Person to allocate the various resource categories, for various reasons there is an increased search for objectivity in Mineral Resource classification (Glacken, Rondon and Levett, 2023).

Drill hole spacing is a critical factor in resource estimation, as it directly affects the accuracy and reliability of the estimated resource. However, determining the optimal drill hole spacing can be challenging, particularly in deposits with complex geology and variable mineralisation. Conditional simulation is a geostatistical technique that can be used to evaluate the impact of different drill hole spacings on resource estimation and classification. It also offers a framework for assessing recoverable resources and the uncertainty associated with them, enabling more informed decision-making and risk mitigation in mine planning and orebody modelling (Emery and Ortiz, 2011; Deraisme and Assibey-Bonsu, 2012; Journel and Kyriakidis, 2004). There are different approaches to drill hole spacing study (Nowak and Leuangthong, 2019; Parker and Dohm, 2014) proposes that for an Indicated resource, the drill hole spacing should be sufficient to predict tonnage, grade and metal on annual production with ±15 per cent relative precision at the 90 per cent confidence level (CI). Per Parker and Dohm's proposal, for a Measured resource, the ±15 per cent relative precision must be achieved on a quarterly production volume.

The paper investigates optimal drill hole spacing aimed at efficient mineral resource classification, reduction of exploration/resource definition costs, and mitigation of orebody risk relating to mineral resource and mineral reserve conversion. The study also aims at optimising costs of grade control drilling.

METHODOLOGY

Two project workflows were applied to address the above study goals. The first workflow involves using a conditional simulation approach to assess the impact of varying drill hole densities. One of the 100 equiprobable simulation realisations was sampled using the elected test drill patterns (10 m × 10 m, 20 m × 10 m, 20 m × 20 m, 40 m × 20 m, 40 m × 40 m and 80 m × 40 m) to produce sets of drill holes. The drill holes sets were imported into Leapfrog Geo for geology interpretation specific to each set of selected drill spacing data. Gold estimates using 3D Ordinary kriging were derived in Datamine Studio RM within these respective geology interpretations and the corresponding sample data set. All the estimates derived in the study were compared using the 10 m × 10 m case as a baseline in deriving the probabilistic error model.

The second workflow involves further empirical reconciliation of the model inventory within the same volume subsequently drilled at a 40 m × 40 m spacing and later infilled to 40 m × 20 m spacing, to further stress-test the efficiency of the probabilistic model recommendations.

Currently, conventional drill spacing for Indicated classification reserves conversion within the extensional vein package bulk stopes at the mine, is at 40 m × 40 m and there is no standardised drill pattern for grade control.

CASE STUDY

Geology

The Invincible South deposit is located in the St Ives gold camp, about 15 km south of Kambalda (Figure 1). Invincible is a relatively recent discovery during the 40 plus years in which gold has been mined at St Ives and is considered to be the largest localised accumulation of gold resources in the area. Two styles of mineralisation are present and are largely controlled by the host lithology. Steep bedding parallel shear veins favour the finer mudstone unit, whilst extensional mostly shallow dipping veins, form in the coarser sediments.

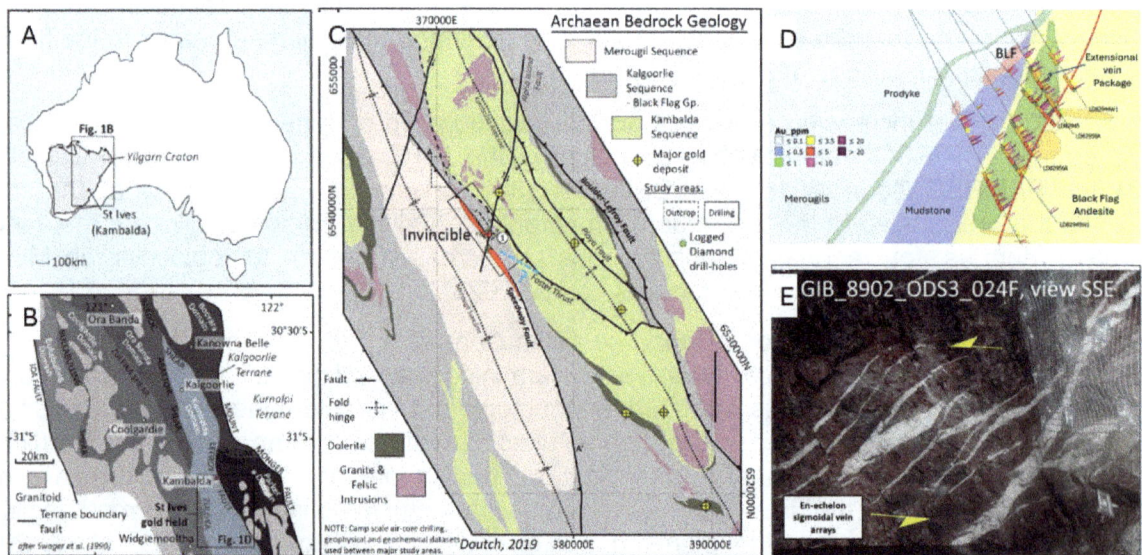

FIG 1 – Location and geology of the St Ives Gold Mine: (a) location of the Yilgarn Craton in Western Australia (after Doutch, 2019); (b) domain subdivisions of the Kalgoorlie Terrane and location of the St Ives Gold Mine (after Swager *et al*, 1990); (c) simplified Archaean Geology map of the study area (after Doutch, 2019); (d) cross-section through the Invincible South orebody; (e) type 2 style of mineralisation exposed in a development face.

Focus on this study is on two geological domains (500 and 520) within the extensional vein package. This style of mineralisation has been mined for the past three years using a bulk mining approach method. As the mine grows deeper, there is a strong motivation to optimise the costs of drilling throughout the exploration-development-production cycle for faster conversion rate to allow year on year reserve conversion and the use of high confidence material to plan and set-up the mine for the future. Because of this, the criticality of attaining an Indicated level of classification was recognised, since this is the lowest confidence level which can be converted to reserves and increased mine life, and thus, contribute to a forecast cash flow via the pre-feasibility and feasibility process.

Workflow

The use of conditional simulation provides the ability to define a dense grid of data, using the parameters of the existing real-world data set, which can then be post-processed in a variety of ways, either to define probability confidence intervals to achieve a resource category confidence, or to look at the drilling required to define error thresholds. The workflow for this study is as follows:

- Real-world data for domains 500 and 520 were validated and composited to 1 m. The drill data density, including mined-out areas, ranges from 10 m × 10 m to 80 m × 80 m with domain 500 being extensively mined underground. Significant portion of the undepleted resources of the type 2 mineralisation is within Domain 520 and its extensions as such most of the work was focused on domain 520.

- Geostatistical analysis was completed on the 1 m composited data and extreme outliers capped. The data was declustered using a 50 m × 50 m × 50 m grid and transformed into the Gaussian space through the Gaussian Anamorphosis modelling and checked for Gaussianity.

- The data was analysed including the spatial continuity from which variograms were modelled in the normal space.

- Conditional Simulation was completed using the Turning Band technique to generate 100 realisations on 1 m × 1 m × 1 m nodes in Isatis software. 500 turning bands were used for the simulation.

- One realisation from the suit of realisations 'Truth' (or benchmark) was then sampled using the elected pseudo drill patterns to produce drill holes. The spacings evaluated were 10 m × 10 m, 20 m × 10 m, 20 m × 20 m, 40 m × 20 m, 40 m × 40 m and 80 m × 40 m. The realisation was chosen to ensure it reflects the key parameters (histogram, variogram, multivariate statistics) of the input data. It was assumed that the pseudo drill holes were optimal and have the same orientation as shown in Figure 2. Figure 3 shows the histogram of the selected realisation. The simulations outputs were validated against the input data, Figure 4 provides some of the simulation checks.

- The drill hole samples were imported into Leapfrog software and geological interpretation and modelling specific to each tested drill set spacing was completed.

- Gold (Au) estimates using 3D Ordinary Kriging (OK) were derived in Datamine within these geology interpretations/models and the corresponding drill hole data set. Appropriate block sizes/supports that reflect the different drill spacing were used to underpin the respective 3D estimates. All six models used the same variograms, which was generated from the original data. It would have also been appropriate to generate different variograms in each drill grid instance, but the differences in continuity were not considered material. The OK models were validated against their own set of input pseudo-drill holes and then each were compared back to the 'Truth' simulation.

- Inventory comparisons on tonnes, grades, and ounces were derived to check the impact of the different drill spacings on the estimates. Estimates from the 10 m × 10 m was considered as the baseline. Additionally, an underground optimisation using the Mineable Shape Optimiser (MSO) software was run at 2.5 g/t using a minimum mining width of 5 m to generate optimised stopes for practical stope design comparisons and to remove any Inventory bias in the comparisons. The results were used for an initial assessment in determining the optimal drill spacing for Indicated classification and grade control.

- A second simulation model was completed on 1 m × 1 m × 1 m nodes using the data set from the 20 m × 10 m and 40 m x 40 m spaced data. Results from the simulations were reblocked to SMU scale (10 m × 10 m × 10 m). The 90/15 rule (Parker and Dohm, 2014; Verly, Postolski and Parker, 2014) was used for the purpose of validation and to achieve a classification confidence.

- Underground Level analysis at 2.5 g/t economic cut-off on 40 m level spacing was completed using the 10 m × 10 m spaced drilling model as a baseline. A 10 per cent probability error (ie a bar representing a 10 per cent probability error) was produced for context purposes.

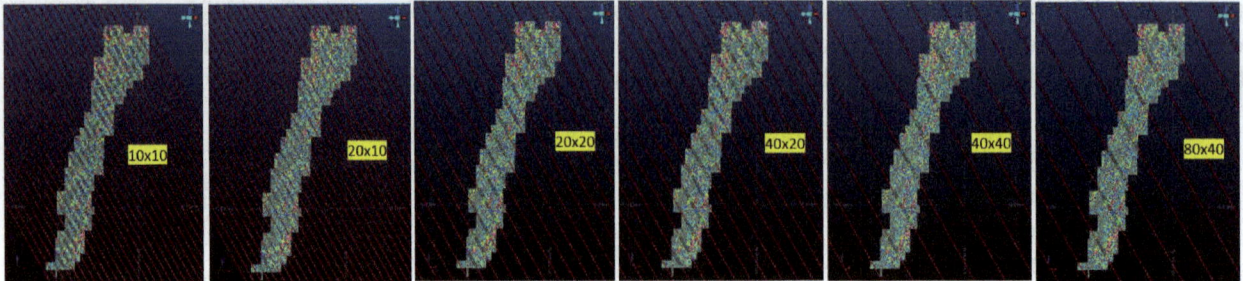

FIG 2 – Cross-section showing one realisation from the simulation output for domain 520. Red traces are the pseudo or equivalent drill holes at the various spacings.

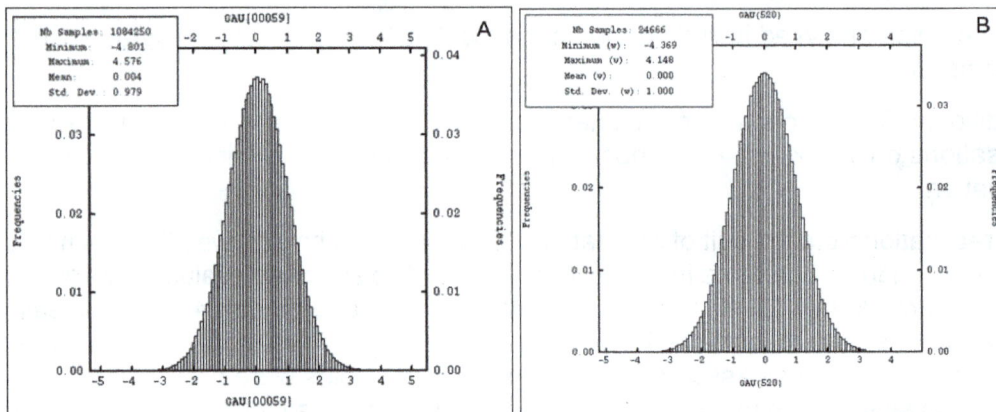

FIG 3 – (a) Histogram of the selected realisation for domain 520; (b) histogram of input data both in the normal space.

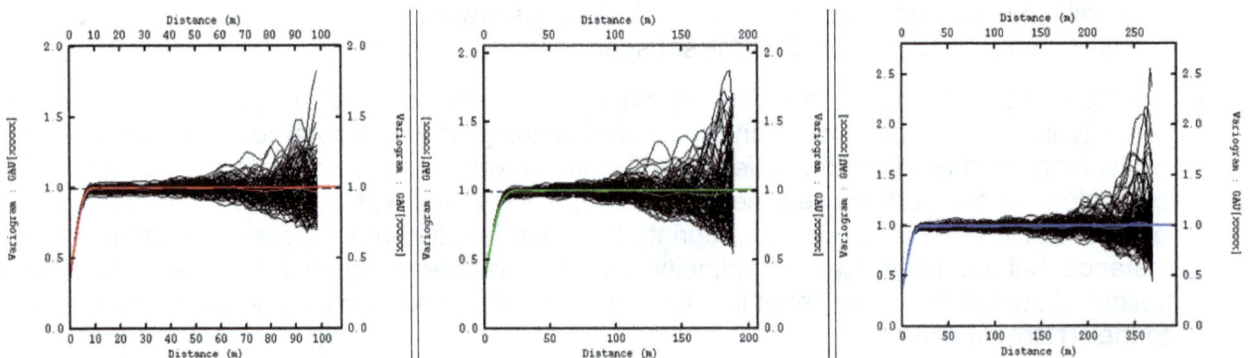

FIG 4 – Validation of simulation results for domain 520, 40 m × 40 m drilling in three directions.

INITIAL INDICATIVE RESULTS

Indicated Resource classification consideration

The effect of different drill hole spacings on resource estimation and classification was analysed by comparing the results from the kriging results. Tables 1 to 4 show the Inventory comparison for the various estimations.

TABLE 1

Inventory comparison at 2.5 g/t economic cut-off within Mineable Shape Optimiser (MSO) shapes between 10 m × 10 m drill spacing and the various drill spacings for domain 500.

Base model (10m x 10m)			Drill hole spacing	Spaced model			Variation			Variation %		
Tonnes (t)	grade (g/t)	metal (oz)		Tonnes (t)	grade (g/t)	metal (oz)	Tonnes (t)	grade (g/t)	metal (oz)	Tonnes (t)	grade (g/t)	metal (oz)
2,403,300	7.0	541,011	20m x 10m	2,519,138	6.4	515,841	115,838	-0.6	-25,170	5%	-9%	-5%
2,403,300	7.0	541,011	20m x 20m	2,527,305	6.1	493,575	124,005	-0.9	-47,436	5%	-13%	-9%
2,403,300	7.0	541,011	40m x 20m	2,241,664	6.0	431,937	-161,636	-1.0	-109,074	-7%	-14%	-20%
2,403,300	7.0	541,011	40m x 40m	1,971,051	6.8	432,427	-432,249	-0.2	-108,584	-18%	-3%	-20%
2,403,300	7.0	541,011	80m x 40m	1,910,798	6.4	396,051	-492,502	-0.6	-144,960	-20%	-9%	-27%

TABLE 2

Inventory comparison at 2.5 g/t economic cut-off within MSO shapes between 40 m × 20 m drill spacing and the 40 m × 40 m drill spacing for domain 500.

40m x 20m grid model			40m x 40m grid model			Variation			Variation %		
Tonnes (t)	grade (g/t)	metal (oz)	Tonnes (t)	grade (g/t)	metal (oz)	Tonnes (t)	grade (g/t)	metal (oz)	Tonnes (t)	grade (g/t)	metal (oz)
2,241,664	6.0	431,937	1,971,051	6.8	432,427	-270,613	0.8	490	-12%	13%	0%

TABLE 3

Inventory comparison at 2.5 g/t economic cut-off within MSO shapes between 10 m × 10 m drill spacing and the various drill spacings for domain 520.

Base model (10m x 10m)			Drill hole Spacing	Spaced models			Variation			Variation %		
Tonnes (t)	grade (g/t)	metal (oz)		Tonnes (t)	grade (g/t)	metal (oz)	Tonnes (t)	grade (g/t)	metal (oz)	Tonnes (t)	grade (g/t)	metal (oz)
3,725,814	4.9	581,030	20m x 10m	3,869,686	4.5	559,684	143,872	-0.4	-21,346	4%	-8%	-4%
3,725,814	4.9	581,030	20m x 20m	4,019,174	4.3	551,895	293,360	-0.6	-29,135	8%	-12%	-5%
3,725,814	4.9	581,030	40m x 20m	3,650,276	4.2	496,956	-75,538	-0.7	-84,074	-2%	-14%	-14%
3,725,814	4.9	581,030	40m x 40m	4,014,428	4.2	543,481	288,614	-0.7	-37,549	8%	-14%	-6%
3,725,814	4.9	581,030	80m x 40m	3,296,133	4.8	504,454	-429,681	-0.1	-76,576	-12%	-2%	-13%

TABLE 4

Inventory comparison at 2.5 g/t economic cut-off within MSO shapes between 40 m × 20 m drill spacing and the 40 m × 40 m drill spacing for domain 520.

40m x 20m grid model			40m x 40m grid model			Variation			Variation %		
Tonnes (t)	grade (g/t)	metal (oz)	Tonnes (t)	grade (g/t)	metal (oz)	Tonnes (t)	grade (g/t)	metal (oz)	Tonnes (t)	grade (g/t)	metal (oz)
3,650,276	4.2	496,956	4,014,428	4.2	543,481	364,152	0.0	46,525	10%	0%	9%

DISCUSSION OF INITIAL RESULTS AND FURTHER WORK

Indicated Resource category

This study incorporates uncertainty or variance for both geology and grade for the respective drilling grids. Indicated Resources provide estimates for both local and global tonnes, grades and ounces. For long-term planning, the global errors are expected to be lower.

Prior to this study, the drill spacing for an Indicated Resource classification for the extensional vein package at Invincible South was 40 m × 20 m. However, Tables 2 and 4 show that, there is no significant ounce variation between the 40 m × 20 m and the 40 m × 40 m grid models for the two domains (ounce variation up to 9 per cent), though they reflect different tonnes and grades above cut-off. These results seem to indicate that 40 m × 40 m drill spacing could be fit for purpose in classifying Indicated CLASS Resource in the area.

Tables 1 and 3 further show the global ounce variation of -20 per cent for domain 500 (40 m × 40 m and 40 m × 20 m) when compared to the baseline model (10 m × 10 m grid). The corresponding global ounces variance for similar drilling in Domain 520 is lower (-6 per cent to -14 per cent). Tables 1 and 3 show material grade variances of up to -14 per cent for both drilling grids (40 m × 40 m and 40 m × 20 m). These global errors are mostly due to smoothing effect of the 3D estimates as a result of block support and information effect differences (Assibey-Bonsu et al, 2024). It should be noted that due to the different drilling configurations, the underlying support/blocks used for the 3D models were different (baseline model blocks (10 m × 10 m × 10 m); 40 m × 40 m grid model used 10 m × 20 m × 10 m blocks). As a result of the smoothing and information effect issues, these results (Tables 1 and 3) were viewed as initial indicative outcomes for the classification work.

The probability modelling analyses in the paper incorporate the impact due to smoothing effect, block supports, and information effect differences for the respective drilling grids. The probability analyses aim at analysing the corresponding recoverable resource models to provide solutions for efficient mineral resource classification.

Probability modelling for Indicated Mineral Resource classification and conversion

The analysis in this section of the paper is a follow-up on the initial case study and the corresponding results above, which showed 40 m × 40 m as possible grid for Indicated CLASS (the smoothing effect etc limitations on the initial results have been discussed in the paper).

Per the methodology/workflow sections, a second simulation model was completed on 1 m × 1 m × 1 m nodes using the data set from the 40 m × 40 spaced data. Results from the simulations were re-blocked to SMU scale (10 m × 10 m × 10 m). The 90/15 rule (Parker and Dohm, 2014) was used for the purpose of modelling a probability classification confidence.

Workflow of probability modelling for Indicated Mineral Resource classification (annual volume)

Using the data from the 40 m × 40 m pseudo grid data, a 100-realisation conditional simulation was completed on 1 m × 1 m × 1 m nodes to quantify the grade and ounce variance under this spacing consideration. The results from the simulation were re-blocked to SMU scale (10 m × 10 m × 10 m). A volume of material within an economic footprint and representing an annual forecast was selected as shown in the white outline in Figure 5. The selection was carefully made to ensure the volume and contained metal align with a typical annual production volume for the bulk extensional package at Invincible. Tonnes, grade, and ounce Inventories were calculated for all the 100 realisations within the selected volumes using 2.5 g/t grade cut-off. The 90 per cent confidence level (CI) and the median value were then derived from the results and used for the 90:15 analyses.

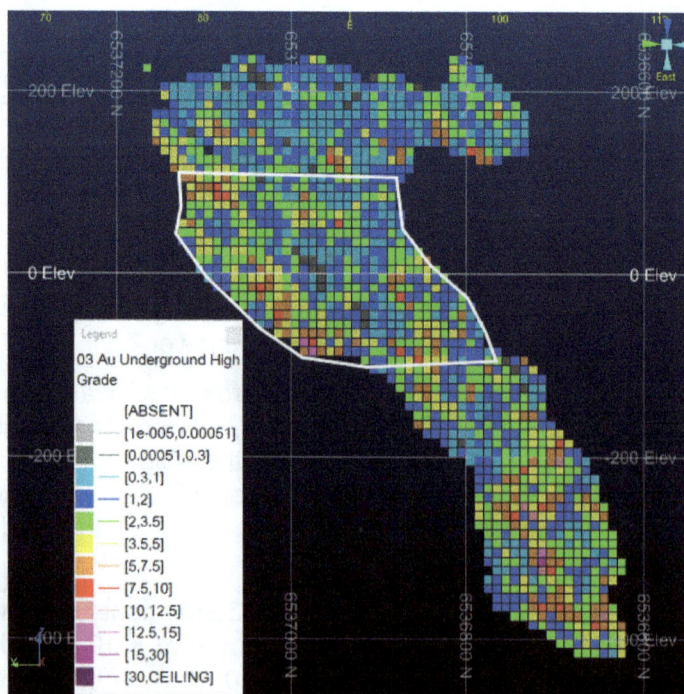

FIG 5 – Sectional view of one of the realisations showing re-blocked simulation using the 40 m × 40 m pseudo drill spaced data (domain 520). White outline is the selected area representing an annual volume used for the 90:15 rule probability model.

Results

The results from the 100 realisations from the simulations on an annual production volume (tonnes, grade and ounce) using the 40 m × 40 m drill data set are shown in Figure 6 and Table 5. Table 5 shows that, there is a 90 per cent probability that the estimated ounces will be within (-13 per cent to +15 per cent) of the median value on an annual volume which suggests that 1 out of 10 annual periods is likely to have a variation within -13 per cent to +15 per cent. The results demonstrate minimal risk and also are within Parker and Dohm's (2014) proposal (ie the 90:15 rule, which states that, for an Indicated Resource, the drill hole spacing should be sufficient to predict tonnage, grade and metal on annual production with ±15 per cent relative precision at the 90 per cent confidence level (CI). The results support using a 40 m × 40 m drilling for an Indicated Resource classification in this area/domain. Similar results were observed for domain 500.

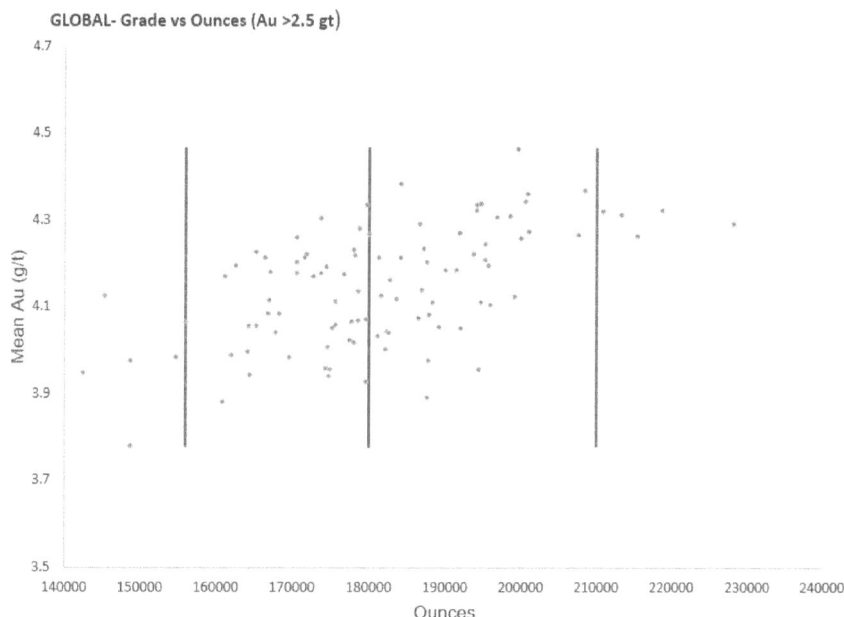

FIG 6 – Plot of the 100 realisations from the simulation using the 40 m × 40 m pseudo drill spaced data. The simulations were selected on an annual volume. Included are the Median ounce (red line) and realisation 90 per cent CI ounce range.

TABLE 5

Probability analysis of annual production volumes at the 90 per cent confidence interval relative to the median.

	Measure	Ounces (koz)	CI relative to median
Annual comparison	5th percentile	156	-13%
	95th percentile	207	15%
	Median	180	

Figure 7 is a visual representation in long section for the 40 m × 20 m and 40 m × 40 m pseudo drill holes and their respective wireframes showing that the lode geometry can be well defined by the 40 m × 40 m spacing for Indicated Resource (compared to the previous 40 m × 20 m Indicated CLASS grid).

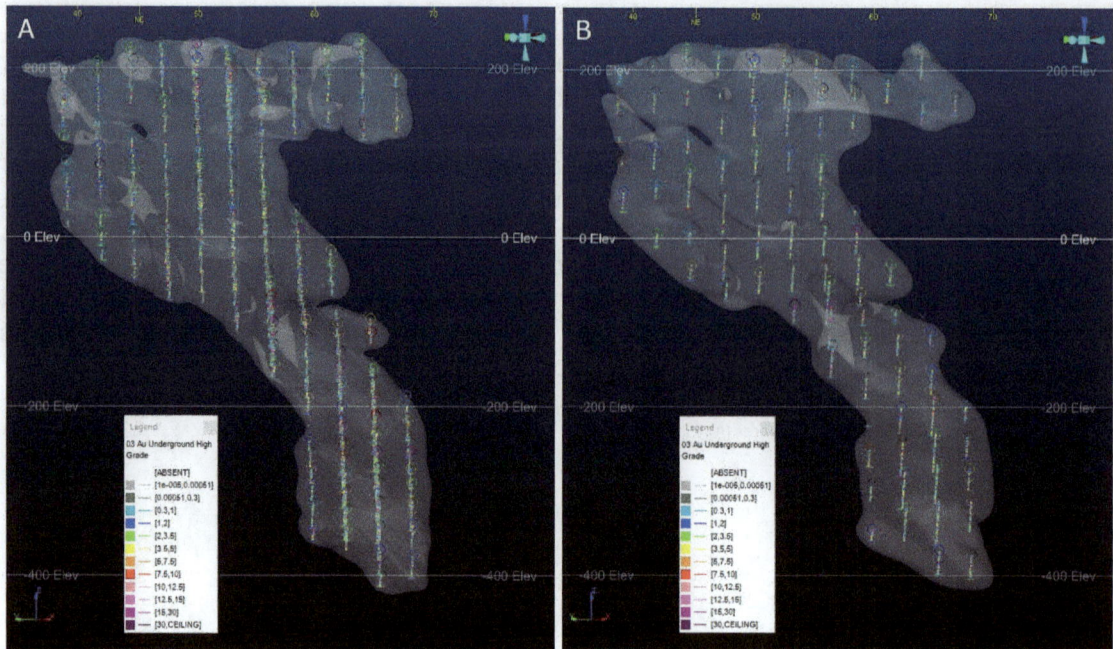

FIG 7 – (a) Long section of domain 520 showing pseudo drill holes at 40 m × 20 m spacing and corresponding wireframe in white; (b) long section of domain 520 showing pseudo drill holes at 40 m × 40 m spacing and corresponding wireframe.

FOLLOW-UP DRILLING TEST OF THE PROBABILITY MODEL

Subsequent to this study, an area within domain 520 on a 40 m × 40 m spacing was infilled to 40 m × 20 m spacing. Table 6 shows the Inventory comparison using a 2.5 g/t cut-off for the follow-up infill work. Table 6 shows that the differences in tonnes, grades and ounces between the estimates from the two-drill hole spacings are not significant (-2 per cent to +3 per cent). The results support the probability model's recommendation for 40 m × 40 m grid to underpin Indicated CLASS classification in the area. The good reconciliation is a further stress-test of the probability model.

TABLE 6

Inventory comparison at 2.5 g/t economic cut-off within MSO shapes within common volume of area drilled on 40 m × 40 m grid and later infill drilled to 40 m × 20 m drill spacing for domain 520.

40m x 20m grid model			40m x 40m grid model			Variation			Variation %		
Tonnes (t)	grade (g/t)	metal (oz)	Tonnes (t)	grade (g/t)	metal (oz)	Tonnes (t)	grade (g/t)	metal (oz)	Tonnes (t)	grade (g/t)	metal (oz)
2,004,556	4.3	275,957	2,071,173	4.2	277,956	66,617	-0.1	1,999	3%	-2%	1%

PROBABILITY MODELLING FOR OPTIMAL GRADE CONTROL DRILL SPACING

Historically, typical drill spacing for grade control at St Ives is 10 m × 10 m. The spacings investigated in this study to assess optimal grade control drilling for the bulk extensional veins were 10 m × 10 m, 20 m × 10 m and 20 m × 20 m.

Inventory comparisons for domains 500 and 520 showed non-material differences in tonnes, grade, and ounces, with variations, ranging -9 per cent to +5 per cent respectively for 10 m × 10 m and the 20 m × 10 m. Table 7 shows the results for domain 520 (-8 per cent to +4 per cent).

TABLE 7

Inventory comparison at 2.5 g/t economic cut-off within MSO shapes between 10 m × 10 m drill spacing and the 20 m × 10 m drill spacing for domain 520.

Base model (10m x 10m)			Drill hole Spacing	20m x 10m grid model			Variation			Variation %		
Tonnes (t)	grade (g/t)	metal (oz)		Tonnes (t)	grade (g/t)	metal (oz)	Tonnes (t)	grade (g/t)	metal (oz)	Tonnes (t)	grade (g/t)	metal (oz)
3,725,814	4.9	581,030	20m x 10m	3,869,686	4.5	559,684	143,872	-0.4	-21,346	4%	-8%	-4%

Further probability simulation work was done for probability error modelling for these initial GC drilling spacing work.

Workflow of probability model for grade control (quarterly volume)

The workflow was similar to the Indicated Mineral Resource probability classification modelling as discussed in the paper. The 20 m × 10 m spacing was used in this case for the simulation and the analysis was done on a quarterly volume basis (white outline block in Figure 8).

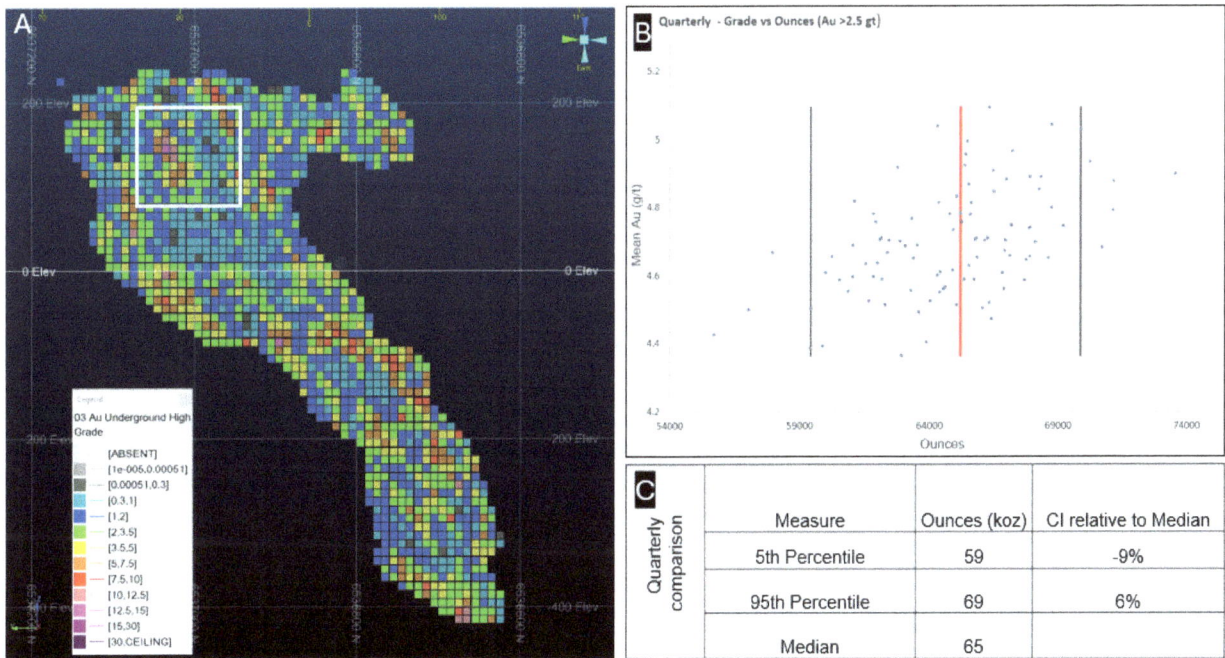

FIG 8 – (a) Long section through domain 520 showing re-blocked simulation results using 20 m × 10 m spaced data; (b and c) quarterly period grade and ounces for the 100 Realisations (point plots). Included are the Median ounce (red line) and Realisation 90 per cent CI ounce range.

Results

Figure 8 shows the results for domain 520 (which were similar to domain 500 results). The results in (Figure 8) confirm the initial results in Table 7.

The quarterly simulation probability analyses for the two domains show that there is a 90 per cent probability that the estimated ounces will be within (-9 per cent and +9 per cent: domain 520 results in Figure 8: -9 per cent/+6 per cent) of the median value on a quarterly volume. These demonstrate that for 1 out of 10 quarterly periods, there is the likelihood of a variance within -9 per cent to +9 per cent. The results indicate that, in both cases, the recommended grade control grid exhibits a reasonable level of grade and geological continuity. The analysis demonstrates minimal risk and also meets the 90:15 rule. The results support using a 20 m × 10 m drilling for grade control spacing or Measured Resource classification.

The work shows that 20 m × 10 m drill hole spacing should be sufficient to predict tonnage, grade, and metal on a quarterly production with ±15 per cent relative precision at the 90 per cent confidence level (CI). This will lead to lower grade control drilling cost together with drilling time savings and will help to alleviate the practical challenge of grade control drilling and production schedules.

Level analysis results

Level analysis was completed on 40 m levels using the block model estimates from the 10 m × 10 m and 20 m × 10 m spacings to analyse the local error and variance. The 10 m × 10 m spacing was used as a baseline for comparison as shown in Figure 9. The results show that, 20 m × 10 m compares well with 10 m × 10 m, and tonnes, grade and ounce are generally within 10 per cent local variation.

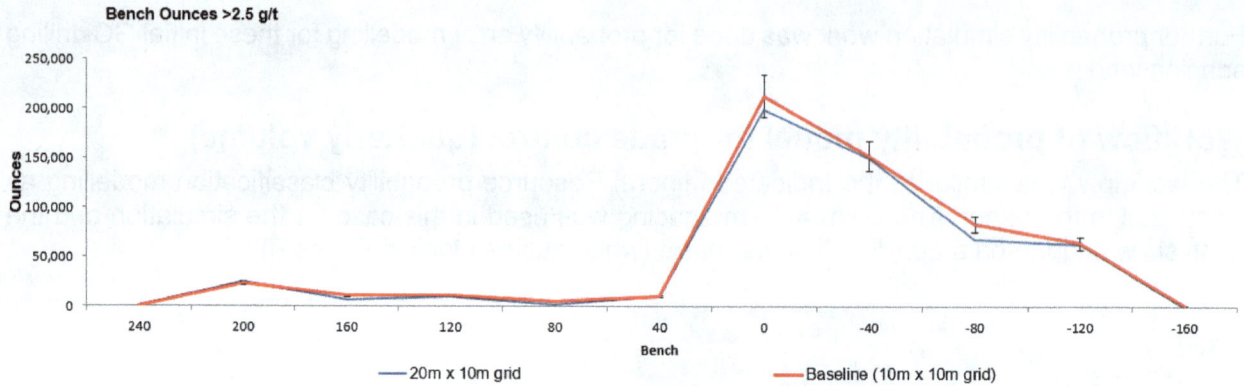

FIG 9 – Ounce level analysis for domain 500 with 10 per cent local error bar.

PRACTICAL CONSIDERATIONS AND COST COMPARISONS FOR INDICATED CLASS CLASSIFICATION (40 m × 40 m VERSUS 40 m × 20 m)

Table 8 shows among other outcomes, the cost and drilling time savings for down-dip extension of domain 520 (Figure 10). Based on the recommendation from the study, five drill holes in red dots were recommended to bring the spacing to 40 m × 40 m. The drilling was completed which resulted in about 300 koz of the Inferred material in the white outline converted to Indicated Resource and further converted to ore reserves based on this probability modelling work. Table 8 demonstrates significant value-adds, including ~$13 million costs savings plus seven months drilling time savings, translating to additional ~300 koz ore reserve conversion at significantly lower reserve conversion cost per ounce. This extends the life of the mine by almost one year.

TABLE 8

Cost and other parameters: comparison between 40 m × 40 m and 40 m × 20 m drilling.

	40 m × 40 m grid	40 m × 20 m grid
Number of drill holes	54	108
Number of metres	27 000 m	54 000 m
Drilling time	7 months/2DD rigs	14 months/2DD rigs
Drilling cost	A$13.2 M	A$26 M
Assay cost	A$97 200	A$194 400
Cost/oz	A$25.7	A$51.5
Ounces converted to Indicated CLASS/Reserves	300 koz	

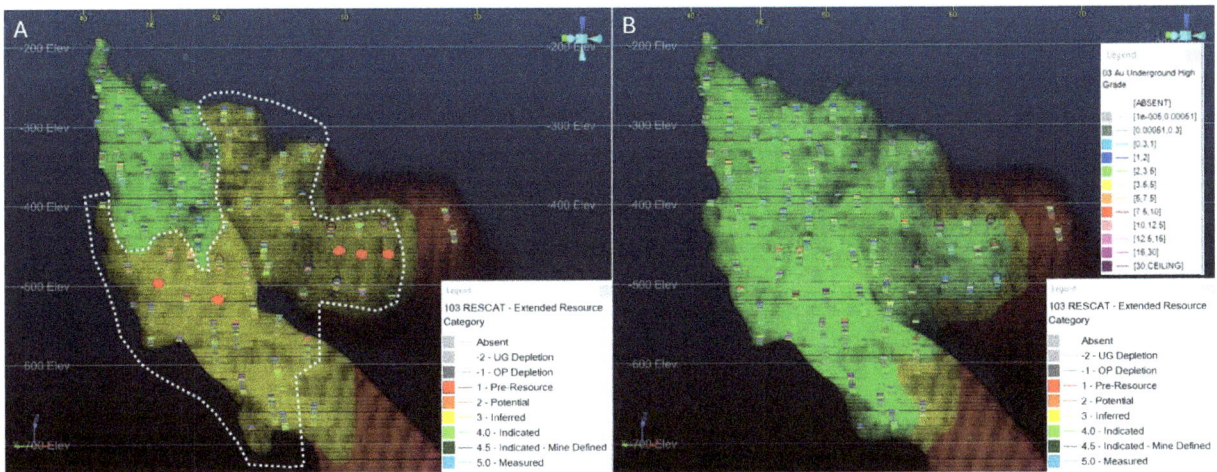

FIG 10 – (a) Long section showing domain 520 extension down-dip block model coloured by Resource Classification. Red dots are additional infill drilling completed. Existing drill holes are coloured by Au grade; (b) final block model showing material upgraded to Indicated.

CONCLUSIONS

The study demonstrates the effectiveness of conditional simulation for optimising drill hole spacing in a gold deposit. The results show that conditional simulation can provide a quantitative measure of the uncertainty associated with different drill hole spacings, allowing for more objective informed decisions about exploration and mining strategies. The use of conditional simulation and probability modelling can lead to significant cost savings, improved resource classification, and increased reserve conversion for life-of mine extension.

The probability error modelling results from the study support a 40 m × 40 m drill spacing to inform an Indicated Resource under the assumption that mineralisation downdip has similar geological analogue as observed in the mined-out area. It is important to ensure optimal drilling angles. The study showed significant value-adds, demonstrating drilling costs savings of ~$13 million as well as seven months drilling time savings. This translates to ~300 koz ore reserve conversion at a significantly lower reserve conversion cost per ounce.

The probability modelling study further shows that a grade control model with a 20 m × 10 m drilling will be sufficient to predict grades, tonnes and ounces accurately enough on a quarterly basis for the bulk mining approach currently being implemented at the mine. The model will be sufficient to mitigate the inherent risk associated with geological and grade variance for grade and ore control drilling at the production stage. While it is understood that local variations will be observed during mining, this study shows that the scale of risk on quarterly basis will be minimised. This will also lead to lower grade control drilling cost together with time savings and will help to alleviate the practical challenge of grade control drilling and production schedules.

ACKNOWLEDGEMENTS

The permission of Gold Fields to publish results from the Invincible South Project is gratefully acknowledged. The authors also acknowledge Tim Goodale for his input in the Leapfrog work and Richard Tully for proofreading and feedback on this paper.

REFERENCES

Assibey-Bonsu, W, Aboagye, M, Appau, K and Muller, C J, 2024. Orebody and mine planning assessment based on alternative recoverable and non-recoverable resource modelling techniques, 12th International Geostatistics Congress, September 2024, Portugal.

Canadian Institute of Mining, Metallurgy and Petroleum (CIM), 2019. CIM Estimation of Mineral Resources and Mineral Reserves Best Practice Guidelines (The CIM guideline) [online]. Available from: <https://mrmr.cim.org/media/1129/cim-mrmr-bp-guidelines_2019.pdf>

Deraisme, J and Assibey-Bonsu, W, 2012. Comparative study of Localized Block Simulations and Localized Uniform Conditioning in the Multivariate case, in *Proceedings of the International Geostatistics Conference*, pp 309–320.

Doutch, D, 2019. Origin, geochemistry, stratigraphic and structural setting of the Archean Invincible gold deposit, St Ives gold camp, Yilgarn Craton, PhD thesis, University of Tasmania.

Emery, X and Ortiz, J, 2011. Two approaches to direct block-support conditional co-simulation, *Computers and Geosciences*, 37(8).

Glacken, I, Rondon, O and Levett, J, 2023. Drill hole spacing analysis for classification and cost optimisation – a critical review of techniques, Mineral Resource Estimation Conference, Perth, Australia.

JORC, 2012. Australasian Code for Reporting of Exploration Results, Mineral Resources and Ore Reserves (The JORC Code) [online]. Available from: <http://www.jorc.org> (The Joint Ore Reserves Committee of The Australasian Institute of Mining and Metallurgy, Australian Institute of Geoscientists and Minerals Council of Australia).

Journel, A G and Kyriakidis, P, 2004. *Evaluation of Mineral Reserves: A Simulation Approach*, Applied Geostatistics Series, Oxford University Press.

Nowak, M and Leuangthong, O, 2019. Optimal drill hole spacing for resource classification, *Mining Goes Digital* (ed: U Mueller), pp 115–124 (Taylor and Francis Group: London).

Parker, H M and Dohm, C E, 2014. Evolution of Mineral Resource classification from 1980 to 2014 and current best practice, Finex 2014 Julius Wernher Lecture, 72 p.

SAMREC, 2016. South African code for the Reporting of Exploration Results, Mineral Resources and Mineral Reserves. (The SAMREC Code) [online]. Available from: <https://www.samcode.co.za/samcode-ssc/samrec>

Swager, C P, Griffin, T J, Witt, W K, Wyche, S, Ahmat, A L, Hunter, W M and McGoldrick, P J, 1990. Geology of the Archaean Kalgoorlie Terrane – an explanatory note.

US Securities and Exchange Commission (SEC), 2021. The SEC SK-1300 Rules [online]. Available from: <https://www.sec.gov/files/rules/final/2018/33-10570.pdf>

Verly, G, Postolski, T and Parker, H M, 2014. Assessing uncertainty with drill hole spacing studies – applications to Mineral Resources, in *Proceedings of the Orebody Modelling and Strategic Mine Planning Symposium 2014*, pp 109–118 (The Australasian Institute of Mining and Metallurgy: Melbourne).

Exploration and geology

Advanced multivariable geological modelling – case study of orogenic Au deposit with multi-generational quartz veins

M Aliaga[1] and M Moore-Roth[2]

1. Senior Geomodeler, Newmont, Colorado Springs CO 80923, USA.
   Email: miguel.aliaga@newmont.com
2. Geosciences Manager, Maptek, Evergreen CO 80439, USA.
   Email: maureen.moore@maptek.com

ABSTRACT

This case study reviews the application of machine learning in modelling complex multi-generational quartz veins of an orogenic gold deposit. Traditional geological modelling methods often rely on subjective interpretations and struggle to incorporate multiple variables, leading to bias and inefficiencies. Machine learning offers a transformative solution by integrating various geological inputs, such as structural data, vein intensity, and mineral associations, into a single multivariable domain characterisation. This approach improves modelling efficiency, allowing geologists to focus on other critical tasks like geological logging, mapping, and resource calculation, supporting a more holistic interpretation.

By leveraging machine learning, geologists can test structural hypotheses and explore multivariable inputs, producing more accurate and realistic geologic models faster than traditional methods. This allows geologists to evaluate how different hypotheses impact downstream interpretations. The case study compares 3D geologic interpretations from both traditional and machine learning methods, highlighting key lessons in modelling the complexity of multigenerational quartz veins in an orogenic Gold deposit.

Chosen for its complexity and multiple generations of quartz veins with strong structural controls, this deposit demands advanced modelling techniques. Applying machine learning enhanced structural analysis and multivariable integration, leading to models that better reflect the intricate relationships between vein structures and mineralisation. This approach respects the geological variability and structural complexity of the deposit, improving prediction reliability and understanding of mineralisation controls within the orogenic gold system. Additionally, adjusting block model resolution and delimiting models to solids or surfaces offers flexibility in optimising model accuracy and resource estimation.

In conclusion, this case study demonstrates how cloud computing and machine learning provide an efficient and adaptable solution for modelling complex deposits. Integrating multiple geological variables and reducing processing time enhances confidence in resource estimation, supports real-time decision-making, and deepens understanding of the deposit's structural framework.

INTRODUCTION

Modelling orogenic gold deposits with complex structural controls and multi-generational quartz veins presents significant challenges in geological interpretation. Traditional geological modelling methods, both explicit and implicit, rely heavily on subjective interpretations, making it difficult to integrate multiple geological variables systematically in the classification of domains. This often results in models that are time-consuming to update and also prone to bias, particularly in structurally complex deposits where vein continuity and mineralisation controls are highly variable.

Machine learning presents a transformative approach to geological modelling. It integrates multivariable data sets—such as lithology, vein intensity, structural data, and mineralogical associations—into a single, comprehensive model (Pym *et al*, 2022). By leveraging computational efficiency and pattern recognition, machine learning enables geologists to explore different structural hypotheses, optimise data-driven interpretations, and reduce model update times from weeks or months to hours. This also enhances model reproducibility but also allows for rapid scenario testing, improving confidence in resource estimation and exploration strategies.

This study applies to a machine learning assisted domain modelling tool designed to incorporate multiple geological variables into a predictive framework. The selected deposit, Merian II, located within the Merian Trend in north-eastern Suriname, was chosen due to its geological complexity, characterised by a structurally controlled orogenic gold system with multiple generations of quartz veins and breccias. The deposit occurs within the Palaeoproterozoic Armina Formation of the Marowijne Supergroup, part of the Guiana Shield, and is influenced by a north-west-trending fold-and-thrust structural setting. Mineralisation is hosted in quartz veins, stockworks, and breccia bodies associated with shear zones and lithological contrasts.

This paper compares traditional geological modelling approaches with machine learning-driven models, evaluating how multivariable integration improves model accuracy, geological continuity, and predictive capabilities. Three different machine learning scenarios were tested, incorporating varying degrees of geological control, from a simple lithological model to a fully integrated structural model. The results demonstrate that incorporating structural controls significantly enhances model reliability, particularly in areas of low data density.

By adopting a machine learning approach, this study highlights key advantages such as faster model generation, reduced subjectivity, and improved geological coherence while also addressing potential challenges, including structural data sensitivity and model reproducibility. The findings contribute to the ongoing evolution of geological modelling techniques, emphasizing the potential of machine learning as a tool for improving exploration efficiency and mineral resource assessment in complex orogenic gold deposits.

GEOLOGY

Regional geology of the Merian Trend

The Merian Trend, located in north-eastern Suriname, lies within the Palaeoproterozoic Guiana Shield, a cratonic province characterised by a complex tectonic history involving multiple orogenic events (Figure 1; Daoust *et al,* 2011). This region is part of a broader framework that includes three major tectono-stratigraphic units:

1. Archean high-grade gneissic complexes in the northern portion of the shield.

2. Palaeoproterozoic greenstone belts, comprising clastic, volcanic, and volcaniclastic sequences, formed between 2.25 and 2.0 Ga.

3. Clastic sequences of the Roraima Formation were deposited at ~1.8 Ga and later intruded by Mesozoic mafic magmatism.

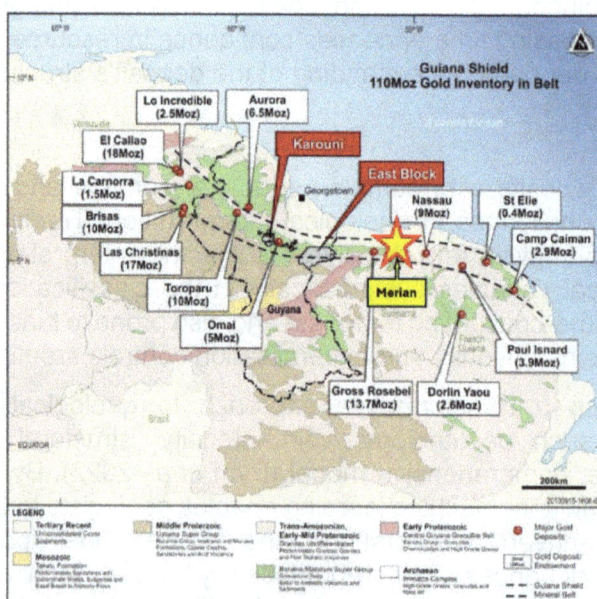

FIG 1 – Main gold deposits of the northern Guiana Shield in Rhyacian supracrustal sequences (Bardoux and Moroney, 2018; modified from Voicu *et al,* 2001).

The Merian Trend is a 25 km-long (Figure 2), north-west-striking mineralised corridor hosted within the Armina Formation of the Marowijne Supergroup. This formation consists of greywackes, mudstones, siltstones, and minor volcaniclastic rocks that have undergone low-grade greenschist metamorphism. Notably, unlike other mineralised regions of the Guiana Shield, no significant igneous intrusions have been directly linked to gold mineralisation within this trend.

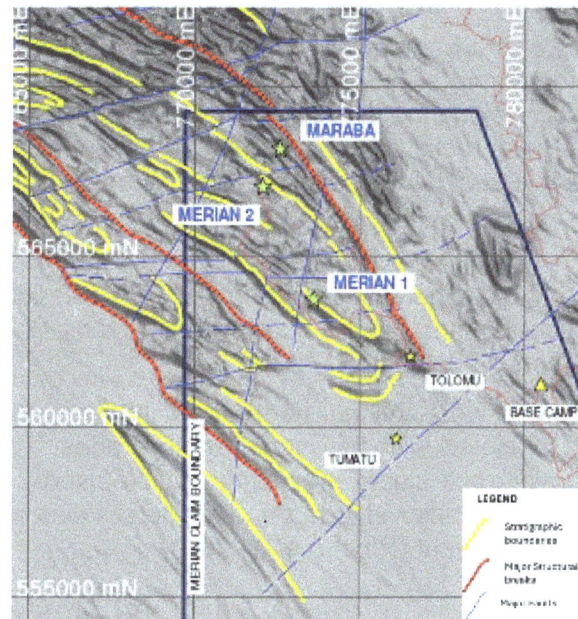

FIG 2 – Merian trend anticline interpreted from airborne magnetic and radiometric data. Background: grey scale reduced to pole horizontal gradient airborne magnetics (Ribeiro *et al*, 2014).

Structural framework

The dominant structural feature in the Merian Trend is the Merian Antiform, a north-west-trending, south-east-plunging fold interpreted from geophysical data, field mapping, and drill core observations. The antiform is accompanied by second-order parasitic folds, creating localised structural complexity. Gold mineralisation is structurally controlled, with mineralised zones primarily associated with:

- North-west-striking faults and shear zones, which acted as fluid conduits during hydrothermal activity. Fold hinges and lithological contacts, were rheological contrasts influence fluid entrapment.

- Subordinate east–west and north-east-striking faults, which locally bound mineralised domains and may represent later structural reactivations.

Mineralisation and alteration

Gold mineralisation within the Merian Trend is classified as epigenetic, mesothermal gold-only, consistent with orogenic gold systems. It is hosted in quartz veins, stockworks, and quartz breccias, commonly associated with pyrite, ankerite, albite, and silica alteration. These hydrothermal events occurred within a fold and thrust structural setting, where deformation and fluid flow were strongly influenced by pre-existing structural features.

The region exhibits a deep lateritic and saprolitic weathering profile, extending 80–100 m below the surface. Beneath the saprolite, a transition zone (saprock) is present, where oxidation decreases, and primary rock textures and structures are partially preserved.

Exploration and prospectivity

Exploration in the Merian Trend has relied on a combination of geochemical surveys, geophysics, and structural mapping, which have been effective in delineating mineralised zones. The trend hosts multiple gold deposits, including Merian I, Merian II, and Maraba (Ribeiro *et al*, 2014).

Local geology of the Merian II deposit

The Merian II deposit is part of the Merian Trend, a north-west-striking, 25 km-long gold-bearing corridor hosted in Palaeoproterozoic siliciclastic rocks of the Armina Formation. The deposit is structurally controlled and occurs along the axial trace of the Merian Antiform, a north-west-trending, south-east-plunging fold. The host rocks primarily consist of highly folded and faulted sandstones, siltstones, and mudstones, subjected to low-grade greenschist metamorphism (Ribeiro *et al*, 2014).

Structural controls

Two distinct structural styles of gold mineralisation are observed at Merian II: Moderately north-east-dipping, north-west-striking sheeted and tabular quartz veins and stockworks. Higher-angle, north-east-dipping irregular quartz breccia bodies.

The '92 shoot', a prominent north-west-striking, north-east-dipping quartz breccia body, forms the core of the deposit and is associated with the highest gold grades. The breccia is composed of quartz, albite, ankerite, and wall rock clasts, displaying multi-episodic brecciation and annealing (Figure 3). This breccia cross-cuts earlier sheeted and irregular veins in the footwall and hanging wall, indicating a later mineralising event (Ribeiro *et al*, 2014).

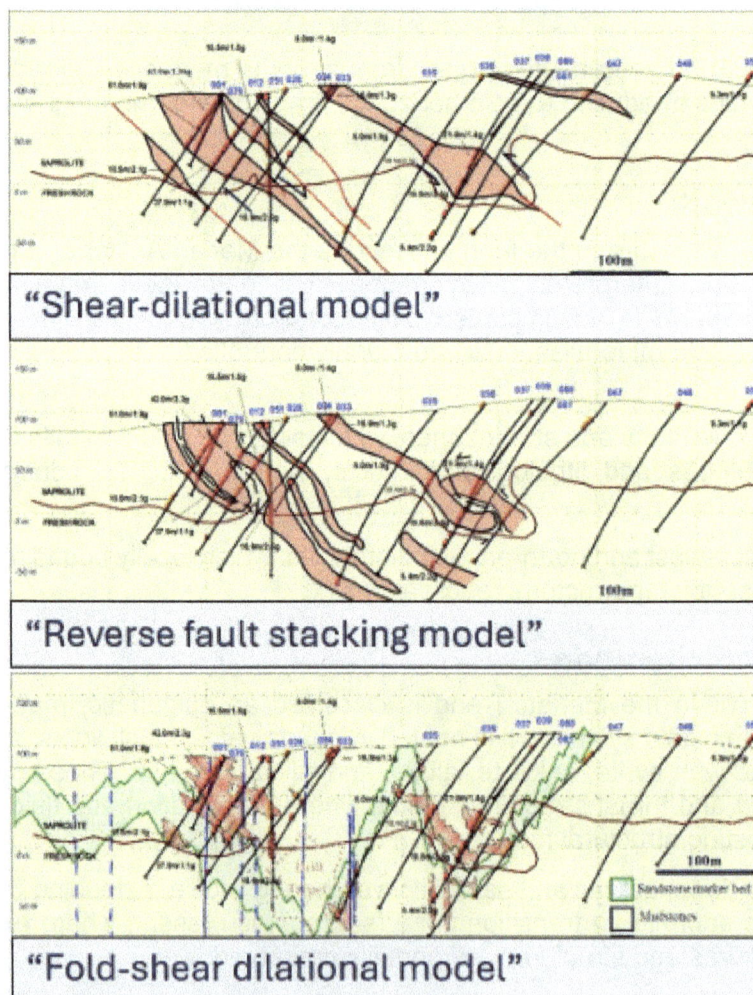

FIG 3 – Several conceptual models (2D sections) were developed based on systematic re-logging and field mapping at the first stages of the Merian II Pit, showing the association between Breccia-Quartz veins and the gold mineralisation (Ribeiro *et al*, 2014).

PROBLEM

Modelling an orogenic gold deposit with multiple generations of veins shaped by geological evolution and structural controls presents a significant challenge. Traditional explicit and implicit modelling techniques often introduce bias due to their dependence on subjective geological interpretation (Figure 4). A major limitation of the current modelling process is its reliance on polylines as an initial input, combined with drill hole data, which makes updating the model difficult. As new data becomes available, it must conform to the original interpretation, restricting flexibility in the modelling process.

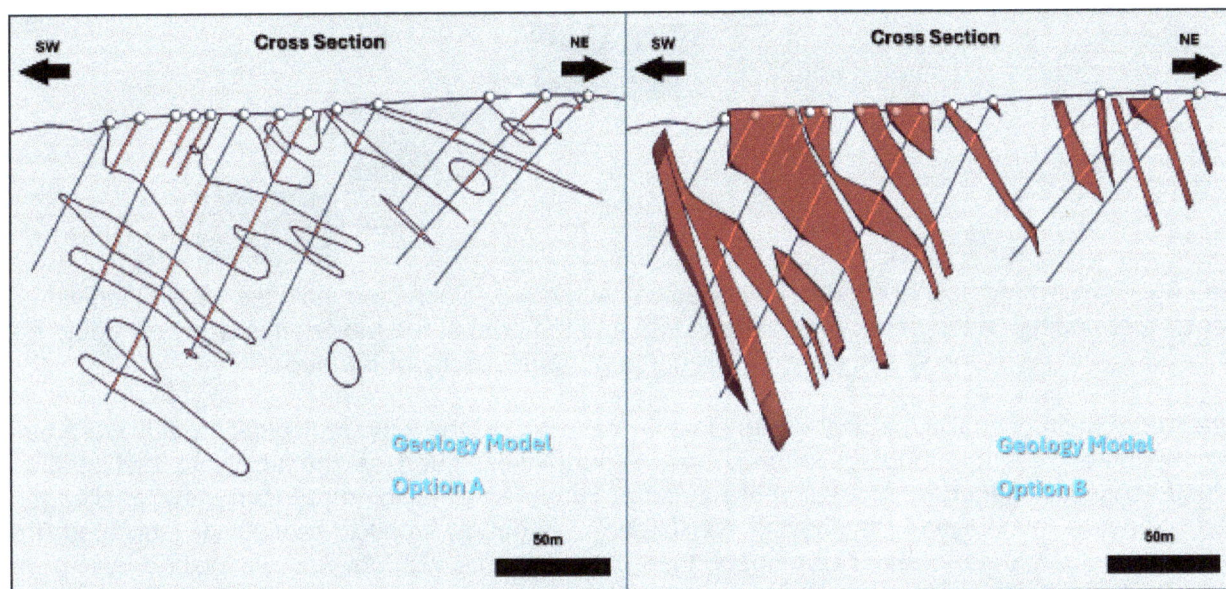

FIG 4 – Examples of geological interpretations using the same data, in this particular case only using drill hole data and geological knowledge from two different geologists with different experience in the same deposit.

Accurately predicting vein behaviour at deeper levels of the deposit using geological mapping and drill hole data remains a core issue. Over the past ten years, various geologists have contributed their interpretations at different stages, many of which were made when data was scarce or when outcrops—such as those now exposed in-pit walls—were not visible. This has led to a collection of interpretations that do not always reflect the current understanding of the deposit.

In addition, current modelling techniques struggle to incorporate systematically collected variables that significantly influence vein system behaviour. Three critical variables—structural data, vein density, and the presence of sulfides associated with mineralisation—are not effectively integrated into traditional models.

GEOLOGICAL MODELLING – CURRENT PROCESS

Over the years, since the initial exploration of the deposit, various techniques have been employed to construct the geological model (Figure 5). These approaches primarily integrated drill hole logging data, with limited geological mapping due to the scarcity of outcroppings. Initially, geological sections were drawn on paper, relying heavily on the interpretations of geologists responsible for the drilling campaigns. As exploration progressed, this early-stage information was digitised, transitioning into explicit modelling techniques. This process involved manually digitising key geological sections, which were then used to construct 3D solids for subsequent resource estimation. However, updating the model with new data followed the same labour-intensive approach as the initial construction, often requiring months of work and the collaboration of multiple geologists.

Geology Model Generation Workflow

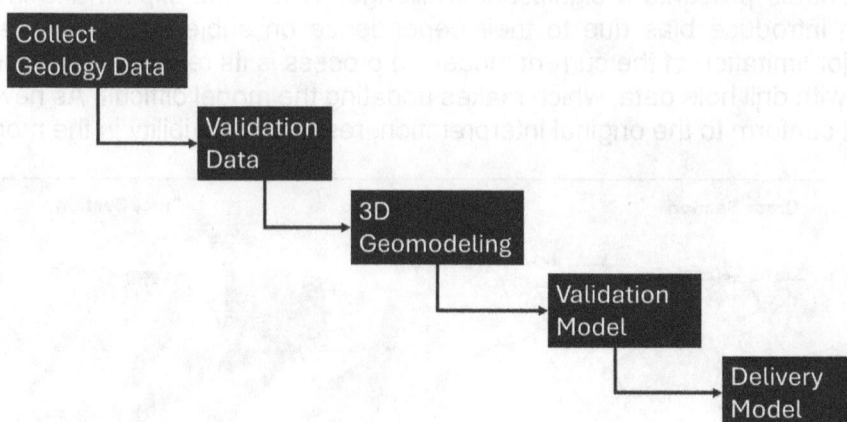

FIG 5 – Current workflow applied to build geological models. Each step includes several individual tasks. Depending on the generation of new data and the goal of the model, this workflow could be changed and optimised regarding the necessity of the deposit.

The introduction of implicit modelling significantly transformed the way geological models were built and updated. This methodology offered new opportunities, such as the ability to test multiple hypotheses and generate models more rapidly. However, it also introduced new challenges, particularly in maintaining geological consistency. Although implicit modelling facilitated the generation of a greater number of models, the quality of some outputs was diminished due to the speed of the process and limited geologic control.

Given the complexity of this deposit, the approach adopted was to integrate explicit interpretations into the implicit modelling environment, using them as guides for future updates. This strategy preserved the original geological interpretations over time, allowing modifications only in areas where new information became available. While this approach ensured continuity and consistency, it also introduced a degree of interpretational bias, limiting the incorporation of alternative hypotheses even when new data suggested potential changes. Additionally, updating models remained a time-consuming process, often taking weeks to complete a single model revision. This balance between preserving geological consistency and allowing for adaptability remains a key challenge in the evolution of geological modelling techniques.

MACHINE LEARNING MODELLING

Mining professionals have applied Deep Learning in orebody modelling through Maptek's DomainMCF (Sullivan, 2022), aiming to leverage neural network-based approaches for orebody knowledge and domain classification. As outlined in Sullivan (2022) the deep learning method begins by constructing a neural network (NN) from pre-coded samples. In this study, we used grade values and structural parameters to build the NN. The algorithm then uses this NN to constrain the interpolation of numerical attributes and generates a 3D block model to represent the data. During the application development processes Maptek used a variety of training data sets to explore how the DomainMCF interpretation and mined-out material.

Generating the geological model using machine learning required adapting to a new interface and workflow. Given that this approach differs from traditional modelling techniques, a structured workflow was established to ensure a comprehensive and reliable model construction. The key steps in the modelling process included the following.

Data acquisition

Collecting relevant geological data, including drill hole logging, geochemical analyses, and structural mapping data. Ensuring data completeness and consistency was a critical first step.

Data processing and validation

This step involves conducting a thorough evaluation of the data set to identify inconsistencies, errors, or missing values. It also involved exploratory data analysis (EDA) to understand relationships between variables such as lithology, quartz vein intensity, sulfide content, and gold mineralisation. Structural data was particularly scrutinised to define major geological domains.

3D domain characterisation block model generation

Implementing machine learning algorithms to integrate multivariable data sets and generate the geological model. Several iterations were performed to refine inputs, with different scenarios tested to assess how various geological controls influenced the output model.

3D domain characterisation block model validation and refinement

The machine learning-generated domain characterisation model was compared against existing implicit models. This included numerical validation—comparing flagged geological units in drill holes with the block model—and visual validation through 3D model review to ensure geological consistency.

Final comparisons and interpretation

Evaluating the differences between the traditional implicit model and the machine learning model, special attention was given to areas of lower data density, where structural control significantly influenced geological continuity. The results were analysed to determine how machine learning could improve geological interpretation and reduce biases inherent in manual modelling processes.

By following this structured workflow, the study successfully demonstrated the potential of machine learning to enhance geological modelling, allowing for faster model updates, improved integration of multiple geological variables, and a more objective approach to interpreting complex gold deposits.

DATA ANALYSIS

Considering that the primary input data originates from drill hole logging (geological descriptions), geochemical analyses (with Au as the key variable of interest), and pit mapping data (particularly structural information), the first crucial step in the modelling process is a thorough evaluation of the input data. This assessment ensures that all information is coherent and reliable before incorporating it into the model.

The initial evaluation focuses on understanding the relationships between different deposit characteristics (Figure 6). For instance, when modelling lithology—specifically breccias—it is essential to determine whether these units exhibit any correlation with other logged variables, such as hydrothermal alteration and vein intensity. This analysis incorporates both recent and historical drill hole data, ensuring that all available information is considered. Additionally, geochemical data is examined, with particular emphasis on Au content, given that this is an orogenic gold deposit. Other elements from the geochemical data set are also reviewed to identify potential correlations with lithology and mineralisation.

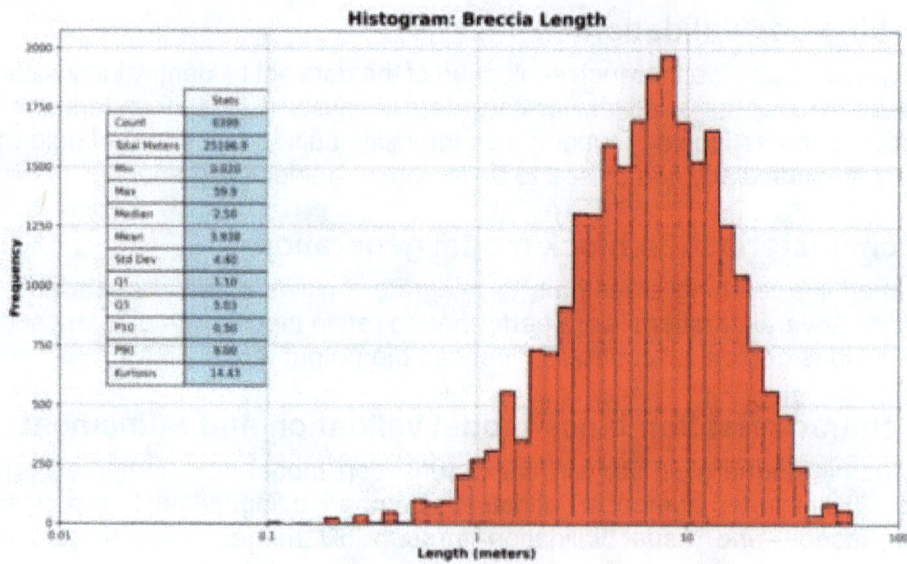

FIG 6 – Histogram displaying the logged interval lengths classified as breccia within the deposit. Understanding the modelled variable is crucial for determining the necessary resolution to capture geological data effectively and for defining optimal machine learning modelling parameters.

Preliminary results indicate a strong correlation between breccia occurrences, vein intensity (regardless of vein type—an initial assumption that requires further validation in future studies), and gold content (Figures 7 and 8). While the correlations are not extremely high, they suggest a certain degree of control over mineralisation. Conversely, hydrothermal alteration appears to be less significant in this specific context and is excluded from further modelling considerations.

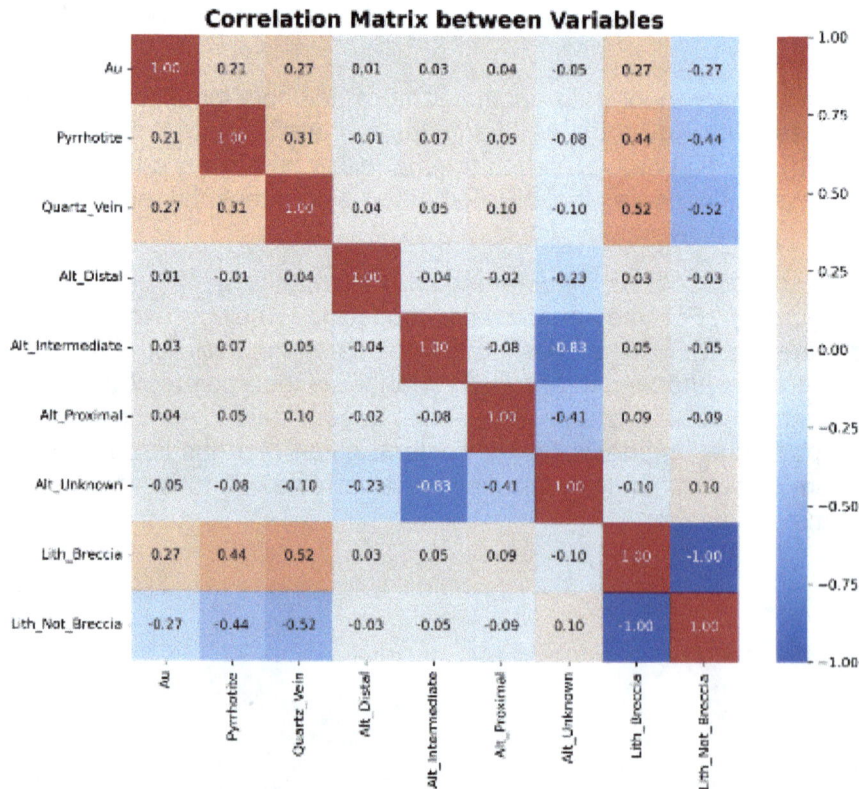

FIG 7 – Correlation matrix representation using available geological logging data, illustrating relationships between breccia and other geological variables. This serves as a foundation for multivariable modelling.

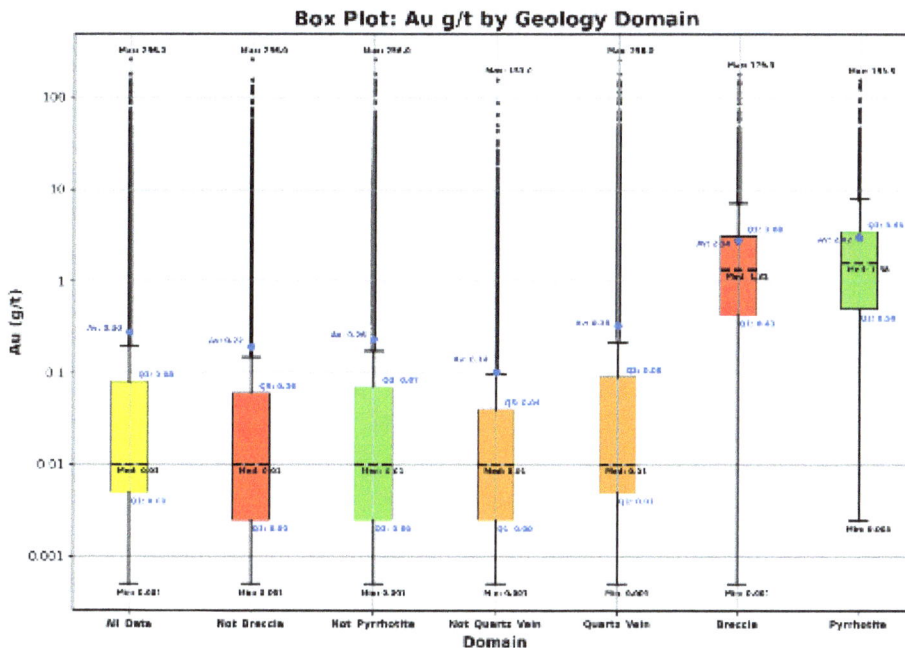

FIG 8 – Gold distribution in relation to geological features: The analysis first presents the full data set, illustrating Au distribution across all data points. It then segments populations based on domains within and outside breccia zones, pyrrhotite, and quartz veins, highlighting the geological controls on economic gold mineralisation.

Structural data is a critical factor in this analysis. A detailed review of the main structural trends reveals that the deposit is influenced by extensional faults associated with a dominant shear zone. This structural framework allows for the definition of five geological domains based on the primary structures. These domains differentiate from the eastern region, where the hanging wall is predominant, from the western region, characterised by the presence of the footwall.

Having established a complete understanding of all relevant variables, it is essential to ensure their integration within the geological model. The goal is to construct a breccia model that accurately incorporates the key geological controls of the deposit, particularly the continuity of breccia bodies. As observed, these bodies exhibit highly irregular thicknesses, with the majority of intercepts measuring less than one metre. Ensuring that the model accounts for this variability is critical to maintaining geological consistency and reliability in resource estimation.

When constructing the geological model, geologists carefully evaluate multiple data inputs. A critical consideration is the length distribution of the geological units to model. By analysing the descriptive statistics shown in Table 1, they gain insights into key parameters such as the minimum interval length, which directly influences the compositing strategy and block size selection. Various composite lengths were evaluated (Figure 9) to identify the most representative configuration for the data set, as reflected in the confusion matrix as seen in Figure 10.

TABLE 1

Different groups with varying composite lengths were tested to ensure the inclusion of logged intervals from the original database inside the trained machine, taking into account the minimum thickness of the unit to be modelled.

Comp group	Minimum modelled unit thickness	Recommended composite length	Recommended parent block size	Recommended sub block size	Percentage of correct predictions
A	4	0.5	8	2	89.19%
B	2	0.25	4	1	92.03%
C	1	0.125	2	0.5	94.33%
D	**0.5**	**0.0625**	**1**	**0.25**	**96%**

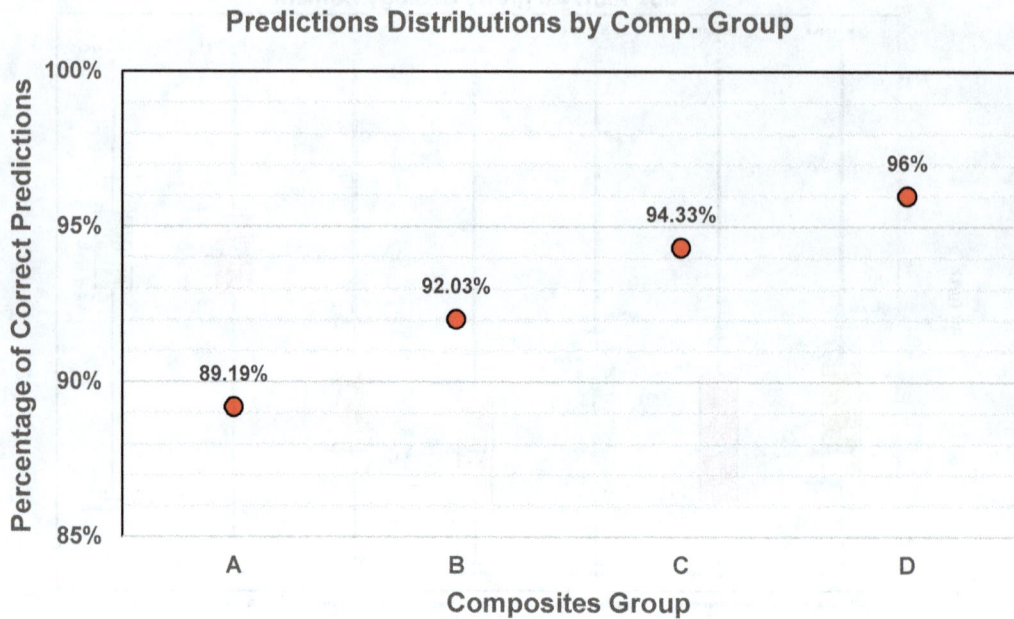

FIG 9 – Scatter plot showing the percentage of correct predictions as a function of composite groups, illustrating that shorter composite lengths—as in Group D—result in higher prediction accuracy.

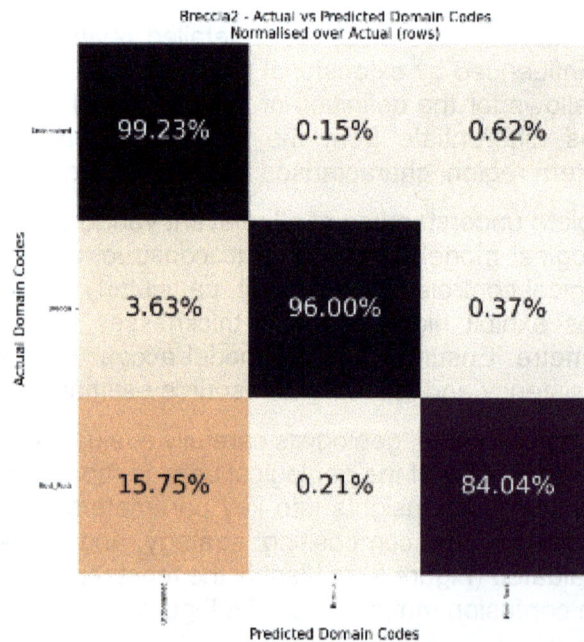

FIG 10 – Confusion matrix generated using an input composite length of 0.0625 m, based on a minimum logged interval length of 0.5 m.

This process ensures the input data is efficiently represented in the training phase of the machine learning workflow. The efficiency of the trained machine is evaluated using the confusion matrix, which measures the percentage of correct positive and negative predictions. As the overall accuracy approaches 100 per cent, it indicates a near-perfect correlation between the input data and the predicted model. However, it is important to note that achieving 100 per cent accuracy is practically impossible in any real-world scenario.

The subsequent step involves defining the most appropriate block model resolution, which should align with the intended mining method—whether open pit or underground. Based on the results, the use of sub-blocking is recommended to ensure that short intervals are accurately captured in the final model.

MACHINE LEARNING TRIALS

During the development of the final model, three different test scenarios were evaluated, each incorporating distinct input variables. First Scenario: The model was built using only breccia data. Second Scenario: The model included breccia data along with gold content. Third Scenario: The model integrated breccia data, gold content, vein intensity, and structural controls.

To assess each scenario's performance, both statistical and visual validations were conducted including a confusion matrix review. Statistically, the logged breccia metres were compared against the intercepts flagged in the drill holes based on the block model results. Visually, the models were reviewed to verify geological consistency—an essential step for validation. Numerically, all three scenarios showed similar statistical outcomes with minimal variation. If these results were compared with traditional implicit modelling techniques, they would likely produce comparable outputs.

However, significant differences were observed in the spatial distribution and geological coherence of the results:

- First scenario: the model failed to establish clear geological patterns in areas with lower data density. While it maintained general trends in zones with high data continuity, the results lacked geological meaning in many cases. Furthermore, at greater distances from data points, the model generated geologically inconsistent features.

- Second scenario: the inclusion of gold as an additional variable improved continuity in some areas, likely compensating for potential inconsistencies in geological logging. This scenario showed better-defined trends in low-data-density zones. However, some inconsistencies remained, particularly in regions where geological understanding was lacking. In this case it gives the geologist the chance to further review geology, to dig deeper into the inconsistencies.

- Third scenario (best performing model): the integration of structural controls resulted in a substantial improvement over the previous scenarios (Figure 11). In low-data-density areas, structural data significantly enhanced continuity, producing results like those in areas with high drill hole density. While vein intensity and gold data supported continuity, the greatest impact came from the structural framework. Intermediate tests using alternative structural inputs revealed high sensitivity to structural orientation, meaning that even slight variations in structural parameters resulted in different model outputs.

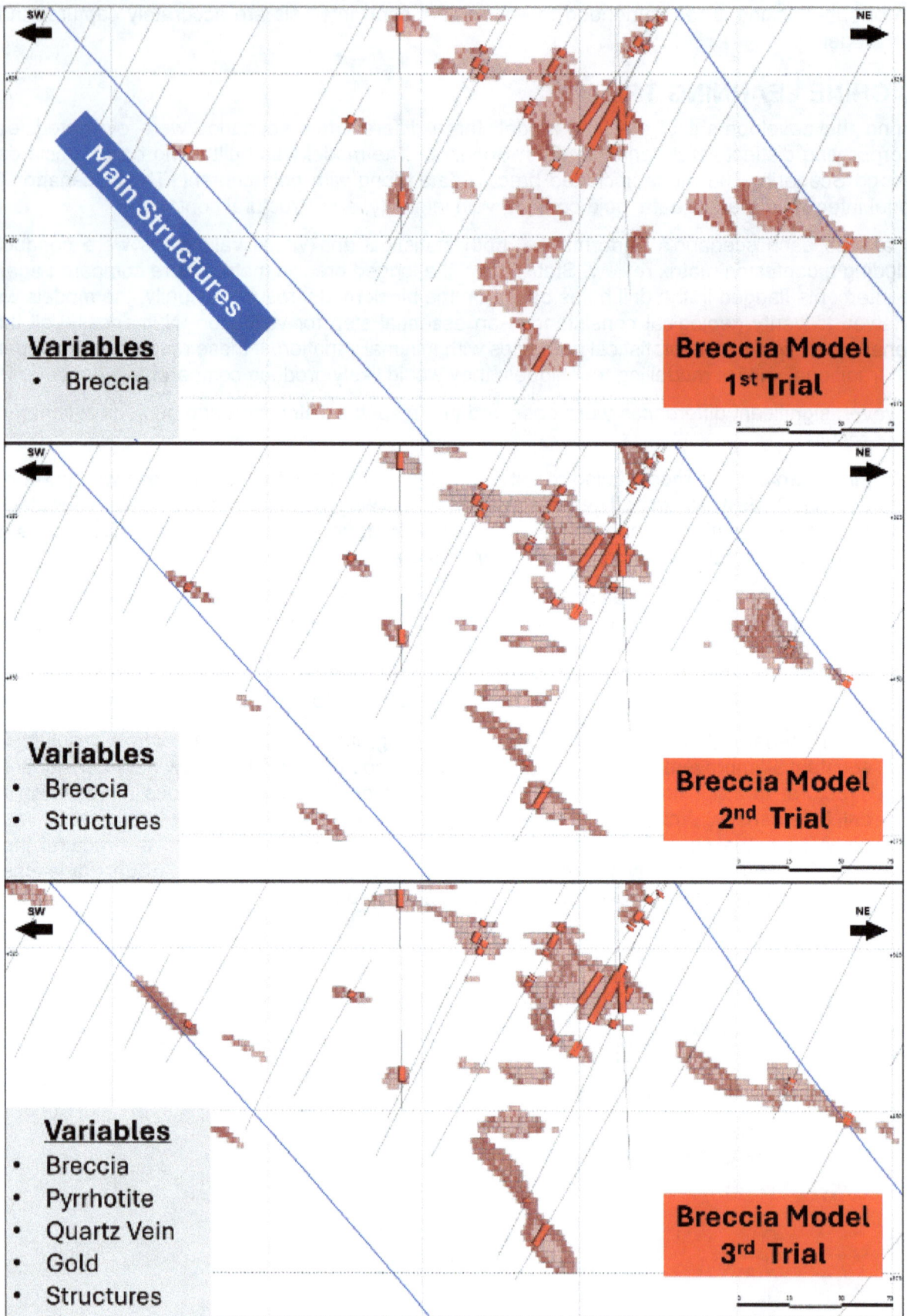

FIG 11 – Breccia model results across different trials, each incorporating varying inputs. The findings demonstrate that increasing geological control yields more geologically consistent and reliable outcomes aligned with deposit knowledge.

VOLUME COMPARISONS

Following the generation of the geologic model, the subsequent step involved validating these results through comparison with the currently adopted official geological model (Figure 12). Additionally, this new model was assessed for its potential application in exploration efforts, particularly in identifying lateral continuity and, more importantly, depth extensions of mineralisation.

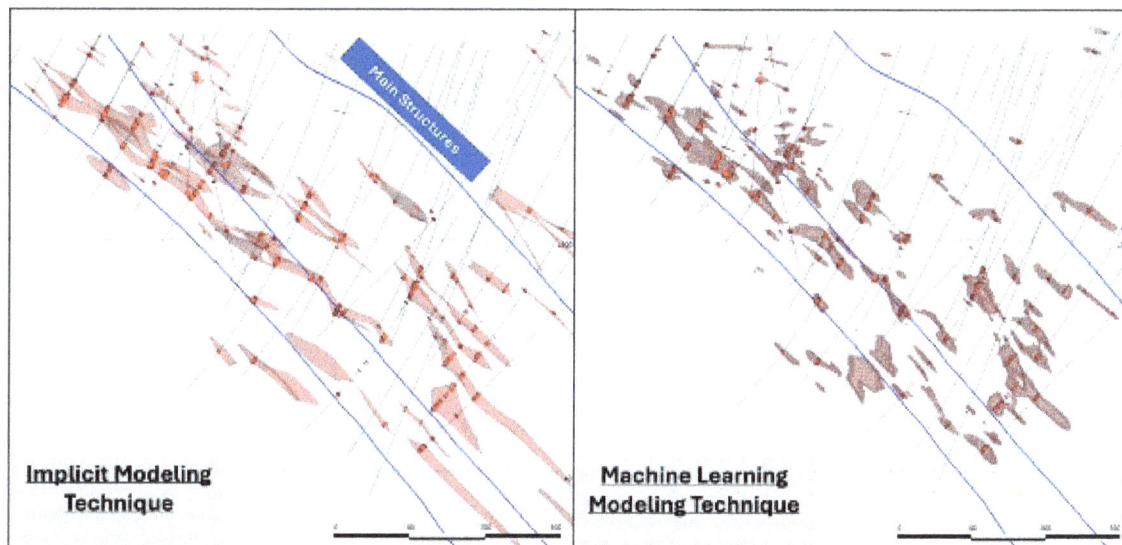

FIG 12 – Comparison of 2D one cross geology section generated using different software: the left panel represents the Implicit Modelling technique, while the right panel applies machine learning. Observed differences are primarily software-driven rather than reflective of geological understanding.

Globally, volume comparisons indicate notable differences in deeper areas and regions with lower drill hole density. In high-data-density areas, no significant differences in volume were observed. However, upon closer inspection, differences in continuity patterns emerged. These variations stem primarily from how continuity is established in the model.

The traditional modelling approach relies heavily on manual selection, where specific intervals are included within the breccia bodies. This introduces potential human bias, which can influence continuity interpretations.

In contrast, the machine learning approach establishes continuity algorithmically, resulting in geometry that is notably different when visualised in 3D. The machine learning generated geometries align more logically with the current geological model and geologic understanding but overcome some of the constraints imposed by implicit modelling techniques.

A critical factor influencing these differences is that the existing implicit model is constrained by its reliance on prior explicit modelling inputs, which were initially developed during the early stages of exploration. As a result, the geometries remain somewhat rigid, limiting the ability to incorporate new geological insights efficiently.

The advantage of this approach is the quick generation and comparison of multiple scenarios, a task that is currently cumbersome and infrequent due to the time required for traditional model updates. Whereas the conventional geological modelling workflow may require several months to incorporate updates and finalise a new model, the machine learning-based approach reduces this time frame to an average of 3–12 hrs for this deposit depending on the block size and volume of data. This efficiency allows for frequent comparisons, helping geologists systematically identify areas that require further refinement or additional geological variables to improve the model's predictive power.

GEOLOGICAL CONTROL

Using machine learning, geological knowledge was incorporated directly into the modelling process. This section outlines the key geological controls applied and demonstrates how the results vary when geological control is introduced. Initially, the model follows a general orientation based on the logged intervals but without considering the specific structural orientations, particularly in areas with complex and highly variable geology. In such cases, defining trend patterns or incorporating other geological variables into the modelling process becomes essential to improving accuracy.

One of the important features of this approach is the ability to perform multivariable modelling, which significantly enhances the model precision. By introducing geological controls, such as structural data and mineral associations, the model can adjust to reflect these influences (Figure 13). The results clearly show how incorporating geological control helps refine the model, particularly in regions with complicated geology. Areas that previously displayed more generalised orientations (Figures 8–11) now exhibit trends that align with geological expectations, highlighting the importance of using multivariable input.

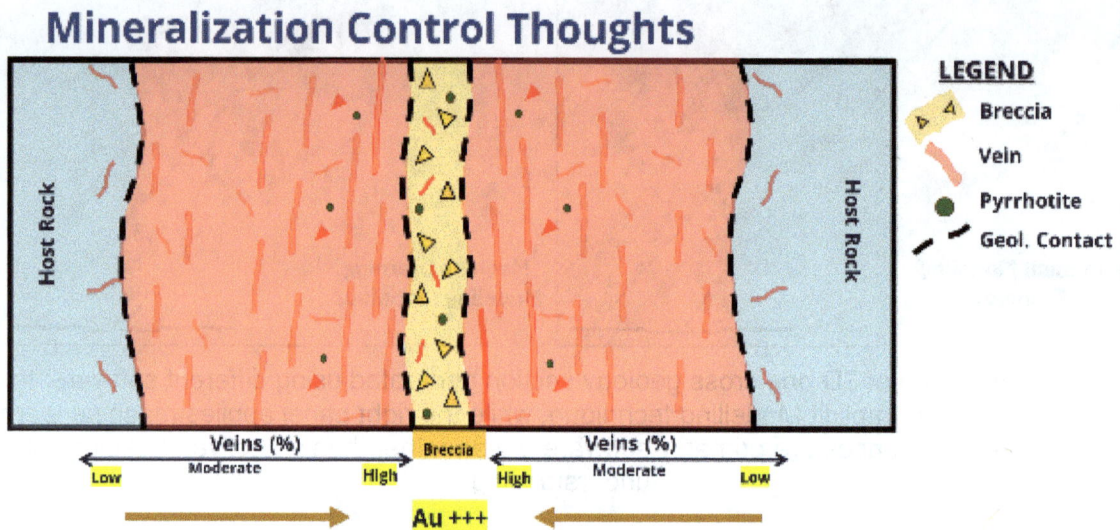

FIG 13 – Schematic representation of the mineralisation controls, summarising an interpretation of the deposit used for machine learning-based modelling. This conceptual framework is embedded in the geological model, integrating structural controls to improve accuracy.

The differences in results when applying geological control are substantial, particularly in challenging geological environments. The ability to observe these variations and fine-tune the model based on geological insight allows for a more robust and realistic representation of the deposit, ultimately improving the model's utility for future decision-making and exploration efforts.

MULTIVARIABLE ANALYSIS

When modelling a specific variable with traditional techniques it is only possible to visually assign intervals within the unit being modelled without fully respecting the factual geological logging. As a result, these models do not always honour the detailed geology of the deposit, leading to inaccuracies. This limitation often results in models that lack precision, especially when trying to account for complex geological controls (Figure 14).

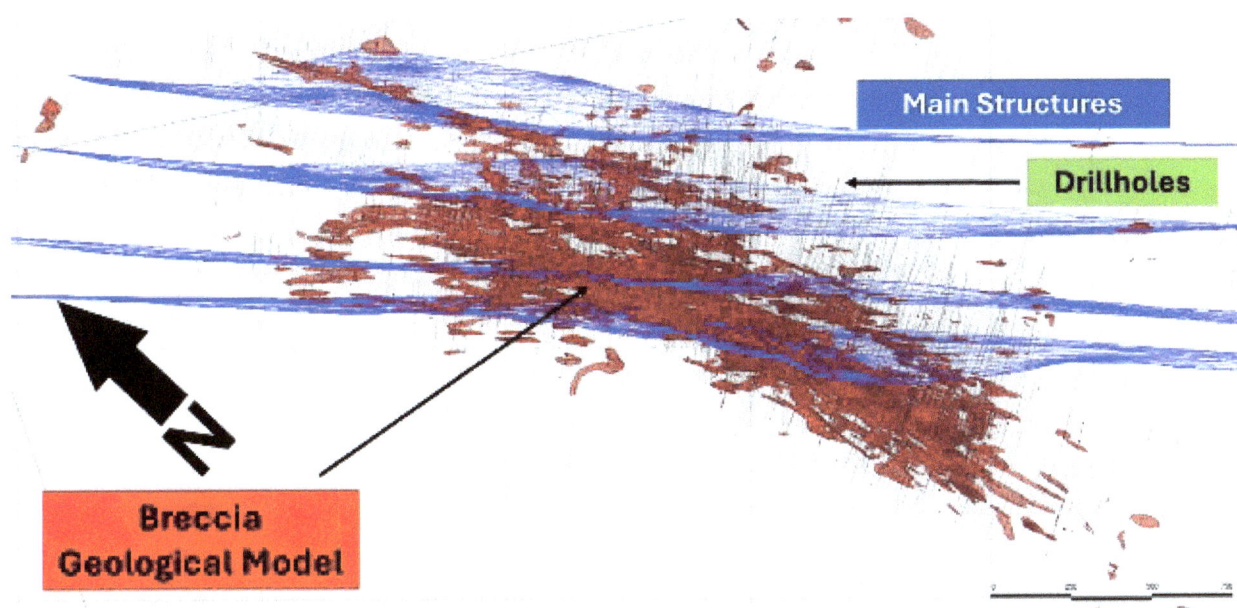

FIG 14 – Final 3D visualisation of the breccia model, integrated with key structures and additional geological variables to define breccia body orientations. The model was executed across the entire modelling domain, making inferences in low-data-density areas while maintaining geologically consistent results.

However, machine learning enables the integration of multiple variables into the modelling process (Figure 15). This allows for the inclusion of additional factors directly into the algorithm, which significantly improves the accuracy of the model. Review of the results shows a notable shift in the orientations generated, as they now reflect the combined influence of all the variables. This multivariable approach delivers greater confidence in the results, as it better represents the true geological conditions of the deposit.

This is particularly important when considering the economic metals in the deposit, as their distribution is often controlled by a combination of many geological factors. By incorporating these variables into the modelling process, machine learning allows for a more accurate reflection of mineralisation controls. This provides a clearer understanding of how different geological elements interact to influence the location and concentration of valuable minerals. This leads to a model that is more geologically sound but also more reliable for future exploration and resource estimation efforts.

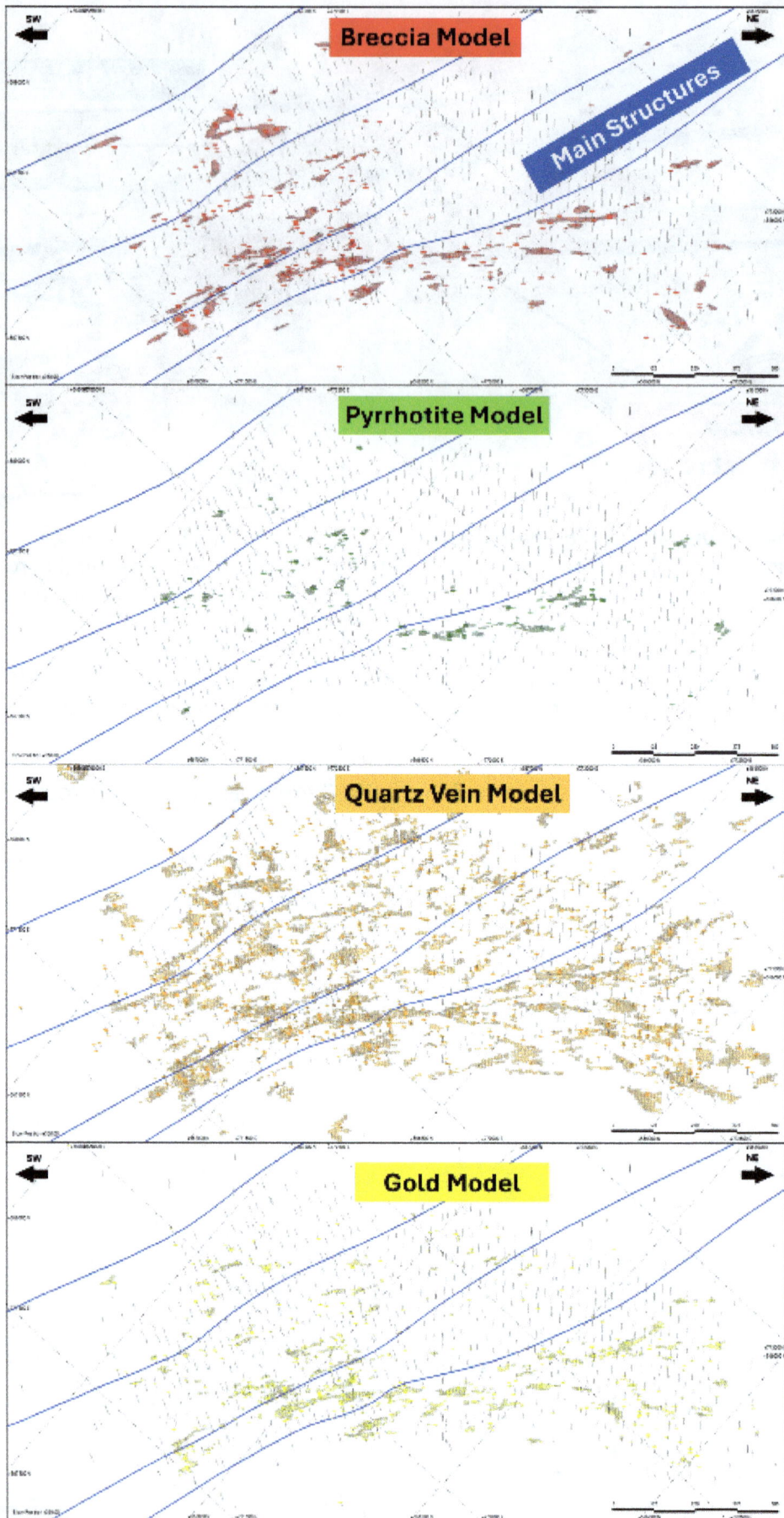

FIG 15 – Multivariable models illustrate the correlations among Breccia, Structures, Pyrrhotite, Quartz Veins, and Gold variables, utilising machine learning for all simulations. Reference: 360 Level, Merian II Pit.

STRUCTURAL ANALYSIS

Incorporating structural data into geological modelling has always been a significant challenge. Often, using structural data to create models can generate inconsistent geometries, particularly when dealing with the inherent complexity of structures in mineral deposits. In many cases, structural elements are not included in the modelling process due to the difficulty of working with them. This complexity is compounded by the need to consider scale—ranging from regional and district-level structures to local or deposit-scale structures—which makes the task even more daunting.

Traditionally, geological models either omit complex structures or struggle to represent them accurately, especially when attempting to model the intricate relationships between structures and mineralisation. However, when using machine learning with structural inputs, the results are notably improved (Figure 16). The model shows a seamless blending of structural information with other variables, allowing for a balanced representation of all contributing factors. What is particularly remarkable is that, unlike with current techniques, the structures do not overpower other variables; instead, they influence the model in a more natural and integrated way.

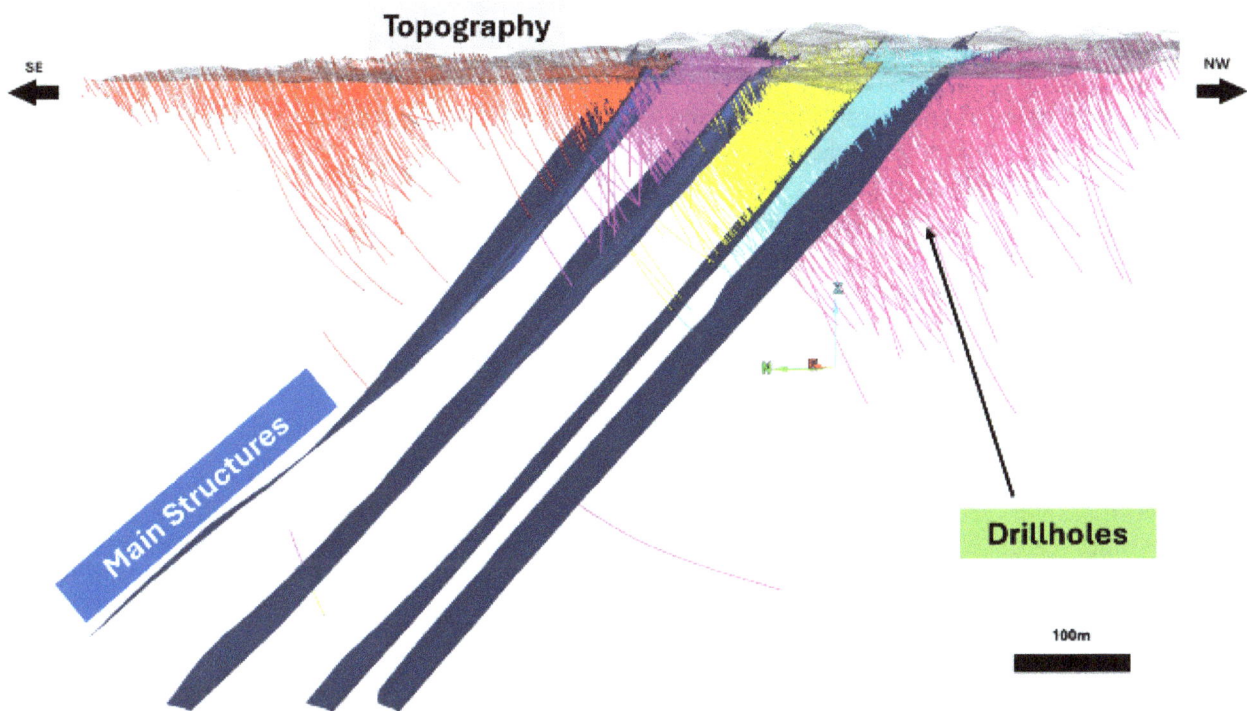

FIG 16 – Structural data is integrated into geological models as an additional variable, interacting with other inputs and contributing weighted influence to model generation.

When using traditional methods, forced orientations are common, where structures are given undue emphasis or appear artificially imposed on the model. With machine learning, however, the model respects the natural complexity of structures while incorporating them alongside other geological controls. By reviewing results with different structural inputs, machine learning offers the ability to rapidly evaluate multiple structural models, which is invaluable for complex deposits. This capability allows for testing different structural hypotheses and provides an additional tool for improving structural understanding by enabling quick comparisons between scenarios.

ADVANTAGES AND DISADVANTAGES

It is important to acknowledge that no geological model is perfect. Debates will always arise regarding whether one model reflects geological understanding better than another. However, this study provides insights into how machine learning-based modelling contrasts with traditional explicit and implicit modelling techniques and highlights its advantages and limitations.

Advantages

The following were advantages identified using machine learning over the traditional methods.

Time efficiency

One of the most significant advantages of using machine learning for geological modelling is the drastic reduction in processing time. In a complex deposit like this, faster model generation reduces bias and also minimises the workload and enables geologists to test multiple hypotheses and geological controls more efficiently.

Data validation

Data validation is equally important for traditional methods and machine learning. Just as in implicit modelling, data integrity remains crucial for building reliable models. Geologists must have a deep understanding of their data set—not only in terms of field geology but also regarding data distribution within tables and 3D space. This step, often referred to as Exploratory Data Analysis (EDA), is mere essential to ensure meaningful modelling results.

Integration of additional variables

Traditional modelling approaches make incorporating multiple geological variables cumbersome, often requiring several preliminary processing steps before they can be incorporated. In contrast, machine learning thrives on integrating large amounts of high-quality data, often identifying previously unseen relationships and improving geological continuity within models. While this does not always guarantee geologically meaningful outputs, there is a clear trend toward generating models with greater continuity and logical spatial distribution.

Challenges and considerations

While machine learning presents clear advantages, it is essential to recognise potential challenges as outlined below.

Reproducibility and geological interpretation

Because machine learning relies on statistical relationships rather than explicit geological interpretation, its models may sometimes fail to capture key geological insights that are well understood by experienced geologists.

Dependence on data quality

While implicit and explicit models also rely on data integrity, machine learning models are highly sensitive to input variations, meaning that poorly curated data sets can result in misleading outputs.

Structural control sensitivity

As observed in the tests performed, the algorithm's response to structural inputs can vary significantly, making structural validation a critical step before integrating such data into machine learning workflows.

Overall, while machine learning introduces a more efficient and flexible approach to geological modelling, it must be applied with careful geological oversight to ensure that models remain geologically meaningful and reliable.

CONCLUSIONS

The application of machine learning to geological modelling of orogenic gold deposits has significantly improved accuracy, efficiency, and adaptability. It enables rapid model updates—cutting processing time from a month to hours—facilitating real-time decision-making and freeing geological teams to focus on exploration and resource estimation. Integrating variables such as structural data, vein intensity, and sulfide mineralisation creates more realistic models, particularly at depth. Unlike traditional methods, machine learning naturally blends complex structural inputs with other data, reducing bias and better reflecting geological relationships. To ensure reliability, rigorous structural validation is essential before incorporating structural data, as poor inputs can skew results. The ability to adjust block model resolution, test structural hypotheses, and delimit models to solids or surfaces enhances flexibility and model consistency across diverse geological settings. While early results show strong performance in data-rich areas, reproducibility in low-density zones remains uncertain and warrants further testing. Overall, machine learning delivers faster, more accurate, and more flexible models, strengthening confidence in mineralisation controls and resource estimation.

ACKNOWLEDGEMENTS

I would like to express my sincere gratitude to Newmont Mining Corporation for the opportunity to publish this paper. In particular, I extend my appreciation to the Exploration – Geoscience Team in the Discovery Support Innovation department and the Exploration Team at Merian for their valuable contributions and geological insights, which played a crucial role in shaping this work. The ability to utilise data from one of Newmont's active mines, with significant future growth potential, was instrumental in the success of this project.

Additionally, I am deeply grateful to Maptek for providing all the essential tools and support that made this project possible. Their GeologyCore and DomainMCF software, which operate within a machine learning environment, enabled the efficient creation of geological models. Furthermore, the Maptek team's extensive training, continuous support, and willingness to provide reference materials were invaluable throughout the process. Their dedication and expertise were key factors in successfully completing this study.

REFERENCES

Bardoux, M and Moroney, R F, 2018. Gold mineralization in the Guiana Shield, Guiana and Suriname, South America: a field trip to the 14th Biennial Society for Geology Applied to Mineral Deposits (SGA) meeting, Geological Survey of Canada, Open File 8351.

Daoust, C, Voicu, G, Brisson, H and Gauthier, M, 2011. Geological setting of the Paleoproterozoic Rosebel Gold District, Guiana Shield, Suriname, *Journal of South American Earth Sciences*, 32(3):222–245. https://doi.org/10.1016/j.jsames.2011.07.001

Pym, F A, Crook, K E, Hetherington, P M and Murphy, M P, 2022. Machine learning in resource geology – why data quality is critical, in *Proceedings of the 12th International Mining Geology Conference 2022*, pp 149–170 (The Australasian Institute of Mining and Metallurgy: Melbourne).

Ribeiro, J W, Schmidt, S, Radjkoemar, S and Anderson, S, 2014. Orogenic Au Deposits of the Merian Trend – Nassau Project, Newmont internal report.

Sullivan, S, 2022. Harness data complexity – how machine learning applies all project data for accurate resource modelling, in *Proceedings of the 12th International Mining Geology Conference 2022*, pp 483–495 (The Australasian Institute of Mining and Metallurgy: Melbourne).

Voicu, G, 2001. Lithostratigraphy, geochronology and gold metallogeny in the northern Guiana Shield, South America: a review, *Ore Geology Reviews*, 18:211–236.

Resource estimation of roll front uranium deposits by using traditional and machine learning methods for Nichols Ranch uranium deposit in Wyoming, USA

A Aydar[1] and K Dagdelen[2]

1. PhD Student, Mining Engineering Department, Colorado School of Mines, Golden CO 80401, USA. Email: arifaydar@mines.edu
2. Professor, Mining Engineering Department, Colorado School of Mines, Golden CO 80401, USA. Email: kdagdelen@mines.edu

ABSTRACT

The fundamental component of resource estimation is domain modelling. However, in such deposits as roll front uranium, they are mostly deposited in sandstone and on the contact of oxidised and reduced rock domains which makes it difficult to model. The Grade–Thickness (GT) as well as contour method is one of the most applied resource estimation techniques in roll-front uranium deposits. However, explicit modelling and GT contouring by using the GT information extracted by drill holes is exceedingly difficult, time consuming and inconsistent.

This research studies domain modelling of roll front uranium mineralisation using the GT values within the radial basis function (RBF) aided implicit modelling framework. It also compares the spatial associations of the RBF determined domain of mineralisation to the domains obtained by GT contours. Then, the block grades within the domain of mineralisation are estimated by using Kriging and selected machine learning models to compare their spatial associations and overall, *in situ* tons and grade estimates with each other and GT contours.

The domain of mineralisation modelled by RBF appears to spatially correlate well with the domain of mineralisation obtained by GT contours. The performance of the selected machine learning models performed quite good with *k*-NN (*k* Nearest Neighbour) having values of Coefficient of Determination (R^2) = 0.792, Root Mean Squared Error (RMSE) = 0.0216 and Mean Absolute Error (MAE) = 0.0048 and the random forest (RF) with R^2=0.751, RMSE=0.0236, and MAE=0.0077. A visual validation of these models, swath plots, grade tonnage curves suggests that the *k*-NN and Ordinary Kriging (OK) results are remarkably close to each other perfectly aligning with the drill hole intersections in terms of grades while RF estimates show significant deviations of higher grades from the other methods and the supporting drill hole information. There is a significant difference between GT *in situ* resource estimates as observed within this study and OK and *k*-NN results were approximately 4 per cent and 1 per cent respectively.

INTRODUCTION

The unique challenges of estimating resources in roll-front uranium deposits, such as those found at the Nichols Ranch ISR (*In situ* Recovery) mine in Wyoming, exemplify the need for advanced estimation methods. Roll-front deposits, characterised by their distinct crescent shapes (Figure 1), complicate traditional estimation methods due to their unusual geometry and the heterogeneous nature of uranium distribution within the orebody. Traditional GT contouring methods, commonly used in these settings, are labour-intensive and subject to considerable error due to the manual drawing and interpretation of contours on two-dimensional maps (Aydar, 2022). This method's reliance on extensive field experience and its time-consuming nature underscores the necessity for more efficient and reliable alternatives.

FIG 1 – Underground mine face of a roll-front uranium (pitchblende) and sub-roll underneath. The rock bolts have a square head with a side length of 0.3 m as a scale. La Sal Mine, Utah. (Courtesy of Travis Boam, Energy Fuels Inc).

Several techniques for modelling roll front uranium deposits have been proposed. Knudsen and Kim (1979) developed variogram models for deposit thickness and grade-thickness (GT) product, fitting experimental variograms with a spherical model. Marbeau and van Brunt (1991) introduced an unfolding technique for uranium roll fronts. Syaeful (2018) applied both Kriging and inverse distance weighting methods to the Rabau Hulu sector uranium deposit in Indonesia. Similarly, Ciputra *et al* (2020) reported a resource model for the same deposit using Ordinary Kriging (OK). Taghvaeenezhad *et al* (2020) advocated for OK in the Khohoumi uranium deposit in Iran, while Yun *et al* (2020) highlighted the benefits of OK for uranium deposits in the Changpai region of the Yangtze River uranium ore field, China, although these deposits are hosted in a fault zone rather than roll fronts. Additionally, Matheron *et al* (1987) proposed a truncated Gaussian simulation (TGS) for sedimentary-hosted deposits, including uranium. Beucher and Renard (2016) further discussed TGS, endorsing its applicability to roll front uranium deposits.

This paper proposes a novel approach to estimate mineral resources for roll front uranium deposits. The proposed approach determines estimation domain by way of implicit modelling using RBF processing of GT data followed by estimation of GT values at block locations using geostatistical and ML techniques. By comparing the results for traditional GT contouring method with both geostatistical Kriging and cutting-edge machine learning (ML) techniques, including *k*-nearest neighbours (*k*-NN) and RF indicate that proposed approach can be effective in mineral resource estimation of roll front uranium deposits. The geostatistical and ML models, augmented by radial basis function (RBF) aided implicit modelling, appears to enhance the accuracy and efficiency of uranium mineral resource estimates. The ML approaches offer substantial advantages over traditional methods, requiring less field experience and providing more consistent results with identical data sets.

Ultimately, this study seeks to assess the effectiveness of more advanced Kriging and ML methodologies applied to block grade estimation within mineralised zones of roll-front uranium deposits compared to traditional GT contouring method. By integrating modern computational techniques with traditional geostatistical practices, this research aims to elevate the standards of precision and reliability in resource estimation. The expected outcomes could significantly influence future approaches to mining exploration and development, particularly in complex geological settings like those of roll- front uranium deposits.

METHODOLOGY

The deposit selected for this study is the Nichols Ranch roll front Uranium deposit located at Casper County, Wyoming, USA. The mine has been operating since 2014 by ISR method which primarily dissolves the Uranium at the host rock by injecting oxidising fluid and the Uranium bearing solution pumped back to processing facility and Uranium oxide (U_3O_8) product which also called 'yellow cake' is produced.

The resource modelling has been performed on 392 drill holes (Figure 2a) with approximately 30 × 30 m spacing which are drilled by rotary drilling method. Downhole geophysical measurements including gamma logging and resistivity at every 1.3 m have been conducted by Energy Fuels Inc which provided to this research. The unit of gamma measurements are counts per second (CPS) which enables us to calculate an equivalent U_3O_8 grade in percentage (% eU_3O_8). The grades with a grade cut-off 0.02 per cent eU_3O_8 were multiplied by thicknesses which equal to GT with 0.2 cut-off. Uranium bearing sandstone formation which is locally called A Sand (Figure 2c) covered by mudstone is nearly horizontal to surface and 30 m in thickness has been divided into ten zones of approximately each 3 m thickness. This mineralised approximately 30 m thick A-Sand formation (Figure 2c) is deposited between approximately 168 m and 198 m below from the current surface (Figure 2b).

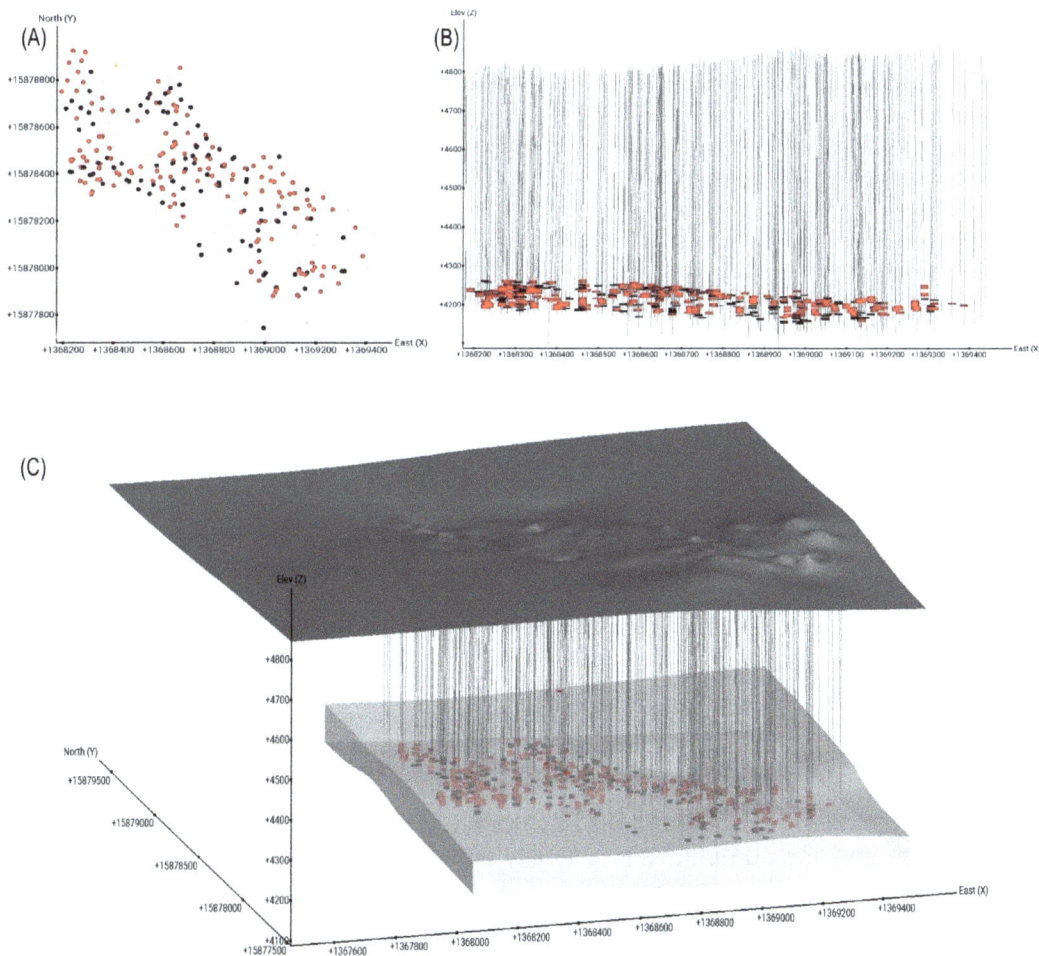

FIG 2 – (a) Plan view of 392 drill holes with GT ≥ 0.2 red circles and GT <0.2 black circles; (b) Looking North view and the black lines represents the drill holes; (c) 3D view of drill holes within light grey coloured mineralised A-Sand domain and the dark grey topography above.

Initially, the GT method have been performed to compare the proposed methodology in this study. The GT values at each zone have been mapped by connecting the same range of GTs with 2.15–

2.75 m width polygons. The GT ranges utilised are 0.2, 0.5, 1, and > 2 with polygons representing the area between these GT ranges. The mapping has been detailed at higher grades which means that the orebody is being thicker at higher and thinner at lower GT values. The average GT of the contoured polygons calculates the resource is multiplied by the polygon area and density factor of 1.17 which results in lbs. of U_3O_8. A density of 16.5 ft^3/short ton (Malensek *et al*, 2022) is used to calculate the density factor.

The resource estimation method proposed in this study involves two steps which are firstly constructing a grade shell by using Radial Basis Function (RBF) to establish the domain of mineralisation and then estimating the 1.5 × 1.5 × 1.5 m block grades within the grade shell representing the domain of mineralisation. Selected block size is smaller than the one expected for kriging, however selecting smaller blocks would increase ML (machine learning) model performance and block sizes are kept consistent for all estimation techniques in this study. The selected estimators are Simple Kriging (SK), OK as traditional geostatistical estimators, and *k*-NN and RF as the machine learning algorithms to be used for estimating the uranium grades.

DATA STATISTICS

The data set used in this study comprises data from 392 drill holes, providing 872 GT values and 4028 composited U_3O_8 grades at 0.3 m intervals for the A sand unit, each with corresponding 'to/from' depth information. Detailed summary statistics for these data are presented in Table 1.

TABLE 1

Summary statistics of GT and U_3O_8 grades.

	Grade U_3O_8	GT
Mean	0.12	0.53
Median	0.10	0.21
Mode	0.07	0.01
Minimum	0.01	0.01
Maximum	0.85	10.16
Range	0.84	10.15
Sum	491	460
Count	4028	872
Standard Deviation	0.094	0.893
Sample Variance	0.009	0.798
Standard Error	0.001	0.030
Skewness	1.966	4.356
Kurtosis	6.216	28.169

To illustrate the distribution of GT values and composited U_3O_8 grades, along with their cumulative percentages, histograms and cumulative probability plots are shown in Figure 3a and 3c. To address the inherent skewness in the original data, GT values and composited U_3O_8 grades were logarithmically transformed. Histograms of these log-transformed values are displayed in Figures 3b and 3D. This transformation normalises the data distributions, making them more suitable for model.

Similarly, the normalised distribution of log-transformed U_3O_8 grades facilitates improved variogram fitting for use in SK and OK models and subsequent variogram modelling and analysis (Figure 4). The resulting normal distribution of log-transformed GT values enhances the stability of variance and improves the robustness of the grade.

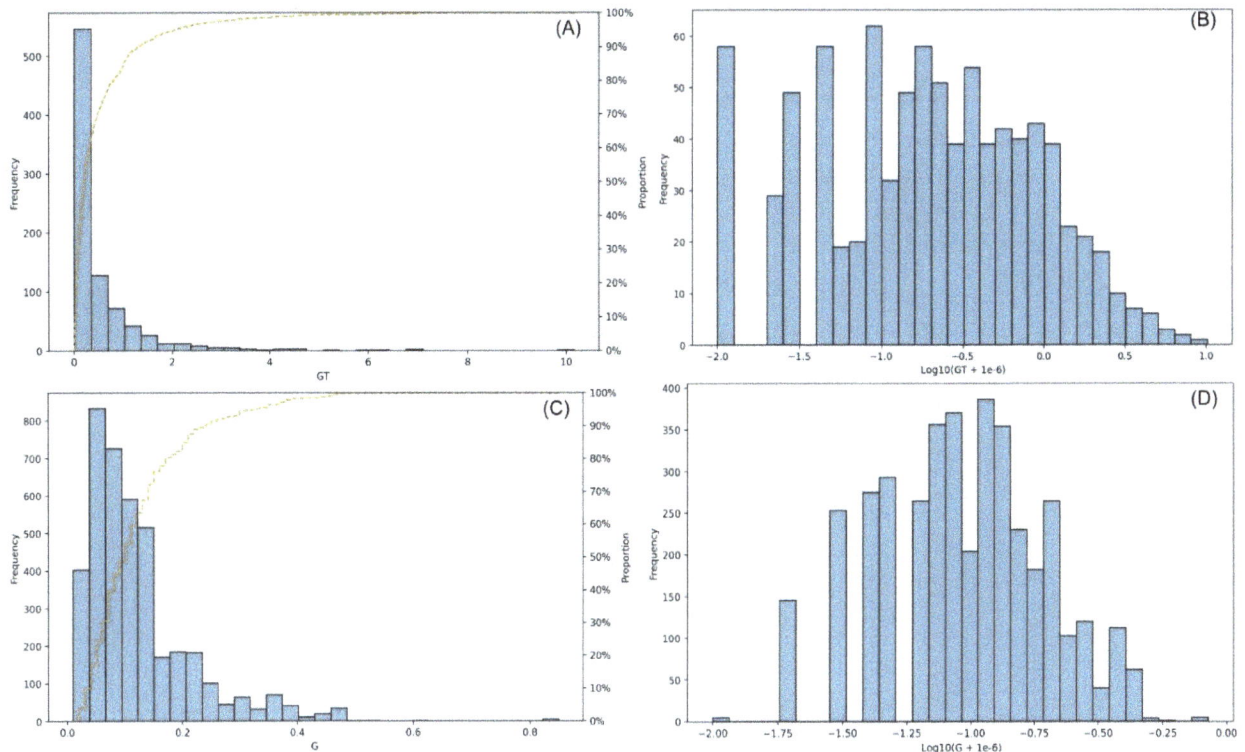

FIG 3 – Histograms show (a) GT values and cumulative probability plot (orange line), (b) log transformed GT values for variogram analysis, (c) Grades as % U_3O_8 of 0.3 m composites and cumulative probability plot (orange line), (d) log transformed grades as % U_3O_8 of 0.3 m composites for variogram analysis and cumulative probability plot (orange).

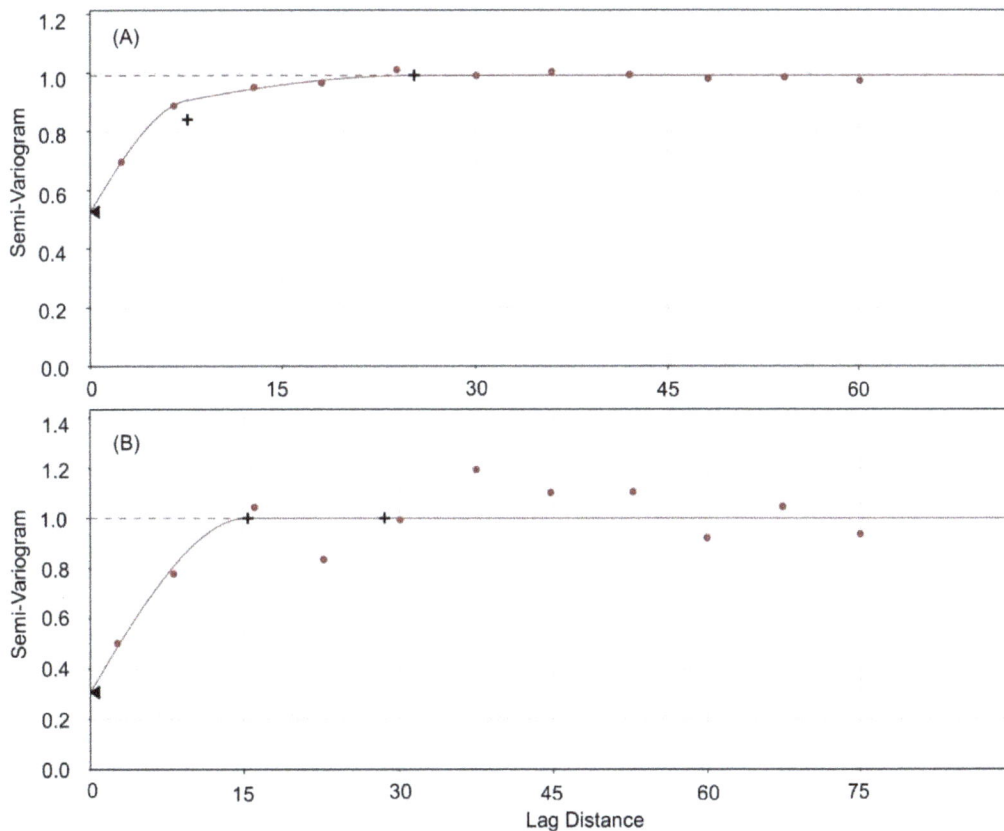

FIG 4 – Semi-Variograms are fitted by double structured Spherical models for (A) log-transformed GT values (>0.2) applied in mineral domain modelling; (B) log-transformed U_3O_8 values (>0.02) utilised for Kriging based grade estimations.

DOMAIN MODELLING

The mineral domain is constructed using a Radial Basis Function (RBF) network, a type of artificial neural network (ANN) with simpler training processes compared to traditional ANNs (Howlett and Jain, 2001). In an RBF network, the output is defined by a radial basis function, $\emptyset$, which is symmetric about a centre point, μ. The function is expressed as $\emptyset(x) = \emptyset(\||x - \mu\||)$, where $\||.\||$ denotes the Euclidean norm, x is the input vector, and μ is the centre vector. In such cases, the Gaussian functions can become an RBF.

The RBF network has three layers (Figure 5): the inputs; the hidden kernel layer; and outputs. The hidden units (centroids or kernels) represent a single radial basis function for each node with an associated centre position and width. Each of the output units represents a weighted summation of the hidden units (Howlett and Jain, 2001).

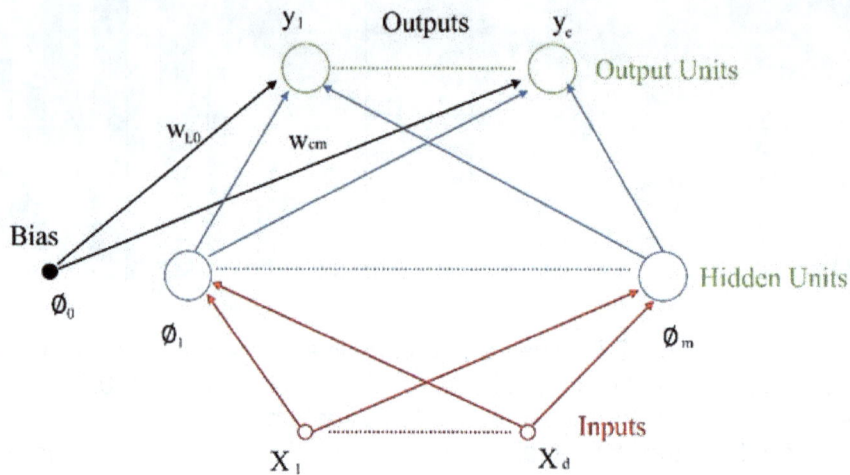

FIG 5 – A radial basis function network.

Estimation domaining is one of the most important steps in the resource estimation and forms the shells of the orebody with respect to geological, mineralogical, alteration and chemical controls of ore forming processes. However, the roll front uranium deposits are formed in mostly sandstone which is porous and enables fluids to transport or conglomerates with the same reason it could form a single geological domain. And the visual determination of sandstone grain size may not give reliable data especially for the rotary or RC type drill holes which can only provide cuttings as samples. Furthermore, the oxidised and reduced information do not form domains because the uranium is not formed uniformly in these domains but formed only as close contacts to the oxide-reduced boundary.

The estimation domain is developed by applying RBF to GT values coming from the mineralised sand unit (A sand) at Nichols Range uranium deposit.

The constructed grade shell, developed using RBF interpolation and 0.3 m composited grade values, is plotted in Figure 6. This grade shell plays a crucial role in delineating the boundaries for the estimation of uranium grades by both geostatistical and machine learning methods, ensuring that the estimations are confined to relevant areas.

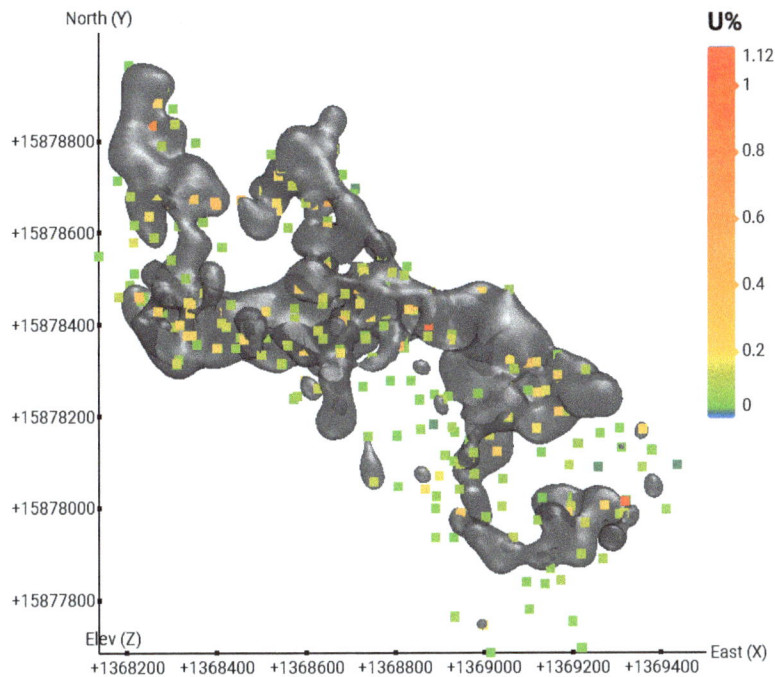

FIG 6 – Plan view of grade shell with % U_3O_8 values of 0.3 m composited samples.

Further, the efficacy of the grade shell in encapsulating high-grade zones is demonstrated through contact analysis between the grade shell and the composited grade values, as presented in Figure 7. This analysis reveals a stark contrast in mean uranium concentrations, with values inside the grade shell averaging 0.14 U_3O_8, compared to just 0.03 U_3O_8 outside. Such a significant drop in grade outside the shell underscores its effectiveness in highlighting areas of uranium mineralisation, confirming the grade shell's role as a critical tool in the targeted estimation of mineral resources.

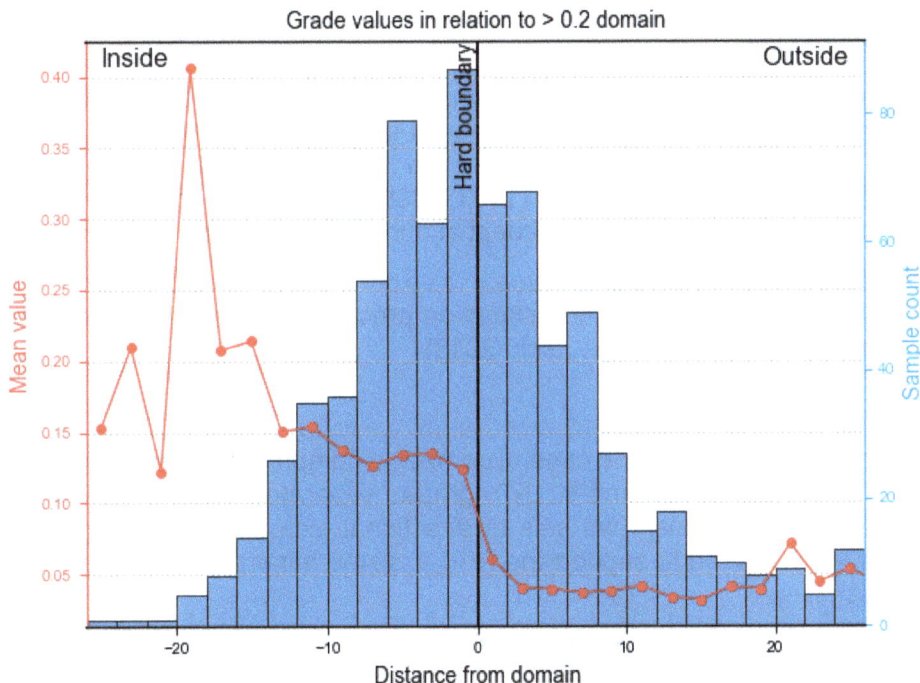

FIG 7 – The histogram shows the mean U_3O_8 values and GT grade shell distance relationship.

ROLL FRONT URANIUM MINERAL RESOURCE ESTIMATION

Traditional method (GT Contour)

The most employed grade estimation technique for roll-front uranium deposits is the GT contour method. This method effectively utilises geometrically consistent contours, representing the grade shell with average grade and area, to delineate uranium grades.

The 2D GT contour method unfolds through a three-step process:

1. GT Value Determination: The GT value for each drill hole is calculated by multiplying the % U_3O_8 grade by the corresponding intersected apparent thickness. For orebodies with regular geometries, the true thickness should be determined. However, in this deposit, the A-sand unit, which primarily hosts uranium mineralisation, is approximately horizontal (Figure 2c), and all drill holes are vertical unless otherwise deviated.

2. Zonal Determination: The deposit is segmented into two-dimensional horizontal or sub horizontal zones, with GT values assigned to specific zones based on their spatial location.

3. Contour Mapping: These GT values are then contoured within each two-dimensional zone. Then these contours representing the average GT's which are then multiplied by the contour area and density factor 1.17 and added each other to result in overall resource. This step demands substantial field expertise to accurately interpret the geological contours and ensure the fidelity of the grade representation.

The Nichols Range mineral resource estimates traditionally were determined by an experienced geologist using GT contouring technique.

Geostatistical methods

The estimation of block GT values can be carried out using many different techniques. Various estimation techniques such as inverse distance weighting (IDW), nearest neighbour, inverse distance power (IDP), and several versions of Kriging are available. In this study, we focused on Simple Kriging (SK) and Ordinary Kriging due to their widespread application and robustness in mineral resource estimation.

SK serves as an estimator characterised by a constant mean m of a variable Z within a domain D (Deutsch and Journel, 1998). The weighting for the SK estimator is derived from a variogram model, which is adjusted to account for spatial dependency in three orthogonal directions, as pioneered by Krige (1951) and further refined by Matheron (1963, 1968). Recognised for its precision, SK is lauded as the optimal unbiased estimation method, adept at minimising the variance of errors in predictions.

OK, on the other hand, recalculates the local mean as a constant within a predefined search ellipsoid, ensuring that the weights assigned in the estimation process sum to one, thereby maintaining the estimator's unbiasedness (Rossi and Deutsch, 2013).

Experimental variograms

Estimation of uranium grades within the mineral domain, a comprehensive variogram model analysis was conducted for both GT values and % U_3O_8 grades separately (Figure 4), and % U_3O_8 grades are composited at 0.3 m intervals. Two sets of directional experimental variograms, aligned with primary directions, were utilised to support the RBF assisted mineral domain definition and SK and OK estimation processes.

For mineral domain definition, GT values were first converted into a 3D format comprising X, Y, Z coordinates and corresponding GT values. Summary statistics for these values are presented in Table 1. Experimental variogram analysis was conducted using a lag distance of 4.6 m to determine the anisotropy directions, identifying a primary anisotropy direction at 226° azimuth. The analysis (Table 2) revealed a nugget effect of 0.53 (Figure 4a, Table 2), reflecting microscale variability or measurement error inherent in the data set. Range parameters were carefully defined for a spherical model, which was selected for its ability to represent the 3D spatial continuity observed in the empirical data. For the major range at 226° azimuth, the first and second structures exhibited ranges of 7.9 m and 25.6 m, respectively. The semi-major range showed ranges of 7.0 m and 8.5 m, while the minor range had ranges of 11.2 m and 14.0 m for the first and second structures, respectively. These parameters capture the anisotropic spatial variability of GT values across different orientations within the deposit. The search ellipsoid was configured with its longest axis aligned at 226° azimuth (25.6 m length, 0° plunge), the shortest axis perpendicular at 136° azimuth (14 m length), and the minor axis in the vertical direction (8.5 m length). This rigorous variogram modelling approach establishes a robust geostatistical framework, accurately capturing the spatial structure of GT values and enhancing the reliability of subsequent resource estimation processes.

TABLE 2

Variogram fit parameters for GT and % U_3O_8 grades.

GT variogram	Variogram type	Major range	Semi-major range	Minor range	Nugget	Azimuth	Dip
First structure	Spherical	7.9 m	7 m	11.2 m	0.53	226°	-90
Second structure	Spherical	25.6 m	8.5 m	14 m			
% U_3O_8 variogram							
First structure	Spherical	15.5 m	4 m	23.5 m	0.3	226°	-90
Second structure	Spherical	29 m	5 m	28 m			

Second set of directional variogram analysis are performed by using 0.3 m composited % U_3O_8 grades. This variogram analysis was used in determination of estimation parameters of the SK and OK models. The parameters selected for the model is critical in capturing the spatial continuity and variability within the deposit is included a lag distance of 6.1 m. With a spherical model, the nugget effect, quantified at 0.3 (Figure 4b, Table 2). Range parameters were carefully defined for two structure spherical model, which was selected for its ability to represent the 3D spatial continuity. For the major range at 226° azimuth, the first and second structures exhibited ranges of 4 m and 5 m, respectively. The semi-major range showed ranges of 23.5 m and 29 m, while the minor range had ranges of 23.5 m and 28 m for the first and second structures, respectively. The search ellipsoid was then configured with its longest axis aligned at 226° azimuth (29 m length, 0° plunge), the shortest axis perpendicular at 136° azimuth (28 m length), and the minor axis in the vertical direction (5 m length). This rigorous variogram modelling approach establishes a robust geostatistical framework, accurately capturing the spatial structure of GT values and enhancing the reliability of subsequent resource estimation processes.

The ore deposit is modelled using cubic blocks with a volume of 1.5 m³, which is significantly smaller than typical block sizes used in similar studies. This smaller block size enhances the representation of the irregularly shaped orebody's boundaries, particularly at the edges of mineralisation, where precision is critical. The mineralisation model was constructed using RBF interpolation. Block centroids, defined by their mid-coordinates and elevations, served as estimation points. Furthermore, these centroids are used for new estimation locations in ML models which is prone to increase ML model accuracy. A total of 101 338 blocks were generated within the mineralised domain. Following SK and OK estimations, the blocks were filtered using a cut-off grade of 0.02 per cent U_3O_8 to delineate the economically viable portions of the deposit.

Machine Learning methods

A generalised sketch of ML model development and prediction is shown in Figure 8. Starting with problem definition which aims to understand the problem and characterise that followed by preparation of all the data with the defined features in the first step. This step also prepares the data into a suitable format. A skewed distribution of data should be rescaled into [0,1] interval by using the normalisation function 'normalise ()' (Erten, 2021; Erten, Yavuz and Deutsch, 2022).

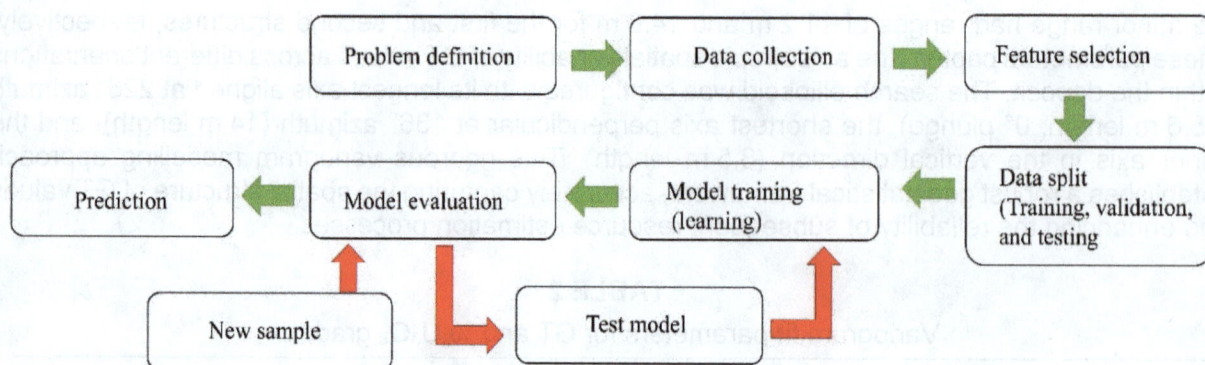

FIG 8 – ML processes steps.

The subsequent phase in our analysis is feature selection, which involves identifying the input parameters that contribute significantly to the learning and prediction processes. For this study, we utilised 0.3 m composited % U_3O_8 grades as our primary features, designated as Easting (X), Northing (Y), and Elevation (Z) for the × inputs, and % U grades as the Y input.

Following the feature selection, the data set is partitioned into two subsets: 80 per cent is allocated for training, and the remaining 20 per cent is reserved for testing. The data splitting is strategically designed to validate the model's performance effectively, ensuring that it can generalise well to new, unseen data. During the model training phase, the algorithm is trained on the 80 per cent training subset. It then proceeds to predict the outcomes on the 20 per cent test subset, allowing for a direct comparison with the true values of these samples. This step is critical as it provides an initial assessment of the model's predictive accuracy.

To further enhance the robustness of our predictive model, a five-fold cross-validation is implemented. This technique involves partitioning the training data set into five equal or nearly equal segments. The model is then trained and validated five times, each with a different segment held out for validation while the remaining segments are used for training. This process not only helps in assessing the model's performance more reliably but also accuracy across different subsets. The average accuracy obtained from these iterations provides a comprehensive view of the model's effectiveness and stability.

The final step in our modelling process is model evaluation, which primarily focuses on hyper-parameter tuning. This involves systematically testing and adjusting the model's parameters to determine the optimal settings that yield the best performance on the validation data. Hyper-parameter tuning is pivotal as it fine-tunes the model to enhance its predictive power and efficiency.

Once the optimum parameters are established, the model is re-operated with these refined settings, following the workflow outlined by the red arrows leading to the 'Test Model' box in Figure 8. This reiteration with the optimised parameters ensures that the model not only performs well on the historical data but is also robust and reliable when applied to new, unseen data set.

This comprehensive approach, from feature selection through to the rigorous testing and optimisation of model parameters, ensures the robustness of our analysis and ensures that the final model is well-equipped to provide accurate predictions. The systematic validation steps, including cross-validation and hyper-parameter tuning, underscore our commitment to achieving the highest possible accuracy and reliability in our predictive modelling efforts.

The performance of the ML models is evaluated by using the RMSE, MAE, and R^2 which all measure the differences between the true and the predicted value of the test data.

RMSE is the squared root of the variance of the estimated errors which measures the closeness of the predicted values to the true values in the test data set. MAE is the average of the absolute error. It is an average deviation on a linear line. The value of the MAE is better as it approaches zero.

Coefficient of determination is the correlation of the predicted and true values of test data. The value of the R^2 approaches to one, the model prediction is better. However, if it is too close to one, it may also mean that the model is overfitting. When the R^2 value is zero, it means that there is no correlation between the true and the predicted values. This value also gets negative values which means the selected ML model is not appropriate and even a horizontal line may estimate better.

After model evaluation step, new sample points which represents the X, Y, Z coordinates of 1.5 × 1.5 × 1.5. Grades are imported to predict their % U_3O_8 values with the tuned and suitable model (with the R^2 values closer to 1).

k-Nearest Neighbours

The *k*-NN is an instance-based learning model that utilises the distance-based similarity functions in regression problems (Cover and Hart, 1967). In other words, *k*-NN assumes that similar things are close to each other. Then, it averages the samples that are close to the estimated sample.

The most important parameter in the *k*-NN in the *k* value which defines the number of closest samples that the algorithm considers. The optimum *k* value can be determined by the hyper-parameter tuning by iterating the model with number of different *k* values and the one with the lowest MSE value is used in model fitting.

The advantages of this method are the algorithm is simple and easy to implement, it is faster than other ML models, can be used in both classification and regression problems. However, as the input data and predictors increase the model gets slower.

The *k*-NN algorithm in regression problems, firstly calculates the distance between new sample point and the *k* nearest samples around, then it calculates the average weighted sum of the *k* nearest sample points for the estimation. For *k*-NN estimation we used *k*=2 (potential for conditional bias) and the resulted R^2 is 0.79 which is the highest among the other machine learning models given in Aydar (2022).

Random forest

Radom forest model similarly works in both classification and regression problems. The RF works with several DTs. Sample selection is random which randomises the trees in RF (Breiman, 2001; Guo *et al*, 2011). RF employs input data (x), trains it and constructs number of DTs (K) and averages the results. RF use training data to increase the diversity of the trees by using a procedure called bagging which creates training data by randomly resampling the data with replacement without making changes or deleting it resulting in increasing the model accuracy (Breiman, 2001).

RF is an advantageous ML method that can be used most of the problems without a long hyper-parameter tuning, without rescaling the data (Breiman, 2001). It is very powerful and has a lower chance of overfitting. However, it gets slower if input data size gets higher. The resultant R-squared of random forest model is 0.75.

NICHOLS RANGE ESTIMATED MINERAL RESOURCE COMPARISON

Validation and comparison of the four-resource model have been performed visually on the cross-sections, by swath plot (Figure 9) and by grade tonnage plot (Figure 10). The swath plot suggests that the *k*-NN and OK model is remarkably similar in terms of average grade by distance and follows a close pattern. The overall resource of these two models is relatively closer than the other two models to the production data provided (Table 3). Similarly, the grade tonnage curves of OK and *k*-NN models are remarkably close to each other.

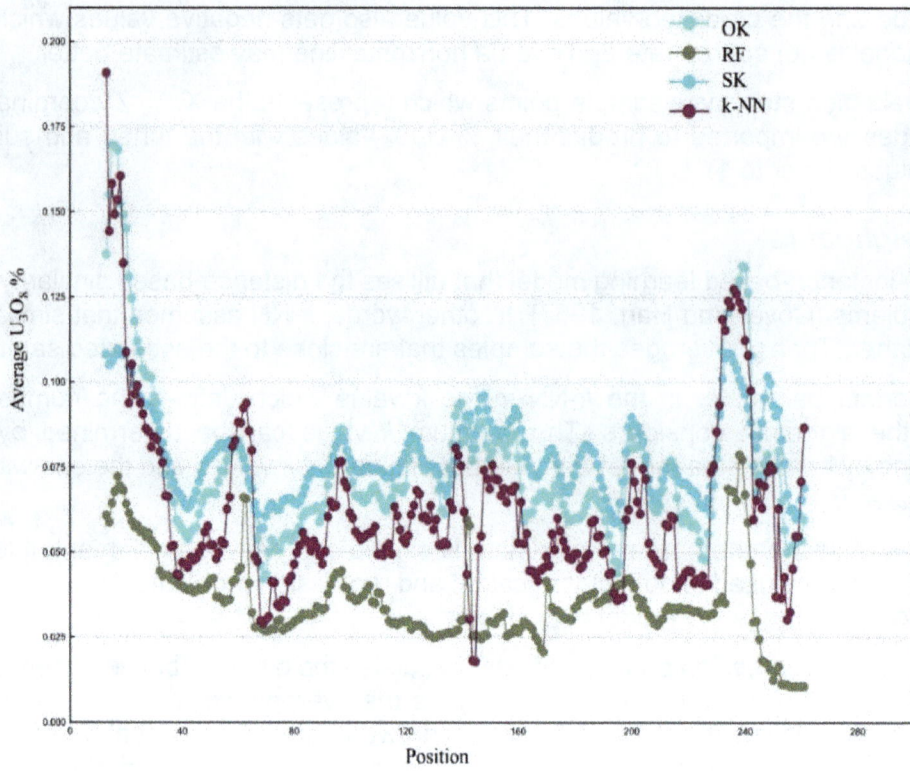

FIG 9 – Swath plot of OK, SK, k-NN and RF resources.

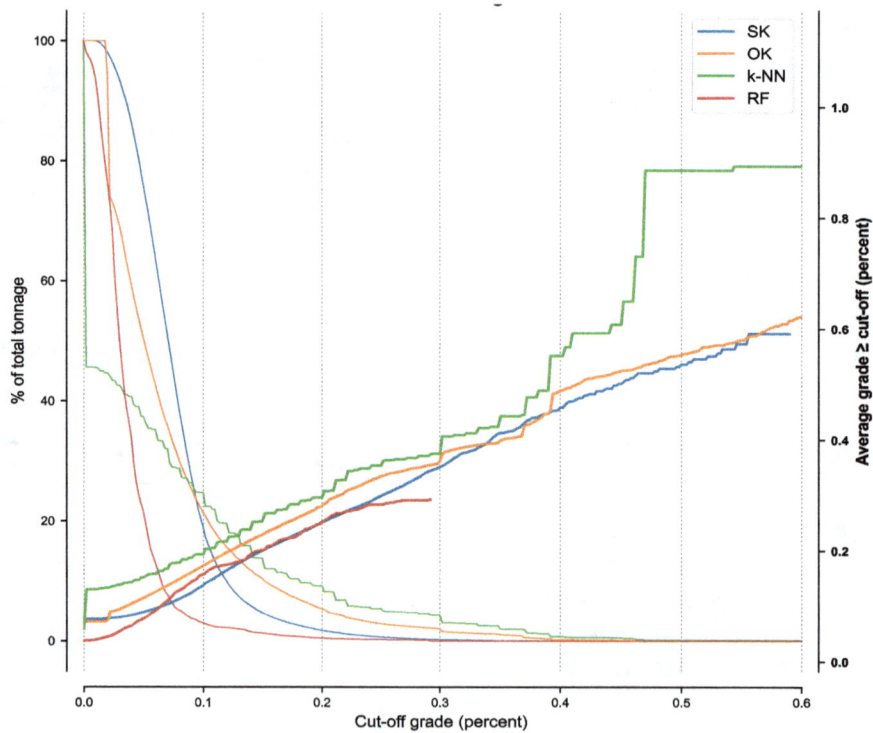

FIG 10 – Grade tonnage curves of four resource estimation models.

TABLE 3

Resource estimation comparison with respect to uranium tonnages at zone 60 and overall deposit. Resources are in metric tons of U_3O_8. *In situ* resource of Company's by GT contour method.

Methods	Models	Zone 60	Overall
Geostatistics	OK	64.8	396.6
	SK	69.2	522.5
GT	GT (this study)	32.7	192.5
Machine Learning	*k*-NN	71.8	409.5
	RF	33.6	190.7
GT	Company's GT	68.5*	413*

One of the good comparisons between GT method and OK and *k*-NN would be the overall grade and tonnage comparison (Table 3) defined in terms of total lbs. of uranium for the area of study. The GT resource model of the company has been produced with 75 per cent accuracy in ISR. Therefore, we compare the estimations with the GT resource model of the Company. Comparison based on total uranium contained in all the zones shows that OK and *k*-NN results are within the 4 per cent and 1 per cent respectively of the Company's GT results. The remarkable difference between the Company's GT model and the GT model prepared by Aydar (2022) and (Table 3) defines the problem depending on the field experience of the resource engineers.

A visual comparison of the resources with the Company's zone 60 GT contour model (a mid-horizontal mineralisation level of 30 m is given in the Table 3. A common feature of these four models is they are spatially associated with the GT model (Aydar, 2022; Aydar and Dagdelen, 2024). The OK model and the *k*-NN model are better in representing the locally enriched grades and more likely to represent the GT interpreted areas. However, this comparison does not fully represent a perfect comparison since the overlay of 2D and the 3D zones are relatively performed. For example, the left and right wing of the GT model do not overlay with the 3D zone 60, however at zone 70 (3 m above horizontal layer), the GT model spatially associate with the 3D model as well.

A comparative analysis of the resource models depicted in Figure 11 reveals that both the *k*-NN model and OK correlate well with the drill hole samples upon visual inspection. It is imperative to recognise the significance of visual validation in mineral resource estimation processes. The ML models, utilising RF, demonstrate an R-squared accuracy of 0.75. Despite this, there is a notable discrepancy in predicting the high-grade Uranium deposits within the drill holes, indicating potential limitations in the model's predictive capability. On the other hand, SK model presents a more accurate representation of Uranium grades. However, the use of a single mean in the kriging estimation leads to a smoothing effect, which potentially diminishes its applicability in reserve calculations or ISR production assessments.

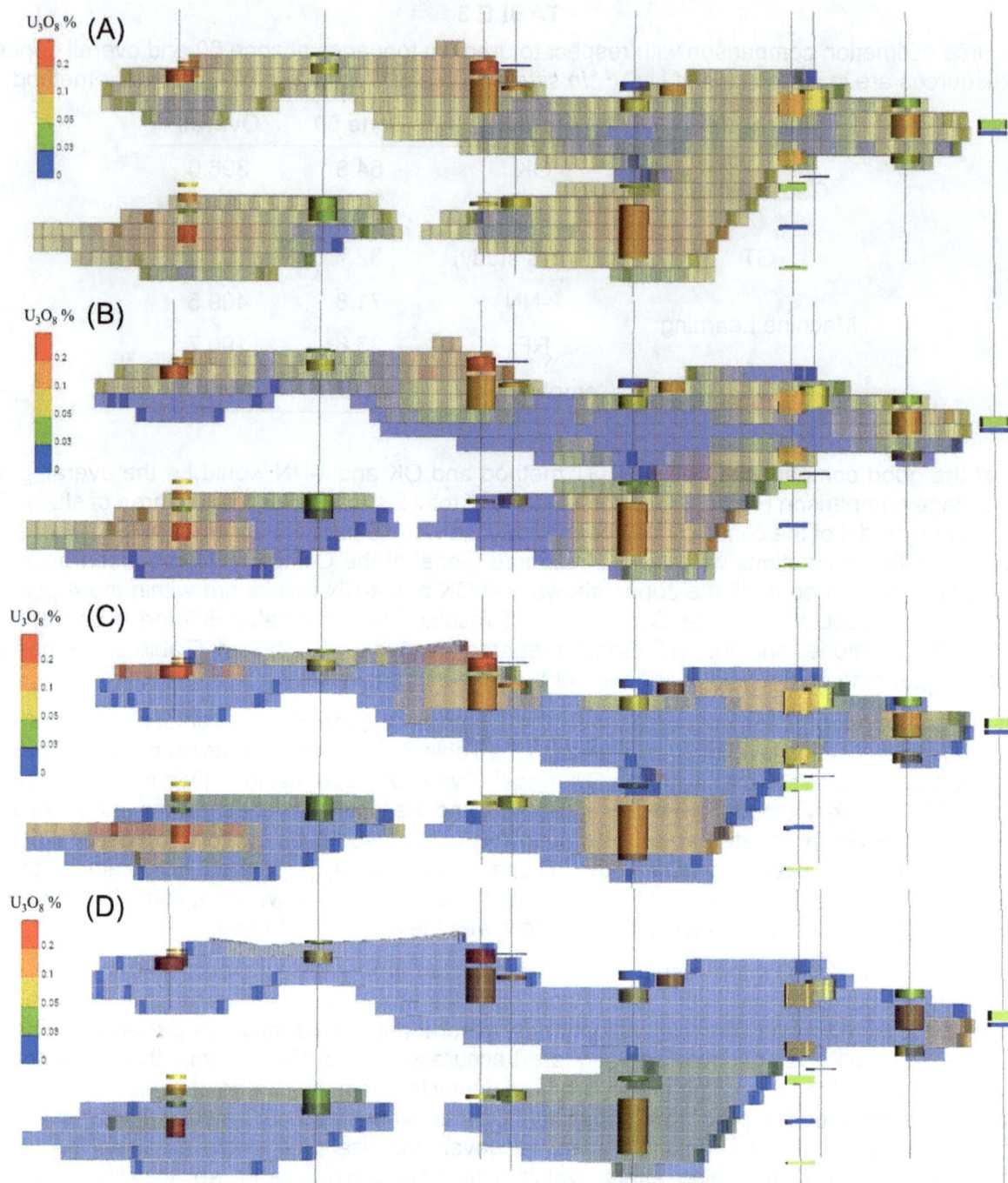

FIG 11 – Cross-sections of 3D blocks and drill hole samples and the block grades estimated by (A) SK, (B) OK, (C) k-NN and (D) RF models.

Overall, both the OK (Figure 11b) and k-NN (Figure 11c) models exhibit robust performance in resource estimation. They accurately reflect the distribution of high and low-grade Uranium, aligning closely with the Uranium yield as determined by the ISR method from this specific section of the Nichols Ranch Uranium Mine.

CONCLUSION

This study provides a comprehensive understanding of the challenges associated with the GT contour model in uranium resource estimation. It highlights the complexities of the method, which is labour intensive, requires extensive field experience, and is time consuming to implement. Moreover, the GT contour method's reliability is questioned as it may yield varying results with each application, underscoring the need for more consistent and efficient methodologies.

Furthermore, this research has successfully utilised RBF to construct a 3D grade shell, effectively defining the mineralised domain within roll-front uranium deposits. This approach significantly enhances the interpretation of uranium rolls traditionally assessed by GT contour methods, offering a more robust framework for resource estimation.

In evaluating the performance of selected machine learning models, k-nearest neighbours (k-NN) demonstrated a promising R^2 value of 0.79, while the RF model exhibited an R^2 of 0.75 during the model training phase. These results indicate a strong predictive capability, underscoring the potential of machine learning in refining resource estimation processes.

The *in situ* resource estimation techniques implemented, specifically k-NN and OK, show deviations of only 1 per cent and 4 per cent respectively when compared to the traditional *in situ* GT contouring resource utilised by the Company at the Nichols Ranch uranium deposit. This minimal variance further validates the effectiveness of these modern approaches in closely approximating the actual recoverable resources.

Moreover, the geospatial estimates produced by the OK and k-NN models align remarkably well with the drill hole intersections, demonstrating that the estimated resources are not only theoretically sound but also practically viable. This alignment is indicative of the models' ability to closely mirror actual uranium production figures.

The methodology advocated in this study for constructing grade shells and estimating grade blocks proves to be significantly more efficient than traditional methods. It offers rapid results, consistent outcomes with repeated applications on the same data, and reduces the dependency on extensive field experience. This streamlined approach not only simplifies the process of resource estimation but also enhances its accuracy and reliability, making it a valuable tool for future applications in uranium mining and other similar mineral resource explorations.

ACKNOWLEDGEMENTS

Arif Aydar was funded by the Turkish Ministry of Education and the General Directorate of Mineral Research and Exploration (MTA) under the 1416 Scholarship Program, which made this research possible. We are grateful to Energy Fuels Inc for providing the drill hole data and supporting this academic research. Special thanks to Travis Boam (Senior Resource Geologist), Bruce Larson (Director of Geology and Land), Gordon Sobering (Senior Mine Engineer), Bernard Bonifas (Director of ISR Operations), and Daniel Kapostasy (Director of Technical Services) for their valuable insights and assistance. A heartfelt thanks to Dr. Rex Bryan for his guidance and expertise in uranium mining and application of geostatistics to Roll Front Uranium Deposits. We would like to thank to Zeynep Ankut, a PhD student at Mining Engineering Department at Colorado School of Mines for reviewing this paper.

REFERENCES

Aydar, A, 2022. Resource estimation of roll front uranium deposits by using traditional and machine learning methods for Nichols Ranch Uranium deposit in Wyoming, MSc Thesis, Colorado School of Mines, Golden, Colorado, USA.

Aydar, A, Dagdelen, K and Bryan, R C, 2024. Overcoming Challenges in Roll-Front Uranium Resource Estimation: RBF and Machine Learning Offer Superior Accuracy and Efficiency, SME Graduate Student Research Poster Contest, Phoenix, Arizona, USA.

Beucher, H and Renard, D, 2016. Truncated Gaussian and derived methods, *Comptes Rendus Geoscience*, 348–7:510–519.

Breiman, L, 2001. Random forests, *Machine learning*, 45(1):5–32.

Ciputra, R, Suharji, S, Kamajati, D and Syaeful, H, 2020. Application of geostatistics to complete uranium resources estimation of Rabau Hulu Sector, Kalan, West Kalimantan.

Cover, T and Hart, P, 1967. Nearest neighbour pattern classification, *IEEE transactions on information theory*, 13(1):21–27.

Deutsch, C V and Journel, A G, 1998. *GSLIB: Geostatistical software library and user's guide* (Oxford University Press: New York).

Erten, G E, 2021. Estimation of Geospatial Data by Using Machine Learning Algorithms, PhD Thesis, Eskişehir Osmangazi Üniversitesi, Eskişehir, Turkey.

Erten, G E, Yavuz, M and Deutsch, C V, 2022. Combination of Machine Learning and Kriging for Spatial Estimation of Geological Attributes, *Natural Resources Research*, 31(1):191–213.

Guo, L, Chehata, N, Mallet, C and Boukir, S, 2011. Relevance of airborne lidar and multispectral image data for urban scene classification using random forests, *ISPRSJ, Photogram, Remote Sensing*, 66:56–66.

Howlett, R J and Jain, L C, 2001. *Radial Basis Function Networks 2* (Springer-Verlag Berlin: Heidelberg).

Knudsen, H P and Kim, Y C, 1979. Development and Verification of Variogram Models in Roll Front Type Uranium Deposits, *Mining Engineers*, pp 1215–1219.

Krige, D, 1951. A statistical approach to some basic mine valuation problems on the Withwaterstand, *SAIMM*, pp 119–139.

Malensek, G A, Mathisen, M B, Collyard, J S, Woods, J L and Brown, P E, 2022. Technical Report on the Nichols Ranch Project, Campbell and Johnson Counties, Wyoming, USA SLR Project No: 138.02544.00001.

Marbeau, J and van Brunt, B, H, 1991. A Geostatistical Study of the North Butte Roll Front Uranium Deposit, Society for Mining, Metallurgy and Exploration, pp 91–102.

Matheron, G, 1963. Principles of geostatistics, *Economic Geology*, 58:1246–1266.

Matheron, G, 1968. Basics of Applied Geostatistics, Mir, Russia.

Matheron, G, Beucher, H, de Fouquet, C, Galli, A, Guerillot, D and Ravenne, C, 1987. Conditional simulation of the geometry of fluvio deltaic reservoirs, SPE, 16753:123–131.

Rossi, M and Deutsch, C, 2013. *Mineral Resource Estimation*, Delray Beach, Florida: Springer.

Syaeful, H, 2018. Geostatistics Application on Uranium Resources Classification: Case Study of Rabau Hulu Sector, Kalan, West Kalimantan, *Eksplorium*, 39–2:131–140.

Taghvaeenezhad, M, Shayestehfar, M, Moarefvand, P and Rezaei, A, 2020. Quantifying the criteria for classification of mineral resources and reserves through the estimation of block model uncertainty using geostatistical methods: a case study of Khoshoumi Uranium deposit in Yazd, Iran, Geosystem Engineering.

Yun, B, PengFei, Z, Jing, Z, WeiHao, K, XiaoCui, L, Ke, C, Lu, S and LinYing, L, 2020. Application of geo-statistics in calculation of a uranium deposit, E3S Web of Conferences 206.

Adaptive search ellipsoid – a new practical algorithm for local anisotropies in mineral resource estimation

A Aydar[1] and K Dagdelen[2]

1. PhD Student, Mining Engineering Department, Colorado School of Mines, Golden CO 80401, USA. Email: arifaydar@mines.edu
2. Professor, Mining Engineering Department, Colorado School of Mines, Golden CO 80401, USA. Email: kdagdelen@mines.edu

ABSTRACT

Current practices in mineral resource estimation challenge resource engineers to analyse locally varying anisotropies more effectively. These anisotropies, driven by local geological conditions, necessitate either sub-domaining of a geological or mineral domain or the use of locally aligned search ellipsoids.

This paper introduces a new algorithm designed to enhance estimation accuracy by systematically addressing and incorporating local anisotropies as part of the estimation process. The algorithm begins by generating directional experimental variograms within a domain at set increments to define anisotropy directions. Once identified, it proceeds by automatically fitting authorized variogram models (Spherical, Exponential or Gaussian) to the experimental variograms, minimising Root Mean Squared Errors (RMSE) to determine the longest axis of global anisotropy. This axis serves as the search radius for filtering local data within a geological neighbourhood. Subsequently, only the filtered local data are used to generate local experimental directional variograms for each estimation locations, followed by auto-fitting and grade estimation using a locally adapted anisotropic variogram.

To test the algorithm, a case study was generated using Walker Lake data set. The case study aimed to estimate unknown values at 725 grid locations from the 470 samples covering the Walker Lake area. The case study with a small sample set showed that more accurate estimation of unknown values at the grid locations is possible for the Walker Lake area if the local variogram anisotropy is determined and used before the estimation of the value at each grid location.

INTRODUCTION

Mineral resource estimation has paramount importance in mining industry that effects decision-making processes in mining investments, mine planning, scheduling even reclamation and environment. Resource estimation starts with gathering quality geological, geochemical, geo-metallurgical and structural data starting from the exploration of a mineral deposit. However, collecting quality data is infinite and requires initial investments on mineral prospects which is finite. Therefore, optimisation of data required to minimise the risk arise from geological uncertainties is necessary. Besides, utilising the available data with the most available technology to reduce the risk is necessary.

The quality of the resource estimation also depends on the suitable interpolation techniques used. Various types of Kriging have been developed since the discovery of the Kriging estimation technique (Krige, 1951; Matheron, 1962, 1963, 1965; Journel and Huijbregts, 1976; Cressie, 1993; Rossi and Deutsch, 2014). However, there are two main risks associated with the Kriging estimation: (1) manual fitting the theoretical models on the semi-variogram; and (2) the local anisotropies within a geological or mineral domain.

As mining software has evolved alongside technological advancements and increased computational power, practitioners have been afforded the capability to visually fit theoretical models. However, visual fitting of variograms can introduce a degree of uncertainty (Isaaks and Srivastava, 1989) and the goodness of fit affects the subsequent estimation workflow. The goodness of fit is measured mostly by the coefficient of determination (CD), RMSE and mean squared error (MSE). To address this issue, contemporary scientific methodologies often employ optimisation techniques to improve the goodness of fit such as SciPy package which can be used to automatically fit variograms, thereby minimising the RMSE or MSE and enhancing model accuracy (Virtanen *et al*, 2020).

In the context of the determining the accurate values in the more complex spatial associations. Figure 1 shows how the using of a single search ellipsoid and local search ellipsoids would change the accuracy of mineral resource estimation. Number of authors have been implemented kriging with a local search (Deutsch and Lewis, 1992; Sullivan, Satchwell and Ferrax, 2007; Xu, 1996). These studies require a locally varying anisotropy (LVA) fields to be constructed by defining the azimuth, dip and plunges in the field. LVA fields can be determined by gradient field which is an image analysis algorithm (Stroetv and Snepvangers, 2005), by moving window (Lillah and Boisvert, 2013), the moment of inertia (Pyrcz and Deutsch, 2014). Additionally, localised uniform kriging technique has been successfully applied by (Abzalov, 2006; Hardtke, Allen and Douglas, 2011) which involves manually adjusting the orientation and dimensions of the search ellipsoid tailored to smaller production panels, thereby refining the kriging process to re-estimate block grades more accurately. Nonetheless, these studies do not utilise the actual similarities of the samples and most of these techniques are limited to availability of surface data.

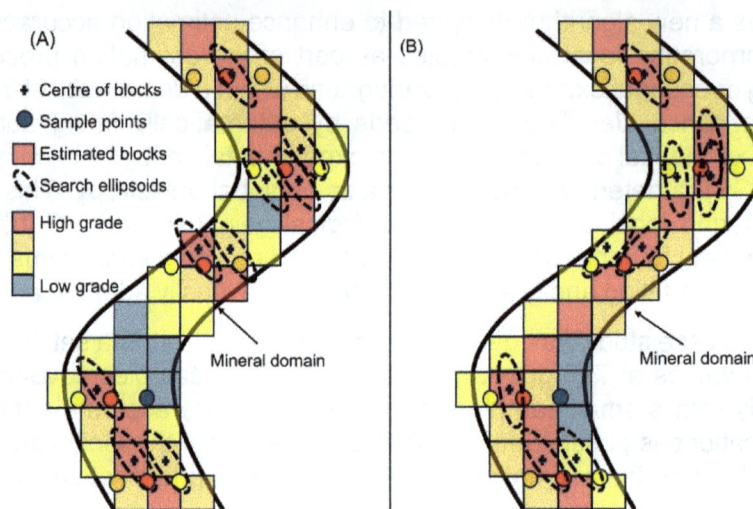

FIG 1 – Illustration of block model within a geologic or mineral domain: (a) ordinary kriging estimation with a single directional search ellipsoid which is calibrated to 135°; (b) using multiple search ellipsoids which adapt to the mineral continuity. The directions of the search ellipsoids determined by LVA fields.

In this paper, we introduce an innovative algorithm which is called Adaptive Search Ellipsoid (ASE) here, designed to enhance the precision of spatial interpolation. ASE is an algorithm methodically calculates local anisotropies at each estimation point, subsequently generating tailored search ellipsoids characterised by individually optimised orientations and dimensions. By allowing for local adaptations in the search strategy and variogram modelling, our algorithm promises to deliver more accurate and reliable estimations in mineral resource estimation and sample interpolation in other branches of science.

IMPLEMENTATION OF ADAPTIVE SEARCH ELLIPSOID

Ordinary Kriging (OK) is widely used estimation technique in mineral resource estimation. However, it is accepted that the OK smooths the estimations, cause incorrect linearities (Figure 1). To address this limitation, OK with LVA field estimations has been proposed. However, generating the LVA field is not directly tied to sample similarities and requires analysing structural trends, pixel surface characteristics, or mineral domain formations. The ASE method aims to reduce estimation errors associated with local anisotropies without requiring additional data during initial resource estimation.

The ASE methodology is straightforward and leverages commercially available Python packages. It operates under classical geostatistical assumptions, including the availability of sample data, second-order stationarity, and the principle that spatial proximity correlates with similarity in attribute values at estimation locations.

The ASE resource estimation process can be summarised in eight steps (Figure 2). The black boxes represent standard OK steps (Steps 1, 2, 3, 4 and 8), which incorporate an auto-variogram fit algorithm. Steps 1–7, including the red boxes outline the novel ASE methodology proposed in this study. The red boxes specifically highlight the ASE algorithm steps, which have been developed and tested in this research.

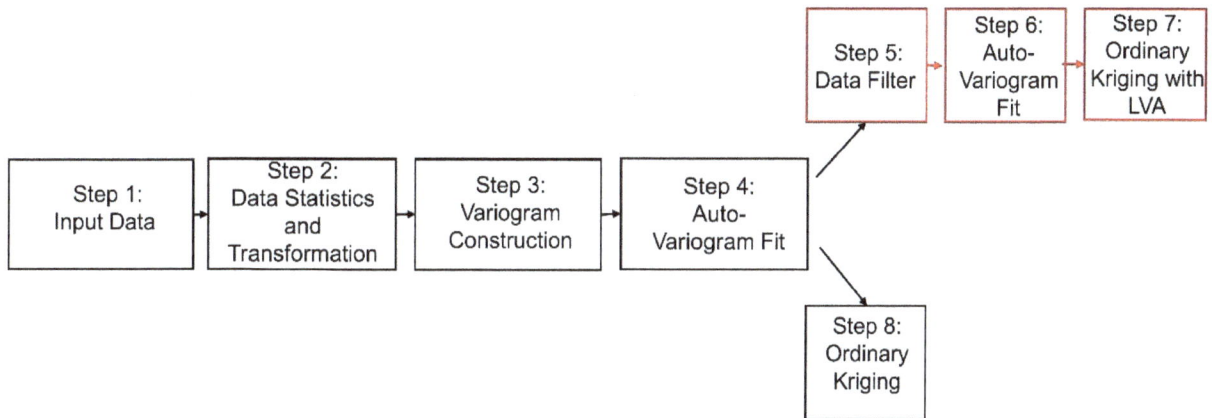

FIG 2 – Information flow chart of the adaptive search ellipsoid algorithm methodology. The red boxes indicate ASE algorithm steps.

The first step in the ASE methodology is to define the input data. For this study, the well-known 2D Walker Lake data set (Isaaks and Srivastava, 1989) is selected. Mineral exploration sample data often exhibit a skewed distribution, necessitating transformation to achieve a normal distribution to create an experimental variogram and theoretical fit which is performed in Step 2. Data transformation should be performed carefully and adapted to specific domain. The step 2 is optional and tailored for the Walker Lake data set only. In Step 3, variogram construction is conducted using equidistant lags with equal-width intervals, a widely adopted configuration. This step is also prone to development with respect to the methods (Mälicke, 2022). The Matheron (1963) variogram estimator, implemented in SciKit-GStat 1.0 (Mälicke, 2022), is employed for this purpose.

Auto-variogram fitting is a critical component of the algorithm, applied in Steps 4 and 6. This process involves fitting three common theoretical models, Spherical, Exponential (Burgess and Webster, 1980), and Gaussian (Journel and Huijbregts, 1976) to directional semi-variograms calculated at 15° intervals with a 30° tolerance angle. The auto-variogram fit utilises the trust-region algorithm in the SciPy package (Virtanen et al, 2020), with increased weights assigned to closer distances. The performance of semi-variogram fits is evaluated using the Root Mean Square Error (RMSE). In Step 4, the optimal directional semi-variogram fit, termed the 'best model fit' in this study, is determined based on the maximum range with the lowest RMSE. The best model fit is characterised by its direction, maximum range, sill, and nugget.

In Step 5, the range of the best model fit serves as the radius of a data-filtering circle. The circle is centred at each estimation location, filtering samples within its bounds. Step 6 replicates the auto-variogram fitting process from Step 4, identifying the best model and the worst model (perpendicular to the best model's azimuth) at each local estimation point. The circle's centre then moves to the next estimation point, and Step 4 is iterated to derive the best and worst model parameters (nugget, sill, range, RMSE) for all estimation locations. Step 6 not only generates the LVA field but also defines the parameters of the local search ellipsoid. This step can also be modified before estimation, as the local best-models have been filtered with respect to the goodness of fit (RMSE > 0.02), ensuring we use only the good models. In this step, external anisotropy directions can also be involved partially. The long axis of the ellipsoid aligns with the optimal range direction (best model), with its length equal to the maximum range, while the short axis corresponds to the worst model's range, perpendicular to the long axis (Vann, Jackson and Bertoli, 2003).

In Step 7, OK is applied at a local scale using the individual search ellipsoid parameters. To enhance accuracy, the best models at each estimation location are selectively used if their RMSE is lower than that of the global directional semi-variogram. For poorly sampled locations, the global best model parameters are applied, or the radius of the filtering circle can be incrementally increased based on RMSE values. In this study, however, all local models are utilised in OK. Anisotropy rotations and back-rotations are also applied at each estimation location based on the best model parameters. Finally, in Step 8, OK is performed using the global semi-directional model parameters and original (untransformed) data to compare results with those obtained from the local models.

AUTOMATIC VARIOGRAM FITTING

A major complication in fitting any variogram model function is that the closer lag classes considerably influence kriging weights and thus, their importance for any type of spatial analysis is greatly enhanced. Though a model may have a good overall fit, when it substantially deviates from fitting the initial few lag classes, it can give inferior kriging results in comparison with a model that is not a good fit overall but honours the initial lags very well; this is principally managed through the range parameter. Other than range, the sill parameter is the only remaining degree of freedom for model fitting. Therefore, in many instances, this undue emphasis on the modelling of closer lag classes will leave the experimental sill, defined simply as sample variance, poorly estimated, especially when the nugget effect is zero. If, however, the nugget is nonzero, an inaccurately estimated sill would influence the nugget-sill ratio and would need to be disposed of from possibly consideration as a valid variogram model. With a well-adjusted range and sill, kriging interpolation may capture the spatial structure of the random field. When kriging employs a pure nugget variogram model, it simply estimates the sample mean, which, although technically right, isn't often practically useful. Thus, the selecting of a variogram model must be done with caution.

Variogram

In the realm of spatial statistics, the variogram serves as a crucial tool for quantifying the spatial correlation between sample points based on their separation distance. Conducting a meaningful analysis of spatially aggregated statistics, it is imperative to define a practical yet robust method for grouping sample data, which can accommodate the inherent variability in data set spacing. The typical approach involves using predefined lag distance classes, within which all point pairs are considered. For each class, the observations are aggregated into a single measure of dissimilarity or semi-variance, denoted by γ This process accounts for the fact that it is exceedingly rare for pairs of points in real-world data sets to be separated by precisely the same distance, thereby necessitating an approximation in the form of distance lags.

The separation distance for each pair of observation points xi, $xi + h$ is calculated meticulously. Various estimators are then employed to compute the semi-variance for these distances, with the Matheron estimator being the most prevalent due to its robustness and simplicity. Proposed by Matheron (1963), this estimator is formally defined for a specific lag distance h as in Equation 1:

$$\gamma(h) = \frac{1}{2N(h)} \sum_{i=1}^{N(h)} [Z(xi) - Z(xi + h)]^2, \tag{1}$$

$N(h)$ represents the number of observation pairs at lag distance h and $Z(x)$ denotes the attribute of interest at location x. Thus, the resulting function is called empirical or experimental variogram.

To effectively model spatial dependencies within a data set, it is imperative to fit a formal mathematical model to the experimental variogram. This essential step facilitates the extraction of critical parameters that elucidate the spatial statistical attributes of the model, potentially extendable to the encompassing random field. These parameters, known as variogram parameters, play pivotal roles in defining the spatial structure and include:

- Nugget: Represents the semi-variance at zero lag distance $h=0$. It encapsulates the variance that remains unexplained by the spatial model, attributed to inherent observational noise, such as measurement inaccuracies or micro-scale variability.

- Sill: Signifies the plateau of the variogram, beyond which the increment in variance ceases, effectively marking the upper limit of the spatial model function. The nugget and sill collectively approximate the total sample variance.

- Effective Range: Denotes the distance at which the variogram value approaches 95 per cent of the sill, beyond which observations are considered statistically independent of each other.

Directional variogram

Construction of a variogram typically assumes isotropy, implying that the spatial relationships among all samples are consistent across all directions. However, this assumption does not hold when the spatial correlation length varies with direction. To address geometric anisotropy, a transformation of the coordinate system is often employed, a method well-documented by Wackernagel (1998). This process begins by conceptualising the geometric locations of the samples as being distributed across a complete 360° space. Commencing from the north (0°), the samples are systematically filtered at incremental angles θ. For each specific angle, the variogram is constructed and subsequently modelled using a theoretical framework tailored to capture the unique spatial dependencies observed at that orientation.

For each directional variogram, only point pairs are considered that are oriented in the direction of the variogram. For two observation locations xi, $xi + h$, the orientation is defined as the angle between the vector u connecting xi and $xi + h$, and a vector along the first-dimension axis: $e= [1,0]$. The cosine of the orientation angle θ can be calculated using Equation 2:

$$\cos(\theta) = \frac{u \circ e}{|e|.|(0,1)|} \tag{2}$$

The directional variogram finally defines an azimuth angle, Equation 2, and a tolerance. Any point pair which deviates less than tolerance from the azimuth is oriented in the direction of the variogram and will be used for estimation. As long as more than one directional variogram is estimated for a data sample, the difference in the estimated variogram parameters describes the degree of anisotropy. In a kriging application, the data sample can now be transformed along the main directions at which the directional variograms differ until the directional variograms do not indicate an anisotropy anymore. The common variogram of the transformed data can be used for kriging, and the interpolated field is finally transformed back.

A geometrical 2D rotation of the given coordinates is performed using Equation 3:

$$X' = X \cdot \cos\theta - Y \cdot \sin(\theta)$$
$$Y' = X \cdot \sin\theta + Y \cdot \cos(\theta) \tag{3}$$

where the $\{X, Y\}$ is the given Easting and Northing coordinates and $\{X', Y'\}$ is the rotated new coordinates (Deutsch and Journel, 1998). A geometrical 3D rotation is explained in (Neufeld and Deutsch, 2005).

Theoretical models

The most common theoretical models used in variogram fitting are Spherical (Burgess and Webster, 1980), Exponential (Journel and Huijbregts, 1976), and Gaussian model (Journel and Huijbregts 1976).

Spherical Model is given in Equation 4:

$$\gamma(h) = \begin{cases} b + C0. \left(1.5. \left(\frac{h}{a}\right) - 0.5. \left(\frac{h}{a}\right)^3\right), h < a, \\ b + C0 \ h \geq a, a := r, \end{cases} \tag{4}$$

where h is the lag distance, and b, $C0$ and a are the variogram model parameters: nugget, sill and range respectively and r is the effective range of 95 percentile of the range a.

Exponential Model is given in Equation 5:

$$\gamma(h) = b + C0. \left[1 - e^\wedge \left(-\frac{h}{a}\right)\right], \qquad a = \frac{r}{3} \tag{5}$$

where h is the lag distance, and b, $C0$ and a are the variogram model parameters: nugget, sill and range respectively and r is the effective range of 95 percentile of the range a. The formula calculating the effective range r is given in the Equation 5.

Gaussian Model is given in Equation 6:

$$\gamma(h) = b + C0. \left[1 - e^{\wedge} \left(- \left(\frac{h}{a} \right)^2 \right) \right], \qquad a = \frac{r}{2} \tag{6}$$

Where h is the lag distance, and b, $C0$ and a are the variogram model parameters, nugget, sill and range respectively and r is the effective range of 95 percentile of the range a. The formula calculating the effective range r is given in the Equation 6.

Fitting a theoretical model

Variogram models can be fitted by using trust region algorithm (Branch, Coleman and Li, 1999), which is a bounded least square algorithm. The objective in variogram model fitting is to minimise the difference between the empirical variogram $\gamma\,emp\,(h)$ and the variogram model $\gamma\,model\,(h;\,A)$ where (h) is the lag distance and A represents the parameters of the model (eg range: a, sill: $C0$, and nugget: b). The fitting can be formulated as a weighted least squares problem in Equation 7:

$$\min A \sum_{i=0}^{n} w(h_i)[\Upsilon emp(h_i) - \Upsilon model(h_i; A)]^2 \tag{7}$$

Unlike global optimisation, the trust region approach focuses on making iterative improvements within a 'trust region' around the current parameter estimate. This region is adjusted based on the success of previous iterations. The ΔA is proposed as an update step and $\Upsilon model(h; \Delta A)$ as model update. If the update reduces the weighted residual sum of squares, accept the $\Delta\theta$ and possibly expand the trust region. If not, reduce the trust region and adjust ΔA. The algorithm iteratively updates A by solving the subproblem defined by the trust region constraints until convergence criteria are met as the changes in the cost function are below a threshold. Ensuring better fit at closer lag distance $w(h) = \frac{1}{1+h}$ is applied. Additionally, the algorithm bounds the values to ensure all the parameters are non-negative.

KRIGING ESTIMATION

Matheron (1963) introduced the Kriging estimator as a linear combination of observed values, formulated to estimate unknown values at new locations is given in Equation 8:

$$\hat{Z}(x0) = \sum_{i=0}^{N} \lambda i\, Z(xi) \tag{8}$$

where $\hat{Z}(x0)$ represents the estimated value at location $x0$, λi are the weights assigned to the observed values $Z(xi)$ at known locations xi and N denotes the number of observed data points. The calculation of the Kriging weights λi ensures that the estimator is unbiased, and the estimation variance is minimised. The condition for unbiasedness (Cressie, 1993) is defined by Equation 9:

$$\sum_{i=1}^{N} \lambda i = 1 \tag{9}$$

To obtain the weights for one unobserved location, a system of equations called the kriging equation system (KES) is formulated in Equation 10, assuming the prediction errors to be 0 (Equation 10) by and substituting Equation 8:

$$E[* (x0) - Z(x0)] = 0 \tag{10}$$

The final kriging, Equation 11, is taken from Montero, Fernández-Avilés and Mateu (2015):

$$\begin{cases} \sum_{j=1}^{N} \lambda_j\, \gamma(xi - xj) + \alpha = \gamma(xi - x0), i = 1, \dots, N \\[2em] \sum_{j=1}^{N} \lambda_j = 1 \end{cases} \tag{11}$$

where α is the Lagrange multiplier to solve the KES. By minimising the prediction variance and requiring the weights to sum to one as in Equation 8, the best linear, unbiased estimation can be obtained.

ALGORITHM

Algorithm 1 explains step by step how to create a LVA field by a given coordinates and sample values and OK with the created LVA field.

Algorithm 1: OK with ASE algorithm

Require: A data set of spatial coordinates and associated values. Transformed the data for better variogram fit if necessary.

1: Initialisation:

Convert azimuth degrees θ_i to radians for computational purposes:

$$radi = \frac{\pi}{180} \times \theta_i$$

2: Directional Variogram Computation:

Compute the directional variogram $V(\theta_i)$ for each azimuth θ_i, given a tolerance, and number of lags:

$V(\theta_i)$=coordinates, values (eg grade), θ_i, 15°, 30°, 10, 'model'

3: Model Definitions:

Define the mathematical formulas for the variogram models:

Spherical Model (Equation 4)

Exponential Model (Equation 5)

Gaussian Model (Equation 6)

4: Parameter Estimation:

For each model type M, estimate parameters b (nugget), $C0$ (sill), and a (range) by minimising the residual sum of squares between the experimental and the modelled semi-variance: (Equation 7).

Calculate the root mean square error (RMSE$_M$) for each model fit to evaluate the accuracy:

$$\text{RMSE}_M = \sqrt{\frac{1}{N}\Sigma(\gamma\,exp - \gamma\,M)^2}$$

5: Fitting Process:

Use the trust region algorithm (Equation 7).

6: Determination of Best Fitting Models:

Determine the best fitting model G based on the lowest RMSE$_M$ and longest range for each azimuth θ_i:

G=arg min(RMSE$_M$–range$_M$)

Filter the data at each estimation location by the model G's range.

7: Local Search Ellipse Definition:

Iterate steps 1 to 6 at each estimation location. This step defines the local search ellipses where the parameters of the best model G at local estimation locations as azimuths, ranges define the direction, and the length of the search ellipses and the perpendicular model's range is the shortest length of the ellipses at each estimation location. Then, use the local searches if the RMSE is smaller than the Global Variogram's RMSE to ensure to use only best fitting variograms, and avoid if the local variogram has not been formed due to lack of samples at a given location. This step of the algorithm can also be detailed with the search strategies given in Vann, Jackson and Bertoli (2003).

8: Ordinary Kriging:

Perform Ordinary Kriging (Equation 8 to Equation 11) at each estimation location by using the individual search ellipses and individual models (from step 7) at each estimation location by considering coordinate rotations in Equation 3.

End Algorithm

CASE STUDY

Data

Walker Lake data set, introduced by Isaaks and Srivastava (1989), is renowned for its comprehensive spatial coverage and consists of two distinct sample sets. The primary set comprises 470 randomly distributed observations, each defined by X and Y coordinates and an associated value (V), representing geospatial data points across the surveyed area.

Additionally, the exhaustive data set contains 78 000 samples, systematically collected on a dense 1 m grid, capturing fine scale spatial variability across the entire study area. To enable direct comparison with results at a coarser spatial resolution, this exhaustive data set was aggregated. Specifically, groups of 100 samples (10 × 10) were averaged to generate a single representative sample per 10 m grid cells to mimic the block grades in mineral resources. This aggregation reduces spatial resolution but facilitates effective comparison of geostatistical estimates at a 10 m scale, aligning with the resolution commonly used in practical geospatial analyses.

Results

Upon completion of the algorithmic procedure (from step 1 through step 8), the optimal filtering radius for further variogram analysis at each estimation location was determined, resulting in specific variogram parameters such as Model Type, Nugget, Sill, Range, and RMSE for each 15° azimuth with a 30° tolerance angle, as detailed in Table 1. These parameters were derived from directional variograms illustrated in Figure 3. Based on the minimisation formula outlined in Step 6 of the algorithm 1, the optimal model was identified as a Gaussian model at 165° with a 76 m range. This selected range facilitated the creation of a circular filter with a radius of 76 m around each estimation point.

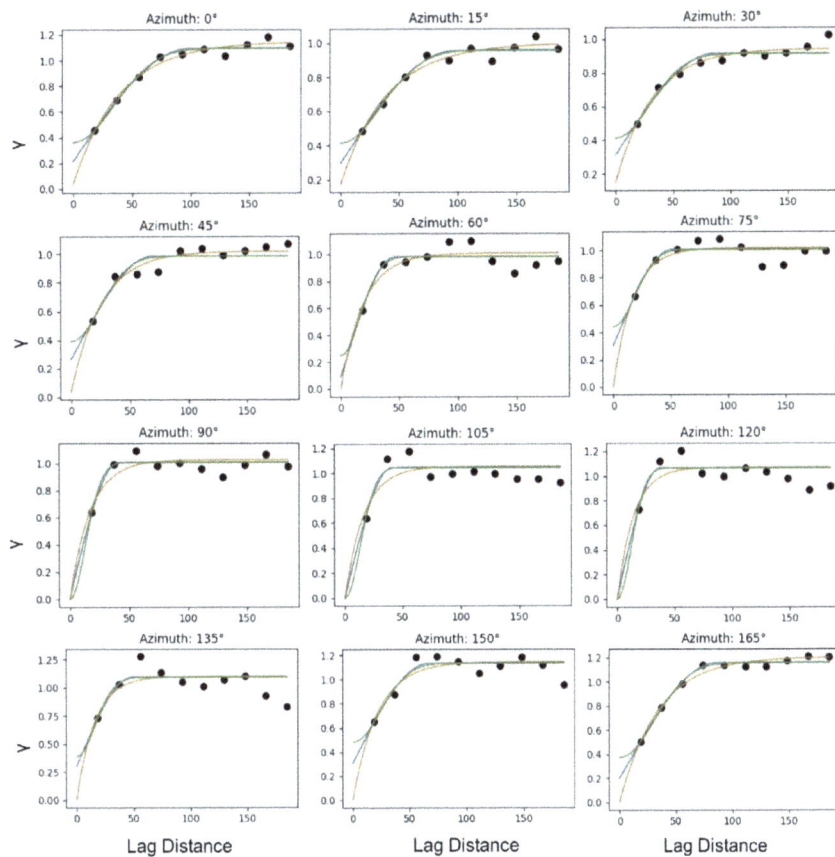

FIG 3 – Automated directional variogram analysis conducted at 15° intervals across all available samples. Theoretical model parameters derived from the analysis are documented in Table 1. Among the models evaluated, the Gaussian model (represented by the green line) at 165° azimuth was identified as the most suitable. Model comparisons are visually represented with different colours: red for Exponential, blue for Spherical, and green for Gaussian models.

Subsequent implementation of the ASE algorithm yielded comprehensive variogram parameters for each azimuth at the various estimation locations. For instance, the location (105, 125) was randomly chosen to demonstrate local application the algorithm 1. The variogram analysis for this location is presented in Figure 4, with the corresponding models and parameters marked with an asterisk (*) in Table 1. Following the guidelines of Step 7 in the algorithm 1, the optimal model for this location was determined to be an Exponential model oriented at 60° azimuth. Figure 5a illustrates the data filtering process and the resultant search ellipse at (105, 125), additionally Figure 5b depicts adjacent search ellipses generated post-algorithm application.

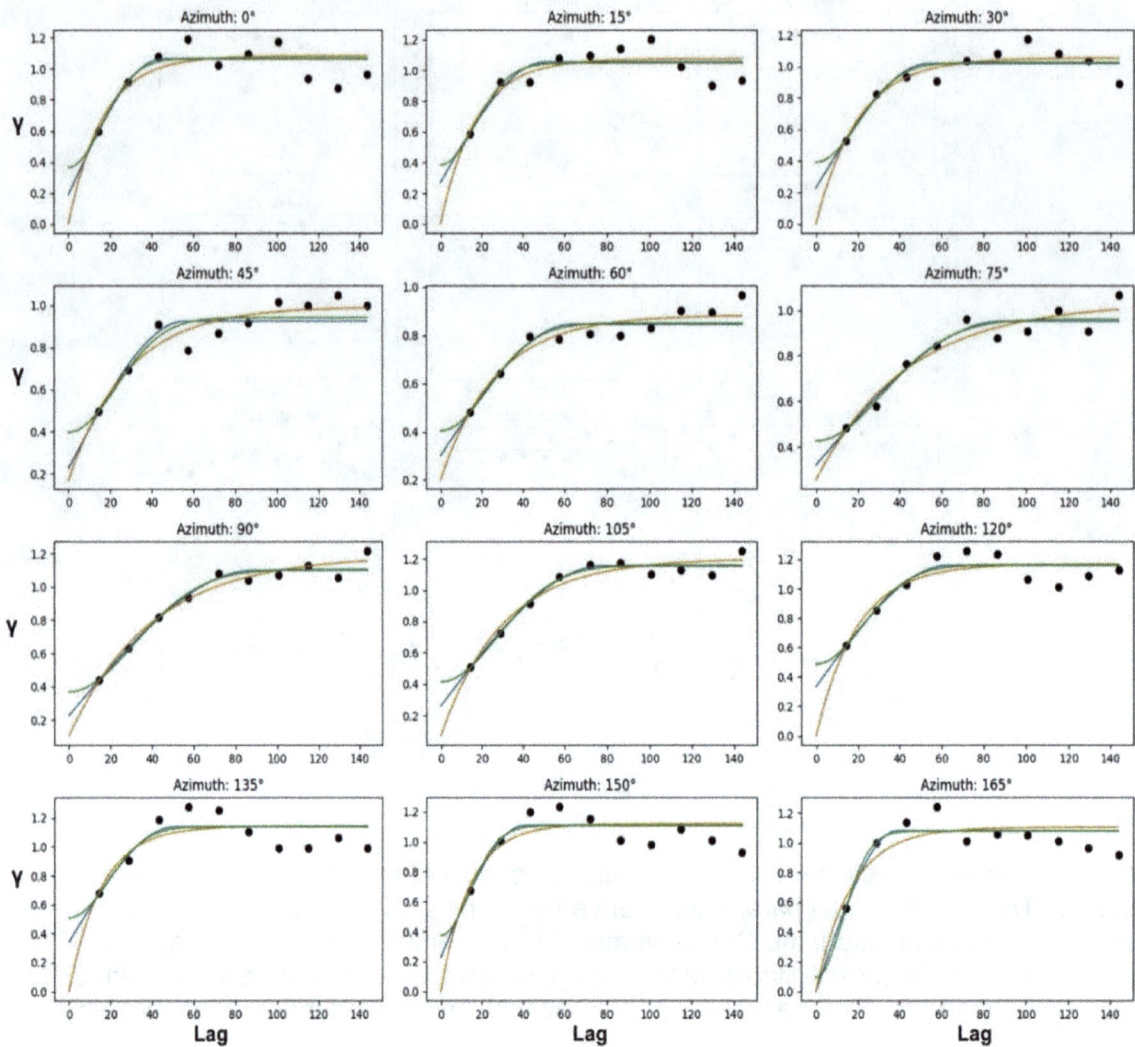

FIG 4 – Automated directional variogram analysis performed at 15° intervals around the point (105, 125), utilising a 76 m radius for data selection. A total of 144 samples were included in this study. Theoretical model parameters derived from the analysis are documented in Table 1 with * columns. Among the models evaluated, the Exponential model (represented by the red line) at 60° azimuth was identified as the most suitable. Model comparisons are visually represented with different colours: red for Exponential, blue for Spherical, and green for Gaussian models.

TABLE 1

Directional variogram model parameters which is used to determine the best model. Azimuth is 165° and Gaussian model is selected as a best model. Model Types are S: Spherical, E: Exponential, and G: Gaussian; Azimut: 0° is North and then clockwise. *Local directional variogram results after locally filtering the data and Azimuth is 60° and Exponential model is selected as a best model.

Azimuth	Model type	Nugget	Sill	Range	RMSE	Nugget*	Sill*	Range*	RMSE*
0	S	0.211	1.093	98.390	0.0379	0.188	1.065	45.0	0.0963
	E	0.030	1.147	39.164	0.0344	0.000	1.084	16.7	0.1081
	G	0.362	1.098	85.771	0.0363	0.365	1.067	39.7	0.0973
15	S	0.295	0.953	96.830	0.0388	0.271	1.055	50.6	0.0932
	E	0.172	0.998	39.791	0.0375	0.001	1.077	17.4	0.0933
	G	0.413	0.959	85.950	0.0367	0.385	1.050	41.1	0.0913
30	S	0.312	0.916	86.373	0.0428	0.229	1.019	54.3	0.0799
	E	0.147	0.945	31.996	0.0304	0.000	1.053	20.2	0.0733
	G	0.413	0.915	73.145	0.0424	0.392	1.025	49.2	0.0775
45	S	0.266	0.986	69.668	0.0627	0.226	0.925	55.9	0.0768
	E	0.037	1.019	25.572	0.0439	0.170	0.994	28.3	0.0555
	G	0.391	0.985	59.435	0.0604	0.398	0.942	55.9	0.0707
60	S	0.090	0.981	47.282	0.0697	0.300	0.846	63.7	0.0529
	E	0.000	1.009	19.660	0.0775	**0.200**	**0.885**	**27.0**	**0.0410**
	G	0.252	0.983	40.935	0.0696	0.404	0.850	56.9	0.0507
75	S	0.304	1.006	51.555	0.0631	0.320	0.951	88.8	0.0534
	E	0.000	1.018	16.773	0.0684	0.255	1.032	43.9	0.0509
	G	0.442	1.007	45.153	0.0633	0.427	0.960	79.0	0.0508
90	S	0.004	1.007	41.115	0.0513	0.227	1.101	89.3	0.0470
	E	0.000	1.026	17.216	0.0690	0.104	1.188	40.1	0.0440
	G	0.001	1.008	32.145	0.0516	0.369	1.106	77.5	0.0455
105	S	0.000	1.043	41.184	0.0813	0.259	1.154	78.2	0.0415
	E	0.000	1.054	16.636	0.1052	0.071	1.201	30.4	0.0530
	G	0.000	1.048	32.808	0.0838	0.413	1.158	68.3	0.0431
120	S	0.001	1.064	37.200	0.0955	0.334	1.160	63.2	0.0760
	E	0.000	1.069	14.300	0.1085	0.000	1.169	19.6	0.0877
	G	0.000	1.065	29.733	0.0967	0.487	1.162	55.3	0.0787
135	S	0.302	1.096	49.059	0.1193	0.344	1.138	50.7	0.1057
	E	0.000	1.100	15.685	0.1265	0.000	1.143	15.6	0.1158
	G	0.390	1.093	39.441	0.1186	0.508	1.140	45.2	0.1082
150	S	0.307	1.131	66.851	0.0737	0.228	1.110	40.8	0.0969
	E	0.000	1.144	21.920	0.0862	0.000	1.120	14.2	0.1089
	G	0.483	1.135	60.745	0.0772	0.372	1.107	34.3	0.0969
165	S	0.199	1.157	86.370	0.0278	0.000	1.079	38.9	0.0899
	E	0.000	1.206	33.975	0.0301	0.000	1.102	16.8	0.1141
	G	**0.374**	**1.161**	**75.904**	**0.0277**	0.093	1.076	31.0	0.0890

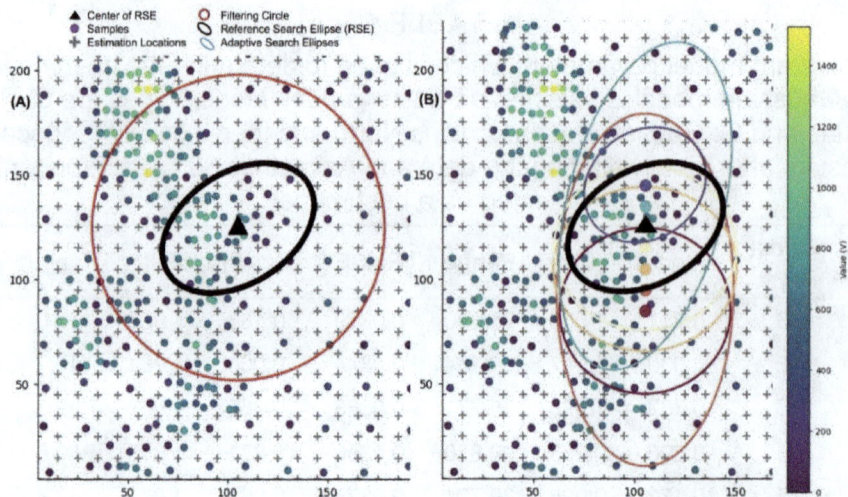

FIG 5 – Walker Lake sample data plots illustrating: (a) A filtering circle (red) with a radius of 76, centred at a randomly selected estimation location (105, 125) along with the resulting search ellipse after filtering the samples at this location; (b) An example of multiple local search ellipses, each depicted in a distinct colour, spanning from (105, 75; 105, 145 at 10 m intervals).

Furthermore, Figure 6 illustrates the LVA field, presenting the RMSEs of theoretical model fits as obtained through the ASE algorithm. This data facilitates the identification of the most precise directions for search ellipses and models at each estimation location. Additionally, it provides detailed parameters, including the nugget, sill, and range values of the optimal model, as well as the range of the model perpendicular to the best direction. Ultimately, 312 Spherical, 212 Exponential, and 201 Gaussian models, each with their specific parameters, were employed in local OK estimation.

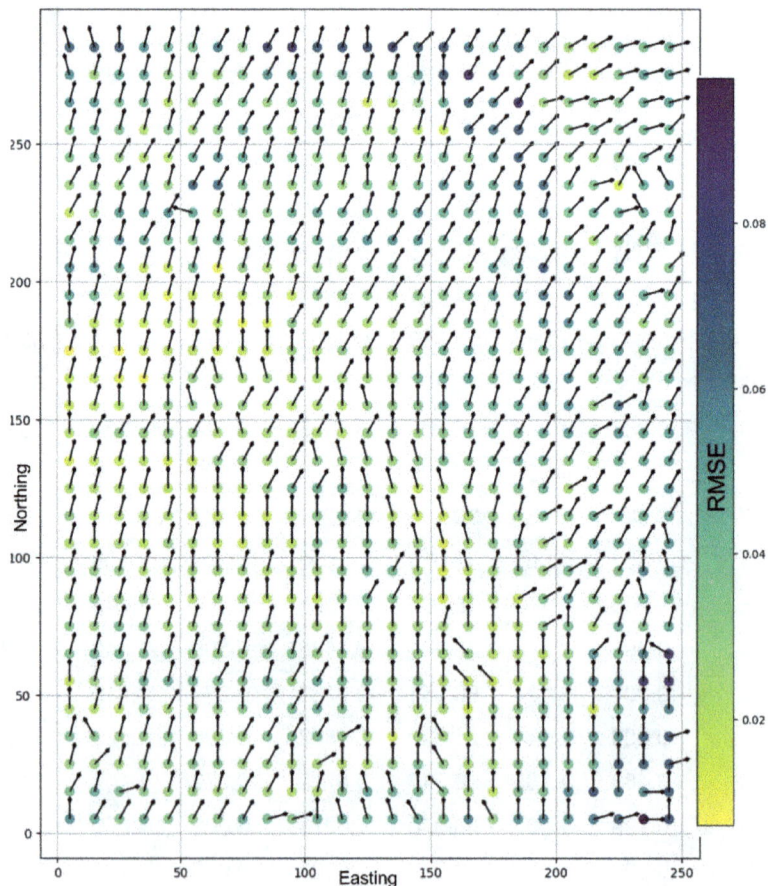

FIG 6 – Locally varying anisotropy field which resulted from adaptive search ellipsoid algorithm. The arrows show the most continuous similarity directions.

MODEL PERFORMANCE

Accuracy of the ASE algorithm has been evaluated by calculating r-squared between exhaustive data, OK Estimation with Single Search Ellipsoid and Estimation with ASE (Figure 7), visually (Figure 8), by plotting the estimated values on the case study, on the cumulative probability plots (Figure 9) and on grade-tonnage curves (Figure 10).

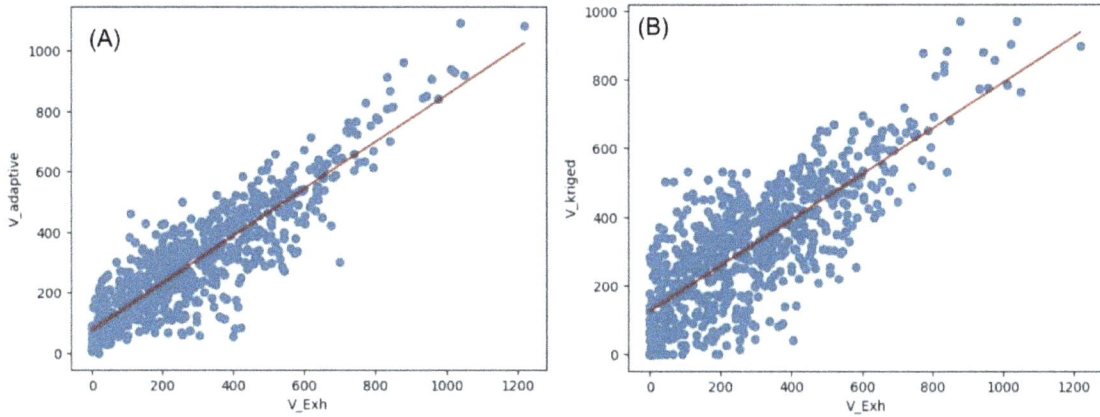

FIG 7 – Estimated V versus Exhaustive V scatter plots illustrating: (a) Estimated by ASE algorithm with an r-squared 0.79; (b) Estimated by a single theoretical model which is given in Figure 4 and Table 1. The r-squared is 0.61.

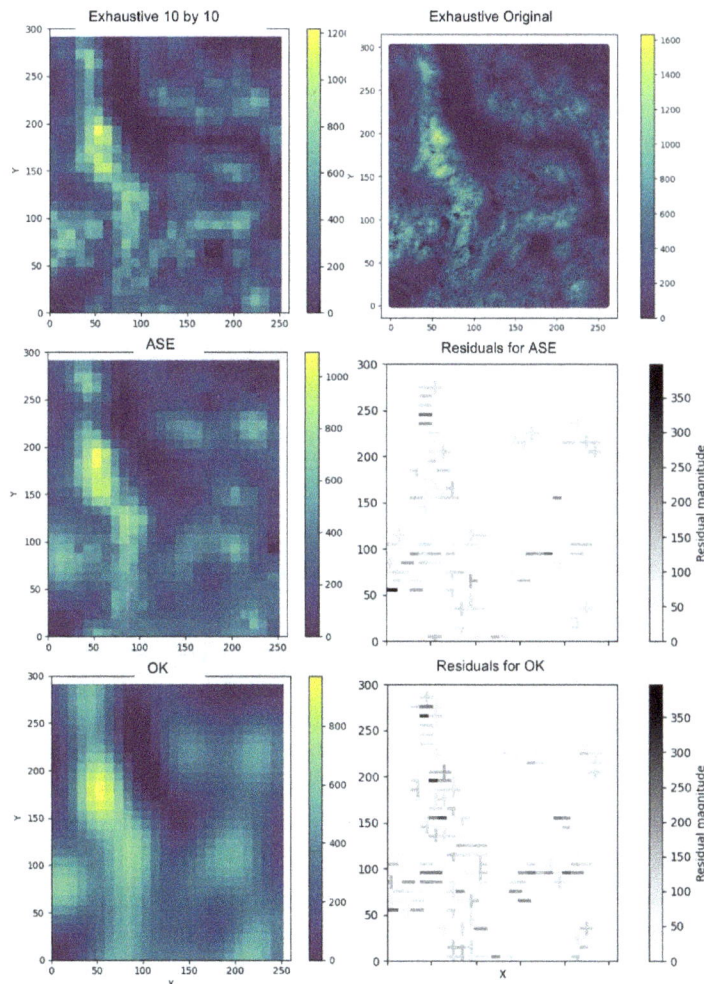

FIG 8 – Walker Lake sample plots, the maps at the top are the exhaustive data both original and the one created by averaging the 1 × 1 samples into 10 × 10 grids, the maps on the middle shows ASE estimation map and residuals at 300 cut-offs, the maps on the bottom shows OK estimation with a single search ellipsoid and the residuals at 300 cut-offs.

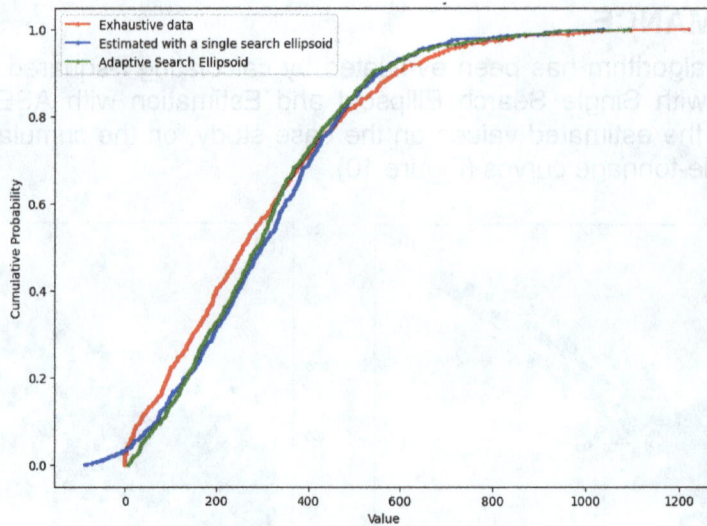

FIG 9 – Cumulative probability plots of exhaustive data, single search ellipsoid estimation and estimation with ASE algorithm.

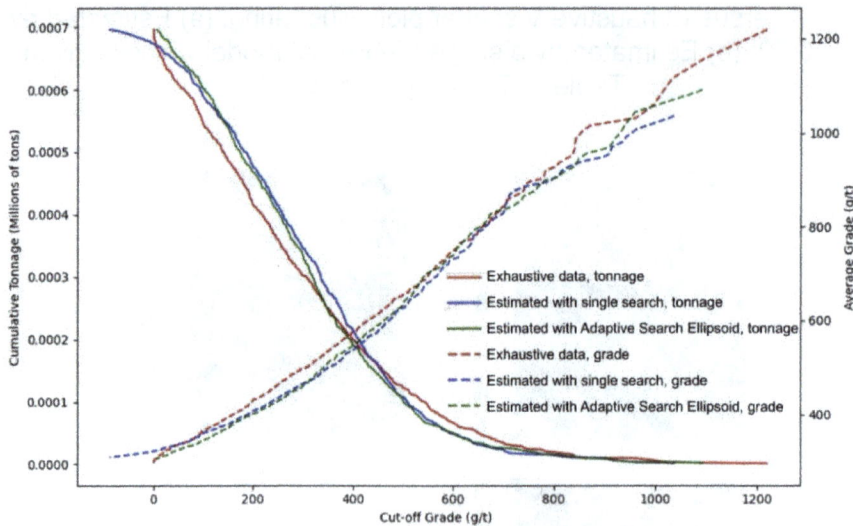

FIG 10 – Grade-Tonnage curves of exhaustive data, estimations with a single search ellipsoid and estimation with ASE algorithm.

Even with the small number of samples (470) and on the 2-dimensional application would express distinctive results in overall estimation. The estimation with ASE results with 79 per cent in r-squared and 61 per cent in r-squared when we use single search ellipsoid (Figure 7).

Visual evaluation of the estimates indicates that the Adaptive Search Ellipsoid (ASE) algorithm outperforms classical estimation methods that rely on a single search ellipsoid, particularly in areas with abrupt grade changes (see Figure 8). The residual maps (Figure 8) further demonstrate that ASE estimation is more accurate than single search ellipsoid estimation. While the ASE algorithm effectively captures higher-grade samples in complex locations, the single search ellipsoid approach tends to smooth estimated grades where grades change abruptly in exhaustive data sets.

The ASE algorithm introduces a novel method for generating LVA fields, utilising only available sample points. This approach aligns with the principles of Kriging with a trend, as demonstrated in previous studies by Eriksson and Siska (2000), Stroetv and Snepvangers (2005), and Lillah and Boisvert (2013). These researchers employed external LVA fields derived through techniques such as moving window analysis, field studies to identify structural trends, and machine learning algorithms. In contrast, our method uniquely automates LVA field creation through directional variogram analysis and automatic variogram fitting, representing a pioneering advancement in geostatistical analysis. This study also highlights the importance of automatic variogram fitting in improving data reproducibility and reducing fitting errors typically associated with manual processes.

The cumulative probability plot (Figure 9) reveals that the ASE estimation trend closely aligns with the exhaustive data set in most cases. Similarly, the grade-tonnage curve (Figure 10) for ASE estimation exhibits trends more consistent with the exhaustive data compared to the single search ellipsoid estimation.

The efficiency of the ASE algorithm could be further improved by incrementally increasing the radius of the data filtering circle while simultaneously narrowing the directional variogram window. Currently, implementing this algorithm on the Walker Lake data set takes approximately one hr on a system with 64 GB RAM and an Intel® Core™ i9-14900HX processor running at 2.20 GHz. Advances in computational technology are expected to facilitate the development of more sophisticated algorithms. These future algorithms could optimise data filtering areas, utilise narrower directional variogram windows, and support more comprehensive theoretical model fits, particularly for large-scale data sets.

Furthermore, a three-dimensional extension of the ASE algorithm is under development. This extension aims to refine the methodologies introduced in this study, offering enhanced insights into spatial variability and anisotropy in geostatistical applications. It will also enable the creation of three-dimensional LVA fields, which are challenging to develop using field studies or machine learning algorithms.

Consequently, the ASE algorithm shows significant promise for improving the accuracy of mineral resource estimation, with far-reaching implications for downstream processes, including mineral reserve calculations, investment decisions, mine planning, scheduling, and reclamation.

CONCLUSION

ASE algorithm introduced in this study offers a robust framework for analysing local variograms by evaluating their specific fitting errors (RMSE) and the associated Kriging estimation inaccuracies. This innovative algorithm facilitates the creation of a LVA field solely based on available primary data, eliminating the reliance on ancillary data sets. Applied to the two-dimensional Walker Lake data set, which comprises 470 samples distributed across 720 estimation points arranged in a 10 × 10 grid representative of block model centre in mineral resource estimation, the ASE algorithm significantly enhances estimation accuracy. This enhancement is demonstrated through improved performance metrics, underscoring the algorithm's effectiveness in geostatistical modelling and its utility in mineral resource estimation.

REFERENCES

Abzalov, M, 2006. Localised uniform conditioning (LUC): A new approach for direct modelling of small blocks, *Mathematical Geology*.

Branch, M A, Coleman, T F and Li, Y, 1999. A subspace, interior, and conjugate gradient method for large-scale bound-constrained minimization problems, *SIAM J Sci Comput*, 21:1–23.

Burgess, T M and Webster, R, 1980. Optimal interpolation and is arithmetic mapping of soil properties, I, The semi-variogram and punctual kriging, *J Soil Sci*, 31:315–331.

Cressie, N, 1993. *Statistics for Spatial Data* (Wiley).

Deutsch, C V and Journel, A G, 1998. *GSLIB, Geostatistical software library and user's guide* (Oxford University Press).

Deutsch, C V and Lewis, R W, 1992. Advances in the practical implementation of indicator geostatistics, in *Proceedings of the 23rd Application of Computers and Operations Research in the Mining Industry Symposium*, pp 169–179.

Eriksson, M and Siska, P P, 2000. Understanding anisotropy computations, *Mathematical Geology,* 32:683–700.

Hardtke, W, Allen, L and Douglas, I, 2011. Localised Indicator Kriging, in 35th APCOM Symposium, Wollongong, NSW.

Isaaks, E H and Srivastava, R, 1989. *An Introduction to Applied Geostatistics*, 561 p (Oxford University Press: New York).

Journel, A G and Huijbregts, C J, 1976. *Mining Geostatistics* (New York: Academic Press).

Krige, D G, 1951. A Statistical Approach to Some Basic Mine Evaluation Problems on the Witwatersrand, *J Chem Metall Min Soc South Africa*, 52:119.

Lillah, M and Boisvert, J, 2013. Inference of 2D and 3D Locally Varying Anisotropy Fields, GeoConvention 2013: Integration.

Mälicke, M, 2022. SciKit-GStat 1.0: a SciPy-flavoured geostatistical variogram estimation toolbox written in Python, *Geosci Model Dev*, 15:2505–2532.

Matheron, G, 1962. Traite de Geostatistique Appliquee, tome 1:111, Paris, France: Editions Technip.

Matheron, G, 1963. Principles of Geostatistics, *Economic Geology*, 58(8):1246–1266.

Matheron, G, 1965. Les Variables Régionalisées et leur Estimation: Une Application de la Théorie des Fonctions Aléatoires aux Sciences de la Nature, Paris, Masson et Cie.

Montero, J-M, Fernández-Avilés, G and Mateu, J, 2015. *Spatial and spatio-temporal geostatistical modeling and kriging* (John Wiley & Sons).

Neufeld, C and Deutsch, C, 2005. Calculating recoverable reserves with uniform conditioning, in *Proceedings of GIS and Spatial Analysis – 2005 Annual Conference of the International Association for Mathematical Geology, IAMG 2005*, pp 1065–1070.

Pyrcz, M, J and Deutsch, C V, 2014. *Geostatistical reservoir modeling* (Oxford University Press).

Rossi, M and Deutsch, C V, 2014. *Mineral Resource Estimation* (Springer).

Stroetv, C and Snepvangers, J, 2005. Mapping curvilinear structures with local anisotropy kriging, *Mathematical Geology*, 37:635–649.

Sullivan, J, Satchwell, S and Ferrax, G, 2007. Grade estimation in the presence of trends; the adaptive search approach applied to the Andina Copper Deposit, Chile, in *Proceedings of the 33rd International Symposium on the Application of Computers and Operations Research in the Mineral Industry*, pp 135–143.

Vann, J, Jackson, S and Bertoli, O, 2003. Quantitative Kriging Neighbourhood Analysis for the Mining Geologist -A Description of the Method with Worked Case Examples, in *Proceedings of the Fifth International Mining Geology Conference*, pp 215–223.

Virtanen, P, Gommers, R, Oliphant, T E, Haberland, M, Reddy, T, Cournapeau, D and Burowski, E, 2020. SciPy 1.0: Fundamental algorithms for scientific computing in Python, *Nature Methods*, 17(3):261–272.

Wackernagel, H, 1998. *Anisotropy*, pp 60–63 (Springer: Heidelberg).

Xu, W, 1996. Conditional curvilinear stochastic simulation using pixel-based algorithms, *Mathematical Geology*, 28:937–949.

Geometallurgical clusters creation in a niobium deposit using Dsclus and hierarchical indicator kriging with trend

J F C L Costa[1], F G F Niquini[2], C L Schneider[3], R M Alcantara[4] and L N Capponi[5]

1. Professor, UFRGS, Porto Alegre, RS, Brazil. Email: jfelipe@ufrgs.br
2. Post Doctoral Researcher, UFRGS, Porto Alegre, RS, Brazil.
   Email: fernanda.gontijo.fn@gmail.com
3. Senior Technologist, CETEM, Rio de Janeiro, RJ, Brazil. Email: claudiol.schneider@gmail.com
4. Geologist, CBMM, Araxa, MG, Brazil. Email: rodrigo.alcantara@cbmm.com
5. Geology Manager, CBMM, Araxa, MG, Brazil. Email: luciano.capponi@cbmm.com

ABSTRACT

Mining in weathered alkaline carbonatite complexes involves the extraction of minerals in areas where geological processes have altered the original composition of the rocks. These complexes are formed by magmatic, hydrothermal, and weathering geological events, which modify the minerals present in the rocks. This geological complexity is reflected in the processing plant, which presents different behaviours in respect to the mine region that is mined. This scenario has motivated the creation of a 3D block model with geometallurgical clusters. To achieve this, four different algorithms were tested: K-Means, Hierarchical Agglomerative Clustering, Dual Space Clustering (Dsclus), and Clustering by Autocorrelation Statistics (Acclus). The first two consider only the multivariate aspects of the data, while the latter take spatial position into account. The Dsclus was the chosen method, once it proved effective in separating zones with similar metallurgical behaviours, while respecting the spatial continuity of the established clusters as well as providing more coherent results with deposit's geology. For the spatial mapping of the defined geometallurgical domains, we propose a workflow based on a well-known geostatistical framework derived from indicator kriging. This method is herein adapted from its original formulation of mapping multiple (K) categories simultaneously to a hierarchical approach, incorporating trends during the estimations in order to well reproduce the circular aspect of the deposit. The results demonstrate the potential of the applied methodologies not only to improve the understanding of the geometallurgical characteristics of the mineral deposit but also to support mine planning and optimise production processes.

INTRODUCTION

Geometallurgy is an increasing field in mining operations and most of its ascent is related to the good results achieved by the companies which applied its concepts. Among several benefits, some are the most pronounced: better use of mineral resources though more accurate definition of ore and waste materials; better management of environmental impacts; increase of the knowledge of the mineralisation and its response in the beneficiation plant. This paper aims to benefit from the latest point, by comprehending the ore and its characteristics in the process.

After data acquisition and quality check of the information available, the next step goes in the direction of understanding how and why different regions in the deposit perform differently in the beneficiation plant. To help in this task, clustering techniques are fundamental tools which, when combined with geological knowledge, can provide valuable results.

The cluster analysis is always made considering samples, usually coming from drill holes or drill dust, but is important to populate the 3D block models with the cluster information, which is used by mining planning. This case study shows a way of fulfill the block model using hierarchical indicator kriging with an additional step, that is the trend model, necessary to better reproduce the deposit's characteristics.

Geological aspects

The world's largest niobium mineral deposits are associated with carbonatite alkaline intrusive rocks, with pyrochlore being the main niobium-bearing mineral. The Araxá niobium mine is related to the Barreiro Carbonatite Alkaline Complex, also known as the Araxá Carbonatite Alkaline Complex, and is recognised as the largest mineral resource and ore reserve of this substance. The main geological processes forming the deposit are:

- Magmatism: Through the intrusion and evolution of a picritic magma and various liquid immiscibility processes, niobium naturally concentrated in the centre of the intrusion, mainly associated with rocks from the petrogenetic series of phoscorites and dolomite carbonatite.

- Hydrothermalism: A late fluid added more niobium to the primary mineralisation and altered the magmatic-origin pyrochlore crystals, replacing calciopyrochlore with bariopyrochlore.

- Tropical weathering: This process altered the carbonate and ferromagnesian minerals of the igneous rocks, leaching the mobile elements and generating a residual concentration of resistant minerals in the weathering zone. In addition to concentrating minerals of significant economic interest, this process also removed important contaminants (carbonates and micaceous minerals), enabling the development of a simplified, low-cost, and highly efficient mineral concentration route.

Other economically interesting substances are associated with this type of mineral deposit, such as phosphate (apatite), iron ore (magnetite), barite, titanium (anatase and ilmenite), and rare earth elements (monazite). Some of these substances are by-products of the pyrochlore and apatite concentration processes, such as magnetite and barite concentrates.

Processing flow sheet

The processing plant consists of three main separation stages that strongly impact the resulting yield and pyrochlore recovery. These are desliming, magnetic separation and flotation. Magnetite, which occurs in the form of relatively large grains, is separated in low intensity magnetic rotating drums. The desliming operation is achieved in a series of hydrocycloning stages after ball milling to reduce the particle sizes to a size that is suitable for efficient flotation concentration of pyrochlore. The flotation plant consists of a relatively complex circuit of rougher, cleaner and recleaner stages that are carried out in closed circuit with each other in order to maximise recovery at the required concentrate grade of Nb_2O_5.

Geometallurgical testing emulates the industrial plant well enough so that it is possible to determine the quality of concentrate that can be achieved from a geometallurgical sample, and the associated yield and recovery.

METHODOLOGY

Database information

The database used in this study contains 26 080 samples isotopically analysed for the following variables: Al_2O_3, Fe_2O_3, Nb_2O_5, P_2O_5 and SiO_2 in the Run-of-mine (ROM), yield, metallurgical recovery, magnetic yield, slimes yield, and Nb_2O_5 and P grades in the flotation concentrate. All flotation tests are made to produce a concentrate inside the specification grade adopted by the company, but some samples do not reach this grade. Figure 1 illustrates the sample positions in the deposit.

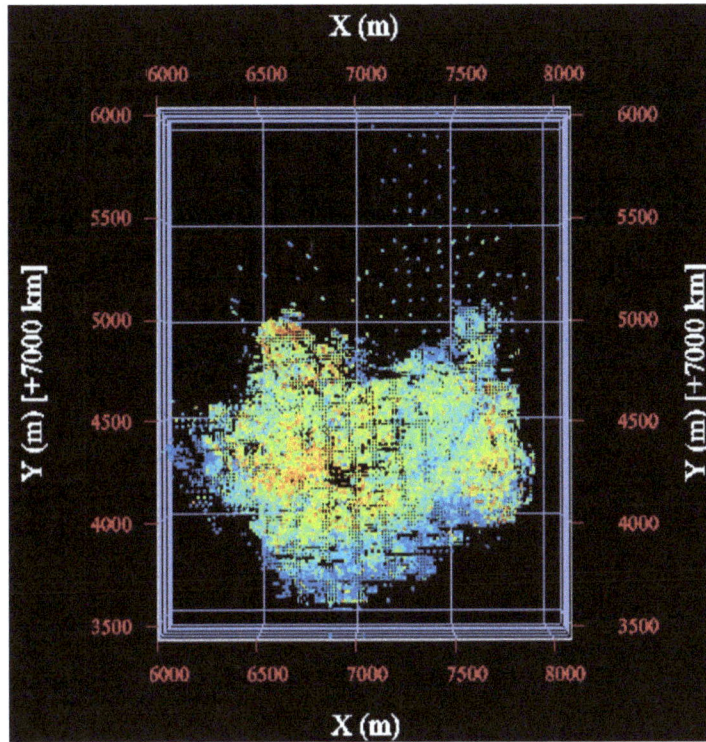

FIG 1 – Samples distribution in the deposit.

One aspect observed in data is the range of variation of the variables: while P grades in concentrate are usually below 1 per cent, the Fe_2O_3 in ROM are usually above 40 per cent. This behaviour harms cluster analysis, which will produce clusters highly influenced by the variables with high magnitude. In order to avoid this problem, a standard scaler was applied to standardise all variables.

Cluster analysis

After standardising the database, the next step was to proceed with the cluster analysis, which provided good results in similar context (Niquini *et al*, 2024). Four techniques were employed in this case study: K-means (MacQueen, 1967), hierarchical clustering (Sokal and Sneath, 1963), dual-space clustering (Martin and Boisvert, 2018) and clustering by autocorrelation statistics (Scrucca, 2005) and the results obtained with each one were evaluated using Pseudo-F (Calinski and Harabasz, 1974) and Spatial Entropy (Martin and Boisvert, 2018) metrics.

The Pseudo-F metric, represented in Equation 1, analyses the heterogeneity between clusters and the homogeneity within the cluster, being the higher the value obtained the better:

$$Pseudo\ F = \frac{(\sum_{i=1}^{k^*} n_i(\bar{x}_{i.}-\bar{x})\prime(\bar{x}_{i.}-\bar{x}))/(k^*-1)}{(\sum_{i=1}^{k^*} \sum_{j=1}^{n_i}(x_{ij}-\bar{x}_{i.})\prime(x_{ij}-\bar{x}_{i.}))/(n-k^*)} \tag{1}$$

Where the index i refers to the group (cluster) number, the index j is related to the sample number in the database, n is the number of samples, $\bar{x}_{i.}$ is the mean vector of group i, $\bar{x}$ is the global mean vector and k is the number of groups evaluated.

Spatial Entropy, represented in Equation 2, measures the spatial structure of the clusters created, without considering any multivariate aspect:

$$H_{Total} = -\sum_{i=1}^{N} \sum_{k=1}^{K} p_{i,k}\ ln\ p_{i,k} \tag{2}$$

where $p_{i,k}$ is the probability of finding a category k near to a local i, considering samples within a given search radius. The smaller the value found, the more homogeneously the clusters are distributed in space.

Figure 2 illustrates the results obtained for these four techniques, for a number of clusters varying from 2 to 5.

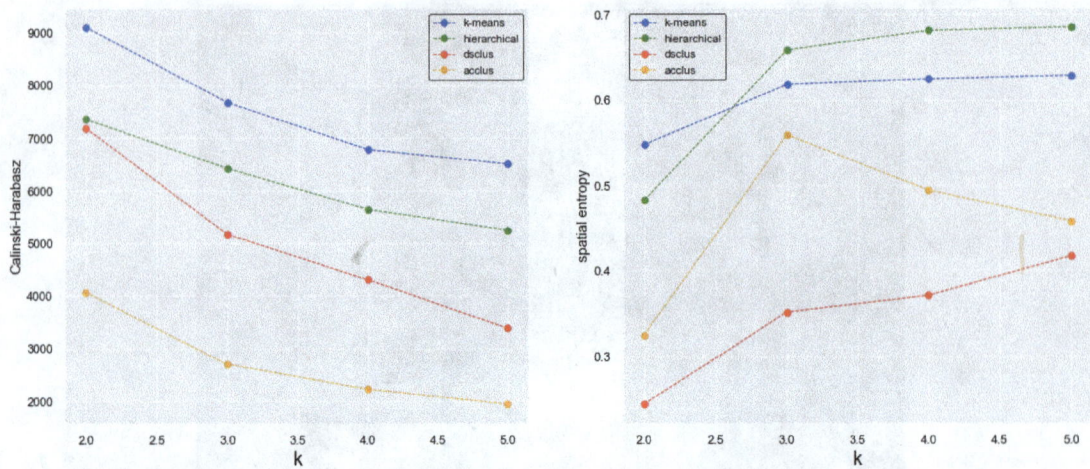

FIG 2 – Pseudo-F (left) and Spatial entropy (right) for different number of clusters and different clustering techniques.

The results obtained showed that, for any number of clusters selected, the K-Means technique (blue line) presented the best statistical results (summarised by the Pseudo-F metric). Analysing the spatial entropy, it was shown a better spatial disposal of the clusters when dual-space clustering (red line) was used, for any number of clusters selected. Figure 3 shows the spatial distribution of the clusters generated when K-means and Dsclus were used. It is remarkable that dual space clustering results resemble more closely the geological behaviour expected, once high variability in short distances is not commonly observed. Those results highlight the spatial cluster techniques superiority in reproducing geological features, even though the multivariate statistical division of samples is not as accurate as the one obtained using K-means.

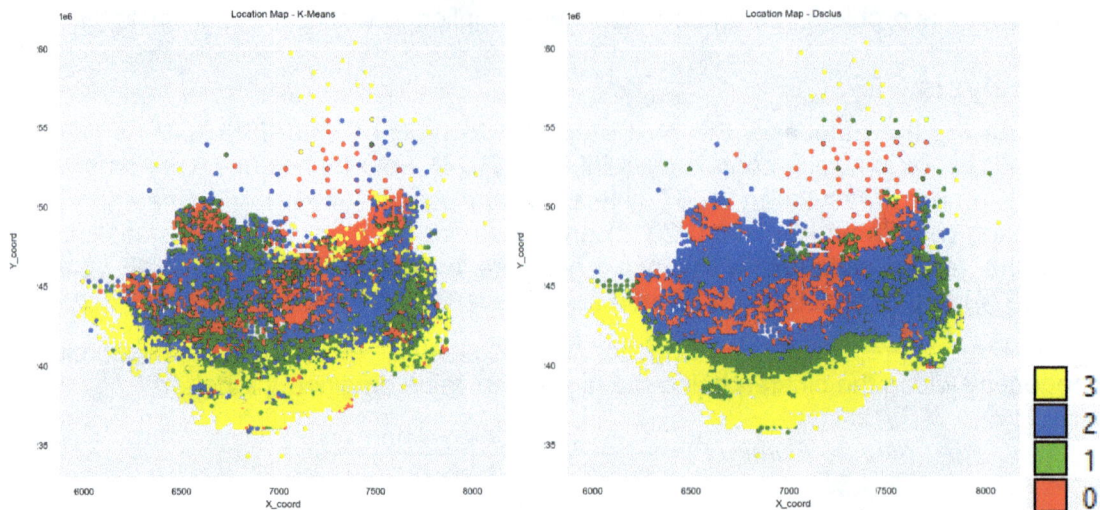

FIG 3 – Spatial distribution obtained when K-means (left) and Dsclus (right) algorithms were used to divide the samples into four clusters.

After choosing the technique, the next step was to decide the ideal number of clusters. This choice was made using the geological knowledge as the key pillar. It is known that adopting only two clusters would oversimplify the complex geology and metallurgical characteristics of the deposit. Dividing data into three domains (Figure 4, left) resulted in merging samples with high and low metallurgical recovery in the same cluster, what is undesirable. By choosing five clusters (Figure 4, right), no differences were perceived between clusters 3 and 4, being the Al_2O_3 grade in ROM the only responsible variable for this split, what has not geological/metallurgical relevance. Thus, it was decided to adopt the results with four clusters (the ones illustrated in the right side of Figure 3), once different metallurgical responses are observed in the groups and the spatial connectivity obtained is adequate for posterior modelling.

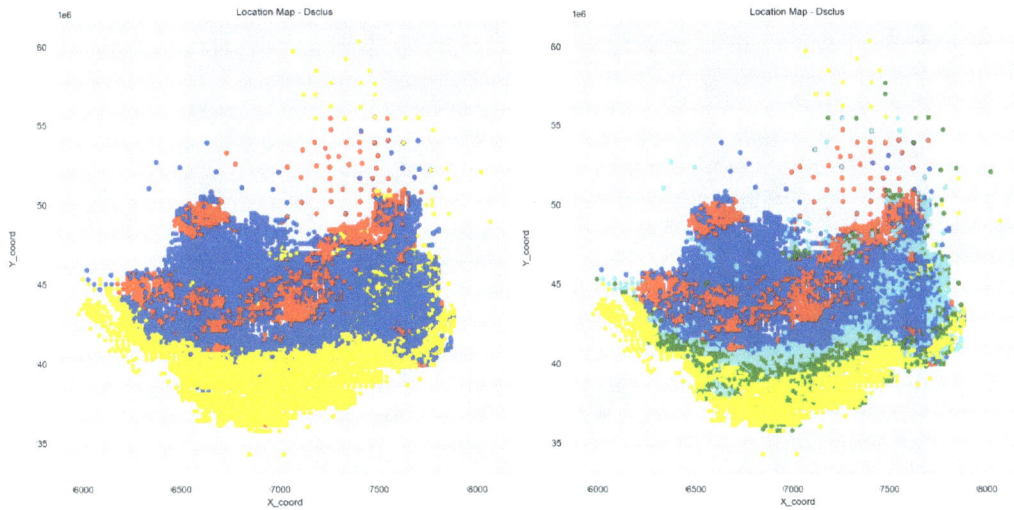

FIG 4 – Scenario with three (left) and five (right) clusters generated with dual space clustering algorithm.

Hierarchical indicator kriging

After defining each sample cluster, the next step was to interpolate this information for all blocks, using hierarchical indicator kriging (Journel, 1982). The first action needed is to create the indicator variables following the logic presented in Equation 3:

$$i(u_\alpha; k) = \begin{cases} 1, if\ cluster = k \\ 0, if\ cluster \neq k \end{cases}, for\ k = 0, 1, 2\ or\ 3 \tag{3}$$

Before starting variography and estimation, there is a need of applying a trend model once it is most probable to observe cluster 3 in the deposit edge and clusters 0 and 2 in the centre, the trend model can capture this regional tendency. Another valuable use of trend models is its use to incorporate qualitative information, as geological maps, aiming at using its information to increase model accuracy. In this case study, cluster 3 is geologically compatible with the bebedourite domain, which surrounds the deposit. It is important to ensure that the block model created reflect this characteristic. So, to better represent the geological features, it was decided to use the geological maps showing the bebedourite area in the trend model creation. By doing that, it is avoided that clusters 0, 1 or 2 appear in the bebedourite area, and, on the other hand, ensures that cluster 3 does not appear in the centre, which is dominated by the dolomitic carbonatites. Figure 5 illustrates the bebedourite points mapped by the geology team, used as secondary information in the trend model. An immediate conversion between bebedourite and cluster 3 was made, so all edge samples were flagged with this code.

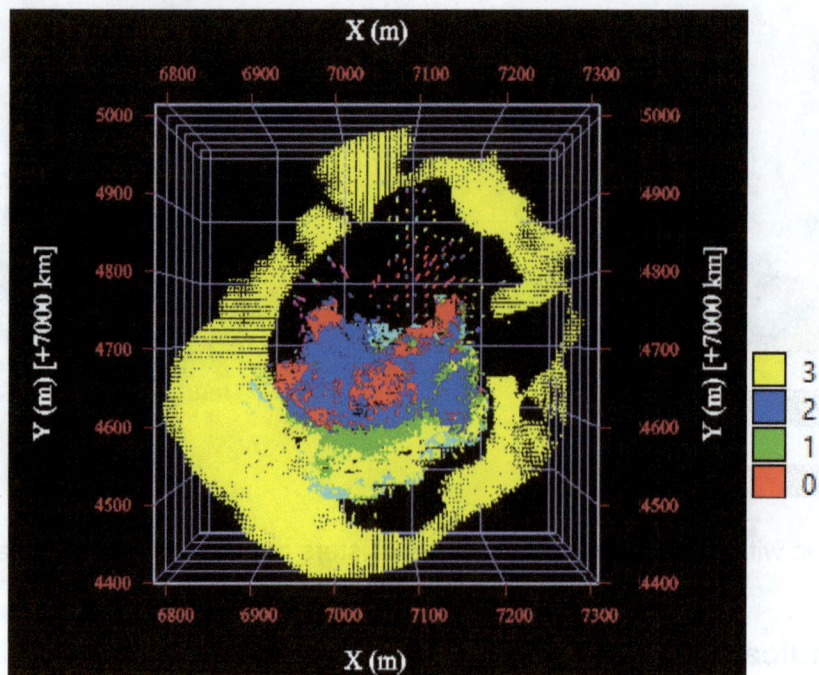

FIG 5 – Bebedourite mapped geologically at the edge of the deposit.

Just to illustrate the results obtained without the trend model, Figure 6 shows the block model coloured by cluster when it was not used. The red lines show bebedourite regions mapped geologically, so cluster 3 should be estimated there. However, due to the proximity of samples which belong to clusters 0 and 1 the forecasts were made taking into account this information, predicting erroneously the cluster category and highlighting the importance of using a trend model.

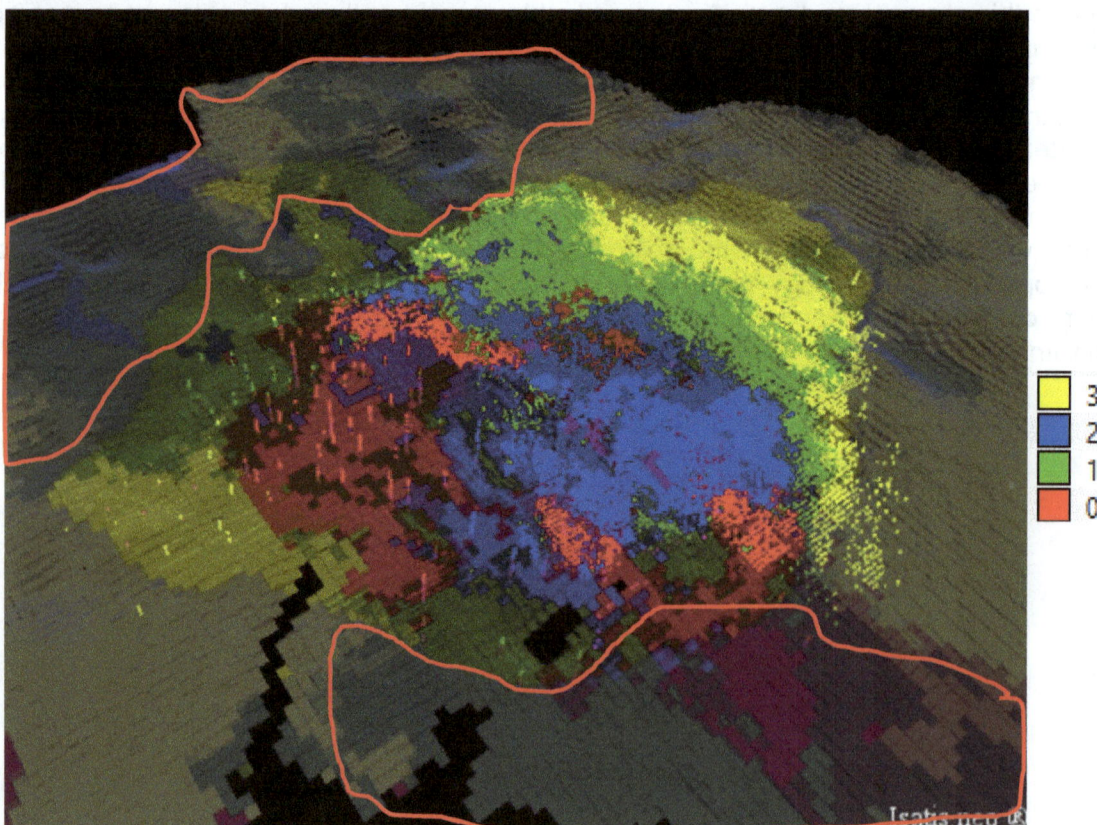

FIG 6 – Block model coloured by cluster when trend model is not used to delimit the cluster 3 area.

After demonstrating the need of adopting a trend model in the estimate, the next step was to build it. The chosen technique was the SPDE (Vergara, Allard and Deassis, 2018), which generated, for each block, its probability of belonging to each cluster, as illustrated in Figure 7, where warmer colours indicate a higher probability of finding the respective cluster at that specific location. After defining the trend probability for each block, it was necessary to define it for each sample in the database, what was done by merging the nearest block trend probabilities for the sample position. Then, the residual for each sample was calculated, using Equation 4 presented here:

$$residual(u_\alpha; k) = i(u_\alpha; k) - trend(u_\alpha; k) \qquad (4)$$

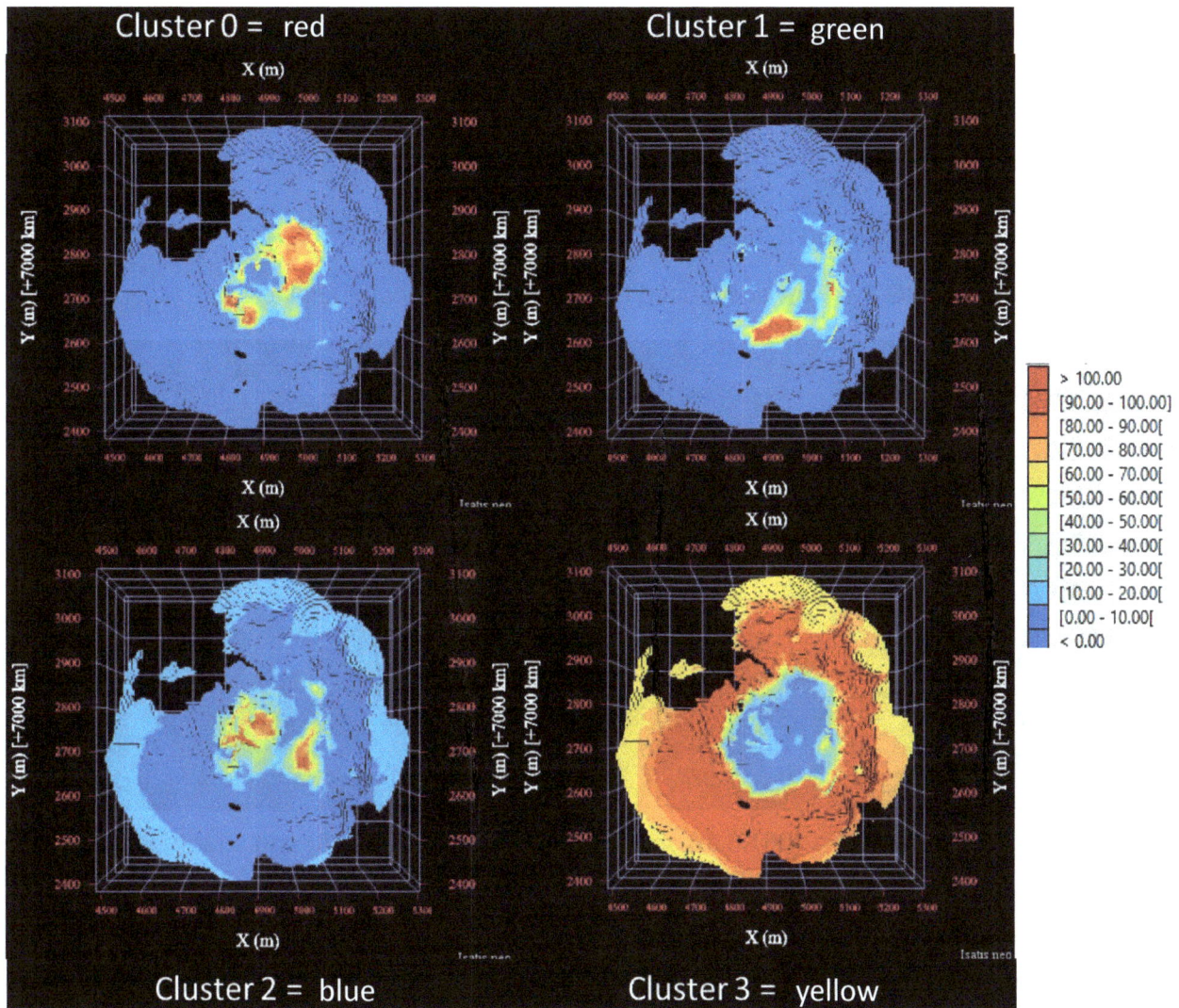

FIG 7 – Trend model created with SPDE, showing the probability of belonging to each cluster.

Figure 8 presents the flow sheet adopted in this study to assign to each block a cluster category. The variograms for each residual variable were calculated and modelled, and the ellipsoids used had dimensions equal to the variogram range in all directions, minimum number of samples to perform kriging equal to two, eight angular sectors and six samples by sector. It is important to remark that, after each residual kriging, the trend model built with SPDE need to be added in each block, to check if the value obtained return a probability equal or higher than 0.5, showing that the block has a significant probability of belonging to a given category. After going through the entire flow sheet, the validations were performed: the proportions of each category were checked to find consistency with the database, the swath plots were calculated, the cross validation was made, together with a visual inspection of the results. Then, it was perceived the need of post processing the results due to 'salt and paper effect' in some regions. This was made using MAPS (Deutsch, 1998) and the results obtained can be visualised in Figure 9.

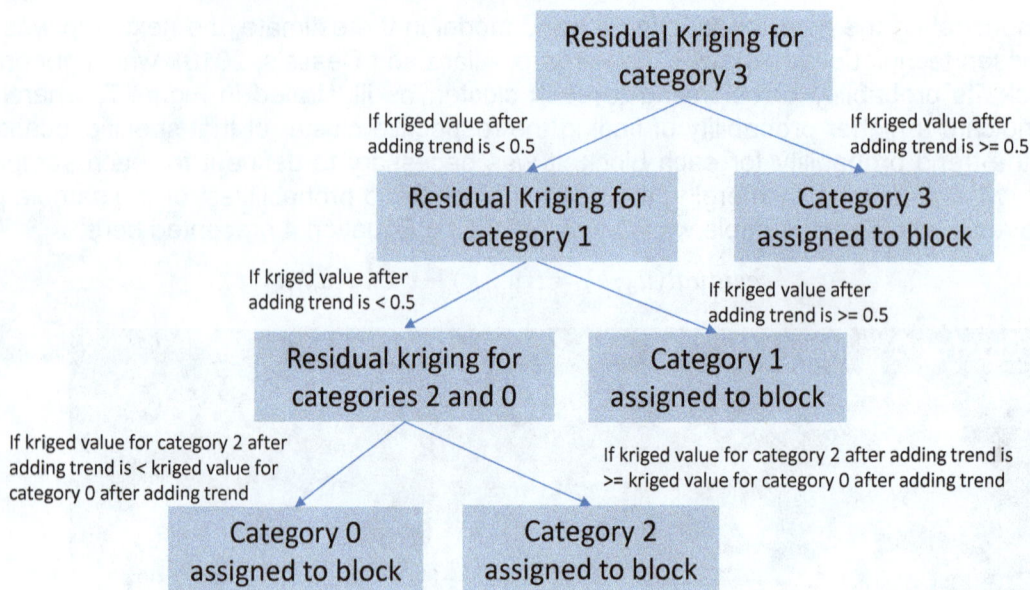

FIG 8 – Flow sheet used for hierarchical indicator kriging.

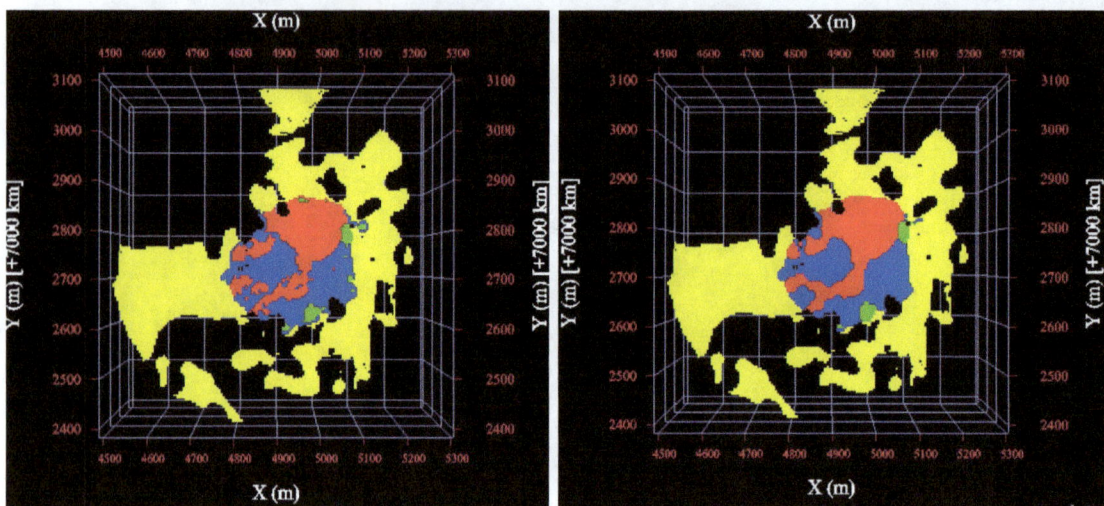

FIG 9 – Vertical section (Z = 1067.6 m) showing the results before (left) and after (right) MAPS post processing applied (Z = 1067.6 m).

RESULTS

To ensure the 3D block model built has geological meaning, the results obtained were evaluated with the aim of correlating the cluster characteristics with lithological and mineralogical aspects. The first step was to analyse the chemical and metallurgical variables' behaviour by cluster, as illustrated in Figure 10. All variables in this figure were standardised to comply with the company's confidentiality requirements, but this does not affect the validity of the conclusions.

Cluster 0 (red) is highly magnetic and have a low percentage of slimes when compared to the other clusters. Cluster 2 (blue) shows the highest mean metallurgical recovery and almost all of its samples achieve the target of Nb_2O_5 in the concentrate. Geologically, the centre zone, where clusters 0 and 2 are located, is predominated by dolomitic carbonatites with subordinated magnetites and phlogopites. Bebedourites and calcitic carbonatites are rarely found in this region. The high magnetic content observed in cluster 0 is probably related to the high presence of magnetites.

Cluster 3 (yellow) shows samples with high percentage of slimes, low mean yield and recovery, and a high number of samples that are unable to achieve the concentration target grade, proving itself as the worse plant feed for the required product. Geologically, these samples belong to a lithological domain composed of silicate rocks from the bebedourite petrogenetic series and small bodies of calcitic carbonatites in the form of ring dykes.

Cluster 1 (green), also have a high percentage of samples which do not achieve the concentrate grade target, but its metallurgical recovery and yield are much better than the observed recoveries in cluster 3. This region is known to be a transition zone between the dolomitic carbonatites to the bebedourites, gradually increasing the presence of phlogopites.

Therefore, it was noted that the geometallurgical model built reflected critical geological features, widely related to the different performances observed when the materials are fed in the processing plant.

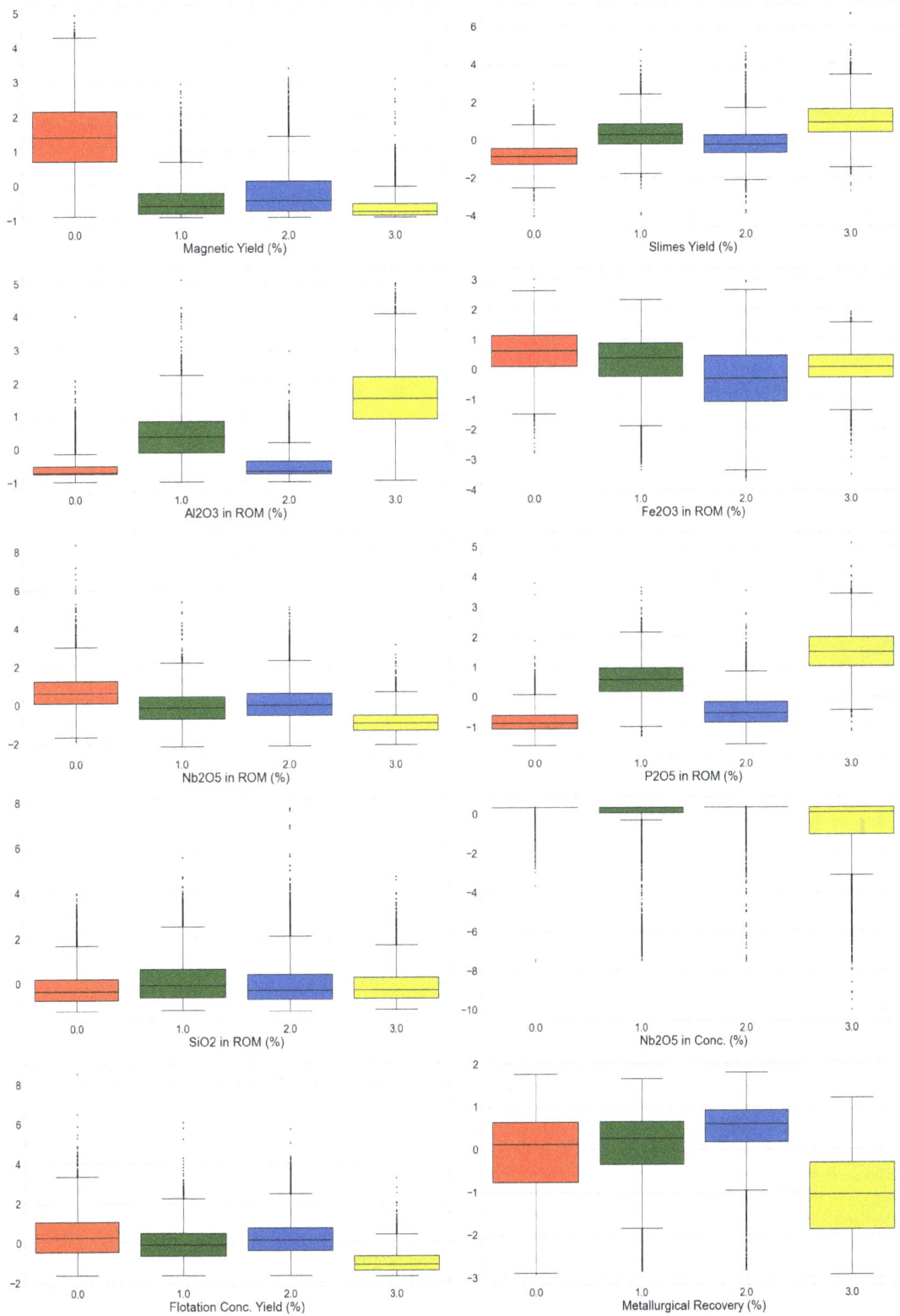

FIG 10 – Input variables box-plots coloured by cluster.

CONCLUSION

This work presents a complete workflow for constructing a block model with geometallurgical clusters, starting from cluster creation using statistical techniques, moving to 3D modelling through hierarchical indicator kriging with trend, and concluding with a geological analysis of the created model, providing key insights for each cluster.

Cluster 0 generates a much larger amount of magnetic material. Knowing this aspect is important to plan the deposition of this material in piles or study the possibility of selling the magnetic concentrates as a by-product, bringing financial benefits to the company. Cluster 3 is marked by a small flotation yield and low recovery, which is very important information for mine planning to avoid long periods of sending only this type of material to the processing plant, having as consequence a small volume of concentrate. This cluster is also marked by the highest percentage of slimes. Cluster 2 usually provides excellent metallurgical recovery and high yield, making it the best ore cluster. Cluster 1 also presents good metallurgical performance, but generates a slightly higher percentage of slimes when compared to clusters 0 and 2, being a point of attention.

REFERENCES

Calinski, T and Harabasz, J, 1974. A dendrite method for *cluster* analysis, *Commun Stat – Theor M*, 3(1):1–27. https://doi.org/10.1080/03610927408827101

Deutsch, C V, 1998. Cleaning categorical variable (lithofacies) realizations with maximum a-posteriori selection, *Computers and Geosciences*, 24(6):551–562.

Journel, A G, 1982. The indicator approach to estimation of spatial data, Proceedings of the 17th APCOM, pp 793–806 (Port City Press: New York).

MacQueen, J, 1967. Some methods for classification and analysis of multivariate observations, in *Fifth Berkeley Symposium on Mathematical Statistics and Probability,* Proceedings, Oakland, 1(14):281–297.

Martin, R and Boisvert, J, 2018. Towards justifying unsupervised stationary decisions for geostatistical modeling: ensemble spatial and multivariate clustering with geomodeling specific clustering metrics, *Comput Geosci*, 120:82–96. https://doi.org/10.1016/j.cageo.2018.08.005

Niquini, F G F, Andrade, I A, Costa, J F C L, Silva, V M and Marcelino, R S, 2024. A workflow to create geometallurgical clusters without looking directly at geometallurgical variables, Minerals Engineering, Volume 222:109171. https://doi.org/10.1016/j.mineng.2024.109171

Scrucca, L, 2005. Clustering multivariate spatial data based on local measures of spatial autocorrelation, *Quademi Del Dipartimento Di Economia,* Finanza e Statistica, Università Di Perugia, 20(1):11.

Sokal, R R and Sneath, P H A, 1963. *Principles of numerical taxonomy* (W H Freeman).

Vergara, R C, Allard, D and Desassis, N, 2018. A general framework for SPDE-based stationary random fields, HAL Open Science, pre-print. https://hal.science/hal-02790087

AI-driven spatial data augmentation for geological modelling and resource estimation

A Gole[1] and S Sullivan[2]

1. Machine Learning Engineer, Maptek Pty Ltd, Adelaide SA 5065.
   Email: arpit.gole@maptek.com.au
2. Technical Lead, Maptek Pty Ltd, Adelaide SA 5065. Email: steve.sullivan@maptek.com.au

ABSTRACT

In real-world data sets, missing values are unavoidable for various reasons. These missing values are typically represented by NaNs, default placeholders, or simply left as blank entries. Depending on the extent of missing data, this can significantly reduce the performance of statistical methods. Additionally, data sets with missing values are incompatible with many machine learning techniques, including random forests, regression models, and neural networks, which rely on the assumption that all features contain complete and relevant information related to the task at hand.

Geological data sets, which capture the 3D representation of a deposit using geological field observations, survey data, drill hole information, and assay grades, are no exception. In geological modelling, it is extremely rare to encounter complete data sets without any missing values. A common but simplistic approach is to exclude observations with missing values altogether. However, when a large portion of the data set contains missing entries, removing those records leads to substantial information loss. This highlights the importance of integrating effective missing data imputation techniques into the data preprocessing workflow—a process that presents several challenges (Rahm and Do, 2000).

Effective imputation methods must account for naturally occurring geological patterns, such as the formation and spatial continuity of rock types, the proportions of different lithologies, and the uncertainty or potential misinterpretation introduced by incomplete data. A method has been developed that captures and reformulates a high level of correlation with the existing geological data. Performance gains using newly imputed data as input to machine learning processes are evaluated using several metrics to provide geological plausibility for the method.

INTRODUCTION

For many reasons, real-world data sets always contain missing values. In diamond drilling, there is always some core loss; an underground drive may only expose part of the orebody for sampling (King, McMahon and Bujtor,1982). Generally, these missing values are represented in a data set by either NaNs, some placeholder value, or even a blank value. In other circumstances, the missing data are ignored completely. Depending on the volume of the missing data, it can degrade the performance of statistical methods. Furthermore, such data sets are incompatible with machine learning techniques like random forest, regression and more, which assume that the entire data set contains numerical values and all the features hold valuable information related to the task.

A geological data set contains a three-dimensional representation of the ore deposit based on geological field observations, survey data, drill hole data and chemical assay elemental grade data. In the case of geological modelling, fully observed data sets without any missing values are a rare occurrence, as shown in Figure 1.

Missing numeric data

Missing domain codes

332476.4	300989.2	541.3	D4	55.9	0.11	9.5	3.3	-0.1		
332476.4	300989.2	538.25	D4	59.4	0.21	7.7	2.6	-0.1		0.0
332476.4	300989.2	535.2	D4	49	0.17	21.7	2.3	-0.1		0.05
332476.4	300989.2	532.15	D4	52.3	0.1	21	1.6	-0.1		0.01
332476.4	300989.2	529.1	D4	51.8	0.16	20.8	1.7	-0.1		0.04
332476.4	300989.2	526.05	D4	41.7	0.1	36.3	0.6	-0.1		0.01
332476.4	300989.2	523	D4	38.6	0.09	41.9	0.7	-0.1		0.01
332476.4	300989.2	519.95	D4	37.8	0.06	42.9	0.3	-0.1		0.02
332476.4	300989.2	516.9	D4	38.7	0.19	38.3	1.3	-0.1		0.03
332476.4	300989.2	514.6	D4	38.7	0.12	38.8	0.4	-0.1		0.01
331042.7	301227.6	578.461		64	0.079	3.89	0.78	0.009	0.037	0.012	0.002
331042.7	301227.6	575.461		64.27	0.092	3	0.71	0.008	0.045	0.008	0.002
331042.7	301227.6	572.461		55.42	0.211	7.66	4.96	0.022	0.048	0.078	0.002
331042.7	301227.6	569.461		61.6	0.13	3.7	1.71	0.012	0.033	0.028	0.004
331042.7	301227.6	566.462		58.52	0.137	5.64	2.93	0.011	0.043	0.057	0.004
331042.8	301227.7	563.462		62.24	0.148	2.27	0.79	0.009	0.023	0.012	0.002
331042.8	301227.7	560.462		54.16	0.13	9.13	6.36	0.011	0.057	0.187	0.008
331042.8	301227.7	557.462		56.74	0.149	6.04	4.07	0.01	0.032	0.099	0.004
331042.8	301227.7	554.462		64.24	0.092	2.21	1.33	0.007	0.033	0.02	0.002
331042.9	301227.7	551.462		42.24	0.173	16.62	12.58	0.013	0.145	0.309	0.014
331042.9	301227.8	548.463		62.45	0.165	3.4	2.41	0.009	0.033	0.061	0.003
331042.9	301227.8	545.463		63.15	0.183	2.87	2.14	0.008	0.04	0.063	0.002
331043	301227.8	542.463		56.3	0.15	7.81	5.95	0.016	0.071	0.175	0.004

FIG 1 – A table with analytical data from the case history with missing data attributes.

A naive solution is to ignore observations with missing values. However, dropping a significant proportion of the data set due to many observations having missing values is a primary source of information loss and hence poor modelling outcomes. It is consequential to keep in mind that geological data collection is a costly and time-consuming process. It becomes vital to impute the missing values in such data sets as a part of the data preprocessing workflow, which poses many challenges (Rahm and Do, 2000). While imputing the missing values, the need for a versatile approach that accounts for naturally occurring geological scenarios is required. Geologically acceptable models create rock shapes that exhibit spatial continuity and maintain proportions of rock type(s) comparable with that observed in source drilling data. Imputing additional data should aim to preserve this behaviour whilst improving certainty or lowering the probability of misinterpretation. This ability to efficiently handle, extract meaningful insights, and discover hidden knowledge from massive volumes of partially complete data becomes critical to any real-world artificial intelligence application in this sphere, as is the case with such applications generally (Lee and Yoon, 2017; Tsai *et al*, 2015).

The motivation and practical value of doing this research is to make faster and smarter uncertainty-based decisions in geological resource modelling—directly from the spatial geological data—especially in the presence of only partially complete information. This way of rapid modelling of geological resources bypasses many time-consuming manual processes that burden the resource geologist, allowing them to focus on interpretation and matters of economic interest rather than rote manual work. To accomplish this aim, the study revolves around the following research questions:

- **RQ 1**: Develop a method to impute the missing lithological domains (categorical rock type) when the chemical assay data (multi-variable continuous numerical data) is present.

- **RQ 2:** Develop a method to impute the missing assay data when the lithological domains (obtained from manual core logging) are present.

- **RQ 3:** Develop a method to impute partially missing assay data when the lithological domains and some assay values are present.

All three missing data scenarios occur with different data collection strategies. Exploration drilling data—logged by geologists for lithological domains and assayed in the lab for elemental composition—is only captured at low spatial density because it is expensive and occurs early on in the economic life of a mine. RC grade control drilling data – although captured at higher density – is not often logged by geologists for lithological domains because of time pressure and the absence of clean core samples – but is assayed in the lab for elemental composition. Blasthole drilling data also is often assayed for elemental composition just prior to mining and is at very high spatial density because these holes host the explosives used to fracture the rock for mining, but no domaining and lithological interpretation occurs on these data. The need for RQ 3 arises when during the initial period of exploration at a mining site the geologists/engineers are interested in certain elements and their assay grades but down the line, they become interested in an increased (or possibly reduced) set of elements and their assay grade values. A justifiable understanding of the above research questions in context with these different types of spatial distributed geological data will enhance the performance of downstream data-driven geological resource modelling methods. An example of this is Maptek's DomainMCF domain modelling solution (Sullivan *et al*, 2019).

In this work, we work towards developing a method to impute missing geological data to have a fully observed geological data set. Quantitatively, the imputed data exhibits a high level of statistical correlation with the existing geological data. Qualitatively, the veracity of the method is assessed using the downstream DomainMCF solution to compare models with and without imputed data in the hands of geologists. These approaches combine to build a sound understanding of the various assumptions and limitations of the technique with suitable geological explanations.

METHODOLOGY

In our context, geological data sets comprise both continuous numerical and categorical data. Additionally, the data set is spatially three-dimensional and is full of complex non-linear spatial interactions that are notably hard to analytically model. Missing data is a wreaking problem in the machine learning paradigm, and the current implementation of DomainMCF at Maptek is no exception. The present technique uses a neural network to implicitly define domain shapes by training on sample data consisting of logged lithological domains and assay grades – with the drawback that every sample needs to have the complete data (Sullivan *et al*, 2019; Kapageridis *et al*, 2021). Quality data is paramount as an input for training the neural network. Using machine learning techniques for imputing the missing values in the data, which then acts as an input to train another machine learning model could lead to a fundamental problem – the data incest problem (Krishnamurthy and Hamdi, 2013). To construct a statistically plausible full data set from a partial data set, let's first briefly discuss categories of missing data and then available and proposed methods for imputing them.

Three categories of missing data (Mislevy, 1991):

1. **Missing at Random (MAR):** Missing data are independent of the unobserved data, but systemically related to values of the observed data. For example, performing selective logging for a rock type based on experience and leaving certain rock types not logged.

2. **Missing Not at Random (MNAR):** Missing data are systemically related to the values of the unobserved data. For example, while drilling, some of the rock gets washed away due to its softness, preventing sampling of that part of the rock type.

3. **Missing Completely at Random (MCAR):** Missing data are independent of both observed and unobserved data. For example, while performing assay, 5 per cent of the samples go missing due to an event not related to anything geological.

The most prevalent imputation techniques include mean, mode, or median substitution in place of the missing value. Since these methods only consider a single feature for imputing the missing value, they do not consider the uncertainty of the imputed value or cross-correlations with other values. This introduces variance and covariance biases into the imputed data set (Arnab, 2017). Other prominent and much-advanced techniques like k-nearest neighbours (kNN Imputation) (Troyanskaya *et al*, 2001) are used for numerical data. Support vector machines (SVM Imputation) (Honghai *et al*, 2005) are used for classification with the ability to use different kernels to reduce computational cost and increase performance, such as Linear, Polynomial, Laplacian, and Radial Basis Function (RBF).

Kriging (Krige, 1953; Matheron, 1962) is a spatial interpolation technique in geostatistics. Kriging is also known as Gaussian process regression and is the primary technique used in the mining context. It has become a go-to method for optimally predicting the values at unknown points in space using nearby data. It uses a variogram to describe the spatial structures present in the data. It can be used to estimate unknown/missing values given nearby completely observed data. Given a set of observation points X and observation values at these locations $Z(x)$, it can be stated that the estimation at an unobserved location $Z^*(x_0)$ is a weighted mean:

$$Z^*(x_0) = \sum_{i=0}^{N} \lambda_i Z(x_i)$$

Here, N is the size of X and λ is the array of weights. To assure unbiasedness in the kriging system, we do this by constrained optimisation, we constrain the weights to sum to 1. A variogram explains how observed points become more dissimilar with distance. At its core, a variogram is only a function of *distance*, although parameters can vary with direction to model anisotropy. Depending upon the stochastic characteristics of the data distribution, different types of kriging apply: simple kriging to independently estimate unknown numerical variables, indicator kriging to estimate the transition probabilities between categories and co-kriging to model and estimate relationships between multiple co-dependent variables. For kriging in general, all weights must be calculated at each estimation location. For a large data set, this means a large matrix system and so solving can be extremely computationally intensive and hence slow.

To aid the machine learning algorithms, we developed a way to impute the missing values in data sets more effectively. The method assumes that the missing data follows specific patterns, such as when the likelihood of missingness is dependent on observed data rather than unobserved information, ie following the MAR principle. This method is highly adaptable and can be applied across various scenarios. It works by iteratively predicting missing values using models based on other available variables in the data set. Notably, the approach can handle different data types, including continuous, unordered categorical, and ordered categorical variables.

The core of this method is to artificially predict the missing data such that it has a similar data distribution as the observed data. It is a parametric multiple imputation method and the ability to tune the model parameters is a complex task that may depend on domain/prior knowledge. Tuning can have a drastic effect on the method's performance. Large data sets will naturally have huge numbers of variables to include in the imputation process. It may not be possible to identify all the meaningful variables to be used during the imputation process. Selection of such variables to include can be guided by various analyses and or based on domain knowledge, but this poses a risk of bias (Collins, Schafer and Kam, 2001). However, the chosen method works by automatically selecting the variables and imputing the values based on the most correlated variables.

In our context, geological data sets comprise both numerical and categorical data. It is 3D point cloud data, with: (a) x, y and z as the spatial coordinates, which are numerical; (b) lithology (also often called stratigraphy in sedimentary geology) as categorical data; and (c) assay data for elemental (or sometimes mineralogical) concentrations which are numerical data. We need evaluation metrics for both numerical and categorical data. *Geological plausibility of the imputed data set is much more than the quantitative statistical analysis of what has been imputed and involves a degree of subjective assessment by experienced geologists.*

The initial selection of the imputed data is based on quantitative analysis using the below metrics on the cross-validation data. Additionally, as part of this study, geologists are asked to visualise/verify the result for the geological plausibility of the imputed data set.

For categorical variables, the evaluation metric is the accuracy:

$$Accuracy = \frac{TP + TN}{TP + TN + FP + FN}$$

where TP is the number of *true positive* predictions, TN is the number of *true negative* predictions, FP is the number of *false positive* predictions, and FN is the number of false negative predictions.

For numerical variables, it is the root mean squared error (RMSE):

$$RMSE = \sqrt{\frac{\sum_{i=1}^{N} (Predicted_i - Actual_i)^2}{N}}$$

Also, kernel density estimate (KDE) plots are plotted for each numerical variable. This helps in the visualisation of the distribution of the numerical values before and after the imputation method is applied to them.

CASE HISTORY

The following case study demonstrates the application of the study method and compares the results of machine learning with and without data imputation.

The Lisheen mine in County Tipperary, Ireland, was mined from 1999 until the mine closure in December 2015 as shown in Figure 2. It produced 22 Mt at 11.6 per cent Zn and 1.9 per cent Pb (Torremans *et al*, 2018). Mineralisation at Lisheen consists of several, largely stratiform, massive sulfide orebodies at or within 30 m of Lower Carboniferous marine carbonates and breccias. The mineralisation is manifest by semi-massive, disseminated, and vein-hosted sulfides in a complex of laterally discontinuous normal faults (Hitzman, 1999).

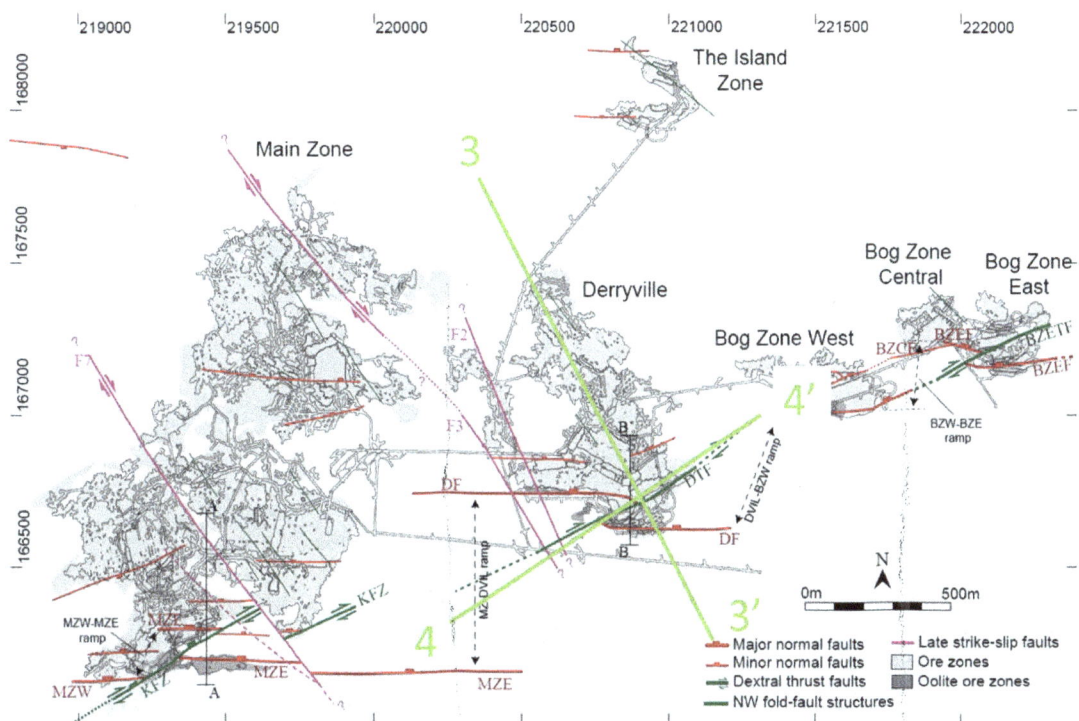

FIG 2 – Structural setting of the Lisheen base metal deposits in Ireland (from Kyne *et al* (2019) modified from Torremans *et al*, 2018).

Previous geological modelling

The Lisheen deposits were investigated using more than 7000 published holes, drilled from surface during exploration and subsequently from underground during mine production. Holes were geologically logged with more than 40 different domain codes being used to describe the host rocks and mineralised intervals. Chemical analysis of mineralised intervals and surrounding host rocks includes determinations for Zn, Pb and a wide suite of related elements. The data from the Lisheen mining operations was released by the government of Ireland as open source data on 27 June 2019 (Department of the Environment, Climate and Communications, 2019).

Modelling during the operation of the Lisheen mines used sectional interpretation of exploration and production drilling and underground channel sampling/mapping observations. Explicit wireframes were created for the footwall and hanging wall of each mineralised horizon. These two surfaces were then combined to make a solid model of each horizon, which was then used to flag an ore code into a 3D block model. Analytical data was used for estimating grades into the block model using inverse distance and ordinary kriging for the numeric attributes of economic interest.

Modelling using machine learning

The origin of the concept of data-driven modelling is discussed in detail by Solomatine, See and Abrahart (2009) and was further updated by Montánsa *et al* (2019). The rapid advance of computing power facilitates computer-based discovery and analysis. In the geological context this allows us to generate a documented workflow which updates the model when the input data changes. This could be a manual process but the most benefit is derived from an automated system which is triggered by a change in state in the underlying data.

This current study built a 3D geological model using the latest commercial machine learning engine, Maptek DomainMCF, which has its origins in collaborative research and development with the mining industry as described in Sullivan *et al* (2019). The details of the data preparation, data validation and the results of the geological modelling process using machine learning are presented in Sullivan (2022).

Modelling using data imputation

A subset of the 7000-hole data set from Lisheen was used for testing the impact of the use of data imputation prior to running the machine learning process. This subset was taken from the SW of the orebody as shown in the study area indicated in Figure 3 and included 1618 exploration and underground production drill holes.

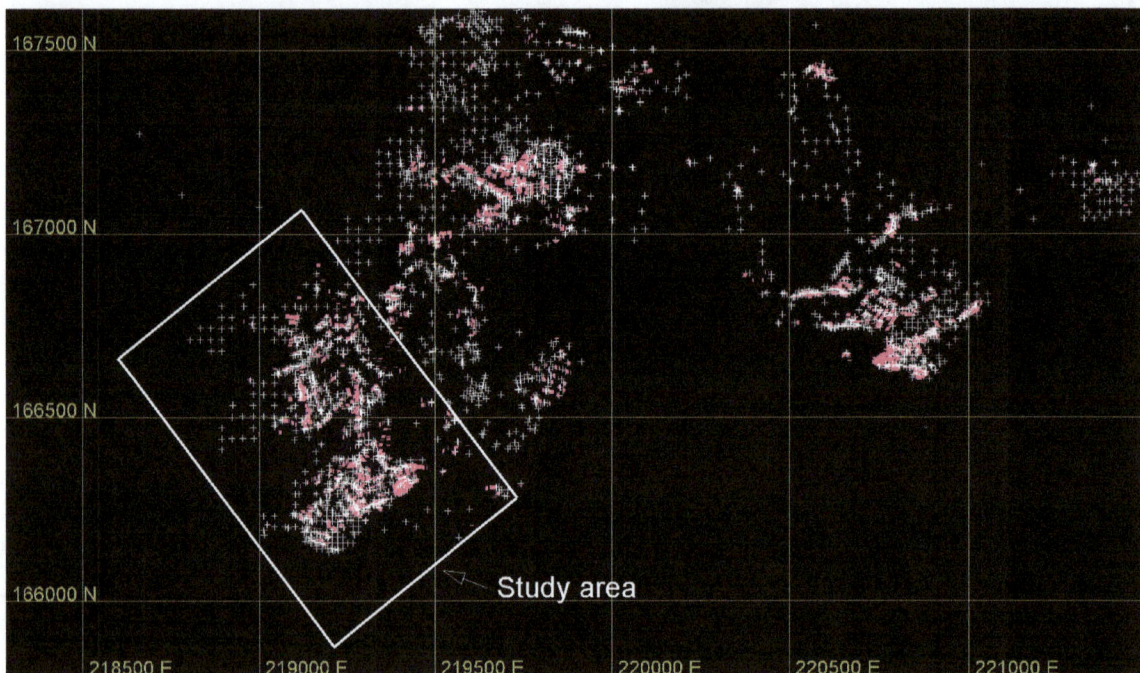

FIG 3 – Lisheen mine data drill hole sample locations with the study area highlighted.

Samples from the subset of drill holes were analysed for Zn, Pb and Fe and reported in per cent (%). Specific gravity (SG) determinations were made from a representative selection of samples and reported in grams per cubic centimetre (g/cm^3). As is the usual practice with semi-massive to massive sulfide orebodies, only drill core samples which have visibly mineralised material are sent for quantitative analysis, saving the cost of analysing barren or waste rock. This resulted in a significant portion of the spatial data points from the drill holes having missing attributes or null values for Zn, Pb, Fe and SG.

A total of 75 159 samples were included in the study area shown in Figure 3. Each data attribute, aside from the spatial coordinates, was well represented but not complete, with missing values of varying proportions as per Table 1.

TABLE 1

Showing the percent of missing data in the case study.

Attribute	Percentage missing
Fe	49.2
Pb	49.9
Zn	49.2
sg	68.8

As discussed earlier, machine learning requires data sets with a complete/full set of attributes for each spatial location. To achieve this, prior to the consideration of data imputation, rows of data with missing domain codes were deleted, as were rows with missing numeric attributes. As can be seen in Table 1, this would result in almost half the observed data being discarded. This scenario was run and used as a base case for comparison with the subsequent imputed scenario.

Before data is uploaded, it needs to be validated. There are hosts of simple rules that can be applied in this process: checking the validity of collar x, y, and z coordinates; checking *from* and *to* sample intervals are not missing or overlapping; and checking that analytical data uses the same units of measurement. Data validation rules are built into DomainMCF so that data going through to the modelling process is as clean as required. This process is critical as the output models will be directly impacted by erroneous input data.

In the Lisheen data, several drill holes were pre-collared through the overburden and were not geologically logged. When composited, these intervals are given the default code of NONE. If unlogged intervals remain in the data-driven process, a domain model will be generated for NONE as per Figure 4a.

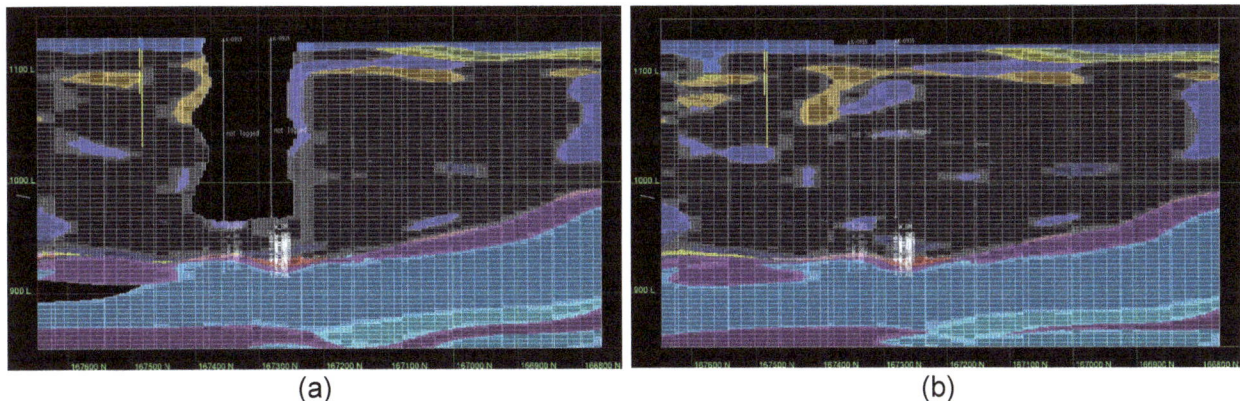

(a) (b)

FIG 4 – (a) A cross-section through the overburden at Lisheen showing the impact of leaving unlogged intervals in the input data to machine learning. (b) The same cross-section shows the impact of the use of imputed data by imputing unlogged intervals.

To remedy the unlogged intervals, the geologist can either manually enter appropriate logging codes based on surrounding drill holes, or delete all unlogged intervals in the input data and allow the machine learning to predict and interpolate relevant domains into the unlogged space shown in Figure 4a. Neither of these options are ideal. Figure 4b shows the impact of applying data imputation to the data missing in Figure 4a. The cross-section through the 3D geology is now complete and a better approximation of the spatial distribution of the domain codes.

Empirically, the result of the domain prediction for the base case: a total of 30 geological domains were predicted from 37 576 samples. This resulted in an overall accuracy of 77 per cent. In the second scenario, data imputation was implemented, allowing all 75 159 samples to be used in the machine-learning process. This scenario was run, and the results of the domain prediction for the imputed data case are as follows: 33 geological domains were predicted, with a strong correlation showing that the method learned well for predictive capabilities. This resulted in an overall accuracy of 84 per cent. We can conclude that using data imputation with this real mine data in the study area allowed an additional three domains to be predicted over the base case. In addition, the accuracy of the domain code prediction was higher when data imputation was included in the modelling process.

Now to compare the distribution of the 37 576 samples before data imputation and for all 75 159 samples after the imputation process. Figure 5 shows the data distribution of the observed data distribution (original data) in blue and the complete data set imputed (imputed data) in red. Assay values are power transformed (Yeo and Johnson, 2000) to improve the interpretability of data distribution by reducing skewness and enhancing the effectiveness of graphical representations to detect the underlying patterns. We can right away see that the data imputation method models the underlying data distribution to a sufficient degree for the assay elements. The general acceptability of the method is taken into account given we only have about 50 per cent of the real data to start the imputation after initial preprocessing.

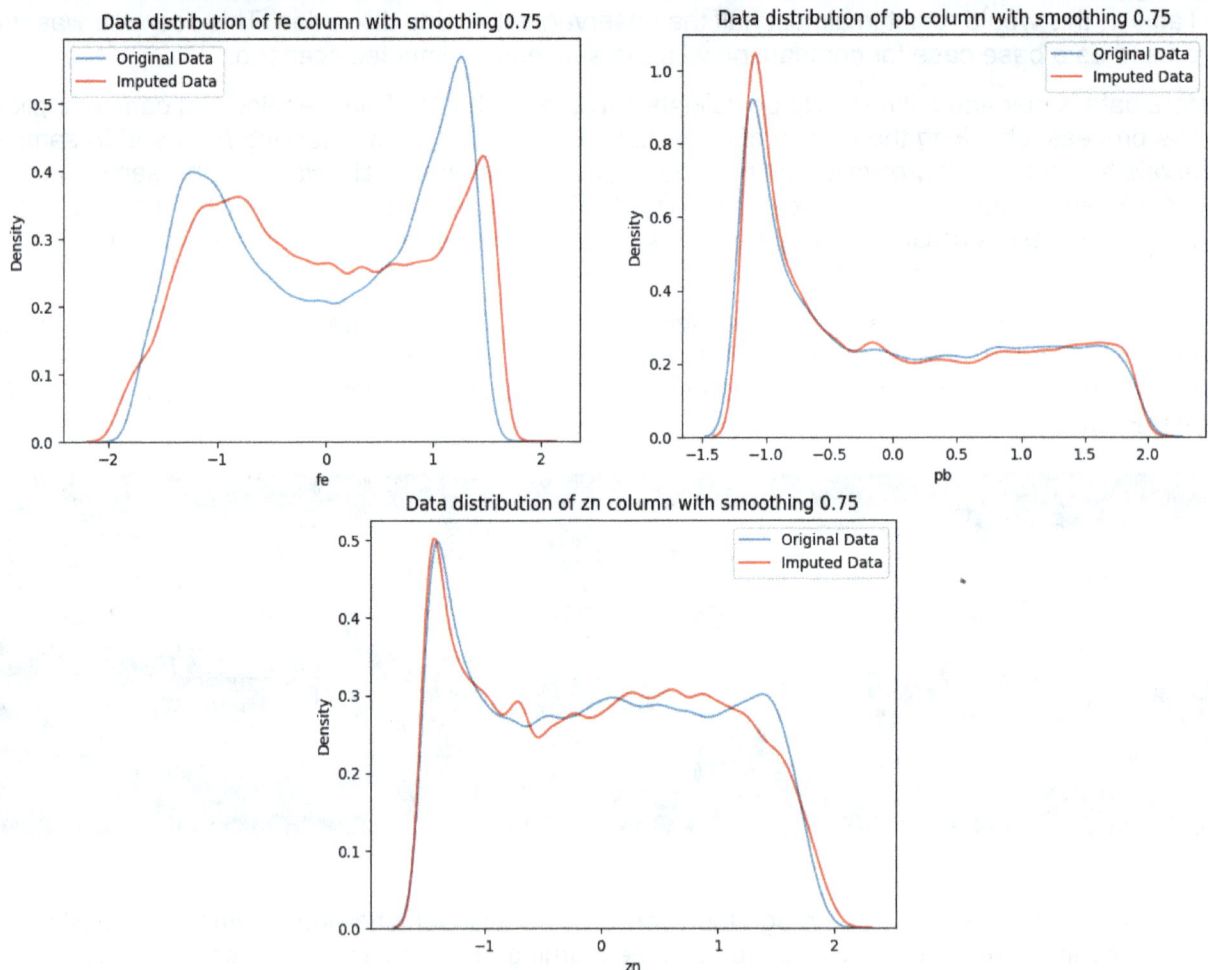

FIG 5 – Showing the data distributions before and after the imputation process by doing the power transformation of the assay values.

CONCLUSION AND FUTURE WORK

The data imputation method discussed is an extremely rapid and data-driven way to produce resource models with uniform and geologically plausible distributions. It requires less human supervision when compared to traditional methods solely relying on domain experts. Incorporating multi-element data allows correlations within these attributes to be analysed and missing domain and numeric data attributes to be imputed with cognisance of the underlying statistical and spatial relationships. Thus, data imputation will be an essential component for improving the outcomes from the implementation of machine learning processes in geological modelling and grade prediction. The documented case history above using real mine data, with all its complexities, shows a significant improvement in the prediction accuracies through the use of data imputation.

Immediate value can be gained in the modelling process by the use of the commercial machine learning engine, Maptek DomainMCF, integrated with the data imputation technique. Because geologists can now use more of the data they have available for rapid resource modelling:

1. Better validation of routine resource modelling can take place.

2. A better and more timely understanding of geological uncertainties can be derived.

3. Many more what-if scenarios can be done in a given amount of time providing a great deal of agility and ability to pivot future decisions.

ACKNOWLEDGEMENTS

The authors acknowledge Maptek for providing the time and approval to prepare and present this paper. They also appreciate the input from Maptek's internal peer review for proofreading and improving the clarity of the paper. Software developers from the DomainMCF team at Maptek are thanked for their ongoing commitment to creating a new environment for geological modelling.

REFERENCES

Arnab, R, 2017. Chapter 1 – Preliminaries and basics of probability sampling, in *Survey Sampling Theory and Applications* (ed: R Arnab), pp 1–21 (Academic Press). https://doi.org/10.1016/B978-0-12-811848-1.00001-7

Collins, L M, Schafer, J L and Kam, C M, 2001. A comparison of inclusive and restrictive strategies in modern missing data procedures, *Psychological Methods*, 6(4):330–351. https://doi.org/10.1037/1082-989X.6.4.330

Department of the Environment, Climate and Communications, 2019. Lisheen Mine Data Release. Available from: <https://www.gov.ie/en/publication/d187c-lisheen-mine-data-release>

Hitzman, M W, 1999. Extensional faults that localize Irish syndiagenetic Zn-Pb deposits and their reactivation during Variscan compression, *Special Publications*, 155:233–245 (Geological Society: London).

Honghai, F, Guoshun, C, Cheng, Y, Bingru, Y and Yumei, C, 2005. A SVM regression-based approach to filling in missing values, in *Knowledge-Based Intelligent Information and Engineering Systems, KES 2005* (eds: R Khosla, R J Howlett and L C Jain), Lecture Notes in Computer Science, 3683:581–587 (Springer: Berlin). https://doi.org/10.1007/11553939_83

Kapageridis, I, Albanopoulos, C, Sullivan, S, Buchanan, G and Gialamas, E, 2021. Application of machine learning to resource modelling of a marble quarry with DomainMCF, *Materials Proceedings*, 5:12. https://doi.org/10.3390/materproc2021005012

King, H F, McMahon, D W and Bujtor, G J, 1982. A guide to the understanding of ore reserve estimation, Supplement to Proceedings No. 281, March 1982 (The Australasian Institute of Mining and Metallurgy: Melbourne).

Krige, D G, 1953. A statistical approach to some basic mine valuation problems on the Witwatersrand, *Operational Research Quarterly*, 4(1):18–18.

Krishnamurthy, V and Hamdi, M, 2013. Mis-information removal in social networks: Constrained estimation on dynamic directed acyclic graphs, *IEEE Journal of Selected Topics in Signal Processing*, 7(2):333–346.

Kyne, R, Torremans, K, Guven, J, Doyle, R and Walsh, J, 2019. 3-D modeling of the Lisheen and Silvermines deposits, County Tipperary, Ireland: Insights into structural controls on the formation of Irish Zn-Pb deposits, *Economic Geology*, 114(1):93–116.

Lee, C H and Yoon, H-J, 2017. Medical big data: Promise and challenges, *Kidney Research and Clinical Practice*, 36(1):3–11. Available from: <https://pubmed.ncbi.nlm.nih.gov/28392994>

Matheron, G, 1962. Treatise on Applied Geostatistics [in French: Traité de géostatistique appliquée], Memoirs of the Bureau of Geological and Mining Research [Mémoires du BRGM], BRGM. Available from: <https://books.google.com.au/books?id=Ej4gzQEACAAJ>

Mislevy, R J, 1991. Review: 'Statistical Analysis with Missing Data' by Roderick J A Little and Donald B Rubin, *Journal of Educational Statistics*, 16(2):150–155. http://www.jstor.org/stable/1165119

Montánsa, F, Chinesta, F, Gómez-Bombarelli, R and Kutz, J, 2019. Data-driven modeling and learning in science and engineering, *Comptes Rendus Mecanique*, 347:845–855.

Rahm, E and Do, H, 2000. Data cleaning: Problems and current approaches, IEEE Data Engineering Bulletin, 23:3–13.

Solomatine, D, See, L and Abrahart, R, 2009. Data-Driven Modelling: Concepts, Approaches and Experiences, in *Practical Hydroinformatics Water Science and Technology Library* (eds: R J Abrahart, L M See and D P Solomatine), vol 68 (Springer: Berlin). https://doi.org/10.1007/978-3-540-79881-1_2

Sullivan, S, 2022. Harnessing data complexity - how machine learning applies all project data for accurate resource modelling, in Proceedings of the AusIMM 12th International Mining Geology Conference, Brisbane, Australia, 22–23 March.

Sullivan, S, Green, C, Carter, D, Sanderson, H and Batchelor, J, 2019. Deep learning – a new paradigm for ore body modelling, AusIMM Mining Geology Conference. Available from: <https://www.maptek.com/pdf/insight/Maptek_Deep_Learning_A_New_Paradigm_for_Orebody_Modelling.pdf>

Torremans, K, Kyne, R, Doyle, R, Güven, J F and Walsh, J J, 2018. Controls on metal distributions at the Lisheen and Silvermines deposits: Insights into fluid-flow pathways in Irish-type Zn-Pb deposits, *Economic Geology*, 113(7):1455–1477.

Troyanskaya, O, Cantor, M, Sherlock, G, Brown, P, Hastie, T, Tibshirani, R, Botstein, D and Altman, R B, 2001. Missing value estimation methods for DNA microarrays, *Bioinformatics*, 17(6):520–525. https://doi.org/10.1093/bioinformatics/17.6.520

Tsai, C-W, Lai, C-F, Chao, H-C and Vasilakos, A V, 2015. Big data analytics: A survey, *Journal of Big Data*, 2(1):21. https://doi.org/10.1186/s40537-015-0030-3

Yeo, I K and Johnson, R A, 2000. A new family of power transformations to improve normality or symmetry, *Biometrika*, 87(4):954–959.

Leveraging machine learning for fast and reliable faulted 3D modelling

A Gole[1] and S Sullivan[2]

1. Machine Learning Engineer, Maptek Pty Ltd, Adelaide SA 5065.
   Email: arpit.gole@maptek.com.au
2. Technical Lead, Maptek Pty Ltd, Adelaide SA 5065. Email: steve.sullivan@maptek.com.au

ABSTRACT

Building the best possible 3D model from observed geological data is critical before grade estimation is performed. For many decades, 3D modelling of faults has been a challenging task. Currently, explicit or implicit modelling techniques are widely used to model faults. Explicit modelling is done manually by geologists based on their knowledge and expertise. The reliability of these models depends on the geologists' abilities and creating them is a time-consuming process (Caumon et al, 2009; Wellmann and Caumon, 2018). On the other hand, implicit modelling takes a more mathematical approach to describe the fault geometry to combine the fault shape and fault frames (Grose et al, 2021; Jessell et al, 2014). Godefroy et al (2018) introduced kinematics with implicit modelling methods and directly applied them to the implicit description of the faulted surfaces.

It is a challenging task to create 3D geological models comprising faults, as they represent a discontinuity in geological feature(s) that are being modelled. Geological discontinuities and their uncertainties can be easily and consistently modelled using machine learning (Sullivan et al, 2019). The underlying architecture of an implicit function has been developed and modified to provide the ability to create geological faulted 3D models in three distinct steps: data preparation; building the fault geometries; and the use of machine learning to generate complex geological faulted 3D models. This approach generates 3D models that fit the supplied data and geological knowledge of the generated fault(s), ie generates a geologically plausible faulted 3D model. The new approach is fast and repeatable compared with existing techniques.

INTRODUCTION

Constructing 3D models is fundamental in any geological exploration and production phase. A critical aspect of this activity is to take all the various geological factors into account. Faulted 3D model construction is a cumbersome task primarily due to the complexity of the underlying geology, the difference in interpretations of this geology and the occurrence of faults. To ease the modelling of these faults, we propose a machine learning approach. This study aims to make the geologist's decision-making process faster and easier through the fast generation of faulted 3D models that accurately represent geological structures including faults.

There are several challenges in creating faulted 3D models. One of these concerns, uncertainties present in the 3D input data of the underlying geology (Wellmann et al, 2010), which may include: (a) inherent randomness, (b) data imprecision or (c) incomplete knowledge. The created 3D model must follow a set of geological criteria to be valid. One example of this is that horizons—the surfaces separating two rock layers—must not intersect themselves (Caumon et al, 2009). Generating faulted 3D models that violate these criteria can result in significant issues, such as the reporting of incorrect volumes of geological units. The ability to accurately model faults by capturing the offsets and their impacts on the underlying geology. Ongoing tectonic stresses can create new faults or modify existing ones. In addition, faults do not necessarily need to terminate at other surfaces (Grose et al, 2021). Figure 1 illustrates an example of a rock volume and demonstrates how the volume changes in response to deformations caused by faults.

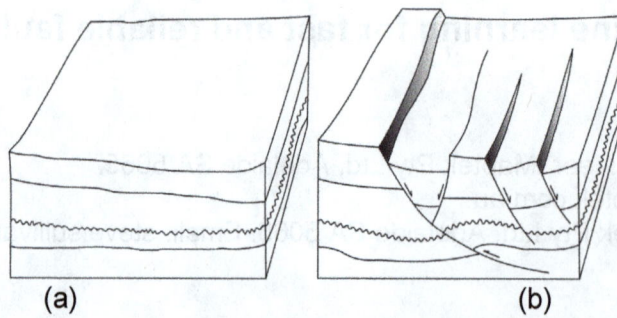

FIG 1 – The figure illustrates a rock volume before and after faulting and displacement, adapted from Zhiwei *et al* (2019). (a) The original rock volume before the fault. (b) The rock volume after faulting, with arrows indicating the direction of movement along the fault plane.

From a geological point of view, faults develop and can be active over a prolonged period of time. Reactivation of these weaknesses in the Earth's crust is commonplace and often there is negligible evidence of how these faults transformed the subsurface into its present configuration. Therefore, the geologist is responsible for evaluating, based on the evidence available, and inferring how faults have shaped the subsurface. This requires identifying faults within the subsurface rock volume being modelled, using patterns of rock discontinuity observed in geological data, to build an understanding of the chronological order of the original geology.

This study aims to address the various challenges facing geological fault modelling and create a faster and iterative ecosystem for creating faulted 3D models. We introduce a neural network deep learning approach to modelling geology, such as creating the faulted model and specifying the influence of faults on one another and the rock layers. We also model how faults cut through the rock layers and take chronological precedence over other fault(s).

METHODOLOGY

The method builds on the existing DomainMCF application, a machine learning approach developed by Maptek (Sullivan *et al*, 2019). It uses raw drill hole data as input to improve the orebody modelling process and capture the spatial continuity in the geological data. DomainMCF builds an artificial neural network (ANN) model to understand the spatial distribution of discrete domain values from a set of samples.

ANNs, such as those developed and deployed in DomainMCF, typically have an architecture as shown in Figure 2. The ANN consists of multiple layers of processing elements (PEs) also known as neurons (McCulloch and Pitts, 1943). There are three types of layers and corresponding PEs—input, hidden and output. PEs from one layer are connected to PEs in the next layer using weighted links known as synapses. PEs transfer the input signal to their outputs using an activation function that differs between the three types of layers.

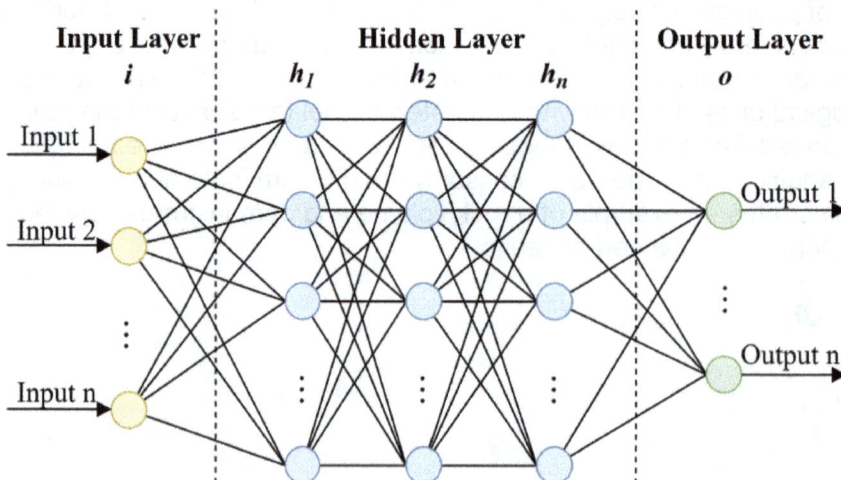

FIG 2 – Illustration of an artificial neural network architecture (Shehzad *et al*, 2021).

In case of DomainMCF, the architecture of the neural network is dynamically decided based on the input data and the fault structure(s) provided by the user, ie the number of hidden layers and PEs per hidden layer are controlled by a process that will find the best configuration based on the user data. For any given sample, the x, y, z coordinates and the information about the fault structure(s) are used as the input to DomainMCF's ANN. Then the ANN is trained to predict the sample domain(s) and, optionally, sample grade(s) and other numeric attributes as the required output. During the inference step, for any given sample location x, y, z and the information about the fault structure(s) provided as an input. As the fault information changes depending on the query location, the predictions obtained for domain and grade are influenced by the faults.

The quality of the ANN trained is directly related to the quality and the quantity of the drilling data supplied to DomainMCF. It is generally accepted that when building a complex 3D model for grade estimation or resource estimation, more samples will be required to train a complex ANN architecture with more PEs and hidden layers, which is essential to capture a geologically complex scenario.

Once the data is supplied and the ANN is trained to understand the quality of the trained model, the following metrics are used.

For categorical variables, the evaluation metric is the accuracy:

$$Accuracy = \frac{TP+TN}{TP+TN+FP+FN}$$

where *TP* is the number of *true positive* predictions, *TN* is the number of *true negative* predictions, *FP* is the number of *false positive* predictions, and *FN* is the number of false negative predictions.

For numerical variables, the root mean squared error (RMSE) is used:

$$RMSE = \sqrt{\frac{\sum_{i=1}^{N} (Predicted_i - Actual_i)^2}{N}}$$

CASE HISTORY

The following case studies demonstrate the application of the study method and compare the results.

The Lisheen mine in County Tipperary, Ireland, was mined from 1999 until the mine closure in December 2015. It produced 22 Mt at 11.6 per cent Zn and 1.9 per cent Pb (Torremans *et al*, 2018). Mineralisation at Lisheen consists of several, largely stratiform, massive sulfide orebodies at or within 30 m of Lower Carboniferous marine carbonates and breccias. The mineralisation is manifest by semi-massive, disseminated, and vein-hosted sulfides in a complex of laterally discontinuous normal faults (Hitzman, 1999).

Bonson, Walsh and Carboni (2012) presented results from studies of the structural geology of three mines in the Lisheen field, which supported a model invoking the existence of normal fault boundaries to be responsible for the localisation of up-fault fluid flow of metal bearing fluids and their entrapment within stratiform dolomite breccias. Normal faults which host these Zn-Pb deposits are generally highly segmented, north-dipping normal faults that offset the Lower Carboniferous succession on the order of 200–400 m. Fault segments tend to strike E-W or ENE-WSW and are arranged in NE-SW oriented, left-stepping, *en échelon* arrays.

A comprehensive 3D synthesis of the base metal mines in Ireland was published by Kyne *et al* (2019). Their research used a combination of traditional section interpretation, wireframing and fault interpolation to define geometric and kinematic links between faulting, structure and mineralisation as shown in Figure 3.

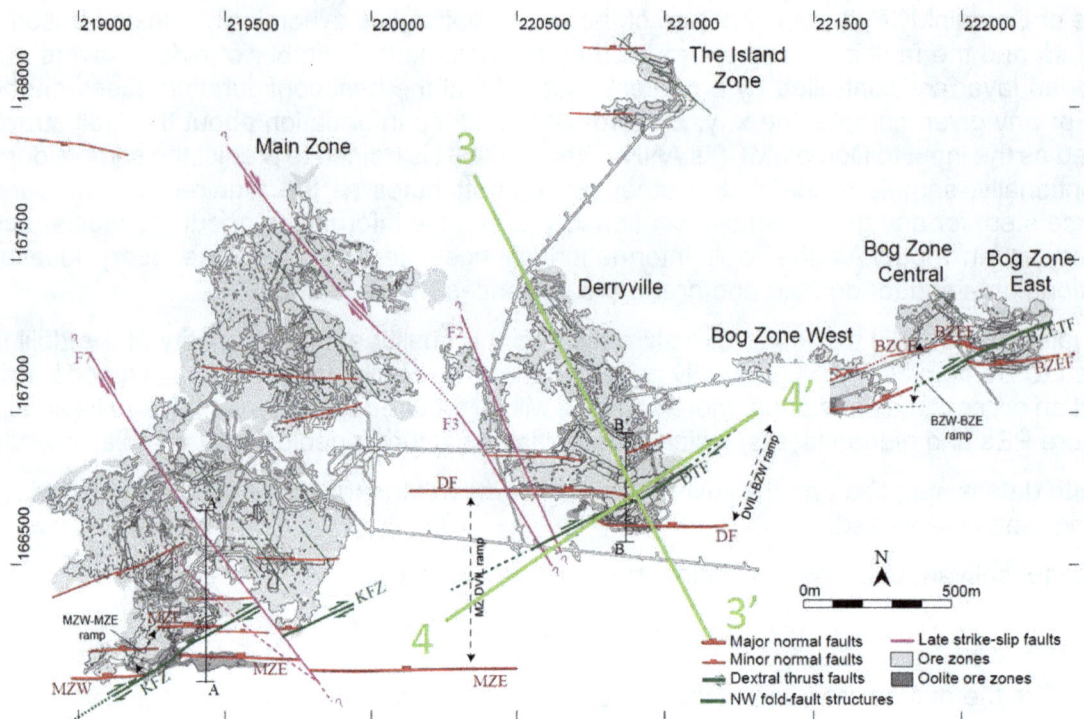

FIG 3 – Structural setting of the Lisheen base metal deposits in Ireland (from Kyne *et al*, 2019 modified from Torremans *et al*, 2018).

Previous geological modelling

The Lisheen deposits contain over 7000 published holes, drilled from the surface during exploration and subsequently from underground during mine production. Holes were geologically logged with more than 40 different domain codes being used to describe the host rocks and mineralised intervals. Chemical analysis of mineralised intervals and surrounding host rocks includes determinations for Zn, Pb and a wide suite of related elements.

Modelling during the operation of the Lisheen mines used sectional interpretation of exploration and production drilling and underground channel sampling/mapping observations. Explicit wireframes were created for the footwall and hanging wall of each mineralised horizon. These two surfaces were then combined to make a solid model of each horizon, which was then used to flag an ore code into a 3D block model.

The mine geologists at Lisheen would update their resource model on a 6 or 12-month cycle. Interpretation and correlation between drill holes on a sectional basis and update of wireframes was a manual process that would take at least a week (Colin Badenhorst, personal communication). The comprehensive 3D modelling and structural synthesis by Kyne *et al* (2019), did not document the time involved in building their models. A plan view from their work is shown in Figure 4.

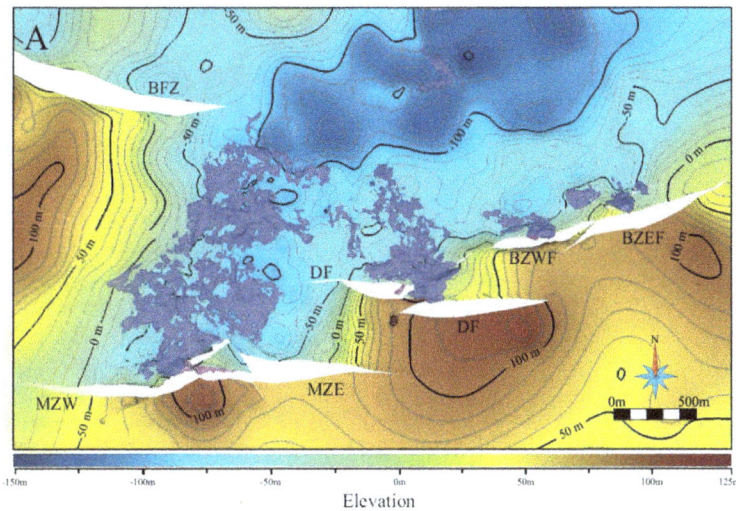

FIG 4 – Plan view of the 3D model generated by Kyne *et al* (2019). The white E-W trending shapes are interpreted faults and are equivalent to the faults marked and annotated in red in Figure 3. The contours represent the elevation above sea level.

Modelling using machine learning

The rapid advance of computing power facilitates computer-based discovery and analysis. In the geological context, this allows us to generate a documented workflow which updates the model when the input data changes. This could be a manual process but the most benefit is derived from an automated system which is triggered by a change in state in the underlying data.

This current study built a 3D geological model using the latest commercial machine learning engine, DomainMCF. The details of the data preparation, data validation and the results of the geological modelling process for one of the Lisheen deposits using machine learning are presented in Sullivan (2022). The DomainMCF system processes the input geological data using multi-threaded cloud computing with cyber secure data transfer for data upload and model download.

Before data is uploaded, it needs to be validated. There are hosts of simple rules that can be applied in this process: checking the validity of collar x, y, and z coordinates; checking *from and to* sample intervals are not missing or overlapping; and checking that analytical data uses the same units of measurement. Data validation rules are built into DomainMCF so that data going through the modelling process is as clean as required. This process is critical as erroneous input data will directly impact the output models.

The machine learning process analyses local and surrounding data, taking into account the orientation, width and other characteristics, and automatically works out where two intervals are not continuous and have been displaced. No manual intervention is required, unlike with grid-based and implicit modelling that wraps a continuous surface between geological data intersections.

Upon generation of any model, it is important to compare it with the input data. For domain codes, this is most easily accomplished by viewing cross-sections in two perpendicular orientations and also in plan view. Data-driven modelling can be brutal as it will honour intervals of inliers and outliers in the data, which otherwise would be glossed or smoothed over during manual interpretation.

Modelled fault breaks are easily visualised using your preferred three-dimensional viewing software. The results of the machine learning process correlate well with interpreted fault models built and observations made during underground mining and mapping at Lisheen as shown in the Figure 5.

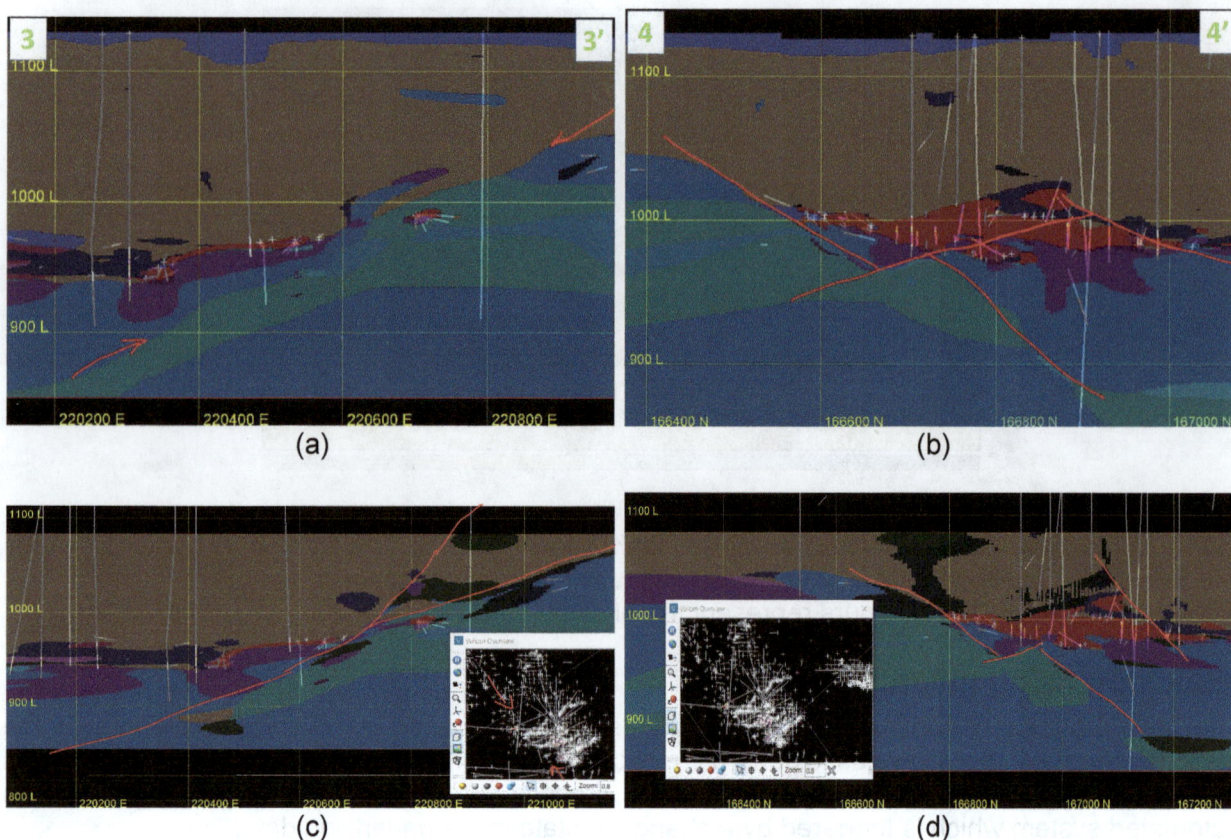

FIG 5 – Cross-sections showing displacement of mineralisation (red domain) across the Derryville thrust. (a) Cross-section 3–3' looking NE (as located on Figure 3) showing displacement of mineralisation (red domain) across the Derryville dextral thrust; (b) Cross-section 4–4' (as located on Figure 3) looking NW showing displacement across the Derryville (north dipping) complicated by interaction with the Derryville south dipping thrust; (c) Cross-section along strike from section 3–3' looking NE (as located on Figure 3) showing displacement of mineralisation (red domain) across the Derryville dextral thrust; (d) Cross-section parallel to section 4–4' (as located on Figure 3) looking NW showing displacement across the Derryville (north dipping) complicated by interaction with the Derryville south-dipping thrust.

A comprehensive review across all sections and plans shows that the DomainMCF data-driven model honours the mineralised data, its surrounding host rocks and displacements along faults, better and faster than the manual interpretation and consequent wireframe models. Empirically, the result of the domain prediction resulted in an overall accuracy of 84.9 per cent. We can right away see that immediate value is gained in the modelling process by using the machine learning approach which is fast and reliable for generating faulted 3D models.

CONCLUSION AND FUTURE WORK

The Lisheen mine data has been subjected to a data-driven modelling process and the resultant models appear superior in mapping the input data compared with a manually controlled process. The correlation between drill data has been derived purely from the input data without bias from the operator. Spatial correlation of domains during the machine learning process has another advantage of adopting the data-driven modelling approach at Lisheen is the speed at which results are generated. The modelling of all mineralised and host rock domains from the raw input data occurred in less than 30 mins. As this is now an automated process, as new data is collected, a model update can be triggered by a change in the data state and the model can always be up to date, no matter the speed or frequency of data collection. Thus machine learning gives geologists more time to analyse results, change settings and run multiple iterations to refine outputs to provide the best possible model(s).

Data-driven geological modelling has not yet been adopted by mainstream mine and resource geologists but as the benefits of its use become widespread, we will look back and wonder why it took so long to embed into our daily processes. Machine learning is not just for traditional mines, as shown in Kapageridis *et al* (2021), which discussed the application of machine learning for domain classification in the quarrying environment. The new approach will become ubiquitous in our world, just as mobile phones and the internet have changed the way we work.

ACKNOWLEDGEMENTS

The authors acknowledge Maptek for providing the time and approval to prepare and present this paper. They also appreciate the input from Maptek's internal peer review for proofreading and improving the clarity of the paper. Software developers from the DomainMCF team at Maptek are thanked for their ongoing commitment to creating a new environment for geological modelling.

REFERENCES

Bonson, C, Walsh, J and Carboni, V, 2012. The role of faults in localising mineral deposits in the Irish Zn-Pb orefield, in *Proceedings of Structural Geology and Resources Symposium*, pp 8–11 (Australian Institute of Geoscientists).

Caumon, G, Collon-Drouaillet, P, de Veslud, C L, Viseur, S and Sausse, J, 2009. Surface-based 3D modeling of geological structures, *Mathematical Geosciences*, 41:927–945.

Godefroy, G, Caumon, G, Ford, M, Laurent, G and Jackson, C A L, 2018. A parametric fault displacement model to introduce kinematic control into modeling faults from sparse data, *Interpretation: Journal of Subsurface Characterization*, 6:B1–B13.

Grose, L, Ailleres, L, Laurent, G, Caumon, G, Jessell, M and Armit, R, 2021. Modelling of faults in LoopStructural 1.0, *Geoscientific Model Development*, 14:6197–6213. https://doi.org/10.5194/gmd-14-6197-2021.

Hitzman, M W, 1999. Extensional faults that localize Irish syndiagenetic Zn-Pb deposits and their reactivation during Variscan compression, *Geological Society, London, Special Publications*, 155:233–245.

Jessell, M, Ailleres, L, De Kemp, E, Lindsay, M, Wellmann, F, Hillier, M and Martin, R, 2014. Next generation three-dimensional geologic modeling and inversion, *Society of Economic Geologists Special Publication*, 18:261–272.

Kapageridis, I, Albanopoulos, C, Sullivan, S, Buchanan, G and Gialamas, E, 2021. Application of machine learning to resource modelling of a marble quarry with DomainMCF, *Materials Proceedings*, 5:12. https://doi.org/10.3390/materproc2021005012

Kyne, R, Torremans, K, Guven, J, Doyle, R and Walsh, J, 2019. 3-D modeling of the Lisheen and Silvermines deposits, County Tipperary, Ireland: Insights into structural controls on the formation of Irish Zn-Pb deposits, *Economic Geology*, 114(1):93–116.

McCulloch, W S and Pitts, W, 1943. A logical calculus of the ideas immanent in nervous activity, *Bulletin of Mathematical Biophysics*, 5:115–133. https://doi.org/10.1007/BF02478259

Shehzad, F, Rashid, M, Sinky, M, Alotaibi, S and Zia, M Y I, 2021. A scalable system-on-chip acceleration for deep neural networks, *IEEE Access*, 9:95412–95426. https://doi.org/10.1109/ACCESS.2021.3094675

Sullivan, S, 2022. Harnessing data complexity – how machine learning applies all project data for accurate resource modelling, in Proceedings of the 12th International Mining Geology Conference (The Australasian Institute of Mining and Metallurgy: Melbourne).

Sullivan, S, Green, C, Carter, D, Sanderson, H and Batchelor, J, 2019. Deep learning – a new paradigm for ore body modelling, in Proceedings of the 11th International Mining Geology Conference (The Australasian Institute of Mining and Metallurgy: Melbourne). Available from: <https://www.maptek.com/pdf/insight/Maptek_Deep_Learning_A_New_Paradigm_for_Orebody_Modelling.pdf>.

Torremans, K, Kyne, R, Doyle, R, Güven, J F and Walsh, J J, 2018. Controls on metal distributions at the Lisheen and Silvermines deposits: Insights into fluid-flow pathways in Irish-type Zn-Pb deposits, *Economic Geology*, 113(7):1455–1477.

Wellmann, F and Caumon, G, 2018. 3-D structural geological models: Concepts, methods and uncertainties, *Advances in Geophysics*, 59:1–121.

Wellmann, J F, Horowitz, F G, Schill, E and Regenauer-Lieb, K, 2010. Towards incorporating uncertainty of structural data in 3D geological inversion, *Tectonophysics*, 490(3):141–151. https://doi.org/10.1016/j.tecto.2010.04.022

Zhiwei, H, Changgui, X, Deying, W, Jian, R, Yubo, L, Shuguang, X and Xin, Z, 2019. Superimposed characteristics and genetic mechanism of strike-slip faults in the Bohai Sea, China, *Petroleum Exploration and Development*, 46(2):265–279.

Enhanced resource domain classification for mineral estimation using geostatistical clustering

J Gonzalez[1], M Liang[2] and I Buitrago[3]

1. GEOVIA Sales Expert Specialist, Dassault Systèmes, Santiago 775000, Chile.
   Email: jose.gonzalez@3ds.com
2. GEOVIA Sales Expert Senior Specialist, Dassault Systèmes, Brisbane Qld 4000.
   Email: min.liang@3ds.com
3. GEOVIA Industry Process Consultant, Dassault Systèmes, Medellin 050023, Colombia.
   Email: isabella.buitrago@3ds.com

INTRODUCTION

This study presents the implementation of machine learning-based methodologies to address classification challenges in the definition of estimation domains, a key process in mineral resource estimation. To achieve this, databases from the Molejón deposit are utilised.

The Molejón deposit is a low-sulfidation epithermal gold deposit located in the Molejón district, Panama. The primary mineralisation of interest consists of gold, with subordinate silver values and low concentrations of molybdenum and copper. Mineralisation is mainly distributed in quartz breccias, concentrated in quartz veins with a strong structural control; however, it can also be found near the surface in saprolites (Laudrum, 1995).

Exploratory data analysis is a fundamental step in mineral resource estimation, culminating in the definition of estimation domains, which are subject to multiple sources of error (Emery, 2019). The definition of these domains is a time-consuming process with a significant manual component, relying on professional judgment to ensure their representativeness, robustness, and validity.

In this context, studying alternative methodologies using machine learning tools becomes relevant, aiming to support informed decision-making by improving the accuracy and/or efficiency in estimation domain definition. This, in turn, enhances the representativeness of the resource model.

Regarding the state-of-the-art in machine learning techniques, there is growing interest in their applications within the mining industry, particularly in resource estimation, geotechnics, and operational control. Specifically, for estimation domain definition, the work of Fustos (2017) is of particular interest, as it explores two geostatistical formalisms applied to clustering techniques to integrate geological knowledge into these algorithms (Romary et al, 2012).

Additionally, the study conducted by Faraj (2021) proposes a workflow for defining estimation domains using hierarchical clustering, emphasizing geology, statistics, and spatial continuity in the process.

For this study, the most suitable approach to support decision-making in estimation domain definition is considered to be unsupervised learning through non-hierarchical clustering techniques. The objective is to evaluate the performance of various algorithms within this category while ensuring the integration of geological knowledge into the process. Furthermore, an alternative methodology capable of incorporating the spatial dependency of regionalised variables into the clustering process is contrasted.

METHODOLOGY

The methodology consisted of three main stages and has been applied to a set of drill hole data. First, an attribute selection phase was conducted, considering spatial continuity and geological relevance. The selection of clustering attributes was justified based on their spatial correlation structures, the number of clusters to be generated, and the chosen geostatistical distance metric. Subsequently, the algorithm was applied, and results were systematically compiled. The final stage involved the validation of estimation domains by comparing the clustering outputs with previous models and interpretations expressed in official reports for reporting resources and reserves (eg NI 43 101).This validation was performed by integrating geological, statistical, and spatial criteria to assess intra- and inter-cluster consistency. Spatial fragmentation was quantified using a percentage-based tolerance range at cluster boundaries. Statistical validation relied on geostatistical similarity measures, including Dunn and Davis-Bouldin indices (Kaufman and Rousseeuw, 1990), as well as descriptive statistical analyses. Geological coherence was assessed based on lithological and temporal consistency within and across clusters to ensure that the geostatistical clustering results aligned with the intrinsic characteristics of the deposit and the estimation domains defined by conventional methodologies.

RESULTS

The implementation of this algorithm requires inputting the Geostatistical Mahalanobis distance matrix (Fustos, 2017) into the computation of the PAM algorithm. To achieve this, the Euclidean distance matrix between object pairs is divided by the direct and cross-variogram matrix.

The construction of both the Euclidean distance matrix and the direct and cross-variogram matrix involves numerical and categorical properties without missing values. The Euclidean distance matrix incorporates lithological domains, spatial coordinates, and the grades of Au, Ag, Cu, and MoS_2, with categorical variables weighted according to data abundance. The direct and cross-variogram matrix is built using spatial coordinates and Au and Ag grades, with Ag being subsampled due to its moderate correlation in Feldspar-Quartz Porphyry. For Euclidean distance calculations, data are scaled to a unit range, while clustering is performed using the Geostatistical Mahalanobis distance as defined by Fustos (2017).

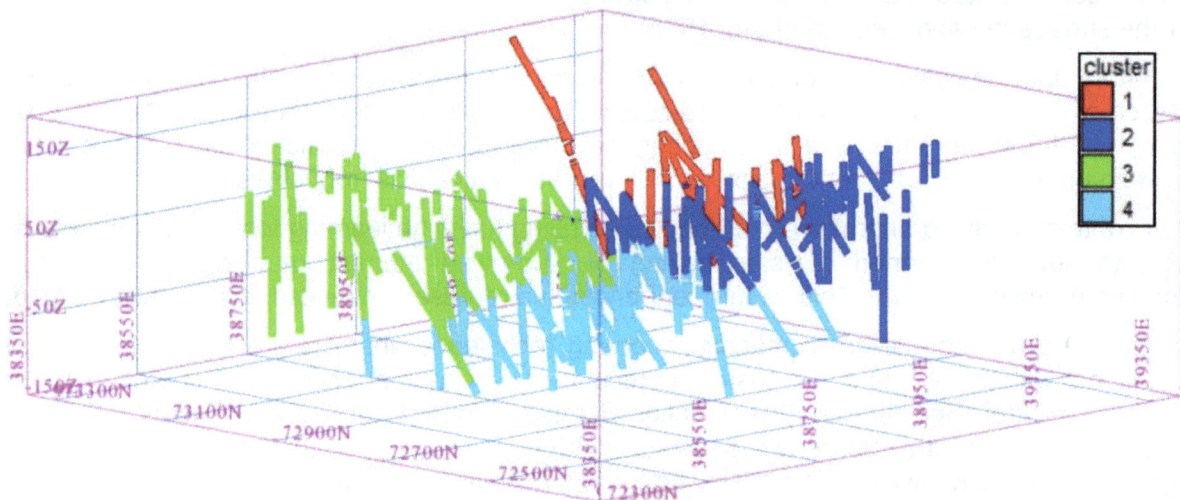

FIG 1 – Spatial display of clusters obtained by PAM algorithm considering the Geostatistical Mahalanobis distance.

The spatial analysis reveals that up to 15.26 per cent of the objects within each cluster exceed the 100 m spacing threshold, with the highest occurrence in Cluster 1, which is characterised by high-grade values. At a global scale, only 2.12 per cent of objects surpass this threshold, indicating a slight fragmentation within the obtained clusters. Additionally, the contact analysis identifies the presence of sharp boundaries between certain clusters, suggesting clear structural differentiations. From a statistical perspective, the clusters exhibit lognormal Au distributions, with significant differences in the mean grades of Au, Ag, and MoS_2. Cluster 2 stands out due to its high Au and Ag grades, whereas the remaining clusters show variations in the concentration of these elements. The low variability in Cu grades suggests that this parameter does not contribute significantly to the analysis.

From a quantitative standpoint and in consideration of performance metrics, the Dunn index (11 × 10^{-4}) and the Davis-Bouldin index (1.23) indicate that, while the clustering method achieves the best intra-group separation and the lowest inter-group dispersion compared to other methods, its overall performance is not optimal. However, the spatial distribution of the clusters largely aligns with the structural patterns of the deposit, with Cluster 2 being the most consistent with expected geological controls. Overall, the segmentation reflects a clustering approach with a degree of structural coherence, although there are opportunities for improvement in differentiating certain lithologies and their relationship with mineralisation grades.

CONCLUSIONS

The Partitioning Around Medoids (PAM) clustering approach, incorporating the Geostatistical Mahalanobis distance function, has proven to be superior to conventional algorithms. Using data from 135 drill holes, early-stage areas of interest in the Molejón deposit were successfully identified, specifically in the NW and SE sectors.

One of the main limitations of this clustering method is its poor definition of boundaries between clusters, as well as its lower selectivity compared to conventional methodologies, particularly in failing to differentiate the Saprolite lithological domain. Despite these shortcomings, the method demonstrates potential for generating representative results in the early stages of exploratory data analysis; however, it is not recommended for direct use in defining estimation domains.

To achieve more reliable estimation domains, further methodological improvements are advised. This includes optimising parameter selection in the clustering process and implementing boundary-smoothing techniques, such as supervised machine learning approaches. Additionally, considering the anisotropy of the deposit and integrating new variables, including missing values, could enhance results.

It is important to note that the clustering method applied, along with its parameters, is specific to this case study. While adjustments can be made for different deposits, the process remains inherently subjective. Generalising clustering algorithms to other types of deposits requires significant effort and represents a promising area for future research.

Ultimately, the objectives of this study have been met, highlighting the need for continued research into methodologies that help reduce subjectivity in mineral resource estimation.

ACKNOWLEDGEMENTS

We would like to express our gratitude to our team from Dassault Systèmes for their pivotal role in enabling the execution of this study and its successful application to a real-world project. Furthermore, we gratefully acknowledge the invaluable guidance and inspiration provided by Professors Xavier Emery and Nadia Mery of the Universidad de Chile in the field of geostatistical innovation.

REFERENCES

Emery, X, 2019. Geoestadística, Universidad de Chile.

Faraj, F, 2021. A Simple Unsupervised Classification Workflow for Defining Geological Domains Using Multivariate Data, *Mining, Metallurgy and Exploration*, pp 1609–1623.

Fustos, R, 2017. Descubrimiento de unidades geometalúrgicas por medio de análisis de conglomerados geoestadístico, Tesis para optar al grado de Doctor en Ingeniería de Minas, Universidad de Chile.

Kaufman, L and Rousseeuw, P, 1990. *Finding Groups in Data: An Introduction to Cluster Analysis*, pp 68–88 (John Wiley & Sons, Inc).

Laudrum, D, 1995. 1994 Summary Report – Geological, Geochemical and Diamond Drilling Molejon Project.

Romary, T, Rivoirard, J, Deraisme, J, Quinoes, C and Freulon, X, 2012. Domaining by clustering multivariate geostatistical data, *Ninth International Geostatistics Congress*, pp 455–466.

Dynamic structural modelling for near real time inputs for geotechnical risk management and optimisation

A Jani[1]

1. Specialist Structural Geologist, Rio Tinto Iron Ore, Perth WA 6000.
   Email: akash.jani@riotinto.com

ABSTRACT

Structural complexities such as folding, faulting, jointing, and shear zones heavily influence the stability of open pit slopes in iron ore mining. As orebodies become more complex and business goals of de-carbonisation and heritage responsibility grows, technology-driven solutions are essential for increasing predictability and adaptability in mining operations. Achieving optimal pit design requires a thorough understanding and modelling of geological structures. However, managing geological uncertainty and variations in orebody geometry remains challenging. Pit face mapping generally employed to provide input into the uncertainty has very long lead times and is high risk. This often results in a reactive approach as mining progresses, leading to unforeseen geotechnical risks or ore sterilisation.

Traditional approaches that rely primarily on drill hole data and resource modelling fall short of capturing the structural detail needed to identify uncertainty and complex structural domains. As such, a pit-scale structural model is necessary for effective slope design and management.

By employing automated 'drone in a box' (DIB) UAVs for data capture and advanced implicit modelling, detailed, local-scale pit models can be created to better address these uncertainties using predictive analysis. Pit face mapping, conducted safely and efficiently using UAVs, provides data for dynamic structural models, which work in near real-time and can integrate orientation data to predict kinematic behaviours. This enhances safety and operational efficiency.

Structural domains informed by this pit mapping, along with resource model data, allow for accurate predictions of rock behaviour under varying extraction conditions. These models are continuously updated as new mapping and drill hole data becomes available, reflecting real-time geology and reducing the risk of encountering unexpected ground conditions, which boosts safety and ore recovery.

Incorporating these technologies, alongside predictive analytics, significantly reduces operational uncertainties, optimises resource extraction, and lowers costs by automating repetitive tasks and reducing human exposure to hazardous areas.

INTRODUCTION

This study focuses on enhancing the stability of open pit slopes in iron ore mining by addressing the challenges of geological uncertainty, structural complexities, and the need for technology-driven solutions. The principal objective is to improve pit design, safety, and ore recovery by utilising automated UAVs ('drone in a box') and advanced implicit modelling. Methods include UAV-based pit face mapping, real-time structural modelling, and predictive analysis. Results show that this approach reduces operational uncertainties, enhances safety, optimises resource extraction, and minimises risks such as unforeseen geotechnical issues and ore sterilisation, thereby lowering costs and improving operational efficiency.

METHODS AND RESULTS

A large part of data collection is done in the pit including pit face mapping and reverse circulation (RC) grade control drilling which feed into the resource and mining models respectively. However, this study largely focuses on remote work techniques and implicit modelling which assist in providing near real time information for structural geological uncertainty.

Structure from motion (SfM) photogrammetry is a commonly adopted technique in mining operations to remotely capture data from inaccessible or high-risk locations, Dey, Roy and Matin (2021). A drone (or UAV) is typically deployed to photograph the highwall, and through computer vision techniques, a 3D mesh is generated. This mesh can then be used for geological interpretation and orientation measurements. Recent advancements in UAV technology have led to the implementation of a 'drone in a box' (DIB) solution, which allows for automated, regular data capture as mining progresses.

An implicit model is set-up in Leapfrog Geo using the latest resource model data including all mapping and drilling data sets. The model build should include a structural surface which can then be used to inform changes in the overall geometry of the stratigraphy in the pit. Any structural uncertainty or gaps can be identified at this stage and flagged for data capture using photogrammetry.

As described by Thiele *et al* (2017), the 3D mesh is interpreted using a combination of automated facet extraction and manual plane picking techniques to identify structural features. The orientation and geometry information gathered from this interpretation is then integrated back into the implicit model to update the structural model, adding explicit controls as needed.

The next step is to address any probabilities of failure based on various kinematic failure mechanisms. This process can be employed directly in the implicit model process using the workflow from the webinar on Seequent (2020). This allows us to assess in near real time any probability of failure based on the latest information available (Figure 1). This process also allows to predict areas of uncertainty or risk based on the current data set and where the highest priority of data capture will be as mining progresses.

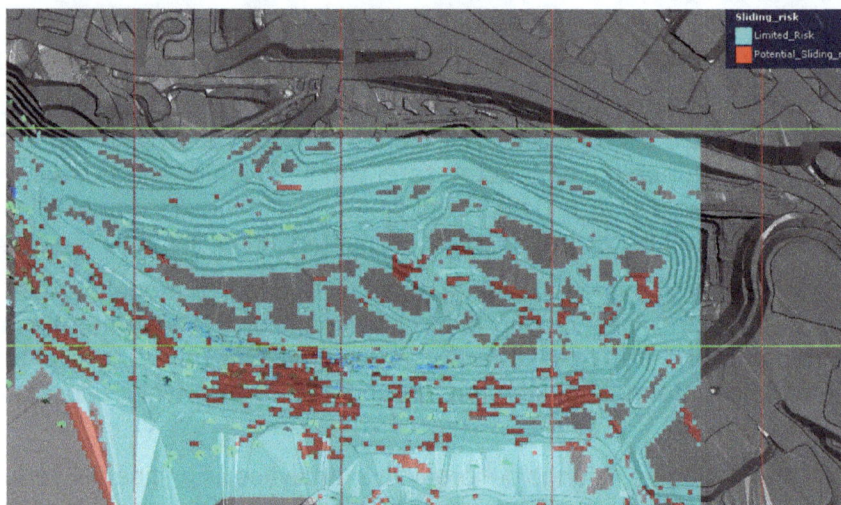

FIG 1 – Areas of pit highlighted based on probability of sliding failure risk where red is potential risk and blue is limited risk.

CONCLUSIONS

Dynamic updates to the structural model provide an efficient way to define and assess geological uncertainty, while also contributing valuable input to geotechnical stability analysis. As a remote method, it eliminates the need for personnel in high-risk environments, minimising disruption to ongoing mining operations. Additionally, it aids in proactively identifying data gaps and potential risks based on the most current data sets. By incorporating probabilistic kinematic failure mechanisms, this approach further assists in predicting potential failure modes. This input is essential for geotechnical risk assessments, as it helps identify and flag areas of high risk and uncertainty.

REFERENCES

Dey, J, Roy, S and Matin, A, 2021. Drone photogrammetry: a structural data gathering tool for open pit mining geotechnics, SRK Mining Services (India) Pvt Ltd, India, University of Calcutta, India.

Seequent, 2020. Improving geotechnical stability analysis using Leapfrog Geo: Edge Partner Webinar, SRK Consulting [online]. Available from: <https://www.seequent.com/improving-geotechnical-stability-analysis-using-leapfrog-geo-edge-partner-webinar-srk-consulting>

Thiele, S T, Grose, L, Cui, T, Cruden, A R and Micklethwaite, S, 2017. Extraction of high-resolution structural orientations from digital data: A Bayesian approach, School of Earth, Atmosphere and Environment, Monash University, Melbourne, Australia.

Resource estimation in domains with soft boundaries

M Liang[1]

1. Senior Mining Consultant, Dassault Systemes – GEOVIA, Brisbane Qld 4000.
   Email: min.liang@3ds.com

ABSTRACT

Accurate resource estimation is essential in mining projects, as it directly influences economic feasibility assessments and operational planning. A fundamental step in this process is the delineation of geological domains, traditionally based on sample data and geological interpretation. Within each domain, resource estimation is commonly performed using methods such as ordinary kriging (OK) or inverse distance weighting (IDW). These methods often simplify domain boundaries as hard, assuming abrupt changes in mineral grades at domain limits. However, in many geological settings, such as porphyry copper deposits, mineralisation transitions gradually between domains, forming soft boundaries. Relying solely on hard boundary assumptions may misrepresent these transitional zones, leading to biased estimations.

This study investigates non-stationary estimation techniques that account for spatial correlations across soft boundaries, improving the accuracy and geological validity of resource models. We evaluate and compare four estimation methods:

1. Ordinary kriging (OK) within domains: Assumes hard boundaries and performs estimation separately within each domain.

2. Ordinary kriging with a non-stationary variogram (OKNS): Introduces a location-dependent variogram, where parameters (range, nugget, and sill) vary with distance to the boundary.

3. Simple kriging with a variant mean (SKVM): Incorporates a local mean that adjusts based on proximity to the boundary, capturing local grade variations.

4. Hybrid approach (OKSKVM): Combines OK in stable areas with SKVM in transition zones, ensuring a balance between numerical accuracy and geological continuity.

The methodologies are applied to a porphyry copper data set, where experimental variograms and contact analysis confirm the presence of a soft boundary between two adjacent domains. Cross-validation results show that OKNS minimises estimation errors, while OKSKVM maintains geological continuity by ensuring a smooth grade transition. These findings highlight the importance of non-stationary models in accurately estimating resources within domains with soft boundaries.

In particular, the study emphasises the critical role of precise grade estimation in the short-term stage of mining, where accurate resource modelling directly informs production scheduling and excavation strategies. Integrating refined estimations into short-term plans ensures optimal extraction sequencing and minimises operational risk, ultimately enhancing both resource recovery and economic performance. The proposed approaches provide a more robust framework for resource evaluation in mining projects, supporting more informed and adaptive short-term planning decisions.

INTRODUCTION

Accurate resource estimation is an essential step in mining projects, as it directly influences economic feasibility assessments and operational planning. A critical aspect of this process is the precise definition of geological and mineral resource domains, which serve as the framework for subsequent grade estimation. These domains are typically defined using a combination of sample data and expert geological interpretation, incorporating lithological, mineralogical, and alteration characteristics, as well as structural features such as faults and folds. Commonly used techniques for domain definition include wireframing, radial basis functions, and machine learning based clustering methods.

Within each geological domain, it is crucial that the spatial distribution of mineral grades is internally consistent to enable reliable estimation. Traditionally, resource estimation is conducted separately within each domain using techniques such as Ordinary Kriging (OK) or Inverse Distance Weighting

(IDW) (Chilès and Delfiner, 2012). These methods generally rely solely on internal samples for variogram modelling and grade interpolation, assuming hard domain boundaries—abrupt transitions in mineralisation at domain limits. However, this assumption often fails to reflect geological reality, particularly in settings where mineralisation transitions gradually across domain boundaries. Ignoring spatial continuity across adjacent domains can introduce significant estimation bias and reduce the reliability of the model.

In many geological settings, such as porphyry copper deposits, mineralisation transitions gradually between domains rather than forming sharp boundaries. These soft boundaries arise due to complex geological processes, such as hydrothermal alteration, mineral mixing, or gradual chemical and physical transformations. Ignoring these transitions can lead to inaccurate resource estimates, as conventional methods may either overestimate or underestimate grades in boundary regions.

The key challenge in soft boundary modelling lies in accurately capturing the grade variations across transitional zones. Traditional hard boundary approaches may misrepresent the true spatial distribution of mineralisation, leading to biased or misleading estimations. To address this limitation, various studies have explored alternative geostatistical techniques (Ortiz and Emery, 2006; Maleki and Emery, 2020; Boisvert and Deutsch, 2008). Besides, non-stationary estimation techniques (Higdon, Swall and Kern, 2022; Liang, Marcotte and Benoit, 2014) account for location dependent spatial structures, allowing for a more geologically realistic and statistically robust resource estimation.

Crucially, in the short-term stage of mining, precise resource modelling is not only desirable but imperative. Grade estimation directly informs production scheduling, digging sequences, and operational tactics. Inaccurate models can lead to inefficient extraction, increased dilution, and suboptimal economic outcomes. Integrating advanced estimation techniques into short-term planning enables mining operations to better align extraction strategies with spatial grade variability, thereby improving ore recovery, reducing waste, and enhancing profitability.

This study investigates the application of Ordinary Kriging and Simple Kriging under both stationary and non-stationary frameworks, with a particular focus on their effectiveness in handling soft boundaries. By comparing these approaches using a porphyry copper case study, we aim to advance resource estimation practices that support more accurate modelling and better-informed decision-making, particularly in the context of short-term mine planning where precision and responsiveness are vital.

METHODOLOGY

To assess the impact of soft boundaries on resource estimation, we conducted a study comparing multiple estimation techniques using a data set of copper grade. The study focuses on two adjacent domains with a soft boundary between them. The geological interpretation of the deposit is beyond the scope of this article; instead, we focus exclusively on the estimation process.

The data set originates from the production stage of the operation, and the estimation model was constructed at the bench scale with a 10 m bench height. Since the grade exhibits minimal variation along elevation, the case is effectively 2D. However, a 3D model was still built because samples locations vary in elevation. To maintain confidentiality, copper grades (%) were transformed before analysis. The data set was then split into 70 per cent for training and 30 per cent for validation.

The estimation methods were implemented using Surpac, a widely used geological and resource modelling software. Block models were created, and the results were evaluated to determine the effectiveness of different estimation techniques in handling soft boundaries.

Estimation approaches

This study evaluates four geostatistical estimation approaches to model mineral grades across geological domains, with particular attention to soft boundary transitions. Each method reflects a different level of complexity in handling spatial continuity and geological realism.

1. Ordinary Kriging within Domains (OK):

Ordinary Kriging was performed independently within each geological domain, assuming that mineral grades are spatially correlated only within, and not across, domain boundaries. This approach treats boundaries as hard and impermeable, meaning no samples from adjacent domains are considered in the estimation process.

2. Ordinary Kriging with Non-Stationary Variogram (OKNS):

Ordinary Kriging with a Non-Stationary Variogram (OKNS) was employed to improve grade estimation across domain boundaries, particularly in transitional zones where mineralisation does not change abruptly. This method assumes stationarity within stable regions—those located sufficiently far from domain boundaries—where traditional Ordinary Kriging (OK) remains valid. However, in boundary-adjacent transition zones, the variogram structure is allowed to vary spatially, accommodating gradual changes in mineral grade continuity.

To achieve this, a location-dependent variogram was introduced, wherein the variogram parameters (range, nugget, and sill) vary as a function of the distance from the block centroid to the nearest domain boundary. For each block i, the distance to the boundary is denoted as d_i, and the variogram parameter at that location is represented as p_i. The spatial interpolation of variogram parameters across the transition zone is defined by:

$$p_i = \begin{cases} p_1 + \frac{1}{2} * (p_2 - p_1) \times \left(1 - \frac{d_i}{d_1}\right), point\ i\ located\ in\ domain\ 1\ with\ width\ d_1 \\ p_2 - \frac{1}{2} * (p_2 - p_1) \times \left(1 - \frac{d_i}{d_2}\right), point\ i\ located\ in\ domain\ 2\ with\ width\ d_2 \end{cases} \tag{1}$$

where p_1 and p_2 are the variogram parameters characteristic of Domain 1 and 2, respectively, and $p_1 < p_2$.

In the transition zone, all available samples that satisfy the selection criteria, such as search radius and minimum/maximum number of samples, are considered in the kriging estimation that is unlike conventional kriging, which typically restricts samples to within-domain boundaries.

3. Simple Kriging with Variant Mean (SKVM):

To better reflect local grade variations near domain boundaries, Simple Kriging with a Variant Mean (SKVM) was implemented. Unlike traditional Simple Kriging, which relies on a constant global mean within each domain, SKVM introduces spatially varying local means based on proximity to the boundary.

Local mean values were computed for groups of samples depending on their distance from the domain boundary, enabling the model to better capture localised grade trends. Simple Kriging was then applied using these location-specific means, allowing the estimation process to adapt to geological transitions within the domain.

Two SKVM configurations were tested:

o SKVMRD (Restricted Domain): Estimation used only the samples from within their respective domains. This approach retained domain integrity while accounting for local variability.

o SKVMG (Global Transition): In transitional zones, all available samples—regardless of domain—were used to capture spatial trends and ensure a smoother grade interpolation across boundaries. In stable zones, however, only domain-specific samples were utilised.

4. Hybrid Approach: Ordinary Kriging in Stable Areas and Simple Kriging with Variant Mean in Transition Areas (OKSKVM):

To combine the strengths of both OK and SKVM, a hybrid approach (referred to as OKSKVM) was implemented:

o In stable areas, Ordinary Kriging was applied using only samples from within each domain, capitalising on stationarity and the relatively consistent grade distributions.

- In transitional zones, the SKVMG variant was used. This allowed the estimation to incorporate local means derived from a broader data set that includes cross-boundary samples, thus promoting geological continuity while maintaining spatial resolution.

Performance evaluation

The performance of each estimation method was evaluated by comparing the estimated values against a testing data set comprising 30 per cent of the original samples, which were withheld from the estimation process to ensure independent validation.

To assess the effectiveness of the proposed techniques, grade estimates were first generated at the testing point locations. Subsequently, the same estimation procedures—with identical parameters— were applied to the block model to evaluate spatial behaviour on a practical mining scale.

Three primary evaluation metrics were employed:

1. Mean Error (ME): Assessed the bias of each method by measuring the average difference between estimated and actual values.

2. Mean Absolute Error (MAE): Quantified overall estimation accuracy, with lower values indicating better predictive performance.

3. Spatial Continuity Analysis: Examined the smoothness and geological realism of grade transitions, particularly across domain boundaries, to determine how well each method preserved expected geological patterns.

Together, these metrics provided a comprehensive evaluation of each approach's ability to model grade distributions in both geologically stable regions and transition zones, allowing for a robust and fair comparison among the tested estimation techniques.

APPLICATION AND RESULTS

Sample data and variogram analysis

This study was carried out on two adjacent geological domains characterised by a soft boundary, where mineralisation transitions gradually rather than abruptly. A total of 271 samples were collected from both exploration and production (blasting) drill holes distributed across the study area.

Sample composites were prepared on a bench-by-bench basis to maintain consistency in vertical sampling resolution. The spatial distribution of the composite samples is presented in Figure 1.

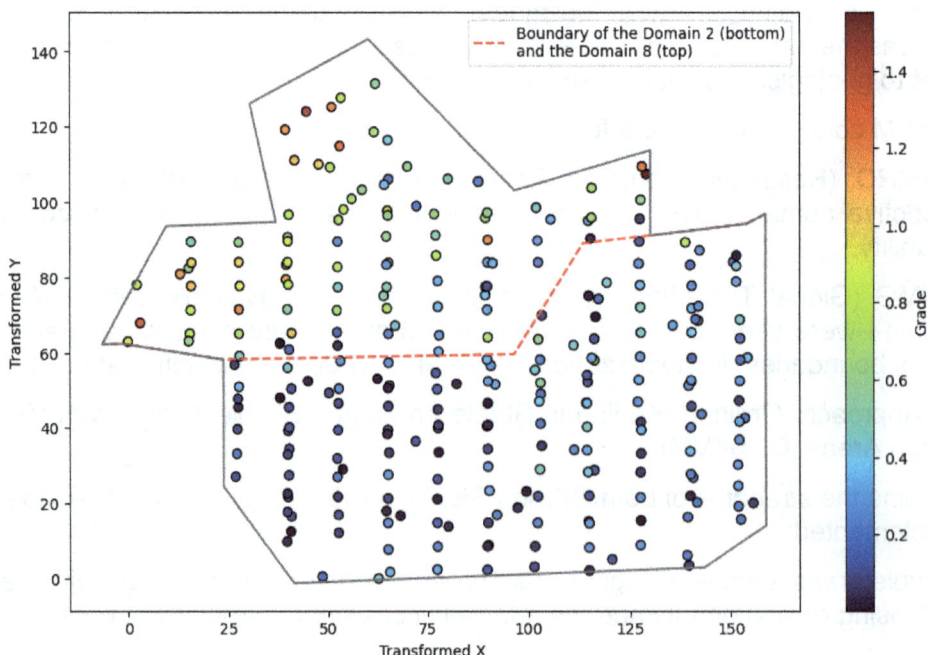

FIG 1 – The sample composites locations in two domains.

To evaluate the spatial continuity of copper grades, experimental variograms were computed separately for each domain. These variograms capture the underlying spatial structure of the grade distribution. Based on the empirical results, theoretical variogram models were fitted to each domain, allowing for more accurate grade interpolation. The fitted variogram models are shown in Figure 2 highlighting differences in spatial correlation characteristics between the two domains.

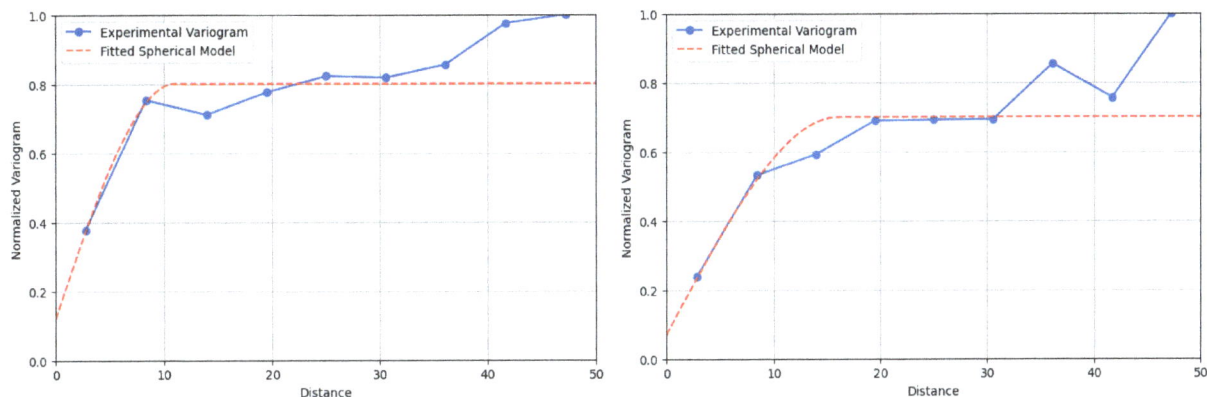

FIG 2 – Experiment variogram of copper for Domain 2 (left), Domain 8 (right) and the model fitted.

Boundary analysis and identification of transition zones

To evaluate the nature of the boundary, a contact plot was generated (Figure 3), which represents the average copper grade as a function of distance from the boundary. This plot revealed a gradual transition in grade values rather than an abrupt change, confirming the presence of a soft boundary. Based on this analysis, the block model was divided into stable zones (regions where grade variations are minimal and relatively uniform) and transition zones (areas near the boundary where grade variations are more pronounced). The transition zone was defined with a width of 10 m on the side of Domain 2 and 20 m on the side of Domain 8, reflecting the asymmetric nature of the boundary gradient as observed in the contact plot.

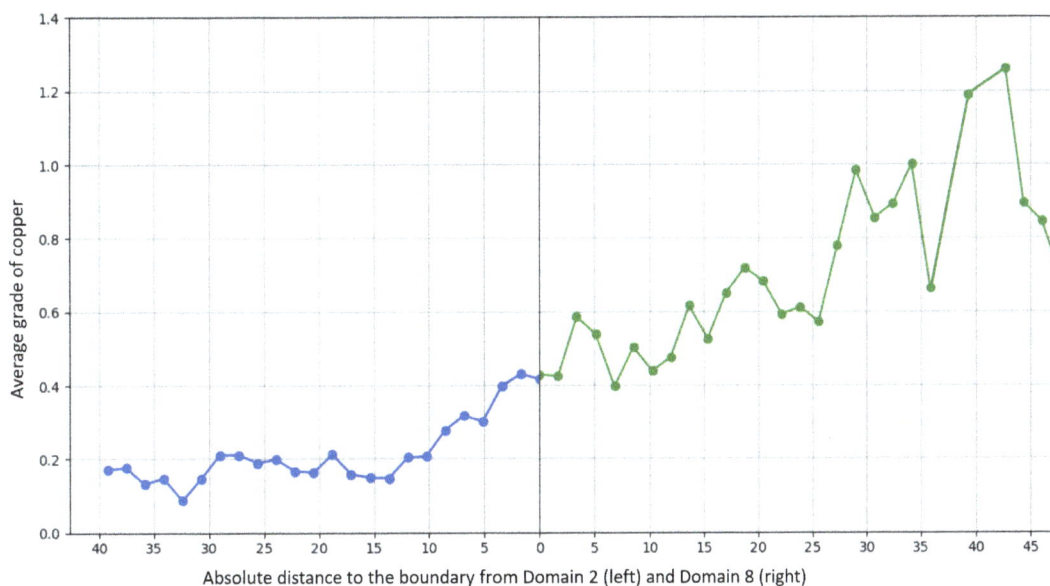

FIG 3 – Contact plot of two domains.

Block model construction and grade estimation

Grade estimation was carried out at both individual testing points and across a full block model to evaluate performance and support short-term mine planning. Testing point estimations were used to quantitatively assess the accuracy of each method, while block model estimations facilitated visualisation and operational application.

A high-resolution bench-scale block model was constructed using a uniform block size of 1 × 1 × 1 m, ensuring detailed spatial resolution suitable for short-term modelling and production planning. The proposed estimation methods (described in the Methodology section) were applied to:

- Predict copper grades for blocks in the model.
- Estimate grades at the exact locations of the testing points for validation purposes.

In all estimation processes, the five closest samples within a search radius equal to the variogram range were used. If fewer than five samples were found within the search radius, a minimum of three samples was used to ensure reliable estimation. Given that the sample data set includes both exploration and production (blasting) drill holes, a sufficient number of samples was available throughout the study area to meet these criteria.

To maintain independence in validation, all testing points were excluded from the data sets used for estimation.

Visualisation and spatial evaluation of the block model outputs were conducted using the 3DEXPERIENCE platform, which enabled integrated analysis, three-dimensional comparison of estimation methods, and qualitative assessment of spatial grade continuity.

Performance evaluation and results

The performance of the five proposed estimation methods was evaluated using both statistical accuracy and geological consistency. Table 1 summarises the descriptive statistics of copper grades from the testing data set and the corresponding estimates. Estimation accuracy was quantified using Mean Error (ME) and Mean Absolute Error (MAE), as presented in Table 2. Additionally, Figure 4 displays scatter plots comparing estimated grades against the testing data, while Figure 5 illustrates the spatial distribution of copper grades in the block model for each method.

TABLE 1

Statistics of copper grade from original testing data and five estimations.

	Domain 8			Domain 2		
	Number	Mean	Variance	Number	Mean	Variance
Original data	28	0.6357	0.1193	54	0.1983	0.0217
OK	28	0.6832	0.0678	54	0.2055	0.0082
OKNS	28	0.6839	0.0659	54	0.2045	0.0083
SKVMRD	28	0.6329	0.0396	54	0.2003	0.0054
SKVMG	28	0.6075	0.0400	54	0.2155	0.0074
OKSKVM	28	0.6472	0.0658	54	0.2162	0.0086

TABLE 2

Copper estimation performance of five cases.

	Globally		Transition region	
	ME	MAE	ME	MAE
OK	0.0313	0.1355	0.0251	0.1387
OKNS	0.0307	0.1262	0.0243	0.1208
SKVMRD	0.0224	0.1359	0.0249	0.1316
SKVMG	0.0303	0.1331	0.0315	0.1229
OKSKVM	0.0364	0.1274	0.0315	0.1227

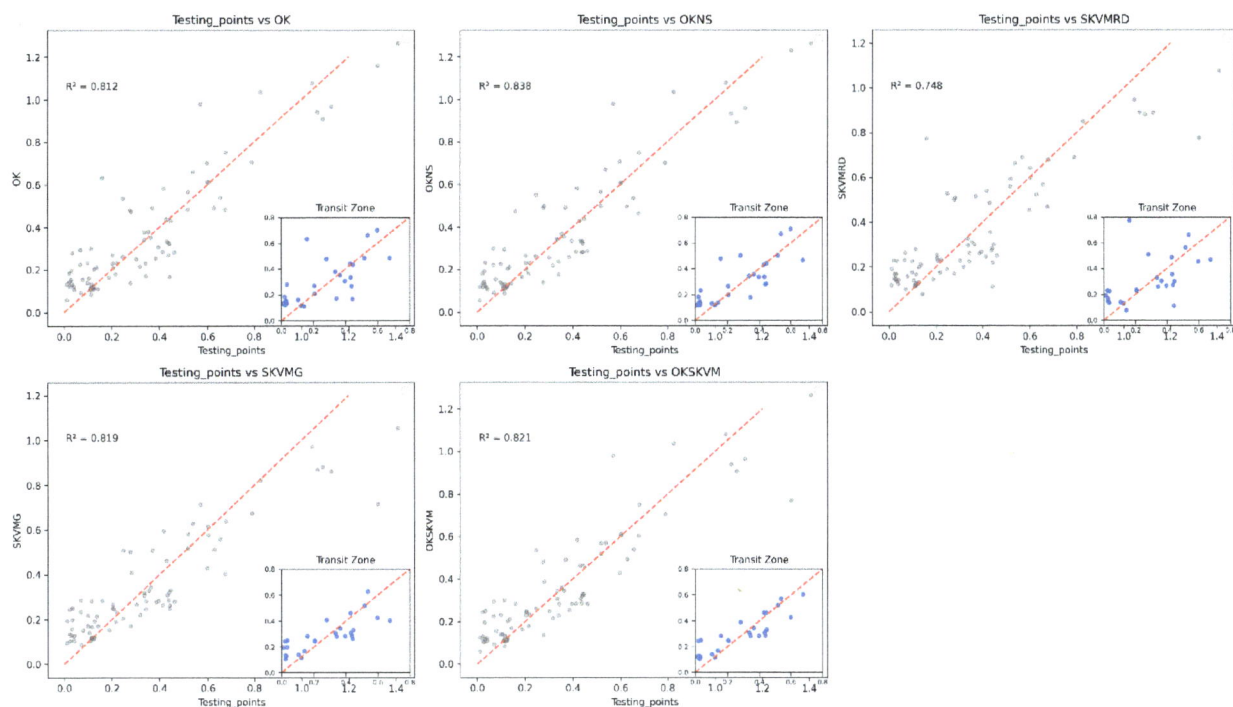

FIG 4 – Scatter plots of estimated copper grade on testing points by OK, OKNS, SKVMRD, SKVMG, and OKSKVM.

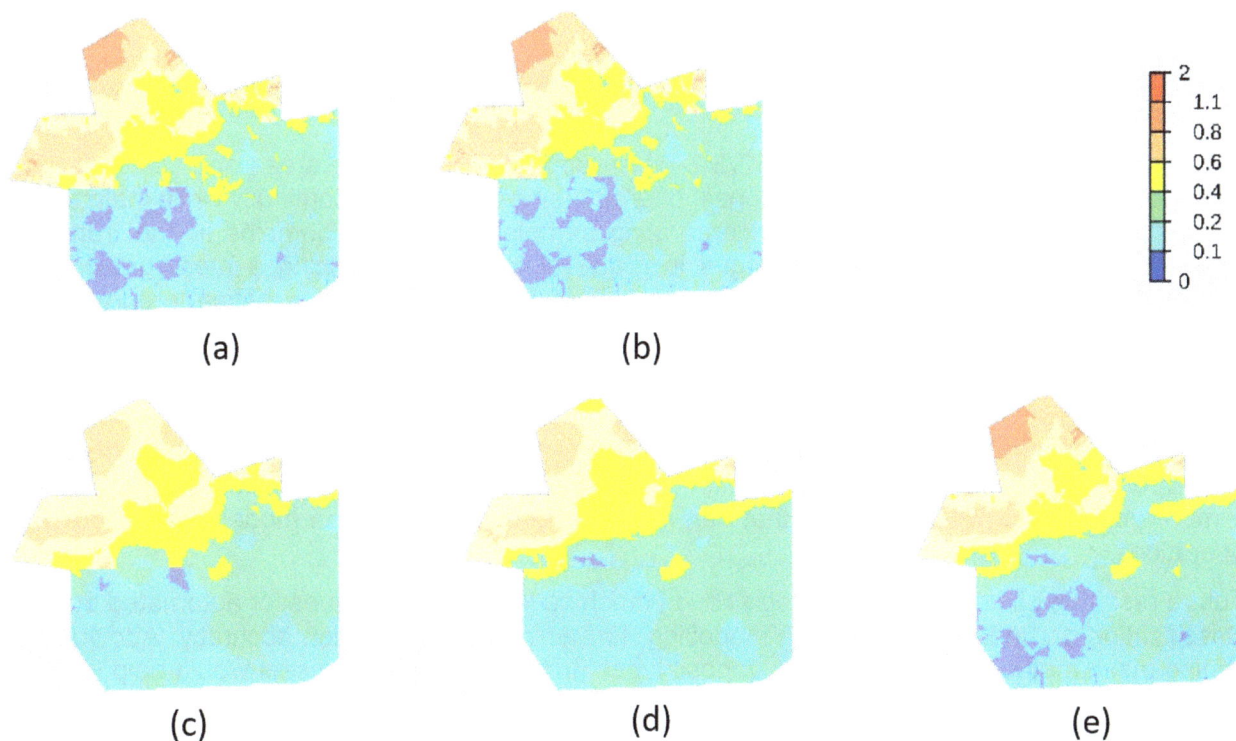

FIG 5 – Estimated copper grade in block model by: (a) OK, (b) OKNS, (c) SKVMRD, (d) SKVMG, and (e) OKSKVM.

Across all methods, ME values remained close to zero, indicating that the estimates were largely unbiased. However, all methods exhibited variance underestimation relative to the original data set—a typical effect of kriging-based techniques due to spatial smoothing. Among the five methods, SKVMRD produced the closest mean values to the original testing data, as it incorporates local means without cross-boundary influence, making it effective in stable zones.

In terms of MAE, OKNS (Ordinary Kriging with a Non-Stationary Variogram) achieved the lowest error, making it the most numerically accurate approach overall. By adjusting variogram parameters (range, nugget, and sill) based on the distance to the domain boundary, OKNS effectively modelled the gradual grade transitions characteristic of soft boundaries. As seen in Figure 4, this approach resulted in estimated values that were more tightly clustered around the actual testing data, particularly in the transition zone, compared to standard OK.

Simple Kriging with a Variant Mean (SKVM) prevented abrupt changes at domain boundaries, ensuring a smoother spatial transition between domains. As a result, MAE values were lower compared to stationary Ordinary Kriging (OK), particularly in areas near the boundary. In the SKVMG case, where data from both domains were incorporated within the transition zone, the method better reflected the natural mineralisation patterns typically observed in deposits with soft boundaries. However, in stable regions far from the boundary, SKVMG produced smoother estimations, leading to performance similar to stationary OK at a global scale.

Overall, Ordinary Kriging with a Transition Variogram (OKNS) is preferred for minimising estimation errors. However, integrating Simple Kriging with a Local Mean (SKVM) in the transition zone further enhances spatial continuity and geological validity, making the hybrid approach (OKSKVM) the most geologically realistic method (the case (e) of Figure 5).

CONCLUSIONS

This study evaluated multiple estimation techniques for handling soft boundaries in resource estimation, focusing on their numerical accuracy and geological consistency. The results indicate that ordinary kriging with a transition variogram (OKNS) provided the best numerical accuracy, minimising estimation errors both globally and in transition zones. By allowing variogram parameters to vary based on distance to the boundary, OKNS effectively captured the gradual grade transition across the soft boundary, reducing bias and improving local accuracy. However, this method resulted in a sharp contrast at the domain boundary, which may not always align with real geological continuity.

In contrast, the hybrid approach (OKSKVM), which applies OK in stable areas and simple kriging with a local mean (SKVM) in transition zones, offered the best geological continuity while maintaining competitive numerical accuracy. By incorporating samples from both domains in the transition region, OKSKVM avoided abrupt changes in grade distribution, producing a more geologically realistic representation of mineralisation patterns. This balance between statistical precision and spatial continuity makes OKSKVM a promising method for soft boundary resource estimation, particularly in deposits where mineralisation transitions gradually across domains.

The findings highlight the importance of considering spatial correlations across soft boundaries rather than relying solely on traditional hard boundary estimation techniques. While OKNS is preferred for minimising numerical estimation errors, integrating SKVM in transition zones provides a more geologically valid approach that can improve resource estimation models used in mining operations.

Future research could explore the integration of machine learning techniques for automated domain classification and adaptive variogram modelling, further refining estimation accuracy. Additionally, testing these methods on different ore deposit types and larger-scale data sets would help validate their broader applicability in the mining industry.

By refining estimation methodologies to better account for soft boundaries, mining operations can achieve more accurate resource models, ultimately leading to better decision-making in mine planning and economic evaluations.

ACKNOWLEDGEMENTS

The author would like to express their sincere gratitude to the anonymous reviewers for their insightful comments and constructive suggestions, which have significantly improved the quality of this manuscript. We also extend our appreciation to Duncan Hall, Mark Simpson and Ralph Smith for their support and insightful discussions throughout this study.

REFERENCES

Boisvert, J and Deutsch, C V, 2008. Kriging in the presence of LVA using Dijkstra's algorithm. Available from: <https://www.ccgalberta.com/ccgresources/report10/2008-110_lva_kriging_dijkstra.pdf>

Chilès, J-P and Delfiner, P, 2012. *Geostatistics: modeling spatial uncertainty* (John Wiley and Sons).

Higdon, D, Swall, J and Kern, J, 2022. Non-stationary spatial modelling, arXiv preprint. arXiv:2212.08043

Liang, M, Marcotte, D and Benoit, N, 2014. A comparison of approaches to include outcrop information in overburden thickness estimation, *Stochastic Environmental Research and Risk Assessment*, 28:1733–1741.

Maleki, M and Emery, X, 2020. Geostatistics in the presence of geological boundaries: Exploratory tools for contact analysis, *Ore Geology Reviews*, 120:103397.

Ortiz, J M and Emery, X, 2006. Geostatistical estimation of mineral resources with soft geological boundaries: a comparative study, *Journal South African Institute Of Mining And Metallurgy*, 106(8):577.

Circular economy in iron ore mining – least squares approach to kaolinite estimation of argillaceous waste material

L J Oliveira[1], D T Ribeiro[2], E K Ferreira[3] and L F B Brito[4]

1. Mining Engineer, Geostatistician, GoGeostats Consulting, Belo Horizonte Minas Gerais 30310370, Brazil. Email: leandro.oliveira@gogeostats.com.br
2. Geologist, Geostatistician PhD, Diniz Consulting, Belo Horizonte Minas Gerais 30000000, Brazil. Email:diniztribeiro@gmail.com
3. Geologist, Federal University of Minas Gerais, Belo Horizonte Minas Gerais 30000000, Brazil. Email: elainekf@gmail.com
4. Chemical Engineer, Circlua, Belo Horizonte Minas Gerais 30000000, Brazil. Email: brunaluizafer@gmail.com

ABSTRACT

Weathered mafic is the primary waste from an iron ore mine (Carajás Province, Brazil). Due to its mineral composition, it can be used as a mineral addition to the cement industry. Analysing the properties of these materials adds value to projects and promotes regional circular economic initiatives, optimising the use of mined resources. This study aimed to infer the mineralogy of weathered mafic, focusing on kaolinite proportions in all samples from a geological database. The normative mineralogical calculation was performed using an alternative method based on Least Squares (LS) to determine the proportions of constituent minerals, minimising the quadratic error between the analysed chemical contents and the chemical contents calculated from the mineralogical norm. The method does not require prior knowledge of the mineralogical sequence, relying solely on the mineralogical context of the analysed samples.

The methodology was validated by comparing X-ray diffraction (XRD) results using the Rietveld method, showing correlations above 0.8 and average relative differences below 5 per cent. After validation, the approach was applied to drill hole databases and block models, proving crucial for predicting kaolinite distribution in mining scheduling. Additionally, absolute density was calculated, which can be used as a weighting variable in mineral resource estimation.

INTRODUCTION

Mining Iron Ore operations, located in the Mineral Province of Carajás, Pará, Brazil, generate waste materials in large quantities, and the disposal of these wastes may lead to economic and environmental problems. These wastes may have characteristics that make them technically and economically viable for use in other industrial sectors. The primary waste material from an iron mine is altered rock derived from basalts and dykes with similar composition. The Parauapebas Formation and Igarapé Cigarra Formation (Meireles et al, 1984) constitute two Neoarchean extrusive units inserted in the metavolcanosedimentary sequence (~2.76 Ga) of the Grão Pará Group, Itacaiúnas Supergroup, being represented by basaltic seeps underlain and overlain by jaspillites, rhyolites, volcanoclastic rocks, and gabbro dykes/sills, thus being inserted in the Serra Norte district in the Mineral Province of Carajás, Pará (Zucchetti, 2007). The weathering product of these rocks, referred to as weathered mafic (DM), predominantly comprises clay minerals (kaolinite) and iron hydroxides (goethite) (Eggleton et al, 1987).

One possibility is to use these wastes as a precursor of supplementary cementitious materials (SCMs) that can partially replace clinker in Portland cement, reducing CO_2 emissions and improving the sustainability of cement production. SCMs generally include industrial by-products such as fly ash, blast-furnace slag and natural materials such as limestone and clays (WBCSD and IEA, 2018).

In the present study, kaolinitic clays are found as a secondary mineral resulting from the weathering of mafic rock compositions. These clay minerals have the potential for SCM production. Mineralogical studies indicate that kaolinitic clays are the primary clay minerals present in these lithologies. The kaolinite content and other clay phases found in decomposed mafic rock, such as halloysite and illite, are key factors in enabling the chemical reactivity of the SCM after thermal processing of the raw material (Fernandez, Martirena and Scrivener, 2011).

When calcined in the 400°C to 800°C temperature range, kaolinite is converted into metakaolinite, which has pozzolanic activity. This means that this calcined material can be added to cement, reacting with hydrated cement product, called portlandite, and forming new cementitious compounds that improve strength and durability properties of cement (Brito and Dweck, 2022). Materials with kaolinite concentrations exceeding 40 per cent can be used as a pozzolanic material, allowing up to 30 per cent of cement to be replaced in mixtures (Ram *et al*, 2023). This substitution makes it possible to reduce clinker production, which is responsible for 8 per cent of CO_2 emissions (International Energy Agency (IEA), 2023).

Thus, it is essential to understand the mineralogical composition of weathered mafic rocks to be used as precursors for pozzolanic materials. However, the test used to determine mineralogy, X-ray diffraction with Rietveld refinement, requires sophisticated equipment and advanced knowledge for interpreting the diffractograms. Therefore, developing mathematical tools to estimate the kaolinite content based on chemical composition results becomes highly important, as these analyses are more widely used and routinely performed in mining operations for ore control and resource estimation.

So, this study aims to evaluate mineral paragenesis to identify the key minerals that produce a precursor of SCM. Normative Mineralogical Calculation (NMC) is the method used to convert chemical compositions in mineral fractions. Other objectives are to select a representative number of samples from the case study database, to quantify their mineral mass fraction using X-ray diffractometry (XRD), and to validate the NMC methodology by comparing these normative mineral fractions and their respective calculated chemical compositions with chemical composition analyses. Once validated, the model will be applied to the entire case study database.

METHODOLOGY

Normative Mineral Calculations are widely used when mineralogy is essential to calibrate the ore or metallurgical plant process. In addition to determining mineralogical proportions, NMC can also be used to calculate mineralogical density (rock density at zero porosity), which, when moisture and porosity saturation are considered, can improve the mass estimate by mineralogy and chemical composition. Ribeiro *et al* (2020) presented a study using NMC for iron ore deposits.

The NMC was calculated using the Least Squares Method (LSM). This method is based on the chemical composition of each mineral and the results of whole-rock chemical analyses. This technique was presented by Herrmann and Berry (2002), who proposed determining the optimal proportion of minerals by minimising the quadratic errors (LSM) between the normative global compositions and the analysed global compositions.

$$Mininum\ Error\ \sum_{i=1}^{n}[Tc_i - Ta_i]^2$$

In the formula above, where Tc represents the calculated normative composition, Ta the analysed composition, i the index of the chemical variable, and n the number of chemical variables, the major components and their respective mineral phases will have the most significant influence on the optimal solution found. Normative Chemistry is calculated by solving the system shown in Figure 1.

MC Matrix (i,j)   X   MP Matrix (j)   =   NC Matrix (i)

Element/Oxide	M1 (Kaolinite)	M2 (Quartz)	M3 (Goetite)	M4 (Anatase)
SiO_2	0.4655	1	0	0
Fe	0	0	0.6285	0
LOI	0.1396	0	0.1010	0
Al_2O_3	0.3950	0	0	0
TiO_2	0	0	0	1

Mineral	Proportion
M1	0.725
M2	0.057
M3	0.200
M4	0.017

Element/Oxide	Normative Chemistry (%)
SiO_2	39.51
Fe	12.60
LOI	12.14
Al_2O_3	28.63
TiO_2	1.70

FIG 1 – Normative chemistry calculation.

The Least Squares Method is applied to minimise the error between normative and analysed chemistry. Figure 2 shows the total squared error calculation.

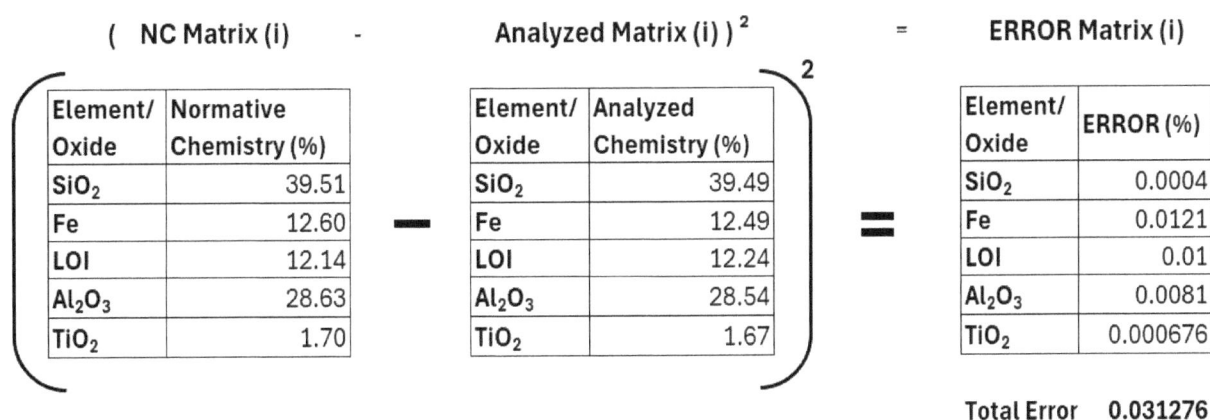

$$(\text{NC Matrix (i)} \quad - \quad \text{Analyzed Matrix (i)})^2 \quad = \quad \text{ERROR Matrix (i)}$$

Element/Oxide	Normative Chemistry (%)
SiO$_2$	39.51
Fe	12.60
LOI	12.14
Al$_2$O$_3$	28.63
TiO$_2$	1.70

Element/Oxide	Analyzed Chemistry (%)
SiO$_2$	39.49
Fe	12.49
LOI	12.24
Al$_2$O$_3$	28.54
TiO$_2$	1.67

Element/Oxide	ERROR (%)
SiO$_2$	0.0004
Fe	0.0121
LOI	0.01
Al$_2$O$_3$	0.0081
TiO$_2$	0.000676

Total Error 0.031276

FIG 2 – Squared errors between normative chemistry and analysed chemistry.

By minimising the total error, the mineral proportion is achieved.

X-ray diffraction (XRD) is a technique widely used for the identification and quantification of crystalline phases. It is based on the constructive interference of monochromatic X-rays diffracted by the periodic atomic planes of a crystal. Each mineral or crystalline compound produces a characteristic diffraction pattern, allowing for phase identification and structural analysis. In this study, XRD analysis was performed using an XRDynamic 500 diffractometer (Anton Paar) equipped with a CuKα radiation source (λ = 1.5418 Å), operating at 40 kV and 50 mA. To enhance data quality and reduce preferred orientation effects, samples were continuously rotated at 30 rev/min during measurement. Data were collected over a 2θ range of 5–70°, with a step size of 0.01° and a scan rate of 1°/min. Phase identification was carried out using the PDF-5+ database from the International Centre for Diffraction Data (ICDD). Quantitative phase analysis were performed using the Rietveld method, implemented in the JADE software. This technique fits a calculated diffraction pattern to the experimental data by refining parameters such as background, unit cell dimensions, peak profiles, scale factors, and atomic coordinates. The accuracy and reliability of the refinement were evaluated through statistical parameters, including the weighted profile R-factor (Rwp), goodness-of-fit (χ^2), and residual curves.

Methodology validation

The validation was carried out by applying the methodology to a synthetic database consisting of the following mineralogical assemblage:

- 25 per cent hematite
- 25 per cent quartz
- 25 per cent kaolinite
- 25 per cent goethite.

The chemical composition of the mineralogy mentioned above is presented in Table 1.

TABLE 1

Synthetic sample with 25 per cent hematite, 25 per cent quartz, 25 per cent kaolinite, 25 per cent goethite.

Sample	Fe (%)	SiO$_2$ (%)	Al$_2$O$_3$ (%)	TiO$_2$ (%)	Mn (%)	LOI (%)	CaO (%)	MgO (%)	K$_2$O (%)	Na$_2$O (%)	P (%)	Stoichiometry (%)
Synthetic	33.2	36.6	9.88	< lim	< lim	6.03	< lim	< lim	< lim	< lim	< lim	100.02

The Least Squares Method was applied in the synthetic sample. The result is presented in Table 2, where the normative calculation results can be compared with the synthetic model compositions. The shaded columns show the normative chemistry results, and the mineralogical proportions are very close to the synthetic samples.

TABLE 2

Synthetic sample with 25 per cent hematite, 25 per cent quartz, 25 per cent kaolinite, 25 per cent goethite.

Sample	Fe (%)	Fe (%) Normative	SiO$_2$ (%)	SiO$_2$ (%) Normative	Al$_2$O$_3$ (%)	Al$_2$O$_3$ (%) Normative	LOI (%)	LOI (%) Normative	Hematite (%)	Quartz (%)	Goethite (%)	Kaolinite (%)
Synthetic	33.21	33.2	36.64	36.63	9.88	9.86	6.03	6.02	24.95	25	25.06	24.99

The graph in Figure 3 shows the correlation between the analysed and the normative chemistry, where it is possible to observe that the obtained compositions closely match, confirming the consistency of the methodology.

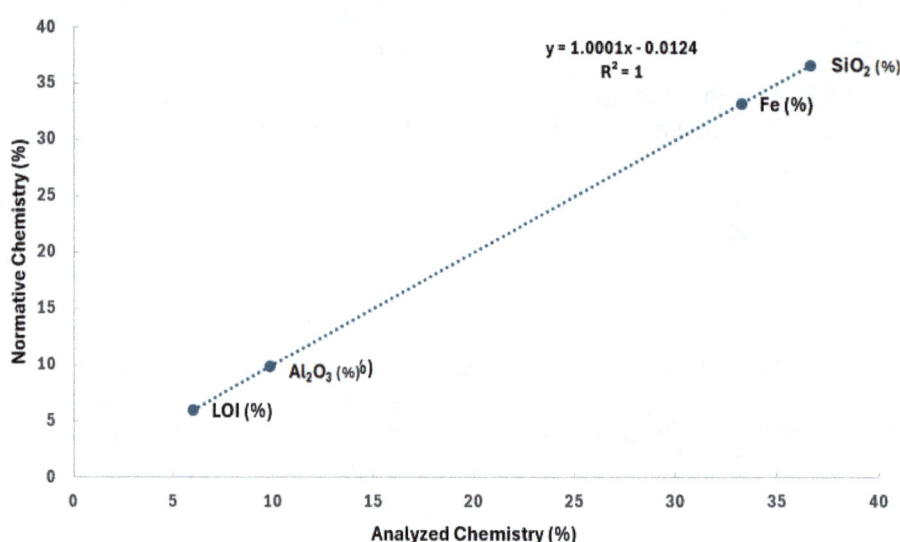

FIG 3 – Squared errors between normative chemistry and analysed chemistry.

The correlation coefficient and the slope of the linear regression line are almost equal to 1. These results demonstrate that the least squares methodology is suitable for determining mineralogy based on chemical analysis results.

In addition to this mathematical validation, defining the main mineral paragenesis is essential to avoid grouping minerals inconsistent with the geological environment.

CASE STUDY

The methodology was developed and applied to a waste from iron ore mining. The project consists of using weathered mafic to produce a precursor of SCMs. The waste lithologies are composed of several minerals. The kaolinite content and other clay phases, such as halloysite and illite, present in decomposed mafic rock are crucial for enabling chemical reactivity.

The database comprises 61 samples and includes complete chemical composition data and the mineralogical phases identified by X-ray diffraction (XRD). Figure 4 displays the diffractogram of one of the analysed samples. A diffractogram is a graphical representation of the intensity of X-rays diffracted by a crystalline material as a function of the diffraction angle (2θ). Each crystalline phase present in the sample generates a specific set of peaks at characteristic positions, defined by the spacing between atomic planes according to Bragg's law. The position (2θ) of each peak reveals the crystallographic structure, while the intensity is related to the abundance and orientation of the crystals. In diffractogram presented, it is possible to see the peaks referring to kaolinite and goethite minerals.

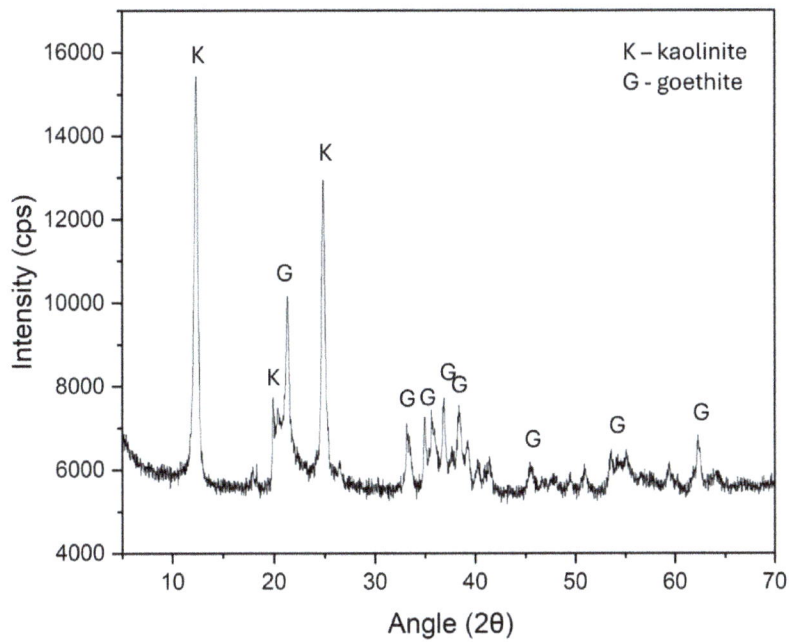

FIG 4 – Example of diffractogram of weathered mafic.

Those samples were used to calibrate the paragenesis and validate the results.

The spatial distribution is shown in Figure 5. The yellow points represent the samples with XRD results.

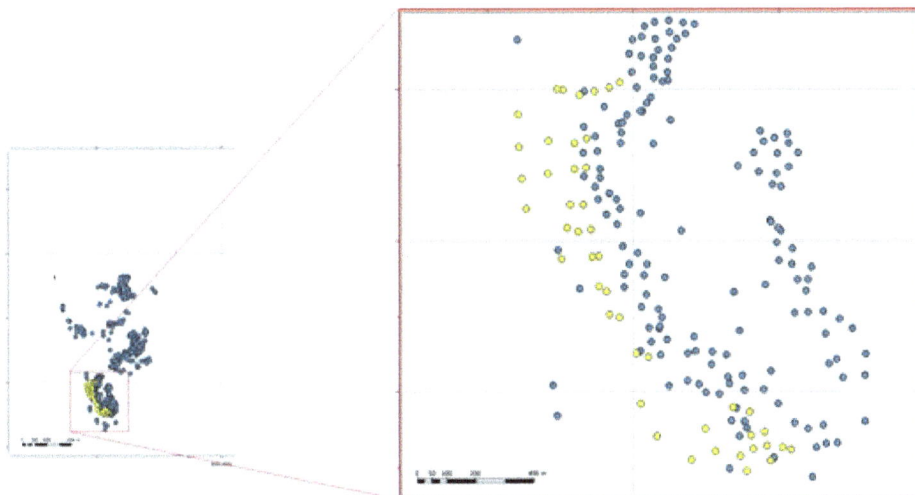

FIG 5 – Yellow points represent the XRD database distribution.

Table 3 shows the elements/oxides univariate statistical analysis.

TABLE 3

Elements/oxides univariate statistical analysis.

Analisys	Element/Oxide	Minimum	Average	Maximum
Chemical Anlalysis	SiO_2	17.80	37.81	66.40
	Fe_2O_3	1.10	20.26	40.24
	LOI	6.79	12.39	14.79
	Al_2O_3	18.44	26.88	34.38
	TiO_2	0.84	2.00	3.12
	CaO	0.01	0.06	0.49
	MgO	0.10	0.27	1.94
	P_2O_5	0.04	0.30	1.97
	Na_2O	0.10	0.15	0.23
	K_2O	0.01	0.34	3.72
	MnO	0.01	0.30	3.80
	Fe	0.77	14.17	28.14
	Mn	0.01	0.23	2.93
	P	0.02	0.13	0.86
Stoichiometry		99.19	100.76	102.29

Table 4 shows the XRD mineralogy univariate statistical analysis.

TABLE 4

Elements/oxides univariate statistical analysis.

Analisys	Mineral	Minimum	Average	Maximum
Mineralogy XRD	Kaolinite	40.60	66.53	87.80
	Quartz	0.00	5.04	50.30
	Microcline	0.00	0.32	19.30
	Muscovite	0.00	0.22	8.70
	Gibsite	0.00	0.74	9.30
	Vermiculite	0.00	0.21	12.60
	Goethite	0.00	22.21	47.90
	Hematite	0.00	0.84	18.80
	Anatase	0.00	1.02	5.00
	Halloisyte	0.00	0.10	5.80
	Illite	0.00	2.64	19.40
	Clorite	0.00	0.02	1.10
Sum Minerals		96.00	100.03	102.00

The exploratory data analysis indicates that there are 12 different minerals (listed in Table 4) present in the collected samples, and the most abundant minerals are:

- **kaolinite** – present in 100 per cent of the samples

- **goethite** – present in 95 per cent of the samples

- **quartz** – present in 41 per cent of the samples

- **anatase** – present in 36 per cent of the samples

- **illite** – present in 26 per cent of the samples.

Those five minerals account for 97.44 per cent of the mineralogy. Figure 6 shows the histogram for those five minerals.

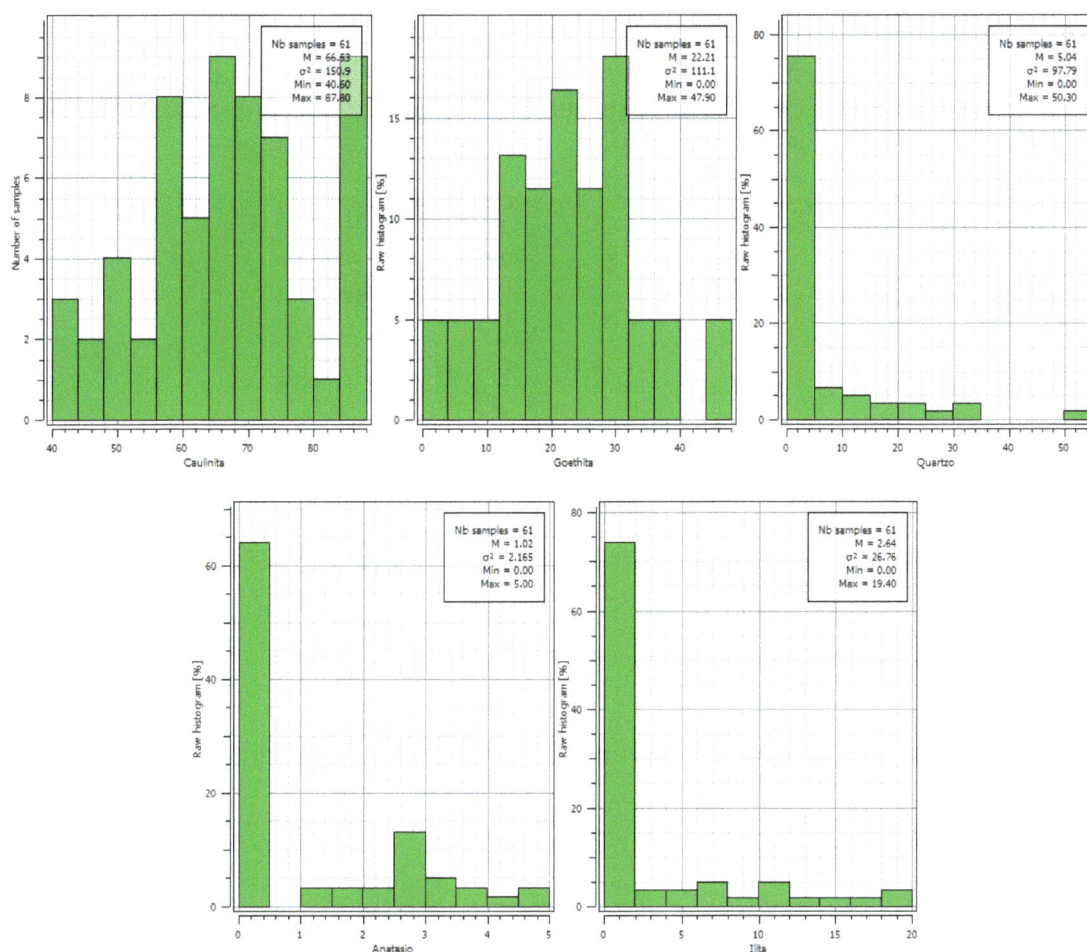

FIG 6 – Histogram kaolinite, goethite, quartz, anatase, and illite.

Normative mineral calculations – NMC

Based on the defined methodology, the mineral proportions were calculated in the database using chemical analysis results alongside the mineralogy obtained by XRD.

A mineralogical matrix was established with the following minerals and their respective compounds, listed in order of importance:

- kaolinite (SiO_2, Al_2O_3, LOI)
- goethite (Fe, LOI)
- quartz (SiO_2)
- illite (SiO_2, Al_2O_3, LOI, MgO, Fe)
- anatase (TiO_2)
- pyrolusite (Mn)
- muscovite (SiO_2, Al_2O_3, K_2O, LOI)
- apatite (P_2O_5, CaO, LOI).

A routine was developed in MS Excel® that allows the selection of minerals to be included in the computation. The routine's output consists of the input database supplemented with columns for the selected minerals, the calculated normative chemistry, and the mineralogical density.

Normative mineral calculations – validation

The input data were compared based on the results of the mineralogical calculation. Figures 7, 8, and 9, respectively, show the correlations between Fe, SiO_2, and Al_2O_3 analysed versus normative. Figure 10 shows the correlation between XRD kaolinite and normative kaolinite.

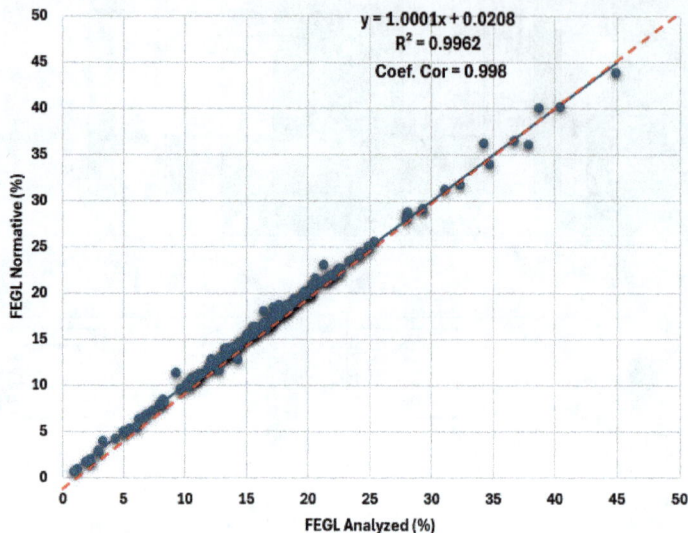

FIG 7 – Correlation between Fe analysed versus Fe normative.

The correlation between normative Fe and analysed Fe is very high, with no scattered values deviating from the trend line.

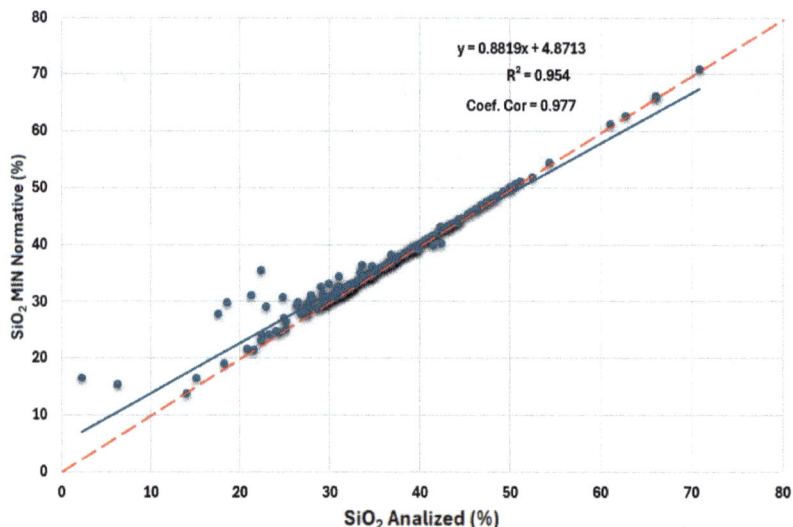

FIG 8 – Correlation between SiO_2 analysed versus SiO_2 normative.

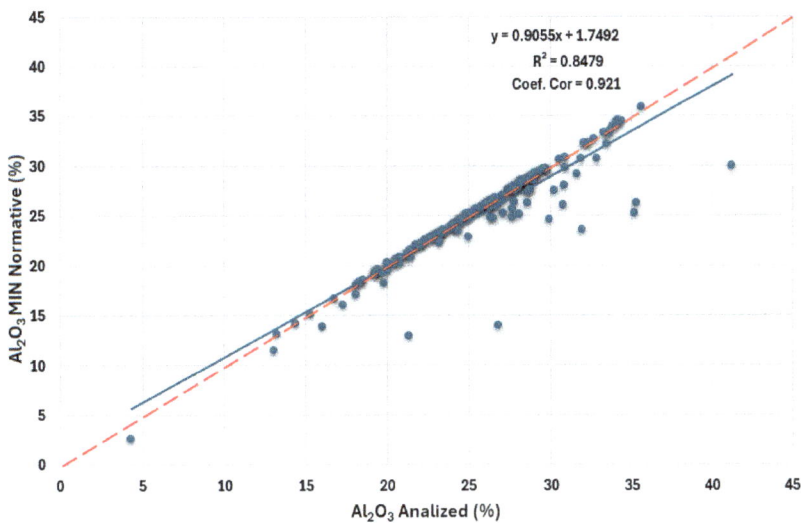

FIG 9 – Correlation between Al_2O_3 analysed versus Al_2O_3 normative.

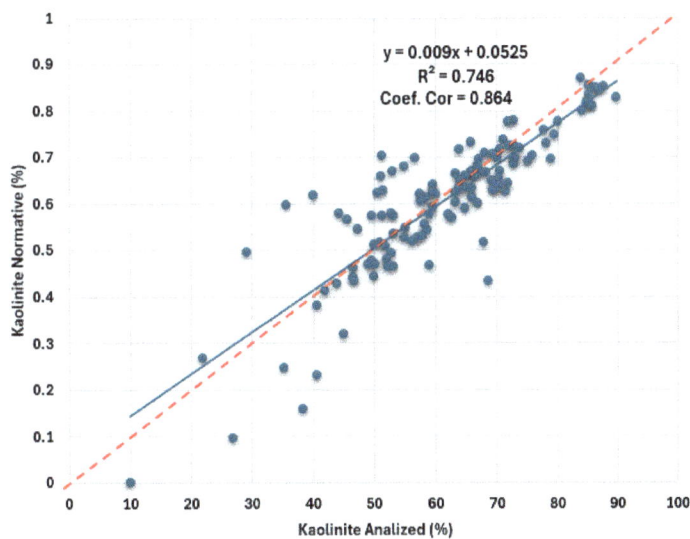

FIG 10 – Correlation between kaolinite XRD versus kaolinite normative.

The mineralogical correlation in Figure 10 is weaker than the chemical correlations in Figures 7, 8 and 9. The mineralogical correlations depend on the precision of the Rietveld methodology and the limitations of normative calculations when dealing with multiple mineral parageneses. New versions of the algorithm are being tested with multiple parageneses. The accuracy of the chemical analysis can be assessed using duplicate samples, which are generally more accurate than XRD methodology.

Table 5 compares the average of the analysed and normative mineralogy so that minor differences can be seen. The results showed that this mafic rock has potential as a precursor for pozzolanic materials, exhibiting a quality that exceeds the required, and making it suitable for partial replacement of cement.

TABLE 5
Average comparison between normative and analysed values.

Source	Kaolinite (%)	Quartz (%)	Gibsite (%)	Goetite + Hematite (%)	Anatase (%)	Ilite (%)
Normative	57.80	4.97	3.86	26.36	1.87	3.31
Analysed (DRX)	57.49	5.57	4.44	26.54	1.45	2.79

The methodology is suitable for mineralogical calculations based on the strong correlation between normative and analysed values and the small differences in the average mineral comparison.

CONCLUSIONS

The results were highly satisfactory. All regression analyses using scatter plots comparing analysed versus normative chemical values, both from XRD and LSM, and between normative mineralogy from XRD and LSM, showed high correlation coefficients and regression line slopes close to 1.

The normative calculation method generally exhibited better correlation indices between the analysed chemical variables and the normative chemistry than those associated with XRD normative chemistry.

Grouping evaluations were also performed, such as one group including quartz and another without it. However, the results did not show any improvement.

The normative mineralogy presented strong results, with correlations above 0.8 and relative differences between the means below 5 per cent. Some localised discrepancies may occur due to differences between the analysed and the input data's XRD chemistry.

This low-cost method is general and applies to all rock types (ore and waste). The quality of the results depends on the representativeness of the selected samples and the QA/QC of the chemical and mineralogical analyses.

Since kaolinite and halloysite have the same chemical composition, halloysite was not included in the normative calculation.

Using Artificial Neural Networks (ANN) for kaolinite prediction yielded highly satisfactory correlations. However, the neural network configuration does not consider coherence between the analysed and ANN-derived normative chemistry.

REFERENCES

Brito, B L F and Dweck, J, 2022. Reuse of kaolinitic waste as a precursor of pozzolanic material, *J Therm Anal Calorim*, 147:6087. https://doi.org/10.1007/s10973-021-10957-2

Eggleton, R A, Foudoulis, C and Varkevisser, D, 1987. Weathering of Basalt: Changes in Rock Chemistry and Mineralogy, *Clays and Clay Minerals*, 35(3):161. https://doi.org/10.1346/ccmn.1987.0350301

Fernandez, R, Martirena, F and Scrivener, K L, 2011. The origin of the pozzolanic activity of calcined clay minerals: A comparison between kaolinite, illite and montmorillonite, *Cement and Concrete Research*, 41:113. https://doi.org/10.1016/j.cemconres.2010.09.013

Herrmann, W and Berry, R F, 2002. MINSQ-a least squares spreadsheet method for calculating mineral proportions from whole rock principal element analyses, *Geochemistry: Exploration, Environment, Analysis*, 2(4):361–368.

International Energy Agency (IEA), 2023. Cement – Analysis. Available from: <https://www.iea.org/energy-system/industry/cement>

Meireles, E M, Hirata, W K, Amaral, A F, Medeiros Filho, C A and Gato, W C, 1984. Geology of the Carajás and Rio Verde sheets, Carajás Mineral Province, State of Pará [in Portuguese: Geologia das folhas Carajás e Rio Verde, Província Mineral de Carajás, Estado do Pará], in Proceedings SBG, Brazilian Congress of Geology, Rio de Janeiro, 5:2164–2174.

Ram, K, Flegar, M, Serdar, M and Scrivener, K, 2023. Influence of Low- to Medium-Kaolinite Clay on the Durability of Limestone Calcined Clay Cement (LC3) Concrete, *Materials*, 16:374. https://doi.org/10.3390/ma16010374

Ribeiro, D, Moraes, I, Kwitko-Ribeiro, R, Braga, D, Spier, C and Santos, P, 2020. From fresh itabirites and carbonates to weathered iron ore: mineral composition, density and porosity of different fresh and altered rocks from the Quadrilátero Ferrífero, Brazil, *Minerals*, 11(1):29.

World Business Council for Sustainable Development (WBCSD) and International Energy Agency (IEA), 2018. Technology Roadmap: Low-carbon transition in the cement industry, International Energy Agency.

Zucchetti, M, 2007. Mafic rocks of the Grão Pará Group and their relationship with the iron mineralization of the N4 and N5 deposits, Carajás, PA [in Portuguese: Rochas máficas do Grupo Grão Pará e sua relação com a mineralização de ferro dos depósitos N4 e N5, Carajás, PA], 165 p, Thesis (Doctorate in Geology) [Tese (Doutorado em Geologia)], Institute of Geosciences, Federal University of Minas Gerais, Belo Horizonte [Instituto de Geociências, Universidade Federal de Minas Gerais, Belo Horizonte].

Models with locally variable spatial continuity using covariance tables

R A Vincenzi[1], J F C L Costa[2], M A Bassani[3] and V C Koppe[4]

1. PhD Candidate, Federal University of Rio Grande do Sul, Porto Alegre 91509900, Brazil. Email: ricardo.vicenzi@gmail.com
2. Professor, Federal University of Rio Grande do Sul, Porto Alegre 91509900, Brazil. Email: jfelipe@ufrgs.br
3. Adjunct Professor, Federal University of Rio Grande do Sul, Porto Alegre 91509900, Brazil. Email: marcel.bassani@ufrgs.br
4. Associate Professor, Federal University of Rio Grande do Sul, Porto Alegre 91509900, Brazil. Email: vkoppe@ufrgs.br

ABSTRACT

Exhaustive secondary data provide valuable insights into local patterns of spatial continuity. The covariance table(s) obtained from the Fast Fourier Transform (FFT) offers a high-performance computational alternative for mapping spatial continuity from exhaustive secondary data. Traditionally, a global covariance table is obtained from the entire grid of secondary data. This global covariance table is then used for estimation/simulation algorithms. The drawback of this approach is that it assumes that the spatial continuity of the variable to be modelled is constant over the area of interest. Many geological formations, such as folded strata, exhibit spatial continuity that changes from one location to another. Using a locally variable spatial continuity model improves the estimates for this type of geological formation. This paper proposes a methodology that considers local spatial continuity in Ordinary Kriging using covariance table(s). First, local covariance table(s) are constructed using FFT applied to sub-areas of the gridded secondary data. Second, local variogram models are fitted automatically based on the experimental points extracted from the local covariance table(s). Finally, Ordinary Kriging is performed using the local variogram models. As a result, each block of the model is estimated using its variogram model. The methodology is applied to a case study using public data. The results show increased precision and accuracy of the estimates due to the use of local variograms.

INTRODUCTION

Geostatistical estimation and simulation in mineral deposits with folds are challenging. For instance, the two-fold limbs usually have different directions of spatial continuity. One alternative for performing geostatistical algorithms in this kind of deposit is to divide the data set into structural domains. In this context, one limb represents one structural domain, which is estimated separately from the other limb, which represents the other domain. Rather than dividing the data set into structural domains, the geoscientist may change the coordinates using an unfolding algorithm. In this case, the model is constructed using unfolded coordinates. A third alternative consists of using local spatial continuity models, where each block is estimated using a different variogram model (Boisvert, 2010; Lillah and Boisvert, 2015; Caixeta, 2020; Li et al, 2024). This approach is known as Local Varying Anisotropy, which is the focus of this paper.

In Earth Science data sets, exhaustive secondary variables are often available. The secondary variable is spatially correlated with the primary variable of interest. In petroleum geostatistics, secondary variables come from seismic surveys. For mining applications, secondary data may come from drone imagery and/or geophysical campaigns (gravimetry and seismic reflection, for instance). With technological advancements, secondary information has become increasingly accessible. When the correlation between the primary and secondary variables is high and linear, the spatial continuities of these variables are similar. In this paper, we explore how to use exhaustive secondary data to obtain the local spatial continuity patterns of the primary variable and incorporate these local patterns into the estimation algorithm.

The secondary variable is usually organised in a regular grid, with equally spaced samples along the X, Y, and Z directions. Quantifying spatial continuity in such data involves calculating experimental variograms in the principal anisotropy directions. The drawback of this method is that, in general, only a few directions are selected for this calculation. A better alternative involves calculating experimental variograms using the Fast Fourier Transform (Marcotte, 1996). The result is a variogram map that informs the variogram value for all distances and orientations. This technique may also be used to calculate experimental covariances and correlograms. When the experimental covariances are calculated, the covariance map is often called a covariance table (Kloeckner *et al*, 2019). The works of Kloeckner *et al* (2019) and Oliveira, Bassani and Costa (2021) used a global covariance table calculated using the entire grid of secondary data. In this work, we investigate using local covariance tables, which are obtained using sub-grids of the secondary data. These subgrids cover subareas that exhibit distinct spatial continuity patterns.

The problem with the FFT-derived covariance tables is that they may not be directly used in geostatistical simulation/estimation algorithms. This problem occurs if the covariance matrices coming from the covariance table are not necessarily positive definite. As a result, the kriging system may not have a unique solution. To overcome this problem, one solution is to correct the covariance table (Yao and Journel, 1998) so that the corrected covariance table is valid; that is, the covariance matrices obtained from this valid covariance table are positive definite. Another alternative for ensuring positive definiteness is to obtain the covariance values from a Linear Model of Regionalization (Goovaerts, 1997). The Linear Model of Regionalization is, by definition, positive definite because it is constructed as a linear combination of known positive definite functions.

The Linear Model of Regionalization is usually obtained by manually fitting the variogram model to the experimental variograms. This approach is practical when only one variogram model is used for the entire area of interest. In our research, each block is assigned its variogram model. In this context, manually fitting a variogram model for each block would be impractical. For this reason, we used an automatic fitting method (Larrondo, Neufeld and Deutsch, 2003) to obtain local variogram models that fit the experimental covariances coming from the local covariance tables — referred to as the locally variable spatial continuity model (LVCM).

The result is a methodology that, with the aid of an exhaustive secondary variable, automatically fits local variogram models and uses this local at kriging estimates. The technique is shown in a case study using synthetic data so that the true data is known. The estimates obtained with the proposed method are compared to those obtained with traditional kriging, which uses only one variogram model per domain.

METHODOLOGY

A covariance table (CT) was created using Marcotte's method (1996), based on the Fast Fourier Transform (FFT) algorithm developed by Cooley and Tukey (1965), implemented in the Python programming language through the NumPy library (Van Der Walt, Colbert and Varoquaux, 2011). After the CT was established, experimental covariances were extracted from the covariance table along pre-defined directions, as illustrated in Figure 1.

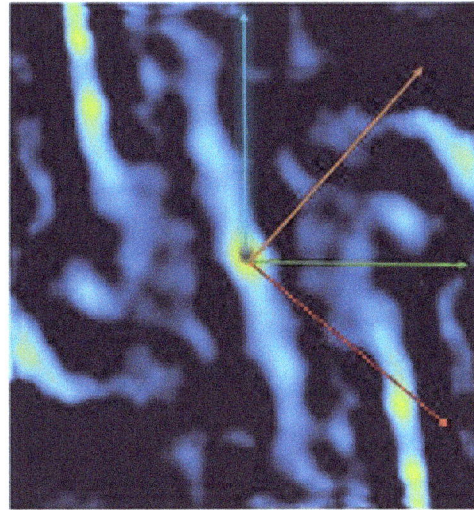

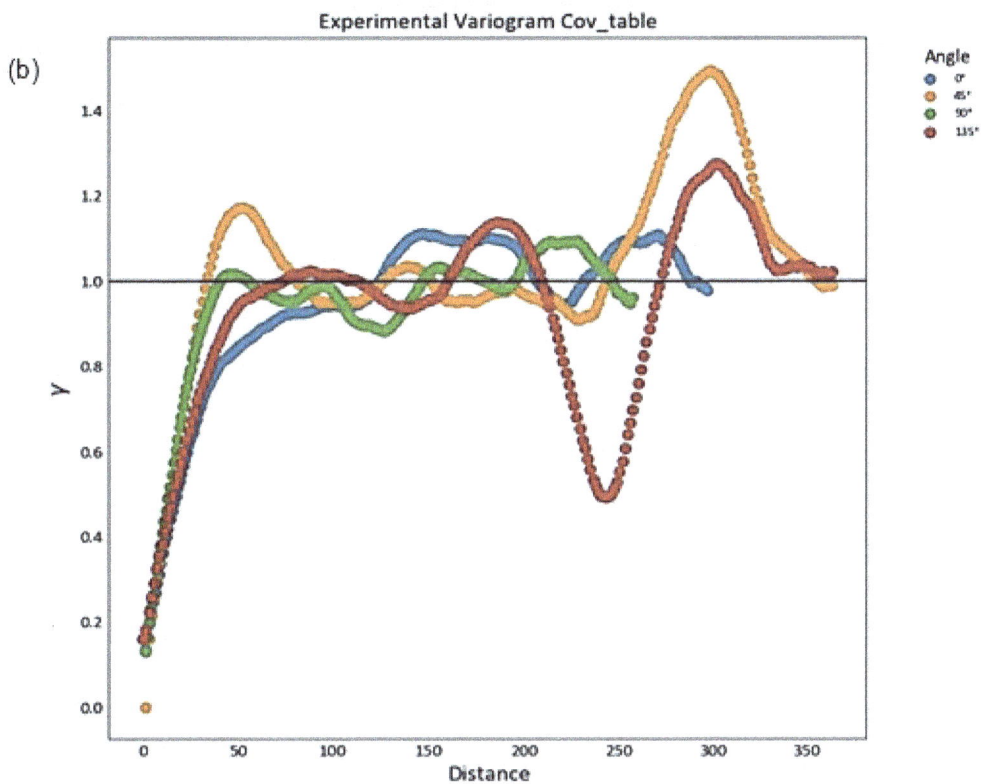

FIG 1 – (a) The CT shows the generation of vectors at predetermined angles; (b) the experimental variograms are shown along vectors from the same CT at every 45°.

The size of the CT is influenced by the Observation Window (OW) used. The OW is the sub-area of the estimation grid used to calculate the CT. A large OW does not capture local variations of spatial continuity, while a small OW may result in a noisy covariance table. A sensitivity analysis is recommended to determine the ideal OW or the method proposed by Yao *et al* (2007) that used the size of the smallest range of the global horizontal variogram of oil and gas deposits (Figure 2).

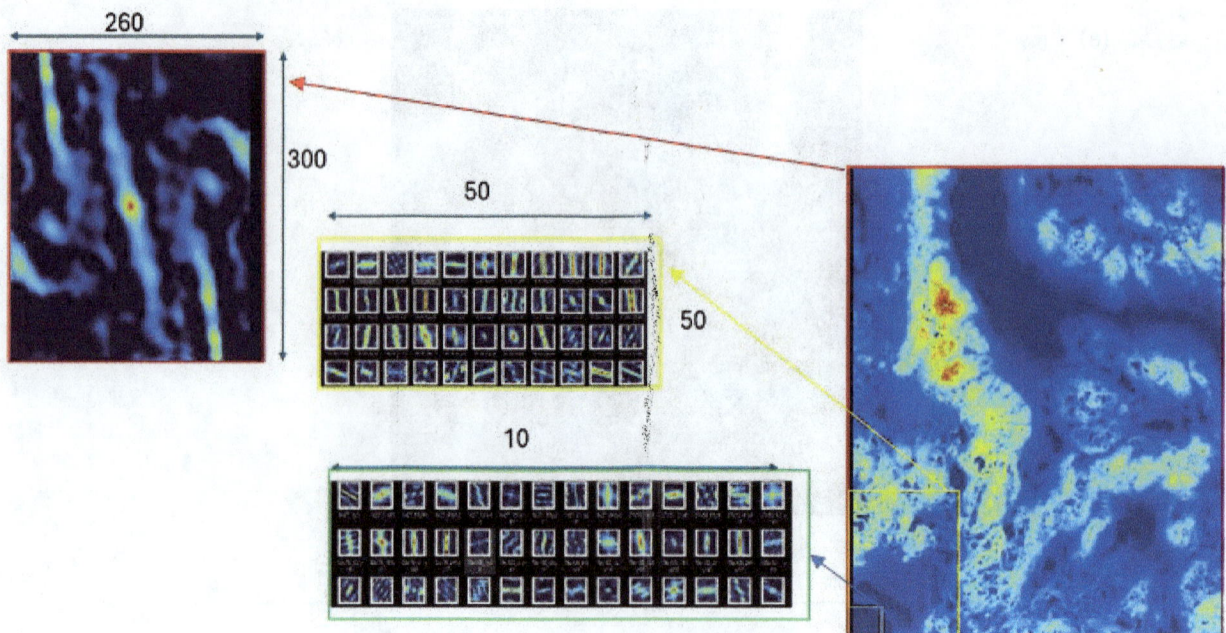

FIG 2 – The figure presents the overlapping covariance windows over the data. In red is the global window; in yellow, a 50 × 50 m window positioned at the origin; and in green, a 10 × 10 m window. Their respective covariance tables accompany each of these windows.

The experimental covariance table is not necessarily positive definite, a condition required for estimation and simulation algorithms based on kriging. For this reason, an automatic variogram fitting method is employed. This automatic fitting approach was previously applied by Machuca-Mory and Deutsch (2009). Larrondo, Neufeld and Deutsch (2003) explained that the goal is to minimise the distance between the variogram model and the experimental variograms through objective functions that randomly vary their parameters and perform two checks. The first check aims to minimise the objective function; any change that increases the objective function is rejected, while changes that decrease it are accepted. The second check ensures that the Linear Model of Regionalization (Journel and Huijbregts, 1978) is positive semi-definite. Changes resulting in a model that is not positive semi-definite are rejected. The local variogram models obtained from the local covariance table(s) are used to estimate the block located at the centre of the OW. The current implementation does not estimate the blocks on the edges of the estimation grid.

The estimation was performed using Ordinary Kriging (*OK*). Each node is estimated independently using a local search strategy (Figure 3).

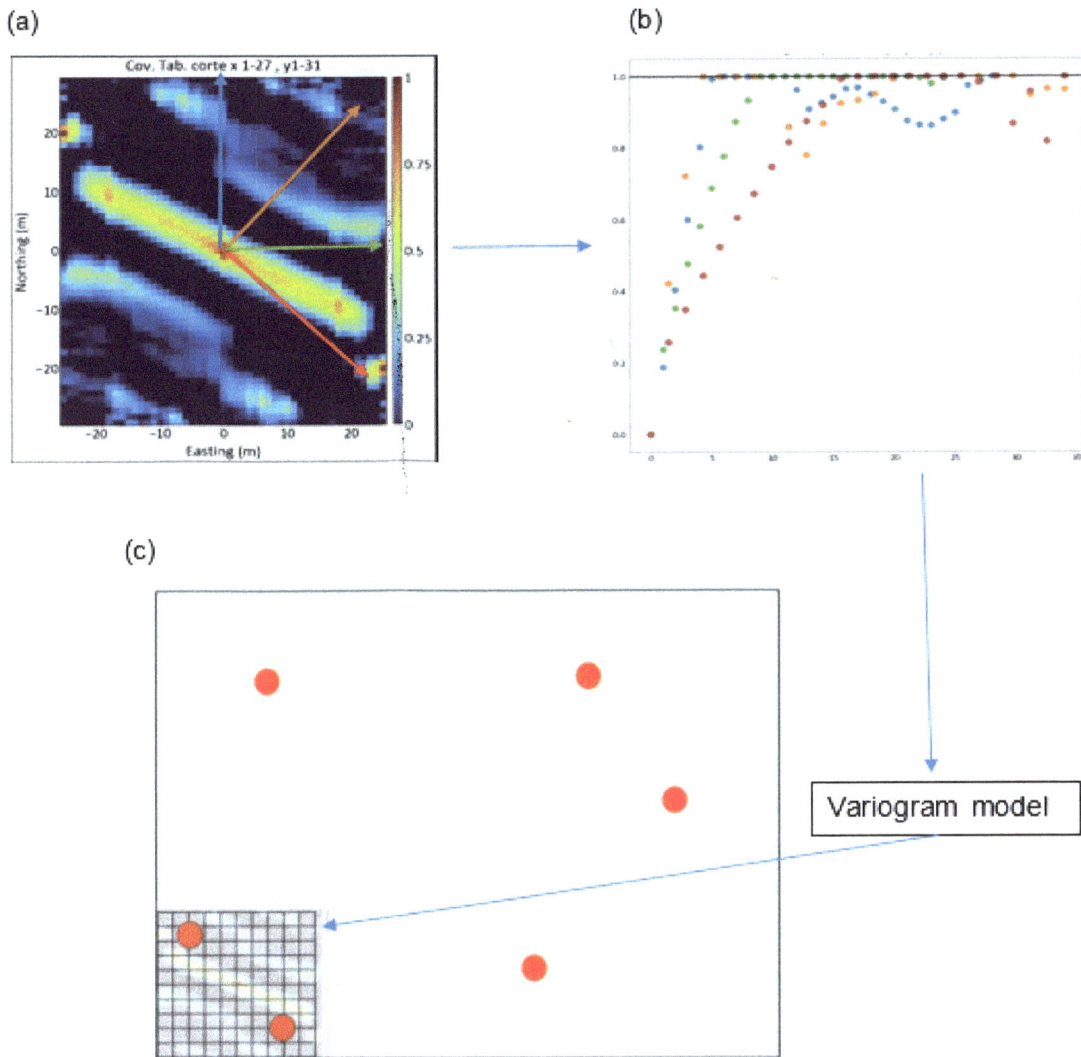

FIG 3 – Illustration of Local OK using LVCM: (a) note the CT transformed into a spatial continuity function; (b) shows its use in the estimation process; (c) with samples in red and the local grid with local estimation in the lower corner.

DATA SET

The Walker Lake data set (Isaaks and Srivastava, 1989) is used in this study. The exhaustive data (Figure 4a) is considered the ground truth and is used to compare the estimates obtained with the different methodologies. The data used for estimation are regularly spaced at every 10 m along the X (East) and Y (North) directions (Figure 4b).

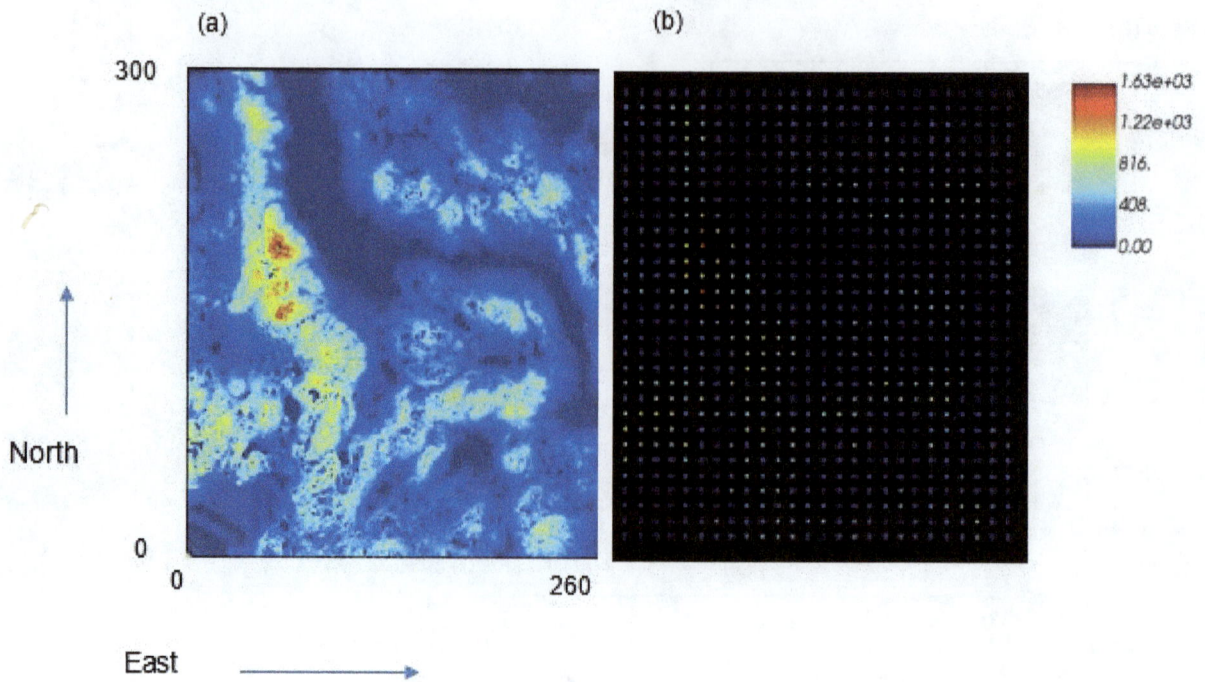

FIG 4 – (a) The exhaustive Walker Lake data set; (b) the regular sampling every 10 units (hard data).

We added random uniform noise of ± 25 per cent to the original exhaustive data (Walker Lake – V values) and used it to mimic secondary data. As the secondary data refers to a variable that is not the variable of interest, adding noise to the exhaustive data of the primary variable is reasonable. The purpose of generating noise is to create a synthetic model of secondary data, which differs sufficiently from reality to simulate the secondary data collection process. The data were renamed *Orig* and *Orig with Noise*, and their distributions, along with a statistical summary of the samples, are presented in Figure 5.

Samples Orig

Orig
Mean: 274.27, Std: 248.26, Var: 61633.96, Median: 214.85

(a)

Comparing Distributions: Orig vs Orig with Noise

Orig
Mean: 277.98, Std: 249.84, Var: 62422.43, Median: 221.25
Orig with noise
Mean: 278.00, Std: 255.89, Var: 65478.06, Median: 217.00

(b)

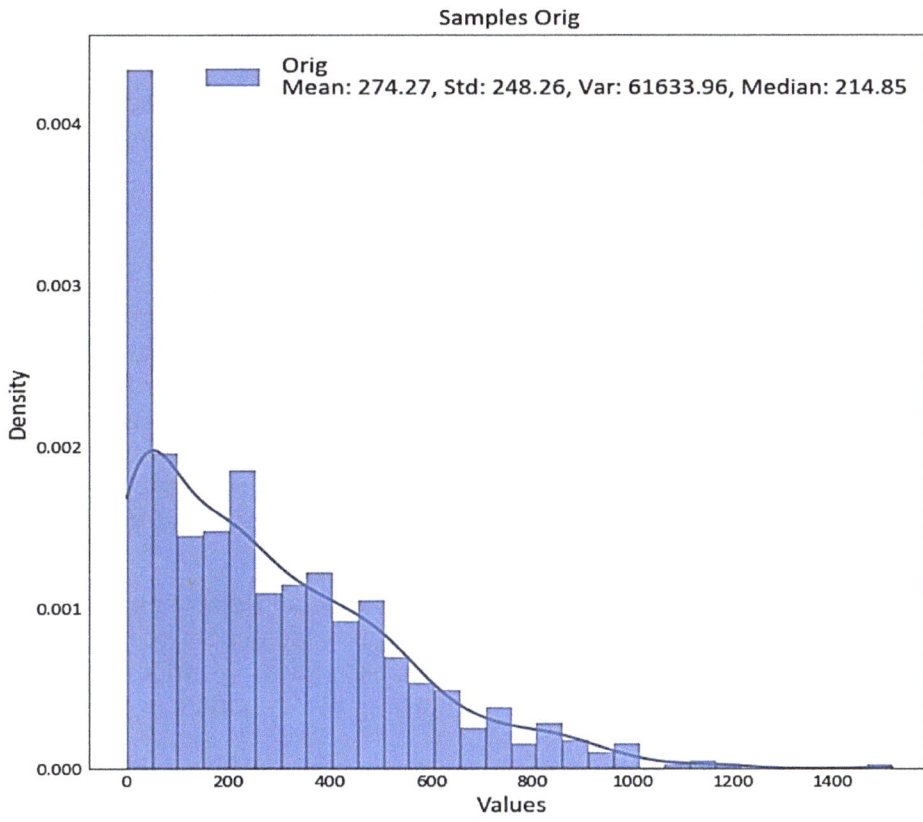

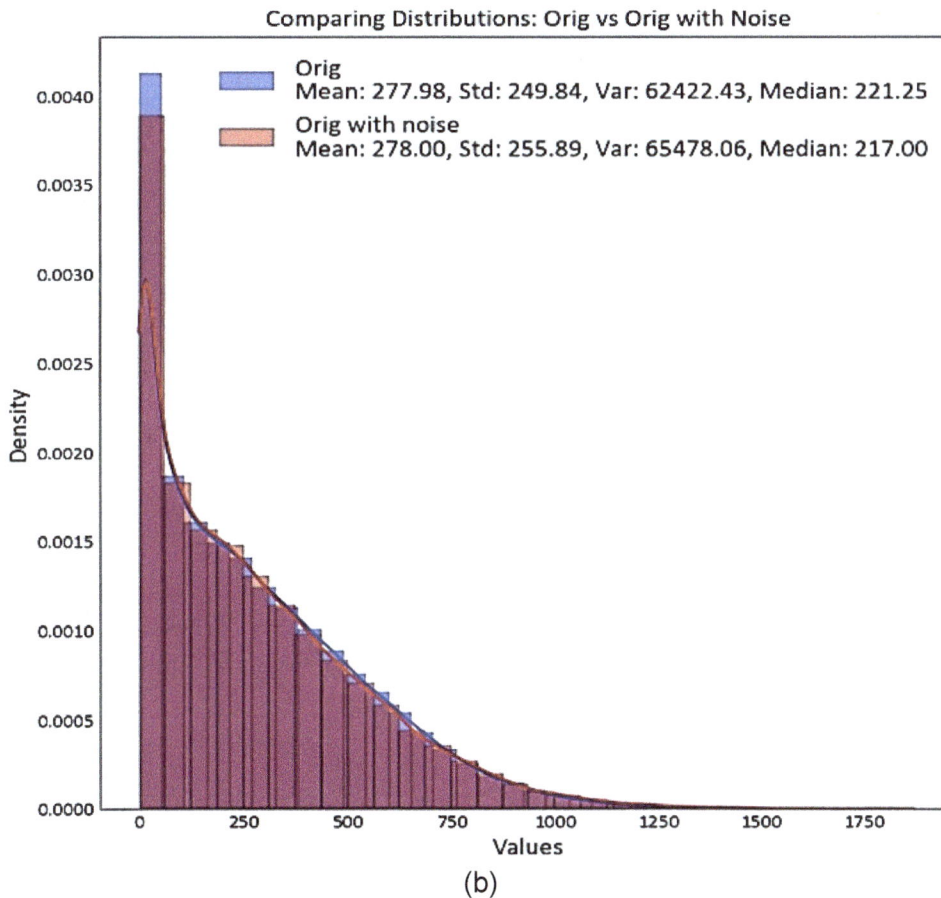

FIG 5 – (a) The distribution of the samples; (b) compares the distributions of the original exhaustive data and the data with noise added.

RESULTS AND DISCUSSIONS

The results show that kriging using a locally variable spatial continuity model (LVCM) achieves better adherence to the data, as demonstrated by swath plot analysis and reduced mean squared error (MSE). Additionally, the method enhances the visual reproduction of the spatial continuity changes in a fully automated process.

Experiment set-up and rationale

The first part of the experiment applied the methodology to regularly sampled data (every 10 m) and used variograms extracted directly from an exhaustive database with no added noise (zero nugget effect). Observation windows (OW) of 26 × 30 units were used, corresponding to one-tenth of the global grid dimensions. This window size was chosen to capture the shortest spatial continuity structures identified in the global variogram. Each window generated a covariance table (CT), and along pre-defined directions (every 45°), experimental variograms were computed. A predefined direction is chosen by the modeller. To improve reproduction, the number of vectors in the CT can be increased. These variograms served as input for the automatic fitting algorithm, which produced the locally variable spatial continuity model (LVCM).

The models were compared in terms of spatial connectivity and statistical behaviour (Figure 6). Models using local variograms exhibited more realistic spatial connectivity compared to the traditional global variogram approach. However, the non-overlapping window approach introduced discontinuity artifacts at window boundaries, reinforcing the importance of overlap for smooth transitions.

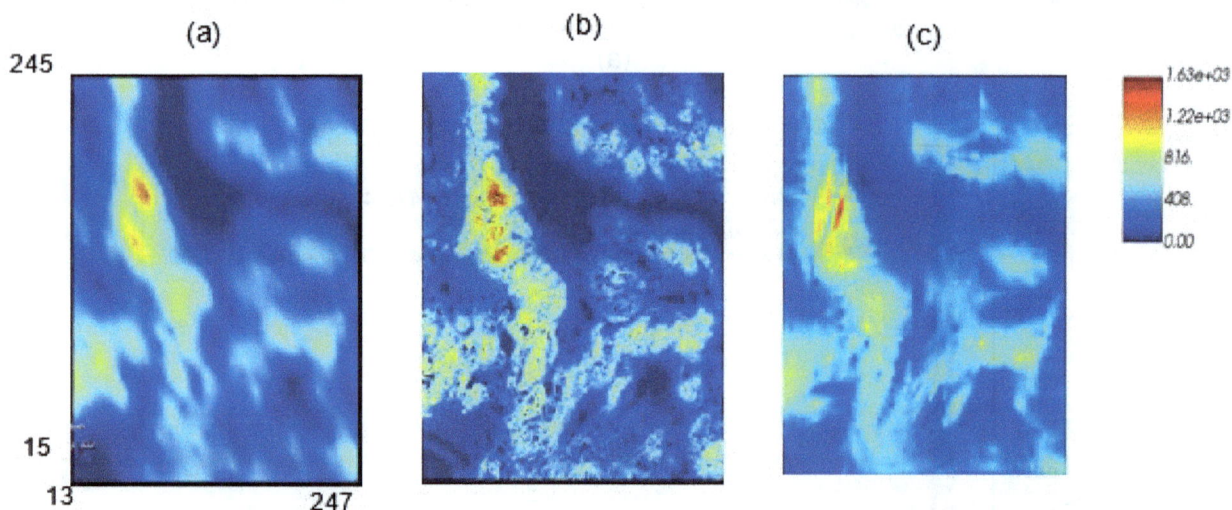

FIG 6 – (a) The traditional kriged model; (b) the original exhaustive Walker Lake database; (c) the OK model using fixed LVCM windows.

Swath plot analysis and statistical performance

Swath plot analysis provided further insights (Figure 7). Estimates obtained with LVCM closely followed the true local section mean, significantly outperforming the traditional OK approach, which produced overly smoothed estimates. This improvement highlights the benefit of capturing local spatial continuity through local variograms.

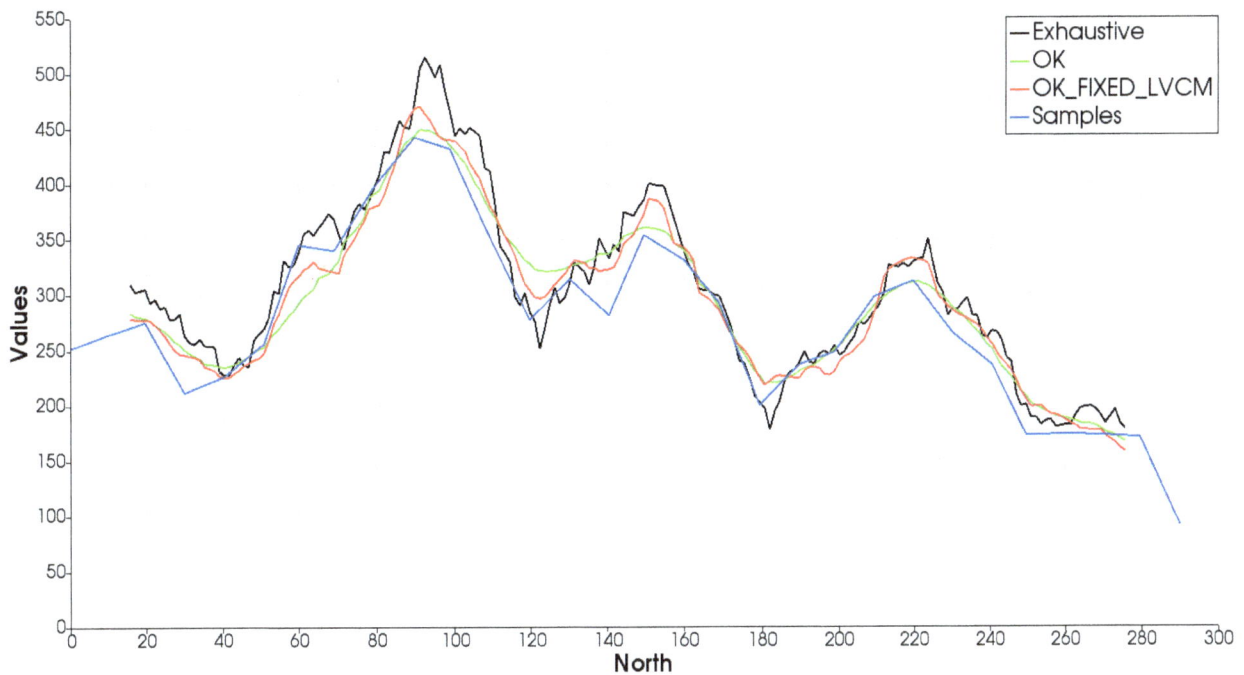

FIG 7 – Swath plot analysis comparing the local continuity method (*OK_Fixed_LVCM*), traditional Kriging (*OK*), samples, and exhaustive data.

Sensitivity to search strategy

The impact of the neighbourhood search strategy was also evaluated, utilising the maximum and minimum number of samples in ordinary kriging, as shown in Table 1.

TABLE 1

All experiments search strategies with maximum (Max) and minimum (Min) number of samples used in ordinary kriging (OK).

Samples	Min	Max
OK	4	12
OK_FIXED_LVCM	4	12
OK_FIXED_LVCM_NOISE	4	12
OK_LVCM	4	12
OK_LVCM _I	8	16
OK_LVCM _II	2	6
OK_LVCM _II_NOISE	2	6

Results compared (OK_LVCM, OK_LVCM _I e OK_LVCM _II) showed that the search strategy had a measurable impact on the estimates (Figure 8). Using too few samples increased variance, while using too many samples introduced over-smoothing due to negative kriging weights. These results emphasise the need to carefully fine-tune search parameters to balance local variability and stability.

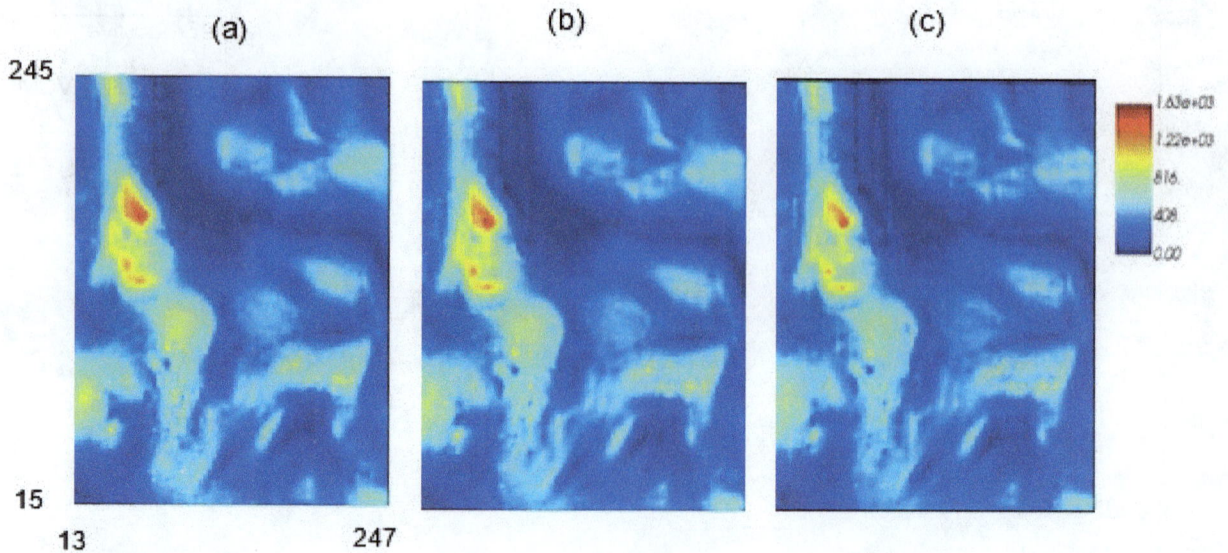

FIG 8 – OK with LVCM using several different search strategies, from fewer data in (a) to more data in (c).

Incorporating secondary data with noise

The methodology was further tested using secondary data with added random uniform noise (±25 per cent) and the best search neighbourhood strategy, with resulted in the minimum MSE presented in Table 2, and the results are shown in Figure 9. Even with noisy secondary data, the local variograms retained their ability to capture realistic spatial continuity patterns. Swath plot analysis (Figures 10 and 11) showed that noise had minimal impact on the spatial continuity representation, and MSE remained lower than with traditional OK.

TABLE 2

Comparison between traditional Kriging and LVCM models in relation to exhaustive data using traditional OK.

	Mean Squared Error (MSE)	Difference (%)
OK	18041	Ref.
OK_FIXED_LVCM	19339	7.2%
OK_FIXED_LVCM_NOISE	19157	6.2%
OK_LVCM	17051	-5.5%
OK_LVCM _I	17860	-1.0%
OK_LVCM _II	16419	-9.0%
OK_LVCM_II_NOISE	16714	-7.4%

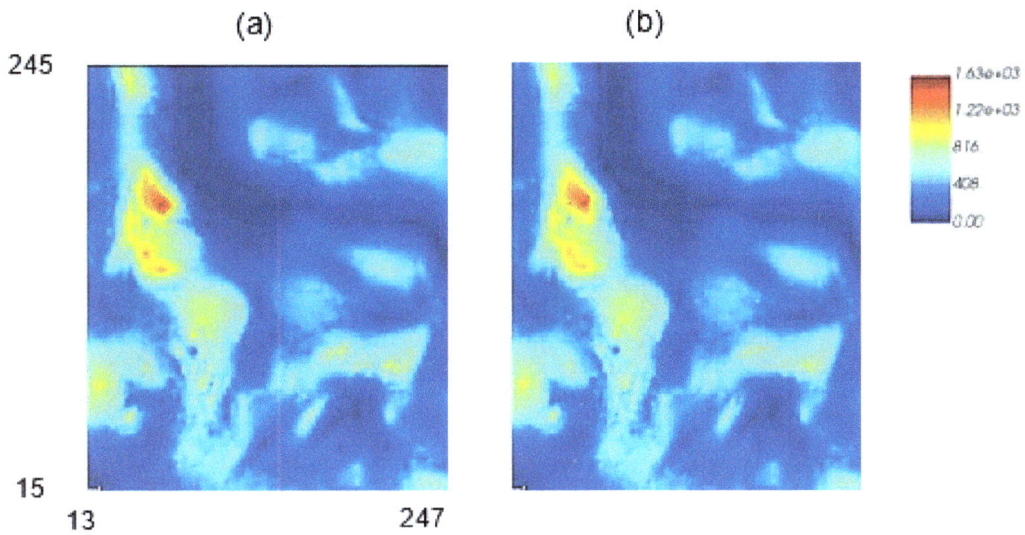

FIG 9 – (a) OK without random uniform noise (OK_LVCM_II); (b) OK with random uniform noise of ±25 per cent (OK_LVCM_II_NOISE).

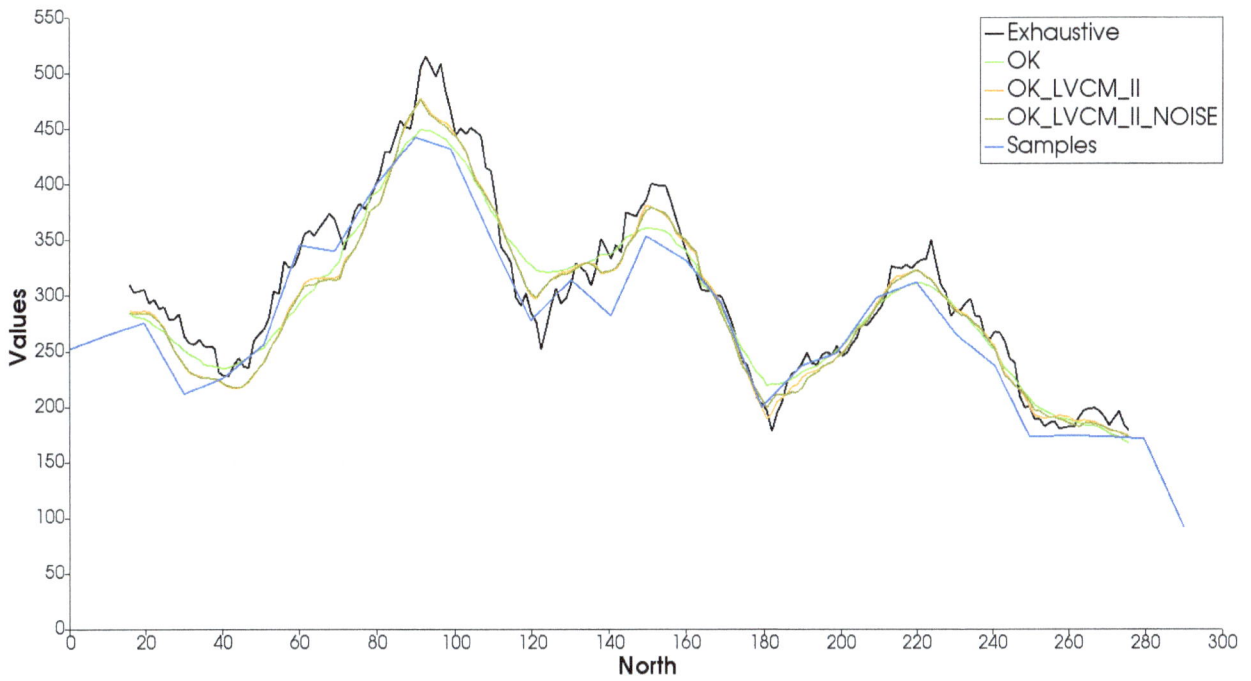

FIG 10 – Swath plot analysis comparing samples of traditional OK (OK) and OK using LVCM with (OK_LVCM_II_NOISE) and without noise (OK_LVCM_II).

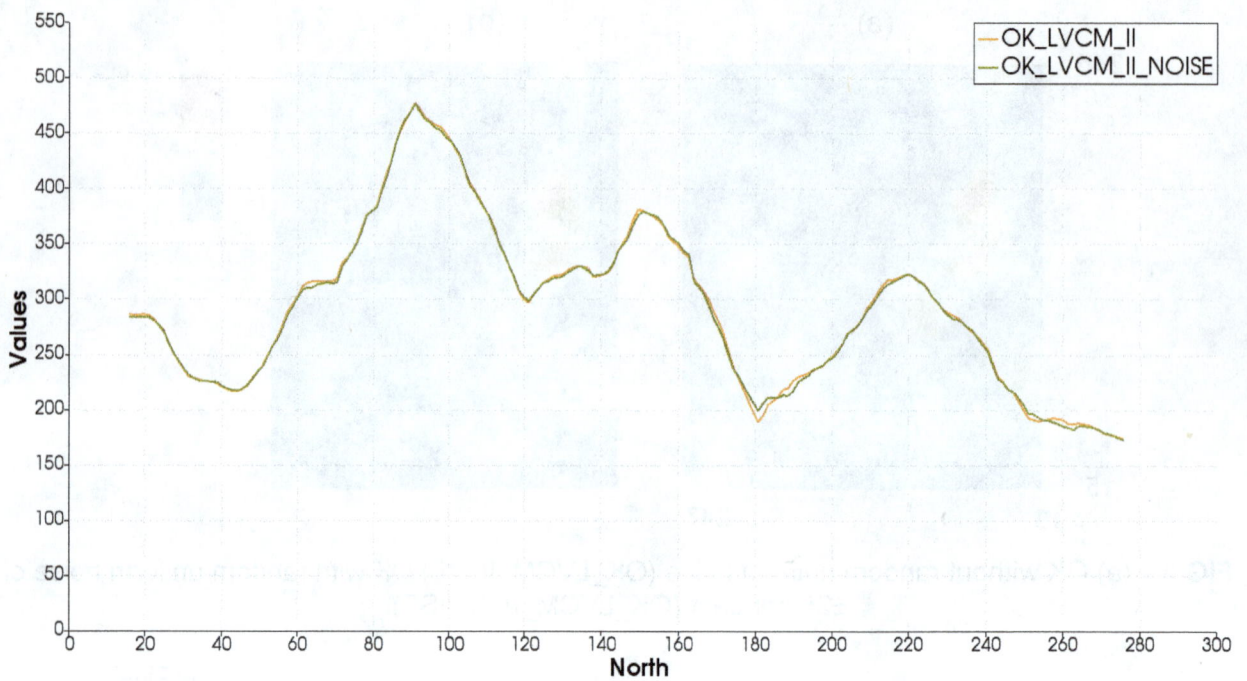

FIG 11 – Swath plot analysis comparing OK using LVCM with (OK_LVCM_II_NOISE) and without noise (OK_LVCM_II).

Summary comparison across all experiments

Finally, Figure 12 compares all tested models — traditional OK, LVCM with different search strategies, and LVCM with and without noise and fixed OW. The results consistently show that methods incorporating local variograms provide more accurate and spatially realistic estimates. Table 2 quantifies these improvements, showing up to 9 per cent reduction in MSE relative to traditional OK. Non-overlapping windows exhibited higher errors and produced artifacts, confirming that window overlapping is essential for maintaining spatial coherence.

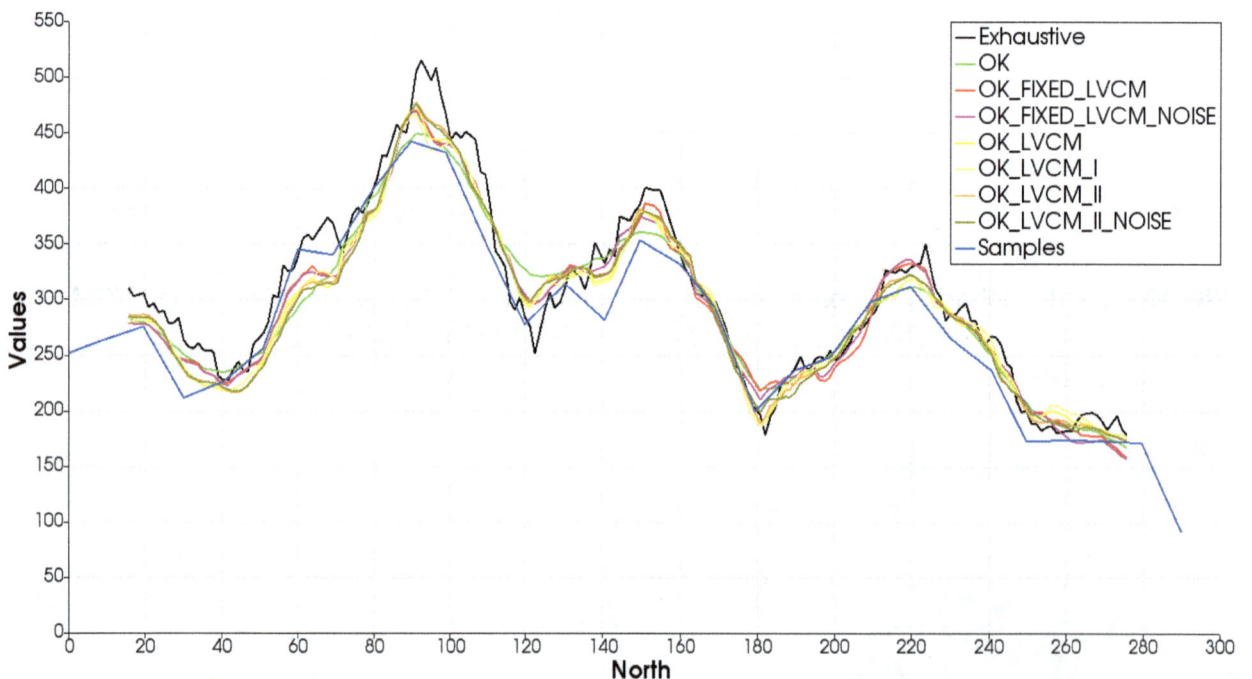

FIG 12 – Swath plot analysis comparing all experiments with samples and exhaustive data.

Note that this study incorporates a set of parameters that remain subject to further optimisation, which is expected to enhance the accuracy and reliability of the results. Key areas for improvement include the refinement of local nugget effect modelling, enhanced variogram fitting, optimisation of the search strategy in relation to local variograms. These advancements will be systematically tested and presented in future studies.

CONCLUSIONS

This study proposes using an exhaustive secondary variable to obtain local variograms and apply these local variograms in geostatistical estimation for complex geological environments. The local variograms are automatically fitted using experimental variograms from a covariance table(s) built with Fast Fourier Transform (FFT). These local variograms are then used in the estimates in Ordinary Kriging (OK).

The technique was applied to a case study. The results show that using local variograms in OK improved the estimates. Compared to the estimates obtained using OK with a single global variogram, the estimates using local variograms are more accurate and precise. The method also highlights the capacity of exhaustive secondary data to inform local patterns of spatial continuity.

ACKNOWLEDGEMENTS

This work was supported by the Mineral Exploration and Mining Planning Laboratory (LPM) at the Federal University of Rio Grande do Sul (UFRGS), Fundação Luiz Englert (FLE), and Petrobras S.A.

REFERENCES

Boisvert, J B, 2010. Geostatistics with Locally Varying Anisotropy, PhD dissertation, University of Alberta, Edmonton, Alberta.

Caixeta, R M, 2020. Contributions to the use of local anisotropies in geostatistics [in Portuguese: Contribuições para o uso de anisotropias locais na geoestatística], PhD dissertation, Federal University of Rio Grande do Sul [Universidade federal do Rio grande do sul].

Cooley, J W and Tukey, J W, 1965. An algorithm for the machine calculation of complex Fourier series, *Math Comput*, 19(90):297–301. https://doi.org/10.2307/2003354.jstor.2003354

Goovaerts, P, 1997. *Geostatistics for Natural Resources Evaluation* (Oxford University Press).

Isaaks, E H and Srivastava, R M, 1989. *An Introduction to Applied Geostatistics* (Oxford University Press).

Journel, A G and Huijbregts, C J, 1978. *Mining geostatistics* (Academic Press, London).

Kloeckner, J, Machado, P L, Rodrigues, A L and Costa, J F, 2019. Covariance table: A fast automatic spatial continuity mapping, *Computers and Geosciences*, 130:94–104. https://doi.org/10.1016/j.cageo.2019.05.001

Larrondo, P F, Neufeld, C T and Deutsch, C V, 2003. VARFIT: A program for semiautomatic variogram modeling, in *Fifth Annual Report of the Centre for Computational Geostatistics*, p 24, University of Alberta, Edmonton.

Li, Z, Zhang, X, Zhu, R, Clarke, K C, Weng, Z and Zhang, Z, 2024. Direct kriging: A direct optimization based model with locally varying anisotropy, *Journal of Hydrology*, 639:131553. https://doi.org/10.1016/j.jhydrol.2024.131553

Lillah, M and Boisvert, J B, 2015. Inference of locally varying anisotropy fields from diverse data sources, *Comput Geosci*, 82(C):170–182. https://doi.org/10.1016/j.cageo.2015.05.015

Machuca-Mory, D F and Deutsch, C V, 2009. Location dependent variograms, in Proceedings of the Eighth International Geostatistics Congress, vol 1, Santiago, Chile.

Marcotte, D, 1996. Fast Variogram Computation with FFT, *Computers and Geosciences*, 22(10):1175–1186. https://doi.org/10.1016/S0098-3004(96)00026-X

Oliveira, C, Bassani, M and Costa, J F, 2021. Application of covariance table for geostatistical modeling in the presence of an exhaustive secondary variable, *Journal of Petroleum Science and Engineering*, 196:108073. https://doi.org/10.1016/j.petrol.2020.108073

Van Der Walt, S, Colbert, S C and Varoquaux, G, 2011. The NumPy array: A structure for efficient numerical computation, *Computing in Science and Engineering*, 13(2):22–30. https://doi.org/10.1109/mcse.2011.37

Yao, T and Journel, A G, 1998. Automatic modeling of (cross) covariance tables using fast Fourier transform, *Math Geol*, 30(6):589–615. https://doi.org/10.1023/A:1022335100486

Yao, T, Calvert, C, Jones, T, Foreman, L and Bishop, G M A, 2007. Conditioning Geologic Models to Local Continuity Azimuth in Spectral Simulation, *Math Geol*, 39:349–354. https://doi.org/10.1007/s11004-007-9082-z

Investment and risk management

Application of fault tree analysis in mining – a review of strategies and advances for risk assessment

M Avilés[1], A Hekmat[2] and I Arroqui[3]

1. Mining Engineering Student, University of Concepcion, Concepción 4030000, Chile.
   Email: maviles2020@udec.cl
2. Assistant Professor, University of Concepcion, Concepción 4030000, Chile.
   Email: ahekmat@udec.cl
3. Mining Engineering Student, University of Concepcion, Concepción 4030000, Chile.
   Email: iarroqui2020@udec.cl

ABSTRACT

Mining is a fundamental component of global economies, facilitating industrial growth and resource availability. However, it remains an inherently high-risk industry, facing significant challenges related to safety and environmental impact. As mining operations reach greater depths and exploit increasingly complex mineral deposits, it becomes essential to implement robust risk assessment methodologies.

Fault Tree Analysis (FTA) is a systematic, graphical method of risk assessment that is used to analyse quantitatively and qualitatively the potential causes of multiple failure scenarios and to understand failure pathways. This review examines the application of FTA in mining contexts, analysing its principles, advantages and challenges to improve safety and sustainability. Its applications cover critical areas such as the management of roof and rib falls in underground mines, the prevention of gas explosions, the assessment of failures in tailings dams and mine hoists, and risk analysis of haul truck-related accidents and conveyor systems. These studies illustrate the value of FTA in risk prioritisation and the guidance of preventive measure design. Nonetheless, applying FTA to large-scale mining systems poses challenges, such as data limitations and the inherent complexity of modelling interconnected risks. Technological advancements, including real-time monitoring systems and artificial intelligence, offer promising opportunities to enhance FTA's accuracy and scope. These innovations allow for more precise modelling of failure probabilities and the integration of dynamic risk factors, ultimately improving decision-making processes.

This article also discusses the potential of combining FTA with other analytical methods, such as Bayesian networks and fuzzy logic, to overcome its limitations and adapt to the complexity of modern mining operations. Recommendations for its practical implementation are provided, alongside an outline of future research directions to further optimise its application and effectiveness in the mining sector.

INTRODUCTION

The mining industry, notorious for its hazardous environments and intricate operations, has persistently sought innovative approaches to enhance safety, efficiency and sustainability. Among the various methodologies employed for risk management, Fault Tree Analysis (FTA) has emerged as a prominent tool for identifying, assessing and mitigating potential failures throughout the mining process. The systematic and structured approach inherent in FTA facilitates the identification of root causes and the analysis of complex failure scenarios, rendering it invaluable in the management of safety and operational risks across a range of mining activities.

This paper explores the application of FTA in diverse mining contexts, highlighting its role in addressing critical safety concerns, including roof and rib falls in underground mines, gas explosions, and failures in tailings dams, mine hoists, and heavy machinery. Furthermore, the paper discusses how FTA has been integrated with advanced methodologies such as fuzzy logic, Bayesian networks, and machine learning to enhance its applicability and address the dynamic and uncertain conditions present in modern mining environments.

As the mining industry transitions towards sustainability, incorporating autonomy and electrification, the integration of FTA with emerging technologies promises a more comprehensive approach to risk management. By improving predictive accuracy and operational efficiency, these advancements provide an opportunity to further enhance safety while reducing environmental impacts. The aim of this paper is to review the current state of FTA applications in mining, to discuss recent innovations, and to propose directions for future research to foster safer, more efficient, and environmentally conscious mining practices.

FUNDAMENTALS OF FTA

FTA is a systematic technique used to analyse the probability of failures in complex systems. The first documented use of FTA was by Bell Telephone Laboratories in 1962 in the context of analysing the safety of the Minuteman missile launch control system (Rausand and Hoyland, 2004). It is based on the graphical representation of events that may lead to an undesired failure, allowing the identification of root causes and the quantification of associated risk, using Boolean logic functions, which integrate primary events with the top event.

FTA uses a deductive approach, ie it starts from an undesired event (top event) and utilises logic gates to represent failure relationships and thereby decompose potential causes. The main elements of the analysis are shown in Table 1.

TABLE 1

Main symbols of a fault tree.

Symbol	Element	Description
	Top Event (TE)	Fault event resulting from the logical combination of the input events, which are operating through the logic gate.
	Intermediate Event (IE)	Contains a description of the event.
	Basic event (BE)	Basic failure or root cause that requires no further development of failure causes. Independent event, used only as the input of a logic gate.
	Undeveloped event (UE)	Event that is not examined further because information is unavailable or because its consequences are insignificant.
	Transfer Event (TE)	The transfer-out symbol indicates that the fault tree is developed further at the occurrence of the corresponding transfer-in symbol.
	AND gate	The output event occurs only if all the input events occur simultaneously.
	OR gate	The output event occurs if any of the input events occur.

FTA symbols are the language that systems use to articulate potential points of failure. They simplify the visualisation of event interactions, and their standardised nature ensures that these diagrams are universally understood, fostering collaboration across industries.

In order to develop an effective FTA it is necessary to define the study's objective. The determination of the top event is the subsequent step, and the elements and conditions that will be included in the study must be defined in order to determine the analysis's scope. The resolution of the analysis is established by determining the level of detail with which each failure will be examined; this may include the consideration of both critical and minor events, as appropriate. The ground rules must be defined, establishing criteria and assumptions for the analysis, such as a deterministic or probabilistic approach, and normal operating conditions. The construction of the fault tree involves the graphical representation of the relationships between faults by means of 'AND' and 'OR' logic gates, connecting basic events and intermediate events. A cut set is defined as a combination of basic events that, if occurring simultaneously, will result in the occurrence of the top event, which is defined as the system failure. A minimal cut set constitutes the smallest possible set of basic events that, when occurring together, directly cause the top event. It is important to note that a minimal cut set cannot be reduced further without losing its ability to trigger the failure. The identification of minimal cut sets is of critical importance in risk assessment, as it facilitates the identification of the most critical failure combinations that need to be addressed to enhance system reliability. Figure 1 presents a generic representation of a fault tree diagram.

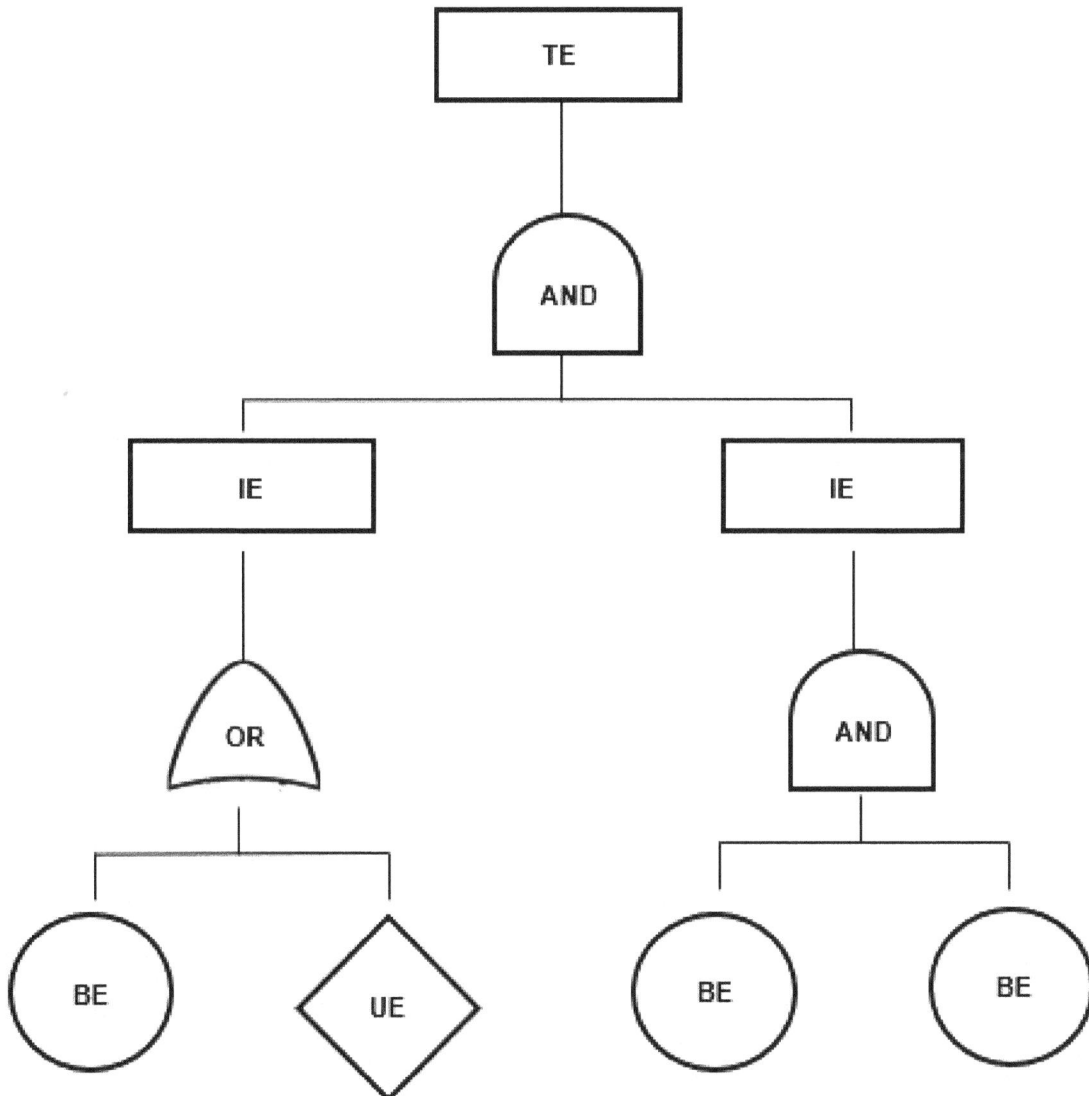

FIG 1 – Basic example of a fault tree diagram.

The quantitative and qualitative analysis of the fault tree requires the calculation of the probability of the top event, using historical data or mathematical models, and the analysis of the contribution of each basic event. Quantitative FTA has been demonstrated to be a highly valuable tool for predicting the future behaviour of systems that have historical failure data available. This information can assist in the development of strategies for preventative maintenance, system redesign, or procedures that mitigate risk. Qualitative FTA is a valuable tool when statistical data is limited or when the objective is to comprehend the potential combinations of events that could result in system failure. Finally, the interpretation and presentation of results is performed, where conclusions are drawn, the most critical failures are identified, and mitigation measures are proposed to improve the safety and reliability of the system.

APPLICATION OF FTA IN THE MINING INDUSTRY

Fault Tree Analysis has emerged as a paramount instrument for risk management in the mining industry. This methodology enables the identification and mitigation of failures across diverse stages of the mining cycle, ranging from geological exploration to mine closure. The implementation of FTA contributes to enhancing safety and efficiency in mining operations.

Haul truck-related fatal accidents in surface coal mining

Zhang, Kecojevic and Komljenovic (2014) conducted an investigation of 12 haul truck-related fatal accidents in surface coal mining in West Virginia using qualitative risk assessment with FTA. Their findings indicated that the two most common root causes are inadequate or improper pre-operational checks and inadequate maintenance. Failure to wear a seat belt and inadequate training were also important contributing factors. A total of eight accidents occurred on haul roads, while ten accidents occurred while the trucks were moving forward. The two most frequently violated provisions of the Code of Federal Regulations were 30 CFR§77.404, which pertains to machinery and equipment; operation and maintenance, and 30 CFR§77.1606, which concerns loading and haulage equipment; inspection and maintenance.

Roof and rib fall in Turkish underground coalmines

Direk (2015) conducted a study that applied the FTA methodology to identify the root causes of roof and rib falls in Turkish underground coalmines. The study's methodology comprised several steps, beginning with data collection and preprocessing, where accident records from 2003 to 2013 at Turkish Hard Coal Enterprise (TTK) – Amasra Hard Coal Institution – were analysed. The subsequent stage entailed the construction of a fault tree, which facilitated the identification of intermediate and basic events contributing to roof and rib falls. The risk quantification stage involved the use of ReliaSoft BlockSim-7 software to calculate the probabilities associated with each basic event and the minimal cut sets. The study determined the root causes by analysing the events that contributed the most to the overall risk. The study identified the occurrence of roof and rib falls in underground coalmines as the primary event, leading to injuries, permanent disabilities, or fatalities. This phenomenon is recognised as one of the most prevalent causes of mining accidents, exerting a substantial impact on worker safety.

The analysis identified several key root causes of roof and rib falls, including the improper use of personal protective equipment (PPE), procedural errors such as non-compliance with safety protocols, and the utilisation of inappropriate tools and equipment. Geological instability, inadequate roof support, and operational factors such as blasting effects and inadequate monitoring were also found to be significant contributors to the incidence of roof and rib falls. The study validates the use of FTA as a valuable tool for the assessment of risks associated with underground mining accidents, and the findings indicate that the lack of proper PPE, procedural errors, and improper tools play a major role in the occurrence of these accidents.

Tailings dam failures

Tailings dams are a critical component of mining infrastructure, as their failure can lead to catastrophic consequences. These consequences may include loss of life, environmental contamination, and severe economic damages. A study by de la Cruz (2017) employed the FTA method to analyse 407 global tailings dam failures, thereby identifying the main causes as follows:

slope slip or slope instability, overflow or surpass, earthquake, leakage, internal erosion (piping), and static liquefaction. The occurrence of a failure event necessitates the presence of three concomitant conditions for the purpose of achieving an uncontrolled release of tailings. Firstly, the material must be in a state of saturation. Secondly, the material must exhibit contractility, defined as a reduction in volume in response to applied stress. Thirdly, the shear stress must exceed the residual strength of the material. The fault trees representativeness of global practice and experience warrants its classification as a generic model. It does not reflect site-specific conditions but rather integrates them with the experiences of numerous different sites (De La Cruz, 2017).

Mine hoists in underground mines

Giraud and Galy (2018) discuss the critical safety concerns of mine hoists, which are integral for both ore extraction and the transportation of miners between underground levels. Given that a moving cage can transport up to 50 or more workers, the safety of these systems is paramount. The study presents two generic fault trees, derived from past accident records, focusing on hoist rope severance and the loss of control of the conveyance. Both scenarios could lead to catastrophic cage crashes at the shaft boundaries, which, though rare, can have devastating consequences.

The fault tree analysis identifies the root causes of such accidents and proposes mitigation strategies. In the case of hoist rope severance, most root causes are attributed to secondary failures of safety mechanisms, such as safety catches designed to secure the cage in the event of rope failure. Particularly those employing the 'Ontario dog' design and a reliable spring system, can contribute to the prevention of crashes if implemented effectively. However, the risk of rope severance remains a significant concern. The article suggests that the implementation of real-time rope tension and condition monitoring, as observed in certain Quebec mines, could serve as a preventative measure against rope failure (Giraud and Galy, 2018).

The study also explores the risk of loss of control of the conveyance, a scenario in which the cage may crash even if the hoist rope remains intact (Giraud and Galy, 2018). This type of accident is mainly caused by failures in the control system, particularly those linked to programmable electronic systems (PES). As automated hoisting machinery becomes more prevalent in mining operations, ensuring the reliability of control systems is paramount in preventing loss of control and safeguarding the safety of miners. While PESs assist in reducing human error, the article underscores that they also introduce novel failure modes that necessitate meticulous management. Giraud and Galy (2018) underscore the significance of enhancing both physical safety mechanisms and electronic control systems to avert major accidents in mine hoisting operations. While technological advancements continue to improve safety, both primary and secondary failures remain challenges that require ongoing attention.

COMBINATION OF FTA WITH OTHER ANALYTICAL METHODS

Fault Tree Analysis has been identified as a valuable tool for identifying and assessing failure pathways in mining operations. However, FTA has notable limitations, particularly when dealing with dynamic scenarios and uncertain conditions. Firstly, FTA is a static method and does not account for time-dependent changes in system conditions. Mining operations involve continuously changing variables (eg equipment wear, evolving geological conditions, and fluctuating environmental factors) that FTA alone may not model effectively. Secondly, FTA's binary classification of events as either 'fail' or 'success' oversimplifies real-world mining risks, which often manifest in degrees, such as partial failures, degraded performance, and near misses.

The combination of FTA and advanced analytical techniques to enhance accuracy and adaptability allows for a more comprehensive risk assessment in complex mining environments.

Fuzzy Logic for handling uncertainty

FTA is contingent upon precise failure probability data, which is frequently inaccessible or indeterminate in mining operations. Incorporating fuzzy logic can address this issue by enabling gradual failure states in substitution for strict binary classifications. This methodology is especially advantageous in evaluating risks where clear failure boundaries are challenging to delineate.

Chrome mine's loading and conveying processes

The study by Yasli and Bolat (2018) applied FTA and a fuzzy approach to assess the risks associated with underground mining, focusing specifically on a chrome mine's loading and conveying processes. The methodology identified all possible undesired events, their root causes, and evaluated them from a common risk perspective without relying on statistical data. This approach is adaptable to various fields and is particularly useful for analysing complex systems with multiple hazard elements.

The top event in fault tree analysis is the occurrence of accidents and incidents in the underground chrome mining process. These events are categorised into two main groups: geological structure-related and human-related causes, which are prevalent in all mining operations. The root causes identified in the study include unsafe working conditions such as dust, fumes, water, mud, inadequate ventilation, harsh environmental factors, and the psychological effects of underground work. The human-related causes encompass errors arising from labour-intensive processes, inadequate safety measures, and insufficient training. These hazardous conditions contribute substantially to occupational accidents and injuries. The study concludes that effective risk management in underground mining necessitates a comprehensive understanding of undesired events and their causes.

The implementation of FTA in conjunction with a fuzzy approach offers a systematic approach to assess and rank risks based on their probability, consequences, and severity. The study emphasises the necessity for preventive measures, including improved ventilation, stricter safety regulations, and better training programmes, with the aim of enhancing occupational health and safety in chrome mining and providing a foundation for future research on accident prevention in the mining sector.

Mine Health and Safety risks in surface mines

A study by Jiskani et al (2022) used the combination of the Z-number concept, fuzzy theory, and FTA, handling the uncertainty arising from lack of complete information and to enhance the reliability of qualitative judgment of experts, to comprehensively analyse the probable causes of Mine Health and Safety (MHS) risks in surface mines, related to machine/equipment, environment factors, and workplace conditions.

The eight fundamental undesired events that represent the minimum conditions necessary to result in an event posing a risk are as follows: blasting incident or uncontrolled initiation, vehicle rollover, ergonomic hazards, dust and explosive fumes, vehicle collision, slips/trips/falls, slope failure, and vibration and noise. However, issues associated with blasting incident, dust and explosive fumes, and vehicle rollover are the most probable undesired events. The most prevalent underlying factors contributing to these events, as identified by the authors, include failure to implement regulations, inattention of the worker, inadequate measures in setting safety perimeter, lack of proper explosive calculations, lack of properly designed roads, inadequate work area cleaning and housekeeping, inadequate skills and ability of workers, the ineffectiveness of management, lack of environment-friendly equipment, and ignoring warnings.

Numerous advantages of this technique include the capacity to represent complicated real-world operating systems, the ability to mitigate analytical uncertainty, and the availability of crisp, fuzzified, and defuzzied reliability parameters (Jiskani et al, 2022).

Fuzzy Polymorphic Bayesian Networks

FTA operates under the assumption of fixed failure probabilities, which may not accurately reflect real-time operational changes. Integrating the FTA with Bayesian Networks (BN) enables the dynamic updating of failure probabilities as new data becomes available. This approach improves predictive accuracy and facilitates adaptive risk assessments based on evolving conditions.

Coalmine gas explosions

Yang, Zhao and Shao (2023) present an advanced methodology that integrates FTA with Fuzzy Polymorphic Bayesian Networks (FPBNs). This integration serves to overcome the limitations of dynamic control, state representation, and uncertainty handling in coalmine automation. The model's innovative approach involves the classification of risk factors into multiple states, alongside the introduction of accuracy corrections to address subjectivity. This integration facilitates real-time probability and risk distribution calculations, thereby providing a valuable technical solution to the challenges posed by gas explosion risk management in coalmines and broadening the new idea of risk assessment from a probability perspective.

A case study of Wangzhuang coalmine demonstrates a 35 per cent probability of a gas explosion accident, underscoring key risk factors such as excessive ventilation resistance, coal spontaneous combustion, and electric sparks, which exhibit the highest sensitivity. The study collected information from 82 coalmine gas explosion accidents in China since 2011, including both traditional coalmines and intelligent coalmines. The study postulates that a coordinated approach can reduce the risks of gas concentration exceeding the limit, the occurrence of ignition sources and gas explosions by 14.6 per cent, 19.2 per cent and 22.9 per cent respectively, providing significant technical support for gas explosion risk management in coalmines.

Firstly, the risk factors of coalmine gas explosions were identified through the implementation of fault tree analysis, and a Bayesian network structure was determined according to the causal logic relationship. The risk factors were then divided into High, Moderate and Low categories through the ALARP (as low as reasonably practicable) criterion, ensuring the Bayesian network's alignment with the actual situation. Secondly, the nodes of the Bayesian network were categorised into root nodes and intermediate nodes. The prior probability of root nodes was determined using the trapezoidal fuzzy theory of seven levels of language variables. To ensure the robustness of the findings, ten experts were invited to conduct field research at the Wangzhuang coalmine, and their subjective influence was corrected from the perspective of their degrees and work experience.

Artificial Neural Network

Coal and gas outbursts in underground mines

Ruilin and Lowndes (2010) conducted a study that integrated FTA with Artificial Neural Networks (ANNs) to assess the risk of coal and gas outbursts in underground mines. The researchers first developed a qualitative FTA for a coal and gas outburst, identifying key intermediate events such as gas properties, physical and mechanical properties of the coal seam, and geological stress conditions. These primary factors were considered essential to the occurrence of an outbursts.

To perform a quantitative assessment, the study applied Boolean algebra to determine minimum intersection sets, focusing on 13 fundamental events from an initial set of 24. Using a backpropagation (BP) algorithm, the model identified eight key parameters, which were then incorporated into the ANN model for risk assessment. The combined FTA-ANN approach achieved a success rate of 87 per cent, demonstrating its potential to improve risk assessment in mining.

The study highlights the complexity of mining environments and emphasises that quantitative analysis may require tailored approaches. While qualitative FTA provides a broad framework, quantitative assessments can be refined by selecting the most relevant risk factors. This highlights the need for innovative methodologies to improve risk prediction and management in mining operations.

ENHANCING FTA WITH ARTIFICIAL INTELLIGENCE AND MACHINE LEARNING

Technological advancements, including real-time monitoring systems and artificial intelligence (AI), present significant opportunities to enhance FTA in mining. Despite the challenge of inadequate historical data, several strategies facilitate the effective integration of AI and machine learning (ML) into FTA. One strategy involves the generation of synthetic data through techniques such as Monte Carlo simulations and physics-based models. These methods enable the prediction of failures without reliance on historical data, making them useful when limited real-world failure records are available.

Another approach is transfer learning, where models trained in other industries, such as aviation or manufacturing, can be adapted to mining operations with small data sets. This allows for leveraging knowledge of general mechanical or structural failures and applying it to mining-specific scenarios.

The integration of real-time data from IoT sensors is also critical. By monitoring conditions such as vibration, temperature, and pressure, sensors can provide current operational data to ML algorithms, thereby providing insights into potential failures without the need for extensive historical records. This helps mining operations to anticipate and address issues proactively.

Additionally, semi-supervised and unsupervised learning techniques can be used when labelled data is scarce. Algorithms such as unsupervised clustering or autoencoders can identify anomalies and failure patterns without the need for labelled failure data, thereby enhancing the model's ability to detect unusual behaviours.

Finally, the integration of expert knowledge is a valuable tool. In situations where data is limited, AI systems can incorporate rules derived from human expertise, thereby improving the accuracy of FTA by combining experienced insights with machine learning models.

FURTHER INVESTIGATION

The transition towards sustainable mining is a critical aspect of modernising the industry to reduce its environmental impact, whilst simultaneously improving efficiency and safety. In recent years, there has been significant progress in the adoption of new technologies, such as automation and electrification, with the aim of minimising environmental impact, enhancing operational efficiency, and safeguarding the well-being of workers in hazardous environments.

Automation plays a central role in reducing human exposure to dangerous conditions within mining sites. The integration of automation unmanned aerial vehicles (UAVs), and autonomous vehicles has become prevalent in tasks such as excavation, drilling, and transportation. This technological transition not only mitigates the risk of accidents but also optimises operational efficiency by enhancing process efficiency, reducing downtime, and improving the precision of various mining activities.

FTA can be a key tool to optimise mining operations that have implemented autonomy in their processes, as is the case with some mining companies in Chile. This approach makes it possible to identify and correct errors, such as the lack of alignment or consistency between the virtual tracks of autonomously operated CAEX equipment and the physical tracks on which they circulate. Such inconsistency can cause serious accidents, as well as losses in production and investment in the equipment.

Conversely, electrification plays a pivotal role in reducing the reliance on fossil fuels. The transition to electric-powered mining equipment, such as electric trucks and drills, contributes to a reduction in greenhouse gas emissions, thereby minimising the environmental impact of mining operations. Electrification also results in quieter operations, leading to a decrease in noise pollution and an enhancement in the quality of life for surrounding communities.

When integrated with risk assessment tools like FTA, these technologies offer a more comprehensive approach to safety and sustainability in mining. FTA can be employed to evaluate the risks associated with the implementation of automation and electrification, ensuring that potential hazards are identified and mitigated before deployment. For instance, FTA can assess the risks of system failures in autonomous machinery or evaluate the safety implications of introducing electric-powered vehicles into underground operations.

Furthermore, the integration of FTA with emerging technologies, such as predictive analytics, artificial intelligence, and machine learning, has the potential to enhance the precision and promptness of risk assessments. These technologies facilitate real-time data collection and analysis, offering insights into equipment performance, environmental conditions, and human factors that contribute to mining risks. As the industry adopts these advancements, further refinement of FTA will be necessary to address the complexities introduced by automation and electrification, ultimately leading to safer, more sustainable, and efficient mining operations.

Further research in this area could explore the synergies between these technologies and risk assessment methodologies, investigating how automation, electrification, and advanced analytics can complement traditional risk management strategies. Additionally, a critical aspect that must be given due consideration is the assessment of the environmental and social impacts of these technological innovations. This is of paramount importance in ensuring that mining operations remain aligned with global sustainability goals, while concurrently maintaining profitability and operational excellence.

CONCLUSIONS

Fault Tree Analysis is a key tool for risk management in the mining industry, allowing for the identification and mitigation of failures across various stages of the mining cycle. Its implementation has proven essential in enhancing safety and operational efficiency. Through studies on haul truck accidents, roof and rib falls in underground mines, tailings dam failures, and hoist system accidents, FTA has identified common root causes such as maintenance failures, improper use of personal protective equipment, and control system failures. However, its static nature and binary event classification limit its ability to address dynamic scenarios and uncertain conditions in the mining environment. To overcome these limitations, integrating FTA with other advanced analytical techniques like fuzzy logic, Bayesian Networks (BN), Fuzzy Polymorphic Bayesian Networks (FPBN), and Artificial Neural Networks (ANNs) enhances risk assessment by addressing uncertainty and dynamic conditions. These methods improve predictive accuracy, allow for real-time updates, and refine risk management, particularly in complex scenarios such as risks related to machine/equipment, environment factors, and workplace conditions, loading and conveying processes, coalmine gas explosions and coal and gas outbursts.

Recent advancements in the fields of artificial intelligence (AI) and machine learning (ML) have rendered enhanced opportunities to augment FTA in the mining industry. The utilisation of techniques such as Monte Carlo simulations and physics-based models facilitates the generation of synthetic data, thereby compensating for the paucity of historical records. The transfer learning paradigm enables the adaptation of models from disparate industries to the mining context, while real-time data from IoT sensors ensures uninterrupted monitoring of operational conditions, thereby facilitating proactive failure prediction. The incorporation of unsupervised learning methods is instrumental in identifying anomalies in scenarios where labelled data is scarce, with the subsequent refinement of models through expert knowledge further enhancing the efficacy of the process. The amalgamation of these technologies with FTA results in a more comprehensive risk assessment, providing insights that are both more accurate and dynamic.

The transition towards sustainable mining has led to the adoption of new technologies, including automation and electrification, to minimise environmental impact and improve operational efficiency. Automation reduces human exposure to hazardous environments, while electrification decreases reliance on fossil fuels, lowering greenhouse gas emissions. Future research should focus on the synergies between emerging technologies and risk management strategies, exploring how automation, electrification, and advanced analytics can support sustainable and efficient mining operations. Additionally, the environmental and social impacts of these innovations must be assessed to ensure that mining operations meet global sustainability goals while maintaining profitability and operational excellence.

ACKNOWLEDGEMENTS

The author wishes to express her sincere gratitude to Universidad de Concepción for providing the academic foundation and resources necessary for the development of this work. Special thanks are extended to Professor Asieh Hekmat for her invaluable guidance, insightful feedback, and continuous support throughout this research. Her expertise and encouragement have been instrumental in shaping this study. The contributions of all co-authors are also greatly acknowledged, as their collaboration and input have been essential in refining the content and direction of this paper.

REFERENCES

De la Cruz, S, 2017. Evaluación del fallo de presas de relaves con el método del árbol de fallos (Evaluation of tailings dam failure using the fault tree method), Master's thesis, Universidad Politécnica de Madrid, Madrid.

Direk, C, 2015. Risk assessment by fault tree analysis of roof and rib fall accidents in an underground hard coal mine, master's thesis, Middle East Technical University, Turkey.

Giraud, L and Galy, B, 2018. Fault tree analysis and risk mitigation strategies for mine hoists, *Safety Science*, 110(A):222–234. https://doi.org/10.1016/j.ssci.2018.08.010

Jiskani, I M, Yasli, F, Hosseini, S, Rehman, A U and Uddin, S, 2022. Improved Z-number based fuzzy fault tree approach to analyze health and safety risks in surface mines, *Resources Policy*, 76. https://doi.org/10.1016/j.resourpol.2022.102591

Rausand, M and Hoyland, A, 2004. *System Reliability Theory Models, Statistical Methods and Applications* (Wiley Series in Probability and Statistics).

Ruilin, Z and Lowndes, I S, 2010. The application of a coupled artificial neural network and fault tree analysis model to predict coal and gas outbursts, *International Journal of Coal Geology*, 84(2):141–152. https://doi.org/10.1016/j.coal.2010.09.004

Yang, J, Zhao, J, and Shao, L, 2023. Risk Assessment of Coal Mine Gas Explosion Based on Fault Tree Analysis and Fuzzy Polymorphic Bayesian Network: A Case Study of Wangzhuang Coal Mine, *Processes*, 11(9):2619. https://doi.org/10.3390/pr11092619

Yasli, F and Bolat, B, 2018. A risk analysis model for mining accidents using a fuzzy approach based on fault tree analysis, *Journal of Enterprise Information Management*, 31(4):577–594. https://doi.org/10.1108/JEIM-02-2017-0035

Zhang, M, Kecojevic, V and Komljenovic, D, 2014. Investigation of haul truck-related fatal accidents in surface mining using fault tree analysis, *Safety Science*, 65:106–117. https://doi.org/10.1016/j.ssci.2014.01.005

MAFMINE 3.1 – integrating cost estimation and ESG constraints for enhanced preliminary economic assessment in mining

C O Petter[1], U V Pinheiro[2], J F G Timm[3], R F D'Arrigo[4], F A C Cardozo[5], H Campos[6], J P Z Oppermann[7], W M Ambrós[8] and R A Petter[9]

1. Professor, Federal University of Rio Grande do Sul, Porto Alegre RS 91509-900, Brazil. Email: cpetter@ufrgs.br
2. Doctoral Student, Federal University of Rio Grande do Sul, Porto Alegre RS 91509-900, Brazil. Email: upiragibe@gmail.com
3. Postdoctoral Student, Federal University of Rio Grande do Sul, Porto Alegre RS 91509-900, Brazil. Email: janainetimm@hotmail.com
4. Doctoral Student, Federal University of Rio Grande do Sul, Porto Alegre RS 91509-900, Brazil. Email: rafael@darrigo.com.br
5. Professor, Federal University of Pampas, Caçapava do Sul RS 96570-000, Brazil. Email: fernando.cantini3@gmail.com
6. Master Student, Federal University of Rio Grande do Sul, Porto Alegre RS 91509-900, Brazil. Email: higor.jose@ufrgs.br
7. Master Student, Federal University of Rio Grande do Sul, Porto Alegre RS 91509-900, Brazil. Email: mineraengenharia.eng@gmail.com
8. Professor, Federal University of Rio Grande do Sul, Porto Alegre RS 91509-900, Brazil. Email: weslei.ambros@ufrgs.br
9. Doctoral Student, Federal University of Rio Grande do Sul, Porto Alegre RS 91509-900, Brazil. Email: rapetter@gmail.com

ABSTRACT

MAFMINE 3.1 is a cost-estimation tool for mining operations, integrating Environmental, Social, and Governance (ESG) considerations. Evolving from the original O'Hara model, MAFMINE 3.1 leverages parametric models to provide preliminary capital expenditure (CAPEX) and operational expenses (OPEX) estimations, achieving an accuracy of 30–50 per cent suited for the Preliminary Economic Assessment (PEA) stage. It supports initial Discounted Cash Flow (DCF) calculations, enabling rapid, order-of-magnitude cost projections with minimal inputs.

This client-server application offers flexible deployment as both a cloud-based tool and a local desktop solution, using ElectronJS to support dual access. Local and synchronised cloud data storage is facilitated through PouchDB, which enables seamless data access in both online and offline modes. For the front end, a multilingual user interface is constructed with Handlebars and GruntJS, making it accessible to a global audience. Mathematical operations for cost and financial modelling are powered by MathJS, which supports quadratic and other complex equations foundational to mining project evaluations. Git version control allows for the ongoing integration of new models and formulas, ensuring adaptability to future needs.

MAFMINE 3.1 also serves as an educational platform for mining engineering students and professionals, offering hands-on experience in cost estimation, financial modelling, and sustainability. ESG considerations, such as water withdrawn from new sources, energy sourcing, and carbon emissions, are incorporated to align with modern standards for responsible resource management. This paper will present MAFMINE 3.1's core functionalities, technical architecture, and practical applications in PEA-level mining assessments, underscoring its role as both a practical tool for cost estimation and an educational resource for sustainable mining practices.

INTRODUCTION

The mining industry relies on complex supply chains and optimisation tools to manage operations efficiently (Leite *et al*, 2019). Concurrent simulation and optimisation models have been developed to address short-term planning challenges, once cost estimation is crucial in mining projects (Fioroni *et al*, 2008). These advancements in modelling and optimisation techniques contribute to more effective decision-making and resource management in the mining industry.

The O'Hara model, developed in 1980, is a methodology used for estimating mining costs, but it is based on foreign economies. To address this limitation, researchers have adapted and optimised the model for specific contexts. In Brazil, it has been adapted for use in national mining projects through the development of MAFMINE software, which incorporates adjustment factors to account for local economic and bureaucratic conditions, to estimate costs in mining projects, specifically in conceptual phases (D'Arrigo, 2012).

MAFMINE evolved from methodologies pioneered by O'Hara (1980) and later refined by institutions such as the Paris School of Mines and the U.S. Bureau of Mines (Camm, 1992). In 2012, the first version of MAFMINE (1.0) was developed in PHP, fully inspired by MAFMO, the software conceived by the Paris School of Mines. In 2017, a major upgrade introduced MAFMINE 3.0, rewritten in JavaScript to ensure cross-browser stability and bilingual support (Portuguese and English). Unlike traditional spreadsheet-based calculations, MAFMINE operates as a client-server application, enabling seamless updates and integration of new industry advancements. A client-server model is model is a computing framework in which a requester (client) orders resources or services to a provider (server), which processes and responds to fulfill them. The core philosophy behind its development was to provide a scalable and adaptable platform that incorporates modern mining technologies, including electrification, automation, and preconcentration.

The tool was designed to bridge the gap between theoretical cost models and real-world mining applications. It achieves this by integrating a robust database of mining costs, sourced from technical studies, feasibility reports, and industry benchmarks. This data-driven approach enhances the accuracy of its parametric models, allowing decision-makers to explore different mining scenarios and optimise project configurations in response to evolving industry trends.

MAFMINE 3.1 employs parametric models for CAPEX and OPEX, utilising equations based on the relationship between key variables such as daily production (tonnes/day), infrastructure, and mineral processing. The model has been validated through benchmarking against mineral feasibility studies and regression analysis on historical data.

Environmental, Social, and Governance (ESG) principles are becoming increasingly significant regarding the transformation and strategic planning of mining projects. An integrated perspective on sustainability goals at strategic, tactical, and operational decision-making levels is essential, requiring that environmental resource allocation be assessed transparently alongside economic objectives. Moreover, incorporating sustainability considerations into the design, operation, and closure of mining projects has become a fundamental prerequisite for ensuring their feasibility. Consequently, the use of sustainability performance indicators and metrics is crucial for any consistent approach to these challenges. While integrating sustainability into quantitative decision-making processes presents significant challenges, mining activities should only proceed if they demonstrably contribute to long-term net positive outcomes for society and the environment (Pimentel, Gonzalez and Barbosa, 2016).

Within this framework, MAFMINE 3.1 serves as a valuable tool for simulating the preliminary feasibility of mining projects by incorporating environmental metrics like Greenhouse Gases (GHG) emissions and water consumption based on the outputs data estimated from the tool's parametric models. The development strategy for this tool is based on the core concept of adapting the parameterisation methodologies to estimate sustainability indices in mining projects (Pinheiro *et al*, 2023). Therefore, MAFMINE 3.1 also serves as an educational platform for mining engineering students and professionals, offering hands-on experience in cost estimation, financial modelling, and sustainability. This paper will present MAFMINE 3.1's core functionalities, technical architecture, and practical applications in Preliminary Economics Assessment level (PEA-level) mining assessments, underscoring its role as both a practical tool for cost estimation and an educational resource for sustainable mining practices.

METHOD

The article is structured into three parts: (i) MAFMINE – Software architecture and functionalities; (ii) ESG Integration; and (iii) case study. The first part, MAFMINE – Software architecture and functionalities, introduces the software, its computational foundation, components, functionalities, and input/output data. The second part, ESG Integration, explains how ESG principles are incorporated into the software's functionalities and how potential environmental impacts are assessed. The third part, case study, presents an analysis of a copper mine, demonstrating how MAFMINE 3.1 can support preliminary economic assessments and estimate potential environmental impacts.

RESULTS AND DISCUSSION

MAFMINE 3.1 – software architecture and functionalities

The parametric models employed by this tool enable the estimation of data with accuracy appropriated for PEA-level mining projects. The software is structured as a distributed system with a client-server model, ensuring accessibility, scalability, and data integrity. The main components include:

- Backend (server-side processing): The backend manages all calculations, database interactions, and business logic. It handles requests from clients, processes parametric models, and generates CAPEX/OPEX estimates.

- Frontend User Interface (UI): A web-based interface that allows users to input project parameters, view cost estimations, and generate reports. The frontend UI is designed for ease of use, providing intuitive navigation and interactive data visualisation.

- Database Management System (DBMS): a structured database that stores mining cost data, project parameters, and environmental metrics. This ensures historical data can be analysed for improved decision-making.

- Cloud-based deployment: MAFMINE is hosted on a cloud infrastructure, enabling remote access, real-time collaboration, and automatic updates. Cloud integration allows seamless scalability as more users and features are added.

- Application Programming Interface (API) integration: the system includes APIs that facilitate integration with external mining software, Internet of Things (IoT) sensors, and financial modelling tools, enhancing interoperability across the mining industry.

The main modules of MAFMINE 3.1 encompass essential aspects of mining and infrastructure, providing comprehensive cost estimations across different project components. The Mine module, applicable to both open pit and underground operations, estimates equipment, development, and infrastructure costs based on key parameters such as production rate, geomechanical conditions, and the distance to energy sources. The Processing Plant module calculates processing costs by considering factors including ore hardness, the selected technological route, and the associated energy and water requirements. Additionally, the Infrastructure module accounts for logistics expenses, worker accommodations, and waste management, ensuring a holistic evaluation of mining project costs.

The intuitive interface enables users to input geological, operational, and environmental data, generating detailed cost reports and Discounted Cash Flow (DCF) analyses. Additionally, MAFMINE 3.1 integrates indexation factors such as the producer price index (PPI) (US Bureau of Labor Statistics, 2022) and the big mac index that is based in the Purchasing Power Parity (PPP) to adjust historical values to current economic conditions.

The client-server application offers a flexible deployment model, functioning both as a cloud-based tool and a local desktop solution, supported by ElectronJS for dual access. Data storage is managed through PouchDB, enabling seamless synchronisation and accessibility in both online and offline modes. The front-end interface, designed with Handlebars and GruntJS, provides multilingual support to accommodate a global user base. Mathematical operations for cost estimation and financial modelling leverage MathJS, which facilitates complex equations, including quadratic functions essential for mining project evaluations. Additionally, Git version control ensures continuous integration of new models and formulas, enhancing adaptability to evolving industry requirements.

MAFMINE 3.1 stands out as both an educational and decision-making tool offering:

- Rapid project evaluation: Provides order-of-magnitude estimates to prioritise early-stage initiatives, reducing investment risks.

- ESG integration: Identifies environmental impacts (eg, water consumption and GHG emissions – CO_2 equivalent) and suggests technological alternatives (eg, fleet electrification).

- Multi-criteria decision-making tools: Features a tool utilising the Analytic Hierarchy Process (AHP) method to support project-related decision-making, along with specific modules for selecting mining methods, main access and ore transport methods.

- Academic support: Used in mining engineering courses, the software simulates real-world scenarios, preparing students for industry challenges.

- Customisable parametric models: future iterations will allow users to modify model parameters and incorporate proprietary cost structures, enabling tailored financial analyses.

Inputs in MAFMINE 3.1 prompt the user to define a production scale as a starting point for estimating costs in three main areas: the mine (open pit and/or underground), the processing plant, and infrastructure. Since all estimates are tied to production scale, empirical rules—first proposed by Taylor in 1977 and later refined by Long in 2009—are commonly used to guide assumptions on capacity and life-of-mine (LOM). The industry's shift toward larger operations is further examined by Wellmer and Drobe (2019).

In each area, subsidiary information is asked, such as topography, distance from energy supply and water source, rock strength, energy supply, more appropriate concentration process, kind of infrastructure for workers and logistics, etc. The mining method, primary access to the underground mine, and ore logistics system are determined using auxiliary multi-criteria decision-making tools included in this version of MAFMINE. These tools rely on classical methodologies—such as Nicholas (1981) or Miller, Pakalnis and Poulin (1995) for mining method selection, and the Vergne (2003) flow chart for access and logistics systems—as described by Cardozo *et al* (2024) and Campos *et al* (2024).

MAFMINE generates model outputs for each input and creates an all-inclusive report, where CAPEX and OPEX and environmental metrics are well detailed. Thus, regarding the environmental metrics, these issues are outputs of the tool, that use data from inputs, mathematical models and environmental factors, as described in following sessions. Figure 1 shows the core module structure of MAFMINE 3.1, emphasising the tool's inputs and outputs.

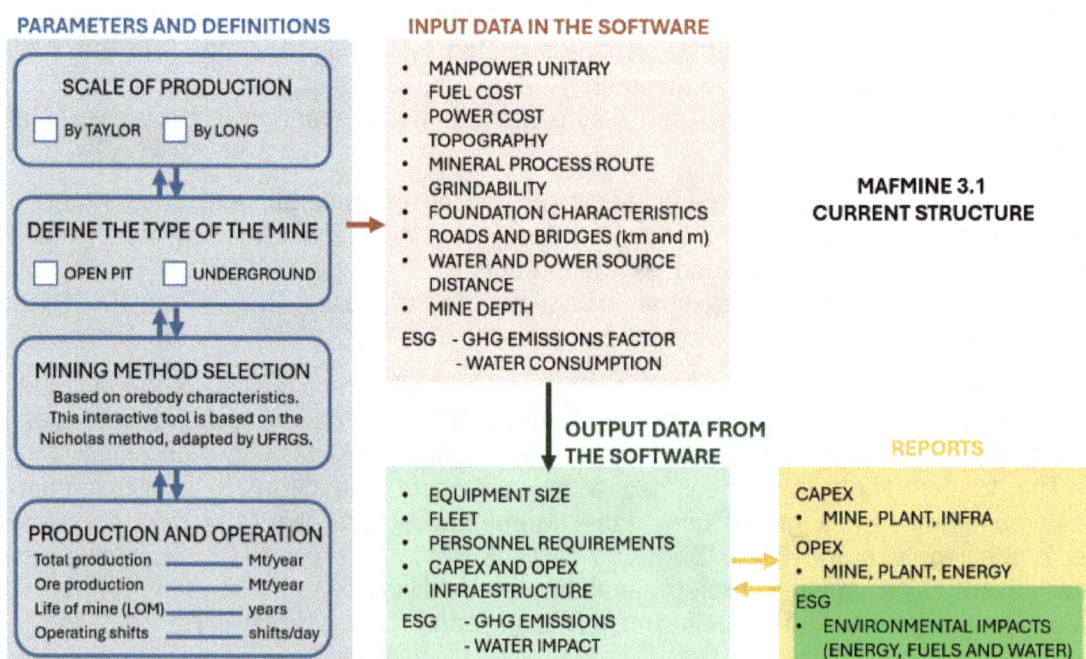

FIG 1 – Core module structure of the software.

ESG integration – environmental metrics

The ESG module of MAFMINE 3.1 was designed to initially assist users in evaluating the environmental impacts of mining operations, particularly GHG emissions and water consumption, providing information for strategic decision-making in mining project assessments. With the increasing demand for sustainable practices and the need to comply with increasingly stringent environmental regulations, this module integrates quantitative assessments of resource consumption and emissions, contributing to more responsible mining aligned with sustainability principles.

The development of the environmental metrics in the MAFMINE 3.1 was based on contemporary challenges in mining, including waste reduction and the implementation of environmentally responsible practices. The industry faces pressure to minimise its environmental impacts, and adopting innovative technologies can bring significant improvements in energy efficiency and emission reductions.

The environmental calculations in MAFMINE 3.1 are parametric, based on mathematical models that allow estimates to be adapted to the specific conditions of each operation. The methodology used enables a flexible and scalable approach, ensuring necessary accuracy in mining feasibility studies and allowing adjustments as new variables and parameters are incorporated.

Among the key metrics provided by the ESG module are:

- Water consumption: allows users to estimate the water efficiency of their operations, per tonnes processed. It considers the water withdrawn from new sources.

- CO_2 equivalent (CO_2e) emissions: calculated based on fuel and energy consumption in operations and processes, using emission factors from the Intergovernmental Panel on Climate Change (IPCC), adjusted for the national energy matrix. For the mine, equipment size is converted into engine power and expected fuel use via parametric curves, which are also applied to estimate energy demand for the plant and site.

- The electrification and automation of the fleet and plant allows to demonstrate the impact in diesel and electricity consumption, altering emission intensity and operational costs.

The selection of the best emission factors for calculating greenhouse gas (GHG) emissions is crucial to obtaining accurate estimates. In the case of this mine located in Brazil, the Global Warming Potential (GWP) standardised by the IPCC were adopted and adjusted according to the national energy matrix. For diesel, a CO_2 equivalent emission factor of 2.68 kg CO_2/L was used, following the guidelines of the Ministry of Science, Technology and Innovation (MCTI, 2023). For CH_4 and N_2O, the emission factors values, when converted to CO_2 equivalent, are extremely low, thus their contribution to the overall emissions is quite negligible. As a result, the total CO_2 equivalent emission factor remains unchanged.

Regarding electricity, the average emission factor of the Brazilian electrical system of 0.0385 t CO_2/MWh was adopted. The Brazilian energy matrix has a significant share of renewable sources, which considerably reduces emissions associated with the processing plant. However, seasonal variations and the use of thermal powerplants can impact these values, making it essential to periodically update the emission factors used (MCTI, 2023). Figure 2 represents a flow chart illustrating the logic behind the calculation of environmental metrics in MAFMINE 3.1.

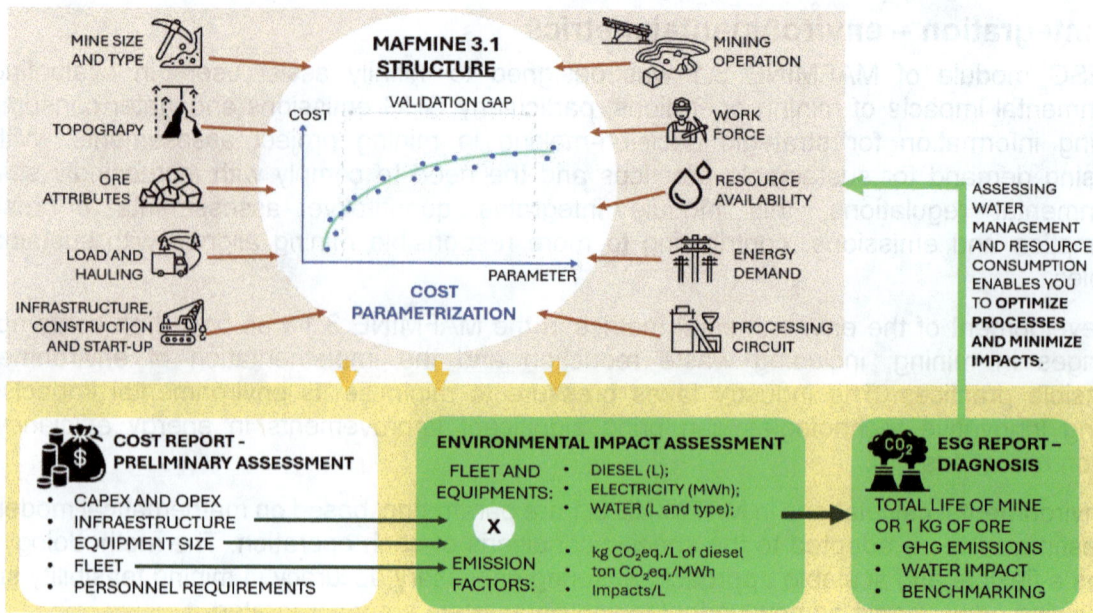

FIG 2 – Schematic flow chart of the integration of ESG in MAFMINE 3.1.

Case study – implementation of MAFMINE 3.1

A case study was conducted to evaluate the applicability of the MAFMINE 3.1 software in the economic modelling of a mineral project, located in a tropical region in northern Brazil. The analysed deposit is situated in a geological context marked by Archean metasedimentary and metavolcanic sequences, with mineralisation hosted in lenticular bodies of copper sulfides, associated with gold and silver. The objective was to simulate the viability of the enterprise by integrating geological data, operational parameters, and indexation factors (PPI and PPP) for monetary and geographical adjustment of the estimated costs.

Figure 3 illustrates the algorithm for estimating mining costs using MAFMINE 3.1. The process involves defining key parameters — such as mineral reserves, rock mass conditions, worker salaries, and the average stripping ratio (which reflects the mine size) — and selecting the mining method (in this case, open pit). The software then applies parametric equations to generate projections for costs, technical data, and environmental metrics.

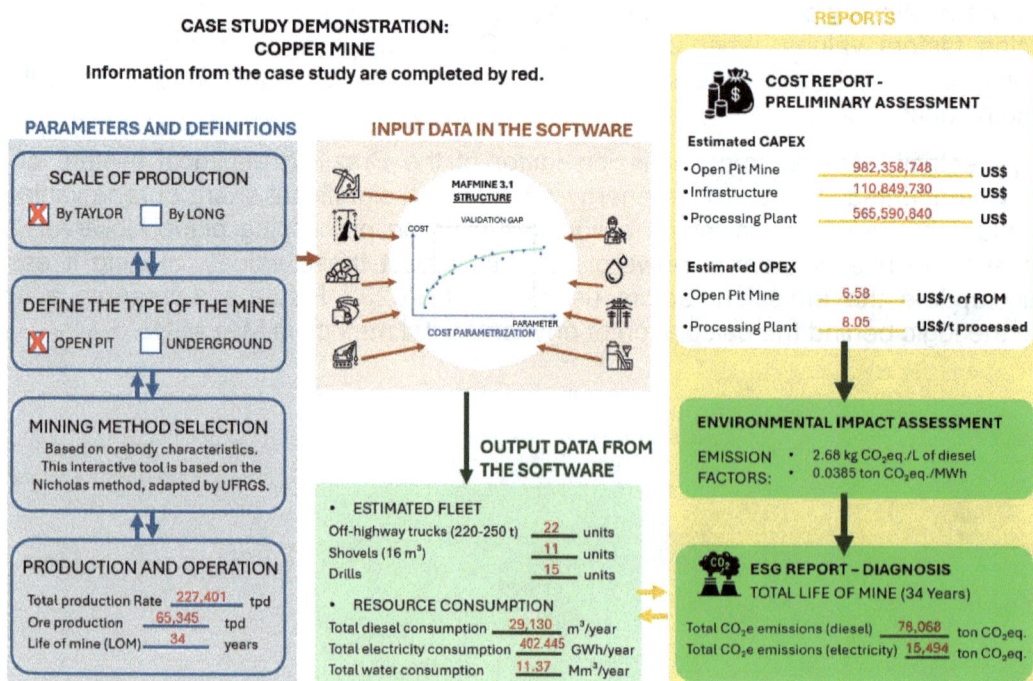

FIG 3 – Schematic flow chart with inputs and outputs data from a simulation in MAFMINE 3.1

It is important to note that some of the input data required by MAFMINE 3.1 can be challenging to obtain during the early stages of a mining project, as they often depend on more advanced techniques applied later in the mineral exploration process. Consequently, the estimates provided are generally preliminary, offering only an order-of-magnitude approximation of the costs. In contrast, the input data presented in Table 1 pertain to a more advanced project, where drilling, sampling and precise geostatistical analysis were carried out. Such situations further highlight the utility of the software as a decision-making tool, as it proposes a marginal cost analysis to guide certain projects, providing a more precise estimation based on the quality of the inputs.

TABLE 1

Input and output parameters from a simulation in MAFMINE 3.1

Parameter		Unit	Input	Output
Base parameters	Mineral Reserves	t	784 300 000	-
	Average stripping ratio	-	2.48	-
	Days in operation per annum	day	350	-
	Ore production rate	tpd	-	65 345
	Total production rate (ore + waste)	tpd	-	227 401
	Life-of-mine (LOM)	year	-	34
	Mine waste bulking	%	50	-
	Road access to mine	Km	20	-
	Access terrain conditions	-	Flat and dry	-
	Ore type	-	Base metals (Cu-Au)	-
	Ore properties	-	WI= 15; 70% <200#	-
	Processing capacity (ROM)	tpd	65 000	-
	Concentrate production	tpd	2400	-
	Tailings disposal method	-	Conventional dam	-
	Distance from mine to plant	km	5	-
Workers' salary	Operation, maintenance and services	USD/d	100	-
	Administration	USD/d	400	-
Workforce	Type of accommodations	-	Daily trips	-
	Workers in processing plant	-	-	229
	Workers in open pit mine	-	-	835
Estimated fleet for open pit mine	Shovels (16 m³)	-	-	11
	Off-highway trucks (220–250 t)	-	-	22
	Drills (10.5 inches diameter)	-	-	15
	Estimated fleet diesel consumption	m³/annum	-	45 596.63
	Operational hrs per annum	h/annum	-	5110
Energy	Energy generation method	-	External supplier[2]	-
	Transmission line length	Km	100	-
	Peak load	kW	-	64 062.70
	Energy consumption	kWh/day	-	1 149 843

Parameter		Unit	Input	Output
Estimated CAPEX	Open pit mine	USD	-	982 358 748
	Infrastructure	USD	-	110 849 730
	Processing plant	USD	-	565 590 840
Estimated OPEX	Open pit mine	USD/t of ROM	-	6.58
	Processing plant	USD/t processed	-	8.05
Emission factors (CO_2e)	Diesel	kg CO_2e/L of diesel	2.68	-
	Electricity	t CO_2e/MWh	0.0385	-
GHG emissions	CO_2e emissions (diesel)	t CO_2e/annum	-	122 198.97
	CO_2e emissions (electricity)	t CO_2e/annum	-	15 494
	Total CO_2e emissions	t CO_2e/annum	-	137 692.97
	Total CO_2e emissions for LOM	t CO_2e/LOM	-	4 681 650.98
Sustainability and mitigation	Potential reduction from fleet electrification	% CO_2e/annum	30%	23.8%
	Biodiesel usage (B20 blend)	% CO_2e/annum	20%	16.6%
	Electrification + biodiesel	% CO_2e/annum	30% / 20%	35%
Resource consumption	Total diesel consumption	m^3/annum	-	45 596.63
	Total electricity consumption	GWh/annum	-	402.445
	Water consumption (withdrawn from new sources)	m^3/t processed	0.5	-
	Total water consumption	Mm^3/annum	-	11.375

Figure 4 illustrates the application interface, which, as mentioned in section 'MAFMINE 3.1 – Software architecture and functionalities', was developed to be simple and intuitive, facilitating the use of the software by professionals in the field and proving instructive in academic subjects related to the mineral sector.

The case study indicates that a mine with an input reserve of 784 300 000 t, an average stripping ratio of 2.48, and 350 operational days per annum is estimated to have an ore production rate of 65 345 tpd and a life-of-mine (LOM) of 34 years. This information is used to estimate a CAPEX of USD 982 million and an OPEX of USD 6.58 per ton of ore mined.

For a base metal mine of this size—with a ROM processing capacity of 65 000 tpd and a concentrate production of 2400 tpd—the processing plant's CAPEX is estimated at USD 565.6 million, with operational costs of USD 8.05 per ton. An additional USD 110.5 million is required for infrastructure such as tailings dam, energy, and water distribution.

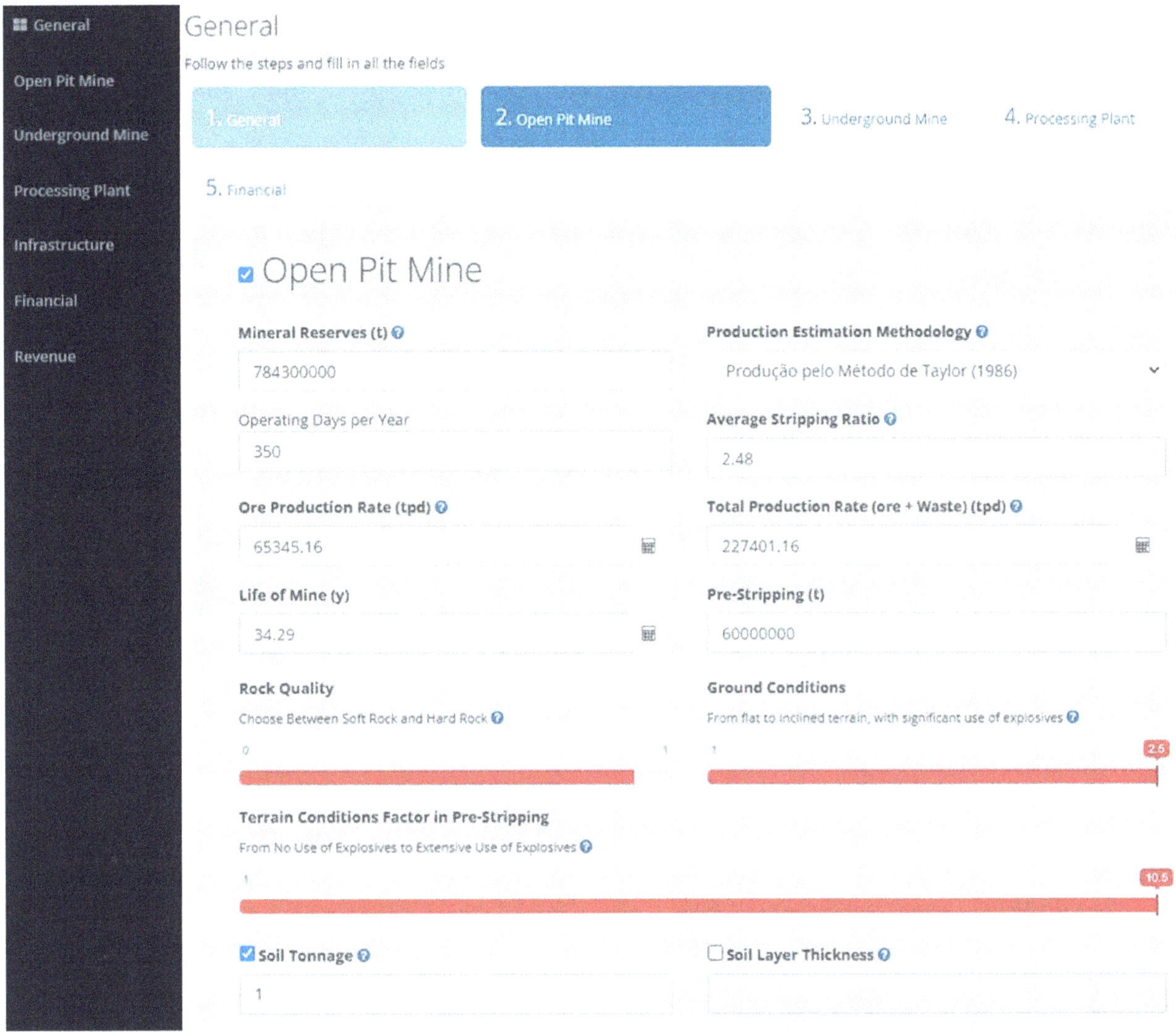

FIG 4 – Open pit page in MAFMINE 3.1 (UFRGS, 2025).

The software estimates that a total of 835 workers is needed for conducting the mine operations and 229 for the processing plant. These numbers include mine operation, equipment maintenance, administration, clerk, engineers and personnel support workers. Regarding the fleet, the estimated requirements include 11 shovels (16 m³), 22 off-highway trucks (220–250 t), and 15 drills (10.5" diameter). Consequently, the estimated fleet diesel consumption is 45 596.63 m³ per annum, with annual operating hours of 5110 hrs per annum.

Regarding potential environmental impacts, the software estimates that CO_2 emissions from diesel amount to 122 198.97 tons CO_2e per annum, while emissions from electricity total 15 494 tons CO_2e per annum. Consequently, the total impact over the life-of-mine (LOM) is estimated at 4 681 560.98 tons CO_2e. Additionally, the software identifies potential strategies to reduce or mitigate these environmental impacts. The first option is partial fleet electrification (30 per cent of the fleet), which could reduce CO_2e emissions by approximately 23.8 per cent per annum. The second option involves substituting diesel with a B20 biodiesel blend, where replacing 20 per cent of diesel with biodiesel could lower CO_2e emissions by approximately 16.6 per cent per annum. The third option combines both strategies – electrification and biodiesel substitution – resulting in an estimated 35 per cent reduction in CO_2e emissions per annum.

The case study demonstrated the software's capacity to deliver estimates for critical project parameters, including production rates, fleet sizing, workforce allocation, capital and operational expenditures, infrastructure investments, and CO_2e emissions. By supporting analyses ranging from preliminary order-of-magnitude estimates to marginal cost evaluations, the software solidified its role as a strategic decision-making tool, particularly during early planning stages. It also identified viable sustainability strategies, such as partial fleet electrification and biodiesel adoption, capable of significantly reducing annual emissions. These features underscore its utility not only in technical validation but also in integrating economic and environmental criteria during the conceptual design of mining projects.

FINAL CONSIDERATIONS

MAFMINE represents a significant advancement in mining project evaluation for educational purposes by providing a fast solution for cost estimation at preliminary level. As the industry continues to adopt digital transformation strategies, the software's integration of ESG considerations and parametric modelling will further enhance its value. Future developments will position MAFMINE 3.1 as a comprehensive platform, bridging the gap between technical feasibility and sustainability.

Thus, it is evident that there is a gap filled with both challenges and opportunities, as there are few studies presenting formal models dedicated to sustainable mining projects. Consequently, following the development and implementation of MAFMINE 3.1, significant results are expected to be achieved. Initially, there is an improvement in environmental assessment, enabling a more comprehensive understanding of the environmental impact of mining projects. This facilitates the identification of areas for improvement and the development of mitigation strategies. Additionally, decision-makers have access to detailed information regarding emissions associated with mining operations, empowering them to take proactive measures to reduce environmental impact and ensure long-term sustainability.

Regarding the future development, the next phase in MAFMINE's evolution is the development of a comprehensive platform for parametric modelling. This initiative aims to expand the tool's capabilities beyond cost estimation to include:

- Advanced ESG modules: the upcoming version will feature a dedicated ESG module that evaluates the socio-environmental trade-offs of mining operations, offering simulations for carbon footprint reduction and water conservation strategies, based on Life Cycle Assessment (LCA) data, providing users with benchmark data for comparison.

- AI-powered decision support: machine learning algorithms will enhance the accuracy of cost predictions by continuously updating the model with real-world mining data.

- Cloud-based collaboration: a web-based interface will facilitate real-time collaboration among engineers, streamlining the decision-making process for mining companies and investors.

ACKNOWLEDGEMENTS

The authors would like to acknowledge the support of the National Council for Scientific and Technological Development (CNPq) under the CNPq/CT-Mineral Call No. 27/2022.

REFERENCES

Camm, T, 1992. Bureau Of Mines: Simplified Cost Models for Prefeasibility Mineral Evaluations, United States Department of the Interior.

Campos, H, Cardozo, F, Petter, C O and Lenz, V M, 2024. Development of a Mining Method Selection Application for Conceptual Mining Projects, in *23rd Mining Seminar Proceedings*, 23:231–241.

Cardozo, F, Campos, H, Petter, C and Lenz, V M, 2024. Mining Method Selection During Conceptual Studies, in *23rd Mining Seminar Proceedings*, 23:242–255.

D'Arrigo, R F, 2012. Model for estimating operating and capital costs in mining projects in the conceptual phase based on the O'Hara model [in Portuguese: Modelo de estimativa de custos operacionais e de capital em projetos de mineração em fase conceitual baseado no modelo de O'Hara], Master's thesis (unpublished), UFRGS, Porto Alegre.

Fioroni, M M, Franzese, L A G, Bianchi, T J, Ezawa, L, Pinto, L R and Miranda, G, 2008. Concurrent simulation and optimization models for mining planning, in *Proceedings of the 2008 Winter Simulation Conference,* pp 759–767.

Leite, J M L G, Arruda, E F, Bahiense, L and Marujo, L G, 2019. Modelling the integrated mine-to-client supply chain: a survey, *International Journal of Mining, Reclamation and Environment*, 34(4):247–293.

Long, K R, 2009. A test and re-estimation of Taylor's empirical capacity–reserve relationship, *Natural Resources Research*, 18(1):57–63.

Miller-Tait, L, Pakalnis, R and Poulin, R, 1995. UBC Mining Method Selection, in *Proceedings of Mine Planning and Equipment Selection*, pp 163–168.

Ministry of Science, Technology and Innovation (MCTI), 2023. National Emissions Registration System, Ministry of Science, Technology and Innovation, Federative Republic of Brazil. Available from: <https://www.gov.br/mcti/pt-br/acompanhe-o-mcti/sirene> [Accessed: 22 February 2025].

Nicholas, D, 1981. Mining Method Selection – A Numerical Approach, *Design and operation of caving and sublevel stoping mine* (ed: D R Stewart), Society of Mining Engineers of the American Institute of Mining, Metallurgical and Petroleum Engineers.

O'Hara, T, 1980. Quick guides to evaluation of orebodies, *CIM Bull*, 2:78–99.

Pimentel, B S, Gonzalez, E S and Barbosa, G N O, 2016. Decision-support models for sustainable mining networks: fundamentals and challenges, *Journal of Cleaner Production*, 112(4):2145–2157.

Pinheiro, U V, Ambros, W M, Petter, C O, Cardozo, F A C, D'Arrigo, R F and Petter, R A, 2023. MAFMINE ESG: Sustainability tool for water management in mining projects [in Portuguese: MAFMINE ESG: Ferramenta de sustentabilidade para gestão de água em projetos de mineração], in *Proceedings of the 7th Symposium on Sustainable Systems,* 2:193–198 (IAHR Publishing).

Taylor, H K, 1977. Mine valuation and feasibility studies, in *Mineral Industry Costs*, pp 1–17 (Northwest Mining Assoc.: Spokane).

UFRGS, 2025. MAFMINE 3.1. Federal University of Rio Grande do Sul, Porto Alegre, Brazil. Available from: <https://www.mafmine.com.br/v3.1/> [Accessed: 26 February 2025].

United States, 1910–1995. *Bureau of Mines Bulletin* (Government Printing Office: Washington DC).

Vergne, J L, 2003. *Hard Rock Miner's Handbook*, 3rd ed.

Wellmer, F-W and Drobe, M, 2019. A quick estimation of the economics of exploration projects – rules of thumb for mine capacity revisited – the input for estimating capital and operating costs, *Boletín Geológico y Minero*, 130(1):7–26.

Mine planning
and operations

Application of causal machine learning in mitigating ore dilution for maximum return in mining

G Asamoah[1], J Liu[2] and R K Asamoah[3]

1. PhD Candidate, University of South Australia, Mawson Lakes SA 5095.
   Email: gloria.asamoah@mymail.unisa.edu.au
2. Associate Professor, Industrial AI Research Centre, University of South Australia, Mawson Lakes SA 5095. Email: jixue.liu@unisa.edu.au
3. Senior Research Fellow, Future Industries Institute, University of South Australia, Mawson Lakes SA 5095. Email: richmond.asamoah@unisa.edu.au

ABSTRACT

The integration and application of causal machine learning models in mining and mineral extraction processes have always been a major problem in the mining industry due to the uncertainties in the geological models such as grade variations, geotechnical issues, mine operations challenges, and metal price fluctuations. Current conventional methods such as the use of stability charts, numerical models and correlation based machine learning are employed in controlling unplanned ore dilution. But the limitations, complexity and computational challenges in the conventional methods result in poor interpretation of outcomes. There are currently contradictory views on the model that are effective for the mining industry in mitigating unplanned ore dilution. Paucity of knowledge, therefore, exists on the specific machine learning model that can be practically applied in mining setting to minimise unplanned ore dilution considering all the dynamic planning parameters, in an efficacious manner for maximum return. The key research questions and hypothesis are:

1. How effective is causal models compared with the other conventional mine planning techniques to obtain high cash flows and maximise output, taking keynote of historical realisations?

2. What fundamental principles limit the transition of most mining companies/operations from the conventional ore dilution control methods to causal models?

3. How can the limitation be eliminated for improved returns?

The objective of this study is to investigate the application of causal model in mitigating unplanned dilution in mining for maximum return while addressing the above research questions. Links to relevant literature will be highlighted.

INTRODUCTION

Currently, there is a high demand for precious (eg gold, silver, and platinum group metals) and critical metals (eg nickel, cobalt and lithium) to address major global converted needs such as high-tech equipment and clean energy systems. This among other challenges has mounted pressure on the mining industry to mine valuable ore, minimise unplanned ore dilution, and increase revenue in an environmentally friendly manner (Henning and Mitri, 2007; Grandell *et al*, 2016; Mooiman, Sole and Dinham, 2016). Therefore, improving the current methods for predicting ore dilution will ensure efficient and cost-effective extraction of high-grade valuable metal ores.

Mining is the extraction of minerals and or metals that are economically viable from the earth, considering the orientation of the geological orebody (Dunbar, 2016). Underground mining is the excavation of ore deposits that are deeper below the earth surface and involves operational activities such as drilling and blasting of ore blocks called stopes, mucking blasted material with Load-Haul-Dump machines (LHDs) and stockpiling (Hamrin, Hustrulid and Bullock, 2001). However, during these activities, ore dilution occurs resulting in the undesirable reduction of value of the ore mined.

Dilution is the contamination that occurs when waste rock (material below the cut-off grade) mixes with valuable ore (material at or greater than the cut-off grade). In underground mining, the categorisation of dilution is planned and unplanned (Jang, 2014). Planned dilution is waste within the ore reserve and included in mineable shapes or stope designs. On the other hand, unplanned dilution occurs when uneconomical material outside the planned stope design walls mixes with the economic ore within the stope boundaries due to stope overbreak or sloughage (Henning and Mitri, 2007). Figure 1 shows the difference between unplanned dilution and planned dilution in underground mine of an ore reserve.

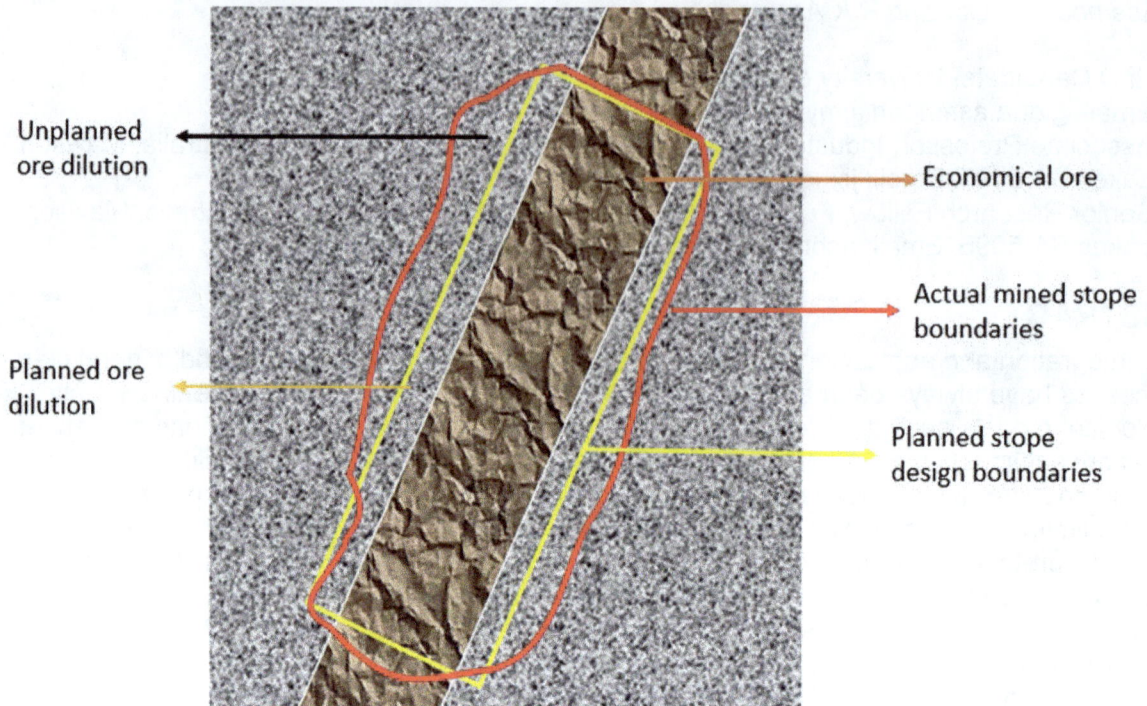

FIG 1 – Schematic diagram showing planned and unplanned ore dilution in an underground stope.

Unplanned ore dilution has been a well-known problem in the mining industry influencing the mine operation cost (Suglo and Opoku, 2012; Whillans, 2018; Delentas, Benardos and Nomikos, 2021). Factors contributing to unplanned ore dilution include drill and blast activities, stope design parameters, and geological and geotechnical factors (Jang, 2014).

In extant literature, the methods employed for predicting ore dilution in mining include empirical, numerical, and correlation-based machine learning.

UNPLANNED ORE DILUTION

Unplanned ore dilution resulting from mine operational activities has been a common problem for years and researchers have employed different methods to solve this problem (Jang, Topal and Kawamura, 2015). Dilution resulting from mine operational activities involves drilling and blasting (Kanchibotla, 2019; Conde and Sanoh, 2022; Himanshu *et al*, 2023; Brantson, Appiah *et al*, 2024; Himanshu *et al*, 2024). After blasting, the ore is mixed with waste (uneconomical part) which is often below the cut-off grade with the valuable mineral of interest resulting in ore dilution (Clark, 1998; Ebrahimi and Eng, 2013; Tommila, 2014; Serdaliyev *et al*, 2022; Temizyürek, 2023; Ming *et al*, 2024). Figure 2 is an example of ore dilution due to overbreak after blast in an underground stope.

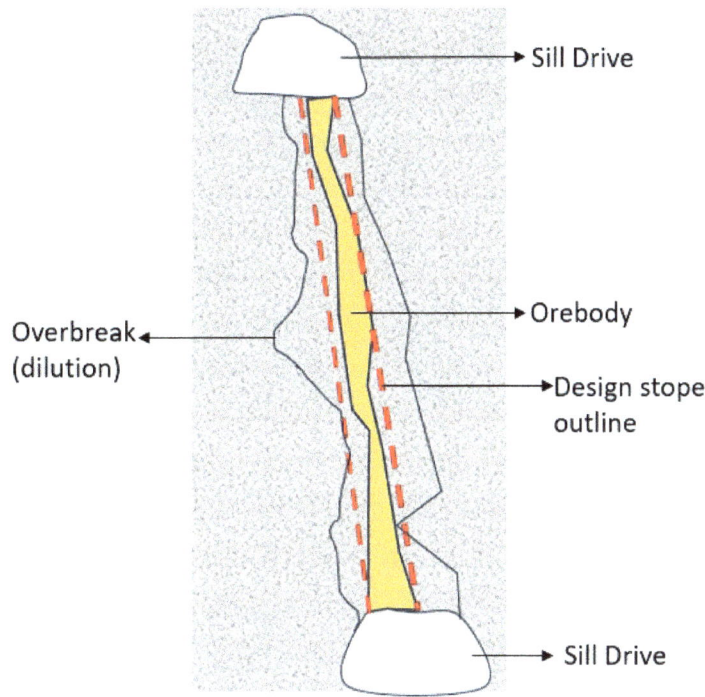

FIG 2 – Dilution due to overbreak in underground stope.

Conventional methods for predicting dilution

The current conventional methods employed to reduce dilution include empirical (Castro and Paredes, 2014; Salgado-Medina *et al*, 2020), numerical (Sun *et al*, 2020; Delentas, Benardos and Nomikos, 2021; Yu *et al*, 2021; Cordova and Gonçalves, 2022) and correlation based machine learning methods (Zhao and Niu, 2020; Bazarbay and Adoko, 2021; Yu, Zheng and Ren, 2021; Korigov, Adoko and Sengani, 2022; Hmoud and Kumral, 2023). Below sub-sections further review these methods, unravelling gaps and the need for advanced methods.

Empirical methods for predicting dilution

The empirical methods employed in predicting dilution involve the use of stability graph (Mawdesley, Trueman and Whiten, 2001). Stability graphs are used for stope designs and estimation of dilution base on stability number (N) and hydraulic radius (HR). The hydraulic radius is the ratio of the area to the perimeter of stope under consideration whiles the stability number is calculated based on empirical factors such as induced stress acting on the stonewall, orientation of stope walls relative to joint planes, gravity and rock quality index (derived from Barton's Q system). The original Mathews stability graph was the first graphical method introduced to assess stability of pillars in underground mines, especially in hard rock mining operations (Stewart, 2005). Figure 3 shows the prediction zones on the Mathews stability graph base on the stability number and hydraulic radius or shape factor. The main idea of Mathews stability graph is that the excavation size relates to how competent a rock mass is, by assessing it stability. However, the graph does not consider the entire excavations but rather the surfaces of the excavations and limited to 50 open stope case histories comprising of depths more than 1000 m (Stewart, 2005).

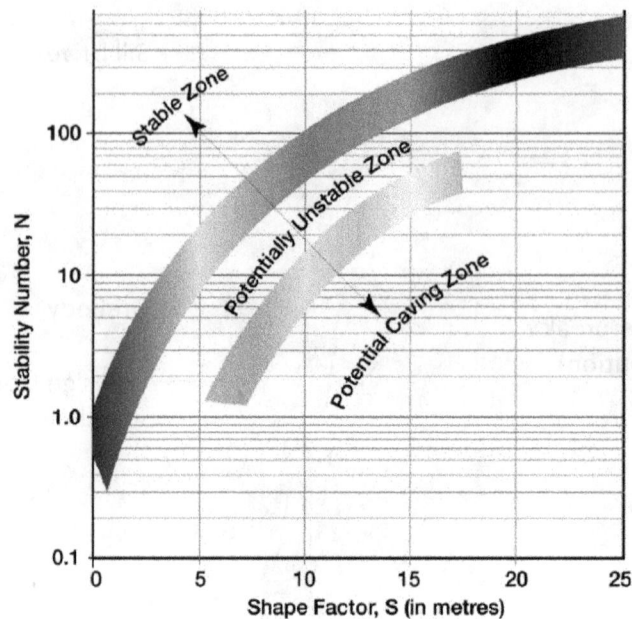

FIG 3 – Original Mathews stability chart (Stewart and Trueman, 2008).

Many researchers have applied stability graphs in extant literature in different ways. For instance, Mawdesley, Trueman and Whiten (2001) extended the original Mathews stability graph considers more database of conditions of rocks and dimensions of stopes. Further analysis was made using logistic regression to explain and optimise the placement of stability zones on the chart. The variables considered in logistic regression analysis include hydraulic radius (shape factor, S), Mathew's stability number and failure zones. Their result showed that logistic regression analysis in conjunction with stability chart are useful in calculating the probability of stability of a particular stope geometry and dimension. Wattimena *et al* (2013) constructed a coal pillar stability chart for a coalmine using logistic regression model to predict the probability of stability of a coal pillar considering its geometry and stress conditions. In their studies, isoprobability contours (lines of equal probability) for stable pillars were developed to assist with the determination of risk associated with the pillar stability. Putra *et al* (2024) research into the integration of Mathews stability chart into the optimisation of stope layout in an underground mine. In their study they considered the use of Mathews stability chart to analyse the stability of stopes in a synthetic block model. The stability conditions were assessed based on the geomechanical data in the block model such as rock mass classification, joint spacing and fault. The outcome of their study showed that, incorporating stability chart in the optimisation process of stope layout yields economic value.

Though the extended Mathews stability graph considered more database of case histories, there are some disadvantages in its application. The first demerit is that the variables considered for predicting the probability of failure of stope does not consider drill and blast variables. The second disadvantage is that the graph is effective in predicting weather a stope is geotechnically stable or unstable but not effective in predicting dilution, because the relationship between stability number and hydraulic radius is stope size dependent (Stewart and Trueman, 2008).

Numerical methods for predicting dilution

In extant literature numerical methods developed for predicting ore dilution involve the use of two or three dimensional numerical simulation of factors causing ore dilution. The factors considered in these simulations include mining depth, *in situ* stress, stope geometry and orientation. Sun *et al* (2020) investigated the draw problem encountered in caving mining method to forecast the ore dilution, based on numerical modelling. They used simulation to estimate the impact of particle flow in stopes on dilution by considering factors such as particle size, drawpoint size and column length. Their numerical simulations results showed that higher draw heights are more prone to ore dilution in a block caving mining method. However, the simulation took 700 hrs real time to finish using high computer memory and processors which shows how numerical analysis can be time consuming. Saeedi *et al* (2010) developed a series of two-dimensional numerical models to analyse the effect of mining depth, *in situ* stress, unsupported roof geometry and dip of coal seam on out of seam dilution in longwall mining method. Their research showed that out of seam dilution in coalmines can be linked to cutting floor and roof fall. The simulation model yielded out of seam dilution of 21.1 per cent compared to the 23 per cent observed in the mine. Also, due to the computer capacity and running time, the panel length and depth for their analysis was restricted to 200 m by 50 m. However, a typical panel length in longwall mining ranges from 200 m to 1000 m and panel depth ranges from 300 m to 1200 m (Smith and Collins, 1985; Isaac and Follington, 1988; Hebblewhite and Holt, 2001; Kelly, Luo and Craig, 2002; Wang *et al*, 2017, 2018, 2024; Khanal *et al*, 2022) highlighting the unrealistic assumptions in numerical modelling. This means that their finding or result cannot be applied in longwall mines of greater length and depth due to computational processing time and cost. Henning and Mitri (2007) employed three-dimensional numerical modelling to analyse the effect of mining depth, *in situ* stress, stope geometry and orientation on ore dilution in blasthole open stoping environment. The result of their studies showed that the reduction of stope height can potentially reduce stope overbreak resulting in dilution. Further, the demerit of the numerical modelling is that it requires strenuous background analysis to be able to determine the right input parameters for the model (Jorquera, Korzeniowski and Skrzypkowski, 2023).

CORRELATION-BASED MACHINE LEARNING METHODS FOR PREDICTING DILUTION

Correlation-based machine learning have been developed by many researchers over the years due to the shortfalls of empirical and numerical used for dilution prediction in dynamic mining environment. Chimunhu *et al* (2024) used Principal Component Analysis – Classification and Regression Tree algorithm (PCA-CART) to predict dilution that assisted long-term production schedules in the prefeasibility stage of mining. They investigated how initial stope designs could be used to establish dilution factors to enhance efficient mine scheduling in underground mining. The dilution factors were estimated by considering variables such as tonnes, grade, stope dimensions, faults and weathering conditions of rocks. The outcome of their study showed that dilution recorded in final stopes was more than initial stopes. Also, dilution was underrated when applied to mine plans. Yu *et al* (2021) also utilised the whale optimisation algorithm in conjunction with Gaussian process to predict dilution and ore loss after blast. Reduction in dilution was more than 50 per cent using their predictive model, which is a significant improvement. Zhao and Niu (2020) focused on the prediction of unplanned dilution using neural network. In addition, Henning and Mitri (2008) investigated a case study on simulation of hanging wall overbreak in stopes to control ore dilution. Their study showed that degree of hanging wall dilution in underground stopes was influenced by using cable bolt for hanging wall reinforcement. The disadvantage of the correlation based machine learning to predict ore dilution is that it uses variables that are statistically related and are in continuous pattern to predict an outcome making causal statement bias.

Therefore, there is still a gap to identify and investigate the causal relationships between mining variables and dilution in mining setting as deposits become more complex and build more robust and reliable models to predict ore dilution. This will also address current challenges with empirical and numerical methods.

SIGNIFICANCE OF CAUSAL MACHINE LEARNING IN MINING COMPARED WITH CONVENTIONAL METHODS IN CONTROLLING ORE DILUTION

Over the years conventional methods focus on the use of stability charts, numerical models and correlation based machine learning methods in predicting ore dilution. Weakness and ineffectiveness of current dilution predictive methods are presented below to further highlight the significance of causal modelling in the mining process and its critical importance for future sustainable mining.

- Application of empirical methods such as the use of stability chart (Mawdesley, Trueman and Whiten, 2001) do not consider other critical variables such as powder factor, blast timing and explosive type (Stewart and Trueman, 2008). These variables can impact on dilution, making empirical approach unable to draw variable-dilution relationships and minimise their negative impact on their dilution.

- Numerical modelling (Heidarzadeh, Saeidi and Rouleau, 2020) does not rely on causation to identify variables impacting on dilution but rather models a subsection of the true probability space with several unrealistic assumptions. For example, in numerical modelling, results from a simulation can show a link between explosive energy and overbreak leading to dilution (An *et al*, 2018). However, as deposit goes deeper and underground conditions dynamically changes, there may be other unknown parameters driving the overbreak which is not currently accounted for in numerical model due to unrealistic assumptions. In addition, numerical methods require strenuous background analysis to be able to determine the right input parameters for the model to predict stope overbreak resulting in high computational cost and complexity (Jorquera, Korzeniowski *et al*, 2023).

- Correlation base machine learning (Petri, 2019) uses the correlation structure in data set to predict dilution and result in bias causal statement, diverting the main cause of a problem in a data set (Klein, El-Assady and Jäger, 2022). Correlation base machine learning are good at identifying patterns in data sets (Zhao and Niu, 2020; Chimunhu *et al*, 2022), but the ability to model mechanisms that drives patterns and also identify the cause and effect relationships is a challenge. For example, there might be correlation between a narrow vein irregular orebody and dilution, as orebody become more narrow and irregular ore dilution increases. However, ore geometry is fixed and will not change, but the adaption of the mining method and blast practices to mine this irregular orebody is critical and causes dilution if not properly managed.

Therefore, it is important to understand the cause and effect of mine planning and drill and blast inputs or variables on ore dilutions. Causal machine learning stem from the understanding of cause and effect that exist between variables where changes in one variable may affect another variable in a data set (Hall, 2000).

Causal models use mathematical models to represent causal relationships and assist in building causal inference from a statistical data (Pearl, 2011). These models have been applied in various disciplines to understand the cause and effect of variable change within a process or a system.

Application of causal models in minimising dilution unplanned ore dilution is important due to the benefits it will bring to the mining industry. The first benefit is that it will increase the financial gain to the mining business by reducing waste for cost efficient operation. This is because correlation-based models can result in financial loss and higher risk when predictions are based on inaccurate feature selections. However, in causal models there is high certainty of prediction, hence the mine is less liable to make financial loss. The second benefit is that mine planning and mine operations can focus on the right areas that requires the right intervention systems, based on the root cause. For example, in a situation of an overbreak, instead of changing the explosive type, the focus can rather be improving the powder factor which may result in more cost-effective and safer approach. Finally, causal models will benefit mining companies to better manage mining sequence during mine scheduling. For example, in a typical underground sublevel open stopping mining sequence where primary stopes are mined and backfilled before secondary stopes are mined, the causal relationship between the mining variables and dilution may change due to different rock properties and parameters employed in blasting the stopes following the mining sequence and direction. The past blasting, mining and backfilling events in the primary stopes can influence the observed dilution outcome for secondary stope. Therefore, understanding the cause and effect relationship between mining variables such as (powder factor, weathering conditions, faults, drill spacing and drill burden) and dilution can assist in predicting and controlling future dilution outcome in secondary stopes.

FACTORS LIMITING THE TRANSITION FROM CONVENTIONAL METHODS OF CONTROLLING ORE DILUTION TO CAUSAL MODELS

Currently causal models have been successfully applied in the field of health science, social science and finance. However, it has not been extensively applied in the mining industry for controlling ore dilution due to the following reasons:

- The high quality of data required for causal analysis. Mining operations faces challenges of missing data and data collection and processing issues.

- Causal machine learning requires expertise who understands causal modelling and are well vexed in knowledge such as statistics and machine learning.

- Causal modelling requires accuracy and can be time consuming.

To eliminate the above factors limiting the application of causal models in the mining industries. Mining companies must do the following:

- Educate and train employees on the benefits of causal models in quality decision-making to solve root cause of problems based on their operational needs.

- Invest in a well-developed structure that enables accurate data collection and processing to prevent missing records.

- The use of automated causal inference approach will increase efficiency and reduce time.

CONCLUSIONS

The application of causal machine learning models in today's mining industry to control unplanned ore dilution is important due to the significant benefits that comes with its implementations. Though there are some factors preventing the use of causal models in the mining industry, overcoming them by training and enlightenment of mining personnels will assist and minimise unplanned ore dilution. There is a need for more causal machine learning model development in the mining industry particularly for mitigating unplanned ore dilution.

ACKNOWLEDGEMENTS

The authors acknowledge University of South Australia and the Australian government for providing funding for this work. The authors also thank the facilities and technical assistance of staff at the Future Industries Institute, University of South Australia.

REFERENCES

An, L, Suorineni, F T, Xu, S, Li, Y-H and Wang, Z-C, 2018. A feasibility study on confinement effect on blasting performance in narrow vein mining through numerical modelling, *International Journal of Rock Mechanics and Mining Sciences*, 112:84–94.

Bazarbay, B and Adoko, A, 2021. A comparison of prediction and classification models of unplanned stope dilution in open stope design, ARMA US Rock Mechanics/Geomechanics Symposium, ARMA.

Brantson, E T, Appiah, T F, Alhassan, I, Dzomeku, G M, Boateng, E O, Takyi, B, Sibil, S, Duodu, E K and Kobi, A K, 2024. A comprehensive review of traditional, modern and advanced presplit drilling and blasting in the mining and construction industries, *Journal of Petroleum and Mining Engineering*, 25(2):87–97.

Castro, R and P, Paredes 2014. Empirical observations of dilution in panel caving, *Journal of the Southern African Institute of Mining and Metallurgy*, 114(6):455–462.

Chimunhu, P, Faradonbeh, R S, Topal, E, Asad, M W A and Ajak, A D, 2024. Development of Novel Hybrid Intelligent Predictive Models for Dilution Prediction in Underground Sub-level Mining, *Mining, Metallurgy and Exploration*, 41:2079–2098.

Chimunhu, P, Topal, E, Ajak, A D and Asad, W, 2022. A review of machine learning applications for underground mine planning and scheduling, *Resources Policy*, 77:102693.

Clark, L M, 1998. Minimizing dilution in open stope mining with a focus on stope design and narrow vein longhole blasting, University of British Columbia.

Conde, F I and O, Sanoh 2022. Analysis and Optimization of Blasting Practices at the Sangaredi Mine, *Journal of Geoscience and Environment Protection*, 10(9):149–169.

Cordova, D P, Zingano, A C and Gonçalves, I G, 2022. Unplanned dilution back analysis in an underground mine using numerical models, *REM-International Engineering Journal*, 75:379–388.

Delentas, A, Benardos, A and Nomikos, P, 2021. Analyzing stability conditions and ore dilution in open stope mining, *Minerals*, 11:1404.

Dunbar, W S, 2016. How mining works, Society for Mining, Metallurgy and Exploration.

Ebrahimi, A and P, Eng 2013. The importance of dilution factor for open pit mining projects, World Mining Congress, Montreal.

Grandell, L, Lehtilä, A, Kivinen, M, Koljonen, T, Kihlman, S and Lauri, L S, 2016. Role of critical metals in the future markets of clean energy technologies, *Renewable Energy,* 95:53–62.

Hall, M A, 2000. Correlation-based feature selection of discrete and numeric class machine learning, in ICML '00: Proceedings of the 17th International Conference on Machine Learning, pp 359–366. https://doi.org/10.5555/645529.657793

Hamrin, H, Hustrulid, W and Bullock, R, 2001. Underground mining methods and applications, *Underground mining methods: Engineering fundamentals and international case studies,* pp 3–14.

Hebblewhite, B and Holt, G, 2001. Regional horizontal movements associated with longwall mining, Coal Mine Subsidence-Current Practice and Issues, Coal Mine Subsidence–Current Practice and Issues, presented at Coal Mine Subsidence-Current Practice and Issues, Maitland, pp 113–122.

Heidarzadeh, S, Shahriyar, A S and Rouleau, A, 2020. Use of probabilistic numerical modeling to evaluate the effect of geomechanical parameter variability on the probability of open-stope failure: a case Study of the Niobec Mine, Quebec (Canada), *Rock Mechanics and Rock Engineering,* 53(3):1411–1431.

Henning, J G and Mitri, H S, 2007. Numerical modelling of ore dilution in blasthole stoping, *International Journal of Rock Mechanics and Mining Sciences,* 44(5):692–703.

Henning, J G and Mitri, H S, 2008. Assessment and control of ore dilution in long hole mining: case studies, *Geotechnical and Geological Engineering,* 26:349–366.

Himanshu, V K, Bhagat, N K, Vishwakarma, A K and Mishra, A K, 2024. *Principles and Practices of Rock Blasting* (CRC Press).

Himanshu, V K, Mishra, A K, Roy, M P and Singh, P K, 2023. *Blasting Technology for Underground Hard Rock Mining* (Springer).

Hmoud, S and Kumral, M, 2023. Spatial Entropy for Quantifying Ore Loss and Dilution in Open pit Mines, *Mining, Metallurgy and Exploration,* 40(6):2227–2242.

Isaac, A and Follington, I, 1988. Geotechnical influences upon longwall mining, *Geological Society, London, Engineering Geology Special Publications,* 5(1):233–242.

Jang, H D, 2014. Unplanned dilution and ore-loss optimisation in underground mines via cooperative neuro-fuzzy network, Curtin University.

Jang, H, Topal, E and Kawamura, Y, 2015. Unplanned dilution and ore loss prediction in longhole stoping mines via multiple regression and artificial neural network analyses, *Journal of the Southern African Institute of Mining and Metallurgy,* 115(5):449–456.

Jorquera, M, Korzeniowski, W and Skrzypkowski, K, 2023. Prediction of dilution in sublevel stoping through machine learning algorithms, 012008, IOP Conference Series: Earth and Environmental Science, IOP Publishing.

Kanchibotla, S S, 2019. Rock Blasting, *SME Mineral Processing and Extractive Metallurgy Handbook,* 347.

Kelly, M, Luo, X and Craig, S, 2002. Integrating tools for longwall geomechanics assessment, *International Journal of Rock Mechanics and Mining Sciences,* 39(5):661–676.

Khanal, M, Qu, Q, Zhu, Y, Xie, J, Zhu, W, Hou, T and Song, S, 2022. Characterization of overburden deformation and subsidence behavior in a kilometer deep longwall mine, *Minerals,* 12(5):543.

Klein, L, El-Assady, M and Jäger, P F, 2022. From Correlation to Causation: Formalizing Interpretable Machine Learning as a Statistical Process, arXiv preprint. arXiv:2207.04969

Korigov, S, Adoko, A C and Sengani, F, 2022. Unplanned dilution prediction in open stope mining: developing new design charts using Artificial Neural Network classifier, *Journal of Sustainable Mining,* 21(2).

Mawdesley, C, Trueman, R and Whiten, W J, 2001. Extending the Mathews stability graph for open–stope design, *Mining Technology,* 110(1):27–39.

Ming, J, Pan, Y, Xie, J, Li, Z and Guo, R, 2024. Study on the law of ore dilution and loss and control strategies under brow line failure in sublevel caving mining, *Scientific Reports,* 14(1):26195.

Mooiman, M B, Sole, K C and Dinham, N, 2016. The precious metals industry: Global challenges, responses and prospects, Metal sustainability: Global challenges, consequences and prospects, pp 361–396.

Pearl, J, 2011. The mathematics of causal relations, Causality and psychopathology: Finding the determinants of disorders and their cures, pp 47–65.

Petri, G, 2019. Predicting the Infinite Dilution Activity Coefficient with Machine Learning,

Putra, D, Karian, T, Sulistianto, B and Heriawan, M N, 2024. Integrating Mathews Stability Chart into the Stope Layout Determination Algorithm, *Mining, Metallurgy and Exploration,* pp 1–14.

Saeedi, G, Shahriar, K, Rezai, B and Karpuz, C, 2010. Numerical modelling of out-of-seam dilution in longwall retreat mining, *International Journal of Rock Mechanics and Mining Sciences,* 47(4):533–543.

Salgado-Medina, L, Núñez-Ramírez, D, Pehovaz-Alvarez, H, Raymundo, C and Moguerza, J M, 2020. Model for dilution control applying empirical methods in narrow vein mine deposits in Peru, Advances in Manufacturing, Production Management and Process Control: Proceedings of the AHFE 2019 International Conference on Human Aspects of Advanced Manufacturing and the AHFE International Conference on Advanced Production Management and Process Control, July 24–28, 2019, Washington DC (Springer).

Serdaliyev, Y, Iskakov, Y, Bakhramov, B and Amanzholov, D, 2022. Research into the influence of the thin ore body occurrence elements and stope parameters on loss and dilution values, *Mining of Mineral Deposits*, 16(4).

Smith, A and R, Collins 1985. The extraction of barrier pillars between adjacent longwall panels at The Durban Navigation Collieries (Pty () Ltd, *Journal of the Southern African Institute of Mining and Metallurgy*, 85(4):125–130.

Stewart, P and R, Trueman 2008. Strategies for minimising and predicting dilution in narrow-vein mines–NVD Method, Australasian Institute of Mining and Metallurgy.

Stewart, P C, 2005. Minimising dilution in narrow vein mines, PhD thesis (unpublished), University of Queensland.

Suglo, R S and S, Opoku 2012. An assessment of dilution in sublevel caving at Kazansi Mine, *International Journal of Mining and Mineral Engineering*, 4(1):1–16.

Sun, H, Jin, A, Elmo, D, Gao, Y and Wu, S, 2020. A numerical based approach to calculate ore dilution rates using rolling resistance model and upside-down drop shape theory, *Rock Mechanics and Rock Engineering*, 53:4639–4652.

Temizyürek, B, 2023. A predictive model using block data for dilution control in open pit mining, University of British Columbia.

Tommila, E, 2014. Mining method evaluation and dilution control in Kittilä mine, Master's thesis (unpublished), Aalto University, Finland.

Wang, P, et al, 2024. Investigation of ground subsidence response to an unconventional longwall panel layout, *International Journal of Coal Science and Technology*, 11(1):68.

Wang, P, Zhao, J, Feng, G and Wang, Z, 2018. Improving stress environment in development entries through an alternate longwall mining layout, *Arabian Journal of Geosciences*, 11:1–17.

Wang, P, Zhao, J, Yoginder, P C and Wang, Z, 2017. A novel longwall mining layout approach for extraction of deep coal deposits, *Minerals*, 7(4):60.

Wattimena, R, Kramadibrata, S, Sidi, I D and Azizi, M A, 2013. Developing coal pillar stability chart using logistic regression, *International Journal of Rock Mechanics and Mining Sciences*, 58:55–60.

Whillans, P T, 2018. Mining Dilution and Mineral Losses, An Underground Operator's Perspective, *Proceedings of the Mining Tech, Santiago, Chile*, 29.

Yu, K, Zheng, C and Ren, F, 2021. Numerical experimental study on ore dilution in sublevel caving mining, *Mining, Metallurgy and Exploration*, 38(1):457–469.

Yu, Z, Shi, X, Zhou, J, Gou, Y, Rao, D and Huo, X, 2021. Machine-learning-aided determination of post-blast ore boundary for controlling ore loss and dilution, *Natural Resources Research*, 30:4063–4078.

Zhao, X and Niu, J A 2020. Method of predicting ore dilution based on a neural network and its application, *Sustainability*, 12(4):1550.

Controlling ore dilution in mining using causality to establish variable relationship

G Asamoah[1], J Liu[2] and R K Asamoah[3]

1. PhD Candidate, University of South Australia, Mawson Lakes SA 5095.
   Email: gloria.asamoah@mymail.unisa.edu.au
2. Associate Professor, Industrial AI Research Centre, University of South Australia, Mawson Lakes SA 5095. Email: jixue.liu@unisa.edu.au
3. Senior Research Fellow, Future Industries Institute, University of South Australia, Mawson Lakes SA 5095. Email: richmond.asamoah@unisa.edu.au

ABSTRACT

Mining industries monitor mining dilution and operating costs as key performance indicators in comparing efficiency of mining operations. These variables are continuously measured by comparing the planned variables to the actuals that were achieved. For instance, key performance variables such as throughput, grade, recovery and cash flow are closely monitored and variance between planned and actual gives indication of process improvement. The effect of dilution manifest in several ways. Firstly, mining dilution can significantly impact metal recovery and grade. For instance, when ore is diluted, the concentration of valuable metals in the ore decreases. The financial viability of a mining operation can be affected through waste processing with further links to poor downstream process performance. Although, data-based upstream mining predictive modelling has gained recognition in recent times, understanding the cause and effect in complex mining variables in industrial data set is still green. The research question that still require further studies is what variables are causing mining dilution. The objective of this study is to establish a reliable relationship among mining variables impacting on dilution. Real industrial data set will be used together with simulations, considering tonnes, grade, equivalent linear overbreak/slough, drill spacing and drill metres.

INTRODUCTION

Dilution is the contamination that occurs when waste rock (material below the cut-off grade) mixes with valuable ore (material at or greater than the cut-off grade). In underground mining, the categorisation of dilution is planned and unplanned (Jang, 2014). Planned dilution is waste within the ore reserve and included in mineable shapes or stope designs. On the other hand, unplanned dilution occurs when uneconomical material outside the planned stope design walls mixes with the economic ore within the stope boundaries due to stope overbreak or sloughage (Henning and Mitri, 2007). Figure 1 shows the difference between unplanned dilution and planned dilution in underground mine of an ore reserve.

Unplanned ore dilution has been a well-known problem in the mining industry influencing the mine operation cost (Suglo and Opoku, 2012; Whillans, 2018; Delentas, Benardos and Nomikos, 2021). Factors contributing to unplanned ore dilution include drill and blast activities, stope design parameters, and geological and geotechnical factors (Jang, 2014).

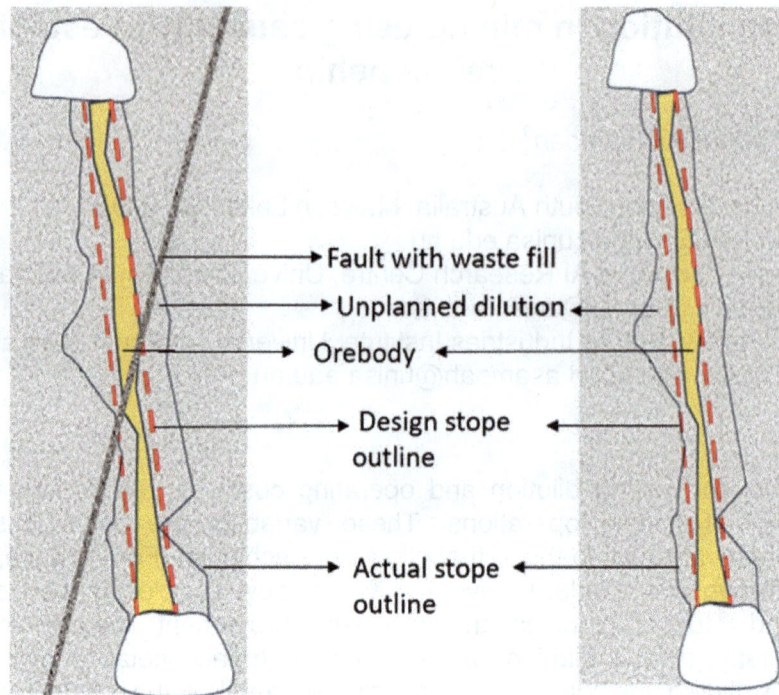

FIG 1 – Schematic diagram showing unplanned ore dilution in an underground stope.

Causal relationship is a link between variables where one variable (cause) influences or drives the occurrence of an outcome (effect). Variables are key parameters or factors that may change under different condition or study for a particular outcome. In this research, the variables are derived from mine production setting where drill and blast activities play role in ore dilution. Examples of these variables include slough, spacing, burden, stope height, stope width, dip and specific gravity. On the other hand, non-causal relationship is where the variables are correlated or associated with the outcome in one way or the other but does not influence the outcome.

Correlation-based machine learning is good at identifying patterns in data sets and able to model mechanisms that drives patterns, but the identification of cause and effect relationships is a challenge. For example, orebody geometry been correlated with dilution does not imply causation. A narrow vein irregular orebody is fixed and will not change, however the adaption of the mining method and blast practices to mine this irregular orebody is critical and causes dilution if not properly managed. Also, in correlation-based machine learning, a shift in the variable distribution can lead to significant misleading results. Given the dilution variable Y and some mining variables $X = (x_1, x_2, \ldots, x_n)$., this problem aims to identify causal relationships between X and Y.

Solving this problem is important due to the financial gain to the mining business upon implementing causal models to minimise dilution. It will also help in reducing waste in mining for cost efficient operation. Correlation-based models can result in financial loss and higher risk when predictions are based on inaccurate feature selections. However, in causal models there is high certainty of prediction, hence the mine is less liable to make financial loss.

DATA COLLECTION AND METHODOLOGY

Data collection and preprocessing

A gold mining company provided operational data for this case study. Due to confidentiality agreement, the name the company cannot be provided. Industrial data set relevant to ore dilution in the underground operation was collected for analysis. The data collected include drill and blast data, Geotech data and geology data. Table 1 shows the mining variables to be collected and used in this study.

TABLE 1

Mining variables.

Variables	Description	Unit
Equivalent linear overbreak/slough	Average overbreak of design stope outline	meters
Stope height	Height of stope in vertical direction	meters
Dip	Angle at which stope is dipping	degrees
Stope width	Distance of stope in horizontal I direction	meters
Tonnes	Actual ore tonnes mined from stope	tonnes
Au	Gold grade	g/t
Drill spacing	Distance between individual drill holes in mining	meters
Drill metres	A measurement unit used in drilling operations	meters
Drill burden	Drill holes perpendicular to the free face	meters

The data was preprocessed to remove outliers using inter-quartile range (IQR), domain knowledge (Dash *et al,* 2023) and missing records. For the multivariable mining data set, scatter plots were generated to help visualise the relationships between variables and identify points that deviate significantly from the trend. With nine initial variables shown in Table 1, the addition of variables to the model was done using stepwise selection and elimination. In stepwise selection and elimination method, variables that are relevant to the model will be maintain whiles irrelevant variable with be eliminated (Pace and Briggs, 2009). From the stepwise selection and elimination, variables such as stope height, stope width and dip were eliminated.

Methodology

Multilinear regression was used to analyse the relationships between the mining variables and dilution. Multilinear regression is an extension of simple linear regression that models the relationship between a dependent variable and two or more independent variables (Grégoire, 2014). It also has the potential to identify causal relationship between variables assuming there is no confoundedness.

Statistical software's such as python or excel was used to calculate the coefficient of regression, the significance of coefficient (p-values) and the goodness fit (R^2). A positive coefficient of the independent variables obtained in the analysis shows that as the values of the independent variable increase the dependent variable (outcome) is expected to increase. Also, a negative coefficient of the independent variable indicates that as the independent variable increase the expected to outcome (dependent variable) will decrease. Further, p-value less than 0.05 shows that the relationship between the independent variable and dependent variable (outcome) is statistically significant, hence the independent variable has an effect on the outcome (Nayebi and Nayebi, 2020).

For multicollinearity, a situation in multilinear regression where two or more independent variables are highly correlated with each other, the associated variables will be eliminated. The high correlation can cause several issues in the regression model, such as difficulties in estimating the relationship between each independent variable and the dependent variable, reduced reliability of the coefficient estimates, and problems with model interpretation, hence requires elimination (Kumari, 2008; Singh, Singh and Paprzycki, 2023).

RESULTS AND DISCUSSION

Multilinear regression with weighting was used to assess the relationship between dilution and mining variables. However, the data set was categorised into different groups before employing the multi linear regression to avoid spurious correlations. These set of groups are geomechanical and design factors and drill and blast factors. Confounding variables such as rock strength was eliminated from the data set before running the regression because rock strength influences both dependent variable and independent variable. Table 2 shows the data categorisation of the data collected.

TABLE 2

Mining variables categorisation.

Dependent variable (effect)	Categorisation		Independent variable (cause)
Dilution	1	Geomechanical condition and design factors	Orebody dip
			Stope height
			Stope width
	2	Drill and blast factors	Spacing
			Burden
			Drill metres
	3	Geotechnical factors	Equivalent linear overbreak/slough
	4	Production	Tonnes
			Grade

Based on dilution analysis collected from 342 rings that were fired from a mining company data collected. The data was categorised into various groups and multilinear regression was run based on the various categorisation and the following p-values in Table 3 was observed to understand is there was relationship between the independent variables and dependent variables.

TABLE 3

Probability values derived from analysis.

Independent variables	P-values
Drill metres	0.8033
Equivalent linear overbreak/slough	0.0005
Tonnes	0.5369
Grade	0.0425
Drill spacing	0.5696

However, after categorising the data, the p-value was observed in the drill parameters was also low showing a relationship between dilution and burden in blast design. The graph in Figure 2 shows the nonlinear relationship between dilution and equivalent linear overbreak.

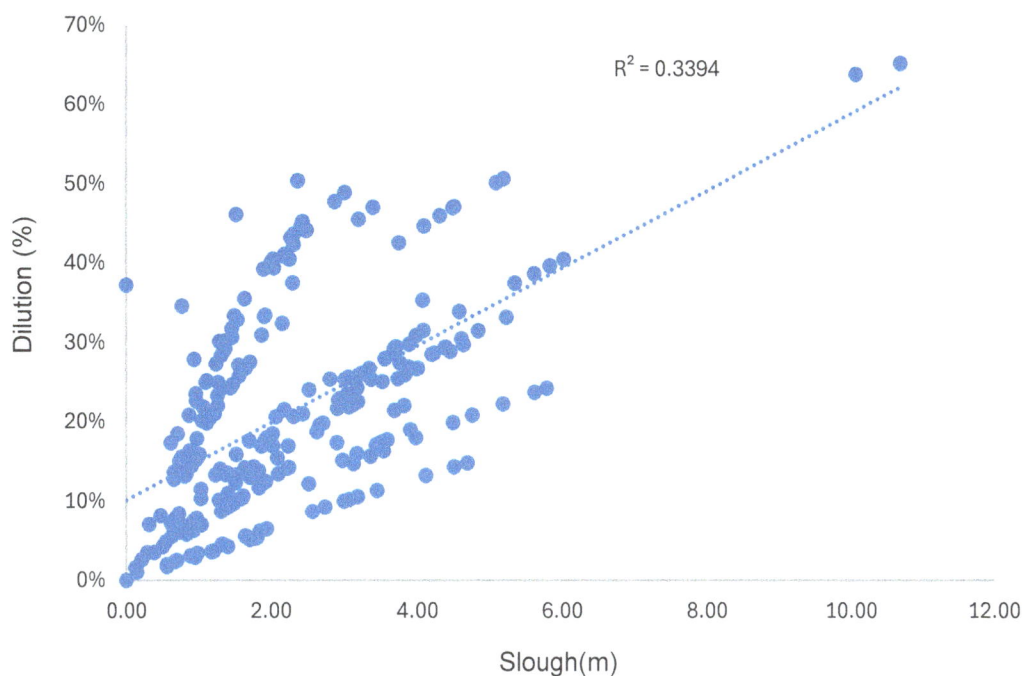

FIG 2 – Non-linear relationship between dilution and equivalent linear overbreak/slough.

CONCLUSIONS

From the multilinear regression analysis, it was observed that the probability value (p-value) of equivalent linear overbreak was the smallest. This indicates that equivalent linear overbreak/slough shows a strong relationship with dilution having p-value less than 0.05, therefore equivalent linear overbreak significantly contributes to dilution. However, drill metres and drill spacing p-values are greater than 0.05 and might not be useful predictor but require the collection of more data to justify.

Future work will involve further investigation of the relationship between dilution and equivalent linear overbreak/slough whether it is direct or indirect. Additional data will be obtained for this work.

ACKNOWLEDGEMENTS

The authors acknowledge University of South Australia and the Australian government for providing funding for this work. The authors also thank the facilities and technical assistance of staff at the Future Industries Institute, University of South Australia.

REFERENCES

Dash, C S K, Behera, A K, Dehuri, S and Ghosh, A, 2023. An outliers detection and elimination framework in classification task of data mining, *Decision Analytics Journal,* 6:100164.

Delentas, A, Benardos, A and Nomikos, P, 2021. Analyzing stability conditions and ore dilution in open stope mining, *Minerals,* 11(12):1404.

Grégoire, G, 2014. Multiple linear regression, *European Astronomical Society Publications Series,* 66:45–72.

Henning, J G and Mitri, H S, 2007. Numerical modelling of ore dilution in blasthole stoping, *International Journal of Rock Mechanics and Mining Sciences,* 44(5):692–703.

Jang, H D, 2014. Unplanned dilution and ore-loss optimisation in underground mines via cooperative neuro-fuzzy network, Curtin University.

Kumari, S S, 2008. Multicollinearity: Estimation and elimination, *Journal of Contemporary Research in Management (JCRM),* 3(1).

Nayebi, H, 2020. Multiple Regression Analysis, Advanced Statistics for Testing Assumed Causal Relationships: Multiple Regression Analysis Path Analysis Logistic Regression Analysis, pp 1–46.

Pace, N L and Briggs, W M, 2009. Stepwise logistic regression, *Anesthesia and Analgesia,* 109(1):285–286.

Singh, P, Singh, S and Paprzycki, M, 2023. Detection and elimination of multicollinearity in regression analysis, *International Journal of Knowledge-based and Intelligent Engineering Systems* (Preprint), pp 1–7.

Suglo, R S and Opoku, S, 2012. An assessment of dilution in sublevel caving at Kazansi Mine, *International Journal of Mining and Mineral Engineering,* 4(1):1–16.

Whillans, P T, 2018. Mining Dilution and Mineral Losses, An Underground Operator's Perspective, Proceedings of the Mining Tech, Santiago, Chile, 29.

Analysis of alternatives to the use of diesel fuel in mining trucks

D Canullan[1], A Anani[2] and S O Adewuyi[3]

1. Masters Student, Department of Mining and Geological Engineering, University of Arizona, Tucson AZ 85719, USA. Email: diegoc@arizona.edu
2. Associate Professor, Department of Mining and Geological Engineering, University of Arizona, Tucson AZ 85721, USA. Email: angelinaanani@arizona.edu
3. Postdoc, Department of Mining and Geological Engineering, University of Arizona, Tucson AZ 85721, USA. Email: sadewuyi@arizona.edu

ABSTRACT

The increasing effects of climate change have resulted in a global agreement on the urgent need for decarbonisation, with governments establishing ambitious emissions reduction targets, such as those set in the 2016 Paris Agreement. The main activities in the mining sector that contribute to greenhouse gas (GHG) emissions are loading and transportation. Minerals are typically transported on large trucks with diesel engines. In order to meet the urgent need to reduce CO_2 emissions in the mining industry, this study examines diesel fuel substitutes for use in mining vehicles. The research analysed the feasibility of using electricity, renewable diesel, and liquefied natural gas (LNG) as alternatives to diesel fuel. We used discrete event simulation (DES) to compare these energy alternatives across various parameters and scenarios. The analysis was performed on data from an open pit mining operation over a 24 hr period. Our results showed that short-term production levels remained largely unaffected by the type of energy used. However, the electric truck scenario required 13 additional trucks to match the production levels of diesel fuels due to the smaller size of electric trucks. LNG trucks demonstrated a significant increase in fuel consumption (+50 per cent) compared to diesel trucks, leading to higher operational costs (+46 per cent). In terms of emissions, electric trucks achieved a 95 per cent reduction in CO_2 emissions, while renewable diesel trucks reduced emissions by 90 per cent. On the other hand, an increase in fuel consumption by LNG trucks compared to diesel trucks resulted in a 30 per cent increase in CO_2 emissions. In addition, simulations with LNG and electric trucks resulted in longer cycle times and queue lengths. The results showed that the assumptions used in this analysis have a major impact on the outcomes, indicating the need for further research as the industry continues to investigate and implement these technologies.

INTRODUCTION

The mining industry is highly dependent on diesel fuel, particularly in open pit operations, where heavy machinery relies on diesel for extraction, transportation, and processing (Cawley *et al*, 2011). This contributes to 40–50 per cent of total emissions, negatively affecting air quality and exacerbating climate change (Bellois, 2022). As global efforts to reduce emissions intensify, the mining sector must transition away from diesel to meet targets like those in the 2016 Paris Agreement, aiming for a 45 per cent reduction in emissions by 2030 and net-zero by 2050 (United Nations, 2015).

Mining and metallurgy contribute about 8 per cent of global carbon emissions, with diesel-powered vehicles, especially those used for loading and hauling, being major contributors (Cox *et al*, 2022). The industry also faces challenges like increasing demand, environmental sustainability concerns, rising operational costs, and technological innovation. To address these, the mining sector must adopt alternative energy sources, digitalisation, and automation (Onifade *et al*, 2023).

Research on alternative energy sources for mining trucks, such as renewable diesel, LNG, and batteries, remains limited. This study aims to analyse the environmental, economic, and operational impacts of these alternatives and develop a simulation model to assess their effectiveness in open pit mining operations.

FUEL TYPES FOR MINE HAULAGE SYSTEMS

Diesel fuel

Diesel is derived from biomass and crude oil, consisting mainly of hydrocarbons (Betiha *et al*, 2018). It is used in high-speed engines due to its low cost, though maintenance can be expensive (Sarkar, 2015). Diesel combustion emits pollutants like carbon dioxide (CO_2) and nitrous oxide (N_2O), which contribute to air pollution and public health issues (Li *et al*, 2024).

Liquefied natural gas (LNG)

LNG is a colourless, non-toxic liquid composed mainly of methane. It has a lower environmental impact compared to diesel, reducing CO_2 emissions by 20–30 per cent and nitrogen oxide emissions by 60–80 per cent. LNG can be used in mining trucks with dual-fuel systems, providing operational flexibility (Bohorquez, 2021).

Renewable diesel

Made from fats and oils, renewable diesel has the same chemical makeup as petroleum diesel. It reduces carbon and nitrogen oxide emissions compared to petroleum diesel, with a 65–90 per cent reduction in carbon intensity. Renewable diesel is compatible with existing mining equipment and infrastructure (Energy Efficiency and Renewable Energy, 2024).

Batteries

Electric fleets powered by batteries offer environmental and health benefits by reducing maintenance, ventilation, and cooling costs. Batteries used in mining equipment are designed to last longer, with safety features that prevent explosions and improve reliability. They can be recharged or swapped out for continuous operation (Epiroc, 2018).

RELEVANT WORK

Although extensive studies have been performed to analyse alternatives fuels for mine haulage systems in the context of fuel consumption, cost, greenhouse gas emissions, and productivity, several limitations and opportunities exist in this domain. The main gap found is that these studies compared alternative energy sources to diesel in an isolated manner limiting the breadth of their analysis. This study addresses this limitation by analysing electricity, renewable diesel, and liquefied natural gas (LNG) as practical alternatives to diesel. Second, several studies do not perform their analysis with data from operating mines. We model the truck-shovel system of an operating mine as discrete event simulation and perform computational experiments to compare different energy sources for haul trucks based on CO_2 emission, energy cost, fuel consumption, and productivity. The succeeding sections detail the methodology used and the results of our analysis.

METHODOLOGY

Arena software is used to develop a model that recreates mine truck routes and their various destinations during operations. The model allows for an in-depth analysis of the mine's transportation and logistics, accounting for key operational parameters like material loading, unloading, fuel loading, shift changes, and waiting times. It supports analysis across different fuel types to evaluate scenarios based on CO_2 emissions, fuel consumption, cycle times, and total fuel costs. The model compares the performance of diesel, LNG, renewable diesel, and battery-powered trucks, providing a comprehensive analysis of each fuel alternative's efficiency and impact.

Data collection

The case study is based on data from an open pit mine in the USA, which operates with five working benches where trucks are loaded with different materials (fines, ore, and waste). The data, collected over two 12 hr shifts, comes from a fleet of 24 trucks with a 265 t capacity, along with two shovel models and one front loader. The trucks follow specific routes starting from the parking area, where they are assigned to one of the five benches for loading. Depending on the material type, trucks are routed to different dumping destinations: ore is taken to DS_ORE1, DS_ORE2, or DS_ORE3, fines to DS_FINES1 to DS_FINES4, and waste to DS_WASTE. Trucks refuel at the parking lot after dumping when their fuel reaches 25 per cent. The operations are planned around two 12 hr shifts.

Data analysis

The study's raw data includes various parameters such as material destinations, haulage distances, travel times, payload data, truck numbers, and model types. The first step in data analysis involved using summary statistics and graphical methods, like scatter plots, to detect outliers, which were removed using domain knowledge and statistical tests. The data was modelled using theoretical distributions, with the Arena Rockwell simulation tool (input analyser) to determine probability distributions. A chi-squared goodness of fit test was performed to validate the selected distribution against the data.

Model logic and construction

The model in Arena software simulates the load-haul-dump operations in the mine. Trucks are assigned to working benches, where they are loaded with material and then travel to designated destinations. The trucks are modelled as entities with attributes such as tonnage, fuel consumption, and travel times. The simulation includes activities like loading, hauling, and dumping, with process modules to model these activities. Resources such as loading equipment and refuelling stations are defined within the simulation, and routes between destinations like the crusher and parking areas are modelled using station modules. Figure 1 shows the logic used for creating the model in the Arena simulation software.

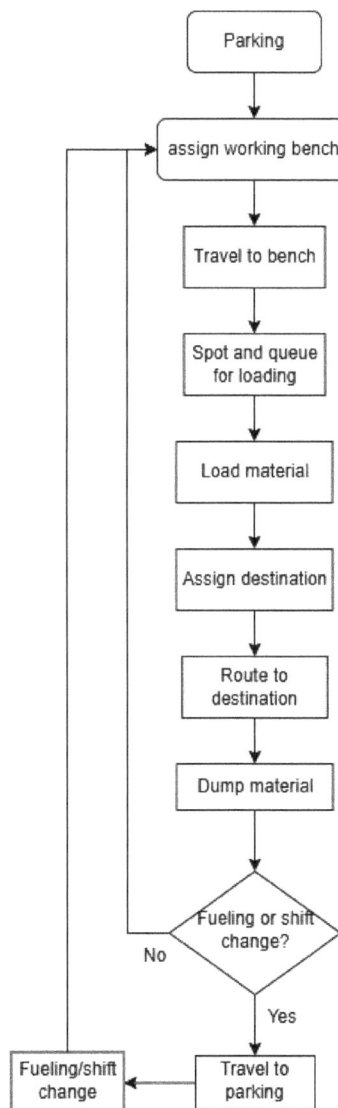

FIG 1 – Flow chart model logic.

Model verification and validation

Model verification involved running the simulation and comparing the preliminary results with actual data from the case study to ensure the model functions correctly. Truck movement animation was also used to confirm the model's behaviour.

For model validation, two key parameters were assessed: daily production and truckloads. The simulation produced 250 899.5 t of production and 954 truckloads, with a 4 per cent difference in production and a 5 per cent difference in truckloads compared to the mine's data. These discrepancies fell within acceptable error margins, confirming that the simulation model is valid and accurately represents the mining system, making it suitable for further use.

Experimental design

Each fuel scenario is simulated independently to analyse its impact, with a base case model developed and validated for further testing (Table 1). The base model is modified to incorporate different experimental scenarios: diesel, liquefied natural gas (LNG), renewable diesel, and battery-powered trucks. For renewable diesel, three CO_2 emission reduction levels are analysed based on literature, with reductions of 75 per cent, 90 per cent, and an average of 82.5 per cent. The input parameters for each scenario include truck capacity, number of trucks, shifts, fuel tank capacity, refuelling time, fuel consumption, CO_2 emissions, and fuel price, with simulations conducted over 24 hrs.

TABLE 1

Scenarios simulation.

Experiment ID	Truck fuel type	Fuel consumption rate	CO₂ emissions
1	Diesel	Base	100%
2	LNG	1.67 × base	Depending on fuel consumption
3	Renewable diesel	1.02 × base	Reduced by 75%
4	Renewable diesel	1.02 × base	Reduced by 82.5%
5	Renewable diesel	1.02 × base	Reduced by 90%
6	Electric	High	Depending on electric consumption
7	Electric	Low	Depending on electric consumption

The CO_2 emissions calculation is based on fuel consumption, assuming that the carbon content fully oxidised into carbon dioxide. The following equation is used to calculate CO_2 emissions resulting from the combustion of fuels (European Environment Agency and European Monitoring and Evaluation Programme (EMEP), 2014):

$$E_{CO2} = 44.011 \cdot \frac{FC}{12.011 + 1.008 r_{HC} + 16.000 r_{OC}} \tag{1}$$

Where:

FC is the fuel consumption in the period considered

r_{HC} is the ratio between hydrogen and carbon atoms

r_{OC} is the ratio between oxygen and carbon atoms (EMEP, 2014)

The diesel scenario serves as the base case, using data from the mine for truck type, number of trucks, fuel tank capacity, refuelling time, and fuel consumption. Table 2 shows the parameters used in this scenario.

TABLE 2

Diesel parameters.

Parameters	Diesel
Truck capacity	265 tons
Number of trucks	24
Shift	2 × 12 hrs
Fuel capacity	1150 gal
Fuelling time	17.5 min average
Fuel consumption	52 gal/hr average
CO_2 emissions	Equation 1, $r_{HC} = 2$, $r_{OC} = 0$
Fuel price	$4.34 USD/gal

In the Liquefied Natural Gas (LNG) scenario, trucks are converted to run exclusively on LNG using a conversion kit (Caterpillar, 2019). This technology allows the simultaneous use of LNG and diesel, but for this study, only LNG is used. Table 3 shows the parameters used in this scenario.

TABLE 3

LNG parameters.

Parameters	LNG
Truck capacity	265 tons
Number of trucks	24
Shift	2 × 12 hrs
Fuel capacity	1150 gal
Fuelling time	17.5 min average
Fuel consumption	86.8 gal/hr average
CO_2 emissions	Equation 1, $r_{HC} = 3.9$, $r_{OC} = 0$
Fuel price	$3.05 USD/gal

In the renewable diesel scenario, trucks require certain modifications to components for proper functionality, but these adjustments do not affect parameters such as payload capacity, number of trucks, fuel tank capacity, or refuelling time. Table 4 presents the parameters for the renewable diesel scenario.

TABLE 4

Renewable diesel parameters.

Parameters	Renewable diesel
Truck capacity	265 tons
Number of trucks	24
Shift	2 × 12 hrs
Fuel capacity	1150 gal
Fuelling time	17.5 min average
Fuel consumption	53.1 gal/hr average
CO_2 emissions	reduced by 75%, 82.5% and 90%
Fuel price	$4.20 USD/gal

For the battery-powered truck scenario, a 265 t capacity truck with a 1.9 MW battery was considered. The fastest charging time of 30 mins was used in the simulation, with an additional 15 min charging time considered to analyse its impact on operations (Parkinson, 2024). Parameters for the battery scenario are summarised in Table 5.

TABLE 5

Battery parameters.

Parameters	Battery
Truck capacity	265 tons
Number of trucks	24
Shift	2 × 12 hrs
Battery capacity	1.9 MW
Charging time	30 min / 1 hr / 2 hr
Electric consumption	3.9 kWh/km / 7.8 kWh/km
CO_2 emissions	3.94×10^{-4} CO_2/kWh
Electricity price	$0.10 USD/kWh

RESULTS AND DISCUSSIONS

The results obtained through discrete event simulation reveal that the choice of energy source for mining haul trucks significantly influences CO_2 emissions, operating costs, and productivity, both in the short-term (24 hrs) and long-term (1 year).

In the short-term analysis, low consumption electric trucks achieved a 95.2 per cent reduction in CO_2 emissions and a 96.7 per cent reduction in energy costs compared to diesel trucks, with less than 1 per cent loss in productivity. These outcomes were consistent at the annual scale, reinforcing the potential of electric haulage as a sustainable long-term strategy. High consumption electric configurations also delivered significant environmental and economic benefits, though with a 3.4 per cent drop in productivity due to higher energy usage and longer charging cycles.

Renewable diesel, specifically in its 90 per cent carbon-reduction form, maintained productivity levels identical to conventional diesel while reducing CO_2 emissions by more than 80 per cent, with no increase in fuel cost. This positions it as a highly effective transitional technology, particularly for operations not yet ready to implement full electrification.

In contrast, LNG consistently underperformed. It led to 69 per cent higher fuel consumption, 47.6 per cent greater CO_2 emissions, and an 18.7 per cent increase in energy costs, while also reducing productivity by 1.3 per cent. These results challenge common assumptions about LNG as a clean alternative and suggest its limited value in decarbonisation strategies for mining.

The simulation also showed that battery charging times above 60 mins significantly reduce operational efficiency, especially in high consumption electric scenarios. This finding underscores the importance of aligning electrification with charging infrastructure planning, operational schedules, and fleet configuration to avoid bottlenecks and productivity loss.

CONCLUSION

Transitioning from diesel to alternative fuels in mining haulage operations is critical for reducing GHG emissions and aligning with global decarbonisation targets. The study highlights that electrification with low consumption systems and fast charging offers the best balance between sustainability and operational performance. Meanwhile, renewable diesel emerges as a practical near-term option, while LNG may not deliver the environmental or economic advantages often attributed to it without broader systemic changes.

Future research should explore hybrid approaches, such as hydrogen fuel cells and trolley-assist systems, to optimise energy efficiency in mining operations. The integration of automation and AI could further enhance sustainability and productivity in the sector.

REFERENCES

Bellois, G, 2022. The impact of climate change on the mining sector, The International Institute for Sustainable Development.

Betiha, M A, Rabie, A M, Ahmed, H S, Abdelrahman, A A and El-Shahat, M F, 2018. Oxidative desulfurization using graphene and its composites for fuel containing thiophene and its derivatives: An update review, *Egyptian Journal of Petroleum*, 27(4):715–730.

Bohorquez, A, 2021. Hauling system in surface mining, National University of Cajamarca, Faculty of Engineering.

Caterpillar, 2019. CAT Dynamic Gas Blending Kit, Caterpillar Inc.

Cawley, T, Wilkinson, S, Pero, L, Parminter, G, Browning, R and McDonald, S, 2011. Analyses of diesel use for mine haul and transport operations, Australian Government, Department of Resources, Energy and Tourism.

Cox, B, Innis, S, Kunz, N C and Steen, J, 2022. The mining industry as a net beneficiary of a global tax on carbon emissions, *Communications Earth and Environment*, 3(17). https://doi.org/10.1038/s43247-022-00346-4

Energy Efficiency and Renewable Energy, 2024. Renewable diesel [online], Alternative Fuels Data Centre, US Department of Energy. Available from: <https://afdc.energy.gov/fuels/renewable-diesel>

Epiroc, 2018. The zero-emission fleet. Available from: <https://www.epiroc.com/en-us/products/electrification-solutions/infrastructure-solutions>

European Environment Agency & European Monitoring and Evaluation Programme (EMEP), 2014. EMEP/EEA air pollutant emission inventory guidebook 2013, Technical report No 12/2013, European Environment Agency.

Li, D, Yu, F, Zhang, R, Zhu, M, Liao, S, Lu, M, Guo, J, Wu, L and Zheng, J, 2024. Real-world greenhouse gas emission characteristics from in-use light-duty diesel trucks in China, *The Science of the Total Environment*, 940:173400.

Onifade, M, Adebisi, J A, Shivute, A P and Genc, B, 2023. Challenges and applications of digital technology in the mineral industry, *Resources Policy*, 85:103978.

Parkinson, G, 2024. Fortescue's 6MW electric vehicle charger stuns the EV and mining industries, *The Driven*. Available from: <https://thedriven.io/2024/09/26/fortescues-6mw-electric-vehicle-charger-stuns-the-ev-and-mining-industries/>

Sarkar, D K, 2015. *Fuels and Combustion*, pp 91–137 (Elsevier eBooks).

United Nations, 2015. The Paris Agreement. Available from: <http://www.un.org/en/climatechange/paris-agreement>

Strategic planning and sequencing optimisation for fully mobile in-pit crushing and conveying (FMIPCC) systems – a novel approach

Y Y Cheng[1], M Nehring[2], M Forbes[3], M S Kizil[4] and P Knights[5]

1. Senior Mining Consultant, Mining Plus, Brisbane Qld 4122. Email: jack.cheng.au@hotmail.com
2. Senior Lecturer, School of Mechanical and Mining Engineering, The University of Queensland, St Lucia Qld 4072. Email: m.nehring@uq.edu.au
3. Senior Lecturer, School of Mathematics and Physics, The University of Queensland, St Lucia Qld 4072. Email: m.forbes@uq.edu.au
4. Associate Professor, School of Mechanical and Mining Engineering, The University of Queensland, St Lucia Qld 4072. Email: m.kizil@uq.edu.au
5. Professor, School of Mechanical and Mining Engineering, The University of Queensland, St Lucia Qld 4072. Email: p.knights@uq.edu.au

ABSTRACT

Truck haulage reliant operations will typically experience higher operating costs due to longer haulage distances. With growing focus on the environmental impacts of mining, particularly greenhouse gas emissions, alternatives like Fully Mobile In-Pit Crusher Conveyor (FMIPCC) are being explored. The feature of this system is that the crusher is fed directly by a digging unit at the working face. A network of conveyor belts then transports material out of pit. However, it requires a robust mine plan to execute due to a loss of flexibility in mine sequencing. At present, there are no commercially available tools to generate and assess strategic mine plans of metalliferous deposit using FMIPCC systems. This research presents one of the first attempts to develop such a tool.

This research introduces a mixed integer programming model for the FMIPCC system to open pit mining. Several key operational constraints are successfully implemented in addition to traditional open pit mining sequencing. The programming approaches have been used to ensure solution times are kept reasonable while ensuring that the unique strategic planning characteristics associated with the implementation of FMIPCC systems are addressed in a novel way. A direct block scheduling approach is used to provide the most meaningful results at a resolution reflective of long-term strategic mine planning. Ultimately the research developed a model that addresses the absence of a planning tool by incorporating essential FMIPCC system features and generating the NPV based on specific constraints.

INTRODUCTION

Conventional large scale surface hard rock mines rely on shovels and trucks for excavation and haulage. As pits become deeper and wider this results in higher operating costs due to the increased haulage distance and a corresponding increase in the consumption of fuel, tyres and labour. As a result of the increased haulage distance with pit depth, energy consumption and the carbon footprint of the operation also expands.

The introduction of conveyors into the open pit operation completely changes the mine configuration and the mine plan when compared to traditional truck and shovel mining. Designing, planning and optimising a strategic plan for a conventional open pit operation is well established. After the block model has been generated using geostatistical approaches (Tolosana-Delgado, Mueller and van den Boogaart, 2019) while also considering selectivity and the size of equipment to be deployed (block resolution), the next step defines the undiscounted ultimate pit limit using the Lerchs and Grossman (1965) or puesdoflow (Hochbaum, 2008) algorithms. Once the undiscounted ultimate pit limit is established the volume of material inside it are further broken down into pushbacks which sets a sequence whereby highest value ore at the least cost is preferred while considering operational constraints such as the equipment to be deployed (Arteaga *et al*, 2014; Arteaga, Nehring and Knights, 2017). Ore and waste materials are then scheduled across mine life to maximise Net Present Value (NPV) while restricted by mining fleet and plant capacity constraints.

While Semi Mobile In-Pit Crusher Conveyor (SMIPCC) systems may be adapted for open pits that have been designed for truck haul, in the case of Fully Mobile In-Pit Crusher Conveyor (FMIPCC) systems, a whole new pit configuration is required. In FMIPCC systems the shovel engages directly with the crusher and thus removes the need for truck haulage. A FMIPCC pit therefore needs to be designed in such a way to account for the rigid sequence required to provide continual conveyor access for mined material to safely and efficiently exit the pit and for this conveyor to be sequentially extended as the pit deepens. Figure 1 shows a comparison between a traditional truck-shovel operation from loading area to dump location (ore or waste) and a FMIPCC system operating in a typical open pit scenario.

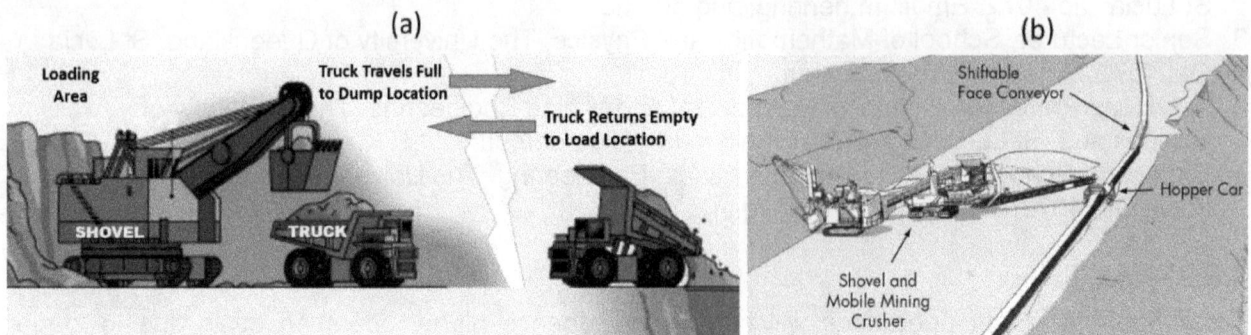

FIG 1 – Illustrations of (a) traditional shovel-truck operation; and (b) fully mobile in-pit crusher conveyor system operating in a typical open pit scenario (Dzakpata *et al*, 2016).

Very limited optimisation tools currently exist to aid the strategic mine planning process for open pit operations that seek to utilise a FMIPCC system. It is within this context that this paper introduces a new mathematical optimisation model for FMIPCC systems to aid the strategic mine planning process for utilisation in large scale open pit metalliferous operations.

THE POTENTIAL OF IN-PIT CRUSHER CONVEYOR SYSTEMS

The potential benefits of using In-pit Crusher Conveyor (IPCC) systems have been well documented. It is worthwhile addressing some of these significant environmental, social and economic benefits in greater detail. An IPCC system is typically powered by electrical energy, which is generally cheaper and cleaner than a diesel powered a truck fleet. Even if the electrical energy used by the IPCC system is generated using thermal coal, the efficiency of an IPCC system is still likely to result in a reduced energy and carbon footprint in comparison to a truck fleet. Alternatively, there are increasing opportunities for the electrical energy to be derived from renewal sources. One example of reducing reliance on diesel is that of a Brazilian iron ore mining operation which is estimated to have reduced diesel consumption of 60 million litres per annum (mL/a) with two installed FMIPCC systems capable of moving 7800 t per hr (t/h) (Raaz and Mentges, 2011). In addition, more of the consumed energy is used in the actual transportation of material with the IPCC system in comparison to truck haulage. It is estimated that conveyors use about 81 per cent of the consumed energy for the transportation of material. This compares to just 39 per cent for trucks (ThyssenKrupp, 2004).

Even though the onset of autonomous haulage is changing labour requirements in many western jurisdictions, another advantage of IPCC systems is the reduced labour requirement in comparison to truck haulage (McCarthy and Cenisio, 2013). The operation of a single truck is estimated to require about seven full-time equivalent people (including operators, maintainers and accounting for relief of personal to cover absences (Dean *et al*, 2015). In addition to the operational financial benefits of a smaller workforce, this also reduces risk from a health and safety perspective.

The operating costs (OPEX) associated with IPCC systems can be significantly lower than truck haulage through a combination of savings associated with reduced diesel consumption and reduced labour requirements (Tutton and Streck, 2009). As a result of this, some estimates suggest that OPEX of a conveyor base system can be as low as one third of a comparable truck haulage system (Dean *et al*, 2015). In addition to reducing the number of trucks, an IPCC system will also generally require less auxiliary equipment (International Mining, 2009). While the OPEX may favour IPCC systems over truck haulage, a greater initial capital expenditure (CAPEX) is generally required to purchase and commission the system. Over time however, the reduced OPEX of an IPCC system should offset the initial greater CAPEX investment. As shown by Nehring *et al* (2018), the lower

OPEX of IPCC systems may allow the viable mining and processing of material to a lower cut-off grade, which in turn results in an increase in the recovery of the resource.

A reduction in the use of haul roads when using conveyors also means that road maintenance and dust mitigation activities such as spraying roads with water are also reduced. Even though some water sprays may still be used for dust suppression in the conveyor system (particularly a transfer locations), overall water consumption is reduced. A further advantage of conveyors is the reduced noise pollution in comparison to truck haulage (Department of Resources, Energy and Tourism (DRET), 2009).

FMIPCC SYSTEM OVERVIEW

IPCC systems are generally split into three different arrangements, with each variation having its own set of advantages and disadvantages for specific operating conditions and deposit type. The three systems are a fixed system, semi mobile system and a fully mobile system. Since this study focuses on the fully mobile system, it is fitting to provide a brief overview of its components and functionality for mine planning purposes.

The FMIPCC system is characterised by several key features. A track mounted crusher/sizer system accompanies the digging unit at the working face. This crusher receives material directly from the digging unit and sizes it appropriately before being discharged onto a conveyor belt. As is the case in traditional open pit mining, the crusher/sizer selection for a FMIPCC system will typically depend on material characteristics including rock strength and fragmentation properties, with crusher throughput rates determined by these factors (Turnbull and Cooper, 2009).

This material then travels via the conveyor network to ultimately end up at the Run-of-mine (ROM) pad if it is ore, or it is discharged via spreaders to construct waste dumps if it is waste material. The FMIPCC system will generally handle both ore and waste materials with a conveyor switch point station activated (located somewhere centrally in the conveyor network) to divert material when switching between ore and waste at the working face. As such, like the traditional shovel-truck open pit mining process, FMIPCC's conveyors (at an angle of inclination of up to 18° to prevent rollback (Alspaugh, 2004)) perform all material movements. Limited truck and shovel application may be necessary for initial development of ramp access.

A FMIPCC system is typically very rigid in the sequencing of material extraction. These sequence constraints not only apply to the system as a whole but also to strip/bench operations (Atchinson and Morrison, 2011). At strip/bench level, this requires shovel movements that start at one end of the bench and continuously move toward the other before moving to the next strip/bench which will thus require a belt move and/or conveyor extension. Since conveyor extensions can take significant amounts of time (typically three weeks), this needs to be considered in equipment availability and utilisation figures used in scheduling. When scheduling, a multi-bench operation can be integrated into one level of mining blocks to feed a single bench conveyor, with the use of *'grasshopper or piggyback conveyors'* to thus simplify the scheduling process. Bench/strip preparation, including drilling and blasting, in both scenarios (single and multi-bench) are conducted sequentially ahead of the digging unit along the bench. The bench conveyors will ultimately feed a main/trunk conveyor which is located in a dedicated ramp that exits the pit.

Ramp access is vitally important for a FMIPCC system in terms of scheduling and access to each level, it also accounts for a significant portion of cost in the system. In a traditional open pit mine plan that utilises shovel-trucks, access ramp design comes into the scheduling phase at a later stage (after pit optimisation is completed) and even then, much of the detail is worked out on an operational level as the pit develops. However, for FMIPCC systems, a dedicated ramp to accommodate the system requires a particular sequence and a significant amount of material movement. This thus needs to be considered when conducting life-of-mine strategic mine planning and scheduling (Hay *et al*, 2019). Since the access ramp is such a key feature in the overall FMIPCC system, this project seeks the development of new mathematical modelling tools that incorporate ramp access requirements into the strategic life-of-mine plan. This is one of the main aspects of FMIPCC planning that distinguishes it from conventional open pit mine planning.

PREVIOUS WORK

The use of conveyor belts to transport mined material instead of traditional truck haulage at large scale was first implemented in Germany during the 1950s. In this case, the motivation for the use of conveyors was to overcome wet and soft ground conditions which made truck haulage difficult (Koehler, 2003; Utley, 2011). As discussed previously, today conveyor based mining can help alleviate many of the environmental, social and economic risks faced by mining operations.

While the German operations of the 1950s could dig the material to a size that was amenable for conveyor handling, the practical application of conveyors into most of the world's hard rock metalliferous operations requires material to be appropriately sized. This necessitates the installation of an in-pit crusher/sizer before conveyor transportation takes place. The selection of an appropriate in-pit crusher/sizer is now a relatively straightforward exercise based on the required size and material characteristics. The original works of Hays (1983) and Huss, Reisler and Almond (1983) form the basis for many of the guides used in today's in-pit crusher/sizer selection practices.

Kesimal (1997), Osanloo (2012) and Paricheh and Osanloo (2019a) provide in-depth insights into the use of conveyors and their applicability to the mining context. To overcome the angle at which conveyors are able to incline before material begins to roll-back on the conveyor, mine planners have the option of designing a designated conveyor slot to take material out of the open pit. This often requires significant additional waste removal. Tutton and Streck (2009) also discussed the resultant waste removal for extracting the conveyor ramp slot but limited discussion from scheduling and optimisation perspectives. The sketch of an open pit with a slot is shown in Figure 2. This has resulted in the development of high-angle conveyors which are able to traverse the slope of a standard pit wall (dos Santos, 1984; Mitchell and Albertson, 1985; dos Santos and Stanisic, 1986; dos Santos, 2016; Liu and Pourrahimian, 2021). These conveyor systems however are yet to achieve the capacities required for large scale bulk material movement.

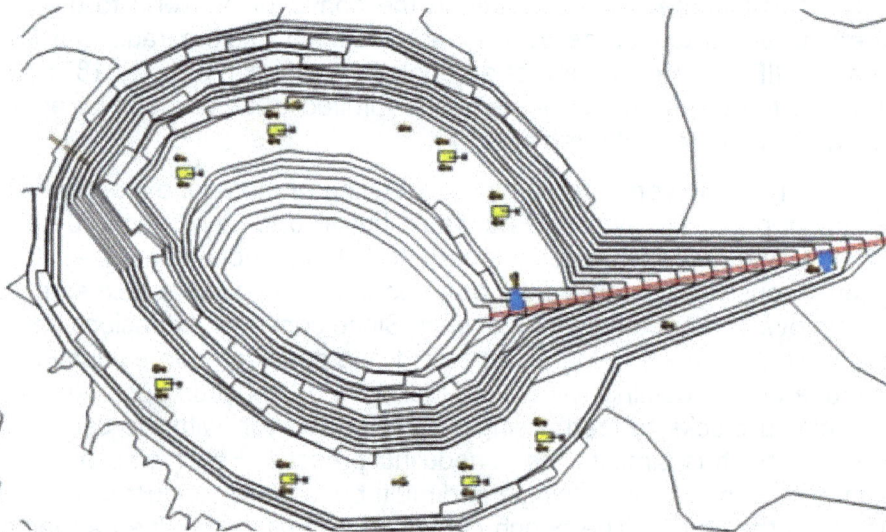

FIG 2 – Conveyor ramp slot location (Tutton and Streck, 2009).

Given the operational capabilities of all the equipment that comprises a FMIPCC system in conjunction with the characteristics of an orebody to be exploited, some researchers have focused on optimising specific aspects of the overall system. Optimal placement of the in-pit crusher within a semi mobile context has been investigated by Roumpos et al (2014), Paricheh and Osanloo (2019b, 2019c) and Abbaspour et al (2019). More recently, studies have focused on the optimal transition point (time and depth) between a truck haulage system and a SMIPCC System (Shamsi and Nehring, 2021).

While some of the traditional open pit planning and optimisation processes to generate the Ultimate Pit Limit (UPL) and production schedule (Nehring et al, 2018) can largely be adapted for SMIPCC systems, it is apparent that there is a significant lack of research and knowledge around mathematical optimisation processes for IPCC systems in general. This lack of research is greatest for FMIPCC systems.

PRODUCTION SCHEDULING OPTIMISATION

Mixed Integer Programming (MIP) is used to provide a mathematical representation of the production scheduling problem for the FMIPCC system. This may then be solved using a specialist solver (such as Gurobi). As such, all subscript notation, sets, decision variables, parameters, objective function and constraints are presented here. Strips to be mined are indexed in the x or y (and z) coordinates depending on the mining direction it follows and are mined in the direction of increasing x coordinates. Other scenarios are run using a similar coordinate set-up, but increasing or decreasing x or y coordinates for different directions. Each strip is mined from the conveyor ramp out in both directions. This is governed by the block precedence constraints.

Sets

B:	set of mine blocks
$T = \{0,1, ..., \|T\| - 1\}$:	set of time periods
S:	set of strips
L:	set of mining levels
$P_b \subset B$:	set of predecessor blocks for $b \in B$. As well as the traditional block access constraints, this encodes the slope of the ramp and the need to mine a strip outward from the ramp.
$B_s \subset B$:	set of blocks in strip $s \in S$
$S_l \subset S$:	set of strips in mining level $l \in L$

Data

c_b:	the value of mining block $b \in B$
ore_b:	the tonnage of ore to be processed in block $b \in B$
$r_s \in B_s$:	the 'ramp' block for strip $s \in S$
$p_s \in S$:	the previous strip to strip s, given the ramp direction
$n_s \in S$:	the next strip to strip s, given the ramp direction
$MineLim$:	the maximum number of blocks that can be mined in a time period
$ProcLim$:	the maximum tonnage of ore that can be processed in a time period
$Discount$:	the NPV discount rate

Variables

$y_{bt} \in \{0,1\}$:	set to 1 if block b has been mined *by the end of* period t. $y_{bt} - y_{b(t-1)}$ is 1 if block b is mined *during* time period t. By definition, $y_{b(-1)} = 0$
$w_{st} \in \{0,1\}$:	set to 1 if strip s is mined in period t. A strip may be mined in more than one time period.
$v_{lt} \in \{0,1\}$:	set to 1 if level l is mined in period t.
$z_s \in \{0,1\}$:	set to 1 if strip s is ever mined.

After defining the required sets, decision variables and parameters, the objective function of the proposed mathematical model, which seeks to maximise NPV over the life of the mine, is defined in Equation 1.

$$Maximise:\ \max \sum_{\substack{b \in B \\ t \in T}} c_b(1 + Discount)^{-t} \tag{1}$$

The related constraints of the proposed model are discussed through Equations 2 to 17, as follows.

$$y_{bt} \geq y_{b(t-1)} \quad \forall\, b \in B, t \in T, t > 0 \tag{2}$$

$$\sum_{b \in B} y_{bt} - y_{b(t-1)} \leq MaxMine \quad \forall\, t \in T \tag{3}$$

$$\sum_{b \in B} ore_b(y_{bt} - y_{b(t-1)}) \leq MaxProc \quad \forall\, t \in T \tag{4}$$

$$y_{bt} \leq y_{b't} \quad \forall\, b \in B, t \in T, b' \in P_b \tag{5}$$

Constraints (Equations 2–5) are standard open pit mine planning constraints. $y_{bt} - y_{b(t-1)}$ is 1 if block b is mined *during* time period t. Equation 3 restricts the mining rate to a predefined tonnage (equivalent blocks) per time period based on the mining equipment capacity. The maximum processing capacity is adhered to using Equation 4. The significant difference for FMIPCC scheduling is the encoded precedent data set that defines the ramp slope and mining strips outward from the ramp into the precedence constraints. Mining capacity and processing capacity are set here to define the maximum capacities of mining and processing.

The section views in Figure 3 demonstrates the wall angle precedent rules for conventional open pit and the dedicated conveyor ramp section in the FMIPCC system. The illustrations compare typical sections of standard precedent for open pit planning and FMIPCC system that required to incorporate the trunk conveyor. Section view A shows that in order to remove the block in the lower level, the corresponding three overlaying blocks (the block directly above, one on the left and one on the right of the targeting block) need to be removed (2D). Section view B shows that two extra corresponding overlaying blocks (towards the side where the out of pit conveyor belt will be placed) need to be removed (2D) for the FMIPCC system to maintain the slope gradient for the conveyor. Thus the FMIPCC conveyor ramp section requires significant amount of material to be removed comparing with the truck and shovel system.

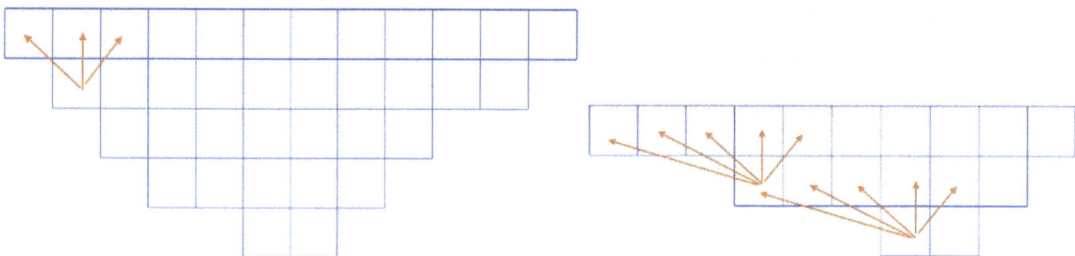

FIG 3 – Section view illustrations for open pit planning and FMIPCC planning. (a) Normal open pit precedent block extraction; (b) FMIPCC block precedent for trunk conveyor.

The diagrams in Figure 3 compare typical cross-sections of standard mining block precedent used in open pit sequence and FMIPCC system's precedence to incorporate the trunk conveyor. The first graph shows that in order to remove the block in the lower level, the corresponding three overlaying blocks need to be removed (2D), while in three dimensions, the overlying five blocks need to be removed. The second graph shows that two extra corresponding overlaying blocks need to be removed (2D) and seven blocks to be removed in three dimensions for the FMIPCC system to maintain an appropriate slope gradient (1 in 4 slope or 14° in the case study) to ensure the conveyor to transporting material at reasonable rate.

$$y_{bt} - y_{b(t-1)} \leq w_{st} \quad \forall\, s \in S, b \in B_s, t \in T \tag{6}$$

$$w_{st} \leq 1 - \sum_{t' \in T, t' > t} w_{p_s t'} \quad \forall\, s \in S, t \in T \tag{7}$$

$$2w_{st} + \sum_{t' \in T, |t-t'| \geq 2} w_{st'} \leq 2 \quad \forall\, s \in S, t \in T \tag{8}$$

$$w_{st} \leq z_s \quad \forall\, s \in S, t \in T \tag{9}$$

$$\sum_{b \in B_s \cup B_{n_s}} y_{b,(|T|-1)} \geq 2z_s \quad \forall\, s \in S \tag{10}$$

$$w_{st} \leq v_{lt} \quad \forall\, l \in L, s \in S_l, t \in T \tag{11}$$

$$v_{lt} + v_{l't} \leq 1 \quad \forall\, l \in L, l' \in L, l' > l + 1, t \in T \tag{12}$$

$$\sum_{l \in L} v_{lt} = 1 \quad \forall\, t \in T \tag{13}$$

$$y_{r_s t} - y_{r_s(t-1)} \leq v_{lt} + v_{(l+1)t} \quad t \in T, l \in L, s \in S_l \tag{14}$$

$$y_{bt} - y_{b(t-1)} \leq y_{r_s t} - y_{r_s(t-2)} \quad t \in T, s \in S, b \in B_s \tag{15}$$

$$y_{r_s t} \leq y_{bt} + 1 - y_{b(|T|-1)} \quad t \in T, s \in S, b \in B_{p_s} \tag{16}$$

$$y_{r_{p_s}(|T|-1)} \leq y_{r_s(|T|-1)} + \sum_{b \in B_{p_s} | r_{p_s} \in P_b} y_{b(|T|-1)} \quad s \in S \tag{17}$$

Constraint (6) activates the w_{st} variables. Constraint (7) ensures that if a strip is mined in a time period, then the previous strip is not mined in any later time period. This forces the strips to be mined in order. Constraint (8) ensures that a strip is mined in at most two time periods, and only in consecutive time periods.

Constraint (9) activates the z_s variables. Constraint (10) ensures at least two blocks are mined in any strip that is mined and its next consecutive strip. This ensures that at least two blocks are mined in the last strip mined in each level. When mining the ramp, it is possible to just one mine one block in a strip.

Constraint (11) activates the v_{lt} variables. Constraint (12) ensures that at most two consecutive levels are mined in each time period.

Constraint (13) forces one reference level for each time period. Constraint (14) ensures the start strips are only on the reference level, or the previous level, in each time period.

Constraint (15) enforces that all blocks in a strip must be mined in the time period the ramp block is mined, or the next time period.

Constraint (16) prevents mining a strip until mining in the previous strip has finished.

Constraint (17) ensures that a minimum of two blocks must be mined in a strip off the ramp.

CASE STUDY

A case study of a copper/gold bearing orebody amenable to open pit mining is used for the purpose of validating and testing the newly developed mathematical model presented earlier. The case study block model was originally derived from the openly available Marvin copper/gold block model (MEA Mine Planning course material). Some block aggregation, re-grading and other alterations have been undertaken for the purpose of simplifying it. While being conceptual in nature, the block model remains appropriate for the purposes of this study. The setting of this mine is a typical remote mining region within Australia. As such, all figures are quoted in Australian Dollars (AUD).

The block model in this context is characterised by nine levels in the vertical ('z') direction, each spanning 60 m, resulting in an overall vertical extent of 540 m. The model extends across 31 blocks in the 'x' direction (each block being 60 m wide) and 30 blocks in the 'y' direction (each block measuring 60 m in length), yielding a total of 8370 blocks. Based on a density of 2.267 t/m³, each 216 000 cubic metre block contains 489 600 t of material. Mineralisation commences at a depth of 60 m below the surface and extends to a depth of 540 m. The highest-grade zone of mineralisation is concentrated between 120 m and 240 m below the surface. Grades exhibit a gradual reduction when moving away from the central core of mineralisation, which is surrounded by a halo of lower-grade material.

Each block in the model is pre-calculated to include information such as three-dimensional coordinates, tonnages, gold and copper grades, and a block value based on equivalent copper grade. This comprehensive data set serves as crucial input for the mathematical optimisation model, providing the necessary parameters to optimise the extraction strategy and maximise the overall economic value of the mining operation.

Planning will take place assuming open pit mining at a pit wall slope angle of 45° for the given block dimension. Typical open pit mining constraints are therefore applicable with the main constraint being that each block can be mined only once the five blocks above have been removed. No other geotechnical issues are assumed to be present to thus allow a fully open sequencing scheme to be used other than those limitations imposed by the IPCC system itself. This includes the development of a ramp/slot to allow an exist path for mined materials. As discussed in the FMIPCC system overview section, continuous bench sequencing is also required which must be reflected in the production schedule. Due to the formation characteristics and dipping of massive hard rock deposit, chosen mining direction and pit progression could significantly influence the values extracted. Mining from various directions from different strips will help to assess the early ore exposure that generating quicker returns.

The mining capacity is set to 13.2 Mt/a (27 blocks). Once ore processing has commenced the processing rate will reach a maximum of 7.8 Mt/a (16 blocks), with the exception of final year. Availability and utilisation of equipment have been considered in the capacity figures provided so to have operational aspects such as the time required for conveyor belt moves and extensions.

A metallurgical recovery of 88 per cent Cu and 88 per cent Au, processing cost of AU$10/t of ore, copper price of AU$8200/t, gold price of AU$68/g, and a discount rate of 10 per cent is applied. After treating all blocks within the data set that are below cut-off grade to the process plant as waste, this reduces the resource to a total of 795 blocks containing copper equivalent mineralisation ranging in grade from 0.05 per cent Cu to 1.28 per cent Cu for a total resource of 389 Mt grading at 0.44 per cent Cu for 1.7 Mt Cu metal.

The initial CAPEX consists of constructing the processing plant and procuring mining equipment and totalling of AU$300 million, which is incurred in the year prior to the commencement of mining and processing. Ore blocks are processed in the same year that they are mined. In this case, considering the initial construction of processing plants and significant stripping to uncover the ore, the first two years' processing will be minimal. No stockpiling is to take place. Extending and key part replacement of conveyor belts systems take place in years six and 11 with the cost of AU$5 million.

RESULTS AND DISCUSSION

The newly developed mathematical model is applied to the case study. The objective is to maximise NPV within mining and mill feed capacities each year and the sequence constraints imposed by the FMIPCC system. The output of the new model is a yearly production and processing schedule. The scheduling results are illustrated in Figure 4, which shows plan views through the orebody at each depth level. The colour legend represents the year in which blocks are scheduled for mining. Since no stockpiling is available, ore blocks that are mined must be processed in the same year.

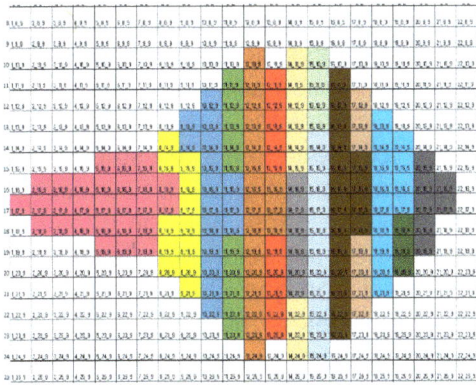

(a)

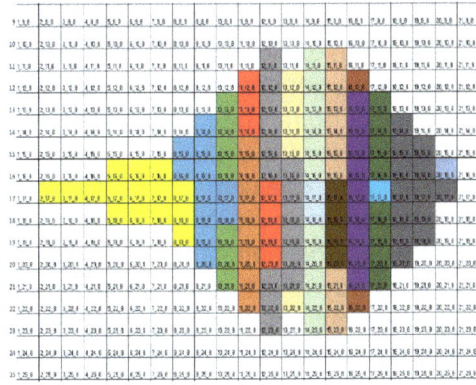

(b)

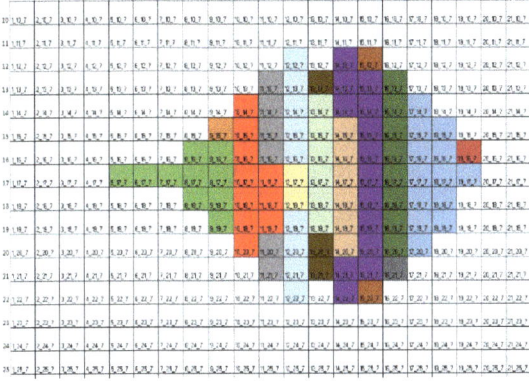

(c)

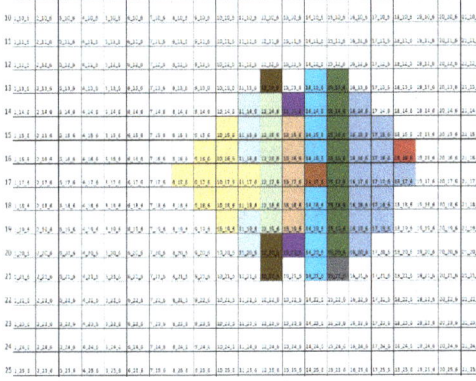

(d)

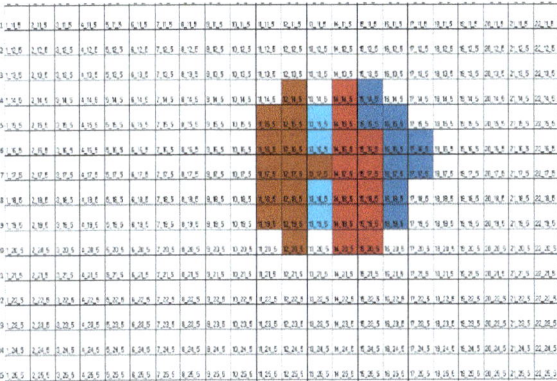

(e)

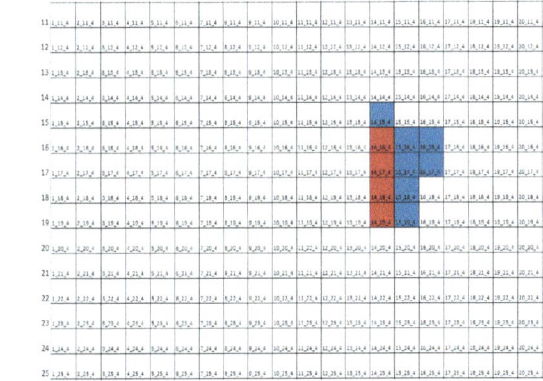

(f)

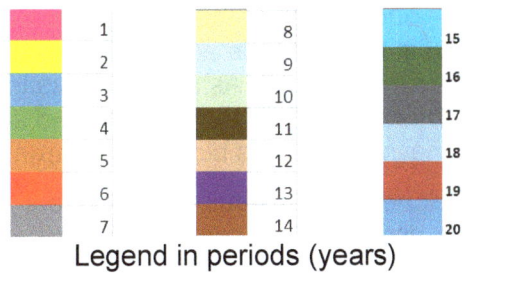

Legend in periods (years)

FIG 4 – Strategic mine plan using FMIPCC – ramp access from west through access y=17.
(a) Level 9 plan view; (b) Level 8 plan view; (c) Level 7 plan view; (d) Level 6 plan view; (e) Level 5 plan view; (f) Level 4 plan view.

Table 1 summarises all 12 cases listing all ramp access scenario results. As discussed in the previous chapter, it has been identified that the optimisation of FMIPCC is not only optimising the mining sequence for the picked ramp, it will also be mining and progressing the pit from different directions to uncover the ore and generate returns early. Imagine a ramp is placed at different locations of the pit and the mining is from a different direction will generate different returns. Therefore, it is worth exploring different mining directions using different ramps to access the mining strips. The 12 cases generated are grouped by ramp access direction and which strip is selected for the ramp to access from. The naming with 'N', 'S', 'W' and 'E' means the access ramp is from north, south, west and east of the pit. The number on the x or y axis represents the strip number that is used for access. As explained, different ramp access from each direction will be assessed. The last two columns summarise the NPV of each case and the solution gaps. Solution gaps indicate the difference between the optimal solution of a problem and the best solution obtained through the model. The accuracy of a solution improves by a reduced solution gap. 0 per cent gap means the solution is proven optimal, for grap greater than 0 per cent, it means there is still uncertainty, and whether the answer is optimal needs to be proved. This has fulfilled one of the original sub-objective on assessing ramp access options from different mining directions to find the optimal solution. Gurobi 10.0.2 version was used for the model.

TABLE 1
Summary of scenarios results.

Case ID	Access direction	Ramp strip	NPV ($ million)	Solution gap
11_y_9 S	South	x=11	1329	5.31%
11_y_9 N	North	x=11	1259	3.57%
12_y_9 S	South	x=12	1360	0.00%
12_y_9 N	North	x=12	1273	1.44%
13_y_9 S	South	x=13	1337	0.00%
13_y_9 N	North	x=13	1257	0.00%
x_15_9 W	West	y=15	1242	1.31%
x_15_9 E	East	y=15	1165	0.00%
x_16_9 W	West	y=16	1270	0.00%
x_16_9 E	East	y=16	1216	0.00%
x_17_9 W	West	y=17	1317	0.00%
x_17_9 E	East	y=17	1236	0.00%

The Gurobi solver was used to solve the newly developed Mixed Integer Programming (MIP) model. The resulting mine plan (Figure 4) shows that access ramps are built before progressing further to each lower depth level. Recall that in this example access is from the left, and in this case the trunk ramp is restricted to be the eighth row of blocks. As such, access to each level is developed before mining of strips can commence. This means that the block immediately above, and the three blocks above and to the left, must be removed to access blocks in the trunk ramp.

In practical applications, the developed model would typically be solved multiple times, considering various candidate trunk ramps to assess the impact of different access strategies. Iterations may involve evaluating options from the left, right, top, and bottom of the diagram, and testing different rows, especially those near the centre of the orebody.

The goal of these iterations is to identify the most optimal trunk ramp configuration that aligns with operational efficiency, resource utilisation, and overall economic objectives. Practical site constraints, such as geological features, infrastructure limitations, or environmental considerations, could further restrict the range of viable options during this selection process.

By systematically exploring different trunk ramp scenarios, the mining operation can fine-tune its strategy, ensuring that the selected access plan is not only mathematically optimal but also practical and feasible within the real-world constraints of the mining site. This iterative approach enhances the adaptability of the model to site-specific conditions, leading to a more robust and effective mine planning solution.

The movement of benches in the system occurs progressively from one side to the other, adhering to the established rule of mining direction. The strip must be completed sequentially before transitioning to the next one in the mining sequence, which could happen that one strip is completed within two continuous periods. This systematic approach ensures an organised and efficient progression through the mining operation, with each strip being progressively addressed before advancing to the subsequent strips.

As stated at the start of this chapter, in order to assess all other ramp access scenarios, another 11 ramp access options have been run through the solver using the same methodology, with the precedent data sets changed to align with the new ramp's starting point. Table 1 summarised all the scenarios' results.

A graphical representation of the corresponding mining and processing totals across the 20-year mine life are provided in Figures 5 and 6 respectively. The planned mining and processing rates are the capacities set in the model. The actual mining and processing rates are the optimisation running results from the solver based on the constraints of the model.

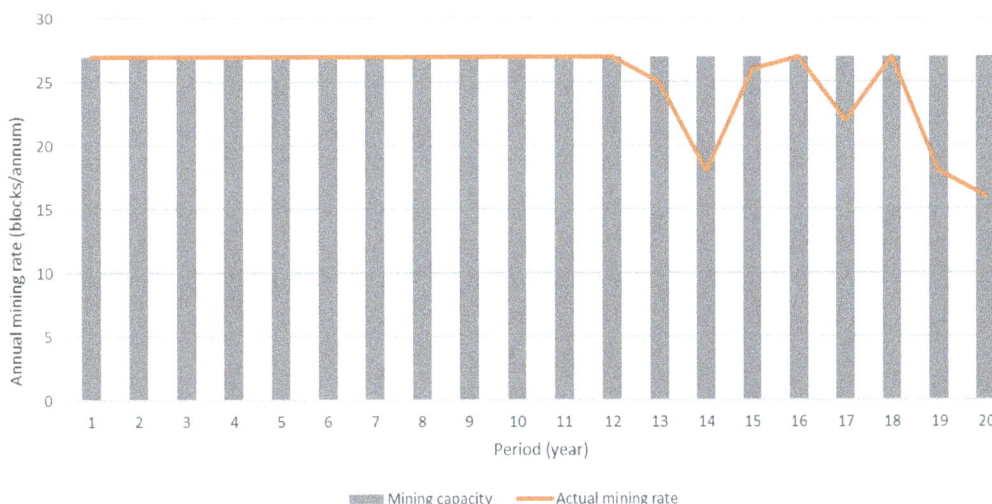

FIG 5 – Mining capacity and actuals.

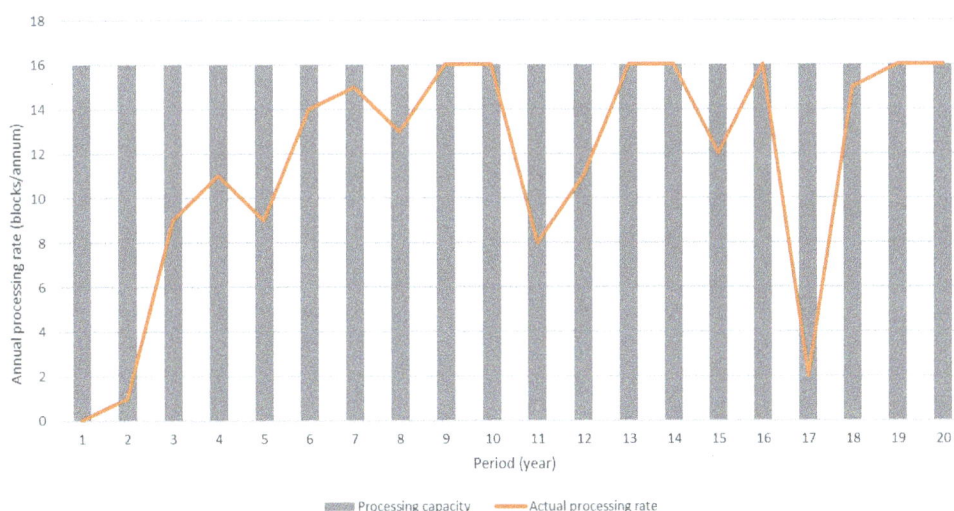

FIG 6 – Processing capacity and actuals.

The mining capacity versus actual rate graph (x_17_9 left case for example) in Figure 5 shows that across the first 12 years' mine life, with the aid of initial striping to uncover the deposit underneath, the scheduled mining rates reach the maximum capacity of 27 blocks a year. In later years when the deposit is mined towards the end of its profitable portion with the consideration of operational constraints of FMIPCC system, the mining rate is lower than installed capacity.

The processing capacity versus actual process rate graph (same x_17_9 left case for example) in Figure 6 shows that in the first two years, due to the nature of the deposit and the focus of initial striping, the scheduled processing rates is minimal. In later years, the processing rates either meet or are close to the designed capacity most of the time with some deficit as stripping is needed to uncover future ore blocks. Apart from the years where ore feed drops, this is a result of waste stripping not being able to keep up with the required removal rate so as to continually gain access to ore. Given that the ability to stockpile was not available in this case, this can be expected and is more likely to be exacerbated due to the very rigorous extraction sequence required by the FMIPCC system.

As shown in Figures 5 and 6, all considered constraints are complied with, including those related to annual mining and processing capacities. Across the 14 year mine life, processing capacities are maintained at or near maximum capacity in each year apart from years 8, 11, 12, 15 and 17, where ore feed drops as the waste stripping not being able to keep up with the required removal rate.

The results that presented for various scenarios are based on the Marvin copper gold block model that is originally developed in Whittle software. This block model was initially employed in case studies to demonstrate ultimate pit limits, mining stage and mine schedule for truck and shovel operations. Thus, the geometry of the orebody including the deposit dip, X-Y dimension extensions and orientation etc are not well-suited for a FMIPCC system application. FMIPCC systems are more suitable for horizontally extensive deposits (in the X-Y dimension) to take the benefits of mining long strips with less conveyor movement and low operating costs.

The solid lines presented in Figure 7 show the cumulative NPV on a yearly basis resulting from the 12 FMIPCC systems' production schedules. Lines in each colour represent different access scenarios for the FMIPCC cases. Capital costs of different systems are brought in at the start of projects, which is year 0 for the discount cash flow calculation.

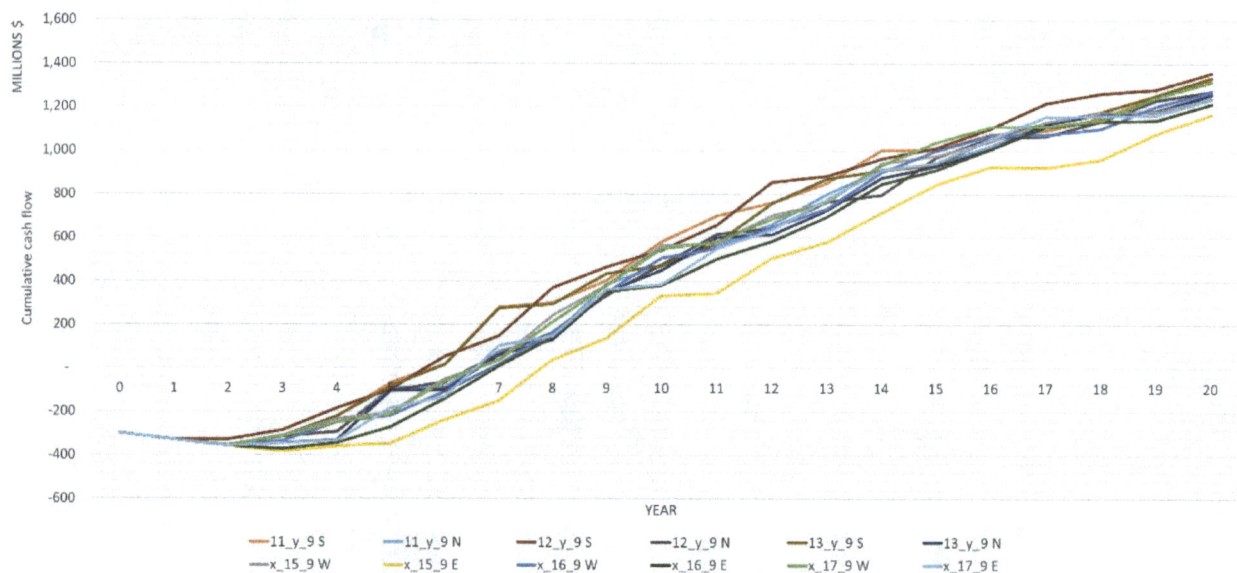

FIG 7 – Cash flow comparison between 12 FMIPCC scenarios.

As shown in Figure 7, the FMIPCC system starts with the high initial CAPEX that results in negative cash flow in the first six years. The operation starts making a positive cumulative cash flow from year six to seven as a result of FMIPCC system's operating cost. Based on the assumptions and parameters used of this case study, an NPV of A\$1.36 billion is generated across the 20-year mine life from one of the 12 FMIPCC case studies.

CONCLUSIONS

This research has investigated the mining sequence of a FMIPCC system in an open pit metalliferous mine context. The investigation encompasses various complex mine sequence constraints and features inherent to the FMIPCC system. The primary objective of the research has been fulfilled through the development of a mathematical model that comprehensively incorporates the essential operational features specific to FMIPCC. This model can serve as a valuable tool for optimising the mining sequence and contribute to a more efficient approach to mine planning in metalliferous open pit contexts. The Marvin block model that used in this research is publicly available and commonly used for demonstration purposes. The model was initially developed for ultimate pit and schedule demonstration for truck and shovel dominant operations. Thus the geometry and grade distribution of the orebody are suitable for this conventional mining method. A horizontally extensive orebody (deposit expands in x and y directions) is deemed to be more suitable for the FMIPCC application.

Direct block scheduling, an alternative to the conventional nested pit shells approach in open pit planning, has been applied in this research to realise FMIPCC operational constraints. The workflow of open pit planning is generally generating the ultimate pit shell, then refining the stage sequence from different pit shells. Mathematical modelling for the mining sequence of FMIPCC will also potentially simplify the ultimate pit shell to mining stage workflow of open pit planning. The MIP model generates the block sequence that takes the discount rate into account for the calculation of NPV. The annual mining and processing rates are close to the designed capacities in the case study contained within this paper with the small deficits being acceptable from a long-term planning perspective, which can be adjusted during short to mid-term planning.

Mathematical formulations have been developed to accommodate major mining constraints in terms of resource, mining sequence and timing. In terms of mining sequence, the constraints encompass FMIPCC movement restriction, horizontal sequence restrictions and strip sequence restrictions that are inherent to FMIPCC systems. This truly customises the results that are generated to the FMIPCC system configuration. As FMIPCC systems require mining sequences in a strip by strip order, direct block scheduling is a good enabler of this feature. MIP together with a Gurobi solver are used to model and solve this complex problem. The FMIPCC system configurations and orientations, as featured in this paper, will change from operation to operation depending on the geology, shape and orientation of the deposit. As shown in the case study, the pit geometry for a FMIPCC system will be different to a truck and shovel mined pit as the pit wall accommodating the straight conveyer layout will restrict how the pit progresses. In addition to this, the mining process is scheduled subject to the operational constraints of the FMIPCC system. The scheduling objective in this case is Net Present Value (NPV).

The conveyor ramp system is a crucial part of scheduling as a means of accessing the deposit for FMIPCC systems and involved significant material movement. Traditional truck and shovel open pit planning doesn't often consider ramps during the ultimate pit shell generation and phase stage assessment, however, this is a critical part of FMIPCC system and is therefore included in the early planning stage of the system. The conveyor ramp's precedent relationships are a key influencing factor on project value. The ramp should be located strategically so as to minimise waste striping and facilitate as early as possible access of the orebody. The time of mining a strip on the same level or starting mining on a new strip in the level below is influenced by the value of the strip of blocks. Twelve potential ramp access locations are assessed within this paper. The proposed ramp locations are strategically chosen from the widest areas of the pit to minimise waste stripping from four cardinal directions: West, East, North and South. The ramps are through the centre/middle of the orebody with an additional potential location being either side of the central slot. The newly developed mathematical model is used to determine the optimal solution between the 12 potential conveyor ramp slot locations. All scheduled sequences have been checked and validated to comply with all modelled constraints.

The suitability of an FMIPCC system for an orebody is dependent on a range of factors. In summary, this research establishes a new approach to schedule FMIPCC system's extraction sequence considering all key operational constraints. The developed model opens the opportunities for evaluating FMIPCC systems for projects. Results that can be generated within reasonable time frames allow more options to be investigated. This in turn enables the rapid evaluation of various potential access ramp locations in order to determine the optimal solutions in terms of NPVs. While IPCC systems are often considered as an alternative to traditional truck and shovel mining for their economic viability, they also provide a significant reduction in the energy intensity of the materials handling system which in turn generates environmental benefits. Therefore, the selection of the system must take into account specific operational, economic and technical considerations as well as social impacts. IPCC systems should not only be treated as an alternative to truck and shovel surface operations from an economic perspective, but they can also potentially add value to operations from an environmental perspective.

RECOMMENDATIONS FOR FURTHER RESEARCH

Since the aim of this study was to provide a preliminary investigation into the development of algorithms suitable to consider the rigorous sequencing constrains inherent to the use of FMIPCC systems, this would naturally lead to undertake a comparison study between the value obtained when extraction of a deposit takes place via a FMIPCC system and one that considers traditional Truck and Shovel. The Marvin block model that used in this research is publicly available and commonly used for demonstration purposes. The model was initially developed for ultimate pit and schedule demonstration for truck and shovel dominant operations. Thus the geometry and grade distribution of the orebody are suitable for this conventional mining method. An extensive orebody (deposit expands in x and y directions and deep in z direction) is deemed to be more suitable for the FMIPCC application. This is because the continuous mining system is more fuel efficient for haulage to the extensive and deep orebody comparing with truck dominant systems, and extensive area favours the conveyor layout in the pit.

Future versions of this model may incorporate the ability to stockpile material. While adding complexity to the mathematical model, this would enable greater flexible around scheduling ore feed to the process plant from a system that is very rigid in its sequence. Additional scheduling flexibility may also be gained by allowing more than one system to operate within a pit. This however, would also introduce a number of system interaction risks that would need constraining.

Exploring different conveyor ramp access according to pit orientation such as zigzag conveyor belts can also be investigated further. This option can potentially further optimise the material stripping based on pit geometry and thus results better returns for the FMIPCC system. A high angle belt conveyor paired with the IPCC system could be another option for steep wall transportation scenarios, this can reduce the significant amount of material removal and associated costs. However, in this case, the capacity bottleneck may be further constrained by the conveyor capacity.

It would also be beneficial to quantify and compare the energy requirements of a FMIPCC system and a traditional truck and shovel system in order to an understanding of the carbon footprint of each. This would allow operations to not only compare mine plans based on traditional financial metrics but also environmental metrics, which are becoming increasingly more important.

REFERENCES

Abbaspour, H, Drebenstedt, C, Paricheh, M and Ritter, R, 2019. Optimum location and relocation plan of semi-mobile in-pit crushing and conveying systems in open-pit mines by transportation problem, *International Journal of Mining, Reclamation and Environment*, 33(5):297–317. https://doi.org/10.1080/17480930.2018.1435968

Alspaugh, M A, 2004. Latest Developments in Belt Conveyor Technology, in Proceedings MINExpo 2004.

Arteaga, F, Nehring, M and Knights, P, 2017. The equipment utilisation versus mining rate trade-off in open pit mining, *International Journal of Mining, Reclamation and Environment*, 31:1–24.

Arteaga, F, Nehring, M, Knights, P and Camus, J, 2014. Schemes of exploitation in open pit mining, in *Proceedings Mine Planning and Equipment Selection 2013*, 1307–1323.

Atchinson, T and Morrison, D, 2011. In-Pit Crushing and Conveying Bench Operations, in *Proceedings Iron Ore Conference*, pp 157–163 (The Australasian Institute of Mining and Metallurgy: Melbourne).

Dean, M, Knights, P, Kizil, MS and Nehring, M, 2015. Selection and planning of fully mobile in-pit crusher and conveyor systems for deep open pit metalliferous applications, in *Proceedings of Third International Future Mining Conference*, pp 219–225 (The Australasian Institute of Mining and Metallurgy: Melbourne).

Department of Resources, Energy and Tourism (DRET), 2009. Airborne contaminants, noise and vibration, Leading Practice Sustainable Development Program for the Mining Industry Handbook, Commonwealth of Australia.

Dos Santos, J A and Stanisic, Z, 1986. In-pit crushing and high angle conveying in a Yugoslavian copper mine, *International Journal of Surface Mining, Reclamation and Environment*, 1(2):97–104. https://doi.org/10.1080/09208118708944108

Dos Santos, J A, 2016. Sandwich belt high angle conveyors coal mine to prep plant and beyond, *Proceedings of the XVIII International Coal Preparation Congress*, pp 111–117 (Springer).

Dos Santos, J, 1984. Sandwich belt high angle conveyors-applications in open pit mining, *Bulk Solids Handling*, 4(1):67–77.

Dzakpata, I, Knights, P F, Kizil, M S, Nehring, M and Aminossadati, S M, 2016. A comparison of the valuable operating time of truck-shovel and in-pit conveyor systems, in *Proceedings of the 2016 Coal Operators' Conference*, pp 347–357.

Hay, E, Nehring, M, Knights, P F and Kizil, M S, 2019. Ultimate pit limit determination for semi mobile in-pit crushing and conveying system: a case study, *International Journal of Mining, Reclamation and Environment*, pp 1–21.

Hays, R M, 1983. Mine planning considerations for in-pit crushing and conveying systems, in *Proceedings of the SME-AIME Fall Meeting* (ed: J Brent Hiskey), pp 3–41 (American Institute of Mining, Metallurgical and Petroleum Engineers: New York).

Hochbaum, D S, 2008. The Pseudoflow Algorithm: A New Algorithm for the Maximum-Flow Problem, *Operations Research*, 56(4):992–1009.

Huss, C, Reisler, N and Almond, R, 1983. Practical and economic aspects of in-pit crushing conveyor systems, in *Proceedings of the SME-AIME Fall Meeting* (ed: J Brent Hiskey), pp 15–31 (American Institute of Mining, Metallurgical and Petroleum Engineers: New York).

International Mining, 2009. IPCC Innovation. Available from: <http://www.infomine.com/publications/docs/International Mining/IMJune2009f.pdf>.

Kesimal, A, 1997. Different types of belt conveyors and their applications in surface mines, *Mineral Resources Engineering*, 6(4):195–219.

Koehler, F, 2003. In-pit crushing system the future mining option, in *Proceedings of the Twelfth International Symposium on Mine Planning and Equipment Selection*, pp 371–376 (The Australasian Institute of Mining and Metallurgy: Melbourne).

Lerchs, H and Grossman, I F, 1965. Optimum design of open-pit mines, *CIM Bulletin*, 58:47–54.

Liu, D and Pourrahimian, Y, 2021. A framework for open-pit mine production scheduling under semi-mobile in-pit crushing and conveying systems with the high-angle conveyor, *Mining*, 1(1):59–79. https://doi.org/10.3390/mining1010005

McCarthy, M and Cenisio, B, 2013. Update on Studies for Implementation of IPCC at the Moatize Coal Project, Cologne: IPCC.

Mitchell, J and Albertson, D, 1985. High angle conveyor offers mine haulage savings, in *Proceedings of the International Materials Handling Conference, BeltCon 3*, 3:9–11 (South African Institute of Materials Handling).

Nehring, M, Knights, P, Kizil, M and Hay, E, 2018. A comparison of strategic mine planning approaches for in-pit crushing and conveying and truck/shovel systems, *International Journal of Mining Science and Technology*, 28(2):205–214.

Osanloo, M, 2012. Future challenges in mining division, are we ready for these challenges? Do we have solid educational program?, in *Proceedings of the 23rd Annual General Meeting of the Society of Mining Professors*, pp 29–39 (Wroclaw University of Technology).

Paricheh, M and Osanloo, M, 2019a. Concurrent open-pit mine production and in-pit crushing–conveying system planning, *Engineering Optimization*, 52(10):1780–1795. https://doi.org/10.1080/0305215X.2019.1678150

Paricheh, M and Osanloo, M, 2019b. How to exit conveyor from an open-pit mine: A theoretical approach, *Proceedings of the 27th International Symposium on Mine Planning and Equipment Selection – MPES 2018*, pp 319–334 (Springer).

Paricheh, M and Osanloo, M, 2019c. A new search algorithm for finding candidate crusher locations inside open pit mines, in *Proceedings of the International Symposium on Mine Planning and Equipment Selection – MPES 2018*, pp 10–25 (Springer).

Raaz, V and Mentges, U, 2011. In pit crushing and conveying with fully mobile crushing plants in regard to energy efficiency and CO_2 reduction, in *Proceedings of in-pit crushing and conveying conference*, pp 21–23.

Roumpos, C, Partsinevelos, P, Agioutantis, Z, Makantasis, K and Vlachou, A, 2014. The optimal location of the distribution point of the belt conveyor system in continuous surface mining operations, *Simulation Modelling Practice and Theory*, 47:19–27.

Shamsi, M and Nehring, M, 2021. Determination of the Optimal Transition Point Between a Truck and Shovel System and a Semi-Mobile In-Pit Crushing and Conveying System, *The Journal of the South African Institute of Mining and Metallurgy*, 121(9):497–504.

ThyssenKrupp, 2004. In-Pit Crushing Continuous Haulage Systems (IPC-CHS), Germany: ThyssenKrupp Fordertechnik.

Tolosana-Delgado, R, Mueller, U and van den Boogaart, K G, 2019. Geostatistics for Compositional Data: An Overview, *Mathematical Geosciences*, 51(4):485–526.

Turnbull, D and Cooper, A, 2009. In-Pit Crushing and Conveying (IPCC) – A Tried and Tested Alternative to Trucks, in Proceedings New Leaders' Conference (The Australasian Institute of Mining and Metallurgy: Melbourne).

Tutton, D and Streck, W, 2009. The application of mobile in-pit crushing and conveying in large, hard rock open pit mines, in Proceedings of the Mining Congress, Niagara.

Utley, R W, 2011. In-pit crushing, *SME Mining Engineering Handbook*, 3rd edn, pp 941–956 (Society for Mining, Metallurgy and Exploration Inc, Englewood).

Reconfiguring the techno-economic-environmental landscape for low-grade ore extraction – a strategic analysis of stope leaching systems

M Gadiile[1], I Onederra[2], M S Kizil[3] and M Nehring[4]

1. PhD candidate, The University of Queensland, St Lucia Qld 4072. Email: m.gadiile@uq.net.au
2. Associate Professor, The University of Queensland, St Lucia Qld 4072.
   Email: i.onederra@uq.edu.au
3. Associate Professor, School of Mechanical and Mining Engineering, The University of Queensland, St Lucia Qld 4072. Email: m.kizil@uq.edu.au
4. Lecturer, The University of Queensland, St Lucia Qld 4072. Email: m.nehring@uq.edu.au

ABSTRACT

As the mining industry confronts escalating pressures to meet global demand for transition metals while upholding rigorous Environmental, Social, and Governance (ESG) standards, the shift towards low-grade and complex orebodies at increasing depth exposes inherent limitations in conventional mining methods. Stope Leaching (SL) is re-emerging as a viable, sustainable alternative to address these complex extraction scenarios, offering improved economic resilience through lower sensitivity to ore grade and a reduced environmental footprint. However, the lack of robust quantitative tools for strategic optimisation significantly hinders comprehensive assessment of SL viability, restricting industry adoption by obscuring key economic and operational insights necessary for informed decision-making. The viability of SL systems depends on complex interactions involving mineralogy, fragmentation, stope geometry, leaching kinetics, and fluid flow dynamics. These parameters introduce spatiotemporal complexities beyond the scope of traditional stoping optimisation methodologies.

This paper identifies the key techno-economic and environmental parameters essential for reliable SL planning and scheduling and proposes a novel, kinetics-informed strategic planning framework tailored specifically to these systems. By incorporating kinetic profiles, solution-handling capacity, and time-evolving operating costs directly into the optimisation model, this dynamic approach ensures critical factors such as diminishing returns and opportunity-cost trade-offs are systematically integrated throughout the mine planning process. The resulting framework enables scenario-based analyses that more accurately capture value generation, enhance operational flexibility, and address sustainability expectations, providing a comprehensive foundation for improved decision-making for SL systems.

INTRODUCTION

The mining industry faces unprecedented challenges as it contends with declining ore grades, with approximately 40 per cent reduction in average copper grades since the early 1990s, increasing mining depths (frequently exceeding 1000 m), and more stringent ESG regulations (Dare-Bryan and Hassanvand, 2023; Estay *et al*, 2023; Farrell and Whitton, 2024; Ghorbani *et al*, 2023; Maybee, Lilford and Hitch, 2023; Rossien, 2020; Wang *et al*, 2019). These compounding pressures expose critical limitations in conventional mining methods, which are progressively constrained either by economic factors or ESG considerations.

Stope leaching, the strategic integration of underground stoping with solution mining techniques, has re-emerged as a promising alternative. This approach involves the systematic circulation of lixiviants through adequately fragmented ore within underground stopes followed by the recovery of metal-bearing pregnant leach solutions (Bahamóndez *et al*, 2016; Boreck *et al*, 1991; Boreck, Lutzens and Speirer, 1990; Dare-Bryan and Hassanvand, 2023; Mousavi and Sellers, 2019; Rossien, 2020; US Department of the Interior, Bureau of Mines, 1994). For copper operations specifically, detailed case studies demonstrate quantifiable advantages: approximately 25 per cent mining capital expenditure (CAPEX) reductions, 49 per cent overall CAPEX savings including processing infrastructure (Dare-Bryan and Hassanvand, 2023), and 33 per cent Net Present Value (NPV) improvement when hybridised with conventional systems as discussed by Mousavi and Sellers (2019). Beyond its operational advantages including reduced energy intensity, lower costs, and minimised surface disturbance as discussed by Estay *et al* (2023), Dare-Bryan and Hassanvand (2023), Mousavi and Sellers (2019), SL offers strategic value by potentially converting previously submarginal resources into reserves, a critical capability given the forecasted temporal disconnect between escalating copper demand and available production capacity through 2035.

Despite these clear advantages, SL implementation faces substantial barriers due to the complex spatiotemporal interactions inherent in SL systems. Prior research has largely focused on subprocess optimisation, lacking integrated frameworks that link system design and metallurgical performance to strategic economic outcomes. Factors such as mineralogy, stope geometry, leach kinetics, and fluid dynamics have been reported to strongly influence value generation but remain inadequately captured by traditional optimisation approaches. Current methodologies often rely on static, recovery-grade-based valuation strategies that overlook the unique dynamics of solution mining processes. This misalignment arises because conventional frameworks were developed for solid ore extraction rather than solution-mining-based recovery systems, which possess distinct time-dependent dynamics, grade-value relationships, and risk profiles.

Consequently, the mining industry currently lacks an integrated, quantitative evaluation framework explicitly tailored for SL implementation. Thus, three critical questions arise:

- Which specific economic parameters and ESG metrics are appropriate for SL's break-even viability yet remain unincorporated into current long-term planning models?

- How do strategic design decisions (eg stope geometry, ore fragmentation methods) interact with leaching kinetics to impact overall metal recovery trajectories over project lifespans?

- What modifications to current optimisation algorithms are required to effectively model SL implementation across diverse geological and mineralogical contexts?

This paper addresses these gaps through a systematic literature analysis and a parametric modelling approach. The essential techno-economic and environmental parameters currently neglected in optimised SL scheduling are identified. Through a critical evaluation of their impacts and systematic comparison to conventional approaches, outcomes from this research provide the foundation for an SL tailored strategic planning framework.

EVOLUTION OF STOPE LEACHING SYSTEMS

Stope leaching has a historical presence dating back to the 1920s, with hard rock applications gaining momentum during the 1970–1980s, primarily as an opportunistic approach for salvaging residual and low-grade ores post conventional mining (Ahlness and Pojar, 1983; Gardner, Johnson and Butler, 1938; Ito, 1976; Van Staden and Laxen, 1989). Figure 1 illustrates the core components and configuration of a typical SL system, including ore access design, lixiviant delivery and containment infrastructure, and stope layout.

The application of SL in hard rock mining has evolved through four distinct, though occasionally overlapping, phases:

- **Early applications** (1920s to 1960s) – Initially adopted as secondary or tertiary recovery system for residual and low-grade ore post conventional mining, primarily in copper mines, with limited systematic design approaches (Ahlness and Pojar, 1983; Gardner, Johnson and Butler, 1938). SL extracted copper from various sources, including old stope filling, low-grade ore in abandoned shrinkage stopes, and crushed, caved ground.

- **Technical advancement period** (1970s to 1980s) – This era was characterised by significant experimentation, including the exploration of various fragmentation approaches and early bioleaching trials. Field experiments helped define fundamental parameters for a range of lixiviant systems, while also exposing critical operational challenges, namely, solution distribution, kinetics, permeability and fragmentation control, all of which had a significant impact on metal recovery rates and economic feasibility (Burton *et al*, 1983; Butler, Ackland and Robinson, 1981; Ito, 1976; Miller, 1986; Rossi, Trois and Visca, 1986; Van Staden and Laxen, 1989).

- **Target research and focused trials** (1990s) – SL evolved into a specialised system for low-grade ore extraction, with research increasingly focused on key technical parameters and system-level optimisation. Core focus areas included bioleaching, solution containment, fragmentation techniques, rock mass characterisation, economic modelling and adaptation of mining methods for SL, though large-scale industry adoption remained limited (Boreck *et al,* 1991; Boreck, Lutzens and Speirer, 1990; Brock, Chomley and Richmond, 1998; Lombardi and Jude, 1994; Miller and Schmuck, 1995; Potts and Webb, 1994; Sand *et al,* 1993; Wang and Dai Yuan, 1993).

- **Decline and re-assessment** (2000s to early 2010s) – Interest in SL declined due to persistent operational challenges, limited innovation, and market conditions favouring alternative techniques such as heap leaching, particularly during periods of elevated commodity prices. However, recent shifts toward environmentally sustainable extraction methods, coupled with technological advancements, have renewed interest in SL as a viable and potentially strategic recovery solution.

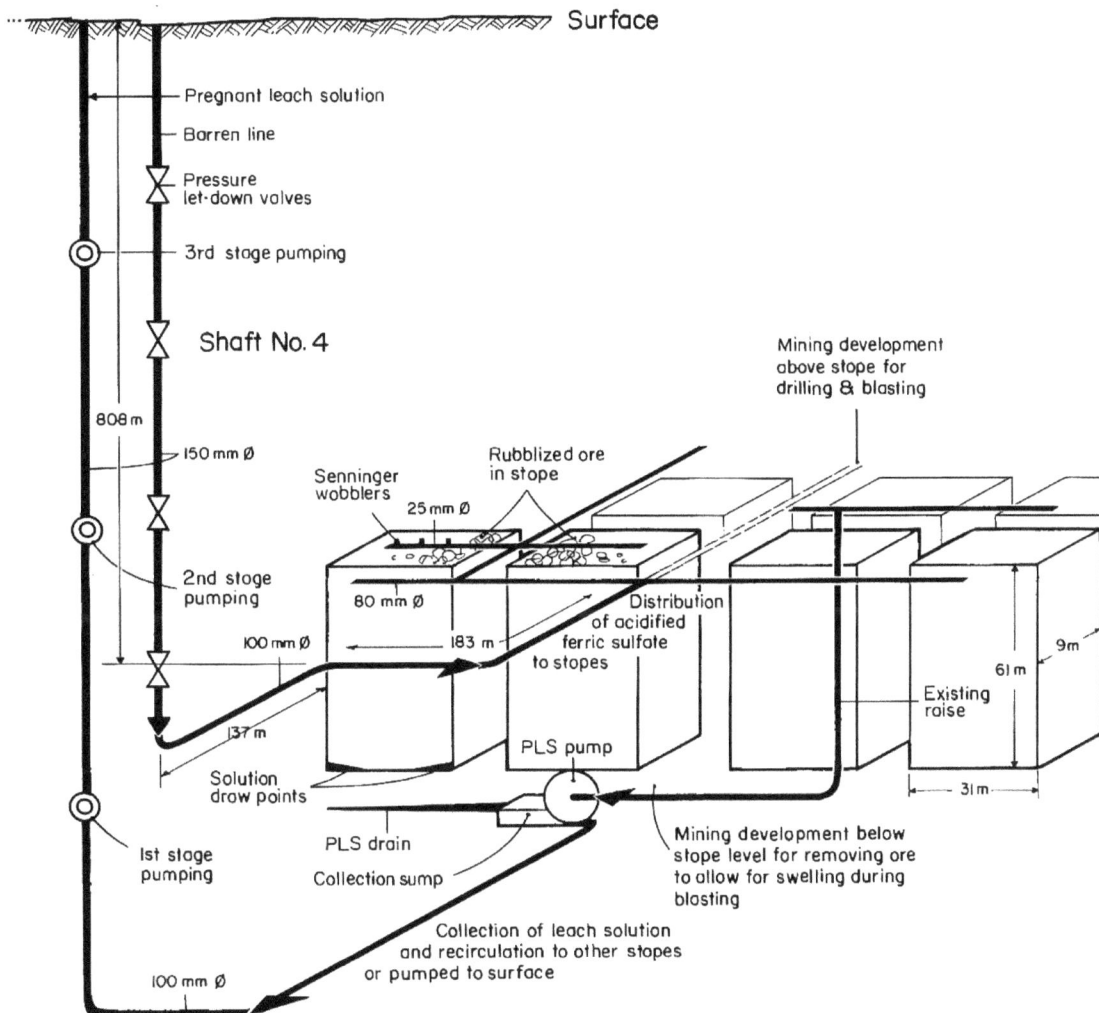

FIG 1 – Generalised schematic of a stope leaching system (adapted from McCready and Sanmugasunderam, 1985).

Analysis of historical case studies reveals four recurring challenges that hindered widespread SL adoption:

- **Insufficient design justification** – Many projects lacked quantitative rationale for system design, often adapting conventional stoping layouts without optimisation for SL-specific performance.

- **Weak linkage between design and outcomes** – Although individual process optimisations were attempted, their strategic economic and operational impacts were rarely evaluated holistically.

- **Overlooked strategic value drivers** – Development efforts remained largely reactive, addressing immediate technical problems rather than proactively enhancing long-term performance.

- **Disjointed integration of geometallurgical insights into mine planning** – The variability of orebodies and their impact on SL performance remained underexplored, further limiting predictive and strategic modelling efforts.

These historical challenges provide essential context for understanding the current resurgence of interest in SL. They underscore the critical knowledge gaps that must be addressed to unlock the full potential of SL in modern mining operations.

ECONOMIC FEASIBILITY AND STRATEGIC PLANNING

Recent studies indicate that, despite lower nominal recoveries (approximately 30 per cent) and slower kinetics, SL can nonetheless achieve profit margins comparable to those of conventional mining due to significantly lower capital intensity and operating costs. Modelling work highlights three primary advantages (Bahamóndez *et al*, 2016; Dare-Bryan and Hassanvand, 2023; Rossien, 2020):

1. **Lower break-even grades** – Suitable for lower-grade deposits that are economically marginal under traditional methods.

2. **Reduced CAPEX requirements** – Particularly beneficial for deep-seated or geologically challenging orebodies.

3. **Lower operating costs (OPEX)** – Due to reduced haulage and comminution requirements.

In parallel, advances in strategic mine planning have begun exploring ways to embed SL into existing scheduling frameworks. For instance, Mousavi and Sellers (2019) proposed an integer programming model that demonstrated potential NPV gains of up to 33 per cent using hybrid mining strategies. The optimisation model also revealed that while higher-grade stopes are typically prioritised for conventional processing, there is no definitive cut-off grade for selecting between conventional processing and SL. This finding highlights the complex interplay of factors beyond ore grade in determining the most economically viable extraction method, challenging traditional grade-based decision-making and the optimisation approaches typically adopted.

The critical limitation becomes evident: existing scheduling frameworks are designed for solid ore extraction, where value ranking is grade-centric and recovery is immediate. They assume a direct, high-grade-to-high-value correlation, a fundamental assumption that does not hold for SL systems. Stope leaching extraction dynamics are governed by delayed, time-dependent leaching kinetics, introducing complex spatiotemporal and mineralogical dependencies. SL's pivotal scheduling decision is not only when to initiate leaching but, more critically, when to terminate the process, balancing diminishing returns against opportunity costs, while accounting for inherent leaching-contingent parameters that influence rate of value generation over time for a given stope. Between these temporal boundaries, both the rate and ultimate extent of recovery are governed exclusively by leaching kinetics, stope fluid flow dynamics, mineralogy, fragmentation, and the rock mass characteristics of the ore column.

THE STOPE LEACHING VALUE GENERATION PROBLEM

In SL operations, the correlation between ore grade and economic value is indirect and modulated by multiple dynamic parameters, including:

- Acid consumption rates, which dictate processing costs and operational efficiency.

- Solution loss dynamics directly impact recoverable metal value and operating costs.

- Spatially variable leaching kinetics, determined by mineralogy and geometallurgical domains.

- Design elements, including stope geometry and fragmentation which affect metal recovery.

- Mineralogy-dependent recovery constraints, which introduce variability in leaching kinetics across different stopes.

Traditional grade-based optimisation approaches can significantly misrepresent actual economic returns, either underestimating or overestimating, by neglecting critical operational factors such as kinetics, acid consumption, solution losses, and other geometallurgical variables that strongly influence value. These parameters significantly affect both the rate of value generation and cumulative economic returns from individual stopes.

Figure 2 illustrates this discrepancy clearly. Despite achieving identical theoretical total recoveries of approximately 40 per cent, variations in stope-level leaching kinetics and operational parameters, particularly acid consumption and solution losses, lead to substantial differences in both the rate and ultimate realised economic value, creating unique, stope-specific effective revenue profiles.

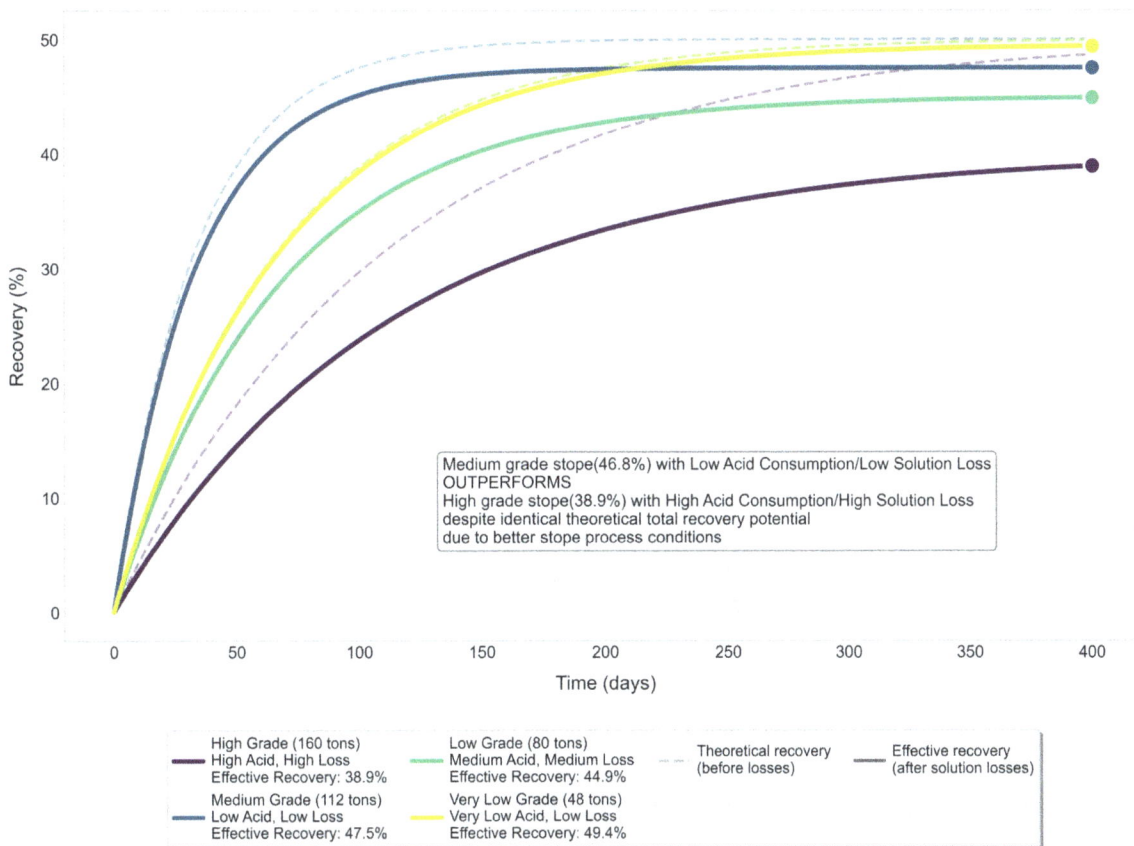

FIG 2 – Stope value ranking methodologies comparison – traditional grade-recovery ranking approach versus geometallurgical-based kinetic-aware evaluation, comparative analysis of recovery profiles.

Figure 3 explicitly highlights this 'value gap', revealing the divergence between conventional valuation methods relying solely on grade-recovery metrics and integrated, kinetics-informed approaches. A key and somewhat counterintuitive insight emerges: a medium-grade stope with favourable acid consumption and minimal solution losses can economically outperform a higher-grade stope burdened by higher acid consumption and greater solution losses. This insight fundamentally challenges conventional grade-centric mine planning, underscoring the need for tailored strategic optimisation approaches for SL systems.

Further reinforcing this concept, Figure 4 illustrates diminishing incremental returns at the stope level as leaching progresses. Extending the leaching phase without considering declining marginal returns significantly increases opportunity costs and can erode overall profitability, shifting strategic objectives from maximising individual stope recovery to balancing incremental stope-level value within the broader system economics. This underscores the necessity to strategically identify optimal termination points rather than the standard approach of targeting total theoretical recovery.

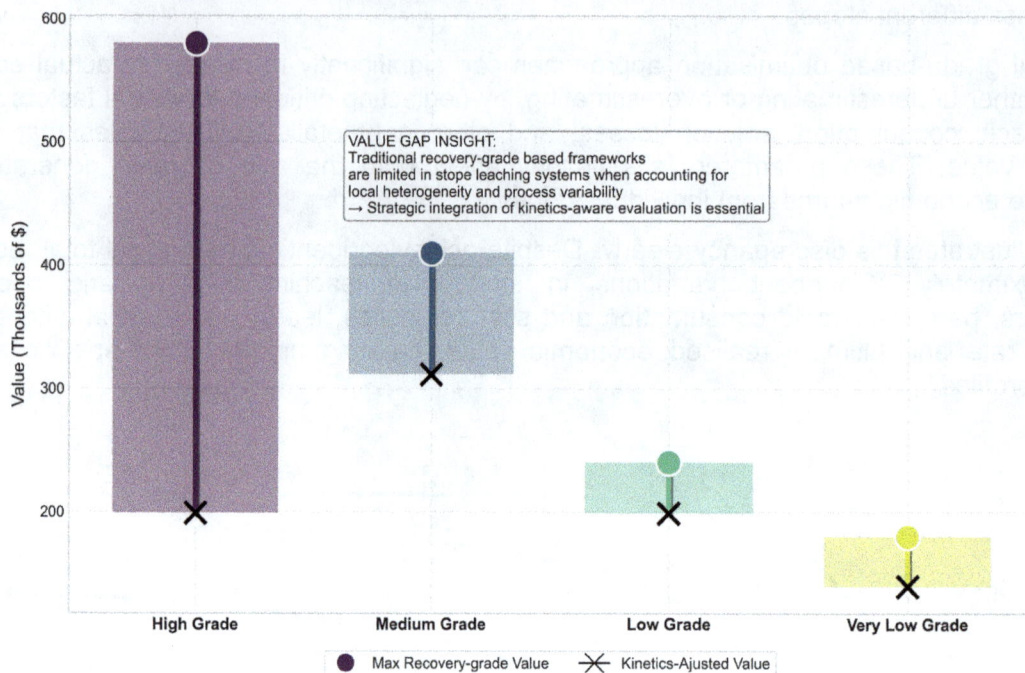

FIG 3 – Stope value ranking methodologies comparison – traditional grade-recovery ranking approach versus geometallurgical-based kinetic-aware evaluation, comparative analysis of stope economic value.

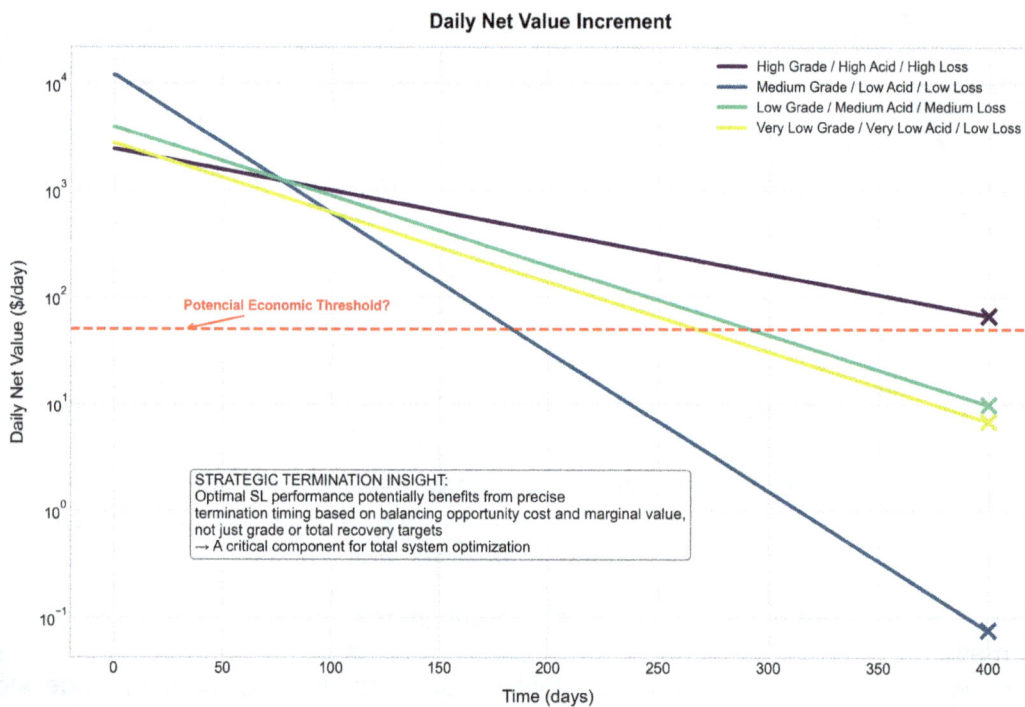

FIG 4 – Diminishing marginal returns in SL – Identifying stope specific optimal termination points for global maximum economic value versus stope total recovery.

Consequently, transitioning from static, grade-based models to dynamic, kinetics-driven optimisation frameworks that explicitly incorporate acid consumption, solution losses, recovery kinetics, and related geometallurgical parameters is essential. This comprehensive approach reduces uncertainty, enhances economic reliability, and facilitates informed, strategic decision-making across the mine life cycle.

Implications for strategic mine planning

In SL, deciding when to stop leaching a stope is as critical as the decision to commence leaching. Extending leaching cycles excessively results in inefficient capital use and diminishing returns, while prematurely halting leaching leads to unrealised recovery potential.

Conventional mine scheduling frameworks prioritise static, recovery-grade-based rankings and fixed cut-off grades, inherently neglecting SL's dynamic recovery processes and opportunity cost trade-offs. Thus, strategic planning models for SL must fundamentally evolve, incorporating these dynamic considerations explicitly.

A critical research gap in current SL optimisation is the absence of an integrated feedback loop linking:

- leaching efficiency

- strategic mine-design decisions

- long-term value optimisation.

The lack of integration between these components results in suboptimal scheduling decisions, undermining long-term profitability. Addressing this gap requires methodologies that move beyond static total recovery-grade-based rankings, incorporating dynamic, kinetics-driven recovery models aligned with SL's inherent spatial-temporal complexities and economic constraints.

- Balancing physical complexity and model practicality to ensure computational feasibility does not compromise accuracy and strategic relevance.

- Reconciling detailed hydrometallurgical data with strategic decision-making, identifying an abstraction level precise enough for accuracy yet manageable enough to remain computationally viable.

Having established these critical limitations in current approaches, the analysis now turns to reverse-engineering the value generation process to identify the critical parameters required to underpin an optimisation framework capable of more accurately capturing the dynamic nature of SL operations.

REVERSE ENGINEERING THE OPTIMISATION FRAMEWORK

To determine the key parameters driving SL decision-making, the objective function is reverse engineered – maximising NPV of the SL system. This is done through a comparative analysis with existing NPV optimisation models, where the NPV is defined by three core components:

1. Capital expenditure (CAPEX).

2. Operating costs (OPEX).

3. Time-dependent revenue streams.

This framework isolates which factors fundamentally alter SL's economic trajectory and require explicit consideration in scheduling and optimisation models.

Capital expenditure consideration

Stope leaching operations follow the same fundamental CAPEX structure as conventional mining but require additional solution management infrastructure that imposes unique scheduling constraints.

Key SL-specific components include:

- **SX/EW facilities** – Essential for solution processing and metal recovery.

- **Swell haulage infrastructure** – Manages fragmented ore movement.

- **Pumping and solution distribution systems** – Dictate solution flow capacity and leaching efficiency.

- **Fragmentation enhancement** – Improves solution penetration and recovery rates.
- **Monitoring and automation technologies** – Enable real-time tracking of leaching performance.

Among these, solution management infrastructure imposes the most significant operational constraint. Pumping capacity and solution handling systems directly determine operation size and the number of stopes that can be leached concurrently, making them a critical input for strategic scheduling models.

Stope leaching requires solution capacity to be treated as a dynamic scheduling parameter. This introduces constraints on:

- **Maximum concurrent active stopes** – Defining the upper limit of simultaneous leaching operations.
- **System-wide solution handling limitations** – Ensuring infrastructure availability aligns with leaching demand.
- **Strategic sequencing dependencies** – Optimising stope activation timing to balance system capacity.

Operating cost dynamics in SL

Stope leaching introduces cost complexities beyond conventional mining, particularly reagent consumption, solution make-up requirements, and solution losses. While economic models incorporate these costs at the financial analysis stage, they are absent from strategic planning frameworks, leading to misaligned cost projections.

Solution loss has a threefold economic impact:

1. Reduces total metal recovery, directly affecting revenue potential.
2. Increases remediation/ESG compliance costs if contamination occurs.
3. Raises reagent and solution replacement expenses, increasing OPEX over time.

These cost variations depend on stope-specific conditions, geometallurgical domains, and leaching cycle stages, creating spatiotemporal cost variability that conventional models do not capture.

Unlike conventional mining, where costs scale with tonnage, SL costs evolve dynamically over time. Leaching efficiency declines, reagent demand shifts, and cumulative operating costs increase with cycle length. The economic viability of a stope is therefore not just a function of grade or size but of time-dependent cost accumulation.

Stope leaching requires a dynamic cost model that integrates:

- **Leaching consumption profiles** – Capturing how reagent demand shifts over time.
- **Solution loss impacts** – Accounting for both recovery efficiency and ESG constraints.
- **Stope-specific cost variability** – Recognising that each stope's economic profile changes throughout the leaching cycle.

Incorporating these factors into scheduling prevents unrealistic cost assumptions and enables optimised sequencing and leaching duration decisions.

Revenue modelling and strategic scheduling implications

Stope leaching revenue modelling fundamentally diverges from conventional mining, where recovery is immediate and value is directly linked to tonnage processed. In SL, extraction follows time-dependent recovery curves with diminishing returns, introducing strategic decision variables absent in conventional optimisation frameworks:

- **Optimal leaching duration** – Defining the economic threshold beyond which continued leaching is no longer viable.

- **Opportunity cost trade-offs** – Allocating resources based on dynamic recovery potential, not static grade rankings.

- **Stope activation sequencing** – Optimising when to initiate and terminate leaching across stopes to balance recovery cycles and infrastructure utilisation.

These factors necessitate a shift from static cut-off grades to dynamic, recovery-driven scheduling models. Unlike traditional frameworks, which optimise based on fixed grade and tonnage constraints, SL scheduling must integrate rate-based recovery dynamics into strategic mine planning. For strategic planning, the level of abstraction is reduced to the stope-bottom PLS fluxes profiles lagged to account for breakthrough time which considers geometric conditions influence of recovery timing.

A kinetics-aware scheduling approach is required, where recovery rate functions as a direct input to optimisation models. This allows for:

- **Opportunity cost evaluation** – Balancing extraction duration against diminishing returns.

- **Resolution variability** – Adapting scheduling models to vary by stope or geometallurgical domain based on required detail.

By incorporating rate-driven scheduling principles, SL optimisation can shift from traditional grade-based models to dynamic, kinetics-informed decision-making, ensuring schedules align with real recovery behaviour rather than static assumptions. The integration of capital expenditure constraints, dynamic operating costs, and time-dependent revenue considerations fundamentally alters how SL systems should be optimised.

Among the factors analysed, solution management capacity constraints, time-evolving operating costs, and diminishing-return revenue profiles emerge as the most critical to accurate SL optimisation. These elements directly dictate stope activation timing, leaching cycle duration, and sequencing dependencies, yet existing models fail to explicitly incorporate key capacity and spatiotemporal constraints required for an accurate representation of SL economics. Capacity constraints related to pumping capacity and distribution limits must be explicitly modelled, as they define the number of stopes that can be leached concurrently and impose fundamental scheduling restrictions. Spatiotemporal parameters such as reagent and lixiviant consumption profiles, solution loss rates, and make-up requirements must be integrated to account for evolving cost structures over the leaching cycle.

Additionally, dissolution rate profiles must be embedded to model time-dependent metal recovery trends, ensuring that economic thresholds for leaching duration are based on actual kinetic behaviour rather than static grade-based assumptions, to potentially offer insights around SL economic cut-off criteria. Incorporating these missing constraints into SL scheduling models is essential to shift to a dynamic, rate-driven optimisation approach that aligns decision-making with the real economic and operational behaviour of SL systems. This analysis demonstrates that treating SL optimisation as an extension of conventional stoping models is inadequate. An innovative approach must integrate kinetics-driven production behaviour, solution-handling limitations, and time-sensitive cost accumulation to ensure SL planning decisions align with real-world recovery behaviours.

In summary, the analysis demonstrates that SL optimisation cannot rely on the static, grade-based approaches typical in conventional stoping models. Instead, a truly dynamic, kinetics aware paradigm is necessary to capture critical parameters, such as solution-handling capacity, time-evolving operating costs, and kinetic-profiles that directly influence scheduling and profitability.

KINETICS-INFORMED STRATEGIC FRAMEWORK FOR STOPE LEACHING SYSTEMS

This section introduces a novel, kinetics-informed strategic planning framework specifically developed for SL operations. In this presentation, a streamlined version of the model is presented to illustrate the key novel logic and temporal dynamics that differentiate this approach from conventional methods. This simplified formulation highlights the core conceptual framework while maintaining analytical clarity. The complete model, which incorporates additional variables and constraints related to geotechnical considerations, development, explicit swell management, leaching management, capacity and resource constraints, adjacency relationships, and detailed operational parameters, will be presented in a forthcoming journal publication. This approach represents a significant advancement beyond traditional optimisation models by explicitly embedding the dynamic, continuous nature of leaching kinetics directly into strategic scheduling, rather than treating stope initiation as a simplistic, binary decision.

The developed framework integrates essential design elements, particularly kinetics-driven stope scheduling and development scheduling, within a unified optimisation approach. While the complete formulation addresses full production logic from development scheduling, this discussion is intentionally streamlined to emphasise its core production-flow logic. This focused presentation clearly show how the model handle core temporal input parameters, which differentiate this framework from traditional approaches, without sacrificing analytical clarity or computational feasibility.

Central to the model is the use of dynamic, stope-specific rate profiles, capturing metal extraction period recoveries, resource utilisation, reagent consumption, and solution-loss dynamics, as direct inputs into NPV optimisation. This dynamic integration allows for a more accurate, robust, and economically realistic evaluation of strategic decisions throughout the mine life cycle. The proposed model addresses four critical limitations in conventional approaches:

1. Static recovery assumptions that fail to capture leaching kinetics, eg diminishing returns, opportunity-cost, and trade-offs between initiating new stopes versus prolonging existing ones.

2. Strategic management of geometallurgical variables that are often non-additive and frequently exhibit non-linear behaviour (Morales *et al,* 2019), the complex nature of these variables often warrants non-linear optimisation approaches.

3. Compatibility with different leaching configurations.

4. Binary treatment of complex, continuous processes.

Sets and indices

$\mathcal{T}:\{t_1, t_2, t_3 \dots, T\}$ Set of time Periods: $t, \tau \in \mathcal{T}, \quad \tau \leq t$

$\mathcal{S}:\{s_1, s_2, s_3 \dots, S\}$ Set of stopes $s \in \mathcal{S}$

Parameters (production and closure)

r: Discounting rate

$\alpha_t: (1 + r)^{-t}$ Discounting factor for period t

p_{sk}: Net cash flow ($/period) from leaching of stope s at elapsed time $k = t - \tau$

v_{sk}: Pre-leaching mining cost ($/period) associated with conditioning stope s

w_{sk}: Swell cash flow ($/period) of stope s at elapsed time $k = t - \tau$

h_{sk}: Closure cost ($/period) for stope s at elapsed time $k = t - \tau$

Decision variables (production and closure)

$y_{s\tau} \in \{0, 1\}$ = 1 if stope s starts production in period τ

$\delta_{st\tau} \in \{0, 1\}$ = 1 if stope s is active in period t, given it started in period $\tau \leq t$

$z_{s\Phi} \in \{0, 1\}$ = 1 if stope s initiates closure in period Φ

$c_{st\Phi} \in \{0, 1\}$: Equals 1 if stope s is in the closure state in period t, having started closure in period $\Phi \leq t$

Objective function (production-focused)

The model seeks to maximise the NPV considering pre-leaching mining, leaching cash flow, and swell material handling, minus closure costs (excluding development terms). Formally:

$$
\max Z = -\underbrace{\sum_{s \in \mathcal{S}} \sum_{t \in \mathcal{T}} \sum_{\tau \leq t} \alpha_t\, v_{sk}\, \delta_{st\tau}}_{\text{Pre-leaching mining cost}}
$$

$$
+ \underbrace{\sum_{s \in \mathcal{S}} \sum_{t \in \mathcal{T}} \sum_{\tau \leq t} \alpha_t\, p_{sk}\, \delta_{st\tau}}_{\text{Leaching cash flow}}
$$

$$
+ \underbrace{\sum_{s \in \mathcal{S}} \sum_{t \in \mathcal{T}} \sum_{\tau \leq t} \alpha_t\, w_{t-\tau}\, \delta_{st\tau}}_{\text{Swell cash flow}}
$$

$$
- \underbrace{\sum_{s \in \mathcal{S}} \sum_{t \in \mathcal{T}} \sum_{\phi \leq t} \alpha_t\, h_{sk}\, c_{st\phi}}_{\text{Closure costs}}
$$

(1)

In this truncated version of the objective function, some of the more intricate components for swell management, pre-leaching mining, and development have been deliberately simplified or entirely omitted to better highlight the core temporal logic embedded within the production framework. This omission is purely for illustrative clarity.

A central feature of this framework is how time-dependent parameters (eg period recovery, acid consumption, and solution loss) are stored in pre-defined arrays and selectively 'activated' only when a stope is scheduled for production. In the objective function cash flow $p_{t-\tau}$ (conceptually) indexes the appropriate recovery at period t for a stope initiated in period τ, ensuring the model always references the correct point along the kinetic profile. This design enables precise alignment between actual leaching behaviour and scheduling decisions, a critical improvement over traditional models, which often rely on static or average parameters that fail to capture dynamic geometallurgical processes.

Additionally, because the non-linear kinetic behaviour is embedded in the input production profiles that essentially trace out the kinetics evolution of the response variables, this approach allows the model to retain its linearity while incorporating the complex underlying logic of leaching behaviour. Thus, the quantitative effect of design is captured in this explicit treatment of input elements, allowing the strategic implications of such design to be evaluated at a strategic level.

The key innovation is that the model accomplishes two critical functions simultaneously: explicitly integrating geometallurgical parameters into strategic planning while embedding non-linear kinetic behaviour into the input data structure, which allows the optimisation model to remain linear (and therefore computationally tractable) while still capturing the complex non-linear leaching dynamics. By integrating these time-indexed arrays, this approach captures essential operational variability, such as consumption variability, resource requirements or diminishing returns in metal recovery over time, within a single optimisation framework. Conventional scheduling methods typically overlook or oversimplify these dynamic aspects, thereby producing schedules that miss crucial nuances of SL. In contrast, the proposed model explicitly accommodates the evolving nature of leaching kinetics, leading to more robust, economically optimal decisions across the mine's life cycle.

This shift from a static to a kinetics-aware paradigm marks a significant departure from legacy scheduling tools, bridging an important gap in stope-level planning for *in situ* leaching operations. By tying operational triggers and rate-dependent behaviours directly into the strategic scheduling horizon, the model provides an analytically rigorous yet practical route to optimise underground SL systems under realistic, time-dependent constraints.

Production constraints

$$
\sum_{\tau \in \mathcal{T}} y_{s\tau} \leq 1 \quad \forall s \in \mathcal{S} \tag{2}
$$

Single initiation – Single Initiation implies that each stope can formally enter its active production phase only once within the strategic planning horizon. However, this constraint strictly governs the strategic production commencement and does not prescribe how the underlying leaching dynamics unfolds. From a modelling standpoint, the time-dependent input parameters (eg period recovery, solution losses, and lixiviant consumption) incorporate the entire evolution of each stope's metallurgical performance under various design configurations. Consequently, once a stope is designated as 'active,' the dynamic nature of leaching, be it cyclical or continuous, is captured through the effective rate profiles embedded in the model's data structures. This design choice ensures that both simple and complex operational leaching strategies are represented within a single, unified strategic scheduling framework. As a result, it remains fully compatible with a wide range of stope leaching configurations, whether they involve intermittent injection and rest cycles, or continuous percolation.

$$\delta_{st\tau} \leq y_{s\tau} \quad \forall(s,t,\tau) \ \tau \leq t \tag{3}$$

Linking active states – A stope is active in period t only if it was started in an earlier or equal period τ.

$$\sum_{\tau \leq t} \delta_{st\tau} \leq 1 \quad \forall s \in \mathcal{S},\ t \in \mathcal{T} \tag{4}$$

Production contiguity – A stope remains active across consecutive periods unless it is terminated, full recovery or early termination by model.

$$(\delta_{st\tau} = 0) \Rightarrow \delta_{s,t+1,\tau} = 0 \quad \forall s,\ t < T,\ \tau \leq t \tag{5}$$

Thus, for each stope, the model enforces a single contiguous production phase. This approach gives the model full autonomy over how far it traverses the kinetic/production profile for a given stope, which makes leaching duration an implicit variable of the model. This allows global NPV optimisation to take priority over local maximisation of individual stope recovery. Additionally, the full model manages precedence relationships between the pre-leaching mining, swell removal, and leaching activation phases, ensuring proper operational sequencing throughout the entire leaching life cycle.

Closure constraints

Unlike conventional mining, where stopes are fully emptied and backfilled, SL inherently leaves behind material, fundamentally altering the post-operation management and cost structure. This operational characteristic, retention of leaching residue within stopes post leaching, along with the potential for early termination and the need to account for closure costs (such as stope neutralisation and rehabilitation cost), necessitates a dedicated closure phase despite the elimination of traditional backfilling requirements. The proposed closure approach captures the post-leaching closure transition state and associated cost profiles that reflect the true economic and environmental obligations of the stopes, including any ongoing remediation liabilities.

$$\delta_{st\tau} - \delta_{st+1\tau} \leq z_{s,t+1} \quad \forall s \in \mathcal{S},\ t \in \mathcal{T} \tag{6}$$

Post-production closure initiation – Closure state transition is governed by the completion of a single contiguous production phase. When a stope that began production in a previous period τ experiences a decline in its production indicator between consecutive periods, this transition formally initiates closure commencement. In mathematical terms, when $\delta_{st\tau} = 1$ and $\delta_{st+1\tau} = 0$, the model enforces $z_{s,t+1} = 1$, ensuring immediate progression to closure activities once production ceases.

$$\sum_{\tau \in \mathcal{T}} z_{s\phi} \leq 1 \quad \forall s \in \mathcal{S} \tag{7}$$

Single closure – Only one closure initiation per stope.

$$c_{st\phi} \leq z_{s\phi} \quad \forall(s,t,\tau,\phi) \tag{8}$$

Valid closure state – A stope may be in closure at t only if closure has been initiated at ϕ.

$$z_{st} \leq \sum_{\tau \leq t} y_{s\tau} \quad \forall (s,t) \tag{9}$$

Closure validity – Closure activities cannot commence without prior production initiation.

$$y_{s,t} = 1 \Rightarrow \sum_{t \geq \Phi} c_{st\Phi} = |\mathcal{C}_s| \quad \forall s \in \mathcal{S}, \ \tau \in \mathcal{T} \tag{10}$$

$$c_{st+1\Phi} \geq c_{st\Phi} \quad \forall s \in \mathcal{S}, \ \tau \in \mathcal{T} \tag{11}$$

Closure continuity – Once closure initiates for a stope, it continues uninterrupted through consecutive periods, and stopes must complete their full closure phase within the planning horizon. The parameter $|\mathcal{C}_s|$ represents the predetermined duration of the closure process for stope s, ensuring that once closure begins, it proceeds for the required number of periods until completion, where $\mathcal{C}_s$ represent a set of closure costs associated with stope s. This approach makes it easier to attach penalty costs associated with environmental remediation linked to liquor loss during both operational and closure phases. In the complete model, operational liquor loss is expressed as a separate objective function term, while closure-related environmental costs are integrated directly into the closure cost structure.

Development, capacity, and geotechnical considerations – The model embeds spatial and operational realism through a hierarchical development framework that governs progressive stope availability.

Where:

$\mathcal{A}^v = \{a_1^v, a_2^v, \dots\}$ Set of shared (interlevel) development activities eg decline

$\mathcal{A}_{\ell_i}^h = \{a_{\ell,1}^h, a_{\ell,2}^h, \dots\}$ Set of intra-level (level-specific) development activities for level ℓ_i eg level access, cross-cuts, liquor ponds of all stopes contained in $\mathcal{S}_\ell$

$Q^v \subset \mathcal{A}^v \times \mathcal{A}^v$, Pairs (a_i^v, a_j^v) Indicating interlevel precedence

$Q_\ell^h \subset \mathcal{A}_\ell^h \times \mathcal{A}_\ell^h$, Pairs $(a_{\ell,i}^h, a_{\ell,j}^h)$ indicating intra-level precedence on level ℓ

$\mathcal{R}_\ell^v \subseteq \mathcal{A}^v$ Interlevel development activities that give access to level ℓ_i

$\mathcal{R}_s^h \subseteq \mathcal{A}_\ell^h$, Intra-level development activities required by stope s

For each level $\ell \in \mathcal{L}$ let $R_\ell^v \subseteq A^v$ denote the set of interlevel developments required to open level ℓ_i. Shared interlevel activities $A^v = \{a_1^v, a_2^v, \dots\}$ (eg declines) and level-specific intra-level activities $A_l^h = \{a_{l,1}^h, a_{l,2}^h, \dots\}$ (eg cross-cuts, ponds) govern progressive access to stopes $s \in S_\ell$ on level $\ell \in L$. Precedence networks are defined by $Q^v \subset A^v \times A^v$ and $Q_\ell^h \subset A_\ell^h \times A_\ell^h$. Required developments to open stope s for production are $R_\ell^v \subseteq A^v$ (interlevel) and $R_s^h \subseteq A_\ell^h$ (intra-level).

Stope-access constraints – Binary variables $d_{a,t}^v$ and $d_{l,a,t}^h$ equal 1 once development activity a is complete by period t. A stope enters production when $y_{s,t} = 1$ and only if:

$$y_{s,t} \leq d_{a,t}^v \forall a \in R_{\ell(s)}^v, \forall t \tag{12}$$

$$y_{s,t} \leq d_{\ell(s),a,t}^h \forall a \in R_s^h, \forall t \tag{13}$$

Thus, the hierarchy $(R_\ell^v \rightarrow R_s^h) \rightarrow y_{s,t}$ ensures progressive stope availability and enforces spatial proximity because prerequisite sets are adjacency-based. Akin to shared interlevel development, the formulation also considers shared intra-level infrastructure that serve multiple stopes.

Optional partial development – Continuous scheduling variables $x_{a,t}^v$ and $x_{\ell,a,t}^h$ ($0 \leq x \leq 1$) permit partial completion when marginal cost exceeds benefit:

$$\sum_{t \in T} x_{a,t}^v \leq 1, \sum_{t \in T} x_{\ell,a,t}^h \leq 1 \tag{14}$$

Production-capacity constraint – Let ϱ_s denote the capacity load (eg required flow) of stope s when active. Installed capacity N_{cap} bounds maximum number of concurrently active stopes:

$$\sum_{s \in S} \delta_{s t \tau} \, \varrho_s \leq N_{\mathrm{cap}} \; \forall t, \tau \in T \tag{15}$$

Detailed adjacency matrices and mutually exclusive geotechnical rules are included in the upcoming journal-length model and are omitted here for brevity.

Discussion and benefits

This production-only formulation incorporates geometallurgical data to capture critical stope-leaching dynamics:

- A single contiguous production phase per stope (with the option for early termination if it becomes suboptimal for global optimisation as determined by the model).

- Time-varying revenue/cost profiles reflecting kinetic behaviour.

- Enforced closure once leaching ends, with closure completed with the planning horizon.

By accounting for diminishing returns later leaching stages (via p_{sk}) and possible solution-loss/reagent-cost penalties, the model identifies economically optimal stope activation and termination. This strategic-level approach maximises overall project NPV while maintaining computational feasibility, well-suited for complex, long-term mine planning scenarios.

CONCLUSIONS

Stope leaching schedules must move beyond static, grade-based planning. By modelling solution-handling limits, time-varying operating costs and kinetics-driven revenues, the proposed formulation captures economic and operational complexity overlooked by conventional methods. It recognises diminishing marginal returns, spatial and temporal cost build-up, and the trade-off between opening new stopes and extending active ones, generating value-oriented plans that remain flexible and consistent with modern sustainability expectations. Priority research directions include:

- **Kinetics-aware stope design and integrated optimisation** – Linking stope geometry with leaching kinetics may establish empirical guidelines for leaching-optimised designs while coupling leaching optimisation with stope design, ventilation and scheduling could advance holistic mine planning.

- **Stochastic optimisation and uncertainty integration** – The proposed formulation offers a solid foundation for stochastic extensions. Key inputs such as solution loss, acid consumption, period recovery, and grade exhibit spatial and temporal variability. Incorporating uncertainty through scenario-based or probabilistic optimisation approaches will improve robustness of planning decisions.

- **Computational and predictive modelling** – Coupling computational fluid dynamics with machine-learning tools such as recurrent neural networks to forecast key stope-scale parameters may enhance both the accuracy and efficiency of strategic optimisation.

- **Tailored fragmentation strategies** – Customised blasting strategies targeting size distributions specifically optimised for leaching kinetics, supported by remote technologies and multi-scale validation, could improve metal liberation and overall kinetics.

REFERENCES

Ahlness, J K and Pojar, M G, 1983. In situ copper leaching in the United States: Case histories of operations, IC 8961. Available from: <https://www.onemine.org/documents/ic-8961-in-situ-copper-leaching-in-the-united-states-case-histories-of-operations>

Bahamóndez, C, Castro, R, Vargas, T and Arancibia, E, 2016. In situ mining through leaching: Experimental methodology for evaluating its implementation and economic considerations, *Journal of the Southern African Institute of Mining and Metallurgy,* 116:689–698.

Boreck, D L, Lutzens, W W and Speirer, R, 1990. Ore leaching in underground stopes, *Mineral Resources Engineering,* 3:31–46.

Boreck, D, Djahanguiri, F, Miller, N, Snodgrass, J and Speirer, R, 1991. Rock mass characterization for designing underground leaching stopes, 11 (Society of Mining Engineers of AIME, Littleton).

Brock, J G, Chomley, J C and Richmond, G D, 1998. Copper leaching at gunpowder, *AusIMM'98 – The Mining Cycle* (The Australasian Institute of Mining and Metallurgy: Melbourne).

Burton, C, Cowman, S, Heffernan, J and Thorne, B, 1983. In-situ bioleaching of sulphide ores at avoca, ireland, Part i – development, characterization and operation of a medium-scale (6000 t) experimental leach site, in *Recent Progress in Biohydrometallurgy* (eds: G Rossi and A E Torma), Associazione Mineraria Sarda, Iglesias, Italy.

Butler, J E, Ackland, M C and Robinson, P C, 1981. Development of in-situ leaching by gunpowder copper limited, Queensland, Australia, SME-AIME Fall Meeting and Exhibit Preprint (Society of Mining Engineers of the American Institute of Mining, Metallurgical and Petroleum Engineers).

Dare-Bryan, P and Hassanvand, A, 2023. Economic and environmental assessment of underground in-situ leaching processes utilising drill and blast to achieve high permeability, *ALTA 2023 – In-Situ Recovery,* pp 106–120.

Estay, H, Díaz-Quezada, S, Arancibia, E and Vargas, T, 2023. Economic assessment of an in situ leaching operation with ore preconditioning using sublevel stoping techniques, *Mining, Metallurgy and Exploration,* 40:493–504.

Farrell, S and Whitton, L, 2024. BHP insights: How copper will shape our future, BHP Insights: how copper will shape our future [online]. Available from: <https://www.bhp.com/news/bhp-insights/2024/09/how-copper-will-shape-our-future>

Gardner, E D, Johnson, C H and Butler, B S, 1938. Copper mining in North America. Available from: <https://www.onemine.org/documents/copper-mining-in-north-america-introduction>

Ghorbani, Y, Nwaila, G T, Zhang, S E, Bourdeau, J E, Cánovas, M, Arzua, J and Nikadat, N, 2023. Moving towards deep underground mineral resources: Drivers, challenges and potential solutions, *Resources Policy,* 80.

Ito, I, 1976. Present status of practice and research works on in-place leaching in Japan, *World Mining and Metals Technology,* 1.

Lombardi, J A and Jude, C V, 1994. Mechanical excavation systems (in three parts), part 2, Stope leach and stope autoclave mineral recovery systems, IC 9420. Available from: <https://onemine.org/documents/ic-9420-mechanical-excavation-systems-in-three-parts-2-stope-leach-and-stope-autoclave-mineral-recovery-systems>

Maybee, B, Lilford, E and Hitch, M, 2023. Environmental, social and governance (ESG) risk, uncertainty and the mining life cycle, *The Extractive Industries and Society,* 14.

McCready, R G L and Sanmugasunderam, V, 1985. The Noranda contract reports on the prefeasibility study of in-place bacterial leaching: A summation. Available from: <https://publications.gc.ca/collections/collection_2019/rncan-nrcan/m38-13/M38-13-85-5-eng.pdf>

Miller, N C and Schmuck, C H, 1995. Use of a tracer for in situ stope leaching solution containment research, RI 9583. Available from: <https://stacks.cdc.gov/view/cdc/10273>

Miller, P C, 1986. Large-scale bacterial leaching of a copper-zinc ore in situ, in Fundamental and applied biohydrometallurgy: Proceedings of the Sixth International Symposium on Biohydrometallurgy (eds: R W Lawrence, R M R Branion and H G Ebner), (Elsevier: Netherlands).

Morales, N, Seguel, S, Cáceres, A, Jélvez, E and Alarcón, M, 2019. Incorporation of geometallurgical attributes and geological uncertainty into long-term open-pit mine planning, *Minerals,* 9.

Mousavi, A and Sellers, E, 2019. Optimisation of production planning for an innovative hybrid underground mining method, *Resources Policy,* 62:184–192.

Potts, T S and Webb, W K, 1994. Mining and in-place bioleaching at gunpowder's mammoth mine, in *Biomine '94. Applications of Biotechnology to the Minerals Industry, International Conference and Workshop* (ed: A M Foundation), Chapter 3.

Rossi, G, Trois, P and Visca, P, 1986. In situ pilot, semi-commercial, bioleaching test at the San Valentino di Predoi Mine (Northern Italy), in Fundamental and applied biohydrometallurgy: Proceedings of the Sixth International Symposium on Biohydrometallurgy (eds: R W Lawrence, R M R Branion and H G Ebner), (Elsevier: Netherlands).

Rossien, M, 2020. Economic modelling and application of in-situ recovery in hard rock mining, in ALTA 2020 – In-Situ Recovery Online (ALTA Metallurgical Services: Melbourne).

Sand, W, Hallmann, R, Rohde, K, Sobotke, B and Wentzien, S, 1993. Controlled microbiological in-situ stope leaching of a sulphidic ore, *Applied Microbiology and Biotechnology,* 40:421–426.

US Department of the Interior, Bureau of Mines, 1994. Stope leaching reduces surface environmental impacts from underground mining. Available from: <https://www.onemine.org/documents/technology-news-no-436-stope-leaching-reduces-surface-environmental-impacts-from-underground-mining>

Van Staden, P J and Laxen, P A, 1989. In-stope leaching with thiourea, *Journal of the Southern African Institute of Mining and Metallurgy,* 89:221–229.

Wang, J and Dai Yuan, N, 1993. In-situ leaching of uranium in China, Technical committee meeting on uranium in situ leaching, pp 129–132. Available from: <https://inis.iaea.org/records/awnjf-r4x20/files/25000366.pdf?download=1>

Wang, S-F, Sun, L-C, Huang, L-Q, Li, X-B, Shi, Y, Yao, J-R and Du, S-L, 2019. Non-explosive mining and waste utilization for achieving green mining in underground hard rock mine in China, *Transactions of Nonferrous Metals Society of China,* 29:1914–1928.

A systematic review of cut-off grade optimisation approaches – application benefits and challenges

J Githiria[1], C Musingwini[2] and B Mutandwa[3]

1. Senior Lecturer, School of Mining Engineering, University of the Witwatersrand, Johannesburg 2050, South Africa. Email: joseph.githiria@wits.ac.za
2. Professor, School of Mining Engineering, University of the Witwatersrand, Johannesburg 2050, South Africa. Email: cuthbert.musingwini@wits.ac.za
3. Lecturer, School of Mining Engineering, University of the Witwatersrand, Johannesburg 2050, South Africa. Email: bright.mutandwa@wits.ac.za

ABSTRACT

Cut-off grade (COG) is a criterion used to distinguish ore from waste material in a mining operation. It directly affects cash flows generated by the operation and consequently, the economic value of the operation which is generally measured using the net present value. Lane's COG approach developed in the 1960s is considered pioneering work for determining optimal COGs and has been applied widely in the mining industry. However, Lane's approach does not account for factors such as commodity price volatility, variable operating costs associated with mining operations and grade uncertainty that are encountered in actual mining practice. This is why research on COG optimisation continues to try and develop approaches that can holistically incorporate all factors affecting COG for realistic application in mining operations. This paper traces modifications to COG optimisation theory since the 1960s obtained from studies that have applied either exact or metaheuristic approaches along the mine value chain and reveals that there has been growth in the adoption of metaheuristic approaches in COG optimisation. This paper also illustrates the benefits and limitations of using these approaches in determining optimal COGs. Relevant articles related to COG optimisation were extracted from the Scopus and Web of Science as primary databases since these databases provide extensive and authoritative coverage of academic research. The articles from both databases were first merged into an Excel data file and duplicates removed before undertaking bibliometric analysis using R-Biblioshiny™ software and then using VOSViewer™ software to visualise the results. The results suggest that the future research directions on COG optimisation should focus on metaheuristic approaches to foster improved adoption by the mining industry.

INTRODUCTION

Mining involves the economic extraction of valuable minerals from the earth's crust using open pit and/or underground mining methods. Economic, geological and operational parameters that govern mining operations include factors such as commodity price, operational cost, and cut-off grade (COG). COG is one of the most important parameters considered when planning and designing mining projects and in mine production scheduling. It is dependent on the reliable estimation of ore grade and commodity price to create a complex relationship as explained in several COG optimisation approaches discussed later in this paper. COG optimisation has widely relied on the work of Lane (1964, 1988).

COG distinguishes ore from waste material for a given orebody, hence affects the economic feasibility of mining projects. The ore mined from open pit and/or underground mining undergoes beneficiation in a series of activities including processing and refining to produce a saleable product. An objective of mining companies is to ensure consistent production of mineral(s) to generate maximum net present value (NPV) throughout the life-of-mine (LOM). The NPV is dependent on interrelated parameters that interact in a complex manner in defining project value. These parameters include COG, commodity price, operational costs, recovery, production capacities, and extraction sequence. Most research studies on optimising COG are deterministic, contrary to what is encountered in actual mining operations, wherein operating parameters change over time in response to factors influencing the global mining industry, thus, requiring the COG to be revised. Generally, the COG is initially determined at the development phase of a mining project and reviewed on a yearly basis to reflect changing economic conditions and variability in geological and operational parameters. However, most of these parameters can change on a daily, weekly or monthly basis, or when new information becomes available, thus, resulting in sub-optimal project value, unless the COG is reviewed at such shorter intervals (Githiria *et al,* 2024). Therefore, the COG policy should be flexible to align with market conditions throughout the LOM. Birch (2018) reviewed 39 mines from different countries, which mined different minerals and illustrated how the mines use different COG approaches to classify their resource as ore or waste. This classification impacts declared Mineral Resources and Mineral Reserves as defined by SAMREC (2016). Arguably, Mineral Resources and Mineral Reserves are among the most significant assets, if not the most important asset for any company operating in the minerals industry because they are the foundation that determines the future economic viability of the company (IASB, 2010; Njowa and Musingwini, 2018; Deloitte, 2023).

The preceding perspectives underpin the importance of COG optimisation and its relationship with Mineral Reserves and mineral project value. Therefore, this paper seeks to highlight the various approaches developed for COG optimisation and their modifications and limitations. It also describes the two common approaches to COG optimisation which are exact and metaheuristic approaches, indicating the applicability and benefits of the different approaches. Exact methods generate optimal solutions but can be limited by computational complexity, especially in real-life large-scale mining projects. Metaheuristic approaches provide near-optimal solutions that address complexities and uncertainties of real-world mining projects. Lastly, through a systematic review and bibliometric analysis, this paper identifies future research directions which show that metaheuristic approaches are more applicable to real-life large-scale mining projects.

SYSTEMATIC REVIEW OF COG OPTIMISATION IN MINING OPERATIONS

A systematic review is a methodology for summarising evidence based on research questions (Tawfik *et al,* 2019). The major steps of a systematic review involve framing research questions, developing selection criteria, searching for articles using existing search engines and databases, appraising the quality of articles by using the selection criteria, interpreting and summarising results (Khan *et al,* 2003; Rys *et al,* 2009; Bello *et al,* 2015; Tawfik *et al,* 2019).

Research problems and solutions in mining are heterogenous and complex, hence the need for systematic reviews of literature to identify research trends, gaps, and recommend future research directions. This includes undertaking systematic reviews of COG optimisation research, given its importance in the mining industry, to enhance understanding of existing research and map future research directions.

Since Lane's pioneering work in the 1960s, only a few literature reviews have been undertaken to identify trends in COG optimisation. For example, Asad, Qureshi and Jang (2016) reviewed COG optimisation models and classified them into four categories namely, break-even, Lane's, stochastic and mathematical models. Biswas, Sinha and Sen (2023b) undertook a review that incorporates enhancements to Lane's COG approach and the application of evolutionary COG algorithms. In this paper a systematic review on COG optimisation was undertaken to highlight advancements and challenges faced in COG optimisation, including benefits to the mining industry from adopting and applying metaheuristic approaches.

COG selection approaches and their impact on the mine value chain

Optimising COG is a critical aspect of mine planning which is used in selecting COGs over the LOM to define a COG policy (Githiria and Musingwini, 2019). A COG policy that can generate higher NPVs tends to use declining COGs over the LOM (Githiria and Musingwini, 2019). Since COG distinguishes ore from waste material, it directly affects the cash flows generated from a mining operation and consequently, the mining project's NPV (Bascetin and Nieto, 2007). Therefore, it is important to optimise COG for any mining operation. Several approaches have been developed for selecting COGs to maximise value along the mine value chain. These approaches can be broadly classified as exact or metaheuristics approaches. In real-life large-scale mining problems COG formulations are generally considered to be non-deterministic polynomial time hard (NP-Hard) for solution by exact methods due to reasons such as:

- Complexity of the problem: COG optimisation involves complex, non-linear, and multi-objective decision-making that must consider parameters affecting mining, processing, market constraints, and geological uncertainties. This complexity makes the problem difficult to solve using exact optimisation techniques.

- Multiple processing streams and polymetallic mines: The COG optimisation problem is more complex to solve for multiple processing streams, stockpiles, or multi-element commodities. For example, the contribution of each metal to the material mined and processed must be considered, thus, increasing the complexity of the problem.

- Predetermined extraction sequence: Lane's algorithm only optimises COG for a predetermined extraction sequence. If the extraction sequence is not known in advance or needs to be optimised simultaneously, the problem becomes NP-Hard.

- Non-linearity and non-convexity: The objective function and constraints involved in COG optimisation are often non-linear and non-convex, which makes the problem computationally challenging, thus, becoming NP-Hard.

- Multiple objectives: Optimising COG while simultaneously considering multiple objectives, such as maximising the NPV, minimising environmental impact, and balancing operational constraints, further increases the complexity of the problem and makes it NP-Hard.

NP-Hard problems are problems where a potential solution can be verified in polynomial time, but finding the solution itself might not be possible in polynomial time. Exact methods cannot solve NP-Hard problems; therefore, resort is then made to use metaheuristic approaches which generate near-optimal solutions for real-life large-scale mining projects.

The development of COG models can be traced back to the 1960s in the published works of Henning (1963), Johnson (1969) and Lane (1964, 1988). Henning (1963) created a break-even COG framework, where the COG is higher during earlier years and lower during later years, reducing to the break-even value that corresponds to the objective of maximising the difference between revenue and cost. The framework confirms that the COG cannot remain constant if the objective is to maximise the NPV. Johnson (1969) applied the Dantzig-Wolfe decomposition method using mixed-integer programming (MIP) to generate dynamic COGs for large mineral deposits, but its solution is not optimal since it has too many constraints such that the blocks in the block model are not completely mined.

Lane (1964, 1988) developed a COG deterministic approach that accounts for economic and geological parameters and production capacities in the calculation of COGs. It uses mathematical derivations to get six COGs, which are then sorted using a sorting algorithm to get an optimum COG. It assumes that a mining operation has three consecutive operational stages namely, mining, processing and refining stages. Three of the six COGs are classified as limiting economic COGs (g_m, g_c, g_r), which refer to mining, processing and refining grades, respectively and are based upon operational costs, commodity price and mining, processing and refining capacities that independently constrain the production throughput. The other three which are determined by assuming that two of the three operational stages are operating in unison at their capacity limits are called balancing COGs (g_{mc}, g_{rc}, g_{mr}) and refer to mining in unison with processing, refining in unison with processing and mining in unison with refining, respectively. The framework uses a grade-tonnage curve and capacities of the production stages and assumes that the production rate of a mining system is determined by production capacities at the three different operational stages.

The Hill of Value (HoV) approach developed by Hall (2014) has been applied in mining economics and strategy optimisation when determining optimal cut-off grades in mining operations. The HoV technique illustrates the relationship between COG selection and project NPV through a hill-shaped surface. It identifies the optimal COG that maximises project's NPV which varies over time as economic conditions change. It also accounts for the opportunity cost of processing lower-grade material that might delay access to higher-grade material.

Mathematical framework of break-even and Lane's COG approaches

The break-even COG is calculated based on economic parameters only, namely, costs and commodity prices. However, the calculation ignores geological variability (grade-tonnage curve), production (mining, processing and refining) capacity constraints and timing of cash flows that must be discounted. Consequently, the COGs calculated using this model are constant over the LOM. Due to its simplicity, several mining operations use this approach to calculate their COG but create sub-optimal mine plans which ignore the reality of commodity price volatility. A challenge encountered when using this formula is deciding on the costs that should be included in the break-even calculations because it is always difficult to justify why specific costs used in the calculations were chosen (Githiria, 2018).

The following steps indicate the procedure involved in a break-even COG model used in a metalliferous mine (Hall, 2014):

1. Formulate a grade-tonnage distribution curve for the whole deposit to be mined.

2. Input the parameters to be used in the COG policy such as the mining capacity (M), processing or milling capacity (C), refining capacity (R), commodity price (s), mining cost (m), processing or milling cost (c), refining cost (r), recovery (y), operational fixed costs per period (f) and the discount rate (d).

3. Determine COG using the break-even COG formula, which determines the grade at which revenue obtained is equal to the cost of producing that revenue (Hall, 2014; Githiria, 2018):

$$\text{Break} - \text{even cut-off grade (g/t)} = \frac{\text{cost (\$/t)}}{\text{commodity price (\$/g)} \times \text{recovery}} \quad (1)$$

4. Determine quantity of ore (q_o) and quantity of waste (q_w), stripping ratio (SR), and average grade ($\bar{g}$) of ore.

5. Determine quantities of material mined (Q_m), ore processed (Q_c), and concentrate refined (Q_r) as follows:

$$Q_m = Q_c\left[1 + \frac{q_w}{q_o}\right] = Q_c[1 + SR] \quad (2)$$

$$Q_c = \frac{Q_m}{(1+SR)} \quad (3)$$

$$Q_r = (Q_c)(\bar{g})(y) \quad (4)$$

6. Determine the cash flow or profit (P_i) for year i, where i is a specific period in a LOM (N):

$$P_i = (s - r)Q_{ri} - cQ_{ci} - mQ_{mi} - f_i \quad (5)$$

Goycoolea *et al* (2020) described how Lane's algorithm optimises production schedules in open pit mines, highlighting its historical context and formal analysis and demonstrated the algorithm's effectiveness in practical mining applications. The mathematical formulation outlined by Lane (1964, 1988) is summarised below, indicating the steps used to determine the COG policy for a mineral deposit. The first two steps resemble the break-even COG approach where they both use the geological and economical parameters to define the mining operation, followed by additional steps outlined as follows:

1. Determine the optimum COG generated in Year i using Equations 6–11; if the initial cash flow V_i is not known set V_i to zero. The optimum COG is determined using a sorting algorithm that sorts COGs generated by Equations 6–11.

$$g_m = \frac{c}{(s-r)*y} \quad (6)$$

$$g_c = \frac{c + \frac{f+d*V}{C}}{(s-r)*y} \quad (7)$$

$$g_r = \frac{c}{\left(s - r - \frac{f+d*V}{R}\right)*y} \quad (8)$$

$$g_{mc} = \frac{\frac{C}{M} - mc(g)}{\left(\frac{mc(g') - mc(g)}{g' - g}\right)} + g \quad (9)$$

$$g_{mr} = \frac{\frac{C}{M} - mr(g)}{\left(\frac{mr(g') - mr(g)}{g' - g}\right)} + g \quad (10)$$

$$g_{cr} = \frac{\frac{C}{M} - cr(g)}{\left(\frac{cr(g') - cr(g)}{g' - g}\right)} + g \quad (11)$$

2. Determine the tonnes of ore and waste, and average grades of the ore associated with the optimum COG. Then, using Equations 2–4 calculate the quantity of material to be mined (Q_m), ore processed or milled (Q_c) and concentrate refined (Q_r). Find the limiting capacity and the remaining LOM from the remaining amount of material to be mined.

3. Determine the yearly profit (P_i) using Equation 5.

4. Compute the NPV using Equation 12 to discount the profits at a given discount rate (d) over the LOM (N).

$$NPV = \sum_{i=1}^{N} \frac{P_i}{(1+d)^i} \quad (12)$$

The value of NPV generated above becomes the second approximation of V (the first was V=0 in Step 1). This is used in the formula used to calculate the optimum COG.

5. Repeat the computation from Step 4 until the value, V, converges.

6. Adjust the grade-tonnage distribution by subtracting the ore tonnes from the grade-tonnage distribution intervals above optimum COG and the waste tonnes from the intervals below the optimum COG in proportionate amounts such that the distribution is not changed.

7. If it is the first iteration and knowing the profits obtained in each year, find the NPV year−by−year by discounting annual profits and go to Step 3. If it is the second iteration, then stop.

8. Use the NPV obtained in Step 5 as the initial NPV for each of the corresponding year for the second iteration.

Lane's COG theory accounts for time value of money by discounting cash flows, making the timing of extraction decisions critical. Hence, the opportunity cost brought about by this approach indicates the potential value that must be foregone when choosing one COG strategy over another. The limiting economic COGs in Lane's theory account for the discount rate assumed (see Equations 6 and 7). These three COGs are then reviewed together with the other balancing COGs to generate the optimal COG.

Modifications to Lane's COG approach

As discussed by Birch (2018), Lane's COG approach has been applied in several mining operations. The adoption of Lane's COG approach by the mining industry is partly due to its capability to generate higher COGs during the first years of operation, hence a higher NPV is obtained, enabling fast recovery of invested capital. However, it can be modified to improve on its wider applicability to account for other mine-specific factors, such as stockpiling and environmental requirements. For example, as Mineral Reserves deplete, the COG declines hence allowing the mine to stockpile potential ore for later processing and this can potentially increase the LOM. Modifications can also be made to overcome some of its deterministic shortcomings such as its assumption that commodity prices and operational costs are constant, yet mining operations face commodity price volatility and operational cost variability. Examples of such modifications and/or adaptations include works reported by Taylor (1972, 1985), Dagdelen (1992, 1993), Whittle and Wharton (1995), King (1999, 2001, 2011), Asad (2002, 2005), Minnitt (2004), Dagdelen and Kawahata (2008), Osanloo, Rashidinejad and Rezai (2008), Gholamnejad (2008, 2009), Githiria (2016), Githiria, Muriuki and Musingwini (2016), King and Newman (2018) and Githiria et al (2024). Other studies that have built on Lane's COG approach include works by Ahmadi and Shahabi (2018), Githiria and Musingwini (2019), Ahmadi and Bazzazi (2019, 2020), Paithankar *et al* (2020), Biswas *et al* (2020), Khan and Asad (2021) and Biswas, Sinha and Sen (2023a). These preceding works demonstrate that Lane's COG approach has been instrumental in laying foundational principles for COG optimisation.

Other COG optimisation approaches

Some COG optimisation studies that have not utilised Lane's COG approach, include Dowd (1976), Krautkraemer (1988), Asad (2007), Cetin and Dowd (2011), Li, Yang and Lu (2012), Thompson and Barr (2014), Moosavi and Gholamnejad (2016), Birch (2016a; 2016b, 2017) and Khan and Asad (2020, 2021). The different approaches used in these studies optimise COGs by incorporating features to account for factors such as:

- Royalty rates, commodity price volatility and cost escalation to reflect real market and operational conditions.

- The reality that mines are generally multi-mineral (or poly-mineral or polymetallic) mines which require COGs to be determined for each mineral whether produced as the main product or by-product.

- The use of equally probable realisations of the ore grade distributions to account for grade uncertainty.

- Simultaneously optimising COG and extraction sequence since COGs are intricately linked to production schedules.

- Multiple processing streams because several destinations exist for mined material (eg stockpile, waste dump, dump leach, heap leach or mill).

The approaches used included formulations based on techniques such as dynamic programming (DP), stochastic programming, MIP and mixed-integer linear programming (MILP). Some of the key findings from these studies included:

- Stochastic price in COG selection depends on the rate of metal price change relative to the discount rate.

- It is important to simultaneously optimise COG and extraction sequence as this results in improved project values.

- Each mining operation reacts differently to adjustments in COGs, thus, underscoring the need for tailored COG approaches to accommodate uncertainties.

- Higher discount rates increase COGs and reduce LOMs, hence optimising solely for maximum NPV decreases overall ore extraction. Therefore, a balanced approach that considers both NPV and longer LOMs is critical for improved resource recovery without risking sterilising mineral resources through high grading.

The preceding examples showed that there are other approaches that can be developed for optimising COG instead of solely relying on Lane's COG approach. This is why efforts have also been directed at exploring the use of heuristic and metaheuristic approaches to COG optimisation.

Heuristic and metaheuristic approach application in mining

Heuristic and metaheuristic methods have sometimes been used as solution methods in studies on COG optimisation. Figure 1 shows a high-level, non-exhaustive classification of commonly used optimisation methods into exact and metaheuristic approaches.

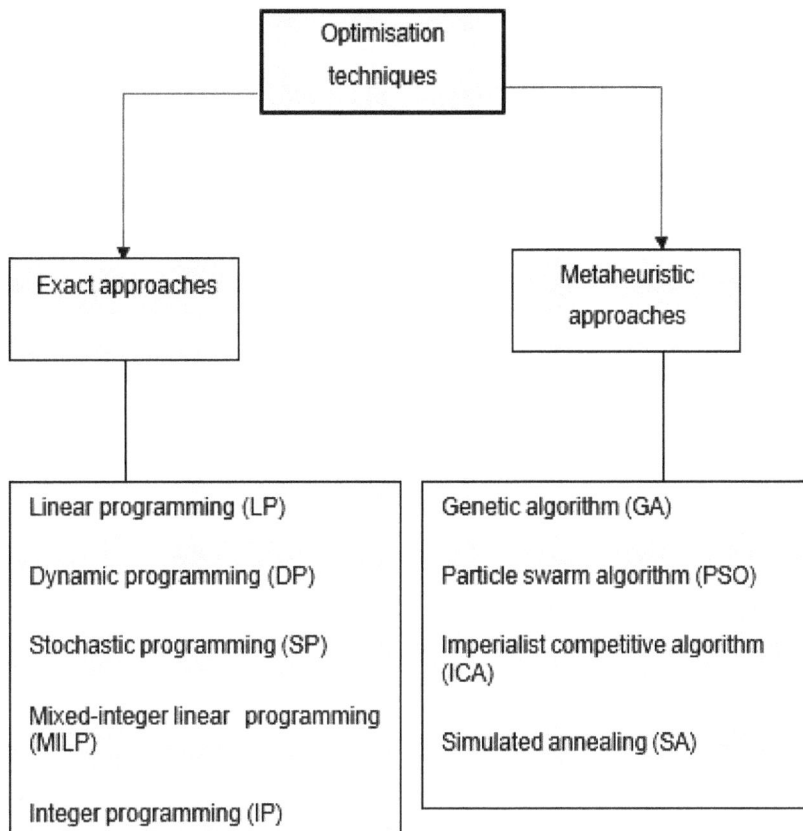

FIG 1 – A non-exhaustive high-level classification of exact and metaheuristic optimisation approaches.

A heuristic approach is a problem-solving method that is structured to solve a complex problem much faster than exact methods. Heuristic techniques rely on cognitive reasoning anchored on induction or analogy as a 'shortcut' way to solving complex problems. Unlike exact methods that guarantee optimal solutions, heuristic methods do not. Metaheuristic approaches are problem-solving techniques that use heuristics to explore the solution space of a complex problem to find an optimal or near-optimal solution and are often inspired by natural processes like human intelligence, evolution, swarm behaviour and genetics (Attea *et al*, 2021). Metaheuristic approaches can solve complex optimisation problems characterised by uncertainty, multiple objectives and non-linear formulations (Tomar, Bansal and Singh, 2023). As such, metaheuristic approaches are useful for solving real-world optimisation problems within reasonable time frames, thus, making them valuable in practical mining applications such as in mine planning and production scheduling.

Dynamic COG optimisation is critical in maximising the economic viability of ore extraction because static COGs can lead to premature mine closures, financial losses, and environmental impacts (Bascetin and Nieto, 2007). COG optimisation has mostly relied on exact approaches but with advancements in computational techniques, metaheuristic approaches have emerged as potential tools for this process. However, despite theoretical advantages of metaheuristic approaches, there is still limited empirical evidence on the extent of their adoption in mining operations.

Studies which have employed metaheuristic algorithms and hybrid approaches have demonstrated the potential of metaheuristic techniques to efficiently find near-optimal dynamic COGs for complex mining scenarios. Exact methods have limitations in handling complexity, non-linearity and multiple objectives inherent in these dynamic COG optimisation problems, hence the increasing use of metaheuristic approaches. Figure 2 shows how the use of metaheuristic approaches to optimise problems has been increasing from 2000 to 2022.

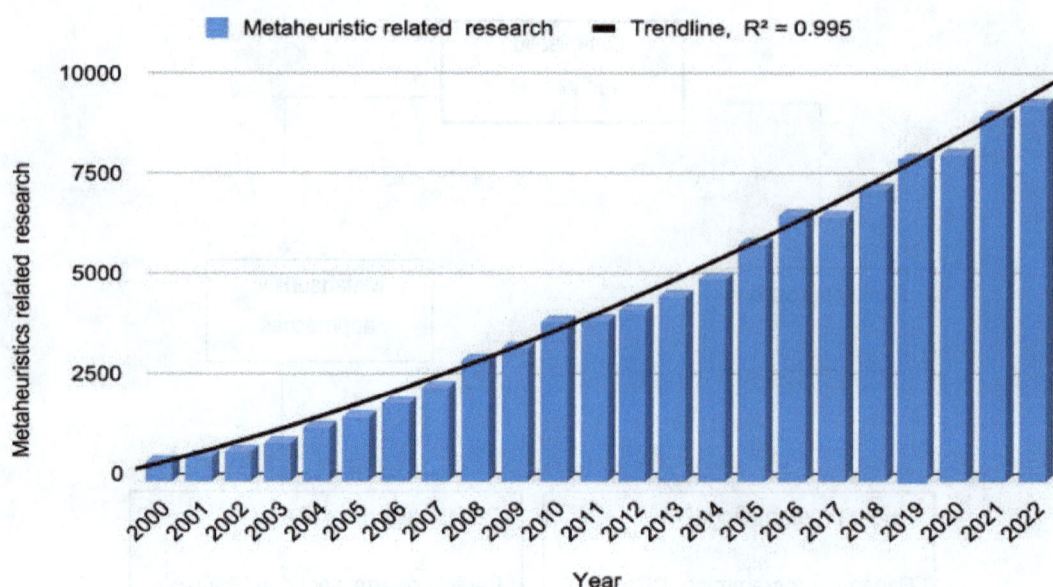

FIG 2 – Metaheuristic related research from 2000 to 2022 (Rajwar, Deep and Das, 2023).

Use of heuristic and metaheuristic approaches in COG optimisation

Despite its wide acceptance in the mining industry, Lane's COG optimisation approach cannot adequately address the complexities and uncertainties in real-world mining scenarios. This has led to other approaches being explored, which include heuristic and metaheuristic approaches as presented in several studies. A non-exhaustive list of such studies includes Cetin and Dowd (2002; 2013; 2016), Ataei and Osanloo (2004), Asad and Dimitrakopoulos (2013a, 2013b), Myburgh, Deb and Craig (2014), Rahimi and Ghasemzadeh (2015), Mohammadi *et al* (2017), Ahmadi and Shahabi (2018), Ahmadi and Bazzazi (2019, 2020), Biswas *et al* (2020), Paithankar *et al* (2020), Paithankar, Chatterjee and Goodfellow (2021), Sotoudeh *et al* (2021), and Biswas, Sinha and Sen (2023a), have used heuristic and/or metaheuristic approaches to optimise COG. Heuristic and/or metaheuristic approaches have been applied in COG optimisation because they tend to generate COGs that are higher than break-even COGs during the early years of a mining operation for faster recovery of invested capital. Then they lower the COGs to break-even COGs during the later years of the mine to maximise extraction of the mineral resource.

Cetin and Dowd (2002) applied a genetic algorithm (GA) to optimise COG for multi-metal mines. Later, Cetin and Dowd (2013) applied a grid search method that could optimise COG for a multi-mineral operation with up to three economic minerals. Cetin and Dowd (2016) used a GA, the grid search method, and DP to optimise production schedules for deposits with three constituent minerals. The study illustrated the significance of applying hybrid stochastic approaches when considering variables in COG optimisation.

Ataei and Osanloo (2004) used a combination of GA and grid search method to optimise COG for multi-metal deposits. Their study supported use of hybrid approaches to COG optimisation.

Asad and Dimitrakopoulos (2013a, 2013b) presented a parametric maximum flow algorithm and heuristic approach to determine optimal COGs based on a stochastic framework. The study illustrated the significance of applying such approaches to account for uncertainty in the supply of ore for a mine.

Myburgh, Deb and Craig (2014) developed a heuristic approach using an evolutionary algorithm combined with both local search and LP algorithms to manage the variation of COGs and permutations of the extraction sequence. The LP algorithm is responsible for determining the optimal material flow-through multiple processing streams as well as stockpile management in the mine. The local search technique is called upon to provide the best production schedule. The approach developed by Myburgh, Deb and Craig (2014) is incorporated in Maptek Evolution® software.

Rahimi and Ghasemzadeh (2015) introduced a metaheuristic algorithm that considers economic, geological and environmental and social aspects to determine optimum COGs. The algorithm demonstrates the importance of balancing various factors in ensuring sustainable mining practices.

Mohammadi *et al* (2017) developed a metaheuristic approach for COG optimisation and production scheduling using the Imperialist Competitive Algorithm (ICA). The approach was used at the Golgohar iron mine in Iran. The study's results suggest that by stockpiling low-grade ore, the concentrate plant can operate at maximum capacity for an additional year beyond the initial plan.

Ahmadi and Shahabi (2018) used a GA to optimise COG. The study compared the results from the GA and Lane's COG optimisation approach and showed that both approaches generated comparable COG profiles over the LOM. However, the GA produced results in much shorter computation time compared to Lane's COG approach, indicating the potential for using metaheuristic approaches.

Ahmadi and Bazzazi (2019, 2020) used the ICA to optimise COGs in open pit mines. Like the work of Mohammadi *et al* (2017), the studies highlighted the advantage of the ICA approach to generate solutions within short computation times.

Biswas *et al* (2020) and Biswas, Sinha and Sen (2023a) introduced a DP and metaheuristic approach to determine optimal COGs in open pit metalliferous deposits. The studies highlighted the significance of COG optimisation in improving recovery and the importance of addressing economic, geological and operational uncertainties in mining.

Paithankar *et al* (2020) introduced a maximum flow algorithm and GA approach to optimise COG, extraction sequence and material flow for a mining complex. Under stochastic conditions, the approach generated higher NPVs and reduced the risk of not meeting production targets compared to commercial software results. Later, Paithankar, Chatterjee and Goodfellow (2021) presented a hybrid solution approach that combines the maximum flow algorithm, GA and Lane's COG optimisation approach to simultaneously optimise the production sequence and dynamic COGs for open pit mining operations. The hybrid approach demonstrated that stockpiling improved the NPV.

Sotoudeh *et al* (2021) used metaheuristic approaches to develop a COG model that integrates pre-concentration systems in underground mining operations to reduce costs and increase profitability. The study demonstrated that this integration improved sustainability by lowering COGs, and reducing processing costs and backfilling requirements, while maintaining high metal production throughput. The approach supports sustainability through maximising the extraction of the mineral resource.

The preceding research studies outlined in this section, collectively underscore the importance of implementing metaheuristic approaches to optimise COGs in mining operations by incorporating diverse factors that enhance sustainability. However, metaheuristic approaches do not always guarantee optimal solutions and employing hybrid approaches can be useful to overcome this shortcoming.

BIBLIOMETRIC ANALYSIS

Bibliometric analysis is typically undertaken to identify key research aspects for mapping future research directions. Bibliometric analysis enables researchers to perform in-depth analysis within a specific research field through pinpointing seminal articles, examining research trends over time, revealing shifts in research focus, identifying new or emerging research techniques, or observe the presence of research collaboration networks among researchers and institutions that create opportunities for partnerships and knowledge exchange. Additionally, bibliometric analysis enables the assessment of the impact of specific studies or methodologies on the specific research field, thereby guiding future research directions (Mutandwa and Musingwini, 2024). In this paper, a bibliometric analysis was conducted in the research field of COG optimisation by collecting data from relevant databases and then analysing the data using relevant software to infer future research directions.

Databases and the analysis tools used

Scopus and Web of Science (WoS) were selected as data sources due to their widespread use as research citation indexes and compatibility with most bibliometric analysis tools (Groote and Raszewski, 2012). However, publication listings from these databases typically overlap by about 43 to 66 per cent due to factors pertaining to subject matter and publication periods (Tabacaru, 2019). Consequently, the listings from both databases must be merged to identify and eliminate duplicates to avoid double-counting of research articles during analysis (Mutandwa and Musingwini, 2024). Various software options are available for bibliometric analysis, such as R-Biblioshiny, VOSviewer, CiteSpace and Gephi. In this paper, R-Biblioshiny was selected because it is freely available from the public domain and is also easy to use.

Data collection

Search strategy

The search strategy involved the use of identified keywords. A combination of keywords was compiled to create a search string for identifying relevant research articles from Scopus and WoS databases. The search string used was as follows: (('cut-off grade' OR 'cut-off grade' or 'cut-off grade' or 'COG policy' or 'economic definition of ore') AND 'optimi*').

Data extraction

The search on Scopus yielded 256 articles after cleaning the data by removing articles and conference papers that were irrelevant to COG optimisation. The same cleaning process was done on the search conducted on WoS which resulted with a total of 119 indexed articles. This process of data search and cleaning was conducted using guidance from the Preferred Reporting Items for Systematic Reviews and Meta-Analysis (PRISMA) approach updated by Page *et al* (2021). The PRISMA approach is a three-step process for records identification, screening and inclusion.

Descriptive analysis

Publication trends

Research publications on COG optimisation were relatively few until the early 2000s as illustrated in Figure 3. Most of the pre-2000 COG optimisation publications appeared in the Australasian Institute of Mining and Metallurgy (AusIMM) journal, with the Journal of the Southern Institute of Mining and Metallurgy and Resources Policy journal indicating a notable increase around 2015. At the time of writing this paper, the Resources Policy appears to be the preferred journal for COG optimisation publications, probably since the journal is a policy-based journal and COG optimisation is part of broader COG policy formulations at mining operations.

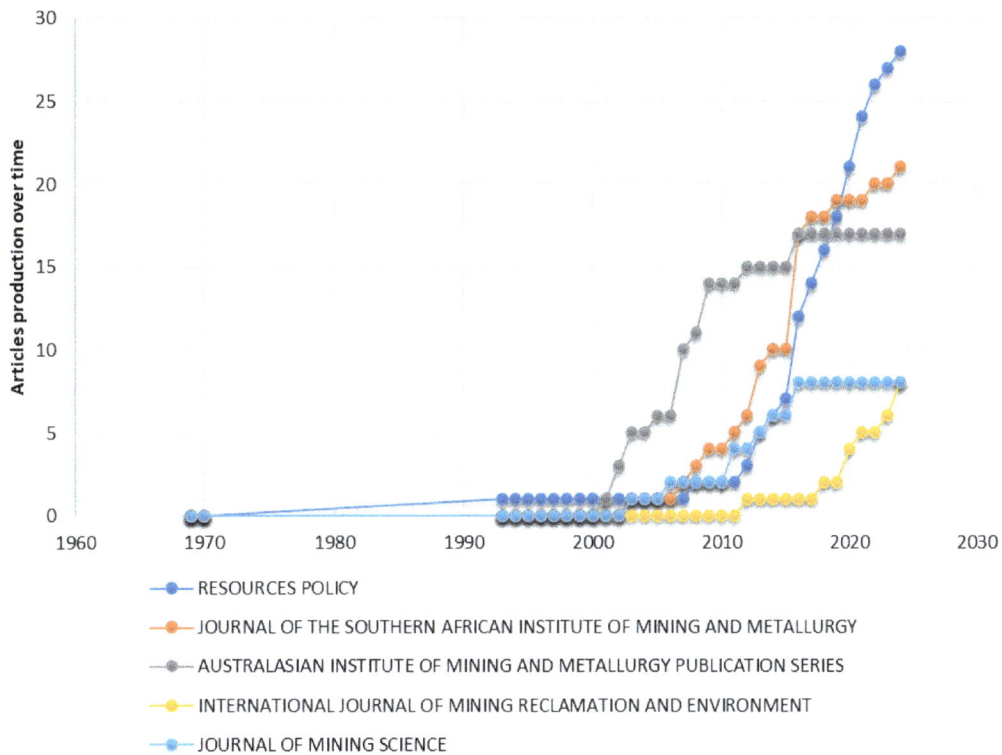

FIG 3 – COG optimisation publications by different journals.

Authorship analysis

Prolific authors frequently influence the research agenda in their field. Ibrahim (2024) posited that identifying these authors facilitates the recognition of the most influential research and trends. In this analysis, ten authors have five or more publications on COG optimisation, with Mohammad Waqar Asad, who is affiliated with Curtin University in Australia, leading the list with 14 articles, as illustrated in Figure 4. Morteza Osanloo, who is affiliated with Amirkabir University of Technology in Iran has the second-largest number of indexed articles, followed by Roussos Dimitrakopoulos and Mustafa Kumral who are both affiliated with McGill University in Canada. Two authors appearing in the top ten from the same institution are Clinton Birch and Joseph Githiria, who are both affiliated with the University of the Witwatersrand in South Africa.

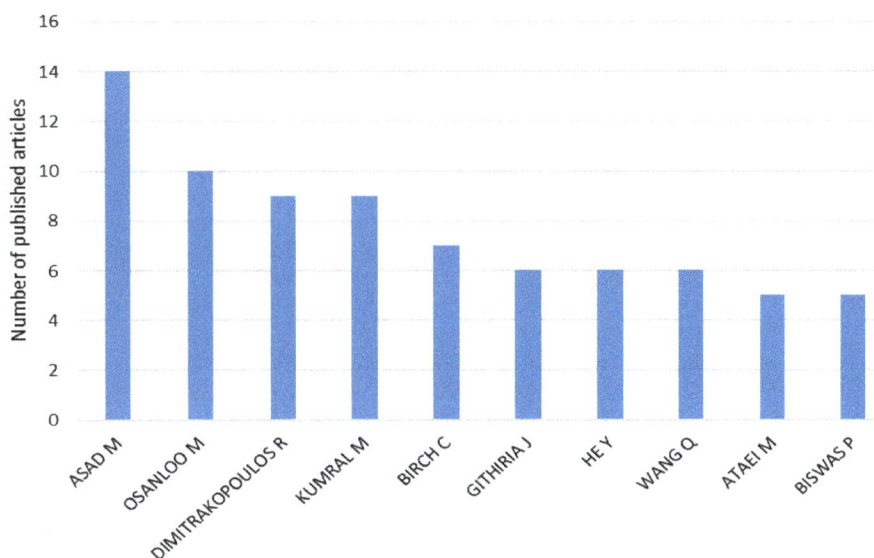

FIG 4 – Top 10 most published authors on COG optimisation.

Citation analysis

For this analysis, normalised total citations and total citations per annum were considered. Normalised total citations is a metric that adjusts for time and variations in citation behaviour, providing a fair comparison of articles. It further offers a more balanced view of an article's influence. Conversely, total citations per annum demonstrate the average number of citations an article receives annually, which highlights its impact comparable on a yearly basis. The article by Ataei and Osanloo (2004), which has the highest number of citations and features among the top ten articles, has the highest number of total citations per annum, despite being published more than 20 years prior to the time of writing of this paper. This may indicate the article's continued relevance and influence on COG optimisation research.

The article by Khan and Asad (2020) which appears as the most recent article in Figure 5, has however, the most citations per annum and ranks third in terms of normalised total citations. This also demonstrates the article's relevance and potential influence in shaping future COG optimisation. It is also important to note that some articles are not open access, and this limits the number of researchers who can access, read, and potentially cite the articles. Moreover, this analysis encompasses only research articles indexed in Scopus and WoS databases, hence there may be some articles with high citation counts beyond those mentioned in this study. Table 1 summarises some of the top cited publications, highlighting some key strengths and shortcomings of models presented in the articles.

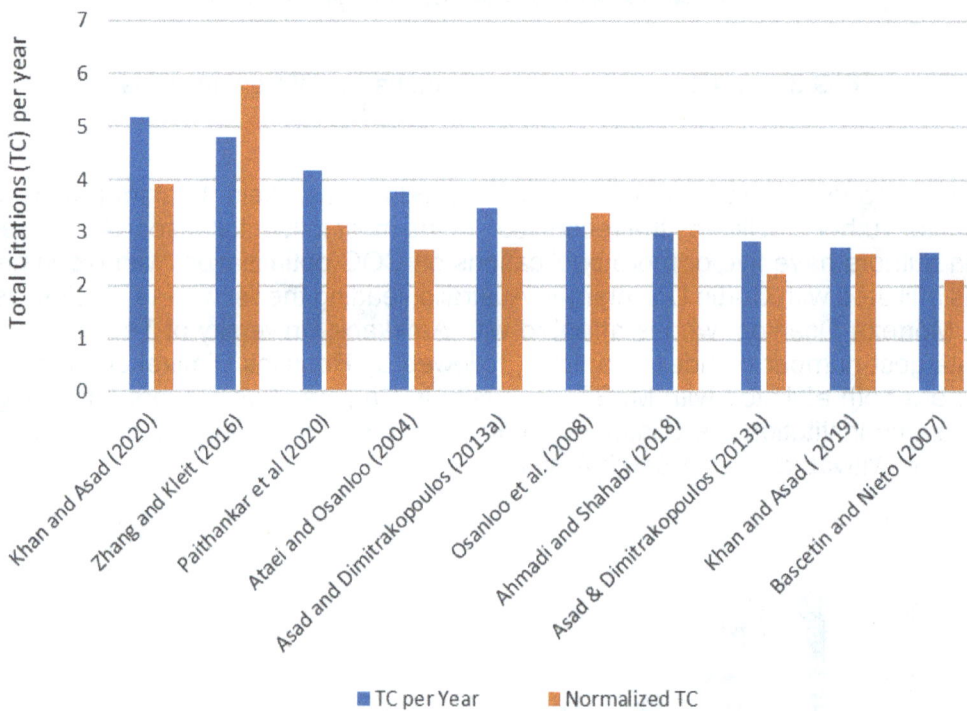

FIG 5 – Total citations per annum and normalised citations per annum.

TABLE 1
Summary of some top-cited publications highlighting their key strengths and weaknesses.

Research	Main subject	Strength(s)	Shortcoming(s)	Characterisation
Khan and Asad (2020)	Formulation of two-stage stochastic linear programming approach	• Dynamic COG policy • Ability to handle uncertainty and generate risk-quantified COG policy.	• Model could be expanded to include other practical aspects, such as environmental and rehabilitation costs.	Mathematical model
Paithankar *et al* (2020)	Stochastic optimisation of production sequence and dynamic COG	• Integrated approach allows for better optimisation of the mining operation.	• Larger or more complex mining problems may pose challenges in terms of solution times and convergence.	Framework that combines a maximum flow algorithm for extraction sequencing and two approaches (GA and Lane's approach) for COG optimisation
Zhang and Kleit (2016)	Optimisation of mining rate considering stockpiling and grade	• Explicit consideration of the impact of stockpiling and how it affects other mining decisions such as COG.	• Model does not account for economies of scale.	Mathematical model
Osanloo, Rashidinejad and Rezai (2008)	How incorporating associated environmental costs into COG optimisation can add value to mining projects.	• Model incorporates costs associated with the management of acid-generating (AG) waste materials and tailings into COG optimisation.	• The iterative nature of the model's calculations, while necessary, may limit its practical application, especially for real-world complex deposits.	Modified model
Ataei and Osanloo (2004)	Optimisation of COGs for multi-metal deposits.	• A hybrid optimisation approach combining GA and the grid search algorithm. • Converges to the optimum solution in a relatively short computational time.	• The optimisation problem being solved is a non-linear programming problem, which can be computationally intensive, especially as the number of metal deposits increases.	A hybrid optimisation approach combining GA and the grid search method

Co-occurrence analysis and thematic mapping

Co-occurrence analysis is used to identify the most common keywords and emerging themes in the literature. To show the conceptual structure of COG optimisation research, a thematic map was constructed in R-Biblioshiny based on authors' keyword co-occurrences to establish initial thematic groupings within the COG optimisation research field. The resulting thematic map is then used to solidify identified networks and compare them in a matrix for comprehensive analysis of co-occurrences to categorise research topics into four distinct clusters which are namely, niche, motor, emerging/declining, and basic themes (Kaiser and Kuckertz, 2024). Niche themes are specialised with limited relevance to the broader research. Motor themes have both a high degree of relevance and development. Basic themes have a high degree of relevance and a low degree of development. Emerging/declining themes have both low relevance and development, for example from Figure 6, Lane's algorithm appears as a declining theme probably due it to being overshadowed by metaheuristic approaches.

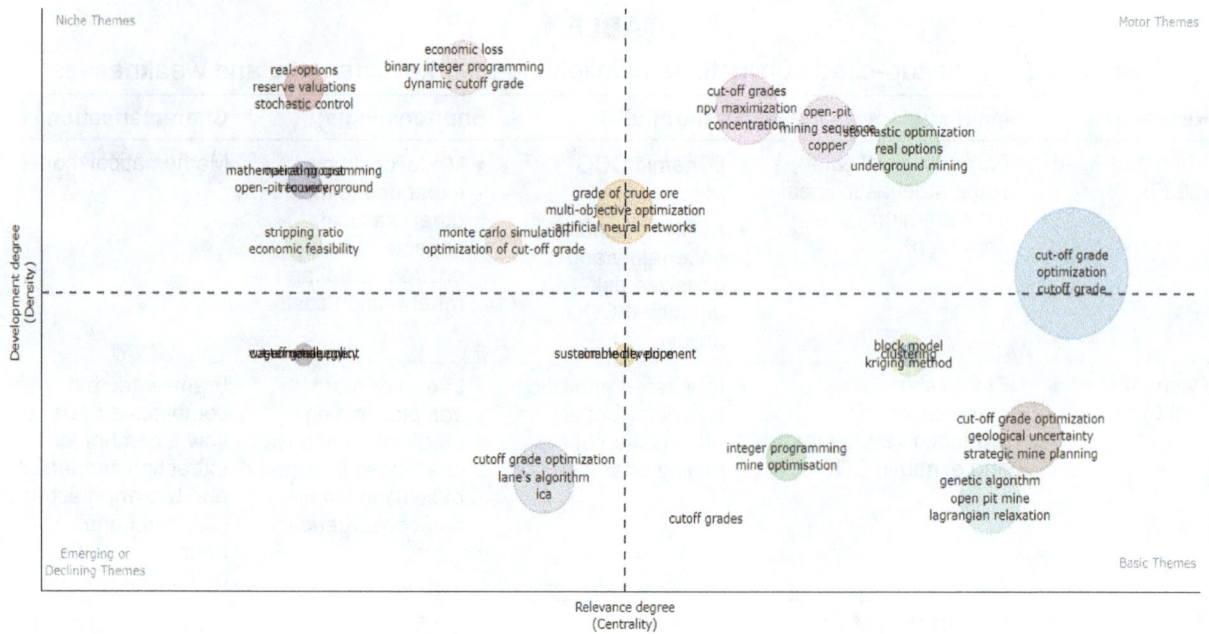

FIG 6 – Thematic analysis of COG optimisation research.

Collaboration analysis

An understanding of the leading authors and their affiliations can facilitate networking and potential collaborations. Researchers frequently seek to collaborate with established experts to enhance the quality and visibility of their work (Meho and Akl, 2025). Most authors who appeared on the list of top ten prolific authors are also present in the collaboration network analysis depicted in Figure 7. However, these authors do not appear to collaborate on GOG optimisation research, except for Mohammad Waqar Asad and Roussos Dimitrakopoulos. The analysed data did not have citation networks to show the interconnections between key papers and this could have assisted in analysing how these prolific authors cite each other's works.

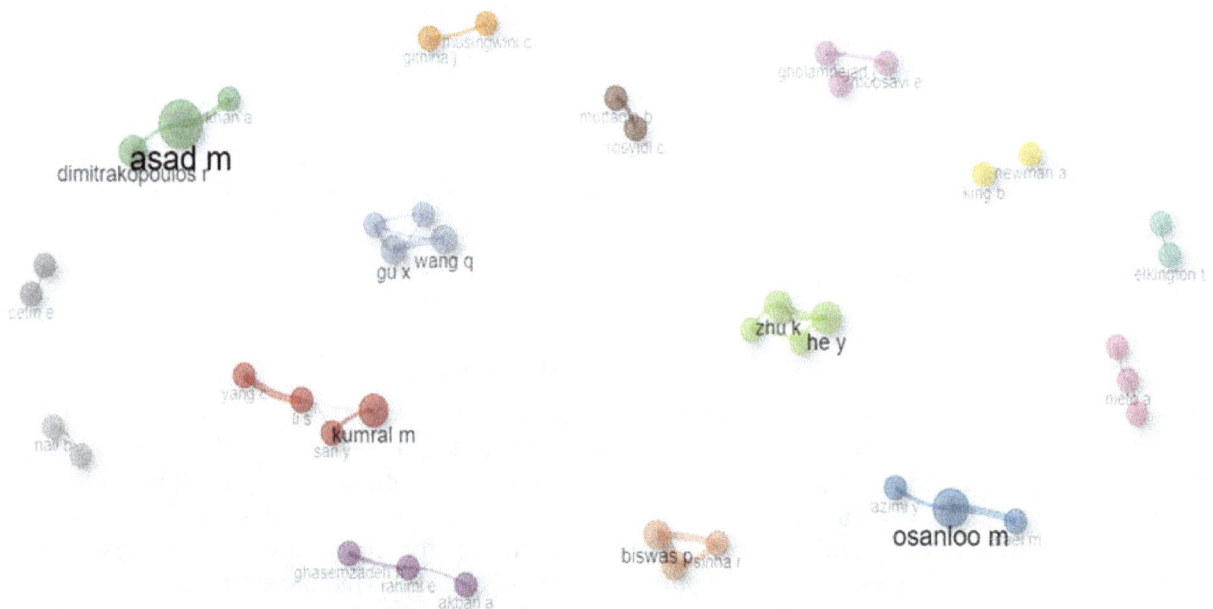

FIG 7 – Collaboration network analysis of COG optimisation research.

CONCLUDING REMARKS AND PROPOSED FUTURE RESEARCH DIRECTIONS

Lane's COG approach, which was developed in the 1960s, is considered pioneering work for determining optimal COGs in the mining industry. However, the approach has some shortcomings which can be overcome by using metaheuristic approaches, hence the growing trend in the use of metaheuristic approaches to solve large-scale real-world COG optimisation problems. This observation was supported by results of a bibliometric analysis on COG optimisation. The bibliometric analysis results suggest that future research directions on COG optimisation should focus on developing more efficient metaheuristic approaches to foster improved adoption by the mining industry.

REFERENCES

Ahmadi, M R and Bazzazi, A A, 2019. Cutoff grades optimisation in open pit mines using meta-heuristic algorithms, *Resources Policy*, 60:72–82. https://doi.org/10.1016/j.resourpol.2018.12.001

Ahmadi, M R and Bazzazi, A A, 2020. Application of meta-heuristic optimisation algorithm to determine the optimal cutoff grade of open pit mines, *Arabian Journal of Geoscience*, 13(5):1–12.

Ahmadi, M R and Shahabi, R S, 2018. Cutoff grade optimisation in open pit mines using genetic algorithm, *Resources Policy*, 55:184–191.

Asad, M W A and Dimitrakopoulos, R, 2013a. Implementing a parametric maximum flow algorithm for optimal open pit mine design under uncertain supply and demand, *Journal of the Operational Research Society*, 64(2):185–197. https://doi.org/10.1057/jors.2012.26

Asad, M W A and Dimitrakopoulos, R, 2013b. A heuristic approach to stochastic cut-off grade optimisation for open pit mining complexes with multiple processing streams, *Resources Policy,* 38:591–597.

Asad, M W A, 2002. Development of generalized cut-off grade optimisation algorithm for open pit mining operations, *Journal of Engineering and Applied Sciences,* 21(2):119–127.

Asad, M W A, 2005. Cut-off grade optimisation algorithm with stockpiling option for open pit mining operations of two economic minerals, *International Journal of Surface Mining, Reclamation and Environment,* 19(3):176–187.

Asad, M W A, 2007. Optimum cut-off grade policy for open pit mining operations through net present value algorithm considering metal price and cost escalation, *Engineering Computations*, 24(7):723–736.

Asad, M W A, Qureshi, M A and Jang, H, 2016. A review of cut-off grade policy models for open pit mining operations, *Resources Policy*, 49:142–152.

Ataei, M and Osanloo, M, 2004. Using a Combination of Genetic Algorithm and the Grid Search Method to Determine Optimum Cutoff Grades of Multiple Metal Deposits, *International Journal of Surface Mining, Reclamation and Environment*, 18(1):60–78. https://doi.org/10.1076/ijsm.18.1.60.23543

Attea, B A, Abbood, A D, Hasan, A A, Pizzuti, C, Al-Ani, M, Özdemir, S and Al-Dabbagh, R D, 2021. A review of heuristics and metaheuristics for community detection in complex networks: current usage, emerging development and future directions, *Swarm and Evolutionary Computation*, 63:100885.

Bascetin, A and Nieto, A, 2007. Determination of optimal cutoff grade policy to optimize NPV using a new approach with optimization factor, *Journal of the Southern African Institute of Mining and Metallurgy*, 107(2):87–94.

Bello, A, Wiebe, N, Garg, A and Tonelli, M, 2015. Evidence-based decision-making 2: systematic reviews and meta-analysis, *Methods in Molecular Biology*, 1281:397–416.

Birch, C, 2016a. Impact of the South African mineral resource royalty on cut-off grades for narrow, tabular Witwatersrand gold deposits, *Journal of the Southern African Institute of Mining and Metallurgy,* 116:237–246.

Birch, C, 2016b. Impact of discount rates on cut-off grades for narrow tabular gold deposits, *Journal of the Southern African Institute of Mining and Metallurgy,* 116:115–122.

Birch, C, 2017. Optimisation of cut-off grades considering grade uncertainty in narrow, tabular gold deposits, *Journal of the Southern African Institute of Mining and Metallurgy,* 117(2):149–156.

Birch, C, 2018. Review of cut-off grade optimisation from Southern African mines, Student assignment-based observations, *Resources Policy*, 56:134–140.

Biswas, P, Sinha, R K and Sen, P, 2023a. A new optimisation model for the selection of optimal cut-off grade using multi-sequential decision algorithm in open-pit metalliferous deposits, *International Journal of Mining, Reclamation and Environment*, 37(9):683–712.

Biswas, P, Sinha, R K and Sen, P, 2023b. A review of state-of-the-art techniques for the determination of the optimum cut-off grade of a metalliferous deposit with a bibliometric mapping in a surface mine planning context, *Resources Policy*, 83:103543. https://doi.org/10.1016/j.resourpol.2023.103543

Biswas, P, Sinha, R K, Sen, P and Rajpurohit, S S, 2020. Determination of optimum cut-off grade of an open-pit metalliferous deposit under various limiting conditions using a linearly advancing algorithm derived from dynamic programming, *Resources Policy*, 66:1–11.

Cetin, E and Dowd, P A, 2002. The use of genetic algorithms for multiple cut-off grade optimisation, in *Proceedings of the 32nd International Symposium on Application of Computers and Operations Research in the Mineral Industry*, pp 769–779.

Cetin, E and Dowd, P A, 2011. Multi mineral cut-off grade optimisation by means of dynamic programming, *Sustainable Production and Consumption of Mineral Resources*, Krakow, Poland.

Cetin, E and Dowd, P A, 2013. Multi-mineral cut-off grade optimisation by grid search, *Journal of the Southern African Institute of Mining and Metallurgy*, 113(8):659–665.

Cetin, E and Dowd, P A, 2016. Multiple cut-off grade optimisation by genetic algorithms and comparison with grid search method and dynamic programming, *Journal of the Southern African Institute of Mining and Metallurgy*, 116(7):681–688.

Dagdelen, K and Kawahata, K, 2008. Value creation through strategic mine planning and cut-off grade optimisation, *Mining Engineering*, 60(1):39–45.

Dagdelen, K, 1992. Cut-off grade optimisation, in *Proceedings of the 23rd International Symposium on Application of Computers and Operations Research in Minerals Industry*, pp 157–165.

Dagdelen, K, 1993. An NPV optimisation algorithm for open pit mine design, in *Proceedings of the 24th International Symposium on Application of Computers and Operations Research in the Mineral Industry*, pp 257–263.

Deloitte, 2023. The Importance of Reliable Mineral Resources and Reserves Reporting: The Bottom Line for Investors, Deloitte [online]. Available from: <https://www.deloitte.com/content/dam/assets-zone1/za/en/docs/services/consulting/2023/za-POV-Resources-Doc-LS-with-bleed.pdf> [Accessed: 10 July 2024].

Dowd, P A, 1976. Application of dynamic and stochastic programming to optimise cut-off grades and production rates, *Transactions of the Institution of Mining and Metallurgy, Section A: Mining Technology*, 85(1):22–31.

Gholamnejad, J, 2008. Determination of the optimum cut-off grade considering environmental cost, *Journal of International Environmental Application and Science*, 3(3):186–194.

Gholamnejad, J, 2009. Incorporation of rehabilitation cost into the optimum cut-off grade determination, *Journal of the Southern African Institute of Mining and Metallurgy*, 108(2):89–94.

Githiria, J and Musingwini, C, 2019. A stochastic cut-off grade optimisation model to incorporate uncertainty for improved project value, *Journal of the Southern African Institute of Mining and Metallurgy*, 119(3):1–12.

Githiria, J, 2016. Cut-off grade optimisation to maximise the net present value using Whittle 4X, *International Journal of Mining and Minerals Engineering*, 7(4):313–327.

Githiria, J, 2018. A stochastic cut-off grade optimisation algorithm, PhD thesis, School of Mining Engineering, University of the Witwatersrand, South Africa.

Githiria, J, Muriuki, J and Musingwini, C, 2016. Development of a computer-aided application using Lane's algorithm to optimise cut-off grade, *Journal of the Southern African Institute of Mining and Metallurgy*, 116(11):1027–1035.

Githiria, J, Musingwini, C, Madahana, M and Khangamwa, G, 2024. Incorporating a machine learning approach for commodity price prediction to generate a cut-off grade optimisation policy under grade uncertainty, *International Journal of Mining, Reclamation and Environment*, pp 1–22. https://doi.org/10.1080/17480930.2024.2420697

Goycoolea, M, Lamas, P, Pagnoncelli, B K and Piazza, A, 2020. Lane's Algorithm Revisited, *Management Science*, 67:3087–3103. https://doi.org/10.1287/MNSC.2020.3685

Groote, S L D and Raszewski, R, 2012. Coverage of Google Scholar, Scopus and Web of Science: A case study of the h-index in nursing, *Nursing Outlook*, 60:391–400. https://doi.org/10.1016/j.outlook.2012.04.007

Hall, B (ed), 2014. *Cut-off grades and optimising the strategic mine plan*, Spectrum Series 20 (The Australasian Institute of Mining and Metallurgy: Melbourne).

Henning, U, 1963. Calculation of cut-off grade, *Canadian Mining Journal*, 84(3):54–57.

Ibrahim, C, 2024. Academic roots: The most prolific authors of bibliometrics, *Jurnal Literasi Perpustakaan dan Informasi, UHO*, 4(2). http://dx.doi.org/10.52423/jlpi.v4i2.49057

International Accounting Standards Board (IASB), 2010. Discussion Paper: Extractive Activities (DP/2010/1). London, United Kingdom, 183 p.

Johnson, T B, 1969. Optimum Production Scheduling, in *Proceedings of the 8th International Symposium Computers and Operations Research*, pp 539–562.

Kaiser, M and Kuckertz, A, 2024. Bibliometrically mapping the research field of entrepreneurial communication: where we stand and where we need to go, *Management Review Quarterly*, 74:2087–2120.

Khan, A and Asad, M W A, 2020. A mathematical programming model for optimal cut-off grade policy in open pit mining operations with multiple processing streams, *International Journal of Mining, Reclamation and Environment*, 34(3):149–158.

Khan, A and Asad, M W A, 2021. A mixed integer programming-based cut-off grade model for open-pit mining of complex poly-metallic resources, *Resources Policy*, 72:102076. https://doi.org/10.1016/j.resourpol.2021.102076

Khan, K S, Kunz, R, Kleijnen, J and Antes, G, 2003. Five steps to conducting a systematic review, *Journal of the Royal Society of Medicine*, 96(3):118–121.

King, B and Newman, A, 2018. Optimising the Cutoff Grade for an Operational Underground Mine, *Interfaces*, 48(4):357–371.

King, B, 1999. Cash flow grades-scheduling rocks with different throughput characteristics, *Proceedings of the Strategic Mine Planning Conference*, pp 1–8.

King, B, 2001. Optimal mine scheduling policies, PhD thesis, Royal School of Mines, Imperial College, London University, UK.

King, B, 2011. Optimal mining practice in strategic planning, *Journal of Mining Science,* 47(2):247–253.

Krautkraemer, J A, 1988. The Cut-Off Grade and the Theory of Extraction, *The Canadian Journal of Economics,* 21(1):146–160.

Lane, K F, 1964. Choosing the optimum cut-off grade, *Colorado School of Mines Quarterly*, 59(4):811–829.

Lane, K F, 1988. *The Economic Definition of Ore: Cut-off Grade in Theory and Practice* (Mining Journal Books: London).

Li, S, Yang, C and Lu, C, 2012. Cut-Off Grade Optimisation Using Stochastic Programming in Open-Pit Mining, *IEEE*, pp 66–69. https://doi.org/10.1109/ICICSE.2012.24

Meho, L I and Akl, E A, 2025. Using bibliometrics to detect questionable authorship and affiliation practices and their impact on global research metrics: A case study of 14 universities, *Quantitative Science Studies*, pp 1–36. https://doi.org/10.1162/qss_a_00339

Minnitt, R, 2004. Cut-off grade determination for the maximum value of a small Wits-type gold mining operation, *Journal of the Southern African Institute of Mining and Metallurgy,* 104(6):277–284.

Mohammadi, S, Kakaie, R, Ataei, M and Pourzamani, E, 2017. Determination of the optimum cut-off grades and production scheduling in multi-product open pit mines using imperialist competitive algorithm (ICA), *Resources Policy*, 51:39–48.

Moosavi, E and Gholamnejad, J, 2016. Optimal extraction sequence modeling for open pit mining operation considering the dynamic cutoff grade, *Journal of Mining Science,* 52(5):956–964.

Mutandwa, B and Musingwini, C, 2024. Insights and future research directions from a bibliometric mapping of studies in stope layout optimisation, *International Journal of Mining, Reclamation and Environment*, 38:577–595. https://doi.org/10.1080/17480930.2024.2325758

Myburgh, C A, Deb, K and Craig, S, 2014. Applying modern heuristics to maximising net present value through cut-off grade optimisation, in *Proceedings of Orebody Modelling and Strategic Mine Planning Symposium*, pp 155–164.

Njowa, G and Musingwini, C, 2018. A framework for interfacing mineral asset valuation and financial reporting, *Resources Policy*, 56:3–15.

Osanloo, M, Rashidinejad, F and Rezai, B, 2008. Incorporating environmental issues into optimum cut-off grades modeling at porphyry copper deposits, *Resources Policy,* 33(4):222–229.

Page, M J, McKenzie, J E, Bossuyt, P M, Boutron, I, Hoffmann, T C, Mulrow, C D, Shamseer, L, Tetzlaff, J M, Akl, E A, Brennan, S E, Chou, R, Glanville, J, Grimshaw, J M, Hróbjartsson, A, Lalu, M M, Li, T, Loder, E W, Mayo-Wilson, E, McDonald, S, McGuinness, L A, Stewart, L A, Thomas, J, Tricco, A C, Welch, V A, Whiting, P and Moher, D, 2021. The PRISMA 2020 statement: an updated guideline for reporting systematic reviews, *BMJ,* 372(71). https://doi.org/10.1136/bmj.n71

Paithankar, A, Chatterjee, S and Goodfellow, R, 2021. Open-pit mining complex optimisation under uncertainty with integrated cut-off grade based destination policies, *Resources Policy*, 70:101875.

Paithankar, A, Chatterjee, S, Goodfellow, R and Asad, M W A, 2020. Simultaneous stochastic optimisation of production sequence and dynamic cut-off grades in an open pit mining operation, *Resources Policy*, 66:1–13.

Rahimi, E and Ghasemzadeh, H, 2015. A new algorithm to determine optimum cut-off grades considering technical, economical, environmental and social aspects, *Resources Policy*, 46:51–63.

Rajwar, K, Deep, K and Das, S, 2023. An exhaustive review of the metaheuristic algorithms for search and optimisation, *Taxonomy, Applications and Open Challenges*, 09 Nov:13187–13257.

Rys, P, Wladysiuk, M, Skrzekowska-Baran, I and Malecki, M T, 2009. Review articles, systematic reviews and meta-analyses: which can be trusted?, *Polskie Archiwum Medycyny Wewnetrznej*, 119(3):148–156. PMID:19514644.

SAMREC, 2016. The South African Code for the Reporting of Exploration Results, Mineral Resources and Mineral Reserves, Johannesburg: Southern African Institute of Mining and Metallurgy.

Sotoudeh, F, Nehring, M, Kizil, M, Knights, P and Mousavi, A, 2021. A novel cut-off grade method for increasing the sustainability of underground metalliferous mining operations, *Minerals Engineering,* 172:107168.

Tabacaru, S, 2019. Web of Science versus Scopus: Journal Coverage Overlap Analysis, Texas A&M University Libraries, pp 1–7. Available from: <https://oaktrust.library.tamu.edu/bitstream/handle/1969.1/175137/Web%20of%20Science%20versus%20Scopus%20Report%202019.pdf?sequence=4&isAllowed=y> [Accessed: 10 January 2025].

Tawfik, G M, Dila, K A S, Mohamed, M Y F, Tam, D N H, Kien, N D, Ahmed, A M and Huy, N T, 2019. A step-by-step guide for conducting a systematic review and meta-analysis with simulation data, *Tropical Medicine and Health*, 47:46.

Taylor, H K, 1972. General background theory of cut-off grades, *Transactions of the Institution of Mining and Metallurgy, Section A: Mining Technology*, 81:160–179.

Taylor, H K, 1985. Cut-off grades – some further reflections, *Transactions of the Institution of Mining and Metallurgy, Section A: Mining Technology*, 96:204–216.

Thompson, M and Barr, D, 2014. Cut-off grade: a real options analysis, *Resources Policy,* 42:83–92.

Tomar, V, Bansal, M and Singh, P, 2023. Metaheuristic Algorithms for Optimisation: A Brief Review, *Engineering Proceedings*, 59(1):238.

Whittle, J and Wharton, C, 1995. Optimising cut-offs over time, in *Proceedings of the 25th International Symposium on the Application of Computers and Mathematics in the Mineral Industries*, pp 261–265.

Zhang, K and Kleit, A N, 2016. Mining rate optimization considering the stockpiling: A theoretical economics and real option model, *Resources Policy*, 47:87–94.

Haulage simulation of open pit mines with ore and waste IPCC

N A Habib[1], E Ben-Awuah[2] and H Askari-Nasab[3]

1. Research Fellow, University of Alberta, Edmonton AB T6G 2R3, Canada.
   Email: nasib@ualberta.ca
2. Associate Professor, Laurentian University, Sadbury ON P3E 2C6, Canada.
   Email: ebenawuah@laurentian.ca
3. Professor, University of Alberta, Edmonton AB T6G 2R3, Canada. Email: hooman@ualberta.ca

ABSTRACT

In-Pit Crushing and Conveying (IPCC) offers a cost-effective and environmentally friendly alternative to traditional diesel truck haulage in mining operations. However, its adoption remains limited across the industry. This study focuses on the implementation of semi-mobile IPCC system within existing or new mines, specifically analysing its impact on production schedules over one year. We developed a haulage simulation model to evaluate the production efficiency of various haulage options: pure truck-shovel, ore IPCC, and a hybrid system combining both IPCC and trucks. The model was verified using operational data from an iron ore mine during the 11th year of its life. Results demonstrate that the hybrid system provides the highest cost savings while closely aligning with optimal production requirements. This research highlights the potential of IPCC technologies to enhance short-term operational efficiency and reduce costs, underscoring the importance of re-evaluating haulage strategies in the mining sector. Our findings advocate for further exploration and implementation of IPCC systems to improve productivity in mining operations.

INTRODUCTION

This study introduces a novel methodology that integrates simulation and optimisation to develop near-optimal short-term mining schedules while incorporating haulage uncertainties. Monthly production schedules are first generated using a Mixed-Integer Linear Programming (MILP) model, which are then fed into a Monte Carlo haulage simulation. This simulation employs probabilistic distributions to represent key operational parameters, including shovel loading time, truck travel and dumping times, as well as failure probabilities for trucks, shovels, and the In-Pit Crushing and Conveying system. While the MILP model effectively constructs optimised short-term schedules for both conventional truck-shovel haulage and IPCC-assisted operations, the Monte Carlo simulation ensures that stochastic variations in haulage operations are adequately captured.

One of the most significant challenges in this research is accurately representing and modelling the uncertainties associated with equipment performance, such as shovel loading and truck travel times, dumping durations, and failure rates. Additionally, achieving a seamless integration between the simulation and optimisation models to efficiently transfer input data from the MILP model within a reasonable time frame poses another key difficulty.

The overarching objective in mining operations is to maximise net present value (NPV) while controlling costs. Mine planning is typically categorised into long-term and short-term phases based on the planning horizon and specific goals. Long-term planning focuses on strategic decisions aimed at optimising NPV across the entire lifespan of the mine. Conversely, short-term planning is more operationally driven, concentrating on tactical decisions such as shovel assignment, grade blending to maintain plant feed quality, and truck dispatching. These short-term plans, which can span monthly, weekly, or even daily time frames, play a crucial role in ensuring alignment with the long-term strategic schedule.

Efficient use of mining equipment is crucial, as haulage costs can account for over 50 per cent of total operating expenses in a truck-shovel operation (Moradi Afrapoli and Askari-Nasab, 2017; Osanloo and Paricheh, 2020). Optimal equipment utilisation can only be achieved by efficiently using all assets to meet the production targets set by the long-term plan. Therefore, the strategic allocation of shovels and trucks in short-term production scheduling is essential to ensure cost control and the achievement of long-term goals.

The majority of contemporary short-term planning models rely on MIP and incorporate explicit precedence constraints. Eivazy and Askari-Nasab (2012) presented a short-term planning model using MIP that incorporates various mining directions and precedence constraints to reduce overall mining expenses, such as processing, haulage, rehandling, and rehabilitation costs. However, by utilising aggregated mining blocks, the model may fall short in optimality, as it neglects specific ore type selection and real-world hauling dynamics. Additionally, the model focuses solely on cost reduction, omitting profit considerations. L'Heureux, Gamache and Soumis (2013) developed a comprehensive mathematical optimisation model for short-term planning, covering operational details for up to three months. The primary goal is to reduce the operational costs of truck and shovel activities, as well as drilling and blasting. This model was successfully applied to scenarios involving up to five shovels, 90 periods, and 132 faces.

Kozan, Liu and Wolff (2013) created a model to manage drilling, blasting, and mining of blocks, along with equipment allocation to these tasks, aiming to minimise the make-span, which is the total time from the start to the end of the schedule. Later, Kozan and Liu (2016) introduced another short-term planning model designed to maximise throughput and minimise equipment idle times during drilling, blasting, and excavation. This model considers equipment capacity, speed, read times, and activity precedence constraints. In their latest work (Liu and Kozan, 2017), they proposed an innovative mine management system that integrates various mathematical models to determine the ultimate pit limit for long-term and medium-term block sequencing. This system also optimises equipment planning using a job-shop scheduling model to enhance mining efficiency. The overall goal of this methodology is to improve mining operations by integrating diverse planning and scheduling models.

Blom, Pearce and Stuckey (2016) and Blom, Burt and Stuckey (2014) introduced an MIP model to generate multiple short-term production schedules. This model aims to optimise equipment and shovel utilisation while considering constraints such as blending requirements, equipment availability, trucking hours, and task precedence relationships. It employs a rolling planning horizon technique and accounts for multiple processing paths.

Thomas, Singh and Krishnamoorthy (2013) and Thomas, Venkateswaran and Krishnamoorthy (2014) tackled the challenge of integrated planning and scheduling within a coal supply chain comprising of several independent mines that must share a limited transportation capacity. Their objectives included minimising total earliness, tardiness, and operational costs while adhering to due dates and transportation constraints. They developed a solution approach based on Lagrangian relaxation, which outperformed traditional MILP models in terms of generating upper and lower bounds and reducing CPU time. Mousavi, Kozan and Liu (2016) developed a comprehensive mathematical model for short-term block sequencing, incorporating constraints like precedence relationships, machine capacity, grade requirements, and processing demands. The goal is to minimise total costs, including rehandling, holding, misclassification, and drop-cut costs. The authors proposed a hybrid solution combining branch and bound with simulated annealing, which can produce solutions with an optimality gap of less than 1 per cent compared to the CPLEX solution when a large neighbourhood search is utilised. A recent optimisation model for short-term open pit planning has been proposed by Nelis and Morales (2022). Their approach maximises profits while meeting operational constraints. Applied to a real copper mine, the model efficiently defines mining cuts and production plans simultaneously. Results demonstrate its effectiveness in generating mining cut configurations quickly, a task that traditionally requires days, completed in less than 15 mins.

Manriquez, González and Nelson (2019) created a short-term planning approach for open pit mining to optimise hierarchical objectives, such as minimising deviations in ore tonnage, plant capacity, metal fines, and shovel movement costs. Using goal programming techniques (weighed sum and hierarchical method), they found both methods produced optimal plans in a copper mine case study, though the model is deterministic and excludes geological uncertainties. Similarly, Upadhyay, Doucette and Askari-Nasab (2022) developed a goal programming-based short-term planning model focusing on optimal shovel allocation, aiming to maximise production and minimise mill grade deviation and shovel movement. Silva-Júnior et al (2023) addressed the complexities of short-term planning in open pit mining operations via a mixed-integer linear goal programming model to optimise truck allocation, routes, and the amount of material transported, aiming to minimise deviations from production targets, chemical grades, and particle sizes, while also reducing the number of trucks required. The study utilises real data from an iron ore mine to validate the approach, demonstrating its effectiveness in enhancing decision-making for truck fleet management and meeting production and quality goals under varying operational scenarios.

Simulation in conjunction with optimisation is widely used in short-term mine planning because simulation can handle uncertainty involved in operations. Ben-Awuah, Pourrahimian and Askari-Nasab (2010) developed a discrete event simulation model to align long-term and short-term mine planning by addressing uncertainties in mining and processing capacities, crusher availability, stockpiling, and blending. This model integrates deterministic long-term plans with dynamic short-term adjustments, enabling planners to evaluate the feasibility and robustness of long-term schedules.

Bodon, Sandeman and Stanford (2011) and Sandeman, Fricke and Bodon (2011) developed simulation optimisation models to maximise tonnes mined and shipped, minimise deviation from quality targets across mine and port stockpiles, and meet blending requirements under constraints of equipment capacity, port capacity, and precedence. Their linear program (LP) integrates optimisation and simulation, offering a more accurate system representation, better solutions, albeit with longer runtime. It facilitates trade-off analysis for capital expenditure and alternative operating practices, including maintenance options. Shishvan and Benndorf (2014, 2016) introduced a stochastic simulation method for optimising short-term production planning in complex continuous mining operations, considering geological uncertainty. Their approach, which combines penalties for production deviation and equipment utilisation in a weighted objective function, helps foresee critical supply and system performance issues. The model utilises geostatistical simulation (20 block model realisations) and discrete event simulation to manage geological and operational uncertainties. This methodology was later applied in industrial case studies by Shishvan and Benndorf (2017).

Torkamani and Askari-Nasab (2015) devised a stochastic discrete event simulation model to analyse truck-shovel material handling and haulage systems in open pit mining. Initially, they formulated an MIP model for optimal truck and shovel allocation across mining faces, integrating these solutions into their simulation framework. Upadhyay and Askari-Nasab (2016, 2017) applied goal-programming to develop a simulation optimisation model for short-term planning in mining. Their approach highlights the integration of proactive decision-making in a dynamic environment, synchronising operational plans with long-term strategies to minimise opportunity costs while optimising production and equipment utilisation. A similar framework has been proposed by Manríquez, Pérez and Morales (2020) to generate an initial schedule and then replicate it to find key performance indicators like equipment utilisation using a discrete event simulation software. Incorporating simulation addresses equipment uncertainty in this otherwise deterministic model, making it applicable across various mining contexts, including open pit mines. However, a limitation of the model is its sole focus on maximising extraction value without considering operational costs in the optimisation process.

Martins, Souza and Assis (2024) presented an integrated simulation and optimisation tool for short-term mine planning. By combining a hierarchical MILP model with discrete event simulation (DES), the tool optimises shift schedules and simulates their execution. The study explored four different scenarios, each with varying priorities of mining objectives such as prioritising ore production, meeting element grade, particle size targets, and balancing waste extraction, showing how various strategies affect production efficiency and plant performance. The findings highlight the benefits of dynamic scheduling and integrated planning in achieving production targets and improving operational performance in open pit mining operations.

Bernardi, Kumral and Renaud (2020) used an ARENA simulation model to compare semi-mobile and fixed IPCC systems for open pit mines, focusing on NPV and production target proximity. For a simplified cone-shaped mine, results showed semi-mobile IPCC generated 10 per cent higher NPV and better met production targets. However, the simplified cost model and mine geometry may not reflect typical mining project complexity. Abbaspour and Drebenstedt (2023) compared various transportation systems in open pit mining, including truck-shovel, fixed IPCC, semi-mobile IPCC (SMIPCC) and fully mobile IPCC (FMIPCC) using a system dynamics modelling. The study introduces a technical index based on system availability, utilisation and power consumption to evaluate the performance over the life-of-mine. The research finds that while the truck-shovel system is generally the most preferred, certain periods favour the FMIPCC system.

Gong *et al* (2023) introduced a near-face stockpile (NFS) mining method that integrates in-pit crushing and conveying (IPCC) with an in-pit near-face stockpile to decouple mining and processing subsystems, enhancing operational flexibility and efficiency. The case study demonstrates that the NFS method leads to a 9.3 per cent increase in NPV and a 20 per cent reduction in head grade deviation compared to traditional methods. This approach aligns with IPCC concepts by enhancing production stability and reducing operational costs through improved equipment utilisation and lower haulage. In a subsequent study, Gong, Moradi Afrapoli and Askari-Nasab (2023) further developed the NFS method, employing discrete event simulation and MILP optimisation to assess its performance. Their case study in an oil sands mine confirms NFS's advantages, demonstrating increased production and reduced transportation costs.

IPCC can be a viable alternative to truck haulage in an era of constantly rising environmental concerns over mining. While IPCC is not a new concept, it is yet to be adopted widely in open pit mines across the world. Several life cycle assessment studies and environmental comparisons (Norgate and Haque, 2013; Erkayaoğlu and Demirel, 2016; Fuming, Qingxiang and Shuzhao, 2015; Bao *et al,* 2023) have found IPCC system to be more ecofriendly compared to pure truck-shovel haulage. Moreover, several economical comparative studies between IPCC and truck-shovel haulage systems found IPCC to be more cost-effective (De Werk, Ozdemir and Kumral, 2017; Nunes, Homero and Giorgio, 2019; Motswaiso and Suglo, 2022). Despite that, mine planning with IPCC has been underexplored. A comprehensive review of short-term mine planning and IPCC by Habib, Ben-Awuah and Askari-Nasab (2023a) shows that mine planning, more specifically short-term mine planning considering IPCC is an almost unexplored area of research. Majority of the IPCC literature, for example, Konak, Onur and Karakus (2007), Taheri, Irannajad and Ataee (2009), Rahmanpour, Osanloo and Adibi (2013), Roumpos, Partsinevelos and Vlachou (2014), Paricheh and Osanloo (2016, 2020), and Paricheh, Osanloo and Rahmanpour (2016, 2017) have been concerned with finding an optimum crusher location and time to install IPCC systems without considering the fact that the optimality of an IPCC system needs to be integrated to the mine plan.

Several methodologies have been developed for simultaneous optimisation of IPCC locations and long-term schedules. Key contributors include Paricheh and Osanloo (2019), Samavati, Essam and Nehring (2020), Shamsi, Pourrahimian and Rahmanpour (2022), Liu and Pourrahimian (2021), Kamrani *et al* (2024), Findlay and Dimitrakapoulos (2024), etc. The primary objective across these studies is to maximise the net present value of the production while leveraging IPCC as the primary mode of material transport. Shamsi and Nehring (2021) analysed scenarios to find the optimal depth for switching from truck-shovel haulage to Semi-Mobile In-Pit Crushing and Conveying (SMIPCC). Using a hypothetical cone-shaped mine with four pushbacks, they found the switch is most economically advantageous at 335 m during the second phase. Gölbaşı and Demirel (2017) developed a simulation algorithm aimed at optimising inspection intervals for mining equipment, with a focus on reducing maintenance costs and maintaining operational efficiency. Using real data from two draglines in a coalmine, the study demonstrated significant cost savings by optimising inspection schedules. The approach is relevant for complex mining operations, like those using IPCC systems, where equipment uptime and cost control are crucial.

The above discussion highlights that short-term planning optimisation using IPCC is a largely unexplored research area. Key decisions regarding IPCC, including optimal crusher location, crusher relocation timing, and conveyor design, are made during the strategic phase of mine planning. Short-term planning must adapt to the installation and movement of crushers and align with long-term strategies to achieve the desired NPV. Furthermore, the uncertainties associated with IPCC, truck and shovel operations add to the complexity of the short-term planning optimisation. To address this issue, the paper introduces and verifies a simulation-optimisation framework designed to optimise short-term planning schedules for open pit mines, accounting for equipment operational uncertainties. This general framework is applicable to mines utilising pure truck-shovel haulage and/or IPCC systems. The framework builds upon and enhances the short-term planning methodology proposed by Habib, Ben-Awuah and Askari-Nasab (2023b) by addressing operational uncertainties. Additionally, it can serve as a comparative tool to determine the performance of IPCC system in comparison with truck-shovel system in terms of haulage cost savings.

PROBLEM DEFINITION

This study aims to develop a robust short-term planning methodology that enhances production schedules by incorporating uncertainties associated with trucks, shovels, and IPCC systems. Rather than relying on deterministic models that may overlook operational variability, this research leverages Monte Carlo simulation to provide a more realistic representation of haulage uncertainties. The simulation model accounts for probability distributions governing truck travel times, shovel loading times, dumping durations, and equipment failures, offering a more comprehensive analysis of production performance under uncertain conditions.

Conventional evaluations of IPCC and truck-shovel haulage systems often disregard inherent operational variabilities, leading to less reliable decision-making. By employing Monte Carlo simulation, this study ensures a more robust assessment of these haulage systems under fluctuating conditions, allowing for the generation of a range of possible outcomes. This probabilistic approach provides deeper insights into how haulage uncertainties influence mine performance.

The key objectives of this research are:

- Simulating various operational scenarios, this study will assess how uncertainties affect haulage costs, production efficiency, and overall revenue.

- Comparing haulage systems under uncertainty: A probabilistic analysis of IPCC and truck-shovel haulage will provide a more thorough evaluation of their performance, moving beyond the limitations of deterministic models.

- Enhancing decision-making for mine planners: The findings from this simulation will support mine planners in making more informed choices regarding IPCC implementation, particularly concerning potential cost savings and revenue optimisation.

ASSUMPTIONS OF THE PROPOSED SIMULATION MODEL

- Probability Distributions: Full and empty haulage speeds, loading/dumping time, bucket capacity and other uncertain variables follow specific probability distributions (eg normal, lognormal, exponential). These distributions are based on recorded historical mining data and are modified for confidentiality reasons.

- Fixed Operational Parameters: Certain parameters, such as route lengths, payload capacities, shovel cycle times and capacities, are assumed to be constant.

- Working Conditions: The working conditions of the mine are assumed to be same for the scenario with IPCC and TS because of the lack of historical mining data with IPCC. As a result, the parameter and variable distributions for trucks and shovels, such as truck speed, loading, and spotting times, are kept the same in both scenarios.

- Failures and Downtime: Equipment failures and downtime are incorporated into the simulation and modelled using statistical distributions based on historical data.

- Steady-State Conditions: The simulation assumes the system reaches steady-state conditions, where the effects of initial transient states are negligible.

- Schedules: Production schedules and haulage demands are fixed across all simulated scenarios and extracted from an MILP optimisation model.

- Mine Layout: The physical layout of the mine, including the road network and dumping locations, is predefined and does not change during the simulation.

- Resource Allocations: The allocation of shovels to mining faces are determined by the MILP model. The ore and waste trucks are separate and locked to respective ore and waste shovels throughout the simulation.

- Waiting time: In this simulation, trucks are not modelled as individual agents or entities. Instead, truck waiting times are generated randomly based on a predetermined distribution. Waiting time is considered for shovel loading only. The M/M/c queuing model, which assumes exponential arrival and service times, has been was used to estimate the waiting time distribution due to the lack of historical waiting time data. The parameters are chosen to reasonably reflect the operational realities.

METHODOLOGY

This paper presents a haulage simulation model designed to evaluate short-term mine planning while accounting for operational uncertainties. Figure 1 provides an overview of the framework. Initially, the block model data for the planning period is utilised to cluster blocks into mining cuts. Shovels are then allocated to these clusters, and a monthly production schedule is generated using a MILP model. The inputs for this model include face IDs, tonnages in each face, costs associated with haulage, mining, and processing, ore price, and the locations of the crusher, conveyor, and waste dumps. The mathematical model is not part of this article. The optimum schedules generated by this model has been used to run the simulation model that captures key haulage dynamics by incorporating statistical distributions for critical parameters such as loading time, travel time, spotting time, and equipment failures.

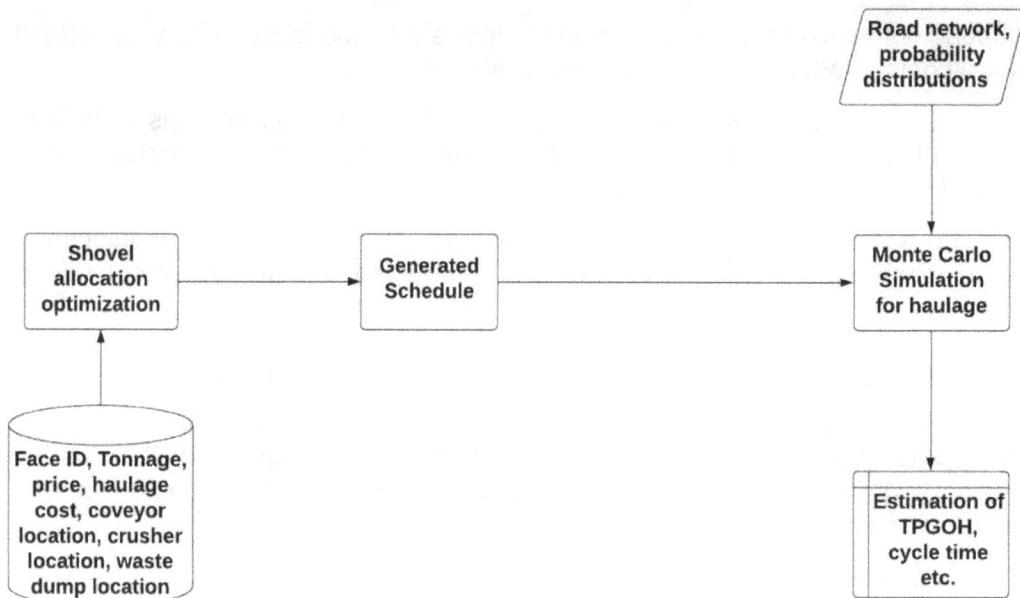

FIG 1 – Outline of the proposed methodology.

The simulation framework utilises detailed road network data and probabilistic inputs to assess haulage performance under varying conditions. By comparing scenarios with IPCC and conventional truck-shovel haulage, the model provides insights into key performance indicators, including tonnes per gross operating hour (TPGOH), truck cycle time, and actual production levels. This approach ensures a more realistic assessment of haulage operations, enabling better decision-making in mine planning.

Simulation logic flow

Figure 2 highlights the simulation logic flow as explained here.

Ore haulage

- **Excavator assignment:** If it is an ore face, an ore shovel is assigned to the scheduled ore face.

- **Mining status check:** The system checks if the face has been mined out.

- **Loading process:** If not mined out, an ore truck positions itself to be loaded by the shovel.

- **Haulage path:**

    o **With IPCC:** The ore truck travels to the in-pit crusher, dumps its load, and returns empty to the shovel.

- o **Without IPCC:** The truck travels to the mill crusher, dumps its load, and returns empty to the shovel.
- **Face reassignment:** If the current ore face is mined out, the simulation checks if all ore faces have been mined out. If not, the shovel is reassigned to the next scheduled ore face.
- **Termination check:**
  - o If all ore faces are mined out, the simulation checks if all waste faces have been mined too.
  - o If both conditions are met, the simulation terminates.
  - o If all waste faces are not mined, the shovel is redirected to waste faces.

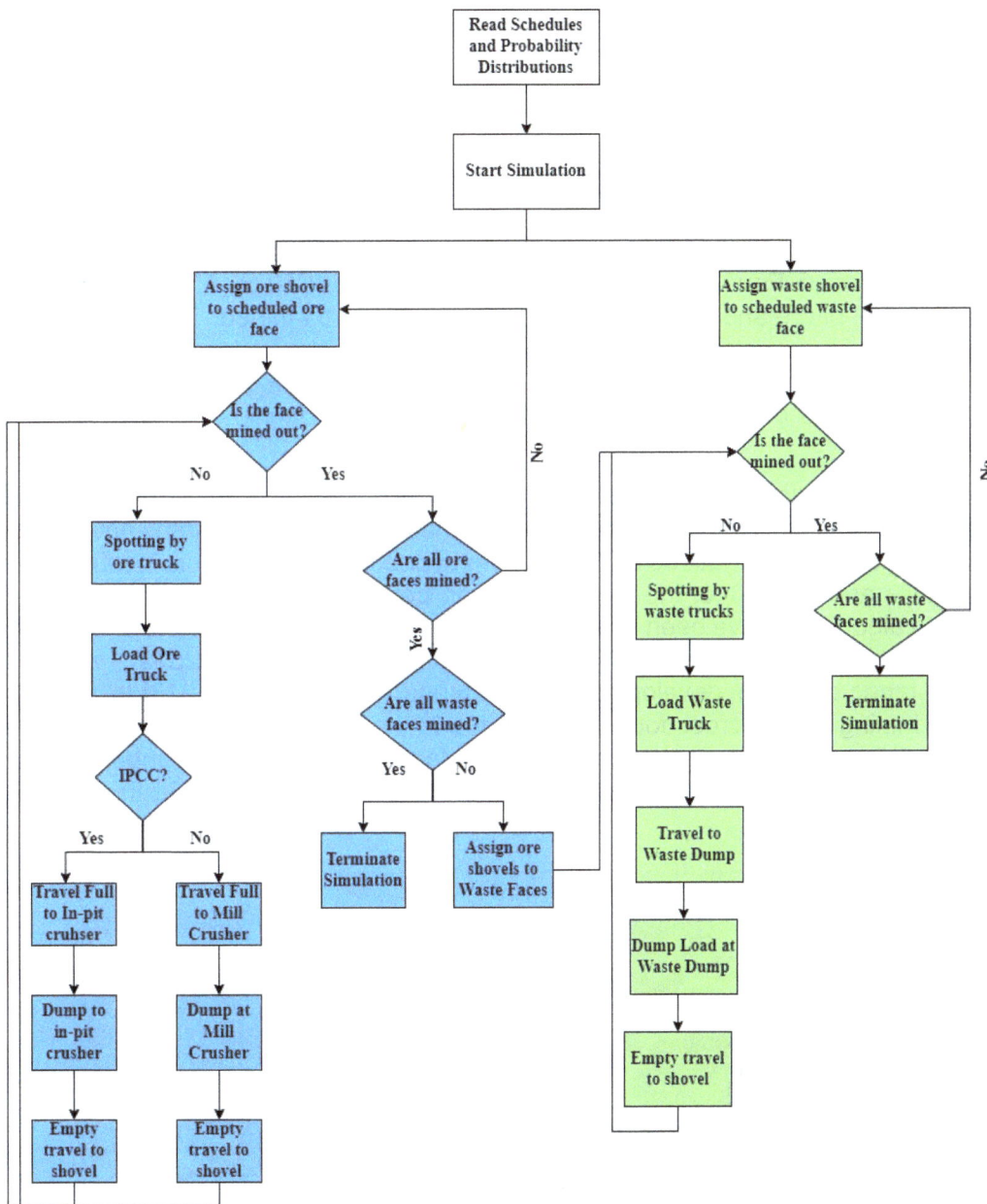

FIG 2 – Haulage simulation flow chart.

Waste haulage

- **Excavator assignment:** Waste excavators are assigned to scheduled waste faces.
- **Mining status check:** The simulation checks if the waste face is mined out.
- **Loading process:**

- o If not mined out, a waste truck positions itself to be loaded by the shovel.
- **Haulage path**:
  - o The loaded waste truck travels to the waste dump, dumps its load, and returns empty to the shovel.
- **Face reassignment**:
  - o If the current face is mined out, the system checks if all waste faces have been mined.
  - o If not, the waste shovel is reassigned to the next waste face.
- **Termination check**:
  - o If all waste faces are mined out, the simulation terminates.

This simulation framework ensures efficient and continuous haulage operations, reflecting the complexities and uncertainties of real mining activities.

CASE STUDY

The case study evaluates two mining scenarios in an iron ore mine: one using the IPCC system for ore and the other with the traditional truck-shovel method, focusing on the short-term schedule for the 11th year of operation. Four benches on elevations 1595 m, 1610 m, 1730 m and 1745 m, with 16 million tonnes (Mt) of ore and 35 Mt of waste are available to be mined. The case study includes one processing plant, one waste dump, and one crusher, which can be located either inside the pit or externally at the plant site. The crusher must process 2700 t per hr, requiring 1.33 Mt per month with two eight-hour shifts daily. The element of interest is the magnetic weight recovery of iron (MWT).

Figure 3 shows the mine layout for the 11th year, with distances from mining faces to the waste dump, crusher, and plant calculated based on the road network. The mine ramps have an 8 per cent grade, and the IPCC scenario requires a 2550 m conveyor belt. The crusher is located on face 3 for the first six months and face 18 for the rest of the year, both at bench 1595. Although several high quality clustering algorithms have been proposed by researchers (Valença Mariz *et al*, 2024; Askari-Nasab, Tabesh and Badiozamani, 2010; Tabesh and Askari-Nasab, 2019), we utilised the hierarchical clustering algorithm developed by Tabesh and Askari-Nasab (2013) to aggregate 4200 blocks into 170 mining faces across four benches due to the method's ease of use and versatility. Figures 4 and 5 illustrate the clustered ore and waste faces, with face IDs numbered, respectively on benches 1595 and 1610. All the faces of the benches on elevation 1730 m and 1745 m are designated as waste.

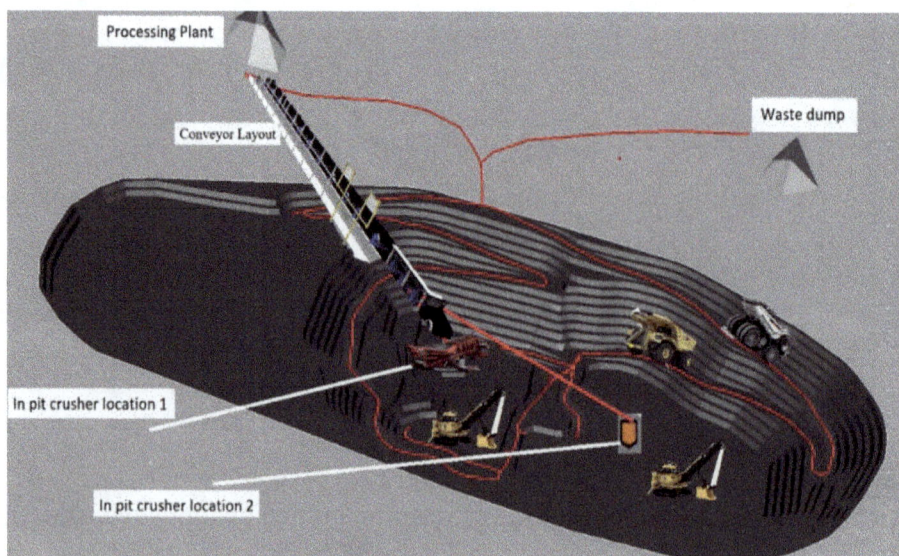

FIG 3 – Mine layout for year 11.

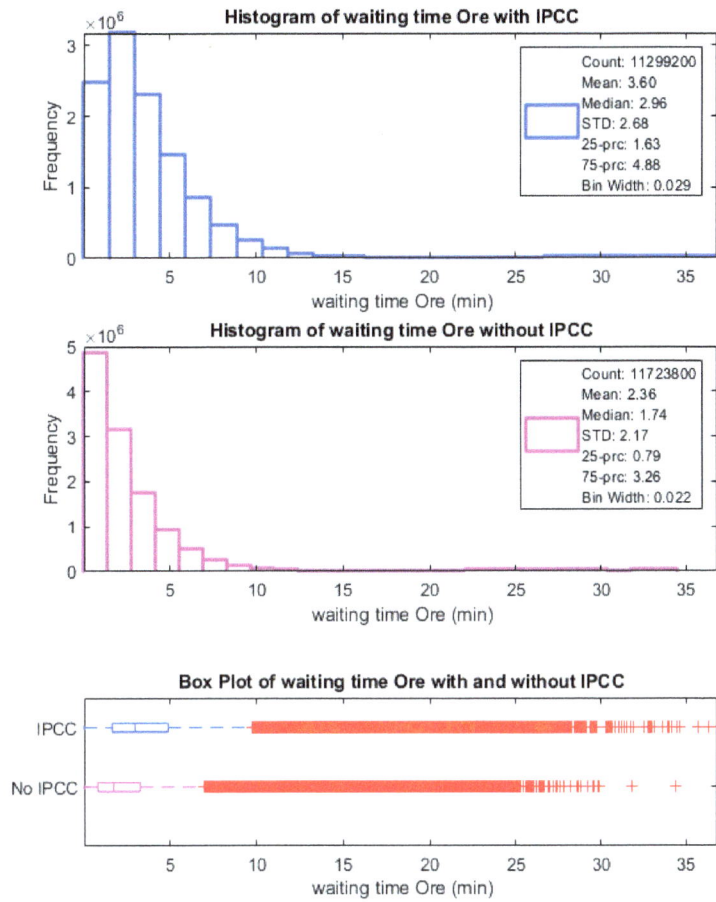

FIG 4 – Comparison of ore truck waiting time.

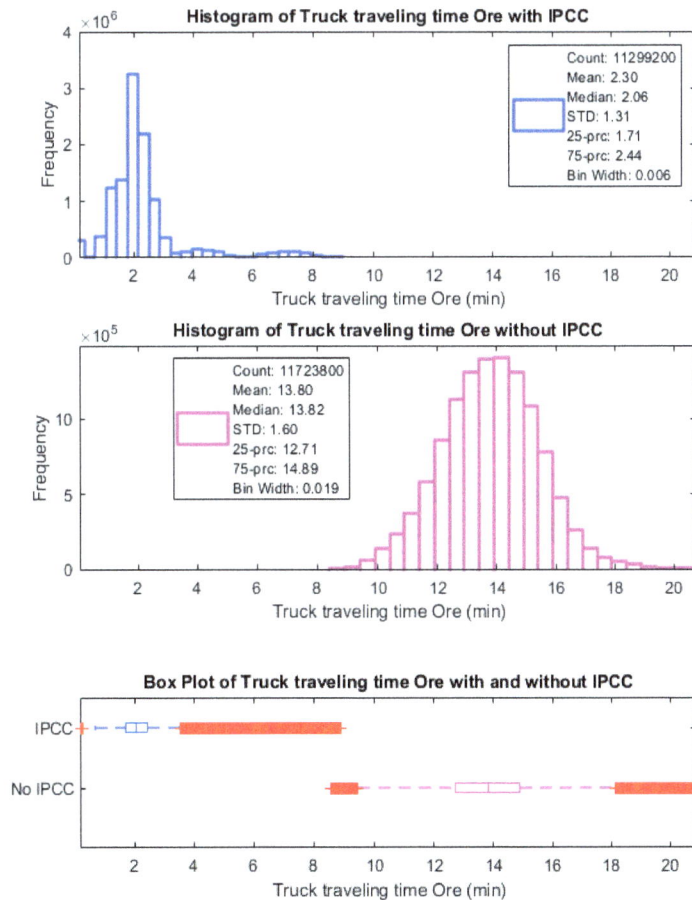

FIG 5 – Comparison of ore truck traveling time.

The mine uses five shovels: two Hit 2500 for ore and three Hitachi 5500Ex for waste. The Hit 2500 shovels have a 12 t bucket capacity with a 22 sec cycle time, while the Hitachi 5500Ex shovels have a 22 t bucket capacity with a 23 sec cycle time. Cat 785C trucks (140 t capacity) are paired with Hit 2500 shovels, and Cat 793C trucks (240 t capacity) are paired with Hitachi 5500Ex shovels.

Equipment breakdowns are incorporated into the haulage simulation using probabilistic modelling. The Monte Carlo Simulation (MCS) framework requires only the probability distribution functions for failure events to be effective. Table 1 outlines the distribution functions used for mean time between failures (MTBF) and mean time to repair (MTTR) in this study. While failure data for trucks and shovels are derived from historical records specific to mining operations, the distribution for conveyor system failures is taken from Londoño, Knights and Kizil (2013), due to the unavailability of real-world IPCC failure data. As explained in Chapter 3, the simulation accounts solely for failures in the conveyor system, excluding potential breakdowns of other IPCC components such as the crusher.

TABLE 1
MTBF and MTTR distribution functions.

Equipment	MTBF (hr)	MTTR (hr)
Hit 2500	WEIB (32, 216)	GAMM (1.4, 1.5)
Hit 5500Ex	WEIB (32, 216)	GAMM (1.4, 1.5)
Cat 785C	WEIB (27, 200)	GAMM (1.4, 1.5)
Cat 793C	WEIB (27, 200)	GAMM (1.4, 1.5)
IPCC conveyor	WEIB (0.714, 31.9)	GAMM (1.4, 1.19)

It is to be noted from (Londoño, Knights and Kizil, 2013) that the conveyor system experiences three distinct types of failures: conveyor take-up, electrical, and mechanical, with MTBF distributions modelled as Weibull (0.48, 36.99), Weibull (0.7, 31.4), and Weibull (0.64, 50.96), respectively. The corresponding MTTRs for these failures follow Gamma distributions, specifically Gamma (1.5, 0.55), Gamma (1.6, 1.1), and Gamma (1.4, 1.7). To simplify analysis, we combined these three failure modes into a single MTBF and MTTR distribution using Monte Carlo simulation. A key assumption in this process was that the failures occur independently. We generated 1000 random values for each of the MTBF and MTTR based on their respective distributions, resulting in 3000 random data points. These were then used to fit the best distribution using ARENA Input Analyzer.

SIMULATION RESULTS

The simulation model integrates data from the optimisation model to effectively simulate ore and waste production, considering operational and equipment uncertainties. It evaluates the haulage performance in both the IPCC and non-IPCC scenarios, utilising a range of variables detailed in Table 2. The statistical properties of these variables, including averages, standard deviations, and other relevant metrics, are derived from 100 replications of the simulation model for each scenario, ensuring robust and reliable performance assessments. The probability distribution and corresponding distribution parameters of the KPIs and truck-shovel failure statistics are generated from the historical data of the iron ore mine. Because of the lack of historical data on IPCC failure, the failure distributions are adopted from Londoño, Knights and Kizil (2013).

TABLE 2

Key performance indicators (KPIs) to be compared.

Variable	Distribution	Variable type
Spotting time (min)	Empirical	
Loading time (min)	Empirical	Independent
Dumping time (min)	Empirical	
Truck speed (km/hr)	Triangular	
Truck traveling time (min)		
Truck cycle time (min)		
Waiting time (min) (Gamma)		Dependent
Ore truck TPGOH (tonne/hr)		
Total production (tonne)		

It is to be noted that the distribution of waiting time has been derived using M/M/C (exponential arrival and service time) queuing model by considering truck speed, distance travelled, number of trucks and shovels. Since, there are two ore and three waste shovels in this case study, the waiting time distribution does not follow an exponential distribution but is better approximated by a Gamma distribution due to the combined effect of multiple servers working in parallel (Bowker, Olkin and Veinott, 2009). The waiting time distributions for ore truck, CAT 785C, are GAMMA (2, 1.18) and GAMMA (2, 1.8) for the scenario without and with IPCC respectively. The distribution of waiting time for the waste truck, CAT 793C, is GAMMA (3,1.62) for both scenarios.

Comparison of independent variables

The independent variables listed in Table 2 comprise truck cycle time. A summary of the comparative statistics for the independent variables is illustrated in Table 3.

A comparison of the individual key performance indicators (KPIs) between the IPCC and non-IPCC scenarios shows that most variables perform identically in both cases. In particular, the average values for ore spotting, loading, and dumping times—as well as those for waste—remain unchanged across scenarios. This uniformity suggests that these operational components are unaffected by whether or not an IPCC system is used. The consistency arises from the assumption that these variables are independent of the haulage system, with the same probability distributions applied in both scenarios.

The confidence intervals (CIs) for the variables in Table 2 are remarkably narrow, indicating a high level of precision in the simulation results. Given that 100 replications were run for each variable, this precision suggests that the simulation model is robust and reliable. For example, the CI for Ore Spotting time (min) under the IPCC scenario is 0.585 ± 0.00015, demonstrating minimal variability around the mean. Similarly, the Ore Loading time (min) has a CI of 3.537 ± 0.0006 for the IPCC scenario, further reinforcing the consistency of the simulation results. This high precision across different variables, whether it is ore Dumping time, or waste loading time, underscores that the simulation outputs are dependable. The narrow CIs imply that the variability in the data is minimal, and the results are not significantly influenced by outliers or anomalies.

TABLE 3

Comparison of independent variables between scenarios.

Variable	Scenario	Mean	Median	St dev	95% confidence interval
Ore spotting time (min)	IPCC	0.58	0.56	0.25	0.58 ± 0.0002
	No IPCC	0.58	0.55	0.25	0.58 ± 0.0001
Ore loading time (min)	IPCC	3.54	3.46	1.03	3.54 ± 0.0006
	No IPCC	3.54	3.46	1.03	3.54 ± 0.0006
Ore dumping time (min)	IPCC	1.35	1.46	0.2	1.35 ± 0.0001
	No IPCC	1.35	1.46	0.2	1.35 ± 0.0001
Ore truck speed (km/hr)	IPCC	36.67	36.34	3.12	36.67 ± 0.002
	No IPCC	36.66	36.34	3.12	36.66 ± 0.002
Waste spotting time (min)	IPCC	0.58	0.54	0.25	0.58 ± 0.0001
	No IPCC	0.58	0.54	0.25	0.58 ± 0.0001
Waste loading time (min)	IPCC	4.51	4.51	1.03	4.51 ± 0.0006
	No IPCC	4.51	4.51	1.03	4.51 ± 0.0006
Waste dumping time (min)	IPCC	1.32	1.46	0.22	1.32 ± 0.0001
	No IPCC	1.32	1.46	0.22	1.32 ± 0.0001
Waste truck speed (km/hr)	IPCC	36.84	36.98	3.49	36.84 ± 0.068
	No IPCC	36.65	36.17	3.02	36.65 ± 0.06

Comparison of dependent variables

Key performance indicators (KPIs) such as truck travel time, waiting time, cycle time, tonnes per gross operating hour (TPGOH), and total production play a crucial role in evaluating haulage system performance. These KPIs are directly influenced by the time spent in truck loading, spotting, dumping, traveling, and queuing. Among them, truck cycle time is particularly important, as it directly impacts the production rate—shorter cycle times typically result in higher output and fewer trucks needed to meet production targets.

The integration of the IPCC system is expected to lower truck cycle times and reduce the overall truck fleet size by decreasing the haul distance. To evaluate haulage efficiency, both TPGOH and total material moved during the planning period are examined. These indicators reflect the system's throughput and productivity. Table 4 presents detailed statistics for these dependent KPIs, providing a comparative analysis of system performance with and without the use of IPCC. All differences in KPI values are measured relative to the scenario without IPCC.

TABLE 4
Comparison of dependent variables.

Variable	Scenario	Mean	Median	St Dev	95% confidence interval
Ore truck traveling time (min)	IPCC	2.30	2.06	1.31	2.30 ± 0.0007
	No IPCC	13.8	13.82	1.61	13.80 ± 0.0009
Difference (%)		83.35	85	18.63	N/A
Waste truck traveling time (min)	IPCC	12.19	12.95	3.84	12.19 ± 0.002
	No IPCC	12.18	12.86	2.85	12.18 ± 0.0014
Difference (%)		-0.05	-0.73	-34.77	N/A
Ore truck waiting time (min)	IPCC	3.60	2.96	2.68	3.6 ± 0.0014
	No IPCC	2.36	1.74	2.17	2.36 ± 0.0013
Difference (%)		-40.34	-70.11	-23.50	N/A
Waste truck waiting time (min)	IPCC	4.86	3.90	3.82	4.86 ± 0.0019
	No IPCC	4.86	3.90	3.82	4.86 ± 0.0019
Difference (%)		0	0	0	N/A
Ore truck cycle time (min)	IPCC	9.52	9.28	1.73	9.52 ± 0.001
	No IPCC	21.63	21.33	2.91	21.63 ± 0.001
Difference (%)		56	55.17	40	N/A
Waste truck cycle time (min)	IPCC	23.43	23.32	5.54	23.43 ± 0.003
	No IPCC	23.44	23.32	5.55	23.44 ± 0.003
Difference (%)		0.05	0	0	N/A
Ore truck TPGOH (tonne/hr)	IPCC	967	956.5	197.35	967 ± 0.12
	No IPCC	376.89	373.34	38.31	376.89 ± 0.022
Difference (%)		-156.57	-156.21	- 415	N/A
Total production (Mt)	IPCC	14.16	14.16	0.14	14.16 ± 0.027
	No IPCC	14.32	14.32	0.09	14.32 ± 0.017
Difference (%)		1.12	1.12	-60	N/A

The traveling, waiting and cycle time of waste trucks remain consistent between IPCC and no IPCC scenarios because of the absence of no waste IPCC system. The ore truck waiting time, however, in the IPCC scenario, is substantially higher, averaging 3.6 mins compared to 2.36 mins in the no IPCC scenario, as depicted by Figure 4. This discrepancy can be attributed to the shorter haul distances in the IPCC scenario. Because trucks in this scenario travel less distance to reach the in-pit crusher, they return to the loading points much quicker than in the no IPCC scenario, where longer haul distances provide more spacing between truck cycles.

The stark reduction in ore cycle time between the scenarios is primarily attributed to the decreased truck travel times and distances facilitated by the implementation of IPCC. In the non-IPCC scenario, the average distance from the ore faces to the mill crusher is 3.1 km. This distance dramatically decreases to just 0.65 km with the integration of an in-pit crusher, illustrating the effectiveness of the IPCC system. This causes the ore truck traveling time to reduce to a mean of 2.29 km/hr with IPCC compared to 13.8 km/hr in the no IPCC scenario. Consequently, the average ore cycle time with IPCC is significantly lower at 9.52 mins, compared to 20.26 mins under the conventional truck-shovel haulage system. Table 4, Figures 5 and 6 also show that the IPCC scenario not only has a lower average traveling time and cycle time but also exhibits less variability, indicating more consistent performance. The two small bumps to the right in the blue histogram on Figure 16 likely indicate that a subset of trucks experienced minor delays, possibly due to maintenance issues or brief stoppages during their routes, causing moderately longer travel times in the IPCC scenario.

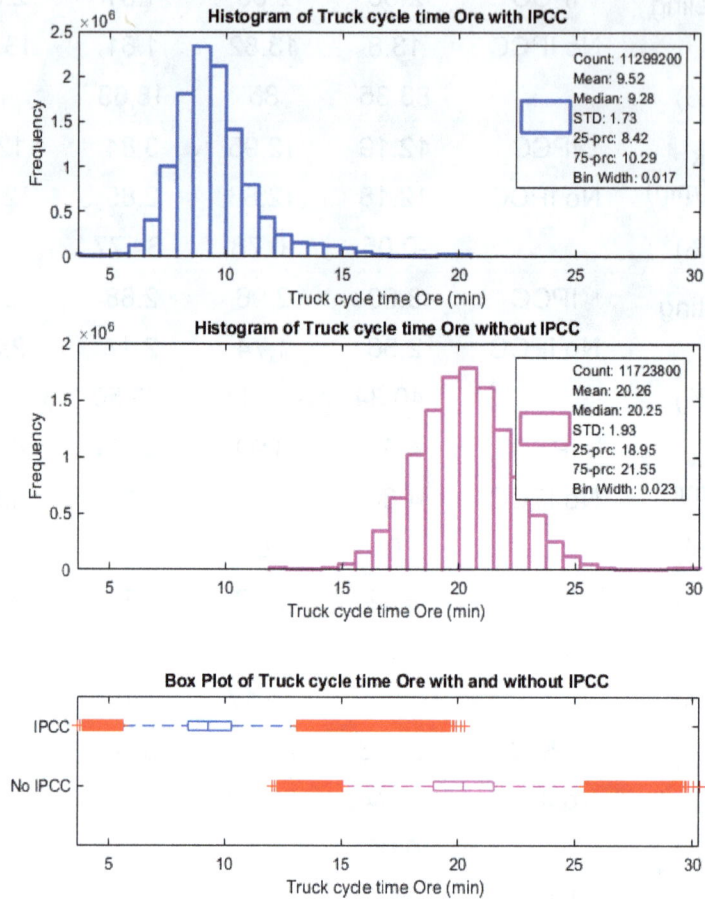

FIG 6 – Ore truck cycle time comparison.

Figures 7 and 8 illustrate the relationship between Tonnage per Hour (TPGOH) and distance to the crusher under IPCC and no IPCC scenarios. In the IPCC scenario, trucks consistently achieve higher TPGOH values, particularly at shorter distances, with values ranging from 1500 to over 3000 t per hr. As distance increases, there is a gradual decline in TPGOH, but the system maintains relatively high productivity due to the continuous operation of the conveyor system, which minimises truck cycle time and maximises throughput. However, the IPCC scenario also shows greater variability in TPGOH, suggesting that while productivity is enhanced, the system introduces additional operational complexities, likely due to IPCC-related failure probabilities. Despite this, four trucks are sufficient to meet the required annual ore tonnage.

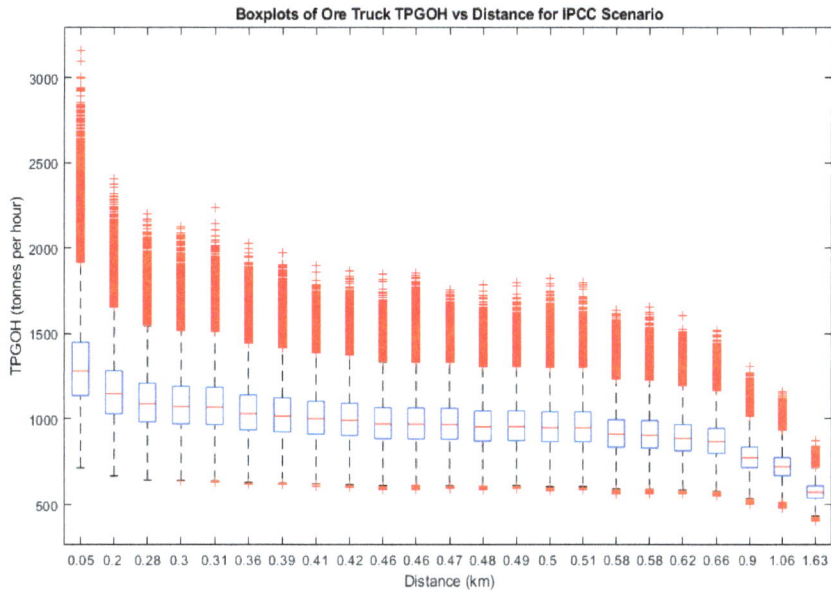

FIG 7 – Box plot of ore truck TPGOH versus distance in the IPCC scenario.

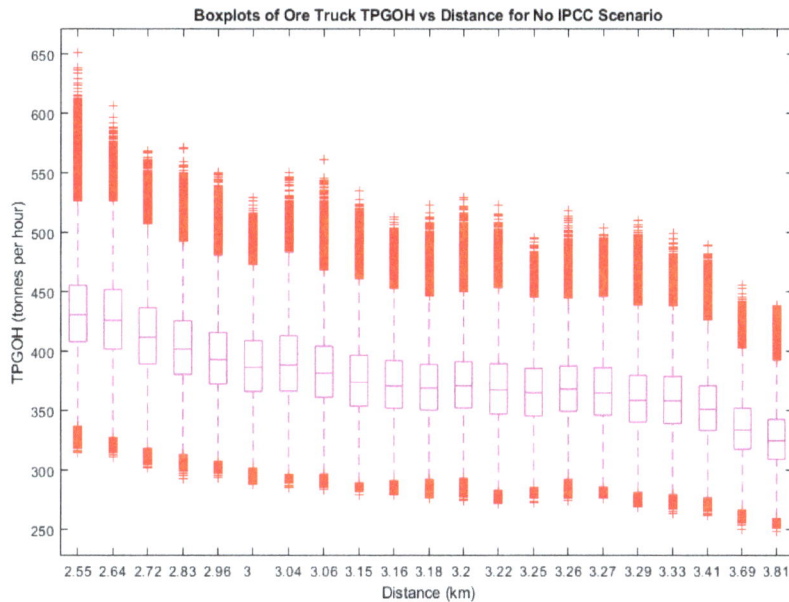

FIG 8 – Box plot of ore truck TPGOH versus distance in the no IPCC scenario.

Conversely, In the no IPCC scenario, TPGOH values are lower, generally between 300 and 600 t per hr, and the decline with distance is more pronounced. While the lower variability indicates more consistent performance, each truck's productivity is reduced. This results in lower overall efficiency despite the steadier output. Consequently, ten trucks are needed to maintain the required production. As a result, more trucks, ten in this case, are needed to maintain the required annual production. This comparison underscores the advantage of IPCC systems in reducing the number of trucks needed to achieve the same production, highlighting its potential for cost savings and operational efficiency in mining operations.

System performance and reliability

When assessing the performance of haulage systems, it is essential to consider not only operational efficiency but also system reliability. This section focuses on a comparative analysis of failure-related statistics for the key components involved in haulage: trucks, shovels, and the IPCC system. For each of these components, the time between failures (MTBF) and time to repair (MTTR) are modelled using Weibull and Gamma distributions, respectively.

The simulation calculates the downtime for each component individually and aggregates them to estimate the total system downtime. This method assumes that failures across trucks, shovels, and conveyors occur independently, with each component subject to breakdowns according to its specific reliability characteristics.

By comparing failure data across the IPCC and no-IPCC scenarios, the analysis aims to highlight the impact of IPCC implementation on system dependability and the variability of operations. Table 5 provides a detailed summary of downtime statistics for the two scenarios, with percentage differences reported relative to the no-IPCC case.

TABLE 5

Comparison of downtime statistics.

Variable	Scenario	Mean	Median	St dev	95% confidence interval
Ore truck downtime (hr)	IPCC	145.58	145.43	14.66	145.58 ± 2.87
	NO IPCC	330.99	330.71	19	330.99 ± 3.72
Difference (%)		56	56	22.84	N/A
Waste truck downtime (hr)	IPCC	445.09	443.23	27.12	445.093 ± 5.317
	NO IPCC	437.63	439.84	26.82	437.63 ± 5.26
Difference (%)		-1.7	-0.77	-1.12	N/A
Ore shovel downtime (hr)	IPCC	288.65	285.08	26.71	288.65 ± 5.23
	NO IPCC	334.91	332.02	22.31	334.91 ± 4.37
Difference (%)		13.81	14.14	-19.72	N/A
Waste shovel downtime (hr)	IPCC	447.12	446.88	22.54	447.12 ± 4.42
	NO IPCC	443.55	442.16	22.51	443.55 ± 4.41
Difference (%)		-0.8	-1.1	-0.13	N/A
IPCC downtime (hr)	IPCC	293.79	289.78	26.73	293.79 ± 5.24

In the case of ore trucks, the IPCC scenario demonstrates a substantial improvement in downtime performance. The average time lost due to failures drops to 145.58 hrs, representing a 56 per cent reduction compared to the 330.99 hrs observed in the no-IPCC scenario. This notable decrease is largely attributed to the shortened haul distances and reduced travel time made possible by the IPCC system. Additionally, the variability in downtime differs significantly between the two cases. The IPCC scenario exhibits a standard deviation of 14.66 hrs, indicating a more consistent and predictable failure pattern. In contrast, the no-IPCC scenario shows a higher standard deviation of 19 hrs, reflecting 23 per cent more variability and less stability in ore truck operations. These differences are clearly illustrated in Figure 9, which emphasises the improved operational reliability introduced by IPCC integration.

However, the inclusion of the IPCC system introduces a new source of downtime. The conveyor system itself experiences an average of 293.79 hrs of downtime, a factor absent in the traditional truck-shovel configuration. This highlights an important trade-off: while IPCC reduces truck-related downtime, it adds its own operational vulnerability.

For waste trucks, the difference in downtime between the two scenarios is minimal. The IPCC scenario reports 445.09 hrs, slightly higher than the 437.63 hrs in the no-IPCC case. Similarly, waste shovel downtime remains largely unchanged, with values of 447.12 hrs and 443.55 hrs for the IPCC and no-IPCC scenarios, respectively. This consistency is expected, as the IPCC system in this case study is implemented exclusively for ore haulage and does not influence waste operations.

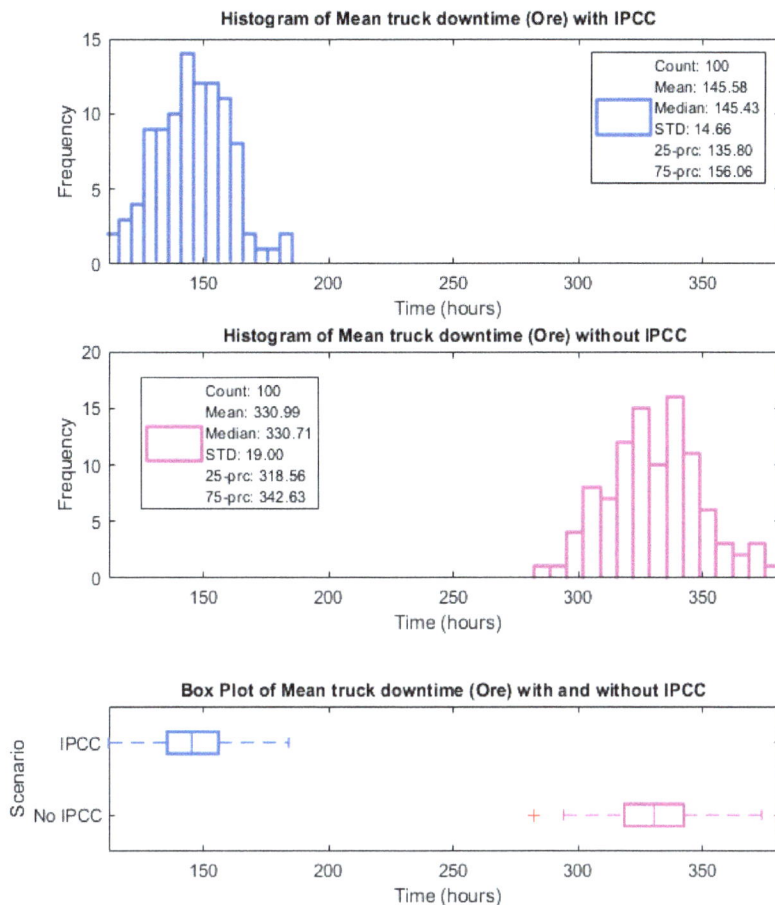

FIG 9 – Comparison of ore truck downtime.

Ore shovel downtime shows moderate improvement with IPCC, dropping to 288.65 hrs compared to 334.91 hrs without IPCC, a reduction of approximately 13.8 per cent. Although less dramatic than the improvement seen with ore trucks, this still reflects a gain in operational efficiency.

Overall, the implementation of IPCC significantly enhances ore truck performance by reducing downtime through shorter travel distances. Nonetheless, it introduces a new layer of downtime risk associated with the IPCC system itself, underscoring the importance of evaluating both benefits and trade-offs when adopting such technologies.

The total repair time in this analysis is the sum of the simulated MTTRs for all components of the haulage system. With the assumption of independent failures for trucks, shovels, and IPCC, this approach generates a reasonable estimate of the mean total repair time for the haulage system as a whole. Figure 10 compares the aggregate repair time between scenarios with and without IPCC. The scatter plot displays the total repair time across 100 simulations for both scenarios. Each blue dot represents a simulation with IPCC, while each red dot represents a simulation without IPCC. The repair time for simulations with IPCC ranges broadly between approximately 1300 to 1550 hrs, while simulations without IPCC show a broader distribution, ranging from about 3700 to 4100 hrs. The scatter plot clearly illustrates that the IPCC scenario results in generally lower and less variable repair times compared to the No IPCC scenario, mainly because of lower truck MTTR and a smaller number of trucks in the fleet.

The IPCC scenario uses four trucks, while the No IPCC scenario uses ten trucks. The larger fleet in the No IPCC scenario distributes the workload, but individual truck failures in the IPCC scenario have a proportionally greater impact on aggregate repair time. Despite the conveyor adding an additional failure component, the IPCC scenario results in substantially lower total mean repair time. The scatter plot highlights the consistent separation between the two scenarios across simulated scenarios.

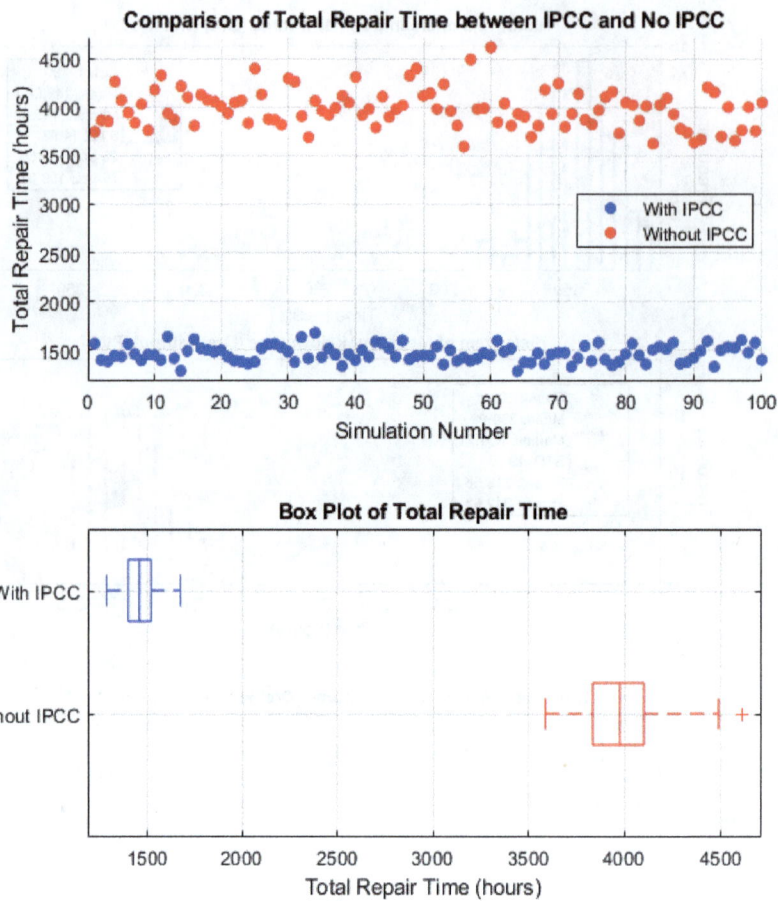

FIG 10 – Aggregate repair time comparison between the scenarios.

The box plot further illustrates these differences. The IPCC scenario has a narrower spread, indicating lower variability in repair time, whereas the No IPCC scenario shows a wider spread with some extreme outliers. This reflects the higher operational redundancy in the No IPCC set-up, despite its overall higher mean time to repair.

Comparison between simulated and optimal production

In this section, simulated production is compared to the optimal production targets set by the MILP for both the IPCC and TS haulage scenarios. The simulation accounts for uncertainties, such as equipment failures and variations in production parameters, which are not considered in the deterministic optimal plan. These uncertainties lead to deviations from the optimal plan in both scenarios, impacting the ability to meet production targets. The comparison aims to assess which haulage system performs better under uncertain conditions. Figure 11 illustrates the production values for both scenarios, highlighting the gap between simulated and optimal production. In the TS scenario, the optimal production target is 15.96 Mt, while the simulated production achieves 14.32 Mt, resulting in a shortfall of 1.64 Mt from the optimal target. This discrepancy underscores the limitations of the TS system, which may stem from longer haulage distances and higher operational inefficiencies.

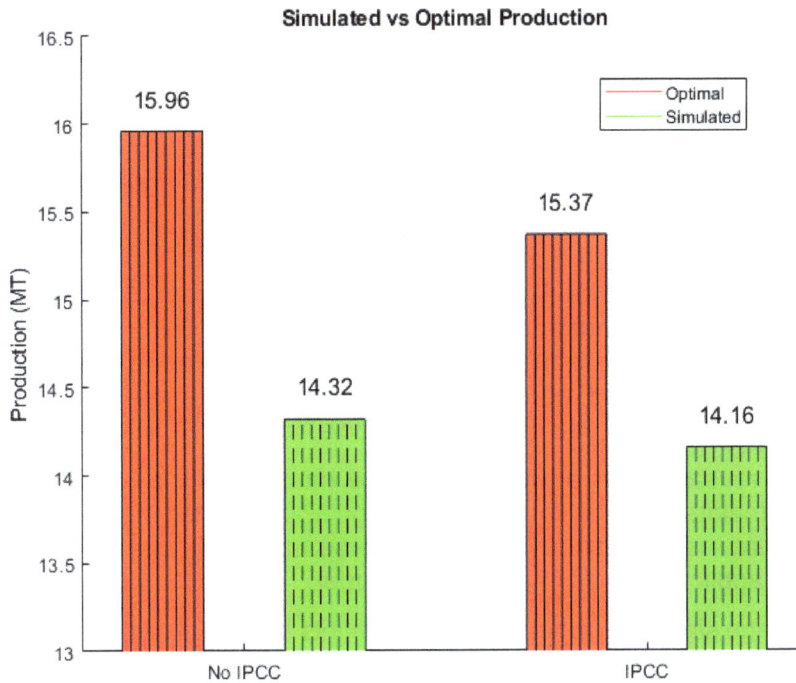

FIG 11 – Simulated versus optimal production.

In the IPCC scenario, the optimal production target is 15.37 Mt, with the simulated production reaching 14.16 Mt. This results in a shortfall of 1.21 Mt from the optimal production target. Although the IPCC system aims to improve efficiency by reducing haulage distance and cycle times, the additional operational complexities and potential downtime introduced by the IPCC system somewhat hinder its ability to reach the optimal production levels.

Figure 12 represents a scatter plot to provide a detailed comparison of the proximity to optimal production for both IPCC and non-IPCC scenarios across 100 simulations. Each point represents the fraction of the optimal production achieved in a specific simulation run. The blue dots indicate the performance of the IPCC scenario, while the red dots represent the performance without IPCC. The black dotted line represents the normalised optimal production target. It is evident from the plot that the IPCC scenario consistently achieves a higher fraction of the optimal production compared to the no IPCC scenario. The IPCC scenario frequently reaches production levels between 0.91 and 0.95 of the optimal production, whereas the no IPCC scenario tends to range between 0.88 and 0.92.

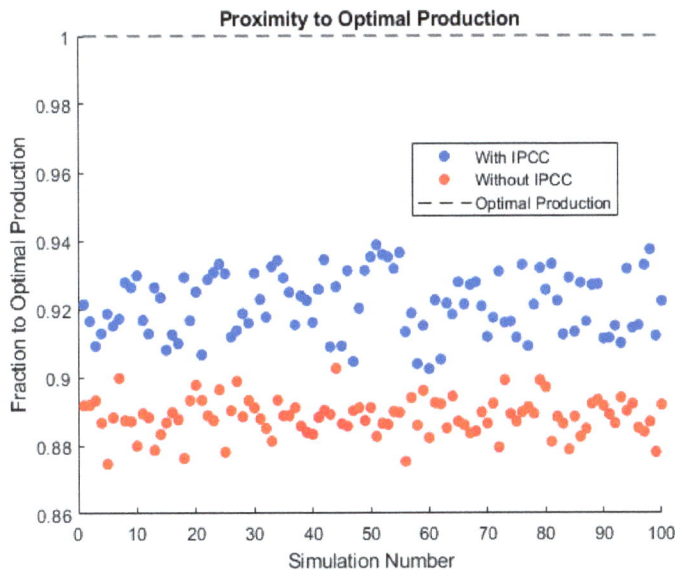

FIG 12 – Proximity to optimal production.

The comparison in Figures 11 and 12 reveals that while both scenarios fall short of their respective optimal production targets, the IPCC scenario has a slightly smaller deviation from the optimal production compared to the TS scenario. This suggests that the IPCC system, despite its operational complexity and potential for increased downtime, outperforms the traditional TS system in maintaining closer proximity to optimal production targets. However, both systems demonstrate the need for further improvements to bridge the gap between simulated and optimal production.

CONCLUSION

This paper presents a detailed haulage simulation framework to evaluate and compare the performance of IPCC and traditional truck-shovel systems in open pit mining. The results highlight that IPCC integration significantly reduces ore truck cycle time and haulage distance, leading to improved operational efficiency and reduced truck requirements. However, these benefits come with added complexity, including increased downtime associated with the IPCC system itself.

While the simulation reveals clear advantages in ore haulage efficiency, it also underscores the importance of robust maintenance strategies to manage IPCC-related failures. As this study focuses on a single case, future research should explore diverse mine settings to validate the findings and examine the broader applicability of the simulation model.

In summary, the haulage simulation offers valuable insights into the trade-offs of implementing IPCC, providing mine planners with a practical tool to assess system performance under uncertainty.

REFERENCES

Abbaspour, H and Drebenstedt, C, 2023. Truck–Shovel vs, In-Pit Crushing and Conveying Systems in Open Pit Mines: A Technical Evaluation for Selecting the Most Effective Transportation System by System Dynamics Modeling, *Logistics,* 7:92.

Askari Nasab, H, Tabesh, M and Badiozamani, M, 2010. Creating Mining Cuts Using Hierarchical Clustering and Tabu Search Algorithms.

Bao, H, Knights, P, Kizil, M and Nehring, M, 2023. Electrification Alternatives for Open Pit Mine Haulage, *Mining,* 3:1–25.

Ben-Awuah, E, Pourrahimian, Y and Askari-Nasab, H, 2010. Hierarchical Mine Production Scheduling using Discrete-Event Simulation, *International Journal of Mining and Mineral Engineering,* 2:137–158.

Bernardi, L, Kumral, M and Renaud, M, 2020. Comparison of fixed and mobile in-pit crushing and conveying and truck-shovel systems used in mineral industries through discrete-event simulation, *Simulation Modelling Practice and Theory,* 103:102100.

Blom, M L, Burt, C N and Stuckey, P J, 2014. A decomposition-based heuristic for collaborative scheduling in a network of open-pit mines, *INFORMS Journal on Computing,* 26:658–676.

Blom, M, Pearce, A and Stuckey, P, 2016. Short-term scheduling of an open-pit mine with multiple objectives, *Engineering Optimization,* 49:1–19.

Bodon, P, Sandeman, T and Stanford, C, 2011. Modeling the mining supply chain from mine to port: A combined optimization and simulation approach, *Journal of Mining Science,* 47:202–211.

Bowker, A H, Olkin, I and Veinott, A F, 2009. Gerald J Lieberman, *Probability in the Engineering and Informational Sciences,* 9:3–26.

De Werk, M, Ozdemir, B and Kumral, M, 2017. Cost analysis of material handling systems in open pit mining: Case study on an iron ore prefeasibility study, *The Engineering Economist,* 62:369–386.

Eivazy, H and Askari-Nasab, H, 2012. A mixed integer linear programming model for short-term open pit mine production scheduling, *Transactions of the Institutions of Mining and Metallurgy, Section A: Mining Technology,* 121:97–108.

Erkayaoğlu, M and Demirel, N, 2016. A comparative life cycle assessment of material handling systems for sustainable mining, *Journal of Environmental Management,* 174:1–6.

Findlay, L and Dimitrakapoulos, R, 2024. Stochastic Optimization for Long-Term Planning of a Mining Complex with In-Pit Crushing and Conveying Systems, *Mining, Metallurgy and Exploration.*

Fuming, L, Qingxiang, C and Shuzhao, C, 2015. A comparison of the energy consumption and carbon emissions for different modes of transportation in open-cut coal mines, *International Journal of Mining Science and Technology,* 25:261–266.

Gölbaşı, O and Demirel, N, 2017. A cost-effective simulation algorithm for inspection interval optimization: An application to mining equipment, *Computers and Industrial Engineering,* 113:525–540.

Gong, H, Moradi Afrapoli, A and Askari-Nasab, H, 2023. Integrated simulation and optimization framework for quantitative analysis of near-face stockpile mining, *Simulation Modelling Practice and Theory*, 128:102794.

Gong, H, Tabesh, M, Moradi Afrapoli, A and Askari-Nasab, H, 2023. Near-face stockpile open pit mining: a method to enhance NP V and quality of the plant throughput, *International Journal of Mining, Reclamation and Environment*, 37:200–215.

Habib, N A, Ben-Awuah, E and Askari-Nasab, H, 2023a. Review of recent developments in short-term mine planning and IPCC with a research agenda, *Mining Technology*, 132:1–23.

Habib, N A, Ben-Awuah, E and Askari-Nasab, H, 2023b. Short-term planning of open pit mines with Semi-Mobile IPCC: a shovel allocation model, *International Journal of Mining, Reclamation and Environment*, 38:236–266.

Kamrani, A, Badiozamani, M M, Pourrahimian, Y and Askari-Nasab, H, 2024. Evaluating the semi-mobile in-pit crusher option through a two-step mathematical model, *Resources Policy*, 95:105113.

Konak, G, Onur, A H and Karakus, D, 2007. Selection of the optimum in-pit crusher location for an aggregate producer, *Journal of the Southern African Institute of Mining and Metallurgy*, 107:161–166.

Kozan, E and Liu, S Q, 2016. A new open-pit multi-stage mine production timetabling model for drilling, blasting and excavating operations, *Transactions of the Institutions of Mining and Metallurgy, Section A: Mining Technology*, 125:47–53.

Kozan, E, Liu, S and Wolff, R, 2013. A short-term production scheduling methodology for open-pit mines, in *Proceedings of the 36th Application of Computers and Operations Research in The Mineral Industry (APCOM) Symposium*, pp 465–473.

L'Heureux, G, Gamache, M and Soumis, F, 2013. Mixed integer programming model for short term planning in open-pit mines, *Transactions of the Institutions of Mining and Metallurgy, Section A: Mining Technology*, 122:101–109.

Liu, D and Pourrahimian, Y, 2021. A Framework for Open-Pit Mine Production Scheduling under Semi-Mobile In-Pit Crushing and Conveying Systems with the High-Angle Conveyor, *Mining*, 1:59–79.

Liu, S Q and Kozan, E, 2017. Integration of mathematical models for ore mining industry, *International Journal of Systems Science: Operations and Logistics*, 6:1–14.

Londoño, J G, Knights, P F and Kizil, M S, 2013. Modelling of In-Pit Crusher Conveyor alternatives, *Mining Technology*, 122:193–199.

Manriquez, F, González, H and Nelson, M V, 2019. Short-term open-pit mine production scheduling with hierarchical objectives, 39th International Symposium on Application of Computers and Operations Research in the Mineral Industry (APCOM 2019), Wroclaw, Poland.

Manríquez, F, Pérez, J and Morales, N, 2020. A simulation–optimization framework for short-term underground mine production scheduling, *Optimization and Engineering*, 21:939–971.

Martins, A G, Souza, M J F and Assis, P S, 2024. An integrated simulation and optimization tool for short-term mining planning problems with different prioritization among competing plant targets, *Computers and Industrial Engineering*, 191:110115.

Moradi Afrapoli, A and Askari-Nasab, H, 2017. Mining fleet management systems: a review of models and algorithms, *International Journal of Mining, Reclamation and Environment*, 33:1–19.

Motswaiso, K and Suglo, R, 2022. Economic evaluation of materials handling systems in a deep open pit mine, International Journal of Mining and Mineral Engineering, *International Journal of Mining and Mineral Engineering*, 13:37–48.

Mousavi, A, Kozan, E and Liu, S Q, 2016. Comparative analysis of three metaheuristics for short-term open pit block sequencing, *Journal of Heuristics*, 22:301–329.

Nelis, G and Morales, N, 2022. A mathematical model for the scheduling and definition of mining cuts in short-term mine planning, *Optimization and Engineering*, 23:233–257.

Norgate, T and Haque, N, 2013. The greenhouse gas impact of IPCC and ore-sorting technologies, *Minerals Engineering*, 42:13–21.

Nunes, R, Homero, J and Giorgio, T, 2019. A decision-making method to assess the benefits of a semi-mobile in-pit crushing and conveying alternative during the early stages of a mining project, *Rem - International Engineering Journal*, 72:285–291.

Osanloo, M and Paricheh, M, 2020. In-pit crushing and conveying technology in open-pit mining operations: a literature review and research agenda, *International Journal of Mining, Reclamation and Environment*, 34:430–457.

Paricheh, M and Osanloo, M, 2016. Determination of the optimum in-pit crusher location in open-pit mining under production and operating cost uncertainties, 6th International Conference on Computer Applications in the Minerals Industries, Istanbul, Turkey.

Paricheh, M and Osanloo, M, 2019. Concurrent open-pit mine production and in-pit crushing–conveying system planning, *Engineering Optimization*, 52:1–16.

Paricheh, M and Osanloo, M, 2020. A New Search Algorithm for Finding Candidate Crusher Locations Inside Open Pit Mines, in *Proceedings of the 28th International Symposium on Mine Planning and Equipment Selection - MPES 2020* (ed: E Topal), pp 10–25 (Springer International Publishing: Cham).

Paricheh, M, Osanloo, M and Rahmanpour, M, 2016. A heuristic approach for in-pit crusher and conveyor system's time and location problem in large open-pit mining, *International Journal of Mining, Reclamation and Environment,* 32:35–55.

Paricheh, M, Osanloo, M and Rahmanpour, M, 2017. In-Pit Crusher Location as a Dynamic Location Problem, *The Southern African Institute of Mining and Metallurgy,* 117:599–607.

Rahmanpour, M, Osanloo, M and Adibi, N, 2013. An approach to determine the location of an in pit crusher in open pit mines, in *Proceedings of 23rd International Mining Congress and Exhibition of Turkey, IMCET 2013,* 1:141–149.

Roumpos, C, Partsinevelos, P and Vlachou, A, 2014. The optimal location of the distribution point of the belt conveyor system in continuous surface mining operations, *Simulation Modelling Practice and Theory,* 47:19–27.

Samavati, M, Essam, D and Nehring, M, 2020. Production planning and scheduling in mining scenarios under IPCC mining systems, *Computers and Operations Research,* 115:104714.

Sandeman, T, Fricke, C and Bodon, P, 2011. Integrating optimization and simulation - A comparison of two case studies in mine planning, Proceedings of the 2010 Winter Simulation Conference.

Shamsi, M and Nehring, M, 2021. Determination of the optimal transition point between a truck and shovel system and a semi-mobile in-pit crushing and conveying system, *Journal of the Southern African Institute of Mining and Metallurgy,* 497–504.

Shamsi, M, Pourrahimian, Y and Rahmanpour, M, 2022. Optimisation of open-pit mine production scheduling considering optimum transportation system between truck haulage and semi-mobile in-pit crushing and conveying, *International Journal of Mining, Reclamation and Environment,* 36:142–158.

Shishvan, M S and Benndorf, J, 2014. Performance optimization of complex continuous mining system using stochastic simulation, Engineering Optimization IV - Proceedings of the 4th International Conference on Engineering Optimization, Lisbon.

Shishvan, M S and Benndorf, J, 2016. The effect of geological uncertainty on achieving short-term targets: A quantitative approach using stochastic process simulation, *Journal of the Southern African Institute of Mining and Metallurgy,* 116:259–264.

Shishvan, M S and Benndorf, J, 2017. Operational Decision Support for Material Management in Continuous Mining Systems: From Simulation Concept to Practical Full-Scale Implementations, *Minerals,* 7:116.

Silva-Júnior, A L, Martins, A G, Pantuza-Jr, G, Cota, L P and Souza, M J F, 2023. Short-term planning of a work shift for open-pit mines: A case study, *Cogent Engineering,* 10:2168172.

Tabesh, M and Askari-Nasab, H, 2013. Automatic Creation of Ore-Selection and Blast Polygons using Clustering Algorithms, Society for Mining, Metallurgy and Exploration (SME) Annual Meeting, Denver, Colorado.

Tabesh, M and Askari-Nasab, H, 2019. Clustering mining blocks in presence of geological uncertainty, *Mining Technology,* 128:162–176.

Taheri, M, Irannajad, M and Ataee, P, 2009. An approach to determine the locations of in-pit crushers in deep open-pit mines, in *Proceedings of the 9th International Multidisciplinary Scientific Geoconference and Expo, SGEM 2009,* pp 341–347.

Thomas, A, Singh, G and Krishnamoorthy, M, 2013. Distributed optimisation method for multi-resource constrained scheduling in coal supply chains, *International Journal of Production Research,* 51:2740–2759.

Thomas, A, Venkateswaran, J and Krishnamoorthy, M, 2014. A resource constrained scheduling problem with multiple independent producers and a single linking constraint: A coal supply chain example, *European Journal of Operational Research,* 236:946–956.

Torkamani, E and Askari-Nasab, H, 2015. A linkage of truck-and-shovel operations to short-term mine plans using discrete-event simulation, *International Journal of Mining and Mineral Engineering,* 6:97–118.

Upadhyay, S P and Askari-Nasab, H, 2016. Truck-shovel allocation optimisation: A goal programming approach, *Transactions of the Institutions of Mining and Metallurgy, Section A: Mining Technology,* 125:82–92.

Upadhyay, S P and Askari-Nasab, H, 2017. Dynamic shovel allocation approach to short-term production planning in open-pit mines, *International Journal of Mining, Reclamation and Environment,* 33:1–20.

Upadhyay, S P, Doucette, J and Askari-Nasab, H, 2022. Short-term production scheduling in open pit mines with shovel allocations over continuous time frames, *International Journal of Mining and Mineral Engineering,* 12:292–308.

Valença Mariz, J L, Peroni, R D L, Silva, R M D A, Badiozamani, M M and Askari-Nasab, H, 2024. A multi-stage constraint programming approach to solve clustering problems in open-pit mine planning, *Engineering Optimization,* 1–24.

Mining project risk analysis of the impact of schedule types of open pit mining on mine value

R Halatchev[1], D Gabeva[2] and A Halatchev[3]

1. Director Mining, AusGEMCO Pty Ltd, Brisbane Qld 4078. Email: ross@ausgemco.com
2. Managing Director, AusGEMCO Pty Ltd, Brisbane Qld 4078. Email: deliana@ausgemco.com
3. Secretary, AusGEMCO Pty Ltd, Brisbane Qld 4078. Email: angel@ausgemco.com

ABSTRACT

Open pit mining enjoys a renaissance nowadays due to the increased demand for base/precious metals and critical minerals. This gave a spur to large amounts of investments, which need to be managed efficiently. A major factor for achieving such management is the development of optimum production schedules, which are the skeleton of any financial model. This specific feature lays down the requirement for the correct understanding of the impact of different production schedule types on mine value and the importance of mitigating the mining project risk over the life-of-mine.

The current paper presents an analysis of the existing types of production schedules in open pit mining and their impact on the mine value. The analysis incorporates a case study to illustrate the impact of the schedule types on the mine value. An author's linear programming optimisation model is used for the generation of schedules, which implements both mine planning principles of ore grade maximisation and waste deferment. The analysis of the schedule types is made with the utilisation of the Cumulative Spatial Graph (CSG) of the mine, which is the most universal tool for mine sequencing and scheduling optimisations. The efficiency of each schedule type is assessed with representative criteria, including the Factor of Waste Deferment, which evaluates the degree of implementation of the principle of waste deferment in the optimum mine sequence. A mining project risk is also used as a criterion to account for the variables exhibiting a stochastic behaviour and contributing to the overall project uncertainty. The risk model is based on a complete discounted cash flow analysis. The Monte Carlo simulation technique is used for modelling the risk of not achieving the planned discounted cash flows over each time step of the cash flows analysis.

INTRODUCTION

The problem of defining the types of production schedules is important because it is not a well-studied aspect of the methodology of open pit mining (Rzhevskii, 1980; Arsentiev, 1961; Kennedy, 1990; Hustrulid and Kuhta, 1998). The recent developments of scheduling algorithms with the methods of Operations Research have already prompted the necessity of defining the schedule types to compare their efficiency. The availability of many scheduling algorithms creates a little chaos about the selection of the best algorithms by users for practical implementations. Unfortunately, the focus in the marketing of these algorithms is placed mostly on their mathematical content, with a simple declaration about their capability to maximise the Net Present Value (NPV) as an objective function of the optimisation model.

The author's recent research achievements in the development of a new approach to open pit production scheduling optimisation have confirmed the importance of defining production schedule types. This is well reflected in the publication of Halatchev (2013), where the concept of using different schedule types for open pit mining and their impact on the mine value was presented briefly. The concept of assessment of the impact of production schedule types in the mine evaluations did not find a place in the fundamental investigations of famous scientists working in the area of Mineral Economics, such as Masse (1962), Matheron (1968), Margolin (1974), and Gentry and O'Neil (1984).

ANALYSIS OF SCHEDULE TYPES AND THEIR IMPACT ON MINE VALUE

Tools of mine sequencing/scheduling

The currently existing types of production schedules in open pit mining can be analysed with the utilisation of the Cumulative Spatial Graph (CSG) of the mine, which is the most informative and universal tool for assessment of the efficiency of pit design in the context of arranging the mine sequence (Arsentiev, 1961). The CSG is usually built with the cumulative function of waste-ore (or metal) quantities of two extreme cases of mine sequence arrangement (Figure 1). The first case is named 'minimum working width sequence (*mnwws*) and it is defined as an arrangement of the mine sequence with a consecutive inclusion of cutbacks (or strips) in the mine exploitation. The second case is named 'maximum working width sequence' (*mxwws*) and it is defined as an arrangement of the mine sequence by mining out consecutively each complete bench within the final pit outlines before the commencement of the lower bench. Both cases of the arrangement of the mine sequence form a Feasible Optimisation Domain (FOD) where the optimum mine sequence (*optims*) is searched. The *optims* can have the shape of a linear piecewise function presented as multiple linear segments, OMNL as shown in Figure 1. Usually, the *mnwws* is the lower sequence of the FOD while the *mxwws* is the upper sequence. The CSG accounts for the spatial aspect of mine sequence arrangement. The time aspect of mine sequence arrangement is accounted for by the Cumulative Time Graph (CTG), which is a derivative of the CSG as it presents the excavation process of any mine sequence. The CTG converts the optimum mine sequence or any other sequence into a schedule. Both graphs are applicable to any type of mineral deposit, mining system, and mining method.

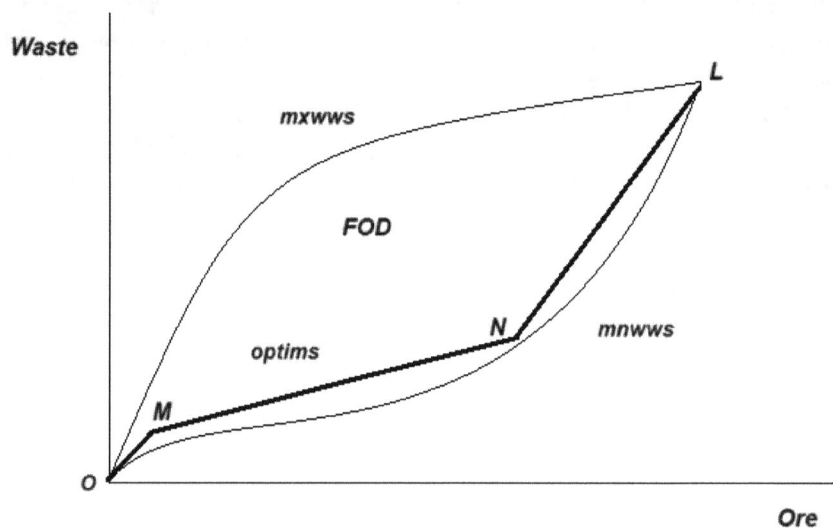

FIG 1 – CSG with optimum mine sequence (*optims*).

The utilisation of the CSG for economic evaluation of mine projects provides evidence about the interval estimate of the NPV related to both extreme cases of mine sequence arrangement. Usually, the *mnwws* has a higher NPV in comparison with the *mxwws,* and both estimates define the interval estimate. The *optims* has an NPV within the interval estimate. The interval estimate of the NPV is due to the time value of money assessed with the project discount rate. If the mineral deposit is presented only by ore with a non-homogeneous grade field, the interval estimate of the NPV will still be valid for such a project. In case of the presence of waste in the mineral deposit, the NPV interval estimate is wider because of the different degrees of implementation of the waste deferment principle for both extreme mine sequences. In other words, the CSG dismisses any concept of a single estimate of the NPV of an open pit mine project, which is an important fact for any economist dealing with the evaluation of mine projects. A theoretical case is possible when the deposit is strongly homogeneous regarding the ore grade and there is no waste. In this case, both sequences *mnwws* and *mxwws* will overlap.

The mine sequencing/scheduling method

The scheduling approach presented here is based on the author's method, which was originated and developed in Australia (Halatchev, 1992, 1993, 1997, 2002, 2005, 2007, 2011, 2015, 2024; Kim and Zhao, 1994; Halatchev and Moustakerov, 1996; Golosinski and Bush, 2000; De Kock, 2007; Wang, Xu and Gu, 2013). The method implements both mine planning principles, such as ore grade maximisation and waste deferment. The method employs a linear programming optimisation model with an objective function reflecting the structure of the NPV of an open pit project on the basis of discounted cash flow (DCF) analysis. The actual level of the implementation of the DCF analysis corresponds to the level of calculating the Earnings Before Interest, Tax, Depreciation and Amortisation (EBITDA). The choice of this level is motivated with by the framework of the VALMIN Code (VALMIN Committee, 2005). The final update of the expression of the objective function for two ore supply flows (basic and secondary ore) is (Halatchev, 2013):

$$Max \sum_{i=0}^{n} d_i \left(1 - R_0\right) \left[\left(S_i - C_i^{ma} - C_i^r \right) \right] M_{b_i} - \left[\left(C_{b_i}^m + C_{b_i}^p + TC_i \right) \left(\alpha_{b_i} \gamma_i \right)^{-1} \right] M_{b_i} -$$

$$\sum_{i=0}^{n} d_i C_{s_i}^m \left(\alpha_{s_i} \right)^{-1} M_{s_i} - \sum_{i=0}^{n} d_i C_{w_i}^m V_i - \sum_{k=1}^{n_m} \sum_{j=1}^{m_k} \sum_{i=0}^{n} d_i h_{kji} NC_{kji} - \sum_{k=1}^{n_m} \sum_{j=1}^{m_k} \sum_{i=0}^{n} d_i u_{kji} DC_{kji} \tag{1}$$

The subscript 'i' in all variables and constants of the model denotes i-th time step of the production scheduling optimisation.

The objective function constants and variables are presented in Tables 1 and 2. The constants of the model constraints are given in Table 3.

TABLE 1
Objective function constants.

Constant	Definition
n	number of time periods to be considered
n_m	number of types of mine equipment
m_k	number of models of equipment per type
d_i	discount factor – $d_i = \left(1+r\right)^{-i}$ where r is the interest rate
S_i	price of payable metal from basic ore
$C_{b_i}^m$	unit operating costs of basic ore mining
$C_{s_i}^m$	unit operating costs of secondary ore mining
$C_{b_i}^p$	unit operating cost of basic ore processing
C_{w_i}	unit operating cost of waste removal
C_i^{ma}	unit marketing cost per payable metal
C_i^r	unit refinery cost per payable metal
R_o	royalty as % of the net revenue
$\alpha_{b_i}, \alpha_{s_i}$	basic and secondary ore grades
γ_i	total recovery of the payable metal
TC_i	time costs
h_{kji}	unit purchase cost of pit capacity of k-th type j-th model of equipment
u_{kji}	unit ownership cost of pit capacity of k-th type j-th model of equipment

TABLE 2
Model variables.

Variable	Definition
M_{b_i}	basic ore metal as a final product
M_{s_i}	secondary ore metal
V_i	waste quantity to be removed
NC_{kji}	new capacity added for k-th type j-th model of equipment
DC_{kji}	capacity decrease for k-th type j-th model of equipment

TABLE 3
Constants of model constraints.

Constant	Definition
$SMBX_i$	cumulative quantity of basic ore metal of the upper-bound basic ore metal function – *mxwws* sequence for i-th production period
$SMBN_i$	cumulative quantity of basic ore metal of the upper-bound basic ore metal function – *mnwws* sequence for i-th production period
$SMBU_i$	maximum cumulative quantity of basic ore metal of both upper-bound basic ore metal functions for i-th production period
$SMBL_i$	minimum cumulative quantity of basic ore metal of both upper-bound basic ore metal functions for i-th production period
$CSMB$	total quantity of basic ore metal
ΔSR_{b_i}	stripping ratio of basic ore metal
$SMSX_i$	cumulative quantity of secondary ore metal of the upper-bound secondary ore metal function – *mxwws* sequence for i-th production period
$SMSN_i$	cumulative quantity of secondary ore metal of the upper-bound secondary ore metal function – *mnwws* sequence for i-th production period
$SMSU_i$	maximum cumulative quantity of secondary ore metal of both upper-bound secondary ore metal functions for i-th production period
$SMSL_i$	minimum cumulative quantity of secondary ore metal of both upper-bound secondary ore metal functions for i-th production period
$CSMS$	total quantity of secondary ore metal
ΔSR_{s_i}	stripping ratio of secondary ore metal
SVU_i	maximum cumulative quantity of waste of both upper-bound waste functions
SVL_i	minimum cumulative quantity of waste of both upper-bound waste functions
CSV	total quantity of waste
C_{kz}^{max}	capacity limit of k-th type and z-th model of production equipment

The revenue in Equation 1 is associated with the metal of basic ore type (BOT) as a marketable product. The secondary ore metal is not treated as a separate final product because the secondary ore type (SOT) goes to stockpiles for later treatment in case of favourable economic conditions. This ore, which is usually named marginal-subgrade material (MSG), has a cut-off grade that is less than the mill-feed cut-off and higher than the break-even cut-off.

The NPV objective function includes variable and fixed operating costs. The variable costs are assigned to different technological processes of mining and processing for any time step of the scheduling horizon. This allows the improvement of the adequacy of the scheduling model. The fixed costs (TC_i) are treated as cost per unit production of ore, and they perform conditionally as variable costs. This approach makes sense only if the annual mill production of the mine varies over the LOM, which leads to the variation of the annual fixed costs. It, however, contradicts the classical definition of fixed costs that are 'independent of throughput' (Gentry and O'Neil, 1984). In case the fixed costs are really independent of the variation of the mill production, then they should not participate in the scheduling optimisation.

An original element of the optimisation model is the introduction of the new capacity (NC_{kji}) and capacity decrease(DC_{kji}) of each type and model of the mine equipment as model variables. This is reflected in the last two terms of the objective function. Both terms manage the location of the optimum sequence in the FOD of the CSG. The mechanism of this type of management of mine sequence is that every increase or decrease of the mining rate has relevant economic consequences. The economic parameters of the mechanism are the unit purchase and ownership costs of each type and model of mine production equipment (eg excavators, trucks, drills) that are related to a unit pit production capacity. These parameters are mine investment parameters, which are an original element of the scheduling model (Hartman and Mutmanski, 2002). They also include the unit purchase and ownership costs of the auxiliary equipment, which are obtained with regard to the pit production capacity (Halatchev, 2013). The mechanism provides a stabilisation of the mining rate as a piecewise linear function, which is determined as a search for the equilibrium between the purchase and ownership costs of the open pit mine production capacity. The fundamental definition of Mine Capacity is provided by Lane (1988). The achievement of mining rate stabilisation makes the production schedule viable for practical execution. This specific feature is better explained with the case study of the paper.

The sequencing/scheduling optimisation model allows two types of schedules to be modelled: a schedule with multi-stage stabilisation of mining rate, and a schedule with single-stage stabilisation of mining rate. Both schedule types are possible to be executed within the FOD of the CSG of the mine. The shape of the FOD predetermines the duration of each stage. Explanations on the schedule types are provided further in this paper.

Constraints

1. Bounds of basic ore metal production

$$\sum_{j=0}^{i} M_{b_j} \leq SMBU_i, \quad i = 0, n-1 \tag{2.1}$$

$$\sum_{j=0}^{i} M_{b_j} \geq SMBL_i, \quad i = 0, n-1 \tag{2.2}$$

$$\sum_{j=0}^{n} M_{b_j} = CSMB \tag{2.3}$$

2. Relationship b/n waste and basic ore metal production

if $SMBX_i \geq SMBN_i$ *then*

$$\Delta SR_{b_i} \sum_{j=0}^{i} M_{b_j} - \sum_{j=0}^{i} V_j \leq \Delta SR_{b_i} SMBL_i + SVL_i, \quad i = 0, n-1 \tag{3.1}$$

elseif SMBX$_i$ < SMBN$_i$ then

$$\Delta SR_{b_i} \sum_{j=0}^{i} M_{b_j} + \sum_{j=0}^{i} V_j \le \Delta SR_{b_i} SMBL_i - SVL_i, \quad i = 0, n-1$$

(3.2)

where: the stripping ratio of basic ore metal is:

$$\Delta SR_{b_i} = \frac{SVU_i - SVL_i}{SMBU_i - SMBL_i}$$

(3.3)

3. Bounds of secondary ore metal production

$$\sum_{j=0}^{i} M_{s_j} \le SMSU_i, \quad i = 0, n-1$$

(4.1)

$$\sum_{j=0}^{i} M_{s_j} \ge SMSL_i, \quad i = 0, n-1$$

(4.2)

$$\sum_{j=0}^{n} M_{s_j} = CSMS$$

(4.3)

4. Relationship b/n waste and secondary ore metal production

if SMSX$_i$ ≥ SMSN$_i$ then

$$\Delta SR_{s_i} \sum_{j=0}^{i} M_{s_j} - \sum_{j=0}^{i} V_j \le \Delta SR_{s_i} SMSL_i + SVL_i, \quad i = 0, n-1$$

(5.1)

elseif SMSX$_i$ < SMSN$_i$ then

$$\Delta SR_{s_i} \sum_{j=0}^{i} M_{s_j} + \sum_{j=0}^{i} V_j \le \Delta SR_{s_i} SMSL_i - SVL_i, \quad i = 0, n-1$$

(5.2)

where: the stripping ratio of secondary ore metal is:

$$\Delta SR_{s_i} = \frac{SVU_i - SVL_i}{SMSU_i - SMSL_i}$$

(5.3)

5. Bounds of waste production

$$\sum_{j=0}^{i} V_j \le SVU_i, \quad i = 0, n-1$$

(6.1)

$$\sum_{j=0}^{i} V_j \ge SVL_i, \quad i = 0, n-1$$

(6.2)

$$\sum_{j=0}^{n} V_j = CSV$$

(6.3)

6. Relationship b/n the capacities of each model of equipment

$$\sum_{j=0}^{i} NC_{kzj} - \sum_{j=0}^{i} DC_{kzi} \ge 0, \quad k \in [1, n_m - 1]; \, z \in [1, m_k]; \, i \in [0, n]$$

(7)

7. Capacity limit of each model of equipment

$$\sum_{i=0}^{n} NC_{kzi} \le C_{kz}^{\max}, \quad k = 1, n_m - 1; z = 1, m_k$$

(8)

8. Definition of the mining rate

$$M_{b_i}(\alpha_{b_i})^{-1} + M_{s_i}(\alpha_{s_i})^{-1} + V_i - \sum_{z=1}^{m_k}\sum_{j=0}^{i} NC_{kzj} + \sum_{z=1}^{m_k}\sum_{j=0}^{i} DC_{kzi} = 0$$

$$k = 1; i \in [0, n]$$

(9)

9. Distribution of new pit capacity among the different types of mine equipment

$$\sum_{z=1}^{m_i}\sum_{j=0}^{i} NC_{izj} - \sum_{z=1}^{m_k}\sum_{j=0}^{i} NC_{kzi} = 0, \quad k \in [2, n_m - 1]; i \in [0, n]$$

(10)

10. Distribution of pit capacity decrease among the different types of mine equipment

$$\sum_{z=1}^{m_i}\sum_{j=0}^{i} DC_{izj} - \sum_{z=1}^{m_k}\sum_{j=0}^{i} DC_{kzi} = 0, \quad k \in [2, n_m - 1]; i \in [0, n]$$

(11)

It is worth noting that constraints (10) and (11) determine the distribution of the new pit capacity and capacity decrease between different types of production equipment in the case that their quantity is equal to or greater than two types ($k \in [2, n_m - 1]$). Types of production equipment can be shovels, loaders, trucks, conveyors, drills, etc. The excavation equipment must be treated conditionally as the first type of equipment in running the optimisation model, as its production predetermines the production of the haulage equipment.

Criteria for strategic scheduling efficiency

The efficiency of the optimum mine sequence in the FOD of the CSG is assessed with representative criteria. A new author's criterion is offered in the paper, which is called the Factor of Waste Deferment (*FWD*). It assesses the degree of implementation of the principle of waste deferment in the optimum sequence (Halatchev, Gabeva and Halatchev, 2024). This is a new mine planning principle dealing with the optimisation of the waste and mining rate schedules by exploring the time value of money (Halatchev, 1993; 2013). The current approach of scheduling optimisations deals mostly with the implementation of the mine planning principle of ore grade maximisation (Golosinski and Bush, 2000). The analytical expression of the Factor of Waste Deferment is as follows:

$$FWD = 100 - FTeC, \%$$

(12)

where: *FTeC* is Factor of Technological Compromise (Halatchev, 1996; 2011):

$$FTeC = \frac{S_O}{S_{FOD}} 100, \%$$

(13)

where:

S_O is the area of the CSG restricted by *optims* and *mnwws*

S_{FOD} area of the FOD.

The *FTeC* assesses the technological potential of the FOD with regard to the mining system used in the pit design.

Another criterion is the Factor of Economic Compromise, which assesses the degree of compromise of using the *optims* with regard to the *mnwws* sequence, also known as the best NPV sequence. The Factor of Economic Compromise (*FEcC*) is assessed with the formula:

$$FEcC = \frac{NPV_{mnwws} - NPV_{optims}}{NPV_{mnwws} - NPV_{mxwws}} 100, \%$$

(14)

where:

NPV_{mnwws} is the NPV of the *mnwws*, which has the best estimate

NPV_{mxwws} NPV of the *mxwws*, which has the worst estimate

NPV_{optims} NPV of the *optims* sequence

The *FEcC* deals with the assessment of the economic efficiency of long-term production scheduling by using the NPV, which is a pure economic criterion.

The integral presentation of the *FEcC* and *FTeC* is the Factor of Total Compromise (*FToC*):

$$FToC = \frac{FEcC\ FTeC}{100}, \%$$

(15)

The above-presented criteria provide exceptionally useful information for the strategic management of surface mining ventures. They can be used for the assessment of the efficiency of any mine design and sequencing/scheduling optimisation model from an economic and technological point of view. The utilisation of these criteria would undoubtedly help mine planning engineers select the most suitable scheduling software.

Case study

The case study is based on a hypothetical large-scale open pit mine exploiting a gold deposit. The mine design comprises 14 cutbacks that define a partial elliptical mining system (Halatchev, 2013). The annual mill production is 5 Mt/a and the LOM horizon is 34 years, which is split into 34 time steps of the scheduling optimisation. Production equipment comprises a fleet of excavators with a production rate of 1800 t/h, availability of 85 per cent, and utilisation of 75 per cent.

The economic input data of scheduling optimisation is summarised in Table 4. The data for the purchase and ownership costs of the mine equipment is summarised in Table 5. It is important to note that the purchase and ownership costs manage the location of the optimum sequence in the CSG, which is transformed into a schedule. Two variants of production scheduling optimisation are defined. Variant 1 imposes a restriction on the open pit mine capacity, while Variant 2 does not impose a restriction. The unit purchase cost of pit capacity is estimated at 2.65 $/t ROM mass while the unit ownership cost of pit capacity is 0.95 $/t ROM mass/a.

TABLE 4

Economic input data of scheduling optimisation.

Parameter	Units*	Value
Price*	$/g	50.00
Royalty	%	5.00
Refinery cost	$/g	4.00
Marketing cost	$/g	2.00
Processing cost	$/t ROM ore	7.80
Ore mining cost	$/t ROM ore	3.37
Waste mining cost	$/t waste	2.99
Time cost	$/t ROM ore	3.54
Others	$/t ROM ore	3.73

(*) price and costs are in US dollars.

TABLE 5

Economic input data of mine equipment.

Equipment	Purchase cost, $/t ROM mass*	Ownership cost, $/t ROM mass/a*
Type: production		
Excavators	1.50	0.50
Trucks	0.80	0.25
Drills	0.20	0.15
Type: auxiliary	0.15	0.05
Total pit capacity	2.65	0.95

(*) all costs are in US dollars.

The computer code 'PITFLOW' developed in Microsoft Visual C++ 6.0 version was used for modelling the schedules with mining rate stabilisation. The computer code uses the ILOG CPLEX Callable Library (version 8.1, 2002, IBM) for solving the production scheduling optimisation model. The code performs a preprocessing of the input data of the scheduling optimisation and calls the optimisation model, which is solved by the Callable Library. The code also does the post-processing of the results and saves them in output files.

It is worth noting that all figures for cumulative spatial graphs and schedules of the study are taken from the author's papers (Halatchev, 2013).

Schedule with variable mining rate

The schedule with variable mining rate (VMR) is based on the transformation of the *mnwws* sequence of the CSG (Figure 2). The *mnwws* is the lower sequence of the FOD. The schedule meets the processing plant demand of ore without the utilisation of stockpiles.

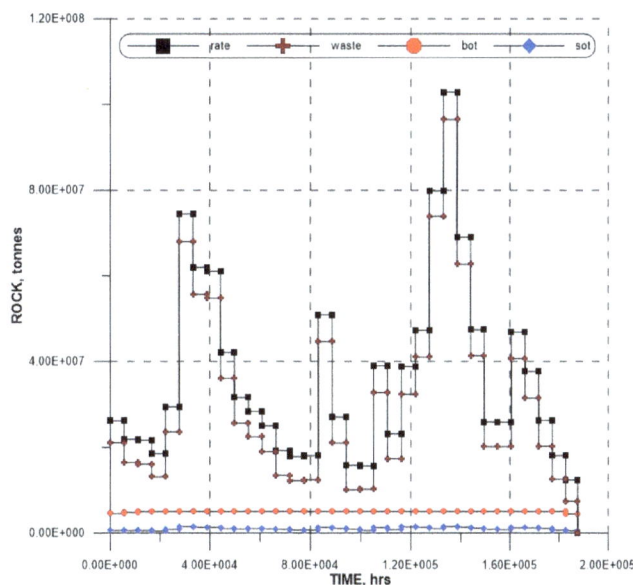

FIG 2 – Production schedule based on *mnwws*.

This schedule has a continuous supply of a fixed quantity of ore from the mine, while the waste removal is a variable quantity. The schedule shows the time distribution of basic ore (BOT), secondary ore (SOT), waste (WST), and mining rate. The analysis of the schedule indicates large fluctuations of the mining rate, which is a summation of waste, basic ore, and secondary ore for each scheduling period. The schedule based on the *mnwws* has the trend of increasing the mining rate over the life-of-mine, which reflects the maximum possible degree of the waste deferment as one of the major mine planning principles in open pit mining.

The estimates of the criteria of the schedule efficiency for this example of a schedule with variable mining rate are as follows: *FEcC* = 0.00 per cent, and *FTeC* = 0.00 per cent. The *FWD* is 100.00 per cent, because the *mnwws* sequence is accepted as an optimum sequence. The NPV estimate is 2728.44 M$. These results are also summarised in Table 6 for the variant of VMR-*mnwws*.

TABLE 6

Results of the schedules comparison

Schedule type	FTeC, %	FEcC, %	FWD, %	NPV*, M $
VMR-mnwws	0.00	0.00	100.00	2728.44
VMR-mxwws	100.00	100.00	0.00	2186.88
SSSMR	36.22	38.34	63.78	2540.89
MSSMR	11.69	7.84	88.31	2687.50
CMR-SP**	68.19	54.75	31.80	2431.92

(*) – NPV calculated with discount rate of 10.00%; (**) – rehandle cost of 1.00$/t is used.

Another variant of the schedule with VMR is shown in Figure 3, which is a schedule transformation of the *mxwws* sequence in the CSG as an upper-bound sequence of the FOD. It also meets the processing plant demand of ore without the utilisation of stockpiles. The mining rate has large fluctuations. The schedule analysis indicates a trend of decreasing the mining rate over the scheduling horizon, which reflects the minimum possible degree of the waste deferment implementation. This explains the difference in the NPV estimates of the schedules based on *mnwws* and *mxwws* sequences.

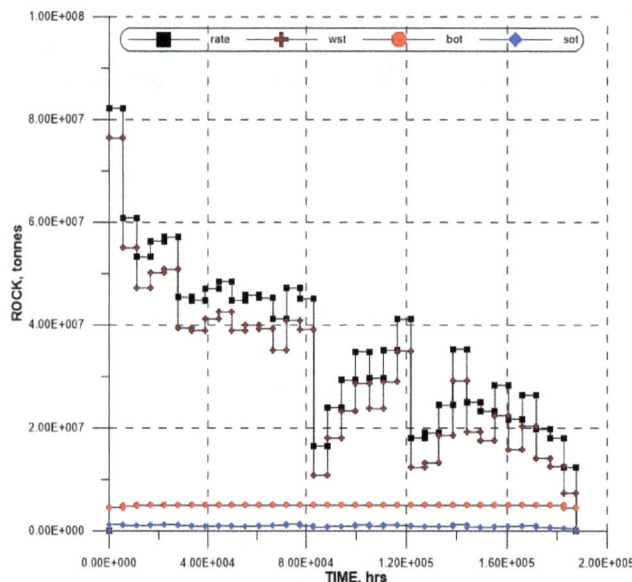

FIG 3 – Production schedule based on *mxwws*.

The estimates of the criteria of the schedule efficiency for this example are as follows: *FEcC* = 100.00 per cent, and *FTeC* = 100 per cent. The *FWD* is 0.00 per cent, because the *mxwws* sequence is accepted as an optimum sequence. The NPV estimate is 2186.88 M$. The estimates are summarised in Table 6 for this variant of VMR-*mxwws*.

Both variants of schedules with VMR can be viable for practical implementation, depending on the degree of fluctuations of the mining rate. Usually, they don't have a practical implementation due to the large fluctuations of the mining rate. This means that their practical realisation would require a very frequent increase and decrease in the size of the excavation equipment fleet, which is a difficult task. Mining is usually conducted in remote areas where the supply of new equipment is a time-consuming task. The schedule of *mnwws* has the highest possible NPV estimate of the FOD, while the schedule of *mxwws* has the minimum NPV estimate because of the impact of the time value of money assessed with the project discount rate.

Schedule with single-stage stabilisation of mining rate

The schedule with a single-stage stabilisation of mining rate (SSSMR) is the schedule of Variant 1 defined in the Case Study section of the paper. The schedule is obtained with the author's scheduling optimisation model, which imposes a restriction on open pit mine capacity. The optimum mining sequence (*optims1*) is shown in the GSG in Figure 4. It is almost a straight line located in the FOD restricted by the *mnwws* and *mxwws* sequences. The *optims1* has 34 nodes corresponding to the number of time steps of the scheduling optimisation.

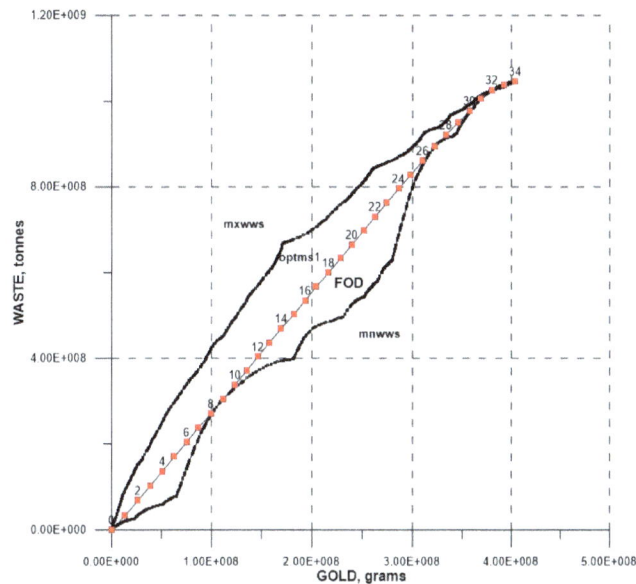

FIG 4 – CSG with optimum mine sequence *optims1*.

The schedule is obtained with the transformation of the optimum sequence (*optims1*) as shown in Figure 5. Its analysis indicates a 100 per cent match of the basic ore supply with the planned mill ore demand without the utilisation of ore stockpiles.

The schedule analysis indicates almost a constant waste removal and mining rate over the scheduling horizon, with the exception of the period of attenuation of the mining operations. There is a dominating period of steady mining rate (38 Mt/a) followed by a few short stages of decreasing the mining rate. With some assumptions, this can be accepted as a schedule with a single-stage stabilisation of mining rate over the LOM. The schedule is viable and can be practically implemented.

The estimates of the criteria of the schedule efficiency for this example of a schedule are as follows: *FEcC* = 38.34 per cent, *FTeC* = 36.22 per cent, and *FWD* = 63.78 per cent (Table 6). The NPV estimate is 2540.89 M$. The estimate of the *FWD* indicates that the optimum schedule has a notable implementation of the principle of waste deferment in the optimum sequence (*optims1*). The NPV estimate is less than the NPV estimate of the *mnwws*-related schedule in Figure 2.

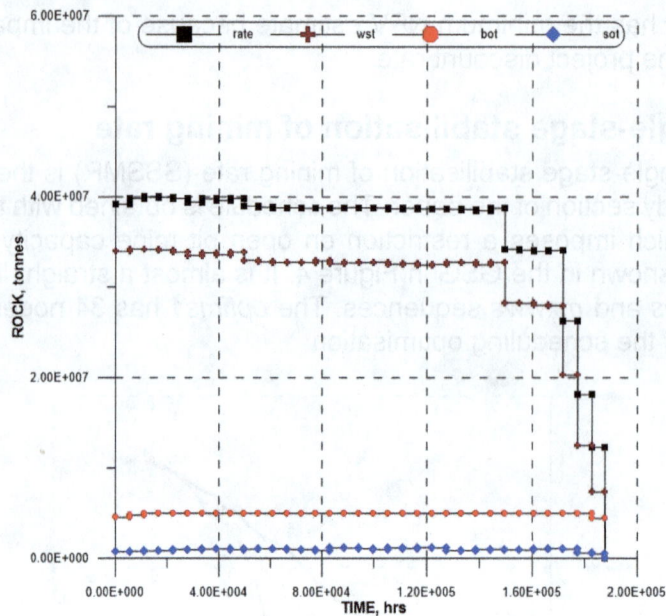

FIG 5 – Schedule with SSSMR based on *optims1*.

Schedule with multi-stage stabilisation of mining rate

The schedule with a multi-stage stabilisation of mining rate (MSSMR) is the schedule of Variant 2 defined in the Case study section of the paper. The schedule is obtained with the optimisation model, which doesn't impose a restriction on open pit mine capacity. The optimum mining sequence (*optims2*) is shown in the GSG (Figure 6). It is a linear stepwise function located in the FOD. The *optims2* has 34 nodes corresponding to the number of time steps of the scheduling optimisation.

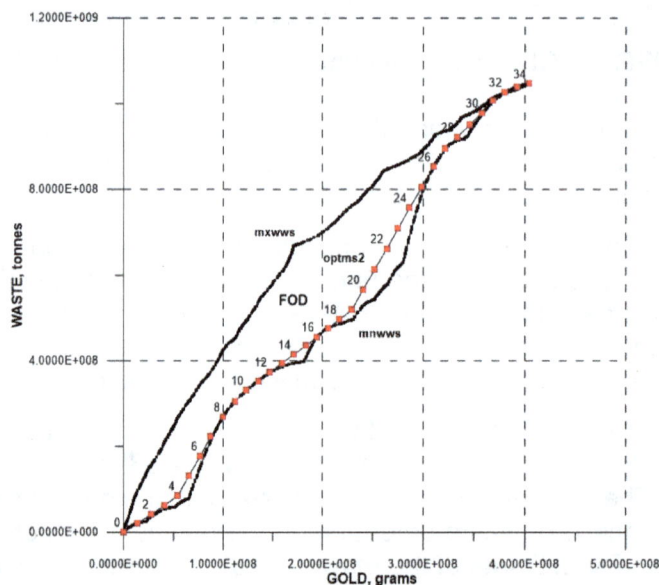

FIG 6 – CSG with optimum mine sequence *optims2*.

The optimum schedule is obtained with the transformation of the optimum sequence (*optims2*) as it is shown in Figure 7. Its analysis indicates a 100 per cent match of the basic ore supply with the planned mill ore demand without the utilisation of ore stockpiles. The schedule has five main stages with constant mining rates. The mining rates are as follows: first stage – 26 Mt/a; second stage – 52 Mt/a; third stage – 26 Mt/a; fourth stage – 54 Mt/a; fifth stage – 34 Mt/a. The mining rate decreases during the last period of attenuation of the mining operations.

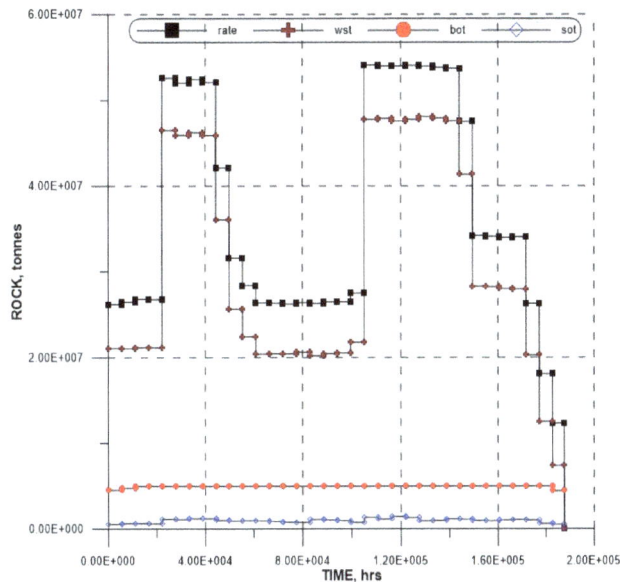

FIG 7 – Schedule with MSSMR based on *optims2*.

The estimates of the criteria of the schedule efficiency for this example of a schedule are as follows: *FEcC* = 7.84 per cent, *FTeC* = 11.69 per cent, and *FWD* = 88.31 per cent (Table 6). The NPV estimate is 2687.50 M$. This schedule has a very high estimate of the *FWD*, which means a very good implementation of the principle of waste deferment in the optimum sequence (*optims2*). The NPV estimate is less than the NPV estimate of the *mnwws*-related schedule, but it is much higher than the NPV of the optimum sequence (*optims1*). The comparison of the schedules of *mnwws* and *optims2* shows that the *optims2* eliminates the fluctuation of the mining rate in Figure 2 and makes the schedule in Figure 7 practically viable.

Schedule with a constant mining rate and using stockpiles

The schedule with constant mining rate and using stockpiles (CMR-SP) is the most popular schedule type in the world mining industry at present, due to the available marketing software. It reflects the simple logic for managing the variable mine ore supply to the processing plant with a constant mining rate by using stockpiles as a buffer between the mine and plant. The variable mine ore supply means surplus or deficit of ore quantity over time. Generally, the schedule with CMR-SP doesn't meet 100 per cent of the planned mill ore rate as a direct mine ore supply, and it requires the management of stockpiles. This is the main difference between the schedule with CMR-SP and the schedule of single-stage or multi-staged stabilisation of mining rate. In other words, the schedule with CMR-SP sets the requirement firstly for working with a constant mining rate and then achievement of the mill ore demand by using stockpiles, while the schedule with SSSMR and MSSMR sets firstly the requirement for meeting the mill ore demand and then searching for the stabilisation of mining rate.

The constant mining rate can be scheduled over the entire life-of-mine (LOM) or for different stages of the pit exploitation where each stage has its constant mining rate achieved by a specific set-up of the mining equipment database and selection of the number of required equipment units. Generally, there is no optimisation procedure for the determination of the mining rate. This is usually a manual procedure or computer algorithm implementation using logical operators, which doesn't optimise the mining rate under the requirement for accurately meeting the planned mill ore rate. An example of such scheduling algorithms is the XPAC software of RPM Global and the SPRY software of Micromine. The methodology behind the schedule of CMR-SP is that it supports the effective utilisation of the available fixed fleet of mine production equipment over a period of time. Mining is usually conducted in remote areas, and working with a schedule without a constant mining rate creates a problem with the management of the delivery of new equipment units to meet the fluctuations in the schedule. The practical implementation of the schedule leads to some small fluctuations of the mining rate due to the reliability of the equipment used as well as the presence of uncertainty in the input data of its modelling. Usually, these fluctuations represent a stationary process.

A graphical illustration of the schedule with CMR-SP is presented in the CSG in Figure 8. It indicates a mining sequence of two stages with a constant mining rate over each stage. The transformation of the sequence into a schedule is shown in Figure 9.

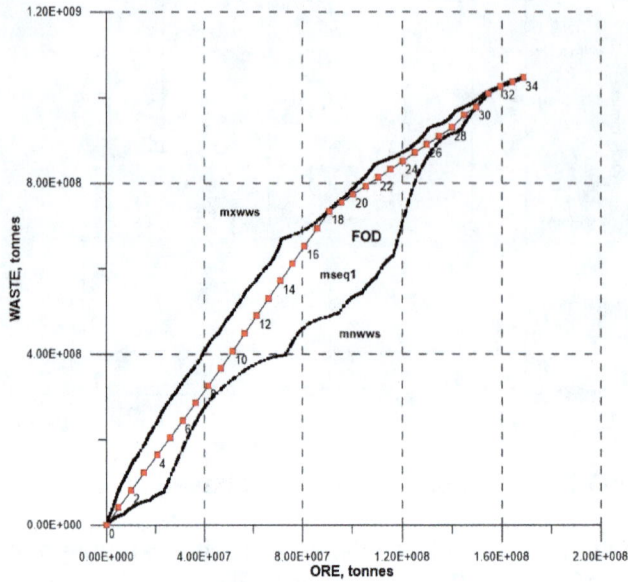

FIG 8 – CSG with sequence of CMR-SP schedule.

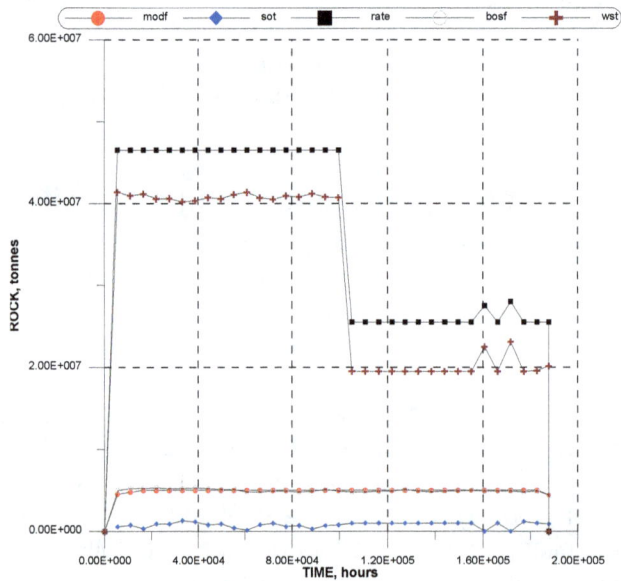

FIG 9 – Production schedule with CMR-SP.

The first stage has a mining rate of 46.2 Mt/a, while the second stage has a rate of 25.4 Mt/a. There are some fluctuations at the last time steps of the schedule because of restrictions on the FOD of the CSG. The schedule is presented with two ore functions: mill ore demand function (*modf*) in red colour, which reflects the planned processing plant rate, and basic ore supply function (*bosf*) in grey colour, which reflects the ore supply from the mine to the plant. The difference between both ore functions determines the surplus and deficit of ore supply to the plant, which is managed with a stockpile. The schedule of ore stockpile inventory is shown in Figure 10 with an ore stockpile balance as cumulative estimates, and ore surplus and deficit as single estimates.

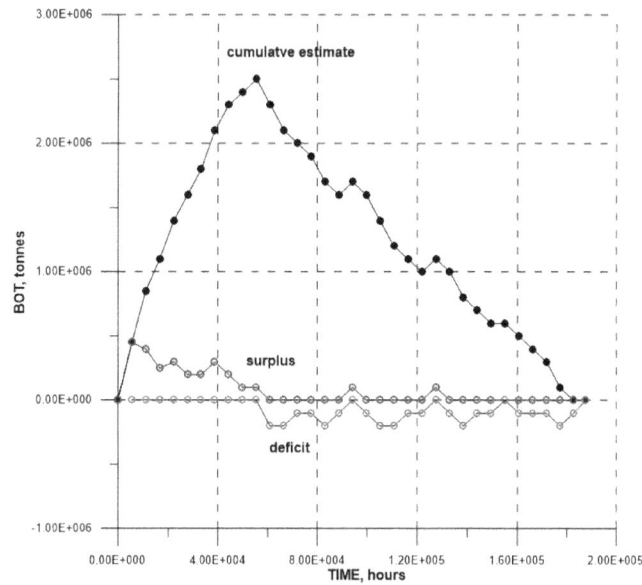

FIG 10 – CSG with optimum mine sequence (*optims*).

The estimates of the criteria of the schedule efficiency are as follows: *FEcC* = 54.75 per cent, *FTeC* = 68.19 per cent, and *FWD* = 31.80 per cent (Table 6). The NPV estimate is 2431.92 M$. The schedule has a low estimate of the *FWD*. The NPV estimate is less than the NPV estimate of the *mnwws*-related schedule as well as the NPV of the SSSMR schedule.

A summary of the results for the criteria of efficiency of all examples of schedule types is presented in Table 6.

MINING PROJECT RISK PROFILES OF SCHEDULE TYPES

Risk model definition

An author's model of mining project risk is used for the analysis of the risk profiles of the schedule types (Halatchev, 2007; Davis, Halatchev, and Potvin, 2007). The model is based on the net operating discounted cash flows of a project, which serve as a measure of mining project risk. The mining project risk is defined as the risk of not achieving positive net operating discounted cash flows at each time step of the Discounted Cash Flow Analysis (DCFA), which is a logical requirement of the mining business. It takes into account the uncertainty of many sets of project variables defined as: $\Re_1$ – set of geological variables; $\Re_2$ – set of mining variables; $\Re_3$ – set of mineral and metallurgical processing variables; $\Re_4$ – set of economic variables; $\Re_5$ – A set of the geomechanical variables of slope stability modelling. Taking into account the above-introduced sets of variables of the DCFA, the mining project risk model is presented as follows (Halatchev, 2007; 2011):

$$R_P(t_i) = \Pr\left\{ ODCF(t_i) < 0 \Big|_{\Re_z = var, \ z=1,2,3,4,5} \right\}, \ \forall i$$

(16)

where:

$ODCF(t_i)$ is the operating discounted cash flow at the time step t_i of DCFA.

The model (Equation 16) defines the risk of not achieving positive net operating discounted cash flows at any time step of the DCFA, given that all sets of variables ($\Re_1, \Re_2, \Re_3, \Re_4, \Re_5$) are treated simultaneously as sets of variables having a stochastic nature. Such a formulation of the mining project risk model means that a strategy for achieving only positive net operating cash flows (ODCF) over the life-of-mine is set by the mining company.

Input data

The mining project risk assessment in this study has been performed with regards to the following parameters: geological variables – gold grade and ore density; mining variables – mining recovery, excavators production rates; mineral processing variables – gold total processing recovery; economic variables – gold price, royalty, marketing cost, processing cost, mining costs for BOT and SOT, waste removal cost, stockpiling cost, and G&A cost; geotechnical variables – excluded from the case study.

An assumption is used for the Gaussian distribution of all variables and the implementation of the Model of Random Quantity. The standard deviations of the gold grade over the scheduling horizon are assessed using the data of the production scheduling as a simplification in comparison with the implementation of stochastic simulation techniques such as Sequential Gaussian Simulation (Ravenscroft, 1992; Souza, Costa and Koppe, 2004). The standard deviations of all variables are assessed as 20 per cent deviation of the mean estimates and using the 6-Sigma rule.

Risk model of the schedule with VMR

The mine project risk profile of the schedule with VMR and *mnwws* sequence is shown in Figure 11. The graph is built with two relationships: ODCFs versus time steps and Risk versus time steps. The analysis of the graph indicates a significant increase of the ODCFs at the beginning of the schedule, followed by a sharp decrease at time steps 6 and 7 where they become negative. This is reflected in the Risk estimates of 92.95 per cent and 95.03 per cent, which are very high. A similar picture is also valid for the time steps 24 and 25, where the risk estimates are 99.98 per cent and 100.00 per cent. For the dominant part of the schedule horizon, the risk estimates are 0.00 per cent. The critical estimates of the Risk can be explained with the analysis of the schedule in Figure 2, where the time steps 6 and 24 are associated with fluctuations of a significant increase of the waste and mining rates.

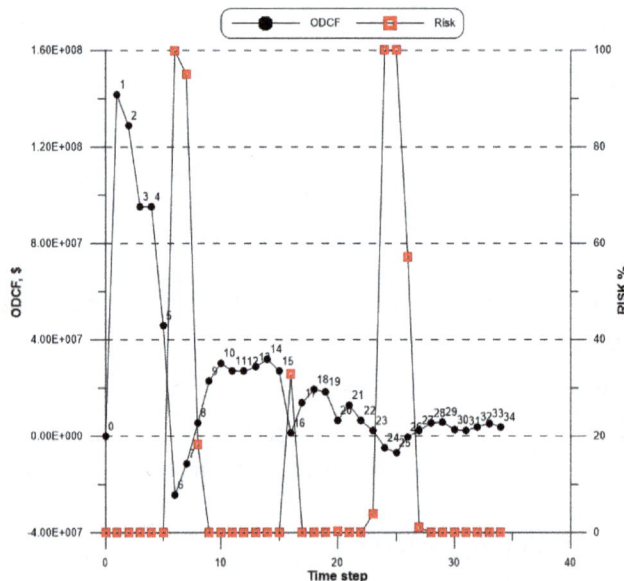

FIG 11 – ODCFs and mining project risk profile of schedule with VMR – *mnwws*.

The mine project risk profile of the schedule with VMR and *mxwws* sequence is shown in Figure 12. The analysis of the graph indicates negative ODCF at the beginning of the schedule horizon (time steps 1 and 2), which leads to a maximum increase of the Risk of 100.00 per cent. After this critical period, the risk estimates are mostly 0.00 per cent due to the positive ODCFs. The maximum Risk estimates can be explained with the analysis of the schedule in Figure 3, which has a maximum waste and mining rate.

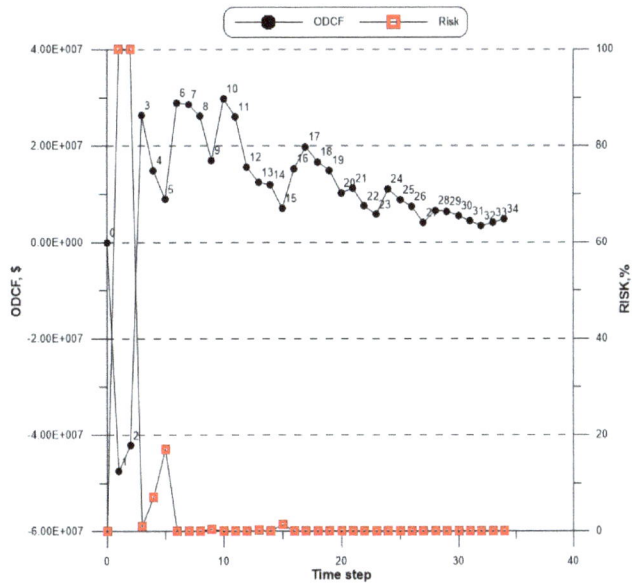

FIG 12 – ODCFs and mining project risk profile of schedule with VMR – *mxwws*.

Risk model of the schedule with SSSMR

The mine project risk profile of the schedule with SSSMR and *optims1* sequence is shown in Figure 13. The analysis of the graph indicates only positive ODCFs over the entire scheduling horizon, and the Risk estimates are equal to 0.00 per cent. This risk profile is associated with Figure 4, presenting the schedule with SSSMR. The mine sequence is a long straight line without any fluctuations of the mining rate.

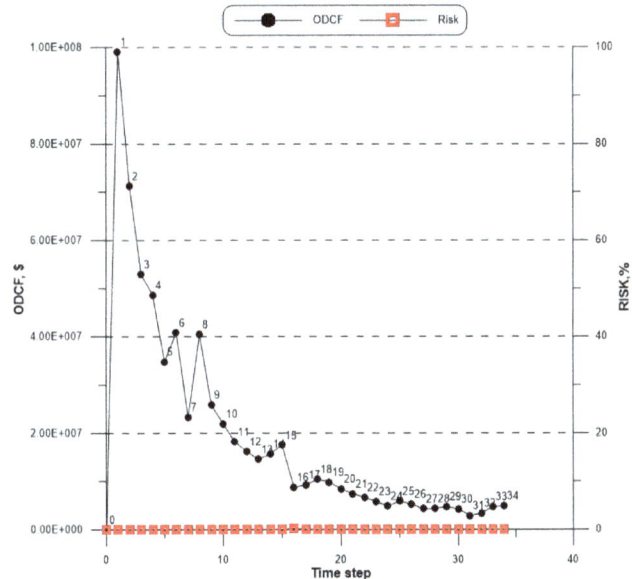

FIG 13 – ODCFs and mining project risk profile of schedule with SSSMR.

Risk model of the schedule with MSSMR

The mine project risk profile of the schedule with MSSMR and *optims2* sequence is shown in Figure 14. The analysis of the graph indicates a significant increase in ODCFs at the time steps 1, 2, 3 and 4, and the Risk is 0.00 per cent for all of them.

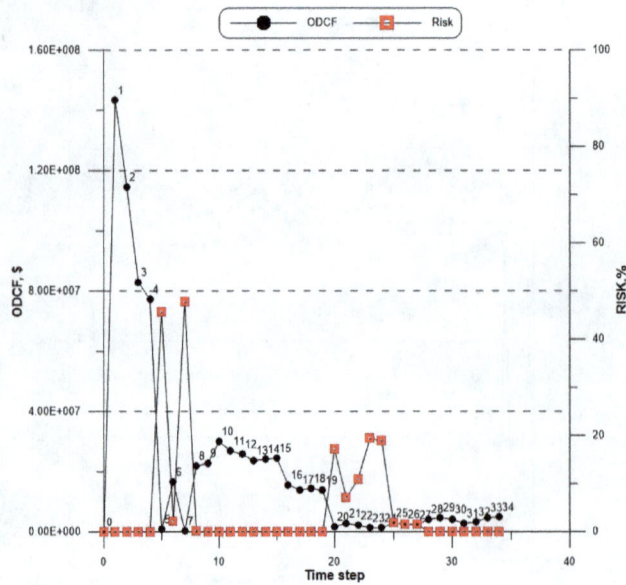

FIG 14 – ODCFs and mining project risk profile of schedule with MSSMR.

At time step 5, the ODCFs get a low estimate, which impacts the increase of the Risk estimate (45.69 per cent). Another increase of the Risk estimates is related to the time steps from 20 up to 24, where they vary within the range from 17.14 per cent up to 19.38 per cent. The rest time steps have Risk estimates of 0.00 per cent. The increase of the Risk at time steps 5 and 20 is due to the *optims2* sequence shown in Figure 6, which reflects the commencement of new phases with a change of the waste and mining rates in a direction of increase.

Risk model of the schedule with CMR-SP

The mine project risk profile of the schedule with CMR-SP is shown in Figure 15. The analysis of the graph indicates only positive ODCFs and Risk estimates of 0.00 per cent.

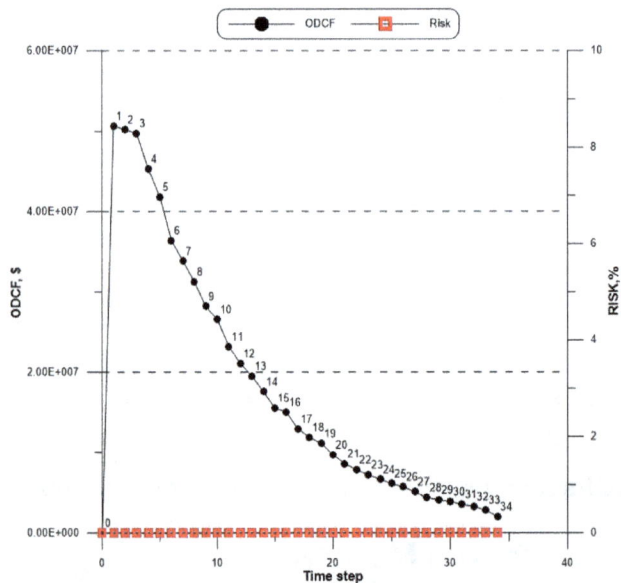

FIG 15 – ODCFs and mining project risk profile of schedule with CMR-SP.

CONCLUSIONS

The results presented in this paper support the following important conclusions:

- Each schedule type has a specific economic potential and impact on the mine project NPV and ore reserve estimates, respectively (see Table 6).

- The schedules with VMR have maximum and minimum NPV possible estimates, but their risk profiles have time steps with a risk of 100.00 per cent due to the fluctuations of mining rates.

- The schedules of SSSMR and MSSMR have the advantage of ignoring the utilisation of ore stockpiles, which imposes additional operating costs on mine production.

- The schedule of MSSMR is characterised by a higher NPV in comparison with the schedule of SSSMR due to a stronger implementation of the mine planning principle of waste deferment, which deals with the time value of money. The schedule of SSSMR has a risk profile of zero risk, while the schedule of MSSMR has some insignificant risk estimates.

- The new criterion introduced as the Factor of Waste Deferment (*FWD*) allows the assessment of the degree of implementation of the principle of waste deferment in any production schedule, and it can be used for the decision-making process of strategic mine planning.

- The schedule of CMR-SP has the disadvantage of using ore stockpiles, which impacts the production costs. The only exception is the case of a compulsory ore blending. This can also be achieved with the schedule of mining rate stabilisation in a more accurate way due to the optimisation procedure. The Risk profile of the schedule is zero.

- The schedules with SSSMR and CMR-SP are good alternatives for achieving a sustainable exploitation of open pit mines because of zero Risk profiles and the design of a single long stage of working without significant fluctuations of mining rate.

- The criteria developed for assessing the efficiency of production scheduling types, including the Factor of Waste Deferment, are sufficiently representative and can be used for the comparison of the currently available scheduling models.

- The schedule types described in the paper are viable for the exploitation of any type of mineral deposits, such as base/precious metals, critical minerals, rare earth, and coal.

ACKNOWLEDGEMENTS

The authors recognise the sponsorship of AusGEMCO Pty Ltd for the paper preparation. The paper reflects the long-term research outcomes of Dr Rossen Halatchev for developing a universal and highly efficient method of open pit production scheduling. The mathematical model of production scheduling optimisation presented in the paper is an outcome from his participation in a research project funded by the Australian Research Council (ARC/SPRIT) Grant #89804477 for the period of 1998–2001.

REFERENCES

Arsentiev, A I, 1961. *Determination of the production rate and limits of open pit mines*, 319 p (Nedra Publisher: Moscow).

Davis, G A, Halatchev, R A and Potvin, Y, 2007. Risk analysis and economic valuation of mining projects, Course notes (Australian Centre for Geomechanics: Perth).

De Kock, P, 2007. A back-to-basics approach of mine strategy formulation, in *Proceedings Sixth International Heavy Minerals Conference 'Back to Basics'*, pp 173–178 (The Southern African Institute of Mining and Metallurgy: Marshalltown).

Gentry, D W and O'Neil, T J, 1984. *Mine investment analysis*, 502 p (AIME: New York).

Golosinski, T and Bush, T, 2000. A comparison of open pit design and production scheduling techniques, *Int Journal of Surface Mining, Reclamation and Environment*, 1(1):53–61.

Halatchev R A, 1996. Factor of Compromise in the assessment of open pit long-term production plans, in *Proceedings Surface Mining '96 Conference*, pp 31–35 (The South African Institute of Mining and Metallurgy: Johannesburg).

Halatchev, R A and Moustakerov, I, 1996. Optimum scheduling of waste and ore production, *Mining Technology Journal*, 78(894):61–64.

Halatchev, R A, 1992. Multi-cut back pit exploitation of gold deposits, Report, Sponsor: Kalgoorlie Consolidated Gold Mines Ltd, Kalgoorlie, WA.

Halatchev, R A, 1993. Stage opencut exploitation of ore deposit, in *Proceedings Applications of Computers in the Mineral Industry Conference* (ed: E Baafi), pp 320–327 (University of Wollongong Printery Service: Wollongong).

Halatchev, R A, 1997. Where four-D continues on, in *Proceedings International Conference 'Optimizing with Whittle,* pp 57–69 (Whittle Programming Pty Ltd: Box Hill).

Halatchev, R A, 2002. The time aspect of the optimum long-term open pit production sequencing, in *Proceedings 30th APCOM Symposium,* pp 133–146 (SME: Littleton, CO).

Halatchev, R A, 2005. A model of discounted profit variation of open pit production sequencing optimization, in *Proceedings 32nd APCOM Symposium,* pp 315–323 (SME: Littleton, CO).

Halatchev, R A, 2007. An approach to variable discount rate modelling of open pit gold mine projects, in *Proceedings 33rd APCOM Symposium,* pp 729–739 (GECAMIN: Santiago, Chile).

Halatchev, R A, 2011. Contemporary criteria of open pit long-term production scheduling, in *Proceedings 35th APCOM Symposium,* pp 265–285 (University of Wollongong Printery Services: Wollongong).

Halatchev, R A, 2013. Owner-Operator versus Contractor Production Scheduling – A Vision for the Effective Exploitation of Australian Gold Resources by Surface Mining, in *Proceedings World Gold 2013 Conference,* pp 255–265 (The Australasian Institute of Mining and Metallurgy: Melbourne).

Halatchev, R A, 2015. The spatial aspect of the optimum long-term open pit production sequencing, in *Proceedings 2015 SME Annual Conference,* pre-print 15–085. (SME: Englewood, Colorado).

Halatchev, R A, Gabeva D A and Halatchev, A R, 2024. The impact of open pit production schedule types on the mine economic value, in *Proceedings International Conference EMMA+2024,* pp 186–199 (Transport Publishing House: Hanoi).

Hartman, H and Mutmanski, J M, 2002. *Introductory Mining Engineering,* 592 p (John Wiley and Sons, Inc: New Jersey).

Hustrulid, W and Kuhta, M, 1998. *Open Pit Mine Planning and Design* (2nd edition), 735 p (Balkema: Rotterdam, Brookfield, VT).

Kennedy, B A, 1990. *Surface Mining* (2nd edition), 1206 p (SME: Littleton, CO).

Kim, Y and Zhao, Y, 1994. Optimum open pit production sequencing – the current state of the art, in *Proceedings SME Annual Meeting,* preprint N-94–224 (American Society for Mining, Metallurgy and Exploration: Littleton).

Lane, K, 1988. *The economic definition of ore – cut-off grades theory and practice,* 149 p (Mining Journal Books Ltd: London).

Margolin, A M, 1974. *Evaluation of mineral deposits, Mathematical methods,* 261 p (Nedra Publishers: Moscow).

Masse, P, 1962. *Optimal investment decisions: rules for action and criteria for choice,* 500 p (Prentice-Hall: Englewood Cliffs, NJ).

Matheron, G, 1968. *Fundamentals of Applied Geostatistics,* 408 p (Mir Publishers: Moscow).

Ravenscroft, P, 1992. Risk analysis for mine scheduling by conditional simulation, *Trans Instn Min Metall (Sect A: Min Industry),* 101:A104–A108.

Rzhevskii, V V, 1980. *Technology and integrated mechanization of open pit mining* (3rd edition), 580 p (Nedra Publishers: Moscow).

Souza, L E, Costa, J F and Koppe, J C, 2004. Uncertainty estimate in resources assessment: a geostatistical contribution, *Natural Resources Research,* 13(1):1–15.

VALMIN Committee, 2005. Code for the Technical Assessment and Valuation of Mineral and Petroleum Assets and Securities for Independent Expert Reports – The VALMIN Code, 2005 edition [online]. Available from: <http://www.valmin.org/valmin_2005.pdf>.

Wang, Q, Xu, X and Gu, X, 2013. Dynamic-programming based model for phase-mining optimization in open-pit metal mines, *Applied Mechanics and Materials,* 316–317:896–901.

Comparison of stochastic versus deterministic open pit mine production schedule optimisation using mixed integer linear programming with conditional orebody simulations

S F Hoerger[1] and K Dagdelen[2]

1. Principal Consultant, Peak View Mine Planning LLC, Englewood CO 80111, USA.
   Email: sfhoerger@gmail.com
2. Professor, Colorado School of Mines, Golden CO 80401, USA. Email: kdagdele@mines.edu

ABSTRACT

Stochastic optimisation creates robust production schedules by solving for mining locations and processing destinations over time to maximise discounted cash flows while accounting for both orebody uncertainty and variability. Conditional simulation inputs provide a model of uncertainty by creating multiple equally likely orebody realisations conditional to existing drill hole information. Each simulated orebody realisation also models the variability of selective mining unit grades.

A new stochastic optimisation model is presented which maximises Net Present Value (NPV) by simultaneously solving for mining locations, process cut-offs, stockpiling, and blending over time. The new model strictly adheres to mine and process constraints for each orebody realisation without relying on arbitrary penalty functions. The model is formulated as a Mixed Integer Linear Program (MILP) with conditionally simulated orebody realisations as the key geostatistical input. The stochastic optimisation model can be easily simplified for deterministic optimisation.

The stochastic optimisation model can also be simplified to allow stochastic evaluation of existing mine plans created via deterministic optimisation or any other method. Stochastic evaluation enables risk analysis to answer questions such as: what annual production ranges are likely? or what NPV uncertainty might be expected?

Year by year production plans, annual cash flows and overall NPV's obtained from stochastic optimisation, deterministic optimisation and stochastic evaluation of open pit mine production schedules are compared using a gold mine case study. Significantly higher NPV is seen from stochastic optimisation with conditional simulation inputs versus deterministic optimisation with Ordinary Kriging inputs. However, when orebody variability and mining selectivity are accurately captured in deterministic geostatistical inputs via an Indicator Kriging, Localised Conditional Simulation or similar process, deterministic optimisation yields long-term open pit mine plans that are very similar to stochastic optimisation solutions.

INTRODUCTION

Mine production scheduling involves determining which parts of the deposit will be mined when and which material will be sent to which destination (Clark and Dagdelen, 2023). An optimum production schedule will maximise an economic objective while satisfying the operation's capacity, blending and sequencing constraints. Many production schedule optimisation methods have been presented (Newman *et al*, 2010; Dimitrakopoulos and Lamghari, 2022) with varying underlying assumptions to achieve solvability for this difficult multi-dimensional problem.

Production scheduling plays a key role in maximising the value of constrained open pit mining complexes. Efficient production schedules defer stripping costs until needed and select mining locations and rates such that scarce processing capacity is utilised most efficiently by balancing the opportunity cost of stockpiling (or wasting) marginally profitable rock against the opportunity cost of slowing the processing of future higher-grade ores. For long life open pit mines, significant value might be created by deferring a stripping campaign by one year or adjusting cut-off grade policies so that the lowest grade 10 per cent of mill feed is replaced by average grade material.

However, production scheduling relies on reliable orebody knowledge. A just-in-time stripping strategy may not deliver value if the process plant runs out of ore due to a model shortfall. Similarly, a cut-off grade optimisation plan may fail if the orebody model does not accurately predict the grade-tonnage distribution that will be encountered. Issues with orebody modelling accuracy and inherent orebody uncertainty can have a significant impact on creating a production schedule that reliably maximises value.

Stochastic production schedule optimisation aims to address both orebody uncertainty and modelling accuracy by using conditional simulation inputs. Conditional simulation provides multiple equally probable orebody models to give a view of uncertainty. The variability contained within each simulated model can provide an accurate representation of the grade tonnage distribution that will be encountered during selective mining.

Stochastic mine planning has been reported to add significant value compared to deterministic mine planning. One well reported stochastic mine planning method has been reported to add 'higher value in production schedules in the order of 25 per cent' (Dimitrakopoulos, 2018). This metaheuristics method (Goodfellow and Dimitrakopoulos, 2016) does not require pre-defined phases and uses penalties to guide solutions that on average satisfy mine and mill capacity constraints. Another study (Menabde et al, 2018) reported 4.1 per cent NPV improvement using stochastic optimisation for a low variability orebody. Their method used Mixed Integer Linear Programming and was constrained to respect constraints on average, but not for each simulated realisation.

The goal of this study is to gain insights into creating increased value from stochastic mine production scheduling and conditional simulations. Can stochastic schedule optimisation be implemented without using arbitrary penalties to minimise constraint deviations? Will increased value still be seen if each simulated realisation is forced to respect process capacity constraints? How much of the value is due to differences in variability/selectivity for deterministic versus stochastic input models? How much of the value is due to deterministic versus stochastic optimisation? Can conditional simulations give uncertainty insights without requiring stochastic optimisation?

GEOSTATISTICAL CONDITIONAL SIMULATIONS FOR MODELLING OREBODY VARIABILITY AND UNCERTAINTY

Conditional simulation, CS, is a common geostatistical technique for accurately modelling the amount of ore and metal which will be recovered by selective mining above one or more cut-off grades. The term 'conditional' means that each point in the model is conditional to the sampled drill hole information and to the other points in the model. The term 'simulation' means that the process creates multiple equally likely outcomes. During the CS process, the data's histogram is the starting point for a Monte Carlo sampling process at each point in the block model and the variogram is used to ensure that each simulated point is conditional to existing drill hole samples and to previously simulated points (Ortiz, 2020). Simulated points can be averaged within blocks to create a simulated block model that reflects the ore control selectivity that will be achieved when selective mining units are defined using future infill drilling information.

For most mine planning software, the most convenient input would be a traditional block model with a single estimated value for each block. Unfortunately, when this type of block model is created, the result may have too much selectivity (ie too much variability) if created by a nearest neighbour method or too little selectivity (ie not enough variability) if created by ordinary kriging with a large number of samples and a high degree of smoothing. Probabilistic estimation methods such as Indicator Kriging (IK) (Journel, 1983) can create a distribution of grades in each block to accurately match the deposit's overall expected ore control selectivity. Modified versions of probabilistic methods, such as Localised Indicator Kriging (Hardtke, Allen and Douglas, 2011), Localised Conditional Simulation (Amihere and Deutsch, 2022) and other methods, can provide a single value for each block while still accurately matching the deposit's overall expected ore control selectivity.

Whether single-value-per-block or probabilistic, estimation methods only give a single orebody block model with no measurement of orebody uncertainty, whereas geostatistical conditional simulations provide multiple equally likely block models. An accurate view of orebody variability is critical for accurate mine planning of selective mining operations. An accurate view of orebody uncertainty is necessary for understanding an operation's risk in providing planned production and cash flow.

Sequential Gaussian Simulation (Rossi and Deutsch, 2014) is implemented in commercially available mining software packages to efficiently create multiple conditional simulation block models, regardless of the data's actual distribution type.

MIXED INTEGER LINEAR PROGRAMMING MODELS FOR MINE PRODUCTION SCHEDULE OPTIMISATION

A Mixed Integer Linear Programming (MILP) model was developed for production scheduling to enable testing of stochastic schedule optimisation versus deterministic optimisation with four main goals. First was to create a stochastic schedule which honours all constraints for each simulated realisation. Second was to avoid use of arbitrarily selected penalties. Third was to use an exact optimiser. Fourth was to create a formulation that could easily be switched between stochastic grade/tonnage inputs and comparable deterministic grade/tonnage inputs. Ideally this could be achieved with efficient solution times too.

One key change to previously defined MILP optimisation formulations (Hoerger *et al,* 1999; Dagdelen and Kawahata, 2008; Dagdelen, 2011) allowed creation of a new formulation that could easily be extended for stochastic schedule optimisation. The extended formulation adheres to all constraints for each orebody simulation without use of arbitrary penalties. Run times are efficient, allowing solutions to be guaranteed to be within a defined tolerance of the best possible solution. The formulation is easily simplified to perform deterministic schedule optimisation allowing for effective comparisons between stochastic and deterministic optimisation and an understanding of deterministic optimisation's sensitivity to alternate methods of creating deterministic grade/tonnage inputs.

The key change versus the previous optimisation formulations was to manage the number of variables by separating the mining flow variables from the destination flow variables—tons mined are tracked from a given panel to the pit exit and process tons by increment are tracked from the pit exit to various destinations. Continuity constraints ensure a balance of mined increment tons with destination tons by increment, but the details of which panel provides increment tons to which destination are not tracked. These details are not necessary to compute a schedule's value and only create an excessive number of variables, particularly for a stochastic formulation.

Stochastic optimisation of mine production schedules

Stochastic production schedule optimisation is implemented as a MILP model for two-stage stochastic programming with recourse (Wolsey, 2021) to determine an optimum mining sequence in terms of mine, layback, panel and time under orebody uncertainty. The first-stage variables are the mining flows, $\mathbf{W}(m,l,p,t)$, which are to be chosen knowing the range of orebody uncertainty defined by the set of conditional simulations but without knowing the specific realisation, k. The second-stage variables are the mine-direct-to-destination flows ($\mathbf{X}$) and the stockpile flows ($\mathbf{Y}$ in, $\mathbf{Z}$ out) which can vary with each simulation, k. This two-stage approach corresponds roughly to making a mining sequence decision based on the grade-tonnage information determined from limited, currently available exploration data while knowing that the final process and stockpile decisions will be made after ore control data reveals a more certain grade tonnage distribution of the material mined.

The variable framework and dimensions (Figure 1) allow for multiple mines, stockpile areas and destinations within a mining complex. Mines are subdivided into predefined laybacks and each layback is split into panels which must be mined in sequential order. Grade/tonnage information for each panel is subdivided into increments which contain similar grade and metallurgical characteristics; each panel's distribution of increments varies from simulation to simulation.

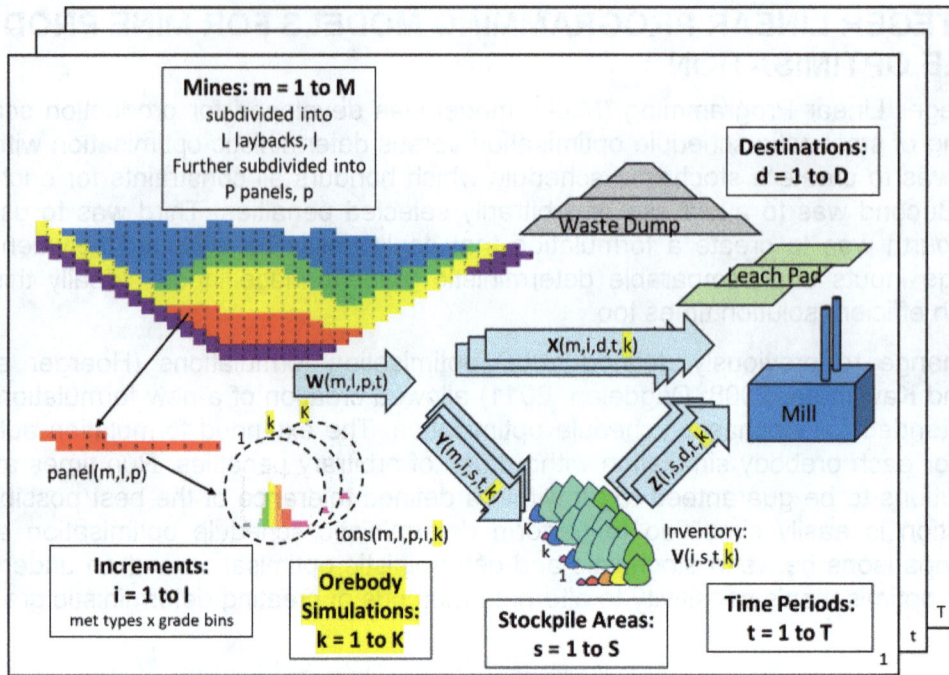

FIG 1 – Key stochastic production scheduling variables and dimensions. For deterministic production scheduling, remove k subscripts.

Objective function and flow constraints

The MILP formulation is set-up to solve for material flows over time that maximise an NPV objective function subject to mining complex constraints for mine and mill tonnages, grades, contained metal and recovered metal and layback sinking rate constraints. The first-stage mining variables (W's) do not have k subscripts, so they are the same for each simulation and are subject to mining capacity constraints for each time period. The second-stage destination and stockpile variables have k subscripts, so they can vary with each simulation, but any constraint involving process or stockpile variables is repeated for each simulation, forcing every simulation to comply with all constraints rather than relying on an arbitrary penalty function to drive compliance or semi-compliance.

Continuity constraints ensure that all material mined is sent to a process destination or stockpile and that tonnages don't get double counted. All flow variables (W, X, Y, Z) and inventory variables (V) must be non-negative to prevent backward flows or negative stockpile inventories.

Maximise Net Present Value =

$$\sum_m \sum_l \sum_p \sum_t pv(t)*MineVal(m,l,p,t)*W(m,l,p,t) +$$

$$1/K * \sum_k \sum_m \sum_i \sum_d \sum_t pv(t)*DestVal(m,i,d,t,k)*X(m,i,d,t,k) +$$

$$1/K * \sum_k \sum_m \sum_i \sum_s \sum_t pv(t)*StockVal(m,i,s,t,k)*Y(m,i,s,t,k) +$$

$$1/K * \sum_k \sum_i \sum_s \sum_d \sum_t pv(t)*ReclaimVal(i,s,d,t,k)*Z(i,s,d,t,k) \tag{1}$$

subject to:

Finite resources ∀ panel (m,l,p):

$$\sum_t W(m,l,p,t) \leq Tons(m,l,p) \tag{2}$$

Mine Continuity ∀ (m,i,t,k):

$$\sum_l \sum_p W(m,l,p,t)*[tons(m,l,p,i,k)/Tons(m,l,p)] = \sum_d X(m,i,d,t,k) + \sum_s Y(m,i,s,t,k) \tag{3}$$

Stockpile Continuity ∀ (i,s,t,k):

$$V(i,s,t,k) - V(i,s,t-1,k) = \sum_m Y(m,i,s,t.k) - \sum_d Z(i,s,d,t,k) \tag{4}$$

Mine Capacity ∀ (m,t):

$$\sum_l \sum_p W(m,l,p,t) \leq MineCapacity(m,t) \tag{5}$$

Layback Capacity $\forall$ (m,l,t):

$$\sum_p \mathbf{W}(m,l,p,t) \leq LaybackCapacity(m,l,t) \tag{6}$$

Process Capacity $\forall$ (d,t,k):

$$\sum_m \sum_i \mathbf{X}(m,i,d,t,k) + \sum_i \sum_s \mathbf{Z}(i,s,d,t,k) \leq DestCapacity(d,t) \tag{7}$$

where the notation $\sum_m$ represents a summation from m=1 to m=M and with the following constants:

pv(t) = 1 / (1 + d)^t: present value factor for discount rate, d

tons(m,l,p,i,k): Orebody tons for each mine, layback, panel, increment, realisation

Tons(m,l,p) = 1/K * $\sum_k \sum_i$ tons(m,l,p,i,k): Total tons for each mine, layback, panel

g(e,i): Element grades (e=1 to E) for each increment

V(i,s,t=0) = specified starting stockpile inventory

Additional formulation details including value coefficient modelling, sequencing constraints and sinking rate constraints are presented in Appendix 1. A simple demonstration case is presented in Appendix 2.

MILP model assumptions

A key model assumption is that partially mined panels must mine the same proportion of tons from each increment in the panel and that density differences from simulation to simulation are not significant, allowing mining constraints to be based on each panel's average total tons, Tons(m,l,p). Also, increments must be tightly defined so that grades for each increment, i, for each element, e, can be treated as a constant, g(e,i). Also note that increment resolution is maintained during stockpile creation and reclaim.

MILP model characteristics

Three key characteristics of this stochastic optimisation model should be emphasised. First, the solution will respect mining and processing constraints for each simulated realisation and no arbitrary penalty functions are needed in the formulation.

Second, cut-off grades are not pre-defined as model inputs; instead, optimum flows for each increment are outputs which can be reported to show cut-off grades. Because the MILP formulation incorporates constraints, grade/tonnage distributions and the opportunity costs associated with deferring cash flows, the MILP solution will reflect a generalisation of Lane's cut-off grade theory (Lane, 1988; Dagdelen, 1992; Rendu, 2014) which incorporates multiple constraints and does not require a pre-defined mining sequence.

Third, as discussed in the introduction to this section, the formulation minimises the number of variables by separating mining variables, $\mathbf{W}$, (which need layback by panel resolution to facilitate cost computations and sinking, tonnage and mine sequencing constraints) from processing, $\mathbf{X}$, and stockpiling, $\mathbf{Y}$, variables (which don't need to know which layback or panel is providing the increment tons). Most critically, this allows for $\mathbf{W}$ variables which don't need a k subscript and keeps the number of $\mathbf{W}$ variables independent of the number of simulations. The MILP formulation was originally implemented in deterministic form as a C program named Peakfinder (Hoerger, 2024) before it was expanded to perform stochastic optimisation (Hoerger and Dagdelen, 2024) and stochastic evaluation (Hoerger and Dagdelen, 2025).

Deterministic optimisation of mine production schedules

The same MILP formulation can be leveraged for deterministic optimisation under the assumption that there is no orebody uncertainty. For deterministic optimisation, there is a single value to quantify the tonnage of each increment in each panel of each layback. The formulation is identical to the stochastic optimisation formulation, except all k subscripts, k summations and 1/K terms are eliminated. Accordingly, the objective function becomes:

Maximise Net Present Value =

$$\sum_m \sum_l \sum_p \sum_t pv(t)*MineVal(m,l,p,t)*\mathbf{W}(m,l,p,t) +$$

$$\sum_m \sum_i \sum_d \sum_t pv(t)*DestVal(m,i,d,t)*\mathbf{X}(m,i,d,t) +$$

$$\sum_m \sum_i \sum_s \sum_t pv(t)*StockVal(m,i,s,t)*\mathbf{Y}(m,i,s,t) +$$

$$\sum_i \sum_s \sum_d \sum_t pv(t)*ReclaimVal(i,s,d,t)*\mathbf{Z}(i,s,d,t) \tag{8}$$

The key input difference for deterministic optimisation is that the grade/tonnage distribution for each panel, tons(m,l,p,i) becomes a fixed input rather than varying by simulation, tons(m,l,p,i,k). Note that this input can come from a single-value-per-block model, such as OK, LIK, LCS or from a probabilistic model which has a distribution of grades for each block such as IK or DK. The input could also come from an accumulation of conditional simulation results (ACS):

$$ACS\ tons(m,l,p,i) = 1/K * \sum_k tons(m,l,p,i,k) \tag{9}$$

Stochastic evaluation for mine production schedule risk analysis

The stochastic optimisation MILP can also be leveraged to use conditional simulation inputs to perform a risk analysis for any specified mining sequence. The formulation is identical to the stochastic optimisation formulation, except all $\mathbf{W}$(m,l,p,t) variables are set to constants based on a user-specified input mining sequence.

Alternatively, the same result could be achieved by performing a deterministic optimisation of the $\mathbf{X}$, $\mathbf{Y}$ and $\mathbf{Z}$ variables for each of the k realisations and reporting the average of the K NPV's. This approach would enable using a large number of simulation realisations without needing to build a MILP problem larger than the deterministic optimisation problem.

Upper bound for mine production schedules (perfect information)

The stochastic optimisation MILP can be expanded to understand the maximum expected value of the orebody if orebody uncertainty was eliminated prior to selecting a mining sequence. This approach is similar to the approach of Froyland *et al* (2018). The formulation would be identical to the two-stage stochastic programming formulation, except that now the mining sequence becomes a second-stage variable, $\mathbf{W}$(m,l,p,t,k), that varies by simulated realisation, and all constraints involving $\mathbf{W}$ variables need to be repeated for each simulation realisation, k. Accordingly, the objective function and first few constraints would be:

Maximise Net Present Value =

$$1/K * \sum_k \sum_m \sum_l \sum_p \sum_t pv(t)*MineVal(m,l,p,t)*\mathbf{W}(m,l,p,t,k) +$$

$$1/K * \sum_k \sum_m \sum_i \sum_d \sum_t pv(t)*DestVal(m,i,d,t,k)*\mathbf{X}(m,i,d,t,k) +$$

$$1/K * \sum_k \sum_m \sum_i \sum_s \sum_t pv(t)*StockVal(m,i,s,t,k)*\mathbf{Y}(m,i,s,t,k) +$$

$$1/K * \sum_k \sum_i \sum_s \sum_d \sum_t pv(t)*ReclaimVal(i,s,d,t,k)*\mathbf{Z}(i,s,d,t,k) \tag{10}$$

subject to:

Finite resources $\forall$ (m,l,p,k):

$$\sum_t \mathbf{W}(m,l,p,t,k) \leq Tons(m,l,p) \tag{11}$$

Mine Continuity $\forall$ (m,i,t,k):

$$\sum_l \sum_p \mathbf{W}(m,l,p,t,k)*[tons(m,l,p,i,k)/Tons(m,l,p)] = \sum_d \mathbf{X}(m,i,d,t,k) + \sum_s \mathbf{Y}(m,i,s,t,k) \tag{12}$$

With similar modifications to all other constraint equations that contain a $\mathbf{W}$ term

Alternatively, the same result could be achieved by performing a deterministic optimisation of the $\mathbf{W}$, $\mathbf{X}$, $\mathbf{Y}$ and $\mathbf{Z}$ variables for each of the k simulated realisations and reporting the average of the k NPV's. This approach would enable the possibility of using a large number of simulations without needing to build a MILP problem larger than the deterministic optimisation problem.

This upper bound (UB-PI) represents the expected NPV of the orebody if perfect orebody information was obtained before choosing the mining sequence. Note that this upper bound is useful for defining a NPV upper bound, but it does not define a specific mining sequence.

Lower bound for mine production schedules (best stochastic evaluation)

Any mining sequence which has a feasible stochastic evaluation will be a lower bound for the stochastic optimisation problem. By creating K mine plans via a deterministic optimisation for each individual simulated realisation, stochastic evaluation can be used to select the deterministic mine plan which creates the highest expected NPV when evaluated against all K simulations. This Best Stochastic Evaluation approach to defining a lower bound (LB-BSE) requires $K+K^2$ deterministic optimisations (for each k, one to create a plan and K to perform the SE), but each of the optimisations is kept to a manageable size equal to the size of the deterministic optimisation problem.

If there is a tight bound between LB-BSE and UB-PI, then there would be limited scheduling value in investigating a full stochastic optimisation. This could occur when there is low orebody uncertainty or when problem constraints such as max sink rate + total mining rate suggest a similar mining schedule for all orebody realisations. In this situation, additional drilling information may not change the mining schedule decisions, but it could still significantly change the expected value and uncertainty around production and profitability metrics.

CASE STUDY – MCLAUGHLIN GOLD MINE

McLaughlin case study – inputs

A case study was performed using data from the McLaughlin gold mine to compare stochastic and deterministic production schedule optimisation techniques. Various geostatistical models were carefully prepared to allow comparisons with different input models with consistent means but different variances.

McLaughlin production scheduling framework and metallurgical scenarios

The McLaughlin mine is an open pit gold mine that was operated from 1984 to 2002. The hot springs style deposit has skewed grade distribution with short variogram range. The gold exploration drilling 20-foot composites were used to create conditional simulations with Hexagon's MinePlan SGS software module. For simplicity in presenting results, nine simulations were used for this study.

Five nested phases for the McLaughlin mine were provided (Aras, Dagdelen and Johnson, 2019) and the first four were split into independent north and south subphases resulting in nine laybacks with up to 62 benches or 31 panels when grouped using two benches per panel. Each panel of each layback was required to not start mining until its inner layback was mined one panel deeper; the final layback was required to wait for both North layback #4 and South layback #4 to be mined one bench deeper. Tonnages for 22 increments defined by gold grade ranges were computed for each panel of each layback for each of several different input models. For case study purposes, destination options were defined as waste dump, leach pad and mill. Four metallurgical scenarios were defined for combinations of with/without stockpiling and with/without leaching.

McLaughlin production scheduling costs and constraints

Key mining parameters were a cost of $1.70/t plus $0.01/t/bench, 10 Mt/a maximum mining capacity, and a maximum of eight 20' benches/a vertical advance for each layback (not counting any panels with less than 49 000 t). Key leaching parameters were a cost of $5/t for 70 per cent recovery, and 2.25 Mt/a maximum leach pad capacity. Key milling parameters were a cost of $12/t for 90 per cent recovery and a 2.25 Mt/a maximum milling capacity. Other important economic parameters were a gold price of $1250/oz and a 12.5 per cent discount rate.

McLaughlin orebody models

A conditional simulation model of the McLaughlin deposit was created using the SGS module of Hexagon's MinePlan software. Each simulated point model (CP) was converted into a selective mining unit (SMU) model (CS) by kriging the simulated points to create a grade for each 25' × 25' × 20' block. For each block, an e-type model was created by averaging the nine SMU simulations—this single-value-per-block model will be called an OK model and shows significantly lower selectivity than the CS model (Figure 2). For each block, the nine SMU simulations were accumulated into a histogram—this probabilistic block model will be called an IK model and shows the same selectivity as the nine CS realisations (Figure 2). These methods of model construction were chosen to allow precise comparability of the different models (Hoerger and Dagdelen, 2024) by ensuring that each block has the same mean value for OK, IK and CS models.

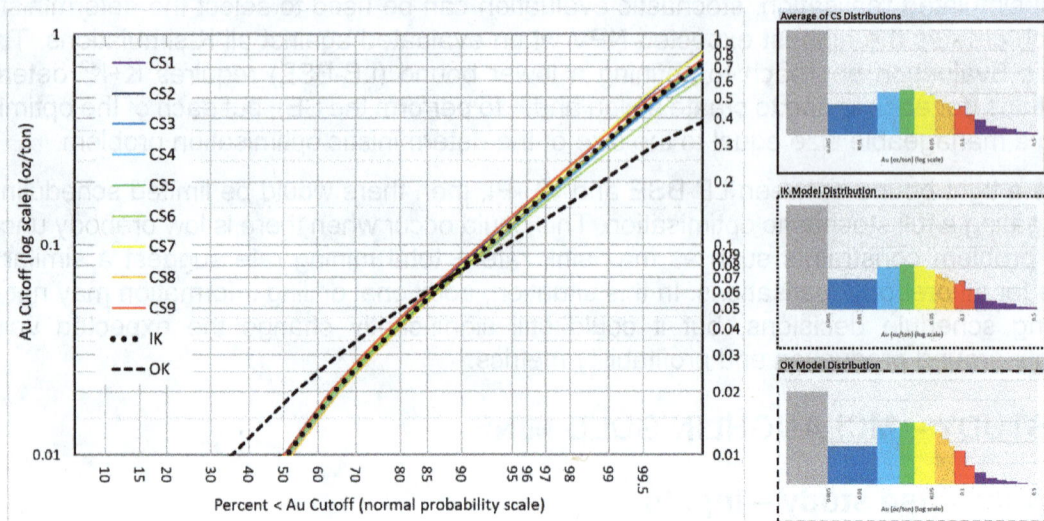

FIG 2 – Selectivity differences between OK models versus CS and IK models.

McLaughlin case study – results

Stochastic versus deterministic(IK) production schedule optimisation – scenario 1

For metallurgical scenario #1, which allows milling, stockpiling and leaching, stochastic production schedule optimisation using CS inputs, SO(CS), was performed and compared to deterministic production schedule optimisation using IK inputs, DO(IK). Both optimisations generated identical mining sequences (same **W**'s: see Figure 3 top three dashboard rows for left versus middle column) as the optimised mining schedules focused on the 'best' laybacks first with each years' 'best' layback constrained by benches/year and total tons mined constrained by mining capacity. The IK model has the same selectivity as the CS models, so, the DO(IK) plan shows the same grade distribution mined as the average of the nine CS models (see Figure 3 third dashboard row for left versus middle column). With the same mining sequence and grade distribution mined for the SO(CS) and DO(IK) plans, the process statistics are also the same (Figure 3 dashboard rows four and five), so planned cash flows and NPV's are also the same (Figure 3 dashboard row six). For this scenario, using deterministic optimisation with inputs that match the CS selectivity gave the same results as stochastic optimisation.

Stochastic evaluation of the DO(IK) plan showed the same results as the stochastic optimisation (ie SE(DO(IK)) graphs would be the same as the left column of Figure 3 dashboard graphs).

Stochastic versus deterministic(OK) production schedule optimisation – scenario 1

Next, deterministic production schedule optimisation was performed for metallurgical scenario #1 using the OK model inputs, DO(OK). Using the OK inputs, deterministic optimisation yields almost an identical mining sequence as stochastic optimisation (Figure 3 top two dashboard rows for right versus left column) as the DO(OK) mining schedule also focuses on the 'best' laybacks first with each year's 'best' layback constrained by benches/year and total tons mined constrained by mining capacity. However, because the OK model has lower selectivity, its grade distribution is smoother (ie more dilution versus CS or IK models) for each year's mined volume (Figure 3 third row). Compared to the CS or IK grade distributions, the smoother OK grade distribution requires pushing the lowest grade mill feed increments to the leach pad and requires pushing the lowest grade leach pad increments to stockpile (Figure 3, row 4). Ultimately, additional dilution tons result in more stockpiled tons and extended mill and leach pad operating lives with lower mill grades leading to lower cash flows and lower planned NPV (Figure 3 rows five, six and seven).

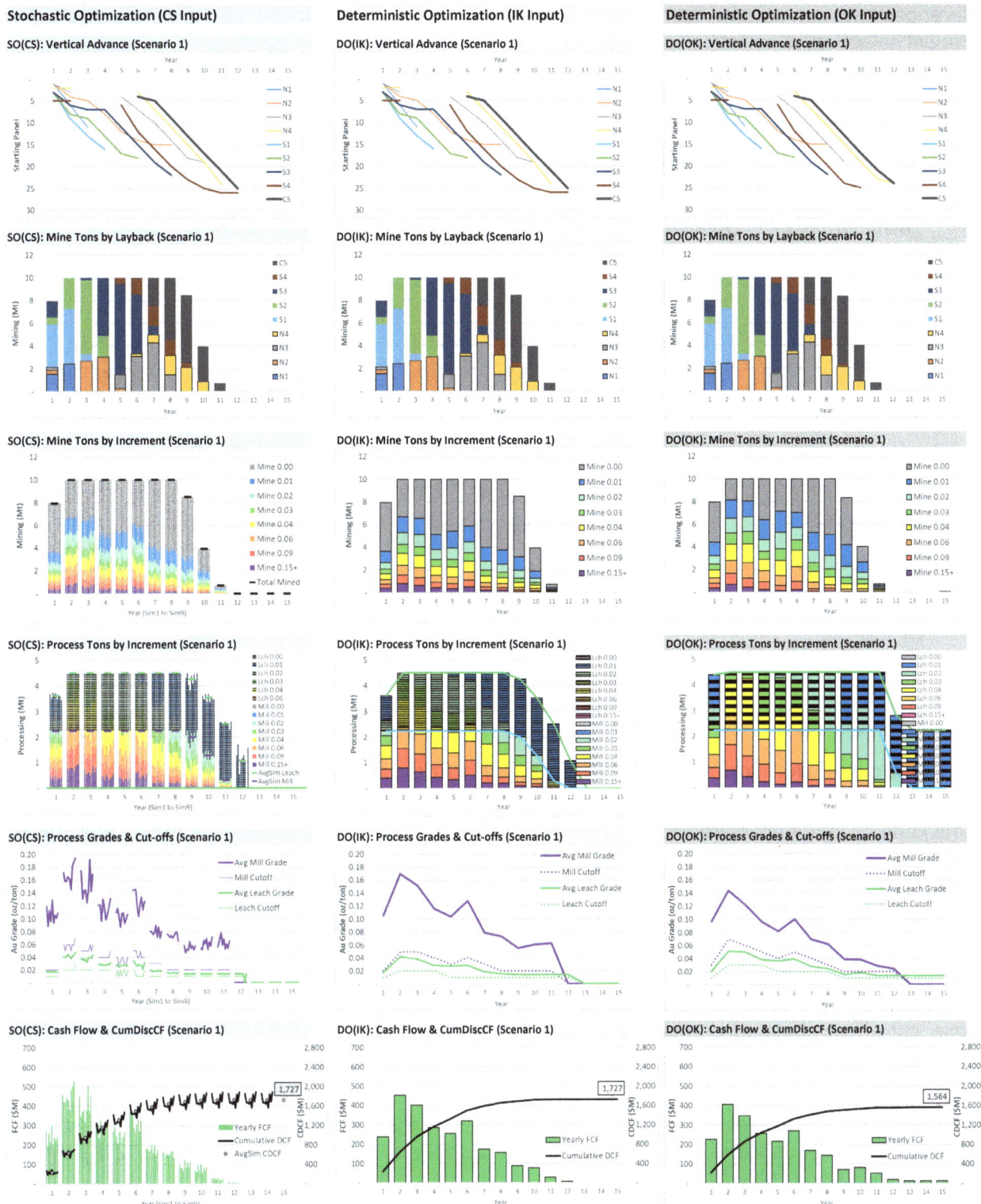

FIG 3 – Production Schedule Dashboards for Scenario 1—Stochastic Optimisation (left column) versus Deterministic Optimisation with IK input (centre column) versus Deterministic Optimisation with OK input (right column): Row 1 mining vertical advance by layback; Row 2 tons mined by layback; Row 3 tons mined by increment; Row 4 tons processed by increment; Row 5 process grades; Row 6 cash flows and cumulative discounted cash flow.

Because the mining sequences were nearly identical, stochastic evaluation of the DO(OK) plan showed the same results as the stochastic optimisation (ie SE(DO(OK)) graphs would be the same as the Figure 3 left column graphs). So, although the planned NPV for DO(OK) is 9.4 per cent lower than the planned NPV for SO(CS), when both SO(CS) and DO(OK) are evaluated consistently using CS models, both show the same realised NPV of $1.727 million (Table 1). For this scenario, differences between deterministic optimisation and stochastic optimisation results were entirely due to model selectivity.

TABLE 1
Stochastic optimisation versus deterministic optimisation with IK and OK inputs.

| Metallurgical Scenario | Prod schedule optimisation | | | | Stochastic evaluation | | |
	Optimisation Method(input)	Planned NPV $M	%diff versus SO(CS)		Fixed Plan	Realised NPV	%diff versus SO(CS)
#1: Mill + Leach	SO(CS)	1726.61	–		SO(CS)	1726.61	–
+ Stockpile	DO(IK)	1727.11	+0.0%		DO(IK)	1726.57	-0.0%
	DO(OK)	1564.02	-9.4%		DO(OK)	1726.53	-0.0%
#2: Mill + Leach	SO(CS)	1707.61	–		SO(CS)	1707.61	–
no Stockpile	DO(IK)	1708.76	+0.1%		DO(IK)	1707.58	-0.0%
	DO(OK)	1526.07	-10.6%		DO(OK)	1689.27	-1.1%
#3: Mill only	SO(CS)	1577.68	–		SO(CS)	1577.68	–
+ Stockpile	DO(IK)	1578.59	+0.1%		DO(IK)	1577.68	+0.0%
	DO(OK)	1324.89	-16.0%		DO(OK)	1570.35	-0.5%
#4: Mill only	SO(CS)	1496.49	–		SO(CS)	1496.49	–
no Stockpile	DO(IK)	1498.54	+0.1%		DO(IK)	1495.72	-0.1%
	DO(OK)	1258.85	-15.9%		DO(OK)	1454.40	-2.8%

Stochastic and deterministic production schedule optimisation – scenario 4

For metallurgical scenario #4, which allows milling, but not stockpiling or leaching, the SO(CS) and DO(IK) mining sequences are nearly identical, but the DO(OK) mining sequences mines at a slower rate. As in scenario 1, the SO(CS) and DO(IK) mining sequences are nearly identical and the input models have the same selectivity, so their planned NPV's and realised NPV's are nearly identical. Because of its similarity to SO(CS), no dashboard graphs are presented for DO(IK).

With fixed mill capacity and no stockpiling option, the less selective OK grade/tonnage distribution requires slower mining to balance the value of the ore milled against the opportunity cost of marginal ore wasted and the opportunity cost of delaying future profits. The slower mining and lower grade above cut-off combine to create a significantly lower planned NPV for the DO(OK) production schedule—15.9 per cent less than the stochastic optimisation plan SO(CS) (Figure 4 middle versus left column).

However, even with slower mining, stochastic evaluation (right column of Figure 4) shows that if the DO(OK) mining sequence is evaluated using the CS models, the realised NPV is only 2.8 per cent lower than SO(CS) (Table 1).

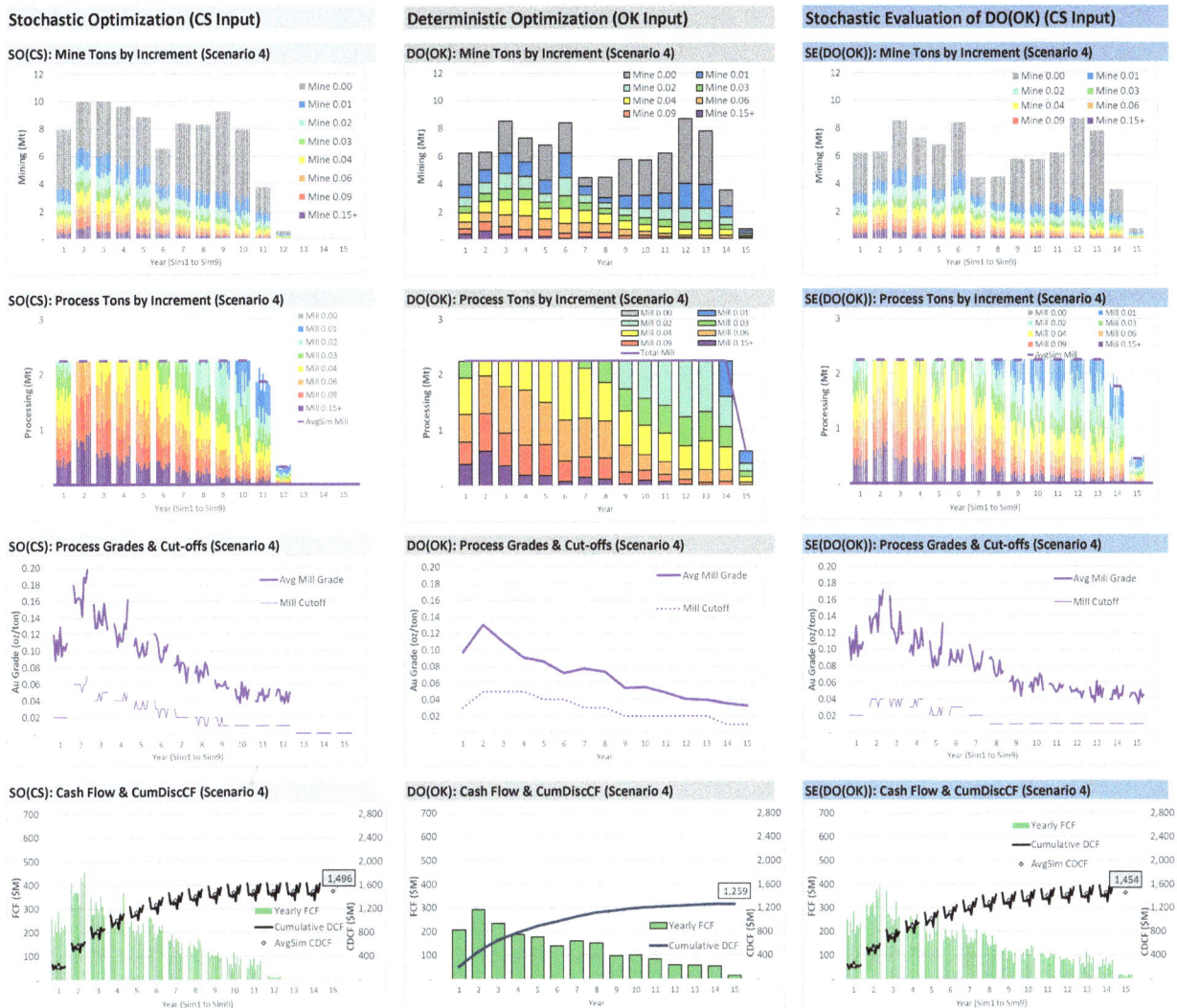

FIG 4 – Production Schedule Dashboards for Scenario 4—Stochastic Optimisation (left column) versus Deterministic Optimisation with OK input (DO(OK)—middle column) versus Stochastic Evaluation of DO(OK) (right column): Row 1 tons mined by increment; Row 2 tons processed by increment; Row 3 process grades; Row 4 cash flows and cumulative discounted cash flow.

Value of perfect information; upper and lower bounds

To examine the value of perfect information, deterministic optimisation was performed for each of the nine simulation realisations as part of the UB-PI process. The average of these nine deterministic optimisations represents an upper bound for the stochastic optimisation. For all four metallurgical scenarios, this perfect information upper bound is very close to the value of stochastic optimisation (Table 2: UB-PI rows shown for scenarios 1 and 4). This indicates that additional information would not significantly change the expected value of the operation. However, with more information, even though the mine plan may not change, uncertainty on the operation's value will be narrowed. In the authors' experience, this is a common paradox in mining: additional information may not change the planned mining sequence, but it will provide more confidence in what metal production and cash flow will actually be delivered when the plan is executed.

Each of the nine plans generated by deterministic optimisation of an individual realisation was then tested against all nine simulations via stochastic evaluation as part of the LB-BSE process. For all four metallurgical scenarios, this lower bound was very close to the NPV produced by stochastic optimisation (Table 2: LB-BSE rows). In this case study, any one of the nine deterministic optimisations would have provided a reasonably tight lower bound for the stochastic optimisation.

TABLE 2
Upper Bound and Lower Bound computations.

Met Scenario		#1: Mill + Leach + Stockpile			#4: Mill only no Stockpile		
			Up Bound	Low Bound		Up Bound	Low Bound
Method(input)	SO(CS)	DO(SimX)	SE(DO(SimX))	SO(CS)	DO(SimX)	SE(DO(SimX))
NPV	SimX	SimX	Avg(Sim1 to 9)	SimX	SimX	Avg(Sim1 to 9)
Sim1	1860	1861	1724	1631	1632	1494
Sim2	1667	1672	1725	1436	1438	1493
Sim3	1672	1679	1725	1452	1461	1491
Sim4	1645	1646	1726	1416	1418	1492
Sim5	1712	1717	1722	1497	1501	1491
Sim6	1565	1569	1725	1341	1350	1489
Sim7	1840	1845	1720	1600	1608	1490
Sim8	1690	1691	1725	1473	1480	1487
Sim9	1888	1892	1723	1623	1637	1488
UB-PI (avg)		1730			1503	
SO(CS) (avg)	1727			1496		
LB-BSE (max)			1726			1494
% of SO(CS)		+0.2%	-0.0%		+0.4%	-0.2%

The demonstrated upper and lower bounds provided an extremely tight bracketing of the stochastic optimisation results. For multi-pit cases where the number of simulation combinations creates a prohibitively large MILP formulation, the bounding processes can be performed with multiple deterministic optimisations with each having a manageable problem size.

McLaughlin case study – discussion

Table 1 shows the full NPV results for all four metallurgical scenarios. In all four scenarios, the SO(CS) and DO(IK) optimisations gave nearly identical mining sequences, planned NPV's and realised NPV's. In this long-term planning case study, the volume mined each year is much greater than the variogram range, so the grade distribution for any given year is fairly similar from realisation to realisation. Compared to a cut-off grade generated by solving the DO(IK) case, cut-offs for any particular simulated realisation only need slight adjustments to meet process capacity constraints. These slight cut-off variations from realisation to realisation add up to little net value impact.

Does stochastic optimisation find a higher value than deterministic optimisation?

In only one metallurgical scenario, #4, SO(CS) gave a significantly different mining sequence than DO(OK). For this scenario, it would be correct to say that the planned NPV is 15.9 per cent lower for deterministic optimisation with OK input versus the planned NPV from stochastic optimisation with CS input. However, when consistently evaluated against the same CS model, the realised NPV difference between the two plans is only 2.8 per cent.

The difference between stochastically and deterministically optimised plans is due to selectivity differences, not optimisation differences. Deterministic optimisation with IK input yields nearly the same plan and value as stochastic optimisation with conditional simulation input. In the other three scenarios, even with the selectivity differences, DO(OK) gave a nearly identical mining sequence to SO(CS). In these cases, planned NPV's are lower when reported based on the lower selectivity OK model, but realised NPV's are nearly identical when mining sequences are consistently evaluated against the same CS models.

Cases where stochastic optimisation might give significantly different mining sequences versus deterministic optimisation include: a) where deterministic input selectivity varies significantly from the conditional simulation selectivity (such as OK input); b) where grade tonnage distributions have little marginal material available to adjust cut-offs up or down to meet capacity constraints (such as the Newsvendor mine in Appendix 2); c) short-term planning cases where mined volume dimensions are within variogram ranges; and d) blending cases where the costs or opportunity costs required to meet constraints dramatically increase for outlier orebody realisations.

Conditional simulations enable risk analysis

Conditional simulations are necessary inputs to provide a risk analysis, whether from stochastic optimisation or from stochastic evaluation of any mine plan created by deterministic optimisation or any other planning approach. Risk analysis can be presented in a variety of formats such as showing every realisation or summarising results in a box plot showing minimums, maximums and quartiles. The grade profile graph in Figure 5 shows an example for communicating how orebody uncertainty will translate to uncertainty in metal mined each year. The box plot graph in Figure 6 shows that the value of a mill only plan (Scenario 4) can be improved with stockpiling (Scenario 3), but the amount of expected NPV increase may be similar in size to the NPV variation due to orebody uncertainty. Additional discussion of using conditional simulations and stochastic evaluation to perform risk analysis can be found in (Hoerger and Dagdelen, 2025).

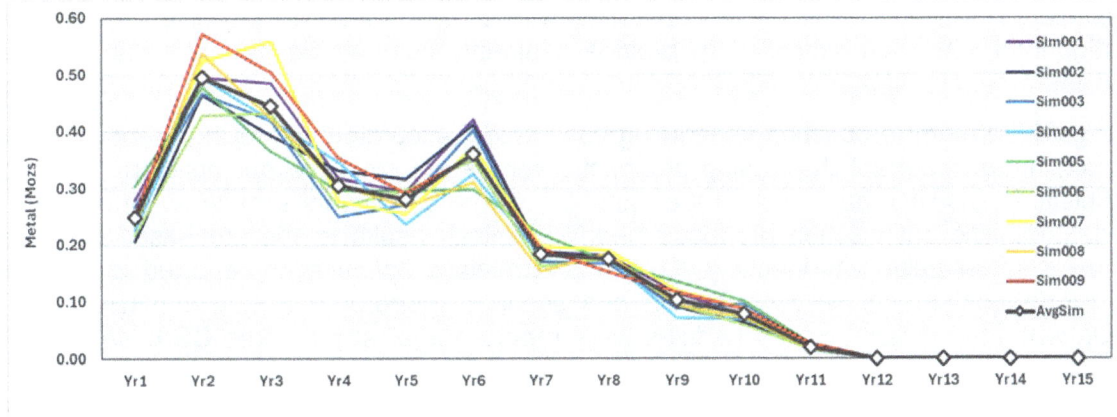

FIG 5 – Scenario 4 uncertainty profile for yearly metal mined ≥ 0.02 oz/ton.

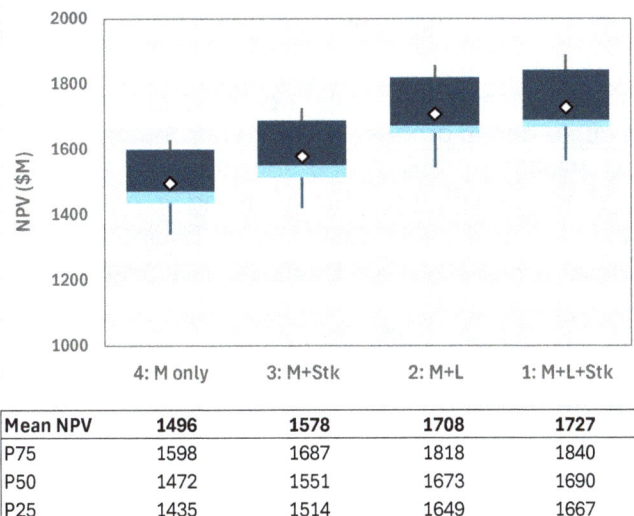

	4: M only	3: M+Stk	2: M+L	1: M+L+Stk
Mean NPV	1496	1578	1708	1727
P75	1598	1687	1818	1840
P50	1472	1551	1673	1690
P25	1435	1514	1649	1667

FIG 6 – NPV uncertainty for four McLaughlin metallurgical scenarios.

Computational notes

The MILP formulations were implemented in a C program named Peakfinder which creates a model file that is solved using IBM's CPLEX. MILP solutions are read back into Peakfinder for summary calculations and export to Excel for graphing. MILP solution tolerances were set to 0.01 per cent to eliminate questions on which optimisations and bounds provided higher or lower values.

The largest scenario, with milling, leaching and stockpiling, created a stochastic optimisation MILP formulation with about 2500 binary variables, 30 000 linear variables, 30 000 equations and 730 000 non-zero coefficients. The corresponding deterministic optimisation MILP formulation was 2500 binary variables, 5500 linear variables, 22 000 equations with 140 000 non-zero coefficients. Using a business laptop computer (1.1 GHz), CPLEX solution times were less than 30 seconds for stochastic optimisation, 10 seconds for deterministic optimisation and 0.5 seconds for stochastic evaluation. Fast solution times are aided by a combination of the sinking rate and cut constraints described in Appendix 1, tuned MILP branch and bound parameters, and the problem formulation which eliminates unnecessary details of increment destinations (ie formulation using $X(m,i,d,t,k)$ rather than $X(m,l,p,i,d,t,k)$).

The proposed bounding processes demonstrated the ability to provide tight bounds using deterministic optimisation of individual conditional simulation realisations without needing to resort to much larger stochastic optimisation problems (Table 2).

CONCLUSIONS

Stochastic optimisation allows production scheduling to incorporate both orebody variability and uncertainty. The algorithm presented allows a MILP implementation of stochastic optimisation where each simulated realisation respects all constraints without use of arbitrary penalties. The stochastic MILP algorithm was then simplified to provide a directly comparable deterministic optimisation method.

For the gold long-term production scheduling case study, accurately modelling orebody variability was much more important than using stochastic versus deterministic optimisation. When using deterministic inputs (IK) that matched the variability of the conditional simulation inputs, the case study deterministic optimisation produced nearly identical results to the stochastic optimisation. When using lower selectivity inputs (OK), the deterministic optimisation reported lower planned NPV's (-9 per cent to -16 per cent) and in one scenario, generated a slower mining schedule; however, when the deterministic schedules were reported against the same CS baseline models, the realised NPV differences were much smaller (0 per cent to -3 per cent).

Even without stochastic optimisation, conditional simulations can still be useful for assessing orebody risk and value impacts. A stochastic evaluation process was demonstrated to show how conditional simulations can provide a risk analysis for any mine sequence. The stochastic evaluation process can measure uncertainty and highlight any difference in variability versus the conditional simulation model.

With multiple orebodies, the stochastic optimisation process presented may become too large for practical MILP solving. An upper bound process (UB-PI) was demonstrated which provides an upper bound to the stochastic optimisation problem and shows the expected value of acquiring perfect information to remove orebody uncertainty prior to fixing a mining sequence. Finally, a best stochastic evaluation process (LB-BSE) was described for finding a good mine plan which may be close to the plan that would be created by a full stochastic optimisation.

APPENDIX 1 – MILP FORMULATION DETAILS

Database for flow value coefficients

A database was used to guide the computation of value coefficients for each possible flow:

$$MineVal(m,l,p,t) = -cM(m,l,p) \tag{13}$$

$$DestVal(m,i,d,t) = -cX(m,d) -cDD(i,d) + \sum_e g(e,i)*yD(e,i,d) *(pr(e,t)-cK(e,i,d))*(1-rD(e,i,d)) \tag{14}$$

$$StockVal(m,i,s,t) = -cS(m,s) \tag{15}$$

$$ReclaimVal(i,s,d,t) = -cR(s,d) -cDS(i,d) + \sum_e g(e,i)*yS(e,i,d)*(pr(e,t)-cK(e,i,d))*(1-rD(e,i,d)) \tag{16}$$

which are calculated from a database of price, cost and recovery constants:

pr(e,t)	Metal price ($/unit of metal) for element e
cM(m,l,p)	Mining cost (to mine exit) ($/ton)
cX(m,d)	Ex-mine hauling cost (from mine exit to destination) ($/ton)
cDD(i,d)	Destination processing cost ($/ton) for direct feed
cDS(i,d)	Destination processing cost ($/ton) for stockpiled feed
cS(m,s)	Stockpiling cost (including haulage from mine exit to stockpile) ($/ton)
cR(s,d)	Reclaiming cost (including haulage from stockpile to destination) ($/ton)
rD(e,i,d)	Royalty (NSR proportion)
cK(e,i,d)	Marketing cost, including smelting, refining, freight ($/unit of metal)
yD(e,i,d)	Yield (recovery) from direct feed
yS(e,i,d)	Yield (recovery) from stockpile reclaim feed

Similar linear constraints can be implemented for minimum tonnages as well as for grades, contained metal and recovered metal:

Example max grade constraint $\forall$ e, d, t, k:

$$\sum_m \sum_i \mathbf{X}(m,i,d,t,k)*[g(e,i)-Gmax(e,d,t)] + \sum_i \sum_s \mathbf{Z}(i,s,d,t,k))*[g(e,i)-Gmax(e,d,t)] \leq 0 \qquad (17)$$

Constraints can also be expressed for groups of mines or groups of destinations, using a group factor, grp(m,mg) or grp(d,dg):

$$\sum_m \sum_l \sum_p \mathbf{W}(m,l,p,t) * grp(m,mg) \leq MineGrpCapacity(mg,t) \; \forall \, mg, t \qquad (18)$$

Sequencing constraints

Sequencing constraints are easiest to implement when the mining flow variables are expressed as the cumulative tonnage proportion mined from each panel by the end of time period t, $\mathbf{U}$(m,l,p,t). This proportion can only increase over time:

$$\mathbf{U}(m,l,p,t-1) \leq \mathbf{U}(m,l,p,t) \; \forall \, m,l,p,t>1 \qquad (19)$$

Then, a binary variable, $\mathbf{O}$(m,l,p,t), can be introduced to allow enforcement of prerequisite ordering. The prerequisite ordering variable $\mathbf{O}$ is defined as 1 if a panel is mined out and 0 if the panel is not mined out, leading to equations defining mined out status:

$$\mathbf{O}(m,l,p,t) \leq \mathbf{U}(m,l,p,t) \; \forall \, (m,l,p,t) \qquad (20)$$

And equations to enforce a preceding panel to be mined out before its successor:

$$\mathbf{O}(m,l,p-1,t) \geq \mathbf{O}(m,l,p,t) \; \forall \, m, l, p>1, t \qquad (21)$$

This equation can be generalised for other prerequisite panels in other laybacks. For open pit mines, besides overlying panels, the most common prerequisites are defined for panels on the same bench of an inner nested layback. Other prerequisite relationships can also be defined for more complex geometries.

Using the $\mathbf{U}$ variables, for the final MILP formulation, all of the equations that include $\mathbf{W}$ variables are algebraically transformed using the following substitution:

$$\mathbf{W}(m,l,p,t) = Tons(m,l,p) * [\mathbf{U}(m,l,p,t) - \mathbf{U}(m,l,p,t-1)] \qquad (22)$$

Sinking rate constraints

Sinking rate constraints can be expressed using the $\mathbf{U}$ variables to limit each layback's number of panels mined per time period:

$$\sum_p (\mathbf{U}(m,l,p,t) - \mathbf{U}(m,l,p,t-1)) \leq MaxPanels(m,l,t) \; \forall \, (m,l,t) \qquad (23)$$

MILP cut constraints

Because the MILP formulation will be solved with a branch and bound algorithm, a large number of binary variables can create long solution times. Although not strictly necessary, the following cuts can speed up the branch and bound algorithm by forcing several other **O** variables to 1 or 0 whenever another **O** variable is set to 1 or 0.

$$O(m,l,p,t) \le O(m,l,p,t+1) \; \forall \; m,l,p,t \; \text{for} \; t<T \; (\text{mined out now} \rightarrow \text{mined out next year}) \qquad (24)$$

$$O(m,l,p+1,t) \le O(m,l,p,t) \; \forall \; m,l,p,t \; \text{for} \; p<P \; (\text{not mined out} \rightarrow \text{postreq not mined out}) \qquad (25)$$

Additional linear constraints

Linear constraints can also be implemented for minimum or maximum grades, contained metal and recovered metal:

Example max grade constraint $\forall$ e, d, t, k:

$$\sum_m \sum_i X(m,i,d,t,k)*[g(e,i)-Gmax(e,d,t)] + \sum_i \sum_s Z(i,s,d,t,k))*[g(e,i)-Gmax(e,d,t)] \le 0 \qquad (26)$$

Constraints can also be expressed for groups of mines or groups of destinations, using a group factor, grp(m,mg) or grp(d,dg) for each mine, m or destination, d:

$$\sum_m \sum_l \sum_p W(m,l,p,t) * grp(m,mg) \le MineGrpCapacity(mg,t) \; \forall \; mg, t \qquad (27)$$

Typically, group factors would be one or zero to include or exclude a mine or destination in a group constraint, but other coefficients can also be used.

APPENDIX 2 – NEWSVENDOR MINE DEMONSTRATION CASE

A simple single panel example was prepared to demonstrate the difference between stochastic and deterministic schedule optimisation. The case was constructed and solved using the MILP models presented in this paper. Additionally, a simple Excel model was created and solved using Excel's built in Evolution Solver and yields the same solution.

Newsvendor demonstration case – inputs

The fictitious Newsvendor mine consists of a 1 gpt Au payable layer at the bottom of a 100 m highwall. The payable layer has an unknown thickness, with uncertainty simulated by nine equally probably thicknesses corresponding roughly to a discrete approximation of a normal distribution with mean of 30 m and standard deviation of 11 m (Figure 7). Orebody width, length and density are such that total payable ore plus overburden adds up to 90 Mt. At an average ore thickness of 30 m, this would equate to 27 Mt of ore and 63 Mt of overburden. If simulation #1 is realised with a thickness of 15.9 m, the 90 Mt mine would contain 14.31 Mt of ore and 75.69 Mt of overburden.

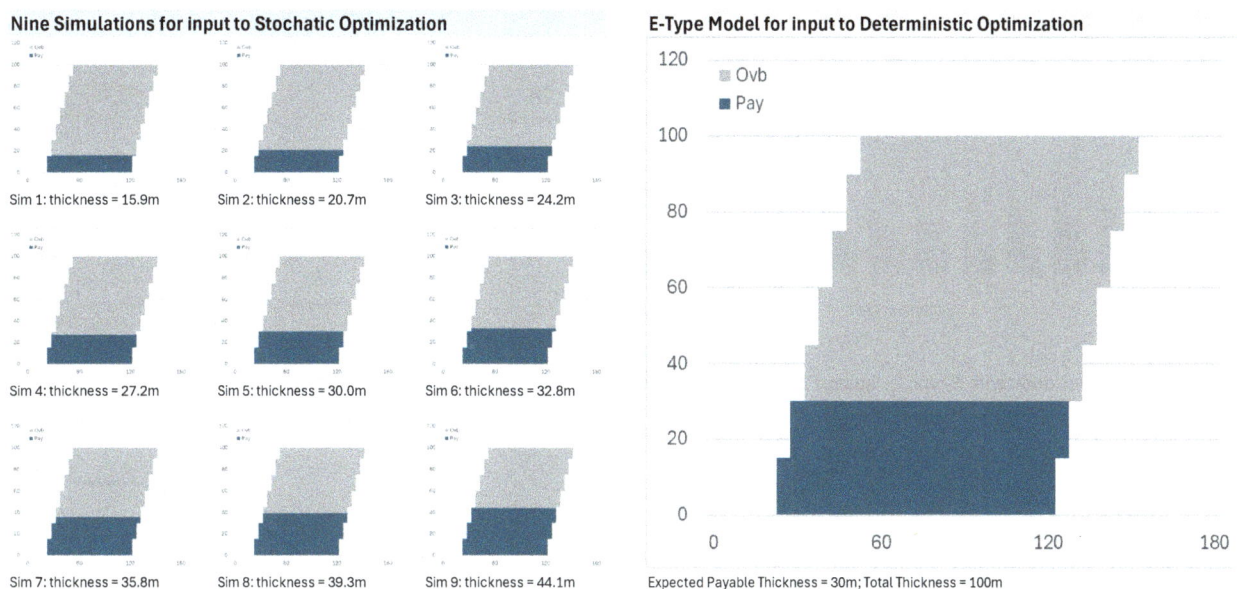

FIG 7 – Newsvendor Stochastic and Deterministic Optimisation Inputs – Orebody cross-section.

Mining assumptions are mining cost of $5/t and no annual capacity constraint. Processing assumptions are milling costs of $20/t for 90 per cent recovery and a maximum mill capacity of 3 Mt/a. No stockpiling is allowed, so any yearly excess of payable ore mined must be sent to the waste dump. Other key assumptions for this example are a discount rate of 10 per cent and a gold price of $80/gram.

Newsvendor demonstration case – results

Stochastic optimisation with simulated inputs – SO(Sim)

For stochastic optimisation, the NPV maximising set of yearly mining and processing tonnages must balance three factors: 1) the opportunity cost of unused mill capacity; 2) the opportunity cost of wasting payable ore; and 3) the opportunity cost of delaying future profits due to slower mining. For purposes of this example, and to match the two-stage stochastic programming assumption that the first-stage variables (ie the yearly total tons mined schedule) will be chosen before orebody uncertainty is revealed, assume that any year's mining does not provide any learning about the payable thickness of future years. This seemingly simple example becomes more complex because the third factor changes over time as the value of future profits decreases. In this example, there are no mill fixed costs, so opportunity cost #1 is zero, while opportunity cost #2 depends on the tail of the ore thickness probability distribution, and opportunity cost #3 decreases over time. The net result is an optimum mining rate that decreases over time to decrease the average amount of payable material wasted over time (Figure 8 top row; Planned NPV = Realised NPV = $590 M). Note that the delay opportunity cost #3 will increase with longer mine life or a higher discount rate.

Deterministic optimisation with e-type inputs – DO(ET)

For deterministic optimisation, using an e-type estimate (ie the expected value of the thickness distribution) of the payable thickness of 30 m, the optimum mine plan is a nine-year mine life with a uniform 10 Mt/a mining and 3 Mt/a processing of payable ore (Figure 8 middle row; Planned NPV = $672 M). Mining and processing are in perfect balance, so the mill operates at full capacity and no payable ore is sent to the waste dump. On paper, this plan looks great. However, as stochastic evaluation will show, any variability in thickness will either result in unused mill capacity or payable ore being sent to the waste dump.

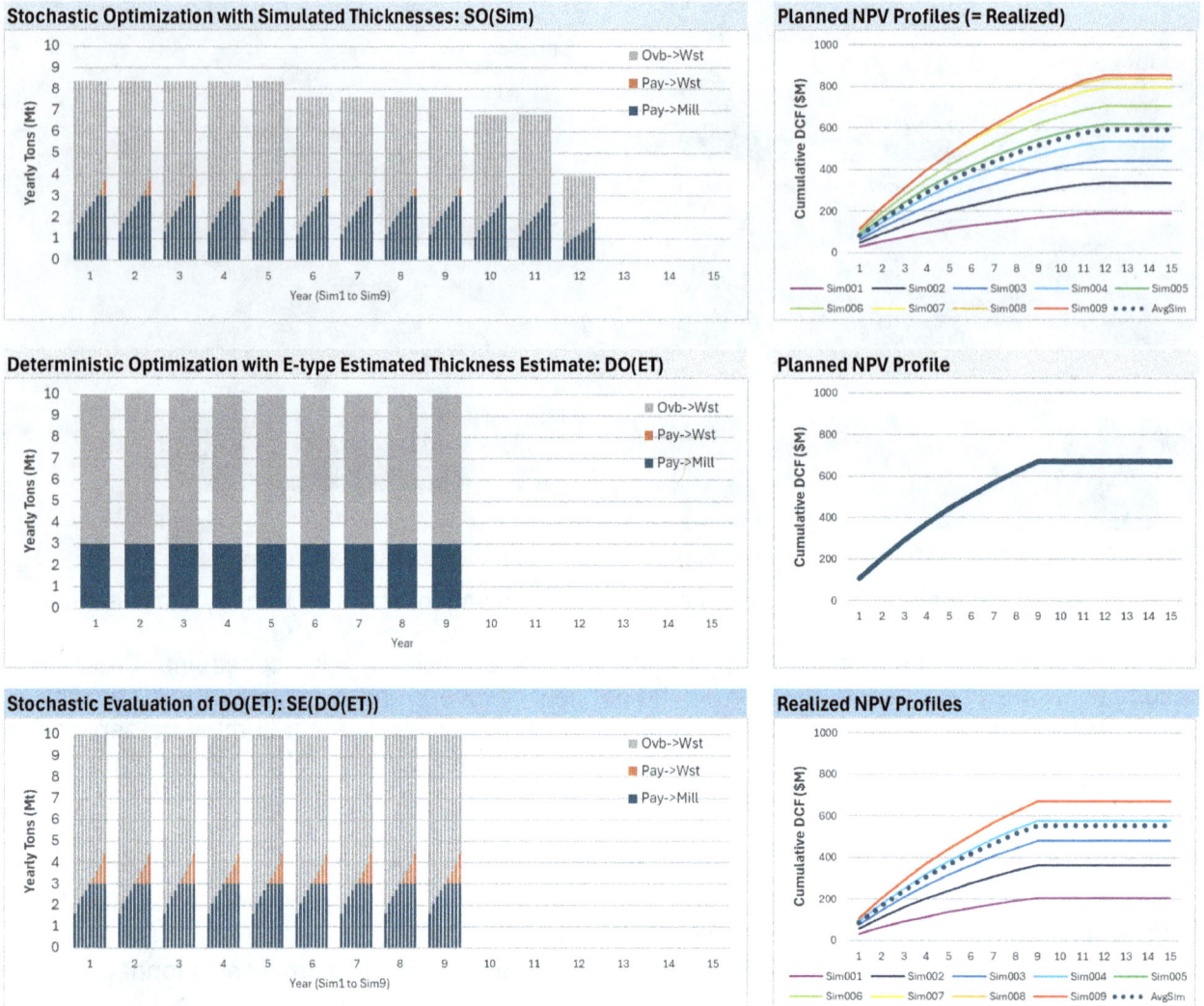

FIG 8 – Newsvendor Mine Stochastic (top row) and Deterministic (middle row) Schedule Optimisation and Stochastic Evaluation of the Deterministic Schedule Optimisation (bottom row).

Stochastic evaluation – SE(DO(ET))

A stochastic evaluation of the deterministically optimised mine plan shows significantly lower realised NPV than the planned NPV of $672 M. Of the nine simulated orebody realisations, four result in mill capacity not being fully utilised and four result in ore in excess of 3 Mt/a being wasted (Figure 8 bottom row; Realised NPV = $554 M; note Sim6-Sim9 realised NPV profiles are all identical with 3 Mt/a ore being sent to the mill and the remainder to waste). Table 3 shows details of the Stochastic Evaluation process for this Newsvendor mine example.

TABLE 3

NPV calculation details for stochastic optimisation, deterministic optimisation, stochastic evaluation and upper and lower bounds for stochastic optimisation NPV.

Calculation Method(input) NPV	SO(CS) SimX	Planned NPV DO(SimX) ET	Realised NPV SE(DO(ET)) SimX	Up Bound DO(SimX) SimX	Low Bound SE(DO(SimX)) Avg(Sim1–9)
Sim1	191	n/a	207	247	269
Sim^2	336	n/a	365	410	399
Sim^3	443	n/a	480	516	476
Sim4	534	n/a	579	599	523
Sim5	619	n/a	671	672	554
Sim6	704	n/a	672	738	578
Sim7	795	n/a	672	804	587
Sim8	837	n/a	672	874	585
Sim9	855	n/a	672	961	568
SO(CS) (avg)	590				
DO(ET)		672			
SE(DO(ET)) (avg)			554		
UB-PI (avg)				647	
LB-BSE (max)					587
% of SO(CS)		+13.8%	-6.1%	+9.6%	-0.5%

Upper bound with perfect information – UB-PI

Performing a deterministic optimisation for each of the nine orebody simulation realisations shows the expected value of perfect information. With perfect information, the mine and mill rates are kept in perfect balance, so the mill operates at full capacity and no payable ore is sent to the waste dump. The average of the nine plans' planned NPV's, creates an upper bound for the NPV of the stochastically optimised plan. The $57 M NPV gap between this upper bound, UB-PI = $647 M, and the stochastically optimised plan, SO(Sim) = $590 M, indicates the expected value of gathering perfect information before making mine plan decisions. Although the expected NPV increases by 10 per cent, the actual value could still be between $247 M if Sim #1 is the actual thickness and $961 M if Sim 9 is the actual thickness. This example parallels mining management situations where additional drilling is done partly to improve mining decisions and partly to remove uncertainty.

Figure 9 illustrates the nine different deterministically optimised mine plans and corresponding discounted cash flow curves resulting from deterministic optimisation of each of the nine simulated orebodies. The mine plans simply consist of mining however many total tons/year are necessary to create 3 Mt/a of payable mill feed, so mining rate and mine life are a simple function of payable thickness and total payable tons. Table 3 shows details of the UB-PI process.

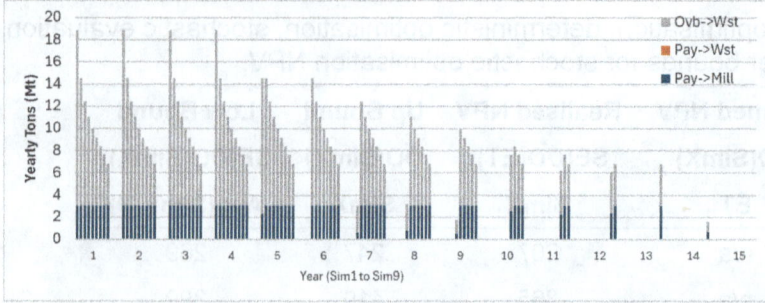

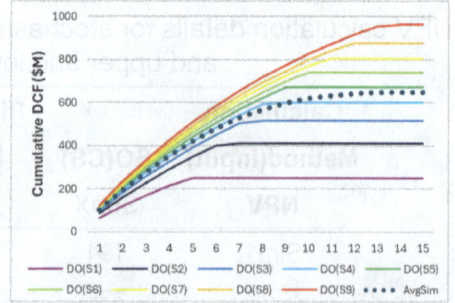

FIG 9 – Determining Stochastic Optimisation Upper Bound if perfect information known before determining mine sequence: UB-PI

Lower bound from best stochastic evaluation – LB-BSE

Each of the deterministic optimisations of individual orebody simulations can be tested to choose the one that yields the highest realised NPV. Stochastic evaluation of each of the nine deterministic plans created during the UB-PI process showed that SE(DO(Sim7)) had the highest realised NPV, $587 M. Using this method, the resulting best stochastic evaluation will be called LB-BSE. For this demonstration case, the LB-BSE had an NPV that was within 1 per cent of the NPV of the Stochastic Optimisation, $590 M.

Figure 10 illustrates stochastic evaluation of the nine deterministic plans created during the UB-PI process to identify a plan whose NPV is nearly as good as the Stochastic Optimisation plan. By using the two bounding processes, it may be possible to create mine plans with NPV nearly as high as the NPV of a Stochastic Optimisation while only working with MILP formulations that are the same size as the Deterministic Optimisation problem. Table 3 shows details of the LB-BSE process.

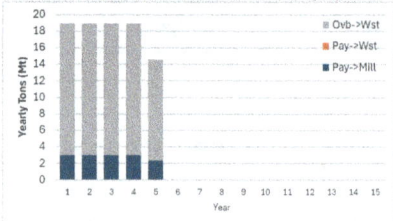
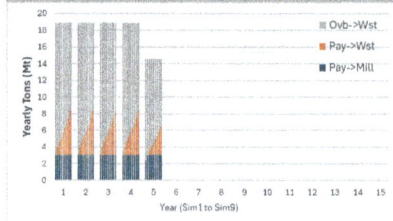
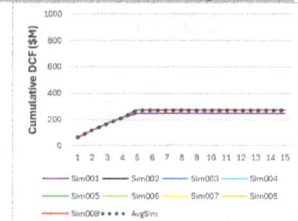

⋮   ⋮   ⋮

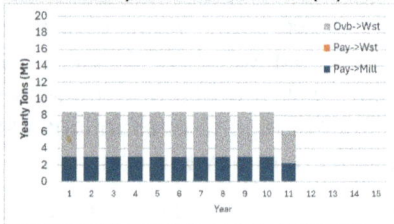
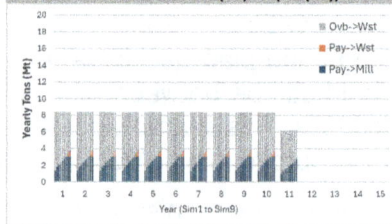
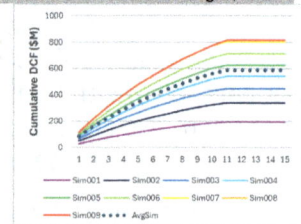

⋮   ⋮   ⋮

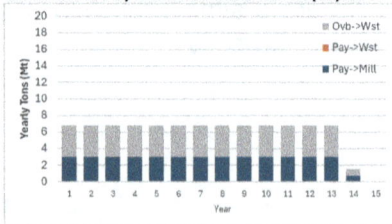
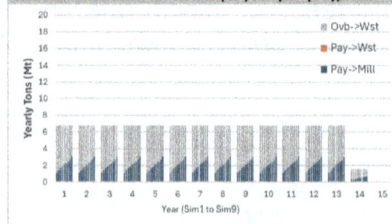
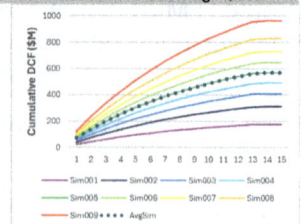

FIG 10 – Determining Stochastic Optimisation Lower Bound (LB-BSE) by stochastic evaluation of each mine sequence shown in Figure 9. SE(DO(Sim7)) provides highest avg NPV of $587 M which provides a lower bound to the stochastic optimisation, SO(CS).

REFERENCES

Amihere, N and Deutsch, C, 2022. Localization of Probabilistic Resource Models, in Geostatistics Lessons (ed: J Deutsch). Available from: <https://geostatisticslessons.com/lessons/localization> [Accessed: 14 Jan 2025].

Aras, C, Dagdelen, K and Johnson, T, 2019. Generating pushbacks using direct block mine production scheduling algorithm, in *Mining Goes Digital* (eds: C Mueller, W Assibey-Bonsu, E Baafi, C Dauber, C Doran, M Jerzy Jaszczuk and O Nagovitsyn), pp 426–436 (Taylor and Francis Group: London).

Clark, L and Dagdelen, K, 2023. Practical Mine Planning and Design, in *SME Surface Mining Handbook* (P Darling), pp 29–67 (SME: Englewood).

Dagdelen, K and Kawahata, K, 2008. Value Creation through Strategic Mine Planning and Cutoff Grade Optimization, *Mining Engineering*, 60(1):39–45.

Dagdelen, K, 1992. Cutoff grade optimization, in *23rd APCOM Proceedings* (ed: Y C Kim), pp 157–165 (SME: Littleton).

Dagdelen, K, 2011. Systems Engineering, in *SME Mining Engineering Handbook* (P Darling), third edition, pp 850–853 (SME: Englewood).

Dimitrakopoulos, R and Lamghari, A, 2022. Simultaneous stochastic optimization of mining complexes – mineral value chains: an overview of concepts, examples and comparisons, *International Journal of Mining, Reclamation and Environment*, 36(3):443–460.

Dimitrakopoulos, R, 2018. Stochastic Mine Planning—Methods, Examples and Value in an Uncertain World, in *Advances in Applied Strategic Mine Planning* (ed: R Dimitrakopoulos), pp 101–115 (Springer: Cham).

Froyland, G, Menabde, M, Stone, P and Hodson, D, 2018. The Value of Additional Drilling to Open Pit Mining Projects, in *Advances in Applied Strategic Mine Planning* (ed: R Dimitrakopoulos), pp 119–138 (Springer: Cham).

Goodfellow, R and Dimitrakopoulos, R, 2016. Global optimization of open pit mining complexes with uncertainty, *Applied Soft Computing*, 40(2016):292–304.

Hardtke, W, Allen, L and Douglas, I, 2011. Localised Indicator Kriging, in *35th APCOM Symposium Proceedings* (eds: E Baafi, R Kininmonth and I Porter), pp 141–147 (The Australasian Institute of Mining and Metallurgy: Melbourne).

Hoerger, S and Dagdelen, K, 2024. Conditional Simulation for Stochastic Production Scheduling of Open Pit Mines, in Geostats 2024–12th International Geostatistics Congress (Springer).

Hoerger, S and Dagdelen, K, 2025. Mine Plan Risk Assessment and Grade Uncertainty Characterization using Geostatistical Conditional Simulation: Gold Mine Case Study, in *Minexchange 2025 SME Annual Conference*, preprint 25–057 (SME: Denver).

Hoerger, S, 2024. MILP Production Scheduling Models for Evaluating Continuous Improvement Projects, in *Minexchange 2024 SME Annual Conference*, preprint 24–058 (SME: Phoenix).

Hoerger, S, Bachmann, J, Criss, K and Shortridge, E, 1999. Long Term Mine and Process Scheduling at Newmont's Nevada Operations, in *28th APCOM Proceedings* (ed: K Dagdelen), pp 739–748 (Colorado School of Mines: Golden).

Journel, A, 1983. Non-parametric estimation of spatial distributions, *Math Geo,* 15(3):445–468.

Lane, K, 1988. *The Economic Definition of Ore—Cut-off Grades in Theory and Practice* (Mining Journal Books: London).

Menabde, M, Froyland, G, Stone, P and Yeates, G, 2018. Mining Schedule Optimisation for Conditionally Simulated Orebodies, in *Advances in Applied Strategic Mine Planning* (ed: R Dimitrakopoulos), pp 91–100 (Springer: Cham).

Newman, A, Rubio, E, Caro, R, Weintraub, A and Eurek, K, 2010. A Review of Operations Research in Mine Planning, *Interfaces*, 40(3):222–245.

Ortiz, J M, 2020. Introduction to sequential Gaussian simulation, Predictive Geometallurgy and Geostatistics Lab Annual Report 2020, pp 7–19 (Queen's University, Montreal).

Rendu, J M, 2014. *An Introduction to Cut-off Grade Estimation*, 2nd edition (SME: Englewood).

Rossi, M and Deutsch, C, 2014. *Mineral Resource Estimation* (Springer: Heidelberg).

Wolsey, L, 2021. *Integer Programming*, 2nd edition (Wiley: Hoboken).

Optimising in-pit crushing and conveying systems for cost-effective open pit mine scheduling – a sensitivity analysis on material throughput and haulage costs

A Kamrani[1], H Askari-Nasab[2] and Y Pourrahimian[3]

1. Postdoctoral fellow, Civil and Environmental Engineering Dept., University of Alberta, Edmonton AB T6G 2H5, Canada. Email: kamrani@ualberta.ca
2. Professor, Civil and Environmental Engineering Dept., University of Alberta, Edmonton AB T6G 2H5, Canada. Email: hooman@ualberta.ca
3. Associate Professor, Civil and Environmental Engineering Dept., University of Alberta, Edmonton AB T6G 2H5, Canada. Email: yashar.pourrahimian@ualberta.ca

ABSTRACT

In open pit mining, in-pit crushing and conveying (IPCC) systems offer significant advantages over traditional truck-shovel (TS) methods, particularly as haul distances grow with increased mine depth. IPCC reduces the need for extensive truck haulage, cutting down on fuel consumption, maintenance, and carbon emissions by transporting material via conveyors. Placing crushers within the pit streamlines material handling and stabilises haulage costs over time, providing a more cost-effective and sustainable operation. This study develops a model to optimise IPCC placement and relocation to minimise haulage costs and maximise Net Present Value (NPV) in long-term mine scheduling. To examine the sensitivity of our model, we vary two key parameters: the tonnage of material processed while the crusher remains in the optimal panel and the dollar-per-ton-kilometer cost of haulage. By adjusting the processed tonnage, we evaluate how throughput levels influence haulage requirements and crusher relocation strategies. Altering the haulage cost, meanwhile, offers insight into how changing market conditions impact the economic viability of crusher relocations and the overall robustness of the model. The methodology utilises a two-stage clustering approach, combining k-medoids and hierarchical algorithms, to define crusher panels and mining cuts precisely, integrating road and conveyor networks. Tested on real mine data, this approach shows significant NPV gains across analysing the scenarios, consistently outperforming traditional TS methods. Our findings highlight that optimal IPCC configurations depend heavily on both material tonnage and haulage costs, underscoring the importance of these factors in sustainable and cost-efficient mining practices. This scalable model provides a flexible solution for modern mine planning, advancing IPCC as a practical alternative to traditional haulage methods.

INTRODUCTION

Material haulage is one of the most energy-intensive and costly components in open pit mining operations. As mines deepen and haulage distances increase, traditional truck and shovel (TS) systems become increasingly inefficient. These systems are not only limited by operational costs and greenhouse gas emissions but also by their reliance on extensive road networks and large fleets of heavy trucks. With the growing demand for sustainability and cost efficiency in mining, alternative haulage methods have garnered significant attention.

One such alternative is the in-pit crushing and conveying (IPCC) system, which offers the potential to reduce haulage distances, lower fuel consumption, and enhance productivity. By crushing materials inside the pit and transporting them via conveyors, IPCC systems can mitigate many of the inefficiencies associated with conventional trucking. Among the various configurations of IPCC systems, semi-mobile solutions are particularly attractive due to their balance of flexibility and cost-effectiveness. These systems can be periodically relocated within the pit to adapt to the spatial and temporal dynamics of long-term mine planning.

Despite their potential advantages, implementing semi-mobile IPCC systems requires careful strategic planning, particularly in terms of their integration with truck-based systems and the long-term scheduling of extraction activities. Furthermore, the economic viability of IPCC systems is highly sensitive to factors such as truck haulage costs, conveyor costs, and the capacity thresholds required for crusher relocation or installation. Capital and operating cost assumptions, in particular, are often chaotic and vary widely across studies and practical implementations, leading to uncertainties in critical decision-making. Consequently, there is a need for a systematic approach that identifies optimal system configurations under different material throughput requirements, while acknowledging how haulage cost factors may affect the project's economic outlook. This paper addresses that need by presenting a comprehensive methodology for optimising IPCC layouts within open pit mines and by providing sensitivity analyses on throughput and hauling costs to highlight robust strategies for cost-effective and flexible mine scheduling.

In several studies, researchers have explored the implementation of IPCC systems to reduce haulage costs and improve operational efficiencies in both small and large-scale mines. In small surface pits of dolomite, mobile crushers were shown to yield lower costs compared to stationary units, primarily by reducing haul and set-up requirements (Klanfar and Vrkljan, 2012). Similar conclusions emerged from investigations on multiseam open cut coal operations, where fully mobile IPCC lowered equivalent unit costs to US$1.22–1.39/t (Londoño, Knights and Kizil, 2013). Analyses of lignite mines incorporating both roads and conveyor ramps further underscored the importance of conveyor design, noting that installation costs of around €3520/m could be offset by long-term savings (Roumpos et al, 2014). Larger metalliferous operations, adopting fully mobile units with capacities of 4000–10 000 t/h, reported capital investments ranging from US$180 million to US$250 million (Dean et al, 2015), though high-level economic assessments indicated notable reductions in truck fleet requirements.

Subsequent work has highlighted the pivotal roles of crusher relocation timing and location in achieving cost savings. In a conical iron ore pit processing 8 kt/d of ore and 12 kt/d of waste, a semi-mobile 907 t/h IPCC required investment outlays of US$48.8 million plus US$1300/m for conveyors yet delivered significant haulage cost reductions (de Werk et al, 2017). Studies on copper operations, including the Sungun Mine in Iran, validated similar trends, showing that crusher relocation (US$1.5 million each move) and conveyor construction (US$3000/m) must be carefully scheduled for optimal benefit (Paricheh, Osanloo and Rahmanpour, 2017, 2018). At Chuquicamata, adding a third gyratory crusher (5600 t/h) at a cost of US$15 million, with US$35 million for installation and about US$2500/m for conveyors, demonstrated how additional IPCC capacity could reduce reliance on trucking (Yarmuch et al, 2017). Further evaluations of simplified copper deposits (Nehring et al, 2018) and the Kahnuj titanium mine (Rahimdel and Bagherpour, 2018) confirmed that both semi-mobile and fully mobile IPCC significantly influence total haulage economics, although capital requirements may reach US$200 million in deeper pits.

Recent investigations have reinforced these findings while examining expanded material handling and practical deployment considerations. A hypothetical copper project employing a 3000 t/h semi-mobile crusher recorded relocation costs of US$1 million and conveyor rates of US$0.20–0.42/t, significantly reducing truck haul distances (Abbaspour, Drebenstedt and Dindarloo, 2018). Similarly, a Brazilian copper–gold mine with a 20 year life found that an IPCC system, rated between 3805 and 2718 t/h, cut total haulage expenditures by 28 per cent (Nunes et al, 2019). Research on large copper porphyries and smaller-scale operations also highlighted cost benefits, indicating that even at lower capacities, IPCC can relieve truck fleet demands, especially if relocation and conveyor charges (eg US$1 million per move, US$0.30/t per level) are optimised (Bernardi, Kumral and Renaud, 2020; Liu and Pourrahimian, 2021). Iron ore mines exhibited similar trends, with semi-mobile crushers around 2700 t/h and conveyor rates of US$0.25/t/km achieving meaningful reductions in truck haulage. In scenarios involving both ore and waste handling, discounted cash flow gains were observed over purely truck-based operations (Kamrani et al, 2024), underscoring that careful planning of crusher configuration and conveyor can offer considerable economic advantages across diverse resource settings.

Kamrani, Pourrahimian and Askari-Nasab (2025) provided a comprehensive framework was developed to optimise the long-term mine scheduling problem while simultaneously integrating road network design and in-pit conveyor configurations. Their two-step mathematical model introduced a novel approach that allowed the strategic placement and relocation of semi-mobile crushers in alignment with material movement and infrastructure development. The study demonstrated significant cost savings and reductions in truck travel distances when comparing IPCC-enabled scenarios against conventional TS systems. While this study provided important insights into the advantages of IPCC systems, their work focused on a fixed set of cost and capacity assumptions.

However, these parameters can vary widely across mines and over time. Therefore, a comprehensive sensitivity analysis is essential to assess the robustness of such models and to provide mine planners with actionable insights under varying operational conditions.

This paper builds upon the foundational work of (Kamrani, Pourrahimian and Askari-Nasab, 2025) and extends it by conducting an extensive scenario-based sensitivity analysis. Specifically, we evaluate how changes in truck haulage costs, conveyor costs, and crusher capacity thresholds affect the overall system performance and haulage cost outcomes. We consider four primary operational configurations: TS only, ore-IPCC, waste-IPCC, and combined ore and waste-IPCC across 32 distinct cost and capacity scenarios. The aim of this study is to provide mine planners and decision-makers with a better understanding of the conditions under which IPCC systems outperform traditional TS methods and how key cost drivers influence strategic planning. The remainder of this paper is organised as follows. Section 2 outlines the methodology, mathematical model and scenario generation process. Section 3 presents the case study and experimental set-up. Section 4 discusses the results, and Section 5 summarises the conclusions and suggests directions for future research while interprets the findings in light of practical planning implications.

METHODOLOGY

The methodology builds upon the first model formulation presented by (Kamrani, Pourrahimian and Askari-Nasab, 2025), which uses a two-step optimisation approach to identify optimal crusher locations and integrate them into the mine schedule while considering haulage network constraints. The methodology begins by optimising a traditional block model to establish the ultimate pit boundary, pushbacks, and a preliminary long-term schedule. From here, each bench-phase within the pushbacks is subdivided into smaller units, storing relevant geological and economic data for every block. Two rounds of clustering then follow. First, a k-medoids clustering method defines 'crusher panels,' each large enough to accommodate the in-pit crusher, its feeder, and conveyor loading areas. Next, a hierarchical clustering algorithm groups blocks into workable 'mining cuts,' ensuring spatial adjacency and uniformity in rock characteristics. In parallel, dedicated road and conveyor ramps are designed for the open pit, allowing for accurate travel-distance measurements by applying a shortest-path algorithm.

After these preparatory steps, two main optimisation models are employed. The first model, aims to select the most cost-effective crusher panel(s) to minimise material transport and installation expenses, accommodating operational configurations that are explained. The second model is a mixed-integer linear program that fine-tunes the long-term production schedule while respecting capacity, grade, and precedence constraints among the mining cuts and crusher panels. It also determines when crushers should be relocated to maximise net present value. Collectively, these models integrate clustering-based block aggregation, in-pit crushing and conveying placement, and operational scheduling, aiming to produce a practical and economically robust open pit mine plan.

Crusher integration and movement

In IPCC-enabled scenarios, semi-mobile crushers are assigned to pre-identified panels called crusher panels, which serve as candidate locations for installation. Once installed, a crusher must remain in a panel until a predefined amount of ore or waste material has been processed. The movement of the crusher is allowed only when the cumulative processed tonnage exceeds the lower bound and does not exceed the upper bound.

Ore and waste crushers operate independently, and their placement is modelled through additional binary variables. Truck haulage is required to move material from the mining face to the crusher location, after which it is transported via conveyor to the destination (eg processing plant or waste dump).

Scenario structure

To understand the effect of economic and operational parameters on the performance of IPCC systems, we constructed a full factorial design involving three key parameters:

- Truck haulage cost: four levels:

Horizontal: [0.1,0.3,0.5,0.7] $/t/km[0.1, 0.3, 0.5, 0.7] \, \$/t/km[0.1,0.3,0.5,0.7]$/t/km

- o Vertical: [0.6,0.8,1.0,1.2] $/t/km[0.6, 0.8, 1.0, 1.2] \, \$/t/km[0.6,0.8,1.0,1.2]$/t/km
- Conveyor haulage cost: four levels:
    - o [0.1,0.3,0.5,0.7] $/t/km[0.1, 0.3, 0.5, 0.7] \, \$/t/km[0.1,0.3,0.5,0.7]$/t/km
- Crusher capacity bounds: two configurations:
    - o Lower bounds: 8 Mt (ore), 22 Mt (waste)
    - o Lower bounds: 16 Mt (ore), 44 Mt (waste)

These combinations result in 4 × 4 × 2 = 324 \times 4 \times 2 = 324 × 4 × 2 = 32 unique scenarios, each defined by a specific set of cost and capacity assumptions.

Each of the 32 scenarios is applied to four broader system configurations to assess how IPCC deployment strategies influence outcomes:

- A: No crusher or conveyor (truck-shovel system only).
- B: One semi-mobile crusher and conveyor for ore.
- C: One semi-mobile crusher and conveyor for waste.
- D: Two semi-mobile crushers and conveyors for both ore and waste.

These configurations reflect varying degrees of IPCC integration, from conventional haulage to full IPCC deployment. In all configurations involving crushers, trucks are still required for short-distance haulage to the in-pit crusher.

Crusher panel optimiser model

To identify the optimal locations for in-pit crushers throughout the mine's operational timeline, it is essential to evaluate material haulage costs between different locations. These costs are computed by multiplying distance matrices by cost coefficients tailored to each scenario. In the base scenario, traditional truck and shovel operations, there is no need to implement the first optimisation model. However, for scenarios that incorporate IPCC, a tailored version of the first model is required.

While this model offers flexibility across different IPCC configurations, technical requirements specific to each material type and scenario must be considered. These include conveyor system components like belts, rollers, drives, pulleys, and more. Each scenario demands a custom formulation with slightly different objectives and constraint sets.

Three primary IPCC configurations are considered:

1. Ore IPCC: A single in-pit crusher handles ore, with conveyors installed on a ramp for ore transport. Waste continues to be hauled by trucks.
2. Waste IPCC: One in-pit crusher is dedicated to waste, using conveyors on a ramp on the same pit side, while ore is trucked out.
3. Ore and Waste IPCC: Two crushers (for ore and waste) are placed on the same crusher panel, using separate conveyors spaced adequately on the same ramp.

The general model designed for the combined Ore and Waste IPCC case is presented in Equations 1 to 7. This version integrates the objective functions and constraints for both ore and waste simultaneously. To adapt the model to single-material scenarios, the *MT* index should be adjusted to only include Ore (O) or Waste (W) material type.

Across all three models, a shared objective component accounts for the cost of installing a crusher at a given panel. This cost may vary based on location, accessibility, and geological factors but is never zero.

$$\min \sum_{p=1}^{P} (f_p \times y_p) + \sum_{MT=\{O,W\}} \sum_{k=1}^{K} \sum_{p=1}^{P} \left[\begin{array}{c} (hoTC \times hoKm_{k,p}) + (veTC \times veKm_{k,p}) \\ + (cc \times cKm_{p,mt}) + cCR_{mt} \end{array} \right] \times z_{k,p,mt} \quad (1)$$

Subject to:

$$\sum_{p=1}^{P} z_{k,p,mt} = do_{k,mt} \ \forall \ k \in \{1,...,K\}, mt \in \{O,W\} \tag{2}$$

$$lM_{p,mt} \times y_p \leq \sum_{k=1}^{K} z_{k,p,mt} \leq uM_{p,mt} \times y_p \ \forall \ p \in \{1,...,P\}, mt \in \{O,W\} \tag{3}$$

$$z_{k,p,mt} \leq do_{k,mt} \times y_p \qquad \forall \ k \in \{1,...,K\}, \ p \in \{1,...,P\}, mt \in \{O,W\} \tag{4}$$

$$\overline{pu} \times \sum_{k=1}^{K} z_{k,p,mt=W} - \overline{mu} \times \sum_{k=1}^{K} z_{k,p,mt=O} \leq 0 \ \forall \ p \in \{1,...,P\}, mt \in \{O,W\} \tag{5}$$

$$z_{k,p,mt} \geq 0 \ \forall \ k \in \{1,...,K\}, \ p \in \{1,...,P\}, mt \in \{O,W\} \tag{6}$$

$$y_p \in \{0,1\} \ \forall \ p \in \{1,...,P\} \tag{7}$$

Where:

$p \in P$	is the index for crusher panels
$k \in K$	is the index for the mining cuts
$mt \in MT$	is the index for the material types being either Ore (O) or Waste (W)
f_p	is the cost of installing crusher in the crusher panel p. It could be different for the crusher panels if they were not chosen within the same bench phase. Otherwise, it has the same value for all the crusher panels
y_p	is a binary decision variable that is equal to one if the crusher is located in crusher panel p, otherwise zero
$hoTC$	is the horizontal unit cost of truck transporting each tonne of material 1 km
$hoKm_{k,p}$	is the horizontal distance between the mining cut k to crusher panel p
$veTC$	is the vertical unit cost of truck transporting each tonne of material 1 km
$veKm_{k,p}$	is the vertical distance between the mining cut k to crusher panel p
cc	is the unit cost of conveyor conveying each tonne of material 1 km towards its destination
$cKm_{p,mt}$	is the distances on the conveyor ramp between the crusher panel p and the material destinations which are mill or waste dump
cCR_{mt}	is the unit cost of crushing one tonnage of ore or waste material
$z_{k,p,mt}$	is a continuous decision variable representing the tonnage of ore or waste from mining cut k sent to crusher panel p
$do_{k,mt}$	is the ore or waste tonnage of mining cut k
$\overline{pu}$	is the average of the upper bounds on the tonnage of ore processing capacity in all the periods
$\overline{mu}$	is the average of the upper bounds on the tonnage of mining capacity in all the periods
$uM_{p,mt}$	is the upper bound on the total ore or waste tonnage milled or dumped during the time that either/both crusher/s is/are located on the crusher panel p
$lM_{p,\ mt}$	is the lower bound on the total ore or waste tonnage milled or dumped during the time that the either/each crusher/s is/are located on the crusher panel p

As mentioned, Equation 1 serves as the objective function of Ore and Waste IPCC scenario which in the given formulation aims at minimising the cost of installing the crusher and transporting ore and waste materials. Equation 2 ensures that all ore and waste tonnages from each mining cut are extracted and assigned to a crusher panel. Equation 3 establishes lower and upper limits for ore and waste, on the total ore tonnage milled or the total waste tonnage dumped during the time that the crusher is located on crusher panel p. Equation 4 sets an upper bound for variable z to ensure the feasibility of the solution. Equation 5 is specific to the Ore and Waste scenario, ensuring that both crushers in the selected optimal crusher panel are fed with an equal amount of material relative to the average upper bound of mining and processing capacities across all periods. Finally, Equations 6 and 7 represent the decision variables' boundaries for z, and y, respectively.

In this study, the scheduling model from Kamrani, Pourrahimian and Askari-Nasab (2025) remains unchanged and is executed only once. Our primary interest lies in assessing how variations in cost and throughput parameters influence haulage costs, rather than examining changes in net present value. Therefore, while the same schedule is used across all 32 scenarios, each operational configuration (eg ore-only in-pit crushing, waste-only, or both) adopts a unique schedule to reflect its respective haulage approach. This maintains a consistent comparative basis yet ensures that the underlying long-term production schedule is internally coherent for each configuration.

CASE STUDY

To evaluate the long-term scheduling performance of semi-mobile IPCC systems under various economic and operational assumptions, a real-size iron ore open pit mine was used as the basis for the case study. The selected mine represents a typical large-scale operation with substantial production volumes, variable topography, and multiple haulage constraints, making it suitable for testing both traditional and IPCC-based haulage strategies.

The mine is divided into 2208 mining cuts, each associated with known tonnage, grade, material type (ore or waste), blocks and spatial coordinates. These cuts are preprocessed into 215 crusher panels using a hybrid approach that combines K-medoids clustering and hierarchical agglomerative clustering (HAC). This panel-based approach significantly reduces the problem size, allowing the optimisation model to handle multi-year schedules and infrastructure placement decisions effectively.

Each panel and cut are defined such that its constituent blocks are spatially contiguous and geologically coherent. The clustering also supports the definition of feasible haulage routes and ensures that crusher panel candidates are positioned in geologically stable zones that can accommodate crusher installation and operation since they are created within a designed pit.

In IPCC-enabled scenarios, the model optimally selects when and where to install the ore and/or waste crusher, subject to tonnage-based movement constraints. The planning horizon spans 20 years, and all results are expressed in cumulative terms unless stated otherwise.

Key economic assumptions include:

- discount rate: 10 per cent

- fixed processing and dumping costs: assumed equal across all scenarios

- truck haulage costs: vary across scenarios based on horizontal and vertical components in $/t/km

- conveyor haulage costs: vary across scenarios in $/t/km

- crusher capacity bounds:
    - upper bounds: 40 Mt (ore), 110 Mt (waste)
    - lower bounds: either 8 Mt/22 Mt or 16 Mt/44 Mt

These cost and capacity settings were combined factorially to produce 32 unique scenarios per system configuration.

The optimisation models for each scenario were implemented in Python, using a combination of MILP formulations and scenario-specific cost inputs. Each of the four system configurations (A, B, C, D) was run across all 32 scenarios, resulting in a total of 128 optimised schedules.

Each model run produced time-series outputs for truck usage, haul distances, tonne-kilometres (TKM), and total haulage cost, which were aggregated and compared across scenarios. These outputs were then analysed to determine which configurations and parameter settings resulted in the most cost-effective solutions and when IPCC systems demonstrated significant advantages over the conventional TS set-up. Figure 1 shows a plan view of the case study with the positioning of the road and conveyor networks, two waste dumps, crusher panels for one bench-phase and mill.

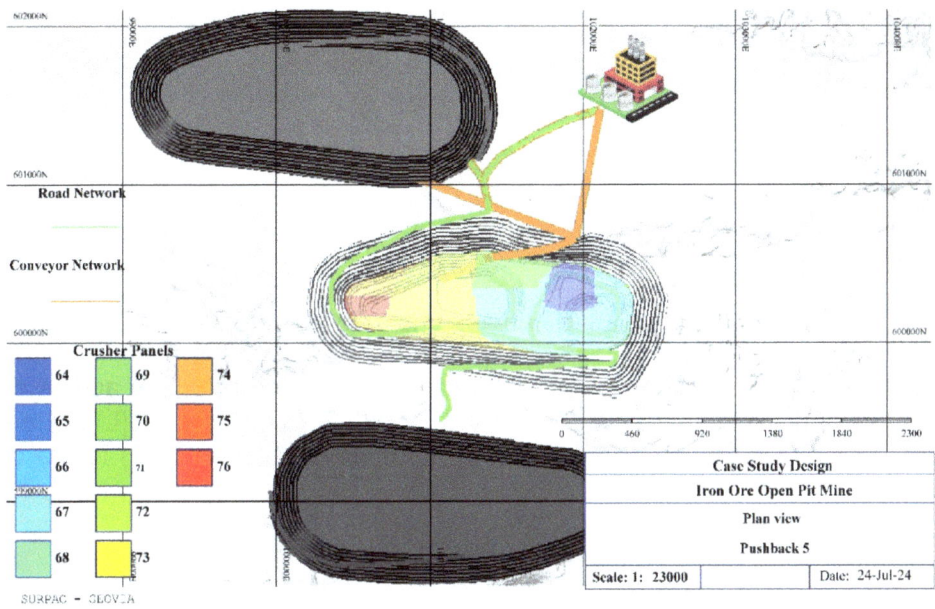

FIG 1 – Plan view of the case study.

DISCUSSION OF RESULTS

In this section we will present the findings of the 128 simulation runs—each representing one of the 32 cost-capacity scenarios under a specific system configuration coded as: CP00 (no IPCC), CP01 (ore IPCC), CP03 (waste IPCC), and CP02 (dual IPCC). Key performance indicators include cumulative haulage costs, ore and waste TKM, and the number of trucks required over the 20 year mine life.

Results

Figure 2 summarises the total cumulative haulage costs (ore + waste) for all scenarios and configurations. As expected, CP00 scenarios consistently incur the highest haulage costs due to reliance on long-distance truck transport. Across all 32 scenarios, CP02 (dual IPCC) achieves the most significant cost reductions, with savings ranging from 25 per cent to over 40 per cent compared to CP00, depending on the cost assumptions.

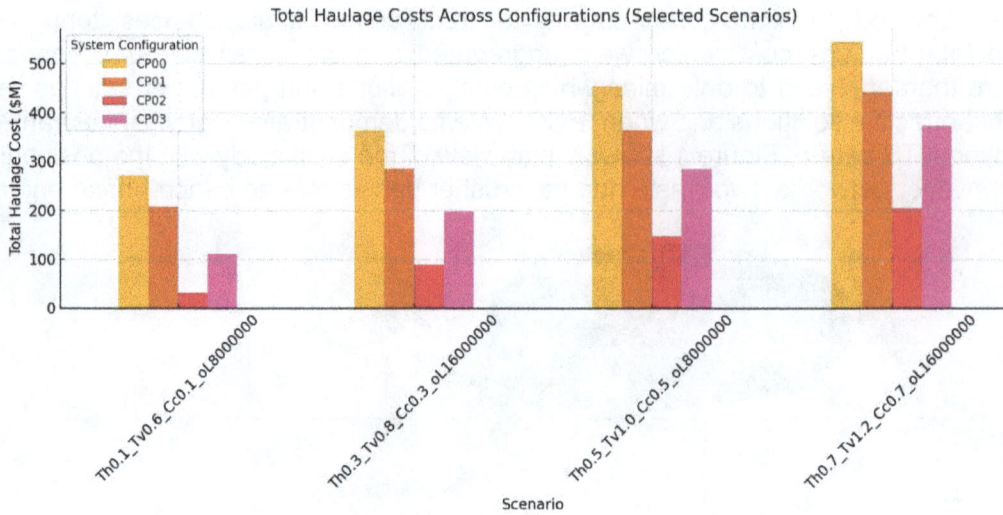

FIG 2 – Total cumulative haulage costs for selected scenarios.

Table 1 provides a detailed breakdown of total haulage cost (in million USD) across configurations and scenarios. Scenario names follow the format Th{H}_Tv{V}_Cc{C}_oL{L}, where {H} is the horizontal truck cost, {V} is vertical truck cost, {C} is conveyor cost, and {L} is ore crusher lower bound.

TABLE 1

Detailed breakdown of total haulage cost across configurations and scenarios.

Scenario	CP00 ($M)	CP01 ($M)	CP03 ($M)	CP02 ($M)
Th0.1_Tv0.6_Cc0.1_oL8000000	272.5	232.1	213.4	188.6
Th0.1_Tv0.6_Cc0.1_oL16000000	272.5	234.8	216.0	192.4
Th0.1_Tv0.6_Cc0.3_oL8000000	272.5	238.7	213.4	188.6
Th0.1_Tv0.6_Cc0.3_oL16000000	272.5	241.4	216.0	192.4
Th0.1_Tv0.6_Cc0.5_oL8000000	272.5	245.2	213.4	188.6
Th0.1_Tv0.6_Cc0.5_oL16000000	272.5	247.8	216.0	192.4
Th0.1_Tv0.6_Cc0.7_oL8000000	272.5	251.6	213.4	188.6
Th0.1_Tv0.6_Cc0.7_oL16000000	272.5	254.2	216.0	192.4
Th0.3_Tv0.8_Cc0.1_oL8000000	309.3	256.1	240.5	213.2
Th0.3_Tv0.8_Cc0.1_oL16000000	309.3	258.9	243.1	216.9
Th0.3_Tv0.8_Cc0.3_oL8000000	309.3	262.6	240.5	213.2
Th0.3_Tv0.8_Cc0.3_oL16000000	309.3	265.3	243.1	216.9
Th0.3_Tv0.8_Cc0.5_oL8000000	309.3	269.0	240.5	213.2
Th0.3_Tv0.8_Cc0.5_oL16000000	309.3	271.7	243.1	216.9
Th0.3_Tv0.8_Cc0.7_oL8000000	309.3	275.5	240.5	213.2
Th0.3_Tv0.8_Cc0.7_oL16000000	309.3	278.2	243.1	216.9
Th0.5_Tv1.0_Cc0.1_oL8000000	340.4	274.8	262.6	231.0
Th0.5_Tv1.0_Cc0.1_oL16000000	340.4	277.5	265.2	234.8
Th0.5_Tv1.0_Cc0.3_oL8000000	340.4	281.3	262.6	231.0
Th0.5_Tv1.0_Cc0.3_oL16000000	340.4	284.0	265.2	234.8
Th0.5_Tv1.0_Cc0.5_oL8000000	340.4	287.8	262.6	231.0
Th0.5_Tv1.0_Cc0.5_oL16000000	340.4	290.5	265.2	234.8

Th0.5_Tv1.0_Cc0.7_oL8000000	340.4	294.2	262.6	231.0
Th0.5_Tv1.0_Cc0.7_oL16000000	340.4	296.9	265.2	234.8
Th0.7_Tv1.2_Cc0.1_oL8000000	394.2	305.2	289.7	258.1
Th0.7_Tv1.2_Cc0.1_oL16000000	394.2	308.0	292.4	262.0
Th0.7_Tv1.2_Cc0.3_oL8000000	394.2	311.7	289.7	258.1
Th0.7_Tv1.2_Cc0.3_oL16000000	394.2	314.4	292.4	262.0
Th0.7_Tv1.2_Cc0.5_oL8000000	394.2	318.2	289.7	258.1
Th0.7_Tv1.2_Cc0.5_oL16000000	394.2	320.9	292.4	262.0
Th0.7_Tv1.2_Cc0.7_oL8000000	394.2	324.6	289.7	258.1
Th0.7_Tv1.2_Cc0.7_oL16000000	394.2	327.3	292.4	262.0

The results show that the inclusion of either ore or waste IPCC (CP01 or CP03) yields substantial haulage cost reductions. However, CP02 generally outperforms both, especially under higher truck cost assumptions and when crusher relocation thresholds are moderate (ie lower bounds of 8 Mt/ 22 Mt).

Figure 3 presents the cumulative TKM for ore and waste materials under different configurations. In CP00, trucks are responsible for all material movement, resulting in the highest TKM values. The use of conveyors in CP01, CP03, and CP02 substantially reduces truck travel distances, with CP02 demonstrating the lowest combined TKM. The reductions are particularly pronounced for waste TKM in CP03 and CP02, highlighting the benefit of applying IPCC to high-tonnage, low-value material like waste. For instance, in high-cost scenarios, waste TKM under CP02 is reduced by over 60 per cent relative to CP00.

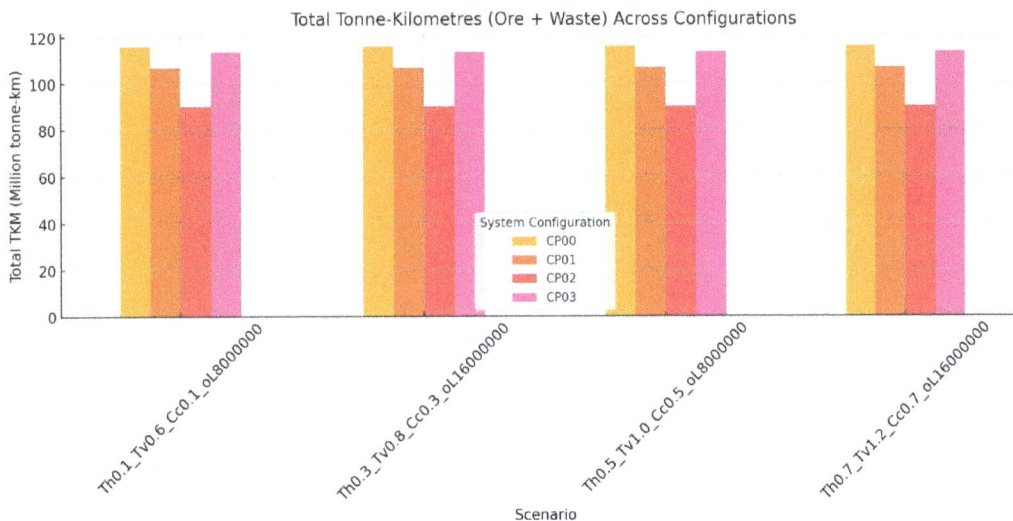

FIG 3 – Cumulative TKM for ore and waste materials under different configurations.

Figure 4 compares the number of ore and waste trucks required over the 20 year horizon. In CP00, truck demand is consistently high due to the need for full-pit haulage. Introducing IPCC systems results in a measurable reduction in truck requirements:

- CP01: Ore truck demand drops significantly, while waste remains unchanged.
- CP03: Waste truck demand drops sharply, especially in low-capacity scenarios.
- CP02: Both ore and waste truck demands are minimised.

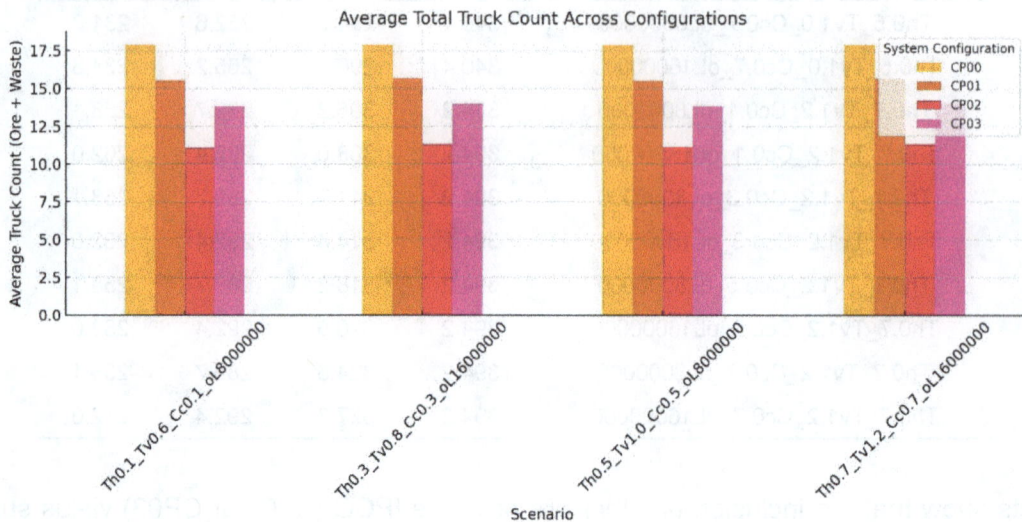

Average Total Truck Count Across Configurations

FIG 4 – compares the number of ore and waste trucks required over the 20 year horizon.

This directly translates to savings in capital expenditure (CAPEX) and maintenance costs, as well as reduced traffic congestion and environmental impact within the pit.

Across all configurations involving crushers, scenarios with the lower capacity threshold (8 Mt ore/ 22 Mt waste) tend to perform better in terms of cost and TKM. The increased frequency of crusher relocation allows the system to remain closer to the active mining face, thereby reducing truck haul distances. However, this benefit comes at the cost of increased logistical complexity and potential downtime during crusher moves. Thus, there is a trade-off between cost efficiency and operational stability that must be evaluated in practice.

While conveyor cost influences overall system cost, its effect is less pronounced than truck haulage cost. Even at the highest tested conveyor rate ($0.7/t/km), IPCC systems remain cost-competitive under medium-to-high truck haulage costs. This suggests that conveyor cost inflation does not significantly undermine the value of IPCC, especially when trucks are expensive to operate.

Figure 3 shows that CP00 consistently produces the highest TKM due to full-distance truck haulage. CP02 significantly reduces TKM, demonstrating the efficiency of combining ore and waste IPCC systems particularly in high-cost scenarios. Also, Figure 4 shows that CP00 requires the largest truck fleet throughout the mine life. The introduction of IPCC systems, especially in CP02, substantially reduces the average number of trucks needed, lowering CAPEX and operational complexity.

Discussion

The results of the sensitivity analysis reveal clear and consistent trends that reinforce the benefits of incorporating semi-mobile in-pit crushing and conveying (IPCC) systems into long-term open pit mine planning. One of the most prominent patterns observed across all scenarios is the increasing effectiveness of IPCC systems as truck haulage costs rise. In scenarios with high horizontal and vertical truck costs (eg Th0.7_Tv1.2), total haulage costs in the TS-only configuration (CP00) escalated dramatically. Conversely, configurations involving conveyors (especially CP02) demonstrated resilience to these increases, primarily due to the reduced dependence on long-distance trucking. These findings imply that in regions where fuel prices are high, haul roads are long and steep, or environmental regulations impose strict emission limits, IPCC becomes not just an alternative but a strategic imperative. The ability to relocate crushers closer to active mining zones helps stabilise haulage costs and enhances system adaptability over time.

Among the three IPCC-enabled configurations, CP02, where both ore and waste are handled by dedicated crusher-conveyor systems, consistently delivered the lowest total haulage cost and truck fleet requirement. The simultaneous reduction in ore and waste TKM demonstrates the operational synergy achieved when both material streams are supported by IPCC infrastructure. While CP01 and CP03 also provide benefits, their advantages are material specific. CP01 is more effective in reducing ore-related TKM and truck use but offers no benefit for waste movement. CP03, conversely, is particularly effective in reducing haulage costs in waste-dominant scenarios. The choice between these configurations should depend on the relative tonnages of ore and waste, crusher relocation feasibility, and haulage profiles in each mine.

The scenarios involving lower tonnage thresholds for crusher relocation (8 Mt for ore and 22 Mt for waste) resulted in better performance across most metrics. These scenarios allowed more frequent repositioning of the crushers, keeping them close to the mining face and minimising truck travel distances. However, frequent relocation introduces logistical challenges. Even if modelled as instantaneous or costless in an optimisation framework, real-world relocations require downtime, workforce planning, and coordination with other operations. Therefore, while lower thresholds may improve cost performance in the model, planners must assess whether their operations can realistically accommodate such frequent movements.

The analysis also shows that while conveyor costs influence the total system cost, they are not as critical a driver as truck haulage costs. Even in scenarios where conveyor costs reach $0.7/t/km, the IPCC-enabled configurations remain economically favourable in medium-to-high truck cost environments. This result is crucial for planners concerned about the capital and operating costs of conveyor systems, as it underscores the robustness of IPCC's value proposition even under less favourable conveyor pricing.

CONCLUSION

This study presented a comprehensive sensitivity analysis of semi-mobile IPCC systems integrated into long-term open pit mine scheduling. Building upon the optimisation framework introduced by Kamrani, Pourrahimian and Askari-Nasab (2025), we evaluated 32 distinct cost and capacity scenarios across four major system configurations: conventional truck-shovel only (CP00), ore IPCC (CP01), waste IPCC (CP03), and dual IPCC for both ore and waste (CP02).

The results demonstrate that IPCC-enabled configurations consistently outperform the conventional truck-shovel system in terms of total haulage cost, tonne-kilometres, and truck fleet requirements. These advantages become particularly pronounced under high truck haulage cost conditions, where IPCC systems, especially CP02, offer substantial cost savings and operational efficiency.

Key findings from the analysis include:

- The dual IPCC system (CP02) yields the greatest cost savings and reductions in truck use, supporting its role as a strategic long-term haulage solution.

- Crusher relocation thresholds significantly affect system performance, with lower thresholds enabling better proximity to active mining zones and reducing haul distances.

- While conveyor costs influence overall haulage expenses, their impact is less critical than that of truck-related costs, affirming the economic resilience of IPCC strategies.

- IPCC integration supports broader sustainability goals, reducing fuel consumption, emissions, and pit traffic congestion.

The results offer guidance for mine planners evaluating the trade-offs between conventional TS and IPCC systems. The detailed scenario analysis provides a decision-support framework to help determine when and where IPCC operational costs become cost-effective, based on-site-specific parameters and market conditions.

Future research may focus on incorporating the second-stage model from Kamrani, Pourrahimian and Askari-Nasab (2025) for relocation timing, scheduling optimisation, modelling downtime and capital costs associated with crusher moves, or integrating stochastic elements to account for material and price uncertainty. Additionally, expanding the framework to consider emissions, energy consumption, and social impact metrics would enhance its applicability in modern sustainability-focused mine planning.

REFERENCES

Abbaspour, H, Drebenstedt, C and Dindarloo, S R, 2018. Evaluation of safety and social indexes in the selection of transportation system alternatives (Truck-Shovel and IPCCs) in open pit mines, *Safety Science*, 108:1–12.

Bernardi, L, Kumral, M and Renaud, M, 2020. Comparison of fixed and mobile in-pit crushing and conveying and truck-shovel systems used in mineral industries through discrete-event simulation, *Simulation Modelling Practice and Theory*, 103:102100. https://doi.org/10.1016/j.simpat.2020.102100

de Werk, M, Ozdemir, B, Ragoub, B, Dunbrack, T and Kumral, M, 2017. Cost analysis of material handling systems in open pit mining: Case study on an iron ore prefeasibility study, *The Engineering Economist*, 62(4):369–386. https://doi.org/10.1080/0013791X.2016.1253810

Dean, M, Knights, P, Kizil, M S and Nehring, M, 2015. Selection and planning of fully mobile in-pit crusher and conveyor systems for deep open pit metalliferous applications, in *Proceedings of the Third International Future Mining Conference*, pp 219–225 (The Australasian Institute of Mining and Metallurgy: Melbourne).

Kamrani, A, Badiozamani, M M, Pourrahimian, Y and Askari-Nasab, H, 2024. Evaluating the semi-mobile in-pit crusher option through a two-step mathematical model, *Resources Policy*, 95:105113. https://doi.org/10.1016/j.resourpol.2024.105113

Kamrani, A, Pourrahimian, Y and Askari-Nasab, H, 2025. Semi-mobile in-pit crushing and conveying vs. truck-shovel systems: Long-term scheduling with road and conveyor networks integration, *Expert Systems with Applications*, 268:126122. https://doi.org/10.1016/j.eswa.2024.126122

Klanfar, M and Vrkljan, D, 2012. Benefits of using mobile crushing and screening plants in quarrying crushed stone, *Min Geoeng,* 36:167.

Liu, D and Pourrahimian, Y, 2021. A Framework for Open pit Mine Production Scheduling under Semi-Mobile In-Pit Crushing and Conveying Systems with the High-Angle Conveyor, *Mining*, 1(1):59–79. https://doi.org/10.3390/mining1010005

Londoño, J G, Knights, P F and Kizil, M S, 2013. Modelling of In-Pit Crusher Conveyor alternatives, *Mining Technology*, 122(4):193–199. https://doi.org/10.1179/1743286313Y.0000000048

Nehring, M, Knights, P F, Kizil, M S and Hay, E, 2018. A comparison of strategic mine planning approaches for in-pit crushing and conveying and truck/shovel systems, *International Journal of Mining Science and Technology*, 28(2):205–214.

Nunes, R A, Junior, H D, de Tomi, G, Infante, C B and Allan, B, 2019. A decision-making method to assess the benefits of a semi-mobile in-pit crushing and conveying alternative during the early stages of a mining project, *Revista Escola de Minas*, 72(2):285–291. https://doi.org/10.1590/0370-44672018720109

Paricheh, M, Osanloo, M and Rahmanpour, M, 2017. In-pit crusher location as a dynamic location problem, *J South Afr Inst Min Metal,* 117:599.

Paricheh, M, Osanloo, M and Rahmanpour, M, 2018. A heuristic approach for in-pit crusher and conveyor system's time and location problem in large open pit mining, *International Journal of Mining, Reclamation and Environment*, 32(1):35–55. https://doi.org/10.1080/17480930.2016.1247206

Rahimdel, M J and Bagherpour, R, 2018. Haulage system selection for open pit mines using fuzzy MCDM and the view on energy saving, *Neural Computing and Applications*, 29(6):187–199. https://doi.org/10.1007/s00521-016-2562-7

Roumpos, C, Partsinevelos, P, Agioutantis, Z, Makantasis, K and Vlachou, A, 2014. The optimal location of the distribution point of the belt conveyor system in continuous surface mining operations, *Simulation Modelling Practice and Theory*, 47:19–27.

Yarmuch, J, Epstein, R, Cancino, R and Peña, J C, 2017. Evaluating crusher system location in an open pit mine using Markov chains, *International Journal of Mining, Reclamation and Environment*, 31(1):24–37. https://doi.org/10.1080/17480930.2015.1105649

Representative measurement using high performance PGNAA to digitalise process feed quality

H Kurth[1] and A Brodie[2]

1. Chief Marketing Officer and Minerals Consultant, Scantech International Pty Ltd, Camden Park SA 5038. Email: h.kurth@scantech.com.au
2. General Manager Marketing, Scantech International Pty Ltd, Camden Park SA 5038. Email: a.brodie@scantech.com.au

ABSTRACT

A major challenge for mineral processing operations is to maximise recovery, minimise operating costs and maintain consistent throughput rates when dealing with a supply of highly variable ore quality from the mine. Ore quality parameters include ore grade, ore hardness, mineralogy, dilution by waste, deleterious components, ore textures affecting liberation, gangue content and quality, and many more. Any or all of these can be variable due to the heterogeneity of orebodies, complexity of ore/waste boundaries and influence of the mining process. Few operators complain of monotonous feed quality and consistent process performance requiring minimal supervision and response requirements. The inability to receive uniform feed material quality is a major reason that mining is unlike manufacturing. This paper uses case studies to demonstrate successful application of real time, representative, composition measurement technology for conveyed plant feed flows to measure, understand and control quality variability to benefit process performance. High performance Prompt Gamma Neutron Activation Analysis (PGNAA) has been proven effective in digitalising conveyed flow quality to enable real time responses to quality variability through ore blending control, ore and waste parcel diversion (bulk sorting), feedback to mining operations and feed forward to process operators. The representative measurement data is used for ore reconciliation and metal accounting applications in multiple commodities. Benefits through preventing unnecessary processing of coarse waste include reductions in GHG emissions, reduced use of consumables and reduced generation of fine tailings per tonne of metal produced. Process feed quality is no longer dictated purely by the mining operations. The ability to measure and control process feed quality has significantly improved operational economics and sustainability in iron ore, base metals and industrial minerals.

INTRODUCTION

Mineral processors are tasked with optimising metal recovery from ore delivered from the mine irrespective of its compositional characteristics. Mining operations schedule ore supply from various sources within the mine and anticipate that the target quality will be delivered to run-of-mine stockpiles or crusher that is suitable for processing. The ore quality feeding the process plant is representatively measured in very few operations and sampling coarse ore representatively requires large and frequent samples and significant resources for sample removal from the flow, sample preparation and analysis in an assay laboratory.

It is well accepted that ore quality variability in process feed has a major effect on metal recoveries with reductions of up to ten to 15 per cent occurring in some commodities (Goodall, 2021). There is an opportunity to manage ore quality to the process plant to maximise recovery, minimise operating costs and maintain consistent throughput rates when dealing with a supply of highly variable ore quality from the mine. Ore quality parameters include ore grade, ore hardness, mineralogy, dilution by waste, deleterious components, ore textures affecting liberation, gangue content and quality, and many more. Any or all of these can be variable due to the heterogeneity of orebodies, complexity of ore/waste boundaries and influence of the mining process. Many of these parameters can be determined representatively and in real time to optimise process performance and plant capacity utilisation. Some parameters such as ore textures are not yet able to be measured or derived representatively in real time from online elemental or mineralogy sensing techniques.

High variability in feed quality can lead to process upsets which reduce metal recovery and few operators complain of monotonous feed quality and consistent process performance requiring minimal supervision and corrective responses. The inability to receive uniform feed material quality is a major reason that mining is unlike manufacturing where inputs are consistent and process performance almost always meets expectations.

A major development over the last 20 to 30 years has been the design and continuous improvement in representative analysis technologies for the Mine to Mill section of a mine's operations. The main technology discussed in this paper is the high-performance prompt gamma neutron activation analysis technique (PGNAA) proven for real time elemental analysis.

PGNAA ELEMENTAL ANALYSIS

PGNAA technology has been adapted to suit analysis of conveyed flows to provide timely, accurate multi-elemental measurement to representatively digitalise conveyed flow quality. Elemental data alone has many benefits, but it can be used to derive mineralogy and ore characteristics that affect process performance, such as ore hardness (Kurth, 2018).

The technique uses a source of neutrons such as Californium-252 which generates the highest neutron emission rate of any radionuclide at suitable neutron energies. Neutrons are thermalised to ensure they are at energies that maximise their chance of 'capture' by elemental nuclei present in the conveyed flow passing through the analyser. The source is typically located under the conveyor for maximum safety as it is well protected by the conveyor structure and extremely unlikely to be dislodged by oversize rocks or surges inflow compared to where it is positioned above the conveyor. The conveyed flow passes through a zone of high neutron flux and the elemental nuclei that do capture neutrons emit unique gamma energy spectral responses that can be sensed by an array of high-performance detectors in the top of the analyser above the conveyed flow as shown in Figure 1. This configuration has proven to provide representative response from the material and minimises the effect of steel cords and chlorine contained in the conveyor belt. Unique measurement capabilities have been developed over time to include direct measurement of elements such as gold at very low concentrations (Balzan et al, 2022).

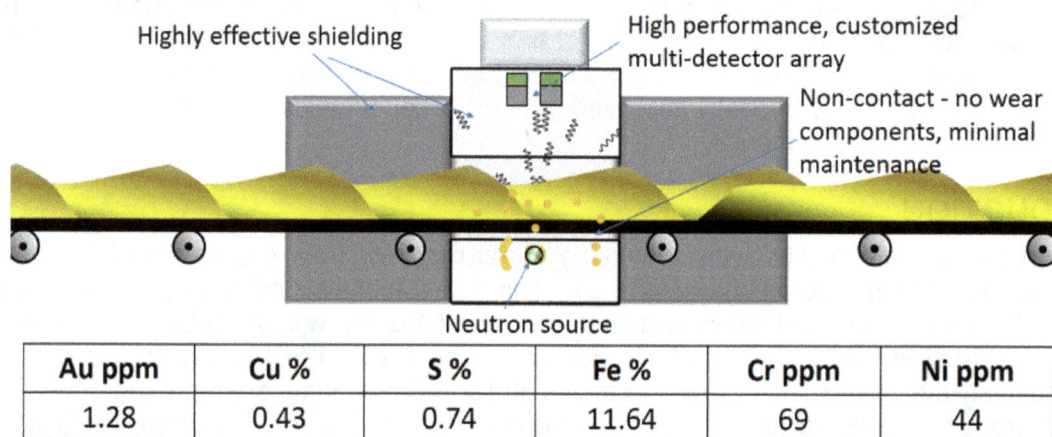

Au ppm	Cu %	S %	Fe %	Cr ppm	Ni ppm
1.28	0.43	0.74	11.64	69	44

FIG 1 – Cross-section through GEOSCAN GOLD high specification PGNAA analyser showing main components and an elemental results for 30 seconds measurement and direct gold measurement over five mins (source: Scantech).

The high performance PGNAA system produces measurements for elements as low as carbon in atomic number and can equally measure multiple elements when high levels of chlorine are present. Chlorine is considered a problem for this technique due to its strong response over multiple gamma energies and causes elemental recognition and resolution limitations in low specification PGNAA systems.

Online analysis technologies such as PGNAA have been extensively used in the cement industry (Harris, Smith and Rossi, 2005), coal industry (Butel et al, 1993), base metals (Arena and McTiernan, 2011; Patel, 2014; Kurth and Balzan, 2017; Balzan et al, 2016; Kurth, 2019; and Nadolski et al, 2018), phosphates (Balzan, Harris and Bauk, 2018), iron ore (Balzan, Beven and Harris, 2015; Balzan et al, 2019; and Matthews and Du Toit, 2011), platinum group metals (Scott et al, 2020), gold

(Balzan *et al*, 2022; Kurth, 2025), manganese (Balzan, Harris and Nel, 2015) and other commodities such as lithium, nickel, chromium (International Mining News, 2025), bauxite and diamonds. Measured elements may not necessarily be those targeted for recovery and could be related to dilution rock types or minerals that negatively affect the process or product quality.

Measurement performance can be optimised through various calibration methods, and these can include static calibration using samples of known composition or dynamic calibration where analyser data is compared to laboratory analysis of samples taken during normal plant operations. Static calibrations may not adequately represent the dynamic conditions present during material flow and therefore mass weighting of results is considered a standard requirement where flow rates may fluctuate or vary randomly. External data from a belt weigher is typically used for tonnage weighting if that source of data is calibrated and trusted. Other external inputs may include moisture measurement so that elemental concentrations can be reported in dry weight percentages as the average composition for each measured increment of flow.

High performance PGNAA is unaffected by particle size, belt speed, mineralogy (no matrix effects), layering or different material compositions, dust or moisture. The technique penetrates up to 550 mm of bed depth and senses the full conveyed cross-section continuously to provide unmatched measurement capability. Data is transmitted directly to the plant control system at the end of each conveyed increment which can vary from 30 seconds to ten mins as needed for the data application. Two mins is a standard frequency for reporting averages for most materials using the high-performance configuration discussed. Figure 2 shows an example of data from an analyser received in a lead-zinc operation each two mins of conveyed ore flow.

Time	Tonnes Analysed	%Pb	%Zn	%Fe	%S	%SiO$_2$	%Al$_2$O$_3$	%CaO	%MgO
12:00:00 AM	52.149	1.144	5.594	11.75	10.85	30.14	5.554	9.400	4.613
12:02:00 AM	51.443	0.354	4.574	11.69	10.60	29.59	5.065	9.438	5.548
12:04:00 AM	51.939	2.138	4.210	12.47	11.80	27.57	4.774	9.555	5.538
12:06:00 AM	52.163	2.231	6.438	12.55	12.02	28.52	5.072	9.576	4.495
12:08:00 AM	51.946	1.910	5.357	12.59	11.94	28.35	4.963	9.861	5.150
12:10:00 AM	51.235	1.560	8.045	11.63	10.98	29.65	5.537	9.714	4.191
12:12:00 AM	50.918	2.045	6.681	11.91	11.27	30.31	5.387	9.381	5.665
12:14:00 AM	51.226	1.722	4.605	11.92	10.75	29.47	4.904	10.186	6.045
12:16:00 AM	50.955	3.312	6.118	11.66	11.00	30.33	5.426	9.385	7.100
12:18:00 AM	51.069	2.053	5.225	12.11	11.52	28.86	5.024	9.252	5.100
12:20:00 AM	51.442	0.280	5.793	12.04	11.07	29.00	5.072	9.615	5.485
12:22:00 AM	51.281	1.746	6.342	12.07	11.40	28.61	4.984	9.439	5.663
12:24:00 AM	51.068	2.524	5.748	12.21	11.52	29.22	5.322	10.004	5.323
12:26:00 AM	51.239	2.708	4.988	12.55	12.00	28.85	4.986	9.280	5.659
12:28:00 AM	51.235	2.724	5.015	12.64	12.09	28.00	5.119	9.774	6.106
12:30:00 AM	51.071	2.207	5.345	12.23	11.93	27.55	5.094	10.305	6.616
12:32:00 AM	50.785	1.269	4.458	11.78	11.54	27.84	4.778	9.581	5.072
12:34:00 AM	50.265	1.540	5.248	11.82	11.37	29.84	5.277	9.208	5.344
12:36:00 AM	50.288	1.411	4.849	12.60	11.87	29.13	5.056	8.990	4.333
12:38:00 AM	24.441	2.158	5.414	12.03	11.63	28.69	4.804	8.718	4.495
12:40:00 AM	38.166	2.965	5.847	10.77	11.19	28.63	5.228	10.707	5.461
12:42:00 AM	47.132	1.984	6.016	11.61	11.29	29.07	5.326	10.112	4.443

FIG 2 – Example of analyser data from two min averages of lead-zinc ore flow received by the process plant, including the tonnage each analysis represents. Elements are user-defined, and the measurement frequency is customised to the application, in this case feed forward control to a dense medium plant. Data is typically date and time stamped as well (source: Scantech).

Figure 3 shows approximately 17 hrs of copper measurement data from an analyser measuring the average grade of each 30 seconds of flow. The analyser was used to understand the variability of ore quality in a large, low-grade copper mine in the Americas. The data demonstrated that expected ore quality was not continuously achieved and that variability was much higher than expected by process staff. Data showing such trends in quality proved useful in analysing process plant response although longer term averages were needed to find correlations with process performance changes.

Cu grade over 17 hours of analysis time

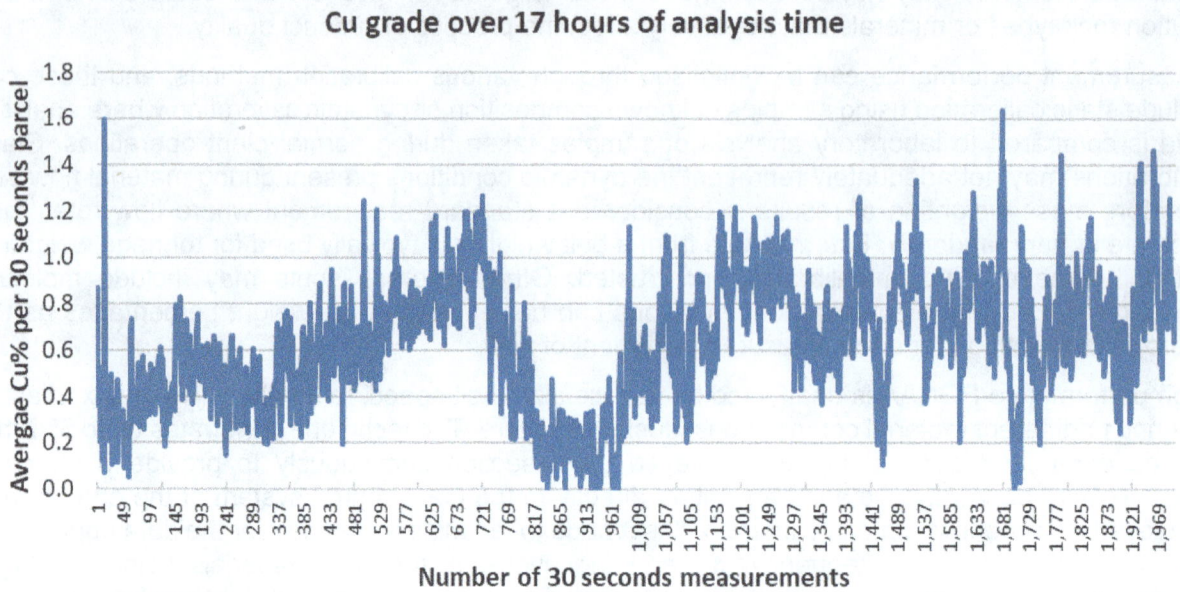

FIG 3 – Copper measurement data showing 30 seconds measurement averages for a copper mine conveyor at 1200 t per hr flow (source: Scantech).

DATA APPLICATIONS

Data received by the plant control system can be used concurrently for many applications. Digitalising conveyed flow quality enables real time responses to quality variability through ore blending control, ore and waste parcel diversion (bulk sorting), feedback to mining operations and feed forward to process operators. The representative measurement data is used for ore reconciliation and metal accounting applications in multiple commodities. Figure 4 shows a simple data integration model.

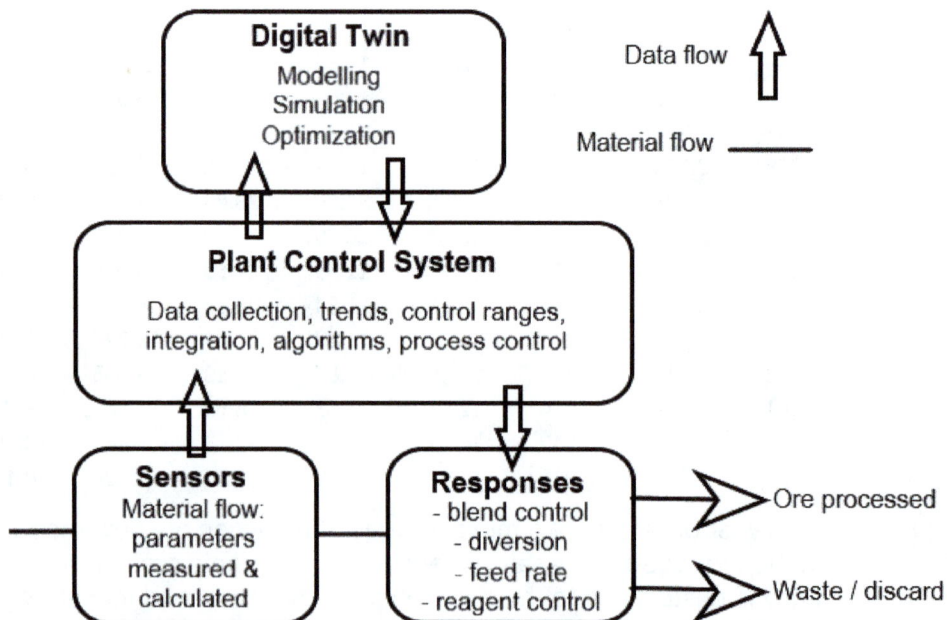

FIG 4 – Simple model showing data flows from mined material flows integrated with plant operations for both real time responses and simulation through digital twin. Data generated from a single location in a process is used for multiple applications concurrently (source: Scantech).

Ore blending

This requires the availability of multiple types of material, either ore grade or ore types, so that ore sources can be proportioned to meet a desired feed quality and consistency. This results in fewer plant upsets and optimised process performance, including improved metal recovery. Blending may include additives or reagents to the ore or to reduce higher contaminant levels. Blending has proven beneficial at Sepon copper-gold operation to control average copper metal content in leach circuit feed based on analyser data (Arena and McTiernan, 2011). Measurement times of five mins, composited over each 30 mins, smoothed highly variable short-term results to minimise blend change frequency. Pyrite addition to the carbonate ore at Sepon was controlled using analyser data (Balzan et al, 2016).

Blending commonly occurs in bulk commodities such as iron ore where different qualities can be blended onto trains or ships to meet an average quality requirement to comply with a desired specification for a customer. Concentrates can be blended in smelter feed and flux material addition controlled through online analysis. This is also the case in limestone feed to raw mills in cement production, sinter basicity control through limestone addition control for blast furnace feed in steelmaking and blending metallurgical coals for coke-making operations.

Increment diversion

Sometimes termed bulk sorting, increment diversion is another capability that relies on accurate and timely measurement of parcels of conveyed material. It can be applied to ores, concentrates, waste flows and other streams where separation using quality parameters may be needed. Examples include the diversion of product quality iron ore to bypass unnecessary beneficiation. Assmang Khumani operations in South Africa diverts approximately one third of its mined ore directly to screening and product stockpiles while lower quality ore is beneficiated through jig circuits (Matthews and du Toit, 2011). The savings in beneficiation cost for the diverted ore are approximately USD 5 million per annum. The diverted material also saves processing emissions of approximately 8 kg of CO_2 per t, or 40 000 t CO_2 e/a. A smaller plant design was possible at lower capital cost than if designed to process all mined ore.

Copper grade in ore can be measured to precisions of 0.02 per cent Cu using high specification PGNAA for 30 second measurement increments. Figure 5 shows an analyser in a bulk sorting application where parcels of 30 seconds are diverted according to their quality.

FIG 5 – GEOSCAN analyser used in a base metals bulk ore sorting plant for 30 seconds parcel diversion (source: Anglo American).

Operations such as New Afton in British Columbia (Nadolski et al, 2018) determined that up to 15 per cent of the mined material could be diverted as waste to avoid unnecessary processing and increase average feed grade. Grade variability in block models is a poor indicator of bulk sorting potential as large blocks are assigned an average grade whereas on conveyor measurement occurs over much smaller parcel sizes in tonnes to tens of tonnes.

Ore grade increases of 5–20+ per cent were achieved at El Soldado copper mine in Chile using bulk ore sorting with 5–20 per cent of mined ore being rejected as waste. Copper ore processing can generate 32 kg CO_2 e/t. Preventing milling and further downstream treatment of this proportion of waste saves up to 20 per cent of GHG emissions, with only minor (up to 2.5 per cent) metal losses. Savings of over 40 000 t CO_2 e/a are achieved at an ore processing rate of 1200 t/h where 14 per cent of the mined ore is discarded at waste. Multiple diversion criteria can be applied and sub-economic material could be diverted to a beneficiation process such as particle sorting.

Pebble circuit flows are ideal for increment diversion to prevent uneconomic material being recycled and unnecessarily consuming mill capacity which is better utilised in processing fresh ore. Waste material measurement may assist in diverting acid generating waste from otherwise clean waste flows.

Feed forward control

Advance notice of process feed quality changes or variability can assist process operators to improve process performance particularly if process sensitivity to changes in feed quality are well understood. Such understanding may result from comprehensive geometallurgical studies. One example is the ability to estimate variations in ore hardness using elements such as silicon and iron and adjust mill feed rate to avoid overloading the mill. Indications of changes in deleterious component concentrations in feed can also trigger process changes.

Reagent addition is proportioned to feed quality, particularly acid-consuming gangue content in a copper leach process feed (Balzan et al, 2016) or for phosphate rock composition variation in phosphoric acid reactor feed. One phosphate plant increased throughput by over 20 per cent as it was discovered through analyser measurement that ore quality was much lower than expected and therefore the plant capacity was being underutilised. The expected two percent improvement was estimated to generate a six-month project payback.

Patel (2014) discusses how analyser measurement data was used to monitor changes in lead-zinc ore feed quality to a dense medium separation plant at Mount Isa Mines are modify density cut points according to composition. Matthews and Du Toit (2011) discuss measuring beneficiation feed to predict upgrade factors for various iron ore types confirmed using analysers on jig product streams and discard flows.

Metal accounting and ore reconciliation

Rarely in any mining operation is the quality of a flow of material able to be measured definitively to the extent it is ideal for feedback to the mine for ore reconciliation to the mine schedule and block model. Measurement data from PGNAA analysis is also suitable for metal accounting for process operations. Analysers are used for elemental balance at the Khumani iron ore operations (Matthews and Du Toit, 2011) because all relevant conveyed flows, including discard flows, are measured to quantify tonnages and grades. Direct gold measurement allows a previously unattainable level of metal accounting to be performed.

CONCLUSIONS

The effectiveness of process control decisions is a function of the quality, timeliness and relevance of sensor data and understanding parameters affecting process performance. High quality, representative data on process feed material composition is not commonly available, particularly in real time on coarse conveyed flows. Data that has proven effective has been generated through:

- High performance sensors customised to data quality and application requirements.
- Sensors designed for high safety, low maintenance, minimal implementation disturbance.
- Sensors proven to perform representative, timely, precise material quality measurement.
- Sensors calibrated through robust techniques involved both static and dynamic processes.

Mining companies that assigned adequate resources to developing their digitalisation strategy, performing data analysis and determining appropriate responses to data have been able to successful integrate sensors into their operations with almost immediate benefit.

Representative, real time measurement data on conveyed flows enables timely responses to ensure that quality management does not finish at the mine. Ore quality can be actively, and automatically in some cases, controlled between the primary crusher and the mill. Over 130 of these analysers have been successfully deployed in the minerals industry in many commodities as discussed in this paper to improve process performance, reduce the amount of waste unnecessarily processed, reduce the fine tailings generated from processing that waste and reduce energy, water and reagent consumption and significantly reduce GHG emissions.

Process feed quality is no longer dictated purely by the mining operations. The ability to measure and control process feed quality has significantly improved operational economics and sustainability in commodities such as iron ore, manganese, chromium and bauxite, base metals (copper, zinc, lead and nickel), precious metals (gold and silver), PGMs, and industrial minerals (phosphate rock, diamonds, lithium, limestone) and coal (metallurgical and thermal).

Benefits of high precision, representative, real time measurement data were initially underestimated in each application discussed in this paper. High resolution data digitalising conveyed ore quality in real time has enabled a better understanding of material variability and created further quality control opportunities. Companies that have adopted high performance PGNAA technology have achieved major improvements and benefits.

ACKNOWLEDGEMENTS

The permission of Scantech International Pty Ltd to publish this paper is gratefully acknowledged as well as various clients and authors mentioned in the paper for their support of the technology and provision of data.

REFERENCES

Arena, T and McTiernan, J, 2011. On-belt analysis at Sepon copper operation, in *Proceedings MetPlant 2011*, pp 527–535 (The Australasian Institute of Mining and Metallurgy: Melbourne).

Balzan, L A, de Paor, A, Doorgapershad, A and Futcher, W, 2022. The end of the rainbow: real time direct gold analysis in run of mine ore at Newcrest's Telfer mine using GEOSCAN analysis, in *Proceedings International Mineral Processing Conference – Asia Pacific 2022*, pp 1140–1149 (The Australasian Institute of Mining and Metallurgy: Melbourne).

Balzan, L A, Swart, E, Gray, L and Kalicinski, M, 2019. Process improvement at Kumba Iron Ore Sishen and Kolomela mines through the use of GEOSCAN on belt analysis equipment, in *Proceedings Iron Ore 2019*, pp 649–659 (The Australasian Institute of Mining and Metallurgy: Melbourne).

Balzan, L, Beven, B J and Harris, A, 2015. GEOSCAN online analyser use for process control at Fortescue Metals Group sites in Western Australia, in *Proceedings Iron Ore 2015*, pp 99–105 (The Australasian Institute of Mining and Metallurgy: Melbourne).

Balzan, L, Harris A and Nel, H, 2015. Adaptation and performance of Geoscan on-belt analysers for manganese ore at Assmang Black Rock, in *Proceedings Africa Australia Technical Mining Conference 2015*, pp 125–130.

Balzan, L, Harris, A and Bauk, Z, 2018. GEOSCAN-M use at a middle eastern phosphate plant, in Proceedings Beneficiation of Phosphates VIII, Cape Town.

Balzan, L, Jolly, T, Harris, A and Bauk, Z, 2016. Greater use of Geoscan on-belt analysis for process control at Sepon copper operation, in Proceedings XXVIII International Mineral Processing Congress (Canadian Institute of Mining, Metallurgy and Petroleum: Quebec).

Butel, D, Howarth, W J, Rogis, J and Smith, K G, 1993. Coal sorting, in *Coal Preparation*, 12:203–214.

Goodall, W, 2021. Understanding what is feeding your process: how ore variability costs money, in *Process Mineralogy Today Blog*, March 10. Available from: <https://minassist.com.au/understanding-what-is-feeding-your-process-how-ore-variability-costs-money/> [Accessed: 26 May 2023].

Harris, A, Smith, K and Rossi, F, 2005. On-belt analysis breakthrough, in *International Cement Review*, October, pp 62–66.

International Mining News, 2025. ERG carrying out chromium ore flow analysis at Donskoy GOK with Scantech GEOSCAN-M. Available from: <https://im-mining.com/2025/02/25/erg-carrying-out-chromium-ore-flow-analysis-at-donskoy-gok-with-scantech-geoscan-m/> [Accessed: 1 March 2025].

Kurth, H and Balzan, L, 2017. Assessing bulk sorting suitability at the New Afton Mine, in *Proceedings Metallurgical Plant Design and Operating Strategies* (MetPlant 2017), pp 315–323 (The Australasian Institute of Mining and Metallurgy: Melbourne).

Kurth, H, 2018. Role of real-time elemental analysis using PGNAA in operational geometallurgy, in Proceedings SAIMM Geometallurgy Conference 2018 (The South African Institute of Mining and Metallurgy: Cape Town).

Kurth, H, 2019. Using real time elemental analysis of conveyed ore flows to improve copper processing, in Proceedings Copper International Conference, 10th edn (Canadian Institute of Mining, Metallurgy and Petroleum, Montreal).

Kurth, H, 2025. Recent developments in representative gold grade measurement of conveyed ore in plant feed, in Proceedings of World Gold 2025.

Matthews, D and du Toit, T, 2011. Validation of material stockpiles and roll out for overall elemental balance as observed in the Khumani iron ore mine, South Africa, in *Proceedings Iron Ore Conference 2011* pp 297–305 (The Australasian Institute of Mining and Metallurgy: Melbourne).

Nadolski, S, Klein, B, Samuels, M, Hart, C J R and Elmo, D, 2018. Evaluation of cave-to-mill opportunities at the New Afton Mine, in *Proceedings of the 50th Annual Canadian Minerals Processors Operators Conference* (eds: B Danyliw, R Cameron and J Zinck), pp 270–281 (Canadian Institute of Mining, Metallurgy and Petroleum, Montreal).

Patel, M, 2014. On-belt elemental analysis of lead-zinc ores using prompt gamma neutron activation analysis, in Proceedings XXVII International Mineral Processing Congress 2014, ch 17 (Gecamin: Santiago, Chile).

Scott, M, Rutter, J, du Plessis, J and Alexander, D, 2020. Operational deployment of sensor technologies for bulk ore sorting at Mogalakwena PGE Mine, in *Proceedings Preconcentration 2020*, pp 169–181 (The Australasian Institute of Mining and Metallurgy: Melbourne).

A comparison of methods for determining weights of criteria in multi-criteria decision-making problems in mine planning

M J Mahase[1], T Tholana[2], C Musingwini[3] and B Mutandwa[4]

1. PhD Candidate, School of Mining Engineering, University of the Witwatersrand, Johannesburg 2001, South Africa. Email: 368976@students.wits.ac.za
2. Senior Lecturer, School of Mining Engineering, University of the Witwatersrand, Johannesburg 2001. Email: tinashe.tholana@wits.ac.za
3. Professor, School of Mining Engineering, University of the Witwatersrand, Johannesburg 2001. Email: cuthbert.musingwini@wits.ac.za
4. Lecturer, School of Mining Engineering, University of the Witwatersrand, Johannesburg 2001. Email: bright.mutandwa@wits.ac.za

ABSTRACT

Decision-making in mine planning is characteristically multi-criteria in nature. Therefore, it requires multi-criteria decision analysis (MCDA) techniques to solve the decision-making problems. MCDA techniques are a subset of multi-criteria decision-making (MCDM) techniques. A bibliometric analysis revealed growth in the use of MCDA techniques in mine planning. This is evident in the increased publication frequency of mine-planning related case studies that have been solved using MCDA techniques, especially post the 1998 global economic meltdown and 2008 global financial crisis. A similar trend can be expected post the COVID-19 global pandemic experienced between 2020 and 2022. Decision-making using MCDA techniques involves several steps. Firstly, decision-making criteria and different feasible solutions (ie alternatives) relevant to the problem are identified. Secondly, weights are assigned to the identified criteria to indicate their relative importance. Thirdly, each alternative is scored against each criterion, and the resulting model is solved to rank the alternatives and identify the most preferred alternative. An important step in MCDA involves assigning weights to the criteria because if this step is done incorrectly, subsequent steps will yield an incorrect solution (ie wrong choice of best alternative). Despite the importance of assigning weights to criteria, not much research has been done on different methods of assigning weights to criteria. Therefore, this paper compared different methods used to assign criteria weights in MCDM and/or MCDA problems in mine planning. The findings from the bibliometric analysis indicated that six criteria weighting methods seem to be the preferred methods in MCDM and/or MCDA mine planning choice decisions. The comparison of the methods to determine criteria weights suggest that future research directions should include the use of hybrid methods and machine learning methods so that uncertainty inherent to MCDM and/or MCDA in mine planning cases can be accounted for and criteria weights determined more reliably.

INTRODUCTION

Uncertainty and global black swan events require mining companies to make both strategic and operational mine planning decisions that are adaptable to change. Flexibility must, therefore, be incorporated in decision-making processes to overcome challenges associated with post-decision changes that often become necessary when new information becomes available, or when conditions change. These changes arise because invariably at the time of making a choice decision in mine planning, the least amount of decision-making information is available, and conditions are not fixed. Companies in the mining sector, like companies in other sectors, encounter such challenges (Ozdemir, 2023). Therefore, it is important to make agile mine planning choice decisions because inflexible decisions can result in significant financial losses during mining operations. Multi-criteria decision-making (MCDM) techniques aid in selecting an 'optimal', 'best' or 'most preferred' alternative. This is achieved by simultaneously evaluating and ranking feasible choices, options, solutions or alternatives in the presence of several decision-making criteria that can sometimes be contradictory, thus, complicating the decision-making process.

MCDM methods have been defined in various ways such as those presented by Zavadskas, Turskis and Kildiene (2014), Jayant and Sharma (2018) and Mirjat et al (2018) and these definitions have a common thread characterising MCDM methods as those dealing with multiple criteria and several alternatives. The MCDM methods are generally divided into multi-attribute decision-making (MADM) methods and multi-objective decision-making (MODM) methods (Zavadskas, Turskis and Kildiene, 2014). MADM methods, which are alternatively called multi-criteria decision analysis (MCDA) methods, select the best alternative from a discrete set of alternatives using multiple and sometimes conflicting criteria, whereas MODM methods use multiple and sometimes conflicting objectives to select the best alternative from a continuous decision space with an infinite number of alternatives. Figure 1 illustrates a generic classification of commonly used MCDM methods.

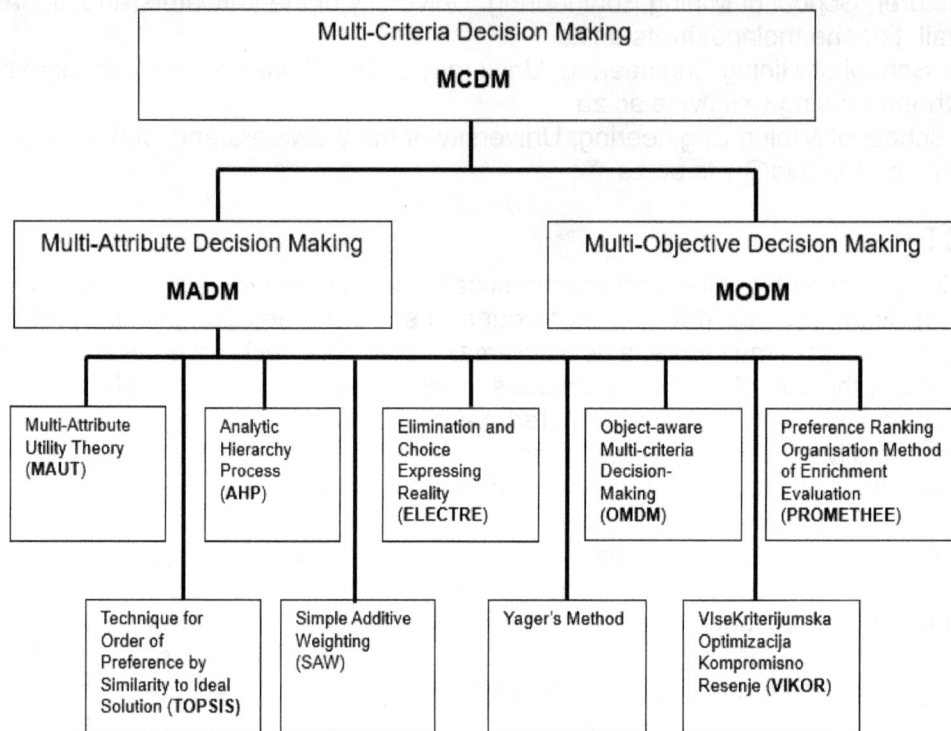

FIG 1 – Generic classification of commonly used MCDM methods (adapted from Mirjat *et al*, 2018).

According to Triantaphyllou (2000) and Musingwini (2010), MCDM and/or MDCA processes involve the following key steps:

- identifying alternatives and decision-making objectives which are then expressed as decision-making criteria for evaluating alternatives relevant to the decision-making problem.

- using a criteria weighting method to assign numerical values to each criterion to indicate the relative importance of each criterion to make the required decision.

- scoring each alternative against each criterion and aggregating the scores together with criteria weights using methods such as those shown in Figure 1, to generate a final set of scores for ranking the alternatives to make the choice decision.

From the preceding description of MCDM steps, it can be concluded that the two steps of identifying criteria and assigning weights to them are critical steps in MCDM and/or MCDA processes, because if these are done incorrectly, then subsequent steps will also be incorrect. According to Belton and Stewart (2002), most literature on MCDM has tended to focus on establishing and ranking different alternatives, while minimal work has gone into analysing methods used to assign weights to criteria more reliably. However, Ayan, Abacıoğlu and Basilio (2023) noted that there has been growing interest among researchers in the past decade in developing criteria weighting methods. This highlights the growing importance of undertaking more research into methods that can more reliably determine weights of criteria in MCDM-type and/or MCDA-type mine planning problems. It is also important to identify the most appropriate MCDM criteria weighting method for a specific need and application (Ayan, Abacıoğlu and Basilio, 2023), hence the need to undertake a bibliometric analysis of criteria weighting methods in different mine planning choice-decision scenarios. Choice decisions typically encountered in mine planning include (Mahase, 2017; Mahase, Musingwini and Nhleko, 2018):

- strategic mine planning option selection from several options given different long-term planning scenarios.

- mine plan selection (eg selecting the best short-term mine plan, medium-term mine plan or long-term mine plan from different mine planning scenarios).

- selecting the optimal mining method from a set of feasible mining methods as part of the mine planning process for new mining projects or expansion brownfield mining projects.

- selecting the optimal location for key mine facilities or infrastructure such as a shaft, waste dump, tailings dam, or plant/processing plant.

- mining equipment selection, including selecting the most appropriate equipment combination from a set of different equipment combinations.

- optimal mining layout selection from among several feasible layout options.

Based on the foregoing perspectives on the importance of research in methods to determine criteria weights, this paper sought to identify and compare methods for determining criteria weights in mine planning choice decisions and undertake a bibliometric analysis of the methods. This would assist in identifying preferred methods for determining criteria weights in MCDM and/or MCDA mine planning choice decisions and in mapping future research directions in this research field.

TRENDS IN THE USE OF MCDM AND/OR MCDA METHODS IN MINE PLANNING CHOICE-DECISIONS

Mahase (2017) noted that there have been occasional upswings in the use of MCDM techniques in the mining industry, such as those observed globally post the 1998 economic meltdown and 2008 Global Financial Crisis (GFC). These upswings are characterised by significant increases in the number of MCDM-type journal papers published between 1997 and 2017. The number of papers show increased use of MCDM techniques in mine planning choice decisions that have long-term implications such as those resulting from the selection of the optimal location of permanent mine infrastructure or choosing a mining method. The global COVID-19 pandemic that started in 2019 in China and seemed to have subsided by the end of 2022, is another example of a global black swan event which can be expected to result in another upswing in the use of MDCM techniques in decision-making. This could again be evident in the number of MCDM-type journal papers published during the aftermath of COVID-19. Figure 2 shows the number of different MCDA case studies and their years of publication. An increase in publication frequency is mainly seen in 2008 and post-2008.

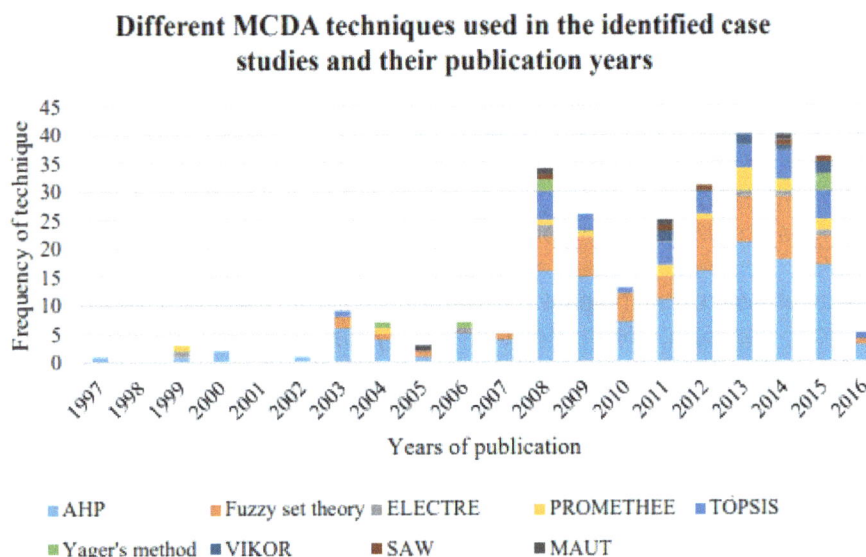

FIG 2 – Number of different MCDA techniques used in different case studies and their publication years (source: Mahase, 2017; Mahase, Musingwini and Nhleko, 2016).

Sitorus, Cilliers and Brito-Parada (2019) also analysed mining and mineral processing MCDM-type case studies and established a similar trend. Figure 3 shows the number of identified case studies and their years of publication. From Figures 2 and 3, there is an apparent marked increase in the use of MCDM techniques post the global 1998 economic meltdown and 2008 GFC.

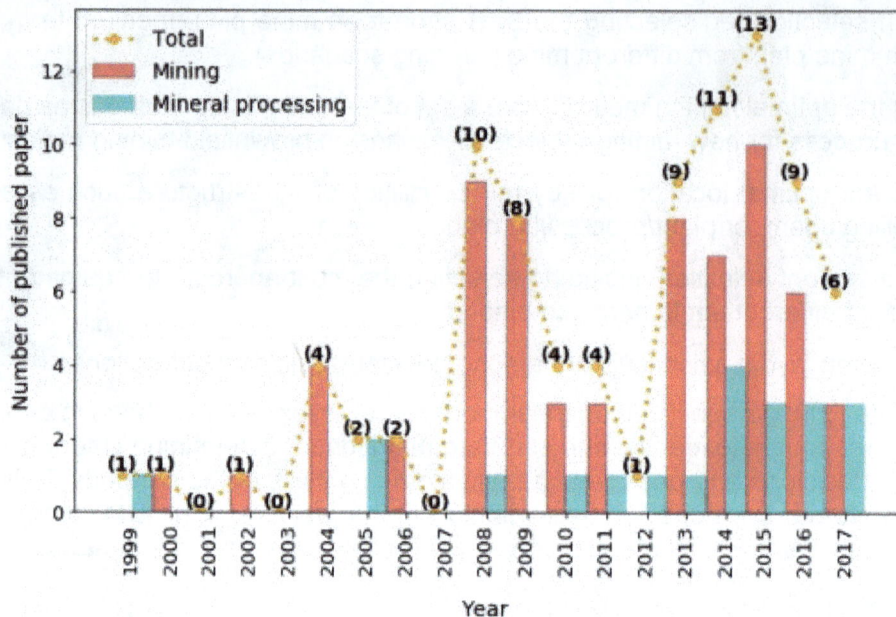

FIG 3 – Number of MCDM-type case study research papers published per annum related to mining and mineral processing (source: Sitorus, Cilliers and Brito-Parada, 2019).

One of the key initial steps in MCDM is to establish decision-making criteria and assign weights to the criteria to reflect their relative importance in the required decision (Musingwini, 2010; Triantaphyllou, 2000). According to Sitorus, Cilliers and Brito-Parada (2019), weights of criteria significantly impact the outcome of the decision hence the assignment of weights to criteria should be carefully executed. Given the importance of reliably determining weights of criteria, this paper therefore, presents work in two complementary areas. Firstly, the paper compares different methods for determining weights of criteria to highlight strengths and weaknesses of each method. Secondly, the paper presents a bibliometric analysis conducted on MCDM-type problems related to mine planning, to enable mapping of future research directions on criteria weight determination, especially with relevance to mine planning choice decisions. Since most mine planning related choice-decision case studies use MCDA methods, this paper focused on analysing MCDA methods and the associated criteria weight determination methods used in mine planning related choice-decision case studies.

MULTI-CRITERIA DECISION-MAKING CRITERIA WEIGHTING METHODS

Odu (2019) and Ayan, Abacıoğlu and Basilio (2023) classified different weighting methods into subjective weighting methods, objective weighting methods, and integrated weighting methods. Subjective weighting methods are based on a decision-maker's opinion. The decision-maker, usually an expert in the field concerned, is given a set of questions often in the form of a questionnaire and is required to provide answers that indicate the weighting or importance of the criteria. This approach, though quite practical, has some degree of subjectivity depending on the respondent's experience and personal preferences. There are challenges associated with using such a subjective approach to assign weights to criteria. Odu (2019) argued that it is difficult for one decision-maker to confidently assign weights to different criteria. However, when more decision-makers are included in the evaluation panel, it becomes more challenging to reach a consensus view on criteria weight assignment, due to differing degrees of subjectivity that each decision-maker has, since different decision-makers are likely to assign different weights to a specific criterion. This could make the process of assigning weights to criteria to be time consuming, hence, the main challenge of MCDA is on determining weights of criteria (Aldian and Taylor, 2005).

In objective weighting methods criteria weights are determined through collecting information on each criterion and this information is processed through mathematical models to ultimately assign weights to criteria (Odu, 2019). In these methods, the opinion or contribution of the decision-maker is not considered. According to Yin (2020), an objective approach to weighting of criteria is suitable when the number of criteria to be considered is relatively large. Objective methods have a limitation of ignoring the importance of human judgement in decision-making (Yin, 2020). In addition, Odu (2019) indicated that each objective approach model has some degree of inherent imperfection. This means that even the most intricate models will have a level of inaccuracy associated with them. Therefore, human interaction is still required to 'sanity check' the results before the decision-making process can be concluded in a more practical way (Odu, 2019).

When subjective and/or objective weighting methods are used in combination to overcome the weaknesses inherent in each of the methods, the combined approach or integrated approach is an integrated weighting method or hybrid weighting method. Some advantages of integrated weighting methods are firstly, that they consider both the decision-maker's opinion and mathematically generated data to assign weights to criteria, and secondly, more data can be easily processed timeously, when numerous criteria are considered (Odu, 2019). The choice of criteria weighting method to use depends on available information and resources, and problem-specific requirements. Table 1 shows a comparison of different methods used to assign weights to criteria in MCDM-type problems.

Not all MCDM criteria weighting methods shown in Table 1 have been applied to mine planning related choice-decision case studies due to the nature of available information on criteria and problem-specific needs of each case study, since mining involves the unique long-term exploitation of non-renewable mineral resources. However, it can be expected that more relevant methods will increasingly be adopted in solving MCDM-type mine planning related choice-decision case studies.

Some of the criteria weight determination methods listed in Table 1 have been used in mine-specific case studies related to mine planning. It can be noted that there is still limited work on MCDA criteria weighting methods and their application to mine planning choice decisions, hence the relevance of this study. Very few authors on MCDM-type mine planning research explicitly write about the criteria weighting methods used to assign criteria weights in their case studies. For example, Rakhmangulov, Burmistov and Osintsev (2024), and Farkas and Hrastov (2021) explicitly discuss criteria weighting methods utilised in their mine planning choice-decision related case studies. Rakhmangulov, Burmistov and Osintsev (2024) used FUCOM to choose criteria, assign weights to the criteria and rank them in order of importance, in the selection of open pit mine dump trucks. A total of 71 criteria were identified and they were aggregated into four groups of criteria, namely technical, technological, environmental, and economic categories. Farkas and Hrastov (2021) used AHP and PROMETHEE to select the optimal final contour design for Tambura quarry located in Croatia. AHP was used to assign weights to criteria and rank them in order of importance. A total of 22 criteria were identified and they were aggregated into five global criteria, namely project parameters, mineral deposit reserve considerations, economic indicators, environmental impacts, and property legal actions. Stevanovic et al (2018) used AHP to select a suitable mining system for the Drmno lignite open pit coalmine in Serbia. Six global criteria were identified, and two alternative mining systems were identified, namely coal exposure using a dragline and coal exposure using a hydraulic shovel. AHP was used to assign weights to criteria and rank alternatives using the identified criteria. The alternative of using coal exposure using a hydraulic shovel was selected as the preferred mining system. Ozdemir (2023) used a hybrid AHP-TOPSIS method to select an optimum mine planning alternative for a chrome mine in Turkey. A group of experts (including geologists, mine planning engineers, investors and operators) was used to identify criteria and alternatives. Thirteen criteria were identified and subsequently aggregated into four global criteria. AHP was used to assign weights to the identified criteria and rank them in order of importance. A total of nine alternatives were identified and TOPSIS was used to rank the identified alternatives. The alternative (A9) with the highest score, which coincidentally also had the highest NPV was ranked as the best alternative, and Ozdemir (2023) concluded that this finding was supported by previous studies on selecting optimum mine planning alternatives.

TABLE 1

A high-level comparison of some commonly used MCDM criteria weighting methods (adapted from: Ayan, Abacıoğlu and Basilio, 2023; Uzhga-Rebrov and Kuleshova, 2023; Odu, 2019).

MCDM criteria weighting method	Classification	Advantage(s)	Disadvantage(s)	Common decision-making application areas
Analytic Hierarchy Process (AHP)	Subjective	Is a trusted MCDM method due to its solid mathematical model. Easy to use, and yields results that are expected by decision-makers. Can be easily applied by using spreadsheets. Has a mechanism for evaluating the degree of inconsistency in criteria weighting, thus, allowing the criteria weighting process to be re-done until inconsistency is below a set threshold.	Not suitable as the number of criteria increases due to the pair-wise comparison process it uses. Less suitable in situations with large numbers of stakeholders. The procedure for processing diverse information obtained from different decision-makers can be challenging.	Engineering; computer science; business; management; energy; environmental science; social science; mathematics; medicine; earth and planetary science; decision science.
Best-Worst Method (BWM)	Subjective	Developed to overcome shortcomings of AHP method. Fewer pair-wise comparisons are required, resulting in fewer inconsistencies during comparison of criteria. More reliable and consistent results compared to AHP, since transitivity relations are not often violated.	Cannot be used in real-world scenarios in which two or more criteria are presented as either best or worst criteria.	Supply chain management; energy; transportation; manufacturing; education; investment; banking; performance evaluation; airline industry; healthcare; communication.
Criteria Importance Through Inter-criteria Correlation (CRITIC)	Objective	Measures the value of each criterion using correlation analysis to improve reliability of weight considerations. Allows for better conflict resolution in a decision-making problem since it considers existing criteria and tries to establish compromise between conflicting criteria. Less computation involved during execution of the method, compared to other objective methods.	Very large weights can be assigned to criteria in the CRITIC method, due to their high standard deviation values. These criteria end up dominating other criteria in the decision-making process.	Business economics; engineering; multi-disciplinary applications; environmental studies.
Criterion Impact Loss (CILOS)	Objective	Eliminates disadvantages of Entropy-based methods. Can be used together with IDOCRIW as an integrated or hybrid weighting method.	Does not consider expert opinion. It is sensitive to data distribution, which may skew the overall decision-making process.	Portfolio optimisation; Used often with IDOCRIW in: (Business economics, Environmental studies, engineering, multi-disciplinary applications).
Full Consistency Method (FUCOM)	Subjective	Requires fewer pair-wise comparisons to be made, hence guarantees better consistency in pair-wise comparisons. Uses a simple algorithm to prioritise criteria by decision-makers. Yields reliable results by obtaining optimal weighting factors with the possibility of validating them.	As a subjective method, it does not consider expert opinion. Not much literature has been published on FUCOM to endorse the validity of the method.	Business economics; environmental studies; engineering.

MCDM criteria weighting method	Classification	Advantage(s)	Disadvantage(s)	Common decision-making application areas
Integrated Determination of Objective Criteria Weights (IDOCRIW)	Objective	Method regarded as a non-biased method. Assigns reliable weights to criteria by combining strengths of CILOS and Entropy method.	Is difficult to apply, especially when handling a relatively large number of criteria. May require data preparation before using, to increase accuracy of the technique.	Business economics; engineering; multi-disciplinary applications; environmental studies.
Level Based Weight Assessment (LBWA)	Subjective	Preserves the structure of a problem, regardless of problem complexity, hence providing a better understanding of the problem. Despite being a subjective method, its simple mathematical architecture can eliminate inconsistent expert preference that is otherwise permitted by BWM and AHP methods. The elasticity coefficient in the method can be used for sensitivity analysis.	It is difficult to achieve high consistency of results when using this method. There is potential subjectivity when defining levels of criteria as global criteria or sub-criteria. Randomness may be introduced during the process of grouping criteria into levels.	Business economics; environmental studies; defence related decision-making; healthcare.
Method based on the Removal Effects of Criteria (MEREC)	Objective	A simple method that does not require complex, detailed computation and can be easily executed.	The method has potential sensitivity to data outliers, which might lead to inaccurate results. Although the method is objective, it still requires subjective input to improve an understanding of the problem.	Business economics; environmental studies; engineering.
Rank Ordered Centroid (ROC) Technique	Subjective	Requires less effort from the decision-maker, compared to other subjective methods. Converts ordinal data into relative weights through mathematical formulations.	High chances of bias associated with the method, since it is subjective.	Engineering; computer science; business management; environmental science; social science; energy; mathematics; medicine; earth and planetary science; decision science.
Rank Reciprocal (RR) Weighting Technique	Subjective	A simple method that is easy to use and implement. Can easily prioritise criteria in order of importance hence easier to determine criteria weights. Requires minimal data since it is subjective.	Can be sensitive to small changes in ranking, leading to inaccurate results. Does not express the absolute importance of each criterion, making results difficult to interpret.	Environmental science; exploration/mining; maintenance strategy of public buildings; tourism development, supplier selection in procurement.
Rank Summed (RS) Weighting Technique	Subjective	Requires less effort from the decision-maker, compared to other subjective methods. Converts ordinal data into relative weights through mathematical formulations.	Associated with high chances of bias since it is subjective.	Environmental science; exploration/mining; maintenance strategy of public buildings; tourism development, project prioritisation.

MCDM criteria weighting method	Classification	Advantage(s)	Disadvantage(s)	Common decision-making application areas
Shannon Entropy	Objective	Weighting of criteria can be assigned even when the decision-maker is partially involved or not involved at all. Accounts for uncertainty in decision-making using random variables in a probability distribution. Weights assigned to criteria are expressed in terms of discrete or continuous probability distributions.	The method's procedure makes it hard to explain or understand any solution obtained through entropy.	Business economics; engineering; multi-disciplinary applications; environmental studies.
Simple Aggregation of Preferences Expressed by Ordinal Vectors-Multi Decision-makers (SAPEVO-M)	Objective	The method increases consistency in decision-making by correcting negative and null criteria weights when process-standardising matrices are integrated.	The method can be difficult to execute. It is resource-intensive and demands resources such as expensive software for its implementation. The method has only been used by a niche community and specialised software, which could result in challenges associated with any form of assistance for any other users.	Defence related decision-making; business economics; engineering; environmental studies.
Simple Multi-Attribute Rating Technique (SMART)	Subjective	Requires fewer resources during implementation because it relies on the decision-maker's expertise. Complexity of the method is generally regarded as being low, hence can be applied without the use of computers (when dealing with small data sets). Enables both relative and absolute weight assignment to be done. It is reliable when handling large data sets.	There is a high chance of bias since the method is a subjective method. There is currently no special software available to support its use and the processing of large data sets.	Engineering; computer science; business; management; environmental science; social science; energy; mathematics; medicine; earth and planetary science; decision science.
Simple Pair-wise Comparison (SPC)	Subjective	It is simple and easy to use. The method breaks down complex problems into simple-pairwise comparisons, making it easy for decision-makers to understand the problem. Can detect small differences between criteria importance.	Requires a relatively large number of comparisons to be made, and this can be time consuming. Potential inconsistency can occur due to the large number of criteria pair-wise comparisons.	Environmental science; exploration/mining; maintenance strategy of public buildings; tourism development.
Simultaneous Evaluation of Criteria and Alternatives (SECA)	Subjective	Consistency in decision-making is achieved by simultaneously evaluating criteria and alternatives. The method is time-efficient, since alternatives and criteria are simultaneously evaluated. Decision-making is holistic since both criteria and alternatives are evaluated simultaneously.	It is difficult to handle complex problems with many criteria and alternatives. There is potential bias associated with the decision-maker dealing with criteria and alternatives. Simultaneous evaluation of criteria and alternatives can overwhelm decision-makers, leading to confusion.	Business economics; environmental studies; defence related decision-making; healthcare.
Stepwise Weight Assessment Ratio Analysis (SWARA)	Subjective	By directly reflecting preferences and experiences of experts, the method enables decision-makers to prioritise criteria without requiring further evaluation/ranking.	The method has inherent bias due to it being a subjective method.	Environmental science; exploration/mining; maintenance strategy of public buildings; tourism development.

MCDM criteria weighting method	Classification	Advantage(s)	Disadvantage(s)	Common decision-making application areas
Swing Weighting Technique (SWING)	Subjective	The method is highly structured hence comprehensible to decision-makers and aids them in understanding the problem and therefore apply weighting methods effectively. A simple, easy to use and transparent method. The method allows decision-makers to easily rank criteria in order of priority.	It is highly dependent on expert judgement and has inherent bias because it is a subjective method. Can be time-consuming when handling a relatively large number of criteria. Can potentially oversimplify a problem reducing the practicality of the solution.	Forecasting; assessing projects; engineering; selecting vendors; business management.

BIBLIOMETRIC ANALYSIS OF MCDA TECHNIQUES IN MINE PLANNING

Bibliometric analysis is conducted to understand past and current research trends to assist in mapping possible future research directions. For example, in a mining context, Mutandwa and Musingwini (2024) used bibliometric analysis to obtain insights on past and current research trends in underground stope optimisation and infer possible future research trends on the subject matter. This is why this paper includes bibliometric analysis work conducted on MCDM criteria weighting research work with a particular focus on mine planning choice decisions, so that future research directions on MCDM criteria weighting in mine planning can be inferred. To conduct the bibliometric analysis, this study utilised R-Biblioshiny™ and VOSviewer software which are freely available from the public domain and are easy to use. The data for analysis was downloaded from Scopus and Web of Science (WOS) as these are the most used databases for sourcing literature for academic research (Mutandwa and Musingwini, 2024). This study also followed the Preferred Reporting Items for Systematic Reviews and Meta-Analyses (PRISMA) guidelines developed by Page et al (2021). The PRISMA guidelines are useful for identifying databases and screening of irrelevant meta-data to compile a more relevant consolidated data set for analysis. In this study, two bibliometric analyses were conducted. The first analysis focused on research papers related to the use of MCDA in mining in general, because as mentioned earlier in the paper, most mine planning related choice-decision case studies have mostly used MCDA methods as a solution approach. The second analysis focused on examining a subset of papers specifically addressing mine planning related choice-decision case studies, so that emerging research trends in MCDA applications to mine planning choice decisions could be mapped.

Application of the PRISMA methodology

The data for the bibliometric analysis was identified, screened and included for analysis as depicted in Figure 4 which shows a generic PRISMA analysis methodology. The keywords were identified from several research papers during a systematic literature review. This informed the search strategy guiding the inclusion and exclusion criteria when screening papers using titles, abstracts, and keywords as was done by Harichandan et al (2022). For consistency, only research papers written in English language were included in the final analysis. Review papers were excluded as these do not address specific case studies. The search for papers had a cut-off date of 2024 since 2025 is still an incomplete year.

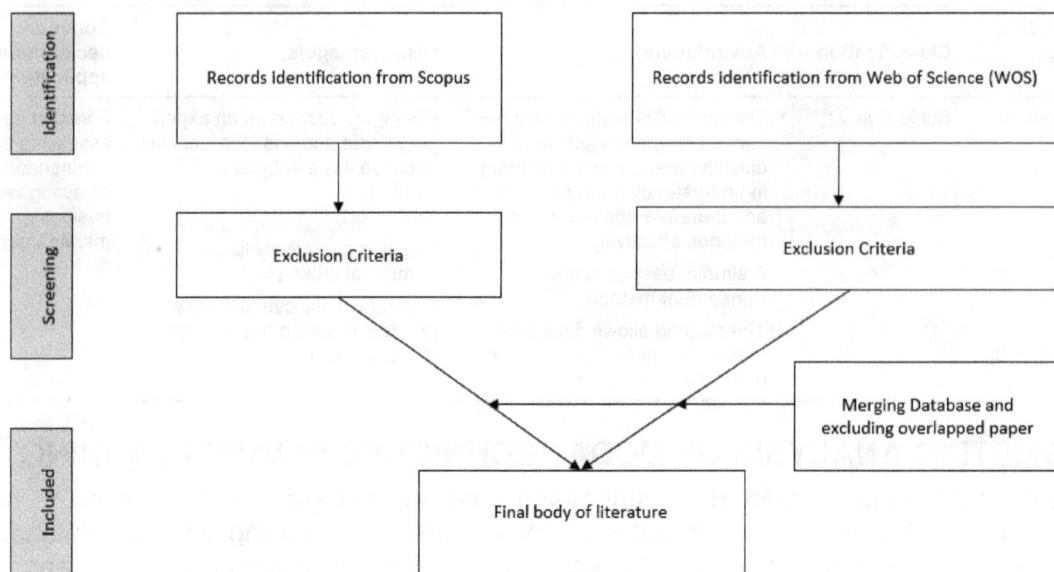

FIG 4 – A generic three-step PRISMA analysis methodology followed in this study (source: Qwabe, Musingwini and Mutandwa, 2024).

Data collection and processing

To obtain relevant data, the following search string was used in Scopus: ("Multi-Criteria Decision Analysis" OR "MCDA" OR "Multi-Attribute Decision Making" OR "MADM" OR "Analytic Hierarchy Process" OR "AHP" OR "Full Consistency Method" OR "FUCOM" OR "Best-Worst Method" OR "BWM" OR "Criteria Importance Through Inter-criteria Correlation" OR "CRITIC" OR "Criterion Impact Loss" OR "CILOS" OR "Integrated Determination of Objective Criteria Weights" OR "IDOCRIW" OR "Level Based Weight Assessment" OR "LBWA" OR "Method based on the Removal Effects of Criteria" OR "MEREC" OR "Rank Ordered Centroid Technique" OR "ROC Technique" OR "Rank Reciprocal Weighting Technique" OR "RR Weighting Technique" OR "Rank Summed Weighting Technique" OR "RS weighting technique" OR "Simple Aggregation of Preferences Expressed by Ordinal Vectors-Multi Decision Makers" OR "SAPEVO-M" OR "Simple Multi-Attribute Rating Technique" OR "SMART" OR "Simple Pair-wise Comparison" OR "SPC" OR "Simultaneous Evaluation of Criteria and Alternatives" OR "SETA" OR "Stepwise Weight Assessment Ratio Analysis" OR "SWARA" OR "Swing Weighting Technique" OR "SWING" OR "Shannon Entropy") AND (" mine planning") AND NOT ("data mining" OR "data mine" OR "text mining" OR "process mining"). The only difference with the search string used in WOS is the replacement of 'AND NOT' with 'NOT'. This search yielded 23 papers in Scopus and 21 papers in WOS. The results suggest that the amount of research on MCDA criteria weighting methods in mine planning choice decisions is still limited.

It was necessary to conduct further search and bibliometric analysis on research papers which used MCDA techniques in mining without limiting to mine planning choice decisions. The following search string was added to the search process: ('MSAHP: An approach to mining method selection') since mining method selection is a component of mine planning.

Again, an almost identical search string was used to search papers indexed in WOS with the only difference being the replacement of 'AND NOT' with 'NOT'. This search string yielded 213 papers in Scopus which reduced to 166 papers after screening out irrelevant and review papers. The WOS search yielded 101 papers, which reduced to 92 papers after screening. These papers are related to the use of MCDA techniques in mining but are not all strictly mine planning focused since the objective was to cover mine planning choice-decision related work or work peripheral to mine planning choice decision-making. The data from Scopus was downloaded as Bibtex files and data from WOS was downloaded as text files to allow the merging and removing duplicates in R-Studio. After merging, 79 duplicate papers were detected and removed resulting in a consolidated file with 179 indexed papers. A file in MS Excel format was then subjected to bibliometric analysis in VOSviewer and R-Biblioshiny.

Network analysis

Figure 5 shows affiliations of authors who published papers in MCDA in mining. Universities in China appear to be the most prominent in the application of MCDA techniques in mining.

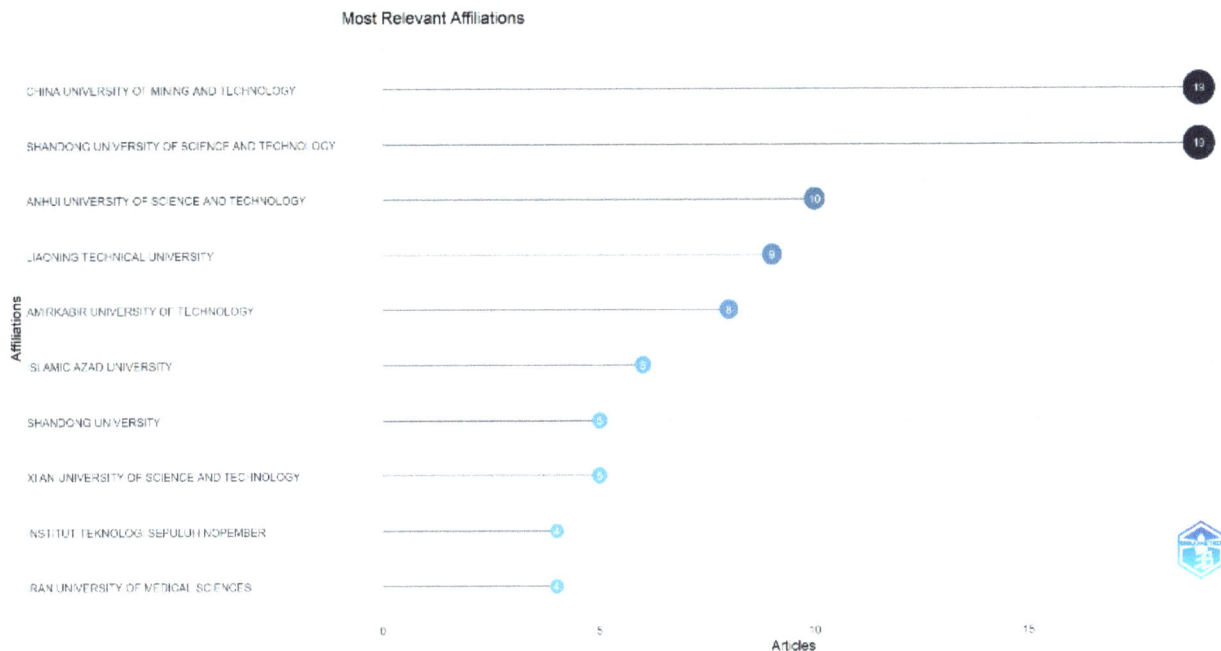

FIG 5 – Most relevant affiliation in using MCDA in mining.

Figure 5 shows that China University of Mining and Technology and the Shandong University of Science and Technology lead on MCDA application in mining, each with 25 relevant papers. Iran ranks second, with significant contributions from the Amirkabir University of Technology, Islamic Azad University, and Iran University of Medical Sciences. Publications from Chinese universities predominantly focus on risk assessment and management in coalmines, as evidenced by works from Fang *et al* (2024), Wang, Sui and Ji (2024) and Hongkai *et al* (2023), among others. Figure 6 shows the most relevant affiliations in using MCDA in mine planning.

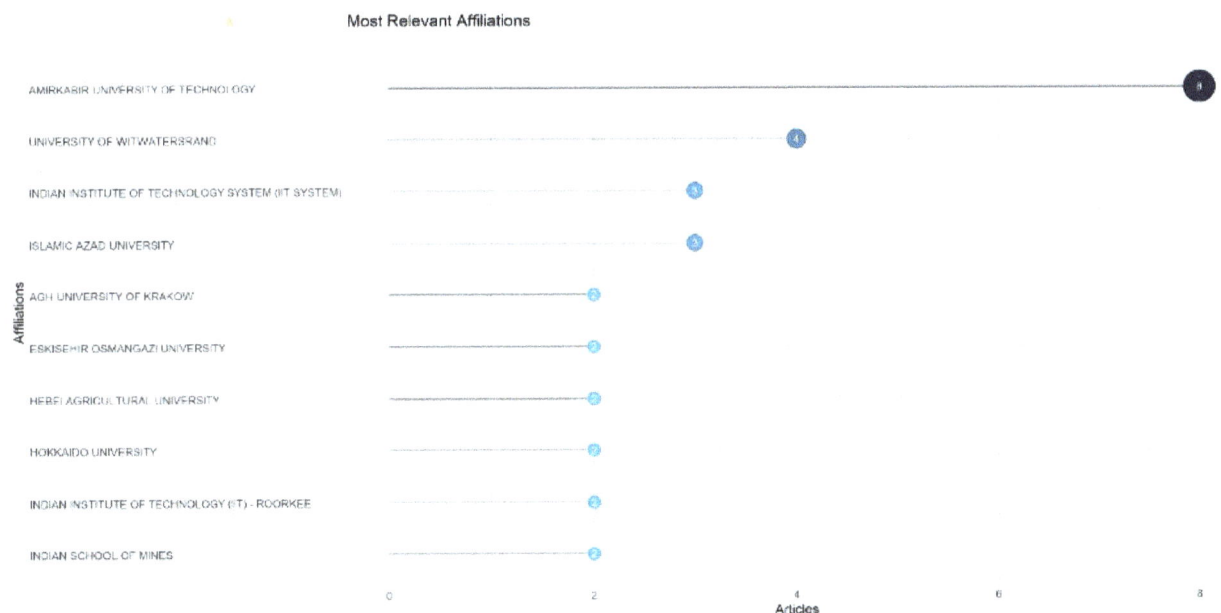

FIG 6 – Most relevant affiliations in using MCDA in mine planning.

The most relevant affiliation of the indexed publications in the mining planning field was Amirkabir University of Technology followed by the University of the Witwatersrand.

Keyword analysis

To get insights into what the research papers focused on, a keyword analysis was done using VOSviewer. Figure 7 shows the network visualisation of the keyword analysis on the use of MCDA techniques in the broader mining sector not specific to mine planning yielded four clusters. The clusters are formed based on their co-occurrences or citations. Clusters represent groups of related items that share the same themes or topics. For example, the green cluster, which has risk assessment as the most commonly appearing keyword, is linked to coal mining. The green cluster shows that the Analytic Hierarchy Process also frequently appeared together with entropy weight, and comprehensive evaluation and weighting methods. The red cluster indicates that the Analytic Hierarchy Process and hierarchy systems frequently appeared together with phrases like decision theory, sensitivity analysis, sustainable development and multi-criteria decision analysis.

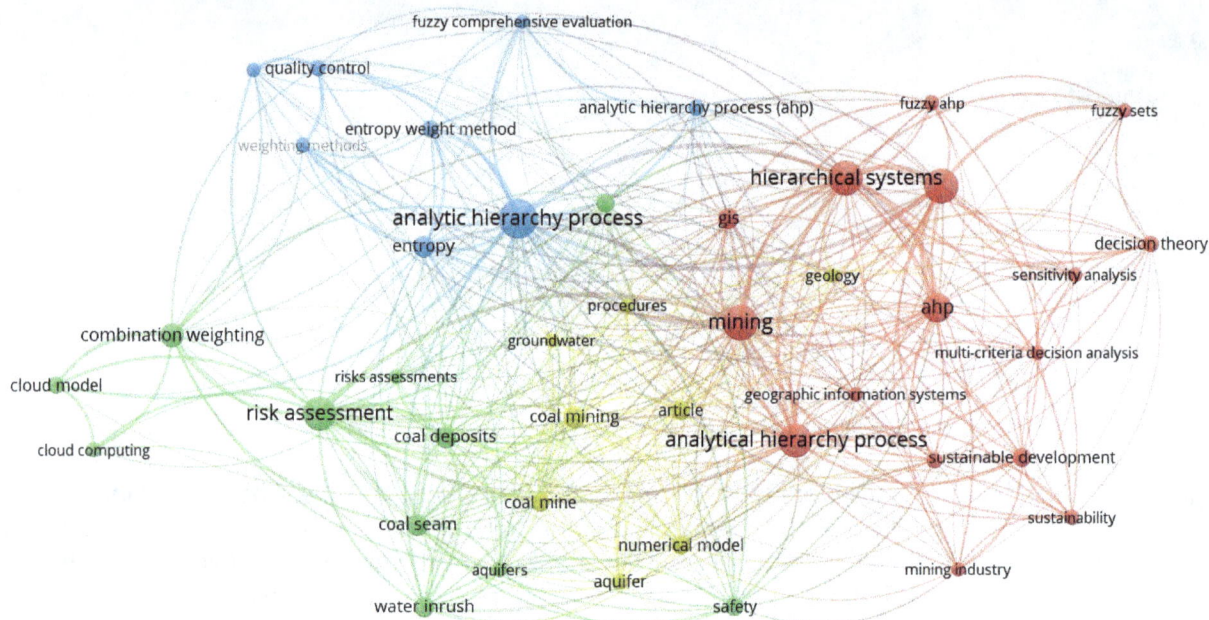

FIG 7 – Network visualisation of keywords by occurrences.

Most frequently used keywords

An analysis of word clouds generated in R-Biblioshiny from 179 research papers on mining that utilised MCDA techniques, compared to those specific to mine planning, revealed some noticeable trends. Figure 8 shows that the most frequently occurring terms in the general mining research papers include 'analytic hierarchy process', 'hierarchical systems', 'decision-making', 'coalmines', and 'risk assessment'. In contrast, papers that focused on the application of MCDA techniques in mine planning emphasise 'mining method selection', 'equipment selection', 'risk assessment', and 'plant location', using the Analytic Hierarchy Process as depicted in Figure 9.

FIG 8 – Network visualisation of keywords by occurrences.

FIG 9 – Word cloud from mining planning specific research papers.

A comparison of the network visualisation (Figure 8) and the two-word clouds (Figure 9) suggest that researchers on mine planning MCDM-type choice-decision research are not explicitly focusing on criteria weighting methods, despite the importance of criteria weighting in MCDM contexts.

Key thematic areas

When using author keywords, a conceptual map with two dimensions was created in R-Biblioshiny. Figure 10 shows the map. The first dimension, which is density, reflects the level of development of themes based on internal connections among keywords. The second dimension, which is centrality, represents the importance of the themes based on external connections among the keywords. The map is divided into four quadrants namely, (i) motor themes (high density and centrality); (ii) basic themes (low density and high centrality); (iii) niche themes (high density and low centrality); and (iv) emerging/declining themes (low density and centrality).

Figure 10 reveals that motor themes on the use of MCDA techniques in mining are on studies related to risk assessment in coal mining, entropy weighting method, improved CRITIC method, grey clustering evaluation and use of the Analytic Hierarchy Process, among others. Basic themes comprised of the following keywords: risk assessment models, risk assess indicator system, evaluation index system and Analytic Hierarchy Process. According to Bretas and Alon (2021), niche themes are from papers on specialised themes, which in Figure 10, are acid mine drainage, coupled weighting method, comprehensive criteria weights, and weighted average method, among others. Lastly, from Figure 10, the quadrant on emerging or declining themes indicates that Delphi analytic hierarchy, fuzzy Delphi analytic, and hierarchy analysis methods are either emerging or declining themes.

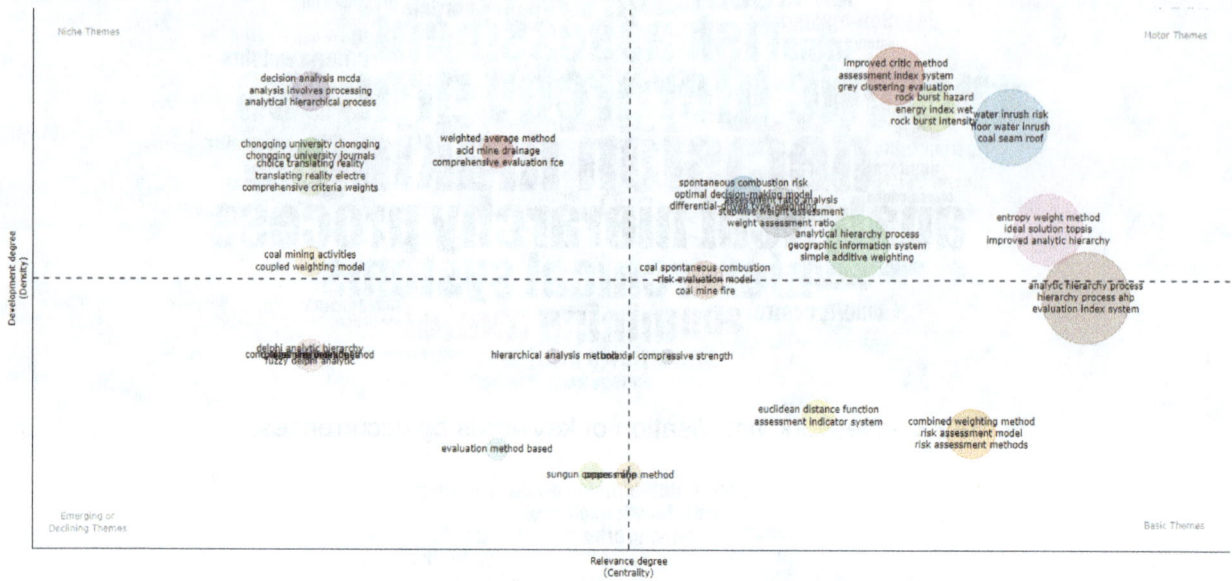

FIG 10 – Thematic map for research papers on the use MCDA techniques in mining.

Figure 11 shows results from analysing author keywords to generate a thematic map for research papers in the field of mine planning. The motor themes consist of papers on the Analytic Hierarchy Process and equipment selection. Mining method selection and optimisation are not located in a specific quadrant indicating that they straddle all four themes.

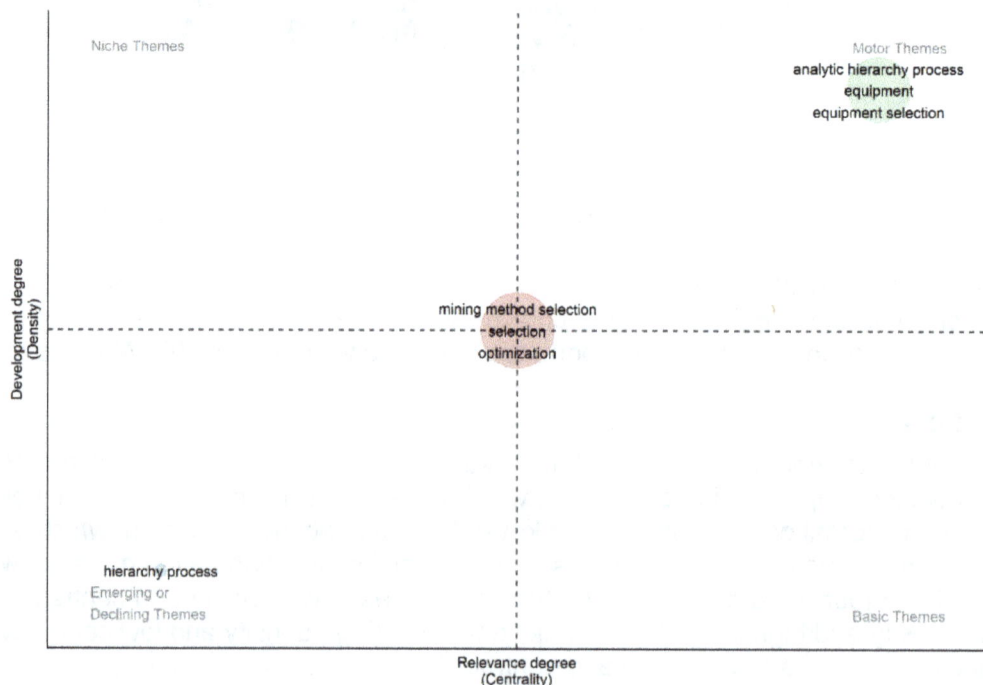

FIG 11 – Thematic map for research papers on the use MCDA techniques in mine planning.

DISCUSSION OF FINDINGS AND IMPLICATIONS FOR FUTURE RESEARCH DIRECTIONS

The results from the bibliometric analysis suggest that the amount of research on MCDA criteria weighting methods in mine planning choice decisions is still limited. This is supported by the bibliometric analysis results (Figures 7 and 10) which showed that the methods 'entropy weighting method, improved CRITIC method, grey clustering evaluation, coupled weighting method, comprehensive evaluation and weighting method, and weighted average method' featured as explicitly mentioned criteria weighting methods applied in mine planning related choice-decision case studies. Therefore, it is important that future mine planning related choice-decision case studies should be explicit on the criteria weighting method used and why it was selected for the specific case in hand.

A comparison of methods for criteria weight determination in MCDM-type choice decision problems indicates that for mine planning MCDA-type problems, there is a gradual shift from using subjective methods to objective methods, when assigning weights to criteria. This is noticeable in the number of objective methods developed between 2019 and 2023 (Ayan, Abacıoğlu and Basilio, 2023; Uzhga-Rebrov and Kuleshova, 2023; Odu, 2019). This can be attributed to advantages that objective weighting methods have over subjective weighting methods, like not being affected by bias (due to not being subjective) and not requiring expert experience, therefore eliminating any potential subjective interference (Ponhan and Sureeyatanapas, 2022). However, objective weighting methods have their own flaws. For example, they are resource intensive and require large data samples to function. In addition, they generally require software for support and are quite complex to understand and apply (Ponhan and Sureeyatanapas, 2022). However, objective weighting methods can be modified to account for uncertainty, for instance, in grey clustering evaluation. In addition, deep learning (a type of machine learning) is being applied to objective methods to increase their reliability. Given the shortcomings of either subjective or objective weighting methods, combined with the challenge of inherent uncertainty in mine planning choice decisions, future research should focus on using hybrid methods and machine learning methods, to account for uncertainty and overcome shortcomings of individual methods. Types of machine learning methods include supervised, unsupervised, semi supervised and reinforcement learning (Sarker, 2021). Under supervised learning, there is classification and regression analysis (Sarker, 2021). Classification is used in predictive modelling, where a model is used to classify data into distinct classes of data. Regression analysis involves using a model on data, to yield a continuous result (Sarker, 2021). Since regression yields a continuous result, it can be considered as one of the methods to account for uncertainty. Regression analysis is used in financial forecasting, cost estimation, trend analysis, among other things. Under unsupervised machine learning, one type of interest is cluster analysis, where closely related data is grouped or clustered to establish an interesting trend in data. These methods under supervised and unsupervised classifications can be used in predictive analysis and intelligent decision-making.

Given that choice decisions require the integration of diverse stakeholder input into the criteria weighting process, it is important to ensure that the selected criteria weights reflect the diverse values and priorities associated with the decision. Also, hybrid weighting methods, whereby both subjective weighting methods (based on expert opinions) and objective weighting methods (based on information from data in mathematical form), become handy in this regard. According to Odu (2019), if criteria weights are determined through objective weighting process, humans should always 'sanity check' the final choice decisions, hence a level of subjectivity is still present. This requires future research to be undertaken on ways to minimise subjectivity for more reliable choice decisions to be made.

Lastly, comparative studies that evaluate the performance of different criteria weighting methods on the same mine planning scenarios can provide valuable insights into the strengths, limitations and applicability of each criteria weighting method. The comparative studies on the mine planning scenarios should use real mine planning data as this will assist in creating deeper methodological or empirical comparisons. Such comparative studies can offer a reflection of work that has been done in a specific field (for instance, assigning of weights in criteria), hence providing further guidance for further research on improving the performance of MCDM methods.

CONCLUSIONS

The assignment of weights to criteria in MCDM and/or MCDA processes is a critical step in mine planning choice decisions because if it is done incorrectly, subsequent steps will yield incorrect decisions. However, existing mine planning choice-decision related research has focused more on the different MCDM and/or MCDA methods with limited attention being paid to criteria weighting methods. Hence, this paper investigated, using a bibliometric analysis approach, research that has been done in mine planning choice decisions but focusing more on criteria weighting methods. The findings indicate that six criteria weighting methods namely, the entropy weighting method, improved CRITIC method, grey clustering evaluation, coupled weighting method, comprehensive evaluation and weighting method, and weighted average method seem to be the preferred methods in MCDM and/or MCDA mine planning choice-decision related case studies. Given that uncertainty is inherent to mine planning processes, this study proposes that future research directions should include the use of hybrid methods and machine learning methods, to account for uncertainty that is inherent to multi-criteria decision-making in mine planning, so that criteria weights can be determined more reliably.

ACKNOWLEDGEMENTS

This paper presented part of the work undertaken by the first author for a PhD study at the University of the Witwatersrand (Wits). The authors would like to acknowledge Wits for supporting them to participate in APCOM 2025 conference.

REFERENCES

Aldian, A and Taylor, M A P, 2005. A consistent method to determine flexible criteria weights for multicriteria transport project evaluation in developing countries, *Journal of the Eastern Asia Society for Transportation Studies*, 6:3948–3963.

Ayan, B, Abacıoğlu, S and Basilio, M P, 2023. A Comprehensive Review of the Novel Weighting Methods for Multi-Criteria Decision-Making, *Information*, 14(285):1–28. https://doi.org/10.3390/info14050285

Belton, V and Stewart, T J, 2002. *Multiple Criteria Decision Analysis: An Integrated Approach*, first edn, 372 p (Dordrecht: Springer Science and Business).

Bretas, V P G and Alon, I, 2021. Franchising research on emerging markets: Bibliometric and content analyses, *Journal of Business Research*, 133:51–65. https://doi.org/10.1016/j.jbusres.2021.04.067

Fang, X, Zhang, K, Hao, M and Wang, Y, 2024. Development and Selection of Fuzzy Variable Weight Vector and Its Application on Prediction for Coal and Gas Outburst of Coal Mining Panels, *International Journal of Fuzzy Systems*, 26:225–238. https://doi.org/10.1007/s40815-023-01590-2

Farkas, B and Hrastov, A, 2021. Multi-criteria analysis for the selection of the optimal mining design solution - a case study on Quarry Tembura, *Energies*, 4(11)3200:1–18. https://www.mdpi.com/1996-1073/14/11/3200

Harichandan, S, Kar, S K, Bansal, R, Mishra, S K, Balathanigaimani, M S and Dash, M, 2022. Energy transition research: A bibliometric mapping of current findings and direction for future research, *Cleaner Production Letters*, 3:100026. https://doi.org/10.1016/j.clpl.2022.100026

Hongkai, S H I, Yun, Q I, Guoen, Z, Xiaoyu, J, Xiang, H E and Yuanhang, S U N, 2023. Risk assessment of coal and gas outburst based on combination weighting and grey clustering, *China Safety Science Journal*, 33(52), https://doi.org/10.16265/j.cnki.issn1003-3033.2023.S1.0593

Jayant, A and Sharma, J, 2018. A comprehensive literature review of MCDM techniques ELECTRE, PROMETHEE, VIKOR and TOPSIS applications in business competitive environment, *International Journal of Current Research*, 10(2):65461–65477. Available from: <https://www.journalcra.com/sites/default/files/issue-pdf/28831.pdf>

Mahase, M J, 2017. Comparison of multi-criteria decision analysis methods in mine planning and related case studies, MSc Research Report (unpublished), University of the Witwatersrand. Available from: <https://hdl.handle.net/10539/25973> [Accessed: 30 November 2020].

Mahase, M J, Musingwini, C and Nhleko, A S, 2016. A survey of applications of multicriteria decision analysis methods in mine planning and related case studies, *The Journal of Southern African Institute of Mining and Metallurgy*, 116(11):1051–1056. Available from: <http://www.scielo.org.za/pdf/jsaimm/v116n11/10.pdf>

Mahase, M J, Musingwini, C and Nhleko, A S, 2018. Using mine planning case studies for empirical estimation of the maximum number of criteria and alternatives for multi-criteria decision making, in *Proceedings of Society of Mining Professors 6th Regional Conference 2018*, pp 293–302.

Mirjat, N H, Uqaili, M A, Harijan, K, Mustafa, M W, Rahman, Md M and Khan, M W A, 2018. Multi-Criteria Analysis of Electricity Generation Scenarios for Sustainable Energy Planning in Pakistan, *Energies*, 11(4):757–790. Available from: <https://www.mdpi.com/1996-1073/11/4/757> [Accessed: 19 May 2022].

Musingwini, C, 2010. Techno-economic optimization of level and raise spacing in Bushveld Complex platinum reef conventional breast mining, *The Journal of The Southern African Institute of Mining and Metallurgy*, 110(8):425–436. Available from: <http://www.scielo.org.za/pdf/jsaimm/v110n8/02.pdf> [Accessed: 10 January 2021].

Mutandwa, B and Musingwini, C, 2024. Insights and future research directions from a bibliometric mapping of studies in stope layout optimisation, *International Journal of Mining, Reclamation and Environment*, 38:577–595. https://doi.org/10.1080/17480930.2024.2325758

Odu, G O, 2019. Weighting methods for multi-criteria decision-making technique, *Journal of Applied Sciences and Environmental Management,* 23(8):1449–1457. Available from: <https://www.ajol.info/index.php/jasem/article/view/189641/178866> [Accessed: 14 February 2021].

Ozdemir, A C, 2023. Use of integrated AHP-TOPSIS method in selection of optimum mine planning for open pit mines, *Archive of Mining Sciences,* 68(1):35–53. https://doi.org/10.24425/ams.2023.144316

Page, M J, McKenzie, J E, Bossuyt, P M, Boutron, I, Hoffmann, T C, Mulrow, C D, Shamseer, L, Tetzlaff, J M, Akl, E A, Brennan, S E, Chou, R, Glanville, J, Grimshaw, J M, Hróbjartsson, A, Lalu, M M, Li, T, Loder, E W, Mayo-Wilson, E, McDonald, S, McGuinness, L A, Stewart, L A, Thomas, J, Tricco, A C, Welch, V A, Whiting, P and Moher, D, 2021. The PRISMA 2020 statement: an updated guideline for reporting systematic reviews, *BMJ 2021,* 372(71):1–9. http://dx.doi.org/10.1136/bmj.n71

Ponhan, K and Sureeyatanapas, P, 2022. A comparison between subjective and objective weighting approaches for multi-criteria decision making: A case of industrial location selection, *Engineering and Applied Science,* 49(6):763–771. Available from: <https://ph01.tci-thaijo.org/index.php/easr/article/view/250477> [Accessed: 02 February 2025].

Qwabe, S, Musingwini, C and Mutandwa, B, 2024. Research trends in mine planning and optimisation techniques identified from a bibliometric mapping and qualitative review of research from selected university-based research entities, in *Proceedings: The Society of Mining Professors 34th Annual General Meeting and Conference,* pp 325–340.

Rakhmangulov, A, Burmistov, K and Osintsev, N, 2024. Multi-criteria system's design methodology for selecting open pit dump trucks, *Sustainability 2024,* 16(823):1–34. https://doi.org/10.3390/info14050285

Sarker, I H, 2021. Machine Learning: Algorithms, Real World Applications and Research Directions, *Springer Nature Computer Science,* 2(160):1–21. Available from: <https://link.springer.com/article/10.1007/S42979-021-00592-X> [Accessed: 02 February 2025].

Sitorus, F, Cilliers, J J and Brito-Parada, P R, 2019. Multi-criteria decision making for the choice problem in mining and mineral processing: Applications and trends, *Expert Systems with Applications,* 121:393–417. https://doi.org/10.1016/j.eswa.2018.12.001

Stevanovic, D, Lekic, M, Krzanovic, D and Ristovic, I, 2018. Application of MCDA in selection of different mining methods and solutions, *Advances in Science and Technology Research Journal,* 12(1):171–180. Available from: <https://yadda.icm.edu.pl/baztech/element/bwmeta1.element.baztech-31959238-7496-44f9-ab86-8f52d1d0e82f> [Accessed: 04 March 2025].

Triantaphyllou, E, 2000. Multi-criteria decision-making methods, in *Multi-Criteria Decision-Making Methods: A Comparative Study,* pp 5–21 (Springer Science +business Media). Available from: <https://link.springer.com/chapter/10.1007/978-1-4757-3157-6_2> [Accessed: 02 November 2020].

Uzhga-Rebrov, O and Kuleshova, G, 2023. A Review and Comparative Analysis of Methods for Determining Criteria Weights in MCDM Tasks, *Information Technology and Management Science,* 26:35–40. https://doi.org/10.7250/itms-2023-0005

Wang, D, Sui, W and Ji, Z, 2024. Fault complexity degree in a coal mine and the implications for risk assessment of floor water inrush, *Geomatics, Natural Hazards and Risk,* 15:2293464. https://doi.org/10.1080/19475705.2023.2293464

Yin, C, 2020. A brief summary of objective weighting methods in MCDM, Linkedin. Available from: <https://www.linkedin.com/pulse/brief-summary-objective-weighting-methods-mcdm-chonghua-yin> [Accessed: 01 March 2023].

Zavadskas, E K, Turskis, Z and Kildiene, S, 2014. State of art surveys of overviews on MCDM/MADM methods, *Technological and Economic Development of Economy,* 20(1):165–179. Available from: <https://www.tandfonline.com/doi/abs/10.3846/20294913.2014.892037> [Accessed: 12 January 2021].

Integrating operational design and data analytics to maximise shovel performance in an open cut coalmine

L C Maia[1], R L Peroni[2], J L V Mariz[3], B T Kuckartz[4] and A F Machado[5]

1. Mining Engineer, Anglo American, Moura Qld 4718. Email: lucianacmaia@hotmail.com
2. Professor, Federal University of Rio Grande do Sul, Porto Alegre, Rio Grande do Sul, Brazil. Email: 00015547@ufrgs.br
3. PhD student, Federal University of Rio Grande do Sul, Porto Alegre, Rio Grande do Sul, Brazil.
4. PhD student, Federal University of Rio Grande do Sul, Porto Alegre, Rio Grande do Sul, Brazil. Email: brukuck@hotmail.com
5. Mining Engineer, Anglo American, Moura Qld 4718. Email: arthurfelicem@hotmail.com

ABSTRACT

Real-time performance solutions are advanced systems that monitor and analyse the actual machine performance, providing instant feedback to operators and planners. While these smart solutions excel at offering real-time feedback, their effectiveness is limited if they operate under a poorly designed plan. The literature agrees that shovel efficiency is influenced by characteristics of the material being loaded, such as fragmentation and bank height. Proper geometric configurations are important to ensure optimal equipment performance. In an open cut coalmine, it may be required multiple shovel cuts to reach the coal seam. The profile of each cut is primarily determined by the thickness of the overburden, the characteristics of the coal seam itself and the shovel size. Using historical data from a shovel monitoring system, this study seeks to correlate bench height with shovel performance metrics in an open cut coalmine in the Bowen Basin. By applying the proposed methodology, the study demonstrates how the height impacts shovel productivity. To conclude, the results of the study indicate that operational performance has a more substantial impact on shovel productivity compared to bench height.

INTRODUCTION

Strip mining is largely employed in horizontal tabular deposits near the surface, such as coal and bauxite deposits. Due to their unique morphology, these deposits are represented as strata, layers, or panels rather than traditional block models (Bassani *et al*, 2025).

One approach for mine design, especially in coalmines, emphasises the integral relationship between material movement and the equipment used. In this context, equipment selection influences the sequence and dimensions of the material blocks to be moved, while the material characteristics impact equipment choice (Kennedy, 1990).

Equipment fleet dimensioning and selection is a base step for operational efficiency improvement and operational cost reduction. There are many haulage alternatives regarding types of equipment to be selected, such as draglines, mobile crushing, haulage conveyors and so on, however the truck and shovel system is still the most used in open pit mines (Lizotte and Bonates, 1986).

In multi-seam mines, it is common to use an integrated mining system where trucks and shovels are combined with draglines for overburden removal. Typically, the truck and shovel system are employed to remove the upper and thinner overburdens within a deposit, while draglines are used to remove the much deeper overburdens that are beyond the working range of shovel operations (Aiken and Gunnett, 1992).

Surface coal mining typically requires the removal of large amounts of overburden for every ton of coal produced. Therefore, loading efficiency plays a pivotal role in increasing production and reducing costs. Electric rope shovels are widely utilised as loading equipment in large open pit coalmines. A typical productive cycle for these shovels includes digging, swinging, dumping, and returning. Profio (1984) highlighted that dig time is the most critical component of cycle time, influenced by several factors such as operator proficiency, the characteristics of the material being loaded (eg fragmentation, bank height, material density, and friction), and the type and condition of the machine.

Since Profio's study, numerous investigations have been conducted on shovel performance. Hendricks, Scoble and Peck (1989) and Khorzoughi and Hall (2016), in the pursuit of correlating the rope shovel performance to the diggability, found that the shovel performance is highly influenced by the operator's skill.

Similarly, Bettens, Grist and Scaree (2022), Yaghini, Hall and Apel (2020) and Patnayak and Tannant (2005) identified operator skill as a significant factor for the shovel productivity. In their study on the Liebherr R996 mining shovel, Bettens, Grist and Scaree (2022) aimed to determine the relative contributions of rock fragmentation, pit geometry, and human factors to variations in excavator productivity. They categorised bench heights into three groups (10 m, 15 m, and 20 m) for each production block, but they concluded that operator skill was the primary driver of productivity. Patnayak and Tannant (2005) also proposed monitoring and assessing the impact of bench height on shovel performance. However, their findings indicated discrepancies between actual digging heights and bench face heights, preventing conclusive results.

Onederra *et al* (2004) observed that flatter regions of the muck pile, which corresponded to smaller dig faces, resulted in reduced production rates. Their manual monitoring and analysis of several blasts revealed significant variability in all components of the loading cycle, with the greatest variation in dig times attributed to muck pile characteristics and operator tactics.

Although past studies have highlighted the importance of operator skill, from a mine planning perspective, this leaves limited room for proactive optimisation of shovel operation.

In an open cut coalmine, it may be required multiple shovel cuts to reach the coal seam. The profile of each cut is primarily determined by the thickness of the overburden, the characteristics of the coal seam itself, and the shovel size. To expose a coal seam across an entire strip, various design scenarios are possible. However, only some specific configurations provide the most cost-effective solution by maximising shovel performance while meeting operational constraints.

Despite its importance, geometric design is often focused on isolated shovel passes without considering their impact on overall strip performance. This disjointed approach can lead to inefficiencies.

Today shovel monitoring systems with high-precision GPS technology provide real-time insights that help operators identify inefficiencies, adjust techniques, and keep performance on track during each cycle. By constantly comparing shovel progress against set targets, these systems ensure better adherence to the mine plan. Additionally, they also offer a chance to assess the performance indicators over a large number of load cycles.

This paper examines the digging productivity of the P&H4100XPB shovel, focusing on the impact of face height on its performance. By analysing historical data from real-time systems, the study aims to identify the most efficient bench height for excavation productivity in an open cut coalmine in the Bowen Basin, Australia's largest coal reserve. The findings are intended to provide actionable insights for mine planners to optimise shovel performance and overall productivity.

The remainder of this study is structured as follows: the Methods section outlines the methods and materials used to make this study possible; the Case Study section applies the proposed methodology to real data, with the aim of investigating shovels performance under different operational configurations; the Results and Discussion section presents the limitations and findings of the experiments; and finally, the Conclusion section summarises the main achievements of this study and proposes future directions.

METHODS

Case study

This study addresses the mining operation in an open cut coalmine located in the Bowen Basin, Central Queensland. Specifically, within the Baralaba Coal Measures. The primary coal seams targeted for mining are labelled A, B, C, D, and E seams, being C and D seams the most economically significant, due to its quality and thickness. While other coal seams exist at greater depths, they are not currently economically viable. The active pits have an average thickness of

250 m and the coal seams dip at angles ranging from 3 to 18° to the west. They all split and coalesce along the strike length of the mining area, exhibiting variations in thickness, quality, and extent.

The mine in study employs conventional strip-mining method, where overburden (waste rock) and coal are mined in strips oriented along the strike of the coal seam. Overburden is removed using a fleet of three draglines, two P&H4100XPB rope shovels and six 500 t class hydraulic excavators. As draglines are the most cost-effective method for overburden removal, they are allocated where there is a demand for the maximum practical volume, considering geotechnical limitations. In practice, draglines are assigned to the bottom two seams. The truck and shovel fleet supports mining by removing the overburden to expose upper seams. Excavators are allocated to the remaining overburden, which is primarily the portion immediately above the coal seam, commonly known as the wedge. Figure 1 illustrates an aerial photo of the mine pit under study.

FIG 1 – Aerial photo of the mine pit under study.

Although the shovels are not the primary machine used for overburden removal, they still play a significant role as they account for approximately 35 per cent of the total overburden movement. Given the importance of shovels in mining operations, there are several systems in place to ensure their optimal performance and contribute to the overall efficiency of the mining process.

For instance, the shovels are equipped with the Argus system by MineWare, an advanced monitoring solution designed for mining operations. The Argus system integrates high precision GPS data to monitor and record real-time payload information, as well as the exact position and performance of the shovel, like bucket fill factors, cycle times, digging conditions and swing angles. This data is stored in a SQL database, enabling detailed analysis of shovel operations and productivity.

From a technical perspective, the system is centred on having an optimal design to ensure an efficient shovel performance. Technical guidelines provided by the mining company to engineers/ planners allow decision-making based on established performance benchmarks.

For instance, the recommended face height for a P&H4100XPB shovel is between 12 and 15 m and no more than 18 m. It is also advised to maintain a consistent face height across the entire dig face and along the dig to ensure operational consistency. Figure 2 illustrates the optimum face height for P&H4100XPB shovels, according to the internal technical standard.

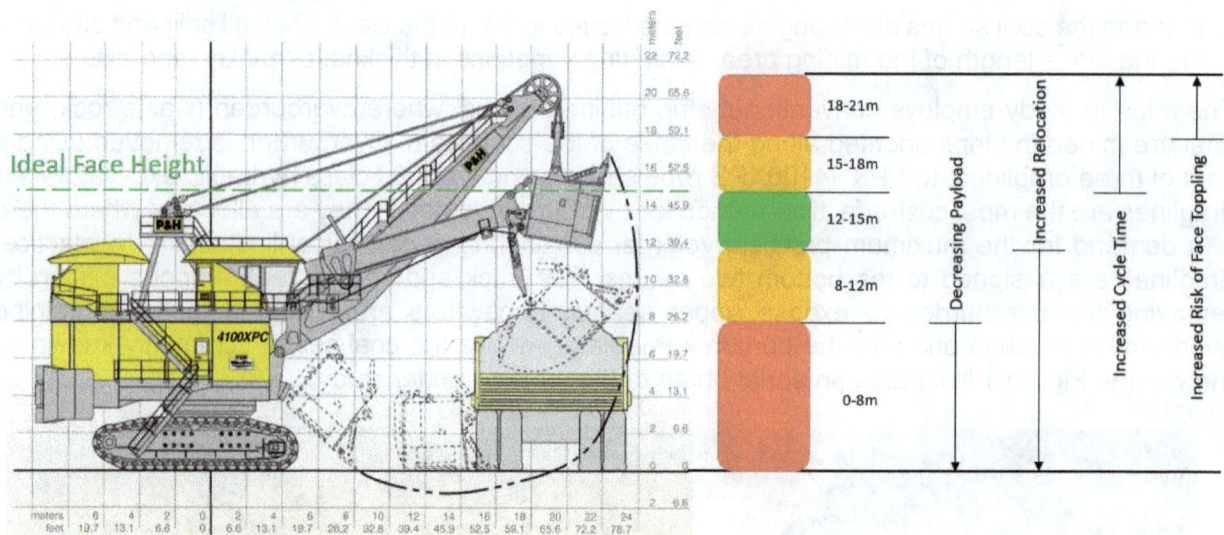

FIG 2 – Face Height recommendation chart for P&H4100XPB shovel.

The methodology employed in this study is composed of three distinct steps, as shown in Figure 3.

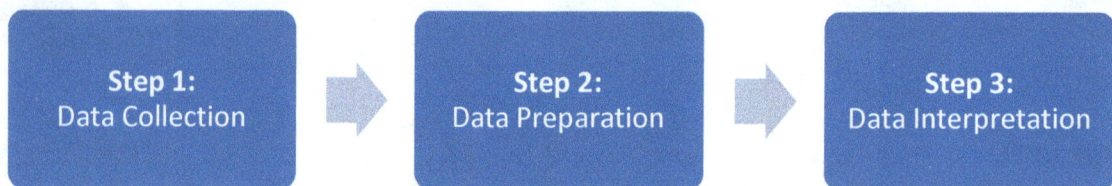

FIG 3 – Flow chart of the methodology used in this study.

Data collection

This study was conducted using shovel performance data obtained from the MineWare's Argus system considering the operation in the studied open cut coalmine. Key metrics include:

- Payload (t), which represents the weight of material in the shovel bucket.

- Phase times, including the fill time (s), swing time (s), dump time (s), and return time (s), which represent the phases of the shovel's operational cycle.

- Cycle time (s), calculated as the sum of the individual phase times.

- Timestamps marking the start time and end time of each cycle.

- End fill height (m), which measures the bucket teeth's position from the ground level at the end of the fill phase, being considered as the measure of bench height in this study.

- Fill_length (m), which measures the distance the bucket teeth travel during the fill.

- GPS coordinate of the machine location.

The data extraction was performed using queries in SQL Server. The analysis covers the period from 01/01/2021 to 09/02/2025 and includes data from two P&H4100XPB shovels operating in one specific pit. Records with less than five satellites were excluded to maintain high precision GPS data.

At the time the analysis was conducted, there were five active pits in the mine, so this study was restricted to data from one of these pits. It includes data from shovels digging the interburden layers of the primary coal seams targeted for mining: A, B, C, D, and E seams. It is important to highlight that the E seam is the least significant for this study as it is primarily targeted for dragline operation.

Data preparation

An exploratory data analysis was conducted to evaluate the variables distribution within the data set. The initial analysis involved examining each numerical variable to determine their mean, standard deviation, minimum and maximum values, quartiles, and statistical outliers, identified using the interquartile range method.

Subsequently, rows containing outliers for the variables payload, fill time, swing time, dump time, return time and end height were identified and eliminated from the data set. This step was important to ensure the reliability of the analysis as outliers can significantly skew results and lead to misleading conclusions.

Then, each dig area was associated with a specific dig design that follows a standard naming convention. To streamline further investigations and filtering, a merge process was executed to associate each cycle with its corresponding dig plan.

Data interpretation

Once the previous steps were performed, equipment performance in each seam was evaluated, allowing for the identification of operational patterns that influence shovel productivity. The initial hypothesis suggested that end_fill_height values would align with bench height. To test this, the data was stratified by seam, and the end_fill_height values across different seams was examined.

A statistical analysis was conducted to further investigate the correlation between height and the operational parameters such as payload, fill time, swing time, dump time, return time, and overall cycle time.

RESULTS AND DISCUSSION

A total of 1 008 454 shovel cycles were extracted from the Argus database. After removing outliers, the data set was reduced to 724 030 records. The Figure 4 shows the 3D view of the database, where each point represents a load cycle and has as attribute the key metrics of that specific load.

FIG 4 – 3D view of the load cycles database.

The results of the descriptive analysis after the outlier removal for each numerical variable are detailed in Table 1.

TABLE 1

Descriptive analysis of the numerical variables.

	mean	std	min	25%	50%	75%	max
Payload (t)	90	17	40	79	91	102	139
Productivity (t/h)	8777	1603	2707	7733	8823	9859	21 660
fill_time (sec)	11.3	3.7	0.3	8.7	11.0	13.7	22.3
swing_time (sec)	10.2	3.3	2.1	7.8	9.5	11.9	20.1
dump_time (sec)	1.8	1.0	1.0	1.0	1.0	3.1	6.1
return_time (sec)	13.9	1.9	8.6	12.7	13.9	15.1	19.5
cycle_time (sec)	37.2	5.4	16.0	33.3	36.7	40.6	63.0
end_fill_height (m)	9.2	1.4	5.5	8.3	9.4	10.1	12.9
fill_length (m)	12.7	3.1	0	10.7	12.4	14.5	45.1

The initial hypothesis suggested that the end_fill_height values would align with the bench height. However, Table 1 shows that the mean end_fill_height is 9.2 per cent less than the P&H4100XPB height of cut (16.8 m) and lies outside the recommended range of 12–15 m according to the design guidelines.

Even when stratifying the data by seam, as shown in Figure 5, the end_fill_height values remain stable across different seams, indicating that seam characteristics do not significantly influence these variations.

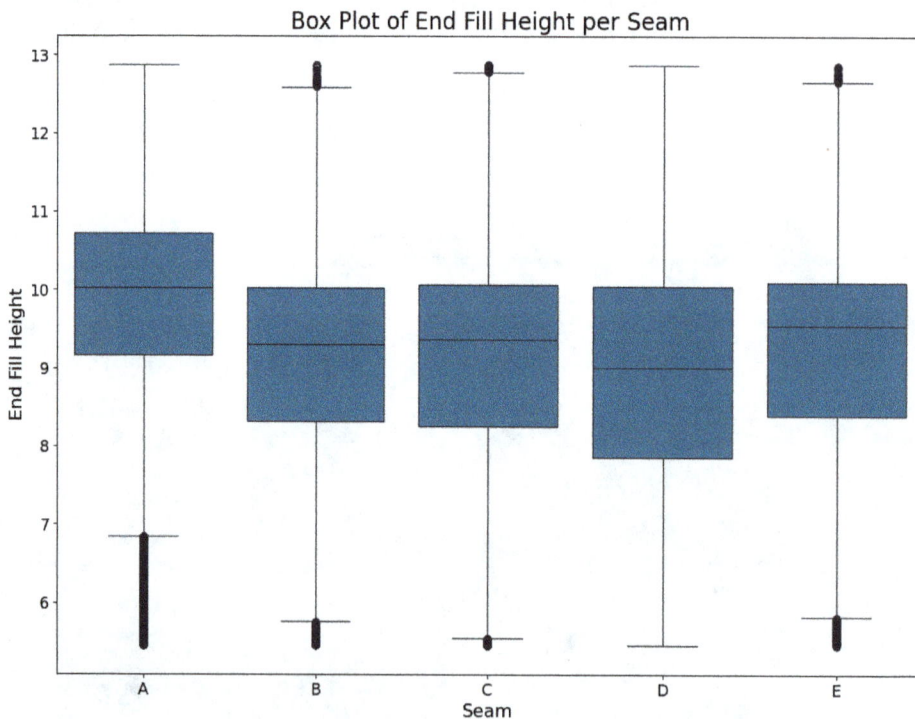

FIG 5 – Box Plot showing the End_fill_height per seam.

As shown in Figure 6, the analysis of the correlation revealed a moderate positive correlation ($r = 0.47$) between fill height and fill time, indicating that higher fill heights are associated with longer fill times, and consequently, longer cycle times. Additionally, a weak positive correlation ($r = -0.34$) was observed between fill height and payload, suggesting that higher fill heights may slightly increase payload. Furthermore, there is a moderate positive correlation ($r = 0.40$) between fill height and fill length. These values collectively indicate that the parameter end_fill_height recorded during the load cycle of the shovels is more closely associated with the digging trajectory than with the bench height, thus being more closely tied to operational performance.

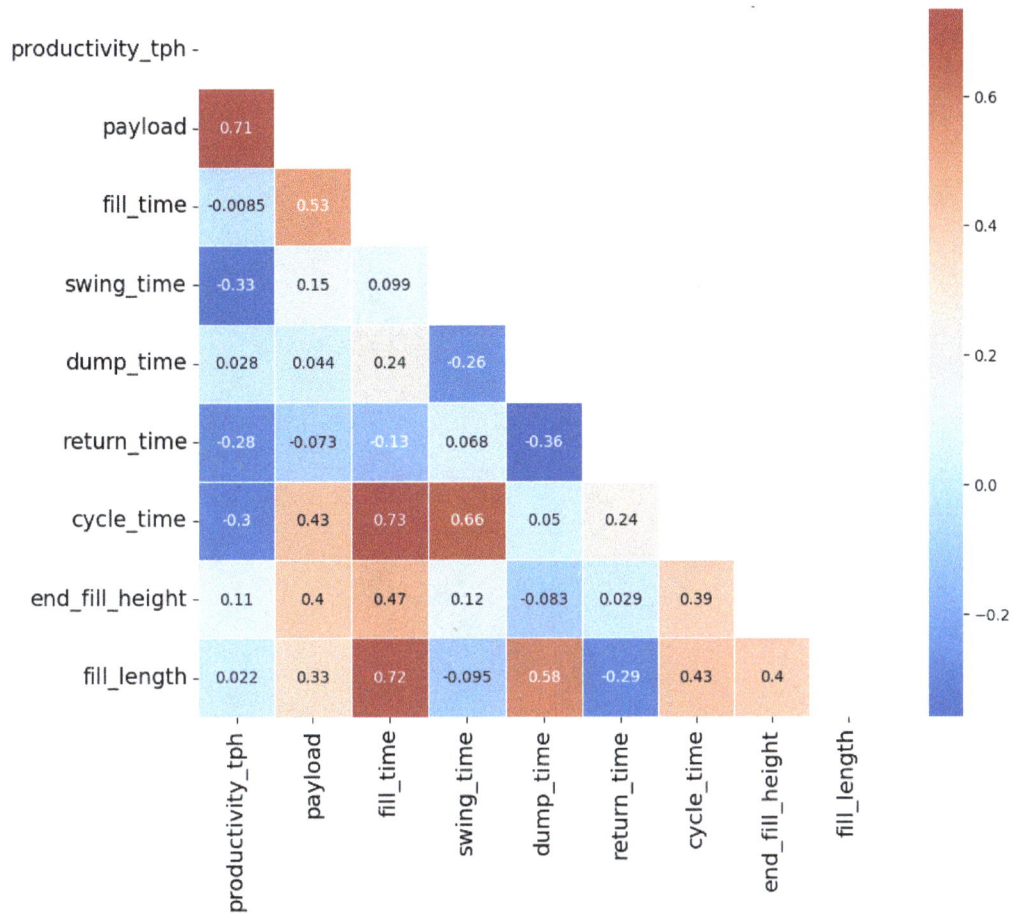

FIG 6 – Correlation matrix, in which cool and heat colours represent negative and positive correlations, respectively.

There are valuable insights to be gained from further analysis of this parameter, though. The correlation matrix shows that end_fill_height has correlations with both payload and fill time, suggesting that this variable significantly impacts productivity. The bar graph illustrated in Figure 7 shows the relationship between end_fill_height categories and shovel productivity.

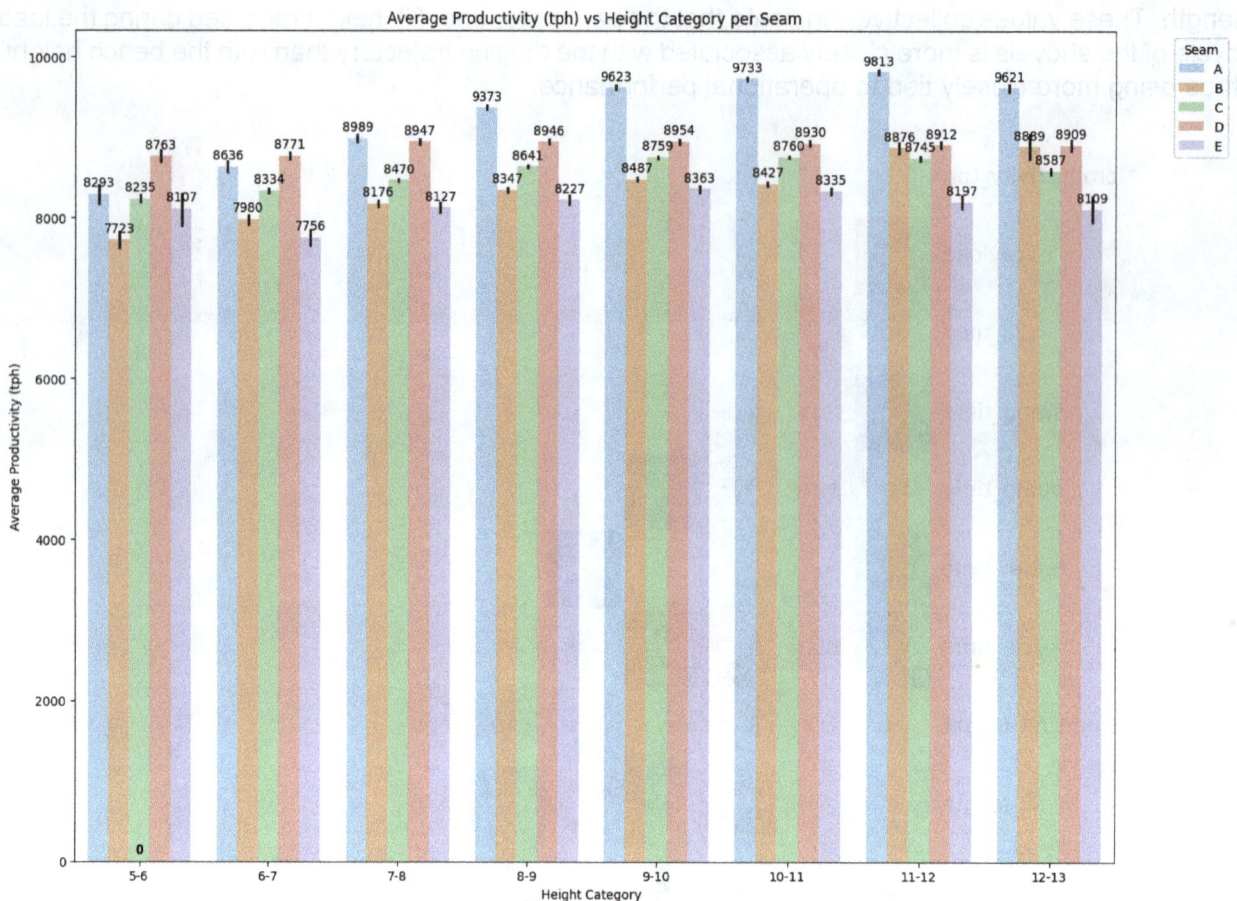

FIG 7 – Average productivity (tph) versus Height Category (m) per seam.

Seam A provides the best shovel productivity rate among all the seams across the height categories, likely due to the characteristics of the rock, which is generally made up of siltstone interbedded with claystone. Contribute to this, the fact that the payload for Seam A is 12 per cent higher than the average. Seam E, on the other hand, exhibits the lowest productivity. This may be explained by the fact that Seam E is the lowest coal seam, typically involving narrow digs. Although draglines are the preferred machines for this seam, the shovel was allocated to the area in the past due to issues with machine availability. This is evident as data from Seam E comprises only 4 per cent of the total data set.

Except for Seam B, all other seams exhibit a sinusoidal behaviour, where the highest performance is observed between the height categories of 9–12 m. For instance, Seam C, which has the largest population in the data set, shows a 2 per cent decrease in the 12–13 m category compared to the 10–11 m category.

Assuming the fill height will always be lower than the bench height, in terms of mine planning and design this data represents an opportunity for the shovel to dig shallower areas without losing efficiency. Typically, an excavator is required to face up for a shovel until the bench height reaches 12 m. However, the data indicates that the shovel can effectively dig at a height of 9 m without compromising its productivity.

When it comes to passes below 8 m, however, the data shows that most seams presented lower performance, supporting the recommendation from the technical guideline.

The hypothesis that end_fill_height values would align with bench height was not supported by the data. Instead, the results imply that operational factors, such as digging trajectory, play a more significant role in shovel productivity. These findings highlights the need to investigate what resources operators and production teams have to control the fill height and, consequently, the cycle time.

CONCLUSIONS

This study analysed historical data from real-time monitoring systems to examine key metrics associated with productivity rates, such as payload and cycle time. The suggested operational pattern was to optimise bench design to maximise shovel productivity, and a hypothesis was formulated that end_fill_height values would align with bench height. The analysis revealed that the parameter end_fill_height is more closely associated with the digging trajectory than with the bench height, emphasizing the importance of operational factors in shovel productivity. The findings corroborate past research, indicating that operational performance plays a more significant role in shovel productivity than bench height.

The study provided valuable insights from the data analysis. Future research should investigate the influence of slope gradient and width on shovel performance metrics. Developing a predictive model to estimate shovel productivity under specific digging conditions will serve as a valuable decision tool for mine planners. Such advancements will further enhance the effectiveness of shovel operations in open cut coalmines.

REFERENCES

Aiken, G E and Gunnett, J W, 1992. Overburden removal in surface mining, in *SME Engineering Handbook*, 2nd edition, ch 6.3, pp 584–619 (Society for Mining, Metallurgy and Exploration: Littleton).

Bassani, M A A, Guimaraes, O R A, Tavares, F H, Cantadori, B, Vicenzi, R, Alves, J L, Mariz, J L V and Peroni, R L, 2025. Linear programming model applied to long-term mine planning in strip mining operations, *Mining, Metallurgy and Exploration*, 42:737–750. https://doi.org/10.1007/s42461-025-01185-5

Bettens, S P, Grist, P and Scaree, P, 2022. How do operators and environment conditions influence the productivity of a large mining excavator, in Proceedings of the SME Annual Meeting.

Hendricks, C, Scoble, M and Peck, J, 1989. Performance monitoring of electric mining shovels, *Transactions of the Institutions of Mining and Metallurgy,* 98:A151–A159.

Kennedy, B A, 1990. Open pit optimization, *Surface Mining*, second edition, illustrated edition, ch 5.3, p 470 (Society for Mining, Metallurgy and Exploration: Littleton). Available from: <https://books.google.com.au/books?hl=en&lr=&id=qJJrYnpT2pYC&oi=fnd&pg=PA470&dq=related:0iXJ6Za6w1oJ:scholar.google.com/&ots=YE4JvNfrop&sig=-brXlRpgCt5ySQFN7QouElHpgGs&redir_esc=y#v=onepage&q&f=false [Accessed: 3 April 2025].

Kennedy, B A, 1990. Surface Coal Mines, Surface Mining, second edition, illustrated edition, ch 5.7, p 495 (Society for Mining, Metallurgy and Exploration: Littleton). Available from: <https://books.google.com.au/books?id=qJJrYnpT2pYC&lpg=PA470&lr&pg=PA495#v=onepage&q&f=false> [Accessed: 3 April 2025].

Khorzoughi, M B and Hall, R, 2016. A study of digging productivity of an electric rope shovel for different operators, *International Journal of Mining and Mineral Engineering*, 7(3):181–209.

Lizotte, Y and Bonates, E, 1986. Truck and shovel dispatching rules assessment using simulation, *Mining Science and Technology*, 5(1):45–58. https://doi.org/10.1016/S0167-9031(87)90910-8

Onederra, I, Brunton, I, Battista, J and Grace, J, 2004. Shot to shovel – understanding the impact of muckpile characteristics and operator proficiency on instantaneous shovel productivity, in *Proceedings EXPLO2004*, pp 205–213.

Patnayak, S and Tannant, D D, 2005. Performance monitoring of electric cable shovels, *International Journal of Surface Mining, Reclamation and Environment*, 19:276–294.

Profio, R L, 1984. Shovel productivity, Harnischfeger Corporation, Milwaukee, Wisconsin, presented at the SME-AIME Annual Meeting, Los Angeles, California, February 26–March 1, 1984.

Yaghini, A, Hall, R A and Apel, D, 2020. Modeling the influence of electric shovel operator performance on mine productivity, *CIM Journal*, 11(1):58–68. Available from: <https://www.researchgate.net/publication/341164481_Modeling_the_influence_of_electric_shovel_operator_performance_on_mine_productivity> [Accessed: 3 April 2025].

Semi-automated fracture characterisation to optimise dimension stone mining

G R S Maior[1], R L Peroni[2], J L V Mariz[3], J T Zagôto[4], D Vale[5] and B T Kuckartz[6]

1. PhD student, Graduate Program Mining, Metallurgical and Materials Engineering – PPGE3M, Federal University of Rio Grande do Sul, Porto Alegre, Rio Grande do Sul, Brazil. Email: gleicon.maior@ifes.edu.br
2. Professor, Graduate Program Mining, Metallurgical and Materials Engineering – PPGE3M, Federal University of Rio Grande do Sul, Porto Alegre, Rio Grande do Sul, Brazil. Email: peroni@ufrgs.br
3. Postdoctoral Fellow, Graduate Program Mining, Metallurgical and Materials Engineering – PPGE3M, Federal University of Rio Grande do Sul, Porto Alegre, Rio Grande do Sul, Brazil. Email: jorge_valenca@hotmail.com
4. Adjunct Professor, Mining Engineering Course, Federal Institute of Espírito Santo, Cachoeiro de Itapemirim, Espírito Santo, Brazil. Email: tessinari@ifes.edu.br
5. Adjunct Professor, Mining Engineering Course, Federal Institute of Espírito Santo, Cachoeiro de Itapemirim, Espírito Santo, Brazil. Email: daniel.vale@ifes.edu.br
6. Postdoctoral Fellow, Graduate Program Mining, Metallurgical and Materials Engineering – PPGE3M, Federal University of Rio Grande do Sul, Porto Alegre, Rio Grande do Sul, Brazil. Email: brukuck@hotmail.com

ABSTRACT

Efficient extraction of dimension stone blocks relies on understanding the structural discontinuities within the rock mass. Accurate characterisation of fractures, faults, and joints allows optimising extraction, minimising losses, and maximising natural resources utilisation. This study introduces a semi-automated approach for detecting and analysing discontinuity families based on structural data, 3D modelling, and field validation. A case study conducted in a dimension stone quarry demonstrated that the proposed method effectively characterised fractures and their influence on block segmentation. Three major discontinuity families with dip angles ranging from 60° to 85° were identified, and their orientation significantly influenced block geometry and extraction efficiency. Fractures presenting steeper dip angles ($\geq 85°$) aligned with natural structural planes, facilitating separation, while those with lower dip angles ($\leq 60°$) led to erratic block shapes, complicating extraction strategies. A progressive removal model of extracted rock slabs was also developed to dynamically visualise the quarry's future evolution. The proposed solution is applicable and scalable, enhancing extraction efficiency and optimising block recovery while integrating geospatial data analysis and structural geology. This approach contributes to mine planning and operational decision-making, highlighting the potential of combining advanced analysis tools with geological insights to modernise the dimension stone sector.

INTRODUCTION

The mining of dimension stones roots back to prehistoric times, when the extraction and use of natural resources were essential to supply the basic needs of the humanity and became integral to human activity. Over time, various types of rocks, further classified as 'dimension stones' or 'ornamental stones', became quite popular in construction and architectural design due to their aesthetic appeal and physical characteristics. Unfortunately, the mining sector of dimension stones still faces problems in operational efficiency and waste management, despite the advances in other areas of the production chain. In fact, a typical dimension stone mine achieves less than 30 per cent recovery rates, which disclose the urgent need for new technologies that can improve block extraction while minimising environmental hazards (Mochi et al, 2011; Vidal, Azevedo, and Castro, 2014; Yurdakul, 2020; Jalalian, Bagherpour and Khoshouei, 2023). The mining of ornamental rocks, as demonstrated in Figure 1, with a variety of geological characteristics, often jeopardises the quality of produced blocks.

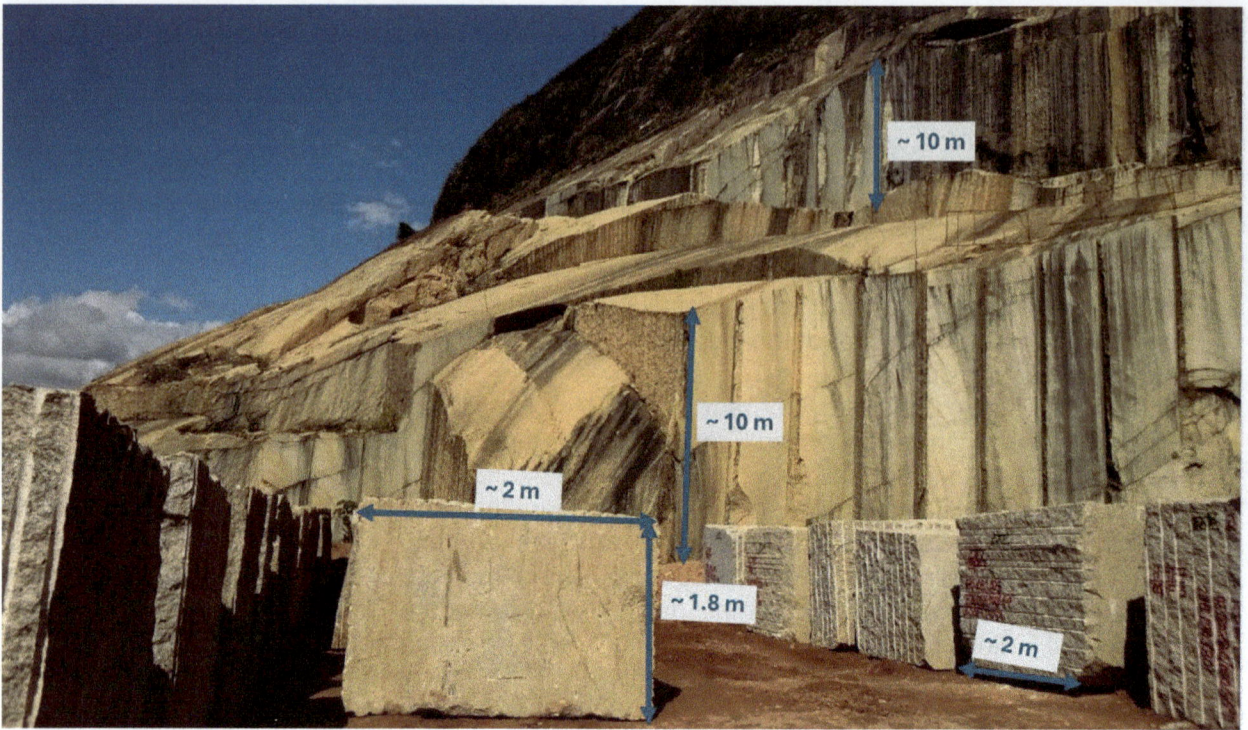

FIG 1 – Dimension stone mining front with evident fractures and different geological characteristics.

As environmental concerns continue to grow and the need to promote rock recovery to increase profits becomes more evident, technological advances become a useful tool to improve the identification and treatment of quarry discontinuities. Geophysical methods, particularly Ground Penetrating Radar (GPR), have been used to detect fractures and other subsurface features. Grandjean and Gourry (1996) showed the use of GPR to identify fractures in dimension stone quarries, helping to preserve economically viable blocks and improving the overall geometry of quarries. In Brazil, Porsani, Sauck and Júnior (2006) used GPR to detect fractures and stress-relief joints in a granite quarry, showcasing the potential of this method to enhance quarrying and reduce cutting costs. Similarly, Selma (2008) observed that GPR is useful to obtain underground 3D images of quarries, helping in structural interpretation and effective resource management.

Before the advent of drones, aerial photogrammetry and GPR, quarry assessment was already undergoing significant transformations. These techniques enable the sketching of realistic 3D models that can be computer processed and then further examined in great detail for structural aspects (Elkarmoty, Bonduá and Bruno, 2020; Riquelme *et al,* 2022). For example, these authors reference Jalalian, Bagherpour and Khoshouei (2023) for computer programs overlaying operations that use photogrammetry processing and structural modelling to reduce waste and maximise the efficiency of operational tasks. The feasibility of these advanced applications is strongly supported by the continuous evolution of computational hardware. Workstations equipped with high-performance GPUs, expanded RAM, and fast SSD storage are now capable of handling and processing large volumes of spatial data—such as dense point clouds and 3D meshes—within practical time frames. This progress has made it possible to integrate photogrammetric models, structural algorithms, and mine planning tools in a seamless and dynamic workflow.

The structural characterisation is complemented by the development of computational algorithms aimed to optimise block recovery in dimension stone quarries. Fernández-de Arriba *et al* (2013) introduced a method for automatically determining the optimal cutting direction, which leads to higher recovery rates and reduced material loss. Thus, the proposed approach aligns with the growing awareness of sustainable mining that considers economic and environmental concerns. Gazi, Silva and Founti (2012) emphasise the need for technologies that improve energy efficiency and reduce environmental impact, urging a more responsible approach to resource utilisation. Geologically heterogeneous deposits and mining operations in brittle rocks pose unique challenges for the development of one of the largest producers and exporters of dimension stones in the world, located in Brazil (Abirochas, 2023).

The detailed characterisation of geological discontinuities, as described by Porsani, Sauck and Júnior (2006), holds considerable potential for reducing costs and improving block recovery rates in Brazilian quarries. Today's biggest challenges are related to the integration of information from different sources, including photogrammetry, GPR, and computational modelling. Rossi, Madonna and Milli (2017), Khoshouei, Jalalian and Bagherpour (2020), and Jalalian, Bagherpour and Khoshouei (2023) demonstrate that a multidisciplinary approach will be essential to overcome these challenges.

METHODOLOGY

This study presents a methodology that integrates structural characterisation with strategic mine planning in ornamental stone quarries, leveraging technologies such as aerial photogrammetry, point cloud analysis and computational analysis. The primary goals are to maximise block recovery, minimise environmental impacts and implement sustainable mining practices. The study area was first surveyed using a DJI Mini drone and processing the acquired data in Agisoft Metashape 2.0 software. Subsequently, analytical software tools such as CloudCompare and Deswik.CAD were employed, specifically using the Deswik 'Extract by Orientation' tool to map structural discontinuities and develop optimised extraction strategies of the deposit according to the geological conditions previously identified. The workflow is organised into four stages, as described herein.

Stage 1 – 3D data acquisition and processing

The first step in this study involved gathering high-resolution 3D data from the quarry using an unmanned aerial vehicle (UAV). A DJI Mavic Mini drone was deployed for aerial photogrammetry, ensuring comprehensive coverage and optimal image overlap. The captured images were then processed in Agisoft Metashape 2.0, generating a dense point cloud and highly detailed topographic contours. These contour lines, spaced at 0.05 m intervals, were exported as vector files and used to create a triangulated surface model in Deswik, providing a detailed representation of the quarry's topography. These contour lines, spaced at 0.05 m intervals, were exported as vector files and used to create a triangulated surface model in Deswik. This high-resolution surface was essential to ensure geometric accuracy for the subsequent application of fracture projection, segmentation, and extraction planning techniques. The 5 cm interval was chosen to balance data density and processing efficiency, allowing detailed topographic modelling without compromising the software's performance in subsequent structural analyses.

Stage 2 – Identification and characterisation of fractures

After building the 3D model, the next step was to identify structural discontinuities in the rock mass. Using the Facets plug-in in CloudCompare, fracture planes were automatically identified based on the coplanarity found in the dense point cloud and resulting in the grouping of discontinuity 'families' according their dip and dip direction. Field measurements taken with a geological compass were used to confirm the orientation of these fractures, validating the automated identification and enabling the continuation of mine planning to optimise block recovery.

Stage 3 – Fracture modelling and orientation-based extraction

With the fracture families identified, the next phase involved modelling their spatial distribution and assessing their impact on extraction. The previously obtained contour lines were imported into Deswik.CAD to create an accurate triangulated surface of the quarry. The structural data was also imported into Deswik.CAD, where the 'Extract by Orientation' tool was used to map the discontinuity families and project their spatial persistence onto the walls. This step facilitated the understanding of the fracture intersections, helping define geological constraints and identify the best extraction zones. The identified main fracture families, characterised by their dip and dip direction, were then integrated into the extraction workflow, improving quarry planning and maximising the cutting efficiency by aligning the mining faces. I t is worth noting that the purpose of this step is not to identify geological wedges or complex structural interactions, but rather to recognise coplanar fracture families and assess how their orientation affects block geometry and extraction efficiency.

Stage 4 – Fracture projection, segmentation, and extraction planning

In the final phase, the fracture families visualised using Deswik.CAD were projected onto the rock mass at 10 m intervals. This value was chosen because, during the fracture family measurements, the persistence of the families could not be fully assessed, and therefore, the 10 m interval was applied arbitrarily. This projection helps visualise how these fractures intersect and how they may influence the geometry of the blocks to be removed, allowing for a better understanding of their impact on extraction. To further refine this analysis, a custom Python script was developed to simulate the continuous removal of the extracted blocks, as shown in Figure 1. This dynamic approach provided more accurate visualisation of the extraction process, offering valuable insights into how the mine's progress according to fracture orientation would behave over time. The integration of the Python script offered some key advantages, such as:

- improved refinement of fracture projection, leading to more accurate block partitioning

- dynamic extraction sequencing that accounts for geological constraints in each mining advance

- optimised partitioning predictions to help mine planning improve block recovery and reduce material loss.

By systematically following these four stages, this methodology provides a comprehensive and scalable approach to characterising discontinuities, modelling fracture behaviour, and improving block extraction strategies in dimension stone quarries. The methodology applied in this research is schematically represented in Figure 2.

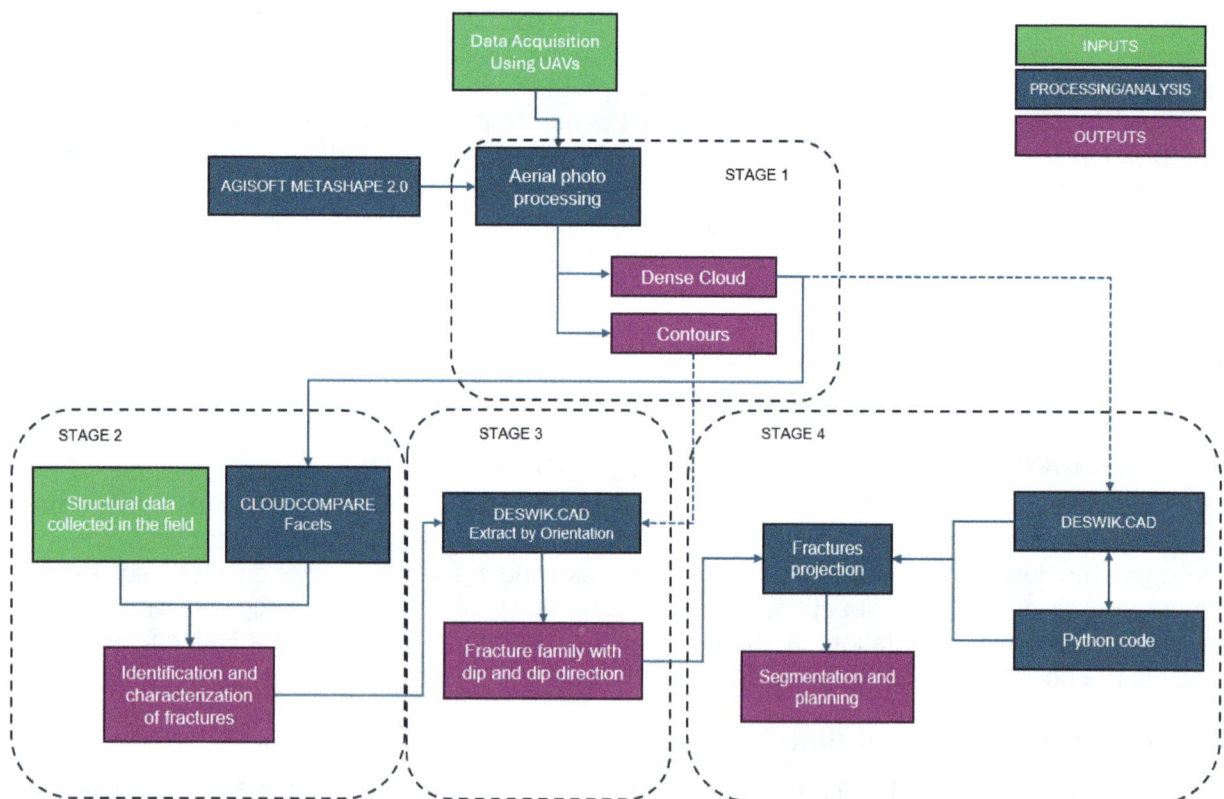

FIG 2 – Flow chart of the methodology applied in this study.

RESULTS

The integrated workflow developed in this study was applied in an active dimension stone quarry, enabling a comprehensive structural analysis of the exposed rock mass. Rather than focusing solely on individual fractures, the approach allowed for the identification of persistent discontinuity families and their geometric relationships, which are crucial for understanding block segmentation patterns. By correlating spatial data with field observations, it was possible to delineate key structural features that govern the extraction potential. The results presented below detail the dominant fracture orientations, their spatial distribution, and their direct impact on block geometry, recovery rates, and planning decisions.

To ensure accuracy and reliability, these results were validated through a comparative analysis with field measurements obtained using a Clar compass. The comparison indicated a strong correlation between the automatically detected fracture orientations and those measured in the field, supporting the method's reliability.

Stereographic analysis in CloudCompare identified three primary fracture families with distinct orientations (see Table 1). Family 1 has an average dip of 81° and a dip direction of 9°. Family 2 shows values of 65° and 340°, respectively, and Family 3 has dip and dip direction values of 82° and 48°. These findings suggest that the main fracture families align with the geological features observed in the study area, providing valuable insights into the structural framework of the exposed outcrop. Figure 3a illustrates the semi-automatically detected discontinuities, represented by different colours, while Figure 3b presents a comparison of the automatically identified fracture Family 1 and its stereographic projection 1. The colours in the stereographic projection of Figure 3b correspond to those of the different planes represented in Figure 3a.

TABLE 1
Main fracture families identified in the quarry.

Family	Average dip (°)	Dip direction (°)	Colour
Family 1	81	9	Orange
Family 2	65	340	Light green
Family 3	82	48	Yellow

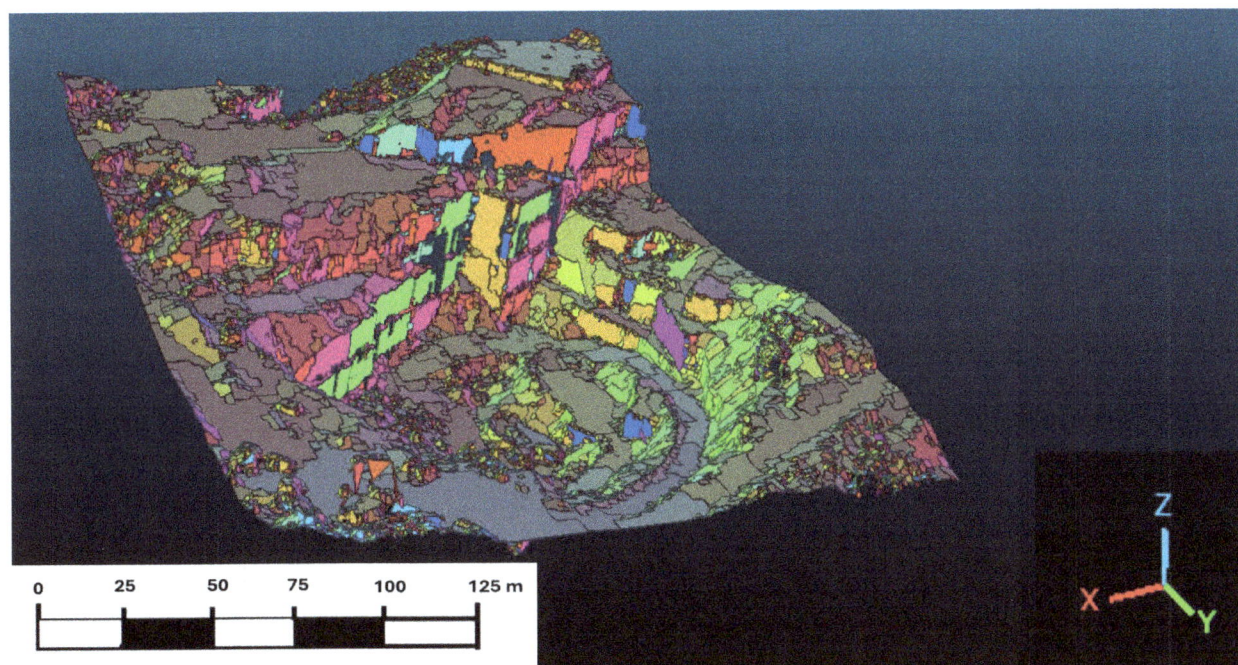

(a)

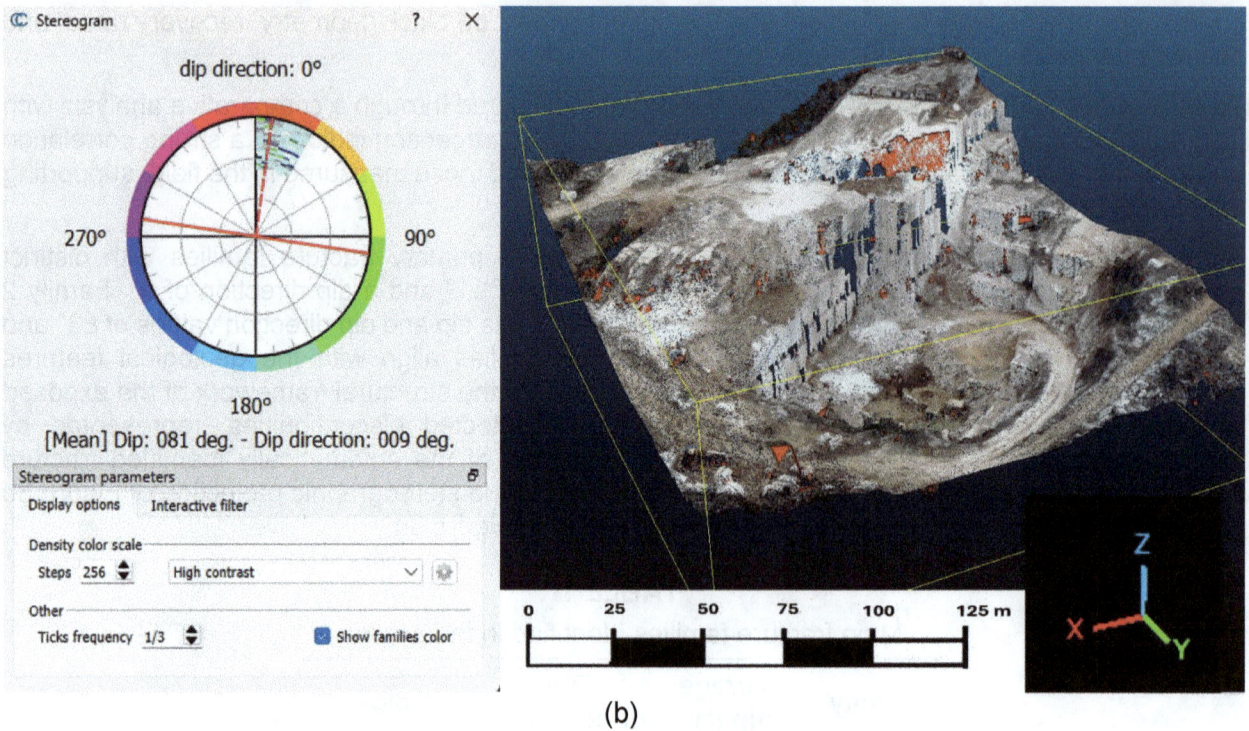

FIG 3 – (a) Geological features identified semi-automatically. Each colour indicates a different identified family; (b) Discontinuities observed in the field and detected through stereographic analysis, showing the average dip and dip direction.

The replication of these fracture families in Deswik enabled the visualisation of structural continuity of the rock mass, helping to identify critical intersection zones between different fracture families. These intersections, if not properly mapped and understood beforehand, can pose challenges to block geometry, often requiring operational adjustments to minimise material loss. Conversely, regions with widely spaced fractures exhibited greater potential for extracting large, regular blocks, thereby improving the overall rock utilisation rate.

Three-dimensional analysis revealed a strong correlation between fracture orientation and block segmentation. Areas where fractures exhibited steeper dips (≥85°) facilitated more efficient extraction, as the rock naturally fragmented along these planes of weakness. In contrast, sectors with lower dips (≤60°) produced less adjustable block geometry, requiring strategic adaptations to optimise rock recovery. These findings align with previous studies, such as those of Fernández-de Arriba *et al* (2013) and Khoshouei, Jalalian and Bagherpour (2023), which emphasise the role of structural characterisation in maximising resource extraction efficiency.

Figure 4 is divided into three parts: a) a general view of the mining front, showing the structural characteristics identified in the rock mass; b) a visualisation of the fracture planes showing their spatial orientations and intersections, illustrating their distribution within the rock mass; c) an enlarged view highlighting the main discontinuity planes, with coloured faces representing fracture families.

The coloured faces (orange, light green and yellow) represent distinct fracture families, each associated with specific dip values and dip direction. A detailed table with the corresponding values for each colour is included in the image to identify each family.

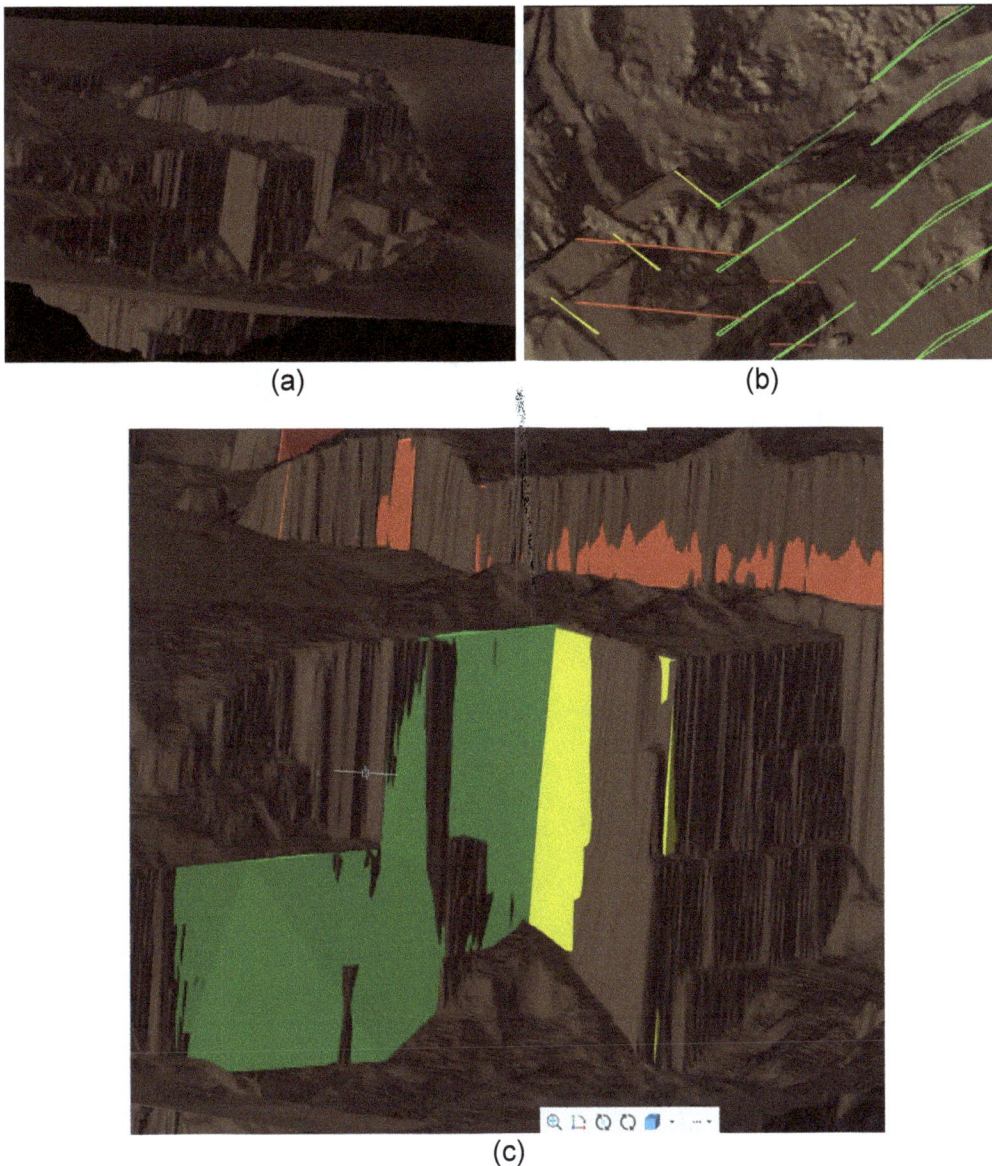

FIG 4 – Discontinuity plans present in the mining front and projections in the rock mass. Family 1 (orange) has a dip of 81° and a Direction dip of 9°, Family 2 (light green) has a dip of 65° and a Direction dip of 340° and Family 3 (yellow) has a dip of 82° and a Direction dip of 48°.

After identifying the fracture families, a Python script code was developed to simulate the progressive removal of the extracted blocks, enabling the visualisation of the mine's progress over time. This approach allows for the evaluation of different extraction scenarios, anticipating the structural impacts of mining and estimating block recovery. The code can be found at: <https://github.com/ GleiconMaior/GleiconMaior/blob/main/Dimension_stones_blocks.ipynb>.

The development of this script became necessary due to the limitation of existing planning software, which lacks specific modules capable of simulating surface removal based on the geometry and particularities of dimension stone mining. Unlike 'conventional' mining, which is based on advances planned through excavated or blasted polygons where cuts follow pre-defined patterns, the extraction of ornamental stones requires a dynamic and adaptive model. This model must consider fracture distribution, block geometry and the sequential removal strategies to optimise the integrity of primary cuts (also called slabs) and maximise the partitioning of final blocks (commercial blocks to be processed into rock plates in the first stage of beneficiation).

Figure 5a shows the superposition of the original surface (in brown) with the surface after the mining advance (in green). The portion removed using the script considers the fracture Families 2 and 3 and, at the very end, forms an edge with family 1 (Figure 5b). The approximate volume of this advance is 1100 m^3 and a potential reduction of material loss can be anticipated, as the visible and detectable discontinuities are only present in the final part of the primary block to be removed (Figure 5c).

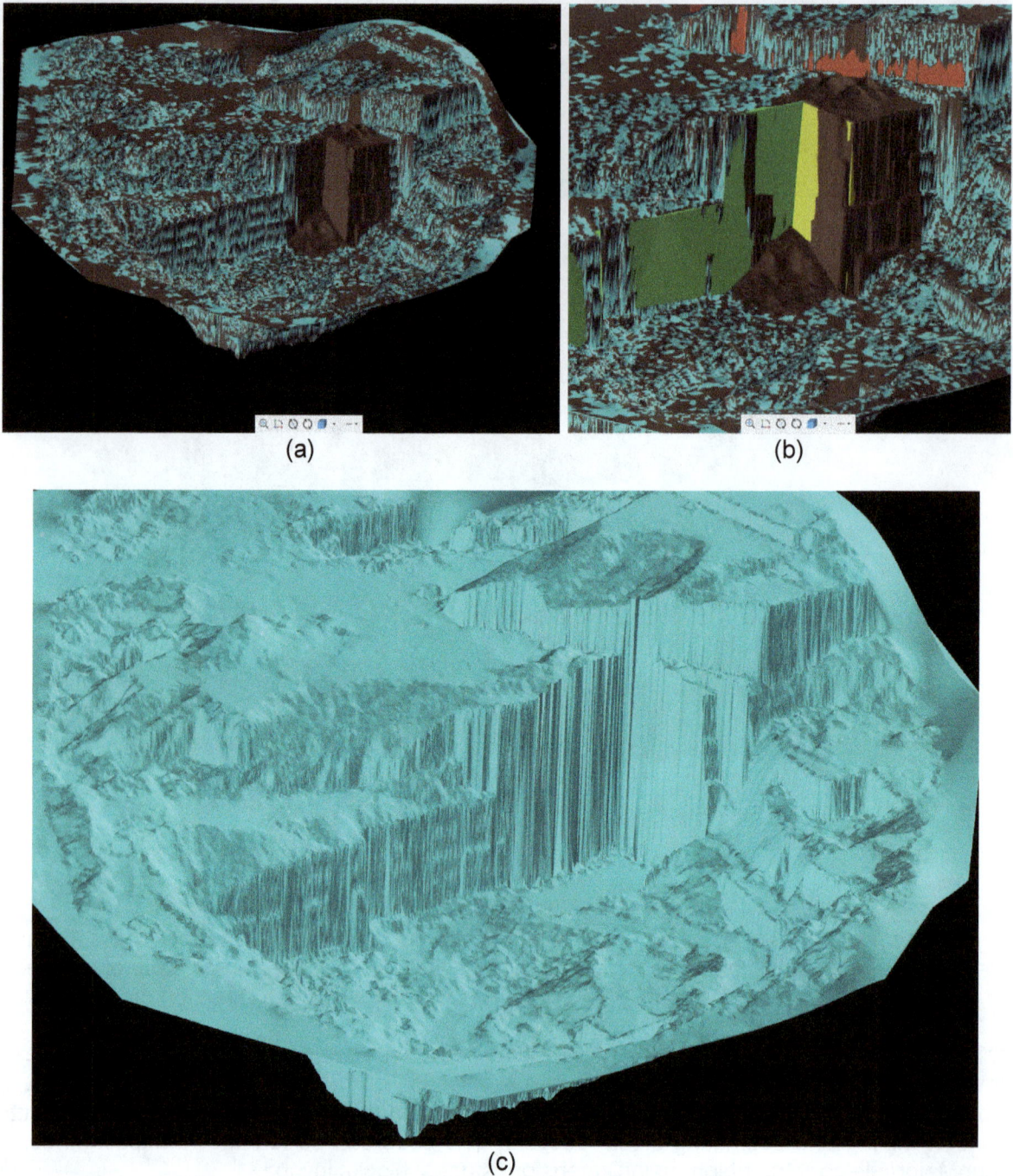

(a)

(b)

(c)

FIG 5 – Blocks removed using Python script after detection of discontinuity families: (a) original rock mass in brown and the surface after removing the first block; (b) representation of the block to be removed with the main fracture families; (c) final topography configuration after removing the mine advance.

With this code, it was possible to:

- simulate the evolution of mining over time based on the identified structural families
- easily evaluate different extraction strategies, optimising mineral recovery
- generate interactive 3D models for analysis and decision-making

- improve the predictability of the extracted block geometry, reducing losses and consequently minimising the waste generated in each cut.

Furthermore, the replication of fracture networks in Deswik facilitated the visualisation of spatial discontinuity distribution and the identification of critical regions where fracture families intersect and interfere with the mining advance. This process allowed for a quantitative assessment of fracture density and potential intersection points, providing insights into areas with higher structural complexity and anticipating potential extraction challenges. By analysing these intersections, it was possible to estimate the proportion of blocks affected by the fracture families, directly informing strategic planning for improved resource recovery. These intersections contribute to the creation of irregular block geometries, demanding planning adjustments to mitigate excessive losses. Conversely, regions with widely spaced discontinuities demonstrated greater potential for the extraction of high-quality, large-sized blocks. The 3D models generated in this study provided a detailed structural framework, enabling the segmentation of extraction blocks based on geological criteria.

Although the methodology has not yet been extensively tested, significant improvements in rock recovery can be anticipated, especially considering that some quarry fronts, such as those in some quartzite mines, currently achieve only a 5 per cent recovery rate. The integration of structural analysis with advanced computational modelling indicated the predominant fracture orientations for the alignment of cutting operations, resulting in:

- more efficient material utilisation by reducing unnecessary waste

- lower energy consumption during extraction, as natural fracture planes facilitate block separation

- optimised cutting strategies that minimise unwanted material loss.

CONCLUSIONS

This study demonstrated the potential of semi-automated fracture characterisation to improve dimension stone mining. By integrating drone surveying, point cloud analysis, stereographic projections, and 3D modelling in Deswik, it was possible to identify and analyse discontinuity families, providing valuable insights for operational planning.

The tools and methodologies used in this study enabled the collection of accurate and reliable data facilitating the design of optimised solutions to address some of the challenges in dimension stone mining. The systematic integration of various approaches highlights the importance of implementing more advanced technologies into strategic planning, thereby enhancing the efficiency and sustainability of mining operations.

Identifying fracture networks in advance enabled detailed assessment of rock mass integrity, allowing for the identification of critical zones where adjustments were needed to minimise rock losses. On the other hand, areas with widely spaced fractures were identified as suitable for extracting larger and more regular primary blocks, enhancing operational efficiency. The study also confirmed that fracture orientation significantly influences extraction: steeply dipping fractures (≥85°) facilitate natural separation, while low-angle fractures (≤60°) require more precise control in cutting and handling.

To further improve predictability and to easily simulate different mining orientations, a Python-based tool was developed to model the progressive removal of blocks. This tool addresses some limitations faced when using commercial planning software applied to dimension stone mine planning and its particularities and enables dynamic scenario analysis, optimising resource recovery and supporting the decision-making process.

Despite the promising results, challenges still remain, particularly in handling large data sets. Future work will focus on integrating machine learning techniques for automated fracture segmentation and refining structural models through hybrid interpolation methods. Additionally, an economic analysis will be conducted to assess the resource utilisation improvements achieved through this approach.

This methodology provides a foundation for modernising dimension stone mining, offering a scalable framework that can be adapted to other operations, driving efficiency and fostering continuous improvement in extraction planning.

ACKNOWLEDGEMENTS

The authors acknowledge the Brazilian CAPES-PROEX program for the funding provided to the post-graduate program PPGE3M, and for also helping support the students with scholarships and providing resources to facilitate ~~support~~ the research.

REFERENCES

Abirochas, 2023. Brazilian Association of the Ornamental Stone Industry, Balance of Brazilian Exports and Imports of Ornamental Stones in 2022 [in Portuguese: Associação Brasileira da Indústria de Rochas Ornamentais, Balanço das Exportações e Importações Brasileiras de Rochas Ornamentais em 2022], Report 05/2023. Available from: <http://www.abirochas.com.br/abirochas-home/> [Accessed: 22 Nov 2024].

Elkarmoty, M, Bonduá, S and Bruno, R, 2020. A 3D optimization algorithm for sustainable cutting of slabs from dimension stone blocks, *Resources Policy*, 65:101533. https://doi.org/10.1016/j.resourpol.2019.101533

Fernández-De Arriba, M, Díaz-Fernández, M E, González-Nicieza, C, Álvarez-Fernández, M I and Álvarez-Vigil, A E, 2013. A computational algorithm for optimization of rock cutting from primary blocks, *Computers and Geotechnics*, 50:29–40. https://doi.org/10.1016/j.compgeo.2012.11.010

Gazi, A, Silva, G and Founti, M A, 2012. Energy efficiency and environmental assessment of a typical dimension stone quarry and processing plant, *Journal of Cleaner Production*, 32:10–21. https://doi.org/10.1016/j.jclepro.2012.03.007

Grandjean, G and Gourry, J C, 1996. GPR data processing for 3D mapping of fractures in a dimension stone quarry (Thassos, Greece), *Journal of Applied Geophysics*, 36:19–30. https://doi.org/10.1016/S0926-9851(96)00029-8

Jalalian, M H, Bagherpour, R and Khoshouei, M, 2023. Environmentally sustainable mining in quarries to reduce waste production and loss of resources using the developed optimization algorithm, *Scientific Reports*, 13:22183. https://doi.org/10.1038/s41598-023-49633-w

Khoshouei, M, Jalalian, M and Bagherpour, R, 2020. The Effect of Geological Properties of Dimension Stones on the Prediction of Specific Energy During Diamond Wire Cutting Operations, *Mining-Geology-Petroleum Engineering Bulletin*, 35(3):17–27. https://doi.org/10.17794/rgn.2020.3.2

Khoshouei, M, Jalalian, M H and Bagherpour, R, 2023. Environmentally sustainable mining in quarries to reduce waste production and resource loss using the developed optimization algorithm, *Scientific Reports*, 13:22183. https://doi.org/10.1038/s41598-023-49633-w

Mochi, S, Nikolayew, D, Ewiak, O and Siegesmund, S, 2011. Optimized extraction of dimension stone blocks, *Environmental Sciences of the Earth*, 63:1911–1924. https://doi.org/10.1007/s12665-010-0825-7

Porsani, J L, Sauck, W A and Júnior, A O S, 2006. GPR for fracture mapping and as a guide for the extraction of ornamental granite from a quarry: A case study from southern Brazil, *Journal of Applied Geophysics*, 58:177–187. https://doi.org/10.1016/j.jappgeo.2005.05.010

Riquelme, A, Martínez-Martínez, J, Martín-Rojas, I, Sarro, R and Rabat, Á, 2022. Contról of natural fractures in historical quarries via 3D point cloud analysis, *Engineering Geology*, 301:106618. https://doi.org/10.1016/j.enggeo.2022.106618

Rossi, F G, Madonna, S and Milli, S, 2017. Facies analysis applied to quarries: a review of the possible use of sedimentology to increase the knowledge and use of ornamental sandstones, *Procedia Engineering*, 208:136–144. https://doi.org/10.1016/j.proeng.2017.11.031

Selma, K, 2008. Photographing layer thicknesses and discontinuities in a dimension stone quarry with 3D GPR visualization, *Journal of Applied Geophysics*, 64:109–114. https://doi.org/10.1016/j.jappgeo.2008.01.001

Vidal, F W H, Azevedo, H C A and Castro, N F, 2014. Dimensional Rock Technology: Research, Mining and Processing; Mineral Technology Center [in Portuguese: Tecnologia de Rochas Ornamentais: Pesquisa, Mineração e Beneficiamento; Centro de Tecnologia Mineral], CETEM/MCTI: Rio de Janeiro, Brazil, 700 p.

Yurdakul, M, 2020. Natural stone waste generation from the perspective of natural stone processing plants: An industrial-scale case study in the province of Bilecik, Turkey, *Journal of Cleaner Production*, 276:123339. https://doi.org/10.1016/j.jclepro.2020.123339

Optimisation of cut-off grade in open pit mines considering spatial and temporal dimensions – a case study

F Manríquez[1], M Valdivia[2], F Pérez[3] and R Montecinos[4]

1. Assistant Professor, Departamento de Ingeniería de Minas, Metalurgia y Materiales, Universidad Técnica Federico Santa María, Santiago, Región Metropolitana, 8940897, Chile. Email: fabian.manriquez@usm.cl
2. Undergraduate student, Departamento de Matemática, Universidad Técnica Federico Santa María, Santiago, Región Metropolitana, 8940897, Chile. Email: marcelo.valdivia@usm.cl
3. Undergraduate student, Departamento de Matemática, Universidad Técnica Federico Santa María, Santiago, Región Metropolitana, 8940897, Chile. Email: felipe.perezav@usm.cl
4. Undergraduate student, Departamento de Matemática, Universidad Técnica Federico Santa María, Santiago, Región Metropolitana, 8940897, Chile. Email: rodrigo.montecinosar@usm.cl

ABSTRACT

Determining cut-off grades throughout a mine's life is essential for strategic open pit mining plans and significantly affects the project's Net Present Value (NPV). Traditional approaches, such as Lane's algorithm and its extensions, have been widely applied, while Direct Block Scheduling (DBS) offers implicit cut-off grade determination over time. However, DBS faces computational challenges in large-scale cases, often requiring heuristics or metaheuristics to mitigate these limitations.

This study presents an optimisation model to determine cut-off grades, integrating spatial and temporal dimensions. Using grade-tonnage distribution curves for predefined bench-phases, the model is formulated as a mixed-integer linear programming problem that maximises the mining plan's NPV. The optimisation identifies break-even and marginal cut-off grades for each bench-phases and period, considering operational constraints like precedence, mining rates, and processing capacities.

Applied to a real-world mine in northern Chile, including additional cut-off grade alternatives per bench-phase led to further NPV increases, demonstrating the model's capacity to unlock greater economic value.

The proposed method solves all scenarios within 2.5 hrs, emphasizing its computational efficiency and practicality for large-scale applications. It positions itself as a viable alternative to DBS, eliminating reliance on heuristics. Additionally, its low computational requirements enable it to optimise mining and processing capacities, provided the mining phases are predefined.

This framework offers a robust, efficient tool for enhancing decision-making in open pit mine planning, balancing economic objectives and operational feasibility.

INTRODUCTION

Open pit mine planning is about finding a profitable sequence for material extraction while meeting operational and technical constraints (Whittle, 1989). This process is structured into strategic (long-term), tactical (medium-term), and operational (short-term) levels (L'Heureux, Gamache and Soumis, 2013), each serving different purposes and information needs.

Strategic planning aims to maximise economic value by deciding ore extraction timing, lifespan, and investment, leading to a production schedule. Tactical scheduling spans up to five years, aiming to meet targets while minimising costs. Operational scheduling manages shorter time frames, optimising mine operations by efficiently allocating equipment.

The planning problem has been addressed using mathematical optimisation techniques to maximise value while ensuring compliance with operational constraints. However, due to its complexity, there is room for further improvement. This work focuses on determining cut-off grades in open pit mining. The next section provides a brief review of the state-of-the-art on cut-off grade optimisation.

RELATED WORK

Determining the optimal cut-off grade over a mine's lifespan is a critical challenge in mining operations, as it significantly impacts the Net Present Value (NPV) of the project (Bascetin and Nieto, 2007). Lane (1964, 1988) introduced a pioneering approach to optimize cut-off grades dynamically, allowing higher-grade ore to be processed earlier, thereby improving the NPV. However, his method is not widely implemented due to its complexity. Various studies have extended Lane's framework to address additional constraints, such as dynamic metal prices (Asad, 2005, 2007), environmental costs (Osanloo, Rashidinejad and Rezai/, 2008), and rehabilitation expenses (Gholamnejad, 2009). Other researchers have developed alternative optimisation techniques, including analytical solutions (Dagdelen, 1993), multi-element considerations (King, 2011), and iterative algorithms for optimising mine, plant, and smelter capacities (Abdollahisharif, Bakhtavar and Anemangely, 2012). More recent efforts have incorporated machine learning and heuristic approaches, such as genetic algorithms combined with nonlinear programming (Azimi and Osanloo, 2011), to further refine cut-off grade optimisation while maximising NPV (Ahmadi, 2018; Ahmadi and Shahabi, 2018).

Despite these advancements, most academic literature still relies on predefined ore-waste classification based on marginal cut-off grades (Fathollahzadeh *et al*, 2021). However, more flexible models allow mining blocks to be dynamically assigned to different destinations, optimising overall mine performance (Kumral, 2012; Moreno *et al*, 2017; Rezakhah, Moreno and Newman, 2020; Rezakhah and Newman, 2020). Although commercial mine planning software, such as Whittle (1999), incorporates Lane's principles, studies have shown that these tools do not always produce optimal solutions (Moreno *et al*, 2017). Key limitations of Lane's original model and its extensions include assumptions of uniform cut-off grades across the ore deposit, constant processing capacities, and the absence of blending constraints, all of which restrict its applicability to complex, multi-zone deposits. Addressing these limitations remains an open challenge in mine planning optimisation.

This paper presents an optimisation model designed to generate long-term mining plans for open pit mines. The model schedules both ore and waste from the bench phases, assuming that the phases have already been defined in a previous stage. It integrates the determination of cut-off grades with the scheduling of the bench phases.

METHODOLOGY

The general methodology applied in this work is as follows. First, the optimisation model is formulated. Next, the computational implementation of the model is carried out. Subsequently, the correct implementation of the model is verified. Then, the model is validated by generating different production plans, using a real mine located in northern Chile as a case study. Finally, the obtained results are presented, followed by the discussion and conclusions.

Optimisation model description

In this section we describe the mixed-integer linear programming (MILP) optimisation model to generate a long-term mine production schedule. The strategic model determines cut-off grades, as well as the material that will be extracted at each bench-phase throughout the scheduling horizon.

Models' inputs

Require technical and economic data, as well as tonnages and grades for each bench-phase of the block model. Defining these phases is outside the scope of this work. We also choose *a priori* a set of potential cut-off grades to apply to each bench-phase.

Models' outputs

Consist of extracted material quantities at each bench-phase throughout the scheduling horizon, along with the corresponding cut-off grades. The extracted material can be sent to the ore processing plant or the waste dump. The strategic model determines one cut-off grade for each bench-phase and period, which classify the material into two categories: ore and waste. It is worth noting that the cut-off grades may vary between periods and bench-phases.

Sets and indexes

$\mathcal{T}, t$ — Set and index of periods

$\mathcal{F}, f$ — Set an index of phases

$\mathcal{B}_f, b$ — Set and index of bench that belong to phase $f \in \mathcal{F}$

$\mathcal{K}, k$ — Set of cut-off grades alternatives

Parameters

r — Annual discount rate (in %)

$v_{f,b,k,t}$ — Profit (in USD/t) when applying the cut-off grade k in bench-phase b, f at period t

M — Material movement capacity (in t/annum)

C — Ore processing capacity (in t/annum)

Decision variables

The principal decision variable is described below:

$x_{f,b,k,t} \in \mathbb{R}_{\geq 0}$ — Extracted tonnage on the bench-phase (b, f) at period t, where we apply the k-th cut-off grade

This decision variable enables the model to select the appropriate cut-off grade (subindex k) to apply across the temporal dimension (subindex t) and spatial dimensions (subindexes f and b).

Objective function

This model aims to maximise the NPV of the production schedule, as shown in Equation 1:

$$\max \sum_{t \in \mathcal{T}} \sum_{f \in \mathcal{F}} \sum_{b \in \mathcal{B}_f} \sum_{k \in \mathcal{K}} \left(\frac{1}{(1+r)^t} \right) v_{f,b,k,t} \cdot x_{f,b,k,t} \tag{1}$$

Constraints description

The optimisation model considers the following constraints:

- Restricts the maximum ore tonnage that can be extracted from a bench-phase in the first period, considering all possible cut-off grades. This maximum corresponds to the total tonnage available in that bench-phase.

- Limits the ore extraction in each period to the total tonnage of the selected cut-off grade in the bench-phase, but only if that specific cut-off grade is chosen by the model.

- Ensures that only one cut-off grade is assigned to each bench-phase in each period.

- Restricts the ore tonnage extracted in periods, considering the remaining tonnage after previous extractions.

- Set the upper extraction limits for total mined material (ore and waste) and processed ore in each period, respectively.

- Set the lower extraction limits for total mined material (ore and waste) and processed ore in each period, respectively.

- Ensures that a bench-phase cannot be fully extracted in a given period unless all material in its face has been removed.

- Enforces precedence between bench-phases.

CASE STUDY

In this case study, a block model of a copper deposit located in northern Chile is used. This block model includes a prior phase definition, meaning that each block belongs to one of these phases.

The phase definition was carried out at an earlier stage and is outside the scope of this study. The mining phases are illustrated in Figure 1.

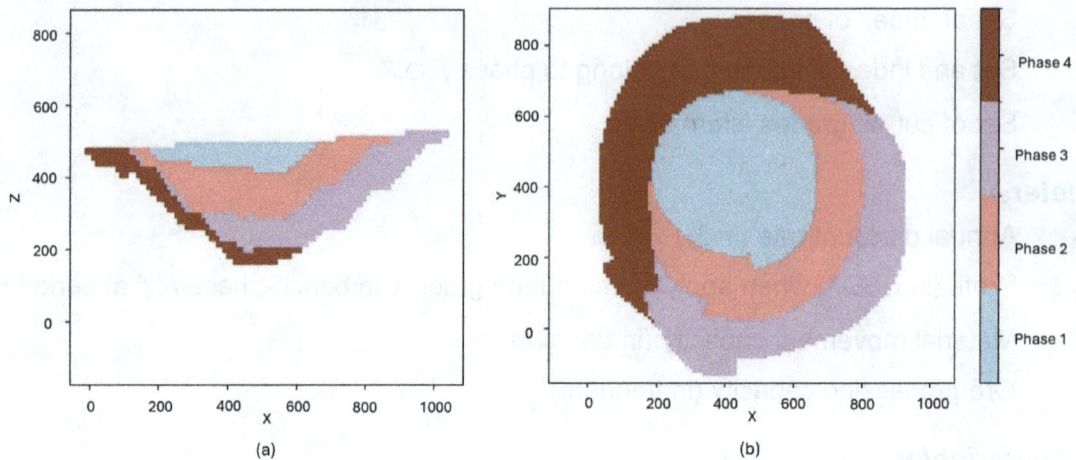

(a)

(b)

FIG 1 – East–west cross-section view (a) and top-down (plant) view (b) of the mining phases in the case study.

Figure 2 presents the mean copper grade in the ore, disaggregated by phase and bench elevation. As shown in the figure, the mean copper grade increases with increasing bench depth in all phases.

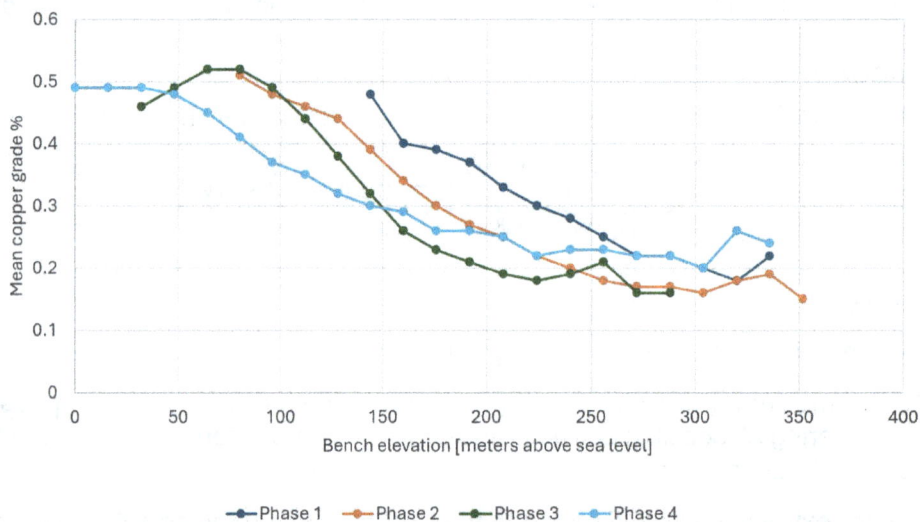

FIG 2 – Mean copper grade in ore by phase and bench elevation.

To verify the proper functioning of the optimisation model described in the previous section, production plans are generated with different mine and plant capacities and varying numbers of cut-off grade alternatives. Specifically, mining plans are compared across two scenarios: one with high mine and plant capacities and another with low mine and plant capacities. The high-capacity scenario is referred to as Scenario A, while the low-capacity scenario is called Scenario B.

The mine and plant capacities used for scenarios A and B are shown in Table 1. For Scenario A, the relationship proposed by Long and Singer (2001) for copper open pit mines was used as an initial approximation to estimate the ore processing capacity. The preliminary estimation of ore reserves was based on the criterion that a block qualifies as ore if its grade is greater than or equal to the marginal cut-off grade. Subsequently, the mine's overall stripping ratio and the previously defined plant capacity were used to determine the total material movement capacity for Scenario A. In Scenario B, the plant capacity from Scenario A was reduced to ensure that, following the ramp-up periods, the ore delivered to the processing plant would fully utilise the installed capacity in each time period. The mine capacity for Scenario B was slightly decreased compared to Scenario A to support this objective.

TABLE 1

Mining and processing capacity for scenarios A and B.

Capacities	Unit	Scenario A	Scenario B
Mining capacity	Mt/annum	22.1	20.0
Ore processing capacity	Mt/annum	11.4	3.5

Within each scenario (A or B), five long-term production plans are computed. Each of these plans differs by the number of cut-off grade alternatives available to the optimisation model to define ore and waste across the mine's phases-benches (spatial dimension) and throughout the planning horizon (temporal dimension).

The candidate cut-off grades are shown in Table 2. The first two cut-off grades used correspond to the marginal cut-off grade and the break-even cut-off grade.

TABLE 2

Cut-off grade alternatives considered in the case study.

Cut-off grade index	Cut-off grade name	Cut-off grade value
1	Marginal cut-off grade	0.1503%
2	Break-even cut-off grade	0.1803%
3	3th cut-off grade	0.4216%
4	4th cut-off grade	0.3000%
5	5th cut-off grade	0.2406%

The marginal cut-off grade is the grade at which the value of sending a block to the processing plant is equal to the value of sending the same block to the waste dump.

The break-even cut-off grade is the grade at which the net benefit of sending a block to the plant is zero. In other words, it is the grade at which revenues balance with the costs of mining, processing, and refining.

In order to directly compare all these mining production plans, the same technical and economic parameters used to evaluate the ore and waste are applied. The economic and technical parameters used in the case study include a mining cost of 2.0 USD per tonne, a processing cost of 10.0 USD per tonne, and a selling and refining cost of 0.25 USD per pound of copper. The copper price was set at 3.8 USD per pound, with a metallurgical recovery of 85.0 per cent. A 10.0 per cent annual discount rate was applied for economic evaluations.

All the schedules presented in this study were obtained on a 3.60 GHz 11th Gen Intel(R) Core™ i7-11700K, with 32 GB RAM, running Windows 11®. The optimisation model was solved using Gurobi Optimiser version 11.0.3.

RESULTS

Results of the mining plans in Scenario A

Table 3 presents the production plan results based on the number of cut-off grade alternatives per bench phase for the scenario with high mine and plant capacities (Scenario A). The base case corresponds to a production plan generated using a single cut-off grade, set equal to the marginal cut-off grade.

TABLE 3

Mining plan results based on the number of cut-off grade alternatives for scenario A.

Number of cut-off grade alternatives	Van [MUSD]	Percentage difference compared to the base case	Mip gap %	Computing time [s]
1	202.6	-	0.0%	308
2	205.7	1.53%	0.0%	1871
3	205.8	1.58%	0.0%	3297
4	206.9	2.12%	0.0%	3112
5	209.9	3.60%	0.0%	5541

It is important to note that the NPV reported in Table 3 does not include the initial investment in the mine and plant. The exclusion of these investments was intentional, as it allows for a direct comparison of the NPV between scenario A and scenario B, isolating the effect on NPV solely due to the scheduling of the bench-phases.

Figure 3 shows the mining production plans for scenario A, comparing the use of a single cut-off grade alternative with the use of five cut-off grade alternatives. From this figure, we observe that both plans exhibit an increasing mean copper grade over the years. This trend is explained by the metal grade distribution in the deposit, as previously shown in Figure 2.

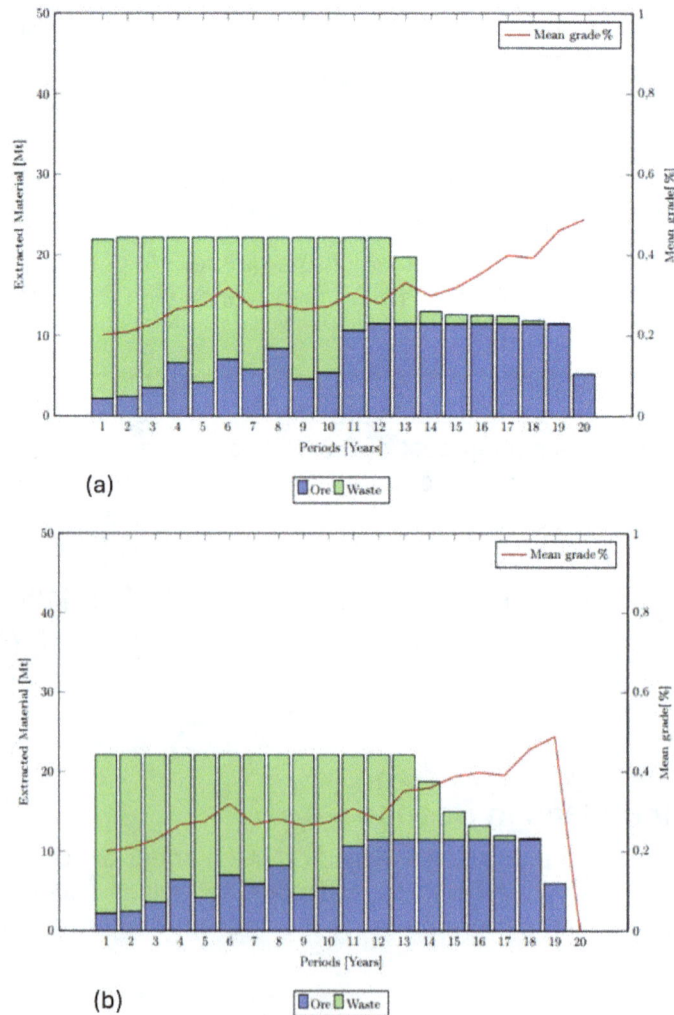

FIG 3 – Mining plan for scenario A using a single cut-off grade alternative (a) and using five cut-off grade alternatives (b).

APCOM 2025 | Perth, Australia | 10–13 August 2025

Results of the mining plans in scenario B

Table 4 presents the production plan results based on the number of cut-off grade alternatives per bench phase for the scenario with low mine and plant capacities (Scenario B). The base case corresponds to a production plan generated using a single cut-off grade, set equal to the marginal cut-off grade.

TABLE 4

Mining plan results based on the number of cut-off grade alternatives for scenario B.

Number of cut-off grade alternatives	Van [MUSD]	Percentage difference compared to the base case	Mip gap %	Computing time [s]
1	38.026	-	0.1	8 462
2	41.471	9.06%	10	3 005
3	44.002	15.72%	10	1 466
4	51.780	36.17%	10	4 442
5	54.081	42.22%	10	1 943

It is important to note that the NPV reported in Table 4 does not include the initial investment in the mine and plant. The exclusion of these investments was intentional, as it allows for a direct comparison of the NPV between scenario A and scenario B, isolating the effect on NPV solely due to the scheduling of the bench-phases.

Figure 4 shows the mining production plans for scenario B, comparing the use of a single cut-off grade alternative with the use of five cut-off grade alternatives. From this figure, we observe that both plans exhibit an increasing mean copper grade over the years. This trend is explained by the metal grade distribution in the deposit, as previously shown in Figure 2.

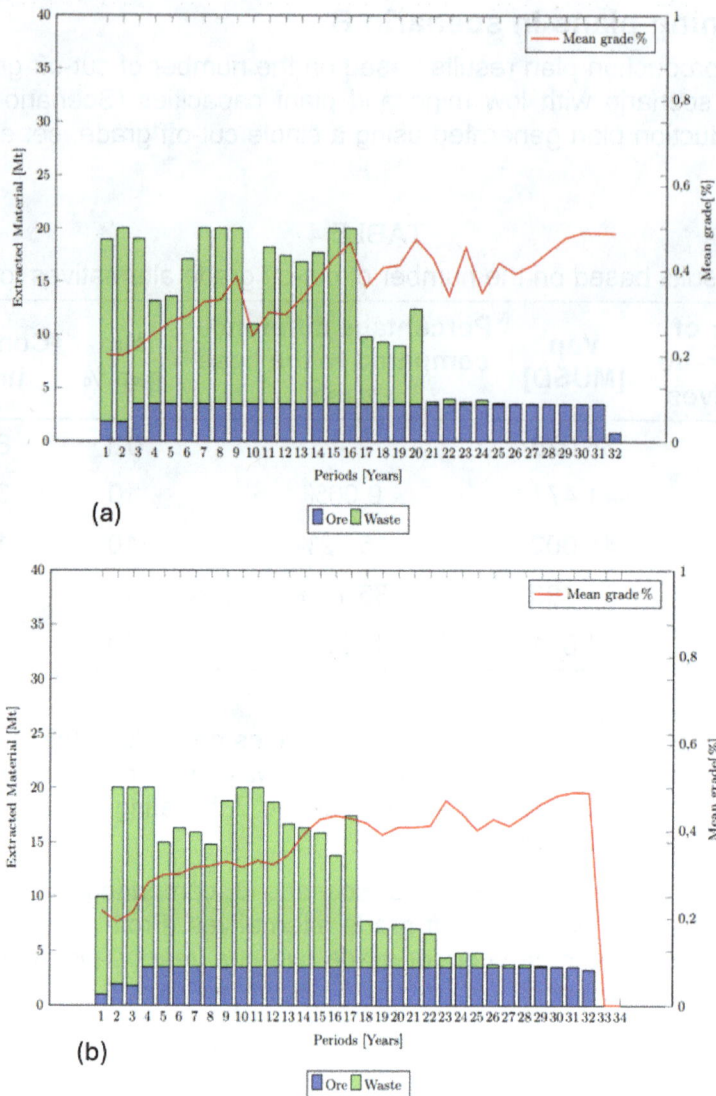

FIG 4 – Mining plan for scenario B with a single cut-off grade alternative.

DISCUSSION

In this section, we analyse and discuss the results presented in the previous section.

It is important to mention that the NPV reported in Tables 3 and 4 does not include the initial investment in the mine and plant. The exclusion of these investments was intentional, as it allows for a direct comparison of the NPV between scenario A and scenario B, isolating the effect on NPV solely due to the scheduling of the bench-phases.

Discussion of the results for the mining plans in Scenario A

As shown in Table 4, as the production plan is generated with a greater number of cut-off grade alternatives per bench phase, the NPV of the plan increases. This result is expected and consistent, as the optimisation model, with a greater number of cut-off grades, can prioritise the processing of higher-grade ore. This, in turn, increases the NPV in these plans, to the detriment of mining plans that do not have the option to prioritise the processing of higher-grade ore.

The mining plan that achieved the highest NPV (209.9 MUSD) was obtained using five cut-off grade alternatives, representing a 3.60 per cent increase compared to the base case, which was generated using a single cut-off grade equal to the marginal cut-off grade.

Analysing the production plans presented in Figures 1 and 2, it can be observed that both mining plans saturate the plant capacity in year 12. This indicates that the chosen mining and plant capacity is not the most suitable for the operating deposit, as mine planners typically aim to saturate the plant in earlier time periods. This is one of the reasons why we computed an additional set of mining plans with lower capacities.

Discussion of the results for the mining plans in Scenario B

As shown in Table 4, as the production plan is generated with a greater number of cut-off grade alternatives per bench phase, the NPV of the plan increases.

Once again, this result is anticipated and aligns with expectations, as the optimisation model, with more cut-off grades, is able to prioritise the processing of higher-grade ore. This leads to an increase in NPV for these plans, compared to mining plans that lack the flexibility to prioritise higher-grade ore processing.

From the analysis of Table 4, it is observed that the NPV of the plans with one and five cut-off grades is 38.026 MUSD and 54.081 MUSD, respectively. This represents a percentage increase of 42.22 per cent. The significant percentage increase in the NPV of the production plan with five cut-off grade alternatives can be explained by the fact that, with a greater number of cut-off grade candidates, the optimisation model has more flexibility to prioritise higher-grade ore for processing. This results in an increase in the average metal grade of the ore sent to the plant. In contrast, when only one cut-off grade is available, the optimisation model cannot prioritise higher-grade ore over lower-grade ore, as there is only a single cut-off grade option.

Discussion of the results for mining plans in Scenarios A and B

In Scenario A, the number of years required to move the total amount of ore and waste is 20 years in the case with one cut-off grade alternative and 19 years in the case with five cut-off grade alternatives.

In Scenario B, the total time required to move the combined amount of ore and waste is 32 years for both the single cut-off grade alternative and the five cut-off grade alternatives.

The Life-of-mine in mine plans of Scenario A are less than the Life-of-mine of mine plans of the scenario B.

From the comparison of the mine production schedule of scenarios A (Figures 1 and 2) and B (Figures 3 and 4) we verify that both mining capacity and ore processing capacity are meet.

As previously mentioned, the percentage difference in NPV between the case with one cut-off grade alternative and the case with five cut-off grade alternatives was 3.60 per cent and 42.22 per cent for the high and low-capacity scenarios, respectively. This indicates that the impact on NPV of having more cut-off grade alternatives to prioritise the processing of higher-grade ore is greater when mine and plant capacities are lower. This is expected, as in the hypothetical extreme case where mine and plant capacities are infinite, the cut-off grade to be applied would be the marginal cut-off grade. In this scenario, the mining plan would consist of just one period (year), meaning the effect of the discount rate would be null, and the objective function would shift to maximising total profit (revenue minus costs).

Since the mine and plant investment is not included in the NPVs of the mining plans, we observe that the NPVs are higher for plans with larger capacities, given the same number of cut-off grades. This is consistent because higher capacities allow for more ore to be processed in the early periods, resulting in increased cash flow during these periods and ultimately yielding a higher NPV compared to plans with lower capacities.

All schedules for Scenario A were solved in under 2 hrs, highlighting the model's computational efficiency and its suitability for large-scale applications. On the other hand, all schedules for Scenario B were completed in under 2.5 hrs, further reinforcing the model's effectiveness. Additionally, as expected, we observed that computation time is influenced by the mine and plant capacities.

The low computational time allows mining engineers to generate mining plans more quickly, facilitating the optimisation of the project's NPV based on the estimated capital costs of the mine and plant, relative to their respective capacities.

CONCLUSIONS

This paper presents an optimisation model designed to generate long-term mining plans for open pit mines. The model schedules both ore and waste from the bench phases, assuming these phases have been defined in a prior stage. It integrates the determination of cut-off grades with the scheduling of the bench phases.

We verify that, when mine and plant capacities remain constant, incorporating additional cut-off grade alternatives per bench-phase increases the NPV.

On the other hand, the impact on NPV improvement from having more cut-off grade alternatives to prioritise the processing of higher-grade ore is more significant when mine and plant capacities are lower.

From the analysis of the case study results, we conclude that the optimisation model effectively prioritises the processing of higher-grade ore when more cut-off grade alternatives are available. As the production plan incorporates more cut-off grade options per bench phase, the NPV increases.

The model described here integrates cut-off grade determination for different phases and benches over time, considering the grade-tonnage distribution for each bench-phase. This improves upon the Lane algorithm, as it and its extensions do not manage bench-phase scheduling or consider the grade-tonnage distribution at the bench-phase level, focusing only on the phase level.

Furthermore, computation times for generating mining plans for the resulting instances are under 2.5 hrs. Generally, computation time depends on mine and plant capacities, the number of phases and benches in each phase, and the specific characteristics of the instance itself.

In conclusion, this model is a robust and efficient tool for optimising decision-making in open pit mine planning.

ACKNOWLEDGEMENTS

The authors express their gratitude to Dr. Gonzalo Nelis for providing the block model used in this case study.

F Manríquez was supported by the Agencia Nacional de Investigación y Desarrollo (ANID) through the Fondecyt Iniciación Project 11250217.

REFERENCES

Abdollahisharif, J, Bakhtavar, E and Anemangely, M, 2012. Optimal cut-off grade determination based on variable capacities in open pit mining, *Journal of the Southern African Institute of Mining and Metallurgy*, 112(12):1065–1069.

Ahmadi, M R and Shahabi, R S, 2018. Cutoff grade optimization in open pit mines using genetic algorithm, *Resources Policy*, 55:184–191.

Ahmadi, M R, 2018. Cutoff grade optimization based on maximizing net present value using a computer model, *Journal of Sustainable Mining*, 17(2):68–75. https://doi.org/10.1016/JJ.SM.2018.04.002

Asad, M W A, 2005. Cut-off grade optimization algorithm for open pit mining operations with consideration of dynamic metal price and cost escalation during mine life, in *Proceedings of the 32nd International Symposium on the Application of Computers and Operations Research in the Mineral Industry*, pp 273–277.

Asad, M W A, 2007. Optimum cut-off grade policy for open pit mining operation through net present value algorithm considering metal price and cost escalation, *International Journal for Computer-Aided Engineering and Software*, 24(7):723–736.

Azimi, Y and Osanloo, M, 2011. Determination of open pit mining cut-off grade strategy using combination of nonlinear programming and genetic algorithm, *Archives of Mining Sciences*, 56(2):189–212.

Bascetin, A and Nieto, A, 2007. Determination of optimal cut-off grade policy to optimise NPV using a new approach with optimisation factor, *Journal of the Southern African Institute of Mining and Metallurgy*, 107(2):87–94.

Dagdelen, K, 1993. An NPV optimisation algorithm for open pit mine design, in *Proceedings of the 24th International Symposium on Application of Computers and Operations Research in the Mineral Industry*, pp 257–263 (CIM).

Fathollahzadeh, K, Asad, M W A, Mardaneh, E and Cigla, M, 2021. Review of Solution Methodologies for Open Pit Mine Production Scheduling Problem, *International Journal of Mining, Reclamation and Environment*, 35(8):564–599. https://doi.org/10.1080/17480930.2021.1888395

Gholamnejad, J, 2009. Incorporation of rehabilitation cost into the optimum cutoff grade determination, *Journal of the Southern African Institute of Mining and Metallurgy*, 109:89–94.

King, B, 2011. Optimal mining practice in strategic planning, *Journal of Mining Science*, 47(2):247–253. https://doi.org/10.1134/S1062739147020110

Kumral, M, 2012. Production planning of mines: Optimisation of block sequencing and destination, *International Journal of Mining, Reclamation and Environment*, 26(2):93–103. https://doi.org/10.1080/17480930.2011.644474

L'Heureux, G, Gamache, M and Soumis, F, 2013. Mixed integer programming model for short term planning in open pit mines, *Mining Technology*, 122(2):101–109. https://doi.org/10.1179/1743286313Y.0000000037

Lane, K F, 1964. Choosing the optimum cut-off grade, *Colorado School of Mines Quarterly*, 59:811–829.

Lane, K F, 1988. *The Economic Definition of Ore: Cut-off Grade in Theory and Practice* (Mining Journal Books: London).

Long, K R and Singer, D A, 2001. A simplified economic filter for open pit mining and heap-leach recovery of copper in the United States, No. 2001-218, US Geological Survey.

Moreno, E, Rezakhah, M, Newman, A and Ferreira, F, 2017. Linear models for stockpiling in open pit mine production scheduling problems, *Eur J Oper Res*, 260:212–221. https://doi.org/10.1016/j.ejor.2016.12.014

Osanloo, M, Rashidinejad, F and Rezai, 2008. Incorporating environmental issues into optimum cut-off grades modeling at porphyry copper deposits, *Resources Policy*, 33:222–229.

Rezakhah, M and Newman, A, 2020. Open pit mine planning with degradation due to stockpiling, *Comp Operations Res*, https://doi.org/10.1016/j.cor.2018.11.009

Rezakhah, M, Moreno, E and Newman, A, 2020. Practical performance of an open pit mine scheduling model considering blending and stockpiling, *Comput Oper Res*, 115:104638. https://doi.org/10.1016/j.cor.2019.02.001

Whittle, J, 1989. The facts and fallacies of open pit design, Whittle Programming Pty Ltd.

Whittle, J, 1999. A decade of open pit mine planning and optimization-the craft of turning algorithms into packages, in *Proceedings of the 28th International Symposium on Application of Computers and Operations Research in the Mineral Industry*, pp 15–24 (Colorado School of Mines).

Mineral processing plant capacity based on geometallurgical block model scheduling

D B Mazzinghy[1], N Morales[2], A Brickey[3], G Nelis[4], J Ortiz[5] and M J F Souza[6]

1. Professor, Department of Mining Engineering, Universidade Federal de Minas Gerais, Belo Horizonte, Minas Gerais 31.270–901, Brazil. Email: dmazzinghy@demin.ufmg.br
2. Professor, Département des Génies Civil, Géologique et des Mines, Polytechnique Montréal, Montreal QC H3T 1J4, Canada. Email: nelson.morales@polymtl.ca
3. Professor, Mining Engineering and Management Department, South Dakota School of Mines and Technology, Rapid City SD 57701, USA. Email: andrea.brickey@sdsmt.edu
4. Professor, Departamento de Ingeniería de Minas, Metalúrgica yMateriales, Universidad Técnica Federico Santa María, Santiago 8940897, Chile. Email: gonzalo.nelis@usm.cl
5. Professor, Camborne School of Mines, University of Exeter, Falmouth, Cornwall TR10 9FE, UK. Email: j.ortiz-cabrera@exeter.ac.uk
6. Professor, Department of Computing, Universidade Federal de Ouro Preto, Ouro Preto, Minas Gerais 35.402–136, Brazil. Email: marcone@iceb.ufop.br

ABSTRACT

The geometallurgy approach seeks to connect different areas, such as geology, mining, processing, finance, and environment, in an integrated workflow to increase the knowledge of the orebody. It has been used successfully to reduce risks in greenfield and brownfield projects. During the development of the greenfield project, the mineral processing plant capacity is usually estimated using a large sample composite by some drill hole intervals from different deposit areas. This sample is often called a 'representative sample', and it is used to perform a pilot plant test to obtain parameters for plant capacity calculation. For example, a greenfield project with 2000 drill hole intervals and chemical composition data may only have 20 comminution tests from composite drill hole intervals. Considering the ore variability, the risk of over or underestimating the plant capacity is high. Nowadays, it is possible to obtain comminution indices during the sample preparation of the drill hole intervals for chemical composition analysis using devices like the Geopyörä Breakage Test (GBT) and Hardness Index Testing (HIT). These devices need a small sample mass, and the tests are done quickly and cheaply. The comminution indices like Impact Breakage Index ($A*b$) and Bond Work Index (BWI) can be converted into specific energy (kWh/t), and the plant capacity per period can be estimated from mine scheduling, considering the plant operational time as a constraint. It is common to have approximately 8000 hrs/annum as plant operational time, depending on the equipment availability. Then, it is possible to find the best time in the Life-of-mine (LOM) to expand the plant to maintain the concentrate quality and production, for example. The methodology presented here can estimate the plant capacity according to the geometallurgical block model scheduling, avoiding future bottlenecks and supporting investment decision-making.

INTRODUCTION

The mining industry is currently changing due to declining ore grades, resulting in a significant increase in the volume of run-of-mine (ROM) to be processed to achieve the same product quantity specifications. It is necessary to understand the orebody deeply to avoid some problems, such as reduced production capacity and out-of-specification products during the life-of-mine (LOM). The heterogeneous nature of orebodies and the variability in physical and chemical properties of the lithologies make mining operations very challenging, and the geometallurgy approach provides a more integrative process. Geometallurgy is an interdisciplinary approach that plays a pivotal role in any project's evaluation or proposed mining operation, and it focuses on spatially characterising the different material types or domains within a deposit, considering their impact on processing, mining performance, and environmental and closure aspects (Global Mining Guidelines Group (GMG), 2025). The geology information from drill holes and the bench metallurgical test results from these drill cores can be used to predict the mineral processing plant performance; also, mine planning can be performed to maximise the net present value (NPV) of the orebody. A challenge during the development of a greenfield project is the determination of the mineral processing plant capacity.

The plant capacity is usually estimated using large sample composites with a defined drill holes interval. This sample is often called a 'representative sample' and it is used to perform a pilot plant test to obtain parameters for plant capacity calculation. It is very common in a greenfield project to have 2000 drill hole intervals with chemical composition data and just only 20 comminution tests from composite drill hole intervals. Considering the ore variability, the risk during the plant capacity estimation is too high, needing more attention to avoid future bottlenecks. For example, consider two blocks with the same grade and same recovery, but with different hardness. The metal per hour produced by each block will be different once the capacity is a function of the specific energy and processing time. Also, the processing cost will vary according to the block hardness with harder blocks requiring more time to be processed. We can conclude from this simple example that the real value of a block cannot be estimated precisely without the specific energy per block (metal per hour). The processing time can modify the mine planning decision-making process and must be considered in the mine schedule optimisation.

The objective of this investigation is to show an open pit mine schedule that incorporates specific energy as a geometallurgical variable block-by-block and how the processing plant capacity can be properly estimated using this approach. The case study developed during this research can be used to illustrate the impact and importance of incorporating geometallurgy into the mine planning process.

OBTAINING COMMINUTION INDICES FROM DRILL HOLES

Currently, it is possible to obtain comminution indices during the sample preparation of the drill hole intervals for chemical composition analysis using geometallurgical devices like the Geopyörä Breakage Test (GBT) and Hardness Index Testing (HIT). These devices need a small sample mass, with tests being cheap and quick. The comminution indices like Impact Breakage Index ($A*b$) and Bond Work Index (BWI) can be converted into specific energy (kWh/t) and then converted in processing time (h). Figure 1 presents the GBT (Bueno et al, 2024) and the HIT (Bergeron et al, 2017) devices.

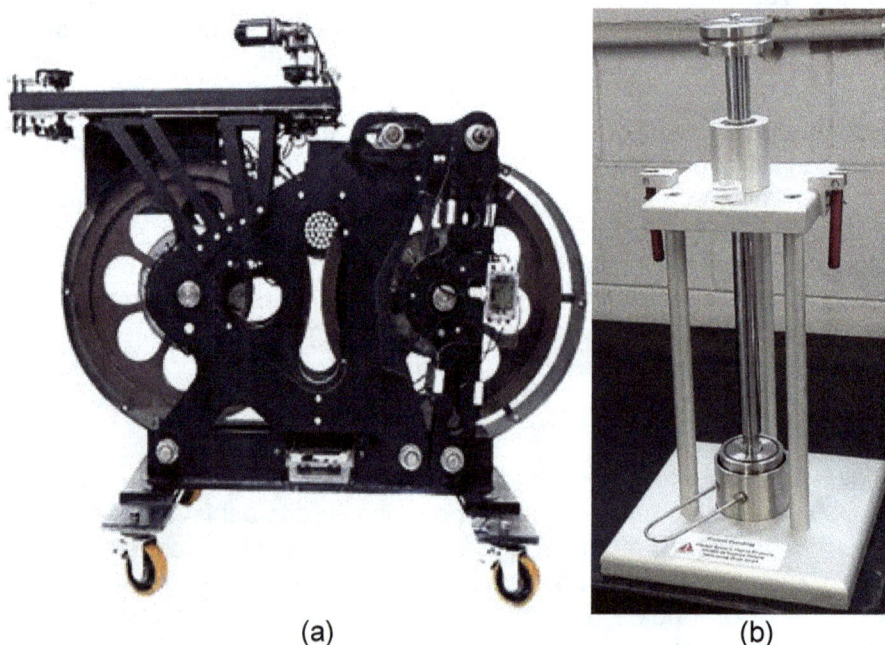

(a)　　　　　　　　　　　　　(b)

FIG 1 – (a) Geopyörä Breakage Test (GBT); and (b) the Hardness Index Testing (HIT).

GBT consists of two counter-rotating wheels that nip and crush the particles with a tightly controlled gap between the rollers. It was designed to obtain comminution indices cheaply and rapidly using a few samples (Bueno et al, 2024). HIT is a low impact device developed for rapid rock-hardness determination at a mine site, allowing the determination of rock hardness variability (Varianemil et al, 2023). The main goal of GBT and HIT is to provide comminution indices to populate orebody models to increase the knowledge about ore hardness variability and support the mine scheduling and decision-making process.

ESTIMATING THE SPECIFIC ENERGY FROM COMMINUTION INDICES

The SMC Test (Steve Morrell Comminution) is a methodology for estimating the specific energy of any comminution circuit (GMG, 2021). We provide an example showing how the specific energy of a SABC (SAG+Ball+Pebble Crusher) circuit can be estimated with the documented methodology. Figure 2 shows a simplified version of the processing flow sheet of the Batu Hijau mine with an SABC circuit (Varianemil *et al*, 2023).

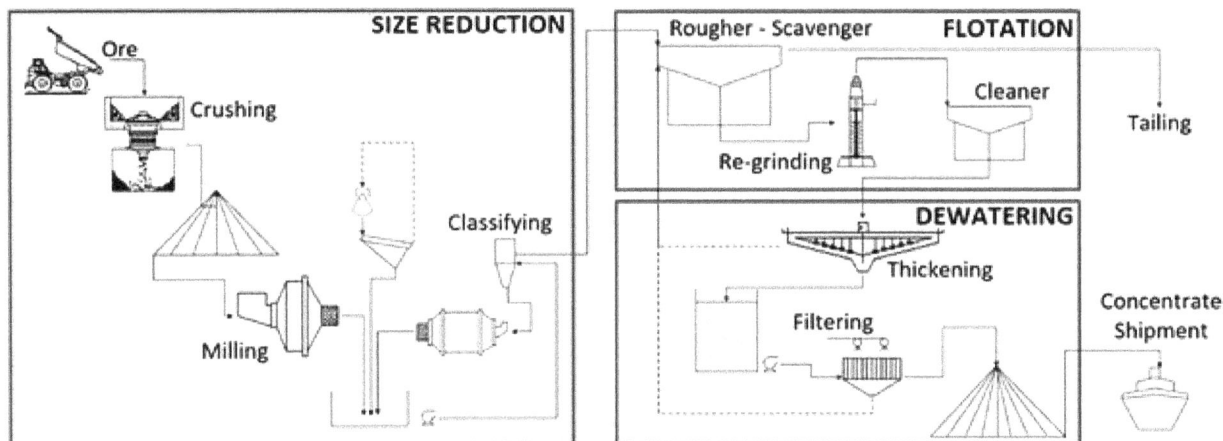

FIG 2 – Batu Hijau process flow sheet showing the SABC circuit (Varianemil *et al*, 2023).

We show the step-by-step calculations necessary to determine the processing time (h) of each block and then populate the block model with the geometallurgical information for use in the mine schedule optimisation process.

Step 1 – Specific energy calculation

The specific energy can be estimated according to Equation 1 (GMG, 2021).

$$E = 4Mia\left(750^{-\left(0,295+\frac{750}{10^6}\right)} - F_{80}^{-\left(0,295+\frac{F_{80}}{10^6}\right)}\right)K_1 + 4Mib\left(P_{80}^{-\left(0,295+\frac{P_{80}}{10^6}\right)} - 750^{-\left(0,295+\frac{750}{10^6}\right)}\right) \tag{1}$$

E	specific energy (kWh/t)
Mia	Working index of the coarse ore fraction
F_{80}	80 per cent passing in the feed of the grinding circuit (µm)
K_1	Pebble mill efficiency factor, being 0.95 when there is pebble recirculation and 1 when there is no pebble recirculation
Mib	Work index of the fine ore fraction
P_{80}	80 per cent passing in the product of the grinding circuit (µm)

Step 2 – Estimating *Mia* and *Mib*

The geometallurgical GBT and HIT devices can provide the comminution indices *A*b* and *BWI*. The *A*b* can be converted into *Mia,* and the *BWI* can be converted into *Mib* using Equations 2 and 3 (Doll, 2025):

$$Mia = 379.4\, A*b^{-0.8} \tag{2}$$

$$Screen\ size\ used\ in\ BWI \begin{cases} 300\mu m \rightarrow Mib = 0.60\, BWI^{1.20} \\ 212\mu m \rightarrow Mib = 0.63\, BWI^{1.21} \\ 150\mu m \rightarrow Mib = 0.69\, BWI^{1.22} \\ 106\mu m \rightarrow Mib = 0.71\, BWI^{1.24} \end{cases} \tag{3}$$

The *BWI* test should be performed using the screen size to produce the P_{80} that will be considered in the mill. Equation 3 shows four screen size models (300 µm, 212 µm, 150 µm, 106 µm). If the P_{80} target in the mill was 150 µm, for example, the screen size used in the BWI test should be the next size above, in this case 212 µm (*BWI@212 µm*).

Step 3 – Estimating *DWi* and F_{80}

The *DWi* can be estimated through the specific gravity SG and the *A*b* using Equation 4 (Doll, 2025):

$$DWi = \frac{100\ SG}{A*b} \tag{4}$$

The F_{80} (mm) of the SAG mill can be estimated considering the close side setting of the primary crusher using Equation 5 (Bailey *et al*, 2009):

$$F_{80} = 0.2\ CSS\ DWi^{0.7} \tag{5}$$

CSS   close side setting (mm)

DWi   Drop Weight Index (kWh/m³)

Step 4 – Estimating the plant throughput

The circuit throughput can be calculated knowing the power available and the specific energy through Equation 6:

$$T = \frac{P}{E} \tag{6}$$

T   is the throughput (t/h)

P   is the mill power available (kW)

E   is the specific energy (kWh/t)

Step 5 – Estimating the processing time

The processing time *PT* (h) can be calculated knowing block mass M_B (t) and the throughput *T* (t/h) using Equation 7:

$$PT = \frac{M_B}{T} \tag{7}$$

Step 6 – Estimating the plant availability time

The total time of the plant operation per annum can be calculated as 7884 hrs, considering 365 days per annum, 24 hrs per day, and 90 per cent operational yield, for example.

OPEN PIT MINE SCHEDULING PROBLEM

We present the two main methodologies of formulating the mine planning problem that were proposed in the 1960s; one presented by Lerchs and Grossmann (1965) and the other by Johnson (1968). However, due to the limitation of hardware and software at the time, only the Lerchs and Grossmann (LG) approach was tractable resulting in its development and becoming the industry standard for many decades. LG model finds the ultimate pit limit that provides the maximum undiscounted revenue at time zero. It can be used to produce nested pits obtained by changing the value of the blocks by applying a revenue factor. Then, the user must find the best pits containing the masses and grades for each period. Nested pit approach serves as an indication of where the highest value sectors are located inside the pit, and therefore, they are used as a guide to define a production schedule. However, there is no guarantee that the production schedule derived from the nested pits is optimal, especially when additional constraints are considered (blending, capacity etc). Nowadays, the pseudoflow algorithm (Hochbaum, 2001) is preferred because it is faster than the LG algorithm. Johnson approach has been proposed to tackle this issue and consists of scheduling the blocks using mixed integer linear programming (MILP). It can obtain an optimum solution, but this method is limited to solving real instances of the problem, which may involve several million blocks. The mine scheduling problem is NP-hard (Deutsch, Dağdelen and Johnson, 2022), and no algorithms solve this problem in polynomial time. Bienstock and Zuckerberg (2009) proposed a new algorithm that, through linear programming relaxations, finds high-quality solutions in instances with millions of variables and constraints of the mine scheduling problem. The BZ algorithm works in multiple steps to produce a mine schedule:

1. Aggregate block scheduling decisions in a column generation procedure using pseudoflow algorithm and then run a very small and fast linear programming optimisation.

2. This result is used for the subsequent round of column generation and this procedure is repeated until reaching the bound.

3. A very fast heuristic is used to give a complete schedule that is close to optimum.

4. Use the bound from BZ to speed up a MILP solver running the full problem.

5. Uses the branch-and-cut to iteratively use BZ to reach the optimum solution (Letelier *et al*, 2020; Minemax, 2025).

Direct block scheduling

In this paper, we will use the direct block scheduling concept to solve the open pit mine scheduling problem. In the literature this kind of problem is known as the Precedence Constrained Production Scheduling Problem (PCPSP) (Espinoza *et al*, 2012). In PCPSP, the economic value for a block can be calculated according to its destination, for example, a processing plant or waste dump, using Equations 8 and 9:

$$Process = M_B \, g \, r \, (S_P - S_c) - M_B(C_P + C_M) \qquad (8)$$

$$Waste = -M_B \, C_M \qquad (9)$$

M_B block mass (t)

g grade

r recovery

S_p selling price ($/t)

S_c selling cost ($/t)

C_p processing costs

C_M mining costs

The optimisation algorithm chooses the best destination for each block based on its value. In this approach, it is not necessary to define cut-off grades or determine before the optimisation if a block is flagged as ore or waste.

We used MiningMath (MM) software in our case study, and the formulation presented here is the one MM used (MiningMath, 2025). Figure 3 presents the simplified flow chart of the MM algorithm.

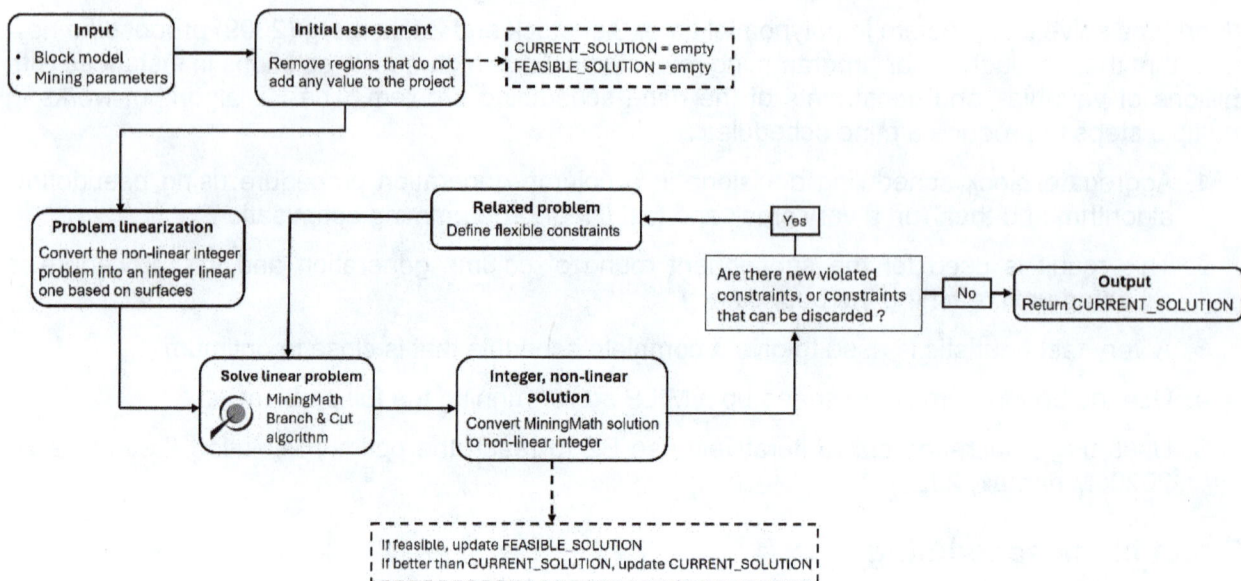

FIG 3 – Simplified flow chart of MiningMath algorithm (MiningMath, 2025).

The first step of the optimisation algorithm is to remove regions that do not add any value to the project, as shown in Figure 4.

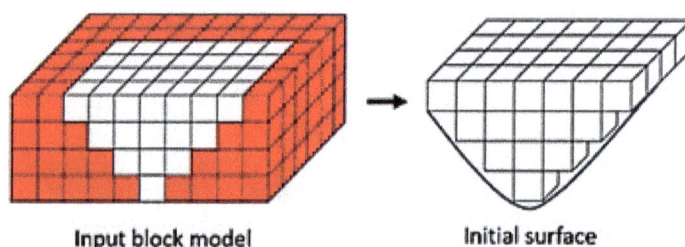

Input block model   Initial surface

FIG 4 – Removing regions that do not add any value to the project (MiningMath, 2025).

The second step of the optimisation algorithm is converting the non-linear integer problem into a linear one based on surfaces (geometry mine constraints), as shown in Figure 5. BW is the bottom width, MW is the mining width, and ADV is the vertical rate of advance.

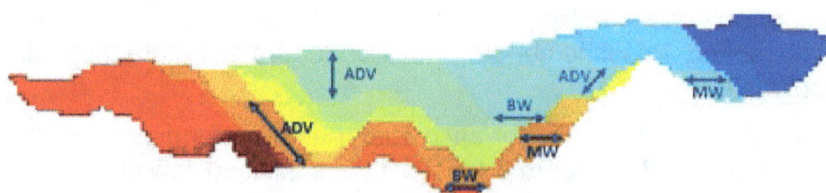

FIG 5 – Example solution with geometric constraints (MiningMath, 2025).

The third step of the optimisation algorithm is converting the linear solution to an integer solution using the branch-and-cut algorithm. This algorithm is more efficient than the branch-and-bound algorithm contained in standard MILP solvers, and it was also fine-tuned for this specific optimisation problem.

Common notation

S number of simulated orebody models considered

s simulation index, $s = 1, ., S$

D number of destinations

d destination index, $d = 1, ., D$

Z	number of levels in the orebody model
z	level index, $z = 1, ., Z$
T	number of periods over which the orebody is being scheduled and also defines the number of surfaces considered
t	period index, $t = 1, ., T$
M	number of cells in each surface, where $M = x \times y$ represents the number of mining blocks in x and y dimensions
c	cell index, $c = 1, ..., M$
G	number of unique destination groups defined. Each group might contain 1, all, or any combination of destinations
g	group index, $g = 1, ..., G$

Objective function

The objective function is the sum of the economic value of blocks mined per period, destination, and simulation. It uses the average result divided by the number of simulations and subtracts the penalties for certain violated restrictions associated with some user-defined parameters (Equation 10):

$$max \ \frac{1}{S} \sum_{s=1}^{S} \sum_{t=1}^{T} \sum_{c=1}^{M} \sum_{z=1}^{Z} \sum_{d=1}^{D} \left(V_{c,t,d,s}^{z} x_{c,t,d}^{z}\right) - p$$

$$c = 1, ..., M, \quad t = 1, ..., T, \quad z = 1, ..., Z, \quad d = 1, ..., D$$

$$x_{c,t,d}^{z} \in \{0,1\}$$

(10)

$V_{c,t,d,s}^{z}$	cumulative discounted economic value of block (c, z) in simulation s, period t and destination d
$x_{c,t,d}^{z}$	simulation-independent binary variable that assumes 1 if block (c, z) is being mined in period t and sent to destination d, and 0 otherwise

The penalties can be calculated using Equation 11:

$$p = \sum_{t=1}^{T} \sum_{g=1}^{G} \left(\overline{\alpha_{t,g}} \left(\sum_{s=1}^{S} \overline{f_{t,g,s}} \right) + \underline{\alpha_{t,g}} \left(\sum_{s=1}^{S} \underline{f_{t,g,s}} \right) + \overline{\beta_{t,g}} \left(\sum_{s=1}^{S} \overline{j_{t,g,s}} \right) + \underline{\beta_{t,g}} \left(\sum_{s=1}^{S} \underline{j_{t,g,s}} \right) \right)$$

(11)

$\overline{f_{t,g,s}}, \underline{f_{t,g,s}}$	continuous variables to penalise sum constraints violated for each period, group of destinations, and simulation. One pair of variables is necessary for each quantifiable parameter modelled block by block whose sum is being constrained. An example would be variables used to control fleet hours spent in different periods or processing time in the plant, groups of destinations, and simulations.
$\overline{j_{t,g,s}}, \underline{j_{t,g,s}}$	continuous variables to penalise average constraints violated for each period, destination, and simulation. One pair of variables is necessary for each quantifiable parameter modelled block by block whose average is being constrained. An example would be variables used to control the average grade of blocks mined in different periods, destination groups, and simulations.
$\overline{\alpha_{t,g}}, \underline{\alpha_{t,g}}$	user-defined weights for variables $\overline{f_{t,g,s}}, \underline{f_{t,g,s}}$ with the same destination group g and period t. The value used is 1 000 000. It can be changed using advanced configuration.
$\overline{\beta_{t,g}}, \underline{\beta_{t,g}}$	user-defined weights for variables with the same destination d and period t. The value used is 1 000 000. It can be changed using advanced configuration.

Surface constraints

Figure 6 shows two surfaces (blue and yellow): a) not crossing each other and respecting the constraint; b) crossing each other and not respecting the constraint.

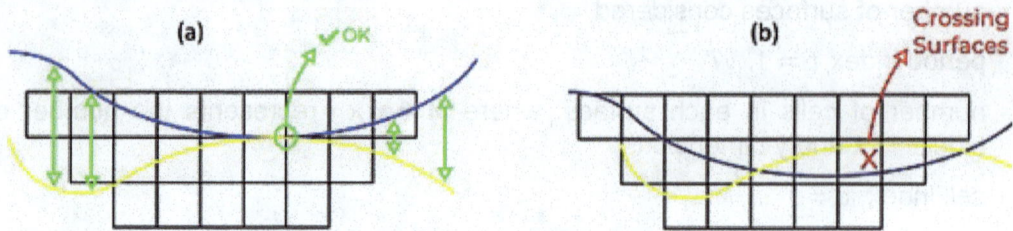

FIG 6 – Surface constraint concept (MiningMath, 2025).

Equation 12 shows the surface constraint:

$$e_{c,t-1} - e_{c,t} \geq 0,$$

$$c = 1, \dots, M, \quad t = 2, \dots, T \tag{12}$$

$$e_{c,t} \in R, \quad t = 1, \dots, T, \quad c = 1, \dots, M$$

$e_{c,t}$ simulation-independent continuous variables associated with each cell c (set of blocks) for each period t, representing cell elevations

Slope constraints

Adjacent elevations on a single surface need to respect a maximum difference. This maximum value will change based on their adjacent direction: x, y, or diagonally (Equations 13, 14, 15):

$$e_{c,t} - e_{x,t} \geq H_x,$$

$$c = 1, \dots, M, \quad t = 1, \dots, T, \quad x \in X_c \tag{13}$$

$$e_{c,t} - e_{y,t} \geq H_y,$$

$$c = 1, \dots, M, \quad t = 1, \dots, T, \quad y \in Y_c \tag{14}$$

$$e_{c,t} - e_{d,t} \geq H_d,$$

$$c = 1, \dots, M, \quad t = 1, \dots, T, \quad d \in D_c \tag{15}$$

H_x, H_y, H_d maximum difference in elevation for adjacent cells in x, y, and diagonal directions

X_c, Y_c, D_c equivalent to H_x, H_y, H_d concept, the sets of adjacent cells, laterally in x, in y, and diagonally, for a given cell c, respectively

Figure 7 shows the adjacent elevations.

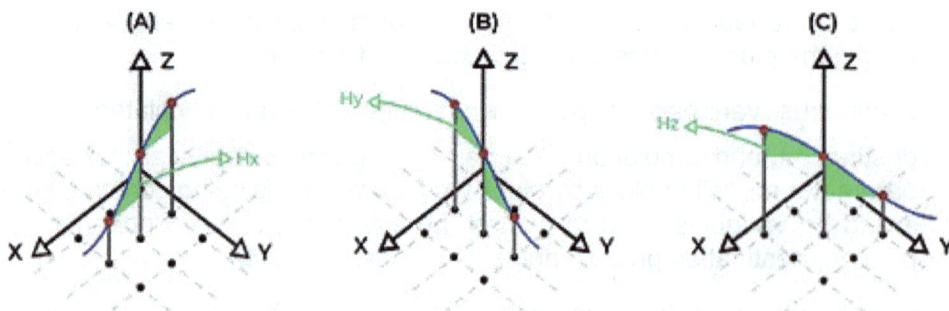

FIG 7 – Adjacent elevations (MiningMath, 2025).

Link constraints

The surfaces define when blocks will be mined. For example, blocks between surfaces associated with period 1 and 2, will be mined in period two. A block is between two surfaces if its centroid is between the two surfaces (Equations 16 and 17):

$$E_c^z \times \sum_{d=1}^{D} x_{c,1,d}^z \geq e_{c,1},$$

$$c = 1, \dots, M, \quad z = 1, \dots, Z$$

(16)

E_c^z elevation of the centroid for a given block (c, z)

$$e_{c,t-1} \geq E_c^z \times \sum_{d=1}^{D} x_{c,t,d}^z \geq e_{c,t},$$

$$c = 1, \dots, M, \quad t = 2, \dots, T, \quad z = 1, \dots, Z$$

(17)

Figure 8 shows the surfaces between two consecutive periods.

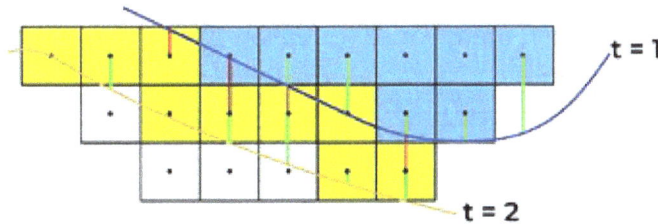

FIG 8 – Surfaces between two consecutive periods (MiningMath, 2025).

Destination constraints

Each mined block can only be sent to one destination (Equation 18):

$$\sum_{d=1}^{D} x_{c,t,d}^z = 1,$$

$$c = 1, \dots, M, \quad t = 1, \dots, T, \quad z = 1, \dots, Z$$

(18)

Mining constraints

For each period and destination group, there is an upper and lower limit of total tonnage to be extracted. Destination groups might be formed by any unique combination of destinations, with 1, many, or all. The sum of the tonnage of mined blocks sent to the same group of destinations in the same period must respect these limits (Equation 19):

$$\sum_{c=1}^{M} \sum_{z=1}^{Z} \sum_{d \in g} T_c^z x_{c,t,d}^z \leq \overline{T_{t,g}},$$

$$t = 1, \dots, T, \quad g = 1, \dots, G$$

(19)

T_c^z tonnage for a given block (c, z)

$\overline{T_{t,g}}$ upper limits in total tonnage to be extracted during period t and destinations in group g

Sum constraints

It is possible to define a certain parameter (ie fleet hours spent, or processing time spent) associated with each mined block to control its sum. The sum of the values of this parameter associated with each mined block must respect lower and upper bounds for each period, destination groups (optional), and simulation (individually or on average). Destination groups might be formed by any unique combination of destinations, with 1, many, or all (Equation 20):

$$F_{t,g,s} \leq \sum_{c=1}^{M} \sum_{z=1}^{Z} \sum_{d \in g} F_{c,d,s}^{z} x_{c,t,d}^{z} + \underline{f_{t,g,s}} - \overline{f_{t,g,s}} \leq \overline{F_{t,g,s}},$$

$$t = 1, \dots, T, \quad g = 1, \dots, G, \quad s = 1, \dots, S \tag{20}$$

$$\overline{f_{t,g,s}}, \underline{f_{t,g,s}} \in \mathbb{R}_{\geq 0}, \quad t = 1, \dots, T, \quad s = 1, \dots, S, \quad d = 1, \dots, D$$

$\underline{F_{t,g,s}}, \overline{F_{t,g,s}}$ lower and upper limits, respectively, in the sum of user-defined parameters to be respected in the period t, destination group g, and simulation s

$F_{c,d,s}^{z}$ the value of user-defined parameter related to a given block (c, z) in destination d and simulation s

Average constraints

It is possible to define a certain parameter (ie grade) associated with each mined block to be controlled on average. This average is weighted by the block's tonnage and by an optional user-defined weight. It must respect lower and upper bounds for each period, destination group (optional), and simulation (individually or on average). Destination groups might be formed by any unique combination of destinations, with 1, many, or all (Equation 21):

$$J_{t,g,s} \leq \frac{\sum_{c=1}^{M} \sum_{z=1}^{Z} \sum_{d \in g} P_{c,t,d,s}^{z} T_{c}^{z} J_{c,s,d}^{z} x_{c,t,d}^{z}}{\sum_{c=1}^{M} \sum_{z=1}^{Z} \sum_{d \in g} P_{c,t,d,s}^{z} T_{c}^{z}} + \underline{j_{t,g,s}} - \overline{j_{t,g,s}} \leq \overline{J_{t,g,s}},$$

$$t = 1, \dots, T, \quad g = 1, \dots, G, \quad s = 1, \dots, S \tag{21}$$

$$\overline{j_{t,g,s}}, \underline{j_{t,g,s}} \in \mathbb{R}_{\geq 0}, \quad t = 1, \dots, T, \quad s = 1, \dots, S, \quad d = 1, \dots, D$$

$\underline{J_{t,g,s}}, \overline{J_{t,g,s}}$ lower and upper limits, respectively, for the average value of the user-defined parameter to be respected in the period t, simulation s, and destination group g

T_{c}^{z} tonnage for a given block (c, z)

$J_{c,s,d}^{z}$ the value of the user-defined parameter of the block (c, z) sent to a destination d in the simulation s

$P_{c,t,d,s}^{z}$ user-defined weight for a block (c, z) in the period t, destination d, and simulation s

Geometric constraints

Surfaces should respect geometric parameters defined by the user, such as minimum bottom width, minimum mining width, minimum mining length, and maximum vertical rate of advance. It is a proprietary constraint not disclosed, and the intuitive idea is show in Equation 22:

$$Geometric\left(e_{c,t}\right) \leq geometric\ restriction,$$

$$c = 1, \dots, M, \quad t = 1, \dots, T, \tag{22}$$

EXPERIMENTS

We used the Marvin block model, which is available for download on the MineLib website (<https://mansci-web.uai.cl/minelib/>). This is a synthetic data set that represents a gold and copper mine (MineLib, 2025).

Table 1 and Figure 9 show the synthetic geometallurgical recovery models for Au and Cu. The main lithologies considered are AvT, GnD, and QzP.

TABLE 1

Geometallurgical models for Au and Cu recoveries.

Lithology	Rec. Au (%)	Rec. Cu (%)
AvT	7.2 ln (Au) + 66.2	3.5 ln (Cu) + 84.2
GnD	5.3 ln (Au) + 62.7	1.2 ln (Cu) + 86.2
QzP	2.4 ln (Au) + 68.1	7.8 ln (Cu) + 88.3

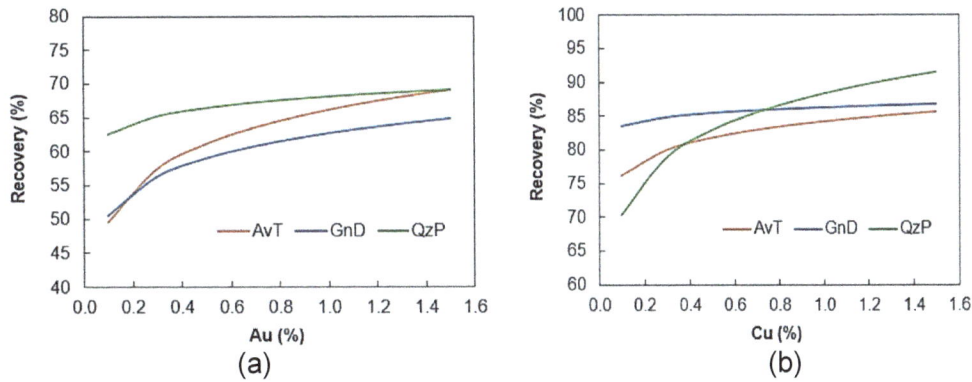

FIG 9 – Geometallurgical recovery models for: (a) Au; and (b) Cu.

Figure 10 shows some variables for the Marvin data set. The Au and Cu grades were original ones from MineLib. Recoveries were calculated through synthetic geometallurgical models using Au and Cu grades as inputs, as presented in Table 1. The comminution indices $A*b$ and BWI were estimated based on real gold and copper mines and imputed into the Marvin data set as synthetic data.

The $A*b$ and BWI were converted to Mia and Mib using Equations 2 and 3. The DWi was estimated using Equation 4. The F_{80} parameter value was estimated using Equation 5, considering the close side setting of the gyratory primary crusher as 150 mm. To calculate the specific energy values of each block, we estimated the F_{80}. We considered $K1$ equal to 0.95 (pebble crusher present) and a P_{80} equal to 150 μm. The throughput (t/h) and processing time (h) were calculated according to Equations 6 and 7, respectively. We considered the mill circuit power available as 28 000 kW.

Table 2 presents the economic parameters considered in this investigation. The values used in the Marvin data set were updated with currently approximated values for Au and Cu mines.

TABLE 2

Economic parameters.

Parameters	Values
Selling Price Au (US$/g)	85.00
Selling Cost Au (US$/g)	1.50
Selling Price Cu (US$/t)	8900.00
Selling Cost Cu (US$/t)	3100.00
Processing Cost ($/t)	10.00
Mining Cost ($/t)	4.00
Discount rate (%)	12%

We considered 20 Mt/annum as the maximum capacity for the plant processing and 60 Mt/annum as the maximum capacity for mine. Considering a processing cost equal to 10 US$/t, and if 40 per cent of this value is the fixed cost of the plant and the other 60 per cent corresponds to the milling stage (SAG and Ball mills), we can have the milling cost vary depending on the ore hardness of each block.

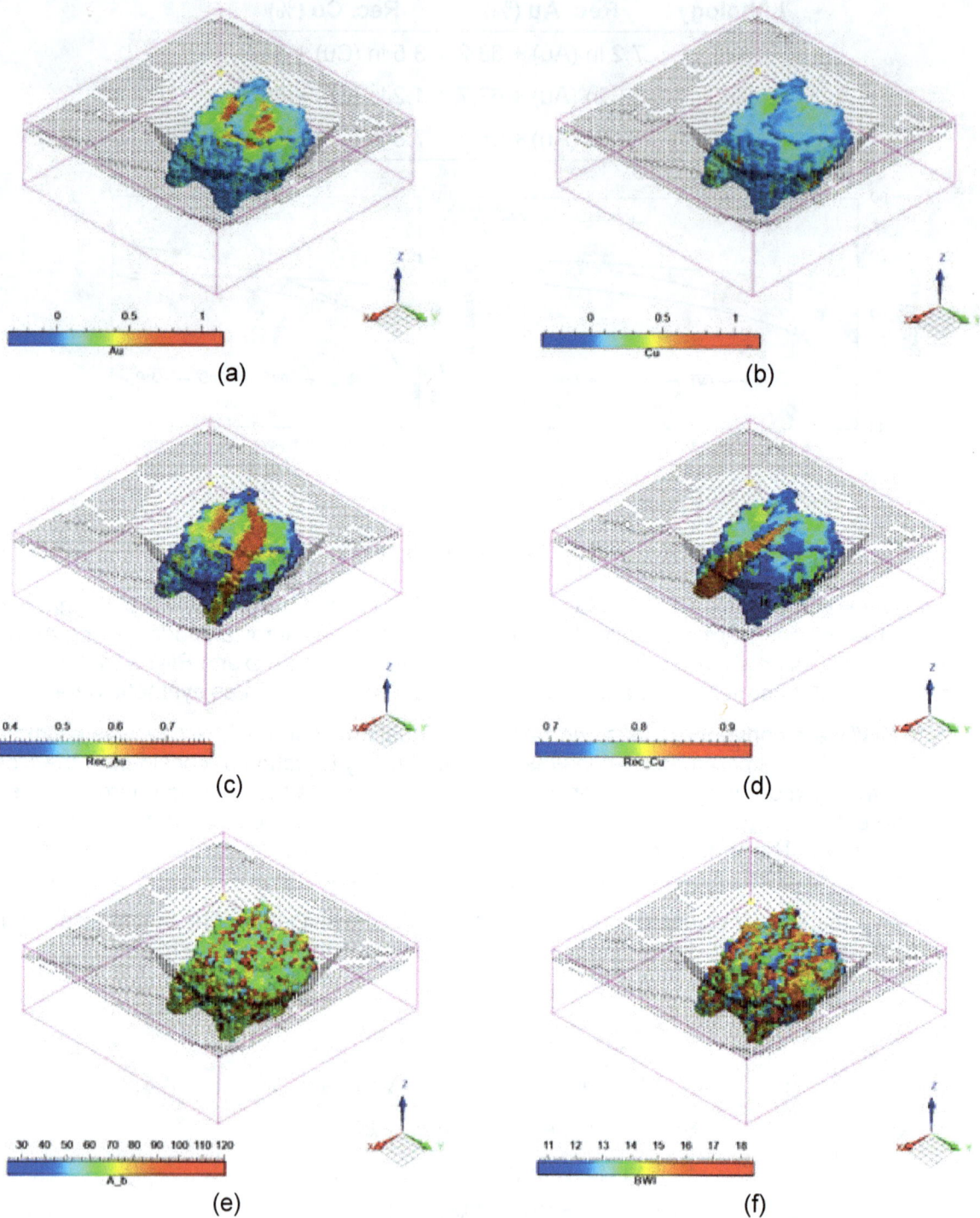

FIG 10 – Marvin data set: (a) Au; (b) Cu; (c) Rec_Au; (d) Rec_Cu; (e) *A*b*; (f) *BWI*.

We produced two scenarios, the first called Marvin and the second called MarvinGeomet. The Marvin scenario considered a fixed processing cost equal 10 US$/t with a consistent specific energy and processing time assigned to all blocks as a base case. The MarvinGeomet scenario considered a variable processing cost according to the ore hardness of each block, and consequently, each block has a different processing time. The maximum processing time considered was 7884 hrs/annum. We compared the processing time of each scenario, even though the processing time of the Marvin scenario was not a constraint in the optimisation. The geometric parameters were the same for both scenarios, minimum mining width equal 60 m, minimum bottom width equal 90 m and maximum vertical rate of advance equal 90 m.

RESULTS AND DISCUSSIONS

The mine scheduling was performed using a Dell Inspiron laptop with an Intel® Core™ i7–10510U CPU@1.80GHz processor, 16 GB of RAM, and a Windows 10 operating system (64-bit). The optimisation time for the two scenarios was less than three mins.

Figure 11 presents the processing and mine capacities, processing time, and net present value for Marvin and MarvinGeomet.

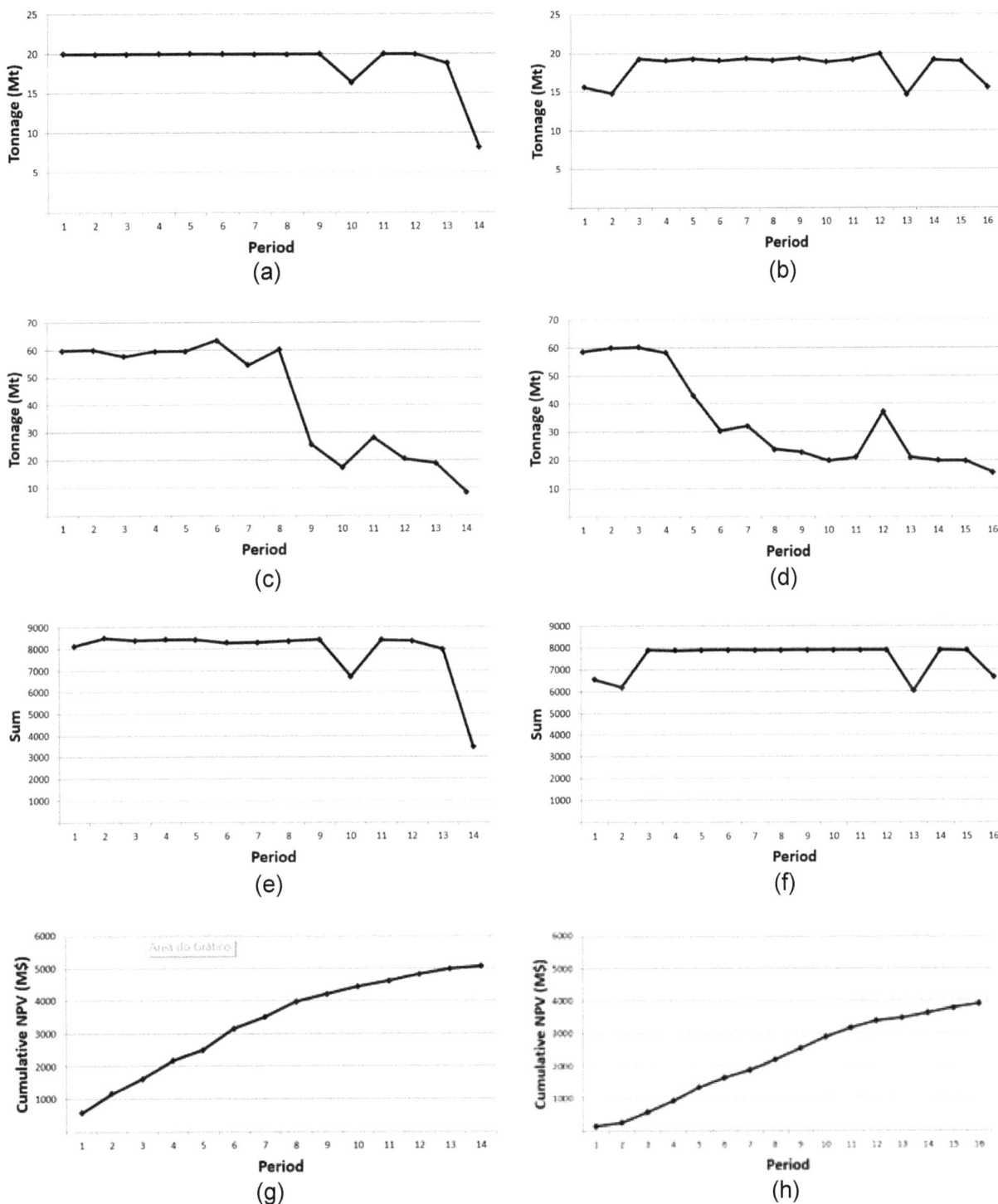

FIG 11 – (Left side) Marvin: (a) processing capacity; (c) mine capacity; (e) processing time; (g) net present value. (Right side) MarvinGeomet: (b) processing capacity; (d) mine capacity; (f) processing time; (h) net present value.

In the Marvin scenario, the processing capacity was respected and was kept at less than 20 Mt/annum. The mine capacity was almost constant until period eight, and then it started to decrease until the end of the mine. The processing time overtook the availability of the plant (7884 hrs/annum), but in this scenario, the processing time was not included as a constraint. This result means that the Marvin scenario will not process the 20 Mt/annum ore target because of a lack of plant time availability. This situation usually happens in mining projects when there is insufficient information about ore hardness variability. The net present value (NPV) for the Marvin scenario was 5055.4 MUS$ with a Life-of-mine (LOM) of 14 years. In the MarvinGeomet scenario, the processing capacity varied during the LOM because the processing time was considered a constraint. The mine capacity decreases during the LOM. The processing time was kept at less than 7884 hrs/annum, and the NPV obtained was 3894.6 MUS$ with an LOM of 16 years. Figure 12 shows periods 4, 8, and 12 for Marvin and MarvinGeomet scenarios.

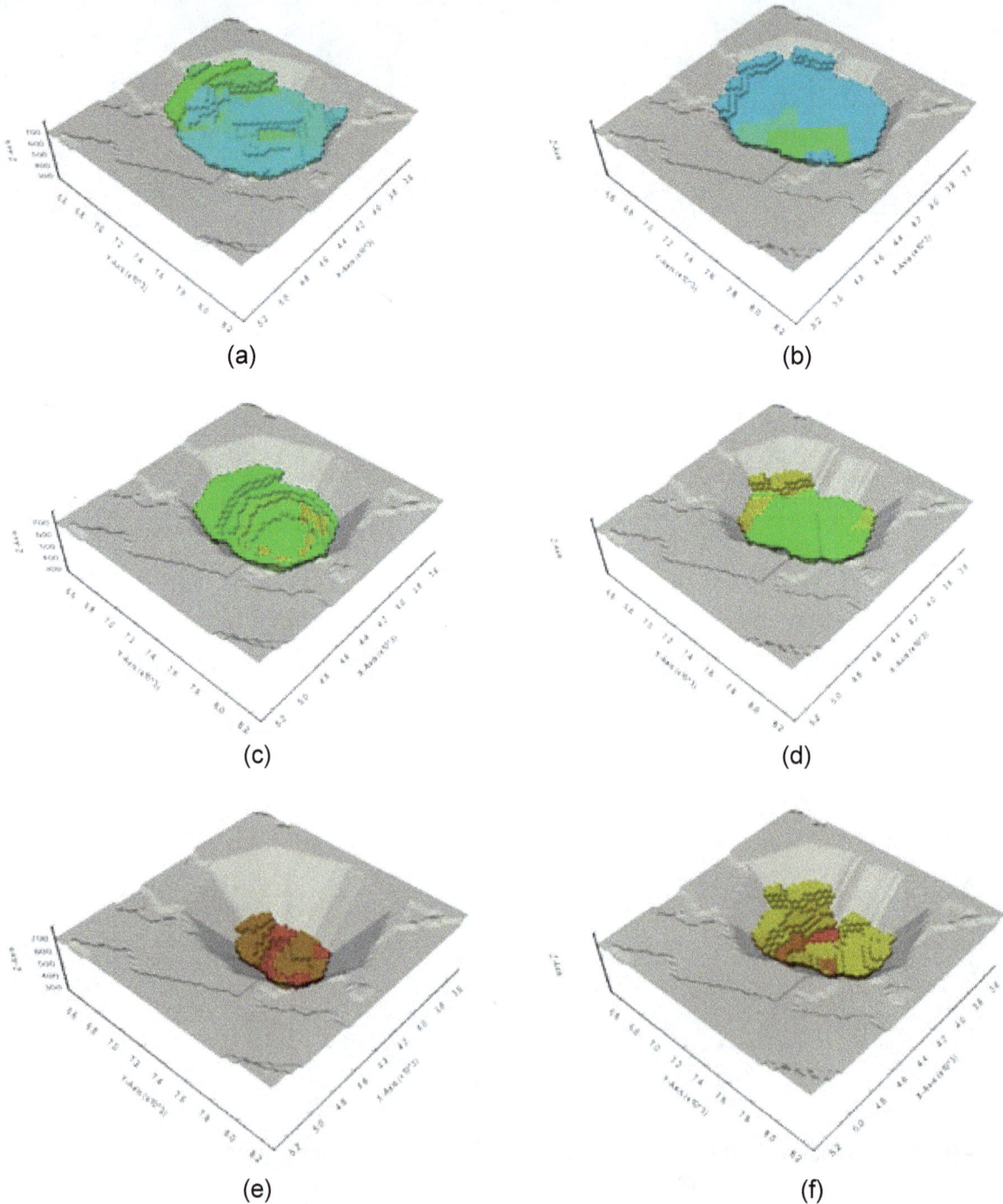

(a)	(b)
(c)	(d)
(e)	(f)

FIG 12 – (Left side) Marvin: (a) period 4; (c) period 8; (e) period 12.
(Right side) MarvinGeomet: (b) period 4; (d) period 8; (f) period 12.

The blocks selected for each period in each scenario were different once the processing cost was fixed in the Marvin scenario, and the processing cost varied due to the ore hardness of each block in the MarvinGeomet scenario. The mine scheduling algorithm must select blocks to maximise the value while respecting the processing time.

CONCLUSIONS

In this article, we showed how to estimate the processing plant capacity based on the ore hardness variability based on geometallurgy characterisation aiming to increase productivity and forecast possible bottlenecks in advance. The comminution indices can be transformed into specific energy and then, into processing time. Using this information, it is possible to have a constraint in the mine schedule optimisation model that considers the maximum operational hours of the processing plant. Then, the mine scheduling algorithm used, MiningMath in this case study, will select blocks to maximise the value while respecting the processing time.

Two scenarios were investigated, the first called Marvin and the second called MarvinGeomet. In the Marvin scenario, the processing capacity was respected (<20 Mt/annum), but the processing time surpassed the availability of the plant (<7884 hrs/annum). In this scenario, the processing time was not included as a constraint. This result means that the Marvin scenario will not produce 20 Mt/annum once there is no processing time available to achieve this production and the plant will be a bottleneck to this project. Without the ore hardness variability of the deposit, it is not possible to forecast the production correctly. The net present value (NPV) for the Marvin scenario was 5055.4 MUS$ with 14 years of Life-of-mine (LOM). In the MarvinGeomet scenario, the processing capacity varied during the LOM because the processing time was considered as a constraint. The time availability of the plant was respected (<7884 hrs/annum) and the NPV obtained was 3894.6 MUS$ with 16 years of LOM. In this scenario, we are close to reality since the mine scheduling optimisation looked for high-value blocks and respected their processing time to feed the plant.

The NPV of Marvin scenario is unrealistic once the processing cost was fixed for all blocks nor considering the ore hardness variability of the orebody. As a result, the NPV of this scenario is considerably higher when compared with the MarvinGeomet scenario. In the MarvinGeomet scenario the processing cost varied block-by-block due the time necessary to process it. This scenario is more realistic and can show future bottlenecks.

For future work, we suggest using this approach in a stochastic mine schedule model and also applying it to other commodities. The specific energy is a part of the milling costs and costs related to wear (balls and liners) should be considered in future studies.

ACKNOWLEDGEMENTS

The first and the last authors would like to thank CNPq (National Council for Scientific and Technological Development - grant 302629/2023) for partially funding the work.

REFERENCES

Bailey, C, Lane, G, Morrell, S and Staples, P, 2009. What Can Go Wrong in Comminution Circuit Design?, In Proceedings of the Tenth Mill Operators' Conference (The Australasian Institute of Mining and Metallurgy: Melbourne).

Bergeron, Y, Kojovic, T, Gagnon, M-d-N and Okono, P, 2017. Applicability of the HIT for Evaluating Comminution and Geomechanical Parameters from Drill Core Samples—The Odyssey Project Case Study, in Proceedings of the COM 2017, Vancouver, Canada.

Bienstock, D and Zuckerberg, M, 2009. A new LP algorithm for precedence constrained production scheduling, Optimisation Online, pp 1–33.

Bueno, M P, Almeida, T, Lara, L and Powell, M, 2024. The Geopyörä Index: A New Instrument for Assessing Comminution and Rock Strength Parameters, in Proceedings of the IMPC, Washington, USA.

Deutsch, M, Dağdelen, K and Johnson, T, 2022. An Open-Source Program for Efficiently Computing Ultimate Pit Limits: MineFlow, Natural Resources Research.

Doll, A, 2025. SMC Test parameters from A×b [online], Linkedin. Available from: <https://www.linkedin.com/in/alex-doll-66b57465/>

Espinoza, D, Goycoolea, M, Moreno, E and Newman, A, 2012. MineLib: A library of open pit mining problems, *Annals of Operations Research*, 206(1):91–114.

Global Mining Guidelines Group (GMG), 2021. The Morrell Method to determine the efficiency of industrial grinding circuits. Available from: <https://gmggroup.org/wp-content/uploads/2024/07/GUIDELINE_The-Morrell-Method-to-Determine-the-Efficiency-of-Industrial-Grinding-Circuits_2021-1.pdf>

Global Mining Guidelines Group (GMG), 2025. Introduction to Geometallurgy. Available from: <https://gmggroup.org/introduction-to-geometallurgy-white-paper/>

Hochbaum, D S, 2001. A new-Old algorithm for minimum-cut and maximum-flow in closure graphs, *Networks: An International Journal*, 37(4):171–193.

Johnson, T B, 1968. Optimum open pit mine production scheduling, PhD thesis, Operations Research Department, University of California, Berkeley.

Lerchs, H and Grossmann, I F, 1965. Optimum Design of Open Pit Mines, *CIM Bulletin*, 58:47–54.

Letelier, O R, Espinoza, D, Goycoolea, M, Moreno, E and Muñoz, G, 2020. Production Scheduling for Strategic Open Pit Mine Planning: A Mixed-Integer Programming Approach, *Operations Research*, 68(5):1425–1444.

Minelib, 2025. Datasets. Available from: <https://mansci-web.uai.cl/minelib/> [Accessed: 5 March 2025].

Minemax, 2025. A Simple Explanation of a Potentially Game-changing New Mine Planning Technology. Available from: <https://www.minemax.com/news/game-changing-new-mine-planning-technology/> [Accessed: 5 March 2025].

MiningMath, 2025. Knowledge Base. Available from: <https://miningmath.com/docs/knowledgebase/theory/algorithm/> [Accessed: 5 March 2025].

Varianemil, D, Kojovic, T, Hakim, D, Dilaga, R and Condori, P, 2023. Ore Hardness Mapping of Batu Hijau Ore Deposit Using the Hardness Index Tester, in Proceedings SAG2023, Vancouver, Canada.

APPENDIX

The pseudo-code of MiningMath optimisation algorithm.

```
INPUT: Block model,
       Mining parameters,
       Optional time limit T
OUTPUT: Excel report summarizing the main results of the optimization,
        Outputs of mining optimization, topography, and pit surfaces in
        .csv format that can also be imported into other mining packages.


EXECUTE initial assessment // Step 1
CREATE problem linearization P // Step 2
SET best_solution to empty
REPEAT // Step 3
    SOLVE P // Optimization engine + proprietary Branch & Cut algorithm
    SET LS to the integer, linear solution of P
    TRANSFORM LS to an integer, non-linear solution RS
    EVALUATE RS
        IF RS is better than best_solution THEN
            SET best_solution to RS
        ENDIF

        IF RS has violated constraints that were unviolated in LS OR
            has constraints that can be discarded/modified THEN
            CREATE new problem linearization P // Step 2
            SET R to TRUE
        ELSE
            SET R to FALSE
        ENDIF
UNTIL R = TRUE OR T has been reached
EXPORT reports and outputs from best_solution
```

Multi-criteria selection of a travelling way development method at a case study conventional underground platinum mine

B Meyer[1], C Musingwini[2] and T Tholana[3]

1. Former Acting Mineral Resources Manager, Sibanye-Stillwater, Marikana East Operations, South Africa; now Group Mineral Resources Manager, CAPM, South Africa. Email: bertom@capm.co.za
2. Professor, School of Mining Engineering, University of the Witwatersrand, Johannesburg 2001, South Africa. Email: cuthbert.musingwini@wits.ac.za
3. Senior Lecturer, School of Mining Engineering, University of the Witwatersrand, Johannesburg 2001, South Africa. Email: tinashe.tholana@wits.ac.za

ABSTRACT

Conventional underground platinum mines on the Bushveld Complex in South Africa, such as the Marikana Operations (Marikana), extract shallow-dipping, narrow, tabular reefs to produce platinum group metals. The reefs are accessed via a network of development openings that include travelling ways, which are inclined tunnels connecting lateral development openings to the upper reef horizon. Marikana had been using the conventional hand-held drill and blast tunnelling method to excavate travelling ways. Marikana then trialled the inverse drop raising tunnelling method and proved its viability in excavating travelling ways, thus, necessitating the selection of a preferred travelling way development method at the mine. The selection is a multi-criteria decision analysis (MCDA) process because the different travelling way development methods must be evaluated by simultaneously considering a set of decision criteria. The Analytic Hierarchy Process (AHP) was selected as the MCDA evaluation technique because of its applicability under conditions of discrete alternatives and criteria. The AHP ranked the inverse drop raising tunnelling method in preference to the conventional hand-held drill and blast tunnelling method by a 7 per cent difference when mined at the standard Marikana travelling way dimensions of 3 m wide, 1.8 m high and 15 m long. Sensitivity analysis subsequently performed in *SuperDecisions*® software supported the robustness of the solution because rank reversal only occurred for the 'Excavation cost' criterion. All the other five criteria showed that the inverse drop raising method remained the preferred travelling way development method. The findings were shared with and endorsed by relevant technical staff at Marikana. This is the first time that a multi-criteria practical selection of a travelling way development method has been done at a conventional underground platinum mine. Therefore, other conventional underground platinum mines can adapt the approach presented in this paper to evaluate other tunnel development methods applicable to their site-specific travelling way development requirements.

INTRODUCTION

The underground mining industry uses several different techniques and technologies to excavate development openings that provide access to production and service areas in the mines. These openings can be primary openings such as shafts or declines, secondary openings such as lateral haulages or tunnels, and tertiary openings such as inclined raises or travelling ways. In South Africa, underground hard rock mines are generally classified either as conventional mines which have traditionally been mined using labour-intensive methods of drilling with hand-held jackhammers, and cleaning stope panels with scraper winches, or mechanised mines that use more modern mining technologies (Egerton, 2004). Conventional underground platinum mines on the Bushveld Complex (BC) extract shallow-dipping, narrow, tabular reefs to produce platinum group metals (PGMs). As Figure 1 shows, conventional underground mining methods are predominantly used to extract PGMs in South Africa. The country is significant in that about 80 per cent of the world's PGM resources and reserves are hosted in the BC in South Africa (Schulte, 2025).

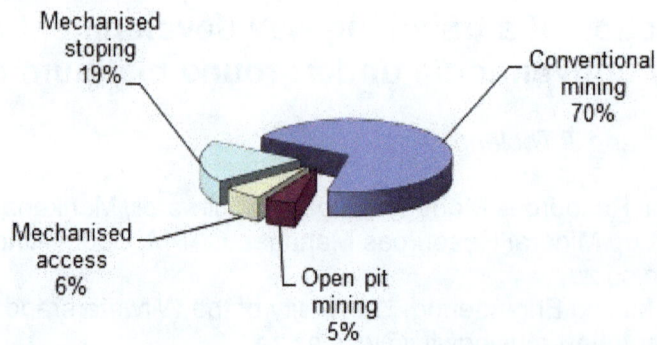

FIG 1 – Distribution of PGM production output by mining method in South Africa (adapted from Pickering, 2007).

The work presented in this paper used Marikana Operations (Marikana) as a case study mining operation, which is one of the conventional underground platinum mining operations on the Western Limb of the BC. Marikana is owned by Sibanye-Stillwater. It is in the North-West Province of South Africa, approximately 80 km west of the country's capital city, Tshwane, and 100 km north-west of the country's commercial capital city, Johannesburg (Sibanye-Stillwater, 2021a). Mining conditions at Marikana resemble those of the other conventional underground platinum mining operations, hence the findings of this study can be generalised to other conventional underground platinum mining operations, but with site-specific adjustments.

The platinum reef horizon in conventional underground platinum mining operations is accessed through a network of development openings which include travelling ways. The travelling ways are inclined tunnels that connect lateral development openings to the reef horizon. Workers use travelling ways to access the reef horizon that contains stoping sections of the mining operation. The travelling ways are also used to provide services required in the stopes. The services include ventilation, compressed air, water, and electrical cables. In a conventional underground platinum mining layout, travelling ways are located between a cross-cut at a lower elevation and a stepover at a higher elevation. Figure 2 illustrates the location of a travelling way relative to other development openings.

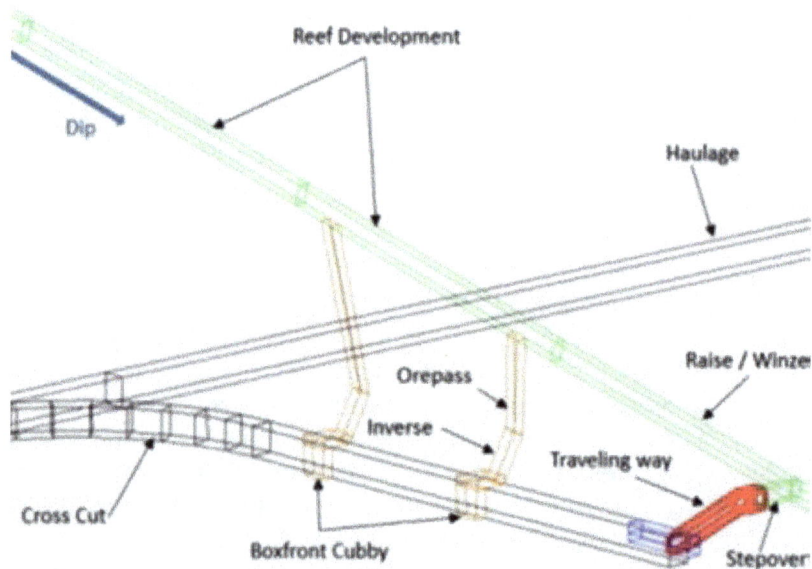

FIG 2 – Isometric view of a development layout depicting the location of a travelling way relative to other development openings at Marikana (Meyer, 2024).

A travelling way is developed from the cross-cut to the reef horizon. The size of travelling ways must enable adequate ventilation flow and accommodate the transportation of equipment and other consumables to and from the stoping areas. At Marikana the standard dimensions of travelling ways are 3 m wide, 1.8 m high and 15 m long.

In a South African context, a travelling way is traditionally excavated at a maximum inclination of 34.5°, so that the reef horizon is intersected in the shortest possible distance, while remaining below a 35° inclination threshold. According to the Mine Health and Safety Council (1996), the 35° limit is set to define an excavation as 'steeply inclined' to comply with the South African Mine Health and Safety Act (MHSA) (1996) which enforces stricter legal requirements and limitations for steeply inclined excavations, mostly of a permanent nature. For example, Sections 7.3.2, 7.5.1, 7.5.2, 7.5.3, 7.7.2, 16.12, 16.13 and 16.14 of the MHSA (1996), regulate the excavation and maintenance of steeply inclined excavations to cater for safety risks related to material transportation, ladderways, and a barrier for adequately closing the inclined excavation. Therefore, travelling ways are mined below the threshold of 35° since they have a temporary use purpose required only during the life of the stope. However, the technical challenge associated with mining at the shallower inclination of 34.5° is that it is more difficult to easily identify shallow-dipping geological structures encountered during excavation and to adequately support them (Meyer, 2024). Operationally, it is also challenging to safely bar down loose rocks in an inclined excavation because when the rocks dislodge, they can easily roll down along the footwall and potentially injure people working in the excavation (Meyer, 2024). These challenges require mining operations to continually trial different tunnelling methods and technologies that can deliver safer travelling way development operations.

At Marikana, the conventional hand-held drill and blast tunnelling method has traditionally been used to excavate travelling ways. However, the inverse drop raising method of excavating travelling ways was also trialled and proved to be viable at Marikana. Since more than one tunnelling method was viable for travelling way development, it became necessary to select an optimal travelling way development method at Marikana. The optimal or preferred tunnelling method should give the best trade-off among factors such as operational efficiency, safety, and cost-effectiveness. As such, the selection process is multi-criteria in nature because the different tunnelling methods must be evaluated by simultaneously considering several factors or decision criteria. The criteria vary across mining sites and are influenced by the site-specific conventional underground mining method being used such as up-dip, down-dip, or breast mining. These criteria include safety, economics, technical requirements, orebody characteristics, excavation dimensions, and support installation requirements. Since different criteria must be considered simultaneously when evaluating different alternatives to select the best alternative from a set of discrete alternatives, the problem is a multi-criteria decision analysis (MCDA) problem that must be solved using MCDA techniques (Balusa and Singam, 2017; Mahase, Musingwini and Nhleko, 2016). MCDA techniques are a sub-category of multi-criteria decision-making (MCDM) techniques. MCDM techniques have recently been used to determine the optimal location of multi-purpose utility tunnels as described in Genger, Luo and Hammad (2021) and Genger, Hammad and Oum (2023), thus, indicating their applicability to the underground mining environment.

SOME TUNNELLING METHODS IN A TRAVELLING WAY DEVELOPMENT CONTEXT

Tunnelling is a process that is applied in various industries including the mining sector. It is undertaken to excavate pathways for infrastructure such as roads, railways, waterways, and pipelines. According to Shah and Rathod (2013), various methods are used to excavate tunnels. These include the following commonly used methods in lateral and inclined tunnels:

- conventional hand-held drill and blast method
- inverse drop raising
- tunnel boring machine (TBM) method
- rail-bound drilling rig tunnel method
- cut-cover method
- immersed tube method
- full-face method
- heading and bench method

- pilot tunnel method
- trenchless tunnelling method.

However, not all the methods can be successfully applied to travelling way development in a conventional underground mining environment. For example, the TBM method is not applicable to Marikana because it is suitable for tunnels with bigger cross-sectional dimensions. The rail-bound drilling rig tunnel method is like the conventional hand-held drill and blast method but requires a drill rig to travel on rails, yet rails are not permitted in travelling ways. Therefore, the method is also not applicable for Marikana travel way development. The cut-over method is applicable to tunnelling excavations located at shallow depths only (Shah and Rathod, 2013). The mining depths at Marikana are at an average depth of around 500 m or more below surface (Sibanye-Stillwater, 2021a), hence the method is not applicable. The immersed tube is a widely used technique in the construction of underwater tunnels (Łotysz, 2010); hence the method is not applicable to Marikana because tunnels are developed through solid ground. The full-face method is used in underground construction projects for continuous excavation and support installation of full-face tunnel cross-sections in difficult ground conditions for both deep and shallow tunnelling depths. The method is not applicable at Marikana where ground conditions are not difficult in travelling way development. The heading and benching method involves excavating the tunnel in two phases. Firstly, the upper part of the tunnel cross-section called a 'heading' is excavated. Thereafter, the bottom part of the excavation called the 'bench' is excavated. The pilot tunnelling method involves the initial excavation of a small diameter 'pilot' hole which is then excavated to the tunnel full-face (Jovičić, 2018). Therefore, both methods are not applicable at Marikana where the travelling ways have small cross-sectional dimensions, precluding excavation as a full-face operation. The trenchless tunnelling method uses horizontal directional drilling techniques and/or TMBs to excavate passageways underwater and underneath built-up areas. TBMs are too big for the standard dimensions of travelling ways at Marikana hence the method is also not applicable for Marikana.

The only applicable methods which are already in use at Marikana, are the conventional hand-held drill and blast, and the inverse drop raising methods. The brief descriptions, advantages, and limitations of these two methods are summarised in Table 1.

TABLE 1

Applicability of tunnelling methods for travelling way development at Marikana (adapted from Meyer, 2024).

Method	Brief description	Advantages	Limitations
Conventional hand-held drill and blast	Method is cyclic and comprises repetitive sequential activities from drilling a full-face tunnel with a pattern of drill holes using drilling equipment, charging the drill holes with explosives, and detonating them to break the rock, cleaning the broken rock and installing support to stabilise the excavation (Pickering, 2004; Shah and Rathod, 2013).	A well-known and widely accepted method that is adaptable to various ground conditions and can navigate changes in tunnel direction.	Short advances are achieved; repeated blasting causes undue fracturing in the hanging wall and footwall and creates an unsmooth floor.
Inverse drop raising	Method is like conventional hand-held drill and blast, except that the drilling that is done is longhole drilling, which results in longer excavation advances from a single blast compared to multiple blasts in conventional hand-held drill and blast that have short advances (Liu and Tran, 2000).	A well-known and widely practised method at Marikana to develop inclined tunnels such as ventilation shafts and orepasses.	Deflection of drill holes can occur at longer drill hole lengths. The drill rigs take up more space compared to drilling equipment used in the conventional hand-held drill and blast method.

These two were, therefore, evaluated for preference through a selection process. Since the selection process is multi-criteria in nature, the next section discusses MCDA techniques.

MULTI-CRITERIA DECISION ANALYSIS TECHNIQUES

Selecting the optimal alternative from a set of alternatives requires multiple factors or decision criteria to be considered simultaneously. These criteria include factors such as technical, financial, environmental, or social requirements. However, some criteria can be contradictory, conflicting, or competing as different stakeholders to a mining project have different and often contradictory expectations. For example, investors expect a project's net present value (NPV) to be maximised, and this is achieved by mining high-grade areas in the early years of the project, leading to a shorter life-of-mine (LOM). On the contrary, governments and surrounding communities expect a longer LOM for increased job security and maximum extraction of the mineral resource endowment, hence some 'trade-off' is required. Criteria are measured in different units making it difficult to configure how to optimally trade-off the competing criteria unless a dimensionless framework is used to overcome the challenge of comparing different units of measure. An additional challenge is that the human brain can easily configure an optimal decision in two-dimensional or three-dimensional spaces when criteria are at most three. Beyond 3D space, decisions require abstract thinking and psychologists have proved that there are general limitations to human abstract thinking, leading to inconsistent judgements being made.

MCDM methods, are a branch of decision-making methods that have been developed to overcome the challenges discussed in the preceding section. MCDM techniques can generally be classified into two sub-categories, namely:

- MCDA techniques that are alternatively called multiple attribute decision-making (MADM) techniques, are used to select the 'best' or 'optimal' alternative from a pre-determined set of discrete alternatives.

- MODM techniques, which are used to select the optimal alternative from a continuous decision space with an infinite number of alternatives.

In this paper, the focus was on MCDA techniques since the tunnelling methods for travelling way development had to be evaluated on discrete sets of alternatives and pre-determined decision-making criteria.

Sample size of experts to decide on number and weights of criteria

MCDA techniques use experts to identify criteria and determine their respective weights. This is generally done by polling different experts to determine the criteria and assign weights. Literature has shown that there is no prescribed number of experts to be used in an MCDA selection process. For example, in different mine planning case studies, Ataei et al (2008) used 17 experts, Naghadehi, Mikaeil and Ataei (2009) used 15 experts and Bazzazi, Osanloo and Karimi (2011) used six experts, respectively. However, it is important to have a panel of experts with specific knowledge of the selection problem at hand to ensure that judgments are practically meaningful and can be trusted.

Number of criteria and alternatives

There is no prescribed limit on the number of criteria and alternatives that can be considered in MCDA techniques. However, it is recommended that the ideal number of criteria should be 'seven-plus-or-minus-two' because of general limitations on human performance in abstract thinking as the human mind will struggle to detect inconsistencies in judgements (Saaty and Ozdemir, 2003). Mahase, Musingwini and Nhleko (2016) analysed a database of mine planning case studies and noted that a combination of a maximum of six criteria and four alternatives tends to produce stable results. Therefore, it is important to perform sensitivity analysis to establish the stability or robustness of a solution obtained from an MCDA process, due to inconsistencies inherent in human judgement in abstract thinking.

Reasons for selecting the Analytic Hierarchy Process (AHP) technique

From a review of several mining-related cases, Sitorus, Cilliers and Brito-Parada (2019) and Mahase, Musingwini and Nhleko (2016) identified the AHP as the most used MCDA technique. AHP is widely used and was selected as the MCDA evaluation method in this paper because of some of its advantages which include its:

- Ability to detect inconsistent judgements and provide an estimate of the degree of inconsistency in the judgements (Coyle, 2004) so that appropriate steps are taken to remedy the inconsistencies until an acceptable level of inconsistency is achieved to proceed with the evaluation.

- Easy application in Microsoft Excel since matrix and vector algebra form the basis of the mathematical framework of the AHP methodology, thus, enabling AHP calculations to be easily performed in Microsoft Excel.

- Formulation as a hierarchal structure which can easily be modelled in decision-making problems of different sizes in mine planning (Zlaugotne *et al*, 2020).

THE AHP

Saaty (1980) developed the AHP methodology, which is based on matrix algebra. Its mathematical framework and main limitation are presented in the next two sections.

Mathematical framework of AHP

The initial step in AHP involves establishing decision criteria for evaluating different alternatives, where n is the total number of criteria. Experts assist in identifying and making pairwise comparisons of the criteria. For two criteria C_i and C_j, the relative weight or importance of C_i over C_j denoted by w_{ij}, is logically an inverse of the relative weight of C_j over C_i (Equation 1):

$$w_{ij} = \frac{1}{w_{ji}}, \forall \, i \neq j \tag{1}$$

Equation 1 represents the reciprocity axiom of the AHP. This means that a criterion is as important as itself (ie $w_{ii} = 1$), hence the diagonal elements of the criteria matrix will always be equal to one. In the second step, the relative weights of the criteria are compiled into a square matrix W of order n, corresponding to the number of criteria as expressed by Equation 2:

$$W = (w_{ij}) \tag{2}$$

The matrix W of criteria weights should satisfy the AHP transitive axiom which says that relative weights are multiplicative. For example, if $C_2 = 2C_1$, and $C_3 = 3C_2$, then Equation 3 must hold:

$$C_3 = 3C_2 = 3 \times (2C_1) = 6C_1 \tag{3}$$

A transitive matrix is referred to as a consistent matrix. However, inconsistencies in judgement arise due to limitations in human abstract thinking when determining the relative importance of criteria. The AHP methodology includes a way of identifying and quantifying the degree of inconsistency. A vector, w, can be determined such that (Equation 4):

$$Ww = \lambda w \tag{4}$$

where w is called an eigenvector of the matrix W and the constant λ is its corresponding eigenvalue. If the matrix W is consistent then Equation 5 must hold:

$$\lambda = n \tag{5}$$

However, when judgements are inconsistent, then Equation 5 cannot hold, and the eigenvector w will satisfy the condition shown by Equation 6:

$$Ww = \lambda_{max}w, \text{where } \lambda_{max} \geq n \tag{6}$$

A Consistency Index, CI, is then calculated from λ_{max} and n using Equation 7:

$$CI = \frac{(\lambda_{max} - n)}{(n - 1)} \tag{7}$$

A Consistency Ratio, *CR*, is then calculated using Equation 8:

$$CR = \frac{CI}{RI}$$ (8)

where *RI* is a corresponding Random Index obtained from Table 2.

TABLE 2

Random index for an n-ordered matrix (Saaty, 1980).

n	1	2	3	4	5	6	7	8	9	10	11	12	13	14	15
RI	0.00	0.00	0.52	0.89	1.11	1.25	1.35	1.40	1.45	1.49	1.51	1.48	1.56	1.57	1.59

Table 2 was generated by simulating samples of random matrices of increasing order and calculating their corresponding simulated *CI*s which are called random indices, *RI*s. It is not possible to have an inconsistent judgement when comparing a criterion to itself or against just one other criterion, so there cannot be any randomness in the judgement, hence matrices of orders 1 and 2 have *RI*s equal to zero. According to Saaty (1980), a *CR* above a threshold of 10 per cent indicates that the judgements are too random to be trusted. In such as case, the pairwise comparison must be redone to reduce the inconsistency in judgements to less than or equal to 10 per cent to proceed with the ranking process.

When the degree of inconsistency in judgements complies with the threshold of 10 per cent, then the score of each alternative per criterion, S_{ij}, is then normalised to eliminate the effect of different units of measure per criterion, hence creating a dimensionless comparison framework that enables clear trade-offs to be made. For *m* alternatives on a criterion, the normalised S_{ij} values denoted by, S_{ij}^{N}, are derived as shown in Equation 9:

$$S_{ij}^{N} = \frac{S_{ij}}{\sum_{i=1}^{m} S_{ij}}$$ (9)

The matrix of normalised scores is finally multiplied by the eigenvector to obtain the aggregated AHP priority score. A decision is then made based on the logic that the higher the AHP priority score for an alternative, the higher it ranks in preference. This is because the AHP is premised on a maximisation framework whereby the criterion with the largest weight is deemed the most important and the alternative with the highest priority score is the most preferred from among the set of alternatives.

Some limitations of AHP

As stated earlier in this paper, AHP has some advantages over other MCDA methods, hence its preference in solving MCDA problems. However, AHP also has some limitations which users need to be aware of, to assist them in their interpretation of results obtained from using it. Some AHP limitations which include those mentioned in Musingwini and Minnitt (2008) are:

- Rank reversal which occurs when the ranking of alternatives changes when some alternatives are either added or removed from the AHP model or when there is a slight change in the weight of a criterion. Saaty and Vargas (2012) noted that rank reversal can be caused by factors such as inconsistency in judgments or sensitivity to minor changes in criteria pairwise comparisons. Sensitivity analysis is usually performed to evaluate the risk of rank reversal and confirm the reliability or robustness of the ranking of the alternatives. Therefore, it was important in this paper to perform a sensitivity analysis.

- When the scale for measuring the relative importance of criteria with respect to each other is changed, say from a scale of 1 to 10 to a scale of 1 to 20, the weight vector may change and cause a change in the ranking of alternatives. In this paper, the scale was kept between 1 and 10 to reduce the degree of subjectivity by the experts.

- AHP only works well if the matrix for the criteria weights is a positive reciprocal matrix, meaning that if a criterion C_i is x times more important than criterion C_j then C_j should be $\frac{1}{x}$ as important as C_i. To overcome this challenge, in this paper the inverses of criteria weights were computed in completing the square matrix of criteria weights, instead of asking experts to estimate reciprocal weights.

- As the number of decision-making criteria increases, the number of pairwise comparisons increases rapidly following a power function, as shown in Table 3. The rapid increase in pairwise comparisons can cloud the judgement process and render the calculations more complex, sometimes generating inconsistent results. This is why in this paper, experts were advised to identify criteria that did not exceed nine, to comply with the *'seven-plus-or-minus-two'* principle proposed by Saaty and Ozdemir (2003).

TABLE 3

Relationship between number of criteria and pairwise comparisons (Kardi, 2006).

Number of criteria	1	2	3	4	5	6	7	n
Number of pairwise comparisons	0	1	3	6	10	15	21	$\frac{n(n-1)}{2}$

RESULTS FROM STEPS FOLLOWED TO DETERMINE THE PREFERRED TUNNELLING METHOD

A workshop to identify and assign weights to criteria for evaluating the different tunnelling methods was conducted on 29 July 2023 as indicated in an official letter from Sibanye-Stillwater (Figure 3).

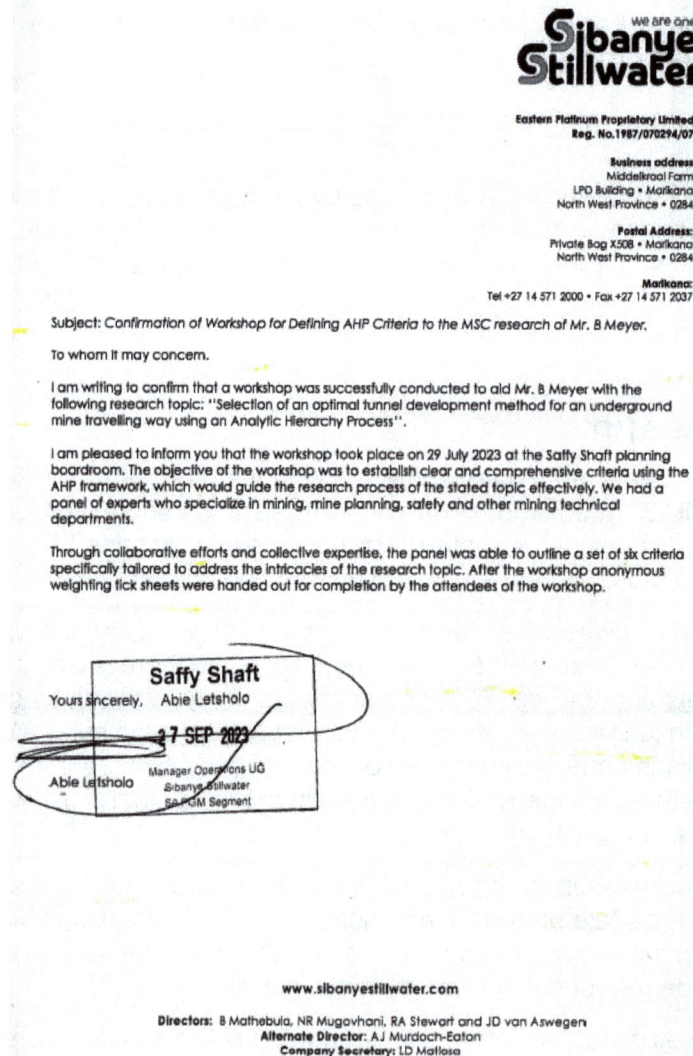

Sibanye Stillwater
we are one

Eastern Platinum Proprietary Limited
Reg. No.1987/070294/07

Business address
Middelkraal Farm
LPD Building • Marikana
North West Province • 0284

Postal Address:
Private Bag X50B • Marikana
North West Province • 0284

Marikana:
Tel +27 14 571 2000 • Fax +27 14 571 2037

Subject: *Confirmation of Workshop for Defining AHP Criteria to the MSC research of Mr. B Meyer.*

To whom it may concern.

I am writing to confirm that a workshop was successfully conducted to aid Mr. B Meyer with the following research topic: "Selection of an optimal tunnel development method for an underground mine travelling way using an Analytic Hierarchy Process".

I am pleased to inform you that the workshop took place on 29 July 2023 at the Saffy Shaft planning boardroom. The objective of the workshop was to establish clear and comprehensive criteria using the AHP framework, which would guide the research process of the stated topic effectively. We had a panel of experts who specialize in mining, mine planning, safety and other mining technical departments.

Through collaborative efforts and collective expertise, the panel was able to outline a set of six criteria specifically tailored to address the intricacies of the research topic. After the workshop anonymous weighting tick sheets were handed out for completion by the attendees of the workshop.

Yours sincerely,

Saffy Shaft
Abie Letsholo

7 SEP 2023

Abie Letsholo

Manager Operations UG
Sibanye-Stillwater
SA PGM Segment

www.sibanyestillwater.com

Directors: B Mathebula, NR Mugovhani, RA Stewart and JD van Aswegen
Alternate Director: AJ Murdoch-Eaton
Company Secretary: LD Matlosa

FIG 3 – Copy of the letter from the mine confirming the workshop (Meyer, 2024).

The workshop involved 12 experts from Marikana who comprised four mining engineers, two safety officers, and six mining technical specialists (Meyer, 2024). The selected experts were experienced in the conventional hand-held drill and blast tunnelling method for travelling way development and had all been previously involved in the trial of the inverse drop raising tunnelling method. Since there is no prescribed size for a panel of experts to be polled, as stated earlier, the panel of experts was chosen based on their knowledge and experience in the two tunnelling methods. Initially, the experts collaboratively identified the criteria to be used for evaluating the two tunnelling methods since they had specialist knowledge in travelling way development. The experts were requested to consider limiting the number of criteria to not more than nine for the reason stated earlier in this paper.

Step-by-step results

Six criteria were identified by the 12 experts, and the hierarchical decision problem was structured as shown in Figure 4. The six criteria are consistent with the findings on the ideal number of MCDA criteria and alternatives, as made by Mahase, Musingwini and Nhleko (2016).

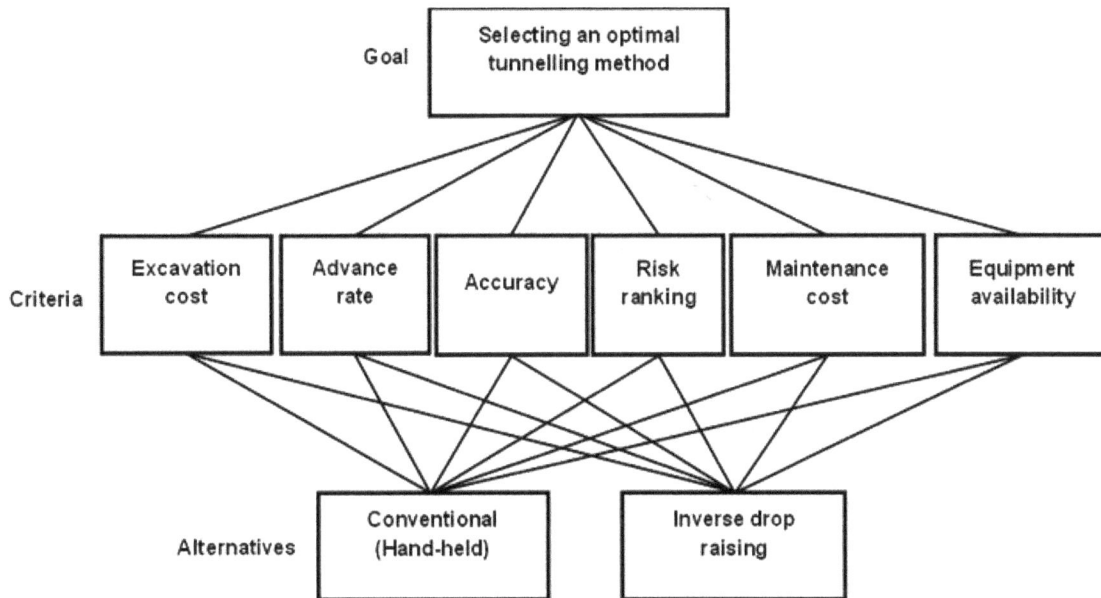

FIG 4 – AHP hierarchical structure depicting the selection problem (Meyer, 2024).

Subsequently, each of the experts anonymously completed a questionnaire to assign weights of one criterion relative to another. Figure 5 shows a sample copy of the anonymous questionnaire completed by one of the experts (Expert 1).

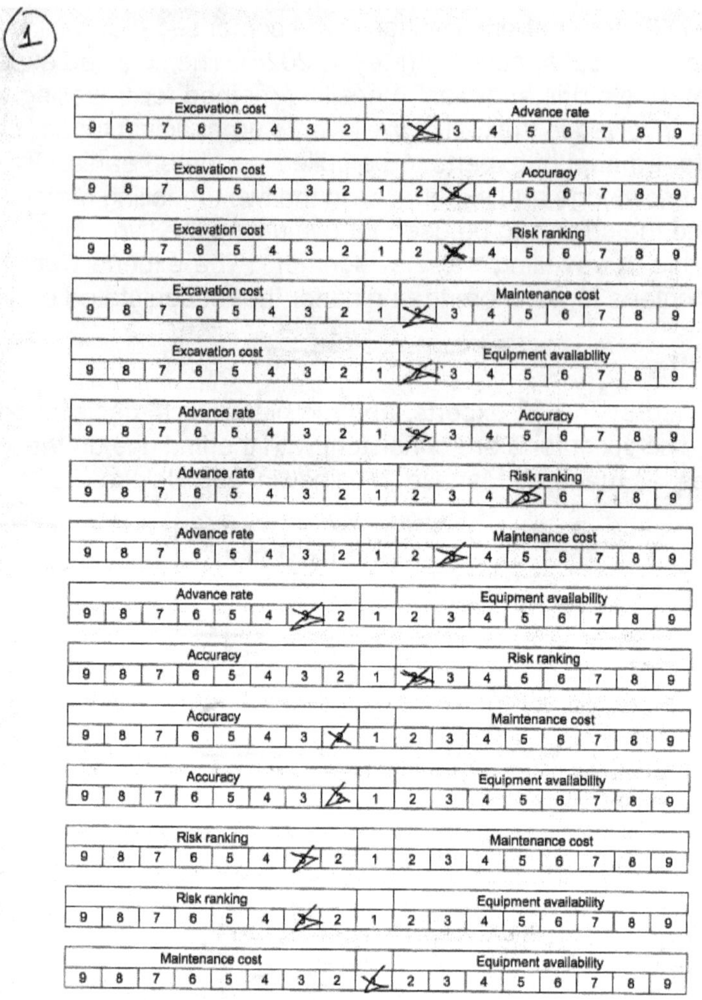

FIG 5 – Sample of a completed anonymous questionnaire.

From Figure 4 for example, Expert 1 ranked the criterion 'Advance rate' to be twice as important as 'Excavation cost'. From the AHP reciprocity axiom, it follows that 'Excavation cost' should be half as important as 'Advance rate' so that a positive reciprocal matrix of criteria weights can be compiled. The weights of the criteria from each of the 12 experts, which were quite comparable to each other probably due to the experts having been involved in both tunnelling methods, were then averaged. The inverses of the averages were then computed to produce a square matrix with diagonal elements being equal to 1 since a criterion is as important as itself. Table 4 shows the aggregated results of the criteria weights as scored by the panel of 12 experts from the questionnaire survey.

TABLE 4

Pairwise comparison of the six criteria (adapted from Meyer, 2024).

Weights	Excavation cost	Advance rate	Accuracy	Risk ranking	Maintenance cost	Equipment availability
Excavation cost	1.000	0.417	0.353	0.238	0.521	0.500
Advance rate	2.400	1.000	0.458	0.332	0.528	0.778
Accuracy	2.835	2.182	1.000	0.350	1.736	2.125
Risk ranking	4.211	3.013	2.857	1.000	4.250	4.000
Maintenance cost	1.920	1.895	0.576	0.235	1.000	1.292
Equipment availability	2.000	1.286	0.471	0.250	0.774	1.000
Column total	14.366	9.793	5.715	2.405	8.809	9.695

The rows in Table 4 indicate the importance of a row criterion relative to column criteria. For example, the last weight in the first row shows that 'Excavation cost' was half as important as 'Equipment availability'. To comply with the reciprocity axiom given by Equation 1, the diagonal weights are equal to one because each criterion is as important as itself, while the entries below the diagonal are approximate reciprocals of the weights above the diagonal, bearing in mind the slight inconsistency in judgements by the experts and rounding-off errors. For example, the last weight in the first column shows that 'Equipment availability' is twice as important as 'Excavation cost'. When excluding the column totals, the 6 × 6 matrix (ie matrix of order $n = 6$) in Table 4 is the matrix, W, as defined previously in Equation 2.

The next step was to divide each entry in Table 5 by its corresponding column sum to normalise the matrix. The normalised entries are shown in Table 5 and should add up to 1.000 since they are normalised values. This was followed by calculating the averages of each row as an eigenvalue entry (ie relative weight of a criterion compared to other criteria). The eigenvalues comprise the eigenvector, w, (ie matrix of relative weights of criteria). Each original row in matrix W was then multiplied by the eigenvector and the result divided by its corresponding eigenvalue to obtain each λ, so that Equations 3, 4 and 5 must hold. The average of all entries was then computed as λ_{max}, which was 6.156 and greater than the order of the matrix of criteria of 6, thus, indicating inconsistency in judgements by the experts as inferred from Equation 6. These steps are illustrated in Table 5.

TABLE 5

Normalised matrix W, its corresponding eigenvector (w) and λ_{max} (adapted from Meyer, 2024).

Weights	Excavation cost	Advance rate	Accuracy	Risk ranking	Maintenance cost	Equipment availability	Eigenvector (w) = matrix of row averages	λ
Excavation cost	0.070	0.043	0.062	0.099	0.059	0.052	*0.064*	6.137
Advance rate	0.167	0.102	0.080	0.138	0.060	0.080	*0.105*	6.025
Accuracy	0.197	0.223	0.175	0.146	0.197	0.219	*0.193*	6.206
Risk ranking	0.293	0.308	0.500	0.416	0.482	0.413	*0.402*	6.267
Maintenance cost	0.134	0.194	0.101	0.098	0.114	0.133	*0.129*	6.171
Equipment availability	0.139	0.131	0.082	0.104	0.088	0.103	*0.108*	6.132
Column total or average	*1.000*	*1.001*	*1.000*	*1.001*	*1.000*	*1.000*	*N/A*	λ_{max} = 6.156

The eigenvector column in Table 5 shows that the experts placed high importance on safety since 'Risk ranking' had a relative weight of 0.402, while 'Excavation cost' had the least importance with a relative weight of 0.064. However, since there was inconsistency in judgements, it was necessary to establish the degree of inconsistency. From Equation 7:

$$CI = (6.156–6) \div (6–1) = 0.031$$

From Equation 8 (since from Table 2, $RI = 1.25$ for n = 6):

$$CR = 0.031 \div 1.25 = 2.50\%$$

The calculated CR was below the 10 per cent threshold, hence the impact of inconsistency of the matrix of criteria was of a negligible level to the AHP process, hence the ranking of alternatives could continue using the estimated criteria weights.

Table 6 shows how each of the two tunnelling methods for travelling way development scored on each of the six criteria based on verified company internal records. The excavation costs were obtained from the Marikana finance department but for company confidentiality and proprietary reasons, the absolute excavation costs were re-based and reported as relative values. For comparability, the advance rates per month were derived from daily schedules of the main activities of the two different methods for a typical 22-shift working month as per the norm in conventional underground platinum mining. The accuracy was obtained by comparing planned dimensions with measured dimensions from the Marikana survey department. The risk ranking was obtained by comparing the two methods using the Sibanye-Stillwater's Safety, Health and Environmental (SHE) Risk Matrix (Sibanye-Stillwater, 2021b), which is a 5 × 5 risk matrix used by Sibanye-Stillwater as part of its standard systematic operational risk management process. The maintenance costs were obtained from the Marikana finance department from records of costs incurred for travelling ways during the entire lifespan of decommissioned travelling ways and then expressed as a cost per metre. The equipment availability was equipment uptime expressed as a percentage of the total scheduled shift time.

TABLE 6

Scores for each method per criterion (adapted from Meyer, 2024).

Criterion	Excavation cost	Advance rate	Accuracy	Risk ranking	Maintenance cost	Equipment availability
Criterion requirement	Minimise	Maximise	Maximise	Minimise	Minimise	Maximise
Conventional hand-held drill and blast	R62 776	22 m/ month	95%	117	R711/m	95.8%
Inverse drop raising	R122 952	22 m/ month	98%	93	R341/m	95.2%
Column totals	*R185 728*	*44 m/ month*	*193%*	*210*	*R1052/m*	*191%*

The two methods (ie alternatives) do not exceed the number of alternatives suggested in the findings by Mahase, Musingwini and Nhleko (2016), hence the AHP process could continue. Using Equation 9, the entries in Table 6 were then normalised by dividing each column entry by its corresponding column total to create Table 7.

TABLE 7

Normalised scores for each method per criterion (adapted from Meyer, 2024).

Criterion	Excavation cost	Advance rate	Accuracy	Risk ranking	Maintenance cost	Equipment availability
Criterion requirement	Minimise	Maximise	Maximise	Minimise	Minimise	Maximise
Conventional hand-held drill and blast	0.338	0.500	0.492	0.557	0.676	0.502
Inverse drop raising	0.662	0.500	0.508	0.443	0.324	0.498
Column totals	*1.000*	*1.000*	*1.000*	*1.000*	*1.000*	*1.000*

The AHP final priority scoring works on a maximisation context. Therefore, it was necessary to convert minimisation criteria requirements to their equivalent maximisation contexts, obtained as complements of the minimisation entry values in Table 7 to produce Table 8.

TABLE 8

Normalised scores for each method per criterion in a maximisation context (adapted from Meyer, 2024).

Criterion	Excavation cost	Advance rate	Accuracy	Risk ranking	Maintenance cost	Equipment availability
Conventional hand-held drill and blast	0.662	0.500	0.492	0.443	0.324	0.502
Inverse drop raising	0.338	0.500	0.508	0.557	0.676	0.498

The matrix in Table 8 was then multiplied by the eigenvector in Table 5 to obtain the final AHP priority scores of 0.464 for the Conventional hand-held drill and blast and 0.537 for the Inverse drop raising. Figure 6 illustrates these findings as percentages. The findings were shared with relevant technical staff at Marikana, and they endorsed the findings.

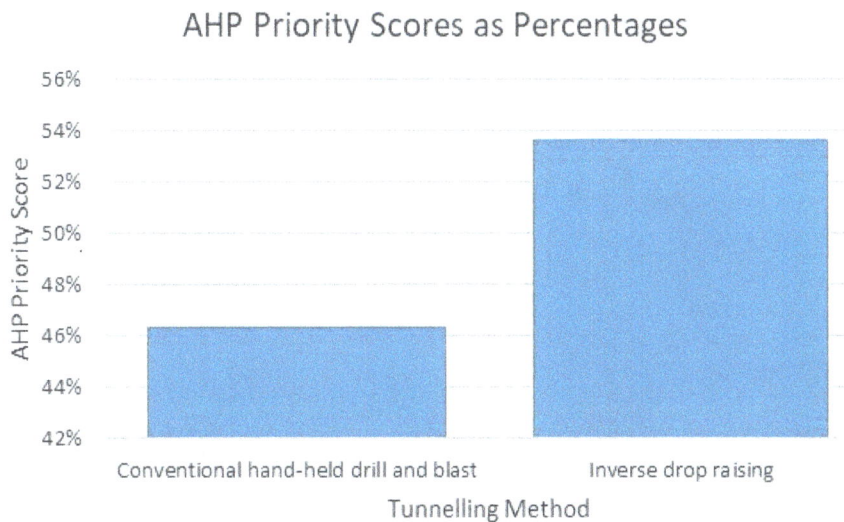

FIG 6 – AHP priority scores for the two tunnelling methods (adapted from Meyer, 2024).

Inferences from the results

Table 9 shows that the inverse drop raising method scored higher than the conventional hand-held drill and blast tunnelling method. The two tunnelling methods differed by approximately 7 per cent. The inverse drop raising tunnelling method is more accurate and requires less maintenance for the temporary use of the excavation. The inverse drop raising tunnelling method also carries a lower risk as experts scored the risk ranking criterion the highest in terms of weight, thus, emphasising the importance of safety at Marikana. Neither of the tunnelling methods had an advantage with respect to advance rate when mining 15 m long travelling ways. However, this might change if longer or shorter travelling ways were to be mined, and such cases will have to be evaluated. Even though the two criteria 'Excavation cost' and 'Maintenance cost' received smaller weights from the panel of experts at the workshop, these were the criteria where the two tunnelling methods differed the most. The implications are that if there is a prolonged drop in PGM commodity prices, Marikana may need to implement budgetary cuts as part of cost-cutting measures. This could result in the inverse drop raising method being less preferred if not done more cost-effectively, because the inverse drop raising tunnelling method is about twice as expensive as the conventional hand-held drill and blast tunnelling method. Therefore, since the inverse drop raising method is the optimal travelling way development method at Marikana as recommended by this study, the mine should investigate ways of executing the method more cost-effectively. Options to consider include either hiring a cheaper contractor, or train in-house mining crews on the inverse drop raising method and purchase own equipment for better cost control.

Sensitivity analysis

As discussed earlier in this paper, the AHP faces a limitation of rank reversal. Therefore, it was necessary to perform sensitivity analysis to confirm the stability or robustness of the solution. Sensitivity analysis could only be done on the criteria only and not on alternatives because only two alternatives had been proven viable for travelling way development at Marikana. This paper utilised the *SuperDecisions*® software because it could be freely downloaded from the public domain for long-term academic research purposes. Figure 7 shows the results of the sensitivity analysis on each of the six criteria.

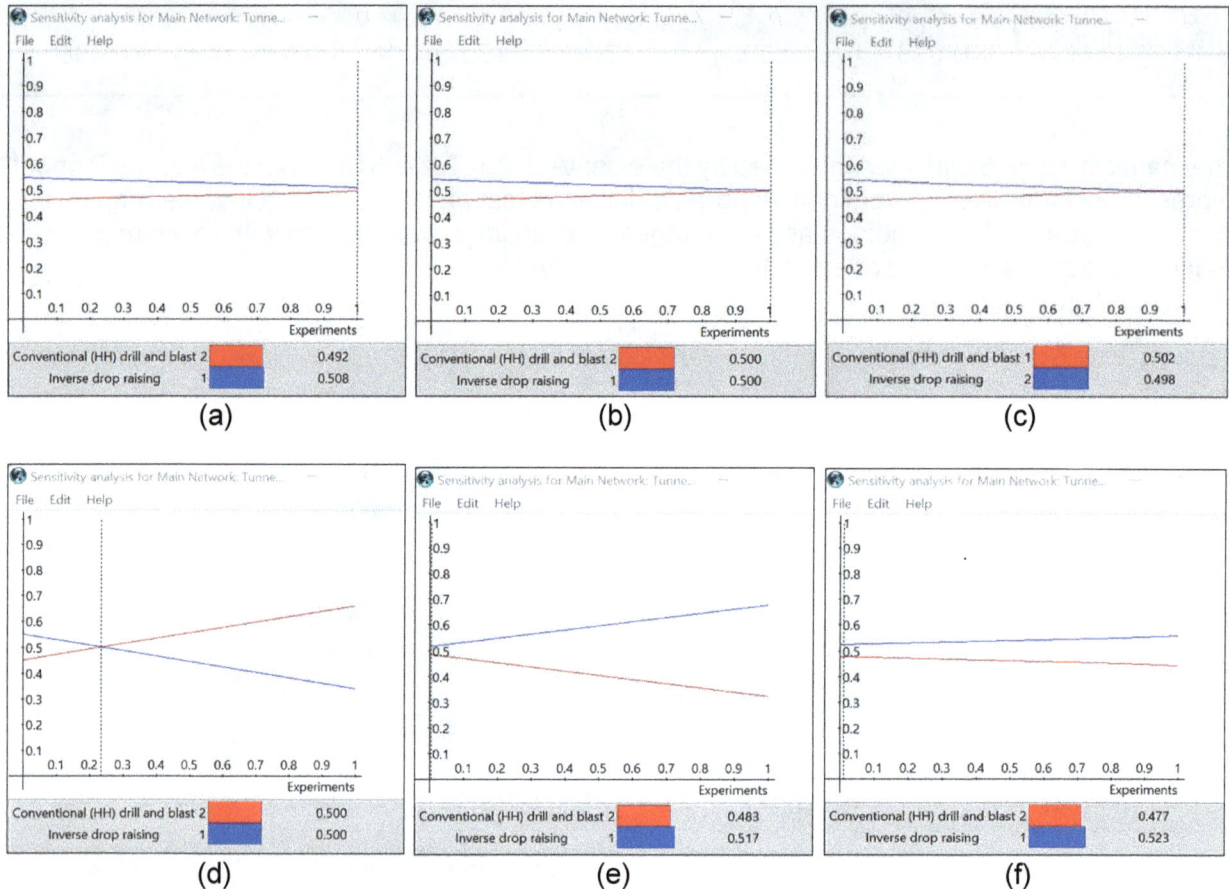

FIG 7 – Sensitivity analysis results of solution to changes in criteria weights: (a) Accuracy; (b) Advance rate; (c) Equipment availability; (d) Excavation cost; (e) Maintenance cost; (f) Risk ranking.

Figure 7 shows that among the six criteria, rank reversal only occurred for the 'Excavation cost' criterion when its weight has changed significantly to be about 0.234 compared to its initial weight of about 0.064. Therefore, the sensitivity results confirm the robustness of the AHP ranking that the inverse drop raising method is the optimal tunnelling method to use for Marikana.

Despite experts having allocated the lowest criterion weight to 'Excavation cost' as shown by the eigenvector in Table 5, the sensitivity of the solution to excavation cost warrants a deeper analysis or explanation. This is due to the sensitivity analysis indicating that 'Excavation cost' is more likely to alter the decision outcome through rank reversal. It is important to re-emphasise that the conventional tunnelling method was done by the mine, while the drop raising method was done by a contractor. The contractor costs are expectedly higher than the mine's own costs (see Table 6) because contractors charge a premium since they must carry the risk associated with the work or any rework that may be required. The maintenance costs reflected in Table 6 at R341/m for the drop raising method and R711/m for the conventional hand-held drill and blast method, are costs that are incurred to keep a travelling way safe and usable during its entire lifespan. The maintenance costs for the conventional hand-held drill and blast method are expectedly higher than for the drop raising method because the conventional method often has overbreak and reduced mining accuracy resulting in more maintenance and equipping needed for staggered floors. In addition, timber is used in the conventional travelling way to create a smooth surface on which scraper winches, and equipment are pulled up. However, timber is not required for the drop raising method as the floor of the excavation is adequately smooth to enable the dragging of equipment directly on the excavated floor, except in rare cases where only minor cement alterations are required to repair the floor during the equipping stage. This saves costs on timber that does not have to be replaced after winches are pulled up the travelling way as the heavy scraper winches damage the timber which then must be replaced. A travelling way excavated using the hand-held method will typically have its timber replaced twice or more during a travelling way's lifespan.

Potential future research work

There is potential future research work that can extend work presented in this paper. For example, there is potential to use hybrid MCDA approaches to solve the choice decision method, such as in the work of Ozdemir (2023) who used a hybrid AHP-TOPSIS method to select an optimum mine planning alternative for a chrome mine in Turkey. AHP was used to assign criteria weights as was done in this paper, while TOPSIS (Technique for Order Preference by Similarity to Ideal Solution) was then used to rank the identified alternatives. Figure 8 in Appendix 1 indicates how such an AHP-TOPSIS hybrid MCDA method was incorporated to add methodological depth in addressing the choice decision problem, where TOPSIS was used to rank the alternatives based on criteria weights generated from AHP. The results of the hybrid AHP-TOPSIS approach confirm the validity of the AHP solution.

Another potential future research area is on exploring the use of other criteria weight determination methods, because as argued by Belton and Stewart (2002), most literature on multi-criteria choice decisions have mostly focused on ranking different alternatives, while minimal work has been done on exploring different criteria weighting methods.

CONCLUSIONS

Conventional underground platinum mining operations on the Bushveld Complex extract shallow-dipping, narrow, tabular reef horizons which are accessed via a network of development openings. The development layout includes a travelling way which is an inclined small cross-sectional area tunnel that connects the lateral development openings to the reef horizon located on an upper elevation. Marikana is one of the mining operations on the Bushveld Complex. It uses two different tunnelling methods to excavate the travelling ways namely the conventional hand-held drill and blast tunnelling and the inverse drop raising tunnelling methods. An exercise was undertaken to rank the preference of the two methods. Since the evaluation is a multi-criteria approach, the Analytic Hierarchy Process (AHP) was used as the evaluation technique due to its several advantages over other MCDA techniques. The inverse drop raising tunnelling method ranked 7 per cent higher than the conventional hand-held drill and blast tunnelling method when mining the standard Marikana travelling ways that are 3 m wide, 1.8 m high and 15 m long. Sensitivity analysis which was subsequently performed, confirmed the stability or robustness of these findings. The sensitivity analysis also indicated that rank reversal only occurred for the 'Excavation cost' criterion and not for the other five criteria. This can be expected despite the 'Excavation cost' being assigned the smallest weighting from the workshop held, because the two tunnelling methods differed the most on this criterion. Therefore, despite the inverse drop raising method being the optimal travelling way development method at Marikana, other cost-effective options must be considered such as hiring a cheaper contractor or training in-house mining crews on the inverse drop raising method and purchase own drop raising equipment for better cost control to enable insourcing of the drop raising function. This is the first time that a multi-criteria practical selection of a tunnelling method for travelling way development in conventional underground platinum mining operations has been done. Therefore, other conventional underground platinum mines can benefit from the findings of this paper by adapting the approach presented to evaluate travelling way tunnel development methods that are applicable to their operating conditions. The conditions, include different travelling way lengths and cross-sectional dimensions as dictated by the mining method being used.

ACKNOWLEDGEMENTS

The work presented in this paper is part of the work conducted by the first author for an MSc Research Report at the School of Mining Engineering, University of the Witwatersrand, Johannesburg, South Africa. At the time of conducting this study, the first author was still employed by Sibanye-Stillwater at Marikana East Operations, South Africa, but is now employed by CAPM, South Africa. The authors would like to express their profound gratitude to the technical staff at Sibanye-Stillwater, Marikana East Operations, for sharing their knowledge and experience and providing resources and support to ensure the successful completion of this study. The authors would also like to acknowledge the University of the Witwatersrand for supporting them to participate in APCOM 2025 conference.

Credit authorship contribution statement

BM: Conceptualisation, Methodology, Formal analysis, Investigation, Resources, Writing – original draft, Writing – review and editing. **CM:** Conceptualisation, Methodology, Writing – review and editing, Supervision. **TT:** Conceptualisation, Methodology, Software, Writing – review and editing.

Declaration of interest

The first author was previously employed at the case study mine where the research work presented in this paper was undertaken.

Funding

Not applicable.

Ethical compliance

The work reported in this paper used human participants. Therefore, it was undertaken with Ethical Clearance approval with Protocol Number EMINN2022/48 granted by the Ethics Committee, School of Mining Engineering, University of the Witwatersrand, Johannesburg, South Africa.

REFERENCES

Ataei, M, Jamshidi, M, Sereshki, F and Jalali, S M, 2008. Mining method selection by AHP approach, *Journal of The Southern African Institute of Mining and Metallurgy*, 108(10):741–749.

Balusa, B C and Singam, J, 2017. Underground mining method selection using WPM and PROMETHEE, *Journal of the Institution of Engineers (India)*, pp 165–171.

Bazzazi, A A, Osanloo, M and Karimi, B, 2011. Deriving preference order of open pit mines equipment through MADM methods: application of modified VIKOR method, *Expert Systems with Applications*, 38(3):2550–2556.

Belton, V and Stewart, T J, 2002. *Multiple Criteria Decision Analysis: An Integrated Approach*, 1st ed, 372 p (Dordrecht: Springer Science and Business).

Coyle, G, 2004. *The Analytic Hierarchy Process (AHP), Practical Strategy: Structured Tools and Techniques*, Open Access Material (Glasgow: Pearson Education Ltd).

Egerton, F M G, 2004. Presidential address: The mechanisation of UG2 mining in the Bushveld Complex, *The Journal of The South African Institute of Mining and Metallurgy*, 104(8):439–450.

Genger, T K, Hammad, A and Oum, N, 2023. Multi-objective optimization for selecting potential locations of multi-purpose utility tunnels considering agency and social lifecycle costs, *Tunnelling and Underground Space Technology*, 140 (2023):105305.

Genger, T K, Luo, Y and Hammad, A, 2021. Multi-criteria spatial analysis for location selection of multi-purpose utility tunnels, *Tunnelling and Underground Space Technology*, 115:104073.

Jovičić, V, 2018. Use of pilot tunnel method to overcome difficult ground conditions in Karavanke tunnel, *Gradjevinski Materijali i Konstrukcije*, 61(1):37–45.

Kardi, T, 2006. Analytic Hierarchy Process (AHP) Tutorial [online]. Available from: <http://people.revoledu.com/kardi/tutorial/AHP/Paired-Comparison.htm> [Accessed: 12 January 2021].

Liu, Q and Tran, H, 2000. Techniques of inverse drop raise blasting and slot drilling, *CIM Bulletin*, 93(1039):45–50.

Łotysz, S, 2010. Immersed tunnel technology: A brief history of its development, *Civil and Environmental Engineering Reports*, 4:97–110.

Mahase, M J, Musingwini, C and Nhleko, S, 2016. A survey of applications of multi-criteria decision analysis methods in mine planning and related case studies, *The Journal of Southern African Institute of Mining and Metallurgy*, 116(11):1051–1056.

Meyer, B, 2024. Selection of an optimal tunnel development method for an underground mine travelling way using an Analytic Hierarchy Process, MSc Research Report, University of the Witwatersrand, South Africa.

Mine Health and Safety Council, 1996. Mine Health and Safety Act (MHSA) No. 29 of 1996 and Regulations, Johannesburg: Mine Health and Safety Council.

Musingwini, C and Minnitt, R C A, 2008. Ranking the efficiency of selected platinum mining methods using the analytic hierarchy process (AHP), in *Proceedings of the Third International Platinum Conference – 'Platinum in Transformation*, pp 319–326. Available from: <https://platinum.org.za/Pt2008/Papers/319-326_Musingwini.pdf> [Accessed: 28 June 2022].

Naghadehi, M Z, Mikaeil, R and Ataei, M, 2009. The application of fuzzy analytic hierarchy process (FAHP) approach to selection of optimum underground mining method for Jajarm Bauxite Mine, Iran, *Expert Systems with Applications*, 36(4):8218–8226. https://doi.org/10.1016/j.eswa.2008.10.006

Ozdemir, A C, 2023. Use of integrated AHP-TOPSIS method in selection of optimum mine planning for open pit mines, *Archive of Mining Sciences*, 68(1):35–53. https://doi.org/10.24425/ams.2023.144316

Pickering, R G B, 2004. The optimization of mining method and equipment, *International Platinum Conference 'Platinum Adding Value'*, pp 4–6 (The South African Institute of Mining and Metallurgy).

Pickering, R G B, 2007. Sandvik Mining and Construction and Narrow Reef Mechanisation, presentation to University of Witwatersrand School of Mining Engineering, 10 May 2007.

Saaty, T L and Ozdemir, M S, 2003. Why the magic number seven plus or minus two, *Mathematical and Computer Modelling*, 38(3–4):233–244.

Saaty, T L and Vargas, L G, 2012. *Models, methods, concepts and applications of the Analytic Hierarchy Process,* 2nd ed (New York: Springer).

Saaty, T L, 1980. *The Analytic Hierarchy Process* (McGraw Hill International: New York).

Schulte, R F, 2025. Platinum-group metals, US Geological Survey, Mineral Commodity Summaries, January 2025. Available from: <https://pubs.usgs.gov/periodicals/mcs2025/mcs2025-platinum-group.pdf> [Accessed: 25 April 2025].

Shah, C J and Rathod, H A, 2013. A review study on methods of tunnelling in hard rocks, *International Journal for Scientific Research and Development*, 1(8):1551–1555.

Sibanye-Stillwater, 2021a. 2021 Mineral Resources and Mineral Reserves Report [online]. Available from: <https://reports.sibanyestillwater.com/2021/download/SSW-RR21.pdf> [Accessed: 28 June 2022].

Sibanye-Stillwater, 2021b. Safety, SA PGM Operations, Procedure, Operational Risk Management, Rustenburg: Sibanye-Stillwater.

Sitorus, F, Cilliers, J J and Brito-Parada, P R, 2019. Multi-criteria decision making for the choice problem in mining and mineral processing: Applications and trends, *Expert Systems with Applications*, 121:393–417.

Zlaugotne, B, Zihare, L, Balode, L, Kalnbalkite, A, Khabdullin, A and Blumberga, D, 2020. Multi-criteria decision analysis methods comparison, *Environmental and Climate Technologies*, 24(1):454–471.

APPENDIX 1

The 6-step solution for the travelling way tunnelling method selection problem in TOPSIS using AHP criteria weights

The TOPSIS method recognises that criteria either maximise benefits or minimise costs, hence the method can be modelled in MS Excel using MAX() and MIN() functions. TOPSIS generates the 'best' solution as the ideal solution that concomitantly maximises benefits while minimising costs of alternatives. It first creates two hypothetical solutions called the positive ideal solution that minimises costs criteria while increasing benefits and the negative ideal solution that increases costs while minimising benefits. The best alternative is the one nearest to the positive ideal solution (S_{i+}) and furthest from the negative ideal solution (S_{i-}) based on either its Euclidean or Geometric Distance.

Criterion	Excavation cost	Advance rate	Accuracy	Risk ranking	Maintenance cost	Equipment availability
Criterion weight from AHP	0.064	0.105	0.193	0.402	0.129	0.108
Alternative						
Conventional hand-held drill and blast	62776	22	95%	117	711	95.80%
Inverse drop raising	122952	22	98%	93	341	95.20%

Normalised Matrix	Excavation cost	Advance rate	Accuracy	Risk ranking	Maintenance cost	Equipment availability
Conventional hand-held drill and blast	0.455	0.707	0.696	0.783	0.902	0.709
Inverse drop raising	0.891	0.707	0.718	0.622	0.432	0.705

Weighted Normalised Matrix	Excavation cost	Advance rate	Accuracy	Risk ranking	Maintenance cost	Equipment availability	S_{i+}	S_{i-}	P_i	Rank
Conventional hand-held drill and blast	0.029	0.074	0.134	0.315	0.116	0.077	0.089	0.028	0.240	2
Inverse drop raising	0.057	0.074	0.139	0.250	0.056	0.076	0.028	0.089	0.760	1

Ideal Solution (max benefits & min costs)	Excavation cost	Advance rate	Accuracy	Risk ranking	Maintenance cost	Equipment availability
Positive ideal solution (V+)	0.029	0.074	0.139	0.250	0.056	0.077
Negative ideal solution (V-)	0.057	0.074	0.134	0.315	0.116	0.076

Criterion weight (W_{ij}) obtained from AHP.

Decision Matrix showing how each alternative scores per each criterion.

Step 1: Use the equation below to calculate cell entries (X_{ij}) for the Normalised Matrix.

$$\overline{X}_{ij} = \frac{X_{ij}}{\sqrt{\sum_{i=1}^{n} X_i^2}}$$

Step 2: Use the equation below to calculate the cell entries (V_{ij}) for the Weighted Normalised Matrix by multiplying each cell entry in the Normalised Matrix with its corresponding criterion weight (W_j).

$$V_{ij} = \overline{X}_{ij} \times W_j$$

Step 3: (i) For each maximisation criterion select the maximum column entry to obtain the ideal best value (V+) entry.

Step 3: (i) For each minimisation criterion select the minimum column entry to obtain the ideal best value (V-) entry.

Step 4: (i) Calculate the Euclidean distance (Si+) from the ideal best values using the equation below.

$$S_i^+ = \left[\sum_{j=1}^{n} \left(V_{ij} - V_j^+ \right)^2 \right]^{0.5}$$

Step 4: (ii) Calculate the Euclidean distance from the ideal worst values using the equation below.

$$S_i^- = \left[\sum_{j=1}^{n} \left(V_{ij} - V_j^- \right)^2 \right]^{0.5}$$

Step 5: Calculate the performance score (P_i) using the equation below.

$$P_i = \frac{S_i^-}{S_i^+ + S_i^-}$$

Step 6: Use the P_i scores to rank the alternatives.

FIG 8 – Hybrid AHP-TOPSIS approach that uses TOPSIS to select the preferred travelling way tunnelling method based on criteria weights generated from AHP.

Towards a system for integrating open pit mine production scheduling stages

P Muke[1], T Tholana[2], C Musingwini[3], M Ali[4] and B Mutandwa[5]

1. PhD Candidate, School of Mining Engineering, University of the Witwatersrand, Johannesburg 2001, South Africa. Email: 1865392@students.wits.ac.za
2. Senior Lecturer, School of Mining Engineering, University of the Witwatersrand, Johannesburg 2001, South Africa. Email: tinashe.tholana@wits.ac.za
3. Professor, School of Mining Engineering, University of the Witwatersrand, Johannesburg 2001, South Africa. Email: cuthbert.musingwini@wits.ac.za
4. Professor, School of Computer Science and Applied Mathematics, University of the Witwatersrand, Johannesburg 2001, South Africa. Email: montaz.ali@wits.ac.za
5. Lecturer, School of Mining Engineering, University of the Witwatersrand, Johannesburg 2001, South Africa. Email: bright.mutandwa@wits.ac.za

ABSTRACT

When optimising production scheduling of open pit mines, there are three main mine production scheduling interconnected stages that are generally considered. These are long-term (LT), medium-term (MT) and short-term (ST) production scheduling stages. Since the early applications of optimisation techniques in mine planning in the 1960s, several exact and/or approximate approaches have been applied to solve the mine production scheduling problem. However, most of these optimisation approaches have been used to optimise either LT, MT or ST production scheduling as standalone stages. This can be detrimental to holistic optimisation of the overall scheduling system as each separately optimised stage can be spatially and/or temporally misaligned with its consecutive upstream or downstream scheduling stage. A misaligned production scheduling system can result in undesirable outcomes such as mineral resource sterilisation or reduced net present value (NPV). However, when production scheduling stages are aligned, the overall production scheduling system can ensure that LT objectives are achieved at MT scheduling horizons, and in turn, MT objectives are also achieved at ST scheduling horizons. This paper used a PRISMA-based bibliometric mapping of production scheduling optimisation approaches and revealed that since the 1960s, the focus has mainly been on optimising the production scheduling stages in isolation, hence future research can focus on integrated approaches. The paper presents a case study example which demonstrates that an integrated LT to MT production scheduling system generates comparable NPVs to when the stages are optimised in isolation, while achieving improved spatial and/or intertemporal alignment between consecutive scheduling stages. Lastly, as part of future research work towards holistic production scheduling optimisation, the paper proposes extensions to the integrated LT to MT production scheduling model by including the ST scheduling stage and feedback loops that enable updating of the production scheduling system from ST to MT and MT to LT when changes occur.

INTRODUCTION

The open pit mine production scheduling problem has garnered significant attention within the field of operations research, leading to the application of several mathematical optimisation techniques (Leite and Dimitrakopoulos, 2014). Mine production scheduling defines the optimal extraction sequence (extraction time and destination) of each material (ore and waste) parcel (Rezakhah, Moreno and Newman, 2020). Therefore, mine production scheduling plays a crucial role in maximising the net present value (NPV) of mining projects and mining operations. Maximum NPV is generated subject to geological, geotechnical, operational, and economic constraints, which ensure that a production schedule is practically feasible (Silva-Júnior *et al*, 2023; Paithankar and Chatterjee, 2019; Boland, Dumitrescu and Froyland, 2008). Geological constraints are factors such as composition, distribution, and characteristics of ore and waste in an orebody (Armstrong *et al*, 2021). These factors influence the geometry of an orebody, grade distribution and spatial location of ore and waste material (Quigley and Dimitrakopoulos, 2019). Geotechnical constraints relate to the structural stability which is important for operational safety and scheduling efficiency (Goodfellow and Dimitrakopoulos, 2013). Operational constraints are factors such as mining width, equipment availability and machine allocation. Economic constraints include factors that affect the economic viability of mining operations (Askari-Nasab and Awuah-Offei, 2009).

Open pit mine production scheduling is divided into interconnected long-term (LT), medium-term (MT), and short-term (ST) scheduling stages to reflect increasing granularity of periods and information (Otto and Musingwini, 2019; Osanloo, Gholamnejad and Karimi, 2008). LT production scheduling typically spans over the life-of-mine (LOM) and aims to maximise NPV over the LOM, with the extraction sequence normally presented on a yearly basis (Jélvez *et al*, 2023; Muke, Nhleko and Musingwini, 2021; Morales *et al*, 2019; Kloppers, Horn and Visser, 2015; Osanloo, Gholamnejad and Karimi, 2008; Akbari, Osanloo and Shirazi, 2008). At the LT horizon, the least amount of information is available, hence uncertainty is higher compared to MT and ST stages (Armstrong *et al*, 2021). The LT production scheduling stage provides an overall scheduling layout that MT and ST schedules must align with (Osanloo, Gholamnejad and Karimi, 2008).

MT mine production scheduling bridges LT and ST production scheduling stages (Otto and Musingwini, 2019). Firstly, MT production scheduling is developed from LT scheduling by translating LT objectives into detailed half-yearly, quarterly or monthly extraction schedules over a horizon of 1–5 years (Osanloo, Gholamnejad and Karimi, 2008). Secondly, it establishes a tactical extraction sequence that minimises operational costs. Azzamouri *et al* (2019) stated that MT production scheduling may include additional detailed information such as stockpiling, which can be used as part of a blending strategy to ensure constant mill feed grade.

ST mine production scheduling is developed from MT production scheduling, which as stated above, would have been developed from LT production scheduling. ST production scheduling focuses on daily, weekly and monthly extraction schedules often over a horizon of 6–12 months (Blom, Pearce and Stuckey, 2019). This stage generates a detailed, actionable production schedule that considers operational activities such as drilling, blasting, loading and hauling (Silva-Júnior *et al*, 2023). The ST production schedule should minimise costs of these activities by maximising equipment and resource utilisation (Silva-Júnior *et al*, 2023). A ST production schedule must be adequately flexible to adapt to changes imposed by factors such as equipment failures or variability in ore grades.

The above explanations indicate that existing LT, MT and ST scheduling approaches attempt to follow an interconnected process, hence it is important to seamlessly integrate the three production scheduling stages. If these stages are optimised as unconnected standalone stages, then such standalone optimisation can be detrimental to holistic optimisation of the overall scheduling system. Consequently, each separately optimised stage can be spatially or temporally misaligned with its consecutive upstream or downstream scheduling stage, resulting in undesirable project outcomes such as:

- mineral resource sterilisation.

- reduced project NPV.

- standalone schedules being locally optimal and not globally optimal leading to missed opportunities from leveraging synergies between scheduling stages.

- greater focus on ST gain at the expense of MT and LT value creation.

- since mining operations have inherent uncertainty, lack flexibility to adapt to unforeseen changes may occur when LT, MT and ST are optimised separately.

Since the early application of optimisation techniques to open pit mine production scheduling in the 1960s, the scheduling stages have generally been optimised as standalone stages in isolation from each other, despite their interconnectedness as stated above. LT production scheduling has been optimised independently from MT and ST schedules and thus fails to systematically inform the MT scheduling stage (Benndorf and Dimitrakopoulos, 2013; Gholamnejad, Lotfian and Kasmaeeyazdi, 2020; Khan, Asad and Topal, 2024; Cutler and Dimitrakopoulos, 2024). Likewise, MT scheduling has been optimised in isolation without following the LT scheduling layout nor informing the ST scheduling stage (Azzamouri *et al*, 2019; Yarmuch and Sepulveda, 2024). ST production scheduling has also been optimised in isolation without following the MT scheduling (Silva-Júnior *et al*, 2023;

Blom, Pearce and Stuckey, 2019; Matamoros and Dimitrakopoulos, 2016). Recently some researchers have attempted to partially integrate the scheduling stages of the production scheduling system. For example, Otto and Musingwini (2019) developed a compliance driver tree (CDT) approach used in conjunction with the root cause analysis technique, to improve spatial and temporal alignment between LT, MT and ST production schedules and mine plan execution, *albeit* manually. Levinson and Dimitrakopoulos (2023) developed a stochastic optimisation framework that links LT and ST scheduling stages, while omitting the MT scheduling stage. Muke (2025) and Muke *et al* (2025) developed a mixed-integer programming (MIP) scheduling model that integrates LT and MT production scheduling stages, while omitting the ST production scheduling stage. These attempts have partially solved the holistic integration of production scheduling stages at open pit mines, hence the need to fully integrate these three scheduling stages. When production scheduling stages are aligned, the overall production scheduling system can ensure that LT objectives are achieved at the MT scheduling horizon, and in turn, the MT objectives are also achieved at ST scheduling horizon.

Based on the foregoing perspectives, this paper uses the Preferred Reporting Items for Systematic Reviews and Meta-Analysis (PRISMA) method to conduct a bibliometric mapping of mine production scheduling optimisation techniques and propose future research directions in this research field. This paper also presents a MIP model for integrating LT to MT production scheduling, which was solved using the genetic algorithm (GA) and the approach was tested on an open pit metal mine block model as a case study. Based on the combined MIP model and GA approach taken, this paper then proposes modifications that can be made to the LT to MT scheduling optimisation model to incorporate the ST scheduling stage, for holistically solving the open pit mine production scheduling system.

OPEN PIT MINE PRODUCTION SCHEDULING OPTIMISATION TECHNIQUES

Techniques for optimising open pit mine production scheduling stages can be categorised as deterministic (or exact) and metaheuristic techniques (Tolouei *et al*, 2020). Deterministic techniques which generate optimal solutions, include methods such as dynamic programming (DP), linear programming (LP), integer programming (IP), MIP, stochastic integer programming (SIP), and stochastic mixed-integer programming (SMIP) (Silva-Júnior *et al*, 2023; Mokhtarian and Sattarvand 2018; Benndorf and Dimitrakopoulos 2013; Lin *et al*, 2012; Newman *et al*, 2010; Dimitrakopoulos and Ramazan 2004). Metaheuristic techniques which generate optimal or near-optimal solutions, include methods such as GA, particle swarm optimisation (PSO), firefly algorithm (FA), ant colony optimisation (ACO), simulated annealing (SA), and bat algorithm (BA) (Gandomi *et al*, 2013; Tolouei *et al*, 2020). Hybrid approaches which combine exact and metaheuristic methods, are sometimes to solve mine production scheduling problems more efficiently. Newman *et al* (2010) gave a comprehensive review of open pit mine production scheduling studies and approaches. Some examples of studies that have used exact, metaheuristic or hybrid methods to solve the open pit mine production scheduling problem are listed in Table 1.

Table 1 indicates a gradual progression from using exact methods to hybrid methods that combine exact and metaheuristic methods to solve the open pit mine production scheduling problem more efficiently. This progression is attributable mainly to exact methods generating solutions that can be impractical when mining blocks that contain heterogeneous material as encountered in real mining conditions, whereas metaheuristic methods can handle variability and uncertainty and are more adaptable to solving real-life large-scale optimisation problems (Armstrong *et al*, 2021). Table 1 also indicates a gradual shift from optimising open pit production scheduling stages as standalone stages in isolation to each other towards holistic optimisation of the stages as interconnected stages. This is why later in this paper, an MIP model that integrates LT and MT open pit mine production schedules solved using GA will be presented.

TABLE 1

Some examples of studies that have solved the open pit mine production scheduling problem.

Author(s)	Method(s) used	Comment(s)
Johnson (1969)	LP	Large multi-period LT production scheduling model was broken down into smaller sub-models that each focus on one period. Constraints considered include mining rate, processing rate, grade target, block precedence and mineral reserve constraints. Model did not fully comply with block extraction precedence constraint resulting in some blocks being suspended above ground level (Gershon, 1983).
Gershon (1983)	MIP	LT mine production scheduling model enhanced the Johnson (1969) study by introducing extra decision variables, to enable partial mining of blocks.
Dagdelen and Johnson (1986)	IP	LT production scheduling model was developed to overcome the shortcoming of the Johnson (1969) study and used a directed graph representation of a block model, where nodes represent blocks and arcs indicate their block precedence relationship and ensured complete extraction of a block in one period.
Denby and Schofield (1994)	GA	Model simultaneously optimises LT production scheduling and pit limits.
Akaike and Dagdelen (1999)	IP	Model used four-dimensional network relaxation and sub-gradient to account for dynamic cut-off grades in LT production scheduling.
Caccetta and Hill (2003)	IP	Model used branch-and-cut to consider different operational constraints such as mill feed capacity, maximum vertical depth, minimum pit bottom width and stockpiles.
Kumral and Dowd (2005)	SA	Model is based on multi-objective and stochastic relaxation to solve the LT production schedule.
Ferland, Amaya and Djuimo (2007)	PSO	Model optimises the LT production scheduling problem.
Boland, Dumitrescu and Froyland (2008)	SMIP	Model incorporates geological uncertainty when solving the LT mine production scheduling problem.
Chicoisne et al (2012)	LP	Model used the knapsack technique to optimise LT production scheduling of a real-sized block model instances in practical time.
Behrang, Hooman and Clayton (2014)	LP	Model used a clustering technique to reduce the number of variables to make the LT production scheduling problem tractable.
Khan and Delius (2015)	MIP and PSO	MIP model optimises LT production schedule and is solved using PSO.
Shishvan and Sattarvand (2015)	ACO	Model based on a max-min ant system to optimise the LT production scheduling problem.
Samavati et al (2018)	LP	Model used the adding cut technique to converge quickly to a near-optimal feasible solution of the LT production scheduling problem.
Ramazan and Dimitrakopoulos (2018)	SIP	Model optimises LT production scheduling by accounting for geological uncertainty through using multiple realisations of simulated orebody models.
Quigley and Dimitrakopoulos (2019)	SMIP	Model optimises ST production scheduling by accounting for mobile equipment allocation, material grade and equipment performance uncertainty.
Gholamnejad, Lotfian and Kasmaeeyazdi (2020)	IP	Model optimises LT production scheduling and reduces the number of active benches per period.
Muke, Nhleko and Musingwini (2021)	MIP and GA	MIP model for optimising LT production scheduling solved using a GA model.
Alipour et al (2022)	IP and GA	Model optimises LT production scheduling.
Muke (2025); Muke et al (2025)	MIP and GA	MIP model solved using GA to improve alignment of LT to MT production schedules.

BIBLIOMETRIC ANALYSIS OF OPEN PIT MINE PRODUCTION SCHEDULING OPTIMISATION TECHNIQUES

Bibliometric analysis is a quantitative and qualitative method for analysing scientific literature, providing insights into trends and identifying emerging areas within a specific research field (Passas, 2024). This paper employed a search methodology based on the PRISMA method to identify, screen and report on research papers published on optimising open pit mine production scheduling.

Search strategy and analysis

Research papers were sourced from Scopus and Web of Science (WoS) databases as these are the most extensively used citation indexes in the global research community and are compatible with most bibliometric analysis tools such as R-Biblioshiny™ and VOSviewer™ (De Groote and Raszewski, 2012; Mutandwa and Musingwini 2024). The following search string was used to find research papers on open pit mine production scheduling: ("open pit mine" OR "open-pit mine") AND ("production scheduling" OR "mine scheduling" OR "mine planning") AND ("optimi*" OR "maximi*" OR "minimi*") AND ("net present value" OR "npv" OR "cost" OR "profit" OR "value") AND ("linear programming" OR "integer programming" OR "mixed integer programming" OR "genetic algorithm" OR "particle swarm" OR "ant colony" OR "exact" OR "deterministic" OR "metaheuristic" OR "heuristic" OR "stochastic" OR "holistic"). A total of 256 publications identified by title, keywords and abstract comprised 174 from Scopus and 82 from WoS. A screening process was conducted using Zotero™, which is an open-source reference management software to manage bibliographic data and related research material (Ray and Ramesh, 2017). Screening in Zotero™ involved merging, removing duplicates and for consistency, discarding papers not written in English and not focused on open pit mine production scheduling. From screening, 180 papers were considered for further analysis.

Five scientific mapping analyses were performed. The first analysis focused on annual science publications of mine production scheduling using R-Biblioshiny™. The second analysis focused on the co-occurrence of keywords using VOSviewer™. The third analysis involved identifying the most relevant authors in the field using R-Biblioshiny™. The fourth analysis consisted of conceptual themes using R-Biblioshiny™. The fifth analysis focused on the quantitative and qualitative analysis of optimisation techniques applied to solve LT, MT and ST production scheduling stages in open pit mining. Figure 1 illustrates the annual publications from the work of Johnson (1969) to a cut-off date of 2024 since 2025 is still an incomplete year.

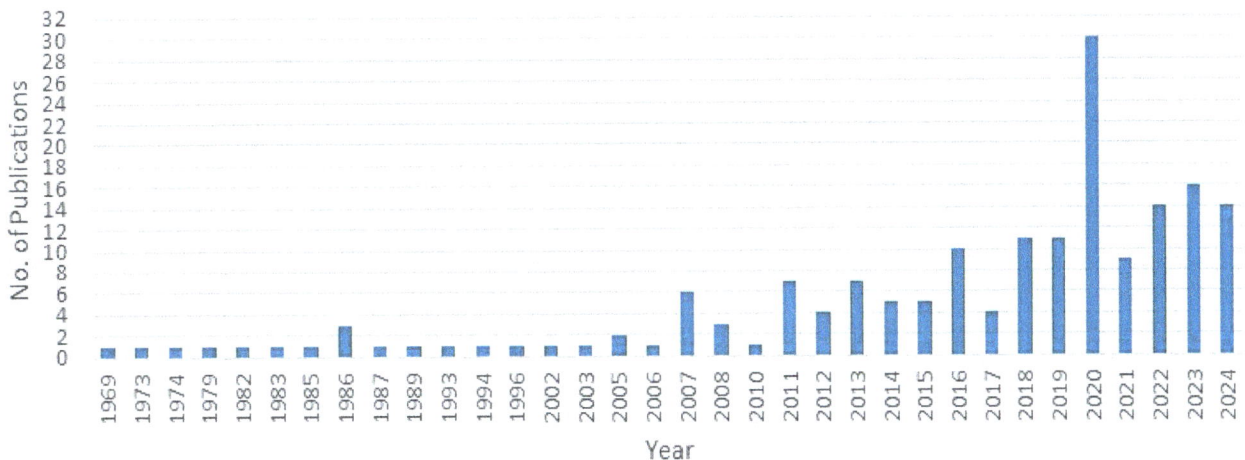

FIG 1 – Annual scientific publications of open pit mine production scheduling solved using optimisation techniques.

Figure 1 shows that from 1969 to 2006, the publication averaged about one article per annum. A surge in publications occurred from 2007 to 2024 averaging about nine papers per annum, reaching a peak of 30 in 2020. The surge indicates growing interest in solving the open pit mine production scheduling optimisation problem.

Franceschet (2009) suggested that clustering is a significant method in bibliometric analysis. The clustering technique organises bibliometric units such as publications based on their correlations that can be visualised as bibliometric maps. These clusters are evident in keyword analysis, specifically co-occurrence analysis, as illustrated in Figure 2, which shows all keywords that appeared at least six times in the co-occurrence analysis to best reflect important keywords. The connections of keywords to 'open pit mining', 'production control', and 'open pit mine' are depicted in red, blue, and green, respectively.

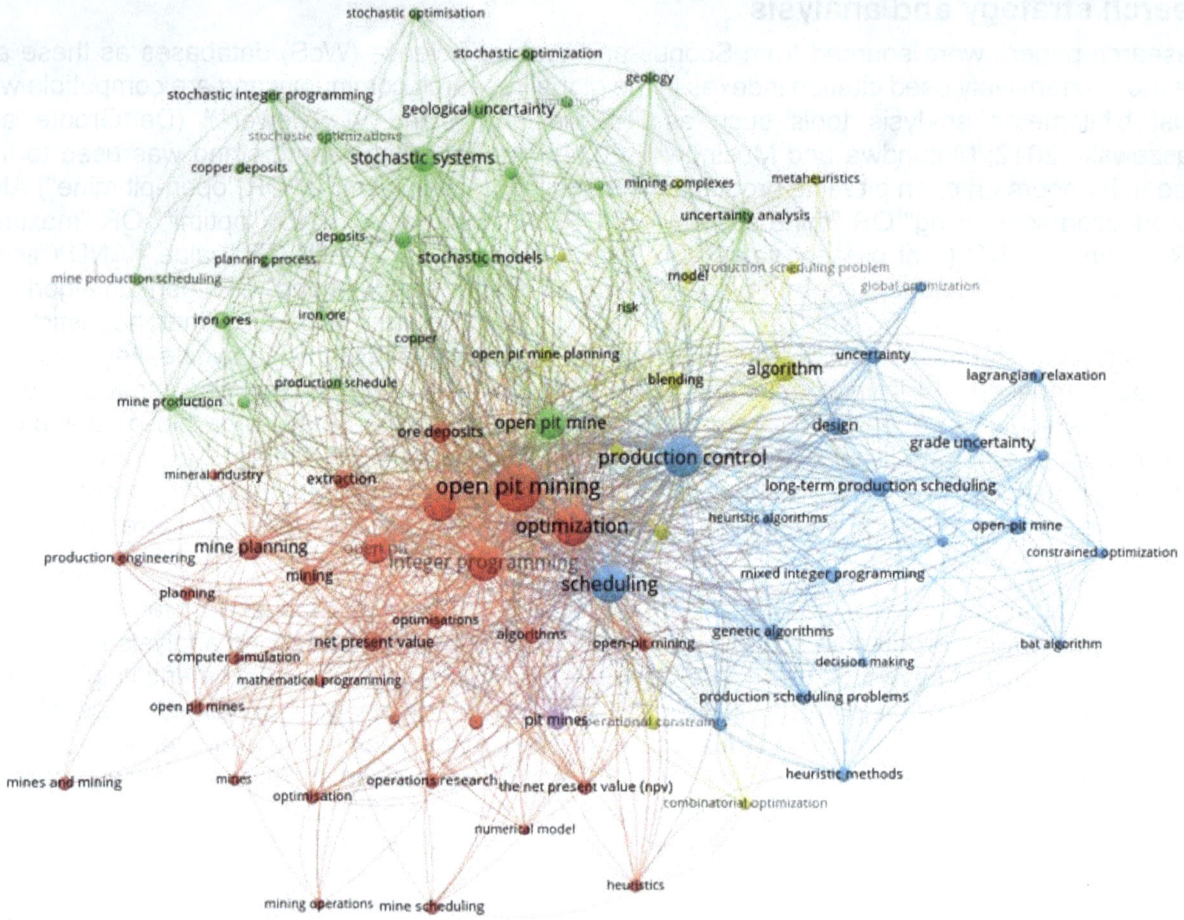

FIG 2 – Co-occurrence of keywords with a minimum threshold of six.

From the keyword list, the term 'long-term' appeared 56 times as 'strategic planning', 'life-of-mine', 'long-term', 'long-range' or 'long-scale'. In contrast, 'medium-term' appeared only twice, linked to 'medium-term' or 'tactical planning', while 'short-term' appeared 16 times, associated with 'short-term' or 'operational planning'. MT scheduling has occurred the least compared to LT and ST scheduling stages, despite its importance in linking LT and ST scheduling, partly due to some publications considering MT as either LT or ST mainly because of different classifications of scheduling horizons by different mines. Therefore, there is a need to standardise the scheduling horizons in the mine planning process. Therefore, this paper proposes that the LT horizon should be over the LOM, MT horizon from 1–5 years, and ST horizon from 1–12 months.

Most relevant authors

A bibliometric analysis of the most contributing authors was done; Figure 3 presents the results.

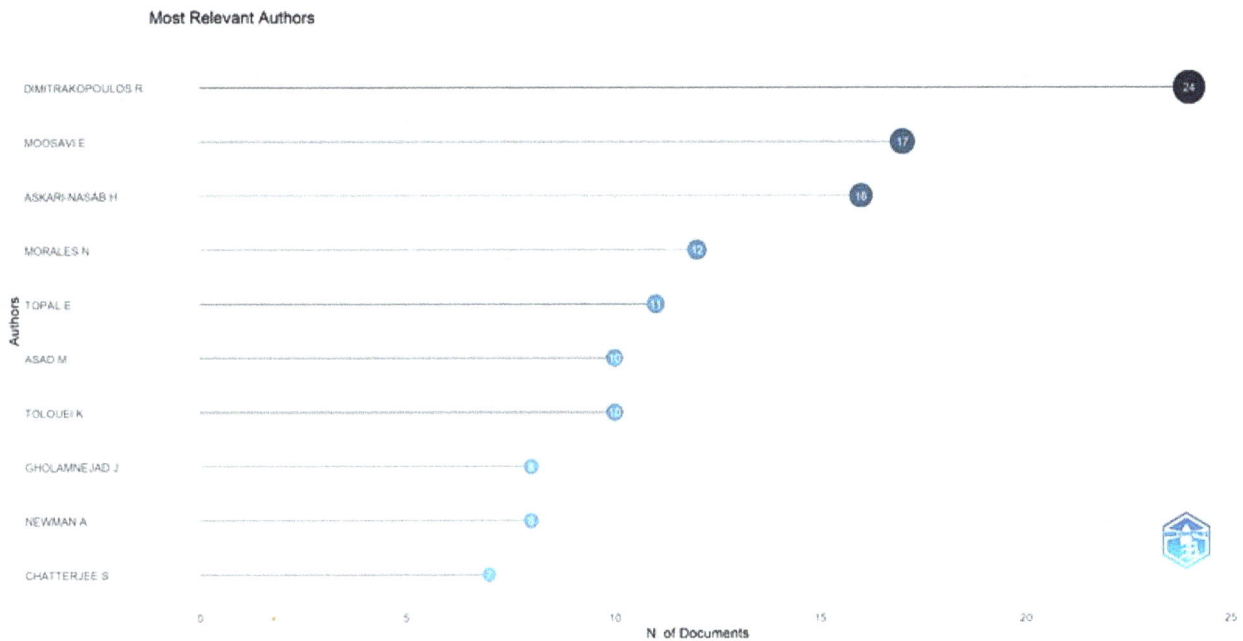

Most Relevant Authors

FIG 3 – Most contributing authors in open pit mine production scheduling publications.

This bibliometric analysis identified R Dimitrakopoulos, E Moosavi, H Askari-Nasab, N Morales and E Topal as the most contributing authors in the field of mine planning with publications ranging from 11 to 24 as shown in Figure 3. Out of the 319 authors identified, 15 authors have at least five publications. Emerging researchers in this field should consider forging collaborations with these identified leading authors.

Key thematic areas

R-Biblioshiny™ generates a thematic map based on prominent keywords. This map is divided into four quadrants, each representing different degrees of relevance and development (maturity of research). Figure 4 shows the conceptual themes in mine production scheduling that can guide authors in identifying focal points for future research. The analysis reveals that basic themes, characterised by low development but high relevance, include topics such as mining complexes, stochastic optimisation, traditional optimisation, simulation, uncertainty algorithms, and global optimisation. Conversely, motor themes, which are both highly developed and highly relevant, encompass topics such as in-pit crushing, conveying systems, artificial intelligence, integer programming, and production control. Niche themes, which are highly developed but less relevant, include particle swarm optimisation, programming models, the Lagrangian-relaxation method, and production scheduling problems. However, there are insufficient keywords to definitively categorise emerging or declining themes.

Out of 180 papers retained for analysis, 160 (ie 89 per cent) relate to LT scheduling, 12 (ie 7 per cent) to ST scheduling, five (ie 3 per cent) to MT scheduling, and only three (ie 2 per cent) relate to integration of production scheduling stages, thus, highlighting the need for future research to focus on holistic integration of production scheduling stages. Figure 5 shows results of a qualitative and quantitative analysis conducted on optimisation techniques applied on LT, MT and ST production schedules.

The tally of optimisation approaches from title, keywords and abstract shows that exact approaches (synonymous with deterministic) are used most in mine production scheduling, followed by stochastic approaches and lastly, metaheuristic approaches to open pit scheduling optimisation. Metaheuristic approaches rank third because they tend to be used to solve seemingly intractable stochastic or exact formulations. Further analysis shows that integrated production scheduling of scheduling stages is an emerging concept as identified in the three papers by Otto and Musingwini (2019, 2020), and Levinson and Dimitrakopoulos (2023). The shortcomings from these three papers were highlighted in the introduction section of this paper. Therefore, this paper presents a MIP model that links LT to MT production scheduling.

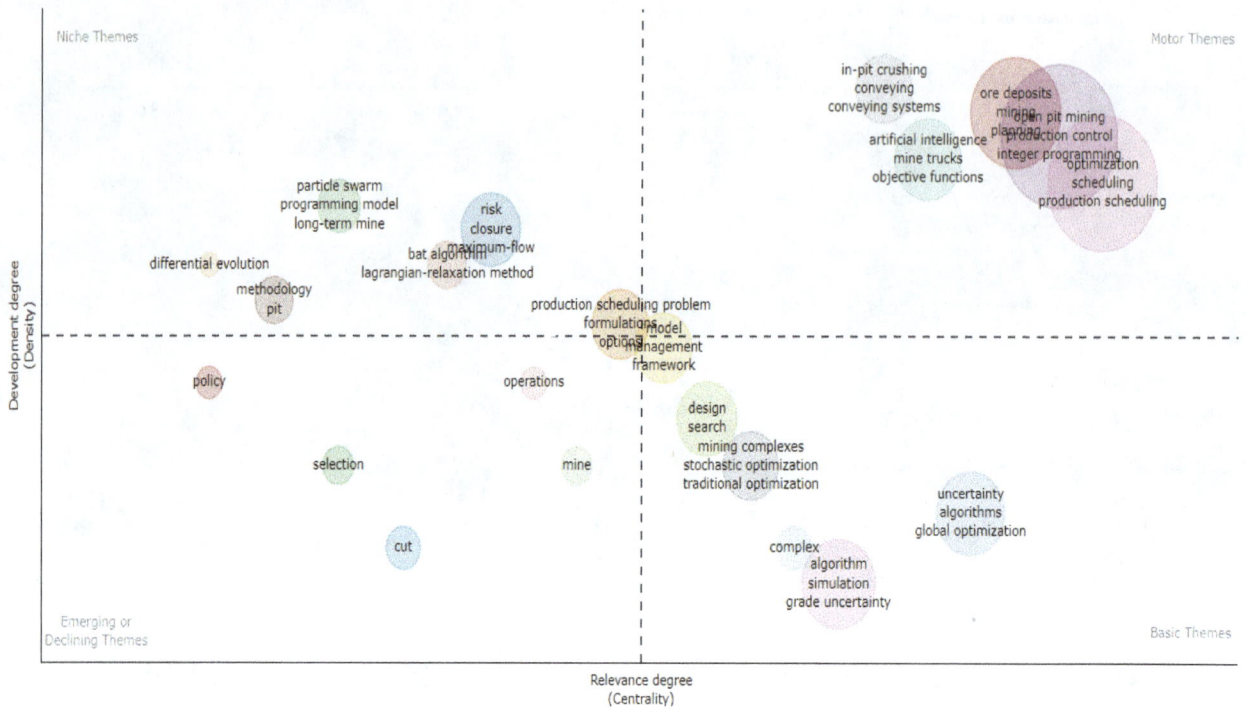

FIG 4 – Conceptual themes in open pit mine production scheduling.

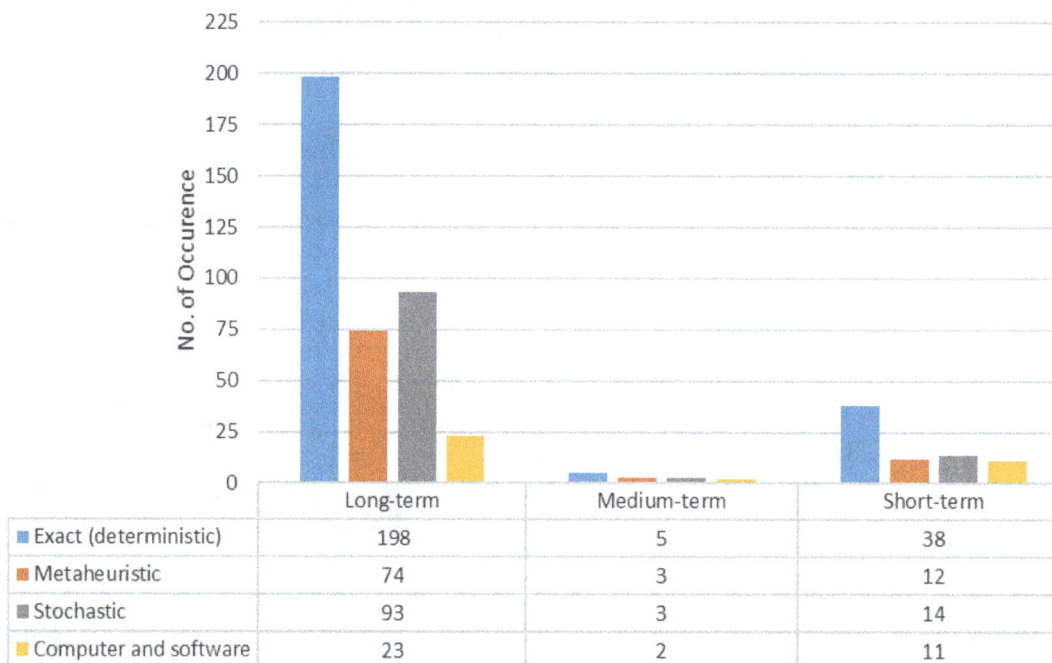

	Long-term	Medium-term	Short-term
■ Exact (deterministic)	198	5	38
■ Metaheuristic	74	3	12
■ Stochastic	93	3	14
■ Computer and software	23	2	11

FIG 5 – Analysis of optimisation techniques for LT, MT and ST production scheduling.

INTEGRATED MINE PRODUCTION SCHEDULING SYSTEM

As mentioned earlier in the paper, open pit mine production scheduling uses a hierarchical and structural scheduling framework that should allow each scheduling stage to feed into its consecutive downstream and upstream stages. Accordingly, this paper presents a MIP model from Muke (2025) and Muke *et al* (2025). The MIP model integrates LT and MT production scheduling stages to maximise NPV, subject to penalties for not meeting targets set for mining rate, processing rate and ore grade. The MIP model was solved using GA, which is a stochastic algorithm. The next sections describe the model formulation and implementation for an open pit metal mine block model as a case study.

Objective function for the integrated production scheduling system

The objective function of the MIP model as expressed by Equation 1 consists of four parts. Part I maximises the discounted cash flow of the production scheduling system for the integrated LT and MT scheduling stages which sums up the MT discounted value into LT discounted value. Parts II and III calculate penalties for deviating from lower- or upper-bounds of mining rates, processing rates and ore grade targets, at LT and MT stages, respectively. Part IV calculates integrated penalties associated with MT mining rate, processing rate and grade target deviations from the LT mining rate, processing rate and grade target, respectively:

$Maximum\ integrated\ Z$

$$
= \underbrace{\sum_{n=1}^{N} \sum_{t=1}^{T^{lt}} \left(\sum_{n=1}^{TNB^{lt}} \sum_{t=1}^{T^{mt}} v_{i,n,t}^{mt} \times X_{i,n,t}^{mt} \right) \times X_{i,n,t}^{lt}}_{Part\ I}
$$

$$
- \underbrace{\sum_{t=1}^{T^{lt}} \left(pc_i^{lt,RM} \times \emptyset_{i,t}^{lt,RM}(x) + pc_i^{lt,RO} \times \emptyset_{i,t}^{lt,RO}(x) + pc_i^{lt,RG} \times \emptyset_{i,t}^{lt,RG}(x) \right)}_{Part\ II}
$$

$$
- \underbrace{\sum_{t=1}^{T^{mt}} \left(pc_i^{mt,RM} \times \emptyset_{i,t}^{mt,RM}(x) + pc_i^{mt,RO} \times \emptyset_{i,t}^{mt,RO}(x) + pc_i^{mt,RG} \times \emptyset_{i,t}^{mt,RG}(x) \right)}_{Part\ III}
$$

$$
- \underbrace{\sum_{t=1}^{T^{mt}} \left(pc_{i,lt}^{mt,RM} \times \emptyset_{i,t,lt}^{mt,RM}(x) + pc_{i,lt}^{mt,RO} \times \emptyset_{i,t,lt}^{mt,RO}(x) + pc_{i,lt}^{mt,RG} \times \emptyset_{i,t,lt}^{mt,RG}(x) \right)}_{Part\ IV}
$$

(1)

Where:

N	is the set of blocks in the block model for a mining location i
n	is the block identifier in the block model $n \in N$
T^{lt}	is the set of LT scheduling periods
t	is the LT scheduling period index $t \in T^{\ell t}$
TNB^{lt}	is the set of blocks in each LT period available for MT scheduling
T^{mt}	is the set of MT scheduling periods
t	is the MT scheduling period index $t \in T^{mt}$
TNB^{mt}	is the set of blocks in each MT period available for ST scheduling

$X_{i,n,t}^{lt}$ is the LT binary variable for each block $n \in N$, mined from mining location i. It takes a value of one if block n is mined in period $t \in T^{lt}$ of LT horizon; zero otherwise.

$X_{i,n,t}^{mt}$ is the MT binary variable for each block $n \in N$ mined from mining location i. It takes a value of one if block n is mined in period $t \in T^{mt}$ of MT horizon; zero otherwise.

The discounted economic value of each block can be calculated at LT and MT scheduling stages using Equations 2 and 3, respectively:

$$v_{i,n,t}^{lt} = [\sum_{n=1}^{N} \sum_{t=1}^{T^{lt}} OT_{i,n,t}^{lt} \times g_{i,n,t} \times r_{i,t} \times (p_{i,t}^{lt} - cs_{i,t}^{lt})]$$

$$\underbrace{\qquad\qquad\qquad\qquad\qquad\qquad\qquad}_{Part\ I}$$

$$- [\sum_{n=1}^{N} \sum_{t=1}^{T^{lt}} (OT_{i,n,t}^{lt} \times cp_{i,t}^{lt}) + (OT_{i,n,t}^{lt} + WT_{i,n,t}^{lt}) \times cm_{i,t}^{lt}]$$

$$\underbrace{\qquad\qquad\qquad\qquad\qquad\qquad\qquad}_{Part\ II}$$

(2)

$$v_{i,n,t}^{mt} = [\sum_{n=1}^{TNB^{lt}} \sum_{t=1}^{T^{mt}} OT_{i,n,t}^{mt} \times g_{i,n,t} \times r_{i,t} \times (p_{i,t}^{mt} - cs_{i,t}^{mt})]$$

$$\underbrace{\qquad\qquad\qquad\qquad\qquad\qquad\qquad}_{Part\ I}$$

$$- [\sum_{n=1}^{TNB^{lt}} \sum_{t=1}^{T^{mt}} (OT_{i,n,t}^{mt} \times cp_{i,t}^{mt}) + (OT_{i,n,t}^{mt} + WT_{i,n,t}^{mt}) \times cm_{i,t}^{mt}]$$

$$\underbrace{\qquad\qquad\qquad\qquad\qquad\qquad\qquad}_{Part\ II}$$

(3)

Equations 4 to 6 are for ensuring temporal alignment, spatial alignment and discounted cash flow alignment from LT to MT schedules:

$$T^{lt}(t) \subseteq T^{mt}(t)$$

(4)

$$TNB_{i,t}^{lt} = \sum_{t=1}^{T^{mt}} TNB_{i,t}^{mt}$$

(5)

$$v_{n,t}^{lt} = \sum_{n=1}^{TNB^{lt}} \sum_{t=1}^{T^{mt}} (v_{n,t}^{mt} \times X_{n,t}^{mt})$$

(6)

$OT_{i,n,t}^{lt}$ and $OT_{i,n,t}^{mt}$ are quantities of ore material contained in block $n \in N$, at location i, in period $t \in T^{\ell t}$ and $t \in T^{mt}$; respectively.

$WT_{i,n,t}^{lt}$ and $WT_{i,n,t}^{mt}$ are quantities of waste material contained in block $n \in N$, at mining location i, in period $t \in T^{\ell t}$ and $t \in T^{mt}$; respectively.

$g_{i,n,t}$ and $g_{i,n,t}$ are grade qualities of ore material in block $n \in N$, at location i, in period $t \in T^{\ell t}$ and $t \in T^{mt}$; respectively.

$r_{i,t}$, and $r_{i,t}$ are mine recoveries of ore material at location i, in period $t \in T^{\ell t}$ and $t \in T^{mt}$; respectively.

$p_{i,t}^{lt}$ and $p_{i,t}^{mt}$ are discounted selling prices of selling prices $P_{i,t}^{lt}$ and $P_{i,t}^{mt}$, respectively, per unit of ore material produced and recovered at mining location i, in period $t \in T^{\ell t}$ and $t \in T^{mt}$; respectively. Where: $p_{i,t}^{lt} = P_{i,t}^{lt}/(1 + d^{lt})^t$ and $p_{i,t}^{mt} = P_{i,t}^{mt}/(1 + d^{mt})^t$.

d^{lt} and d^{mt} are economic discounted rates in periods $t \in T^{\ell t}$ and $t \in T^{mt}$, respectively.

$cs_{i,t}^{lt}$ and $cs_{i,t}^{mt}$ are discounted selling costs of selling costs $CS_{i,t}^{lt}$ and $CS_{i,t}^{mt}$, respectively, for selling unit ore material from location i, in period $t \in T^{\ell t}$ and $t \in T^{mt}$; respectively. Where: $cs_{i,t}^{lt} = CS_{i,t}^{lt}/(1 + d^{lt})^t$ and $cs_{i,t}^{mt} = CS_{i,t}^{mt}/(1 + d^{mt})^t$.

$cp_{i,t}^{lt}$ and $cp_{i,t}^{mt}$ are discounted processing costs of processing costs $CP_{i,t}^{lt}$ and $CP_{i,t}^{mt}$, respectively, for processing unit ore material from location i, in period $t \in T^{\ell t}$ and $t \in T^{mt}$; respectively. Where: $cp_{i,t}^{lt} = CP_{i,t}^{lt}/(1 + d^{lt})^t$ and $cp_{i,t}^{mt} = CP_{i,t}^{mt}/(1 + d^{mt})^t$.

$cm_{i,t}^{lt}$ and $cm_{i,t}^{mt}$ are discounted mining costs of mining cost $CM_{i,t}^{lt}$ and $CM_{i,t}^{mt}$, respectively, for mining unit material (ore and waste) from location i, in period $t \in T^{\ell t}$ and $t \in T^{mt}$; respectively. Where: $cm_{i,t}^{lt} = CM_{i,t}^{lt}/(1 + d^{lt})^t$ and $cm_{i,t}^{mt} = CM_{i,t}^{mt}/(1 + d^{mt})^t$.

$pc_i^{lt,RM}$ is the discounted mining penalty cost of mining penalty cost $PC_{i,t}^{lt,RM}$, for deviating from lower-bound mining rates $L_i^{lt,RM}$, or upper-bound mining rate $U_i^{lt,RM}$, at location i, in period $t \in T^{\ell t}$. Where: $pc_i^{lt,RM} = PC_{i,t}^{lt,RM}/(1 + d^{lt})^t$.

$pc_i^{mt,RM}$ is the discounted mining penalty cost of mining penalty cost $PC_{i,t}^{mt,RM}$, for deviating from lower-bound mining rates $L_i^{mt,RM}$, or upper-bound mining rate $U_i^{mt,RM}$, at location i, in period $t \in T^{mt}$. Where: $pc_i^{mt,RM} = PC_{i,t}^{mt,RM}/(1 + d^{mt})^t$.

$pc_i^{lt,RO}$ is the discounted processing penalty cost of processing penalty cost $PC_{i,t}^{lt,RO}$, for deviating from lower-bound processing rates $L_i^{lt,RO}$, or upper-bound processing rate $U_i^{lt,RO}$, at location i, in period $t \in T^{\ell t}$. Where: $pc_i^{lt,RO} = PC_{i,t}^{lt,RO}/(1 + d^{lt})^t$.

$pc_i^{mt,RO}$ is the discounted processing penalty cost of processing penalty cost $PC_{i,t}^{mt,RO}$, for deviating from lower-bound processing rates $L_i^{mt,RO}$, or upper-bound processing rate $U_i^{mt,RO}$ at location i, in period, $t \in T^{mt}$. Where: $pc_i^{mt,RO} = PC_{i,t}^{mt,RO}/(1 + d^{mt})^t$.

$pc_i^{lt,RG}$ is the discounted grade target penalty cost of grade target penalty cost $PC_{i,t}^{lt,RG}$, for deviating from lower-bound grade target rates $L_i^{lt,RG}$, or upper-bound grade target $U_i^{lt,RG}$ at location i, in period $t \in T^{\ell t}$. Where: $pc_i^{lt,RG} = PC_{i,t}^{lt,RG}/(1 + d^{lt})^t$.

$pc_i^{mt,RG}$ is the discounted grade target penalty cost of grade target penalty cost $PC_{i,t}^{mt,RG}$, for deviating from lower-bound grade target rates $L_i^{mt,RG}$, or upper-bound grade target $U_i^{mt,RG}$, at location i, in period $t \in T^{mt}$. Where: $pc_i^{mt,RG} = PC_{i,t}^{mt,RG}/(1 + d^{mt})^t$.

$L_i^{lt,RM}$ and $L_i^{mt,RM}$ are lower-bound mining rates at mining location i, in periods $t \in T^{\ell t}$ and $t \in T^{mt}$, respectively.

$U_i^{lt,RM}$ and $U_i^{mt,RM}$ are upper-bound mining rates at mining location i, in periods $t \in T^{\ell t}$ and $t \in T^{mt}$, respectively.

$\emptyset_{i,t}^{lt,RM}(x)$ and $\emptyset_{i,t}^{mt,RM}(x)$ are mining rate penalty functions at location i, during periods $t \in T^{\ell t}$ and $t \in T^{mt}$, respectively. These mining rate penalty functions are defined using Equations 7 and 8, respectively:

$$\emptyset_{i,t}^{lt,RM}(x) = max\{0, -f_t^{lt,RM}(x) + L_{i,t}^{lt,RM}, f_t^{lt,RM}(x) - U_{i,t}^{lt,RM}\} \tag{7}$$

$$\emptyset_{i,t}^{mt,RM}(x) = max\{0, -f_t^{mt,RM}(x) + L_{i,t}^{mt,RM}, f_t^{mt,RM}(x) - U_{i,t}^{mt,RM}\} \tag{8}$$

Where $f_t^{lt,RM}(x) = RM_{i,t}^{lt}$ and $f_t^{mt,RM}(x) = RM_{i,t}^{mt}$.

$RM_{i,t}^{lt}$ and $RM_{i,t}^{mt}$ are the reference continuous variables representing the actual mining rates achieved at location i, during period $t \in T^{lt}$ and $t \in T^{mt}$, respectively.

$\emptyset_{i,t}^{lt,RO}(x)$ and $\emptyset_{i,t}^{mt,RO}(x)$ are processing rate penalty functions at location i, during periods $t \in T^{\ell t}$ and $t \in T^{mt}$, respectively. These processing rate penalty functions are defined using Equations 9 and 10, respectively:

$$\emptyset_{i,t}^{lt,RO}(x) = max\{0, -f_t^{lt,RO}(x) + L_{i,t}^{lt,RO}, f_t^{lt,RO}(x) - U_{i,t}^{lt,RO}\} \tag{9}$$

$$\emptyset_{i,t}^{mt,RO}(x) = max\{0, -f_t^{mt,RO}(x) + L_{i,t}^{mt,RO}, f_t^{mt,RO}(x) - U_{i,t}^{mt,RO}\} \tag{10}$$

Where $f_t^{lt,RO}(x) = RO_{i,t}^{lt}$ and $f_t^{mt,RO}(x) = RO_{i,t}^{mt}$.

$RO_{i,t}^{lt}$ and $RO_{i,t}^{mt}$ are the reference continuous variables representing the actual processing rates achieved at location i, during period $t \in T^{lt}$ and $t \in T^{mt}$, respectively.

$\emptyset_{i,t}^{lt,RG}(x)$ and $\emptyset_{i,t}^{mt,RG}(x)$ are grade target penalty functions at location i, during periods $t \in T^{\ell t}$ and $t \in T^{mt}$, respectively. These grade target penalty functions are defined using Equations 11 to 12, respectively:

$$\emptyset_{i,t}^{lt,RG}(x) = max\{0, -f_t^{lt,RG}(x) + L_{i,t}^{lt,RG}, f_t^{lt,RG}(x) - U_{i,t}^{lt,RG}\} \tag{11}$$

$$\emptyset_{i,t}^{mt,RG}(x) = max\{0, -f_t^{mt,RG}(x) + L_{i,t}^{mt,RG}, f_t^{mt,RG}(x) - U_{i,t}^{mt,RG}\} \tag{12}$$

Where $f_t^{lt,RG}(x) = RG_{i,t}^{lt}$ and $f_t^{mt,RG}(x) = RG_{i,t}^{mt}$.

$RG_{i,t}^{lt}$ and $RG_{i,t}^{mt}$ are the reference continuous variables representing the actual grate target achieved at location i, during period $t \in T^{lt}$ and $t \in T^{mt}$, respectively.

Integrated penalties associated with MT mining rate, processing rate and ore grade target deviations from the LT mining rate, processing rate and ore grade target, are calculated by Equations 13 to 15, respectively:

$$\emptyset_{i,t,lt}^{mt,RM}(x) = max\{0, -f_{t,lt}^{mt,RM}(x) + L_{i,t}^{lt,RM}, f_{t,lt}^{mt,RM}(x) - U_{i,t}^{lt,RM}\} \tag{13}$$

$$\emptyset_{i,t,lt}^{mt,RO}(x) = max\{0, -f_{t,lt}^{mt,RO}(x) + L_{i,t}^{lt,RO}, f_{t,lt}^{mt,RO}(x) - U_{i,t}^{lt,RO}\} \tag{14}$$

$$\emptyset_{i,t,lt}^{mt,RG}(x) = max\{0, -f_{t,lt}^{mt,RG}(x) + L_{i,t}^{lt,RG}, f_{t,lt}^{mt,RG}(x) - U_{i,t}^{lt,RG}\} \tag{15}$$

$\emptyset_{i,t,lt}^{mt,RM}(x)$, $\emptyset_{i,t,lt}^{mt,RO}(x)$ and $\emptyset_{i,t,lt}^{mt,RG}(x)$ are integrated penalty functions of MT mining rate, processing rate and ore grade target for deviating from actual LT mining rate, processing rate and grate target achieved at location i, respectively, during period $t \in T^{mt}$, for each $t \in T^{lt}$.

Where $f_{t,lt}^{mt,RM}(x) = RM_{i,t,lt}^{mt}$, $f_{t,lt}^{mt,RO}(x) = RO_{i,t,lt}^{mt}$ and $f_{t,lt}^{mt,RG}(x) = RG_{i,t,lt}^{mt}$ are integrated LT and MT functions.

$RM_{i,t,lt}^{mt}$, $RO_{i,t,lt}^{mt}$ and $RG_{i,t,lt}^{mt}$ are integrated reference continuous variables measuring the MT mining rate, processing rate and ore grade target deviations from the LT mining rate, processing rate and ore grade target achieved at location i, during period $t \in T^{mt}$, for each $t \in T^{lt}$.

Integrated production scheduling constraints

The objective function expressed by Equation 1 is subject to the constraints following described in the next sub-sections.

Integrated scheduling mining rate constraint

Equations 16 and 17 ensure the integration of LT and MT mining rate constraints. Mining rates constraints are limitations imposed on the amount of material including ore and waste extracted at location i, during period $t \in T^{lt}$ and $t \in T^{mt}$; respectively:

$$RM_{i,t,lt}^{mt} = f_{t,lt}^{mt,RM}(x) = \sum_{t=1}^{T^{lt}} \sum_{n=1}^{TNB^{lt}} \sum_{t=1}^{T^{mt}} \left(OT_{i,n}^{mt} + WT_{i,n}^{mt}\right) \times X_{n,t}^{mt} \tag{16}$$

Where:

$$RM_{i,t}^{mt} = f_t^{mt,RM}(x) = \sum_{n=1}^{TNB^{lt}} \sum_{t=1}^{T^{mt}} \left(OT_{i,n}^{mt} + WT_{i,n}^{mt}\right) \times X_{n,t}^{mt} \tag{17}$$

Integrated scheduling processing rate constraint

Equations 18 and 19 ensure the integration of LT and MT processing rate constraints. Processing rate constraints are limitations imposed on the quantity of ore extracted at location i, during period $t \in T^{lt}$ and $t \in T^{mt}$; respectively:

$$RO_{i,t,lt}^{mt} = f_{t,lt}^{mt,RO}(x) = \sum_{t=1}^{T^{lt}} \sum_{n=1}^{TNB^{lt}} \sum_{t=1}^{T^{mt}} OT_{i,n}^{mt} \times X_{n,t}^{mt} \tag{18}$$

Where:

$$RO_{i,t}^{mt} = f_t^{mt,RO}(x) = \sum_{n=1}^{TNB^{lt}} \sum_{t=1}^{T^{mt}} OT_{i,n}^{mt} \times X_{n,t}^{mt} \tag{19}$$

Integrated scheduling grade target constraint

Equations 20 and 21 ensure the integration of LT and MT ore grade target constraints. Ore grade target constraints are limitations imposed on the quality of ore material extracted at location i, during period $t \in T^{lt}$ and $t \in T^{mt}$; respectively:

$$RG_{i,t,lt}^{mt} = f_{t,lt}^{mt,RG}(x) = \sum_{t=1}^{T^{lt}} \sum_{n=1}^{TNB^{lt}} \sum_{t=1}^{T^{mt}} \frac{\left(OT_{i,n}^{mt} \times g_{n,c}\right) \times X_{i,t}^{mt}}{OT_{i,n}^{mt} \times X_{i,t}^{mt}} \tag{20}$$

Where:

$$RG_{i,t}^{mt} = f_t^{mt,RG}(x) = \sum_{n=1}^{TNB^{lt}} \sum_{t=1}^{T^{mt}} \frac{\left(OT_{i,n}^{mt} \times g_{n,c}\right) \times X_{n,t}^{mt}}{OT_{i,n}^{mt} \times X_{n,t}^{mt}} \tag{21}$$

Integrated scheduling Mineral Reserves constraint

Equation 22 ensures the integration of LT, MT and ST Mineral Reserves constraint at location i, from period $t \in T^{lt}$ to $t \in T^{mt}$:

$$N = \sum_{t=1}^{T^{lt}} \sum_{t=1}^{T^{mt}} \sum_{\tau=1}^{T^{st}} TNB_{i,\tau}^{st} \tag{22}$$

Integrated block precedence constraint

Equation 23 ensures the block precedence constraint is embedded in the integrated scheduling system regardless of the scheduling stage at mining location i, during period $t \in T^{lt}$ and $t \in T^{mt}$:

$$9X_{ljk}^{lt,mt} - \sum_{r=1}^{T^{lt,mt}} (X_{l-1,j-1,k+1}^r + X_{l-1,j,k+1}^r + X_{l-1,j+1,k+1}^r + X_{l,j-1,k+1}^r + X_{l,j,k+1}^r + X_{l,j+1,k+1}^r \tag{23}$$
$$+ X_{l+1,j-1,k+1}^r + X_{l+1,j,k+1}^r + X_{l+1,j+1,k+1}^r) \leq 0, \forall l,j,k$$

Equation 23 represents a 1:9 pattern which requires the removal of nine blocks, r directly situated above a block. Where indices l, j and k are 3D coordinates of a block.

An additional operational constraint is applied to the MT schedule, defined by the number of benches bh along the k direction to be mined in each MT scheduling period. The number of benches to mine in each MT schedule is constrained by Equation 24:

$$\sum_{bh=bt=1}^{bb} \left(RM_{bh_k}^{mt} \right) \leq RM_{i,t}^{mt}; \; bh_k \in [bt, bb], \forall k, t \tag{24}$$

Where:

bt is the top bench of MT blocks from the surface

bb is the bottom bench of MT blocks in period t

$RM_{bh_k}^{mt}$ is the tonnage of material (ore and waste) on bench bh_k for the MT schedule in period t

The maximum number of benches is attained when the cumulative tonnage satisfies the MT mining rate constraint.

Since the generic block model structure of an open pit mine can be represented as a framework of chromosomes, it makes the GA a suitable algorithm to solve the production scheduling problem (Muke, Nhleko and Musingwini, 2021). The MIP model was solved using GA, which was coded in Python because it is a freely available open-source programming language that can handle large data sets and complex computations such as those encountered in open pit mine production scheduling. Figure 6 illustrates the GA framework followed in solving the MIP production scheduling model.

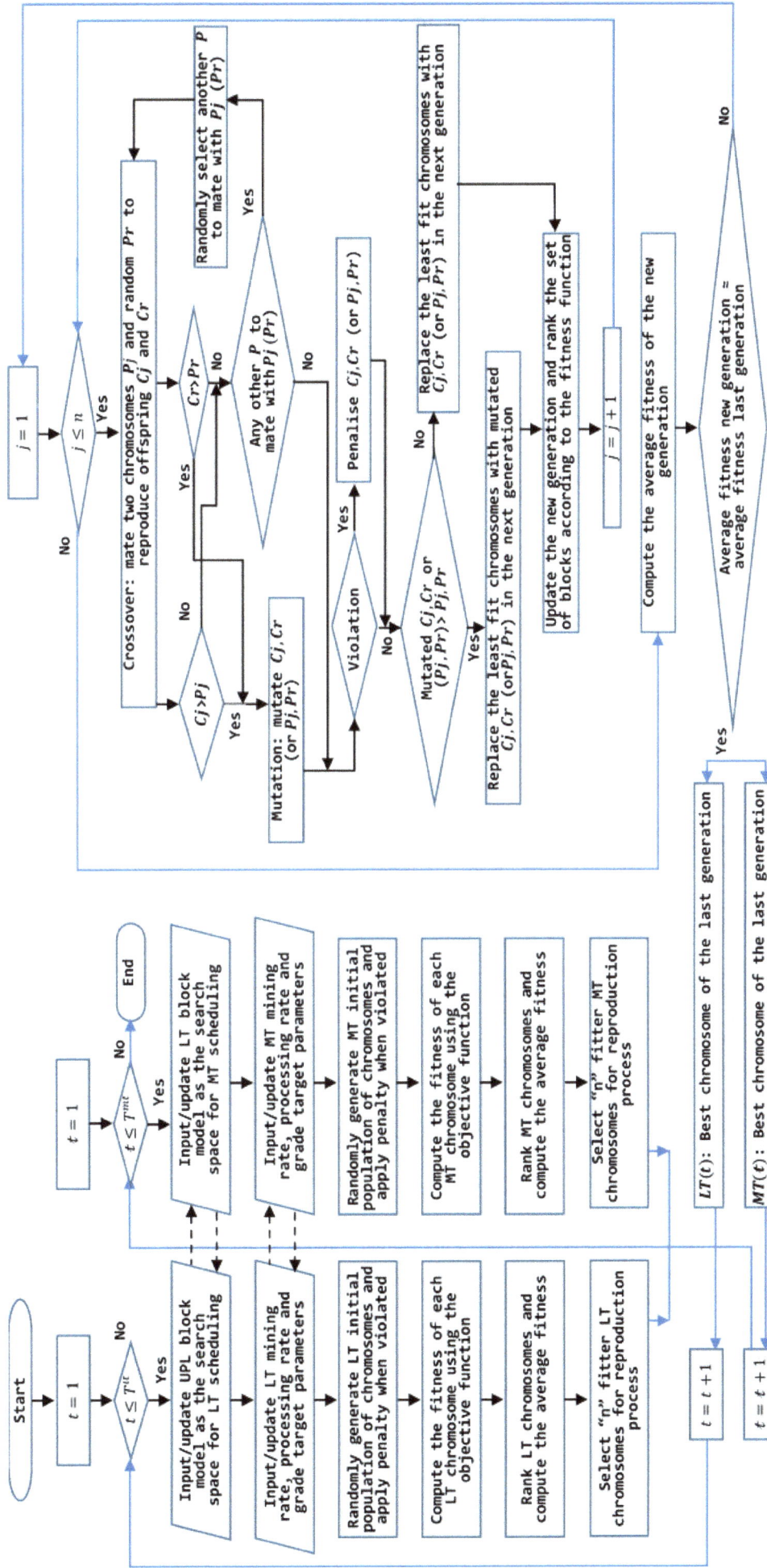

FIG 6 – Framework illustrating the GA approach followed in solving the MIP production scheduling model (Muke, 2025; Muke *et al*, 2025).

OPEN PIT METAL MINE CASE STUDY APPLICATION

To test the combined MIP model and GA approach, this paper used the KD block model data set as a case study. The KD block model was downloaded from *MineLib*, which is a library of publicly available mining data sets for research experiments. The data includes a block model and precedence files providing the basic information about the KD block model. The block model file provides information for each block including block identifier, block coordinates, block tonnage, block economic value, block destination and block grade of copper. The precedence file provides information regarding precedence relationships of blocks. Table 2 describes *MineLib's* input parameters for the KD block model.

TABLE 2

Description of input parameters of the KD block model.

Parameter	Description	Value
N	Total number of blocks	14 153
Γ_n	Maximum number of block predecessors of block	25
d	Discounted rate	0.15
$L_i^{lt,RM}$	Lower-bound mining rate for LT scheduling stage	-
$U_i^{lt,RM}$	Upper-bound mining rate for LT scheduling stage	Unlimited
$L_i^{lt,RO}$	Lower-bound processing rate for LT scheduling stage	-
$U_i^{lt,RO}$	Upper-bound processing rate for LT scheduling stage	10 000 000
$L_i^{mt,RM}$	Lower-bound mining rate for MT scheduling stage	-
$U_i^{mt,RM}$	Upper-bound mining rate for MT scheduling stage	Unlimited
$L_i^{mt,RO}$	Lower-bound processing rate for MT scheduling stage	-
$U_i^{mt,RO}$	Upper-bound processing rate for MT scheduling stage	5 000 000
$g_{i,n}$	Grade target of copper (Cu)	0.317 Cu%
$CM_{i,t}^{lt}, CM_{i,t}^{mt}$ and $CM_{i,\tau}^{st}$	Unit mining cost	0.75 $/t

Implementation of the combined MIP model and GA approach determined blocks that must be mined in every LT and MT scheduling period, simultaneously. Firstly, it determined the time in years, and secondly, the time in half-years when blocks should be mined to maximise NPV. Integrated scheduling enabled the spatial and temporal alignment between LT and MT schedules over the first five years of operation because the MT horizon spans up to five years. Blocks that are mined in MT half-year periods are always mined in the corresponding LT year period. The sum of two consecutive half-year periods of the MT schedule results in one year period of the LT schedule. Figure 7 shows the integrated LT and MT production scheduling with spatial compliance from LT Period 1 to MT Period 1 and MT Period 2.

The integrated LT and MT production schedule shown in Figure 7 complies with the block precedence constraint presented in Table 2. A block is mined either in the same period or a subsequent period to satisfy the precedence constraint. The integrated schedule also complied with hard constraints which should not be violated, namely LT and MT mining rate limits and mineral reserve constraints. Processing rate limits and ore grade constraints are soft constraints as they are dependent on the quantity and quality of available ore blocks. Table 3 summarises results of the combined MIP model and GA approach applied on the KD block model.

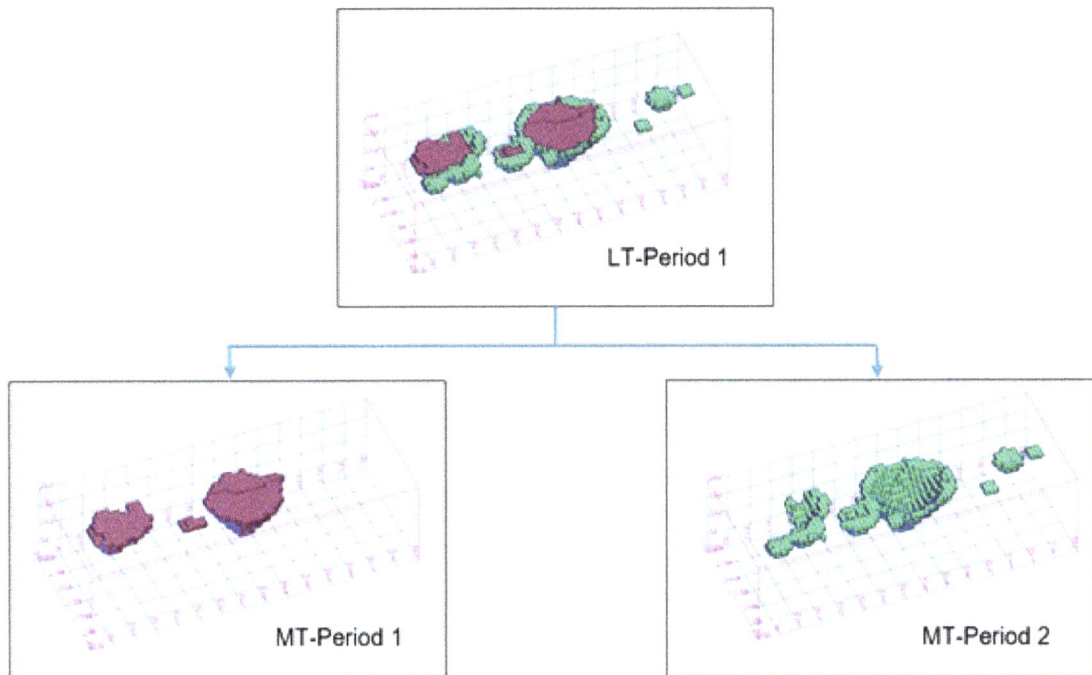

FIG 7 – Integrated LT and MT production scheduling with spatial compliance from LT Period 1 to MT Period 1 and MT Period 2.

The integrated LT and MT schedule solution was optimised over six years. The first five years aligned LT and MT scheduling stages and the sixth year for the LT schedule. The optimisation of the isolated LT schedule generated US$389.17 million NPV. The integrated LT and MT schedule generated US$397.27 million NPV, which is 2.08 per cent higher than isolated LT schedule. This demonstrates that integrated scheduling can generate higher NPVs but also improves the spatial and temporal compliance between scheduling stages. *MineLib* provides also a benchmark LT production scheduling solution of the KD block model, which has US$406.87 million NPV using LP relaxation of the production scheduling model that was solved using the TopoSort algorithm. The isolated LT solution is 4.35 per cent less, but when compared with the integrated schedule solution the gap is reduced to 2.36 per cent. Nevertheless, integrated LT and MT schedule ensures that LT objectives are achievable at MT horizons. Another run of the combined MIP and GA approach on the KD block model may generate a different result since GA is stochastic.

Table 4 shows the scheduling compliance of the isolated and integrated LT and MT production scheduling results. Mine production compliance is a temporal and/or spatial metric for measuring the total actual material mined, both in and out of the schedule, as a percentage of the scheduled material between scheduling stages. Equations 25 and 26 by Otto and Musingwini (2020) were used to calculate the compliance ratios:

$$LT \ to \ MT \ production \ compliance = \frac{Total \ actual \ MT \ material \ mined}{Total \ LT \ material \ scheduled} \tag{25}$$

$$LT \ to \ MT \ cash \ flow \ compliance = \frac{Total \ actual \ MT \ cash \ flow}{Total \ LT \ cash \ flow} \tag{26}$$

TABLE 3

Integrated LT and MT production scheduling results of the KD block model.

Period [Year] t	No. blocks		Total tonnage [T]		Ore tonnage [T]		Waste tonnage [T]		Average grade [Cu%]		Cash flow [US$ mil]	
	$TNB^{lt}_{i,t}$ LT	$TNB^{mt}_{i,t}$ MT	$RM^{lt}_{i,t}$ LT	$RM^{mt}_{i,t}$ MT	$RO^{lt}_{i,t}$ LT	$RO^{mt}_{i,t}$ MT	$RW^{lt}_{i,t}$ LT	$RW^{mt}_{i,t}$ MT	$RG^{lt}_{i,t}$ LT	$RG^{mt}_{i,t}$ MT	$V^{lt}_{n,t}$ LT	$V^{mt}_{n,t}$ MT
1	2400	1200	36 903 540	18 364 380	17 135 460	8 684 340	19 768 080	9 680 040	0.82	0.81	105.28	52.01
		1200		18 539 160		8 451 120		10 088 040		0.82		53.18
2	2400	1200	37 290 120	18 644 220	14 273 520	8 991 720	23 016 600	9 652 500	0.81	0.79	83.73	53.90
		1200		18 645 900		5 281 800		13 644 100		0.85		29.83
3	2400	1200	37 592 460	18 694 500	13 918 500	7 449 600	23 673 960	11 244 900	0.86	0.84	92.72	48.76
		1200		18 897 960		6 468 900		12 429 060		0.90		43.95
4	2400	1200	37 939 980	18 888 420	13 243 020	5 604 480	24 696 960	13 283 940	0.97	0.97	102.40	41.03
		1200		19 051 560		7 638 540		11 413 020		0.97		61.37
5	2400	1200	38 375 400	19 148 340	18 453 900	9 133 260	19 921 500	10 015 080	0.86	0.86	128.67	63.53
		1200		19 227 060		9 320 640		9 906 420		0.85		65.14
6	2153		34 618 560		19 153 020		15 465 540		0.77		117.55	
Total / Avg	14 153	12 000	222 720 060	188 101 500	96 177 420	77 024 400	126 542 640	111 077 100	0.84	0.87	630.35	512.70

NPV [US$ mil] at 15% discount rate for integrated LT and MT scheduling — 397.27

TABLE 4

Scheduling compliance of the isolated and integrated LT and MT production scheduling results on KD.

Period t [year]	t $\left[\frac{year}{2}\right]$	Isolated LT results achieved $RM_{i,t}^{lt}$ [tonne]	$RO_{i,t}^{lt}$ [tonne]	$V_{n,t}^{lt}$ [$mil]	Isolated MT results achieved $RM_{i,t}^{mt}$ [tonne]	$RO_{i,t}^{mt}$ [tonne]	$V_{n,t}^{mt}$ [$mil]	Isolated LT and MT scheduling compliance ratio RM [%]	RO [%]	V [%]	Integrated LT and MT scheduling compliance ratio RM [%]	RO [%]	V [%]
1	1	36 903 540	17 135 460	105.28	12 706 257	6 937 290	46.37	114.77	134.95	146.83	100	100	100
	2				29 647 935	16 187 012	108.21						
2	3	37 290 120	14 273 520	83.73	22 226 403	7 897 824	45.93	99.34	92.22	91.43	100	100	100
	4				14 817 602	5 265 216	30.62						
3	5	37 592 460	13 918 500	92.72	16 997 806	7 169 001	52.89	100.48	114.46	126.76	100	100	100
	6				20 775 097	8 762 113	64.64						
4	7	37 939 980	13 243 020	102.40	26 192 434	10 128 129	83.67	106.21	117.66	125.70	100	100	100
	8				14 103 618	5 453 608	45.05						
5	9	38 375 400	18 453 900	128.67	29 044 037	12 246 008	74.38	108.12	94.80	82.58	100	100	100
	10				12 447 444	5 248 289	31.88						
Total or Average		**188 101 500**	**77 024 400**	**513**	**198 958 637**	**85 294 493**	**583.64**	**105.77**	**110.73**	**113.81**	**100**	**100**	**100**

The results of the isolated production schedule show misalignment between LT and MT scheduling stages, averaging 105.77 per cent, 110.73 per cent and 113.81 per cent for total material mined, ore mined, and cash flow generated, respectively. The integrated LT and MT production scheduling model generated better scheduling compliance compared to the isolated optimisation of these scheduling stages. In the integrated model, LT and MT schedules share the same constraints and performance metrics, such as total material mined, ore mined, and cash flow, because MT production scheduling is both quasi-strategic and quasi-operational. However, an additional constraint, consisting of the number of active benches, was added to the MT production scheduling stage to improve the practicability of the schedule in addressing the ST operational constraint. Therefore, the model can be extended to integrate ST production scheduling by adding operational constraints such as machine allocation, mining width, and resource utilisation, and using feedback loops to update MT and LT scheduling stages with operational changes when they occur.

CONCLUSION

A total of 180 papers downloaded from Scopus and WoS databases were used for bibliometric analysis using R-Biblioshiny™ and VOSviewer™. The PRISMA-based bibliometric analysis revealed that since the first application of optimisation techniques in 1969 for solving the open pit mine production scheduling problem, the focus has mainly been on optimising LT (89 per cent) followed by ST (7 per cent) then MT (3 per cent) stages *albeit* in isolation. The MT scheduling stage occurred the least compared to LT and ST scheduling stages probably due to inconsistent definition of the time frame for the MT scheduling horizon. The analysis also revealed that exact approaches are commonly applied even though there is a gradual shift towards stochastic and metaheuristic approaches to enable the incorporation of large numbers of constraints and uncertainty inherent in the scheduling process.

This paper presented a MIP model that integrated LT and MT production scheduling stages and the model was solved using GA. Application of the approach *MineLib*'s KD block model, generated 2.08 per cent higher NPV compared to the isolated optimisation of LT production scheduling. Integrated production scheduling ensured 100 per cent compliance between LT and MT production scheduling stages. However, isolated LT and MT scheduling had an average scheduling misalignment of 105.77 per cent, 110.73 per cent and 113.81 per cent for total material mined, ore mined, and cash flow generated, respectively. In addition to the benefit of improving compliance, several other benefits such as avoiding mineral resource sterilisation were highlighted in the paper. Therefore, future research should focus on using techniques that integrate LT, MT and ST production schedules.

ACKNOWLEDGEMENTS

This paper presented part of the work undertaken by the first author for a PhD study at the University of the Witwatersrand. The authors acknowledge the University of the Witwatersrand for supporting them to participate in APCOM 2025 conference.

REFERENCES

Akaike, A and Dagdelen, K, 1999. Strategic production scheduling method for an open pit mine, in *Proceedings of the 28th Application of Computers and Operations Research in The Mineral Industry*, pp 729–738.

Akbari, A, Osanloo, M and Shirazi, M, 2008. Ultimate pit limit determination through minimizing risk costs associated with price uncertainty, *Gospodarka Surowcami Mineralnymi - Mineral Resources Management*, 24(4):157–170.

Alipour, A, Khodaiari, A A, Jafari, A and Moghaddam, R T, 2022. An integrated approach to open pit mines production scheduling, *Resources Policy*, 75:102459.

Armstrong, M, Lagos, T, Emery, X and Homem-de-Mello, T, 2021. Adaptive open pit mining planning under geological uncertainty, *Resources Policy*, 72:102086.

Askari-Nasab, H and Awuah-Offei, K, 2009. Open pit optimisation using discounted economic block values, *Mining Technology*, 118(1):1–12.

Azzamouri, A, Fenies, P, Fontane, F and Giard, V, 2019. Modelling the tactical decisions for open pit mines [online]. Laboratory for Analysis and Modeling of Systems for Decision Support, Dauphine Université Paris. Available from: <https://hal.science/hal-02158146v1/file/cahier_380.pdf> [Accessed: 2 November 2024].

Behrang, K, Hooman, A N and Clayton, D, 2014. A linear programming model for long-term mine planning in the presence of grade uncertainty and a stockpile, *International Journal of Mining Science and Technology*, 24:451–459.

Benndorf, J and Dimitrakopoulos, R, 2013. Stochastic long-term production scheduling of iron ore deposits: Integrating joint multi-element geological uncertainty, *Journal of Mining Science*, 49(1):68–81.

Blom, M, Pearce, A R and Stuckey, P J, 2019. Short-term planning for open pit mines: a review, *International Journal of Mining, Reclamation and Environment*, 33(5):318–339.

Boland, N, Dumitrescu, I and Froyland, G, 2008. A Multistage Stochastic Programming Approach to Open Pit Mine Production Scheduling with Uncertain Geology [online]. ResearchGate – Mining Engineering, 33 p. Available from: <https://www.researchgate.net/publication/228947110_A_multistage_stochastic_programming_approach_to _open_pit_mine_production_scheduling_with_uncertain_geology> [Accessed: 5 November 2024].

Caccetta, L and Hill, S P, 2003. An application of branch and cut to open pit mine scheduling, *Journal of Global Optimization*, 27:349–365.

Chicoisne, R, Espinoza, D, Goycoolea, M, Moreno, E and Rubio, E, 2012. A new algorithm for the open pit mine scheduling problem, *Operations Research*, 60(3):517–528.

Cutler, J and Dimitrakopoulos, R, 2024. Joint stochastic optimisation of open pit mine production scheduling with ramp design, *International Journal of Mining, Reclamation and Environment*, 38(6):480–495.

Dagdelen, K and Johnson, T, 1986. Optimum open pit mine production scheduling by Lagrangian parameterization, in *Proceedings 9th APCOM Symposium*, pp 127–141 (The American Institute of Mining, Metallurgical and Petroleum Engineers, Incorporated (AIME)).

De Groote, S L and Raszewski, R, 2012. Coverage of Google Scholar, Scopus and Web of Science: A case study of the h-index in nursing, *Nursing Outlook*, 60(6):391–400.

Denby, B and Schofield, D, 1994. Open pit design and scheduling by use of genetic algorithms, *Transactions of the Institution of Mining and Metallurgy*, 103:A21–A26.

Dimitrakopoulos, R and Ramazan, S, 2004. Uncertainty-based production scheduling in open pit mining, *Society for Mining, Metallurgy and Exploration*, 316:106–112.

Ferland, J A, Amaya, J and Djuimo, M S, 2007. Application of a particle swarm algorithm to the capacitated open pit mining problem, in *Autonomous Robots and Agents* (eds: S C Mukhopadhyay and G S Gupta), pp 127–133 (Springer: Berlin Heidelberg).

Franceschet, M, 2009. A cluster analysis of scholar and journal bibliometric indicators, *Journal of the American Society for Information Science and Technology*, 60(10):1950–1964.

Gandomi, A H, Yang, X-S, Talatahari, S and Alavi, A H, 2013. Metaheuristic Algorithms in Modeling and Optimization, *Metaheuristic Applications in Structures and Infrastructures*, pp 1–24.

Gershon, M, 1983. Optimal mine production scheduling: evaluation of large scale mathematical programming approaches, *International Journal of Mining Engineering*, 1:315–329.

Gholamnejad, J, Lotfian, R and Kasmaeeyazdi, S, 2020. A practical, long-term production scheduling model in open pit mines using integer linear programming, *The Journal of the Southern African Institute of Mining and Metallurgy*, 120(12):665–670.

Goodfellow, R and Dimitrakopoulos, R, 2013. Algorithmic integration of geological uncertainty in pushback designs for complex multiprocess open pit mines, *Mining Technology*, 122(2):67–77.

Jélvez, E, Ortiz, J, Varela, N M, Askari-Nasab, H and Nelis, G, 2023. A Multi-Stage Methodology for Long-Term Open pit Mine Production Planning under Ore Grade Uncertainty, *Mathematics*, 11(18):3907.

Johnson, T B, 1969. Optimum open pit production scheduling, 1st edn, University of California, Berkeley: Defense Technical Information Center.

Khan, A and Delius, C N, 2015. Application of particle swarm optimisation to the open pit mine scheduling problem, in *Proceedings of the 12th International Symposium Continuous Surface Mining*, pp 195–212 (Springer International Publishing).

Khan, A, Asad, A M W and Topal, E, 2024. A heuristic method for production scheduling of an open pit mining operation, *International Journal of Mining, Reclamation and Environment*, 38(4):293–305.

Kloppers, B, Horn, C and Visser, J, 2015. Strategic and tactical requirements of a mining long-term plan, *Journal of the Southern African Institute of Mining and Metallurgy*, 115(6):515–521.

Kumral, M and Dowd, P, 2005. A simulated annealing approach to mine production scheduling, *Journal of the Operational Research Society*, 56(8):922–930.

Leite, A and Dimitrakopoulos, R, 2014. Stochastic optimization of mine production scheduling with uncertain ore/metal/waste supply, *International Journal of Mining Science and Technology*, 24(6):755–762.

Levinson, Z and Dimitrakopoulos, R, 2023. Connecting planning horizons in mining complexes with reinforcement learning and stochastic programming, *Resources Policy*, 86(1):104–136.

Lin, M-H, Tsai, J-F, Yu, C-S and Hu, Y-C, 2012. A Review of Deterministic Optimization Methods in Engineering and Management, *Mathematical Problems in Engineering*, 2012(1):756023.

Matamoros, M E V and Dimitrakopoulos, R, 2016. Stochastic short-term mine production schedule accounting for fleet allocation, operational considerations and blending restrictions, *European Journal of Operational Research*, 255(3):911–921.

Mokhtarian, M and Sattarvand, J, 2018. Integration of commodity price uncertainty in long-term open pit mine production planning by using an imperialist competitive algorithm, *Journal of the Southern African Institute of Mining and Metallurgy*, 118(2):165–172.

Morales, N, Seguel, S, Cáceres, A, Jélvez, E and Alarcón, M, 2019. Incorporation of Geometallurgical Attributes and Geological Uncertainty into Long-Term Open pit Mine Planning, *Minerals*, 9(2):108.

Muke, P, 2025. A dynamic long-term and medium-term integrated open pit mine production scheduling system based on the genetic algorithm, PhD Thesis (unpublished) submitted to the University of the Witwatersrand, South Africa.

Muke, P, Nhleko, A and Musingwini, C, 2021. A genetic algorithm model for optimising long-term open pit mine production scheduling, in *Proceedings of the Application of Computers and Operations Research in the Minerals Industries (APCOM 2021)*, pp 1–18 (The Southern African Institute of Mining and Metallurgy).

Muke, P, Tholana, T, Musingwini, C and Ali, M, 2025. A genetic algorithm for temporal and spatial alignment of long- and medium-term mine production scheduling for open pit mines, submitted to *Resources Policy*.

Mutandwa, B and Musingwini, C, 2024. Insights and future research directions from a bibliometric mapping of studies in stope layout optimisation, *International Journal of Mining, Reclamation and Environment*, 38(8):577–595.

Newman, A M, Rubio, E, Caro, R, Weintraub, A and Eurek, K, 2010. A review of operations research in mine planning, *Interfaces*, 40(3):222–245.

Osanloo, M, Gholamnejad, J and Karimi, B, 2008. Long-term open pit mine production planning: A review of models and algorithms, *International Journal of Mining, Reclamation and Environment*, 22(1):3–35.

Otto, T and Musingwini, C, 2019. A spatial mine-to-plan compliance approach to improve alignment of short- and long-term mine planning at open pit mines, *The Journal of the Southern African Institute of Mining and Metallurgy*, 119:253–259.

Otto, T and Musingwini, C, 2020. A compliance driver tree (CDT) based approach for improving the alignment of spatial and intertemporal execution with mine planning at open pit mines, *Resources Policy*, 69:1–9.

Paithankar, A and Chatterjee, S, 2019. Open pit mine production schedule optimization using a hybrid of maximum-flow and genetic algorithms, *Applied Soft Computing Journal*, 81:105507.

Passas, I, 2024. Bibliometric Analysis: The Main Steps, *Encyclopedia*, 4(2):1014–1025.

Quigley, M and Dimitrakopoulos, R, 2019. Incorporating geological and equipment performance uncertainty while optimising short-term mine production schedules, *International Journal of Mining, Reclamation and Environment*, 34(5):362–383.

Ramazan, S and Dimitrakopoulos, R, 2018. Stochastic optimisation of long-term production scheduling for open pit mines with a new integer programming formulation, in *Advances in Applied Strategic Mine Planning*, pp 139–153 (Springer International Publishing).

Ray, A K and Ramesh, D, 2017. Zotero: Open Source Citation Management Tool for Researchers, *International Journal of Library and Information Studies*, 7(3):238–245.

Rezakhah, M, Moreno, E and Newman, A, 2020. Practical performance of an open pit mine scheduling model considering blending and stockpiling, *Computers and Operations Research*, 115:104638.

Samavati, M, Essam, D, Nehring, M and Sarker, R, 2018. A new methodology for the open pit mine production scheduling, *Omega*, 81:169–182.

Shishvan, M S and Sattarvand, J, 2015. Long term production planning of open pit mines by ant colony optimization, *European Journal of Operational Research*, 240:825–836.

Silva-Júnior, A L, Martins, A G, Pantuza Jr, G, Cota, L P and Souza, M J F, 2023. Short-term planning of a work shift for open pit mines: A case study, *Cogent Engineering*, 10(1):2168172.

Tolouei, K, Moosavi, E, Bangian, A H, Afzal, P and Aghajani, B A, 2020. A comprehensive study of several meta-heuristic algorithms for the open pit mine production scheduling problem considering grade uncertainty, *Journal of Mining and Environment*, 11(3):721–736.

Yarmuch, J L and Sepulveda, G, 2024. Application of Mining Width-Constrained Open Pit Mine Production Scheduling Problem to the Medium-Term Planning of Radomiro Tomic Mine: A Case Study, *Mining, Metallurgy and Exploration*, 41(2):681–693.

Operational design of ramps in open pit mines considering switchbacks through the assistance of mathematical optimisation – a case study

P Nancel-Penard[1,2], F Manríquez[3], R Figueroa[4], M Lobos[5] and M Núñez[6]

1. Researcher, Delphos Mine Planning Laboratory, Department of Mining Engineering, Faculty of Physical and Mathematics Sciences, University of Chile, Santiago, Chile. Email: pierre.nancel@amtc.uchile.cl
2. Researcher, Advanced Mining Technology Center, University of Chile, Tupper 2007, Santiago, Chile. Email: pierre.nancel@amtc.uchile.cl
3. Assistant Professor, Departamento de Ingeniería de Minas, Metalurgia y Materiales, Universidad Técnica Federico Santa María, Santiago, Región Metropolitana, 8940897, Chile. Email: fabian.manriquez@usm.cl
4. Undergraduate student, Departamento de Ingeniería de Minas, Metalurgia y Materiales, Universidad Técnica Federico Santa María, Santiago, Región Metropolitana, 8940897, Chile. Email: rodrigo.figueroaol@sansano.usm.cl
5. Undergraduate student, Departamento de Ingeniería de Minas, Metalurgia y Materiales, Universidad Técnica Federico Santa María, Santiago, Región Metropolitana, 8940897, Chile. Email: matias.lobos@usm.cl
6. Undergraduate student, Departamento de Ingeniería de Minas, Metalurgia y Materiales, Universidad Técnica Federico Santa María, Santiago, Región Metropolitana, 8940897, Chile. Email: martin.nuneze@sansano.usm.cl

ABSTRACT

Open pit mine design is a vital component of strategic mining operations, traditionally carried out through a labour-intensive 'traditional design' approach. This study introduces a two-step methodology to streamline the creation of operational mine designs while incorporating 180° ramp turns, or 'switchbacks.' The first step employs integer linear programming to generate a block-supported mine design, where the layout is represented using discrete blocks. In the second step, this preliminary design is transformed into an operational mine design, referred to as the 'aided design.' CAD software assists mine designers in replicating the ramp paths derived from the block-supported design. Remarkably, the aided design achieves outcomes comparable to those of a human expert, with minimal percentage differences in value. This demonstrates the method's effectiveness in maintaining design quality while significantly reducing the time and resources needed for mine planning. By lowering reliance on individual expertise, this approach frees resources to explore various design scenarios, such as optimal switchback placement and surface ramp configurations, allowing for better-informed decision-making. Although the methodology offers substantial advantages, the conclusions are case-specific and depend on factors such as the block model, pushback geometry, economic parameters, and technical constraints. Nonetheless, this study highlights the potential of combining optimisation tools and CAD software to enhance the efficiency and consistency of open pit mine design processes. The proposed methodology not only simplifies the design process but also opens new possibilities for analysing diverse mine configurations. The semi-automatic design positions itself as a valuable tool for mine planning, delivering substantial advantages to mine designers.

INTRODUCTION

Open pit mine design is a vital component of strategic mining operations, traditionally carried out through a labour-intensive 'traditional design' approach.

Open pit mine design involves planning haul roads, known as ramps, which allow high-capacity trucks to transport material from loading fronts within the mine to designated destinations. These include waste dumps for overburden, primary crushers for ore processing, and stockpiles for temporary ore storage. This study specifically considers in-pit ramps, which are the haul roads located within the mine.

An open pit mine design considers the following operational parameters: berm width, ramp width, bench height maximum ramp gradient, bench face angle, inter-ramp angle (IRA), and overall slope angle. In-pit ramps can be either spiral or feature 180° turns, commonly known as switchbacks. Switchbacks require more space to allow trucks to navigate the turns safely. For further details on the manual design of switchbacks, refer to Hustrulid, Kutcha, and Martin (2013). Couzens (1979) mentions that it is desirable to avoid the use of switchbacks in a pit due to the following causes: tend to slow down traffic; cause greater tire wear; cause various maintenance problems; and probably pose more of a safety hazard than do spiral roads (vision problems, machinery handling, among others). Also, switchbacks affect the cost and safety of the haulage truck's operation (Thompson, 2011). Indeed, designing a switchback impacts the global angle. Considering the same ramp width, and the same pit height H_p, the global angle is lower in the switchback design ($\alpha_{g,1} > \alpha_{g,2}$). Therefore, the quantity of waste to be extracted is greater in the slope face where the switchback is located, which leads to an additional stripping cost.

Usually, the ramp design process is carried out by a mine designer using CAD software. This engineer applies his experience and judgment to find a ramp design that is as close as possible to the pit obtained in the previous stage.

The ultimate pit problem is a fundamental component of open pit mine planning, aiming to determine the pit outline that maximises the economic value of extracted material while adhering to slope and precedence constraints. Traditionally solved using the Lerchs-Grossmann algorithm (Lerchs and Grossmann, 1965), modern approaches have introduced more efficient methods, such as the pseudo-flow algorithm. This algorithm, based on network flow theory, offers improved computational performance and scalability, particularly for large block models (Hochbaum and Chen, 2000). Tools incorporating these techniques, such as Whittle® software (version 4.7.3100, 2025, Dassault Systèmes, GEOVIA), have become standard in the industry for generating optimal final pit designs. Other algorithms, such as the Floating Cone method (Pana, 1966), along with subsequent improvements by David, Dowd, and Korobov (1974) and Wright (1999), have also been proposed to solve the ultimate pit limit problem.

The ramp design process has a series of disadvantages:

- it is a trial-and-error process.
- it involves considerable time (between days and weeks).
- the result depends on the person who carries out the designs.
- it is not known if the obtained design is the best possible or how far it is from the best possible.

In practice, the mine designer looks for a feasible solution. However, this solution is not necessarily the optimal one. The open pit design is expected to be iterative (Deswik, 2018). The design may need several attempts to come up with a satisfactory final design. The mine designer will get faster and develop better designs with practice and experience.

This study introduces a two-step methodology to streamline the creation of operational mine designs while incorporating 180° ramp turns, or 'switchbacks.' This paper contributes by extending the integer linear programming model proposed by Nancel-Penard et al (2019) to account for switchbacks in open pit ramp design, ensuring sufficient space is allocated for each switchback. The output of the model—a block-support ramp layout—serves as a guideline for planners to develop an operational ramp design.

RELATED WORK

Several studies have addressed the problem of open pit ramp design. Early approaches, such as Dowd and Onur (1992), introduced heuristic algorithms to minimise profit loss by generating spiral ramps without switchbacks. More recent methods leverage integer linear programming models, as seen in Morales, Nancel-Penard, and Parra (2017) and Nancel-Penard et al (2019), which optimise ramp placement while considering economic impact. Espejo, Nancel-Penard, and Morales (2019) extended this approach by automating operational ramp designs based on block-bench optimisation. Additional contributions, such as Yarmuch et al (2020), formulated ramp design as a binary linear programming problem, though limitations remain regarding economic profit integration and geometric fidelity. More recent works explore non-linear optimisation (Kaykov *et al*, 2023) and automated methodologies for integrating ramp design into mine planning (Morales, Nancel-Penard, and Espejo, 2023). More recently, Sanhueza *et al* (2025) models in-pit haulage network design as a modified Steiner tree problem.

The contribution of this paper is to present a case study that applies an extension of the integer linear programming model described in Nancel-Penard *et al* (2019), which generates open pit ramp design at block support considering switchbacks. The results of the model serve as a guide for the planner in developing an operational design.

METHODOLOGY

This section presents a detailed description of the general methodology for generating an operational open pit design using mathematical optimisation. A central component of this methodology is the construction of a block-support ramp, which serves as a critical intermediate step. A brief overview of the integer linear programming model used to generate this block-support ramp within the overall methodological framework is also presented.

Optimisation model description

This article extends the integer linear programming model presented in Nancel-Penard *et al* (2019) to explicitly incorporate switchbacks in the construction of an open pit ramp at block support. The optimisation model used the concepts of candidates block for in-pit ramp design, spiral paths and switchback paths, which are described herein.

Candidate blocks for in-pit ramp design

Figure 1a illustrates a block model and the final pit, with its contour outlined in black. Grey blocks represent those within the pit, while white blocks indicate those outside.

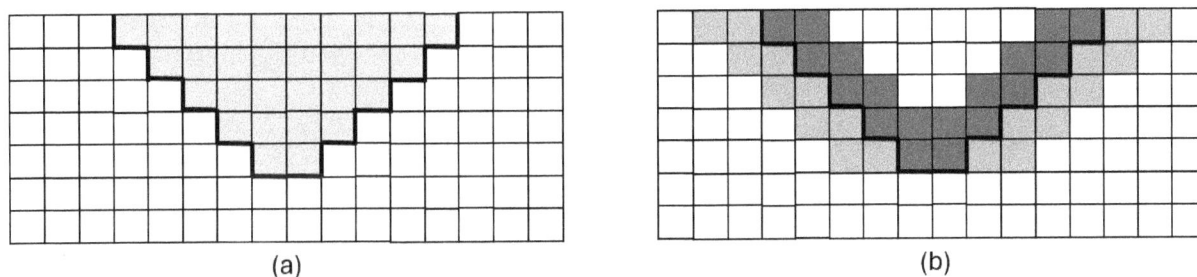

(a)   (b)

FIG 1 – (a) Block model and final pit. The final pit contour is outlined in black. Light grey blocks represent those within the pit, while white blocks indicate those outside. (b) Candidate blocks for ramp design. This set includes dark grey blocks, which are inside the final pit and close to its contour, and light grey blocks, which are outside the pit but near its contour.

Based on the final pit boundary at the block scale, an internal and external width (measured in number of blocks) is defined. This width determines the set of candidate blocks for ramp design.

Figure 1b shows the candidate blocks for ramp design. This set includes dark grey blocks, which are inside the final pit and close to its boundary, and light grey blocks, which are outside the pit but near its boundary. In this figure, an internal buffer of two blocks and an external buffer of two blocks are considered.

For the optimisation of block-suport open pit ramp, the concept of pre-computed spiral paths from Morales, Nancel-Penard, and Parra (2017) is recalled here.

Spiral paths

A purely spiral in-pit ramp is composed of segments called 'spiral paths,' which are computed within the set of candidate blocks for ramp construction. Each spiral path represents a section of the ramp at a specific bench. It consists of blocks at the same bench, along with an additional block at the bench immediately below, ensuring connectivity between benches. Each spiral path is designed to maintain a slope that does not exceed a predefined maximum slope, ensuring a smooth ramp design.

Switchback paths

To consider a ramp with switchbacks (180° turns), special segments must be created to represent them. The generation of these new paths, referred to as switchback paths is explained here. A switchback path requires more space than a spiral path to allow haulage trucks to safely make a 180° turn. The goal is to generate switchback paths with a lower slope, α_{sw}, compared to the slope of the spiral paths, α_{sp}. As the slope decreases, the length of the switchback paths increases.

The connection of two switchback paths that turn in opposite directions (clockwise and anticlockwise) on consecutive benches results in the design of a switchback. Then, the mine designer can use this additional space to properly design the switchback when smoothing the design obtained by the integer linear programming model.

Thus, the integer linear programming model used in this study selects and connects the spiral and switchbacks paths to construct a continuous in-pit ramp that links all benches of the final pit while maximising the total profit at the block scale.

Decision variables, objective function and constraints

Two different decision variables are defined. First, the binary variable y_b determines whether a block b from the considered block model is extracted. The second variable, x_{ki}, is a binary variable that determines whether the i-th path at bench k is selected to be part of the ramp. In the integer linear programming model, the blocks comprising the selected ramp path are extracted.

The objective function maximises the overall profit of the ramp design. It corresponds to the profit of all extracted blocks.

The following constraints are considered within the integer linear programming model:

- Constraint 1 ensures connectivity between paths, meaning that each selected path at bench k is connected to the selected path at bench k−1.

- Constraint 2 specifies that there can be at most one path per bench.

- Constraint 3 prevents the extraction of any block for which all preceding blocks have not been extracted first.

- Constraint 4 ensures that the blocks belonging to a selected path with a specified ramp width are extracted.

- Constraint 5 ensures that for each selected path, all blocks between the ramp and the slope beneath the ramp are extracted.

- Constraint 6 prevents the extraction of blocks directly below the ramp blocks and those located between the ramp and the slope at level k-1.

- Constraint 7 guarantees that the ramp connects to the lowest bench of the pit.

- Constraint 8 prevents an unconnected path from being an eligible path.

- Constraint 9 imposes that connected paths turning in opposite directions (clockwise and anticlockwise) must be switchback paths.

It is important to note that implementing the Constraint 9 requires both spiral and switchback paths to have additional attributes: first, the direction of turn (clockwise or anti-clockwise), and second, the type of path (spiral or switchback). These constraints ensure the use of switchback paths in the construction of switchbacks, thus providing sufficient space to design a switchback in the next step of the methodology: the generation of the operational design.

The proposed integer linear programming model allows for specifying the starting block at the pit surface. If the starting block for the ramp is not defined, the processing time increases significantly due to the computation of numerous additional paths, which substantially expands the number of possible path connections.

In the traditional approach, the placement of a switchback in the ramp design is typically determined by the engineer. Therefore, the proposed integer linear programming model enables the user to specify between which benches the optimisation should generate a switchback. This is accomplished by constructing paths in opposite directions exclusively between the designated bench pairs.

Alternatively, the integer linear programming model also supports determining the location of switchbacks and the starting point of the block-support ramp automatically. This approach involves solving an instance with a larger number of decision variables compared to the case in which the user specifies the switchback locations, thereby resulting in significantly higher computational times.

General methodology description

The proposed methodology has the following sequential steps:

1. Compute the final pit using traditional methodology (Lerchs and Grossmann, 1965), using the technical and economic parameters, along with slope precedence, based on the overall global angle of the pit.

2. Define the set of candidate blocks along the contour of the final pit, determining the internal and external widths in terms of the number of blocks.

3. Define the maximum slope for both the spiral and switchback paths.

4. Compute the set of paths within the candidate blocks designated for ramp design.

5. Apply the integer linear programming model to generate the open pit ramp with block support, using the final pit obtained in step 1 and the paths generated in step 4 as input.

6. Generate an operational open pit design using CAD software, based on the open pit ramp with block support obtained in the previous step. The planner uses as a guide the open pit ramp to generate an operational open pit ramp design.

Therefore, the proposed assisted methodology was evaluated by comparing its performance with that of the traditional approach. Specifically, the aided design was compared to several manual designs based on global slope angles, considering percentage differences in extracted tonnage and overall profit.

CASE STUDY

The case study employs a block model with blocks of 40 [m] × 40 [m] × 20 [m]. The block model represents a massive-type copper deposit. The total number of blocks are 34 200. The economic parameters considered include a mining cost of 2.2 USD per tonne and a processing cost of 10.0 USD per tonne. The metallurgical recovery is set at 90 per cent, while the copper price is assumed to be 3.0 USD per pound. Additionally, a selling and refining cost of 0.5 USD per pound is applied.

The open pit design is based on a bench height of 20 m and a berm width of 8.5 m. The ramps are designed with a width of 40 m to accommodate equipment access. Slope stability considerations include a bench face angle of 68.4°, an inter-ramp angle of 50.6°, and an overall slope angle of 45.0°.

Table 1 summarises the results of the final pit for the case study.

TABLE 1

Results of the final pit for the case study.

Result	Final pit	Unit
Total tonnage	528.27	Mt
Ore tonnage	332.12	Mt
Waste tonnage	196.15	Mt
Stripping ratio	0.59	-
Average copper grade in ore	0.41	%
Tonnage of copper fine contained in ore	1.36	Mt
Ore profit	2652.95	MUSD
Waste profit	-431.53	MUSD
Total profit	2221.42	MUSD

RESULTS AND DISCUSSIONS

Manual designs

Based on the final pit contour, three designs—labelled A, B, and C—are manually generated. The main differences among these designs lie in the location of the initial ramp points on the surface and the positioning of the switchbacks. Figure 2 presents the manual designs associated with the case study.

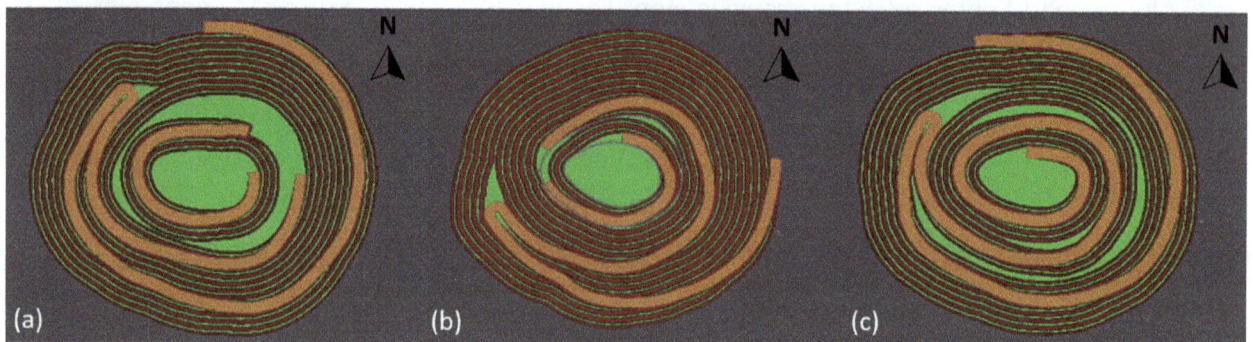

FIG 2 – Plant view of the manual designs: A, B and C for the case study.

Assisted design

The application of the integer linear programming model, along with the open pit design parameters from the case study, results in a block-supported ramp, as shown in Figure 3a. Based on the shape of the block supported ramp, a human designer using a CAD software obtains a smoothed assisted design shown in Figure 3b.

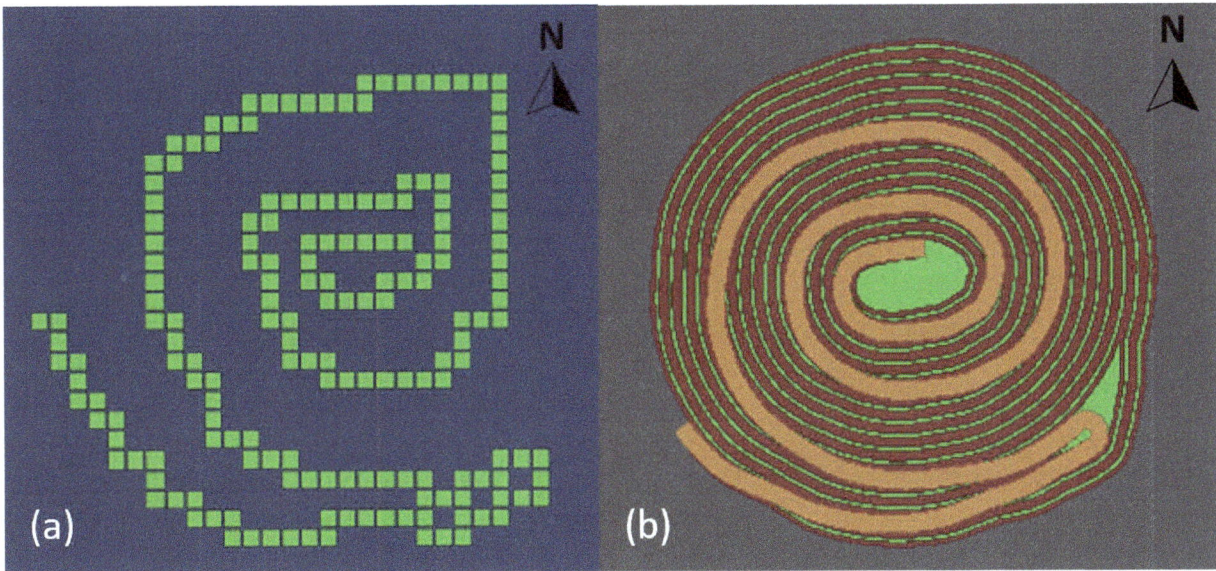

FIG 3 – Plant view (from above) of the smoothed assisted design (a) and block-supported ramp (b).

Figure 4 compares the block-supported ramp (in blue) with the assisted design (in orange). From this figure, it is evident that the human designer has made slight adjustments to the assisted design, deviating it slightly from the block-supported ramp. Figure 4 visually confirms the coherence between the location of the block-supported ramp and the assisted design.

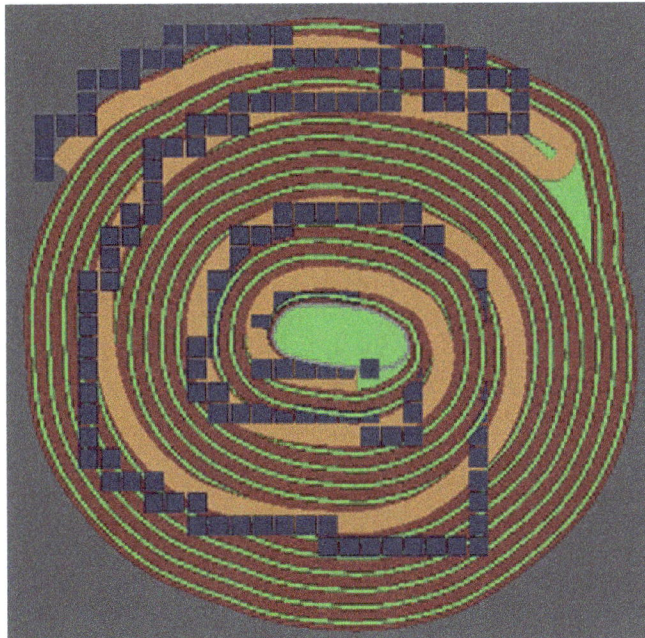

FIG 4 – Assisted design (in orange) associated with the block-supported ramp (in blue), plant view from below.

Comparison of manual and assisted designs

Table 2 reports the percentage differences in tonnage and profit for the manual and assisted designs compared to the final pit.

TABLE 2

Summary of results for the manual and assisted designs.

Design	Percentage difference in total tonnage compared to the final pit	Percentage difference in total profit compared to the final pit
Manual design A	26.59%	-7.76%
Manual design B	15.82%	-5.08%
Manual design C	-1.91%	-4.25%
Assisted design	-4.22%	-4.18%

Note that the manual designs A, B, and C were obtained sequentially, with manual design A created first, followed by B, and finally C. The goal of each new design was to reduce the percentage deviation in total tonnage and total profit. Through this iterative process, the human designer learned from previous designs, refining each subsequent attempt to achieve a better outcome. As a result, the manual designs were progressively improved.

From the manual designs, the manual design C is the one with less deviation in terms of total tonnage and total profit. In terms of deviation of profit, the assisted design obtains practically the same profit deviation of the manual design C. However, in term of tonnage deviation, the assisted design obtains a large deviation (-4 to 22 per cent) in comparison with the manual design C (-1 to 91 per cent). This is because the human designer had to make slight adjustments to try to follow the trajectory of the block-supported design. These deviations imply deviations in terms of ore and waste tonnage.

Table 3 presents the overall angles associated with the final pit, manual designs, and assisted design. These results verifies that both the manual designs and the assisted design comply with the 45° overall angle imposed by the final pit.

TABLE 3

Overall slope angles (in degrees) associated with the final pit, manual designs, and assisted design.

Design/slope	North slope	South slope	East slope	West slope	Average
Final pit	40.6	45.00	45.00	45.00	43.9
Manual design A	44.94	44.82	45.6	44.18	44.9
Manual design B	44.82	44.82	44.94	44.40	44.7
Manual design C	45.12	45.06	44.76	44.88	45.0
Assisted design	45.06	43.10	45.06	44.94	44.5

CONCLUSIONS

Remarkably, in the case study, the aided design achieves outcomes comparable to those of a human expert, with minimal percentage differences in value. This demonstrates the method's effectiveness in maintaining design quality in the case study while significantly reducing the time and resources needed for mine planning. By lowering reliance on individual expertise, this approach frees resources to explore various design scenarios, such as optimal switchback placement and surface ramp configurations, allowing for better-informed decision-making. Although the methodology offers substantial advantages, the conclusions are case-specific and depend on factors such as the block model, pushback geometry, economic parameters, and technical constraints. Nonetheless, this study highlights the potential of combining optimisation tools and CAD software to enhance the efficiency and consistency of open pit mine design processes. The proposed methodology not only simplifies the design process but also opens new possibilities for analysing diverse mine configurations. The semi-automatic design positions itself as a valuable tool for mine planning, delivering substantial advantages to mine designers.

The integer linear programming model is capable of automatically determining both the location of switchbacks and the starting point of the block-support ramp. However, this automated approach significantly increases the number of decision variables, leading to substantially higher computational times compared to scenarios where switchback locations are predefined by the user. To address this challenge, future research will explore the application of metaheuristics and/or constraint programming techniques to improve computational efficiency.

ACKNOWLEDGEMENTS

The authors would like to thank the Advanced Mining Technology Centre (AMTC) of the University of Chile, through Basal Fund AFB230001.

F Manríquez was supported by the Agencia Nacional de Investigación y Desarrollo (ANID) through the Fondecyt Iniciación Project 11250217.

All figures presented in this article are original and were created by the authors. No material has been reproduced from other sources.

REFERENCES

Couzens, T R, 1979. Aspects of production planning: Operating layout and phase plan, in *Open Pit Mine Planning and Design* (eds: W A Hustrulid and J Crawford) pp 217–232 (Society of Mining Engineers).

David, M, Dowd, P A and Korobov, S, 1974. Forecasting departure from planning in open pit design and grade control, in 12th Symposium on the application of computers and operations research in the mineral industries (APCOM), 2:F131–F142).

Deswik, 2018. Guidelines and considerations for open pit designers [online]. Available from: <https://www.deswik.com/news/guidelines-considerations-openpit-designers> [Accessed: 17 August 2023].

Dowd, P and Onur, P, 1992. Optimizing Open Pit Design and Sequencing, in 23rd International Symposium of Application of Computers and Operations Research.

Espejo, N, Nancel-Penard, P and Morales, N, 2019. A procedure to generate optimized ramp designs using mathematical programming, in Proceedings of the 39th international symposium on Application of Computers and Operations Research in the Mineral Industry (APCOM), Wroclaw. https://doi.org/10.1201/9780429320774

Hochbaum, D S and Chen, A, 2000. *Performance analysis and best implementations of old and new algorithms for the open pit mining problem, Operations Research*, 48(6):894–914.

Hustrulid, W A, Kutcha, M and Martin, R, 2013. Geometrical Considerations, in Open Pit Mine Planning and Design, Leiden, the Netherlands, pp 290–408 (Taylor and Francis).

Kaykov, D, Terziyski, D, Arsova, K and Stajić, M, 2023. A non-linear optimization approach to in-pit haul road design, *Sustainable Extraction and Processing of Raw Materials Journal*, 4(4)47–57. https://doi.org/10.58903/u17190831

Lerchs, H and Grossmann, I F, 1965. Optimum design of open pit mines, *Canadian Mining and Metallurgical Bulletin*, 58(633):47–54.

Morales, N, Nancel-Penard, P and Espejo, N, 2023. Development and analysis of a methodology to generate operational open pit mine ramp designs automatically, *Optimization and Engineering*, 24(2):711–741.

Morales, N, Nancel-Penard, P and Parra, A, 2017. An integer linear programming model for optimizing open pit ramp design, in Proceedings of the 38th International Symposium on the Applications of Computers and Operations Research in the Mineral Industry (APCOM), Golden, Colorado. https://doi.org/10.1145/2465506

Nancel-Penard, P, Morales, N, Parra, A and Widzyk-Capehart, E, 2019. Profit-optimal design of ramps in open pit mining, *Archives of Mining Sciences*, 64(2):399–413. https://10.24425/ams.2019.128691

Pana, M T, 1966. The simulation approach to open pit design, in Short Course and Symposium on Computers and Computer Applications in Mining and Exploration, vol 2, College of Mines, University of Arizona.

Sanhueza, E, Yarmuch, J L, Nancel-Penard, P and Gainza, N, 2025. A novel approach for the design of minimum-cost open pit haulage network, *International Journal of Mining, Reclamation and Environment*, pp 1–20. https://doi.org/10.1080/17480930.2024.2444880

Thompson, R J, 2011. Design, Construction and Maintenance of Haul Roads, in *SME Mining Engineering Handbook* (ed: P Darling), pp 957–976 (Society for Mining, Metallurgy and Exploration).

Wright, E A, 1999. Moving Cone II—a simple algorithm for optimum pit limits design, in *Proceedings of the 28th Symposium on the Application of Computers and Operations Research in the Mineral Industries (APCOM)*, pp 367–374.

Yarmuch, J L, Brazil, M, Rubinstein, H and Thomas, D A, 2020. Optimum ramp design in open pit mines, *Computers and Operations Research*, 115:104739. https://doi.org/10.1016/j.cor.2019.06.013

Short-term grade uncertainty and its impact on operational dig-limit definitions

G Nelis[1], C Aguilera[2], A Campos[3] and N Morales[4]

1. Assistant Professor, Universidad Técnica Federico Santa María, Santiago, Chile. Email: gonzalo.nelis@usm.cl
2. Postgraduate Student, Universidad Técnica Federico Santa María, Santiago, Chile. Email: constanza.aguilera@usm.cl
3. Undergraduate Student, Universidad Técnica Federico Santa María, Santiago, Chile. Email: arleth.campos@sansano.usm.cl
4. Associate Professor, Polytechnique Montréal, Montréal, QC, Canada. Email: nelson.morales@polymtl.ca

ABSTRACT

Short-term mine planning plays a crucial role in determining the destination of all extracted material to align with long-term strategic goals and meet production targets. Compared to other planning horizons, short-term planning benefits from a higher volume of available geological data. However, this advantage is offset by the daily operational challenges posed by the variability of the deposit. Fluctuations in grade, rock type, and geometallurgical performance can significantly affect the achievement of production targets and reduce the expected revenue of the operation.

In this study, we analyse the impact of grade uncertainty on a critical step within the grade control workflow: the dig-limit definition. This step incorporates operational constraints related to shovel selectivity to generate a feasible short-term destination plan. While manual methods are commonly employed for this task, several optimisation-based approaches have been proposed to enhance efficiency and accuracy. Using an optimisation model, we demonstrate that grade uncertainty can lead to profit deviations of up to 20 per cent, depending on the scenario, in a real-world case study. Additionally, by evaluating the variability of profit through a deterministic model, we highlight a significant gap between the expected value derived from a traditional grade control workflow and the actual realised profit, particularly in low-grade benches. We also show that a most-probable destination assignment based on the optimal dig-limits outcomes for each grade simulation is virtually identical to the deterministic dig-limit based on the estimated scenario, with no advantages in terms of profit or variability. Our findings underscore the importance of integrating grade uncertainty into short-term planning optimisation models to improve decision-making and maximise operational profitability.

INTRODUCTION

Open pit mining relies on the accurate definition of dig-limits to maximise ore recovery while minimising mining dilution. Dig-limits establish the boundaries between different materials and their respective destinations, directly influencing resource utilisation, production target fulfillment, and overall mine profitability. However, defining optimal dig-limits is a complex task due to the interplay of geological uncertainty, operational constraints, and economic considerations.

Traditional methods for dig-limit definition primarily rely on block model estimates and predefined cut-off grades. In many operations, short-term planning teams manually delineate dig-limits, incorporating operational constraints imposed by loading equipment, ie the minimum selectivity size of the shovel. While this manual approach ensures operational feasibility, it is often time-consuming and prone to suboptimal results, leading to increased waste handling, ore loss, and dilution. These inefficiencies can significantly impact operational performance and overall financial returns.

To address these challenges, researchers have proposed various optimisation techniques and algorithms to automate dig-limit definition. Early studies focused on heuristic approaches to generate mining polygons that maximise profitability—defined in terms of the economic value of material assignment—while also minimising penalties related to polygon shape to enhance operational feasibility (Neufeld, Norrena and Deutsch, 2003; Norrena, Neufeld and Deutsch, 2002; Norrena and Deutsch, 2001).

Isaaks, Treloar and Elenbaas (2014) further explored heuristic techniques, incorporating a minimum mining width constraint to ensure operational viability. Ruiseco, Williams and Kumral (2016) introduced a genetic algorithm metaheuristic to optimise dig-limits, which was later extended to determine optimal equipment size (Ruiseco and Kumral, 2017) and combined with deep learning techniques to improve computational efficiency (Williams et al, 2021).

Deutsch (2017) proposed an optimisation model coupled with a branch-and-bound algorithm to systematically solve the dig-limit problem, complemented by a post-processing stage using simulated annealing to refine solutions and reduce computational time. Similarly, Vasylchuk and Deutsch (2019) developed a specialised heuristic to improve dig-limit determination, emphasizing the need to incorporate geological variability to define more robust short-term limits.

A significant advancement in this field came from Sari and Kumral (2018), who introduced the first Mixed Integer Linear Programming (MILP) model capable of solving real-world dig-limit problems optimally within practical runtimes using commercial solvers. Their set-covering formulation represented the minimum selectivity size of a mining shovel with a covering element, ensuring that each block was covered by at least one frame while assigning each frame to a single destination. The application of this model across multiple benches demonstrated a 6.5 per cent increase in profitability compared to manual designs. This approach was later extended by Hmoud and Kumral (2024), enabling the simultaneous consideration of multiple destinations.

More recently, Nelis et al (2025) proposed an alternative set-covering formulation that incorporated capacity and blending constraints, along with destination-specific selectivity requirements. Their study also benchmarked computational efficiency against previous approaches, demonstrating faster convergence to optimal solutions while reporting a 7 per cent improvement in profitability over manual designs.

While significant progress has been made in optimising dig-limit definition, limited research has explored the influence of grade uncertainty on the resulting limits. Although short-term block models benefit from high-density blasthole or grade control sampling, data quality, especially with blastholes, can sometimes be limited, leading to potential errors in short-term grade estimations. At the same time, the sampling pattern can be incomplete, which can also induce errors in the estimated grade. These errors can propagate through the planning process, affecting dig-limit accuracy and, ultimately, operational performance.

To tackle the lack of information in short-term horizons, geostatistical simulations have been proposed to better represent the variability of block grades (Deutsch, Magri and Norrena, 2000; Magri and Ortiz, 2000; Verly, 2005). Such simulations have been widely applied in short-term scheduling (de Carvalho and Dimitrakopoulos, 2024; Jewbali and Dimitrakopoulos, 2018; Matamoros and Dimitrakopoulos, 2016; Quigley and Dimitrakopoulos, 2020; Shishvan and Benndorf, 2016) and related problems, such as mining cut definition (Nelis, Morales and Jelvez, 2023; Tabesh and Askari-Nasab, 2019). Geostatistical simulation has proven to be an effective tool to better evaluate mineral reserves in the presence of uncertainty (Journel and Kyriakidis, 2004).

Additionally, blast movement introduces another layer of uncertainty in dig-limit definition. The displacement of material due to blasting can alter the expected grade distribution, further complicating destination assignment. Simulations of blast movement have been shown to be an effective tool for assessing its impact on dig-limits (Hmoud and Kumral, 2022), and when combined with grade variability models, they enable a more comprehensive understanding of uncertainty (Hmoud and Kumral, 2024).

This paper investigates the impact of grade uncertainty on dig-limit definition by applying an exact optimisation approach to a real-world data set. We generate multiple grade scenarios from actual blasthole data and analyse the resulting variability in dig-limits. The study quantifies the influence of grade uncertainty on dig-limit location, expected ore tonnage, and overall profitability.

METHODOLOGY

In this study, a Mixed Integer Linear Programming (MILP) approach is employed to automatically define dig-limits across multiple grade realisations obtained through traditional geostatistical simulations. This section outlines the tools and methods used in our research, with a specific focus on the optimisation model and its application to short-term dig-limit definition.

Dig-limit optimisation

To optimise dig-limits, we implemented the model proposed by Nelis *et al* (2025), which, based on our literature review, demonstrates higher computational efficiency compared to other exact optimisation approaches. This advantage is crucial, as it enables the rapid calculation of dig-limits across multiple scenarios, facilitating a more robust evaluation of grade uncertainty. A comprehensive validation of this optimisation model, and comparisons against previous exact approaches and manual procedures for the dig-limit definition problem, can be found in (Nelis *et al*, 2025).

The model is formulated as a set-covering problem, where the fundamental covering element—also referred to as the structural element—is designed to be compatible with the size and shape of the mining shovel. Traditionally, square elements are used to represent the minimum mining width, but alternative shapes may be more appropriate depending on the specific mining operation and equipment configurations. The set-covering constraint ensures that all assigned destinations comply with operational feasibility requirements. In this approach, block destinations are adjusted to conform to the shape of the structuring element while simultaneously maximising overall profitability.

The optimisation model defines two primary decision variables:

1. Destination Assignment – Each block within the bench is assigned a destination, such as processing, stockpiling, or waste dumping, to maximise economic value.

2. Validity of Structuring Elements – A structuring element is considered *valid* if all the blocks it covers share the same destination.

To enforce an operationally feasible dig-limit definition, the model applies a covering constraint, ensuring that every block is covered by at least one valid structuring element. Since the structuring elements adhere to the minimum selectivity size requirements, this constraint guarantees that the destination assignments comply with operational constraints.

Additionally, the model incorporates several key operational constraints:

- Capacity Constraints – These constraints ensure that the total material sent to each processing destination does not exceed available capacity. For example, blocks assigned to the processing plant must comply with constraints related to maximum tonnage or available milling hours.

- Blending Constraints – To maintain product quality, the model imposes constraints on average grade composition. For example, the concentration of contaminants can be controlled to optimise metallurgical recovery, and stockpiles can be managed to maintain the expected head grade.

- Destination-Specific Selectivity – Since different destinations may be extracted by different shovels with varying selectivity capabilities, the model allows for different minimum selectivity sizes for each destination. This flexibility enables the optimisation of shovel allocation across different mining faces and better adhering to real mining conditions.

By integrating these constraints, the MILP model provides a structured and automated approach to dig-limit optimisation, ensuring operational feasibility while maximising profitability.

Formally, lets define as $\mathcal{B}$ the set of blocks, $\mathcal{D}$ the set of destinations, and $\mathcal{F}_d$ the set of structuring elements compatible with destination $d \in \mathcal{D}$. If a structuring element $f \in \mathcal{F}_d$ covers a block $b \in \mathcal{B}$, we denote that $b \in \mathcal{B}_f$. $\mathcal{F}_{db}$ is the set of all structuring elements $f \in \mathcal{F}_d$ such that SMU $b \in \mathcal{B}_f$. Each structuring element $f \in \mathcal{F}_d$ has a size k_f, which represents the number of SMUs covered by the element.

The destination assignment is constrained by a set of additive resources $\mathcal{C}$, such as tonnage, metal content, milling hours, etc, w_{cbd} defines the utilisation of resource c by SMU b if sent to destination

d. $\overline{W_{cd}}$ and $\underline{W_{cd}}$ represent the upper and lower limits of resource c in destination d. Similarly, Q is the set of relevant ore grades, and g_{qb} is the grade of element q in block b.

$\overline{G_{qd}}$ and G_{qd} represent the upper and lower limits for the average grade allowed in each destination d, respectively. Finally, p_{bd} is the profit obtained if block $b \in \mathcal{B}$ is assigned to destination $d \in \mathcal{D}$.

Binary decision variable x_{bd} is 1 if SMU b is sent to destination d, and 0 otherwise. Binary variable v_{fd} is 1 if all blocks in the structuring element f are sent to the destination d and 0 otherwise. A summary of the mathematical notation used in the article is displayed in Table 1.

TABLE 1

Mathematical notation.

Notation	Definition
$\mathcal{B}$	Set of blocks
$\mathcal{D}$	Set of destinations
$\mathcal{F}_d$	Set of structuring elements for destination $d \in \mathcal{D}$
$\mathcal{B}_f$	Set of blocks covered by element $f \in \mathcal{F}_d$
$\mathcal{F}_{db}$	Set of elements $f \in \mathcal{F}_d$ such that $b \in \mathcal{B}_f$
$\mathcal{C}$	Set of resources
Q	Set of grades
k_f	Size of element $f \in \mathcal{F}_d$
w_{cbd}	Utilisation of resource $c \in \mathcal{C}$ if block $b \in \mathcal{B}$ is assigned to destination $d \in \mathcal{D}$
$\overline{W_{cd}}$	Maximum limit of resource $c \in \mathcal{C}$ in destination $d \in \mathcal{D}$
$\underline{W_{cd}}$	Minimum limit of resource $c \in \mathcal{C}$ in destination $d \in \mathcal{D}$
g_{qb}	Grade of element $q \in Q$ in block $b \in \mathcal{B}$
$\overline{G_{qd}}$	Maximum limit of the average grade $q \in Q$ in destination $d \in \mathcal{D}$
$\underline{G_{qd}}$	Minimum limit of the average grade $q \in Q$ in destination $d \in \mathcal{D}$
p_{bd}	Profit of block $b \in \mathcal{B}$ if assigned to destination $d \in \mathcal{D}$
x_{bd}	Decision variable. 1 if block $b \in \mathcal{B}$ is assigned to destination $d \in \mathcal{D}$
v_{fd}	Decision variable. 1 if all blocks $b \in \mathcal{B}_f$ are assigned to destination $d \in \mathcal{D}$

The model is shown in Equations 1 to 8:

$$\max \quad \sum_{b \in \mathcal{B}} \sum_{d \in \mathcal{D}} x_{bd} p_{bd} \tag{1}$$

$$\text{s.t.} \quad v_{fd} \leq \frac{1}{k_f} \sum_{b \in \mathcal{B}_f} x_{bd} \qquad \forall\, d \in \mathcal{D}, f \in \mathcal{F}_d \tag{2}$$

$$\sum_{d \in \mathcal{D}} \sum_{f \in \mathcal{F}_{db}} v_{fd} \geq 1 \qquad \forall b \in \mathcal{B} \tag{3}$$

$$\sum_{d \in \mathcal{D}} x_{bd} = 1 \qquad \forall b \in \mathcal{B} \tag{4}$$

$$W_{cd} \leq \sum_{b \in \mathcal{B}} x_{bd} w_{cbd} \leq \overline{W_{cd}} \qquad\qquad \forall d \in \mathcal{D}, c \in \mathcal{C} \qquad (5)$$

$$\underline{G_{qd}} \sum_{b \in \mathcal{B}} x_{bd} w_{cbd} \leq \sum_{b \in \mathcal{B}} x_{bd} w_{cbd} g_{qb} \leq \overline{G_{qd}} \sum_{b \in \mathcal{B}} x_{bd} w_{cbd} \qquad \forall d \in \mathcal{D}, c \in \mathcal{C}, q \in \mathcal{Q} \qquad (6)$$

$$x_{bd} \in \{0,1\} \qquad\qquad \forall b \in \mathcal{B}, d \in \mathcal{D} \qquad (7)$$

$$v_{fd} \in \{0,1\} \qquad\qquad \forall f \in \mathcal{F}_d, d \in \mathcal{D} \qquad (8)$$

Equation 1 is the objective function and represents the profit obtained by assigning all SMUs to a destination. Equation 2 checks if the structuring element $f \in \mathcal{F}_d$ is valid. Equation 3 ensures that each block is covered by at least one valid structuring element. Equation 4 forces that every block in the bench must be assigned to a destination. Equation 5 is the capacity constraint for each destination and Equation 6 is a blending constraint Finally, Equations 7 and 8 impose that both decisions variables are binary.

Impact of grade variability

To analyse the impact of grade variability on dig-limit definitions, we employ a traditional geostatistical simulation framework for continuous variables. Specifically, a Sequential Gaussian Simulation (SGS) algorithm is used to generate multiple (N) grade realisations based on blasthole sample data. This approach allows us to capture the inherent grade uncertainty arising from an incomplete sampling pattern.

It is important to note that this study focuses solely on grade variability due to limited sampling, without considering additional uncertainty introduced by blast movement. However, the methodology proposed by Hmoud and Kumral (2022) could be integrated to simulate the impact of blast movement on grade variability. Furthermore, a combined approach incorporating both sampling uncertainty and blast-induced variability, as suggested by Hmoud and Kumral (2024), could provide a more comprehensive assessment of dig-limit uncertainty.

Each generated grade scenario represents a different possible realisation of the orebody, where individual block grades vary between simulations. Consequently, the economic value and optimal destination assignment for each block differ across realisations. For each simulation, we apply the optimisation model described in the previous section to determine the corresponding optimal dig-limits. This type of optimisation outcome is commonly referred to as the 'Perfect Knowledge' solution, as it assumes full and accurate knowledge of the actual deposit grades, something that is not available in real mining operations. By comparing these multiple optimised dig-limit solutions, we conduct a statistical analysis to quantify the impact of grade uncertainty on key operational parameters, including ore tonnage and profit variability.

In addition to analysing grade simulation scenarios, we also evaluate a scenario based on traditional grade estimation techniques. Using the same blasthole data set, an ordinary kriging interpolation is performed to generate an estimated block model. This kriged model is then used as input for the same optimisation algorithm, simulating the workflow typically applied in conventional grade control operations.

In the context of stochastic mine planning, this kriging-based approach is referred to as the 'Deterministic Solution', as it relies on a single, fixed grade estimate rather than a probabilistic representation of geological uncertainty. By comparing the deterministic dig-limit solution with those derived from multiple simulated grade realisations, we quantify the potential biases and risks associated with relying solely on traditional estimation techniques.

Figure 1 provides a graphical summary of the methodology employed in this study, illustrating the sequence of steps used to generate grade scenarios, apply the optimisation model, and compare outcomes across different estimation techniques.

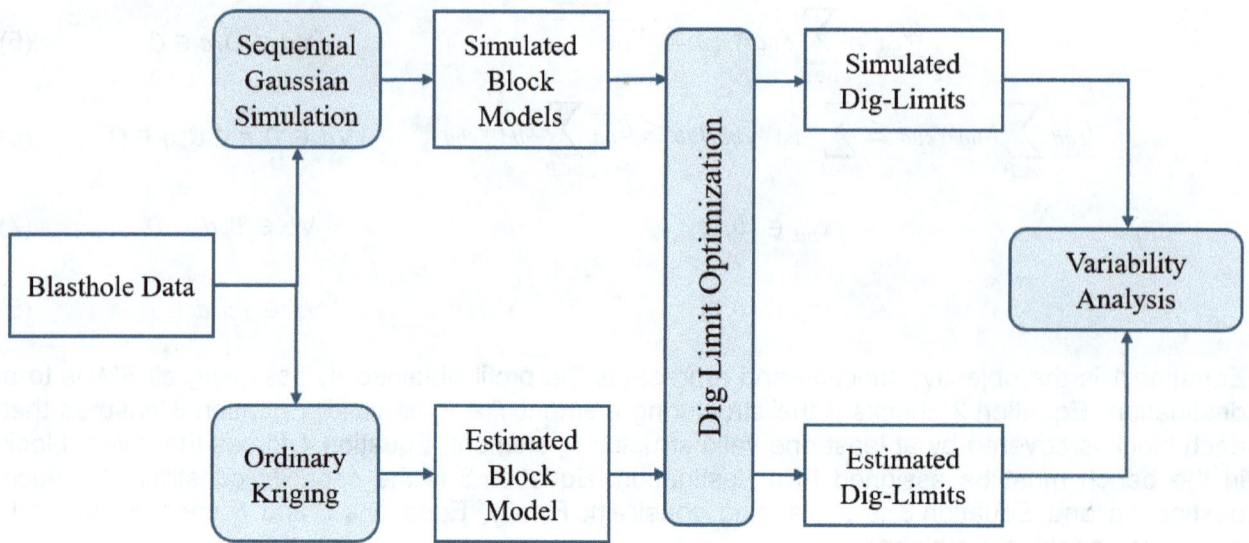

FIG 1 – Methodology workflow.

CASE STUDY

This study utilises real blasthole data from a low-grade porphyry copper deposit located in Central Chile. The data set consists of 162 copper samples, collected in a pseudo-regular grid, with an average grade of 0.3 per cent Cu. To account for grade uncertainty, Sequential Gaussian Simulation (SGS) was applied to generate 100 grade realisations, each representing a possible scenario of the orebody's true grade distribution. Additionally, an ordinary kriging (OK) interpolation was performed to produce a single estimated grade model. This kriged scenario represents the traditional grade control approach typically used in operational mining environments. Figure 2 shows the blasthole data, a simulated scenario (highlighting the expected higher local variability compared to deterministic estimation) and the kriging estimation, which smooths grade variations due to the nature of kriging interpolation.

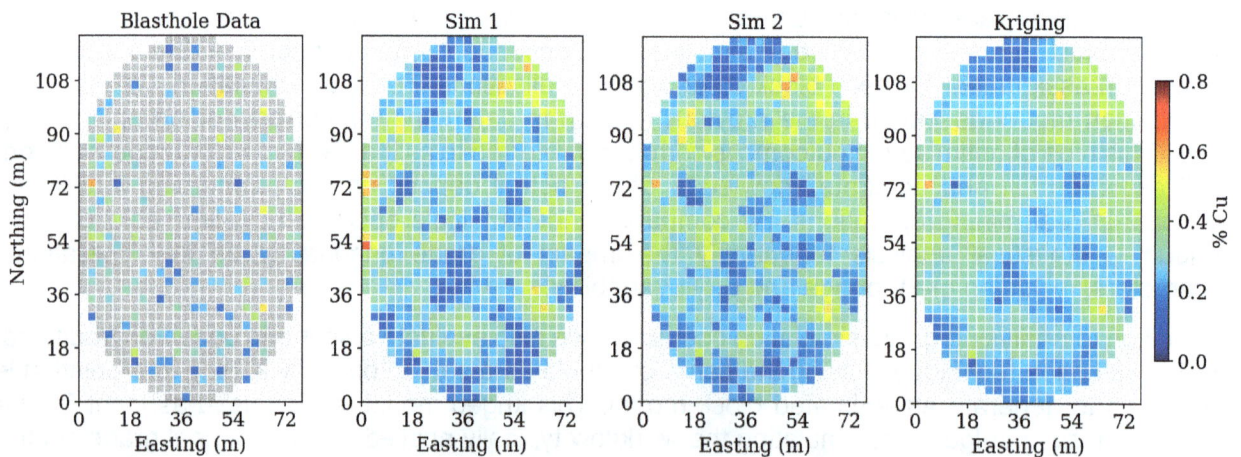

FIG 2 – Blasthole data and simulations.

Two destinations are considered: a processing plant to send the ore, and a waste dump. The minimum selectivity size for this case study was defined as a 2 × 2 square for the plant, and a 3 × 3 square for the waste dump, since waste regions are mined with larger equipment. Table 2 shows the economic parameters used in this case study to obtain the profit for each block and destination. The cut-off grade in this case is 0.28 per cent Cu.

TABLE 2

Economic parameters.

Parameter	Plant	Waste dump
Copper Price (US$/lb)	2.65	2.65
Processing Cost (US$/t)	11	0
Recovery	90%	0%
Selling and Refinement Cost (US$/lb)	0.48	0
Mining Cost (US$/t)	1.0	1.0

The optimisation model described in the Methodology section was implemented using Python 3.11 and Gurobipy 12.0 as the solver. The computational experiments were conducted on a Ryzen 7950X CPU with 64GB of RAM. The optimisation model was able to solve all scenarios in under two seconds.

RESULTS

We first analyse the variability of the dig-limits obtained across different scenarios. Figure 3 presents the dig-limit definitions for two simulated grade scenarios and the deterministic solution obtained from the kriging-based estimation. Differences in the placement of dig-limits are noticeable, particularly in areas where the block grades are close to the cut-off grade. This is expected since minor variations in grade can lead to shifts in the classification of blocks as ore or waste.

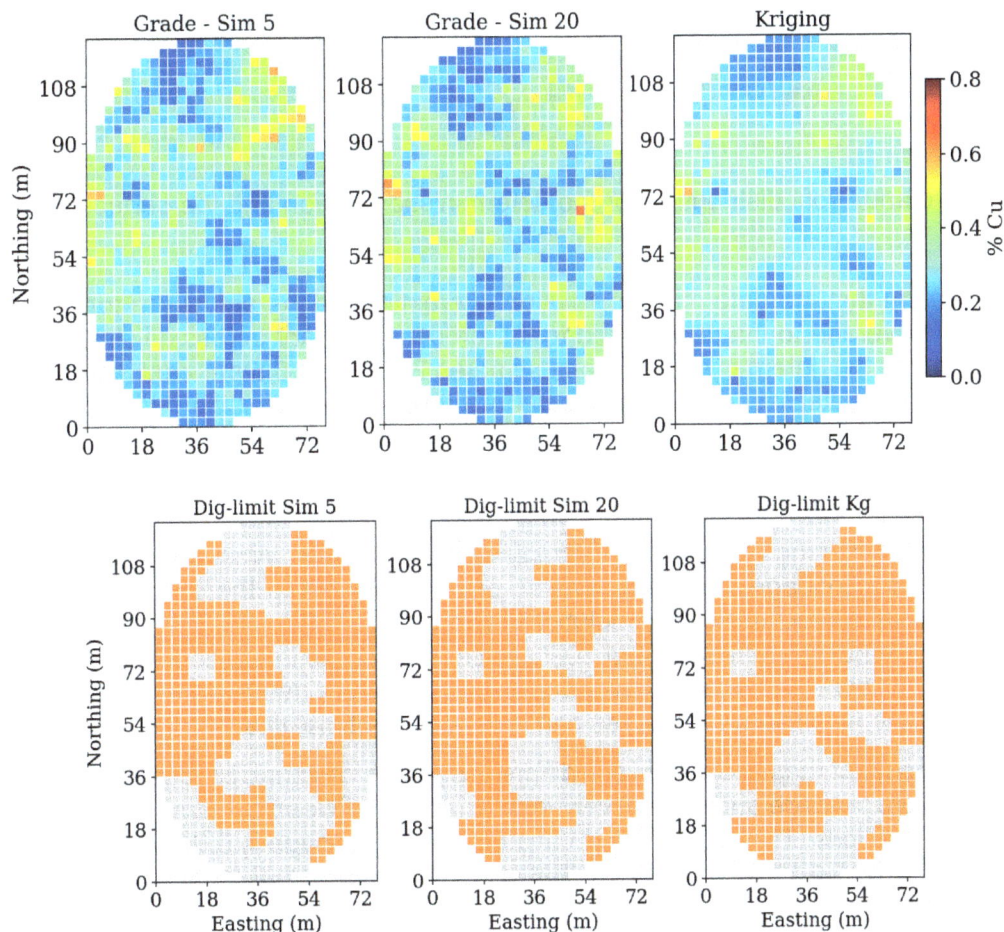

FIG 3 – Dig-limits and grades for two simulations and the deterministic scenario. For the dig-limit definitions, orange blocks are assigned to the Plant, and grey blocks to the Waste Dump.

Certain regions of the bench, however, exhibit consistency across all three cases. The western region shows a persistent presence of ore, while the northern and southern regions are consistently classified as waste. In contrast, the central region shows significant variability, with assigned destinations changing across different scenarios. This indicates that this area is more sensitive to grade uncertainty, which affects the final dig-limit definition.

A notable trend observed in these scenarios is that the deterministic solution classifies more blocks as ore compared to the simulations. If these dig-limits were applied across all scenarios, this would lead to an overestimation of ore tonnage and an increase in mining dilution. To quantify this effect, we analyse the distribution of ore tonnage across all simulated scenarios (perfect knowledge solutions) and compare it with the deterministic solution in Figure 4.

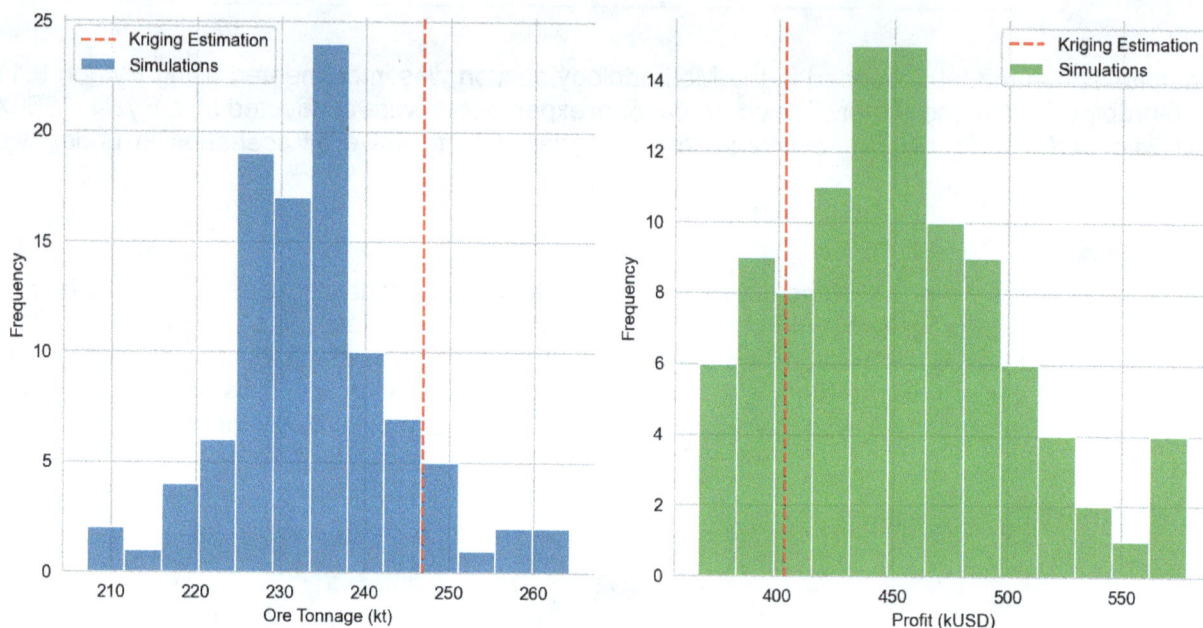

FIG 4 – Ore and Profit distribution for the deterministic solution and the perfect knowledge solutions.

The results confirm that the deterministic solution overestimates the actual ore tonnage present in the deposit. The average tonnage assigned to the processing plant across all simulated scenarios is 233.9 kt, whereas the deterministic solution estimates 246.8 kt, a 5.5 per cent increase. Moreover, in 90 out of 100 simulated scenarios, the actual ore tonnage is lower than what the deterministic solution predicts. This suggests that the kriging-based estimation tends to systematically overestimate ore tonnage, which, if applied in an operational setting, would result in a higher proportion of waste material reaching the processing plant.

In contrast, when examining profitability, a different trend emerges. While the deterministic solution predicts a higher ore tonnage, it underestimates profit compared to the simulated scenarios. The average profit for the perfect knowledge solutions is 451.2 kUS$, whereas the deterministic solution achieves only 402.8 kUS$, representing a 10.7 per cent decrease in expected profit. Out of the 100 simulations, only 16 cases resulted in lower actual profit than the deterministic estimate, reinforcing that traditional grade estimation techniques can lead to suboptimal financial performance.

An interesting aspect of these results is the different degrees of variability observed in ore tonnage and profit. While ore tonnage shows relatively low variability between scenarios, profitability exhibits much greater fluctuations. The coefficient of variation for ore tonnage is 0.04, whereas for profit, it is more than double at 0.10. This suggests that even relatively small changes in the destination assignment of blocks can lead to significant differences in overall financial performance.

To illustrate this, we examine the range of results obtained across the 100 scenarios. The worst-performing scenario yields a profit of 365.5 kUS$, while the best scenario achieves 578.8 kUS$, a difference of 213 kUS$ (58.8 per cent). In comparison, the variation in ore tonnage is less pronounced, ranging from 207.2 kt to 264.2 kt, a 27.5 per cent difference. These findings highlight the importance of accurately defining dig-limits, as even minor misclassifications can lead to substantial financial outcomes.

In a real mining operation, the deterministic solution derived from kriging would typically be used to define the final dig-limits, without considering geological uncertainty. To evaluate how well this approach performs, we apply these deterministic dig-limits across all 100 simulated grade scenarios and analyse the profit distribution obtained (Figure 5).

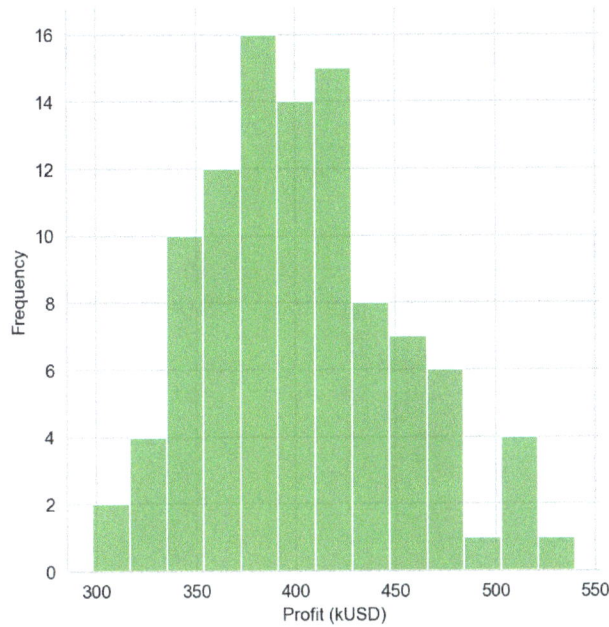

FIG 5 – Profit distribution of the deterministic solution evaluated over 100 scenarios.

The results show that while the average profit achieved using deterministic dig-limits remains 402 kUS$, in line with its original estimate, the variability in actual profits is substantial. This means that while, on average, the deterministic solution aligns with the expected profit, individual outcomes can differ significantly, leading to potential misestimations of actual financial performance in the mining operation.

To further investigate this, we compare the profit gap between the deterministic dig-limits and the perfect knowledge solutions in each scenario in Figure 6. The scenarios are sorted by the perfect knowledge solution profitability for clarity.

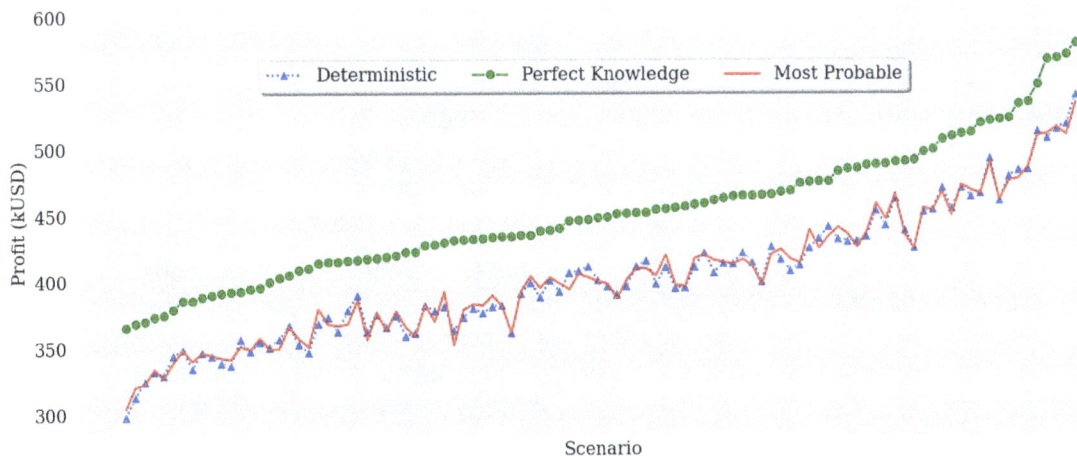

FIG 6 – Profit for Perfect Knowledge, Deterministic and Most Probable solutions across 100 scenarios.

In all scenarios, the perfect knowledge solution performs better than the deterministic solution as expected. The gap between the two solutions varies depending on the scenario. In the best case, the difference is 27.1 kUS$ (6.9 per cent), while in the worst case, it reaches 72.47 kUS$ (20.1 per cent). The discrepancy tends to be more pronounced in scenarios with lower overall profitability, highlighting the critical importance of an adequate sampling strategy, particularly for low-grade deposits. A more robust sampling approach could help reduce uncertainty, bringing operational outcomes closer to the perfect knowledge solutions.

Finally, we leverage the 100 simulated scenarios to analyse how frequently each block is classified as ore or waste. Figure 7 presents a probability map indicating the frequency with which each block is assigned to the processing plant. Using this information, we construct an alternative 'most probable' dig-limit definition, where each block is classified based on the most frequent destination assignment across all scenarios. This definition could be useful to try to capture additional profit leveraging the knowledge from each simulation.

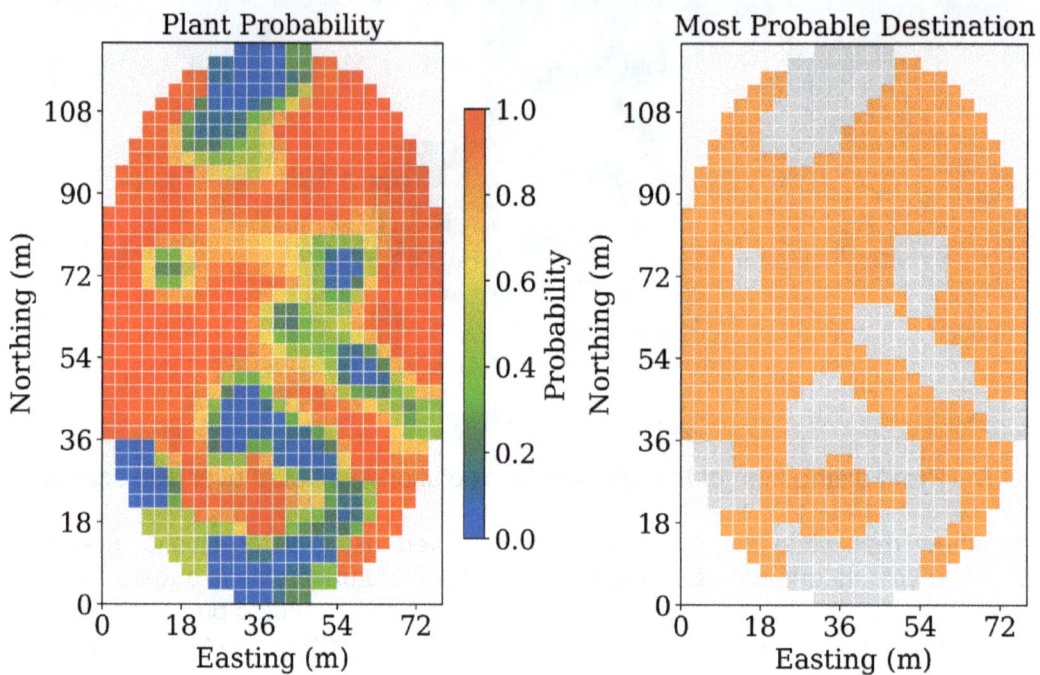

FIG 7 – Probability map for ore classification and Most Probable destination.

While this approach provides additional insights, it also presents limitations. In this case study, the most probable dig-limit definition achieves an average profit of 403 kUS$, only 1 kUS$ higher than the deterministic solution—a negligible improvement. Furthermore, in some scenarios, the most probable dig-limit performs worse than the deterministic approach, as shown in Figure 6.

Another drawback of this method is that it does not inherently guarantee operational feasibility. Figure 7, highlights areas where the minimum required selectivity size is not met, meaning that manual adjustments would be necessary to correct these inconsistencies. This suggests that while probability-based dig-limit definitions can provide additional guidance, they are not a standalone solution and would require further refinement to ensure practical applicability.

Overall, the findings of this case study underscore the substantial impact of grade uncertainty on short-term planning decisions. The significant gap observed between the deterministic and stochastic solutions reinforces the need for more sophisticated approaches to dig-limit optimisation. By integrating stochastic models into mine planning, it is possible to capture some of the value lost when relying on traditional estimation techniques, thereby improving both economic performance and operational feasibility.

CONCLUSIONS

This study provides a statistical analysis of the variability in dig-limit definitions resulting from grade uncertainty. The results highlight that grade uncertainty significantly influences the optimal dig-limit decisions, leading to variability in both ore tonnage and expected profitability in the short-term. This variability can directly impact the ability to meet mining targets and, as a result, must be managed to achieve long-term operational goals.

The traditional grade control workflow, which relies on a deterministic estimation of block grades (often through methods like kriging), can produce a high range of possible outcomes when uncertainty is considered. In such cases, relying on these conventional techniques can lead to underestimations of ore tonnage, higher dilution, and reduced profitability over short-term planning horizons. These results underscore the fact that, even in operations with relatively dense sampling strategies, grade uncertainty remains a significant issue that can affect financial results.

In this work, the most-probable destination assignment, based on the set of dig-limits derived from the grade simulation, shown no advantages over the deterministic dig-limit definition based on the kriged scenario. This shows the limitations of deterministic dig-limit optimisation models in tackling uncertainty in short-term planning.

To address this uncertainty, the study suggests two potential avenues. First, the integration of novel stochastic optimisation models could provide more robust solutions, accounting for the inherent uncertainties in grade distribution. These models would enable mining operations to define more flexible and profitable dig-limits, helping to capture additional value and mitigate the risks of over- or underestimating ore tonnage.

Alternatively, enhancing sampling strategies could also be a valuable approach, especially in low-grade deposits. By improving the precision of the geological data, planners and geologist could reduce uncertainty and better align the estimated ore with the actual deposit characteristics. In this context, investing in better sampling techniques may be essential to fully capture the profit potential of these deposits.

ACKNOWLEDGEMENTS

This work was funded by the National Agency of Research and Development (ANID Chile) through the Fondecyt de Iniciación Project 11230022.

REFERENCES

de Carvalho, J P and Dimitrakopoulos, R, 2024. Simultaneous shovel allocation and grade control decisions for short-term production planning of industrial mining complexes – an actor-critic approach, *International Journal of Mining, Reclamation and Environment*, 38(1):53–78. https://doi.org/10.1080/17480930.2023.2247196

Deutsch, C, Magri, E and Norrena, K, 2000. Optimal grade control using geostatistics and economics: methodology and examples, *SME Transactions*, 308:43–52.

Deutsch, M, 2017. A branch and bound algorithm for open pit grade control polygon optimization, in *Proceedings of the 38th International Symposium on the Applications of Computers and Operations Research in the Mineral Industry (APCOM)*, 8 p.

Hmoud, S and Kumral, M, 2022. Effect of Blast Movement Uncertainty on Dig-Limits Optimization in Open pit Mines, *Natural Resources Research*. https://doi.org/10.1007/s11053-021-09998-z

Hmoud, S and Kumral, M, 2024. Risk-Based Optimization of Post-Blast Dig-Limits Incorporating Blast Movement and Grade Uncertainties with Multiple Destinations in Open pit Mines, *Natural Resources Research*. https://doi.org/10.1007/s11053-024-10428-z

Isaaks, E, Treloar, I and Elenbaas, T, 2014. Optimum dig lines for open pit grade control, in *Proceedings of 9th International Mining Geology Conference*, pp 425–432 (The Australasian Institute of Mining and Metallurgy: Melbourne).

Jewbali, A and Dimitrakopoulos, R, 2018. Stochastic Mine Planning—Example and Value from Integrating Long- and Short-Term Mine Planning Through Simulated Grade Control, Sunrise Dam, Western Australia, in *Advances in Applied Strategic Mine Planning*, pp 173–189 (Springer International Publishing). https://doi.org/10.1007/978-3-319-69320-0_13

Journel, A G and Kyriakidis, P C, 2004. *Evaluation of Mineral Reserves* (Oxford University Press: New York). https://doi.org/10.1093/oso/9780195166941.001.0001

Magri, E and Ortiz, J, 2000. Estimation of economic losses due to poor blast hole sampling in open pits, in *Geostatistics 2000, Proceedings of the 6th International Geostatistics Congress*, pp 732–741.

Matamoros, M E V and Dimitrakopoulos, R, 2016. Stochastic short-term mine production schedule accounting for fleet allocation, operational considerations and blending restrictions, *European Journal of Operational Research*, 255(3):911–921. https://doi.org/10.1016/j.ejor.2016.05.050

Nelis, G, Morales, N and Jelvez, E, 2023. Optimal mining cut definition and short-term open pit production scheduling under geological uncertainty, *Resources Policy*, 81:103340. https://doi.org/10.1016/j.resourpol.2023.103340

Nelis, G, Morales, N, Estay, R, Manriquez, F, Vivar, P and Morales, C, 2025. A Novel Optimization Model for the Dig-Limit Definition Problem in Open Pit Mines with Multiple Destinations, *Resources Policy*, 102:105510. https://doi.org/10.1016/j.resourpol.2025.105510

Neufeld, C, Norrena, K and Deutsch, C, 2003. Semi-Automatic Dig Limit Generation, in *Center for Computational Geostatistics Annual Report Papers* (1):1–23. Available from: <papers2://publication/uuid/5155E453-EB0A-4808-9452-B86CAA180583>

Norrena, K and Deutsch, C, 2001. Automatic Determination of Dig Limits Subject to Geostatistical, Economic and Equipment Constraints, in *Center for Computational Geostatistics Annual Report Papers*, pp 1–18. Available from: <papers2://publication/uuid/8E3B6A0A-4131-4F32-8D36-406B0312C5C4>

Norrena, K, Neufeld, C and Deutsch, C, 2002. An Update on Automatic Dig Limit Determination, in *Center for Computational Geostatistics Annual Report Papers*, pp 1–17. Available from: <papers2://publication/uuid/FF1B87D4-7799-4EEB-837D-2067AF429CDC>

Quigley, M and Dimitrakopoulos, R, 2020. Incorporating geological and equipment performance uncertainty while optimising short-term mine production schedules, *International Journal of Mining, Reclamation and Environment*, 34(5):362–383. https://doi.org/10.1080/17480930.2019.1658923

Ruiseco, J R and Kumral, M, 2017. A Practical Approach to Mine Equipment Sizing in Relation to Dig-Limit Optimization in Complex Orebodies: Multi-Rock Type, Multi-Process and Multi-Metal Case, *Natural Resources Research*, 26(1):23–35. https://doi.org/10.1007/s11053-016-9301-8

Ruiseco, J R, Williams, J and Kumral, M, 2016. Optimizing Ore–Waste Dig-Limits as Part of Operational Mine Planning Through Genetic Algorithms, *Natural Resources Research*, 25(4):473–485. https://doi.org/10.1007/s11053-016-9296-1

Sari, Y A and Kumral, M, 2018. Dig-limits optimization through mixed-integer linear programming in open pit mines, *Journal of the Operational Research Society*, 69(2):171–182. https://doi.org/10.1057/s41274-017-0201-z

Shishvan, M S and Benndorf, J, 2016. The effect of geological uncertainty on achieving short-term targets: A quantitative approach using stochastic process simulation, *Journal of the Southern African Institute of Mining and Metallurgy*, 116(3):259–264. https://doi.org/10.17159/2411-9717/2016/v116n3a7

Tabesh, M and Askari-Nasab, H, 2019. Clustering mining blocks in presence of geological uncertainty, *Mining Technology: Transactions of the Institute of Mining and Metallurgy*, 128(3):162–176. https://doi.org/10.1080/25726668.2019.1596425

Vasylchuk, Y V and Deutsch, C V, 2019. Optimization of Surface Mining Dig Limits with a Practical Heuristic Algorithm, *Mining, Metallurgy and Exploration*, 36(4):773–784. https://doi.org/10.1007/s42461-019-0072-8

Verly, G, 2005. Grade Control Classification of Ore and Waste: A Critical Review of Estimation and Simulation Based Procedures, *Mathematical Geology*, 37(5):451–475. https://doi.org/10.1007/s11004-005-6660-9

Williams, J, Singh, J, Kumral, M and Ramirez Ruiseco, J, 2021. Exploring Deep Learning for Dig-Limit Optimization in Open pit Mines, *Natural Resources Research*, 30(3):2085–2101. https://doi.org/10.1007/s11053-021-09864-y

AMC's Hill of Value® – modelling sequential decisions to transfer strategy optimisation into an operational mine plan

T M Pelech[1] and F Grobler[2]

1. Senior Mining Consultant, AMC Consultants, West Perth WA 6005.
   Email: tpelech@amcconsultants.com
2. Principal Consultant, AMC Consultants, West Perth WA 6005.
   Email: fgrobler@amcconsultants.com

ABSTRACT

Mine strategy optimisation techniques such as AMC's Hill of Value® (HoV®) use numerical modelling of an operation to rapidly explore a wide range of strategic options. This is achieved by testing and flexing variables such as the cut-off grade, throughput rates, mining methods, scheduling sequences and more to understand which of these decision variables are strategically important. The output of the HoV® process is an optimal set of decision variables, guided by one or more objectives (eg NPV, strategy resilience, mine life) which can be converted into an operational mine plan. It enables a high number of strategic options to be rapidly assessed for long-term planning guidance, but at the expense of tactical resolution.

Once an optimal strategy is chosen and committed to for a long-term plan, tactical deviations or modifications from this plan may be required to minimise short-term disruptions and achieve better operational alignment. Ideally, value-accretive tactical deviations could be undertaken without compromising the overall strategy. This is attainable with confidence if the appropriate level of detail is modelled for the strategy selection.

A robust strategy selection process requires that any potential tactical choices fall within its bounds, albeit at a higher resolution. This will enable the promotion of tactical modifications that create value while constraining decisions that destroy it. To account for these tactical deviations within a strategy, they must also be represented in the numerical model.

Tactical decision-making becomes difficult to model numerically when multiple sequential decisions are possible. This is especially the case for underground mines, where the sequential decisions of mine scheduling, capital commitment, access development, and production phase of selecting and extracting the ore, stockpiling and processing typically exist. Each of these sequential decisions can have an upstream and downstream impact on the overall mine value chain.

This paper investigates a practical method for capturing high resolution sequential decisions in a strategy optimisation model and enables these to coherently transfer to the operational mine plan.

INTRODUCTION

There are numerous methods, products and approaches available to help determine a strategic mine plan. A common approach, Strategy Selection, requires hundreds of hours of engineering work to produce a relatively small set of detailed schedules. Following a simple NPV comparison, the best option is typically selected to move forward to execution. At the other end of the spectrum, Strategy Optimisation, is a more algorithmic approach, allowing an indefinite number of potential options to be rapidly generated, tested and optimised. One of the more well-known strategy optimisation techniques is AMC's Hill of Value® process (HoV®), which applies numerical modelling to explore a wide range of strategic options. This paper focuses on the practical application of HoV® in ensuring the final strategy selection is truly the most desirable outcome, rather than simply an NPV maximum and is also resilient to various potential tactical deviations. This paper also examines how sequential decisions, changing economic conditions and managerial flexibility can be effectively integrated into a comprehensive strategy optimisation, ensuring that strategic decisions are aligned with operational realities.

LITERATURE REVIEW

Real options value approach

Elkington and Gould (2012) outlined the limitations of relying solely on Net Present Value (NPV) for project assessment in mine planning. Traditional NPV-only optimisation approaches often overlook the benefits of tactical decision-making and managerial flexibility, which we define as a manager's ability to adapt the execution of a strategy in response to changing circumstances, uncertainty, or new opportunities. It mentioned that NPV maximisation tends to encourage features such as deferred waste/development, expedited extraction of high-grade ore and high production rates. These features may not always align with optimal strategic decisions. It emphasised that NPV does not solely account for the dynamic nature of mining operations and the various uncertainties and risks involved. As a result in practice, it is common for mine operators to intuitively adjust their strategy execution to compensate for the flaws inherent in NPV maximisation. Examples of this include mining and processing 'incremental' (below cut-off grade) material to fill the processing plant in periods of low ore availability or delaying closure due to perceived potential exploration success. In some cases, these strategy deviations are vindicated as every manager would hope, when a significant increase in the product price occurs or a large new ore source is discovered.

To address the perceived limitation of NPV-only assessment, Elkington and Gould (2012) suggest the adoption of Real Options Valuation (ROV) as an alternative approach. In their paper, ROV incorporates managerial flexibility and allows dynamic responses to changes in commodity prices, enabling a more comprehensive evaluation of mining projects. In their case study results, it is suggested that reducing processing capacity, increasing cut-off grade and significantly increasing use of stockpiles (particularly for material deemed sub-economic at the time of decision-making) may be optimal compared to the NPV-maximised case.

The practical demonstration of ROV by Elkington and Gould (2012) is captured in a limited example. As they have noted, it lacks scalability for longer time periods and larger numbers of potential strategic options. These are common features for full-scale mining project optimisations. ROV's practical application to larger mine optimisations may also be limited without implementing a different style of algorithm as the number of variables increases.

Strategy optimisation approach

The book 'Cut-Off Grades and Optimising the Strategic Mine Plan' by Hall (2014) provides a detailed exploration of various methodologies and principles related to the derivation and application of cut-off grades and strategic mine planning. It emphasises the importance of integrating cut-off grade determination into the overall mine planning framework to maximise value and achieve corporate goals. The book outlines a strategy optimisation approach that involves a multi-dimensional analysis, taking into account various parameters such as geology, production capacities, and economic conditions.

The Strategy Optimisation approach is a broad definition and involves evaluating various options and scenarios early in the planning process without mandating a specific algorithmic approach. Some algorithms suggested include:

- exhaustive (brute force) calculation and comparison
- genetic algorithms
- dynamic programming
- linear and mixed integer programming.

Strategy Optimisation emphasises the need for a flexible and multi-dimensional approach to test various uncertain parameters (eg product price, operating cost, mining rates) and variables (mining sequences, strategic options etc) to maximise value and importantly, minimise risk in mining operations. It is aligned with the reality outlined by Elkington and Gould (2012), where mine operators intuitively understand that the best NPV does not always correlate with the best strategy. The Strategy Optimisation approach reinforces the value of trade-offs in various risk measures or other objectives against lost NPV.

The book advocates for the use of the Hill of Value®, which traditionally shows the change in NPV by varying cut-off grades and processing throughput rates. The numerical modelling behind the namesake chart includes the parameters, variables and relationships mentioned above. However, this traditional visualisation of the Hill of Value® becomes less useful for higher-dimensional optimisations that may be necessary to properly test the resilience of certain strategies. Furthermore, the generation of a Hill of Value® chart generally assumes that sub-optimal solutions must be generated, to show the relative gains of the optimal solution as part of the solution space. This assumes some form of exhaustive calculation and comparison, which, similar to the limitations of the ROV approach, can become computationally onerous or limiting for larger mine optimisation problems that attempt to account for higher resolution, potentially including tactical deviations.

Nested stopes and constraint-based scheduling

Jetmore and Roos (2022) introduce a technique similar to Strategy Optimisation. It enables rapid testing of strategic underground mine schedules with nested cut-off stope designs and constraint-based scheduling. Jetmore and Roos (2022) leverage the Theory of Constraints to isolate bottleneck mining activities and simplify the mine scheduling process to ease the computational requirements for scheduling. This allows a high number of schedules to be assessed for optimality, at the expense of schedule resolution.

To streamline the scheduling process, several key simplifications are made:

- Omitting Secondary Activities: Example activities such as ventilation, ground support installation, and backfilling are omitted to focus on production-driven schedules, usually driven by ore tonnage constraints.

- Development: Level and ramp development is assumed to be in place in time for production.

- Ignoring Stope Sequence: The design and sequencing of stopes are not detailed, the average tonnages and grades for aggregated groups of stopes are used instead.

- Simplified Cost Model: Used to adjust mining costs relative to mine size, usually based on assumed fixed and variable component splits.

- Backfill Assumptions: Similar to development, it is assumed that backfill would cure in time for production.

- Simplified Stope Geometries: Used to speed up the inventory generation process.

To determine an optimal solution, this simplified approach also uses elements of the Hill of Value® and graphs of incremental return by cut-off. This approach allows for quick, simple and reliable scheduling, enabling engineers to analyse various scenarios and make informed strategic decisions. A key issue with this approach relative to the ROV method is that with a simplified schedule, the cost or benefit of tactical deviations and managerial flexibility cannot be accounted for. Furthermore, it is relatively more difficult to transfer a simplified schedule into an executable mine plan. For most if not all strategic optimisation approaches, at least a small amount of additional work is required in adding detail to a strategic schedule to transfer it to an executable plan, but for the simplified approach, this amount of additional work is relatively much higher.

Dynamic economics approach

Deucker, Phillips, and Kato (2021) present a specific method for optimising mining operations on an activity scale resolution. This is a key step in moving towards accounting for managerial flexibility and tactical deviations within a strategy.

In the Dynamic Economics approach, each stope activity in an underground mining schedule is optimised by selecting a stope design based on a cut-off grade. It is also an option to entirely exclude the stope. Varied stope cut-off increments are tested until the highest cash flow schedule output is found. For each stope, any related development or other activities can also be excluded from the schedule based on a dynamic economics model, which ensures that cash flow is optimised for all tasks, not just the stope itself. Costs are modelled for both fixed (by period) and variable (by ore tonne) cost inputs across each schedule period.

This approach highlights the challenges of determining the optimal mining strategy when dealing with marginal material and the shortfalls of a simplified approach. Traditional cut-off grade approaches are applied as averages over entire mining zones, such as in Jetmore and Roos (2022), and may miss intricacies of each sequential mining decision that potentially impacts the ultimate economic outcome for each activity. The activity scale resolution in Dynamic Economics addresses this issue by implementing economic logic within the schedule, allowing for optimal economic decisions for each schedule activity.

The Dynamic Economic case study outlined by Deucker, Phillips, and Kato (2021) demonstrates the potential improvement in NPV by using an activity-based valuation approach, showing significant increases in NPV compared to traditional methods. The detailed nature of this approach computationally limits its ability to assess the indefinite number (for some operations, tens or hundreds of millions) of potential variable combinations that are possible. Without exploring these potential combinations, it is arguable that although it may achieve a mathematical optimum for a limited model, the approach is unable to explore the potential deviations and resilience of a larger strategy that can be insightful.

Improvements can be made to this approach by implementing multiple objective trade-offs. This is especially the case in terms of risk management. Finding the appropriate balance between simplicity and resolution is required to adequately assess the millions of potential combinations of schedule variables that may be present with larger multi-objective problems in a reasonable time frame.

The Strategy Optimisation approach outlined in Hall (2014) emphasises the need for a multi-dimensional analysis to maximise value and minimise risk and suggests some applicable computational methods. The simplified scheduling approach shown by Jetmore and Roos (2022) demonstrates a method to assess a large number of potential outcomes, at the expense of schedule resolution. Conversely, the activity-based optimisation presented by Deucker, Phillips, and Kato (2021) provides a more precise approach to strategic mine planning, demonstrating significant improvements in NPV. A combination of elements of these approaches can be devised by implementing more sophisticated risk evaluation and multi-objective optimisation, especially to account for managerial flexibility and tactical deviations as will be shown in this paper.

In summary, the existing literature highlights the importance of incorporating managerial flexibility and tactical decision-making into mine planning, moving beyond the limitations of NPV-only assessments. None of the reviewed literature have shown a practical method of achieving this type of optimisation.

MULTI-OBJECTIVE OPTIMISATION AND ANALYSIS

Multi-Objective optimisation can be used to find the highest value and the most resilient strategies by comparing two or more objectives and identifying the desired trade-off region and corresponding decision variables. It is shown in this paper to be very useful in allowing corporate and operational decision-makers to select their optimal strategies based on trade-offs encompassing features that are difficult to model in a simple NPV objective such as social/other obligations and the lost opportunity to discover new ore sources. Importantly, the output of the HoV® process allows decision-makers to understand the consequences of certain sub-optimal strategies that might otherwise be given inappropriate weight in a traditional strategy selection process.

The Non-dominated Sorting Genetic Algorithm III (NSGA-III) described in Deb and Jain (2014) is used as the optimisation algorithm in this paper. It is an evolutionary algorithm, designed for solving multi-objective problems and an extension of the well-known NSGA-II algorithm, which is widely used across various software platforms for multi-objective optimisation. NSGA-III introduces some enhancements to address the challenges associated with multiple objective optimisation problems. By allowing the algorithm access to a certain range of model variables and giving it target objective variables to read from, the algorithm iterates through a process of mutating input variable combinations to converge to optimum sets for each objective.

Post-processing techniques, or data analysis is then undertaken on the NSGA-III output for the Hill of Value® process. This analysis allows mine management to gain valuable insights from the optimisation process. Most importantly is the definition of a Pareto Front (or Optimum Frontier), the line of optimal trade-off combinations between two or more objectives. Along this edge, all of the additional model variable data can be pulled out of the results and analysed to see how those model parameters are changing along the Front. This kind of insight is very valuable to strategy execution managers who need to understand the consequences of changing any of their operational settings, such as mining rates, cut-off grade, processing rates, and inventory sets. This kind of analysis allows managers to understand the importance of each of these potential parameters in the context of achieving their strategy, allowing them greater tactical or managerial flexibility to achieve those outcomes in practice (in the case where they do need to change some settings). The delivered Hill of Value® strategy with these enhancements included is now far more nuanced and useful than the conventional output based on a simple 3D surface chart.

CASE STUDIES

The case studies outlined for this paper are shown in Table 1. They will be used to illustrate the methods currently used in the Hill of Value® process to prove the resilience of potential strategies. They demonstrate the application of an advanced Strategy Optimisation approach combining all the above literature with a multi-objective optimisation algorithm. All numerical values (NPV etc) have been modified in this paper to obscure sensitive information of the actual operating mine sites.

TABLE 1
Case studies.

Case	Mining method	Nominal process throughput	Commodity
Mine X	Stoping	4.5 Mtpa (multiple mines)	Polymetallic
Mine Y	Stoping	3 Mtpa (single mine)	Pb–Zn–Ag
Mine Z	Narrow vein stoping	0.7 Mtpa (multiple sources)	Au–Cu

For the strategy optimisation of Mine X, there was particular interest in the sensitivity to product price. This posed a similar problem to that raised by Elkington and Gould (2012), whereby the mine operators knew they would be able to make changes to any strategy executed if there was a significant change in metal prices received.

The traditional approach to mine strategy optimisation as shown in Hall (2014) and Jetmore and Roos (2022), wherein the cut-off grade is applied uniformly across mining zones, does not allow for the resolution to account for managerial flexibility if metal prices change. Traditionally and in those approaches, a price sensitivity is carried out on a static inventory and schedule. This does not allow for changes in the mine design and inventory that would occur if for example the metal prices change, and it is determined that some of that inventory is no longer economically viable. A mine-wide (or by mining zone) cut-off grade application treats all stopes equally, regardless of how much additional cost is required to extract an individual stope. These additional costs could include, for example of additional development along strike or at depth for simplified mining zones that reach across multiple levels. However, in practice, the decisions to commit to development for each stope is a sequential decision that is made by management at the time of execution. It can therefore be changed up until the time of execution (or even thereafter), and this ability to change at shorter notice should be more accurately accounted for when testing sensitivity to economic conditions.

Mine Y sought to investigate the benefit of some infrastructure options while also minimising overall capital expenditure, including closure costs. This kind of assessment is considered a strategic option that should be easily tested with a simple schedule and NPV analysis as shown by Jetmore and Roos (2022) or a simple Real Options Valuation. However, the underlying complications become apparent when looking at the numerous combinations of options to modify cut-off grades and the changes in mining rates/productivities when the benefits of new infrastructure are made available. This can be otherwise examined as a sequential decision problem, where the impacts of new infrastructure go beyond an immediate cost and productivity benefit, but also impacts how the cut-off grade strategy could be applied. Potentially new, unexpected solution combinations may become apparent when mining zone productivity assumptions are modified in conjunction with cut-off grades while also attempting to minimise capital expenditure. Considering this as a simple ROV problem doesn't account for these relevant additional dimensions. It must be re-cast as a multi-objective trade-off problem that cannot be solved with simple algorithms.

Mine Z simply wanted a cut-off grade strategy to be used as an input into their tactical mine plans and budgeting process. As with many Strategy Optimisation studies, maximising NPV was the stated goal. However, being a narrow-vein gold operation with complex geology, mine-life extensions had been experienced nearly every year for the previous ten years by discovering and developing new ore sources. Committing to an NPV maximising cut-off grade that leads to a shorter mine life and thereby diminishing future opportunities to extend the mine life appears illogical to the dynamic and flexible manager. In these types of cases, additional constraints may also exist where the mine must operate for a certain amount of time to fulfill specific corporate, social or contractual obligations. It may be that the maximum NPV solution is not actually the best solution after all. This again becomes a multi-objective trade-off problem that requires decision-makers to understand the range of potential outcomes and decide on the best trade-off for their own site-specific scenario.

KEY METHODS AND STEPS

Strategy Optimisation requires a somewhat customised approach for each mining operation. Ultimately the mining model to be optimised must be able to represent an operation and its base schedule, to an appropriate level of accuracy and include any of the additional operational states that are desired to be explored. There are numerous approaches to this, and they all share the similarity that a Mineral Resource model is converted to a set of mineable shapes, which is then transferred into a schedule. For the case study results in this paper, many of the concepts from previous literature are integrated to enable exploration of the mine optimisation problem space. A short-list of the concepts that enable this problem space exploration are as follows:

- Concepts of relative valuation of real options (Elkington and Gould, 2012) and strategy selection.

- Strategy Optimisation and the Hill of Value® (Hall, 2014).

- Simplified constraint-based scheduling (Jetmore and Roos, 2022):

  o But only simplified to the appropriate level required to capture consequential tactical deviations and managerial flexibility.

- Elements of Dynamic Economics (Deucker, Phillips and Kato, 2021).

- Multi-objective optimisation techniques, specifically the NSGA-III algorithm (Deb and Jain, 2014).

The introduction of tactical deviations after a strategy has been committed to presents a high risk to its success. It is therefore prudent to consider any of these potential deviations during the optimisation process. This is primarily undertaken by proving the resilience of the optimal strategy during the optimisation process, by testing it against changing conditions (eg commodity prices) and importantly, the input of managerial opinions and potential future actions. The chosen 'optimal' strategy may change depending on these additional inputs and opinions, so it is important to test the strategy's level of resilience when selecting it.

Multi-objective optimisation techniques can be used to find an optimal and resilient strategy, allowing trade-offs between one or more objectives in addition to Net Present Value. However, to prove the resilience of a strategy against tactical deviations, the model inputs must first be adequately defined.

In addition to implementation of the above concepts, the following critical steps are undertaken for this.

Derive important relationships

Many mine sites will have data available for the current operational conditions, for example the metallurgical recoveries derived from most common combination of minerals produced at the mine, costs reported to stock exchanges, and data demonstrating the overall productivity of the mine. The information that is probably not available, is the data representing conditions that are different from the current or historical settings. This missing data represents the region of the optimisation problem space that must be modelled based on either benchmarked data, known physical or observed relationships (such as the Six-Tenths rule for scaling costs according to equipment capacity, Lanz and Seabrook, 2012) and first-principles derivations. It is important that, rather than having a single data point available that represents only one operating point, numerical relationships (eg derived regression functions) are built into the optimisation model that consider all of the operating points that should be tested. The exact nature of important relationships will be unique to every mine site, and an extended period of consultation is required to determine their appropriate makeup.

Parameterise mining inventories

The mining inventory contains an array of all the tasks in a schedule, sometimes grouped into larger strategic blocks or zones for simplification. Each of the tasks, or strategic blocks contains attributes for all the development, ore tonnages, average grades, backfill, and any other cost or revenue related mining attributes required to produce a representative schedule that can be assessed by NPV or any other desired measure.

To maintain simplicity in the schedule but still capture the important features of Dynamic Economics as in Deucker, Phillips, and Kato (2021), a parametric mining inventory generation is implemented in this paper, based on directional dependencies and a Revenue Factor. In this paper, the Revenue Factor is multiplicative factor applied to the product prices and allows a simple profit assessment for individual stopes to determine whether they should be included in the inventory set denoted by that Revenue Factor.

This approach, in contrast to Dynamic Economics, allows the optimisation algorithm to select the most valuable inventory under any economic environment which has already been pre-processed and is fit for a mining zone resolution cut-off grade optimisation. In some cases, it is also possible for a Revenue Factor denoted inventory to be specified based on an economic scenario. This ensures changes in the mine inventory are considered when mine economics models and relationships are varied during the optimisation. This approach results in a more accurate optimisation and risk assessment tool when compared to a simple sensitivity analysis, which does not change the inventory, only the economic conditions.

As in Dynamic Economics, all sequential decisions prior to the stoping operation may be excluded if the stope itself does not cover the required costs. All development is therefore assessed in this manner, prior to optimising the cut-off grade of a stope shape.

The following section describes the logic used to derive lateral and vertical development meters and to generate calibrated optimal inventories for varying price scenarios. This captures, at a high level, the relationship between increased revenue, and an increased economic argument to expend development costs at higher price points.

The development inventory shown in Figure 1 is generally categorised into the following types:

- RAMP – For declines and inclines (ramps) including any associated ramp stockpiles, or infrastructure.
- STRIKE – Footwall development for transverse stoping zones and ore drives for longitudinal stoping zones.
- ACC – For level access between the ramp and orebody strike development and any additional stockpiles, cuddies or level infrastructure drives.
- VERT – Vertical development including vent rises, orepasses and escapeways. Does not include the access development required to reach the vertical rise.
- OP – For operating development, including slot drives and cross-cuts for transverse stoping zones.
- Stope – the smallest unit of production ore in an underground stoping mine.

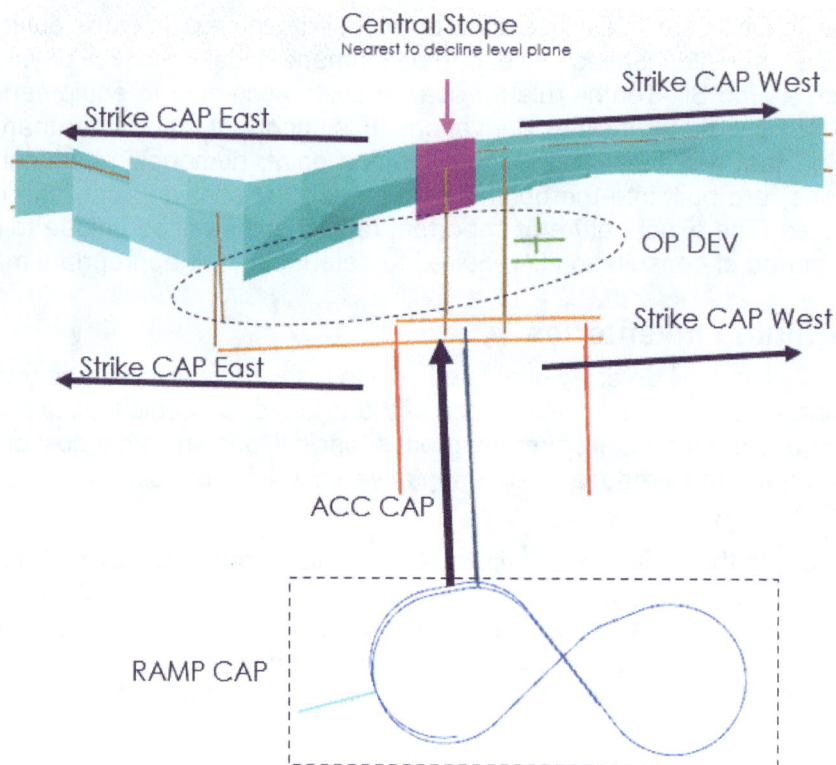

FIG 1 – Plan view of levels in a stoping mine illustrating capital and operating development meter derivation.

As a simplification for bulk mining operations, development ore tonnage is derived from any development completed inside stope shapes. However, this can be overridden by designed development inventories for more selective mining methods, such as in a narrow vein mine.

Inclusion of stoping material beyond economic development extents can lead to HoV® output cut-off grade values for the mine zone to be higher than economically justified, being penalised by the inclusion of negative cash flow material. To minimise this, a method of filtering out increments with likely negative cash flow is implemented to generate the Revenue Factor denoted inventory. Optimised development strike and stoping inventories are generated by maximising the value of each level by testing each incremental stope along strike according to the following equation:

$$Incremental\ Strike\ Value\ =\ Estimated\ Revenue\ -\ Estimated\ Cost$$

Estimated Revenue is determined using metal prices, dilution and recovery assumptions.

The value along strike is calculated directionally from the access, along the strike of the orebody as described in Figure 2. The cumulative value is calculated moving from the access point, towards the outer extents along strike. The Optimal Level Strike is identified where the cumulative value along strike is at the maximum. More complex orebodies do require more complicated frameworks, particularly when there is less of a relationship between required development and strike length of the orebody.

Any development and stoping material beyond the optimal strike will be excluded from the inventory set at this price point and cut-off grade.

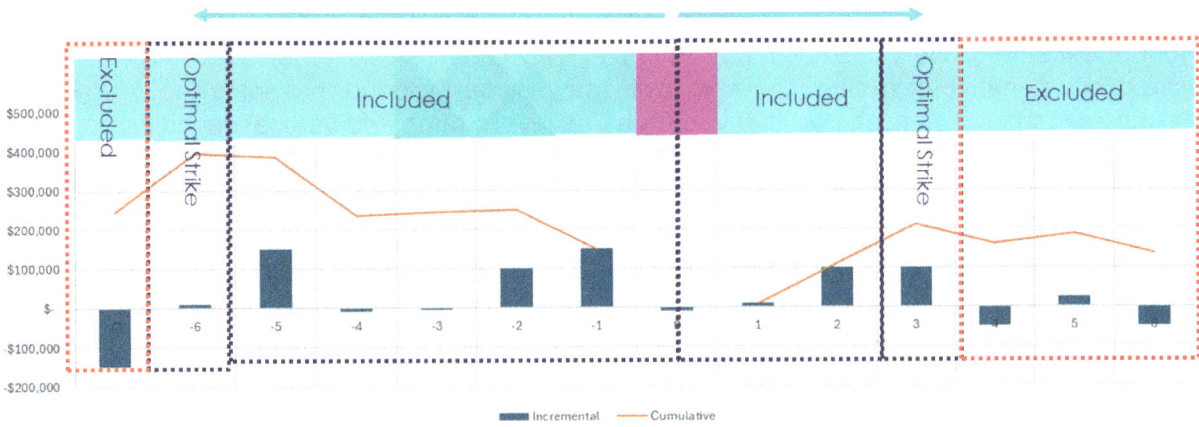

FIG 2 – Section view of a stoping level illustrating logic in determining optimal strike for the inventory.

The same logic used to determine the optimal inventory strike can also be used to determine the optimal inventory vertical extents of the Ramp (either incline or decline). The input design ramp must be directional, indicating an incline or decline.

Using the optimal strike values for each level, the incremental value of each level can be determined using the following equation:

$$\text{Incremental RL Value} = \sum \text{Optimal Level Strike Values} - \text{Capital Cost}$$

Where, the Optimal Level Strike Values are the maximum of the cumulative Strike values for the level, in each direction along the orebody, from the access.

The Capital Cost is the sum of the cost for the RAMP, ACC, and VERT categories. An assumed unit cost for development per metre of advance is used, derived from each operation's data or benchmarks.

The optimal vertical extents of an inventory set can be determined by calculating the cumulative value for each additional level, similar to the process of identifying the optimal strike. The process of identifying the Optimal Level Strike Value is visualised in Figure 3, progressing down a decline.

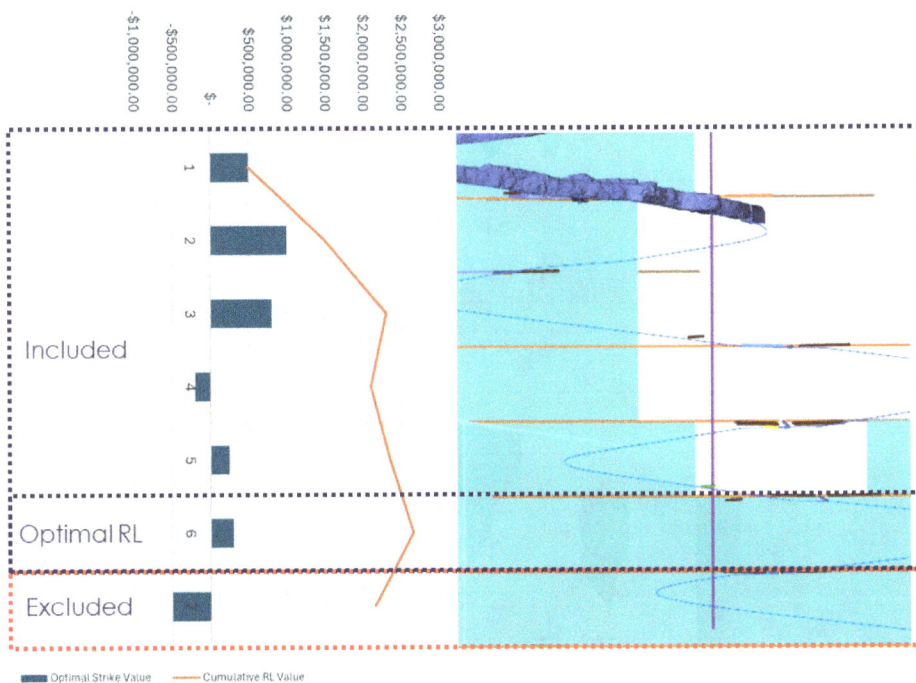

FIG 3 – Diagram illustrating logic in determining optimal RL for the inventory.

The cumulative value accounts for intermediary levels, allowing the algorithm to find optimal extents beyond the first negative value level. For example, even if optimal strike for a level is 0 m and the level exhibits negative cash flow, the next level may still be included, providing that it covers the previous level's negative cash flow (derived from that level's ramp and vertical development costs).

With this method, the mining inventory generation process allows a flexible parametric relationship to economic inputs, allowing inventories to be modified based on changes in metal price and mining cost. An optimal mining inventory can be generated for each of the price scenarios to be tested. (eg metal prices of -20 per cent, -10 per cent, Base, +10 per cent, +20 per cent). Each inventory set can then be designated with a Revenue Factor (RF) attribute to identify the price scenario used to generate it. These RF attributes can be used to filter mining zone inputs into the HoV® model, selecting a set depending on the price scenario, or even allow the evolutionary algorithm to select an optimal inventory during the optimisation loop.

This process of assessing capital commitment, access development, and production (selecting and extracting ore), is a sequential decision process where each decision impacts both upstream and downstream steps. This parametric inventory generation for underground mines allows the capture of sequential decisions for each Revenue Factor assessed in a discrete format for ease of computation during the optimisation.

CASE RESULTS AND DISCUSSION

Mine X was most interested in the optimal strategy's resilience against the change in economic conditions. A chart has been developed (Figure 4) to map potential options against two pricing scenarios, High and Reduced. The blue line represents the potential strategic options optimised against a high price, indexed to a case number along the x-axis. In this example, the x-axis corresponds to changes in strategy required to increase the mine life as this was another desirable property of the final strategy. Each data series is based on a different inventory set, with a different Revenue Factor, meaning the available inventory and mine life do not directly align, hence the need to index to the case number. The red line represents the same strategies if a metal price reduction occurs. The y-axis is normalised to the maximum value of NPV for each case, so that the cases are directly comparable to each other and any direct effect of the price on the NPV is excluded (unlike the results of a simple sensitivity analysis). The mining inventory for the price changes is also different, generated according to the method described in the section Parameterise Mining Inventories, it is another key differentiator to a simple sensitivity analysis.

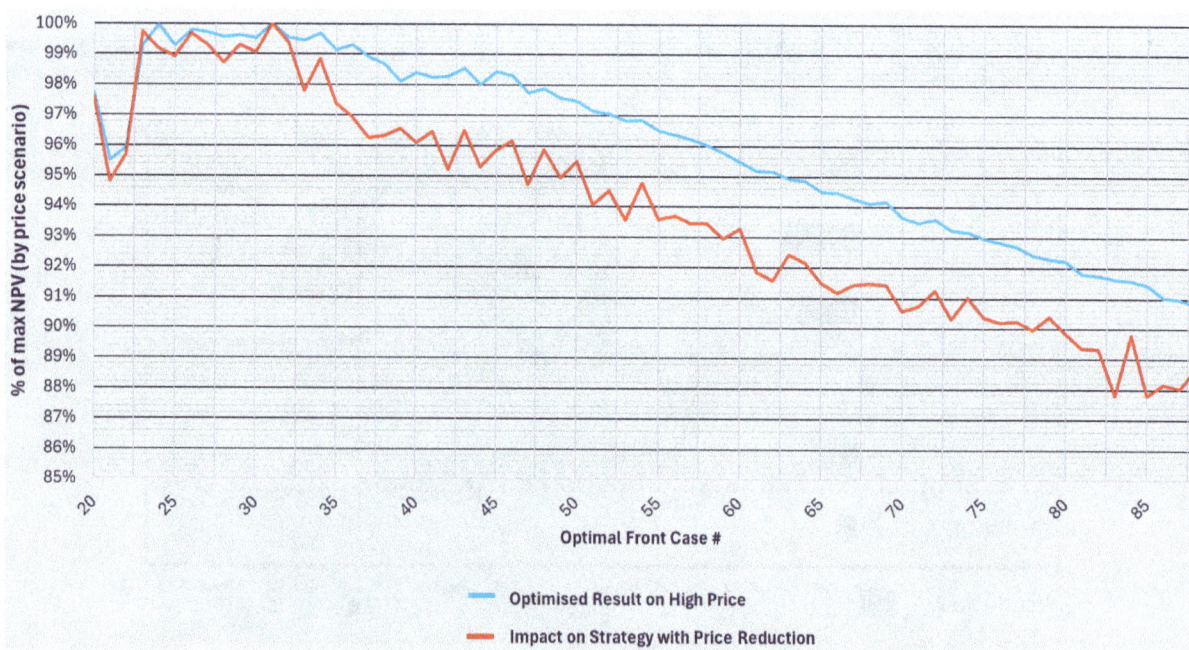

FIG 4 – Mine X strategy NPV resilience to price change.

The maximum value around case #31 is the best strategy for both price settings. However, there is very little difference in effectiveness of the selected strategy from anywhere between cases #20 to #35 indicating that all of these cases are resilient to price changes, even with the associated changes to the mining inventory based on economic conditions.

Choosing a strategy between the most resilient range of cases leads to the best outcomes. However, if a strategy is desired that reflects the operational settings in higher case numbers (for this example, yielding longer mine lives), the consequences of a price change on those cases can also be seen, understood and navigated by management.

Mine Y exhibited thousands of viable strategy combinations when three infrastructure options were considered in conjunction with other variables. Many more than could be considered with a simple ROV or similar approach. In Figure 5, the three colours blue, green and orange represent the three infrastructure options, Base Case, New Vent Shaft and Decline Bypass respectively. When implementing these infrastructure options, there is a new set of resulting combinations of capital cost, updated productivity and cut-off grades yielding the various data points on an NPV versus Capex expenditure chart. This approach enables testing of numerous different strategies, accounting for the sequential decision of capex cost, productivity changes, and the impact on optimal cut-off grade strategy. A selection can also be made to align with more than one goal, in this case by maximising NPV while also minimising capital expenditure. The Calibration Case line on the chart represents the NPV of the mine plan currently in execution, for a comparison of the relative benefit of each trade-off combination.

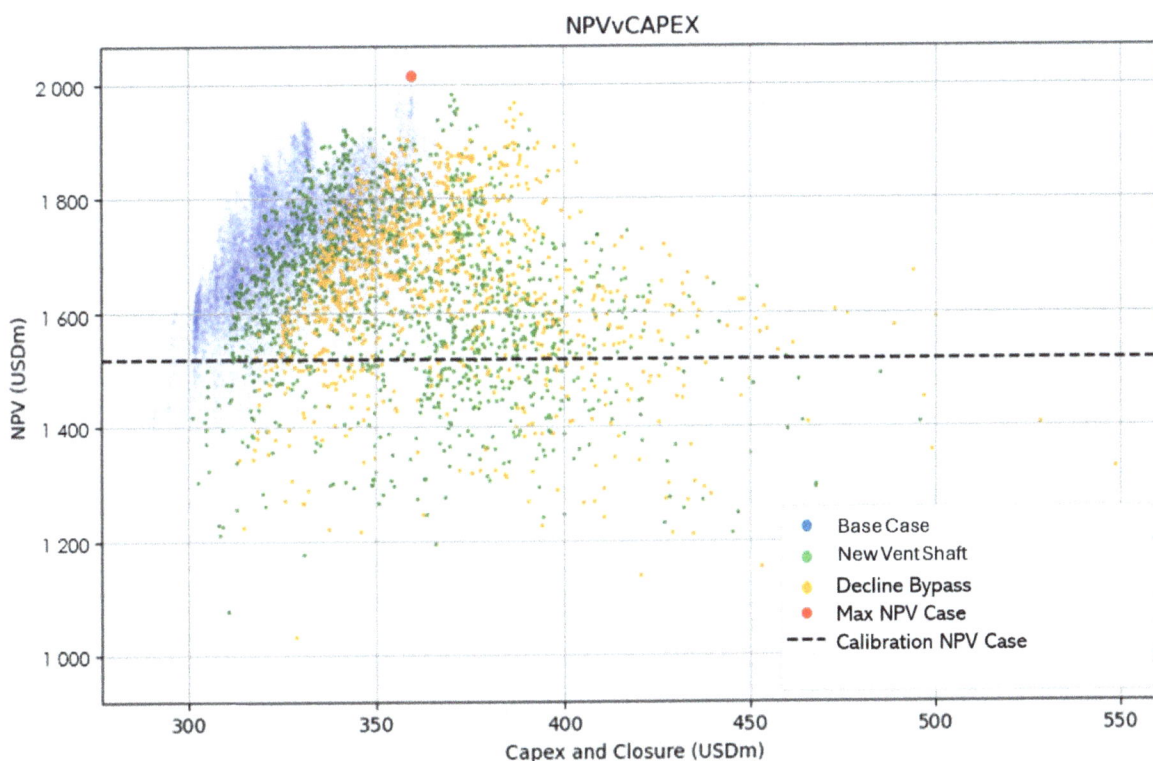

FIG 5 – Mine Y Multi-objective trade-off – maximised NPV versus minimised Capex.

In this study (Mine Y), the Base Case exhibited the lowest capital costs and the highest NPV, however there is significant overlap between the options and very few potential combinations around the NPV peak regions.

Mine Z sought a simple cut-off grade optimisation to use as an input into its tactical planning process. However, as shown in Figure 6 the NPV maximised case resulted in a significant shortening of the expected mine life compared to the current mine plan, denoted with a star on the chart.

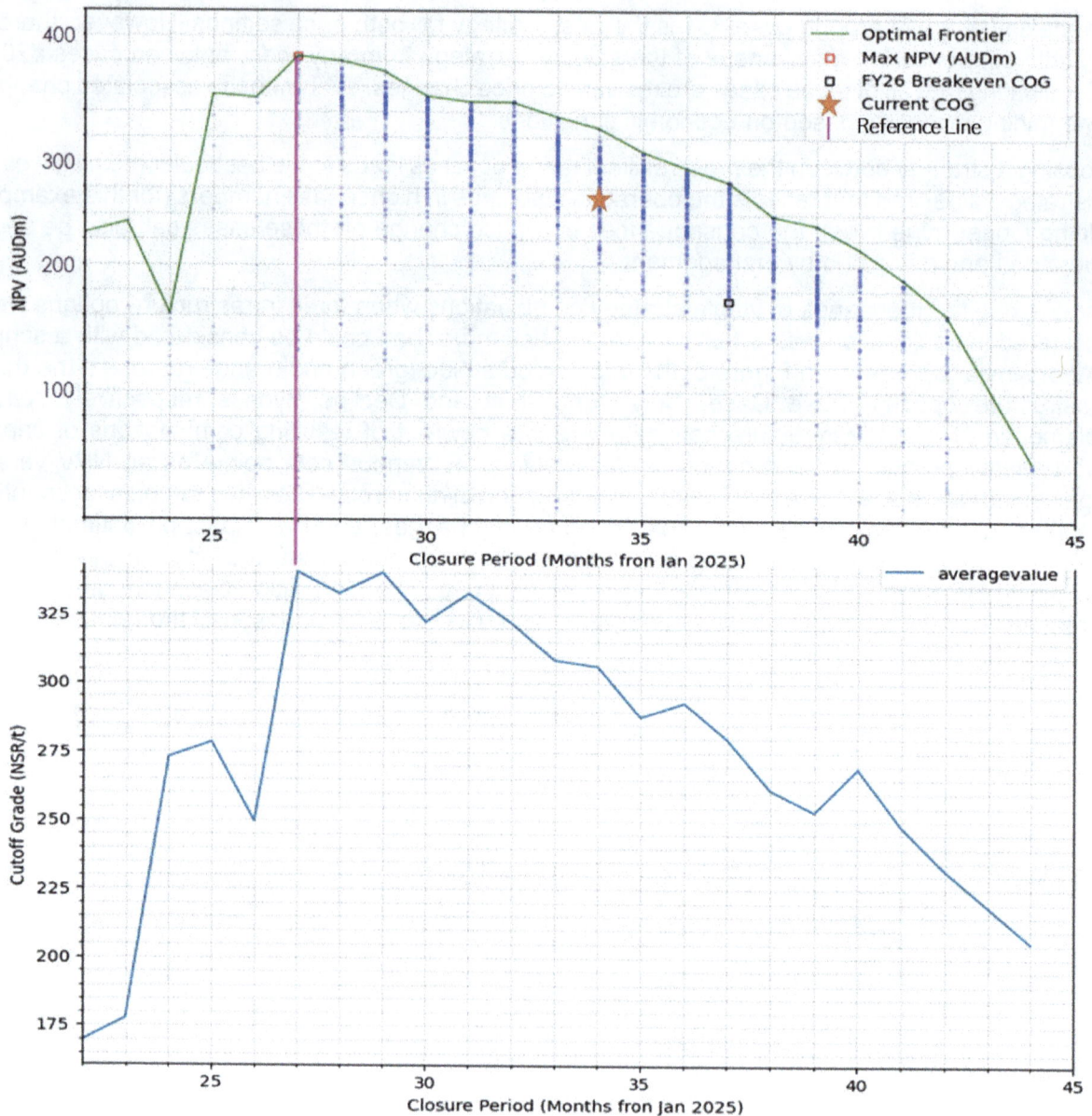

FIG 6 – Mine Z – Multi-objective trade-off – Optimal Cut-off grade and NPV versus Mine Life.

The multi-objective optimisation and resulting trade-off map between NPV and mine life allows the management to account for other factors at play, which may include social obligations or potential future inclusions of undiscovered material. The next step of a study such as this would be to attempt to value the potential of any undiscovered resources, and their potential strategic impact.

This results map can be used to improve the NPV of the mine when compared to the traditional method of deriving cut-off grades (Breakeven cut-off grade, also shown on the chart) (Hall, 2014). It can also be used by management to navigate a more nuanced strategy that includes their flexibility and desire to allow more time to discover new resources. The result of this optimisation enables management to decide on a cut-off grade policy along the optimal frontier, which is not necessarily at the maximum NPV. A time limit can then be set on waiting for the results of planned resource exploration programmes, with the right reserved to revert to an NPV maximum setting if those exploration programmes are not fruitful.

The composite chart in Figure 6 shows the corresponding average cut-off grade (expressed as Net Smelter Return) that corresponds to each point along the Optimal Frontier, linked up by the reference line shown in pink.

CONCLUSIONS AND FUTURE WORK

The application of AMC's Hill of Value® in mine strategy optimisation provides a robust framework for making well-informed strategic decisions. By incorporating sequential decisions into the inputs of the optimisation model, the resilience of the strategy is enhanced. Value-accretive tactical deviations are less likely to compromise the overall strategy, and important trade-offs can be visualised and discussed easily.

The case studies presented in this paper demonstrate the practical benefits of this methodology, highlighting its potential to improve the strategic resilience, in particular where sequential mining decisions are possible, and ultimately economic outcomes of mining operations. Mine X was able to test numerous strategic options in response to economic changes, including inventory sets that vary according to economic viability under various conditions, and not just the applied stope cut-off grades.

Mine Y investigated a detailed trade-off between different infrastructure options, and contrasting the varied costs, and productivity benefits, with sequentially dependent optimal cut-off grade settings. The multi-objective solutions map also allows decision-makers to select between a combination of minimising capex outlays and maximising NPV.

Mine Z has been provided with a NPV and mine life trade-off map, allowing decision-makers to account for the likely possibility of discovering more ore sources and the impacts of mine closure. It gives them the ability to pivot to a higher NPV strategy once all their other objectives are satisfied. This result demonstrates that a more detailed strategy selection is possible with this kind of information, as decision-makers can also see the relative cost to (reduction in) NPV of any decisions alternate to the maximum NPV option.

Finally, further development of this approach is required with near-term targets of:

- Investigation of the application of Dynamic Economics within the optimisation loop, which may require a new optimisation algorithm.

- Investigation of higher resolution scheduling techniques without sacrificing human effort and losing the algorithmic ability to test an indefinite number of cases.

ACKNOWLEDGEMENTS

The authors acknowledge AMC Consultants for supporting this publication.

REFERENCES

Deb, K and Jain, H, 2014. An Evolutionary Many-Objective Optimization Algorithm Using Reference-Point-Based Nondominated Sorting Approach, Part I: Solving Problems With Box Constraints, *IEEE Transactions on Evolutionary Computation*, 18(4):577–601.

Deucker, E, Phillips, F and Kato, J, 2021. Maximising asset value through optimal block by block decisions on marginal material, in *Proceedings Underground Operators Conference 2021*, 18 p (The Australasian Institute of Mining and Metallurgy: Melbourne).

Elkington, T and Gould, J, 2012. Real Option Value Optimisation, in *Proceedings Project Evaluation Conference 2012*, pp 35–40 (The Australasian Institute of Mining and Metallurgy: Melbourne).

Hall, B (ed), 2014. *Cut-Off Grades and Optimising the Strategic Mine Plan*, Spectrum Series 20 (The Australasian Institute of Mining and Metallurgy: Melbourne).

Jetmore, K and Roos, C, 2022. Strategic Underground Mine Planning Through Nested Stopes and Constraint-Based Scheduling, SME Annual Meeting (Society for Mining, Metallurgy and Exploration).

Lanz, T and Seabrook, W, 2012. Operating Cost Estimation, in *Cost Estimation Handbook*, 2nd edn, Monograph 27 (The Australasian Institute of Mining and Metallurgy: Melbourne).

Analysing the haul road design from the simulation perspective – theoretical and real improvements

R L Peroni[1], J L V Mariz[2], E G O Neto[3], D J Souza[4] and A Moradi Afrapoli[5]

1. Full Professor, Mineral Research and Mine Planning Laboratory, Universidade Federal do Rio Grande do Sul, Porto Alegre RS, Brazil. Email: peroni@ufrgs.br
2. Post-doctorate, Mineral Research and Mine Planning Laboratory, Universidade Federal do Rio Grande do Sul, Porto Alegre RS, Brazil Email: jorge_valenca@hotmail.com
3. Mining Engineer, Mineral Research and Mine Planning Laboratory, Universidade Federal do Rio Grande do Sul, Porto Alegre RS, Brazil. Email: eduardo_goneto@hotmail.com
4. Mining Engineer, Mineral Research and Mine Planning Laboratory, Universidade Federal do Rio Grande do Sul, Porto Alegre RS, Brazil. Email: dayssouza11@gmail.com
5. Assistant Professor, Sustainable Intelligent Mining Laboratory, University of Kentucky, Lexington KY, USA. Email: ali.moradi@uky.edu

ABSTRACT

In mining operations where trucks are the primary means of moving ore to processing plants and waste to dump sites, the efficiency of the production flow system is heavily influenced by the quality of haul roads. A thorough understanding of the requirements for constructing durable haul roads and predicting their behaviour, including deterioration patterns and maintenance frequency over their lifespan, is essential. Haul road quality directly impacts the performance of the transportation system, which can account for 45–55 per cent of mining costs. While equipment performance charts from suppliers offer insights into key indicators, such as truck speed, cycle time, and fuel consumption, these often assume deterministic conditions with a constant road quality. However, challenges arise when road design varies along its length or when poor design, inadequate drainage, and accelerated deterioration disrupt initial performance assumptions. Additionally, most mines feature road networks shared by multiple equipment types, including production and maintenance vehicles and light-duty traffic, leading to frequent interferences. Discrete-event simulation was applied to model these complexities in a phosphate mine in Brazil, providing a realistic view of fleet performance under varying conditions. Considering a theoretical base case, whose productivity was calculated according to the equipment's rimpull curves, and stochastic scenarios that address the impact of different rolling resistances (RR), it was verified that only the increase in RR promoted a significant loss in productivity of up to 5 per cent from RR values of 2 per cent to 6 per cent in the simulated scenarios. By comparing scenarios with and without best practices in haul road construction, the advantages of well-maintained roads and quantify the impact of road deterioration on equipment performance over time were demonstrated. This approach highlights the importance of maintaining steady road quality to achieve optimal transportation system efficiency, underscoring the benefits of sound haul road practices in minimising costs and maximising operational stability.

INTRODUCTION

Building and maintaining haul roads is a key aspect of production logistics in mining. Adopting a careful approach at this stage of planning can lead to significant beneficial impacts, including financial savings, reduction of maintenance frequency and fuel consumption, reduction of greenhouse gas emission, increase in equipment productivity and availability. According to Thompson and Visser (2003), mining haul roads have historically been built on an empirical basis, which does not provide adequate economic and technical results. Truck haulage costs can account to more than 50 per cent of the total operating costs in surface mines, depending on the circumstances; therefore, planners should be aware of the risks they are assuming under these conditions (Thompson, 2011). However, another conceptual approach has emerged, establishing that the success of a mining haul road is directly connected to the execution of a project subdivided into components such as geometry, structure, functionality, and maintenance (Deslandes and Dickerson, 1989; Thompson and Visser, 1999).

Thompson and Visser (2006) suggest that road design must be approached holistically, meaning that if one component of the design is deficient, the others will not perform optimally. The first step is the geometric design, which is associated with road features that aim to provide overall safe and efficient driving. Once the haulage equipment is defined, space requirements and performance limitations arise, hence the impact of the road design on vehicle's controllability must be regarded. Therefore, aspects such as vertical alignment, horizontal alignment, safety berms, and drainage components are fundamental in road development over varying topography. The recommended longitudinal gradient for in-pit mining roads is 10 per cent, ensuring safe trafficability, leading to a reduced average fuel consumption and exiting the pit as fast as possible, according to truck manufacturers performance specifications. Even though a constant gradient is ideal to avoid frequent gear shifts and a lost in productivity, real roads generally present multiple gradients, which also contribute to higher maintenance costs and premature replacement of truck's components (Thompson and Visser, 1999; Thompson, Peroni and Visser, 2019).

Structural design refers to the hidden part of the road, composed of layers of appropriate materials capable to provide the support needed to the design, where the base layer thickness is determined according to the imposed vehicular mass on the pavement. The primary objective is to protect weaker subgrade layers and ensure that pavement deformation is minimal, thus avoiding excessive maintenance interventions. Starting with the subgrade, characterised by the *in situ* material on which the road is designed, other layers such as the base and the wearing course are dimensioned using the California Bearing Ratio (CBR) as a methodology. The CBR is an adaptation for mine haul roads of a method widely used in public highways, which consists of tests to determine the percentage ratio between the pressure exerted by a piston compacting a sample under study and the pressure exerted by the piston, under the same conditions, on a standard sample with a CBR value of 100 per cent. Therefore, the material to be used in the layers must undergo testing, and based on the material's indices, along with the known maximum wheel load applied by the loaded truck, it is possible to determine the required thickness for each structural layer (Hustrulid, Kuchta and Martin, 2013).

Furthermore, the wearing course is also part of the functional design, where the selected material must possess characteristics that ensure low rolling resistance, dust control, traction, reduced vibration, and consistent vehicle speed, thereby prioritising transportation economy (Thompson and Visser, 1999; Thompson, Peroni, and Visser, 2019). According to Thompson and Visser (1999), the chosen material may be a combination of different types, as long as it falls within an acceptable or ideal range, defined by specified intervals between the shrinkage product and the grading coefficient. The shrinkage product (Sp) is determined by the linear shrinkage of the selected material, which is approximately half the value of its plasticity index, multiplied by the percentage of material passing through a 0.425 mm sieve. The grading coefficient (Gc), by its turn, is determined based on the percentage of material passing through the 26.5 mm, 2 mm, and 4.75 mm sieves (Thompson and Visser, 2000).

The magnitude of the force transmitted through the truck components on irregular roads is proportional to the square of the speed at which the equipment hits the irregular sections and the gross vehicle weight (GVW), corroborating the need to reduce road roughness in order to increase the productivity and component durability. In this context, rolling resistance (RR) represents the traction force required for a truck to overcome the retardation effect between its tires and the road surface (Tannant and Regensburg, 2001; Thompson, Peroni and Visser, 2019). This parameter is often expressed in terms of percentage of road gradient (RG), so Table 1 depicts an estimation of rolling resistance given the road condition.

TABLE 1

Estimation of rolling resistance based on road condition (Thompson, Peroni and Visser, 2019).

Road conditions	Rolling resistance (%)
Strong layers and well-constructed, compacted roads with low tire penetration	2
Intermediate strength layers, compacted roads, well-constructed and frequently maintained, with minimal (<25 mm) tire penetration	2–3
Weak layers or wearing material, tire penetration 25–50 mm, rutted and poorly maintained	3–5
Weak layers or surface material, tire penetration 50–100 mm, rutted and poorly maintained	5–8

Therefore, rolling resistance is a fundamental parameter for calculating the effective inclination that must be overcome by the equipment, in addition to the road gradient itself, thus resulting in the calculation of the total resistance (TR), as shown in Equation (1):

$$TR(\%) = RR \pm RG \tag{1}$$

where TR represents the total resistance (%), RR is the rolling resistance, and RG is the road gradient, which is positive or negative for the uphill and downhill sections, respectively (Tannant and Regensburg, 2001; Soofastaei et al, 2016; Thompson, Peroni and Visser, 2019; Vera-Burau et al, 2023).

Therefore, the use of methodologies capable of answering questions about mining operations, haul road design and fleet performance plays a key role in mining, especially for questions that would be difficult to analyse from an algebraic perspective. In this regard, Discrete Event Simulation (DES) is a methodological approach used to model and analyse systems in which changes occur at discrete points in time. Its main characteristics include:

- Event-based: Unlike continuous simulation, DES events that occur at specific time instants.

- Discrete states: The system transitions between distinct states in response to scheduled events.

- Simulation clock: Maintains control over the temporal sequence of events.

- Flexibility and customisation: It can be adapted to different domains and scenarios, including production processes, traffic management and communication network simulations.

- Statistical modelling: Uses statistical distributions to represent uncertainties inherent to the simulated systems (Pooch and Wall, 1992).

Pooch and Wall (1992) state that the scientific methodology for applying a simulation system consists of four main phases:

1. Planning: Involves defining the problem, analysing system characteristics, estimating required resources, and setting objectives, thus ensuring a clear understanding of the system and its influencing factors.

2. Modelling: Focuses on creating a system mathematical representation, acquiring and processing necessary data, in addition to preparing the model for computer execution, hence translating real-world problems into a structured computational model.

3. Validation: Assess the model's accuracy and reliability, ensuring that it accurately represents the real-world system and runs as intended.

4. Application: Encompasses running simulations, analysing the results, and formulating recommendations for problem-solving, which can be further improved after implementation and new rounds of application.

In open pit mining, DES approaches are often used to optimise haulage operations and evaluate their main characteristics, allowing the prediction of parameters such as truck efficiency, optimum fleet size, fuel consumption, and greenhouse gas emission. In this sense, different variables can be assessed as statistical distributions, including tonnage and loading time, equipment speed, and road resistance, thus generating distinct scenarios to verify their impact on productivity (Chaowasakoo et al, 2017; Moradi Afrapoli and Askari-Nasab, 2019; Alamdari et al, 2022). Fleet management systems can be single-staged, in which the dispatching process does not consider production constraints or goals, or multi-staged, where the main problems to be solved are the shortest path (Elbrond and Soumis, 1987; Temeng, Otuonye and Frendewey, 1997), the production optimisation and fleet allocation (Koenigsberg, 1958; White and Olson, 1986; Soumis, Ethier and Elbrond, 1989) and the real-time truck dispatching (Lizotte, Bonates and Leclerc, 1987; Topal and Ramazan, 2010). Figure 1 depicts a cyclic truck-shovel operation (left) and a configuration of multiple trucks for one shovel (right).

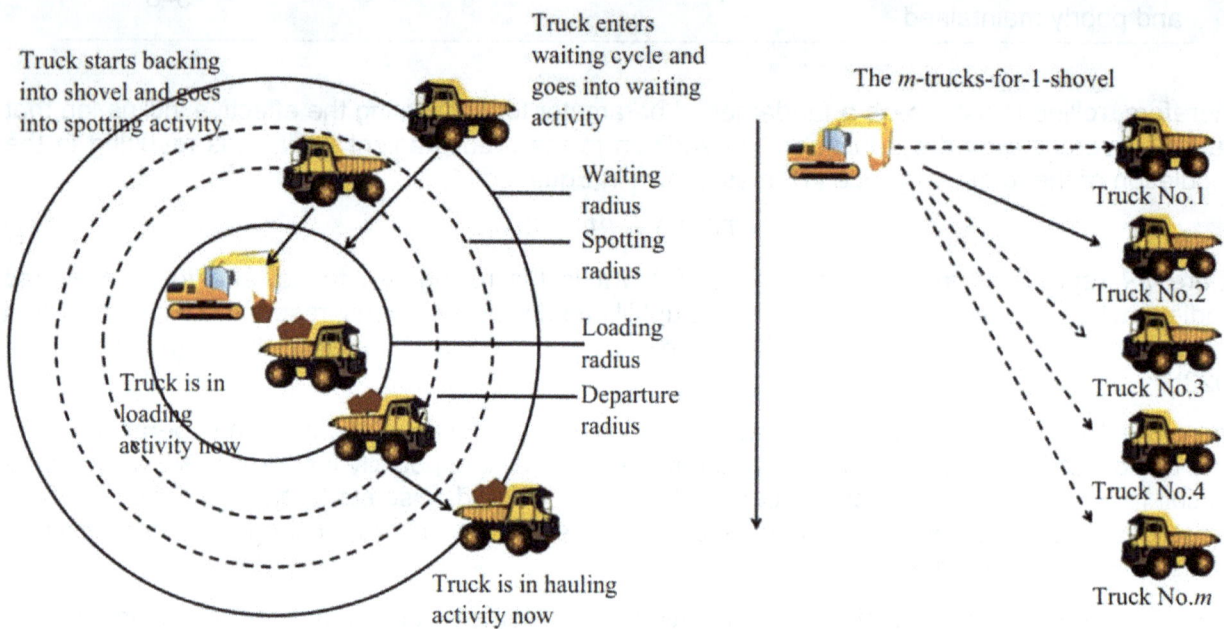

FIG 1 – A cyclic truck-shovel operation (left) and a configuration of multiple trucks for one shovel (right) (Modified from Chaowasakoo et al, 2017).

Therefore, Discrete Event Simulation stands out as an essential tool in mining, allowing the simulation of different scenarios and anticipating impacts before implementing real changes in the studied systems. It facilitates reliable decision-making and resource optimisation to maximise efficiency and reduce operational costs. Assessing the impact that different rolling resistances would have on the productivity of a phosphate mine in Brazil is the main analysis approached by this study. To demonstrate this effect, a haul road stretch connecting a mining front and a crusher was modelled and had its characteristics changed for confidentiality reasons. Scenarios were simulated considering different RR values, where an increase in RR would cause a negative impact on the truck distribution velocity curves. A theoretical base case scenario was proposed to enable comparison and discussion of the results.

METHODOLOGY

The methodology adopted in this study encompasses four stages, which address the modelling of the haul road studied; the definition of the theoretical base scenario relying on the equipment rimpull curves; the generation of different stochastic scenarios through DES; and the comparison of scenarios, culminating in statistical analysis and discussion of results. The equipment configuration is comprised by a single excavator and a fleet tested of up to ten trucks. The selected truck model is the CAT 777, with a nominal payload of 91 t, as the equipment is already used in the operation.

All simulations were performed using ARENA software v.16.20 – Student Mode (Figure 2), covering a 30-day period with three shifts and fuelling per day, so that each truck had to stop if it was crossing a parking lot near the mining face or crushing area within a 25 min and 30 min window, respectively, before the start of each new shift. The output of the simulations were the excavator utilisation rate (%), average waiting time for loading (min), and number of cycles per month, allowing the calculation of monthly productivity (kt) from the product of the number of cycles by the truck's payload. In all scenarios, the loading and dumping times were arbitrarily set as 5 min and 1.1 min, respectively. The road sections that provide access to the mining front and crushing area, which were not discretised into road stretches, also had fixed travel times assigned. Therefore, it takes 4.3 min for trucks to reach the excavator and 5 min to pass through the mining front when loaded, respectively, and then it takes 3.3 min for trucks to reach the crusher and 2.6 min to pass through the crushing area, respectively. This study disregarded any possibility of trucks reaching speeds below 5 km/h or above 50 km/h, which means that all cases assume a constant speed along the stretches with the same longitudinal grade and rolling resistance conditions.

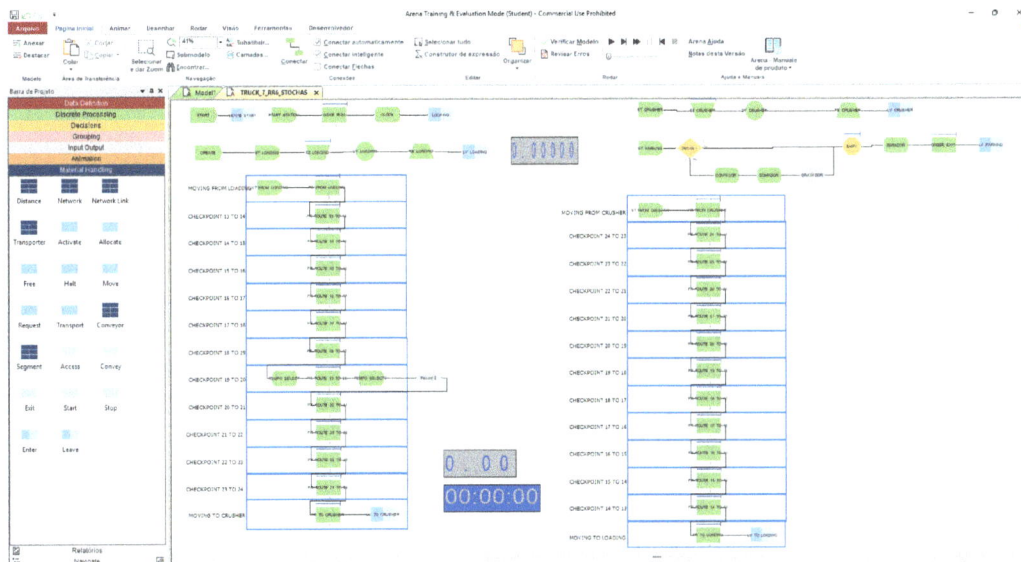

FIG 2 – Arena software interface.

Stage 1

The topographic data were obtained through surveys using an unmanned aerial vehicle (UAV) and modelled by the mining company, which provided the mine model in the form of contour lines. From this model, the road selected for study had the longitudinal gradients analysed. Then, the chosen road was segmented into sections (road stretches) according to the different grades verified, so that the influence of each stretch on overall productivity could be analysed independently.

Stage 2

To evaluate the theoretical speeds of the trucks on each road stretch, the equipment's rimpull curves were used for the uphill and downhill stretches, considering a RR of 3 per cent as the real situation in the mine. In these charts, the vertical dashed lines represent the loaded (L) and empty (E) conditions. The use of these charts is based on the visual plotting analysis of vertical or horizontal lines by the user. Figure 3 shows the CAT 777 truck's rimpull (left) and retarding (right) curves for the uphill and downhill roads, respectively. For example, with a total gross vehicle weight of 160 t and a total resistance of 5 per cent, the rimpull would result in approximately 9000 kg, which limits the speed up to 28 km/h.

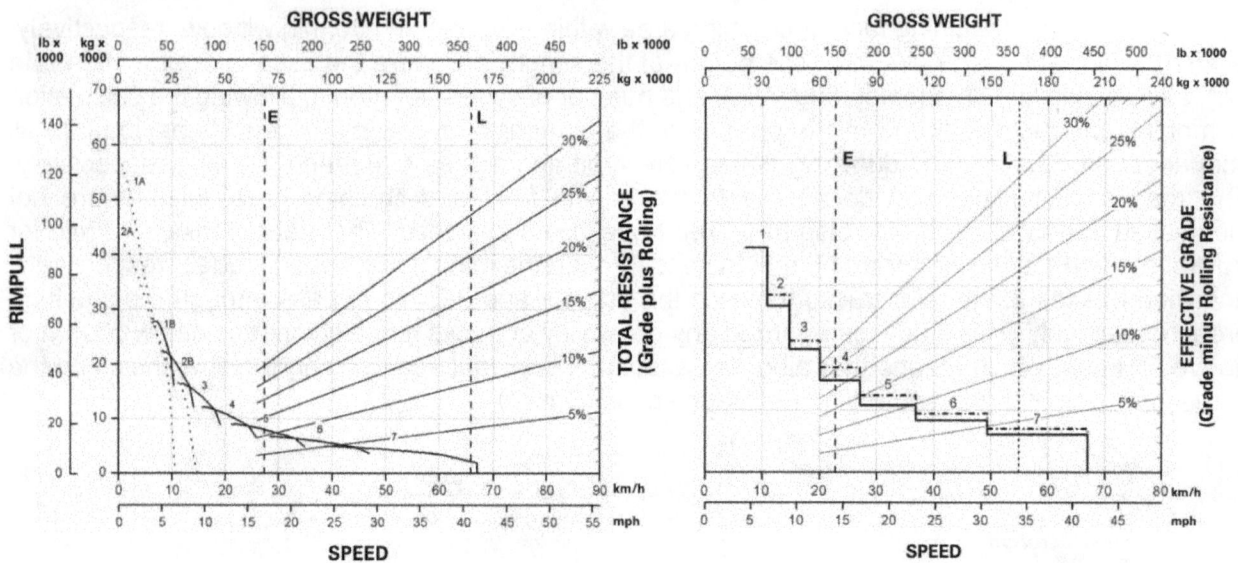

FIG 3 – CAT 777 truck's rimpull and retarding curves for the uphill and downhill roads, respectively (modified from Caterpillar, 2019).

Furthermore, to assess the ideal number of trucks in the fleet operating on the studied road under the previously mentioned conditions, scenarios with these deterministic speeds were run in ARENA, considering fleets with five to ten trucks. Although using software that allows stochastic analysis, as all speeds on the different road sections were deterministic, these scenarios were also deterministic. Therefore, the scenario was elected as the theoretical base case, combining reasonable assumptions of utilisation rate, average waiting time, and productivity.

Stage 3

Considering the ideal number of trucks in the fleet according to the results of Stage 2, to incorporate uncertainty into the assessment, statistical distributions were built from real truck data provided by the mining company. This data covers two weeks and comprises information such as truck coordinates, speed, and status, which can be loaded or empty, resulting in 40 727 rows after data treatment procedures to eliminate inconsistencies. Hence, the data were subdivided considering the truck status for each road stretch, thus the histograms of these speeds were fitted to statistical distributions to allow incorporation of speed uncertainty into the simulation. Considering a probabilistic approach, for each stretch, these speed distributions were sampled randomly for each truck cycle, which may result in different cycle times for the same truck in each scenario being evaluated. Furthermore, to comply with the mining company's safety standards and avoid assigning very low speeds to trucks on the stretches, the fitted distributions were truncated on the predetermined minimum and maximum speeds of 5 and 50 km/h, respectively.

To evaluate the impact of different rolling resistances on productivity, different scenarios were simulated in which the truck speed distributions were shifted by 4 km/h when they were loaded and by 3 km/h when empty, gaining speed when the RR decreased or losing speed when it increased, although the histograms' minimum and maximum values remain the same. Five stochastic scenarios were carried out considering rolling resistances from 2 per cent to 6 per cent, in which the original data curves were used in the stochastic scenario with RR of 3 per cent. To determine the ideal number of simulations to be performed in each stochastic scenario, initial runs with 100 realisations were performed and the average accumulated productivity per month (kt) was plotted for each realisation. Therefore, a threshold was determined at which all simulations stabilised and the final experiments were performed considering this optimal number of realisations.

Stage 4

At this stage, the results of the theoretical base case and the stochastic scenarios were compared in terms of the average metrics and a boxplot of the productivity per month. Furthermore, the results were discussed and future directions are proposed. Figure 4 displays a flow chart with a summary of the stages proposed in this methodology.

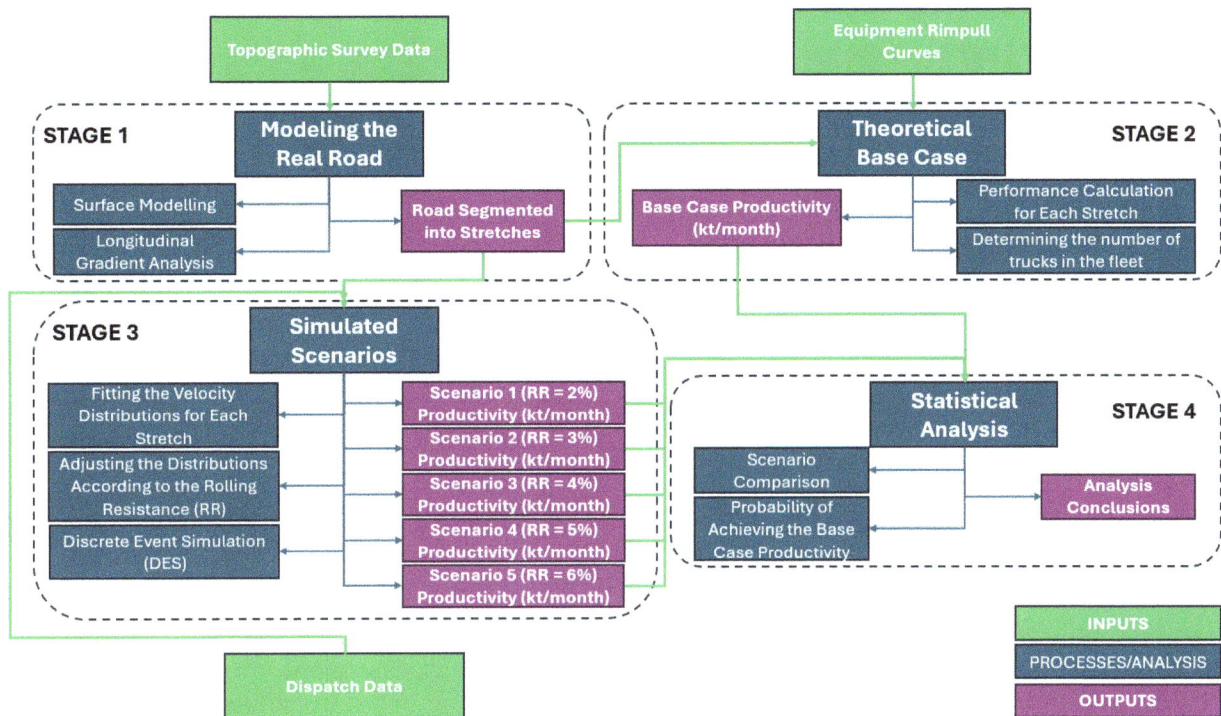

FIG 4 – Flow chart with the stages proposed in this methodology.

RESULTS AND DISCUSSION

The four stages of the methodology were executed in chronological sequence using data from a phosphate mine located in Brazil.

Stage 1

Based on the information collected from the contour lines representing the mine's topography, focusing on the road networks used by trucks to travel between the mining faces and the crusher, a 2.7 km road section was selected to observe the grade differences along the route, as well as their lengths. For this purpose, centrelines were design over the road surface, which had previously been converted into a digital terrain model (DTM). Along the road, it was observed that dividing it into segments at every gradient variation would introduce a level of complexity difficult to accommodate in the simulation study. Therefore, 11 road segments were created, prioritising gradient consistency by smoothing out some variations occurring over very short distances. Thus, the longitudinal gradient of the road ranges between -2 per cent and 5 per cent, considering the longitudinal profile of the road from the mining face to the crusher.

After defining these stretches, 11 polygons were created to encompass the entire road surface for each section as can be visualised in Figure 5. This approach was implemented to filter dispatch data and produce distribution curves derived from truck speed information.

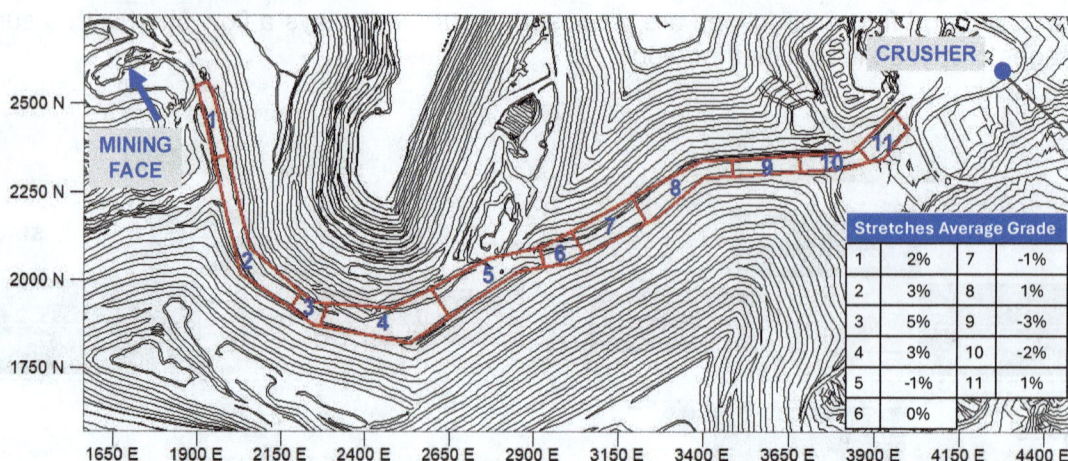

FIG 5 – Road stretches overlaid on the topography represented by contour lines.

Stage 2

Each road segment had a longitudinal gradient to which the rolling resistance, set at 3 per cent, was added. Using the total value obtained from this, the speed for both loaded and empty state was determined based on the rimpull curves of the CAT 777 truck.

Based on the velocities and distances of the segments, the times taken by the truck to travel each road segment were input into the model built in Arena software. Therefore, the time values were constant, and from this, it was possible to establish the variation in the truck fleet to determine the ideal quantity. Six scenarios were created, differentiated by the variation in the number of trucks on the fleet, ranging from five to ten. Table 2 presents some metrics collected after running the scenarios, which simulate a 30 day period of production.

TABLE 2

Results of replications of deterministic scenarios that fleet truck size was tested.

Fleet truck size	Loading utilisation rate	Average queue time for loading (minutes)	Number of cycles	Production (kt)
5	77.1%	0.2	6658	605.9
6	91.7%	0.5	7917	720.4
7	94.8%	4.5	8188	745.1
8	95.8%	9.3	8278	753.3
9	96.9%	14.0	8368	761.5
10	97.9%	18.6	8457	769.6

The fleet of seven trucks was chosen considering all the conditions presented in Table 2 as it demonstrated a reasonable utilisation rate for the excavator combined with low queue times while delivering the desired production. From eight trucks onward, the time for the last truck to complete its shift change became too tight, and scenarios considering fleets with nine and ten trucks would have trucks that could not complete a shift change.

Stage 3 and 4

Using the same foundation as the model built in Stage 2, in Stage 3 the constant round trip times of the trucks were replaced by probabilistic distribution curves constructed for each segment based on the real dispatch data of this operation. Therefore, each segment has its own curve for the loaded truck's travel and a second curve for the empty truck's travel. These curves are represented by multiple distributions such as normal, beta, and Weibull. Different types of distributions occur because each segment has distinct trafficability conditions. Additionally, factors such as curves, other slower of stopped equipment on the road, and variations in load weight also contribute to speed variations within a segment.

Thus, based on the theoretical base case, five scenarios were constructed for joint analysis, where the rolling resistance varies between 2 per cent and 6 per cent, with the value of 3 per cent assigned to the original data set. On the other hand, the other scenarios involved a shift in the curve by 4 km/h for the loaded status and 3 km/h for the empty status. Each scenario was executed with 100 replications, a number chosen based on the stabilisation of the cumulative average production (Figure 6).

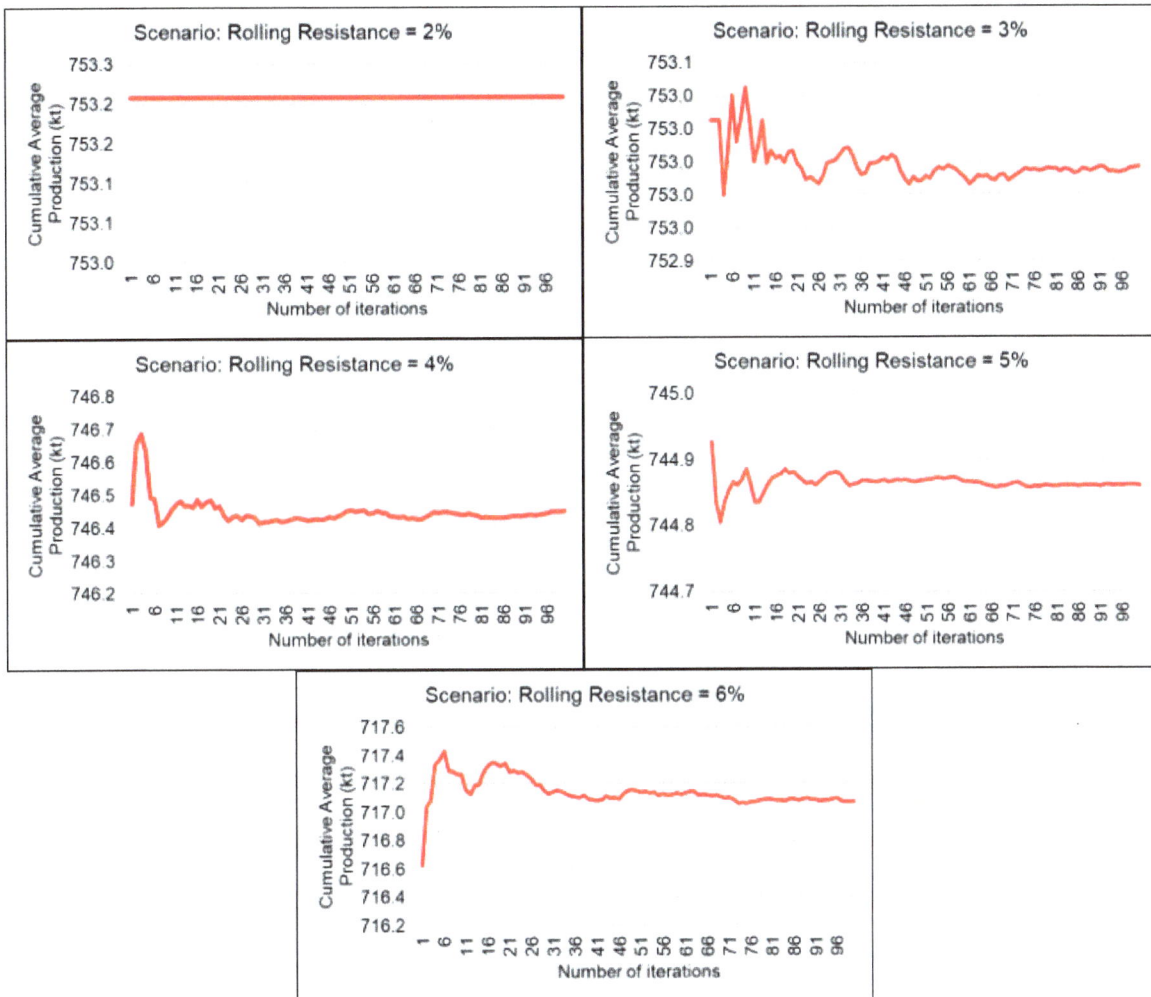

FIG 6 – Graphs showing the cumulative average as a function of the number of simulated iterations.

The simulated scenarios, in addition to the base case, have their results presented in Table 3, which shows the averages of the replications for loading utilisation rate, average queue time for loading, number of cycles, and production.

TABLE 3

Results of average of iterations from stochastic scenarios varying the rolling resistance.

Mode	Rolling resistance	Loading utilisation rate	Average queue time for loading (minutes)	Number of cycles	Production (kt)
Deterministic	3%	94.8%	4.5	8188	745.1
Stochastic	2%	95.8%	6.4	8277	753.2
	3%	95.8%	5.8	8275	753.0
	4%	95.0%	4.9	8203	746.4
	5%	94.8%	3.4	8185	744.9
	6%	91.3%	2.2	7880	717.1

The deterministic case exhibits behaviour that diverges from the stochastic cases for 2 per cent and 3 per cent of RR, as the theoretical velocities input into the fixed-value model are lower than the average velocities found in the distributions.

Productivity decreased as RR increased considering the truck needs to reduce its speed to travel on a more irregular surface, consequently completing fewer cycles within during each shift. The average queue times and the utilisation of the loaded equipment also show a tendency to decrease with the increase in RR. Although it might seem beneficial for queue times to decrease, this does not apply in these cases, as the travel time between loading and dumping points tends to increase, increasing the distances between trucks and reducing the likelihood of queues. On the other hand, the excavator tends to experience more idle time.

Figure 7 shows a boxplot graph that illustrates the statistics of the replications for each scenario in terms of monthly production, followed by Table 4.

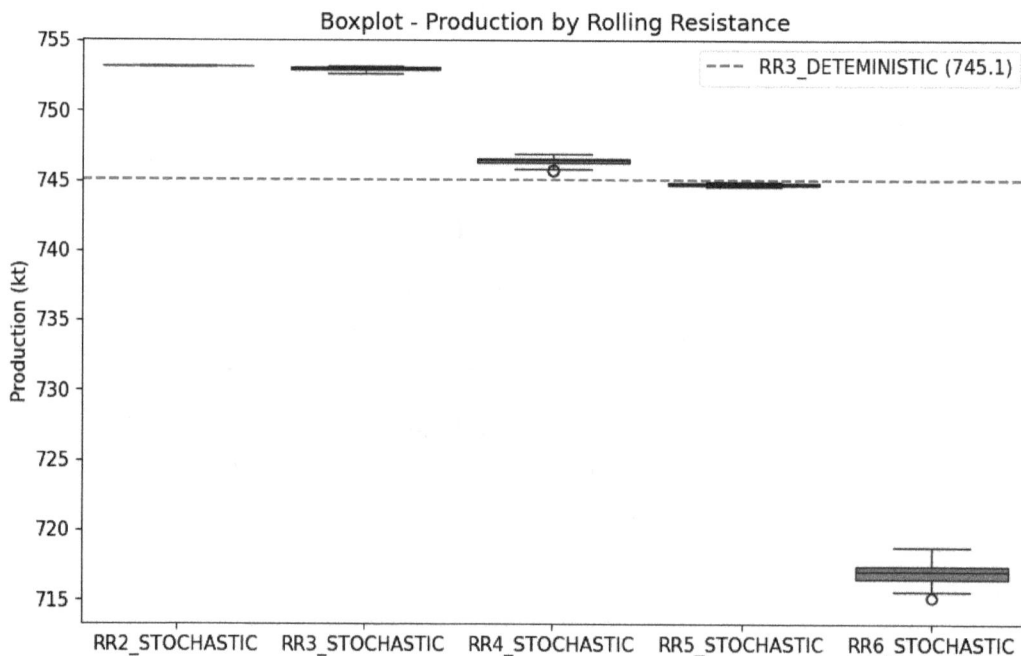

FIG 7 – Boxplot for each scenario varying the rolling resistances for stochastic situation plus a line of deterministic scenario.

TABLE 4

Statistics from iterations by each scenario analysing monthly production in tonnes.

	RR2	RR3	RR4	RR5	RR6
Minimum	753 207	752 661	745 836	744 562	715 169
Quartile 1	753 207	752 934	746 200	744 744	716 625
Quartile 2	753 207	753 025	746 473	744 835	717 080
Quartile 3	753 207	753 116	746 587	744 926	717 535
Maximum	753 207	753 207	747 019	745 017	718 809

It can be observed that, as the rolling resistance (RR) increases, the productivity variance also increases, resulting in spreader boxplots. For the scenarios with lower RR values, the speeds sampled in each stretch were very close to the upper limit of the distributions (speed limit of 50 km/h), thus reducing the variability. On the other hand, as the distributions were shifted to regions with lower speeds, the amplitude of the sampled values increased, resulting in spreader boxplots. This behaviour highlights the importance of maintaining the rolling resistance of haul roads at acceptable levels, otherwise poor-quality roads can significantly reduce fleet productivity. In addition, high RR values are associated with reduced equipment and component life, although this analysis is beyond the scope of this study.

CONCLUSION

This study emphasises the importance of roads as fundamental infrastructure for the success of mine planning. The increase in rolling resistance over time is inevitable; however, a comprehensive approach to road design can help mitigate these effects. This work provides a significant analysis of various aspects of road design, as well as the use of discrete event simulation as a tool in a practical case.

A deterministic approach to variables in road design may fail to account for uncertainties, potentially leading to operational losses in mining. The comparison between the deterministic and stochastic cases demonstrates that a theoretical approach does not always align with real-world conditions. The study revealed that as rolling resistance increases, mine productivity tends to decrease due to the performance of trucks. Comparing the average production of the 3 per cent RR scenario (real data for distribution curves) with the 6 per cent RR scenario shows a production drop of nearly 5 per cent. Additionally, as rolling resistance increases, the variability in truck performance may also rise.

For future studies, a more detailed approach to distribution curves could yield more representative results. Furthermore, other variables could be simulated using distribution curves. While rolling resistance in this study was treated as fixed for each scenario, an approach where this parameter varies over time could also be implemented.

REFERENCES

Alamdari, S, Basiri, M H, Mousavi, A and Soofastaei, A, 2022. Application of machine learning techniques to predict haul truck fuel consumption in open pit mines, *J Min Environ*, 13(1):69–85.

Caterpillar, 2019. Caterpillar Performance Handbook, vol 49, (Caterpillar: Peoria).

Chaowasakoo, P, Seppälä, H, Koivo, H and Zhou, Q, 2017. Digitalization of mine operations: Scenarios to benefit in real-time truck dispatching, *Int J Min Sci Technol*, 27(2):229–236.

Deslandes, J V and Dickerson, A W, 1989. A new concept for mine haul route surface, in *Proceedings of the International Symposium on Off-Highway Haulage in Surface Mines*, pp 247–254 (Balkema: Rotterdam).

Elbrond, J and Soumis, F, 1987. Towards integrated production planning and truck dispatching in open pit mines, *Int J Surf Min Reclam Environ*, 1(1):1–6.

Hustrulid, W, Kuchta, M and Martin, R, 2013. *Open Pit Mine Planning and Design*, e-book (CRC Press).

Koenigsberg, E, 1958. Cyclic queues, *J Oper Res Soc*, 9(1):22–35.

Lizotte, Y, Bonates, E and Leclerc, A, 1987. A design and implementation of a semi-automated truck/shovel dispatching system, in *Proceeding of the 20th APCOM* (eds: I C Lemmer, L Wade, H Schaum, J R Cutland, F A G M Camisani-Calzolari and R W O Kersten), pp 377–387 (South African Institute of Mining and Metallurgy: Johannesburg).

Moradi Afrapoli, A and Askari-Nasab, H, 2019. Mining fleet management systems: a review of models and algorithms, *Int J Min Reclam Environ*, 33(1):42–60.

Pooch, U W and Wall, J A, 1992. *Discrete Event Simulation: A Practical Approach*, 312 p (CRC Press: New York).

Soofastaei, A, Aminossadati, S M, Arefi, M M and Kizil, M S, 2016. Development of a multi-layer perceptron artificial neural network model to determine haul trucks energy consumption, *Int J Min Sci Technol*, 26(2):285–293.

Soumis, F, Ethier, J and Elbrond, J, 1989. Evaluation of the new truck dispatching in the Mount Wright mine, in *Proceeding of the 21st APCOM*, pp 674–682 (Society of Mining Engineers: Littleton).

Tannant, D D and Regensburg, B, 2001. *Guidelines for Mine Haul Road Design*, 401 p (University of British Columbia: Vancouver).

Temeng, V A, Otuonye, F O and Frendewey, J O, 1997. Real-time truck dispatching using a transportation algorithm, *Int J Surf Min Reclam Environ*, 11(4):203–207.

Thompson, R J and Visser, A T, 1999. Management of unpaved road networks on opencast mines, *Transp Res Rec*, 1652(1):217–224.

Thompson, R J and Visser, A T, 2000. The functional design of surface mine haul roads, *Journal of the South African Institute of Mining and Metallurgy*, 100(3):169–180.

Thompson, R J and Visser, A T, 2003. Mine haul road maintenance management systems, *Journal of the South African Institute of Mining and Metallurgy*, 103(5):303–312.

Thompson, R J and Visser, A T, 2006. The Impact of Rolling Resistance on Fuel, Speed and Costs, *HME: Continuous Improvement Case Studies*, pp 68–75.

Thompson, R J, 2011. Design, Construction and Maintenance of Haul Roads, in *SME Mining Engineering Handbook* (ed: P Darling), 3rd edn, 10.6:957–976.

Thompson, R J, Peroni, R L and Visser, A T, 2019. *Mining Haul Roads: Theory and Practice*, 245 p (CRC Press: London).

Topal, E and Ramazan, S, 2010. A new MIP model for mine equipment scheduling by minimizing maintenance cost, *Eur J Oper Res*, 207(2):1065–1071.

Vera-Burau, A, Álvarez-Ramirez, D, Sanmiquel, L and Bascompta, M A, 2023. Comparison of the fuel consumption and truck models in different production scenarios, *Appl Sci*, 13(9):5769.

White, J W and Olson, J P, 1986. Computer-based dispatching in mines with concurrent operating objectives, *Min Eng*, 38(11):1045–1054.

A mathematical model for production scheduling from uncertain resources to run-of-mine stockpiles and processing plant in the sub-level caving method

J Qu[1], C Xu[2] and P Dowd[3]

1. PhD student, The University of Adelaide, Adelaide SA 5005. Email: junlin.qu@adelaide.edu.au
2. Associate professor, The University of Adelaide, Adelaide SA 5005.
   Email: chaoshui.xu@adelaide.edu.au
3. Professor, The University of Adelaide, Adelaide SA 5005. Email: peter.dowd@adelaide.edu.au

ABSTRACT

Traditional sub-level caving (SLC) production scheduling models primarily focus on optimising draw control while overlooking the integration of post-extraction ore flow, including stockpiling and processing constraints. This omission can potentially lead to inefficiencies in ore blending, increased dilution, and suboptimal resource utilisation. To address this limitation, this study develops a mixed-integer programming (MIP) model that explicitly considers ore flow across the entire mining value chain. The proposed approach integrates a 3D block model, enabling more detailed constraint formulation and precise production sequencing compared to conventional 3D models. Furthermore, the model incorporates deleterious element constraints and stockpiling strategies, improving ore quality control and minimising contamination penalties. The model is implemented using the Gurobi optimiser and validated through a case study of an iron oxide copper-gold (IOCG) deposit. Results demonstrate that the proposed method enhances scheduling flexibility, increases mine operational efficiency, and improves ore quality management while maximising net present value (NPV).

INTRODUCTION

Sub-level caving (SLC) is a widely adopted underground mining method due to its high production capacity and low production cost. However, effective production scheduling in SLC is challenging due to constrained blasting conditions, unpredictable ore flow, and ore-waste mixing, which complicate ore recovery and grade control. Traditional scheduling models primarily focus on draw control optimisation, determining the sequence and timing of block extraction. While these models improve production efficiency, they often fail to integrate post-extraction ore flow, such as stockpile management and processing constraints, which play a crucial role in maintaining ore quality and minimising economic losses.

A well-structured production schedule is essential for optimising equipment utilisation, minimising costs, and ensuring consistent ore recovery. Various scheduling approaches have been developed, including manual planning, heuristic algorithms, and exact optimisation methods. While exact methods such as mixed-integer programming (MIP) can provide optimal solutions, many existing MIP-based models for SLC scheduling focus only on optimising the draw order of blocks without considering how extracted ore moves through the entire mining system, from the drawpoint to stockpiles and processing plants. This lack of integration can result in inefficient blending, excessive rehandling, and potential processing penalties due to high contaminant levels. Moreover, production scheduling, stockpiling, and blending in mining operations, which should function as an integrated system within the production process, are often studied in isolation. This fragmented approach undermines the overall effectiveness of the mining value chain.

To address these challenges, this study presents a mathematical production scheduling model that explicitly incorporates post-extraction ore flow management with production scheduling. Unlike conventional models, the proposed approach employs a 3D block model instead of a 2D representation, enabling more accurate constraint formulation at the production ring level. Furthermore, the model integrates stockpiling strategies and deleterious element constraints, optimising ore quality control and enhancing scheduling flexibility and applicability to real-world mining operations. By considering the production scheduling, stockpiling and blending, the model is capable of considering the ore flow-through the mining value chain, thereby improving the effectiveness and overall coherence of decision-making in mining operations.

The remainder of this paper is structured as follows: Section 2 reviews relevant literature on mathematical production scheduling models for SLC and the integration of stockpiling in scheduling optimisation. Section 3 presents the formulation of the proposed MIP model, which optimises production scheduling while accounting for ore flow, contamination penalties, and blending constraints. Section 4 describes the case study and data set used for model validation. Section 5 presents the results and demonstrates that the proposed approach improves NPV and ore quality control, reducing penalty costs associated with excessive contaminant levels.

LITERATURE REVIEW

Mathematical production scheduling models for SLC

Research on production scheduling for sub-level caving (SLC) mines remains relatively limited due to the complex constraints associated with this mining method. These constraints make scheduling more challenging compared to open pit mining and other underground mining methods. Despite the challenges, relevant research has been undertaken, particularly in the Kiruna mine, one of the world's largest underground mines, located in northern Sweden.

One of the earliest attempts to optimise underground mine production scheduling was made by Williams, Smith and Wells (1973), who introduced a linear programming (LP) approach for generating detailed production schedules in a sub-level stoping copper mine. Their model enabled planners to efficiently create initial schedules and make necessary adjustments, significantly improving upon traditional manual scheduling methods. A key advantage of mathematical programming is its ability to determine an optimal schedule that maximises NPV or minimises costs while ensuring adherence to operational constraints. In contrast, manual planning methods and heuristic algorithms, although capable of producing practical schedules, cannot guarantee optimality (Riddle, 1977).

Since 1994, several researchers have explored the use of MIP for SLC production scheduling, primarily focusing on minimising deviations from production targets. Almgren (1994) developed a mathematical model that used machine placement as the fundamental mining entity, incorporating both long-term and short-term scheduling requirements for the Kiruna mine. However, the model had limited practical applicability due to its unrealistic assumption that each block would be fully mined within a single time period.

To address the limitations of heuristic scheduling approaches, Kuchta, Newman and Topal (2004) introduced an MIP model aimed at generating optimal production schedules for SLC mines. A major advantage of this formulation was the reduction in the number of decision variables, which significantly lowered deviations from planned production, less than 5 per cent, compared to 10–20 per cent deviation observed in manual scheduling approaches. Building upon this work, Newman and Kuchta (2007) sought to enhance model tractability by incorporating machine allocation based on sequencing rules and analysing worst-case performance scenarios after data aggregation. Later, Newman and Martinez (2011) extended this model by integrating both long-term and short-term scheduling constraints. Using an optimisation-based decomposition algorithm, the refined model minimised deviations from pre-planned monthly production targets, improving overall scheduling accuracy.

More recently, Shenavar, Ataee-pour and Rahmanpour (2020) developed an MIP-based model to optimise SLC production scheduling with the objective of maximising NPV. Their model incorporated key technical and operational constraints, including production capacity, sub-level geometry, and mine development considerations. To enhance computational efficiency, the block model was preprocessed using the floating stope algorithm, which defined the ultimate mine boundary and reduced the number of decision variables. However, like most SLC scheduling models, their approach relied on a 2D block model, which does not fully account for mine development in 3D.

Building on these advancements, Khazaei and Pourrahimian (2023) proposed a mixed-integer linear programming (MILP) framework for long-term production scheduling optimisation in SLC mines. Their model aimed to maximise NPV while incorporating key operational constraints, including mine development activities, mining and processing capacities, continuous mining requirements, active mining units, grade blending, and sequencing. Similar to Shenavar, Ataee-pour and Rahmanpour

(2020), the work of Khazaei and Pourrahimian, emphasised the need for more detailed scheduling at the production ring level, as each mining unit consists of multiple production rings across the width of the orebody. Integrating a 3D block model at this level could further enhance the adaptability and applicability of the model to real-world mining operations.

Production scheduling optimisation with stockpiling and ore blending

Stockpiles play a crucial role in ore quality management by serving as both a buffer for operational variability and a blending mechanism, particularly for controlling levels of deleterious elements in the concentrate. However, the non-linearity introduced by stockpiling complicates its integration into traditional production scheduling models. Despite its importance, relatively few studies have examined the dynamic impact of stockpile blending on production scheduling, especially in the presence of deleterious elements. Most production planning models that incorporate post-extraction ore flow as part of the value chain have been developed for open pit mining, where stockpiling and blending are essential to meet grade and tonnage targets.

In 2012, Bley et al (2012) developed an MIP model for open pit mining to generate production schedules while managing material flow-through stockpiles. The proposed model combined multiple formulations with an aggressive branching scheme and a branch-and-bound algorithm to mitigate the complexity introduced by quadratic constraints.

Building on this work, Moreno et al (2017) proposed several approximate linear models that incorporated stockpiles into open pit mine planning. Their approach introduced the L-average, L-bound, and K-buckets models, which defined average stockpile grades, minimum grade constraints, and segmented stockpile compositions, respectively. However, these models did not fully account for the complexities of stockpile blending and material mixing.

In the same year, Tabesh and Askari-Nasab (2017) introduced a non-linear model to estimate stockpile grades and control the head grade of material sent to processing plants. To improve computational efficiency, the non-linear formulation was transformed into a linear model using piecewise linearisation and solved using MILP techniques. A key limitation of existing stockpiling studies is that most models consider only a single element of interest, failing to account for the effects of multiple elements on ore blending and processing.

Recently, Rezakhah, Moreno and Newman (2020) extended the L-average bound model to incorporate multiple metals in stockpile blending. Different from previous studies, the author regarded gold as a copper-equivalent mineral, simplifying the model and accelerating solution time. However, in cases where deleterious elements must be strictly controlled, converting contaminants into equivalent elements may increase model complexity and affect grade and tonnage control.

Despite these advancements, existing SLC scheduling models largely overlook the integration of post-extraction ore flow, leading to inefficiencies in blending, stockpile management, and contaminant control. This study addresses these gaps by:

- Explicitly incorporating ore flow into SLC scheduling, improving stockpiling strategies and blending optimisation.

- Employing a 3D block model, allowing for more precise constraint formulation at the ring level.

- Integrating deleterious elements in modelling, improving ore quality control and minimising processing penalties.

By addressing these aspects, this study enhances mine planning flexibility, improves operational efficiency, and optimises ore quality control in SLC operations.

METHODOLOGY

Model formulation

Notation

Indices and sets

$i \in \{1, \ldots, I\}$	Index for stopes within the mining layout in the horizontal direction.
$k \in \{1, \ldots, K\}$	Index for stopes within the mining layout in the vertical direction.
$d \in \{1, \ldots, D\}$	Index for rings in stope i,j in the longitudinal direction.
$j \in \{1, \ldots, J\}$	Index for stockpile ID in the mining value chain with a total of J stockpiles.
$t \in \{1, \ldots, T\}$	Index for time periods in the production schedule.
$e \in \{1, \ldots, E\}$	Index for elements representing attributes of gold, copper, and uranium.

Parameters

I	Number of stopes within the level in the horizontal direction.
K	Number of levels in the vertical direction.
D	Number of rings in stope i,j in longitudinal direction.
J	Total number of stockpiles in the mining value chain.
T	Number of time periods in the production schedule.
E	Number of elements in scheduling models to be considered.
V_e^t	Revenue; generated by selling the final product from element e in period t.
MC_+^t	Upper bound on the mining capacity in period t.
MC_-^t	Lower bound on the mining capacity in period t.
PC_+^t	Maximum tonnage allowed to be sent to plant in period t.
PC_-^t	Minimum tonnage allowed to be sent to plant in period t.
L^t	The maximum number of active levels available for mining in period t.
Q^t	The maximum number of active rings available for mining in period t.
$g_{i,k,d}^e$	Average grade of element e in the ring indexed i,k,d.
$w_{i,k,d}$	The ore tonnage of ring indexed i,k,d.
wbR_j^t	The ore tonnage status of the stockpile indexed i,k,d before implementing reclamation in period t.
waR_j^t	The ore tonnage status of the stockpile indexed i,k,d after implementing reclamation in period t.
$gbR_j^{t,e}$	The average grade of element e in the stockpile indexed i,k,d before implementing reclamation in period t.
$gaR_j^{t,e}$	The average grade of element e in the stockpile indexed i,k,d after implementing reclamation in period t.
$wDev_{stope,+}^t$	Continuous variable; the excess over the stope production tonnage upper bound in period t.
$wDev_{stope,-}^t$	Continuous variable; the shortfall from the stope production tonnage lower bound in period t.

$gDev_{p,+}^{t,e}$	Continuous variable; the grade excess of the metal e sent to the processing plant over the grade upper bound in period t.
$gDev_{p,-}^{t,e}$	Continuous variable; the metal e grade (the ore sent to the processing plant) shortfall from the lower bound in period t.
$wDev_{p,+}^{t}$	Continuous variable; the ore tonnage (the ore sent to the processing plant) excess over the ore tonnage upper bound in period t.
$wDev_{p,-}^{t}$	Continuous variable; the shortfall from the ore tonnage lower bound of the ore sent to the processing plant in period t.
$gb_{p,+}^{t,e}$	Upper bound on the allowable grade of element e dispatched to the processing plant p in period t.
$gb_{p,-}^{t,e}$	Lower bound on the allowable grade of element e dispatched to the processing plant p in period t.
$wb_{stope,+}^{t}$	Upper bound on the allowable tonnage produced from stopes in period t.
$wb_{stope,-}^{t}$	Lower bound on the allowable tonnage produced from stopes in period t.
$wc_{stope,+}^{t}$	per-unit penalty cost for the upper ore tonnage target deviation of the ore tonnage produced from the stopes in period t.
$wc_{stope,-}^{t}$	per-unit penalty cost for the lower target deviation of the ore tonnage produced from the stopes in period t.
$wc_{p,+}^{t}$	per-unit penalty cost for the upper ore tonnage target deviation (the ore dispatched to the processing plant) in period t.
$wc_{p,-}^{t}$	per-unit penalty cost for the lower ore tonnage target deviation (the ore dispatched to the processing plant) in period t.
$gc_{p,+}^{t,e}$	per-unit penalty cost in terms of the upper grade target deviation on the element e sent to the processing plant in period t.
$gc_{p,-}^{t,e}$	per-unit penalty cost in terms of the upper grade target deviation on the element e sent to the processing plant in period t.
$CostO_{ikd}^{t}$	Cost per tonne in terms of mining for stope i,k,d in period t.
$CostR_{j}^{t}$	Cost per tonne in terms of reclaiming from stockpile j in period t.
h	The financial discount rate.

Decision Variables

$O_{ikd,p}^{t}$	Continuous variable: the tonnage of ore extracted from ring i,k,d to the processing plant in period t.
$O_{ikd,j}^{t}$	Continuous variable: the tonnage of ore extracted from ring i,k,d to the stockpile j in period t.
R_{j}^{t}	Continuous variable: the tonnage of ore reclaimed from the stockpile j and sent to the processing plant in period t.
$A_{i,k,d}^{t}$	Binary variable: representing the ring i,k,d is being mined in period t.

Objective function

$$\max \left\{ \sum_{t=1}^{T}\sum_{i=1}^{I}\sum_{k=1}^{K}\sum_{d=1}^{D}\sum_{j=1}^{J}\sum_{e=1}^{E}\left[O_{ikd,p}^{t} \times g_{i,k,d}^{e} \times \left(V_e^t - CostO_{ikd}^t \right) + R_j^t \times gbR_j^{t,e} \times \left(V_e^t - CostR_j^t \right) \right. \right.$$
$$\left. - O_{ikd,j}^t \times CostO_{ikd}^t \right] \times (1+h)^{-t} \right] \, part1 -$$

$$\sum_{t=1}^{T}\left(wc_{stope,+}^t \times wDev_+^{t,stope} + wc_{stope,-}^t \times wDev_-^{t,stope} \right) \times (1+h)^{-t} \, part \, 2 - \tag{1}$$

$$\sum_{t=1}^{T}\sum_{i=1}^{I}\sum_{k=1}^{K}\sum_{d=1}^{D}\sum_{j=1}^{J}\sum_{e=1}^{E}\left[\left(gc_{p,+}^{t,e} \times wDev_{p,+}^t + gc_{p,-}^{t,e} \times wDev_{p,-}^t \right) \times \left(O_{ikd,p}^t + O_{ikd,j}^t \right) \right.$$
$$\left. + \left(wc_{p,+}^t \times gDev_{p,+}^{t,e} + wc_{p,-}^t \times gDev_{p,-}^{t,e} \right) \times (1+h)^{-t} \right] part3 \right\}$$

Constraints

$$MC_-^t \le O_{ikd,p}^t + O_{ikd,j}^t \le MC_+^t \qquad\qquad \forall t\epsilon\{1,\dots,T\} \tag{2}$$

$$PC_-^t \le O_{ikd,p}^t + R_{j,p}^t \le PC_+^t \qquad\qquad \begin{array}{l}\forall i\epsilon\{1,\dots,I\}, \forall k\epsilon\{1,\dots,K\}, \\ \forall d\epsilon\{1,\dots,D\}, \forall j\epsilon\{1,\dots,J\}\end{array} \tag{3}$$

$$\sum_{t=1}^{T}\left(O_{ikd,p}^t + O_{ikd,j}^t \right) \le w_{i,k,d} \qquad\qquad \begin{array}{l}\forall i\epsilon\{1,\dots,I\}, \forall k\epsilon\{1,\dots,K\}, \\ \forall d\epsilon\{1,\dots,D\}\end{array} \tag{4}$$

$$\sum_{t=1}^{T} R_{j,p}^t \le wbR_j^t \qquad\qquad \forall j\epsilon\{1,\dots,J\} \tag{5}$$

$$waR_j^t = wbR_j^t - R_{j,p}^t \qquad\qquad \forall j\epsilon\{1,\dots,J\} \tag{6}$$

$$waR_j^t = wbR_j^{t+1} \qquad\qquad \forall j\epsilon\{1,\dots,J\}, \forall e\epsilon\{1,\dots,E\} \tag{7}$$

$$gbR_j^{t,e} = gaR_j^{t-1,e} \qquad\qquad \begin{array}{l}\forall t\epsilon\{2,\dots,T\}, \forall j\epsilon\{1,\dots,J\}, \\ \forall e\epsilon\{1,\dots,E\}\end{array} \tag{8}$$

$$gaR_j^{t,e} = \frac{gbR_j^{t,e} \times wbR_j^t - R_{j,p}^t \times gbR_j^{t,e}}{wbR_j^t - R_{j,p}^t} \qquad\qquad \begin{array}{l}\forall t\epsilon\{1,\dots,T\}, \forall j\epsilon\{1,\dots,J\}, \\ \forall e\epsilon\{1,\dots,E\}\end{array} \tag{9}$$

$$\sum_{i=1}^{I}\sum_{k=1}^{K}\sum_{d=1}^{D}\left(O_{ikd,p}^t + O_{ikd,j}^t \right) - wDev_{stope,+}^t \le wb_{stope,+}^t \qquad \forall j\epsilon\{1,\dots,J\} \tag{10}$$

$$\sum_{i=1}^{I}\sum_{k=1}^{K}\sum_{d=1}^{D}\left(O_{ikd,p}^t + O_{ikd,j}^t \right) + wDev_{stope,-}^t \ge wb_{stope,-}^t \qquad \forall t\epsilon\{1,\dots,T\}, \forall j\epsilon\{1,\dots,J\} \tag{11}$$

$$\sum_{i=1}^{I}\sum_{k=1}^{K}\sum_{d=1}^{D}\left(O_{ikd,p}^t + R_{j,p}^t \right) - wDev_{p,+}^t \le wb_{p,+}^t \qquad \forall t\epsilon\{1,\dots,T\}, \forall j\epsilon\{1,\dots,J\} \tag{12}$$

$$\sum_{i=1}^{I}\sum_{k=1}^{K}\sum_{d=1}^{D}\left(O_{ikd,p}^t + R_{j,p}^t \right) + wDev_{p,-}^t \ge wb_{p,-}^t \qquad \begin{array}{l}\forall t\epsilon\{1,\dots,T\}, \forall j\epsilon\{1,\dots,J\}, \\ \forall e\epsilon\{1,\dots,E\}\end{array} \tag{13}$$

$$\sum_{i=1}^{I}\sum_{k=1}^{K}\sum_{d=1}^{D}\sum_{e=1}^{E}\left(g_{i,k,d}^e \times O_{ikd,p}^t + gbR_j^{t,e} \times R_{j,p}^t \right) - gb_{p,+}^{t,e} \qquad \begin{array}{l}\forall t\epsilon\{1,\dots,T\}, \forall j\epsilon\{1,\dots,J\}, \\ \forall e\epsilon\{1,\dots,E\}\end{array} \tag{14}$$
$$\times \left(O_{ikd,p}^t + R_{j,p}^t \right) - gDev_{p,+}^{t,e} \le 0$$

$$\sum_{i=1}^{I}\sum_{k=1}^{K}\sum_{d=1}^{D}\sum_{e=1}^{E} gb_{p,-}^{t,e} \times (O_{ikd,p}^{t} + R_{j,p}^{t}) - (g_{i,k,d}^{e} \times O_{ikd,p}^{t} + gbR_{j}^{t,e} \times R_{j,p}^{t}) - gDev_{p,-}^{t,e} \leq 0 \qquad \forall t\epsilon\{1,\dots,T\}, \forall j\epsilon\{1,\dots,J\} \qquad (15)$$

$$A_{i,k,d}^{t} \leq A_{i+n,k-1,d}^{t} \qquad \forall t\epsilon\{1,\dots,T\}, \forall i\epsilon\{1,\dots,I\}, \forall k\epsilon\{1,\dots,K\}, \forall d\epsilon\{1,\dots,D\} \qquad (16)$$

$$A_{i,k,d}^{t} \leq A_{i-1,k,d}^{t} \qquad \forall t\epsilon\{1,\dots,T\}, \forall i\epsilon\{1,\dots,I\}, \forall k\epsilon\{1,\dots,K\}, \forall d\epsilon\{1,\dots,D\} \qquad (17)$$

$$A_{i,k,d}^{t} \leq A_{i,k,d-1}^{t} \qquad \forall t\epsilon\{1,\dots,T\}, \forall i\epsilon\{1,\dots,I\}, \forall k\epsilon\{1,\dots,K\}, \forall d\epsilon\{1,\dots,D\} \qquad (18)$$

$$A_{i,k,d}^{t} \leq A_{i-1,k,d+1}^{t} \qquad \forall t\epsilon\{1,\dots,T\}, \forall i\epsilon\{1,\dots,I\}, \forall k\epsilon\{1,\dots,K\}, \forall d\epsilon\{1,\dots,D\} \qquad (19)$$

$$A_{i,k,d}^{t} \leq A_{i,k-1,d}^{t} \qquad \forall t\epsilon\{1,\dots,T\}, \forall i\epsilon\{1,\dots,I\}, \forall k\epsilon\{1,\dots,K\}, \forall d\epsilon\{1,\dots,D\} \qquad (20)$$

$$\sum_{i=1}^{I} A_{i,k,d}^{t} \leq L^{t} \qquad \forall t\epsilon\{1,\dots,T\}, \forall k\epsilon\{1,\dots,K\}, \forall d\epsilon\{1,\dots,D\} \qquad (21)$$

$$\sum_{i=1}^{I}\sum_{k=1}^{K}\sum_{d=1}^{D} A_{i,k,d}^{t} \leq Q^{t} \qquad \forall t\epsilon\{1,\dots,T\} \qquad (22)$$

The objective function maximises the discounted net profit derived from directly sending ore from the stopes to the processing plant and reclaiming ore from the stockpile while deducting penalties associated with tonnage and grade violations beyond the specified soft bounds. Equations 2 and 3 impose hard constraints that restrict the minimum and maximum extraction and processing capacities for each period. In this model, hard constraints represent strict limits that cannot be violated, whereas soft constraints allow decision variables to deviate from the prescribed limits but impose penalty costs proportional to the degree of violation. Equations 4 and 5 ensure that the total tonnage of ore extracted from production rings and reclaimed from stockpiles does not exceed their respective reserves. Equations 6–9 ensure the accurate delivery of the grade and tonnage status in stockpiles from period to period. If the stockpile is empty initially, the wbR_{j}^{t} and $gbR_{j}^{t,e}$ are equal to the tonnage and ore grade sent to the stockpile in this period. Otherwise, the initial tonnage and parameter should be considered as the waR_{j}^{0} and $gaR_{j}^{0,e}$, and then calculate with $O_{ikd,j}^{1}$ as the result of wbR_{j}^{1} and $gbR_{j}^{1,e}$. It is also noticeable that the frequency of calculating the grade and tonnage status in stockpiles depends on the times of stockpiling and reclamation in the value chain, which makes the proposed approach suitable to the model characterised by generating real-time optimisation strategy and the system equipped with ore-tracking technologies.

Equations 10–13 are flexible constraints for the total tonnage extracted and processed in each period. Bounds $wDev_{stope,+}^{t}$, $wDev_{stope,-}^{t}$, $wDev_{p,+}^{t}$, and $wDev_{p,-}^{t}$ are used as buffers to allow tonnage deviations but financially penalise the excess and shortfall in the objective function for producing a more stable production schedule. $wb_{stope,+}^{t}$, $wb_{stope,-}^{t}$, $wb_{p,+}^{t}$, and $wb_{p,+}^{t}$ are the two pairs of soft bounds for ore tonnage mined and sent to the processing plant in period t, respectively.

Likewise, Equations 14 and 15 impose flexible constraints to stabilise fluctuations in ore grade dispatched to the processing plant, ensuring it remains within an acceptable range. Given the characteristics of SLC, a downward mining method, Equation 16 enforces vertical access constraints, specifying that a minimum number of overlying sublevel stopes must be extracted before mining of the underneath sublevel stopes can commence. Similarly, Equation 17 defines horizontal access constraints, requiring a minimum number of blocks to be extracted to gain access to a specific stope. This constraint also governs the mining sequence within a sub-level. Considering the ring production sequence within a stope, shown in Figure 1, Equation 18 defines the longitudinal access constraints within a stope. This constraint also governs the mining sequence within a stope. Equation 19 defines that for any two adjacent stopes in a level, ring d in the stope i must be extracted before the corresponding rings in the stope $i + 1$. Equation 20 ensures that during the mining process of rings i, k, d, the production process of ring $i, k - 1, d$ have completed. Equations 21 and 22 defines the maximum number of active levels and rings in each period.

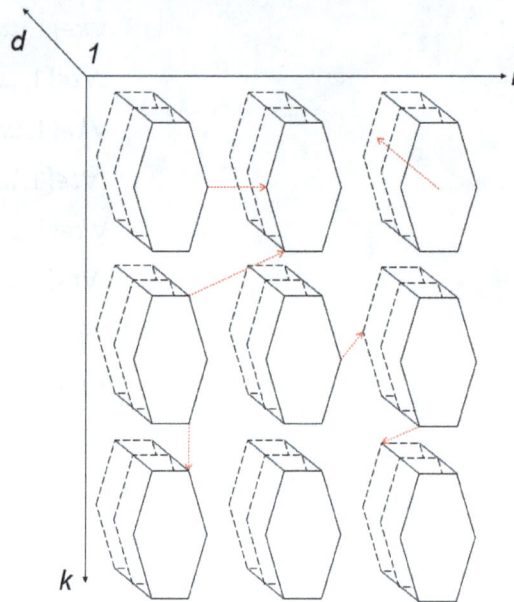

FIG 1 – Schematic representation of indexes and constraints (the figure is for illustration only).

Unlike most existing SLC production scheduling models, which formulate constraints based on a lateral 2D block model, the proposed MIP model considers the production scheduling in the ring level. Traditional 2D-based scheduling approaches often overlook equipment availability constraints and charging operation delays, which can impact the feasibility of the schedule. By incorporating 3D operational constraints, the proposed model enables a more detailed and practical scheduling strategy, improving operational efficiency.

CASE STUDY – IRON OXIDE COPPER-GOLD DEPOSIT

Data from an operating mine

The proposed MIP model is validated using a case study of an IOCG deposit. This MIP framework presented in section 3 can produce a mining schedule considering multiple elements, SLC operational features based on a 3D block model, and the ore flow in the whole mining value chain, including blending and stockpiling. The goal is to maximise the NPV of the mining operation.

The main copper-uranium-gold minerals in the deposit are chalcopyrite, bornite, and uraninite. Using the available geological information, one mineralised domain is interpreted and modelled, which contains 27 Mt of ore, with average grades of Cu, Au and U at 1.4 per cent, 0.71 g/t and 160 g/t, respectively. The mine design incorporates 25 m sub-level spacing, 5 m wide production drives at a spacing of 15 m (centre to centre) and a standard SLC ring design. Figure 2 show the grade distributions of copper, gold, and uranium. These distributions reveal the heterogeneous nature of the orebody, which suggests the need for ore blending in the mining system to stabilise the mill feed grades and optimise the production schedules.

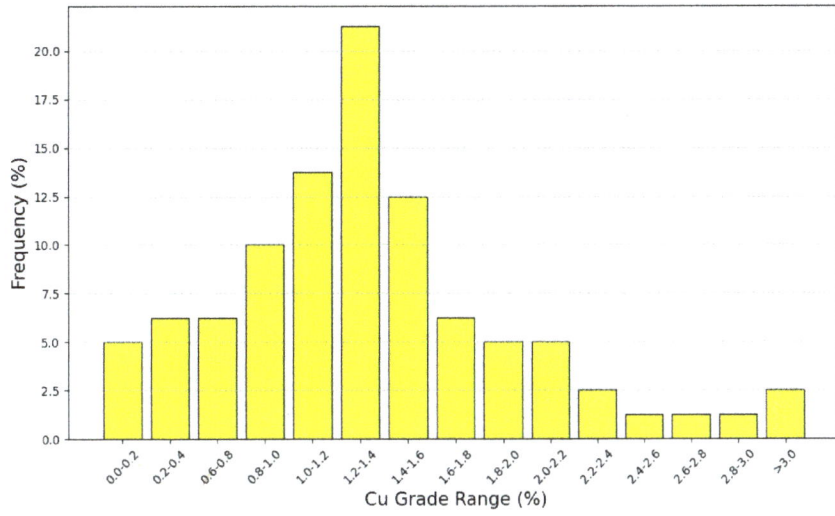

(a)

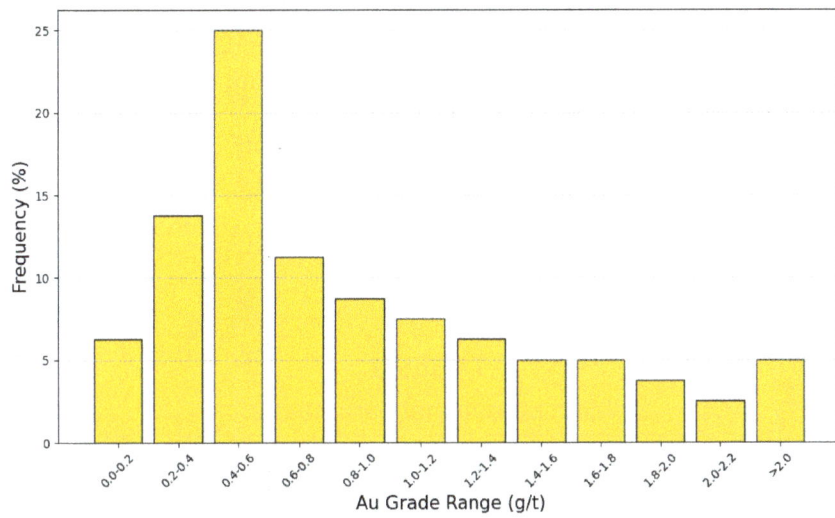

(b)

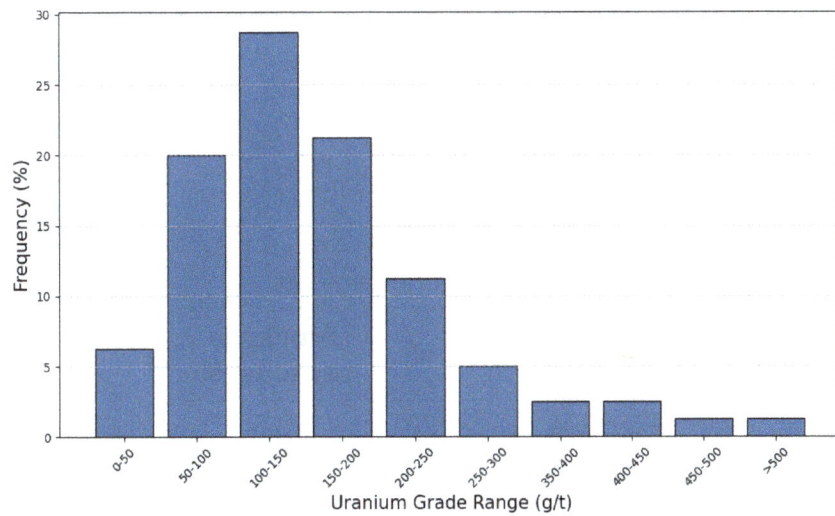

(c)

FIG 2 – (a) Copper grade distribution of all stopes in the stope layout; (b) gold grade distribution of all stopes in the stope layout; (c) uranium grade distribution of all stopes in the stope layout.

From this mineralised domain, we consider a total of four levels and four stopes in each level. A total *in situ* ore tonnage is 3.52 Mt and average grades of Cu, Au, and U are 1.35 per cent, 0.71 g/t and 160 g/t, respectively. Each stope is subdivided into five rings, each ring contains 44 000 t of ore, and such a process generates a total of 80 rings. The economic parameters used in this study are listed in Table 1. Mine operational parameters and stockpile grade arrangements are listed in Tables 2 and 3, respectively. In Table 3, the assumed Cu, Au and U grades of the reclaimed material from each stockpile are also listed. To demonstrate the benefit of integrating ore flow in the whole value chain, the proposed approach is compared to a conventional method that does not consider the ore flow-through the stockpiles, and the feed tonnage is fixed to 352 000 t each period.

TABLE 1
Mine economic parameters.

Economic parameter	
Copper selling price, $/t	8300
Gold selling price, $/g	51.3
Mining costs, $/t	44.16
Processing costs, $/t	15.13
Reclamation costs, $/t	1.29
Gold recovery, %	65
Copper recovery, %	84
Uranium recovery, %	16
Discount factor, %	10

TABLE 2
Mine operational parameters.

Operational parameters	Min	Max
Mining capacity, tonnes	300 000	450 000
Processing capacity, tonnes	300 000	350 000
Feed tonnage, tonnes	352 000	352 000
Copper feed grade, %	1.0	1.5
Uranium feed grade, g/t	0	140
Gold feed grade, g/t	0.9	1.0

TABLE 3
Stockpiling arrangements.

Stockpile	U, g/t	Stockpile capacity, Mt
Stockpile 1	<120	0.5
Stockpile 2	<300	0.5
Stockpile 3	<400	0.5

Three stockpiles with different acceptable ranges of uranium for stockpiling and blending are modelled to make the ore grade and tonnage acceptable to the processing plant. Due to the consideration of multiple elements and polarised grade distribution in some rings, acceptable uranium grade in stockpiles is divided into three different but continuous ranges, and the model will generate the acceptable grade for the other two elements. This definition aims to test the model's flexibility and adaptability.

Computational results

The optimisation system described in Section 3 was applied to the case discussed above, with ten time intervals modelled, each corresponding to one calendar year. The proposed MIP model comprised 137 225 constraints, 31 570 continuous variables, and 8855 integer variables, reflecting its comprehensive treatment of ore flow and operational constraints. In contrast, the conventional model was substantially less complex, with only 18 260 constraints, 11 695 continuous variables, and 8845 integer variables, reflecting its limited scope focused solely on draw sequencing.

The models were solved using the Gurobi Optimiser (Gurobi Optimisation, LLC, 2024) within a Python-based environment, ensuring efficient and accurate optimisation for this complex scheduling problem. The computation time required for each run was 334.19 seconds (337.43 work units) on a desktop equipped with an Intel (R) Core™ i7–4790 CPU @ 3.60 GHz and 16GB RAM. The models were solved to a 6 per cent optimality gap.

Figure 3 and Table 4 illustrate the dynamic ore flow patterns across the mine value chain produced by the proposed model, revealing how blending and stockpiling are leveraged to balance processing requirements and ore variability. As shown in Table 4, over the ten production periods, the total produced ore tonnage exhibits moderate fluctuations, ranging from approximately 300 000 to 440 000 t. As reserves are progressively depleted, total ore flow approaches the lower operational bound by period 10.

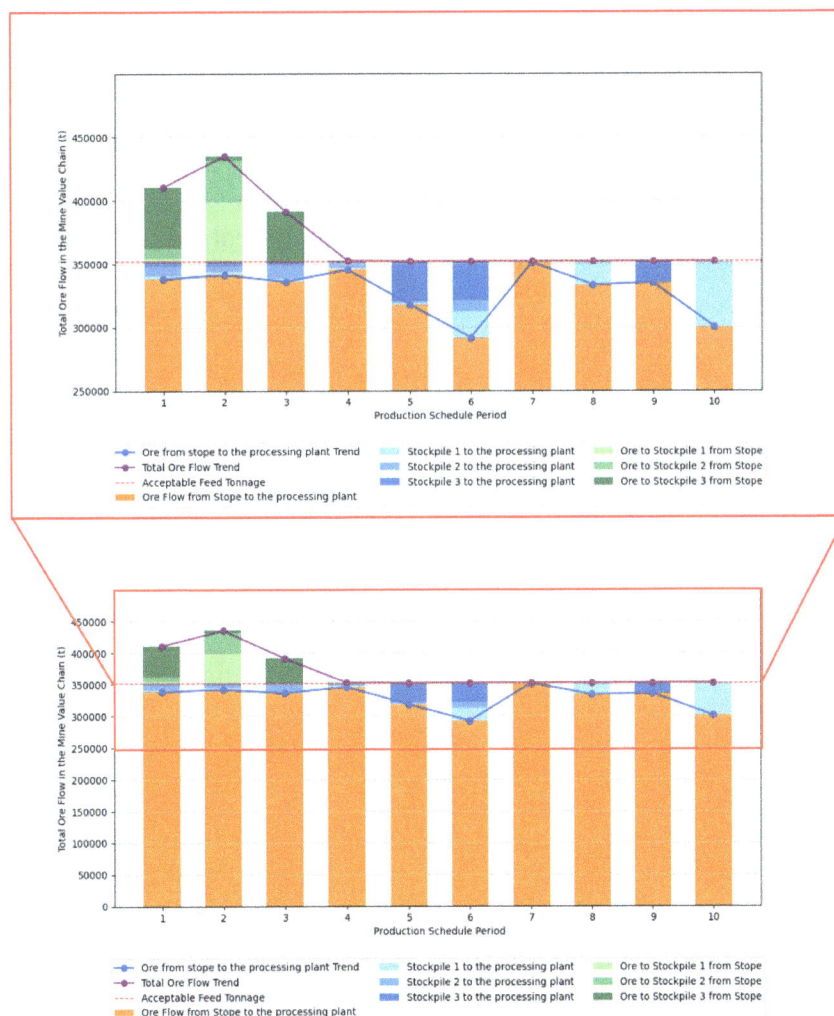

FIG 3 – Schematic representation of the total ore flow-through the mine value chain generated by the proposed models.

TABLE 4
Ore flow summary table.

Time	U grade mined, g/t	Tonnage mined	Tonnage sent to the processing plant	Total tonnage reclaimed	Average blended U fed to the mill, g/t	Reduction in U from blending, g/t
1	152	440 000	352 000	12 398	140	12
2	164	423 864	352 000	10 991	140	24
3	184	374 631	352 000	16 451	140	44
4	142	345 589	352 000	6848	140	2
5	155	317 584	352 000	34 413	140	15
6	171	300 000	352 000	60 401	140	31
7	140	350 673	352 000	1326	140	0
8	148	332 982	352 000	19 017	140	8
9	147	334 674	352 000	17 325	140	7
10	164	300 000	352 000	52 000	140	24

Initially, underground stopes serve as the dominant ore source. However, starting in period 4, a notable increase in the proportion of the ore reclaimed from stockpiles and delivered to the processing plant is observed. This transition demonstrates the model's strategic foresight in leveraging early-stage extraction (periods 1 to 3), during which it proactively increases extraction to build-up stockpile buffers. These buffers, in turn, provide greater operational flexibility in later stages, particularly for regulating multi-element feed grades. Stockpile reclamation volumes peak in periods 6 and 10, highlighting their role as buffers that stabilise both throughput and grade consistency. This strategic deployment of reclaimed ore becomes especially valuable when extraction sequencing constraints limit immediate access to desirable grade material.

Different uranium thresholds across the three stockpiles induce distinct reclamation behaviours. For instance, Stockpile 3, with its higher allowable U content, serves as a dedicated sink for uranium-rich material, mitigating the risk of U-grade violations. Simultaneously, the model maintains the copper and gold grade consistency required by the processing facility.

To highlight the impact of blending strategies, Table 4 compares uranium grades at different stages. The data clearly show a consistent reduction in uranium grade, from mined levels exceeding 150 g/t to a uniform feed grade of 140 g/t, demonstrating the model's effectiveness in managing deleterious elements through time-phased blending. As the same orebody is mined in both cases, the reduction in uranium feed grade is purely attributable to the optimised schedule, which incorporates mining, stockpiling, and blending decisions to reduce the variability of uranium content in the feed. This scheduling flexibility enables the model to bring the uranium level under the pre-defined acceptable range, thereby minimising financial penalties.

Figure 4 presents the ore flow generated by the conventional model. Without stockpiling mechanisms, the model struggles to maintain stable feed grades due to rigid extraction sequencing and polarised ring grades. Unlike the proposed approach, which achieves smoother and constraint-compliant grade profiles, the conventional model exhibits pronounced volatility in uranium content, resulting in substantial penalty costs. This contrast underscores the proposed model's key advantage, its ability to decouple extraction and processing through intermediate stockpile control.

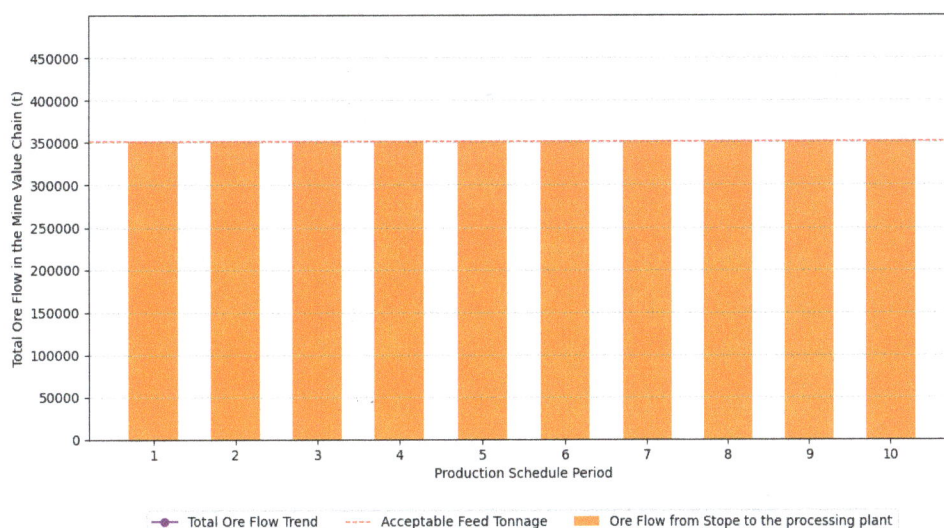

FIG 4 – Schematic representation of the total ore flow-through the mine value chain generated by the conventional model.

The proposed model achieves an optimal NPV of approximately $137 million, representing a 29 per cent increase over the $97 million attained using the conventional model. This improvement is primarily driven by strategic blending that eliminates uranium-related penalties ($64 million) and significantly reduces gold and copper violations. While the proposed model incurs minor penalties on Au and completely mitigates the penalty on U and Cu, it effectively manages grade distributions through proactive stockpiling and blending. In contrast, the conventional model, constrained by direct dispatch without blending, suffers a $40 million NPV loss and demonstrates inferior feed grade stability.

This discrepancy is mainly due to the lack of a buffering mechanism to manage ore grade before dispatch to the processing plant, as evidenced by a $64 million uranium grade penalty cost and $3.6 and $4.5 million copper and gold grade penalty costs, respectively, in the conventional model. In contrast, the proposed model effectively manages element grades sent to the plant and thus achieves higher NPV and a more stable feed grade.

The results demonstrated by Table 5 clearly reveal a significant increase in NPV when stockpiling is included. Since the conventional model does not account for blending, it attempts to construct a series of optimal linear tonnage combinations from accessible rings to meet plant constraints, resulting in only marginal violations of Cu and Au grade bounds. However, this comes at the cost of a $64 million uranium grade penalty due to the lack of a buffering and blending mechanism in the conventional model, which fails to control the variability in deleterious element content. The proposed model significantly decreases the cost of uranium feed grade penalty by storing and blending the ore in the stockpiles and simultaneously seeks a stable Cu and Au feed grade. From the perspective of generating an optimal ore flow strategy, the proposed MIP model significantly outperforms its conventional counterpart.

TABLE 5

NPV, undiscounted operational costs and revenues of the optimal production schedules.

	Proposed MIP model	Conventional model
Cu revenue ($ million)	329	329
Au revenue ($ million)	109	109
Cost ($ million)	213	211
U penalty costs ($ million)	0	64
Cu penalty ($ million)	0	3.6
Au penalty ($ million)	3.5	4.5
NPV ($ million)	137	97

Table 6 further highlights that the proposed model, through dynamic stockpile blending, achieves an average uranium feed grade of 140 g/t, which is approximately 10 per cent lower than in the conventional model, while maintaining desirable Cu (1.28 per cent) and Au (0.88 g/t) feed grades. This balance across the three elements showcases the model's capacity to simultaneously control both deleterious and valuable metals. The gold feed grade falls below the target lower bound only in periods 9 and 10. In comparison, although the conventional model achieves higher average copper and gold feed grades, it fails to meet the upper and lower bounds for all elements in more periods, which is not financially optimal. More importantly, the proposed model successfully controls the uranium feed grade within acceptable limits across all periods, which is an outcome not achieved by the conventional model. The results presented in Table 6 further demonstrate the proposed optimisation system's capability to mitigate the negative impact of uranium contamination while deriving an optimal mining, stockpiling, and blending strategy.

TABLE 6

Feed grades to the processing plant for both the proposed and conventional model.

time	Consider stockpiling and blending			Without stockpiling and blending		
	Cu %	Au g/t	U g/t	Cu %	Au g/t	U g/t
1	1.50	0.90	140	1.31	0.91	140
2	1.34	0.94	140	1.47	0.90	163
3	1.31	0.90	140	1.26	0.90	170
4	1.50	0.90	140	1.25	0.95	140
5	1.07	0.90	140	1.30	0.96	198
6	1.07	0.92	140	1.53	0.94	176
7	1.34	0.97	140	1.37	1.03	129
8	1.29	0.90	140	1.08	0.75	149
9	1.35	0.84	140	1.64	1.14	140
10	1.04	0.62	140	1.19	0.79	164
average	1.28	0.88	140	1.34	0.93	157

Figures 5 to 7 illustrate the grade discrepancies between material extracted from the stopes and that delivered to the processing plant for copper, gold, and uranium, respectively. As depicted in Figure 5, the proposed model achieves a notably more stable copper feed grade profile across all periods. This outcome is particularly significant given that copper constitutes the primary source of revenue in the operation. Accordingly, the model prioritises maintaining the copper feed grade within its specified upper and lower bounds. For example, in period 4, the model avoids a grade violation by strategically diverting high-grade ore to the stockpile, thereby preserving compliance. In contrast, the conventional model fails to meet the upper bound constraints in periods 6 and 9, incurring penalty costs.

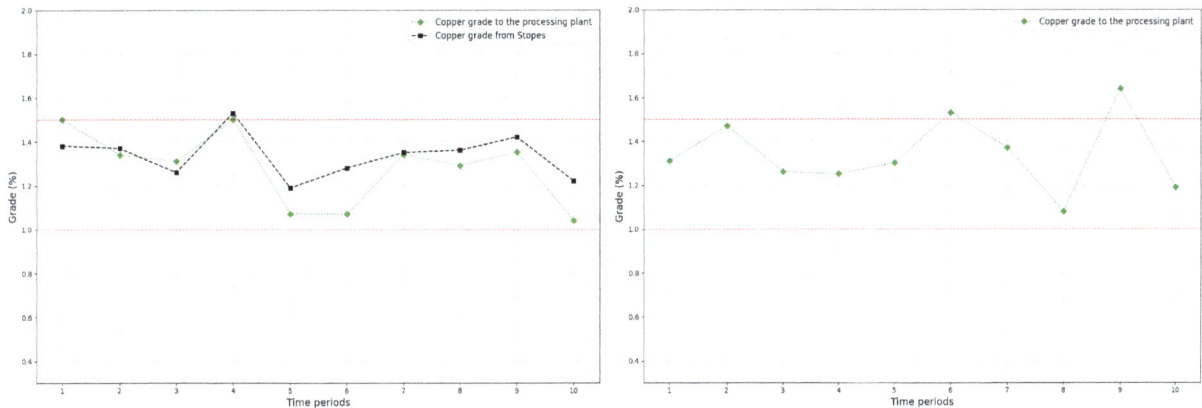

FIG 5 – Grade gap between copper grade extracted from stopes and sent to the processing plant (MIP (left) versus conventional model (right)).

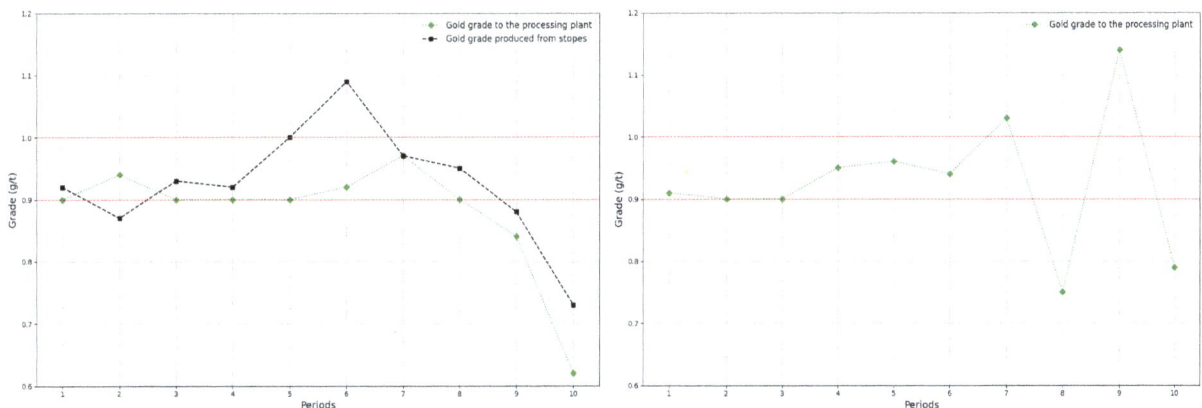

FIG 6 – Grade gap between gold grade extracted from stopes and sent to the processing plant (MIP (left) versus conventional model (right)).

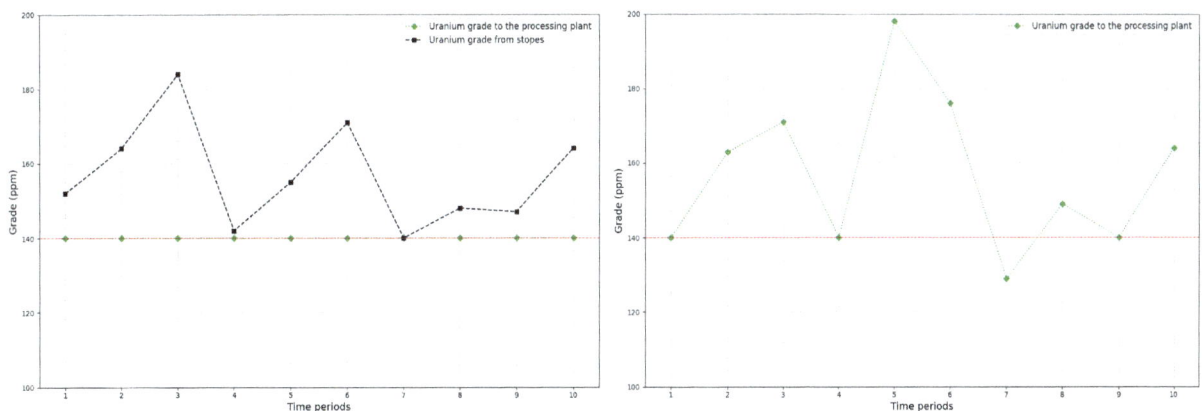

FIG 7 – Grade gap between Uranium grade extracted from stopes and sent to the processing plant (MIP (left) versus conventional model (right)).

Figure 6 presents the gold grade gap and again shows reduced volatility under the proposed MIP-based model. Importantly, the acceptable grade range for gold is extremely narrow, just 0.1 g/t, which makes strict compliance particularly challenging. The model demonstrates its effectiveness by avoiding penalties in periods 2 and 6, while deferring unavoidable violations to the final periods. These late-stage penalties are less detrimental in economic terms due to the application of a discount rate, which diminishes their present value. This strategic timing reflects the model's ability to balance constraint satisfaction with economic optimisation.

Finally, Figure 7 shows a significant reduction in fluctuations of the uranium feed grade under the proposed model. In this case, the proposed model successfully maintains uranium grades within the plant's acceptable range across all periods, entirely avoiding penalty costs. Notably, this performance is achieved without compromising the stability of copper and gold feed grades. This comprehensive balance across multiple elements underscores the model's distinctive capability to harmonise feed grade requirements in multi-element ore systems, something conventional models are structurally ill-equipped to achieve. When considered alongside the ore flow chart (Figure 3) and the NPV analysis (Table 5), these results demonstrate that the proposed model consistently delivers superior grade control and economic outcomes, even in the face of inherent grade heterogeneity.

CONCLUSIONS

In this study, an MIP model was developed for production scheduling in SLC operations, with a particular focus on ore flow, stockpiling, and blending across the full mining value chain. By incorporating a 3D block model, the approach allows more detailed scheduling decisions at the ring level and better reflects the practical constraints faced in underground mining operations.

The case study on an IOCG deposit demonstrated that integrating post-extraction ore flow into the scheduling process can significantly improve production outcomes. The computational results demonstrate that the proposed model not only achieves a 29 per cent higher NPV than the conventional model ($137 million versus $97 million), but also significantly improves the stability of the feed grades for copper, gold, and uranium. Specifically, by incorporating three distinct stockpiles with defined uranium grade thresholds, the model enables flexible blending strategies that reduce uranium penalty costs by $64 million while maintaining consistent Cu and Au grades within acceptable bounds.

One of the key advantages of the proposed model is its ability to balance grade targets of multiple minerals by using multiple stockpiles with different grade ranges. This makes it possible to blend ore more effectively and respond to variations in stope output, which is often a challenge in SLC operations due to the variability of ore grades and limited control during draw. The results also show that blending strategies not only help maintain stable feed grades to the processing plant but also improve the flexibility of the production schedules.

Although the model performs well in the current case, further improvements could be made by integrating real-time ore tracking systems and dynamic stockpile monitoring. These enhancements would allow for more responsive scheduling and better control of feed quality over time. Additionally, exploring the interaction between stockpiles designed for different purposes, such as controlling deleterious elements or maintaining valuable metal grades, could further improve operational flexibility.

In summary, the proposed model provides a practical tool for improving production scheduling in SLC mines. By considering ore flow and blending as part of the scheduling process, it offers a more integrated and realistic approach that can support better decision-making and improved economic outcomes in underground mining projects.

ACKNOWLEDGEMENTS

This work was supported by the Australian Research Council Integrated Operations for Complex Resources Industrial Transformation Training Centre (No. IC190100017) and funded by universities, industry, and the Australian Government.

REFERENCES

Almgren, T, 1994. An Approach to Long Range Production and Development Planning with Application to the Kiruna Mine, Sweden, PhD thesis, Luleå University of Technology, Sweden.

Bley, A, Boland, N, Froyland, G and Zuckerberg, M, 2012. Solving mixed integer nonlinear programming problems for mine production planning with stockpiling, Optimization Online. Available from: <http://www.optimization-online.org/DB_HTML/2012/11/3674.html>

Gurobi Optimization, LLC, 2024. Gurobi Optimizer Reference Manual. Available from: <https://www.gurobi.com>

Khazaei, S and Pourrahimian, Y, 2023. An Innovative Optimization Model for Long-term Production Scheduling of Sublevel Caving Mines Using Mixed Integer Linear Programming, MOL Report Eleven.

Kuchta, M, Newman, A and Topal, E, 2004. Implementing a Production Schedule at LKABs Kiruna Mine, *Interfaces*, 34:124–134.

Moreno, E, Rezakhah, M, Newman, A and Ferreira, F, 2017. Linear models for stockpiling in open pit mine production scheduling problems, *European Journal of Operational Research*, 260(1):212–221.

Newman, A M and Kuchta, M, 2007. Using aggregation to optimize long-term production planning at an underground mine, *Eur J Oper Res*, 176:1205–1218.

Newman, A M and Martinez, M A, 2011. A solution approach for optimizing long- and short-term production scheduling at LKAB's Kiruna mine, *Eur J Oper Res*, 211:184–197.

Rezakhah, M, Moreno, E and Newman, A, 2020. Practical performance of an open pit mine scheduling model considering blending and stockpiling, *Computers and Operations Research,* 115:104638.

Riddle, J M, 1977. A dynamic programming solution of a block-caving mine layout, in *Proceedings of the 14th International Symposium on the Application of Computers and Operations Research in the Mineral Industry (APCOM)*, pp 767–780 (Society for Mining, Metallurgy and Exploration Inc, Colorado).

Shenavar, M, Ataee-pour, M and Rahmanpour, M, 2020. A New Mathematical Model for Production Scheduling in Sub-level Caving Mining Method, *Journal of Mining and Environment*, 11(3):765–778.

Tabesh, M and Askari-Nasab, H, 2017. Linearized Stockpile Modeling for Long-Term Open pit Production Planning, *Mining Optimization Laboratory*, 1.780:1.

Williams, J K, Smith, L and Wells, P M, 1973. Planning of underground copper mining, in *Proceedings of the 10th International Symposium on the Application of Computers and Operations Research in the Mineral Industry (APCOM)*, pp 251–254.

Advancing long-term open pit mine planning with quantum computing – concepts, reformulations and implementation

A Quelopana[1], B Keith[2] and J Canales[3]

1. Professor, Departamento de Ingeniería de Sistemas y Computación, Universidad Católica del Norte, Antofagasta 1270709, Chile. Email: aldo.quelopana@ucn.cl
2. Professor, Departamento de Ingeniería de Sistemas y Computación, Universidad Católica del Norte, Antofagasta 1270709, Chile. Email: brian.keith@ucn.cl
3. Data Scientist and Quantum Algorithm Researcher, CoreDevX, Santiago 7510838, Chile. Email: javiera.canales@coredevx.com

ABSTRACT

Over the years, advances in long-term open pit mine planning have significantly optimised ore extraction and transport to metallurgical plants. However, as additional aspects, such as uncertainty, more stages in the mineral value chain, and greater detail within them are integrated, model formulations have become increasingly complex. This complexity has heightened the demand for computational resources, often leading researchers to prioritise efficiency over precision in their proposals. The advent of quantum computing presents a promising approach to addressing these limitations, offering computational power that has shown remarkable success across diverse fields. With the growing availability of quantum computers and simulators, this work presents the basis of some preliminary findings from an ongoing Chilean project exploring quantum technology applications in the long-term mine planning. This paper introduces key concepts in quantum computing, the necessary mathematical reformulations, and a proof-of-concept implementation integrating the Ultimate Pit Limit (UPL) problem with the Quantum Approximate Optimisation Algorithm (QAOA) via two case studies: one executed on a real quantum computer and another on a quantum simulator. These case studies not only reveal current technological limitations but also provide insights into the potential of quantum approaches for addressing mine-planning challenges.

INTRODUCTION

Mine planning involves devising a schedule for ore extraction that maximises the project's net present value (NPV) while satisfying a broad array of operational constraints (Campos, Arroyo and Morales, 2018). Traditionally, mine planning is divided into hierarchical levels—long, medium, and short-terms (L'Heureux, Gamache and Soumis, 2013). This work focuses on the long-term dimension, which is central to defining the ultimate geometry of the mine by selecting which blocks to mine and establishing their extraction timing. In doing so, long-term plans set the stage for more detailed tactical and operational decisions (Rahnema, Amirmoeini and Moradi Afrapoli, 2023).

Mining operations adopt different extraction techniques based on the geologic characteristics of the deposit. For deposits that are economically viable and relatively close to the surface, open pit mining is often the method of choice. This method stands out due to its cost efficiency, as it minimises the need for complex excavation and tunnelling and typically requires less overburden removal compared to underground methods. Open pit mining also benefits from economies of scale, making it particularly attractive for processing lower-grade ores (Darling, 2011).

A critical aspect of mine planning is the reliance on models that predict the orebody characteristics. When these predictions are based on methods like kriging or inverse distance weighting, the approach is considered deterministic. However, such estimation techniques can be problematic, especially when high-grade ore is next to waste rock. In these cases, the smoothing inherent in the estimation process tends to underrepresent high-grade zones and overrepresent lower-grade material (Yamamoto, 2008). This misrepresentation can lead to substantial losses in potential NPV—studies suggest that more accurate models could boost project value by 10 per cent to 25 per cent (Ramazan and Dimitrakopoulos, 2013; Dimitrakopoulos and Lamghari, 2022). To address these inaccuracies, conditional simulation methods such as Turning Bands Simulation and Sequential Gaussian Simulation have been introduced (Paravarzar, Emery and Madani, 2015). These techniques generate multiple equiprobable orebody scenarios, each reflecting the essential statistical and spatial properties of the deposit (Maleki et al, 2022). This paradigm shift has paved the way for stochastic mine planning, which develops extraction schedules that perform robustly across a suite of plausible geological scenarios rather than being optimised for a single, deterministic case (Nelis, Morales and Widzyk-Capehart, 2019).

The challenge of mine planning, however, extends beyond geological uncertainty. When planning spans the entire mineral value chain—from extraction to mineral processing and metallurgy—the complexity increases dramatically. Historically, optimisation efforts have concentrated on individual stages of the process, sometimes attempting to coordinate between them in hopes of achieving synergistic benefits. Yet, optimising each stage in isolation does not guarantee that the overall system will be operated at its full potential (Skyttner, 2001). More recent studies have introduced the concept of operational modes in mineral processing, where a set of key parameters (for example, metal recovery rates and processing capacities) defines distinct plant configurations (Quelopana and Navarra, 2024). By discretising the range of possible configurations, these operational modes allow metallurgical plants to adapt more flexibly to feed variability (Navarra, Rafiei and Waters, 2017). In practical terms, this means that a plant can be preconfigured to handle different types of ore blends more efficiently, and mining operations can receive clearer guidelines on the quantity and quality of material to deliver (Quelopana et al, 2023). The integration of these plant process models into long-term mine planning frameworks represents a significant advance in aligning the entire value chain.

Given the multifaceted nature of these problems, conventional exact optimisation methods—which aim for a global optimum—have struggled to keep pace with real-world complexities. This limitation has driven the adoption of metaheuristic techniques. Although metaheuristics, such as Particle Swarm Optimisation (Khan, 2018), Simulated Annealing (Levinson and Dimitrakopoulos, 2020), Tabu Search (Senecal and Dimitrakopoulos, 2020), and Variable Neighbourhood Descent (Lamghari, Dimitrakopoulos and Ferland, 2014), often yield high-quality solutions, they do not guarantee an optimum (Quelopana et al, 2023). Moreover, these methods typically involve a range of tuneable parameters whose settings can dramatically affect solution quality and computational effort. In practice, the process of parameter tuning itself may become an optimisation problem, adding another layer of difficulty to the overall task (Agrawal, 2021).

In response to these computational challenges, alternative approaches are being explored. One promising avenue is the application of artificial intelligence (AI) techniques in mine planning (Noriega and Pourrahimian, 2022; Mariz et al, 2024). AI methods, including reinforcement learning (Sutton and Barto, 2018) and neural networks (Goodfellow, Bengio and Courville, 2016), have begun to emerge as tools that can complement traditional optimisation strategies. While these AI-driven approaches show potential for handling the inherent uncertainties and dynamic nature of mine planning, they are still in an early stage of development. Their effectiveness is often contingent on fine-tuning and the capacity to emulate decision-making processes without necessarily seeking the global optimum (Russell and Norvig, 2021).

Given the inherent complexity of mine planning—arising from both geological uncertainty and the need to coordinate a chain of interdependent processes—innovative strategies that move beyond traditional methods are essential. A central difficulty is that the underlying mine-sequencing (open pit production-scheduling) problem is NP-hard: the number of feasible extraction sequences grows exponentially with the number of blocks, and no polynomial-time algorithm is known that guarantees the global optimum in the general case (Dagdelen and Johnson, 1986; Bienstock and Zuckerberg, 2010). Practitioners therefore resort to decomposition techniques, heuristics and meta-heuristics, yet these still struggle when instances become large or when stochastic constraints are introduced.

As part of an ongoing research initiative, this paper investigates the potential of quantum computing to address the inherent complexity of mine planning—particularly the intractability that arises in large-scale optimisation under geological and operational constraints. The study focuses on the Ultimate Pit Limit (UPL) problem, presenting a proof of concept that integrates this foundational model with the Quantum Approximate Optimisation Algorithm (QAOA) (Blekos et al, 2024). Alongside a conceptual overview of relevant quantum principles and mathematical reformulations, the paper highlights current technological limitations while underscoring the feasibility of hybrid quantum-classical approaches. Although the UPL formulation adopted here is deterministic and does not yet reflect the full stochastic dynamics or interconnected decisions across the mineral value chain, it marks a crucial first step. This work lays the groundwork for future extensions toward more realistic models that embrace NP-hard complexity and leverage emerging quantum capabilities.

The UPL problem is widely regarded as a cornerstone in open pit mine planning, delineating the maximum economically extractable volume of material within a deposit. It determines the ultimate shape of the pit, directly influencing downstream decisions such as resource allocation, scheduling, and long-term investment strategies (Campos, Arroyo and Morales, 2018). Given its strategic significance and the combinatorial nature of the decisions involved, the UPL problem has been the subject of intensive research for decades. Accurately solving it is essential not only for short-term profitability but also for ensuring the sustainable development of the mining operation over its entire life cycle.

While previous research in quantum computing has shown promise in related fields—such as scheduling (Scherer *et al*, 2021; Yonaga *et al*, 2022; Kurowski *et al*, 2023; Schworm *et al*, 2023, 2024), optimisation (Wang, Kim and Suresh, 2023; Abbas *et al*, 2024), and other industrial applications (Bayerstadler *et al*, 2021)—no studies have yet applied this technology to stochastic mine planning. Some results of this branch of research were submitted to a computational and operational research audience, with details tailored to that field of study; hence, this paper aims to further elaborate on the work for a mining and metallurgical audience.

QUANTUM COMPUTING

Quantum computing, rooted in the principles of superposition and entanglement, offers a fundamentally novel approach to solving problems that are computationally intensive for classical systems (Benenti *et al*, 2019). Unlike traditional computers that operate on binary bits, quantum computers utilise quantum bits (qubits), which can exist in multiple states simultaneously (Salm *et al*, 2020). This inherent parallelism enables the exploration of vast solution spaces, making quantum computing a promising tool for optimisation tasks in fields such as mine planning, where complex problem-solving is essential (Shishvan and Sattarvand, 2015).

This technology represents a paradigm shift in computational science. By exploiting unique quantum mechanical phenomena, quantum computers process information in fundamentally different ways compared to conventional systems. The ability to leverage quantum parallelism opens up transformative potential for tackling problems that would otherwise be computationally prohibitive, thereby complementing the advanced integration of plant process models and AI methods in modern mine planning strategies.

Current quantum systems can boast architectures with over 1000 qubits (Castelvecchi, 2023), yet practical applications are often limited by challenges such as error correction and maintaining quantum coherence (Córcoles *et al*, 2019; Filippov, Maniscalco and García-Pérez, 2024). Some of the commercialised architectures are based on the quantum properties of atomic nuclei in dissolved molecules, where potent magnetic fields align nuclear spins and microwaves manipulate these spins to perform quantum operations. Despite these advancements, the number of effectively usable qubits remains a bottleneck, restricting the complexity of the problems that these systems can presently address.

In light of these technological limitations, quantum simulators have become a critical component in the field (D'Urbano, Angelelli and Catalano, 2023). They enable researchers to assess the quantum advantage of various methods and to experiment with quantum algorithms before deploying them on actual hardware. By generating empirical evidence on performance differences and characterising errors like noise, these simulators accelerate the refinement of quantum algorithms (D'Urbano, Angelelli and Catalano, 2023). This development is crucial for devising hybrid solutions that integrate quantum computing with traditional approaches, ultimately offering new avenues to address the computational challenges inherent in mine planning.

Quantum optimisation

Quantum optimisation encompasses a broad spectrum of problem types, each demanding customised algorithmic approaches depending on the structure of the decision variables and constraints (Abbas *et al*, 2024). A key category involves discrete optimisation problems, where decision variables take binary or integer values. These are often formulated as Quadratic Unconstrained Binary Optimisation (QUBO) problems, a standard representation that can be mapped onto quantum systems using techniques from statistical physics (Lucas, 2014). In such mappings, the problem is expressed as a Hamiltonian—an energy function whose lowest energy state (the *ground state*) corresponds to the optimal solution. This representation enables the use of quantum algorithms like quantum annealing, the QAOA, or other variational quantum algorithms to search for high-quality solutions to complex, NP-hard problems such as those found in scheduling, logistics, or long-term mine planning (Barahona, 1982).

Beyond discrete optimisation, continuous optimisation problems represent another essential category—where variables can take values over continuous domains and the problems may be convex (eg, quadratic programming) or non-convex (eg, nonlinear resource allocation) in nature (Abbas *et al*, 2024). In these settings, quantum algorithms aim to accelerate key computational steps, such as solving systems of linear equations or semidefinite programs, by leveraging advanced quantum techniques like quantum singular value transformation and block encoding (Abbas *et al*, 2024). These techniques can reduce runtime complexities under certain assumptions, offering potential speedups over classical solvers.

Among first-order methods, which rely primarily on gradient information, the Multiplicative Weights Update (MWU) framework has emerged as a versatile approach for handling large-scale convex optimisation, particularly in online or iterative settings (Arora, Hazan and Kale, 2005). In contrast, second-order methods, such as Interior Point Methods (IPMs), incorporate curvature information (ie., second derivatives) to more efficiently navigate the feasible region. IPMs are widely used in fields like operations research and engineering design to solve constrained optimisation problems. In these methods, a key computational bottleneck is the Newton step, which involves solving a large linear system. Quantum computing has been proposed as a means to accelerate this step—for instance, by using quantum linear system solvers like the Harrow-Hassidim-Lloyd (HHL) algorithm—to improve scalability in complex optimisation settings (Rebentrost *et al*, 2019).

A hybrid category emerges in the form of mixed-integer programming, which combines both discrete and continuous variables. Such problems are particularly challenging because they merge the combinatorial complexity of discrete choices with ability to make more nuanced choices with continuous optimisation (Abbas *et al*, 2024). Quantum-inspired strategies in this domain often involve reformulating the problem—using techniques such as penalty methods, decomposition, or hybrid branch-and-bound algorithms—to isolate the most computationally intensive components. These components can then be tackled with quantum-classical algorithms, including methods analogous to the Alternating Direction Method of Multipliers (ADMM) (Gambella and Simonetto, 2020), potentially reducing overall solution times even if full fault-tolerant quantum computation remains on the horizon.

Overall, the range of quantum optimisation techniques—from exact methods that require error-corrected devices to near-term variational and hybrid approaches—demonstrates the potential for significant speedups in solving complex optimisation problems while also highlighting the importance of matching the algorithm to the underlying structure of each problem class. As mentioned in the introduction, this study is focused on the UPL problem, which falls within the domain of discrete optimisation. Given that QUBO formulations are the most common approach in this field, the problem will be reformulated accordingly and solved using the QAOA algorithm, a widely used method for addressing this type of problem.

Quadratic unconstrained binary optimisation (QUBO)

The QUBO formulation represents a mathematical framework for expressing optimisation problems using binary variables, where all constraints are embedded into the objective function via penalty terms. Although QUBO problems can be framed as either maximisation or minimisation tasks, quantum computing applications typically require them to be cast as minimisation problems. This characteristic allows QUBO to seamlessly integrate constraints within a single quadratic expression, thereby avoiding the need for explicit constraint handling—a feature that aligns well with the operational paradigms of quantum systems (Punnen, 2022).

The intrinsic advantages of QUBO further underscore its suitability for quantum computing. The binary nature of its variables maps directly onto qubits, enabling straightforward representation in quantum hardware (Warren, 2013). Moreover, the quadratic interactions effectively model the relationships between variables, a capability that can be exploited using quantum entanglement

(Glover *et al*, 2022). This formulation's versatility is evidenced by its ability to recast numerous NP-hard optimisation problems, thereby establishing a robust framework for a wide range of applications (Chatterjee, Bourreau and Rančić, 2024; Lucas, 2014; Chapuis *et al*, 2017). In practical domains such as the UPL problem, QUBO can be employed to encode complex operational constraints—like precedence relationships between mining blocks—into a structure amenable to solution via quantum algorithms, such as the QAOA.

The QUBO problem formulation is displayed in Equations 1 and 2:

Objective Function:

$$\text{Min}_x f(x) = x^T Q x = \sum_{i=1}^{N} \sum_{j=1}^{N} q_{ij} x_i x_j \tag{1}$$

Constraint:

$$x_i \in \{0,1\}, \forall i \in \{1, \dots, n\} \tag{2}$$

Problem Details:

- n is the number of binary decision variables.
- x is an n-dimensional vector of binary variables.
- Q is an $n \times n$ symmetric matrix.
- q_{ii} encodes the interaction cost or weight between the binary variables x_i and x_j.
- The Q matrix includes both linear and quadratic components:
  - Diagonal entries q_{ii} correspond to linear terms.
  - Off-diagonal entries q_{ij} ($i \neq j$) correspond to pairwise interactions.

Because of the Q matrix can be symmetric, an equivalent way of writing the objective function is given in Equation 3. The matrix is then supplied to the QAOA algorithm to find the optimum.

$$\text{Min}_x f(x) = \sum_{i=1}^{n} q_{ii} x_i + \sum_{i<j}^{n} q_{ij} x_i x_j \tag{3}$$

Reformulation the UPL problem as a QUBO

This section begins by presenting the classical UPL formulation. This optimisation model aims to maximise overall profit from mining blocks, subject to critical precedence constraints, as shown in Equations 4–6. The parameters for this problem are:

Parameters:

- N is the total number of blocks in the model
- V_i is the net value of block i, computed as revenue minus cost
- P_i is the set of blocks that must be mined prior to block i
- x_i is a binary decision variable, where x_i = 1 indicates that block i is mined, and x_i = 0 otherwise

Objective Function:

$$\max Z = \sum_{i=1}^{N} V_i x_i \tag{4}$$

Constraints:

- Precedence Constraint: Each block i can be mined only if every block j in P_i has already been mined. Formally,

$$x_i \leq x_j, \forall j \in P_i \tag{5}$$

Binary Constraint:

$$x_i \in \{0,1\}, \forall i = 1,2, \dots, N \tag{6}$$

This formulation ensures the correct mining order while maximising total profit in the open pit scenario.

A way to solve the UPL using quantum algorithms involves transforming its formulation into a minimisation QUBO problem. For this purpose, Equation 4 is replaced with Equation 7:

$$\min f(x) = -Z = - \sum_{i=1}^{N} V_i x_i \tag{7}$$

The constraints in the UPL formulation, represented as $x_i \leq x_j$ (Equation 5), are integrated into the objective function through the penalty term $\lambda(x_i - x_i x_j)$ (Glover et al, 2022), where λ is a positive constant penalising any violation of the precedence requirements. In general form, precedence constraints can be modelled as shown in Equation 8:

$$\lambda \sum_{i=1}^{N} \sum_{j \in P_i} (1 - x_j) x_i \tag{8}$$

By combining the minimisation term (Equation 7) with the precedence constraint (Equation 8), the resulting objective function is presented in Equation 9:

$$\min f(x) = -Z = - \sum_{i=1}^{N} V_i x_i + \lambda \sum_{i=1}^{N} \sum_{j \in P_i} (1 - x_j) x_i \tag{9}$$

To derive the Q matrix, Equation 9 is expanded to match the canonical QUBO form (Equation 3):

$$\min f(x) = -Z = - \sum_{i=1}^{N} V_i x_i + \lambda \sum_{i=1}^{N} \sum_{j \in P_i} x_i - \lambda \sum_{i=1}^{N} \sum_{j \in P_i} x_i x_j \tag{10}$$

Rewriting:

$$\min f(x) = -Z = \sum_{i=1}^{N} (-V_i + \lambda |P_i|) x_i - \lambda \sum_{i=1}^{N} \sum_{j \in P_i} x_i x_j \tag{11}$$

Based on this correspondence, the Q matrix entries are identified as indicated in Equations 12 and 13:

$$q_{ii} = -V_i + \lambda \sum_{j \in P_i} 1 \tag{12}$$

$$q_{ij}, q_{ji} = -\frac{\lambda}{2} \tag{13}$$

Quantum approximate optimisation algorithm (QAOA)

QAOA is a hybrid quantum-classical approach that iteratively refines parameterised quantum states to approximate solutions for combinatorial optimisation problems (Farhi, Goldstone and Gutmann, 2014). In this scheme, a quantum circuit encodes potential solutions, while a classical optimiser tunes the circuit parameters to either minimise or maximise the specified objective. By exploiting quantum interference and entanglement, QAOA can traverse the search space more efficiently than some purely classical algorithms, making it particularly suitable for QUBO-based problems. Its potential advantages have been explored in various domains, including portfolio management, scheduling, and supply chain logistics, and early work indicates the possibility of a measurable quantum speedup or resource efficiency gain (Kurowski et al, 2023; Pirnay et al, 2024).

A high-level description of the QAOA workflow is shown in Algorithm 1. It begins by initialising the system in a uniform superposition of all candidate solutions. Next, two operators—the cost operator and the mixer operator—are alternately applied across p layers, each governed by a set of adjustable parameters $\vec{\gamma} = (\gamma_1, \gamma_2, \ldots, \gamma_p)$ and $\vec{\beta} = (\beta_1, \beta_2, \ldots, \beta_p)$. In this context, the loop index k in Algorithm 1 denotes the current QAOA layer, with each layer applying a fixed pair (γ_k, β_k). After the final round of these transformations, measurements yield a bitstring which is then scored via the objective function. A classical optimiser updates the parameters, aiming to improve the result on subsequent iterations. This hybrid loop repeats until termination criteria are met, returning the bitstring that achieves the lowest (or highest) observed cost value.

ALGORITHM 1
Quantum Approximate Optimisation Algorithm (QAOA).

1:	**function** QAOA(C,p)	Cost function C and depth p as input
2:	Initialise parameters $\vec{\gamma} = (\gamma_1, \gamma_2, \ldots, \gamma_p)$ and $\vec{\beta} = (\beta_1, \beta_2, \ldots, \beta_p)$	
3:	Prepare uniform superposition state: $$\|\psi_0\rangle = \frac{1}{\sqrt{2^n}} \sum_{z \in \{0,1\}^n} \|z\rangle$$	
4:	**for** k = 1 to p **do**	
5:	Apply cost unitary: $U_C(\gamma_k) = e^{-i\gamma_k C}$	
6:	Apply mixer unitary: $U_M(\beta_k) = e^{-i\beta_k B}$, where $B = \sum_{i=1}^{n} \sigma_x^i$	
7:	**end for**	
8:	Measure quantum state to obtain bitstring z	
9:	Evaluate cost function $C(z)$	
10:	Update $\vec{\gamma}$ and $\vec{\beta}$ using classical optimisation	
11:	**return z**	Bitstring with the lowest cost $C(z)$
12:	**end function**	

QAOA has shown impressive adaptability in a variety of practical settings, making it a leading candidate for tackling intricate optimisation challenges. Applications ranging from wireless computer vision (Li *et al*, 2020), and protein folding (Chandarana *et al*, 2023) to machine learning model training (Date, Arthur and Pusey-Nazzaro, 2021) and quantum finance (Canabarro *et al*, 2022) highlight its ability to handle problems that are computationally demanding for classical methods. This versatility, coupled with the algorithm's compatibility with near-term quantum hardware, makes QAOA an attractive option for real-world deployment. In industry, QAOA has already been employed for tasks like vehicle routing (Azad *et al*, 2022), telecommunications network optimisation (Cui *et al*, 2022), and clustering (Moussa *et al*, 2022), with promising outcomes. Although current quantum devices still face size and noise constraints, ongoing research results indicate that QAOA could eventually become a key component of quantum-enhanced optimisation efforts.

Case studies

Two illustrative case studies were conducted to demonstrate the quantum-based approach to the UPL problem. One utilised an actual quantum computer, and the other employed a quantum simulator. Both are proof-of-concept examples focused on identifying the optimal solution.

Case study 1 – 3-qubit quantum computer

The first case used a 3-qubit desktop quantum computer, Triangulum SpinQ (for more info: <https://www.spinquanta.com/products-services/triangulum>), which relies on nuclear magnetic resonance (NMR). Because the computer supports only three qubits, a problem consisting of three blocks was designed (Figure 1), with their values representing economic returns in USD.

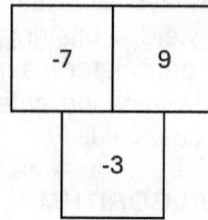

FIG 1 – Blocks for case study 1.

Applying Equations 12 and 13, the matrix Q for this problem, with λ=500, is:

$$Q = \begin{pmatrix} 7 & 0 & -250 \\ 0 & -9 & -250 \\ -250 & -250 & 1003 \end{pmatrix}$$

The SpinQit library (provided by the hardware manufacturer) was used to implement the QAOA algorithm, combined with the COBYLA classical optimiser for tuning the parameters $\vec{\gamma}$ and $\vec{\beta}$. The approach converged to the optimal solution [0, 1, 0], indicating that only the top-right block should be mined. Consequently, case study 1 yields an objective value of USD -9.0, which corresponds to a profit of USD 9.00.

Case study 2 – 3D ore deposit via quantum simulation

The second case study involves a notional 3D copper deposit containing 32 blocks, extracted via open pit mining. Blocks with a grade at or above a specified cut-off undergo through a combined Leaching and Solvent Extraction (SX)/Electrowinning (EW) process, while lower-grade blocks incur only mining costs. The value of each block v_b, is computed using Equations 14 and 15:

If grade ≥ cut-off:

$$v_b = m_b \times r \times d - t_b \times \left(c_{mining} + c_{processing}\right) \tag{14}$$

Otherwise:

$$v_b = -t_b \times c_{mining} \tag{15}$$

where m_b is the metal content, r is the recovery rate, d is the metal price, t_b is the block tonnage, c_{mining} is the mining cost, and $c_{processing}$ is the processing cost. Table 1 summarises these parameters for the studied data set. Precedence constraints reduced the problem to 20 blocks. Simulations were conducted on a quantum simulator from CoreDevX <https://coredevx.com/site/QSimulator200/> that supports up to 30 qubits and allows Python programming via Qiskit <https://www.ibm.com/quantum/qiskit>.

TABLE 1

Mining and Processing Parameters for the data set used in case study 2.

Parameter	Value
Number of Blocks	4 × 4 × 2
Block Weight (tons)	20 000
Metal Price (US$/lb)	4.31
Recovery (%)	90
Mining Cost (US$/ton)	22
Processing Cost (US$/ton)	24
Cut-Off Grade (%)	0.35

QAOA was again applied, with COBYLA serving as the classical optimiser. Due to memory limitations, Qiskit was unable to simulate more than 20 qubits under the iterative requirements of QAOA. However, since the problem does not include the full set of 32 blocks (4 × 4 × 2) — as some blocks can never be extracted due to precedence constraints (Equation 5) — only 20 blocks were considered (one block per qubit). After constructing the 20 × 20 Q matrix (with λ = 10 000 000 and nine precedence relationships), an optimal solution was obtained. This solution was independently verified using the Gurobi Optimiser <https://www.gurobi.com/>, and both approaches yielded the same optimal result. The final solution's value, which was US$2 245 184, is illustrated by a 3D pit design figure (Figure 2).

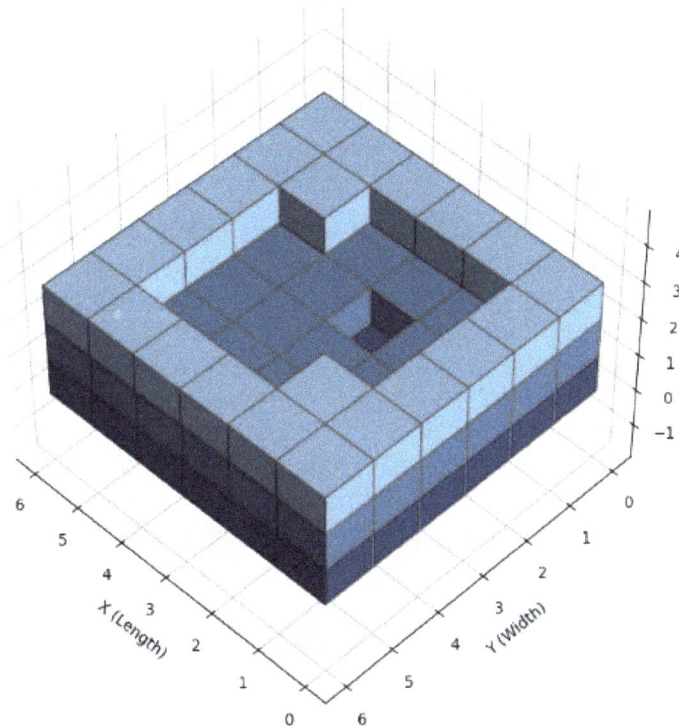

FIG 2 – Ultimate pit outline for the second case study.

Discussion

The successful application of QAOA to solve the UPL problem highlights the potential of quantum optimisation in mine planning. The algorithm consistently reached optimal solutions in both case studies while honouring precedence constraints. Although the UPL problem is simpler than more complex mining scenarios (such as stochastic mine planning), it remains a fundamental starting point in open pit mining upon which higher-level decisions rely (Newman *et al*, 2010).

Despite this success, the first case study clearly illustrates the limitations of quantum hardware in terms of qubit availability. The exponential growth of the state space, tied to the number of blocks, remains a major hurdle (Córcoles *et al*, 2019; Filippov, Maniscalco and García-Pérez, 2024). However, quantum computing is advancing rapidly due to the efforts of the quantum research community. Recent developments in hardware improvements and error mitigation (D'Urbano, Angelelli and Catalano, 2023) suggest that current constraints may be temporary, paving the way for future scalability. Additionally, recent announcements about new quantum chips, such as Willow by Google and Majorana by Microsoft, indicate continuous progress in the field.

The second case study highlighted notable memory challenges when scaling quantum optimisation to industrial-sized block models in quantum simulators. One approach to addressing this computational bottleneck involves reducing the dimensionality of the Q matrix, following the method proposed by Lewis and Glover (2017). They introduced a preprocessing technique based on five rules, which has demonstrated promising results. Another approach involves leveraging tensor network methods (Markov and Shi, 2008; Schollwöck, 2011; Orús, 2014). By decomposing large quantum states into smaller, interconnected tensors, these techniques have shown promise in efficiently simulating quantum systems (Markov and Shi, 2008; Villalonga *et al*, 2019; Gray and Kourtis, 2021). Applying tensor network concepts to mine planning could facilitate meaningful testing of quantum methodologies on larger, more representative models while awaiting further advancements in quantum hardware.

The results reported here suggest that quantum techniques could be valuable in addressing these and other mining optimisation challenges. Although the technology is still evolving, recent progress indicates steady movement towards the level of maturity required for broader industrial adoption.

CONCLUSIONS

This paper presents what appears to be the first instance of using quantum optimisation to solve the UPL problem in mine planning. A QUBO-based formulation combined with QAOA effectively determined optimal mining boundaries while respecting precedence constraints. Key concepts of quantum computing were introduced, mathematical reformulations were presented, and successful outcomes in two test cases validated the theoretical foundation of this approach as a proof of concept.

Implementation on a quantum simulator demonstrates both the potential and present-day limitations of quantum computing in large-scale mine planning. Although the method consistently found correct solutions for the test cases, significant memory requirements exposed a gap between current quantum hardware capacity and the computational demands of industrial mining applications. In particular, running the approach in Python/Qiskit encountered memory constraints, underscoring the need for preprocessing the Q matrix to reduce its dimensions and/or employing advanced simulation techniques—such as tensor network models—that may enable simulations of up to 100 qubits on available hardware. These barriers do not negate the value of the framework, as continued development (eg, through other platforms or forthcoming methods from the literature) may mitigate existing constraints, especially when quantum devices become more capable.

The methodology established here can serve as a foundation for future advancements in quantum optimisation for mine planning. More complex mining optimisation challenges may be addressed in subsequent research as hardware evolves. Going forward, hybrid quantum-classical approaches are expected to help bridge current technological gaps, ensuring that the mining industry is ready to exploit new hardware developments as they arise.

REFERENCES

Abbas, A, Ambainis, A, Augustino, B, Bärtschi, A, Burhman, H, Coffrin, C, Cortiana, G, Dunjko, V, Egger, D J, Elmegreen, B G, Franco, N, Fratini, F, Fuller, B, Gacon, J, Gonciulea, C, Gribling, S, Gupta, S, Hadfield, S, Heese, R, Kircher, G, Kleinert, T, Koch, T, Korpas, G, Lenk, S, Marecek, J, Markov, V, Mazzola, G, Mensa, S, Mohseni, N, Nannicini, G, O'Meara, C, Peña-Tapia, E, Pokutta, S, Proissl, M, Rebentrost, P, Sahin, E, Symons, B C B, Tornow, S, Valls, V, Woerner, S, Wolf-Bauwens, M L, Yard, J, Yarkoni, S, Zechiel, D, Zhuk, S and Zoufal, C, 2024. Challenges and opportunities in quantum optimisation, *Nature Reviews Physics*, 6:718–735. https://doi.org/10.1038/s42254-024-00770-9

Agrawal, T, 2021. *Hyperparameter optimization in machine learning: make your machine learning and deep learning models more efficient* (Apress). https://doi.org/10.1007/978-1-4842-6579-6

Arora, S, Hazan, E and Kale, S, 2005. Fast algorithms for approximate semidefinite programming using the multiplicative weights update method, in *46th Annual IEEE Symposium on Foundations of Computer Science (FOCS 2005)*, pp 339–348. https://doi.org/10.1109/SFCS.2005.35

Azad, U, Behera, B K, Ahmed, E A, Panigrahi, P K and Farouk, A, 2022. Solving vehicle routing problem using quantum approximate optimisation algorithm, *IEEE Transactions on Intelligent Transportation Systems*, 24:7564–7573. https://doi.org/10.1109/TITS.2022.3172241

Barahona, F, 1982. On the computational complexity of Ising spin glass models, *Journal of Physics A: Mathematical and General*, 15(10):3241–3253. https://doi.org/10.1088/0305-4470/15/10/028

Bayerstadler, A, Becquin, G, Binder, J, Botter, T, Ehm, H, Ehmer, T, Erdmann, M, Gaus, N, Harbach, P, Hess, M, Klepsch, J, Leib, M, Luber, S, Luckow, A, Mansky, M, Mauerer, W, Neukart, F, Niedermeier, C, Palackal, L, Pfeiffer, R, Polenz, C, Sepulveda, J, Sievers, T, Standen, B, Streif, M, Strohm, T, Utschig-Utschig, C, Volz, D, Weiss, H and Winter, F, 2021. Industry quantum computing applications, *EPJ Quantum Technology*, 8(1). https://doi.org/10.1140/epjqt/s40507-021-00114-x

Benenti, G, Casati, G, Rossini, D and Strini, G, 2019. *Principles of quantum computation and information: a comprehensive textbook*, (World Scientific).

Bienstock, D and Zuckerberg, M, 2010. Solving LP relaxations of large-scale precedence constrained problems, in *Proceedings of the 12th International Conference on Integer Programming and Combinatorial Optimization (IPCO XII)* (eds: M Jünger and P Mutzel), pp 1–14 (Springer).

Blekos, K, Brand, D, Ceschini, A, Chou, C H, Li, R H, Pandya, K and Summer, A, 2024. A review on quantum approximate optimisation algorithm and its variants, *Physics Reports*, 1068:1–66. https://doi.org/10.1016/j.physrep.2023.06.001

Campos, P, Arroyo, C and Morales, N, 2018. Application of optimised models through direct block scheduling in traditional mine planning, *Journal of the Southern African Institute of Mining and Metallurgy*, 118:381–386. https://doi.org/10.17159/2411-9717/2018/v118n4a8

Canabarro, A, Mendonça, T M, Nery, R, Moreno, G, Albino, A S, de Jesus, G F and Chaves, R, 2022. Quantum finance: a tutorial on quantum computing applied to the financial market, *arXiv preprint*. https://doi.org/10.48550/arXiv.2208.04382

Castelvecchi, D, 2023. Underdog technologies gain ground in quantum-computing race, *Nature*, 614:400–401. https://doi.org/10.1038/d41586-023-00278-9

Chandarana, P, Hegade, N N, Montalban, I, Solano, E and Chen, X, 2023. Digitized counterdiabatic quantum algorithm for protein folding, *Physical Review Applied*, 20:014024. https://doi.org/10.1103/PhysRevApplied.20.014024

Chapuis, G, Djidjev, H N, Hahn, G and Rizk, G, 2017. Finding maximum cliques on a quantum annealer, in *Proceedings of the Computing Frontiers Conference,* pp 63–70. https://doi.org/10.1145/3075564.3075575

Chatterjee, Y, Bourreau, E and Rančić, M J, 2024. Solving various NP-hard problems using exponentially fewer qubits on a quantum computer, *Physical Review A,* 109:052441. https://doi.org/10.1103/PhysRevA.109.052441

Córcoles, A D, Kandala, A, Javadi-Abhari, A, McClure, D T, Cross, A W, Temme, K, Nation, P D, Steffen, M and Gambetta, J M, 2019. Challenges and opportunities of near-term quantum computing systems, *Proceedings of the IEEE,* 108:1338–1352. https://doi.org/10.1109/JPR O C, 2020.2991508

Cui, J, Xiong, Y, Ng, S X and Hanzo, L, 2022. Quantum approximate optimisation algorithm based maximum likelihood detection, *IEEE Transactions on Communications*, 70:5386–5400. https://doi.org/10.1109/TCOMM.2022.3184220

D'Urbano, A, Angelelli, M and Catalano, C, 2023. The significance of classical simulations in the adoption of quantum technologies for software development, in *International Conference on Product-Focused Software Process Improvement*, pp 60–67 (Springer). https://doi.org/10.1007/978-3-031-49269-3_6

Dagdelen, K and Johnson, T B, 1986. Optimum open pit mine production scheduling, *Proceedings of the 19th Symposium on the Application of Computers and Operations Research in the Mineral Industry (APCOM)*, pp 127–142.

Darling, P (ed), 2011. *SME Mining Engineering Handbook*, 3rd edn (Englewood: Society for Mining, Metallurgy and Exploration).

Date, P, Arthur, D and Pusey-Nazzaro, L, 2021. QUBO formulations for training machine learning models, *Scientific Reports*, 11:10029. https://doi.org/10.1038/s41598-021-89461-4

Dimitrakopoulos, R and Lamghari, A, 2022. Simultaneous stochastic optimisation of mining complexes – mineral value chains: an overview of concepts, examples and comparisons, *International Journal of Mining, Reclamation and Environment*, 36:443–460. https://doi.org/10.1080/17480930.2022.2065730

Farhi, E, Goldstone, J and Gutmann, S, 2014. A quantum approximate optimisation algorithm, *arXiv preprint*. https://doi.org/10.48550/arXiv.1411.4028

Filippov, S N, Maniscalco, S and García-Pérez, G, 2024. Scalability of quantum error mitigation techniques: from utility to advantage, *arXiv preprint*. https://doi.org/10.48550/arXiv.2403.13542

Gambella, C and Simonetto, A, 2020. Multiblock ADMM heuristics for mixed-binary optimisation on classical and quantum computers, *IEEE Transactions on Quantum Engineering*, 1:3102022. https://doi.org/10.1109/TQE.2020.3033139

Glover, F, Kochenberger, G, Hennig, R and Du, Y, 2022. Quantum bridge analytics I: a tutorial on formulating and using QUBO models, *Annals of Operations Research*, 314:141–183. https://doi.org/10.1007/s10479-022-04634-2

Goodfellow, I, Bengio, Y and Courville, A, 2016. *Deep learning*, (Cambridge: MIT Press).

Gray, J and Kourtis, S, 2021. Hyper-optimised tensor network contraction, *Quantum*, 5:410. https://doi.org/10.22331/q-2021-03-15-410

Khan, A, 2018. Long-term production scheduling of open pit mines using particle swarm and bat algorithms under grade uncertainty, *Journal of the Southern African Institute of Mining and Metallurgy*, 118(4):361–368. https://doi.org/10.17159/2411-9717/2018/v118n4a5

Kurowski, K, Pecyna, T, Slysz, M, Różycki, R, Waligóra, G and Węglarz, J, 2023. Application of quantum approximate optimisation algorithm to job shop scheduling problem, *European Journal of Operational Research*, 310(2):518–528. https://doi.org/10.1016/j.ejor.2023.03.013

L'Heureux, G, Gamache, M and Soumis, F, 2013. Mixed integer programming model for short term planning in open pit mines, *Mining Technology*, 122:101–109. https://doi.org/10.1179/1743286313Y.0000000037

Lamghari, A, Dimitrakopoulos, R and Ferland, J, 2014. A variable neighbourhood descent algorithm for the open pit mine production scheduling problem with metal uncertainty, *Journal of the Operational Research Society*, 65:1305–1314. https://doi.org/10.1057/jors.2013.81

Levinson, Z and Dimitrakopoulos, R, 2020. Simultaneous stochastic optimisation of an open pit gold mining complex with waste management, *International Journal of Mining, Reclamation and Environment*, 34:415–429. https://doi.org/10.1080/17480930.2019.1621441

Lewis, M and Glover, F, 2017. Quadratic unconstrained binary optimization problem preprocessing: theory and empirical analysis, *Networks*, 70(2):79–97. https://doi.org/10.1002/net.21751

Li, J, Alam, M, Saki, A A and Ghosh, S, 2020. Hierarchical improvement of quantum approximate optimisation algorithm for object detection, in *21st International Symposium on Quality Electronic Design (ISQED)*, pp 335–340 (IEEE). https://doi.org/10.1109/ISQED48828.2020.9136973

Lucas, A, 2014. Ising formulations of many NP problems, *Frontiers in Physics*, 2(5). https://doi.org/10.3389/fphy.2014.00005

Maleki, M, Mery, N, Soltani-Mohammadi, S, Khorram, F and Emery, X, 2022. Geological control for in-situ and recoverable resources assessment: a case study on Sarcheshmeh porphyry copper deposit, Iran, *Ore Geology Reviews*, 150. https://doi.org/10.1016/j.oregeorev.2022.105133

Mariz, J L V, Badiozamani, M M, Peroni, R d L and Silva, R M d A, 2024. A critical review of bench aggregation and mining cut clustering techniques based on optimisation and artificial intelligence to enhance the open pit mine planning, *Engineering Applications of Artificial Intelligence*, 133. https://doi.org/10.1016/j.engappai.2024.108334

Markov, I L and Shi, Y, 2008. Simulating quantum computation by contracting tensor networks, *SIAM Journal on Computing*, 38:963–981. https://doi.org/10.1137/050644756

Moussa, C, Wang, H, Bäck, T and Dunjko, V, 2022. Unsupervised strategies for identifying optimal parameters in quantum approximate optimisation algorithm, *EPJ Quantum Technology*, 9:11. https://doi.org/10.1140/epjqt/s40507-022-00131-4

Navarra, A, Rafiei, A A and Waters, K, 2017. A systems approach to mineral processing based on mathematical programming, *Canadian Metallurgical Quarterly*, 56:35–44. https://doi.org/10.1080/00084433.2016.1261501

Nelis, G, Morales, N and Widzyk-Capehart, E, 2019. Comparison of different approaches to strategic open pit mine planning under geological uncertainty, in *Proceedings of the 27th International Symposium on Mine Planning and Equipment Selection – MPES 2018* (eds: E Widzyk-Capehart, A Hekmat and R Singhal), pp 95–105 (Cham: Springer International Publishing).

Newman, A, Rubio, E, Caro, R and Weintraub, A, 2010. A review of operations research in mine planning, *Interfaces*, 40:222–245. https://doi.org/10.1287/inte.1090.0492

Noriega, R and Pourrahimian, Y, 2022. A systematic review of artificial intelligence and data-driven approaches in strategic open pit mine planning, *Resources Policy*, 77:102727. https://doi.org/10.1016/j.resourpol.2022.102727

Orús, R, 2014. A practical introduction to tensor networks: matrix product states and projected entangled pair states, *Annals of Physics*, 349:117–158. https://doi.org/10.1016/j.aop.2014.06.013

Paravarzar, S, Emery, X and Madani, N, 2015. Comparing sequential Gaussian and turning bands algorithms for cosimulating grades in multielement deposits, *Minerals Engineering*, 347:84–93. https://doi.org/10.1016/j.crte.2015.05.008

Pirnay, N, Ulitzsch, V, Wilde, F, Eisert, J and Seifert, J P, 2024. An in-principle super-polynomial quantum advantage for approximating combinatorial optimisation problems via computational learning theory, *Science Advances*, 10:eadj5170. https://doi.org/10.1126/sciadv.adj5170

Punnen, A P, 2022. *The Quadratic Unconstrained Binary Optimization Problem: Theory, Algorithms and Applications*, (Springer International Publishing). https://doi.org/10.1007/978-3-031-04520-2

Quelopana, A and Navarra, A, 2024. Incorporating operational modes into long-term open pit mine planning under geological uncertainty: an optimisation combining variable neighbourhood descent with linear programming, *Mining, Metallurgy and Exploration*. https://doi.org/10.1007/s42461-024-01052-9

Quelopana, A, Ordenes, J, Araya, R and Navarra, A, 2023. Geometallurgical detailing of plant operation within open pit strategic mine planning, *Processes*, 11:381. https://doi.org/10.3390/pr11020381

Rahnema, M, Amirmoeini, B and Moradi Afrapoli, A, 2023. Incorporating environmental impacts into short-term mine planning: a literature survey, *Mining*, 3(1):10. https://doi.org/10.3390/mining3010010

Ramazan, S and Dimitrakopoulos, R, 2013. Production scheduling with uncertain supply: a new solution to the open pit mining problem, *Optimization and Engineering*, 14:361–380. https://doi.org/10.1007/s11081-012-9186-2

Rebentrost, P, Lloyd, S, Schuld, M, Petruccione, F and Wossnig, L, 2019. Quantum gradient descent and Newton's method for constrained polynomial optimisation, *New Journal of Physics*, 21(7). https://doi.org/10.1088/1367-2630/ab2a9e

Russell, S and Norvig, P, 2021. *Artificial Intelligence: A Modern Approach*, 4th edn (Pearson Education).

Salm, M, Barzen, J, Breitenbücher, U, Leymann, F, Weder, B and Wild, K, 2020. The NISQ analyzer: automating the selection of quantum computers for quantum algorithms, in *Symposium and Summer School on Service-Oriented Computing*, pp 66–85 (Springer). https://doi.org/10.1007/978-3-030-64846-6_5

Scherer, A, Guggemos, T, Grundner-Culemann, S, Pomplun, N, Prüfer, S and Spörl, A, 2021. On-call operator scheduling for satellites with Grover's algorithm, in *Computational Science – ICCS 2021* (eds: M Paszynski, D Kranzlmüller, V V Krzhizhanovskaya, J J Dongarra and P M A Sloot), pp 803–810 (Springer). https://doi.org/10.1007/978-3-030-77980-1_2

Schollwöck, U, 2011. The density-matrix renormalisation group in the age of matrix product states, *Annals of Physics*, 326:96–192. https://doi.org/10.1016/j.aop.2010.09.012

Schworm, P, Wu, X, Klar, M, Glatt, M and Aurich, J C, 2024. Multi-objective quantum annealing approach for solving flexible job shop scheduling in manufacturing, *Journal of Manufacturing Systems*, 72:142–153. https://doi.org/10.1016/j.jmsy.2023.11.015

Schworm, P, Wu, X, Wagner, M, Ehmsen, S, Glatt, M and Aurich, J C, 2023. Energy supply scheduling in manufacturing systems using quantum annealing, *Manufacturing Letters*, 38:47–51. https://doi.org/10.1016/j.mfglet.2023.09.005

Senecal, R and Dimitrakopoulos, R, 2020. Long-term mine production scheduling with multiple processing destinations under mineral supply uncertainty, based on multi-neighbourhood Tabu search, *International Journal of Mining, Reclamation and Environment*, 34:459–475. https://doi.org/10.1080/17480930.2019.1595902

Shishvan, M S and Sattarvand, J, 2015. Long-term production planning of open pit mines by ant colony optimisation, *European Journal of Operational Research*, 240:825–836. https://doi.org/10.1016/j.ejor.2014.07.040

Skyttner, L, 2001. *General systems theory – ideas and applications* (Singapore: World Scientific Publishing Co, Pte, Ltd).

Sutton, R S and Barto, A G, 2018. *Reinforcement learning: an introduction*, 2nd edn (Cambridge: MIT Press).

Villalonga, B, Boixo, S, Nelson, B, Henze, C, Rieffel, E and Biswas, R, 2019. A flexible high-performance simulator for verifying and benchmarking quantum circuits implemented on real hardware, *NPJ Quantum Information*, 5:1–16. https://doi.org/10.1038/s41534-019-0196-1

Wang, Y, Kim, J E and Suresh, K, 2023. Opportunities and challenges of quantum computing for engineering optimisation, *Journal of Computing and Information Science in Engineering*, 23(6). https://doi.org/10.1115/1.4062969

Warren, R H, 2013. Adapting the traveling salesman problem to an adiabatic quantum computer, *Quantum Information Processing*, 12:1781–1785. https://doi.org/10.1007/s11128-012-0490-8

Yamamoto, J, 2008. Estimation or simulation? That is the question, *Computational Geosciences*, 12:573–591. https://doi.org/10.1007/s10596-008-9096-8

Yonaga, K, Miyama, M, Ohzeki, M, Hirano, K, Kobayashi, H and Kurokawa, T, 2022. Quantum optimisation with Lagrangian decomposition for multiple-process scheduling in steel manufacturing, *ISIJ International*, 62(9):1874–1880. https://doi.org/10.2355/isijinternational.ISIJINT-2022-019

Mine operations management – the path to digital connected mining operations

S Rahangdale[1] and K Chhabra[2]

1. Industry Consultant, Dassault Systèmes, Singapore. Email: sagar.rahangdale@3ds.com
2. Industry Consultant Manager, Dassault Systèmes, Perth WA 6000.
   Email: kriti.chhabra@3ds.com

ABSTRACT

Mining companies must use every possible strategy to manage the pressure to produce more. Yet many mine companies are missing out on significant opportunities to boost the productivity of their mines and processing plant operations.

Many industries have achieved a level of execution discipline well beyond what the mining industry is experiencing today. For example, some mines still capture operations data manually, often writing information down on paper before transferring it to an Excel spreadsheet that they then send out by email. It builds organisational and technological silos, where important information may be held in one area when it could be of great use to someone in another department, but they don't even know it exists.

This paper defines the path for connected, digital mine operations using Mine Operations Management (MOM).

Executing digital transformation through mine operations management

Digital transformation requires the real world to be reflected in a digital world – reproducing products, plans and processes in digital form.

The domain where digital and physical process converge can be managed by MOM solution– Mine Operations Management. While many mining companies have elements of MOM, only a few are just starting to adopt the ISA-95 architecture found in manufacturing. It serves middle domain to bridge the gap between business systems and physical operations.

MOM solution provides a unified data platform and allows users to connect directly into equipment's and real time controls as well as ERP, geology, planning and engineering applications to automate data collection and display it all within one platform.

MOM enables superior work management through increased visibility by managing Material Tracking and Reconciliation from pit to port, managing the work and workforce either in mine or processing plants. It also provides control over performance with comprehensive monitoring of operations KPI, Asset or Equipment Performance monitoring with Maintenance capabilities.

INTRODUCTION

The mining industry is a fundamental pillar of the global economy, providing essential raw materials for construction, manufacturing, energy production, and technological advancements. Despite its significance, the industry faces persistent challenges, including fluctuating commodity prices, operational inefficiencies, environmental concerns, and the need for technological adaptation to enhance productivity and sustainability.

As the industry navigates evolving regulatory frameworks and market demands, companies that invest in digital transformation and energy-efficient solutions will be best positioned to drive productivity while minimising environmental impact.

This paper intended to provide details on Mine Operation Management (MOM) offerings as it supports digital transformation by enabling different departments and functions to use single collaborative unified platform for different functionalities.

DIGITAL TRANSFORMATION IN MINING

Although the concept of digital transformation has existed for some time, there is no one-size-fits-all approach or widespread adoption within the industry. However, the aim of any new initiatives must be to address below business drivers and then subsequently identify the areas and use cases and tie them to business strategy:

- enhancing operational efficiency

- connected operations visibility

- data-driven decision-making

- meeting regulatory pressures

- enabling remote operations

- adapting to changing market demands.

Mine sites are facing important challenges regarding the adoption and development of innovative concepts, in contrast, the manufacturing sector, which shares similarities with mining in terms of operations and execution, has been at the forefront of embracing digital transformation through initiatives like smart factories, IIOT, Industry 4.0, and connected factories.

Digital transformation requires the real world to be reflected in a digital world – reproducing products, plans and processes in digital form. This domain where these worlds converge is orchestrated by the critical element of a digital transformation strategy – a MOM solution.

MOM software represents an evolution and extension of the manufacturing execution system (MES), Manufacturing Operations Management from manufacturing industry with specific use cases from mining operations. In a modern mine sites or processing plants, a system for process orchestration with predictive insight is mandatory. This system must seamlessly collaborate with Geologist, Mine Planner, Surveyor, Plant Managers, Supervisors, Maintenance Team for effective usage of solution across the value chain. Thus, the future of digital transformation in mining operations largely hinges on the future of MOM.

MOM technology needs to continue evolve in ways that make digital transformation accessible to a broader spectrum of mining companies. The MOM solution will accelerate digital transformation for the agile, innovative mining companies because it enables low-risk, high-reward adoption of specific MOM functionality in step-by-step approach.

The MOM technology has to be more modular in approach and offering to support the step wise or systematic approach for covering the digital transformation.

The comprehensive digital transformation strategy involves IIOT, IT-OT convergence, Virtual Twin and MOM. The following section outlines the concept of MOM (Mine Operations Management) and its key components.

MINE OPERATIONS MANAGEMENT

The Mine Operations Management solution offers a comprehensive portfolio of end-to-end applications spanning the entire mining value chain, as illustrated in Figure 1. The applications help manage production performance in real-time and ensure compliance with the plan and allows to capitalise on operational data and produce actionable insights for faster and informed strategic decision-making.

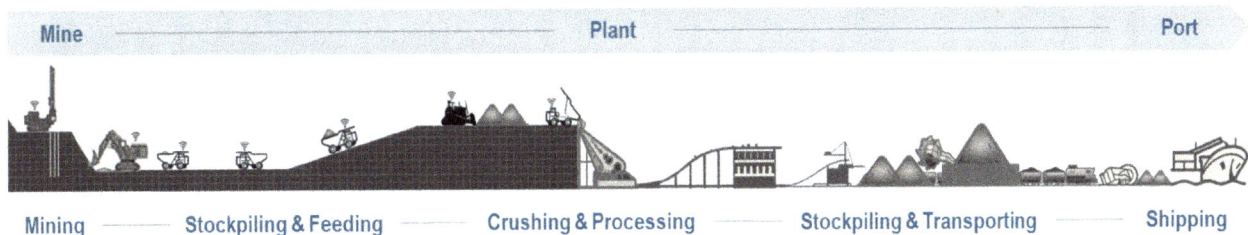

FIG 1 – MOM coverage.

MOM manages the entire process, starting from the extraction phase when the ore is dug and moved to the next location, all the way through to when it is transported to the port or stored. In this way, MOM covers both mine sites and processing plants, covering the following key functions:

- managing the complex material movement data
- manage the master data such geological data, block model data from material balance calculation
- capture movement transactions
- stockpile management with inventory in out reporting
- capture surveys
- end of month material reconciliation
- sampling data capture
- provide instructions for manufacturing the product on the shop floor
- manage actual production activities, equipment and processes
- collect production data and provide access to other functions for planning and quality purposes
- manage machine/asset maintenance
- track machine performance
- provide instructions
- verify that operators are properly certified and trained
- short interval control plan versus actual
- shift logs
- quality checks and analysis.

MOM functionality can be categorised in different sections with common platform for master data management and reporting and analytics.

To maximise benefits and adopt a modular, step-by-step approach, a unified platform that covers all use cases is an ideal solution. This reduces the need for excessive integrations within individual functions, allowing greater focus on connecting with other business systems, devices, and assets, ultimately leading to a more streamlined architecture across the IT landscape.

MOM components

The comprehensive MOM solution is a combination of COTS/OOB packages and specific functions, tailored to meet the mine's unique needs and requirements, addressing the unique challenges. At a broader level, the MOM solution can be categorised into three components as shown in Figure 2. Material Reconciliation consolidates production data and monitors material flow in real time to maintain alignment with physical inventory and minimise discrepancies. Operation control manages and tracks tasks, resources, and execution to ensure on-plan performance, and Assets Performance optimises equipment utilisation, schedules maintenance, and drives uptime and efficiency.

FIG 2 – MOM modules.

Material tracking and reconciliation

The reconciliation solution is used to manage actual material tracking and inventory reporting in proactive and systematic way. It helps standardise the procedure for performing month end reconciliation across the company in an open and documented fashion. This simplified process helps easily identify and analyse variance to calibrate the mine plan.

The solution provides different features to capture the movement, integrate with FMS, historian etc for the data form mine sites activities and synchronise the data with material envelope for accurate reporting of inventory.

Operational control

Within the operational control, we can focus on the following two essential functions that ensure both tasks and people are aligned for seamless mine operations execution:

1. Work Order/Task Management: Establishes a standardised workflow for generating, assigning, and tracking work orders, ensuring tasks are executed efficiently, accurately and with complete visibility.

2. Workforce Management: Real-time workforce performance tracking and centralised labour management across mine and processing plants.

The complete operational control solution targets the digital way of executing work with screens and configurable workflow targeting production/processing plant technicians, operators, supervisors to assist in their day-to-day job while capturing necessary information for operational enhancements. To illustrate how the solution adds value in day-to-day operations, here are a few example use cases:

- Shift Log – provides feature and options for shift supervisor to collect data from different areas and track.

- Production/Operation Execution: Manage coordination of production and other activities and tracks task execution.

- Directed real-time control, verified data collection, Tracks task execution and KPIs.

- Digital work instructions: Enable employees to efficiently perform tasks by providing guidance and work instructions (Paperless Operations).
- Electronically tracks time and attendance for all employees.
- Labor related KPIs and integration with ERP and HR applications.

Asset performance and plant monitoring

A plant monitoring solution offers real-time visibility and work-in-progress reports by collecting data from various equipment and correlating it with operational data. It enables data analysis through advanced analytics and reporting tools, providing valuable insights into plant performance. By utilising MOM (Manufacturing Operations Management) integration technology and leveraging IIoT (Industrial Internet of Things), data from different devices and equipment can be seamlessly integrated into the MOM system for continuous monitoring and improved operational efficiency.

Asset Maintenance: The maintenance solution offers features to manage machine or asset maintenance effectively. It allows for the planning of preventive or scheduled maintenance activities and supports the execution of both preventive and reactive maintenance digitally. This includes providing digital work instructions, tracking spare part inventory availability, assigning maintenance tasks, and monitoring their progress to ensure timely completion and optimal asset performance. Maintenance solution in MOM, can be integrated to EAM Enterprise Asset Management solution to get asset related data.

Data management

Master Data: Manage the master data required from different use case at one place example dig locations, measurements, material compounds, equipment etc data can be managed directly in MOM or automate from source system such as Geology Model, ERP, etc.

Transactional Data: Transactional data is primarily composed of two elements: direct data entry or workflow execution within the MOM system, and data integrated from other systems and devices. As mentioned earlier, continuous data exchange is essential for various use cases, such as FMS (Fleet Management System) integration for material movement data, and equipment data for OEE (Overall Equipment Efficiency) or performance analysis.

Since the MOM system supports a wide range of use cases across different departments within the mining value chain, it is crucial to integrate with various other systems to ensure seamless data flow and enable effective decision-making.

MOM systems adhere to ISA-95 standards for integration between business systems and operational systems. At a high level, they must integrate directly with instrumentation and real-time control solutions (ISA-Level 2), as well as with geology, planning, and engineering applications (ISA-Level 3). Additionally, they need to connect with business systems like ERP and HRM (ISA-Level 4). This integration is achieved through secure architecture and the use of communication protocols such as OPC UA, lightweight MQTT messaging, Web Services and other industry-standard protocols that are best suited for the mining operational environment.

Leveraging data from across the mine value chain, the MOM solution can also support advanced use cases, such as emissions reporting and the prediction of abnormal material movement, by utilising advanced analytics capabilities.

CONCLUSIONS

From initial mine site activities to processing plants, a fully integrated modular MOM will help a mining operation achieve operational excellence by connecting and standardising key business processes. It will also ensure that processes are handed off seamlessly between different business disciplines. The modular MOM technology, with its flexible data modelling and unified data management, enables rapid scalability and reduces upgrade risks. It accelerates the ability to achieve connected operations, integrate IIoT, and adopt other future advancements related to MOM, ensuring long-term adaptability and growth.

Dynamic on-bench loading location optimisation of explosive trucks for short-term mine planning

M Samaei[1], R Shirani Faradonbeh[2], P Klaric[3] and R Suhane[4]

1. PhD Candidate, Minerals, Energy and Chemical Engineering, WA School of Mines (WASM), Kalgoorlie WA 6430. Email: samaei.masoud@curtin.edu.au
2. Lecturer, Minerals, Energy and Chemical Engineering, WA School of Mines (WASM), Kalgoorlie WA 6430. Email: roohollah.shiranifaradonbeh@curtin.edu.au
3. Principal, Drill and Blast, Fortescue, Perth WA 6000. Email: paul.klaric@fortescue.com
4. Senior Mining Engineer, Fortescue, Perth WA 6000. Email: rahul.suhane@fortescue.com

ABSTRACT

The mining industry is progressing steadily toward operation optimisation, with drill-and-blast operations presenting significant challenges in this transition. Many mining processes still require optimisation especially for mine sites with a substantial geographic spread. One such process, which consumes substantial time and resources, involves traveling distance for charging blastholes using Mobile Processing Units (MPUs) and re-loading MPUs at facilities which may not be in proximity to the charging location. This study introduces an innovative programming tool designed to optimise the number of MPUs and their routes within a pit, taking into account factors such as explosive types and MPU capacities based on the provided mine plan inputs. Additionally, the tool employs a Genetic Algorithm (GA) and satellite road networking mapping to determine the optimal on-bench loading locations for multiple pits, minimising travel distances between MPUs, pits, on-bench loading areas, and magazines. The tool was tested over four consecutive six month periods of the mine plan for Fortescue, demonstrating significant reductions in travel distances, thus providing the potential to minimise operational costs.

INTRODUCTION

The mining industry consistently seeks to reduce operational costs and enhance efficiency to maximise profitability while adopting more sustainable practices. In the past decade, a key focus area has been reducing transportation costs across various operational processes, while transportation costs can account for up to 50 per cent of total operating expenses in mining, amounting to millions of dollars annually. These efforts can be classified into several categories, including the adoption of alternative energy sources such as hydrogen and electric-powered vehicles (Guerra et al, 2020; Issa et al, 2023), the development of energy recovery systems (ERSs) (Terblanche et al, 2019), improvements to haul road surfaces (Bodziony and Patyk, 2024), transitions between transportation modes (eg road, rail, and monorail) (Ryabko and Gutarevich, 2021), path optimisation (Brazil et al, 2015; Yardimci and Karpuz, 2019), and optimising fleet dispatch locations (Tomy, Seiler and Hill, 2024; Moradi-Afrapoli and Askari-Nasab, 2020). Most of the mentioned fields require time and cost for research and development, while categories such as path and dispatching location optimisation could be used to achieve advantages faster.

Mobile Processing Units (MPUs) are essential in open pit mining, as they are designed to manufacture or blend bulk explosives directly at the blasting site (Worksafe Western Australia, 2024). These specialised vehicles are reloaded with bulk explosives precursors either at the magazine location or through On-Bench Loading (OBL). The distances between the magazine, OBL sites, and pits significantly affect drill and blast operations, which in turn affect overall operational efficiency and mining profitability. Despite the critical role of drill and blast operations—often a bottleneck in mining workflows—optimisation strategies for MPU dispatching and path planning have received little attention in the literature. Given that drill and blast efficiency directly impacts downstream mining and processing activities, further research in this area is essential (Samaei et al, 2025). Open pit Mine Scheduling Problems (OPMSPs), as discussed in Chicoisne et al (2012), have primarily focused on managing block-related tasks such as mining scheduling (determining extraction sequences), dump scheduling (allocating material to waste dumps or processing plants), and haulage route planning (optimising transportation routes). However, the role of auxiliary equipment such as drilling rigs and MPUs has been largely overlooked in these studies, as confirmed by the literature review conducted by the authors.

In small-scale operations, a single MPU may be sufficient, and optimising its operation would have a minimal impact on cost and efficiency. However, in large-scale operations with multiple, geographically spread pits, the number of MPUs and their reloading frequency increase significantly, creating logistical and cost-related challenges. This study identifies a gap in evaluating the operational costs of MPUs and exploring strategies to mitigate them. Therefore, the research focuses on developing a system for short-term mine planning that determines the required number of MPUs and calculates the cumulative distances they must travel from each pit to the magazine location. To enhance efficiency without relocating the magazine, optimising mid-term haulage between the magazine and the pits should be prioritised over simply selecting a median-based location. However, since the ideal location for a magazine may change over time—and relocating it is both costly and disruptive—establishing an OBL area as a temporary feeding point for MPUs presents a more practical alternative. Positioned just before blasting, the OBL site allows for real-time adjustments, reduces travel distances for MPUs, and improves operational flexibility. Its strategic placement is crucial for enhancing drill and blast efficiency, ensuring safety, and optimising the preparation and loading of explosives into blastholes.

By integrating these considerations into a mine planning framework, this study aims to develop an optimisation approach that determines the most effective dynamic OBL locations for groups of pits in different time frames, ultimately improving cost efficiency and operational effectiveness in large-scale mining operations.

STUDY AREA AND DATA

This study utilises short-term planning data from the Solomon Hub mines (see Figure 1), a major iron ore operation within Fortescue company's portfolio. The Solomon Hub consists of six mining areas: Firetail North (FTN), Firetail South (FTS), Valley of the Kings (VOK), Valley of the Queens (VOQ), Trinity (TRN), and FRD (Fredericks). The Solomon Hub is situated in the Pilbara region of Western Australia, within the mineral-rich Hamersley Ranges. Its strategic location, in proximity to Fortescue's rail infrastructure, enables efficient transport of iron ore to Port Hedland for export, reinforcing the hub's role as a major contributor to Fortescue's total production.

FIG 1 – Location of the case study, Solomon, located in the Western Hub (Fortescue, 2015).

Fortescue's presence in the region dates to the early 2000s, when extensive exploration identified significant iron ore deposits. Mining operations officially commenced in 2012, marking a pivotal expansion in Fortescue's growth. The Solomon Hub's development has been implemented in phases: the Firetail deposit began production in May 2013, followed by Kings Valley in December 2013. In 2019, Fortescue approved the development of Queens Valley, with an investment of A$417 million, and the mine became operational in 2022. This phased expansion reflects Fortescue's strategic approach to scaling production while integrating technological advancements such as autonomous equipment and innovative ore processing facilities.

The system developed in this study consists of two primary components. The first component focuses on determining the number of required MPUs for each mine, considering every blasting pattern and delivery amount, as well as the distances MPUs travel within each pit.

The following data was utilised in this phase:

- Blasting period date.
- Pit name.
- Types of explosives used.
- Explosives volume.
- Number of MPUs deployed.
- Volume capacity of MPUs for each explosive type.
- Average burden during the time frame.
- Average spacing during the time frame.
- Number of blastholes within the time frame for each explosive type.

The second component focuses on optimising the location of the OBL area for each time frame. This involves determining whether MPUs should load explosives from the OBL or the magazine, as well as identifying the most efficient routes for material transport. To achieve this, the following data was considered in this phase:

- Number of required MPUs.
- Distance travelled by MPUs within each pit.
- Location of pits and magazine.
- Available routes between pits and the magazine.

By integrating these two components, the system aims to enhance the efficiency of MPU allocation and optimise explosive loading locations, ultimately improving the overall drill and blast operation.

METHODOLOGY

Two algorithms were developed for the system's components. The first algorithm was a heuristic approach specifically designed for this project, making it a novel solution that does not rely on any existing methodology. It was tailored to determine the required number of MPUs for each type of explosive, considering factors such as the type of explosive, vehicle capacity, blast pattern hole configuration, explosive volume per hole, burden, and spacing.

When a drill pattern is unavailable, the algorithm estimates distances using average burden, spacing, and the assumed ratio of the pattern. However, if a specific blast pattern is provided, it generates paths tailored to that pattern and further optimises routes by adjusting path directions. This optimisation aims to minimise tramming distances and reduce overall project time.

The second phase of the project employs a Genetic Algorithm (GA) to determine the optimal location for the OBL area. GA is an optimisation algorithm inspired by natural selection and evolution, widely used for complex problems where traditional methods may struggle (Samaei et al, 2022). Originally developed by John Holland in the 1970s, GA belongs to the family of evolutionary algorithms and simulates biological processes such as reproduction, mutation, and survival of the fittest. They are particularly effective for problems with large search spaces, including engineering design, scheduling, and machine learning parameter tuning, where finding an efficient solution is crucial (Sampson, 1976).

GAs follow a structured yet adaptive process. They begin with an initial population of randomly generated solutions, typically represented as strings or chromosomes (eg binary codes or real numbers). Each solution's fitness is evaluated, and the fittest individuals are selected—using methods like roulette wheel selection or tournament selection—for reproduction. During this process, crossover (combining segments of two parent solutions) and mutation (randomly modifying solution elements) introduce diversity, preventing the algorithm from converging prematurely on local optima. Over successive generations, the population evolves as weaker solutions are eliminated and stronger ones move toward an optimal or near-optimal outcome. This iterative process continues until a termination criterion is met, such as reaching a maximum number of generations or achieving a satisfactory fitness level, making GAs a powerful tool for adaptive problem-solving (Sampson, 1976). Figure 2 shows the flow chart of the GA algorithm adjusted for OBL location optimisation.

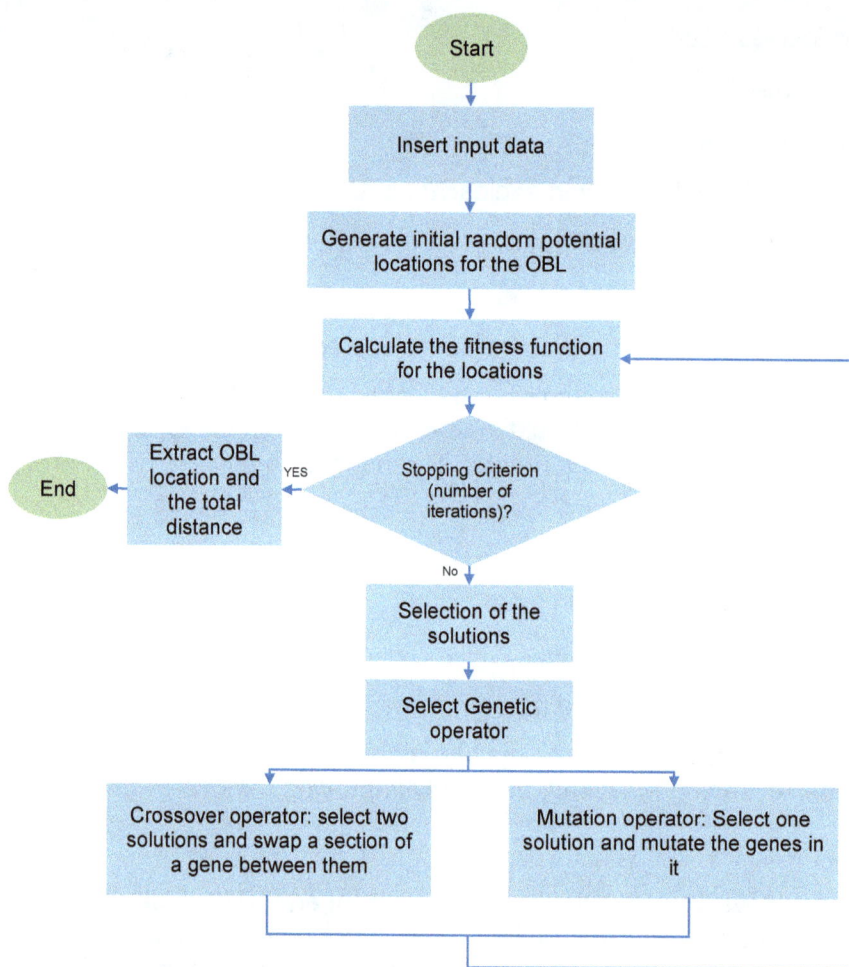

FIG 2 – Flow chart of the Genetic Algorithm (GA).

RESULTS AND DISCUSSION

As mentioned in the previous section, the system starts with calculating the number of required MPUs for each type of explosive and, subsequently, the in-pit distances for the given time frame. The code is fed with a six-month time period to calculate the overall distances. The in-pit distance is not only the distance between holes; the distance that each MPU takes to reach the starting hole and, after discharge, to arrive at the pit entrance is also important. Hence, it is essential to consider the distance between the pit entrance and the starting point for charging, as this can significantly impact the total distance compared to measuring only the distances between holes. If actual patterns are provided to the system, in-pit distances will be calculated based on the given pattern by the system. Otherwise, the system will generate a schematic pattern using available information, such as burden, spacing, and the probable pattern's dimensional ratio.

The image below depicts the approximate path for the given pattern. Since path generation is not the primary focus, minor errors can be overlooked. Additionally, such errors may occur across all patterns and pits but have little impact on estimating the OBL location. Each colour presents the holes that could be charged by a fully loaded MPU and the distance from the bench entrance to the starting and end holes.

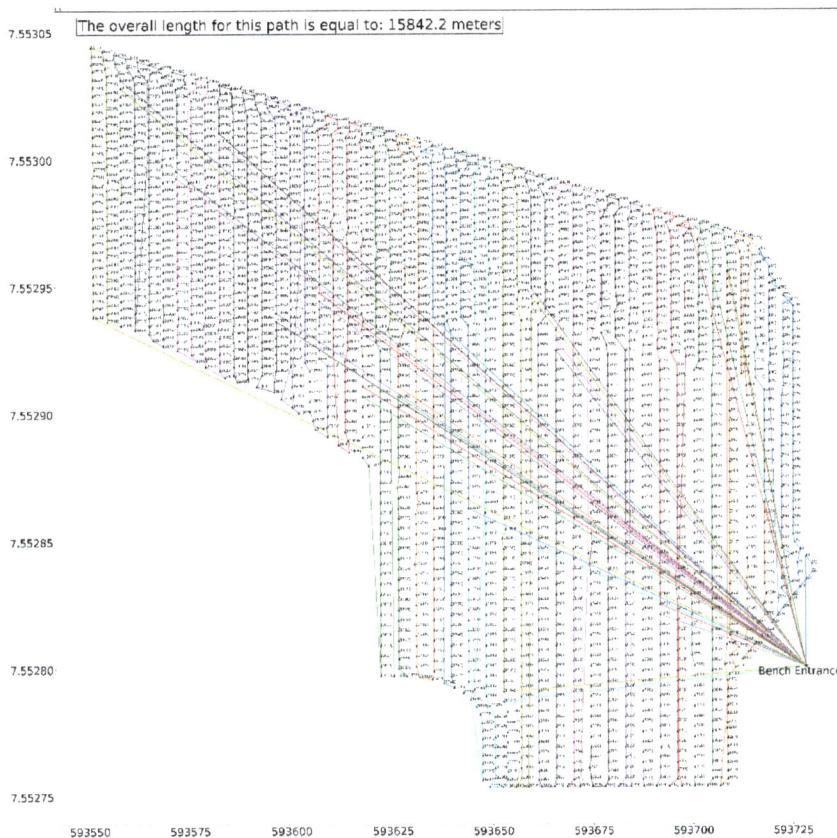

FIG 3 – Approximate path of MPUs on a blasting pattern consisting the bench entrance, the location of the first hole to be loaded and the last hole before the vehicle discharge.

After calculating the overall distances within a pit for the time frame, the second stage is to find the most optimum OBL location for the hub. As mentioned in the previous section, a GA was used to find the location. The most important part of the GA is the definition of an appropriate fitness function to the algorithm in which the algorithm tries to minimise or maximise it. In the current study, the aim was to minimise the overall distance of MPUs commuting in a six-month period. Therefore, the Equation 1 is defined as the fitness function.

$$OTO = \sum_{1}^{n} POD_n \tag{1}$$

While,

$$
\begin{aligned}
POD = \ &[2 * \text{ number of MPUs} \\
&* \text{ distance between entrance point and OBL}] \\
&+ [\text{overall distance within a pit that MPUs travel} \\
&- \text{ distance between holes}]
\end{aligned}
\tag{2}
$$

The fitness function was defined as the overall distance from a pit entrance to the on-bench loading location (OTO), which is equal to the pit to OBL location distance (POD) from each of the pits. In the current study, six pits are available; therefore, $OTO = POD_1 + POD_2 + POD_3 + POD_4 + POD_5 + POD_6$ tried to find the best location for OBL in which OTO is minimum.

Another criterion defined for the algorithm was the search area. It is evident that only certain points can serve as OBL locations, while others may be unsuitable due to factors such as topography or proximity to facilities. In this study, only a 30 m wide corridor parallel to existing routes was considered as a potential OBL location. However, future studies should focus on developing a more comprehensive set of criteria for determining suitable OBL locations. GA has several hyperparameters which can influence the final results. The optimum values of these parameters were acquired using a trial-and-error procedure. Table 1 shows the obtained values.

TABLE 1

Applied hyperparameters for the GA.

Max iteration	1000
Population size	200
Mutation probability	0.1
Crossover probability	0.8
Crossover type	Uniform

As shown in Figure 4, the model convergence graph shows an appropriate model development process. The graph shows that the algorithm could decrease the distance by moving the OBL location to the potential locations and, after almost 700 iterations, reached the most optimum solution.

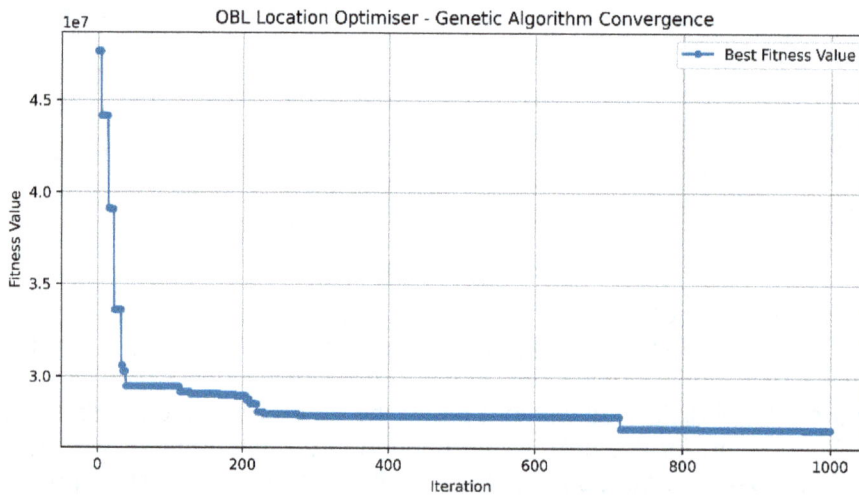

FIG 4 – GA convergence graph.

Finally, the results of the code for four consecutive six month periods are shown in Figure 5. The analysis of OBL locations over different time frames reveals that the optimal position remained relatively stable during the initial periods, shifting slightly towards the right-side pits before moving significantly in the final time frame. This stability suggests that mining operations in the VOQ pit were significantly higher compared to the other pits, and that haulage distances and logistical factors did not change considerably during the first 18 months, maintaining the efficiency of the chosen OBL location. However, in the last six month period, the suggested location of the OBL moved closer to the FTN and FTS pits, indicating a substantial shift in operational conditions. This relocation was likely driven by factors such as increased activity in these pits or a limiting of production at the VOQ pit. The connectivity between pits and the OBL, as represented by haulage paths, further supports the hypothesis that a different set of pits became dominant in the operation. This shift suggests a planned transition in production focus, emphasising the need for dynamic optimisation of the OBL location to adapt to evolving mining conditions.

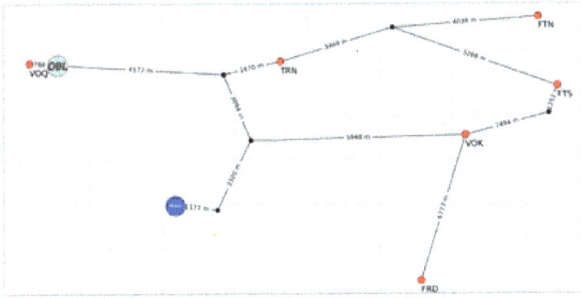

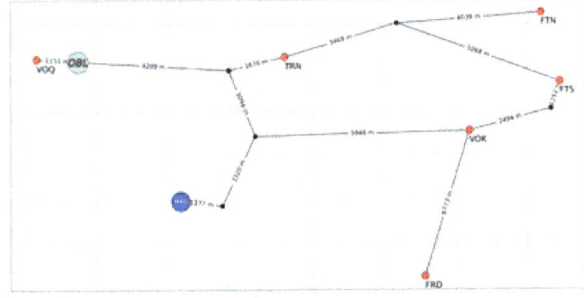

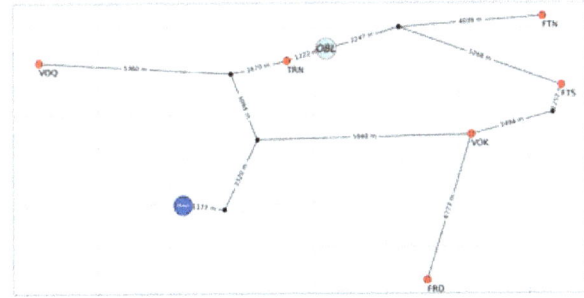

- VOQ is better to use the OBL
- VOK is better to use the Magazine
- FTN is better to use the OBL
- FRD is better to use the Magazine
- TRN is better to use the OBL
- FTN is better to use the Magazine

(a)

- VOQ is better to use the OBL
- VOK is better to use the Magazine
- FTN is better to use the OBL
- FRD is better to use the Magazine
- TRN is better to use the OBL
- FTN is better to use the Magazine

(b)

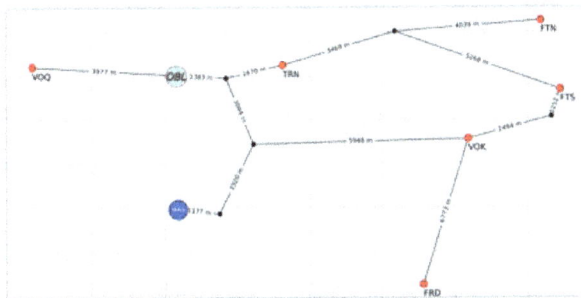

- VOQ is better to use the OBL
- VOK is better to use the OBL
- FTN is better to use the OBL
- FRD is better to use the OBL
- TRN is better to use the OBL
- FTN is better to use the OBL

(c)

- VOQ is better to use the OBL
- VOK is better to use the Magazine
- FTN is better to use the OBL
- FRD is better to use the Magazine
- TRN is better to use the OBL
- FTN is better to use the OBL

(d)

FIG 5 – The optimised OBL location for different time frames: (a) 1–6 months; (b) 7–12 months; (c) 13–18 months; (d) 19–24 months (distances are aerial).

In the next step, the generated images of the optimised OBL locations and haulage routes were overlaid onto satellite imagery, providing a real-world comparison of the computed paths with actual terrain conditions. The analysis reveals that, for 12–18 month time frames, the proposed OBL location may not be feasible due to discrepancies between the real road network and the routes considered in the optimisation process; however, close proximity to the calculated location in the nearest available location could be used as an OBL location. This misalignment highlights the necessity for improving accuracy in future studies, ensuring that the optimisation framework accounts for actual road constraints, topographical limitations, and potential operational restrictions. Incorporating high-resolution road data, terrain models, and real-time operational constraints could significantly enhance the robustness and applicability of the optimisation model.

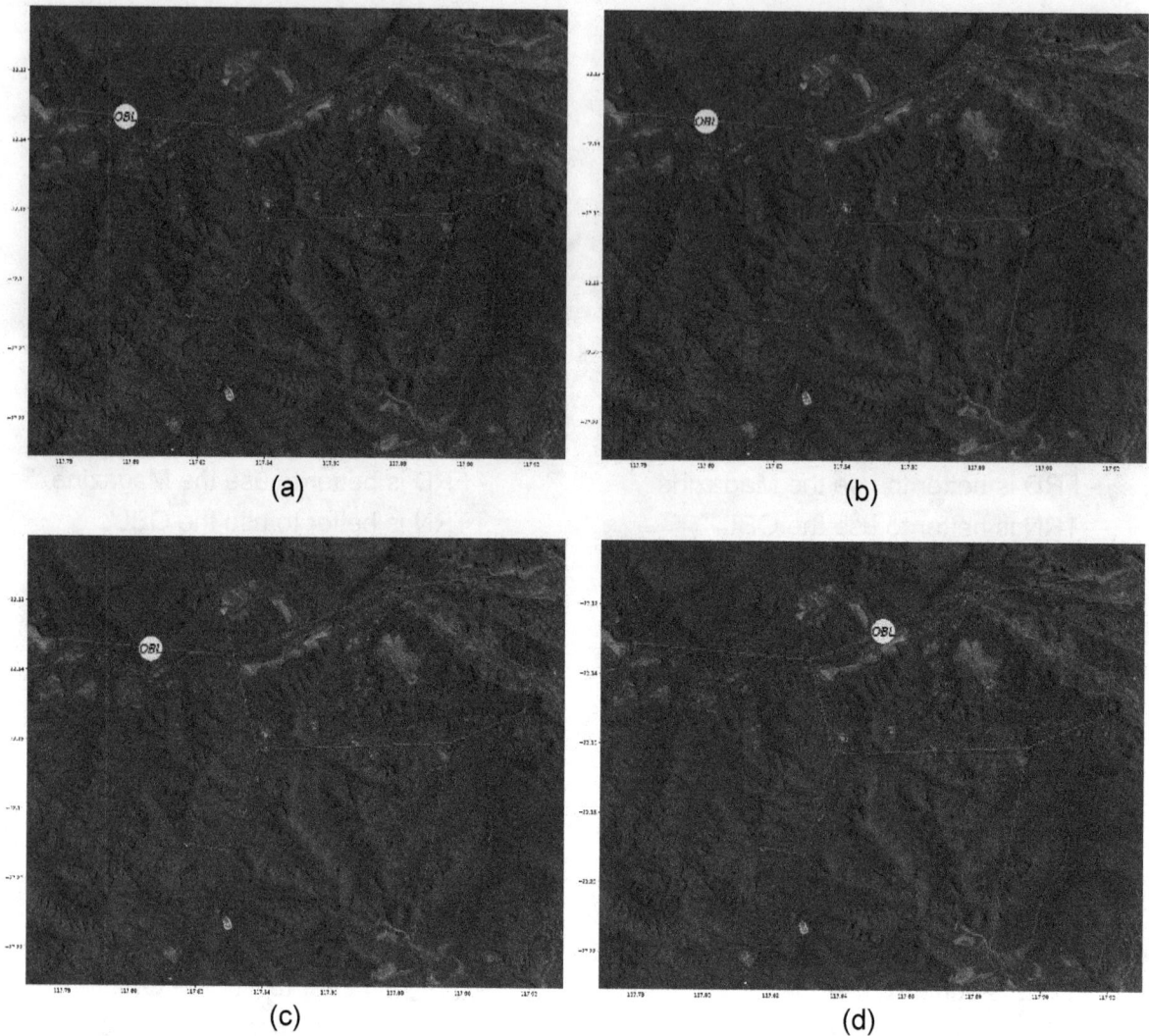

FIG 6 – The optimised OBL location on satellite images for different time frames presenting the feasibility of the locations based on the topographical constraints: (a) 1–6 months; (b) 7–12 months; (c) 13–18 months; (d) 19–24 months.

CONCLUSIONS

This study successfully developed an optimisation system for OBL locations in open pit mining, integrating a novel in-pit distance calculator and a GA to enhance drill-and-blast operations. The results demonstrate that this approach effectively reduces the total distance travelled by MPUs, leading to a significant improvement in operational efficiency. By dynamically optimising OBL locations over multiple planning periods, the system has the potential to minimise costs and reduce operational time, ultimately enhancing the profitability of mining operations. The reductions in MPU travel distances not only lower fuel consumption but also decrease the need for vehicles and operators, further contributing to cost savings. Despite the promising results, this area of research has received limited attention in the literature, highlighting the need for further investigation. Existing studies on mine scheduling and fleet management often overlook the impact of OBL locations on drill-and-blast operations. Given the substantial cost implications of MPU travel distances, this optimisation approach should be explored further to establish more refined decision-making frameworks for OBL placement.

Future improvements to this system could focus on incorporating high-resolution road network data, real-time operational constraints, and terrain models to enhance accuracy. Additionally, integrating machine learning techniques alongside GA could further optimise OBL locations based on evolving mining conditions. As mining operations continue to move towards full automation, the implementation of such optimisation tools will play a critical role in maximising efficiency, reducing costs, and improving overall operational sustainability.

ACKNOWLEDGEMENT

The authors would like to express their sincere gratitude to Fortescue for providing the necessary data and operational insights that made this research possible. We also extend our appreciation to the WA School of Mines (WASM), Curtin University, for their academic support and resources throughout this study. The collaboration between WASM and Fortescue has been instrumental in advancing this research on drill-and-blast optimisation. Finally, we acknowledge the contributions of all individuals who provided valuable feedback and technical discussions that helped refine this work.

REFERENCES

Bodziony, P and Patyk, M, 2024. The Influence of the Mining Operation Environment on the Energy Consumption and Technical Availability of Truck Haulage Operations in Surface Mines, *Energies*, 17:2654.

Brazil, M, Grossman, P A, Rubinstein, J H and Thomas, D A, 2015. Demonstrating efficiency gains from installing truck turntables at crushers, arXiv preprint. arXiv:1511.02443

Chicoisne, R, Espinoza, D, Goycoolea, M, Moreno, E and Rubio, E, 2012. A new algorithm for the open pit mine production scheduling problem, *Operations research*, 60:517–528.

Fortescue, 2015. 2015 Annual Report. Available from: <https://cdn.fortescue.com/docs/default-source/announcements-and-reports/150824-annual-financial-results-(fy2015)-presentation.pdf?sfvrsn=7d6f4bd_1>

Guerra, C F, Reyes-Bozo, L, Vyhmeister, E, Caparrós, M J, Salazar, J, Godoy-Faúndez, A, Clemente-Jul, C and Verastegui-Rayo, D, 2020. Viability analysis of underground mining machinery using green hydrogen as a fuel. *International Journal of Hydrogen Energy*, 45:5112–5121.

Issa, M, Ilinca, A, Rousse, D R, Boulon, L and Groleau, P, 2023. Renewable energy and decarbonization in the Canadian mining industry: Opportunities and challenges, *Energies*, 16:6967.

Moradi-Afrapoli, A and Askari-Nasab, H, 2020. A stochastic integrated simulation and mixed integer linear programming optimisation framework for truck dispatching problem in surface mines, *International Journal of Mining and Mineral Engineering*, 11:257–284.

Ryabko, K and Gutarevich, V, 2021. Substantiation of performance indicators of mine monorail locomotives, *Gornye nauki i tekhnologii (Mining Science and Technology; Russian)*, 6:136–143.

Samaei, M, Faradonbeh, R S, Topal, E and Goodwin, J, 2025. Path Optimisation For Drill Rigs: A Step Toward Full Automation In Mining, ROC 2025, Brisbane.

Samaei, M, Massalow, T, Abdolhosseinzadeh, A, Yagiz, S and Sabri, M M S, 2022. Application of soft computing techniques for predicting thermal conductivity of rocks, *Applied Sciences*, 12:9187.

Sampson, J R, 1976. Adaptation in natural and artificial systems (John H Holland), Society for Industrial and Applied Mathematics.

Terblanche, P J, Kearney, M P, Nehring, M and Knights, P F, 2019. Potential of on-board energy recovery systems to reduce haulage costs over the life of a deep surface mine, *Mining Technology*, 128:51–64.

Tomy, M, Seiler, K M and Hill, A J, 2024. MCTS Based Dispatch of Autonomous Vehicles under Operational Constraints for Continuous Transportation, *20th International Conference on Automation Science and Engineering (CASE)*, 4104–4111 (IEEE).

Worksafe Western Australia, 2024. Mobile Processing Units: Code of Practice. Available from: <https://www.worksafe.wa.gov.au/publications/mobile-processing-units-code-practice>

Yardimci, A G and Karpuz, C, 2019. Shortest path optimization of haul road design in underground mines using an evolutionary algorithm, *Applied Soft Computing*, 83:105668.

A robust scheduling framework for short-term planning of underground mining operations using constraint programming – a Canadian case study

R Shahin[1], M Gamache[2] and G Pesant[3]

1. Postdoctoral Fellow Researcher, Department of Mathematics and Industrial Engineering, Polytechnique Montréal, Québec H3T 0A3, Canada. Email: rezaa.shahin.1992@gmail.com
2. Professor, Department of Mathematics and Industrial Engineering, Polytechnique Montréal, Québec H3T 0A3, Canada. Email: michel.gamache@polymtl.ca
3. Professor, Department of Computer Engineering and Software Engineering, Polytechnique Montreal, Quebec H3T 0A3 Canada. Email: gilles.pesant@polymtl.ca

ABSTRACT

Underground mining presents significant challenges due to escalating costs and reduced profit margins, particularly as extraction reaches greater depths. These operations follow a cyclical workflow, comprising sequential tasks—such as drilling, charging, blasting, and loading—that rely on specialised equipment to progress each stage efficiently. Given the complexity and dependency of each step, effective scheduling of tasks and equipment is crucial for optimising resource management, ensuring operational efficiency, and maintaining profitability. This study introduces a robust scheduling framework tailored specifically for these operations, with a practical application to a Canadian underground mine. By leveraging Constraint Programming, the framework addresses both production and development scheduling across a one-month planning horizon, accounting for the multifaceted nature of the problem. The study models and solves the scheduling problem through three distinct approaches, each capturing varying degrees of certainty and complexity: (1) a deterministic model where task processing times are fully known and fixed (2) a deterministic model incorporating uncertain processing times, and (3) a two-stage multi-scenario stochastic model where task processing times are generated by a probability distribution, enabling multiple scenarios being solved simultaneously. The stochastic approach is particularly beneficial in handling the inherent uncertainties of the problem, allowing for adaptability to variable real-world conditions. Through a comparative analysis of the three models, this research underscores the value of accounting for uncertainty, revealing that the stochastic model not only achieves an optimal scheduling outcome but also produces schedules that are significantly more resilient and adaptable to unexpected disruptions. By demonstrating the effectiveness of stochastic scheduling in enhancing operational robustness, this framework provides a pathway for improving decision-making in underground mining, offering a scalable solution for resource-intensive and high-risk.

INTRODUCTION

Underground mining is characterised by high costs, with profit margins decreasing as extraction depths increase. A major determinant of efficiency and profitability in these operations is the coordination of mobile machinery. Currently, machinery fleet scheduling is predominantly manual (Aalian, Pesant and Gamache, 2023), relying on outdated methods and tools that fail to meet the increasing demands of modern operations (Aalian, Gamache and Pesant, 2024). This highlights the pressing need for improved system-level coordination to ensure the long-term profitability of mining operations.

In underground mining, activities are generally divided into two main categories: development and production. Development activities, which do not generate immediate financial returns, are mostly performed in waste rock areas to provide access to economically valuable ore deposits (Aalian, Pesant and Gamache, 2023). In contrast, production activities are carried out in valuable rock formations to extract ore material from areas referred to as stopes (Campeau and Gamache, 2020). The mine production and development scheduling problem, a critical and com- plex component of mining operations, entails selecting and sequencing activities for development and extraction to meet specific goals. In the scheduling phase, tasks include assigning mining equipment to different activities, determining the start times for each activity, and sequencing these activities. Short-term scheduling is done on a shift basis over a planning horizon of one to two weeks or one month, depending on the case study under investigation.

Mathematical programming methods, and in recent years constraint programming methods, have enabled the development of mathematical models for short-term planning of mining operations. These methods make it possible to efficiently construct work schedules for one or more shifts that take into account all operational constraints. However, implementing these schedules is difficult due to the many uncertainties related to mining operations: equipment breakdown, large variation in the duration of mining activities due to different geological characteristics from one site to another, variation in traffic on the underground mine transport network depending on the sites exploited, and so on. To overcome this problem, short-term planning models must integrate these uncertainties on the duration of activities into optimisation models in order to construct more robust schedules.

For nearly ten years, efficient communication systems have been available in underground mines, allowing the real-time transfer of information and various data on ongoing operations. This information can now be stored and analysed in order to optimise the progress of operations.

The primary objective of this study is to explore whether more robust short-term schedules can be developed for underground mining operations. Robustness, in this context, refers to the ability of a schedule to maintain feasibility and performance despite variations in task processing times. To achieve this, we integrate uncertainty considerations into the scheduling framework by introducing a two-stage stochastic formulation alongside a deterministic model with a certain degree of uncertainty, scheduling the tasks in a more 'conservative' approach. The effectiveness of these approaches is assessed by analysing their impact on schedule stability and operational efficiency under different levels of conservatism. Furthermore, two primary approaches are used to address this problem: (1) proactive and (2) reactive. The proactive approach assumes that all relevant information is known in advance, operating within a static environment where the solver has complete knowledge of future events throughout the planning horizon. In contrast, the reactive approach involves a dynamic setting in which data is gradually disclosed to the solver as the planning horizon progresses. In this study, we adopt the proactive approach.

This study contributes a CP model-based framework to aid industry decision-makers in scheduling development and production tasks. We propose two models in this study. The first model addresses uncertainty in task processing times under deterministic conditions (fixed times) and six degrees of conservatism (worst-case scenarios). The second model is a two-stage stochastic formulation, using probability distributions to solve multiple scenarios simultaneously, and its results are compared to deterministic outcomes. Finally, we generate nine instances, along with 50 scenarios per instance that include probabilistic task times. Machine sequences from the two models—deterministic with conservatism, and stochastic—are then solved using the newly generated processing times and subsequently analysed and compared to assess their performance.

In the remainder of the paper, we present a literature review in Section 2 and outline the problem description and mathematical formulation in Section 3. Section 4 covers the methodology of the study, followed by design of experiments and results in Sections 5 and 6, respectively. Finally, we conclude in Section 7.

LITERATURE REVIEW

This section reviews the existing literature on the task scheduling problem within the mining industry. The studies are categorised based on the type of mine analysed. Specifically, the literature is divided into two primary groups: research focusing on open pit mines and studies examining underground mines. Each of these categories is further classified based on the methodological approach adopted, distinguishing between works that address the problem deterministically and those that consider stochastic methods. It is noteworthy that the literature identifies three distinct levels of scheduling: long-term, medium-term, and short-term. Given that this study aims to schedule underground mining tasks within a short-term framework, our focus is limited to the short-term planning horizon for task scheduling as discussed in the literature. The search strategy to find target paper is explain in Online Appendix (see the following link for the online appendix file: <https://drive.google.com/file/d/1IAkB ZeYJD75xoJquRF3MUkYiHeKo65wj/view?usp=sharing>) part A. In the following, we review studies for open pit and underground mines, respectively.

Deterministic open pit mine

Menezes and dos Santos Correa (2022) presented a short-term scheduling model for open pit mining operations. The model along with a heuristic effectively balanced resource utilisation, production targets, and logistical constraints. Nelis and Morales (2022) developed an optimisation model that simultaneously addresses the scheduling and configuration of mining cuts in short-term open pit planning to maximise profit. The proposed model successfully integrated mining cut configuration and short-term scheduling, producing practical, high-value solutions.

Manríquez, González and Morales (2023) developed a LP formulation that optimised short-term open pit mine production scheduling, while considering multiple objectives, such as maximising equipment productivity, ensuring compliance with ore extraction targets, and managing shovel movements and stockpile utilisation. The model successfully generated balanced and robust short-term production schedules that outperformed traditional single-objective approaches. Silva-Junior et al (2023) addressed short-term planning for open pit mines, focusing on the efficient allocation and dynamic routing of a heterogeneous truck fleet. The paper aimed to develop a reliable short-term planning model that efficiently allocates trucks and schedules routes to meet production, quality, and particle size targets while minimising the use of resources, such as the number of trucks required. Habib, Ben-Awuah and Askari-Nasab (2024) introduced a LP model designed for short-term planning in open pit mines that incorporates the use of a Semi-Mobile In-Pit Crushing and Conveying (IPCC) system. The model sought to maximise profit by balancing the cost of mining, haulage, and ore processing against revenue from ore sales. Finally, Rahnema, Grenon and Moradi Afrapoli (2024) introduced a novel LP model tailored for short-term open pit mine production scheduling that explicitly integrates environmental considerations, specifically GHG emissions. The primary objective was to develop a scheduling model that maximises profit while reducing GHG emissions.

Stochastic open pit mine

Nelis, Morales and Jelvez (2023) introduced a novel LP model for short-term production scheduling in open pit mines while accounting for geological uncertainty. The objective of this study was to define mining cuts and develop a short-term production schedule that maximises profit and minimises deviation from production targets. The finding showed that the integration of uncertainty into short-term scheduling significantly enhanced robustness of results and target compliance compared to deterministic methods. Martins, Souza and Assis (2024) integrated a LP model with a Discrete Event Simulation (DES) to optimise and simulate short-term mining planning in open pit mines. The objective was to develop a robust and adaptive tool for short-term mining planning that maximises productivity and minimises deviations from plant production and quality targets. Their model could facilitate efficient scheduling while considering the real-time dynamics of mining operations. Both and Dimitrakopoulos (2020) developed a stochastic optimisation model that jointly addressed short-term mine production scheduling and fleet management. The primary objective was to maximise metal production and overall profit while minimising operational disruptions. The results showed that they could reduce operational costs by 3.1 per cent. Navarro Torres et al (2020) also combined a LP model with DES to optimise short-term open pit mining. operations. The main goal was to optimise the allocation of shovels and trucks in short-term open pit mine operations to minimise deviations The integrated model provided a comprehensive decision support system for short-term mine planning.

Deterministic underground mine

Campeau and Gamache (2020) developed a LP model for short-term planning in underground gold mines, addressing the complexity of scheduling and resource management. The main goal was to maximise ore extraction while ensuring efficient use of resources and maintaining adherence to operational constraints. Campeau and Gamache (2022) introduced a LP model that integrated short-term underground mine scheduling. Their model helped to significantly reduced dependency on manual planning and ensured that activities are optimally allocated. Campeau and Gamache (2022) introduced a novel approach to short-term underground mine planning by using CP to optimise resource allocation and scheduling. The primary goal was to create an efficient scheduling tool that can optimise short-term mine planning, balancing resource use while maximising profitability and ensuring operational constraints are met. The CP model demonstrated superior performance compared to a traditional LP model. Ayaburi *et al* (2024) introduced two integrated LP models for optimising short-term production scheduling in underground mines, to develop integrated scheduling models that maximise NPV while ensuring compliance with ventilation and heat management requirements.

Amoako, Chowdu and Brickey (2023) presented a LP formulation designed to improve short-term production scheduling in underground mining by minimising deviations from medium-term production forecasts. The authors offered a robust tool for handling unforeseen operational challenges, ensuring that short-term schedules remain as aligned as possible with long-term strategic goals. Astrand, Johansson and Zanarini (2020) introduced a novel approach to short-term underground mine scheduling, combining CP with a Large Neighbourhood Search (LNS) method. Gliwan and Crowe (2022) developed a LP formulation for weekly scheduling in underground gold mines. The main objective was to maximise the total gold mine, while adhering to transportation and processing constraints within the mine. The paper demonstrated that an optimisation model can enhance productivity. The goal was to optimise the scheduling of mobile machines, minimising overall makespan while considering safety constraints.

Stochastic underground mine

Manríquez, Pérez and Morales (2020) developed a MILP model integrated with DES to enhance the adherence of short-term production schedules in underground mines, ensuring these schedules are executable under real-world conditions. Their iterative approach improved schedule adherence without negatively impacting NPV, though NPV was not the primary optimisation objective. Aalian, Pesant and Gamache (2023) presented a CP model designed to optimise short-term underground mine planning. This model introduced a more efficient scheduling method compared to manual planning, which traditionally relies heavily on a planner's expertise and experience. Aalian, Gamache and Pesant (2024) developed a model for short-term scheduling of production activities that accounts for the uncertainty in activity durations. The research considered two approaches: a scenario-based model and a confidence-constraint model, both of which generate robust schedules that are less susceptible to delays and operational disruptions. The confidence-constraint model and scenario-based approach outperformed traditional deterministic models by producing schedules that had more robust results when evaluated under simulated scenarios.

Linear Programming versus Constraint Programming

In the context of the mining industry, LP is widely employed to tackle task scheduling problems by optimising resource allocation under given constraints. On the other hand, Constraint Programming (CP) is a powerful technique used to solve task scheduling problems, particularly in complex scenarios with multiple constraints. In the mining industry, CP is especially valuable for handling scheduling challenges where traditional optimisation methods may struggle, providing a more robust approach to meeting operational constraints and objectives.

Previous studies have highlighted the effectiveness of CP in addressing scheduling challenges (Laborie *et al*, 2018). CP is an exact method used for modelling and solving combinatorial optimisation problems, and it has found extensive application in fields such as planning, scheduling, transportation, and automated systems scheduling (Laborie *et al*, 2018). CP achieves high computational efficiency through constraint propagation, which narrows the domains of variables, and by employing advanced search strategies that minimise the search space (Pesant, 2014). Additionally, CP provides more flexible and intuitive formulations by offering a rich set of variable types, functions, and global constraints (Kanet, Ahire and Gorman, 2004). This flexibility allows for more compact.

models with fewer decision variables and constraints compared to other mathematical programming methods, such as LP (Aalian, Gamache and Pesant, 2024). These advantages make CP well-suited for addressing large-scale scheduling problems. In the context of underground mining, CP's comprehensive functions can more effectively model mining-specific constraints in short-term scheduling compared to other optimisation techniques (Aalian, Gamache and Pesant, 2024). Therefore, this study adopts a CP formulation to model and resolve the problem under consideration.

Our contribution

Recent progress has demonstrated the superior performance of CP models in comparison to traditional LP approaches (Campeau and Gamache, 2022; Laborie *et al*, 2018). Consistent with these advancements, our research employs CP to formulate and address the scheduling problems encountered in underground mining complexes. Furthermore, to our knowledge, this study is the first to handle a data set that simultaneously schedules both development and production activities. Furthermore, the model presented in this study is based on the model of (Aalian, Pesant and Gamache, 2023), while our model accounts for the time required by each piece of equipment to travel between various locations within the mine, a factor that is not addressed in (Aalian, Pesant and Gamache, 2023). In addition, we formulate a two-stage stochastic variant of the problem, enabling the simultaneous resolution of multiple scenarios and thereby producing more robust outcomes. In Aalian, Pesant and Gamache (2023), the authors address two separate instances of the problem: one focused on production tasks and the other on development tasks. Their case study represents a specific scenario in which production activities take place during the day shift, while development tasks are carried out during the night shift. In contrast, our approach considers a more complex situation where both production and development tasks occur during both day and night shifts. This adds to the problem's complexity by requiring certain pieces of equipment to be shared between production and development tasks, while also incorporating the time needed for equipment to travel between different sites.

This study presents the following contributions. First, we introduce a framework that serves as a decision-making tool, helping decision-makers in scheduling both production and development tasks, thus offering a holistic approach to underground mine planning. Second, to validate our framework, we implement it in a Canadian underground mining scenario to assess its effectiveness. Third, the proposed model is analysed using two distinct versions: (1) a deterministic version with fixed task processing times and incorporating six degrees of conservatism in task processing times; (2) a two-stage stochastic formulation that considers multiple scenarios simultaneously, following the probability distribution. Then, we conduct a thorough comparison of these versions, illustrating the practical advantages for decision-makers. Both models are evaluated across nine distinct instances to obtain a comprehensive assessment of the makespan under various conditions. Fourth, a total of 50 scenarios are generated, with task processing times modelled based on a probability distribution. The generated task processing times are subsequently applied to the machine sequences obtained from the two models to assess the results and conduct a comparative analysis.

PROBLEM DESCRIPTION AND MATHEMATICAL MODEL

In this section we describe our problem, highlighting its key characteristics, and formulate its mathematical model.

Problem description

Underground mining operations can be divided into two main types of activities: development and production. For each development or production cycles to be completed, certain number of tasks should be performed. Development tasks are carried out in waste rock, which holds no economic value, in order to gain access to ore deposits of financial interest. Conversely, production tasks occur in economically valuable rock within areas known as stopes. These mining activities are performed cyclically at various worksites designated for these tasks. Figure 1 illustrates the development and production cycles, with their respective activities arranged in a sequence-dependent order. Figure 1a illustrates the development cycle. Figure 1b illustrates the production cycle.

Each task is executed by a specific machine. Table 1 outlines the machine types required for each task. Multiple machines are available for each activity type. Each machine is considered a unitary resource, capable of performing only one task at a time. Various underground mining methods are used to extract deep mineral deposits. In our case study, the mine employs the longhole stoping method, a widely used technique that is effective for large, steeply dipping ore deposits with a tabular shape.

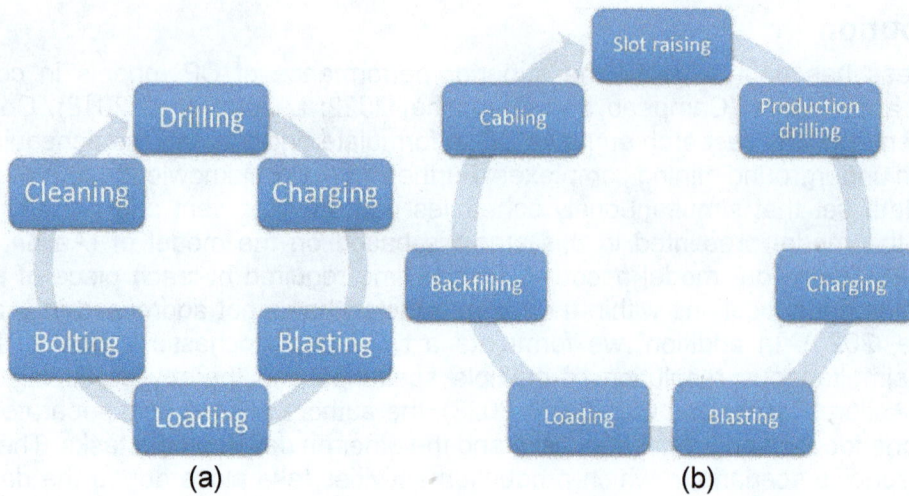

FIG 1 – Steps to be taken for development and production in a mine: (a) Steps for development; (b) Steps for production.

TABLE 1

Description of required machines for each task.

Tasks	Machines
Drilling	Drilling rigs
Charging	Anfo loader
Loading	Scooptram
Bolting	Bolter
Cleaning	Scooptram
Cabling	Cabling machine
Slot raising	Raise borer

Mathematical model

In this section, we present the methodology employed to tackle the problem under consideration. We outline the CP formulation model we developed to solve the problem, detailing the key components and structure of the optimisation model. A CP-based optimisation model has been developed for scheduling in underground mining, incorporating the specific operational requirements of underground mining processes. In this model, interval variables are used to represent activities. An interval variable represents a time interval during which an activity occurs, where the start time, end time, and duration are decision variables within the scheduling problem. More formally, an interval variable a is characterised by three key attributes: a start time $s(a)$, an end time $e(a)$, and a size (or duration) $\delta(a) = e(a) - s(a)$. These values are non-negative integers, ensuring that $s(a)$, $e(a)$ $\in N$ and $e(a) \geq s(a)$. The flexibility of interval variables allows the scheduling model to determine the best possible allocation of activities over time while satisfying operational constraints.

Additionally, interval variables can be optional, meaning that their inclusion in the final schedule is subject to optimisation decisions. An optional interval variable b can either be present in the schedule as an active interval or completely omitted. Mathematically, an optional interval variable is defined as $b \in \{\emptyset\} \cup \{(s, e)|s, e \in N, e \geq s\}$. Each optional interval variable is associated with a Boolean status, indicating whether the corresponding activity is included in or excluded from the schedule. In the problem to be solved, each activity must be assigned to at most one machine among a set of machines capable of performing it. To model this allocation, we use optional interval variables, allowing the solver to select the most suitable resource dynamically. By leveraging optional interval variables, the model can represent different assignment possibilities while ensuring that an activity is either scheduled on exactly one machine or excluded if necessary. Moreover, Table 2 presents the notations of the proposed model.

TABLE 2
Notations of the model.

Notation	Description
Sets	
J	Set of activities
M	Set of machines
S	Set of scenarios
M_j	Set of eligible machines to perform activity j
A_j	Set of activities that must occur after activity j
B	Set of blast activities
Parameters	
p_j	Processing time of activity j
BC	Blast calendar during which only blasting is permitted
Variables	
$Y_{j,s}$	Interval variable for activity j in scenario s
$X_{j,m,s}$	Optional interval variable to perform activity j using machine m in scenario s
$K_{m,s}$	Sequence variable for machine m in scenario s

Resource allocation is represented through a set of interval variables known as a sequence variable, which ensures that activities within the sequence do not overlap in time.

Additionally, the developed CPO model incorporates several functions and constraints, which are discussed in detail below:

- ***endOf***: This function retrieves the end value of an interval variable if it is defined, otherwise it returns zero. For example, it considers an interval variable representing a task starting at 2 and ending at 5. Applying endOf to this variable would return 5. If the interval is optional and absent, the function would return 0. In our model, this function is particularly useful when dealing with optional tasks whose presence depends on scheduling decisions, allowing us to handle absent tasks seamlessly in temporal constraints.

- ***alternative***: This function enforces that among a given set of optional interval variables, exactly one must be present if the main interval variable is present, and the selected one must have the same start and end times as the main interval. For example, suppose two optional tasks, task A and task B, are available. If the main task is scheduled, then exactly one of task A or task B must be chosen to occur, with identical start and end times as the main task. If task A is selected and scheduled from time 3 to 6, task B must remain absent, and *vice versa*. This constraint is used to model mutually exclusive choices where the execution timing must be synchronised.

- ***noOverlap***: This function is used when a set of interval variables, defined by a sequence variable, must be scheduled without any overlap. It is particularly useful when tasks require exclusive use of a resource, such as a machine that can only process one task at a time. For example, if tasks A and B share the same machine, the *noOverlap* constraint ensures that task A must finish before task B starts (or *vice versa*), preventing both tasks from running simultaneously. In our context, this constraint ensures resource feasibility by modelling unary capacity resources, where each machine can handle only one task at a time without interruption.

- **endBeforeStart**: This function enforces a strict precedence relationship between two interval variables. It ensures that if both tasks are present in the schedule, the first task must finish before the second one begins. This constraint is used when there is a strict dependency between tasks, such as when task B cannot begin until task A is fully completed. For instance, if task A finishes at time 5, then task B can only start at time 5 or later. This is particularly useful in workflow processes in underground mining. Additionally, **endBeforeStart** guarantees a minimal temporal gap between consecutive activities when necessary, which is critical in tightly coupled operational processes such as sequential underground operations.

- **forbidExtent**: This function prevents an interval variable from overlapping with a forbidden region, which is represented using a step function. A step function is a piecewise constant function that defines availability over time by assigning values to specific time intervals. Typically, the function takes the value 1 when an interval is allowed and 0 when it is forbidden. The **forbidExtent** constraint ensures that an interval variable cannot overlap with any time period where the step function has a value of 0. As a result, the interval variable must either finish before the forbidden region starts or begin after it ends. For example, suppose task A cannot occur during a blast time window from 8 to 10, which is defined as a forbidden region where the step function takes a value of 0. The **forbidExtent** function ensures that task A must either finish before 8 or start after 10, preventing it from being scheduled within the restricted period. In our model, **forbidExtent** is primarily used to avoid scheduling activities during blast periods, enhancing safety and operational realism.

- **sameSequence**: This function ensures that a sequence of interval variables maintains the same relative order across different scenarios. It is particularly useful in stochastic scheduling models, where activity durations vary according to probability distributions. In our model, the *sameSequence* function is employed to construct a unique sequence of activities for each machine while accounting for different scenarios. Each scenario is generated by randomly sampling activity durations from predefined distribution functions. The constraint ensures that, despite variations in task durations across scenarios, the relative order of activities on each machine remains unchanged. This guarantees consistency in the scheduling structure while allowing the model to handle uncertainty in processing times effectively. Enforcing the same sequence across scenarios significantly reduces the combinatorial complexity of the stochastic model by preventing reordering of activities between scenarios, which would otherwise require extensive branching in the solution tree.

The objective function of the model is:

$$Minimize\ Max \sum_{s \in S} \sum_{j \in J} \left(\left(endOf(Y_{j,s}) \right) / S \right) \tag{1}$$

Subject to the following constraints:

$$presenceOf(X_{j,m,s}) = 0 \qquad \forall j \in J, m \in M \setminus M_j, s \in S \tag{2}$$

$$alternative(Y_{j,s}, X_{j,m,s} \mid m \in M_j \qquad \forall j \in J, s \in S \tag{3}$$

$$noOverlap(s_{m,s}, D) \qquad \forall m\ in\ M, s \in S \tag{4}$$

$$endBeforeStart(Y_{j,s}, Y_{i,s}) \qquad \forall j \in J, i \in A_j, s \in S \tag{5}$$

$$forbidExtent(Y_{j,s}, B) \qquad \forall j \in J \setminus B, s \in S \tag{6}$$

$$SameSequence(K_{m,s}, K_{m,s'}) \qquad \forall m \in M, s, s' \in S, s \neq s' \tag{7}$$

Objective (1) of the CP model aims to minimise the makespan (ie the time corresponding to the end time of the activity ending last). Constraints (2) ensures that each activity is assigned to eligible machine type. Constraints (3) ensure that only one optional variable is selected for an interval variable, meaning that only one machine (with the appropriate type) is assigned to perform a specific activity. Constraints (4) prohibit the simultaneous use of machines, ensuring that each machine is allocated to only one activity at a time. Constraints (5) account for the order in which activities must be executed at a specific site to take into account the precedence constraints in the development and production cycles. Constraints (6) ensure that blasting activities occur only during designated blast windows. Constraints in (7), referred to as $SameSequence$, ensure that the interval variables representing tasks in both scenarios one and two retain the same relative order. For instance, if task a precedes task b in scenario one, then its counterpart a' must also precede b' in scenario two. The model provided represents the stochastic version of the problem, designed for two-stage stochastic experiments. However, transitioning to the deterministic version is straightforward. To achieve this, we relax the stochastic constraint (refer to Constraint 7) and reduce the number of scenarios to one. By doing so, the model becomes simplified and deterministic, allowing for a solution based on a single scenario.

As discussed earlier, two models will be tested in this study:

- The deterministic model, denoted $\mathcal{P}_{1,n}^{\alpha}$ uses a single scenario. In this scenario, the duration of each activity is based on the average duration of this type of activity observed from historical data. Moreover, we integrated different degrees of conservatism to simulate the worst case scenarios. The value of α represents a certain level of conservatism where $\alpha = 0$ is equivalent to no conservatism (which follows the average task processing time from historical data) and $\alpha = 1$ is the most conservative one since the duration of all task is set to its maximum value. Finally, n represents the instance number for which the model is being solved. For instance, when referring to the results of $\mathcal{P}_{1,6}^{\alpha=0.75}$, it indicates that the discussion pertains to the first model applied to instance six, with a conservatism degree of 0.75. In Section 5, we will elaborate on the method used to compute α.

- The second model, denoted as $\mathcal{P}_{2,n}$ represents the stochastic model. In this model, task processing times follow a probability distribution, and ten scenarios are solved simultaneously. This number is particularly chosen as it was suggested in the literature by (Aalian, Pesant and Gamache, 2023). Across these ten scenarios, the machine sequence remains the same, while each scenario features distinct task processing times.

METHODOLOGY

In this section, we outline the methodological steps employed in this study. Broadly, the methodology is divided into three main steps, referred to as 'Step 1', 'Step 2', and 'Step 3' as illustrated in Figure 2. The details of each step are presented in the corresponding subsections.

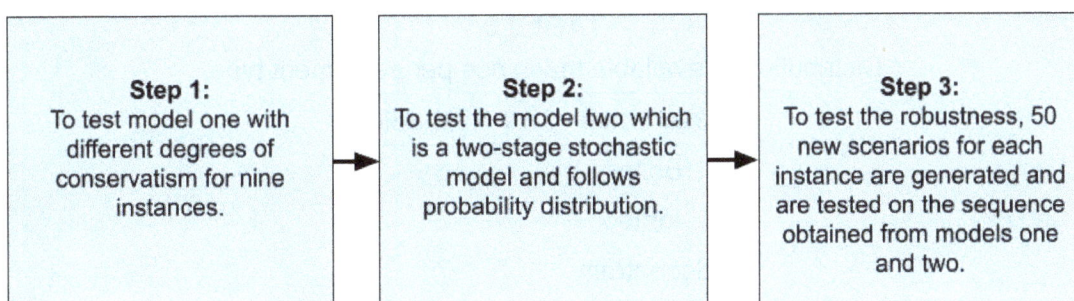

| Step 1: To test model one with different degrees of conservatism for nine instances. | Step 2: To test the model two which is a two-stage stochastic model and follows probability distribution. | Step 3: To test the robustness, 50 new scenarios for each instance are generated and are tested on the sequence obtained from models one and two. |

FIG 2 – Methodological steps followed in this study.

Step 1

In the first step, we evaluate our initial model ($\mathcal{P}_{1,n}^{\alpha}$) to address the problem, considering various configurations. Specifically, we test nine different instances, ranging from $\mathcal{P}_{1,1}^{\alpha}$ to $\mathcal{P}_{1,9}^{\alpha}$. Additionally, we incorporate different degrees of conservatism to construct worst-case scenarios. In this study, we consider seven conservatism levels: 0, 0.75, 0.8, 0.85, 0.9, 0.95, and 1. Therefore, 63 configurations are generated and tested in step 1. The process used to generate these instances is and the methodology for determining and applying these conservatism levels in relation to α is thoroughly discussed in Section 5.

Step 2

In the second step, we evaluate our second model, which is a two-stage stochastic model. This model is also tested on nine different instances; however, in each test, ten scenarios are solved simultaneously. For instance, the second model for $Instance$ 1 is denoted as $\mathcal{P}_{2,1}$. It is important to note that α is not included as an index in this model, as its stochastic nature follows a probability distribution. In this step, after obtaining the results from the second model, they are compared with those of the first model to facilitate a comprehensive comparative analysis.

Step 3

In the third and final step, we generate 50 new scenarios for each instance, resulting in a total of 450 newly created scenarios that follow a probability distribution for each type of equipment within a specified range for each instance. We then apply the machine sequences obtained from both models under different configurations and test them against these newly generated scenarios. In other words, the machine sequences remain fixed (as determined by the two models), while the new task processing times from the new scenarios are evaluated based on these sequences. This test will help us understand the robustness and adaptability of the machine sequences derived from both models when subjected to previously unseen scenarios. By analysing the performance of these sequences under new conditions, we can assess the stability and effectiveness of each model's decisions in handling variability and conservatism.

PARAMETER AND EXPERIMENTAL SETTINGS

In this section, we present a detailed analysis of the experimental design and parameter setting. For our tests, nine instances are created. For each instance, the two models will be tested. Table 3 lists the types and quantities of machines available for use, essential for resource allocation and scheduling decisions across different tasks. Table 4 details for the first instance, the processing time for each activity, specifying both average and maximum durations, as well as the calculated delays at six conservatism levels, previously define as α (0.75, 0.8, 0.85, 0.9, 0.95, and 1), resembling worst case scenarios. These settings reflect potential fluctuations in task durations, providing a realistic framework to account for delays that may arise in practice. In particular, when α is set to 0.75, it results in an adjusted processing time of 23.5 mins for the mucking task (see Table 4). This value is calculated as the average processing time plus the 0.75 per cent of maximum delay which is 10 for the task (16 + (0.75 × 10) = 23.5). Varying levels of conservatism provide deeper insights into the model's behaviour and allow us to assess its sensitivity to input variations (Shahin *et al*, 2024a, 2023).

TABLE 3

Distribution of available machines per equipment type.

Machines	# Available
Bolter	6
Jumbo	3
Scooptram	7
Anfo loader	4
Cable	5
Slot raise	1
Drill	4
Truck	7

TABLE 4
Distribution of task processing times with respect to potential delays for *Instance* 1.

Activity	Min	Average	Max	Max delay	0.75	0.80	0.85	0.90	0.95	1
Development										
Mucking	14.00	16.00	26.00	10.00	23.50	24.00	24.50	25.00	25.50	26.00
Bolting	36.00	43.00	72.00	29.00	64.75	66.20	67.65	69.10	70.55	72.00
Clean face	3.00	5.00	9.00	4.00	8.00	8.20	8.40	8.60	8.80	9.00
Drilling	16.00	24.00	33.00	9.00	30.75	31.20	31.65	32.10	32.55	33.00
Charging	5.00	8.00	18.00	10.00	15.50	16.00	16.50	17.00	17.50	18.00
Production										
Haulting	49.68	69.00	117.30	48.30	105.23	107.64	110.06	112.47	114.89	117.30
Cleaning	43.20	60.00	102.00	42.00	91.50	93.60	95.70	97.80	99.90	102.00
Cable	64.08	89.00	151.30	62.30	135.73	138.84	141.96	145.07	148.19	151.30
Paste fill	4.32	6.00	10.20	4.20	9.15	9.36	9.57	9.78	9.99	10.20
Rock fill	47.52	66.00	112.20	46.20	100.65	102.96	105.27	107.58	109.89	112.20
Mass	58.32	81.00	137.70	56.70	123.53	126.36	129.20	132.03	134.87	137.70
Charging	18.00	25.00	42.50	17.50	38.13	39.00	39.88	40.75	41.63	42.5
Drilling	102.24	142.00	241.40	99.40	216.55	221.52	226.49	231.46	236.43	241.40
Re-drilling	21.60	30.00	51.00	21.00	45.77	46.80	47.85	48.90	49.95	51.00
Slot raising	13.68	19.00	32.30	13.30	28.98	29.64	30.31	30.97	31.64	32.30

In addition to the initial task processing times provided in Table 4, referred to as *Instance* 1, we created eight additional instances to introduce varying ranges for the execution times of each task. Specifically, we adjusted the 'range' of task durations (minimum, average, and maximum) in order to generate diverse processing time distributions, thereby enhancing the comprehensiveness of our experiments. These adjustments facilitate better estimation when addressing the underlying problem. Table 5 provides a summary of differences between different instances generated, while the tables for other eight instances are presented in online Appendix B.

TABLE 5
Summary of differences between Instance 1 to 9.

# of Instance	Description
1	Taken from historical data
2	Move the average of #1 to left by one unit
3	Move the average of #1 to right by one unit
4	Move the average of #2 to left by one unit
5	Move the average of #2 to right by one unit
6	The range of tasks in #1 is reduced by a factor of 0.5
7	The range of tasks in #1 is reduced by a factor of 0.75
8	The range of tasks in #1 is increased by a factor of 1.25
9	The range of tasks in #1 is increased by a factor of 1.5

Turning our focus on the experimental design, the results are structured into three main parts to comprehensively illustrate the model's performance under different assumptions and configurations. The first part examines the outcomes from model $\mathcal{P}_{1,n}^{\alpha=0}$, where all task processing times are assumed to be fixed and certain. Here, task durations are based on 'average' processing time across all instances (from $Instance$ 1 to $Instance$ 9), providing a baseline assessment of scheduling efficiency and resource utilisation when uncertainties are absent. This deterministic evaluation offers insights into how the model performs under ideal conditions, helping us understand the baseline operational feasibility. Moreover, we evaluate six levels of conservatism for each instance. For example, in the case of $Instance$ 1, we consider values ranging from $\mathcal{P}_{1,1}^{\alpha=0}$ to $\mathcal{P}_{1,1}^{\alpha=1}$, where α takes the values 0, 0.75, 0.8, 0.85, 0.9, 0.95, and 1. The case of $\alpha = 0$ indicates that the average processing time for each task is used, as outlined in Table 4, whereas higher levels of α lead to a more conservative system.

In the second part, we present the results from the two-stage stochastic model, denoted $\mathcal{P}_2$ where ten scenarios are solved simultaneously to assess performance under full stochastic conditions. Here, here-and-now variables represent decisions that are fixed across all scenarios, made before uncertainty is realised. In contrast, wait-and-see variables allow the model to adjust decisions after uncertainties unfold, adapting to specific conditions in each scenario. This division enables the model to make foundational scheduling and resource allocation choices upfront, while retaining flexibility for scenario-specific adjustments. It is important to note that in this model, task processing times follows the probability distribution between the 'minimum' and 'maximum' values for each task that are presented in Table 4 for $Instance$ 1. This use of probability distributions ensures a more realistic representation of conservatism, enabling the model to capture the inherent variability in task processing times and improve the robustness of the scheduling and resource allocation decisions. This test is also implemented for $Instance$ 2 through $Instance$ 9.

Finally, in the third part, 50 scenarios are created in which task processing times adhere to the probability distribution. For each scenario, we maintain the sequence of tasks on each machine obtained from the optimal solution of the two models but we adjust the processing duration of each task and recalculate the value of the objective function in order to analyse the variation of the makespan.

For the experiments, the CP models are implemented in IBM ILOG CPLEX Optimisation Studio version 12.10 and solved using the Constraint Programming Optimiser. All experiments were conducted on a computer with an Intel® Core™ i7–1165G7 CPU @ 2.80GHz processor and 16GB of RAM.

EXPERIMENTAL RESULTS

In this section, we analyse the results obtained for the three steps (parts) as described in Sections 4 and 5.

Regarding the findings from the first part ($\mathcal{P}_{1,n}^{\alpha}$), the results are detailed in Table 6. It is noteworthy that the value of the objective function decreases from approximately 1400 to 1380 and 1360 for configurations $\mathcal{P}_{1,1}^{\alpha=0}$, $\mathcal{P}_{1,2}^{\alpha=0}$ and $\mathcal{P}_{1,4}^{\alpha=0}$, respectively, while it increases for $\mathcal{P}_{1,3}^{\alpha=0}$ and $\mathcal{P}_{1,5}^{\alpha=0}$, being approximately 1420 and 1440, respectively. The trend for reduction in $\mathcal{P}_{1,2}^{\alpha=0}$ and $\mathcal{P}_{1,4}^{\alpha=0}$ can be attributed to the reduction in the average task processing time by one and two units for $Instance$ 2 and $Instance$ 4, respectively, as indicated in Table 5. Conversely, the observed increase in the objective function for $\mathcal{P}_{1,3}^{\alpha=0}$ and $\mathcal{P}_{1,5}^{\alpha=0}$ is justified by the rise in the average task processing time for $Instance$ 3 and $Instance$ 5 by one and two units, as also evidenced in Table 5. It is worth noting that all solutions listed in Table 6 achieved optimality (0 per cent gap) within the initial three seconds of the experiment.

TABLE 6

Distribution of results across *Instance* 1 to *Instance* 9 for the second model.

Ins. No.	$\mathcal{P}_{1,n}^{\alpha=0}$	$\mathcal{P}_{1,n}^{\alpha=0.75}$	$\mathcal{P}_{1,n}^{\alpha=.8}$	$\mathcal{P}_{1,n}^{\alpha=0.85}$	$\mathcal{P}_{1,n}^{\alpha=.9}$	$\mathcal{P}_{1,n}^{\alpha=0.95}$	$\mathcal{P}_{1,n}^{\alpha=1}$
$n=1$	1400.08	2135.05	2184.10	2233.05	2282.10	2331.05	2380.10
$n=2$	1380.10	2115.05	2164.10	2213.05	2262.10	2311.10	2360.10
$n=3$	1420.10	2155.13	2204.14	2253.17	2302.22	2351.17	2400.14
$n=4$	1360.10	2095.09	2144.14	2193.09	2242.14	2291.05	2340.10
$n=5$	1440.10	2175.05	2224.10	2273.09	2322.10	2371.09	2420.14
$n=6$	1375.98	1761.52	1787.23	1812.86	1838.57	1864.28	1890.10
$n=7$	1343.04	1936.96	1976.66	2016.19	2055.82	2095.43	2135.05
$n=8$	1235.04	2277.44	2347.04	2416.56	2486.00	2555.52	2625.05
$n=9$	1160.01	2442.48	2528.00	2613.44	2698.96	2784.48	2870.10

Analysing the results for $\mathcal{P}_{1,6}^{\alpha=0}$ through $\mathcal{P}_{1,9}^{\alpha=0}$, we observe a consistent decline in the objective function values, which decrease from 1400 to approximately 1375, 1343, 1235, and 1160 for *Instance* 6 to *Instance* 9, respectively (refer to Table 6). Although the ranges of task processing times are reduced for *Instance* 6 and *Instance* 7 (with reduction factors of 0.5 and 0.75, respectively) and increased for *Instance* 8 and *Instance* 9 (with increment factors of 1.25 and 1.5, respectively), the objective function values exhibit a general downward trend across these instances. This phenomenon can be attributed to the methodology employed in obtaining the results: the average processing time for each task is used. Typically, the average processing time lies closer to the lower boundary of the given range rather than the upper boundary, rather than representing the median value of the range. For instance, consider the 'clean face' task. Table 7 presents the details of processing times for this task in the original configuration (*Instance* 1) and for *Instance* 6 through *Instance* 9, where the ranges are modified. Notably, the maximum processing time required for the task decreases from 9 to 7 and 8 time units in *Instance* 6 and *Instance* 7, respectively, when the range is reduced by factors of 0.5 and 0.75. Conversely, the maximum time increases to 10 and 11 time units in *Instance* 8 and *Instance* 9, respectively, due to the increment factors of 1.25 and 1.5. Nevertheless, across all instances, the average processing time for the 'clean face' task is consistently lower compared to the original configuration. This trend holds true for the majority of tasks, thereby explaining the overall reduction in objective function values observed across these instances.

TABLE 7

Distribution of clean face processing times across five different instances.

Instances	Minimum	Average	Maximum	Range factor	Object of corresponding Instance
Instance 1	3.00	5.00	9.00	1.00	1400.00
Instance 6	4.00	4.67	7.00	0.50	1375.00
Instance 7	3.50	4.38	8.00	0.75	1343.00
Instance 8	2.50	3.54	10.00	1.25	1235.00
Instance 9	2.00	3.00	11.00	1.50	1160.00

Shifting our attention to the outcomes of the first part with certain degrees of conservatism, the findings are presented in the different column of Table 6. It can be observed that entries $\mathcal{P}_{1,1}^{\alpha=0.75}$ through $\mathcal{P}_{1,1}^{\alpha=1}$ correspond to the results for *Instance* 1 under varying levels of conservatism. Similarly, analogous results are reported for *Instance* 2 through *Instance* 9, denoted as $\mathcal{P}_{1,2}^{\alpha=0.75}$ through $\mathcal{P}_{1,2}^{\alpha=1}$, and continuing in the same pattern up to $\mathcal{P}_{1,9}^{\alpha=0.75}$ through $\mathcal{P}_{1,9}^{\alpha=1}$ for the rest of instances.

In general, it can be observed that the results for each instance with any level of conservatism are consistently higher than the outcomes obtained under '0' conservatism. This observation indicates that incorporating uncertainty leads to an increase in the objective function. Furthermore, the additional task processing time caused by higher levels of conservatism does not affect the solving time, which remains minimal throughout this part, nor does it influence the optimality gap, which remains at 0 per cent. When comparing the instances, a similar trend to that identified in part one. For example, *Instance* 2 and *Instance* 4 exhibit lower objective functions compared to the baseline setting (*Instance* 1) across various levels of conservatism. In contrast, *Instance* 3 and *Instance* 5 display higher objective functions. The reduction in objective functions for *Instance* 2 and *Instance* 4 can be attributed to a leftward shift in their respective ranges by one and two units, respectively. Conversely, the increase in objective functions for *Instance* 3 and *Instance* 5 is explained by a rightward shift in their ranges by one and two units, respectively.

For *Instance* 6 and *Instance* 7, the results show a reduction in the objective functions by 17.52 per cent and 9.32 per cent, respectively, for configurations $\mathcal{P}_{1,6}^{\alpha=0.75}$ and $\mathcal{P}_{1,7}^{\alpha=0.75}$, when compared to configuration $\mathcal{P}_{1,1}^{\alpha=0.75}$. Although both instances demonstrate a decrease in their objective functions relative to the baseline setting, *Instance* 7 has a higher objective function than *Instance* 6. This difference can be explained by the reduction factor applied, which is 0.75 for *Instance* 7 compared to 0.5 for *Instance* 6. Following the same reasoning, *Instance* 8 and *Instance* 9 exhibit higher objective functions. Specifically, the objective functions for configurations $\mathcal{P}_{1,8}^{\alpha=0.75}$ and $\mathcal{P}_{1,9}^{\alpha=0.75}$ are 6.62 per cent and 14.38 per cent greater, respectively, compared to $\mathcal{P}_{1,1}^{\alpha=0.75}$. Finally, the worst-case scenarios are presented in the rows where the degree of conservatism is equal to one, specifically in $\mathcal{P}_{1,1}^{\alpha=1}$, $\mathcal{P}_{1,2}^{\alpha=1}$, and continuing through $\mathcal{P}_{1,9}^{\alpha=1}$ for *Instance* 9.

Regarding the results presented in the second part ($\mathcal{P}_{2,n}$), they are summarised in Table 8. In this analysis, nine instances with varying ranges were examined based on a probability distribution, generated between minimum and maximum values for each task, to provide a more accurate estimation of the objective functions under different conditions. A two-stage stochastic model was employed in this part, where the solver tries to find the sequence that will best satisfy ten scenarios where the duration of the tasks is randomly generated according to the task distribution functions. This approach ensures that the sequence of machines remains consistent across the ten scenarios within each instance, even though task processing times differ among the scenarios. Such a methodology aids in obtaining more robust results across various instances.

TABLE 8
Distribution of results across Instance 1 to Instance 9 for the second model.

Config.	$\mathcal{P}_{2,1}$	$\mathcal{P}_{2,2}$	$\mathcal{P}_{2,3}$	$\mathcal{P}_{2,4}$	$\mathcal{P}_{2,5}$	$\mathcal{P}_{2,6}$	$\mathcal{P}_{2,7}$	$\mathcal{P}_{2,8}$	$\mathcal{P}_{2,9}$
Obj.	1712.60	1648.34	1724.87	1645.63	1727.03	1547.08	1643.73	1721.18	1723.43

The maximum solution time limit for this model was set to eight hrs (28 800 seconds). Upon reaching this time limit, the model records the best solution found, along with the corresponding optimality gap percentage. Due to the increased complexity arising from the two-stage stochastic programming approach and the simultaneous resolution of ten scenarios, an approximate gap of 40 per cent was observed across different instances. The CPO was able to find a feasible solution within 10 secs, though with a gap of approximately 57 per cent. Over the next 15 mins, the gap was reduced to around 40 per cent, but no further improvement in the solution was achieved before the time limit expired. A short discussion of this integrity gap will be made later when analysing the results of part three of our methodology.

The results for are represented by $\mathcal{P}_{2,1}$, demonstrate reductions in $\mathcal{P}_{2,2}$ and $\mathcal{P}_{2,4}$, while increases are noted in $\mathcal{P}_{2,3}$ and $\mathcal{P}_{2,5}$. Furthermore, reductions in the objective functions are observed in $\mathcal{P}_{2,6}$ and $\mathcal{P}_{2,7}$ compared to $\mathcal{P}_{2,1}$. Among these, the reduction in $\mathcal{P}_{2,6}$ is more pronounced due to the application of a 0.5 factor, whereas a 0.75 factor contributed to the reduction in $\mathcal{P}_{2,7}$. Conversely, the objective function increased in $\mathcal{P}_{2,8}$ and $\mathcal{P}_{2,9}$ relative to $\mathcal{P}_{2,1}$.

Finally, in the third part, we generate 50 new scenarios for each instance, helping us to have 450 newly generated scenarios in total. These scenarios follow probability distribution in given ranges for each instance (minimum and maximum task processing time for each task in each instance). Then, the sequence of machines that we stored from parts one and two are taken, while the task processing time of the newly generated scenarios for each instance is taken into account for the model. This would help us to have a good understanding of the robustness of our results.

After storing the results, the average of the objective function is calculated for each configuration. For instance, in the case of $\mathcal{P}_{1,1}^{\alpha=0}$, its machine sequence was tested with 50 scenarios, yielding 50 results. The average of these 50 objective function values is then computed and is shown in Table 9, row $n' = 1$ for column $\mathcal{P}_{1,1}^{\alpha=0}$, while the original result from model one is presented in row $n = 1$. This procedure is repeated for the other configurations. The results are presented in Table 9. The column labelled with different values for $n' \in 1,2,\ldots,9$ provides the results for the insertion objective function. From the results, it is evident that, in each instance, the values in the n' row are very close, and in some cases, identical. This can be attributed to the fact that the machine sequencing originates from configurations with optimal assignment (0 per cent gap in the obtained solution). In *Instance* 1, which consists of eight configurations (from $\mathcal{P}_{1,1}^{\alpha=0}$ to $\mathcal{P}_{2,1}$), the n' row values are approximately 1679, with only slight variations in the decimal places. A similar pattern is observed in other instances. However, the deviations become more noticeable in *Instance* 8 and *Instance* 9, where the differences amount to a few units rather than just decimals.

TABLE 9

Distribution of results of newly generated scenarios that are inserted in different configurations across instances.

Ins. No.	$\mathcal{P}_{1,n}^{\alpha=0}$	$\mathcal{P}_{1,n}^{\alpha=0.75}$	$\mathcal{P}_{1,n}^{\alpha=0.8}$	$\mathcal{P}_{1,n}^{\alpha=0.85}$	$\mathcal{P}_{1,n}^{\alpha=0.9}$	$\mathcal{P}_{1,n}^{\alpha=0.95}$	$\mathcal{P}_{1,n}^{\alpha=1}$	$\mathcal{P}_{2,1}$
$n = 1$	1400.08	2135.05	2184.1	2233.05	2282.1	2331.05	2380.1	1625.06
$n' = 1$	1679.03	1679.02	1679.32	1679.02	1679.02	1679.02	1679.03	1679.04
$n = 2$	1380.1	2115.05	2164.1	2213.05	2262.1	2311.05	2360.1	1617.2
$n' = 2$	1629.72	1629.38	1629.43	1629.42	1631.9	1629.43	1629.4	1629.37
$n = 3$	1420	2155.13	2204.14	2253.17	2302.22	2351.17	2400.14	1698.95
$n' = 3$	1701.77	1701.81	1701.81	1701.8	1701.8	1701.8	1701.81	1701.79
$n = 4$	1360	2095.09	2144.14	2193.09	2242.14	2291.05	2340.1	1657.85
$n' = 4$	1632.2	1632.16	1632.16	1632.16	1632.16	1632.18	1632.20	1632.2
$n = 5$	1440	2175.05	2224.1	2273.09	2322.1	2371.09	2420.14	1735.04
$n' = 5$	1703.13	1702.67	1702.66	1702.64	1703.07	1702.62	1702.62	1702.66
$n = 6$	1375.98	1761.52	1787.23	1812.86	1838.57	1864.28	1890.1	1545.68
$n' = 6$	1544.79	1544.79	1544.81	1544.81	1544.82	1544.79	1544.8	1544.77
$n = 7$	1343.04	1936.96	1976.66	2016.19	2055.82	2095.43	2135.05	1637.06
$n' = 7$	1616.5	1616.51	1616.55	1616.51	1616.51	1616.52	1616.52	1616.45
$n = 8$	1235.04	2277.44	2347.04	2416.56	2486	2555.52	2625.05	1766.54
$n' = 8$	1743.26	1740.56	1740.58	1740.58	1743.26	1743.44	1740.57	1740.56
$n = 9$	1160.01	2442.48	2528	2613.44	2698.96	2784.48	2870.1	1787.15
$n' = 9$	1820.7	1818.22	1819.01	1818.86	1818.45	1818.21	1818.22	1818.22

The task processing times in the n' row align with the probability distribution specific to the range of each instance. For example, the value in row $\mathcal{P}_{1,1}^{\alpha=0}$, under the $n' = 1$ row, corresponds to the range specified for *Instance* 1 and represents the average of 50 repetitions using the machine sequence of $\mathcal{P}_{1,1}^{\alpha=0}$. A similar process occurs for $\mathcal{P}_{1,1}^{\alpha=0.75}$, where the probability distribution range is the same as that of *Instance* 1 (similar to $\mathcal{P}_{1,1}^{\alpha=0}$), but with the sequence of $\mathcal{P}_{1,1}^{\alpha=0.75}$ over 50 repetitions. It is worth noting that although the individual task processing times and machine sequences differ between $\mathcal{P}_{1,1}^{\alpha=0}$ and $\mathcal{P}_{1,1}^{\alpha=0.75}$, both configurations utilise optimal machine assignments and follow the same 50-repetition approach. This consistency explains why the results in the n' row is nearly identical.

An analysis of the results presented in Table 9 yields the following observation. Considering the outcomes for $\mathcal{P}_{1,1}^{\alpha=0}$, the reported original objective function value is approximately 1400. This indicates that when no conservatism is incorporated into the model while addressing *Instance* 1, the makespan of the final executed task is 1400. However, when this value is compared with the corresponding entry in the $n' = 1$ row, the makespan increases to 1679. This suggests that the initial makespan of 1400 is overly optimistic for scheduling purposes. Furthermore, examining the original objective function of the same model with a conservatism level of one ($\mathcal{P}_{1,1}^{\alpha=1}$), the makespan is observed to be approximately 2380, whereas the corresponding value in the $n' = 1$ row is around 1679. This implies that a conservatism level of one results in an excessively pessimistic estimation. Overall, the expected makespan of 1679 appears to fall within the range between conservatism levels (α) of 0 and 0.75, based on the results obtained for the first model in *Instance* 1. A similar pattern can be observed across other instances as well. Moreover, examining the results obtained from the second model, as presented in the last column of the table where various values of n are considered, it is evident that this model yields results that are the closest to those obtained for n'. For example, in *Instance* 1, the result obtained from the second model is 1625.06, whereas its counterpart ($n' = 1$) is 1679.04, reflecting an approximate deviation of 3.32 per cent. Conversely, when comparing the results of the first model for *Instance* 1, the deviation increases to nearly 29 per cent (see the results of $\mathcal{P}_{1,1}^{\alpha=1}$). A similar trend is observed across other instances, reinforcing the robustness of the results produced by the second model. It is worth noting that the second model successfully solved ten scenarios simultaneously. However, as previously mentioned, we observed that the optimality gap was approximately 40 per cent. This observation prompted us to solve each of the ten scenarios individually, store the results, and compare them with the outcomes obtained from solving all ten scenarios simultaneously. When addressing each scenario separately, we achieved the optimal solution for each case. The results indicate that the optimal solution for *Instance* 1 across the ten scenarios falls within the range of 1543 to 1677, corresponding to the scenarios with the minimum and maximum makespan, respectively. Consequently, the solution derived from the stochastic model, which considers all ten scenarios simultaneously, should be within this range. However, due to high symmetry, the solver is unable to close the gap. Despite this significant optimality gap, the results obtained in Step 3 of the methodology, as reported in Table 9, demonstrate that the solution achieved after 15 mins remains highly effective.

To deepen the analysis of the optimality gap, we introduce a new test case referred to as *scenario 11*. This scenario is constructed by selecting, for each task, the minimum processing time observed across scenarios 1 through 10. For example, if the task 'mucking' has varying durations across the initial ten scenarios, scenario 11 assigns to it the shortest of these values. This approach is applied uniformly across all tasks, resulting in a configuration that represents the most favourable (optimistic) processing times among the existing scenarios. Scenario 11 is then solved independently, and its makespan is stored. Subsequently, the makespan derived from scenario 11 is adopted as a lower bound for the stochastic model. This choice is justified by the fact that scenario 11 reflects the most optimistic configuration possible, and as such, its makespan represents the minimum achievable under ideal conditions. Employing this value as a lower bound aids the solver by reducing the symmetry within the problem space, thereby potentially accelerating convergence. Implementing this strategy resulted in a reduction of the optimality gap from approximately 40 per cent to 32 per cent. Therefore, it can be concluded that the inclusion of a lower bound could help speed up the process and reduce the gap.

To better understand the similarities in the $n' = 1$ row for each instance, we conducted a deeper analysis by examining the concept of the *critical path*. The critical path in project management and scheduling refers to the sequence of tasks or activities that determines the shortest possible duration required to complete a project (Liu and Hu, 2021). Any delay in the activities on the critical path directly impacts the overall project timeline, as these activities have no flexibility (or slack) in their scheduling (Liu and Hu, 2021). To know about calculation of the critical path, please refer to online Appendix C. After performing the computation to identify the critical path, we concluded that the critical paths across the instances are highly similar, which further explains the closeness of the values in the $n' = 1$ row for each instance.

CONCLUSION

Traditional manual scheduling methods are no longer adequate to meet the demands of modern mining projects, given the high financial stakes and operational challenges. Effective scheduling is crucial not only for maximising productivity and ensuring safety, but also for optimising the use of substantial investments and improving the overall profitability of mining projects. This study addressed the critical problem of scheduling development and production tasks in underground mining operations, where inefficiencies can lead to significant cost overruns and reduced profitability. The complexity of these operations, combined with their uncertain environment, underscores the importance of efficient scheduling and resource management.

Our study makes several key contributions to tackle these challenges effectively. First, we presented a CP model that simultaneously schedules development and production tasks for underground mines. Second, a proactive approach is proposed that incorporates task duration uncertainty into the optimisation model, which is a significant advancement over traditional deterministic models. By simulating task delays and accounting for variability, our approached enhances the robustness and reliability of the resulting schedules. To demonstrate the efficacy of our methodology, we solved the scheduling problem under both deterministic conditions, where task processing times are known and fixed, and under multiple degrees of conservatism, simulating worst-case scenarios. Finally, we introduced a two-stage stochastic formulation, which simultaneously addresses multiple scenarios to generate more robust schedules that are better equipped to handle real-world uncertainties. By comparing the outcomes of the stochastic model with the deterministic versions, we provided valuable insights into the benefits of using a proactive approach.

Practical implication

The results of this study demonstrate significant opportunities for improving operational efficiency through advanced scheduling techniques. Specifically, the practical implications highlighted below provide actionable insights for optimising mining processes and addressing the challenges posed by uncertain task processing times:

1. The adoption of sophisticated scheduling models, such as CP, could be essential for managing task allocation for shorter computation times, particularly when considering uncertain conditions.

2. By leveraging our CP model, mine planners can mitigate delays and ensure the optimal use of resources. This can improve the reliability of daily operations and provide a framework for responding swiftly to unanticipated disruptions.

3. The developed model's ability to incorporate uncertainty equips mining operations with a robust tool for risk management. By simulating different scenarios, planners can anticipate the impact of potential delays and adjust plans proactively.

4. The use of stochastic modelling offers schedules that are more resilient to disruptions. By generating solutions that account for variability, operations are better positioned to handle the unpredictable nature of underground mining.

Limitation and future research

There are several limitations that present opportunities for future research and enhancement. Addressing these limitations could lead to more robust and practical optimisation models that better reflect the realities of underground mining operations.

One major limitation of our approach is the large optimality gap observed in the stochastic method. This indicates that the current method struggles to consistently identify solutions that are close to optimal within a reasonable computational time. As a result, exploring alternative stochastic methods that may be better suited for implementation in CPO becomes a promising direction for future research.

Another area for improvement lies in the lack of a systematic literature review that can categorise papers based on the employed methodology and show cases the improvement of algorithms over years to address the problem at hand (Shahin *et al*, 2024b).

REFERENCES

Aalian, Y, Gamache, M and Pesant, G, 2024. Short-term underground mine planning with uncertain activity durations using constraint programming, *Journal of Scheduling*, pp 1–17.

Aalian, Y, Pesant, G and Gamache, M, 2023. Optimization of short-term underground mine planning using constraint programming, in 29th International Conference on Principles and Practice of Constraint Programming, Schloss Dagstuhl-Leibniz-Zentrum für Informatik.

Amoako, R, Chowdu, A and Brickey, A, 2023. A deviation-minimization approach to short-term underground mine schedule optimization, *Mining, Metallurgy and Exploration*, 40(5):1749–1765.

Astrand, M, Johansson, M and Zanarini, A, 2020. Underground mine scheduling of mobile machines using constraint programming and large neighborhood search, *Computers and Operations Research*, 123:105036.

Ayaburi, J, Swift, A, Brickey, A, Newman, A and Bienstock, D, 2024. Optimizing ventilation in medium-and short-term mine planning, *Optimization and Engineering*, pp 1–26.

Both, C and Dimitrakopoulos, R, 2020. Joint stochastic short-term production scheduling and fleet management optimization for mining complexes, *Optimization and Engineering*, 21(4):1717–1743.

Campeau, L-P and Gamache, M, 2020. Short-term planning optimization model for underground mines, *Computers and Operations Research*, 115:104642.

Campeau, L-P and Gamache, M, 2022. Short-and medium-term optimization of underground mine planning using constraint programming, *Constraints*, 27(4):414–431.

Campeau, L-P, Gamache, M and Martinelli, R, 2022. Integrated optimisation of short-and medium-term planning in underground mines, *International Journal of Mining, Reclamation and Environment*, 36(4):235–253.

Gliwan, S E and Crowe, K, 2022. A network flow model for operational planning in an underground gold mine, *Mining*, 2(4):712–724.

Habib, N A, Ben-Awuah, E and Askari-Nasab, H, 2024. Short-term planning of open pit mines with semi-mobile IPCC: A shovel allocation model, *International Journal of Mining, Reclamation and Environment*, 38(3):236–266.

Kanet, J J, Ahire, S L and Gorman, M F, 2004. Constraint programming for scheduling, in *Handbook of Scheduling: Algorithms, Models and Performance Analysis* (ed: J Y-T Leung), pp 47-1–47-21 (CRC Press: Boca Raton).

Laborie, P, Rogerie, J, Shaw, P and Vilím, P, 2018. IBM ILOG CP optimizer for scheduling: 20+ years of scheduling with constraints at IBM/ILOG, *Constraints*, 23:210–250.

Liu, D and Hu, C, 2021. A dynamic critical path method for project scheduling based on a generalised fuzzy similarity, *Journal of the Operational Research Society*, 72(2):458–470.

Manríquez, F, González, H and Morales, N, 2023. Short-term open pit production scheduling optimizing multiple objectives accounting for shovel allocation in stockpiles, *Optimization and Engineering*, 24(1):681–707.

Manríquez, F, Pérez, J and Morales, N, 2020. A simulation–optimization framework for short-term underground mine production scheduling, *Optimization and Engineering*, 21:939–971.

Martins, A G, Souza, M J F and Assis, P S, 2024. An integrated simulation and optimization tool for short-term mining planning problems with different prioritization among competing plant targets, *Computers and Industrial Engineering*, 191:110115.

Menezes, G C and Santos Correa, J dos, 2022. Model and algorithms applied to short-term integrated programming problem in mines, *Resources Policy*, 79:102950.

Navarro Torres, V F, Mateus, G R, Martins, A G, Carneiro, W and Chaves, L S, 2020. Integrated optimization and simulation models for short-term open pit mine planning, *Journal of the Southern African Institute of Mining and Metallurgy*, 120(11):617–626.

Nelis, G and Morales, N, 2022. A mathematical model for the scheduling and definition of mining cuts in short-term mine planning, *Optimization and Engineering*, 23(1):233–257.

Nelis, G, Morales, N and Jelvez, E, 2023. Optimal mining cut definition and short-term open pit production scheduling under geological uncertainty, *Resources Policy*, 81:103340.

Pesant, G, 2014. A constraint programming primer, *EURO Journal on Computational Optimization*, 2:89–97.

Rahnema, M, Grenon, M and Moradi Afrapoli, A, 2024. Sustainable open pit mining through GHG-conscious short-term production scheduling, *International Journal of Mining, Reclamation and Environment*, pp 1–22.

Shahin, R, Beaulieu, M, Bélanger, V and Cousineau, M, 2024a. Optimizing strategic personal protective equipment stockpiling: An assessment using Morris sensitivity analysis, *Engineering Proceedings*, 76(1):29.

Shahin, R, Hosteins, P, Pellegrini, P and Vandanjon, P-O, 2023. A full factorial sensitivity analysis for a capacitated flex-route transit system, in *Proceedings of the 8th International Conference on Models and Technologies for Intelligent Transportation Systems (MT-ITS)*, pp 1–6 (IEEE).

Shahin, R, Hosteins, P, Pellegrini, P, Vandanjon, P-O and Quadrifoglio, L, 2024b. A survey of flex-route transit problem and its link with vehicle routing problem, *Transportation Research Part C: Emerging Technologies*, 158:104437.

Silva-Junior, A L, Martins, A G, Pantuza-Jr, G, Cota, L P and Souza, M J F, 2023. Short-term planning of a work shift for open pit mines: A case study, *Cogent Engineering*, 10(1):2168172.

Geotechnology applied to quality control in bauxite mining in Brazil

I Soares[1], R Radtke[2], A Pina[3], K Gomes[4], B Gomes[5] and S Farias[6]

1. Engineer, Hydro, Paragominas, Pará 68628-486, Brazil. Email: ismael.soares@hydro.com
2. Geologist, Hydro, Paragominas, Pará, 68628-486, Brazil. Email: ricardo.souza@hydro.com
3. Geologist, Hydro, Paragominas, Pará, 68628-486, Brazil. Email: acacio.pina.neto@hydro.com
4. Coordinator, Hydro, Paragominas, Pará, 68628-486, Brazil.
   Email: keila.pam.gomes@hydro.com
5. Manager, Hydro, Paragominas, Pará, 68628-486, Brazil. Email: bruno.gomes@hydro.com
6. Senior Manager, Hydro, Paragominas, Pará, 68628-486, Brazil. Email: silvia.farias@hydro.com

ABSTRACT

The Hydro Bauxite Mine in Paragominas, Pará, Brazil, faces the daily challenge of monitoring the quality of mining operations. This task is undertaken by trained technicians who assess the boundary between waste and ore throughout the day across various shifts. However, the current process does not include a qualitative evaluation of monitored activities, which hampers the efficiency and reporting of these operations. To address this limitation, the solution involves the adoption of geotechnology. The ArcGIS system was introduced to innovate the quality control methodology using the Field Maps and Survey123 applications. Field Maps provides technicians with access to geological model data and mining plans on an offline map, while all field controls were digitised using Survey123 forms. These forms include fields for identification, geographic coordinates, qualitative evaluations, photographs, and comments. The implementation of this new process has demonstrated significant efficiency gains, reducing the time required for data entry to approximately 3 mins, compared to an estimated 30 mins with the previous method. This represents a 90 per cent reduction in time, resulting in increased productivity and enhanced control over mining activities. Additionally, the recorded data is stored in ArcGIS Online's cloud platform, enabling the creation of a standardised, secure, and traceable database. The integration of geological model data accessible in the field, qualitative assessments, photographs, and georeferenced information enhances the efficiency and accuracy of technicians' decision-making processes, contributing to improved management of quality control activities within the mine.

INTRODUCTION

One major challenge in the mining industry is guaranteeing the maximisation of mineral resources by extracting as much of the mineable deposit as possible with minimal dilution and loss of ore. This makes the operation economically viable and ensures rational mining, extending the life of the mine.

To achieve this, mining and geology technicians play a crucial role in evaluation and technical guidance during mineral extraction, being responsible for Quality Control activities.

Quality Control ensures efficiency in mine operations and the optimal use of the ore layer. Its main functions include controlling dilution and contamination of the ore by identifying lithological transitions and boundaries between waste and ore; monitoring ore quality through sample collection during the extraction process stages; inspecting and managing ore stockpiles; following the mining plan; and preparing daily reports with field information to guide management in making strategic and operational decisions.

In this context, the use of technologies focused on spatial and geographical information, that is, geotechnologies, is applied in all stages of conception and operation of a mining project (Rocha et al, 2023).

Through the Geographic Information System (GIS), it is possible to collect information, integrate different databases, analyse, and map data from mineral exploration, mine planning and operation, to environmental recovery and management (Choi, Baek and Park, 2020).

Geospatial data and GIS are the foundation of the mineral industry, as the sector has a crucial relationship with the Earth's surface, both above and below ground. Furthermore, mine operation is an inherently spatial activity where the use of GIS applications is indispensable, particularly in evaluating important operational indicators, digitising processes, reducing risks, improving performance in blasting and ore loading, and mapping the orebody (du Plessis, Grobler and Mashimbye, 2022).

Therefore, the use of geotechnologies and GIS applied to Quality Control activities can bring significant benefits to mining companies. Conversely, not using this technology can lead to manual workflows, inefficiencies in the process, untraceable and even lost information.

Thus, this study aimed to implement the use of geotechnologies in the operational activities of the Quality Control team at the bauxite mine of Mineração Paragominas S.A. (MPSA), a company of the Norsk Hydro S.A. group, located in the northern region of Brazil (Figure 1).

FIG 1 – Aerial view of the administrative and industrial areas of Mineração Paragominas S.A.

METHODOLOGY

Study area

The study area is the MPSA mining area, located in the municipality of Paragominas, in the north-western region of the State of Pará (PA), Northern Brazil. It consists of two bauxite deposits on contiguous plateaus, referred to as Miltonia 3 (M3) and Miltonia 5 (M5). The geographic coordinates of the mine are 2° 59' 51" S; 47° 21' 13" W, as shown in Figure 2.

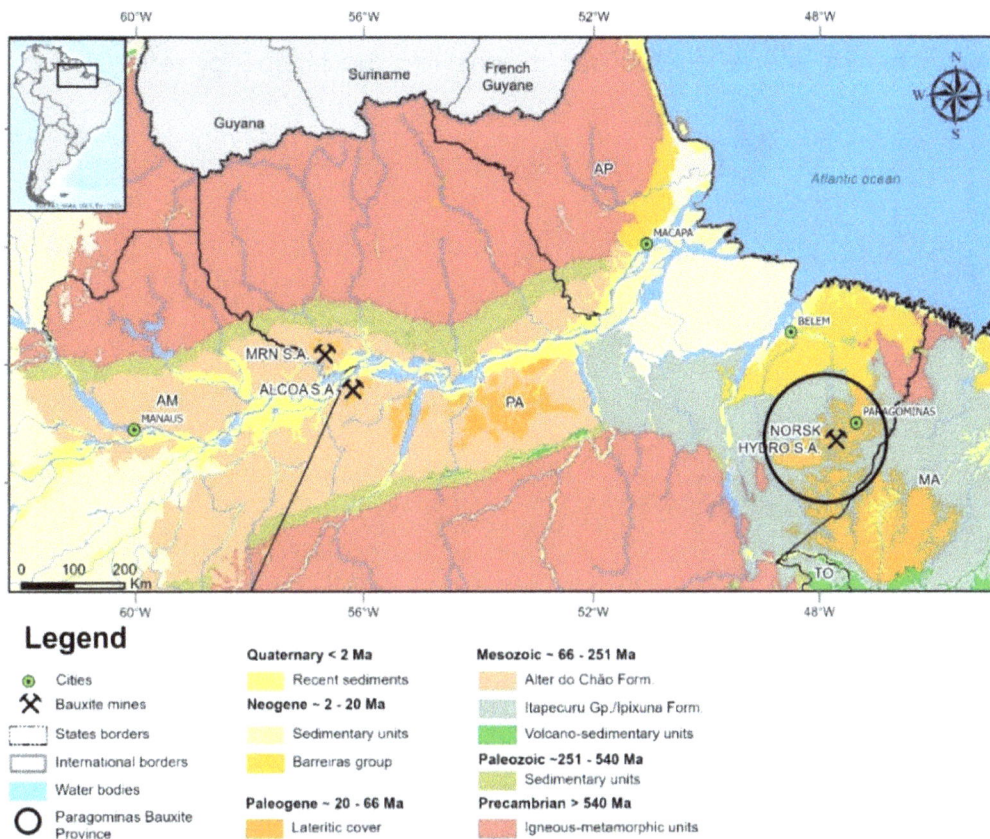

FIG 2 – Simplified geological map of northern Brazil showing the distribution of the main lithostratigraphic units.

Mine operation

Due to the geometric and spatial characteristics of the deposits, which occur on flat plateaus with the mineralised body in a tabular and horizontal shape, the most suitable mining method is strip mining.

This method essentially involves mining the deposit in successive strips, with the overburden from one strip being deposited in the pit resulting from the mining of the previous strip (Figure 3). This minimises the transportation distance for the overburden, and the area degraded by mining can be gradually rehabilitated shortly after extraction, thus accelerating environmental recovery.

FIG 3 – Bauxite extraction using the strip-mining method with a surface miner.

In summary, the operation consists of the following steps: First, the vegetation is removed, and a layer of soil (typically called Organic Soil or Topsoil) is separately mined for rehabilitation purposes. Then, the overburden layer is excavated, stacked, and stored. At this point, the bauxite is mined and loaded onto trucks to be transported to the beneficiation plant. After bauxite removal, the overburden layer returns to the pit and is levelled before receiving the topsoil for rehabilitation processes (Figure 4).

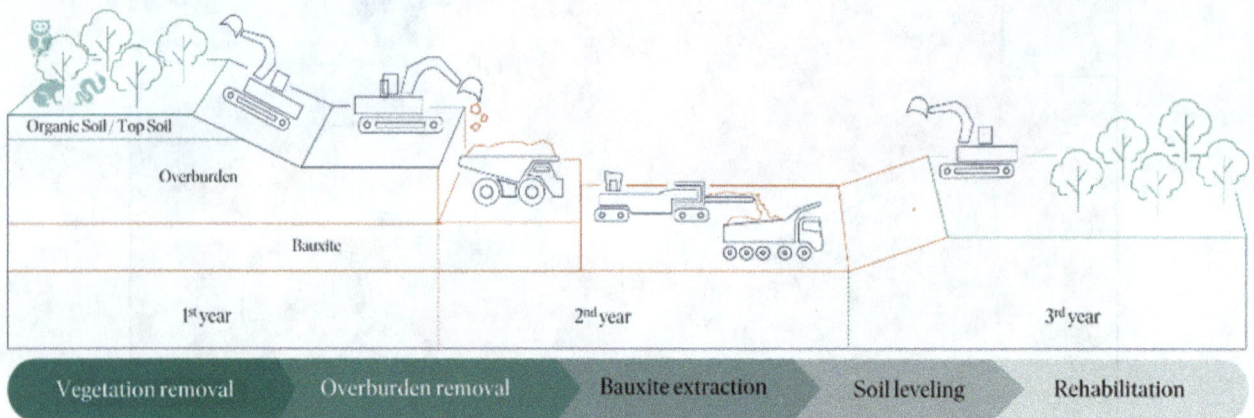

FIG 4 – Operation flow at the bauxite mine in Paragominas.

The MPSA's operations comprise an integrated production system that includes mining, beneficiation, and the transport of bauxite pulp through a 244 km-long pipeline that crosses seven municipalities in the State of Pará (PA), mostly underground. Along the pipeline route, there is also a 236 km-long power transmission line that supplies the plant with the required electricity. A Pumping Station (PS2) is located in Tomé-Açu/PA. The pipeline and the transmission line transport the bauxite pulp to the municipality of Barcarena, where it is used as the raw material to produce alumina in Hydro Alunorte S.A., another company of the Norsk Hydro S.A. group (Brandt Meio Ambiente, 2006).

Quality control

The Quality Control team at MPSA is responsible for a crucial role in monitoring mining operations, overseeing activities at different mining fronts throughout the operation of three mines located on the Miltonia 3 plateau and one mine on the Miltonia 5 plateau.

This team works in 12 hr shifts, covering an average of 130 km per shift, with the goal of providing support to the mining and dispatch operations (Figure 5). The main activities include:

- Monitoring of overburden removal.
- Marking of waste and ore contact boundaries.
- Control of the release of operational reserves.
- Update and management of mining fronts.
- Control of ore edge recovery.
- Management of stockpile operations.
- Sampling of density and pit floor.
- Control of the thickness of mined ore.
- Control of ore loss.
- Preparation of daily shift reports.

FIG 5 – The Quality Control team providing guidance to operators and assessing the deposit.

Initially, the technical team recorded operational information through manual notes on printed forms, filled out throughout the shift with data and observations related to the activities carried out at the mine. At the end of the day, this information was digitally transcribed and sent as reports to managers and other team members. However, this manual process limited immediate access to the collected data, making it difficult to obtain real-time information about the mine's operation, geological anomalies, and other relevant aspects of the extraction process.

Given this scenario, it became evident that there was a need to modernise and automate the reporting process, ensuring data traceability, the creation of a structured database, and the efficient integration of results with management and other strategic sectors.

Geotechnology and GIS

To meet the need for modernisation, automation, and optimisation of Quality Control activities, it was essential to use tools capable of integrating and providing data from various sources. In this context, geotechnologies and GIS provided the necessary tools to create an integrated system, enabling the availability of offline geological data and automating field data collection. In summary, the use of GIS optimised the flow of Quality Control information during bauxite extraction and enhanced decision-making in mining operations.

The system was developed on the ArcGIS Online Platform, using the Field Maps, Survey123, Map Viewer and ArcGIS Dashboards applications (Figure 6). An intelligent, unified, collaborative, and automated solution was created to empower the Quality Control technicians by providing access to geological model information, optimising field data collection, and delivering essential information to decision-makers. Additionally, the system ensures real-time data agility and traceability through interactive management dashboards that display key Quality Control indicators.

FIG 6 – ArcGIS tools and applications employed in developing the solution.

RESULTS

Digitalisation

To optimise and digitalise the Quality Control data collection at MPSA, an interactive map of the mine was developed using the Field Maps application from the ArcGIS platform. This map incorporates all the current structures of the operation, including access points and the location of mining fronts. To represent the different activities monitored by technicians during the work shift, vector layers of the point type were created, where each layer corresponds to a specific activity and is linked to a digital form with the same requirements as the manual forms previously used.

The system allows navigation through the map and data recording even in offline mode. Once an internet connection is reestablished, the collected data is automatically synchronised with an online database.

In this way, the technician can register information about an operational activity directly on the map and at the corresponding location, entering the data as georeferenced points with specific symbols for each activity. At the end of the shift, all the information recorded throughout the day is automatically updated in the central database (Figure 7).

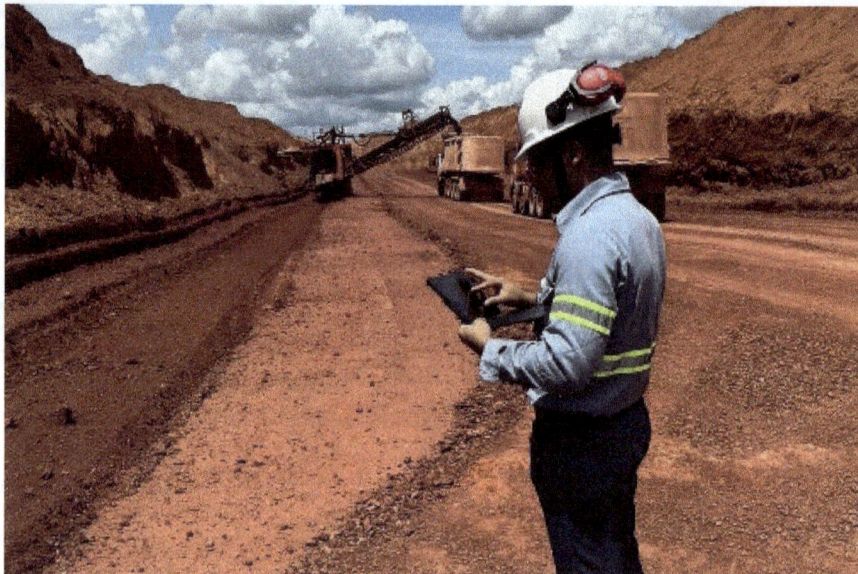

FIG 7 – Quality Control technician monitoring the mining operation and entering data through Field Maps.

In addition to Field Maps, the use of the Survey123 application, also from the ArcGIS platform, was implemented. Unlike Field Maps, which is based on a mapping environment, Survey123 focuses on data collection through structured forms, ensuring the capture of geographic coordinates even in offline environments.

With Survey123 tool, activities such as monitoring the thickness of the mined ore, controlling ore losses, and marking the contact between waste and ore were digitised, improving both accuracy and efficiency in monitoring the operational aspects of the mine (Table 1).

TABLE 1

Application used for each Quality Control activity.

Quality control activities	App ArcGIS	
	Field Maps	Survey123
Monitoring of overburden removal	X	
Marking of waste and ore contact		X
Control of operational reserve release	X	
Updating and management of mining fronts	X	
Control of ore edge recovery	X	
Stockpile operations management	X	
Density and pit bottom sampling		X
Control of mined ore thickness		X
Control of ore loss		X
Preparation of daily shift reports	X	

The digitisation of field activities has resulted in a significant improvement in operational efficiency. Previously, the collected data had to be manually transcribed and shared individually via email, a process that took approximately 30 mins per shift. With the implementation of the new digital system, the entry of information related to Quality Control activities has been substantially optimised, reducing the registration time to about 3 mins.

Furthermore, after synchronisation, the data becomes immediately accessible in real-time, enabling its visualisation, analysis, and interaction through ArcGIS Dashboards (Figure 8), which provides greater agility in decision-making and monitoring of operations.

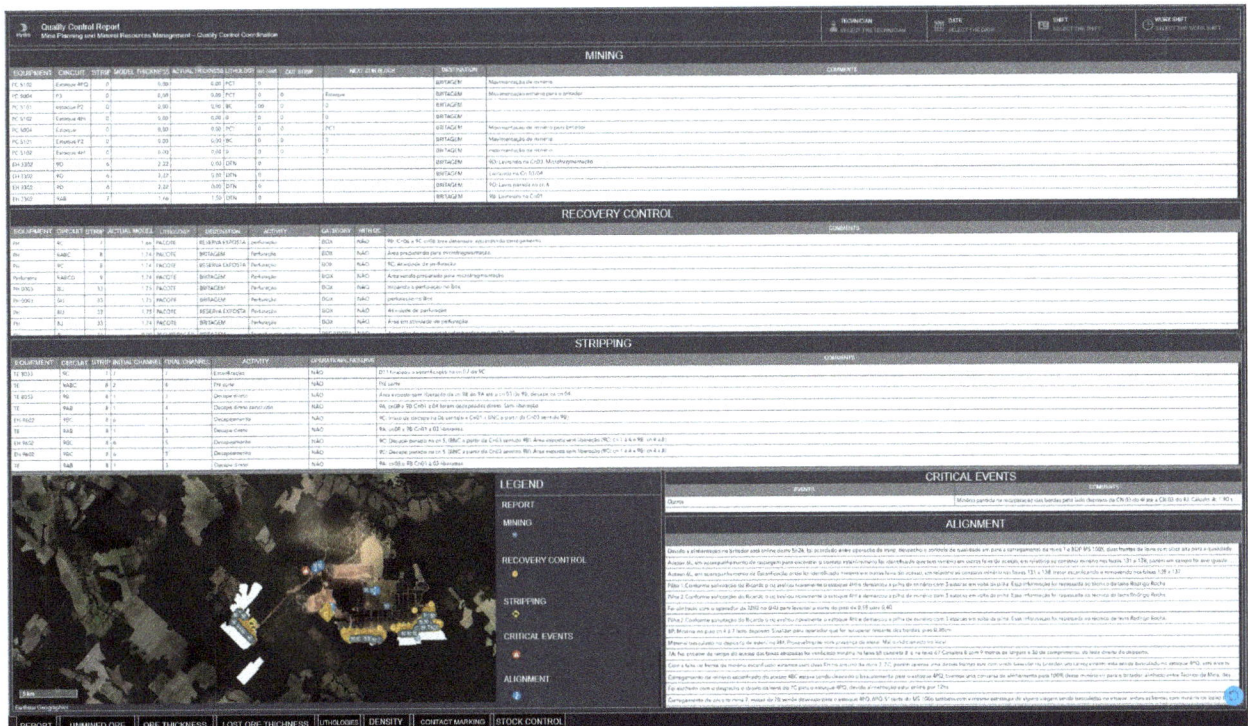

FIG 8 – Dashboards for monitoring data collected by the Quality Control team.

Geological model

The application of geotechnology through the ArcGIS platform enabled the integration of chemical quality and ore thickness data directly into the field map used by Quality Control. Previously, this information was limited to specialised software for manipulating geological models, making it difficult to access in the field. With the new approach, this data is now available offline on the digital map, allowing technicians to consult it immediately during operations.

This integration was made possible through the volumetric calculation of the mining front polygons in Datamine software, followed by the editing and incorporation of the variables into ArcGIS Pro. As a result, the volumetric polygons were incorporated into the operational Quality Control map, making geological information accessible in real-time to support decision-making in the field (Figure 9).

FIG 9 – Illustrative schematic of the geological model accessible in the field through Field Maps.

ROM reconciliation

The implementation of the digitalisation of Quality Control activities and the incorporation of additional geological model information began in April 2024, providing greater autonomy to field technicians. As a result, decision-making related to the initiation and cessation of mining fronts became more precise, directly enhancing the guidance of mine operations.

The primary indicator used to evaluate the effectiveness of these changes was ROM (Run-of-mine) reconciliation, which involves comparing the ore mass predicted by the geological model with the actual mined mass. This is measured through topographic surveys and expressed as a ratio of the surveyed volume to the estimated ore. The higher this ratio, the more efficient and sustainable the mining operation was.

In 2024, a 3.7 per cent increase in annual ROM reconciliation was observed compared to 2023, following the implementation of operational and procedural improvements within the Quality Control team. Although this positive outcome coincides with the changes made, it is important to emphasise that ROM reconciliation is a multifactorial indicator, influenced by a range of variables throughout the production chain — from the geological model to topographic measurement processes. Therefore, it is not methodologically sound to attribute this improvement solely to the actions of the Quality Control team. However, it is undeniable that effective mining control — particularly the accurate positioning and closure of mining fronts based on geological information — has a direct impact on reconciliation accuracy. Thus, it can be concluded that the more assertive performance of the Quality Control technicians contributed significantly to the observed result.

CONCLUSIONS

The use of geotechnologies in mining has become an essential practice, encompassing activities from mineral exploration to mining operations. However, the implementation of new control mechanisms, process automation, the structuring of a traceable database, the optimisation of report management, and the expansion of information access for the Quality Control team presented considerable challenges—overcome through the adoption of Geographic Information Systems (GIS).

The ArcGIS platform proved to be a versatile, intuitive, and highly customisable solution, enabling the full transition of field activities to an integrated digital environment. The digitalisation of processes allowed for the automation of operational reports and controls, optimising approximately two hrs per day in the routine of quality control technicians. In terms of governance, the management of performance indicators (KPIs) began to be carried out directly through a centralised database on ArcGIS Online, promoting greater traceability, standardisation, and agility in decision-making.

Additionally, the geological model became available in real time for field consultation via mobile devices integrated with Field Maps, which significantly increased the accuracy of mining guidance.

Therefore, the results demonstrate that the use of geotechnologies can be broadly applied in the mining industry, regardless of ore type or operational objectives, directly contributing to increased efficiency, precision, and sustainability in mining processes.

ACKNOWLEDGEMENTS

We would like to express our sincere gratitude to Norsk Hydro – Mineração Paragominas for the opportunity to develop a system utilising geotechnologies to optimise the Quality Control processes. Our deepest thanks go to the Management of Planning and Mineral Resources team for their partnership and continuous support throughout the development of the system. Their guidance and encouragement were invaluable.

A special thanks to the Quality Control team, whose active involvement was crucial in the system's development. Their inputs and feedback ensured that the system became not only practical and user-friendly but also intelligent and effective in meeting the operational needs. Without their dedication and collaboration, this project would not have been as successful.

REFERENCES

Brandt Meio Ambiente – Casaverdehidrosam, 2006. Environmental Impact Study of Bauxite Mining and Processing by Mineração Vera Cruz S/A [in Portuguese: Estudo de Impacto Ambiental de Lavra e Beneficiamento de Bauxita pela Mineração Vera Cruz S/A], MRCParagominas, Brandt Meio Ambiente – Casaverdehidrosam, Brazil.

Choi, Y, Baek, J and Park, S, 2020. Review of GIS-Based Applications for Mining: Planning, Operation and Environmental Management, *Applied Sciences*, 10(7):2266. https://doi.org/10.3390/app10072266

du Plessis, H, Grobler, H and Mashimbye, C, 2022. A review of GISc education, its value and use in the mining and exploration industries, *South African Journal of Geomatics*, 11(1):113–129. https://doi.org/10.4314/sajg.v11i1.9

Rocha, J, Gomes, E, Boavida-Portugal, I, Viana, C M, Truong-Hong, L and Phan, A T (eds), 2023. *GIS and Spatial Analysis*, Rijeka: IntechOpen. https://doi.org/10.5772/intechopen.100705

Development of a simulation-integrated mine road quality index (MRQI) for efficient dispatching operations in surface mines

S N Topal[1], A Elshahawy[2] and O Gölbaşı[3]

1. Student, Middle East Technical University, Ankara 06690, Turkiye. Email: nil.topal@metu.edu.tr
2. Sales Export Specialist, Noksel Steel Pipe, Ankara 06690, Turkiye.
   Email: ahmedelshahawy@gmail.com
3. Associate Professor Dr, Middle East Technical University, Ankara 06690, Turkiye.
   Email: golbasi@metu.edu.tr

ABSTRACT

Haul roads are critical to the safety, efficiency, and productivity of surface mining operations. The deterioration of haul road functionality, whether partial or complete, can result in significant challenges such as production delays, increased wear on machinery, and induced safety risks, all of which may disrupt or even halt dispatching activities. In many cases, the construction of mine roads relies on experience-based practices rather than standardised methodologies. The absence of systematic and standardised approaches in haul road construction contributes to the emergence of various operational problems, as the frequency and severity of uncertainties increase. These issues often appear as potholes, slip cracks, and uneven surfaces on the road.

This study aims to address these challenges by developing a standardised framework, the Mine Road Quality Index (MRQI), designed to assess and improve the quality of mine roads. The methodology integrates fuzzy logic, fault tree analysis, and discrete event simulation to analyse and prioritise the uncertainty factors that influence road quality and performance. To achieve this, a Fuzzy Fault Tree Analysis (FFTA) integrated with expert panel is conducted first to evaluate 17 key uncertainty factors in five main categories: structural design uncertainties, functional design uncertainties, management uncertainties, geometric design uncertainties, and geological uncertainties. These factors are assessed based on their alignment with both planned and actual road conditions, as well as their frequency and severity in contributing to road failures and impacting production rates.

Constructed FFTA allows the development of a Discrete Event Simulation (DES) model for a more in-depth evaluation of uncertainties affecting road quality and for exploring road improvement strategies by integrating the MRQI into a dispatch algorithm. This integrated approach offers a systematic approach to simulate and predict road performance across various scenarios, enabling improved operational outcomes.

INTRODUCTION

Surface mines rely heavily on haul roads, which are essential for ensuring the safety, efficiency, and productivity of mining operations. These roads are critical for the transportation of extracted materials, directly influencing the cost, performance, and safety of mining activities. High-quality haul roads contribute to reduced fuel consumption, improved equipment longevity, and minimised maintenance expenses. However, poorly designed or inadequately maintained roads can lead to increased operational costs, vehicle damage, and safety hazards. Despite their significance, the development and upkeep of mine roads often rely on empirical practices rather than standardised frameworks, resulting in inconsistencies and vulnerabilities in road performance.

A key challenge in haul road management lies in addressing the various uncertainties that affect their functionality. Structural uncertainties, such as fluctuations in subgrade strength or variations in applied loads, can result in uneven surfaces and deformations. Functional issues, like inconsistencies in surface material quality or changes due to weather conditions, impact road durability and vehicle traction. Management-related uncertainties, including irregular traffic patterns, fluctuating costs, and inconsistent maintenance schedules, further complicate road maintenance. Geometric factors, such as road gradients, sight distances, and braking requirements, also influence safety and efficiency. Additionally, geological uncertainties, such as clay content and subsurface discontinuities, can compromise road stability and longevity.

Given these challenges, there is a critical need for a systematic and data-driven approach to evaluating and improving haul road quality. This study integrates expert survey techniques, fuzzy logic, fault tree analysis (FTA), and simulation modelling to develop a comprehensive Mine Road Quality Index (MRQI). Expert surveys are employed to systematically assess the impact and significance of uncertainties, assigning weighted values based on the experience and expertise of respondents. Fuzzy logic-based fault tree analysis (FFTA) enables the quantification of uncertainty relationships, providing a structured methodology to evaluate risk factors and failure probabilities. Finally, simulation modelling allows for a dynamic assessment of how uncertainties evolve over time, offering insights into their cumulative effects on road performance.

By leveraging these advanced analytical techniques, this research aims to establish a standardised and quantitative framework for haul road assessment. The MRQI will serve as a decision-making tool to optimise haul road design, maintenance, and management, ultimately contributing to safer, more efficient, and cost-effective mining operations.

BACKGROUND INFORMATION

The quality of haul roads is a critical factor influencing the safety, efficiency, and overall productivity of surface mining operations. Extensive research has explored the impact of road conditions on mining performance, highlighting various aspects such as geometric design, material properties, maintenance strategies, and fuel efficiency.

For instance, Posada-Henao, Sarmiento-Ordosgoitia and Correa-Espinal (2023) demonstrated that road gradients exceeding 5 per cent significantly increase fuel consumption, underscoring the importance of optimising geometric design. Alegre, Peroni and Aquino (2024) further examined the role of road width and curvature, showing their impact on operational costs and sustainability. Kubler (2024) emphasised that optimised ramp gradients and smoother transitions reduce fuel consumption, reinforcing the importance of well-planned haul road designs. Similarly, Strack (2015) reviewed Australian haul road design practices, concluding that inadequate road structures result in higher rolling resistance, increased fuel usage, and accelerated vehicle wear.

In addition to geometric considerations, road surface conditions play a major role in haul road quality. Ngwangwa and Heyns (2014) introduced artificial neural networks (ANNs) to detect surface defects and classify road roughness, providing a novel method for real-time quality assessment. Thompson *et al* (2006) highlighted the importance of integrating sensor-based monitoring systems with maintenance protocols to optimise haul road performance and minimise costs. Similarly, Terziyski (2024) analysed open pit road construction methodologies, stressing the necessity of drainage systems and adequate surface thickness to prevent erosion and degradation over time.

Despite these contributions, most existing studies address individual factors affecting haul road performance but fail to offer a standardised framework for quantifying road uncertainties comprehensively. A key research gap lies in the lack of a structured methodology that integrates geometric, structural, functional, management, and geological uncertainties into a unified decision-making model. Additionally, previous studies do not adequately assess how these uncertainties dynamically evolve over time and impact mining efficiency.

To bridge this gap, researchers have explored methodologies such as fault tree analysis (FTA), fuzzy logic, and simulation-based modelling. FTA has been extensively used in reliability engineering to evaluate failure probabilities in complex systems (Rausand and Høyland, 2004). Its integration with fuzzy logic provides a means to model uncertainties in data-limited environments (Zadeh, 1965). Furthermore, discrete event simulation (DES) allows for the real-time assessment of failure probabilities, predicting how uncertainties influence road performance over time (Banks *et al*, 2010). Cheliyan and Bhattacharyya (2018) emphasised the role of expert elicitation in system reliability assessment, highlighting the benefits of expert-driven methodologies in uncertainty modelling. In a related study, Tekbey (2023) developed an uncertainty assessment model for deviations between long- and short-range production plans in surface metal mines using fuzzy logic fault tree analysis (FFTA) and DES, demonstrating the effectiveness of integrating expert insights with probabilistic modelling.

Although these methodologies have advanced the study of uncertainty in mining operations, a comprehensive predictive model for haul road quality assessment remains lacking. This study aims to address this deficiency by developing the Mine Road Quality Index (MRQI)—a structured framework that incorporates expert survey data, fuzzy logic-based FTA, and simulation modelling to quantify haul road uncertainties systematically. Unlike previous research, this study evaluates the cumulative effects of geometric, structural, functional, management, and geological uncertainties, enabling a proactive approach to haul road design, maintenance, and performance optimisation.

By integrating expert assessments, real-world data, and predictive modelling, this study moves beyond conventional road assessment techniques. Specifically, it focuses on geometric uncertainties (stopping distance, sight distance, diggability, road dimensions, and slope), structural uncertainties (applied load and subgrade strength variability), functional uncertainties (surface material, particle size distribution, and traction and adhesion), management uncertainties (traffic volume, cost variability, maintenance scheduling, and equipment fleet diversity), and geological uncertainties (dominant formation, presence of clay, water infiltration, and subsurface discontinuities). Figure 1 illustrates the categorisation of these uncertainty factors.

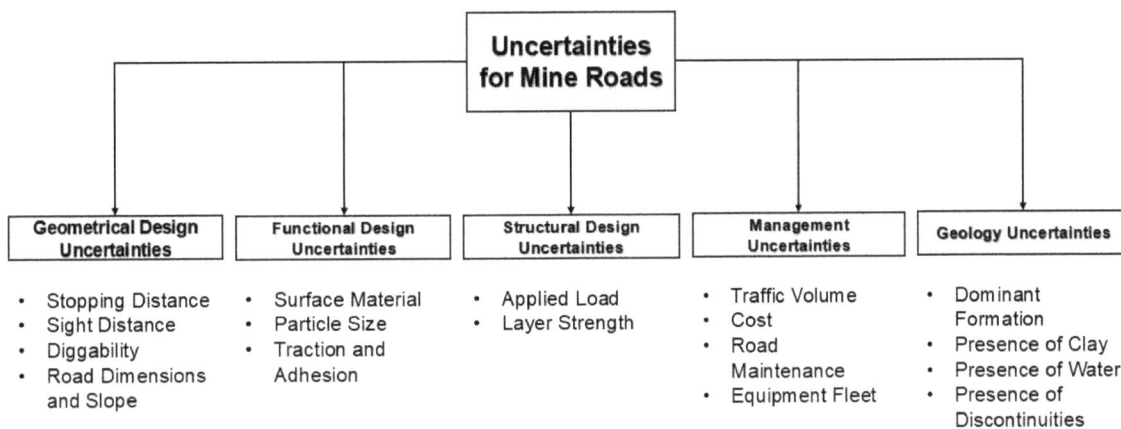

FIG 1 – Uncertainty categories.

The integration of these factors into a unified framework represents a significant contribution to literature, offering practical insights for safer, more efficient, and cost-effective mining operations. By shifting from a reactive to a proactive haul road management strategy, this study enhances the industry's ability to anticipate and mitigate road worsening risks, ultimately improving long-term operational sustainability.

METHODOLOGY

This study aims to develop a Mine Road Quality Index (MRQI) by integrating expert survey, fuzzy logic, fault tree analysis, and simulation modelling. The methodology is structured into three main stages: (i) expert survey and data collection, (ii) fuzzy logic-based fault tree analysis for uncertainty assessment, and (iii) simulation model development for system behaviour evaluation. Figure 2 represents the methodology of the study.

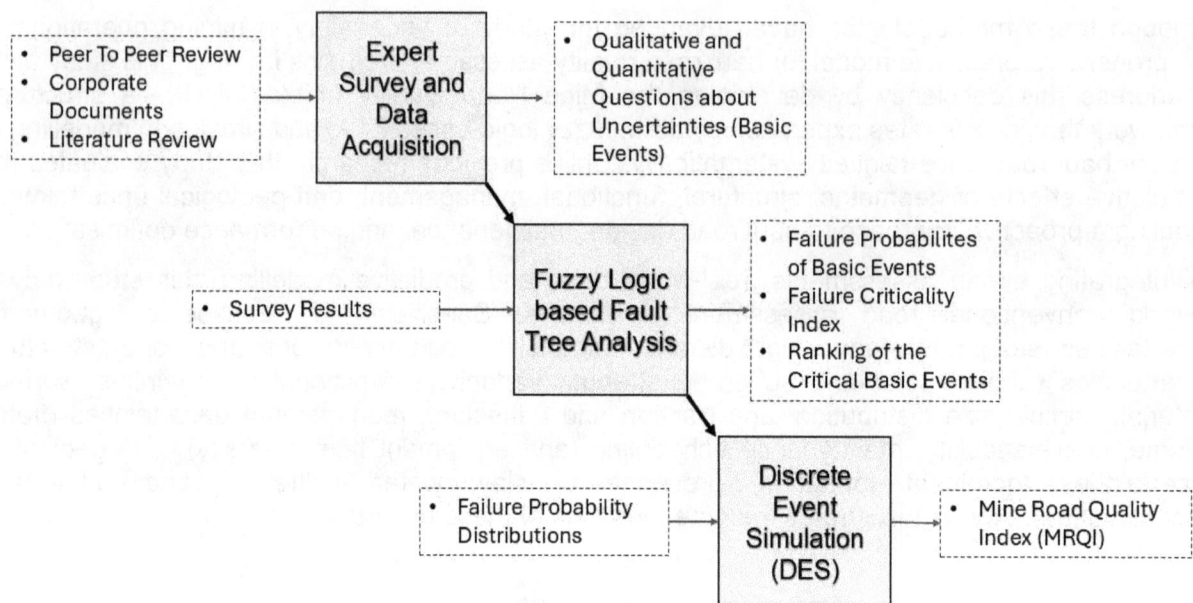

FIG 2 — Methodology of the study.

EXPERT SURVEY AND DATA COLLECTION

The first step in data collection involves conducting an expert survey to identify and evaluate uncertainties affecting mine road quality. Expert elicitation is a crucial method when failure probability data is scarce, as it provides a structured approach to assess system reliability and prioritise influencing factors (Cheliyan and Bhattacharyya, 2018). The survey serves as a foundation for fuzzy logic evaluation, fault tree analysis, and simulation modelling by systematically gathering knowledge on the significance and impact of uncertainties.

The survey and collected data are based on peer-to-peer review, corporate reports and literature review. It consists of five distinct questions for each identified uncertainty. Two of these are qualitative questions aimed at understanding how uncertainties differ between the design and construction phases. The remaining three focus on quantitatively assessing the frequency of issues arising from uncertainties and their impact on production losses. At this stage, since real-world expert responses are not available, the number of participants, their backgrounds, and their experience-based opinions have been based on hypothetical data. This ensures that the simulation framework can still be developed and validated before integrating actual field data.

FUZZY LOGIC-BASED FAULT TREE ANALYSIS

Fuzzy logic method provides a robust framework for modelling systems characterised by uncertainty and imprecision, using linguistic variables and approximate reasoning rather than strict binary classifications. Its flexible structure enables the formulation of rules that closely resemble human decision-making processes (MathWorks, 2022). The first step in fuzzy logic-based fault tree analysis (FFTA) is constructing an expert weighting table based on participants' backgrounds. This weighting system ensures that expert opinions are evaluated proportionally to their experience and qualifications. Table 1 presents a comparative analysis based on the age and professional background data of 520 mining engineers who graduated from mining engineering programs and are currently working in the mining sector (Tekbey, 2023). For instance, individuals with less than five years of experience are assumed to be under the age of 30.

TABLE 1

Background study of mining engineers.

Title	Attributes		
	Experience in years	Age in years	Weight
Senior Mine (Planning) Manager	≥ 25	> 50	5
Mine (Planning) Manager	20–25	40–50	4
Mine Planning Superintendent	10–20	35–40	3
Mine Planning Chief/ Senior Mine Planning Engineer	5–10	30–35	2
Mine Planning Engineer	<5	<30	1

After calculating the weighted average, participants are required to complete a questionnaire where they provide responses on a seven-point scale: Very Low, Low, Mildly Low, Medium, Mildly High, High, and Very High. Within the questionnaire, participants will specify their minimum, most expected, and maximum values on a scale from 1 to 10. For example, for a senior mine planning engineer, Very High category might have a minimum value of 7, a most expected value of 8, and a maximum value of 10. The result of the questionnaire will give the fuzzy sets which can be later used to transform linguistic variables into fuzzy numbers.

At this stage, the responses corresponding to the seven-point linguistic scale will be valuable in understanding the probability of failure of uncertainties (basic events). Participants will provide linguistic responses to each basic event and its associated sub-questions. These linguistic evaluations will then be transformed into fuzzy numbers using Equation 1 (Cheliyan and Bhattacharyya, 2018).

$$M_i = \sum_{j=1}^{N_e} A_{ij} w_j \tag{1}$$

Where:

i	$(1, 2, ..., N)$
N	is the number of Basic Events (Uncertainties)
N_e	is the number of participants of the expert survey
w_j	is the weighting score of the expert j
A_{ij}	is the linguistic expression of the corresponding basic event given by the expert j (It will be given in the numerical example section)
M_i	is the resultant fuzzy number of BEi

In the study conducted by Tekbey (2023), the fuzzy numbers obtained from the Equation 1 were converted into Fuzzy Possibility Scores (FPS). This variable represents the possibility of the corresponding basic event. At this stage, the Left and Right Fuzzy Ranking Method was employed, as it is one of the most widely used approaches for such assessments:

$$FPS = \frac{\mu_r + (1 - \mu_l)}{2} \tag{2}$$

where $\mu_l = \frac{1-a}{1+b-a}$ and $\mu_r = \frac{c}{1+c-b}$ for a triangular fuzzy set number of Fn = (a,b,c).

The generated fuzzy possibility value can be converted into fuzzy probability using the formula proposed in the study by Cheliyan and Bhattacharyya (2018). Equation 3 can be represented as follows:

$$P(X) = \begin{cases} \frac{1}{10^k} \; for \; FPS \neq 0 \\ 0, \; for \; FPS = 0 \end{cases} \; where \; k = 2.301 \sqrt[3]{(\frac{1-FPS}{FPS})} \qquad (3)$$

As explained in the steps, the failure probability of basic events can be calculated using this method. Consequently, the prioritisation ranking of basic events contributing to the top event in the fault tree analysis can be determined.

Fault Tree Analysis (FTA) is a systematic and deductive technique used to identify the root causes of system failures by mapping the logical relationships between various events that lead to an undesired outcome. It is commonly represented as a diagram that starts with a 'top event,' typically a failure, and branches down into various contributing events through logical gates like 'AND' and 'OR'.

In the context of this project, FTA is employed to model and analyse the uncertainties affecting mine road quality. By constructing a fault tree, the key uncertainty factors, including structural, functional, management, geometric, and geological aspects—are systematically examined to understand how they interact and contribute to road performance failures.

Through this analysis, FTA not only highlights the most critical factors that reduce mine road quality but also offers insights into how these factors can be moderated to improve road performance. By integrating FTA results with the simulation framework, this project aims to develop a strong and data-driven approach to prioritise interventions and optimise mine road management strategies. The constructed fault tree is presented in Figure 3.

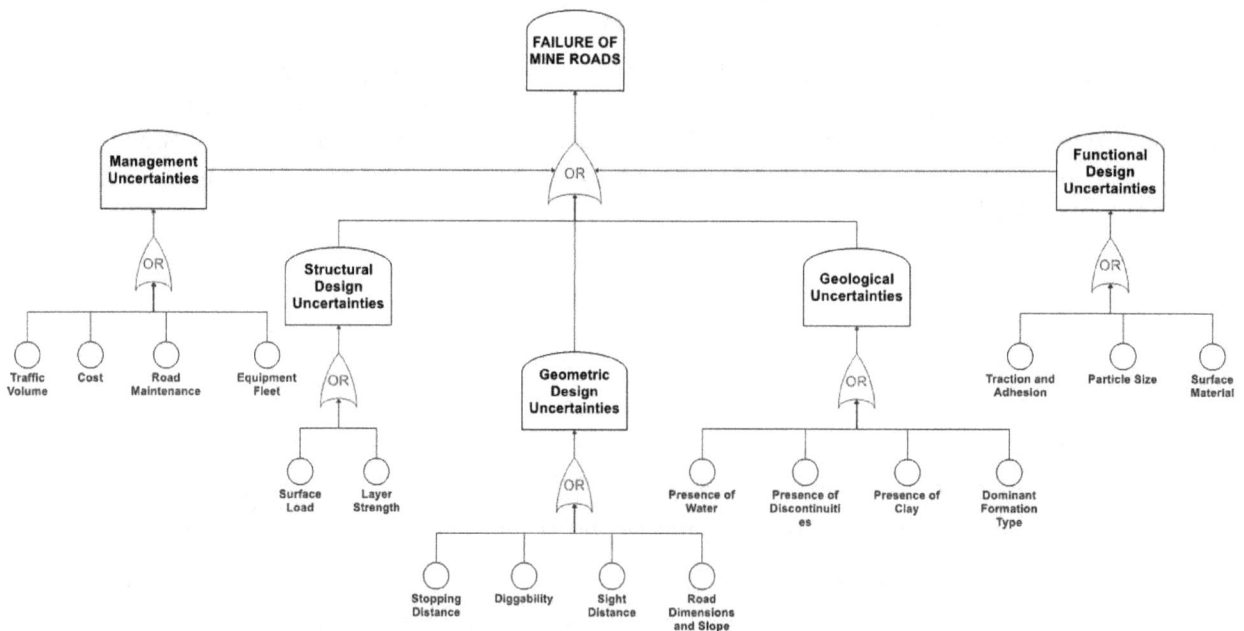

FIG 3 – Fault tree for failure of mine roads.

SIMULATION ALGORITHM DEVELOPMENT PROCESS

Discrete event simulation is a technique used to model systems that undergo changes in state variables at a discrete set of time points. These models can be static—representing the system at a specific moment—or dynamic, capturing its evolution over time, which aligns well with analysing time-dependent mining operations (Soofastaei *et al*, 2016). To evaluate and improve mine road quality, a structured simulation model is developed to analyse how various uncertainties impact road performance over time. Mine haul roads are subject to multiple uncertainty factors, such as structural design, functional design, geometric design, management and geology, all of which can lead to operational inefficiencies and increased maintenance costs. This study employs fuzzy logic-based fault tree analysis (FFTA) and discrete event simulation (DES) modelling to systematically assess the probabilistic relationships between these uncertainties and their potential consequences.

Given that mine road quality is influenced by a combination of qualitative and quantitative factors, expert opinions play a crucial role in defining the degree of impact of each uncertainty. A structured expert survey is conducted to gather information on how these uncertainties evolve across different project stages, from design to operational phases.

To further analyse the temporal and stochastic nature of these uncertainties, a simulation model is developed. The primary objective is to provide an assessment of uncertainty-driven variations in road quality, helping decision-makers to enhance road performance.

At this stage, the study focuses on structuring the simulation algorithm steps and defining the input parameters, variables, and boundary conditions. Future work will involve implementing the Mine Road Quality Index (MRQI) calculation and validating the model's predictive accuracy through case studies. The proposed framework presents a promising approach to quantifying road quality uncertainties.

Variables, probability density functions (PDFs), and sets within the scope of the simulation algorithm are shown below Tables 2 and 3.

TABLE 2
Simulation model variables.

Variables	Description
FO_i	Failure occurrence time of uncertainty i
s	Simulation number
RD_i	Road downtime duration of uncertainty i
s_t	Target simulation number
OC^{direct}	Weekly operational cost
UC_i	Unit downtime cost of uncertainty i
Dm^{in}	The minimum remaining duration for the failure to occur

Sets/PDFs	Description
t_t	Target simulation time
$f(x)_i$	Probability density function of road failure occurrence time of uncertainty i
$r(x)_i$	Probability density function of road downtime time of uncertainty i

TABLE 3
Simulation model equations.

	Description	Equation
ND_i	The new remaining duration for failure to occur of uncertainty i after one is failed	$FO_i - Dm^{in}$
$OC_i^{indirect}$	Additional road maintenance cost of uncertainty i	$RD_i \times (UC_i)$
t_a	Active simulation time	$FO_i + RD_i.$

The following section provides an overview of the simulation steps.

- In the developed simulation model, a discrete event simulation (DES) approach has been utilised, which models system behaviour as a sequence of distinct events occurring at specific points in time. The inputs provided to the algorithm are derived from probability density functions (PDFs), indicating that the system does not operate with fixed values but rather simulates outcomes using randomly generated values. This highlights the stochastic structure of the model, where uncertainties are dynamically represented. Additionally, in this framework, time progresses internally, meaning the system advances through discrete time steps based on event occurrences.

- In this algorithmic model, time progresses in weekly increments. At the beginning of the first iteration, a random value is assigned from $f(x)_i$ to each uncertainty. This process determines the problem occurrence time, FO_i, for each uncertainty, establishing that failure events are dynamically generated. Once all uncertainties have been assigned a time value, the algorithm identifies the smallest value, Dm^{in}, among them which represents the next failure event to occur. Additionally, the system determines which specific uncertainty is responsible for this failure, allowing the model to proceed accordingly in the simulation process.

- Each uncertainty is assigned to a separate branch, allowing the model to process different types of failures independently. Once the earliest problem occurrence time is identified and the corresponding uncertainty is determined, the algorithm activates the relevant branch, directing the simulation flow towards the specific failure type. From this point forward, all calculations and assessments will proceed within the active branch.

- In the active branch, a random mine road downtime value, RD_i, is assigned from $g(x)_i$ corresponding to the failure occurrence time. Within the algorithm, the mine road repair time is assumed to be equal to the assigned downtime, meaning that once a failure occurs, the system will remain non-operational for the duration of the downtime before resuming normal function.

- To update the failure occurrence times of inactive uncertainties, the algorithm recalculates their new remaining time by subtracting the active uncertainty's problem occurrence time from their previously assigned values which is represented by $FO_i - Dm^{in}$. For example, if failure events associated with the first and second uncertainties are expected to occur in weeks 3 and 4, respectively, the algorithm will follow the branch corresponding to the failure in week 3. The time until the next failure will be updated as (4–3) = 1 week which is denoted as the new problem occurrence duration, ND_i, for inactive uncertainties. For the active uncertainty, reassignment of the problem occurrence time can be done by calling a new value from the corresponding probability function, $f(x)_i$.

- The time increments in the simulation follow a structured process where the system progresses in weekly steps, ensuring that all failure-related calculations for each uncertainty are completed before advancing to the next iteration. At each time step, arrays storing the total downtime, associated additional operational costs, and the cumulative impact on overall mining performance are updated. Within this framework, direct costs, OC^{direct}, remain fixed as predetermined operational expenses, whereas indirect costs, $OC_i^{indirect}$, fluctuate based on the severity and frequency of mine road failures. It can be calculated as $RD_i \times (UC_i)$. These values are continuously tracked and stored for each uncertainty.

- The simulation cycle continues until the active simulation time, t_a, reaches the target simulation time, t_t. At this stage, the active time can be found as $FO_i + RD_i$. The target simulation time must be sufficiently large to encompass the occurrence frequencies of all basic events (uncertainties) embedded within the algorithm. Therefore, the model allows for the observation of all possible failure scenarios. In cases where the computed active time does not yet match the target simulation time ($t_a \neq t_t$), the algorithm enters a looping process and iterates continuously until the target is reached.

NUMERICAL EXAMPLE

In this section, the numerical implementation of the study, whose theory and framework were established above, will be conducted. In the first part, a weight will be assigned based on the participants' backgrounds. Then, fuzzy sets for the participants will be created using a questionnaire. The responses given for each basic event, along with the participants' weight scores, will subsequently form the aggregated fuzzy numbers. This process is called fuzzy aggregation (Tekbey, 2023). Then, using the defuzzification method, the FPS (fuzzy possibility score) for the basic events will be obtained. The fault tree method will be used to convert possibility into probability, thereby obtaining the FCI% (Failure Criticality Index) corresponding to the basic events. The final part of the numerical example section consists of the inputs and outputs used in the discrete event simulation.

Since real-world data is not yet available, a hypothetical expert data set has been created, representing age, years of experience in the mining sector, and education level as seen in Table 4. The weighting score calculations are made by using Table 1.

TABLE 4
Weighting score calculations of the participants.

Title	Experience (years)	Age (years)	Title weight	Experience weight	Age weight	Total weight
Mine Planning Superintendent	15	43	3	3	4	10
Senior Mine Manager	30	58	5	5	5	15
Senior Mine Planning Engineer	12	40	2	3	3	8
Mine Manager	24	50	4	4	4	12
Mine Planning Engineer	6	28	1	2	3	6
Mine Planning Superintendent	18	48	3	3	4	10
Mine Manager	20	45	4	4	4	12
Senior Mine Manager	35	60	5	5	5	15
Mine Planning Chief	10	32	2	3	2	7
Senior Mine Planning Engineer	7	33	1	2	2	5

To determine the weighted influence of each expert, the total score assigned to an expert is divided by the sum of all expert scores. Thus, the weighted influence of each participant is determined as follows in the Table 5.

TABLE 5
Result of the weighting calculations.

Expert ID	E1	E2	E3	E4	E5	E6	E7	E8	E9	E10
Total score	10	15	8	12	6	10	12	15	7	5
Weighted average	0.1	0.15	0.08	0.12	0.06	0.1	0.12	0.15	0.07	0.05

The weighted influence distribution reveals that E2 and E8 have the highest impact, each contributing 15 per cent to the total weighting. In contrast, E5 and E10, exhibit the lowest influence, collectively accounting for 11 per cent. This weighting data will be useful to calculate aggregated triangular fuzzy number of the basic events.

The fuzzy membership function is a key component in fuzzy logic, mapping input linguistic variables to a membership value between 0 and 1. Essentially, it translates fuzzy linguistic sets into numerical fuzzy outputs. Each linguistic variable requires a corresponding membership function, which can be derived from a structured questionnaire.

For this study, participants will be asked to fill out the linguistic variables. As the fuzzy numbers used are triangular, experts will provide minimum, maximum, and most expected values on a scale from one to ten. The answers of the linguistic variables are presented in Table 6.

TABLE 6
Membership function results.

Participant ID		VH	H	MH	M	ML	L	V L
	Minimum	7	6	4	3	2	1	1
1	Most expected	8	7	5	4	3	2	2
	Maximum	9	8	6	5	4	3	2
	Minimum	8	7	5	4	3	2	1
2	Most expected	9	8	6	5	4	3	2
	Maximum	10	9	7	6	5	4	3
	Minimum	9	8	6	5	3	2	1
3	Most expected	10	9	7	6	4	3	2
	Maximum	10	9	8	7	5	4	3
	Minimum	7	7	5	3	2	1	1
4	Most expected	8	8	6	4	3	2	2
	Maximum	9	9	7	5	4	3	3
	Minimum	8	7	6	4	3	2	1
5	Most expected	9	8	7	5	4	3	2
	Maximum	10	9	8	6	5	4	3
	Minimum	9	8	7	5	4	2	1
6	Most expected	10	9	8	6	5	3	2
	Maximum	10	9	8	7	6	4	3
	Minimum	8	7	5	3	2	1	1
7	Most expected	9	8	6	4	3	2	2
	Maximum	10	9	7	5	4	3	3
	Minimum	7	7	5	4	2	1	1
8	Most expected	8	8	6	5	3	2	2
	Maximum	9	9	7	6	4	3	3
	Minimum	8	7	6	5	3	2	1
9	Most expected	9	8	7	6	4	3	2
	Maximum	10	9	8	7	5	4	3
	Minimum	7	6	5	3	2	1	1
10	Most expected	8	7	6	4	3	2	2
	Maximum	9	8	7	5	4	3	3

As stated in the fuzzy logic-based fault tree analysis section, the questionnaire will give the fuzzy sets which can be found by conducting descriptive statistics of the data. The fuzzy sets can be seen in Table 7.

TABLE 7
Fuzzy sets.

Linguistic term	Fuzzy set
Very High	(0.79, 0.87, 1.00)
High	(0.7, 0.8, 0.88)
Mildly High	(0.56, 0.66, 0.75)
Medium	(0.4, 0.5, 0.6)
Mildly Low	(0.27, 0.37, 0.47)
Low	(0.16, 0.26, 0.36)
Very Low	(0.1, 0.2, 0.3)

In Table 8, the basic event descriptions and their corresponding linguistic responses are given.

TABLE 8
Linguistic responses for a corresponding basic event.

Basic event ID	Basic event description	Exp 1	Exp 2	Exp 3	Exp 4	Exp 5	Exp 6	Exp 7	Exp 8	Exp 9	Exp 10
BE1	Surface load	M	M	ML	MH	MH	ML	ML	M	H	MH
BE2	Layer strength	MH	MH	H	M	H	H	VH	H	M	M
BE3	Traffic volume	H	H	MH	VH	MH	H	M	VH	H	M
BE4	Cost	H	VH	VH	M	MH	H	MH	H	M	VH
BE5	Road maintenance	H	H	VH	M	M	VH	MH	M	H	MH
BE6	Equipment fleet	VH	H	H	H	VH	M	MH	VH	M	MH
BE7	Surface material	MH	VH	M	MH	H	MH	H	MH	VH	H
BE8	Particle size	H	M	M	VH	MH	H	MH	VH	H	M
BE9	Traction and adhesion	MH	H	MH	VH	M	MH	H	MH	VH	M
BE10	Stopping distance	H	VH	H	MH	M	M	VH	MH	H	M
BE11	Sight distance	H	M	MH	VH	H	MH	M	MH	VH	H
BE12	Diggability	H	VH	M	MH	MH	H	M	VH	MH	M
BE13	Road dimensions and slope	VH	MH	M	VH	H	MH	M	MH	H	VH
BE14	Dominant formation type	H	VH	M	MH	MH	H	M	VH	H	M
BE15	Presence of clay	H	H	MH	VH	M	MH	H	VH	MH	M
BE16	Presence of discontinuities	MH	M	VH	MH	MH	H	MH	H	M	VH
BE17	Presence of water	H	VH	MH	M	M	H	VH	MH	H	M

The aggregated fuzzy numbers, fuzzy possibility score and its corresponding probabilities are represented in Table 9. As seen, BE15 and BE3 can be interpreted as the two basic events with the highest probability of failure.

TABLE 9

Fuzzy possibility scores and corresponding probabilities.

BE_i	Aggregated fuzzy numbers (a,b,c)	FPS	$P(X_i)$
1	(0.42, 0.52, 0.62)	0.52	0.0056
2	(0.60, 0.70, 0.79)	0.68	0.0165
3	(0.65, 0.75, 0.85)	0.73	0.0219
4	(0.64, 0.74, 0.84)	0.72	0.0206
5	(0.59, 0.69, 0.79)	0.67	0.0155
6	(0.65, 0.75, 0.85)	0.73	0.0218
7	(0.63, 0.73, 0.82)	0.71	0.0191
8	(0.62, 0.71, 0.81)	0.69	0.0175
9	(0.62, 0.72, 0.82)	0.70	0.0185
10	(0.62, 0.72, 0.82)	0.70	0.0184
11	(0.59, 0.69, 0.78)	0.67	0.0152
12	(0.62, 0.71, 0.81)	0.69	0.0177
13	(0.61, 0.70, 0.80)	0.69	0.0169
14	(0.63, 0.72, 0.82)	0.70	0.0187
15	(0.66, 0.75, 0.85)	0.73	0.0222
16	(0.59, 0.69, 0.78)	0.67	0.0153
17	(0.62, 0.72, 0.82)	0.70	0.0184

The probability outputs obtained from the fault tree are used as inputs for the ReliaSoft Analytical Fault Tree and Simulation Fault Tree tools. Using the PDF equations, the probability occurrence of sub-events and the top event is calculated which is the difference in the mine roads design phase and mine roads operating phase. The PDF equations and failure probability of sub-events and top event can be seen here:

$D_1 = +R_{Surface Load} \cdot R_{Layer Strength}$

$D_2 = +R_{Traffic Volume} \cdot R_{Cost} \cdot R_{Road Maintenance} \cdot R_{Equipment Fleet}$

$D_3 = +R_{Traction and Adhesion} \cdot R_{Particle Size} \cdot R_{Surface Material}$

$D_4 = +R_{Stopping Distance} \cdot R_{Sight Distance} \cdot R_{Diggability} \cdot R_{Road Dimensions and Slope}$

$D_5 = +R_{Dominant Formation Type} \cdot R_{Presence of Clay} \cdot R_{Presence of Discontinuities} \cdot R_{Presence of Water}$

$D_6 = +D_5 D_4 D_3 D_2 D_1$

Statistically, the total failure probability is calculated as 0.2612, indicating a 26.12 per cent chance of system failure. Among the sub-event categories, D5 has the highest failure probability (0.0725), while D1 has the lowest (0.0220), suggesting that factors under D5 (Presence of Clay Problems) have the most significant impact on road quality.

Table 10 presents basic events based on their fuzzy probability scores. However, it is essential to evaluate them within the context of the overall system behaviour, considering how each event's failure contributes to the total system failure. To address this, the failure criticality index (FCI) is calculated using the ReliaSoft – Simulation Fault Tree tool and it is presented below Table. This index represents the relative frequency with which a component's failure leads to the failure of the entire system (Tekbey, 2023).

TABLE 10

Failure probability of sub-events and top event.

Basic event ID, BE_i	Basic event	Sub-event categories	Failure probability of sub events	Failure probability of the top event
BE1	Surface load	D1	0.0220	
BE2	Layer strength			
BE3	Traffic volume			
BE4	Cost	D2	0.0775	
BE5	Road maintenance			
BE6	Equipment fleet			
BE7	Surface material			
BE8	Particle size	D3	0.0542	
BE9	Traction and adhesion			0.2612
BE10	Stopping distance			
BE11	Sight distance	D4	0.0665	
BE12	Diggability			
BE13	Road dimensions and slope			
BE14	Dominant formation type			
BE15	Presence of clay	D5	0.0725	
BE16	Presence of discontinuities			
BE17	Presence of water			

Table 11 presents the Failure Criticality Index (FCI) for various basic events, ranking them based on their relative impact on system failure. The most critical events include the presence of clay (10.36 per cent), stopping distance (9.46 per cent), traction and adhesion (7.66 per cent), and presence of discontinuities (7.66 per cent), indicating that soil properties, road traction, and stopping distance significantly contribute to failures. In contrast, the least critical events, including surface load (2.70 per cent), particle size (2.70 per cent), and layer strength (3.60 per cent), show minimal impact.

TABLE 11
Failure criticality index and rankings of the sub-events.

Basic event ID, BE_i	Basic event categories	FCI (%)	FCI based ranking	Sub-event categories	FCI (%)
BE1	Surface load	2.70	16	D1	6.30
BE2	Layer strength	3.60	15		
BE3	Traffic volume	6.76	5	D2	23.9
BE4	Cost	4.95	10		
BE5	Road maintenance	5.41	9		
BE6	Equipment fleet	6.76	5		
BE7	Surface material	6.31	8	D3	16.7
BE8	Particle size	2.70	16		
BE9	Traction and adhesion	7.66	3		
BE10	Stopping distance	9.46	2	D4	25.7
BE11	Sight distance	4.95	10		
BE12	Diggability	6.76	5		
BE13	Road dimensions and slope	4.50	13		
BE14	Dominant formation type	4.50	13	D5	27.5
BE15	Presence of clay	10.36	1		
BE16	Presence of discontinuities	7.66	3		
BE17	Presence of water	4.95	10		

Analysing the sub-event categories as presented in Figure 4, D5 (27.5 per cent) has the highest contribution to failures, emphasizing the significance of ground conditions, while D4 (25.7 per cent) highlights the importance of stopping distance, sight distance, and diggability. D2 (23.9 per cent) and D3 (16.7 per cent) also play substantial roles, whereas D1 (6.30 per cent) has the least impact.

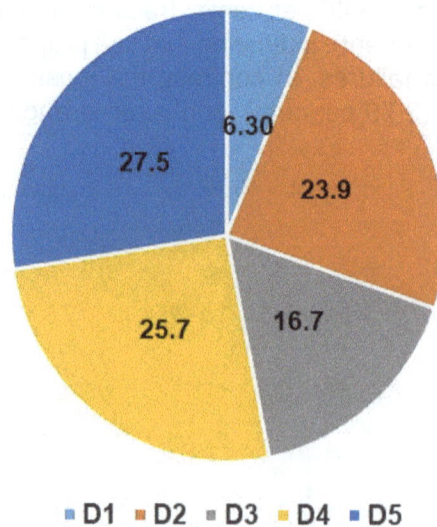

Figure 4 – Pie Chart of Sub-Events based on FCI %.

Therefore, it can be concluded that the failure of mine roads can be attributed to geological uncertainties, functional design uncertainties, management uncertainties, geometrical design uncertainties, and structural design uncertainties, and should be addressed during the simulation phase.

Table 12 presents the inputs to be used in the simulation. Each basic event has been considered separately within the simulation model. As stated in the Simulation Algorithm Development Process Section, system behaviour is represented as sequential events occurring at specific times. Probability distribution functions (PDFs) are used as inputs, ensuring that outcomes are generated stochastically rather than being fixed. Time progresses in weekly increments, and each uncertainty is assigned a random occurrence time, determining when a failure event will take place. The algorithm identifies the earliest failure event and its corresponding uncertainty, dynamically directing the simulation process. Each uncertainty is handled through separate branches, allowing independent assessment of different failure types. Once a failure occurs, the model continues calculations and evaluations accordingly.

TABLE 12

Simulation inputs.

Failure of mine roads simulation inputs	
Item, unit	**Value**
Surface load problem occurrence frequency, week	TRD(1,3,9)
Layer strength problem occurrence frequency, week	TRD(5,8,10)
Traffic volume problem occurrence frequency, week	TRD(3,4,9)
Cost problem occurrence frequency, week	TRD(1,7,11)
Road maintenance problem occurrence frequency, week	TRD(2,3,4)
Equipment fleet problem occurrence frequency, week	TRD(7,11,12)
Surface material problem occurrence frequency, week	TRD(4,11,12)
Particle size problem occurrence frequency, week	TRD(2,3,5)
Traction and adhesion problem occurrence frequency, week	TRD(7,15,19)
Stopping distance problem occurrence frequency, week	TRD(4,5,13)
Sight distance problem occurrence frequency, week	TRD(3,6,15)
Diggability problem occurrence frequency, week	TRD(10,11,12)
Road dimensions and slope problem occurrence frequency, week	TRD(12,13,16)
Dominant formation type problem occurrence frequency, week	TRD(2,6,10)
Presence of clay problem occurrence frequency, week	TRD(1,2,7)
Presence of discontinuities problem occurrence frequency, week	TRD(3,5,8)
Presence of water problem occurrence frequency, week	TRD(2,4,6)
Operational cost, ($/week)	TRD(50000, 75000, 100000)
Road downtime frequency (week)	TRD(0.25, 3, 5)

Table 13 presents the simulation outputs related to the failure of mine roads. It consists of failure types in terms of indirect costs, direct costs, and overall operational expenditure (OPEX). Within the scope of the study on the Mine Road Quality Index (MRQI), these results offer a preliminary understanding of the economic implications of different uncertainties affecting haul road performance.

TABLE 13

Simulation outputs.

	Failure of mine roads simulation outputs		
	Average indirect cost ($)	**Average direct cost ($)**	**Average total OPEX ($)**
Surface load	7538	74 986	82 524
Layer strength	2675	75 628	78 303
Traffic volume	1282	74 352	75 634
Cost	2555	73 946	76 502
Road maintenance	4953	74 347	79 300
Equipment fleet	15 363	75 763	91 126
Surface material	14 818	75 914	90 731
Particle size	34 000	74 916	108 916
Traction and adhesion	15 433	74 866	90 299
Stopping distance	2666	81 989	84 057
Sight distance	1305	75 173	72 867
Diggability	2605	76 601	79 206
Road dimensions and slope	4901	74 674	79 575
Dominant formation type	2674	75 095	77 770
Presence of clay	1284	74 999	76 283
Presence of discontinuities	2549	75 092	77 641
Presence of water	4918	75 340	80 258

Notably, the highest indirect costs are associated with particle size ($34 000), equipment fleet issues ($15 363), and traction and adhesion failures ($15 433), highlighting the significant role of material properties and vehicle-road interactions in maintenance expenses and operational efficiency.

RESULTS AND DISCUSSIONS

The foundation for the Mine Road Quality Index (MRQI) framework has been established, offering valuable insights into the factors influencing haul road quality in surface mining operations. By integrating fuzzy logic, fault tree analysis (FTA), and discrete event simulation (DES), this study has systematically explored the uncertainties affecting haul road performance and identified key areas for improvement. While MRQI has not yet been fully developed, the methodology presented in this study provides a structured approach that will guide its future formulation. Future studies will focus on refining the MRQI framework and validating it through real-world case studies, making it a promising tool for optimising haul road quality in mining operations. The findings discussed below focus on failure probabilities, criticality indices, and the overall impact of uncertainties on road quality, serving as a steppingstone for future indexing studies.

The fault tree analysis (FTA) revealed that the total failure probability of the haul road system is 26.12 per cent, indicating a significant risk of road performance degradation. Among the sub-event categories, D5 (Geological Uncertainties) had the highest probability of failure (0.0725), followed by D4 (Geometric Design Uncertainties) (0.0665), D2 (Management Uncertainties) (0.0775), and D3 (Functional Design Uncertainties) (0.0542). The least critical category was D1 (Structural Design

Uncertainties) with a failure probability of 0.0220. These results highlight the importance of addressing geological and geometric design factors, which have the most substantial impact on road quality.

The Failure Criticality Index (FCI) further emphasised the relative contribution of each basic event to system failure. The presence of clay (BE15) emerged as the most critical factor, with an FCI of 10.36 per cent, followed by stopping distance (BE10) at 9.46 per cent, and traction and adhesion (BE9) at 7.66 per cent. These findings suggest that soil properties, road traction, and stopping distance are the primary drivers of road failures. In contrast, factors such as surface load (BE1) and particle size (BE8) had minimal impact, with FCIs of 2.70 per cent and 2.70 per cent, respectively.

The high failure probability and criticality of geological uncertainties underscore the importance of addressing subsurface conditions in haul road design and maintenance. The presence of clay (BE15) and discontinuities (BE16) were identified as the most significant contributors to road instability. Clay, in particular, can lead to reduced road strength and increased susceptibility to water infiltration, which exacerbates road degradation. These findings align with previous research by Terziyski (2024), who emphasised the need for adequate drainage systems and surface thickness to mitigate the effects of geological uncertainties.

Geometric design factors, particularly stopping distance (BE10) and sight distance (BE11), were also identified as critical contributors to road failures. Poorly designed road gradients and inadequate sight distances can lead to increased braking requirements, higher fuel consumption, and safety hazards. These results are consistent with the findings of Posada-Henao, Sarmiento-Ordosgoitia and Correa-Espinal (2023), who demonstrated that road gradients exceeding 5 per cent significantly increase fuel consumption and operational costs. The study by Kubler (2024) further supports the importance of optimising ramp gradients and transitions to reduce fuel consumption and improve road performance.

Management-related factors, such as traffic volume (BE3) and road maintenance schedules (BE5), were also found to significantly impact road quality. Irregular traffic patterns and inconsistent maintenance schedules can lead to accelerated road wear and increased downtime. The integration of sensor-based monitoring systems, as suggested by Thompson et al (2006), could help mitigate these issues by providing real-time data on road conditions and enabling proactive maintenance.

Functional design factors, including surface material quality (BE7) and traction and adhesion (BE9), were identified as critical contributors to road performance. Inconsistent surface material quality can lead to uneven road surfaces and reduced vehicle traction, increasing the risk of accidents and equipment damage. The findings of Ngwangwa and Heyns (2014) support the use of artificial neural networks (ANNs) for real-time road condition assessment, which could help identify and address functional design issues more effectively.

While structural design uncertainties had the lowest failure probability, they still play a role in road performance. Factors such as surface load (BE1) and layer strength (BE2) can lead to road deformation and uneven surfaces if not properly managed. These findings highlight the need for standardised construction practices and the use of high-quality materials to ensure road durability.

The discrete event simulation (DES) model provided valuable insights into the temporal and stochastic nature of road uncertainties. By simulating various failure scenarios, the model enabled the prediction of road performance over time and the identification of optimal maintenance strategies. The simulation outputs revealed that uncertainties related to the equipment fleet ($91 126 OPEX), surface material ($90 731 OPEX), and traction and adhesion ($90 196 OPEX) resulted in the highest total operational expenditures. These findings underscore the importance of targeted interventions to address high-cost uncertainties and optimise haul road performance in surface mining operations.

CONCLUSIONS

To sum up, this study explores the factors influencing haul road quality in surface mining operations by addressing a critical gap in the literature. A comprehensive literature review was conducted, revealing a lack of a standardised framework for assessing mine road quality. Based on the information gained from the review, 17 key uncertainties affecting haul road performance were identified, covering geological, structural, geometric, functional, and management-related factors.

To prioritise uncertainties, a hypothetical expert survey was designed, and the responses were analysed using the fuzzy fault tree analysis (FFTA) method. The results indicated that among all uncertainties, the presence of clay under geological factors exhibited the highest failure probability. Furthermore, the overall system failure (top-event) was calculated as 26.12 per cent. When analysed in terms of Failure Criticality Index (FCI%), the presence of clay (10.36 per cent) ranked as the most critical uncertainty, followed by stopping distance (9.46 per cent).

To further analyse the impact of the uncertainties dynamically, the outputs of the fuzzy fault tree analysis were integrated into a discrete event simulation (DES) model. This simulation framework allows for the stochastic representation of haul road performance over time by modelling failure events as discrete occurrences. The model operates on a probabilistic basis, assigning random failure times to each uncertainty and simulating their impact on road conditions and maintenance requirements. The simulation results highlighted that the highest indirect cost was associated with particle size ($34 000), followed by uncertainties related to equipment fleet ($15 363).

ACKNOWLEDGEMENTS

The authors would like to acknowledge the Scientific and Technological Research Council of Türkiye (TUBITAK), Grant No: 1919B012339507 for TUBITAK 2009-A, for financial support.

The authors would like to thank Yunus Kardoğan for his technical contributions to creating the Expert Survey.

REFERENCES

Alegre, D A G, Peroni, R L and Aquino, E R, 2024. The impact of haul road geometric parameters on open pit mine strip ratio, *Mining Science*, 78:34–52.

Banks, J, Carson, J S, Nelson, B L and Nicol, D M, 2010. *Discrete-event system simulation*, 5th ed (Pearson).

Cheliyan, D and Bhattacharyya, B, 2018. Expert elicitation for system reliability assessment: A review, *Reliability Engineering and System Safety*, 176:110–123.

Kubler, K A C, 2024. Optimisation of off-highway truck fuel consumption through mine haul road design, *International Journal of Mining Engineering*, 62:22–40.

MathWorks, 2022. *What Is Fuzzy Logic?*, Available from: <https://www.mathworks.com/help/fuzzy/what-is-fuzzy-logic.html> [Accessed: 30 April 2025].

Ngwangwa, H M and Heyns, P S, 2014. Application of an ANN-based methodology for road surface condition identification on mining vehicles and roads, *Journal of Terramechanics*, 53:59–74.

Posada-Henao, J J, Sarmiento-Ordosgoitia, I and Correa-Espinal, A, 2023. Effects of road slope and vehicle weight on truck fuel consumption, *Sustainability*, 15(3):1934.

Rausand, M and Høyland, A, 2004. *System reliability theory: Models, statistical methods and applications*, 2nd edn (Wiley-Interscience).

Soofastaei, A, Aminossadati, S M, Kizil, M S and Knights, P, 2016. A discrete-event model to simulate the effect of truck bunching due to payload variance on cycle time, hauled mine materials and fuel consumption, *International Journal of Mining Science and Technology*, 26(5):745–752. https://doi.org/10.1016/j.ijmst.2016.03.012

Strack, A L, 2015. A review on Australian mine haul road design procedures, *University of Southern Queensland Publications*, 12:105–120.

Tekbey, T B, 2023. An uncertainty assessment modeling for deviations between long- and short-range production plans at surface metal mines, Master's thesis, Middle East Technical University.

Terziyski, D, 2024. Optimization of technological schemes for construction of open mine roads, *Bulgarian Academy of Sciences Journal*, 77:5–22.

Thompson, R J, Visser, A T, Heyns, P S and Hugo, D, 2006. Mine road maintenance management using haul truck response measurements, *Mining Technology*, 115(2):84–92.

Zadeh, L A, 1965. Fuzzy sets, *Information and Control*, 8(3):338–353.

Towards responsible mining – ESG-integrated multi-objective optimisation in sublevel stoping production scheduling

W Ullah[1], M Nehring[2], M S Kizil[3] and P Knights[4]

1. PhD student, School of Mechanical and Mining Engineering, The University of Queensland, St Lucia Qld 4072. Email: g.ullah@student.uq.edu.au
2. Senior Lecturer, School of Mechanical and Mining Engineering, The University of Queensland, St Lucia Qld 4072. Email: m.nehring@uq.edu.au
3. Associate Professor, School of Mechanical and Mining Engineering, The University of Queensland, St Lucia Qld 4072. Email: m.kizil@uq.edu.au
4. Professor, School of Mechanical and Mining Engineering, The University of Queensland, St Lucia Qld 4072. Email: p.knights@uq.edu.au

ABSTRACT

The increasing focus on sustainable practices in mining operations has created a demand for comprehensive production scheduling models that incorporate Environmental, Social and Governance (ESG) factors. This research presents a novel multi-objective optimisation model specifically developed for production scheduling in sublevel stoping mining operations, targeting sustainability challenges within the industry. The model addresses three primary objectives: maximising Net Present Value (NPV), minimising the socio-environmental effects of carbon emissions and ground vibration. Using NSGA-II, a Pareto front of optimal solutions was generated, allowing for a comprehensive evaluation of trade-offs between the objectives. The study used conceptual data to model the production of 200 stopes, optimising NPV, carbon emissions and Peak Particle Velocity (PPV). Three solutions from the Pareto front were analysed. The solution with the highest NPV achieved A\$311.78 million, while the most environmentally favourable option reduced carbon emissions to 13 551 t. Solution C, chosen as the compromise option, delivered an NPV of A\$310.19 million (0.51 per cent below the highest), carbon emissions of 18 890 t (12.3 per cent lower than the maximum) and a PPV of 331 mm/s (13.8 per cent below the peak value). Applying NSGA-II to this production scheduling problem provides a flexible and powerful tool for decision-makers to evaluate a wide range of trade-offs and select an optimal solution that aligns with their corporate and environmental objectives. By considering both the financial and non-financial outcomes, mining companies can make more informed choices that contribute to long-term sustainability while still achieving strong profitability.

INTRODUCTION

Mining is essential for sustaining global economic development, providing the raw materials required for construction, manufacturing, energy production and numerous consumer goods. Minerals like iron, copper, gold and other valuable resources are the backbone of modern industry, powering infrastructure, technology and even green energy solutions such as solar panels and electric vehicles (Pouresmaieli *et al*, 2023). The extraction and processing of these minerals enable advancements in various industries, which in turn support the well-being and progress of societies worldwide (Shiquan *et al*, 2022). As the demand for these critical resources continues to rise, mining operations must adapt to produce them efficiently while addressing the Environmental, Social and Governance (ESG) challenges that accompany large-scale extraction (Ullah *et al*, 2023).

The extraction of valuable minerals from the earth involves various mining methods, depending on many factors, including location, depth, type of ore, etc. One such method is underground mining, extracting minerals beneath the surface. Sublevel stoping is a widely used technique in underground mining, particularly for steeply dipping orebodies (Sotoudeh *et al*, 2020). It involves drilling and blasting ore in horizontal sections, or stopes, while allowing the country rock material to remain in place. Sublevel stoping is known for its high recovery rates and cost-effectiveness, but like any mining method, it comes with environmental and social risks, such as ground instability, emissions and disturbances to local communities (Sotoudeh *et al*, 2020; Ullah *et al*, 2023).

In modern mining operations, there is growing recognition of the need to balance economic outcomes with ESG considerations. ESG factors encompass a company's responsibility to reduce its environmental footprint, foster positive social impacts and adhere to ethical governance practices (Mitchell, 2023). In mining, this means minimising environmental damage, such as carbon emissions and habitat degradation, ensuring the safety and well-being of workers and local communities and complying with regulations and industry standards. Integrating ESG considerations into mining operations is increasingly vital as stakeholders, including investors, regulators and the public, demand more transparency and accountability from mining companies (Fikru et al, 2024).

The motivation for integrating ESG factors into production scheduling arises from the complex trade-offs between maximising profit and minimising environmental and social side effects. Traditionally, production schedules in mining were focused solely on optimising economic returns, with little attention given to the long-term impacts on the environment or surrounding communities (Mirzehi and Afrapoli, 2024). However, as sustainability becomes a key priority across industries, mining companies face increasing pressure to adopt more responsible practices (Taylor and Connellan, 2024). By incorporating ESG into production schedules, mining companies can make more informed decisions that balance profitability with their broader responsibilities to the environment and society. They can also improve their social license to operate and meet evolving regulatory demands (Minerals Council of Australia, 2021).

There is a noticeable lack of research that simultaneously considers both ESG and economic objectives in sublevel stoping. Recently, multi-objective optimisation techniques have become more prominent due to their flexibility. These techniques effectively solve real-world challenges, making them ideal for balancing economic and sustainability goals in complex mining scenarios.

This research addresses the challenge above by introducing a new approach, a novel multi-objective optimisation model designed to improve production scheduling in sublevel stoping while incorporating key ESG factors and NPV objectives. The model focuses on three main goals: maximising NPV, minimising carbon emissions and minimising ground vibrations. Utilising the Non-dominated Sorting Genetic Algorithm II (NSGA-II) helps decision-makers balance the trade-offs between economic, environmental and social objectives. This novel model offers a practical, scalable solution for mining companies, contributing to the growing body of knowledge on sustainable mining practices and supporting the broader goals of sustainable development.

PRODUCTION SCHEDULING IN SUBLEVEL STOPING

Production scheduling in sublevel stoping has been a critical area of research due to its impact on operational efficiency and profitability. Traditionally, studies have focused on optimising the extraction sequence to maximise ore recovery while minimising costs and operational delays. Early works primarily utilised linear programming (LP) and mixed-integer programming (MIP) techniques to develop schedules that ensure a smooth flow of materials and efficient use of equipment (Gholamnejad, Lotfian and Kasmaeeyazdi, 2020). Researchers began using more advanced optimisation techniques like heuristics and metaheuristics as computational capabilities grew. Genetic algorithms (GA), particle swarm optimisation (PSO) and simulated annealing (SA) became popular due to their ability to solve complex and large-scale production scheduling problems. These models generally emphasise economic objectives, such as maximising NPV, minimising costs and improving resource utilisation, but often lack consideration of environmental or social factors, which are increasingly relevant in modern mining practices.

Williams, Smith and Wells (1972) first developed an LP model to optimise production scheduling in sublevel stoping. Their method focused on reducing fluctuations in ore production but assumed that the material at each level was uniform, which did not accurately reflect the mining method's selectivity. Trout (1995) was one of the pioneers in applying MIP to maximise NPV in production scheduling for sublevel stoping mining operations at Mt Isa, Australia. Introducing a fill mass exposure, Nehring (2006) and Nehring and Topal (2007) improved the model. Little, Nehring and Topal (2008) further enhanced the efficiency of Trout's models by reducing solution time by decreasing the number of variables. To overcome some of the drawbacks of Nehring's (2011) model for the planning and scheduling of sublevel stoping procedures, Basiri (2018) created an innovative approach. A stochastic integer program with two stages was developed by Furtado e Faria,

Dimitrakopoulos and Pinto (2022). This study aimed to satisfy geotechnical limitations while managing the risk of missing production targets, maximising discounted revenues and minimising development expenses.

A mixed-integer linear programming (MILP) formulation is used in the mathematical programming framework presented by Appianing, Ben-Awuah and Pourrahimian (2023). This system explicitly tackles the difficulties associated with integrated open stope development and production scheduling by integrating backfilling, operational level management and stope extraction duration control. An overview of the production scheduling optimisation for sub-level stoping was presented by Sotoudeh *et al* (2020) in the context of a review of the advancement of production schedule optimisation. Sotoudeh *et al* (2023) introduced a new mathematical model to optimise production scheduling in underground mining with a pre-concentration system. This model helps schedule and sequence stopes in sublevel stoping. The study showed that a pre-concentration system could boost economic benefits by increasing the NPV.

However, these studies primarily focused on economic objectives, such as maximising NPV, minimising costs and improving resource utilisation, without addressing environmental or social factors. This gap in research highlights the need for integrating sustainability considerations in production scheduling models, particularly in light of increasing regulatory and societal pressures on mining operations.

PREVIOUS WORK INTEGRATING ESG IN MINING AND OPTIMISATION

ESG considerations have become increasingly important in mining operations due to growing awareness of environmental impacts and social responsibilities (Maybee, Lilford and Hitch, 2023). The mining industry faces significant pressure to operate sustainably, minimising adverse environmental effects such as carbon emissions, habitat destruction and ground vibrations while ensuring positive social contributions and maintaining ethical governance practices (Fu *et al*, 2024; Onifade *et al*, 2024; Ullah *et al*, 2023). Despite the importance of ESG in mining, only limited research has integrated these factors into optimisation models for production scheduling.

Recent studies have begun to investigate the integration of ESG into mining operations, with a primary focus on reducing carbon emissions and enhancing community relations. Smith and Brooks (2018) was among the first to study how social performance issues can be included in long-term planning for mineral resources in South Africa. They explained the key rules of the Anglo-American 'Social Way' framework, which helps manage social responsibilities. The study conducted by Nehring and Knights (2024) looked at the methods for methodically identifying, evaluating and integrating ESG risks and opportunities into the initial stages of planning and design for natural resource projects. Fikru *et al* (2024) investigated rating variances based on business size and place of origin using sustainalytics' ESG risk ratings for Ecuador's expanding mining sector. Numerous observations from this study could guide future research. Most of the research that has been done so far is still scattered and does not have a complete plan for balancing economic, environmental and social goals in scheduling sublevel stoping operations.

MULTI-OBJECTIVE OPTIMISATION TECHNIQUES IN MINING OPERATIONS

Multi-objective optimisation has become a powerful tool for balancing conflicting objectives in mining operations. Multi-objective optimisation methods, such as the Non-dominated Sorting Genetic Algorithm II (NSGA-II) and the Pareto-based approaches, allow for simultaneous optimisation of economic, environmental and social goals. These techniques have been applied across various mining problems, particularly in production scheduling, where multiple objectives often conflict. For instance, Wang *et al* (2018) developed a multi-objective optimisation model for the Huogeqi Copper Mine in China to use mineral resources more effectively for sustainable development. The model aims to maximise both economic profit and resource efficiency, using the NSGA-II algorithm to find the best solutions. Foroughi *et al* (2019) formulated a multi-objective integer programming model to enhance mine planning and production scheduling in sublevel stoping mining. The model uses the NSGA-II to generate solutions to maximise NPV and metal recovery.

Despite their effectiveness, multi-objective optimisation techniques are still underutilised in sublevel stoping mining operations, especially regarding ESG integration. Most multi-objective models focus on traditional economic and operational goals, such as maximising ore recovery or minimising costs. These multi-objective techniques provide a promising avenue for addressing the trade-offs between financial returns and sustainability in mining, yet further research is needed to fully integrate ESG factors into the production scheduling of sublevel stoping operations.

Problem description and the optimisation model/model formulation

This work presents a multi-objective optimisation model designed explicitly for the sublevel stoping mining method, with an emphasis on including ESG considerations. The research aims to analyse the trade-offs between ESG objectives in mining operations. The NSGA-II algorithm evaluates these competing objectives, effectively balancing them to generate optimal solutions. The detailed structure and formulation of the optimisation model, including its parameters and constraints, are discussed in the following sections.

Indices

S	set of stopes in the model
T	set of time periods

Parameters

g_s	ore grade of stope, g / t
r^t	ore recovery at time period, %
p^t	metal price at time period, $ / t
m^t	mining cost at time period, $ / t
n^t	processing cost at time period, $ / t
o_s	ore tonnage of stope, t
i	discount rate, %
PPV	peak particle velocity, mm / s
Q	charge weight / day, kg
D	distance from blast to monitoring point, m
k, l and m	experimental constants depending on the blast geometry and rock type, generally, $0.5 \leq l \leq 1$ and $-1.5 \leq m \leq -1$
a	scaling coefficient
b	decay coefficient
E	number of equipment types used
Δ_e	energy consumption per tonne by the e type of equipment, PJ
C_Δ	corresponding carbon emission factor of the energy type consumed by the e type of equipment, tCO_2 / J
K	number of rock types
P_1	rated power of the drilling rigs, kW
B_1	average number of boreholes drilled by the drilling rigs
μ_1	drilling efficiency of drilling rigs, m / h
$C_{electricity}$	the carbon emission factor of electric energy, tCO_2 / kWh
L_k^1	unit consumption of explosives for the same type of rock mass for preparation blasting, kg / m³
L_k^2	unit consumption of explosives for the same type of rock mass for ore blasting, kg / m³

α	the proportion of preparatory work in the whole sublevel stoping mine, %
$C_{explosive}$	the carbon emission factor for industrial explosives used in mines, tCO_2/t
C_{V-D-A}	amount of greenhouse gases released during the pressurisation, drainage and ventilation procedures used to treat a unit cube of rock mass, tCO_2/t
v, d, a	quantity of ventilator, drainage pump and compressor varieties, m^3/s?
P_v, P_d, P_a	corresponding operating power of a certain kind of fan, drain pump and compressor, kW
n_v, n_d, n_a	quantity of a certain kind of ventilator, drainage pump and compressor units in operation
t_v, t_d, t_a	average daily operating hours of a certain ventilator, drainage pump and compressor, h
Q_{day}	total daily ore and waste rock production in the sublevel stoping mine, t / day
Q_{day}^1	daily average rock mass extracted in a sublevel stoping mine using compressed air equipment, t / day
H	number of LHDs (load, haul, dump)
P_h	rate power of LHDs, w
λ	power ratio coefficient of an engine under no-load and heavy load
t_h	average round-trip time of the LHDs, s
C_{diesel}	the carbon emission factor of diesel, tCO_2/J
α_{diesel}	diesel combustion conversion efficiency, %
V_h	bucket capacity for a certain kind of diesel LHD used in the mine, m^3
k_s	full bucket coefficient of the LHD, %
P_c	working power of certain types of crusher, kW
t_c	average daily working time of a certain type of crusher, h
ρ	Average density of ore, t/m^3
Q_{day}^3	daily average amount ore crushed, t / day
R, U	types of mixers and pumps used in the process of mixing and pumping backfilling
P_r, P_u	rated power of each type of mixer and pump, kW
t_r, t_u	average working time of mixer and pump to complete a workflow, h
n_r, n_r	number of mixers and pumps for each type
V_r, V_u	treatment volume of the mixer and pump in their respective single workflow time, t
off_s	set of all stopes which should be offset from stope s
cb_s	set of all stopes that share common blocks with stope s
ext_s	set of all stopes which have practical extraction levels with stope s
adj_s	set of all stopes that are adjacent to stope s
tpt	set of time periods including all time periods up to and the current period t
OH_u	maximum ore handling tonnage limit for time period t
OH_l	minimum ore handling tonnage limit for time period t
BF	backfill supply limit in time period t
U_t	maximum contained metal tonnage target in time period t

U_l minimum contained metal tonnage target in time period t

Variables

$$e_{s,t} = \begin{cases} 1 \text{ if extraction from stope s is schedule in time period t} \\ \quad\quad\quad\quad\quad 0 \text{ otherwise} \end{cases}$$

Objective functions

The optimisation model incorporates three distinct objective functions to address various aspects of sublevel stoping. The first objective is to maximise the NPV, ensuring the highest possible financial return from the mining operations. The second objective focuses on minimising carbon emissions, targeting the reduction of environmental impact by limiting the greenhouse gases produced during the energy-intensive mining process. Lastly, the third objective aims to minimise ground vibrations, which helps mitigate the social and environmental disturbances caused by mining activities, particularly in areas near communities or sensitive ecosystems. These three objectives reflect a balanced approach to optimising economic gains while considering environmental sustainability and social responsibility.

Economic objective

The objective function is defined as the discounted revenue by Sari and Kumral (2023) using Equation 1, which takes into account several important variables. It considers the stope's tonnage, mining and processing expenses, metal price, ore grade and ore recovery. When taken as a whole, these components offer a thorough assessment of the economic return, considering the quantity and quality of the ore, related costs and the metal's market value:

$$\text{NPV} = \sum_{s=1}^{S}\sum_{t=1}^{T}\left[\frac{(g_s.\ r^t.p^t - m^t - n^t).o_s}{(1+i)^t}\right] e_{s,t} \tag{1}$$

Social objective

Blast-induced ground vibrations in and around mines have become a significant social and environmental concern. This issue requires careful attention and management from the mining industry's perspective. As local councils impose stricter regulations, vibration monitoring has become crucial to mining operations. Ground vibrations can be effectively controlled by adjusting the location and orientation of the point of interest relative to the blast site (Garai, Agrawal and Mishra, 2023). The United States Bureau of Mines (USBM) introduced the first significant equation for calculating PPV in 1959. Additionally, various researchers and institutions have proposed modified predictors (as reported in Prashanth and Nimaje, 2018; including Langefors and Kihlstrom, 1963; Ambraseys and Hendron, 1968; Bureau of Indian Standards, 1973; Ghosh and Daemen, 1983). The blast waves from individual holes combine to produce a PPV is calculated by the Equation 2 (Cardu, Coragliotto and Oreste, 2019) and the goal is to minimise this PPV during the blasting process:

$$\text{PPV} = \sum_{s=1}^{S}\sum_{t=1}^{T} k\, Q^l D^m\, e_{s,t} \tag{2}$$

Environmental objective

The Intergovernmental Panel on Climate Change (IPCC) (Ashford *et al*, 2006) suggests using the carbon emission coefficient method for calculating carbon emissions, with two approaches: *'top-down'* and *'bottom-up'*. The *'top-down'* method estimates emissions by measuring energy use and applying carbon emission factors, while the *'bottom-up'* method measures emissions directly from each piece of equipment. Due to the complex nature of mining operations, the *'top-down'* method has been adopted in this study to assess carbon emissions for sublevel stoping operations (Wang and Zhou, 2022). The goal was to reduce emissions from key activities like drilling, blasting, air compression, drainage, ventilation, transportation, processing and backfilling. Equation 3 (Ren *et al*,

2023; Ulrich, Trench and Hagemann, 2020) outlines the objective function for minimising total carbon emissions from these processes:

$$\text{Carbon emission} = \sum_{s=1}^{S} \sum_{t=1}^{T} \left(a. g_s^b + \sum_{e=1}^{E} \Delta_e. C_\Delta \right). o_s. e_{s,t} \tag{3}$$

Modelling carbon emissions during the rock drilling

The primary source of carbon emissions during drilling comes from the energy consumption of drilling equipment. Underground drilling includes roadway development and ore production blastholes, with drill rigs (Jumbos and production drill rigs) being the primary tools. The energy used for drilling blastholes depends on the machine's power and the time spent breaking the rock, which is affected by rock properties, the number of holes and the drilling length. Since machine power is known, the following formula calculates the carbon emissions per tonne of rock mass from drilling operations (Ren *et al*, 2023):

$$C_{\text{drilling}} = \sum_{k=1}^{K} P_1 \frac{B_1}{\mu_1} C_{electricity} \tag{4}$$

Modelling carbon emissions during the blasting process

Blasting-related carbon emissions mainly result from industrial explosives and are calculated by dividing the total explosive consumption by the volume or tonnage of blasted rock (kg/m³ or kg/t). In sublevel stoping, blasting involves excavation, preparation and stoping. Excavation and preparation, being more difficult due to limited free surfaces, require more explosives per unit than stoping. The average explosive consumption for different rock types is determined using a rock general coefficient. The model for calculating carbon emissions per cubic metre of rock mass during blasting is as follows (Ren *et al*, 2023):

$$C_{\text{blasting}} = \sum_{k=1}^{K} \frac{\left[L_k^1. \alpha + L_k^2. (1 - \alpha) \right]. C_{explosive}}{1000} \tag{5}$$

Modelling carbon emissions during the ventilation, drainage and air compression processes

Carbon emissions from underground mine ventilation, drainage and compressed air systems result from power consumption, which increases as the mining area grows. The continuous operation of the primary fan, drainage pump and compressor is essential for mine safety. To calculate carbon emissions per unit of ore, the ratio of daily power consumption to the total daily ore and waste rock production is used, since these systems are vital for safety but not directly tied to ore quantity. Carbon emissions from air compression per unit of rock mass are determined by dividing daily power use by the average daily rock mined (Ren *et al*, 2023):

$$C_{V-D-A} = \frac{\left(\sum_{v=1}^{V} P_v. n_v. t_v + \sum_{d=1}^{D} P_d. n_d. t_d \right). C_{electricity}}{Q_{day}. 1000} + \frac{\sum_{a=1}^{A} P_a. n_a. t_a. C_{electricity}}{Q_{day}^1. 1000} \tag{6}$$

Modelling carbon emissions during the transportation process

Loading and hauling are essential in mining, where underground loaders, trucks, or Load-Haul-Dump (LHD) carriers transport ore to processing areas and remove waste. Carbon emissions during transportation are complex due to the changing stope position, scope and varying distances travelled by equipment. Power consumption also fluctuates based on factors like load type (heavy or no-load) and terrain (uphill or downhill). The transport distance parameter is used to calculate the average round-trip time, disregarding slope and vehicle performance. For mining trucks, the power ratio coefficient, λ, is set at 0.91 when the truck is either empty or loaded (Zhang *et al*, 2020). LHDs, which run on diesel, have a specific carbon emission model for moving a unit of rock mass (Ren *et al*, 2023):

$$C_{\text{transportation}} = \sum_{h=1}^{H} \frac{P_h(1+\lambda).t_h.C_{diesel}}{2.\alpha_{diesel}.V_h.k_s} \tag{7}$$

Modelling carbon emissions during the crushing process

Crushing is a crucial step in mining, as it breaks down ore to extract valuable minerals. This process reduces ore size, preparing it for further stages like mineral separation. In underground mining, crushers like jaw, gyratory, or impact types are typically used for primary crushing. Heavy machinery, often powered by electricity or diesel, is involved, leading to carbon emissions from fossil fuel combustion or electricity generation. The energy source, such as coal, natural gas, or renewables, significantly affects the carbon footprint. The following formula can be used to calculate carbon emissions per tonne of ore processed:

$$C_{\text{crushing and grinding}} = \sum_{k=1}^{K} \frac{P_c.t_c.\rho}{Q_{day}^3.1000} C_{electricity} \tag{8}$$

Modelling carbon emissions during the backfilling process

The backfilling process in sublevel stoping is essential for improving safety, providing ground support and maximising resource use. In this method, the ore is extracted in horizontal slices from a vertical orebody and the resulting voids are filled with materials like cemented fill, hydraulic fill, or waste rock mixtures. Carbon emissions during backfilling come from the significant energy used in producing and transporting the backfill material. The formula for calculating carbon emissions per tonne of backfill is as follows (Ren *et al*, 2023):

$$C_{filling} = \sum_{r=1}^{R} \frac{P_r.t_r.n_r.C_{electricity}}{V_r} + \sum_{u=1}^{U} \frac{P_u.t_u.n_u.C_{electricity}}{V_u} \tag{9}$$

Constraints

Sublevel stoping involves several geotechnical and sequential constraints that must be carefully managed to ensure both safety and efficiency. Geotechnical constraints relate to the stability of the rock mass, requiring proper ground support systems and careful planning to prevent collapse or rockfalls. Sequential constraints, on the other hand, involve the order in which different parts of the mine are accessed and extracted, ensuring that the mining sequence follows a logical and safe progression. These constraints impact the design, scheduling and overall strategy of the mining operation (Foroughi *et al*, 2019):

$$\sum_{t=1}^{T} e_{s,t} + \sum_{t=1}^{T} e_{s',t} \le 1 \ \forall \ s|s' \in off_s \tag{10}$$

$$\sum_{t=1}^{T} e_{s,t} + \sum_{t=1}^{T} e_{s',t} \le 1 \ \forall \ s|s' \in cb_s \tag{11}$$

$$\sum_{t=1}^{T} e_{s,t} + \sum_{t=1}^{T} e_{s',t} \le 1 \ \forall \ s|s' \in ext_s \tag{12}$$

$$e_{s,t} + e_{s',t} \le 1 \ \forall s,t|s' \in adj_s \tag{13}$$

$$\sum_{t' \in tpt} e_{s,t'} + \sum_{s' \in adj_s} e_{s',t} \le 2 \ \forall \ s, t \tag{14}$$

$$\sum_{s=1}^{S} o_s.e_{s,t} \leq OH_u \qquad (15)$$

$$\sum_{s=1}^{S} o_s.e_{s,t} \geq OH_l \qquad (16)$$

$$\sum_{s=1}^{S} d_{fv}\, e_{s,t} \leq BF \qquad (17)$$

$$\sum_{s=1}^{S} o_s.g_s.mr^t.e_{s,t} \leq U_t \; \forall t \qquad (18)$$

$$\sum_{s=1}^{S} o_s.g_s.mr^t.e_{s,t} \geq L_t \; \forall t \qquad (19)$$

Constraint (10) prohibits the formation of offset stopes, which are stopes directly above one another and of the same size. This helps prevent the creation of vertical planes between backfilled stopes, reducing the risk of material failure. Constraint (11) ensures that only one stope among those sharing at least one block can be produced at a time, preventing overlapping stopes and ensuring that a stope is in only one phase at any given time. Constraint (12) prevents the selection of adjacent stopes unless they share common drawpoint levels, ensuring practical and functional drawpoint levels. Constraint (13) limits the total extraction variables to never exceeding one, preventing adjacent stopes from being created simultaneously, thereby avoiding large voids that could compromise geotechnical stability. Constraint (14) restricts production to only one adjacent stope after backfilling, effectively managing strain within the mine. Constraints (15), (16) and (17) regulate material removal and backfill delivery based on the handling system's capacity and a predetermined limit. Finally, constraints (18) and (19) help maintain consistent plant feed by ensuring that metal remains within upper and lower bounds, reducing grade fluctuations in each production period.

SOLUTION METHOD – MULTI-OBJECTIVE OPTIMISATION

Multi-objective optimisation (MOO) models deal with problems with multiple conflicting objectives, making it necessary to find solutions that balance trade-offs between objectives. Two primary techniques exist to solve multiple objective optimisation problems: classical methods and evolutionary algorithms. Classical methods like the weighted sum approach combine the objectives into a single scalar function by assigning weights to each objective. While this approach is simple, it depends heavily on the proper selection of weights, which can be challenging. Additionally, this method may miss parts of the Pareto front, especially when the objective functions are non-convex. Another traditional approach is the ε-constraint method, where one objective is optimised while others are treated as constraints. However, both methods have limitations in exploring the full diversity of possible solutions (Ullah and Nehring, 2021).

Evolutionary algorithms (EAs) have gained prominence in solving MOO models because they can handle complex, nonlinear and non-convex relationships between objectives without needing specific assumptions about the shape of the solution space. Some popular evolutionary algorithms include Particle Swarm Optimisation (PSO) and Genetic Algorithms (GA). These methods generate a population of solutions and iteratively improve them through operations like selection, crossover and mutation. Multi-Objective Evolutionary Algorithms (MOEAs), such as the Strength Pareto Evolutionary Algorithm (SPEA) and Multi-Objective Genetic Algorithm (MOGA), build on these concepts to handle multiple objectives efficiently by maintaining a diverse population of solutions and using Pareto dominance to guide the search process (Coello, Lamont and van Veldhuizen, 2007).

Among the many MOEAs, the NSGA-II is one of the most effective and widely used methods. NSGA-II uses a fast non-dominated sorting approach to classify solutions based on their Pareto dominance and applies a crowding distance mechanism to maintain diversity in the population. Its ability to explore various trade-off solutions makes it ideal for MOO problems where conflicting objectives must be balanced. The current model seeks to maximise the NPV, minimise carbon emissions and minimise ground vibration in sublevel stoping mining operations, so NSGA-II is chosen as the solution method. Its robust performance and ability to generate a well-distributed Pareto front make it well-suited to address the complexities and conflicting objectives of the mining optimisation problem (Deb *et al*, 2000).

CASE STUDY

To evaluate the effectiveness of the proposed approach, the optimisation model was tested on a conceptual copper deposit comprising 200 stopes. Each stope had a volume of 2700 m^3, with dimensions of 30 m in length, width and height (30 m × 30 m × 30 m). This configuration allowed for a detailed simulation of the mining operation. The parameters involved in this model, including economic, operational and mining-related factors, were crucial for accurately assessing the outcomes. A detailed summary of these parameters, such as stope dimensions, extraction costs and metal prices, can be found in Table 1 providing a comprehensive view of the inputs used for this analysis.

TABLE 1

The list of parameters regarding the case study.

Parameter	Unit	Value
Recovery	%	95
Average iron price	$/t	90
Fixed extraction cost	$/stope	1 400 000
Variable extraction cost	$/t	20
Fixed backfill cost	$/stope	1 200 000
Variable backfill cost	$/$m^3$	15
Discount rate	%	10
Ore production capacity in each period	t	150 000
Minimum contained metal tonnage target per period	kg	60 000
Maximum contained metal tonnage target per period	kg	90 000
Backfill availability per time period	m^3	40 000
Backfilling, waste and ore density	t/m^3	2.1, 3 and 5.2
Mine life	period	200

Table 2 presents the technical parameters associated with various rock masses, each characterised by distinct lithologies in the sublevel stoping gold-copper mine. These parameters include critical data such as the drilling length and the number of drill holes, both of which were carefully calculated for each unit cube of rock mass. These measurements are essential for accurately modelling the drilling requirements and performance across different rock types, ensuring efficiency and safety in the mining process.

TABLE 2
Technical parameters of different rock masses.

Common rock mass types	Average drill hole length m/m³	Average number of drill holes
(Orebody) Skarn	0.83	5.0
(Orebody) Marble	0.83	5.0
(Wall rock) Quartz diorite porphyrite	0.94	5.0
(Wall rock) Diorite	0.94	5.4

Table 3 provides detailed information on the technical specifications of the equipment used in tunnelling and drilling operations. This includes the features of the drilling rigs and tunnelling machinery, such as their power, capacity and performance metrics, which are essential for assessing the effectiveness and operational efficiency of the mining activities. Together, these tables offer a comprehensive overview of the technical aspects critical to the success of the sublevel stoping mining method.

TABLE 3
Technical parameters of the tunnelling and drilling rigs.

Drill type	Model of drill	Nominal power (kW)	Rock breaking efficiency (m/h)	Equipment size (m)	Equipment weight (t)
Tunnelling	Huatai HT82	62	30	11 × 1.45 × 2.08	10.0
Deep-hole drilling	Huatai HT72	62	60	9.05 × 1.45 × 2.08	11.5

The primary explosive used in the mining process is Ammonium Nitrate/Fuel Oil (ANFO), with a carbon emission factor of 0.189 t CO_2 per tonne. Table 4 shows the average explosive consumption for different rock types during various blasting operations in the mine.

TABLE 4
Explosive blasting parameters.

Rock types	Solid coefficient of rock	Unit explosive for preparatory work (kg/m³)	Unit explosive for ore blasting (kg/m³)
Skarn	8~10	1.62~1.89	1.49
Marble	10~12	1.89~2.11	1.58

During the mine's production process, approximately 3500 t of ore and 350 t of waste rock were produced daily, with an average density of 3200 kg/m³. Compressed air equipment powered 75 per cent of the operations. A variable speed fan, running 24 hrs daily, reduced energy consumption by 40 per cent. Two operational tables shared the same drainage pump, while others remained on standby during the three-hour average workday. The air compressor operated continuously from 8:00 to 16:00 and cycled between operations during other hours. Tables 5, 6 and 7 provide details on the mine fan, drainage pump and air compressor.

TABLE 5

Fan data.

Type of fan	Operating capacity (kW)	Number of working devices	Fan air volume (m³/min)	Static pressure (Pa)	Fan speed (rev/min)
K40–6-no 14	30	1	984~2064	150~695	960
K45–6-no 14	45	1	1434~2718	500~959	980
FCDZ -6-no 22	370	3	2400~7600	750~2750	990
K40–4-no 12	37	1	882~1926	242~1118	1450

TABLE 6

Drainage pump data.

Type of pump	Operating capacity (kW)	Number of working devices	Pumping capacity (m³/h)	Fan speed (rev/min)
200D43x6	300	1	280	1480
MD280–65x7	630	2	280	1480
MD280–43x5	250	2	280	1480
MD280–65x9	800	2	280	1480

TABLE 7

Air compressor data.

Type of compressor	Operating capacity (kW)	Number of working devices	Operating time (h)	Rated exhaust pressure (MPa)	Nominal volume flow (m³/min)
TS325–400	300	8	8	0.7	41.8
TS325–400	300	3	16	0.7	61.7

It is likely that only one type of LHD is used throughout the whole process for moving the same rock pile. The LHD has an average round-trip time of 200 seconds and its diesel engine is 35 per cent efficient. Table 8 displays the data of LHDs utilised in mining operations.

TABLE 8

LHD data.

Type of LHDs	Rated power (kW)	Number of working devices	Bucket capacity (m³)	Full-bucket coefficient
WJ-1.5	63	8	1.50	1.12
WJ-0.75	58	4	0.75	1.09
WJ-1	58	7	1.00	1.10

Backfilling is crucial in sublevel stoping mines, utilising high-pressure piston pumps and mixer equipment. Table 9 provides detailed specifications for the machinery used in the backfilling process.

TABLE 9

Backfilling the equipment parameter table.

Equipment	Number of types	Number of working devices	Rated power (kW)	Operating time (h)
Mixer	SJ6x6	2	30	16
	SJ6x8	1	30	16
Pump	80ZBYL-450	7	90	8
	150ZJ-I-A70	1	200	8
	100ZJ-I-A50	3	90	8
	100ZJ-I-A50	2	55	8

Carbon emissions from ore processing plants in mining originate from energy-intensive operations, such as crushing, grinding and refining. Table 10 provides the detailed specifications of the equipment used in these processes.

TABLE 10

Crushing and Grinding equipment parameter table.

Equipment	Model	Feeding size (mm)	Rated power (kW)	Capacity (t/h)	Operating time (h)
Crushers	PE600x900	500	75	140	16
	PE1200x1500	1020	250	800	16
Grinder	Φ2.4x10	25	570	30	16
	Φ3.4x7.5	25	1000	60	16

RESULTS AND DISCUSSIONS

The multi-objective production scheduling problem in sublevel stoping, which aims to maximise NPV, minimise carbon emissions and minimise ground vibration, was successfully addressed using the NSGA-II. These three objectives represent critical considerations for the mining industry: economic profitability (through NPV maximisation), environmental sustainability (by lowering carbon emissions) and social responsibility (by minimising ground vibration to reduce the impact on nearby communities and structures). Each objective carries challenges and requires careful balancing, as they often conflict. Actions that boost economic returns may lead to higher environmental or social costs.

NSGA-II, a robust MOO optimisation algorithm known for generating diverse trade-off solutions, was employed to address these competing priorities. The algorithm's parameters are provided in Table 11, which outlines the various configurations used to fine-tune the performance of the optimisation process. Table 12 presents the range of values for each objective function, offering insights into how well the algorithm balanced the competing goals of maximising NPV, minimising carbon emissions and minimising ground vibration.

TABLE 11

Parameters for NSGA-II.

Parameter	Value
Population size	100
Total number of iterations	10 000
Crossover probability	0.9
Mutation probability	0.8

TABLE 12

Summary statistics of objectives.

	NPV ($)	Carbon emissions (tonnes CO_2)	Peak particle velocity (mm/s)
Minimum output	$215 685 000	13 551	219
Maximum output	$311 784 000	21 530	384
Mean	$263 734 500	17 540	301

Figure 1 visually represents the Pareto front obtained through the NSGA-II process, illustrating the optimal trade-offs between the three objectives. The Pareto front comprises a set of non-dominated solutions, meaning no solution is universally better across all objectives. For example, a solution that maximises NPV may lead to higher carbon emissions or increased ground vibration, while a solution that minimises ground vibration may result in lower economic returns. These non-dominated solutions are classified as Pareto optimal, where improving one objective can only come at the expense of another.

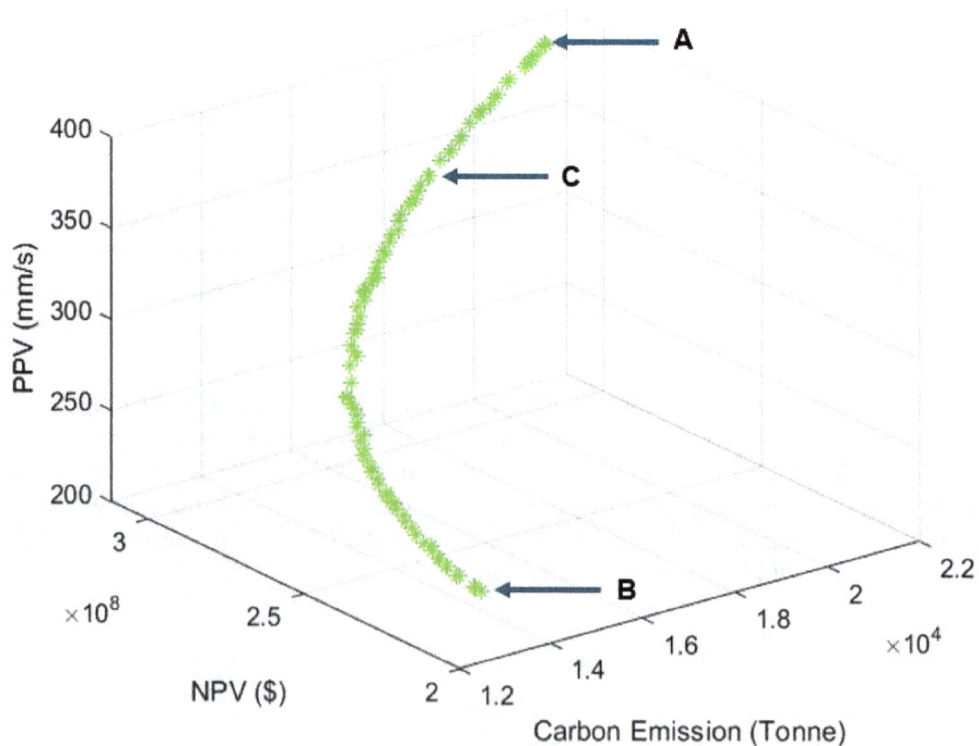

FIG 1 – Pareto front.

In multi-objective optimisation problems like this, it is critical to acknowledge that without a clear preference from the decision-maker, no solution within the Pareto front can be considered inherently superior. Each solution presents a different compromise between the objectives and it remains essential for the decision-maker to select one solution for implementation based on their priorities. This selection often involves evaluating the trade-offs in light of external factors, such as regulatory requirements, community concerns and long-term business goals.

In practice, the decision-maker must carefully assess the available solutions and choose one that best aligns with the mining operation's objectives. Whether the primary focus is on profitability, environmental impact, or social responsibility, the selected solution must balance these considerations in a way that meets the needs of stakeholders. Notably, the convergence of all solutions to the Pareto front highlights the effectiveness of NSGA-II in addressing the complex, conflicting goals in sublevel stoping, providing a robust framework for decision-making in real-world applications.

In this study, the production scheduling of sublevel stoping is optimised based on three critical objectives: maximising Net NPV, minimising carbon emissions and minimising ground vibration, represented by PPV. Using an MOO approach with NSGA-II, three notable solutions emerge from the Pareto front: solutions A, B and C, each representing different trade-offs between the economic, environmental and social objectives. Table 13 represents the summary statistics of three solutions A, B and C.

TABLE 13

Summary statistics of three solutions.

Solution	NPV ($)	Carbon emissions (tonnes CO_2)	Peak particle velocity (mm/s)
Solution A	$311 784 000	21 530	384
Solution B	$215 695 000	13 551	219
Solution C	$310 187 000	18 890	331

Solution A represents the extreme case of maximising NPV, achieving an impressive value of A$311.78 million. However, this economic gain comes at the cost of higher environmental and social impacts. The carbon emissions for solution A are 21 530 t, which is significantly higher than the other solutions. Additionally, solution A generates a high ground vibration level, with a PPV of 384 mm/s. While this solution is optimal from an economic perspective, it does not align well with the environmental and social objectives of minimising carbon footprint and reducing the negative impact of vibration on surrounding areas. The high PPV could result in potential risks to nearby structures and communities, as ground vibrations from blasting operations are a known cause of environmental disturbances and social concerns. Thus, while solution A maximises profitability, it fails to address the pressing need for sustainable and socially responsible mining practices.

Solution B, on the other hand, represents the opposite end of the Pareto front, prioritising environmental and social factors. It achieves the lowest carbon emissions at 13 551 t and a significantly lower PPV of 219 mm/s. However, these improvements in social and ecological performance are accompanied by a considerable reduction in NPV, A$215.69 million. This solution reflects a strong commitment to reducing the environmental impact and improving the quality of life for surrounding communities by mitigating the adverse effects of ground vibrations. However, the economic trade-off is substantial, with the lower NPV potentially affecting the financial feasibility of the mining operation. Solution B presents a favourable option for mining companies and stakeholders seeking to balance profitability with corporate social responsibility. However, the lower financial returns may not meet the expectations of all decision-makers.

Solution C offers a balanced compromise between the extremes of solutions A and B. It achieves an NPV of A$310.18 million, close to the maximum value seen in solution A, while significantly reducing carbon emissions and ground vibration compared to the first solution. The carbon emissions in solution C are 18 890 t, which is a notable reduction from solution A and the PPV is 331 mm/s, lower than that for solution A but slightly higher than that for solution B. This middle-ground approach allows for substantial economic gains while addressing environmental and social concerns. By balancing the three objectives, solution C is a practical and sustainable choice for real-world implementation.

Practically, a mine plan around solution C is recommended for implementation. It achieves a near-optimal NPV while significantly reducing carbon emissions and ground vibration. This compromise ensures that the mining operation remains financially viable, with acceptable environmental and social impacts. Solution C exemplifies how multi-objective optimisation can lead to solutions that do not sacrifice one objective for another but instead strike a balance, contributing to the overall sustainability of the mining operation while maximising economic returns. Given the growing importance of ESG factors in mining, solution C represents a forward-thinking approach that aligns with the industry's shift towards more sustainable practices.

STOPE SEQUENCING

Stope sequencing is crucial in optimising the production schedule of sublevel stoping mining operations. The arrangement and extraction sequence of stopes directly influence a mining project's economic outcomes, environmental footprint and operational efficiency. In the context of this research, stope sequencing is analysed for three distinct solutions: A, B and C, which are derived from the multi-objective optimisation model focused on maximising NPV, minimising carbon emissions and minimising PPV. The stope sequences of these solutions reflect the trade-offs between these three objectives.

Solution A's stope sequencing (Figure 2) is heavily driven by economic considerations, focusing on maximising NPV. The stope sequence prioritises high-value areas of the orebody, extracting these regions as early as possible to generate the highest immediate cash flows. This sequencing strategy often leads to a more aggressive extraction plan, aiming to maximise ore recovery in the shortest time. However, this rapid approach may result in higher energy consumption due to the concentrated drilling and blasting activities, leading to increased carbon emissions. Additionally, the sequencing in solution A may cause greater ground disturbance, as reflected by the higher PPV values, due to the intensive and closely spaced blasting events. The focus on economic gains in this solution comes with the cost of greater environmental and social impacts.

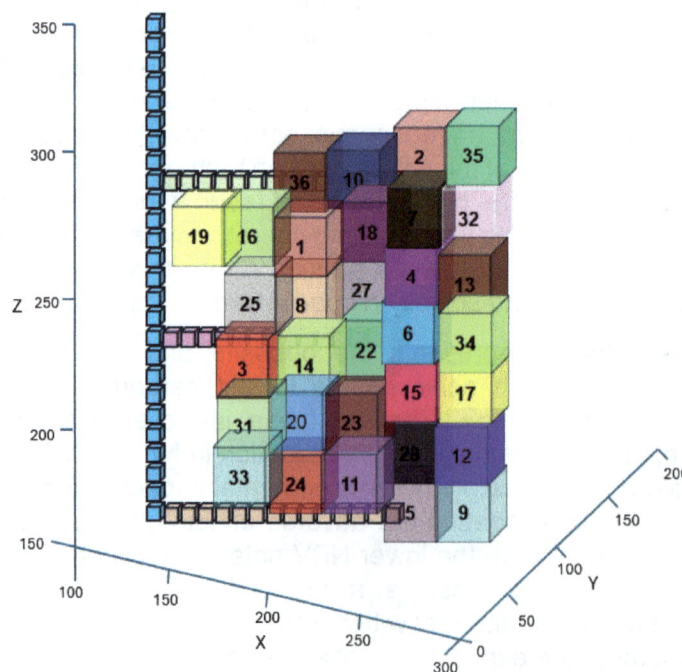

FIG 2 – Stope sequencing for solution A.

Solution B's stope sequencing (Figure 3), in contrast, is designed with a strong focus on minimising environmental and social impacts, particularly carbon emissions and ground vibrations. The sequencing in this solution adopts a more cautious approach, extracting stopes to reduce the frequency and intensity of blasting. This typically involves selecting stopes less likely to generate high vibrations or require intensive energy use, leading to a more distributed and less aggressive extraction plan. The result is a stope sequence that produces significantly lower carbon emissions and minimises the impact of vibrations on nearby infrastructure and communities. However, this approach may delay the extraction of high-value stopes, contributing to this solution's lower NPV. The stope sequencing in solution B demonstrates how prioritising environmental and social factors can result in a slower but more sustainable mining process.

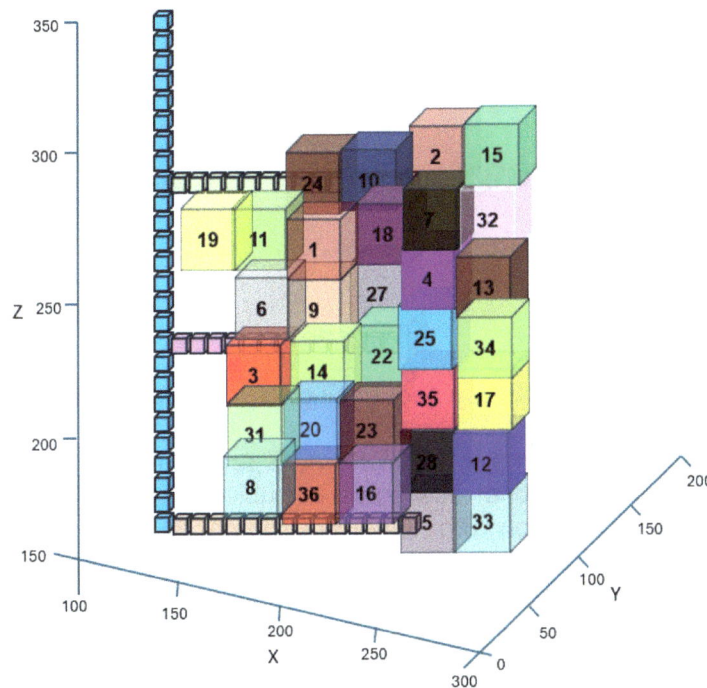

FIG 3 – Stope sequencing for solution B.

Solution C's stope sequencing (Figure 4) represents a balanced approach, aiming to achieve a compromise between the economic and environmental objectives. The sequencing plan in solution C combines elements from both the aggressive and cautious approaches seen in solutions A and B. High-value stopes are extracted earlier in the schedule, ensuring that the NPV remains close to its maximum potential. At the same time, careful consideration is given to minimising the environmental impact during critical phases of the operation. The stope sequencing in this solution avoids over-concentration of blasting activities, distributing them in a way that limits excessive ground vibration and reduces carbon emissions. This balanced sequencing strategy enables the organisation to sustain robust economic performance while minimising its environmental impact and complying with acceptable social standards.

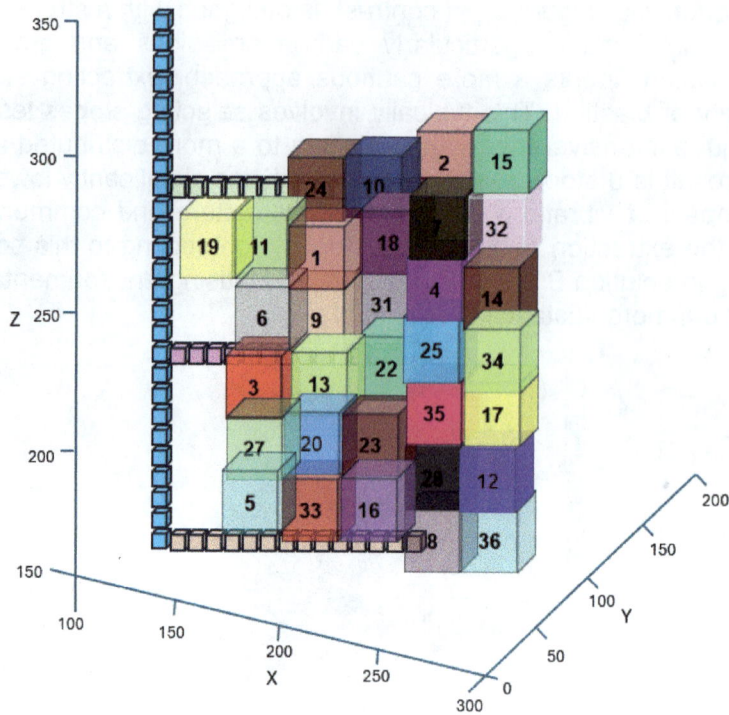

FIG 4 – Stope sequencing for solution C.

Overall, the stope sequencing patterns in solutions A, B and C reveal how different optimisation priorities can lead to varied operational strategies in sublevel stoping mining. While solution A focuses on rapid, high-value extraction with higher environmental costs, solution B adopts a more conservative, eco-friendly approach that reduces the mining project's immediate financial returns. Solution C offers a well-rounded stope sequencing that balances the economic, environmental and social aspects, making it the most suitable choice for practical implementation in modern mining operations.

CONCLUSIONS

This research presented a comprehensive multi-objective production scheduling model for sublevel stoping, aimed at optimising three critical objectives: maximising NPV, minimising carbon emissions and minimising ground vibrations. The model, solved using the NSGA-II algorithm, successfully produced a Pareto front that enables decision-makers to evaluate trade-offs between financial, environmental and social metrics. The results offer insights into how different production strategies impact these objectives and provide a clear path for implementing sustainable mining practices.

The study showcased the practical application of multi-objective optimisation in a real-world mining context by exploring a range of optimal solutions. The outcomes demonstrate the variability in NPV, carbon emissions and ground vibration across different production strategies. A clear trend emerges when considering the percentage differences between the extreme solutions in the Pareto front and the chosen compromise solution: a balanced approach to mining operations can lead to substantial environmental benefits with only marginal reductions in financial performance. For instance, the NPV difference between the highest and lowest economic solutions is significant, with the largest NPV at A\$311.78 million and the smallest at A\$215.69 million. However, the selected compromise solution achieves an NPV of A\$310.19 million, representing only a 0.51 per cent reduction from the maximum NPV while maintaining a 43.81 per cent higher NPV than the most environmentally conservative solution. This small economic concession is a compelling argument for adopting a more sustainable mining approach, especially considering the substantial environmental benefits of this trade-off.

The reduction in carbon emissions and ground vibration was much more pronounced on the environmental front. The difference in carbon emissions between the extreme solutions spans from 13 551 t to 21 530 t, with the compromise solution producing 18 890 t emissions. This represents a 12.26 per cent reduction in carbon emissions compared to the highest-emission scenario, achieving nearly optimal financial results. This demonstrates the ability to implement meaningful environmental improvements without sacrificing profitability, showcasing the potential for mining operations to meet carbon reduction targets while remaining economically viable.

Regarding ground vibration, the results further highlighted the effectiveness of the compromise solution. The maximum PPV observed was 384 mm/s, while the minimum was 219 mm/s. The compromise solution achieved a PPV of 331 mm/s, resulting in a 13.8 per cent reduction from the highest vibration level. This is particularly important in mining operations near populated or sensitive areas, where high-ground vibrations can lead to structural damage and social resistance. By reducing the PPV to this level, the operation can enhance community relations and minimise the risk of damaging nearby infrastructure while maintaining economic strength.

The percentage differences between the various objectives emphasise the potential for mining operations to integrate sustainability without sacrificing their financial goals. The model presented in this research highlights that a balanced approach that marginally reduces economic returns while significantly lowering carbon emissions and ground vibration can lead to a more sustainable and socially responsible mining practice. This is especially critical in today's mining industry, where ESG factors are becoming central to operational decision-making and long-term business strategy.

The application of NSGA-II to this production scheduling problem provides a flexible and powerful tool for decision-makers. It enables them to evaluate a wide range of trade-offs and select an optimal solution that aligns with their corporate and environmental objectives. By considering both the financial and non-financial outcomes, mining companies can make more informed choices that contribute to long-term sustainability while still achieving strong profitability.

In conclusion, this research demonstrated the use of MOO, which enables mining companies to pursue meaningful reductions in carbon emissions and ground vibrations while maintaining nearly optimal financial outcomes. The chosen compromise solution, with its small financial trade-off and significant environmental benefits, represents the future of sustainable mining. The results of this research underscore the importance of integrating economic, environmental and social factors into mining operations to achieve sustainable development in the industry. The mining sector can contribute to global sustainability goals through careful planning and advanced optimisation techniques while remaining economically competitive.

Further research could include additional environmental and social factors, such as water usage and worker safety, into the optimisation model to enhance the sustainability of mining operations. Incorporating real-world data from operational mines would allow for more accurate simulations and validation of the model. Future studies could also investigate using alternative algorithms or hybrid approaches to improve computational efficiency and solution quality.

REFERENCES

Appianing, E J A, Ben-Awuah, E and Pourrahimian, Y, 2023. Life-of-mine optimization for integrated open stope development and production scheduling using a mixed-integer linear programming framework, *Mining Technology: Transactions of the Institutions of Mining and Metallurgy*, 132:106–120. https://doi.org/10.1080/25726668.2023.2182285

Ashford, P, Baker, J A, Clodic, D, Devotta, S, Godwin, D, Harnisch, J, Irving, W, Jeffs, M, Kuijpers, L, McCulloch, A, Peixoto, R D A, Uemura, S, Verdonik, D P, Kenyon, W G, Rand, S and Woodcock, A, 2006. Emissions of fluorinated substitutes for ozone depleting substances, *Industrial Processes and Product Use*, 3.

Basiri, Z, 2018. Stopes Layout and Production Scheduling Optimization in Sublevel Stoping Mining, MS thesis, University of Alberta.

Cardu, M, Coragliotto, D and Oreste, P, 2019. Analysis of predictor equations for determining the blast-induced vibration in rock blasting, *Int J Min Sci Technol*, 29:905–915. https://doi.org/10.1016/j.ijmst.2019.02.009

Coello, C A C, Lamont, G B and van Veldhuizen, D A, 2007. *Evolutionary Algorithms for Solving Multi-Objective Problems* (Springer).

Deb, K, Agrawal, S, Pratap, A and Meyarivan, T, 2000. A Fast Elitist Non-dominated Sorting Genetic Algorithm for Multi-objective Optimization: NSGA-I I, in *International Conference on Parallel Problem Solving from Nature*, pp 849–858.

Fikru, M G, Avila-Santamaria, J J, Soria, R, Logan, A and Romero, P P, 2024. Evaluating ESG risk ratings of mining companies: What are lessons for Ecuador's developing mining sector?, *Resources Policy*, 94. https://doi.org/10.1016/j.resourpol.2024.105133

Foroughi, S, Hamidi, J K, Monjezi, M and Nehring, M, 2019. The integrated optimization of underground stope layout designing and production scheduling incorporating a non-dominated sorting genetic algorithm (NSGA-II), *Resources Policy*, 63:101408. https://doi.org/10.1016/j.resourpol.2019.101408

Fu, C, Yu, C, Guo, M and Zhang, L, 2024. ESG rating and financial risk of mining industry companies, *Resources Policy*, 88. https://doi.org/10.1016/j.resourpol.2023.104308

Furtado e Faria, M, Dimitrakopoulos, R and Pinto, L C L, 2022. Integrated stochastic optimization of stope design and long-term underground mine production scheduling, *Resources Policy*, 78. https://doi.org/10.1016/j.resourpol.2022.102918

Garai, D, Agrawal, H and Mishra, A K, 2023. Impact of orientation of blast initiation on ground vibrations, *Journal of Rock Mechanics and Geotechnical Engineering*, 15:255–261. https://doi.org/10.1016/j.jrmge.2022.03.012

Gholamnejad, J, Lotfian, R and Kasmaeeyazdi, S, 2020. A practical, long-term production scheduling model in open pit mines using integer linear programming, *The Journal of the Southern African Institute of Mining and Metallurgy*, 120. https://doi.org/10.17159/2411

Little, J, Nehring, M and Topal, E, 2008. A new mixed-integer programming model for mine production scheduling optimisation in sublevel stope mining, *Australian Mining Technology Conference 2008*, pp 157–172.

Maybee, B, Lilford, E and Hitch, M, 2023. Environmental, Social and Governance (ESG) risk, uncertainty and the mining life cycle, *Extractive Industries and Society*. https://doi.org/10.1016/j.exis.2023.101244

Minerals Council of Australia, 2021. ESG Change for the better.

Mirzehi, M and Moradi Afrapoli, A, 2024. Sustainable long-term production planning of open pit mines: An integrated framework for concurrent economical and environmental optimization, *Resources Policy*, 94. https://doi.org/10.1016/j.resourpol.2024.105131

Mitchell, P D, 2023. Top 10 business risks and opportunities for mining and metals in 2024, EY.

Nehring, M and Knights, P, 2024. A Systems Engineering Approach to Incorporate ESG Risks and Opportunities in Early-Stage Mine Design and Planning, *Mining*, 4:546–566. https://doi.org/10.3390/mining4030031

Nehring, M and Topal, E, 2007. Production schedule optimisation in underground hard rock mining using mixed integer programming, in *Proceedings of the Project Evaluation Conference 2007*, pp 169–175 (The Australasian Institute of Mining and Metallurgy: Melbourne).

Nehring, M, 2006. Stope Sequencing and Optimisation in Underground Hardrock Mining, BEng thesis, The University of Queensland, Brisbane.

Nehring, M, 2011. Integrated Production Schedule Optimisation for Sublevel Stoping Mines, PhD thesis, The University of Queensland, Brisbane.

Onifade, M, Zvarivadza, T, Adebisi, J A, Said, K O, Dayo-Olupona, O, Lawal, A I and Khandelwal, M, 2024. Advancing toward sustainability: The emergence of green mining technologies and practices, *Green and Smart Mining Engineering*, 1:157–174. https://doi.org/10.1016/j.gsme.2024.05.005

Pouresmaieli, M, Ataei, M, Nouri Qarahasanlou, A and Barabadi, A, 2023. Integration of renewable energy and sustainable development with strategic planning in the mining industry, *Results in Engineering*, 20. https://doi.org/10.1016/j.rineng.2023.101412

Prashanth, R and Nimaje, D S, 2018. Estimation of peak particle velocity using soft computing technique approaches: a review, *Noise and Vibration Worldwide*. https://doi.org/10.1177/0957456518799536

Ren, G, Wang, W, Wu, W, Hu, Y and Liu, Y, 2023. Carbon Emission Prediction Model for the Underground Mining Stage of Metal Mines, *Sustainability*, 15. https://doi.org/10.3390/su151712738

Sari, Y A and Kumral, M, 2023. Stope Sequencing Optimization for Underground Mines Through Chance-Constrained Programming, *Min Metall Explor*, 40:1737–1748. https://doi.org/10.1007/s42461-023-00821-2

Shiquan, D, Amuakwa-Mensah, F, Deyi, X and Yue, C, 2022. The impact of mineral resource extraction on communities: How the vulnerable are harmed, *Extractive Industries and Society*, 10. https://doi.org/10.1016/j.exis.2022.101090

Smith, G L and Brooks, L, 2018. Incorporation of the socio-cultural dimension into strategic long-term planning of mineral assets in South Africa, *J South Afr Inst Min Metall*, 118:331–336. https://doi.org/10.17159/2411-9717/2018/v118n4a1

Sotoudeh, F, Nehring, M, Kizil, M, Knights, P and Mousavi, A, 2023. A New Mathematical Programming Formulation for Production Scheduling Optimisation of Sublevel Stoping Operations in the Presence of Pre-concentration Systems, *Min Metall Explor*, 40:2255–2267. https://doi.org/10.1007/s42461-023-00843-w

Sotoudeh, F, Nehring, M, Kizil, M, Knights, P and Mousavi, A, 2020. Production scheduling optimisation for sublevel stoping mines using mathematical programming: A review of literature and future directions, *Resources Policy*, 68:101809. https://doi.org/10.1016/j.resourpol.2020.101809

Taylor, A and Connellan, C, 2024. What does it mean to be a responsible mining and metals player in 2024?, White & Case.

Trout, L P, 1995. Underground mine production scheduling using mixed integer programming, in Proceedings of the 25th International Application of Computers and Operations Research in the Minerals Industry (APCOM) Symposium (The Australasian Institute of Mining and Metallurgy: Melbourne).

Ullah, G M W and Nehring, M, 2021. A multi-objective mathematical model of a water management problem with environmental impacts: An application in an irrigation project, *PLoS One*, 16:1–16. https://doi.org/10.1371/journal.pone.0255441

Ullah, G M W, Nehring, M, Kizil, M and Knights, P, 2023. Environmental, Social and Governance Considerations in Production Scheduling Optimisation for Sublevel Stoping Mining Operations: a Review of Relevant Works and Future Directions, *Min Metall Explor*. https://doi.org/10.1007/s42461-023-00869-0

Ulrich, S, Trench, A and Hagemann, S, 2020. Greenhouse gas emissions and production cost footprints in Australian gold mines, *J Clean Prod*, 267. https://doi.org/10.1016/j.jclepro.2020.122118

Wang, G and Zhou, J, 2022. Multiobjective Optimization of Carbon Emission Reduction Responsibility Allocation in the Open-Pit Mine Production Process against the Background of Peak Carbon Dioxide Emissions, *Sustainability*, 14. https://doi.org/10.3390/su14159514

Wang, X, Gu, X, Liu, Z, Wang, Q, Xu, X and Zheng, M, 2018. Production process optimization of metal mines considering economic benefit and resource efficiency using an NSGA-II model, *Processes*, 6:228. https://doi.org/10.3390/pr6110228

Williams, J K, Smith, L and Wells, P M, 1972. Planning of Underground Copper Mining, in *Proceedings of the 10th International Application of Computers and Operations Research in the Minerals Industry (APCOM) Symposium*, pp 251–254.

Zhang, Z, Song, G, Chen, J, Zhai, Z and Yu, L, 2020. Development of a simplified model of speed-specific vehicle-specific power distribution based on vehicle weight for fuel consumption estimates, *in Transportation Research Record*, pp 52–67 (SAGE Publications Ltd). https://doi.org/10.1177/0361198120947415

Fleet management system in PT Indo Muro Kencana

H Utama[1], H T Wibowo[2] and L Silitonga[3]

1. MAusIMM, Junior Manager Mining, Mining Department, PT Indo Muro Kencana, Central Kalimantan 73961, Indonesia. Email: heru.utama@imkgold.co.id
2. General Manager Operation, PT Indo Muro Kencana, Central Kalimantan 73961, Indonesia. Email: hendro.triwibowo@imkgold.co.id
3. Manager, Mining Department, PT Indo Muro Kencana, Central Kalimantan 73961, Indonesia. Email: lasher.silitonga@imkgold.co.id

ABSTRACT

PT Indo Muro Kencana (IMK) is a surface gold mine located in Mount Muro area, Murung Raya, Central Kalimantan province, Indonesia. The mining operation is conducted by using multiple open pits. Mining activity is operated by using combination of excavator and articulated dump truck (ADT). ADT with capacity of 40 t and 60 t are operated to deliver ore to mineral processing plant and to discard waste material to waste dump. To begin with, the objective of this paper is to explain condition of IMK before implementation of fleet management system (FMS) and IIS (Integrated Information System). Secondly, the purpose of this paper is to elucidate daily practice of FMS in IMK site. IMK is located in remote area, hilly topography and limited communication network (limited wi-fi and GSM signal). Correspondingly, ore hauling distance from pit to ore processing plant is approximately 8 km. At this point, it is recommended that FMS is required to track location and activity of mining units. In addition, it is required due to remote area location and to anticipate security issue due ore hauling. Currently, MineTrack VHF/UHF system (a product from SNCTechnologies, part of SatNetCom) is operated as FMS in IMK. It is operated by using existing mining radio frequency and facility (existing mining radio towers). Furthermore, daily practice of MineTrack VHF/UHF system as FMS is combined with IIS. Combination of MineTrack VHF/UHF system and IIS are used by mining operation engineer and mine production engineer to track location, activities of mining units, to get raw production data and to create mine production report. In conclusion, it can be seen that implementation of MineTrack VHF/UHF system as FMS and combined with IIS is applicable for medium-scale mining operation that is located in remote area with hilly topography and operated in multiple pit operation. This system is affordable for medium-scale mining operation that is located in remote area with potential security and social issue (non-technical aspects of mining operation). Also, it is claimed that this system is a low cost of operational cost and investment.

INTRODUCTION

PT Indo Muro Kencana (IMK) is a surface gold mining operation that is located in Mount Muro, Murung Raya, Central Kalimantan province, Indonesia. Mining operation is conducted in multiple open pit by using combination of excavator and articulated dump truck (ADT). Excavators with various capacities are operated to achieve production target. ADT with capacity of 40 t and 60 t are operated to deliver ore to the mineral processing plant and to discard waste material to the waste dump area. Type of ADT are Volvo A40G, Volvo A40F (with capacity of 40 t) and Volvo A60H (with capacity of 60 t). Additionally, waste dump is acronym to describe area to dump of barren rock (waste material). Instead, Rom pad stockpile is acronym to define ore stockpile at processing plant area. It is located at crusher area (mineral processing plant) and mining office. List of mining equipment in IMK is shown in Table 1.

TABLE 1

Mining equipment.

Unit	Type	Capacities (payload)	Number of units
Excavator	Excavator Volvo 750BL	4.5 bcm	3
Excavator	Excavator Caterpillar Cat 349	3.6 bcm	1
Excavator	Excavator Volvo 480D	3.5 bcm	6
Excavator	Excavator PC 400	2.6 bcm	2
ADT Hauler	ADT Volvo A40G/F	40 t	45
ADT Hauler	ADT Volvo A60H	60 t	14

Figure 1 explains the mining process and system in IMK site. It shows mining and transportation system from mining area to the crushing plant (ore) and to waste dump (waste). In addition, Figure 2 shows mining equipment during mining activity (loading and hauling).

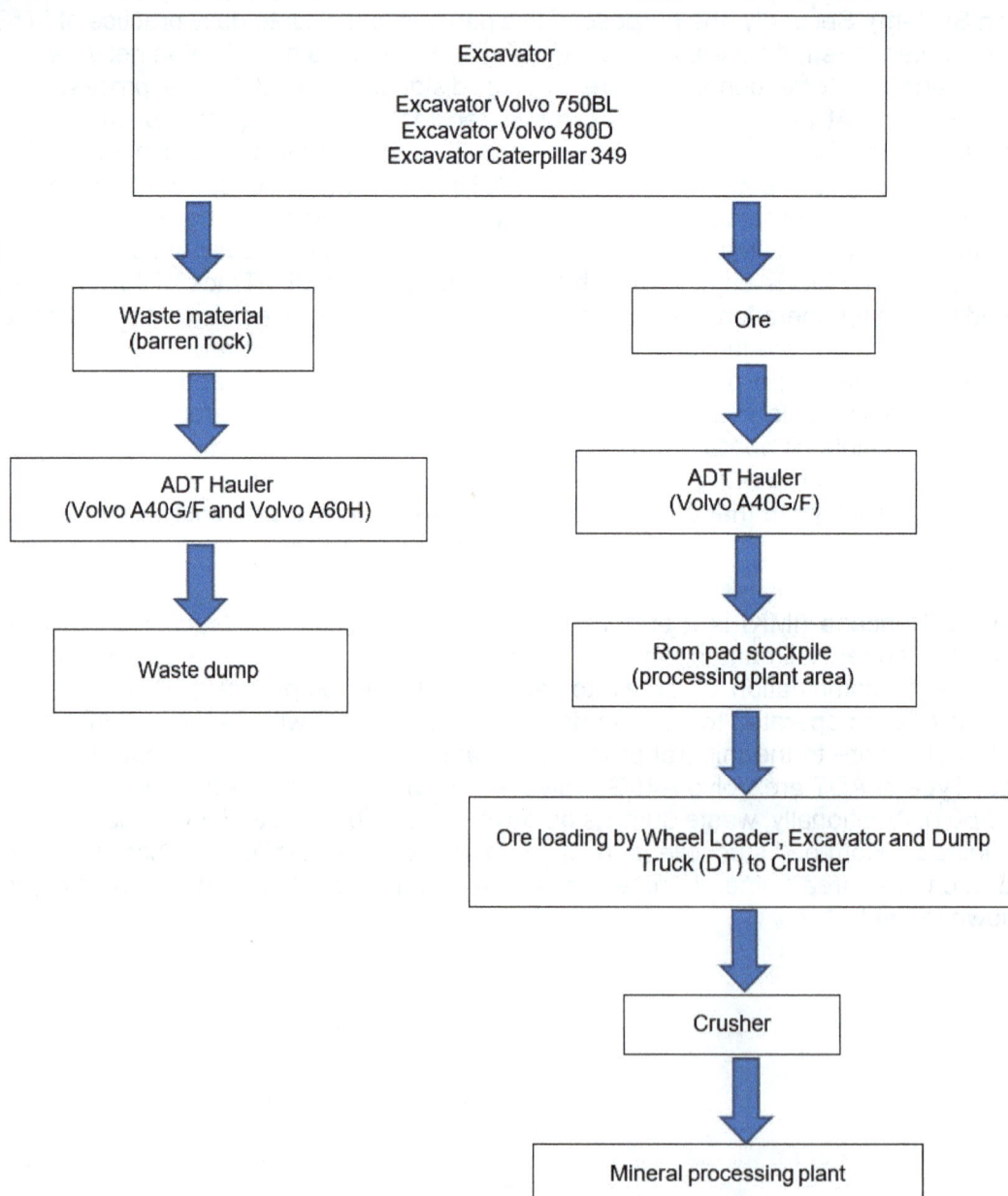

FIG 1 – Mining process and system in IMK.

FIG 2 – Mining equipment.

DAILY PRACTICE OF MINING ENGINEERING

Mining operation engineering and mine production engineer

Initially, Utama and Heryadi (2024) state that mine operation engineering is one of sub-section in mining operation department of PT IMK. It is consisting of mining engineers that are responsible to conduct mining operation daily practice, such as fleet management system, manpower management (mining operator), and implementation of mine plan (daily and weekly plan). In addition, this sub-section is responsible to control mining operation (ore mined, waste mined and pit development). Correspondingly, these mining engineers are responsible to conduct mining operation daily reporting and to manage fleet management system. At this point, daily practice of mining operation engineering is conducted by mining operation engineer (mining department) and mine production reporting activity is conducted by production engineer of Mine Geo Services (MGS) department. On this occasion, previous daily practice of mining operation engineering and mine production reporting before implementation of FMS and post-implementation of FMS will be explained. Furthermore, the combination of MineTrack VHF/UHF system as FMS and IIS (Integrated Information System) will be discussed.

Daily practice before implementation of FMS and IIS

Previous daily practice (2018–2023)

Daily practice of mining operation engineering and mine production reporting before implementation of FMS and IIS is described in Table 2.

TABLE 2

Daily practice before implementation of FMS and IIS.

Item	Description	Comment
Mining method	Surface mining operation: open pit gold mining. Multiple pit operation.	Excavator and Articulated Dump Truck (ADT) 2 shift per day.
Location	Remote area. Ore hauling distance from active pit to ore processing plant = 8 km.	Mining location in remote area. Hills topography. Potential for social issue and safety issue. Non-technical aspects of mining operation.
Pre start check for mining unit	Paper-based form for daily pre start	Paper consuming. Potential for data delay, data error and missing information.
Mining unit (production line up)	Written manually on board, updated manually by mine operation engineer.	Time consuming. Potential delay for operation in order to start mining on 6:00 am
Manpower (operator) management	Line-up of mining manpower (written manually) on white board, updated manually by mine operation engineer.	Time consuming. Potential delay for operation in order to start mining on 6:00 am
Technical issue	Network availability in remote area. Technical issue due to communication.	Limited wi-fi availability. Insufficient of GSM network.
Technical issue	Crowed condition of radio communication during reporting.	Radio is primary communication instrument for mining.
Security issue	Ore stealing (ore theft) during ore hauling. Ore hauling: interaction of ore hauling ADT with local people, illegal miners.	Non-technical aspects of mining operation: security issue during ore hauling from pit to Rom area (ore stealing, burglary).
Social issue	Mining operation: interaction of mining activity with local people (local villagers), illegal miners and artisanal mining.	Non-technical aspects of mining operation: social issue and safety issue.
Reporting of production	Raw production data was reported from pit using radio communication to base control on hourly basis.	Base control is noted on paper-based report. Potential for missing information.
Production data	Raw production data from checker (reported by using radio communication)	Data input (manual) by base control officer. Extra time required to reconciled (production data) with mining operation engineer.
Mine production report	Data process by mine production engineer	Delay in sending production report to mining management.

Proposed mining improvement (2023–2024)

Previously, a mining improvement program was proposed in order to improve daily practice of mining operation and production reporting. Initially, the reasons of this proposed mining improvement were to reduce paper consumption (for pre start check for mining unit), to improve mining operation daily practice (to promote digitalisation in mining operation) and also to improve daily practice of mine production report. Secondly, next reason of this proposed mining improvement is because there is an expectation for mining operation to track unit location, operator's activities, to get raw production data and to generate production report to mining operation management. Moreover, another reason of this proposed improvement is to anticipate potential security issue and social issue (non-technical aspects of mining operation) during ore hauling from active pit to Rom pad stockpile.

Furthermore, proposed mining improvement (before implementation of FMS and IIS) is shown in Table 3. At this point, it is proposed that implementation of FMS and IIS as a mining improvement.

This improvement is a collaboration project between mining operation engineer (mining department), mine production engineer (mine geo services department) and information technology team (administration department) of PT Indo Muro Kencana. It could also be said that this mining improvement will improve previous daily practice of mining operation engineering and will improve daily practice of production engineering in term of mine production reporting. Also, it is projected that this improvement will reduce paper consumption for pre-start checklist of mining unit. Correspondingly, it is predicted that crowded radio situation due overused of radio communication during verbal reporting will be reduced significantly.

TABLE 3

Proposed mining improvement (before implementation of FMS and IIS).

Item	Proposed mining improvement	Comment
FMS	FMS to support mining operation department and mine production engineer (MGS department).	MineTrack VHF/UHF system as FMS
IIS	IIS. Collaboration of IT team (administration department), mine operation engineer (mining department) and production engineer (MGS department).	System for mining operation engineer and mine production engineer
Digitalisation	Improvement in daily practice of mining operation engineering.	Line up visualisation of mining unit and mining operator arrangement will be shown on big screen LCD monitor in mining operation shelter.
Digitalisation	Improvement in pre start unit by using mobile phone to reduce the use of paper	On line pre start of mining unit (excavator, ADT, light vehicle).

Post-FMS and IIS implementation (2024–2025)

Mine Track VHF/UHF system as fleet management system

Right now, MineTrack VHF/UHF system is operated as the existing FMS in IMK site. In addition, MineTrack VHF/UHF system is operated in IMK by using existing mining radio frequency and facility (use existing mining radio towers without wi-fi, internet and GSM communication). In general, Post-FMS implementation in PT IMK is shown in Table 4.

Initially, additional information of MineTrack VHF/UHF system is described in Table 5. Secondly, architecture system of MineTrack LongHaul (MineTrack VHF/UHF system) with MDT Gen 2 is shown in Figure 3. Moreover, MineTrack VHF/UHF system in PT Indo Muro Kencana, hardware and main hardware are explained in Figure 4. Next, it can be seen that operator interface (Android rugged tablet) and activity cycle of MineTrack UHF/VHF system are shown in Figures 5 and 6 respectively. Furthermore, MineTrack application and web application for mining operator of Indo Muro Kencana is shown in Figure 7. In this case, installation of FMS at mining equipment (ADT) PT IMK is shown in Figure 8.

Next, it could also be said that daily practice of MineTrack VHF/UHF as fleet management system in IMK is combined with IIS. It is noted that combination of MineTrack VHF/UHF system and IIS is applied by mining operation engineer (mining department) and mine production engineer (MGS department) in order to track unit location, operator's activities, to obtain raw production data and also to generate production report to mining management.

On the other hand, from point of view of the IT network infrastructure, it is found that not all features of MineTrack VHF/UHF system can be implemented on-site. The main reason of this is because GSM network and wi-fi facilities in-pit area are not appropriate to support full features of MineTrack VHF/UHF system. As a result, live streaming of MineTrack cannot be performed.

TABLE 4

Post-FMS implementation in PT IMK.

Item	Description	Comment
System implemented	MineTrack VHF/UHF system	VHF/UHF frequencies to send data
Support facility	Operated by using existing radio frequency and facility	Existing radio towers. Limited internet wi-fi and gsm communication facility
System structure	MineTrack LongHaul with MDT Gen 2 (mobile data terminal)	Refer to Figure 3 – MineTrack LongHaul with MDT Gen 2 and Figure 4 – MineTrack VHF/UHF system in PT Indo Muro Kencana
Operations	Android MDT (mobile data terminal), MineTrack Client	The MineTrack client application is an application used by operators to interact with the system that will produce data in forms: position, activity, status, reason, login, logoff
Hardware	Operator interface: Android rugged tablet (MDT Gen 2)	1.1 GHz processor, 1 GB RAM, 8 GB Flash ROM, NFC reader
Software application	MineTrack Client	Software of MDT Gen 2
Inputs	MineTrack client Android	Select the login button. Operator taps the RFID card. The login status with the RFID number value will appear on the screen. MineTrack Mobile will send the operator's login information to the server.
Outputs (basic data)	Basic data To track location and activities of mining units	Basic data generated: unit location, unit status (ready, production, breakdown, delay, standby)
Outputs (tracking)	Tracking of unit activity	Loading, queuing, hauling, traveling, dumping
Outputs (ore hauling)	To track location, activities of ore hauling and to anticipate security issue	To monitor security issue due ore hauling; ore hauling distance from pit to process plant area = 8 km
Outputs (login and logoff)	Login and logoff tracking	To monitor working hours. To monitor operator's performance/attitude
Outputs (safety issue)	Overspeed or under speed alarm	Safety issue. To monitor operator's attitude
Optimisation	Tracking: type of material	Ore and waste. Soil, clay, transition rock, fresh rock
Dispatch decision-making	Raw data from mine tracking to create a report to monitor operator's performance	Mining dashboard for mining operation superintendent to monitor and to evaluate mining performance.
	To review productivity of mine production	To increase the accuracy of decision-making for mining operation management.
Performance assessment	To review daily operator's performance	Input, review and consideration data for mining operation management
	Production summary (daily mining operation performance)	
Mine production reporting	Raw production data from MineTrack VHF/UHF	Raw data is processed using IIS

Item	Description	Comment
Technical challenges	Log in and log off for mining operator	Relatively new technology and new habit for mining operator of PT IMK.
	Habit: operator's attitude, discipline (routine activity) and human error	Mining supervisor and mine operation engineer need to monitor about operator's attitude and discipline
Technical challenges	Radio communication and MineTrack VHF/UHF system	Possibility of existing radio towers and MineTrack facility in mining area have potential electricity problem (due to extreme weather).
		It is possible that during this technical problem, some area in hauling road (from active pit to process plant) are not covered by MineTrack VHF/UHF system and radio communication.
Technical challenges (related to potential social issue and security issue)	Radio towers and MineTrack facilities in mining area	Radio towers and MineTrack facilities have potential interaction with local people and illegal miners.
		Potential security issue, safety issue and social issue. Non-technical aspects of mining operation.
Post-FMS implementation	Production gains	Mining production data is up to date with actual mine production (at around one hr from real time)
Post-FMS implementation	Reduction in reporting error (reconciled)	Error in reporting = 3–5%
Post-FMS implementation	Time saving; reduction in paper-based report and reduction paper consumption	Mine production report is released faster than previous practice; at around three hrs in advance from previous daily practice

TABLE 5
Additional information of MineTrack VHF/UHF system.

Item	Description	Comment
General overview	MineTrack VHF/UHF system is recommended to apply in mining environments without a clear line of site due to mountainous or heavily forested terrain.	MineTrack UHF/VHF system is the solution for GPS tracking that uses VHF/UHF frequency bands for mobile radio. It uses TDMA technology which provides faster data transmission.
Communication network	Fix tower. wi-fi radio. VHF or UHF base station. Pole (weighbridge)	MineTrack VHF/UHF system consists of a base transmitter, repeater units, client receiver and server. uses VHF/UHF frequencies and compatible with existing radio facilities.
Application system of MineTrack VHF/UHF system	Server, MineTrack dispatch, MineTrack web, MineTrack report, MineTrack dashboard, MineTrack weighbridge	
Hardware of MineTrack VHF/UHF system at the mining unit	Mobile data terminal (MineTrack client Android), Reveon UHF or VHF tracker, GPS antenna, UHF/VHF antenna, RFID tag	Hardware (operator interface): Android rugged tablet (MDT Gen 2)

ARCHITECTURE SYSTEM WITH MDT GEN 2

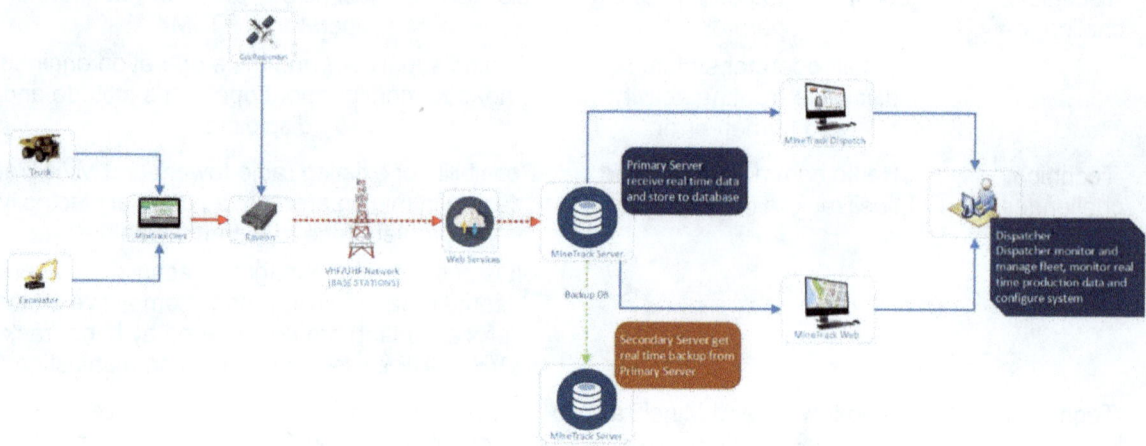

FIG 3 – MineTrack LongHaul with MDT Gen 2.

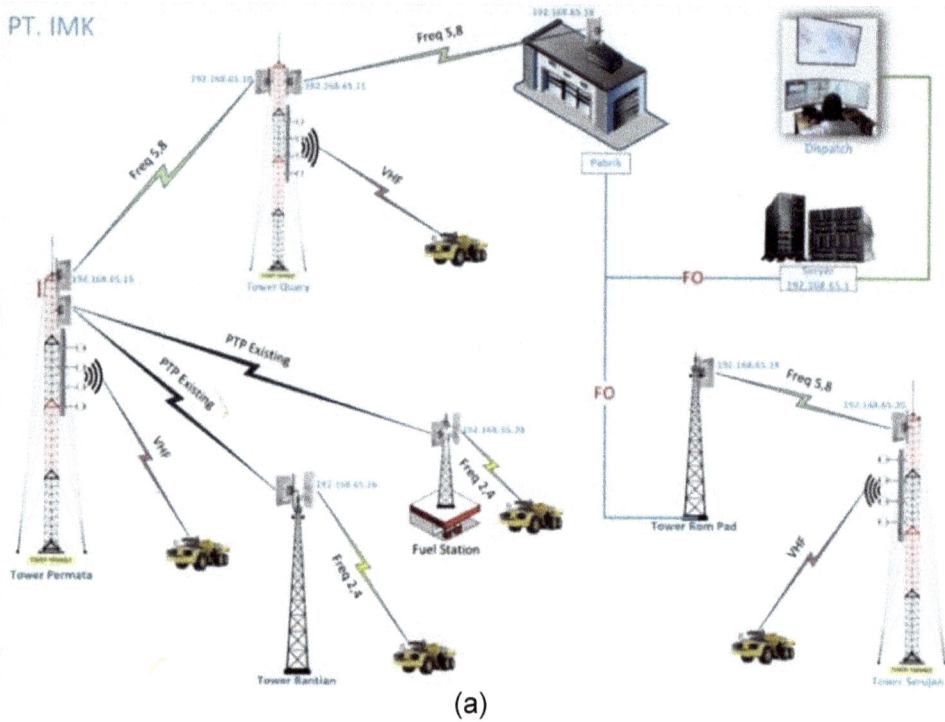

(a)

(b)

Radio UHF/VHF GPS Tracker : ST–GR-V or ST-GR-U

MOBILE DATA TERMINAL (MDT) GENERASI 2 (ANDROID)

(c)

FIG 4 – MineTrack VHF/UHF system in PT Indo Muro Kencana (a); (b) hardware; and (c) main hardware.

FIG 5 – Operator interface: Android rugged tablet.

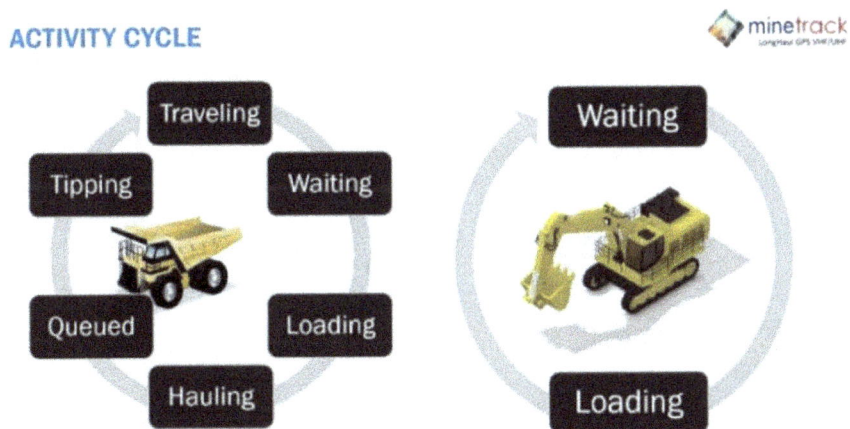

FIG 6 – Activity cycle of MineTrack UHF/VHF system.

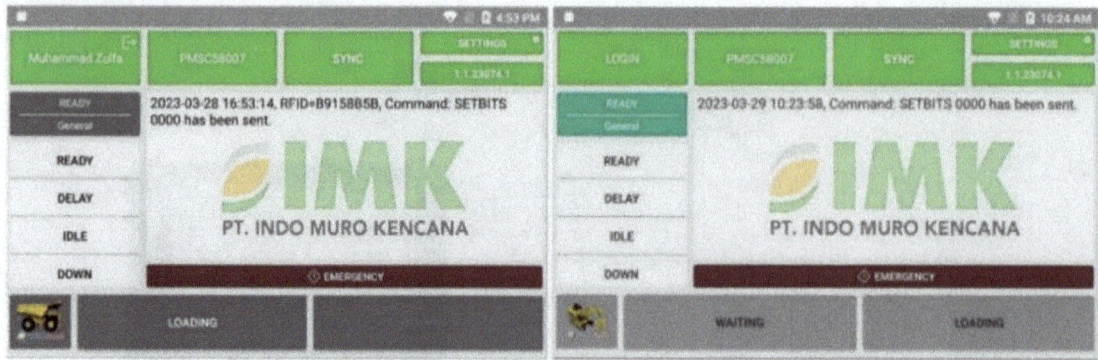

(a)

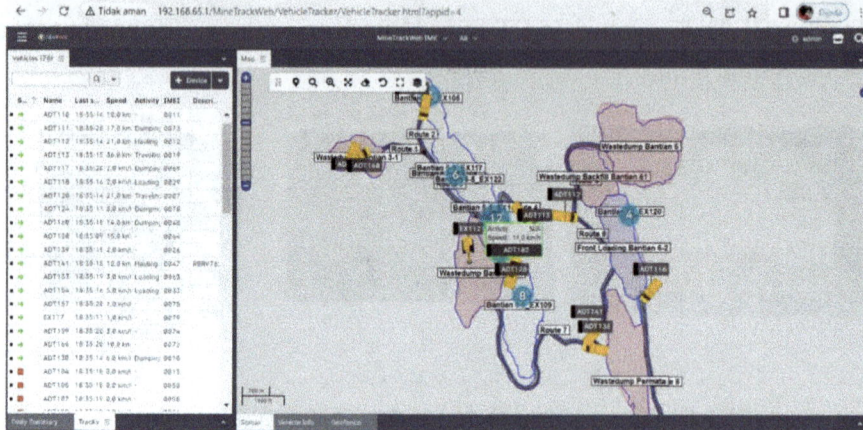

(b)

FIG 7 – MineTrack client application (a) and web application (b) for mining operator PT Indo Muro Kencana.

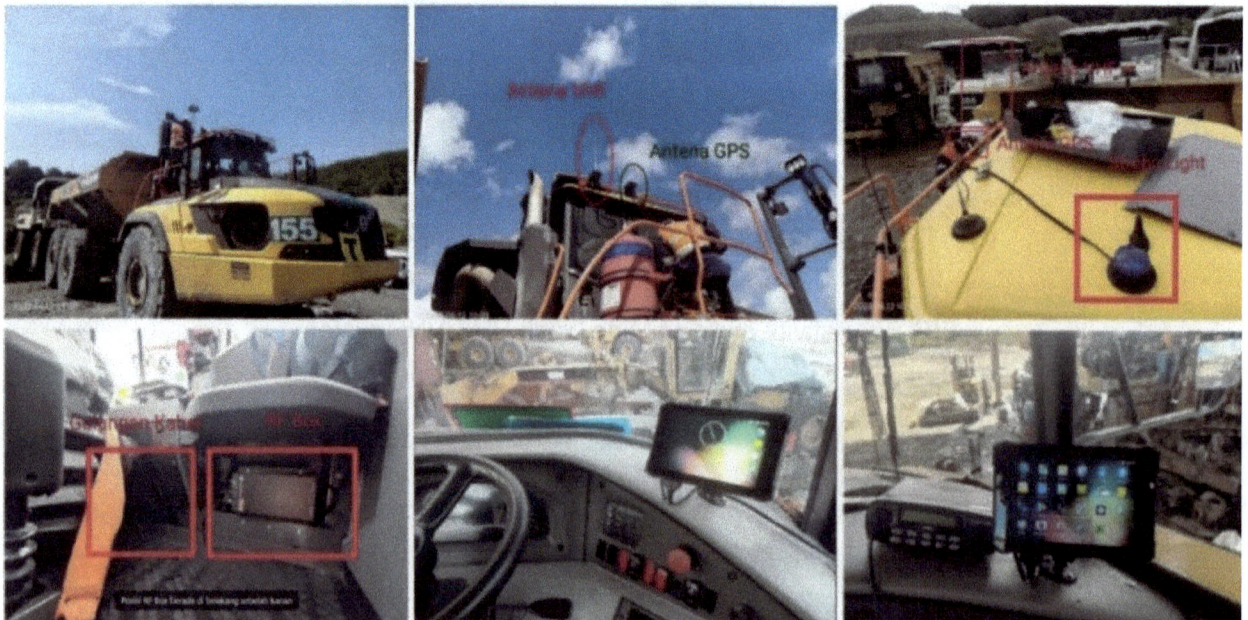

FIG 8 – Installation at mining equipment PT Indo Muro Kencana.

IIS (Integrated Information System)

At this point, post-IIS implementation in PT IMK is described in Table 6. On this occasion, IIS is a system that created by collaboration of IT team (administration department), mine operation engineering team (mining department) and production engineering team (MGS department).

TABLE 6

Post-IIS implementation in PT IMK.

Item	Description(s)	Comment(s)
System implemented	IIS	Integrated Information System
Support facility	Basic data (raw date) from MineTrack VHF/UHF system	MineTrack VHF/UHF system operated by using existing mining radio frequency and facility.
Digitalisation in mining operation	Mobile phone application. Digitalisation of mine equipment pre start	IIS is available for a mobile phone application. Reduce paper consumption
IIS for mining operation department	Effectiveness of mine production information is improved significantly.	IIS as a digitalisation improvement is give clear information to mining supervisor and mine operator.
		Before shift start, information about line up mining unit, mining operator allocation and position of mining unit is shown on big screen LCD monitor on mining shelter. Operational delay can be reduced significantly.
Dashboard of IIS	Dashboard of IIS as a module to monitor	To monitor: daily fuel consumption, hour metre (HM), operator's line up, operator's performance, mine production summary and unit status (breakdown, ready, standby unit).
Dashboard of IIS	For mining operation management	To monitor mining performance, to increase the accuracy of decision-making for mining management.
Dashboard of IIS and performance assessment	To review daily operator's performance. production summary (daily mining operation performance).	Input, review and consideration data for mining operation management.
IIS and mine production reporting	Mine production reporting by combination of raw production data from MineTrack VHF/UHF system and IIS.	Raw production data from MineTrack VHF/UHF system. Raw data is processed by using IIS.
Technical challenges	Log in and log off for mining operator.	Relatively new technology and new habit for mining operator of PT IMK.
	Habit: operator's attitude, discipline (routine activity) and human error.	Mining supervisor and mine operation engineer need to monitor about operator's attitude and discipline.
Technical challenges	Radio communication and MineTrack VHF/UHF system	Existing radio towers and MineTrack facility in mining area have potential electricity problem (technical problem): some area in hauling road (from pit to process plant) are not covered by MineTrack VHF/UHF system and radio communication.
Post-IIS and FMS implementation	Production gains	Mine production data is up to date with actual mine production (at around one hr from real time production)
Post-IIS and FMS implementation	Reduction in reporting error (reconciled)	Error in reporting = 3–5%

Post-IIS and FMS implementation	Time saving. Reduction in paper-based report and reduction paper consumption.	Mine production report is released faster than previous practice. At around three hrs in advance from previous daily practice.

Dashboard of IIS for PC (desktop) is shown in Figure 9. In addition, IIS is used as a software for mining operation engineer to conduct daily practice and mine production engineer in order to create production report. Next, data processing using IIS is conducted by production engineer in order to create a mine production report. Additionally, dashboard of IIS and data processing by using IIS is shown in Figure 10.

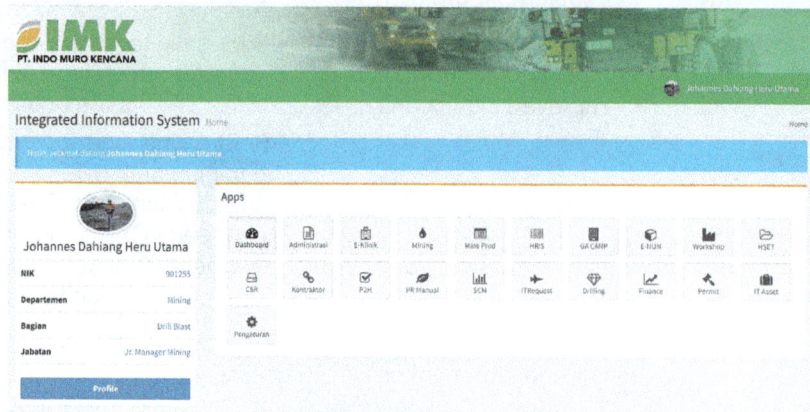

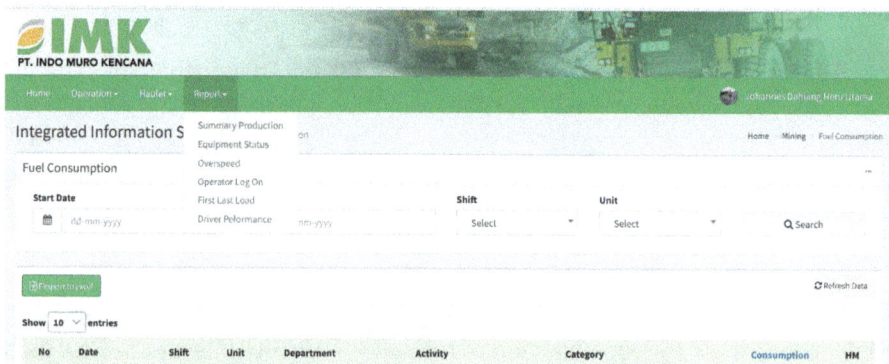

FIG 9 – Dashboard of IIS.

FIG 10 – IIS (data processing).

Furthermore, it is important to note that IIS is also available for mobile phone application. It seems that this application is easy to operate for mining operator. It could also be said that IIS mobile application is useful to conduct pre start check for mining unit. As a result, information related to unit condition, especially unsafe condition, substandard unit condition and equipment damage can be immediately reported to base control (control room) and also informed to mechanics (mobile maintenance plant department). Next, mobile application for IIS is shown in Figure 11.

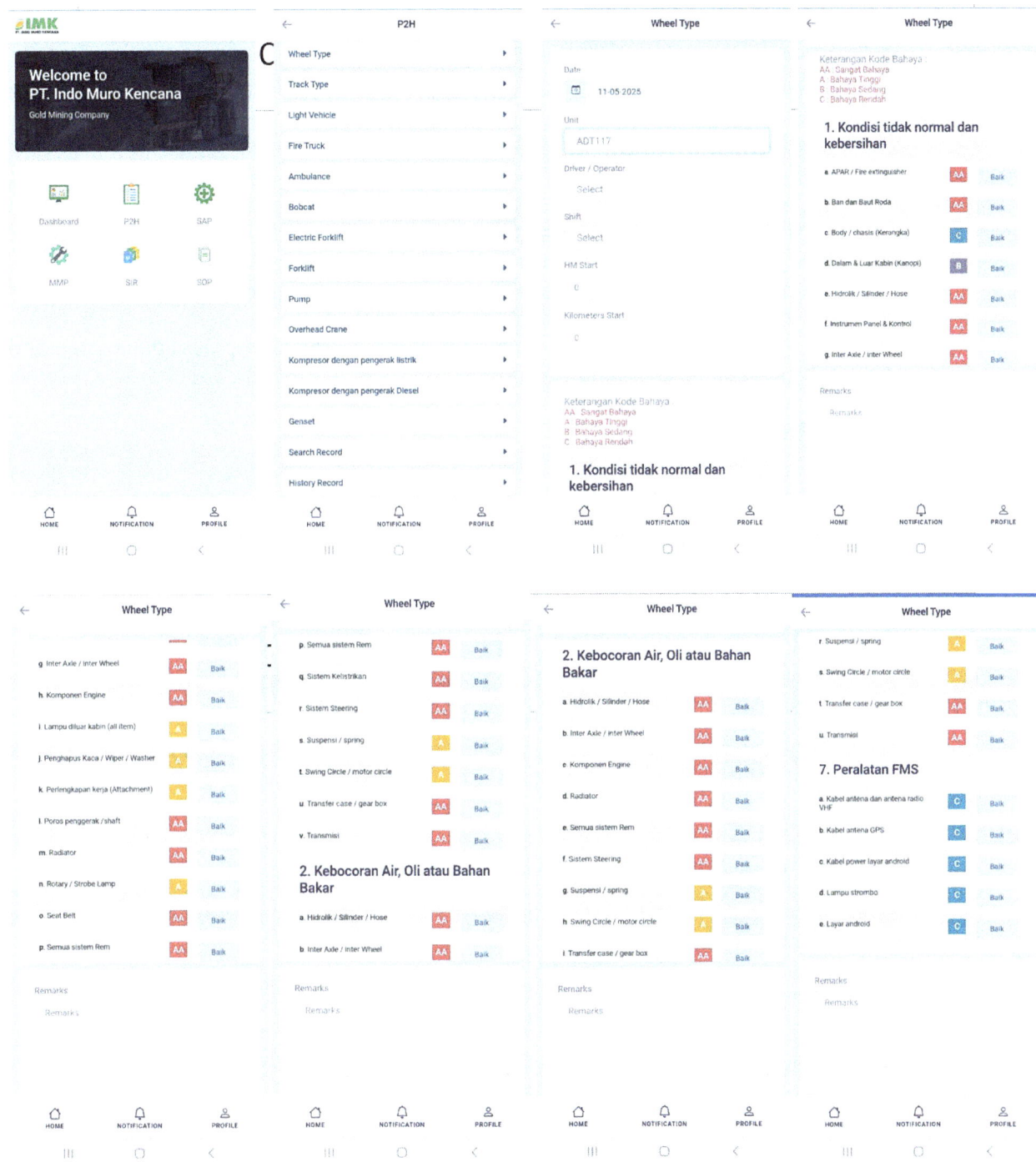

FIG 11 – Mobile phone application for IIS.

CONCLUSIONS

At present, MineTrack VHF/UHF system (a product from SNCTechnologies, part of SatNetCom) is used as fleet management system in IMK site. Initially, the main reason of this is because MineTrack VHF/UHF system is applicable to operate in remote area location and it is applicable with current radio communication facility (VHF/UHF frequency) in IMK site. Additionally, it is clear that post-FMS implementation, there is a significant reduction of paper-based report (reduction in paper consumption for mining operation). Correspondingly, post-implementation of MineTrack VHF/UHF system and IIS, it is claimed that paper usage for mine operation purpose is reduced significantly. Next, it can be said that mining production data (post-FMS implementation) is up to date with actual mine production (at around one hr from real time). Also, there is a significant reduction in reporting error and data reconciled. It could also be said, percentage of error in reporting is reduced significantly to at around 3–5 per cent. Furthermore, it can be seen that mine production report is released faster than previous practice. It is released by mine production engineer at around three hrs in advance from previous daily practice of mine reporting.

In conclusion, it clear that implementation and combination of MineTrack VHF/UHF system as FMS and IIS (Integrated Information System) is applicable for medium scale mining operation, especially mining operation that is located in remote area with hills topography and operated in multiple pit operation. In addition, it is possible that FMS is required in IMK site in order to anticipate potential security and social issue during ore hauling, to track unit location and to monitor mine operator's activities. Furthermore, it is important to note that the application of MineTrack VHF/UHF system as FMS in IMK is combined with the application of IIS (Integrated Information System) in order to track unit's location, to monitor operator's performance and to create mine production report. In other words, it is claimed that this system is a low cost of operational cost, low investment and is affordable for medium scale gold mining operation that is located in remote area with potential security and social issue (non-technical aspects of mining operation).

ACKNOWLEDGEMENTS

The authors greatly acknowledge the managements of PT Indo Muro Kencana (PT IMK) who gives permission to utilise the data used in this paper.

The authors would like to express appreciation to Mining operation department as a team to perform daily practice of FMS (fleet management system) and IIS (Integrated Information System) in PT IMK.

The authors also would like to express appreciation to production engineering team of Mine Geo Services (MGS) department as a team to conduct daily practice of FMS and IIS in PT IMK.

The authors would like to express appreciation to IT team (Information Technology team, Administration department) as a team to perform daily practice of FMS and IIS in PT IMK.

The authors acknowledge the SNC Technologies who gives consent to utilise the data of MineTrack VHF/UHF system (a product from SNCTechnologies, part of SatNetCom) used in this paper.

REFERENCE

Utama, H and Heryadi, R, 2024. Future skills and workforce evolution – training and skills development for mining operation engineer and drill and blast engineer in surface mineral mine PT Indo Muro Kencana, in *Proceeding of the International Future Mining Conference 2024*, pp 73–90 (The Australasian Institute of Mining and Metallurgy: Melbourne).

Spatial data management system applied to the stability of underground excavations

F Vardanega[1], R L Peroni[2], J L V Mariz[3] and B T Kuckartz[4]

1. Mining Engineer, Deswik, Belo Horizonte MG 30330–160, Brazil.
   Email: fabio.vardanega@outlook.com
2. Professor, Universidade Federal do Rio Grande do Sul, Porto Alegre RS 91509–900, Brazil.
   Email: peroni@ufrgs.br
3. Post-Doctorate, Universidade Federal do Rio Grande do Sul, Porto Alegre RS 91509–900, Brazil. Email: jorge_valenca@hotmail.com
4. Post-Doctorate, Federal University of Rio Grande do Sul, Porto Alegre RS 91509–900, Brazil. Email: brukuck@hotmail.com

ABSTRACT

Overbreak and underbreak factors have a major influence on the economic assessment and ground stability of underground mining operations. The overbreak, geometrically represented by the Equivalent Linear Overbreak Slough (ELOS), can be estimated by a combination of the stope's geomechanical characterisation (rock mass quality) and geometric information (dimension, shape, and position), according to Matthew's stability graph. However, the quality of this estimation is highly dependent on the accuracy and the level of detail of the input information used to perform the calculation. It also relies on the number of available samples to build and calibrate the graph for each site or specific region of the mine. This work proposes to use a spatial data management system to assist in the collection and storage of the information required to calculate the dilution factor and extend the use of the stored data to enhance the collaboration between teams of the technical services departments. The system can improve the processes by increasing productivity, guaranteeing that more analyses can be carried out in the same period by integrating the data collection and analysis into pre-existing routines. It can also provide greater reliability, by making the analyses auditable and reproducible, reducing the risk of human error. It can increase efficiency, by allowing more detailed analyses to be performed without significant additional effort. Collaboration is also increased by facilitating access and sharing the information within the different teams involved in planning the excavations, allowing filtering results by region, level, panel, support systems (cabled versus unsupported stopes) or any other relevant spatial reference available on the data set. Furthermore, since the data is continuously stored in a spatial system, it allows new information obtained in the future to be correlated with the historical data, to improve and anticipate the excavations' behaviour, according to operational and local conditions observed in the past by similarity.

INTRODUCTION

In the context of underground mines, Wang (2004) states that both overbreak and stope stability can be determined by aspects such as rock mass characteristics, stope dimensions and geometry, *in situ* stress, blasting parameters, presence of geological structures, and stope exposure time. Nevertheless, addressing this problem through empirical approaches such as Matthew's stability graph (Potvin, 1988; Clark, 1998) has become popular due to its simplicity, although it may ignore relevant influencing factors on dilution and stability such as exposure time, hanging wall geometry, blasting powder factor, and stresses (Wang, 2004).

Stability in underground environments

The Matthew's stability graph empirical method, according to Potvin (1988), considers three aspects related to the excavation of a rock mass, which, in turn, are subdivided into five components:

1. Rock mass characteristics:

   o Block size, represented by the ratio between the Rock Quality Designation (RQD) and the number of joint sets (J_n), ranging between 0.1 and 400.

- Joint orientation adjustment factor (B), ranging between 0.2 and 1, which can be multiplied by the shear strength of jointed rocks factor, calculated by the ratio between the joint roughness number (J_r) and the joint alteration number (J_a), with values that can vary between 0.02 and 4.

2. Induced stress:

- Compressive stress factor (A), ranging between 0.1 and 1.

3. Physical conditions of the problem:

- Gravity factor (C), with values that can vary between 2 and 8.

- Stope size and shape factor, represented by the hydraulic radius (HR), calculated as the ratio between the area and the perimeter of the analysed stope face.

The first four components mentioned can be combined to generate the modified stability number (N'), calculated as Equation 1:

$$N' = \left(\frac{RQD}{J_n}\right) \times \left(\frac{J_r}{J_a}\right) \times B \times A \times C \tag{1}$$

Thus, as shown in Figure 1, the modified stability number (N'), in the Y axis, and the hydraulic radius, in the X axis, compound the graph. Each observation is then plotted according to those components and classified by its stability condition. This allows for defining regions of the graph with a greater propensity for stability or caving, and a transition zone of instability (greyed in the figure) can also be highlighted (Potvin, 1988). Furthermore, the radius factor, calculated as half of the maximum harmonic radius of a surface, closely approximates the hydraulic radius and, therefore, it is believed that it can be used in the stability graph method (Milne, Pakalnis and Lunder, 1996).

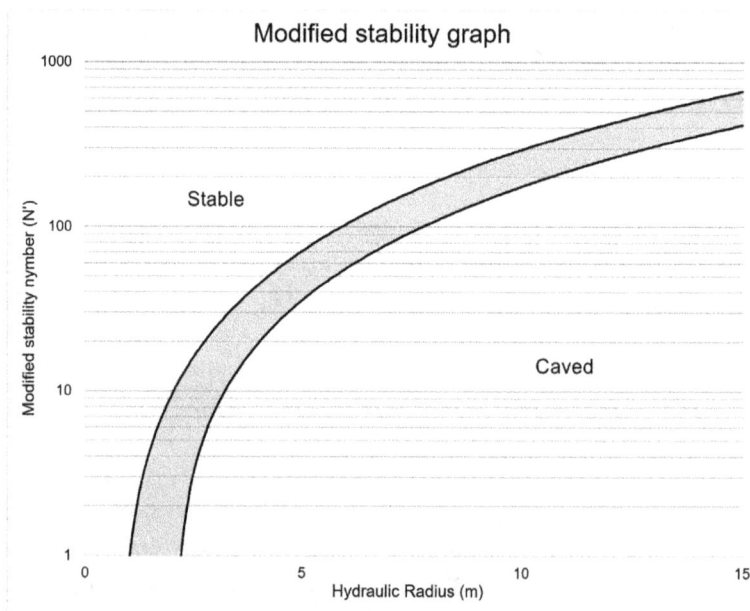

FIG 1 – Modified Matthew's stability graph (modified from Potvin, 1988).

One way to classify the stability condition and estimate the dilution of a stope is through the concept of Equivalent Linear Overbreak Slough (ELOS), calculated as the ratio between the overbreak volume and the area of the analysed stope face, according to Equation 2 (Clark, 1998):

$$ELOS = \frac{Volume\ of\ slough\ from\ stope\ surface(m^3)}{Stope\ height(m) \times Wall\ strike\ lenght(m)} \tag{2}$$

Using the ELOS concept allows for easier interpretation of dilution when compared to volumetric measurements, also allowing the development of a classification scheme in which ELOS values determine the boundaries between stability zones.

Numerical methods such as finite element methods (FEM) (Lima, Guimarães and Gomes, 2024) and boundary element methods (BEM) (Panji *et al*, 2016) are also a common approach to tackle these problems, which are used to determine the induced stress around excavations numerically. Despite their potential, numerical methods may be inaccurate in evaluating the influence of features such as rock mass strength, exposure time, and blasting properties, depending strongly on the input data. The FEM approach discretises the studied domain into small finite elements (rectangles, in 2D, or prisms, in 3D), which allows the modelling of nonlinear and heterogeneous material properties. However, the definition of the external limits of the domain may lead to discretisation errors, in addition to the processing time required to perform the calculations, which may be intensive. On the other hand, in the BEM approach, only the limits of the problem are defined and discretised, so there is better modelling of the external limits of the domain under analysis, reducing discretisation errors and processing time. However, this method should only be used in homogeneous and linear materials (Wang, 2004).

To mitigate the limitations of the mentioned methods, machine learning (ML) approaches have been employed to predict dilution and stability in underground mines from complex data sets, mainly in a supervised machine learning context, which approaches these tasks as classification or regression problems from labelled data. In general, the original data set is segmented into training and testing data sets, where ML-based models are calibrated and adjusted to the distributions of the variables present in the training data set, and only then is their ability to predict new data evaluated on the testing data set. Due to the strong dependence of models on the data set and the chosen partitions, it is necessary to employ statistical techniques such as cross-validation to reduce the possibility of overfitting or underfitting models, situations in which they would be unable to adequately predict the behaviour of new data (Mariz *et al*, 2024). Among the ML-based models successfully used in these tasks, it is possible to mention neural networks (He *et al*, 2024), random forest (Liu *et al*, 2024), and support vector machine (Jorquera, Korzeniowski and Skrzypkowski, 2023).

Data management

Currently, society and industry are growing in connectivity and, consequently, in the amount of data generated. Despite the increase in information availability, allowing for more informed decision-making, it also produces some challenges, like the ones highlighted by Acquire (2024):

- Increase in data volume and sources: people, equipment, and sensors generate a large amount of data all the time, which makes it difficult to process and use.

- Realtime reports: increase in the amount and frequency of reports to fulfill legal requirements or support decision-making processes, highlighting the need to organise the data that serves as the basis for such reports.

- Siloed data: using spreadsheets or isolated data sets, potentially with duplicated versions, makes it impractical to access and use information reliably.

- Data quality: manual data gathering and filling processes increase the human error risk, eroding the quality and reliability of such data sources.

In a report, Seequent (2019) highlights a 2017 survey in which 83 per cent of respondents identified data management as one of the top five issues facing their organisations. The way companies deal with data management in day-to-day operations contains several weaknesses, despite the relevance of this topic. Examples include some of the questions raised by Herbele (2020) when analysing the use of spreadsheets as a source of information for monitoring production in a mining company:

- Are the inserted data the most up-to-date?

- How much time does it take to find the correct file?

- Is the spreadsheet information reliable?

- Who modified the data in the spreadsheet? Should they have permission for that?

In addition to using spreadsheets as a data source, another critical feature of information management is how the files are stored and organised. Using shared network folder structures, cloud-sharing tools, or even emails as reference locations, the time spent by geologists and engineers in data management may surpass 20 per cent of their available hours, which consumes a large amount of the effort that could be dedicated to solve issues of their areas of expertise. The data format and integrity, when found, also harm the productivity of these professionals, causing geologists, for example, to spend 10 per cent of their weeks just performing data conversion and adaptation tasks (Seequent, 2019).

Benefits of using databases

Oracle (2020) defines a database as 'an organised collection of structured information – or data – usually stored in a computer system.' It aims to promote data integration, cataloguing, governance, and security, combining and provisioning information from different areas and defining rules and responsibilities for accessing and changing stored data (SAP, 2025).

The main differences between using a database and spreadsheets are how the information is stored and manipulated, the definition of data access permission, and the amount of data that can be stored. While spreadsheets are designed for use by a single user or a small group of users with basic needs for storing and manipulating information, databases allow for simultaneous access by multiple users and can contain a much larger amount of information (Oracle, 2020).

According to Oracle (2020), a spatial or graph database is a type of database that 'stores data in terms of entities or the relationships between entities'. Cândido (2023), in turn, defines geospatial data as 'information that describes a location and the geographic characteristics of a given place'. Therefore, a spatial database can be understood as a data set based on entities with 3D components. Information management with spatial databases has become crucial in the mining industry, where the core business assets are mineral deposits and, therefore, 3D entities.

Additionally, storing information like mineral inventory, mine design, mining blocks, and mine plans in a single database allows for reporting and data monitoring, ensuring accuracy and auditability and maintaining stakeholder confidence (Acquire, 2024).

Study proposal

This study aims to evaluate using a spatial database system to provide better information for underground stability assessment by integrating survey and geomechanics jobs and creating a procedure to update the results when a reassessment is required. The results of an experiment using the database were compared to a traditional method for calculating the parameters of Matthew's stability graph, which uses only the final as-built stope solid to reconcile against the planned stope. Later, the modified stability number of those results was reassessed to consider a new structural joint. In this regard, the number of graphical dots and the overall shape of the modified stability graph were compared between the two approaches and after the reassessment. Also, the timing required to apply the proposed processes was recorded to evaluate the methodologies' effectiveness.

The remainder of this study is as follows: a methodology section outlines the experiments and their assumptions and describes the input data; a data analysis section presents and analyses the graph results and timings from both experiments; a conclusion section summarises the main findings of this study and proposes directions for future studies.

METHODOLOGY

This methodology aims to test the benefits of incorporating a spatial database into the framework for underground stability assessment, based on Matthew's stability graph empirical method. Two experiments were conducted to verify the ease of running and updating the reconciliation analysis, relying on a synthetic database built for this study:

- Experiment 1: Incorporates the reconciliation process into the surveyors' routine of updating the underground excavations as-built, bringing greater detail to the analysis by generating a new record in the graph for each scan.

- Experiment 2: Simulates the process of updating the reconciliation analysis when new information that may change the previously calculated results is available.

The flow chart in Figure 2 shows the processes and data involved in the preparation and execution of each experiment, as well as the analysis conducted based on their results. The green boxes represent data (files or tridimensional entities), yellow boxes represent processes, and white boxes represent the data sources (whether the data comes directly from the spatial database or external files). The reconciliation solids, one of the main products of the first experiment, are written to the database for reporting purposes. These solids also act as input and output for the second experiment, where they are downloaded from the system to recalculate their modified stability number and are uploaded to the database to update the reports.

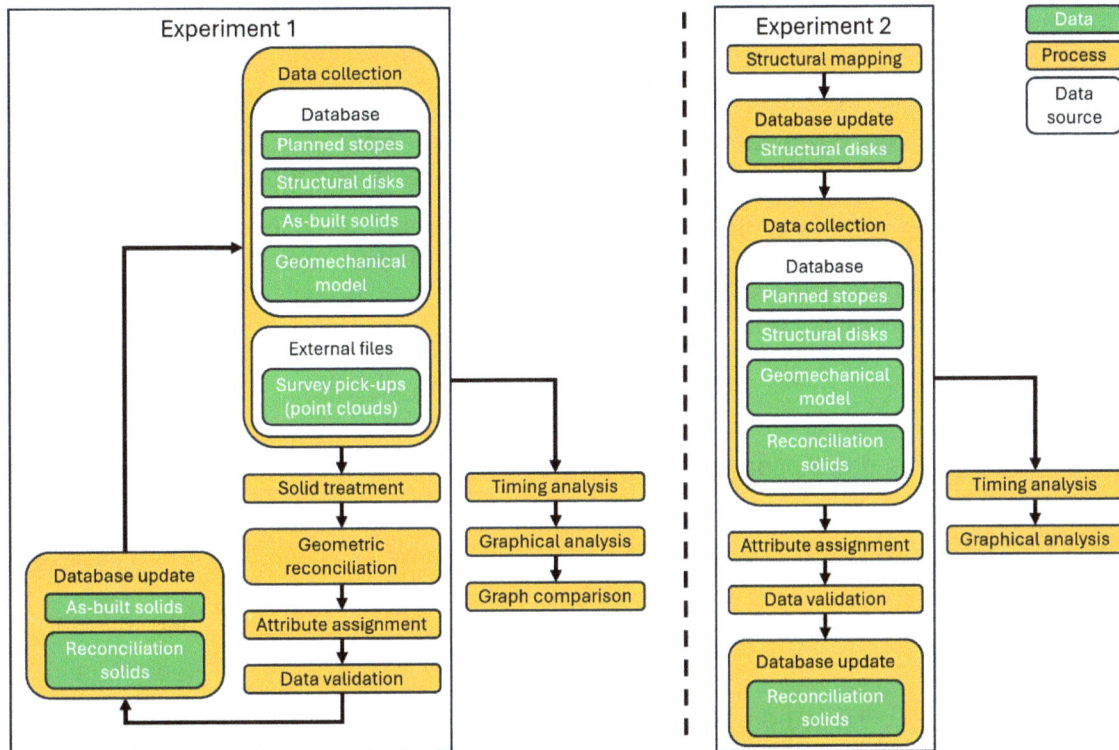

FIG 2 – Flow chart that depicts the two experiments conducted in this study and the analysis performed based on their results.

In both experiments, Deswik.MDM (Mining Data Management), version 2025.1.1 (by Deswik®), a spatial database, was used to organise and centralise the data for the study. Besides the tridimensional entities, the system can store and control other files, like the geomechanical block model. Deswik.CAD, version 2024.2.1926 (by Deswik®), a CAD (Computer Aided Design) package, was used to execute the remaining processes of each experiment. Based on the updated information in the database, the graphical analysis and comparison were carried out in Power BI Desktop™, version 2.140.1205.0 (by Microsoft®), using APIs (Application Programming Interface) provided by Deswik.MDM to automatically update the reports. The experiments used a 64.0 GB RAM (Random Access Memory) computer with an 11th gen Intel® Core™ i7–11850H of 2.5 GHz processor. The details of the data set, the system configuration, and the procedure of each experiment are outlined herein.

Data set and system configuration

The synthetical data set prepared for this study was based on an underground mine using the sublevel stoping method. It consisted of:

- one oredrive as-built solid.
- one planned stope solid.
- five incremental point cloud scans, representing five blasts of the planned stope.

- one geomechanical block model containing the compressive rock strength (σ_c) and the RQD/Jn * Jr/Ja information as fields.

- four structural disks, representing mapped joints in the rock mass.

- one additional structural disk, representing a joint mapped after the completion of the mining of that stope.

- ninety (90) reconciliation solids, representing previously reconciled stopes utilising the same methodology as proposed in this study, ie incremental reconciliation for each scan.

- thirty (30) reconciliation solids, representing previously reconciled stopes utilising the traditional method, ie single reconciliation for each stope.

The geomechanical block model was stored in a file manager inside the system, which ensures that its latest version is always downloaded to the user's computer after login. Because of that, no time was necessary to find and download the geomechanical model during the experiments.

The five-point clouds were *.las format files and were an input provided by the user during the execution of the first experiment. The additional structural disk was created inside the CAD environment in preparation for the second experiment. The remaining entities were stored in different categories inside Deswik.MDM, and contained a set of attributes to facilitate access to the data and reporting, as outlined in Table 1.

TABLE 1

Attributes description and assignment to each type of date (category).

Attribute	Purpose	Categories
Level	Identification of mine level	As-built solids Planned stope Reconciliation solids
Name	Heading or stope identification	As-built solids Planned stope Reconciliation solids
Joint strike	Strike of each joint to calculate the modified stability number factors	Structural joints
Joint dip	Dip of each joint to calculate the modified stability number factors	Structural joints
Joint dip direction	Dip direction of each joint to calculate the modified stability number factors	Structural joints
ELOS	Overbreak measurement for reporting	Reconciliation solids
RF	Radius factor for reporting	Reconciliation solids
Stability assessment	Classification based on ELOS values for reporting	Reconciliation solids
N'	Modified stability number for reporting	Reconciliation solids
Location	Hanging wall or footwall identification	Reconciliation solids

The stability assessment classification was calculated in the reconciliation results based on the ELOS, following range values that can vary from site to site:

- ELOS < 0.75 m: Stable

- 0.75 m ≤ ELOS < 1.2 m: Unstable

- ELOS ≥ 1.2 m: Caving.

The calculation of the factors for the modified stability number (N') considered the following assumptions:

- The main induced stress (σ_i) direction is vertical and can be calculated by the depth (Z) of the stope, the average density (d) of the overlying rock, and the gravity acceleration (g) by Equation 2

- The relative elevation of the stope is calculated in CAD and has a value of -652.6 m

- The average density of the overlying rock is 2.8 t/m^3

- The gravity acceleration is 10 m/s^2

- A conversion factor was applied to calculate the induced stress in MPa

- The stope's hanging wall face is the most prone to failure

- The stope face's dip (θ) was calculated during the process

- Factors A, B, and C were calculated based on equations translated from Potvin's abacus

- Factor A was calculated based on the Equation 3

- Factor C was calculated based on Equation 4, considering the slabbing and gravity fall modes.

$$\sigma_i = -Z \times d \times g \tag{2}$$

$$A = \begin{cases} 0.1, \dfrac{\sigma_c}{\sigma_i} < 2 \\ 0.1125 \ x \left(\dfrac{\sigma_c}{\sigma_i}\right) - 0.125, \dfrac{\sigma_c}{\sigma_i} < 10 \\ 1, \dfrac{\sigma_c}{\sigma_i} \geq 10 \end{cases} \tag{3}$$

$$C = 8 - 7 \times Cos(\theta) \tag{4}$$

To conduct the experiments, three workflows were also configured in the system. Each workflow contained a set of input and output categories, a template CAD file to organise the downloaded data, and a process map, a user-customisable Deswik.CAD tool to automate repetitive tasks and processes, which ensures the correct field naming and calculation according to the standards for the project. The data is filtered from the database using attributes and spatial coordinates specified in each workflow. After the data selection, the system creates, downloads, and opens a Deswik.CAD file on the user's computer, and the user can start the work. When the workflow is completed, Deswik.MDM validates the data, preventing invalid information (malformed entities or required attributes with missing values) from being written to the database.

As Deswik.MDM records the time taken to run the workflow and Deswik.CAD records the duration of each process map command, those metrics were also used to evaluate the experiments and the proposed methodology.

The details of the workflows will be elucidated in the description of each experiment.

Experiment 1

Experiment 1 considered the recurrent execution of the as-built solids update routine five times, which was possible due to the availability of the same number of point cloud files containing the incremental scans of the same stope, as demonstrated in Figure 3. The workflow for this routine allows the surveyor to filter the planned stope solid by its unique 'name' attribute, and the as-built solids are filtered by 'level', allowing direct access to development headings and other stope solids close to the survey scan. The structural disks are filtered according to their coordinates around the stope solid.

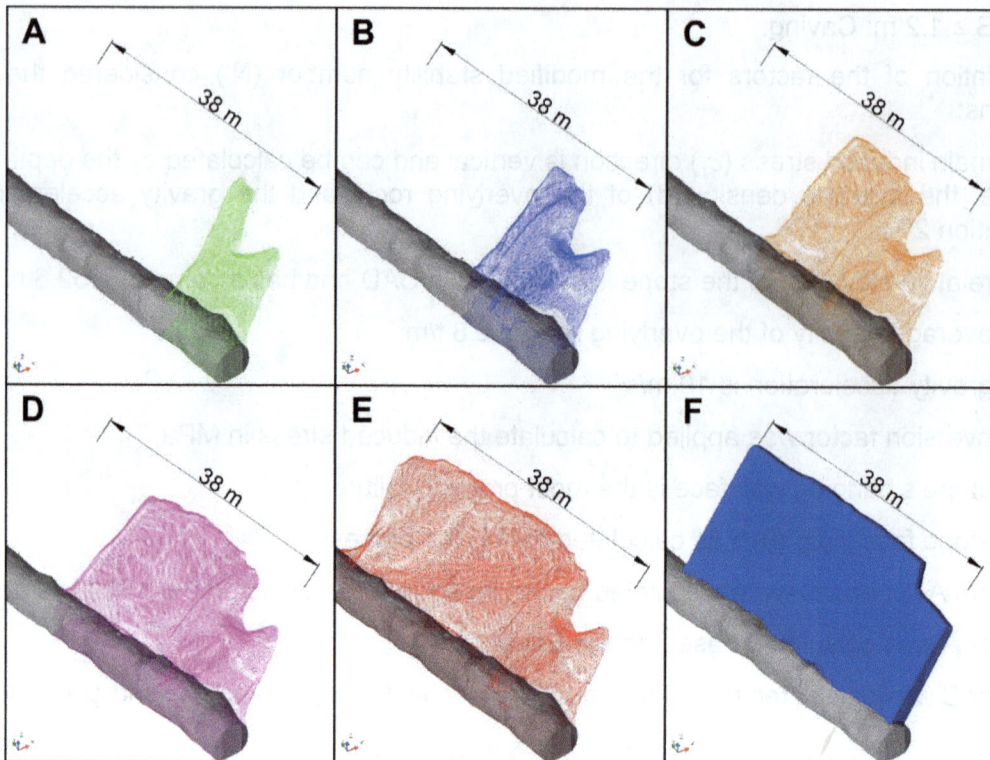

FIG 3 – The development heading solid (grey) is present in all pictures. (a–e) show the evolution of the point clouds from the five subsequent scans; (f) shows the planned stope solid.

Firstly, the point cloud of the survey scan is imported into CAD space and properly identified with attributes. Then, using the process map, it generates and treats the as-built solid. This includes the solid creation based on the cleaned point cloud and the overlaps removal between the current solid and previous ones (stope and development headings).

With the as-built solid created, the user can run the geometric reconciliation between it and the planned stope solids using Deswik's stope and development reconciliation tool. All the as-built solids of the stope sharing the same value for the 'name' attribute are united to be reconciled against the planned stope. Also, the user is asked to split the planned solid (cut the spatial entity) in the position of the latest blast so that the planned stope has the same dimension as the scanned solid and underbreak is not overcalculated. During the reconciliation process, the tool also calculates the radius factor and the ELOS, and this information is assigned to the reconciliation solids.

During the attribute assignment process, the planned stope solid is interrogated against the geomechanical block model and, in conjunction with the geometric evaluation between the stope's hanging wall face and the structural features present on the surroundings of the excavation, the modified stability number (N') is calculated. The database is then updated with the new as-built solid and its reconciled results, which allows the routine to start again for the next scan, using the previous as-built solids as part of the inputs.

After the execution of each interaction, the additional effort required to include the reconciliation analysis in the survey routine was quantified by the processing time of the following three 'process map' commands:

1. Modified stability number calculation: the interrogation of the planned stope solids against the geomechanical block model and the calculation of the modified stability number factors, especially the interaction between the joints and the stope face.

2. Preparation of solids: the process of cutting the planned stope solid in the same position as the latest blast to avoid underbreak overcalculation and the union of all incremental as-built solids for the stope.

3. Geometric reconciliation: the execution of the stope and development reconciliation tool, generating the reconciliation solids, calculating the ELOS and radius factor values, and assignment of reporting attributes, like the stability assessment, into the results.

A Power BI™ report reproducing the stability graph based on the reconciliation solids information was used to analyse the details provided by the proposed methodology and compare them with how the graph would look after the usual approach of evaluating the geometric reconciliation only when the stope is fully mined (single information dot for each stope). The report contains controls for calibrating two second-degree polynomial curves representing the transition between the stable–unstable–caving areas.

Experiment 2

Experiment 2 aims to test the proposed workflow for updating the modified stability number in planned stopes and reconciliation solids when new information is available in the spatial database after the work conducted in experiment 1 is completed. A new preparation step was performed before this experiment, which comprehended the inclusion into the system of a new structural feature corresponding to a new joint mapped by the geotechnical team using a specific workflow.

Then, the data collection process for this scenario occurs entirely throughout the system, using the third workflow, where the user can select the recently added entity and filter all the planned stope shapes, structural disks, and reconciliation solids around it by appropriate spatial bounds.

A CAD file and process map commands similar to those used in experiment 1 are used to recalculate the modified stability number in the planned stope solids and update this information in the corresponding reconciliation results, which is accomplished by matching the 'name' attribute between the entities. Then, the database is updated with the new values.

Following the completion of experiment 2, the time required for the update was evaluated using the same methods applied in experiment 1, and the stability graph report was analysed to understand the impact of the new information on the curves' calibration.

DATA ANALYSIS

The next two sections present and discuss the results obtained in experiments 1 and 2, considering the application of the proposed methodology.

Experiment 1

Table 2 demonstrates the timing records for each iteration of experiment 1, in minutes. The 'Workflow time' column accounts for the entire workflow run, including data selection, file creation and download, all the processing time inside CAD, data validation, file upload, and database update. The 'Reconciliation processes time' column shows the additional effort required to perform the geometric reconciliation in the same routine as the as-built update.

TABLE 2

Timing records for workflow execution and reconciliation processes (modified stability number calculation, solids preparation, and reconciliation).

Iteration	Workflow time (min)	Reconciliation processes time (min)
1	4	1.22
2	5	1.42
3	5	1.35
4	6	1.78
5	7	1.95
Average	5.4	1.55

As expected, the time required to complete the workflow increases between each iteration, ranging from 4–7 mins, mainly because of the processing time to clean up the point cloud and create and treat the as-built solid due to the entities' size. The same behaviour is noted regarding the reconciliation processes, ranging from 1.22–1.95 mins (1:13–1:57 mins), as the geometric reconciliation tool involves many calculations and Boolean operations to produce the results.

Besides that, adding an average of 1.55 mins (1:33 mins) to execute reconciliation is a small effort compared to the value these analyses can bring to the business, which is compensated by the time saved by using a spatial database to get the required data quickly and in a reliable format: only 5.4 mins (5:24 mins) on average for a complete as-built update and geometric reconciliation process is a small processing time when compared to what is usually seen in the industry.

It's important to note that the workflow timing also depends on factors such as the number of entities stored in the database, the amount of data matching the selection, and the network quality between the system server and the user's computer. As the number of entities increases, more computational effort is required to filter the database and compile the CAD file, while the network quality is a key factor in the download and upload speeds.

Regarding the details in the stability assessment provided by the proposed methodology, Figure 5 compares the stability graph when a single analysis is available for each stope (A) and when the analysis is performed for each scan (B), where each dot represents one incremental solid inside of each stope. The green dashed line represents the transition between the theoretical stable-unstable areas, and the red dashed line represents the transition between the unstable-caving zones, which can be adjusted as more data is recorded. The red ellipse highlights the data generated by the iterations of experiment 1, which will be relevant when analysing the results of experiment 2.

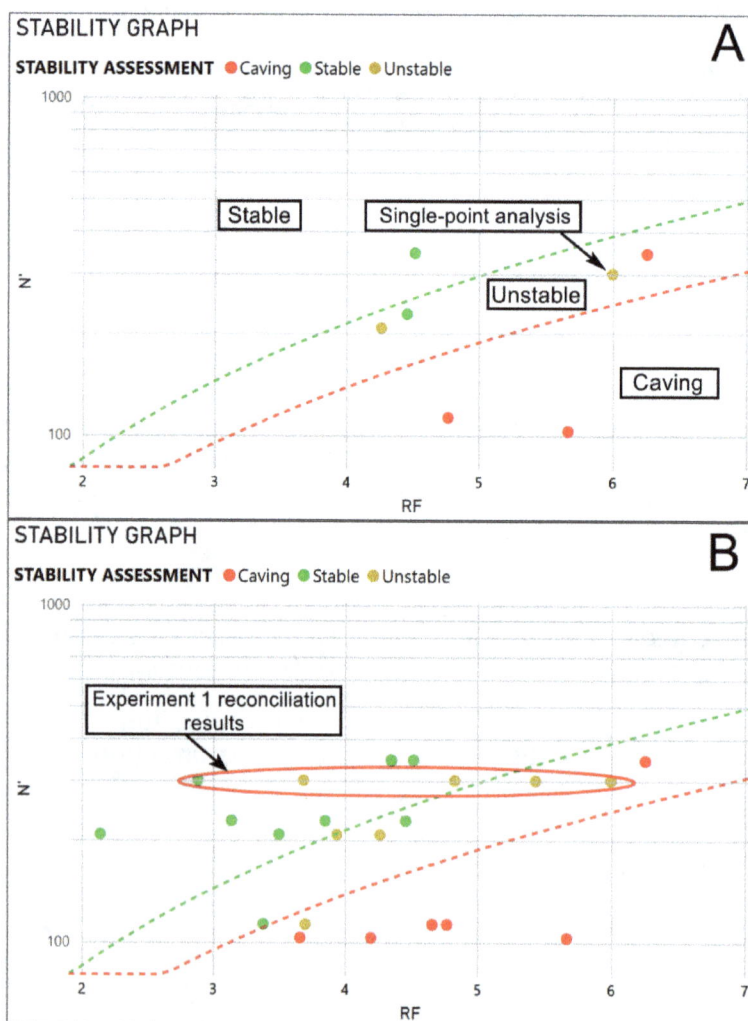

FIG 5 – Comparison between the stability graph with a single reconciliation for each stope (a); and the proposed methodology (b).

For the first graph, the information is focused only on the higher radius factors (final stope configuration), as the incremental excavations within the same stope are not captured. With the proposed methodology, the left-most point of each stope represents the initial scan, and the new points are recorded to the right as the mining progresses, sometimes transitioning from stable (green markers) to unstable (yellow) or even caving (red) conditions. With that, there are three to four times more information dots (depending on the number of survey scans), and a wider range of radius factors is covered, which can help anticipate a premature failure of the stope according to the conditions of the rock after each blast within the stope. The additional details also allow a better understanding of the rock mass behaviour and support requirements, bringing a more precise calibration of the transition curves.

The link between the reporting package and the database system also enables individualising the stability graph analysis into as many details as available. For example, having the mine area, level, and some information about supporting (whether the stope was cabled or not) assigned to the reconciliations results allows the graph to be filtered by any combination of these attributes and calibrating a set of curves for each characteristic.

Experiment 2

Figure 6 shows a view in the MDM's 3D space, where the new mapped joint, added to the database after the survey as-built update and reconciliation analyses, can be seen in the middle of the stope solid (represented by a light grey disk). Also, a bounding box measuring 60 m in each direction was used as a spatial bound to filter other joints, planned stopes, and reconciliation results, allowing for quick identification and access to the surrounding entities.

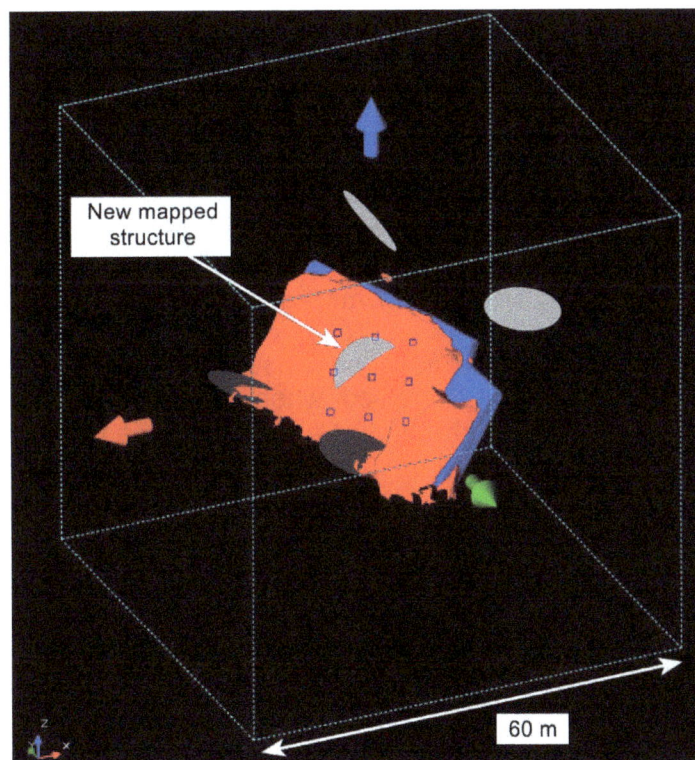

FIG 6 – The new mapped joint in the middle of the stope and a 60 m × 60 m × 60 m spatial bounding box used to filter planned stopes (in blue), reconciliation solids (in red), and closer joints in the MDM preview.

With this filtering method, the reconciliation update workflow had a total execution time of 2 mins, including the time to recalculate the modified stability number of the stope and the attribute assignment on the reconciliation solids. Also, as highlighted in Figure 7, after the inclusion of the new mapped discontinuity and execution of the workflow, the graphical report showed a decrease in the modified stability number (N') from 302 to 164 in all the dots for the reconciled stope (circled by the ellipse in both parts of Figure 7).

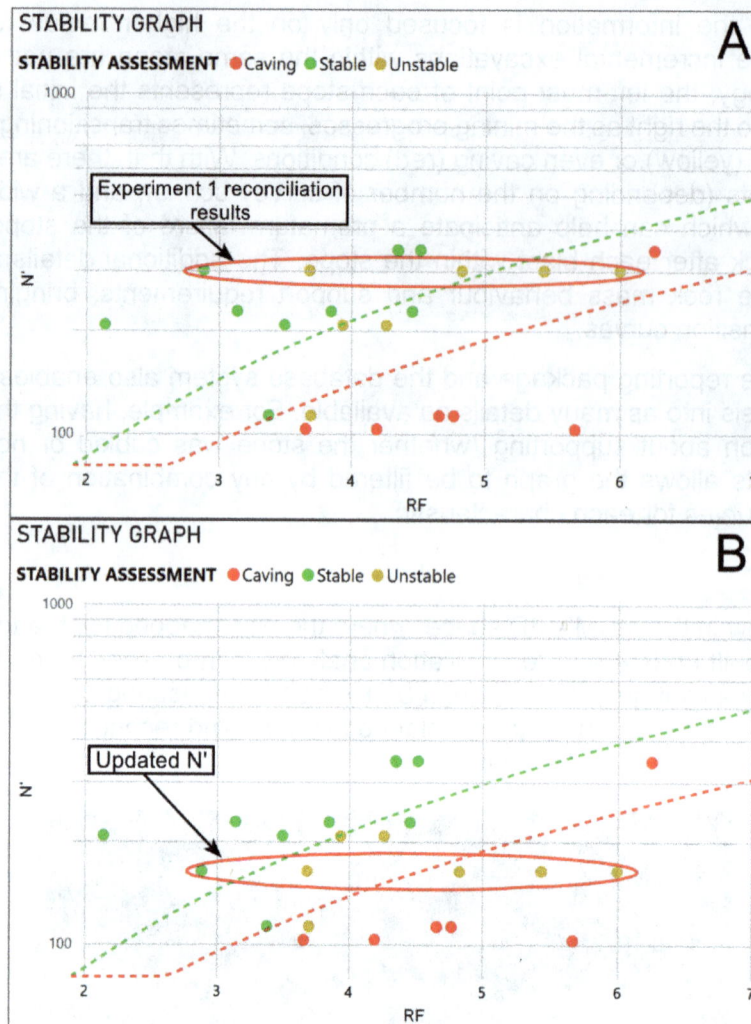

FIG 7 – Comparison between the stability graph before (a); and after the update of the modified stability number based on a new joint mapped after the stope excavation (b).

The ability to run such updates means that some behaviours, potentially interpreted as weird due to the lack of information, can later be corrected when new data is available, improving the geomechanics capabilities on predicting the dilution and defining the proper supporting and mining strategies by implementing this methodology.

CONCLUSIONS

This study presented a methodology for combining the stope reconciliation analysis into an as-built solids update routine using a spatial database as a central location for data and workflow management. The results showed an additional effort of only 1.55 mins (1:33 mins) on average in processing time for each stope scan treatment, producing three to four times more data for the stability assessment analysis.

The methodology also provided a better coverage of analysis regarding the radius factor, ensuring that the reconciliation would be carried out with the smallest possible increments (constrained by the operational characteristics and surveyors' efficiency). The spatial bounds filtering method, used in experiment 2, proved to be a valuable tool for updating past analyses based on new information, contributing to a better understanding of the rock mass behaviour.

The characteristics of an organised spatial database can save the technical services teams significant time usually spent searching for files and correcting data, allowing each person to focus on their expertise. In the example presented in this study, the geomechanics can employ their time to analyse the field data and calibrate the stability graphs instead of searching for the latest as-built or planned stope in a shared network or email.

The ability to reference the database into reporting packages allowed for faster and more reliable reports, as they can be kept up to date with minimal effort. The standards validation on attribute names and values also ensures that the reports will not be broken on the next update.

Additional studies can be conducted to improve the understanding of spatial database capabilities for the mining industry in general and the technical services department in particular:

- Extensive time comparison between the work carried out at a site before and after the implementation of a spatial database system.

- Improve the stability assessment by including other variables mentioned in the literature, like supporting, blasting, filling etc.

- Apply the stability assessment in a wider mine area and test the effectiveness of attribute filtering to create separate graphs for each region.

ACKNOWLEDGEMENTS

The authors gratefully acknowledge the Federal University of Rio Grande do Sul, Deswik, and all advisors and colleagues who supported our work.

REFERENCES

Acquire, 2024. Top 5 information management hurdles for the mining and mineral resources industry [online]. Available from: <https://www.acquire.com.au/newsroom/top-5-information-management-challenges-for-the-mining-and-mineral-resources-industry/> [Accessed: 11 February 2025].

Cândido, D H, 2023. Artigo: Tendências da gestão de dados geoespaciais na mineração [online]. *Mundo Geo,* Available from <https://mundogeo.com/2023/12/14/artigo-tendencias-da-gestao-de-dados-geoespaciais-na-mineracao/> [Accessed: 11 February 2025].

Clark, L M, 1998. Minimizing dilution in open stope mining with a focus on stope design and narrow vein longhole blasting, MSc Thesis, University of British Columbia, Vancouver.

He, B, Li, J, Armaghani, D J, Hashim, H, He, X, Pradhan, B and Sheng, D, 2025. The deep continual learning framework for prediction of blast-induced overbreak in tunnel construction, *Expert Syst Appl*, 264:125909.

Herbele, J M C, 2020. Desenvolvimento de metodologia para consolidação de indicadores de desempenho para controle e gestão na mineração: Um estudo de caso, Msc thesis, Universidade Federal do Rio Grande do Sul, Porto Alegre.

Jorquera, M, Korzeniowski, W and Skrzypkowski, K, 2023. Prediction of dilution in sublevel stoping through machine learning algorithms, *IOP Conf Ser Earth Environ Sci*, 1189:012008.

Lima, M P, Guimarães, L J N and Gomes, I F, 2024. Numerical modeling of the underground mining stope stability considering time-dependent deformations via finite element method, *REM Int Eng, J,* 77(3):e230080.

Liu, Y, Li, A, Wang, S, Yuan, J and Zhang, X, 2024. A feature importance-based intelligent method for controlling overbreak in drill-and-blast tunnels via integration with rock mass quality, *Alex Eng J,* 108:1011–1031.

Mariz, J L V, Ferraz, T S G, Lima, M P and Silva, R M A, 2024. Uso de Aprendizado de Máquina Supervisionado e Algoritmo Genético na predição da diluição em minas subterrâneas, in *Proceedings XI Fórum de Mineração* (Universidade Federal de Pernambuco: Recife).

Milne, D M, Pakalnis, R C and Lunder, P J, 1996. Approach to the quantification of hanging-wall behaviour, *The Institute of Materials, Minerals and Mining*, University of British Columbia, Vancouver.

Oracle, 2020. O que é um banco de dados? [online]. Available from <https://www.oracle.com/br/database/what-is-database/> [Accessed: 15 February 2025].

Panji, M, Koohsari, H, Adampira, M, Alielahi, H and Marnani, J A, 2016. Stability analysis of shallow tunnels subjected to eccentric loads by a boundary element method, *J Rock Mech Geotech Eng*, 8(4):480–488.

Potvin, Y, 1988. Empirical open stope design in Canada, PhD Thesis, University of British Columbia, Vancouver.

SAP, 2025. O que é gerenciamento de dados? [online]. Available from <https://www.sap.com/brazil/products/technology-platform/what-is-data-management.html> [Accessed: 11 February 2025].

Seequent, 2019. The data management challenge [online], Seequent. Available from: <https://www.seequent.com/community/research-reports/the-data-management-challenge/> [Accessed: 11 February 2025].

Wang, J, 2004. Influence of stress, undercutting, blasting and time on open stope stability and dilution, PhD thesis, University of Saskatchewan, Saskatoon.

Remote monitoring of slurry pumps in mining and mineral processing – enhancing reliability and performance through digitalisation

A Varghese[1,5], J Xiang[2], P Mikkonen[3], E Lessing[4], G M Hassan[6] and A Karrech[7]

1. Product Head – Pumps Digital, Metso, Perth WA 6110. Email: alan.varghese@metso.com
2. Performance Center Manager, Metso, Changsha, Hunan Province, China.
   Email: jinhua.xiang@metso.com
3. Vibration Expert, Metso, Espoo 02230, Finland. Email: petri.mikkonen@metso.com
4. Vice President Engineering – Pumps, Metso, Espoo 02230, Finland.
   Email: evert.lessing@metso.com
5. PhD Student, University of Western Australia, Perth WA 6009.
   Email: alan.varghese@research.uwa.edu.au
6. Senior Lecturer, University of Western Australia, Perth WA 6009.
   Email: ghulam.hassan@uwa.edu.au
7. Professor, University of Western Australia, Perth WA 6009. Email: ali.karrech@uwa.edu.au

ABSTRACT

Mining and mineral processing operations depend heavily on slurry pumps for the hydraulic transportation of solids. These pumps operate under harsh conditions that place extreme stress on the pumps. Frequent pump failures not only disrupt operations but can also lead to significant financial losses. The challenges are further compounded by remote and often isolated locations of mines, coupled with shortage of skilled labour, and the necessity for continuous performance despite varying ore conditions. The rise of digitalisation technologies provides a strategic solution to these industry-wide issues. This paper explores the remote monitoring of slurry pumps through advanced digital tools, technologies, and expert services that enable continuous tracking of pump condition and performance. Pumps Condition Monitoring Service (PCMS) is introduced as a practical and scalable entry point into pump digitalisation, demonstrating how real-time data, wireless sensors, cloud-based platforms, and expert-driven analysis can transform pump maintenance from reactive to proactive. Illustrative case studies from copper, iron, and steel operations highlight the tangible benefits of this approach, including improved reliability, reduced downtime, and more efficient maintenance planning. The paper also presents insights from monitoring 46 pumps globally over a one-year period, summarising common faults, alarms, corrective actions, challenges, and key lessons learned. These findings demonstrate how digital solutions not only enhance pump performance and reliability but also contribute to operational efficiency and cost reduction across mining sites. Through these results, the paper underscores the critical role of digitalisation and remote monitoring in modernising pump maintenance practices and building more resilient, sustainable mining operations.

INTRODUCTION

Slurry pumps are critical components in mining and mineral processing operations, where they facilitate the hydraulic transport of solids across various stages of production. These pumps often operate in extremely demanding conditions. Frequent breakdowns not only result in costly unplanned downtime but also compromise the overall efficiency and safety of operations. Customer challenges with slurry pumps in mining include strict environmental, safety, and social regulations, declining ore grades, difficulties in predicting failures, lost production due to unplanned maintenance, inventory and maintenance planning, labour and expertise shortages, and increasing energy consumption and costs. Figure 1 provides a visual example of a slurry pump in operation at a mine site.

In recent years, the mining sector has seen a strong push towards digital transformation, driven by the urgency to reduce environmental impact, improve operational efficiency, and ensure long-term sustainability. Digitalisation in this context refers to the use of computerised or connected digital devices and digitalised data to reduce costs, enhance business productivity and transform mining practices (Barnewold and Lottermoser, 2020). The digitalisation of mining practices is being accelerated by technological advancements and the increasing availability of smart, connected equipment. These developments are largely influenced by goals related to electrification, supply chain resilience, and Environmental, Social, and Governance (ESG) frameworks. Beyond improving safety and reducing environmental risks, digital technologies offer an opportunity to lower operational costs and boost overall productivity.

FIG 1 – Example of slurry pump operating at a mine site.

In this paper, the traditional maintenance methods currently used in the industry are compared against remote monitoring and proactive data-driven maintenance, which is rising due to digitalisation as an enabler. Data-driven maintenance utilises advanced digital tools and technologies to connect equipment, such as wireless sensors, cloud-based platforms, and expert services, to their condition and performance. It introduces the concept of Pump Condition Monitoring Services (PCMS) as a practical step towards this transformation for pumps. Furthermore, the study presents findings from the monitoring of 46 pumps over one-year period, including frequently observed faults, case studies, limitations encountered, and future considerations for advancing digital maintenance practices in the mining sector.

DIGITALISATION AS AN ENABLER

Digitalisation is a key enabler for reliability. Typically, in pump performance analysis, data collection is the first step. This is often done manually by asking operators and maintenance personnel on-site about changes to the process, any abnormal conditions, and potential system issues. However, the responses can be vague, with answers like 'nothing has changed,' leading to insufficient, incomplete, or out-of-calibration data. This results in low-quality assessments that address quick fixes without delving into a thorough root cause analysis. Consequently, the quality of equipment performance is compromised. This is where data-driven performance analysis shows clear added value by bridging gaps and starting the analysis on more solid ground.

Traditional maintenance methods for pumps typically involve reactive or preventive maintenance. Reactive maintenance focuses on repairing faults after a breakdown, restoring non-operating assets to desired conditions. The general concept is to operate assets until they break and then fix them. The challenge with this approach is the random timing and inefficient operation of non-optimal assets that are beyond their useful life, adding costs to operations. Preventive maintenance, on the other hand, is calendar-based scheduled maintenance at fixed intervals. It is planned and routine maintenance intended to ensure continuous operation of assets, prevent time-based failures, reduce unexpected breakdowns, and extend asset life. However, the challenge here is premature repairs, leading to substantial cost additions from underutilised equipment or parts that could still be used. The machine may continue to operate without issues (Karki *et al*, 2022).

Digitalisation and reliability are closely intertwined, forming the foundation for data-driven maintenance and asset operation. Through digital technologies, maintenance evolves beyond traditional methods to encompass a comprehensive approach that includes monitoring, predictive analytics, diagnostics, troubleshooting, optimisation, reporting, commissioning, and asset modernisation. This integrated strategy enables informed decision-making and timely interventions. Reliable asset performance, enabled by digital tools, contributes to process stability, enhanced resource efficiency, improved safety, and a reduced environmental footprint. By leveraging accurate and continuous data, mining operations are empowered to better measure, communicate, and progress toward operational excellence and sustainability goals, transforming insights into tangible value.

PCMS – A PRACTICAL STARTING POINT FOR PUMP DIGITALISATION

Pumps Condition Monitoring Service (PCMS) provides a practical, efficient and cost-effective way to start the digitalisation journey for pumps. One of the most immediate benefits of PCMS is enhanced visibility into pump health and performance through remote monitoring, addressing the key challenge of overlooked maintenance needs in traditional reactive maintenance. This lack of visibility can lead to a cascade of issues, ranging from safety risks (Varghese *et al*, 2023) to excessive consumption of labour, materials, and resources, ultimately inflating operational costs. Additionally, unplanned pump failures can severely impact production and lead to costly downtime.

PCMS enhances operational transparency by enabling early detection of developing faults, long before they escalate into major failures. By delivering insights into pump health and performance, PCMS enables a shift from reactive to proactive maintenance strategies. A wide range of condition monitoring technologies are available in the market such as vibration analysis, temperature sensing, ultrasound, electrical measurements and infrared thermography to deliver a comprehensive assessment of pump condition. The growing availability of affordable, wireless sensors, especially for vibration and temperature, has made PCMS more accessible and efficient. These smart sensors eliminate the need for manual data collection and deliver insights within minutes.

PCMS includes installation of sensors, connectivity through a gateway, and parametrising the cloud monitoring and analytics platform. Figure 2 presents an example of wireless sensors installed on a slurry pump. The monitored parameters include triaxial vibration and surface temperature readings, which enable the detection of various issues such as bearing defects, gear defects, overheating, inadequate lubrication, looseness, unbalance, misalignment, rubbing, mechanical strain, cavitation, shaft bend, base defects, and electrical faults. Figures 3 and 4 illustrate data analytics examples for bearing defects and insufficient lubrication, respectively.

FIG 2 – Wireless sensors installed on slurry pump in gold squares.

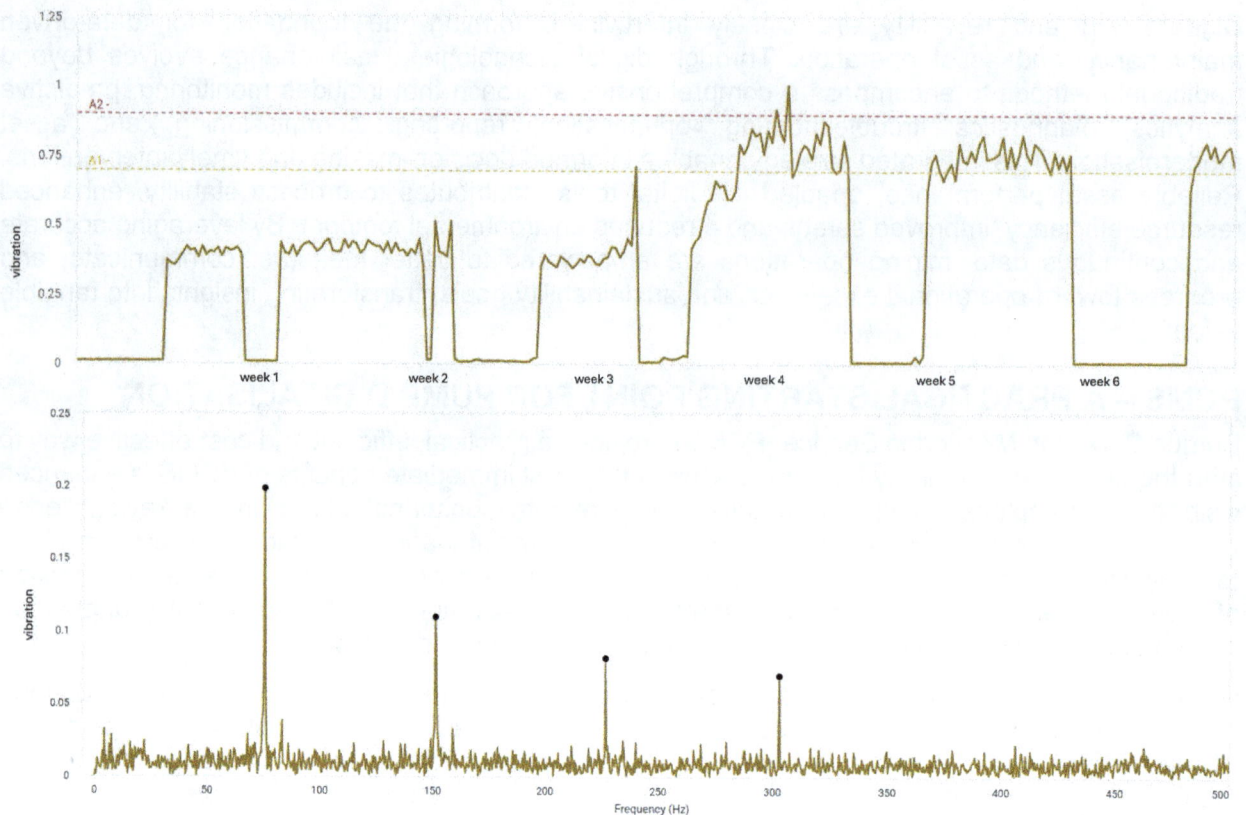

FIG 3 – Typical bearing defect detection.

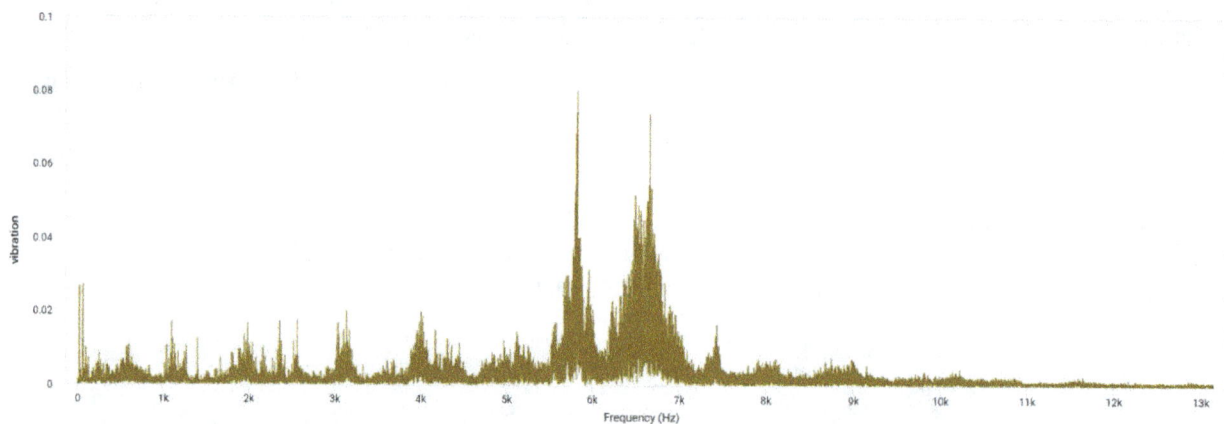

FIG 4 – Typical insufficient lubrication.

Once deployed, PCMS enables continuous tracking of pump operating conditions through remote performance centres where dedicated monitoring experts configure threshold alarms and continuously assess data for any deviations. Upon detection of an anomaly, the issue is validated, diagnosed, and resolved with actionable recommendations tailored to the asset's criticality and operating environment. A comparison of traditional pump diagnostics versus PCMS is presented in Figure 5 that showcases significant savings of over 90 hrs with better quality results. This approach allows proactive management of pump conditions, optimising operations and extending pump life.

Collaboration is key to success. While many pump manufacturers now offer some form of PCMS, its effectiveness depends heavily on collaboration among all stakeholders, including OEMs, operations teams, and maintenance departments. While dashboards and analytics tools are valuable starting points for understanding pump behaviour, the real value lies in transforming insights into timely actions and long-term improvements. It is this synergy between digital tools and human expertise that ultimately drives performance gains and sustainable operational excellence.

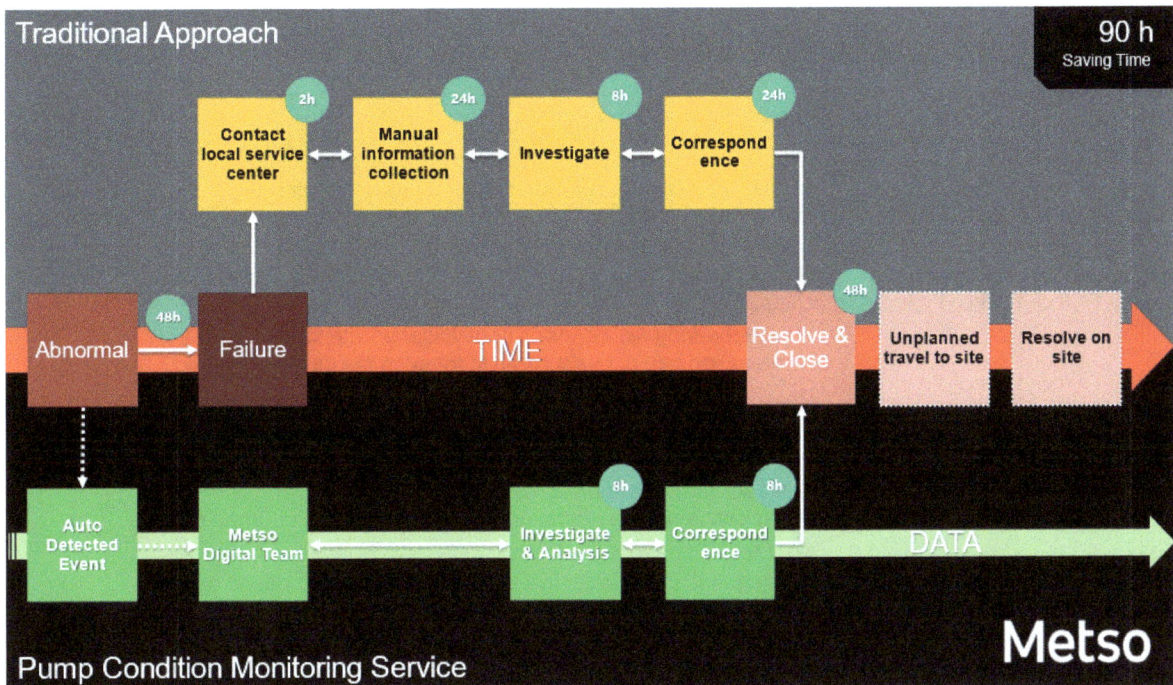

FIG 5 – Traditional versus pumps condition monitoring service approach.

CASE STUDIES AND RESULTS

Case studies have been presented to demonstrate the tangible benefits of implementing PCMS (remote monitoring) and to showcase how digitalisation can significantly enhance the reliability and efficiency of pump operations. The first example focuses on a copper mine site, where the Pump Condition Monitoring System (PCMS) was implemented on a single pump application. This pump was closely monitored and compared against a non-monitored pump. The results clearly illustrate substantial operational improvements and cost savings driven by data, monitoring, and timely interventions. The second case study involves a group of pumps at a global iron and steel facility that were recently integrated into the PCMS. Remarkably, within just 48 hrs of installation and monitoring, several performance and reliability issues were uncovered, ranging from mechanical misalignments to process-related inefficiencies.

Furthermore, Table 1 highlights common faults identified by remote monitoring 46 pumps over one-year period globally across various industrial sites. The result includes corrective actions and savings that can generate from remote monitoring. These early insights enabled prompt corrective actions, validating the rapid impact and value of condition-based monitoring. In addition to the benefits, several challenges and lessons learned throughout the implementation of PCMS have been identified as key areas for future improvement. Together, these studies underscore how digital tools and proactive strategies are reshaping maintenance practices, reducing unplanned downtime, and contributing to more sustainable and cost-effective operations. By leveraging advanced monitoring technologies, sites have achieved significant improvements in pump performance, reliability, and overall operational efficiency.

TABLE 1

Common faults, alarms and recommended actions identified through PCMS.

Identified faults	Alarms reported	Description – actions and success	Sensor data
Inadequate lubrication	Motor (11) Pump (7)	Typically, lubrication is carried out within days of notification, often resulting in a return to baseline condition in many cases. A key advantage of this approach is the high visibility and prompt response it enables. However, it has been noted that untimely action can result in complete bearing failure. Additionally, water ingress or contamination can act as a catalyst for such failures.	Vibration temperature
Bearing faults	Gearbox (20) Motor (13) Pump (8)	A reduction of over 8 hrs of unplanned downtime was achieved in the majority of cases. This enabled maintenance to be carried out in a planned manner, with resources such as spare parts and labour readily available, effectively mitigating operational risk. Following repairs, equipment performance has returned to baseline operation. In one notable case, proactive contingency planning based on risk analysis allowed operators to prepare nearly a month in advance, ensuring a timely and efficient response to the event.	Vibration temperature
Overheating	8	Bearings can overheat due to excessive loads, improper lubrication (either over or insufficient), contamination, and incorrect installation (Detweiler, 2011). If the equipment is not instrumented, low visibility can prevent early detection, leading to sudden failures. We have observed over eight cases of overheating and recommended appropriate actions. Sometimes, overheating occurs momentarily, such as right after installation, requiring close observation to ensure the issue is resolved.	Temperature
Cavitation	5	Cavitation in centrifugal pumps can arise from several factors, one of the most common being insufficient inlet pressure. This may result from a low liquid level in the supply tank or leaks within the suction piping, both of which reduce the pressure at the pump inlet (Morris and Gutierrez, 2024). We identified five such cases where cavitation was linked to these issues. In each instance, process recommendations were provided, including adjustments to tank levels and updates to control logic to ensure stable and adequate inlet pressure.	Vibration
Foreign materials	2	In two instances, foreign materials such as mill scats or balls were found to pass through the pumps due to upstream equipment failures (eg screens). These foreign materials can cause premature damage to the pumps. One such instance was identified and corrected within 12 hrs of detection.	Vibration
Looseness	3	Fastening bolts can loosen over time, making periodic inspections essential. With remote monitoring, these checks can be better aligned with predicted risk, allowing for smarter prioritisation of maintenance activities. Instances have been detected where motor or pump mountings became loose and were subsequently readjusted.	Vibration
Suction pipe vibration transmission	2	Vibrations from nearby process systems can be transferred to the pump. If these vibrations fall within the pump's critical speed range, they can be amplified, leading to premature failure. Two instances of suction pipe vibration transmission have been detected and notified for correction.	Vibration

Identified faults	Alarms reported	Description – actions and success	Sensor data
Misalignment – belt over tension or slipping	8	Majority of the pumps we currently monitor use V-belt transmission. It is commonly noted in these cases that belts maybe over tensioned or become loose over the period and the pulleys are misaligned. Pumps that have operated for extended periods without regular inspection or monitoring often show signs of neglect. Ensuring proper belt tension and pulley alignment is critical for reliable and efficient pump operation.	Vibration
Unbalance	2	Two cases of unbalanced pump impeller were identified causing issues of free rotation	Vibration
Rotor damage	1	One case of electric motor rotor damage was detected and notified to maintenance	Vibration
Gear teeth high load, cracked or broken	7	Seven cases have been reported where gear teeth were found to be operating under high load conditions or were cracked or completely broken. Pumps with gearbox drives typically experience higher load conditions, making them particularly beneficial to monitor. Continuous monitoring in these applications provides early detection of abnormal load patterns or gear deterioration, allowing timely maintenance actions and preventing costly unplanned downtime.	Vibration

Remote monitoring of single pump at a copper mine

The case study of a pump application at a copper mine site demonstrates how a proactive approach significantly improved pump performance, life, reliability, and overall operational efficiency. These approaches include selecting the right pump for the application, implementing remote monitoring, taking timely service actions, and maintaining high visibility of operating conditions. These measures have resulted in substantial savings in operating costs, improved slurry pumping efficiency, and water savings.

Over the course of a year, four campaigns were conducted, each yielding positive results. Initially, the site used a 300 mm inlet rubber-lined pump, which was oversized and inefficient for the duty. To understand pump performance, data such as flow, pressure, and density were collected and monitored. Upon reviewing this data, it became evident that a smaller, 250 mm inlet pump would be more suitable. Following the installation of this correctly sized pump, power and current levels dropped significantly due to improved efficiency.

During the first campaign, pump efficiency and water utilisation efficiency were substantially higher compared to previous pumps due to the right-sized pump for the application. However, the pump's lifespan was similar to the previous model due to an unexpected failure of upstream screens, which allowed mill balls to fall through and cause early wear and damage. This issue was addressed in subsequent campaigns. In the second campaign, closer operational control was implemented, allowing the pump to achieve over 2500 hrs of operation, showcasing a marked improvement in reliability. In the third campaign, switching to a metal liner addressed the mill scats issue, extending the pump's operational life beyond 3000 hrs.

Additionally, the introduction of a Pump Condition Monitoring System (PCMS) provided enhanced visibility into pump operations by identifying potential issues early. Common issues like motor bearing wear, lubrication problems, and pump bearing problems were detected early. This early detection allowed for a proactive maintenance approach, preventing unexpected downtime and maintaining high operational reliability.

Figure 6 summarises the outcomes of these campaigns. Pump efficiency was calculated by averaging flow and pressure data collected remotely over the campaign period and plotting them on the pump curve. Over time, the lifespan of the pumps has been extended from 1200 to 3135 hrs, ensuring more reliable operation. Compared to the previous pump, operating costs have been reduced by up to 32 per cent. Pump efficiency has increased from 49 per cent to 69 per cent, while water utilisation efficiency has improved from 70 per cent to 83 per cent. These improvements illustrate the tangible value of a data-driven, proactive maintenance strategy in demanding environments.

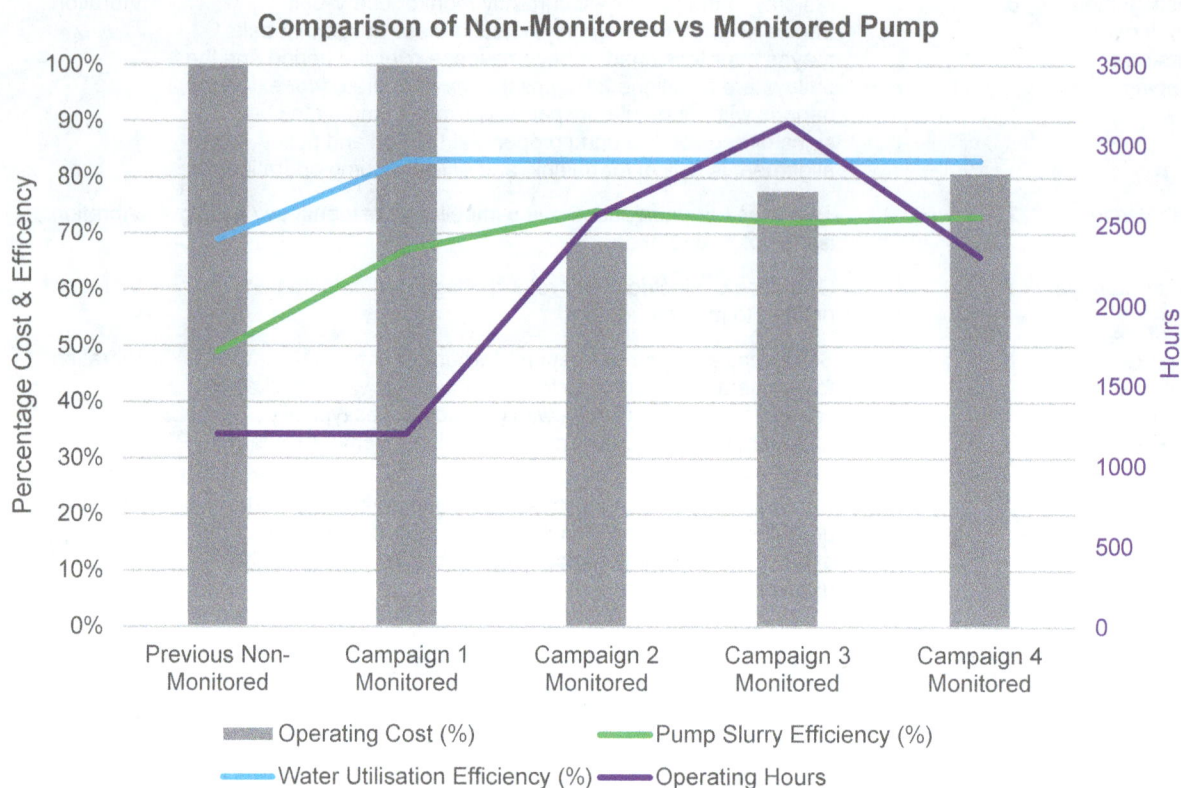

FIG 6 – Comparison of non-monitored versus monitored pump.

Remote monitoring of pump fleet at a global iron and steel facility

In this case study, a global iron and steel manufacturer implemented PCMS for several of their existing pumps on-site. These pumps operate in high-temperature applications and are critical to the process operation. Within 48 hrs of installation and monitoring, numerous optimisations opportunities were identified, ranging from simple to complex issues.

The initial findings included misalignment issues with pulleys, belts being either loose or over-tensioned, instances of cavitation, and vibrations transmitted through suction pipes to the pumps. These issues are detailed in Table 1. Following the initial 48-hour period, a baseline operational mode was established, and a comprehensive plan for corrective actions was developed.

The corrective actions included adjusting belt tensions on all monitored pumps, using laser alignment to ensure proper alignment within acceptable tolerances, raising the level in the tank, and updating the programming logic to prevent cavitation. Additionally, certain operational conditions were modified to ensure more reliable pump performance. These measures resulted in significantly improved pump stability, efficiency, and safety, leading to enhanced savings and operational efficiency for the plant.

Challenges and learnings

During the implementation of PCMS, several challenges were encountered that highlight the importance of both technical and human factors in the success of digital, data-driven maintenance approaches. One of the primary technical issues was maintaining stable connectivity, especially in remote mining locations. These areas often face limited telecommunications infrastructure, making reliable internet access a challenge. In some cases, such as in Chile, specific local requirements necessitated the use of Narrowband IoT (NB-IoT) cellular providers. Where cellular connectivity was not viable, alternative solutions like Starlink satellite internet were explored to ensure consistent data transmission.

Another issue involved physical damage or loss of sensors. During routine pump maintenance, sensors were sometimes accidentally damaged by tools, misplaced, or removed without being reinstalled, often due to limited training or awareness among on-site personnel. For instance, when pumps were sent off-site for repairs, sensors were removed during disassembly but occasionally forgotten during reassembly. Figure 7 provides examples of sensors that are damaged or lost during maintenance activities.

FIG 7 – Examples of damaged or lost condition monitoring sensors.

Infrastructure reliability also plays a vital role in sustaining effective PCMS. The stability, speed, and completeness of data transmission are critical to maintaining a consistent monitoring program. There were cases where gateways failed intermittently, often requiring firmware updates or replacements to restore full functionality and access the latest features.

Equally important is the mindset and engagement of key stakeholders. Effective collaboration, open communication, and a clear feedback loop are essential for translating system insights into meaningful action. While anomaly detection is a major step forward, its impact is only realised when prompt corrective actions are taken, and the outcome is communicated back to close the loop. Without this feedback, the potential value of the monitoring system is left unrealised.

Nonetheless, these hurdles are largely intermittent and manageable when compared to the significant value that the service can deliver. The insights gained through continuous, remote monitoring empower mining and industrial operations to shift from reactive to proactive maintenance, reducing unplanned downtime, enhancing safety, and optimising overall pump performance.

CONCLUSION AND FUTURE WORK

Slurry pumps are vital to mining and mineral processing operations for hydraulic transport of solids through harsh and demanding environments. These conditions subject pumps to intense stress, often leading to frequent failures that disrupt production and incur high financial costs. However, the rise of digital technologies presents a powerful opportunity to mitigate these challenges and improve overall asset performance.

This paper examined the remote condition monitoring of slurry pumps using PCMS, which leverages advanced digital tools, wireless sensors, cloud platforms, and expert analysis to monitor pump health and performance. PCMS was introduced as an effective entry point into the broader digitalisation journey for pumping systems. Its benefits were highlighted through case studies at copper, iron, and steel operations, demonstrating measurable improvements in reliability, early fault detection, and operational efficiency.

Alongside the benefits, the paper identified commonly observed pump faults, associated alarms, corrective actions, and key learnings from monitoring 46 pumps over a year. While challenges such as sensor damage, connectivity limitations, infrastructure reliability, and stakeholder engagement were encountered, they are largely manageable and far outweighed by the long-term value of the service. Addressing these challenges requires targeted training, stronger maintenance practices, and structured change management to ensure smooth integration and sustained collaboration across operational teams.

Looking ahead, future development focuses on scaling PCMS across larger pump fleets, minimising false positives, and automating the diagnosis of recurring faults. These improvements will enhance the efficiency of expert monitoring teams and unlock the full potential of data-driven maintenance.

Ultimately, the adoption of PCMS and similar technologies marks a fundamental shift from reactive to predictive and prescriptive maintenance practices. It positions mining operations not only to improve pump reliability and performance but also to achieve greater sustainability, cost control, and resilience in an increasingly demanding environment.

ACKNOWLEDGEMENTS

The authors gratefully acknowledge the valuable support and collaboration from numerous individuals and organisations who contributed to this research. Special thanks are extended to Metso and University of Western Australia for their ongoing partnership and guidance. We also express our appreciation to the Pumps Business Line and Digital Solutions team at Metso for providing access to critical data and technical insights. Lastly, we sincerely thank the global pumps sales and service teams at Metso for their field support and contributions throughout the study.

REFERENCES

Barnewold, L and Lottermoser, B G, 2020. Identification of digital technologies and digitalisation trends in the mining industry, *International Journal of Mining Science and Technology*, 30(6):747–757. https://doi.org/10.1016/j.ijmst.2020.07.003

Detweiler, W H, 2011. Common causes and cures for roller bearing overheating, SKF USA Inc. Available from: <http://www.maintenanceresources.com/ReferenceLibrary/Bearings/Common.htm>

Karki, B R, Basnet, S, Xiang, J, Montoya, J and Porras, J, 2022. Digital maintenance and the functional blocks for sustainable asset maintenance service – A case study, *Digital Business*, 2(2):100025. https://doi.org/10.1016/j.digbus.2022.100025

Morris, G and Gutierrez, D, 2024. Understanding and mitigating cavitation in pumps. Available from: <https://www.pumpsandsystems.com/cavitation-101> [Accessed: 24/03/2025].

Varghese, A, Martins, S, Lessing, E, Hassan, G M and Karrech, A, 2023. Slurry Pump Safety Considerations in Mineral Processing Plants, MetPlant 2023. Adelaide, Australia.

Incorporating materials supply in strategic mine planning

D K Walker[1], E Y Baafi[2] and S Kiridena[3]

1. Principal Underground Mine Planning, BMA Technical Services, Brisbane Qld 4000.
   Email: david.walker5@bhp.com
2. Honorary Professor, University of Wollongong, Wollongong NSW 2500.
   Email: ebaafi@uow.edu.au
3. Senior Lecturer, University of Wollongong, Wollongong NSW 2500.
   Email: skiriden@uow.edu.au

ABSTRACT

As an underground coalmine continues to expand away from its surface infrastructure bases, the criticality of good supply chain management increases. The simplest way to slacken the logistical supply chain and to lessen strain on the supply system is to move the surface infrastructure closer to the productive faces.

There is currently no mechanism for underground coalmine strategic mine planners to identify potential future logistical supply chain bottlenecks within a mine plan and address them proactively. By identifying any logistics constraints as early as possible, the best opportunity to rectify the problem at the least expense is realised.

Circumstances will prevent sufficient relief to logistics strain by locating portals closer to productive activities as the mine expands over time. Logistics strain will continue to increase with the mine expansion until there is a point where the productive activities are heavily influenced mainly by bottlenecks and the rate of production begins to deteriorate due to either materials supply not being able to keep up with the desired rate of expansion or personnel not being able to arrive quickly enough to operate the machinery at its utilisation rate.

The paper discusses an approach to strategically identify logistics bottlenecks and the impacts that coalmine planning parameters might have on these at any point in time throughout a life-of-a mine plan. The developed system uses a suite of unique algorithms that are designed to 'bolt onto' existing mine plans with the XPAC mine scheduling software package developed by RPM Global (<https://rpmglobal.com/product/xpac/>). The developed system identifies at a strategic level the number of material delivery loads required to maintain planned productivity for a mining operation. It also strategically identifies logistics bottlenecks and the impacts that mine planning parameters might have on these at any point in time throughout the life-of-mine plan.

INTRODUCTION

Underground coal mining logistics is typically regarded as all activities undertaken to support production (Hanslovan and Visovsky, 1984). Underground coal mining logistics for the purposes of this paper is limited to the areas between the portal access of the mine and the production faces and includes:

- The transportation of the workforce to the production districts or other areas where support of such production is required.

- The transportation of materials to all necessary areas either using purpose-built machines, load-haul-dump (LHD) units and trailer configurations or flat top rail networks. This typically includes:

  o conveyor componentry, including structure, belts, rollers and idlers, and drive-head construction materials

  o primary and secondary support material

  o longwall equipment between change-outs

  o longwall consumables for the duration of extraction within a block

  o ventilation control device construction materials

- o production support machinery
- o pipework for transport of water
- o electrical reticulation and cabling and supporting infrastructure such as switch rooms, transformers, and district circuit breakers.

There is currently no mechanism for mine management and strategic planners to identify potential future logistical supply chain bottlenecks within a mine plan and address them proactively (Walker *et al*, 2020). A mine plan is the combination of the physical design, machine paths or sequence, and productivities for a particular mining operation. The general assumption is that any mine plan can be supported logistically at a strategic level and that the tactical level can deal with any unforeseen challenges on a day-to-day basis. The impact of this assumption will depend on the size of the operation itself. If a logistics bottleneck is missed because the planning does not consider it, the larger mine will be first to encroach the logistics breaking strain threshold. By only tactically identifying and logistics breaking strain the mine may delay the inevitable productivity decrease for a short period of time but will likely create a new long-term shortfall in productive capacity compared to the original mine plan or an expensive recovery to reactively address the bottleneck created by logistics.

Consider a large underground coalmine. Consumables/supplies and personnel enter and exit through an adit, drift or through a shaft. Over time the active mine working will migrate away from these portals. With an adit through a sub-crop or outcrop the mine will always move away from the portal entrance. A centralised drift or shaft may migrate away from the shaft only to return closer to the shaft again after several years to commence a new domain in a new direction.

Centralised shaft or drift

As can be seen in Figure 1, operations with a shaft or cross measure drift operations can be seen to migrate away from the surface access portals as domains EP1, NP1 and WP1 migrate away from the East Portal, North Portal, and West Portal, respectively.

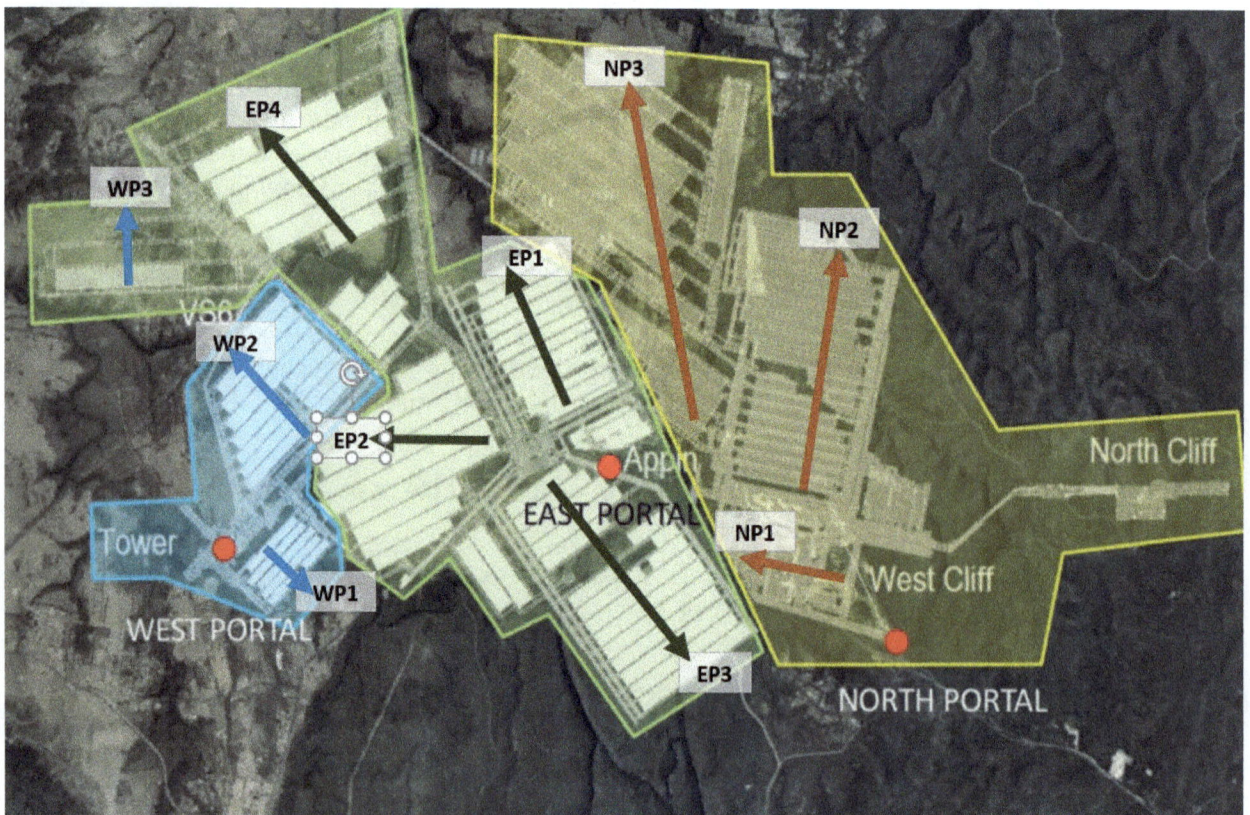

FIG 1 – Shaft and cross measure drift portals (in red) for a large coal mining complex Appin Mine highlighting the migration away and return to the portals for active operations (underlay map source: Young (2017)).

At the completion of these domains, the logistics maintaining this supply chain of personnel and consumables are at their farthest point, then relaxation can occur of the supply chain if the next active workings commence closer to the shaft or drift portal as per EP2 from EP1, EP3 from EP2, and NP3 from NP2. Therefore, in a shaft or drift portal context if there were no other limitations to extraction then the active mine workings would radiate out somewhat like Figure 2.

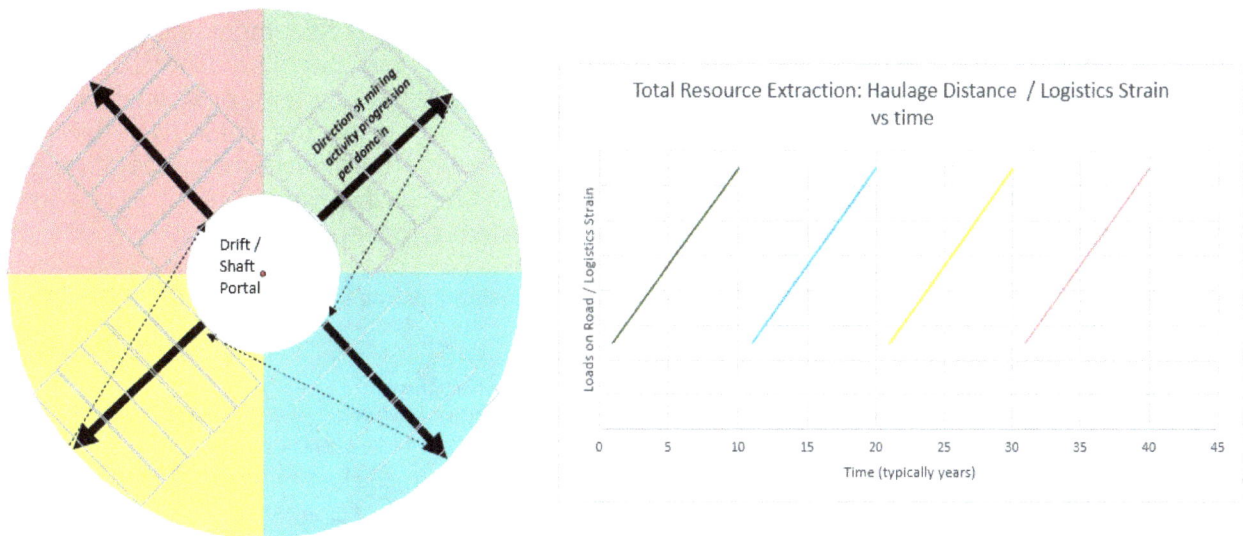

FIG 2 – The relationship between domain changes for a centralised shaft or portal for production activity progression and haulage distance/logistics strain.

Figure 2 is a simplified example of an unrestricted resource with homogenous quality and conditions. What can be seen is four mining domains. Each domain has progressive production panels (eg longwall panels) radiating away from the central shaft or drift. As the longwalls continue to progress outwards, longwall tonnes per development metre decrease until the operation will reach a crossover point beyond the panel operating cost minima which is where the cost of continuing to radiate out in that domain is more expensive than developing the next domain much closer to the shaft/drift. This is the trigger point to progress to the next domain and so on.

Figure 2 forms the basis of any strategic mine plan when analysing the priority of extraction between domains. Mining conditions are not unrestricted nor are conditions and quality in coal ever homogenous. This relationship remains valid between a centralised shaft/drift and the fluctuation logistics strain of the operation. It is also important to point out that the logistics strain of the operation is not dissimilar to the highest ventilation load of an underground which is never the last longwall panel in a domain. Rather is the most inbye location of a production district where all mining activities are still wholly within that domain. After this time development of a new domain commences where at least productive activity is closer to the shaft/drift and therefore reducing total logistics strain.

Logistics breaking strain

Circumstances may therefore prevent relief to logistics strain by sinking shafts or drifts closer to productive activities as a mine expands. Logistics strain will continue to increase with the expansion until there is a point where the productive activities bottleneck transfers to logistics and the rate of production begins to deteriorate due to either materials supply not being able to keep up or personnel not being able to arrive quickly enough to operate the machinery at the utilisation required. For the purposes of this paper such an event is known as the logistics breaking strain.

Logistics breaking strain can occur intermittently. In certain circumstances the number of supplies and the number of personnel travelling to and from the productive area may need to increase due to geotechnical changes requiring more supplies and specialist personnel needing to be delivered, then once through this phase logistics strain returns to normal. Consider logistics strain over a timeline as shown in Figure 3. Multiple unique events may occur at any stage which accelerate the mine towards logistics breaking strain. For example, more supplies coupled with high productivity and falling machine reliability.

More units, more problems

The natural way to fix a fall in productivity particularly in development is to increase the amount of development units. But if the operations do not know that logistics breaking strain has been encountered, significant diseconomies of scale will likely occur. More development units will require more personnel in transit, more supplies to differing areas of the mine and more distractions of the operational oversight to manage to the point where a reduction in units increases or at least matches productivity. If the operation is blind to logistics breaking strain being the reason for declining productivity, then the operation will eventually reach a point of untenability by driving operating and sustaining capital costs up to address phantom bottlenecks, when all that may be required is a 'hard reset' of the operation and a rebuild until productivity losses are experienced.

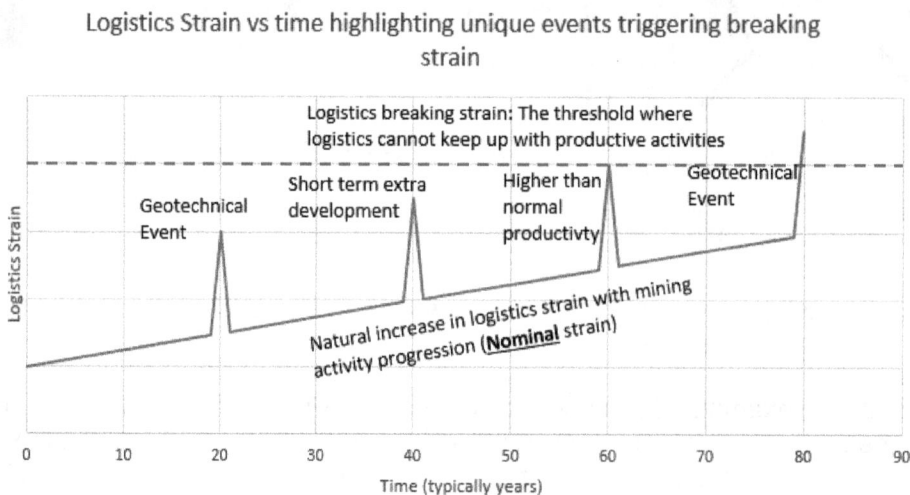

FIG 3 – A simplistic example of unique events increasing logistics strain for a small period before returning to normal.

LOGISTICS SIMULATION SYSTEM

The proposed logistics simulation system should be able to quickly and easily 'bolt on' to existing life-of-mine schedules and not affect these in any way. There are three computer systems which were used to develop the logistics simulation system. The analysis starts with ProgeSoft's ProgeCAD where a mine design is drafted up and mine reserves and physicals are calculated. These are input to Runge Pincock Minarco (RPM) XPAC software where these physicals are combined with machine productivities to generate a schedule and a mine plan. This is traditionally where the mine plan ends, and the financial analysis would typically begin. Iterations between the mine design, the schedule and financial analysis to optimise value would be undertaken as shown in Figure 4.

However, the planning cycle must consider the supply chain and therefore design and schedule the mine to alleviate/defer logistics strain for as long as possible. A discrete simulation software, FlexSim is introduced within this part of the planning process for two reasons. Firstly, it will be a used to study the effect of the schedule on the supply chain, introduced as a pass/fail check and to transfer between XPAC and the Financial Analysis to ensure the eventual value proposition is more realistic as shown in Figure 5. Secondly as FlexSim provides discrete event simulation, the materials flow is studied for peak events at a 'microscopic level' to determine if a solution can be found for relief of logistics strain that can be transferred back up to the macroscopic (XPAC) level which could increase the threshold of logistics breaking strain. In Figure 5, ProgeCAD is the mine design software where all mine geometry is constructed and then reserves are run ready to export into XPAC scheduling software. This includes mains headings, gate roads, longwalls and/or pillar extraction panels. Essentially the software demarcates between development (first workings) and longwall or pillar extraction (second workings).

Current Mine Planning Cycle for Value Determination

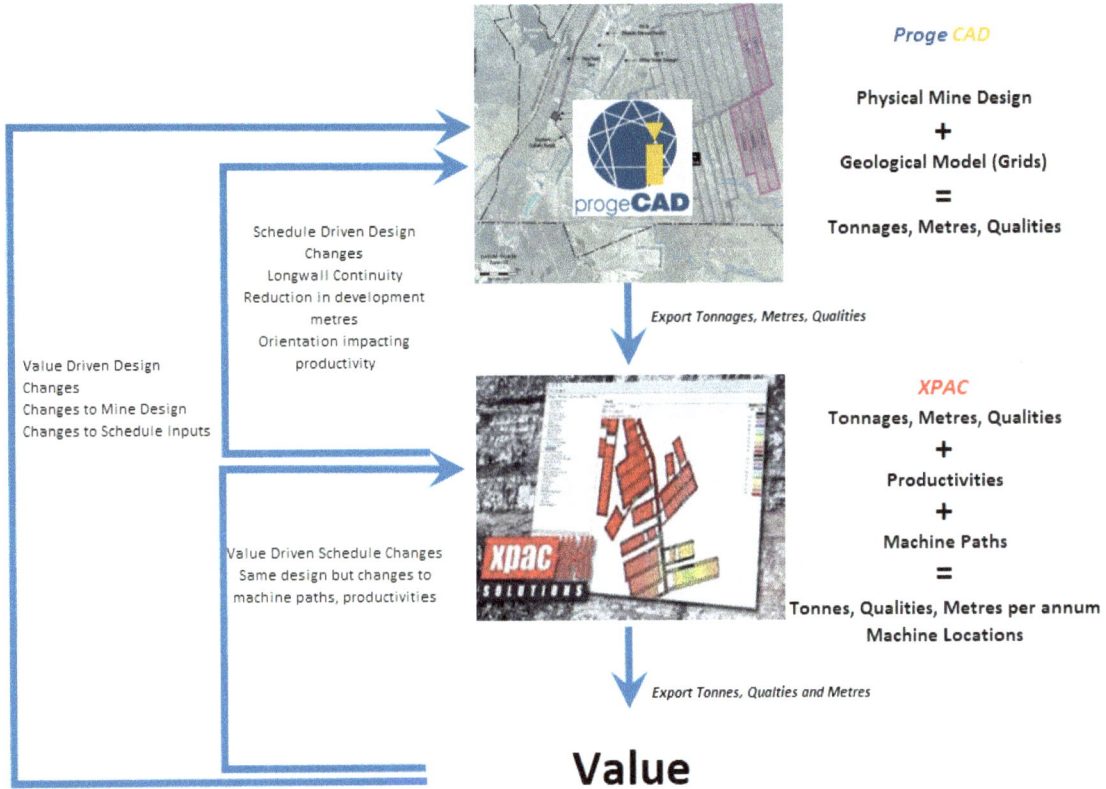

FIG 4 – The traditional planning cycle and software used for value determination.

Mine Planning Cycle for Value Determination proposed in this research

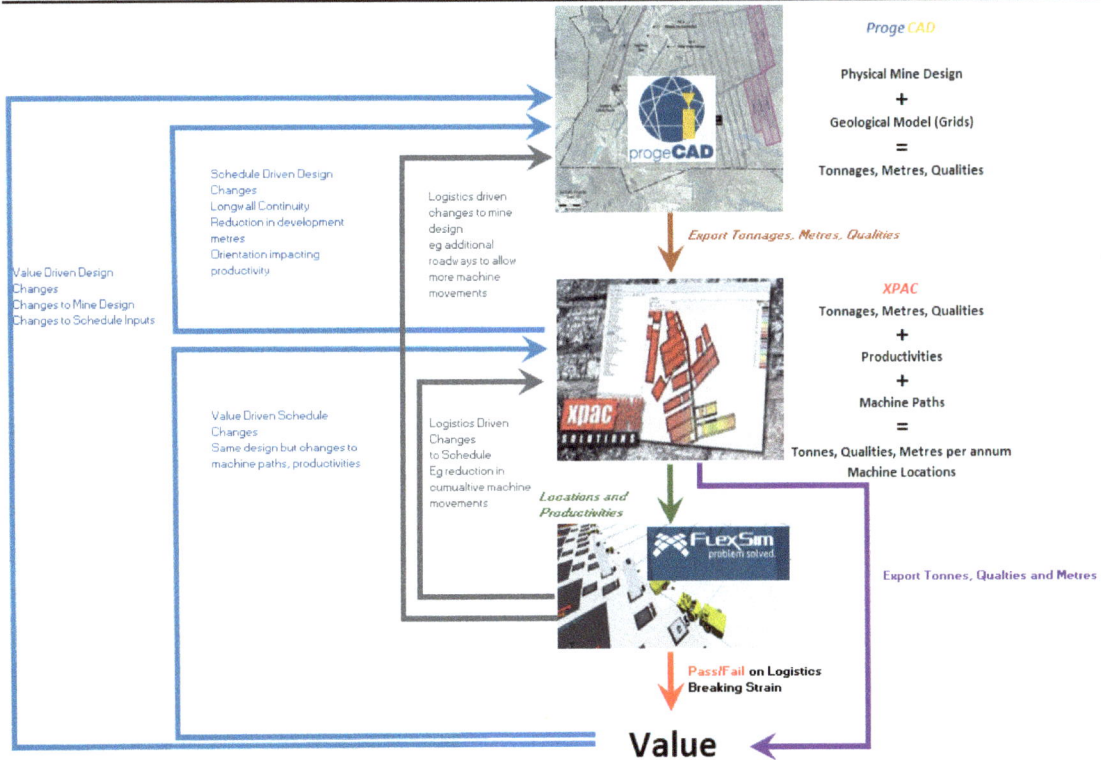

FIG 5 – The insertion of FlexSim between schedule output and financial evaluation.

Strategic logistics model

The developed strategic logistic model uses a suite of unique algorithms, designed to bolt onto 'Runge Pincock Minarco XPAC scheduling software. The entire logistics strategic identification model is depicted in Figure 6.

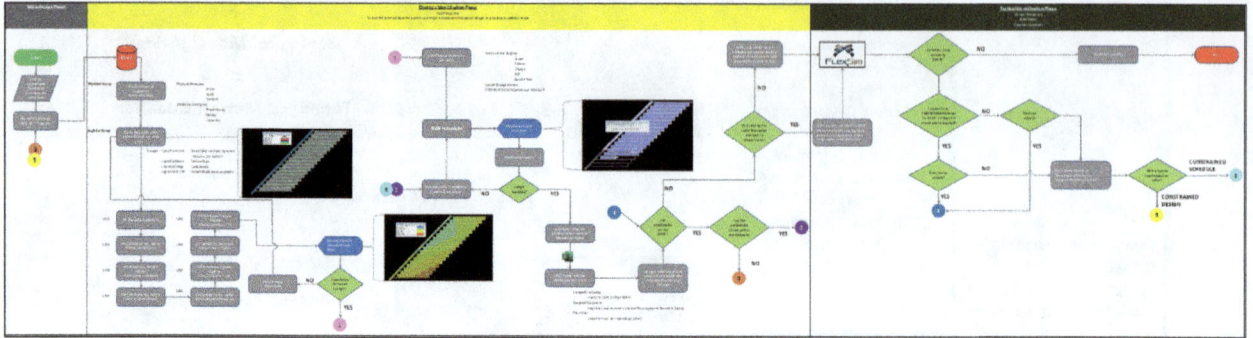

FIG 6 – Flow chart of logistics bottleneck identification and rectification.

The model:

- Estimates logistics strain due to migration away from and return to a portal.

- Estimates the logistics strain due to migration away from a portal in a general direction.

- Explains the unique events that contribute to logistics strain.

- Predicts the mechanisms for how logistics strain reaches breaking strain.

- Explains the impact on logistics strain by adding more delivery LHDs to the system (more units – more problems).

Typical model outcome is depicted in Figure 7 showing the 12-week average loads on road is just below six loads. This implies if loads on road for 12 weeks around the peak requirement day is maintained and is less than the logistic breaking strain, there will be sufficient materials delivered and stockpiled to meet production.

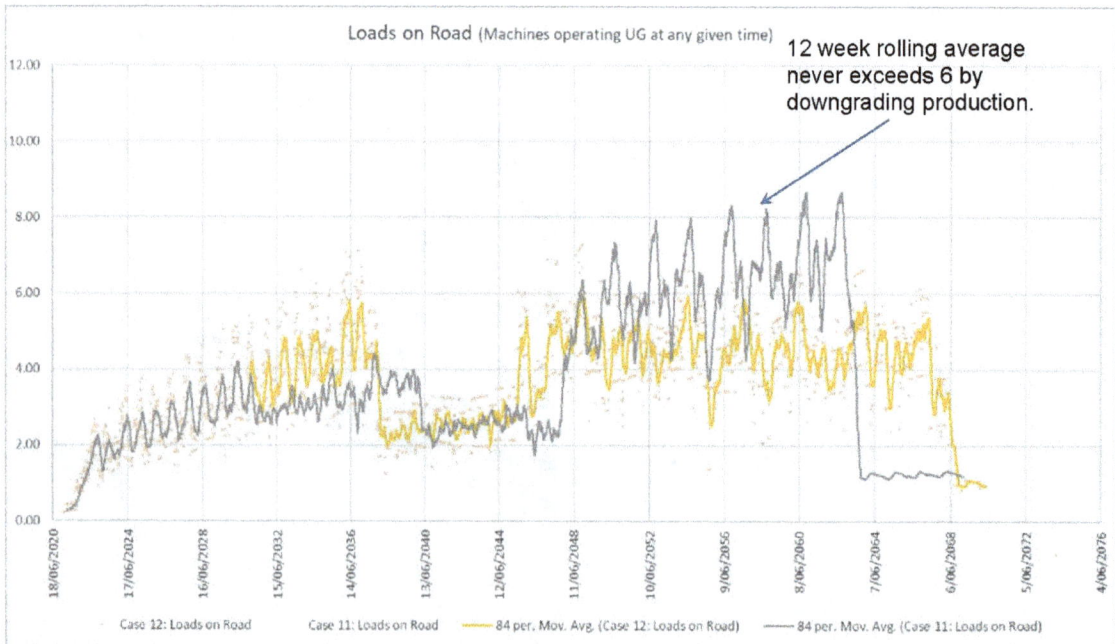

FIG 7 – Loads on road – never exceeds six loads on road for the 12-week rolling average.

CONCLUDING REMARKS

Some mine sites within Australia experience significant logistics delay, particularly mines within the NSW Southern Coalfields. Older mines will sink new shafts and drifts closer to workings. However, there is no formal system to identify when exactly a mine should sink a shaft or drifts or how to financially justify such an expense.

A strategic system has been developed to identify early logistics bottlenecks and therefore give the mine planner the best opportunity to proactively address unacceptable logistics strain. This system allows speedy sensitivity analysis to either remove or defer logistics breaking strain. In the case that neither deferral or removal of logistics breaking strain can occur, the system is able to identify the gap and schedules are run that are downgraded to reflect the future reality. This allows the planner and strategic decision-makers to understand the revised value of the mine plan that will closely reflect reality.

REFERENCES

Hanslovan, J J and Visovsky, R G 1984. *Logistics of Underground Coal Mining*, Energy Technology Review series, issue 91 (Noyes Publications).

Walker, D, Harvey, C, Baafi, E, Porter, I and Kiridena, S, 2020. Benchmarking Study of Underground Coal Mine Logistics, ACARP Report C28021.

Young, W, 2017. Complexity in integration high voltage and ventilation systems, NSW Environment and Planning.

Separation of fine-grained rutile and zircon by flotation – a study of sensitive flotation chemistry with considerations for operational circuits

G Wren[1] and G Senanayake[2]

1. MAusIMM(CP), Senior Metallurgist Project Development, Talison Lithium, Greenbushes WA 6254. Email: george.wren@talisonlithium.com
2. Associate Professor in Extractive Metallurgy, Murdoch University, Murdoch WA 6150. Email: g.senanayake@murdoch.edu.au

ABSTRACT

Flotation is extensively used to concentrate various sulfide ores including copper, lead, zinc, gold and nickel. It is not as common for oxide minerals; however, it is used for a range of commodities including rare earths, lithium, and tin. Less common are applications in mineral sands, which are usually processed with physical techniques. Some mineral sand deposits are challenging for the conventional approach and alternative techniques including flotation can provide an advantage. One example is the separation of rutile and zircon at Sierra Rutile with poor performance reported from attempts to use electrostatics on fine grained particles, below 150 µm. To improve the separation, a flotation system was developed, incorporating an amine collector, starch depressant and fluoride activator.

This zircon flotation system has been considered for Australian fine-grained deposits such as the WIM150 resource in Victoria, however little discussion of the underlying flotation chemistry is available. In this work, experimental results are presented to demonstrate that the system relies on sensitive surface and solution chemistry. The analysis includes fitting adsorption isotherm models using two nonlinear least square regression methods: orthogonal distance and a generalised reduced gradient algorithm (Microsoft Excel Solver add in – GRG Nonlinear). Curves prepared with both methods are compared, and the physical interpretation discussed. This computational approach provided a significant contribution to characterising the sensitive collector adsorption process that involves the formation of multilayers that for zircon, can be extensive.

This work highlights the benefit oxide flotation can provide and acknowledges the challenge of operating a complex flotation circuit. It demonstrates that successful separation relies on good control over a sensitive chemical process. Some practical operational considerations are also provided, including the requirement for good water quality, automated reagent preparation and appropriate strategies to monitor reagents in flotation process streams. These considerations are relevant to the flotation of many different oxide minerals and are likely to assist operators to develop strategies for good consistent flotation performance.

INTRODUCTION

Zircon is a commodity with strong industrial demand for use as an opacifier and colouring agent for tiles and ceramics, zirconium chemicals, refractories and others (Elsner, 2013). Review of United States Geological Survey (USGS) reports (2016, 2019, 2021, 2022, 2024) shows zircon reserves have reduced from 78 Mt to 39 Mt (Table 1) and Australian producers have assessed the potential of developing alternative resources, such as the WIM deposits in the Murray Basin region of Victoria (Wren, 2024). Table 2 indicates the size of nine WIM style deposits: over nine Bt of ore containing 381 Mt of heavy mineral (HM). The HM has elevated zircon content, around 15 per cent, suggesting a significant resource.

TABLE 1

Australian and global Zirconium reserves 2015 to 2023.

Location	Zirconium reserves (kt)			
	2015	2018	2020	2023
Australia	51 000	42 000	42 000	55 000
Australia JORC compliant	NR	NR	13 000	20 000*
World total	77 700	73 000	63 555	74 000
World total (Australia – JORC)	77 700	73 000	34 555	39 000

*For Australia, JORC compliant reserves were 20 million tonnes.
Some values not reported shown as NR (USGS 2016, 2019, 2021 and 2024).

TABLE 2

Some fine-grained WIM style deposits with estimates for heavy mineral (HM) and zircon.

Deposit	Ore tonnes (millions)	HM grade (%)	HM tonnes (millions)
WIM 50	710	3.5	24.8
WIM 100	162	3.0	4.9
WIM 150	452	2.41	10.9
WIM 200	430	7.2	31.0
WIM 250	1270	5.3	67.3
Avon Bank	705	5.4	38.1
Donald	4780	3.7	177
Goschen	28	2.6	0.7
Central Goschen	520	5.0	26.0
Total	**9057**	**4.2**	**381**

(Sources: Olshina and Van Kann, 2012; Astron Resources, nd).

The WIM deposits were defined by the corporate body Conzinc Riotinto of Australia Exploration, or CRAE (Iluka, 2024) but were not successfully developed as 'extraction and milling of the ore has proven too costly and difficult due to the fine-grained nature of EL3330' (Olshina and Van Kann, 2012; Astron Resources, nd). The conventional process used to generate heavy mineral concentrates uses gravity, magnetic and electrostatic techniques to produce mineral products for the market, or downstream processing. Electrostatic separators have poor efficiency when treating fine grained particles, evident in the treatment of process streams at the Sierra Rutile mine (Davies, Keila and Wonday, 1994). The operation attempted to separate zircon/rutile concentrates finer than 100 mesh (150 µm) using electrostatics with little success. This was attributed to the distribution of heavy minerals, which was below the size range where the conventional electrostatic technique works efficiently. The issue was addressed by developing a flotation system that used dodecyl amine acetate (DDA), starch and fluoride to separate the minerals into zircon and rutile products. By using DDA flotation, fine grained rutile was recovered as a marketable concentrate and the potential for zircon production demonstrated.

More recent work applied the system to a sample from the fine-grained WIM100 deposit. It achieved a reasonable separation, however tight control of test procedures was required to demonstrate repeatability (Wren, 2024). This indicates a sensitive system, as subtle changes to solution and surface chemistry influence the separation which introduces a significant risk, or problem for an industrial scale application. These interactions are important and depend on reagents, conditions and minerals, but are poorly understood with little information available.

The work presented in this paper was completed to advance the understanding of the sensitive surface and solution chemistry that underlies the separation of zircon and rutile by DDA flotation. It compares adsorption isotherm models with experimental results, identifying the most suitable equation for a given system and considering the physical interpretation.

RELATED WORK

Separation of zircon and rutile by flotation

Before examining the adsorption behaviour of DDA on pure mineral samples, it is worth considering a related application that has been used in industry (Sierra Rutile) to separate fine grained rutile and zircon by flotation. Figure 1 presents results as kinetic curves from two flotation tests completed on a sample from WIM100 (Vic, Australia) at different pH targets: 3.0 and 7.0 (Wren, 2024). All other conditions were constant. At pH 3.0 zircon, indicated by Zr, is fast floating achieving a high recovery that exceeded 90 per cent. Rutile and ilmenite have slow kinetics with Ti recovery around 3.0 per cent. These conditions permit good separation of zircon from the titanium minerals rutile and ilmenite.

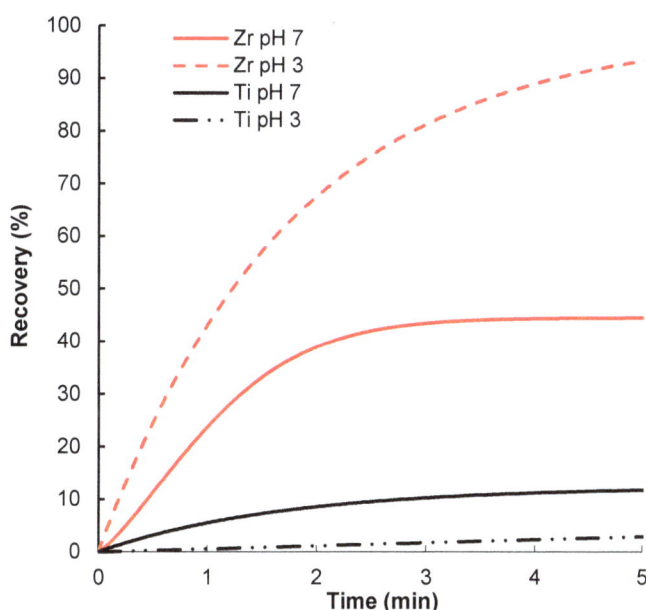

FIG 1 – Comparison of batch flotation test results at pH 7.0 and pH 3.0 for zircon (Zr) and the titanium minerals ilmenite and rutile (Ti). Reagent additions: 900 g t^{-1} NaF, 2000 g t^{-1} of corn starch, and 125 g t^{-1} of Armac C at 20°C (from Wren, 2024).

Increasing pH to 7 resulted in poor selectivity: titanium and iron recovery increased (>10.0 per cent), and recovery of zircon reduced to around 40 per cent. These results clearly highlight a sensitive system that requires careful pH control for consistent performance. Other variables such as temperature, reagent concentrations, conditioning procedure, pulp density and conditioning intensity have also been examined using batch flotation tests (Wren, 2024). The results show that careful control of all parameters is important for consistent, efficient reagent adsorption and flotation performance.

Adsorption models

Selective mineral flotation relies on the adsorption of surfactants at specific minerals, creating hydrophobic surfaces that can attach to an air bubble. For sensitive flotation systems such as the separation of zircon and rutile by DDA, starch and NaF, a good understanding of the underlying solution and surface chemistry will assist operators to achieve consistent and efficient flotation performance.

One relevant consideration is the adsorption process, there are many well-known adsorption isotherms that are used to model different systems, and the selection of an appropriate equation is critical for characterising adsorption systems. This work compares experimental isotherms with three different models: Langmuir, Brunauer, Emmett and Teller (BET), and Langmuir Freundlich. All are common models used to describe and characterise adsorption systems.

Langmuir

The Langmuir model assumes adsorption forms a monolayer only, adsorption sites are all homogenous, adsorption energy is constant, and no lateral interactions are present between adsorbed molecules. The model is described by Equation 1, where q_e is the adsorbed amount at equilibrium, S_T the maximum adsorption capacity, K_L is the Langmuir constant and C_e the adsorbates equilibrium concentration (Kalam et al, 2021):

$$q_e = S_T \frac{K_L\, C_e}{1 + K_L\, C_e} \tag{1}$$

BET

The BET model is the basis for the standard used to estimate particle surface area. It assumes multilayer adsorption occurs due to Van der Waal's forces and applies to type II isotherms (disperse, nonporous or macro porous solids) and type IV isotherms (mesoporous solids with pore diameter between 2 nm and 50 nm). It assumes inaccessible pores are not detected and has been modified and applied to describe systems with a liquid adsorbate to estimate the monolayer adsorption capacity and liquid phase saturation concentration (Beltran, Pignatello and Teixido, 2016; Brunauer, Emmett and Teller, 1938; Ebadi, Jafar and Anvar, 2009; ISO 9277, 2010). Equation 2 provides one form of the BET equation, S_M is BET maximum monolayer adsorption capacity, K_{BET} the BET constant, C_S and C_e the BET adsorbate monolayer saturation concentration and equilibrium concentration, respectively (Beltran, Pignatello and Teixido, 2016):

$$q_e = \frac{S_M K_{BET} C_e}{(C_S - C_e)\left(1 + (K_{BET} - 1)\frac{C_e}{C_S}\right)} \tag{2}$$

Langmuir Freundlich

The Freundlich model is used to describe adsorption on heterogenous surfaces with varied activities and applies to multilayer adsorption. It is described by Equation 3, where K_f is the Freundlich constant and N_F the Freundlich heterogenous parameter. The value of N indicates favourability of adsorption, if $1 < n < 10$ the surface was heterogenous, and adsorption easily occurred:

$$q_e = K_f\, C_e^{\frac{1}{N_F}} \tag{3}$$

A combination of the Langmuir and Freundlich models is in Equation 4. Known as the Langmuir Freundlich isotherm, this model describes adsorption on a heterogenous surface. Adsorption is dominated by the Langmuir model for low concentrations and the Freundlich for higher concentrations. The parameters are the Langmuir-Freundlich maximum adsorption capacity (q_{MLF}) the heterogenous parameter (MLF) with value >0, <1 and K_{LF} is the Langmuir Freundlich equilibrium constant:

$$q_e = \frac{q_{MLF}(K_{LF}C_e)^{MLF}}{1 + (K_{LF}C_e)^{MLF}} \tag{4}$$

Tools for fitting adsorption models to experimental data

Various methods are available to fit adsorption models to experimental data. These include plotting equations in the linear form, linear regression analysis and nonlinear techniques. Beltran, Pignatello and Teixido (2016) discussed the limitations of ordinary least squares regression, outlining that the approach is inadequate for systems with three or more parameters and can lead to misleading results. To achieve a reasonable model fit, nonlinear least squares (NLLS) regression can be used. NLLS methods aim to minimise a function (U), with an iterative process to define the model parameters. U is the sum of squared residuals, or the difference between the experimental adsorption ($q_{exp,i}$) and calculated adsorption ($q_{calc,i}$):

$$U = \sum_{i=1}^{n} w_i \left(q_{exp,i} - q_{calc,i} \right)^2 \tag{5}$$

In this work, two different NLLS methods were used to define model parameters. The first used the generalised reduced gradient method, a standard generalised reduced gradient algorithm from the Microsoft Excel Solver add in (GRG Nonlinear). The second method used a tool developed by Beltran, Pignatello and Teixido (2016) 'ISOT_calc' and parameter optimisation macro 'Ref_GN_LM'. This approach will minimise the residuals in both the aqueous (C_e) and solid phases (q), a technique known as orthogonal distance regression:

$$U_d = \sum_{i=1}^{n} \left\{ \left(\frac{\Delta C_{e,i}}{C_{e,exp,i}} \right)^2 + \left(\frac{\Delta q_i}{q_{exp,i}} \right)^2 \right\} \tag{6}$$

The tool provides ten common predefined adsorption models and has provision for a user defined model. It was tested using three different systems with a good fit achieved in all examples.

EXPERIMENTAL

Materials

High purity samples of zircon and rutile were prepared from a WIM 100 ore sample, a detailed procedure has been provided in other work (Wren, 2024). Milliq water was used in all experiments. AR grade dodecyl amine acetate (DDA) was supplied by Sigma Aldrich/Merck.

The 4-chloro-7-nitrobenzofurazan (chloro-NBD), suitable for fluorescence, AR grade methanol and AR grade sodium bicarbonate were sourced from Sigma Aldrich/Merck for use in derivatisation of DDA solution samples prior to analysis with HPLC. The HPLC mobile phase consisted of two parts: water with AR grade acetic acid and pH adjusted to 4.5 using AR grade sodium hydroxide, and acetonitrile suitable for HPLC.

Methods

Adsorption experiments

Experimental adsorption isotherms were prepared based on the procedure described by Choi and Whang (1963). All experiments were completed with 20 mL of solution added to a 50 mL conical flask. A magnetic stirrer was added to the flask along with 2.0 g of zircon or rutile sample. Sufficient agitation was applied to ensure that the mineral samples were suspended for the duration of the test. Adsorption experiments were consistently agitated for two mins and then allowed to equilibrate for 2 hrs before filtration and analysis. The selection of two hrs equilibration time was made following initial experiments that demonstrated equilibrium at one hr.

Determination of residual DDA

Residual dodecylamine was measured using a similar approach to the procedure described by Hao et al (2004). The method used for preparation of standards is provided in the appendix. A similar procedure was used for filtered experimental samples and a description of the method follows.

A 1 mL aliquot of DDA solution (standard or filtered experimental sample) was mixed with 1 mL of methanol and 2 mL of Chloro-NBD/sodium carbonate solution before derivatisation at 70°C for 1 hr. This reaction was completed in a water bath using submerged 4 mL vials. After the reaction was complete particles were removed with a 0.45 µm Teflon syringe filter. The reactant was found to interact with the filter media. To avoid experimental error, each filter was primed with 1 mL of sample and set aside for 30 seconds before being flushed with another 1 mL of sample. After flushing, the syringe was set aside for a further 30 seconds before the final 1.5 mL of sample was delivered to the vial for HPLC analysis. To ensure good precision, standards were prepared for each batch of experimental samples.

Analysis was completed using HPLC fitted with a photo diode array detector (PDA). Mobile phase was an 80:20 (v/v) mixture of acetonitrile and water. The water was 40 mM acetic acid with pH adjusted to 4.5 using dilute NaOH. A Zorbax Eclipse XDB C18 150 mm × 4.6 mm, 5 µm column was used with 100 µL injection and mobile phase at 2.0 mL min^{-1}. Using this procedure, the dodecylamine acetate which had been labelled by derivatisation with Chloro-NBD was detected after 5.1 to 5.2 mins of retention time at 470 nm. All samples were analysed for 7 mins to ensure peaks were found.

DISCUSSION

DDA adsorption on zircon at pH 3

Experimental isotherms for DDA adsorption on zircon were prepared and compared with the Langmuir isotherm. This model, described in Equation 1, applies to systems that have monolayer adsorption on homogenous sites with constant adsorption energy and no lateral interactions. Both methods, GRG nonlinear and orthogonal distance (OR) were used to fit the Langmuir model to the experimental results. The isotherms are provided graphically in Figure 2 and equation parameters and measure of fit are in Table 3.

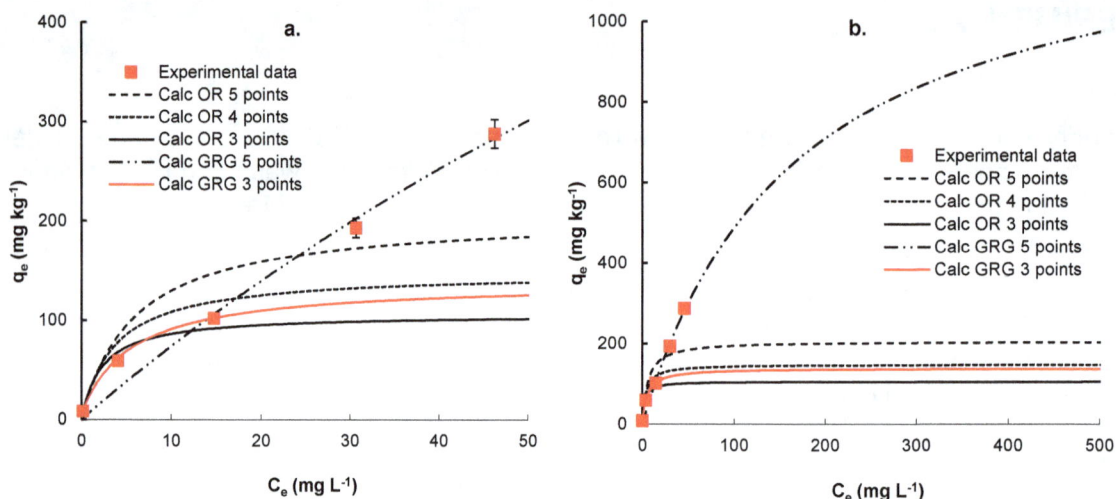

FIG 2 – Adsorption isotherm for DDA at low (a) and high (b) equilibrium concentrations, on zircon at pH 3.0 in milli-q water showing experimental and calculated values for the Langmuir isotherm with parameters defined by two methods: orthogonal regression (OR) and GRG nonlinear (excel solver). Different isotherms are shown for calculated values with parameters defined by considering experimental results for the three lowest initial DDA concentrations, the four lowest initial DDA concentrations, and all data points. Error bars are included for the experimental results in (a) (from Wren, 2024).

TABLE 3

Langmuir equation parameters for DDA on zircon at pH 3.0 and measure of fit for five iterations of isotherms prepared by defining parameters for different concentration ranges and regression methods (from Wren, 2024).

Initial DDA concentration of experimental data points used to define model parameters (mg L^{-1})	Regression method	Langmuir parameters[a]		Measure of fit		
		S_T (mg kg^{-1})	K_L (L mg^{-1})	MWSE	RMSE	R^2
1.00, 10.0, 25.0	GRG	139	0.19	N/A	N/A	0.99
1.00, 10.0, 25.0	OR	106	0.43	0.01	0.20	N/A
1.00, 10.0, 25.0, 50.0	OR	148	0.27	0.06	0.51	N/A
1.00, 10.0, 25.0, 50.0, 75.0	OR	206	0.17	0.11	0.73	N/A
1.00, 10.0, 25.0, 50.0, 75.0	GRG	1.30×10^3	6.0×10^{-3}	N/A	N/A	0.97

With the GRG nonlinear method, a poor fit was achieved when all five experimental data points were included. Reducing the number of experimental data points to consider only the three lowest initial concentrations (C_I) of DDA (C_I: 1 mg L^{-1}, 10 mg L^{-1} and 25 mg L^{-1}) significantly improved the result with the correlation coefficient (R^2) increasing from 0.97 to 0.99. Using orthogonal distance regression to minimise the sum of squares showed a similar trend, with poor fit using all five data points and an improved result when the three experimental results for the lowest initial DDA concentration were considered. This confirms the trend identified using the GRG nonlinear method and suggests the Langmuir model only applies when the initial DDA concentration is around or below 25 mg L^{-1}. It is likely that at higher initial concentrations multilayer formation occurs, therefore the experimental system no longer displays Langmuir adsorption character. The Langmuir model contains a parameter for maximum monolayer adsorption capacity (S_T), and for the best fitting example (three lowest concentrations only, GRG nonlinear method) S_T = 139 mg kg^{-1}. When adsorption exceeds this limit multilayer formation and lateral interactions are expected to occur.

Analysis of the Langmuir results suggested that DDA adsorption on zircon initially forms a monolayer however as DDA concentration increases, multilayer formation and lateral interactions become significant. The BET model (Equation 2) has been used to characterise multilayer formation for gas and liquid adsorbate, providing information about monolayer adsorption capacity and multilayer formation. Both methods were used to fit the BET model to the experimental isotherm (DDA adsorption at pH 3) with a better result using the GRG method. This is seen as a chart (Figure 3) with additional data provided in Table 4 around measure of fit and model parameters. The parameter S_M is the BET monolayer adsorption capacity and with the GRG approach, the value is 130 mg kg^{-1}, very close to the estimate for maximum adsorption capacity using the Langmuir model (139 mg kg^{-1}, data points for three lowest initial concentrations NLLS regression by GRG).

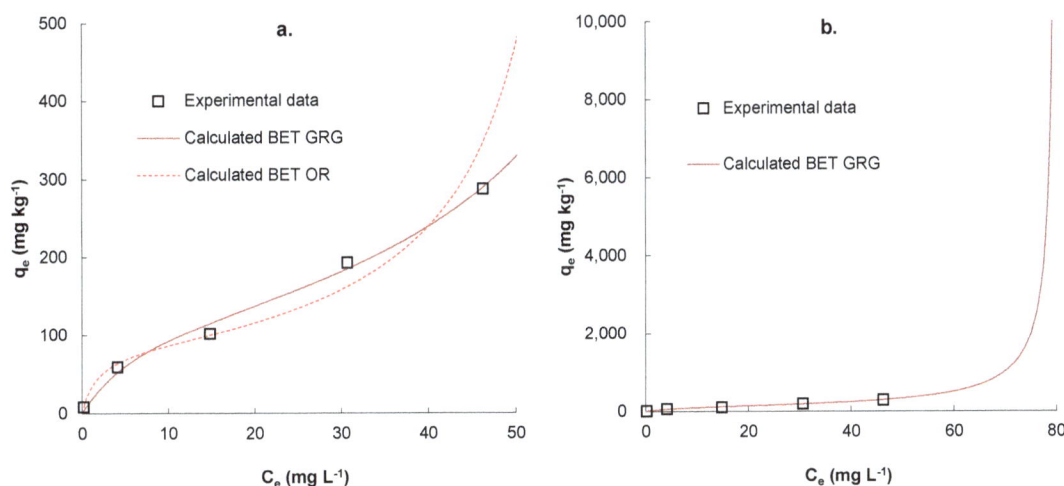

FIG 3 – Adsorption isotherm for DDA at low (a) and high (b) equilibrium concentrations, on zircon at pH 3.0 in milli-q water showing experimental values and calculated values for the BET model with parameters defined using orthogonal regression and GRG (from Wren, 2024).

TABLE 4

Comparison of fit and model parameters for BET model (DDA adsorption on zircon at pH 3.0).

NLLS regression method	Measure of fit	Model parameters
Generalised reduced gradient	R^2 = 0.99[a]	S_M = 130 mg kg^{-1} K_{BET} = 11.3 L mg^{-1} C_S = 80.2 mg L^{-1}
Orthogonal distance regression	MWSE = 5.0 × 10^{-3} RMSE = 0.17	S_M = 81.7 mg kg^{-1} K_{BET} = 36.1 L mg^{-1} C_S = 60.4 mg L^{-1}

R^2: correlation coefficient. MWSE: mean weighted square error. RMSE: root mean square error.

DDA adsorption on zircon at pH 7

Additional experiments were completed to characterise DDA adsorption on zircon at neutral pH. A photograph is provided in Figure 4 which shows six flasks containing zircon and amine solution at different initial concentrations (C_I) after the procedure was terminated (C_I: 1.00 mg L^{-1}, 10.0 mg L^{-1}, 25.0 mg L^{-1}, 50.0 mg L^{-1}, 75.0 mg L^{-1} and 100 mg L^{-1}). The initial DDA concentration was increased from left to right and the photo clearly shows increasing initial DDA concentration corresponds to an increase in adsorption density. The lowest concentration at 1.00 mg L^{-1} (flask 7) has no flocculation and a small amount of mineral agglomerating on the side of the flask and floating on the surface of the solution. As the concentration of DDA is increased more flotation is observed and agglomerates form with flocculation from flask 10 (CI 50.0 mg L^{-1} DDA), a strong indication of multilayer adsorption.

FIG 4 – Photograph of flasks containing DDA solution and zircon after agitating and equilibrating at pH 7.0. Initial DDA concentration increases from left to right: 1.00 mg L^{-1} (flask 7), 10.0 mg L^{-1} (flask 8), 25.0 mg L^{-1} (flask 9), 50.0 mg L^{-1} (flask 10), 75.0 mg L^{-1} (flask 11) and 100 mg L^{-1} (flask 12) (from Wren, 2024).

As described for DDA adsorption on zircon at pH 3, nonlinear least squares regression was used to define parameters for the Langmuir and BET models with the experimental isotherms obtained for DDA adsorption on zircon at pH 7. The results are provided in Figure 5 (Langmuir), Figure 6 (BET) and Table 5 (measure of fit and model parameters). As for pH 3, DDA adsorption on zircon at a neutral pH only fits the Langmuir model at lower concentrations. Figure 5 shows a good fit when only the three lowest initial DDA concentrations are considered, with the Langmuir model estimating monolayer adsorption (S_T) at 203 mg kg^{-1} (Table 5).

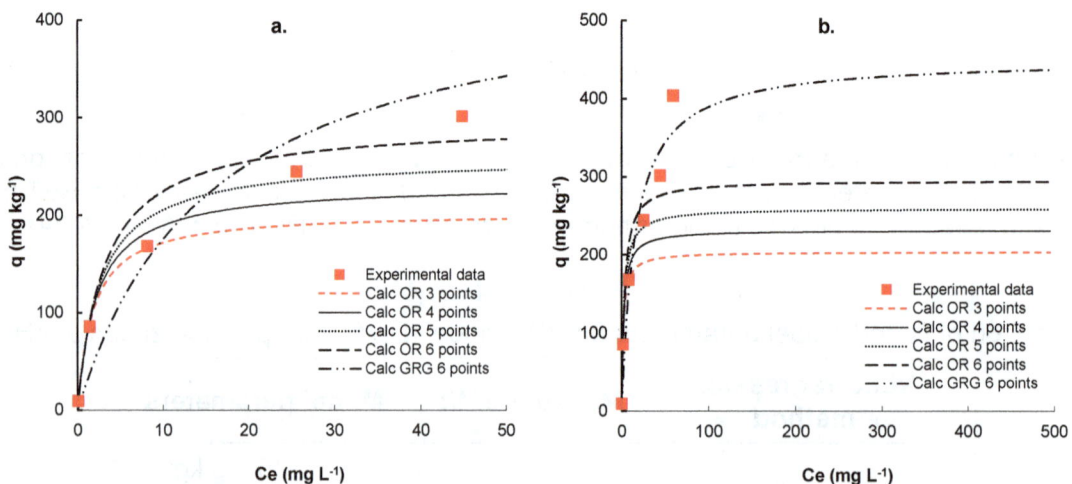

FIG 5 – Adsorption isotherm for DDA at low (a) and high (b) equilibrium concentrations on zircon at pH 7.0 in milli-q water showing experimental and calculated values for the Langmuir isotherm with parameters defined by two methods: orthogonal regression (OR) and GRG nonlinear (excel solver). Different isotherms are shown for calculated values with parameters defined by considering experimental results for the three lowest initial DDA concentrations, the four lowest initial DDA concentrations, the five lowest initial DDA concentrations and all data points (from Wren, 2024).

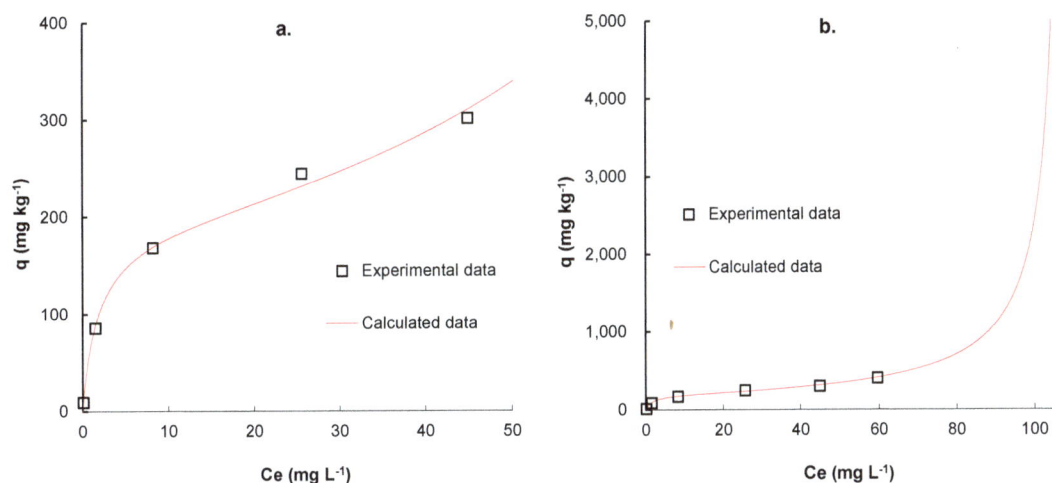

FIG 6 – Adsorption isotherm for DDA at low (a) and high (b) equilibrium concentrations, on zircon at pH 7.0 in milli-q water showing experimental values and calculated values for the BET model with parameters defined using orthogonal regression (from Wren, 2024).

TABLE 5

Comparison of fit and model parameters for the Langmuir and BET model (DDA on zircon at pH 7.0) (from Wren, 2024).

Model	NLLS regression method	C_i of data points included in regression (mg L⁻¹)	Measure of fit	Model parameters
Langmuir	Orthogonal distance	1.00, 10.0, 25.0	MWSE[a] = 2.0×10^{-3} RMSE[b] = 7.3×10^{-2}	$S_T = 203$ mg kg⁻¹ $K_L = 0.54$ L mg⁻¹
	Orthogonal distance	1.00, 10.0, 25.0, 50	MWSE = 1.4×10^{-2} RMSE = 0.24	$S_T = 232$ mg kg⁻¹ $K_L = 0.45$ L mg⁻¹
	Orthogonal distance	1.00, 10.0, 25.0, 50.0, 75	MWSE = 2.6×10^{-2} RMSE = 0.25	$S_T = 260$ mg kg⁻¹ $K_L = 0.38$ L mg⁻¹
	Orthogonal distance	1.00, 10.0, 25.0, 50.0, 75.0, 100	MWSE = 4.9×10^{-2} RMSE = 0.31	$S_T = 296$ mg kg⁻¹ $K_L = 0.32$ L mg⁻¹
	Generalised reduced	1.00, 10.0, 25.0, 50.0, 75.0, 100	$R^2 = 0.87$[c]	$S_T = 451$ mg kg⁻¹ $K_L = 6.4 \times 10^{-2}$ L mg⁻¹
BET	Orthogonal distance	1.00, 10.0, 25.0, 50.0, 75.0, 100	MWSE = 1.0×10^{-3} RMSE = 5.5×10^{-2}	$S_M = 185$ mg kg⁻¹ $K_{BET} = 64.9$ L mg⁻¹ $C_S = 108$ mg L⁻¹

MWSE: mean weighted square error. RMSE: root mean square error. R^2: correlation coefficient.

The BET equation achieved a very good fit to the experimental data (Figure 6), predicting monolayer adsorption at 185 mg kg⁻¹, in good agreement with the Langmuir model (203 mg kg⁻¹). The BET model also predicts extensive multilayer formation, with adsorption continually increasing with equilibrium DDA concentration. This is consistent with the experimental observations, including the photograph in Figure 4.

DDA adsorption on rutile

The Langmuir Freundlich equation (Equation 4) was found to provide a good fit to the experimental isotherm prepared for the adsorption of DDA on rutile. This model combines the Langmuir equation to describe monolayer adsorption at low concentrations and at higher concentrations, the Freundlich equation to predict multilayer adsorption. The results are provided for adsorption of DDA on rutile at pH 7 in Figure 7 with experimental results compared to the Langmuir Freundlich isotherm (LF) and Langmuir isotherm. Both models provide a good fit, however the mean weighted square error (MWSE) and root mean square error (RMSE) show that LF provides a better description of the adsorption system than the Langmuir model alone.

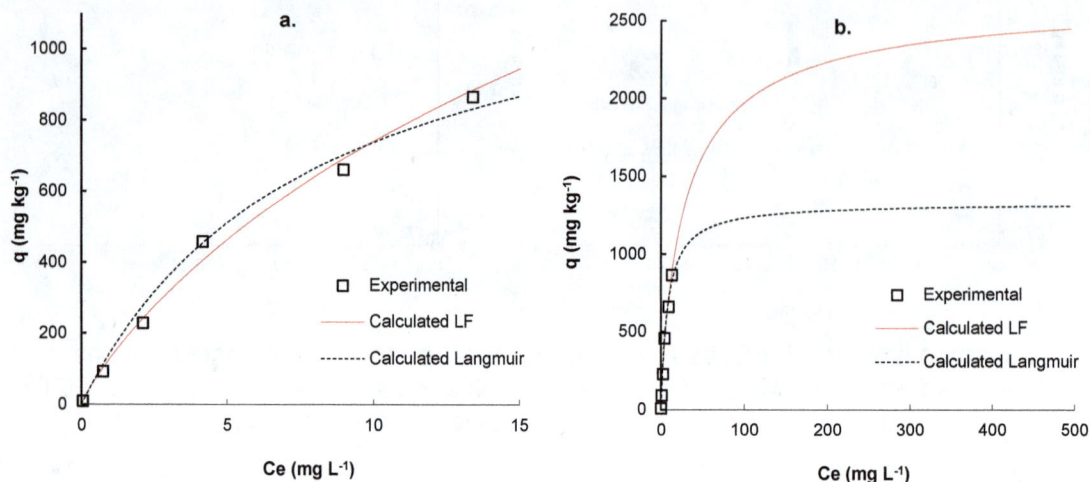

FIG 7 – Adsorption isotherm for DDA at low (a) and high (b) equilibrium concentrations, on rutile at pH 7.0 in milli-q water showing experimental values and calculated values for the Langmuir Freundlich (LF) isotherm with parameters defined using orthogonal regression. The Langmuir isotherm is provided for comparison (from Wren, 2024).

At pH 3.0, DDA adsorption on rutile was best described by the LF isotherm, provided in Figure 8, with model parameters and measure of fit in Table 6. As for DDA on rutile at pH 7.0, the LF model provides a good fit to all experimental points and predicts multilayer formation with an eventual saturation point, or maximum adsorption.

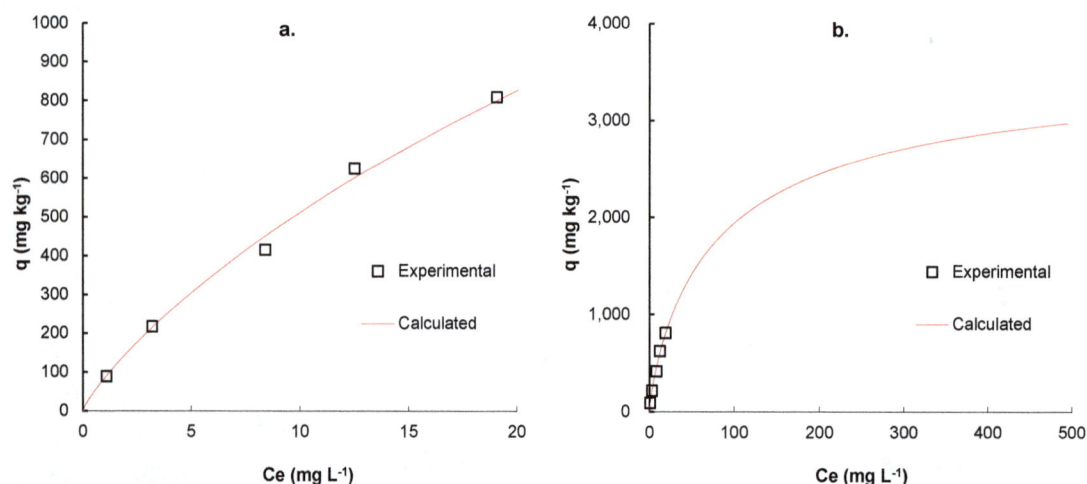

FIG 8 – Adsorption isotherm for DDA at low (a) and high (b) equilibrium concentrations, on rutile at pH 3.0 in milli-q water showing experimental values and calculated values for the Langmuir Freundlich isotherm with parameters defined using orthogonal distance regression (from Wren, 2024).

TABLE 6

Comparison of fit and model parameters for the adsorption of DDA on rutile and pH 3 and 7. Values for two models are presented, Langmuir Freundlich and Langmuir (from Wren, 2024).

Model	Adsorption pH	MWSE[a]	RMSE[b]	Model parameters
Langmuir Freundlich	7	8.0×10^{-3}	0.15	$S_T = 2.66 \times 10^3$ mg kg^{-1} $K_LF = 0.03$ $MLF = 0.87$
	3	1.9×10^{-3}	9.7×10^{-2}	$S_T = 3.65 \times 10^3$ mg kg^{-1} $K_LF = 0.01$ $MLF = 0.85$
Langmuir	7	2.5×10^{-2}	0.22	$S_T = 1.33 \times 10^3$ mg kg^{-1} $K_L = 0.12$ L mg^{-1}
	3	5.5×10^{-3}	0.12	$S_T = 1.61 \times 10^3$ mg kg^{-1} $K_L = 0.05$ L mg^{-1}

MWSE: mean weighted square error. RMSE: root mean square error.

CONCLUSION AND CONSIDERATIONS FOR COMMERCIAL APPLICATION

This work has contributed to characterising a sensitive system, demonstrating DDA adsorption on zircon forms extensive multilayers, which increase when the pH is raised from 3 to 7. For this system, a good description is provided by the BET model, and at low DDA concentrations, the Langmuir model. The Langmuir model fails to provide a good fit when adsorption exceeds the monolayer saturation capacity, estimated at 139 mg kg^{-1} (Langmuir) and 130 mg kg^{-1} (BET). The BET model also predicts the formation of extensive DDA multilayers on zircon as more adsorbate is added to the system.

The Langmuir Freundlich model provided the best description for DDA adsorption on rutile, predicting multilayer formation with an eventual saturation, or maximum adsorption. Comparing DDA adsorption on rutile with zircon shows two systems capable of forming multilayers, however these are more extensive for zircon.

Two nonlinear least square regression tools were used to fit adsorption models to experimental isotherms: the generalised reduced gradient method provided in the solver add in for Microsoft Excel and an orthogonal distance method developed by Beltran, Pignatello and Teixido (2016). Both methods found reasonable values for the model parameters however for zircon at pH 3.0 the GRG method provided a better fit.

The work considered three parameters related to the DDA flotation system: mineral type (zircon or rutile), DDA concentration and pH. For successful separation in a laboratory or operational environment many other parameters need to be considered, including the presence of other minerals, the relative quantity of different minerals, NaF concentration, amylose (starch) concentration and preparation, water quality, pulp density, temperature, conditioning intensity and time. All these parameters will interact in a flotation system and for consistent good performance, tight control is required.

This includes: (i) water quality, a consistent supply of quality water will be an asset; (ii) reagent concentrations and preparation of starch, this should follow an automated procedure. Changes in mineral assemblage are likely to affect performance, requiring tight control of blending and proactive testing to inform operators of required process changes. Automated systems that use X-Ray fluorescence to monitor process stream character are available for flotation plants and should be considered a requirement for operations. A good system can be configured to adjust reagent concentrations based on real time changes in the character of feed and product streams. In addition, routine sampling and analysis to monitor flotation reagent concentrations will assist operators to make informed process decisions, mitigating variability to help achieve consistent performance.

These aspects are critical for good operational results with an oxide flotation circuit, such as zircon/rutile separation with DDA, and should be considered good practice. At the project development stage, environmental and economic implications related to the DDA flotation system should also be considered. This includes assessment of project value (flotation based compared with the conventional physical approach), reagent recycling and waste management.

REFERENCES

Astron Resources, nd. The Donald Mineral Sand Project, Astron Resources. Available from: <http://www.astronlimited.com.au/projects-operations/donald-mineral-sands.aspx> [Accessed: 26/7/17].

Beltran, J, Pignatello, J and Teixido, M, 2016. ISOT_Calc: A versatile tool for parameter estimation in sorption isotherms, *Computers and Geosciences*, 94:11–17.

Brunauer, S, Emmett, P and Teller, E, 1938. Adsorption of gases in multimolecular layers, *Journal of the American Chemical Society*, 60(2):308–319.

Choi, H S and Whang, K U, 1963. Surface properties and floatability of zircon, *Transactions of the Canadian Mining and Metallurgical Society (CIM Bulletin)*, 56:466–468.

Davies, J P, Keila, A K and Wonday, S, 1994. Development and operation of zircon flotation at Sierra Rutile Limited, in *Proceedings of 10th Industrial Minerals International Congress*, pp 160–172.

Ebadi, A, Jafar, S and Anvar, K, 2009. What is the correct form of BE isotherm for modeling liquid phase adsorption, *Adsorption*, 15:65–73.

Elsner, H, 2013. Zircon – Insufficient supply in the future?, DERA Rohstoffinformationen, The German Mineral Resources Agency (Deutsche Rohstoffagentur; DERA), Federal Institute for Geosciences and Natural Resources (Bundesanstalt für Geowissenschaften und Rohstoffe; BGR). Available from: <https://www.zircon-association.org/assets/files/KnowledgeBank/rohstoffinformationen-14.pdf> [Accessed: 25 March 2021].

Hao, F, Lwin, T, Bruckard, W and Woodcock, J, 2004. Determination of aliphatic amines in mineral flotation liquors and reagents by high-performance liquid chromatography after derivatization with 4-chloro-7-nitrobenzofurazan, *Journal of Chromatography A*, 1055:77–85.

Iluka Resources, 2024. WIM100 Mineral Resource Estimate Update. Available from: <https://www.iluka.com/media/t5nctvdr/wim100-mineral-resource-estimate-update.pdf> [Accessed: 22 July 2025].

International Organization for Standardization (ISO), 2010. ISO 9277.2010 – Determination of the specific surface area of solids by gas adsorption – BET method ISO 9277:2010(E), International Organization for Standardization (ISO).

Kalam, S, Abu-Khamsin, S, Kamal, M and Shirish, P, 2021. Surfactant adsorption isotherms: a review, *ACS omega*, 6(48):32342–32348.

Olshina, A and Van Kann, M, 2012. GSV TR2012/1 – Heavy Mineral Sands in the Murray Basin of Victoria, Geological Survey of Victoria Technical Record 2012/1.

United States Geological Survey (USGS), 2016. Zirconium and hafnium, Mineral Commodity Summaries 2016, Mineral Commodity Summaries. Available from: <https://apps.usgs.gov/minerals-information-archives/mcs/mcs2016.pdf> [Accessed: 25 March 2021].

United States Geological Survey (USGS), 2019. Zirconium and hafnium, Mineral Commodity Summaries 2019, Mineral Commodity Summaries. Available from: <https://apps.usgs.gov/minerals-information-archives/mcs/mcs2019.pdf> [Accessed: 25 March 2019].

United States Geological Survey (USGS), 2021. Zirconium and hafnium, Mineral Commodity Summaries 2021, Mineral Commodity Summaries. Available from: <https://pubs.usgs.gov/periodicals/mcs2021/mcs2021.pdf> [Accessed: 25 March 2021].

United States Geological Survey (USGS), 2022. Zirconium and hafnium, Mineral Commodity Summaries 2022 – Nitrogen. Available from: <https://pubs.usgs.gov/periodicals/mcs2022/mcs2022.pdf> [Accessed: 27 January 2023].

United States Geological Survey (USGS), 2024. Zirconium and hafnium, Available from: <https://pubs.usgs.gov/periodicals/mcs2024/mcs2024.pdf> [Accessed: 02 May 2024].

Wren, G D, 2024. Separation of fine-grained zircon and rutile: a study of flotation chemistry and kinetics, PhD thesis, Murdoch University.

APPENDIX – METHOD USED FOR A MINE DETERMINATION BY DERIVATISATION AND ANALYSIS WITH HPLC

Stage	Detail	Reagents
1 Prepare calibration solutions	Prepare 25 mL of 1000 mg L^{-1} (0.10% w/v) DDA (0.025 g) in ultra-pure water for calibration standards. Store in 50 mL vial with a cap.	• AR grade dodecylamine acetate (DDA) • Ultrapure water
	Prepare six standards: • 15.0 mg L^{-1} • 12.0 mg L^{-1} solution • 9.00 mg L^{-1} solution n • 6.00 mg L^{-1} solution • 1.00 mg L^{-1} solution • 0.10 mg L^{-1} solution	• DDA calibration solutions • Ultrapure water
2 Prepare chloro-NBD / bicarbonate solution	Prepare chloro-NBD solution: • Dissolve 0.2 g (0.4) of chloro-NBD in 25 mL of methanol, added by pipette to a 50 mL flask • Add 0.84 (1.68) g of $NaHCO_3$ and make up to 50 mL with MilliQ water. Adjusted to pH 3.0 with H_2SO_4 or at neutral pH	• 4-chloro-7-nitrobenzofurazan (chloro-NBD) • Methanol • $NaHCO_3$ • Ultrapure water
3 Prepare solutions for derivatisation	Add 1 mL methanol to 1 mL DDA calibration solution	• DDA calibration solutions • Methanol
	Add 2 mL chloro-NBD solution to methanol/DDA/water solution	• Calibration solutions (methanol/water) • Chloro-NBD bicarbonate solution
4 Derivatisation	Derivatise the amine by heating at 70°C for 60 mins and cooling	• Calibration solutions with Chloro-NBD bicarbonate solution
5 Prepare HPLC eluent	Prepare dilute (5% w/w) sodium hydroxide solution	• Ultrapure water • Sodium hydroxide
	Prepare solution of water buffered in 40 mM acetic acid	• Ultrapure water • Acetic acid
	Adjust pH of 40 mM acetic acid solution to 4.5 with dilute sodium hydroxide solution	• Water buffered with acetic acid • Dilute sodium hydroxide solution
6 Analyse with HPLC	Elute at 2.0 mL min^{-1} in C-18 column (150 mm long, 3.9 mm ID)	• Acetonitrile • Ultrapure water buffered with acetic acid

Digital twins for stockyard reclamation optimisation

S Zhao[1], T F Lu[2], L Statsenko[3], H Assimi[4] and C Garcia[5]

1. Research Fellow, University of Adelaide, Adelaide SA 5005. Email: shi.zhao@adelaide.edu.au
2. Associate Professor, University of Adelaide, Adelaide SA 5005.
   Email: tien-fu.lu@adelaide.edu.au
3. Associate Professor, University of South Australia, Adelaide SA 5000.
   Email: larissa.statsenko@unisa.edu.au
4. Research Fellow, University of Adelaide, Adelaide SA 5005.
   Email: hirad.assimi@adelaide.edu.au
5. System Engineer, Eka CTRM Solutions Pty Ltd, Adelaide SA 5000.
   Email: chris.garcia@eka1.com

ABSTRACT

Run-of-Mine (ROM) stockpiles are commonly used as buffers and quality control units between the mine and the processing plant. Its quality control function is achieved by stacking ores of varying quality selectively into either a new stockpile or onto an existing stockpile. The decision on where to stack the ore follows a rule of thumb, each stockpile contains ore of a similar quality, and the incoming material should not significantly change the overall quality of the stockpile. Reclamation operations blend stacked materials, further reducing quality variation. Thereby, a stockpile is considered to have a uniform quality distribution, and its quality is saved as weighted percentages of the elements of interest in most of current stockpile management systems. This paper identifies the reason why the efficiency and effectiveness of such systems are lower than expected and proposes a new management strategy, named proactive reclaiming. The key idea behind this strategy is to use 3D stockpile models for quality calculations, store the results at corresponding 3D locations, and then utilise such quality embedded models to optimise the reclaiming sequence for multiple loaders. The objective of the optimisation is to achieve the required quantity and quality combinations with smaller margins, while also reducing the reclaiming time.

INTRODUCTION

Ore excavated from the Earth is usually a mixture of deleterious impurities that substantially reduce its market value for three main reasons. First, some impurities, such as cadmium, mercury, lead, and antimony, are detrimental to the environment and to human health when present above natural level. Second, some impurities, such as fluoride and uranium, may introduce significant problems to processing plants and even cause equipment damage. Lastly, they have adverse impacts on the quality of the produced products. For example, bismuth can make copper cathode brittle (Fountain, 2013), while phosphorus can result in a similar embrittling effect on steel (Dub, Dub and Makarycheva, 2006). Due to these factors, copper smelters impose financial penalties for the presence of deleterious impurity elements in copper concentrates (Lane *et al*, 2016). Similar penalties also apply in iron ore export (Pownceby *et al*, 2019). Consequently, producers are keen to control deleterious impurities and maintain a low-grade variability throughout the entire downstream process to ensure that delivered product meet with quality requirements.

ROM stockpiles are the simplest storage facilities for blasted ore between the mine(s) and the primary crushing plant, typically formed by haul truck dumps. When crushing is required, the stockpiled ore is picked up by Front End loaders (FELs) and transported by haul trucks to the crushing plant. Additionally, ROM stockpiles play a crucial role in downstream processing, because stockpile blending is widely recognised in the mining industry as an effective method to adjust the quality of delivered material and reduce variations in quality. Currently, at most ROM stockyards, blending is achieved through selectively stacking ores with different qualities to different locations or stockpiles, allowing materials to mix naturally during reclamation.

Many researchers focus on optimise the stacking operations. Everett (1996) introduced a stress vector to describe the deviations of four quality elements for iron ore and aimed to minimise the increment in the stress vector in each stockpile. A new term 'attenuation', which is the standard deviation of reclaimed materials divided by the standard deviation of the same size stacked materials was added to the stress vector. A small attenuation indicates a good reduction in short-term grade variability (Jupp, Howard and Everett, 2013). Statsenko and Melkoumian (2014) proposed a decision support system for stacking operations. They developed a parametric model for a real stockpile and used vector equations to describe the quality, variances, and quantity. Trouchina and Topal (2020) proposed a linear model to obtain more homogenous iron ore grade for a ROM stockpile through optimised stacking strategies. Xie, Neumann and Neumann (2021), proposed a nonlinear model for large-scale ROM stockpile stacking problem. They used the differential evolution (DE) algorithm with assistance of a fitness function to find optimised stacking sequences. The experimental results indicated that the optimised solutions could make better predication for long-term optimisation problem.

The above-mentioned publications can be categorised as selectively stacking algorithms. There is clear evidence that these algorithms improve stockpile blending. However, integrating these methods into real stockpile management systems remains challenging. The primary limitation is that the most accurate quality assay results of the orebody is often unavailable during the stacking process. Chemical analysis of multiple elements from sampled ore typically lags behind handling (loading, hauling and stacking) operations by approximately 24 to 48 hrs. Consequently, using inaccurate or outdated quality data in these algorithms may lead to inefficient or ineffective stacking sequences.

Proactively reclamation refers to reclaiming materials from different locations or stockpiles to meet the required quantity and quality specifications. It overcomes the above-mentioned limitation because reclaiming operations are less time-critical than stacking operations. There is typically a buffer of 1 to 2 days before reclamation begins after ore is stacked. Some published works also aims to optimise the proactively reclamation process. Lu and Myo (2010) described an algorithm to fulfill quality grade target with minimum reclaimer overall movement. Assimi et al (2022) used greedy and ant colony optimisation algorithms to meet the required quantity and quality combinations with minimum reclaiming time. In both studies, stockpiles are modelled as triangular prisms and then further divided into voxels to represent quality variations within a stockpile. The voxel-based 3D stockpile models enable a more precise representation of ore quality distribution because the quality of mined ore varies over time and location due to natural geological heterogeneity. Thus, allowing for more complex decision-making in the reclamation process. However, such 3D voxel models are not able to be updated continuously. Additionally, in literature, most current stockpile management systems use the weighted average grade (WAG) method to calculate the quality. The quality of a stockpile is determined by the weighted average of the quality of all individual dumps, with an assumption that it is uniformly distributed across the stockpile. This assumption differs from real-world situations and restricts proactive reclamation strategies.

Three-dimensional stockpile models are essential for optimising proactively reclamation. These models could serve as placeholders for the lagging quality assay results if they can be generated continuously. For instance, to create a 3D dump model for each truck dumping event and associate each model with the quality when the assay results become available. If a model is considered as a layer of a stockpile, a 3D stockpile model will consist of multiple layers, each corresponding to different dumping events. Such multi-layered, quality embedded stockpile model enhances spatial quality representation within the stockpile, enabling better-informed reclamation decisions. To create such 3D models, Zhao et al (2013) discussed an algorithm that converts a bucket wheel reclaimer into a mobile scanning device to measure stockpile profiles. This approach was soon surpassed by aerial surveys, which proved to be faster and more cost-effective. Nowadays, drone photogrammetry and airborne-LiDAR systems are widely used for 3D stockpile modelling (Alsayed and Nabawy, 2023). However, most 3D models generated from measurements are primarily used for volume estimation at moment. No published work has been found in literature on integrating these measurement models into stockpile management systems for blending optimisation. One possible reason is that no system can continuously scan stockpiles 24/7, limiting the ability to track real-time shape changes. Another challenge is the large volume of data and additional workload. Building 3D models from measurements is a time-consuming process and still requires human participation at some stages.

To summarise, there is a need to convert current selectively stacking algorithms into proactively reclaiming algorithms for either short-term or long-term blending optimisation. The translation requires a fast 3D stockpile modelling technique capable of recording real-time stockpile shape changes. Additionally, a user-friendly interface is essential to facilitate seamless integration of modelling and optimisation systems into existing stockyard/stockpile management systems. The work presented here is a part of a translation project funded by the Premier's Research and Industry Fund (PRIF) Research Consortia Program. A primary objective of this project is to demonstrate that the stockpile modelling techniques owned by translation partner, Quor Eka and multiple optimisation algorithms developed from another PRIF project, A RP5, can be integrate into an existing stockpile/stockyard management system, thereby improving the stockpile blending efficiency and effectiveness. For this purpose, a new module has been added to the optimisation system from the A RP5 project. The module communicates with the two backup databases provided by our end-user partner, BHP, and a 3D stockyard management system (InSight CM) developed by Quor Eka.

OPTIMISATION SYSTEM

The optimisation system integrates real-world data to build accurate, multi-layered 3D stockpile models, calculates the quality distributions within each 3D stockpile model using quality assay results, and generates optimised proactive reclaiming sequences for FELs. In this project, it is designed as a command-and-control centre that communicates across different systems/platforms. It comprises two key modules, the communication module, which exchange data across different systems/platforms, and the optimisation module, which generates optimised reclaiming sequences using the quality embedded 3D stockpile models. Both modules are programmed in Python. Figure 1 illustrates data and message flow of the optimisation system.

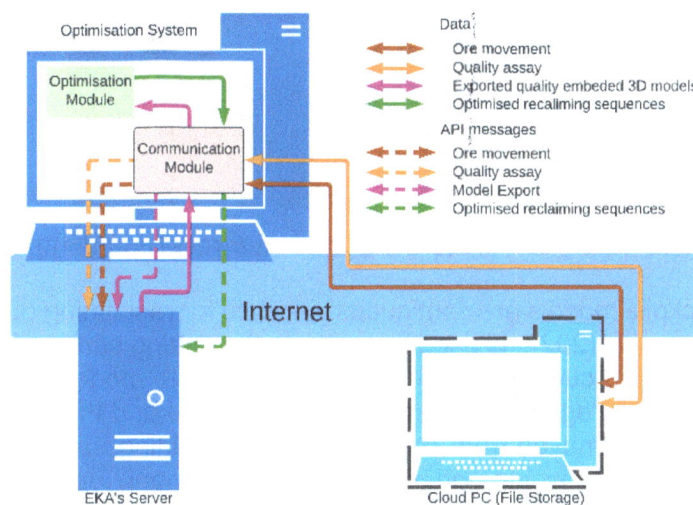

FIG 1 – Optimisation system design.

Generate models using data

To simulate data exchange between the optimisation system and existing stockyard management systems, a secured cloud computer is used as the SQL server. Raw backups from two stockyard management systems, one managing underground stopes to the in-pit stockyard translations and the other handling the in-pit to ROM stockyard translation, are restore into the MS SQL Management Studio. Eka's system, InSite CM, has been modified to import new stockyard, which is derived from real measurement data and vehicle models for environmental visualisation and user interpretation. Additionally, new communication protocols have been defined in InSite CM to receive messages from the communication system. The communication module queries the SQL server to extract event ID, time stamps, vehicle locations, loading/dumping locations, and scheduled stacking/reclaiming events from specific tables in the database. The time interval between each query is fixed at 5 mins to mimic real-time updates in the databases. If an update is detected, the retrieved data will be converted into corresponding API (Application Programming Interface) messages and then sent to InSite CM, which processes the messages and update the 3D models accordingly. Additionally, the communication module extracts the quality assay results (a separated file) on daily basis and send to InSite CM through API messages.

Truck dump messages are used to build and update the 3D stockpile models during stacking operations. The tonnage, truck orientation (heading) at the dump, dumping location (geo-positioning information), the dump height above the ground, together with the material properties, such as density and repose angles, are used to generate a dumping vector for a dump event. After receiving a truck dump message, InSite CM creates a 3D dump model, typically within 3 to 5 seconds. These 3D dump models serve as placeholders for delayed quality assay results. In parallel, the communication module also sends out ore movement messages. These messages enable InSite CM to track the source of each dump, identifying which underground stope the material originated from. Since the samples taken for quality assay are also collected from the same underground stope, it is possible to associate the quality of a 3D dump model with the corresponding assay result. A stockpile contains multiple dumps mined from different stopes. As a result, the quality of each dump varies. Consequently, a stockpile model in InSite CM may consist multiple layers, each with different material quality. Thus, the quality distribution can be analysed and visualised effectively, providing insights into material composition and variability for blending processes. Figure 2 shows such 3D stockpile modes generated by InSite CM.

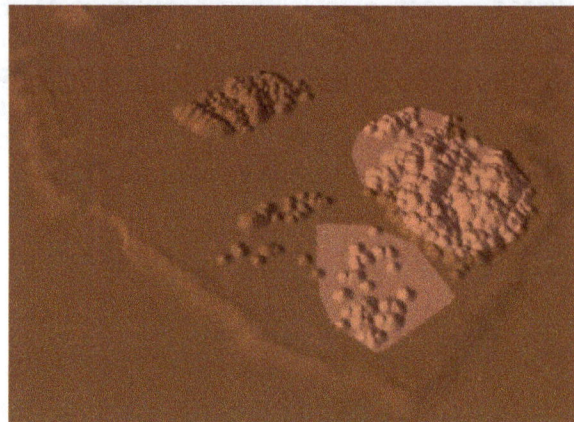

FIG 2 – 3D stockpile modes generated after receiving dump events.

Three-dimensional stockpile models are continuously update for reclaiming operations. A kinematics model proposed by Zhao *et al* (2022) is used to generate the cutting surface for the bucket of a FEL. The intersection between the cutting surface and a 3D stockpile model defines the new stockpile surface after excavation. The region between the cutting surface and the original stockpile model is considered as the material removed from the stockpile. The cutting surface is fully integrated into InSight CM. Ideally, InSight CM would update stockpile models only upon receiving a FEL loading message. However, not all excavation locations for FELs are saved in the database and it is difficult to identify these locations from vehicle trajectory. The current solution to these issues is to trigger the update process for every loader position message received from the communication module. If an intersection is found, the model is updated. Otherwise, no update occurs. The authors are now developing an algorithm to filter out significant errors in the loader orientation data. Once this filter is integrated into the system, the update can be visualised in InSight CM. Figure 3 illustrates such an updating process using MATLAB scripts.

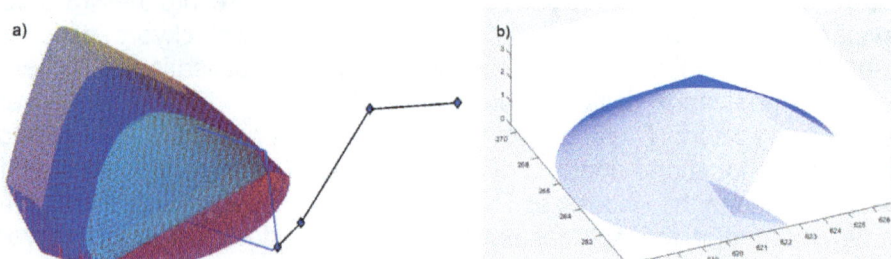

FIG 3 – Update 3D models during reclaiming operations: (a) materials removed by a loader in three continuous excavations; (b) The updated stockpile models. The surface is represented by point clouds for better illustration purpose.

Using model for optimisation

Three-dimensional stockpile models generated from computer simulation are then partitioned into reclaiming steps called voxels. In this study, a voxel is considered as the smallest quantity and quality unit for reclaiming and optimisation. For instance, a voxel is a sickle-shaped cut as shown in Figure 3a. The volume of each voxel is calculated using double integral. If a voxel contains materials from different layer, the volume from each layer is calculated separately. The material quality is determined based on the volume and the quality metrics of each layer. Both 3D stockpile models, voxels and their quality are then exported as separated files for optimisation.

The optimisation module developed in A RP5 has been evaluated using a synthesis data set. The results obtained from the deterministic and randomised greedy algorithms demonstrated promising improvements in achieving the required the quantity and quality combinations with optimised reclaiming time (Assimi et al, 2021). The source code was updated to for the optimisation module. After the update, it was further validated using the real stockpile measurement data before integrating into the optimisation system. Four Comma-Separated Values (CSV) files generated from the measurement data are used to initialise the optimisation module. The Node.csv file contains location, quality, and quantity information for each voxel. The Precedence.csv file defines the precedence relationships among nodes, guiding the order in which reclamation operations should be executed first. The machine.CSV define the safety operation region for multiple FELs and the Cost.csv defines the travelling cost from among nodes. All these CSV files are generated using computer programs.

Model reconciliation

Using computer simulation technique allow 3D models to be generated and updated continuously. However, the modelling accuracy is primarily limited by two factors. First, position and orientation accuracy. The geo-positioning system for on-site vehicles relies on differential Global Positioning System (GPS) technique, which provides 3D accuracy within a range of 3–5 m. Additionally, truck headings saved in the database are only directions of travel since all trucks only have one GPS antenna installed. When trucks move slowly or are stationary, these headings become unreliable. Meanwhile, although most of loaders have at least two GPS antenna installed, the roll and pitch angles derived from GPS measurements are still prone to errors. Second, the time constraint in representing real-world complexities. During a dump event, a truck tends to move forward slightly when the dump body reaches its maximum height. Rock segregation and collapse also occurs during stacking and reclaiming operations. Accurately simulating these processes requires significant computational resources and time, making real-time or near-real-time modelling challenging.

Maintaining the accuracies of computer simulation models is a crucial role for the optimisation system. The reconciliation process ensure that the simulation models remain aligned with real-world stockpiles and continue to provide reliable outputs for reclaiming optimisation. This process is achieved by using the measurement models to update and refine the simulation models. Because both models use the same local reference frame, there is no need to align them using point registration algorithms. However, it is still necessary to detection the boundaries of a stockpile from both models and use the boundary points as features for the reconciliation. To improve the efficacy, image processing techniques are applied to detect the boundaries.

To generate an 2D image, x, y and z coordinates in a point cloud file are converted into 2D grid coordinates (matrices) separately. Matrix X contain i rows and each row is a copy of all unique x-values, where i is the total number of unique y-values. Matrix Y contains j columns and each column is a copy of all unique y-values, where j is the total number of unique x-values. Together, matrix X and Y from a structured 2D grids. Matrix Z stores the corresponding heights to a pair of $(X_{i,j}, Y_{i,j})$. As a result, all three 2D matrices have the same dimensions. Matrix Z is transforms into an intuitionistic fuzzy set (IFS) Z_F, where values range from 0 to 1 for edge detection. A value of 1 represents a full membership in in Z_F, while a value of 0 represents no membership. A fuzzy logic-based boundary detection algorithm described by Chaira and Ray (2008) is applied to Z_F.

This algorithm adds a new parameter called intuitionistic fuzzy divergence (IFD) to measure the difference between two IFSs. A floating 3 × 3 mask is applied to Z_F and divide it into non-overlapping image windows/IFSs. Additionally, the algorithm generates 16 preset templates. For each image window, the IFDs between the window and all 16 preset templates are computed. The most appropriate IFD, determined by the max–min relationship, is selected and used to replace the corresponding values in the image window. A fuzzy divergence matrix will be produced after the entire process. After applying a threshold, the divergence matrix is converted into a binary edge image. The key limitation of this algorithm is that the preset templates and the threshold need to be manually adjusted to achieve the optimal results. The detected boundaries will be filtered to extract the most outside edges of the stockpile. Because a value in Matrix Z is associated with a pair of x and y coordinates, the true boundary of a stockpile can be detected through mapping the final binary image to Matrix X and Y. Figure 4 illustrates the boundary detection result for a measurement mode.

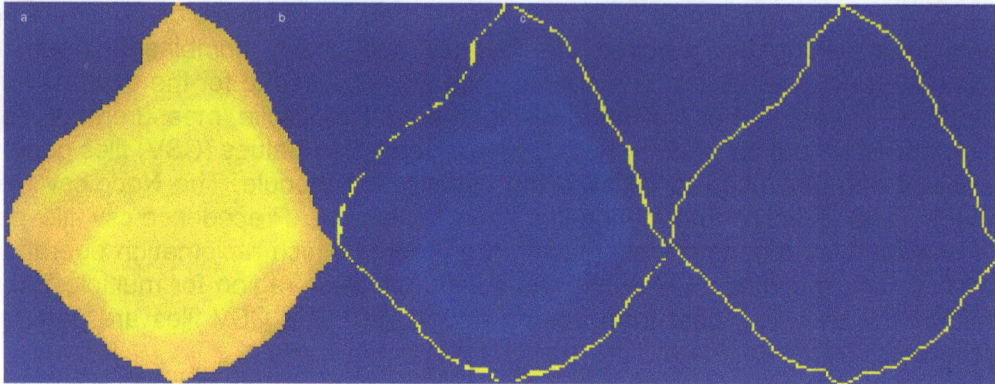

FIG 4 – Detect stockpile boundaries for reconciliation: (a) A 3D stockpile model; (b) Boundary detected using the fuzzy logic; (c) The true stockpile boundaries after filtering out unnecessary points.

The reconciliation process is achieved through aligning the boundary points for both models along the x and y axes. Once aligned, the maximum height ($z_{j,k}^{com(n)}$) at a given (x_j, y_k) coordinate in the computer simulation model, which represents a point on the top surface of the outermost layer, the nth layer, is replaced with the corresponding z-value ($z_{j,k}^{mea}$) from the measurement model. An extra step is needed for the layers under the outmost layer (i^{th} layer and $i = 1, 2, ..., n-1$) if $z_{i,j}^{mea} <$ $z_{i,j}^{com(n)}$. The height of the top surfaces of these layers is determined by Equation 1:

$$\begin{cases} z_{j,k}^{com(i)} = z_{j,k}^{com(i)} - \left(z_{i,j}^{com(n)} - z_{i,j}^{mea}\right) \ if \ z_{j,k}^{com(i)} - \left(z_{i,j}^{com(n)} - z_{i,j}^{mea}\right) \geq 0 \\ z_{j,k}^{com(i)} = 0 \ if \ z_{j,k}^{com(i)} - \left(z_{i,j}^{com(n)} - z_{i,j}^{mea}\right) < 0 \end{cases} \tag{1}$$

CONCLUSION AND FUTURE WORK

This paper discussed the benefits of using 3D stockpile models to enhance the quality control and blending efficiency for ROM stockpiles. To demonstrate the benefits, the authors describe a system that is able to mimic a real operational environment using backup databases from a mining site. The ultimate objective for the demonstration is to validate that the use of simulation stockpile models generated from actual operational data will enable more predictable quality control and optimised blending. This system described here is expected to serve as a foundation for further development, such as integration into existing stockyard/stockpile management systems for automated decision-making in mining operations.

Significant progress has been achieved in developing such a demonstration system. The future work mainly focused on exporting simulation models from Insight CM, generating required input files for the optimisation module using Python program, and displaying the optimised reclaiming sequences on InSight CM.

ACKNOWLEDGEMENTS

This research has been supported by the South Australian Government through the Premier's Research and Industry Fund (PRIF) Research Consortia Program 'Unlocking Complex Resources through Lean Processing'.

REFERENCES

Alsayed, A and Nabawy, M R, 2023. Stockpile volume estimation in open and confined environments: a review, *Drones*, 7(8):537.

Assimi, H, Koch, B, Garcia, C, Wagner, M and Neumann, F, 2021. Modelling and optimization of run-of-mine stockpile recovery, Proceedings of the 36th Annual ACM Symposium on Applied Computing.

Assimi, H, Koch, B, Garcia, C, Wagner, M and Neumann, F, 2022. Run-of-mine stockyard recovery scheduling and optimisation for multiple reclaimers, Proceedings of the 37th ACM/SIGAPP Symposium on Applied Computing.

Chaira, T and Ray, A K, 2008. A new measure using intuitionistic fuzzy set theory and its application to edge detection, *Applied Soft Computing*, 8(2):919–927.

Dub, V, Dub, A and Makarycheva, E, 2006. Role of impurity and process elements in the formation of structure and properties of structural steels, *Metal Science and Heat Treatment*, 48(7):279–286.

Everett, J E, 1996. Iron ore handling procedures enhance export quality, *Interfaces*, 26(6):82–94.

Fountain, C, 2013. The whys and wherefores of penalty elements in copper concentrates, in *Proceedings of MetPlant 2013*, pp 502–518 (The Australasian Institute of Mining and Metallurgy: Melbourne).

Jupp, K, Howard, T and Everett, J, 2013. Role of pre-crusher stockpiling for grade control in iron ore mining, *Applied Earth Science*, 122(4):242–255.

Lane, D J, Cook, N J, Grano, S R and Ehrig, K, 2016. Selective leaching of penalty elements from copper concentrates: A review, *Minerals Engineering*, 98:110–121.

Lu, T-F and Myo, M T R, 2010. Optimization of reclaiming voxels for quality grade target with reclaimer minimum movement, in 11th International Conference on Control Automation Robotics and Vision.

Pownceby, M I, Hapugoda, S, Manuel, J, Webster, N A and MacRae, C M, 2019. Characterisation of phosphorus and other impurities in goethite-rich iron ores–Possible P incorporation mechanisms, *Minerals Engineering*, 143:106022.

Statsenko, L and Melkoumian, N S, 2014. Modeling blending process at open pit stockyards: A Northern Kazakhstan Mining Company case study, Mine Planning and Equipment Selection: Proceedings of the 22nd MPES Conference, Dresden, Germany, 14th–19th October 2013.

Trouchina, O P and Topal, E, 2020. Effective Methods to Reduce Grade Variability in Iron Ore Mine Operations, Proceedings of the 28th International Symposium on Mine Planning and Equipment Selection-MPES.

Xie, Y, Neumann, A and Neumann, F, 2021. Heuristic strategies for solving complex interacting large-scale stockpile blending problems, 2021 IEEE Congress on Evolutionary Computation (CEC).

Zhao, S, Lu, T-F, Koch, B and Hurdsman, A, 2013. Dynamic modelling of 3D stockpile for life cycle management through sparse range point clouds, *International Journal of Mineral Processing*, 125:61–77.

Zhao, S, Lu, T-F, Statsenko, L, Koch, B and Garcia, C, 2022. A framework for near real-time ROM stockpile modelling to improve blending efficiency, *Journal of Engineering, Design and Technology*, 20(2):497–515.

Sustainable practices

Modelling blasting vibration in sustainable mine planning using machine learning techniques – a case study

M Aghdamigargari[1], S Avane[2], A Anani[3] and S O Adewuyi[4]

1. PhD student, Department of Mining and Geological Engineering, University of Arizona, Tucson AZ 85721, USA. Email: mehriaghdami@arizona.edu
2. PhD student, Department of Mining and Geological Engineering, University of Arizona, Tucson AZ 85721, USA. Email: sylvesteravane@arizona.edu
3. Associate Professor, Department of Mining and Geological Engineering, University of Arizona, Tucson AZ 85721, USA. Email: angelinaanani@arizona.edu
4. Postdoc, Department of Mining and Geological Engineering, University of Arizona, Tucson AZ 85721, USA. Email: sadewuyi@arizona.edu

ABSTRACT

Traditional mine planning has primarily focused on economic and operational objectives, often overlooking the significant impacts of mining activities on surrounding communities. Today, it is essential for mining operations to assess their activities against sustainable mining criteria to ensure environmental and social responsibility. Blasting, an indispensable part of mining, poses substantial challenges to nearby communities due to its adverse effects, such as vibration, noise, and fly rock. This research addresses these impacts by developing a predictive framework to assess the effects of blasting, focusing on vibrations. By integrating machine learning models with data from real mining operations, this study aims to quantify and predict Peak Particle Velocity (PPV) as a key measure of blast-induced vibration in different blasting schedules. Initial efforts involve collecting data on blasting parameters such as total charge and distance to train and validate the predictive model. In this case study, a data set comprising 373 blasting events was analysed using a Deep Neural Network (DNN) model. The results revealed a strong negative correlation between PPV and scaled distance. These findings highlight the potential to embed the developed model into production scheduling, promoting sustainable mine planning that enhances community well-being while optimising Net Present Value (NPV).

INTRODUCTION

Blasting is a crucial operation in mining projects, playing a key role in breaking rock masses to facilitate excavation. However, it also generates ground vibrations that can negatively impact nearby structures, the environment, and local communities. Predicting these vibrations before blasting can aid in designing blasts and minimising adverse effects. The USBM empirical equation (Equation 1) has long been the primary traditional method for estimating Peak Particle Velocity (PPV) (Jansrud, 2024). In this equation, the parameters K and b are determined using various regression techniques based on blasting data sets that include PPV and scaled distance (SD), as defined in Equation 2. Here, W represents the maximum charge weight per delay, and D denotes the distance between the blast site and monitoring points.

$$PPV = K(SD)^b \tag{1}$$

$$SD = (D/W)^{1/2} \tag{2}$$

While the USBM empirical equation provides a straightforward initial estimation of blast-induced vibrations, its accuracy is often limited in complex scenarios due to its reliance on only a few parameters. As a result, it may not always yield optimal predictions. In recent years, machine learning (ML) techniques have emerged as powerful alternatives for PPV prediction, offering enhanced accuracy by incorporating additional factors such as geological conditions, which traditional methods overlook (Nguyen *et al*, 2020; Pradeep, Chandrahas and Fissha, 2024).

Key parameters influencing blast-induced ground vibrations include the distance between the blast site and monitoring points, maximum charge weight per delay, burden, spacing, and blasthole depth. Among ML techniques, Artificial Neural Networks (ANNs), particularly the feed-forward back-propagation neural network (BPNN), have been widely adopted due to their ability to capture complex nonlinear relationships between input variables and PPV (Roy, 2005; Khandelwal and Singh, 2009; Murmu, Maheshwari and Verma, 2018; Nguyen *et al*, 2020).

Das, Sinha and Ganguly (2019) developed an ANN model with 15 inputs utilising 248 data samples collected from three coalmines with varying geo-mining conditions. The model was trained using 70 per cent of data set, and it achieved a correlation coefficient of 0.96 between the predicted and measured PPV values for a new set of data, significantly outperforming the traditional empirical model, which had a correlation coefficient of 0.63. This highlights the superior predictive accuracy of ANNs compared to conventional methods. However, other statistical metrics, such as coefficient of determination (R^2), root mean square error (RMSE), and mean absolute error (MAE), were not addressed in this study, and it is unknown whether their proposed model will predict blasting vibration in a complex geological structure.

Integrating multiple machine learning algorithms can also improve predictive performance. Hosseini *et al* (2023) used ANN with Extreme Gradient Boosting (XGBoost) to introduce an ensemble modelling approach for PPV estimation. The effectiveness of the base models was assessed using various validation metrics, including the R^2, RMSE, MAE, variance accounted for (VAF), and overall accuracy. The results demonstrated that the ensemble model provided higher prediction accuracy compared to the best-performing individual models.

In recent years, in addition to ANNs, various other machine learning algorithms, such as Support Vector Machine (SVM), K-Nearest Neighbours (KNN), and Random Forest (RF), have been implemented and compared for vibration prediction (Chandrahas *et al*, 2022; Nguyen, Bui and Drebenstedt, 2023).

Among various types of neural networks, the multi-layer perceptron (MLP) is a form of feed-forward that employs backpropagation algorithm and serves as the foundation for deep learning, enhancing the performance of ANNs. MLP employs the backpropagation algorithm for training, making it a powerful supervised learning technique capable of handling nonlinear problems effectively. When an MLP network incorporates multiple hidden layers and deep learning techniques, it is referred to as a deep neural network (DNN). A DNN consists of at least three layers: an input layer, one or more hidden layers, and an output layer. In the input layer, neurons function as receptors, transmitting information to the hidden layers. The hidden layer neurons process and learn from the data by performing calculations and adjusting the connection weights between them. These weighted values are then passed to the output layer, where the final predictions are generated and displayed (Brownlee, 2018).

Nguyen *et al* (2021) applied a hybrid DNN framework combined with several nature-inspired optimisation algorithms (Harris Hawks Optimisation Algorithm (HHOA), Whale Optimisation Algorithm (WOA), and Particle Swarm Optimisation (PSO)) to predict vibration in an open pit coalmine. Using 229 blasting events and two hidden layers in MLP, the hybrid models outperformed individual DNNs, with the HHOA-DNN achieving the highest accuracy of R^2 = 0.930, MSE = 2.361 for the testing, and R^2 = 0.941, MSE = 1.540 for the training data set. The study demonstrated the importance of features such as explosive charge, monitoring distance, and time delay.

In another study, Wang *et al* (2022) developed a long short-term memory (LSTM) model to predict the full waveform of blast-induced vibrations. Unlike models that focus only on PPV, this approach focuses on time series, allowing to captured temporal and spatial complexities. Using 20 simulated and real-world blasting events, they demonstrated the LSTM model's broader applicability and improved accuracy in representing vibration duration and frequency. Jansrud (2024) demonstrated the superiority of a DNN model over the USBM empirical model in predicting PPV. The DNN model was trained on 9724 samples of data points from a mining company, and evaluation was conducted using three statistics criteria, R^2, MSE, and MAE. The network they used consists of eight neurons in the input layer corresponding to eight input features of the data set including Site, Blast ID, Scaled Distance, Distance, Maximum Instantaneous Charge, Blast Direction, Time frame, Groundwater presence (binary). It also had three hidden layers and one output layer to predict PPV. The new idea improved predictive accuracy by over 70 per cent compared to the industry standard.

In this regard, DNNs approaches have revolutionised the prediction of blast-induced vibrations by offering greater accuracy and incorporating a wider range of features compared to traditional methods. Integrating these models into mining operations can significantly improve social and environmental safety, ensure regulatory compliance, and enhance operational efficiency. This study applies DNNs to an open pit mine as a new case study, providing insights into their applicability in diverse mining environments.

The remainder of this paper is organised as follows: The next section outlines the research methodology employed in this study. The results and discussion section presents the model findings and comparative analysis. The final section provides conclusions and explores future research opportunities.

METHODOLOGY

The data for blast-induced ground vibration was collected from a large copper mine in Arizona, United States. After data cleaning and preprocessing, records from 373 blasting events were retained, including key parameters such as hole depth, distance, number of holes, total charge, maximum instantaneous charge, and charge weight per delay. Ground vibration measurements were recorded using Mini Seis III Pro devices. Based on this data, the scaled distance was calculated, and the aforementioned parameters, along with scaled distance, were used as input variables for a ML model to predict PPV as the output.

Figure 1 outlines the main steps of this study. A Deep Neural Network (DNN) with two hidden layers was developed, and key hyperparameters were tuned to maximise the R^2 score. The tuning process explored the following options: number of neurons in the first hidden layer [16, 32], second hidden layer [8, 16], learning rate [0.001, 0.01], activation functions ['relu', 'tanh', 'sigmoid'], and optimisers ['adam', 'sgd', 'adagrad', 'rmsprop'].

K-fold cross-validation with k=5 was employed to increase the generalisation ability of the obtained model. However, it caused a higher complexity to the model too. The network consisted of seven neurons in the input layer (corresponding to the seven input variables) and one neuron in the output layer. 80 per cent of the data set was allocated for training and the rest for testing the ability of network to predict PPV. The network was trained in a supervised manner using the backpropagation algorithm to optimise a multi-layer feedforward network. Before training begins, the data underwent preprocessing, which includes normalising both input and output values to improve model performance and stability. Additionally, the top 1 per cent and bottom 1 per cent of the data were removed from the analysis to eliminate potential outliers and enhance model accuracy.

At the next step, the USBM empirical equation was applied as a second method for PPV prediction. In this approach, least squares regression was used to estimate the unknown parameters b and K, fitting the model to the data. Finally, the performance of both methods was compared to evaluate their predictive accuracy.

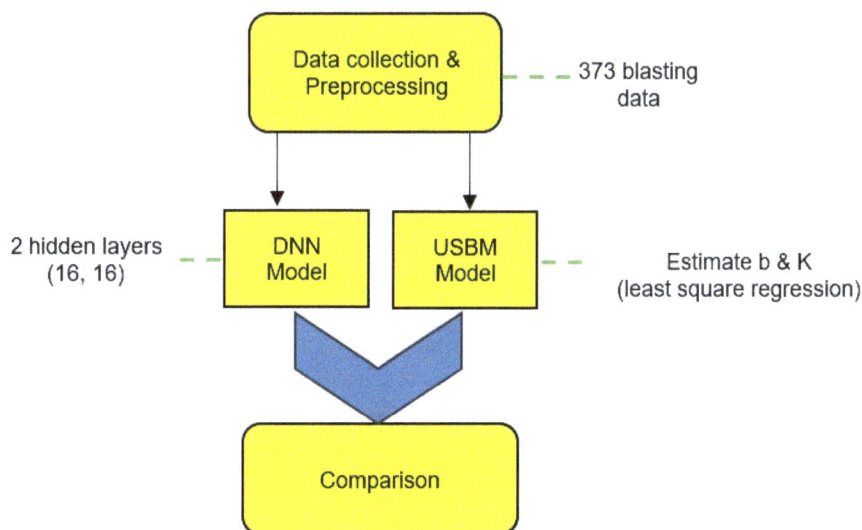

FIG 1 – Research methodology.

RESULTS AND DISCUSSION

The correlation matrix in Figure 2 shows a visual representation of the relationships between different variables in the data set. The colour intensity reflects the strength of the correlation, with the scale on the right indicating that red shades signify positive correlations, blue shades represent negative correlations, and light grey indicates little to no correlation. The matrix highlights the following key relationships:

- PPV and distance/scaled distance exhibit a strong negative correlation (-0.83, -0.86), represented by the dark blue colour. This indicates that as scaled distance increases, PPV tends to decrease, and *vice versa*.

- PPV and charge weight per delay/Max instantaneous charge show a slight positive association, suggesting that larger explosive charges tend to slightly increase Max PPV.

- PPV's correlation with other variables is weak, as evidenced by the light colours, implying that PPV has minimal association with most other features in the data set.

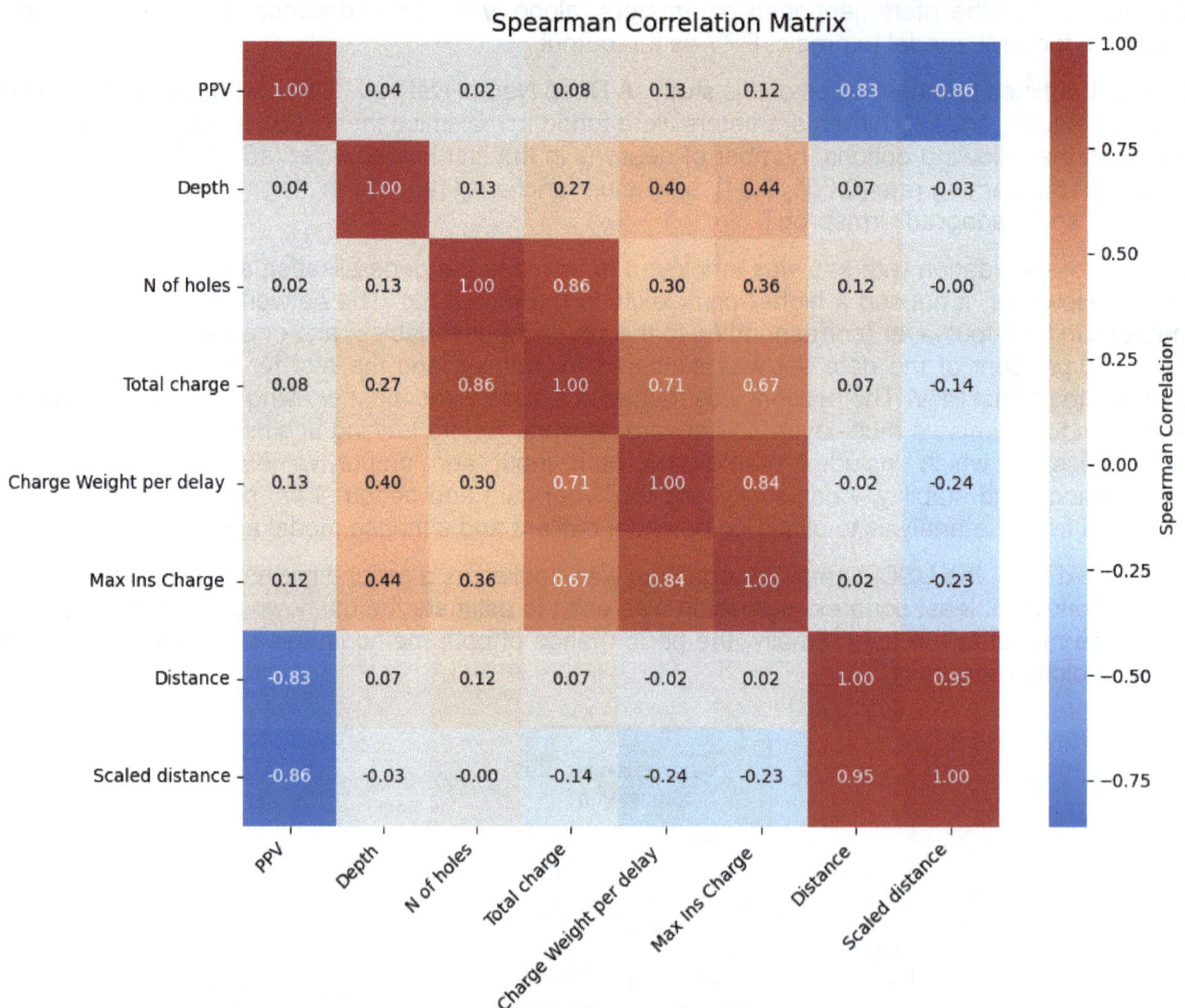

FIG 2 – Feature correlation matrix.

Applying cross-validation and tunning hyperparameters in 150 epochs the best DNN model was obtained with the following parameters:

- Hidden 1 neurons: 16
- Hidden 2 neurons: 16
- Learning rate: 0.01

- Activation function: sigmoid
- Optimiser: rmsprop.

The scatter plot in Figure 3 presents the predicted PPV versus true PPV for the best DNN model for both training and testing data, suggesting that the model is capturing the relationship between input features and PPV reasonably well. However, some scatter around the ideal y = x line indicates prediction errors and the possibility for improvement in model accuracy. The best DNN model achieved R^2 values of 0.76 on the training data and 0.72 on the testing data.

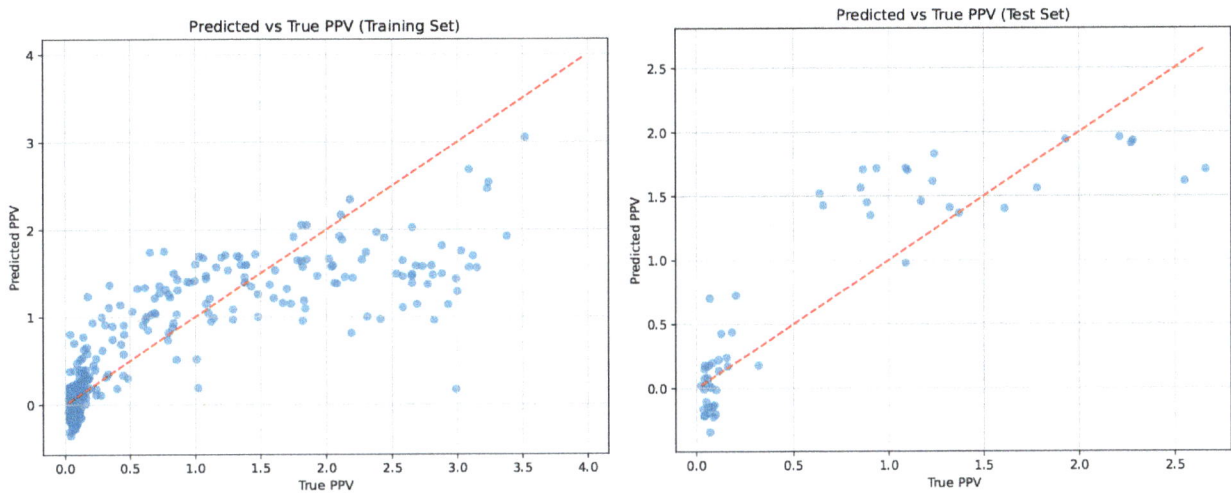

FIG 3 – Predicted PPV versus True PPV for the DNN approach.

In the second phase, implementing the empirical method using the same data set, we tried to fit a curve between the scale distance and PPV using least square regression which resulted in Equation 3:

$$PPV=16 \ (SD)^{-1.14} \tag{3}$$

where K = 16 and b = 1.14.

Finally, we employed various statistical metrics to assess both methods and compare their performance in predicting PPV. The results are displayed in Table 1.

TABLE 1

Evaluation results of the DNN model compared to USBM model.

Evaluation metrics	DNN model	USBM empirical equation
Root Mean Square Error (RMSE)	0.4041	0.5813
Mean Absolute Error (MAE)	0.3066	0.3346
R^2	0.7208	0.5074

From the table, it can be observed that the DNN method outperforms USBM across all three statistical measures, indicating its superior predictive capability. The lower errors for DNN suggest improved accuracy and reliability in capturing the underlying patterns of the data set. Also, an increase in the R^2 value indicates that the DNN model outperforms the empirical equation and can explain more than 72 per cent of the variability in this blast vibration data.

CONCLUSIONS

This research established a predictive framework to evaluate the impact of blast-induced vibration, a key environmental and social consequence of mining operations. By utilising DNN, the study revealed a strong inverse relationship between PPV and scaled distance in the 373 blasting data assessed. Also, the result of comparative evaluation highlights the effectiveness of deep learning-based approaches in enhancing prediction accuracy compared to traditional methods like USBM. However, the model developed in this study is not the most accurate for predicting PPV. By incorporating more data, we can improve the R^2 value and reduce errors, leading to a more robust model capable of capturing a broader range of data. Additionally, in future work, this predictive model can be integrated with mine planning optimisation to promote socially sustainable mining practices, avoiding blasting patterns that could incur into environmental issues and ultimately enhancing the relationship between mining companies and local communities.

REFERENCES

Brownlee, J, 2018. Deep learning for time series forecasting: predict the future with MLPs, CNNs and LSTMs in Python, Machine Learning Mastery.

Chandrahas, N S, Choudhary, B S, Teja, M V, Venkataramayya, M S and Prasad, N K, 2022. **XG** boost algorithm to simultaneous prediction of rock fragmentation and induced ground vibration using unique blast data, *Applied Sciences*, 12(10):5269.

Das, A, Sinha, S and Ganguly, S, 2019. Development of a blast-induced vibration prediction model using an artificial neural network, *Journal of the Southern African Institute of Mining and Metallurgy*, 119(2):187–200.

Hosseini, S, Pourmirzaee, R, Armaghani, D J and Sabri Sabri, M M, 2023. Prediction of ground vibration due to mine blasting in a surface lead–zinc mine using machine learning ensemble techniques, *Scientific Reports*, 13(1):6591.

Jansrud, G, 2024. Enhancing Prediction of Blast-Induced Ground Vibrations through Machine Learning, Master's thesis, UiT Norges arktiske universitet.

Khandelwal, M and Singh, T N, 2009. Prediction of blast-induced ground vibration using artificial neural network, *International Journal of Rock Mechanics and Mining Sciences*, 46(7):1214–1222.

Murmu, S, Maheshwari, P and Verma, H K, 2018. Empirical and probabilistic analysis of blast-induced ground vibrations, *International Journal of Rock Mechanics and Mining Sciences*, 103:267–274.

Nguyen, H, Bui, X N and Drebenstedt, C, 2023. Machine Learning Algorithms for Data Enrichment: A Promising Solution for Enhancing Accuracy in Predicting Blast-Induced Ground Vibration in Open pit Mines, Inżynieria Mineralna.

Nguyen, H, Bui, X N, Tran, Q H, Nguyen, D A, Hoa, L T T, Le, Q T and Giang, L T H, 2021. Predicting blast-induced ground vibration in open pit mines using different nature-inspired optimization algorithms and deep neural network, *Natural Resources Research*, 30:4695–4717.

Nguyen, H, Drebenstedt, C, Bui, X N and Bui, D T, 2020. Prediction of blast-induced ground vibration in an open pit mine by a novel hybrid model based on clustering and artificial neural network, *Natural Resources Research*, 29(2):691–709.

Pradeep, T, Chandrahas, N S and Fissha, Y, 2024. A principal component-enhanced neural network framework for forecasting blast-induced ground vibrations, *J Civ Hydraul Eng*, 2(4):206–219.

Roy, P P, 2005. *Rock blasting: effects and operations* (CRC Press).

Wang, Y, Zheng, G, Li, Y and Zhang, F, 2022. Full waveform prediction of blasting vibration using deep learning, *Sustainability*, 14(13):8200.

Reducing a quarry's greenhouse gas emissions through air deck implementation

R Heryadi[1,2], H Utama[3] and L Aprisko[4]

1. COO, PT Suma Yogara Sejahtera, South Jakarta 12810, Indonesia.
   Email: rudy.heryadi@breny.my.id
2. Lecturer, PGRI Kalimantan University, Banjarmasin, South Kalimantan 70122, Indonesia.
3. Junior Mine Manager, PT Indo Muro Kencana, Murung Raya, Central Kalimantan, Indonesia.
   Email: heru.utama@imkgold.co.id
4. Drill and Blast Engineer, PT Karimun Granit, Karimun Island, Indonesia.
   Email: leoaprisko7@gmail.com

ABSTRACT

Generally, drilling and blasting are the most common methods for rock breaking in the quarry to get the desired fragmentation size. Several efforts have been made to reduce greenhouse gas (GHG) emissions in blasting operations at the quarry, especially through drilling and blasting design optimisation. One simple method for the drilling and blasting process that can increase blasting energy efficiency is the blasting method using air decking. By applying air decking, the use of explosives can be reduced, and proportionally, the environmental impact of the explosives usage can also be reduced. This research aims to determine how much GHG emissions can be reduced by air decking in one of the limestone quarries in Central Sulawesi, Indonesia. The air deck is installed at the bottom of the blasthole and is 0.3 m long. The blasthole has a 89 mm diameter, the total number of holes is 144, the average depth is 3.6 m, and the type of explosive used is ANFO. GHG emission calculations were carried out using emission factors in blasting activities and several input materials in the blasting process. The calculation result shows that estimated GHG emission reduction in quarries can be attained significantly, especially when compared with conventional quarry standards and those that have made energy-efficiency efforts in mining activities in their quarries. The total reduction of ANFO is 216 kg, which correlates with 283.6 kg CO_2-eq of GHG emission reduction per blast. Approximately five blasts every month are conducted; therefore, it can reduce the GHG emissions to 1418 kg CO_2-eq per month. Reducing GHG emissions by implementing an air deck does not sacrifice the blasting result, whereas in this experiment, the blasting recovery obtained is 91 per cent. The GHG emissions reduction potential from implementing an air deck in the limestone quarry in this research is approximately 7.5 per cent.

INTRODUCTION

Blasting in the mining industry for coal, minerals, and quarries is needed, especially for breaking rocks to get the products. During the mining operation, caution is required during drilling and blasting operations. Blasting activities are preceded by drilling blastholes, filling explosives into the blastholes, and continuing with stemming, tie-up, and shot firing. The commercial explosives used for blasting operations can generate a high carbon footprint or Greenhouse gas (GHG) emission, especially from Ammonium Nitrate mixed with Fuel Oil (ANFO), which forms a large percentage of commercial explosives (Heryadi, Utama and Prasmoro, 2024). The issue of reducing the carbon footprint from mining activity is becoming the main concern for most mining players in the world. ANFO explosives normally consist of 96 per cent Ammonium Nitrate (AN) and 4 per cent fuel oil (FO) (Suppajariyawat et al, 2019), and when an explosion occurs, water, Nitrogen (N_2), and Carbon Dioxide (CO_2) are formed (Gilmartin, 2020). CO_2 is one of the GHGs that create global warming and is the most important in the atmosphere (Lecksiwilai et al, 2016). It occurs naturally and is a product of human activities, including burning fossil fuels.

In quarries or mining in general, the explosives must be used efficiently to ensure optimum fragmentation, good floor conditions, and face stability. There are various ways to use explosives efficiently, and one method is by implementing air decking. Air decking implementation can provide more benefits than just reducing explosive consumption. Other advantages of using an air deck/bottom deck besides saving on using explosives are reduced vibration, minimal back break, reduced fly rock distance, and minimised toe (Chiappetta, 2010). Figure 1a shows the application of an air deck in a blasthole, and Figures 1b and 1c show the type of bottom and top air deck under the brand name Sysdeck™Blastfren.

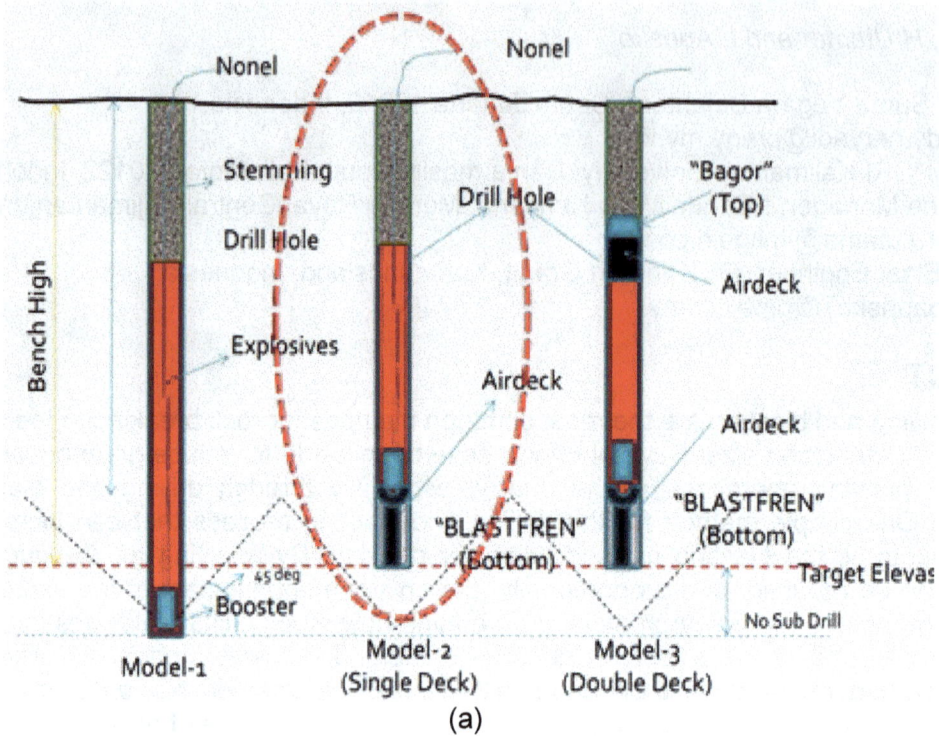

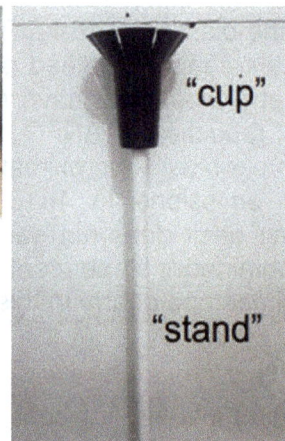

FIG 1 – (a) Air deck in blasthole; (b) top deck; (c) bottom deck.

The application of air decking, aside from saving AN usage, can also increase the fragmentation quality. The application of air decking can reduce bouldering by up to 80 per cent. Shovel loading efficiency can increase by 20 to 40 per cent on sandstone-type materials, and explosives costs can be reduced by 10 to 35 per cent, depending on the type of material (Jhanwar and Jethwa, 2000).

So far, various studies on air decking implementation have only been pertinent to its benefit for explosives saving and other results, such as vibration reduction, fly rock reduction, and minimal toe formation at the floor (Chiappetta, 2010; Masda Rohal Sadiq, 2021; Monjezi et al, 2022). Air decks can also improve fragmentation with fewer explosives, where the average fragmentation and digging rate improve. Vibration is one of the environmental impacts caused by blasting activity, and air deck application gave satisfactory results in reducing the vibration by up to 33 per cent (Chiappetta, 2010). Various studies have addressed blasting vibration as the main environmental impact (Hidayat, Puspitasari and Tantina, 2011; Ramadhan, Adnyano and Purnomo, 2020; Yang, Zhang and Wang, 2022). Other studies related to air deck implementation are summarised in Table 1. Another benefit of using an air deck where the carbon footprint or life cycle GHG emission can be reduced has never been studied, especially in a quarry. Only one article from Gilmartin (2020) ever discussed the potential benefit of an air deck in quarry operation, where the discussion emphasises the potential reduction of carbon footprint in blasting activity when an air deck is applied.

TABLE 1

Summary of the research pertinent to the air decking implementation in blasting activities.

No	Author(s)	Title/topic	Finding/research result
1	Hayat, Alagha and Ali (2019)	Air deck in surface mining operations	A comparison of the position of the air deck in the blasthole based on the research results is made between the middle, top, and bottom positions (for soft to hard rock types). Engineers must design air decks starting from the lowest recommended length. No discussion on other potential benefits of air decks to the environment.
2	Monjezi *et al* (2022)	Comparison of bottom and top air-deck applications to improve blasting operations in Gypsum and iron mines (Iran)	The application of an air deck in both the Top and Bottom sections shows more or less the same advantages. There is a discussion on vibration and fly rock as an environmental impact, where the top deck provides better results to reduce environmental impact (vibration and fly rock)—no discussion on potential carbon footprint reduction.
3	Jhanwar (2012)	Investigation of the effect of air deck on blasting performance in open pit/surface mines in India	Applying an air deck for open pit mines is superior to conventional blasting, increasing efficiency in coalmines in several parameters. No discussion on environmental impact was found.
4	Chiappetta (2010)	New blasting techniques to eliminate sub-grade drilling, improve fragmentation, reduce explosive consumption, and reduce ground vibrations	Using bottom decking at the North Nevada mine. No problems were encountered with the blasting results using the bottom deck. There is a reduction in environmental impact in the form of ground vibration by 33 per cent. There is no discussion or result related to a reduction in carbon footprint.
5	Zarei *et al* (2022)	Air deck to improve surface blasting performance	Air deck is applied in the bottom blasthole on medium-strength rock (UCS 75, density 2.7 g/cc, limestone). Applying air decking (single) in several blasting zones shows better/lower MFS results than conventional blastholes without air decks (18–20% reduction). Material cost per bcm can be reduced from USD 0.115 to USD 0.105 (10.7% reduction)—no environmental impact discussion in this article.

As summarised in Table 1, many studies of air deck application have been conducted. However, the significant contribution of air decks to reducing GHG emissions is still lacking. This study is strategically important, especially as an additional effort to reduce GHG emissions through simple and inexpensive methods in blasting operations; therefore, this study aims to determine how much GHG emissions can be reduced by air decking in the quarry. The blasting activity was conducted in a limestone quarry in North Morowali, Central Sulawesi, Indonesia, belonging to PT Gita Perkasa Mineralindo (PT GPM).

METHOD

There are several steps in calculating the reduction of GHG emission potential in a limestone quarry. The steps taken are: (1) Explosives consumption calculation and (2) Calculating GHG emission reduction from explosives consumption.

Explosives consumption calculation

Consumption of the explosives (ANFO) can be determined during the design stage. The blasting design parameter is shown in Table 2.

TABLE 2

Limestone quarry blasting design.

Blasting design parameter	Unit	With air deck*	Conventional**
Blasthole diameter	mm	89	89
Burden	m	3	3
Spacing	m	4	4
Subdrill	m	0	0
Stemming	m	1	1
Average depth	m	3.6	3.6
Average air deck length (bottom deck)	m	0.3	0

* The actual blasting is carried out with 100% air deck.

** Blast design if blasting is carried out without an air deck.

Figure 2 shows the blasting area description where the elevation was 75 m before blasting, and after blasting, the elevation was 72 m. There were 144 blastholes, and actual blasting was carried out using a 100 per cent bottom air deck with a 0.3 m length. The air deck was in cup form with the brand name Sysdeck™Blastfren, which comes with a PVC pipe stand. Figure 1c shows the type of air deck utilised.

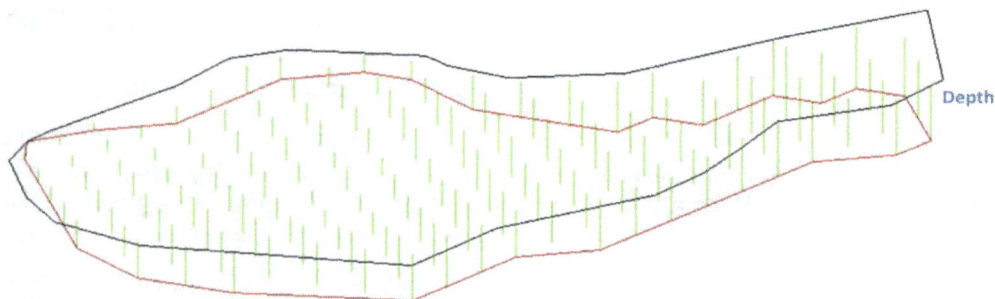

Depth

FIG 2 – Blasting area, initial elevation was 75 m, and final elevation was 72 m.

GHG emission reduction calculation

The life cycle of GHG emissions was based on its life cycle. However, the comprehensive data on the GHG emission of ANFO was based on the previous GHG emission data ratio from Climateleaders.org (2020). The life cycle GHG emission ratio of ANFO explosives from the reference was directly multiplied by the total number of ANFO used to obtain the total life cycle GHG emission using Equation 1. The total life cycle GHG emission reduction was then calculated by subtracting ANFO's GHG emission in conventional blasting from ANFO's life cycle GHG emission with the air deck. Table 3 provides ANFO's life cycle GHG emission from the Climateleaders.org (2020) used in the calculation.

TABLE 3

Assumed value of life cycle GHG emission of ANFO used in the GHG emission reduction calculation.

Stage of explosives use life cycle	GHG emission (tonnes-CO$_2$eq/tonne-ANFO)
Manufacturing of AN	1.6
Transportation of AN	0.1
Blasting	0.2

$$TLC_{GHG\ ANFO} = LC_{GHG\ ANFO\ R} \times m_{ANFO} \tag{1}$$

Where:

$TLC_{GHG\ ANFO}$ — is the total GHG emission produced from blasting activity in tonnes CO$_2$-eq

$LC_{GHG\ ANFO\ R}$ — is the ratio of GHG emission of ANFO from blasting activity in tonnes CO$_2$-eq/tonne-ANFO

m_{ANFO} — is the quantity of ANFO in tonnes used during blasting in the limestone quarry

The total GHG emission of ANFO from blasting using an air deck and conventional blasting without an air deck can be calculated using Equation 1, and the reduction of life cycle GHG emission of blasting activity using an air deck can be found using Equation 2.

$$R_{LC\ GHG\ ANFO} = TLC_{GHG\ ANFO\ conventional} - (TLC_{GHG\ ANFO\ air\ deck} + TLC_{air\ deck}) \tag{2}$$

Where:

$R_{LC\ GHG\ ANFO}$ — is the total GHG emission reduction when using air deck in tonne-CO$_2$-eq

$TLC_{GHG\ ANFO\ conventional}$ — is life cycle GHG emission produced when blasting using ANFO conducted without an air deck in tonne CO$_2$-eq

$TLC_{GHG\ ANFO\ air\ deck}$ — is life cycle GHG emission produced when blasting using ANFO conducted with an air deck in tonne CO$_2$-eq

$TLC_{airdeck}$ — is life cycle GHG emission of the air deck set in tonne CO$_2$-eq

The GHG emission calculation considers the GHG emission from using the cup (PP plastic) and a 1-inch diameter Stand (PVC). The total weight of the cup plus standpipe was 120 g per set, and the total weight per blast was only 16.2 kg. By taking the highest GHG emission from PVC at 7.83 kg CO$_2$-eq per kg plastic (PP only 3.576 kg CO$_2$-eq per kg plastic) (Climatiq, 2021; Liang and Yu, 2023), the total GHG emission per blast was 0.1268 t CO$_2$-eq.

RESULT AND DISCUSSION

ANFO consumption and blasting result

The ANFO consumption, both actual with the air deck and conventional, is displayed in Table 4. The actual recovery and volume blasted are also shown in Table 4.

TABLE 4

ANFO consumption and blasted material obtained.

Description	Unit	Value
Actual ANFO consumed using the air deck	tonnes	1.763
ANFO consumed in conventional blasting	tonnes	1.979
Saving of ANFO	tonnes	0.216
Theoretical blasted material volume	m³	4815
Actual blasted material volume	m³	4377
Actual blasted material mass	tonnes	8754
Blasting recovery percentage	%	91

It is obvious that the blasting result using a bottom air deck displayed a similar result with conventional blasting; this is in line with several previous studies, which stated that there is no reduction in the quality of fragmentation but instead increases its quality (Melnikov *et al,* 1978; Jhanwar and Jethwa, 2000; Jhanwar, 2011, 2012; Masda Rohal Sadiq, 2021). In this article, the blasting activity with an air deck only measured AN consumption; other parameters, such as average fragmentation measurement using software, ground vibration, and fly rock distance, have never been measured. Since the quarry location is quite remote and no other communities are nearby (an isolated island in North Morowali, Central Sulawesi), the concern for ground vibration affecting communities is negligible. The most important thing for a limestone quarry is the fragmentation size that the rock crusher feeding line can accommodate. The blasting recovery results at 91 per cent are quite satisfying, as this value is confirmed by other studies where the blasting recovery is generally within the acceptable range above the minimum 90 per cent recovery targeted (Hakim *et al,* 2016; Libriyon and Kopa, 2019). Figure 3 shows the picture of the before and after blasting, where it was clear that the fragmentation result was good (visually).

FIG 3 – (left) Before blast picture; (right) after blast picture.

Based on visual observation, Figure 3 indicates no failed blasts. There was a decent heave as a result, though the ANFO charge was reduced due to the implementation of the air deck.

Life cycle GHG emission reduction

The GHG emission reduction can be calculated based on the actual ANFO consumption and theoretical calculation when blasting is conducted without an air deck, multiplied by the life cycle GHG emission ratio for ANFO from Table 3. Figure 4 displays the life cycle GHG emission of blasting when an air deck is used versus the life cycle GHG emission reduction of blasting without an air deck.

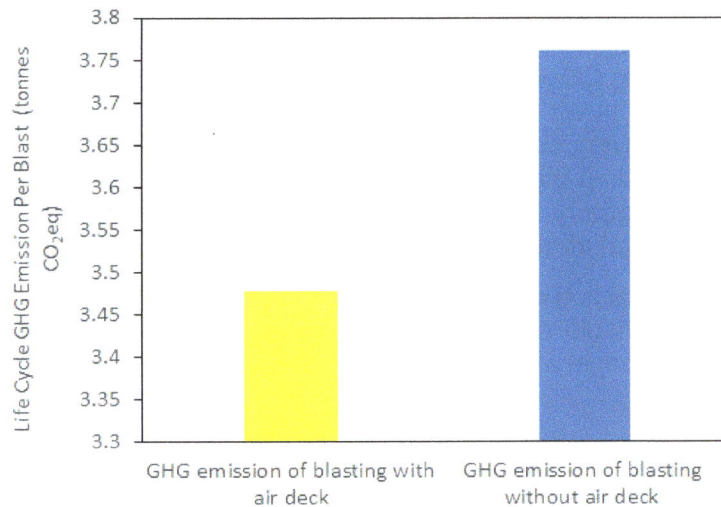

FIG 4 – Life cycle GHG emission comparison of blasting with and without an air deck.

From Figure 4, the GHG emissions of blasting per every blasting activity with air deck and without air deck are 3.4765 t CO_2-eq and 3.3761 t CO_2-eq, respectively. The life cycle GHG emission reduction per blast is 0.2836 t CO_2-eq or 283.6 kg CO_2-eq. On average, the blasting activity is carried out five times per month. By assuming that the blasting activity is conducted five times per month, the life cycle GHG emission reduced is at 1418 kg CO_2-eq per month, and annually, it can reduce the life cycle GHG emission up to 17 016 kg CO_2-eq or 17.016 t CO_2-eq. This annual GHG emission potential to be reduced is equivalent to approximately three passenger vehicle emissions annually (US EPA, 2024). At the quarry site, there are approximately three units of Light Vehicles (LVs), and by assuming that the GHG emission of LVs at the site is the same as that of passenger vehicles emissions, implementation of an air deck in blasting operation can offset the GHG emission of LVs at the quarry site. Annual production of limestone aggregates from the limestone quarry of PT GPM is approximately 500 000 t, and assuming that the quarry's Life cycle GHG emission is 3.13 kg CO_2-eq per t limestone rock product (Kittipongvises, 2017), the total GHG emission per annum is 1565 t CO_2-eq. Life cycle GHG emission reduction in limestone quarry from blasting activity is at 1.09 per cent of the total life cycle GHG emission from limestone quarry activity, while from blasting activity itself, the GHG emission reduction is at 7.5 per cent. Nevertheless, reducing GHG emissions should not only be done through drilling and blasting; a combined effort targeting every section and aspect of quarry operation is needed to achieve a greater reduction in GHG emissions, hence improving the quarry operation's sustainability by using fewer resources.

Reducing GHG emissions from blasting activity through air deck application at the bottom blasthole can be considered the simplest alternative to reduce GHG emissions from blasting operation, where at least 7.5 per cent of GHG emissions can be reduced. The deeper the hole depth, the longer the air deck is applied, and there should be a higher reduction in GHG emission obtained (Jhanwar and Jethwa, 2000; Hayat, Alagha and Ali, 2019; Masda Rohal Sadiq, 2021; Monjezi et al, 2022). Implementing an air deck for every blasting activity is not just for the limestone quarry operation; other types of quarry and mineral mining should consider air deck as one of the methods to help reduce the GHG emissions in their operation. The longer the air deck is applied, the more explosives can be reduced, ultimately reducing GHG emissions and increasing the operation's sustainability.

CONCLUSIONS

Evaluating the usage of air decks in limestone quarry operations shows promising results in reducing GHG emissions. However, the GHG emission reduction is quite small compared with the overall emission from the limestone quarry itself. However, starting from a simple way through air deck application in drilling and blasting operation, further efficiency to reduce carbon footprint in limestone quarry can be conducted and targeted per stage or section until all sections in the quarry operation achieve their carbon footprint.

The total reduction of ANFO is 216 kg, which correlates with 283.6 kg CO_2-eq of GHG emission reduction per blast. Approximately five blasts every month are conducted; therefore, it can reduce the GHG emissions to 1418 kg CO_2-eq per month or 17.016 t of CO_2-eq per annum. Reducing GHG emissions by implementing an air deck does not sacrifice the blasting result, whereas in this experiment, the blasting recovery obtained is 91 per cent. The GHG emissions reduction potential from implementing an air deck in this blasting activity is approximately 7.5 per cent. From the total GHG emission of limestone quarry PT GPM, approximately a 1.09 per cent reduction of life cycle GHG emission can be obtained from blasting activity.

The use of air decks for all blasting operations is not limited to limestone quarries; other quarry and mineral mining operations should also consider using air decks to lower GHG emissions. More explosives can be used with longer air decks, lowering GHG emissions and improving the operation's sustainability.

REFERENCES

Chiappetta, R F, 2010. NEW Blasting Technique PowerDeck™ Concept, Allentown, Pennsylvania: Blasting Analysis International, Inc.

Climateleaders.org, 2020. Improving Efficiency and Reducing GHG Emissions, Climateleaders.org. Available from: <https://www.climateleaders.org.au/case-studies/incitec-pivot-1/> [Accessed: 5 February 2021].

Climatiq, 2021. Emission Factor: PP (polypropylene), Climatiq. Available from: <https://www.climatiq.io/data/emission-factor/8f6dd6e1-31a3-43a3-b58e-5b7a21e10685> [Accessed: 9 March 2025].

Gilmartin, S, 2020. Air decking and the environment, *Quarry Magazine*. Available from: <https://www.quarrymagazine.com/air-decking-and-the-environment/> [Accessed: 9 March 2025].

Hakim, R N, Nurhakim, K and Ridha, A, 2016. Split Rock and Cutting Drill for Stemming Material in Rock Breaking Activities with Blasting [in Indonesian: Batu Split dan Cutting Bor untuk Material Stemming Dalam Kegiatan Pemberaian Batuan Dengan Peledakan], *Info-Teknik*, 17(2):263–272.

Hayat, M B, Alagha, L and Ali, D, 2019. Air Decks in Surface Blasting Operations, *Journal of Mining Science*, 55(6):922–929. https://doi.org/10.1134/S1062739119066307

Heryadi, R, Utama, J D H and Prasmoro, A V, 2024. Cost and fragmentation prediction of ammonium nitrate fuel oil versus bulk emulsion explosives, *IOP Conference Series: Earth and Environmental Science*, 1422(1). https://doi.org/10.1088/1755-1315/1422/1/012001

Hidayat, L, Puspitasari, R and Tantina, 2011. Analisis sensitivitas sebagai faktor penting dalam suatu pengambilan keputusan investasi, *Jurnal Ilmiah Ranggagading*, 11(2):134–140 [in Indonesian].

Jhanwar, J C and Jethwa, J L, 2000. The use of air decks in production blasting in an open pit coal mine, *Geotechnical and Geological Engineering*, 18:269–287.

Jhanwar, J C, 2011. Theory and Practice of Air-Deck Blasting in Mines and Surface Excavations: A Review, *Geotechnical and Geological Engineering*, pp 651–663. https://doi.org/10.1007/s10706-011-9425-x

Jhanwar, J C, 2012. Investigation into the influence of air-decking on blast performance in opencast mines in India: A study, *Blasting in Mining – New Trends*, pp 105–110. https://doi.org/10.1201/b13739-18

Kittipongvises, S, 2017. Assessment of environmental impacts of limestone quarrying operations in Thailand, *Environmental and Climate Technologies*, 20(1):67–83. https://doi.org/10.1515/rtuect-2017-0011

Lecksiwilai, N, et al, 2016. Net Energy Ratio and Life cycle greenhouse gases (GHG) assessment of bio-dimethyl ether (DME): produced from various agricultural residues in Thailand, *Journal of Cleaner Production*, 134(Part B):523–531. https://doi.org/10.1016/j.jclepro.2015.10.085

Liang, Q and Yu, L, 2023. Assessment of carbon emission potential of polyvinyl chloride plastics, *E3S Web of Conferences*, 393:0–3. https://doi.org/10.1051/e3sconf/202339301031

Libriyon, D P and Kopa, R, 2019. Evaluasi geometri peledakan terhadap fragmentasi batuan hasil peledakan digging Time alat gali muat dan recovery peledakan di Pit B PT, Darma Henwa Tbk Bengalon Coal Project Kalimantan Timur, *Bina Tambang*, 5(1):200–212. Available from: <http://ejournal.unp.ac.id/index.php/mining/article/view/107636> [in Indonesian].

Masda Rohal Sadiq, 2021. Implementasi bottom air deck dan expand pattern secara terintegrasi dalam rangka optimalisasi penggunaan bahan peledak di Pit South Pinang PT Kaltim Prima Coal, *Indonesian Mining Professionals Journal*, 3(1):17–30 [in Indonesian].

Melnikov, N V, et al, 1978. Blasting methods to improve rock fragmentation, *Acta Astronautica*, 5(11–12):1113–1127. https://doi.org/10.1016/0094-5765(78)90014-0

Monjezi, M, et al, 2022. Comparison and application of top and bottom air decks to improve blasting operations, *AIMS Geosciences*, 9(1):16–33. https://doi.org/10.3934/geosci.2023002

Ramadhan, L, Adnyano, A I A and Purnomo, H, 2020. Analisis Perbandingan Metode Air Deck Dan Non Air Deck Terhadap Ground Vibration Hasil Peledakan Di, *Perhapi*, 01(02):151–162 [in Indonesian].

Suppajariyawat, P, et al, 2019. Classification of ANFO samples based on their fuel composition by GC–M S and FTIR combined with chemometrics, *Forensic Science International*, 301:415–425. https://doi.org/10.1016/j.forsciint.2019.06.001

US Environmental Protection Agency (US EPA), 2024. Greenhouse Gas Emissions from a Typical Passenger Vehicle, US Environmental Protection Agency. Available from: <https://www.epa.gov/greenvehicles/greenhouse-gas-emissions-typical-passenger-vehicle> [Accessed: 9 March 2025].

Yang, X, Zhang, Y and Wang, J, 2022. Blasting Vibration Monitoring and a New Vibration Reduction Measure, *Journal of Engineering and Technological Sciences*, 54(1). https://doi.org/10.5614/j.eng.technol.sci.2022.54.1.12

Zarei, M, et al, 2022. The use of air decking techniques for improving surface mine blasting, *Arabian Journal of Geosciences*, 15(19):1–12. https://doi.org/10.1007/s12517-022-10826-8

The effect of social license on the exploitation of mines using machine learning algorithm

S Kohanpour[1], E Moosavi[2,3] and M Zakeri Niri[4]

1. Department of Mining Engineering, Electronic Campus, Islamic Azad University, Tehran, Iran. Email: sama.pars.2020@gmail.com
2. Department of Petroleum and Mining Engineering, South Tehran Branch, Islamic Azad University, Tehran, Iran. Email: se_moosavi@iau.ac.ir
3. Research Center for Modeling and Optimization in Science and Engineering, South Tehran Branch, Islamic Azad University, Tehran, Iran. Email: se.moosavi@yahoo.com
4. Department of Civil, Eslamshahr, Islamic Azad University, Eslamshahr, Iran. Email: zakeriiau@gmail.com

ABSTRACT

Optimum use of natural resources and sustainable development are critical and challenging issues in today's world. One of the important economic sectors that is highly dependent on natural resources is the mining sector. Exploitation of mines as one of the important factors of economic growth and development of different regions has always attracted the attention of many researchers and policy makers. In the meantime, one of the major challenges is obtaining a social license (SL) to exploit mines, which can have profound effects on the process of extracting and using mineral resources. SL to Operate means obtaining the consent and support of the local community and various stakeholders to carry out mining activities. This license is not only limited to obtaining consent through government laws and regulations, but also includes gaining the trust and cooperation of the local community, non-governmental organisations and other relevant stakeholders. Failure to obtain a social permit can lead to delays in mining projects, increased costs, and even a complete stop of mining operations. The study employs machine learning algorithms to analyse extensive mining operations and community interaction data sets, including support vector machine (SVM), and random forests (RF). These algorithms not only facilitate the identification of underlying patterns but also enable the prediction of potential future impacts, thereby providing a robust analytical tool for stakeholders. These algorithms not only facilitate the identification of underlying patterns but also enable the prediction of potential future impacts, thereby providing a robust analytical tool for stakeholders. This research tries to provide a predictive model by analysing available data and identifying influential patterns that can help mining companies in obtaining SL and improving extractive processes. The RF demonstrates better performance compared to the SVM, with an accuracy of 88 per cent and an F1 score of 86 per cent, based on the evaluation metrics. Therefore, RF is a more suitable option for predicting SL acquisition.

INTRODUCTION

Social License to Operate (SLO) is described as an informal, unspoken social agreement with society or a specific social group that allows an extraction or processing operation to enter a community, begin its activities, and continue operating. Various scholars have discussed this concept, with Jim Cooney being recognised as a pioneer in 1997 (Franks and Cohen, 2012).

From April 2013 to December 2018, over 60 per cent of social conflicts were related to social-environmental issues (Defensoría del Pueblo, 2018). These disputes, which focus on extraction activities, have impacted companies operating in the country, as local communities fail to see the benefits stemming from these operations. Jaskoski (2014) noted that the most intense conflicts between communities and companies were driven by complaints that companies did not fulfill promised benefits and by the lack of sufficient community involvement in the distribution of these benefits.

Given the complexities and multiplicity of factors affecting social licensing, using traditional methods to analyse and predict these factors may be ineffective. In the meantime, machine learning algorithms, as a powerful and modern tool, can play an important role in analysing data and predicting stakeholder behaviour patterns. Machine learning can help improve the decision-making process and enhance the level of cooperation between mining companies and local communities by analysing big data and identifying hidden patterns (Jamarani *et al,* 2024).

As highlighted by Moffat and Zhang (2014), there has been limited research on the factors that influence host communities' acceptance of mining projects. This study seeks to address that gap by exploring the key elements that determine the granting of social licenses to operate within the mining regions of Peru. To identify these factors, we examined the perspectives of stakeholders regarding two mining companies, which serve as representatives of the large-scale mining sector. These companies operate in distinct geographic locations, each characterised by different population structures and socio-economic conditions (Sícoli, 2016).

Research on the impact of social licensing on mining exploitation using machine learning algorithms is important and necessary in several ways. Unsustainable exploitation of mineral resources can lead to environmental degradation and a decrease in the quality of life of local communities. By obtaining social licensing and better understanding the needs and concerns of the local community, sustainable development and conservation of natural resources can be contributed to (Komnitsas, 2020). Dissatisfaction and opposition from the local community can lead to project suspension, increased costs, and reduced productivity. By identifying and managing the factors affecting social consent, tensions can be reduced and cooperation between mining companies and local communities can be increased.

Hormozgan province, with its rich mineral resources and special geographical location, is one of the important regions of Iran in the field of mining activities. Environmental and social issues in this region are of particular importance, and the results of this research can help to better manage these challenges (Hedayat Allah, Ahmadi Kahanali and Sharyari, 2023). Using machine learning algorithms to analyse social and environmental issues in the mining sector is a novel and efficient approach that can lead to more accurate and applicable results. This approach can be used as a model for other similar research in different fields.

Considering these studies and the importance of the subject, the main issue of this research is to investigate and analyse the impact of social licensing on mining exploitation in Hormozgan Province using machine learning algorithms. This research seeks to answer the question of how to identify the factors affecting social licensing, using existing data and machine learning methods, and provide predictive models to improve extraction processes and increase cooperation with the local community. The results of this research can greatly help policymakers and mining managers in making better decisions and improving the level of cooperation with the local community as a model. This study employs a qualitative exploratory approach, selecting two mining companies as case studies to develop a model that identifies key factors influencing the granting of SLO in Hormozgan. This research design is justified by the need for an in-depth analysis of the complex interactions between various stakeholders, as understanding these relationships requires a detailed exploration of the phenomenon. Consequently, the study investigates the interactions of local communities with issues surrounding the granting of SLO, exploring the dynamics within each company's context and examining the causes and potential outcomes of these relationships.

IMPACT OF SOCIAL LICENSING ON MINING EXPLOITATION

As one of the key factors in the success or failure of mining projects, social licensing refers to the acceptance and approval of mining activities by the local community and stakeholders. In areas where mines are located close to local communities, there are concerns about the environmental and social impacts of mining. Therefore, obtaining a social license for mining companies means building trust and confidence among community members. Transparency and effective communication with the local community are key factors in obtaining this license, and mining companies must provide accurate information about their activities and impacts. A mining license ensures that mining activities are carried out in compliance with laws and regulations. In Hormozgan Province, obtaining these licenses means complying with environmental and social standards. The licensing process includes submitting extraction plans, assessing environmental and social impacts, and respecting the rights of local communities. Managing potential conflicts with local communities and holding consultation meetings with local residents are key challenges in this process. Continuous monitoring of the implementation of permit conditions and monitoring of environmental and social impacts help maintain a balance between industrial development and environmental protection, leading to sustainable development of the region.

METHODOLOGY

The main objective of this study is to investigate the impact of social licensing on mining exploitation in Hormozgan province using machine learning algorithms. The data of this study are divided into social data, economic data, and data. Data collection method in this study is to design and implement structured surveys among the local community in the mining areas of Hormozgan province. The surveys will include questions about the attitudes, behaviours, and expectations of the local community from mining activities. Conducting in-depth interviews with key people in the local community, mining company managers, and local experts to collect qualitative data, collecting information related to the economic performance of mining companies from financial reports, economic databases, and other reliable sources, and collecting data related to environmental impacts from environmental reports available in mining companies and the Environmental Protection Organisation. Also, the use of satellite images and remote sensing data to analyse changes in vegetation and environmental impacts was important in this study.

The algorithm used in this research is the selection of appropriate machine learning algorithms such as Random Forest (RF) and Support Vector Machine (SVM) for data analysis. The materials, equipment and standards used in this research are statistical and machine learning software such as Python, R, and other specialised software for data analysis, remote sensing data, questionnaires and survey tools. This research method, considering the purpose, type of data and implementation method, comprehensively and accurately examines the impact of social licensing on mining exploitation in Hormozgan province using machine learning algorithms and provides practical and operational results.

Random forest

Random Forest is a type of supervised machine learning algorithm that can be used for both classification and regression machine learning. This method is an ensemble learning technique that predicts using a combination of classifiers and improves model performance. The random forest algorithm consists of several random trees for specified subsets of data and uses averaging to help improve prediction accuracy.

RF uses a majority voting method to obtain prediction results through bootstrapping and a feature random selection strategy. The RF flow chart is shown in Figure 1. RF performs well in regression and classification tasks, offering robustness, anti-overfitting ability, and efficient parallel processing capabilities. Bootstrapping is the process of training different subsets of the data set simultaneously using different decision trees, ensuring that each decision tree in the RF model is unique, thus increasing the diversity and generalisation ability of the model (Josso *et al*, 2023). As the number of decision trees in the random forest increases, the misclassification rate decreases sharply before stabilising. When the number of trees reaches around 500, the mis-classification rate stabilises. Therefore, the number of trees is set to 500 for training the random forest model, and the number of features for classification nodes was determined to be the square root of the total number of features.

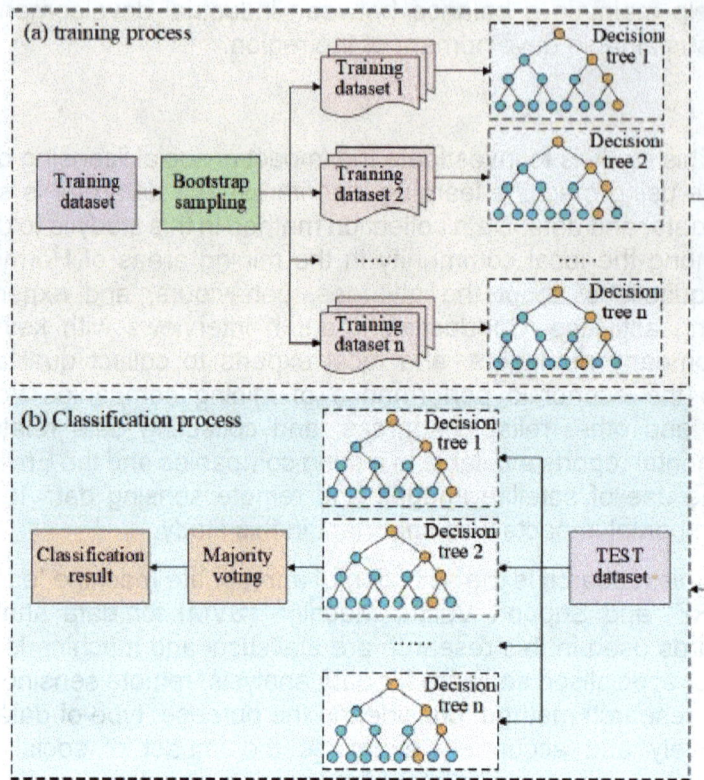

FIG 1 – Random forest algorithm.

Support vector machine

SVM is another supervised algorithm method that is commonly used for classification problems. To use SVM, raw data is defined as scattered points in an n-dimensional space. (n is the number of features available for the data). The main goal of this algorithm is to create a hyperplane or decision boundary by which different data in a data set can be separated. Figure 2 illustrates the support vectors, which are the data points that define the optimal hyperplane in the SVM model.

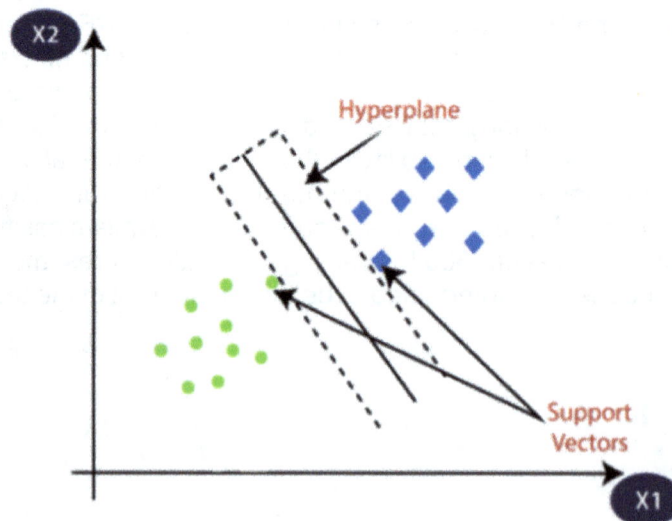

FIG 2 – The data points as SVM determining the hyperplane (García-Gonzalo *et al*, 2016).

Case study

Compared to other regions, Hormozgan province faces unique challenges in obtaining social permits due to its proximity to the sea and sensitive ecosystems. The main differences include the greater importance of environmental impacts and water resource management. In other regions, economic impacts may be more important. However, there are also similarities; for example, in all regions, effective communication with the local community and the use of new technologies can help improve social permits. Cultural and social differences also play an important role in obtaining social permits in some regions. This research is based on two initial assumptions derived from preliminary observations. There are significant differences between the two cases:

- Metallic and non-metallic mines represent two distinct cases within the mining industry, each exhibiting different relationships with the communities where they operate.

- The extraction activities of these companies are geographically situated at the outer limits of the country: Metallic mines and non-metallic mines is located in various locations within Hormozgan. Due to their geographic locations, the needs of each community may differ, influenced by cultural variations.

There are also similarities that justify analysing both classifications within the same study:

- Both categories began operations in early 1998s with private capital, and their shares are a mix of local and foreign investment.

- Both categories are considered significant players in Hormozgan mining sector during the 2006–2022 period.

Mining exploitation in Hormozgan requires going through legal procedures and obtaining the necessary permits from relevant institutions. These steps include geological studies, applying for an exploration permit, environmental assessments, and finally obtaining an exploitation permit. An important aspect of social permits is interaction and cooperation with the local community, which includes consulting with local residents and creating shared benefits. The limitations and impacts on each environmental parameter are as shown in Table 1.

TABLE 1
Limits and impacts on each environmental parameter.

Destructive effects	Value	Beneficial effects	Value
Very high negative impact	-5	Very high beneficial effect	5
Significantly damaging impact	-4	Significant beneficial effect	4
Moderately damaging impact	-3	Moderate beneficial effect	3
Slightly damaging impact	-2	Low beneficial effect	2
Very slightly damaging impact	-1	Very low beneficial effect	1

DATA ANALYSIS

Following the validation of the initial assumptions, interview responses were systematically categorised into three analytical groups. This classification served multiple purposes: to identify recurring themes across interviews, to compare varying individual perspectives within each case study, and to analyse the frequency and potential interconnections between categories. By doing so, the study aimed to pinpoint the critical factors that influence communities in granting a social license to operate for mining companies.

Mining effects

As an initial step in developing a model for granting social licenses, individuals were consulted about their perceptions of the impacts of mining activities in their communities. Their responses highlighted both positive and negative effects. On the positive side, mining was linked to economic growth, social development, corporate social responsibility initiatives, and employment opportunities. Conversely, negative impacts included environmental contamination, infrastructure damage, insufficient social investment, and declining benefits due to reduced mining activity (Figure 3).

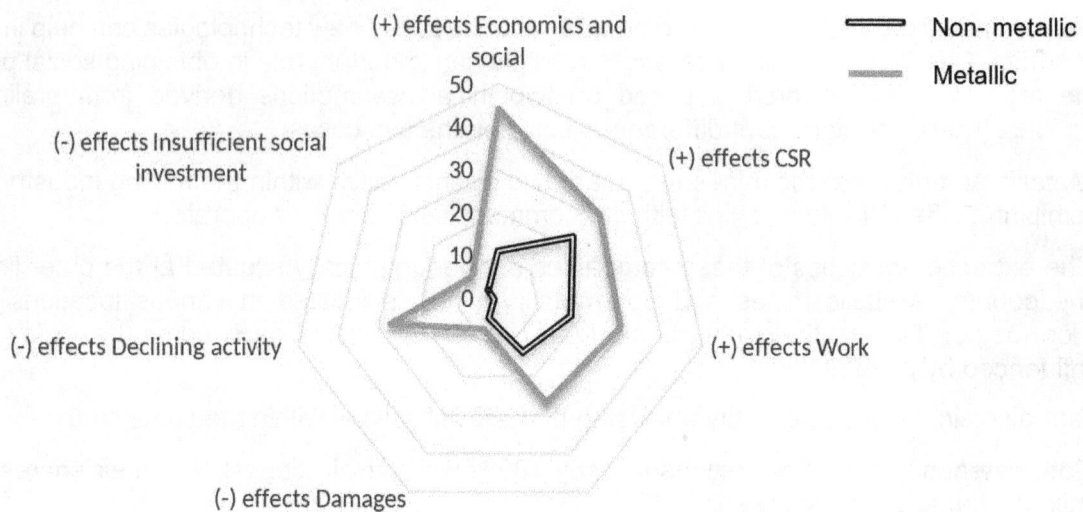

FIG 3 – Mining effects answers frequencies.

In metallic mine, two contrasting effects were most frequently mentioned. On one hand, respondents in urban areas emphasised economic and social benefits stemming from mining operations. On the other hand, concerns about environmental contamination—particularly water pollution—were predominant among negative perceptions:

- There have been improvements in economic indicators due to the mining tax revenue, leading to better road infrastructure, electricity, and sanitation. These are clear advancements because there is more financial inflow (Frente de Defensa representative).

- People claim—though I don't have concrete data—that the levels of heavy metals in drinking water have risen (university student).

- Some residents say, 'I don't know, but my water looks bluish, cloudy, with strange particles.' That's what we've been hearing (municipality official).

In non-metallic mine, the most frequently mentioned benefits were linked to the company's corporate social responsibility initiatives. Similar to the metallic case, concerns about pollution were raised; however, interviewees struggled to pinpoint specific environmental damages directly caused by the company.

- The company actively engages with local districts, addressing their specific needs and demands, particularly those within the mine's direct area of influence (advisor to the Mayor).

- All the pollution is towards the wind (taxi driver).

Conflicts

To further explore the factors influencing the granting of social licenses, interviewees were asked about challenges and conflicts between mining companies and local communities. The responses were categorised under 'Conflicts,' differentiating various issues such as water-related concerns, socio-economic disparities linked to mining, corruption, failure to meet commitments by both companies and the government, and the presence of foreign entities all of which held varying degrees of significance across the two case studies (Figure 4).

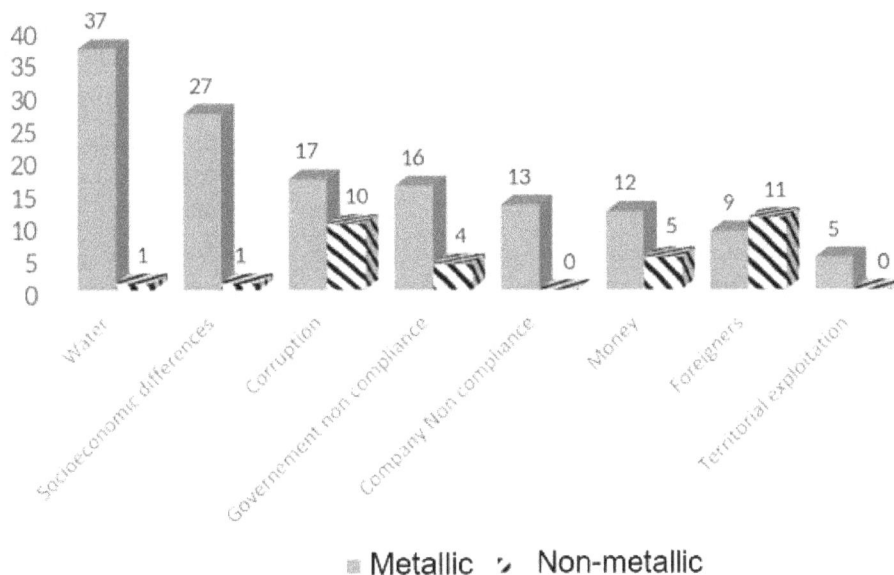

Metallic Non-metallic

FIG 4 – Sources of conflicts answers frequencies.

In metallic mine, the most frequently cited issues were water-related concerns, socio-economic inequalities, corruption, and governmental noncompliance.

The 'Water' subcategory encompassed concerns about water scarcity, fears of contamination due to mining operations, and the absence of a reliable government authority to ensure water quality:

- The company's primary responsibility should be ensuring access to clean water before engaging in mining activities. It would be valuable to ask whether they provide potable water to the local community (City Council Member).

- The company's primary responsibility should be ensuring access to clean water before engaging in mining activities. It would be valuable to ask whether they provide potable water to the local community (City Council Member).

The category of 'socio-economic differences' encompasses responses highlighting disparities in the treatment of local residents by individuals associated with the company, including engineers, their spouses, and mining workers. It also reflects concerns about income inequalities and the historical context of the community:

- Engineers perceive themselves as belonging to a superior social class, which hinders the establishment of equal and respectful relationships with the local community. (Municipality Official).

The 'Corruption' category highlights concerns regarding bribery of government officials by mining companies and the mismanagement of public funds by authorities:

- Unfortunately, the operational model of mining companies in Hormozgan follows a vicious cycle dividing communities, marginalising local populations, and bribing influential community leaders (Frente de Defensa representative).

The 'Government Noncompliance' subcategory reflects interviewees' concerns that the government fails to mediate effectively between companies and communities and does not properly oversee mining operations:

- The Central Government rarely acts as an effective mediator, and, most importantly, remote communities are often neglected (university student).

In non-metallic, conflicts are predominantly associated with corruption and the employment of non-local workers. Corruption concerns stem from allegations that mining companies resolve disputes through financial incentives rather than fostering genuine community engagement:

- Regrettably, the community relations team of the mining company often resorts to bribing local leaders and community board members, using monetary incentives to settle issues instead of addressing them substantively (university professor, referring to a mining project).

The issue of 'foreigners' pertains to concerns regarding the mining company's preference for hiring workers from outside the local community rather than employing residents. This practice is perceived as creating social tensions and limiting economic opportunities for locals:

- Some workers come from other regions and assume they can integrate seamlessly. But their behaviours and attitudes do not align with the local community (advisor to the Mayor).

Government participation

A key stakeholder identified in the analysis is the government, represented by various agencies and administrative levels. Institutional representatives perceive themselves as facilitators in fostering trust between mining companies and local communities by providing information, mediating disputes, and overseeing corporate practices. However, they also acknowledge persistent governmental shortcomings in addressing community needs, ensuring effective engagement between mining operations and local populations, and implementing initiatives that enhance the overall quality of life:

- Local governments are ineffective, and the royalty funds, along with other resources generated by the company, are either mismanaged or embezzled, as is evident in certain cases' (Osinergmin official).

The dynamics between companies, government entities, and community stakeholders varied significantly based on the specific context of each case.

RESULTS

The results of the machine learning algorithms are presented in this section. These results include accuracy, recall accuracy, positive predictive accuracy, and F1 score for each algorithm (Table 2).

TABLE 2

Performance comparison of different algorithms.

Methods	F1	Precision	Recall	Accuracy
RF	86%	87%	86%	88%
SVM	81%	82%	80%	83%

RF performs well and is close to SVM. These algorithms have provided better results due to their ability to handle complex and nonlinear data. SVM shows the lowest accuracy, which may be due to its sensitivity to tuning parameters or the need for more optimisation data.

CONCLUSION

This research has shown that social licensing plays an important role in the successful and sustainable exploitation of mines. Emphasizing positive engagement with the local community and compliance with environmental and social requirements can lead to reduced operational barriers and increased productivity. Mines that have been able to gain the consent of the local community and manage their environmental impacts have been significantly more successful in obtaining social licensing. These findings highlight the importance of implementing comprehensive environmental and social management programs. The role of policymakers and government institutions in regulating and supporting mining activities is also very important. The development and implementation of stricter laws and regulations to protect the environment and encourage companies to use new and environmentally friendly technologies can have positive effects on mining exploitation. Continuous engagement with the local community and transparency in providing information can also lead to increased trust and support from the local community.

Finally, the use of green and advanced technologies can help improve efficiency and reduce negative environmental impacts. New technologies in waste management, renewable energy, and smart monitoring systems can lead to process optimisation and increased productivity of mining operations. Together, these measures can help achieve sustainable exploitation and increase local community satisfaction.

REFERENCES

Defensoría del Pueblo, 2018. Social Conflict Report N°178 [in Spanish: Reporte de Conflictos Sociales], December 2018, Office of the Deputy for the Prevention of Social Conflicts and Governance [Adjuntía para la Prevención de Conflictos Sociales y la Gobernabilidad]. Available from: <https://www.defensoria.gob.pe/wp-content/uploads/2019/01/Conflictos-Sociales-N°-178-Diciembre-2018.pdf>

Franks, D and Cohen, T, 2012. Social license in design: constructive technology assessment within a mineral research and development institution, Brisbane (Australia), Center for Social Responsibility in Mining, Sustainable Minerals Institute, University of Queensland.

García-Gonzalo, E, Fernández-Muñiz, Z, Garcia, N, Paulino, J, Sánchez, A and Menéndez, M, 2016. Hard rock Stability Analysis for Span Design in Entry-Type Excavations with Learning Classifiers, *Materials*, 9:531. https://doi.org/10.3390/ma9070531

Hedayat Allah, N, Ahmadi Kahanali, R and Sharyari, M, 2023. A futurology of the sustainable development process of Hormozgan Province, Iran, *Futures*, 154:103248. https://doi.org/10.1016/j.futures.2023.103248

Jamarani, A, Haddadi, S, Sarvizadeh, R, Haghi Kashani, M, Akbari, M and Moradi, S, 2024. Big data and predictive analytics: A systematic review of applications, *Artificial Intelligence Review*, 57. https://doi.org/10.1007/s10462-024-10811-5

Jaskoski, M, 2014. Environmental licensing and conflict in Peru's mining sector: a path-dependent analysis, *World Dev*, 64:873–883. https://doi.org/10.1016/j.worldev.2014.07.010

Josso, P, Hall, A, Williams, C, Le Bas, T, Lusty, P and Murton, B, 2023. Application of Random-Forest Machine Learning Algorithm for Mineral Predictive Mapping of Fe-Mn Crusts in the World Ocean, *Ore Geol Rev*, 162:105671.

Komnitsas, K, 2020. Social License to Operate in Mining: Present Views and Future Trends, Resources, 9:79. https://doi.org/10.3390/resources9060079

Moffat, K and Zhang, A, 2014. The paths to social licence to operate: an integrative model explaining community acceptance of mining, *Resource Policy*, 39:61–70. https://doi.org/10.1016/j.resourpol.2013.11.003

Sícoli, C, 2016. Factors determining the social license to operate in Peru: the case of large-scale mining [in Spanish: Factores que determinan la licencia social para operar en Perú: el caso de la gran minería], PhD thesis (unpublished), Barcelona, Spain: Polytechnic University of Catalonia, Department of Business Organization [España: Universitat Politècnica de Catalunya, Departament d'Organització d'Empreses].

Haul truck intelligence system for effective decarbonisation

P McBride[1] and A Soofastaei[2]

1. Australian Director, Cascadia Scientific, Sydney NSW 2000.
   Email: pmcbride@cascadiascientific.com
2. Australian Performance Assurance Manager, Cascadia Scientific, Sydney NSW 2000.
   Email: asoofastaei@cascadiascientific.com

ABSTRACT

Mining haul trucks are responsible for up to 75 per cent of mine site emissions, posing a critical challenge for decarbonising open pit operations. While zero-emission technologies offer long-term solutions, they remain economically and logistically constrained. This study introduces an AI-powered optimisation framework integrating high-precision fuel monitoring, real-time diagnostics, and machine learning analytics to improve fuel efficiency. Deployment across four mine sites reduced fuel consumption by 6.2 per cent to 10.3 per cent, surpassing Australia's 4.9 per cent CO_2 reduction mandate. Key interventions included operator feedback systems, predictive maintenance, and haul route optimisation. The system demonstrated a return on investment within six months and integrated seamlessly with existing fleet management software. These findings underscore the framework's scalability, cost-effectiveness, and alignment with sustainability and operational goals, positioning it as a viable bridge to zero-emission haulage.

INTRODUCTION

This study introduces a practical and scalable solution to one of the mining sector's most critical decarbonisation challenges: accurately measuring haul truck fuel consumption using high-precision flowmeters; visualising fuel usage; identifying potential savings opportunities; forecasting future fuel consumption; and optimising fuel efficiency through artificial intelligence. Deployment across four mining operations achieved fuel consumption reductions ranging from 6.2 per cent to 10.3 per cent, with an average return on investment realised in under six months. Key outcomes included measurable enhancements in operator performance, predictive maintenance scheduling, and haul road condition management. These results exceeded Australia's Safeguard Mechanism requirement of a 4.9 per cent reduction in CO_2 emissions and demonstrated strong technological feasibility and economic viability under real-world mining conditions. Notably, the AI-driven system integrates seamlessly with existing fleet management platforms, enabling rapid adoption without requiring extensive modifications to legacy infrastructure.

The mining industry is at a critical crossroads as it faces intensifying pressure to reduce greenhouse gas (GHG) emissions while maintaining operational efficiency to meet the soaring global demand for critical minerals essential to modern infrastructure and technology. The need to align with ambitious decarbonisation targets and climate policies has placed the sector under heightened scrutiny, necessitating the adoption of innovative, data-driven operational models that ensure continued economic viability, productivity, and supply chain security. Among the most significant contributors to carbon emissions in open pit mining, haul trucks consume between 15 and 45 million litres of diesel per site annually (MCA, 2023), making them a substantial source of Scope 1 emissions. Therefore, the imperative for scalable, immediately deployable emissions reduction strategies is paramount to reducing environmental impact while maintaining operational resilience and financial sustainability.

The economic burden of haul truck fuel consumption is equally significant, with diesel expenditures comprising 18 per cent to 25 per cent of total mining operational costs (MinEx, 2023). With fuel price volatility, increasingly stringent carbon taxation mechanisms, and escalating emissions regulations, mining enterprises must embrace cost-effective fuel optimisation strategies that offer tangible economic benefits while significantly curbing carbon footprints. Additionally, stakeholder expectations and ESG (Environmental, Social, and Governance) compliance requirements push the industry toward greater transparency, accountability, and sustainability, reinforcing the need for advanced fleet management solutions and intelligent fuel efficiency interventions.

While zero-emission haulage technologies, including battery-electric and hydrogen-powered trucks, present a promising long-term solution, their broad implementation remains constrained by technological, economic, and infrastructural limitations. Most mining operations are in remote regions with inadequate grid infrastructure and insufficient hydrogen refuelling networks, impeding near-term large-scale electrification and alternative powertrain deployment. Furthermore, high capital expenditures, charging constraints, and energy storage limitations present formidable barriers to immediate fleet-wide adoption. Consequently, the industry must prioritise short-term, high-impact operational improvements that deliver immediate emissions reductions, enhance fuel efficiency, and optimise fleet performance while concurrently establishing a strategic foundation for the long-term transition to zero-emission mining logistics.

This study introduces a comprehensive, data-driven framework to bridge the gap between existing diesel-dependent haul truck operations and future zero-emission technologies. It proposes an AI-powered optimisation strategy designed to enhance fuel efficiency, minimise emissions, and generate actionable operational insights. Through the integration of high-precision telemetry, real-time fuel monitoring, and machine learning-driven analytics, this research presents a scalable, cost-effective approach that allows mining operations to maximise the efficiency of current fleets while gradually integrating next-generation haulage solutions. This methodology aligns with Australia's decarbonisation objectives, which mandate a 43 per cent reduction in emissions below 2005 levels by 2030 while contributing to the broader commitment of achieving net-zero carbon neutrality by 2050 (Clean Energy Regulator, 2023).

The proposed AI-enabled framework leverages real-time sensor data, predictive maintenance analytics, and advanced operational modelling to deliver intelligent, adaptive recommendations for optimising haul truck operations, reducing idle time, refining route efficiency, and improving operator performance. By incorporating machine learning algorithms and real-time analytics, the system ensures that fuel efficiency enhancements are continuously monitored, refined, and tailored to each mine site's unique operational conditions. This leads to substantial reductions in fuel wastage and carbon intensity per unit of ore transported. Furthermore, this approach extends equipment lifespan, reduces maintenance costs, and enhances overall fleet sustainability, reinforcing its role as a practical, scalable emissions reduction strategy.

Beyond the environmental and regulatory advantages, AI-driven haul truck optimisation offers a strong economic value proposition, delivering mining enterprises a clear return on investment (ROI) through reduced fuel expenditures, extended maintenance intervals, and increased fleet utilisation rates. The system mitigates unexpected downtime and costly mechanical failures, ensuring compliance with evolving emissions regulations while maintaining long-term operational efficiency. The findings of this study underscore the transformative impact of machine learning and real-time data analytics in revolutionising mining fleet logistics, demonstrating that technology-driven optimisation can serve as an immediate catalyst for decarbonisation while preserving economic stability.

As the mining sector accelerates toward net-zero emissions objectives, AI-enhanced operational intelligence emerges as a crucial enabler in transitioning from diesel-powered fleets to electrified and hydrogen-based haulage systems. This research establishes that machine learning-powered optimisation strategies are indispensable for achieving near-term and long-term emissions reduction goals, ensuring that mining enterprises remain technologically advanced, operationally resilient, and environmentally responsible. Ultimately, integrating data-centric decision-making, automation, and AI-driven predictive analytics will drive the next generation of sustainable mining fleet management, setting a new benchmark for efficiency, compliance, and environmental leadership in the sector.

Haul truck emission sources

The Haul Truck Emission Sources graph (Figure 1) illustrates the key contributors to greenhouse gas emissions in mining haulage operations. The largest source of emissions is engine combustion, accounting for approximately 50 per cent of total emissions, primarily due to the combustion of diesel fuel to generate power. Idle emissions contribute 20 per cent, often resulting from inefficient haul cycle planning or excessive waiting times. Other significant sources include tire wear (10 per cent) and brake wear (8 per cent), which release particulate matter and contribute to environmental degradation. Aerodynamic drag (7 per cent) impacts fuel efficiency, particularly in high-speed haulage scenarios, while rolling resistance (5 per cent) stems from road conditions and tire performance, further affecting fuel consumption (Soofastaei *et al*, 2022).

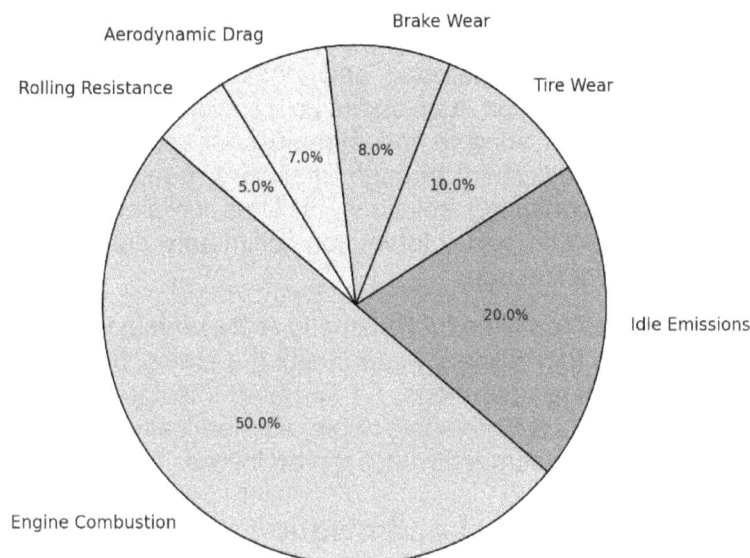

FIG 1 – Haul truck emission sources.

Understanding these emission sources allows for targeted optimisation strategies, such as operator performance enhancements, predictive maintenance, and haul road management, to mitigate environmental impact while improving fuel efficiency.

Challenges in fuel efficiency monitoring

Accurately monitoring and optimising fuel efficiency in haul truck operations remains a significant challenge due to the inherent limitations of conventional assessment methodologies. Traditionally, fuel efficiency evaluations have relied on Engine Control Module (ECM) estimates, which, despite their widespread adoption, exhibit accuracy deviations of up to 12 per cent (Soofastaei *et al*, 2022). This margin of error results in substantial inefficiencies, leading to missed opportunities for optimisation and unquantified excess fuel consumption.

Empirical analyses indicate that between 5 per cent and 10 per cent of total fuel consumption in mining haulage operations is attributable to inefficiencies that remain undetected under conventional monitoring systems (Paraszczak and Fytas, 2022). These inefficiencies stem from various operational factors, including suboptimal operator behaviours, inconsistencies in payload distribution, fluctuating haul road conditions, and performance disparities across engine types. Reliance on ECM data alone fails to capture high-resolution, real-time fuel consumption metrics, thereby hindering the implementation of targeted, data-driven interventions designed to minimise fuel waste and enhance operational efficiency.

Extensive field studies involving over 100 haul trucks have revealed a transparency deficit of approximately 94 per cent between reported and actual fuel consumption, exposing systemic deficiencies in existing fuel accounting methodologies. This disparity highlights the lack of granularity and precision in current fuel reporting frameworks, which impedes efforts to ensure accurate fuel tracking, financial accountability, and regulatory emissions compliance. Without robust, high-fidelity fuel monitoring infrastructure, decision-makers lack the empirical foundation necessary to drive substantial improvements in fuel efficiency, perpetuating inefficiencies that exert direct financial, environmental, and operational burdens on mining enterprises.

Additional challenges arise from inconsistencies in fuel distribution logging, inaccuracies in refuelling event records, and integration gaps within legacy fleet management systems, all of which exacerbate fuel monitoring deficiencies. The absence of real-time fuel telemetry integration within traditional fleet monitoring solutions creates blind spots in operational visibility, preventing the timely implementation of strategic fuel conservation measures. Moreover, external variables such as climatic conditions, altitude variations, and haul road degradation significantly influence fuel consumption rates, yet these factors remain inadequately accounted for in static ECM-based assessments.

To overcome these challenges, there is an urgent need for real-time, AI-enhanced fuel monitoring systems that integrate high-precision fuel flow metres, advanced telematics, and machine learning-driven analytics. These technologies provide a comprehensive, data-rich alternative that enhances fuel tracking accuracy, operational transparency, and predictive insights for fleet optimisation. By shifting from reactive, post-consumption fuel audits to proactive, continuous fuel management frameworks, mining operations can achieve measurable cost reductions, improve sustainability initiatives, and optimise haulage efficiency through intelligent, adaptive decision-making models. Furthermore, these advanced monitoring solutions facilitate predictive maintenance strategies, allowing for early anomaly detection and intervention, ultimately contributing to enhanced fleet reliability and extended equipment lifespan.

Integrating AI-driven analytics and real-time fuel telemetry represents a transformative advancement in mining logistics. It ensures that fuel efficiency improvements are continuously refined and adapted to the evolving conditions of mining operations. By adopting intelligent, data-driven solutions, the mining sector can transition towards a more sustainable, economically efficient, and environmentally responsible future while maintaining competitiveness in an increasingly regulated global landscape.

Proposed solution – AI-driven fuel optimisation

This study presents a machine learning-integrated haul truck intelligence framework, engineered to revolutionise fuel optimisation through high-resolution fuel telemetry, real-time vehicular diagnostics, and predictive analytics. By employing AI-driven optimisation algorithms, this system enables quantification, diagnosis, and systematic mitigation of inefficiencies throughout the haul cycle, improving fleet-wide fuel efficiency, minimising operational costs, and reducing carbon emissions.

The proposed framework implements a multi-layered data acquisition and analytical model, synthesizing real-time fuel flow telemetry, engine diagnostics, GPS-based geospatial tracking, and predictive analytics to construct a high-fidelity, dynamic representation of haul truck energy consumption. Unlike conventional monitoring methodologies that primarily depend on historical Engine Control Module (ECM) estimates, which are often subject to inaccuracies, this AI-driven system delivers granular, real-time insights into fuel consumption patterns, operator behaviour, haul road conditions, and mechanical performance anomalies. These capabilities facilitate proactively identifying and remedying inefficiencies before they escalate into excessive fuel wastage or premature component failures.

Key components of the AI-driven optimisation framework

- **Real-time fuel flow telemetry** – Captures and processes high-resolution fuel consumption data with a margin of error as low as **±0.5 per cent**, a significant improvement over traditional Engine Control Module (ECM) estimates, which typically exhibit error rates of up to **±12 per cent**. This enhanced precision enables the early detection of inefficiencies and deviations from optimal performance metrics, forming a critical foundation for effective optimisation.

- **AI-powered predictive analytics** – This system employs advanced machine learning algorithms to detect fuel inefficiencies, predict maintenance requirements, and generate adaptive, data-informed interventions. It is designed to achieve high sensitivity levels and can identify deviations as small as **2 per cent to 3 per cent** in fuel consumption and operational parameters, ensuring timely and precise corrective actions.

- **Operator performance analytics** – Monitors and assesses driving patterns, throttle modulation, braking efficiency, and idling behaviours, providing real-time feedback for performance optimisation.

- **Haul road condition monitoring** – This deploys GPS tracking, topographic modelling, and terrain analysis algorithms to correlate road conditions with fuel inefficiencies, allowing for real-time route optimisation and proactive road maintenance scheduling.

- **Automated anomaly detection** – Identifies irregularities in engine diagnostics, fuel system operations, and vehicular telemetry, enabling pre-emptive maintenance actions that mitigate unexpected downtime and prevent unnecessary fuel inefficiencies.

Scalability and economic viability

The empirical findings of this study demonstrate that the AI-driven fuel optimisation framework is not only technologically robust but also economically viable. By enhancing fuel efficiency, reducing maintenance expenditures, and extending the operational lifespan of critical mining equipment, this approach yields a measurable return on investment (ROI) within six months. The framework is highly scalable across diverse mining environments, accommodating variations in fleet sizes, geological conditions, and operational constraints. Additionally, seamless integration with existing mine fleet management systems ensures minimal operational disruptions while enhancing interoperability with existing digital infrastructure.

Long-term sustainability and operational resilience

Beyond immediate fuel efficiency improvements, this AI-powered framework is a strategic enabler for long-term sustainability by aligning mining operations with global decarbonisation initiatives. The framework enhances operational resilience by continuously refining fuel consumption models, operational workflows, and predictive maintenance strategies and positions mining enterprises for a seamless transition toward low-emission haulage solutions, including battery-electric and hydrogen-powered trucks.

This study underscores the transformational role of AI-driven fuel optimisation in modern mining logistics, illustrating how intelligent, data-driven decision-making is fundamental to achieving sustainable, economically viable, and resilient mining operations. As the industry navigates evolving regulatory landscapes and increasing cost pressures, AI-enhanced fuel optimisation emerges as an indispensable tool for securing long-term efficiency, sustainability, and competitiveness in large-scale mining operations.

METHODOLOGY

Data collection and measurement approach

This study deployed high-precision fuel metering systems across multiple mining sites to capture real-time fuel consumption data, significantly enhancing accuracy over traditional estimation techniques (Thompson and Visser, 2023). These metering systems, rigorously validated against Original Equipment Manufacturer (OEM) specifications, achieved ±0.1 per cent measurement accuracy through K-Factor linearisation adjustments (ISO/IEC, 2017). By integrating advanced telemetry and high-frequency fuel flow monitoring, the system facilitated fine-grained analysis of fuel usage patterns, aligning with industry best practices in mining fleet analytics (GMG, 2023).

The study incorporated a multi-modal data acquisition framework, synchronising real-time telemetry, geospatial positioning, and engine diagnostics to ensure a holistic view of operational efficiency (Australian Standards, 2023). GPS and elevation data, processed via differential correction algorithms, provided centimetre-level accuracy, adhering to mining survey calibration protocols (Bilski and Paweł, 2022). Additionally, engine performance metrics were captured via direct CAN bus integration, ensuring fidelity to OEM-defined operational standards (SAE International, 2023). Payload weights were recorded using calibrated weighing systems to enhance data robustness, while road condition assessments followed established mining haul road evaluation methodologies (ISO, 2023; Thompson and Visser, 2023).

To maximise operational feasibility, the system was designed to seamlessly integrate with existing legacy Fleet Management Systems (FMS), such as Modular Mining Dispatch, Wenco, and Caterpillar MineStar. Data streams from the fuel metres, GPS, and diagnostics modules were aggregated via standard communication protocols (eg TCP/IP, CANopen) and mapped into pre-existing data schemas used by the site's operational software. This allowed real-time fuel data, equipment health indicators, and spatial telemetry to be visualised through the mine's standard dashboards without requiring significant modifications to infrastructure or workflow. The interoperability framework ensured minimal disruption to day-to-day operations, enabling rapid deployment and operator acceptance.

The study's data set encompassed a statistically significant sample size across multiple mining operations, providing sufficient analytical power to detect meaningful operational variations while controlling for site-specific variables. This comprehensive sampling approach ensured robust model training and validation, with cross-validation techniques employed to verify predictive accuracy.

Analytics framework

The study employed a machine learning-driven analytics pipeline to conform to rigorous data science methodologies applicable to mining operations (GMG, 2023). This approach incorporated industry-specific operational technology standards, ensuring seamless integration with mining telemetry systems (ISO, 2023).

Data preprocessing and quality assurance

A rigorous data preprocessing pipeline was implemented to refine input data sets, incorporating automated anomaly detection, error correction, and statistical validation techniques (Australian Standards, 2023). This protocol ensured high data fidelity, eliminating measurement inconsistencies and reducing outlier-induced biases in model training. The resulting validated data set met industry benchmarks for data integrity in mining technology systems.

Machine learning framework

This research developed a predictive machine learning framework to optimise haul truck energy efficiency, leveraging a structured cycle of model design, training, intervention execution, impact evaluation, and iterative retraining. This cyclical approach enabled continuous refinement of fuel efficiency insights, driving sustainable operational improvements.

Model development and training

The study employed gradient-boosted decision trees, a machine learning architecture well-suited for nonlinear predictive modelling in industrial settings. The model was trained on an extensive multi-variable data set, incorporating:

- haul cycle duration and total distance travelled
- payload tonnage and vertical displacement
- operator behavioural metrics
- truck-specific fuel efficiency parameters
- geospatial road condition factors.

The primary target variable was fuel consumption per haul cycle, with a focus on minimising predictive error and quantifying key performance determinants.

Feature importance analysis and optimisation

Following training, the model underwent post-hoc interpretability analysis, identifying the most influential predictive variables affecting fuel efficiency. Key insights revealed:

- Operator driving techniques significantly impact energy consumption.
- Road surface variability is a dominant factor in fuel inefficiencies.
- Truck-specific performance deviations warrant targeted maintenance interventions.

Intervention design and deployment

Upon identifying inefficiencies, targeted intervention protocols were developed and implemented, focusing on:

- Operator optimisation programs emphasise throttle control, braking efficiency, and synchronisation of shift timing.

- Predictive maintenance scheduling, detecting anomalies indicative of incipient mechanical failures, thus preventing excessive fuel consumption.

- Haul road infrastructure enhancements, optimising surface conditions to minimise rolling resistance and maximise haul cycle efficiency.

Impact assessment and iterative model refinement

Intervention outcomes were systematically evaluated through quantitative performance assessments, measuring the effectiveness of fuel optimisation strategies. The post-intervention data set was reintegrated into the machine learning framework, triggering an adaptive learning cycle that improved predictive accuracy and operational insights.

By embedding machine learning-driven optimisation into mining haulage operations, this methodology delivers a scalable, data-informed strategy for fuel efficiency enhancement, leading to substantial reductions in emissions and cost savings within the mining sector.

RESULTS

Operational efficiency enhancements

A rigorous analysis of implementation data from multiple study sites identified key pathways for emissions reduction, validated through controlled trials and advanced statistical methodologies. This study systematically assessed operator performance optimisation, predictive maintenance interventions, and haul road management strategies, each contributing to quantifiable reductions in fuel consumption and carbon emissions.

Operator performance optimisation

Machine learning-driven behavioural analytics revealed that operator driving patterns significantly influenced fuel efficiency variability. The deployment of real-time feedback mechanisms and structured training programs resulted in measurable improvements in fuel consumption, supporting prior research on performance-based optimisation in mining operations.

The integration of operator-specific performance tracking led to an observed fuel savings range of 5–8 per cent across all evaluated sites, with the most proficient operators achieving sustained reductions of up to 12 per cent.

Notable behavioural modifications contributing to these gains included:

- Optimised gear shifting protocols, minimising unnecessary fuel consumption.

- Refined acceleration and braking patterns, reducing energy losses.

- Consistent implementation of adaptive load-aware driving techniques.

Predictive maintenance impact

Incorporating high-precision fuel monitoring systems enabled early anomaly detection, identifying signs of performance deterioration up to 200 operational hours before traditional maintenance triggers. The predictive modelling framework pinpointed underperforming haul trucks, facilitating targeted maintenance interventions to address the following:

- Obstructed fuel filters, restricting optimal combustion.

- Deteriorated fuel injectors, leading to inefficient fuel utilisation.

- Drivetrain component wear exacerbates rolling resistance and energy expenditure.

Impact of AI-based optimisation on maintenance intervals

Table 1 highlights the significant improvements in component lifespan achieved through AI-driven predictive maintenance strategies. Before optimisation, maintenance schedules were based on fixed-interval servicing, often leading to premature part replacements or unexpected failures. With AI-based optimisation, maintenance intervals were extended by 30–60 per cent, depending on the component, by leveraging real-time performance monitoring and anomaly detection. For instance, fuel injector servicing intervals increased from 500 to 700 operating hrs (a 40 per cent extension), while brake pad replacements were extended from 600 to 900 hrs (a 50 per cent increase). These improvements resulted in reduced downtime, lower maintenance costs, and increased fleet availability, demonstrating the effectiveness of predictive analytics in optimising equipment reliability while minimising unnecessary servicing and resource consumption (Soofastaei *et al*, 2022).

TABLE 1

Maintenance interval comparison.

Component	Pre-optimisation interval (operating hrs)	Post-optimisation interval (operating hrs)	Extended lifespan (%)
Fuel injectors	500	700	40.0
Engine oil	250	400	60.0
Brake pads	600	900	50.0
Tires	1000	1300	30.0
Air filters	400	600	50.0
Transmission system	800	1100	37.5

Mining sites implementing predictive maintenance protocols informed by fuel efficiency metrics realised:

- A 12 per cent reduction in maintenance costs due to proactive servicing.

- An 8 per cent increase in fleet availability, minimising downtime.

- A high correlation ($r = 0.87$) between fuel efficiency degradation trends and mechanical failure probabilities, reinforcing the predictive validity of the machine learning framework.

Haul road management optimisation

Assessments of road condition impacts on fuel consumption facilitated the development of adaptive, data-driven haul road maintenance schedules, yielding an additional 3–5 per cent efficiency improvement across study sites. Real-time road surface monitoring and strategic grading interventions resulted in:

- A 15 per cent increase in haul road maintenance efficiency, optimising resource utilisation.

- Enhanced road geometry configurations, reducing rolling resistance and excessive fuel use.

Comparative implementation pathways

This study evaluated the relative effectiveness of emissions reduction strategies across multiple mining sites, benchmarking their outcomes against established industry performance standards (see Table 2) (GMG, 2023):

- **Mine A:** Focused on operator-centric performance enhancements, leveraging behavioural analytics and dynamic feedback systems. The findings confirmed the long-term benefits of data-driven operator training, aligning with research on efficiency improvement frameworks (Dindarloo and Siami-Irdemoosa, 2023).

- **Mine B:** Prioritised predictive maintenance integration, utilising real-time engine diagnostics and condition-based monitoring. Results supported best practices in AI-enhanced predictive maintenance, underscoring the strategic value of fuel efficiency-informed servicing schedules (Thompson and Visser, 2023; Soofastaei *et al,* 2022).

- **Mine C:** Achieved the most significant emissions reduction (15 per cent) through comprehensive haul road management interventions. Continuous surface monitoring and grade optimisation significantly enhanced fuel efficiency, validating the correlation between road quality and fleet energy consumption (Thompson and Visser, 2023).

- **Mine D:** Implemented a multi-faceted strategy, incorporating operator performance training, predictive maintenance, and road condition optimisation. Although achieving a 9 per cent reduction in emissions, findings suggest that site-specific targeted interventions, as demonstrated in Mines A and C, may yield higher initial efficiency gains.

TABLE 2
Summary of emission reduction outcomes across study sites.

Site	Fleet size	Baseline emissions (tCO_2/annum)	Reduction achieved	Primary optimisation pathway
Mine A	12	24 500	12%	Operator performance
Mine B	8	16 800	8%	Maintenance optimisation
Mine C	15	31 200	15%	Road management
Mine D	10	20 900	9%	Combined approach

These findings substantiate the scalability and cost-effectiveness of machine learning-driven fuel optimisation strategies, providing a viable framework for sustainable emissions reductions and operational cost efficiency in the mining sector.

DISCUSSION

Key determinants of implementation success

The analysis of implementation outcomes identified several critical factors influencing the success of the fuel optimisation framework. These insights, corroborated by industry research and empirical evidence from operational sites, provide a foundation for best practices in mining fleet efficiency enhancement (MinEx, 2023).

Data integrity and measurement fidelity

The efficacy of optimisation strategies was intrinsically linked to the precision and reliability of data acquisition systems. The ability to capture high-resolution telemetry data with minimal error margins was paramount in ensuring accurate fuel consumption analytics and operational decision-making. These findings align with established protocols for mining technology deployment (GMG, 2023). Inconsistent or incomplete data sets posed a substantial risk to performance assessments, reinforcing the necessity for continuous calibration and rigorous sensor validation.

System interoperability and integration efficiency

Seamless integration with existing mine automation systems and operational workflows emerged as a crucial determinant of sustainable implementation. Sites with full interoperability between AI-driven analytics and legacy fleet management software demonstrated greater scalability, user adoption, and operational continuity. These findings align with contemporary research on digital transformation strategies in mining, underscoring the significance of adaptive infrastructure and modular system design (Paraszczak and Fytas, 2022).

Operator involvement and performance feedback systems

Human factors played an essential role in the sustained impact of implementation. The effectiveness of operator engagement models and real-time performance feedback loops was strongly correlated with long-term behavioural modifications and efficiency gains. Operators who received automated feedback on fuel usage trends exhibited consistent improvements in haul cycle optimisation and fuel conservation behaviours. This outcome is congruent with findings from human-centred mining performance studies, which highlight the necessity of structured training programs and AI-driven intervention mechanisms (Dindarloo and Siami-Irdemoosa, 2023). Additionally, management endorsement and implementation of transparent performance metrics proved essential in fostering a culture of continuous operational improvement (MCA, 2023).

CHALLENGES IN IMPLEMENTATION

Despite the measurable benefits of AI-driven optimisation, several technical and organisational challenges impeded seamless adoption. These barriers were consistent across multiple study sites and aligned with challenges documented in digitalisation initiatives within the mining sector (GMG, 2023).

Resistance to organisational and technological change

Introducing AI-driven decision-support systems encountered initial resistance from operational teams, necessitating structured change management frameworks. Effective deployment required phased rollout strategies, comprehensive user training, and demonstrable performance benefits, ensuring alignment with workforce expectations and operational needs. These findings are consistent with industry best practices in mining technology integration (MinEx, 2023).

Technical barriers to system compatibility

Integrating advanced machine learning models with mine operational systems posed considerable technical hurdles. Many mining operations relied on heterogeneous and outdated fleet management software, necessitating custom API development and extensive reconfiguration to achieve complete data interoperability. These challenges reflect broader industry concerns regarding legacy infrastructure limitations in mining digital transformation (CSIRO, 2023).

Scalability and data governance complexities

The deployment of large-scale AI-driven fuel monitoring and analytics systems introduced challenges related to data volume processing, computational efficiency, and real-time analytics scalability. Sites lacking structured data governance frameworks struggled to ensure data reliability and mitigate inconsistencies in fuel telemetry streams. To address this, the study emphasised the necessity of automated data validation pipelines and machine-learning-driven anomaly detection. These methodologies align with established best practices in mining data architecture and operational intelligence (Bilski and Paweł, 2022). Furthermore, maintaining a long-term focus on continuous improvement initiatives was crucial in preventing regression in optimisation gains, reinforcing insights from studies on sustainable operational advancements in mining (Thompson and Visser, 2023).

CONCLUSIONS

This study demonstrates that integrating high-resolution fuel monitoring and machine learning-driven analytics delivers immediate and substantial reductions in greenhouse gas emissions from mining haul truck operations. Empirical evidence gathered across multiple deployment sites indicates scope one emissions reductions ranging from 6.2 per cent to 10.3 per cent, significantly exceeding the regulatory benchmark established by Australia's Safeguard Mechanism at 4.9 per cent (Clean Energy Regulator, 2023). Crucially, these reductions were achieved without compromising operational productivity, reaffirming the synergy between environmental responsibility and economic efficiency in the mining sector.

The findings of this study underscore the efficacy of machine learning-augmented optimisation strategies in identifying inefficiencies that traditional monitoring methodologies fail to capture

(Soofastaei *et al,* 2022). The study identified three primary pathways as key drivers of emissions mitigation:

1. Operator performance enhancement through real-time behavioural analytics and structured training programs.

2. Predictive maintenance execution, utilising machine learning models for proactive fault detection and asset management.

3. Dynamic haul road management, optimising terrain interactions to minimise rolling resistance and improve fuel efficiency.

As the Department of Industry, Science, and Resources (2023) projected, these targeted interventions align with and exceed industry benchmarks, reinforcing their potential for widespread adoption across large-scale mining operations.

Furthermore, the study highlights critical success factors determining these interventions' scalability and long-term effectiveness. The results emphasise integrating technological advancements with structured change management strategies. The highest-performing sites demonstrated strong operator engagement, systematic implementation methodologies, and adherence to best practices in operational efficiency enhancement (Dindarloo and Siami-Irdemoosa, 2023). This evidence reinforces that sustainable technology adoption requires a well-defined ecosystem of stakeholder collaboration, performance accountability, and continuous professional development.

Future research directions should explore integrating this machine learning framework with autonomous haulage systems, investigate the transferability of these optimisation models to underground mining environments, and develop predictive algorithms that can facilitate the transition from diesel to zero-emission technologies by modelling energy requirements under varying operational conditions.

Looking ahead, this data-driven framework provides a strategic roadmap for mining operations to meet short-term regulatory mandates while simultaneously positioning themselves for long-term decarbonisation initiatives. Insights derived from this study will be crucial in guiding investment decisions related to next-generation haulage solutions, including adopting zero-emission fleet technologies. Most importantly, the efficiency gains from short-interval control optimisation offer mining enterprises a cost-effective, validated pathway toward sustainability, operational resilience, and regulatory compliance.

ACKNOWLEDGEMENTS

The author acknowledges the collaboration of the mining operations participating in this study and their willingness to share data and insights to advance industry knowledge in this critical area. Cascadia Scientific has approved the publication of this paper.

REFERENCES

Australian Standards, 2023. AS 5327-2023 – Mining Equipment Monitoring and Data Collection, Standards Australia, Sydney.

Bilski, P and Paweł, W, 2022. Automated assessment of haul road condition using machine learning methods, *International Journal of Mining Science and Technology*, 32(2):271–280.

Clean Energy Regulator, 2023. Safeguard Mechanism Reforms 2023–2030, Australian Government, Canberra.

CSIRO, 2023. Mining Technology Roadmap 2023: Pathways to Sustainable Operations, Commonwealth Scientific and Industrial Research Organisation, Canberra.

Department of Industry, Science and Resources, 2023. Resources and Energy Major Projects: 2023, Commonwealth of Australia. Available from: <https://www.industry.gov.au/publications/resources-and-energy-major-projects-2023>

Dindarloo, S R and Siami-Irdemoosa, E, 2023. Applied machine learning in mining operations: a comprehensive review, *International Journal of Mining Science and Technology*, 33(1):13–25.

Global Mining Guidelines Group (GMG), 2023. Guideline for the Implementation of Autonomous Systems in Mining, ver 2, GMG, Montreal.

International Organisation for Standardisation (ISO), 2023. ISO 23725:2023 – Mining Operations and Equipment Monitoring Requirements, ISO, Geneva.

International Organisation for Standardisation/International Electrotechnical Commission (ISO/IEC), 2017. ISO/IEC 17025:2017 – General Requirements for Testing and Calibration Laboratories Competence, ISO, Geneva.

Minerals Council of Australia (MCA), 2023. Climate Action Plan: 2023 Progress Report, MCA, Canberra.

MinEx, 2023. Surface Mining Operational Efficiency Guidelines, MinEx, Wellington.

Paraszczak, J and Fytas, K, 2022. Opportunities for fuel consumption and carbon emission reduction in mining operations, *Mining Technology*, 131(4):254–266.

SAE International, 2023. J1939-71 Vehicle Application Layer, Society of Automotive Engineers, Warrendale.

Soofastaei, A, Aminossadati, S M, Kizil, M S and Knights, P, 2022. Development of an artificial intelligence model for energy efficiency improvement in surface mines, *Journal of Cleaner Production*, 326:129328.

Thompson, R J and Visser, A T, 2023. Mine haul road maintenance management systems: recent developments, *Mining Technology,* 132(2):89–102.

Author index